LearnSmart™ Study Modules

Powered by Connect™ Biology, McGraw-Hill LearnSmart™ provides students with a GPS (Guided Path to Success). It is an adaptive diagnostic tool based on artificial intelligence that constantly assesses a student's knowledge of the course material. Sophisticated diagnostics adapt to each student's individual knowledge base, and vary the questions to determine what the student doesn't know, knows but has forgotten, and how best to match and improve their knowledge level. Students actively learn the required concepts, and instructors can get specific LearnSmart™ reports to monitor overall progress.

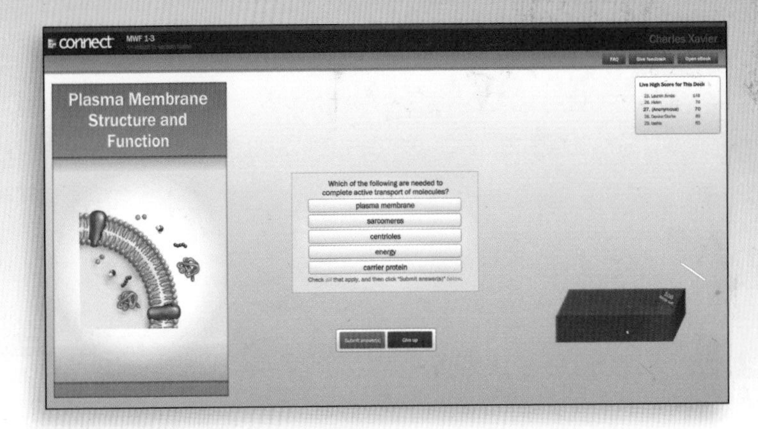

Learn Fast. Learn Easy. Learn Smart.

Learn more at www.mhlearnsmart.com.

My Lectures—Tegrity

McGraw-Hill Tegrity Campus™ records and distributes lectures with just a click of a button. Students can view anytime/anywhere via computer, iPod, or mobile device. It indexes as it records PowerPoint presentations and anything shown on the computer, so students can use keywords to find exactly what they want to study.

Instructor Resources

Connect™ Biology provides easy access to the following resources:

- Enhanced image PowerPoints with editable art
- Lecture PowerPoints with animations
- Animation PowerPoints
- Labeled and unlabeled Jpeg files of art, photos, and tables from the textbook

Chapter	Enhanced Image PPTs (includes photos, and editable art)	Lecture PPTs with Animations	Animation PowerPoints	Labeled Jpeg Images	Base Art Image Files (.jpgs, no labels or leader lines)
All Chapters	Enhanced Image PPTs (707,634 KB)	Lecture Animation PPTs (649,609 KB)	Animation PPTs (1,64,060 KB)	Labeled Images (859,337 KB)	Base Images (599,793 KB)
Ch01	Ch. 1 Enhanced Image PPTs (23,977 KB)	Ch. 1 Lecture Animation PPTs (14,860 KB)	There are no Animation PPTs correlated to this chapter.	Ch. 1 Labeled Images (28,669 KB)	Ch. 1 Base Images (21,333 KB)
Ch02	Ch. 2 Enhanced Image PPTs (9,907 KB)	Ch. 2 Lecture Animation PPTs (6,012 KB)	Ch. 2 Animation PPTs (6,794 KB)	Ch. 2 Labeled Images (15,054 KB)	Ch. 2 Base Images (8,805 KB)
Ch03	Ch. 3 Enhanced Image PPTs (16,605 KB)	Ch. 3 Lecture Animation PPTs (14,220 KB)	Ch. 3 Animation PPTs (1,833 KB)	Ch. 3 Labeled Images (18,346 KB)	Ch. 3 Base Images (14,607 KB)

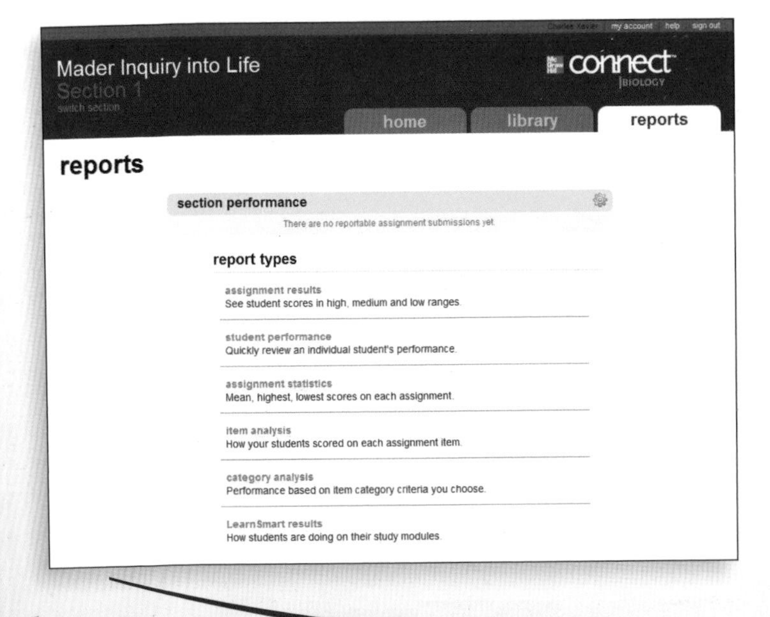

Powerful Reporting Solutions

Connect™ Biology offers detailed reporting so instructors can quickly assess how students are doing in regards to overall class performance, individual assignments, and each question.

All practice and test bank questions are tagged to the textbook by chapter, section, topic, and Bloom's level of difficulty, and aligned with the learning outcomes in the textbook to aid these reporting features.

Inquiry into Life

Thirteenth Edition

Sylvia S. Mader

Jeffrey A. Isaacson
Nebraska Wesleyan University

Kimberly G. Lyle-Ippolito
Anderson University

Andrew T. Storfer
Washington State University

The McGraw·Hill Companies

INQUIRY INTO LIFE, THIRTEENTH EDITION

Published by McGraw-Hill, a business unit of The McGraw-Hill Companies, Inc., 1221 Avenue of the Americas, New York, NY 10020.

Some ancillaries, including electronic and print components, may not be available to customers outside the United States.

This book is printed on acid-free paper.

1 2 3 4 5 6 7 8 9 0 DOW/DOW 1 0 9 8 7 6 5 4 3 2 1 0

ISBN 978–0–07–340344–1
MHID 0–07–340344–X

Vice President & Editor-in-Chief: *Marty Lange*
Vice President, EDP: *Kimberly Meriwether-David*
Publisher: *Janice Roerig-Blong*
Executive Editor: *Michael S. Hackett*
Director of Development: *Kristine Tibbetts*
Senior Developmental Editor: *Rose M. Koos*
Senior Marketing Manager: *Tamara Maury*
Senior Project Manager: *Jayne L. Klein*
Lead Production Supervisor: *Sandy Ludovissy*
Senior Media Project Manager: *Jodi K. Banowetz*
Senior Designer: *Laurie B. Janssen*
(USE) Cover Image: *©Gettyimages*
Senior Photo Research Coordinator: *Lori Hancock*
Photo Research: *Evelyn Jo Johnson*
Art Studio: *Electronic Publishing Services Inc., NYC*
Compositor: *Electronic Publishing Services Inc., NYC*
Typeface: *10/12 Palatino*
Printer: *R. R. Donnelley*

Library of Congress Cataloging-in-Publication Data

Mader, Sylvia S.
Inquiry into life / Sylvia S. Mader. — 13th ed. / Kimberly G. Lyle-Ippolito, Jeffrey A. Isaacson, Andrew T. Storfer.
p. cm.
Includes index.
ISBN 978–0–07–340344–1 — ISBN 0–07–340344–X (hard copy : alk. paper) 1. Biology—Textbooks. I. Title.

QH308.2.M363 2011
570—dc22

2009021386

www.mhhe.com

Brief Contents

Contents

Readings

Bioethical Focus

Ecology Focus

Health Focus

Science Focus

What Sets Mader Apart?

The Human Approach

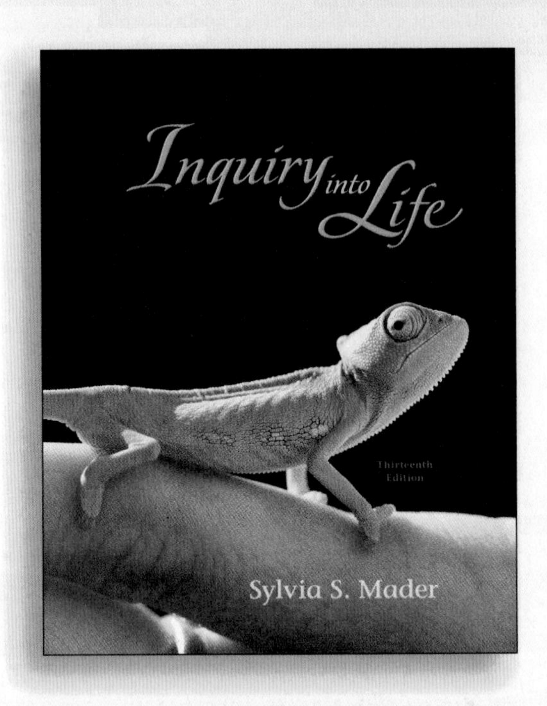

Inquiry into Life was originally developed to reach out to science-shy students. Dr. Mader and her colleagues all believed in teaching science from a human perspective and sought human applications that would make the material more relevant to the student. After teaching for several years, Dr. Mader developed a methodology that enabled most students to learn biology, and from those teaching experiences, she developed the first edition of *Inquiry into Life.*

Biology has changed rapidly since *Inquiry into Life* was first published. New findings, ideas, and concepts have emerged over the years, and each edition of *Inquiry into Life* has blended the classic with the new to create a book that is both time-tested and current.

Mader Writing Style

Well-known for its clarity and simplicity, the Mader writing style makes the content accessible to students. Mader's writing appeals to students because it meets them where they are and helps them understand the concepts with its clear "take-home messages."

"The book is readable, clear and concise. I rarely have students asking me to explain material that they have read—this has not been the case with other texts."
Sandra Devenny
Delaware County Community College

"This book is written very well for a non-science student."
Jason Jennings
Southwest Tennessee Community College

"The greatest strong point of this book is its straight forward approach to difficult topics that are presented in an easily readable format. Mader integrates good examples and illustrations to support the text."
William Sproat
Walters State Community College

"This is probably the hardest material in the book, and it was delivered in clear and concise language."
Francesca Catalano
American Public University System

Relevancy and Applications

Inquiry into Life covers the entire field of general biology to provide the fundmental principles and the application of those principles to human concerns.

As with previous editions, the central theme of *Inquiry into Life* is understanding the workings of the human body and how humans fit into the world of living things. This is accomplished through the features highlighted below.

New Applying the Concepts and Applying the Concepts Revisited

Each chapter begins with a short, thought-provoking vignette that relates chapter material to a real-life situation, and questions at the end of the chapter encourage students to reconsider the situation in light of the chapter concepts just learned.

Applying the Concepts

Fifty is the new 30! In 2008, several famous women, including Madonna,Sharon Stone, and Michelle Pfeiffer turned 50. All of the media hype suggests that today's 50-year-old women are as healthy as 30-year-old women used to be. What health changes have occurred since 1958 when these women were born? In 1960, the life expectancy at birth in the United States was 69.7 years. In 2006, the last year with available statistics, the average life expectancy at birth was 78.1—a gain of 8.4 years. Many factors affect life expectancy, including reduction in infant mortality, improvement ... r nutrition. But changes in medi- ... example, open heart and bypass ... ized axial tomography) scans and ... were invented in the 1970s. New ... the 1980s. There has also been ... vironment and thereby reducing ... of the Environmental Protection ... hat cleaner air from 1978 to the ... ths to our life expectancy. With ... rhaps soon we will read headlines

Applying the Concepts **[Revisited]**

Although there have been many advances made in medicine and public health, much of the responsibility for good health as we age rests with choices made by the individual. Madonna is a "fitness fanatic" who exercises daily and eats a macrobiotic diet. Sharon Stone exercises and abstains from all caffeine and alcohol.

1. What good health habits have you incorporated into your routine that will extend your health span?
2. What bad habits should you try to eliminate from your life now, before they cause damage that will age you?

Readings

Inquiry into Life has four types of application readings.

Health Focus readings review procedures and technology that can contribute to our well-being.

Ecology Focus readings show how the concepts of the chapter can be applied to ecological concerns.

Science Focus readings describe how experimentation and observations have contributed to our knowledge about the living world.

Bioethical Focus readings describe modern situations that call for value judgments and challenge students to develop a point of view.

See page vii for a complete listing of Focus reading topics.

Health Focus

Vegetarians: Where Do You Get Your Protein?

There are approximately 5 million adult vegetarians in the United States. Although definitions vary, most abstain from eating red meat, ... complementation, the idea that foods with insufficient levels of one or more essential amino acids needed to be ingested at the ... proteins in plant food such as cereals may be harder to digest than animal proteins, vegetarians may need to include a higher percent-

Ecology Focus

Fill 'Er Up—with Algae?

Algae are at the base of many food chains. But researchers are now using algae to produce fuel for cars! In 2007, a San Francisco ... these genes turned off, the algae actually make more oil. Algae-based biofuels are considered ... LiveFuels, are trying to produce oil from algae using sunlight. GreenFuel Technologies grows algae in sealed, transparent tubes exposed to

Science Focus

How Cells Talk to One Another

All organisms are able to sense and respond to specific signals in their environment. A bacterium that has taken up residence in ... occasion, or one tissue may need to perform one of its various functions only at particular times. In plants, external signals, such as ... sites around the body. For example, the pancreas releases a hormone called insulin, which is transported in blood vessels to the

Bioethical Focus

Just a Snip, Please: Testing Hair for Drugs

Imagine you have just started your first real full-time job, with health insurance, a pension plan, and other benefits. The first day, after you settle into your cubicle, a manager comes by and asks for a sample of your hair. "We normally just snip a bit off in the back, where it will hardly show," he says, smiling. When you ask why, he replies that it is company policy to screen employees for drug use every six months.

Tests that detect illicit drugs or their metabolic products in urine or saliva are of limited value, because most of these chemicals are secreted by the body for only a few days. In contrast, many illicit drugs and/or their metabolites are incorporated into growing hair shafts throughout the body, where they remain indefinitely. Especially in the first 1.5 inches of hair growth from the scalp, each 0.5 inch is considered to represent 30 days' worth of growth (and thus, potential drug use). Hair from anywhere on the body can be used, although the growth of body hair is usually slower, so the time of any drug use cannot be determined. It generally takes four to five days from the time a drug is taken into the body until it begins to appear in hair.

Many commercial laboratories now offer hair testing. In general, these labs contend that they can distinguish between environmental exposure to a particular drug—from being in the vicinity of someone smoking marijuana, for example—and actual drug use by an individual. In recent years, several court decisions have supported the idea that hair testing can accurately distinguish actual drug use from such "passive" exposure. This has led to the marketing of several shampoos for cleaning or "detoxifying" the hair shafts, but these may not be effective. And even if they were, a lab could conceivably test hair for common contaminants expected to be found in everyone's hair. If these contaminants were not found, this could be used as evidence that a person has attempted to hide prior drug usage.

Even if one accepts that hair testing is an accurate way to prove that a person has used drugs, is it ethical for businesses to require their employees to undergo such tests? Does it matter what type of job a person has? For example, would it be more difficult to argue against mandatory drug testing for school bus drivers than for stockbrokers? And for those of you still living with your parents, how upset would you be to find out that your mom or dad had slipped into your bedroom at night, snipped off a small bit of hair from the back of your head, and mailed it to one of several companies that now offer hair testing to the general public?

Form Your Own Opinion

1. The Fourth Amendment to the U.S. Constitution guards against "unreasonable searches and seizures" by the government. Do you believe that the types of drug testing described here are unconstitutional?
2. In which of the following additional situations would you support mandatory drug testing: To test airline pilots for hallucinogens? To test high school athletes for steroids? To test NBA players for marijuana?
3. Could you imagine any circumstance where you might want to test your own children for drug use?

What Sets Mader Apart?

Vivid and Captivating Illustrations

Mader's hallmark illustration program continues to be a foundation of *Inquiry into Life*, helping students visualize biology and allowing instructors to easily incorporate dynamic art in lecture or as part of assignments.

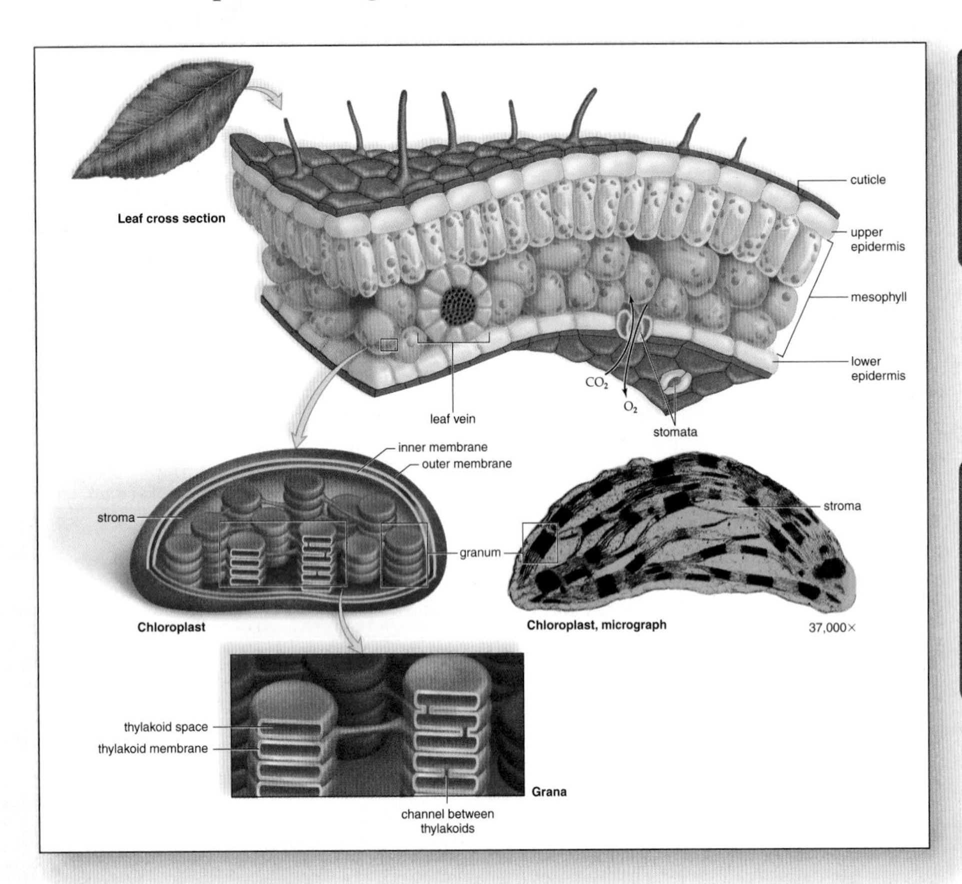

Multi-Level Perspective

Illustrations depicting complex structures show macroscopic and microscopic views to help students connect the two levels.

Combination Art

Drawings of structures are paired with micrographs to give students the best of both perspectives: the realism of photos and the explanatory clarity of line drawings.

Animal cells

plasma membrane

nucleus

6.6 μm

In an isotonic solution, there is no net movement of water.

6.6 μm

In a hypotonic solution, water enters the cell, which may burst (lysis).

6.6 μm

In a hypertonic solution, water leaves the cell, which shrivels (crenation).

Plant cells

nucleus

central vacuole

chloroplast

25 μm

In an isotonic solution, there is no net movement of water.

cell wall

25 μm

In a hypotonic solution, the central vacuole fills with water, turgor pressure develops, and chloroplasts are seen next to the cell wall.

plasma membrane

40 μm

In a hypertonic solution, the central vacuole loses water, the cytoplasm shrinks (plasmolysis), and chloroplasts are seen in the center of the cell.

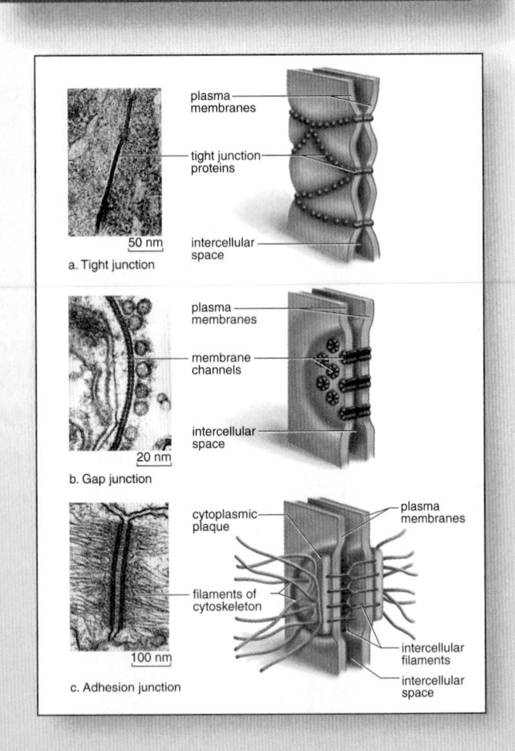

Process Figures

These figures break down processes into a series of smaller steps and organize them in an easy-to-follow format.

carrier protein
Outside
Inside

1. Carrier has a shape that allows it to take up 3 Na^+.

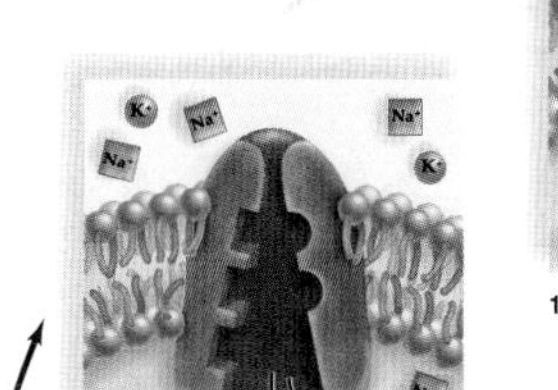

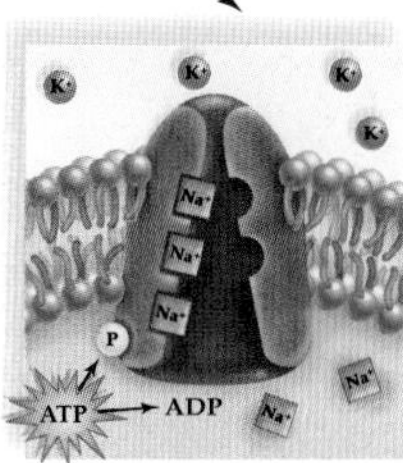

2. ATP is split, and phosphate group attaches to carrier.

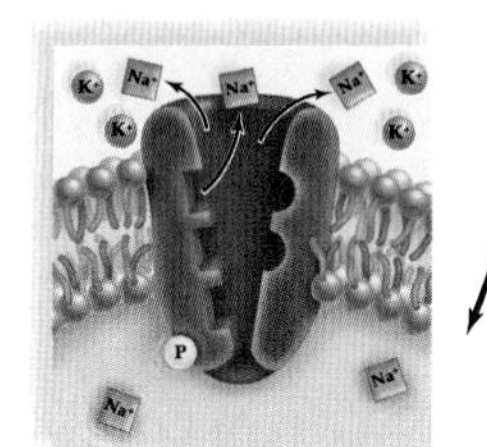

3. Change in shape results and causes carrier to release 3 Na^+ outside the cell.

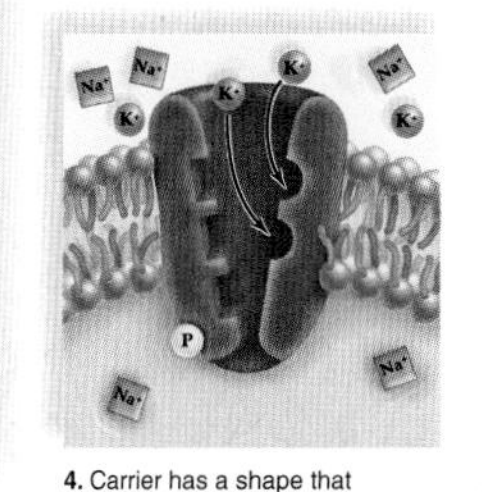

4. Carrier has a shape that allows it to take up 2 K^+.

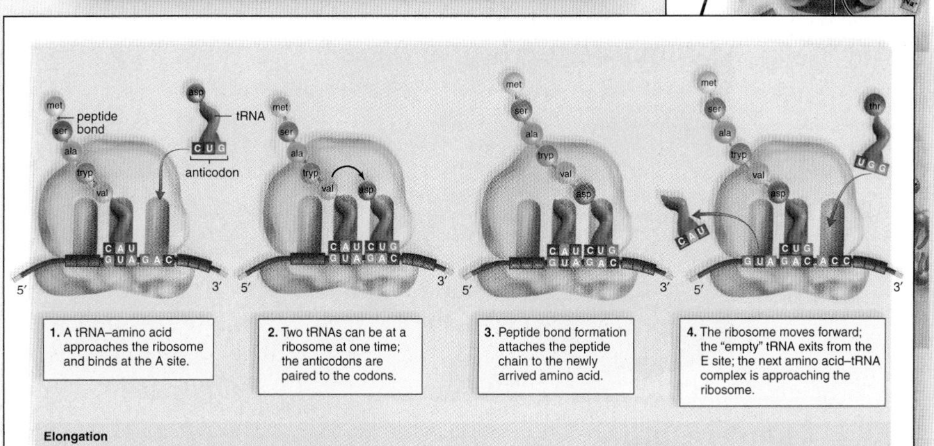

1. A tRNA–amino acid approaches the ribosome and binds at the A site.

2. Two tRNAs can be at a ribosome at one time; the anticodons are paired to the codons.

3. Peptide bond formation attaches the peptide chain to the newly arrived amino acid.

4. The ribosome moves forward; the "empty" tRNA exits from the E site; the next amino acid–tRNA complex is approaching the ribosome.

Elongation

Icons

Icons help orient the student.

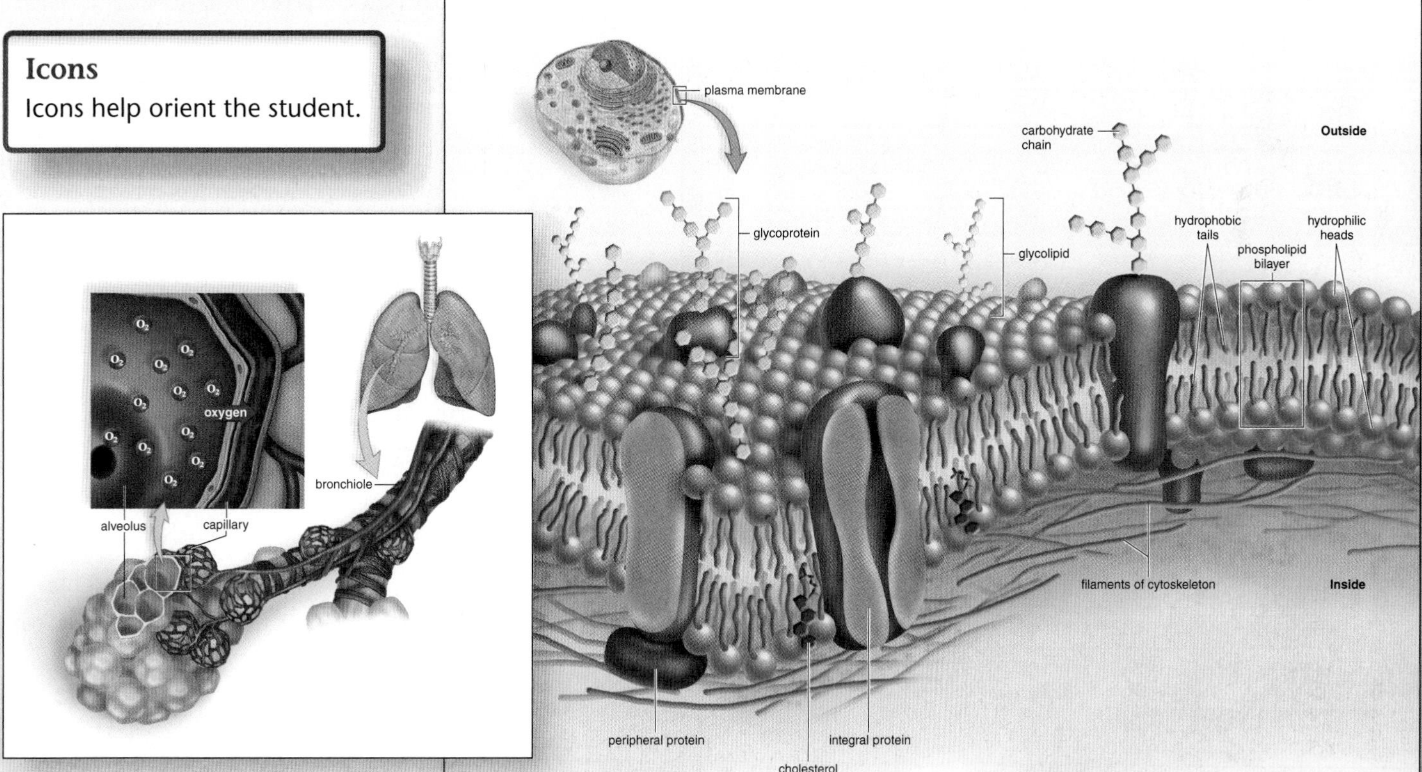

What Sets Mader Apart?

The Mader Learning System

Inquiry into Life continues to have a strong pedagogical framework that helps students excel in their study of biology.

Chapter Outline

The chapter outline provides the overall organization of the chapter.

Meet
dics
eart
hree
levi-
l.

The cardiovascular system, which includes the heart and blood vessels, transports oxygen, nutrients, and wastes to and from the tissues. Diseases of this system, such as atherosclerosis, are a major cause of death in the more developed countries of the world. If you live in the United States, you have about a one in three chance of dying of heart disease, and if you add in all the other conditions that can affect the blood vessels, your odds are greater than 50:50 of eventually developing cardiovascular disease.

Tim Russert was no doubt aware that he needed to be vigilant about cardiovascular disease; he was on cholesterol-lowering drugs and had passed a heart stress test just a few weeks before his death. And yet his heart ceased functioning at a relatively young age. Even more alarming, two out of three cardiovascular-related deaths occur without any prior diagnosis. However, there are ways to reduce your risk of cardiovascular disease, and these are discussed later in this chapter.

Cardiovascular System

CHAPTER OUTLINE

213

New Learning Outcomes

Learning goals are provided at the beginning of each section to clearly outline expectations.

216 Part III Maintenance of the Human Body

12.2 The Human Heart

LEARNING OUTCOMES

1. Name the major components of the heart, including the four chambers and four valves.
2. Trace the path of blood through the heart and lungs.
3. Describe the intrinsic and extrinsic control of the heartbeat.

The **heart** is a cone-shaped, muscular organ about the size of a fist (Fig. 12.3). It is located between the lungs directly behind the sternum (breastbone) and is tilted so that the apex (the pointed end) is oriented to the body's left. The major portion of the heart, called the **myocardium,** consists largely of cardiac muscle tissue. The muscle fibers of the myocardium are branched and tightly joined to one another. The heart lies within the **pericardium,** a thick, serous membrane that secretes a small quantity of lubricating liquid. The inner surface of the heart is lined with endocardium, a membrane composed of connective tissue and endothelial tissue.

Internally, a wall called the septum separates the heart into a right side and a left side (Fig. 12.4). The heart has four chambers. The two upper, thin-walled atria (sing., **atrium**) are located above the two lower, thick-walled **ventricles.** The ventricles pump the blood to the lungs and the body.

The heart also has four valves that direct the flow of blood and prevent its backward movement. The two valves that lie between the atria and the ventricles are called the **atrioventricular valves.** These valves are supported by strong fibrous strings called **chordae tendineae.** The chordae, which are attached to muscular projections of the ventricular walls, support the valves and prevent them from inverting when the heart contracts. Tl
right side is called th
flaps, or cusps. The va
(or mitral) valve becau
valves, between the v
the **semilunar valves,**
pulmonary semilunar
and the pulmonary t
between the left ventr

left subclavian artery
left common carotid artery
brachiocephalic artery
superior vena cava
aorta
left pulmonary artery
pulmonary trunk
left pulmonary veins
right pulmonary artery
right pulmonary veins
left atrium
left cardiac vein
right atrium
right coronary artery
left ventricle
right ventricle
inferior vena cava
apex
a.

superior vena
right coronary artery
inferior vena cava
left cardiac vein
right cardiac vein
b.

Figure 12.3 External heart anatomy.
a. The venae cavae and the pulmonary trunk are attached to the right side of the heart. The aorta and the pulmonary veins are attached to the left side of the heart. **b.** The coronary arteries and cardiac veins pervade cardiac muscle. The coronary arteries bring oxygen and nutrients to cardiac cells, which derive no benefit from blood coursing through the heart. The cardiac veins drain blood into the right atrium.

Path of Blood Through the Heart

By referring to Figure 12.4, we can trace the path of blood through the heart in the following manner:

- The superior vena cava and the inferior vena cava, which carry O_2-poor blood that is relatively high in carbon dioxide, enter the right atrium.
- The right atrium sends blood through the tricuspid valve to the right ventricle.
- The right ventricle sends blood through the pulmonary semilunar valve into the pulmonary trunk and through the two **pulmonary arteries** to the lungs.
- Four **pulmonary veins,** which carry O_2-rich blood, enter the left atrium.
- The left atrium sends blood through the bicuspid (mitral) valve to the left ventricle.
- The left ventricle sends blood through the aortic semilunar valve into the aorta to the rest of the body.

From this description, it is obvious that O_2-poor blood never mixes with O_2-rich blood, and that blood must go through the lungs in order to pass from the right side to the left side of the heart. In fact, the heart is a double pump because the right ventricle sends blood into the lungs and the left ventricle sends blood into the rest of the body. Because the left ventricle has the harder job of pumping blood to the entire body, its walls are thicker than those of the right ventricle, which pumps blood a relatively short distance to the lungs. In a person with an average heart rate of 70 beats per minute, the output of the left ventricle is about 5.25 liters of blood per minute, which is about equal to the total amount of blood in the body.

The pumping of the heart sends blood out under pressure into the arteries. Because the left side of the heart is the stronger pump, blood pressure is greatest in the aorta. Blood pressure then decreases as the cross-sectional area of arteries and then arterioles increases. The **pulse** is a wave effect that passes down the walls of arteries when the aorta expands and then recoils with each ventricular contraction. The arterial pulse can be used to determine the heart rate, and a weak or "thready" pulse may indicate a weak heart or low blood pressure.

CHECK YOUR PROGRESS

1. Name each blood vessel and heart chamber that blood passes through on its journey through the heart and lungs.
2. If the left ventricle was not able to pump blood properly, what effect might this have on the lungs?

New Check Your Progress

Questions are provided at the end of chapter sections to help students assess their understanding of the key concepts.

232 Part III Maintenance of the Human Body

SUMMARIZING THE CONCEPTS

12.1 The Blood Vessels

There are three types of blood vessels:

- Arteries (and arterioles) take blood away from the heart.
- Capillaries, where exchange of substances with the tissues occurs.
- Veins (and venules) take blood to the heart.

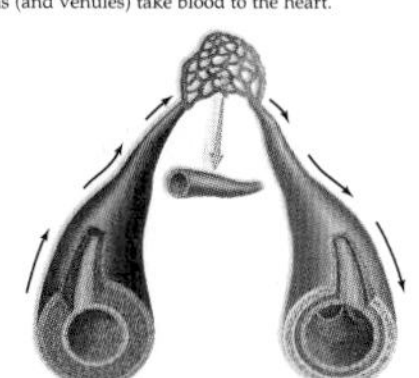

12.2 The Human Heart

The heart has a right and left side and four chambers:

- On the right side, the venae cavae deliver O_2-poor blood from the body into the right atrium, and the right ventricle pumps it via the pulmonary trunk into the pulmonary circuit.
- On the left side, the pulmonary veins bring O_2-rich blood from the lungs into the left atrium, and the left ventricle pumps it through the aorta into the systemic circuit.
- Blood from the atria passes through atrioventricular valves into the ventricles. Blood leaving the ventricles passes through semilunar valves.
- During the cardiac cycle, the SA node (cardiac pacemaker) initiates the heartbeat by causing the atria to contract. The AV node conveys the stimulus to the ventricles, causing them to contract.
- The heart sounds, "lub-dup," are due to the closing of the atrioventricular valves, followed by the closing of the semilunar valves.

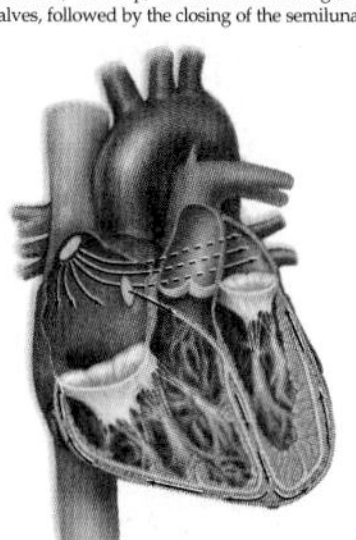

12.3 The Vascular Pathways

The cardiovascular system is divided into the pulmonary circuit and the systemic circuit:

- In the pulmonary circuit, the pulmonary trunk from the right ventricle of the heart and the two pulmonary arteries take O_2-poor blood to the lungs, and four pulmonary veins return O_2-rich blood to the left atrium of the heart.
- In the systemic circuit, O_2-rich blood is pumped from the left ventricle into the aorta, which branches off to form arteries going to specific organs. Eventually, arteries divide into arterioles and capillaries, and capillaries lead to venules, which join to form veins. The vein that carries O_2-poor blood to the vena cava most likely has the same name as the artery that

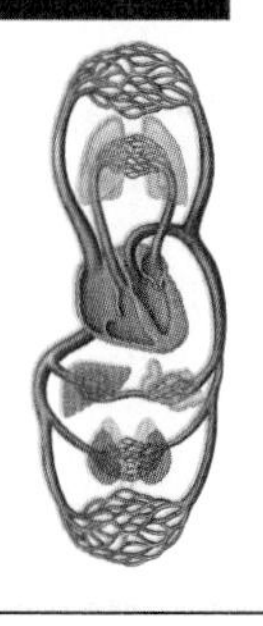

Chapter Summary

The summary is organized according to the major sections in the chapter and helps students review the important concepts and topics.

Chapter 12 Cardiovascular System 233

TESTING YOURSELF

Choose the best answer for each question.

1. ____ lie between ____ and ____.
   a. Arteries, veins, capillaries c. Veins, arteries, capillaries
   b. Capillaries, arteries, veins d. None of these are correct.
2. Gas (oxygen and carbon dioxide) exchange occurs across the ____ of the ____.
   a. veins, lungs c. arteries, tissues
   b. capillaries, tissues d. All of these are correct.
3. The myocardium is made of
   a. muscle.
   b. epithelium.
   c. connective tissue.
   d. None of these are correct
4. The average adult heart rate is about ____ beats per minute.
   a. 120 c. 45
   b. 100 d. 70
5. The "lub," or first heart sound, is produced by closing of the
   a. aortic semilunar valve.
   b. pulmonary semilunar valve.
   c. tricuspid valve.
   d. bicuspid valve.
   e. both AV valves.
6. Label the following ECG wave chart.

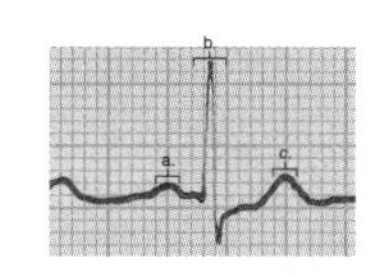

7. All arteries in the body contain O_2-rich blood with the ex[…] tion of the
   a. aorta.
   b. pulmonary artery.
   c. renal artery.
   d. coronary arteries.
8. Which of the following assists in the return of venous blo[…]

10. Label the following diagram of the systemic circuit.

11. What is the type of cell that becomes any of the formed elements in blood?
   a. multipotent stem cells c. lymphoid stem cells
   b. myeloid stem cells d. All of these are correct.
12. Which association is incorrect?
   a. white blood cells—infection fighting
   b. red blood cells—blood clotting
   c. red blood cells—hemoglobin
   d. platelets—blood clotting
13. A decrease in lymphocytes would result in problems associated with

234 Part III Maintenance of the Human Body

UNDERSTANDING THE TERMS

agranular leukocyte 224, albumin 222, anemia 224, aneurysm 228, angina pectoris 228, angioplasty 229, aorta 220, arteriole 214, artery 214, atherosclerosis 228, atrioventricular bundle 218, atrioventricular valve 216, atrium 216, AV (atrioventricular) node 218, basophil 224, blood pressure 221, capillary 214, cardiac cycle 218, cardiac pacemaker 218, chordae tendineae 216, clotting 225, coronary artery 220, dendritic cell 224, diastole 218, diastolic pressure 221, electrocardiogram (ECG) 219, embolus 228, endothelium 214, eosinophil 224, erythropoietin 224, fibrin 225, fibrinogen 225, formed element 222, granular leukocyte 224, heart 216, heart attack 228, hemoglobin 222, hepatic portal system 221, hepatic portal vein 221, hepatic vein 221, hypertension 231, inferior vena cava 220, lymph 227, lymphatic capillaries 227, lymphocyte 224, macrophage 224, megakaryocyte 225, monocyte 224, mononuclear cell 224, myocardium 216, neutrophil 224, pericardium 216, plaque 228, plasma 222, platelet 225, prothrombin 225, prothrombin activator 225, pulmonary artery 217, pulmonary circuit 220, pulmonary vein 217, pulse 217, Purkinje fibers 218, red blood cell (erythrocyte) 222, SA (sinoatrial) node 218, semilunar valve 216, serum 226, stem cell 226, stent 229, stroke 228, superior vena cava 220, systemic circuit 220, systole 218, systolic pressure 221, thrombin 225, thromboembolism 228, thrombus 228, tissue fluid 227, total artificial heart (TAH) 229, valve 215, varicose veins 215, vein 214, ventricle 216, venule 215, white blood cell (leukocyte) 224

THINKING CRITICALLY

1. A few specialized tissues do not contain any blood vessels, including capillaries. Can you think of two or three? How might these tissues survive without a direct blood supply?
2. Assume your heart rate is 70 beats per minute (bpm), and each minute your heart pumps 5.25 liters of blood to your body. Based on your age to the nearest day, about how many times has your heart beat so far, and what volume of blood has it pumped?
3. Examine the following abnormal ECG tracings. Compared with the normal tracing in Testing Yourself Question 6 or Figure 12.6*b*, what is your best guess as to what type of problem each patient may have? What type of device can be implanted in a patient's heart to correct some abnormalities of the heart, and how does it generally work?

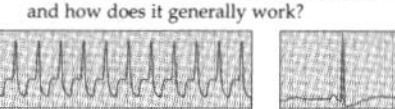

4. There are not enough living hearts available to meet the need for heart transplants. What are some of the major problems a manufacturer would have to overcome when attempting to design an artificial heart that will last for years inside a patient's body?

INQUIRY INTO LIFE WEBSITE

The companion website for *Inquiry into Life* provides a wealth of information organized and integrated by chapter. You will find practice tests, animations, videos, and much more that will complement your learning and understanding of general biology.

http://www.mhhe.com/maderinquiry13

Quiz Questions

Challenging quiz questions close each chapter. In addition, a page-referenced ***Understanding the Terms*** section reinforces the scientific vocabulary used in the chapter, and critical-thinking questions encourage students to apply what they have learned.

Inquiry into Life Website

www.mhhe.com/maderinquiry13

Students can access additional practice tests, animations, flashcards, and much more to complement their learning.

What's *New* in the Thirteenth Edition?

The thirteenth edition of *Inquiry into Life* continues the tradition of being time-tested and current. In this new edition, you will find:

- Extensive revisions to the Genetics, Evolution and Diversity, and Behavior and Ecology chapters
- New pedagogical tools that help students assess their understanding and make their learning relevant
- Integrated media package that includes study and assessment tools to directly support the learning outcomes in the book

Content Changes

Genetics—In light of the rapidly changing world of genetics, genomics, and biotechnology, and the impact these changes are having in our world, the entire genetics section has been reorganized and rewritten. These chapters build a foundation for recent discoveries in the field and engage students with topics currently in the news.

Evolution and Diversity—Based on up-to-date molecular data, chapters have been updated to incorporate new taxonomic relationships and phylogenetic reconstructions of organisms.

Behavior and Ecology—The modern evolutionary-ecological approach to the study of behavior, and more up-to-date discussions of biodiversity value and conservation have been included in this new edition.

See pages xviii and xix for a detailed list of content changes made to this new edition of *Inquiry into Life*.

New Pedagogical Tools

Learning Outcomes outline expectations.

Check Your Progress questions assess understanding.

Practice and test questions in McGraw-Hill Connect™ are aligned with these learning outcomes to help instructors measure student comprehension.

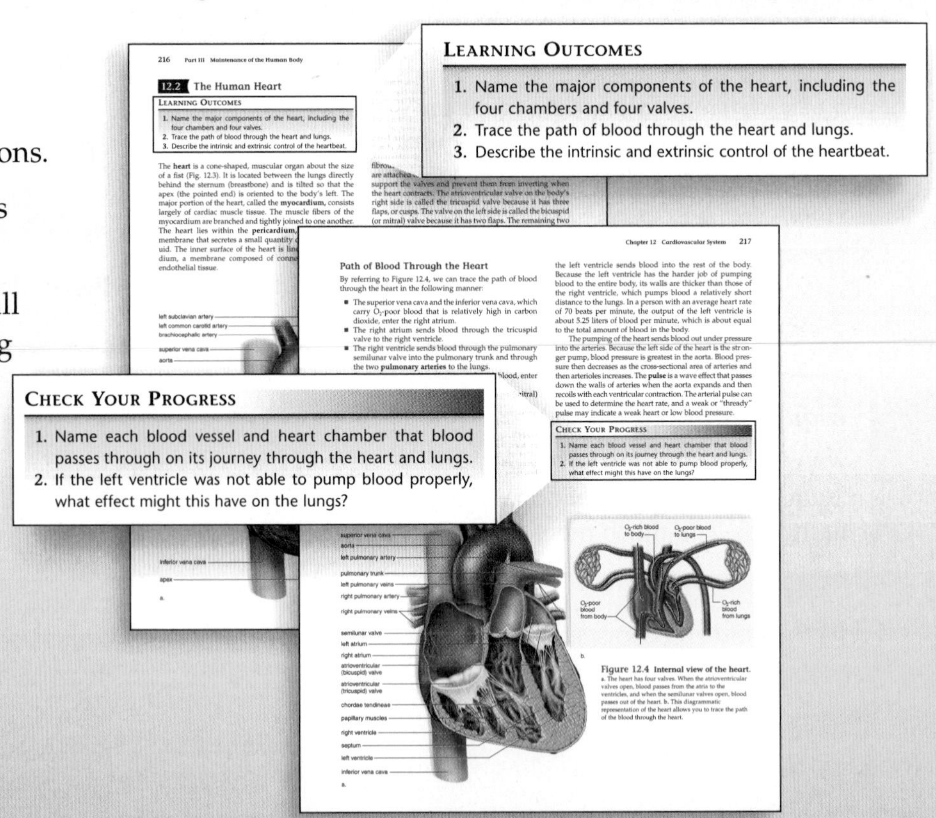

Integrated Media Package

McGraw-Hill Connect™ Biology offers the following in one, easy-to-use interface.

- **Practice Bank Questions with Rich Interactive Questions**
  - **Coverage of Key Concepts** Book-specific questions covering the key concepts have been created using four new, engaging tools: labeling, sequencing, classification, and composition, in addition to standard multiple choice and true/false questions.
  - **Powerful Lecture Enhancement** Instructors can use the zoom feature in the labeling and sequencing questions to pinpoint an exact area of the figure during lecture.
  - **Unmatched Flexibility** Nearly every aspect of the interactive questions can be customized—labels, hints, feedback, and more.
- **Test Bank Questions**
  - All practice and test bank questions are tagged to the textbook by chapter, figure, section, topic, and Bloom's level of difficulty, and aligned with the learning outcomes.
- **Animations with Quizzing**

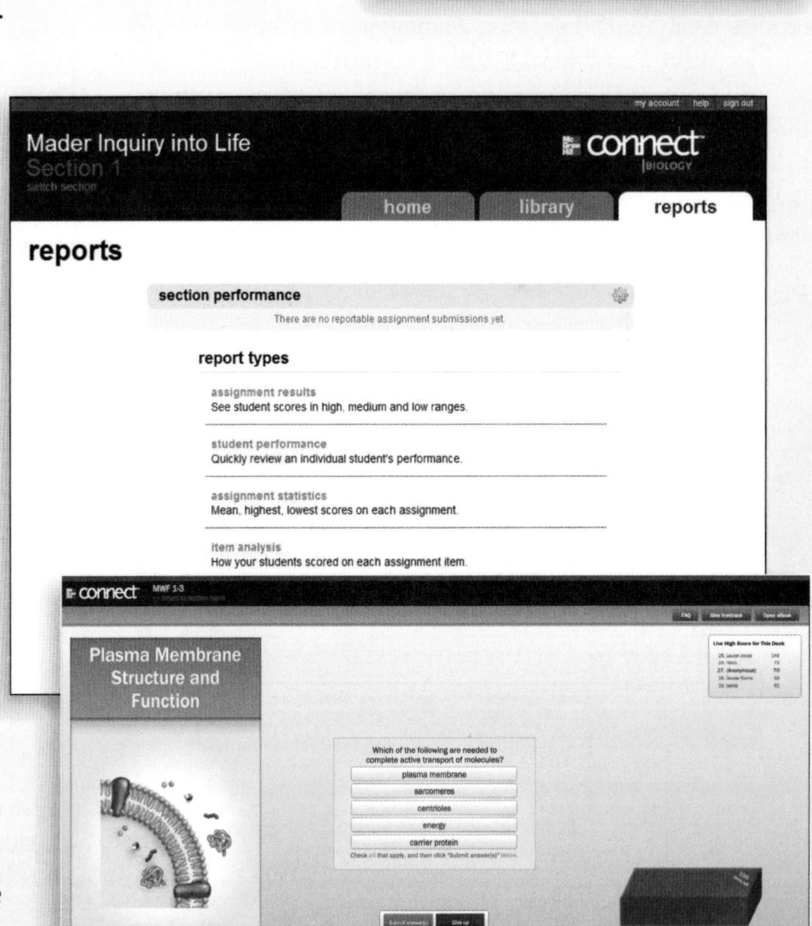

- **Impressive Reporting Solutions** With Connect's detailed reporting, instructors can quickly assess how students are doing in regards to overall class performance, individual assignments, and each question.
- **LearnSmart™** is a diagnostic learning tool
- **Tegrity Campus™** records and distributes lectures
- **Lecture PowerPoints with Embedded Animations**
- **FlexArt and Jpeg Files of Art, Photos, and Tables from the Textbook**
- **Videos Correlated by Textbook Chapters**
- **McGraw-Hill ConnectPlus™ Biology** provides students with all the advantages of Connect™ Biology, plus access to an **eBook** that includes animations and inline assessments.

See pages xvi and xvii for more information.

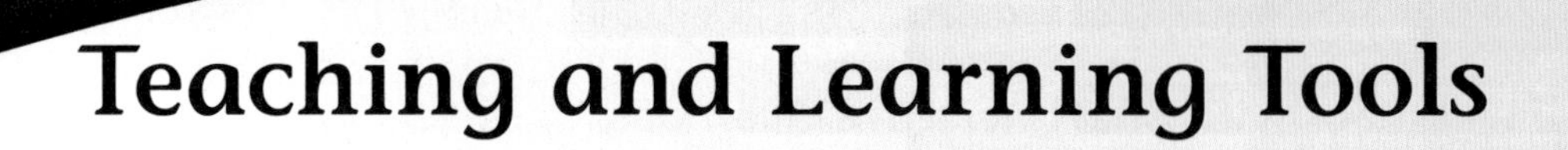

Teaching and Learning Tools

McGraw-Hill Connect™ Biology

www.mhhe.com/maderinquiry13

McGraw-Hill Connect™ Biology provides online presentation, assignment, and assessment solutions. It connects your students with the tools and resources they'll need to achieve success.

With Connect™ Biology you can deliver assignments, quizzes, and tests online. A robust set of questions and activities are presented and aligned with the textbook's learning outcomes. As an instructor, you can edit existing questions and author entirely new problems. Track individual student performance—by question, assignment, or in relation to the class overall—with detailed grade reports. Integrate grade reports easily with Learning Management Systems (LMS), such as WebCT and Blackboard. And much more.

ConnectPlus™ Biology provides students with all the advantages of Connect™ Biology, plus 24/7 online access to an eBook. This media-rich version of the book is available through the McGraw-Hill Connect™ platform and allows seamless integration of text, media, and assessments.

To learn more, visit

www.mcgrawhillconnect.com

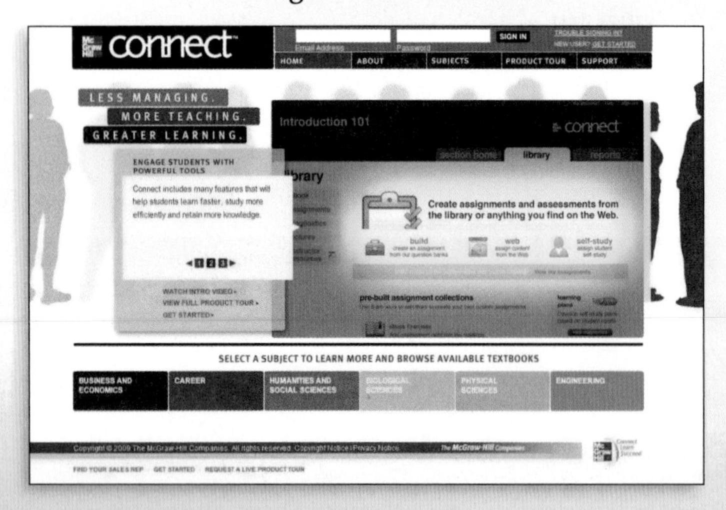

LearnSmart™

LearnSmart™ is available as an integrated feature of McGraw-Hill Connect™ Biology and provides students with a GPS (**G**uided **P**ath to **S**uccess) for your course. Using artificial intelligence, LearnSmart™ intelligently assesses a student's knowledge of course content through a series of adaptive questions. It pinpoints concepts the student does not understand and maps out a personalized study plan for success. This innovative study tool also has features that allow instructors to see exactly what students have accomplished, and a built-in assessment tool for graded assignments.

Visit the site below for a demonstration.

www.mhlearnsmart.com

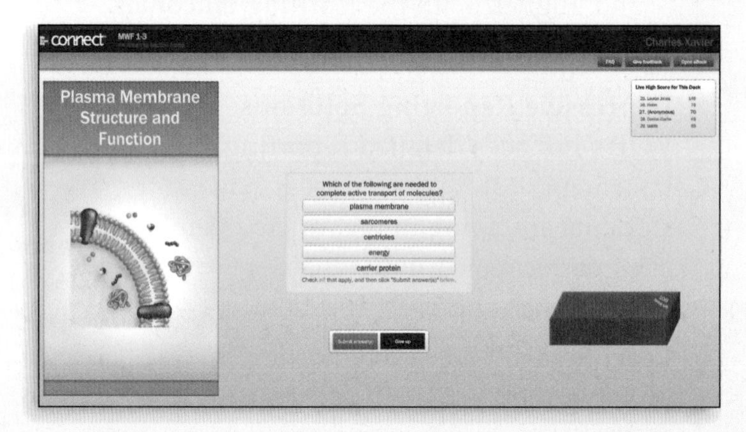

Computerized Test Bank

A comprehensive bank of test questions is provided within a computerized test bank powered by McGraw-Hill's flexible electronic testing program EZ Test Online. EZ Test Online allows you to create paper and online tests or quizzes in this easy to use program! A new tagging scheme allows you to sort questions by Bloom's difficulty level, topic, and section. Imagine being able to create and access your test or quiz anywhere, at any time, without installing the testing software. Now, with EZ Test Online, instructors can select questions from multiple McGraw-Hill test banks or author their own, and then either print the test for paper distribution or give it online.

Presentation Tools

Everything you need for outstanding presentations in one place!

www.mhhe.com/maderinquiry13.

- *FlexArt Image PowerPoints*—including every piece of art that has been sized and cropped specifically for superior presentations, as well as labels that can be edited and flexible art that can be picked up and moved on key figures. Also included are tables, photographs, and unlabeled art pieces.
- *Lecture PowerPoints with Animations*—animations illustrating important processes are embedded in the lecture material.
- *Animation PowerPoints*—animations only are provided in PowerPoint.
- *Labeled Jpeg Images*—Full-color digital files of all illustrations that can be readily incorporated into presentations, exams, or custom-made classroom materials.
- *Base Art Image Files*—unlabeled digital files of all illustrations.

Chapter	Enhanced Image PPTs (includes photos, and editable art)	Lecture PPTs with Animations	Animation PowerPoints	Labeled Jpeg Images	Base Art Image Files (.jpgs, no labels or leader lines)
All Chapters	Enhanced Image PPTs (707,634 KB)	Lecture Animation PPTs (649,609 KB)	Animation PPTs (1,64,060 KB)	Labeled Images (859,337 KB)	Base Images (599,793 KB)
Ch01	Ch. 1 Enhanced Image PPTs (23,977 KB)	Ch. 1 Lecture Animation PPTs (14,860 KB)	There are no Animation PPTs correlated to this chapter.	Ch. 1 Labeled Images (28,669 KB)	Ch. 1 Base Images (21,333 KB)
Ch02	Ch. 2 Enhanced Image PPTs (9,907 KB)	Ch. 2 Lecture Animation PPTs (6,012 KB)	Ch. 2 Animation PPTs (6,794 KB)	Ch. 2 Labeled Images (15,054 KB)	Ch. 2 Base Images (8,805 KB)
Ch03	Ch. 3 Enhanced Image PPTs (16,605 KB)	Ch. 3 Lecture Animation PPTs (14,220 KB)	Ch. 3 Animation PPTs (1,833 KB)	Ch. 3 Labeled Images (18,346 KB)	Ch. 3 Base Images (14,607 KB)

Presentation Center

In addition to the images from your book, this online digital library contains photos, artwork, animations, and other media from an array of McGraw-Hill textbooks that can be used to create customized lectures, visually enhanced tests and quizzes, compelling course websites, or attractive printed support materials.

My Lectures—Tegrity

Tegrity Campus™ records and distributes your class lecture, with just a click of a button. Students can view anytime/anywhere via computer, iPod, or mobile device. It indexes as it records your PowerPoint presentations and anything shown on your computer so students can use keywords to find exactly what they want to study. Tegrity is available as an integrated feature of McGraw-Hill Connect™ Biology or as standalone.

Instructor's Manual

The instructor's manual contains chapter outlines, lecture enrichment ideas, and critical thinking questions.

Laboratory Manual

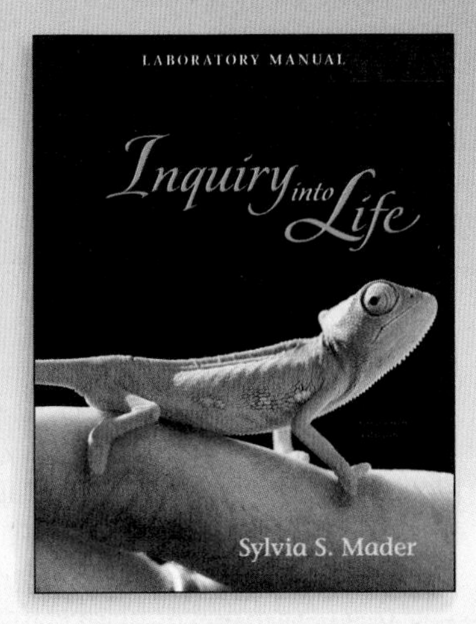

The *Inquiry into Life Laboratory Manual* is written by Dr. Sylvia Mader. With few exceptions, each chapter in the text has an accompanying laboratory exercise in the manual. Every laboratory has been written to help students learn the fundamental concepts of biology and the specific content of the chapter to which the lab relates, as well as gain a better understanding of the scientific method.

Companion Website

www.mhhe.com/maderinquiry13

The Mader: *Inquiry into Life* companion website allows students to access a variety of free digital learning tools that include:

- Chapter-level quizzing with pretest and posttest
- Bio Tutorial animations
- Vocabulary flashcards
- Virtual Labs

Biology Prep, also available on the companion website, helps students to prepare for their upcoming coursework in biology. This website enables students to perform self assessments, conduct self-study sessions with tutorials, and perform a post-assessment of their knowledge in the following areas:

- Introductory Biology Skills
- Basic Math Review I and II
- Chemistry
- Metric System
- Lab Reports and Referencing

McGraw-Hill: Biology Digitized Video Clips

ISBN (13) 978-0-312155-0
ISBN (10) 0-07-312155-X

McGraw-Hill is pleased to offer an outstanding presentation tool to text adopting instructors—digitized biology video clips on DVD! Licensed from some of the highest-quality science video producers in the world, these brief segments range from about five seconds to just under three minutes in length and cover all areas of general biology from cells to ecosystems. Engaging and informative, McGraw-Hill's digitized videos will help capture students' interest while illustrating key biological concepts and processes such as mitosis, how cilia and flagella work, and how some plants have evolved into carnivores.

Detailed List of Content Changes in *Inquiry into Life,* Thirteenth Edition

Part I Cell Biology

Every chapter-opening story and several Focus readings have been revised to be more current and relevant to students.

- Chapter 4, *Membrane Structure and Function,* now includes a discussion on gating of channel proteins and a new Science Focus on how cells talk to each other.
- Chapter 5, *Cell Division,* includes information on the control of the cell cycle and cancer, and proto-oncogenes and tumor suppressor genes. The discussion of cell division in prokaryotes has been moved from this chapter to the microbiology chapter (28).
- Chapter 6, *Metabolism: Energy and Enzymes,* has been revised to discuss cellular respiration and metabolism as it relates to humans and their diet.

Part II Plant Biology

- Chapter 9, *Plant Organization and Function,* has been significantly revised to include new photos, new terminology, and new organization. Plant organs (roots, stems, and leaves) as well as the different types of plant tissues that make up these organs are discussed in more detail. Students are shown the journey that cotton takes from being grown in the field to being sold as T-shirts. Plus a new Science Focus examines how the science of tree rings applies to historical buildings.

Part III Maintenance of the Human Body

- Chapter 11, *Human Organization,* introduces the issue of the risks and benefits of exposure of human skin to UV light through a new Health Focus reading, and a new Bioethical Focus probes the issue of testing hair for drugs. Updates for currency include new information on warts caused by human papillomavirus and some recently approved treatments for acne.
- Chapter 12, *Cardiovascular System,* has been revised to include information on automatic external defibrillators, anemias and their common causes, heart valve disease and its treatment, recent problems with stents, and a new image of an actual heart from a patient that has undergone coronary bypass surgery. The Health Focus on Prevention of Cardiovascular Disease has been updated to include recent information on omega-3 fatty acids, as well as the effect of alcohol on the heart.
- Chapter 13, *Lymphatic and Immune Systems,* includes a new chapter-opening vignette describing the case of a 3-year-old boy with X-linked agammaglobulinemia, to introduce the importance of the immune system. Updates for timeliness include a comparison of some specific types of vaccines, such as conjugate vaccines used to immunize children against *Haemophilus influenzae* type B. A new monoclonal antibody approved for treatment of allergies is discussed, and a new image shows the appearance of a positive TB skin test.
- Chapter 14, *Digestive System and Nutrition,* begins with the scenario of an obese man who had gastric bypass surgery to help students focus on both the anatomy and functions of the digestive system, as well as the problem of eating disorders. A new Health Focus reading about vegetarianism clarifies some recent information about how protein requirements can be met on a diet without meat. Other updates include an expanded discussion of heartburn and gastroesophageal reflux disease, as well as a new paragraph about Crohn's disease.
- Chapter 16, *Urinary System and Excretion,* begins with an opening vignette that illustrates the relatively common instance of a baby born with polycystic kidney disease, and thus the critical role of the kidneys in homeostasis is emphasized. Updates include an expanded discussion of diuretics to include their non-therapeutic use by bodybuilders and other athletes, and the section on hemodialysis has been clarified and expanded to include a brief discussion of the merits of home dialysis.

Part IV Integration and Control of the Human Body

- Chapter 17, *Nervous System,* includes a new Science Focus reading that provides some possible answers about the biological function(s) of sleep, and considers the possible functions of dreaming. Other updates include a new paragraph on how certain venoms, poisons, and toxins (like Botox) can affect the activity of neurotransmitters, and information on new treatments for Alzheimer and Parkinson disease.
- Chapter 18, *Senses,* discusses new research studies on the use of gene therapy to treat a type of congenital blindness, as well as the development of an artificial retina.
- Chapter 19, *Musculoskeletal System,* includes a new Bioethical Focus that asks students to consider whether physical modifications, such as the prosthetic legs, constitute an unfair advantage in sports. Updates include information on the stages of fracture healing, herniated intervertebral disks, the effects of tetanus toxin on muscle contraction, the appropriateness of creatine phosphate as a dietary supplement for athletes, and the use of an electromyogram to measure muscle activity.

Part V Continuance of the Species

- Chapter 21, *Reproductive System,* opens with a new vignette that explores some of the issues raised by the case of Nadya Suleman, the "Octomom." The section on the male reproductive system has been updated to include some of the latest studies on the possible benefits of circumcision. The Control of Reproduction section has been updated to include information on the vaginal contraceptive ring. The information on sexually transmitted diseases has been expanded to consider cultural factors that may play a role in HIV transmission, new information about antiretroviral drugs, a recent study on the link between human papillomavirus (HPV) and oral cancer, as well as the controversy over vaccinating teenage girls against HPV.

In light of the rapidly changing world of genetics, genomics, and biotechnology, and the impact these changes are having in our world, the entire genetics section has been reorganized and rewritten. These chapters build a foundation for recent discoveries in the field and engage the students with topics currently in the news.

- Chapter 23, *Patterns of Gene Inheritance,* focuses on inheritance at the gene level. It begins with Mendel's laws and then moves to human pedigree analysis and genetic disorders, including both autosomal dominant and recessive genetic disorders. This increases student understanding by linking inheritance concepts directly with their application to human diseases. More complex inheritance patterns and environmental influences on gene expression are included as well as a new Science Focus reading on the genetics of taste. The artwork has also been updated.
- Chapter 24, *Chromosomal Basis of Inheritance,* focuses on human inheritance at the level of the chromosome. It begins with a discussion of sex-linked genes with the corresponding pedigree analysis and genetic disorders. It also includes a discussion of how linked genes on chromosomes are inherited and common chromosome number and structure disorders.
- Chapter 25, *DNA Structure and Control of Gene Expression,* presents an overall view of how DNA controls phenotype. It lays a foundation with DNA structure and replication, including the contribution of Rosalind Franklin to the elucidation of the double helical structure of DNA. It then covers RNA structure and function, transcription, and translation. The chapter includes an updated section on how gene expression is regulated in both prokaryotes and eukaryotes. The chapter concludes with a section on gene mutations and their consequences. This includes a discussion of cancer as a failure in regulation of gene expression with a new section on the role of signaling pathways, proto-oncogenes, and tumor suppressor genes.
- Chapter 26, *Biotechnology and Genomics,* incorporates some of the latest advancements in genetically modified organisms, genomic sequencing, comparative genomics, and functional genomics. Examples include a more detailed description of PCR that is accessible to an introductory biology student, new figures to describe DNA fingerprinting, a description of use of microarrays for genetic disease testing, and a brief discussion of bioinformatics and the use of computers to analyze genetic/genomic data.

Part VI Evolution and Diversity

- Chapter 27, *Evolution of Life,* begins with a new vignette that discusses MRSA (methicillin-resistant staph aureus). This vignette brings relevancy to biology as the important issue of evolution of antibiotic resistance is discussed. In addition, new taxonomic relationships have been incorporated, and several figures describing these relationships have been added or revised.
- Chapter 28, *Microbiology,* has been significantly revised to reflect new developments. Timely discussions of recent infectious disease outbreaks have been included. A new Health Focus reading compares the development of antibiotics and probiotics, reflecting our increasing awareness of how microbes can be harmful, but also beneficial to human health. The section on different metabolic groups of archaea has also been expanded, to emphasize the uniqueness of this domain of life. The sections on protists and fungi have been reorganized with more focus on diversity and two new figures have been added to emphasize shared characteristics. New information has been added to recognize various efforts to control mosquito populations in countries where malaria is present, as well as to update students on the recent progress towards a malaria vaccine. In the section on fungi, the chytrids are now briefly described as a major group, and additional information has been added about poisonous and hallucinogenic mushrooms, as well as the possible role of Stachybotrys in "sick building syndrome." The section of viruses has also been updated, especially with regard to recent influenza outbreaks and vaccinations for chickenpox and shingles.
- Chapters 29 and 30, *Plants* and *Animals: Part I,* have been updated to incorporate new taxonomic relationships and phylogenetic reconstructions of plants (Ch. 29) and animals (Ch. 30) based on the most up-to-date molecular data. Several new figures describing these relationships have been incorporated.
- Chapter 31, *Animals: Part II,* has been updated to incorporate new taxonomic relationships and phylogenetic reconstructions of vertebrates based on the most up-to-date molecular data. The section on human evolution has been updated with the latest information, including new ideas about the original evolution of hominids, the discovery of new fossils, and a discussion of *Homo floresiensis.* It also brings in new ideas about the distribution of modern humans, and contrasting ideas about how we originally colonized the planet. New figures describing the evolutionary relationships of hominids have been incorporated.

Part VII Behavior and Ecology

- Chapter 32, *Behavioral Ecology,* has undergone substantial revisions. While it begins with a classical approach to studying animal behavior, the later sections discuss the modern evolutionary-ecological approach to studying behavior and incorporate modern ideas on mating systems, mate choice, optimal foraging, and altruism. New examples of studies will help instructors better convey the concepts to students, with new figures of studies that describe these concepts.
- Chapter 36, *Conservation Biology,* has been significantly revised and incorporates sections on nature reserve design, a new section on sustainability, and more up-to-date discussions of biodiversity value and conservation. New information has been incorporated on reasons for biodiversity loss, including emerging infectious diseases. Several new figures have been included to demonstrate concepts of nature reserve design, how loss of particular species can affect others in the food chain, new data on global warming, how habitat conservation designed for protecting one species can help many others, and new advancements in renewable resources.

Acknowledgments

It would be impossible for a single individual to publish a book of the quality of *Inquiry into Life*. I want to thank Jeff Isaacson, Kimberly Lyle-Ippolito, and Andrew Storfer. We met at the onset of the project to discuss my overall vision for *Inquiry into Life*, and together we formulated a plan for developing the thirteenth edition. Their unique and specialized backgrounds, in addition to their current experiences in the classroom, have been extremely helpful in shaping this edition.

It has been a pleasure to work with McGraw-Hill publishing: Michael Hackett, my editor; Tamara Maury, my marketing manager; Rose Koos, my developmental editor; and Jayne Klein, my project manager.

The design of the book was managed by Laurie Janssen, and the illustration program was carried to completion by the artists at Electronic Publishing Services. Evelyn Jo Johnson and Lori Hancock did a superb job of securing quality photographs.

Textbook writing is very time consuming, and I have always appreciated the continued patience and encouragement of my family. By now my children are grown, but they never fail to keep up with how I am progressing with my work. My husband, Arthur Cohen, himself a biology teacher, gives me considerable support each day and discusses with me the details of my work. His input is invaluable to me.

360° Development Process

McGraw-Hill's 360° Development Process is an ongoing, never-ending, education-oriented approach to building accurate and innovative print and digital products. It is dedicated to continual large-scale and incremental improvement, driven by multiple user feedback loops and checkpoints. This is initiated during the early planning stages of our new products, intensifies during the development and production stages, then begins again after publication in anticipation of the next edition.

This process is designed to provide a broad, comprehensive spectrum of feedback for refinement and innovation of our learning tools, for both student and instructor. The 360° Development Process includes market research, content reviews, course- and product-specific symposia, accuracy checks, and art reviews. We appreciate the expertise of the many individuals involved in this process.

Textbook Reviewers

Andrew Baldwin, *Mesa Community College*
David A. Battigelli, *Ivy Tech Community College*
Donald R. Baud, *University of Memphis*
Thomas Buettner, *Southern Illinois University*
Emily B. Carlisle, *Pearl River Community College*
Sara Carlson, *University of Akron*
Francesca Catalano, *American Public University System*
Sandra Devenny, *Delaware County Community College*
Donald Dorfman, *Monmouth University*
Michael J. Farabee, *Estrella Mountain Community College*
Tiara J. Harms, *Pensacola Junior College*
Reba Harrell, *Hinds Community College*
Daniece Harris-Williams, *Hinds Community College, Rankin Campus*
Jason Jennings, *Southwest Tennessee Community College*
Gail I. Jones, *Texas Christian University*
Diann Jordon, *Alabama Sate University*
Mary Jane Keleher, *Salt Lake Community College*
Cynthia J. Kincer, *Wytheville Community College*
Todd A. Kostman, *University of Wisconsin Oshkosh*
Jerome Krueger, *South Dakota State University*
Todd C. Martin, *Metropolitan Community College, Blue River*
Mary Delores McCright, *Texarkana College*
Melinda Neal, *Cowley County Community College*
Meredith Somerville Norris, *University of North Carolina, Charlotte*
Ashley Ramer, *University of Akron*
David A. Resnik, *Johnson & Wales University*
Jean Revie, *South Mountain Community College*
Darryl Ritter, *Okaloosa–Walton College*
Stephanie Songer, *North Georgia College and State University*
L. D. Spears, *John A. Logan Community College*
William Sproat, *Walters State Community College*
Martha W. Stratton, *Tennessee State University*
Mike Taylor, *Santiago Canyon College*
Mohammed Rafique Uddin, *LeMoyne–Owen College*
Pete van Dyke, *Walla Walla Community College*
Kathy Webb, *Bucks County Community College*
Hilda Wells, *Hinds Community College*
Cindy White, *University of Northern Colorado*
Lura C. Williamson, *University of New Orleans*
James Wise, *Hampton University*
Ted Zerucha, *Appalachian State University*

Ancillary Authors

Test Bank: Joy Brookshire, *Kennesaw State University;* Kimberly Lyle-Ippolito, *Anderson University;* Stephanie Songer, *North Georgia College and State University*

Connect™ Content: Raymond Burton, *Germanna Community College*

Website Quizzes: Dan Matusiak, *St. Charles Community College*

Lecture and Image PowerPoints: Brenda Leady, *University of Toledo*

Media Asset Correlations: Jennifer Burtwistle, *Northeast Community College*

Instructor's Manual: Kimberly Lyle-Ippolito, *Anderson University*

LearnSmart™

Authors: Patrick Galliart, *North Iowa Area Community College;* Janet Forrest Kent, *Valdosta State University;* Karen Meisch, *Austin Peay State University*

Reviewers: Raymond Burton, *Germanna Community College;* Murad Odeh, *South Texas College*

The Study of Life

Applying the Concepts

Biodiversity—the diversity of life on Earth is astonishing. More than 1.75 million species have been described, and scientists estimate that as many as 5–30 million species may exist. Yet all living things—from bacteria to trees to dogs to mosquitoes to birds to humans—share certain common characteristics. Why? Living things are similar because all forms of life can be traced back to an ancient common ancestor, the very first living thing.

Because of this common ancestry, humans and plants, although quite different in appearance, share a common genetic code and basic life processes. Their many obvious differences, on the other hand, have been shaped by millions of years of evolution.

In this chapter, you will learn how basic characteristics of life can account for the similarities and differences among the world's species and how this biodiversity is organized. You will also learn how the Earth's biosphere that all species inhabit is organized. Currently, the Earth's biodiversity is being threatened by human activities, and as many as half the world's species may become extinct in the next 100 years. Why should we care? Living things depend on one another in complex ways, and thus the loss of one species can lead to the loss of other species that depend on it, including those species that directly affect humans. We will revisit the importance of biodiversity at the end of the chapter.

Chapter Outline

Figure 1.1 Living things on planet Earth.

If aliens ever visit our corner of the universe, they will be amazed at the diversity of life on our planet. They will also note that it is the only known place where oxygen is present in large quantities in the atmosphere. This is one of the criteria for the diversity of modern life as we know it.

1.1 The Characteristics of Life

Learning Outcomes

1. Describe the characteristics of living things.
2. Place the levels of biological organization in a hierarchy.

Life. Except for the most desolate and forbidding regions of the polar ice caps, planet Earth is teeming with life. Without life, our planet would be nothing but a barren rock hurtling through space. The variety of life on Earth is staggering, and human beings are a part of it. So are butterflies, trees, birds, mushrooms, and giraffes (Fig. 1.1). The variety of living things ranges in size from bacteria, much too small to be seen by the naked eye, all the way up to giant sequoia trees that can reach heights of 100 meters or more.

The diversity of life seems overwhelming, and yet all living things have certain characteristics in common. Taken together, these characteristics give us insight into the nature of life and help us distinguish living things from nonliving things. Living things generally (1) are organized, (2) acquire materials and energy, (3) reproduce, (4) respond to stimuli, (5) are homeostatic, (6) grow and develop, and (7) have the capacity to adapt to their environment. Let us consider each of these characteristics in order.

Living Things Are Organized

Living things can be organized in a hierarchy of levels (Fig. 1.2). In trees, humans, and all other organisms, atoms join together to form molecules, such as DNA molecules that occur within cells. A **cell** is the smallest unit of life, and some organisms are single-celled. In multicellular organisms, a cell is the smallest structural and functional unit. For example, a human nerve cell is responsible for conducting electrical impulses to other nerve cells. A **tissue** is a group of similar cells that perform a particular function. Nervous tissue is composed of millions of nerve cells that transmit signals to all parts of the body. Several tissues then join together to form an **organ.** The main organ that receives signals from nerves is the brain. Organs then work together to form an **organ system.** In the nervous system, the brain sends messages to the spinal cord, which in turn sends them to body parts through spinal nerves. Complex **organisms** such as trees and humans are a collection of organ systems. The human body contains a digestive system, a cardiovascular system, and several other systems in addition to the nervous system.

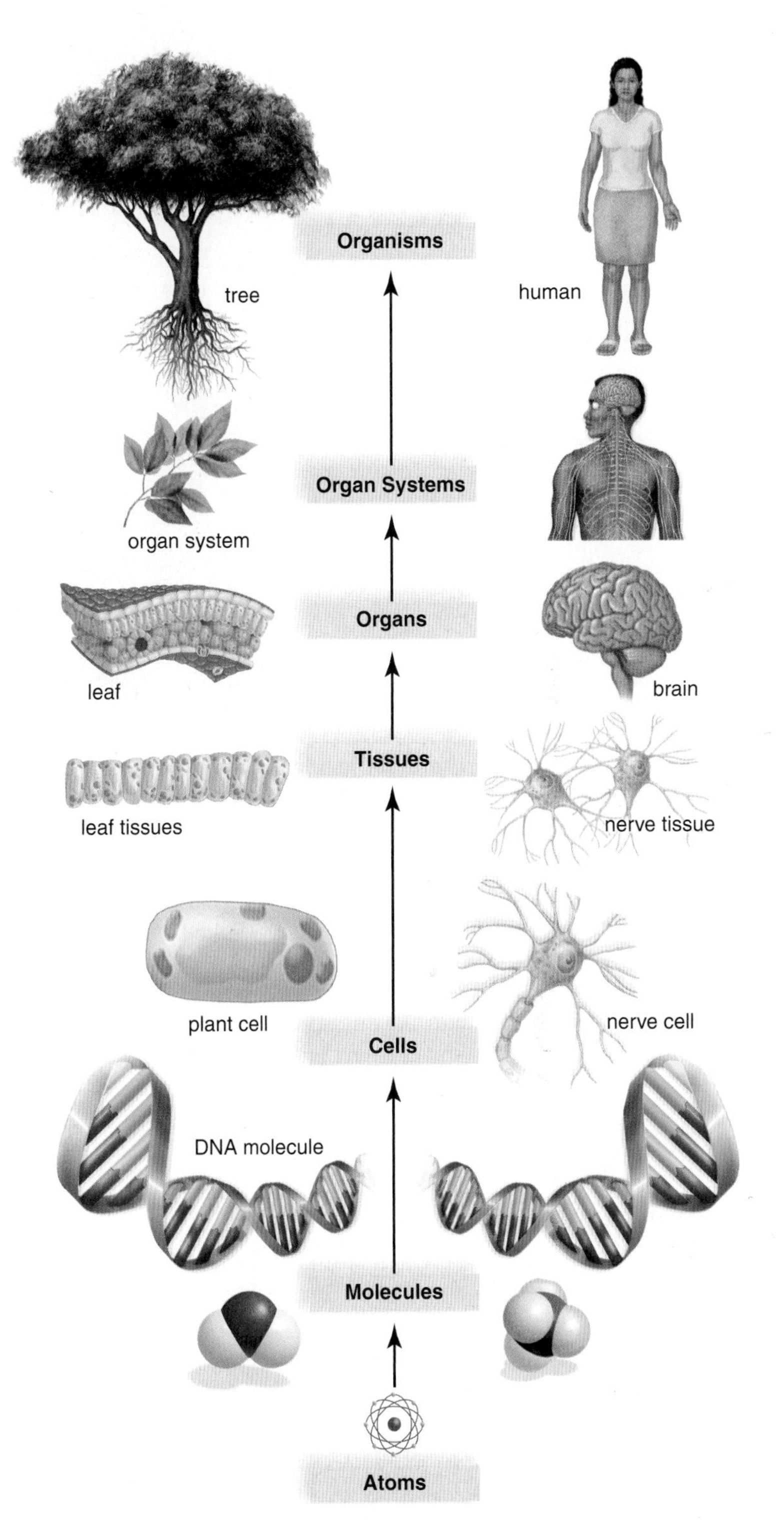

Figure 1.2 Levels of biological organization. At the lowest level of organization, atoms join to form molecules (including DNA) that are found in cells. Cells, the smallest units of life, differ in size, shape, and function. Tissues are groups of cells that work together to serve the same function. Organs, which are composed of different tissues, make up the organ systems of complex organisms. Each organ and organ system performs functions that keep the organism alive and well.

Living Things Acquire Materials and Energy

Living things need an outside source of materials and energy to maintain their organization and carry on life's other activities. Plants, such as trees, use carbon dioxide, water, and solar energy to make their own food. Human beings and other animals acquire materials and energy by eating food.

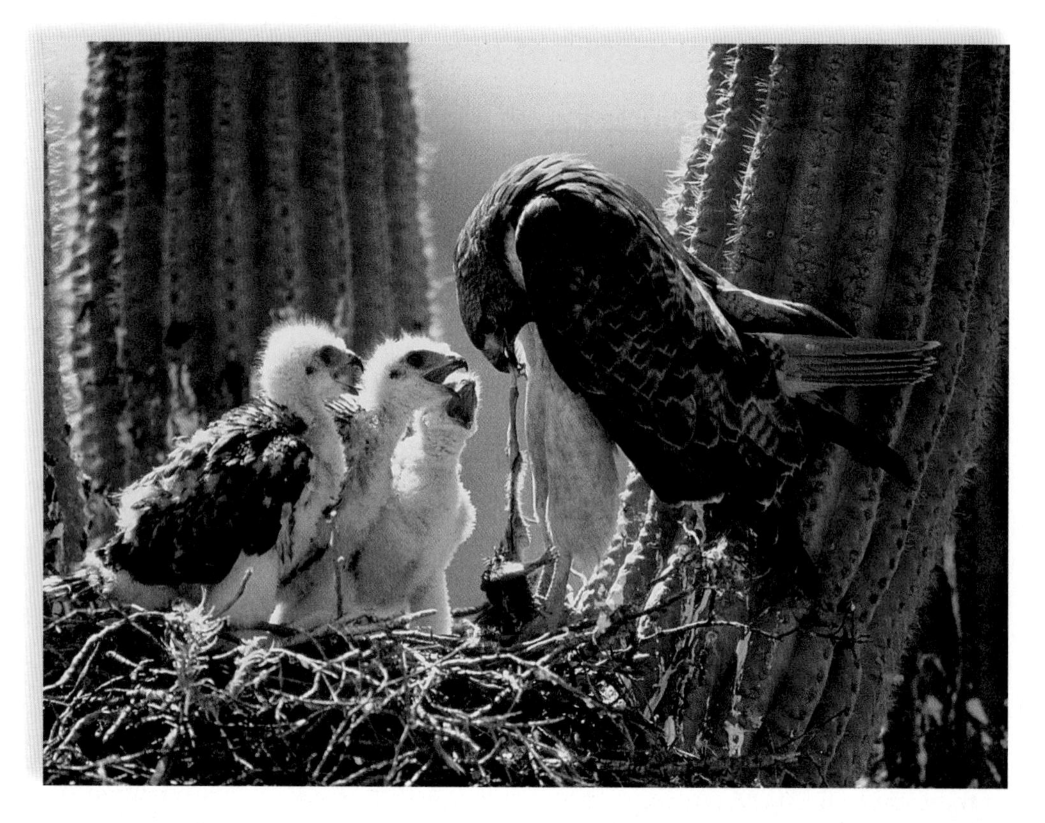

Figure 1.3 Living things acquire materials and energy, and they reproduce.
A red-tailed hawk has captured a rabbit, which it is feeding to its young.

Food provides nutrient molecules, which cells use as building blocks or for **energy**—the capacity to do work. Cells use energy from nutrient molecules to carry out everyday activities. Some nutrient molecules are broken down completely to provide the necessary energy to carry out synthetic reactions, such as building proteins. Thus, cells need energy to do work, and it takes work to maintain the organization of a cell as well as an organism. Above, a red-tailed hawk feeds its young to provide the energy they need (Fig. 1.3).

Most living things can convert energy into motion. Self-directed movement, as when we decide to stand from a chair, is even considered by some to be one of life's characteristics.

Living Things Reproduce

Life comes only from life. The information encoded in **genes** within each individual's DNA contains instructions that direct organismal **reproduction**—that is, to make more of itself. Before reproduction occurs, DNA is replicated so that exact copies of genes are produced. DNA contains the hereditary information that directs cellular functions and life's activities in general. Unicellular organisms reproduce asexually simply by dividing. The new cells have the same genes and structure as the single parent. Multicellular organisms most often reproduce sexually. Each parent, male and female, contributes roughly one-half the total number of genes to its offspring, which does not resemble either parent exactly.

Living Things Respond to Stimuli

Living things respond to external stimuli, often by moving toward or away from a stimulus, such as the smell of food. Movement in animals, including humans, is dependent upon their nervous and musculoskeletal systems. Other living things use a variety of mechanisms in order to move. The leaves of plants track the passage of the sun during the day, and when a houseplant is placed near a window, hormones help its stem bend to face the sun.

The movement of an organism, whether self-directed or in response to a stimulus, constitutes a large part of its **behavior.** Behavior is largely directed toward minimizing injury, acquiring food, and reproducing.

Living Things Are Homeostatic

Homeostasis means "staying the same." Actually, the internal environment of an organism stays *relatively* constant. For example, human body temperature fluctuates slightly throughout the day. Also, the body's ability to maintain a normal internal temperature is somewhat dependent on the external temperature—we will die if the external temperature becomes too hot or cold.

One of the major purposes of this text is to show how all the organ systems of the human body help maintain homeostasis. The digestive system provides nutrient molecules; the cardiovascular system transports them throughout the body; and the urinary system rids blood of wastes. The nervous and endocrine systems coordinate the activities of the other systems.

Living Things Grow and Develop

Growth, recognized by an increase in the size of an organism and often in the number of cells, is a part of development.

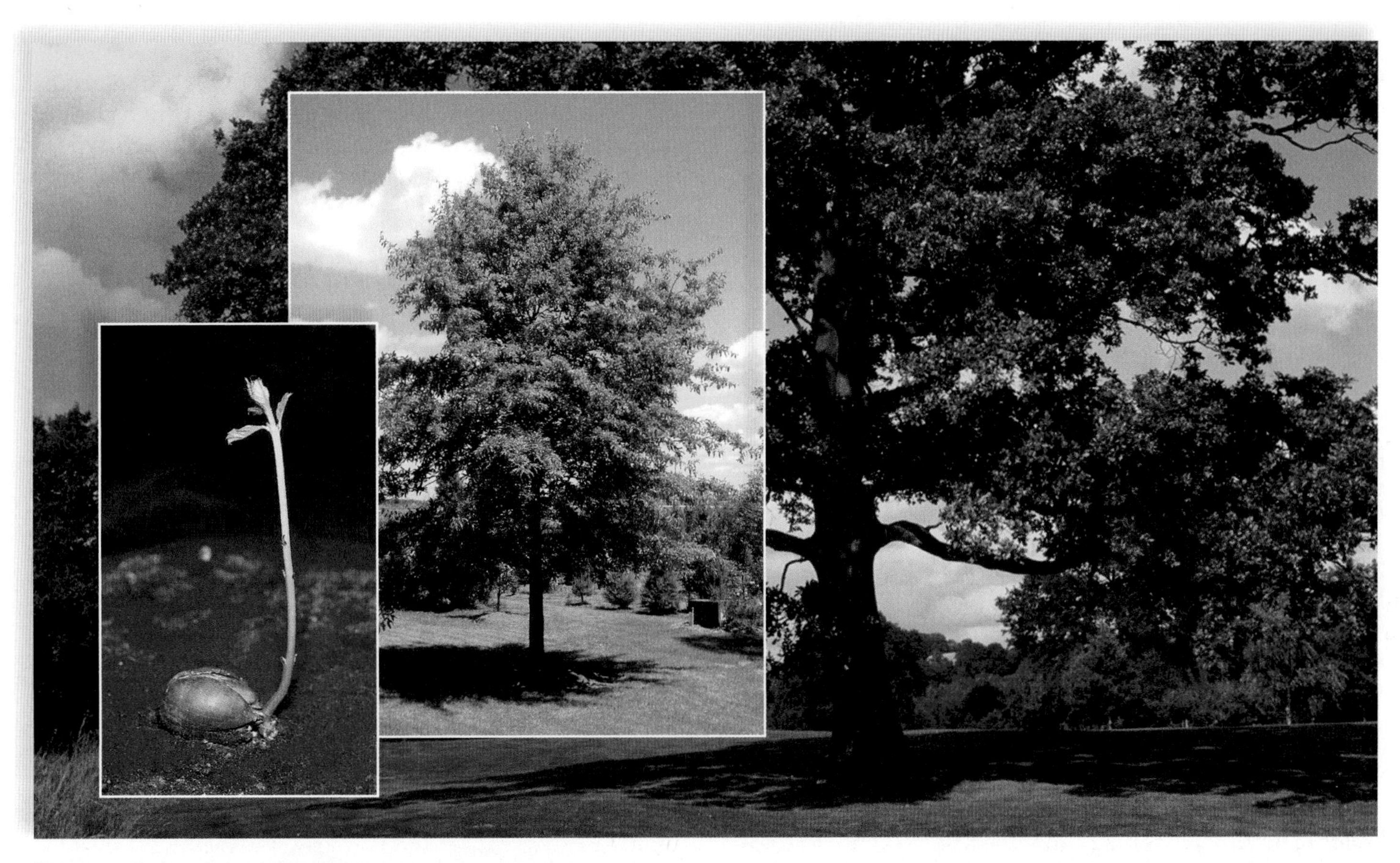

Figure 1.4 Living things grow and develop.
A small acorn gives rise to a very large oak tree through the process of growth and development.

All organisms undergo some form of development. Figure 1.4 illustrates that an acorn progresses to a seedling before it becomes an adult oak tree.

In humans, **development** includes all the changes that take place between conception and death. First, the fertilized egg develops into a newborn, and then a human goes through childhood, adolescence, adulthood, and aging. Development also includes repair that takes place following an injury.

Living Things Have the Capacity to Adapt

Throughout the nearly 4 billion years that life has been on Earth, the environment has constantly been changing. For example, glaciers that once covered much of the world's surface 10,000–15,000 years ago have since receded, and many areas that were once covered by ice are now habitable. On a smaller scale, a hurricane or fire could drastically change the landscape in an area quite rapidly.

As the environment changes, some individuals of a **species** (a group of organisms that can successfully interbreed and produce fertile offspring) may possess certain features that make them better suited to the new environment. We call such features **adaptations.** For example, consider the hawk in Figure 1.3, which catches and eats rabbits. A hawk, like other birds, can fly because it has hollow bones, which can be considered an adaptation. Similarly, its strong feet can take the shock of a landing after a hunting dive, and its sharp claws can grab and hold onto prey.

Individuals of a species that are better adapted to their environment tend to live longer and produce more offspring than other individuals. This differential reproductive success, called **natural selection,** results in changes in the characteristics of a population (all the members of a species within a particular area) through time. That is, adaptations that result in higher reproductive success will increase in frequency in a population across generations. This change in the frequency of traits in populations and species is called **evolution.**

Evolution explains both the unity and diversity of life. As stated at the beginning of this chapter, all organisms share the same basic characteristics of life because we all share a common ancestor—the first cell or cells—that arose nearly 4 billion years ago. During the past 4 billion years, the Earth's environment has changed drastically, and the diversity of life has been shaped by the evolutionary responses of organisms to these changes.

Check Your Progress

1. What are the characteristics that help define life?
2. Explain why organisms share these characteristics.

1.2 The Classification of Living Things

Learning Outcome

1. Describe how living things are classified.

Since life is so diverse, it is helpful to have a classification system to group organisms according to their similarities.

Systematics is the discipline of identifying and classifying organisms according to specific criteria. Each type of organism is placed in a **species, genus, family, order, class, phylum, kingdom,** and finally **domain.**

DOMAIN ARCHAEA

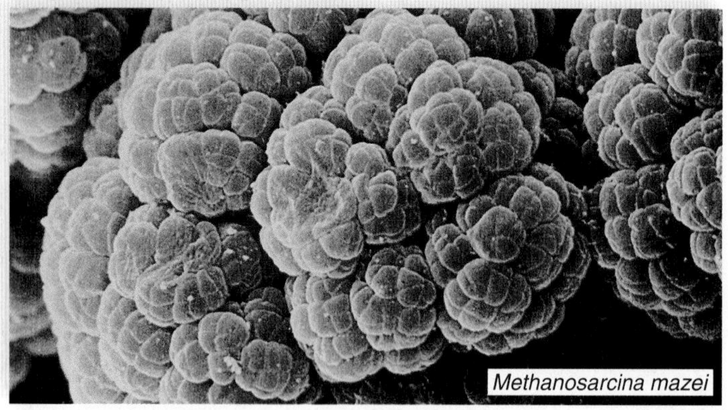

a. Archaea are capable of living in extreme environments.

DOMAIN BACTERIA

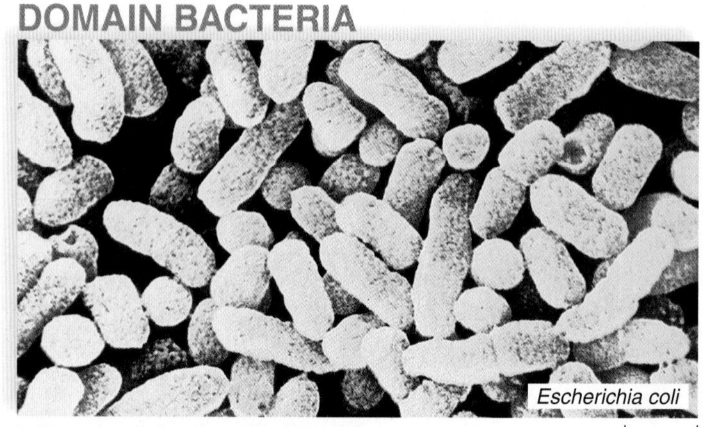

b. Bacteria are found nearly everywhere.

DOMAIN EUKARYA

Kingdom	Organization	Type of Nutrition	Representative Organisms	
Protista	Complex single cell, some multicellular	Absorb, photosynthesize, or ingest food	paramecium, euglenoid, slime mold, dinoflagellate	Protozoans, algae, water molds, and slime molds
Fungi	Some unicellular, most multicellular filamentous forms with specialized complex cells	Absorb food	black bread mold, yeast, mushroom, bracket fungus	Molds, yeasts, and mushrooms
Plantae	Multicellular form with specialized complex cells	Photosynthesize food	moss, fern, pine tree, nonwoody flowering plant	Mosses, ferns, nonwoody and woody flowering plants
Animalia	Multicellular form with specialized complex cells	Ingest food	sea star, earthworm, finch, raccoon	Invertebrates, fishes, reptiles, amphibians, birds, and mammals

c. Eukaryotes are divided into four kingdoms.

Figure 1.5 Pictorial representation of the three domains of life.
Archaea (**a**) and bacteria (**b**) are both prokaryotes but are so biochemically different that they are not believed to be closely related. **c.** Eukaryotes are biochemically similar but structurally dissimilar. Therefore, they have been categorized into four kingdoms. Many protists are unicellular, but the other three kingdoms are characterized by multicellular forms.

Domains

Domains are the largest classification category. Biochemical and genetic evidence tells us that there are three domains: **Archaea, Bacteria,** and **Eukarya** (Fig. 1.5). Both domain Archaea and domain Bacteria contain unicellular **prokaryotes,** which lack the membrane-bounded nucleus found in the cells of **eukaryotes** in domain Eukarya. The genes of eukaryotes are found in the nucleus, which thereby controls the cell.

Prokaryotes are structurally simple (Fig. 1.5*a, b*) but metabolically complex. The *Archaea* live in aquatic environments that lack oxygen or are too salty, too hot, or too acidic for most other organisms. Perhaps these environments are similar to those of the primitive Earth, and archaea represent the first cells that evolved. *Bacteria* are found almost anywhere—in the water, soil, and atmosphere, as well as on our skin and in our digestive tracts. Although some bacteria cause diseases, others are beneficial, both environmentally and commercially. For example, bacteria can be used or engineered to develop new medicines, to clean up oil spills, or to help purify water in sewage treatment plants.

Kingdoms

Systematists are in the process of deciding how to categorize archaea and bacteria into kingdoms. The *eukaryotes* are currently classified into at least four kingdoms with which you may be familiar (Fig. 1.5*c*). **Protists** (kingdom Protista) range from unicellular to a few multicellular organisms. Some can make their own food (photosynthesizers), while others must ingest their food. Because of the great diversity among protists, however, many biologists have split the Protista into several kingdoms. Among the **fungi** (kingdom Fungi) are the familiar molds and mushrooms that, along with bacteria, help decompose dead organisms. **Plants** (kingdom Plantae) are well known as multicellular photosynthesizers, while **animals** (kingdom Animalia) are multicellular and ingest their food.

Other Categories

The other classification categories are phylum, class, order, family, genus, and species. Each classification category is more specific than the one preceding it. For example, the species within one genus share very similar characteristics, while those within the same kingdom share only general characteristics. Modern humans are the only living species in the genus *Homo,* but many different types of animals are in the animal kingdom (Table 1.1). To take another example, all species in the genus *Pisum* (pea plants) look quite similar, while the species in the plant kingdom can be quite different, as is evident when we compare grasses to trees.

Systematics helps biologists make sense out of the bewildering variety of life on Earth because organisms are classified according to their presumed evolutionary relationships. Organisms placed in the same genus are the most closely related, and those in separate domains are the most distantly related. Therefore, all eukaryotes are more closely related to one another than they are to bacteria or archaea. Similarly, all animals are more closely related to one another than they are to plants. As more is learned about evolutionary relationships among species, systematic relationships are changed. Systematists are even now making observations and performing experiments that will soon result in changes in the classification system adopted by this text.

TABLE 1.1 Classification of Humans

Classification Category	Characteristics
Domain Eukarya	Cells with nuclei
Kingdom Animalia	Multicellular, motile, ingestion of food
Phylum Chordata	Dorsal supporting rod and nerve cord
Class Mammalia	Hair, mammary glands
Order Primates	Adapted to climb trees
Family Hominidae	Adapted to walk erect
Genus *Homo*	Large brain, tool use
Species *Homo sapiens**	Body proportions of modern humans

* To specify an organism, you must use the full binomial name, such as *Homo sapiens.*

Scientific Names

Within the broad field of systematics, **taxonomy** is the assignment of a binomial, or two-part name, to each species. For example, the scientific name for human beings is *Homo sapiens,* and for the garden pea, *Pisum sativum.* The first word is the genus to which the species belongs, and the second word is the species name. (Note that both words are in italics, but only the genus name is capitalized.) The genus name can be used alone to refer to a group of related species. Also, a genus can be abbreviated to a single letter if used with the species name (e.g., *P. sativum*).

Scientific names are in a common language—Latin—and biologists use them universally to avoid confusion. Common names, by contrast, tend to overlap and are often in a particular language.

Check Your Progress

1. What are the main criteria for classification of organisms into domains and kingdoms?
2. Why do we assign species to a hierarchical classification system (e.g., kingdom, phylum, class, etc.)?

1.3 The Organization of the Biosphere

Learning Outcomes

1. Describe how life is organized on the planet.
2. Discuss how humans influence ecosystems.

The organization of life extends beyond the individual to the **biosphere,** the zone of air, land, and water on the Earth where living organisms are found. Individual organisms belong to a **population,** all the members of a species within a particular area. All the different populations in the same area make up a **community,** in which organisms interact among themselves and with the physical environment (soil, atmosphere, etc.), thereby forming an **ecosystem.**

Figure 1.6 depicts a grassland inhabited by populations of rabbits, mice, snakes, hawks, and various types of plants. These populations exchange gases with and give off heat to the atmosphere. They also take in water from and give off water to the physical environment. In addition, the populations interact with each other by forming food chains in which one population feeds on another. For example, mice feed on plants and seeds, snakes feed on mice, and hawks feed on rabbits and snakes. The interactions among the various food chains make up a food web.

Ecosystems are characterized by chemical cycling and energy flow, both of which begin when photosynthetic plants, algae, and some bacteria take in solar energy and inorganic nutrients to produce food in the form of organic nutrients. The gray arrows in Figure 1.6 represent chemical cycling—chemicals move from one species to another in a food chain, until with death and decomposition, inorganic nutrients are returned to living plants once again. The yellow to red arrows represent energy flow. Energy flows from the sun through plants and other members of the food chain as one species feeds on another. With each transfer, most energy is lost as heat. Eventually, all the energy taken in by photosynthesizers has dissipated into the atmosphere. Because energy flows and does not cycle, ecosystems could not stay in existence without a constant input of solar energy and the ability of photosynthesizers to absorb it.

Climate largely determines where different ecosystems are found around the globe. For example, deserts are found in areas of almost no rain, grasslands require a minimum amount of rain, and forests generally require much rain. The two most biologically diverse ecosystems—tropical rain forests and coral reefs—occur where solar energy is most abundant. Coral reefs, which are found just offshore from the continents and islands of the Southern and Northern Hemispheres, are built up from the skeletons of sea animals called corals. Inside the tissues of corals are tiny, one-celled protists that carry on photosynthesis and provide food to their hosts. Reefs provide a habitat for many other animals, including jellyfish, sponges, snails, crabs, lobsters, sea turtles, moray eels, and some of the world's most colorful fishes (Fig. 1.7).

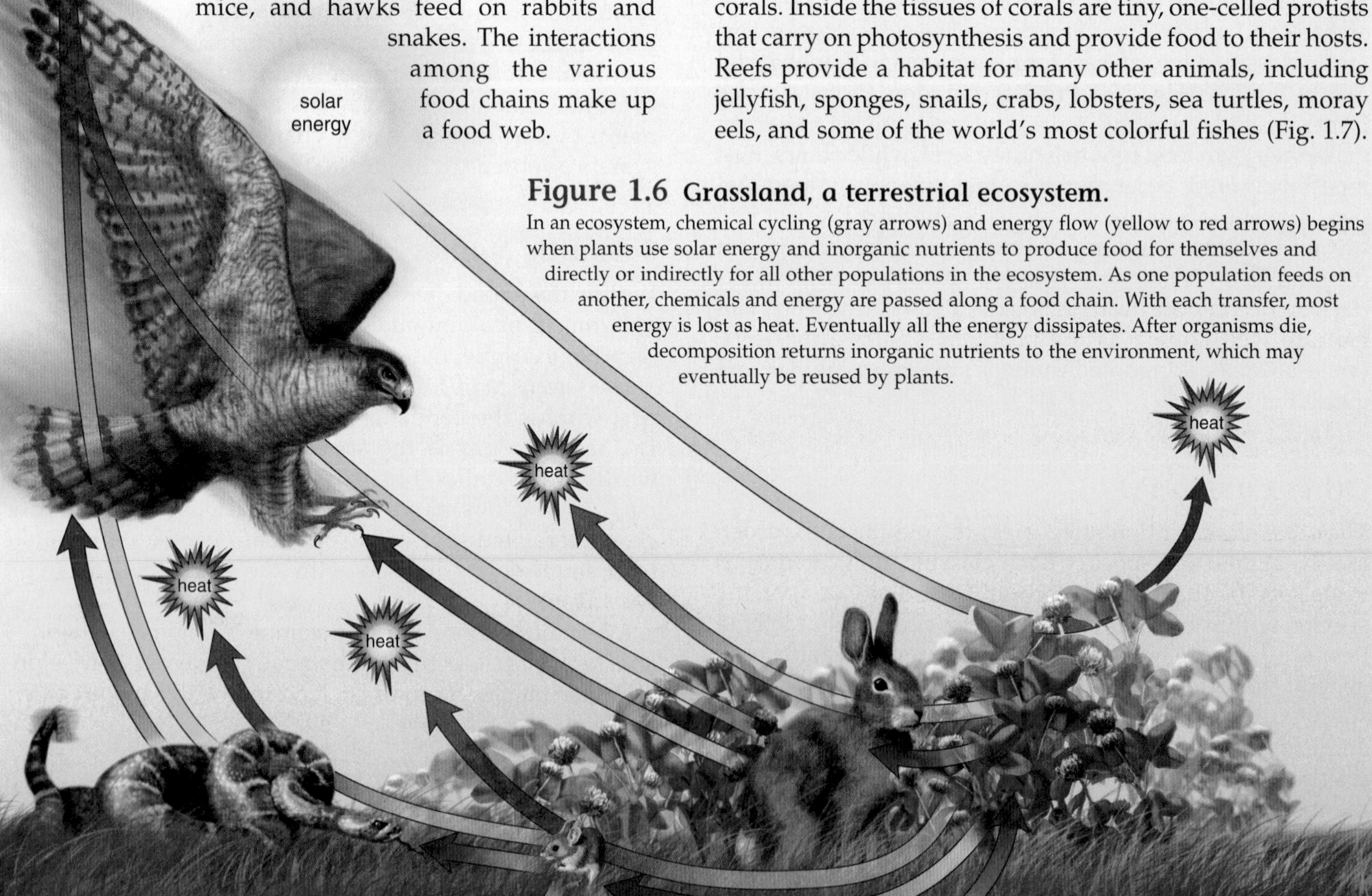

Figure 1.6 Grassland, a terrestrial ecosystem. In an ecosystem, chemical cycling (gray arrows) and energy flow (yellow to red arrows) begins when plants use solar energy and inorganic nutrients to produce food for themselves and directly or indirectly for all other populations in the ecosystem. As one population feeds on another, chemicals and energy are passed along a food chain. With each transfer, most energy is lost as heat. Eventually all the energy dissipates. After organisms die, decomposition returns inorganic nutrients to the environment, which may eventually be reused by plants.

The Human Species

The human species tends to modify existing ecosystems for its own purposes. Humans clear forests or grasslands in order to grow crops; later, they build houses on what was once farmland; and finally, they convert small towns into cities. As coasts are developed, humans send sediments, sewage, and other pollutants into the sea.

Like tropical rain forests, coral reefs are severely threatened as the global human population increases in size. Some reefs are 50 million years old, and yet in just a few decades, human activities have destroyed or severely degraded nearly 25% of all coral reefs and another third are threatened. At this rate, nearly three-quarters could be destroyed within 50 years. Similar levels of destruction have occurred in tropical rain forests.

It has long been clear that human beings depend on healthy ecosystems for food, medicines, and various raw materials. We are only now beginning to realize that we depend on them even more for the services they provide. Just as chemical cycling occurs within ecosystems, so do ecosystems keep chemicals cycling throughout the entire biosphere. The dynamics of ecosystems ensure that the environmental conditions of the biosphere are suitable for the continued existence of humans. In turn, many biologists believe that ecosystems cannot function properly unless they remain biologically diverse.

Biodiversity

Biodiversity encompasses the total number of species, the variability of their genes, and the ecosystems in which they live. The present biodiversity of our planet has been estimated at as high as 5–30 million species, and so far, about 1.75 million have been identified and named. Extinction is the permanent loss of a species or larger taxonomic group. Human activites are causing the extinction of about 400 species per day. For example, several species of fishes have all but disappeared from the coral reefs of Indonesia and along the African coast because of overfishing. Many biologists are alarmed about the present rate of extinction and believe it has surpassed the rates of the five mass extinctions that have occurred during our planet's history. The dinosaurs became extinct during the last mass extinction, 65 million years ago.

Most biologists agree that a primary bioethical issue of our time is conservation of biodiversity. Just as a native fisherman who assists in overfishing a reef is doing away with his own food source, so we as a society are contributing to the destruction of our home, the biosphere. If instead we adopt an ethic that conserves the biosphere, we are helping to ensure the continued existence of its species.

Check Your Progress

1. How are ecosystems characterized?
2. How might human activities disrupt basic ecosystem processes?

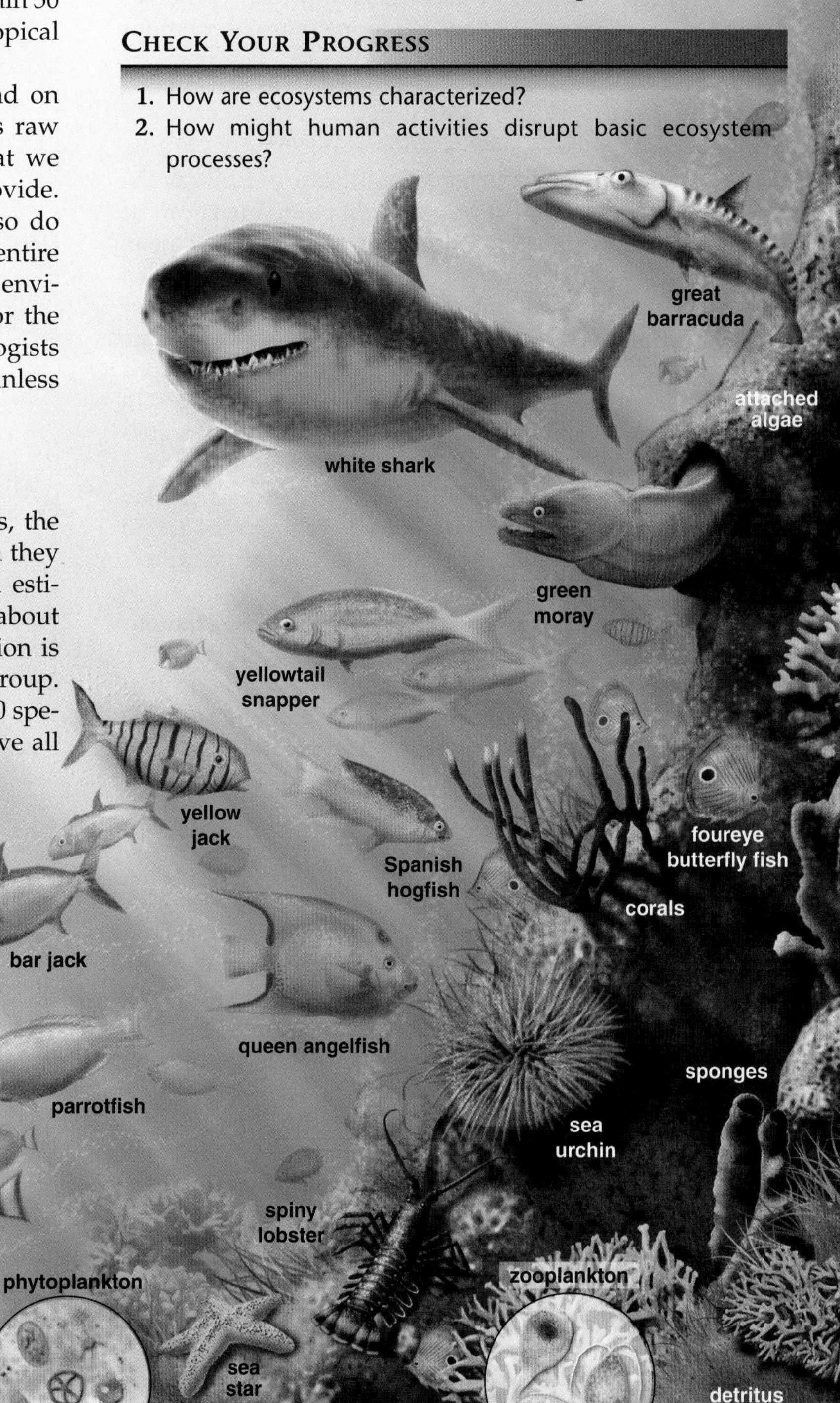

Figure 1.7 Coral reef, a marine ecosystem. Coral reefs, a type of ecosystem found in tropical seas, contain many diverse forms of life, a few of which are shown here. Coral reefs are now threatened because of certain human activities. Saving such biodiversity is one of our greatest modern-day challenges.

1.4 The Process of Science

Learning Outcomes

1. Formulate a hypothesis.
2. List the steps in conducting a scientific experiment.
3. Interpret a controlled study.

Biology is the scientific study of life. Biologists can be found almost anywhere studying life-forms. Often, they perform experiments, either in the field (outside the laboratory) or in the laboratory.

Biologists—and all scientists—generally test hypotheses using the scientific method based on previous observations (Fig. 1.8).

Observation

Scientists ultimately are curious about nature and how the world works. They believe that natural **phenomena** can be understood more fully by observing and studying them. Scientists use all their senses to make **observations.** We can observe with our noses that dinner is almost ready and our ears that a piano needs tuning. Scientists also extend the ability of their senses by using instruments. For example, a microscope enables us to see objects too small to view with the naked eye. Finally, scientists may expand their understanding even further by taking advantage of the knowledge and experiences of other scientists. For instance, they may look up past studies in scientific journals, or they may attend scientific meetings and hear presentations from others researching similar topics.

Chance alone can help a scientist get an idea. A famous case pertains to penicillin. When examining a petri dish, Alexander Fleming observed an area around a mold that was free of bacteria. Upon investigating, Fleming found that the mold produced an antibacterial substance he called penicillin. This observation led Fleming to think that perhaps penicillin would be useful in humans.

Hypothesis

After making observations and gathering knowledge about a phenomenon, a scientist uses inductive reasoning. **Inductive reasoning** occurs whenever a person uses creative thinking to combine isolated facts into a cohesive whole. In this case, the scientist comes up with a **hypothesis,** a tentative explanation for the natural event. The scientist presents this hypothesis as a falsifiable statement.

Past experiences and observations, no matter what they might be, most likely influence the formation of a hypothesis. But a scientist only considers hypotheses that can be tested. Moral and religious beliefs, while very important to our lives, differ between cultures and through time, and are not testable scientifically.

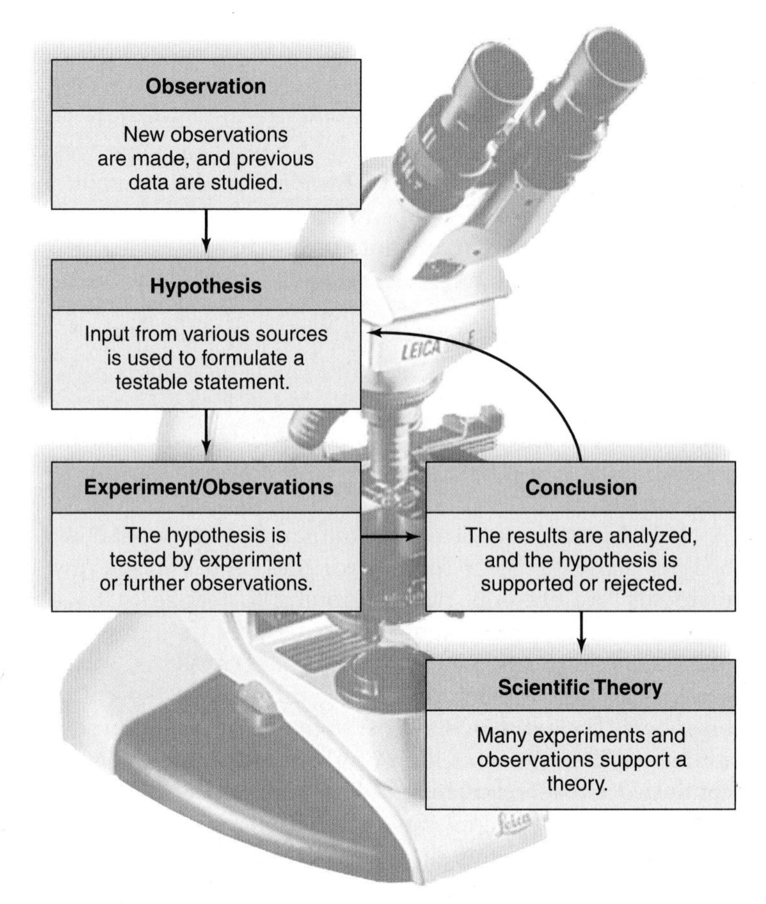

Figure 1.8 Flow diagram for the scientific method. On the basis of new and/or previous observations, a scientist formulates a hypothesis. The scientist then tests the hypothesis by making further observations and/or performing experiments. The resulting new data either support or do not support the hypothesis. The return arrow indicates that a scientist often chooses to retest the same hypothesis or to test an alternative hypothesis. Conclusions from many different but related experiments may lead to the development of a scientific theory. For example, studies of development, anatomy, and fossil remains all support the theory of evolution.

Experiment/Further Observations

Testing a hypothesis involves either conducting an experiment or making further observations. To determine how to test a hypothesis, a scientist uses deductive reasoning. **Deductive reasoning** involves "if, then" logic. For example, a scientist might reason, *if* organisms are composed of cells, *then* microscopic examination of any part of an organism should reveal cells. We can also say that the scientist has made a **prediction** that the hypothesis can be supported by doing microscopic studies. Making a prediction helps a scientist know what to do next.

Further observations may be used to test the hypothesis. Most often, however, scientists perform experiments. The manner in which a scientist intends to conduct an experiment is called the **experimental design.** A good experimental design ensures testing is specific and that results will be meaningful. An experiment should always include a control group. A control group, or simply the **control,** goes through all the steps of an experiment but lacks the factor, or is not exposed to the factor, being tested. For example, in a drug trial the treatment group receives the actual drug being tested while the control group receives a placebo (sugar pill).

In their experiments, scientists often use a **model,** a representation of an actual subject. For example, scientists can use computer modeling to suggest future ways human activities may affect global climate change, or they may use mice as a model, instead of humans, when doing cancer research. When a medicine is effective in mice, researchers still need to test it in humans before making recommendations or using it in widespread treatment. Thus, results obtained using models are considered a hypothesis until the experiment can be performed using the actual subject.

Data

The results of an experiment are referred to as the **data.** Data should be observable and objective, rather than subjective or based on opinion. Mathematical data are often displayed in the form of a graph or table. Many studies rely on statistical data. Let's say an investigator wants to know if eating onions can prevent women from getting osteoporosis (weak, porous bones). The scientist conducts a survey asking women about their onion-eating habits and then correlates these data with the condition of their bones. Other scientists critiquing this study would want to know: How many women were surveyed? How old were the women? What were their exercise habits? What proportion of the diet consisted of onions? And what criteria were used to determine the condition of their bones? Should the investigators conclude that eating onions does protect a woman from osteoporosis, other scientists would want to know that the relationship did not occur simply by chance. The larger the sample size (in this case, the number of women studied) and the larger the effect (in this case, strength of bones in women eating more onions compared to those eating fewer onions), the less likely that the result is due to chance alone. Statistical tests are used to evaluate how likely it is that the results of an experiment are due to chance. And in the end, statistical data of this sort would only be suggestive until we know of some ingredient in onions that has a direct biochemical or physiological effect on bone strength. Therefore, scientists are skeptics who always pressure one another to keep investigating a particular topic.

Conclusion

Scientists must analyze the data to reach a **conclusion** as to whether the hypothesis is supported or not. Because science progresses, the conclusion of one experiment can lead to the hypothesis for another experiment, as represented by the return arrow in Figure 1.8. Sometimes, results that do not support one hypothesis can help a scientist formulate another hypothesis to be tested. Scientists report their findings in scientific journals, subject to critical review by their peers, so that their methodology and data are available to other scientists. Experiments and observations must be repeatable—that is, any scientist who repeats an experiment must get the same results, or else the data are suspect. This scientific process is continuous, and in biology scientists are discovering new things all the time.

Scientific Theory

The ultimate goal of science is to understand the natural world in terms of **scientific theories,** concepts that join together well-supported and related hypotheses. In ordinary speech, the word *theory* refers to a speculative idea. But in science, a theory is supported by a broad range of observations, experiments, and data and is thus similar to a law, such as the "theory" of gravity.

Some of the basic theories of biology are as follows:

Theory	Concept
Cell	All organisms are composed of cells, and new cells only come from preexisting cells.
Homeostasis	The internal environment of an organism stays relatively constant—within a range that is protective of life.
Gene	Organisms contain coded information that dictates their form, function, and behavior.
Ecosystem	Organisms are members of populations, which interact with each other and with the physical environment within a particular locale.
Evolution	A change in the frequency of traits that affect reproductive success in a population or species across generations.

When we use the word "theory" in everyday terms, we often mean a guess or a hunch. However, in scientific terms, a theory is supported by observation and repeated empirical investigation. So, a theory in science is akin to a "law," such as gravitational theory or atomic theory. The theory of evolution is the unifying concept of biology because it pertains to many different aspects of living things. The theory of evolution enables scientists to understand the history of life; the diversity of living things; and the anatomy, physiology, and development of organisms—even their behavior.

The theory of evolution has been a very fruitful theory scientifically, meaning that it has helped scientists generate new hypotheses. Because this theory has been supported by so many observations and experiments for over 100 years, some biologists refer to the **principle** of evolution, a term that refers to theories generally accepted by an overwhelming number of scientists. Other biologists even prefer to use the term **law** instead of principle.

Check Your Progress

1. What is a scientific hypothesis?
2. Describe the steps of the scientific method.

A Controlled Study

Most investigators do controlled studies in which some groups receive a treatment (experimental groups), while other groups receive no treatment (control). To ensure that the outcome is due to the **experimental variable** (independent variable) alone, all other conditions between the groups in the experiment are controlled to be identical. The result is called the **response variable** (or dependent variable) because it is due to the experimental treatment.

Experimental Variable (Independent Variable)	Response Variable (Dependent Variable)
Factor of the experiment being tested	Result or change that occurs due to the experimental variable

In the following study, researchers performed an experiment in which nitrogen fertilizer is the experimental variable and crop yield is the response variable. Nitrogen fertilizer in the short run long has been known to enhance yield and increase food supplies. However, excessive nitrogen application can cause pollution by adding toxic levels of nitrates to water supplies. Also, applying nitrogen fertilizer year after year may alter soil properties such that crop yields may decrease instead of increase. A solution is to let the land remain unplanted for several years until the soil recovers naturally.

An alternative to nitrogen fertilizers is the use of legumes, plants such as peas and beans, that provide a home for bacteria that convert atmospheric nitrogen to a form usable by plants. The bacteria live in nodules on the roots (Fig. 1.9). The products of photosynthesis move from the leaves to the root nodules. In turn, the nodules supply the plant with nitrogen compounds the plant can use to make proteins.

Numerous legume crops can be rotated (planted every other season) with any number of cereal crops. The nitrogen added to the soil by the legume crop increases cereal crop yield. The particular rotation used depends on the location, climate, and market demand.

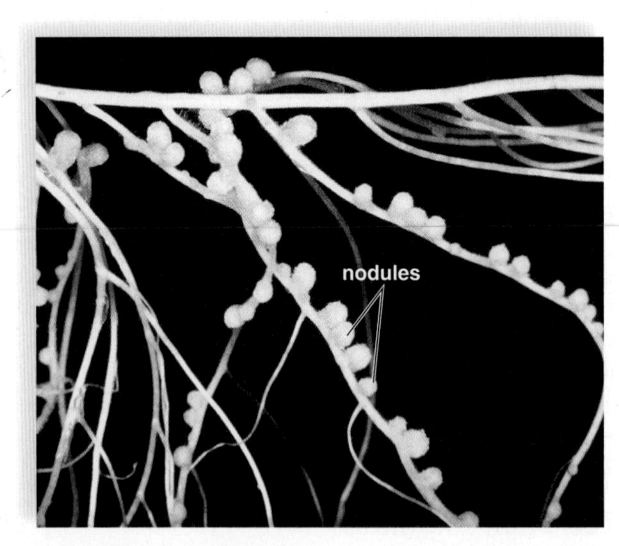

Figure 1.9 Root nodules.
Bacteria that live in nodules on the roots of legumes, such as pea plants, convert nitrogen in the soil to a form that plants can use to make proteins and other nitrogen-containing molecules.

The Experiment

The pigeon pea plant is a legume with a high rate of atmospheric nitrogen conversion. This plant is widely grown as a food crop in India, Kenya, Uganda, Pakistan, and other subtropical countries. Researchers formulated the hypothesis that rotating pigeon pea and winter wheat crops would be a reasonable alternative to using nitrogen fertilizer to increase the yield of winter wheat.

HYPOTHESIS: A pigeon pea/winter wheat rotation will cause winter wheat production to increase as well as or better than the use of nitrogen fertilizer.

PREDICTION: Wheat production (biomass) following the growth of pigeon peas will surpass wheat biomass following nitrogen fertilizer treatment.

In this study, the investigators decided on the following experimental design (Fig. 1.10*a*):

Control Pots

- Winter wheat was planted in pots of soil that received no fertilization treatment (i.e., no nitrogen fertilizer and no preplanting of pigeon peas).

Test Pots

- Winter wheat was grown in clay pots in soil treated with nitrogen fertilizer equivalent to 45 kilograms (kg)/hectare (ha).
- Winter wheat was grown in clay pots in soil treated with nitrogen fertilizer equivalent to 90 kg/ha.
- Pigeon pea plants were grown in clay pots in the summer. The pigeon pea plants were then tilled into the soil, and winter wheat was planted in the same pots.

To ensure a controlled experiment, the conditions for the control pots and the test pots were identical with the exception of the treatments noted. The plants were exposed to the same environmental conditions and watered equally. During the following spring, the wheat plants were dried and weighed to determine total wheat production (biomass) in each of the pots.

The Results After the first year, wheat biomass was higher in certain test pots than in the control pots (Fig. 1.10*b*). Specifically, test pots with 45 kg/ha of nitrogen fertilizer (*orange*) had only slightly more wheat biomass production than the control pots, but test pots that received 90 kg/ha treatment (*green*) had nearly twice the biomass production of the control pots. Thus, application of nitrogen fertilizer had the potential to increase wheat biomass. To the surprise of investigators, wheat production following summer planting of pigeon peas did not result in as high a biomass production as the control pots.

CONCLUSION: The hypothesis is not supported. Wheat biomass following the growth of pigeon peas is not as high as that obtained with nitrogen fertilizer treatments.

Figure 1.10 Pigeon pea/winter wheat rotation study.
a. The experiment involves control pots that receive no fertilization and three types of test pots: test pots that received 45 kg/ha of nitrogen; test pots that received 90 kg/ha of nitrogen; and test pots in which pigeon peas rotated with winter wheat. **b.** The graph compares wheat biomass for each of three years. Wheat biomass in test pots that received the most nitrogen fertilizer declined, while wheat biomass in test pots with pigeon pea/winter wheat rotation increased dramatically.

Continuing the Experiment

The researchers decided to continue the experiment using the same design and the same pots as before, to see if the buildup of residual soil nitrogen from pigeon peas would eventually increase wheat biomass. They formed a new hypothesis.

HYPOTHESIS: A sustained pigeon pea/winter wheat rotation will eventually cause an increase in winter wheat production.

PREDICTION: Wheat biomass following two years of pigeon pea/winter wheat rotation will surpass wheat biomass following nitrogen fertilizer treatment.

The Results After two years, the yield following 90 kg/ha nitrogen treatment (*green*) was not as much as it was the first year (Fig. 1.10*b*). As predicted, wheat biomass following summer planting of pigeon peas (*brown*) was the highest of all treatments, suggesting that buildup of residual nitrogen from pigeon peas had the potential to provide fertilization for winter wheat growth.

CONCLUSION: The hypothesis was supported. At the end of two years, the yield of winter wheat following a pigeon pea/winter wheat rotation was better than for the other types of pots.

The researchers continued their experiment for still another year. After three years, winter wheat biomass production had decreased in the control pots and in the pots treated with nitrogen fertilizer. Pots treated with nitrogen fertilizer still had increased wheat biomass production compared with the control pots, but not nearly as much as pots

Bioethical Focus

The Pros and Cons of DDT

The widespread use of DDT began in 1945 to control agricultural pests and mosquitoes. DDT is extremely toxic to insects—both those that benefit crops and those that destroy them. DDT is inexpensive, kills most pests, and provides lasting protection because it breaks down slowly. DDT is believed to have helped save millions of human lives. The military used it during World War II to control body lice and thus typhus fever, and the World Health Organization (WHO) used it to control mosquitoes and consequently the malaria they carry.

However, subsequent studies showed that DDT (and other pesticides) can become concentrated in the food chain. DDT is known to have led to the decline of birds of prey, such as the bald eagle, because it causes thinning of eggshells. Adults then inadvertently crush the eggs when attempting to incubate them, killing the embryos before they hatch.

Because of public concerns about DDT's effects on the environment and possible long-term effects on human health, the U.S. Environmental Protection Agency banned DDT in 1972. Since then, bald eagles have recovered to the point that they are no longer on the endangered species list. In 2004, the United Nations passed the international Persistent Organic Pollutants treaty to reduce DDT in the environment by permitting its use only under strict regulations.

Still, DDT is the most cost-effective chemical for preventing malaria because it kills the mosquitoes that carry the disease. According to WHO, malaria causes 300 million acute illnesses and over 1 million deaths each year in the tropics, subtropics, and portions of the Middle East. Researchers state in the *Journal of Infectious Diseases*, "Today, DDT is still needed for malaria control. If the pressure to abandon this effective insecticide continues, millions of additional malaria cases world-wide [will result]." So, should we continue to cut back on the use of DDT because of its effects on animals and people? Or should we consider allowing its use to help control malaria and prevent some of the many deaths that occur each year?

Form Your Own Opinion

1. Present regulations ban the manufacture and use of DDT in many, but not all, countries. Should it be banned everywhere? Why or why not?
2. Under what circumstances might the use of DDT be acceptable? Explain.
3. Currently in the U.S., the insecticide malathion is widely used to control mosquitoes that may carry West Nile virus. Although malathion breaks down more quickly than DDT, some evidence suggests it can be harmful to amphibians and fish. What data are needed before continued widespread use of malathion?

following summer planting of pigeon peas. Compared to the first year, wheat biomass increased almost fourfold in pots having a pigeon pea/winter wheat rotation (Fig. 1.10*b*). The researchers suggested that the soil had been improved by the organic matter, as well as by the addition of nitrogen from the pigeon peas, and published their results in a scientific journal.[1]

Ecological Importance of This Study

This study showed that the use of a legume, namely pigeon peas, to improve the soil produced a far better yield than the use of a nitrogen fertilizer over the long haul. Legumes are plants that house bacteria capable of converting atmospheric nitrogen into a form a plant can use. When the pigeon pea plants were turned over into the soil, the winter wheat plants could make use of this nitrogen.

Rotation of crops, as was done in this study, is an important part of organic farming. Sometimes there are adverse economic results when farmers switch from chemical-intensive to organic farming practices. But, because the benefit of making the transition may not be realized for several years, some researchers advocate that the process be gradual. This study suggests, however, that organic farming will eventually be beneficial both for farmers and the environment!

[1] Bidlack, J. E., Rao, S. C., and Demezas, D. H. 2001. Nodulation, nitrogenase activity, and dry weight of chickpea and pigeon pea cultivars using different *Bradyrhizobium* strains. *Journal of Plant Nutrition* 24:549–60.

Check Your Progress

1. Describe the scientific process used in the pigeon pea experiment.
2. Design an experiment that includes control and experimental groups and outline the difference between the independent and dependent variables. When would the results of the experiment support versus refute a hypothesis?

1.5 Science and Social Responsibility

Learning Outcome

1. Discuss the costs and benefits of technology.

Many scientists are engaged in fields of study that sometimes seem remote from our everyday lives. Other scientists are interested in using the findings of past and present scientists to produce a product or develop a technique that directly affects our lives. The application of scientific knowledge for a practical purpose is called **technology.** For example, virology, the study of viruses and molecular chemistry, led to the discovery of new drugs that extend the life span of people who have HIV/AIDS and a vaccine to prevent smallpox. Cell biology research has allowed physicians to develop various cancer treatments.

Most technologies have benefits but also drawbacks. Research has led to modern agricultural practices that help to feed the burgeoning human population. However, the use of nitrogen fertilizers leads to water pollution, and the use of pesticides, as you may know, kills not only pests but also other types of organisms.

Who should decide how, and even whether, a technology should be put to use? Making value judgments is not a part of science. Ethical and moral decisions must be made by all people. Therefore, the responsibility for how to use scientific technology must reside with people from all walks of life, not with scientists alone. Presently, we need to decide whether we want to stop producing bioengineered organisms that may be harmful to the environment. Also, through gene therapy, we are developing the ability to cure diseases and the possibility of altering the genes of our offspring. Perhaps one day we might even be able to clone ourselves. Should we do these things? So far, as a society, we continue to believe in the sacredness of human life, and therefore we have passed laws against human cloning. Even if the procedure is perfected, we may also continue to rule against human cloning.

The Bioethical Focus in this chapter presents the complexities of banning the use of DDT. Each of us must wrestle with this and the other bioethical issues discussed in this text, such as stem cell research, in our everyday lives and hopefully make decisions that are beneficial to society overall.

Check Your Progress

1. Technological advances from scientific discoveries can help people but may also harm the environment. Give an example of such a discovery and discuss the pros and cons.

Applying the Concepts [Revisited]

At the beginning of the chapter, it was brought to your attention that the Earth's biodiversity is being threatened by human activities. Specifically, while our use of technology benefits the quality of our lives by, for example, improving health care, it also causes harm to the environment through pollution, loss of habitat for wildlife, and other damage. An example of this is illustrated with the use of DDT, which controls mosquitoes that carry diseases like malaria, but also was thought to have caused bird declines. Our increase in use of the world's resources also has resulted in the destruction of coral reefs and tropical rain forests, areas rich in biodiversity. As we move through this text, you will have the tools to better understand the influence of humans on the environment, the biodiversity on Earth, and choices that we can make to balance our needs with the needs of all other species.

1. Given what you know about how the Earth's biodiversity is classified, which species do you think are most susceptible to human-caused environmental degradation and why?
2. One of the biggest bioethical issues we face in the twenty-first century is the loss of biodiversity, much of which is due to human activities, such as development and agriculture. These activities benefit human society by providing shelter, food, and other necessities of life, and yet they also threaten many of the world's species with extinction. Given the benefits of development to humans, why should we conserve biodiversity?

SUMMARIZING THE CONCEPTS

1.1 The Characteristics of Life

Evolution accounts for both the diversity and the unity of life we see about us. All organisms share the following characteristics of life:

- Living things are organized. The levels of biological organization extend as follows: atoms and molecules ⟶ cells ⟶ tissues ⟶ organs ⟶ organ systems ⟶ organisms ⟶ populations ⟶ communities ⟶ ecosystems. In an ecosystem, populations interact with one another and with the physical environment.
- Living things take materials and energy from the environment; they need an outside source of nutrients.
- Living things reproduce; they produce offspring that resemble themselves.
- Living things respond to stimuli; they react to internal and external events.
- Living things are homeostatic; internally they stay just about the same despite changes in the external environment.
- Living things grow and develop; during their lives, they change—most multicellular organisms undergo various stages from fertilization to death.
- Living things have the capacity to adapt to a changing environment.

1.2 The Classification of Living Things

- Living things are classified according to evolutionary relationships into ever-more-inclusive categories: species, genus, family, order, class, phylum, kingdom, and domain.
- Species in different domains are only distantly related; species in the same genus are very closely related.
- Each species has a Latin scientific name that consists of the genus and species name. Both names are italicized, but only the genus is capitalized, as in *Homo sapiens*.

1.3 The Organization of the Biosphere

- Living things generally belong to a population, defined as all members of a single species in a particular area.
- Populations interact with each other within a community and with the physical environment, forming an ecosystem.
- Ecosystems are characterized by chemical cycling and energy flow, which begin when a photosynthesizer becomes food (organic nutrients) for an animal.
- Food chains tell who eats whom in an ecosystem. As one population feeds on another, the energy dissipates, but nutrients do not.
- Eventually, inorganic nutrient molecules are decomposed and return to photosynthesizers, which use them plus solar energy to produce more food.

- Human activities have totally altered many ecosystems and are putting stress on most of the others. Coral reefs and tropical rain forests, for example, are quickly disappearing.
- The health of the biosphere is essential to the future continuance of the human species.

1.4 The Process of Science

- When studying the natural world, scientists use the scientific method as the following steps:
  1. Observations, along with previous data, are used to formulate a hypothesis.
  2. New observations and/or experiments are carried out to test the hypothesis. Experimental designs should include a control group.
  3. The experimental and observational results are analyzed, and the scientist decides whether the results support the hypothesis or prove it false.
  4. Hypotheses can be supported and modified based on new experiments because science is always open to change.
- Several related conclusions resulting from similar scientific experiments may allow scientists to arrive at a theory, such as the cell theory, the gene theory, or the theory of evolution. The theory of evolution is a unifying theory of biology.

1.5 Science and Social Responsibility

- Science does not consider moral or ethical questions. It is up to all of us to decide how the various technologies that grow out of basic science should be used or regulated.

TESTING YOURSELF

Choose the best answer for each question.

1. Which sequence represents the correct order of increasing complexity in living systems?
   a. cell, molecule, organ, tissue
   b. organ, tissue, cell, molecule
   c. molecule, cell, tissue, organ
   d. cell, organ, tissue, molecule
2. Classification of organisms reflects
   a. similarities.
   b. evolutionary history.
   c. Neither a nor b is correct.
   d. Both a and b are correct.
3. A population is defined as
   a. the number of species in a given geographic area.
   b. all of the individuals of a particular species in a given area.
   c. a group of communities.
   d. all of the organisms in an ecosystem.
4. The ultimate source of energy in any ecosystem is (are)
   a. the sun.
   b. green plants.
   c. fungi.
   d. animals.
5. Which are the most biologically diverse ecosystems?
   a. deserts and rain forests
   b. rain forests and coral reefs
   c. deserts and coral reefs
   d. None of these are correct.
6. What is the unifying theory in biology that explains the relationships of all living things?
   a. ecology
   b. evolution
   c. biodiversity
   d. taxonomy
7. Which sequence exhibits an increasingly more-inclusive scheme of classification?
   a. kingdom, phylum, class, order
   b. phylum, class, order, family
   c. class, order, family, genus
   d. genus, family, order, class
8. The best explanation as to why "big, fierce carnivores (animals that eat other animals) are rare" would be the fact that
   a. energy flows and dissipates.
   b. energy is stored in living systems.
   c. nutrients are cycled.
   d. nutrients flow.
9. Features that make an organism suited to its way of life are called
   a. ecosystems.
   b. populations.
   c. adaptations.
   d. None of these are correct.

For questions 10–13, match the statements with the correct terms.

a. photosynthetic ability
b. absorb food
c. ingest food
d. all populations within a given area
e. all the interactions of all organisms with each other and with the environment in a given area

10. Animals
11. Plants
12. Ecosystem
13. Community
14. What's the difference between a test group and a control group?
    a. There is no real difference because both are part of a study.
    b. The test group is testing something, but the control group is not testing anything.
    c. The test group is exposed to the factor being tested, but the control group is not exposed.
    d. The control group proves the results true, but the test group does not.
    e. All of these are correct.
15. In the controlled study of nitrogen fertilizers and legumes (see pages 12–13),
    a. pigeon peas and winter wheat were treated similarly; for example, winter wheat was kept inside to stabilize the temperature.
    b. no pigeon peas were grown in the control pots, and no fertilizer was added.
    c. there were two test groups; therefore, there were two test pots.
    d. the investigators were surprised that the control group didn't show better results.
    e. All of these are correct.

UNDERSTANDING THE TERMS

adaptation 5
animal 7
Archaea 7
Bacteria 7
behavior 4
biodiversity 9
biology 10
biosphere 8
cell 3
class 6
community 8
conclusion 11
control 11
data 11
deductive reasoning 10
development 5
domain 6
ecosystem 8
energy 4
Eukarya 7
eukaryote 7
evolution 5
experimental design 10
experimental variable 12
family 6
fungi 7
gene 4
genus 6
homeostasis 4
hypothesis 10
inductive reasoning 10
kingdom 6
law 11
model 11
natural selection 5
observation 10
order 6
organ 3
organism 3
organ system 3
phenomenon 10
phylum 6
plant 7
population 8
prediction 10
principle 11
prokaryote 7
protist 7
reproduction 4
response variable 12
scientific theory 11
species 5
systematics 6
taxonomy 7
technology 14
tissue 3

THINKING CRITICALLY

1. How can evolution explain both the unity and diversity of life?
2. What is the definition of life? Explain in your own words how scientists believe life originally evolved.
3. You are a scientist working at a pharmaceutical company and have developed a new cancer medication that has the potential for use in humans. Outline a series of experiments, including the use of a model, to test whether the cancer medication works.

INQUIRY INTO LIFE WEBSITE

The companion website for *Inquiry into Life* provides a wealth of information organized and integrated by chapter. You will find practice tests, animations, videos, and much more that will complement your learning and understanding of general biology.

http://www.mhhe.com/maderinquiry13

PART I

Cell Biology

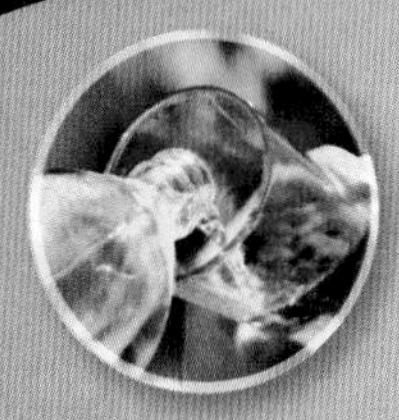

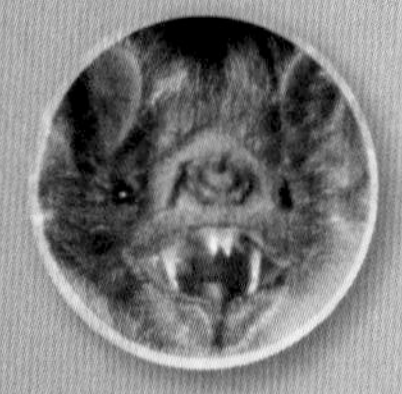

The Molecules of Cells

Applying the Concepts

For years it was common knowledge that you should drink eight glasses of water a day for proper hydration. You still need to stay hydrated, but now researchers are questioning where the number eight in the eight glasses a day came from. Apparently there was never any scientific research study to back up the recommendation that each individual needs to drink eight glasses of water per day. So where did the number come from? Some scientists think it started from a misreading of a report from the Food and Nutrition Board of the National Research Council. The report recommended drinking one milliliter of water for every calorie of food eaten. That would amount to 64–80 ounces of water per day—eight to ten glasses. The next line of the report, however, states that most of this water can be obtained in prepared foods. Regardless of how it started, Americans became addicted to drinking water! Nutrition researchers know that one number does not fit all people or all situations. The amount of water you need to stay hydrated depends on a number of conditions, including the surrounding temperature, how much you exercise, how much you sweat, and body size. You need to replace daily the water that you lose through breathing, sweating, and waste elimination. Drinking water is only one way to get hydrated. Other beverages such as juices and milk provide water. You can also count caffeinated drinks in your total fluid intake now, although less caffeine is still better than more. Foods also contain water. Fruits, vegetables, soups, and dairy all contain more than 80% water! So eating these foods can supply part of your hydration needs. Unless you are sick or an endurance athlete, then your thirst should be an adequate indicator of when you need fluid. The choice of fluid is up to you!

Water is an interesting molecule with unusual characteristics. In this chapter you will learn how the unique properties of water make it essential to the existence of life on Earth.

Chapter Outline

2.1 Basic Chemistry

Learning Outcomes

1. Define and give examples of matter.
2. Describe the structure of an atom, including the subatomic particles, their charges, and location.
3. Know how the periodic table is organized.
4. Explain how isotopes differ.
5. Discuss beneficial and harmful uses for radiation.

Everything—including the book you're holding, the chair you're sitting on, the water you drink, and the air you breathe—is composed of matter. **Matter** refers to anything that takes up space and has mass. Matter has many diverse forms, but it can only exist in three distinct states: solid, liquid, and gas.

All matter, both nonliving and living, is composed of certain basic substances called **elements.** An element is a substance that cannot be broken down to simpler substances with different properties by ordinary chemical means. (A property is a physical or chemical characteristic, such as density, solubility, melting point, and reactivity.) Only 92 naturally occurring elements serve as the building blocks of all matter. Other elements have been "human-made" and are not biologically important.

Earth's crust as well as all organisms are composed of elements, but they differ as to which elements are predominant (Fig. 2.1). Only six elements—carbon, hydrogen, nitrogen, oxygen, phosphorus, and sulfur—are basic to life and make up about 95% of the body weight of organisms. The acronym CHNOPS helps us remember these six elements. The properties of these elements are essential to the uniqueness of cells and organisms. Other elements, including sodium, potassium, calcium, iron, and magnesium are also important to living things.

Atomic Structure

In the early 1800s, the English scientist John Dalton championed the atomic theory, which says that elements consist of tiny particles called **atoms.** An atom is the smallest part of an element that displays the properties of the element. An element and its atom share the same name. The atomic symbol is composed of one or two letters, which stands for this name. For example, the symbol H means a hydrogen atom, the symbol Cl stands for chlorine, and the symbol Na (for *natrium* in Latin) is used for a sodium atom.

Physicists have identified a number of subatomic particles that make up atoms. The three best-known subatomic particles are positively charged **protons,** uncharged **neutrons,** and negatively charged **electrons.** Protons and neutrons are located within the nucleus of an atom, and electrons move about the nucleus. Figure 2.2 shows the arrangement of the subatomic particles in a helium atom, which has only two electrons. In Figure 2.2*a*, the stippling shows the probable

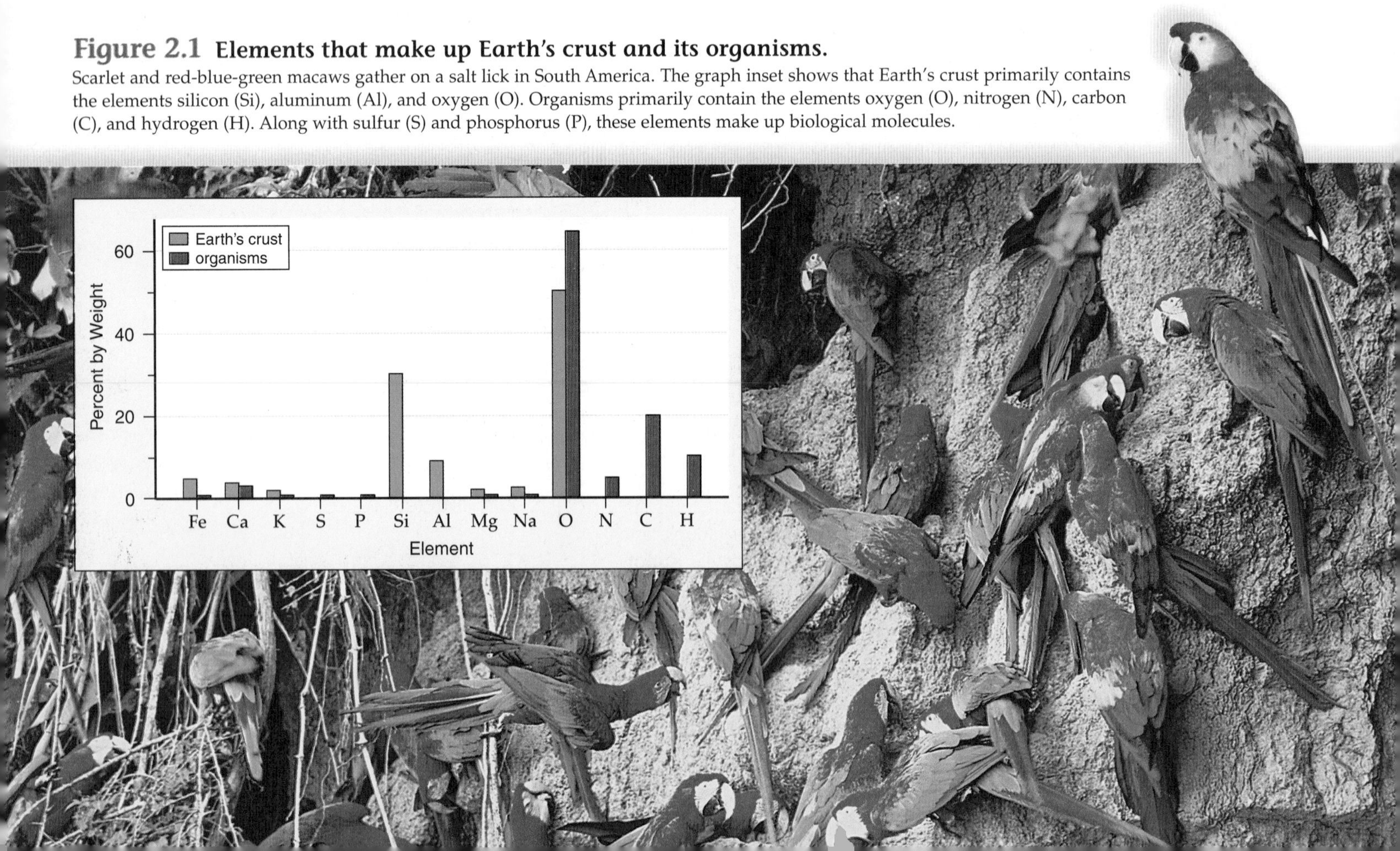

Figure 2.1 Elements that make up Earth's crust and its organisms.
Scarlet and red-blue-green macaws gather on a salt lick in South America. The graph inset shows that Earth's crust primarily contains the elements silicon (Si), aluminum (Al), and oxygen (O). Organisms primarily contain the elements oxygen (O), nitrogen (N), carbon (C), and hydrogen (H). Along with sulfur (S) and phosphorus (P), these elements make up biological molecules.

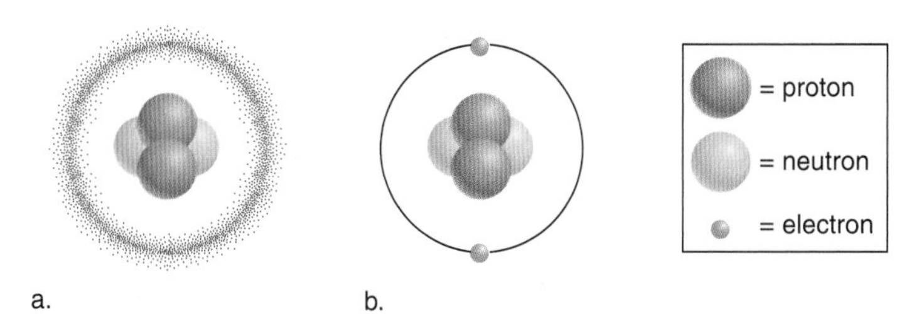

Subatomic Particles			
Particle	**Electric Charge**	**Atomic Mass**	**Location**
Proton	+1	1	Nucleus
Neutron	0	1	Nucleus
Electron	−1	~0	Electron orbital

c.

Figure 2.2 Model of helium (He).
Atoms contain subatomic particles called protons, neutrons, and electrons. Protons and neutrons are within the nucleus, and electrons are outside the nucleus. **a.** The stippling shows the probable location of the electrons in the helium atom. **b.** A circle termed an electron orbital represents the average location of an electron. **c.** The electric charge and the atomic mass units of the subatomic particles vary as shown.

Periods ↓ / Groups →	I	II	III	IV	V	VI	VII	VIII
	1 **H** 1.008							2 **He** 4.003
	3 **Li** 6.941	4 **Be** 9.012	5 **B** 10.81	6 **C** 12.01	7 **N** 14.01	8 **O** 16.00	9 **F** 19.00	10 **Ne** 20.18
	11 **Na** 22.99	12 **Mg** 24.31	13 **Al** 26.98	14 **Si** 28.09	15 **P** 30.97	16 **S** 32.07	17 **Cl** 35.45	18 **Ar** 39.95
	19 **K** 39.10	20 **Ca** 40.08	31 **Ga** 69.72	32 **Ge** 72.59	33 **As** 74.92	34 **Se** 78.96	35 **Br** 79.90	36 **Kr** 83.60

Figure 2.3 A portion of the periodic table.
In the periodic table, the elements, and therefore atoms, are arranged in the order of their atomic numbers and placed in groups (vertical columns) and periods (horizontal rows). All the atoms in a particular group have certain chemical characteristics in common. These four periods contain the elements that are most important in biology. The complete periodic table is in Appendix D.

location of electrons, and in Figure 2.2*b*, the circle represents an electron orbital, the average location of electrons.

The concept of an atom has changed greatly since Dalton's day. If an atom could be drawn the size of a football field, the nucleus would be like a gumball in the center of the field. The electrons would be tiny specks whirling about in the upper stands. Most of an atom is empty space. We can only indicate the orbital where the electrons are expected to be most of the time. In our analogy, the electrons might very well stray outside the stadium at times.

All atoms of an element have the same number of protons. This is called the **atomic number.** The number of protons in the nucleus makes each atom unique. The atomic number is often written as a subscript to the lower left of the atomic symbol.

Each atom has its own specific mass. The **atomic mass** of an atom is essentially the sum of its protons and neutrons. Protons and neutrons are assigned one atomic mass unit each. Electrons are so small that their mass is considered zero in most calculations (Fig. 2.2*c*). The term *atomic mass* is used, rather than *atomic weight,* because mass is constant while weight changes according to the gravitational force of a body. The gravitational force of Earth is greater than that of the moon. Therefore, substances weigh less on the moon even though their mass has not changed. The atomic mass is often written as a superscript to the upper left of the atomic symbol. For example, the carbon atom can be noted in this way:

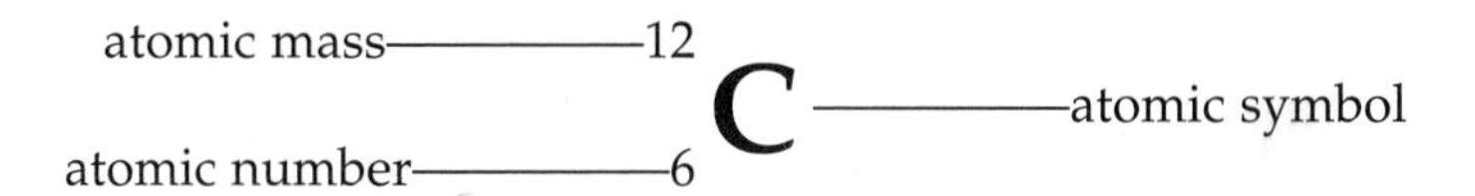

The Periodic Table

Once chemists discovered a number of the elements, they began to realize that even though each element consists of a different atom, certain chemical and physical characteristics recur, or showed periodicity. The periodic table was constructed as a way to group the elements, and therefore atoms, according to these characteristics. Notice in Figure 2.3 that the periodic table is arranged according to increasing atomic number. The vertical columns in the table are groups. The horizontal rows are periods, which cause each atom to be in a particular group. For example, all the atoms in group VII react with one atom at a time, for reasons we will soon explore. The atoms in group VIII are called the noble gases because they are inert and rarely react with another atom. Helium and krypton are noble gases.

In Figure 2.3 the atomic number is above the atomic symbol and the atomic mass is below the atomic symbol. The atomic number tells you the number of positively charged protons. It therefore also tells you the number of negatively charged electrons if the atom is electrically neutral. To determine the number of neutrons, subtract the number of protons from the atomic mass, and take the closest whole number. The periodic table in Figure 2.3 shows only some of the elements. A complete table is available in Appendix D at the end of the book.

Check Your Progress

1. Why would the predominant elements in Earth's crust differ from those present in living organisms?
2. Draw a calcium atom including the protons, neutrons, and electrons.
3. The next element under Kr in the periodic table is Xe. How many electrons does it have in its outer orbital?

Isotopes

Isotopes are atoms of the same element that differ in the number of neutrons. Isotopes have the same number of protons, but they have different atomic masses. Because the number of protons gives an atom its identity, changing the number of neutrons affects the atomic mass but not the name of the atom. For example, the element carbon has three common isotopes:

$$^{12}_{6}C \qquad ^{13}_{6}C \qquad ^{14}_{6}C^{*}$$

*radioactive

Carbon 12 has six neutrons, carbon 13 has seven neutrons, and carbon 14 has eight neutrons. Unlike the other two isotopes of carbon, carbon 14 is unstable. It changes over time into nitrogen 14, which is a stable isotope of the element nitrogen. As carbon 14 decays, it releases various types of energy in the form of rays and subatomic particles, and therefore it is a **radioactive isotope.** Today, biologists use radiation to date objects, create images, and trace the movement of substances.

Low Levels of Radiation

The chemical behavior of a radioactive isotope is essentially the same as that of the stable isotopes of an element. This means that you can put a small amount of radioactive isotope in a sample and it becomes a tracer or tag by which to detect molecular changes.

The importance of chemistry to medicine is nowhere more evident than in the many medical uses of radioactive isotopes. Specific tracers are used in imaging the body's organs and tissues. For example, after a patient drinks a solution containing a minute amount of radioactive ^{131}I, it becomes concentrated in the thyroid—the only organ to take up iodine. A subsequent image of the thyroid indicates whether it is healthy in structure and function (Fig. 2.4*a*). Positron emission tomography (PET) is a way to determine the comparative activity of tissues. Radioactively labeled glucose, which emits a subatomic particle known as a positron, is injected into the body. The radiation given off is detected by sensors and analyzed by a computer. The result is a color image that shows which tissues took up glucose and are metabolically active (Fig. 2.4*b*). A PET scan of the brain can help diagnose a brain tumor, Alzheimer disease, epilepsy, or whether a stroke has occurred.

High Levels of Radiation

Radioactive substances in the environment can harm cells, damage DNA, and cause cancer. When Marie Curie was studying radiation, its harmful effects were not known, and she and many of her co-workers developed cancer. The release of radioactive particles following a nuclear power plant accident can have far-reaching and long-lasting effects on human health. However, the effects of radiation can also be put to good use (Fig. 2.5). Radiation from radioactive

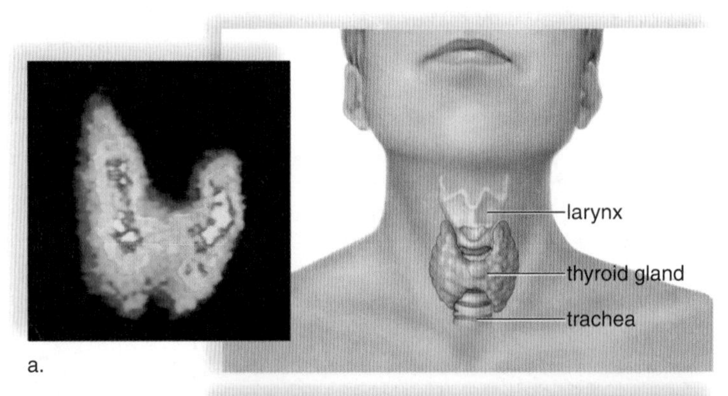

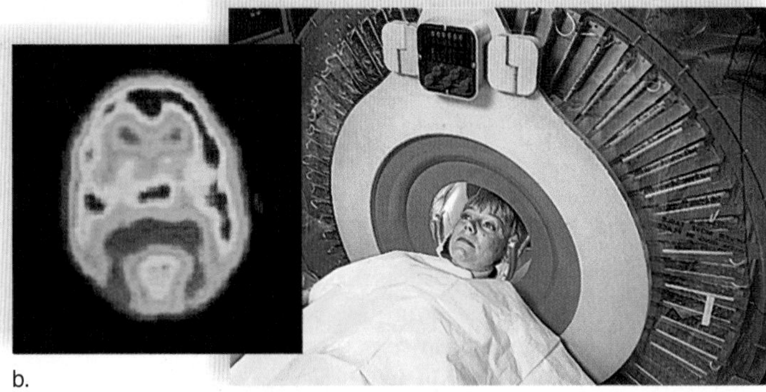

Figure 2.4 Low levels of radiation.
a. The missing area in this thyroid scan (*upper left*) indicates the presence of a tumor that does not take up the radioactive iodine. **b.** A PET (positron emission tomography) scan reveals which portions of the brain are most active (yellow and red colors).

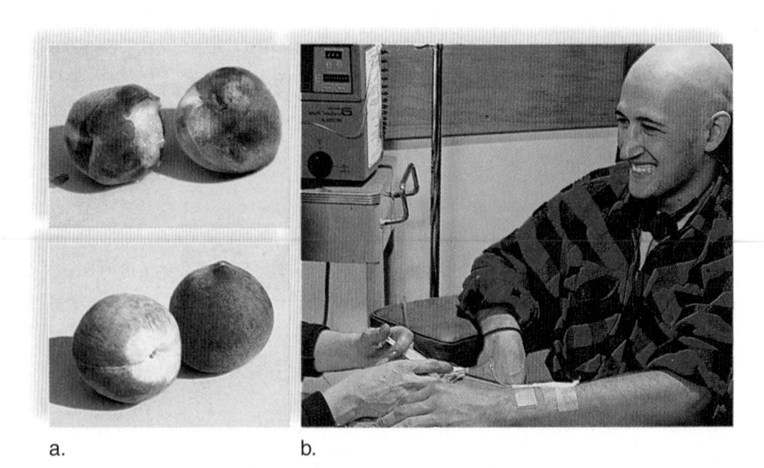

Figure 2.5 High levels of radiation.
a. Radiation kills bacteria and fungi. Irradiated peaches (*bottom*) spoil less quickly and can be kept for a longer length of time. **b.** Physicians use radiation therapy to kill cancer cells.

isotopes has been used for many years to sterilize medical and dental products. Following the terrorist attacks of 9/11, radiation was used to sterilize the U.S. mail and other packages to free them of possible pathogens, such as anthrax spores. The ability of radiation to kill cells is often applied to cancer cells. Targeted radioisotopes can be introduced into the body so that the subatomic particles emitted destroy only cancer cells, with little risk to the rest of the body. X-rays, another form of high-energy radiation, can be used for medical diagnosis.

Electrons

In an electrically neutral atom, the positive charges of the protons in the nucleus are balanced by the negative charges of electrons moving about the nucleus. Various models in years past have attempted to illustrate the precise location of electrons. Figure 2.6 uses the Bohr model, which is named after the physicist Niels Bohr. The Bohr model is useful, but today's physicists tell us it is not possible to determine the precise location of any individual electron at any given moment.

In the diagrams in Figure 2.6, the energy levels (also termed electron orbitals) are drawn as concentric rings about the nucleus. For atoms up through number 20 (i.e., calcium), the first orbital (closest to the nucleus) can contain two electrons. Thereafter, each additional orbital can contain eight electrons. For these atoms, each lower level is filled with electrons before the next higher level contains any electrons.

The sulfur atom, with an atomic number of 16, has two electrons in the first orbital, eight electrons in the second orbital, and six electrons in the third, or outer, orbital. Revisit the periodic table (see Fig. 2.3), and note that sulfur is in the third period. In other words, the horizontal row tells you how many orbitals an atom has. Also note that sulfur is in group VI. The group tells you how many electrons an atom has in its outer orbital.

If an atom has only one orbital, the outer orbital is complete when it has two electrons. Otherwise, atomic orbitals follow the octet rule, which states that the outer orbital is most stable when it has eight electrons. As mentioned previously, atoms in group VIII of the periodic table are called the noble gases because they do not ordinarily react. Atoms with fewer than eight electrons in the outer orbital react with other atoms in such a way that after the reaction, each has a stable outer orbital. Atoms can give up, accept, or share electrons in order to have eight electrons in the outer orbital.

Check Your Progress

1. How can radiation be both beneficial and harmful at the same time?
2. In Figure 2.6, why does carbon have only two orbitals while phosphorus and sulfur have three orbitals?
3. What is the difference between oxygen 16 and oxygen 18?

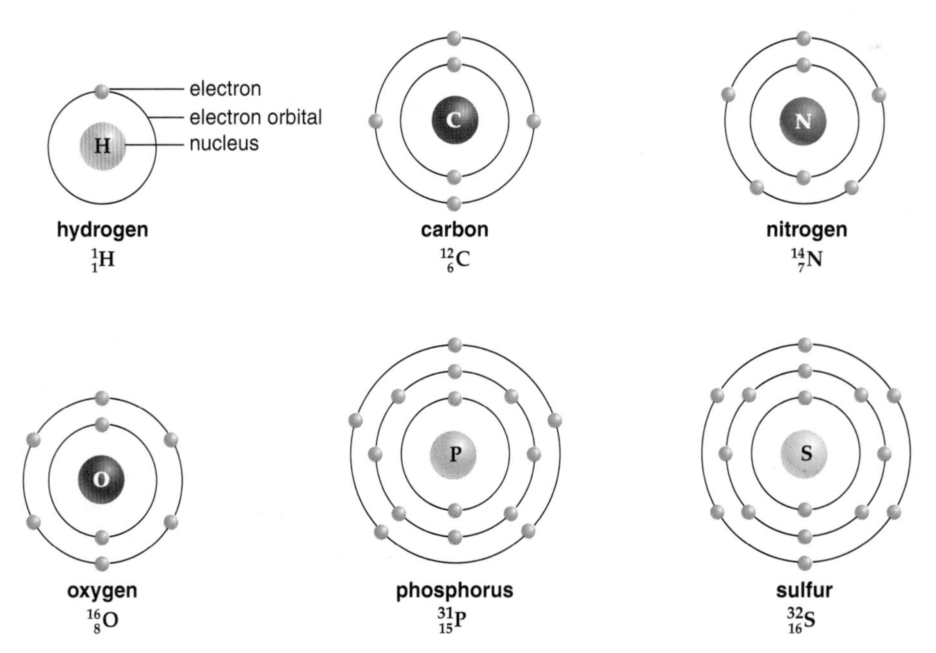

Figure 2.6 Bohr models of atoms.
Electrons orbit the nucleus at particular energy levels (electron orbitals). The first orbital contains up to two electrons. Each orbital thereafter can contain up to eight electrons as long as we consider only atoms with an atomic number of 20 or below. Each orbital is filled before electrons are placed in the next orbital.

2.2 Molecules and Compounds

LEARNING OUTCOMES

1. Define a molecule and a compound.
2. Compare and contrast ionic and covalent bonds.
3. Explain why water is a polar molecule and how this enables the formation of hydrogen bonds.

Atoms, except for noble gases, routinely bond with one another. A **molecule** is formed when two or more atoms bond together. For example, oxygen does not exist in nature as a single atom, O. Instead, two oxygen atoms are joined to form a molecule of oxygen, O_2. When atoms of two or more different elements bond together, the product is called a **compound.** Water (H_2O) is a compound that contains atoms of hydrogen and oxygen. We can also speak of molecules of water because a molecule is the smallest part of a compound that still has the properties of that compound.

Electrons possess energy, and the bonds that exist between atoms also contain energy. Organisms are directly dependent on chemical-bond energy to maintain their organization. When a chemical reaction occurs, electrons shift in their relationship to one another, and energy may be given off or absorbed. This same energy is used to carry on our daily lives.

Ionic Bonding

Ions form when electrons are transferred from one atom to another. For example, sodium (Na), with only one electron in its third orbital, tends to be an electron donor (Fig. 2.7*a*). Once it gives up this electron, the second orbital, with eight electrons, becomes its outer orbital. Chlorine (Cl), on the other hand, tends to be an electron acceptor. Its outer orbital has seven electrons, so if it acquires only one more electron, it has a completed outer orbital. When a sodium atom and a chlorine atom come together, an electron is transferred from the sodium atom to the chlorine atom. Now both atoms have eight electrons in their outer orbitals.

This electron transfer, however, causes a charge imbalance in each atom. The sodium atom has one more proton than it has electrons. Therefore, it has a net charge of +1 (symbolized by Na^+). The chlorine atom has one more electron than it has protons. Therefore, it has a net charge of −1 (symbolized by Cl^-). Such charged particles are called **ions.** Sodium (Na^+) and chloride (Cl^-) are not the only biologically important ions. Some, such as potassium (K^+), are formed by the transfer of a single electron to another atom. Others, such as calcium (Ca^{2+}) and magnesium (Mg^{2+}), are formed by the transfer of two electrons.

Ionic compounds are held together by an attraction between negatively and positively charged ions called an **ionic bond.** When sodium reacts with chlorine, an ionic compound called sodium chloride (NaCl) results. Sodium chloride is a salt, commonly known as table salt because it is used to season our food (Fig. 2.7*b*). Salts can exist as a dry solid, but when salts are placed in water, they release ions as they dissolve. NaCl separates into Na^+ and Cl^-. Ionic compounds are most commonly found in this dissociated (ionized) form in biological systems because these systems are 70–90% water.

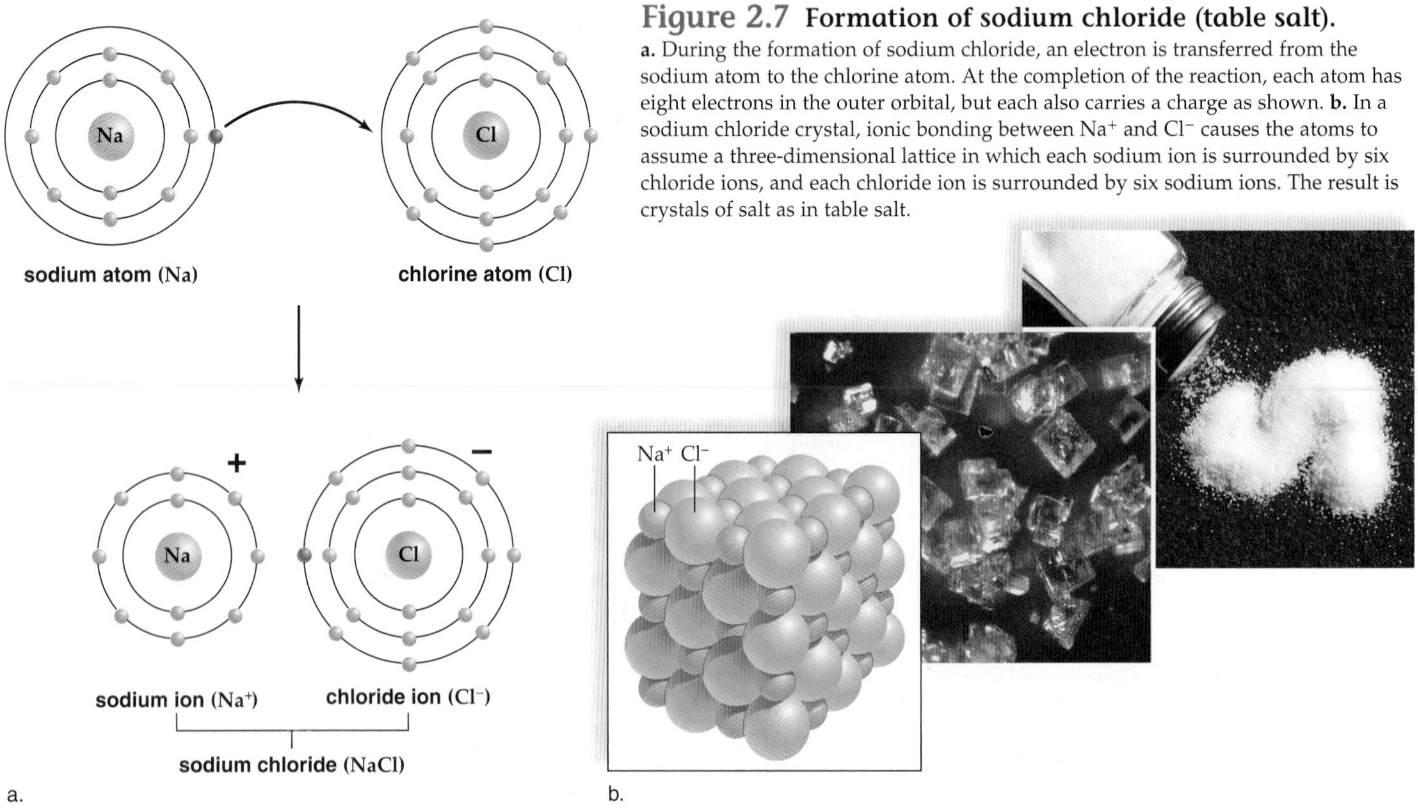

Figure 2.7 Formation of sodium chloride (table salt). **a.** During the formation of sodium chloride, an electron is transferred from the sodium atom to the chlorine atom. At the completion of the reaction, each atom has eight electrons in the outer orbital, but each also carries a charge as shown. **b.** In a sodium chloride crystal, ionic bonding between Na^+ and Cl^- causes the atoms to assume a three-dimensional lattice in which each sodium ion is surrounded by six chloride ions, and each chloride ion is surrounded by six sodium ions. The result is crystals of salt as in table salt.

Covalent Bonding

A **covalent bond** results when two atoms share electrons in such a way that each atom has an octet of electrons in the outer orbital. In a hydrogen atom, the outer orbital is complete when it contains two electrons. If hydrogen is in the presence of a strong electron acceptor, it gives up its electron to become a hydrogen ion (H^+). But if this is not possible, hydrogen can share with another atom and thereby have a completed outer orbital. For example, one hydrogen atom will share with another hydrogen atom. Their two orbitals overlap, and the electrons are shared between them (Fig. 2.8*a*). Because they share the electron pair, each atom has a completed outer orbital.

A more common way to symbolize that atoms are sharing electrons is to draw a line between the two atoms, as in the structural formula H—H. In a molecular formula, the line is omitted and the molecule is simply written as H_2.

Sometimes, atoms share more than one pair of electrons to complete their octets. A double covalent bond occurs when two atoms share two pairs of electrons (Fig. 2.8*b*). To show that oxygen gas (O_2) contains a double bond, the molecule can be written as O═O.

It is also possible for atoms to form triple covalent bonds, as in nitrogen gas (N_2), which can be written as N≡N. Single covalent bonds between atoms are quite strong, but double and triple bonds are even stronger.

Shape of Molecules

Structural formulas make it seem as if molecules are one-dimensional, but actually molecules have a three-dimensional shape that often determines their biological function. Molecules consisting of only two atoms are always linear, but a molecule such as methane with five atoms (Fig. 2.8*c*) has a tetrahedral shape. Why? Because, as shown in the ball-and-stick model, each bond is pointing to the corners of a tetrahedron (Fig. 2.8*d, left*). The space-filling model comes closest to the actual shape of the molecule. In space-filling models, each type of atom is given a particular color—carbon is always black and hydrogen is always off-white (Fig. 2.8*d, right*).

The shapes of molecules are necessary to the structural and functional roles they play in living things. For example, hormones have specific shapes that allow them to be recognized by the cells in the body. Antibodies combine with disease-causing agents, like a key fits a lock, to keep us well. Similarly, homeostasis is maintained only when enzymes have the proper shape to carry out their particular reactions in cells.

Nonpolar and Polar Covalent Bonds

When the sharing of electrons between two atoms is fairly equal, the covalent bond is said to be a nonpolar covalent bond. All the molecules in Figure 2.8, including methane

Electron Model	Structural Formula	Molecular Formula
H H	H—H	H_2

a. Hydrogen gas

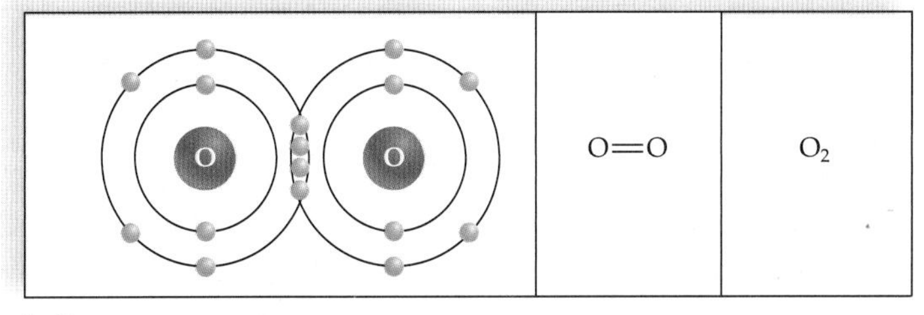

b. Oxygen gas

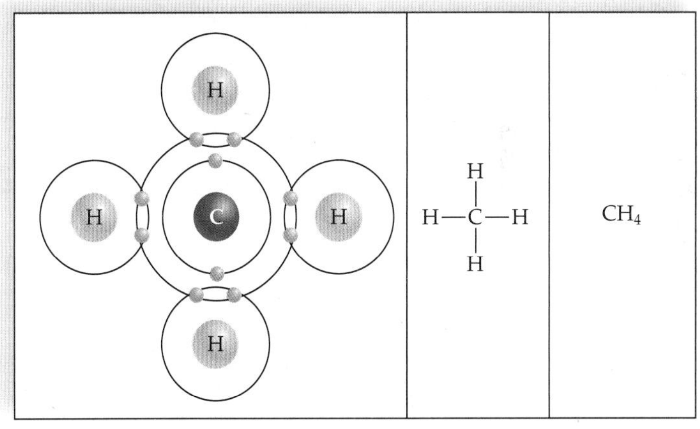

c. Methane

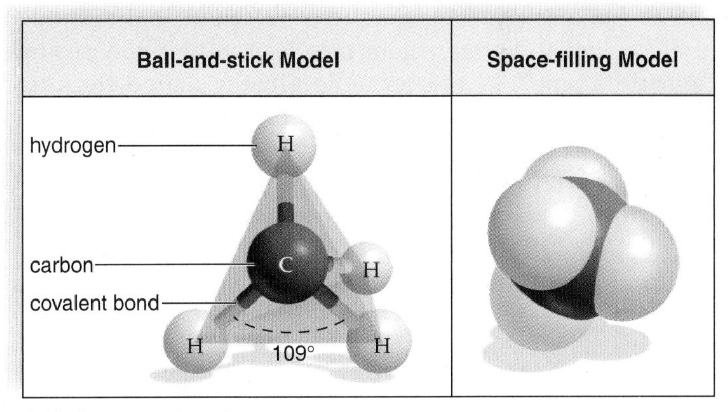

d. Methane–continued

Figure 2.8 Covalently bonded molecules.
In a covalent bond, atoms share electrons, allowing each atom to have a completed outer orbital. **a.** A molecule of hydrogen (H_2) contains two hydrogen atoms sharing a pair of electrons. This single covalent bond can be represented in any of these three ways. **b.** A molecule of oxygen (O_2) contains two oxygen atoms sharing two pairs of electrons. This results in a double covalent bond. **c.** A molecule of methane (CH_4) contains one carbon atom bonded to four hydrogen atoms. **d.** When carbon binds to four other atoms, as in methane, each bond actually points to one corner of a tetrahedron. Ball-and-stick models and space-filling models are three-dimensional representations of a molecule—in this case, methane.

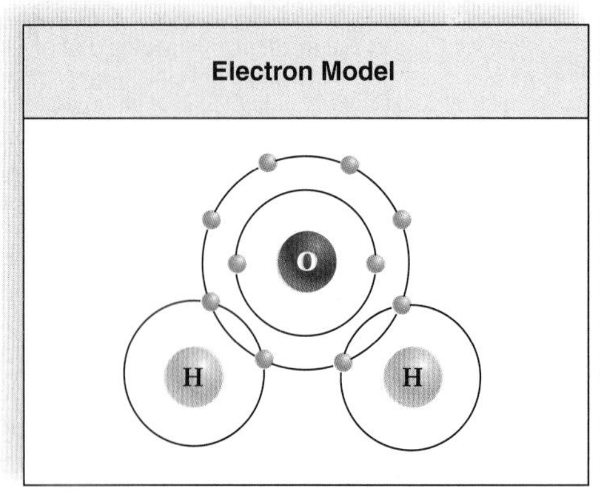

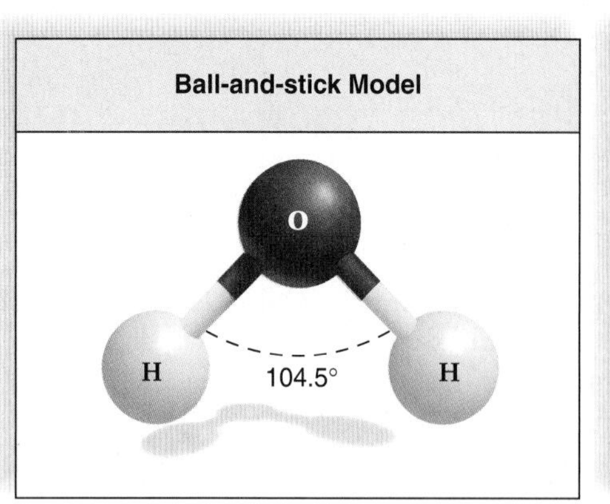

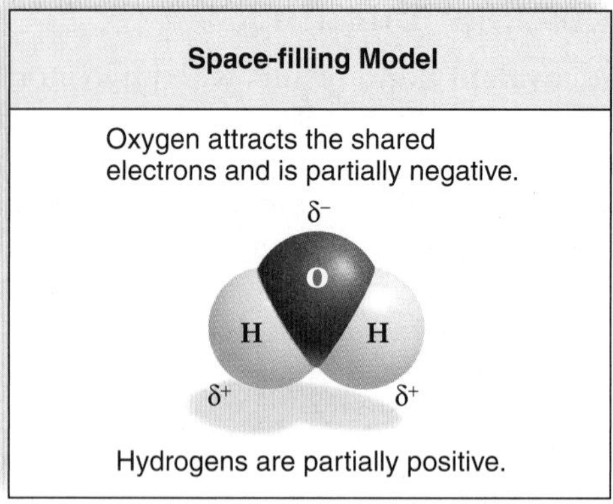

a. Water (H_2O)

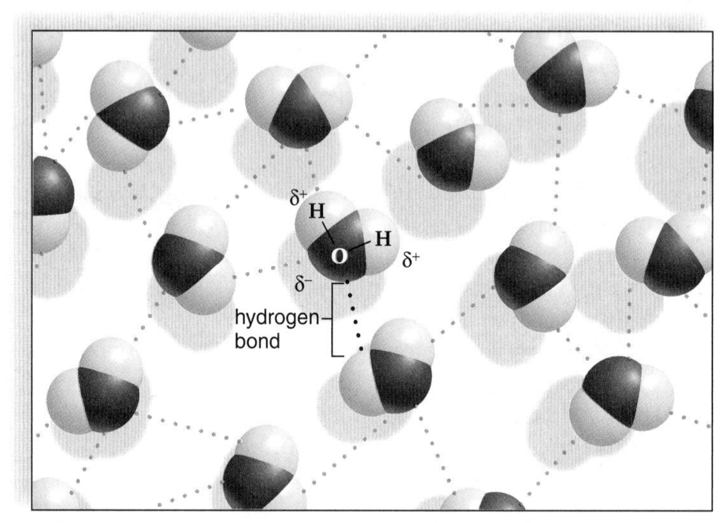

b. Hydrogen bonding between water molecules

Figure 2.9 Water molecule.
a. Three models for the structure of water. The electron model does not indicate the shape of the molecule. The ball-and-stick model shows that the two bonds in a water molecule are angled at 104.5°. The space-filling model also shows the V shape of a water molecule. **b.** Hydrogen bonding between water molecules. A hydrogen bond is the attraction of a slightly positive hydrogen to a slightly negative atom in the vicinity. Each water molecule can hydrogen-bond to four other molecules in this manner. When water is in its liquid state, some hydrogen bonds are forming and others are breaking at all times.

(CH_4), are nonpolar. In the case of water (H_2O), however, the sharing of electrons between oxygen and each hydrogen is not completely equal. The attraction of an atom for the electrons in a covalent bond is called its electronegativity. The larger oxygen atom, with the greater number of protons, is more electronegative than the hydrogen atom. The oxygen atom can attract the electron pair to a greater extent than each hydrogen atom can. It may help to think of electronegativity as where the electron pair chooses to "spend its time." In a water molecule, the shared electron pair spends more time around the nucleus of the oxygen atom than around the nucleus of the hydrogen atom. This causes the oxygen atom to assume a slightly negative charge (δ^-), and it causes the hydrogen atoms to assume slightly positive charges (δ^+). The unequal sharing of electrons in a covalent bond creates a polar covalent bond. In the case of water, the molecule itself is a polar molecule (Fig. 2.9*a*).

Hydrogen Bonding

Polarity within a water molecule causes the hydrogen atoms in one molecule to be attracted to the oxygen atoms in other water molecules (Fig. 2.9*b*). This attraction, although weaker than an ionic or covalent bond, is called a **hydrogen bond.** Because a hydrogen bond is easily broken, it is often represented by a dotted line. Hydrogen bonding is not unique to water. Many biological molecules have polar covalent bonds involving an electropositive hydrogen and usually an electronegative oxygen or nitrogen. In these instances, a hydrogen bond can occur within the same molecule or between different molecules.

Although a hydrogen bond is more easily broken than a covalent bond, many hydrogen bonds taken together are quite strong. Hydrogen bonds between cellular molecules help maintain their proper structure and function. For example, hydrogen bonds hold the two strands of DNA together. When DNA makes a copy of itself, each hydrogen bond easily breaks, allowing the DNA to unzip. On the other hand, the hydrogen bonds acting together add stability to the DNA molecule. As we shall see, many of the important properties of water are the result of hydrogen bonding.

Check Your Progress

1. Is carbon dioxide (CO_2) a molecule, a compound, or both? Nitrogen gas (N_2)?
2. A carbon atom (C) is capable of forming four covalent bonds. Why would you expect that?
3. Why would you expect any bond between H and another atom (other than itself) to be a polar bond?
4. Why is the H in a water molecule said to be partially positive and the O partially negative instead of just positive or negative?

2.3 Chemistry of Water

Learning Outcomes

1. Describe the unique properties of water and the advantages of these properties for life.
2. Be able to define an acid and a base and use the pH scale.
3. Recognize the importance of buffers to living organisms.

The first cell(s) evolved in water, and all living things are 70–90% water. Water is a polar molecule, and water molecules are hydrogen-bonded to one another (see Fig. 2.9*b*). Due to hydrogen bonding, water molecules cling together. Without hydrogen bonding between molecules, water would change from a solid to liquid state at –100°C and from a liquid to gaseous state at –91°C. This would make most of the water on Earth steam, and life unlikely. But because of hydrogen bonding, water is a liquid at temperatures typically found on Earth's surface. It melts at 0°C and boils at 100°C. These and other unique properties of water make it essential to the existence of life.

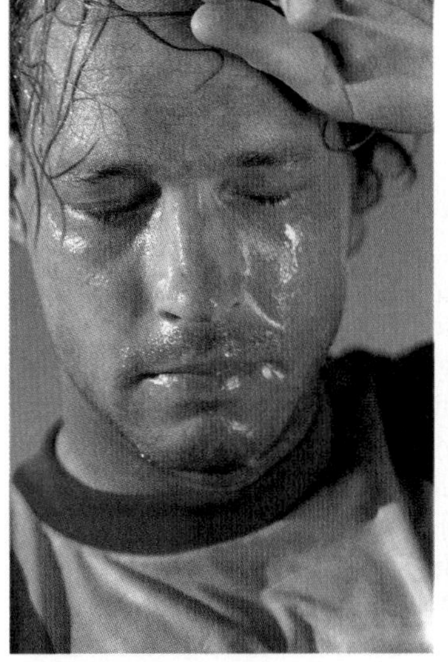

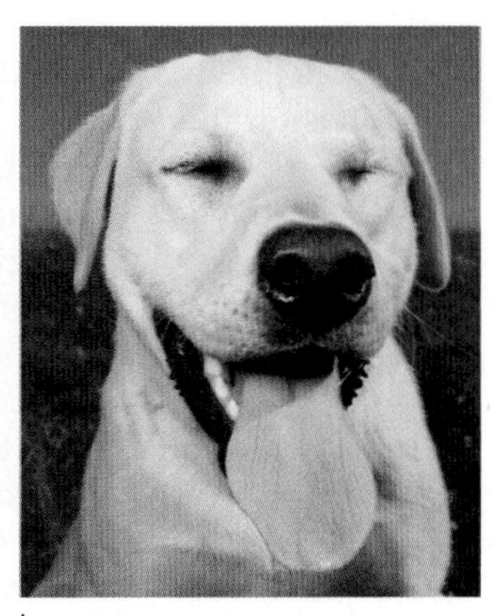

Figure 2.10 The advantage of water's high heat of vaporization.
a. When body temperature increases, sweat is produced by glands in the dermal layer of the skin. The evaporation of the water off your skin aids in cooling your body. **b.** Dogs cannot sweat like humans do. Instead a dog pants. As air passes over the moist tongue, the water evaporates cooling the body.

Properties of Water

Water has a high heat capacity. A **calorie** is the amount of heat energy needed to raise the temperature of 1 gram (g) of water 1°C. In comparison, other covalently bonded liquids require input of only about half this amount of energy to rise 1°C in temperature. The many hydrogen bonds that link water molecules help water absorb heat without a great change in temperature.

Converting 1 g of the coldest liquid water to ice requires the loss of 80 calories of heat energy. Water holds onto its heat, and its temperature falls more slowly than that of other liquids. This property of water is important not only for aquatic organisms but also for all living things. Because the temperature of water rises and falls slowly, organisms are better able to maintain their normal internal temperatures and are protected from rapid temperature changes.

Water has a high heat of vaporization. Converting 1 g of the hottest water to a gas requires an input of 540 calories of heat energy. Water has a high heat of vaporization because hydrogen bonds must be broken before water boils and water molecules vaporize—that is, evaporate into the environment. Water's high heat of vaporization gives animals in a hot environment an efficient way to release excess body heat (Fig. 2.10). When an animal sweats, or gets splashed, body heat is used to vaporize the water, thus cooling the animal.

Because of water's high heat capacity and high heat of vaporization, temperatures along coasts are moderate. During the summer, the ocean absorbs and stores solar heat, and during the winter, the ocean releases it slowly. In contrast, the interior regions of continents can experience severe changes in temperature.

Water is a solvent. Due to its polarity, water facilitates chemical reactions, both outside and within living systems. It dissolves a great number of substances. A solution contains dissolved substances, which are then called **solutes.** When ionic salts—for example, sodium chloride (NaCl)—are put into water, the negative ends of the water molecules are attracted to the sodium ions, and the positive ends of the water molecules are attracted to the chloride ions. This causes the sodium ions and the chloride ions to separate, or dissociate, in water:

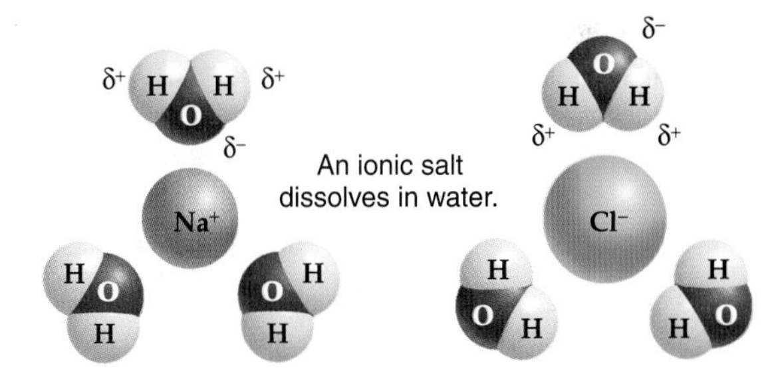

Water is also a solvent for larger molecules that contain ionized atoms or are polar molecules.

Those molecules that can attract water are said to be **hydrophilic.** When ions and molecules disperse in water, they move about and collide, allowing reactions to occur. Nonionized and nonpolar molecules, such as oil, that cannot attract water are said to be **hydrophobic.**

Water molecules are cohesive and adhesive. Cohesion is apparent because water flows freely, and yet water molecules do not separate from each other. They cling together because of hydrogen bonding. Water exhibits adhesion because its positive and negative poles allow it to adhere to polar surfaces. Cohesion and adhesion allow water to fill a tubular vessel. Therefore, water is an excellent transport system, both outside of and within living organisms.

Unicellular organisms rely on external water to transport nutrient and waste molecules, but multicellular organisms often contain internal vessels through which water transports nutrients and wastes. For example, the liquid portion of our blood, which transports dissolved and suspended substances throughout the body, is 90% water.

Cohesion and adhesion also contribute to the transport of water in plants. The roots of plants are anchored in the soil, where they absorb water, but the leaves are uplifted and exposed to solar energy. How is it possible for water to rise to the top of even very tall trees? A plant contains a system of vessels that reaches from the roots to the leaves. Water evaporating from the leaves is immediately replaced with water molecules from the vessels. Because water molecules are cohesive, a tension is created that pulls a water column up from the roots. Adhesion of water to the walls of the vessels also helps prevent the water column from breaking apart.

Water has a high surface tension. The stronger the force between molecules in a liquid, the greater the surface tension. As with cohesion, hydrogen bonding causes water to have a high surface tension. This property makes it possible for humans to skip rocks on water. The water strider, a common insect, can even walk on top of a pond without breaking the surface.

Frozen water (ice) is less dense than liquid water. As liquid water cools, the molecules come closer together. They are densest at 4°C, but they are still moving about, bumping into each other (Fig. 2.11). At temperatures below 4°C, including at 0°C when water is frozen, the water forms a regular crystal lattice that is rigid and more open since the water molecules are no longer moving about. For this reason water expands as it freezes, which is why cans of soda burst when placed in a freezer or why frost heaves make northern roads bumpy in the winter. It also means that ice is less dense than liquid water, and therefore ice floats on liquid water.

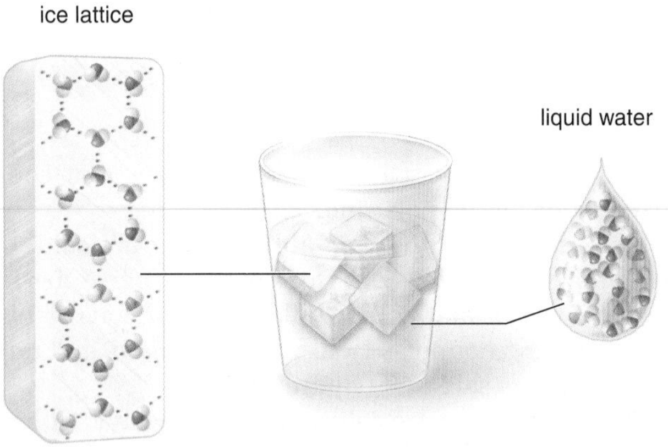

If ice did not float on water, it would sink, and ponds, lakes, and perhaps even the ocean would freeze solid. This would make life impossible in the water and also on land. Instead, bodies of water always freeze from the top down. When a body of water freezes on the surface, the ice acts as an insulator to prevent the water below it from freezing. This protects aquatic organisms so that they can survive the winter. As ice melts in the spring, it draws heat from the environment, helping to prevent a sudden change in temperature that might be harmful to life.

ice lattice

liquid water

Figure 2.11 Ice floats on water.
Remarkably, water is more dense at 4°C than at 0°C. Most substances contract when they solidify. But water expands when it freezes because the water molecules in ice form a lattice in which the hydrogen bonds are farther apart than in liquid water.

Acids and Bases

When water ionizes, it releases an equal number of hydrogen ions (H^+) and hydroxide ions (OH^-):

$$\underset{\text{water}}{H{-}O{-}H} \rightleftharpoons \underset{\text{hydrogen ion}}{H^+} + \underset{\text{hydroxide ion}}{OH^-}$$

Only a few water molecules at a time dissociate, and the actual number of H^+ and OH^- is very small (1×10^{-7} moles/liter).[1]

Acidic Solutions (High H^+ Concentrations)

Lemon juice, vinegar, tomatoes, and coffee are all acidic solutions. What do they have in common? **Acids** are substances that release hydrogen ions (H^+)[2] when they dissociate in water. For example, hydrochloric acid (HCl) is an important acid that dissociates in this manner:

$$HCl \longrightarrow H^+ + Cl^-$$

Because dissociation is almost complete, HCl is called a strong acid. If hydrochloric acid is added to a beaker of water, the number of hydrogen ions (H^+) increases greatly.

Basic Solutions (Low H^+ Concentrations)

Baking soda and antacids are common basic solutions familiar to most people. **Bases** are substances that either take up hydrogen ions (H^+) or release hydroxide ions (OH^-). For example, sodium hydroxide (NaOH) is an important base that dissociates in this manner:

$$NaOH \longrightarrow Na^+ + OH^-$$

Because dissociation is almost complete, sodium hydroxide is called a strong base. If sodium hydroxide is added to a beaker of water, the number of hydroxide ions increases.

[1] In chemistry, a mole is defined as the amount of matter that contains as many objects (atoms, molecules, ions) as the number of atoms in exactly 12 g of ^{12}C.

[2] A hydrogen atom contains one electron and one proton. A hydrogen ion has only one proton, so it is often simply called a proton.

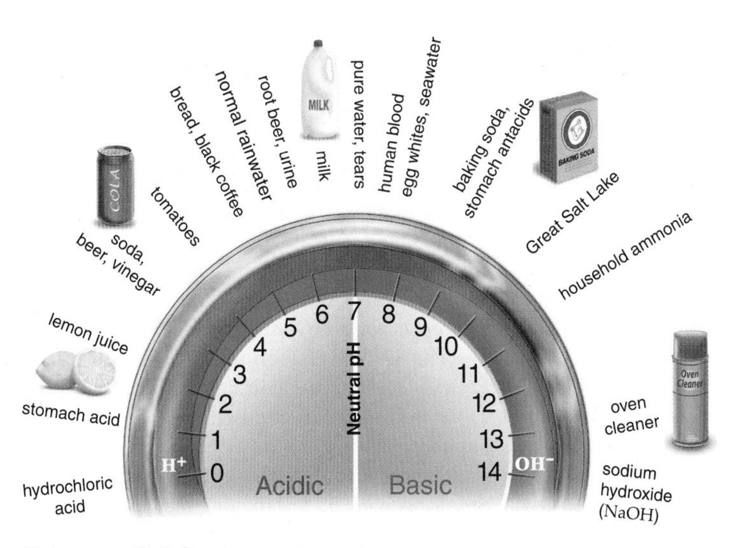

Figure 2.12 The pH scale.
The dial of this pH meter indicates that pH ranges from 0 to 14, with 0 the most acidic and 14 the most basic. pH 7 (neutral pH) has equal amounts of hydrogen ions (H^+) and hydroxide ions (OH^-). An acidic pH has more H^+ than OH^-, and a basic pH has more OH^- than H^+.

pH Scale

The **pH scale** is used to indicate the acidity or basicity (alkalinity) of solutions.[3] The pH scale (Fig. 2.12) ranges from 0 to 14. A pH of 7 represents a neutral state in which the hydrogen ion and hydroxide ion concentrations are equal. A pH below 7 is an acidic solution because the hydrogen ion concentration [H^+] is greater than the hydroxide concentration [OH^-]. A pH above 7 is basic because [OH^-] is greater than [H^+]. As we move down the pH scale from pH 14 to pH 0, each unit has ten times the H^+ concentration of the previous unit. As we move up the scale from 0 to 14, each unit has ten times the OH^- concentration of the previous unit.

The pH scale was devised to eliminate the use of cumbersome numbers. For example, the possible hydrogen ion concentrations of a solution are on the left in the following listing, and the pH is on the right:

[H^+] (moles per liter)		pH
0.000001	$= 1 \times 10^{-6}$	6
0.0000001	$= 1 \times 10^{-7}$	7
0.00000001	$= 1 \times 10^{-8}$	8

To further illustrate the relationship between hydrogen ion concentration and pH, consider the following question. Which of the pH values listed indicates a higher hydrogen ion concentration [H^+] than pH 7, and therefore would be an acidic solution? A number with a smaller negative exponent indicates a greater quantity of hydrogen ions than one with a larger negative exponent. Therefore, pH 6 is an acidic solution.

Buffers and pH

A **buffer** is a chemical or a combination of chemicals that keeps pH within normal limits. Many commercial products, such as Bufferin, shampoos, or deodorants, are buffered as an added incentive for us to buy them. Buffers resist pH changes because they can take up excess hydrogen ions (H^+) or hydroxide ions (OH^-).

In animals, the pH of body fluids is maintained within a narrow range, or else health suffers. The pH of our blood when we are healthy is always about 7.4—that is, just slightly basic (alkaline). If the blood pH drops to about 7, acidosis results. If the blood pH rises to about 7.8, alkalosis results. Both conditions can be life threatening. Normally, pH stability is possible because the body has built-in mechanisms to prevent pH changes. Buffers are the most important of these mechanisms. For example, carbonic acid (H_2CO_3) is a weak acid that minimally dissociates and then re-forms in the following manner:

$$\underset{\text{carbonic acid}}{H_2CO_3} \underset{\text{re-forms}}{\overset{\text{dissociates}}{\rightleftharpoons}} H^+ + \underset{\text{bicarbonate ion}}{HCO_3^-}$$

Blood always contains a combination of some carbonic acid and some bicarbonate ions. When hydrogen ions (H^+) are added to blood, the following reaction occurs:

$$H^+ + HCO_3^- \longrightarrow H_2CO_3$$

When hydroxide ions (OH^-) are added to blood, this reaction occurs:

$$OH^- + H_2CO_3 \longrightarrow HCO_3^- + H_2O$$

These reactions prevent any significant change in blood pH.

CHECK YOUR PROGRESS

1. What is the difference between water's high heat capacity and high heat of vaporization?
2. Would you expect charged molecules (either positively or negatively charged) to be hydrophilic or hydrophobic?
3. Explain why solution with a pH of 6 contains more H^+ than a solution with a pH of 8.
4. Why is a weakly dissociating acid/base and not a strongly dissociating one a better buffer?

[3] pH is defined as the negative log of the hydrogen ion concentration [H^+]. A log is the power to which ten must be raised to produce a given number.

Ecology Focus

Are Fish in the Stream Near Your House Taking Antidepressants?

In 1999–2000 the U.S. Department of the Interior's Geological Survey conducted a "reconaissance survey" of pharmaceuticals, hormones, and other organic waste products present in U.S. streams. This was the first nationwide survey of its type. The U.S. Geological Survey sampled 139 streams in 30 states. They chose streams they expected to be contaminated, so the results are not representative of all the streams in the United States. They tested for 95 different contaminants. What they found was shocking. Of the 139 streams tested, 80% had one or more contaminants present. The median number of contaminants per stream was seven, although one stream had 38 contaminants present. They found 82 different contaminants, including:

- triclosan, the antimicrobial disinfectant found in hand soap;
- estradiol, found in hormone replacement pills;
- norethisterone, found in birth control pills;
- ciprofloxacin, an antibiotic;
- codeine, an analgesic; and
- fluoxetine, an antidepressant.

The concentration of the contaminants was generally very low and very few exceeded the drinking water guidelines. However, many of these contaminants do not have established drinking water guidelines. There is little to nothing known about the adverse effects of many of these chemicals in drinking water. Additionally, there is little known about the long-term effects of exposure to very low concentrations of many of these compounds.

Endocrine disruptors such as reproductive hormones are the exception. Even low levels of endocrine disruptors adversely affect aquatic organisms and humans. These agents mimic hormones within the body which act at very low concentrations. The U.S. Geological Survey published the results of experiments in which male minnows were exposed to wastewater effluent (Fig. 2A). These male minnows exhibited an endocrine disruptive response. Endocrine disruptors affect development, reproduction, and any other processes mediated by hormones.

It has also been proposed that long-term exposure to low levels of antibiotics in drinking water may increase the incidence of antibiotic resistance in bacteria. This could become a problem for both humans and livestock.

The Office of National Drug Control Policy suggests that you not flush outdated or unused medication down the toilet. They suggest mixing medication with coffee grounds or kitty litter and then sealing it in an empty can or bag before disposal. This will prevent the drugs from being diverted into the water supply or into the hands of unauthorized users. Medication should only be flushed if the label says it can.

Discussion Questions

1. Investigate whether your community has a pharmaceutical take-back program.
2. Each community is required to publish a water quality report by July 1. Obtain a report for your community and discuss the results with your class.

Figure 2A Effects of organic wastewater contamination.
In these aquaria, male fathead minnows were exposed to wastewater effluent for 28 days to determine the potential effects of organic contaminates on growth and development.

2.4 Organic Molecules

Learning Outcomes

1. Distinguish inorganic from organic molecules.
2. Define a functional group.
3. Describe how monomers are joined to form polymers.

Inorganic molecules constitute nonliving matter, but even so, inorganic molecules such as salts (e.g., NaCl) and water play important roles in living things. The molecules of life, however, are organic molecules. **Organic molecules** always contain carbon (C) and hydrogen (H). The chemistry of carbon accounts for the formation of the very large variety of organic molecules found in living things. A carbon atom has four electrons in the outer orbital. In order to achieve eight electrons in the outer orbital, a carbon atom shares electrons covalently with as many as four other atoms, as in methane (CH_4). A carbon atom can share with another carbon atom, and in so doing, a long hydrocarbon chain can result:

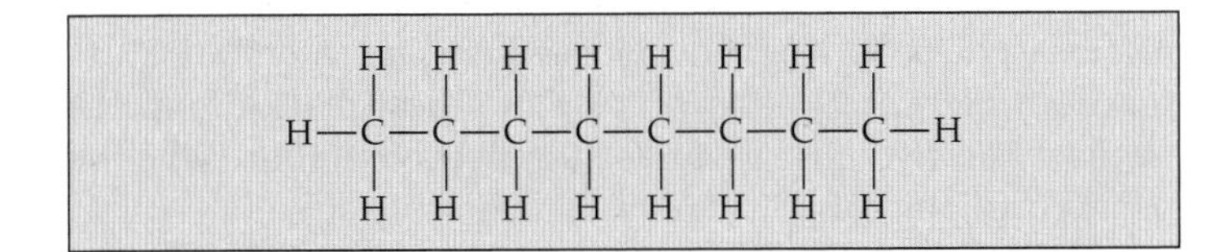

As we shall see, the chain can turn back on itself to form a ring compound. So-called functional groups can be attached to carbon chains. A **functional group** is a particular cluster of atoms that always behaves in a certain way. One functional group of interest is the acidic (carboxyl) group —COOH because it can give up a hydrogen (H^+) and ionize to $—COO^-$.

Many molecules of life are macromolecules—that is, they contain many molecules joined together. A **monomer** (*mono*, one) is a simple organic molecule that exists individually or can link with other monomers to form a **polymer** (*poly*, many). The polymers in cells form from monomers as follows:

Polymer	Monomer
carbohydrate (e.g., starch)	monosaccharide
protein	amino acid
nucleic acid	nucleotide

Aside from carbohydrates, proteins, and nucleic acids, the other organic molecules in cells are lipids. You are very familiar with carbohydrates, lipids, and proteins because certain foods are known to be rich in these molecules, as illustrated in Figure 2.13. The nucleic acid DNA makes up our genes, which are hereditary units that control our cells and the structure of our bodies.

Cells have a common way of joining monomers to build polymers. During a **dehydration reaction,** an —OH (hydroxyl group) and an —H (hydrogen atom), the equivalent of a water molecule, are removed as the reaction proceeds (Fig. 2.14*a*). To degrade polymers, the cell uses a **hydrolysis reaction,** in which the components of water are added (Fig. 2.14*b*).

Check Your Progress

1. Why are the molecules of life called organic molecules?
2. What do dehydration and hydrolysis reactions have in common?

Figure 2.13 Common foods.
Carbohydrates, such as bread, potatoes, and pasta, are digested to sugars. Lipids, such as oils and butter, are digested to glycerol and fatty acids. Proteins, such as meat, are digested to amino acids. Cells use these subunit molecules to build their own macromolecules.

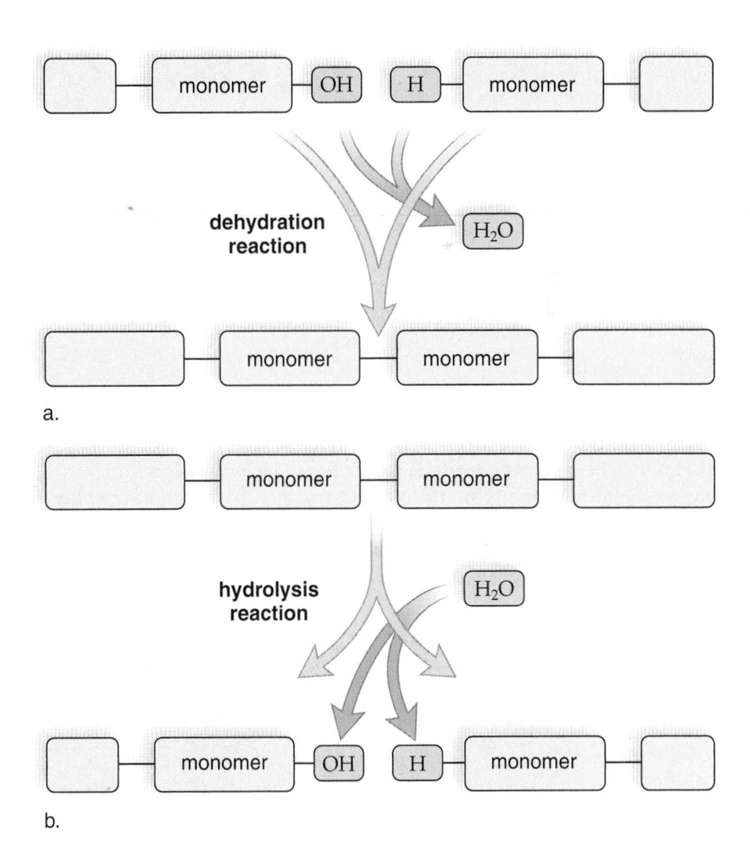

Figure 2.14 Synthesis and degradation of polymers.
a. In cells, synthesis often occurs when monomers join (bond) during a dehydration reaction (removal of H_2O). **b.** Degradation occurs when the monomers in a polymer separate during a hydrolysis reaction (addition of H_2O).

2.5 Carbohydrates

Learning Outcomes

1. Recognize the structure of a carbohydrate.
2. Compare and contrast different types of carbohydrates.

Carbohydrates first and foremost function for quick fuel and short-term energy storage in all organisms, including humans. Carbohydrates play a structural role in woody plants, bacteria, and animals such as insects. In addition, carbohydrates on cell surfaces are involved in cell-to-cell recognition, as we will learn in Chapter 4.

Carbohydrate molecules are characterized by the presence of the atomic grouping H—C—OH, in which the ratio of hydrogen atoms (H) to oxygen atoms (O) is approximately 2:1. Since this ratio is the same as the ratio in water, the name "hydrates of carbon" seems appropriate.

Simple Carbohydrates

If the number of carbon atoms in a molecule is low (from three to seven), then the carbohydrate is a simple sugar, or **monosaccharide.** The designation **pentose** means a 5-carbon sugar, and the designation **hexose** means a 6-carbon sugar. **Glucose,** a hexose, is blood sugar (Fig. 2.15). Our bodies use glucose as an immediate source of energy. Other common hexoses are fructose, found in fruits, and galactose, a constituent of milk. These three hexoses (glucose, fructose, and galactose) all occur as ring structures with the molecular formula $C_6H_{12}O_6$. The exact shape of the ring differs, as does the arrangement of the hydrogen (—H) and hydroxyl (—OH) groups attached to the ring.

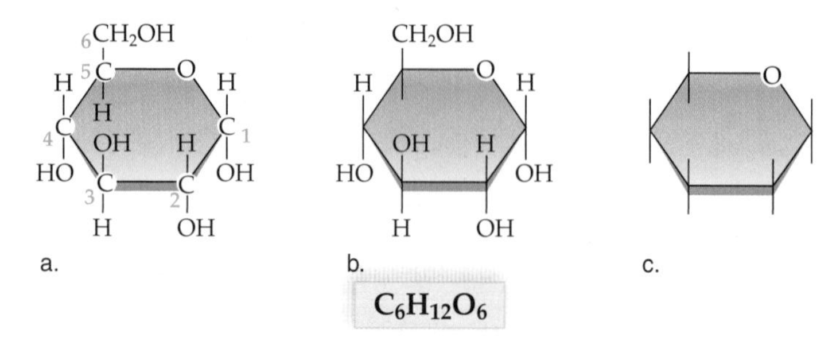

Figure 2.15 Three ways to represent the structure of glucose.
$C_6H_{12}O_6$ is the molecular formula for glucose. The *far left* structure (**a**) shows the carbon atoms, but the middle structure (**b**) does not show the carbon atoms. The *far right* structure (**c**) is the simplest way to represent glucose. Note that in ***a*** and ***b***, each carbon has an attached H and OH group. Those groups are assumed in *c*.

A **disaccharide** (*di*, two; *saccharide*, sugar) contains two monosaccharides that have joined during a dehydration reaction. Figure 2.16 shows how the disaccharide maltose forms when two glucose molecules bond together. Note the position of this bond. Our hydrolytic digestive juices can break this bond, and the result is two glucose molecules. (Think about the sweet taste of malted milk balls!) When glucose and fructose join, the disaccharide sucrose forms. Sucrose is another disaccharide of special interest because we use it at the table to sweeten our food. We acquire the sugar from plants such as sugarcane and sugar beets. You may also have heard of lactose, a disaccharide found in milk. Lactose is glucose combined with galactose. Some people are lactose intolerant because they cannot break down lactose. This leads to unpleasant gastrointestinal symptoms when they drink milk.

Polysaccharides

Long polymers such as starch, glycogen, and cellulose are **polysaccharides** that contain many glucose subunits.

Starch and Glycogen

Starch and **glycogen** are ready storage forms of glucose in plants and animals, respectively. Some of the polymers in starch are long chains of up to 4,000 glucose units. Starch has fewer side branches, or chains of glucose that branch off from the main chain, than does glycogen, as shown in Figures 2.17 and 2.18. Flour, which we use for baking and usually acquire by grinding wheat, is high in starch, and so are potatoes.

After we eat starchy foods such as potatoes, bread, and cake, glucose enters the bloodstream. The liver stores glucose as glycogen. In between meals, the liver releases glucose so that the blood glucose concentration is always about 0.1%.

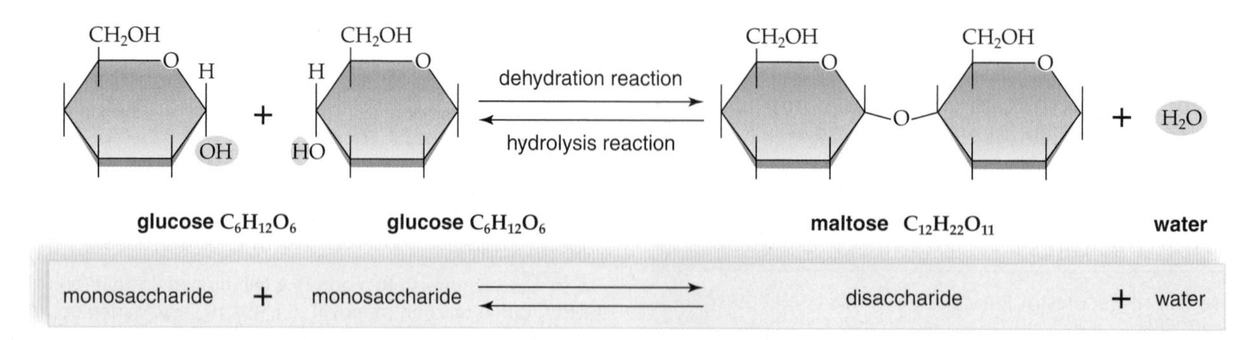

Figure 2.16 Synthesis and degradation of maltose, a disaccharide.
Synthesis of maltose occurs following a dehydration reaction when a bond forms between two glucose molecules and water is removed. Degradation of maltose occurs following a hydrolysis reaction when this bond is broken by the addition of water.

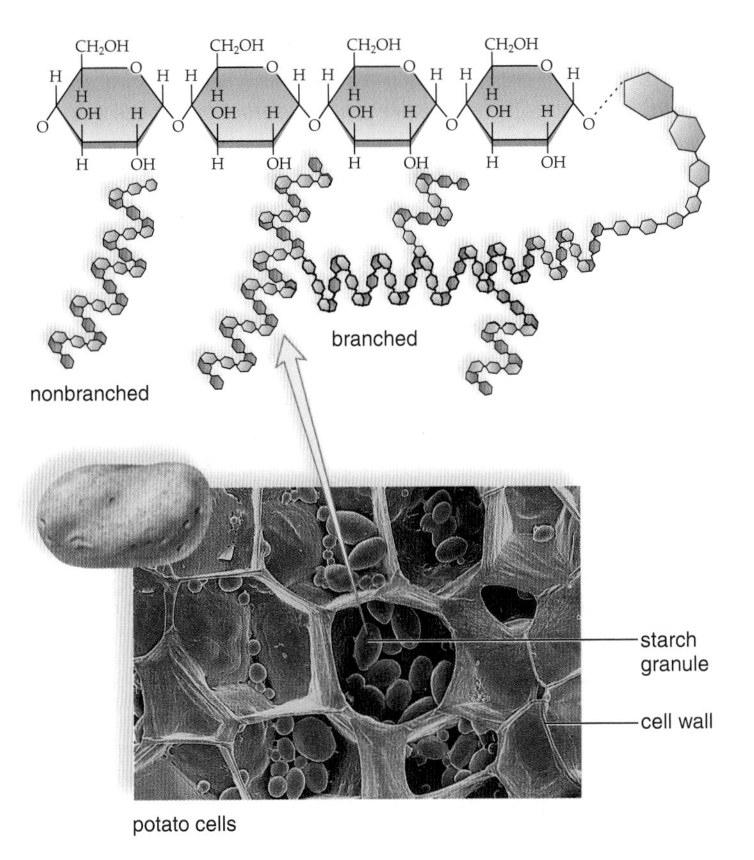

Figure 2.17 Starch structure and function.
Starch is composed of chains of glucose molecules. Some chains are branched, as indicated. Starch is the storage form of glucose in plants. The electron micrograph shows starch granules in potato cells. When we eat starch-containing foods such as corn, potatoes, bread, white rice, and pasta, glucose enters our bloodstream.

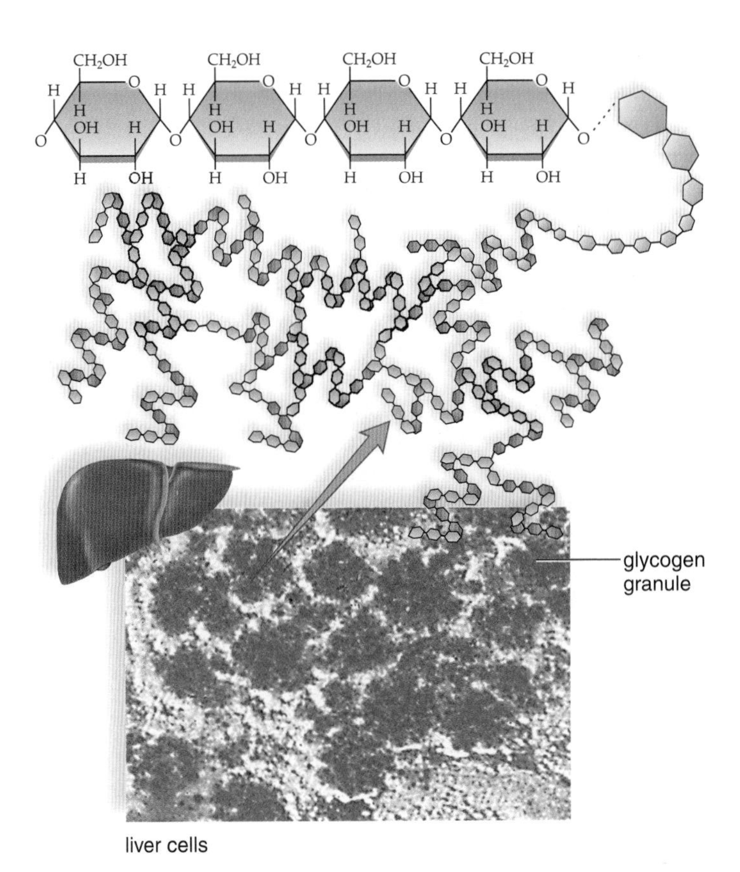

Figure 2.18 Glycogen structure and function.
Glycogen is a highly branched polymer of glucose molecules. Glycogen is the storage form of glucose in animals. The electron micrograph shows glycogen granules in liver cells. Muscle cells also store glycogen.

Cellulose

Some types of polysaccharides function as structural components of cells. The polysaccharide **cellulose** is found in plant cell walls, and this accounts, in part, for the strong nature of these walls.

In cellulose (Fig. 2.19), the glucose units are joined by a slightly different type of linkage than that found in starch or glycogen. (Observe the alternating up/down position of the oxygen atoms in the linked glucose units.) While this might seem to be a technicality, it is significant because we are unable to digest foods containing this type of linkage. Therefore, cellulose largely passes through our digestive tract as fiber, or roughage. Medical scientists believe that fiber in the diet is necessary to good health, and some have suggested it may even help prevent colon cancer.

Chitin, which is found in the exoskeleton (shell) of crabs and related animals, is another structural polysaccharide. Scientists have discovered that chitin can be made into a thread and used as a suture material.

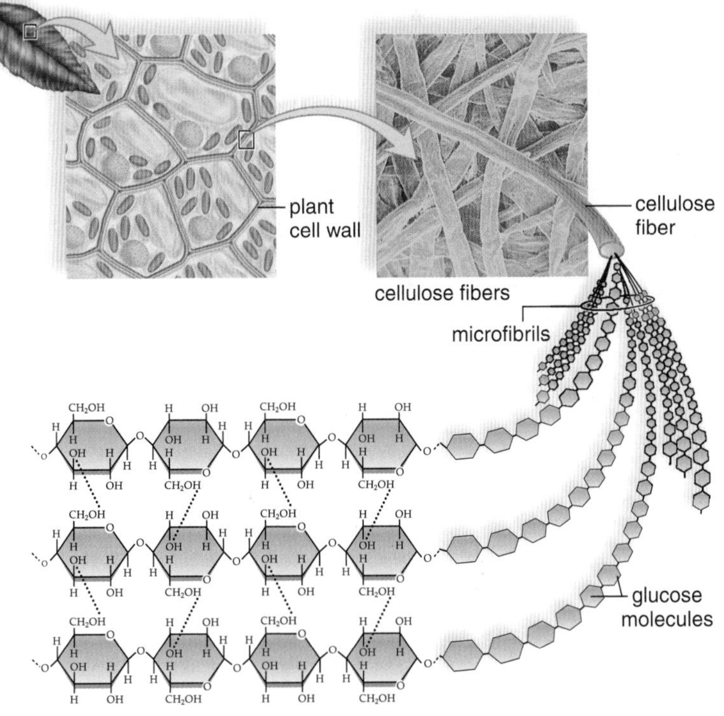

Figure 2.19 Cellulose structure and function.
In cellulose, the linkage between glucose molecules is slightly different from that in starch or glycogen. Plant cell walls contain cellulose, and the rigidity of these cell walls permits nonwoody plants to stand upright as long as they receive an adequate supply of water.

Check Your Progress

1. What structural element do all carbohydrates have in common?
2. Why is starch in plants a source of glucose for our bodies but cellulose in plants is not?

2.6 Lipids

LEARNING OUTCOMES

1. Describe the structure of the various lipids.
2. List the functions lipids play in our bodies.

Lipids contain more energy per gram than other biological molecules, and fats and oils function as energy storage molecules in organisms. Phospholipids form a membrane that separates the cell from its environment and forms its inner compartments as well. The steroids are a large class of lipids that includes, among others, the sex hormones.

Lipids are diverse in structure and function, but they have a common characteristic: they do not dissolve in water. Lipids are hydrophobic.

Fats and Oils

The most familiar lipids are those found in fats and oils. **Fats,** which are usually of animal origin (e.g., lard and butter), are solid at room temperature. **Oils,** which are usually of plant origin (e.g., corn oil and soybean oil), are liquid at room temperature. Fat has several functions in the body: it is used for long-term energy storage, it insulates against heat loss, and it forms a protective cushion around major organs.

Fats and oils form when one glycerol molecule reacts with three fatty acid molecules (Fig. 2.20). A fat molecule is sometimes called a **triglyceride** because of its three-part structure, and the term *neutral fat* is sometimes used because the molecule is nonpolar.

Emulsification

Emulsifiers can cause fats to mix with water. The emulsifiers contain molecules with a nonpolar end and a polar end. The molecules position themselves about an oil droplet so that their nonpolar ends project inward and their polar ends project outward. Now the droplet disperses in water, which means that **emulsification** has occurred:

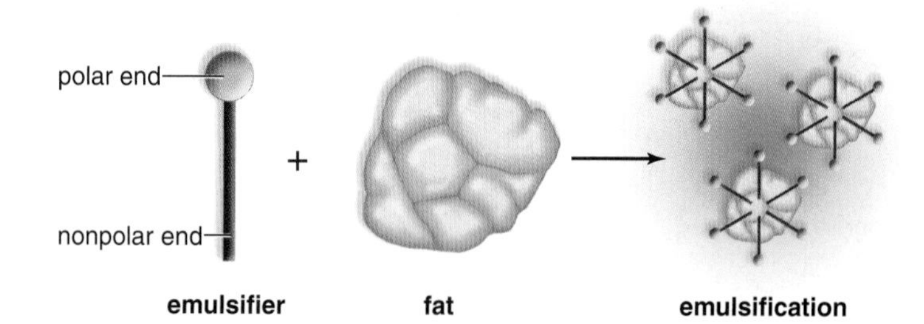

Emulsification takes place when dirty clothes are washed with soaps or detergents. It explains why some salad dressings are uniform in consistency (emulsified) while others separate into two layers. Also, prior to the digestion of fatty foods, fats are emulsified by bile in the intestines. The gallbladder stores bile for use when a meal is eaten, and a person who has had the gallbladder removed may have trouble digesting fatty foods.

Saturated and Unsaturated Fatty Acids

A **fatty acid** is a hydrocarbon chain that ends with the acidic group —COOH (see Fig. 2.20). Most of the fatty acids in cells contain 16 or 18 carbon atoms per molecule, although smaller ones with fewer carbons are also known.

Fatty acids are either saturated or unsaturated. **Saturated fatty acids** have no double covalent bonds between carbon atoms. The carbon chain is saturated, so to speak, with all the hydrogens it can hold. Saturated fatty acids account for the solid nature at room temperature of fats such as lard and butter. **Unsaturated fatty acids** have double bonds between carbon atoms wherever the number

glycerol + 3 fatty acids ⇄ (dehydration reaction → ; ← hydrolysis reaction) fat molecule + $3\ H_2O$ (3 water molecules)

Figure 2.20 Synthesis and degradation of a fat molecule.
Fatty acids can be saturated or unsaturated. Saturated fatty acids have no double bonds between carbon atoms, whereas unsaturated fatty acids have some double bonds (colored yellow) between carbon atoms. When a fat molecule (triglyceride) forms, three fatty acids combine with glycerol, and three water molecules are produced.

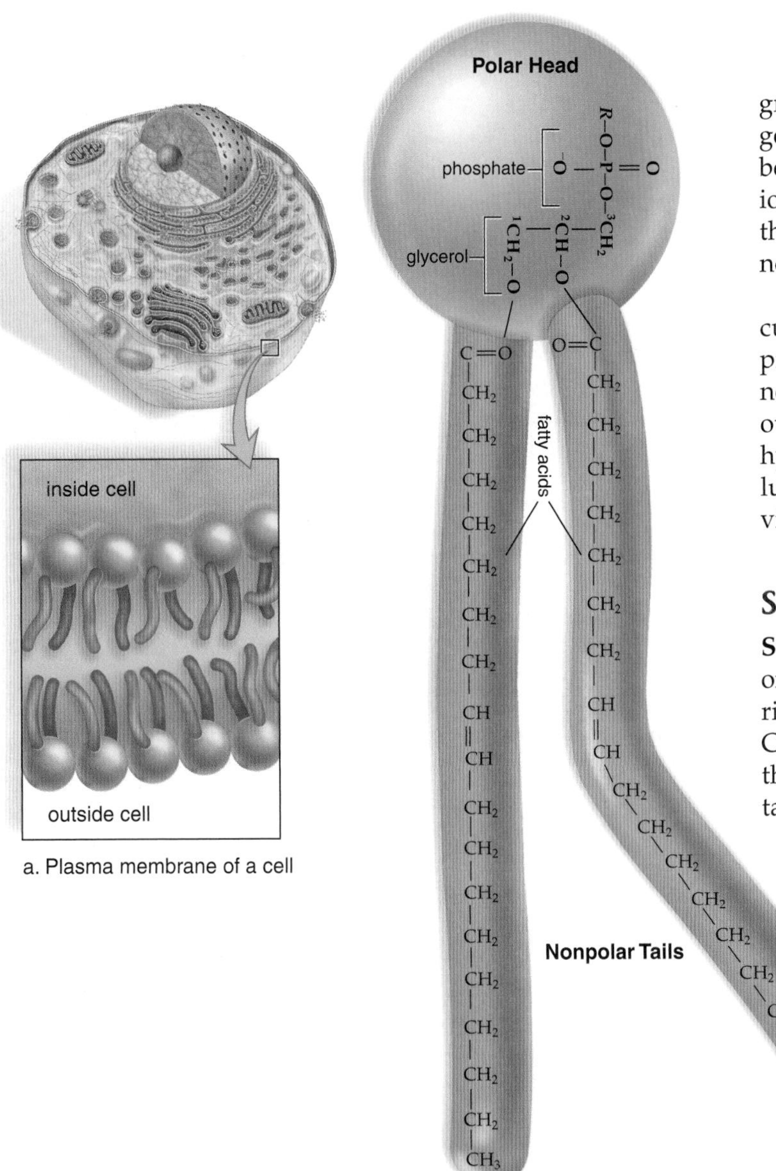

Figure 2.21 Phospholipids form membranes.
a. Phospholipids arrange themselves as a bilayer in the plasma membrane that surrounds cells. **b.** Phospholipids are constructed like fats, except that in place of the third fatty acid, they have a polar phosphate group. The bilayer structure forms because the polar (hydrophilic) head is soluble in water, whereas the two nonpolar (hydrophobic) tails are not. The polar heads interact with the inside and outside of the cell while the nonpolar tails interact with each other.

of hydrogens is less than two per carbon atom. Unsaturated fatty acids account for the liquid nature of vegetable oils at room temperature. Hydrogenation, the chemical addition of hydrogen to vegetable oils, converts them into a solid. This type of fat, called *trans fat,* is often found in processed foods.

Phospholipids

Phospholipids, as their name implies, contain a phosphate group. Essentially they are constructed like fats, except that in place of the third fatty acid, there is a polar phosphate group or a grouping that contains both phosphate and nitrogen. These molecules are not electrically neutral, as are fats, because the phosphate and nitrogen-containing groups are ionized. They form the so-called polar (hydrophilic) head of the molecule, while the rest of the molecule becomes the nonpolar (hydrophobic) tails (Fig. 2.21).

Phospholipids illustrate that the chemistry of a molecule helps determine its function. Phospholipids are the primary components of cellular membranes. They spontaneously form a bilayer in which the hydrophilic heads face outward toward watery solutions and the tails form the hydrophobic interior. Plasma membranes separate extracellular from intracellular materials and thus are absolutely vital to the form and function of a cell.

Steroids

Steroids have a backbone of four fused carbon rings. Each one differs primarily by the arrangement of the atoms in the rings and the type of functional groups attached to them. Cholesterol is a steroid formed by the body that also enters the body as part of our diet. Cholesterol has several important functions. It is a component of an animal cell's plasma membrane and is the precursor of several other steroids, such as bile salts and the sex hormones testosterone and estrogen (Fig. 2.22).

Now we know that a diet high in saturated fats, trans fats, and cholesterol can cause fatty material to accumulate inside the lining of blood vessels thereby reducing blood flow. The Health Focus on page 36 discusses which sources of carbohydrates, fats, and proteins are recommended for inclusion in the diet.

Check Your Progress

1. List the two types of lipid molecules found in the plasma membranes of animal cells.
2. How does the presence of a double bond in an unsaturated fatty acid affect whether that substance is a solid or liquid?

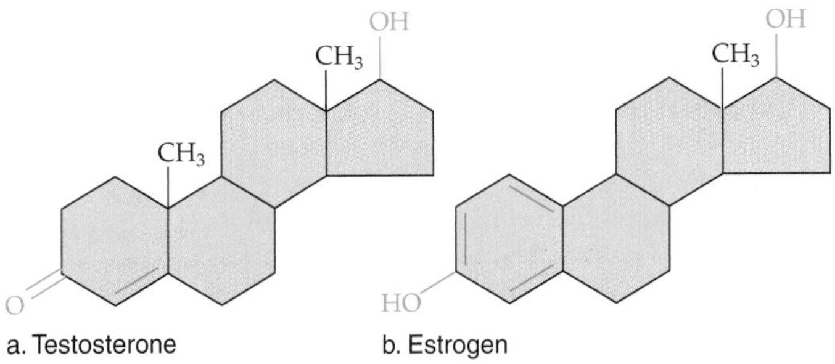

a. Testosterone b. Estrogen

Figure 2.22 Steroids.
All steroids have four adjacent rings, but their attached groups differ. The effects of (**a**) testosterone and (**b**) estrogen on the body largely depend on the difference in the attached groups (shown in blue).

Health Focus

A Balanced Diet

Everyone agrees that we should eat a balanced diet, but just what is a balanced diet? The U.S. Department of Agriculture (USDA) released a new food pyramid in April of 2005 (Fig. 2B). The new pyramid contains vertical bands showing that all types of foods are needed for a balanced diet. Some bands are wider than others because you need more of some foods and less of others in your daily diet.

Progressing up the pyramid, the bands get smaller. The bands at the base represent food with little or no added fat or sugar. The narrower top area stands for foods with more added sugars and fats—which should be eaten more sparingly. For example, apple pie may contain fruit, but it also contains fat and sugar and should be eaten in smaller quantities than raw apples.

Carbohydrates

Carbohydrates (sugars and polysaccharides) are the quickest, most readily available source of energy for the body. Complex carbohydrates, such as those in whole-grain breads and cereals, are preferable to simple carbohydrates, such as candy and ice cream, because they contain dietary fiber (nondigestible plant material), plus vitamins and minerals. Insoluble fiber has a laxative effect, and soluble fiber combines with the cholesterol in food and prevents cholesterol from exiting the intestinal tract and entering the blood. However, researchers have found that the starch in potatoes and processed foods, such as white bread and white rice, leads to a high blood glucose level just as simple carbohydrates do. Researchers ask, Is this why many adults are now coming down with type 2 diabetes?

Fats

We have known for many years that saturated fats in animal products contribute to the formation of deposits called plaque, which clog arteries and lead to high blood pressure and heart attacks. Even more harmful than naturally occurring saturated fats are the so-called trans fats, created artifically using vegetable oils. Trans fats are partially hydrogenated to make them semisolid. Trans fats are found in shortenings, solid margarines, and some processed foods.

Notice that the new food pyramid advocates an intake of certain liquid oils. These oils contain monounsaturated and polyunsaturated fatty acids, which researchers have found are protective against the development of cardiovascular disease.

Other Nutrients

Red meat is rich in protein, but it is usually also high in saturated fat; therefore, fish and chicken are preferred sources of protein. Also, a combination of rice and *legumes* (a group of plants that includes peas and beans) can provide all of the amino acids the body needs to build cellular proteins.

Nutritionists agree that eating fruits and vegetables is beneficial. At the very least, they provide us with the vitamins we need in our diet.

Discussion Questions

1. List the number and types of fruits and vegetables you eat in a typical day. Are you getting the recommended amounts of these important foods? Do you have variety in the types of fruits and vegetables you consume?
2. Why are some fats "good" for you and some "bad" for you? What is the difference in these fats?
3. Chemically, what is the difference between a "whole-grain" carbohydrate and a simple carbohydrate?

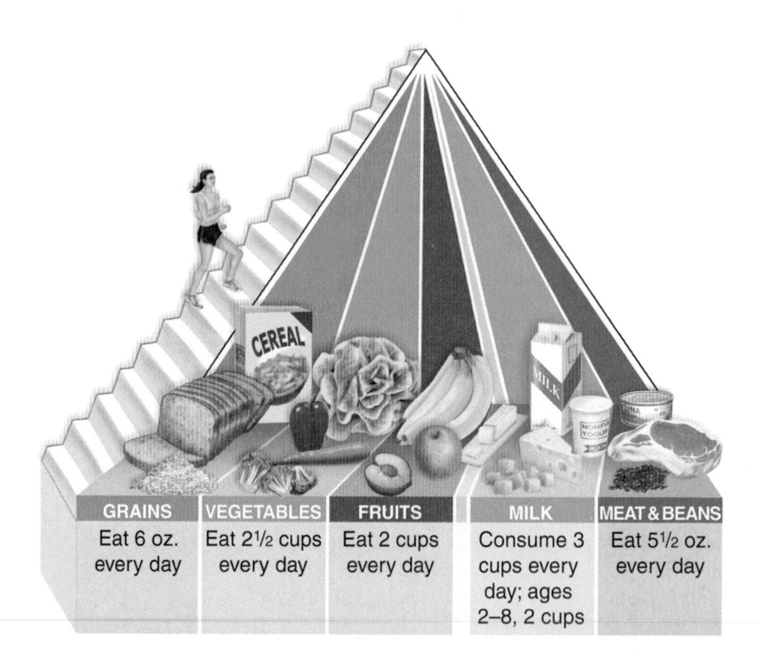

Figure 2B Food guide pyramid.
The United States Department of Agriculture (USDA) developed this pyramid as a guide to better health. The different widths of the food group bands suggest how much food a person should choose from each group. The six different colors illustrate that foods from all groups are needed each day for good health. The orange represents grains; the green represents vegetables; the red represents fruits; the yellow represents oils; the blue represents milk; and the purple represents meat and beans. The wider base is supposed to encourage the selection of foods containing little or no solid fats or added sugars. The person climbing the pyramid represents the balance between foods and physical activity.

2.7 Proteins

Learning Outcomes

1. Describe the monomer unit of a protein and how monomer units are assembled into peptides.
2. Explain the primary, secondary, tertiary, and quarternary structure of a protein and describe the relationship between protein structure and function.

Proteins are polymers composed of amino acid monomers. An amino acid has a central carbon atom bonded to a hydrogen atom and three functional groups. The name of the molecule is appropriate because one of these groups is an amino group (—NH_2) and another is an acidic group (—COOH). The third group is called an *R* group, and **amino acids** differ from one another by their *R* group.The *R* group varies from having a single carbon to being a complicated ring structure (Fig. 2.23).

Name	Structural Formula	*R* Group
alanine (Ala)	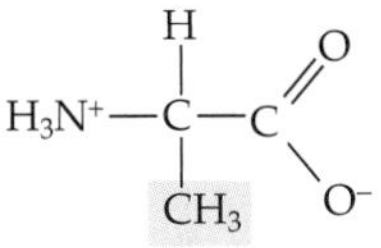 $H_3N^+—C(H)(CH_3)—C(=O)O^-$	*R* group has a single carbon atom
valine (Val)	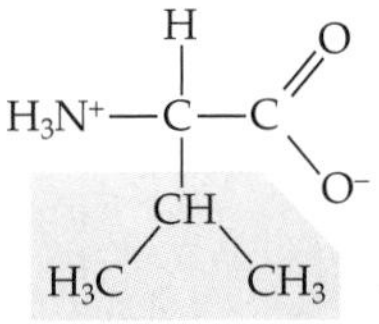 $H_3N^+—C(H)(CH(H_3C)(CH_3))—C(=O)O^-$	*R* group has a branched carbon chain
cysteine (Cys)	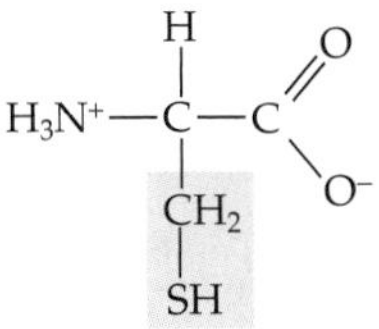 $H_3N^+—C(H)(CH_2—SH)—C(=O)O^-$	*R* group contains sulfur
phenylalanine (Phe)	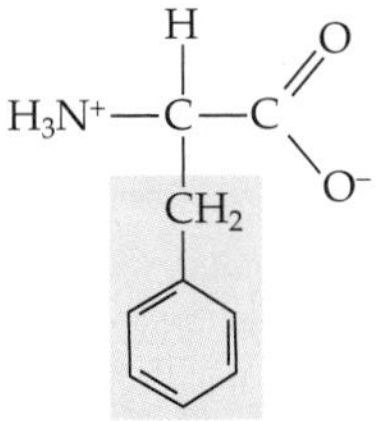 $H_3N^+—C(H)(CH_2—C_6H_5)—C(=O)O^-$	*R* group has a ring structure

Figure 2.23 Representative amino acids.
Amino acids differ from one another by their *R* group. The simplest *R* group is a single hydrogen atom (H). *R* groups (blue) containing carbon vary as shown.

Proteins perform many functions. Proteins such as keratin, which makes up hair and nails, and collagen, which lends support to ligaments, tendons, and skin, are structural proteins. Some proteins are enzymes. Enzymes are necessary contributors to the chemical workings of the cell, and therefore of the body. **Enzymes** speed chemical reactions. They work so quickly that a reaction that normally takes several hours or days without an enzyme takes only a fraction of a second with an enzyme. Many hormones, messengers that influence cellular metabolism, are also proteins. The proteins actin and myosin account for the movement of cells and the ability of our muscles to contract. Some proteins transport molecules in the blood. Hemoglobin is a complex protein in our blood that transports oxygen. Antibodies in blood and other body fluids are proteins that combine with foreign substances, preventing them from destroying cells and upsetting homeostasis.

Proteins in the plasma membrane of cells have various functions: some form channels that allow substances to enter and exit cells; some are carriers that transport molecules into and out of the cell; and some are enzymes.

Peptides

Figure 2.24 shows that a synthesis reaction between two amino acids results in a dipeptide and a molecule of water. A **polypeptide** is a single chain of amino acids. The bond that joins any two amino acids is called a **peptide bond.** The atoms associated with a peptide bond—oxygen (O), carbon (C), nitrogen (N), and hydrogen (H)—share electrons in such a way that the oxygen has a partial negative charge (δ^-) and the hydrogen has a partial positive charge (δ^+):

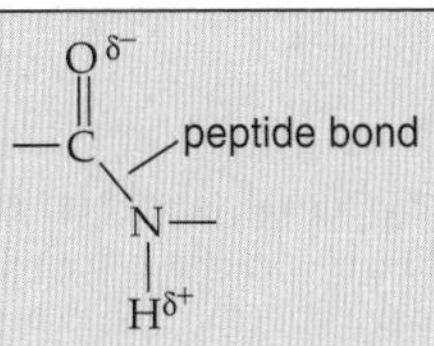

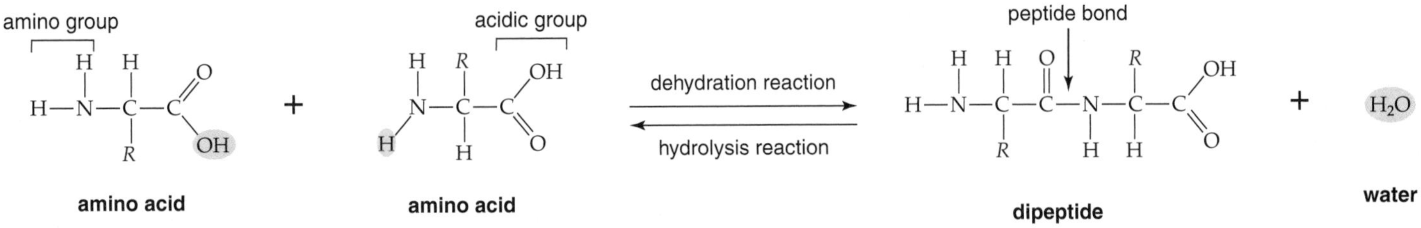

$O^{\delta-}$
—C—N— (peptide bond)
$H^{\delta+}$

Figure 2.24 Synthesis and degradation of a dipeptide.
Following a dehydration reaction, a peptide bond joins two amino acids, and a water molecule is released. Following a hydrolysis reaction, the bond is broken with the addition of water.

Visual Focus

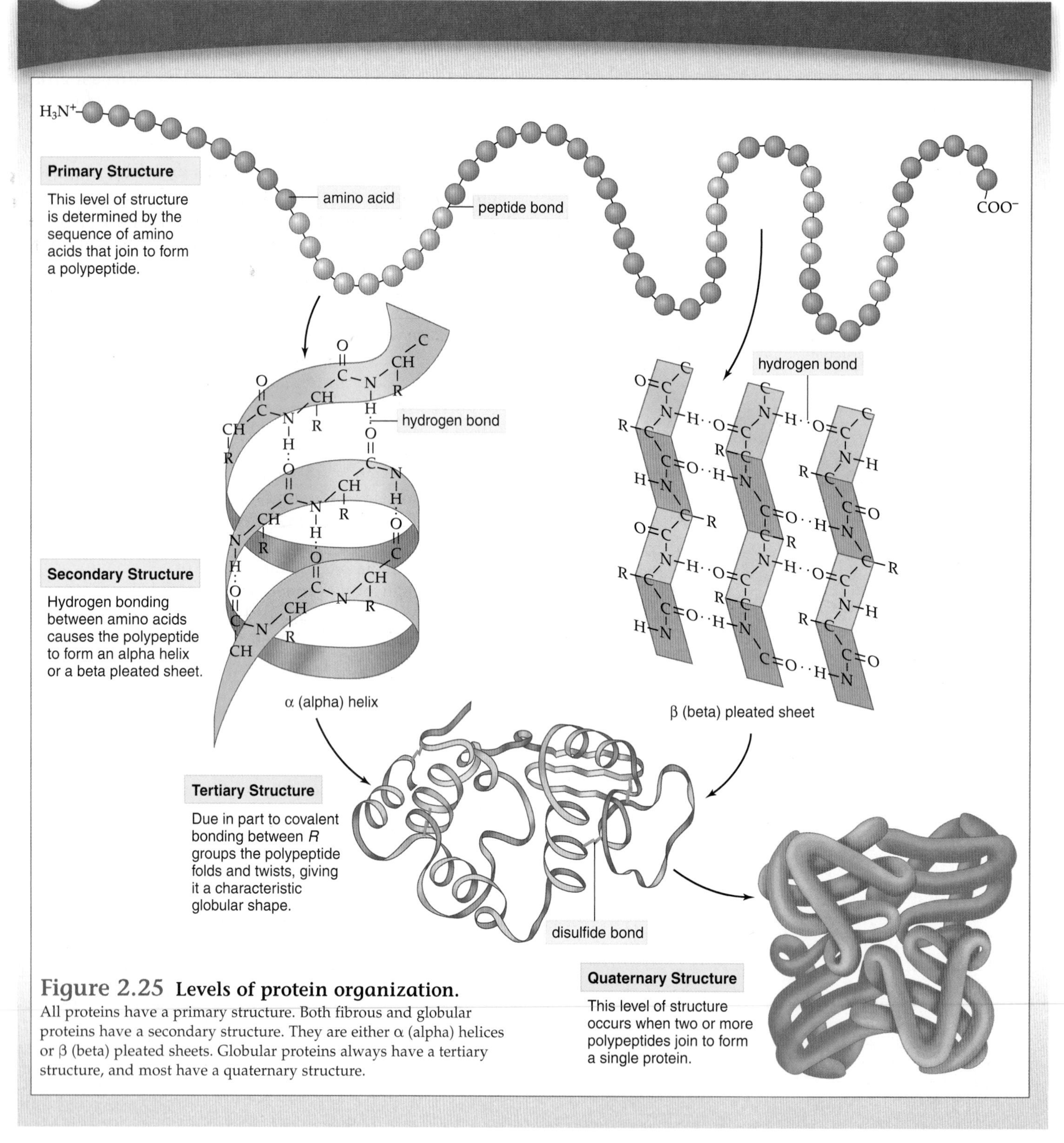

Figure 2.25 Levels of protein organization. All proteins have a primary structure. Both fibrous and globular proteins have a secondary structure. They are either α (alpha) helices or β (beta) pleated sheets. Globular proteins always have a tertiary structure, and most have a quaternary structure.

Therefore, the peptide bond is polar, and hydrogen bonding is possible between the C=O of one amino acid and the N—H of another amino acid in a polypeptide.

Levels of Protein Organization

Proteins have at least three levels of structural organization. Some can have four levels (Fig. 2.25). The first level, called the *primary structure,* is the linear sequence of the amino acids joined by peptide bonds. Polypeptides can be quite different from one another. Think of a polysaccharide like a necklace that contains a single type of "bead," namely glucose. Then polypeptide "necklaces" make use of 20 different possible types of "beads," namely amino acids. Each particular polypeptide has its own sequence of amino acids.

Therefore, each polypeptide differs by the sequence of its *R* groups.

The *secondary structure* of a protein comes about when the polypeptide takes on a certain orientation in space. A coiling of the chain results in an α (alpha) helix, or a right-handed spiral, similar to a spiral staircase. Or a folding of the chain results in a β (beta) pleated sheet similar to a hand-held fan. Hydrogen bonding between peptide bonds holds the shape in place.

The *tertiary structure* of a protein is its final three-dimensional shape. In muscles, myosin molecules have a rod shape ending in globular (globe-shaped) heads. In enzymes, the polypeptide bends and twists in different ways. Invariably, the hydrophobic portions are packed mostly on the inside, and the hydrophilic portions are on the outside where they can make contact with water. The tertiary shape of a polypeptide is maintained by various types of bonding between the *R* groups; covalent, ionic, and hydrogen bonding all occur. One common form of covalent bonding between *R* groups is a disulfide (S—S) linkage between two cysteine amino acids.

Some proteins have only one polypeptide, and others have more than one polypeptide, each with its own primary, secondary, and tertiary structures. In proteins with multiple polypeptide chains, these separate polypeptides are arranged to give such proteins a fourth level of structure, termed the *quaternary structure.* Hemoglobin is a complex protein having a quaternary structure. Most enzymes also have a quaternary structure. Thus, proteins can differ in many ways, such as in length, sequence, and structure. Each individual protein is chemically unique.

The final shape of a protein is very important to its function. As we will discuss in Chapter 6, for example, enzymes cannot function unless they have their usual shape. When proteins are exposed to extremes in heat and pH, they undergo an irreversible change in shape called **denaturation.** For example, we are all aware that adding acid to milk causes curdling and that heating causes egg white, which contains a protein called albumin, to coagulate. Denaturation occurs because the normal bonding between the *R* groups has been disturbed. Once a protein loses its normal shape, it is no longer able to perform its usual function. Researchers hypothesize that an alteration in protein organization is related to the development of Alzheimer disease and Creutzfeldt-Jakob disease (the human form of "mad cow" disease).

Check Your Progress

1. List some of the functions of proteins.
2. How does the structure of the monomer unit of proteins result in the name amino acid?
3. What is responsible for holding the primary structure of a protein together? The secondary structure? The tertiary structure?

2.8 Nucleic Acids

Learning Outcomes

1. Describe the structure of DNA and RNA.
2. Explain the role of ATP in the cell.

The two types of nucleic acids are **DNA (deoxyribonucleic acid)** and **RNA (ribonucleic acid).** The discovery of the structure of DNA has had an enormous influence on biology and on society in general. DNA stores genetic information in the cell and in the organism. Further, the cell replicates and transmits this information when it reproduces and when an organism reproduces. We now not only know how genes work, but we can manipulate them. The science of biotechnology is largely devoted to altering the genes in living organisms.

Structure of DNA and RNA

Both DNA and RNA are polymers of nucleotides. Every **nucleotide** is a molecular complex of three types of subunit molecules—phosphate (phosphoric acid), a pentose sugar, and a nitrogen-containing base:

Nucleotide structure

The nucleotides in DNA contain the sugar deoxyribose and the nucleotides in RNA contain the sugar ribose. This difference accounts for their respective names (Table 2.1). There are four different types of bases in DNA: **adenine (A), thymine (T), guanine (G),** and **cytosine (C).** The base can have two rings (adenine or guanine) or one ring (thymine or cytosine). In RNA, the base **uracil (U)** replaces the base thymine. These structures are called bases because their presence raises the pH of a solution.

TABLE 2.1 DNA Structure Compared With RNA Structure

	DNA	RNA
Sugar	Deoxyribose	Ribose
Bases	Adenine, guanine, thymine, cytosine	Adenine, guanine, uracil, cytosine
Strands	Double stranded with base pairing	Single stranded
Helix	Yes	No

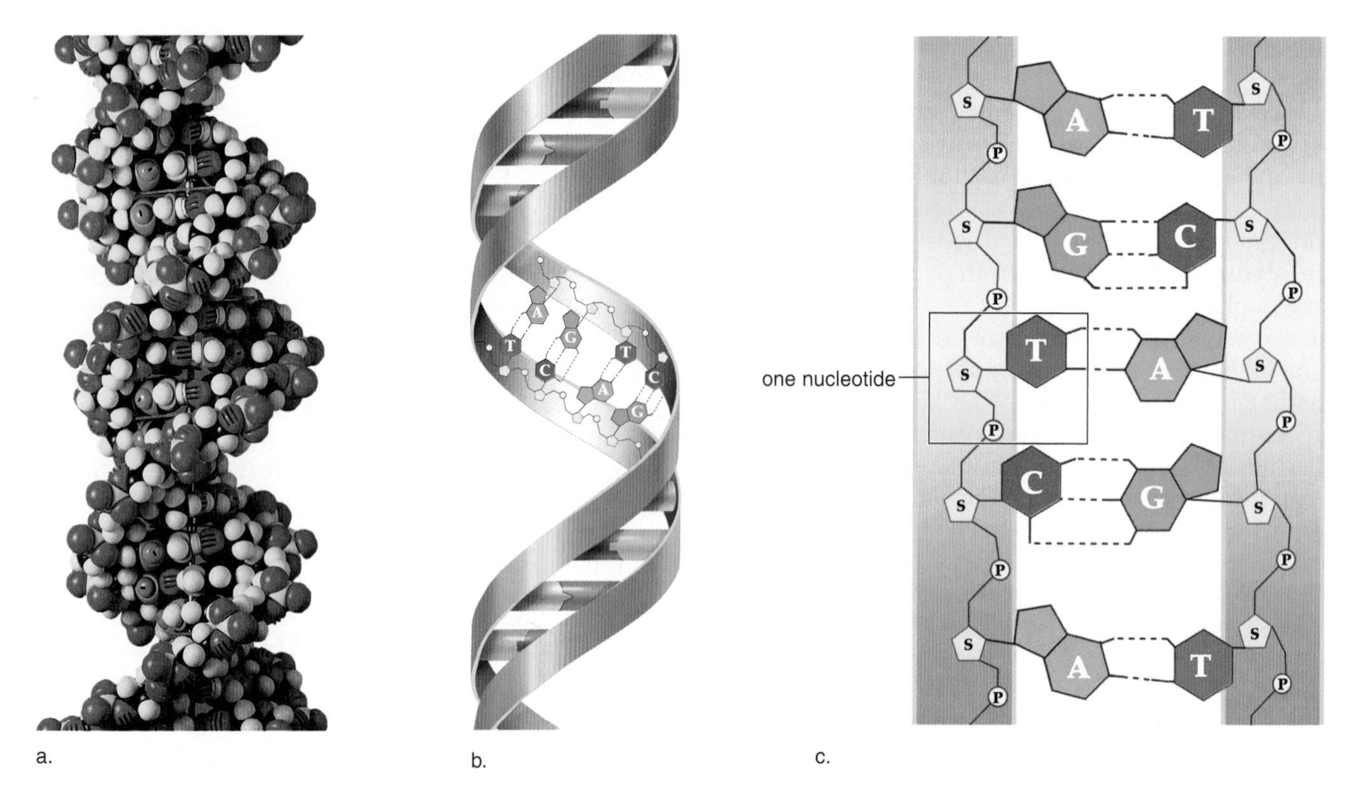

Figure 2.26 Overview of DNA structure.
The structure of DNA is absolutely essential to its ability to replicate and to serve as the genetic material. **a.** Space-filling model of DNA's double helix. **b.** Complementary base pairing between strands. **c.** Ladder configuration. Notice that the uprights are composed of alternating phosphate and sugar molecules and that the rungs are complementary nitrogen-containing paired bases. The heredity information stored by DNA is the sequence of its bases, which determines the primary structure of the cell's proteins.

The nucleotides form a linear molecule called a strand, which has a backbone made up of alternating phosphates and sugars with the bases projecting to one side of the backbone. The nucleotides and their bases occur in a definite order. After many years of work, researchers now know the sequence of all the bases in human DNA—the human genome. This breakthrough is expected to lead to improved genetic counseling, gene therapy, and medicines to treat the causes of many human illnesses.

DNA is double stranded, with the two strands twisted about each other in the form of a **double helix** (Fig. 2.26*a, b*). In DNA, the two strands are held together by hydrogen bonds between the bases. When unwound, DNA resembles a ladder. The uprights (sides) of the ladder are made entirely of the alternating phosphate and sugar molecules, and the rungs of the ladder are made only of complementary paired bases. Thymine (T) always pairs with adenine (A), and guanine (G) always pairs with cytosine (C). Complementary bases have shapes that fit together (Fig. 2.26*c*).

Complementary base pairing allows DNA to replicate in a way that ensures the sequence of bases will remain the same. The base sequence of some sections of DNA contains a code that specifies the sequence of amino acids in the proteins of the cell.

RNA is single stranded and is formed by complementary base pairing with one DNA strand. There are several types of RNA. One type of RNA, mRNA or messenger RNA, carries the information from the DNA strand to the ribosome where it is translated into the sequence of amino acids specified by the DNA.

ATP (Adenosine Triphosphate)

In addition to being the monomers of nucleic acids, nucleotides have other metabolic functions in cells. When adenosine (adenine plus ribose) is modified by the addition of three phosphate groups instead of one, it becomes **ATP (adenosine triphosphate),** an energy carrier in cells. Glucose is broken down in a step-wise fashion so that the energy of glucose is converted to that of ATP molecules. ATP molecules serve as small "energy packets" suitable for supplying energy to a wide variety of a cell's chemical reactions. ATP can be said to be the energy "currency" (like paper money) of the cell. Reactions in the cell that need energy "spend" ATP.

Bioethical Focus

Blue Gold

Environmentalists believe that the world is running out of clean drinking water. Over 97% of the world's water is salt water found in the oceans. Salt water is unsuitable for drinking without expensive desalination. Of the fresh water in the world, most is locked in frozen form in the polar ice caps and glaciers and therefore unavailable. This leaves only a small percentage in groundwater, lakes, and rivers that could be available for drinking, industry, and irrigation. However, some of that water is polluted and unsuitable.

Water has always been the most valuable commodity in the Middle East, even more valuable than oil. But as fresh water becomes limited and the world's population grows, the lack of sufficient clean water is becoming a worldwide problem.

The combination of increasing demand and dwindling supply has attracted global corporations who want to sell water. Water is being called the "blue gold" of the twenty-first century, and an issue has arisen regarding whether the water industry should be privatized. That is, could water rights be turned over to private companies to deliver clean water and treat wastewater at a profit, similar to the way oil and electricity are handled? Private companies have the resources to upgrade and modernize water delivery and treatment systems, thereby conserving more water. However, opponents of this plan claim that water is a basic human right required for life, not a need to be supplied by the private sector. In addition, a corporation can certainly own the pipelines and treatment facilities, but who owns the rights to the water? For example, North America's largest underground aquifer, the Ogallala, covers 175,000 square miles under several states in the southern Great Plains. If water becomes a commodity, do we allow water to be taken away from people who cannot pay in order to give it to those who can?

Form Your Own Opinion

1. Do you agree that the water industry should be privatized? Why or why not?
2. Is access to clean water a "need" or a "right"? If it is a right, who pays for that right?
3. Since water is a shared resource, everyone believes they can use water, but few people feel responsible for conserving it. What can you do to conserve water?

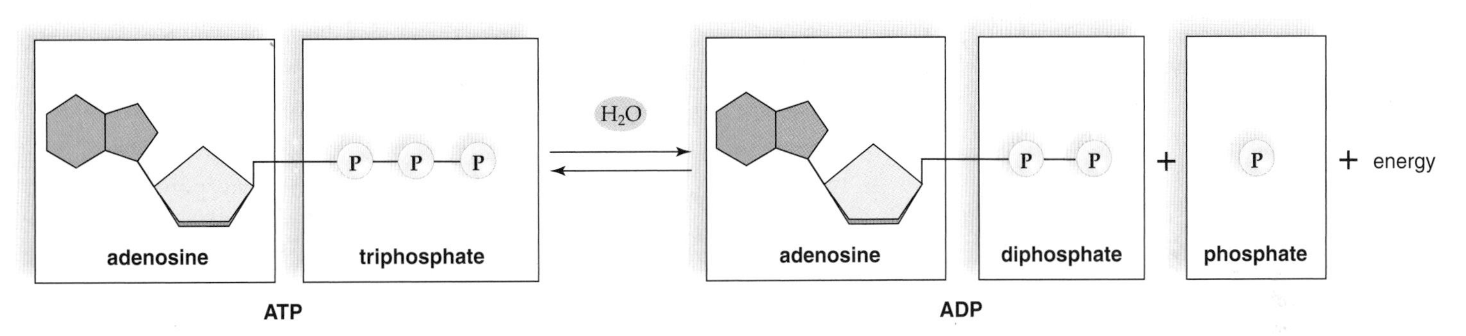

Figure 2.27 ATP reaction.
ATP, the universal energy "currency" of cells, is composed of adenosine and three phosphate groups (called a triphosphate). When cells require energy, ATP undergoes hydrolysis, producing ADP+Ⓟ, with the release of energy.

ATP is a high-energy molecule because the last two phosphate bonds are unstable and easily broken. In cells, the terminal phosphate bond usually is hydrolyzed, leaving the molecule **ADP (adenosine diphosphate)** and a molecule of inorganic phosphate Ⓟ (Fig. 2.27). The cell uses the energy released by ATP breakdown to synthesize macromolecules such as carbohydrates and proteins. Muscle cells use the energy for muscle contraction, and nerve cells use it for the conduction of nerve impulses. After ATP breaks down, it is rebuilt by the addition of Ⓟ to ADP. Notice in Figure 2.27 that an input of energy is required to re-form ATP.

Check Your Progress

1. How is information stored in DNA?
2. How is energy stored in ATP?

Applying the Concepts [Revisited]

Living things, including humans, are composed of 70–90% water. As discussed in this chapter, the unique properties of water enable it to transport substances in the blood, keep you cool, and dissolve other molecules.

1. How does the polarity of water help it to carry nutrients and waste in your blood? What kind of substances would you expect to dissolve in the water in your blood? What kinds of substances would not dissolve? How might those be transported in the blood?
2. You can now consume coffee for part of your water needs, although too much is not recommended. Why is coffee not the equivalent of water? Why is a soft drink not the equivalent of water?

SUMMARIZING THE CONCEPTS

2.1 Basic Chemistry

- Both living and nonliving things are composed of matter consisting of elements. The acronym CHNOPS stands for the most significant elements found in living things: carbon, hydrogen, nitrogen, oxygen, phosphorus, and sulfur.
- Elements contain atoms, and atoms are composed of subatomic particles. Protons and neutrons in the nucleus determine the atomic mass of an atom. The atomic number indicates the number of protons and the number of electrons in electrically neutral atoms. Protons have positive charges, neutrons are uncharged, and electrons have negative charges. Isotopes are atoms of a single element that differ in their numbers of neutrons. Radioactive isotopes have many uses, including serving as tracers in biological experiments and medical procedures.
- The number of electrons in the outer orbital determines the reactivity of an atom. The first orbital is complete when it is occupied by two electrons. In atoms up through calcium, number 20, every orbital beyond the first one is complete with eight electrons. The octet rule states that atoms react with one another in order to have a completed outer orbital.

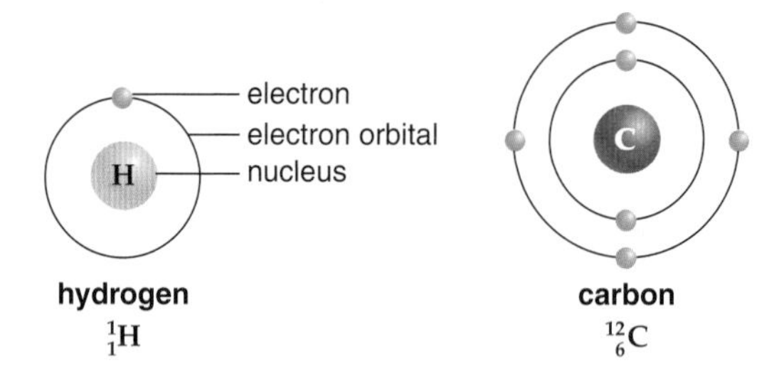

2.2 Molecules and Compounds

- Ions form when atoms lose or gain one or more electrons to achieve a completed outer orbital. An ionic bond is an attraction between oppositely charged ions. A covalent bond is one or more shared pairs of electrons.
- When carbon combines with four other atoms, the resulting molecule has a tetrahedral shape. The shape of a molecule is important to its biological role.
- In polar covalent bonds, the sharing of electrons is not equal. One of the atoms exerts greater attraction for the shared electrons than the other, and a slight charge on each atom results. A hydrogen bond is a weak attraction between a slightly positive hydrogen atom and a slightly negative oxygen or nitrogen atom within the same or a different molecule. Hydrogen bonds help maintain the structure and function of cellular molecules.

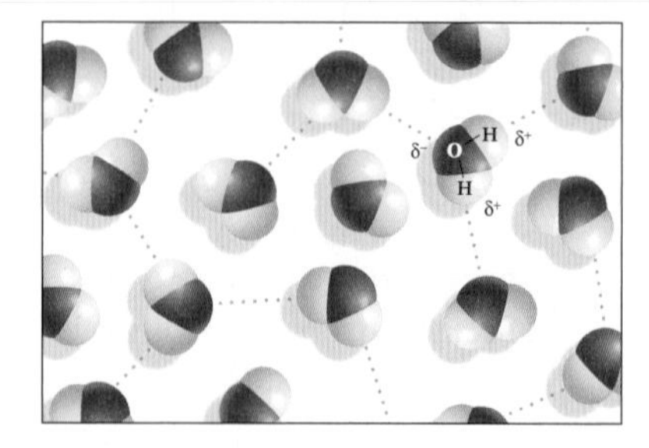

2.3 Chemistry of Water

- Water is a polar molecule. Its polarity allows hydrogen bonding to occur between water molecules. Water's polarity and hydrogen bonding account for its unique properties. These features allow living things to exist and carry on cellular activities.
- A small fraction of water molecules dissociate to produce an equal number of hydrogen ions and hydroxide ions. Solutions with equal numbers of H^+ and OH^- are termed neutral. In acidic solutions, there are more hydrogen ions than hydroxide ions. These solutions have a pH less than 7. In basic solutions, there are more hydroxide ions than hydrogen ions. These solutions have a pH greater than 7. Cells are sensitive to pH changes. Biological systems often contain buffers that help keep the pH within a normal range.

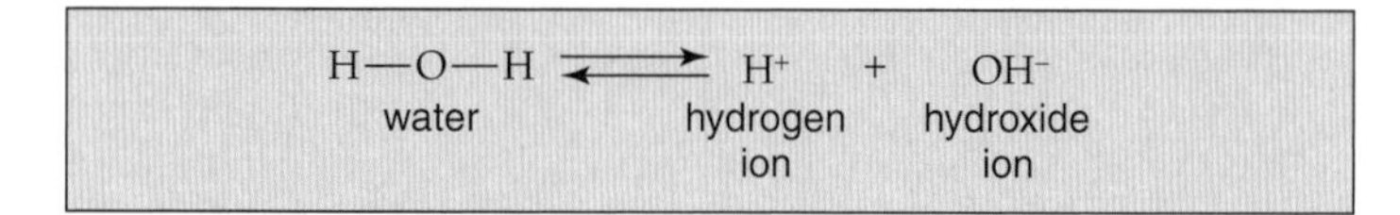

2.4 Organic Molecules

- The chemistry of carbon accounts for the chemistry of organic compounds. Carbohydrates, lipids, proteins, and nucleic acids are macromolecules with specific functions in cells.
- Polymers are formed by the joining together of monomers. Long-chain carbohydrates, proteins, and nucleic acids are all polymers. For each bond formed during a dehydration reaction, a molecule of water is removed. For each bond broken during a hydrolysis reaction, a molecule of water is added.

2.5 Carbohydrates

- Glucose is the 6-carbon sugar most utilized by cells for "quick" energy. Sugar monomers join to form polysaccharides. Plants store glucose as starch, and animals store glucose as glycogen. Humans cannot digest cellulose, which forms plant cell walls.

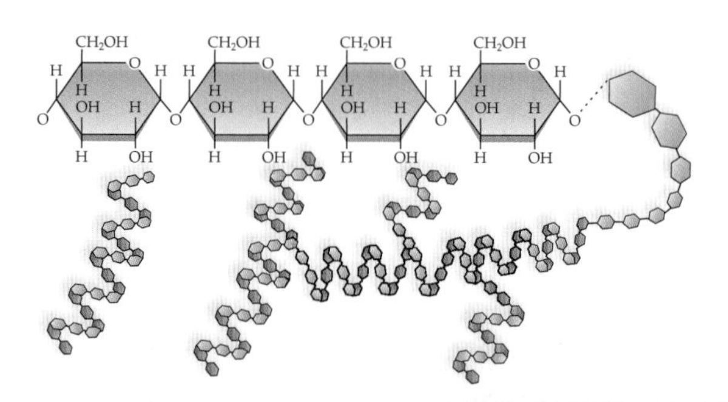

2.6 Lipids

- Lipids are varied in structure and function. Fats and oils, which function in long-term energy storage, contain glycerol and three fatty acids. Fatty acids can be saturated or unsaturated. Plasma membranes in cells contain phospholipids that have a polarized end. Certain hormones are derived from cholesterol, a complex ring compound.

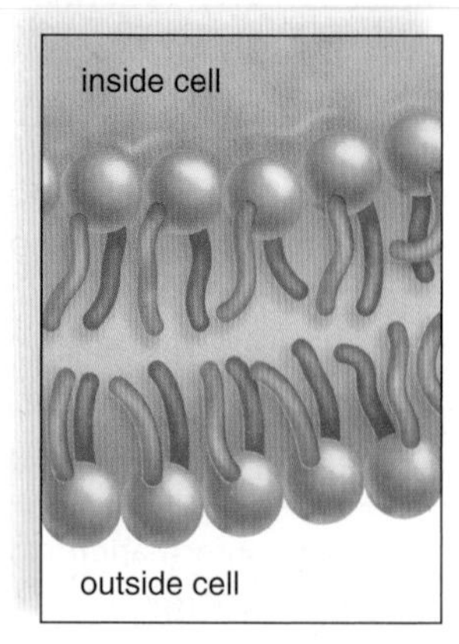

2.7 Proteins

- Proteins have numerous functions in cells. Some are enzymes that speed chemical reactions. The primary structure of a polypeptide is its own particular sequence of the possible 20 types of amino acids. The secondary structure is often an alpha (α) helix or a beta (β) pleated sheet. Tertiary structure occurs when a polypeptide bends and twists into a three-dimensional shape. A protein can contain several polypeptides, and this accounts for a possible quaternary structure.

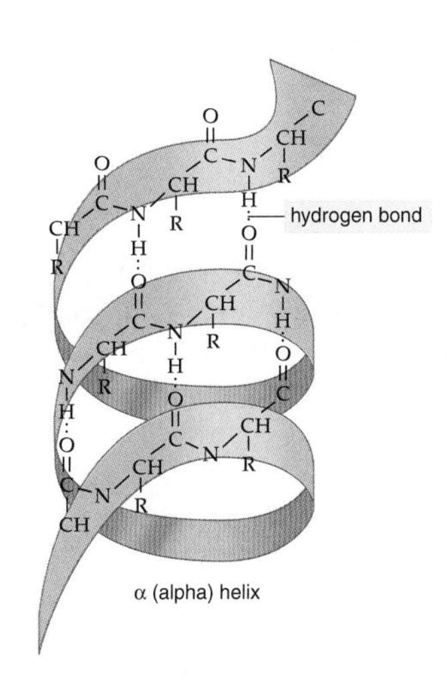

2.8 Nucleic Acids

- Nucleic acids are polymers of nucleotides. Each nucleotide has three components: a sugar, a base, and phosphate (phosphoric acid). DNA, which contains the sugar deoxyribose, is the genetic material that stores information for its own replication and for the sequence of amino acids in proteins. DNA, with the help of RNA, specifies protein synthesis.
- ATP, with its unstable phosphate bonds, is the energy currency of cells. Hydrolysis of ATP to ADP+ Ⓟ releases energy that the cell uses to do metabolic work.

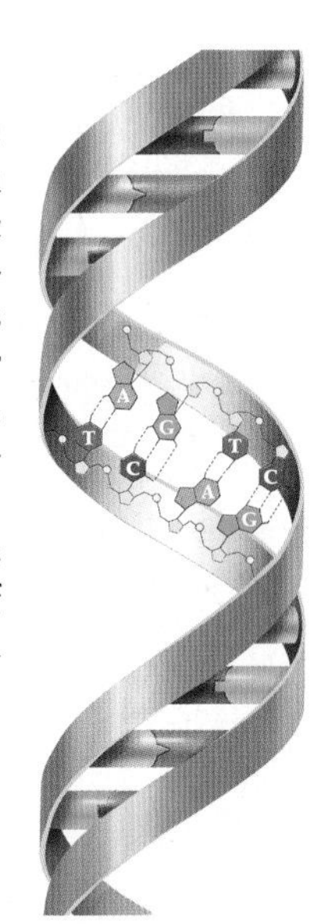

TESTING YOURSELF

Choose the best answer for each question.

1. Which of these ions is important for acid-base balance?
   a. K^+ c. Cl^-
   b. Na^+ d. H^+
2. Which of these offers the most accurate information concerning the shape of a molecule?
   a. electron-dot formula c. molecular formula
   b. structural formula d. space-filling model
3. An example of a hydrogen bond would be
   a. the bond between a carbon atom and a hydrogen atom.
   b. the bond between two carbon atoms.
   c. the bond between sodium and chlorine.
   d. the bond between two water molecules.
4. Covalent bonding between *R* groups in proteins is associated with the ________________ structure.
   a. primary c. tertiary
   b. secondary d. secondary and tertiary
5. The polysaccharide found in plant cell walls is
   a. glucose. c. cellulose.
   b. starch. d. maltose.
6. Ionic bonding involves
   a. loss of electrons.
   b. gain of electrons.
   c. attraction of opposite charges.
   d. All of these are correct.
7. Phosphates can be found in
   a. DNA. c. glucose.
   b. fatty acids. d. Both a and b are correct.
8. Which of these cannot be considered an organic molecule?
   a. table salt c. DNA
   b. glucose d. enzyme
9. Removal of the gallbladder may cause difficulty in digesting
   a. fats. c. sugars.
   b. proteins. d. water.
10. ________________ is the precursor of ________________.
    a. Estrogen, cholesterol
    b. Cholesterol, glucose
    c. Testosterone, cholesterol
    d. Cholesterol, testosterone and estrogen
11. An example of a dehydration reaction would be
    a. building amino acids from proteins.
    b. building lipids from fatty acids.
    c. breaking down glucose into starch.
    d. breaking down starch into glucose.
12. Radioactive elements differ in their
    a. number of protons. c. number of neutrons.
    b. atomic number. d. number of electrons.
13. Nucleotides
    a. contain sugar, a nitrogen-containing base, and a phosphate molecule.
    b. are the monomers for fats and polysaccharides.
    c. join together by covalent bonding between the bases.
    d. are found in DNA, RNA, and proteins.
14. The bond joining two adjacent amino acids is called
    a. a peptide bond.
    b. a hydrogen bond.
    c. an ionic bond.
    d. All of these are correct.
15. Which of the following pertains to an RNA nucleotide and not to a DNA nucleotide?
    a. contains the sugar ribose
    b. contains a nitrogen-containing base
    c. contains a phosphate molecule
    d. becomes bonded to other nucleotides by dehydration
16. Saturated fatty acids and unsaturated fatty acids differ in
    a. the number of carbon-to-carbon bonds.
    b. their consistency at room temperature.
    c. the number of hydrogen atoms present.
    d. All of these are correct.

17. Which of the following is not one of the elements commonly found in living organisms?
    a. glucose
    b. hydrogen
    c. carbon
    d. nitrogen

18. A solution at pH 6 contains ________ than a solution at pH 4.
    a. 10 H^+ ions more
    b. 100 H^+ ions more
    c. 10 H^+ ions less
    d. 100 H^+ ions less

UNDERSTANDING THE TERMS

acid 28
adenine (A) 39
ADP (adenosine diphosphate) 41
amino acid 37
atom 20
atomic mass 21
atomic number 21
ATP (adenosine triphosphate) 40
base 28
buffer 29
calorie 27
carbohydrate 32
cellulose 33
compound 24
covalent bond 25
cytosine (C) 39
dehydration reaction 31
denaturation 39
disaccharide 32
DNA (deoxyribonucleic acid) 39
double helix 40
electron 20
element 20
emulsification 34
enzyme 37
fat 34
fatty acid 34
functional group 31
glucose 32
glycogen 32
guanine (G) 39
hexose 32
hydrogen bond 26
hydrolysis reaction 31
hydrophilic 27
hydrophobic 27
inorganic molecule 30
ion 24
ionic bond 24
isotope 22
lipid 34
matter 20
molecule 24
monomer 31
monosaccharide 32
neutron 20
nucleotide 39
oil 34
organic molecule 30
pentose 32
peptide bond 37
phospholipid 35
pH scale 29
polymer 31
polypeptide 37
polysaccharide 32
protein 37
proton 20
radioactive isotope 22
RNA (ribonucleic acid) 39
saturated fatty acid 34
solute 27
starch 32
steroid 35
thymine (T) 39
triglyceride 34
unsaturated fatty acid 34
uracil (U) 39

THINKING CRITICALLY

1. Why can we use potatoes as a food source but not wood?
2. Since proteins are composed of the same limited number of amino acids, why are they all so different?
3. Given the following ions, can you determine the number of electrons normally present in the outer orbital of each atom?
    a. K^+
    b. Ca^{2+}
    c. F^-
    d. N^{3-}

INQUIRY INTO LIFE WEBSITE

The companion website for *Inquiry into Life* provides a wealth of information organized and integrated by chapter. You will find practice tests, animations, videos, and much more that will complement your learning and understanding of general biology.

http://www.mhhe.com/maderinquiry13

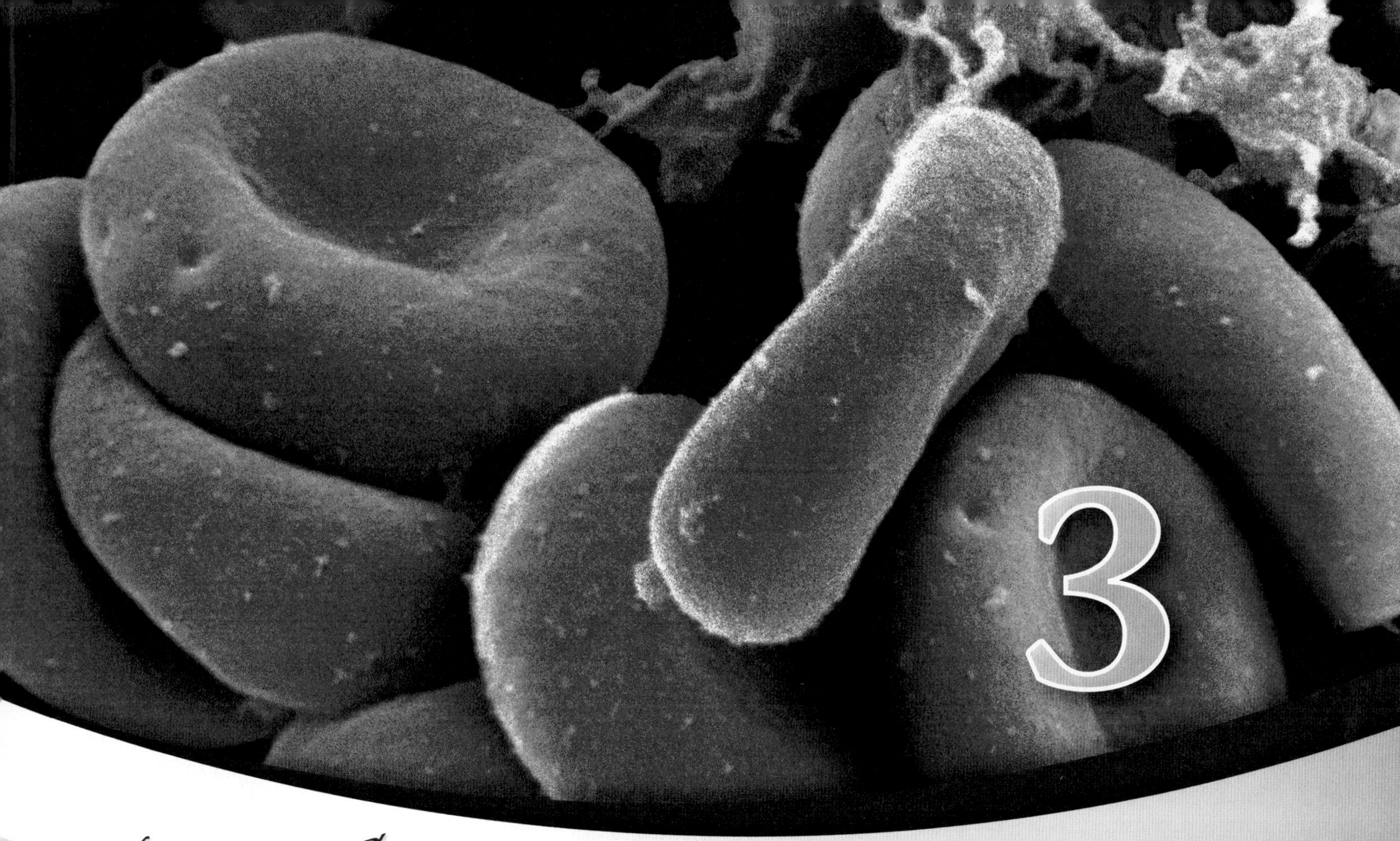

Cell Structure and Function

Applying the Concepts

One of the characteristics used to distinguish prokaryotic from eukaryotic cells is the presence of a nucleus. However, mature red blood cells (also called erythrocytes) in mammals including humans are unusual in that they do not contain a nucleus—or many of the other specialized cellular structures called organelles that will be discussed in this chapter. Red blood cells come from stem cells in the bone marrow. During maturation, the part of the cell containing the nucleus is pinched off and then destroyed by cells of the immune system. The other organelles are destroyed internally.

Mature erythrocytes are much smaller than other cells in the body and are shaped like biconcave disks with a depression in the middle. The lack of a nucleus and organelles enables these red blood cells to be incredibly flexible, so they can squeeze through tiny capillaries in the tissues where they release their oxygen load. For example, a red blood cell is about 8 micrometers (μm) in diameter, while some of the capillaries in the brain are only 2 μm in diameter. In order to enter these capillaries, the red blood cell stretches into a long, thin shape by rearranging its cytoskeleton, which is responsible for the cell shape.

The loss of the nucleus and organelles also enables a red blood cell to hold more hemoglobin, necessary to carry oxygen. The cell is packed full of hemoglobin—approximately 280 million molecules in a single red blood cell! This does mean, however, that a red blood cell cannot repair itself or divide. A red blood cell circulates in the blood for about 120 days before it is removed from the circulation by immune cells within the liver and spleen.

In this chapter, we will discuss the characteristics of a typical animal cell. But most of the cells in the human body differentiate into specialized cells like a red blood cell—so there really is no typical animal cell!

Chapter Outline

3.1 The Cellular Level of Organization

Learning Outcomes

1. Describe the two categories of cells.
2. Define cell theory.
3. Understand why actively metabolizing cells must be small.

The cell marks the boundary between the nonliving and the living. The molecules that serve as food for a cell and the macromolecules that make up a cell are not alive, and yet the cell is alive. The cell is the structural and functional unit of an organism, the smallest structure capable of performing all the functions necessary for life. Thus, the answer to what life is must lie within the cell. The smallest living organisms are unicellular, while larger organisms are multicellular—that is, composed of many cells.

Cells can be classified as either prokaryotic or eukaryotic. Prokaryotic cells do not contain the membrane-enclosed structures found in eukaryotic cells. Therefore, eukaryotic cells are thought to have evolved from prokaryotic cells (see Section 3.4). Prokaryotic cells are exemplified by the bacteria.

The diversity of cells is illustrated by the many types in the human body, such as muscle cells and nerve cells. But despite a variety of forms and functions, human cells contain the same components. The basic components that are common to all eukaryotic cells, regardless of their specializations, are the subject of this chapter. Viewing these components requires a microscope. The Science Focus on page 53 introduces you to the microscopes most used today to study cells. Electron microscopy and biochemical analyses have revealed that eukaryotic cells actually contain tiny, specialized structures called **organelles.** Each organelle performs specific cellular functions.

Today, we are accustomed to thinking of living things as being constructed of cells. But the word cell didn't enter biology until the seventeenth century. Antonie van Leeuwenhoek of Holland is now famous for making his own microscopes and observing all sorts of tiny things that no one had seen before. Robert Hooke, an Englishman, confirmed Leeuwenhoek's observations and was the first to use the term "cell." The tiny chambers he observed in the honeycomb structure of cork reminded him of the rooms, or cells, in a monastery.

Over 150 years later—in the 1830s—the German microscopist Matthias Schleiden stated that plants are composed of cells. His counterpart, Theodor Schwann, stated that animals are also made up of living units called cells. This was quite a feat, because aside from their own exhausting examination of tissues, both had to take into consideration the studies of many other microscopists. Rudolf Virchow, another German microscopist, later came to the conclusion that cells don't suddenly appear; rather, they come from preexisting cells. Think about how we reproduce. The sperm fertilizes an egg and a human being develops from a resulting cell, called a zygote.

The **cell theory** states that all organisms are made up of basic living units called cells, and that all cells come only from previously existing cells. Today, the cell theory is a basic theory of biology.

Cell Size

Cells are quite small. A frog's egg, at about 1 millimeter (mm) in diameter, is large enough to be seen by the human

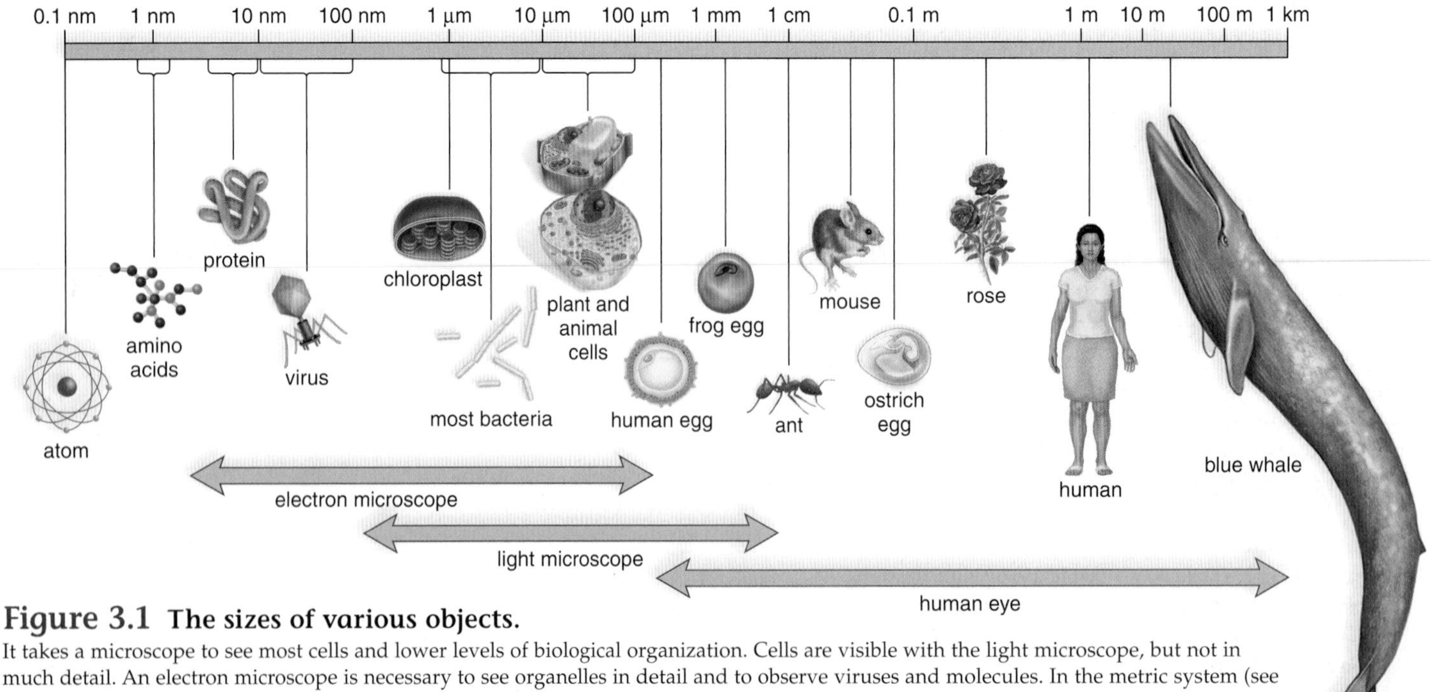

Figure 3.1 The sizes of various objects.
It takes a microscope to see most cells and lower levels of biological organization. Cells are visible with the light microscope, but not in much detail. An electron microscope is necessary to see organelles in detail and to observe viruses and molecules. In the metric system (see Appendix C), each higher unit is ten times greater than the preceding unit. (1 meter = 10^2 cm = 10^3 mm = 10^6 μm = 10^9 nm.)

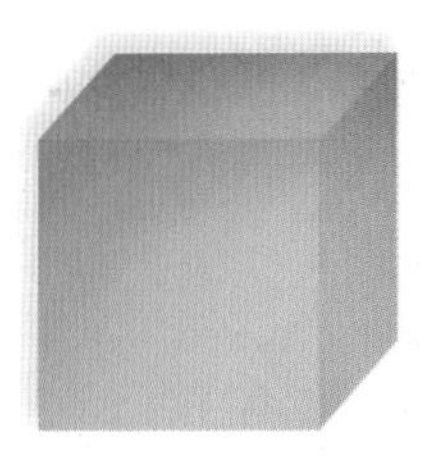

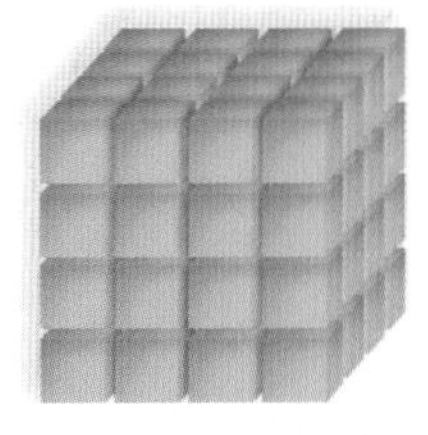

Figure 3.2 Surface-area-to-volume relationships. All three have the same volume, but the group on the right has four times the surface area.

eye. But most cells are far smaller than 1 mm. Some are even as small as 1 micrometer (μm)—one thousandth of a millimeter. Cell inclusions and macromolecules are smaller than a micrometer and are measured in terms of nanometers (nm). Figure 3.1 outlines the visual range of the eye, light microscope, and electron microscope. The discussion of microscopy in the Science Focus explains why the electron microscope allows us to see so much more detail than the light microscope does.

The fact that cells are so small is a great advantage for multicellular organisms. Nutrients, such as glucose and oxygen, enter a cell, and wastes, such as carbon dioxide, exit a cell at its surface. Therefore, the amount of surface area affects the ability to get material into and out of the cell. A large cell requires more nutrients and produces more wastes than a small cell. But, as cells get larger in volume, the proportionate amount of surface area actually decreases. For example, for a cube-shaped cell, the volume increases by the cube of the sides (height × width × depth), while the surface area increases by the square of the sides and number of sides (height × width × 6). If a cell doubles in size, its surface area only increases fourfold while its volume increases eightfold. Therefore, small cells, not large cells, are likely to have an adequate surface area for exchanging nutrients and wastes. As Figure 3.2 demonstrates, cutting a large cube into smaller cubes provides a lot more surface area per volume.

Most actively metabolizing cells are small. The frog's egg is not actively metabolizing. But once the egg is fertilized and metabolic activity begins, the egg divides repeatedly without growth. These cell divisions restore the amount of surface area needed for adequate exchange of materials. Further, cells that specialize in absorption have modifications that greatly increase their surface-area-to-volume ratio. For example, the columnar epithelial cells along the surface of the intestinal wall have surface foldings called microvilli (sing., microvillus) that increase their surface area.

Check Your Progress

1. Do you expect a chicken's egg to be actively metabolizing?
2. Are the cells that make up your body classified as prokaryotic or eukaryotic?

3.2 Prokaryotic Cells

Learning Outcome

1. Describe the structures and functions of the various parts of a prokaryotic cell.

Cells can be classified by the presence or absence of a nucleus. **Prokaryotic cells** lack a membrane-bounded nucleus. The domains Archaea and Bacteria consist of prokaryotic cells. Prokaryotes generally exist as unicellular organisms (single cells) or as simple strings and clusters. Many people think of germs when they hear the word bacteria, but not all bacteria cause disease. In fact, most bacteria are beneficial and are essential for other living organisms' survival.

Plasma Membrane and Cytoplasm

All cells are surrounded by a **plasma membrane** that consists of a phospholipid bilayer in which some protein molecules are embedded:

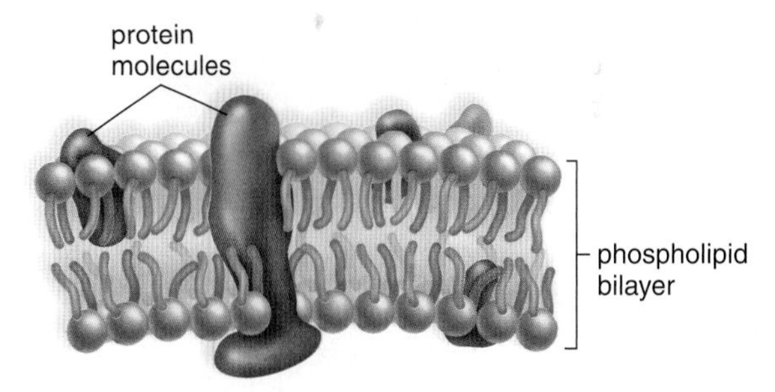

The plasma membrane is a living boundary that separates the living contents of the cell from the nonliving surrounding environment. Inside the cell is a semifluid medium called the **cytoplasm.** The cytoplasm is composed of water, salts, and dissolved organic molecules. The plasma membrane regulates the entrance and exit of molecules into and out of the cytoplasm.

Bacterial Anatomy

Figure 3.3 illustrates the main features of bacterial anatomy. The **cell wall,** located outside of the plasma membrane, contains peptidoglycan, a complex molecule that is unique to bacteria and composed of chains of disaccharides joined by peptide chains. The cell wall protects the bacteria. Some antibiotics, such as penicillin, interfere with the synthesis of peptidoglycan. In some bacteria, the cell wall is further surrounded by a **capsule** and/or a gelatinous sheath called a **slime layer.** Some bacteria have long, very thin appendages called flagella (sing., **flagellum**), which are composed of subunits of the protein flagellin. The flagella rotate like propellers, allowing the bacterium to move rapidly in a fluid medium. Some bacteria also have **fimbriae,** which are short appendages that help them attach to an appropriate surface. The capsule and fimbriae often give pathogenic bacteria increased ability to cause disease.

Prokaryotes have a single chromosome (loop of DNA and associated proteins) located within a region of the

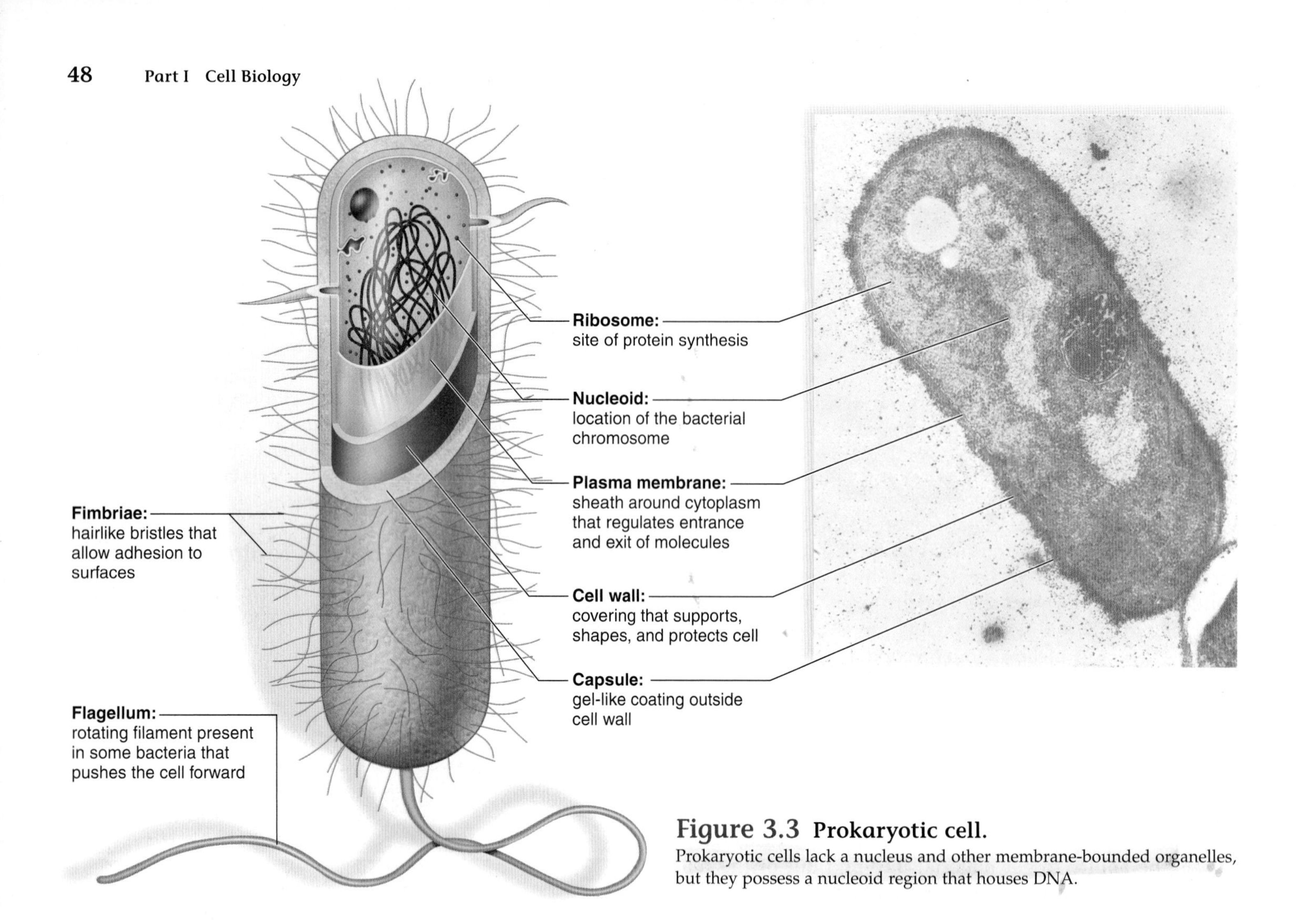

Figure 3.3 Prokaryotic cell.
Prokaryotic cells lack a nucleus and other membrane-bounded organelles, but they possess a nucleoid region that houses DNA.

cytoplasm called the **nucleoid.** The nucleoid is not bounded by a membrane. Many prokaryotes also have small accessory rings of DNA called **plasmids.** The cytoplasm has thousands of **ribosomes** for the synthesis of proteins. The ribosomes of prokaryotic organisms are smaller and structurally different from those of eukaryotic cells, which makes ribosomes a good target for antibacterial drugs. In addition, the photosynthetic cyanobacteria have light-sensitive pigments, usually within the membranes of flattened disks called **thylakoids.**

Although prokaryotes are structurally simple, they are much more metabolically diverse than eukaryotes. Many of them can synthesize all their structural components from very simple, even inorganic molecules. Indeed, humans exploit the metabolic capability of bacteria by using them to produce a wide variety of chemicals and products. Prokaryotes also have adapted to living in almost every environment on Earth. In particular, archaea have been found living under conditions that would not support any other form of life—for example, in water at temperatures above boiling. Archaeal membranes have unique membrane-spanning lipids that help them survive in extremes of heat, pH, and salinity. Table 3.1 compares the major structures of prokaryotes (archaea and bacteria) with those of eukaryotes.

Check Your Progress

1. What is the function of the plasma membrane?
2. Make a table of each of the bacterial structures in Figure 3.3 and their function.
3. How do the bacterial structures shown in Figure 3.3 allow bacteria to survive as unicellular organisms?

TABLE 3.1 Comparison of Major Structural Features of Archaea, Bacteria, and Eukaryotes

	Archaea	Bacteria	Eukaryotes
Cell wall	Usually present, no peptidoglycan	Usually present, with peptidoglycan	Sometimes present, no peptidoglycan
Plasma membrane	Yes	Yes	Yes
Nucleus	No	No	Yes
Membrane-bounded organelles	No	No	Yes
Ribosomes	Yes	Yes	Yes, larger than prokaryotic

3.3 Eukaryotic Cells

Learning Outcome

1. Describe the structure and function of the organelles within eukaryotic cells.

Eukaryotic cells are structurally very complex. The principal distinguishing feature of eukaryotic cells is the presence of a nucleus, which separates the chromosomes from the cytoplasm of the cell. In addition, eukaryotic cells possess a variety of other organelles, many of which are surrounded by membranes. Animals, plants, fungi, and protists are all composed of eukaryotic cells.

Cell Walls

Some eukaryotic cells have a permeable but protective cell wall in addition to a plasma membrane. Many plant cells have both a primary and a secondary cell wall. A main constituent of a primary cell wall is cellulose molecules. Cellulose molecules form fibrils that lie at right angles to one another for added strength. The secondary cell wall, if present, forms inside the primary cell wall. Such secondary cell walls contain lignin, a substance that makes them even stronger than primary cell walls. The cell walls of some fungi are composed of cellulose and chitin, the same type of polysaccharide found in the exoskeleton of insects. Algae, members of the kingdom Protista, contain cell walls composed of cellulose.

Organelles of Animal and Plant Cells

Originally the term organelle referred to only membranous structures. However, it has come to mean any well-defined subcellular structure that performs a particular function (Table 3.2). By analogy, a eukaryotic cell can be thought of as a factory. Just as all the assembly lines of a factory operate at the same time, so all the organelles of a cell function simultaneously. Raw materials enter a factory where different departments turn them into various products. In the same way, the cell takes in chemicals, and then the organelles process them. The factory must also get rid of waste, and the cell performs that function as well.

If we limit our discussion to just animal and plant cells, both animal cells (Fig. 3.4) and plant cells (Fig. 3.5) contain mitochondria. Only plant cells have chloroplasts and only animal cells have centrioles. In the illustrations throughout this text, note that each of the organelles has an assigned color.

Check Your Progress

1. What is the role of a cell wall in eukaryotes?
2. Define an organelle.

TABLE 3.2 Eukaryotic Structures in Animal Cells and Plant Cells

Structure	Composition	Function
Cell wall (plant cells only)	Contains cellulose fibrils	Support and protection
Plasma membrane	Phospholipid bilayer with embedded proteins	Defines cell boundary; regulates molecule passage into and out of cells
Nucleus	Nuclear envelope, nucleoplasm, chromatin, and nucleoli	Storage of genetic information; synthesis of DNA and RNA
Nucleoli	Concentrated area of chromatin, RNA, and proteins	Ribosomal subunit formation
Ribosomes	Protein and RNA in two subunits	Protein synthesis
Endoplasmic reticulum (ER)	Membranous flattened channels and tubular canals	Synthesis and/or modification of proteins and other substances, and distribution by vesicle formation
Rough ER	Network of folded membranes studded with ribosomes	Folding, modification, and transport of proteins
Smooth ER	Network of folded membranes having no ribosomes	Various; lipid synthesis in some cells
Golgi apparatus	Stack of small membranous sacs	Processing, packaging, and distribution of proteins and lipids
Lysosomes (animal cells only)	Membranous vesicle containing digestive enzymes	Intracellular digestion
Vacuoles and vesicles	Membranous sacs of various sizes	Storage of substances
Peroxisomes	Membranous vesicle containing specific enzymes	Various metabolic tasks
Mitochondria	Inner membrane (cristae) bounded by an outer membrane	Cellular respiration
Chloroplasts (plant cells only)	Membranous grana bounded by two membranes	Photosynthesis
Cytoskeleton	Microtubules, intermediate filaments, actin filaments	Shape of cell and movement of its parts
Cilia and flagella (cilia are rare in plant cells)	9 + 2 pattern of microtubules	Movement of cell
Centriole (animal cells only)	9 + 0 pattern of microtubules	Formation of basal bodies

Figure 3.4 Animal cell anatomy.

Micrograph of an insect cell and drawing of a generalized animal cell. See Table 3.2 for a description of these structures, along with a listing of their functions.

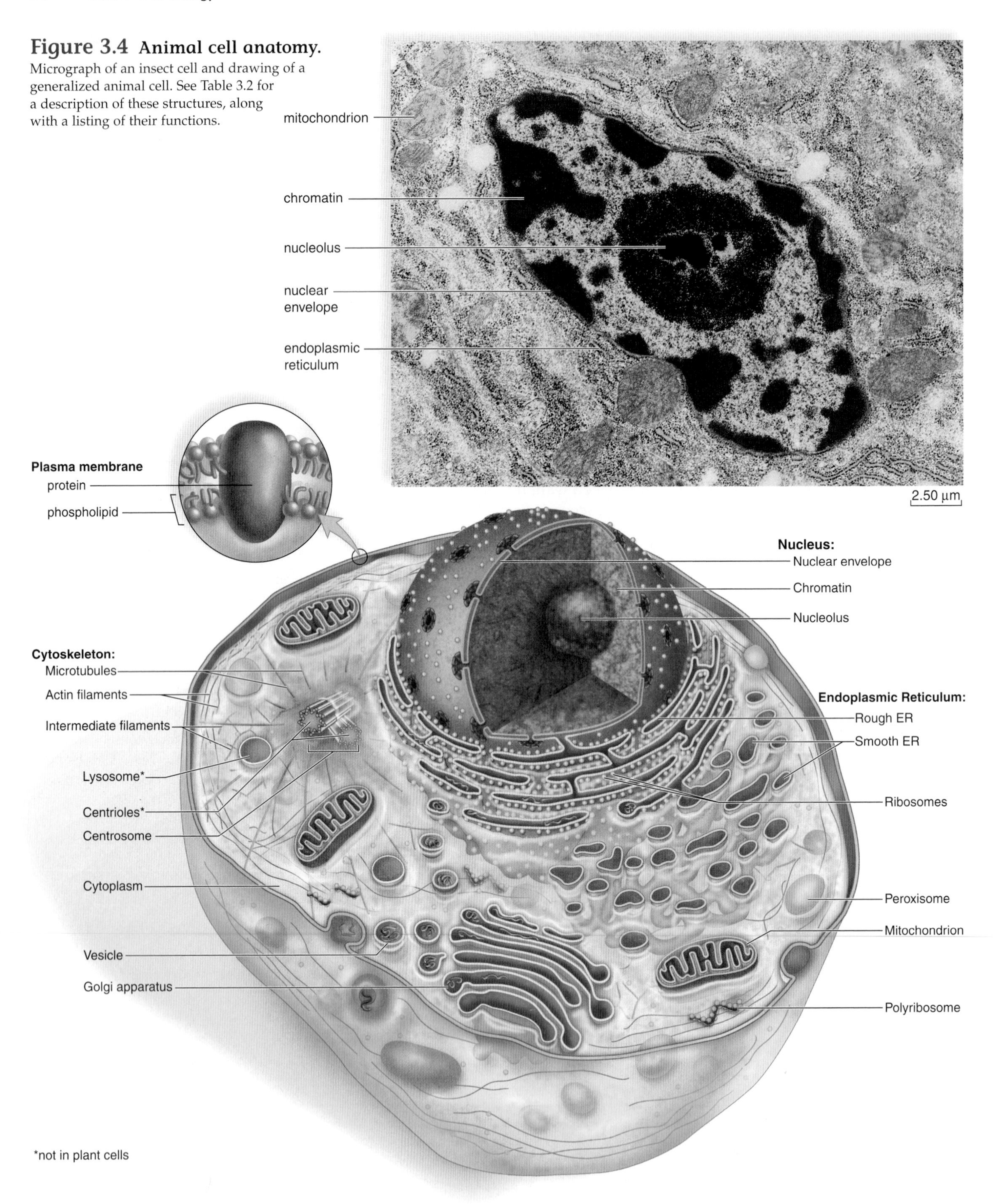

peroxisome
mitochondrion
nucleus
ribosomes
central vacuole
plasma membrane
cell wall
chloroplast
1 μm

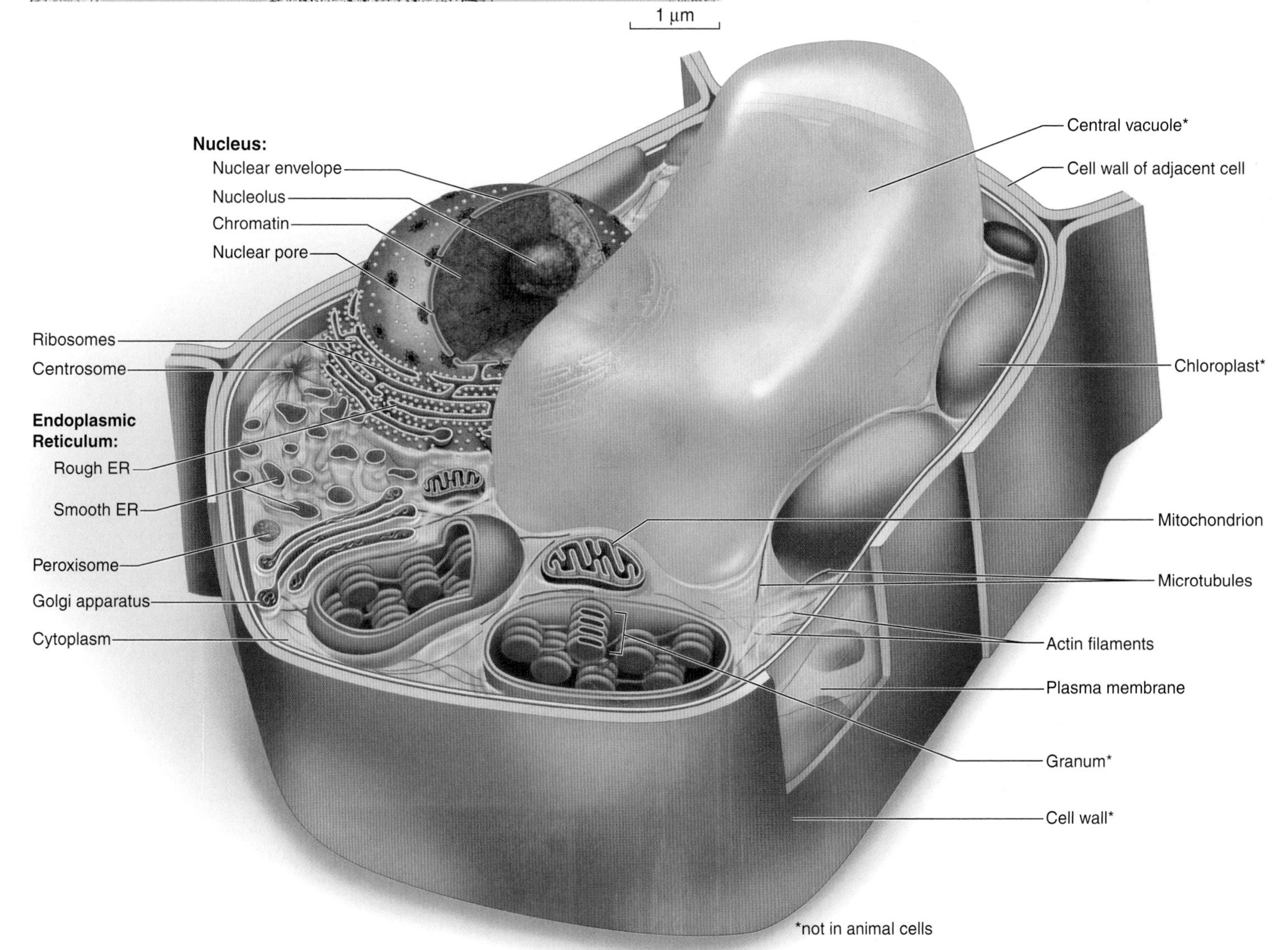

Figure 3.5 Plant cell anatomy.

Micrograph of a young plant cell and drawing of a generalized plant cell. See Table 3.2 for a description of these structures, along with a listing of their functions.

The Nucleus

The **nucleus,** which has a diameter of about 5 μm, is a prominent structure in the eukaryotic cell. In photographs or slides it looks like a dark golf ball within the cell. The nucleus is of primary importance because it stores the genetic material, DNA. In the factory analogy, the nucleus is the head office. DNA governs the characteristics of the cell and its metabolic functioning. Every cell in an individual contains the same DNA, but in each cell type, certain genes are turned on and certain others are turned off.

When you look at the nucleus, even in an electron micrograph, you cannot see a DNA molecule (Fig. 3.6). You can see **chromatin,** which consists of DNA and associated proteins. Chromatin in most eukaryotic cells is not one continuous strand. It is divided into long, threadlike structures called **chromosomes.** Human cells contain 46 chromosomes. During most of the cell's lifetime, chromosomes are dispersed and look grainy. But when the cell is ready to undergo cell division, chromosomes undergo coiling and become highly condensed. Chromosomes are immersed in a semifluid medium called the **nucleoplasm.** A difference in pH between the nucleoplasm and the cytoplasm suggests that the nucleoplasm has a different composition.

When you look at an electron micrograph of a nucleus, you may see one or more regions that look darker than the rest of the chromatin. These are nucleoli (sing., **nucleolus**), where another type of RNA, called ribosomal RNA (rRNA), is produced and where rRNA joins with proteins to form the subunits of ribosomes described in the next section.

The nucleus is separated from the cytoplasm by a double membrane known as the **nuclear envelope,** which is continuous with the endoplasmic reticulum (also discussed next). The nuclear envelope has **nuclear pores** of sufficient size (100 nm) to permit the bidirectional transport of proteins and ribosomal subunits.

Interestingly, during cell division, the nuclear envelope completely disappears and the contents of the nucleus are mixed with the cytoplasm. Following cell division, the nuclear envelope re-forms around the chromosomes, and the other contents of the nucleus are imported through the nuclear pores. Regulation of transport through the nuclear pores can control events within the nucleus or the entire cell.

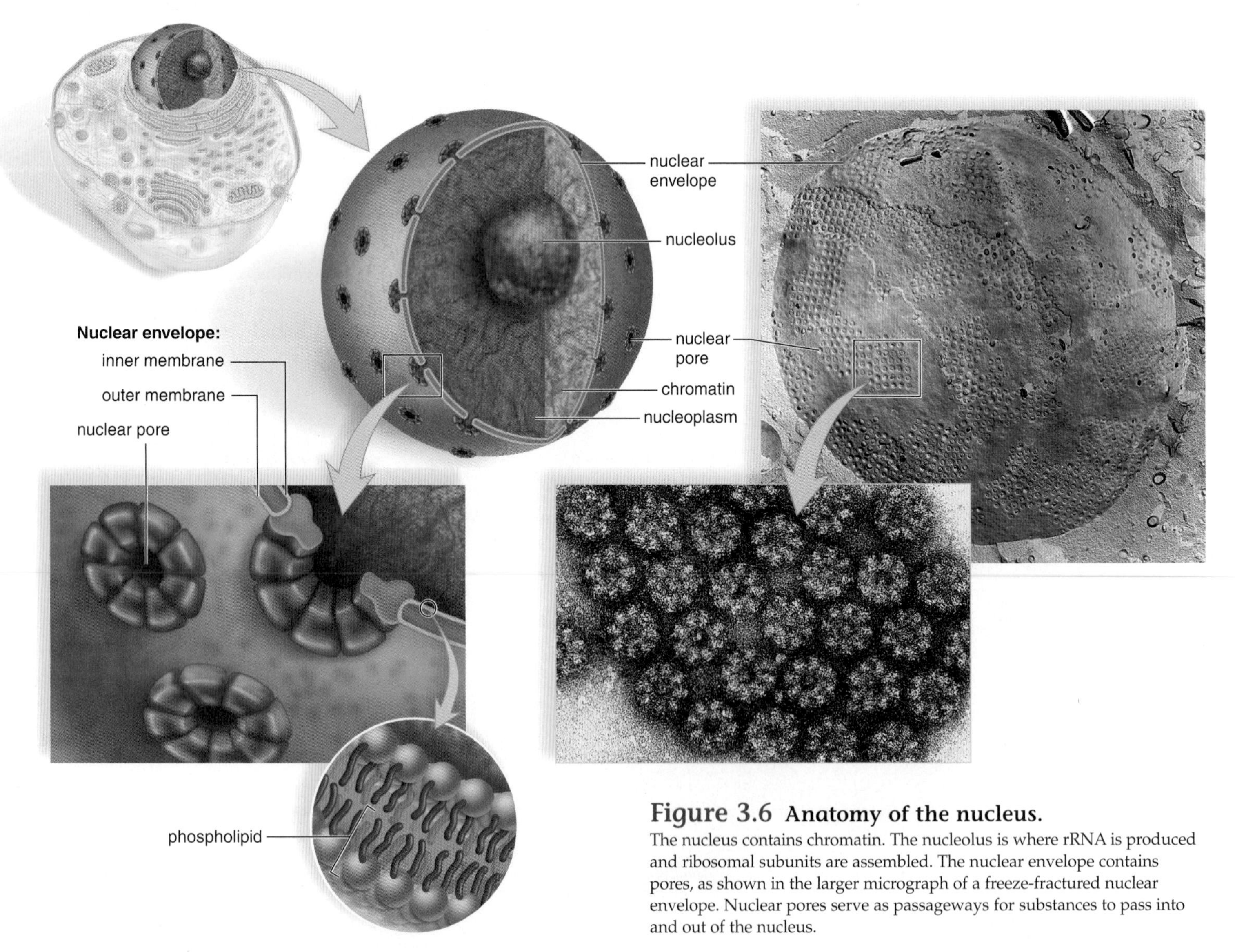

Figure 3.6 Anatomy of the nucleus.
The nucleus contains chromatin. The nucleolus is where rRNA is produced and ribosomal subunits are assembled. The nuclear envelope contains pores, as shown in the larger micrograph of a freeze-fractured nuclear envelope. Nuclear pores serve as passageways for substances to pass into and out of the nucleus.

Science Focus

Microscopy Today

Today, scientists use two types of microscopes: light microscopes and electron microscopes. Figure 3A depicts one type of light microscope and two types of electron microscopes, along with micrographs obtained with each one.

The most commonly used light microscope is the compound light microscope. "Compound" refers to the presence of more than one lens. In a compound light microscope, light rays pass through a specimen such as a blood smear, a tissue section, or even living cells. The light rays are brought to a focus by a set of glass lenses and the resulting image is then viewed. Specimens are often stained to provide viewing contrast.

In an electron microscope, electrons rather than light rays are brought to focus by a set of electromagnetic lenses, and the resulting image is projected onto a viewing screen or photographic film. Electron microscopes produce images of much higher magnification than may be obtained with even the best compound light microscope. The ability to distinguish two points as separate points (resolution) is much greater with electron microscopes. The greater resolving power is due to the fact that electrons travel at a much shorter wavelength than do light rays. Live specimens, however, cannot be viewed since the specimen must be dried out. Two commonly used types of electron microscopes are transmission and scanning.

A transmission electron microscope (TEM) may magnify an object's image over 200,000 times, compared to approximately 1,000 times with a compound light microscope. Also, a TEM has a much greater ability to make out detail. Much of our knowledge of the internal structures of cells comes from with a TEM.

The magnifying and resolving powers of a typical scanning electron microscope (SEM) are not as great as with a TEM although they are still greater than those of a compound light microscope. A SEM provides a striking three-dimensional view of the surface of an object.

A picture obtained using a compound light microscope is sometimes called a photomicrograph, and a picture resulting from the use of an electron microscope is called a transmission or scanning electron micrograph, depending on the type of microscope used. Color may be added to electron micrographs, such as those in Figure 3A, using a computer.

Discussion Questions

1. What distortions or artifacts might be seen in electron microscopy when viewing a dried, nonliving specimen?
2. What happens to an image if the magnification is increased without increasing the resolution?
3. Why are electron micrographs always in black and white (or artificially colored)?

Figure 3A Diagram of microscopes with accompanying micrographs.
The compound light microscope and the transmission electron microscope provide an internal view of an organism. The scanning electron microscope provides an external view of an organism.

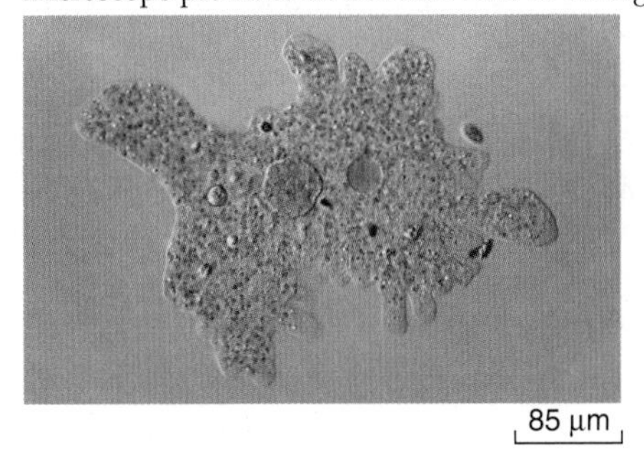

amoeba, light micrograph, stained with dye

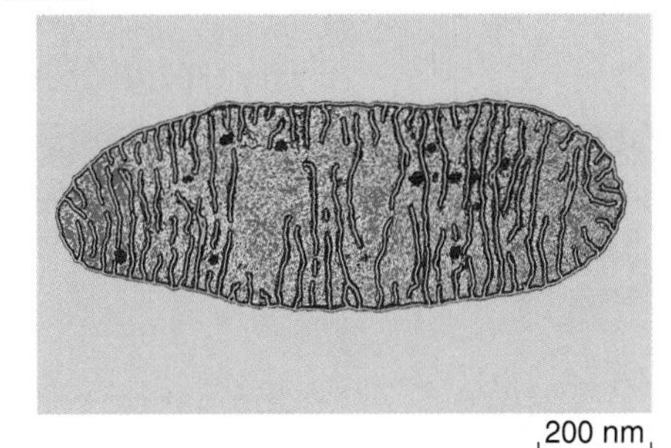

mitochondrion, TEM, artificially colored

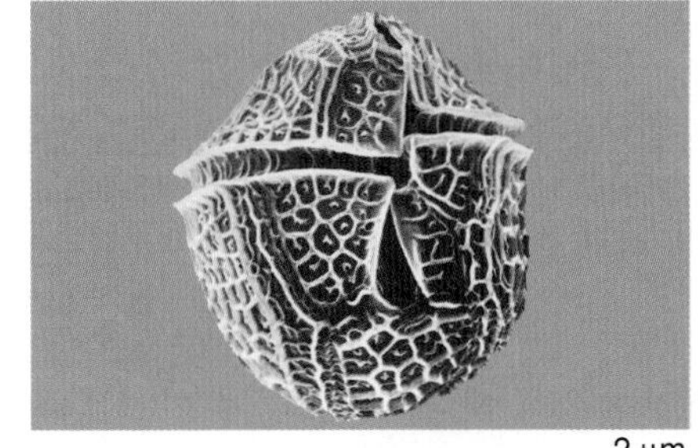

dinoflagellate, SEM, artificially colored

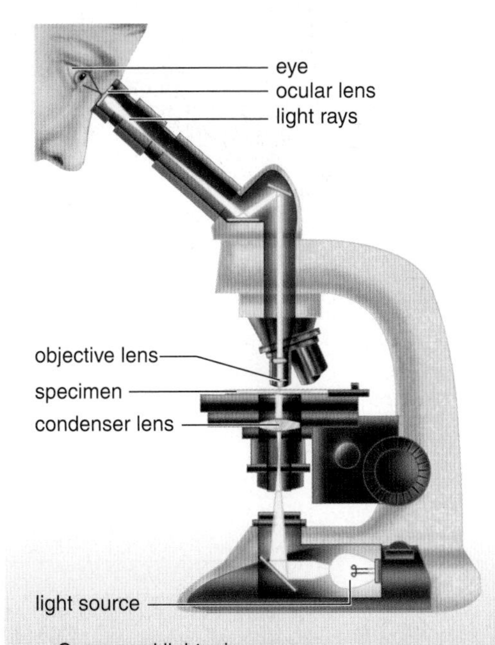

a. Compound light microscope

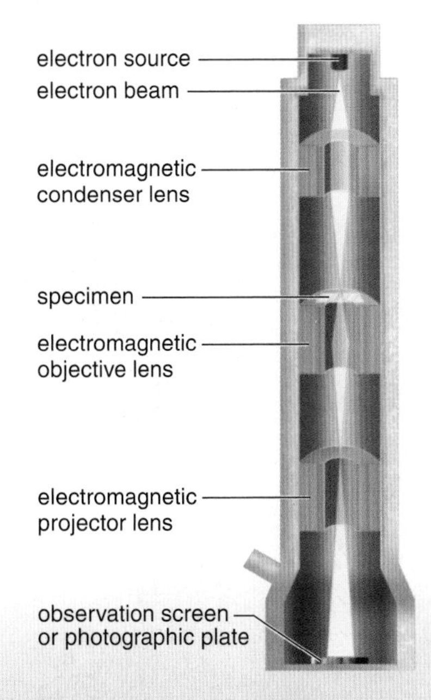

b. Transmission electron microscope

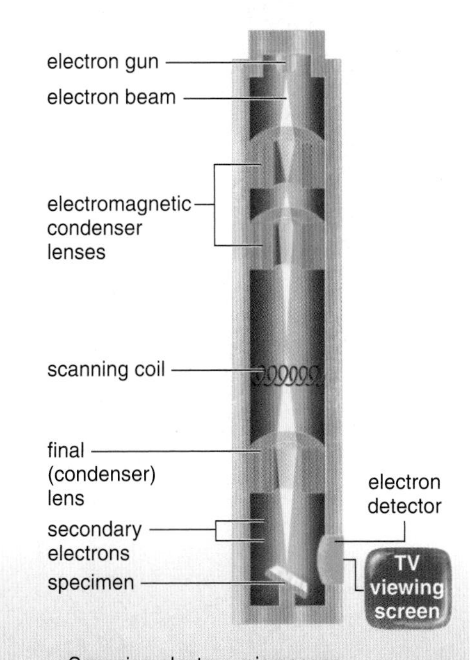

c. Scanning electron microscope

Ribosomes

Ribosomes are responsible for the synthesis of proteins using messenger RNA as a template. Ribosomes are composed of two subunits, called "large" and "small" because of their sizes relative to each other. Each subunit is a complex of unique ribosomal RNA (rRNA) and protein molecules. Ribosomes can be found individually in the cytoplasm, as well as in groups called **polyribosomes** (several ribosomes associated simultaneously with a single mRNA molecule). Ribosomes can also be found attached to the endoplasmic reticulum, a membranous system of sacs and channels discussed in the next section. Proteins synthesized at ribosomes attached to the endoplasmic reticulum have a different destination from that of proteins synthesized at ribosomes free in the cytoplasm.

The Endomembrane System

The endomembrane system consists of the nuclear envelope, the endoplasmic reticulum, the Golgi apparatus, and several **vesicles** (tiny membranous sacs). Continuing the factory analogy, the endomembrane system is essentially the transportation and product-processing section of the cell. This system compartmentalizes the cell so that particular enzymatic reactions are restricted to specific regions. Organelles that make up the endomembrane system are connected either directly or by transport vesicles.

The Endoplasmic Reticulum

The **endoplasmic reticulum (ER),** a complicated system of membranous channels and sacs (flattened vesicles), is physically continuous with the outer membrane of the nuclear envelope. Rough ER is studded with ribosomes on the side of the membrane that faces the cytoplasm (Fig. 3.7). As proteins are synthesized on these ribosomes they pass into the interior of the ER, where processing and modification begin. Proteins synthesized here are destined for the membrane of the cell or to be secreted from the cell.

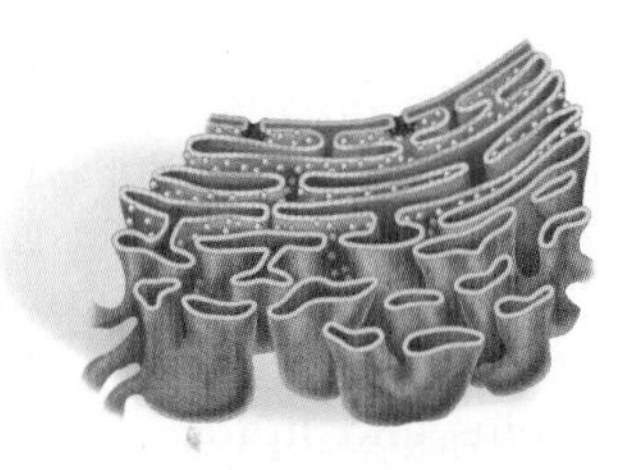

Proper folding, processing, and transport of proteins are critical to the functioning of the cell. For example, in cystic fibrosis a mutated plasma membrane channel protein is retained in the endoplasmic reticulum because it is folded incorrectly. Without this protein in its correct location, the cell is unable to regulate the transport of the chloride ion, resulting in the various symptoms of the disease.

Smooth ER, which is continuous with rough ER, does not have attached ribosomes. Smooth ER synthesizes the phospholipids that occur in membranes and performs various other functions depending on the particular cell. In the testes, it produces testosterone, and in the liver, it helps detoxify drugs. In muscle cells, the smooth ER stores calcium ions. Regardless of any specialized function, smooth ER also forms vesicles in which products are transported to the Golgi apparatus.

The Golgi Apparatus

The **Golgi apparatus** is named for Camillo Golgi, who discovered its presence in cells in 1898. The Golgi apparatus consists of a stack of three to twenty slightly curved sacs

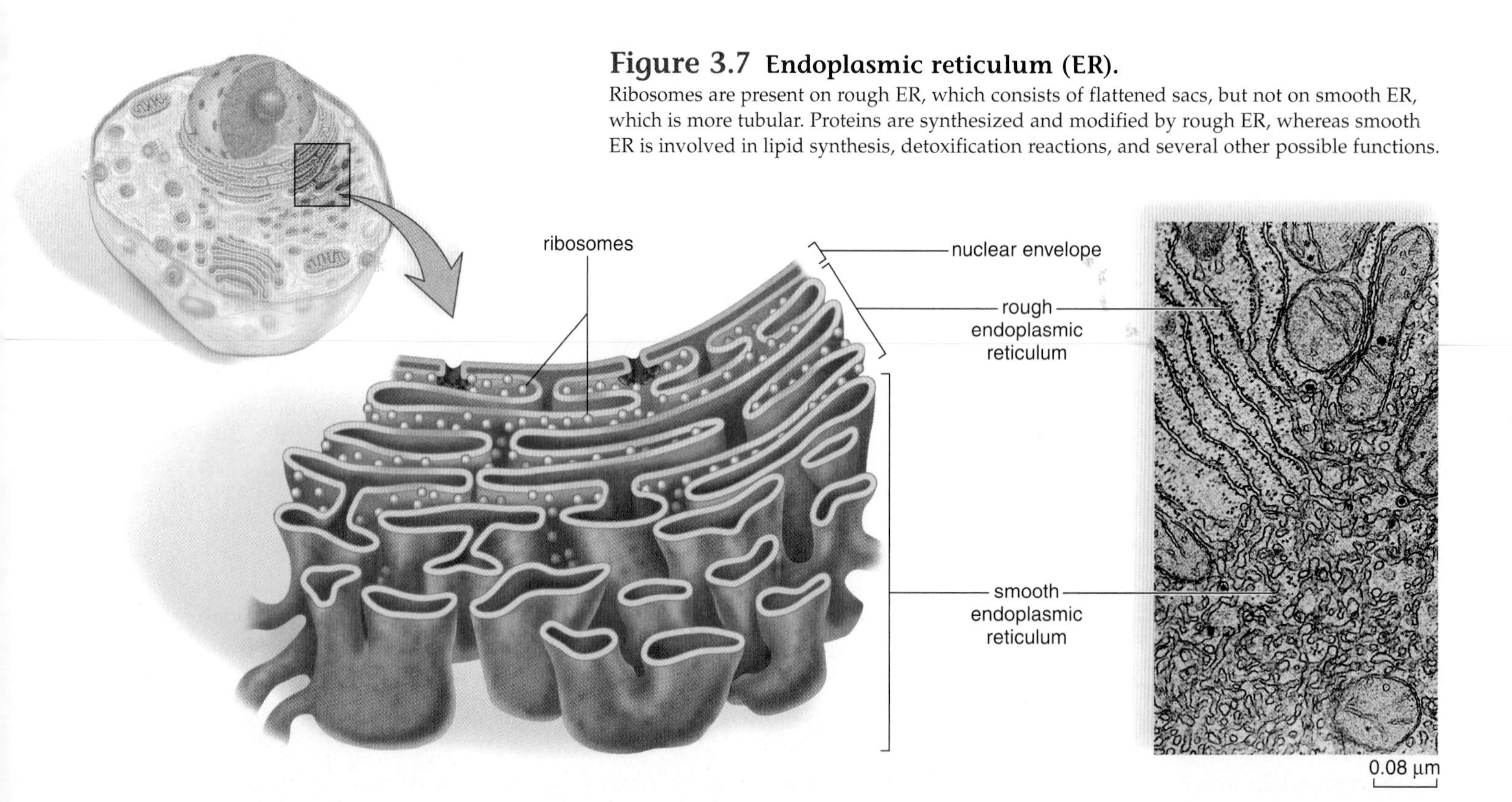

Figure 3.7 Endoplasmic reticulum (ER).
Ribosomes are present on rough ER, which consists of flattened sacs, but not on smooth ER, which is more tubular. Proteins are synthesized and modified by rough ER, whereas smooth ER is involved in lipid synthesis, detoxification reactions, and several other possible functions.

whose appearance can be compared to a stack of pancakes (Fig. 3.8). The Golgi apparatus is referred to as the post office of the cell because it collects, sorts, packages, and distributes materials such as proteins and lipids. In animal cells, one side of the stack (the inner face) is directed toward the ER, and the other side of the stack (the outer face) is directed toward the plasma membrane. Vesicles can frequently be seen at the edges of the sacs.

The Golgi apparatus receives proteins and also lipid-filled vesicles that bud from the ER. These molecules then move through the Golgi from the inner face to the outer face. How this occurs is still being debated. According to the maturation saccule (small sac) model, the vesicles fuse to form an inner face saccule, which matures as it gradually becomes a saccule at the outer face. According to the stationary saccule model, the molecules move through stable saccules from the inner face to the outer face by shuttle vesicles. It is likely that both models apply, depending on the organism and the type of cell.

During their passage through the Golgi apparatus, proteins and lipids can be modified before they are repackaged in secretory vesicles. Secretory vesicles proceed to the plasma membrane, where they discharge their contents. This action is termed **secretion.**

The Golgi apparatus is also involved in the formation of lysosomes, vesicles that contain proteins and remain within the cell. How does the Golgi apparatus direct traffic—in other words, what makes it direct the flow of proteins to different destinations? It now seems that proteins made at the rough ER have specific molecular tags that serve as "zip codes" to tell the Golgi apparatus whether they belong inside the cell in some membrane-bounded organelle or in a secretory vesicle.

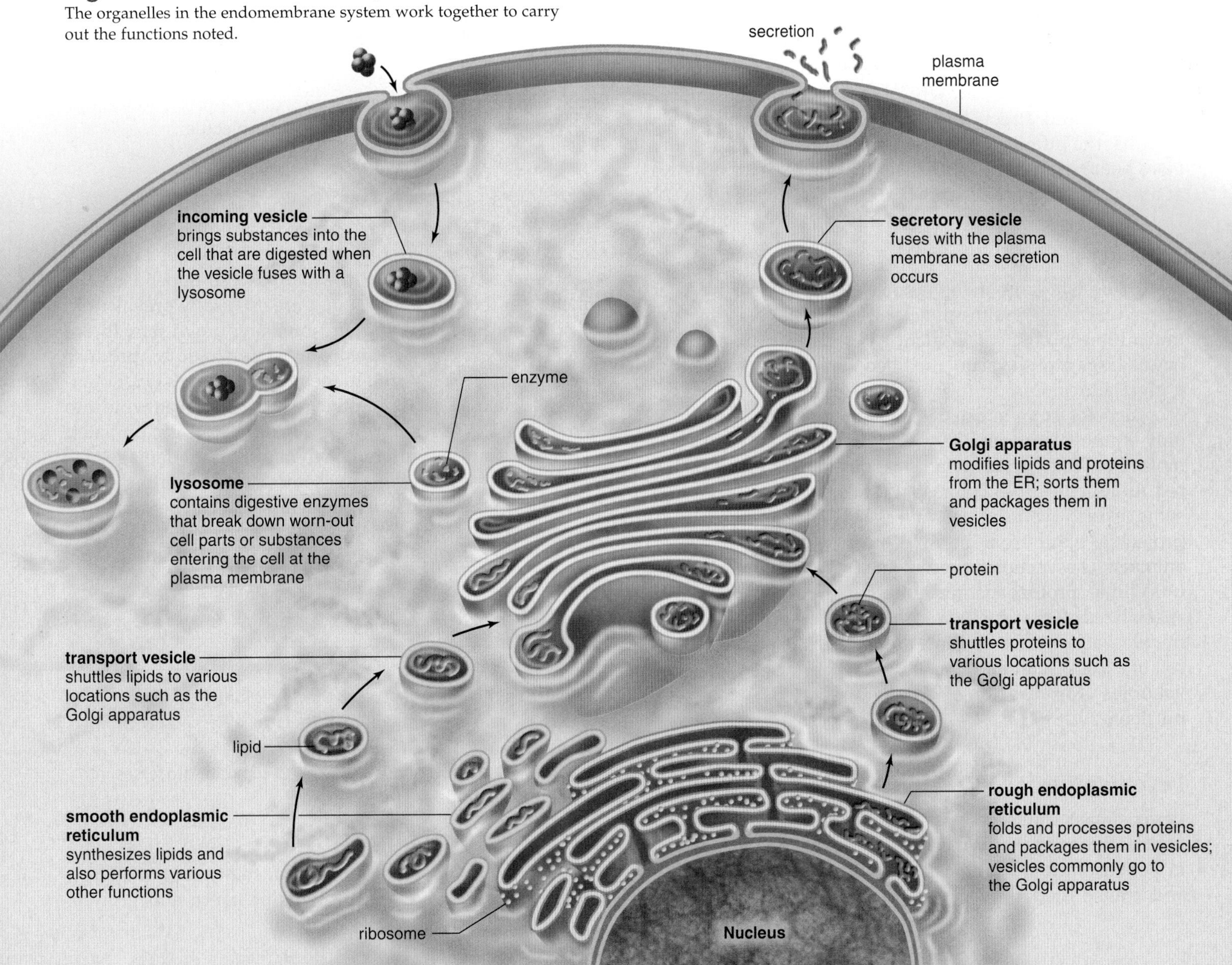

Figure 3.8 Endomembrane system.
The organelles in the endomembrane system work together to carry out the functions noted.

Lysosomes

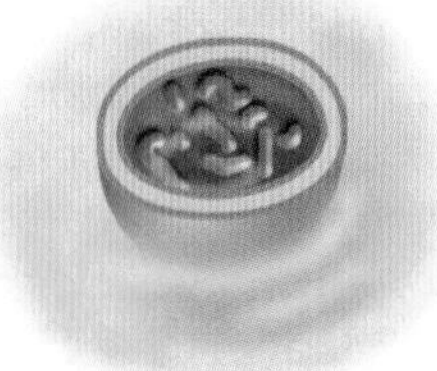

Lysosomes are membrane-bounded vesicles produced by the Golgi apparatus. Lysosomes contain hydrolytic digestive enzymes. Lysosomes are the garbage disposals of the factory.

Sometimes macromolecules are brought into a cell by vesicle formation at the plasma membrane (see Fig. 3.8). When a lysosome fuses with such a vesicle, its contents are digested by lysosomal enzymes into simpler subunits that then enter the cytoplasm. Some white blood cells defend the body by engulfing pathogens via vesicle formation. When lysosomes fuse with these vesicles, the bacteria are digested. Even parts of a cell are digested by its own lysosomes (called autodigestion). For example, the finger webbing found in the human embryo is later dissolved by lysosomes so that the fingers are separated.

Lysosomes contain many enzymes for digesting all sorts of molecules. Occasionally, a child inherits the inability to make a lysosomal enzyme, and therefore has a lysosomal storage disease. For example, in Tay Sachs disease, the cells that surround nerve cells cannot break down a particular lipid, which then accumulates inside lysosomes and affects the nervous system. At about six months, the infant can no longer see and, then, gradually loses hearing and even the ability to move. Death follows at about three years of age.

Vacuoles

A **vacuole** is a large membranous sac. A vacuole is larger than a vesicle. Although animal cells have vacuoles, they are much more prominent in plant cells. Typically, plant cells have a large central vacuole so filled with a watery fluid that it gives added support to the cell (see Fig. 3.5).

Vacuoles store substances. Plant vacuoles contain not only water, sugars, and salts, but also pigments and toxic molecules. The pigments are responsible for many of the red, blue, or purple colors of flowers and some leaves. The toxic substances help protect a plant from herbivorous animals. The vacuoles present in unicellular protozoans are quite specialized. They include contractile vacuoles for ridding the cell of excess water and digestive vacuoles for breaking down nutrients.

Peroxisomes

Peroxisomes, similar to lysosomes, are membrane-bounded vesicles that enclose enzymes (Fig. 3.9). However, the enzymes in peroxisomes are synthesized by cytoplasmic ribosomes and transported into a peroxisome by carrier proteins.

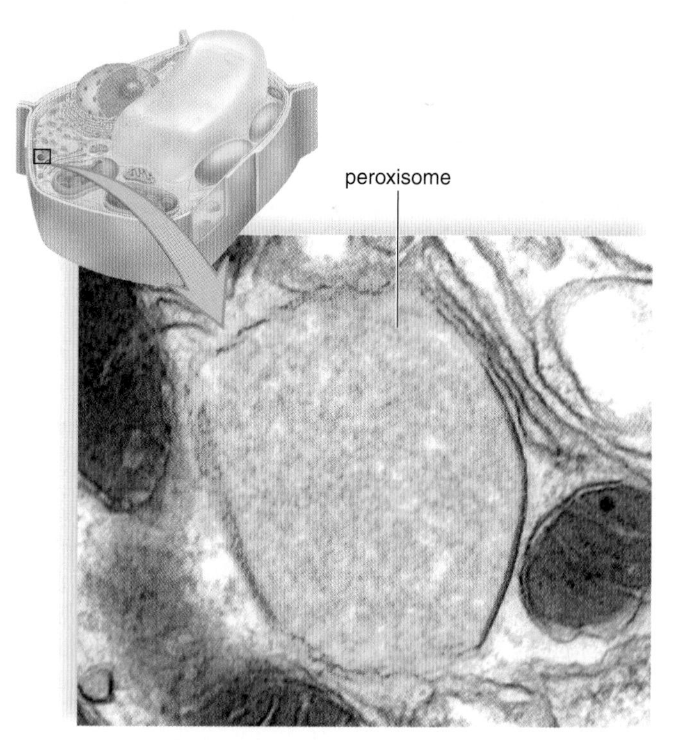

Figure 3.9 Peroxisomes.
Peroxisomes contain one or more enzymes that can oxidize various organic substances. Peroxisomes also contain the enzyme catalase, which breaks down the hydrogen peroxide (H_2O_2) that builds up after organic substances are oxidized.

Typically, peroxisomes contain enzymes whose action results in hydrogen peroxide (H_2O_2), a toxic molecule:

$$RH_2 + O_2 \longrightarrow R + H_2O_2$$
$$(R = \text{remainder of molecule})$$

Hydrogen peroxide is immediately broken down to water and oxygen by another peroxisomal enzyme called catalase.

The enzymes present in a peroxisome depend on the function of the cell. Peroxisomes are especially prevalent in cells that are synthesizing and breaking down fats. In the liver, some peroxisomes break down fats and others produce bile salts from cholesterol. In the movie *Lorenzo's Oil,* Lorenzo's cells lacked a carrier protein to transport an enzyme into peroxisomes. As a result, long-chain fatty acids accumulated in his brain and he suffered from neurological damage.

Plant cells also have peroxisomes. In germinating seeds, they oxidize fatty acids into molecules that can be converted to sugars needed by the growing plant. In leaves, peroxisomes can carry out a reaction that is opposite to photosynthesis—the reaction uses up oxygen and releases carbon dioxide.

CHECK YOUR PROGRESS

1. What is the role of the nucleus?
2. What is the role of the ribosomes within the cell?
3. What organelles would you expect to be especially prominent in an actively secreting cell?
4. Why is it advantageous for a cell to separate processes into different compartments?

Energy-Related Organelles

Life is possible only because of a constant input of energy. Organisms use this energy for maintenance and growth.

Chloroplasts and mitochondria are the two eukaryotic membranous organelles that specialize in converting energy to a form the cell can use. **Chloroplasts** use solar energy to synthesize carbohydrates, and carbohydrate-derived products are broken down in mitochondria (sing., **mitochondrion**) to produce ATP molecules, as shown in the following diagram:

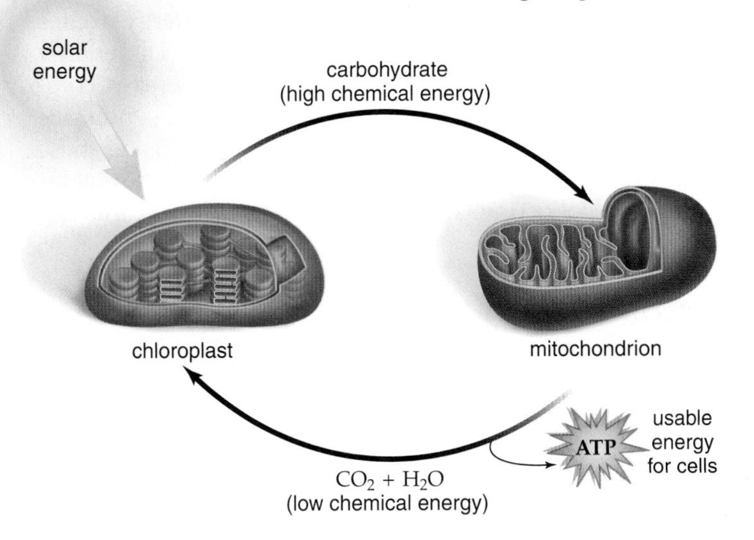

This diagram shows that chemicals recycle between chloroplasts and mitochondria, but energy flows from the sun through these organelles to ATP. When cells use ATP as an energy source, energy dissipates as heat. Life could not exist without a constant input of solar energy.

Only plants, algae, and cyanobacteria are capable of carrying on **photosynthesis** in this manner:

solar energy + carbon dioxide + water ⟶ carbohydrate + oxygen

Plants and algae have chloroplasts (Fig. 3.10), while cyanobacteria carry on photosynthesis within independent thylakoids. Solar energy is the ultimate source of energy for most cells because nearly all organisms, either directly or indirectly, use the carbohydrates produced by photosynthesizers as an energy source.

Many organisms carry on **cellular respiration,** the process by which the chemical energy of carbohydrates is converted to that of ATP (adenosine triphosphate), the common energy carrier in cells. All organisms, except prokaryotes, complete the process of cellular respiration in mitochondria. Cellular respiration can be represented by this equation:

carbohydrate + oxygen ⟶ carbon dioxide + water + energy

Here, *energy* is in the form of ATP molecules. When a cell needs energy, ATP supplies it. The energy of ATP is used for all energy-requiring processes in cells.

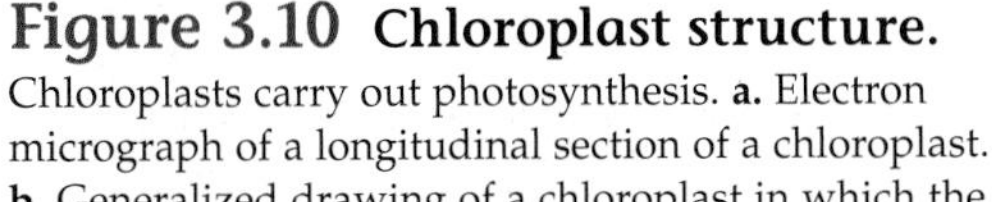

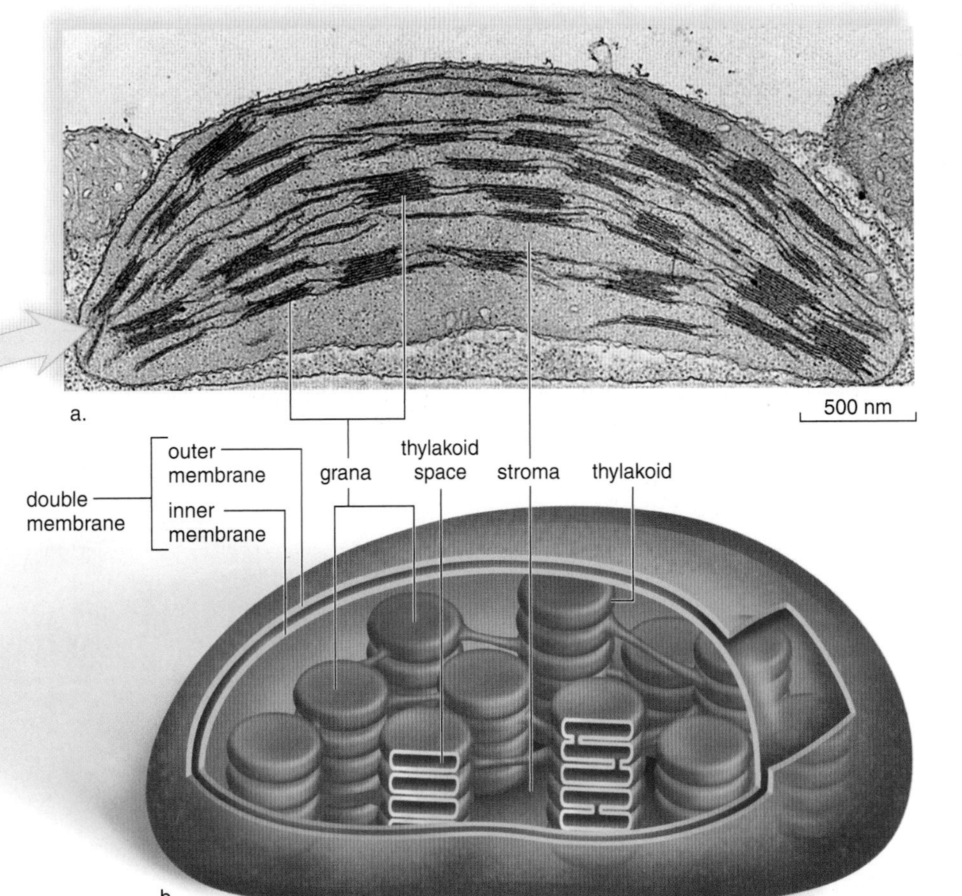

Figure 3.10 Chloroplast structure. Chloroplasts carry out photosynthesis. **a.** Electron micrograph of a longitudinal section of a chloroplast. **b.** Generalized drawing of a chloroplast in which the outer and inner membranes have been cut away to reveal the grana, each of which is a stack of membranous sacs called thylakoids. In some grana, but not all, it is obvious that thylakoid spaces are interconnected.

Chloroplasts

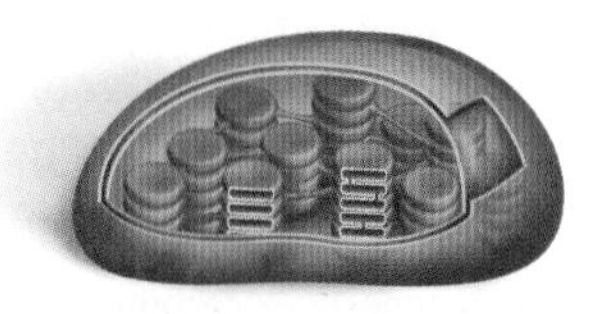

Plant and algal cells contain chloroplasts, the organelles that allow them to use solar energy to produce organic molecules. Chloroplasts are about 4–6 μm in diameter and 1–5 μm in length. They belong to a group of organelles known as plastids. Among the plastids are also the *amyloplasts*, common in roots, which store starch, and the *chromoplasts*, common in leaves, which contain red and orange pigments. A chloroplast is green because it contains the green pigment chlorophyll. A typical plant cell in a leaf may contain approximately 50 chloroplasts. Chloroplasts divide by splitting in two in a manner similar to how bacteria divide.

A chloroplast is bounded by two membranes that enclose a fluid-filled space called the **stroma** (see Fig. 3.10). The stroma contains a single circular DNA molecule as well as ribosomes. The chloroplast contains its own genome and makes some, but not all, of its own proteins. The others are encoded by nuclear genes and imported from the cytoplasm. A membrane system within the stroma is organized into interconnected flattened sacs called thylakoids. In certain regions, the thylakoids are stacked up in structures called grana (sing., **granum**). There can be hundreds of grana within a single chloroplast (see Fig. 3.10). Chlorophyll, which is located within the thylakoid membranes of grana, captures the solar energy needed to enable chloroplasts to produce carbohydrates. The solar energy excites an electron within the chlorophyll molecule. The energy from that excited electron is used to make high energy compounds. The stroma contains the enzymes that synthesize carbohydrates from carbon dioxide and water using these high energy compounds. The chloroplast is a tiny solar panel for collecting sunlight and a factory for making carbohydrates.

Mitochondria

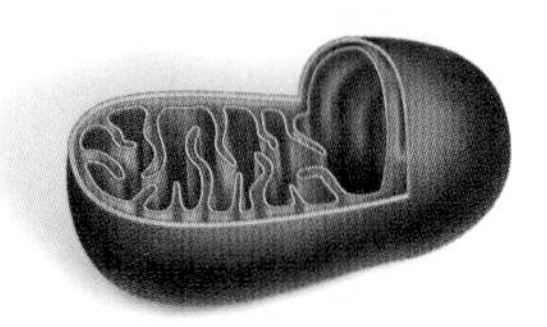

All eukaryotic cells, including those of plants and algae, contain mitochondria. This means that plant cells contain *both* chloro-

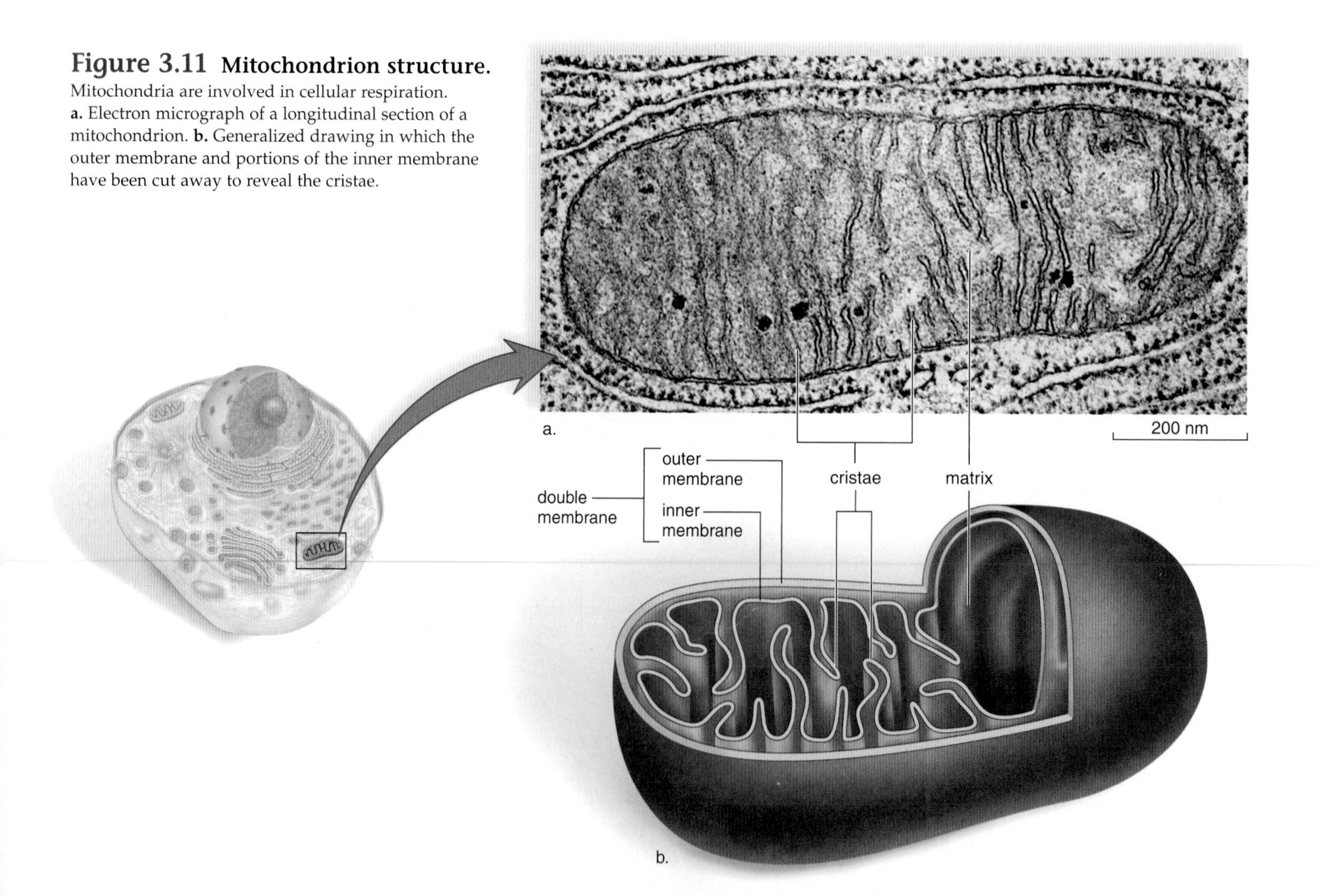

Figure 3.11 Mitochondrion structure. Mitochondria are involved in cellular respiration. **a.** Electron micrograph of a longitudinal section of a mitochondrion. **b.** Generalized drawing in which the outer membrane and portions of the inner membrane have been cut away to reveal the cristae.

plasts and mitochondria. Mitochondria are usually 0.5–1.0 μm in diameter and 2–5 μm in length. Mitochondria are the power plants of the cell. Substrates broken down in the cytoplasm are transported into the mitochondria and converted into ATP to be used by the cell for its various needs. Mitochondria are also involved in cellular differentiation and cell death. It is thought that mitochondria may play a role in aging.

Mitochondria, like chloroplasts, divide by splitting in two. Mitochondria are also bounded by a double membrane (Fig. 3.11). In mitochondria, the inner fluid-filled space is called the **matrix.** As with chloroplasts, mitochondria contain their own circular DNA chromosome and encode some, but not all, of their own proteins. The matrix contains ribosomes, and enzymes that break down carbohydrate products, releasing energy to be used for ATP production.

The inner membrane of a mitochondrion invaginates to form **cristae.** Cristae provide a much greater surface area to accommodate the protein complexes and other participants that produce ATP.

The number of mitochondria per cell can vary considerably. Some cells only have one mitochondrion, while other cells may have thousands. Tissues that need large amounts of energy have more mitochondria per cell than tissues with lower energy demand. Mutations in the mitochondrial DNA usually affect high-energy-demand tissues such as the eye, central nervous system, and muscles.

In humans, all mitochondria come from the maternal line through the egg. The father's sperm does not contribute any mitochondria to the offspring. This has made mitochondrial DNA studies useful for population genetic studies. An interesting example of this is the search for "mitochondrial Eve" supporting the Out of Africa theory for human evolution.

Check Your Progress

1. How are chloroplasts and mitochondria alike?
2. Do mitochondria and chloroplasts provide *opposite functions* (one works and the other reverses what the first one did) or *sequential functions* (one works and then the other works) in energy acquisition? Why?

The Cytoskeleton

The protein components of the **cytoskeleton** interconnect and extend from the nucleus to the plasma membrane in eukaryotic cells. Prior to the 1970s, scientists believed that the cytoplasm was an unorganized mixture of organic molecules. Then, high-voltage electron microscopes, which can penetrate thicker specimens, showed instead that the cytoplasm is highly organized. The technique of immunofluorescence microscopy identified the makeup of the protein components within the cytoskeletal network (Fig. 3.12).

The cytoskeleton contains actin filaments, intermediate filaments, and microtubules, which maintain cell shape and allow the cell and its organelles to move. Therefore, the cytoskeleton is often compared with the bones and muscles of an animal. However, the cytoskeleton is dynamic, especially because its protein components can assemble and disassemble as appropriate.

Actin Filaments

Actin filaments (formerly called microfilaments) are long, extremely thin, flexible fibers (about 7 nm in diameter) that occur in bundles or meshlike networks. Each actin filament contains two chains of globular actin monomers twisted about one another in a helical manner.

Actin filaments play a structural role when they form a dense, complex web just under the plasma membrane, to which they are anchored by special proteins. They are also seen in the microvilli that project from intestinal cells, and their presence accounts for the formation of pseudopods (false feet), extensions that allow certain cells to move in an amoeboid fashion.

How are actin filaments involved in the movement of the cell and its organelles? They interact with **motor molecules,** which are proteins that can attach, detach, and reattach farther along an actin filament. For example, in muscle cells the motor molecule myosin pulls actin filaments along in this way using the energy of ATP. Myosin has both a head and a tail. The tails of several muscle myosin molecules are joined to form a thick filament, while the heads interact with ATP and the actin filament.

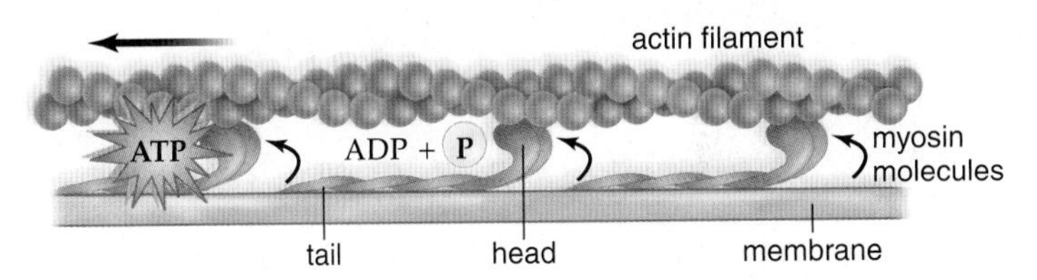

During animal cell division, the two new cells form when actin, in conjunction with myosin, pinches off the cells from one another.

Intermediate Filaments

Intermediate filaments (8–11 nm in diameter) are intermediate in size between actin filaments and microtubules and perform a structural role in the cell. They are a ropelike assembly of fibrous polypeptides, but the specific type varies according to the tissue. Some intermediate filaments support the nuclear envelope, whereas others support the plasma membrane and take part in the formation of cell-to-cell junctions. In skin cells, intermediate filaments, made of the protein keratin, give great mechanical strength. They, too, are dynamic structures.

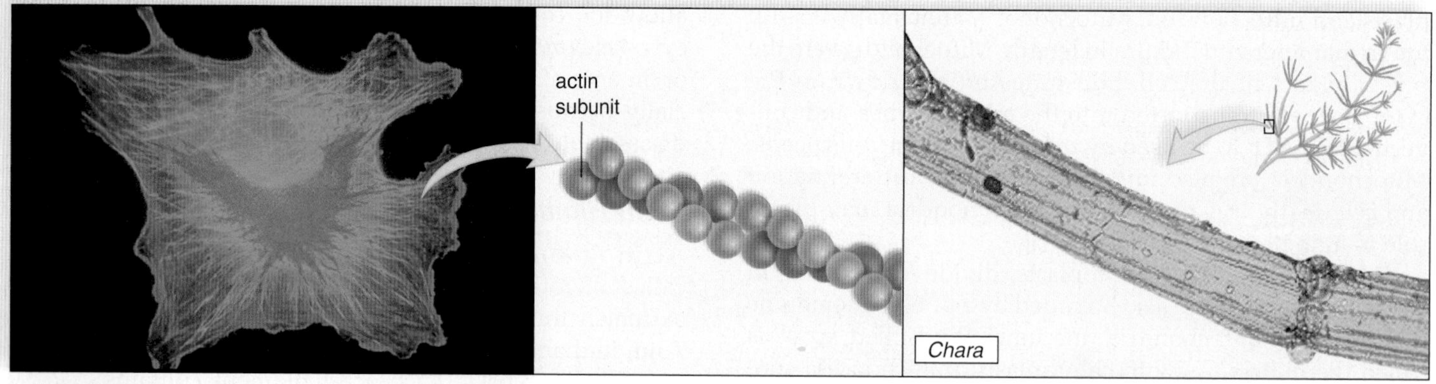

a. Actin filaments

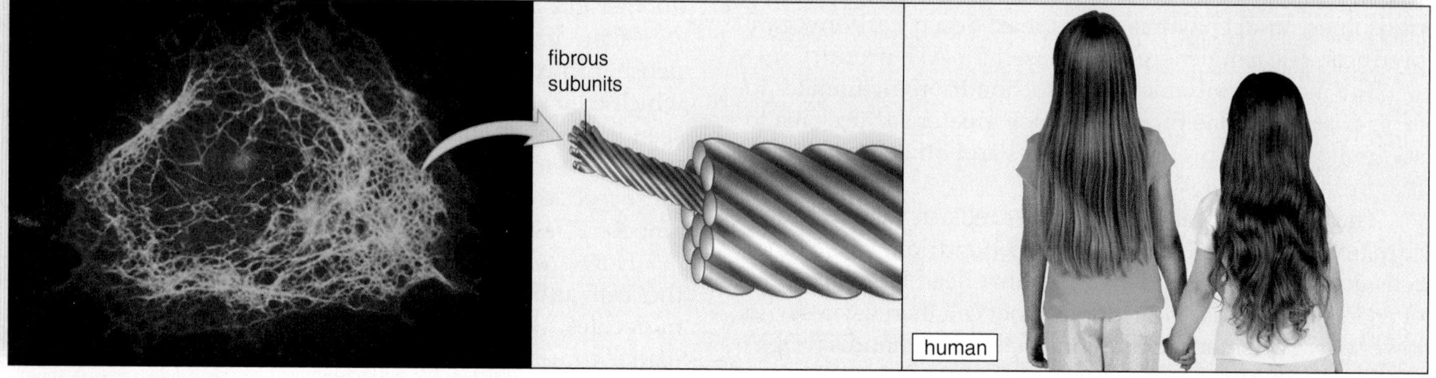

b. Intermediate filaments

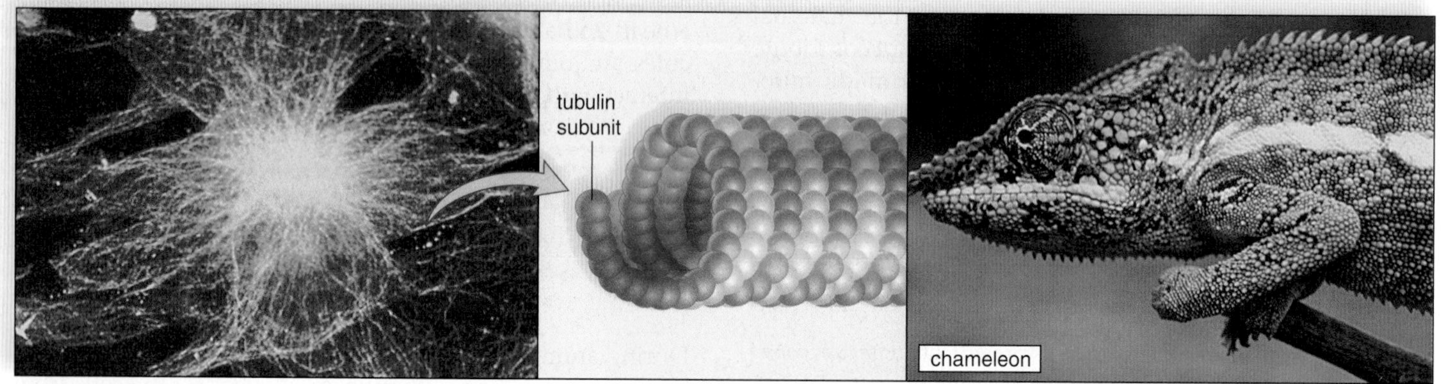

c. Microtubules

Figure 3.12 The cytoskeleton.
The cytoskeleton maintains the shape of the cell and allows its parts to move. Three types of protein components make up the cytoskeleton. They can be identified in cells by using a special fluorescent technique that detects only one type of component at a time. **a.** *Left to right:* Fibroblasts in animal tissue have been treated so that actin filaments can be microscopically detected. The drawing shows that actin filaments are composed of a twisted double chain of actin subunits. The giant cells of the green alga *Chara* rely on actin filaments to move organelles from one end of the cell to another. **b.** *Left to right:* Fibroblasts in an animal tissue have been treated so that intermediate filaments can be microscopically detected. The drawing shows that fibrous proteins account for the ropelike structure of intermediate filaments. Human hair is strengthened by the presence of intermediate filaments. **c.** *Left to right:* Fibroblasts in an animal tissue have been treated so that microtubules can be microscopically detected. The drawing shows that microtubules are hollow tubes composed of tubulin subunits. The skin cells of a chameleon rely on microtubules to move pigment granules around so that they can take on the color of their environment.

Microtubules

Microtubules are small, hollow cylinders about 25 nm in diameter and 0.2–25 μm in length.

Microtubules are made of the globular protein tubulin, which is of two types, called α and β. Microtubules have 13 rows of tubulin dimers, surrounding what appears in electron micrographs to be an empty central core.

The regulation of microtubule assembly is controlled by a microtubule organizing center. In most eukaryotic cells, the main microtubule organizing center is in the **centrosome,** which lies near the nucleus. Microtubules radiate from the centrosome, helping to maintain the shape of the cell and acting as tracks along which organelles can move. Whereas the motor molecule myosin is associated with actin filaments, the motor molecules kinesin and dynein are associated with microtubules:

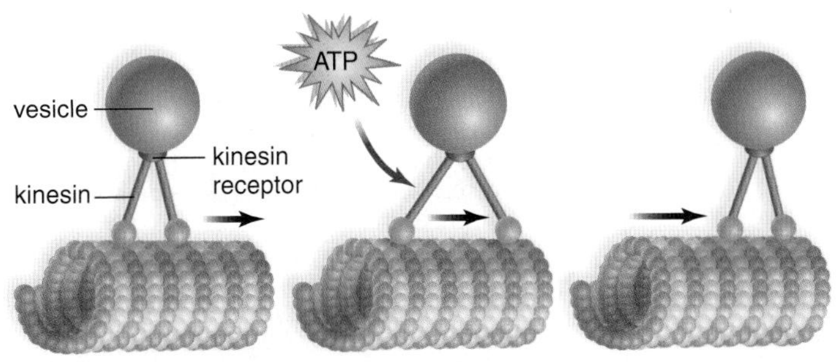

Before a cell divides, microtubules disassemble and then reassemble into a structure called a spindle. The spindle apparatus attaches to the chromosomes and ensures that they are distributed in an orderly manner. It also participates in dividing the cell in half. At the end of cell division, the spindle disassembles, and microtubules reassemble once again into their former array.

Centrioles

Both plant and animal cells contain centrosomes, the major microtubule organizing center for the cell. But in animal cells, a centrosome contains two centrioles lying at right angles to each other. The centrioles may be involved in the process of microtubule assembly and disassembly. **Centrioles** are short cylinders of microtubules with a 9 + 0 pattern of microtubule triplets—that is, a ring having nine sets of microtubule triplets, with none in the middle (Fig. 3.13).

Before an animal cell divides, the centrioles replicate such that the members of each pair are again at right angles to one another (see Fig. 3.13). Then, each pair becomes part of a separate centrosome. During cell division, the centrosomes move apart and may function to organize the mitotic spindle.

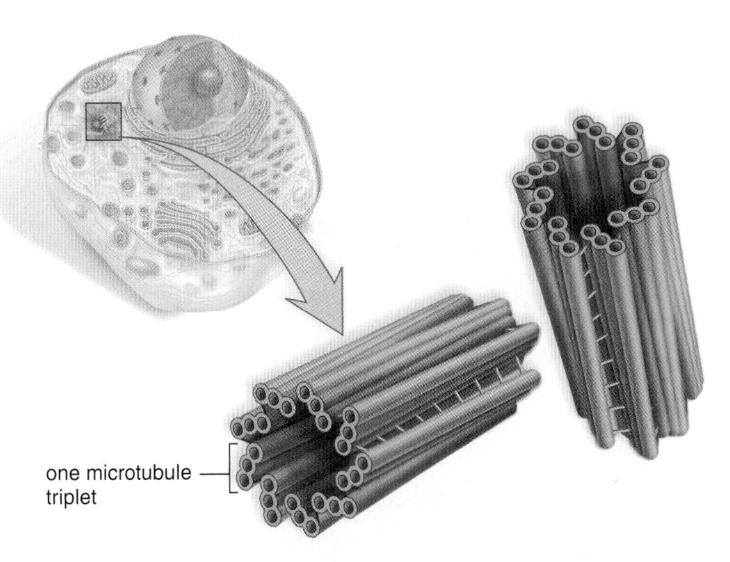

Figure 3.13 Centrioles.
In a nondividing animal cell, a single pair of centrioles lies in the centrosome located just outside the nucleus. Just before a cell divides, the centrioles replicate, producing two pairs of centrioles. During cell division, centrioles in their respective centrosomes separate so that each new cell has one centrosome containing one pair of centrioles.

Cilia and Flagella

Cilia (sing., **cilium**) and flagella are hairlike projections that can move either in an undulating fashion, like a whip, or stiffly, like an oar. Cells that have these organelles are capable of movement. For example, unicellular organisms called paramecia move by means of cilia, whereas sperm cells move by means of flagella. In the human body, the cells that line our upper respiratory tract have cilia that sweep debris trapped within mucus back up into the throat, where it can be swallowed or ejected. This action helps keep the lungs clean.

In eukaryotic cells, cilia are much shorter than flagella, but they have a similar construction. Both are membrane-bounded cylinders. The cylinders are composed of nine microtubule doublets arranged in a circle around two central microtubules. Therefore, they have a 9 + 2 pattern of microtubules. Cilia and flagella move when the microtubule doublets slide past one another (Fig. 3.14).

Check Your Progress

1. How are actin filaments, intermediate filaments, and microtubules alike? How are they different?
2. How does the structure of cilia and flagella differ from that of centrioles?
3. Study Figure 3.14 on page 62. Then explain how cilia and flagella move.

Visual Focus

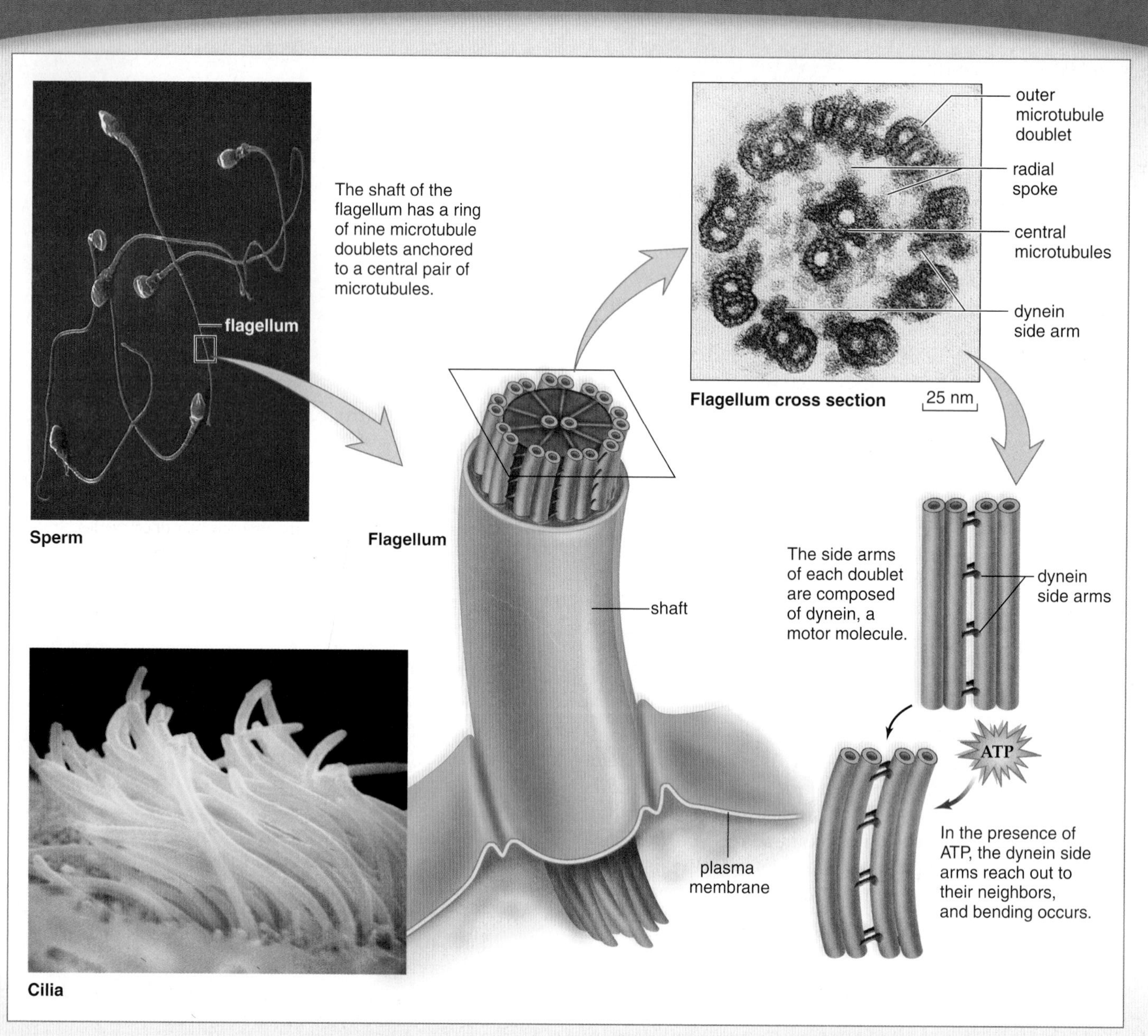

Figure 3.14 Structure of a flagellum or cilium.
The shaft of a flagellum (or cilium) contains microtubule doublets whose side arms are motor molecules that cause the projection to move. Sperm have flagella. Without the ability of sperm to move to the egg, human reproduction would not be possible. Cilia cover the surface of the cells of the respiratory system where they beat upward to remove foreign matter.

3.4 Origin and Evolution of the Eukaryotic Cell

Learning Outcomes

1. Identify the first cells and how eukaryotic cells came to be.
2. Describe the evidence for the endosymbiotic theory.

The fossil record, which is based on the remains of ancient life, suggests that the first cells were prokaryotes. Therefore, scientists believe that eukaryotic cells evolved from prokaryotic cells. Both bacteria and archaea are candidates, but biochemical data suggest that eukaryotes are more closely related to the archaea. The eukaryotic cell probably evolved from a prokaryotic cell in stages. Invagination of the plasma membrane might explain the origin of the nuclear envelope and such organelles as the endoplasmic reticulum and the Golgi apparatus. Some believe that the other organelles could also have arisen in this manner.

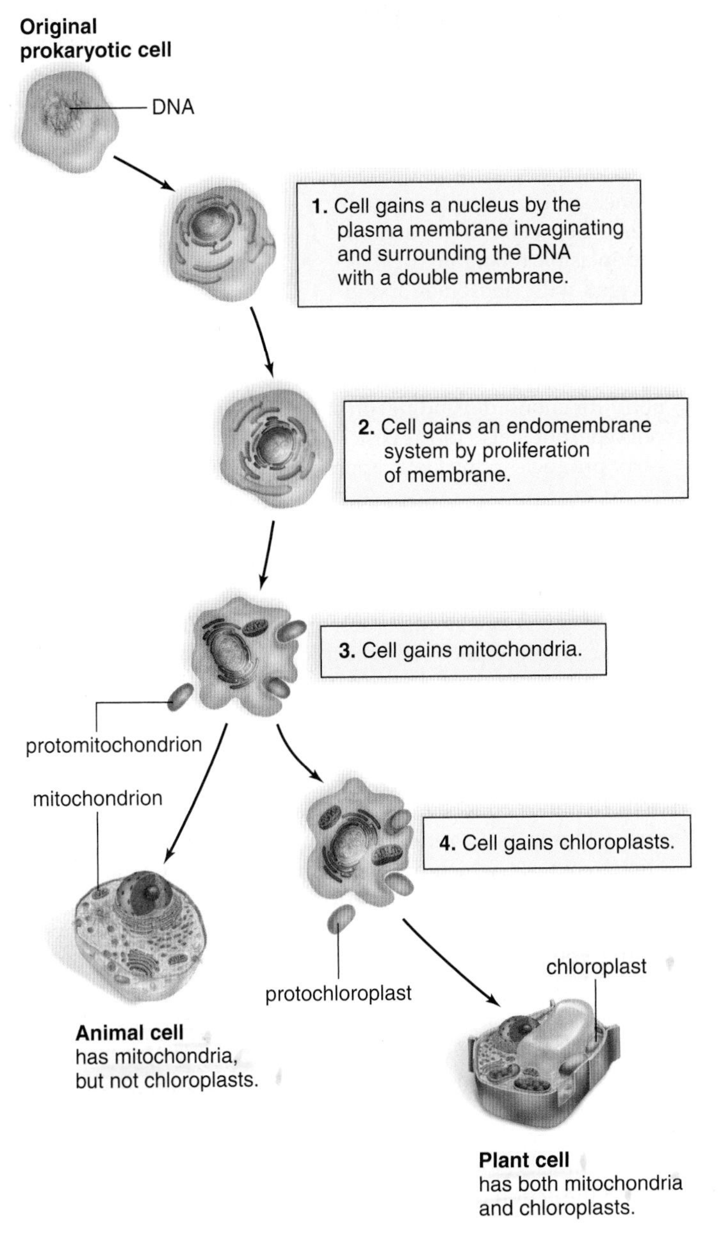

Figure 3.15 Origin of organelles.
Invagination of the plasma membrane could have created the nuclear envelope and an endomembrane system. The endosymbiotic theory suggests that mitochondria and chloroplasts were once independent prokaryotes that took up residence in a eukaryotic cell.

Another, more interesting hypothesis for the origin of some organelles has been put forth. Observations in the laboratory indicate that an amoeba infected with bacteria can become dependent upon them. Some investigators believe mitochondria and chloroplasts are derived from prokaryotes that were taken up by a much larger cell (Fig. 3.15). Perhaps mitochondria were originally aerobic heterotrophic bacteria, and chloroplasts were originally cyanobacteria. The eukaryotic host cell would have benefited from an ability to utilize oxygen or synthesize organic food when, by chance, the prokaryote was taken up and not destroyed. After the prokaryote entered the host cell, the two would have begun living together cooperatively. This proposal is known as the **endosymbiotic theory** (*endo-*, in; *symbiosis*, living together). Some of the evidence supporting this hypothesis is as follows:

1. Mitochondria and chloroplasts are similar to bacteria in size and in structure.
2. Both organelles are bounded by a double membrane—the outer membrane may be derived from the engulfing vesicle, and the inner one may be derived from the plasma membrane of the original prokaryote.
3. Mitochondria and chloroplasts contain a limited amount of genetic material and divide by splitting. Their DNA (deoxyribonucleic acid) is a circular loop like that of prokaryotes.
4. Although most of the proteins within mitochondria and chloroplasts are now produced by the eukaryotic host, they do have their own ribosomes and they do produce some proteins. Their ribosomes resemble those of prokaryotes.
5. The RNA (ribonucleic acid) base sequence of the ribosomes in chloroplasts and mitochondria also suggests a prokaryotic origin of these organelles.

It is also possible that the flagella of eukaryotes are derived from an elongated bacterium with a flagellum that became attached to a host cell. However, the flagella of eukaryotes are constructed differently from those of modern bacteria.

Check Your Progress

1. What evidence suggests that chloroplasts and mitochondria were once independently living prokaryotes?

Applying the Concepts Revisited

The cells that are part of your body are all eukaryotic animal cells. Yet each cell type specializes in ways that enable it to carry out its particular function within the body, similar to the red blood cells discussed earlier. Some cells have more of certain organelles than other cells because of their specialization. For example, cells that specialize in secreting protein products have more extensive endoplasmic reticulum and Golgi apparatus.

1. If a cell specializes in destroying invaders, what organelle(s) would you expect it to have more of?
2. How do you think plant cells are specialized to perform their functions?
3. Why are prokaryotic cells not specialized?

SUMMARIZING THE CONCEPTS

3.1 The Cellular Level of Organization

- All organisms are composed of cells, the smallest units of living matter.
- Cells are capable of self-reproduction, and new cells come only from preexisting cells.
- Cells must remain small in order to have an adequate ratio of surface-area-to-volume for exchange of molecules with the environment.

3.2 Prokaryotic Cells

- All cells have a plasma membrane consisting of a phospholipid bilayer with embedded proteins. The membrane regulates the movement of molecules into and out of the cell. The inside of the cell is filled with a fluid called cytoplasm.
- Prokaryotic cells do not have a nucleus, but they do have a nucleoid that is not bounded by a nuclear envelope. They also lack most of the other organelles that compartmentalize eukaryotic cells.
- Prokaryotic cells, as exemplified by the bacteria and archaea, are structurally less complex than eukaryotic cells but metabolically very diverse.

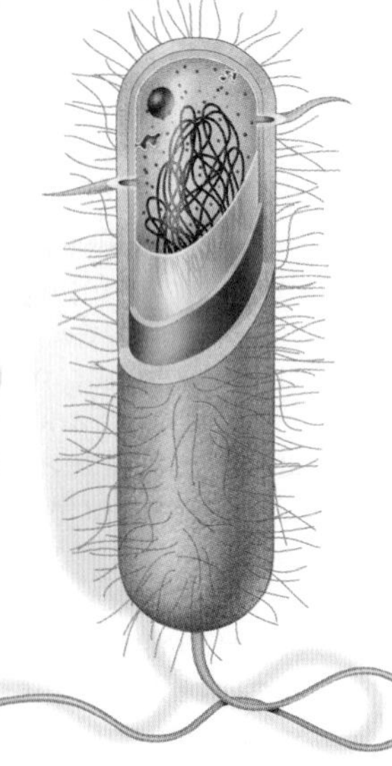

3.3 Eukaryotic Cells

- Eukaryotic cells are very complex. Animals, plants, fungi, and protists are all examples of eukaryotes.
- Some eukaryotic cells contain cell walls for protection.
- Eukaryotic cells contain organelles, which are subcellular structures that perform a particular function.
- The nucleus of eukaryotic cells is bounded by a nuclear envelope containing pores. These pores serve as passageways between the cytoplasm and the nucleoplasm. Within the nucleus, the chromatin is a complex of DNA and protein. Chromatin is divided into separate structures called chromosomes.
- The nucleolus is a special region of the chromatin where rRNA is produced and where proteins from the cytoplasm gather to form ribosomal subunits.
- Ribosomes are organelles that function in protein synthesis. They can be bound to the endoplasmic reticulum (ER) or can exist within the cytoplasm singly or in groups called polyribosomes.
- The endomembrane system includes the ER, the Golgi apparatus, the lysosomes, and other types of vesicles and vacuoles. The endomembrane system compartmentalizes the cell. The rough endoplasmic reticulum (RER) is covered with ribosomes and is involved in the folding, modification, and transport of proteins. The smooth ER has various metabolic functions depending on the cell type, but it also forms vesicles that carry products to the Golgi apparatus.
- The Golgi apparatus processes proteins and repackages them into lysosomes, which carry out intracellular digestion, or into vesicles for transport to the plasma membrane or other organelles.
- Vacuoles are large storage sacs, and vesicles are smaller ones. The large single plant cell vacuole not only stores substances but also lends support to the plant cell.
- Peroxisomes contain enzymes that oxidize molecules by producing hydrogen peroxide, which is subsequently broken down.
- Cells require a constant input of energy to maintain their structure. Chloroplasts capture the energy of the sun and carry on photosynthesis, which produces carbohydrates. Carbohydrate-derived products are broken down in mitochondria to produce ATP. The mitochondria are considered the energy powerhouse of the cell. Only plants, algae, and cyanobacteria contain chloroplasts.
- The cytoskeleton contains actin filaments, intermediate filaments, and microtubules. These maintain cell shape and allow the cell and its organelles to move. Actin filaments interact with motor molecules such as myosin in muscle cells. Microtubules are present in centrioles, cilia, and flagella. In the cytoplasm they serve as tracks along which vesicles and other organelles move due to the action of specific motor molecules.

3.4 Origin and Evolution of the Eukaryotic Cell

- The first cells were probably prokaryotic cells. Eukaryotic cells most likely arose from prokaryotic cells in stages.
- Biochemical data suggest that eukaryotic cells are closer evolutionarily to the archaea than to the bacteria.
- The nuclear envelope most likely evolved through invagination of the plasma membrane, but mitochondria and chloroplasts may have arisen through endosymbiotic events.

TESTING YOURSELF

Choose the best answer for each question.

1. Proteins are produced
   a. in the cytoplasm or the ER.
   b. in the nucleus.
   c. in the Golgi apparatus.
   d. None of these are correct.
2. Cell walls are found in _____ and contain _____.
   a. animals, chitin
   b. animals, cellulose
   c. plants, chitin
   d. plants, cellulose
3. Which structure is characteristic of prokaryotic cells?
   a. nucleus
   b. mitochondria
   c. nucleoid
   d. chloroplast
4. Ribosomes are found in what type of cells?
   a. animal
   b. plant
   c. bacterial
   d. All of these are correct.
5. _____ are (is) produced by the smooth ER.
   a. Proteins
   b. DNA
   c. Lipids
   d. Ribosomes
6. Lysosomes function in _____
   a. protein synthesis.
   b. processing and packaging.
   c. intracellular digestion.
   d. All of these are correct.
7. Which of these could you see with a light microscope?
   a. atom
   b. proteins
   c. amino acid
   d. None of these are correct.

For questions 8–12, match the structure to the function in the key.

Key:

a. movement of cell
b. transport of proteins
c. photosynthesis
d. ribosome formation
e. folds and processes proteins

8. Chloroplasts
9. Flagella
10. Golgi apparatus
11. Rough ER
12. Nucleolus
13. Which of these sequences depicts a hypothesized evolutionary scenario?
    a. cyanobacteria ⟶ mitochondria
    b. Golgi ⟶ mitochondria
    c. mitochondria ⟶ cyanobacteria
    d. cyanobacteria ⟶ chloroplast
14. The products of photosynthesis are
    a. carbohydrates and oxygen.
    b. oxygen and water.
    c. carbon dioxide and water.
    d. carbohydrates and water.
15. Label these parts of the cell that are involved in protein synthesis and modification. Give a function for each structure.

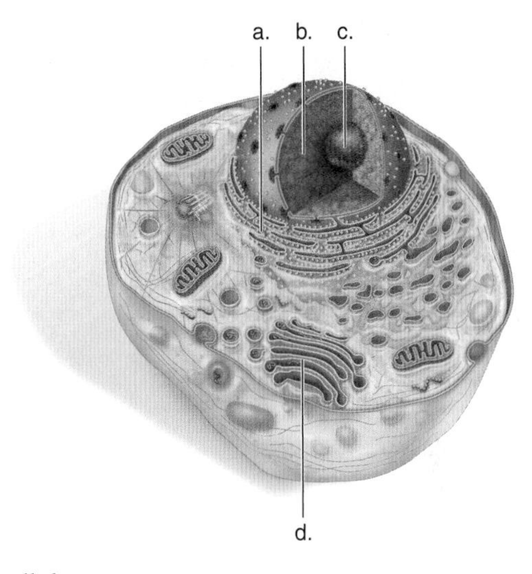

16. The cell theory states:
    a. Cells form as organelles and molecules become grouped together in an organized manner.
    b. The normal functioning of an organism does not depend on its individual cells.
    c. The cell is the basic unit of life.
    d. Only eukaryotic organisms are made of cells.
17. The small size of cells is best correlated with
    a. the fact that they are self-reproducing.
    b. their prokaryotic versus eukaryotic nature.
    c. an adequate surface area for exchange of materials.
    d. their vast versatility.
18. Which of the following structures would be found in both plant and animal cells?
    a. centrioles
    b. chloroplasts
    c. cell wall
    d. mitochondria
    e. All of these are found in both types of cells.

UNDERSTANDING THE TERMS

actin filaments 59
capsule 47
cell theory 46
cellular respiration 57
cell wall 47
centriole 61
centrosome 61
chloroplast 57
chromatin 52
chromosome 52
cilium (pl., cilia) 61
cristae 59
cytoplasm 47
cytoskeleton 59
endoplasmic reticulum (ER) 54
endosymbiotic theory 63
eukaryotic cell 49
fimbriae 47
flagellum (pl., flagella) 47
Golgi apparatus 54
granum (pl., grana) 58
intermediate filaments 59
lysosome 56
matrix 59
microtubules 61
mitochondrion 57
motor molecule 59
nuclear envelope 52
nuclear pore 52
nucleoid 48
nucleolus (pl., nucleoli) 52
nucleoplasm 52
nucleus 52
organelle 46
peroxisome 56
photosynthesis 57
plasma membrane 47
plasmid 48
polyribosome 54
prokaryotic cell 47
ribosome 48
secretion 55
slime layer 47
stroma 58
thylakoid 48
vacuole 56
vesicle 54

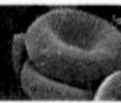

THINKING CRITICALLY

1. In the 1958 movie *The Blob,* a giant, single-celled alien creeps and oozes around, attacking and devouring helpless humans. Why couldn't there be a real single-celled organism as large as the Blob?
2. Calculate the surface-area-to-volume ratio of a 1-mm cube and a 2-mm cube. Which has the smaller ratio?
3. Are centrioles required for microtubule assembly and disassembly?

INQUIRY INTO LIFE WEBSITE

The companion website for *Inquiry into Life* provides a wealth of information organized and integrated by chapter. You will find practice tests, animations, videos, and much more that will complement your learning and understanding of general biology.

http://www.mhhe.com/maderinquiry13

4

Applying the Concepts

Have you ever bitten into a hot pepper and had the sensation that your mouth is on fire? Your eyes water and you are in real pain! The feelings of heat and pain are due to a membrane protein in your sensory nerves. Capsaicin is the chemical in chili peppers that binds to a channel protein in specialized sensory nerve cell endings called nociceptors (*noci-* means hurtful). One of the important functions of a membrane is to control what molecules move into and out of the cell and when they move. This particular channel protein, when activated, allows calcium ions to flow into the cell. In addition to capsaicin, other things such as acidic pH, heat, electrostatic charges, and a variety of chemical agents can activate this channel protein. Once activated by any of these signals, the response is the same. The channel opens, calcium ions flow into the cell, and the nociceptor sends a signal to the brain—*pain*! What is the quickest way to alleviate the pain? You have to get that channel closed by removing the capsaicin! Some people drink cold water, but that does very little other than cool down your mouth. Capsaicin is lipid-soluble and does not dissolve in water. However drinking milk, or eating rice or bread usually help. If you are a true "chili head," you know that if you survive the first bite, the next bite is easier. That is because within minutes, you become desensitized to the pain. But whatever you do—don't rub your eyes after handling hot peppers!

In this chapter we will discuss the various functions of proteins embedded in the membranes of your cells and how the membrane controls what enters and leaves the cell. We will also describe how cells "talk" to each other through signals sent to receptor proteins in the cell membrane.

Membrane Structure and Function

Chapter Outline

4.1 Plasma Membrane Structure and Function

Learning Outcomes

1. Describe the fluid-mosaic model of membrane structure.
2. List the various types of integral proteins in the membrane and their functions.

The plasma membrane separates the internal environment of the cell from the external environment. It regulates the entrance and exit of molecules into and out of the cell. In this way, it helps the cell and the organism maintain a steady internal environment. The plasma membrane is a phospholipid bilayer in which protein molecules are either partially or wholly embedded. The phospholipid bilayer has a fluid consistency, comparable to that of light oil. The proteins are scattered either just outside or within the membrane. Therefore, they form a *mosaic* pattern. This description of the plasma membrane is called the **fluid-mosaic model** of membrane structure (Fig. 4.1).

The hydrophilic (water-loving) polar heads of the phospholipid molecules face the outside and inside of the cell where water is found, and the hydrophobic (water-fearing) nonpolar tails face each other (see Fig. 4.1). Cholesterol is another lipid found in animal plasma membranes. Related steroids are found in the plasma membranes of plants. Cholesterol stiffens and strengthens the membrane, thereby helping to regulate its fluidity.

The proteins in a membrane consist of peripheral proteins and integral proteins. Peripheral proteins are associated with only one side of the plasma membrane. Peripheral proteins on the inside of the membrane are often held in place by cytoskeletal filaments. In contrast, integral proteins span the membrane, and can protrude from one or both sides. They are embedded in the membrane, but they can move laterally, changing their position in the membrane.

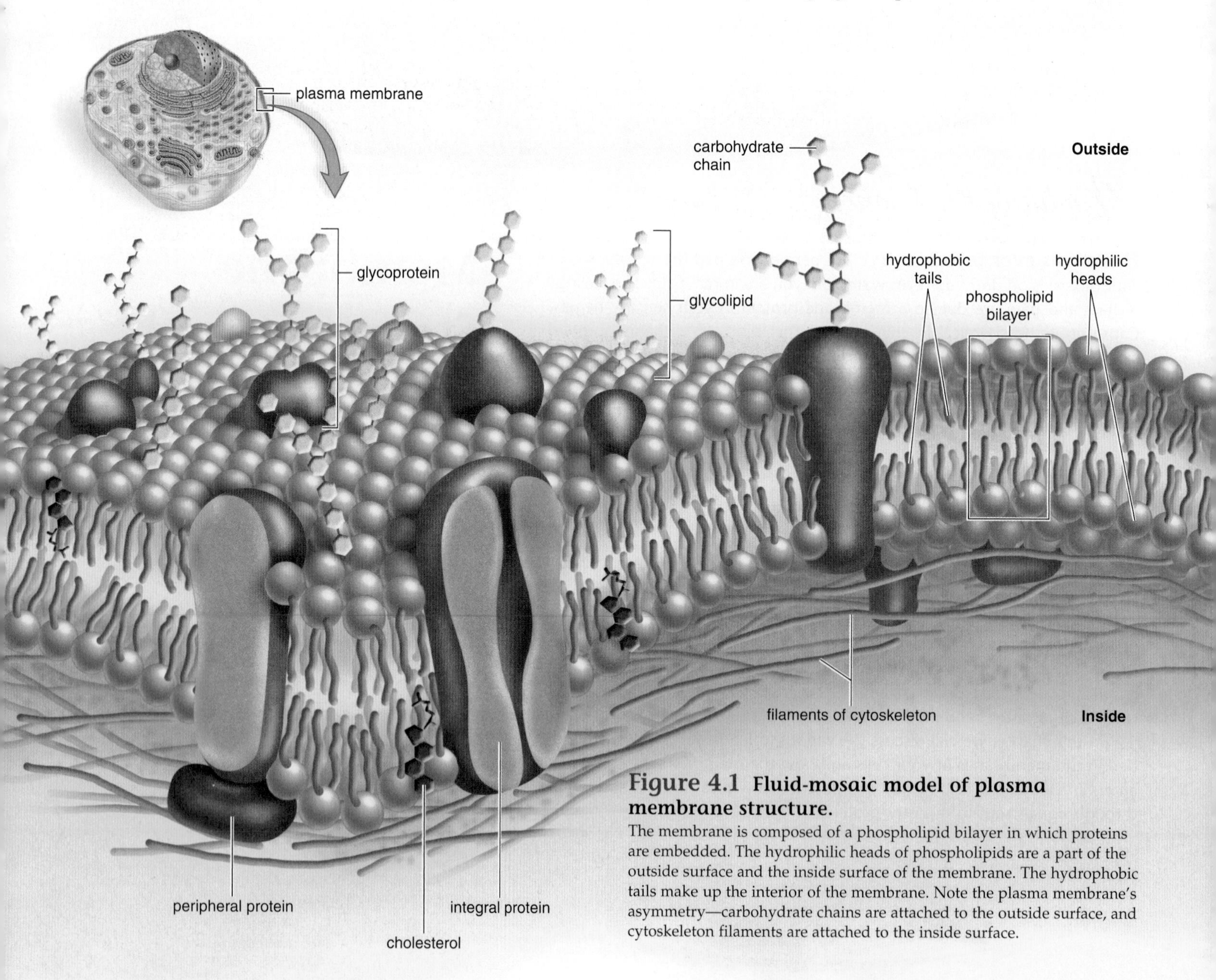

Figure 4.1 Fluid-mosaic model of plasma membrane structure.
The membrane is composed of a phospholipid bilayer in which proteins are embedded. The hydrophilic heads of phospholipids are a part of the outside surface and the inside surface of the membrane. The hydrophobic tails make up the interior of the membrane. Note the plasma membrane's asymmetry—carbohydrate chains are attached to the outside surface, and cytoskeleton filaments are attached to the inside surface.

Both phospholipids and proteins can have attached carbohydrate (sugar) chains. If so, these molecules are called **glycolipids** and **glycoproteins,** respectively. Since the carbohydrate chains occur only on the outside surface and peripheral proteins occur asymmetrically on one surface or the other, the two halves of the membrane are not identical.

Functions of the Proteins

The plasma membranes of various cells and the membranes of various organelles each have their own unique collections of proteins. The peripheral proteins often have a structural role in that they help stabilize and shape the plasma membrane. They may also function in signaling pathways. The integral proteins largely determine a membrane's specific functions. Integral proteins can be of the following types:

Channel proteins are involved in the passage of molecules through the membrane. They have a channel that allows a substance to simply move across the membrane (Fig. 4.2*a*). For example, a channel protein allows hydrogen ions to flow across the inner mitochondrial membrane. Without this movement of hydrogen ions, ATP would never be produced. Channel proteins may contain a gate that must be opened by the binding of a specific molecule to the channel.

Carrier proteins are also involved in the passage of molecules through the membrane. They combine with a substance and help it move across the membrane (Fig. 4.2*b*). For example, a carrier protein transports sodium and potassium ions across a nerve cell membrane. Without this carrier protein, nerve conduction would be impossible.

Cell recognition proteins are glycoproteins (Fig. 4.2*c*). Among other functions, these proteins help the body recognize when it is being invaded by pathogens so that an immune reaction can occur.

Receptor proteins have a shape that allows a specific molecule to bind to it (Fig. 4.2*d*). The binding of this molecule causes the protein to change its shape and thereby bring about a cellular response. The coordination of the body's organs is totally dependent on such signal molecules. For example, the liver stores glucose after it is signaled to do so by insulin.

Enzymatic proteins carry out metabolic reactions directly (Fig. 4.2*e*). The integral membrane proteins of the electron transport chain carry out the final steps of aerobic respiration. Without the presence of enzymes, some of which are attached to the various membranes of the cell, a cell would never be able to perform the metabolic reactions necessary to its proper function.

The peripheral proteins often have a structural role in that they help stabilize and shape the plasma membrane.

Check Your Progress

1. What type of organic molecule is responsible for the fluidity of the plasma membrane?
2. How do channel proteins differ from carrier proteins?

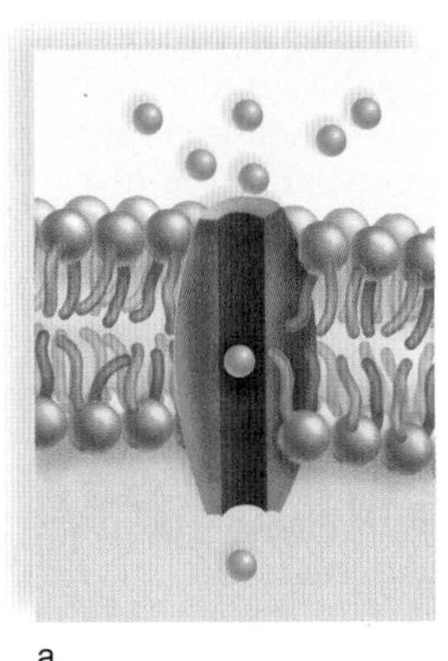

Channel Protein
Allows a particular molecule or ion to cross the plasma membrane freely. Cystic fibrosis, an inherited disorder, is caused by a faulty chloride (Cl^-) channel; a thick mucus collects in airways and in pancreatic and liver ducts.

a.

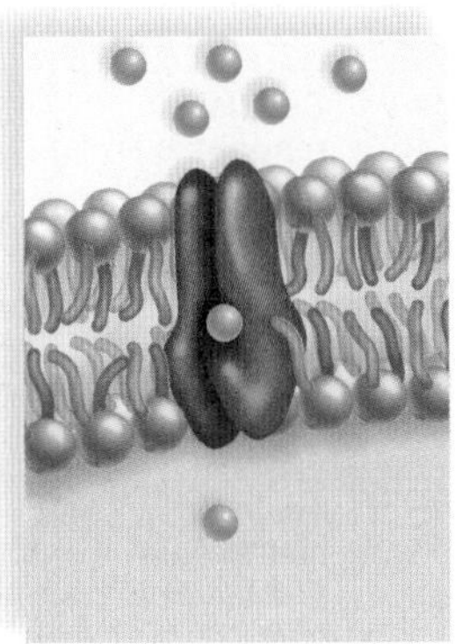

Carrier Protein
Selectively interacts with a specific molecule or ion so that it can cross the plasma membrane. The family of GLUT carriers transfers glucose in and out of the various cell types of the body. Different carriers respond differently to blood levels of glucose.

b.

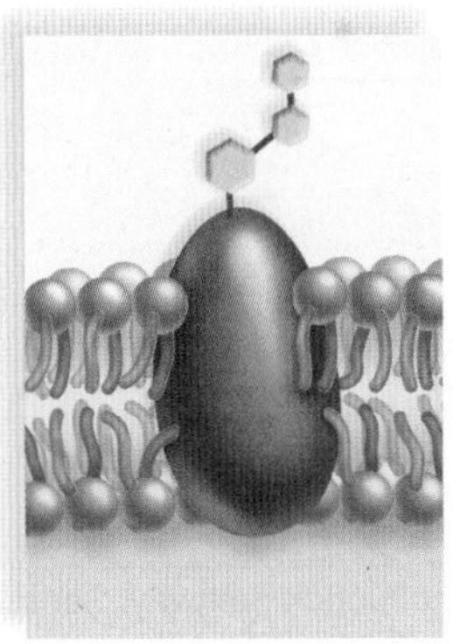

Cell Recognition Protein
The MHC (major histocompatibility complex) glycoproteins are different for each person, so organ transplants are difficult to achieve. Cells with foreign MHC glycoproteins are attacked by white blood cells responsible for immunity.

c.

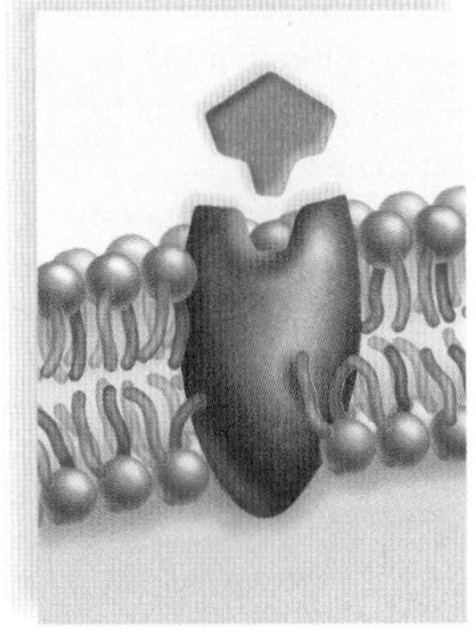

Receptor Protein
Shaped in such a way that a specific molecule can bind to it. Some types of dwarfism result not because the body does not produce enough growth hormone, but because the plasma membrane growth hormone receptors are faulty and cannot interact with growth hormone.

d.

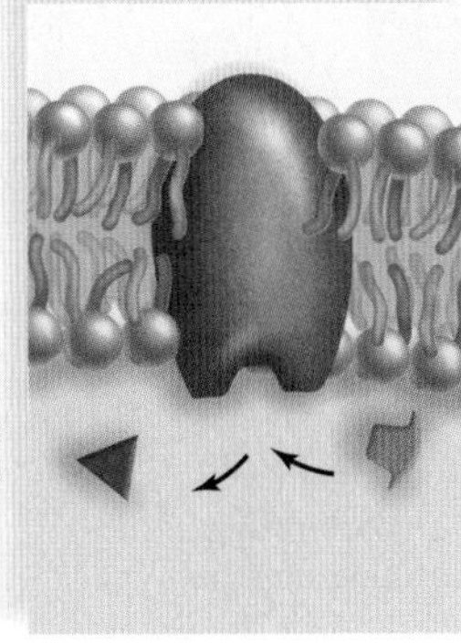

Enzymatic Protein
Catalyzes a specific reaction. The membrane protein, adenylate cyclase, is involved in ATP metabolism. Cholera bacteria release a toxin that interferes with the proper functioning of adenylate cyclase, which eventually leads to severe diarrhea.

e.

Figure 4.2 Examples of membrane protein diversity. These are some of the functions performed by integral proteins found in the plasma membrane.

How Cells Talk to One Another

All organisms are able to sense and respond to specific signals in their environment. A bacterium that has taken up residence in your body is responding to signaling molecules when it finds food and escapes immune cells in order to stay alive. Signaling helps bread mold on stale bread in your refrigerator detect the presence of an opposite mating strain and begin its sexual life cycle. Similarly, the cells of an embryo are responding to signaling molecules when they move to specific locations and assume the shape and perform the functions of specific tissues (Fig. 4A*a*). In the newborn, signaling is still required because the functions of a specific tissue may be necessary only on occasion, or one tissue may need to perform one of its various functions only at particular times. In plants, external signals, such as a change in the amount of light, tells them when it is time to resume growth or flower. Internal signaling molecules enable plants to coordinate the activities of roots, stems, and leaves.

Cell Signaling

The cells of a multicellular organism "talk" to one another by using signaling molecules, sometimes called chemical messengers. Some messengers are produced at a distance from a target tissue and, in animals, are carried by the circulatory system to various sites around the body. For example, the pancreas releases a hormone called insulin, which is transported in blood vessels to the liver, and thereafter, the liver stores glucose as glycogen. Failure of the liver to respond appropriately results in a medical condition called diabetes. Growth factors act locally as signaling molecules and cause cells to divide. Overreacting to growth factors can result in a tumor characterized by unlimited cell division. The importance of cell signaling causes much research to be directed toward understanding the intricacies of the process.

We have learned that cells respond to only certain signaling molecules. Why? Because they must bind to a receptor protein, and cells have receptors for only certain signaling molecules. Each cell has receptors for numerous signaling molecules and often the final response is due to a summing up of all the various signals received. These molecules tell a cell what it should be doing at the moment, and without any signals, the cell dies.

Signaling not only involves a receptor protein, it also involves a pathway called a transduction pathway and a response. To understand the process, consider an analogy. When a TV camera (the receptor) is shooting a scene, the picture is converted to electrical signals (transduction pathway) that are understood by the TV in your house and are converted to a picture on your screen (the response). The process in cells is more complicated because each member of the pathway can turn on the activity of a number of other proteins. As shown in Figure 4A*b*, the cell response to a transduction pathway can be a change in the shape or movement of a cell, the activation of a particular enzyme, or the activation of a specific gene.

Discussion Questions

1. What happens if a cell is missing a receptor for a particular signaling molecule?
2. How does the binding of one type of signaling molecule sometimes result in multiple cellular responses?
3. Propose a mechanism whereby a mutated growth factor receptor results in a cell that grows out of control.

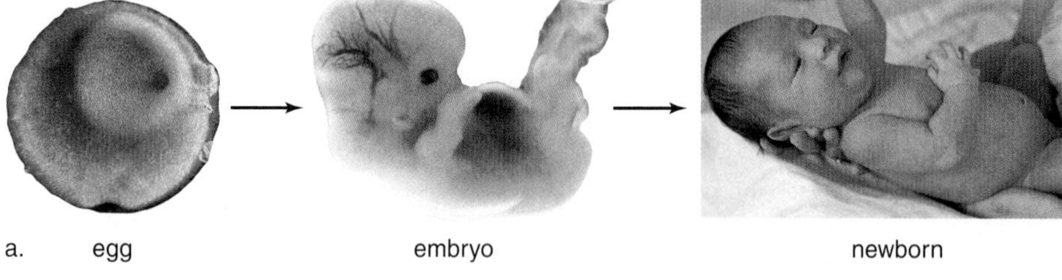

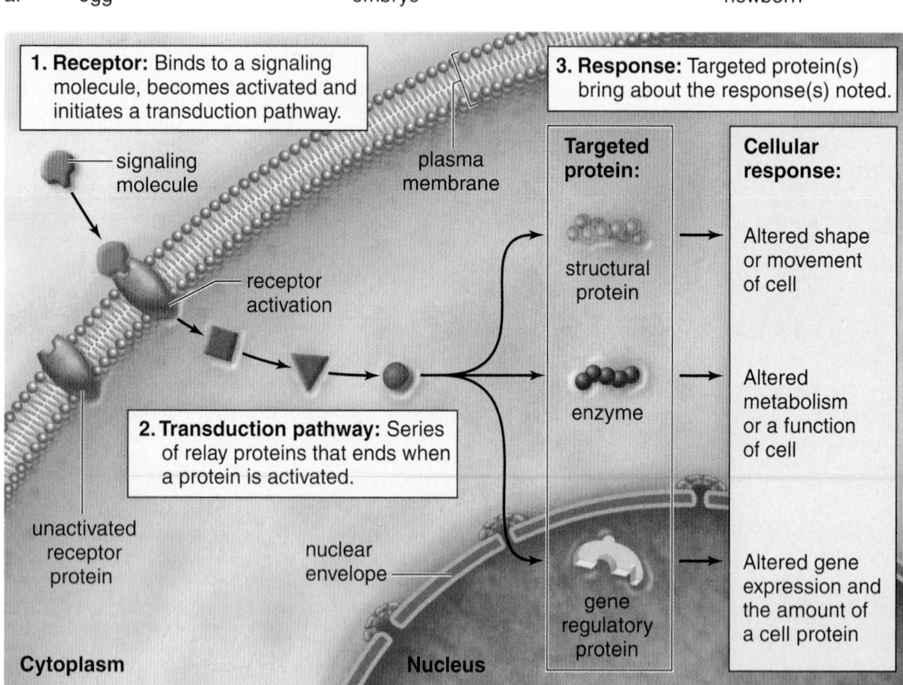

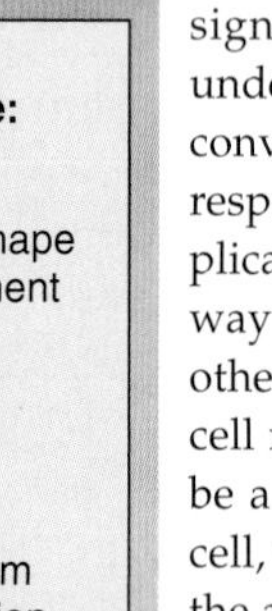

Figure 4A Cell signaling.
a. The process of signaling helps account for the transformation of an egg into an embryo and then an embryo into a newborn. **b.** The process of signaling involves three steps: binding of the signaling molecule, transduction of the signal, and response of the cell depending on what type protein is targeted.

4.2 The Permeability of the Plasma Membrane

Learning Outcomes

1. Define the term differentially permeable.
2. Predict the movement of molecules in diffusion and osmosis.
3. Explain how molecules and ions get into and out of cells.

The plasma membrane regulates the passage of molecules into and out of the cell. This function is critical because the life of the cell depends on maintenance of its normal composition. The plasma membrane can carry out this function because it is **differentially permeable,** meaning that certain substances can move across the membrane while others cannot.

Table 4.1 lists, and Figure 4.3 illustrates, which types of molecules can freely (i.e., passively) cross a membrane and which may require transport by a carrier protein and/or an expenditure of energy. In general, water and small, noncharged molecules, such as carbon dioxide, oxygen, glycerol, and alcohol, can freely cross the membrane. They are able to slip between the hydrophilic heads of the phospholipids and pass through the hydrophobic tails of the membrane. These molecules are said to go "down" their **concentration gradient** as they move from an area where their concentration is high to an area where their concentration is low. Some molecules are able to go "up" their concentration gradient, or move from an area where their concentration is low to an area where their concentration is high, but this requires energy.

Water actually uses two methods to enter the cell. In most cells, it diffuses across the plasma membrane—but in some cells, there are special water channels called aquaporins that allow larger amounts of water to cross the membrane.

Large molecules and some ions and charged molecules are unable to freely cross the membrane. They can cross the plasma membrane through channel proteins, with the assistance of carrier proteins, or in vesicles. A channel protein forms a pore through the membrane that allows molecules of a certain size and/or charge to pass. Carrier proteins are specific for the substances they transport across the plasma membrane—for example, sodium ions, amino acids, or glucose.

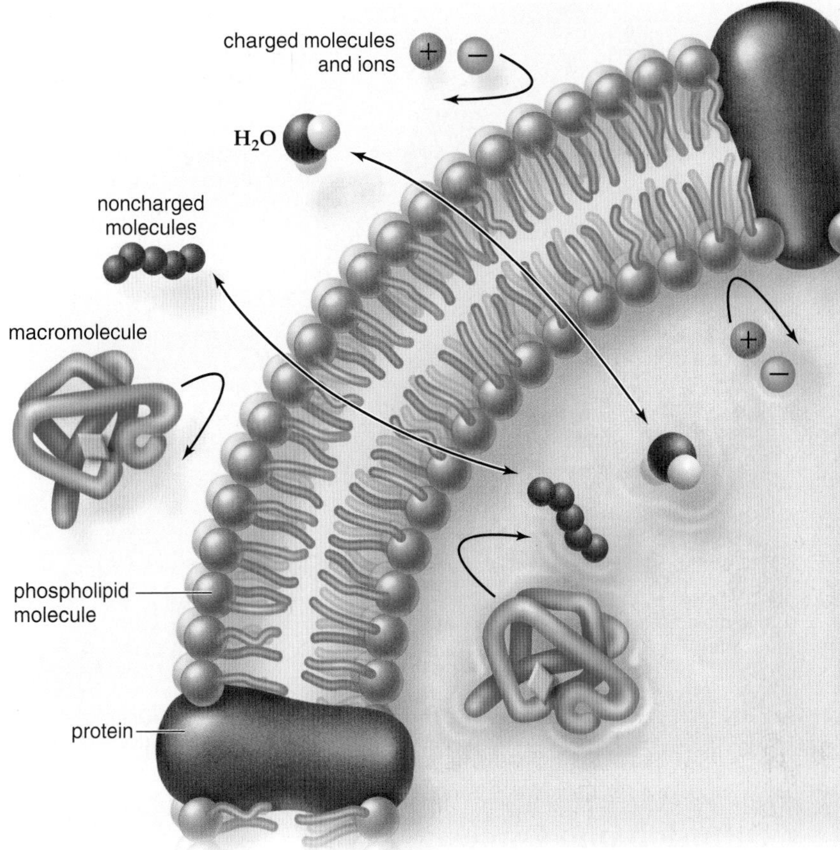

Figure 4.3 How molecules cross the plasma membrane.
Molecules that can diffuse across the plasma membrane are shown with long back-and-forth arrows. Substances that cannot diffuse across the membrane are indicated by the curved arrows.

Vesicle formation is another way a molecule can exit a cell (called exocytosis) or enter a cell (called endocytosis). This method of crossing a plasma membrane is reserved for macromolecules or even larger materials, such as a virus.

Check Your Progress

1. What characteristics of a molecule determine whether or not it can cross the membrane?

TABLE 4.1 Passage of Molecules Into and Out of the Cell

	Name	Direction	Requirement	Examples
Energy Not Required	Diffusion	Toward lower concentration	Concentration gradient	Lipid-soluble molecules, water, and gases
	Facilitated transport	Toward lower concentration	Channels or carrier and concentration gradient	Some sugars and some amino acids
Energy Required	Active transport	Toward higher concentration	Carrier plus energy	Sugars, amino acids, and ions
	Exocytosis	Toward outside	Vesicle fuses with plasma membrane	Macromolecules
	Endocytosis	Toward inside	Vesicle formation	Macromolecules

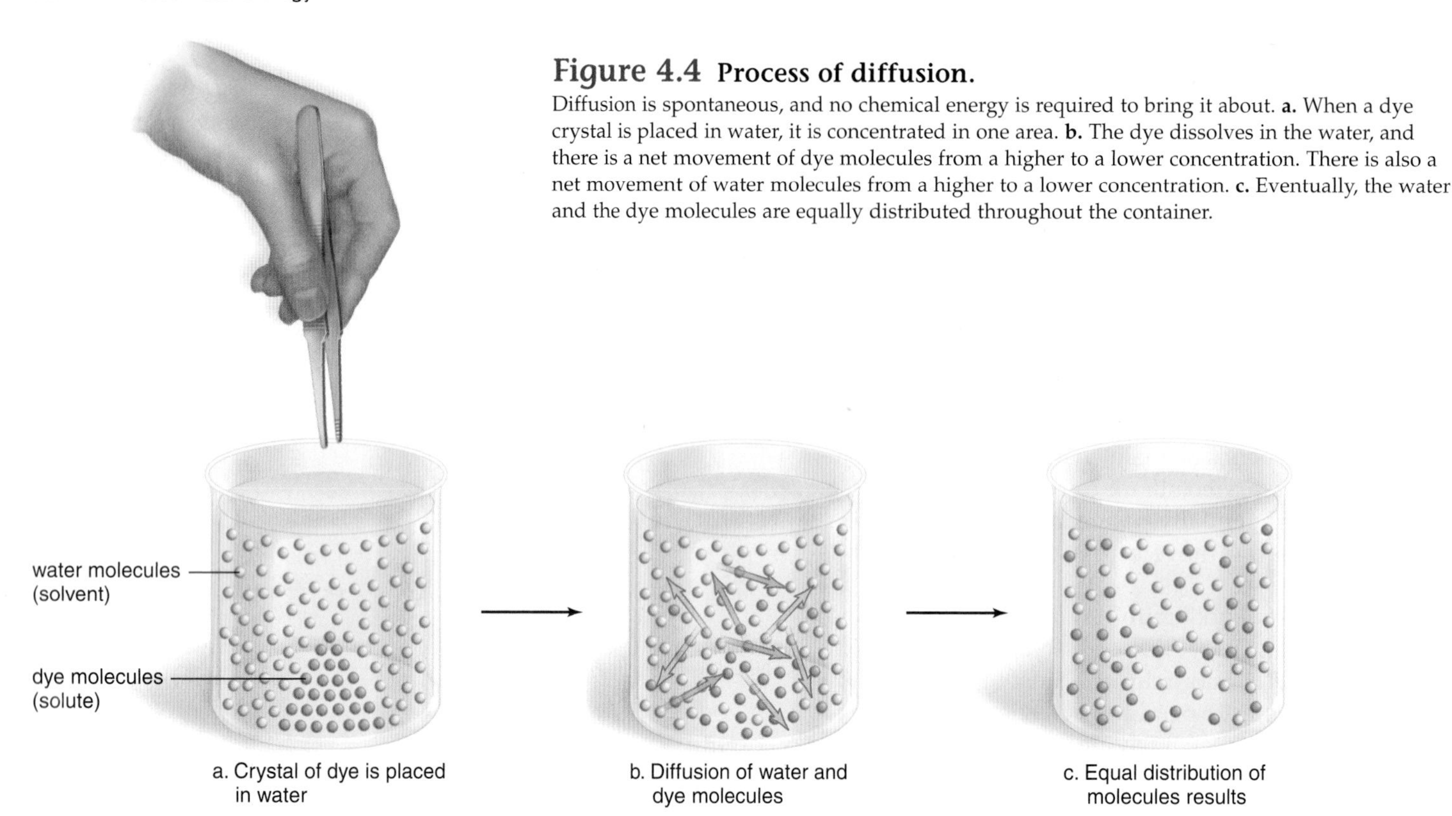

Figure 4.4 Process of diffusion.
Diffusion is spontaneous, and no chemical energy is required to bring it about. **a.** When a dye crystal is placed in water, it is concentrated in one area. **b.** The dye dissolves in the water, and there is a net movement of dye molecules from a higher to a lower concentration. There is also a net movement of water molecules from a higher to a lower concentration. **c.** Eventually, the water and the dye molecules are equally distributed throughout the container.

Diffusion and Osmosis

Diffusion is the movement of molecules from a higher to a lower concentration—that is, down their concentration gradient—until equilibrium is achieved and they are distributed equally. Diffusion is a physical process that can be observed with any type of molecule. For example, when a crystal of dye is placed in water (Fig. 4.4), the dye and water molecules move in various directions, but their net movement, which is the sum of their motion, is toward the region of lower concentration. Eventually, the dye is dissolved in the water, resulting in equilibrium and a colored solution.

A solution contains both a solute, usually a solid, and a **solvent,** usually a liquid. In this case, the solute is the dye and the solvent is the water molecules. Once the solute and solvent are evenly distributed, their molecules continue to move about, but there is no net movement of either one in any direction.

The chemical and physical properties of the plasma membrane allow only a few types of molecules to enter and exit a cell simply by diffusion. Gases can diffuse through the lipid bilayer. This is the mechanism by which oxygen enters cells and carbon dioxide exits cells. Also, consider the movement of oxygen from the alveoli (air sacs) of the lungs to the blood in the lung capillaries (Fig. 4.5). After inhalation (breathing in), the concentration of oxygen in the alveoli is higher than that in the blood. Therefore, oxygen diffuses into the blood.

Several factors influence the rate of diffusion. Among these factors are temperature, pressure, electrical currents, and molecular size. For example, as temperature increases, the rate of diffusion increases.

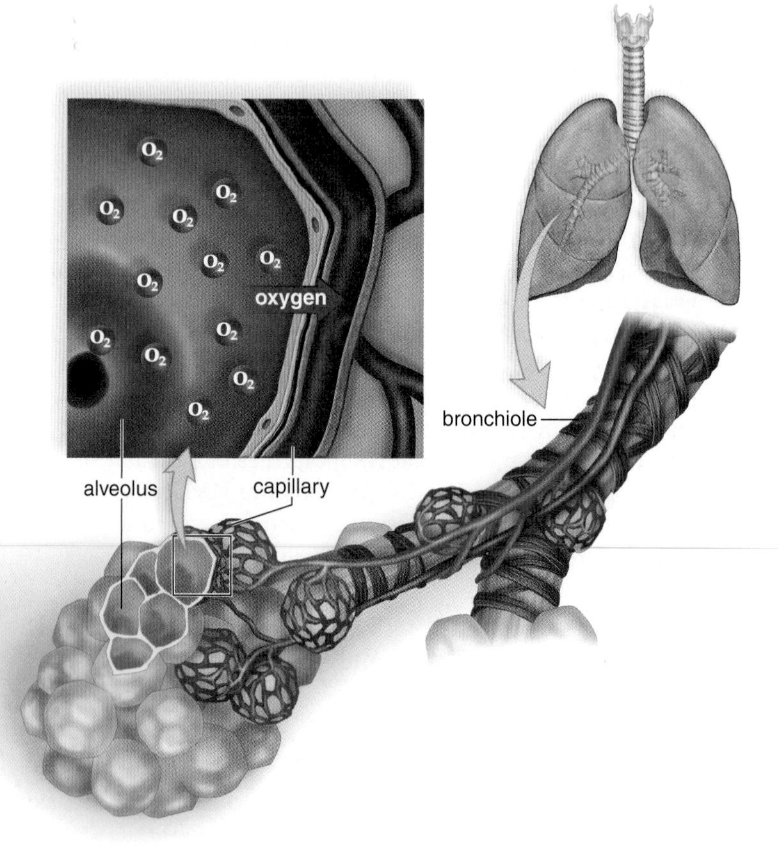

Figure 4.5 Gas exchange in lungs.
Oxygen (O_2) diffuses into the capillaries of the lungs because there is a higher concentration of oxygen in the alveoli (air sacs) than in the capillaries.

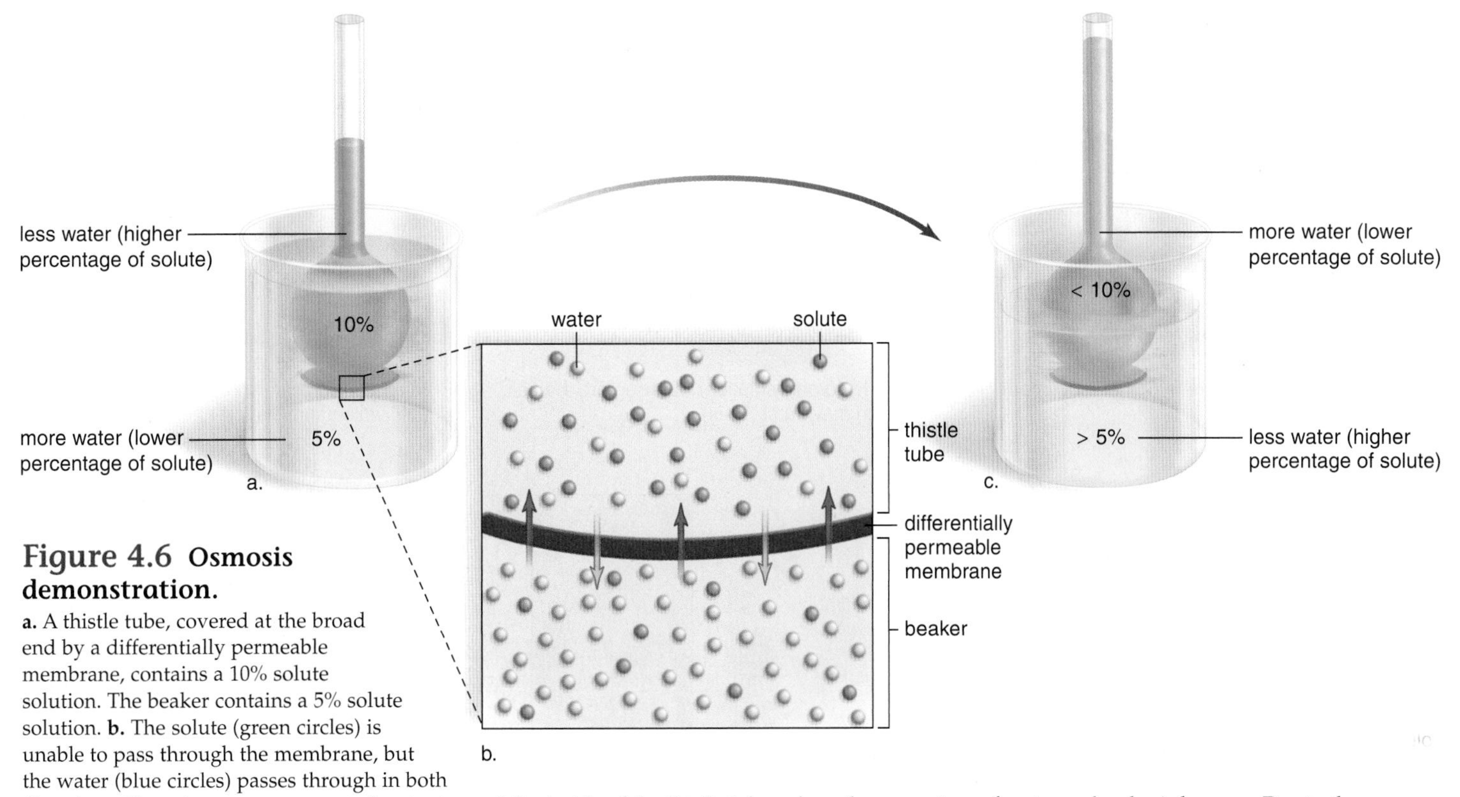

Figure 4.6 Osmosis demonstration.
a. A thistle tube, covered at the broad end by a differentially permeable membrane, contains a 10% solute solution. The beaker contains a 5% solute solution. **b.** The solute (green circles) is unable to pass through the membrane, but the water (blue circles) passes through in both directions. There is a net movement of water toward the inside of the thistle tube, where the percentage of water molecules is lower. **c.** Due to the incoming water molecules, the level of the solution rises in the thistle tube.

Osmosis

The diffusion of water across a differentially permeable membrane due to concentration differences is called **osmosis.** To illustrate osmosis, a thistle tube containing a 10% solute solution[1] is covered at one end by a differentially permeable membrane and then placed in a beaker containing a 5% solute solution (Fig. 4.6). The beaker has a higher concentration of water molecules (lower percentage of solute), and the thistle tube has a lower concentration of water molecules (higher percentage of solute). Diffusion always occurs from higher to lower concentration. Therefore, a net movement of water takes place across the membrane from the beaker to the inside of the thistle tube.

The solute does not diffuse out of the thistle tube. Why not? Because the membrane is not permeable to the solute. As water enters and the solute does not exit, the level of the solution within the thistle tube rises (Fig. 4.6*c*). In the end, the concentration of solute in the thistle tube is less than 10%. Why? Because there is now less solute per unit volume of solution. And, the concentration of solute in the beaker is greater than 5% because there is now more solute per unit volume.

Water enters the thistle tube due to the osmotic pressure of the solution within the thistle tube. **Osmotic pressure** is the pressure that develops in a system due to osmosis.[2] In other words, the greater the possible osmotic pressure, the more likely it is that water will diffuse in that direction. Due to osmotic pressure, water is absorbed by the kidneys and taken up by capillaries in the tissues. Osmosis also occurs across the plasma membrane.

Isotonic Solution In the laboratory, cells are normally placed in **isotonic solutions**—that is, the solute concentration and the water concentration both inside and outside the cell are equal, and therefore there is no net gain or loss of water (Fig. 4.7). The prefix *iso-* means "the same as," and the term tonicity refers to the osmotic pressure or tension of the solution. A 0.9% solution of the salt sodium chloride (NaCl) is known to be isotonic to red blood cells. Therefore, intravenous solutions medically administered usually have this tonicity. Terrestrial animals can usually take in either water or salt as needed to maintain the tonicity of their internal environment. Many animals living in an estuary, such as oysters, blue crabs, and some fishes, are able to cope with changes in the salinity (salt concentrations) of their environment. Their kidneys, gills, and other structures help them do this.

Hypotonic Solution Solutions that cause cells to swell, or even to burst, due to an intake of water are said to be **hypotonic solutions.** The prefix *hypo-* means "less than" and refers to a solution with a lower concentration of solute (higher concentration of water) than inside the cell. If a cell is placed in a hypotonic solution, water enters the cell. The net movement of water is from the outside to the inside of the cell.

Any concentration of a salt solution lower than 0.9% is hypotonic to red blood cells. Animal cells placed in such a solution expand and sometimes burst or lyse due to the buildup of pressure. The term **cytolysis** is used to refer to disrupted cells. Hemolysis, then, is disrupted red blood cells.

The swelling of a plant cell in a hypotonic solution creates **turgor pressure.** When a plant cell is placed in a

[1] Percent solutions are grams of solute per 100 ml of solvent. Therefore, a 10% solution is 10 g of sugar with water added to make 100 ml of solution.

[2] Osmotic pressure is measured by placing a solution in an osmometer and then immersing the osmometer in pure water. The pressure that develops is the osmotic pressure of a solution.

Animal cells

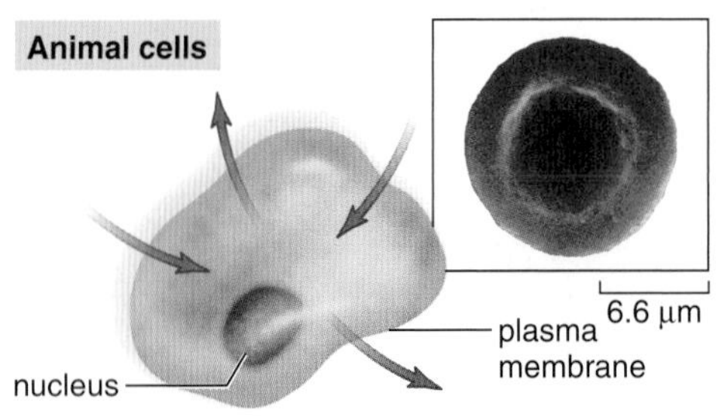

In an isotonic solution, there is no net movement of water.

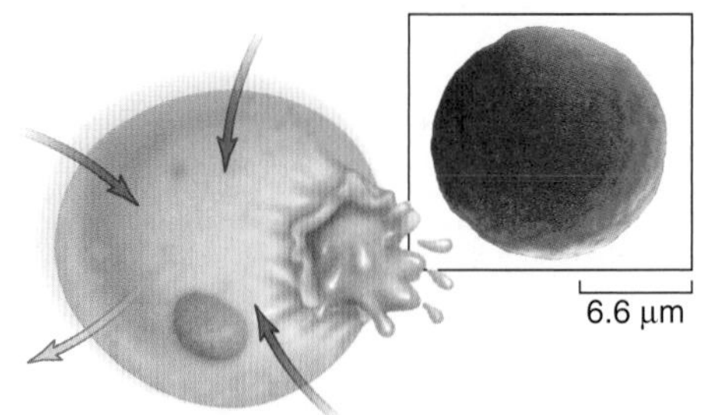

In a hypotonic solution, water enters the cell, which may burst (lysis).

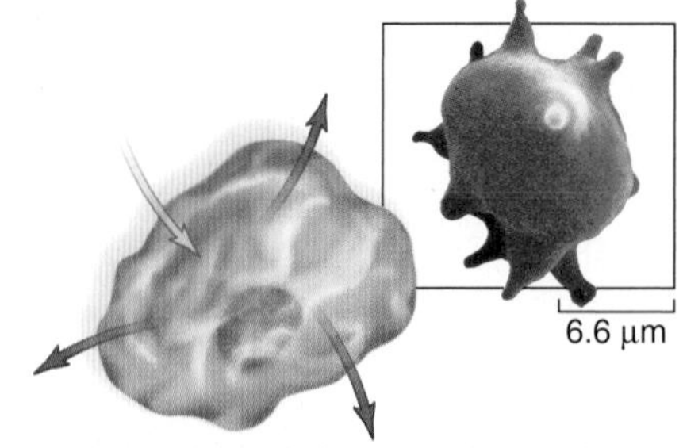

In a hypertonic solution, water leaves the cell, which shrivels (crenation).

Plant cells

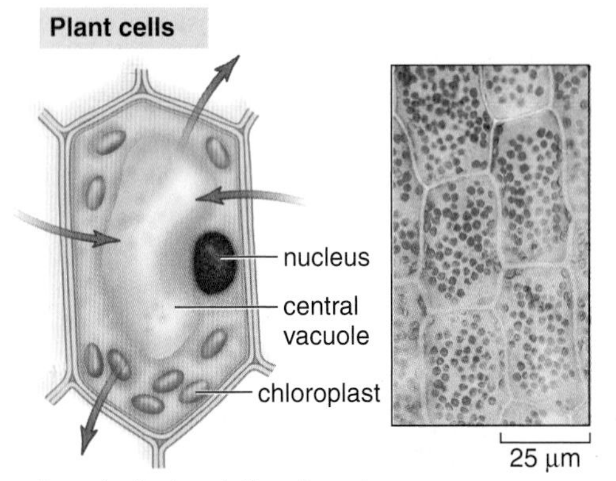

In an isotonic solution, there is no net movement of water.

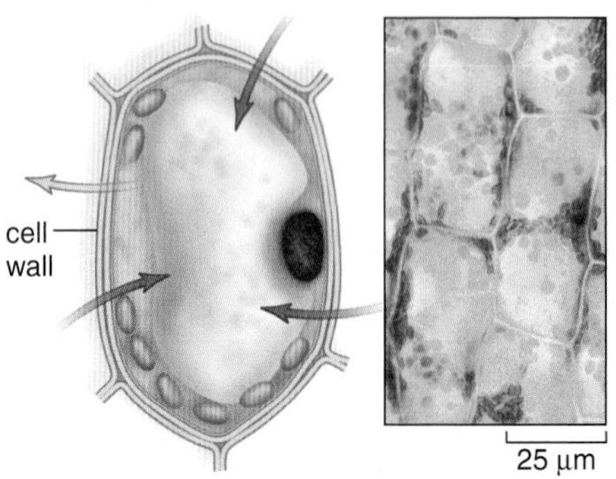

In a hypotonic solution, the central vacuole fills with water, turgor pressure develops, and chloroplasts are seen next to the cell wall.

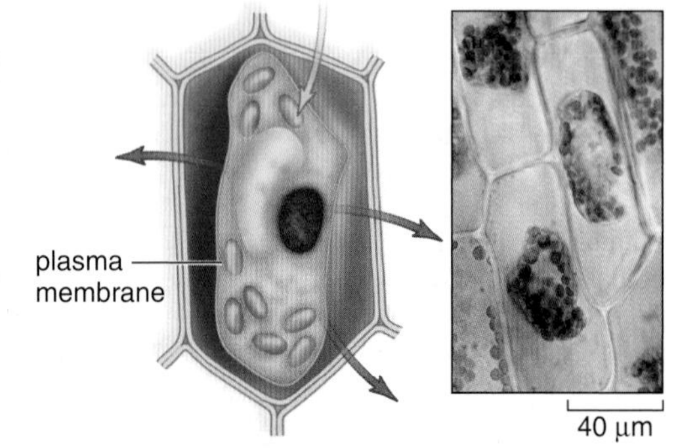

In a hypertonic solution, the central vacuole loses water, the cytoplasm shrinks (plasmolysis), and chloroplasts are seen in the center of the cell.

Figure 4.7 Osmosis in animal and plant cells.
The arrows indicate the movement of water molecules. To determine the net movement of water, compare the number of arrows that are taking water molecules into the cell with the number that are taking water out of the cell. In an isotonic solution, a cell neither gains nor loses water; in a hypotonic solution, a cell gains water; and in a hypertonic solution, a cell loses water.

hypotonic solution, the cytoplasm expands because the large central vacuole gains water and the plasma membrane pushes against the rigid cell wall. The plant cell does not burst because the cell wall does not give way. Turgor pressure in plant cells is extremely important to the maintenance of the plant's erect position. If you forget to water your plants, they wilt due to decreased turgor pressure.

Organisms that live in fresh water have to prevent their internal environment from becoming hypotonic. Many protozoans, such as paramecia, have contractile vacuoles that rid the body of excess water. Freshwater fishes have well-developed kidneys that excrete a large volume of dilute urine. Even so, they have to take in salts at their gills. Even though freshwater fishes are good osmoregulators, they would not be able to survive in either distilled water or a marine environment.

Hypertonic Solution Solutions that cause cells to shrink or shrivel due to loss of water are said to be **hypertonic solutions.** The prefix *hyper-* means "more than" and refers to a solution with a higher percentage of solute (lower concentration of water) than the cell. If a cell is placed in a hypertonic solution, water leaves the cell. The net movement of water is from the inside to the outside of the cell.

Any concentration of a salt solution higher than 0.9% is hypertonic to red blood cells. If animal cells are placed in this solution, they shrink. The term **crenation** refers to the shriveling of a cell in a hypertonic solution. Meats are sometimes preserved by salting them. The salt kills any bacteria present because it makes the meat a hypertonic environment.

When a plant cell is placed in a hypertonic solution, the plasma membrane pulls away from the cell wall as the large central vacuole loses water. This is an example of **plasmolysis,** shrinking of the cytoplasm due to osmosis. The dead plants you may see along a salted roadside died because they were exposed to a hypertonic solution during the winter. Also, when salt water invades coastal marshes due to storms or human activities, coastal plants die. Without roots to hold the soil, it washes into the sea, thereby losing many acres of valuable wetlands.

Marine animals cope with their hypertonic environment in various ways that prevent them from losing water. Sharks increase or decrease the urea in their blood until their blood is isotonic with the environment. Marine fishes and other types of animals excrete salts across their gills. Have you ever seen a marine turtle cry? It is ridding its body of salt by means of glands near the eye.

Check Your Progress

1. Explain what it means to say that molecules move "down" their concentration gradient.
2. What is the difference between osmosis and diffusion?
3. If you suspended red blood cells in a solution containing 1.5% NaCl, what would happen to them?

Transport by Carrier Proteins

The plasma membrane impedes the passage of all but a few substances. Yet, biologically useful molecules are able to enter and exit the cell at a rapid rate because of carrier proteins in the membrane. Carrier proteins are specific. Each can combine with only a certain type of molecule or ion, which is then transported through the membrane. Scientists do not completely understand how carrier proteins function, but after a carrier combines with a molecule, the carrier is believed to undergo a change in shape that moves the molecule across the membrane. Carrier proteins are required for both facilitated transport and active transport (see Table 4.1).

Facilitated Transport

Facilitated transport explains the passage of such molecules as glucose and amino acids across the plasma membrane even though they are not lipid-soluble. The passage of glucose and amino acids is facilitated by their reversible combination with carrier proteins, which in some manner transport them through the plasma membrane. These carrier proteins are specific. For example, various sugar molecules of identical size might be present inside or outside the cell, but glucose can cross the membrane hundreds of times faster than the other sugars.

A model for facilitated transport (Fig. 4.8) shows that after a carrier has assisted the movement of a molecule to the other side of the membrane, it is free to assist the passage of other similar molecules. Neither diffusion nor facilitated transport requires an expenditure of energy (use of ATP) because the molecules are moving down their concentration gradient in the same direction they tend to move anyway.

Active Transport

During **active transport,** molecules or ions move through the plasma membrane, accumulating either inside or outside the cell. For example, iodine collects in the cells of the thyroid gland; glucose is completely absorbed from the gut by the cells lining the digestive tract; and sodium can be almost completely withdrawn from urine by cells lining the kidney tubules. In these instances, molecules have moved to the region of higher concentration, exactly opposite to the process of diffusion.

Both carrier proteins and an expenditure of energy are needed to transport molecules against their concentration gradient. In this case, chemical energy, usually in the form of ATP, is required for the carrier to combine with the substance to be transported. Therefore, it is not surprising that cells involved primarily in active transport, such as kidney cells, have a large number of mitochondria near membranes where active transport is occurring.

Proteins involved in active transport are often called pumps because, just as a water pump uses energy to move water against the force of gravity, proteins use energy to move a substance against its concentration gradient. One type of pump that is active in all animal cells, but is especially associated with nerve and muscle cells, moves sodium ions (Na^+) to the outside of the cell and potassium ions (K^+) to the inside of the cell. These two events are linked, and the carrier protein is called a **sodium-potassium pump.** A change in carrier shape after the attachment of a phosphate group, and again after its detachment, allows the carrier to combine alternately with sodium ions and potassium ions (Fig. 4.9). The phosphate group is donated by ATP when it is broken down enzymatically by the carrier. The sodium-potassium pump results in both a solute concentration gradient and an electrical gradient for these ions across the plasma membrane.

The passage of salt (NaCl) across a plasma membrane is of primary importance to most cells. The chloride ion (Cl^-) usually crosses the plasma membrane because it is

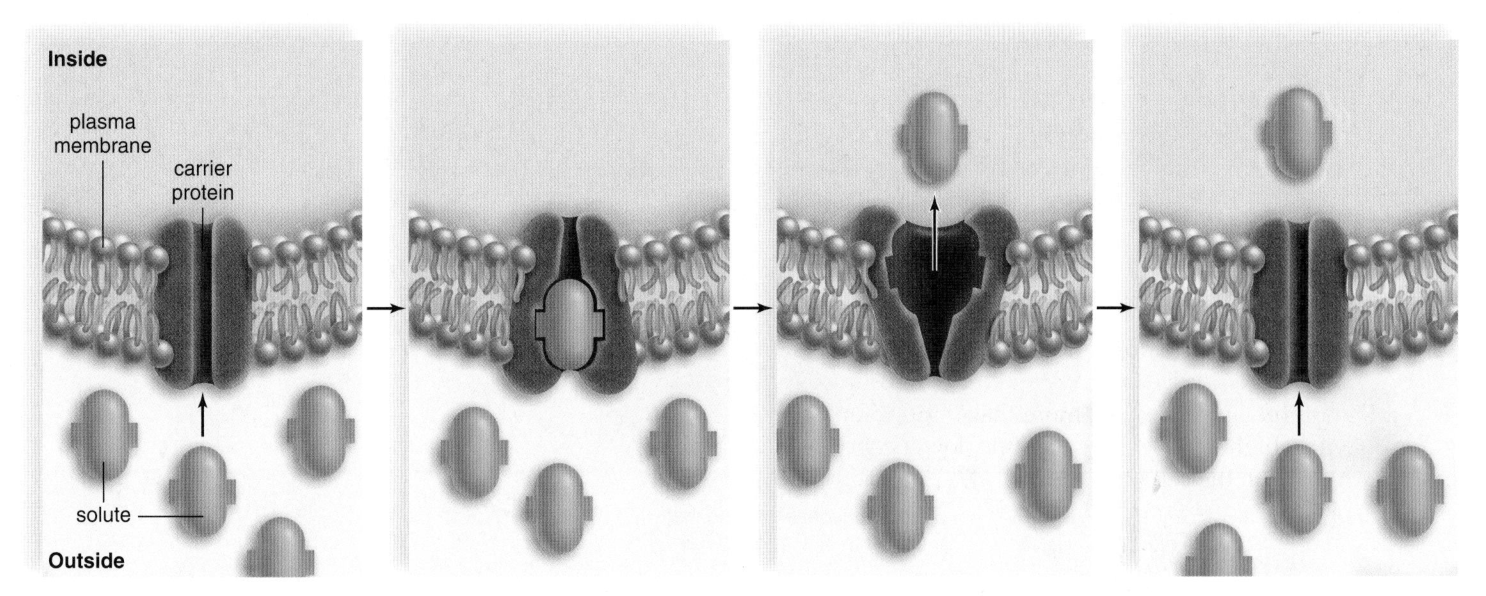

Figure 4.8 Facilitated transport.
During facilitated transport, a carrier protein speeds the rate at which the solute crosses the plasma membrane toward a lower concentration. Note that the carrier protein undergoes a change in shape as it moves a solute across the membrane.

attracted by positively charged sodium ions (Na^+). First sodium ions are pumped across a membrane, and then chloride ions simply diffuse through channels that allow their passage.

As noted in Figure 4.2*a*, the genetic disorder cystic fibrosis results from a faulty chloride channel. In cystic fibrosis, Cl^- transport is reduced, and so is the flow of Na^+ and water. Researchers believe that the lack of water causes the mucus in the bronchial tubes and pancreatic ducts to be abnormally thick, thus interfering with the function of the lungs and pancreas.

Check Your Progress

1. Compare and contrast facilitated diffusion with active transport.
2. Why does the chloride ion require a channel to cross the membrane?

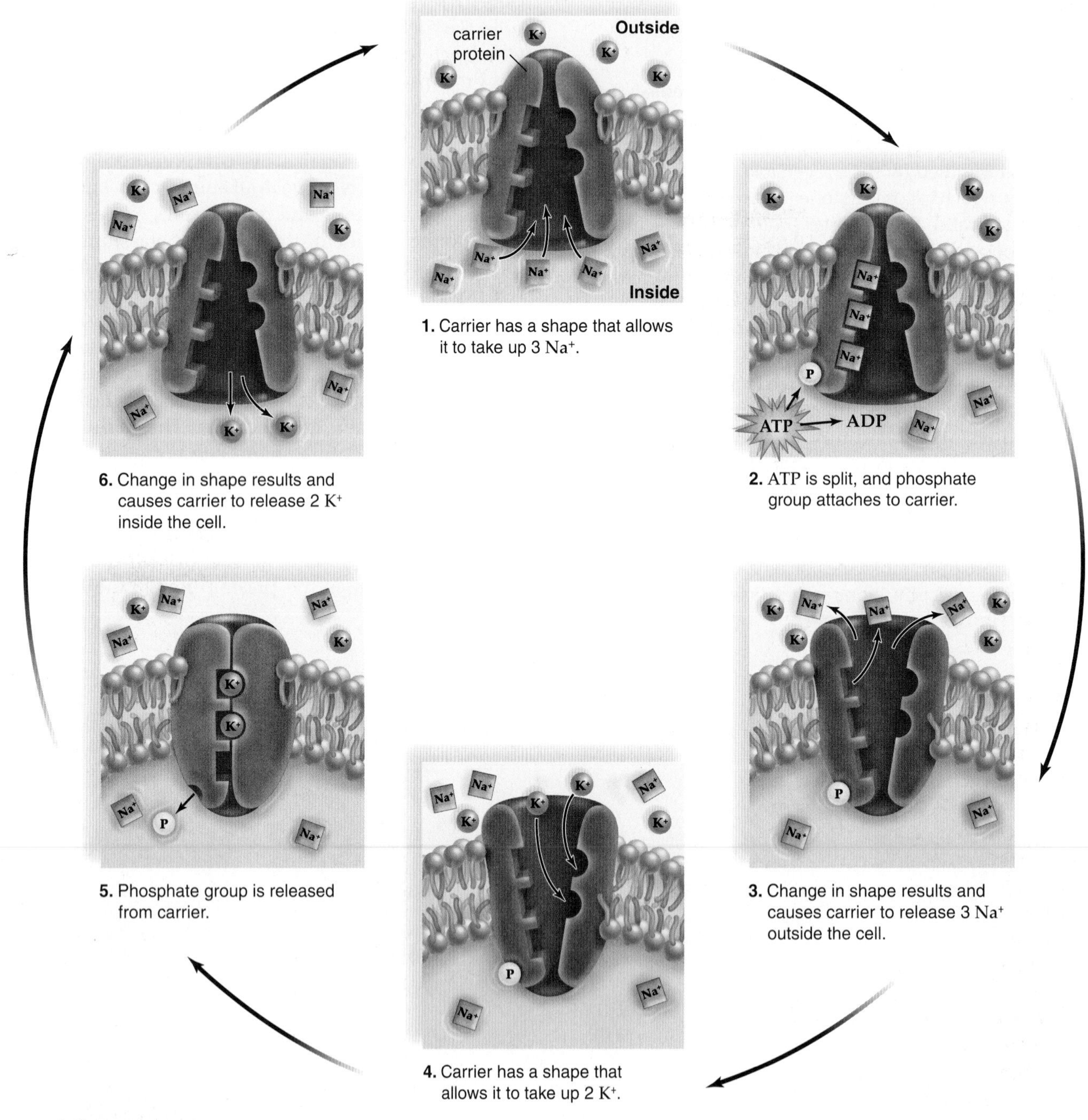

Figure 4.9 The sodium-potassium pump.
The same carrier protein transports sodium ions (Na^+) to the outside of the cell and potassium ions (K^+) to the inside of the cell because it undergoes an ATP-dependent change in shape. Three sodium ions are carried outward for every two potassium ions carried inward. Therefore, the inside of the cell is negatively charged compared to the outside.

Vesicle Formation

How do macromolecules such as polypeptides, polysaccharides, or polynucleotides enter and exit a cell? Because they are too large to be transported by carrier proteins, macromolecules are transported into and out of the cell by vesicle formation. Vesicle formation is called membrane-assisted transport because membrane is needed to form the vesicle. Vesicle formation requires an expenditure of cellular energy, but an added benefit is that the vesicle membrane keeps the contained macromolecules from mixing with molecules within the cytoplasm. Exocytosis is a way substances can exit a cell, and endocytosis is a way substances can enter a cell.

Exocytosis

During **exocytosis,** a vesicle fuses with the plasma membrane as secretion occurs (Fig. 4.10). Hormones, neurotransmitters, and digestive enzymes are secreted from cells in this manner. The Golgi apparatus often produces the vesicles that carry these cell products to the membrane. Notice that during exocytosis, the membrane of the vesicle becomes a part of the plasma membrane, which is thereby enlarged. For this reason, exocytosis can be a normal part of cell growth. The proteins released from the vesicle adhere to the cell surface or become incorporated in an extracellular matrix.

Cells of particular organs are specialized to produce and export molecules. For example, pancreatic cells produce digestive enzymes or insulin, and anterior pituitary cells produce growth hormone, among other hormones. In these cells, secretory vesicles accumulate near the plasma membrane, and the vesicles release their contents only when the cell is stimulated by a signal received at the plasma membrane. A rise in blood sugar, for example, signals pancreatic cells to release the hormone insulin. This is called regulated secretion, because vesicles fuse with the plasma membrane only when it is appropriate to the needs of the body.

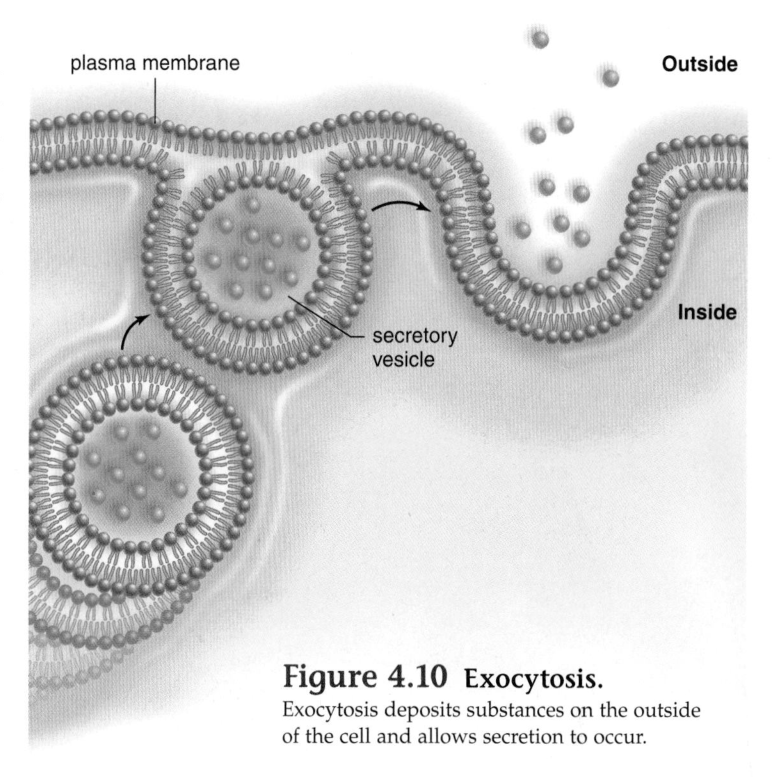

Figure 4.10 Exocytosis.
Exocytosis deposits substances on the outside of the cell and allows secretion to occur.

Endocytosis

During **endocytosis,** cells take in substances by vesicle formation. A portion of the plasma membrane invaginates to envelop the substance, and then the membrane pinches off to form an intracellular vesicle. Endocytosis occurs in one of three ways, as illustrated in Figure 4.11. Phagocytosis transports large substances, such as viruses, and pinocytosis transports small substances, such as macromolecules, into cells. Receptor-mediated endocytosis is a special form of pinocytosis.

Phagocytosis When the material taken in by endocytosis is large, such as a food particle or another cell, the process is called **phagocytosis.** Phagocytosis is common in unicellular organisms such as amoebas (Fig. 4.11*a*). It also occurs in humans. Certain types of human white blood cells are amoeboid—that is, they are mobile like an amoeba, and they are able to engulf debris such as worn-out red blood cells or viruses. When an endocytic vesicle fuses with a lysosome, digestion occurs. We will see that this process is a necessary and preliminary step toward the development of immunity to bacterial diseases.

Pinocytosis **Pinocytosis** occurs when vesicles form around a liquid or around very small particles (Fig. 4.11*b*). Blood cells, cells that line the kidney tubules or the intestinal wall, and plant root cells all use pinocytosis to ingest substances.

Whereas phagocytosis can be seen with the light microscope, an electron microscope is required to observe pinocytic vesicles, which are no larger than 0.1–0.2 μm. Still, pinocytosis involves a significant amount of the plasma membrane because it occurs continuously. The loss of plasma membrane due to pinocytosis is balanced by the occurrence of exocytosis, however.

Receptor-Mediated Endocytosis **Receptor-mediated endocytosis** is a form of pinocytosis that is quite specific because it uses a receptor protein shaped so that a specific molecule, such as a vitamin, peptide hormone, or lipoprotein, can bind to it (Fig. 4.11*c*). The receptors for these substances are found at one location in the plasma membrane. This location is called a coated pit because there is a layer of protein on the cytoplasmic side of the pit. Once formed, the vesicle becomes uncoated and may fuse with a lysosome. When empty, used vesicles fuse with the plasma membrane, and the receptors return to their former location.

Receptor-mediated endocytosis is selective and much more efficient than ordinary pinocytosis. It is involved in uptake and also in the transfer and exchange of substances between cells. Such exchanges take place when substances move from maternal blood into fetal blood at the placenta, for example.

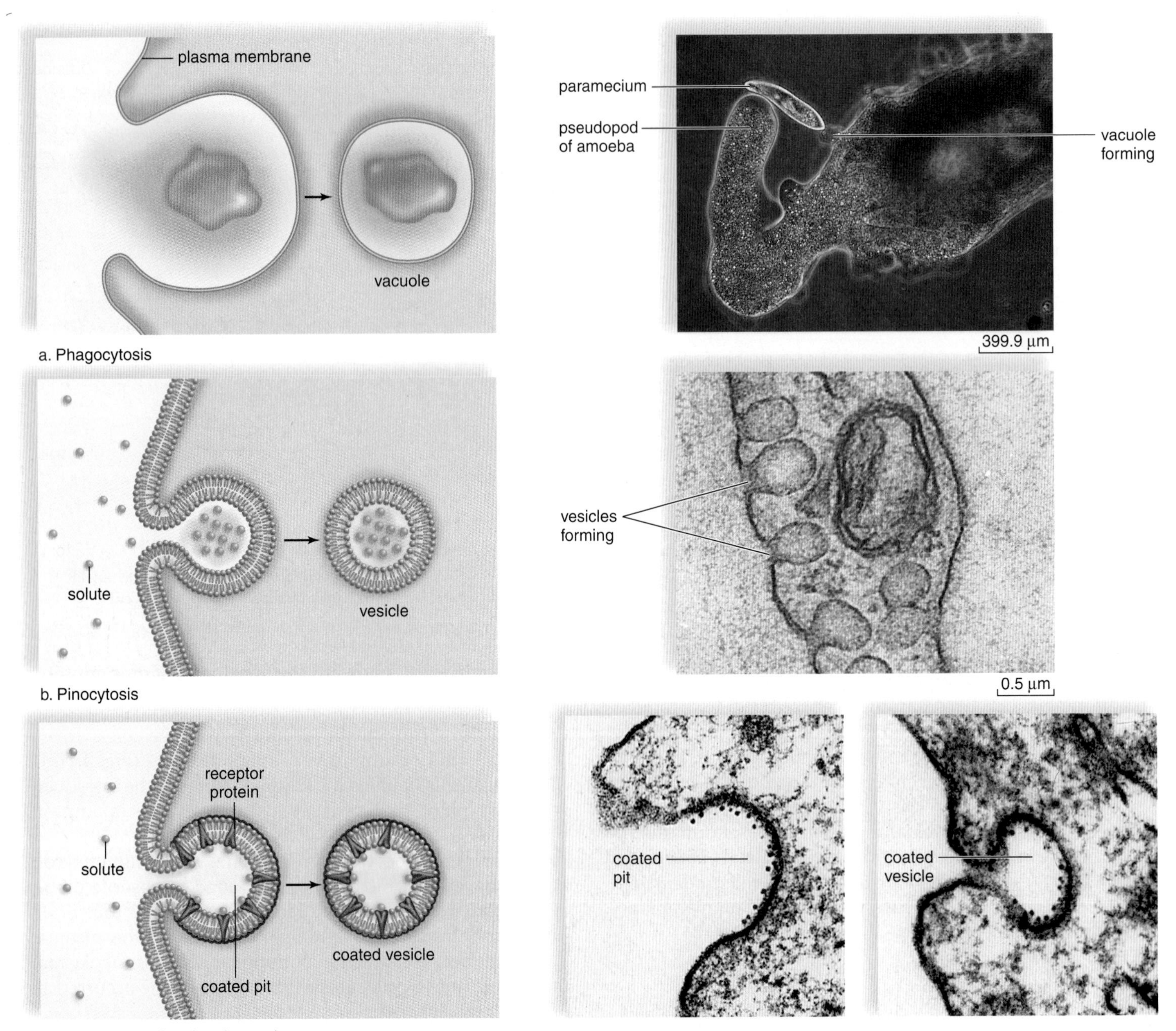

Figure 4.11 Three methods of endocytosis.
a. Phagocytosis occurs when the substance to be transported into the cell is large. Amoebas ingest by phagocytosis. Digestion occurs when the resulting vacuole fuses with a lysosome. **b.** Pinocytosis occurs when a macromolecule such as a polypeptide is transported into the cell. The result is a vesicle (small vacuole). **c.** Receptor-mediated endocytosis is a form of pinocytosis. Molecules first bind to specific receptor proteins, which migrate to or are already in a coated pit. The vesicle that forms contains the molecules and their receptors.

The importance of receptor-mediated endocytosis is demonstrated by a genetic disorder called familial hyper-cholesterolemia. Cholesterol is transported in the blood by a complex of lipids and proteins called low-density lipoprotein (LDL). Ordinarily, body cells take up LDL when LDL receptors gather in a coated pit. But in some individuals, the LDL receptor is unable to properly bind to the coated pit, and the cells are unable to take up cholesterol. Instead, cholesterol accumulates in the walls of arterial blood vessels, leading to high blood pressure, occluded (blocked) arteries, and heart attacks.

Check Your Progress

1. Where is the vesicle produced in exocytosis? Where does it end up?
2. Compare and contrast the three methods of endocytosis.

Applying the Concepts Revisited

In the case of the particular channel described in the opening of this chapter, calcium ions enter the cell and initiate a cascade of events that results in the sensation of pain. A different type of voltage-gated ion channel also allows calcium ions to enter the cell. In cells with this type of channel, muscular contraction or neuron excitation occurs.

1. Why could there be different responses to the same ion crossing the membrane?
2. Drugs known as calcium channel blockers are used to decrease blood pressure (hypertension). How do you think these drugs act?
3. Capsaicin is also sold as a topical pain reliever for arthritis patients. What information in the opening story gives you a clue as to why capsaicin can relieve pain?

SUMMARIZING THE CONCEPTS

4.1 Plasma Membrane Structure and Function

- According to the fluid-mosaic model of the plasma membrane, a lipid bilayer is fluid and has the consistency of light oil. The hydrophilic heads of phospholipids form the inner and outer surfaces, and the hydrophobic tails form the interior.
- Proteins within the membrane are the mosaic portion. The peripheral proteins often have a structural role in that they help stabilize and shape the plasma membrane. They may also function in signaling pathways. The integral proteins form channels, act as carriers to move substances across the membrane, function in cell recognition, act as receptors, and carry on enzymatic reactions.
- Carbohydrate chains are attached to some of the lipids and proteins in the membrane. These are glycolipids and glycoproteins.

4.2 The Permeability of the Plasma Membrane

- Some substances, such as gases and water, freely cross a plasma membrane, while others—particularly ions, charged molecules, and macromolecules—have to be assisted across.
- Passive ways of crossing a plasma membrane (diffusion and facilitated transport) do not require an expenditure of chemical energy (ATP). Active ways of crossing a plasma membrane (active transport and vesicle formation) do require an expenditure of chemical energy.
- Lipid-soluble compounds, water, and gases simply diffuse across the plasma membrane from the area of higher concentration to the area of lower concentration.
- The diffusion of water across a differentially permeable membrane is called osmosis. Water moves across the membrane into the area of lower water (higher solute) content. When cells are in an isotonic solution, they neither gain nor lose water; when they are in a hypotonic solution, they gain water; and when they are in a hypertonic solution, they lose water.
- Some molecules are transported across the membrane by carrier proteins that span the membrane. During facilitated transport, a carrier protein assists the movement of a molecule down its concentration gradient. No energy is required.
- During active transport, a carrier protein acts as a pump that causes a substance to move against its concentration gradient. Energy in the form of ATP molecules is required for active transport to occur.
- Larger substances can exit and enter a membrane by exocytosis and endocytosis. Exocytosis involves secretion. Endocytosis includes phagocytosis and pinocytosis. Receptor-mediated endocytosis, a type of pinocytosis, makes use of receptor molecules in the plasma membrane and a coated pit, which pinches off to form a vesicle.

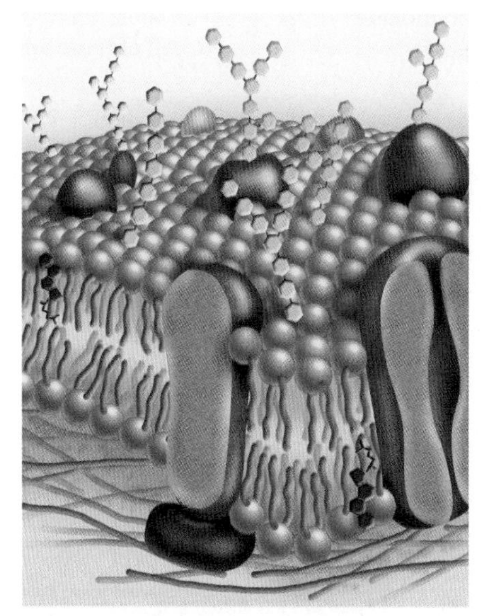

TESTING YOURSELF

Choose the best answer for each question.

1. Energy is used for
   a. diffusion.
   b. osmosis.
   c. active transport.
   d. None of these are correct.
2. When mixed together, water is a ____, and dye is a ____.
   a. solute, solvent
   b. solvent, solvent
   c. solute, solute
   d. solvent, solute
3. Na^+ movement across a plasma membrane occurs by
   a. diffusion.
   b. endocytosis.
   c. active transport.
   d. None of these are correct.
4. Carrier proteins are ____ in their action.
   a. specific
   b. not specific
   c. involved in diffusion
   d. None of these are correct.
5. Which of these are methods of endocytosis?
   a. phagocytosis
   b. pinocytosis
   c. receptor-mediated
   d. All of these are correct.
6. An isotonic solution has the/a ____ concentration of water as the cell and the/a ____ concentration of solute as the cell.
   a. same, same
   b. same, different
   c. different, same
   d. different, different

7. The mosaic part of the fluid-mosaic model of membrane structure refers to
   a. the phospholipids.
   b. the proteins.
   c. the glycolipids.
   d. the cholesterol.
8. The carbohydrate chains on lipids and proteins are found
   a. on the inside of the membrane.
   b. on the outside of the membrane.
   c. on both sides of the membrane.
9. A coated pit is associated with
   a. diffusion.
   b. osmosis.
   c. receptor-mediated endocytosis.
   d. pinocytosis.
10. Exocytosis involves
    a. fusion with the plasma membrane.
    b. fusion with the nucleus.
    c. fusion with the mitochondria.
    d. fusion with a ribosome.
11. When mitochondria are found near the surface of kidney cells, the cells are probably
    a. using active transport.
    b. using facilitated transport.
    c. pumping a molecule across the membrane.
    d. allowing a molecule to diffuse into the cell.
    e. Both a and c are correct.
12. Which of these passes through a plasma membrane by way of diffusion?
    a. carbon dioxide
    b. oxygen
    c. water
    d. All of these are correct.
13. A phospholipid molecule has a head and two tails. The tails are found
    a. at the surfaces of the membrane.
    b. in the interior of the membrane.
    c. spanning the membrane.
    d. where the environment is hydrophilic.
    e. Both a and b are correct.
14. Besides phospholipids, the other lipid molecule that is vital to an animal's plasma membrane is
    a. cholesterol.
    b. glycogen.
    c. triglycerides.
    d. glycerol.
15. When a cell is placed in a hypotonic solution,
    a. solute exits the cell to equalize the concentration on both sides of the plasma membrane.
    b. water exits the cell toward the area of lower solute concentration.
    c. water enters the cell toward the area of higher solute concentration.
    d. solute exits and water enters the cell.
    e. Both c and d are correct.
16. When a cell is placed in a hypertonic solution,
    a. solute exits the cell to equalize the concentration on both sides of the plasma membrane.
    b. water exits the cell toward the area of higher solute concentration.
    c. water enters the cell toward the area of higher solute concentration.
    d. solute exits and water enters the cell.
    e. Both a and c are correct.

UNDERSTANDING THE TERMS

active transport 75
carrier protein 69
cell recognition protein 69
channel protein 69
concentration gradient 71
crenation 74
cytolysis 73
differentially permeable 71
diffusion 72
endocytosis 77
enzymatic protein 69
exocytosis 77
facilitated transport 75
fluid-mosaic model 68
glycolipid 69
glycoprotein 69
hypertonic solution 74
hypotonic solution 73
isotonic solution 73
osmosis 73
osmotic pressure 73
phagocytosis 77
pinocytosis 77
plasmolysis 74
receptor-mediated endocytosis 77
receptor protein 69
sodium-potassium pump 75
solvent 72
turgor pressure 73

THINKING CRITICALLY

1. When a signal molecule such as a growth hormone binds to a receptor protein in the plasma membrane, it stays on the outside of the cell. How might the inside of the cell know that the signal has bound?
2. Some antibiotics interfere with the formation of the bacterial cell wall, thus weakening the cell wall. How might this cause a bacterium to be killed?
3. How could a channel protein regulate what enters the cell?

INQUIRY INTO LIFE WEBSITE

The companion website for *Inquiry into Life* provides a wealth of information organized and integrated by chapter. You will find practice tests, animations, videos, and much more that will complement your learning and understanding of general biology.

http://www.mhhe.com/maderinquiry13

Applying the Concepts

Cell Division

Chapter Outline

In August 2008, actress Christina Applegate, 36, revealed to shocked fans that she had been diagnosed with breast cancer. Every year over 192,000 American women are diagnosed with breast cancer. Christina's mother is a breast and cervical cancer survivor, and Christina tested positive for a *BRCA1* (breast cancer 1, early onset) gene mutation, linked to both breast and ovarian cancer. Women with a mutation in either the *BRCA1* or *BRCA2* gene are three to seven times more likely to be diagnosed with breast cancer than those without a mutation. Several medical approaches, including increased surveillance and prophylactic surgeries, are available for those who test positive for a mutated *BRCA* gene. Christina had been undergoing routine mammograms since age 30, so the cancer was caught at an early stage. Still, Christina made the decision to have a double mastectomy, removing both breasts even though the cancer was only in one, instead of undergoing longer-term treatments such as chemotherapy. She stated that it was a tough decision, but she wanted to be sure that the cancer was completely gone.

Cancer results from a failure in control of the cell cycle, the orderly set of stages that result in cell division. *BRCA1* is a tumor suppressor gene. As we shall see in this chapter, tumor suppressor genes prevent the formation of cancer. When these genes are mutated, cancer can develop.

5.1 Cell Increase and Decrease

Learning Outcomes

1. Explain the two processes that change the number of cells in the body.
2. Describe the stages of the cell cycle and what occurs in each stage.

Cell division which increases the number of **somatic cells** (body cells) occurs throughout our lives. Each of us started out as a single-celled fertilized egg, but now we are composed of trillions of cells. Even now, your body is producing thousands of new red blood cells, skin cells, and cells that line your respiratory and digestive tracts. Also, if you suffer an injury, cell division will repair it.

Apoptosis, programmed cell death, decreases the number of cells. Both cell division and apoptosis are normal parts of growth and development. Apoptosis occurs during development to remove unwanted tissue—for example, the tail of a tadpole disappears as it matures into a frog. In humans, the fingers and toes of the embryo are initially webbed, but are normally freed from one another later in development as a result of apoptosis. Apoptosis also plays an important role in preventing cancer. An abnormal cell that could become cancerous will often die via apoptosis, thus preventing a tumor from developing.

The Cell Cycle

Cell division is a part of the cell cycle. The **cell cycle** is an orderly set of stages that take place between the time a cell divides and the time the resulting cells also divide.

The Stages of Interphase

As Figure 5.1*a* shows, most of the cell cycle is spent in **interphase.** This is the time when a cell carries on its usual functions, which are dependent on its location in the body. It also gets ready to divide: it grows larger, the number of organelles doubles, and the amount of DNA doubles. For mammalian cells, interphase lasts for about 20 hours, which is 90% of the cell cycle.

Interphase is divided into three stages: the G_1 stage occurs before DNA synthesis, the S stage includes DNA synthesis, and the G_2 stage occurs after DNA synthesis. Originally G stood for the "gaps" that occur before and after DNA synthesis during interphase. But now that we know growth occurs during these stages, the G can be thought of as standing for growth.

During the G_1 stage, a cell doubles its organelles (such as mitochondria and ribosomes), and it accumulates the materials needed for DNA synthesis. Some cells, such as nerve and muscle cells, typically do not complete the cell cycle and are arrested. These cells are said to have entered a G_0 stage.

During the S stage, DNA replication occurs. At the beginning of the S stage, each chromosome is composed of one DNA molecule, which is called a chromatid. At the end of this stage, each chromosome consists of two sister chromatids that have identical DNA sequences. Another way of expressing this is to say that DNA replication has resulted in duplicated chromosomes.

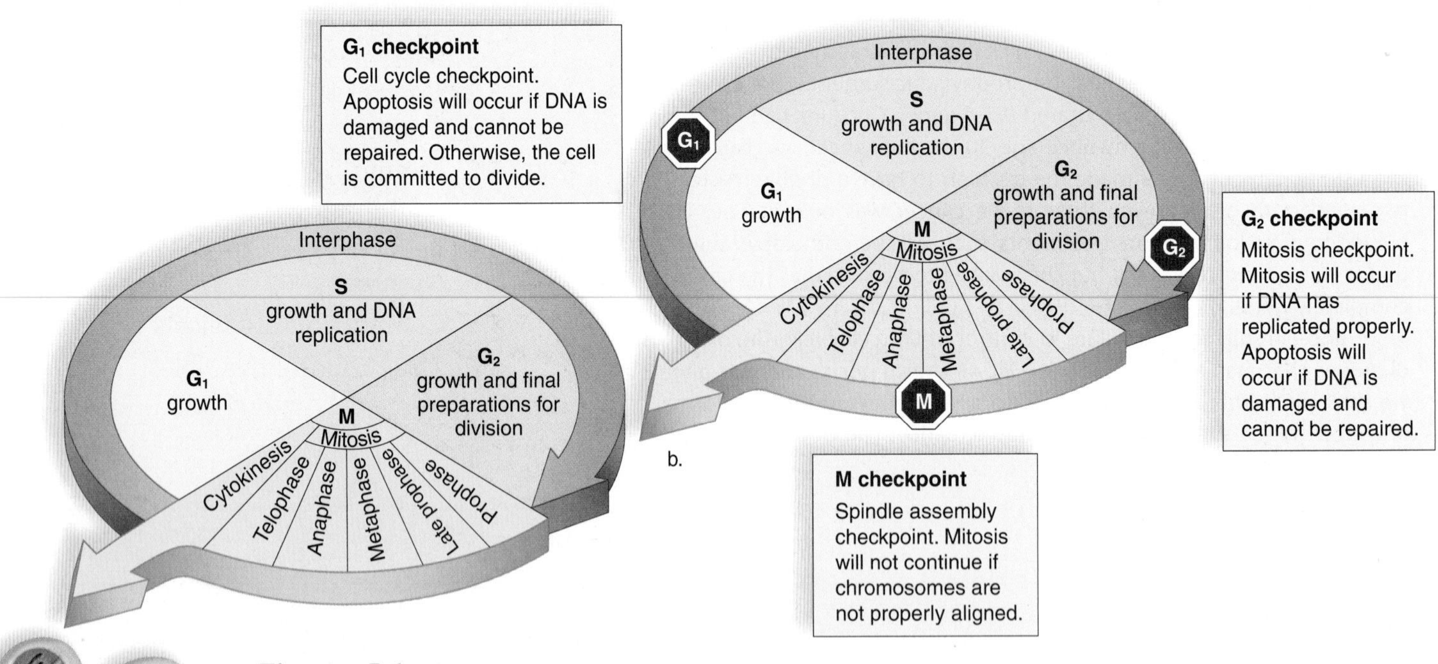

Figure 5.1 The cell cycle.
a. Cells go through a cycle that consists of four stages: G_1, S, G_2, and M. Each stage is characterized by a major activity.
b. The cell cycle may stop at any of three checkpoints if necessary.

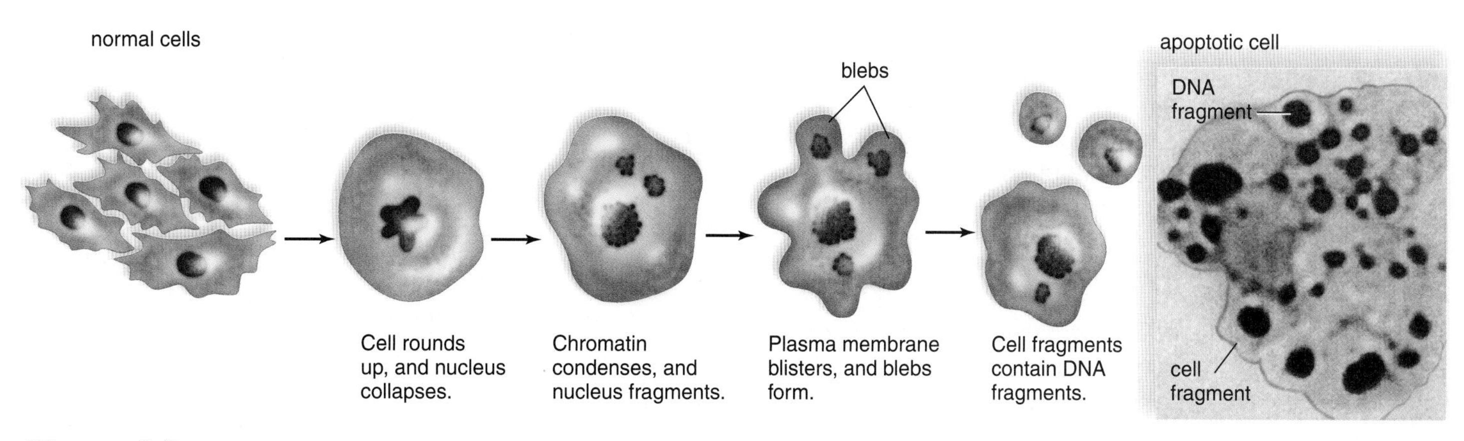

Figure 5.2 Apoptosis.
Apoptosis is a sequence of events that results in a fragmented cell. The fragments are engulfed by white blood cells.

During the G_2 stage, the cell synthesizes the proteins needed for cell division, such as the protein found in the spindle apparatus, described in section 5.3.

The Mitotic Stage

Following interphase, the cell enters the M (for mitotic) stage. This stage not only includes **mitosis**, the division of the nucleus and genetic material, but also **cytokinesis**, the division of the cytoplasm, if it occurs. During mitosis, as we shall see, the sister chromatids of each chromosome separate, becoming daughter chromosomes that are distributed to two daughter nuclei. When cytokinesis is complete, two daughter cells that are identical to the mother cell are present. Mammalian cells usually require only about four hours to complete the mitotic stage.

Apoptosis

During apoptosis, the cell progresses through a typical series of events that bring about its destruction (Fig. 5.2). The cell rounds up and loses contact with its neighbors. The nucleus fragments, and the plasma membrane develops blisters. Finally, the cell fragments, and its bits and pieces are engulfed by white blood cells.

A remarkable finding of the past few years is that cells routinely harbor the enzymes, now called caspases, that bring about apoptosis. The enzymes are ordinarily held in check by inhibitors, but they can be unleashed either by internal or external signals. There are two sets of caspases. The first set, the "initiators," receive the signal to activate the second set, the "executioners," which then activate the enzymes that dismantle the cell. For example, executioners turn on enzymes that tear apart the cytoskeleton and enzymes that chop up DNA.

Check Your Progress

1. Draw the cell cycle, and label the major activity that occurs during each stage.
2. What is apoptosis and how is it beneficial to the body?

5.2 Control of the Cell Cycle and Cancer

Learning Outcomes

1. Distinguish between internal and external controls of the cell cycle.
2. Describe the checkpoints for the cell cycle.
3. Differentiate between the role of proto-oncogenes and tumor suppressor genes in cancer.

Eukaryotic cells have evolved a complex system for regulation of the cell cycle. The cell cycle is controlled by both internal and external signals. The internal signals ensure that the stages follow one another in the normal sequence and that each stage is properly completed before the next stage begins. The external signals tell the cell whether or not to divide.

The events of the cell cycle must occur in the correct order, even if the steps take longer than normal. The red stop signs in Figure 5.1*b* represent three checkpoints when the cell cycle possibly stops. Researchers have identified proteins called **cyclins** that increase and decrease as the cell cycle continues. The appropriate cyclin has to be present for the cell to proceed from the G_1 stage to the S stage and from the G_2 stage to the M stage.

The first checkpoint during G_1 allows the cell to determine whether conditions are favorable to begin the cell cycle. The cell needs to assess whether there are building blocks available for duplication of the DNA and if the DNA is intact. DNA damage can stop the cell cycle at the G_1 checkpoint. In mammalian cells, the p53 protein stops the cycle at the G_1 checkpoint when DNA is damaged. First, the p53 protein attempts to initiate DNA repair, but if that is not possible, it brings about apoptosis.

The cell cycle stops at the G_2 checkpoint if DNA has not finished replicating. This prevents the initiation of the M stage before completion of the S stage. Also, if DNA is damaged, stopping the cell cycle at this checkpoint allows time for the damage to be repaired. If repair is not possible, apoptosis occurs.

Another cell cycle checkpoint occurs during the mitotic (M) stage. The cycle stops if the chromosomes are not going to be distributed accurately to the daughter cells.

These checkpoints are critical for preventing cancer development. A damaged cell should not complete mitosis, but instead should undergo apoptosis.

Mammalian cells tend to enter the cell cycle only when stimulated by an external factor. Growth factors are hormones that are received at the plasma membrane. These signals set into motion the events that result in the cell entering the cell cycle. For example, when blood platelets release a growth factor, skin fibroblasts in the vicinity are stimulated to finish the cell cycle so an injury can be repaired.

Proto-oncogenes and Tumor Suppressor Genes

Carcinogenesis, the development of cancer, is a multi-stage process involving disruption of normal cell division and behavior. Genetic changes in the cell occur during several of the stages of cancer development. During initiation of cancer, a mutation occurs in the DNA of the cell. This mutation could be inherited from a parent, the result of an error in replication, or induced by an environmental agent. Later, as the altered cell grows and divides to become a population of cells, further mutations occur. When these mutations are examined, many are found to occur in two different categories of genes: proto-oncogenes and tumor suppressor genes. **Proto-oncogenes** encode proteins that promote the cell cycle and prevent apoptosis. They are often likened to the gas pedal of a car because they cause cells to continue through the cell cycle. **Tumor suppressor genes** encode proteins that stop the cell cycle and promote apoptosis. They are often likened to the brakes of a car because they inhibit cells from progressing through the cell cycle.

When proto-oncogenes mutate, they become cancer-causing genes, called **oncogenes**. For example, the protein product of the *RAS* proto-oncogene is part of a signal transduction pathway, one of a series of relay proteins (Fig. 5.3). When a signaling molecule, such as a growth factor, binds to a receptor on the cell surface, the Ras protein is activated. Activated Ras protein conveys a message to the nucleus resulting in gene expression that brings about cell division. If the *RAS* proto-oncogene is mutated to become an oncogene in such a way that the Ras protein is always activated, no growth factor needs to bind to the cell for the altered Ras protein to send the signal to divide. An altered Ras protein is found in approximately 25% of all tumors.

When tumor suppressor genes mutate, their products no longer inhibit the cell cycle. *p53* is a tumor suppressor gene. The p53 protein is such an important protein for regulating the cell cycle that almost half of all human cancers have a mutation in the *p53* gene.

Check Your Progress

1. Why might a cell stop at each of the three cell cycle checkpoints?
2. Explain how oncogenes and mutated tumor-suppressor genes act in opposite ways but both result in cancer.

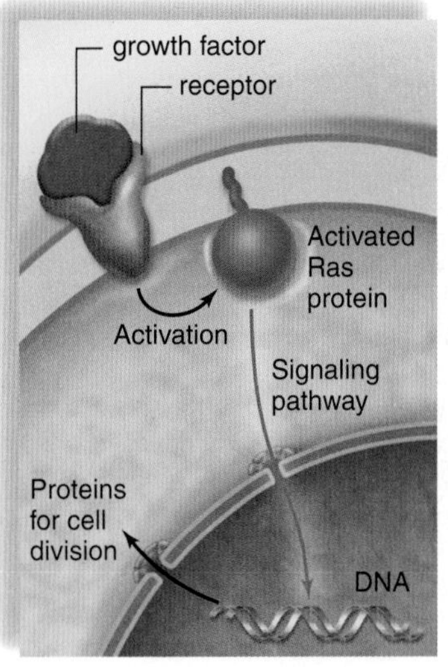

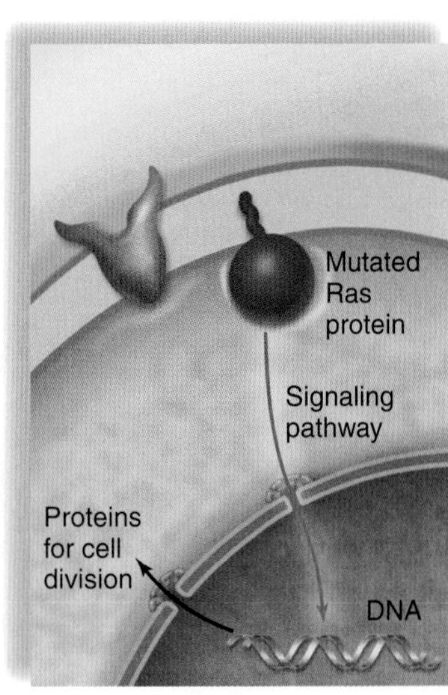

Figure 5.3 Activity of Ras protein.
a. When a growth factor binds to its receptor on the cell surface, Ras protein, part of a signal transduction pathway, is activated and relays a signal to the nucleus to express genes whose products promote cell division. **b.** When a mutation occurs in the *RAS* proto-oncogene such that the Ras protein is always activated, it sends a signal for cell division without the presence of a growth factor.

5.3 Maintaining the Chromosome Number

Learning Outcomes

1. Explain the purpose and the process of mitosis.
2. Compare and contrast mitosis in plant and animal cells.

Eukaryotic chromosomes are composed of chromatin, a combination of both DNA and protein. Some of these proteins are concerned with DNA and RNA synthesis, but a large proportion, termed histones, seem to play primarily a structural role. A human cell contains at least 2 m of DNA. Yet all of this DNA is packed into a nucleus that is about 5 μm in diameter. The histones are responsible for packaging the DNA so that it can fit into such a small space. When a eukaryotic cell is not undergoing division, the chromatin is dispersed or extended. This makes the DNA available for RNA synthesis. At the time of cell division, chromatin coils, loops, and condenses into a highly compacted form.

Each species has a characteristic chromosome number. For instance, human cells contain 46 chromosomes, corn has 20 chromosomes, and the crayfish has 200! This number is called the **diploid (2n) number** because it contains two (a pair) of each type of chromosome. Humans have 23 pairs of chromosomes. One member of each pair originates from the mother, and the other member originates from the father. In Figure 5.4, the blue chromosomes are inherited from one parent, and the red are inherited from the other.

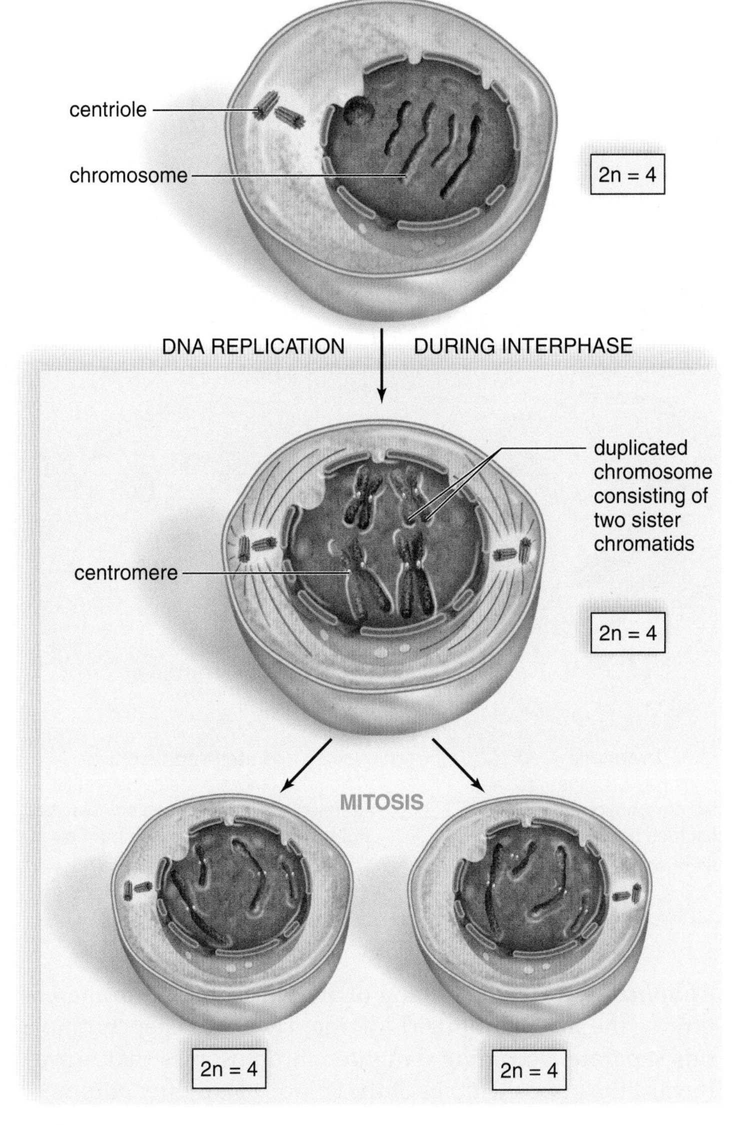

Figure 5.4 Mitosis overview.
Following DNA replication during interphase, each chromosome in the parental nucleus is duplicated and consists of two sister chromatids. During mitosis, the centromeres divide and the sister chromatids separate, becoming daughter chromosomes that move into the daughter nuclei. Therefore, daughter cells have the same number and kinds of chromosomes as the parental cell. (The blue chromosomes were inherited from one parent, and the red chromosomes were inherited from the other.)

Half the diploid number, called the **haploid (n) number** of chromosomes, contains only one of each kind of chromosome. In the life cycle of humans, only sperm and eggs have the haploid number of chromosomes.

Overview of Mitosis

Mitosis is nuclear division in which the chromosome number stays constant. A 2n nucleus divides to produce daughter nuclei that are also 2n (see Fig. 5.4). It would be possible to diagram this as 2n ⟶ 2n. Before nuclear division takes place, DNA replication occurs, duplicating the chromosomes. A duplicated chromosome is composed of two sister chromatids held together in a region called the **centromere. Sister chromatids** are genetically identical—they contain the same DNA sequences. At the completion of mitosis, each chromosome consists of a single chromatid:

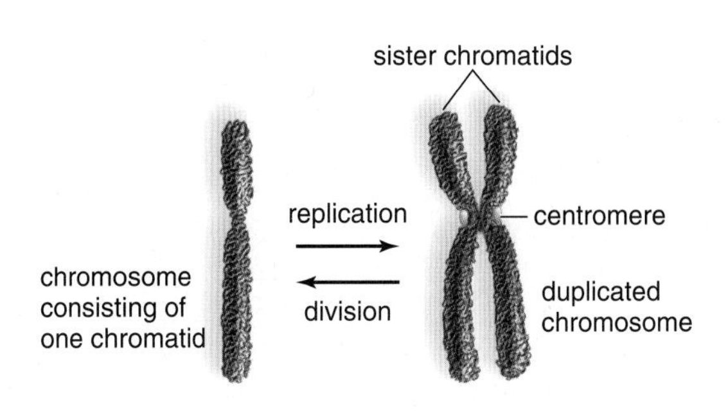

Figure 5.4 gives an overview of mitosis; for simplicity, only four chromosomes are depicted. (*In determining the number of chromosomes, it is only necessary to count the number of independent centromeres.*) During mitosis, the centromeres divide and then the sister chromatids separate, becoming **daughter chromosomes.** Therefore, each daughter nucleus gets a complete set of chromosomes and has the same number of chromosomes as the parental cell. This makes the daughter cells genetically identical to each other and to the parental cell.

Mitosis is the type of nuclear division that occurs when tissues grow or when repair occurs. Following fertilization, the zygote begins to divide mitotically, and mitosis continues during development and the life span of the individual.

Mitosis in Detail

Mitosis is nuclear division that produces *two daughter nuclei, each with the same number and kinds of chromosomes as the parental nucleus.*

During mitosis, a **spindle** brings about an orderly distribution of chromosomes to the daughter cell nuclei. The spindle contains many fibers, each composed of a bundle of microtubules. Microtubules are able to disassemble and assemble. The centrosome, which is the main microtubule organizing center of the cell, divides during late interphase. It is believed that centrosomes are responsible for organizing the spindle. In animal cells each centrosome contains a pair of barrel-shaped organelles called **centrioles** and an **aster,** which is an array of short microtubules that radiate from the centrosome. The fact that plant cells lack centrioles suggests that centrioles are not required for spindle formation.

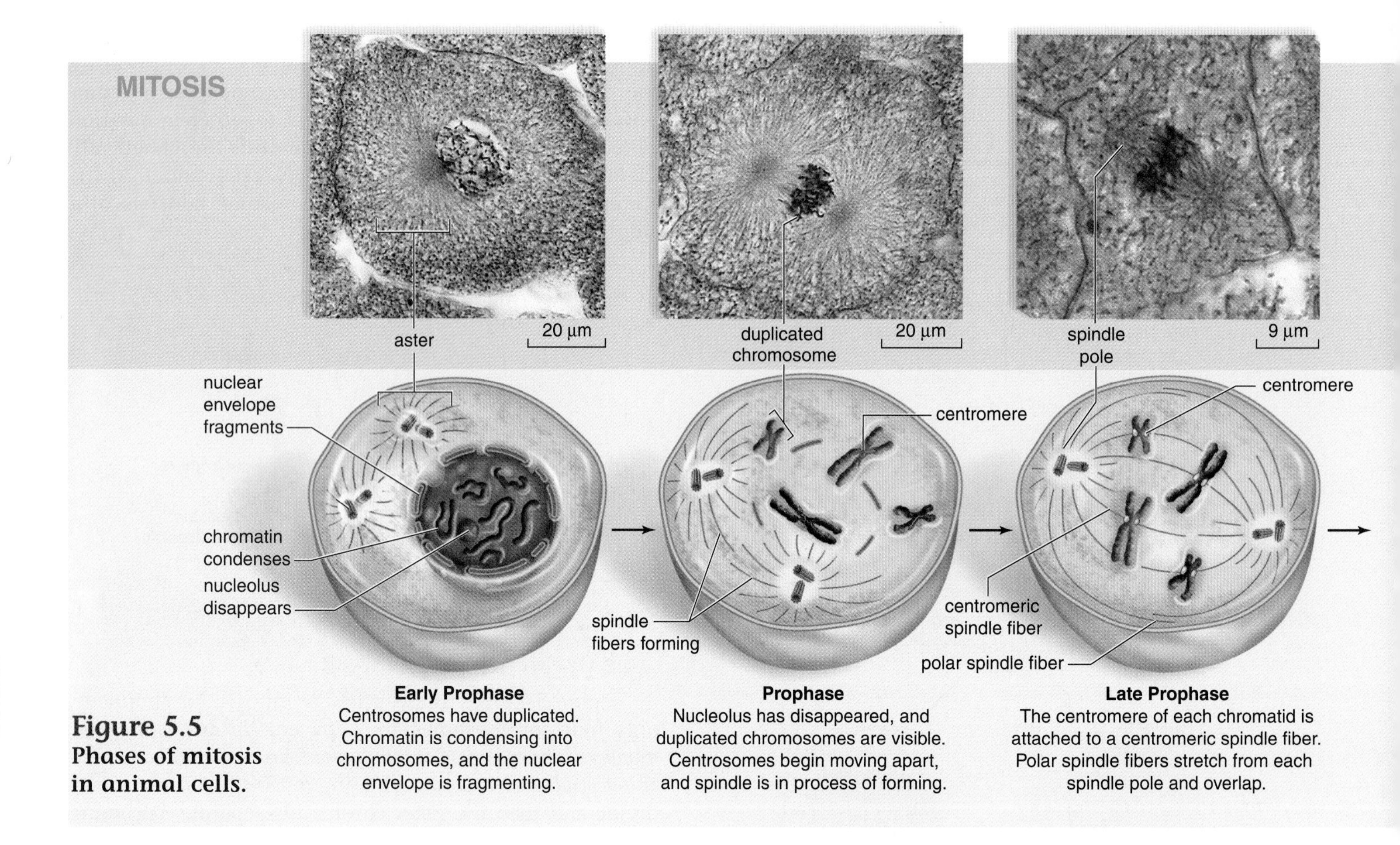

Figure 5.5 Phases of mitosis in animal cells.

Mitosis in Animal Cells

Mitosis is a continuous process that is arbitrarily divided into four phases for convenience of description: prophase, metaphase, anaphase, and telophase (Fig. 5.5).

Prophase It is apparent during early **prophase** that cell division is about to occur. The centrosomes begin moving away from each other toward opposite ends of the nucleus. Spindle fibers appear between the separating centrosomes as the nuclear envelope begins to fragment, and the nucleolus begins to disappear.

The chromatin condenses and the chromosomes are now visible. Each is duplicated and composed of sister chromatids held together at a centromere. The spindle begins forming during late prophase, and the chromosomes become attached to the spindle fibers. Their centromeres attach to fibers called centromeric (or kinetochore) fibers. As yet, the chromosomes have no particular orientation, moving first one way and then the other.

Metaphase By the time of **metaphase,** the fully formed spindle consists of poles, asters, and fibers. The **metaphase plate** is a plane perpendicular to the axis of the spindle and equidistant from the poles. The chromosomes attached to centromeric spindle fibers line up at the metaphase plate during metaphase. Polar spindle fibers reach beyond the metaphase plate and overlap.

Anaphase At the beginning of **anaphase,** the centromeres uniting the sister chromatids divide. Then the sister chromatids separate, becoming daughter chromosomes that move toward the opposite poles of the spindle. Daughter chromosomes have a centromere and a single chromatid.

What accounts for the movement of the daughter chromosomes? First, the centromeric spindle fibers shorten, pulling the daughter chromosomes toward the poles. Second, the polar spindle fibers push the poles apart as they lengthen and slide past one another.

Telophase During **telophase,** the spindle disappears, and nuclear envelope components reassemble around the daughter chromosomes. Each daughter nucleus contains the same number and kinds of chromosomes as the original parental cell. Remnants of the polar spindle fibers are still visible between the two nuclei.

The chromosomes become more diffuse once again, and a nucleolus appears in each daughter nucleus. Cytokinesis is under way, and soon there will be two individual daughter cells, each with a nucleus that contains the diploid number of chromosomes.

Mitosis in Plant Cells

As with animal cells, mitosis in plant cells permits growth and repair. A particular plant tissue called meristematic

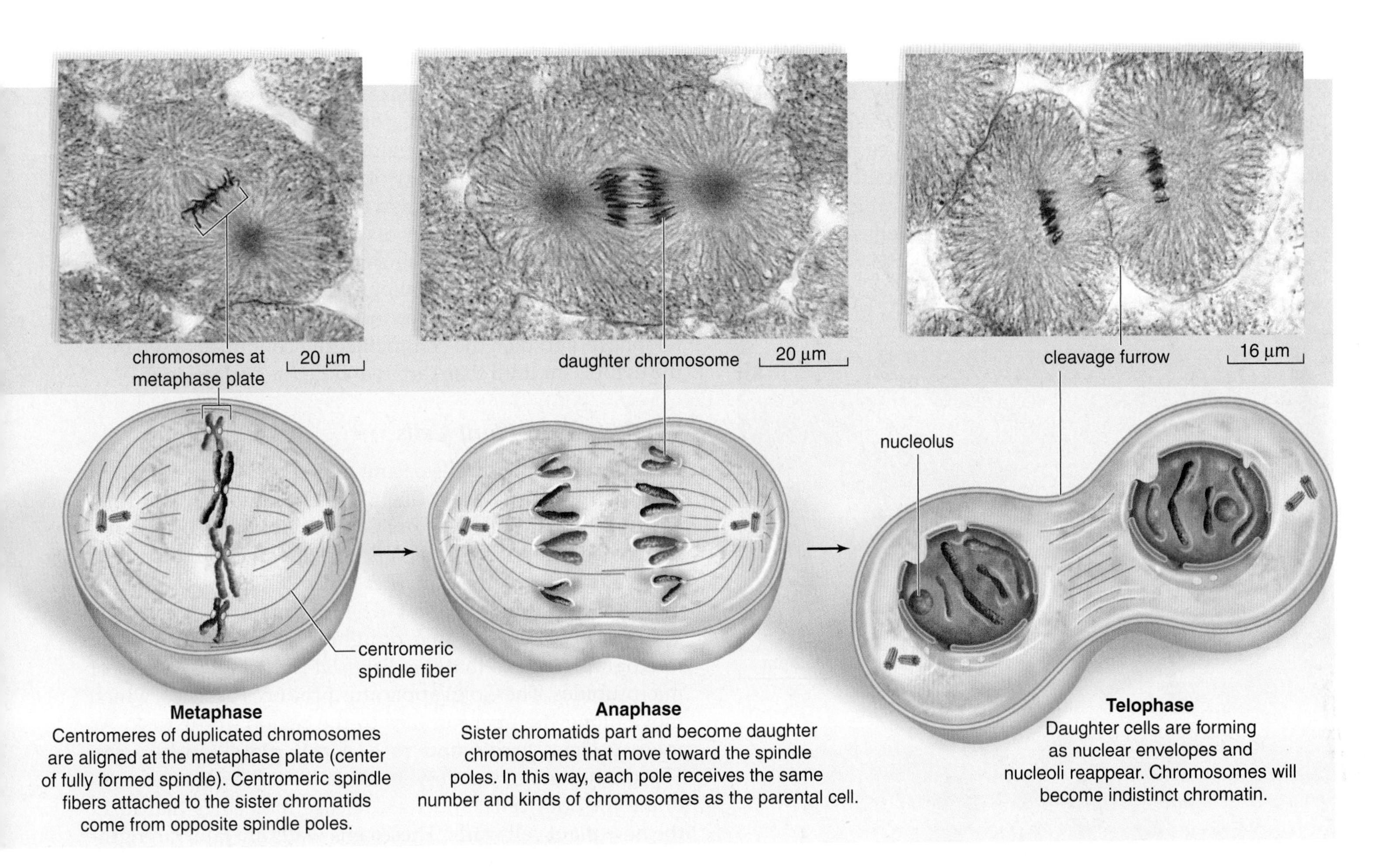

tissue retains the ability to divide throughout the life of a plant. Meristematic tissue is found at the root tip and also at the shoot tip of stems. Lateral meristematic tissue accounts for the ability of trees to increase their girth each growing season.

Figure 5.6 illustrates mitosis in plant cells. Exactly the same phases occur in plant cells as in animal cells. Although plant cells have a centrosome and spindle, there are no centrioles or asters during cell division. The spindle still brings about the distribution of the chromosomes to each daughter cell.

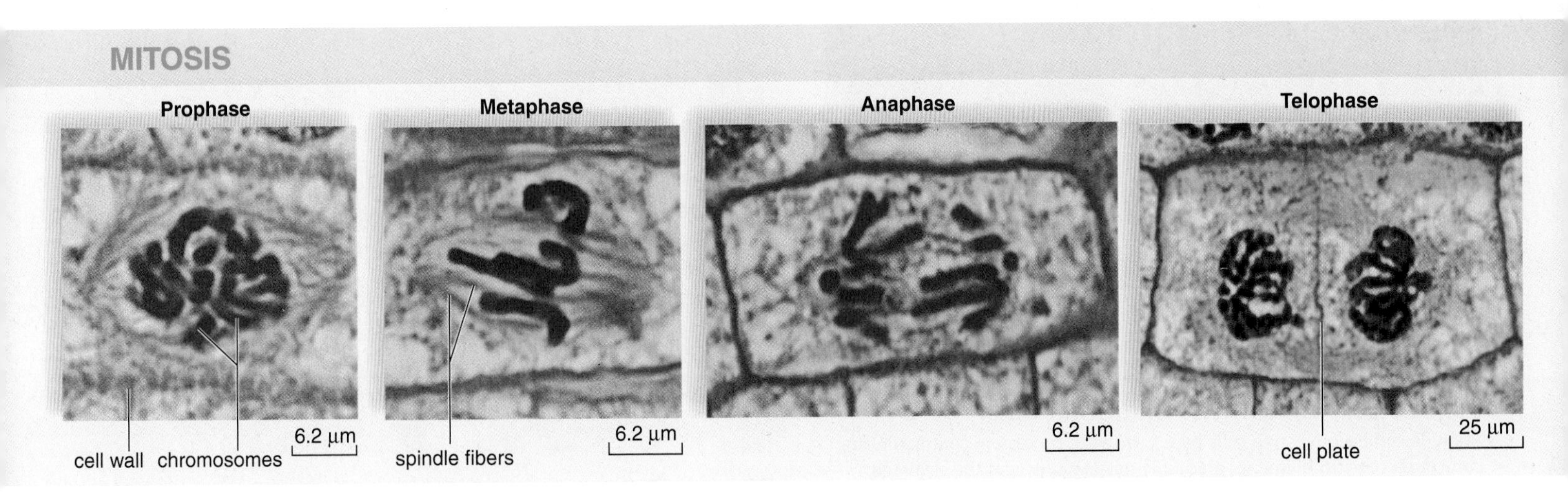

Figure 5.6 Phases of mitosis in plant cells.
Note the presence of the cell wall and the absence of centrioles and asters. Even so, plant cells have a spindle that brings about the movement of the chromosomes and distributes them so that the two daughter nuclei will each have the same number of chromosomes.

Cytokinesis in Animal and Plant Cells

Cytokinesis, or cytoplasmic cleavage, usually accompanies mitosis, but they are separate processes. Division of the cytoplasm begins in anaphase and continues in telophase but does not reach completion until just before the next interphase. By that time, the newly forming cells have received a share of the cytoplasmic organelles that duplicated during the previous interphase.

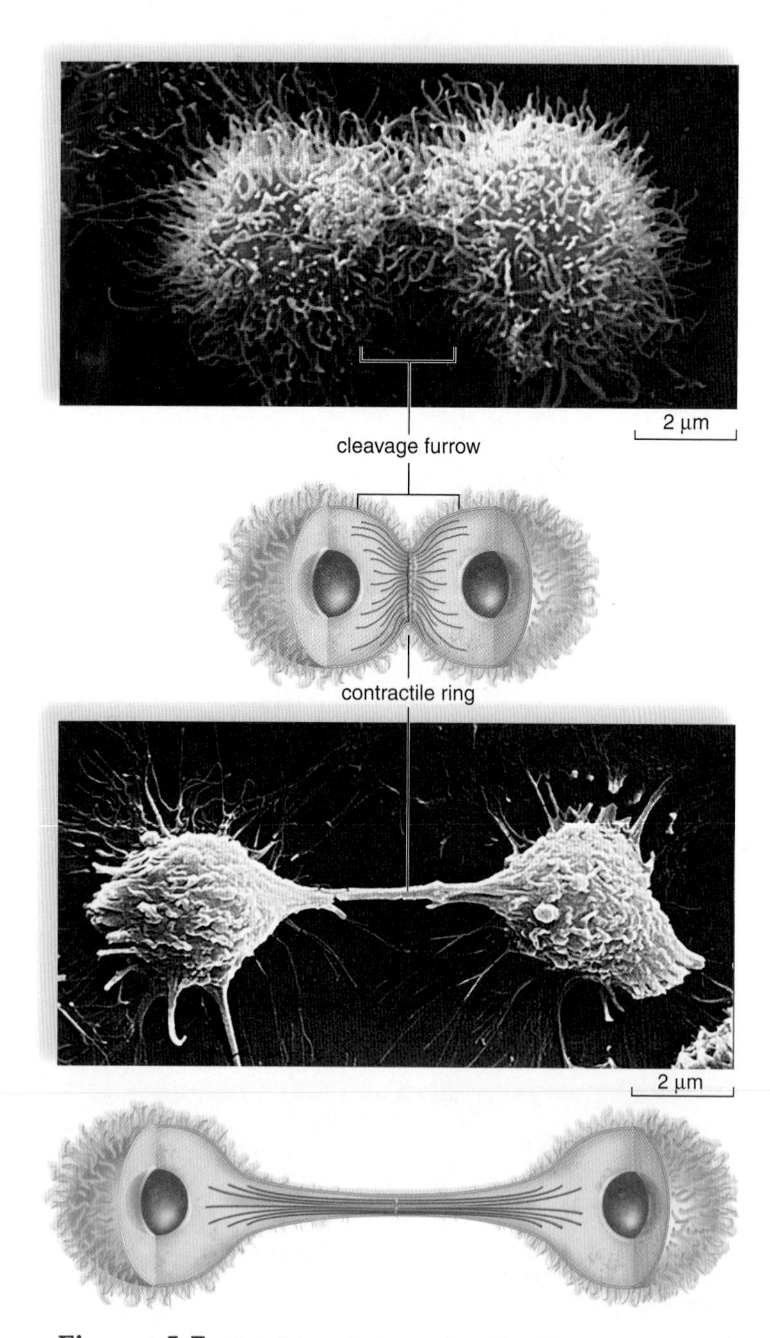

Figure 5.7 Cytokinesis in animal cells.
A single cell becomes two cells by a furrowing process. A contractile ring composed of actin filaments gradually gets smaller, and the cleavage furrow pinches the cell into two cells.

Cytokinesis in Animal Cells

As anaphase draws to a close in animal cells, a **cleavage furrow,** which is an indentation of the membrane between the two daughter nuclei, begins to form. The cleavage furrow deepens when a band of actin filaments, called the contractile ring, slowly forms a constriction between the two daughter cells. The action of the contractile ring can be likened to pulling a drawstring ever tighter about the middle of a balloon, causing the balloon to constrict in the middle.

A narrow bridge between the two cells can be seen during telophase, and then the contractile ring continues to separate the cytoplasm until there are two daughter cells (Fig. 5.7).

Cytokinesis in Plant Cells

Cytokinesis in plant cells occurs by a process different from that seen in animal cells (Fig. 5.8). The rigid cell wall that surrounds plant cells does not permit cytokinesis by furrowing. Instead, cytokinesis in plant cells involves building new cell walls between the daughter cells.

Cytokinesis is apparent when a small, flattened disk appears between the two daughter plant cells. Electron micrographs reveal that the disk is at right angles to a set of microtubules. The Golgi apparatus produces vesicles, which move along the microtubules to the region of the disk. As more vesicles arrive and fuse, a **cell plate** can be seen. The membrane of the vesicles completes the plasma membrane for both cells, and they release molecules that form the new plant cell walls. These cells walls are later strengthened by the addition of cellulose fibrils.

Check Your Progress

1. Before replication, how many chromatids does a chromosome contain? After replication?
2. Draw each of the stages of mitosis beginning with four chromosomes.
3. Compare and contrast mitosis in plant and animal cells.

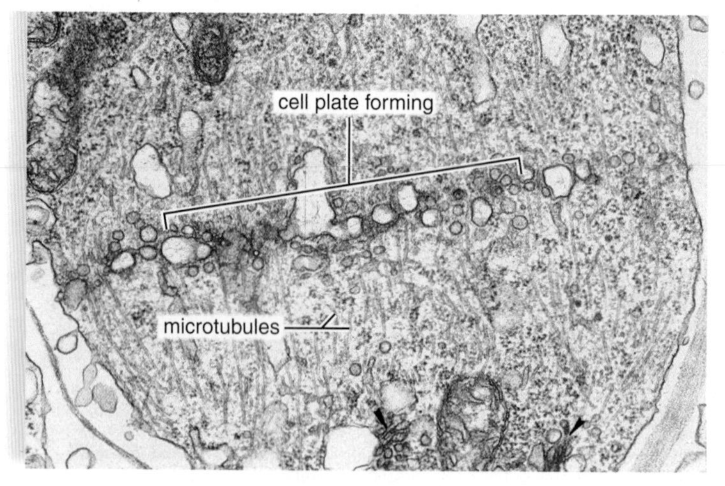

Figure 5.8 Cytokinesis in plant cells.
During cytokinesis in a plant cell, a cell plate forms midway between two daughter nuclei and extends to the plasma membrane.

5.4 Reducing the Chromosome Number

Learning Outcomes

1. Explain the purpose of meiosis.
2. Describe the stages of meiosis.

Meiosis occurs in any life cycle that involves sexual reproduction. **Meiosis** reduces the chromosome number in such a way that the daughter nuclei receive only one of each kind of chromosome. The process of meiosis ensures that the next generation of individuals will have only the diploid number of chromosomes and a combination of traits different from that of either parent.

Overview of Meiosis

At the start of meiosis, the parental cell has the diploid number of chromosomes. In Figure 5.9, the diploid number is 4, which you can verify by counting the number of centromeres. Notice that meiosis requires two cell divisions, meiosis I and meiosis II, and that the four daughter cells have the haploid number of chromosomes. DNA replication occurs prior to meiosis I.

Recall that when a cell is 2n, or diploid, the chromosomes occur in pairs. For example, the 46 chromosomes of humans occur in 23 pairs. In Figure 5.9, there are two pairs of chromosomes. The members of a pair are called **homologous chromosomes** or **homologues.** Homologues are the same size but are indicated by different colors—the blue chromosomes are inherited from one parent, and the red are inherited from the other.

Meiosis I

During meiosis I, the homologous chromosomes come together and line up side by side. This so-called **synapsis** results in an association of four chromatids that stay in close proximity during the first two phases of meiosis I.

Because of synapsis, there are pairs of homologous chromosomes at the metaphase plate during meiosis I. Notice that only during meiosis I is it possible to observe paired chromosomes at the metaphase plate. When the members of these pairs separate, each daughter nucleus receives one member of each pair. Therefore, each daughter cell now has the haploid number of chromosomes, as you can verify by counting its centromeres.

Meiosis II and Fertilization

Replication of DNA does not occur between meiosis I and meiosis II because the chromosomes are still duplicated (composed of two sister chromatids). During meiosis II, the centromeres divide and the sister chromatids separate, becoming daughter chromosomes that are distributed to daughter nuclei. In the end, each of four daughter cells has the haploid number of chromosomes, and each chromosome consists of one chromatid.

In some life cycles, such as that of humans (see Fig. 5.15), the daughter cells mature into **gametes** (sex cells—sperm and egg) that fuse during fertilization. **Fertilization** restores the diploid number of chromosomes in a cell that will develop into a new individual. If the gametes carried the diploid instead of the haploid number of chromosomes, the chromosome number would double with each fertilization.

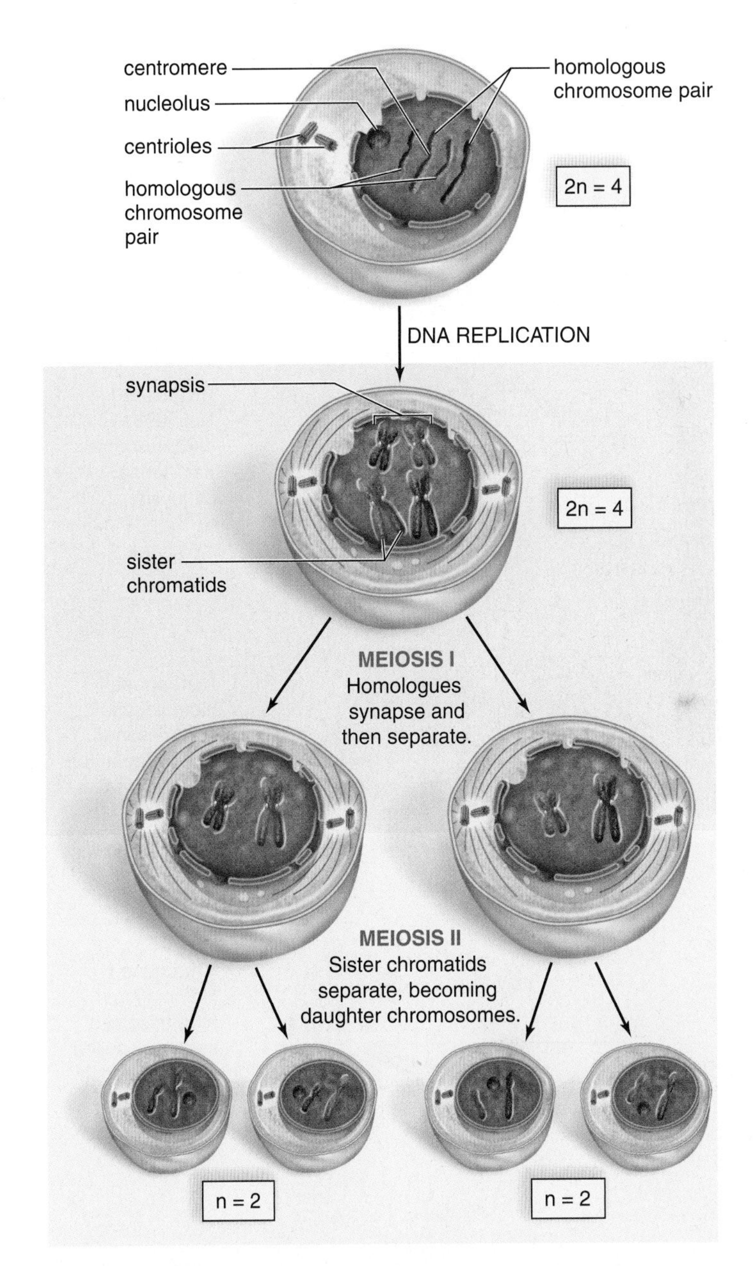

Figure 5.9 Overview of meiosis.
Following DNA replication, each chromosome is duplicated. During meiosis I, the homologous chromosomes pair during synapsis and then separate. During meiosis II, the centromeres divide and the sister chromatids separate, becoming daughter chromosomes that move into the daughter nuclei.

Meiosis in Detail

Meiosis, which requires two nuclear divisions, *results in four daughter nuclei, each having one of each kind of chromosome and therefore half the number of chromosomes as the parental cell.*

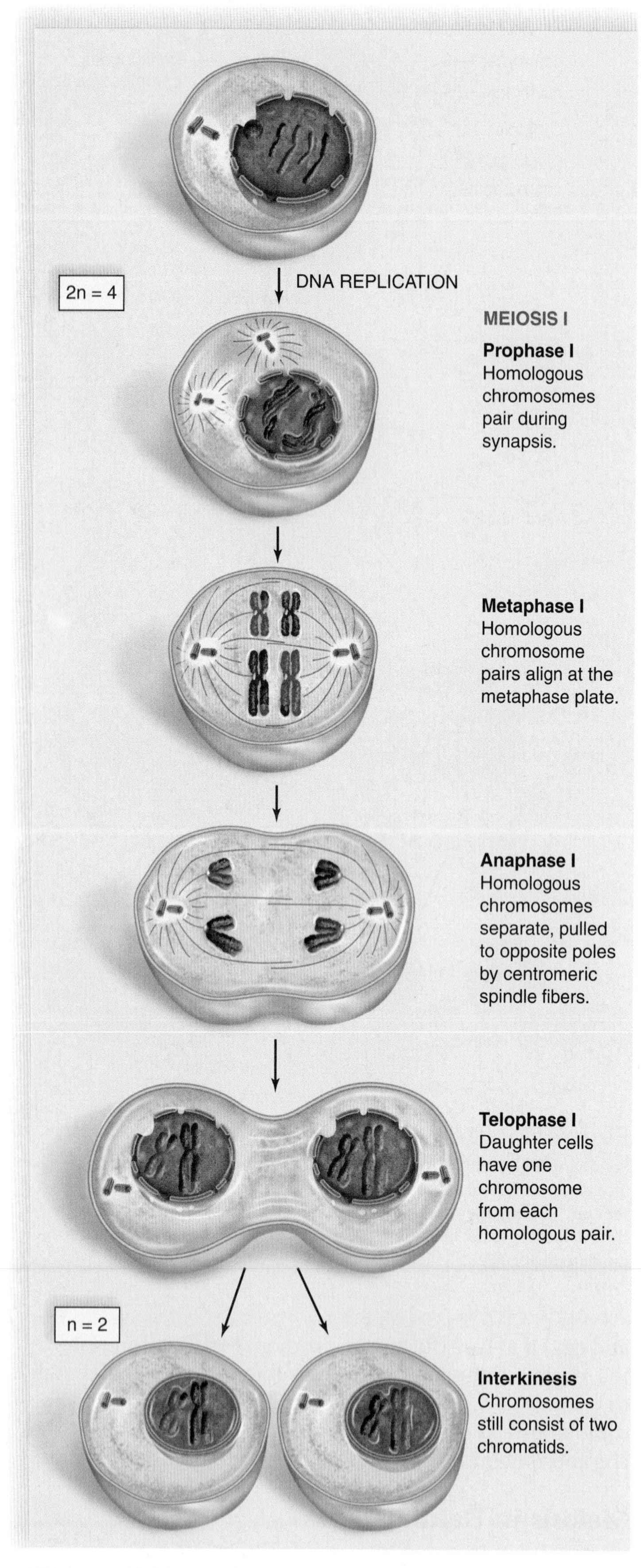

Figure 5.10 Meiosis I in an animal cell.
The exchange of color between nonsister chromatids represents crossing-over. Notice that the daughter cells have only 2 centromeres compared to the parental cell's 4.

First Division

Meiosis I has four phases, called prophase I, metaphase I, anaphase I, and telophase I. The phases of meiosis I for an animal cell are diagrammed in Figure 5.10. Meiosis helps ensure that **genetic variation** of the parental genes occurs through two key events: crossing-over and independent assortment of chromosome pairs. Because the members of a homologous pair can carry slightly different instructions for the same genetic trait, these events are significant. For example, one homologue may carry instructions for brown eyes, while the corresponding homologue may carry instructions for blue eyes. Fertilization ensures the offspring will have different combinations of genes from either parent.

Prophase I In prophase I, synapsis occurs, and then the spindle appears while the nuclear envelope fragments and the nucleolus disappear. Figure 5.11 shows that during synapsis, the homologous chromosomes come together and line up side by side. Now an exchange of genetic material may occur between the *nonsister* chromatids of the homologues. This exchange is called **crossing-over.** Crossing-over means that the chromatids held together by a centromere are no longer identical. As a result of crossing-over, the daughter cells receive chromosomes with recombined genetic material. In Figures 5.10 and 5.11, crossing-over is represented by an exchange of color. Because of crossing-over, first the nonsister chromatids, and then the chromosomes they give rise to, have a different combination of genes than the parental cell.

Metaphase I and Anaphase I During metaphase I, the homologues align at the metaphase plate. Depending on how they align, the maternal or paternal member of each

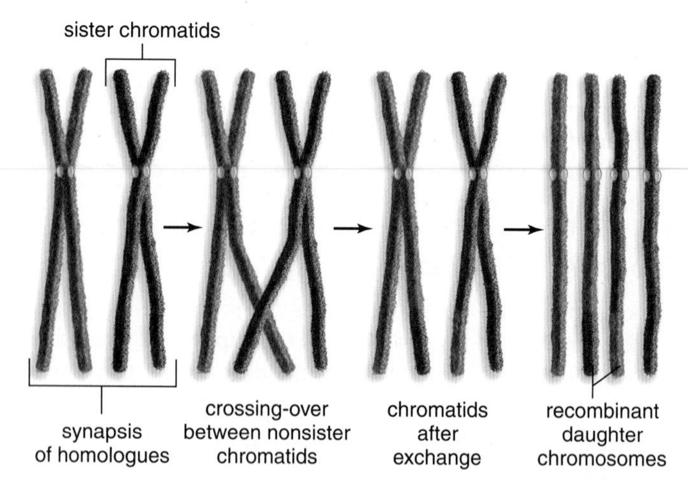

Figure 5.11 Synapsis and crossing-over.
During prophase I, *from left to right,* duplicated homologous chromosomes undergo synapsis when they line up with each other. During crossing-over, nonsister chromatids break and then rejoin. The two resulting daughter chromosomes will have a different combination of genes than they had before.

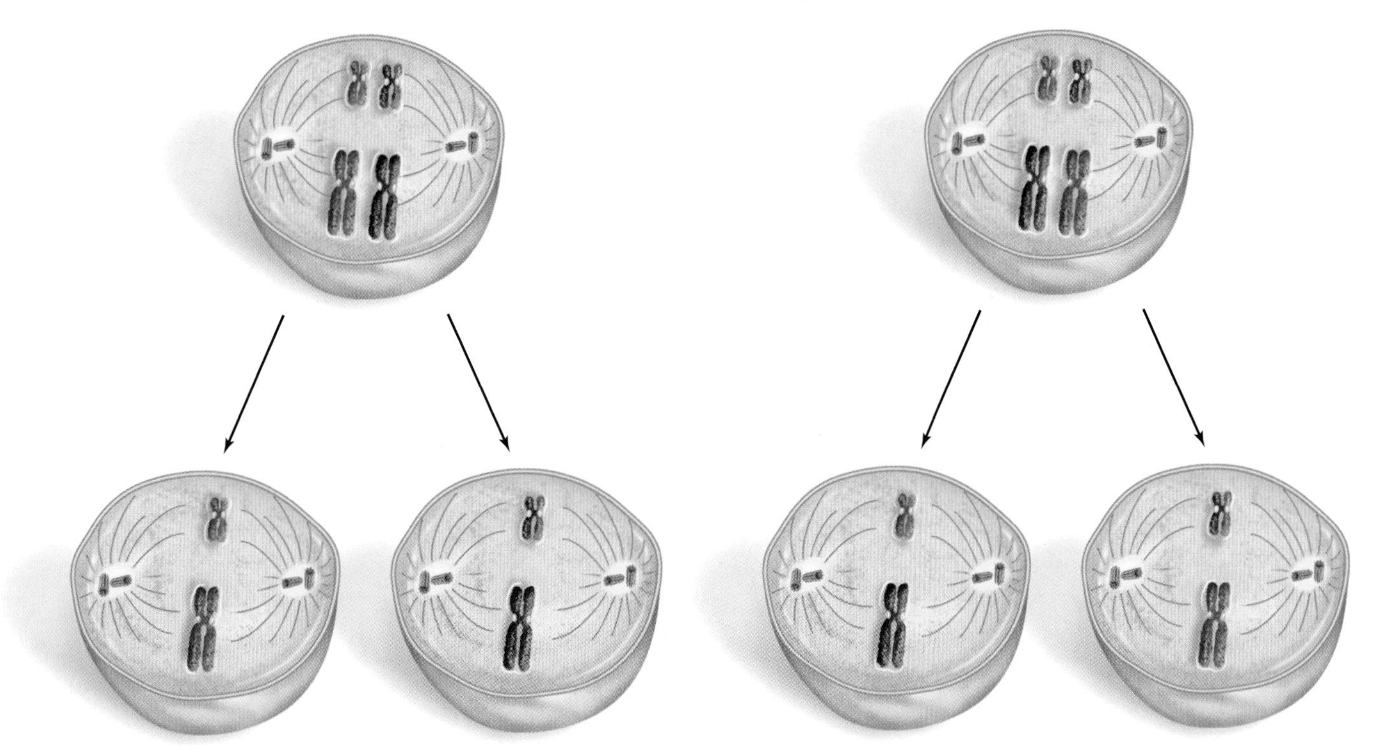

Figure 5.12 Independent assortment.
Two possible orientations of homologous chromosome pairs at the metaphase plate are shown for metaphase I. Each of these will result in daughter nuclei with a different combination of parental chromosomes (independent assortment). In a cell with two pairs of homologous chromosomes, there are 2^2 possible combinations of parental chromosomes in the daughter nuclei.

pair may be oriented toward either pole. **Independent assortment** occurs when these homologous pairs separate from each other during anaphase I, generating cells with different combinations of maternal and paternal chromosomes. Notice that each chromosome still consists of two sister chromatids. Figure 5.12 shows two possible orientations for a cell that contains only two pairs of chromosomes (2n = 4). Gametes from the first cell, but not the other, will have two paternal or two maternal chromosomes. The gametes from the other cell will have different combinations of maternal and paternal chromosomes. When all possible orientations are considered for a cell containing two pairs of chromosomes, the result will be 2^2, or four, possible combinations of maternal and paternal chromosomes in the resulting gametes from this cell. In humans, where there are 23 pairs of chromosomes, the number of possible chromosomal combinations in the gametes is a staggering 2^{23}, or 8,388,608. And this does not even consider the genetic recombinations that are introduced due to crossing-over. Together, prophase I, metaphase I, and anaphase I introduce the genetic variation that helps ensure that no two offspring have the same combination of genes as the parents.

Telophase I In some species, a telophase I phase occurs at the end of meiosis I. If so, the nuclear envelopes re-form, and nucleoli reappear. This phase may or may not be accompanied by cytokinesis, which is separation of the cytoplasm.

Interkinesis The period of time between meiosis I and meiosis II is called **interkinesis.** No replication of DNA occurs during interkinesis. Why is this appropriate? Because the chromosomes are already duplicated.

Second Division

Phases of meiosis II for an animal cell are diagrammed in Figure 5.13. At the beginning of prophase II, a spindle appears while the nuclear envelope disassembles and the nucleolus disappears. Each duplicated chromosome attaches to the spindle, and then they align at the metaphase plate during metaphase II. During anaphase II, sister chromatids separate, becoming daughter chromosomes that move into the daughter nuclei. In telophase II, the spindle disappears as the nuclear envelope re-forms.

During the cytokinesis that can follow meiosis II, the plasma membrane furrows to produce two complete cells, each of which has the haploid number of chromosomes. Because each cell from meiosis I undergoes meiosis II, there are four daughter cells altogether.

The Importance of Meiosis

Meiosis produces haploid cells. One of the hallmarks of meiosis is the possibility of genetic variation: the haploid cells produced are no longer identical to the diploid parent cell from which they came.

Variation occurs in two ways. First, during prophase I, crossing-over between nonsister chromatids of the paired homologous chromosomes rearranges genes, with the result that the sister chromatids of each homologue may no longer be identical. Second, the independent assortment of chromosomes during anaphase I means that the gametes produced by meiosis II may have different combinations of chromosomes than the parental cell.

Upon fertilization, the combining of chromosomes from genetically different gametes, even if the gametes are from a single individual (as in many plants), helps ensure

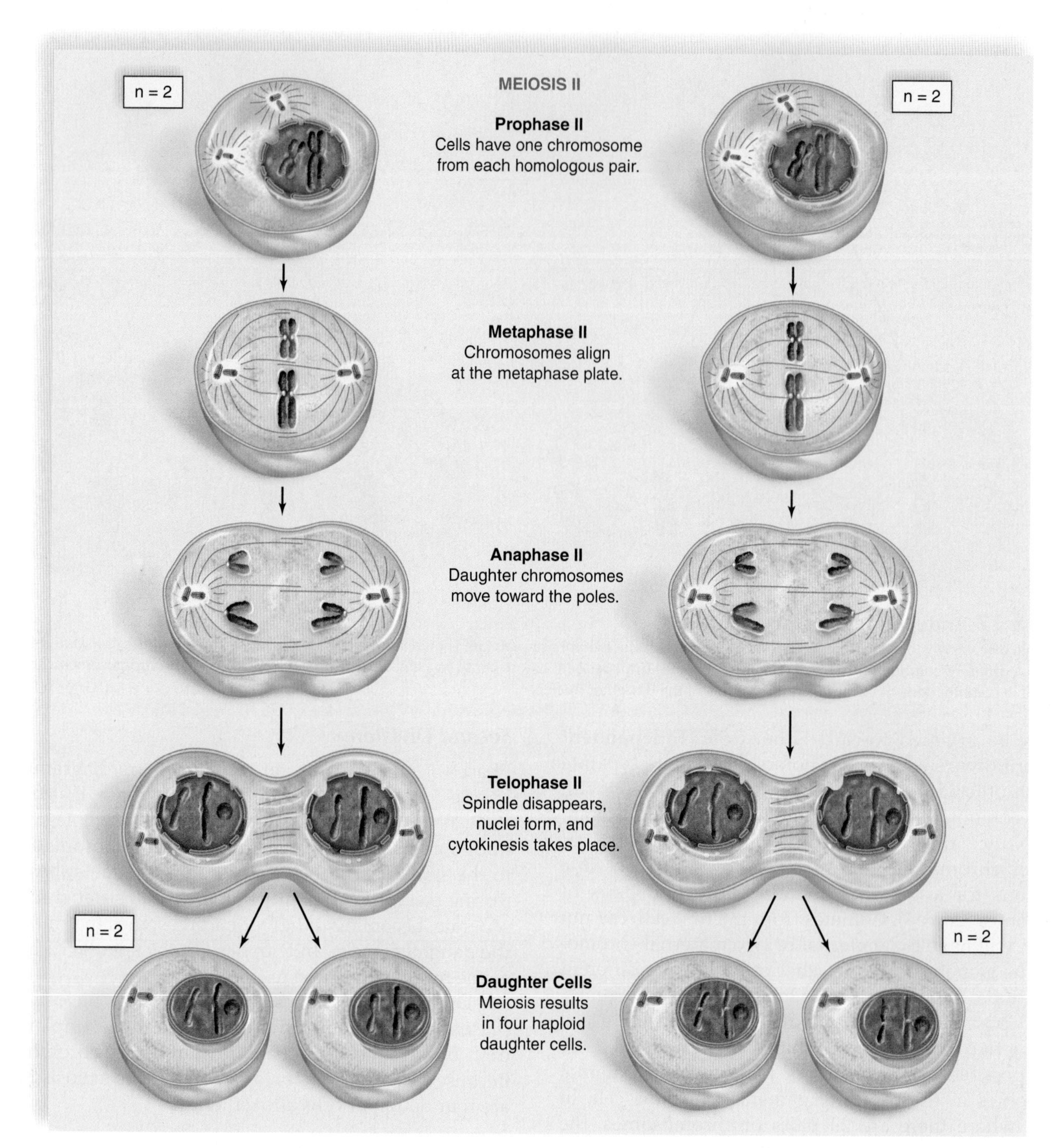

Figure 5.13 Meiosis II in an animal cell.
During meiosis II, sister chromatids separate, becoming daughter chromosomes that are distributed to the daughter nuclei. Following meiosis II, there are four haploid daughter cells. Comparing the number of centromeres in each daughter cell with the number in the parental cell at the start of meiosis I (Fig. 5.10) verifies that each daughter cell is haploid.

that offspring are not identical to their parents. This genetic variability is the main advantage of sexual reproduction.

The staggering amount of genetic variation achieved through meiosis is particularly important to the long-term survival of a species because it increases genetic variation within a population. If the environment changes, genetic variability among offspring introduced by sexual reproduction may be advantageous. Under the new conditions, some offspring may have a better chance of survival and reproductive success than others in a population.

Check Your Progress

1. Why are there two divisions in meiosis?
2. Why isn't DNA replicated between meiosis I and meiosis II?
3. Draw meiosis I and II starting with four chromosomes (two pairs of homologous chromosomes).
4. How do crossing-over and independent assortment introduce genetic variation?

5.5 Comparison of Meiosis with Mitosis

LEARNING OUTCOME

1. Compare and contrast meiosis and mitosis.

Figure 5.14 compares meiosis to mitosis. Notice the following differences:

- DNA replication takes place only once prior to either meiosis or mitosis. However, meiosis requires two nuclear divisions, while mitosis requires only one nuclear division.
- Meiosis produces four daughter nuclei, and following cytokinesis, there are four daughter cells. Mitosis followed by cytokinesis results in two daughter cells.
- The four daughter cells following meiosis are haploid and have half the chromosome number as the parental cell. The daughter cells following mitosis have the same chromosome number as the parental cell.
- The daughter cells resulting from meiosis are not genetically identical to each other or to the parental cell. The

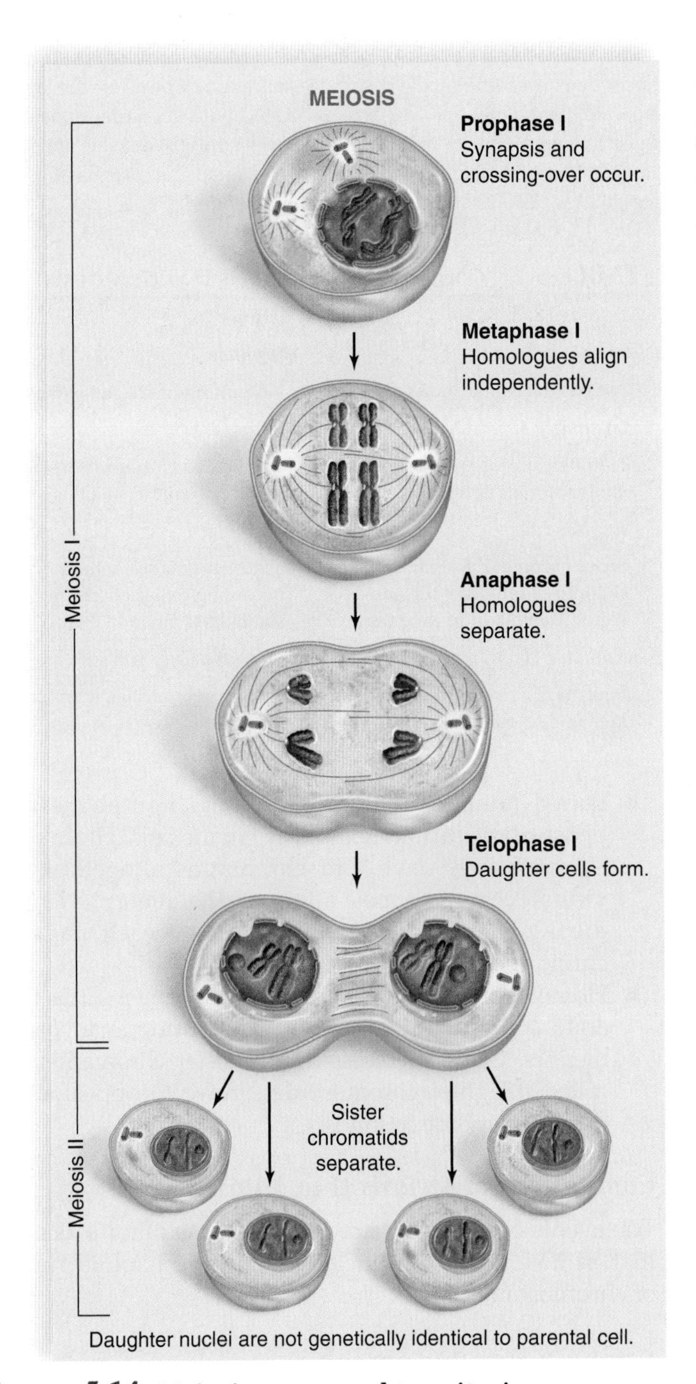

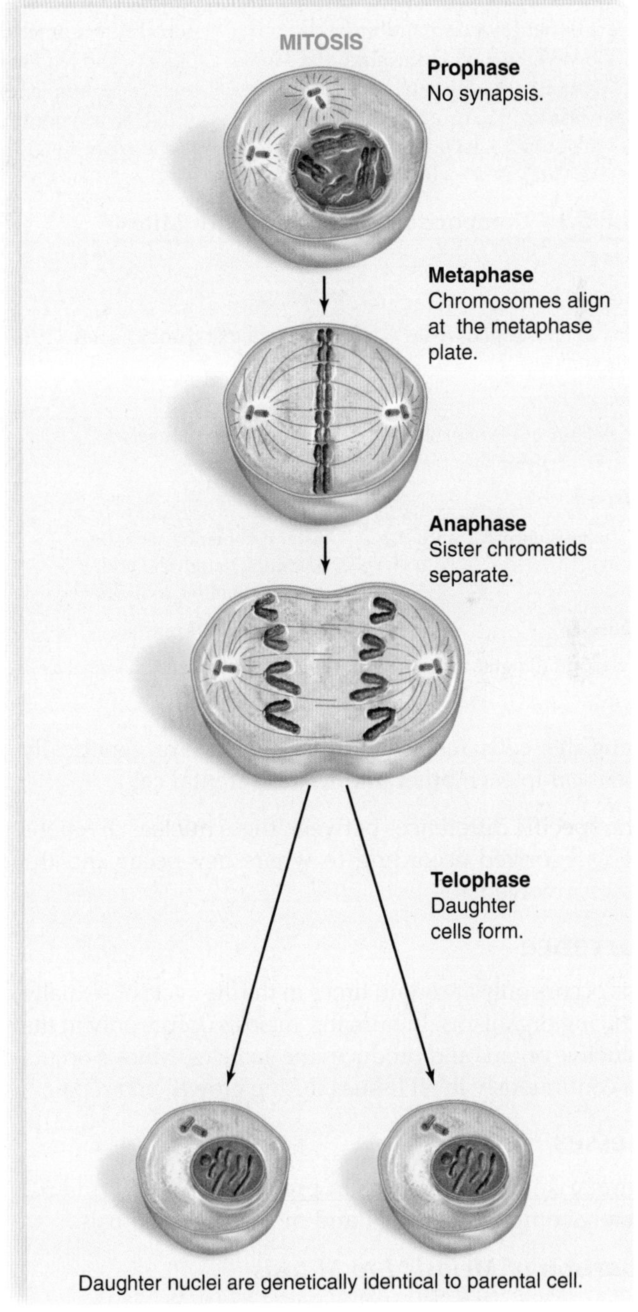

Figure 5.14 Meiosis compared to mitosis.
Compare metaphase I of meiosis to metaphase of mitosis. In meiosis metaphase I, homologues are paired at the metaphase plate, but in mitosis metaphase, all chromosomes align individually at the metaphase plate. Individual homologues separate during meiosis anaphase I, so that the daughter cells are haploid. (Blue chromosomes are from one parent and red from the other; exchange of color represents crossing-over.)

Bioethical Focus

Delaying Childbirth

More and more women are delaying childbirth until their thirties and forties. From 1990 to 2006, the number of births to women aged 35–39 increased 57%. In the same time period, the number of births to women aged 40–44 increased 70%. Why such an increase in older mothers? Some women are focusing on their education and getting established in a career. Other women are marrying later or experiencing divorce and remarriage. Also, the availability of birth control methods has made it easier to delay childbirth.

However, delaying the start of a family is not without risks. For some women, increasing age results in reduced fertility, and others may not be able to get pregnant at all. Women in their forties have one-and-a-half times the miscarriage rate of mothers in their twenties. Older women tend to experience more health complications, such as high blood pressure and diabetes, during pregnancy as well. But there are also risks to the baby, including low birth weight or chromosomal abnormalities.

The risk of a chromosomal abnormality is higher for the child of an older mother because all of a female's eggs are arrested in prophase I when she is a fetus. After 40 years, the chromosomes in those eggs become "sticky" and tend to not separate during meiosis. For example, the risk of having a child with Down syndrome, caused by three copies of chromosome 21, is approximately 1 in 100 for a 40-year-old mother compared to 1 in 1,250 for a 20-year-old. There is also evidence that the father's age may have a role in Down syndrome. Other abnormalities may increase with increasing paternal or maternal age.

Form Your Own Opinion

1. Do the advantages of delaying childbirth outweigh the disadvantages?
2. If delaying pregnancy means increased health and fertility problems, should insurance companies pay for the extra costs associated with such pregnancies?
3. Should doctors strongly discourage couples from trying to get pregnant after they reach a certain age?

TABLE 5.1 Comparison of Meiosis I with Mitosis

Meiosis I	Mitosis
Prophase I	*Prophase*
Pairing of homologous chromosomes	No pairing of chromosomes
Metaphase I	*Metaphase*
Homologous duplicated chromosomes at metaphase plate	Duplicated chromosomes at metaphase plate
Anaphase I	*Anaphase*
Homologous chromosomes separate.	Sister chromatids separate, becoming daughter chromosomes that move to the poles.
Telophase I	*Telophase/Cytokinesis*
Two haploid daughter cells	Two daughter cells, identical to the parental cell

TABLE 5.2 Comparison of Meiosis II with Mitosis

Meiosis II	Mitosis
Prophase II	*Prophase*
No pairing of chromosomes	No pairing of chromosomes
Metaphase II	*Metaphase*
Haploid number of duplicated chromosomes at metaphase plate	Duplicated chromosomes at metaphase plate
Anaphase II	*Anaphase*
Sister chromatids separate, becoming daughter chromosomes that move to the poles.	Sister chromatids separate, becoming daughter chromosomes that move to the poles.
Telophase II	*Telophase/Cytokinesis*
Four haploid daughter cells, not identical to parental cell	Two daughter cells, identical to the parental cell

daughter cells resulting from mitosis are genetically identical to each other and to the parental cell.

The specific differences between these nuclear divisions can be categorized according to when they occur and the processes involved.

Occurrence

Meiosis occurs only at certain times in the life cycle of sexually reproducing organisms. In humans, meiosis occurs only in the reproductive organs and produces the gametes. Mitosis occurs almost continuously in all tissues during growth and repair.

Processes

To summarize the differences in process, Tables 5.1 and 5.2 separately compare meiosis I and meiosis II to mitosis.

Comparison of Meiosis I to Mitosis

Notice that these events distinguish meiosis I from mitosis:

- Homologous chromosomes pair and undergo crossing-over during prophase I of meiosis, but not during mitosis.
- Paired homologous chromosomes align at the metaphase plate during metaphase I in meiosis. These paired chromosomes have four chromatids altogether. Individual chromosomes align at the metaphase plate during metaphase in mitosis. They each have two chromatids.
- Homologous chromosomes (with centromeres intact) separate and move to opposite poles during anaphase I in meiosis. Centromeres split, and sister chromatids, now called daughter chromosomes, move to opposite poles during anaphase in mitosis.

Comparison of Meiosis II to Mitosis

The events of meiosis II are just like those of mitosis except that in meiosis II, the nuclei contain the haploid number of chromosomes.

Check Your Progress

1. In what ways are meiosis I and mitosis different?
2. In what ways are meiosis II and mitosis similar?

5.6 The Human Life Cycle

LEARNING OUTCOMES

1. Describe the human life cycle in terms of haploid and diploid cells.
2. Explain the process of gamete production in both males and females.

The human life cycle requires both meiosis and mitosis (Fig. 5.15). A haploid sperm (n) and a haploid egg (n) join at fertilization, and the resulting **zygote** has the full, or diploid (2n), number of chromosomes. During development of the fetus before birth, mitosis keeps the chromosome number constant in all the cells of the body. After birth, mitosis is involved in the continued growth of the child and repair of tissues at any time. As a result of mitosis, each somatic cell in the body has the same number of chromosomes.

Spermatogenesis and Oogenesis in Humans

In human males, meiosis is a part of **spermatogenesis,** which occurs in the testes and produces sperm. In human females, meiosis is a part of **oogenesis,** which occurs in the ovaries and produces eggs (Fig. 5.16). Further description of the human reproductive system can be found in Chapter 21.

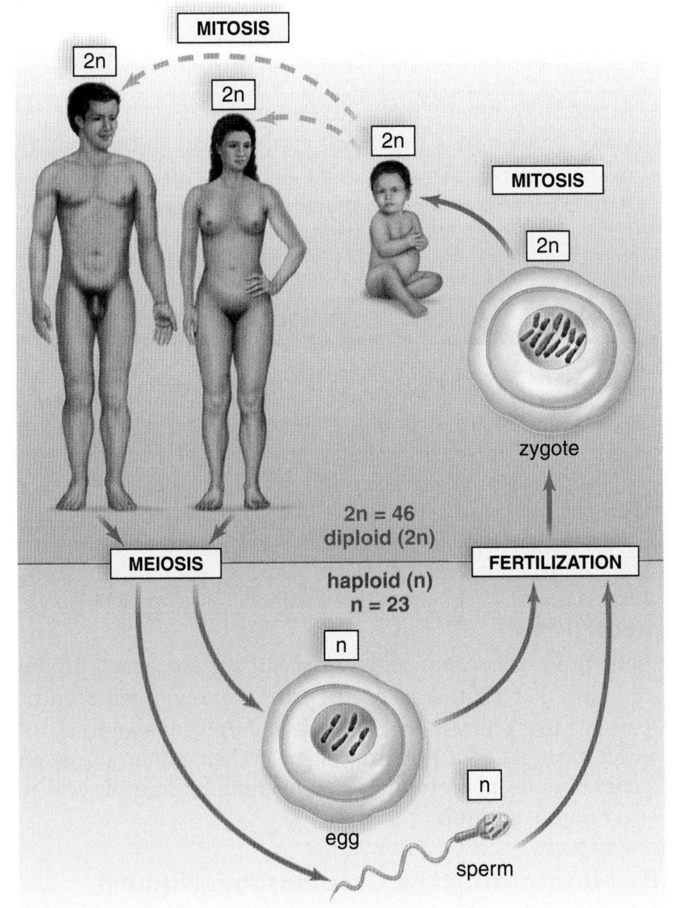

Figure 5.15 Life cycle of humans.
Meiosis in males is a part of sperm production, and meiosis in females is a part of egg production. When a haploid sperm fertilizes a haploid egg, the zygote is diploid. The zygote undergoes mitosis as it develops into a newborn child. Mitosis continues throughout life during growth and repair.

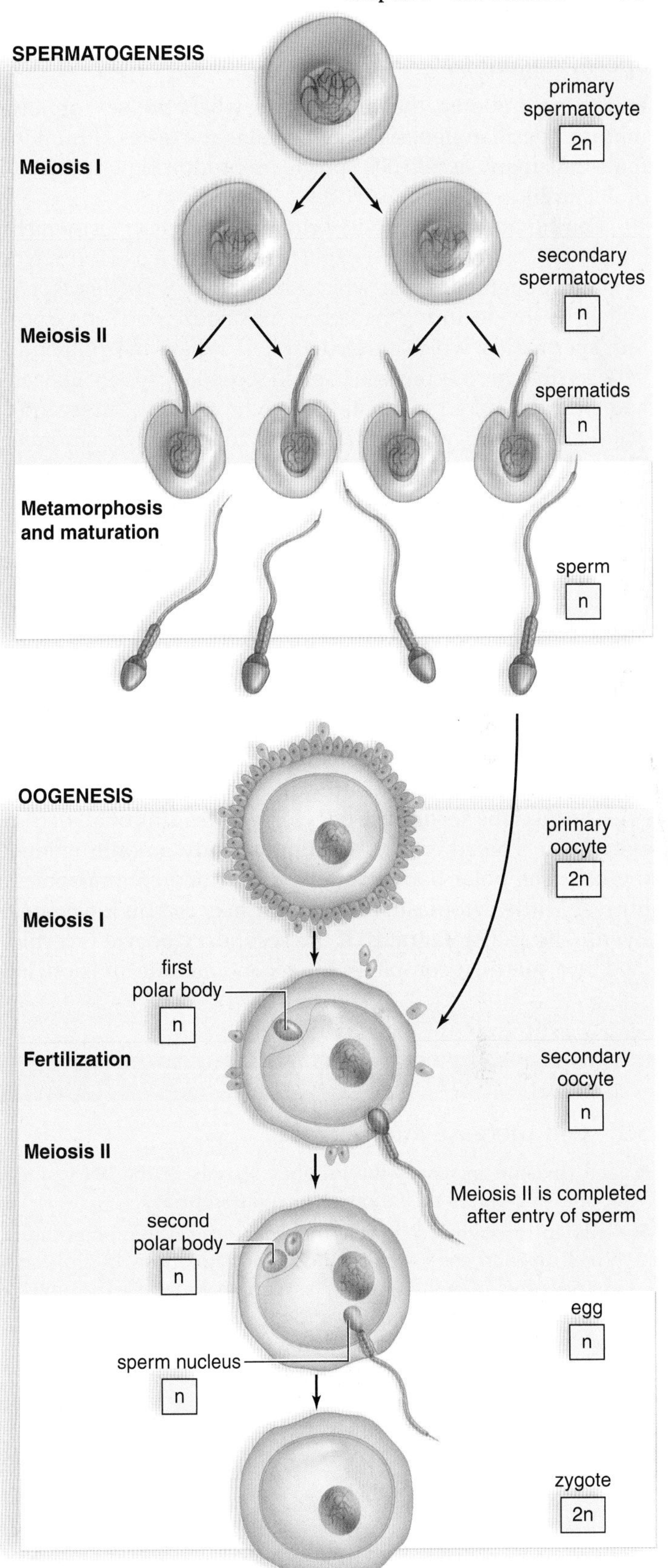

Figure 5.16 Spermatogenesis and oogenesis in mammals.
Spermatogenesis produces four viable sperm, whereas oogenesis produces one egg and at least two polar bodies. In humans, both the sperm and the egg have 23 chromosomes each. Therefore, following fertilization, the zygote has 46 chromosomes.

Spermatogenesis

Following puberty, the time of life when the sex organs mature, spermatogenesis is continual in the testes of human males. As many as 300,000 sperm are produced per minute, or 400 million per day.

During spermatogenesis, primary spermatocytes, which are diploid, divide during the first meiotic division to form two **secondary spermatocytes,** which are haploid. Secondary spermatocytes divide during the second meiotic division to produce four **spermatids,** which are also haploid. What's the difference between the chromosomes in haploid secondary spermatocytes and those in haploid spermatids? The chromosomes in secondary spermatocytes are duplicated and consist of two chromatids, while those in spermatids consist of only one chromatid. Spermatids then mature into **sperm (spermatozoa).**

Oogenesis

Meiosis in human females begins in the fetus. In the ovaries of the fetus, all of the primary oocytes (diploid cells) begin meiosis but become arrested in prophase I. Following puberty and the initiation of the menstrual cycle, one primary oocyte begins to complete meiosis. It finishes the first meiotic division as two cells, each of which is haploid, although the chromosomes are still duplicated. One of these cells, termed the **secondary oocyte,** receives almost all of the cytoplasm. The other is the first **polar body,** a nonfunctioning cell. The polar body contains duplicated chromosomes but very little cytoplasm and may or may not divide again. Eventually it disintegrates. If the secondary oocyte is fertilized by a sperm, it completes the second meiotic division, in which it again divides unequally, forming an **egg** and a second polar body. The chromosomes of the egg and sperm nuclei then join to form the 2n zygote. If the secondary oocyte is not fertilized by a sperm, it disintegrates and passes out of the body with the menstrual flow.

Check Your Progress

1. What cells in the human body are haploid? What role do these cells play in the life cycle?
2. How many sperm are produced from one primary spermatocyte? How many eggs from one oocyte?

Applying the Concepts Revisited

Approximately 10–15% of the women who are diagnosed with breast cancer have a hereditary form of the disease. This means that they inherited a genetic mutation that increases their risk of developing cancer. A genetic mutation does not guarantee that they will develop cancer nor does it determine when or where they may develop cancer, if they do. Many of these mutations occur in proto-oncogenes or tumor suppressor genes. In Christina Applegate's case, she inherited a mutated *BRCA1* gene. The *BRCA1* gene is a tumor suppressor gene whose protein product is involved in DNA repair.

1. How might a mutated *BRCA1* gene increase the risk for cancer?
2. Experts recommend that you test for your *BRCA1/BRCA2* gene status if you have a significant family history of breast and/or ovarian cancer. Why would family history be important in determining whether or not you need the test?

Summarizing the Concepts

5.1 Cell Increase and Decrease

- Cell division increases the number of cells in the body, and apoptosis reduces this number when appropriate.
- Cells go through a cell cycle that includes (1) interphase and (2) cell division, consisting of mitosis and cytokinesis. Interphase, in turn, includes G_1 (growth as certain organelles double), S (DNA synthesis), and G_2 (growth as the cell prepares to divide). Cell division occurs during the mitotic stage (M) when daughter cells receive a full complement of chromosomes.

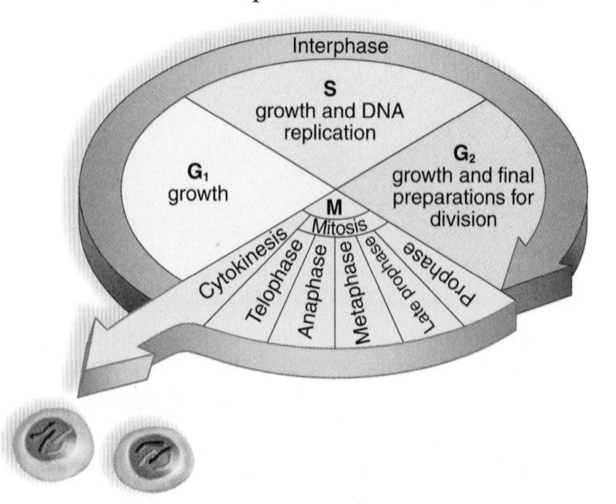

5.2 Control of the Cell Cycle and Cancer

- Internal and external signals control the cell cycle, which can stop at any of three checkpoints: in G_1 prior to the S stage, in G_2 prior to the M stage, and near the end of mitosis. DNA damage is one reason the cell cycle stops. The p53 protein is active at the G_1 checkpoint, and if DNA is damaged and can't be repaired, this protein initiates apoptosis. During apoptosis, enzymes called caspases bring about destruction of the nucleus and the rest of the cell.
- Both proto-oncogenes and tumor suppressor genes produce proteins that control the cell cycle. Proto-oncogenes encode proteins that promote the cell cycle and prevent apoptosis. *RAS* is an example of a proto-oncogene. When tumor suppressor genes mutate, their products no longer inhibit the cell cycle. *p53* is a tumor suppressor.

5.3 Maintaining the Chromosome Number

- Each species has a characteristic number of chromosomes. The total number is the diploid number, and half this number is the haploid number. Among eukaryotes, cell division involves nuclear division and division of the cytoplasm (cytokinesis).

- Replication of DNA precedes cell division. The duplicated chromosome is composed of two sister chromatids held together at a centromere. During mitosis, the centromeres divide, and daughter chromosomes go into each new nucleus.

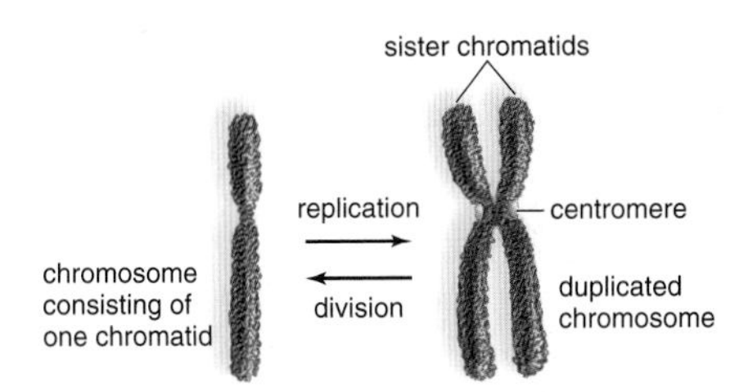

- Mitosis has the following phases: prophase, which includes early prophase, when chromosomes have no particular arrangement, and late prophase, when the chromosomes are attached to spindle fibers; metaphase, when the chromosomes are aligned at the metaphase plate; anaphase, when the chromatids separate, becoming daughter chromosomes that move toward the poles; and telophase, when new nuclear envelopes form around the daughter chromosomes and cytokinesis is well under way.
- Cytokinesis differs between plant and animal cells. Plant cells form a cell plate between the daughter cells, while animal cells develop a cleavage furrow that separates the daughter cells.

5.4 Reducing the Chromosome Number

- Meiosis occurs in any life cycle that involves sexual reproduction. The end result of meiosis is daughter cells with the haploid number of homologous chromosomes. In some life cycles, the daughter cells become gametes, and upon fertilization, the offspring have the diploid number of chromosomes, the same as their parents.
- Crossing-over and independent assortment of chromosomes during meiosis I ensure genetic variation in daughter cells.
- Meiosis utilizes two nuclear divisions. During meiosis I, homologous chromosomes undergo synapsis, and crossing-over between nonsister chromatids occurs. When the homologous chromosomes separate during meiosis I, each daughter nucleus receives one member from each pair of chromosomes. Therefore, the daughter cells are haploid.
- Distribution of daughter chromosomes derived from sister chromatids during meiosis II then leads to a total of four new cells, each with the haploid number of chromosomes.

5.5 Comparison of Meiosis with Mitosis

- Figure 5.14 contrasts the phases of mitosis with the phases of meiosis.

5.6 The Human Life Cycle

- The human life cycle involves both mitosis and meiosis. Mitosis ensures that each somatic cell has the diploid number of chromosomes.
- Meiosis is a part of spermatogenesis and oogenesis. Spermatogenesis in males produces four viable sperm, while oogenesis in females produces one egg and two polar bodies. Oogenesis does not go on to completion unless a sperm fertilizes the secondary oocyte.

TESTING YOURSELF

Choose the best answer for each question.

1. DNA synthesis occurs during
   a. interphase.
   b. prophase.
   c. metaphase.
   d. telophase.
2. Mitosis involves
   a. gametes.
   b. somatic cells.
   c. Both a and b are correct.

In questions 3–5, match the part of the diagram of the cell cycle to the statements provided.

3. DNA replicates.
4. Organelles double.
5. Cytokinesis occurs.

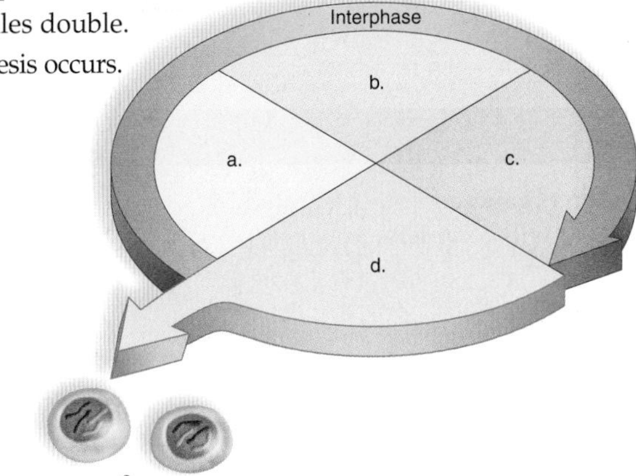

6. Which is not one of the stages of mitosis?
   a. anaphase
   b. telophase
   c. interphase
   d. metaphase
7. Where is the checkpoint for DNA damage?
   a. G_1
   b. S
   c. G_2
   d. M

In questions 8–12, match the part of the diagram that follows to the appropriate statement.

8. Meiosis II
9. Primary spermatocyte
10. Sperm
11. Secondary spermatocyte
12. Spermatids

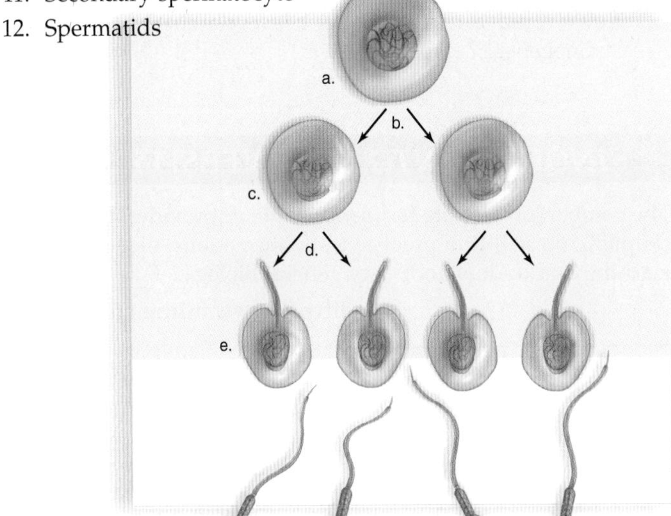

13. Programmed cell death is called
    a. mitosis.
    b. meiosis.
    c. cytokinesis.
    d. apoptosis.
14. Which helps to ensure that genetic diversity is maintained?
    a. independent assortment
    b. crossing-over
    c. genetic recombination
    d. All of these are correct.
15. When mutated, a proto-oncogene
    a. becomes an oncogene.
    b. prevents the cell from going through the cell cycle.
    c. pushes the cell through the cell cycle.
    d. causes apoptosis.
    e. Both a and c are correct.

For questions 16–17, match the descriptions that follow to the terms in the key. Answers may be used more than once.

Key:

a. centriole
b. microtubule
c. chromosome
d. centromere

16. Point of attachment for sister chromatids
17. Found at a pole in the center of an aster
18. If a parental cell has 14 chromosomes prior to mitosis, how many chromosomes will the daughter cells have?
    a. 28
    b. 14
    c. 7
19. Crossing-over occurs
    a. between sister chromatids of the same chromosome.
    b. between chromatids of homologous chromosomes.
    c. during mitosis.
    d. during meiosis II.
20. Plant cells
    a. do not have centrioles.
    b. have a cleavage furrow like animal cells do.
    c. form a cell plate by fusion of several vesicles.
    d. do not have the four phases of mitosis: prophase, metaphase, anaphase, and telophase.
    e. Both a and c are correct.

UNDERSTANDING THE TERMS

anaphase 86
apoptosis 82
aster 85
cell cycle 82
cell plate 88
centriole 85
centromere 85
cleavage furrow 88
crossing-over 90
cyclin 83
cytokinesis 83
daughter chromosome 85
diploid (2n) number 84
egg 96
fertilization 89
gamete 89
genetic variation 90
haploid (n) number 85
homologous chromosome (homologue) 89
independent assortment 91
interkinesis 91
interphase 82
meiosis 89
metaphase 86
metaphase plate 86
mitosis 83
oncogenes 84
oogenesis 95
polar body 96
prophase 86
proto-oncogenes 84
secondary oocyte 96
secondary spermatocyte 96
sister chromatid 85
somatic cell 82
sperm (spermatozoa) 96
spermatid 96
spermatogenesis 95
spindle 85
synapsis 89
telophase 86
tumor suppressor genes 84
zygote 95

THINKING CRITICALLY

1. Why is crossing-over much more likely to occur during meiosis rather than mitosis?
2. How would a nonfunctional cell cycle checkpoint lead to carcinogenesis?
3. From an evolutionary standpoint, why is sexual reproduction advantageous for the continuation of a species?

INQUIRY INTO LIFE WEBSITE

The companion website for *Inquiry into Life* provides a wealth of information organized and integrated by chapter. You will find practice tests, animations, videos, and much more that will complement your learning and understanding of general biology.

http://www.mhhe.com/maderinquiry13

Metabolism: Energy and Enzymes

Applying the Concepts

Vampire bats were inspiration for the many legends about vampires. Vampire bats do exist, but they are not as frightening as legends would have you believe. Vampire bats are not large animals and they do not normally attack humans. The body of the common vampire bat, *Demodus rotundus,* is about the size of a human thumb with a wing span of only eight inches. These bats normally feed on cattle, horses, and pigs. Also, vampire bats do not "suck" the blood of their animal host. Instead, they make small cuts in the skin of the animal and lap up the blood that flows from the injury. It has been known for years that vampire bat saliva has an amazing ability to dissolve blood clots, allowing the blood to continue to flow from the wound while the bat feeds. Fibrin is an insoluble protein that forms the basis of a blood clot. Fibrin is dissolved by the enzyme plasmin, which circulates in the blood in an inactive form called plasminogen. Another enzyme activates plasminogen, which is converted to plasmin, and then dissolves the fibrin of the clot. Researchers have discovered an enzyme in the saliva of the vampire bat that is 150 times more potent at activating plasminogen and thus dissolving clots than any known drug. This enzyme may one day be used to treat victims of ischemic stroke, caused when a clot blocks blood supply to the brain.

This chapter describes the general characteristics and functions of enzymes, and how enzymes function in the flow of energy and metabolism. Each enzyme in our body is responsible for one, unique metabolic reaction, as illustrated by the vampire bat enzyme which is responsible for converting plasminogen into plasmin.

Chapter Outline

6.1 Life and the Flow of Energy

Learning Outcomes

1. Explain why energy flows through an ecosystem.
2. State the two laws of thermodynamics and how these laws apply to cells.

Energy is the ability to do work or bring about a change. In order to maintain their organization and carry out metabolic activities, cells as well as organisms need a constant supply of energy. This energy allows living things to carry on the processes of life, including growth, development, locomotion, metabolism, and reproduction.

The majority of organisms get their energy from organic nutrients produced by photosynthesizers (algae, plants, and some bacteria). Therefore, life on Earth is ultimately dependent on solar energy.

Forms of Energy

Energy occurs in two forms: kinetic and potential. **Kinetic energy** is the energy of motion, as when a ball rolls down a hill or a moose walks through grass. **Potential energy** is stored energy—its capacity to accomplish work is not being used at the moment. The food we eat has potential energy because it can be converted into various types of kinetic energy. Food is specifically called **chemical energy** because it contains energy in the chemical bonds of organic molecules. When a moose walks, it has converted chemical energy into a type of kinetic energy called **mechanical energy** (Fig. 6.1).

Two Laws of Thermodynamics

Figure 6.1 illustrates the flow of energy in a terrestrial ecosystem. Plants capture only a small portion of solar energy. When plants photosynthesize and then make use of the food they produce, some energy dissipates as heat. Still, there is enough remaining to sustain a moose and the other organisms in an ecosystem. As living things metabolize nutrient molecules, all the captured solar energy eventually dissipates as heat. Therefore, energy flows through an ecosystem. It does not recycle. Two laws of thermodynamics explain why energy flows in ecosystems and in cells.

The first law of thermodynamics—the law of conservation of energy—states that energy cannot be created or destroyed, but it can be changed from one form to another.

When leaf cells photosynthesize, they use solar energy to form carbohydrate molecules from carbon dioxide and water. (Carbohydrates are energy-rich molecules, while carbon dioxide and water are energy-poor molecules.) Not all of the captured solar energy is used to form carbohydrates; some becomes heat:

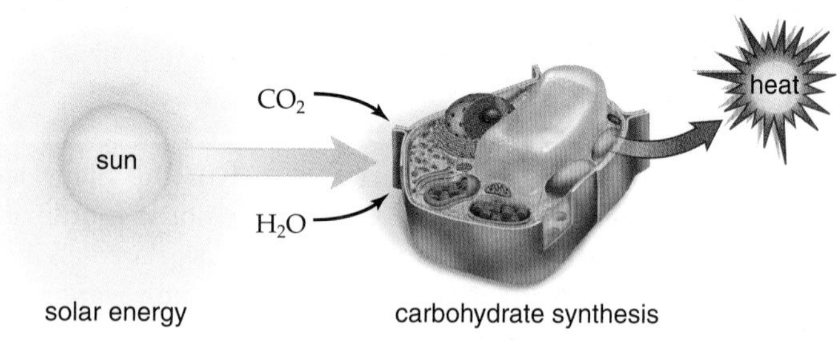

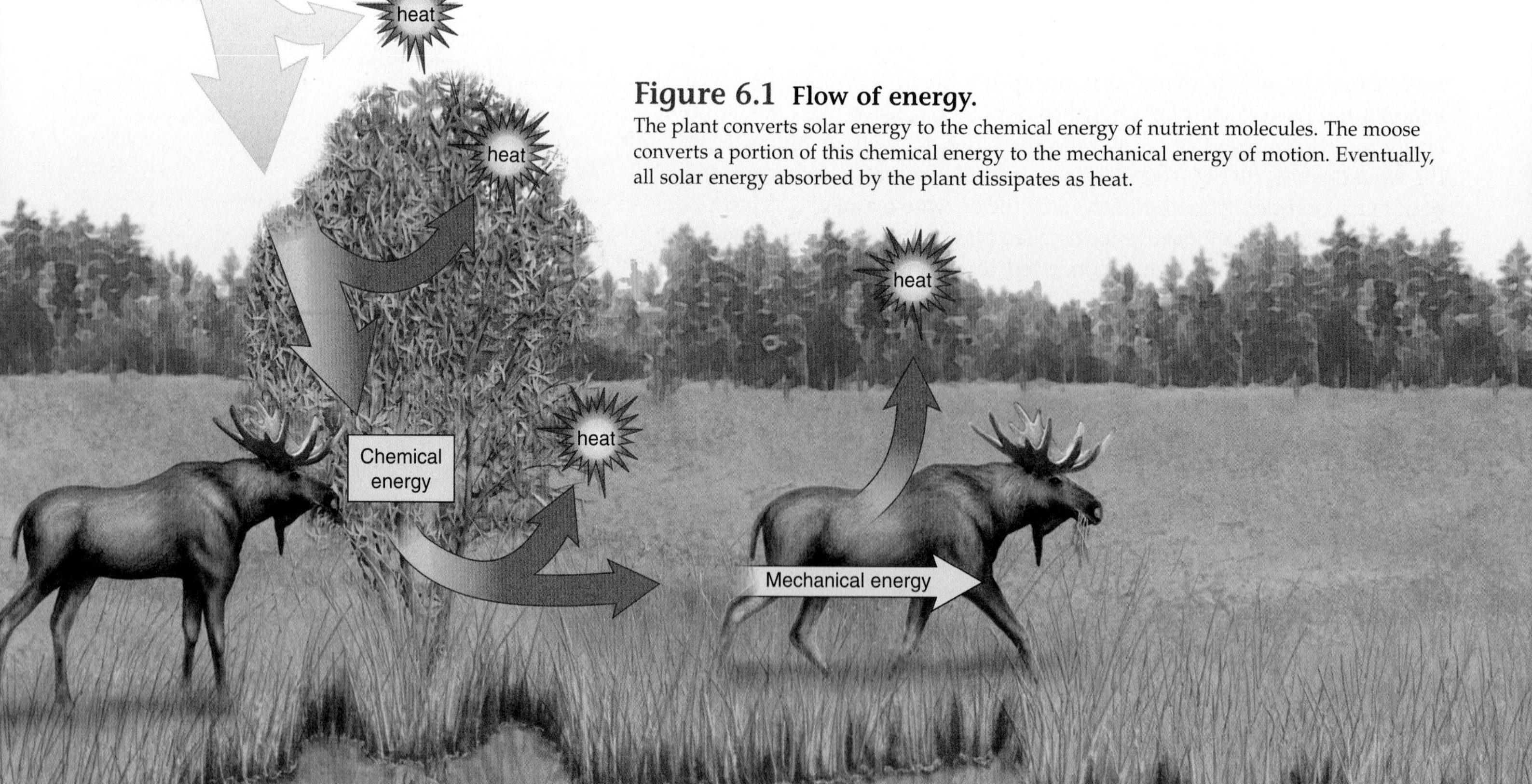

Figure 6.1 Flow of energy. The plant converts solar energy to the chemical energy of nutrient molecules. The moose converts a portion of this chemical energy to the mechanical energy of motion. Eventually, all solar energy absorbed by the plant dissipates as heat.

Obviously, plant cells do not create the energy they use to produce carbohydrate molecules. That energy comes from the sun. Is any energy destroyed? No, because heat is also a form of energy. Similarly, a moose uses the energy derived from carbohydrates to power its muscles. And as its cells use this energy, none is destroyed, but some becomes heat, which dissipates into the environment:

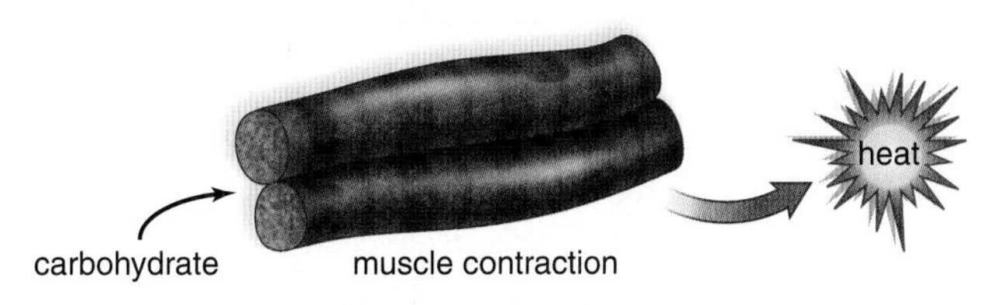

The second law of thermodynamics, therefore, applies to living systems.

The second law of thermodynamics states that energy cannot be changed from one form to another without a loss of usable energy.

Some of the solar energy taken in by the plant and some of the chemical energy within the nutrient molecules taken in by the moose become heat. When heat dissipates into the environment, it is no longer usable—that is, it is not available to do work. With transformation upon transformation, eventually all usable forms of energy become heat that is lost to the environment. Heat that dissipates into the environment cannot be captured and converted to one of the other forms of energy.

As a result of the second law of thermodynamics, no process requiring a conversion of energy is ever 100% efficient. Much of the energy is lost in the form of heat. In automobiles, the gasoline engine is between 20% and 30% efficient in converting chemical energy into mechanical energy. The majority of energy is obviously lost as heat. Cells are capable of about 40% efficiency, with the remaining energy given off to their surroundings as heat.

Cells and Entropy

The second law of thermodynamics can be stated another way: Every energy transformation makes the universe less organized and more disordered. The term **entropy** is used to indicate the relative amount of disorganization. Since the processes that occur in cells are energy transformations, the second law means that every process that occurs in cells always does so in a way that increases the total entropy of the universe. Then, too, any one of these processes makes less energy available to do useful work in the future.

Figure 6.2 shows two processes that occur in cells: glucose breakdown and hydrogen ion diffusion. The second law of thermodynamics tells us that glucose tends to break apart into carbon dioxide and water. Why? Because glucose is more organized, and therefore less stable, than its breakdown products. Also, hydrogen ions on one side of a membrane tend to move to the other side unless they are prevented from doing so. Why? Because when they are distributed randomly, entropy has increased. As an analogy, you know from experience that

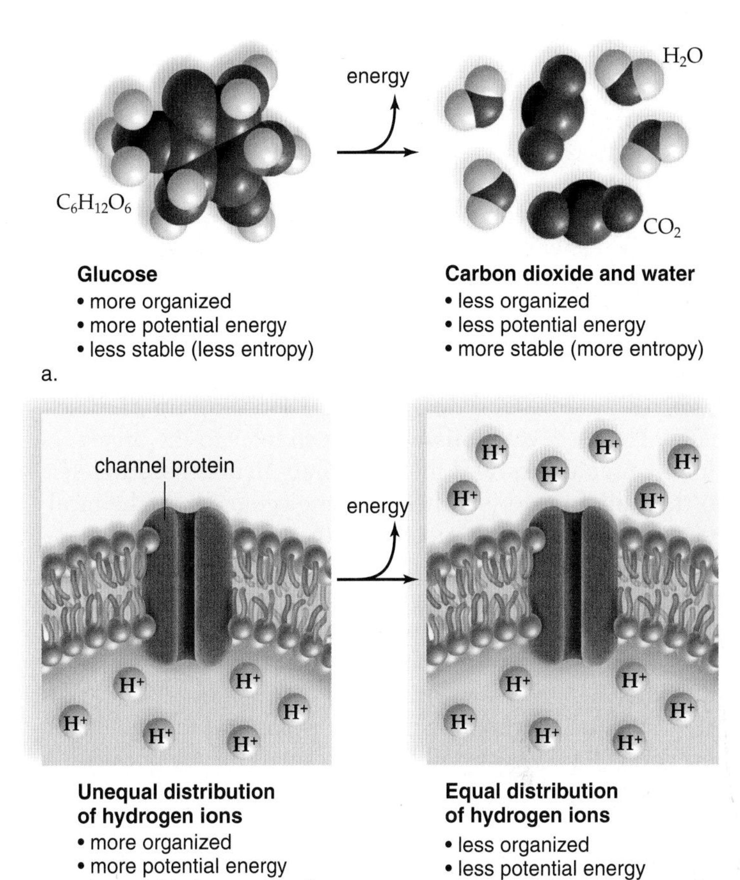

Figure 6.2 Cells and entropy.
The second law of thermodynamics tells us that (**a**) glucose, which is more organized, tends to break down to carbon dioxide and water, which are less organized. **b.** Similarly, hydrogen ions (H^+) on one side of a membrane tend to move to the other side so that the ions are randomly distributed. Both processes result in a loss of potential energy and an increase in entropy.

a neat room is more organized but less stable than a messy room, which is disorganized but more stable. How do you know a neat room is less stable than a messy room? Consider that a neat room always tends to become more messy.

On the other hand, you know that some cells can make glucose out of carbon dioxide and water, and all cells can actively move ions to one side of the membrane. How do they do it? These cellular processes obviously require an input of energy from an outside source. This energy ultimately comes from the sun. Living things depend on a constant supply of energy from the sun because the ultimate fate of all solar energy in the biosphere is to become randomized in the universe as heat. A living cell is a temporary repository of order purchased at the cost of a constant flow of energy.

Check Your Progress

1. Why is energy not recyclable?
2. If a cell becomes more organized, what has become less organized in the universe in order to fulfill the second law of thermodynamics?

6.2 Energy Transformations and Metabolism

Learning Outcomes

1. Be able to determine whether a chemical reaction will go forward as written.
2. Describe how ATP is used in metabolism.

Cellular **metabolism** is the sum of all the chemical reactions that occur in a cell. A significant part of cellular metabolism involves the breaking down and the building up of molecules. The term **catabolism** is used to refer to the breaking down of molecules, and the term **anabolism** is used to refer to the building up (synthesis) of molecules. In a chemical reaction, **reactants** are substances that participate in a reaction, while **products** are substances that form as a result of a reaction. For example, in the reaction $A + B \longrightarrow C + D$, A and B are the reactants while C and D are the products. How do you know whether this reaction will go forward as written?

Using the concept of entropy, it is possible to state that a reaction will go forward if it increases the entropy of the universe. But in cell biology, we do not usually wish to consider the entire universe. We simply want to consider a particular reaction. In such instances, cell biologists use the concept of free energy instead of entropy. **Free energy** is the amount of energy available—that is, energy that is still "free" to do work—after a chemical reaction has occurred. The change in free energy after a reaction occurs is calculated by subtracting the free energy content of the reactants from that of the products. A negative result means that the products have less free energy than the reactants, and the reaction will go forward. In our reaction, if C and D have less free energy than A and B, the reaction will "go."

Exergonic reactions are spontaneous and release energy, while **endergonic reactions** require an input of energy to occur. In the body, many reactions, such as protein synthesis, nerve impulse conduction, or muscle contraction, are endergonic, and they are driven by the energy released by exergonic reactions. ATP is a carrier of energy between exergonic and endergonic reactions.

ATP: Energy for Cells

ATP (adenosine triphosphate) is the common energy currency of cells. When cells require energy, they "spend" ATP. The more active the organism, the greater the demand for ATP. However, the amount on hand at any one moment is minimal because ATP is constantly being generated from **ADP (adenosine diphosphate)** and a molecule of inorganic phosphate Ⓟ (Fig. 6.3). A cell is assured of a supply of ATP, because glucose breakdown during cellular respiration provides the energy for the buildup of ATP in mitochondria. Only 39% of the free energy of glucose is transformed to ATP; the rest is lost as heat.

There are many biological advantages to the use of ATP as an energy carrier in living systems. ATP is a common and universal energy currency because it can be used in many different types of reactions. Also, when ATP is converted to energy, ADP, and Ⓟ, the amount of energy released is sufficient for a particular biological function, and little energy is wasted. In addition, ATP breakdown can be

adenosine triphosphate

P — P — P

Energy from exergonic reactions (e.g., cellular respiration)

ATP

Energy for endergonic reactions (e.g., protein synthesis, nerve impulse conduction, muscle contraction)

ADP + P

P — P + P

adenosine diphosphate + phosphate

a.

b. 2.25×

Figure 6.3 The ATP cycle.
a. In cells, ATP carries energy between exergonic reactions and endergonic reactions. When a phosphate group is removed by hydrolysis, ATP releases the appropriate amount of energy for most metabolic reactions. **b.** In order to produce light, a firefly breaks down ATP.

coupled to endergonic reactions in such a way that it minimizes energy loss.

Structure of ATP

ATP is a nucleotide composed of the nitrogen-containing base adenine and the 5-carbon sugar ribose (together called adenosine) and three phosphate groups. ATP is called a "high-energy" compound because of the energy stored in the chemical bonds of the phosphates. Under cellular conditions, the amount of energy released when ATP is hydrolyzed to ADP + Ⓟ is about 7.3 kcal per mole.[1]

Coupled Reactions

In **coupled reactions,** the energy released by an exergonic reaction is used to drive an endergonic reaction. ATP breakdown is often coupled to cellular reactions that require an input of energy. Coupling, which requires that the exergonic reaction and the endergonic reaction be closely tied, can be symbolized like this:

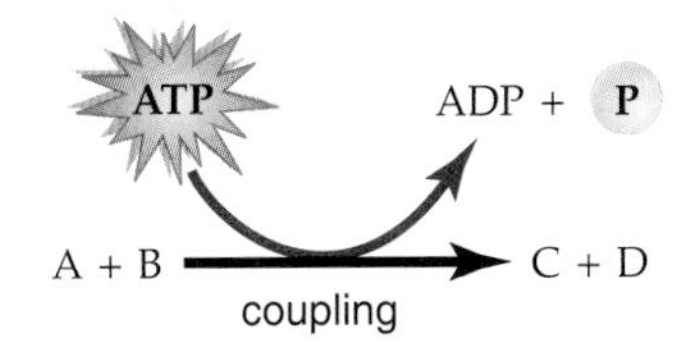

[1] A mole is the number of molecules present in the molecular weight of a substance (in grams).

Notice that the word *energy* does not appear following ATP breakdown. Why not? Because this energy was used to drive forward the coupled reaction. Figure 6.4 shows that ATP breakdown provides the energy necessary for muscular contraction to occur. The transfer of energy is not complete, and some energy is lost as heat.

Function of ATP

In living systems, ATP can be used for:

Chemical work ATP supplies the energy needed to synthesize macromolecules (anabolism) that make up the cell, and therefore the organism.

Transport work ATP supplies the energy needed to pump substances across the plasma membrane.

Mechanical work ATP supplies the energy needed to permit muscles to contract, cilia and flagella to beat, chromosomes to move, and so forth.

In most cases, ATP is the immediate source of energy for these processes.

Check Your Progress

1. Is an anabolic reaction more likely to be exergonic or endergonic? To have a positive or negative value for ΔG?
2. How is ATP like currency (money)?

Figure 6.4 Coupled reactions.
Muscle contraction occurs only when it is coupled to ATP breakdown.

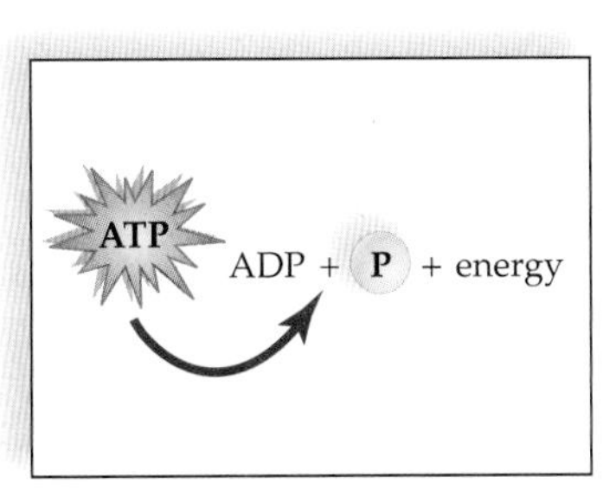

a. ATP breakdown is exergonic.

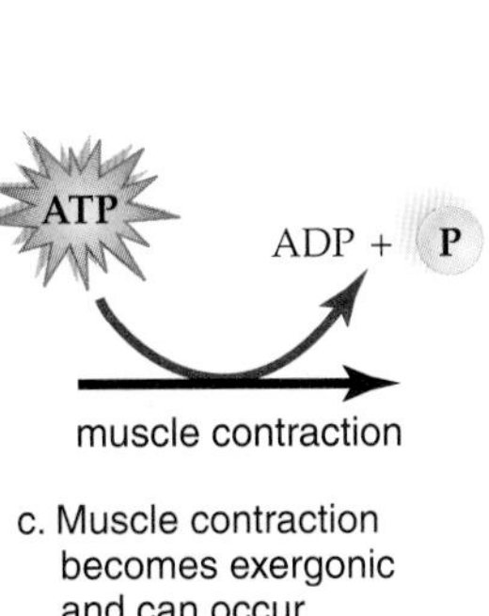

b. Muscle contraction is endergonic and cannot occur without an input of energy.

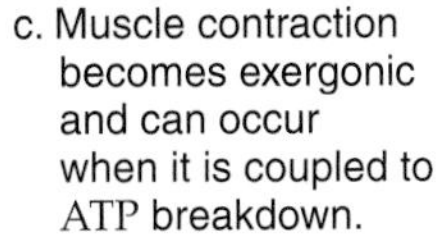

c. Muscle contraction becomes exergonic and can occur when it is coupled to ATP breakdown.

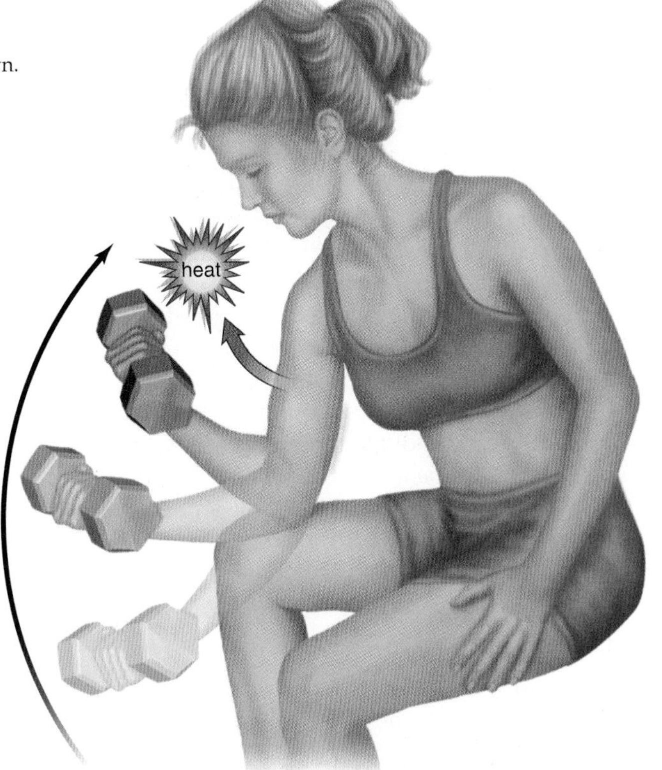

6.3 Enzymes and Metabolic Pathways

LEARNING OUTCOMES

1. List the components of a metabolic pathway, and state why such pathways are useful.
2. Describe how an enzyme binds its substrate.
3. Explain how an enzyme increases the rate of a reaction, and list factors that affect enzyme speed.

Reactions do not occur haphazardly in cells. They are usually part of a **metabolic pathway,** a series of linked reactions. Metabolic pathways begin with a particular reactant and terminate with an end product. Some metabolic pathways are cyclical, regenerating the starting material. While it is possible to write an overall equation for a pathway as if the beginning reactant went to the end product in one step, actually many specific steps occur in between. In the pathway, one reaction leads to the next reaction, which leads to the next reaction, and so forth in an organized, highly structured manner. This arrangement makes it possible for one pathway to lead to several others, because various pathways have several molecules in common. Also, metabolic energy is captured and utilized more easily if it is released in small increments rather than all at once.

A metabolic pathway can be represented by the following diagram:

$$A \xrightarrow{E_1} B \xrightarrow{E_2} C \xrightarrow{E_3} D \xrightarrow{E_4} E \xrightarrow{E_5} F \xrightarrow{E_6} G$$

In this diagram, the letters A – F are reactants, and the letters B – G are products in the various reactions. In other words, the products from the previous reaction become the reactants of the next reaction. The letters E_1 – E_6 are enzymes.

An **enzyme** is a protein that functions to speed a chemical reaction. (Some RNA molecules, called ribozymes, can also speed chemical reactions.)

An enzyme is a catalyst—it participates in the chemical reaction, but is not used up by the reaction. Note that the enzyme does not determine whether the reaction goes forward; that is determined by the free energy of the reaction. Enzymes simply increase the rate of the reaction.

The reactants in an enzymatic reaction are called the **substrates** for that enzyme. In the first reaction, A is the substrate for E_1, and B is the product. Now B becomes the substrate for E_2, and C is the product. This process continues until the final product (G) forms. Any one of the molecules (A – G) in this linear pathway could also be a substrate for an enzyme in another pathway. The presence or absence of an active enzyme determines which reaction takes place. A diagram showing all the possibilities would be highly branched.

Energy of Activation

Molecules frequently do not react with one another unless they are activated in some way. In the lab, for example, in the absence of an enzyme, activation is very often achieved by heating a reaction flask to increase the number of effective collisions between molecules. The energy that must be added to cause molecules to react with one another is called the **energy of activation (E_a).** Even though the reaction will go forward (ΔG is negative), the energy of activation must be overcome. The burning of firewood is a very exergonic reaction, but firewood in a pile does not spontaneously combust. The input of some energy, perhaps a lit match, is required to overcome the energy of activation.

Figure 6.5 shows E_a when an enzyme is not present compared to when an enzyme is present, illustrating that enzymes lower the amount of energy required for activation to occur. Nevertheless, the addition of the enzyme does not change the end result of the reaction. Notice that the energy of the products is less than the energy of the reactants. This results in a negative ΔG, so the reaction will go forward. But the reaction will not go at all unless the energy of activation is overcome. Without the enzyme, the reaction rate will be very slow. By lowering the energy of activation, the enzyme increases the rate of the reaction.

Sugar in your kitchen cupboard will break down to carbon dioxide and water because the energy of the products (carbon dioxide and water) is much less than the free energy of the reactant (sugar). However, the rate of this reaction is so slow that you never see it. If you eat the sugar, the enzymes in your digestive system greatly increase the speed at which the sugar is broken down. However, the end result (carbon dioxide and water) is still the same.

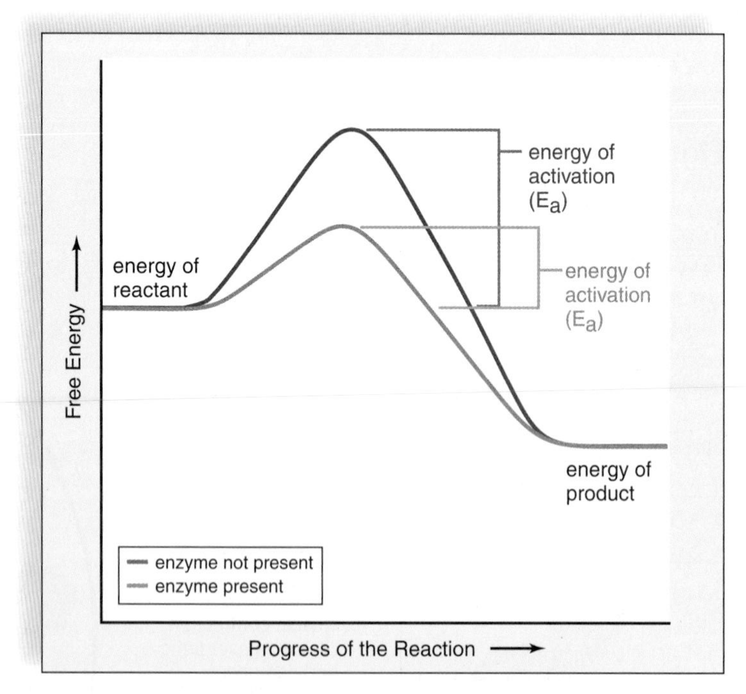

Figure 6.5 Energy of activation (E_a). Enzymes speed the rate of reactions because they lower the amount of energy required for the reactants to react. Even reactions like this one, in which the energy of the product is less than the energy of the reactant (ΔG is negative), speed up when an enzyme is present.

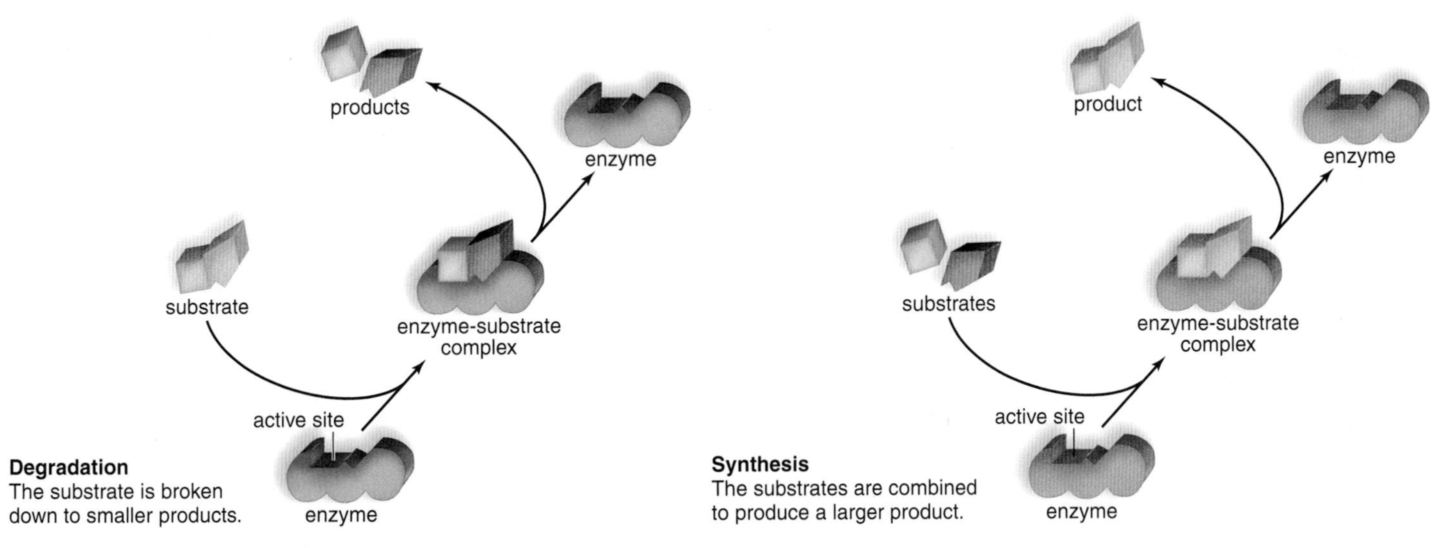

Figure 6.6 Enzymatic action.
An enzyme has an active site where the substrates and enzyme fit together in such a way that the substrates react. Following the reaction, the products are released, and the enzyme is free to act again. The enzymatic reaction can result in the degradation of a substrate into multiple products (catabolism) or the synthesis of a product from multiple substrates (anabolism).

How Enzymes Function

The following equation, which is illustrated in Figure 6.6, is often used to indicate that an enzyme forms a complex with its substrate:

S	+	E	⟶	ES	⟶	E	+	P
substrate		enzyme		enzyme-substrate complex				product

In most instances, only one small part of the enzyme, called the **active site,** complexes with the substrate(s). It is here that the enzyme and substrate fit together, seemingly like a key fits a lock. However, cell biologists now know that the active site undergoes a slight change in shape in order to accommodate the substrate(s). This is called the **induced fit model** because the enzyme is induced to undergo a slight alteration to achieve optimum fit (Fig. 6.7).

The change in shape of the active site facilitates the reaction that now occurs. After the reaction has been completed, the product(s) is released, and the active site returns to its original state, ready to bind to another substrate molecule. Only a small amount of enzyme is actually needed in a cell because enzymes are not used up by the reaction.

Some enzymes do more than simply complex with their substrate(s); they participate in the reaction. For example, trypsin digests protein by breaking peptide bonds. The active site of trypsin contains three amino acids with *R* groups that actually interact with members of the peptide bond—first to break the bond and then to introduce the components of water. This illustrates that the formation of the enzyme-substrate complex is very important in speeding up the reaction.

Every reaction in a cell requires that its specific enzyme be present. Because enzymes complex only with their

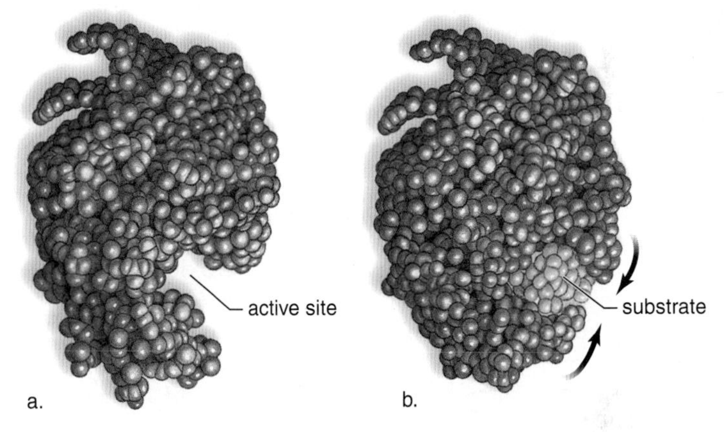

Figure 6.7 Induced fit model.
These computer-generated images show an enzyme called lysozyme that hydrolyzes its substrate, a polysaccharide that makes up bacterial cell walls. **a.** Shape of enzyme when no substrate is bound to it. **b.** After the substrate binds, the shape of the enzyme changes so that hydrolysis can better proceed.

substrates, they are often named for their substrates with the suffix *-ase,* as in the following examples:

Substrate	Enzyme
Lipid	Lipase
Urea	Urease
Maltose	Maltase
Ribonucleic acid	Ribonuclease
Lactose	Lactase

Check Your Progress

1. What are the advantages of having a particular chemical reaction (such as the breakdown of sugar) proceed via a metabolic pathway instead of in just one reaction?
2. Why is the induced fit model more accurate for enzyme-substrate interaction than the lock-and-key model?

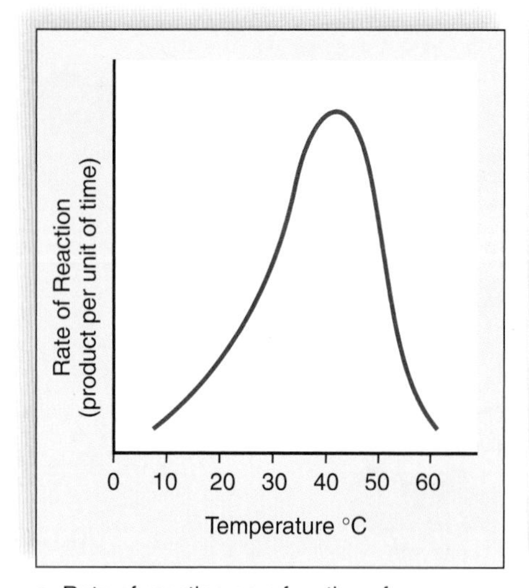

a. Rate of reaction as a function of temperature.

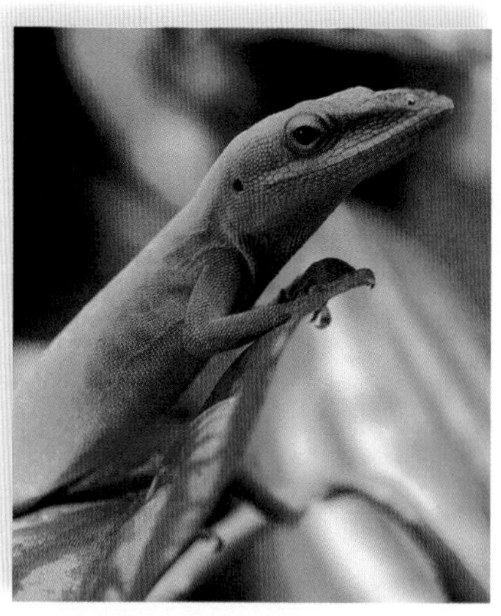

b. Body temperature of ectothermic animals often limits rates of reactions.

c. Body temperature of endothermic animals promotes rates of reactions.

Figure 6.8 The effect of temperature on rate of reaction.
a. Usually, the rate of an enzymatic reaction doubles with every 10°C rise in temperature. This enzymatic reaction is maximum at about 40°C. Then it decreases until the reaction stops altogether, because the enzyme has become denatured. **b.** The body temperature of ectothermic animals, which require an environmental source of heat, often limits rates of reactions. **c.** The body temperature of endothermic animals, which generate heat through their own metabolism, promotes rates of reaction.

Factors Affecting Enzymatic Speed

Enzymatic reactions proceed quite rapidly. Consider, for example, the breakdown of hydrogen peroxide (H_2O_2) as catalyzed by the enzyme catalase: $2\,H_2O_2 \longrightarrow 2\,H_2O + O_2$. The breakdown of hydrogen peroxide can occur 600,000 times a second when catalase is present! In order to achieve maximum product per unit time, there should be enough substrate to fill the enzyme's active sites most of the time. In addition to substrate concentration, the rate of an enzymatic reaction can also be affected by environmental factors (temperature and pH) or by cellular mechanisms, such as enzyme activation, enzyme inhibition, and cofactors.

Substrate Concentration

Generally, enzyme activity increases as substrate concentration increases because there are more collisions between substrate molecules and the enzyme. As more substrate molecules fill active sites, more product results per unit time. But when the enzyme's active sites are filled almost continuously with substrate, the enzyme's rate of activity cannot increase any more. The maximum rate has been reached.

Temperature and pH

As the temperature rises, enzyme activity increases (Fig. 6.8). This occurs because higher temperatures cause more effective collisions between enzyme and substrate. However, if the temperature rises beyond a certain point, enzyme activity eventually levels out and then declines rapidly because the enzyme is **denatured.** An enzyme's shape changes during denaturation, due to the loss of secondary and tertiary structure, and then it can no longer bind its substrate(s) efficiently. As the temperature decreases, enzyme activity decreases.

Each enzyme also has a preferred pH at which the rate of the reaction is highest. Figure 6.9 shows the preferred pH for the enzymes pepsin and trypsin. At this pH value, these enzymes have their normal configurations. The globular structure of an enzyme is dependent on interactions, such as hydrogen bonding, between *R* groups (see Fig. 2.23). A change in pH can alter the ionization of *R* groups and disrupt normal

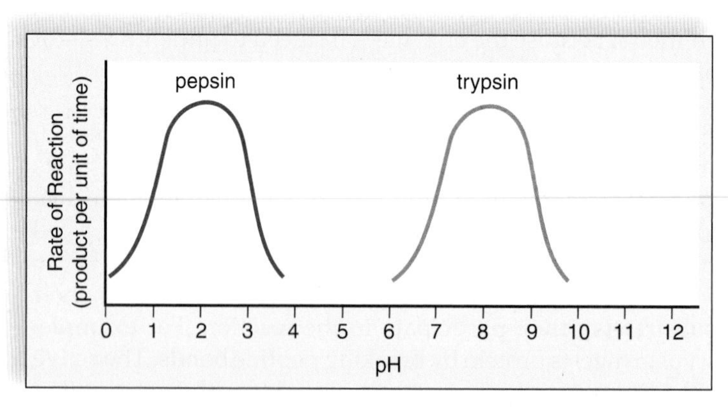

Figure 6.9 The effect of pH on rate of reaction.
The preferred pH for pepsin, an enzyme that acts in the stomach, is about 2, while the preferred pH for trypsin, an enzyme that acts in the small intestine, is about 8. At the preferred pH an enzyme maintains its shape so that it can bind with its substrates.

interactions, and under extreme pH conditions, denaturation eventually occurs. If the enzyme's shape is altered, it is then unable to combine efficiently with its substrate.

Enzyme Activation

Not all enzymes are needed by the cell all the time. Genes can be turned on to increase the concentration of an enzyme in a cell or turned off to decrease the concentration. But enzymes can also be present in the cell in an inactive form. Activation of enzymes occurs in many different ways. Some enzymes are covalently modified by the addition or removal of phosphate groups. An enzyme called a kinase adds phosphates to proteins, as shown below, and an enzyme called a phosphatase removes them. In some proteins, adding phosphates activates them; in others, removing phosphates activates them. Enzymes can also be activated by cleaving or removing part of the protein, or by associating with another protein or cofactor.

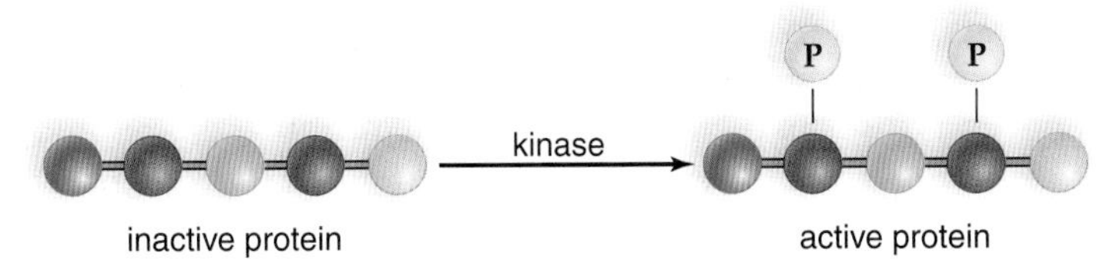

Enzyme Inhibition

Enzyme inhibition occurs when the substrate is unable to bind to the active site of an enzyme. The activity of almost every enzyme in a cell is regulated by feedback inhibition. In the simplest case, when there is plenty of product, it binds to the enzyme's active site, and then the substrate is unable to bind. As the product is used up, inhibition is reduced, and more product can be produced. In this way, the concentration of the product is always kept within a certain range.

Most metabolic pathways in cells are regulated by a more complicated type of feedback inhibition (Fig. 6.10). In these instances, the end product of an active pathway binds to a site other than the active site of the first enzyme. The binding changes the shape of the active site so that the substrate is unable to bind to the enzyme, and the pathway shuts down (inactive). Therefore, no more product is produced.

Poisons are often enzyme inhibitors. Cyanide is an inhibitor for an essential enzyme (cytochrome *c* oxidase) in all cells, which accounts for its lethal effect on humans. Penicillin is an *antimicrobial agent* that blocks the active site of an enzyme unique to bacteria. Therefore, penicillin is a poison for bacteria.

Enzyme Cofactors

Many enzymes require an inorganic ion or an organic, but nonprotein, helper to function properly. The inorganic ions are metals such as copper, zinc, or iron. These helpers are called **cofactors.** The organic, nonprotein molecules are called **coenzymes.** These cofactors assist the enzyme and may even accept or contribute atoms to the reactions.

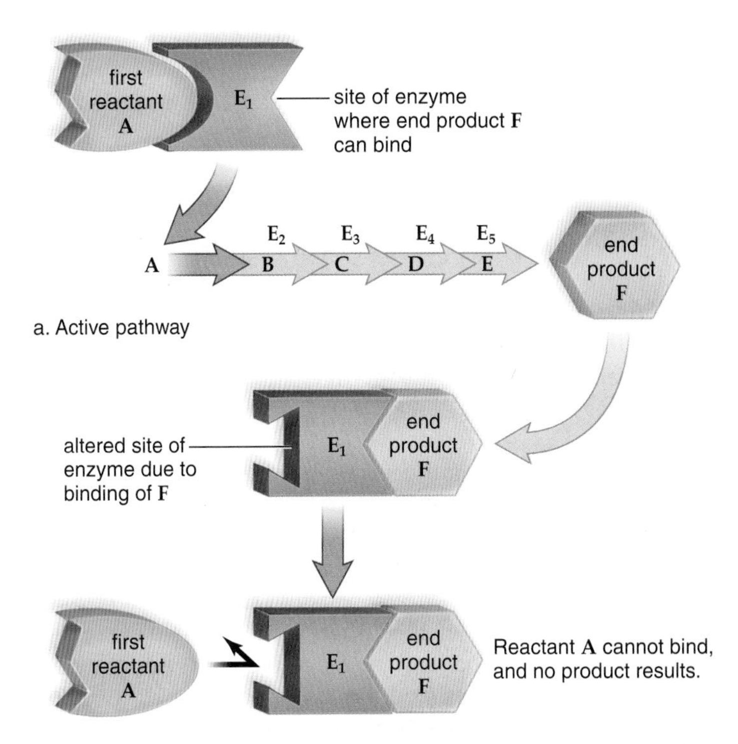

Figure 6.10 Feedback inhibition.
a. In an active pathway, the first reactant (A) is able to bind to the active site of enzyme E_1. **b.** Feedback inhibition occurs when the end product (F) of the metabolic pathway binds to the first enzyme of the pathway—at a site other than the active site. This binding causes the active site to change its shape. Now reactant A is unable to bind to the enzyme's active site, and the whole pathway shuts down.

Vitamins are often components of coenzymes. **Vitamins** are relatively small organic molecules that are required in trace amounts in our diet and in the diets of other animals for synthesis of coenzymes that affect health and physical fitness. The vitamin becomes a part of the coenzyme's molecular structure. For example, the vitamin niacin is part of the coenzyme NAD, and riboflavin (B_2) is part of the coenzyme FAD.

A deficiency of any one of these vitamins results in a lack of the coenzyme and therefore a lack of certain enzymatic actions. In humans, this eventually results in vitamin-deficiency symptoms. For example, niacin deficiency results in a skin disease called pellagra, and riboflavin deficiency results in cracks at the corners of the mouth.

Check Your Progress

1. What factors could denature an enzyme and cause it to cease functioning?
2. What is the advantage to inhibiting a metabolic pathway by feedback inhibition?

6.4 Oxidation-Reduction and Metabolism

Learning Outcomes

1. Show that the overall equations for photosynthesis and cellular respiration are oxidation-reduction reactions.
2. Draw a diagram to illustrate that the metabolic reactions within chloroplasts and mitochondria are intimately related.
3. Explain how proteins and fats in addition to glucose can be used as mitochondrial fuel.

When oxygen (O) combines with a metal such as iron or magnesium (Mg), oxygen receives electrons and becomes an ion that is negatively charged. The metal loses electrons and becomes an ion that is positively charged. When magnesium oxide (MgO) forms, it is appropriate to say that magnesium has been oxidized. On the other hand, oxygen has been reduced because it has gained negative charges (i.e., electrons). Today, the terms oxidation and reduction are applied to many reactions, whether or not oxygen is involved. Very simply, **oxidation** is the loss of electrons, and **reduction** is the gain of electrons. In the reaction $Na + Cl \longrightarrow NaCl$, sodium has been oxidized (loss of electron), and chlorine has been oxidized (gain of electrons). Because oxidation and reduction go hand-in-hand, the entire reaction is called a **redox reaction**.

The terms oxidation and reduction also apply to covalent reactions in cells. In this case, however, oxidation is the loss of hydrogen atoms ($e^- + H^+$), and reduction is the gain of hydrogen atoms. Notice that when a molecule loses a hydrogen atom, it has lost an electron, and when a molecule gains a hydrogen atom, it has gained an electron. This form of oxidation-reduction is exemplified in the overall equations for photosynthesis and cellular respiration.

Chloroplasts and Photosynthesis

The chloroplasts in plants capture solar energy and use it to convert water and carbon dioxide into a carbohydrate. Oxygen is a by-product that is released (Fig. 6.11 *left*). The overall equation for photosynthesis can be written like this:

$$\text{energy} + \underset{\text{carbon dioxide}}{6\,CO_2} + \underset{\text{water}}{6\,H_2O} \longrightarrow \underset{\text{glucose}}{C_6H_{12}O_6} + \underset{\text{oxygen}}{6\,O_2}$$

This equation shows that during photosynthesis, hydrogen atoms are transferred from water to carbon dioxide as glucose forms. In this reaction, therefore, carbon dioxide has been reduced and water has been oxidized. It takes energy to reduce carbon dioxide to glucose, and this energy is supplied by solar energy. Chloroplasts are able to capture solar energy and convert it to the chemical energy of ATP, which is used along with hydrogen atoms to reduce carbon dioxide.

The reduction of carbon dioxide to form a mole of glucose stores 686 kcal in the chemical bonds of glucose. This is the energy that living things utilize to support themselves only because carbohydrates (and other nutrients) can be oxidized in mitochondria.

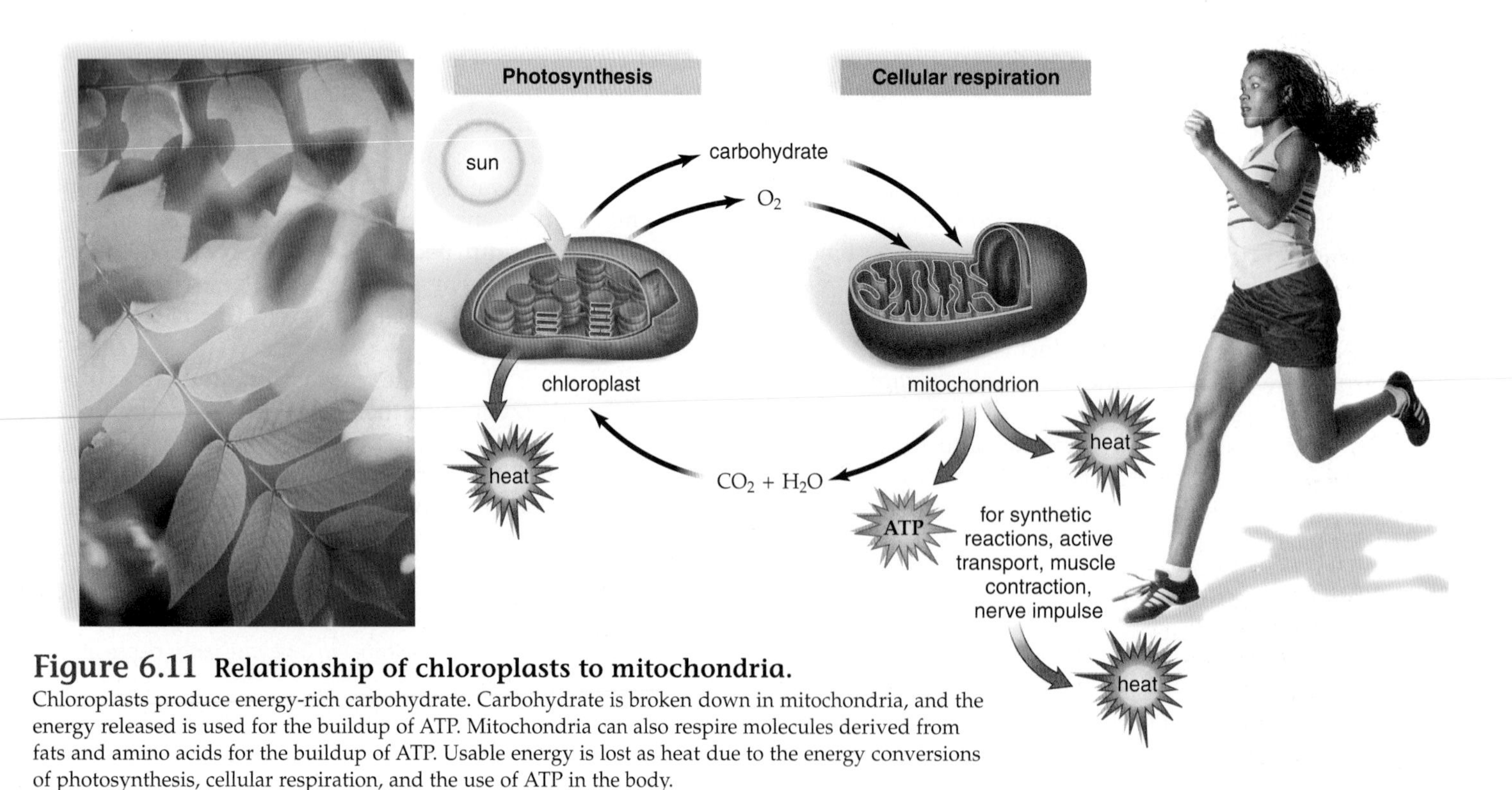

Figure 6.11 Relationship of chloroplasts to mitochondria.
Chloroplasts produce energy-rich carbohydrate. Carbohydrate is broken down in mitochondria, and the energy released is used for the buildup of ATP. Mitochondria can also respire molecules derived from fats and amino acids for the buildup of ATP. Usable energy is lost as heat due to the energy conversions of photosynthesis, cellular respiration, and the use of ATP in the body.

Mitochondria and Cellular Respiration

Mitochondria, present in both plants and animals, oxidize carbohydrates and use the released energy to build ATP molecules (Fig. 6.11 *right*). Cellular respiration therefore consumes oxygen and produces carbon dioxide and water, the very molecules taken up by chloroplasts. The overall equation for cellular respiration is the opposite of the one we used to represent photosynthesis:

$$\underset{\text{glucose}}{C_6H_{12}O_6} + \underset{\text{oxygen}}{6\,O_2} \longrightarrow \underset{\text{carbon dioxide}}{6\,CO_2} + \underset{\text{water}}{6\,H_2O} + \text{energy}$$

In this reaction, glucose has lost hydrogen atoms (been oxidized), and oxygen has gained hydrogen atoms (been reduced). When oxygen gains electrons, it becomes water. The complete oxidation of a mole of glucose releases 686 kcal of energy, and some of this energy is used to synthesize ATP molecules. If the energy within glucose were released all at once, most of it would dissipate as heat instead of some of it being used to produce ATP. Instead, cells oxidize glucose step by step. The energy is gradually stored and then converted to that of ATP molecules, which is used in animals in the many ways listed in Figure 6.11.

Figure 6.11 shows us very well that chloroplasts and mitochondria are involved in a cycle. Carbohydrate produced within chloroplasts becomes a fuel for cellular respiration in mitochondria, while carbon dioxide released by mitochondria becomes a substrate during photosynthesis in chloroplasts. These organelles are involved in a redox cycle because carbon dioxide is reduced during photosynthesis and carbohydrate is oxidized during cellular respiration. Note that energy does not cycle between the two organelles; instead, it flows from the sun through each step of photosynthesis and cellular respiration until it eventually becomes unusable heat as ATP is used by the cell.

Cellular Respiration and Humans

Human beings, like all eukaryotic organisms, are involved in the cycling of molecules between chloroplasts and mitochondria. Our food is derived from plants or we eat other animals that have eaten plants. Also, we take in oxygen released by plants. Nutrients from our food and oxygen enter our mitochondria, which produce ATP (Fig. 6.12). Without a supply of energy-rich molecules, ultimately derived from plants, we could not produce the ATP molecules needed to maintain our bodies. On the other hand, our mitochondria release carbon dioxide and water. The carbon dioxide is exhaled and it enters the atmosphere where it is accessible to plants once again.

So far we have discussed only the use of glucose as a mitochondrial fuel, but hopefully not much of your diet is actually glucose. Our food, such as a slice of pizza with cheese and pepperoni on top, consists of carbohydrates, fats, and proteins. These macromolecules are broken down to simpler molecules in our digestive tract. Starch in the crust

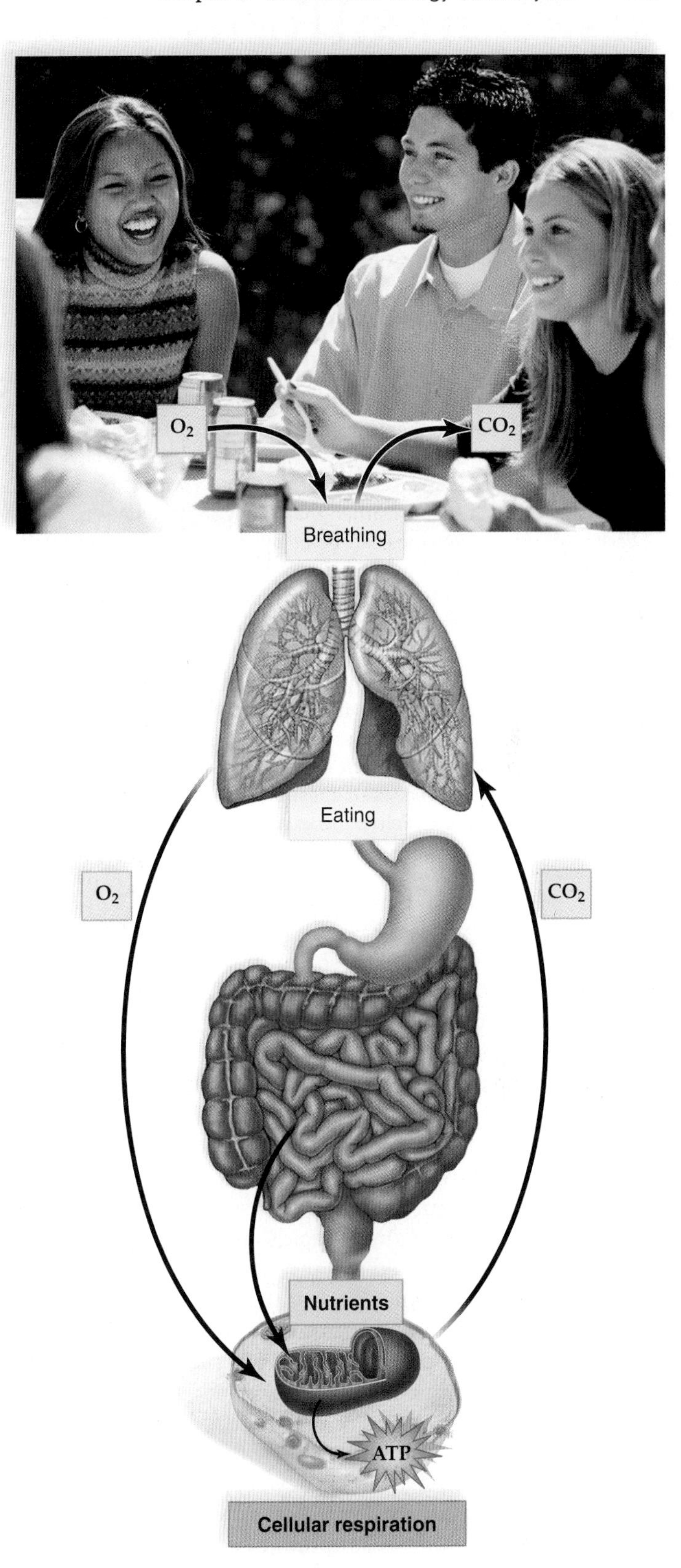

Figure 6.12 Relationship between breathing, eating, and cellular respiration.
The O_2 we inhale and the nutrients resulting from the digestion of our food are carried by the bloodstream to our cells where they enter mitochondria and undergo cellular respiration. Following cellular respiration, the ATP stays in the cell, but the CO_2 is carried to the lungs for exhalation.

Eat Your Enzymes!

You have been told to eat your fruits and vegetables, but have you ever been told to eat your enzymes? Enzyme therapy is a booming business. According to manufacturer's advertisements, "enzyme therapy can decrease inflammation and increase energy and overall well-being." Other manufacturers claim certain enzyme supplements can improve or even cure serious diseases. But opponents of enzyme therapy are becoming more and more vocal about the dangers of this type of alternative medicine.

Medical science has learned that taking certain digestive enzymes can be beneficial for people who have genetic defects. For example, people with cystic fibrosis take pancreatin, a mix of pancreatic enzymes, to assist their digestion. But for other people, opponents of enzyme therapy believe that purchasing supplemental enzymes is a waste of money. They say that other enzymes would be broken down by the stomach acid and the digestive enzymes in the small intestine and would never be able to help patients. Opponents also caution that just because something is "natural" doesn't mean it cannot be harmful. Worse, they believe, is that some people fail to seek traditional medical attention for serious ailments because they believe these enzymes will heal them—claims that are unsubstantiated by medical research.

Proponents of enzyme therapy strongly believe that the goal of medicine should be to keep people healthy instead of only to treat disease. They believe everyone could benefit from extra enzymes, not just people who have defects. Aging and poor diets contribute to a need for extra enzymes. Cooking food destroys the natural enzymes in the food, so all of us are lacking in these enzymes. Enzyme therapy proponents claim that because enzyme therapies are labeled "alternative," funding is unavailable for the types of studies that need to be done. And enzyme therapists can point to many satisfied customers who feel much better after taking enzyme supplements. Just because traditional medicine doesn't currently understand how enzyme therapy works doesn't mean that it won't move from alternative to traditional medical practice in a few years.

Form Your Own Opinion

1. Should manufacturers of enzyme supplements be required to complete controlled medical research studies before selling their product?
2. If taking enzyme supplements helps you feel better, should you be prevented from taking them just because they are not traditional medicine?
3. If someone chooses enzyme therapy for a serious medical problem and later dies, can/should the enzyme therapist and/or manufacturer be sued for treating this patient in a nontraditional manner? Keep in mind that traditional medicine does not have a 100% cure rate, either.

of a pizza is digested to glucose, but the fat is broken down to glycerol and fatty acids, and proteins are hydrolyzed to amino acids. Glycerol, fatty acids, and amino acids enter cellular respiration at different places in the cellular respiration pathway, as discussed in Chapter 7. Fatty acids are a concentrated source of energy compared to a carbohydrate because each fatty acid has a long chain of reduced carbon atoms. Only when we are on a diet that restricts carbohydrates and fats do our bodies then use amino acids as an energy source. When amino acids are metabolized, the amino groups are removed in a process called **deamination**. The amino group becomes part of urea, the primary excretory product of humans. The remainder of the amino acid can enter cellular respiration at various places in the pathway, depending on the length of its carbon skeleton. When we are literally starving and have exhausted stored supplies of glycogen and fat, the body will withdraw protein from our muscles and use it as an energy source.

If we were to study all the various metabolic steps involved in the cellular respiration of glucose, fatty acids, and amino acids, you would soon realize that certain substrates recur along the way. This allows the cell to switch metabolic gears and even switch to building up molecules rather than breaking them down. A good example is when you get fat from eating too much ice cream, cakes, and pies. A particular breakdown product of glucose can be used to form fatty acids, which then join with glycerol and the product fat gets stored at various locations in your body.

Check Your Progress

1. How are the overall equations for photosynthesis and cellular respiration redox reaction?
2. Draw and explain a diagram that shows that chloroplasts and mitochondria are involved in a cycling of materials and a redox cycle.
3. How do humans participate in the cycle described in question number 2?
4. What three types of nutrients can be used in mitochondria to build up ATP molecules?

Applying the Concepts Revisited

The clotting cascade involves several enzymes that circulate in the blood. When a blood vessel is damaged, platelets release prothrombin activator, an enzyme that converts the plasma protein prothrombin into thrombin. Thrombin, in turn, is an enzyme that converts fibrinogen into fibrin, forming the basis of the clot. When the damage has been repaired, the plasma protein plasminogen is converted into plasmin, which destroys the fibrin.

1. Why would it be advantageous to have preformed but inactive enzymes for the clotting cascade present in the blood?
2. Some drugs inhibit blood clotting. How might these drugs work in the body?

SUMMARIZING THE CONCEPTS

6.1 Life and the Flow of Energy

- Two energy laws are basic to understanding energy-use patterns in cells and ecosystems. The first law states that energy cannot be created or destroyed but can only be changed from one form to another. The second law of thermodynamics states that one usable form of energy cannot be converted into another form without loss of usable energy. Therefore, every energy transformation makes the universe less organized.
- As a result of these laws, we know that the entropy of the universe is increasing and that only a constant input of energy maintains the organization of living things.

6.2 Energy Transformations and Metabolism

- The term metabolism encompasses all the chemical reactions occurring in a cell. Catabolism involves breaking down reactants, whereas anabolism involves building up products.
- Considering individual reactions, only those that result in products that have less usable energy than the reactants go forward. Such reactions, called exergonic reactions, release energy.
- Endergonic reactions, which require an input of energy, occur in cells only because it is possible to couple an exergonic process with an endergonic process. For example, glucose breakdown is an exergonic metabolic pathway that drives the buildup of many ATP molecules. These ATP molecules then supply energy for cellular work.

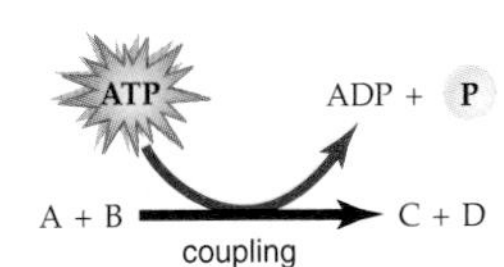

- ATP goes through a cycle of constantly being built up from, and then broken down to, ADP + Ⓟ.

6.3 Enzymes and Metabolic Pathways

- A metabolic pathway is a series of reactions that proceed in an orderly, step-by-step manner. Each reaction requires a specific enzyme.
- Reaction rates increase when enzymes form a complex with their substrates. Generally, enzyme activity increases as substrate concentration increases. Once all active sites are filled, the maximum rate has been achieved.
- Any environmental factor, such as temperature or pH, affects the shape of a protein, and therefore also affects the ability of an enzyme to do its job.

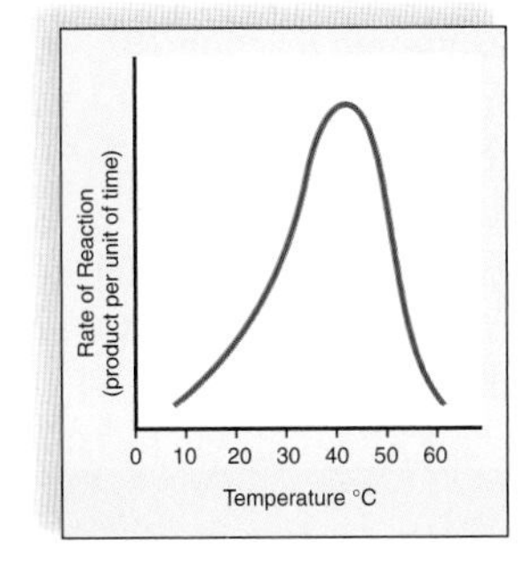

- Cellular mechanisms regulate enzyme quantity and activity. The activity of most metabolic pathways is regulated by feedback inhibition. Many enzymes have cofactors or coenzymes that help them carry out a reaction.

6.4 Oxidation-Reduction and Metabolism

- The overall equation for photosynthesis is the opposite of that for cellular respiration. Both processes involve oxidation-reduction reactions. Redox reactions are a major way in which energy is transformed in cells.
- During photosynthesis, carbon dioxide is reduced to glucose, and water is oxidized. Glucose formation requires energy, and this energy comes from the sun. Chloroplasts capture solar energy and convert it to the chemical energy of ATP molecules, which are used along with hydrogen atoms to reduce carbon dioxide to glucose.
- During cellular respiration, glucose is oxidized to carbon dioxide, and oxygen is reduced to water. This reaction releases energy, which is used to synthesize ATP molecules in all types of cells.
- Energy flows through all living things. Photosynthesis is a metabolic pathway in chloroplasts that transforms solar energy to the chemical energy within carbohydrates, and cellular respiration is a metabolic pathway completed in mitochondria that transforms this energy into that of ATP molecules. Eventually, the energy within ATP molecules becomes heat.

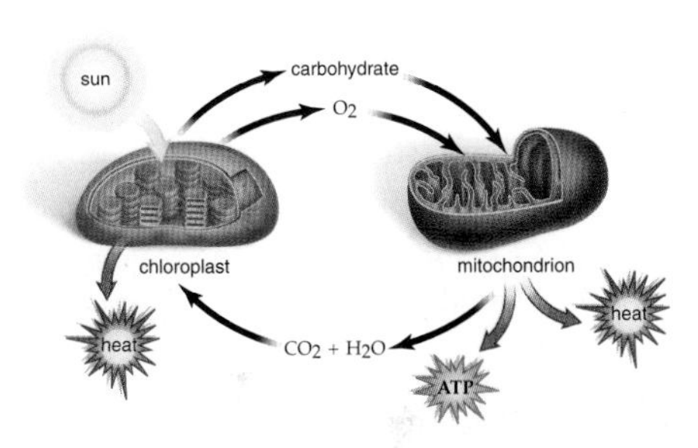

- Cellular respiration is an aerobic process that requires oxygen and gives off carbon dioxide. Cellular respiration involves oxidation. The air we inhale contains the oxygen, and the food we digest after eating contains the carbohydrate glucose needed for cellular respiration.
- Besides carbohydrates, fats and proteins can be used to provide ATP via cellular respiration.
- Anabolism can use molecules directly from our food or substrates from other metabolic pathways to synthesize new molecules.

TESTING YOURSELF

Choose the best answer for each question.

1. The fact that energy transformation is never 100% efficient is stated in the _____ law of thermodynamics.
   a. first
   b. second
   c. third
   d. None of these are correct.
2. ATP is a
   a. modified protein.
   b. modified amino acid.
   c. nucleotide.
   d. modified fat.
3. The amount of energy needed to get a chemical reaction started is known as the
   a. starter energy.
   b. activation energy.
   c. reaction energy.
   d. product energy.

In questions 4–7, match the statements concerning enzyme action to the appropriate graph.

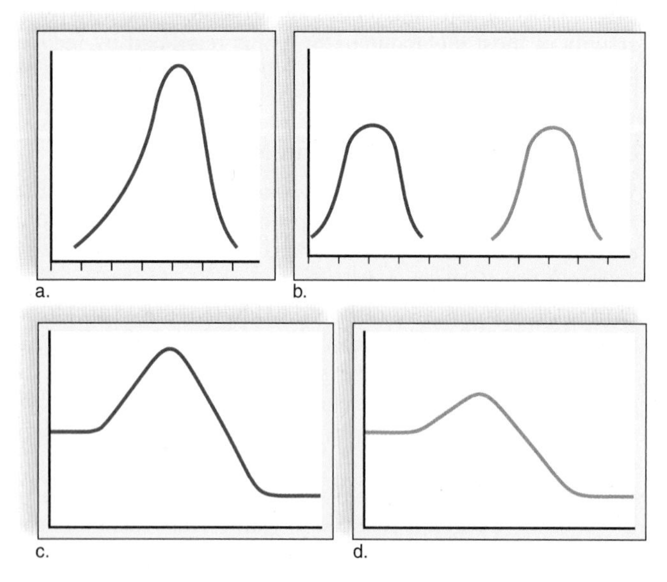

4. Progress of reaction in the absence of an enzyme
5. Enzymatic action with varying temperature
6. Progress of reaction in the presence of an enzyme
7. Enzymes with varying pH optima
8. Enzymatic action is sensitive to changes in
   a. temperature.
   b. pH.
   c. substrate concentration.
   d. All of these are correct.
9. The end products of photosynthesis are
   a. CO_2 and H_2O.
   b. O_2 and CO_2.
   c. O_2 and H_2O.
   d. O_2 and carbohydrates.
10. Which has more potential energy: glucose or CO_2 and H_2O?
    a. glucose
    b. CO_2 and H_2O
    c. Both are the same because reactants and products are equal.
    d. Neither is correct.
11. ____ is the loss of electrons, and ____ is the gain of electrons.
    a. Oxidation, reduction
    b. Reduction, oxidation
    c. Neither a nor b is correct.
    d. Need more information to answer.
12. Entropy is a term used to indicate the level of
    a. usable energy.
    b. disorganization.
    c. enzyme action.
    d. None of these are correct.
13. Enzymes catalyze reactions by
    a. bringing the reactants together.
    b. lowering the activation energy.
    c. Both a and b are correct.
    d. Neither a nor b is correct.
14. If the reaction A + B $\longrightarrow$ C + D + energy occurs in a cell,
    a. the reaction is exergonic.
    b. an enzyme could still speed the reaction.
    c. ATP is not needed to make the reaction go.
    d. A and B are reactants; C and D are products.
    e. All of these are correct.
15. The active site of an enzyme
    a. is identical to that of any other enzyme.
    b. is the part of the enzyme where its substrate can fit.
    c. can be used over and over again.
    d. is not affected by environmental factors such as pH and temperature.
    e. Both b and c are correct.

UNDERSTANDING THE TERMS

active site 105
ADP (adenosine diphosphate) 102
anabolism 102
ATP (adenosine triphosphate) 102
catabolism 102
chemical energy 100
coenzyme 107
cofactor 107
coupled reactions 103
deamination 110
denatured 106
endergonic reaction 102
energy 100
energy of activation (E_a) 104
entropy 101
enzyme 104
enzyme inhibition 107
exergonic reaction 102
free energy 102
induced fit model 105
kinetic energy 100
mechanical energy 100
metabolic pathway 104
metabolism 102
oxidation 108
potential energy 100
product 102
reactant 102
redox reaction 108
reduction 108
substrate 104
vitamin 107

THINKING CRITICALLY

1. Entropy is increased when nutrients breakdown so why are enzymatic metabolic pathways required for cellular respiration? (Hint: see Figure 6.5)
2. Why would you expect glucose storage as glycogen to be an energy-requiring process?
3. If photosynthesis and cellular respiration are reverse equations, then why can't mitochondria carry on photosynthesis?

INQUIRY INTO LIFE WEBSITE

The companion website for *Inquiry into Life* provides a wealth of information organized and integrated by chapter. You will find practice tests, animations, videos, and much more that will complement your learning and understanding of general biology.

http://www.mhhe.com/maderinquiry13

Cellular Respiration

Chapter Outline

Applying the Concepts

The 113th running of the Boston Marathon took place on April 20, 2009. Over 22,000 runners completed the race including 29 participants in wheelchairs. A marathon consists of 42.195 kilometers (26 miles 385 yards) and Deriba Merga, of Ethiopia, was the first place men's finisher with a time of 2:08:42. The first place women's finisher was Salina Kosgei, of Kenya, with a time of 2:32:416. Running a marathon requires endurance running as well as some short periods of sprinting. Among primates, endurance running is unique to humans. Endurance running is defined as running many kilometers over a long time period using aerobic metabolism. In aerobic runs, athletes run at slower rates than during a sprint. This enables muscle cells to use oxygen in order to completely break down glucose, producing more ATP, a high energy molecule used for muscle contraction, than otherwise. The breakdown of glucose with oxygen to produce carbon dioxide and water in the cytoplasm and mitochondria of the cell is called cellular respiration. Anaerobic running, in contrast, involves running short sprints as fast as possible. Without oxygen (anaerobic), glucose cannot be broken down completely. It is changed into lactate, which is responsible for that muscle burn we sometimes feel after strenuous exercise. Following anaerobic running, you continue to breathe heavily until you obtain enough oxygen to dispose of the lactate.

In this chapter, we will discuss the metabolic pathways of cellular respiration that allow the energy within a glucose molecule to be converted into ATP.

7.1 Overview of Cellular Respiration

Learning Outcomes

1. Write the overall equation for cellular respiration.
2. Describe the role of electron carriers in respiration.
3. List the phases of cellular respiration and indicate where they occur in a cell.

Cellular respiration is the release of energy from molecules such as glucose accompanied by the use of this energy to synthesize ATP molecules. Cellular respiration is an aerobic process that requires oxygen (O_2) and gives off carbon dioxide (CO_2). It usually involves the complete breakdown of glucose as shown here:

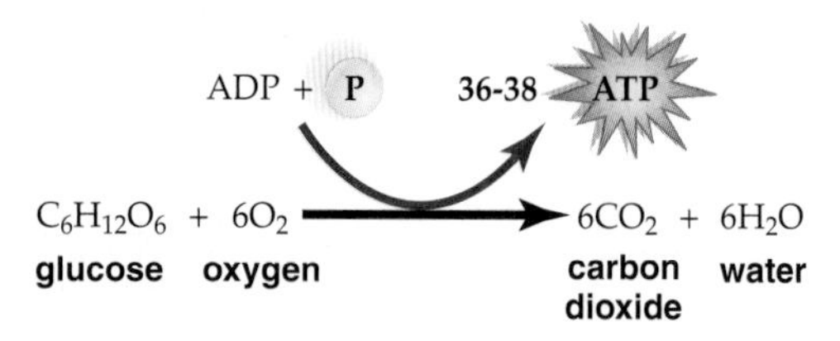

As glucose is broken down, ATP is built up, and this is why the ATP reaction is drawn using a curved arrow above the glucose reaction arrow. The breakdown of one glucose molecule results in 36 or 38 ATP molecules. This represents about 40% of the potential energy within a glucose molecule. The rest of the energy dissipates. This conversion is more efficient than many others. For example, only about 25% of the energy within gasoline is converted to the motion of a car.

Why does cellular metabolism convert the energy within the bonds of glucose to energy within the bonds of ATP? ATP contains just about the amount of energy required for most cellular reactions, such as joining one amino acid to another during protein synthesis or joining actin to myosin during muscle contraction. This energy is released by a relatively simple procedure: the removal of a single phosphate group. The cell uses the reverse reaction for building up ATP again.

NAD^+ and FAD

Even though we can write an overall equation for cellular respiration it involves many individual reactions, each one catalyzed by its own enzyme. Certain of these enzymes utilize the coenzyme **NAD^+ (nicotinamide adenine dinucleotide)** as an electron carrier. Coenzymes help an enzyme do its job and may even participate in the reaction. In this instance, NAD^+ receives two electrons as the substrate is oxidized. You'll recall from Chapter 6 that each electron is received by NAD^+ as part of a hydrogen atom. A hydrogen atom consists of a hydrogen ion (H^+) and an electron (e^-). As shown in Figure 7.1, NAD^+ receives 2 e- and 2 H^+ to give NADH + H^+.

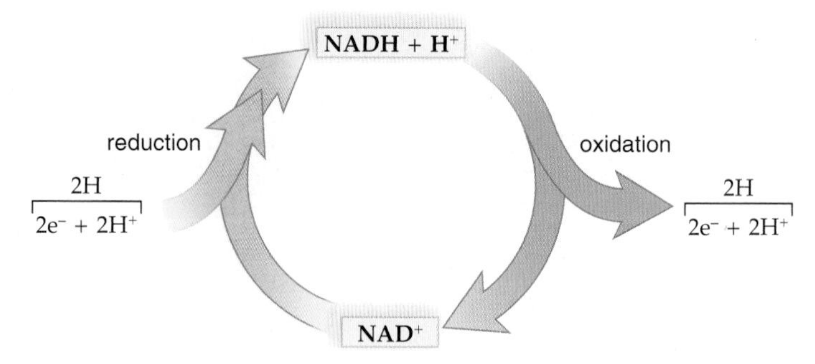

Figure 7.1 The NAD^+ cycle.
The coenzyme NAD^+ accepts two hydrogen atoms (H^+ + e^-), and NADH + H^+ results. When NADH passes on electrons, NAD^+ results. Only a small amount of NAD^+ need be present in a cell, because each NAD^+ molecule is used over and over again.

FAD (flavin adenine dinucleotide) is another frequently used electron carrier as a coenzyme. When FAD accepts 2 e- and 2 H^+, $FADH_2$ results. NAD^+ and FAD are analogous to electron shuttle buses. They pick up electrons at specific enzymatic reactions in either the cytoplasm or the matrix of the mitochondria and carry these high-energy electrons to an electron transport chain in the cristae of the mitochondria, where they drop them off. The empty NAD^+ or FAD is then free to go back and pick up more electrons.

Phases of Cellular Respiration

The metabolic pathways of cellular respiration allow the energy within a glucose molecule to be released in a stepwise fashion so that it can be coupled to ATP production. Cells would lose a tremendous amount of energy if glucose breakdown occurred all at once—much of that energy would become nonusable heat. These reactions have been organized into four phases (Fig. 7.2). The first phase, glycolysis, takes place outside the mitochondria and does not utilize oxygen. The other phases take place inside the mitochondria, where oxygen is the final acceptor of electrons.

- **Glycolysis** is the breakdown of glucose ($C_6H_{12}O_6$) to two molecules of pyruvate, a C_3 molecule. Oxidation by removal of electrons (e^-) and hydrogen ions (H^+) provides enough energy for the immediate buildup of two ATP.
- During the **preparatory (prep) reaction,** pyruvate is oxidized to a C_2 acetyl group carried by CoA (coenzyme A), and CO_2 is removed. Since glycolysis ends with two molecules of pyruvate, the prep reaction occurs twice per glucose molecule.
- The **citric acid cycle** is a cyclical series of oxidation reactions that give off CO_2 and produce one ATP. The citric acid cycle used to be called the Krebs cycle in honor of the man who worked out most of the steps. The cycle is

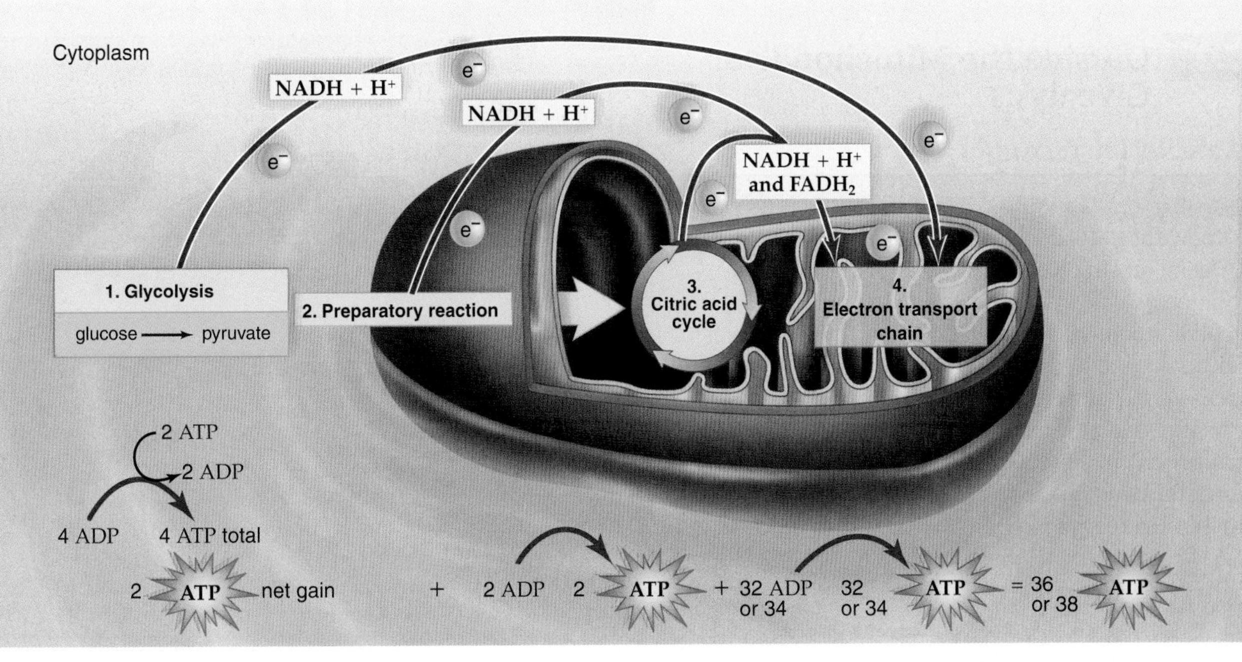

Figure 7.2 The four phases of complete glucose breakdown.
The complete breakdown of glucose consists of four phases. (1) Glycolysis in the cytoplasm produces pyruvate, which enters mitochondria if oxygen is available. (2,3) The preparatory reaction and the citric acid cycle that follow occur inside the mitochondria. (4) Also inside mitochondria, the electron transport chain receives the electrons that were removed from glucose breakdown products. The result of glucose breakdown is a maximum of 36 or 38 ATP, depending on the particular cell.

now named for citric acid (or citrate), the first molecule in the cycle. The citric acid cycle turns twice because two acetyl CoA molecules enter the cycle per glucose molecule. Altogether, the citric acid cycle accounts for two immediate ATP molecules per glucose molecule.

- The **electron transport chain** is a series of membrane-bound carriers that pass electrons from one carrier to another. High-energy electrons are delivered to the chain, and low-energy electrons leave it (Fig. 7.3). The electron transport chain is like stair steps. As something bounces down a flight of stairs it loses potential energy. Similarly, as electrons pass through the carriers from a higher-energy to a lower-energy state, energy is released and used for ATP synthesis. The electrons from one glucose molecule passing down the electron transport chain result in a maximum of 32 or 34 ATP, depending on certain conditions.

In cellular respiration, the low-energy electrons are finally received by O_2, which then combines with H^+ and becomes water.

Pyruvate is a pivotal metabolite in cellular respiration. If oxygen is not available to the cell, **fermentation** occurs in the cytoplasm instead of continued **aerobic** (with oxygen) cellular respiration. During fermentation, pyruvate is reduced to lactate or to carbon dioxide and alcohol, depending on the organism. As we shall see in section 7.4 fermentation results in a net gain of only two ATP per glucose molecule.

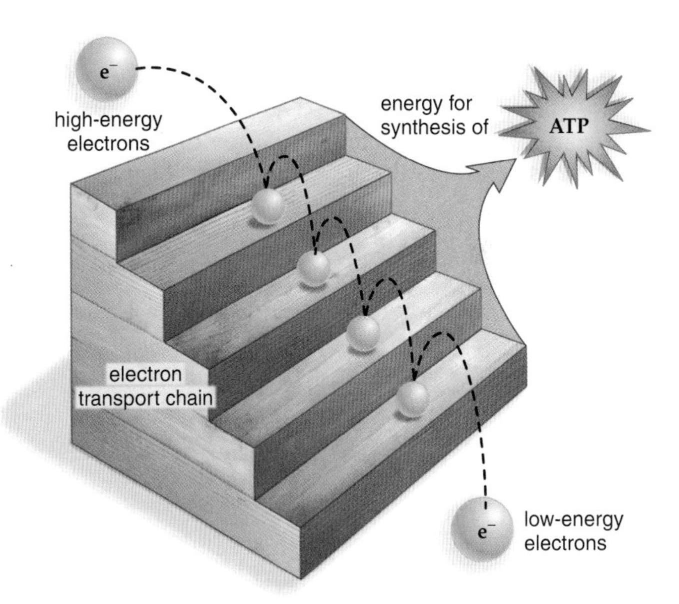

Figure 7.3 Electron transport chain.
High-energy electrons are delivered to the chain. As they pass from carrier to carrier, energy is released and used for ATP production.

Check Your Progress

1. What reactant does the CO_2 product come from?
2. Which phases in cellular respiration occur in the cytoplasm? In the mitochondria?
3. Which phase of cellular respiration results in the most ATP?

7.2 Outside the Mitochondria: Glycolysis

Learning Outcomes

1. List the inputs and outputs for glycolysis.
2. Know the cellular location of glycolysis.
3. Define substrate-level ATP synthesis.

Glycolysis, which takes place within the cytoplasm, is the breakdown of glucose to two pyruvate molecules (Fig. 7.4*a*). Since glycolysis occurs universally in all organisms, it most likely evolved before the citric acid cycle and the electron transport chain. Glycolysis mostly likely evolved when environmental conditions were anaerobic and before cells had mitochondria. This may be why glycolysis does not require oxygen and occurs in the cytoplasm.

Energy-Investment Steps

As glycolysis begins, two ATP are used to activate glucose, and the molecule that results splits into two C_3 molecules **(G3P, glyceraldehyde 3-phosphate),** each of which has an attached phosphate group. From this point on, each C_3 molecule undergoes the same series of reactions (Fig. 7.4*b*).

Energy-Harvesting Steps

Oxidation of G3P occurs by the removal of hydrogen atoms ($H^+ + e^-$). The hydrogen atoms are picked up by NAD^+, and $NADH + H^+$ results. Later, NADH will pass electrons on to the electron transport chain. Oxidation of G3P and subsequent substrates results in four high-energy phosphate groups, which are used to synthesize four ATP. This is **substrate-level ATP synthesis,** in which an enzyme passes a high-energy phosphate to ADP, and ATP results:

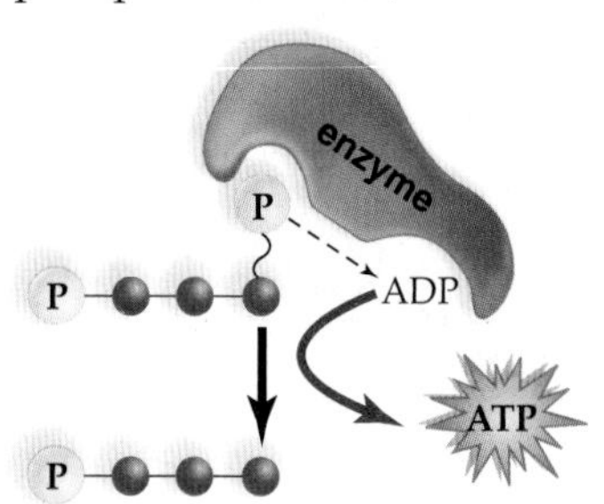

Subtracting the two ATP that were used to get started, glycolysis yields a net gain of two ATP (Fig. 7.4*b*).

When oxygen is available, the end product, pyruvate, enters the mitochondria, where it undergoes further breakdown. If oxygen is not available, pyruvate can be used for fermentation as described in section 7.4. For each glucose that enters glycolysis, two ATP, two $NADH + H^+$, and two pyruvate are formed.

Check Your Progress

1. What occurs when G3P is oxidized? What molecule receives the electrons?
2. How many total ATP are made during the energy-harvesting steps of glycolysis? Why is there only a net gain of two ATP?

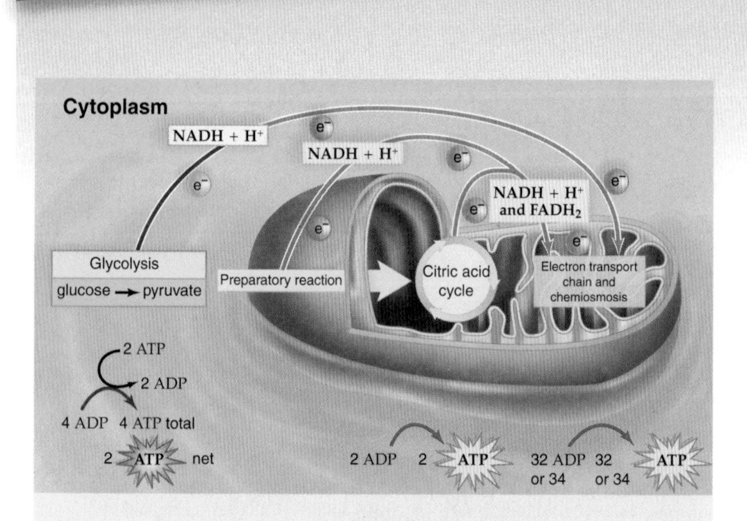

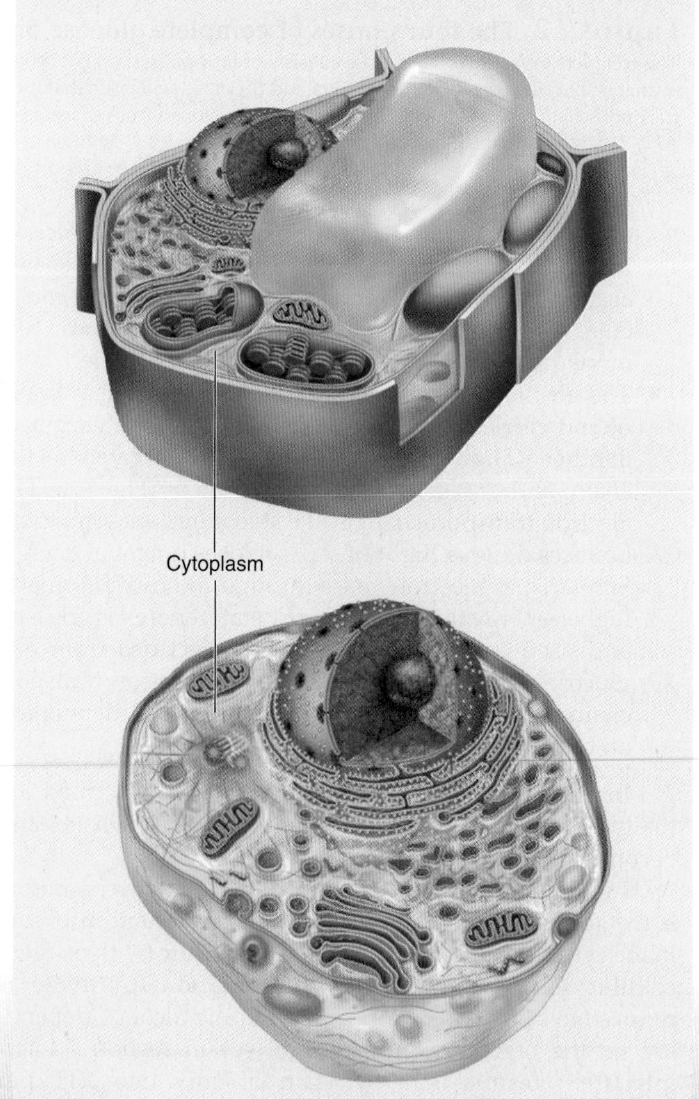

a.

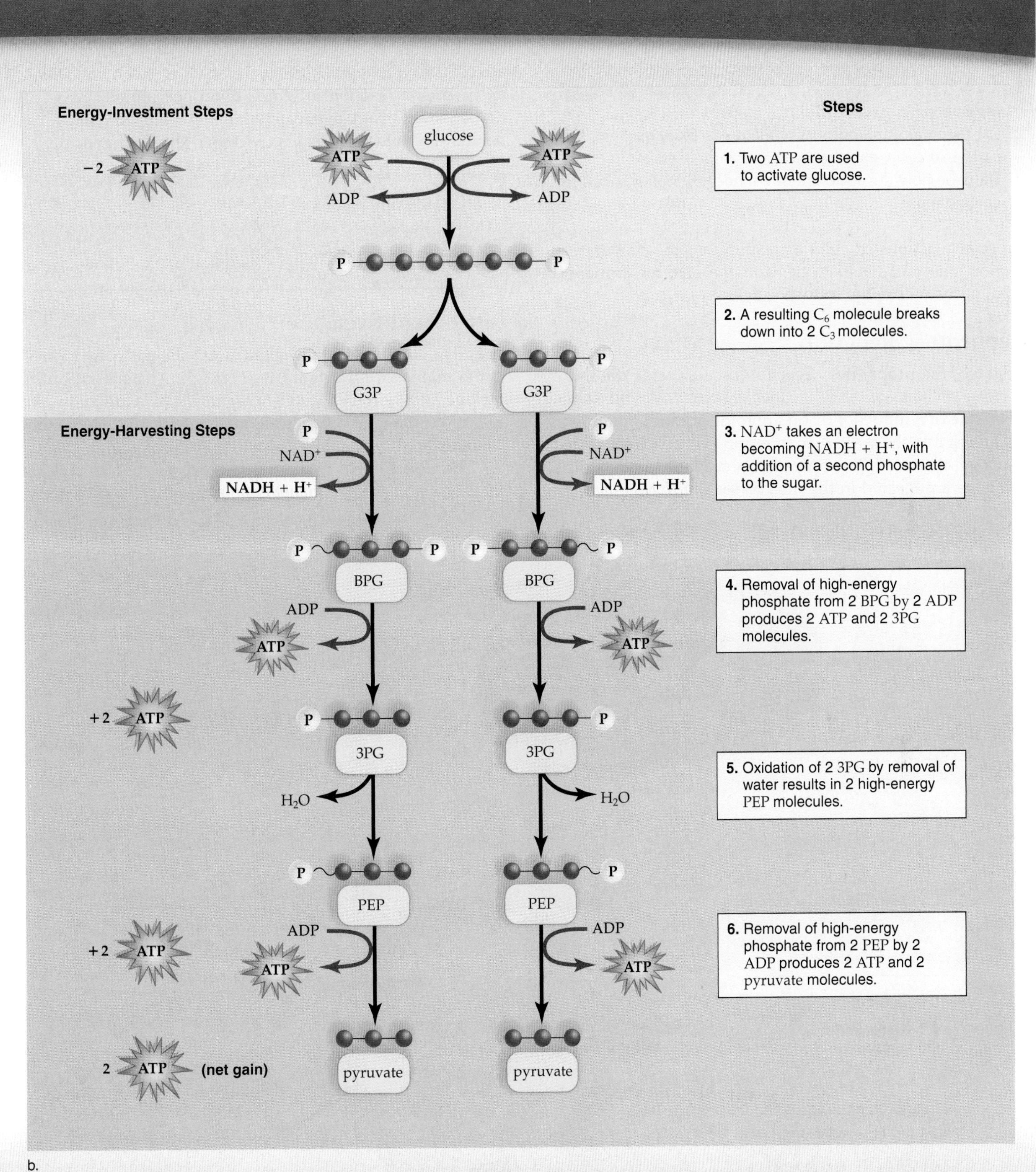

b.

Figure 7.4 Glycolysis.
a. Glycolysis takes place in the cytoplasm of almost all cells. **b.** This pathway begins with glucose and ends with two pyruvate molecules. There is a gain of two $NADH + H^+$ and a net gain of two ATP from glycolysis.

7.3 Inside the Mitochondria

Learning Outcomes

1. Know the precise location of the last two phases of cellular respiration.
2. List the inputs and outputs of the preparatory reaction, the citric acid cycle, and the electron transport chain.
3. Describe how the structure of a mitochondrion is suited to chemiosmosis.

The final reactions of cellular respiration, the preparatory reaction, the citric acid cycle, and the electron transport chain, occur within the mitochondria.

Preparatory Reaction

As stated, the preparatory reaction occurs inside the mitochondria. Where specifically does it occur? As you know, the **cristae** of a mitochondrion are folds of inner membrane that jut out into the **matrix,** an innermost compartment filled with a gel-like fluid. The preparatory reaction and the citric acid cycle are located in the matrix (see Fig. 7.2).

The preparatory (prep) reaction is so called because it produces the molecule that can enter the citric acid cycle. In this reaction, pyruvate is converted to a C_2 *acetyl group* attached to *coenzyme A (CoA),* and CO_2 is given off. This is an oxidation reaction in which hydrogen atoms ($H^+ + e^-$) are removed from pyruvate by NAD^+ and $NADH + H^+$ results. This reaction occurs twice per glucose molecule:

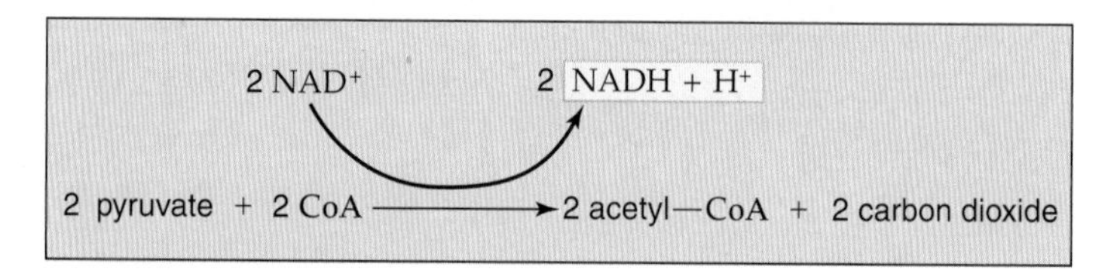

Citric Acid Cycle

The citric acid cycle is a cyclical metabolic pathway located in the matrix of mitochondria (Fig. 7.5). At the start of the citric acid cycle, the C_2 acetyl group carried by CoA joins with a C_4 molecule, and a C_6 citrate molecule results. (Note that the citric acid cycle will turn twice per glucose molecule.)

The CoA returns to the preparatory reaction to pick up another acetyl group. During the citric acid cycle, each acetyl

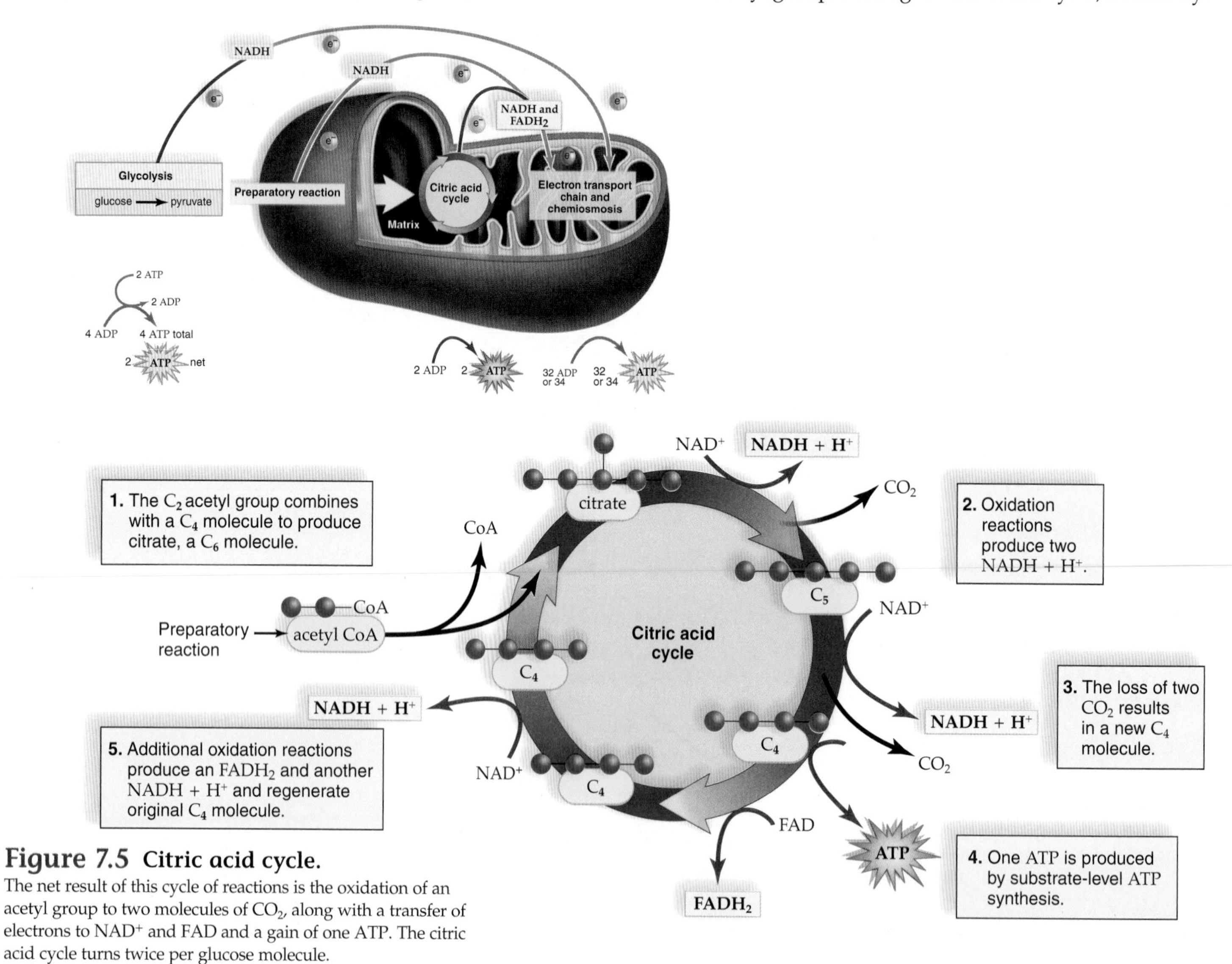

Figure 7.5 Citric acid cycle.
The net result of this cycle of reactions is the oxidation of an acetyl group to two molecules of CO_2, along with a transfer of electrons to NAD^+ and FAD and a gain of one ATP. The citric acid cycle turns twice per glucose molecule.

group received from the preparatory reaction is oxidized to two CO_2 molecules. As the reactions of the cycle occur, oxidation is carried out by the removal of hydrogen atoms ($H^+ + e^-$). In three instances, $NADH + H^+$ results, and in one instance, $FADH_2$ is formed.

Substrate-level ATP synthesis is also an important event of the citric acid cycle. In substrate-level ATP synthesis, an enzyme passes a high-energy phosphate to ADP, and ATP results.

Because the citric acid cycle turns twice for each original glucose molecule, the inputs and outputs of the citric acid cycle per glucose molecule are as follows:

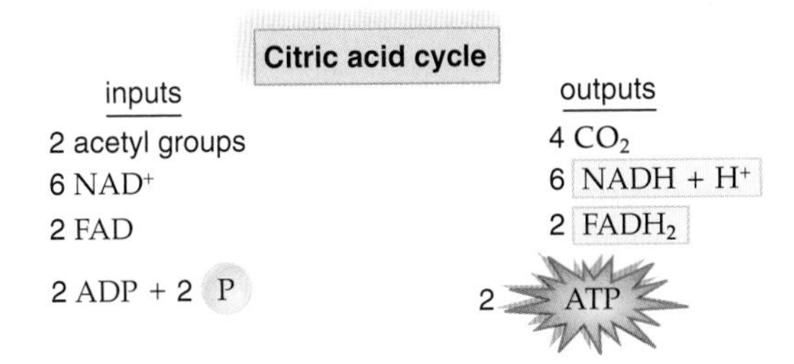

The six carbon atoms originally located in the glucose molecule have now become CO_2. The preparatory reaction produces two CO_2, and the citric acid cycle produces four CO_2 per glucose molecule.

Electron Transport Chain

The electron transport chain located in the cristae of the mitochondria is a series of carriers that pass electrons from one to the other. Some of the electron carriers of the system are called **cytochrome** molecules. Cytochromes are a class of iron-containing proteins important in redox reactions.

Figure 7.6 is arranged to show that high-energy electrons enter the system and low-energy electrons leave the system. Notice that when NADH gives up its two electrons to the chain, it becomes NAD^+ and 2 H^+ remain. Similarly, when $FADH_2$ gives up two electrons to the chain, it becomes FAD and 2 H^+ remain. The next carrier gains the electrons and is reduced. Then each of the carriers, in turn, becomes reduced and then oxidized as the electrons move down the system.

As the electrons pass from one carrier to the next, energy that will be used to produce ATP molecules is captured and stored as a hydrogen ion gradient (see Fig. 7.7). Oxygen receives the energy-spent electrons from the last of the carriers. After receiving electrons, oxygen combines with hydrogen ions and forms water.

Once NADH has delivered electrons to the electron transport chain, the NAD^+ that results can pick up more hydrogen atoms. In like manner, once $FADH_2$ gives up electrons to the chain, the FAD that results can pick up more hydrogen atoms. The recycling of coenzymes and ADP increases cellular efficiency because it does away with the necessity to synthesize NAD^+ and FAD every time.

Check Your Progress

1. List the inputs and outputs of the preparatory reaction.
2. List the inputs and outputs of the citric acid cycle.
3. How are NAD^+ and FAD like a shuttle bus?

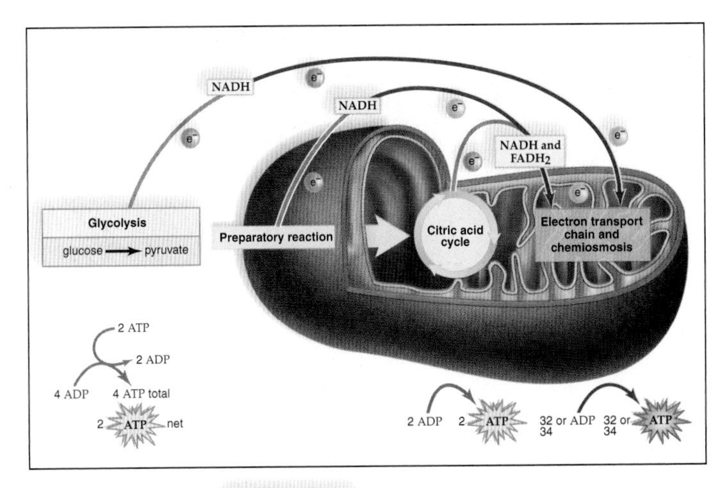

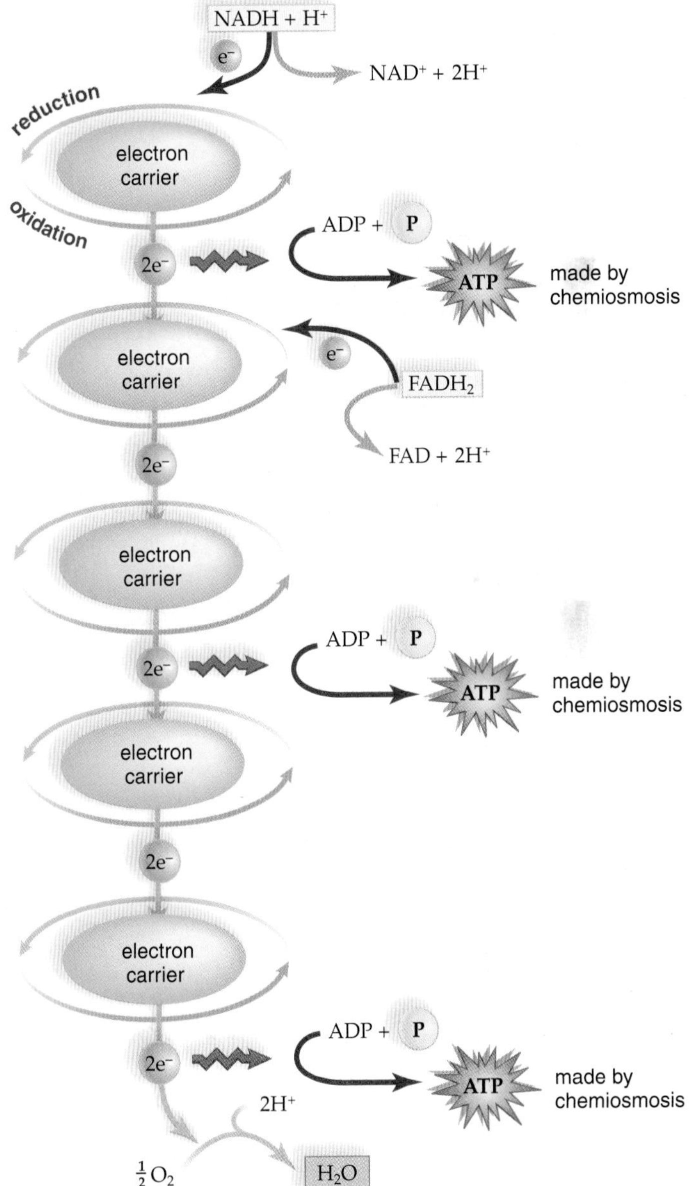

Figure 7.6 The electron transport chain.
NADH and $FADH_2$ bring electrons to the electron transport chain. As the electrons move down the chain, energy is captured and used to form ATP. For every two electrons that enter by way of NADH, two to three ATP result. For every two electrons that enter by way of $FADH_2$, one to two ATP result. Oxygen, the final acceptor of the electrons, becomes a part of water.

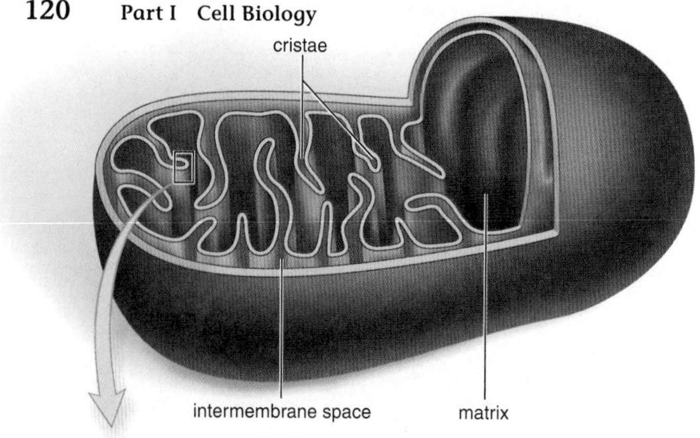

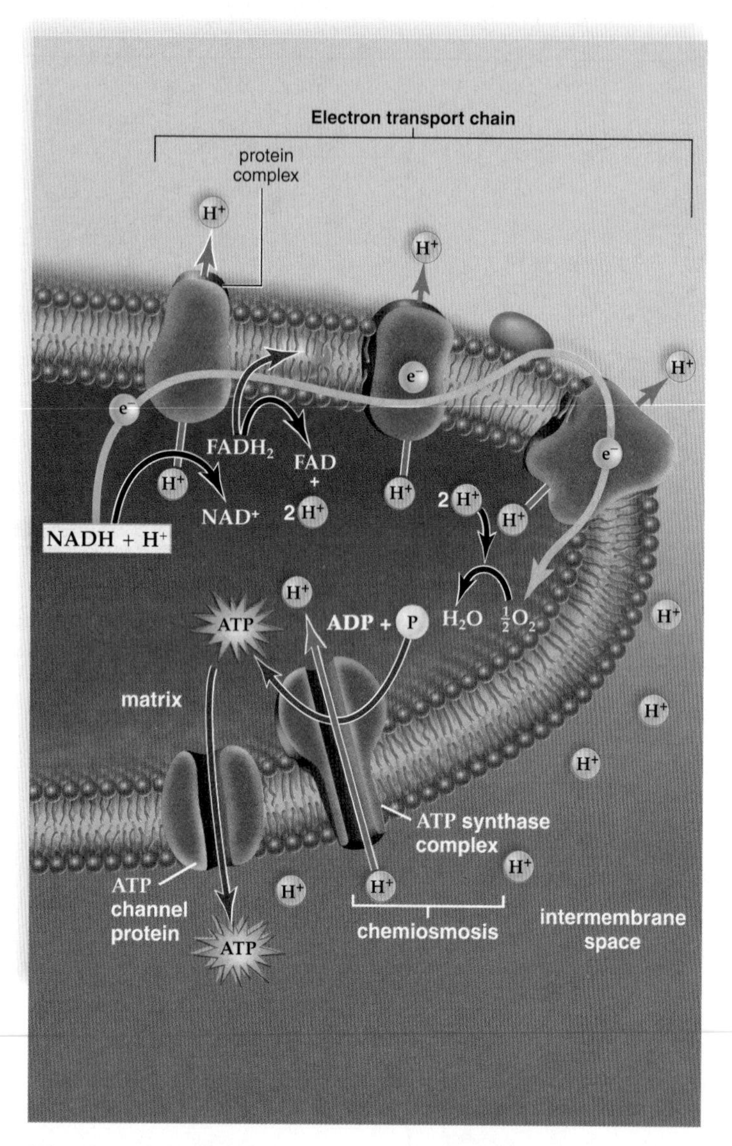

Figure 7.7 Organization and function of cristae in mitochondria.

The electron transport chain is located in the cristae. As electrons move from one protein complex to the other, hydrogen ions (H^+) are pumped from the mitochondrial matrix into the intermembrane space. As hydrogen ions flow down a concentration gradient from the intermembrane space into the matrix, ATP is synthesized by the enzyme ATP synthase. ATP leaves the matrix by way of a channel protein.

Organization of Cristae

The electron transport chain is located within the cristae of the mitochondria. The cristae increase the internal surface area of a mitochondrion, thereby increasing the area devoted to ATP formation. Figure 7.6 is a simplified overview of the electron transport chain. Figure 7.7 shows how the components of the electron transport chain are arranged in the cristae.

We have been stressing that the carriers of the electron transport chain accept electrons, which they pass from one to the other. What happens to the hydrogen ions from NADH + H^+? The complexes in the cristae use the energy released by electrons as they move down the electron transport chain to pump H^+ from the mitochondrial matrix into the space between the outer and inner membrane of a mitochondrion. This space is called the **intermembrane space.** The pumping of H^+ into the intermembrane space establishes an unequal distribution of H^+ ions; there are many H^+ in the intermembrane space and few in the matrix of a mitochondrion. This means that the intermembrane space is positively charged in relation to the matrix as well as more acidic.

The cristae also contain an *ATP synthase complex.* The H^+ ions flow through an ATP synthase complex from the intermembrane space into the matrix. The flow of H^+ through an ATP synthase complex brings about a change in shape, which causes the enzyme ATP synthase to synthesize ATP from ADP + Ⓟ . Mitochondria produce ATP by **chemiosmosis,** so called because ATP production is tied to an electrochemical gradient, namely the unequal distribution of H^+ across the cristae. Once formed, ATP molecules pass into the cytoplasm.

Chemiosmosis is similar to using water behind a dam to generate electricity. The pumping of H^+ out of the matrix into the intermembrane space is like pumping water behind the dam. The floodgates of the dam are like the ATP synthase complex. When the floodgates are open, the water rushes through, generating electricity. In the same way, H^+ rushing through the ATP synthase complex is used to produce ATP.

Energy Yield from Cellular Respiration

Figure 7.8 calculates the maximum ATP yield for the complete breakdown of glucose to CO_2 and H_2O. Per glucose molecule, there is a net gain of two ATP from glycolysis, which takes place in the cytoplasm. The citric acid cycle, which occurs in the matrix of mitochondria, accounts for two ATP per glucose molecule. This means that a total of four ATP are formed by substrate-level ATP synthesis outside the electron transport chain.

The remaining 32-34 ATP are produced by the electron transport chain and chemiosmosis. Per glucose molecule, ten NADH and two $FADH_2$ deliver electrons to the electron transport chain. ATP synthesis can be measured by experiments where intact mitochondria are suspended in a solution

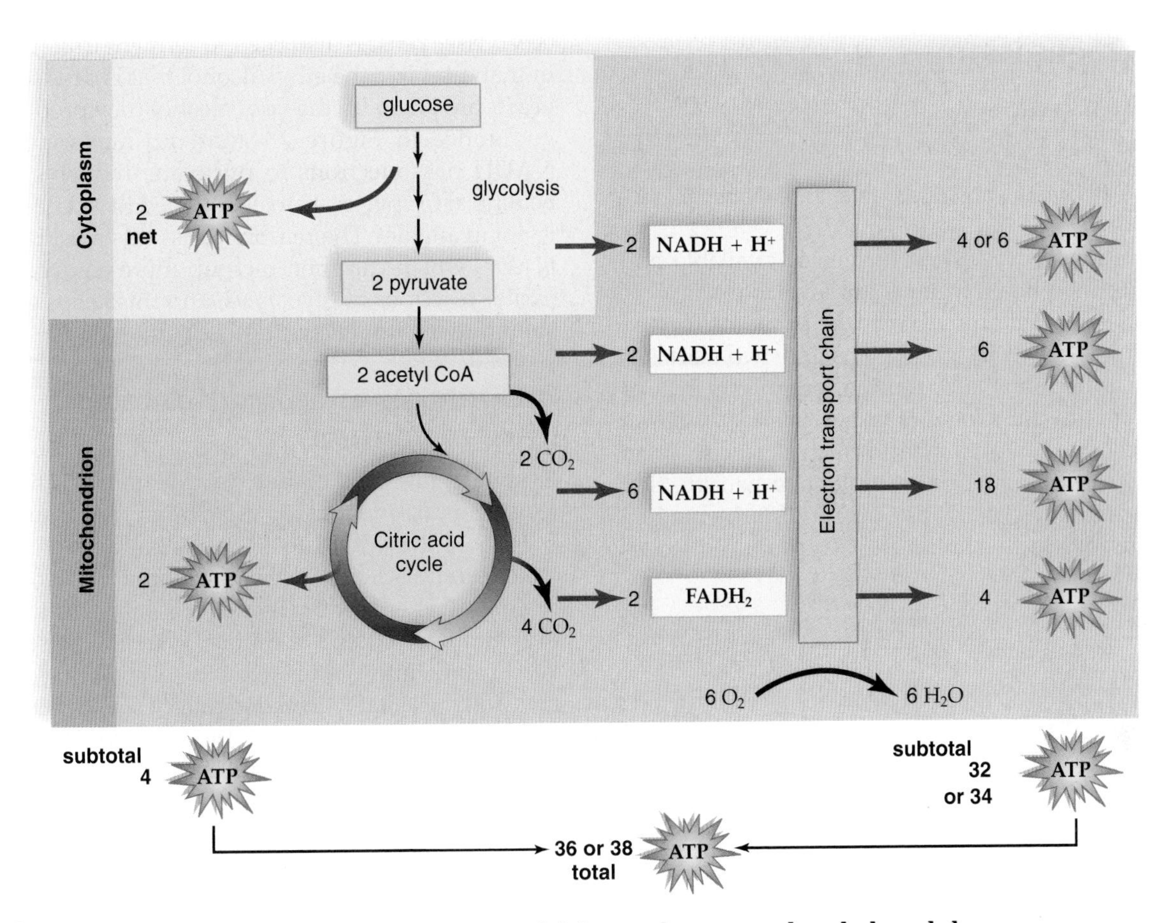

Figure 7.8 Accounting of the maximum energy yield per glucose molecule breakdown.
Substrate-level ATP synthesis during glycolysis and the citric acid cycle accounts for four ATP. Chemiosmosis accounts for a maximum of 32 or 34 ATP, depending on the shuttle mechanism involved for transporting cytoplasmic NADH + H^+ into the mitochondrion. The maximum total of ATP is therefore 36 or 38 ATP.

containing oxygen. Such experiments yield between two and three ATP for the two electrons delivered to the electron transport chain by each NADH, and between one and two ATP for the two electrons delivered by each $FADH_2$. Researchers are unable to determine more accurate values because ATP and oxygen are being consumed by other reactions in the mitochondria, and researchers cannot purify the electron transport chain and study it in isolation. Why the difference between NADH and $FADH_2$? Figure 7.6 shows the reason for this difference: $FADH_2$ delivers its electrons to the transport chain lower in the chain than does NADH, and therefore these electrons cannot account for as much ATP production.

These calculations are for NADH formed *inside* the mitochondria by the citric acid cycle. What about the ATP yield of NADH generated *outside* the mitochondria by the glycolytic pathway? In some cells, NADH cannot cross mitochondrial membranes, but a transfer mechanism allows its electrons to be delivered to the electron transport chain inside the mitochondria. The cost to the cell is one ATP for the two electrons transferred from outside to inside the mitochondrion. Since there are two NADH formed during glycolysis, this reduces the overall count of ATP produced, in some cells, to four instead of six ATP (see Fig. 7.8).

Efficiency of Cellular Respiration

It is interesting to calculate how much of the energy in a glucose molecule eventually becomes available to the cell. The difference in energy content between the reactants (glucose and O_2) and the products (CO_2 and H_2O) is 686 kcal. An ATP phosphate bond has an energy content of 7.3 kcal, and if 36 of these are produced during glucose breakdown, 36 phosphates are equivalent to a total of 263 kcal. Therefore, 263/686, or 39%, of the available energy is usually transferred from glucose to ATP. The rest of the energy dissipates in the form of heat.

Check Your Progress

1. Why is the intermembrane space more acidic than the matrix?
2. How is the ATP synthase complex like the gate of a dam?
3. Why do the electrons carried by $FADH_2$ result in fewer ATP than do the electrons carried by NADH?

7.4 Fermentation

Learning Outcomes

1. Explain when fermentation occurs.
2. Describe the advantages and disadvantages of fermentation.

Cellular respiration occurs when oxygen is available and fermentation occurs when oxygen is not available (Fig. 7.9). In human cells, like other animal cells, the pyruvate formed by glycolysis accepts two hydrogen ions and two electrons and is reduced to lactate. Other types of organisms instead produce alcohol with the release of CO_2. Bacteria vary as to whether they produce an organic acid, such as lactate, or an alcohol and CO_2. Yeasts are good examples of organisms that generate ethyl alcohol and CO_2 as a result of fermentation. When yeast is used to leaven bread, the CO_2 produced makes bread rise. Yeast is also used to ferment fruit juice into wine. In that case, it is the ethyl alcohol that is desired. Eventually, yeasts are killed by the very alcohol they produce.

Notice in Figure 7.9 that during fermentation, two NADH pass electrons to pyruvate, reducing it. Why is it beneficial for pyruvate to be reduced to lactate when oxygen is not available? The reason is that this reaction regenerates NAD^+, which can then pick up more electrons during the earlier reactions of glycolysis, and this keeps glycolysis and substrate-level ATP synthesis going.

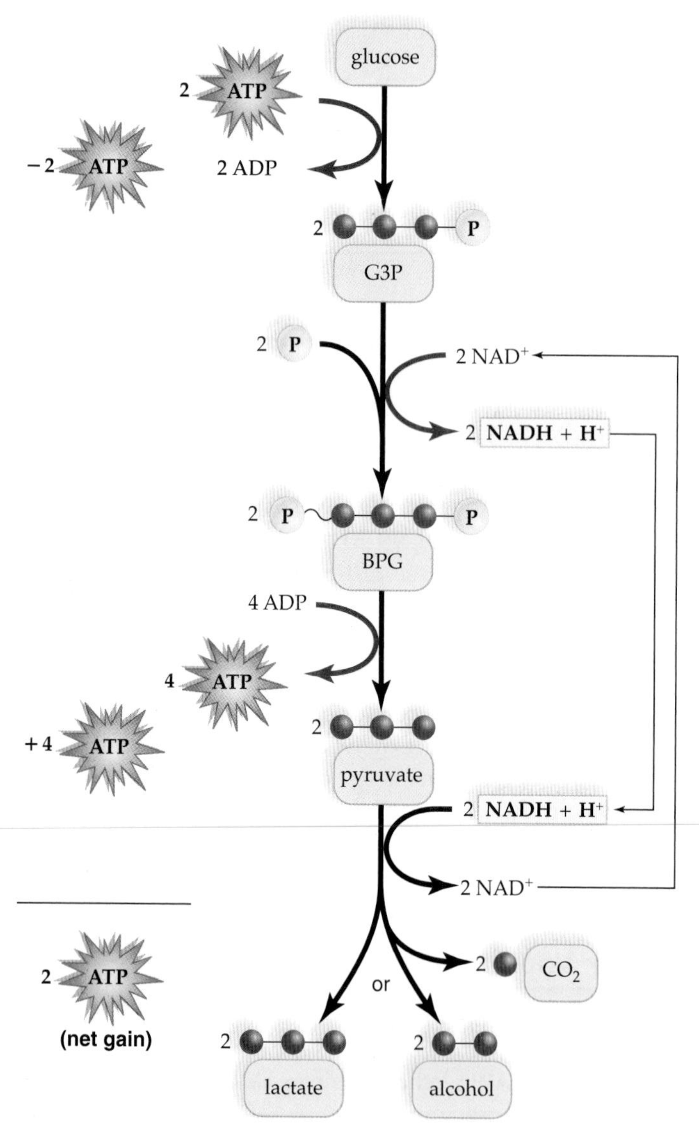

Figure 7.9 Fermentation.
Fermentation consists of glycolysis followed by a reduction of pyruvate by NADH + H^+. The resulting NAD^+ returns to the glycolytic pathway to pick up more hydrogen atoms.

Advantages and Disadvantages of Fermentation

Despite its low yield of only two ATP, fermentation is essential to humans. It can provide a rapid burst of ATP, and thus muscle cells more than other cells are apt to carry on fermentation. When our muscles are working vigorously over a short period of time, as when we run, fermentation is a way to produce ATP even though oxygen is temporarily in limited supply.

Lactate, however, is toxic to cells. At first, blood carries away all the lactate formed in muscles. But eventually lactate begins to build up, changing the pH and causing the muscles to "burn." When we stop running, our bodies are in **oxygen debt,** as signified by the fact that we continue to breathe very heavily for a time. Recovery is complete when the lactate is transported to the liver, where it is reconverted to pyruvate. Some of the pyruvate is broken down completely, and the rest is converted back to glucose.

Energy Yield of Fermentation

Fermentation produces only two ATP by substrate-level ATP synthesis. These two ATP represent only a small fraction of the potential energy stored in a glucose molecule. As noted earlier, complete glucose breakdown during cellular respiration results in a maximum of 36 to 38 ATP. Therefore, following fermentation, most of the potential energy a cell can capture from the respiration of a glucose molecule is still waiting to be released.

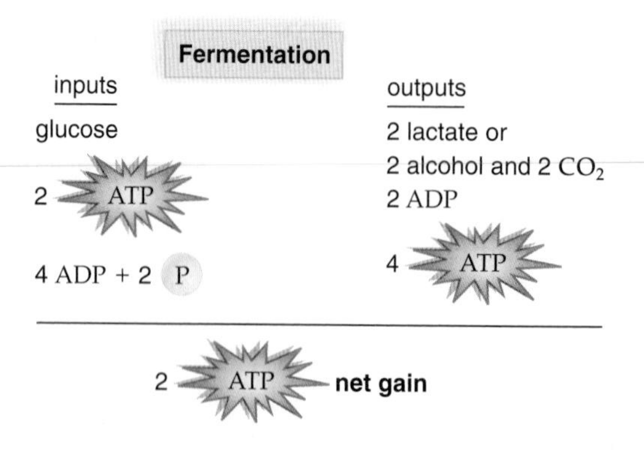

Check Your Progress

1. What do you suppose would happen if the cell ran out of NAD^+?
2. What chemical is responsible for that muscle burn you feel during heavy exercise?

Hibernation: A Long Sleep with No Eats!

When food becomes scarce during the winter months, some animals go into a period of dormancy known as hibernation. Animals such as raccoons, skunks, bats, chipmunks, and bears hibernate. Hibernation periods usually begin in early October and end in March or April. Bears have a different method of hibernation than smaller animals. Smaller animals drop their body temperature, enter a comatose state, and slow their heart rate and respirations to almost imperceptible levels. However, they must wake periodically to eat and drink, and rid themselves of waste. Bears will drop their heart rate to approximately one-fifth of normal and their body temperature only five to nine degrees below normal. They are not comatose. American black bears can go as long as 100 days without eating or drinking, or eliminating waste by urination or defecation. How is it possible to go this long without taking in food? What do hibernating bears use for their energy source? In preparation for this long period of inactivity, bears may gain up to 30 pounds of fat per week. The stored fat is their energy source. Fat yields almost twice the energy of carbohydrates or protein. Fat is converted into acetyl CoA, which enters the citric acid cycle and is oxidized to carbon dioxide and water. While carbon dioxide is exhaled, the water is used for metabolic needs.

What is fascinating to scientists is that bears do not develop any of the toxic side effects of utilizing fat as a sole energy source as humans would. Bears also do not undergo muscle wasting as humans would during such prolonged inactivity. Nor do they suffer from osteoporosis. Researchers are very interested in the medical applications of hibernation. NASA is especially interested in how to maintain muscle strength and bone mass in space under inactive conditions.

Discussion Questions

1. Why is fat a good energy source?
2. What advantage do bears have over a small animal like a squirrel when it comes to hibernation during the winter?

Applying the Concepts Revisited

During aerobic exercise, oxygen is available for the muscles to completely break down glucose to carbon dioxide and water. During anaerobic exercise, oxygen is not available and the muscles produce lactic acid.

1. Of the four phases of cellular respiration, which one requires oxygen?
2. Some bacteria produce lactic acid as the final product of glucose breakdown, but this is not the case in muscle tissue. How does the human body metabolize lactic acid produced during anaerobic exercise?

SUMMARIZING THE CONCEPTS

7.1 Overview of Cellular Respiration

- During cellular respiration, glucose is oxidized to CO_2 and H_2O. Four phases are required for glucose to be metabolized to carbon dioxide and water. During oxidation of substrates, electrons are removed along with hydrogen ions (H^+).

7.2 Outside the Mitochondria: Glycolysis

- Glycolysis, the breakdown of glucose to two pyruvates, is a series of enzymatic reactions that occur in the cytoplasm.

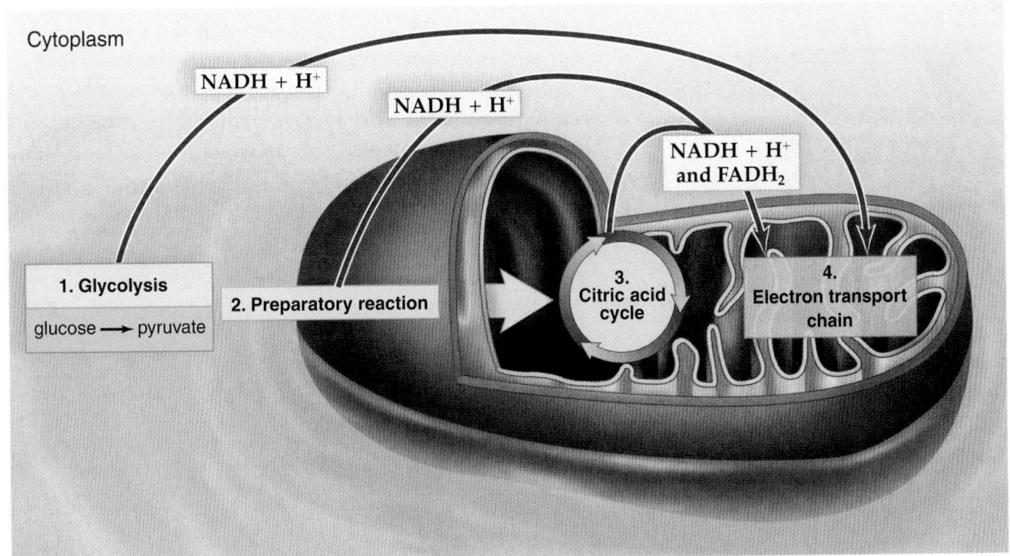

- Oxidation by NAD^+ releases enough energy immediately to give a net gain of two ATP by substrate-level ATP synthesis. Two NADH + H^+ are formed.

7.3 Inside the Mitochondria

- Pyruvate from glycolysis enters a mitochondrion, where the preparatory reaction takes place. During this reaction, oxidation occurs as CO_2 is removed. NAD^+ is reduced, and CoA receives the C_2 acetyl group that remains. Since the reaction must take place twice per glucose, two NADH + H^+ result.

- The acetyl group enters the citric acid cycle, a series of reactions located in the mitochondrial matrix. Complete oxidation follows, as two CO_2, three NADH + H^+, and one $FADH_2$ are formed. The cycle also produces one ATP. The entire cycle must turn twice per glucose molecule.
- The final stage of glucose breakdown involves the electron transport chain located in the cristae of the mitochondria. The electrons received from NADH and $FADH_2$ are passed down a chain of electron carriers until they are finally received by O_2, which combines with H^+ to produce H_2O. As the electrons pass down the chain, ATP is produced.
- The carriers of the electron transport chain are located in protein complexes on the cristae of the mitochondria. Each protein complex receives electrons and pumps H^+ into the intermembrane space, setting up an electrochemical gradient. When H^+ ions flow down this gradient through the ATP synthase complex, energy is released and used to form ATP molecules from ADP and Ⓟ. This is ATP synthesis by chemiosmosis.
- To calculate the total number of ATP produced per glucose molecule, consider that for each NADH + H^+ formed inside the mitochondrion, three ATP are produced. Each molecule of $FADH_2$ results in the formation of only two ATP because the electrons enter the electron transport chain at a lower energy level than the electrons delivered by NADH. In cells that cannot transfer NADH across the mitochondrial membrane, the ATP counts are reduced by two because the electrons from NADH generated in the cytoplasm during glycolysis must be transferred from outside to inside the mitochondria using two ATP.
- Of the maximum number of 36 or 38 ATP formed by cellular respiration, four are produced outside the electron transport chain: two by glycolysis and two by the citric acid cycle. The rest are produced by the electron transport chain.

7.4 Fermentation

- Fermentation involves glycolysis, followed by the reduction of pyruvate by NADH + H^+ to either lactate or alcohol and CO_2. The reduction process regenerates NAD^+ so that it can accept more electrons during glycolysis.
- Although fermentation results in only two ATP, it still serves a purpose: in humans, it provides a quick burst of ATP energy for short-term, strenuous muscular activity. The accumulation of lactate puts the individual in oxygen debt because oxygen is needed to completely metabolize lactate to CO_2 and H_2O.

TESTING YOURSELF

Choose the best answer for each question.

For questions 1–5, match the letters in the diagram to the appropriate statement.

1. Preparatory reaction
2. Electron transport chain
3. Glycolysis
4. Citric acid cycle
5. NADH + H^+ and $FADH_2$

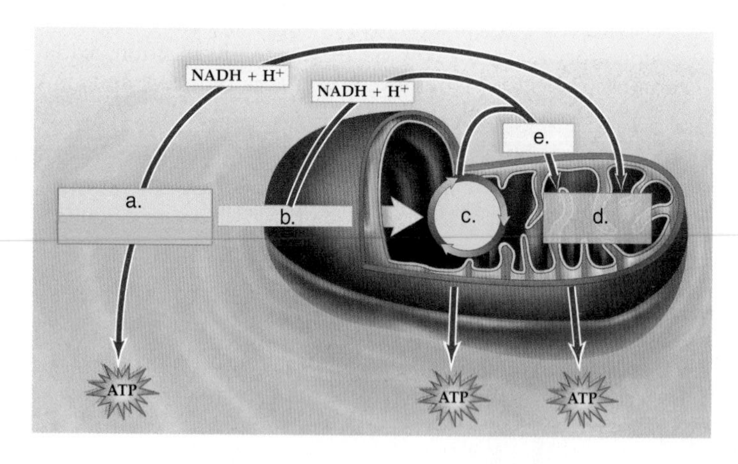

6. Which type of human cells carry on the most fermentation?
   a. fat
   b. muscle
   c. nerve
   d. bone
7. Which phase of cellular respiration occurs in the cytoplasm?
   a. glycolysis
   b. citric acid cycle
   c. preparatory reaction
   d. electron transport chain
8. Cellular respiration cannot occur without
   a. sodium.
   b. oxygen.
   c. lactic acid.
   d. All of these are correct.
9. Mitochondria produce ATP by
   a. glycolysis.
   b. fermentation.
   c. chemiosmosis.
   d. All of these are correct.
10. An ATP synthase complex can be found in the
    a. cytoplasm.
    b. mitochondria.
    c. Both a and b are correct.
    d. Neither a nor b is correct.
11. The correct order of these processes would be
    a. citric acid cycle, glycolysis, prep reaction.
    b. glycolysis, prep reaction, citric acid cycle.
    c. prep reaction, citric acid cycle, glycolysis.
    d. None of these are correct.
12. Which of the following is needed for glycolysis to occur?
    a. pyruvate
    b. glucose
    c. NAD^+
    d. ATP
    e. All of these are needed except a.
13. Which of the following is *not* a product, or end result, of the citric acid cycle?
    a. carbon dioxide
    b. pyruvate
    c. NADH + H^+
    d. ATP
    e. $FADH_2$

14. How many ATP molecules are produced by the oxidation of one molecule of NADH via the electron transport chain?
    a. 1 to 2
    b. 2 to 3
    c. 36 to 38
    d. 8 to 10
15. What is the name of the process that adds the third phosphate to an ADP molecule using the flow of hydrogen ions across a membrane?
    a. substrate-level ATP synthesis
    b. fermentation
    c. reduction
    d. chemiosmosis
16. Which are possible products of fermentation?
    a. lactic acid
    b. alcohol
    c. CO_2
    d. All of these are correct.
17. The metabolic process that produces the most ATP molecules is
    a. glycolysis.
    b. the citric acid cycle.
    c. the electron transport chain.
    d. fermentation.
18. Substrate-level ATP synthesis takes place during
    a. glycolysis and the citric acid cycle.
    b. the electron transport chain and the prep reaction.
    c. glycolysis and the electron transport chain.
    d. the citric acid cycle and the prep reaction.

UNDERSTANDING THE TERMS

aerobic 115
cellular respiration 114
chemiosmosis 120
citric acid cycle 114
cristae 118
cytochrome 119
electron transport chain 115
FAD (flavin adenine dinucleotide) 114
fermentation 115
G3P (glyceraldehyde 3-phosphate) 116
glycolysis 114
intermembrane space 120
matrix 118
NAD^+ (nicotinamide adenine dinucleotide) 114
oxygen debt 122
preparatory (prep) reaction 114
pyruvate 115
substrate-level ATP synthesis 116

THINKING CRITICALLY

1. Bacteria do not have mitochondria, and yet they contain an electron transport chain. On what membrane could this be located?
2. Rotenone is a broad-spectrum insecticide that inhibits the electron transport chain. Why might it be toxic to humans?
3. Explain the benefit of a stepwise breakdown of glucose during cellular respiration rather than a rapid breakdown.

INQUIRY INTO LIFE WEBSITE

The companion website for *Inquiry into Life* provides a wealth of information organized and integrated by chapter. You will find practice tests, animations, videos, and much more that will complement your learning and understanding of general biology.

http://www.mhhe.com/maderinquiry13

PART II

Plant Biology

Photosynthesis

Chapter Outline

Applying the Concepts

Taking a walk in the woods in the fall when the leaves are turning colors can be an enjoyable and relaxing form of exercise. Interestingly, the same process that causes leaves to change colors in the fall is also involved in the ripening of fruits such as apples and pears—and the end product of the process may have health benefits as well! Leaves contain several types of pigments, including the green chlorophylls and the yellow to red carotenoids. These pigments absorb the solar energy that the plant utilizes to carry out photosynthesis. In the fall when lower temperatures signal a change in the seasons, the supply of water and nutrients to the leaves declines and chlorophyll begins to degrade. Now, the carotenoids, which were formerly masked by chlorophyll become visible allowing us to enjoy the change of color. Researchers have discovered that when a fruit ripens and its skin changes color, chlorophyll is degraded in the same manner as in leaves. The degradation of chlorophyll produces molecules called non-fluorescent chlorophyll catabolites (NCCs) that become concentrated in the skin. When researchers examined the properties of these NCCs, they discovered that they were antioxidants! Antioxidants stabilize free radicals, dangerous molecules that otherwise damage the DNA and proteins of a cell. Several health problems, including cancer and heart disease, are thought to be promoted by free radicals. So the breakdown of chlorophyll provides us with colorful woodlands and can also give us a healthy dose of antioxidants when we eat fruits.

In this chapter we will see how pigments are involved in the process of photosynthesis. The solar energy they capture is used by the plant to make its food and this food sustains the plant and any organisms that feed on the plant.

8.1 Overview of Photosynthesis

Learning Outcomes

1. Define autotroph and heterotroph.
2. Understand the critical role of photosynthesis for all organisms on Earth.
3. Write the overall chemical equation for photosynthesis.
4. Describe the process of photosynthesis in terms of two sets of reactions that take place in a chloroplast.

Photosynthesis converts solar energy into the chemical energy of a carbohydrate. Photosynthetic organisms, including plants, algae, and cyanobacteria, are called **autotrophs** because they produce their own food (Fig. 8.1). Each year, photosynthesizing organisms produce approximately 170 billion metric tons of carbohydrates. No wonder photosynthetic organisms are able to sustain themselves and all other living things on Earth.

With few exceptions, it is possible to trace any food chain back to plants and algae. In other words, producers, which have the ability to synthesize carbohydrates, feed not only themselves but also consumers, which must take in preformed organic molecules. Collectively, consumers are called **heterotrophs.** Both autotrophs and heterotrophs use organic molecules produced by photosynthesis as a source of building blocks for growth and repair and as a source of chemical energy for cellular work.

Pigments allow photosynthetic organisms to capture solar energy and this is the "fuel" that makes photosynthesis possible. Most photosynthetic organisms contain **chlorophyll,** the pigment that gives them a green color. However, the green of chlorophyll can be masked by other pigments. The **carotenoids** give photosynthesizing cells a yellow to red color. In addition, the phycobilins give red algae their red color and cyanobacteria a bluish color.

Flowering Plants as Photosynthesizers

Although many different organisms can photosynthesize, we will limit our discussion to the flowering plants. Portions of the plant, particularly the leaves, contain chlorophyll and other pigments that enable the plant to carry on photosynthesis. The leaf of a flowering plant contains mesophyll tissue in which cells are specialized for photosynthesis (Fig. 8.2). The raw materials for photosynthesis are water

a. *Oscillatoria* 100×

b. Kelp

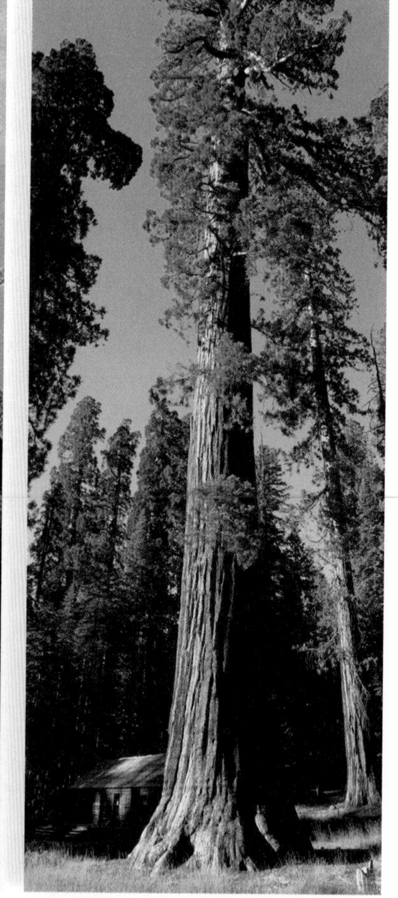

c. Sequoia

Figure 8.1 Photosynthetic organisms. Photosynthetic organisms include **a.** cyanobacteria such as *Oscillatoria,* which are a type of bacterium; **b.** algae such as kelp, which typically live in water and can range in size from microscopic to macroscopic; and **c.** plants such as the sequoia, which typically live on land.

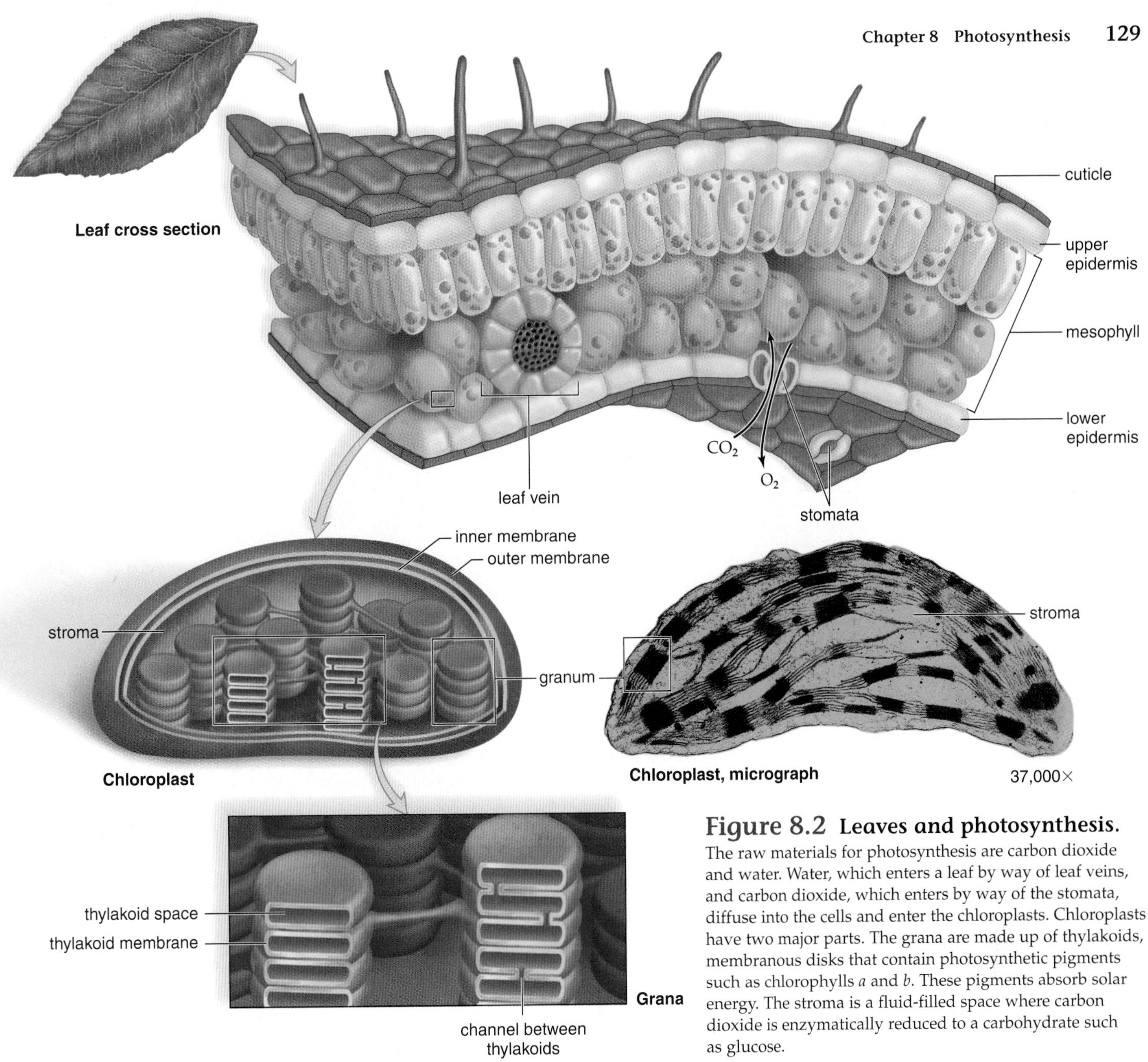

Figure 8.2 **Leaves and photosynthesis.** The raw materials for photosynthesis are carbon dioxide and water. Water, which enters a leaf by way of leaf veins, and carbon dioxide, which enters by way of the stomata, diffuse into the cells and enter the chloroplasts. Chloroplasts have two major parts. The grana are made up of thylakoids, membranous disks that contain photosynthetic pigments such as chlorophylls *a* and *b*. These pigments absorb solar energy. The stroma is a fluid-filled space where carbon dioxide is enzymatically reduced to a carbohydrate such as glucose.

and carbon dioxide. The roots of a plant absorb water, which then moves in vascular tissue up the stem to a leaf by way of the leaf veins. Carbon dioxide in the air enters a leaf through small openings called **stomata** (sing., stoma). After entering a leaf, carbon dioxide and water diffuse into the cells and enter the chloroplasts, the organelles that carry on photosynthesis.

A double membrane surrounds a chloroplast and its fluid-filled interior, which is called the stroma. A different membrane system within the stroma forms flattened sacs called thylakoids. In some places, thylakoids are stacked to form grana (sing., granum), so called because they looked like piles of seeds to early microscopists. The space of each thylakoid is connected to the space of every other thylakoid within a chloroplast, thereby forming a continuous inner compartment within chloroplasts called the thylakoid space.

Chlorophyll and other pigments that are part of a thylakoid membrane are capable of absorbing solar energy. This is the energy that drives photosynthesis. The stroma, filled with enzymes, is where carbon dioxide is first attached to an organic compound and is then reduced to a carbohydrate. Therefore, it is proper to associate the absorption of solar energy with the thylakoid membranes making up the grana, and to associate the reduction of carbon dioxide to a carbohydrate with the stroma of a chloroplast.

Human beings, and indeed nearly all organisms, release carbon dioxide into the air. This is some of the same carbon dioxide that enters a leaf through the stomata and is converted to carbohydrate. Carbohydrate, in the form of glucose, is the chief energy source for most organisms.

Check Your Progress

1. What is the source of energy for carbohydrate production in plants?
2. Where does photosynthesis take place in a plant?

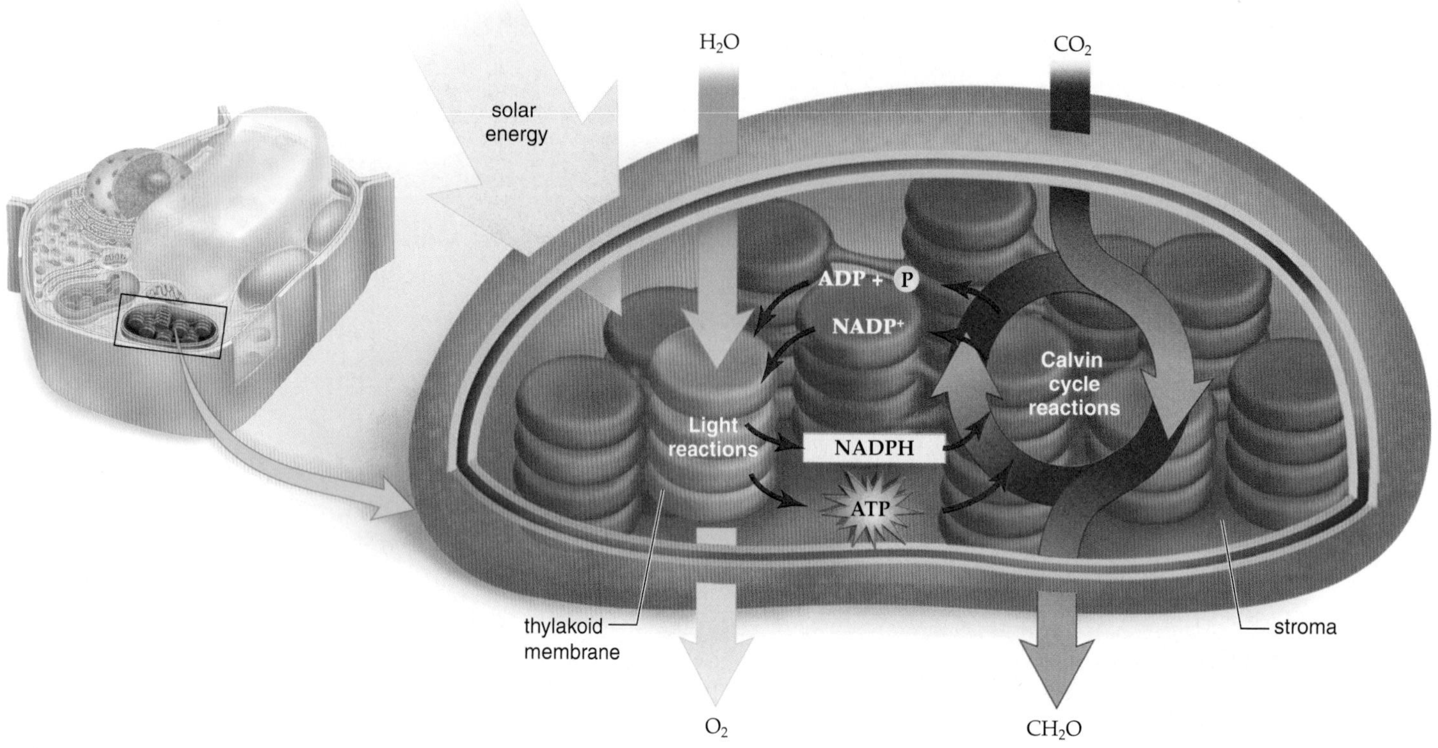

Figure 8.3 Overview of photosynthesis.
The process of photosynthesis consists of the light reactions and the Calvin cycle reactions. The light reactions, which produce ATP and NADPH, occur in the thylakoid membrane. These molecules are used in the Calvin cycle reactions in the stroma to reduce carbon dioxide to a carbohydrate.

Photosynthetic Reaction

For convenience, the overall equation for photosynthesis is sometimes simplified in this manner:

$$6\,CO_2 + 6\,H_2O \xrightarrow{\text{solar energy}} C_6H_{12}O_6 + 6\,O_2$$

This equation shows glucose and oxygen as the products of photosynthesis.

The oxygen given off by photosynthesis comes from water. This was proven experimentally by exposing plants first to CO_2 and then to H_2O that contained an isotope of oxygen called heavy oxygen (^{18}O). Only when heavy oxygen was a part of water did this isotope appear in O_2 given off by the plant. Therefore, O_2 released by chloroplasts comes from H_2O, not from CO_2.

The equation also shows that during photosynthesis, CO_2 gains hydrogen atoms and becomes a carbohydrate. Reduction of any molecule requires energy and during photosynthesis this energy is provided by the sun.

The overall chemical equation for photosynthesis tells us why plants are so important to our lives. Plants provide us with food in the form of carbohydrates and they produce the oxygen that we breathe and use for cellular respiration.

Two Sets of Reactions

An overall equation for photosynthesis tells us the beginning reactants and the end products of the pathway. But much goes on in between. The word photosynthesis suggests that the process requires two sets of reactions: *photo,* which means light, refers to the reactions that capture solar energy, and *synthesis* refers to the reactions that produce carbohydrate.

The two sets of reactions are called the **light reactions** (light-dependent reactions) and the **Calvin cycle reactions** (light-independent reactions) (Fig. 8.3). We will see that the coenzyme **NADP+ (nicotinamide adenine dinucleotide phosphate)** carries hydrogen atoms from the light reactions to the Calvin cycle reactions. When NADP+ accepts hydrogen atoms, it becomes NADPH. Also, ATP carries energy between the two sets of reactions.

Check Your Progress

1. Write the overall equation for photosynthesis.
2. Name the two sets of reactions involved in photosynthesis.

8.2 Solar Energy Capture

Learning Outcomes

1. Relate the photosynthetic pigments to the radiant energy they absorb.
2. Explain the role of the noncyclic electron pathway and the cyclic electron pathway.
3. Describe the organization of the thylakoid and how this organization is critical to the production of ATP during photosynthesis.

During the light reactions, the pigments within the thylakoid membranes absorb solar energy. Solar energy (radiant energy from the sun) can be described in terms of its wavelength and its energy content. Figure 8.4 lists the different types of radiant energy, from the shortest wavelength, gamma rays, to the longest, radio waves. White, or *visible,* light is only a small portion of this spectrum.

Visible Light

Visible light itself contains various wavelengths of light. When it is passed through a prism, we see all the different colors that make up visible light. (Actually, it is our brains that interpret these wavelengths as colors.) The colors in visible light range from violet (the shortest wavelength) to indigo, blue, green, yellow, orange, and red (the longest wavelength). The energy content is highest for violet light and lowest for red light.

Only about 42% of the solar radiation that hits Earth's atmosphere ever reaches the surface of Earth, and most of this radiation is within the visible-light range. Higher-energy wavelengths are screened out by the ozone layer in the atmosphere, and lower-energy wavelengths are screened out by water vapor and carbon dioxide (CO_2) before they reach Earth's surface. Both the organic molecules within organisms and certain life processes, such as vision and photosynthesis, are adapted to the solar radiation that is most prevalent in the environment.

The pigments found within most types of photosynthesizing cells, the chlorophylls *a* and *b* and the carotenoids, are capable of absorbing various portions of visible light. The absorption spectrum for these pigments is shown in Figure 8.5. Both chlorophyll *a* and chlorophyll *b* absorb violet, indigo, blue, and red light better than the light of other colors. Because green light is reflected and only minimally absorbed, leaves appear green to us. The yellow or orange carotenoids are able to absorb light in the violet-blue-green range. These pigments and others become noticeable in the fall when chlorophyll breaks down and the other pigments are uncovered. Photosynthesis begins when the pigments within thylakoid membranes absorb solar energy.

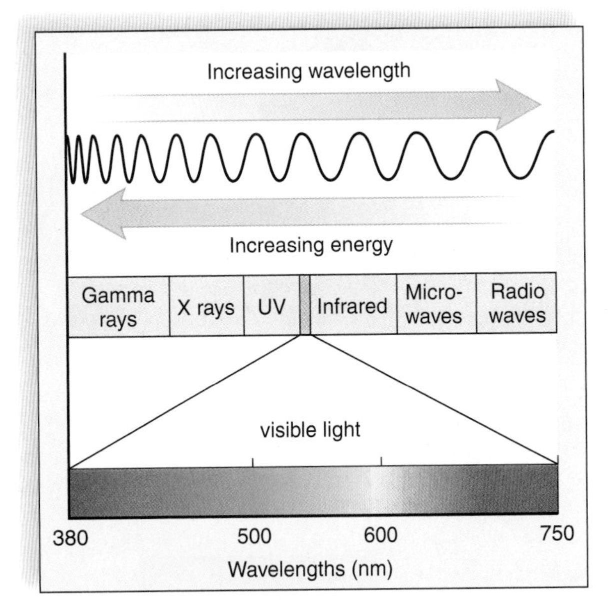

Figure 8.4 The electromagnetic spectrum.
The electromagnetic spectrum extends from the very short wavelengths of gamma rays through the very long wavelengths of radio waves. Visible light, which drives photosynthesis, is expanded to show its component colors. The components differ according to wavelength and energy content.

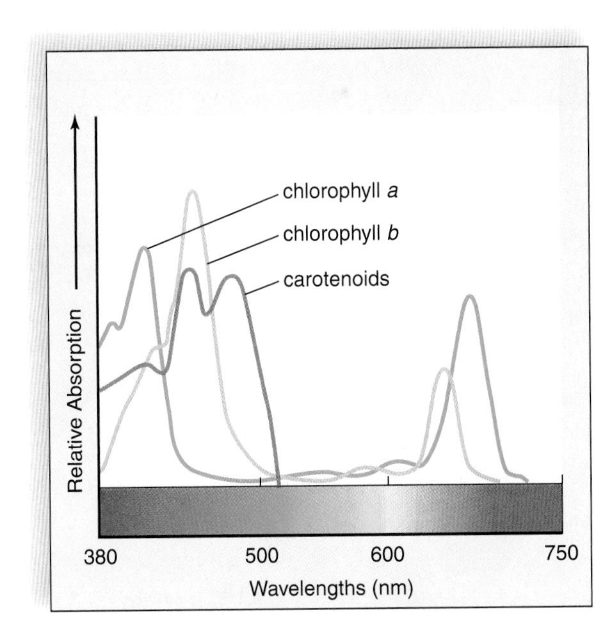

Figure 8.5 Photosynthetic pigments and photosynthesis.
The photosynthetic pigments in chlorophylls *a* and *b* and the carotenoids absorb certain wavelengths within visible light. This is their absorption spectrum. Notice that they do not absorb green light. That's why the leaves of plants appear green to us. You observe the color that is not absorbed by the leaf.

Light Reactions

The light reactions that occur in the thylakoid membrane consist of two electron pathways called the noncyclic electron pathway and the cyclic electron pathway. Both electron pathways produce ATP, but only the noncyclic pathway also produces NADPH.

Noncyclic Electron Pathway

The **noncyclic electron pathway** is so named because the electron flow can be traced from water to a molecule of $NADP^+$ (Fig. 8.6). This pathway uses two photosystems, called photosystem I (PS I) and photosystem II (PS II). The photosystems are named for the order in which they were discovered, not for the order in which they participate in the photosynthetic process. A **photosystem** consists of a pigment complex (molecules of chlorophyll *a*, chlorophyll *b*, and the carotenoids) and an electron acceptor within the thylakoid membrane. The pigment complex serves as an "antenna" for gathering solar energy.

The noncyclic pathway begins with photosystem II. The pigment complex absorbs solar energy, which is then passed from one pigment molecule to another until it is concentrated in a particular pair of chlorophyll *a* molecules called the *reaction center*. Electrons (e^-) in the reaction center chlorophyll become so energized that they escape from the reaction center and move to a nearby electron acceptor.

Photosystem II would disintegrate without replacement electrons. These replacement electrons are provided by water, which splits, releasing oxygen to the atmosphere. Many organisms, including plants and even ourselves, use this oxygen. The hydrogen ions (H^+) stay in the thylakoid

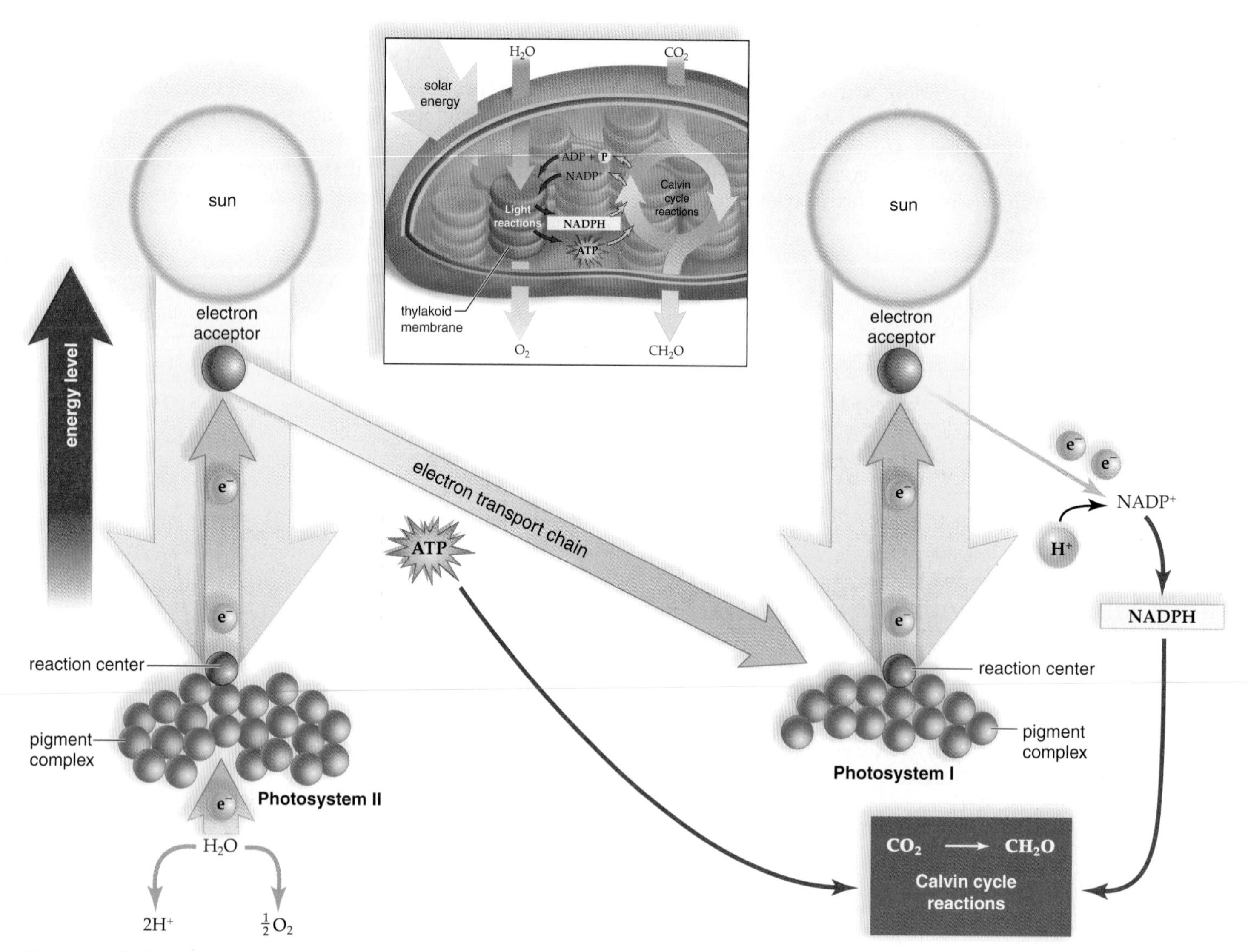

Figure 8.6 Noncyclic electron pathway: Electrons move from water to $NADP^+$.
Energized electrons (taken from water, which splits, releasing oxygen) leave photosystem II and pass down an electron transport chain, leading to the formation of ATP. Energized electrons (replaced by photosystem II) leave photosystem I and pass to $NADP^+$, which then combines with H^+, becoming NADPH.

space and contribute to the formation of a hydrogen ion gradient. As shown in the overall chemical reaction for photosynthesis, water is used up and oxygen is produced.

The electron acceptor that received the energized electrons from the reaction center then sends those electrons down an electron transport chain a series of carriers that pass electrons from one to the other. As the electrons pass from one carrier to the next, the energy that is released is used to move hydrogen ions (H^+) from the stroma into the thylakoid space, forming a hydrogen ion gradient. When these hydrogen ions flow down their electrochemical gradient through **ATP synthase** complexes, ATP production occurs, as will be described shortly. Notice that this ATP will be used in the Calvin cycle reactions in the stroma to reduce carbon dioxide to a carbohydrate.

Similarly, when the photosystem I pigment complex absorbs solar energy, energized electrons leave its reaction center and are captured by a different electron acceptor. After going through the electron transport chain, the electrons from photosystem II are now low-energy electrons. These electrons are used to replace those lost by photosystem I. The electron acceptor in photosystem I passes its electrons to $NADP^+$ molecules. Each $NADP^+$ accepts two electrons and a H^+ to become a reduced form of the molecule—that is, NADPH. This NADPH will also be used by the Calvin cycle reactions in the stroma to reduce carbon dioxide to a carbohydrate.

Cyclic Electron Pathway

The **cyclic electron pathway** (Fig. 8.7) begins when the photosystem I pigment complex absorbs solar energy that is passed from one pigment to the other until it is concentrated in a reaction center. As with photosystem II, electrons (e^-) become so energized that they escape from the reaction center and move to nearby electron acceptor molecules.

This time, instead of the electrons moving on to $NADP^+$, energized electrons (e^-) taken up by an electron acceptor are sent down an electron transport chain. As the electrons pass from one carrier to the next, their energy becomes stored as a hydrogen (H^+) gradient. Again, the flow of hydrogen ions down their electrochemical gradient through ATP synthase complexes produces ATP.

The spent electrons return to photosystem I after the electron transport chain. This is how photosystem I receives replacement electrons and why this electron pathway is called cyclic. It is also why the cyclic pathway produces ATP but does not produce NADPH.

In plants, the reactions of the Calvin cycle can use any extra ATP produced by the cyclic pathway because the Calvin cycle reaction requires more ATP molecules than NADPH molecules. Also, other enzymatic reactions, aside from those involving photosynthesis, are occurring in the stroma and can use this ATP. It's possible that the cyclic flow of electrons is used alone when carbohydrate is not being produced. At this time, there would be no need for NADPH, which is produced only by the noncyclic electron pathway.

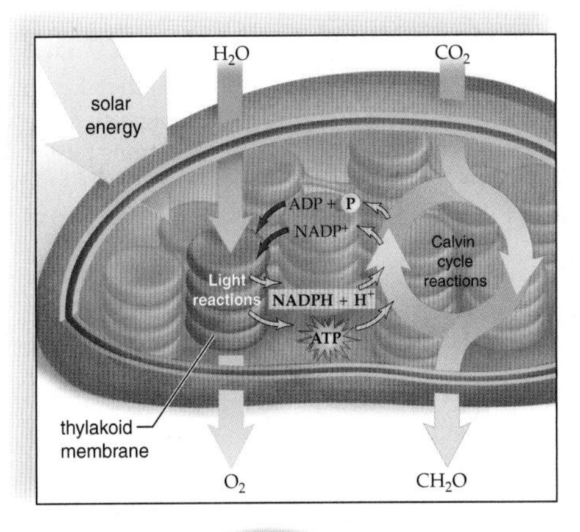

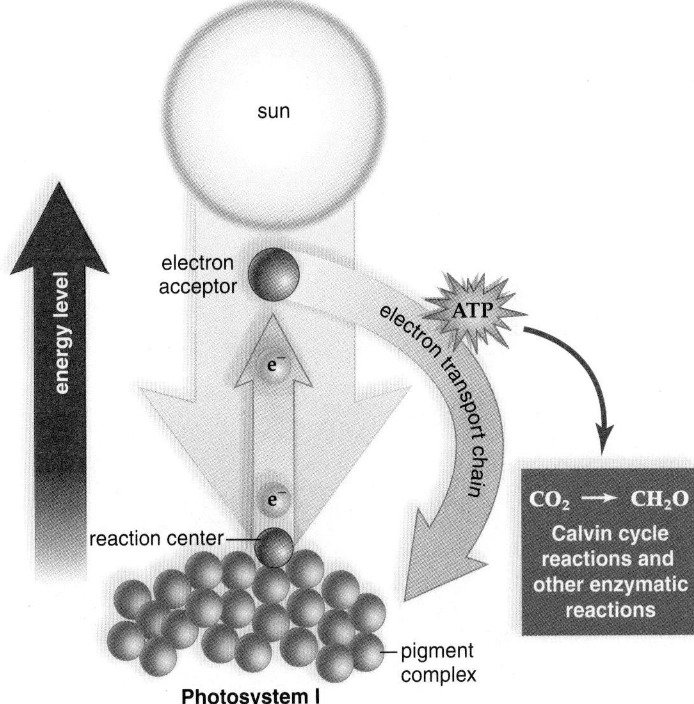

Figure 8.7 Cyclic electron pathway: Electrons leave and return to photosystem I.
Energized electrons leave the photosystem I reaction center and are taken up by an electron acceptor, which passes them down an electron transport chain before they return to photosystem I. Only ATP production occurs as a result of this pathway.

Check Your Progress

1. With reference to Figure 8.5, the carotenoids absorb what type of radiant energy?
2. Why do leaves appear green to us?
3. Relate the production of both NADPH and ATP to the involvement of both photosystems in noncyclic photosynthesis.
4. Relate the production of only ATP to the involvement of only one photosystem in cyclic photosynthesis.

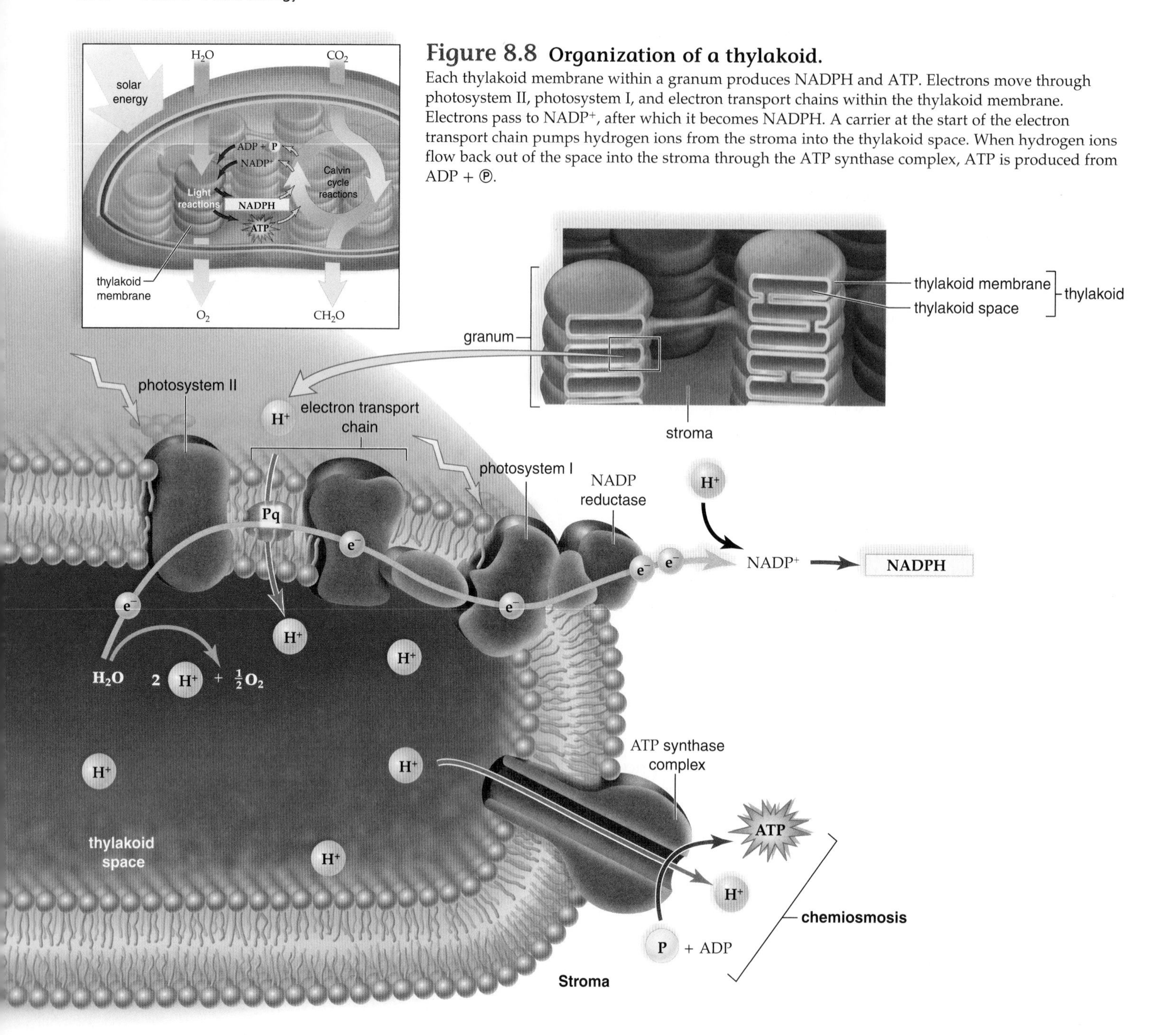

Figure 8.8 Organization of a thylakoid.
Each thylakoid membrane within a granum produces NADPH and ATP. Electrons move through photosystem II, photosystem I, and electron transport chains within the thylakoid membrane. Electrons pass to $NADP^+$, after which it becomes NADPH. A carrier at the start of the electron transport chain pumps hydrogen ions from the stroma into the thylakoid space. When hydrogen ions flow back out of the space into the stroma through the ATP synthase complex, ATP is produced from ADP + Ⓟ.

The Organization of the Thylakoid Membrane

As we have discussed, the following molecular complexes are present in the thylakoid membrane (Fig. 8.8):

Photosystem II, which consists of a pigment complex and an electron acceptor molecule, receives electrons from water, which splits, releasing oxygen.

The electron transport chain carries electrons from photosystem II to photosystem I, and pumps H^+ from the stroma into the thylakoid space.

Photosystem I, which also consists of a pigment complex and an electron acceptor molecule, is adjacent to NADP reductase, which reduces $NADP^+$ to NADPH.

The ATP synthase complex crosses the thylakoid membrane and contains an interior channel and a protruding ATP synthase, an enzyme that joins ADP + Ⓟ.

ATP Production

The thylakoid space acts as a reservoir for hydrogen ions (H^+). First, each time oxygen is removed from water, two H^+ remain in the thylakoid space. Second, as the electrons move from carrier to carrier along the electron transport chain, the electrons give up energy, which is used to pump H^+ from the stroma into the thylakoid space. This is like pumping water behind a dam. The end result is that there are more H^+ in the thylakoid space than in the stroma. Because H^+ is charged, this is an electrochemical gradient. Not only are there are more hydrogen ions, but there are more positive charges within the thylakoid space than the stroma. When a channel is opened in the ATP synthase complex, H^+ flows from the thylakoid space into the stroma, just like water flowing from behind a dam. The flow of H^+ from high to low concentration across the thylakoid membrane provides the energy that allows the ATP synthase enzyme to enzymatically produce ATP from ADP + Ⓟ. This method of producing ATP is called chemiosmosis because ATP production is tied to the establishment of an H^+ gradient.

Check Your Progress

1. What part of a thylakoid contains the photosystems, electron transport chain, and the ATP synthase complex?
2. Why is the H^+ gradient across a thylakoid membrane referred to as a storage of energy?

8.3 Calvin Cycle Reactions

Learning Outcomes

1. Describe the three major steps of the Calvin cycle.
2. Relate the product of the Calvin cycle to the other molecules found in a plant and therefore to heterotrophs.

The Calvin cycle reactions follow the light reactions. The Calvin cycle is a series of reactions that produce carbohydrate before returning to the starting point once more (Fig. 8.9). Therefore, this set of reactions is called a cycle. The cycle is named for Melvin Calvin, who, with colleagues, used the radioactive isotope ^{14}C as a tracer to discover the reactions making up the cycle.

This series of reactions uses carbon dioxide from the atmosphere to produce carbohydrate. How does carbon

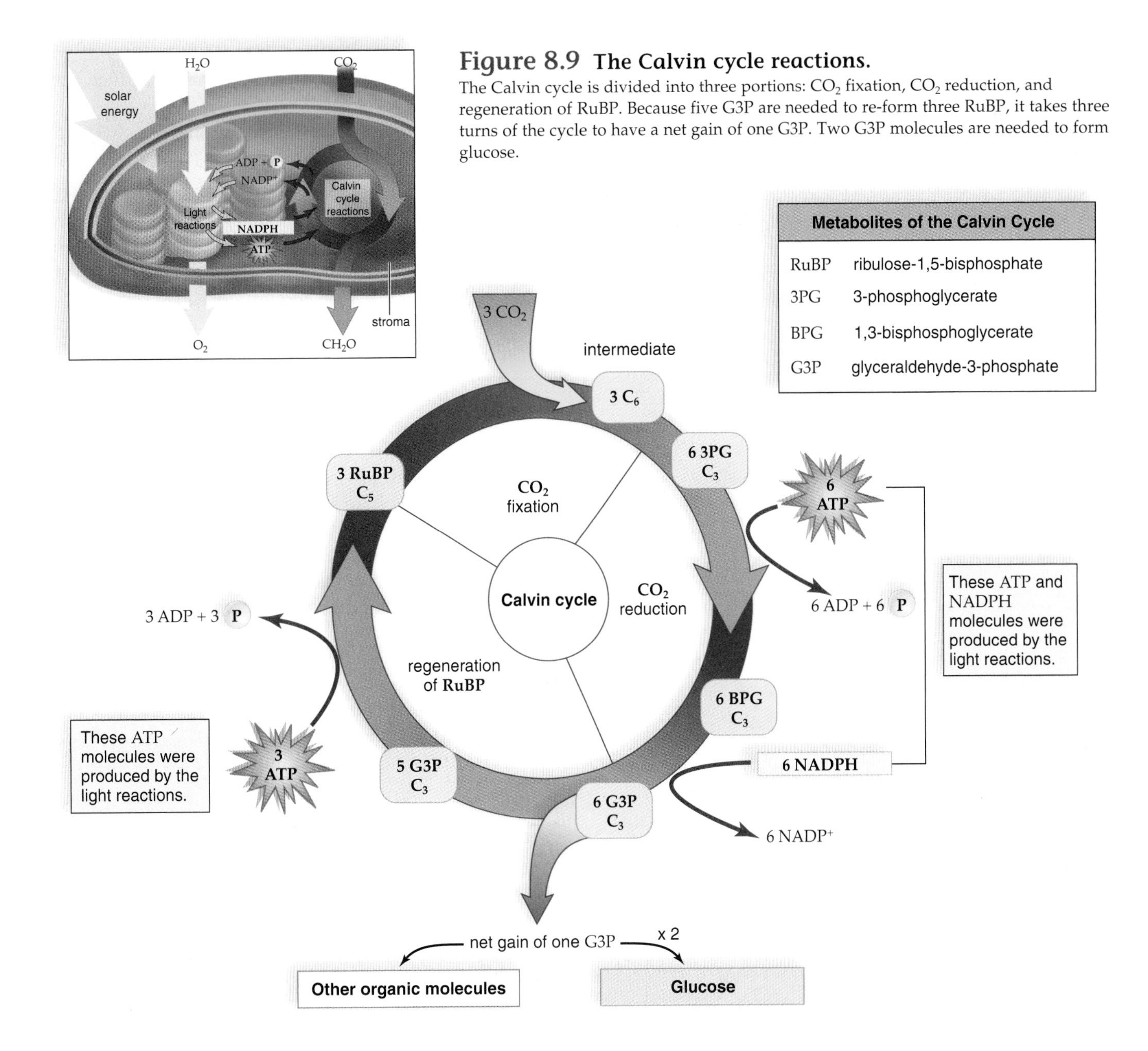

Figure 8.9 The Calvin cycle reactions. The Calvin cycle is divided into three portions: CO_2 fixation, CO_2 reduction, and regeneration of RuBP. Because five G3P are needed to re-form three RuBP, it takes three turns of the cycle to have a net gain of one G3P. Two G3P molecules are needed to form glucose.

dioxide get into the atmosphere? Humans and most other organisms take in oxygen from the atmosphere and release carbon dioxide to the atmosphere. The Calvin cycle includes (1) carbon dioxide fixation, (2) carbon dioxide reduction, and (3) regeneration of **RuBP** (ribulose-1, 5-bisphosphate), the starting material of the cycle.

Fixation of Carbon Dioxide

Carbon dioxide (CO_2) fixation is the first step of the Calvin cycle. During this reaction, three molecules of carbon dioxide from the atmosphere are attached to three molecules of RuBP, a 5-carbon molecule. The result is three 6-carbon molecules. Each 6-carbon molecule then splits in half, forming a total of six 3-carbon molecules. This first 3-carbon molecule is called 3-phosphoglycerate or 3PG.

The enzyme that speeds this reaction, called RuBP carboxylase, is a protein that makes up about 20% to 50% of the protein content in chloroplasts. The reason for its abundance may be that it is unusually slow (it processes only a few molecules of substrate per second compared to thousands per second for a typical enzyme), and so there has to be a lot of it to keep the Calvin cycle going.

Reduction of Carbon Dioxide

Each of two 3PG molecules undergoes reduction to G3P in two steps. The first step converts 3PG into 1,3-bisphosphoglycerate (BPG) using ATP, and the second reaction converts BPG into glyceraldehyde-3-phosphate (G3P) using NADPH.

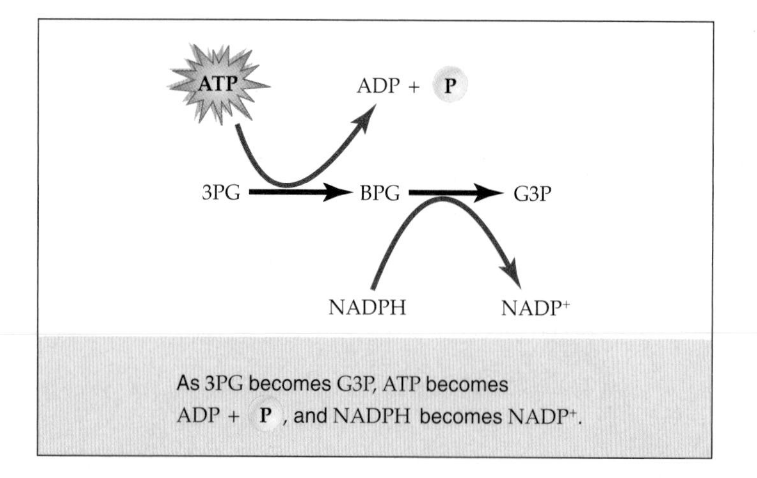

As 3PG becomes G3P, ATP becomes ADP + P, and NADPH becomes NADP+.

This sequence of reactions uses some ATP and NADPH from the light reactions. These reactions also result in the reduction of carbon dioxide to a carbohydrate because $R—CO_2$ has become $R—CH_2O$. Energy and electrons are needed for this reduction reaction, and they are supplied by ATP and NADPH.

Regeneration of RuBP

Notice that the Calvin cycle reactions in Figure 8.9 are multiplied by three because it takes three turns of the Calvin cycle to allow one G3P to exit. Why? Because, for every three turns of the Calvin cycle, five molecules of G3P are used to re-form three molecules of RuBP and the cycle continues. Notice that 5×3 (carbons in G3P) $= 3 \times 5$ (carbons in RuBP):

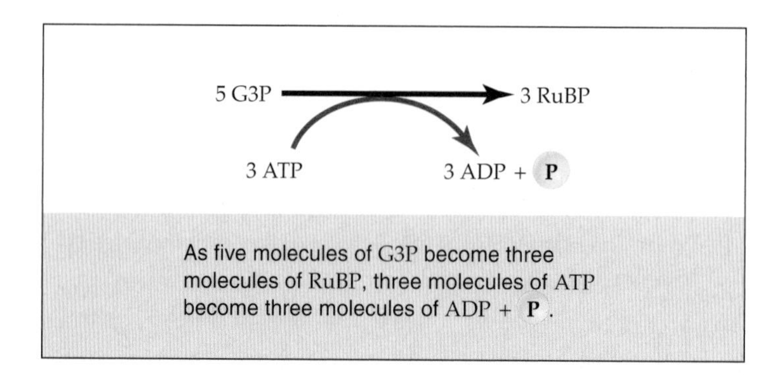

As five molecules of G3P become three molecules of RuBP, three molecules of ATP become three molecules of ADP + P.

This reaction also uses some of the ATP produced by the light reactions.

The Importance of the Calvin Cycle

G3P (glyceraldehyde-3-phosphate) is the product of the Calvin cycle that can be converted to other molecules a plant needs. Compared to animal cells, algae and plants have enormous biochemical capabilities. They use G3P for glucose, sucrose, starch, cellulose, fatty acid, and amino acid synthesis.

Glucose phosphate is one of the organic molecules that results from G3P metabolism. Glucose is the molecule that plants and animals most often metabolize to produce the ATP molecules they require for their energy needs.

Glucose phosphate can be combined with fructose (and the phosphate removed) to form sucrose, the molecule plants use to transport carbohydrates from one part of the plant to the other.

Glucose phosphate is also the starting point for the synthesis of starch and cellulose. Starch is the storage form of glucose. Some starch is stored in chloroplasts, but most starch is stored in roots. Cellulose is a structural component of plant cell walls and becomes fiber in our diet because we are unable to digest it.

A plant can use the hydrocarbon skeleton of G3P to form fatty acids and glycerol, which are combined in plant oils. We are all familiar with corn oil, sunflower oil, or olive oil used in cooking. Also, when nitrogen is added to the hydrocarbon skeleton derived from G3P, amino acids are formed.

Check Your Progress

1. Describe how carbon dioxide is fixed and then reduced to a carbohydrate.
2. How many turns of the cycle does it take to produce one glucose molecule? Why?
3. Explain how G3P is a pivotal molecule in plant metabolism.

8.4 Alternative Pathways for Photosynthesis

Learning Outcome

1. Differentiate between C_3, C_4, and CAM photosynthesis.

Plants are able to live under all sorts of environmental conditions, and one reason is that various modes of photosynthesis have evolved (Fig. 8.10).

C_3 Photosynthesis

The leaves of C_3 plants have a particular structure and a different means of fixing CO_2 compared with C_4 plants. In **C_3 plants,** such as wheat, rice, and oats, mesophyll cells are in parallel layers. The bundle sheath cells around the plant veins do not contain chloroplasts (Fig. 8.11*a*). This structure exposes the cells containing the Calvin cycle to the incoming CO_2. CO_2 is fixed by RuBP carboxylase of the Calvin cycle, and the first detectable molecule following fixation is a 3-carbon molecule (see Fig. 8.9). Unfortunately, RuBP carboxylase cannot only bind CO_2, it can also bind with O_2. When the enzyme binds oxygen, it undergoes a nonproductive, wasteful reaction called photorespiration because it uses oxygen and releases carbon dioxide. Photorespiration makes C_3 photosynthesis an inefficient way to produce carbohydrate when the oxygen concentration rises in leaf space. This can happen when the weather is hot and dry because this type of weather leads to the closing of stomata in order to conserve water.

C_4 Photosynthesis

In **C_4 plants,** such as sugarcane and corn, the mesophyll cells are arranged in concentric rings around the bundle sheath cells, which also contain chloroplasts (Fig. 8.11*b*). In the mesophyll cells, CO_2 is initially fixed by forming a 4-carbon molecule. (The formation of this molecule accounts for the terms C_4 plant and C_4 photosynthesis.) The 4-carbon molecule releases CO_2 to the Calvin cycle and carbohydrate production follows.

Because of the need to transport molecules from where CO_2 is fixed to where the Calvin cycle is located, it would appear that the C_4 pathway would be little utilized among plants. However, when the weather is hot and dry, C_4 plants have an advantage. This is because when the stomata close in order to conserve water and oxygen increases in leaf air space, RuBP carboxylase is not exposed to this oxygen in C_4 plants and photorespiration does not occur. Instead in C_4 plants, the carbon dioxide is delivered to the Calvin cycle, which is located in bundle sheath cells that are sheltered from the leaf air spaces.

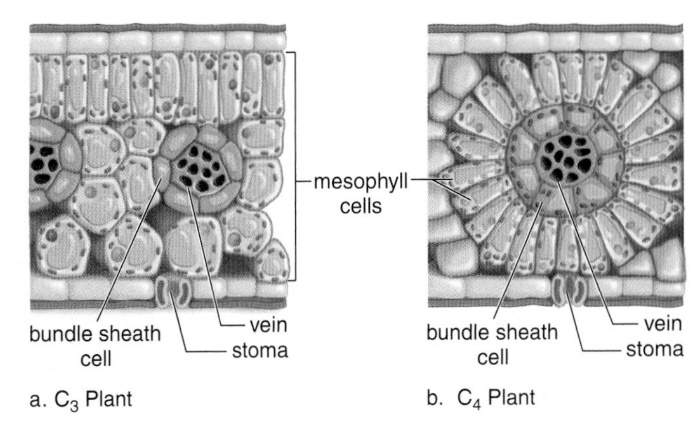

Figure 8.11 C_3 and C_4 plant leaf cell arrangement. **a.** C_3 plants contain mesophyll cells in parallel layers. The bundle sheath cells do not contain chloroplasts. **b.** C_4 plants contain mesophyll cells arranged in concentric rings around chlorophyll containing bundle sheath cells.

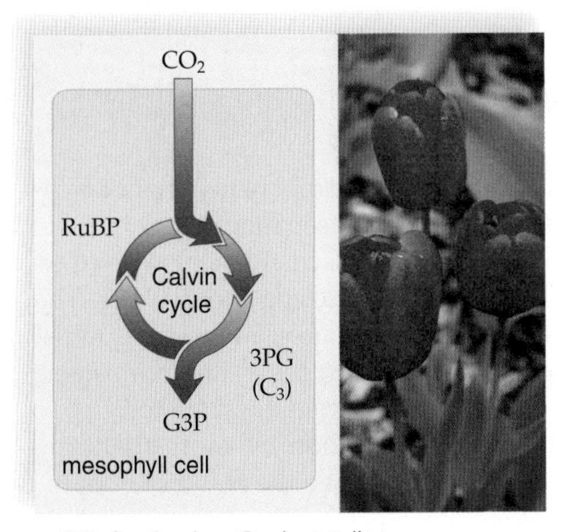

a. CO_2 fixation in a C_3 plant, tulip

b. CO_2 fixation in a C_4 plant, corn

c. CO_2 fixation in a CAM plant, pineapple

Figure 8.10 Alternative pathways for photosynthesis.
Photosynthesis can be categorized according to how, where, and when CO_2 fixation occurs. **a.** In C_3 plants, CO_2 is taken up by the Calvin cycle directly in mesophyll cells. **b.** C_4 plants from a C_4 molecule in mesophyll cells prior to releasing CO_2 to the Calvin cycle in bundle sheath cells. **c.** CAM plants fix CO_2 at night, forming a C_4 molecule.

Bioethical Focus

The New Rice

There is a global food crisis according to international government leaders who met in Japan in July of 2008. The world's population continues to grow, and grain is being diverted for biofuels and to feed livestock. The world's farmers are not keeping up with the demand for wheat, sorghum, maize, and rice. But according to the International Rice Research Institute (IRRI) in Los Baños, Phillipines, two-thirds of the world's poorest people subsist primarily on rice (Fig. 8A). So a shortage of rice affects world hunger more than the other grains. More rice is being consumed than grown, and world rice stockpiles have been declining since 2001. In order to keep up with demand, the world must grow 50 million tons more rice per year than it did in 2005. This requires an increase globally of 1.2% per year.

Figure 8A Rice to feed the world.
Rice provides food for two-thirds of the world's poorest people. New technologies may increase the production of rice.

Flooding is one cause of decreased rice production. Most varieties of rice will die if they are submerged for three or more days. However, one variety, called flood resistant (FR), could survive even if submerged for three weeks. A single gene, called the *Sub1A* gene, appeared to confer a significant degree of submergence tolerance to the FR variety. IRRI scientists were able to introduce the gene into commercial rice strains via hybridization, and four varieties are currently in field trials.

China plants approximately 57% of its rice crop in hybrid varieties. Hybrid rice produces approximately 20% more per acre than the traditional inbred varieties. The hybrids developed by the Chinese are best suited for growth in the tropics and are not effective in temperate zones. In addition, the rice lacks in flavor compared to the inbred varieties.

One of the more long-term research programs with rice is trying to convert rice from a C_3 into a C_4 plant! C_4 plants are 50% more efficient at turning sunlight into food than C_3 plants. There are two aspects of a C_4 plant that would need to be introduced into rice. The first would be the actual enzymes involved in fixing carbon dioxide. The second would be the leaf anatomy of the mesophyll and bundle sheath cells. Researchers have succeeded in introducing the appropriate enzymes from maize, a C_4 plant, into rice plants. Without the proper leaf anatomy, however, there is no guarantee the cloned enzymes will increase the efficiency of photosynthesis.

Form Your Own Opinion

1. If the poorest people in the world go hungry due to the lack of grain, how much difference would it make to solving world hunger if you reduced your meat and fuel consumption?
2. Are genetically modified organisms the only way to feed the world's growing population?

When the weather is moderate, C_3 plants ordinarily have the advantage, but when the weather becomes hot and dry, C_4 plants have the advantage, and we can expect them to predominate. In the early summer, C_3 plants such as Kentucky bluegrass and creeping bent grass predominate in lawns in the cooler parts of the United States, but by midsummer, crabgrass, a C_4 plant, begins to take over.

CAM Photosynthesis

Another alternative pathway, called the **CAM pathway,** has also evolved because of environmental pressures. CAM stands for crassulacean-acid metabolism. Crassulaceae is a family of flowering succulent plants that live in warm, arid regions of the world. CAM photosynthesis is prevalent among most succulent plants that grow in desert environments, including the cacti.

While the C_4 pathway separates components of photosynthesis by location, the CAM pathway separates them in time. CAM plants fix CO_2 into a 4-carbon molecule at night, when they can keep their stomata open without losing much water. The 4-carbon molecule is stored in large vacuoles in their mesophyll cells until the next day. During the day, the 4-carbon molecule releases the CO_2 to the Calvin cycle within the same cell.

Plants are capable of fixing carbon by more than one pathway. These pathways appear to be the result of adaptation to different climates.

Check Your Progress

1. Name some plants that use a method of photosynthesis other than C_3 photosynthesis.
2. Explain why C_4 photosynthesis is advantageous in hot, dry conditions.

8.5 Photosynthesis Versus Cellular Respiration

Learning Outcomes

1. Write the overall chemical equation for photosynthesis and cellular respiration.
2. Understand that the processes of photosynthesis and cellular respiration are not the opposite of each other.

Both plant and animal cells carry on cellular respiration, but animal cells cannot photosynthesize. Plants, algae, and cyanobacteria are capable of photosynthesis. The organelle for cellular respiration is the mitochondrion, while the organelle for photosynthesis is the chloroplast. (Cyanobacteria do not have organelles and thus do not have chloroplasts, although they do have thylakoids.) Photosynthesis is the building up of glucose, while cellular respiration is the breaking down of glucose. Figure 8.12 compares the two processes.

The following overall chemical equation for photosynthesis is the opposite of that for cellular respiration. The reaction in the forward direction represents photosynthesis, and the word *energy* stands for solar energy. The reaction in the opposite direction represents cellular respiration, and the word *energy* then stands for ATP:

$$\text{energy} + 6\,CO_2 + 6\,H_2O \underset{\text{cellular respiration}}{\overset{\text{photosynthesis}}{\rightleftarrows}} C_6H_{12}O_6 + 6\,O_2$$

Both photosynthesis and cellular respiration are metabolic pathways within cells, and therefore consist of a series of reactions that the overall equation does not indicate. Both pathways, which use an electron transport chain located in membranes, produce ATP by chemiosmosis. Both also use an electron carrier; photosynthesis uses $NADP^+$, and cellular respiration uses NAD^+.

Both pathways utilize the following reaction, but in opposite directions. For photosynthesis, read the reaction from left to right; for cellular respiration, read the reaction from right to left:

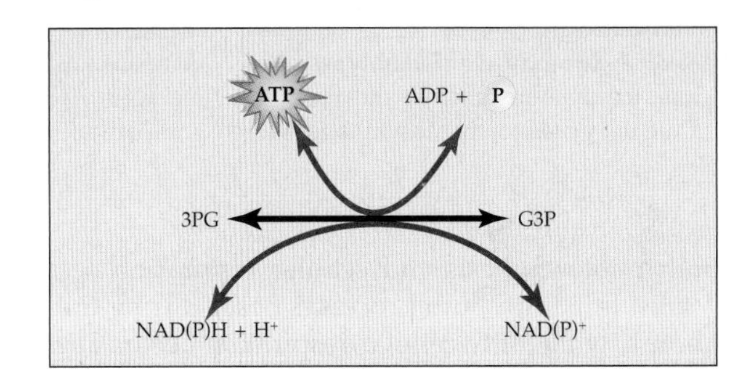

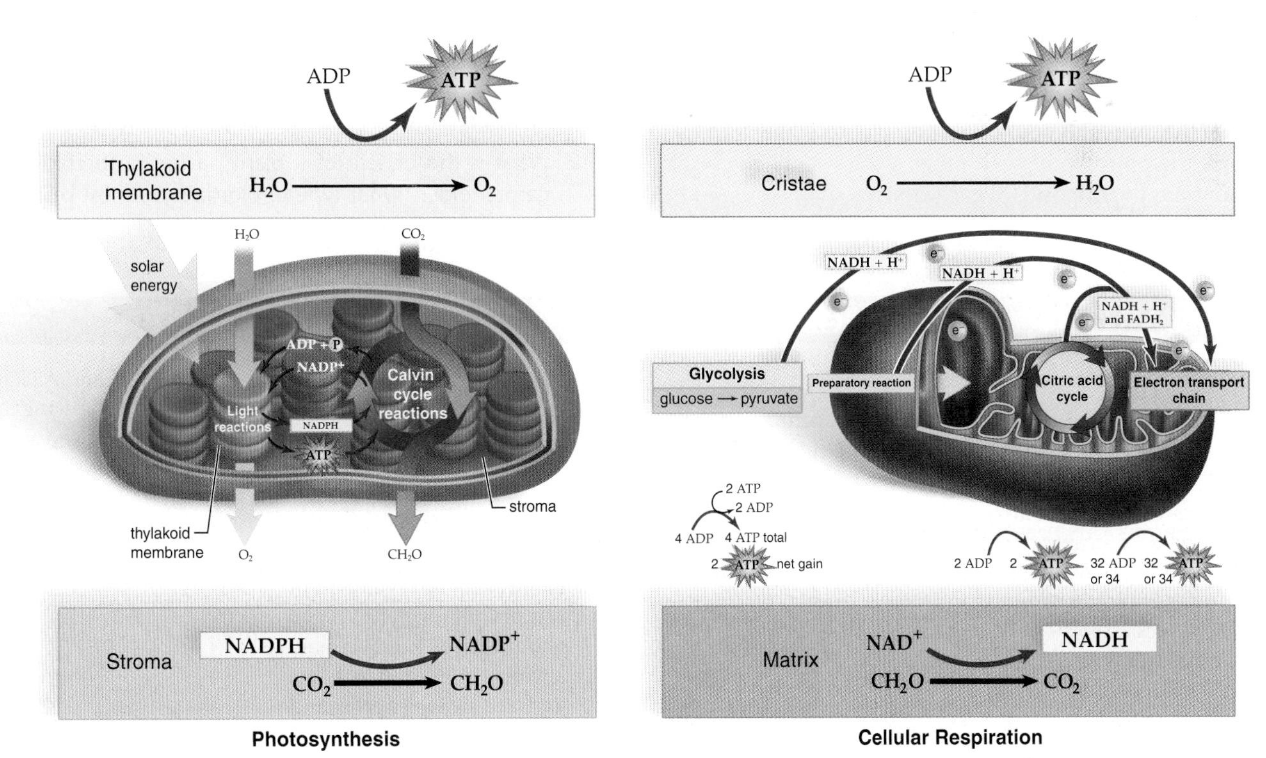

Figure 8.12 Photosynthesis versus cellular respiration.
Both photosynthesis and cellular respiration have an electron transport chain located within membranes, where ATP is produced. Both processes have enzyme-catalyzed reactions located within the fluid interior of respective organelles. In photosynthesis, hydrogen atoms are donated by NADPH + H^+ when CO_2 is reduced in the stroma of a chloroplast. During cellular respiration, NADH forms when glucose is oxidized in the cytoplasm and glucose breakdown products are oxidized in the matrix of a mitochondrion.

Fill 'Er Up—with Algae?

Algae are at the base of many food chains. But researchers are now using algae to produce fuel for cars! In 2007, a San Francisco company called Solazyme received a $2 million grant from the National Institute of Standards and Technology to develop crude oil from algae. In January 2008, Solazyme announced a development agreement with Chevron to produce algae-based fuel for a diesel car. The process they developed does not depend on the photosynthetic ability of the algae. It uses genetically modified algae grown in large vats. Instead of being exposed to sunlight, the algae are fed sugar that they convert into various types of oil. Different types of algae produce different types of oil. Because the algae are grown in the dark, they turn off the genes for photosynthesis. With these genes turned off, the algae actually make more oil.

Algae-based biofuels are considered "green" fuels. After the oil is harvested, the remaining waste can be used as fertilizer. Algae can be grown on marginally productive, even desert, lands and do not require fresh water. And if the fuel is spilled—it is biodegradable!

The U.S. government's National Renewable Energy Laboratory worked on algae as a source of biofuel for almost 20 years, beginning in the 1970s, but was never able to make the venture profitable. However, with the increase in costs for diesel, many other companies as well as the federal government are again working on algae fuels. Two other companies, GreenFuel Technologies and LiveFuels, are trying to produce oil from algae using sunlight. GreenFuel Technologies grows algae in sealed, transparent tubes exposed to sunlight. LiveFuels is growing algae in an open pond in Southern California.

As diesel costs continue to rise, algae fuel may soon become economically feasible. Then the question will become—which method will grow algae the fastest?

Discussion Questions

1. If algae-based biofuel reduces our dependence on foreign oil, how much cheaper does it need to be before we will switch?
2. What other costs might be involved in switching to biofuel besides just the production costs of the fuel?

Both photosynthesis and cellular respiration occur in plant cells. In plants, both processes occur during the daylight hours, while only cellular respiration occurs at night. During daylight hours, the rate of photosynthesis exceeds the rate of cellular respiration, resulting in a net increase and storage of glucose. The stored glucose is used for cellular metabolism, which continues during the night.

Check Your Progress

1. Why are the processes of photosynthesis and cellular respiration not the opposite of each other if the overall chemical reactions are the opposite of each other?

Applying the Concepts Revisited

As shown in Figure 8.5, the pigments in plant leaves have different absorption patterns and this gives plants a more efficient overall absorption pattern for photosynthesis. The absorption of solar energy in the form of light rays provides the energy needed for photosynthesis.

1. Why are pigments needed for photosynthesis?
2. What is the benefit of a plant having both chlorophylls and carotenoids? What type of pigment do most plants lack?

Summarizing the Concepts

8.1 Overview of Photosynthesis

- Photosynthesis is absolutely essential for the continuance of life because it supplies the biosphere with food, which is used as a source of building blocks and energy.
- Chloroplasts carry on photosynthesis. During photosynthesis, solar energy is converted to chemical energy within carbohydrates. A chloroplast contains two main portions: the stroma and membranous grana made up of thylakoid sacs. Chlorophyll and other pigments within the thylakoid membrane absorb solar energy, and enzymes in the fluid-filled stroma reduce CO_2.

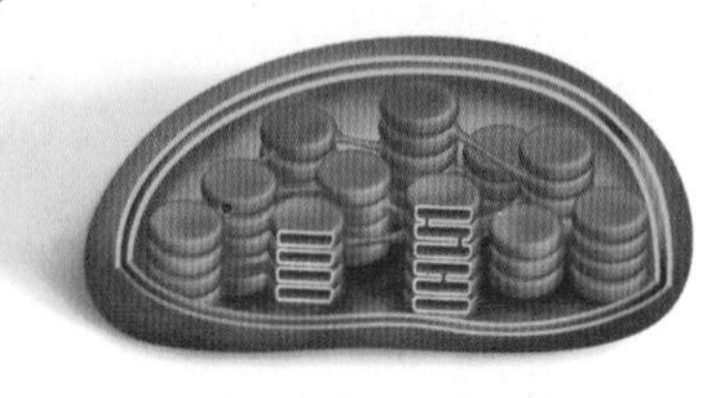

- The overall equation for photosynthesis is $6CO_2 + 6H_2O \longrightarrow C_6H_{12}O_6 + 6O_2$. Photosynthesis consists of two reactions: the light reactions and the Calvin cycle reactions. The Calvin cycle reactions, located in the stroma, use NADPH and ATP to reduce carbon dioxide. These molecules are produced by the light reactions located in the thylakoid membranes of the grana, after chlorophyll captures the energy of sunlight.

8.2 Solar Energy Capture

- Photosynthesis uses solar energy in the visible-light range. Chlorophylls *a* and *b* and the carotenoids largely absorb violet, indigo, blue, and red wavelengths and reflect green wavelengths. This causes leaves to appear green to us.

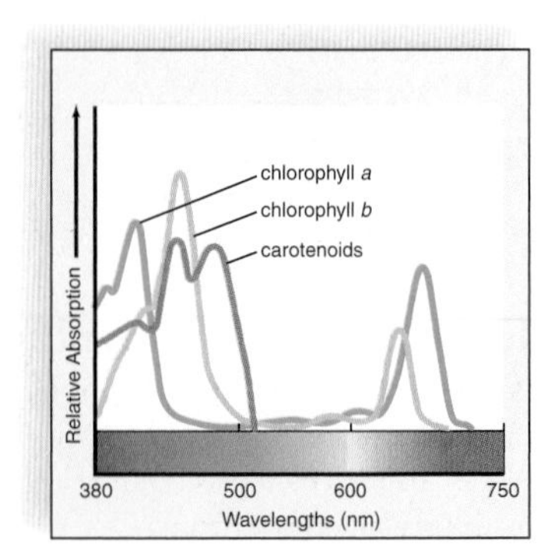

- Photosynthesis begins when pigment complexes within photosystem I and photosystem II absorb radiant energy. In the noncyclic electron pathway, electrons are energized in photosystem II before they enter an electron transport chain. Electrons from H_2O replace those lost in photosystem II. As the electrons pass through the electron transport chain, they help establish a hydrogen ion gradient. The electrons energized in photosystem I pass to $NADP^+$, which becomes NADPH. Electrons from the electron transport chain replace those lost by photosystem I.

- In the cyclic electron pathway of the light reactions, electrons energized by the sun leave photosystem I and enter an electron transport chain that produces a hydrogen ion gradient. Then the energy-spent electrons return to photosystem I.
- The hydrogen ion gradient across the thylakoid membrane is used to synthesize ATP using an ATP synthase enzyme complex.

8.3 Calvin Cycle Reactions

- The ATP and NADPH made in thylakoid membranes pass into the stroma, where carbon dioxide is reduced during the Calvin cycle reactions.
- Carbon dioxide is attached to a 5-carbon molecule named RuBP by the enzyme RuBP carboxylase. The resulting 6-carbon molecule splits into two molecules of 3-carbons, each called 3PG. ATP and NADPH from the light reactions are then used to reduce 3PG to G3P.
- G3P is used to synthesize various molecules, including carbohydrates such as glucose.

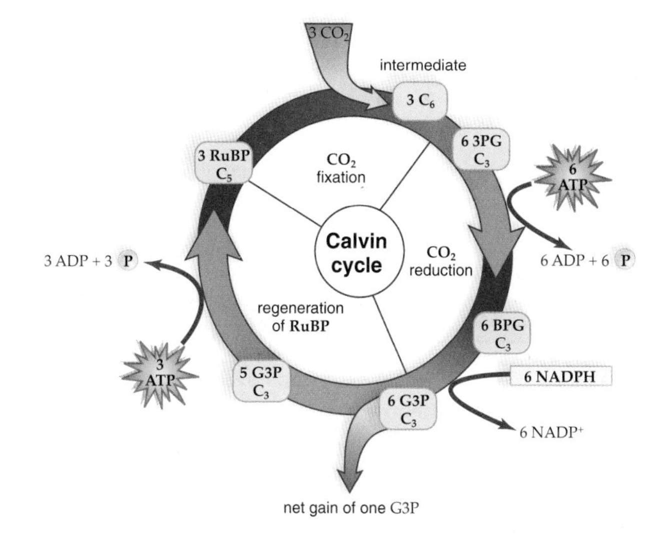

8.4 Alternative Pathways for Photosynthesis

- In C_3 plants, the first molecule observed following carbon dioxide fixation is a 3-carbon molecule. The structure of these plants allows RuBP carboxylase to bind with oxygen during photorespiration.
- Plants utilizing C_4 photosynthesis have a different construction than plants using C_3 photosynthesis. C_4 plants fix CO_2 in mesophyll cells and then deliver the CO_2 to the Calvin cycle in bundle sheath cells. Now, O_2 cannot compete for the active site of RuBP carboxylase when stomata are closed due to hot and dry weather. This represents a partitioning of pathways in space: carbon dioxide fixation occurs in mesophyll cells, and the Calvin cycle occurs in bundle sheath cells.
- CAM plants fix carbon dioxide at night when their stomata remain open. This represents a partitioning of pathways in time: carbon dioxide fixation occurs at night, and the Calvin cycle occurs during the day.

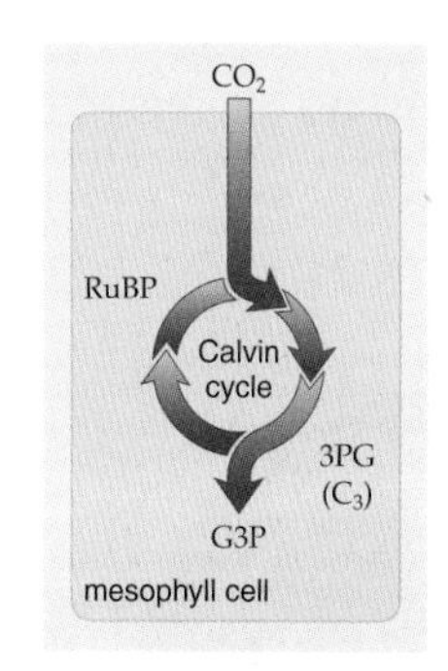

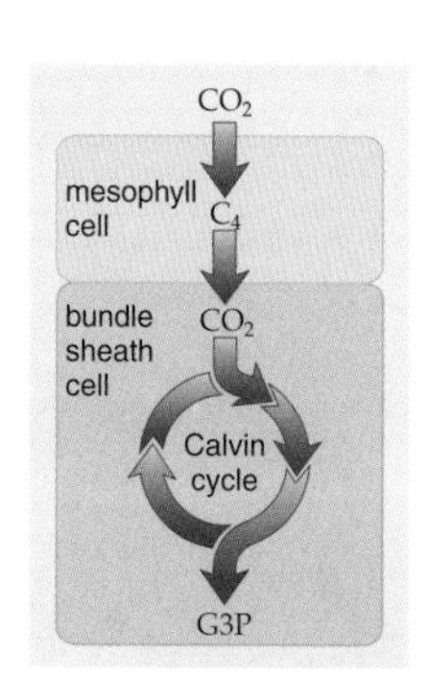

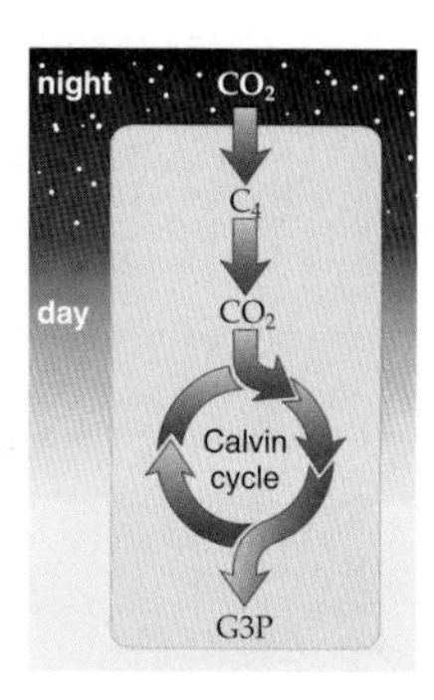

8.5 Photosynthesis Versus Cellular Respiration

- Both photosynthesis and cellular respiration utilize an electron transport chain and ATP synthesis. However, photosynthesis reduces CO_2 to a carbohydrate. The oxidation of H_2O releases O_2. Cellular respiration oxidizes carbohydrate, and CO_2 is given off. Oxygen is reduced to water.

TESTING YOURSELF

Choose the best answer for each question.

1. Photosynthetic activity can be found in
   a. plants.
   b. bacteria.
   c. algae.
   d. All of these are correct.
2. Cellular respiration and photosynthesis both
   a. use oxygen.
   b. produce carbon dioxide.
   c. contain an electron transport chain.
   d. occur in the chloroplast.
3. The noncyclic electron pathway, but not the cyclic pathway, generates
   a. 3PG.
   b. chlorophyll.
   c. ATP.
   d. NADPH.
4. In plants, G3P is used to form
   a. fatty acids.
   b. amino acids.
   c. starch.
   d. All of these are correct.

For questions 5–9, match each definition with a type of plant in the key.

Key:

a. C_3 plants
b. C_4 plants

5. Contain chloroplasts in bundle sheath cells
6. Mesophyll cells arranged in parallel
7. Rice
8. Corn

9. Can be inefficient
10. In the absence of sunlight, plants are not able to engage in the Calvin cycle due to a lack of
    a. ATP.
    b. oxygen.
    c. NADPH.
    d. Both a and c are correct.
11. Disrupting the flow of H^+ across the thylakoid membrane will also disrupt
    a. CO_2 production.
    b. water production.
    c. ATP production.
    d. None of these are correct.
12. In respiration, electrons are carried by ______, while in photosynthesis they are carried by ______.
    a. $NADP^+$, NAD^+
    b. ATP, ADP
    c. NAD^+, $NADP^+$
    d. ATP, FAD
13. Which light rays are used for photosynthesis?
    a. light having long wavelengths
    b. light having short wavelengths
    c. violet, indigo, and blue light
    d. ultraviolet light
    e. Both b and c are correct.
14. Label a, b, c, d, and e in the following diagram of a chloroplast.

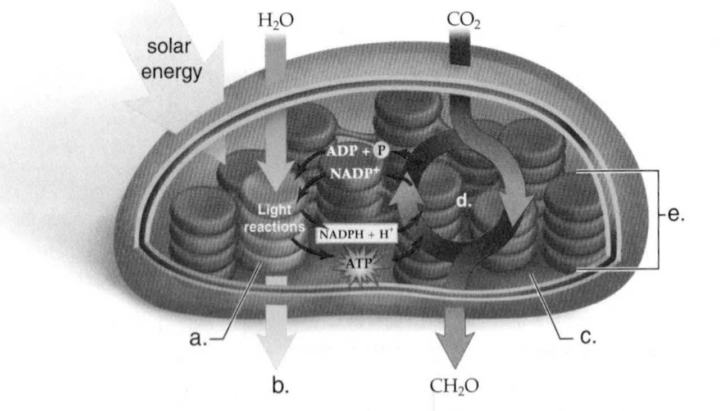

15. The oxygen given off by photosynthesis comes from
    a. H_2O.
    b. glucose.
    c. CO_2.
    d. RuBP.
16. The final acceptor of electrons during the noncyclic electron pathway is
    a. photosystem I.
    b. ATP.
    c. photosystem II.
    d. $NADP^+$.
    e. water.
17. Which of these should not be associated with an electron transport chain?
    a. chloroplasts
    b. protein complexes
    c. movement of H^+ into the thylakoid space
    d. formation of ATP
    e. the RuBP complex
18. Chemiosmosis depends on
    a. protein complexes in the thylakoid membrane.
    b. a difference in H^+ concentration between the thylakoid space and the stroma.
    c. ATP breaking down to ADP + Ⓟ.
    d. the action spectrum of chlorophyll.
    e. Both a and b are correct.

UNDERSTANDING THE TERMS

ATP synthase 133
autotroph 128
C_3 plant 137
C_4 plant 137
Calvin cycle reactions 130
CAM pathway 138
carbon dioxide (CO_2) fixation 136
carotenoids 128
chlorophyll 128
cyclic electron pathway 133
heterotroph 128
light reactions 130
$NADP^+$ (nicotinamide adenine dinucleotide phosphate) 130
noncyclic electron pathway 132
photosynthesis 128
photosystem 132
RuBP (ribulose-1, 5-bisphosphate) 136
stomata (sing., stoma) 129

THINKING CRITICALLY

1. Why are broad, thin leaves advantageous for photosynthesis?
2. Based on what you know about pigments, why do some people who live in very hot, sunny climates wear white?
3. Why would you expect the electron transport chain in chloroplasts to pump H^+ into the thylakoid space instead of into the stroma?

INQUIRY INTO LIFE WEBSITE

The companion website for *Inquiry into Life* provides a wealth of information organized and integrated by chapter. You will find practice tests, animations, videos, and much more that will complement your learning and understanding of general biology.

http://www.mhhe.com/maderinquiry13

Plant Organization and Function

Applying the Concepts

Plants have evolved mechanisms to protect themselves from the damaging effects of the environment and predators. In nonwoody plants, a thin, outer layer of epidermal cells secretes a waxy coating called a cuticle that is impermeable to water. The cuticle covers the surfaces of the leaves and aerial portions of the plant. Cacti have thick cuticles to protect against water loss in the dry desert. Seashore plants have thick cuticles to protect against salt in seawater spray. The epidermis of a plant can also produce hairs. In windy habitats, plant hairs reduce evaporation by breaking up the wind flow across the plant surface. In cold climates, plant hairs protect the living tissue from freezing. Plant hairs can also inhibit herbivores attempting to feed on the plant. This inhibition is a result of hair size, prickliness, and taste. Plants have also evolved antiherbivory chemical defenses against predators. For example, nicotine, made in the roots and stored in the leaves of tobacco plants, is toxic to insects. Caffeine from the leaves and fruits of some plants can kill certain insects feeding on the plant. In spite of these defenses, insect herbivory causes considerable loss of plant tissue.

In this chapter we will describe the structure of flowering plants. We will describe roots, stems, and leaves and how each is specialized to carry out its particular function in the plant.

Chapter Outline

9.1 Plant Organs

Learning Outcomes

1. Define an organ and list the vegetative and reproductive organs of plants.
2. Explain the function and structure of roots, stems, and leaves.

From cacti living in hot deserts to water lilies growing in a pond, the flowering plants, or **angiosperms,** are extremely diverse. But despite their great diversity in size and shape, flowering plants share many common structural features. Most flowering plants possess a root system and a shoot system (Fig. 9.1). The **root system** simply consists of the roots, while the **shoot system** consists of the stem and leaves. Just as in animals, an **organ** is defined as a structure that contains different tissues and performs one or more specific functions. The roots, stems, and leaves are the vegetative organs common to plants. Flowers, seeds, and fruits are structures involved in reproduction.

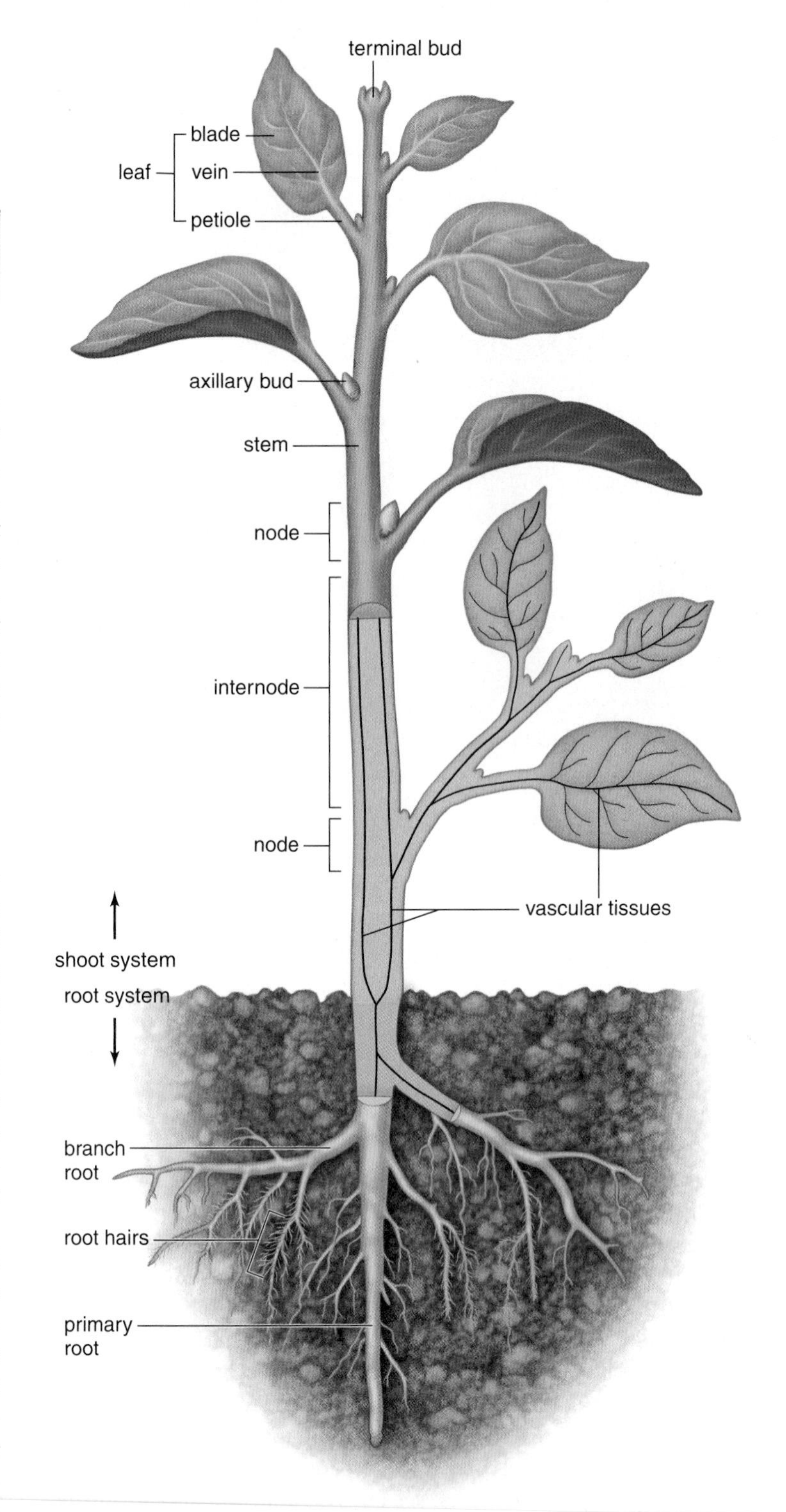

Figure 9.1 Organization of a plant body.
The body of a plant consists of a root system and a shoot system. The shoot system contains the stem and leaves, two types of plant vegetative organs. Axillary buds can develop into branches or flowers, the reproductive structures of a plant. The root system is connected to the shoot system by vascular tissue (brown lines) that extends from the roots to the leaves.

Roots

The root system in the majority of plants is located underground. The depth and distribution of plant roots depends on the type of plant, the timing and amount of rainfall, and the soil composition. For example, plants growing in deserts tend to have deeper roots than those growing in temperate grasslands. A common misconception is that the size and distribution of tree roots reflect the aboveground trunk and branches of the tree. In reality, 90% of a tree's roots are located within 1 m of the surface. However, a tree's roots extend out much farther than the crown of the tree. On average, a tree's roots will extend two to four times the diameter of the aboveground portion of the tree.

The extensive root system of a plant anchors it in the soil and gives it support (Fig. 9.2*a*). The root system absorbs water and minerals from the soil for the entire plant. The cylindrical shape of a root allows it to penetrate the soil as it grows and to absorb water from all sides. The absorptive capacity of a root is also increased by its many branch (lateral) roots and root hairs located in a special zone near a root tip. Root hairs, which are projections from epidermal root-hair cells, are so numerous that they increase the absorptive surface of a root tremendously. In a classic study completed in 1937, H. J. Dittmer counted the number of root hairs in a single, 20-inch-high rye plant. He counted 13,800,000 root hairs! Root-hair cells are constantly being replaced, so this same size rye plant may form about 100 million new root-hair cells every day. A plant roughly pulled out of the soil will not fare well when transplanted. This is because small lateral roots and root hairs are torn off. Transplantation is more apt to be successful if you take a part of the surrounding soil along with the plant, leaving as much of the branch roots and the root hairs intact as possible.

Roots have still other functions. In certain plants, the roots are modified for food storage. For example, branch roots form storage organs for carbohydrates in yams and sweet potatoes. Roots can also store water. Some plants in the pumpkin family store large amounts of water in their roots. Roots also produce hormones that stimulate the growth of stems and coordinate their size with the size of the root. It is most efficient for a plant to have root and stem sizes that are proportional.

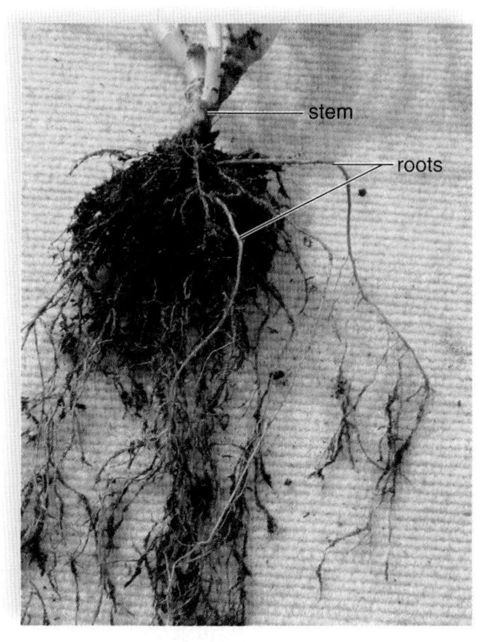

a. Root system, geranium

b. Shoot system, geranium

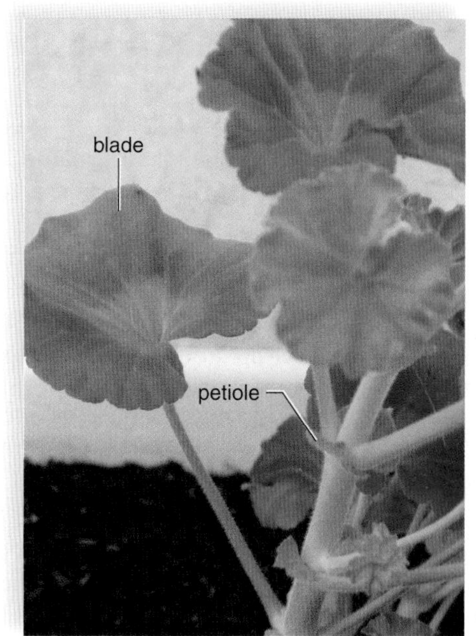

c. Leaves, geranium

Figure 9.2 Vegetative organs of a eudicot.
a. The root system anchors the plant and absorbs water and minerals. **b.** The shoot system consists of a stem and its branches, which support the leaves and transport water and organic nutrients. **c.** The leaves, which are often broad and flat, carry on photosynthesis.

Stems

The shoot system of a plant is composed of the stem, the branches, and the leaves. A **stem,** the main axis of a plant, terminates in tissue that allows the stem to elongate and produce leaves (Fig. 9.2*b*). A stem can sometimes expand in girth as well as length. As trees grow taller each year, they accumulate woody tissue that adds to the strength of their stems. Most stems are upright and support leaves in such a way that each leaf is exposed to as much sunlight as possible. A **node** occurs where leaves are attached to the stem, and an **internode** is the region between the nodes (see Fig. 9.1). The presence of nodes and internodes is the characteristic used to identify a stem, even if it happens to be an underground stem. In some plants, the nodes of horizontal stems can be used to asexually produce new plants.

In addition to supporting the leaves, a stem has vascular tissue that transports water and minerals from the roots through the stem to the leaves. Vascular tissue also transports the organic products of photosynthesis, usually in the opposite direction. **Xylem** cells consist of nonliving cells that form a continuous pipeline for water and mineral transport, while **phloem** cells consist of living cells that join end to end for organic nutrient transport. Stems may have functions other than transport. In some plants (e.g., cactus), the stem is the primary photosynthetic organ. In succulent plants, the stem is a water reservoir. Tubers are underground horizontal stems that store nutrients.

Leaves

Leaves are the major part of a plant that carries on photosynthesis, a process that requires water, carbon dioxide, and sunlight. Leaves receive water from the root system by way of the stem. Plant leaves that are broad and flat have the maximum surface area for the absorption of carbon dioxide and the collection of solar energy needed for photosynthesis. Also, unlike stems, leaves are almost never woody. With few exceptions, their cells are living, and the bulk of a leaf contains tissue specialized to carry on photosynthesis. The wide portion of a foliage leaf is called the **blade,** and the stalk that attaches the blade to the stem is the **petiole** (Fig. 9.2*c*).

The size, shape, color, and texture of leaves are highly variable. These characteristics are fundamental in plant identification. The leaves of some aquatic duckweeds may be less than 1 mm in diameter, while some palms may have leaves that exceed 6 m in length. The shapes of leaves can vary from cactus spines to deeply lobed oak leaves. Leaves can exhibit a variety of colors, from many shades of green to deep purple. The texture of leaves varies from smooth and waxy like a magnolia to coarse like a sycamore. Not all leaves are foliage leaves. Some are specialized to protect buds, attach to objects (tendrils), store food (bulbs), or even capture insects.

The upper acute angle between the petiole and stem is the leaf axil, where an **axillary bud** (or lateral bud) originates. This bud may become a branch or a flower. Plants that lose their leaves every year are called **deciduous.** This is in contrast to most of the gymnosperm plants (conifers), which usually retain their leaves for two to seven years. These plants are called **evergreens.**

Check Your Progress

1. Why is it more accurate to say that a stem has nodes and internodes than to define a stem as being aboveground?
2. How are roots uniquely adapted to their function? Stems? Leaves?

9.2 Cells and Tissues of Plants

Learning Outcomes

1. Explain the function of meristematic tissue.
2. Describe the tissue types in plants.

Figure 1.2 told us that plants have the same levels of biological organization as animals. As in animals, a tissue is composed of specialized cells that perform a particular function. In this section we are going to be looking at plant tissues and in that way we will have the opportunity to discuss the specialized cells within a plant. We will see that simple tissues are made up of a single cell type, while complex tissues contain several different cell types. All the tissue types in a plant arise from **meristematic tissue.** Meristematic tissue allows a plant to grow its entire life because it retains cells that ever have the ability to divide and produce more tissues. All plants can grow their entire lives. Even a 5,000 year-old tree is still growing!

Meristematic Tissue

Meristematic tissue is present in a shoot tip and a root tip where it is called apical meristem. **Apical meristem** causes an increase in length called *primary growth.* In addition to apical meristems, monocots, particularly grasses, have a type of meristem called intercalary meristem, which allows them to regrow lost parts. Intercalary meristems occur at the base of nodes and leaf blades, and account for why grass can so readily regrow after being grazed by a cow or cut by a lawn mower.

Apical meristem continually produces three types of primary meristem, and these develop into the three types of specialized primary tissues in the body of a plant: Protoderm gives rise to epidermal tissue; ground meristem produces ground tissue; and procambium produces vascular tissue. The functions of these three specialized tissues include:

1. **Epidermal tissue** forms the outer protective covering of a plant.
2. **Ground tissue** fills the interior of a plant.
3. **Vascular tissue** transports water and nutrients in a plant and provides support.

Primary growth is responsible for herbaceous, or nonwoody stems. In contrast, wood forms in plants that have lateral meristems or cambium in their roots and stems. This allows the width of the root and stem or trunk to increase over time. This process is called secondary growth.

Epidermal Tissue

As mentioned, the primary meristem tissue called protoderm gives rise to epidermal tissue. The entire body of both nonwoody (herbaceous) and young woody plants is covered by a layer of **epidermis,** which in most plants contains a single layer of closely packed epidermal cells. The walls of epidermal cells that are exposed to air are covered with a waxy **cuticle** to minimize water loss. The cuticle also protects against bacteria and other organisms that might cause disease.

In roots, certain epidermal cells have long, slender projections called **root hairs** (Fig. 9.3*a*). The hairs increase the surface area of the root for absorption of water and minerals.

On stems, leaves, and reproductive organs, epidermal cells produce hairs called **trichomes** that have two important functions: protecting the plant from too much sun and conserving moisture. Sometimes trichomes, particularly glandular ones, help protect a plant from herbivores by producing a toxic substance. Under the slightest pressure, the stiff trichomes of the

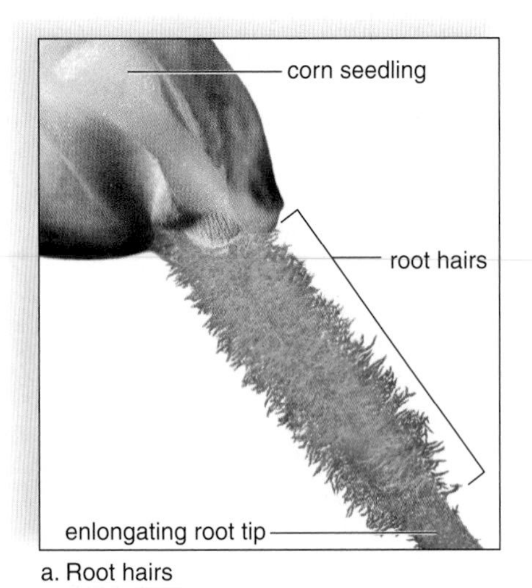

a. Root hairs

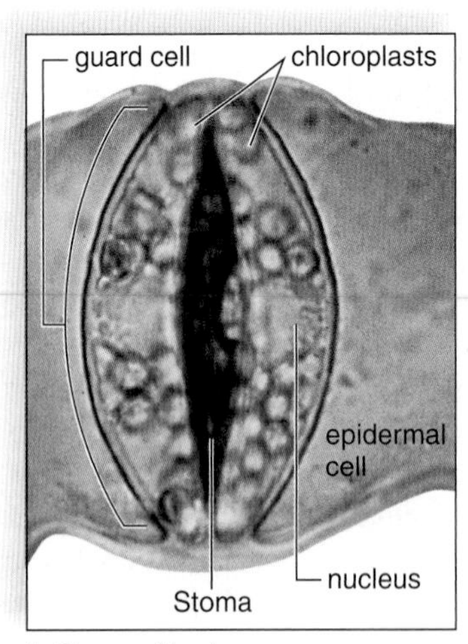

b. Stoma of leaf

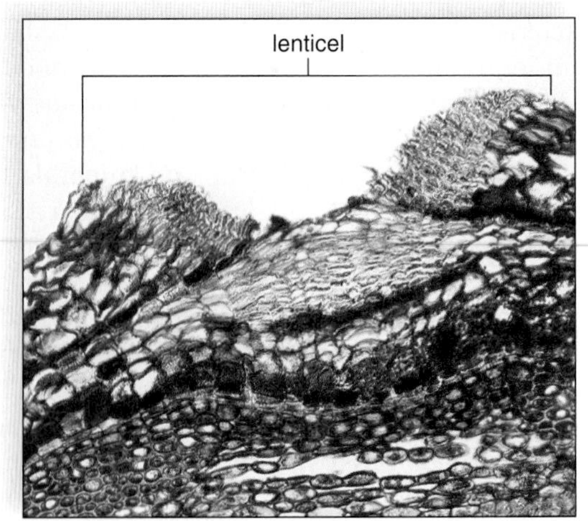

c. Cork of older stem

Figure 9.3 Modifications of epidermal tissue.
a. Root epidermis has root hairs to absorb water. **b.** Leaf epidermis contains stomata (sing., stoma) for gas exchange. **c.** Periderm includes cork and cork cambium. Lenticels in cork are important in gas exchange.

stinging nettle lose their tips, forming "hypodermic needles" that inject an intruder with a stinging secretion.

A waxy cuticle reduces the gas exchange in leaves and so leaves contain specialized cells called **guard cells** and microscopic pores called stomata (sing., stoma). Guard cells, which are epidermal cells with chloroplasts, surround stomata (Fig. 9.3*b*). When the stomata are open, gas exchange is possible but water loss also occurs.

In older woody plants, the epidermis of the stem is replaced by **periderm,** which is composed primarily of boxlike **cork** cells. At maturity, cork cells can be sloughed off, and new cork cells are made by a meristem called cork cambium (Fig. 9.3*c*). As the new cork cells mature, they increase slightly in volume, and their walls become encrusted with suberin, a lipid material, so that they are waterproof and chemically inert. These nonliving cells protect the plant and make it resistant to attack by fungi, bacteria, and animals. Some cork tissues are commercially used for bottle corks and other products.

The cork cambium overproduces cork in certain areas of the stem surface. This causes ridges and cracks to appear. These areas where the cork layer is thin and the cells are loosely packed are called **lenticels.** Lenticels are important in gas exchange between the interior of a stem and the air.

Ground Tissue

Ground meristem is the primary meristem that produces ground tissue. Ground tissues form the bulk of a plant and fill the space between the epidermal and the vascular tissue. Most of the photosynthesis and carbohydrate storage takes place in ground tissue. Ground tissue is also responsible for producing hormones, toxins, pigments, and other specialized chemicals. These cell types are present in ground tissue: parenchyma, collenchyma, and sclerenchyma cells.

Parenchyma cells are the most abundant and correspond best to the typical plant cell (Fig. 9.4*a*). These are the least specialized of the cell types and are found in all the organs of a plant. Parenchyma cells can divide and give rise to more specialized cells, as when roots develop from stem cuttings placed in water. They have relatively thin walls, are alive at maturity, and may contain chloroplasts and carry on photosynthesis, or they may contain colorless plastids that store the products of photosynthesis. A juicy bite from an apple yields mostly storage parenchyma cells.

Collenchyma cells are like parenchyma cells except they have thicker primary walls (Fig. 9.4*b*). The thickness is irregular with the corners of the cell being thicker than other areas. Collenchyma cells often form bundles just beneath the epidermis and give flexible support to immature regions of a plant body. The familiar strands in celery stalks (leaf petioles) are composed mostly of collenchyma cells.

Sclerenchyma cells have thick secondary cell walls (an additional cell wall layer produced between the primary wall and the plasma membrane) impregnated with **lignin,** which is a highly resistant organic substance that makes the walls tough and hard (Fig. 9.4*c*). If we compare a cell wall to reinforced concrete, cellulose fibrils would play the role of steel rods, and lignin would be analogous to the cement. Most sclerenchyma cells are dead at maturity. Their primary function is to support the mature regions of a plant. Tracheids and vessel elements are sclerenchyma cells that not only function in support but transport water. Two other types of sclerenchyma cells are fibers and sclereids. Fibers are long and slender, and may be grouped in bundles that are sometimes commercially important. Hemp fibers can be used to make rope, and flax fibers can be woven into linen. Flax fibers, however, are not lignified, which is why linen is soft. Sclereids, which are shorter than fibers and have variable shapes, are found in seed coats and nutshells. Sclereids, or "stone cells," are responsible for the gritty texture of pears. The hardness of nuts and peach pits is also due to sclereids.

Vascular Tissue

Primary meristem tissue called procambium produces vascular tissue. There are two types of vascular (transport) tissue. Xylem transports water and minerals from the roots to the leaves, and phloem transports sugar and other organic compounds, including hormones, throughout the plant. Both xylem and phloem are considered complex tissues

Figure 9.4 Ground tissue cells.

a. Parenchyma cells are the least specialized of the plant cells. **b.** The walls of collenchyma cells are much thicker than those of parenchyma cells. **c.** Sclerenchyma cells have very thick walls and are nonliving at maturity—their primary function is to give strong support.

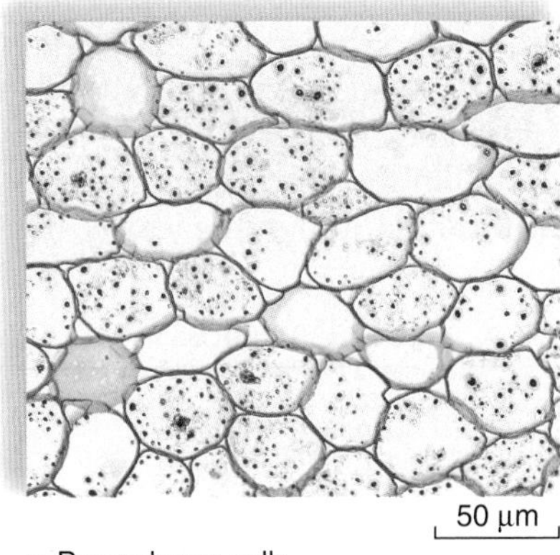

a. Parenchyma cells

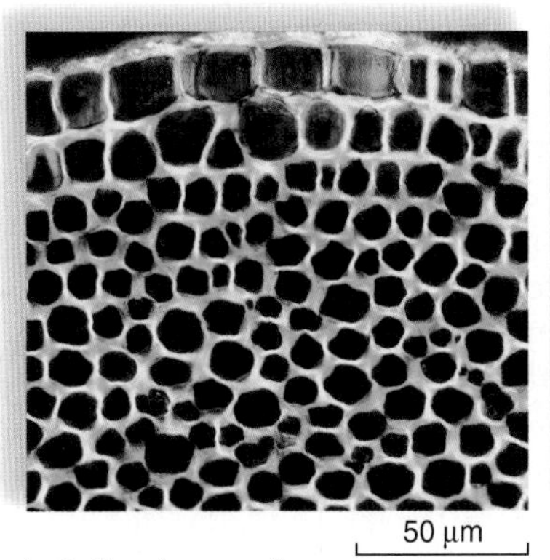

b. Collenchyma cells

c. Sclerenchyma cells

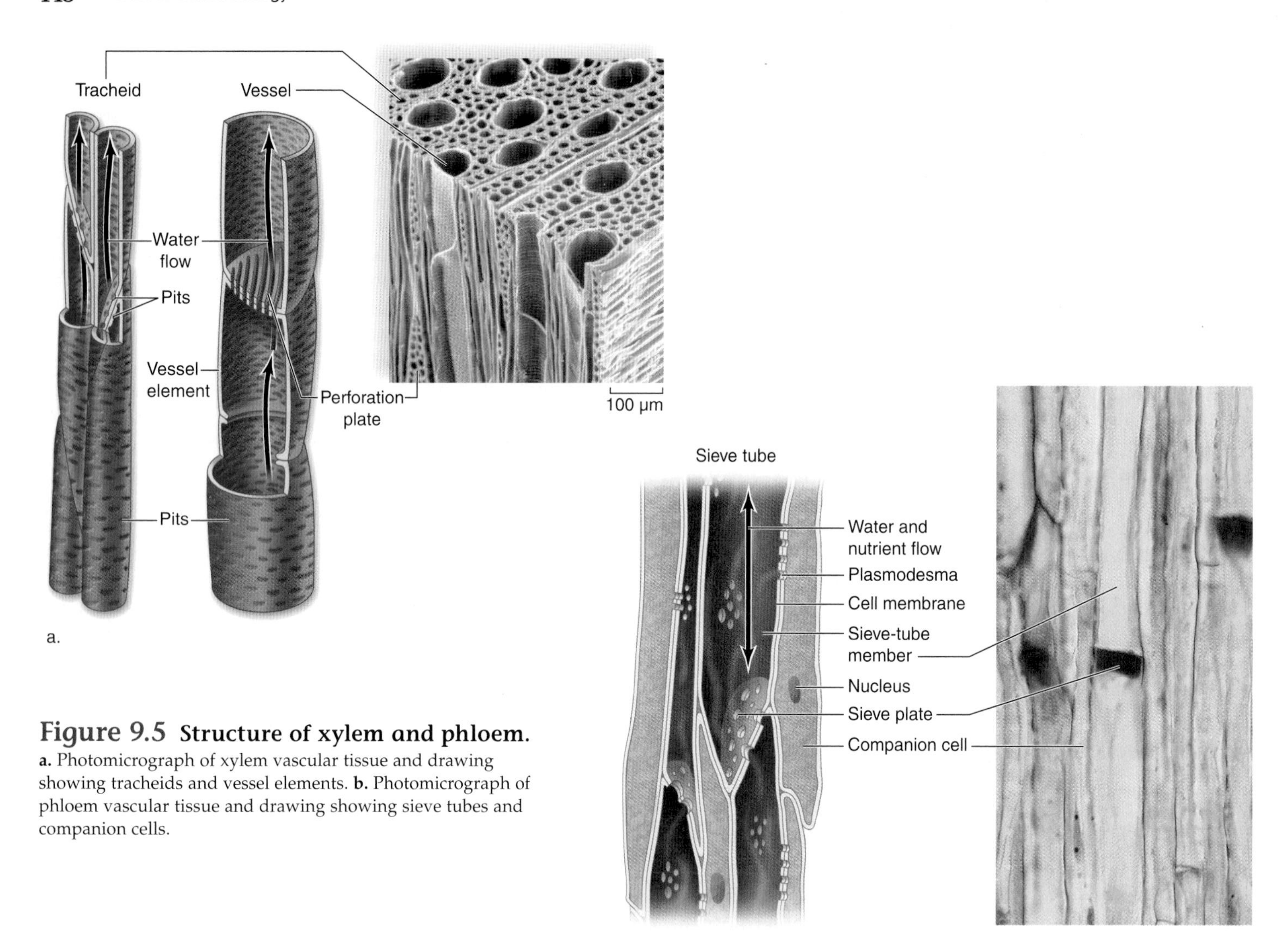

Figure 9.5 Structure of xylem and phloem. **a.** Photomicrograph of xylem vascular tissue and drawing showing tracheids and vessel elements. **b.** Photomicrograph of phloem vascular tissue and drawing showing sieve tubes and companion cells.

because they are composed of two or more kinds of cells. Xylem contains two types of conducting cells: tracheids and vessel elements (Fig. 9.5*a*). Both types of conducting cells are hollow and nonliving, but the **vessel elements** are shorter and wider. Vessel elements have plates with perforations in their end walls and are arranged to form a continuous vessel for water and mineral transport. The elongated **tracheids,** with tapered ends, form a less efficient means of transport, but water can move across the end walls and side walls because there are pits, or depressions, where the secondary wall does not form. In addition to vessel elements and tracheids, xylem contains parenchyma cells that store various substances. Xylem also contains fibers that lend support.

The conducting cells of phloem are **sieve-tube members** arranged to form a continuous sieve tube (Fig. 9.5*b*). Sieve-tube members contain cytoplasm but no nuclei. The term *sieve* refers to a cluster of pores in the end walls, collectively called a sieve plate. Each sieve-tube member has a **companion cell,** which does have a nucleus. The two are connected by numerous **plasmodesmata,** strands of cytoplasm extending from one sieve tube member to another, through the sieve plate. The nucleus of the companion cell controls and maintains the life of both cells. The companion cells are also believed to be involved in phloem's transport function.

It is important to realize that vascular tissue (xylem and phloem) extends from the root through stems to the leaves and vice versa (see Fig. 9.1). In the roots, the vascular tissue is located in the **vascular cylinder.** In the stem, it forms **vascular bundles,** and in the leaves, it is found in **leaf veins.**

Check Your Progress

1. What type of tissue in a plant gives rise to all the other types and allows plants to grow their entire lives?
2. Give a general function for epidermal tissue, ground tissue, and vascular tissue in a plant.
3. What cell types are found in ground tissue and vascular tissue? Give a function for each cell type.

9.3 Monocot Versus Eudicot Plants

Learning Outcome

1. Differentiate between monocots and eudicots.

Flowering plants are divided into two groups, depending on the number of **cotyledons,** or seed leaves, in the embryonic plant (Fig. 9.6). Most cotyledons emerge, grow larger, and become green when the seed germinates. Some plants have one cotyledon, and are known as monocotyledons, or **monocots.** Other embryos have two cotyledons, and are known as eudicotyledons, or **eudicots.**

The vascular (transport) tissue is organized differently in monocots and eudicots. In the monocot root, vascular tissue occurs in a ring. In the eudicot root, phloem, which transports organic nutrients, is located between the arms of xylem, which transports water and minerals, and has a star shape. In the monocot stem, the vascular bundles, which contain vascular tissue surrounded by a bundle sheath, are scattered. In a eudicot stem, the vascular bundles occur in a ring.

Leaf veins are vascular bundles within a leaf. Monocots exhibit parallel venation, and eudicots exhibit netted venation, which may be either pinnate or palmate. Pinnate venation means that major veins originate from points along the centrally placed main vein, and palmate venation means that the major veins all originate at the point of attachment of the blade to the petiole:

Adult monocots and eudicots have other structural differences, such as the number of flower parts and the number of apertures (thin areas in the wall) of pollen grains. The flower parts of monocots are arranged in multiples of three, and the flower parts of eudicots are arranged in multiples of four or five. Eudicot pollen grains usually have three apertures, and monocot pollen grains usually have one aperture.

Although the division between monocots and eudicots may seem of limited importance, it does in fact affect many aspects of their structure. The eudicots are the larger group and include some of our most familiar flowering plants—from dandelions to oak trees. The monocots include grasses, lilies, orchids, and palm trees, among others. Some of our most significant food sources are monocots, including rice, wheat, and corn.

Check Your Progress

1. Compare the following in monocots and eudicots: number of cotyledons, leaf venation, and flower parts.

	Seed	Root	Stem	Leaf	Flower
Monocots	One cotyledon in seed	Root xylem and phloem in a ring	Vascular bundles scattered in stem	Leaf veins form a parallel pattern	Flower parts in threes and multiples of three
Eudicots	Two cotyledons in seed	Root phloem between arms of xylem	Vascular bundles in a distinct ring	Leaf veins form a net pattern	Flower parts in fours or fives and their multiples

Figure 9.6 Flowering plants are either monocots or eudicots.
Five features are used to distinguish monocots from eudicots: the number of cotyledons in the seed; the arrangement of vascular tissue in roots, stems, and leaves; and the number of flower parts.

9.4 Organization of Roots

LEARNING OUTCOMES

1. List the zones in the root involved in primary growth.
2. Describe the anatomy of eudicot and monocot roots.
3. Identify different types of roots and root specializations.

The longitudinal section of a eudicot root shown in Figure 9.7 reveals zones where cells are in various stages of differentiation as primary growth occurs. The root apical meristem is in the region protected by the **root cap.** Root cap cells have to be replaced constantly because they are worn away by rough soil particles as the root grows. The primary meristems are located in the zone of cell division, which

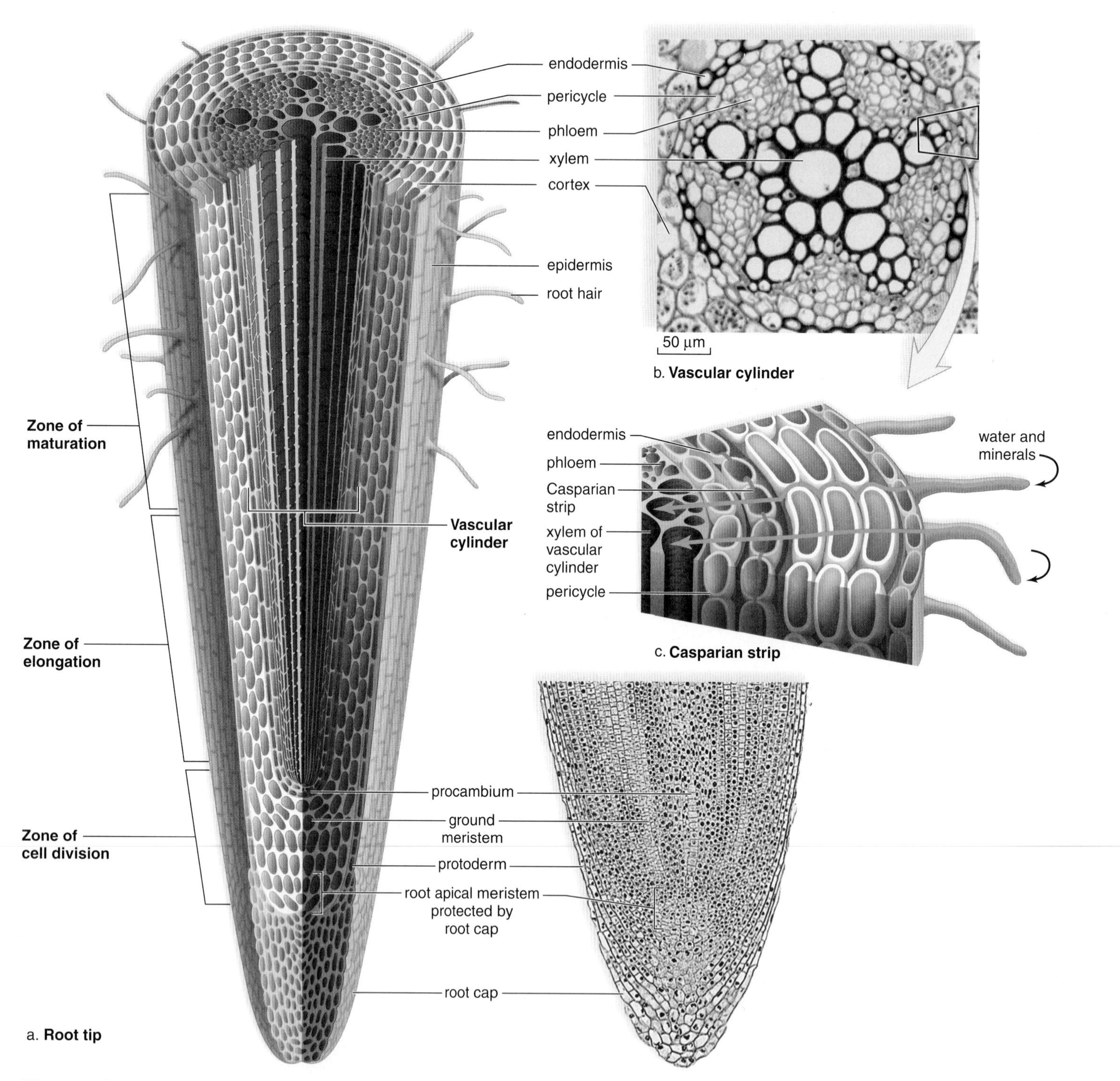

Figure 9.7 Eudicot root tip.
a. The root tip is divided into three zones, best seen in a longitudinal section such as this. **b.** The vascular cylinder of a eudicot root contains the vascular tissue. Xylem is typically star-shaped, and phloem lies between the points of the star. **c.** Water and minerals either pass between cells, or they enter root cells directly. In any case, because of the Casparian strip, water and minerals must pass through the cytoplasm of endodermal cells in order to enter the xylem. In this way, endodermal cells regulate the passage of minerals into the vascular cylinder.

continuously provides cells to the zone of elongation above by mitosis. In the zone of elongation, the cells lengthen as they become specialized. The zone of maturation, which contains fully differentiated cells, is recognizable by the presence of root hairs.

Anatomy of a Eudicot Root

Figure 9.7*a, b* also shows a cross section of a root at the zone of maturation. These specialized tissues are identifiable:

Epidermis The epidermis, which forms the outer layer of the root, consists of only a single layer of cells. The majority of epidermal cells are thin-walled and rectangular, but in the zone of maturation, many epidermal cells have root hairs that project as far as 5–8 mm into the soil.

Cortex Moving inward, next to the epidermis, large, thin-walled parenchyma cells make up the **cortex,** a type of ground tissue. These irregularly shaped cells are loosely packed, so that water and minerals can move through the cortex without entering the cells. The cells contain starch granules, and the cortex functions in food storage.

Endodermis The **endodermis** is a single layer of rectangular cells that forms a boundary between the cortex and the inner vascular cylinder. The endodermal cells fit snugly together and are bordered on four sides by a layer of impermeable lignin and suberin known as the **Casparian strip** (Fig. 9.7*c*). This strip prevents the passage of water and mineral ions between adjacent cell walls. The two sides of each cell that contact the cortex and the vascular cylinder respectively remain permeable. Therefore, the only access to the vascular cylinder is through the endodermal cells themselves, as shown by the arrows in Figure 9.7*c*.

Vascular tissue The **pericycle,** the first layer of cells within the vascular cylinder, can become meristematic and start the development of branch roots (Fig. 9.8). The main portion of the vascular cylinder contains xylem and phloem. The xylem appears star-shaped in eudicots because several arms of tissue radiate from a common center (see Fig. 9.7). The phloem is found in separate regions between the arms of the xylem.

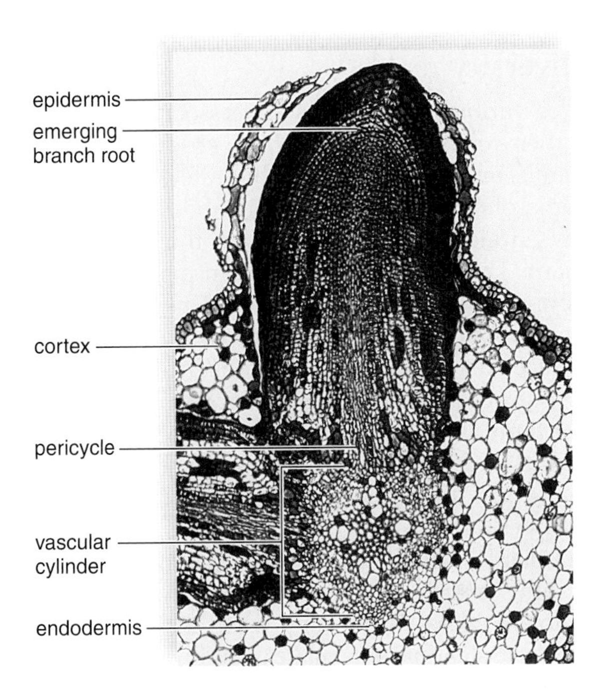

Figure 9.8 Branching of eudicot root.
This cross section of a willow shows the origination and growth of a branch root from the pericycle.

Anatomy of Monocot Roots

Monocot roots have the same growth zones as eudicot roots, but they do not undergo secondary growth as many eudicot roots do. Also, the organization of their tissues is slightly different. A monocot root contains **pith,** a type of ground tissue, which is centrally located. The pith is surrounded by a vascular ring composed of alternating xylem and phloem bundles (Fig. 9.9).

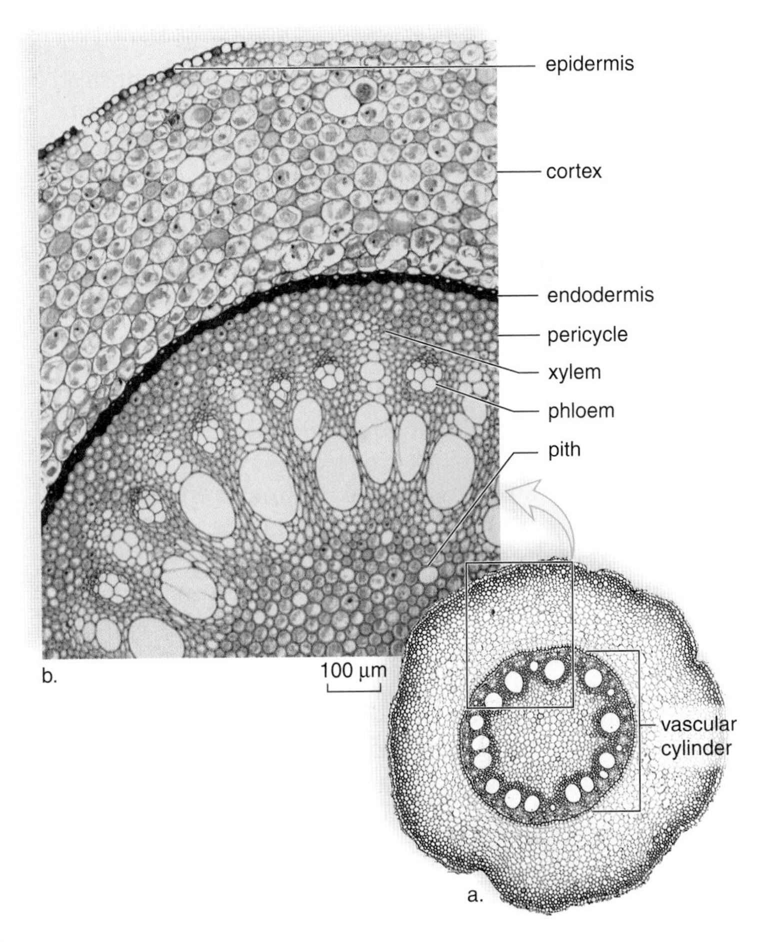

Figure 9.9 Monocot root.
a. This overall cross section of a monocot root shows that a vascular ring surrounds a central pith. **b.** The enlargement shows the exact placement of various tissues.

CHECK YOUR PROGRESS

1. How does a plant control what enters the vascular bundles?
2. Contrast eudicot and monocot roots.

Root Diversity

Roots have various adaptations and associations to better perform their functions of anchorage, absorption of water and minerals, and storage of carbohydrates.

In some plants, notably eudicots, the first or **primary root** grows straight down and remains the dominant root of the plant. This so-called **taproot** can penetrate the soil from a few centimeters to a reported 35 m for a mesquite tree. Lateral roots grow from the taproot to provide additional surface area for water absorption. The taproot is often fleshy and stores food (Fig. 9.10*a*). Carrots, beets, and turnips have taproots that we consume as vegetables. Sweet potato plants don't have taproots, but they do have roots that expand to store starch. We call these storage areas sweet potatoes.

Other plants, notably monocots, have no single, main root, but rather a large number of slender roots. These grow from the lower nodes of the stem when the first (primary) root dies. These slender roots make up a **fibrous root system** (Fig. 9.10*b*). Fibrous root systems grow relatively close to the soil surface where they form dense mats. Many grasses have fibrous root systems that strongly anchor the plants to the soil.

Root Specializations

When roots develop from organs of the shoot system instead of the root system, they are known as **adventitious roots.** Some adventitious roots emerge above the soil line, as in corn plants, where their main function is to help anchor the plant. If so, they are called prop roots (Fig. 9.10*c*). Other examples of adventitious roots are those found on horizontal stems or the rootlets at the nodes of climbing English ivy. As the vines climb, the rootlets attach the plant to any available vertical structure.

Black mangroves live in swampy water and have pneumatophores, root projections that rise above the water and acquire oxygen for cellular respiration (Fig. 9.10*d*). Some plants have poorly developed roots or no roots at all because minerals and water are supplied by other mechanisms. **Epiphytes** are "air plants." They do not grow in soil but on larger plants, which give them support, but no nutrients. Some epiphytes have roots that absorb moisture from the atmosphere, and many catch rain and minerals in special pockets at the base of their leaves.

Plants such as dodders and broomrapes are parasitic on other plants. Their stems have rootlike projections called haustoria (sing., haustorium) that grow into the host plant and make contact with vascular tissue from which they extract water and nutrients (Fig. 9.10*e*).

Two symbiotic relationships assist roots in taking up mineral nutrients. In the first type, legumes (soybeans and alfalfa) have roots infected by nitrogen-fixing *Rhizobium* bacteria. These bacteria can fix atmospheric nitrogen (N_2) by breaking the $N \equiv N$ bond and reducing nitrogen to NH_4^+ for incorporation into organic compounds. The bacteria live

a. Taproot

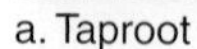

b. Fibrous root system

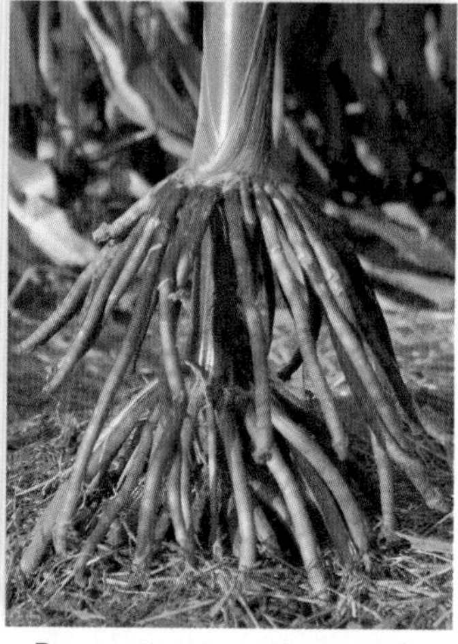

c. Prop roots, a type of adventitious root

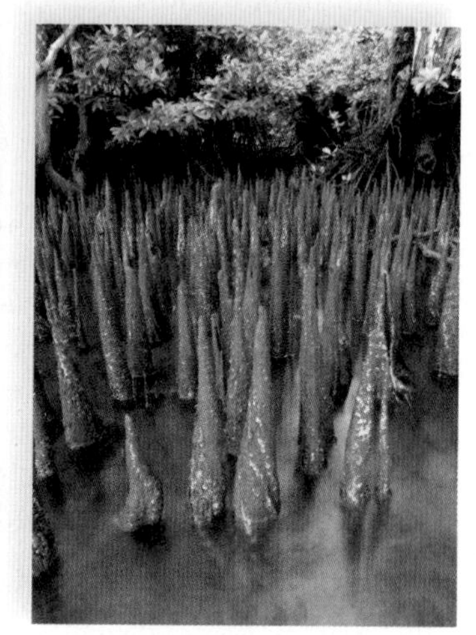

d. Pneumatophores of black mangrove trees

e. Dodder

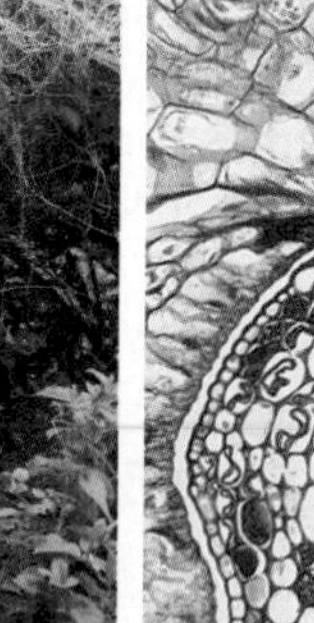

Figure 9.10 Root diversity and specialization.
a. A taproot may have branch roots in addition to a main root. **b.** A fibrous root has many slender roots with no main root. **c.** Prop roots are specialized for support. **d.** The pneumatophores of a black mangrove tree allow it to acquire oxygen even though it lives in swampy water. **e.** *Left:* Dodder is a parasitic plant consisting mainly of orange-brown twining stems. (The green in the photograph is the host plant.) *Right:* Haustoria are rootlike projections of the stem that tap into the host's vascular system.

in **root nodules** and are supplied with carbohydrates by the host plant (Fig. 9.11). The bacteria, in turn, furnish their host with nitrogen compounds.

The second type of symbiotic relationship, called a mycorrhizal association, involves fungi and almost all plant roots (Fig. 9.12). Only a small minority of plants do not have **mycorrhizae,** sometimes called fungus roots. Ectomycorrhizae form a mantle that is exterior to the root, and they grow between cell walls. Endomycorrhizae can penetrate cell walls. In any case, the fungus increases the surface area available for mineral and water uptake and breaks down organic matter, releasing nutrients the plant can use. In return, the root furnishes the fungus with sugars and amino acids. Plants are extremely dependent on mycorrhizae. Orchid seeds, which are quite small and contain limited nutrients, do not germinate until a mycorrhizal fungus has invaded their cells. Nonphotosynthetic plants, such as Indian pipe, use their mycorrhizae to extract nutrients from nearby trees. Plants without mycorrhizae are usually limited as to the environment in which they can grow.

Check Your Progress

1. Why would humans consume roots as food?
2. List several specializations of roots. Which of the three main functions of a root is involved in each specialization?

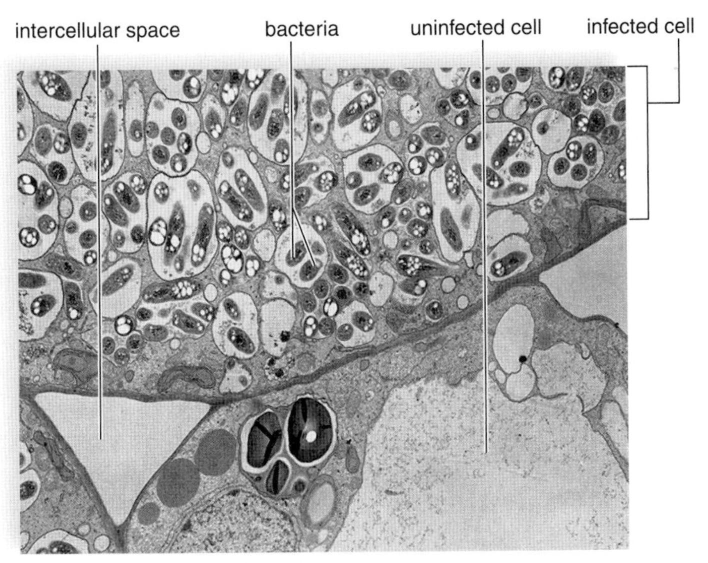

Figure 9.11 Root nodules.
a. Nitrogen-fixing bacteria live in nodules on the roots of plants, particularly legumes. **b.** In infected nodule cells, bacteria fix atmospheric nitrogen and make reduced nitrogen available to a plant. The plant passes carbohydrates to the bacteria.

Figure 9.12 Mycorrhizae.
Top: Experimental results show that plants grown with mycorrhizae (two plants on right) grow much larger than a plant (left) grown without mycorrhizae. *Bottom:* Ectomycorrhizae on red pine roots.

9.5 Organization of Stems

Learning Outcomes

1. Diagram the sections of a woody twig.
2. Describe the anatomy of eudicot and monocot stems.
3. Explain the secondary growth of stems.
4. List some stem modifications.

The anatomy of a woody twig ready for next year's growth illustrates the organization of a stem (Fig. 9.13). The **terminal bud** contains the shoot tip protected by bud scales, which are modified leaves. Leaf scars and bundle scars mark the location of leaves that have dropped. Dormant axillary buds that can give rise to branches or flowers are also found here. Each spring when growth resumes, bud scales fall off and leave a scar. You can tell the age of a stem by counting these groups of bud scale scars because there is one for each year's growth.

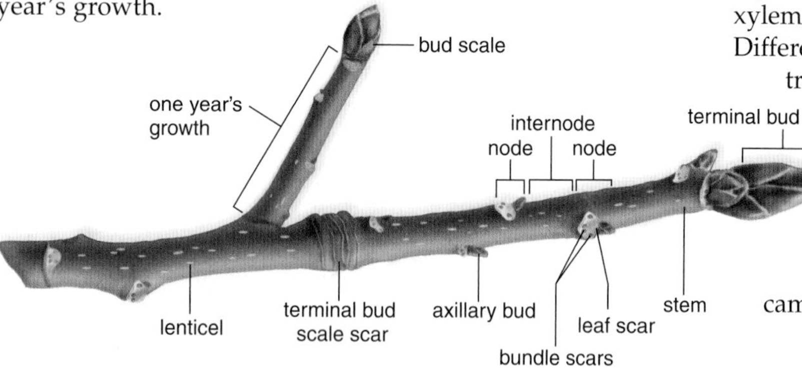

Figure 9.13 Woody twig.
The major parts of a stem are illustrated by a woody twig collected in winter.

When growth resumes, the apical meristem at the shoot tip produces new cells that elongate and thereby increase the length of the stem (primary growth). The shoot apical meristem is protected within the terminal bud, where immature leaves called leaf primordia (sing., primordium) envelop it (Fig. 9.14*a*). The portion of stem between nodes is an internode. As a stem grows, the internodes increase in length.

As described on page 146, three specialized types of primary meristem develop from a shoot apical meristem (Fig. 9.14*b*). These primary meristems contribute to the length of a shoot. The protoderm, the outermost primary meristem, gives rise to epidermis. The ground meristem produces two tissues composed of parenchyma cells. The parenchyma tissue in the center of the stem is the pith, and the parenchyma tissue between the epidermis and the vascular tissue is the cortex.

The procambium, shown as an orange strand of tissue in Figure 9.14*a*, produces the first xylem cells, called primary xylem, and the first phloem cells, called primary phloem. Differentiation continues as certain cells become the first tracheids or vessel elements of the xylem within a vascular bundle. The first sieve-tube members of a vascular bundle do not have companion cells and are short-lived. (Some live only a day before being replaced.) Mature vascular bundles contain fully differentiated xylem, phloem, and a lateral meristem called **vascular cambium.** Vascular cambium is discussed more fully later in this section.

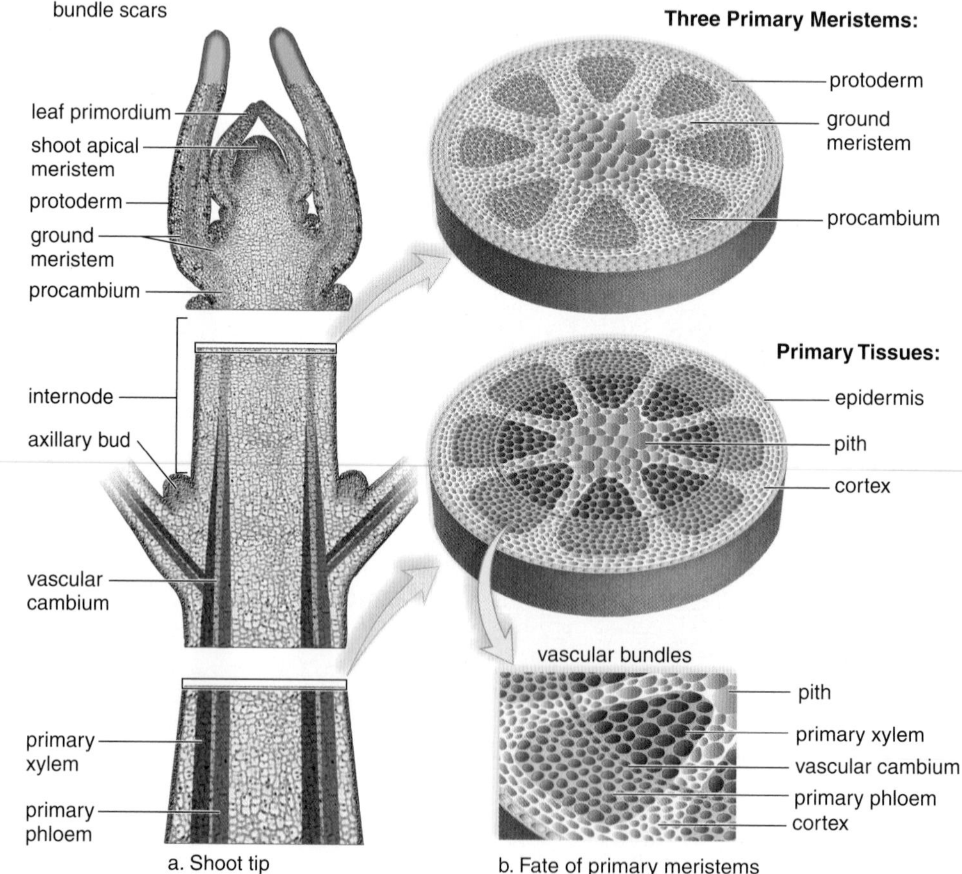

Figure 9.14 Shoot tip and primary meristems.
a. The shoot apical meristem within a terminal bud is surrounded by leaf primordia. **b.** The shoot apical meristem produces the primary meristems. Protoderm gives rise to epidermis; ground meristem gives rise to pith and cortex; and procambium gives rise to vascular tissue, including primary xylem, primary phloem, and vascular cambium.

Herbaceous Stems

Mature nonwoody stems, called **herbaceous stems,** exhibit only primary growth. The outermost tissue of herbaceous stems is the epidermis, which is covered by a waxy cuticle to prevent water loss. In each distinctive vascular bundle, xylem is typically found toward the inside of the stem, and phloem is found toward the outside.

In a herbaceous eudicot stem, such as a sunflower, the vascular bundles are arranged in a distinct ring that separates the cortex from the central pith, which stores water and the products of photosynthesis (Fig. 9.15). The cortex is sometimes green and carries on photosynthesis. In a monocot stem such as corn, the vascular bundles are scattered throughout the stem, and often there is no well-defined cortex or well-defined pith (Fig. 9.16).

Check Your Progress

1. What do each of the primary meristems give rise to in the stem?
2. Compare and contrast eudicot and monocot stems.

Figure 9.15 Herbaceous eudicot stem.

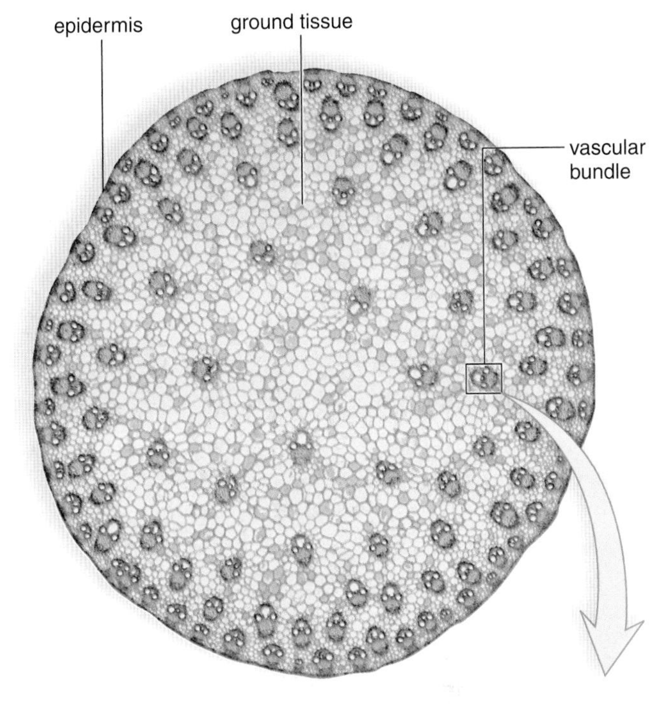

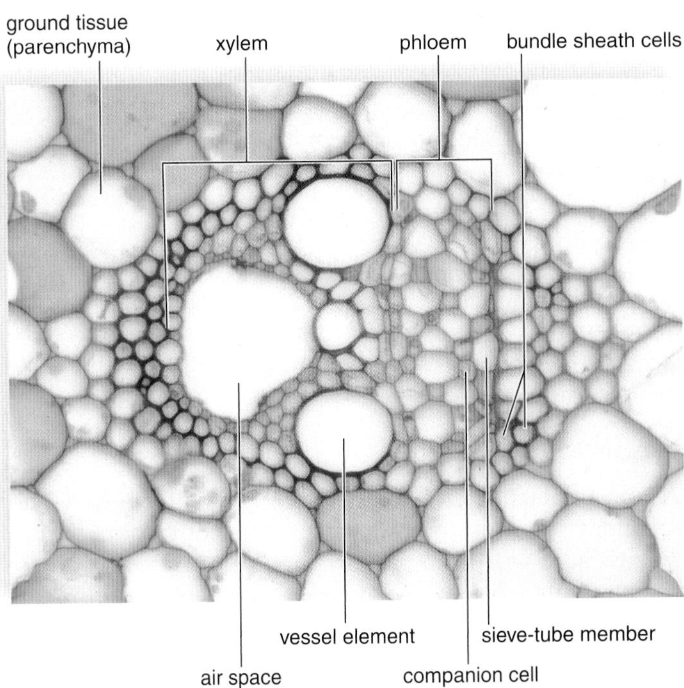

Figure 9.16 Monocot stem.

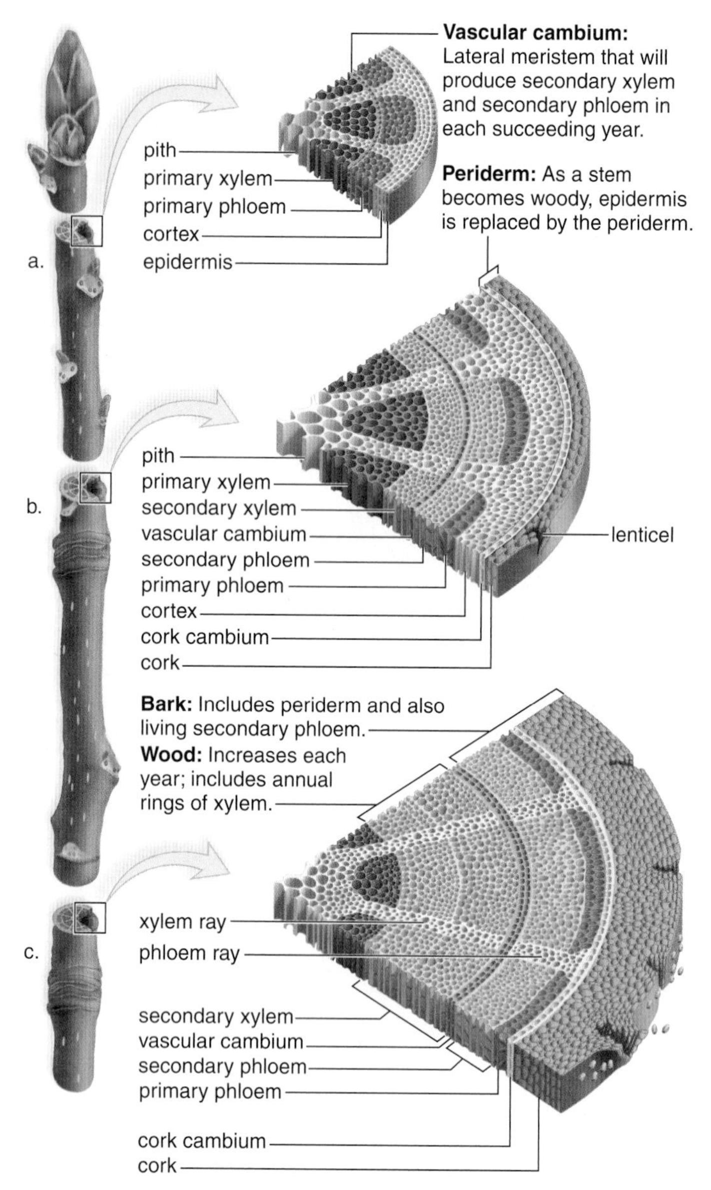

Figure 9.17 Secondary growth of stems.
a. Diagram showing a eudicot herbaceous stem just before secondary growth begins. **b.** Secondary growth has begun, and periderm has replaced the epidermis. Vascular cambium produces secondary xylem and secondary phloem each year. **c.** In a two-year-old stem, the primary phloem and cortex have disappeared, and only the secondary phloem (within the bark) produced by vascular cambium will be active that year. Secondary xylem builds up to become the annual rings of a woody stem.

Woody Stems

A woody plant, such as an oak tree, has both primary and secondary tissues. Primary tissues are those new tissues formed each year from primary meristems right behind the shoot apical meristem. Secondary tissues develop during the first and subsequent years of growth from lateral meristems: vascular cambium and cork cambium. Primary growth occurs in all plants. *Secondary growth,* which occurs only in conifers and woody eudicots, increases the girth of trunks, stems, branches, and roots.

Trees and shrubs undergo secondary growth due to activities of the vascular cambium (Fig. 9.17). In herbaceous plants, vascular cambium is present between the xylem and phloem of each vascular bundle. In woody plants, the vascular cambium develops to form a ring of meristem that divides parallel to the surface of the plant, and produces new xylem and phloem each year. Eventually, a woody eudicot stem has an entirely different organization from that of a herbaceous eudicot stem. A woody stem has no distinct vascular bundles and instead has three distinct areas: the bark, the wood, and the pith. Vascular cambium lies between the bark and the wood.

You will also notice in Figure 9.17*c* the xylem rays and phloem rays that are visible in the cross section of a woody stem. Rays consist of parenchyma cells that permit lateral conduction of nutrients from the pith to the cortex as well as some storage of food. A phloem ray is actually a continuation of a xylem ray. Some phloem rays are much broader than other phloem rays.

Bark

The **bark** of a tree contains both periderm (cork, cork cambium, and a single layer of cork cells filled with suberin) and phloem. Although secondary phloem is produced each year by vascular cambium, phloem does not build up from season to season. The bark of a tree can be removed. However, doing so is very harmful because without phloem organic nutrients cannot be transported.

Cork cambium develops beneath the epidermis. When cork cambium first begins to divide, it produces tissue that disrupts the epidermis and replaces it with cork cells. Cork cells are impregnated with suberin, a waxy layer that makes them waterproof but also causes them to die. This is protective because now the stem is less edible. But an impervious barrier means that gas exchange is impeded except at lenticels, which are pockets of loosely arranged cork cells not impregnated with suberin.

Wood

Wood is secondary xylem that builds up year after year, thereby increasing the girth of trees. In trees that have a growing season, vascular cambium is dormant during the winter. In the spring, when moisture is plentiful and leaves require much water for growth, the secondary xylem contains wide vessel elements with thin walls. In this so-called *spring wood,* wide vessels transport sufficient water to the growing leaves. Later in the season, moisture is scarce, and the wood at this time, called *summer wood,* has a lower proportion of vessels (Fig. 9.18). Strength is required because the tree is growing larger and summer wood contains numerous thick-walled tracheids. At the end of the growing season, just before the cambium becomes dormant again, only heavy fibers with especially thick secondary walls may develop. When the trunk of a tree has spring wood followed by summer wood, the two together make up one year's growth, or an **annual ring.** You can tell the age of a tree by counting the annual rings (Fig. 9.19*a*). The outer annual rings, where transport occurs, are called sapwood.

In older trees, the inner annual rings, called the heartwood, no longer function in water transport. The cells become plugged with deposits such as resins, gums, and other substances that inhibit the growth of bacteria and fungi. Heartwood may help support a tree, although some trees stand erect and live for many years after the heartwood has rotted away. Figure 9.19*b* shows the layers of a woody stem in relation to one another.

The annual rings are important not only in telling the age of a tree, but also in serving as a historical record of tree growth. For example, if rainfall and other conditions were extremely favorable during a season, the annual ring may be wider than usual. The Science Focus on page 161 describes another use for annual tree rings.

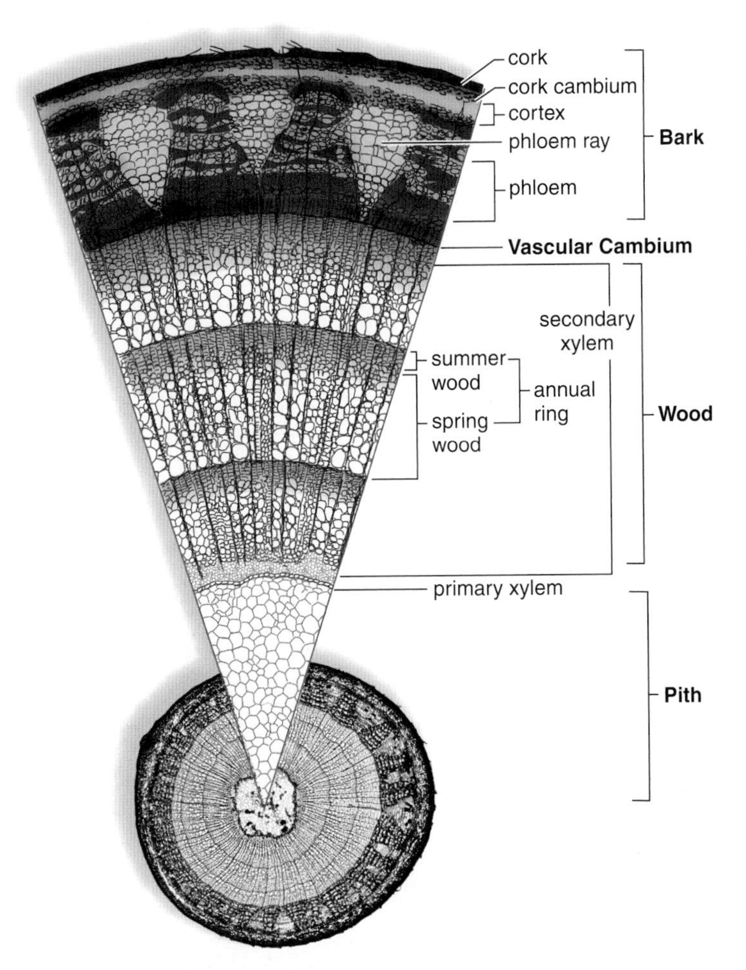

Figure 9.18 Three-year-old woody twig.
The buildup of secondary xylem in a woody stem results in annual rings, which tell the age of the stem. The rings can be distinguished because each one begins with spring wood (large vessel elements) and ends with summer wood (smaller and fewer vessel elements).

Woody Plants Is it advantageous to be woody? With adequate rainfall, woody plants can grow taller than herbaceous plants and increase in girth because they have adequate vascular tissue to support and service their leaves. However, it takes energy to produce secondary growth and prepare the body for winter if the plant lives in the temperate zone. Also, woody plants need more defense mechanisms because a long-lasting plant that stays in one spot is likely to be attacked by herbivores and parasites. Then, too, trees don't usually reproduce until they have grown for several seasons, by which time they may have succumbed to an accident or disease. In certain habitats, it is more advantageous

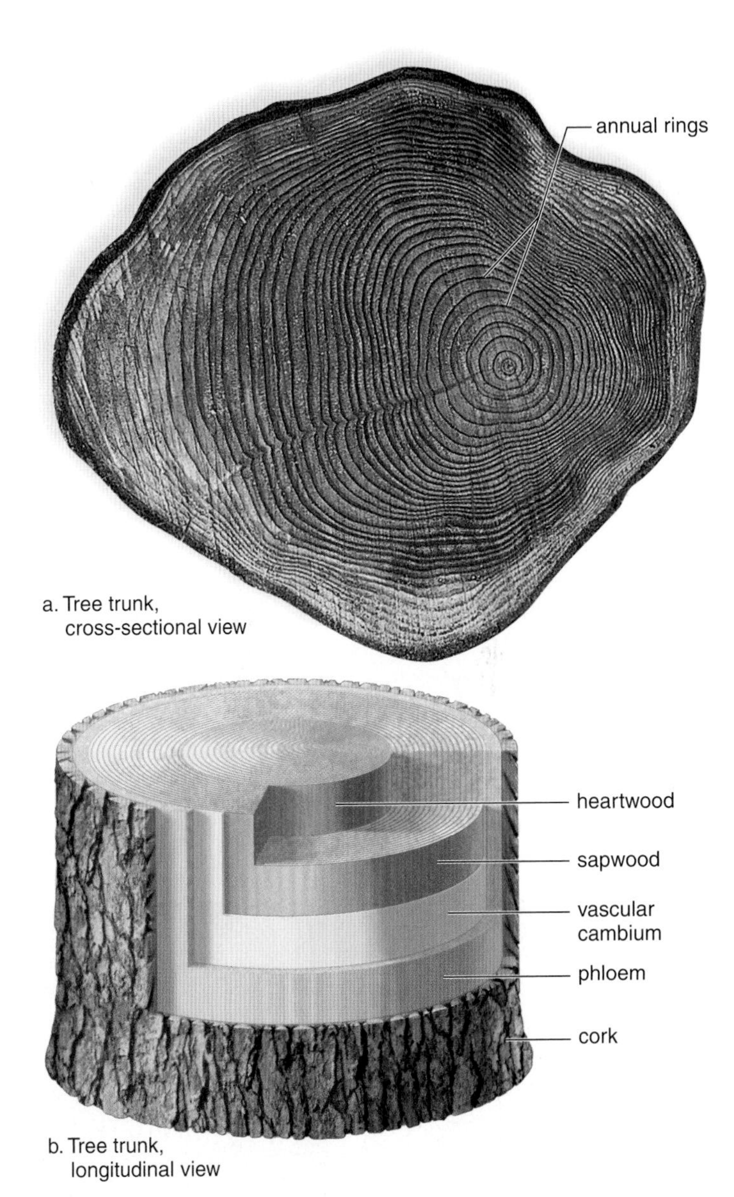

Figure 9.19 Tree trunk.
a. A cross section of a 39-year-old tree trunk. The xylem within the darker heartwood is inactive. The xylem within the lighter sapwood is active. **b.** The relationship of bark, vascular cambium, and wood is retained in a mature stem. The pith has been buried by the growth of layer after layer of new secondary xylem.

for a plant to put most of its energy into simply reproducing rather than being woody.

Stem Diversity

Stems exist in diverse forms, or modifications (Fig. 9.20). Aboveground horizontal stems, called **stolons** or runners, produce new plants where nodes touch the ground. The strawberry plant is a common example of this type of stem, which functions in vegetative reproduction.

Aboveground vertical stems can also be modified. For example, cacti have succulent stems specialized for water storage, and the tendrils of grape plants (which are stem branches) allow them to climb. Morning glory and its relatives have stems that twine around support structures. Such tendrils and twining shoots help plants expose their leaves to the sun.

Underground horizontal stems, called **rhizomes,** may be long and thin, as in sod-forming grasses, or thick and fleshy, as in irises. Rhizomes survive the winter and contribute to asexual reproduction because each node bears a bud. Some rhizomes have enlarged portions called tubers, which function in food storage. Potatoes are tubers. The eyes are buds that mark the nodes.

Corms are bulbous underground stems that lie dormant during the winter, just as rhizomes do. They have thin, papery leaves and a thick stem. They also produce new plants the next growing season. Gladiolus corms are referred to as bulbs by laypersons, but the botanist reserves the term *bulb* for a structure composed of thick modified leaves attached to a short vertical stem. An onion is a bulb.

Humans make use of stems in many ways. The stem of the sugarcane plant is a primary source of table sugar. The spice cinnamon and the drug quinine are derived from the bark of *Cinnamomum verum* and various *Cinchona* species, respectively.

Check Your Progress

1. Why are there two portions to every ring in a tree's annual rings?
2. Which parts of a woody stem are still living?
3. A potato is an enlargement of a stem. How could you tell that a potato is a stem?

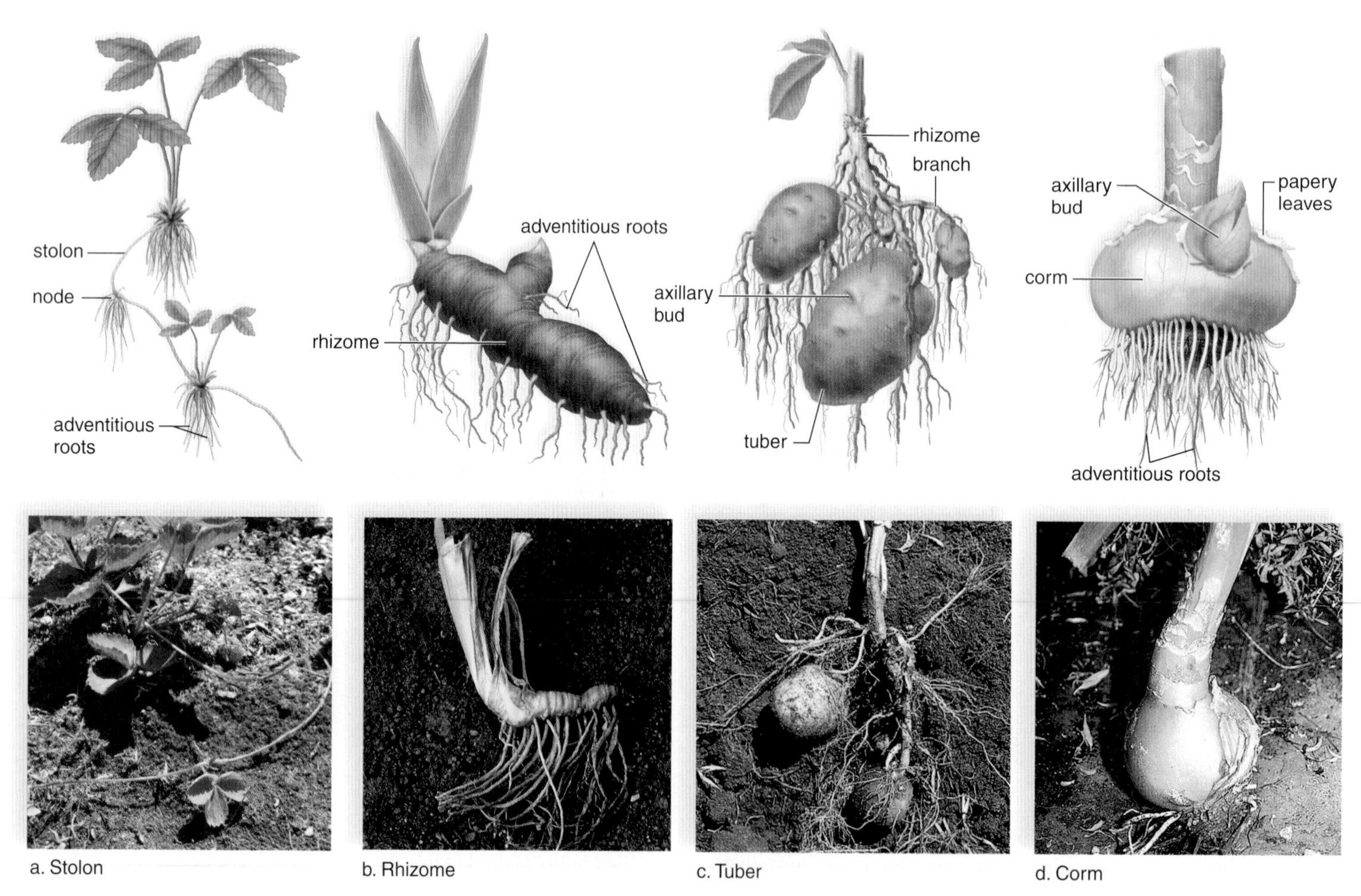

Figure 9.20 Stem diversity.
a. A strawberry plant has aboveground horizontal stems called stolons. Every other node produces a new shoot system. **b.** The underground horizontal stem of an iris is a fleshy rhizome. **c.** The underground stem of a potato plant has enlargements called tubers. We call the tubers potatoes. **d.** The corm of a gladiolus is a thick stem covered by papery leaves.

Ecology Focus

Where Do T-shirts Come From?

Americans love cotton fabrics—from sheets to towels to T-shirts. In fact, cotton is the number one clothing fiber in the United States. And cotton fibers come from the farm!

The cotton plant has been around for at least 7,000 years. The oldest pieces of cotton cloth, dated 7,000 years old, were found in a cave in Mexico. Although the cotton plant grows wild in several places around the world, India is thought to be the first place where it was cultivated. Ancient Egyptian, Chinese, and Indian civilizations were adept at spinning cotton as well as weaving and dyeing elaborate textiles from the yarn. Arab traders introduced cotton to Europe in 800 A.D. and by 1500 A.D. cotton had spread around the world.

Today, China, followed by India and then the United States, are the world's leaders in cotton production. Brazil, Pakistan, Uzbekistan, and Turkey also produce large cotton crops. In the southern United States, the cotton belt extends from Virginia to California. Cotton has one of the longest growing seasons, from five to six months, depending on the weather. Seedlings emerge five to ten days after the seeds are planted. Flower buds appear in approximately five to seven weeks. Cotton flowers are usually self-pollinated, although bees and other insects can cross-pollinate the flowers. Following pollination, the flower petals fall off and the ovaries develop into pods called cotton bolls (Fig. 9A). Cotton fibers, which develop on the surface of the seeds within a boll, are composed of cells with thick walls of cellulose. Fibers are actually individual cells that can grow up to 2 inches in length. An average boll may contain 500,000 fibers, and there may be 100 bolls per plant. The bolls split open approximately two months after the plant blooms and the air dries the white fibers for harvest. In the United States, cotton is harvested mechanically. Stripper harvesters remove the entire boll from the plant, while spindle pickers pull the cotton from the open bolls.

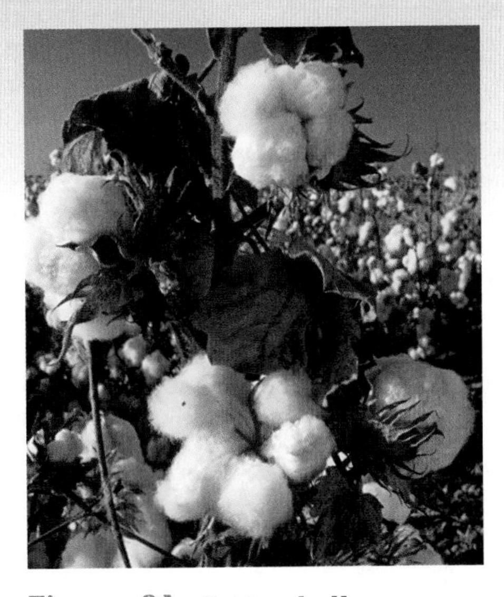

Figure 9A Cotton boll. Cotton is such a valuable worldwide crop it is sometimes called "white gold."

From the field, the cotton is first dried and cleaned before it goes to the cotton gin. The cotton gin was invented in 1793 by Eli Whitney. He watched workers separating the fibers from the seeds by hand and invented a machine that would do the same work 50 times faster. Today's cotton gin uses circular saws to pull the fibers off the seeds and through small holes while leaving the seeds behind. The total length of fibers from one cotton boll may exceed 300 miles! The fibers, which are now called lint, are then compressed into bales weighing approximately 500 lb each.

Next the cotton is graded as to quality based on the color, length of fiber, fineness of the fibers, and other characteristics. Because cotton is not perishable it can be stored in warehouses or sold immediately at market. Of the cotton that is sold at market, more than half is used to make clothing. The rest is used for home furnishings and industrial purposes.

In the textile factory, the bales are opened and blended together by automated machinery. Carding machines then align the fibers into a thin web that is drawn through a funnel to produce a soft, ropelike strand. Other machines stretch the strands and twist them to obtain the needed-size yarn. The yarns are wound around tubes in preparation for the fabric-weaving step. Modern mills use high-speed weaving machines to produce fabrics from the various yarns. These fabrics then undergo several finishing steps including dyeing, printing, and finishing. Finishing can change the look and feel of the fabric and add characteristics such as flame retardation, wrinkle resistance, and shrinkage control.

One T-shirt requires approximately 8 oz of cotton lint to produce. At that conversion rate, one bale of cotton would produce 1,000 T-shirts. But cotton does not just produce fibers. It also produces seeds. Five percent of the seeds are saved for replanting and the rest are used whole or crushed. Whole cotton seeds are fed to livestock. Crushed seeds are used to produce oil, for salad dressings and margarine, and meal for food products.

For the 2007–2008 growing season, the United States produced 19.2 million bales of cotton. China produced 35.8 million bales, while the rest of the world produced 64.3 million bales. That is a lot of T-shirts!

Discussion Questions

1. If someone asked you where T-shirts come from, what would you say?
2. Not all clothing is made from plant products. Give examples of clothing that is derived from animal products.
3. What was the total number of bales of cotton produced in the 2007–2008 growing season? How many T-shirts could that much cotton make? If the total population of the world is approximately 6.7 billion, how many T-shirts per person does that much cotton make?

9.6 Organization of Leaves

Learning Outcomes

1. Diagram the structure of a leaf.
2. List some leaf modifications.

Leaves are the organs of photosynthesis in vascular flowering plants. Just as a solar panel collects sunlight, so does a leaf. As mentioned earlier, a leaf usually consists of a flattened blade and a petiole connecting the blade to the stem. The blade may be single or composed of several leaflets. Externally, it is possible to see the pattern of the leaf veins, which contain vascular tissue. Leaf veins have a net pattern in eudicot leaves and a parallel pattern in monocot leaves (see Fig. 9.6).

Figure 9.21 shows a cross section of a typical eudicot leaf of a temperate zone plant. At the top are layers of epidermal tissue that often bear trichomes, protective hairs sometimes modified as glands that secrete irritating substances. These features may prevent the leaf from being eaten by insects. The epidermis characteristically has an outer, waxy cuticle that helps keep the leaf from drying out. The cuticle also prevents gas exchange because it is not gas permeable. The epidermal layers contain openings called stomata that allow carbon dioxide to move into a leaf and oxygen to move out of the leaf. Water loss also occurs at stomata, but each stoma has two guard cells that regulate its opening and closing, and stomata close when the weather is hot and dry.

The body of a leaf is composed of **mesophyll** tissue. Most eudicot leaves have two distinct regions: **palisade mesophyll,**

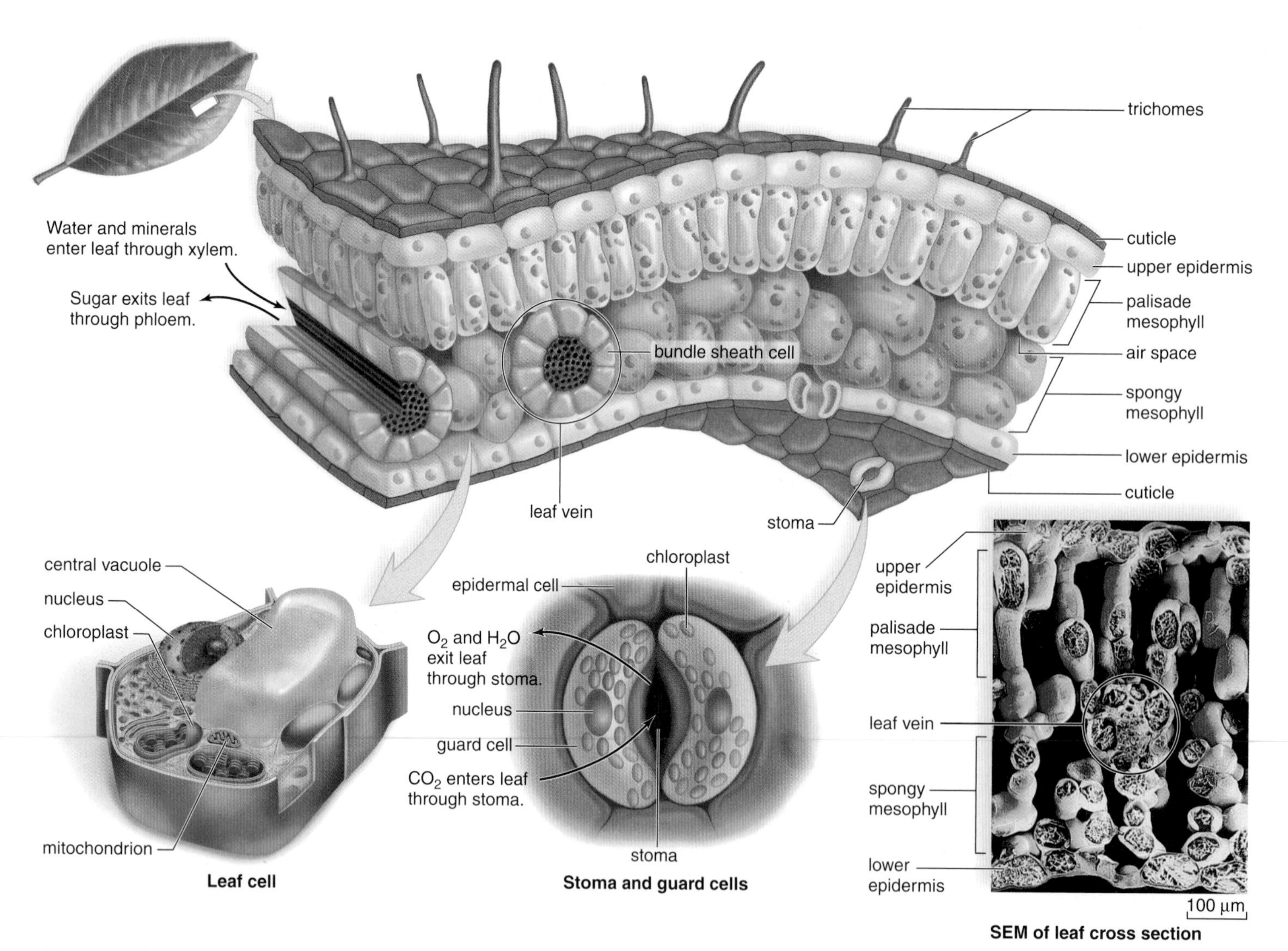

Figure 9.21 Leaf structure.
Photosynthesis takes place in the mesophyll tissue of leaves. The leaf is enclosed by epidermal cells covered with a waxy layer, the cuticle. Trichomes (leaf hairs) are also protective. The veins contain xylem and phloem for the transport of water and solutes. A stoma is an opening in the epidermis that permits the exchange of gases.

Lincoln Slept Here

Nestled in the beautiful hills near Hodgenville, Kentucky, is Sinking Spring Farm, home to the Abraham Lincoln Birthplace National Historic Site. Within the memorial, visitors can see a 13- by 17-ft log cabin at the site of his birth (Fig. 9B). The National Park Service has conceded that it really isn't the log cabin Lincoln was actually born in, calling it a "symbolic 19th century Kentucky log cabin." But when a documentary was being made about Lincoln, filmmakers wanted to be sure it really wasn't the birthplace cabin. They called in a team of scientists led by dendrochronologist Henri Grissino-Mayer, from the University of Tennessee in Knoxville.

What does a dendrochronologist do? The science of tree-ring dating is called *dendrochronology,* derived from two Greek words: *dendron* for tree and *chronos* for time. Tree-ring dating takes advantage of the secondary growth pattern of trees. Every growth season, trees grow in width. In a cross section of a tree that grows in a temperate climate such as the United States, an alternating light and dark concentric ring represents one year of growth. The light part of the ring is growth that occurs during the spring and early summer. The dark part marks the end of that season's growth. Environmental conditions, such as rainfall, temperature, and amount of sunlight, influence the width of the rings. For example, in dry years the rings are much thinner, while in years with plenty of rainfall, the rings are much thicker. Trees of the same species growing in the same area will have a similar pattern of rings. Dendochronologists match the pattern of thick and thin rings from living trees with the pattern of rings in logs taken from old homes. This allows them to count back from the present using the living tree and construct a time line, thereby dating when the tree that produced the log was cut down. Of course, knowing when a tree was cut down does not mean that it was used in that year to build a house. However, it could *not* have been used for construction before that year!

Figure 9B Lincoln's "Birthplace" Cabin.
Within the Abraham Lincoln Birthplace National Historic Site memorial, visitors can see a "symbolic 19th century Kentucky log cabin."

Grissino-Mayer was asked to authenticate Lincoln's birthplace cabin. His team discovered that the logs used to build the cabin had been cut in the 1840s or 1850s. Lincoln was born in 1809. It turns out that the logs were actually taken from multiple cabins and used to construct this cabin in the 1890s. Not only could Lincoln not have been born in this log cabin, he could not have slept there either!

Discussion Questions

1. Why does tree-ring dating require a living tree to establish a chronology?
2. Why would it be important to use trees of the same species grown in the same region for dating?
3. Why is there only one light/dark ring in temperate climate trees? Would this be the same pattern seen in tropical trees?

containing elongated cells, and **spongy mesophyll,** containing irregular cells bounded by air spaces. The parenchyma cells of these layers have many chloroplasts and carry on most of the photosynthesis for the plant. The loosely packed arrangement of the cells in the spongy layer increases the amount of surface area for gas exchange.

Leaf Diversity

The blade of a leaf can be simple or compound (Fig. 9.22*a*). A simple leaf has a single blade in contrast to a compound leaf, which is divided in various ways into leaflets. For example, a magnolia tree has simple leaves, and a buckeye tree has compound leaves. In pinnately compound leaves, the leaflets occur in pairs, as in a black walnut tree, while in palmately compound leaves, all of the leaflets are attached to a single point, as in a buckeye tree. Plants such as the mimosa have bipinnately compound leaves, with leaflets subdivided into even smaller leaflets.

Leaves can be arranged on a stem in three ways: alternate, opposite, or whorled (Fig. 9.22*b*). The American beech has alternate leaves; the maple has opposite leaves, being attached to the same node; and bedstraw has a whorled leaf arrangement, with several leaves originating from the same node.

Leaves are adapted to environmental conditions. Shade plants tend to have broad, wide leaves, and desert plants tend to have reduced leaves with sunken stomata. The leaves of a cactus are the spines attached to the succulent (fleshy)

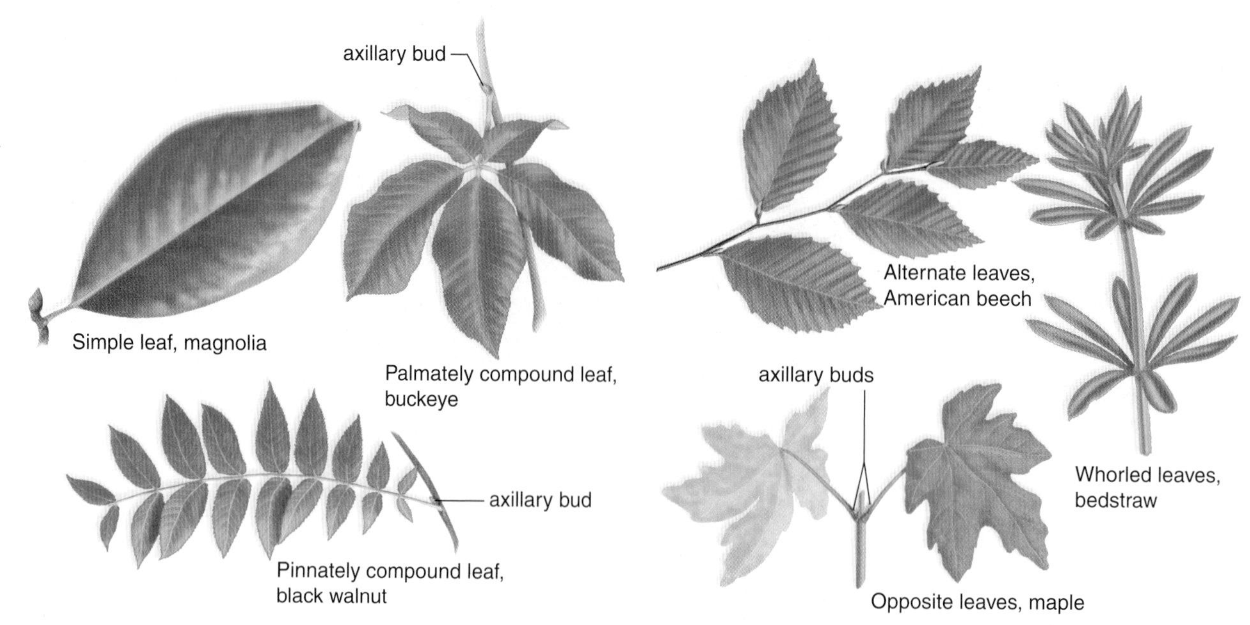

Figure 9.22 Classification of leaves.
a. Leaves are simple or compound, being either pinnately compound or palmately compound. Note the one axillary bud per compound leaf. **b.** The leaf arrangement on a stem can be alternate, opposite, or whorled.

stem (Fig. 9.23*a*). Other succulents have leaves adapted to hold moisture.

An onion bulb is composed of leaves surrounding a short stem. In a head of cabbage, large leaves overlap one another. The petiole of a leaf can be thick and fleshy, as in celery and rhubarb. Leaves of climbing plants, such as those of peas and cucumbers, are modified into tendrils that can attach to nearby objects (Fig. 9.23*b*).

Some plants are carnivores! Their leaves are specialized for catching insects. A sundew has sticky trichomes that trap insects and others that secrete digestive enzymes. The Venus flytrap has hinged leaves that snap shut and interlock when an insect triggers sensitive trichomes that project from inside the leaves (Fig. 9.23*c*). Insectivorous plants commonly grow in marshy regions, where the supply of soil nitrogen is severely limited. The digested insects provide the plants with a source of organic nitrogen.

Check Your Progress

1. The body of the leaf contains what two regions? How are they different?
2. How are leaves adapted to their environment?

a. Cactus

b. Cucumber

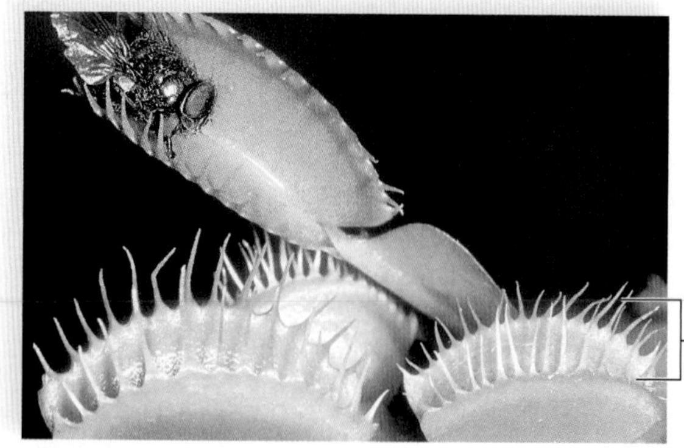

c. Venus flytrap

Figure 9.23 Leaf diversity.
a. The spines of a cactus are modified leaves that protect the fleshy stem from animal predation. **b.** The tendrils of a cucumber are modified leaves that attach the plant to a physical support. **c.** The modified leaves of the Venus flytrap serve as a trap for insect prey. When triggered by an insect, the leaf snaps shut. Once shut, the leaf secretes digestive juices that break down the soft parts of the insect's body.

9.7 Uptake and Transport of Nutrients

Learning Outcomes

1. Explain the movement of water in a plant according to the cohesion-tension model of xylem transport.
2. Explain the movement of organic nutrients in a plant according to the pressure-flow model of phloem transport.

In order to produce a carbohydrate, a plant requires carbon dioxide from the air and water from the soil. One of the major functions of roots is to supply the plant with water as well as minerals. Humans make use of a plant's ability to take minerals from the soil. We depend on them for our basic supply of such minerals as calcium, to build bones and teeth, and iron, to help carry oxygen to our cells. Minerals such as copper and zinc are cofactors for the functioning of enzymes.

Water and minerals are transported through a plant in xylem, while the products of photosynthesis are transported in phloem.

Water Uptake and Transport

Water and minerals enter a plant at the root, primarily through the root hairs. Osmosis of water and diffusion of minerals aid the entrance of water and minerals from the soil into the plant. Eventually, however, a plant uses active transport to bring minerals into root cells. From there, water, along with minerals, moves across the tissues of a root until it enters xylem.

Once water enters xylem, it must be transported upward to the top of a tree. This can be a daunting task. Water entering root cells creates a positive pressure called root pressure that tends to push xylem sap upward. Atmospheric pressure can support a column of water to a maximum height of approximately 10.3 m. However, some trees can exceed 90 m in height, so other factors must be involved in causing water to move from the roots to the leaves. Figure 9.24 illustrates the accepted model for the transport of water and, therefore, minerals in a plant. It is called the **cohesion-tension model.**

Cohesion-Tension Model of Xylem Transport

Recall that the vascular tissue of xylem contains the hollow conducting cells called tracheids and vessel elements (see Fig. 9.5). The vessel elements are larger than the tracheids, and they are stacked one on top of the other to form a pipeline that stretches from the roots to the leaves. It is an open pipeline because the vessel elements have no end walls, just perforation plates, separating one from the other. The tracheids, which are elongated with tapered ends, form a less obvious means of transport. Water can move across the end and side walls of tracheids because of pits, or depressions, where the secondary wall does not form.

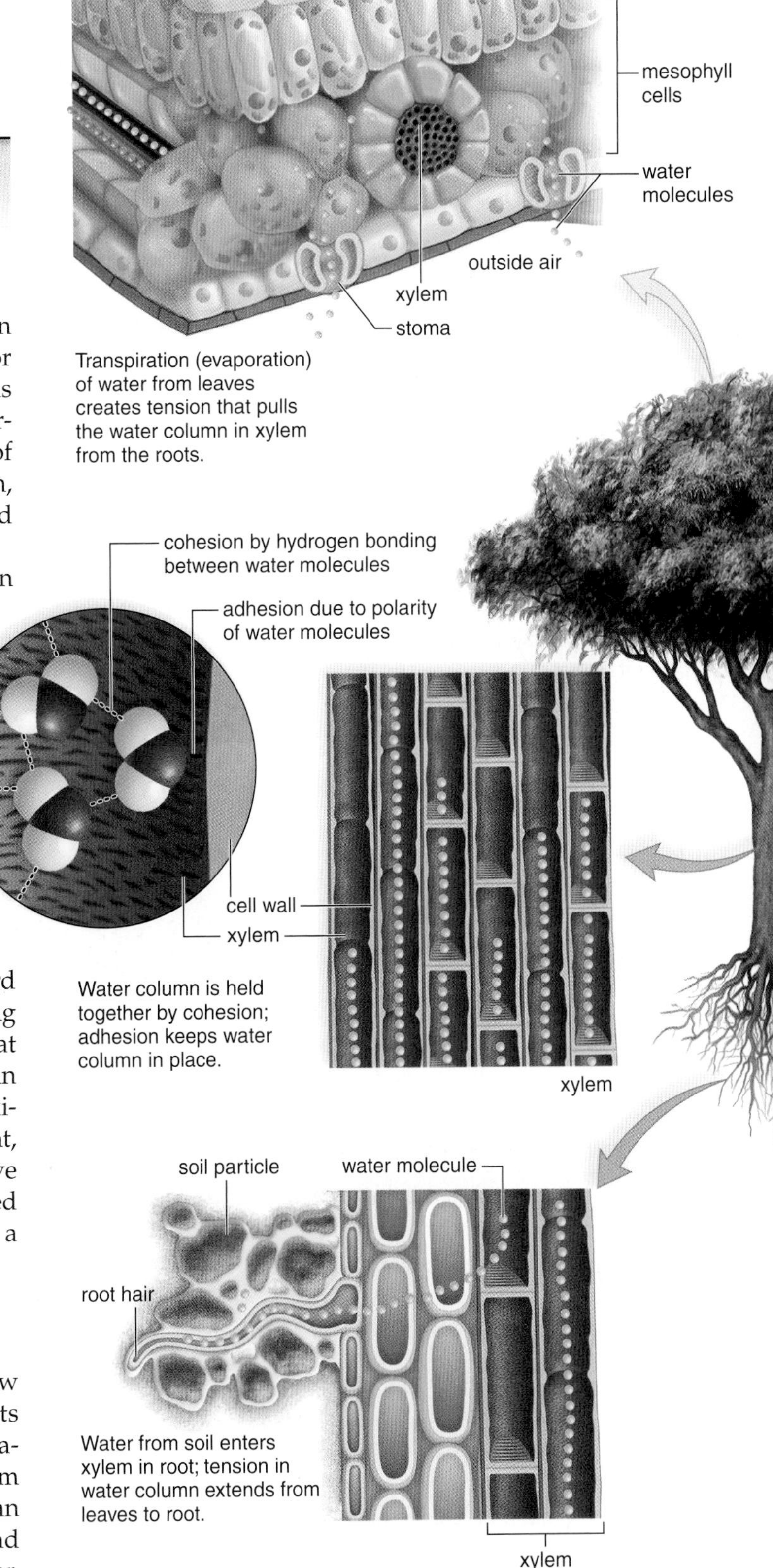

Figure 9.24 Cohesion-tension model of xylem transport.
Tension created by evaporation (transpiration) at the leaves pulls water along the length of the xylem from the root hairs to the leaves.

Explanation of the Model Unlike animals, which rely on a pumping heart to move blood through their vessels, plants utilize a passive, not active, means of transport to move water in xylem. The cohesion-tension model of xylem transport relies on the properties of water (see section 2.3). The term *cohesion* refers to the tendency of water molecules to cling together. Because of hydrogen bonding, water molecules interact with one another, forming a continuous water column in xylem from the leaves to the roots that is not easily broken. *Adhesion* refers to the ability of water, a polar molecule, to interact with the molecules making up the walls of the vessels in xylem. Adhesion gives the water column extra strength and prevents it from slipping back.

Why does the continuous water column move passively upward? Consider the structure of a leaf. When the sun rises, stomata open, and carbon dioxide enters a leaf. Within the leaf, the mesophyll cells—particularly the spongy layer—are exposed to the air, which can be quite dry. Water now evaporates from mesophyll cells. Evaporation of water from leaf cells is called **transpiration.** At least 90% of the water taken up by the roots is eventually lost by transpiration. This means that the total amount of water lost by a plant over a long period of time is surprisingly large. A single *Zea mays* (corn) plant loses somewhere between 135 and 200 liters of water through transpiration during a growing season.

The water molecules that evaporate are replaced by other water molecules from the leaf veins. In this way, transpiration exerts a driving force—that is, a *tension*—which draws the water column up in vessels from the roots to the leaves. As transpiration occurs, the water column is pulled upward, first within the leaf, then from the stem, and finally from the roots.

Tension can reach from the leaves to the root only if the water column is continuous. What happens if the water column within xylem is broken, as by cutting a stem? The water column "snaps back" in the xylem vessel, away from the site of breakage, making it more difficult for conduction to occur. This is why it is best to maximize water conduction by cutting flower stems under water. This effect has also allowed investigators to measure the tension in stems. A device called the pressure bomb measures how much pressure it takes to push the xylem sap back to the cut surface of the stem.

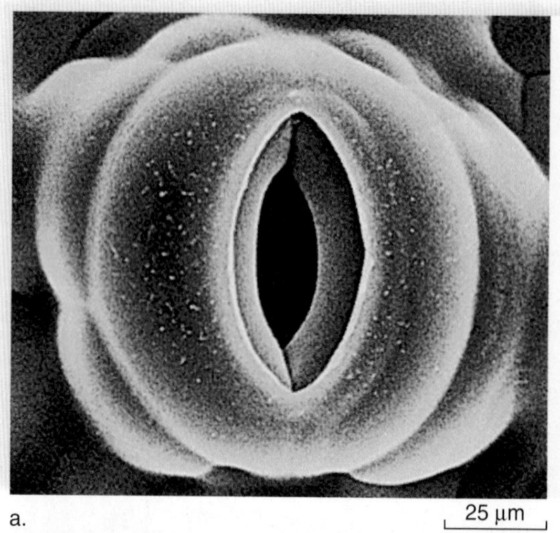

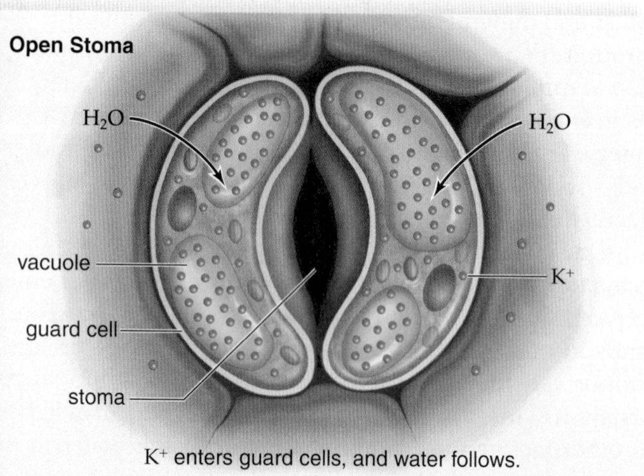

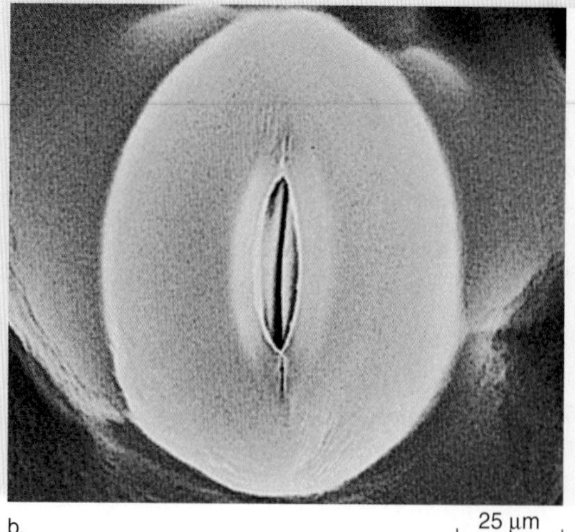

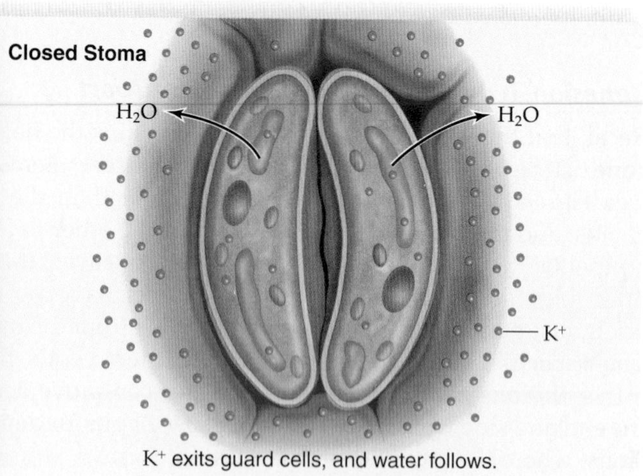

Figure 9.25 Opening and closing of stomata.
a. A stoma opens when turgor pressure increases in guard cells due to the entrance of K^+ followed by the entrance of water.
b. A stoma closes when turgor pressure decreases due to the exit of K^+ followed by the exit of water. (SEM artificially colored.)

There is an important consequence to the way water is transported in plants. When a plant is under water stress, the stomata close. Now the plant loses little water because the leaves are protected against water loss by the waxy cuticle of the upper and lower epidermis. When stomata are closed, however, carbon dioxide cannot enter the leaves, and plants are unable to photosynthesize. (CAM plants are a notable exception. See p. 138.) Photosynthesis, therefore, requires an abundant supply of water so that the stomata remain open and allow carbon dioxide to enter.

Opening and Closing of Stomata

Each stoma, a small pore in the leaf epidermis, is bordered by two guard cells. When water enters the guard cells and turgor pressure increases, the stoma opens. When water exits the guard cells and turgor pressure decreases, the stoma closes. Notice in Figure 9.25 that the guard cells are attached to each other at their ends and that the inner walls are thicker than the outer walls. When water enters, a guard cell's radial expansion is restricted because of cellulose microfibrils in the walls, but lengthwise expansion of the outer walls is possible. When the outer walls expand lengthwise, they buckle out from the region of their attachment, and the stoma opens.

Since about 1968, it has been clear that potassium ions (K^+) accumulate within guard cells when stomata open. In other words, active transport of K^+ into guard cells causes water to follow by osmosis and stomata to open.

If plants are kept in the dark, stomata open and close about every 24 hours, just as if they were responding to the presence of sunlight in the daytime and the absence of sunlight at night. This means that some sort of internal biological clock must be keeping time. Circadian rhythms (behaviors that occur nearly every 24 hours) and biological clocks are areas of intense investigation at this time. Other factors that influence the opening and closing of stomata include temperature, humidity, and stress.

Check Your Progress

1. What is the difference between cohesion and adhesion? How is each involved in moving water up a plant?
2. Define transpiration. How is it involved in moving water up a plant?
3. What does osmosis have to do with the opening and closing of stomata?

Organic Nutrient Transport

Not only do plants transport water and minerals from the roots to the leaves, but they also transport organic nutrients to the parts of plants that need them. This includes young leaves that have not yet reached their full photosynthetic potential, flowers that are in the process of making seeds and fruits, and the roots, whose location in the soil prohibits them from carrying on photosynthesis.

Role of Phloem

As long ago as 1679, Marcello Malpighi suggested that bark is involved in translocating sugars from leaves to roots. He observed the results of removing a strip of bark from around a tree, a procedure called **girdling.** If a tree is girdled below the level of the majority of its leaves, the bark swells just above the cut, and sugar accumulates in the swollen tissue. We know today that when a tree is girdled, the phloem is removed, but the xylem is left intact. Therefore, the results of girdling suggest that phloem is the tissue that transports sugars.

Radioactive tracer studies with carbon 14 (^{14}C) have confirmed that phloem transports organic nutrients. When ^{14}C-labeled carbon dioxide (CO_2) is supplied to mature leaves, radioactively labeled sugar is soon found moving down the stem into the roots. It's difficult to get samples of sap from phloem without injuring the phloem, but this problem is solved by using aphids, small insects that are phloem feeders. The aphid drives its stylet, a sharp mouthpart that functions like a hypodermic needle, between the epidermal cells, and sap enters its body from a sieve-tube member (Fig. 9.26). If the aphid is anesthetized using ether, its body can be carefully cut

a. An aphid feeding on a plant stem

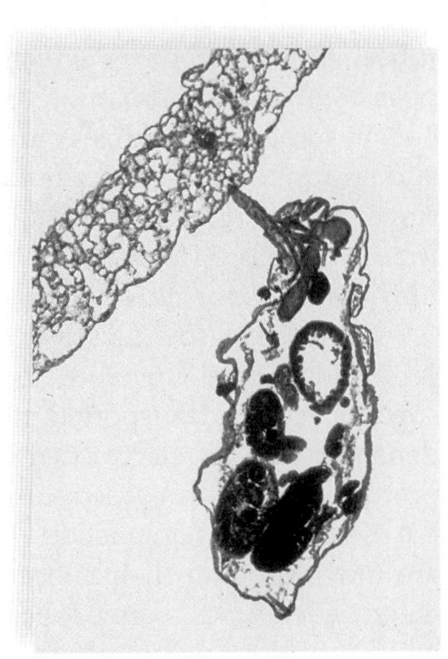

b. Aphid stylet in place

Figure 9.26 Acquiring phloem sap. Aphids are small insects that remove nutrients from phloem by means of a needlelike mouthpart called a stylet. **a.** Excess phloem sap appears as a droplet after passing through the aphid's body. **b.** Micrograph of a stylet in plant tissue. When an aphid is cut away from its stylet, phloem sap becomes available for collection and analysis.

away, leaving the stylet. Phloem can then be collected and analyzed. The use of radioactive tracers and aphids has revealed that the movement of nutrients through phloem can be as fast as 60–100 cm per hour and possibly up to 300 cm per hour.

Pressure-Flow Model of Phloem Transport

The **pressure-flow model** is the current explanation for the movement of organic materials in phloem (Fig. 9.27). Consider the following experiment in which two bulbs are connected by a glass tube. The left-hand bulb contains solute at a higher concentration than the right-hand bulb. Each bulb is bounded by a differentially permeable membrane, and the entire apparatus is submerged in distilled water:

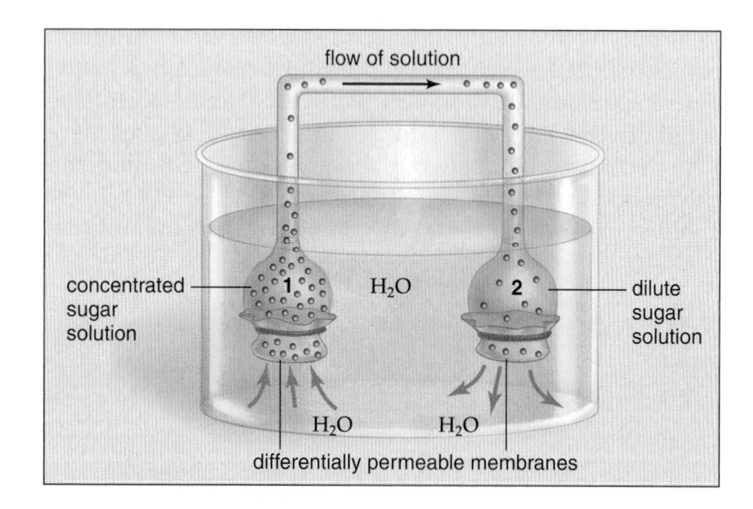

Distilled water flows into the left-hand bulb to a greater extent because it has the higher solute concentration. The entrance of water creates a positive *pressure,* and water *flows* toward the second bulb. This flow not only drives water toward the second bulb, but it also provides enough force for water to move out through the membrane of the second bulb—even though the second bulb contains a higher concentration of solute than the distilled water.

In plants, sieve tubes are analogous to the glass tube that connects the two bulbs. Sieve tubes are composed of sieve-tube members, each of which has a companion cell. It is possible that the companion cells assist the sieve-tube members in some way. The sieve-tube members align end to end, and strands of plasmodesmata (cytoplasm) extend through sieve plates from one sieve-tube member to the other. Sieve tubes, therefore, form a continuous pathway for organic nutrient transport throughout a plant.

During the growing season, photosynthesizing leaves are producing sugar. Therefore, they are a **source** of sugar. This sugar is actively transported into phloem. Again, transport is dependent on an electrochemical gradient established by a proton pump, a form of active transport. Sugar is carried across the membrane in conjunction with hydrogen ions (H^+), which are moving down their concentration gradient. After sugar enters sieve tubes, water follows passively by osmosis. The buildup of water within sieve tubes creates the positive pressure that starts a flow of phloem contents. The roots (and other

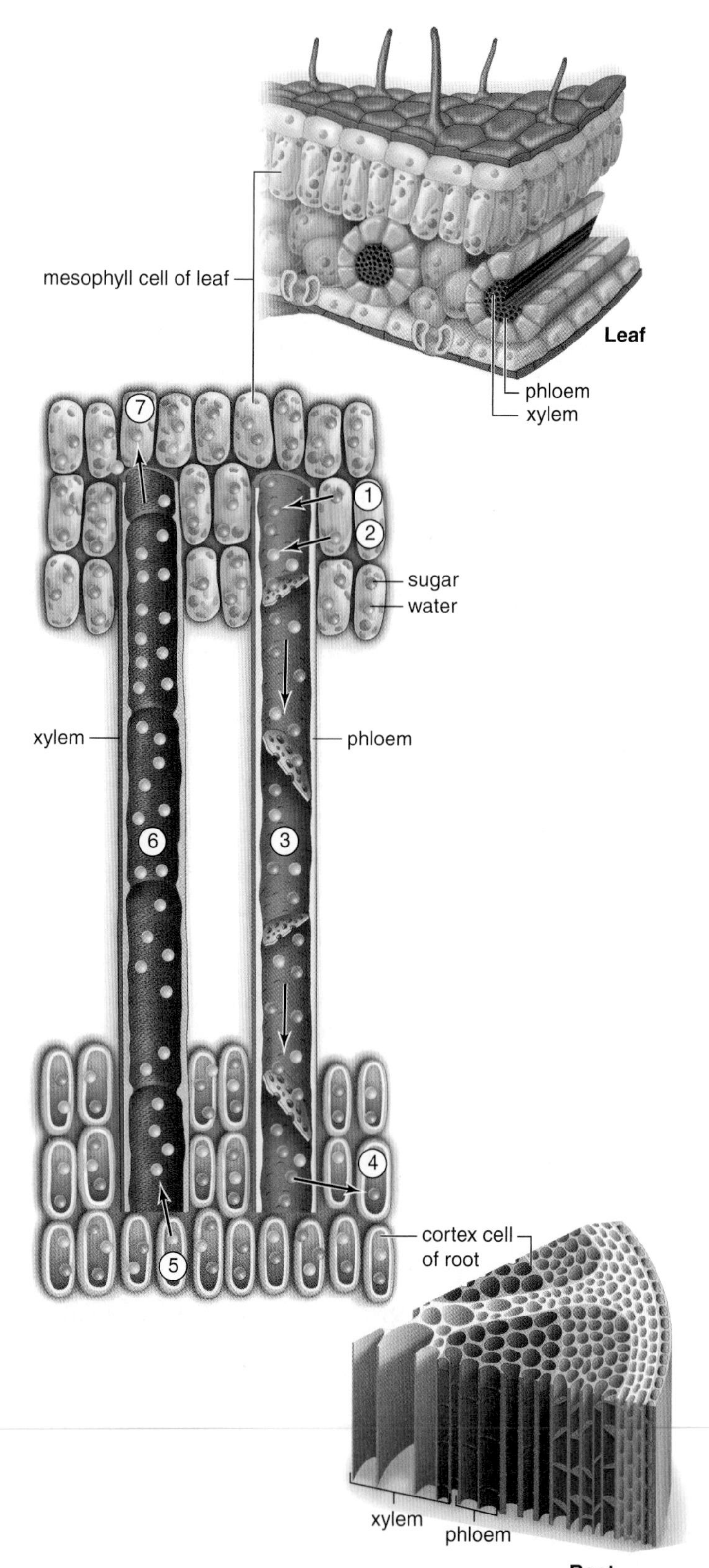

Figure 9.27 Pressure-flow model of phloem transport.
At a source, ① sugar is actively transported into sieve tubes. ② Water follows by osmosis. ③ A positive pressure causes phloem contents to flow from the source to a sink. At a sink, ④ sugar is actively transported out of sieve tubes, and cells use it for cellular respiration. Water exits by osmosis. ⑤ Some water returns to the xylem, where it mixes with more water absorbed from the soil. ⑥ Xylem transports water to the mesophyll of the leaf. ⑦ Most of the water is transpired, but some is used for photosynthesis, and some reenters the phloem by osmosis.

growth areas) are a **sink** for sugar, meaning that they are removing sugar and using it for cellular respiration. After sugar is actively transported out of sieve tubes, water exits phloem passively by osmosis (which reduces pressure at the sink) and is taken up by xylem, which transports water to leaves, where it is used for photosynthesis. Now, phloem contents continue to flow from the leaves (source) to the roots (sink).

The pressure-flow model of phloem transport can account for any direction of flow in sieve tubes if we consider that the direction of flow is always from source to sink. For example, recently formed leaves can be a sink, and they will receive sucrose until they begin to maximally photosynthesize.

Check Your Progress

1. Explain how osmosis is involved in moving organic nutrients in the plant.
2. Does phloem always transport organic nutrients from the leaves to the roots? Why or why not?

Applying the Concepts Revisited

The roots, stems, and leaves have structural and chemical adaptations to protect the plant from the physical elements and attack by other organisms such as insects and bacteria. Interestingly, cottonwood trees developed a way to defend themselves against one predator (beavers) that actually ended up aiding another one of their predators (beetles). If a beaver cuts down a cottonwood tree, the tree will resprout and the new shoots have a high concentration of a chemical that the beavers find offensive. However, the caterpillar of a leaf beetle that feeds on the cottonwood tree eats this compound and stores it in its own body to use as a defensive compound against ants, which attack it!

1. How do woody plants protect themselves?
2. Would adaptations to protect a plant from the physical elements (wind, temperature, etc.) differ from adaptations to protect a plant from an herbivore? Why or why not?

Summarizing the Concepts

9.1 Plant Organs

- A flowering plant has three vegetative organs. Roots anchor a plant, absorb water and minerals, and store the products of photosynthesis. Stems support leaves, conduct materials to and from roots and leaves, and help store plant products. Leaves are specialized for gas exchange, and they carry on most of the photosynthesis in the plant.

9.2 Cells and Tissues of Plants

- A plant has the ability to grow its entire life because it possesses meristematic (embryonic) tissue.
- Apical meristems are located at or near the tips of stems and roots, where they increase the length of these structures. This increase in length is called *primary growth.* The apical meristems continually produce three types of meristem: protoderm, ground meristem, and procambium.
- The entire body of both nonwoody (herbaceous) and young woody plants is covered by a layer of epidermis, which in most plants contains a single layer of closely packed epidermal cells. The walls of epidermal cells that are exposed to air are covered with a waxy cuticle to minimize water loss.
- Ground tissue forms the bulk of a plant and contains parenchyma, collenchyma, and sclerenchyma cells. Parenchyma cells are thin-walled and capable of photosynthesis when they contain chloroplasts. Collenchyma cells have thicker walls for flexible support. Sclerenchyma cells are hollow, nonliving support cells with secondary walls fortified by lignin.
- Vascular tissue consists of xylem and phloem. Xylem contains two types of conducting cells: vessel elements and tracheids. Xylem transports water and minerals. In phloem, sieve tubes are composed of sieve-tube members, each of which has a companion cell. Phloem transports sugar and other organic compounds, including hormones.

9.3 Monocot Versus Eudicot Plants

- Flowering plants are divided into monocots and eudicots according to the number of cotyledons in the seed; the arrangement of vascular tissue in roots, stems, and leaves; and the number of flower parts.

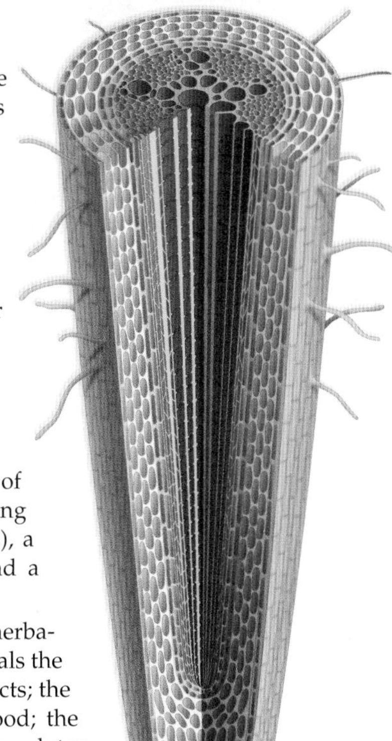

9.4 Organization of Roots

- A root tip has a zone of cell division (containing the primary meristems), a zone of elongation, and a zone of maturation.
- A cross section of a herbaceous eudicot root reveals the epidermis, which protects; the cortex, which stores food; the endodermis, which regulates the movement of minerals; and the vascular cylinder, which is composed of vascular tissue.
- In the vascular cylinder of a eudicot, the xylem appears star-shaped, and the phloem is found in separate regions, between the arms of the

xylem. In contrast, a monocot root has a ring of vascular tissue with alternating bundles of xylem and phloem surrounding the pith.

- Roots are diversified. Taproots are specialized to store the products of photosynthesis. A fibrous root system covers a wider area. Prop roots are adventitious roots specialized to provide increased anchorage.
- Roots enter into symbiotic relationships with bacteria and fungi.

9.5 Organization of Stems

- The activity of the shoot apical meristem within a terminal bud accounts for the primary growth of a stem. A terminal bud contains internodes and leaf primordia at the nodes. When stems grow, the internodes lengthen.
- In a cross section of a nonwoody eudicot stem, epidermis is the outermost layer of cells, followed by cortex tissue, vascular bundles in a ring, and an inner pith. Monocot stems have scattered vascular bundles, and the cortex and pith are not well defined.
- Secondary growth of a woody stem is due to activities of the vascular cambium, which produces new xylem and phloem every year, and cork cambium, which produces new cork cells when needed. Cork, a part of the bark, replaces epidermis in woody plants.
- In a cross section of a woody stem, the bark is composed of all the tissues outside the vascular cambium: secondary phloem, cork cambium, and cork. Wood consists of secondary xylem, which builds up year after year and forms annual rings.

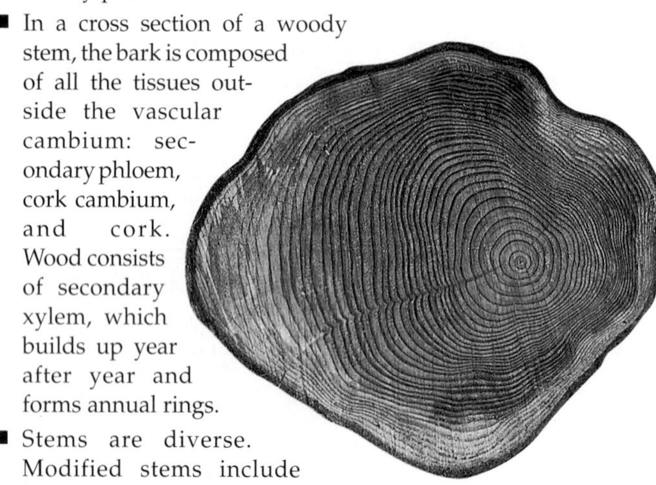

- Stems are diverse. Modified stems include horizontal aboveground and underground stems, corms, and some tendrils.

9.6 Organization of Leaves

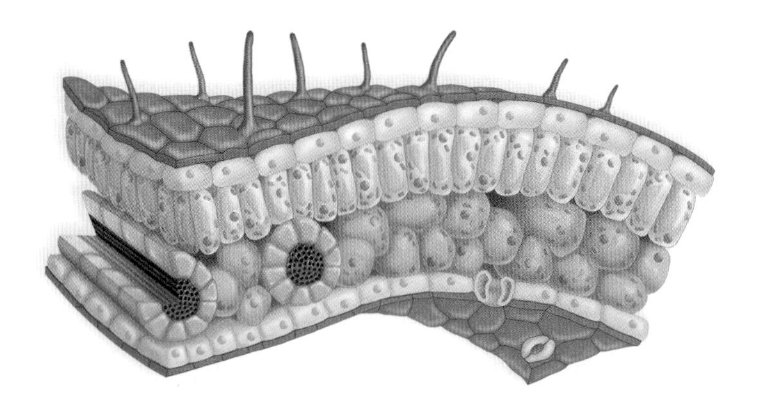

- The bulk of a leaf is mesophyll tissue bordered by an upper and lower layer of epidermis. The epidermis is covered by a cuticle and may bear trichomes. Stomata tend to be in the lower layer. Vascular tissue is present within leaf veins.
- Leaves are diverse. The spines of a cactus are leaves. Other succulents have fleshy leaves. An onion is a bulb with fleshy leaves, and the tendrils of peas are leaves. The Venus flytrap has leaves that trap and digest insects.

9.7 Uptake and Transport of Nutrients

- Water transport in plants occurs within xylem. The cohesion-tension model of xylem transport states that transpiration (evaporation of water at stomata) creates tension, which pulls water upward in xylem. This method works only because water molecules are cohesive. Transpiration and carbon dioxide uptake occur when stomata are open.
- Stomata open when guard cells take up potassium (K^+) ions and water follows by osmosis. Stomata open because the entrance of water causes the guard cells to buckle out.
- Transport of organic nutrients in plants occurs within phloem. The pressure-flow model of phloem transport states that sugar is actively transported into phloem at a source, and water follows by osmosis. The resulting increase in pressure creates a flow, which moves water and sugar to a sink.

TESTING YOURSELF

Choose the best answer for each question.

1. _____ forms the outer protective covering of a plant.
   a. Ground tissue
   b. Root tissue
   c. Epidermal tissue
   d. Vascular tissue
2. Which of these is an incorrect contrast between monocots (stated first) and eudicots (stated second)?
   a. one cotyledon—two cotyledons
   b. leaf veins parallel—net veined
   c. vascular bundles in a ring—vascular bundles scattered
   d. flower parts in threes—flower parts in fours or fives
   e. All of these are correct contrasts.

3. Which of these cells in a plant is apt to be nonliving?
   a. parenchyma
   b. collenchyma
   c. sclerenchyma
   d. epidermal
   e. guard cells
4. Root hairs are found in the zone of
   a. cell division.
   b. elongation.
   c. maturation.
   d. apical meristem.
   e. All of these are correct.
5. Insect-eating plants are common in regions where ____ is in limited supply.
   a. water
   b. CO_2
   c. nitrogen
   d. oxygen
6. Stomata are open when
   a. guard cells are filled with water.
   b. potassium ions move into the guard cells.
   c. turgor pressure increases.
   d. All of these are correct.
7. Annual rings are the number of
   a. internodes in a stem.
   b. rings of vascular bundles in a monocot stem.
   c. layers of xylem in a stem.
   d. bark layers in a woody stem.
   e. Both b and c are correct.
8. What role do cohesion and adhesion play in xylem transport?
   a. Like transpiration, they create a tension.
   b. Like root pressure, they create a positive pressure.
   c. Like sugars, they cause water to enter xylem.
   d. They create a continuous water column in xylem.
9. The sugar produced by mature leaves moves into sieve tubes by way of ____, while water follows by ____.
   a. osmosis, osmosis
   b. active transport, active transport
   c. osmosis, active transport
   d. active transport, osmosis
10. In a cross section of which of the following structures would you find scattered vascular bundles?
    a. monocot root
    b. monocot stem
    c. eudicot root
    d. eudicot stem
11. Which of the following is considered a stem?
    a. potato
    b. sweet potato
    c. cabbage
    d. carrot

For questions 12–15, match the definitions with the structure in the key.

Key:

a. stoma
b. node
c. xylem
d. parenchyma cell

12. Major constituent of ground tissue
13. Region where leaves attach to stem
14. Opens when transpiration occurs
15. Contains tracheids

UNDERSTANDING THE TERMS

adventitious root 152
angiosperms 144
annual ring 157
apical meristem 146
axillary bud 145
bark 156
blade 145
Casparian strip 151
cohesion-tension model 163
collenchyma 147
companion cell 148
cork 147
cork cambium 156
cortex 151
cotyledon 149
cuticle 146
deciduous 145
endodermis 151
epidermal tissue 146
epidermis 146
epiphyte 152
eudicot 149
evergreen 145
fibrous root system 152
girdling 165
ground tissue 146
guard cell 147
herbaceous stem 155
internode 145
leaves (sing., leaf) 145
leaf vein 148
lenticel 147
lignin 147
meristematic tissue 146
mesophyll 160
monocot 149
mycorrhizae 153
node 145
organ 144
palisade mesophyll 160
parenchyma 147
pericycle 151
periderm 147
petiole 145
phloem 145
pith 151
plasmodesmata (sing., plasmodesma) 148
pressure-flow model 166
primary root 152
rhizome 158
root cap 150
root hair 146
root nodule 153
root system 144
sclerenchyma 147
shoot system 144
sieve-tube member 148
sink 167
source 166
spongy mesophyll 161
stem 145
stolon 158
taproot 152
terminal bud 154
tracheid 148
transpiration 164
trichome 146
vascular bundle 148
vascular cambium 154
vascular cylinder 148
vascular tissue 146
vessel element 148
wood 156
xylem 145

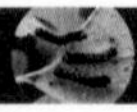

THINKING CRITICALLY

1. List all the specific tissue types in roots, stems, and leaves where you would expect to find parenchyma cells.
2. Why is it more accurate to define a stem as containing nodes and internodes than as being upright?
3. Compare and contrast the transport of water and organic nutrients in plants with the transport of blood in humans.

INQUIRY INTO LIFE WEBSITE

The companion website for *Inquiry into Life* provides a wealth of information organized and integrated by chapter. You will find practice tests, animations, videos, and much more that will complement your learning and understanding of general biology.

http://www.mhhe.com/maderinquiry13

Applying the Concepts

It is nice to be able to eat a slice of cold watermelon on a very hot summer day—without being bothered by the seeds. Seedless watermelons, once a novelty, have become a common convenience. But how are seedless watermelons produced?

Seeded watermelons have a diploid number (2n) of 22 chromosomes. Genetic manipulation of a 2n plant can produce a 4n plant with 44 chromosomes. When a 2n plant (pollen parent) is crossed with a 4n plant (seed parent), the resulting fruit has 3n seeds. The plant that germinates from this 3n seed is sterile (unable to produce seeds) because meiosis cannot proceed as usual. Hence the seedless watermelon!

Although seedless watermelons do not produce viable seeds, they develop small white "seeds." These are actually soft, tasteless seed coats that can be eaten with the watermelon. Obviously, these seed coats can't be used to start next year's crop and the 3n seeds needed for the next crop must be produced anew by crossing 2n and 4n plants. Expense, time, and energy are required all to save us the inconvenience of spitting out watermelon seeds!

Our story of seedless watermelons reminds us that we have the capability of genetically modifying plants and other organisms, often simply called GMOs. Unlike seedless watermelons, there is controversy surrounding the potential health effects of eating GM (genetically modified) foods. Nonetheless, humans modify plants all the time.

In this chapter, you will learn how normal flowering plant reproduction occurs, along with seed production and development, to better understand how plants operate under natural conditions. You will also learn how humans can modify plants.

Plant Reproduction and Responses

Chapter Outline

10.1 Sexual Reproduction in Flowering Plants

Learning Outcomes

1. Show that flowering plants exhibit an alternation of generations even though they produce two types of spores and two types of gametophytes.
2. Label only the reproductive parts of a flower and describe the function of each part.
3. In flowering plants, diagram and describe the development of male and female gametophytes and the development of the sporophyte.

In contrast to animals, plants have two multicellular stages in their life cycle; therefore, their life cycle is called an **alternation of generations.** In this life cycle, a diploid sporophyte alternates with a haploid gametophyte (see Fig. 29.2):

- The 2n **sporophyte** produces haploid **spores** by meiosis. The spores divide by mitosis to become gametophytes.
- The n **gametophyte** produces gametes by mitosis. Upon fertilization, the cycle returns to the 2n sporophyte.

Now that you have an overview of the plant life cycle, we will examine how it pertains to the flowering plant life cycle.

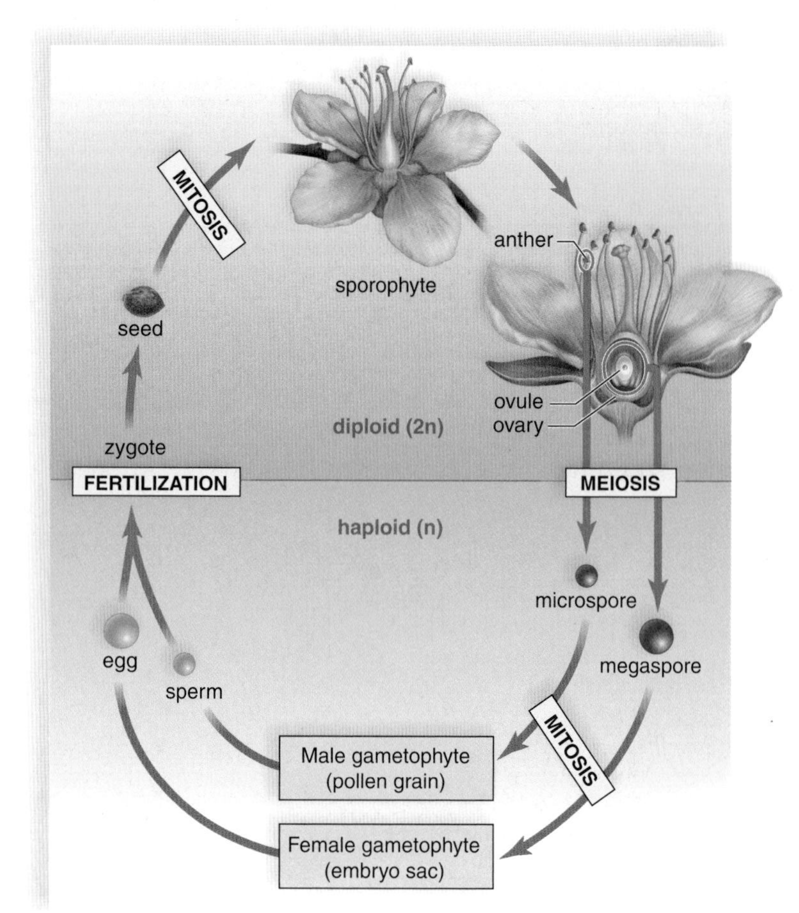

Figure 10.1 Alternation of generations in flowering plants.
The sporophyte bears flowers. The flower produces microspores within anthers and megaspores within ovules by meiosis. A megaspore becomes a female gametophyte, which produces an egg within an embryo sac, and a microspore becomes a male gametophyte (pollen grain), which produces sperm. Fertilization results in a seed-enclosed zygote and stored food.

Flowering Plant Life Cycle

In flowering plants, the sporophyte is dominant and it is the generation that bears flowers (Fig. 10.1). A **flower** is a reproductive structure and it produces two types of spores: microspores and megaspores. A **microspore** undergoes mitosis and becomes a pollen grain, which is the male gametophyte. Meanwhile, the **megaspore** has undergone mitosis to become a microscopic embryo sac, which is the female gametophyte. The female gametophyte is retained within the flower.

A pollen grain is either windblown or carried by an animal to the vicinity of the embryo sac. At maturity, a pollen grain contains two nonflagellated sperm that travel down a pollen tube to the embryo sac. Upon fertilization, the structure surrounding the embryo sac develops into a **seed.** The seeds are enclosed by a **fruit,** which aids in dispersing the seeds. When a seed germinates, a new sporophyte emerges and develops into a mature organism.

Adaptation to a Land Environment

The life cycle of flowering plants is adapted to a land existence because all stages of the life cycle are protected from drying out. For example, the microscopic gametophytes develop within the sporophyte. Pollen grains are not released until they have a thick wall, and they are carried by wind or an animal (usually an insect) to another flower where they develop a pollen tube that carries the sperm to the egg. Following fertilization, the seed coat protects the embryo until conditions favor regrowth.

Flowers

The flower is unique to angiosperms. The evolution of the flower was a major factor leading to the success of angiosperms, which now number over 240,000 described species. Aside from producing the spores and protecting the gametophytes, flowers often attract pollinators. Flowers also produce the fruits that enclose the seeds.

A flower develops in response to environmental signals such as day length (see section 10.4). In many plants, a shoot apical meristem suddenly stops producing leaves and starts producing a flower enclosed within a bud. In other plants, axillary buds develop directly into flowers.

A typical flower has four whorls of modified leaves attached to a receptacle at the end of a flower stalk. The receptacle bearing a single flower is attached to a structure called a peduncle, while the structure that bears one of several flowers is called a pedicel. Figure 10.2 shows the following structures of a typical flower:

1. The **sepals,** which are the most leaflike of all the flower parts, are usually green, and they protect the bud as the flower develops within. Sepals can also be the same color as the flower petals.
2. An open flower next has a whorl of **petals,** whose color accounts for the attractiveness of many flowers. The size, shape, and color of petals are attractive to a specific pollinator. Wind-pollinated flowers may have no petals at all.

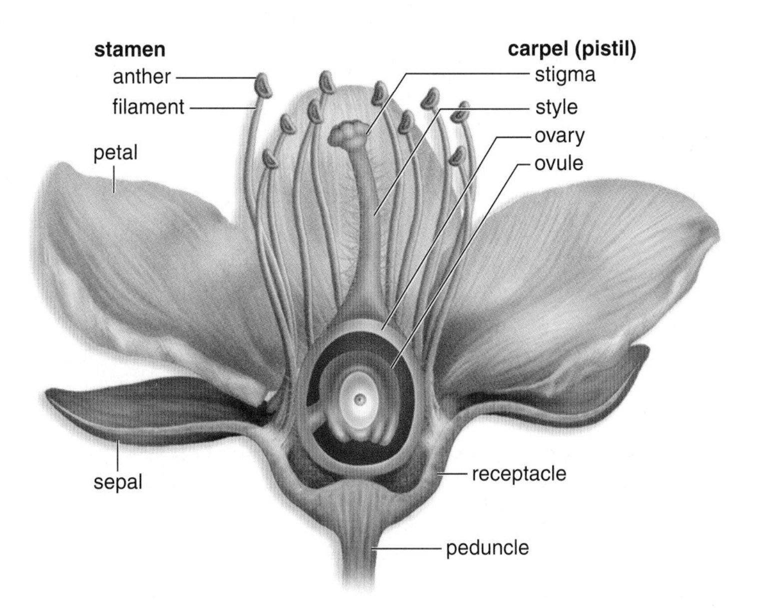

Figure 10.2 Anatomy of a flower.

A complete flower has all flower parts: sepals, petals, stamens, and at least one carpel.

3. **Stamens** are the "male" portion of the flower. Each stamen has two parts: the **anther,** a saclike container, and the **filament,** a slender stalk. Pollen grains develop from the microspores produced in the anther.
4. At the very center of a flower is the **carpel** (also called the pistil), a vaselike structure that represents the "female" portion of the flower. A carpel usually has three parts: the **stigma,** an enlarged sticky knob; the **style,** a slender stalk; and the **ovary,** an enlarged base that encloses one or more ovules.

In monocots, flower parts occur in threes and multiples of three. In eudicots, flower parts are in fours or fives and multiples of four or five (Fig. 10.3).

A flower can have a single carpel or multiple carpels. Sometimes several carpels are fused into a single structure, in which case this compound ovary has several chambers containing **ovules,** each of which will house a megaspore. For example, an orange develops from a compound ovary, and every section of the orange is a chamber.

Not all flowers have sepals, petals, stamens, and carpels. Those that do, such as the flower in Figure 10.2, are said to be *complete,* and those that do not are said to be *incomplete.* Flowers that have both stamens and carpels are called perfect (bisexual) flowers. Those with only stamens or only carpels are imperfect (unisexual) flowers. If staminate and carpellate (pistillate) flowers are on one plant, the plant is monoecious (one house) (Fig. 10.4). If staminate and carpellate flowers are on separate plants, the plant is dioecious (two houses).

Check Your Progress

1. In flowering plants, which generation produces two types of spores and what happens to the spores?
2. List only the reproductive parts of a flower and give a function for each part mentioned.

a. Daylily

b. Festive azalea

Figure 10.3 Monocot versus eudicot flowers.

a. Monocots, such as daylilies, have flower parts usually in threes. In particular, note the three petals and three sepals. **b.** Azaleas are eudicots. They have flower parts in fours or fives; note the five petals of this flower. (p = petal; s = sepal.)

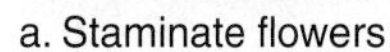

a. Staminate flowers

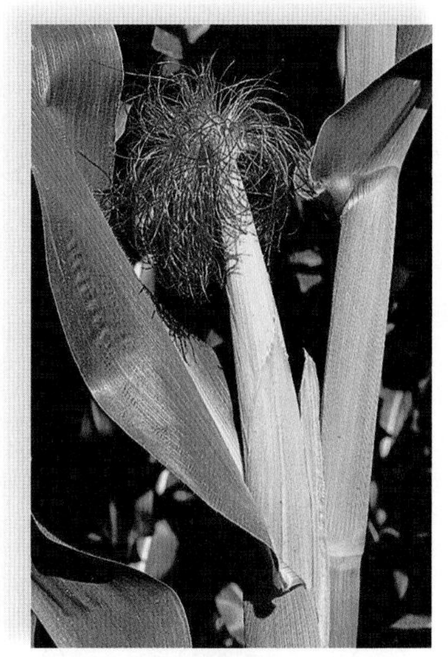

b. Carpellate flowers

Figure 10.4 Corn plants are monoecious.

A corn plant has (**a**) clusters of staminate flowers and (**b**) clusters of carpellate flowers. Staminate flowers produce the pollen that is carried by wind to the carpellate flowers, where an ear of corn develops.

Life Cycle of Flowering Plants

In plants, the sporophyte produces haploid spores by meiosis. The haploid spores grow and develop into haploid gametophytes, which produce gametes by mitotic division.

Flowering plants are heterosporous, meaning that they produce microspores and megaspores. Microspores become mature male gametophytes (sperm-bearing pollen grains), and megaspores become mature female gametophytes (egg-bearing embryo sacs).

Development of Male Gametophyte

Microspores are produced in the anthers of flowers (Fig. 10.5). An anther has four pollen sacs, each containing many **microsporocytes** (microspore mother cells). A microsporocyte undergoes meiosis to produce four haploid microspores. In each, the haploid nucleus divides mitotically, followed by unequal cytokinesis, and the result is two cells enclosed by a finely sculptured wall. This structure, called the **pollen grain,** is at first an immature **male gametophyte** that consists of a tube cell and a generative cell. The larger tube cell will eventually produce a *pollen tube.* The smaller generative cell divides mitotically either now or later to produce two sperm. Once these events take place, the pollen grain has become the mature male gametophyte.

Development of Female Gametophyte

The ovary contains one or more ovules. An ovule has a central mass of parenchyma cells almost completely covered by layers of tissue called integuments except where there is an opening, the micropyle. One parenchyma cell enlarges to become a **megasporocyte** (megaspore mother cell), which undergoes meiosis, producing four haploid megaspores (Fig. 10.5). Three of these megaspores are nonfunctional. In a typical pattern, the nucleus of the functional megaspore divides mitotically until there are eight nuclei in the **female gametophyte.**

Figure 10.5 Life cycle of flowering plants. *Development of gametophytes (far page):* A pollen sac in the anther contains microspore mother cells, which produce microspores by meiosis. A microspore develops into a pollen grain, which germinates and has two sperm. An ovule in an ovary contains a megaspore mother cell, which produces a megaspore by meiosis. A megaspore develops into an embryo sac containing seven cells, one of which is an egg. *Development of sporophyte (this page):* A pollen grain contains two sperm by the time it germinates and forms a pollen tube. During double fertilization, one sperm fertilizes the egg to form a diploid zygote, and the other fuses with the polar nuclei to form a triploid (3n) endosperm cell. A seed contains the developing sporophyte embryo plus stored food.

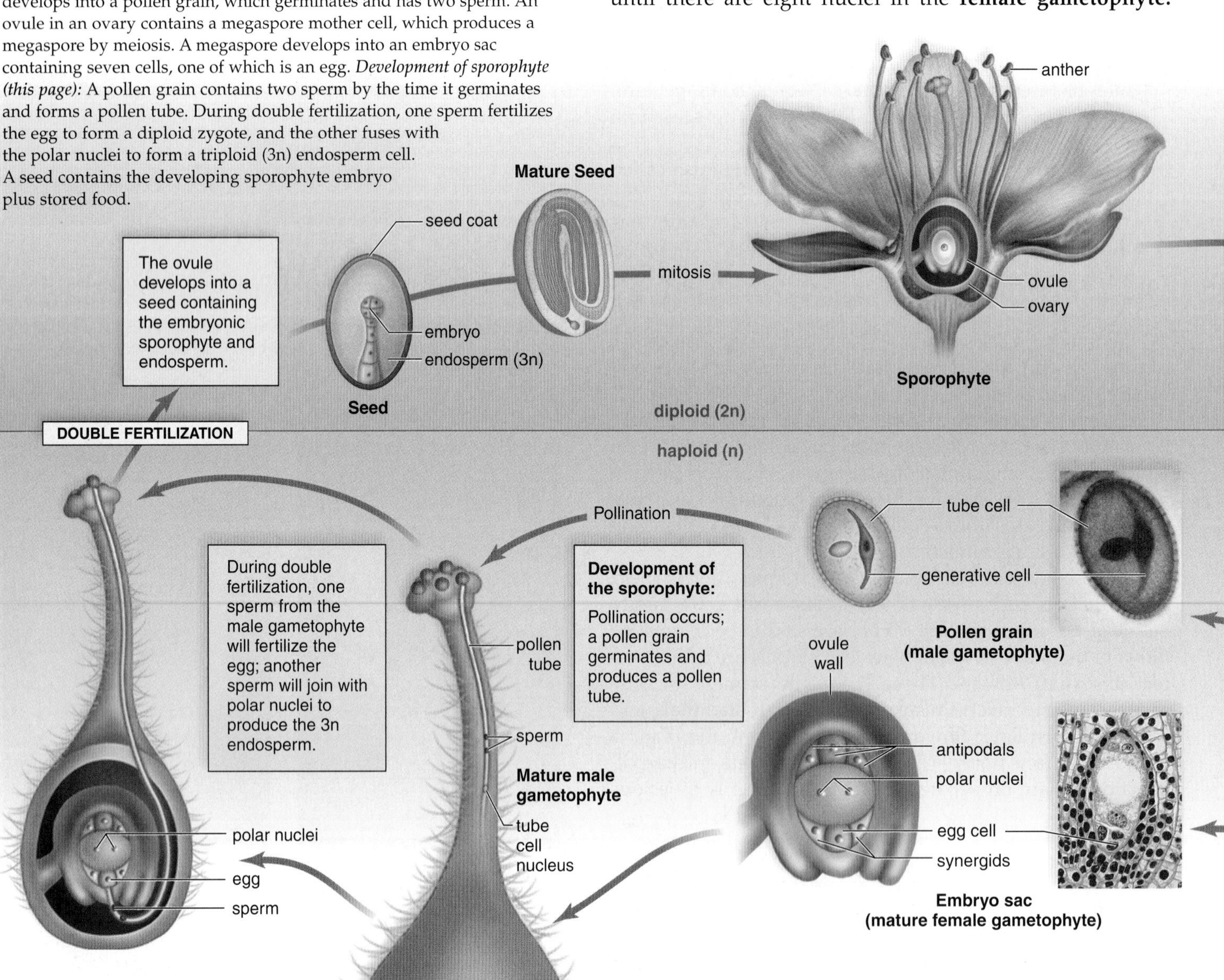

When cell walls form later, there are seven cells, one of which is binucleate. The female gametophyte, also called the **embryo sac,** consists of these seven cells:

1 egg cell;
2 synergid cells;
1 central cell containing two polar nuclei; and
3 antipodal cells

Development of Sporophyte

The walls separating the pollen sacs in the anther break down when the pollen grains are ready to be released (Fig. 10.6). **Pollination** is simply the transfer of pollen from an anther to the stigma of a carpel. Self-pollination occurs if the pollen is from the same plant, and cross-pollination occurs if the pollen is from a different plant of the same species.

Angiosperms often have adaptations to foster cross-pollination. For example, the carpels may mature only after the anthers have released their pollen. As discussed in the Ecology Focus on page 180, cross-pollination may also be brought about with the assistance of a particular pollinator. If a pollinator goes from flower to flower of only one type of plant, cross-pollination is more likely to occur in an efficient manner. Color, odor, and secretion of nectar are ways that pollinators are attracted to plants. Over time, certain pollinators have become adapted to reach the nectar of only one type of flower. In the process, pollen is inadvertently picked up and taken to another plant of the same type.

When a pollen grain lands on the stigma of the same species, it germinates, forming a pollen tube (see Fig. 10.5). The germinated pollen grain, containing a tube cell and two sperm, is the mature male gametophyte. As it grows, the pollen tube passes between the cells of the stigma and the style to reach the micropyle, a pore of the ovule. Now **double fertilization** occurs. One sperm nucleus unites with the egg nucleus, forming a 2n zygote, and the other sperm nucleus migrates and unites with the polar nuclei of the central cell, forming a 3n endosperm cell. The zygote divides mitotically to become the **embryo,** a young sporophyte, and the endosperm cell divides mitotically to become the endosperm. **Endosperm** is the tissue that will nourish the embryo and seedling as they develop.

Figure 10.6 Pollen grains.
a. Cocksfoot grass releasing pollen. **b.** Pollen grains of Canadian goldenrod. **c.** Pollen grains of pussy willow. The shape and pattern of pollen grain walls are quite distinctive, and experts can use them to identify the genus, and sometimes even the species, that produced a particular pollen grain. Pollen grains have strong walls resistant to chemical and mechanical damage. Therefore, they frequently become fossils.

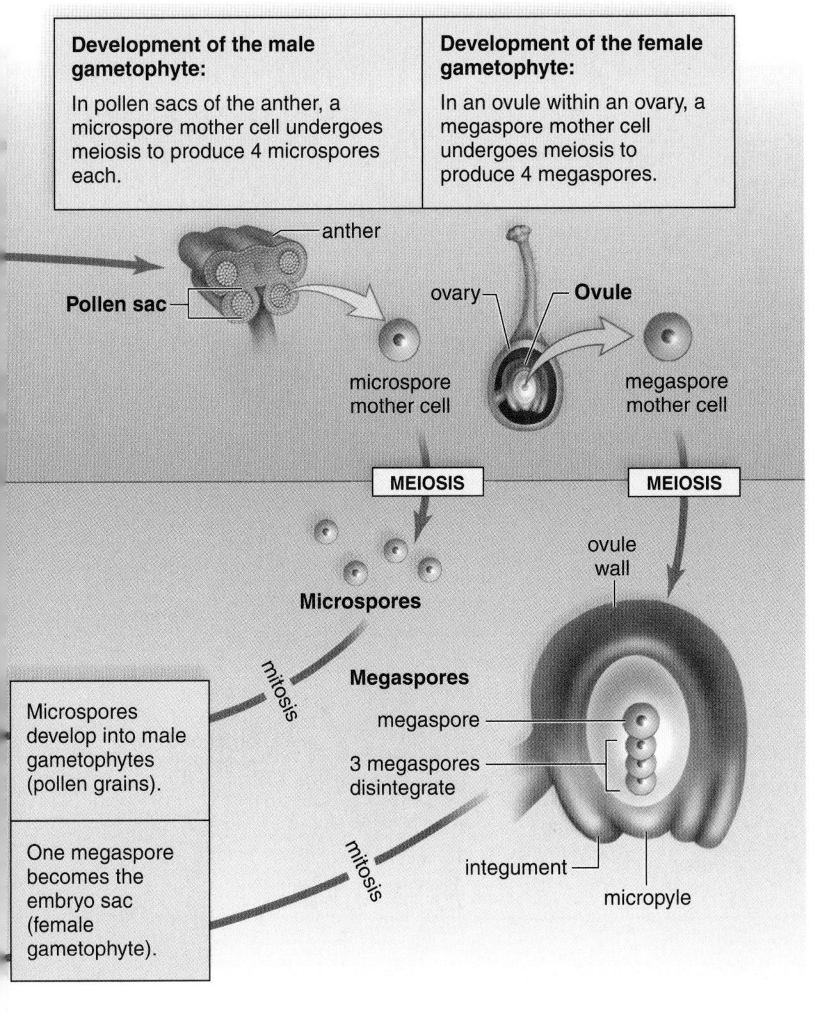

Check Your Progress

1. Why are flowering plants said to be "heterosporous"?
2. What happens in an ovule from the time it contains megaspores to the time it is a seed?
3. Describe the process of double fertilization.

10.2 Growth and Development

Learning Outcomes

1. Trace the development of a eudicot embryo and compare the function of its cotyledons to that of a cotyledon in monocots.
2. Classify and give examples of different types of fruits.
3. Label seed structure and describe germination and dispersal.

Development of the Eudicot Embryo

The endosperm cell, shown in Figure 10.7*a* divides to produce the endosperm tissue, shown in Figure 10.7*b*. The zygote also divides, but asymmetrically. One of the resulting cells is small, with dense cytoplasm. This cell is destined to become the embryo, and it divides repeatedly in different planes, forming a ball of cells. The other, larger cell also divides repeatedly, but it forms an elongated structure called a suspensor, which has a basal cell. The suspensor pushes the embryo deep into the endosperm tissue. The suspensor then disintegrates as the seed matures.

During the globular stage, the embryo is a ball of cells (Fig. 10.7*c*). The root-shoot axis of the embryo is already established at this stage because the embryonic cells near the suspensor will become a root, while those at the other end will ultimately become a shoot.

The embryo has a heart shape when the **cotyledons,** or seed leaves, appear (Fig. 10.7*d*). As the embryo continues to enlarge and elongate, it takes on a torpedo shape (Fig. 10.7*e*). Now the root tip and shoot tip are distinguishable. The shoot apical meristem in the shoot tip is responsible for aboveground growth, and the root apical meristem in the root tip is responsible for underground growth. In a mature embryo, the epicotyl is the portion between the cotyledon(s) that contributes to shoot development (Fig. 10.7*f*). The hypocotyl is that portion below the cotyledon(s) that contributes to stem development. The radicle contributes to root development. Now, the embryo is ready to develop the two main parts of a plant: the shoot system and the root system.

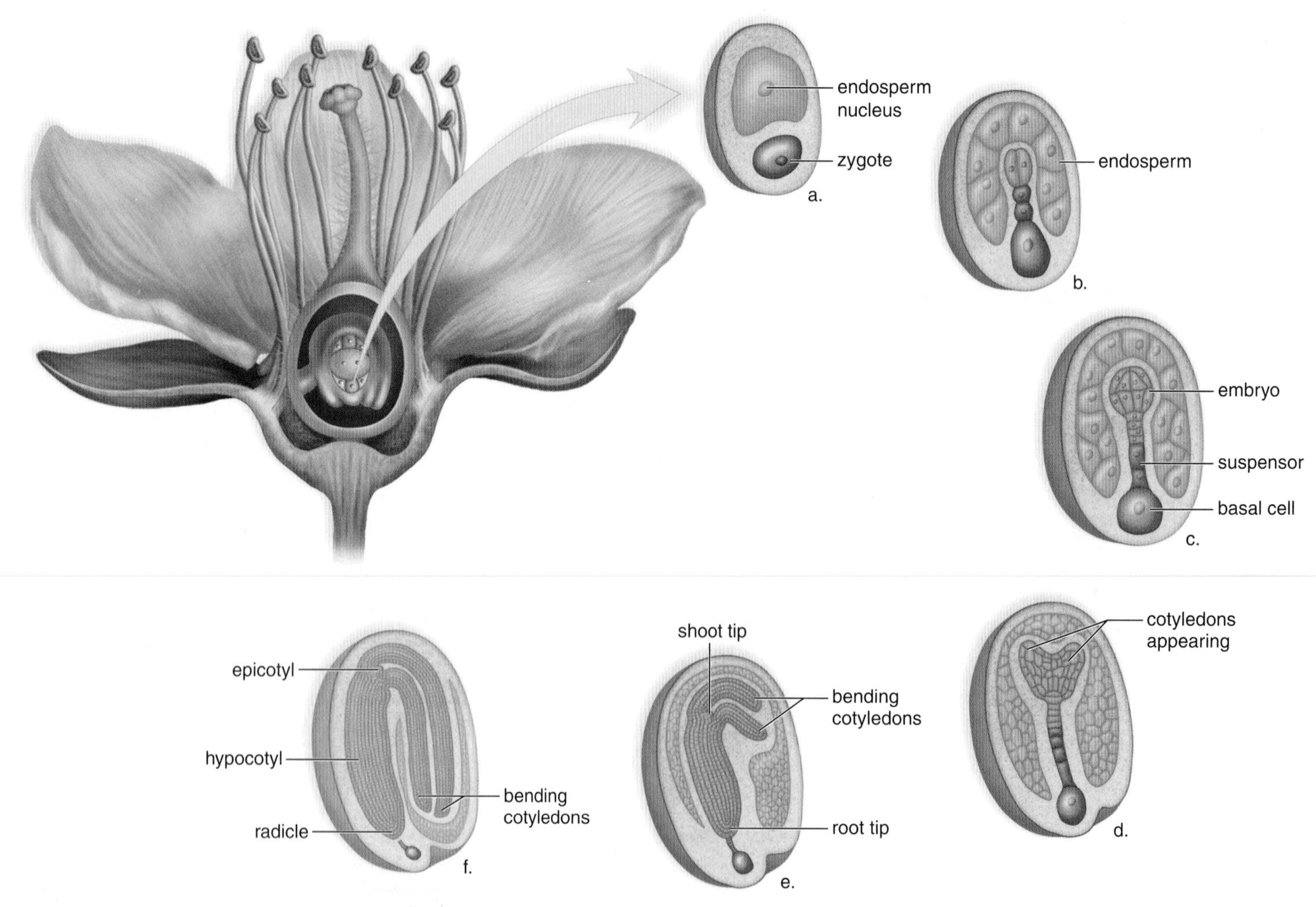

Figure 10.7 Development of a eudicot embryo.
a. The unicellular zygote lies beneath the endosperm nucleus. **b., c.** The endosperm is a mass of tissue surrounding the embryo. The embryo is located above the suspensor. **d.** The embryo becomes heart shaped as the cotyledons begin to appear. **e.** There is progressively less endosperm as the embryo differentiates and enlarges. As the cotyledons bend, the embryo takes on a torpedo shape. **f.** The embryo consists of the epicotyl, the hypocotyl, and the radicle.

The cotyledons are quite noticeable in a eudicot embryo and may fold over. As the embryo develops, the wall of the ovule becomes the seed coat.

Monocot Versus Eudicot Embryos

Whereas eudicots have two cotyledons, monocots have only one cotyledon. In monocots, the cotyledon rarely stores food. Instead, it absorbs food molecules from the endosperm and passes them to the embryo. In eudicots, the cotyledons usually store the nutrient molecules that the embryo uses. Therefore, the endosperm disappears because it has been taken up by the two cotyledons.

Figures 10.9 and 10.10, on pages 178–179, contrast the structure of a bean seed (eudicot) and a corn kernel (monocot).

Check Your Progress

1. What part of a plant embryo will contribute to aboveground growth and which part will contribute to belowground growth?
2. What is the function of endosperm in monocots versus eudicots?

Fruit Types and Seed Dispersal

A fruit—derived from an ovary and sometimes other flower parts—protects and helps disperse offspring. As a fruit develops, the ovary wall thickens to become the pericarp (Fig. 10.8*a*). The pericarp can have as many as three layers: exocarp, mesocarp, and endocarp.

Simple fruits are derived from a simple ovary of a single carpel or from a compound ovary of several fused carpels. A pea pod is a simple fruit. At maturity, the pea pod breaks open on both sides of the pod to release the seeds. Peas and beans are legumes. A *legume* is a fruit that splits along two sides when mature.

Like legumes, cereal grains of wheat, rice, and corn are *dry fruits.* Sometimes, these grainlike fruits are mistaken for seeds because a dry pericarp adheres to the seed within. These dry fruits are indehiscent, meaning that they don't split open. Humans gather grains before they are released from the plant and then process them to acquire their nutrients.

In some simple fruits, the mesocarp becomes fleshy. When ripe, *fleshy fruits* often attract animals and provide them with food (Fig. 10.8*b*). Peaches and cherries are examples of fleshy fruits that have a hard endocarp. This type of endocarp protects the seed so it can pass through the digestive system of an animal and remain unharmed. In a tomato, the entire pericarp is fleshy. If you cut open a tomato, you see several chambers because the flower's carpel is composed of several fused carpels.

Among fruits, apples are an example of an *accessory fruit* because the bulk of the fruit is not from the ovary, but from the receptacle. Only the core of an apple is derived from the ovary. If you cut an apple crosswise, it is obvious that an apple, like a tomato, came from a compound ovary with several chambers.

Compound fruits develop from several individual ovaries. For example, each little part of a raspberry or blackberry is derived from a separate ovary. Because the flower had

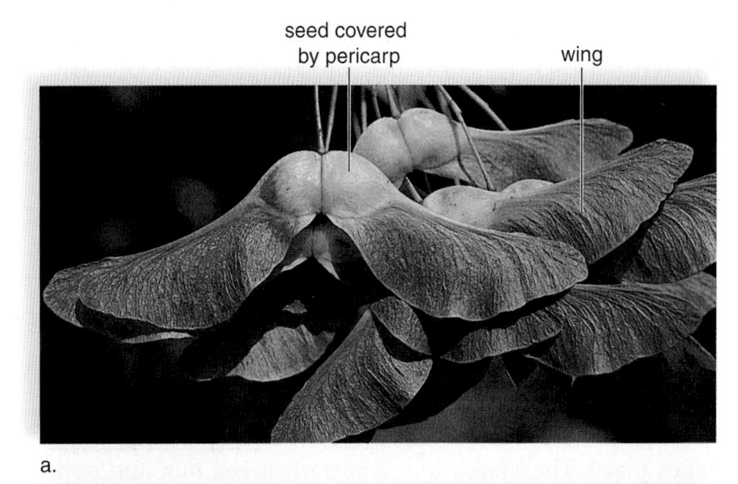

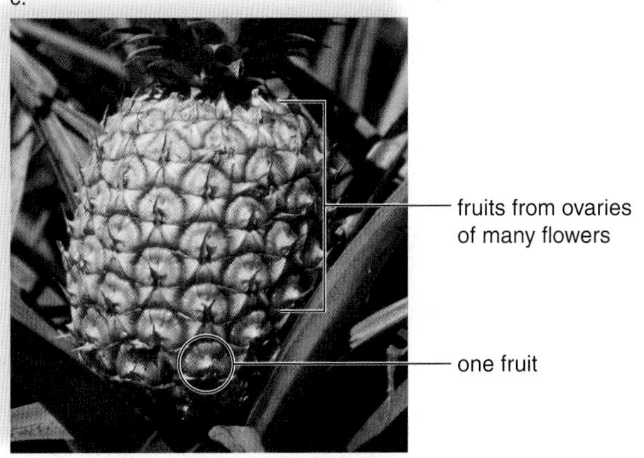

Figure 10.8 Structure and function of fruits.
a. Maple trees produce a winged fruit. The wings rotate in the wind and keep the fruit aloft. **b.** Strawberry plants produce an accessory fruit. Each "seed" is actually a fruit on a fleshy, expanded receptacle. (A slug is snacking on this strawberry.) **c.** Like strawberries, raspberries are aggregate fruits. Each "berry" is derived from an ovary within the same flower. **d.** A pineapple is a multiple fruit derived from the ovaries of many flowers.

many separate carpels, the resulting fruit is called an *aggregate fruit* (Fig. 10.8*c*). The strawberry is also an aggregate fruit, but each ovary becomes a one-seeded fruit called an achene. The flesh of a strawberry is from the receptacle. In contrast, a pineapple comes from many different carpels that belong to separate flowers. As the ovaries mature, they fuse to form a large, *multiple fruit* (Fig. 10.8*d*).

Dispersal of Seeds

For plants to be widely distributed, their seeds have to be dispersed—that is, distributed, preferably long distances, from the parent plant.

Plants have various means of ensuring that dispersal takes place. The hooks and spines of clover, bur, and cocklebur attach to the fur of animals and the clothing of humans. Birds and mammals sometimes eat fruits, including the seeds, which are then defecated (passed through the digestive tract with the feces) some distance from the parent plant. Squirrels and other animals gather seeds and fruits, which they bury some distance away.

Some plants have fruits with trapped air or seeds with inflated sacs that help them float in water. The fruit of the coconut palm, for example, which can be dispersed by ocean currents, may land hundreds of kilometers away from the parent plant. Many plants have seeds with structures such as woolly hairs, plumes, and wings that adapt them for dispersal by wind. The seeds of an orchid, however, are so small and light that they need no special adaptation to carry them far away. The somewhat heavier dandelion fruit uses a tiny "parachute" for dispersal. The winged fruit of a maple tree, which contains two seeds, has been known to travel up to 10 kilometers from its parent (see Fig. 10.8*a*). A touch-me-not plant has seed pods that swell as they mature. When the pods finally burst, the ripe seeds are hurled out.

Germination of Seeds

Following dispersal, seeds **germinate**—that is, they begin to grow so that a seedling appears. Some seeds do not germinate

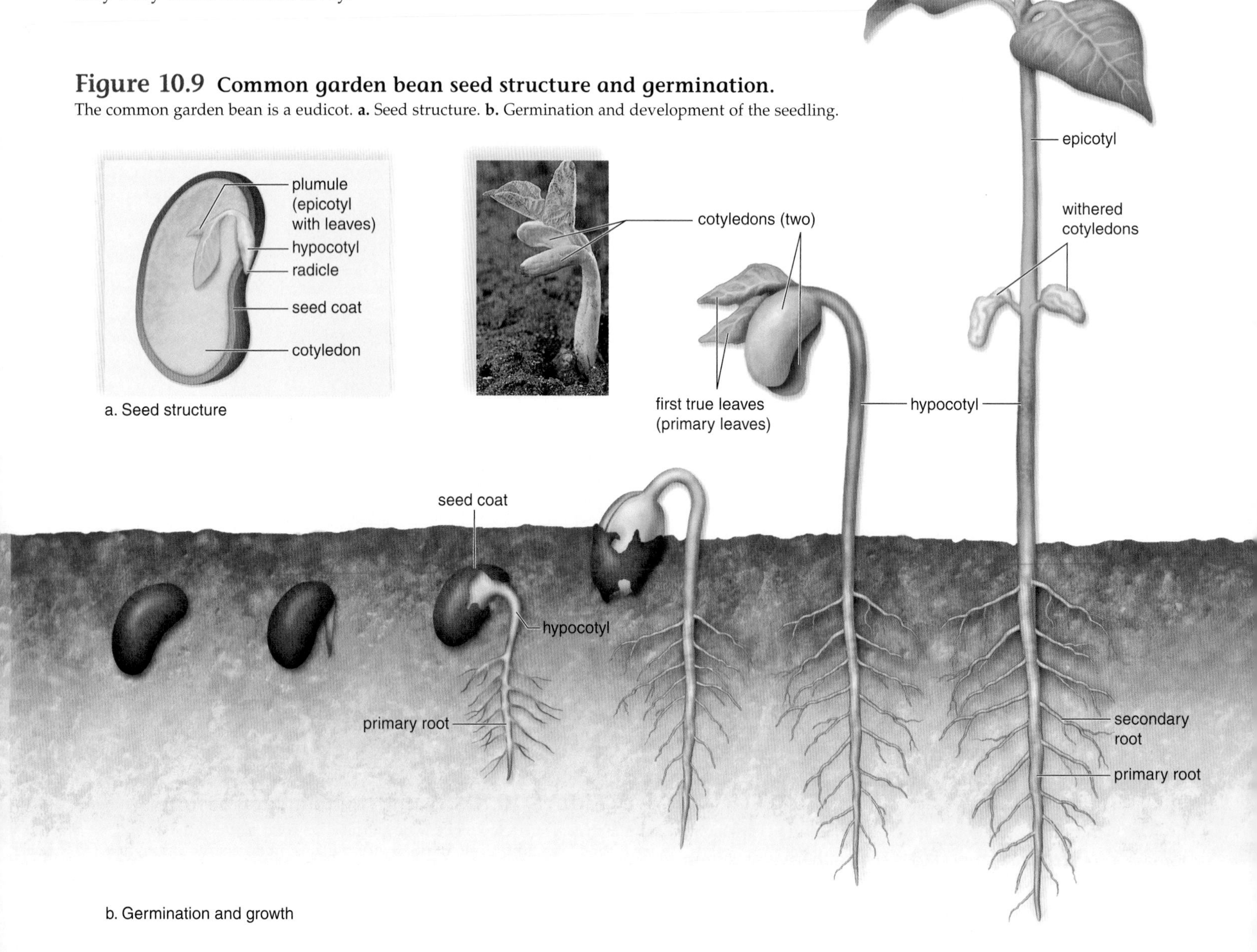

Figure 10.9 Common garden bean seed structure and germination. The common garden bean is a eudicot. **a.** Seed structure. **b.** Germination and development of the seedling.

until they have been dormant for a period of time. For seeds, **dormancy** is the time during which no growth occurs, even though conditions may be favorable for growth. In the temperate zone, seeds often have to be exposed to a period of cold weather before dormancy is broken. In deserts, germination does not occur until there is adequate moisture. This requirement helps ensure that seeds do not germinate until the most favorable growing conditions are present. Germination takes place if there is sufficient water, warmth, and oxygen to sustain growth.

Germination requires regulation, and both inhibitors and stimulators are known to exist. Fleshy fruits (e.g., apples and tomatoes) contain inhibitors so that germination does not occur until the seeds are removed and washed. In contrast, stimulators are present in the seeds of some temperate-zone woody plants. Mechanical action may also be required. Water, bacterial action, and even fire can act on the seed coat, allowing it to become permeable to water. The uptake of water causes the seed coat to burst.

Eudicot Versus Monocot Seed Germination

As mentioned, eudicot embryos have two cotyledons that have absorbed the endosperm. The cotyledons supply nutrients to the embryo and seedling and eventually shrivel and disappear. If the two cotyledons of a bean seed are parted, you can see a rudimentary plant (Fig. 10.9*a*). The epicotyl bears young leaves and is called a **plumule.** As the eudicot seedling emerges from the soil, the shoot is hook shaped to protect the delicate plumule. The hypocotyl becomes part of the stem, and the radicle develops into the roots.

In monocots, the endosperm is the food-storage tissue, and the single cotyledon does not have a storage role. Corn is a monocot, and its kernels are actually fruits, and therefore, the outer covering is the pericarp (Fig. 10.10). The plumule and radicle are enclosed in protective sheaths called the coleoptile and the coleorhiza, respectively. The plumule and the radicle burst through these coverings during germination.

Check Your Progress

1. Give examples of plant adaptations for seed dispersal by wind.
2. Why don't seeds germinate immediately after dispersal?
3. Describe the differences between eudicot and monocot seed germination.

Figure 10.10 Corn kernel structure and germination.
A corn plant is a monocot. **a.** Grain structure. **b.** Germination and development of the seedling.

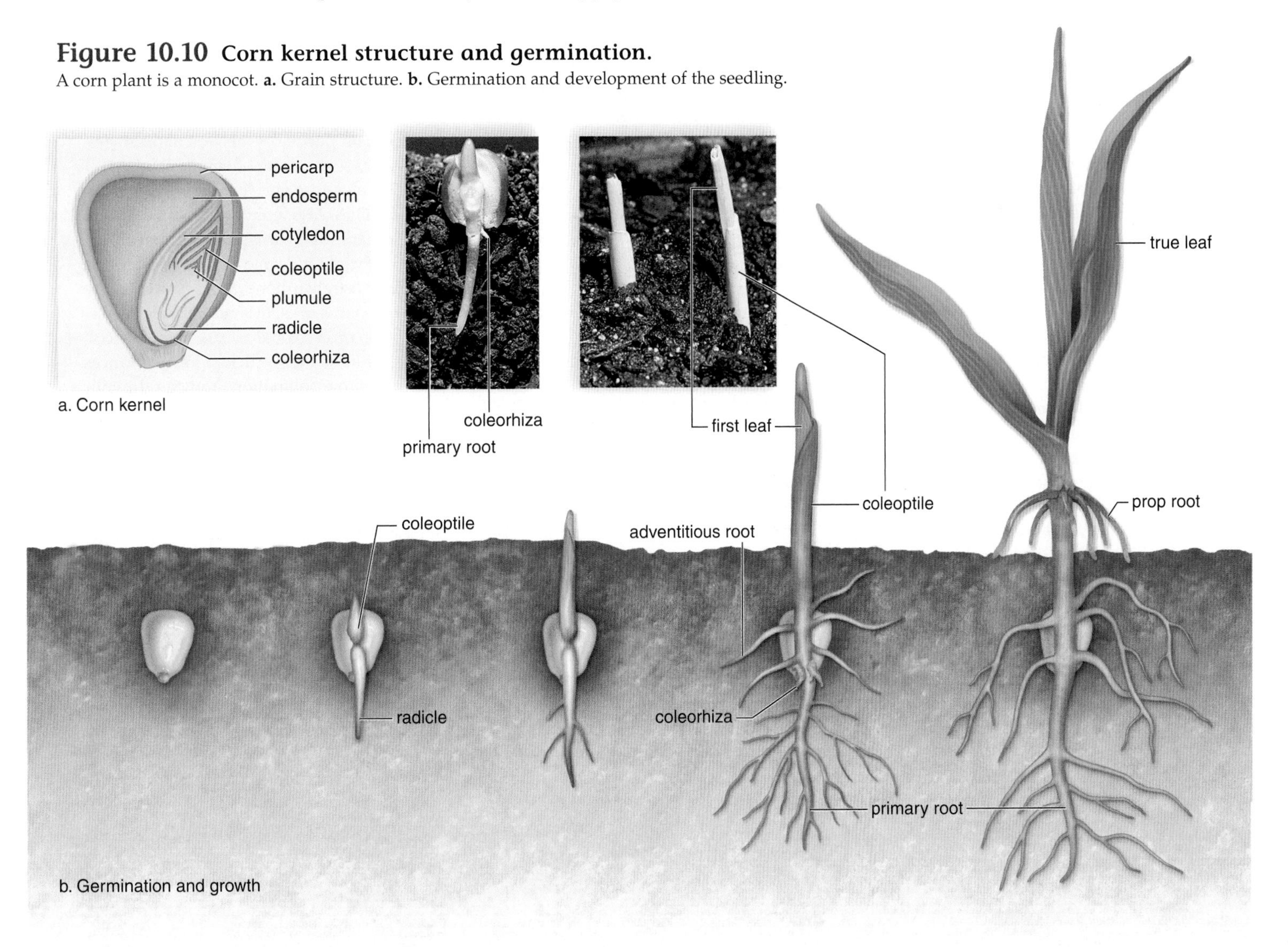

Ecology Focus

Pollinators and You

Bees, butterflies, and other pollinators play a critical role in our world. Many agricultural crops are dependent upon insects to carry their pollen between flowers. Only then can these plants produce a fruit, which becomes food for humans and/or farm animals. Thus, pollinators make possible the fruits, vegetables, and even the coffee we enjoy every day.

Plants and their pollinators have a symbiotic relationship. In return for pollination and fertilization of flowers, the flower provides the pollinator with food, often a sweet liquid called nectar. Often, the flower and pollinator are well suited to one another. Flowers are often colors a specific pollinator species can see and are open at times their pollinators are active. In some cases, the structure of some species of flowers is an exact match to the mouthparts of a particular species of pollinator!

Important pollinators include a variety of insects as well as birds and mammals, particularly bats. The best-known pollinator is the honeybee (Fig. 10A*a*). Beekeepers maintain around 2.4 million colonies of the European honeybee in the United States. Wild honeybee colonies also occur. Current estimates value honeybee pollination at approximatly $5 billion in Canada and $15 billion in the United States per year. It is also estimated that about one-third of human nutrition is due to bee pollination.

Unlike humans, bees can see ultraviolet light. Bee-pollinated flowers often have ultraviolet shadings, called nectar guides, which highlight the reproductive portion of the flower. The nectar guides point to a narrow floral tube large enough for the bee's feeding apparatus, but too small for other insects to reach the nectar. Bee-pollinated flowers are sturdy and irregular in shape because they often have a landing platform, where the bee can alight. Here, the bee brushes up against the anther and stigma as it moves toward the floral tube. Pollen clings to its hairy body, and the bee then stores the pollen in pollen baskets on the third pair of its legs. Bees feed pollen to their larvae.

Butterflies have a weak sense of smell, so their flowers tend to be odorless. The flower has bright colors, even red, because butterflies, unlike bees, can see the color red. Unable to hover, butterflies need a place to land. Composite flowers (composed of a compact head of numerous individual flowers) are especially favored by butterflies because these flowers provide a flat landing platform. Each flower has a long, slender floral tube, accessible to the long, thin butterfly proboscis (Fig. 10A*b*).

a.

b.

Figure 10A Two types of pollinators.
a. A bee-pollinated flower is a color other than red (bees can't see this color) and has a landing platform where the reproductive structures of the flower brush up against the bee's body. **b.** A butterfly-pollinated flower is often a composite, containing many individual flowers. The broad expanse provides room for the butterfly to land, after which it lowers its proboscis into each flower, in turn.

Today, there are more than 240,000 described species of flowering plants and over 900,000 described species of insects. This diversity suggests that the success of angiosperms has contributed to the success of insects, and vice versa.

In recent years, however, populations of bees and other pollinators have been declining worldwide. Consequently, some plants are endangered because they have lost their normal pollinator. Research suggests that the decline in pollinator populations has been caused by a variety of factors, including pollution, habitat loss, and emerging diseases.

Recently, declines of honeybees in particular have become a global concern. In 2006, commercial beekeepers in the United States began reporting rapid declines in their honeybee colonies. Scientists later named this phenomenon "colony collapse disorder" (CCD) due to its severity. Although honeybee losses have been common in the past, CCD has alarmed beekeepers and farmers alike due to the combination of rapid and widespread losses, along with failure of bees to return to the hive. Current reports suggest CCD affects bee colonies in 35 of the 50 U.S. states. A recent survey of 13% of U.S. honeybee colonies suggests losses increased by 14% between 2007 and 2008.

Possible causes of CCD include pesticides, parasites and diseases, stress, poor nutrition, and a lack of genetic diversity among bees. These factors may act together, with impacts magnified relative to single causes.

Although the actual cause of CCD remains a mystery, two serious emerging diseases of honeybees are caused by mites. The tracheal mite, a stowaway from Europe, lives in the breathing tubes of adult honeybees and sucks their blood, causing adult bees to become disoriented and weak. The other mite, called the varoa mite, from Asia, is an external parasite that feeds on pupae, larvae, and adult stages, and can cause the death of entire colonies.

Mites alone are unable to explain the number of declines. Also recently implicated in honeybee declines is a virus that may have been accidentally imported into the United States via infected Australian honeybees. However, the role of this virus in CCD is currently debated.

Today, professional farmers rent bee colonies and have them trucked long distances to ensure crop pollination. You can do your part to help sustain honeybee populations by planting pollen- and nectar-producing plants and by limiting the use of insecticides on your lawn and garden.

In general, it is important to note that pollinators cannot be exchanged on a one-for-one basis because they are often specific to certain plants. Therefore, each species of pollinator should be conserved.

Discussion Questions

1. Why do you think pollinators are often very specific to the plants they pollinate?
2. If a flower is white and exhibits a strong odor, do you think its pollinator species visits during the day or night? Explain your answer.
3. What could be done to counteract the possible causes of CCD?

10.3 Asexual Reproduction

Learning Outcomes

1. Discuss how asexual reproduction differs from sexual reproduction.
2. Describe how plants are propagated in tissue culture.
3. Explain how genetic engineering can be used to improve plant traits.

In asexual reproduction, there is only one parent, instead of two as in sexual reproduction. Because plants contain nondifferentiated meristem tissue, they routinely reproduce asexually by vegetative propagation. For example, complete strawberry plants will grow from the nodes of stolons (see Fig. 9.20*a*) and irises will grow from the nodes of rhizomes (see Fig. 9.20*b*). White potatoes are actually portions of underground stems, and each eye is a bud that will produce a new potato plant if it is planted with a portion of the swollen tuber. Sweet potatoes are modified roots. They can be propagated by planting sections of the root. You may have noticed that the roots of some fruit trees, such as cherry and apple trees, produce "suckers," small plants that can grow into new trees.

Many types of plants are now propagated from stem cuttings. The discovery that the plant hormone auxin can cause roots to develop from a stem, as discussed in section 10.4, has expanded the list of plants that can be propagated from stem cuttings.

Propagation of Plants in Tissue Culture

Tissue culture is the growth of a *tissue* in an artificial liquid or solid *culture* medium. Tissue culture now allows botanists to breed a large number of plants from somatic tissue and to select any that have superior hereditary characteristics. These and any plants with desired genotypes can be propagated rapidly in tissue culture.

Plant cells are **totipotent,** which means that an entire plant can be produced from most plant cells. The only exceptions are the plant cells that lose their nuclei (sieve-tube members) or that are dead at maturity (xylem, sclerenchyma, and cork cells). Commercial micropropagation methods now produce thousands, even millions, of identical seedlings in a small vessel. One favorite such method is meristem culture. If the correct proportions of hormones are added to a liquid medium, many new shoots will develop from a single shoot tip. When these are removed, more shoots form. Because the shoots are genetically identical, the adult plants that develop from them, called clonal plants, all have the same traits. Another advantage to meristem culture is that meristem, unlike other portions of a plant, is virus-free. Therefore, the plants produced are also virus-free. (The presence of plant viruses weakens plants and makes them less productive.)

When mature plant cells, as opposed to meristem cells, are used to grow entire plants, enzymes are often used to digest the cell walls of a small piece of tissue, usually mesophyll tissue from a leaf, and the result is cells without walls, called **protoplasts** (Fig. 10.11*a*). The protoplasts regenerate a new cell wall (Fig. 10.11*b*) and begin to divide, forming aggregates of cells and then a callus (Fig. 10.11*c,d*). These clumps of cells can be manipulated to produce **somatic embryos** (asexually produced embryos) (Fig. 10.11*e*). It's possible to produce millions of somatic embryos at once in large tanks called bioreactors. This is done for certain vegetables, such as tomato, celery, and asparagus, and for ornamental plants, such as lilies, begonias, and African violets. Somatic embryos that are encapsulated in a protective hydrated gel (and sometimes called artificial seeds) can be shipped anywhere. A mature plant develops from each

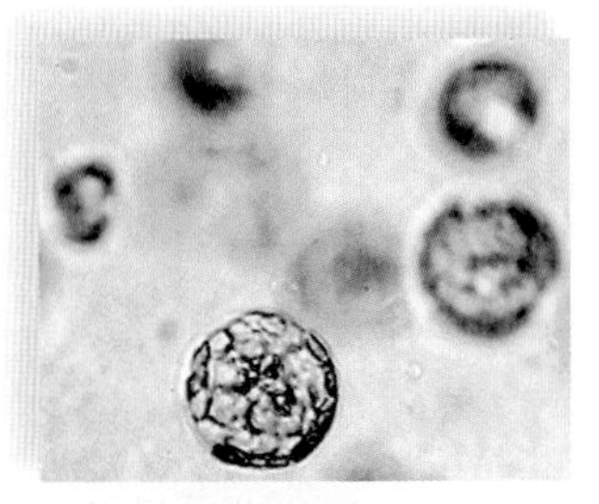
a. Protoplasts

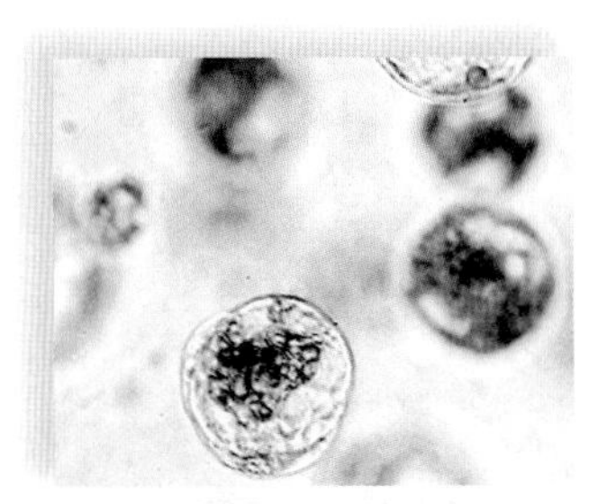
b. Cell wall regeneration

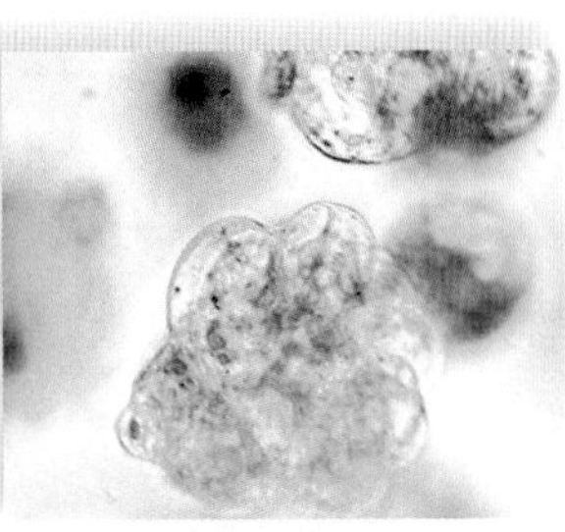
c. Aggregates of cells

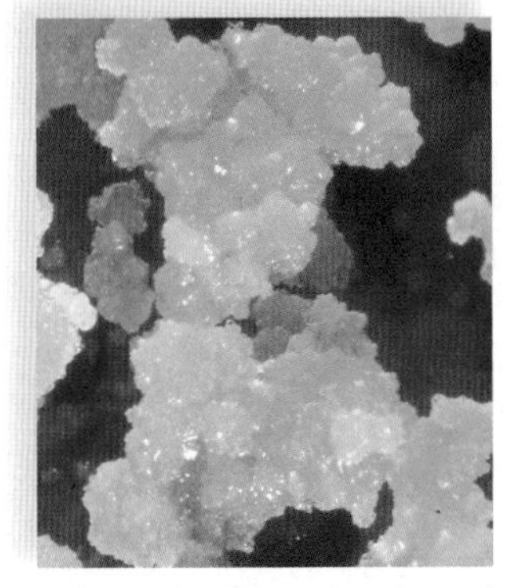
d. Callus (undifferentiated mass)

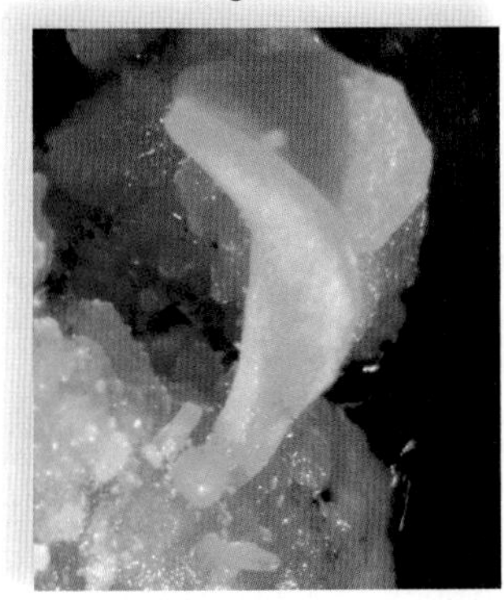
e. Somatic embryo

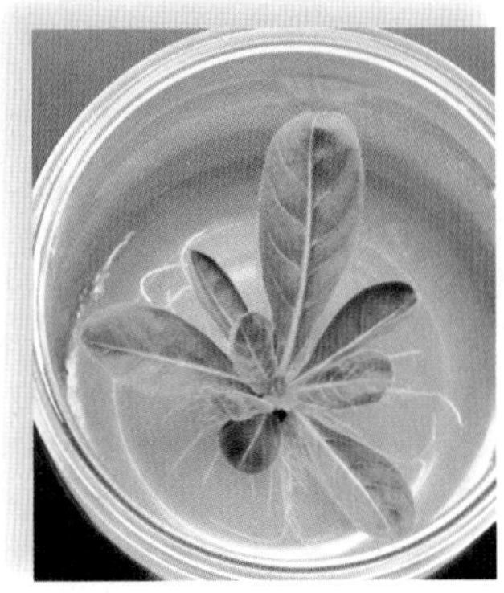
f. Plantlet

Figure 10.11 Tissue culture of plants.
a. When plant cell walls are removed by digestive enzyme action, the resulting cells are called protoplasts. **b.** Cell walls regenerate, and cell division begins. **c.** Cell division produces an aggregate of cells. **d.** An undifferentiated mass, called a callus, forms. **e.** Somatic cell embryos appear. **f.** The embryos develop into plantlets that can be transferred to soil for growth into adult plants.

somatic embryo (Fig. 10.11*f*). Plants generated from the somatic embryos vary somewhat because of mutations that arise during the production process. These so-called somaclonal variations are another way to produce new plants with desirable traits.

Recall that pollen grains are produced in the anthers of a flower. Anther culture is a direct way to produce a line of plants whose homologous chromosomes have the same genes. Anther culture involves mature anthers that are cultured in a medium containing vitamins and growth regulators. The haploid tube cells within the pollen grains divide, producing proembryos consisting of as many as 20 to 40 cells. Finally, the pollen grains rupture, releasing haploid embryos. The experimenter can now generate a haploid plant, or chemical agents can be added that encourage chromosomal doubling. The resulting plants are diploid, and the homologous chromosomes carry the same genes.

The culturing of plant tissues has led to a technique called cell suspension culture. Rapidly growing calluses are cut into small pieces and shaken in a liquid nutrient medium so that single cells or small clumps of cells break off and form a suspension. These cells will produce the same chemicals as the entire plant. For example, cell suspension cultures of *Cinchona ledgeriana* produce quinine, and those of *Digitalis lanata* produce digitoxin.

Genetic Engineering of Plants

Traditionally, **hybridization,** the crossing of different varieties of plants or even species, was used to produce plants with desirable traits. Hybridization, followed by vegetative propagation of the mature plants, generated a large number of identical plants with these traits. Today, it is possible to directly alter the genes of organisms and, in that way, produce new varieties with desirable traits.

Tissue Culture and Genetic Engineering

Protoplasts growing in a tissue culture medium can be genetically altered. A foreign gene isolated from any type of organism—plant, animal, or bacteria—is placed in the tissue culture medium. High-voltage electric pulses are then used to create pores in the plasma membrane so that the foreign gene enters the cells. In one such procedure, a gene for the production of the firefly enzyme luciferase was inserted into tobacco protoplasts, and the adult plants glowed when sprayed with the substrate luciferin.

Unfortunately, the regeneration of cereal grains from protoplasts has been difficult. As a result, other methods are used to introduce DNA into plant cells with intact cell walls. Today, it is possible to use a gene gun to bombard a callus (undifferentiated mass of cells) with DNA-coated microscopic metal particles. Then genetically altered somatic embryos develop into genetically altered adult plants.

Many plants, including corn and wheat varieties, have been genetically engineered by these methods. Such plants are called **transgenic plants** (or **genetically modified plants**) because they carry a foreign gene and have new and different traits. Figure 10.12 shows two types of transgenic plants. In another technique, foreign DNA is inserted into the plasmid of the bacterium *Agrobacterium,* which normally infects plant cells. The plasmid then contains *recombinant DNA* because it has genes from different sources, namely those of the plasmid and also the foreign genes of interest. When a bacterium with a recombinant plasmid infects a plant, the recombinant plasmid enters the cells of the plant and can incorporate into the plant's DNA.

Agricultural Plants with Improved Traits

Corn and cotton plants, in addition to soybean and potato plants, have been engineered to be resistant to either herbicides or insect pests. Some corn and cotton plants have

a. Herbicide-resistant soybean plants

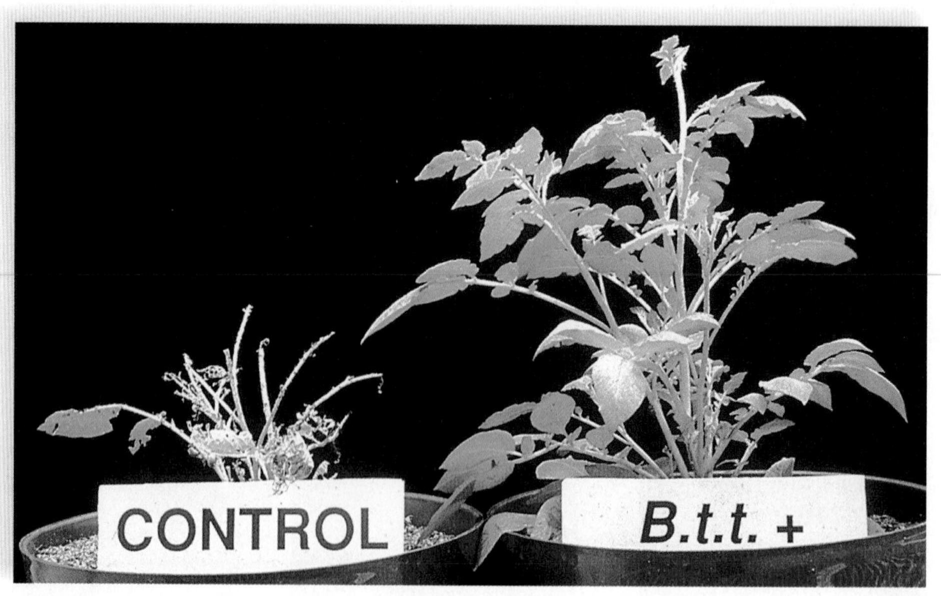

b. Nonresistant potato plant

c. Pest-resistant potato plant

Figure 10.12 Transgenic plants.

a. These soybean plants have been given a gene that causes them to be resistant to a particular herbicide. **b.** This potato plant is not resistant to a pest and does poorly when exposed to pests. **c.** This potato plant is resistant to a pest and grows beautifully.

Transgenic Crops of the Future	
Improved Agricultural Traits	
Herbicide resistant	Wheat, rice, sugar beets, canola
Salt tolerant	Cereals, rice, sugarcane, canola
Drought tolerant	Cereals, rice, sugarcane
Cold tolerant	Cereals, rice, sugarcane
Improved yield	Cereals, rice, corn, cotton
Modified wood pulp	Trees
Disease protected	Wheat, corn, potatoes
Improved Food-Quality Traits	
Fatty acid/oil content	Corn, soybeans
Protein/starch content	Cereals, potatoes, soybeans, rice, corn
Amino acid content	Corn, soybeans

a. Desirable traits

b. Salt-intolerant Salt-tolerant

Figure 10.13 Transgenic crops of the future.
a. Transgenic crops of the future include those with improved agricultural or food quality traits. **b.** A salt-tolerant plant has been engineered. The plant to the left does poorly when watered with a salty solution, but the engineered plant to the right is tolerant of the solution. The development of salt-tolerant crops would increase food production in the future.

been developed that are both insect and herbicide resistant. According to the not-for-profit group International Service for the Acquisition of Agri-biotech Applications (ISAAA), the global planted area of biotech crops has increased by more than 50-fold, from 4.2 million acres in six countries since 1996 to 282.4 million acres in 23 countries in 2007. This group predicts that the dramatic growth in commercialization of biotech crops seen during the first decade will be surpassed in the second decade.

If crops are resistant to a broad-spectrum herbicide and weeds are not, then the herbicide can be used to kill the weeds. When herbicide-resistant plants were planted, weeds were easily controlled, less tillage was needed, and soil erosion was minimized. However, wildlife and native plants declined because increased amounts of herbicides were used. Another concern is the potential transfer of herbicide-resistant genes from the modified crop plant to a related wild, "weedy" species through natural hybridization.

Crops with other improved agricultural and food-quality traits are desirable (Fig. 10.13). Irrigation, even with fresh water, inevitably leads to salinization of the soil, which reduces crop yields. Thus, the development of salt-tolerant crops would increase yield on such land. A salt-tolerant tomato will soon be field tested. First, scientists identified a gene coding for a channel protein that transports Na^+ into a vacuole, preventing it from interfering with plant metabolism. Then the scientists used the gene to engineer tomato plants that overproduce the channel protein. The modified plants thrived despite being watered with a salty solution. Salt- and also drought- and cold-tolerant cereals, rice, and sugarcane might help provide enough food for a growing world population.

Potato blight is the most serious potato disease in the world. About 150 years ago, it was responsible for the Irish potato famine that caused the death of millions of people. By placing a gene from a naturally blight-resistant wild potato into a farmed variety, researchers have now made potato plants that are no longer vulnerable to a range of blight strains.

Some progress has also been made in increasing the food quality of crops. Soybeans have been developed that mainly produce monounsaturated fatty acid, a change that may improve human health. These altered plants also produce acids that can be used as hardeners in paints and plastics. The necessary genes were derived from *Vernonia* and castor bean seeds and were transferred into the soybean DNA.

Other types of genetically engineered plants are also expected to increase productivity. Stomata might be altered to take in more carbon dioxide or lose less water. A team of Japanese scientists is working on introducing the C_4 photosynthetic cycle into rice (see the Bioethical Focus in Chapter 8). Unlike C_3 plants, C_4 plants do well in hot, dry weather. These modifications would require a more complete reengineering of plant cells than the single-gene transfers that have been done so far.

Some people have expressed health and environmental concerns regarding the growing of transgenic crops, as discussed in the Bioethical Focus at the end of this chapter.

Commercial Products

Single-gene transfers have allowed plants to produce various products, including human hormones, clotting factors, and antibodies. One type of antibody made by corn can deliver radioisotopes to tumor cells, and another made by soybeans may be developed to treat genital herpes.

The tobacco mosaic virus has been used as a vector to introduce a human gene into adult tobacco plants in the field. (Note that this technology bypasses the need for tissue culture completely.) Tens of grams of α-galactosidase, an enzyme needed for the treatment of a human lysosomal storage disease, were harvested per acre of tobacco plants. And it took only 30 days to get tobacco plants to produce antibodies to treat non-Hodgkin's lymphoma after being sprayed with a genetically engineered virus.

Check Your Progress

1. List two ways in which plants can reproduce asexually.
2. How are plants grown in tissue culture?
3. Discuss two examples of how and why plants are genetically engineered.

10.4 Control of Growth and Responses

Learning Outcomes

1. Explain the importance of plant hormones.
2. Match the types of plant hormones with their function.
3. Discuss how plants respond to stimuli.

Plants respond to environmental stimuli such as light, gravity, and seasonal changes, usually by a change in their pattern of growth. Hormones are involved in these responses.

Plant Hormones

Plant hormones are small organic molecules produced by the plant that serve as chemical signals between cells and tissues. Currently, the five commonly recognized groups of plant hormones are auxins, gibberellins, cytokinins, abscisic acid, and ethylene. Other chemicals, some of which differ only slightly from the natural hormones, also affect the growth of plants. These and the naturally occurring hormones are sometimes grouped together and called plant growth regulators.

Plant hormones bring about a physiological response in target cells after binding to a specific receptor protein in the plasma membrane. Figure 10.14 shows how the hormone auxin brings about elongation of a plant cell, a necessary step toward differentiation and maturation.

Auxins

The most common naturally occurring **auxin** is indoleacetic acid (IAA). It is produced in shoot apical meristem and is found in young leaves, flowers, and fruits. Therefore, you would expect auxin to affect many aspects of plant growth and development.

Effects of Auxin Auxin is present in the apical meristem of a plant shoot, where it causes the shoot to grow from the top, a phenomenon called **apical dominance**. Only when the terminal bud is removed, deliberately or accidentally, are the axillary buds able to grow, allowing the plant to branch (Fig. 10.15). Pruning the top (apical meristem) of a plant generally achieves a fuller look because of increased branching of the main body of the plant.

The application of a weak solution of auxin to a woody cutting causes adventitious roots to develop more quickly than they would otherwise. Auxin production by seeds also promotes the growth of fruit. As long as auxin is concentrated in leaves or fruits rather than in the stem, leaves and fruits do not fall off. Therefore, trees can be sprayed with auxin to keep mature fruit from falling to the ground.

How Auxins Work When a plant is exposed to unidirectional light, auxin moves to the shady side, where it binds to receptors and activates an ATP-driven pump that transports hydrogen ions (H^+) out of the cell (see Fig. 10.14). The acid environment weakens cellulose fibrils, and activated enzymes further degrade the cell wall. Water now enters the cell, and the resulting increase in turgor pressure causes the cell to

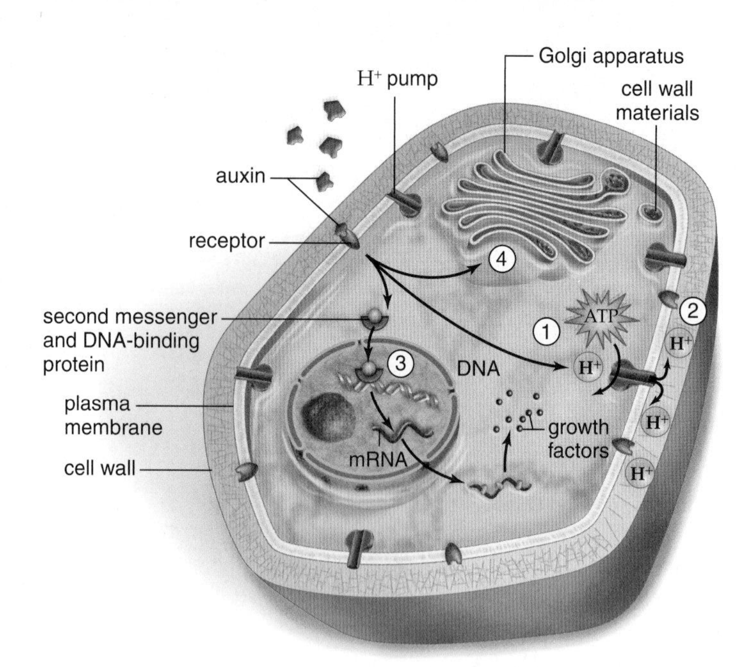

Figure 10.14 Auxin mode of action.
(1) The binding of auxin to a receptor stimulates an H^+ pump, moving hydrogen ions into the cell wall. (2) The resulting acidity causes the cell wall to weaken, and water enters the cell. (3) At the same time, a second messenger stimulates production of growth factors, and (4) new cell wall materials are synthesized. The result of all these actions is that the cell elongates, causing the stem to elongate.

elongate and the stem to bend toward the light. This growth response to unidirectional light is called **phototropism**. Besides phototropism, auxins are also involved in **gravitropism**, in which roots curve downward and stems curve upward in response to gravity, as discussed later in this section.

Gibberellins

Gibberellins were discovered in 1926 while a Japanese scientist was investigating a fungal disease of rice plants called

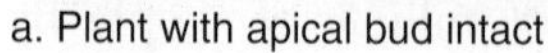

a. Plant with apical bud intact

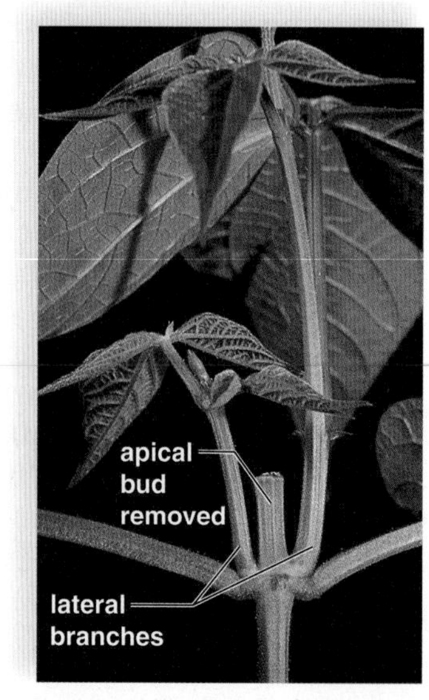

b. Plant with apical bud removed

Figure 10.15 Apical dominance.
a. Auxin in the apical bud inhibits axillary bud development, and the plant exhibits apical dominance. **b.** When the apical bud is removed, lateral branches develop.

"foolish seedling disease." The plants elongated too quickly, causing the stem to weaken and the plant to collapse. The fungus infecting the plants produced an excess of a chemical called gibberellin, named after the fungus *Gibberella fujikuroi.* It wasn't until 1956 that gibberellic acid was isolated from a flowering plant rather than from a fungus. Sources of gibberellin in flowering plant parts are young leaves, roots, embryos, seeds, and fruits.

Gibberellins are growth-promoting hormones that bring about elongation of the resulting cells. We know of about 70 gibberellins, and they differ chemically only slightly. The most common of these is gibberellic acid, GA_3 (the subscript designation distinguishes it from other gibberellins).

Effects of Gibberellins When gibberellins are applied externally to plants, the most obvious effect is stem elongation (Fig. 10.16). Gibberellins can cause dwarf plants to grow, cabbage plants to become 2 m tall, and bush beans to become pole beans.

During dormancy a plant does not grow, even though conditions may be favorable for growth. The dormancy of seeds and buds can be broken by applying gibberellins, and research with barley seeds has shown how GA_3 influences the germination of seeds. Barley seeds have a large, starchy endosperm that must be broken down into sugars to provide energy for the embryo to grow. It is hypothesized that after GA_3 attaches to a receptor in the plasma membrane, calcium ions (Ca^{2+}) combine with a protein. This complex is believed to activate the gene that codes for amylase. Amylase then acts on starch to release the sugars needed for the seed to germinate. Gibberellins have been used to promote barley seed germination for the beer brewing industry.

Cytokinins

Cytokinins were discovered as a result of attempts to grow plant tissue and organs in culture vessels in the 1940s. It was found that cell division occurred when coconut milk (a liquid endosperm) and yeast extract were added to the culture medium. Although the effective agent or agents could not be isolated, they were collectively called cytokinins because, as you may recall, cytokinesis means division of the cytoplasm. A naturally occurring cytokinin was not isolated until 1967. Because it came from the kernels of maize (*Zea*), it was called zeatin.

Effects of Cytokinins The cytokinins promote cell division. These substances, which are derivatives of adenine, one of the purine bases in DNA and RNA, have been isolated from actively dividing tissues of roots and also from seeds and fruits. A synthetic cytokinin, called kinetin, also promotes cell division.

Researchers have found that **senescence** (aging) of leaves can be prevented by applying cytokinins. When a plant organ, such as a leaf, loses its natural color, it is most likely undergoing senescence. During senescence, large molecules within the leaf are broken down and transported to other parts of the plant. Senescence does not always affect the entire plant at once. For example, as some plants grow taller, they naturally lose their lower leaves. Not only can cytokinins prevent the death of leaves, but they can also

Figure 10.16 Effect of gibberellins.
The cabbage plants on the right were treated with gibberellins. The cabbage plants on the left were not treated. Gibberellins are often used to promote stem elongation in economically important plants, but the exact mode of action remains unclear.

initiate leaf growth. Axillary buds begin to grow despite apical dominance when cytokinin is applied to them.

Researchers are well aware that the ratio of auxin to cytokinin and the acidity of the culture medium determine whether a plant tissue forms an undifferentiated mass, called a callus, or differentiates to form roots, vegetative shoots, leaves, or floral shoots. Researchers have reported that chemicals called oligosaccharins (chemical fragments released from the cell wall) are effective in directing differentiation. They hypothesize that reception of auxin and cytokinins, which leads to the activation of enzymes, releases these fragments from the cell wall.

Abscisic Acid

Abscisic acid (ABA) is produced by any "green tissue" with chloroplasts, monocot endosperm, and roots. Abscisic acid was once thought to function in **abscission,** the dropping of leaves, fruits, and flowers from a plant. But even though the external application of abscisic acid promotes abscission, researchers no longer believe this hormone functions naturally in this process. Instead, they think the hormone ethylene, discussed next, brings about abscission.

Effects of Abscisic Acid Abscisic acid is sometimes called the stress hormone because it initiates and maintains seed and bud dormancy and brings about the closure of stomata when a plant is under water stress (Fig. 10.17). Dormancy has begun when a plant stops growing and prepares for adverse conditions (even though conditions at the time are favorable for growth). For example, researchers believe that abscisic acid moves from leaves to vegetative buds in the fall, and thereafter, these buds are converted to winter buds. A winter bud is covered by thick, hardened scales. A reduction in the level of abscisic acid and an increase in the level of gibberellins are believed to break seed and bud dormancy. Then seeds germinate, and buds send forth leaves.

Ultimately abscisic acid causes potassium ions (K^+) to leave guard cells. Thereafter, the guard cells lose water, and stomata close.

Ethylene

Ethylene is a gas that can move freely in the air. Like the other hormones studied, ethylene works with other hormones to bring about certain effects.

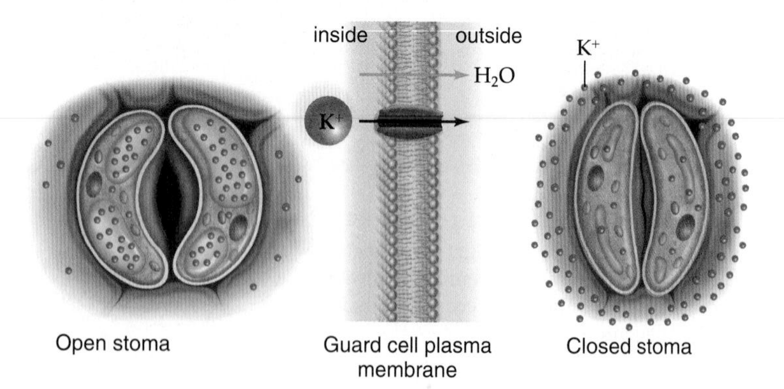

Figure 10.17 Abscisic acid control of stoma closing.
K^+ is concentrated inside the guard cells, and the stoma is open (*left*). With the loss of K^+, water (H_2O) exits the guard cells, and the stoma closes (*right*).

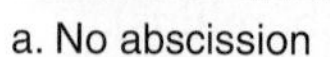

Figure 10.18 Functions of ethylene.
a. Normally, there is no abscission when a holly twig is placed under a glass jar for a week. **b.** When an ethylene-producing ripe apple is also under the jar, abscission of the holly leaves occurs. **c.** Similarly, ethylene given off by this one ripe tomato will cause the others to ripen.

Effects of Ethylene Ethylene is involved in abscission. Low levels of auxin and perhaps gibberellin in the leaf, compared to the stem, probably initiate abscission. But once the process of abscission has begun, ethylene stimulates certain enzymes, such as cellulase, which cause leaf, fruit, or flower drop (Fig. 10.18*a,b*). Cellulase hydrolyzes cellulose in plant cell walls.

In the early 1900s, it was common practice to prepare citrus fruits for market by placing them in a room with a kerosene stove. Only later did researchers realize that ethylene, an incomplete combustion product of kerosene, was ripening the fruit (Fig. 10.18*c*). Because it is a gas, ethylene can act from a distance. A barrel of ripening apples can induce the ripening of a bunch of bananas, even if they are in different containers. Ethylene is released at the site of a plant wound due to physical damage or infection (which is why one rotten apple spoils the whole bushel).

Check Your Progress

1. What are plant hormones?
2. What are the functions of auxins, gibberellins, cytokinins, abscicic acid, and ethylene?

Plant Responses to Environmental Stimuli

Environmental signals determine the seasonality of growth, reproduction, and dormancy in plants. Plant responses are strongly influenced by such environmental stimuli as light, day length, gravity, and touch. The ability of a plant to respond to environmental signals fosters the survival of the plant and the species in a particular environment.

Plant responses to environmental signals can be rapid, as when stomata open in the presence of light, or they can take some time, as when a plant flowers in season. Despite their variety, most plant responses to environmental signals are due to growth and sometimes differentiation, brought about at least in part by particular hormones.

Plant Tropisms

Plant growth toward or away from a directional stimulus is called a **tropism.** Tropisms are due to differential growth—one side of an organ elongates faster than the other, and the result is a curving toward or away from the stimulus. The following three well-known tropisms were each named for the stimulus that causes the response:

phototropism	growth in response to a light stimulus
gravitropism	growth in response to gravity
thigmotropism	growth in response to touch

Growth toward a stimulus is called a positive tropism, and growth away from a stimulus is called a negative tropism. For example, in positive phototropism, stems curve toward the light, and in negative gravitropism, stems curve away from the direction of gravity (Fig. 10.19). Roots, of course, exhibit positive gravitropism.

The role of auxin in the positive phototropism of stems has been studied for quite some time. Because blue light, in particular, causes phototropism to occur, researchers believe that a yellow pigment related to the vitamin riboflavin acts as a photoreceptor for light. Following reception, auxin migrates from the bright side to the shady side of a stem. The cells on that side elongate faster than those on the bright side, causing the stem to curve toward the light. Negative gravitropism of stems occurs because auxin moves to the lower part of a stem when a plant is placed on its side. Figure 10.14 explains how auxin brings about elongation.

a.

b.

Figure 10.19 Tropisms.
a. In positive phototropism, the stem of a plant curves toward the light. This response is due to the accumulation of auxin on the shady side of the stem. **b.** In negative gravitropism, the stem of a plant curves away from the direction of gravity 24 hours after the plant was placed on its side. This response is due to the accumulation of auxin on the lower side of the stem.

Flowering

Flowering is a striking response in angiosperms to environmental seasonal changes. In some plants, flowering occurs according to the **photoperiod,** which is the ratio of the length of day to the length of night over a 24-hour period. Plants can be divided into three groups:

1. **Short-day plants** flower when the day length is shorter than a critical length. (Examples are cocklebur, poinsettia, and chrysanthemum.)
2. **Long-day plants** flower when the day length is longer than a critical length. (Examples are wheat, barley, clover, and spinach.)
3. **Day-neutral plants** do not depend on day length for flowering. (Examples are tomato and cucumber.)

Experiments have shown that the length of continuous darkness, not light, controls flowering. For example, the cocklebur does not flower if a suitable length of darkness is interrupted by a flash of light. In contrast, clover does flower when an unsuitable length of darkness is interrupted by a

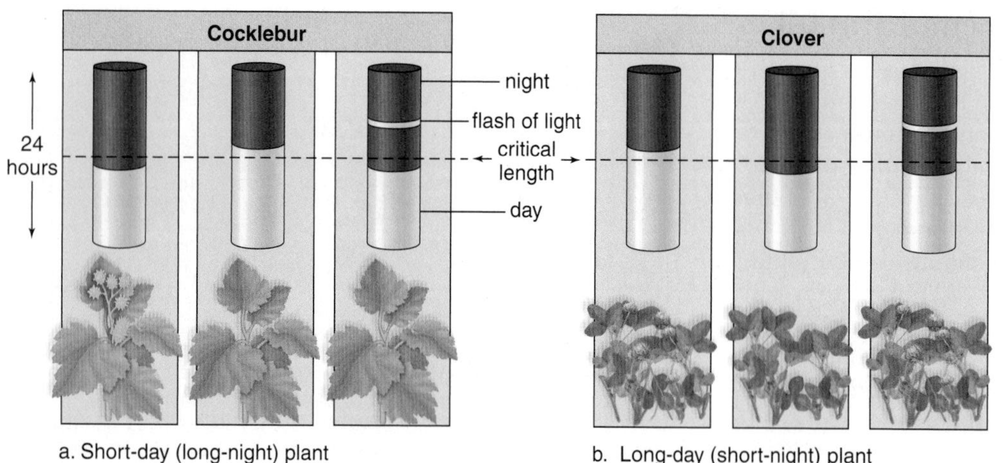

Figure 10.20 Photoperiodism and flowering.
a. Short-day plant. When the day is shorter than a critical length, this type of plant flowers. The plant does not flower when the day is longer than the critical length. It also does not flower if the longer-than-critical-length night is interrupted by a flash of light. **b.** Long-day plant. The plant flowers when the day is longer than a critical length. When the day is shorter than a critical length, this type of plant does not flower. However, it does flower if the slightly longer-than-critical-length night is interrupted by a flash of light.

flash of light (Fig. 10.20). (Interrupting the light period with darkness has no effect on flowering.)

Phytochrome and Plant Flowering

If flowering is dependent on day and night length, plants must have some way to detect these periods. This appears to be the role of **phytochrome,** a blue-green leaf pigment that alternately exists in two forms—P_r and P_{fr}:

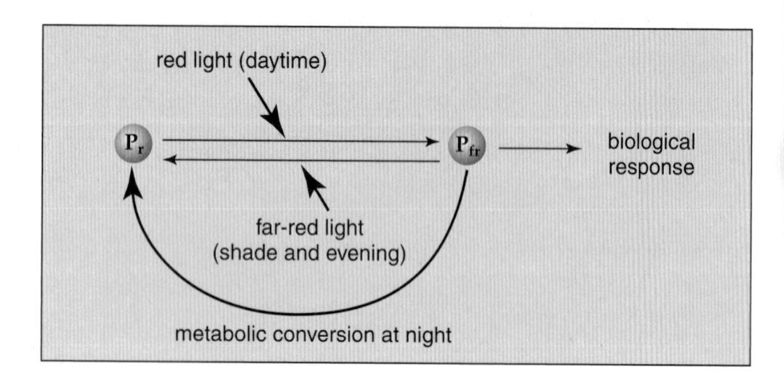

Direct sunlight contains more red light than far-red light; therefore, P_{fr} (far red) is apt to be present in plant leaves during the day. In the shade and at sunset, there is more far-red light than red light. Therefore, P_{fr} is converted to P_r (red) as night approaches. There is also a slow metabolic replacement of P_{fr} by P_r during the night.

It is possible that phytochrome conversion is the first step in a signaling pathway that results in flowering. A flowering hormone has never been discovered.

Other Functions of Phytochrome

The $P_r \longrightarrow P_{fr}$ conversion cycle has other functions. The presence of P_{fr} indicates to seeds of some plants that sunlight is present and conditions are favorable for germination. Such seeds must be only partly covered with soil when planted. Following germination, the presence of P_r indicates that stem elongation may be needed to reach sunlight. Seedlings that are grown in the dark etiolate—that is, the stem increases in length, and the leaves remain small (Fig. 10.21). Once the seedling is exposed to sunlight and P_r is converted to P_{fr}, the seedling begins to grow normally—the leaves expand, and the stem branches.

Figure 10.21 Phytochrome control of growth pattern.
a. If red light is prevalent, as it is in bright sunlight, normal growth occurs. These effects are due to phytochrome. **b.** If far-red light is prevalent, as it is in the shade, etiolation occurs.

Check Your Progress

1. Describe two different types of plant growth tropism.
2. What is the difference between short-day and long-day plants?
3. Describe the potential role of phytochrome in plant flowering.

Applying the Concepts Revisited

At the beginning of this chapter, you learned that genetic manipulation will produce a 3n seedless watermelon. You have also learned that it is possible to genetically modify plants by using tissue culture techniques and various ways of giving them foreign genes. Some of these plants become GM (genetically modified) foods. Although it is economically beneficial to produce GM foods, some are concerned about their health effects.

1. Would you support more intensive scientific testing of the health effects of GM foods even if it resulted in higher food costs?
2. Do you think all GM foods should be labeled at the supermarket? in restaurants?

Bioethical Focus

Transgenic Crops

Transgenic plants (genetically modified plants, or genetically modified organisms [GMOs]) may allow crop yields to keep up with the ever-increasing worldwide demand for food. And some of these plants have the added benefit of requiring less fertilizer and/or pesticides, which can be harmful to human health and the environment.

But some scientists believe that transgenic crops pose their own threat to the environment, and many activists believe that GMOs are themselves dangerous to our health. Studies have shown that wind-carried pollen can cause transgenic crops to hybridize with nearby weedy relatives, raising concern that some weeds may become uncontrollable pests. Or perhaps a toxin produced by transgenic crops could hurt other species. Many researchers are conducting tests to see if this might occur. Also, although transgenic crops have not caused any illnesses in humans, some scientists concede the possibility that people could be allergic to a transgene's protein product. Such concerns led to a massive recall in September 2000 that pulled about 2.8 million boxes of Taco Bell taco shells from grocery stores after unapproved genetically modified corn was found in the product.

Already, transgenic plants must be approved by the Food and Drug Administration before they are considered safe for human consumption, and they must meet certain Environmental Protection Administration standards. Some people believe safety standards for transgenic crops should be further strengthened, while others fear stricter standards will compromise food production. Another possibility is to retain the current standards but require clear labeling so buyers can choose whether or not to eat genetically modified foods.

Form Your Own Opinion

1. If plant pollen is transferred by an insect rather than by the wind, do you think there would still be a risk of GMOs hybridizing with weedy relatives?
2. What are the advantages and disadvantages of genetically engineering a plant versus using artificial breeding methods to develop new traits?
3. If a food label states that the food has been genetically modified, do you think most people would still buy the product? Why or why not?

SUMMARIZING THE CONCEPTS

10.1 Sexual Reproduction in Flowering Plants

- Flowering plants have an alternation of generations life cycle.
- Flowers borne by the sporophyte produce microspores and megaspores by meiosis. Microspores develop into a male gametophyte, and megaspores develop into the female gametophyte. Gametophytes produce gametes by mitosis.
- Flowers contain: sepals that form an outer whorl; petals, which are the next whorl; and stamens that form a whorl around the base of at least one carpel. The carpel, in the flower's center, consists of a stigma, style, and ovary, which contains ovules.
- Anthers contain microsporocytes, each of which divides meiotically to produce four haploid microspores. Each microspore divides mitotically to produce a two-celled pollen grain—a tube cell and a generative cell. The generative cell later divides mitotically to produce two sperm cells.
- The pollen grain is the male gametophyte. After pollination, the pollen grain germinates, and as the pollen tube grows, sperm cells travel to the embryo sac.
- Each ovule contains a megasporocyte, which divides meiotically to produce four haploid megaspores, only one of which survives. This megaspore divides mitotically to produce the female gametophyte (embryo sac), which usually has seven cells.
- Flowering plants undergo double fertilization. One sperm nucleus unites with the egg nucleus, forming a 2n zygote, and the other unites with the polar nuclei of the central cell, forming a 3n endosperm cell. The zygote becomes the sporophyte embryo, and the endosperm cell divides to become endosperm tissue.

10.2 Growth and Development

- Prior to seed formation, the zygote undergoes growth and development to become an embryo.
- The seeds enclosed by a fruit contain the embryo (hypocotyl, epicotyl, plumule, radicle) and stored food (endosperm and/or cotyledons).
- Following dispersal, a seed germinates.

10.3 Asexual Reproduction

- Many flowering plants reproduce asexually (e.g., buds give rise to entire plants, or roots produce new shoots).
- Asexual reproduction allows for clonal propagation in tissue culture. This micropropagation is now commercialized—cell suspension cultures can produce chemicals of medical importance, and plant tissue culture allows genetic engineering. One goal of genetic engineering is to improve crop yields or food quality.

10.4 Control of Growth and Responses

- Plant hormones are chemical signals involved in plant growth and responses to environmental stimuli.
- There are five commonly recognized groups of plant hormones:
  1. Auxins cause apical dominance, the growth of adventitious roots, and two types of tropisms—phototropism and gravitropism.
  2. Gibberellins promote stem elongation and break seed dormancy so that germination occurs.
  3. Cytokinins promote cell division, prevent senescence of leaves, and along with auxin, influence differentiation of plant tissues.
  4. Abscisic acid initiates and maintains seed and bud dormancy and closes stomata under water stress.
  5. Ethylene causes abscission of leaves, fruits, and flowers. It also causes some fruits to ripen.
- Environmental signals play a significant role in plant growth and development.

- Tropisms are growth responses toward or away from unidirectional stimuli. When a plant is exposed to light, auxin moves laterally from the bright to the shady side of a stem. Then, cells on the shady side elongate, and the stem bends toward the light. Similarly, auxin is responsible for negative gravitropism and growth upward, opposite gravity.
- Flowering is a striking response to seasonal changes. Phytochrome, a plant pigment that responds to daylight, is believed to be part of a biological clock system that in some unknown way brings about flowering. Phytochrome has various other functions, such as seed germination, leaf expansion, and stem branching.
- Short-day plants flower when days are shorter (nights are longer) than a critical length, and long-day plants flower when days are longer than a critical length. Some plants are day-neutral.

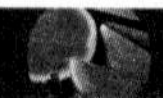

TESTING YOURSELF

Choose the best answer for each question.

1. Stigma is to carpel as anther is to
   a. sepal.
   b. stamen.
   c. ovary.
   d. style.
2. _____ always promotes cell division.
   a. Auxin
   b. Phytochrome
   c. Cytokinin
   d. None of these are correct.

For questions 3–6, match the definitions with the structure in the key.

Key:

a. flower
b. megaspore
c. gibberellin
d. style
e. senescence

3. Develops into the female gametophyte
4. Growth hormone that promotes elongation
5. Aging in plants
6. Produces seeds enclosed by fruits
7. Double fertilization refers to the formation of a _____ and a _____.
   a. zygote, zygote
   b. zygote, pollen grain
   c. zygote, megaspore
   d. zygote, endosperm
8. Which is the correct order of the following events: (1) megaspore becomes embryo sac, (2) embryo formed, (3) double fertilization, (4) meiosis?
   a. 1, 2, 3, 4
   b. 4, 1, 3, 2
   c. 4, 3, 2, 1
   d. 2, 3, 4, 1
9. In an accessory fruit, such as an apple, the bulk of the fruit is from the
   a. ovary.
   b. style.
   c. pollen.
   d. None of these are correct.
10. Ethylene stimulates the action of _____ to produce _____.
    a. auxins, root formation
    b. gibberellins, flower formation
    c. cellulase, flower formation
    d. cellulase, flower dropping
11. Label the following diagram of a flower.

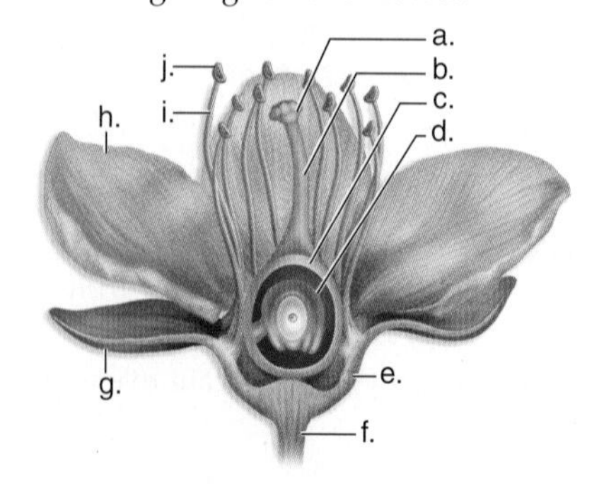

12. Label the following diagram of alternation of generations in flowering plants.

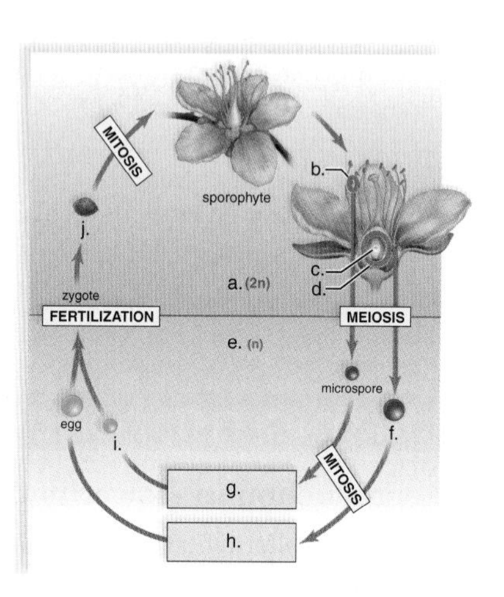

13. Short-day plants
    a. are the same as long-day plants.
    b. are apt to flower in the fall.
    c. do not have a critical photoperiod.
    d. will not flower if a short day is interrupted by bright light.
    e. All of these are correct.
14. Which of the following plant hormones causes plants to grow in an upright position?
    a. auxin
    b. gibberellins
    c. cytokinins
    d. abscisic acid
    e. ethylene
15. How is the megaspore in the plant life cycle similar to the microspore? Both
    a. have the diploid number of chromosomes.
    b. become an embryo sac.
    c. become a gametophyte that produces a gamete.
    d. are necessary for seed production.
    e. Both c and d are correct.

UNDERSTANDING THE TERMS

abscisic acid (ABA) 186
abscission 186
alternation of generations 172
anther 172
apical dominance 184
auxin 184
carpel 173
cotyledon 176
cytokinin 185
day-neutral plant 187
dormancy 179
double fertilization 175
embryo 175
embryo sac 175
endosperm 175
ethylene 186
female gametophyte 174
filament 173
flower 172
fruit 172
gametophyte 172
genetically modified plant 182
germinate 178
gibberellin 184
gravitropism 184
hybridization 182
long-day plant 187
male gametophyte 174
megaspore 172
megasporocyte 174
microspore 172
microsporocyte 174
ovary 173
ovule 173
petal 172
photoperiod 187
phototropism 184
phytochrome 188
plant hormone 184
plumule 179
pollen grain 174
pollination 175
protoplast 181
seed 172
senescence 185
sepal 172
short-day plant 187
somatic embryo 181
spore 172
sporophyte 172
stamen 172
stigma 173
style 173
tissue culture 181
totipotent 181
transgenic plant 182
tropism 187

THINKING CRITICALLY

1. Under which environmental conditions would it be advantageous for a plant to carry out asexual reproduction?
2. Why was it advantageous for plants to evolve to have just one type of pollinator rather than having many different types of pollinators?
3. If someone offered you unlimited funding to genetically engineer a plant in order to benefit humankind, what idea would you pursue and why?

INQUIRY INTO LIFE WEBSITE

The companion website for *Inquiry into Life* provides a wealth of information organized and integrated by chapter. You will find practice tests, animations, videos, and much more that will complement your learning and understanding of general biology.

http://www.mhhe.com/maderinquiry13

PART III

Maintenance of the Human Body

Human Organization

Applying the Concepts

The human body is amazingly complex. At the cellular level, metabolic reactions must occur in a tightly controlled and highly regulated manner, so that all the biological functions take place as needed. DNA must be synthesized and genes must be transcribed. Enzymes must produce thousands of products when and where they are needed. These processes are complicated enough in single-celled organisms, but most of the cells that comprise multicellular species like humans have specialized functions as a part of tissues, organs, or organ systems. In the next several chapters, we will examine how these organs and systems carry out their functions.

Interestingly, even though all humans belong to the same species, individuals can have significant variations in the way that their body systems function. Outwardly, one of the most obvious differences between humans is skin color. A study published in 2005 showed that although many genes are involved in producing all the nuances of skin color found in humans, a mutation in a single, "master" gene seems to be very important in determining this characteristic. By studying people with different skin colors, the investigators found that a change of a single adenine (A) to thymidine (T) in the gene coding for a protein of unknown function was the major factor in determining darker or lighter skin. This may have made it easier for different human populations to adapt to varying levels of sun exposure. So even though the entire body is incredibly complicated as a whole, a particular feature can be strongly influenced in a relatively simple manner!

Chapter Outline

11.1 Types of Tissues

Learning Outcomes

1. List the four major types of tissues found in the human body, and describe the location and major features of each.
2. Describe the functions of each tissue type, including the roles of specific cell types and cellular modifications within each tissue.

Recall the biological levels of organization described in Chapter 1. Cells are composed of molecules; a tissue has similar types of cells; an organ contains several types of tissues; and several organs make up an organ system. In this chapter, we consider the tissue, organ, and organ system levels of organization.

A **tissue** is composed of similarly specialized cells that perform a common function in the body. The tissues of the human body can be categorized into four major types:

Epithelial tissue covers body surfaces and lines body cavities.
Connective tissue binds and supports body parts.
Muscular tissue moves the body and its parts.
Nervous tissue receives stimuli, processes that information, and conducts nerve impulses.

Cancers are classified according to the type of tissue from which they arise. **Carcinomas,** the most common type, are cancers of epithelial tissue. The Health Focus later in this chapter describes one type of carcinoma, the melanoma. Sarcomas are cancers arising in muscle or connective tissue, especially bone or cartilage. Leukemias are cancers of the blood cells, and lymphomas are cancers that originate in lymph nodes. The chance of developing cancer in a particular tissue shows a positive correlation to the rate of cell division. Epithelial cells reproduce at a high rate, and 2,500,000 new blood cells appear each second. Therefore, carcinomas and leukemias are common types of cancers.

Epithelial Tissue

Epithelial tissue, also called epithelium, consists of tightly packed cells that form a continuous layer. Epithelial tissue covers surfaces and lines body cavities. Epithelial tissue has numerous functions in the body. Usually, it has a protective function, but it can also be modified to carry out secretion, absorption, excretion, and filtration.

On the external surface, epithelial tissue protects the body from injury, drying out, and possible invasion by microbes such as bacteria and viruses. On internal surfaces, modifications help epithelial tissue carry out both its protective and specific functions. Epithelial tissue secretes mucus along the digestive tract and sweeps up impurities from the lungs by means of cilia (sing., cilium). It efficiently absorbs molecules from kidney tubules and from the intestine because of minute cellular extensions called microvilli (sing., **microvillus**).

A **basement membrane** usually joins an epithelium to underlying connective tissue. We now know that the basement membrane consists of glycoprotein secreted by epithelial cells and collagen fibers that belong to the connective tissue.

Epithelial tissue is classified according to the shape of cell it is composed of (squamous, cuboidal, or columnar) and on the number of layers in the tissue (Fig. 11.1). **Squamous epithelium** is characterized by flattened cells and is found lining the lungs and blood vessels. **Cuboidal epithelium** contains cube-shaped cells and lines the kidney tubules. **Columnar epithelium** has cells resembling rectangular pillars or columns, with nuclei usually located near the bottom of each cell. This epithelium lines the digestive tract. Ciliated columnar epithelium lines the oviducts, where it propels the egg toward the uterus, or womb.

Classification is also based on the number of layers in the tissue. *Simple* epithelium has a single layer of cells, whereas *stratified* epithelium has layers of cells piled one on top of another. The walls of the smallest blood vessels, called **capillaries,** are composed of simple squamous epithelium. The permeability of this single layer of cells allows exchange of substances between the blood and tissue cells. Simple cuboidal epithelium lines kidney tubules and the cavities of many internal organs. Stratified squamous epithelium lines the nose, mouth, esophagus, anal canal, and vagina. The outer layer of skin is also stratified squamous epithelium, but the cells have been reinforced by keratin, a protein that provides strength.

When an epithelium is *pseudostratified,* it appears to be layered, but true layers do not exist because each cell touches the basement membrane. The lining of the windpipe, or trachea, is pseudostratified ciliated columnar epithelium. A secreted covering of mucus traps foreign particles, and the upward motion of the cilia carries the mucus to the back of the throat, where it either may be swallowed or expelled. Smoking can cause a change in mucous secretion and inhibit ciliary action, and the result is a chronic inflammatory condition called bronchitis.

When an epithelium secretes a product, it is said to be glandular. A **gland** can be a single epithelial cell, as are the mucus-secreting goblet cells within the columnar epithelium lining the digestive tract, or a gland can contain many cells. Glands that secrete their product into ducts are called exocrine glands, and those that secrete their product into the bloodstream are called endocrine glands. The pancreas is both an exocrine gland, because it secretes digestive juices into the small intestine via ducts, and an endocrine gland, because it secretes insulin into the bloodstream.

Junctions Between Epithelial Cells

The cells of a tissue can function in a coordinated manner when the plasma membranes of adjoining cells interact. The junctions between cells help cells function as a tissue. A **tight junction** forms an impermeable barrier because adjacent plasma membrane proteins actually join, producing a zipperlike fastening

Visual Focus

Figure 11.1 Epithelial tissue.
Certain types of epithelial tissue—squamous, cuboidal, and columnar—are named for the shapes of their cells. They all have a protective function in addition to other specific functions.

Pseudostratified, ciliated columnar
- lining of trachea
- sweeps impurities toward throat

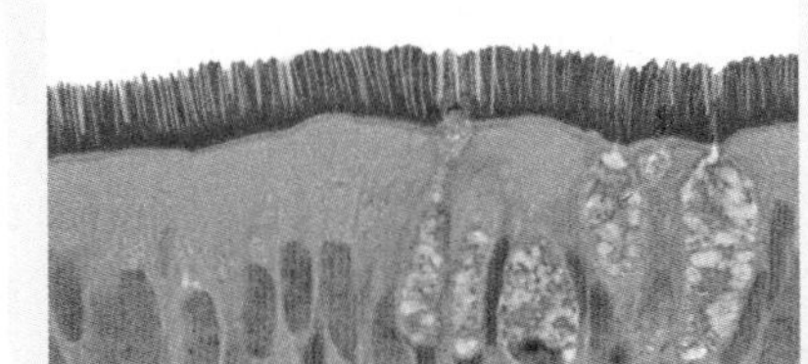

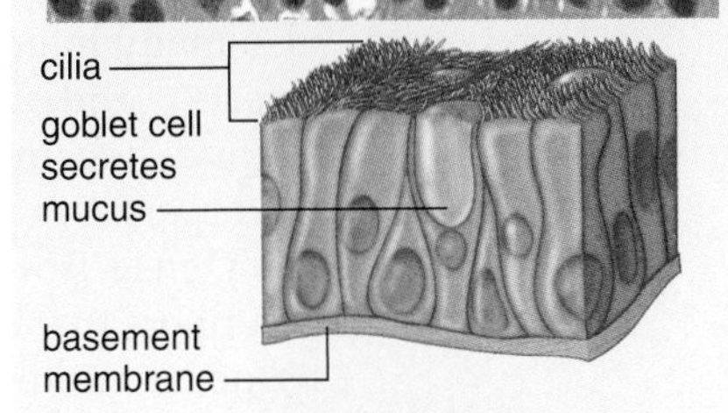

Simple squamous
- lining of lungs, blood vessels
- protects

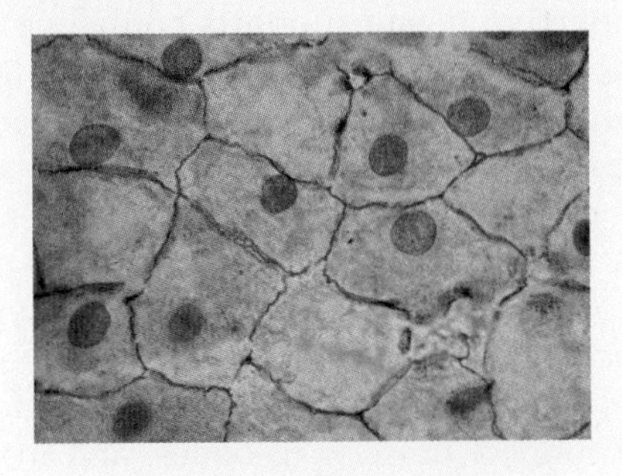

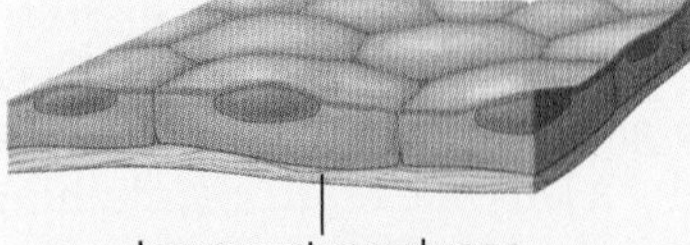

Stratified squamous
- skin (epidermis)
- lining of nose, mouth, esophagus, anal canal, vagina
- protects

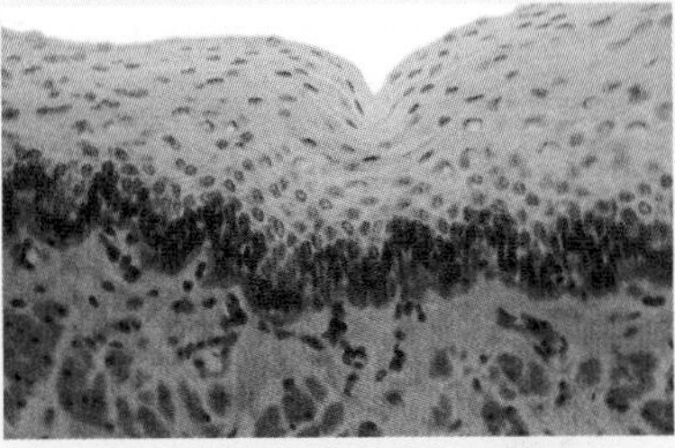

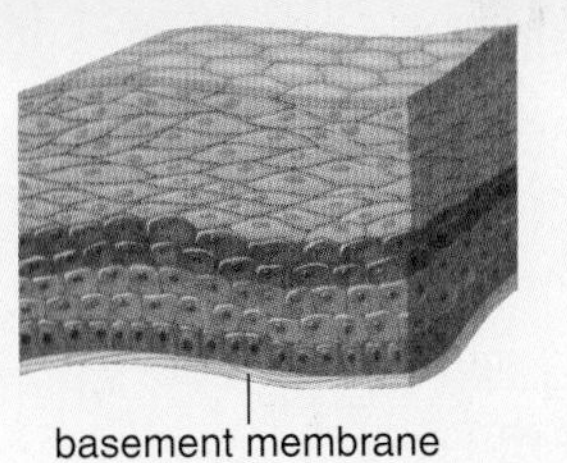

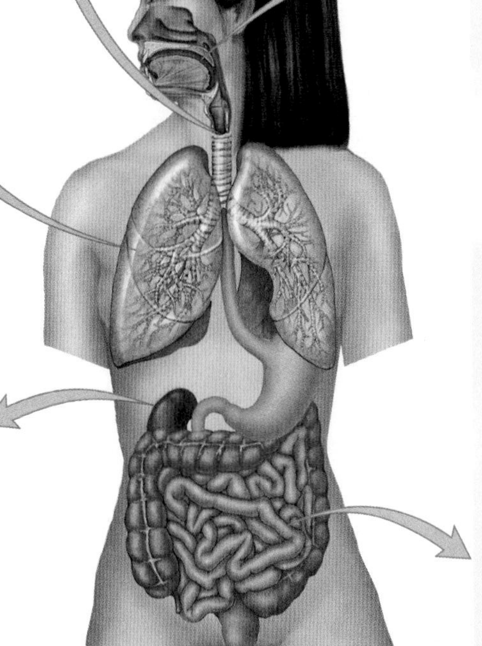

Simple cuboidal
- lining of kidney tubules, various glands
- absorbs molecules

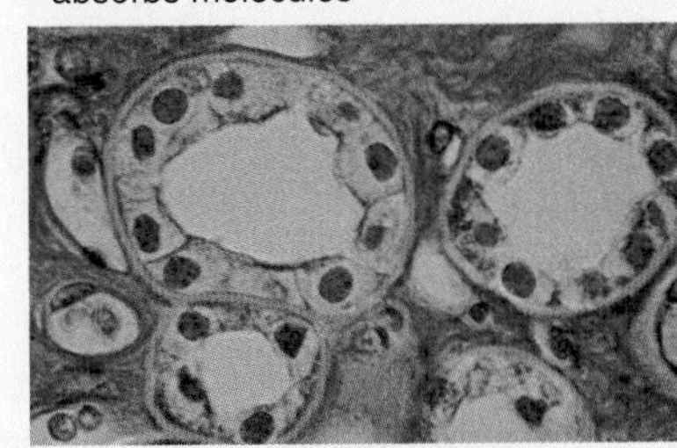

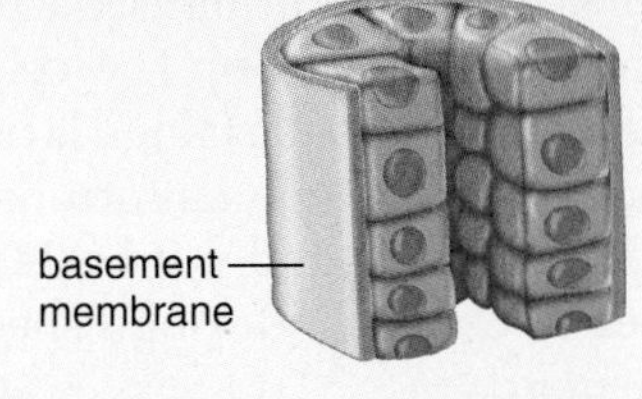

Simple columnar
- lining of small intestine, oviducts
- absorbs nutrients

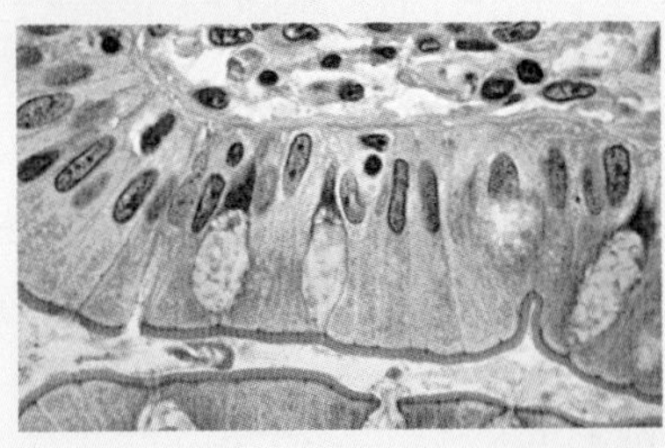

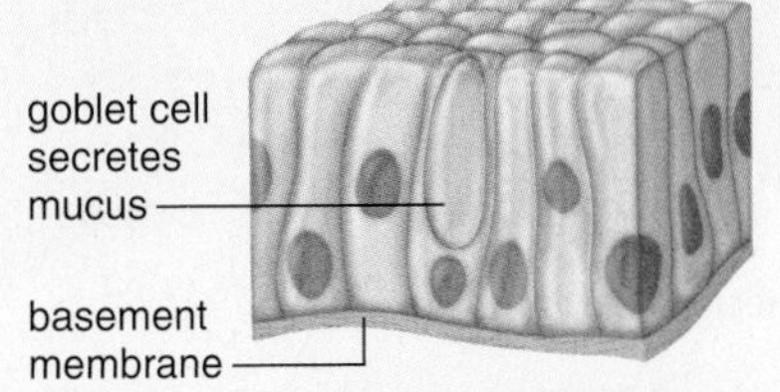

(Fig. 11.2*a*). In the intestine, the gastric juices stay out of the body, and in the kidneys, the urine stays within kidney tubules because epithelial cells are joined by tight junctions.

A **gap junction** forms when two adjacent plasma membrane channels join (Fig. 11.2*b*). This lends strength, but it also allows ions, sugars, and small molecules to pass between the two cells. In an **adhesion junction** (also called a desmosome), the adjacent plasma membranes do not touch but are held together by intercellular filaments firmly attached to cytoplasmic plaques (Fig. 11.2*c*). Adhesion junctions act like rivets or "spot welds" that enhance the strength of a tissue, such as the skin.

Check Your Progress

1. What are the main functions of epithelial tissues?
2. By what two characteristics are epithelial tissues usually classified?
3. What are the unique functions of tight, gap, and adhesion junctions?

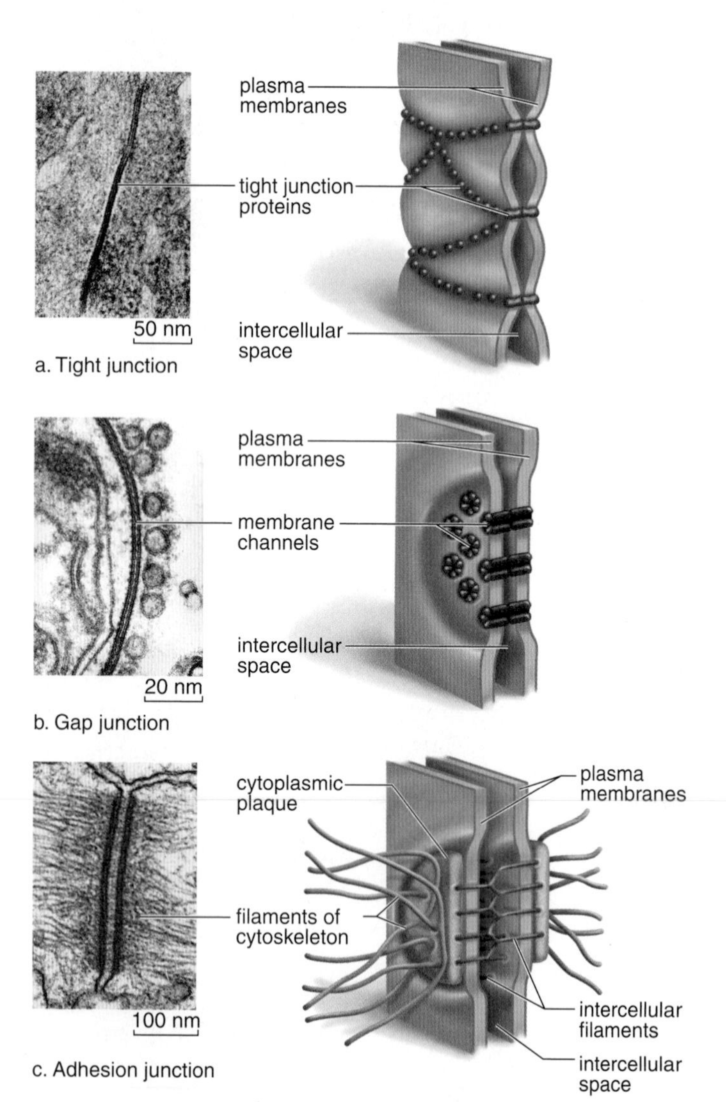

Figure 11.2 Junctions between epithelial cells.
Epithelial tissue cells are held tightly together by (**a**) tight junctions that hold cells together; (**b**) gap junctions that allow materials to pass from cell to cell; and (**c**) adhesion junctions that increase tissue strength.

Connective Tissue

Connective tissue binds organs together, provides support and protection, fills spaces, produces blood cells, and stores fat. As a rule, connective tissue cells are widely separated by a **matrix,** consisting of a noncellular material that varies in consistency from solid to jellylike to fluid. A nonfluid matrix may have fibers of three possible types. White **collagen fibers** contain collagen, a protein that gives them flexibility and strength. **Reticular fibers** are very thin collagen fibers that are highly branched and form delicate supporting networks. Yellow **elastic fibers** contain elastin, a protein that is not as strong as collagen but more elastic.

Loose Fibrous and Dense Fibrous Tissues

Loose fibrous connective tissue supports epithelium and also many internal organs (Fig. 11.3*a*). Its presence in lungs, arteries, and the urinary bladder allows these organs to expand. It forms a protective covering enclosing many internal organs, such as muscles, blood vessels, and nerves.

Dense fibrous connective tissue contains many collagen fibers that are packed together (Fig. 11.3*b*). This type of tissue has more specific functions than does loose connective tissue. For example, dense fibrous connective tissue is found in **tendons,** which connect muscles to bones, and in **ligaments,** which connect bones to other bones at joints.

Both loose fibrous and dense fibrous connective tissues have cells called **fibroblasts** located some distance from one another and separated by a jellylike matrix containing white collagen fibers and yellow elastic fibers.

Adipose Tissue and Reticular Connective Tissue

In **adipose tissue** (Fig. 11.3*c*), the fibroblasts enlarge and store fat (thus, they are called adipocytes). The body uses this stored fat for energy, insulation, and organ protection. Adipose tissue is found beneath the skin, around the kidneys, and on the surface of the heart. **Reticular connective tissue** forms the supporting meshwork of lymphoid tissue in lymph nodes, the spleen, the thymus, and the bone marrow. All types of blood cells are produced in red bone marrow, but a certain type of lymphocyte (T lymphocyte) completes its development in the thymus. The lymph nodes are sites of lymphocyte responses (see Chapter 13).

Cartilage

Cartilage is a specialized form of dense fibrous connective tissue, which most commonly forms the smooth surfaces that allow bones to slide against each other in joints. Cartilage is found in many other locations, however. The cells in cartilage lie in small chambers called lacunae (sing., **lacuna**), separated by a matrix that is solid yet flexible. Unfortunately, because this tissue lacks a direct blood supply, it heals very slowly. There are three types of cartilage, distinguished by the main type of fiber in the matrix.

Hyaline cartilage (Fig. 11.3*d*), the most common type of cartilage, contains only very fine collagen fibers. The matrix has a white, translucent appearance. Hyaline cartilage is

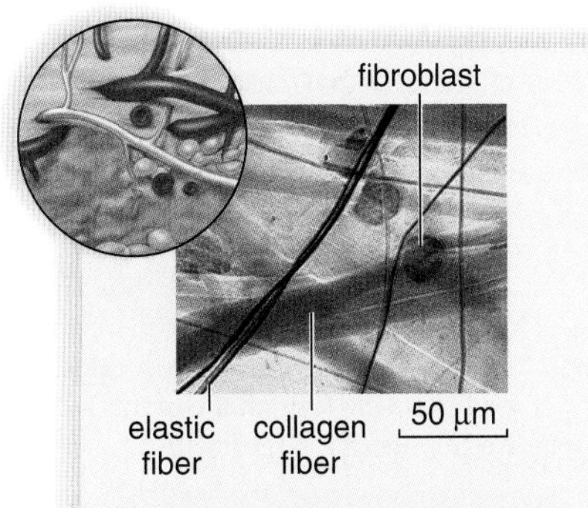

Loose fibrous connective tissue
- has space between components.
- occurs beneath skin and most epithelial layers.
- functions in support and binds organs.

a.

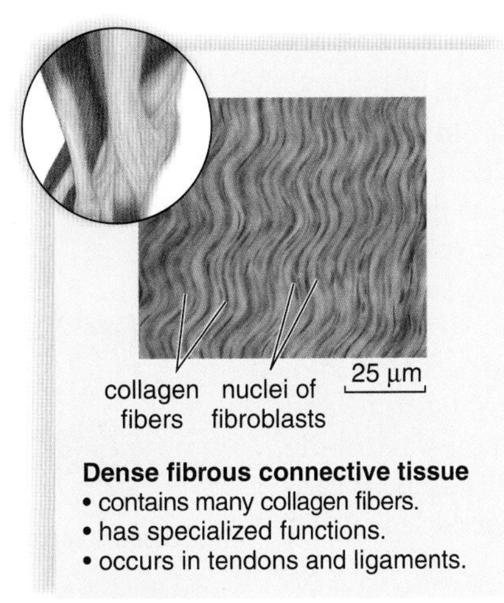

Dense fibrous connective tissue
- contains many collagen fibers.
- has specialized functions.
- occurs in tendons and ligaments.

b.

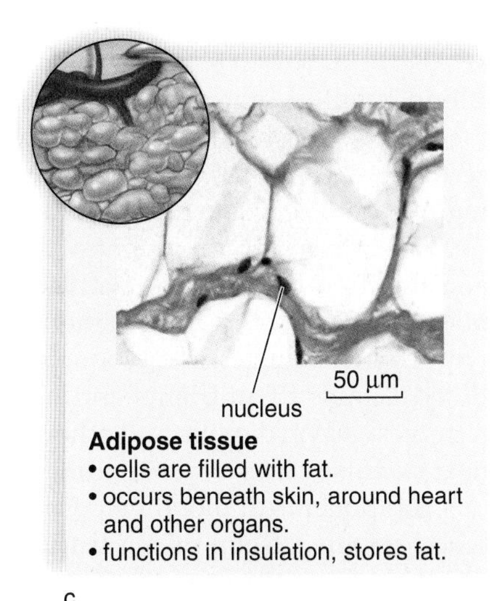

Adipose tissue
- cells are filled with fat.
- occurs beneath skin, around heart and other organs.
- functions in insulation, stores fat.

c.

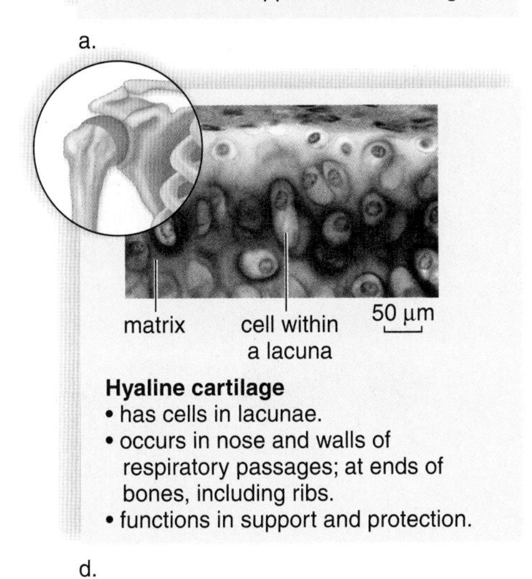

Hyaline cartilage
- has cells in lacunae.
- occurs in nose and walls of respiratory passages; at ends of bones, including ribs.
- functions in support and protection.

d.

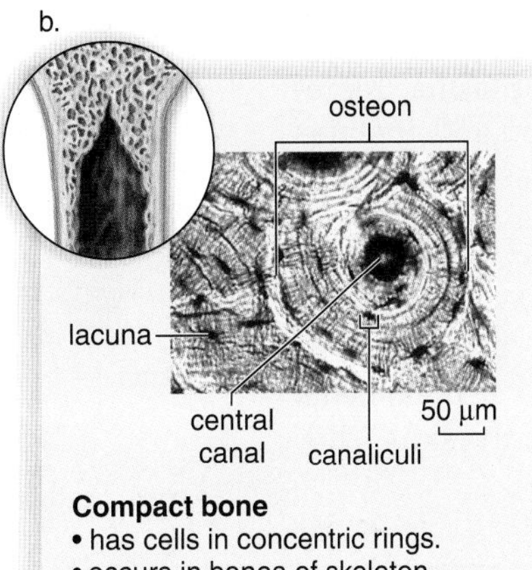

Compact bone
- has cells in concentric rings.
- occurs in bones of skeleton.
- functions in support and protection.

e.

Figure 11.3 Connective tissue examples.
a. In loose fibrous connective tissue, cells called fibroblasts are separated by a jellylike matrix, which contains both collagen and elastic fibers. **b.** Dense fibrous connective tissue contains tightly packed collagen fibers for added strength. **c.** Adipose tissue cells (adipocytes) have nuclei pushed to one side because the cells are filled with fat. **d.** In hyaline cartilage, the flexible matrix has a white, translucent appearance. **e.** In compact bone, the hard matrix contains calcium salts. Concentric rings of cells in lacunae form an elongated cylinder called an osteon. An osteon has a central canal that contains blood vessels and nerve fibers.

found in the nose and at the ends of the long bones and the ribs, and it forms rings in the walls of respiratory passages. The fetal skeleton is also made of this type of cartilage. Later, the cartilaginous fetal skeleton is replaced by bone.

Elastic cartilage has more elastic fibers than hyaline cartilage. For this reason, it is more flexible and is found, for example, in the framework of the outer ear.

Fibrocartilage has a matrix containing strong collagen fibers. Fibrocartilage is found in structures that withstand tension and pressure, such as the pads between the vertebrae in the backbone and the wedges in the knee joint.

Bone

Bone is the most rigid connective tissue. It consists of an extremely hard matrix of inorganic salts, notably calcium salts, deposited around protein fibers, especially collagen fibers. The inorganic salts give bone rigidity, and the protein fibers provide elasticity and strength, much as steel rods do in reinforced concrete.

Compact bone makes up the shaft of a long bone (Fig. 11.3*e*). It consists of cylindrical structural units called osteons. The central canal of each osteon is surrounded by rings of hard matrix. Bone cells are located in spaces called lacunae between the rings of matrix. In the central canal, nerve fibers carry nerve impulses and blood vessels carry nutrients that allow bone to renew itself. Nutrients can reach all of the bone cells because they are connected by thin processes within canaliculi (minute canals) that also reach to the central canal. The shaft of long bones such as the femur (thigh bone) has a hollow chamber filled with bone marrow, a site where blood cells develop.

The ends of a long bone contain **spongy bone,** which contains numerous bony bars and plates, separated by irregular spaces. Although lighter than compact bone, spongy bone is still designed for strength. Just as braces are used for support in buildings, the solid portions of spongy bone follow lines of stress. See section 19.1 to learn more about bone and cartilage.

Check Your Progress

1. What do loose and dense fibrous connective tissue have in common? How are they different?
2. List the three major types of cartilage and one body region where each is found.
3. What characteristics of bone make it different from all other types of connective tissue?

Blood

Blood is unlike other types of connective tissue in that the matrix (i.e., plasma) is not made by the cells. In fact, some scientists do not classify blood as connective tissue, suggesting instead a separate category called vascular tissue.

The internal environment of the body consists of blood and the fluid between the body's cells. The systems of the body help keep the composition and chemistry of blood within normal limits, and blood, in turn, creates tissue fluid. Blood transports nutrients and oxygen to tissue fluid and removes carbon dioxide and other wastes. It helps distribute heat and also plays a role in fluid, ion, and pH balance. Various components of blood help protect us from disease, and blood's ability to clot prevents fluid loss.

If blood is transferred from a person's vein to a test tube and prevented from clotting, it separates into two layers (Fig. 11.4*a*). The upper, liquid layer, called **plasma,** represents about 55% of the volume of whole blood and contains a variety of inorganic and organic substances dissolved or suspended in water (Table 11.1). The lower layer consists of red blood cells (erythrocytes), white blood cells (leukocytes), and blood platelets (thrombocytes) (Fig. 11.4*b*). Collectively, these are called the formed elements, and they represent about 45% of the volume of whole blood. Formed elements are manufactured in the red bone marrow of the skull, ribs, vertebrae, and ends of the long bones.

The **red blood cells** are small, biconcave, disk-shaped cells without nuclei. The presence of the red pigment hemoglobin makes the cells red and, in turn, makes the blood red. Hemoglobin is composed of four units. Each contains the protein globin and a complex, iron-containing structure called heme. The iron forms a loose association with oxygen, and in this way, red blood cells transport oxygen.

White blood cells differ from red blood cells in that they are usually larger, have a nucleus, and without staining would appear translucent. When smeared onto microscope slides and stained, the nuclei of white blood cells typically look bluish (Fig. 11.4*b*). White blood cells fight infection in a number of different ways. Some white blood cells are phagocytic and engulf infectious pathogens, while others are responsible for the adaptive immunity that develops after an individual is exposed to various pathogens or toxins, either through natural infection or by vaccination. One important feature of acquired immunity is the production of antibodies, molecules that combine with foreign substances to inactivate them.

Platelets are not complete cells. Rather, they are fragments of large cells present only in bone marrow. When a blood vessel is damaged, platelets help to form a plug that seals the vessel, and injured tissues release molecules that stimulate the clotting process.

For more details on red blood cells, white blood cells, and platelets, see section 12.4 and Chapter 13.

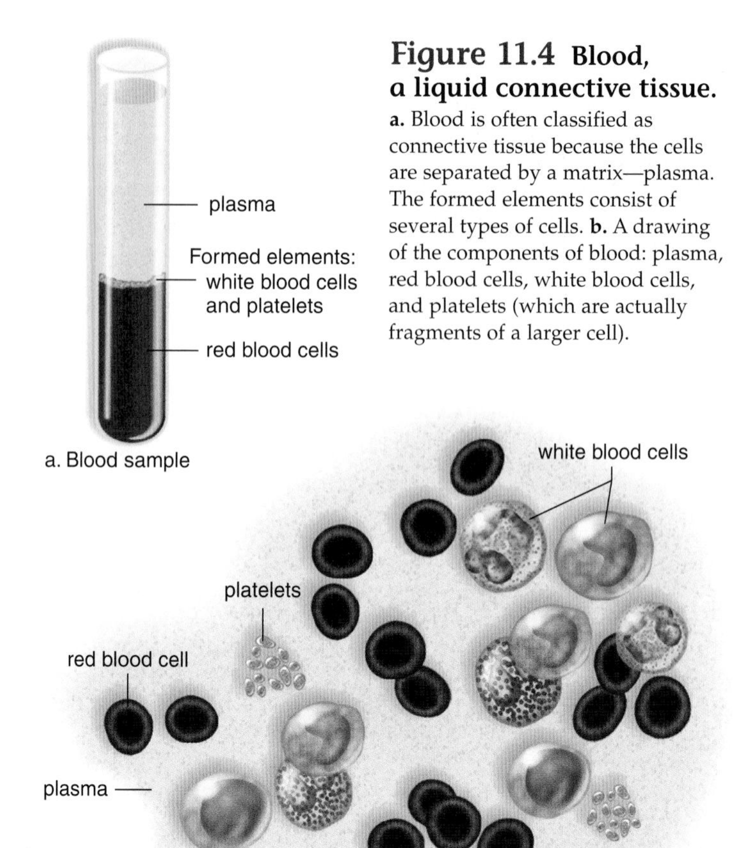

Figure 11.4 Blood, a liquid connective tissue. **a.** Blood is often classified as connective tissue because the cells are separated by a matrix—plasma. The formed elements consist of several types of cells. **b.** A drawing of the components of blood: plasma, red blood cells, white blood cells, and platelets (which are actually fragments of a larger cell).

TABLE 11.1 Components of Blood Plasma

Water (90–92% of total)	
Solutes (8–10% of total)	
Inorganic ions (electrolytes)	Na^+, Ca^{2+}, K^+, Mg^{2+}, Cl^-, HCO_3^-, HPO_4^{2-}, SO_4^{2-}
Gases	O_2, CO_2
Plasma proteins	Albumin, globulins, fibrinogen, transport proteins
Organic nutrients	Glucose, lipids, phospholipids, amino acids, etc.
Nitrogen-containing waste products	Urea, ammonia, uric acid
Regulatory substances	Hormones, enzymes

Check Your Progress

1. What are the main components of blood?
2. Contrast the functions of red blood cells, white blood cells, and platelets.
3. How is blood similar to, yet different from, other types of connective tissue?

Muscular Tissue

Muscular tissue is composed of cells called muscle fibers. Muscle fibers contain actin filaments and myosin filaments, whose interaction accounts for movement. The three types of muscle tissue are skeletal, smooth, and cardiac.

Skeletal muscle (Fig. 11.5*a*) is usually attached by tendons to the bones of the skeleton, and when it contracts, body parts move. Contraction of skeletal muscle is under voluntary control. Skeletal muscle fibers are cylindrical and quite long—sometimes they run the length of the muscle. They arise from fusion of several cells, resulting in one fiber with multiple nuclei. The fibers have alternating light and dark bands that give them a **striated** appearance.

Smooth muscle is so named because the cells lack striations. The spindle-shaped fibers, each with a single nucleus, form layers in which the thick middle portion of one cell is opposite the thin ends of adjacent cells. Consequently, the nuclei form an irregular pattern in the tissue (Fig. 11.5*b*). Smooth muscle is not under voluntary control, and therefore is said to be involuntary. Smooth muscle is found in the walls of viscera (intestine, stomach, and other internal organs) and blood vessels.

Cardiac muscle (Fig. 11.5*c*) is found only in the walls of the heart. Cardiac muscle combines features of both smooth muscle and skeletal muscle. Like skeletal muscle, it has striations, but the contraction of the heart is involuntary. Cardiac muscle cells also differ from skeletal muscle cells in that they usually have a single, centrally placed nucleus. The cells are branched and seemingly fused one with the other, and the heart appears to be composed of one large interconnecting mass of muscle cells. Actually, cardiac muscle cells are separate and individual, but they are bound end to end at **intercalated disks,** areas where folded plasma membranes between two cells contain adhesion junctions and gap junctions. These areas promote the flow of electrical current when the heart muscle contracts by allowing ions to flow freely between cells. See chapters 12 and 9, respectively, for more information about cardiac and skeletal muscle.

Check Your Progress

1. What are three types of muscle tissue?
2. Besides body movement, what other essential functions are performed by muscle tissue?

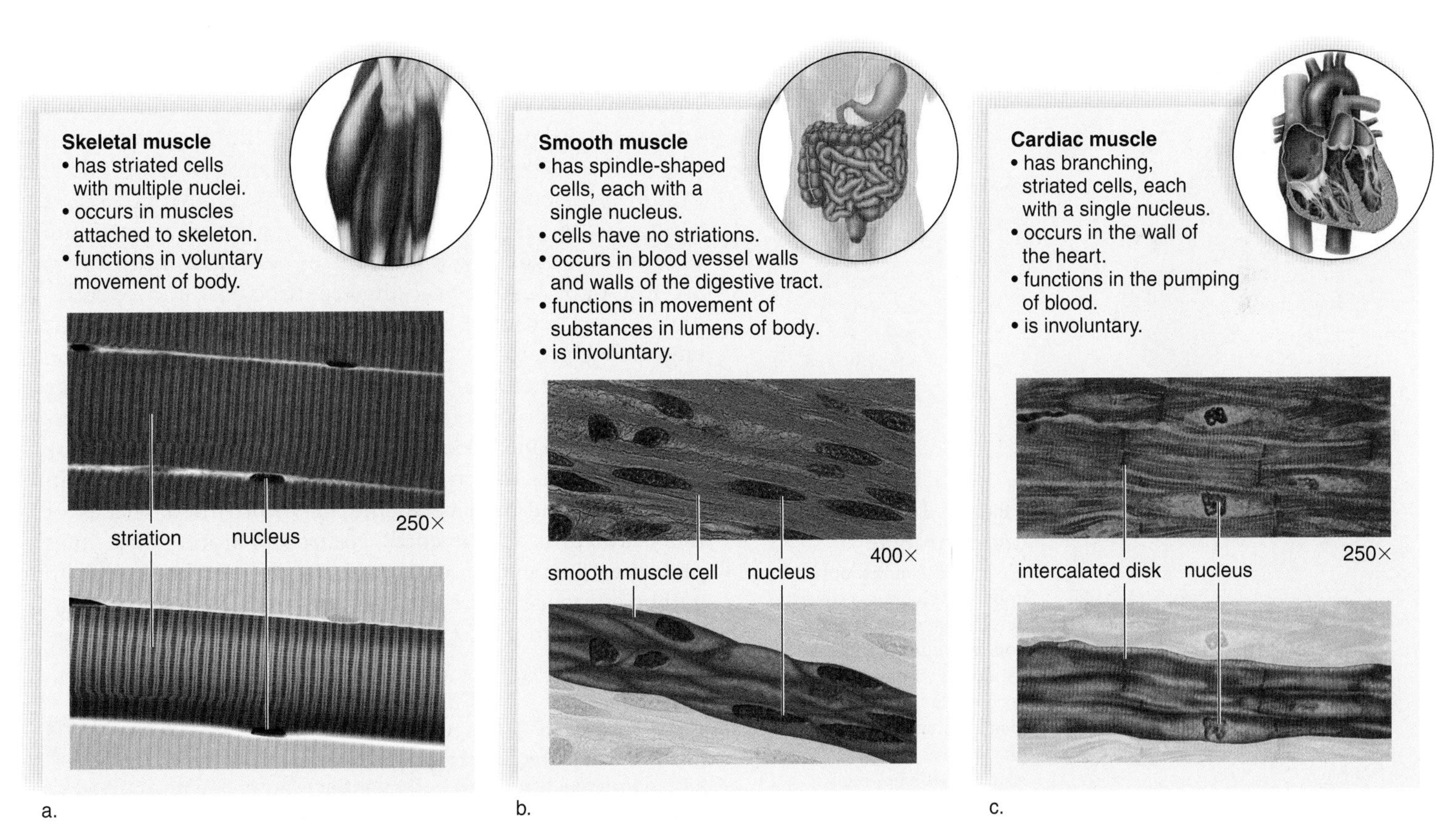

Figure 11.5 Muscular tissue.
a. Skeletal muscle is voluntary and striated. **b.** Smooth muscle is involuntary and nonstriated. **c.** Cardiac muscle is involuntary and striated. Cardiac muscle cells branch and fit together at intercalated disks.

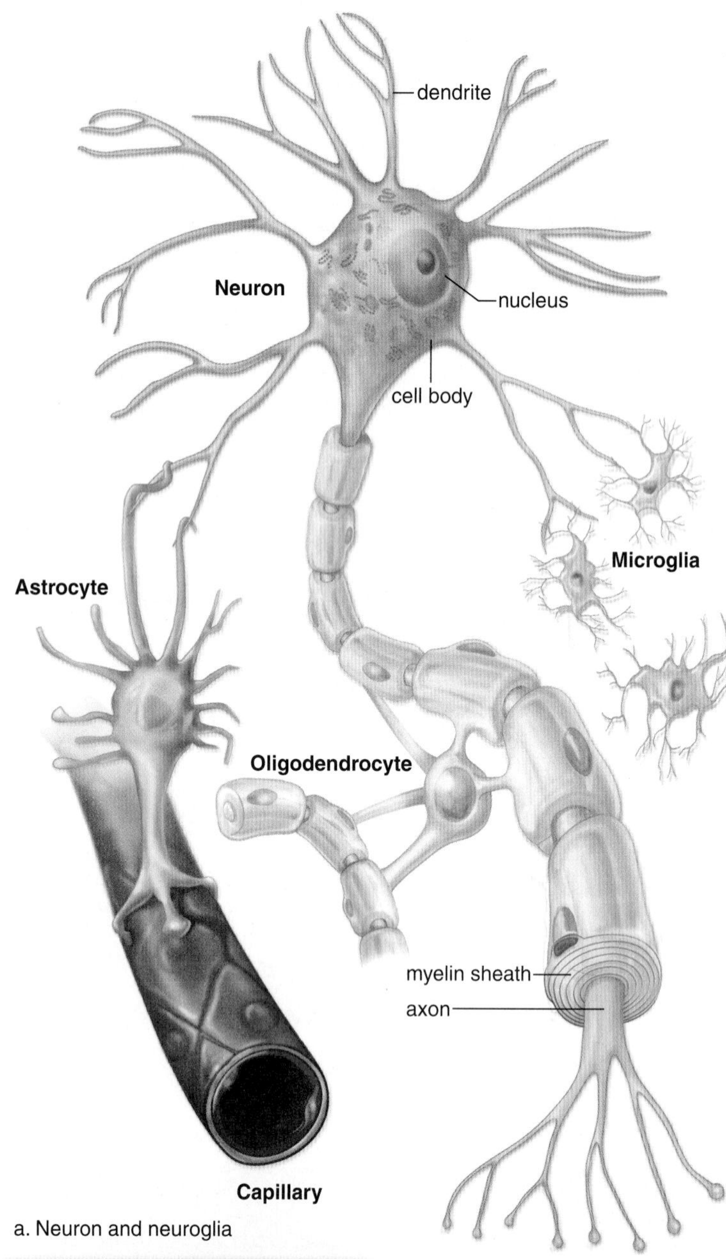

a. Neuron and neuroglia

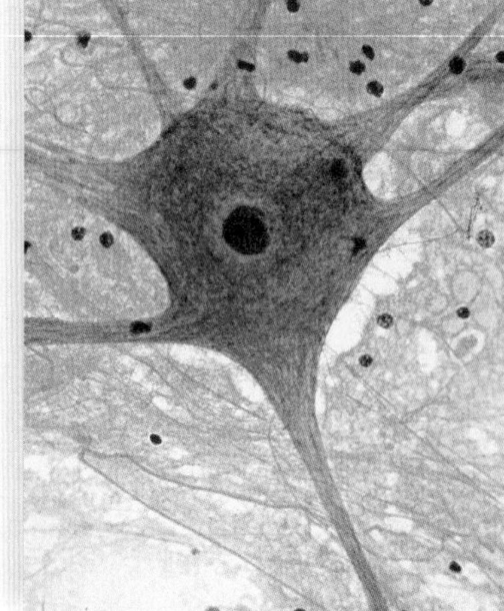

b. Micrograph of a neuron

Figure 11.6 Neurons and neuroglia.
a. Neurons conduct nerve impulses. Neuroglia consist of cells that support and nourish neurons and have various functions. Microglia become mobile in response to inflammation and phagocytize debris. Astrocytes lie between neurons and a capillary. Therefore, nutrients entering neurons from the blood must first pass through astrocytes. Oligodendrocytes form the myelin sheaths around fibers in the brain and spinal cord. **b.** A neuron cell body as seen with a light microscope.

Nervous Tissue

Nervous tissue, which contains nerve cells called neurons (Fig. 11.6*a*), is present in the brain and spinal cord. A **neuron** is a specialized cell that has three parts: a cell body, dendrites, and an axon (Fig. 11.6*a, b*). The cell body contains the major concentration of the cytoplasm and the nucleus of the neuron. A dendrite is a process that conducts signals toward the cell body. An axon is a process that typically conducts nerve impulses away from the cell body. Long axons are covered by myelin, a white fatty substance, which increases the speed of nerve impulses. The term *fiber*[1] is used here to refer to an axon along with its myelin sheath if it has one. Outside the brain and spinal cord, fibers bound by connective tissue form **nerves.**

The nervous system has just three functions: sensory input, integration of data, and motor output. Nerves conduct impulses from sensory receptors to the spinal cord and the brain where integration occurs. The phenomenon called sensation occurs only in the brain, however. Nerves also conduct nerve impulses away from the spinal cord and brain to the muscles and glands, causing them to contract and secrete, respectively. In this way, a coordinated response to the stimulus is achieved.

Neuroglia

In addition to neurons, nervous tissue contains neuroglia. **Neuroglia** are cells that outnumber neurons nine to one and take up more than half the volume of the brain (Fig. 11.6*a*). Although the primary function of neuroglia is to support and nourish neurons, research is currently being conducted to determine how much they directly contribute to brain function. The four types of neuroglia in the brain are microglia, astrocytes, oligodendrocytes, and ependymal cells. Microglia, in addition to supporting neurons, engulf bacterial and cellular debris. Astrocytes provide nutrients to neurons and produce a hormone known as glia-derived growth factor, which has potential as a treatment for Parkinson disease and other diseases caused by neuron degeneration. Oligodendrocytes form myelin sheaths. Ependymal cells line the fluid-filled spaces of the brain and spinal cord. Neuroglia don't have long processes, but even so, researchers are now beginning to gather evidence that they do communicate among themselves and with neurons! To learn more about the nervous system, see Chapter 17.

Check Your Progress

1. Name the three parts of a neuron, and define nerve fiber.
2. What are the three functions of the nervous system?
3. List four types of neuroglial cells, and review the function of each.

[1]In connective tissue, a fiber is a component of the matrix; in muscle tissue, a fiber is a muscle cell; in nervous tissue, a fiber is an axon and its myelin sheath.

11.2 Body Cavities and Body Membranes

Learning Outcomes

1. Know the two main cavities of the human body and how each can be further divided.
2. Compare and contrast the location(s) and function(s) of the different types of body membranes.

The human body is divided into two main cavities: the ventral cavity and the dorsal cavity (Fig. 11.7*a*). The ventral cavity, which is called a **coelom** during development, becomes divided into the thoracic, abdominal, and pelvic cavities (the latter two are sometimes grouped together as the abdominopelvic cavity). The thoracic cavity contains the right and left lungs and the heart. The thoracic cavity is separated from the abdominal cavity by a horizontal muscle called the diaphragm. The stomach, liver, spleen, gallbladder, and most of the small and large intestines are in the upper portion of the abdominal cavity. The pelvic cavity contains the rectum, the urinary bladder, the internal reproductive organs, and the rest of the large intestine. Males have an external extension of the abdominal wall, called the scrotum, containing the testes.

The dorsal cavity also has two parts: the cranial cavity within the skull contains the brain; the vertebral canal, formed by the vertebrae, contains the spinal cord.

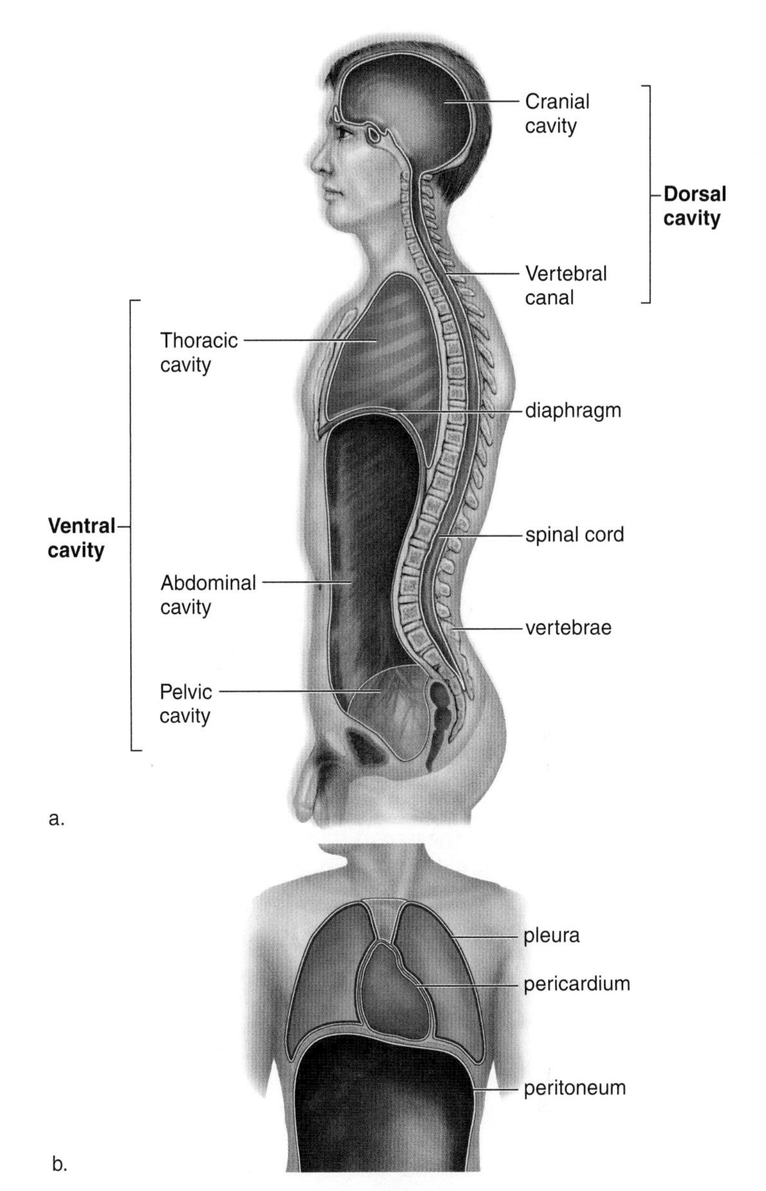

Figure 11.7 Mammalian body cavities.
a. Side view. The dorsal (toward the back) cavity contains the cranial cavity and the vertebral canal. The brain is in the cranial cavity, and the spinal cord is in the vertebral canal. In the ventral (toward the front) cavity, the diaphragm separates the thoracic cavity and the abdominal cavity. The heart and lungs are in the thoracic cavity, and most other internal organs are in the abdominal cavity. **b.** Types of serous membranes.

Body Membranes

Body membranes line cavities and the internal spaces of organs and tubes that open to the outside.

Mucous membranes line the tubes of the digestive, respiratory, urinary, and reproductive systems. They are composed of an epithelium overlying a loose fibrous connective tissue layer. The epithelium contains goblet cells that secrete mucus. This mucus ordinarily protects the body from invasion by bacteria and viruses. More mucus is secreted and expelled when a person has a cold and has to blow her/his nose.

Serous membranes, which line the thoracic and abdominal cavities and cover the organs they contain, are also composed of epithelium and loose fibrous connective tissue. They secrete a watery fluid that keeps the membranes lubricated. Serous membranes support the internal organs and compartmentalize the large thoracic and abdominal cavities. This helps hinder the spread of any infection.

Serous membranes have specific names according to their location (Fig. 11.7*b*). The pleurae (sing., pleura) line the thoracic cavity and cover the lungs; the pericardium covers the heart; the peritoneum lines the abdominal cavity and covers its organs. A double layer of peritoneum, called mesentery, supports the abdominal organs and attaches them to the abdominal wall. Peritonitis is a potentially life-threatening infection of the peritoneum that may occur if an inflamed appendix bursts before it is removed, or if the digestive tract is perforated for any other reason.

Synovial membranes, composed only of loose connective tissue, line freely movable joint cavities. They secrete synovial fluid that lubricates the cartilage at the ends of the bones so that they can move smoothly in the joint cavity.

The **meninges** are membranes within the dorsal cavity. They are composed only of connective tissue and serve as a protective covering for the brain and spinal cord.

Check Your Progress

1. Which cavities are separated by the diaphragm?
2. What is the function of the fluids produced by various body membranes?

11.3 Organ Systems

LEARNING OUTCOMES

1. Describe the general function of each body system.
2. Provide several examples of how different organ systems work closely together.

Figure 11.8 illustrates the organ systems of the human body. It should be emphasized that just as organs work together in an organ system, so do organ systems work together in the body. In some cases, it is arbitrary to assign a particular organ to one system when it also assists the functioning of many other systems.

Integumentary System

The **integumentary system** contains skin, which is made up of two main types of tissue: the epidermis is composed of stratified squamous epithelium, and the dermis is composed of fibrous connective tissue. The integumentary system also includes nails, located at the ends of the fingers and toes (collectively called the digits), hairs, muscles that move hairs, the oil and sweat glands, blood vessels, and nerves leading to sensory receptors. Besides having a protective function, skin also synthesizes vitamin D, collects sensory data, and helps regulate body temperature.

Cardiovascular System

In the **cardiovascular system,** the heart pumps blood and sends it under pressure into the blood vessels. While blood is moving throughout the body, it distributes heat produced by the muscles. Blood transports nutrients and oxygen to the cells and removes their waste molecules, including carbon dioxide. Despite the movement of molecules into and out of the blood, it has a fairly constant volume and pH. The blood is also a route by which cells of the immune system can be distributed throughout the body.

Lymphatic and Immune Systems

The **lymphatic system** consists of lymphatic vessels (which transport lymph), lymph nodes, and other lymphatic (lymphoid) organs. This system protects the body from disease by purifying lymph and storing lymphocytes, the white blood cells responsible for adaptive immunity. Lymphatic vessels absorb fat from the digestive system and collect excess tissue fluid, which is returned to the cardiovascular system.

The **immune system** consists of all the cells in the body that protect us from disease, especially those caused by infectious agents. The lymphocytes, in particular, belong to this system.

Digestive System

The **digestive system** includes the mouth, esophagus, stomach, small intestine, and large intestine (colon), along with associated organs such as the teeth, tongue, salivary glands, liver, gallbladder, and pancreas. This system receives food and digests it into nutrient molecules, which can enter the body's cells. The nondigested remains are eventually eliminated.

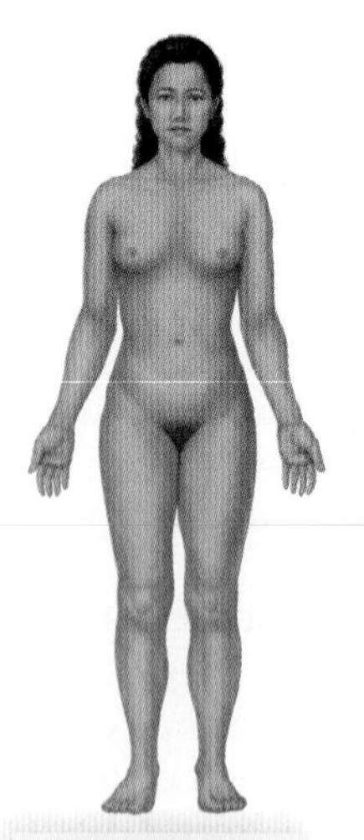

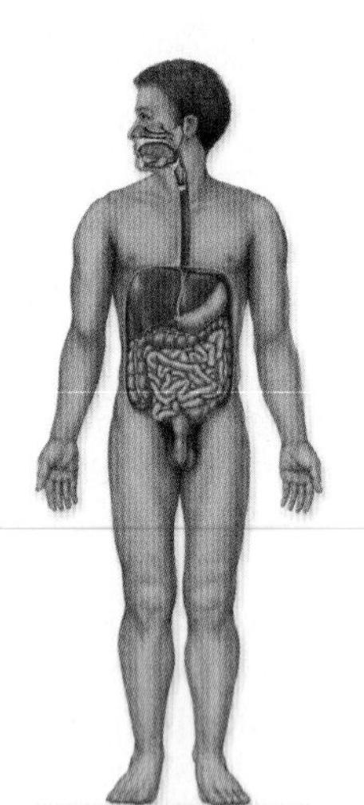

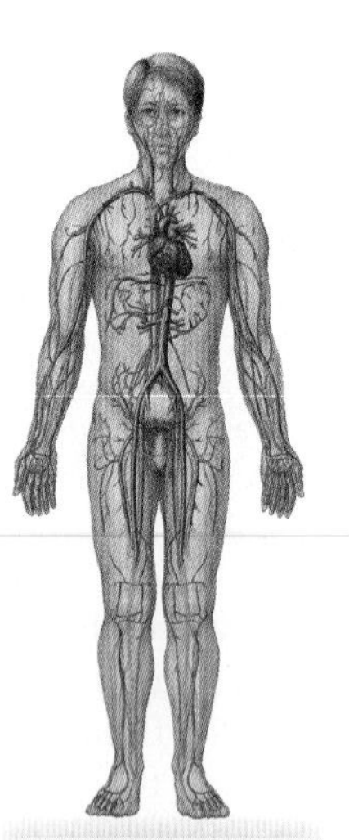

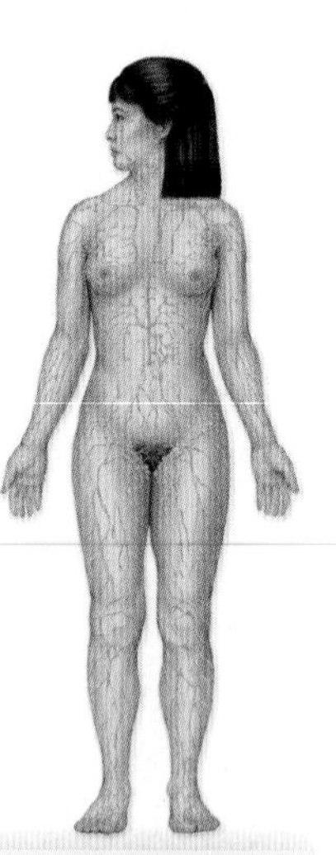

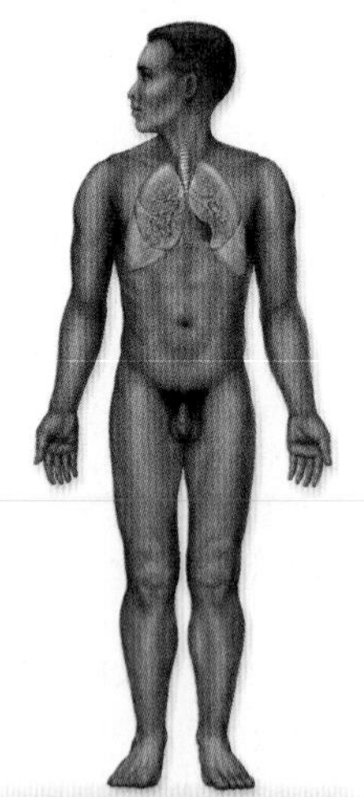

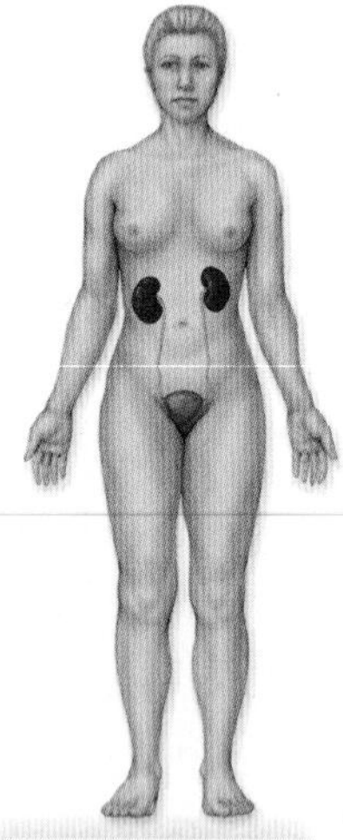

Figure 11.8 Organ systems of the body.

Respiratory System

The **respiratory system** consists of the lungs and the tubes that take air to and from them. The respiratory system brings oxygen into the body for cellular respiration and removes carbon dioxide from the body at the lungs, restoring pH.

Urinary System

The **urinary system** contains the kidneys, the urinary bladder, and the tubes that carry urine. This system rids the body of metabolic wastes, and helps regulate the fluid balance and pH of the blood.

Skeletal System

The bones of the **skeletal system** provide a scaffolding that helps hold and protect body parts. For example, the skull forms a protective encasement for the brain. The skeleton also helps move the body because it serves as a place of attachment for the skeletal muscles. It stores minerals, and it produces blood cells within the bone marrow.

Muscular System

In the **muscular system,** skeletal muscle contraction maintains posture and accounts for the movement of the body and its parts. Cardiac muscle contraction results in the heartbeat. The walls of internal organs contract due to the presence of smooth muscle.

Nervous System

The **nervous system** consists of the brain, spinal cord, and associated nerves. The nerves conduct nerve impulses from sensory receptors to the brain and spinal cord where integration occurs. Nerves also conduct nerve impulses from the brain and spinal cord to the muscles and glands, allowing us to respond to both external and internal stimuli.

Endocrine System

The **endocrine system** consists of the hormonal glands, which secrete chemical messengers called hormones. Hormones have a wide range of effects, including regulating cellular metabolism, regulating fluid and pH balance, and helping us respond to stress. Both the nervous and endocrine systems coordinate and regulate the functioning of the body's other systems. The endocrine system also helps maintain the functioning of the male and female reproductive organs.

Reproductive System

The **reproductive system** has different organs in the male and female. The male reproductive system consists of the testes, other glands, and various ducts that conduct semen to and through the penis. The testes produce sex cells called sperm. The female reproductive system consists of the ovaries, oviducts, uterus, vagina, and external genitals. The ovaries produce sex cells called eggs, or oocytes. When a sperm fertilizes an oocyte, an offspring begins to develop.

Check Your Progress

1. Name several cell types found in each system.
2. Discuss three examples of how different organ systems work together.

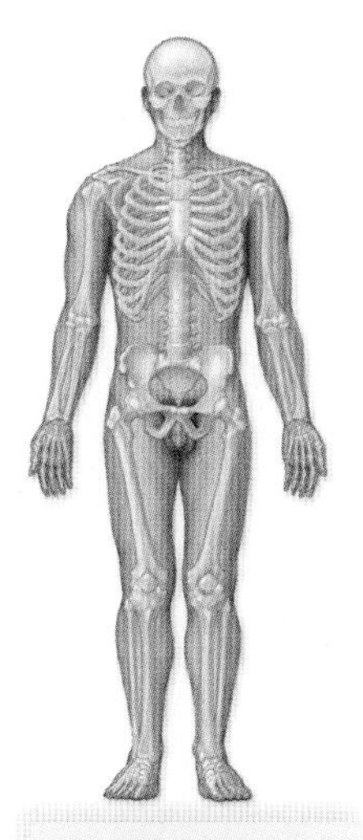

Skeletal system
- supports the body.
- protects body parts.
- helps move the body.
- stores minerals.
- produces blood cells.

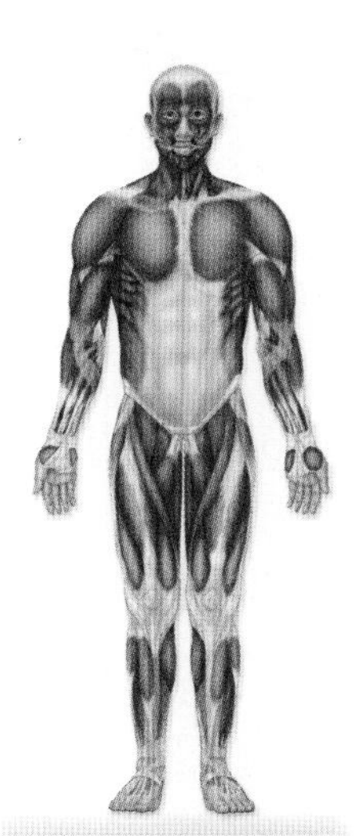

Muscular system
- maintains posture.
- moves body and internal organs.
- produces heat.

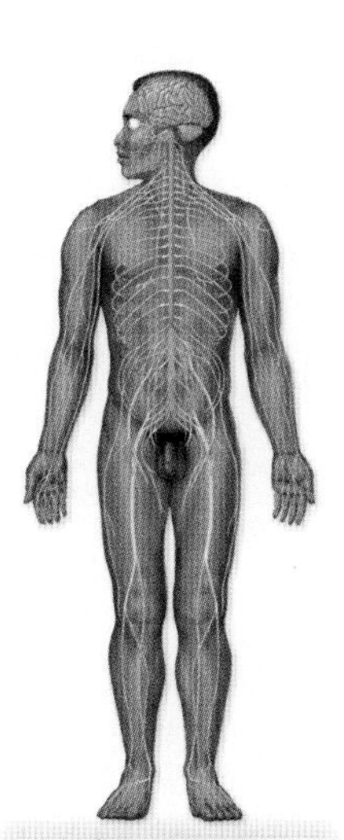

Nervous system
- receives sensory input.
- integrates and stores input.
- initiates motor output.
- helps coordinate organ systems.

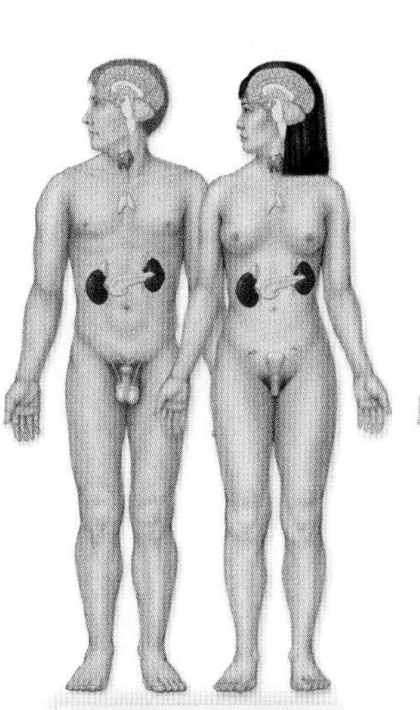

Endocrine system
- produces hormones.
- helps coordinate organ systems.
- responds to stress.
- helps regulate fluid and pH balance.
- helps regulate metabolism.

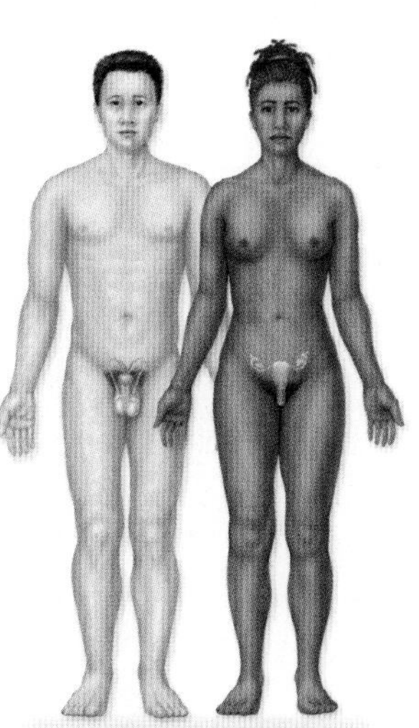

Reproductive system
- produces gametes.
- transports gametes.
- produces sex hormones.
- in females, nurtures and gives birth to offspring.

Figure 11.8 Organ systems of the body—continued

11.4 Integumentary System

Learning Outcomes

1. Identify the two main regions of skin, and how these are distinguished from the subcutaneous layer.
2. Describe the function(s) of each of the major accessory organs of the skin.

The skin and its accessory organs (nails, hair, oil glands, and sweat glands) are collectively called the integumentary system. Skin protects underlying tissues from physical trauma, pathogen invasion, and water loss. It also helps regulate body temperature. The skin even synthesizes vitamin D with the aid of ultraviolet radiation. Skin also contains sensory receptors, which help us to be aware of our surroundings and to communicate through touch.

Regions of the Skin

The **skin** has two regions: the epidermis and the dermis (Fig. 11.9). A subcutaneous layer is present between the skin and any underlying structures, such as muscle or bone. The subcutaneous layer is not a part of the skin.

The **epidermis** is made up of stratified squamous epithelium. New cells derived from *basal cells* become flattened and hardened as they are pushed to the surface by cells forming underneath them. Hardening takes place because the cells produce keratin, a waterproof protein. A thick layer of dead keratinized cells, arranged in spiral and concentric patterns, forms fingerprints and footprints. Specialized cells in the epidermis called **melanocytes** produce melanin, the main pigment responsible for skin color. Different amounts of melanin in the skin provide protection from damage by excessive UV radiation, while allowing sufficient UV activation of vitamin D (see the Health Focus later in this chapter).

The **dermis** is a region of fibrous connective tissue beneath the epidermis. The dermis contains collagen and elastic fibers. The collagen fibers are flexible but offer great resistance to overstretching. They prevent the skin from being torn. The elastic fibers maintain normal skin tension but also stretch to allow movement of underlying muscles and joints. (The number of collagen and elastic fibers

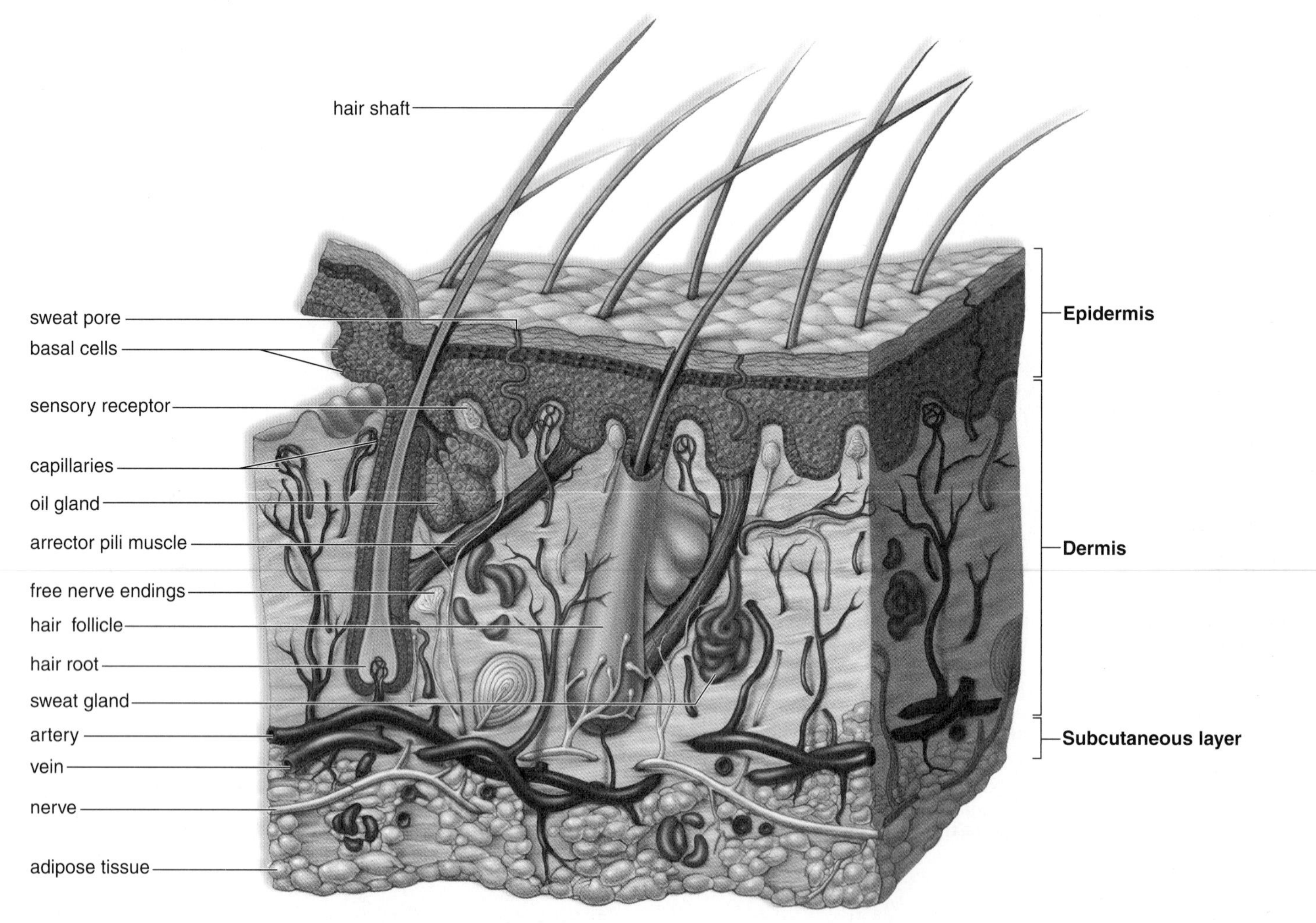

Figure 11.9 Human skin anatomy.
Skin consists of two regions, the epidermis and the dermis. A subcutaneous layer lies below the dermis.

decreases with exposure to the sun, causing the skin to become less supple and more prone to wrinkling.) The dermis also contains blood vessels that nourish the skin. When blood rushes into these vessels, a person blushes, and when there is minimal blood in them, the skin turns "blue."

Sensory receptors are specialized free nerve endings in the dermis that respond to external stimuli. There are sensory receptors for touch, pressure, pain, and temperature. The fingertips contain the most touch receptors, and these add to our ability to use our fingers for delicate tasks.

The **subcutaneous layer** is composed of loose connective tissue and adipose tissue, which stores fat. Fat is a stored source of energy in the body. Adipose tissue also helps thermally insulate the body from either gaining heat from the outside or losing heat from the inside.

Warts are small areas of skin proliferation caused by the human papillomavirus. Though they can occur at any age, nongenital warts most commonly occur between the ages of 12 and 16. Though sometimes embarrassing, these are generally harmless and disappear without treatment—more than half do so within two years. Warts that cause cosmetic disfigurement or are painful (such as plantar warts on the foot) can be surgically removed, frozen with liquid nitrogen, or treated with various pharmaceutical compounds. One example is cantharidin, an extract from the blister beetle. Genital warts are discussed in Chapter 21.

Accessory Organs of the Skin

Nails, hair, and glands are structures of epidermal origin, even though some parts of hair and glands are largely found in the dermis.

Nails are a protective covering of the distal part of the digits. Nails can help pry things open or pick up small objects. Nails grow from special epithelial cells at the base of the nail in the portion called the nail root. These cells become keratinized as they grow out over the nail bed. The visible portion of the nail is called the nail body. The cuticle is a fold of skin that hides the nail root. The whitish color of the half-moon-shaped base, or lunula, results from the thick layer of cells in this area (Fig. 11.10).

Hair follicles are in the dermis and continue through the epidermis where the hair shaft extends beyond the skin. Epidermal cells form the root of hair, and their division causes a hair to grow. The cells become keratinized and dead as they are pushed farther from the root. Interestingly, chemical substances in the body such as illicit drugs and byproducts of alcohol metabolism are incorporated into growing hair shafts, where they can be detected by laboratory tests (see the Bioethical Focus, page 210).

When we are scared or cold, contraction of the arrector pili muscles attached to hair follicles may cause the hairs to "stand on end" and goose bumps to develop. The purpose of this phenomenon in humans is unknown, although it may help fur-covered mammals to look larger, as well as to keep warm by trapping air within the fur. The color of hair is due to the variable presence of different forms of melanin. In general, the more melanin, the darker the hair. A decreasing

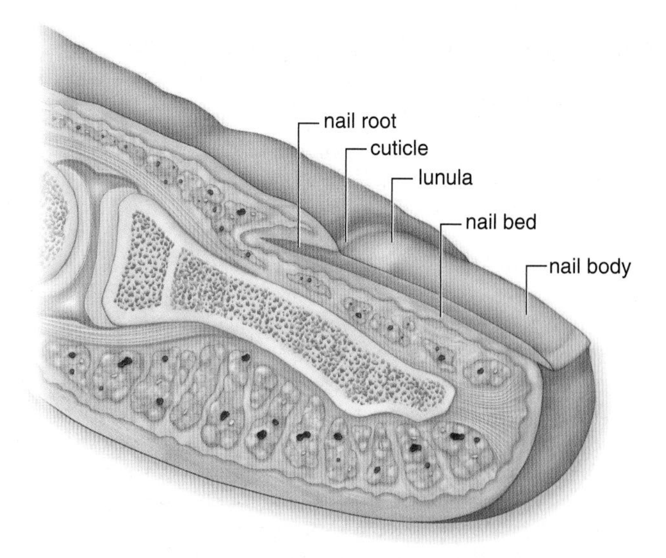

Figure 11.10 Nail anatomy.
Cells produced by the nail root become keratinized, forming the nail body.

melanin content with age results in the various shades of gray hair. White hair contains little or no melanin. Hair loss is most often age-related, but also may occur following many types of illness, especially of the endocrine system (see Chapter 20).

Each hair follicle has one or more **oil glands,** which secrete sebum, an oily substance that lubricates the hair within the follicle and the skin itself. If the oil glands fail to discharge, the secretions collect and form "whiteheads" or "blackheads." The color of blackheads is due to oxidized sebum. Acne is an inflammation of the oil glands that most often occurs during adolescence. Hormonal changes during this time cause the oil glands to become more active, especially in those areas that have the highest density of oil glands, such as the face, chest, shoulders, and back. Most medications for acne work by drying out and unclogging blocked pores and/or by killing bacteria that can infect the oil glands. In 2005, the FDA approved Zeno, a hand-held medical device that kills bacteria that cause some types of acne by heating the skin to 120 degrees for 2.5 minutes.

The skin of an average adult also has about 2 to 4 million **sweat glands.** A sweat gland is a coiled tubule within the dermis that straightens out near its opening, or pore. Some sweat glands open into hair follicles, but most open onto the surface of the skin. Sweat glands play a role in modifying body temperature. When body temperature starts to rise, sweat glands become active. Sweat absorbs body heat as it evaporates. Once body temperature lowers, sweat glands are no longer active.

Check Your Progress

1. Contrast the structure and function of skin with that of the mucous and serous membranes described in section 11.2.
2. List three accessory organs of the skin, and discuss the major function(s) of each.

UV Rays: Too Much Exposure or Too Little?

The sun is the major source of energy for life on Earth. Without the sun, most organisms would quickly die out. But the sun's energy can also be damaging. In addition to visible light, the sun emits ultraviolet (UV) radiation, which has a shorter wavelength (and thus a higher energy) than visible light. Based on wavelength, UV radiation can be grouped into three types: UVA, UVB, and UVC, but only UVA and UVB reach the Earth's surface. Scientists have developed a UV index to determine how powerful the solar rays are in different U.S. cities. In general, the more southern the city, the higher the UV index and the greater the risk of skin cancer. Maps indicating the daily risk levels for various U.S. regions can be viewed at http://www.weather.gov/os/uv/.

Exposure to UV radiation can damage skin cells, leading to the pain and redness characteristic of a sunburn. Tanning occurs when melanin granules increase in keratinized cells at the surface of the skin as a way to prevent further damage by UV rays. For many years, it was believed that most of the damaging effects of natural sun exposure, including sunburn, cataracts, and some forms of skin cancer, were caused by UVB rays. However, recent studies have elevated awareness that UVA rays, which penetrate more deeply into the skin, can induce premature aging of the skin, as well as a particular type of skin cancer called (malignant) melanoma. Both types of rays can induce skin cancer by damaging DNA present in skin cells, resulting in mutations that can cause the cell to divide at an abnormally high rate. UVA rays have also been shown to suppress the function of immune system cells residing in the skin, which may help cancer cells to avoid destruction once they appear.

It is estimated that more than 1 million Americans will be diagnosed with skin cancer in 2008, making it the most commonly diagnosed type of cancer in the United States. The most deadly form is melanoma, which occurs in adolescents and young adults as well as in older people. If detected early, over 95% of patients survive at least five years, but if the cancer cells have already spread throughout the body, only 10% to 20% can expect to live this long. Melanoma is most common in fair-skinned persons, particularly if they had several severe sunburns as children. It affects pigmented cells and often has the appearance of an unusual mole. But unlike a mole that is circular and confined, melanoma moles often look like spilled ink spots (Fig. 11A). Any moles that become malignant are removed surgically. If the cancer has spread, chemotherapy and a number of other treatments are also available. In March 2007, the USDA approved a new vaccine to treat melanoma in dogs, which is the first time a vaccine has been approved as a treatment for any cancer in animals or humans. Clinical studies of the canine vaccine, conducted at the Animal Medical Center in New York City, demonstrated significantly improved survival, even in dogs with advanced disease. Similar trials are under way to test such a vaccine for use in humans.

So how can we protect ourselves against damage from the sun? Sunscreens generally do a better job of blocking UVB than UVA rays. In fact, the SPF, or sun protection factor printed on sunscreen labels, refers only to the degree of protection against UVB. Most sunscreens don't provide as much protection against UVA, and people may have a false sense of security because they are avoiding sunburn, so they spend more time in the sun. Some sunscreens do a better job of blocking UVA—look for those that contain zinc oxide, titanium dioxide, avobenzone, or Mexoryl SX. Because no sunscreen is 100% effective at blocking all UV rays, other recommended measures include minimizing sun exposure between 10 A.M. and 4 P.M. (when the UV rays are most intense), and wearing protective clothing, hats, and sunglasses. Interestingly, tanning salons use lamps that emit UVA rays that are two to three times more powerful than the UVA rays emitted by the sun. Because of the potential damage to deeper layers of skin, most medical experts recommend avoiding indoor tanning salons altogether.

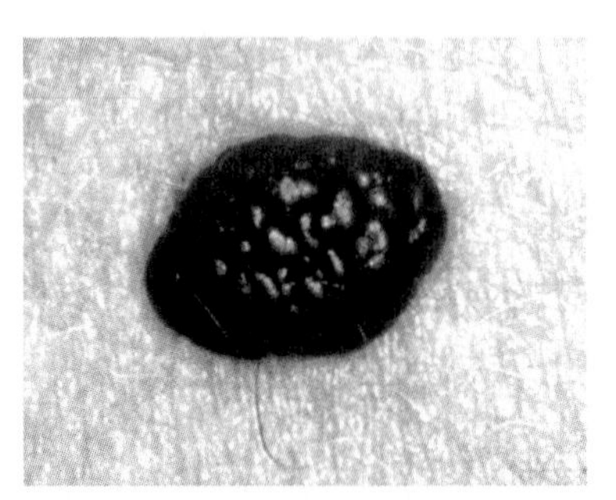
a. Normal mole

b. Melanoma

Figure 11A Skin cancer.
a. Normal mole tends to be symmetrical, with even edges and color. **b.** Malignant melanoma results from a proliferation of pigmented cells. Warning signs include a change in the shape, size or color of a normal mole, as well as itching, tenderness, or pain.

So if UV light is potentially damaging, why haven't all humans developed the more protective dark skin that is common to humans living in tropical regions? It turns out that vitamin D is only produced in the body when UVB rays interact with a form of cholesterol found mainly in the skin. This "sunshine vitamin" serves several important functions in the body, including keeping bones strong, boosting the immune system, and reducing blood pressure. Certain foods also contain vitamin D, but it can be difficult to obtain sufficient amounts through diet alone. Therefore, in more temperate areas of the planet, lighter-skinned individuals have the advantage of being able to synthesize sufficient vitamin D. Interestingly, dark-skinned people living in such regions may be at increased risk for vitamin D deficiency. A 2006 study published in the journal *Science* made the case that people of African descent living in the United States might be more susceptible to tuberculosis because of their lower vitamin D levels.

So how much sun exposure is enough in temperate regions of the world? During the summer months, an average fair-skinned person will synthesize plenty of vitamin D after exposure to 10–15 minutes of midday sun. During winter months, however, anyone living north of Atlanta probably receives too few UV rays to stimulate vitamin D synthesis, and therefore they must fulfill their requirement through their diet (see section 14.4 for more specific dietary recommendations).

Discussion Questions

1. Is it hard for you to imagine that pale, fair skin was once considered the most desirable look for Caucasian women? Why do you think tanning has become so popular?
2. In addition to the increased risk of cancer, skin damage from sun exposure is known to make the skin age faster and become more wrinkled. Which is more important to you—having a "healthy-looking" tan now, or preserving your skin's health later in life? Why?
3. If caught early, melanoma is rarely fatal. What are some factors that could delay detection of a melanoma?

11.5 Homeostasis

Learning Outcomes

1. Define homeostasis, and describe why it is essential to living organisms.
2. Distinguish between positive and negative feedback mechanisms, and list specific examples of each in the human body.
3. Discuss the roles of organ systems in maintaining homeostasis, and the health consequences if homeostasis is disrupted.

Homeostasis is the maintenance of a relatively constant internal environment by an organism, or even by a single cell. Even though external conditions may change dramatically, internal conditions stay within a narrow range. For example, regardless of how cold or hot it gets, the temperature of the body stays around 98.6°F (37°C). Even if you consume an acidic food such as yogurt (with a pH around 4.5), the pH of your blood is usually about 7.4, and even if you eat a candy bar, the amount of sugar in your blood stays between 0.05% and 0.08%.

It is important to realize that internal conditions are not absolutely constant. They tend to fluctuate above and below a particular value. Therefore, the internal state of the body is often described as one of *dynamic* equilibrium. If internal conditions change to any great degree, illness results. This makes the study of homeostatic mechanisms medically important.

Negative Feedback

Negative feedback is the primary homeostatic mechanism that keeps a variable close to a particular value, or set point. A homeostatic mechanism has at least two components: a sensor and a control center (Fig. 11.11). The sensor detects a change in internal conditions. The control center then directs a response that brings conditions back to normal again.

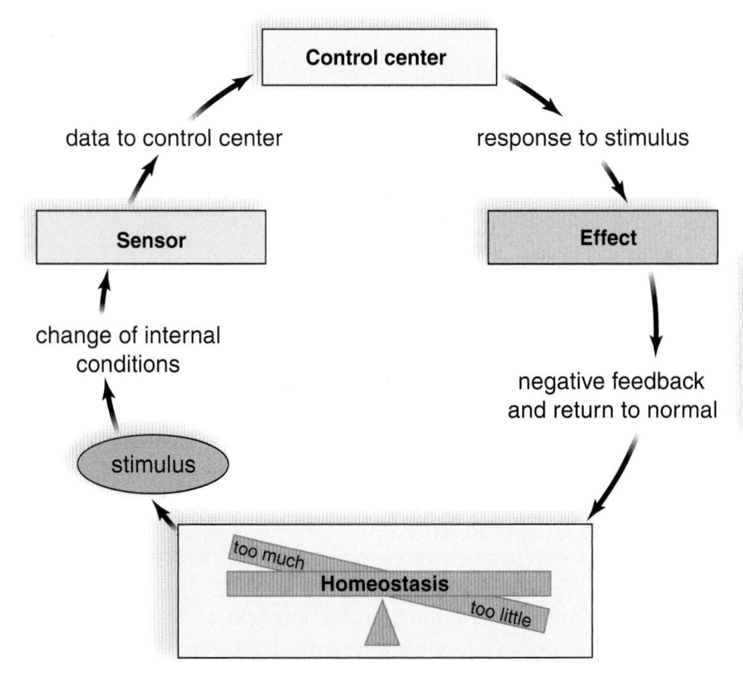

Figure 11.11 Negative feedback mechanism. The sensor and control center of a feedback mechanism work together to keep a variable close to a particular value.

Now, the sensor is no longer activated. In other words, a negative feedback mechanism is present when the output of the system dampens the original stimulus.

Let's take a simple example. When the pancreas detects that the blood glucose level is too high, it secretes insulin, a hormone that causes cells to take up glucose. Now the blood sugar level returns to normal, and the pancreas is no longer stimulated to secrete insulin.

Mechanical Example

A home heating system is often used to illustrate how a more complicated negative feedback mechanism works (Fig. 11.12).

You set the thermostat at, say, 68°F. This is the *set point*. The thermostat contains a thermometer, a sensor that detects

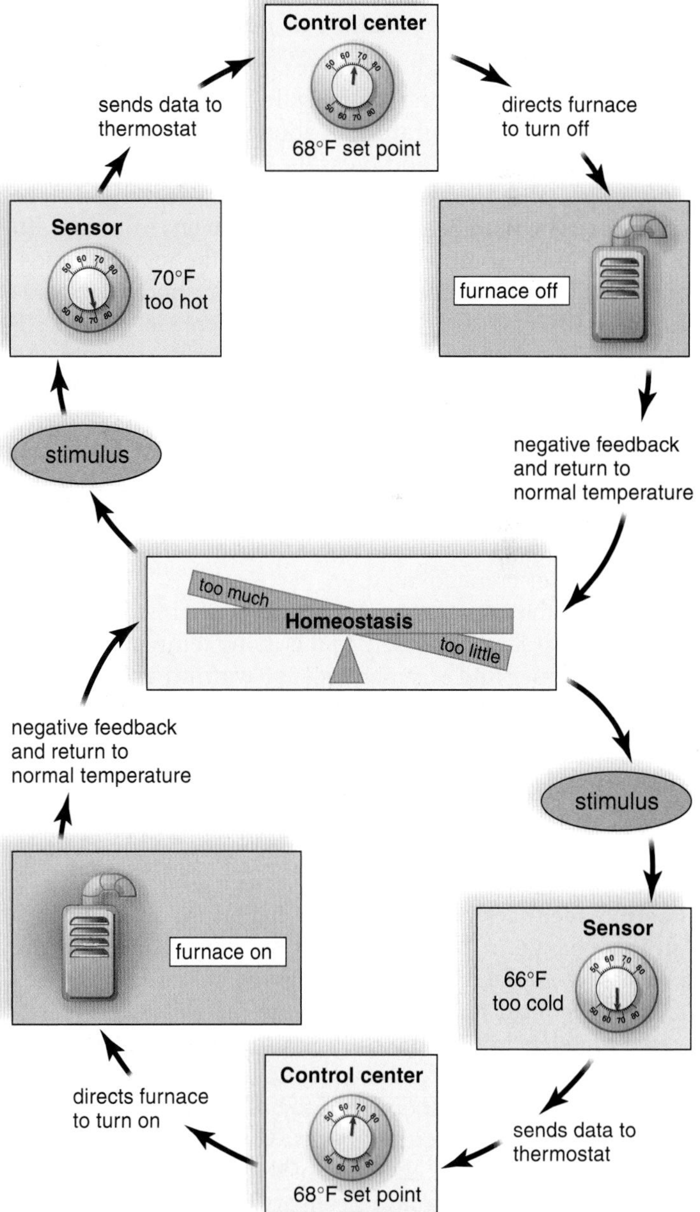

Figure 11.12 Complex negative feedback mechanism. When a room becomes too warm, negative feedback allows the temperature to return to normal. A contrary cycle, in which the furnace turns on and gives off heat, returns the room temperature to normal when the room becomes too cool.

when the room temperature is above or below the set point. The thermostat also contains a control center. It turns the furnace off when the room is too hot and turns it on when the room is too cold. When the furnace is off, the room cools, and when the furnace is on, the room warms. In other words, typical of negative feedback mechanisms, there is a fluctuation above and below normal.

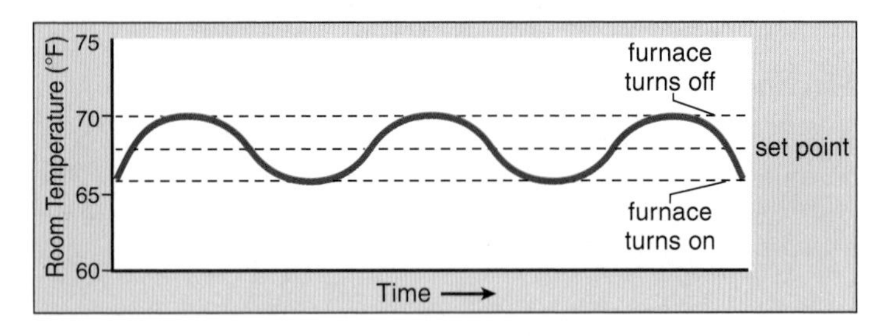

Human Example: Regulation of Body Temperature

The sensor and control center for body temperature are located in a part of the brain called the hypothalamus. When the body temperature is above normal, the control center directs the blood vessels of the skin to dilate (Fig. 11.13). This allows more blood to flow near the surface of the body, where heat can be lost to the environment. In addition, the nervous system activates the sweat glands, and the evaporation of sweat helps lower body temperature. Gradually, body temperature decreases to 98.6°F (37°C). When the body temperature falls below normal, the control center (via nerve impulses) directs the blood vessels of the skin to constrict. This conserves heat. If body temperature falls even lower, the control center sends nerve impulses to the skeletal muscles, and shivering occurs. Shivering generates heat, and gradually body temperature rises to 98.6°F. When the temperature rises to normal, the control center is inactivated.

Notice that a negative feedback mechanism prevents change in the same direction; that is, body temperature does not get warmer and warmer because warmth brings about a change toward a lower body temperature. Also, body temperature does not get colder and colder because a body temperature below normal brings about a change toward a warmer body temperature.

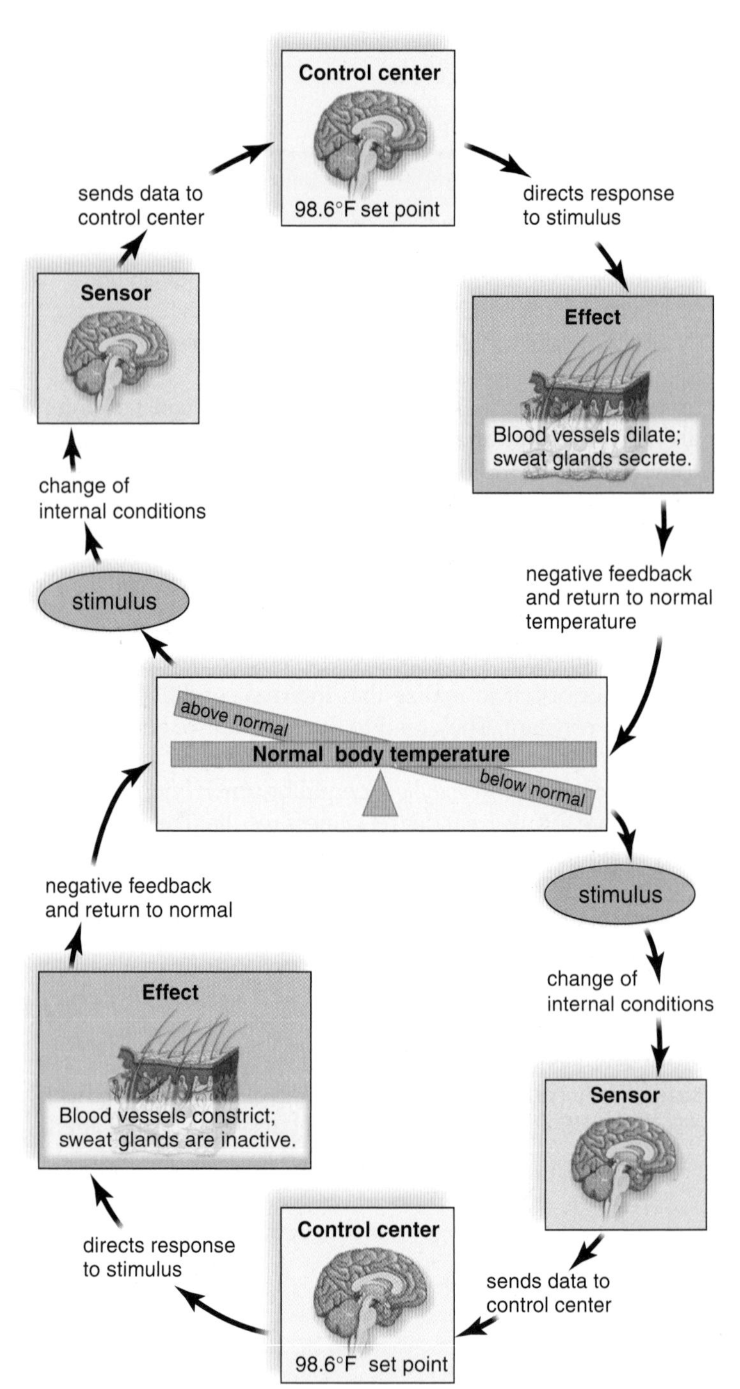

Figure 11.13 Regulation of body temperature.
Normal body temperature is maintained by a negative feedback system.

Positive Feedback

Positive feedback is a mechanism that brings about an ever greater change in the same direction. One example is the process of blood clotting, during which injured tissues release chemical factors that activate platelets. These activated platelets initiate the clotting process, and also release factors that stimulate further clotting (see section 12.4).

Positive feedback loops tend to be involved in processes that have a definite cutoff point. Consider that when a woman is giving birth, the head of the baby begins to press against the cervix, stimulating sensory receptors there. When nerve impulses reach the brain, the brain causes the pituitary gland to secrete the hormone oxytocin. Oxytocin travels in the blood and causes the uterus to contract. As labor continues, the cervix is ever more stimulated, and uterine contractions become ever stronger until birth occurs.

Check Your Progress

1. What is the primary homeostatic mechanism that keeps most internal conditions in the body close to a set value?
2. How do positive feedback systems differ from those that use negative feedback?

Homeostasis and Body Systems

All systems of the body contribute toward maintaining homeostasis.

The Transport Systems

The cardiovascular system conducts blood to and away from capillaries, where exchange of gases, nutrients, and wastes occurs. Tissue fluid, which bathes all the cells of the body, is refreshed when molecules such as oxygen and nutrients move into tissue fluid from the blood and when carbon dioxide and wastes move from tissue fluid into the blood (Fig. 11.14).

The lymphatic system is an accessory to the cardiovascular system. Lymphatic capillaries collect excess tissue fluid, which is returned via lymphatic vessels to the cardiovascular system. Lymph nodes are sites where the immune system responds to invading microorganisms.

The Maintenance Systems

The respiratory system adds oxygen to and removes carbon dioxide from the blood. It also plays a role in regulating blood pH because removal of CO_2 causes the pH to rise, just as CO_2 retention helps to lower the pH. The digestive system takes in and digests food, providing nutrient molecules that enter the blood to replace the nutrients that are constantly being used by the body cells. The liver, an organ that assists the digestive process by producing bile, also plays a significant role in regulating blood composition. Immediately after glucose enters the blood, any excess is removed by the liver and stored as glycogen. Later, the glycogen can be broken down to replace the glucose used by the body cells. In this way, the glucose composition of the blood remains relatively constant. The liver also removes toxic chemicals, such as ingested alcohol and other drugs from the blood. The liver makes urea, a nitrogenous end product of protein metabolism. Urea and other metabolic waste molecules are excreted by the kidneys, which are a part of the urinary system. Urine formation by the kidneys is extremely critical to the body, not only because it rids the body of unwanted substances, but also because urine formation offers an opportunity to carefully regulate blood volume, salt balance, and pH.

The Support Systems

The integumentary, skeletal, and muscular systems protect the internal organs. In addition, the integumentary system produces vitamin D, while the skeletal system stores minerals and produces the blood cells.

The Control Systems

The nervous system and the endocrine system work together to control other body systems so that homeostasis is maintained. We have already seen that in negative feedback mechanisms, sensory receptors send nerve impulses to control centers in the brain, which then rapidly direct effectors to become active. Effectors can be muscles or glands. Muscles bring about an immediate change. Endocrine glands secrete hormones that bring about a slower, more lasting change that keeps the internal environment relatively stable.

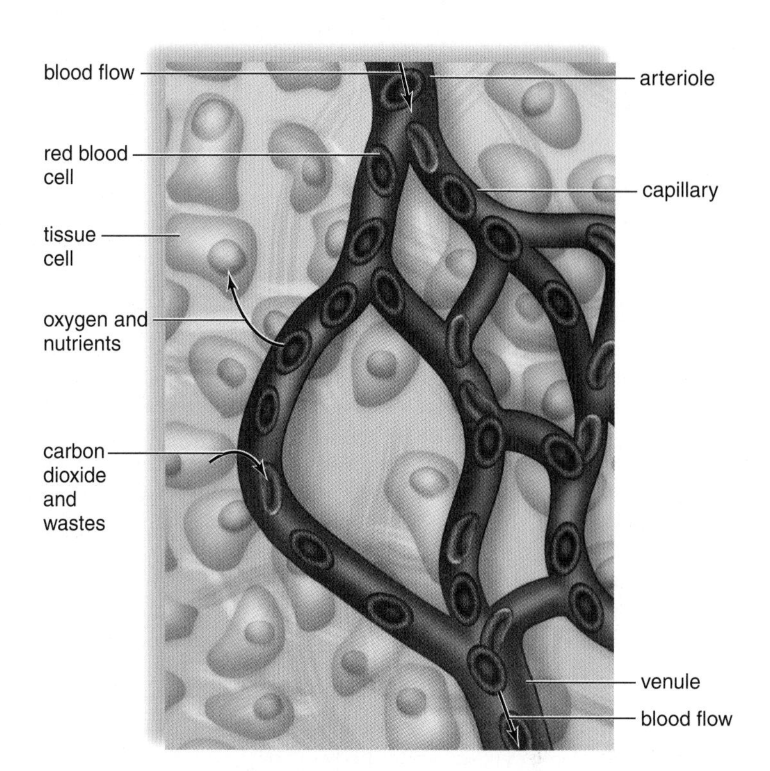

Figure 11.14 Regulation of tissue fluid composition.
Cells are surrounded by tissue fluid, which is continually refreshed because oxygen and nutrient molecules constantly exit, and carbon dioxide and waste molecules continually enter the bloodstream.

Disease

Disease is an abnormality in the body's normal processes that significantly impairs normal function. When homeostasis fails, the body (or part of the body) no longer functions properly. For example, different types of cancer can hinder the ability of virtually any body system to perform its normal homeostatic function. A malignant cancer often becomes **systemic,** meaning that it affects the entire body or at least several organ systems. Other diseases, including certain types of infections or degenerative diseases such as arthritis or Alzheimer disease, are more **localized** to a specific part of the body. Diseases may also be classified on the basis of their severity and duration. **Acute diseases,** such as poison ivy dermatitis or influenza, tend to occur suddenly and generally last a short time, although some can be life threatening. **Chronic diseases,** such as multiple sclerosis or AIDS, tend to develop slowly and last a long time, even for the rest of a person's life, unless an effective cure is available.

Check Your Progress

1. Describe how several body systems can interact to maintain homeostasis.
2. List several specific diseases that result when a particular body system fails to perform its function.

Bioethical Focus

Just a Snip, Please: Testing Hair for Drugs

Imagine you have just started your first real full-time job, with health insurance, a pension plan, and other benefits. The first day, after you settle into your cubicle, a manager comes by and asks for a sample of your hair. "We normally just snip a bit off in the back, where it will hardly show," he says, smiling. When you ask why, he replies that it is company policy to screen employees for drug use every six months.

Tests that detect illicit drugs or their metabolic products in urine or saliva are of limited value, because most of these chemicals are secreted by the body for only a few days. In contrast, many illicit drugs and/or their metabolites are incorporated into growing hair shafts throughout the body, where they remain indefinitely. Especially in the first 1.5 inches of hair growth from the scalp, each 0.5 inch is considered to represent 30 days' worth of growth (and thus, potential drug use). Hair from anywhere on the body can be used, although the growth of body hair is usually slower, so the time of any drug use cannot be determined. It generally takes four to five days from the time a drug is taken into the body until it begins to appear in hair.

Many commercial laboratories now offer hair testing. In general, these labs contend that they can distinguish between environmental exposure to a particular drug—from being in the vicinity of someone smoking marijuana, for example—and actual drug use by an individual. In recent years, several court decisions have supported the idea that hair testing can accurately distinguish actual drug use from such "passive" exposure. This has led to the marketing of several shampoos for cleaning or "detoxifying" the hair shafts, but these may not be effective. And even if they were, a lab could conceivably test hair for common contaminants expected to be found in everyone's hair. If these contaminants were not found, this could be used as evidence that a person has attempted to hide prior drug usage.

Even if one accepts that hair testing is an accurate way to prove that a person has used drugs, is it ethical for businesses to require their employees to undergo such tests? Does it matter what type of job a person has? For example, would it be more difficult to argue against mandatory drug testing for school bus drivers than for stockbrokers? And for those of you still living with your parents, how upset would you be to find out that your mom or dad had slipped into your bedroom at night, snipped off a small bit of hair from the back of your head, and mailed it to one of several companies that now offer hair testing to the general public?

Form Your Own Opinion

1. The Fourth Amendment to the U.S. Constitution guards against "unreasonable searches and seizures" by the government. Do you believe that the types of drug testing described here are unconstitutional?
2. In which of the following additional situations would you support mandatory drug testing: To test airline pilots for hallucinogens? To test high school athletes for steroids? To test NBA players for marijuana?
3. Could you imagine any circumstance where you might want to test your own children for drug use?

Applying the Concepts Revisited

Obviously the human body consists of an incredibly complex system of molecules, cells, tissues, and organs functioning together. As introduced in Chapter 2, each of these components is produced from information encoded in our genes, which are made of DNA.

1. How might a single gene product (i.e., a protein) specifically influence the development of each of the following human characteristics?
   a. skin color
   b. height
   c. weight
   d. male fertility
2. In what specific layer of the skin would melanin be found? Considering the function of melanin, why does it make biological sense that the melanocytes would be in this layer?

SUMMARIZING THE CONCEPTS

11.1 Types of Tissues

Human tissues are categorized into four groups:

- Epithelial tissue covers the body and lines its cavities. The different types of epithelial tissue (squamous, cuboidal, and columnar) can be simple, stratified, or pseudostratified and have cilia or microvilli. Epithelial cells sometimes form glands that secrete either into ducts or into the blood.
- Connective tissues, in which cells are separated by a matrix, often bind body parts together. Loose fibrous connective tissue supports epithelium and encloses organs. Dense fibrous connective tissue, such as that of tendons and ligaments, contains closely packed collagen fibers. Adipose tissue stores fat. Both cartilage and bone provide support for other tissues, but the matrix for cartilage is more flexible than that for bone, which contains calcium salts. Blood is a connective tissue in which the matrix is a liquid.
- Muscular tissue is of three types. Both skeletal and cardiac muscle are striated. Both cardiac and smooth muscle are involuntary. Skeletal muscle is found in muscles attached to bones, and smooth muscle is found in internal organs.
- Nervous tissue has one main type of conducting cell, the neuron, and several types of supporting neuroglia. Each neuron has a cell body, dendrites, and an axon. Axons are specialized to conduct nerve impulses.

11.2 Body Cavities and Body Membranes

The internal organs occur within cavities.

- The thoracic cavity contains the heart and lungs.
- The abdominal and pelvic cavities contain organs of the digestive, urinary, and reproductive systems, among others.

Four types of membranes line body cavities and the internal spaces of organs.

- Mucous membrane lines the tubes of the digestive system,
- Serous membrane lines the thoracic and abdominal cavities and covers the organs they contain.
- Synovial membranes line movable joint cavities.
- Meninges cover the brain and spinal cord.

11.3 Organ Systems

The organ systems of the human body can be grouped according to their major function(s):

- The digestive, cardiovascular, lymphatic, respiratory, and urinary systems perform processing and transport functions that maintain the normal conditions of the body.
- The immune system protects the body against infectious disease.
- The skeletal and muscular systems support the body and permit movement.
- The nervous system receives sensory input and directs muscles and glands to respond.
- The endocrine system produces hormones, some of which influence the functioning of the reproductive system.
- The integumentary system serves protective functions and also makes vitamin D, collects sensory data, and helps regu-late body temperature.

11.4 Integumentary System

- The integumentary system is composed of skin and the accessory organs (nails, hair, oil glands, and sweat glands).
- The main functions of the integumentary system are to protect underlying tissues while providing a barrier to pathogen invasion and water loss. Skin also contains sensory receptors.
- Human skin has two regions: (1) the epidermis contains basal cells that produce new epithelial cells, which become keratinized as they move toward the surface; and (2) the dermis, a largely fibrous connective tissue, contains epidermally derived glands and hair follicles, nerve endings, and blood vessels. Sensory receptors for touch, pressure, temperature, and pain are also in the dermis.

11.5 Homeostasis

- The body's internal environment consists of blood and tissue fluid.
- Homeostasis is the maintenance of a relatively constant internal environment, mainly by two mechanisms: (1) negative feedback mechanisms keep the environment relatively stable (when a sensor detects a change above or below a set point, a control center brings about an effect that reverses the change and brings conditions back to normal again); and (2) positive feedback mechanisms bring about rapid change in the same direction as the stimulus.

TESTING YOURSELF

Choose the best answer for each question.

1. A grouping of similar cells that perform a specific function is called a
   a. sarcoma. b. membrane. c. tissue. d. None of these are correct.
2. The smallest blood vessels are called
   a. veins. b. arteries. c. arterioles. d. capillaries.
3. Tight junctions are associated with
   a. connective tissue. b. adipose tissue. c. cartilage. d. epithelium.
4. A reduction in red blood cells would cause problems with
   a. fighting infection. b. carrying oxygen. c. blood clotting. d. None of these are correct.
5. Which choice is true of both cardiac and skeletal muscle?
   a. striated b. single nucleus per cell c. multinucleated cells d. involuntary control
6. The skeletal system functions in
   a. blood cell production. b. mineral storage. c. movement. d. All of these are correct.
7. Which system plays the biggest role in fluid balance?
   a. cardiovascular
   b. urinary
   c. digestive
   d. integumentary
8. Label the diagram of body cavities on the right.

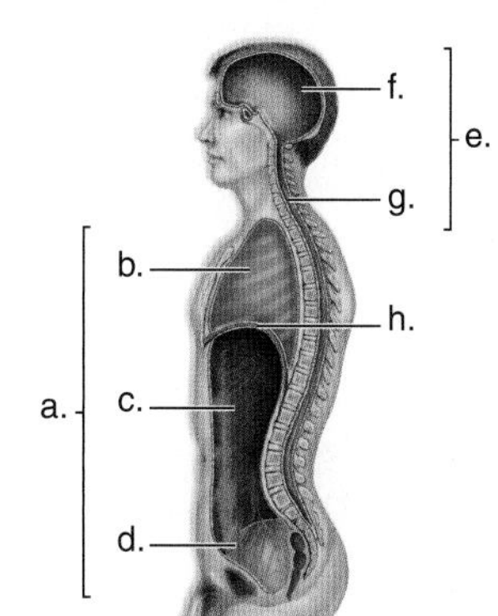

9. Which of the following is a function of skin?
   a. temperature regulation
   b. manufacture of vitamin D
   c. collection of sensory input
   d. protection from invading pathogens
   e. All of these are correct.
10. Which of these is involved in storing energy?
    a. epidermis b. dermis c. subcutaneous layer d. None of these are correct.
11. Which of these correctly describes a layer of the skin?
    a. The epidermis is simple squamous epithelium in which hair follicles develop and blood vessels expand when we are hot.
    b. The subcutaneous layer lies between the epidermis and the dermis. It contains adipose tissue, which keeps us warm.
    c. The dermis is a region of connective tissue that contains sensory receptors, nerve endings, and blood vessels.
    d. The skin has a special layer, still unnamed, in which there are all the accessory structures such as nails, hair, and various glands.
12. Without melanocytes, skin would
    a. be too thin.
    b. lack nerves.
    c. lack color.
    d. None of these are correct.
13. Label the diagram of human skin on the right.

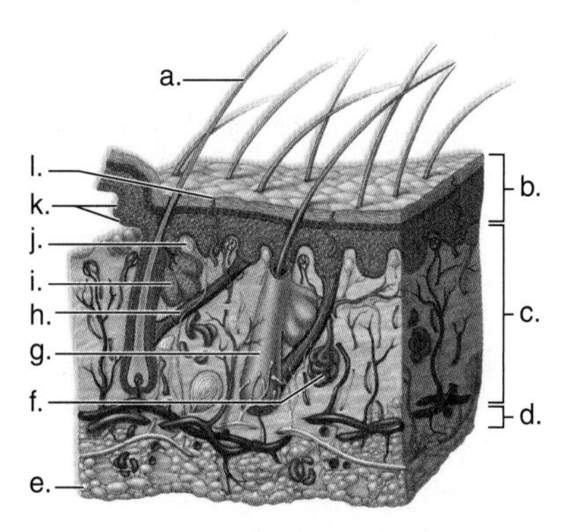

14. The correct order for homeostatic processing is
    a. sensory detection, control center, effect brings about change.
    b. control center, sensory detection, effect brings about change in environment.
    c. sensory detection, control center, effect causes no change in environment.
    d. None of these are correct.
15. Which allows rapid change in one direction and does not achieve stability?
    a. homeostasis
    b. positive feedback
    c. negative feedback
    d. All of these are correct.
16. Which of the following is an example of negative feedback?
    a. Air conditioning goes off when room temperature lowers.
    b. Insulin decreases blood sugar levels after eating a meal.
    c. Heart rate increases when blood pressure drops.
    d. All of these are examples of negative feedback.

UNDERSTANDING THE TERMS

acute disease 209
adhesion junction 196
adipose tissue 196
basement membrane 194
blood 198
bone 197
capillary 194
carcinoma 194
cardiac muscle 199
cardiovascular system 202
cartilage 196
chronic disease 209
coelom 201
collagen fiber 196
columnar epithelium 194
compact bone 197
connective tissue 196
cuboidal epithelium 194
dense fibrous connective tissue 196
dermis 204
digestive system 202
disease 209
elastic cartilage 197
elastic fiber 196
endocrine system 203
epidermis 204
epithelial tissue 194
fibroblast 196
fibrocartilage 197
gap junction 196
gland 194
hair follicle 205
homeostasis 207
hyaline cartilage 196
immune system 202
integumentary system 202
intercalated disk 199
lacuna 196
ligament 196
localized 209
loose fibrous connective tissue 196
lymphatic system 202
matrix 196
melanocyte 204
meninges 201
microvillus 194
mucous membrane 201
muscular system 203
muscular tissue 199
nail 205
negative feedback 207
nerve 200
nervous system 203
nervous tissue 200
neuroglia 200
neuron 200
oil gland 205
plasma 198
platelet 198
positive feedback 208
red blood cell 198
reproductive system 203
respiratory system 203
reticular connective tissue 196
reticular fiber 196
serous membrane 201
skeletal muscle 199
skeletal system 203
skin 204
smooth muscle 199
spongy bone 197
squamous epithelium 194
striated 199
subcutaneous layer 205
sweat gland 205
synovial membrane 201
systemic 209
tendon 196
tight junction 194
tissue 194
urinary system 203
wart 205
white blood cell 198

THINKING CRITICALLY

1. In what way(s) is blood like a tissue?
2. Which of these homeostatic mechanisms in the body are examples of positive feedback, and which are examples of negative feedback? Why?
    a. The adrenal glands produce epinephrine in response to a hormone produced by the pituitary gland in times of stress; the pituitary gland senses the epinephrine in the blood and stops producing the hormone.
    b. As the bladder fills with urine, pressure sensors send messages to the brain with increasing frequency signaling that the bladder must be emptied. The more the bladder fills, the more messages are sent.
    c. When you drink an excess of water, specialized cells in your brain as well as stretch receptors in your heart detect the increase in blood volume. Both signals are transmitted to the kidneys, which increase the production of urine.
3. Homeostatic systems in the body can be categorized as transport systems, maintenance systems, support systems, and control systems. What organ systems (from section 11.3) could be placed into more than one of these categories?

INQUIRY INTO LIFE WEBSITE

The companion website for *Inquiry into Life* provides a wealth of information organized and integrated by chapter. You will find practice tests, animations, videos, and much more that will complement your learning and understanding of general biology.

http://www.mhhe.com/maderinquiry13

12 Cardiovascular System

Chapter Outline

Applying the Concepts

On June 13, 2008, Tim Russert, the longtime host of NBC's *Meet The Press,* collapsed in his office. Co-workers called 911, and paramedics arrived four minutes later. After determining that 58-year-old Russert's heart wasn't beating, they began CPR and attempted to defibrillate his heart three times, with no success. Less than an hour later, the much-respected television personalilty and journalist was pronounced dead at a local hospital.

The cardiovascular system, which includes the heart and blood vessels, transports oxygen, nutrients, and wastes to and from the tissues. Diseases of this system, such as atherosclerosis, are a major cause of death in the more developed countries of the world. If you live in the United States, you have about a one in three chance of dying of heart disease, and if you add in all the other conditions that can affect the blood vessels, your odds are greater than 50:50 of eventually developing cardiovascular disease.

Tim Russert was no doubt aware that he needed to be vigilant about cardiovascular disease; he was on cholesterol-lowering drugs and had passed a heart stress test just a few weeks before his death. And yet his heart ceased functioning at a relatively young age. Even more alarming, two out of three cardiovascular-related deaths occur without any prior diagnosis. However, there are ways to reduce your risk of cardiovascular disease, and these are discussed later in this chapter.

12.1 The Blood Vessels

Learning Outcomes

1. List the three main types of blood vessels, their structural features, and their major functions.
2. Identify the type of blood vessel in which exchange takes place. Explain what is being "exchanged" and why.

The cardiovascular system has three types of blood vessels: the **arteries** (and arterioles), which carry blood away from the heart to the capillaries; the **capillaries,** which permit exchange of material with the tissues; and the **veins** (and venules), which return blood from the capillaries to the heart.

Like other tissues, the blood vessels require oxygen and nutrients, and therefore the larger ones have blood vessels in their own walls.

The Arteries

An arterial wall has three layers (Fig. 12.1*a*). The inner layer is a simple squamous epithelium called **endothelium** with a connective tissue basement membrane that contains elastic fibers. The middle layer is the thickest layer and consists of smooth muscle that can contract to regulate blood flow and blood pressure. The outer layer is fibrous connective tissue near the middle layer, but it becomes loose connective tissue at its periphery. The largest artery in the human body is the aorta. It is approximately 25 mm wide and carries O_2-rich blood from the heart to other parts of the body.

Smaller arteries branch off from the aorta, eventually forming a large number of arterioles. **Arterioles** are small arteries just visible to the naked eye, averaging about 0.5 mm in diameter. The inner layer of arterioles is endothelium. The middle layer is composed of some elastic tissue but mostly of smooth muscle with fibers that encircle the arteriole. When these muscle fibers are contracted, the vessel has a smaller diameter (is constricted). When these muscle fibers are relaxed, the vessel has a larger diameter (is dilated). Whether arterioles are constricted or dilated affects blood pressure. The greater the number of vessels dilated, the lower the blood pressure.

As we will see in section 12.5, diseases affecting the arteries, especially those supplying the heart and brain, are the leading cause of death in the United States.

The Capillaries

Capillaries join arterioles to venules (Fig. 12.1*b*). Capillaries are extremely narrow—only 8–10 μm wide—and have thin walls composed only of a single layer of endothelium with

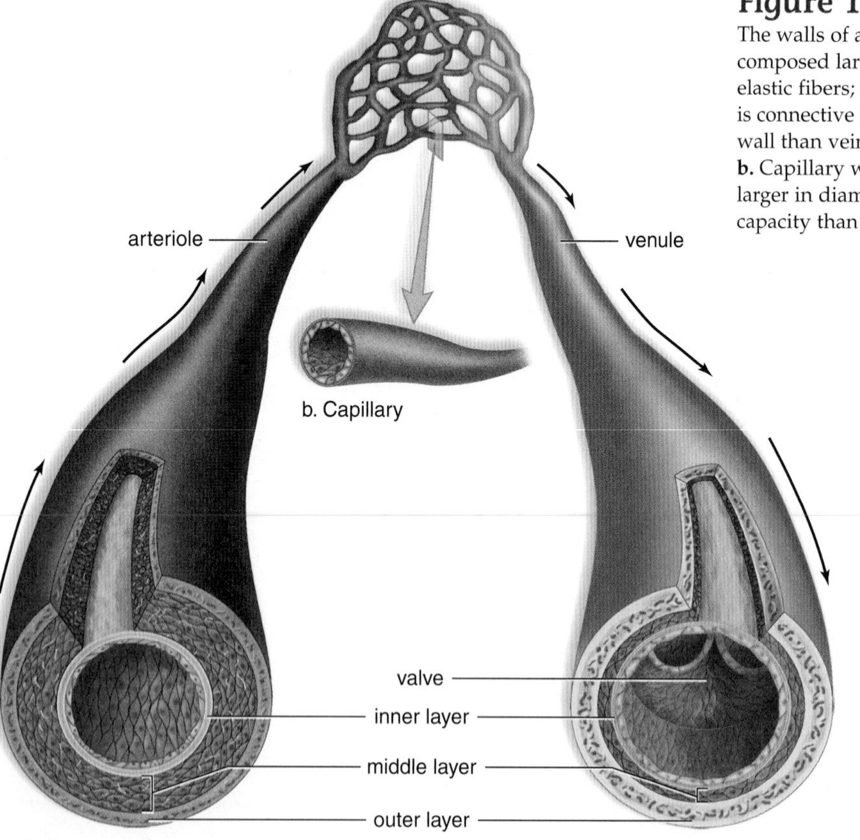

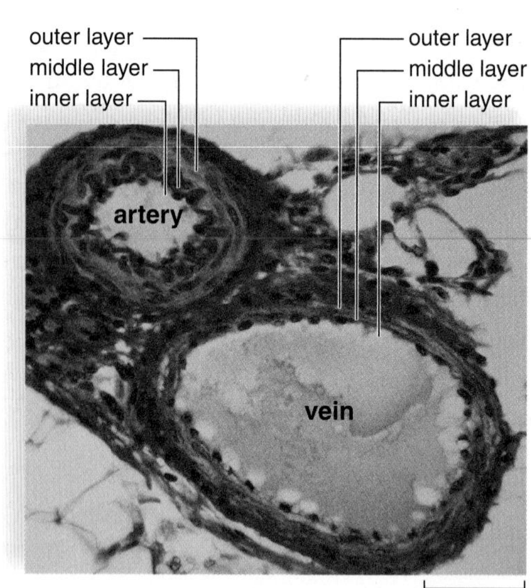

Figure 12.1 Blood vessels.
The walls of arteries and veins have three layers. The inner layer is composed largely of endothelium, with a basement membrane that has elastic fibers; the middle layer is smooth muscle tissue; the outer layer is connective tissue (largely collagen fibers). **a.** Arteries have a thicker wall than veins because they have a larger middle layer than veins. **b.** Capillary walls are one-cell-thick endothelium. **c.** Veins are generally larger in diameter than arteries, so collectively, veins have a larger holding capacity than arteries. **d.** Light micrograph of an artery and a vein.

a basement membrane. Although each capillary is small, they form vast networks. Their total surface area in a human is about 6,000 square meters (m^2). Capillary beds (networks of many capillaries) are present in nearly all regions of the body. Consequently, a cut to almost any body tissue draws blood. One region of the body that is nearly capillary-free is the cornea of the eye, so that light can pass through. Therefore, the cells of the cornea must obtain nutrients by diffusion from the tears on the outside surface, and from the aqueous humor on the inside surface (see Chapter 18).

Capillaries play a very important role in homeostasis because an exchange of substances takes place across their thin walls. Oxygen and nutrients, such as glucose, diffuse out of a capillary into the tissue fluid that surrounds cells. Wastes, such as carbon dioxide, diffuse into the capillary. Some water also leaves a capillary. Any excess is picked up by lymphatic vessels, as discussed in section 12.4 and in Chapter 13. The relative constancy of tissue fluid is absolutely dependent upon capillary exchange.

Because capillaries serve the cells, the heart and the other vessels of the cardiovascular system can be thought of as the means by which blood is conducted to and from the capillaries. Only certain capillary beds are completely open at any given time. For example, after eating, the capillary beds that serve the digestive system are mostly open, and those that serve the muscles are mostly closed. Each capillary bed has an arteriovenous shunt that allows blood to go directly from the arteriole to the venule, bypassing the bed (Fig. 12.2). Contracted precapillary sphincter muscles prevent the blood from entering the capillary vessels.

A decreased blood flow to the brain after meals has also been offered as an explanation for "postprandial somnolence," or the sleepiness that many people feel after eating. However, recent evidence suggests that the blood supply to the brain is maintained under most physiological conditions, including ingestion of a heavy meal, and that hormones released by the digestive tract may instead be the culprit.

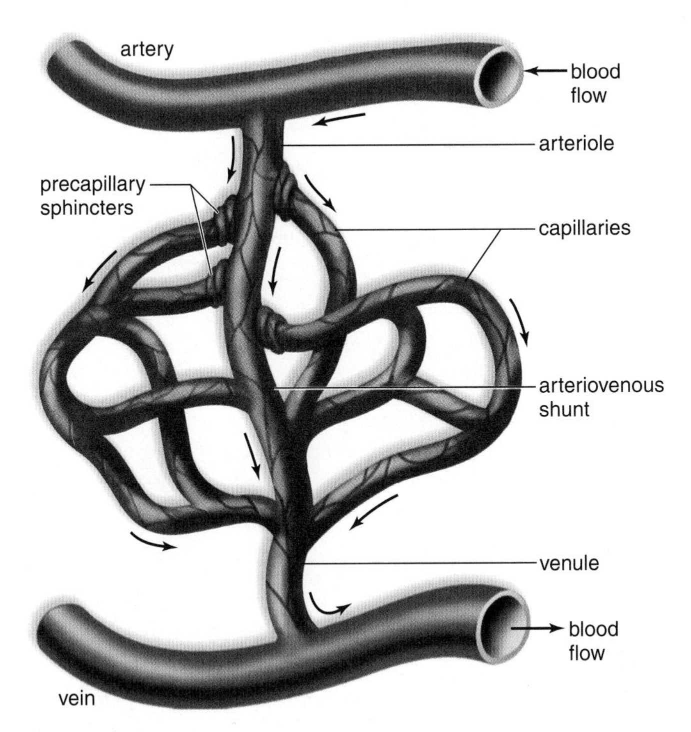

Figure 12.2 Anatomy of a capillary bed.
A capillary bed forms a maze of capillary vessels that lies between an arteriole and a venule. When precapillary sphincter muscles are relaxed, the capillary bed is open, and blood flows through the capillaries. When sphincter muscles are contracted, blood flows through a shunt that carries blood directly from an arteriole to a venule. As blood passes through a capillary in the tissues, it gives up its oxygen (O_2). Therefore, blood goes from being O_2-rich in the arteriole (red color) to being O_2-poor in the vein (blue color).

The Veins

Veins and venules take blood from the capillary beds to the heart. First, the **venules** (small veins) drain blood from the capillaries and then join to form a vein. The walls of veins (and venules) have the same three layers as arteries, but there is less smooth muscle and connective tissue (see Fig. 12.1*c*). Therefore, the wall of a vein is thinner than that of an artery.

Veins often have **valves,** which allow blood to flow only toward the heart when open and prevent blood from flowing backward when closed. Valves are found in the veins that carry blood against the force of gravity, especially the veins of the lower limbs. Unlike blood flow in the arteries and arterioles, which is kept moving by the pumping of the heart, blood flow in veins is primarily due to skeletal muscle contraction. If these valves become damaged by disease or through the normal wear-and-tear of aging, blood may begin pooling in the veins, causing them to enlarge and be visible as **varicose veins.** These most commonly occur in the lower legs of older individuals. Varicose veins of the anal canal are known as hemorrhoids.

Because the walls of veins are thinner, they can expand to a greater extent (see Fig. 12.1*d*). At any one time, about 70% of the blood is in the veins. In this way, the veins act as a blood reservoir. If blood is lost due to hemorrhaging, nervous stimulation causes the veins to constrict, providing more blood to the rest of the body. The largest veins in the body are the superior vena cava (20 mm wide) and the inferior vena cava (35 mm wide). These veins deliver O_2-poor blood into the heart.

Check Your Progress

1. How is blood flow controlled in each of the three major types of blood vessels?
2. How do O_2 and CO_2 move across capillary walls?

12.2 The Human Heart

Learning Outcomes

1. Name the major components of the heart, including the four chambers and four valves.
2. Trace the path of blood through the heart and lungs.
3. Describe the intrinsic and extrinsic control of the heartbeat.

The **heart** is a cone-shaped, muscular organ about the size of a fist (Fig. 12.3). It is located between the lungs directly behind the sternum (breastbone) and is tilted so that the apex (the pointed end) is oriented to the body's left. The major portion of the heart, called the **myocardium,** consists largely of cardiac muscle tissue. The muscle fibers of the myocardium are branched and tightly joined to one another. The heart lies within the **pericardium,** a thick, serous membrane that secretes a small quantity of lubricating liquid. The inner surface of the heart is lined with endocardium, a membrane composed of connective tissue and endothelial tissue.

Internally, a wall called the septum separates the heart into a right side and a left side (Fig. 12.4). The heart has four chambers. The two upper, thin-walled atria (sing., **atrium**) are located above the two lower, thick-walled **ventricles.** The ventricles pump the blood to the lungs and the body.

The heart also has four valves that direct the flow of blood and prevent its backward movement. The two valves that lie between the atria and the ventricles are called the **atrioventricular valves.** These valves are supported by strong fibrous strings called **chordae tendineae.** The chordae, which are attached to muscular projections of the ventricular walls, support the valves and prevent them from inverting when the heart contracts. The atrioventricular valve on the body's right side is called the tricuspid valve because it has three flaps, or cusps. The valve on the left side is called the bicuspid (or mitral) valve because it has two flaps. The remaining two valves, between the ventricles and their attached vessels, are the **semilunar valves,** whose flaps resemble half-moons. The pulmonary semilunar valve lies between the right ventricle and the pulmonary trunk. The aortic semilunar valve lies between the left ventricle and the aorta.

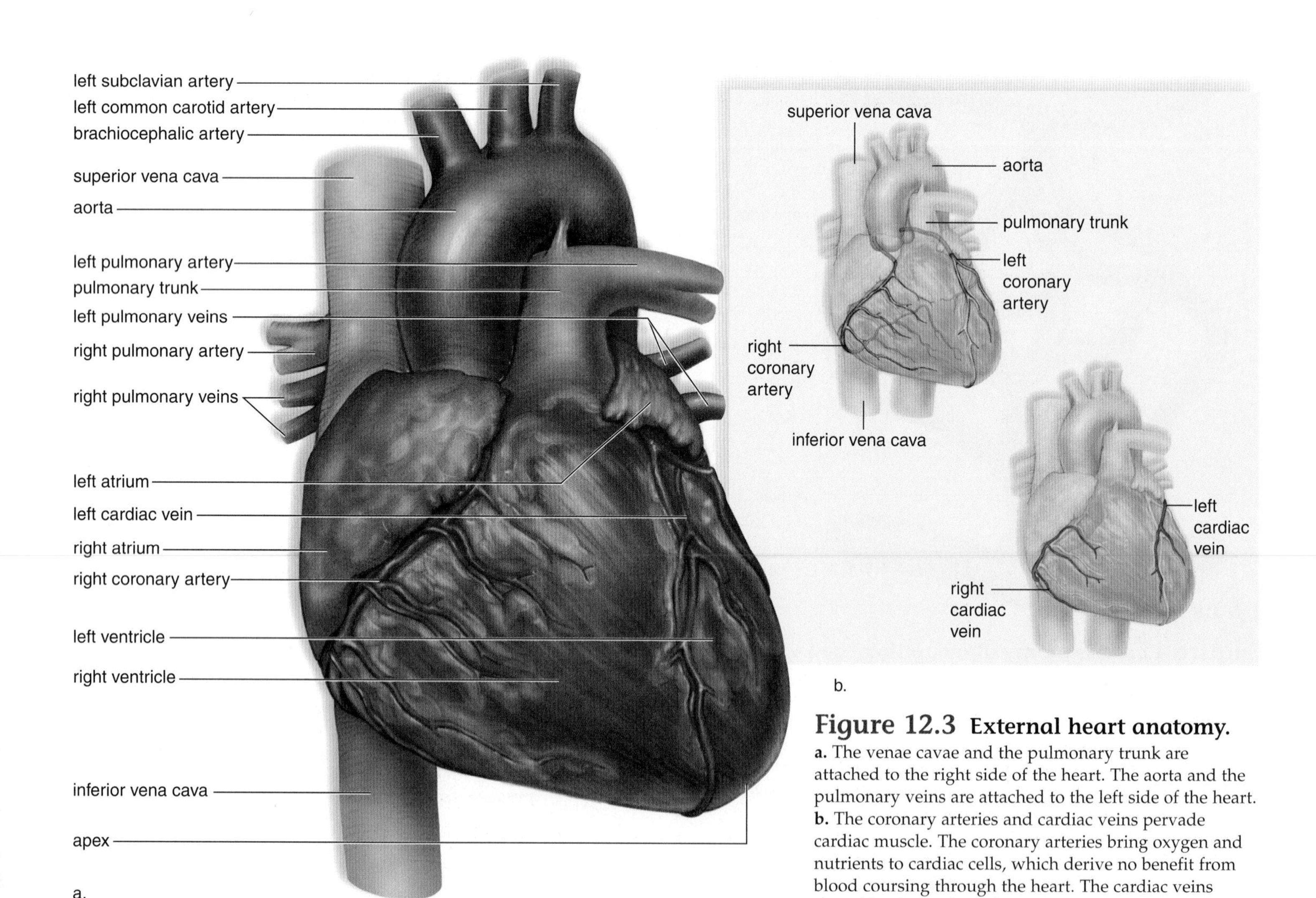

Figure 12.3 External heart anatomy. **a.** The venae cavae and the pulmonary trunk are attached to the right side of the heart. The aorta and the pulmonary veins are attached to the left side of the heart. **b.** The coronary arteries and cardiac veins pervade cardiac muscle. The coronary arteries bring oxygen and nutrients to cardiac cells, which derive no benefit from blood coursing through the heart. The cardiac veins drain blood into the right atrium.

Path of Blood Through the Heart

By referring to Figure 12.4, we can trace the path of blood through the heart in the following manner:

- The superior vena cava and the inferior vena cava, which carry O_2-poor blood that is relatively high in carbon dioxide, enter the right atrium.
- The right atrium sends blood through the tricuspid valve to the right ventricle.
- The right ventricle sends blood through the pulmonary semilunar valve into the pulmonary trunk and through the two **pulmonary arteries** to the lungs.
- Four **pulmonary veins,** which carry O_2-rich blood, enter the left atrium.
- The left atrium sends blood through the bicuspid (mitral) valve to the left ventricle.
- The left ventricle sends blood through the aortic semilunar valve into the aorta to the rest of the body.

From this description, it is obvious that O_2-poor blood never mixes with O_2-rich blood, and that blood must go through the lungs in order to pass from the right side to the left side of the heart. In fact, the heart is a double pump because the right ventricle sends blood into the lungs and the left ventricle sends blood into the rest of the body. Because the left ventricle has the harder job of pumping blood to the entire body, its walls are thicker than those of the right ventricle, which pumps blood a relatively short distance to the lungs. In a person with an average heart rate of 70 beats per minute, the output of the left ventricle is about 5.25 liters of blood per minute, which is about equal to the total amount of blood in the body.

The pumping of the heart sends blood out under pressure into the arteries. Because the left side of the heart is the stronger pump, blood pressure is greatest in the aorta. Blood pressure then decreases as the cross-sectional area of arteries and then arterioles increases. The **pulse** is a wave effect that passes down the walls of arteries when the aorta expands and then recoils with each ventricular contraction. The arterial pulse can be used to determine the heart rate, and a weak or "thready" pulse may indicate a weak heart or low blood pressure.

Check Your Progress

1. Name each blood vessel and heart chamber that blood passes through on its journey through the heart and lungs.
2. If the left ventricle was not able to pump blood properly, what effect might this have on the lungs?

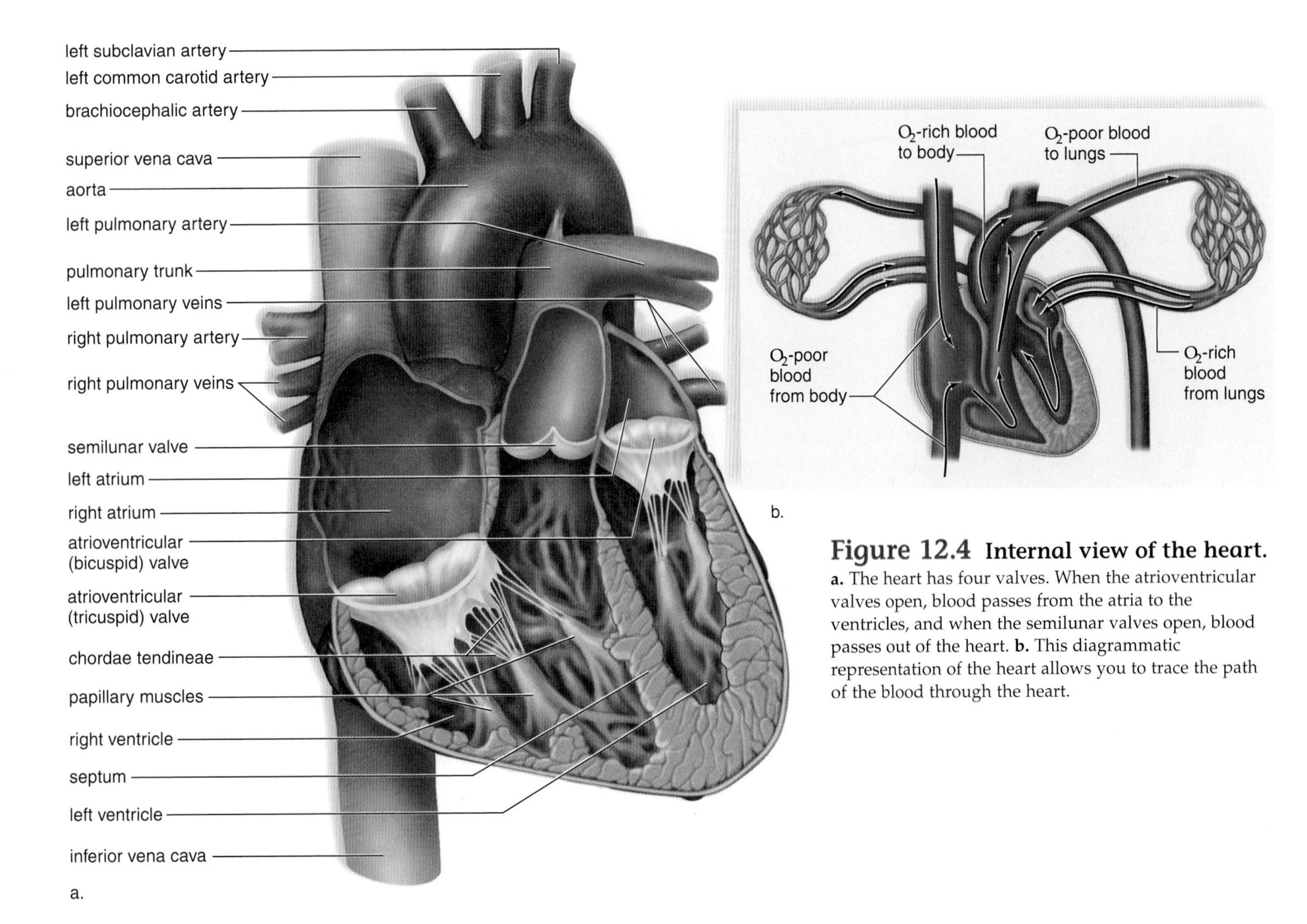

Figure 12.4 Internal view of the heart. **a.** The heart has four valves. When the atrioventricular valves open, blood passes from the atria to the ventricles, and when the semilunar valves open, blood passes out of the heart. **b.** This diagrammatic representation of the heart allows you to trace the path of the blood through the heart.

The Heartbeat

Each heartbeat is called a **cardiac cycle** (Fig. 12.5). When the heart beats, first the two atria contract at the same time; then the two ventricles contract at the same time. Then all of the chambers relax. The word **systole** refers to contraction of heart muscle, and the word **diastole** refers to relaxation of heart muscle. The heart contracts, or beats, about 70 times a minute, and each heartbeat lasts about 0.85 second, apportioned as follows:

Cardiac Cycle		
Time	**Atria**	**Ventricles**
0.15 sec	Systole	Diastole
0.30 sec	Diastole	Systole
0.40 sec	Diastole	Diastole

A normal adult rate at rest can vary from 60 to 80 beats per minute, which adds up to about 2.5 million beats in a lifetime!

When the heart beats, the familiar "lub-dup" sound occurs. The longer and lower-pitched "lub" is caused by vibrations occurring when the atrioventricular valves close due to ventricular contraction. The shorter and sharper "dup" results when the semilunar valves close due to back pressure of blood in the arteries. A heart murmur, or a slight whooshing sound after the "lub," is most commonly due to blood flowing back through an ineffective mitral valve (see section 12.5).

Intrinsic Control of Heartbeat

The rhythmic contraction of the atria and ventricles is due to the intrinsic (or internal) conduction system of the heart that is made possible by the presence of nodal tissue, a unique type of cardiac muscle. Nodal tissue, which has both muscular and nervous characteristics, is located in two regions of the heart. The **SA (sinoatrial) node** is located in the upper dorsal wall of the right atrium. The **AV (atrioventricular) node** is located in the base of the right atrium very near the septum (Fig. 12.6*a*). The SA node initiates the heartbeat and sends out an excitation impulse every 0.85 second. This causes the atria to contract. When impulses reach the AV node, a slight delay allows the atria to finish contraction before the ventricles begin to contract. The signal for the ventricles to contract travels from the AV node through specialized cardiac muscle fibers called the **atrioventricular bundle** (AV bundle) before reaching the numerous and smaller **Purkinje fibers.**

The SA node is called the **cardiac pacemaker** because it usually keeps the heartbeat regular. If the SA node fails to work properly, the heart still beats due to impulses generated by the AV node. But the beat is slower (40 to 60 beats

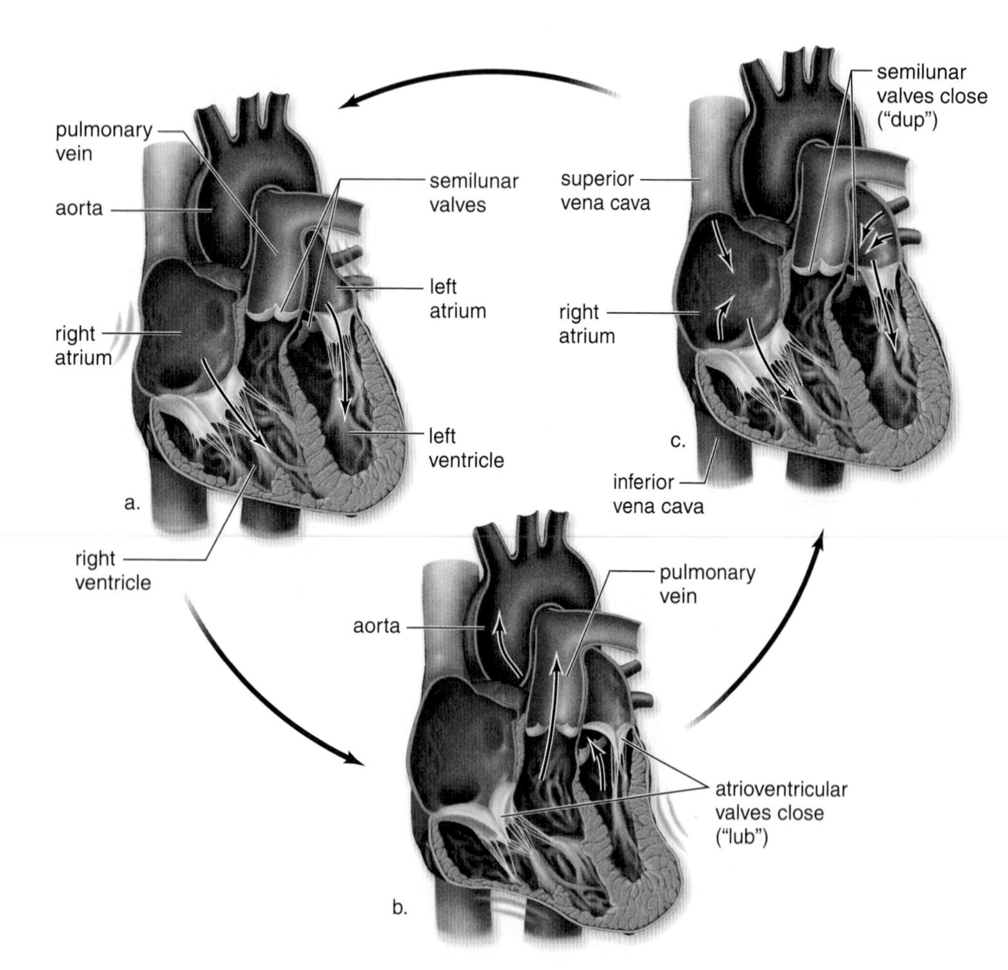

Figure 12.5 Generation of heart sounds during the cardiac cycle. **a.** When the atria contract, the ventricles are relaxed and filling with blood. **b.** When the ventricles contract, the atrioventricular valves close, preventing blood from flowing back into the atria and producing the "lub" sound of the heartbeat. **c.** After the ventricles contract, the "dup" sound of the heartbeat results from the closing of the semilunar valves to prevent arterial blood from flowing back into the ventricles.

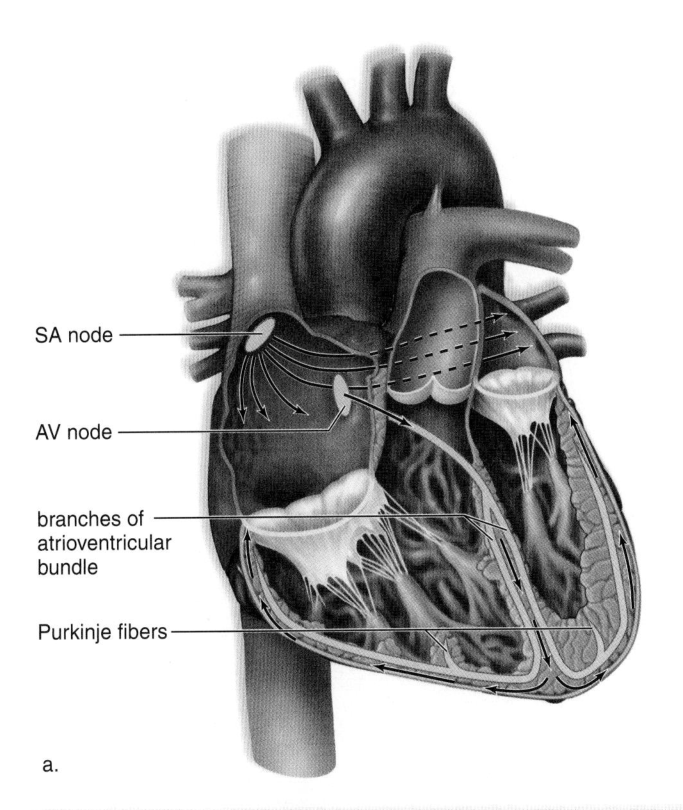

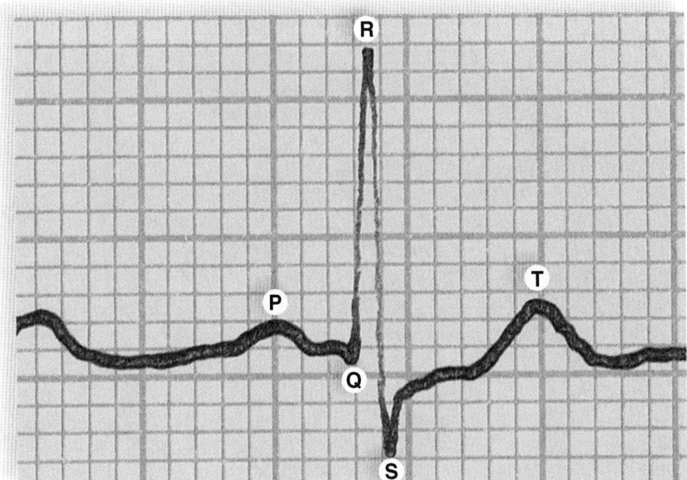

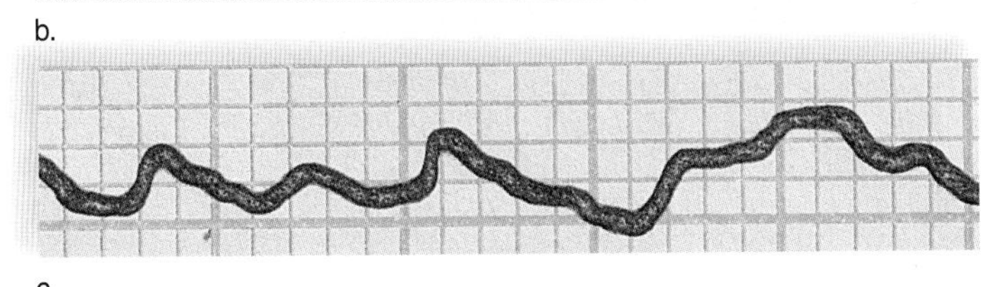

Figure 12.6 Conduction system of the heart.
a. The SA node sends out a stimulus (black arrows), which causes the atria to contract. When this stimulus reaches the AV node, it signals the ventricles to contract. Impulses pass down the two branches of the atrioventricular bundle to the Purkinje fibers, and thereafter the ventricles contract. **b.** A normal ECG usually indicates that the heart is functioning properly. The P wave occurs just prior to atrial contraction; the QRS complex occurs just prior to ventricular contraction; and the T wave occurs when the ventricles are recovering from contraction. **c.** Ventricular fibrillation produces an irregular electrocardiogram due to irregular stimulation of the ventricles.

per minute). To correct this condition, an artificial pacemaker may be implanted that gives an electrical stimulus to the heart every 0.85 second.

Extrinsic Control of Heartbeat

The body also has extrinsic (or external) ways to regulate the heartbeat. A cardiac control center in the medulla oblongata, a portion of the brain that controls internal organs, can alter the beat of the heart by way of the autonomic system, a portion of the nervous system. The autonomic system has two subdivisions: the parasympathetic division, which promotes functions of a resting state, and the sympathetic division, which brings about responses to increased activity or stress. The parasympathetic division decreases SA and AV nodal activity when we are inactive, and the sympathetic system increases these nodes' activity when we are active or excited.

The hormones epinephrine and norepinephrine, which are released by the adrenal medulla (inner portion of the adrenal gland), also stimulate the heart. When we are frightened, for example, the heart pumps faster and stronger due to sympathetic stimulation and because of the release of epinephrine and norepinephrine.

The Electrocardiogram

An **electrocardiogram (ECG)** is a recording of the electrical changes that occur in the myocardium during a cardiac cycle. Body fluids contain ions that conduct electrical currents, and therefore the electrical changes in the myocardium can be detected on the skin's surface. Electrodes placed on or near the chest are connected by wires to an instrument that detects and records the myocardium's electrical changes. Figure 12.6*b* depicts the electrical changes during a normal cardiac cycle.

When the SA node triggers an impulse, the atrial fibers produce an electrical change called the P wave. The P wave indicates that the atria are about to contract. After that, the QRS complex signals that the ventricles are about to contract. The electrical changes that occur as the ventricular muscle fibers recover produce the T wave.

Various types of abnormalities can be detected by an electrocardiogram. One of these, called ventricular fibrillation, causes uncoordinated contraction of the ventricles (Fig. 12.6*c*). Ventricular fibrillation is of special interest because it can be caused by an injury or drug overdose. Once the ventricles are fibrillating, they must be defibrillated by applying a strong electrical current for a short period of time. Then the SA node may be able to reestablish a coordinated beat. Many public places now have automatic external defibrillators (AEDs), which are small devices that can be used to determine whether a person is suffering from ventricular fibrillation—and if so, administer an appropriate electrical shock to the chest.

Check Your Progress

1. What specifically causes the sounds of the heartbeat?
2. Why is it important for the speed and strength of heart contractions to be regulated both intrinsically and extrinsically?

12.3 The Vascular Pathways

Learning Outcomes

1. Construct a simple diagram showing the flow of blood from the heart through all major parts of the body.
2. Describe the factors that affect blood pressure in arteries, capillaries, and veins.

The cardiovascular system includes two circuits (Fig. 12.7). The **pulmonary circuit** circulates blood through the lungs, and the **systemic circuit** serves the needs of body tissues.

The Pulmonary Circuit

Blood from all regions of the body first collects in the right atrium and then passes into the right ventricle, which pumps it into the pulmonary trunk. The pulmonary trunk divides into the right and left pulmonary arteries, which branch as they approach the lungs. The arterioles take blood to the pulmonary capillaries, where gas exchange occurs. Blood then passes through the pulmonary venules, which lead to the four pulmonary veins (two from each lung) that enter the left atrium. Because blood in the pulmonary arteries is O_2-poor but blood in the pulmonary veins is O_2-rich, it is not correct to say that all arteries carry blood high in oxygen and all veins carry blood low in oxygen. It is just the reverse in the pulmonary circuit.

The Systemic Circuit

The systemic circuit includes the major arteries and veins shown in Figure 12.8. The largest artery in the systemic circuit is the **aorta,** and the largest veins are the **superior vena cava** and the **inferior vena cava.** The superior vena cava collects blood from the head, the chest, and the arms, and the inferior vena cava collects blood from the lower body regions. Both enter the right atrium. The aorta and the venae cavae serve as the major pathways for blood in the systemic circuit.

The path of systemic blood to any organ in the body begins in the left ventricle. For example, trace the path of blood to and from the legs in Figure 12.8:

left ventricle—aorta—common iliac artery—femoral artery—leg capillaries—femoral vein—common iliac vein—inferior vena cava—right atrium

Notice that, when tracing blood, you need only mention the aorta, the proper branch of the aorta, the region, and the vein returning blood to the vena cava. In most instances, the artery and the vein that serve the same region are given the same name (Fig. 12.8).

The **coronary arteries** (see Fig. 12.3) serve the heart muscle itself. (The heart is not nourished by the blood in its own chambers.) The coronary arteries are the first branches off the aorta. They originate just above the aortic semilunar valve, and they lie on the exterior surface of the heart, where they divide into diverse arterioles. Because they have a very small diameter, the coronary arteries may become clogged, as discussed in section 12.5. The coronary capillary beds join

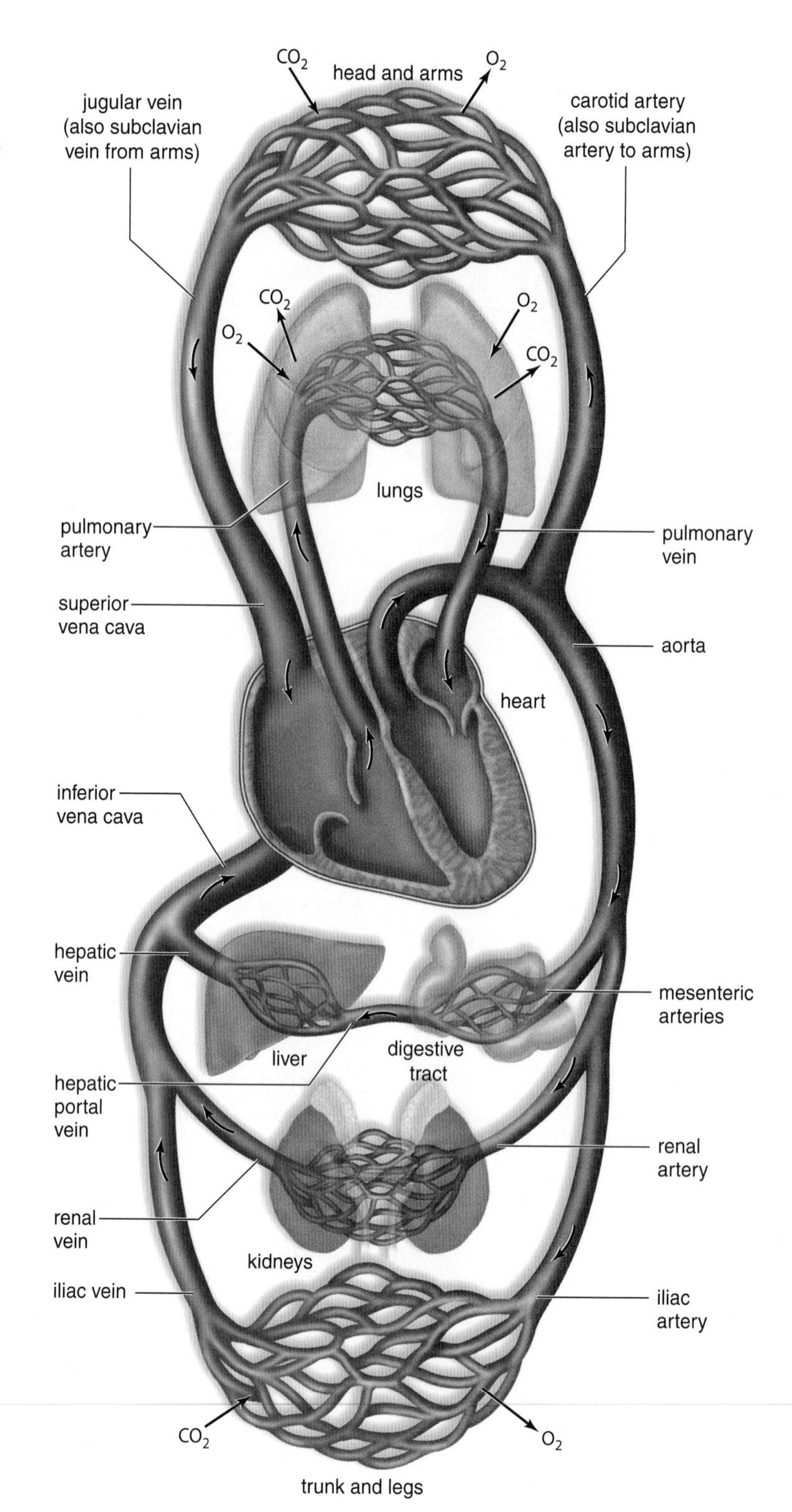

Figure 12.7 Path of blood.
This symbolic and not-to-scale drawing shows the path of blood in the pulmonary and systemic circuits. The pulmonary arteries and veins take blood from the right (blue) to the left (red) side of the heart. Tracing blood from the digestive tract to the right atrium in the systemic circuit involves the hepatic portal vein, the hepatic vein, and the inferior vena cava. The blue-colored vessels carry O_2-poor blood, and the red-colored vessels carry O_2-rich blood; the arrows indicate the direction of blood flow.

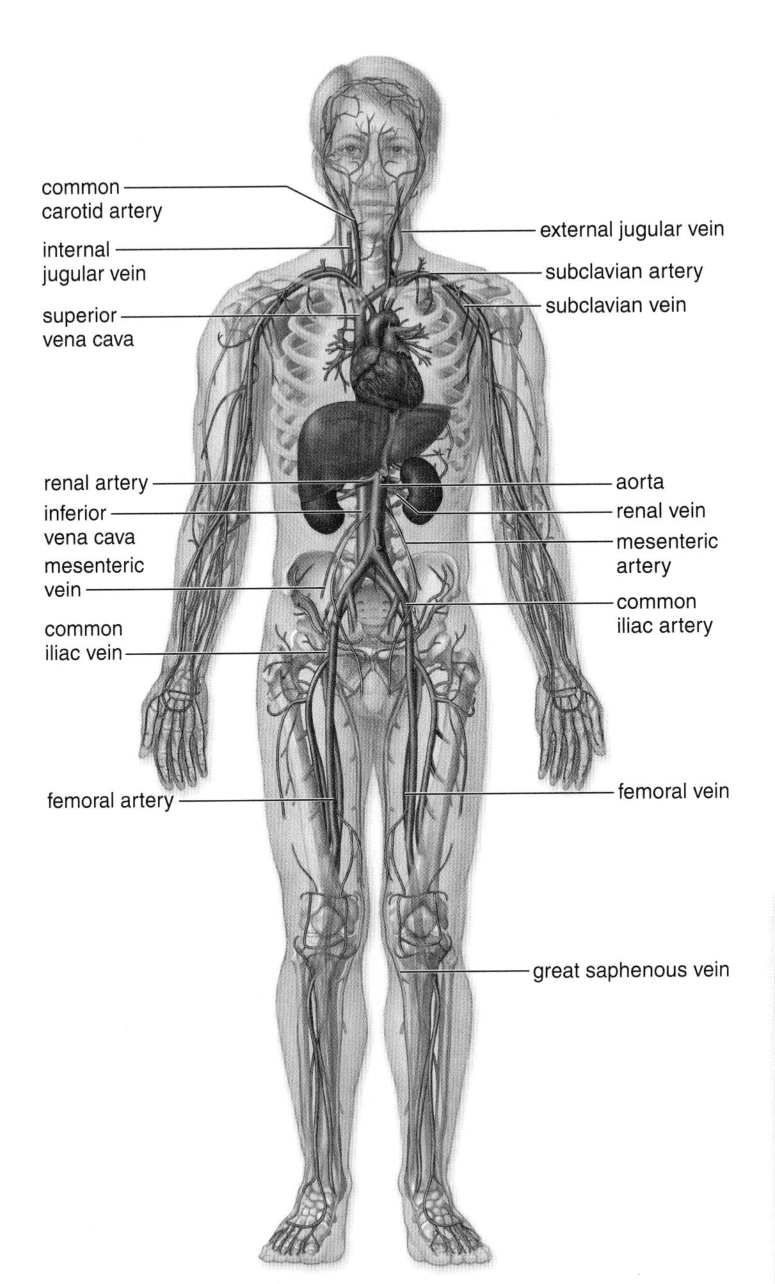

Figure 12.8 Major arteries (red) and veins (blue) of the systemic circuit.
This representation of the major blood vessels of the systemic circuit shows how the systemic arteries and veins are arranged in the body. The superior and inferior venae cavae take their names from their relationship to which organ?

to form venules. The venules converge to form the cardiac veins, which empty into the right atrium.

A *portal system* in blood circulation begins and ends in capillaries. One such system, the **hepatic portal system,** is associated with the liver. Capillaries that occur in the villi of the small intestine pass into venules that join to form the **hepatic portal vein.** This vein carries the blood to a set of capillaries in the liver, an organ that monitors the makeup of the blood (see Fig. 12.7). The **hepatic vein** leaves the liver and enters the inferior vena cava.

Blood Pressure

When the left ventricle contracts, blood is forced into the aorta and then into other systemic arteries under pressure. **Systolic pressure** results from blood being forced into the arteries during ventricular systole, and **diastolic pressure** is the pressure in the arteries during ventricular diastole.

As blood flows from the aorta into the arteries and arterioles, blood pressure falls. Also, the difference between systolic and diastolic pressure gradually diminishes. In the capillaries, blood flow is slow and fairly even. This may be related to the very high total cross-sectional area of the capillaries (Fig. 12.9). It has been calculated that if all the blood vessels in a human were connected end to end, the total distance would reach around Earth at the equator two times! Most of this distance would be due to the large number of capillaries.

Human **blood pressure** can be measured with a sphygmomanometer, which has a pressure cuff that determines the amount of pressure required to stop the flow of blood through an artery. Blood pressure is normally measured on the brachial artery, an artery in the upper arm. Today, automated manometers are often used to take one's blood

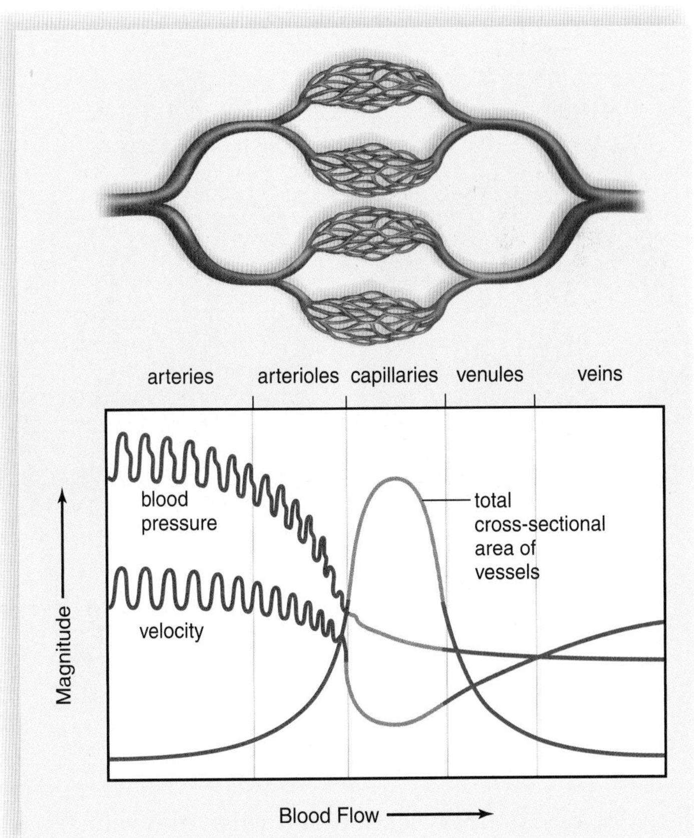

Figure 12.9 Blood velocity and blood pressure.
In capillaries, blood is under minimal pressure and has the least velocity. Blood pressure and velocity drop off because capillaries have a greater total cross-sectional area than arterioles.

pressure instead. Blood pressure is expressed in millimeters of mercury (mm Hg). A blood pressure reading consists of two numbers that represent systolic and diastolic pressures, respectively—for example, 120/80. Blood pressure that is too low or—more commonly—too high can be a significant health problem (see section 12.5).

Blood pressure in the veins is low and by itself is an inefficient means of moving blood back to the heart, especially from the limbs. When skeletal muscles near veins contract, they put pressure on the veins and the blood they contain. Valves prevent the backward flow of blood in veins, and therefore muscle contraction is sufficient to move blood toward the heart (Fig. 12.10). During long periods of sitting or other inactivity, clots may form in the deep veins of the legs. Because these can cause serious problems, especially if they break free and become lodged in the lungs, take frequent stretch breaks on long car or plane trips.

Check Your Progress

1. What arteries carry O_2-poor blood?
2. Through what major blood vessels does blood flow to the lungs and other parts of the body and return to the heart?
3. What causes blood to flow in arteries? In veins?
4. Why is systolic blood pressure higher than diastolic?

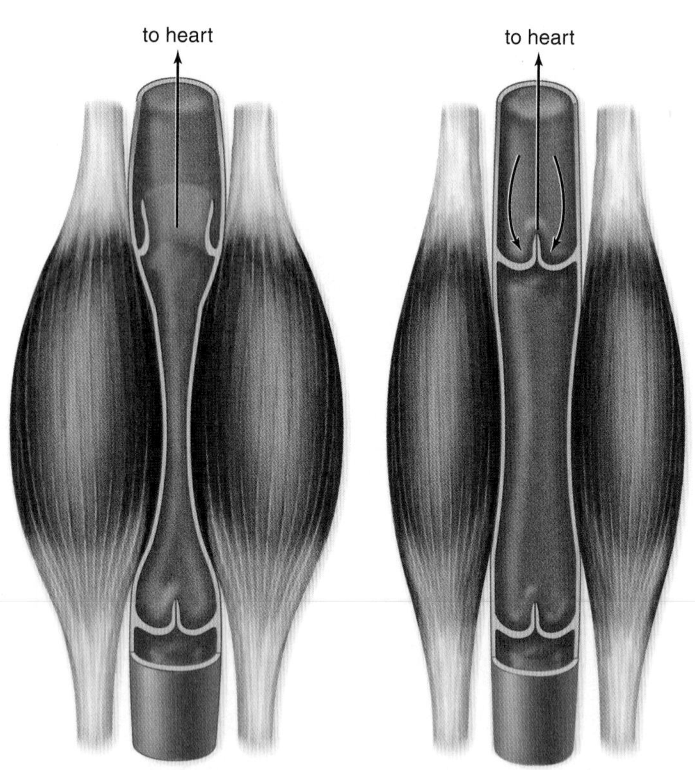

Figure 12.10 Cross section of a valve in a vein. **a.** Pressure on the walls of a vein, exerted by skeletal muscles, increases blood pressure within the vein and forces the valve open. **b.** When external pressure is no longer applied to the vein, blood pressure decreases, and back pressure forces the valve closed. Closure of the valve prevents the blood from flowing in the opposite direction.

12.4 Blood

Learning Outcomes

1. List the main types of cells in blood and their functions.
2. Identify the major molecular and cellular events that result in a blood clot.
3. Define capillary exchange, and describe the two major forces involved.
4. Discuss the role of blood in health and disease.

If blood is collected from a person's vein into a test tube and prevented from clotting, it separates into two layers (Fig. 12.11*a*). The lower layer consists of the **formed elements,** and the upper layer is **plasma,** the liquid portion of blood. The formed elements consist of red blood cells, white blood cells, and blood platelets.

Blood has transport functions, regulatory functions, and protective functions. In addition to nutrients and wastes, blood also transports hormones. Blood helps regulate body temperature by dispersing body heat and helps regulate blood pressure because the plasma proteins contribute to the osmotic pressure of blood. Buffers in blood help maintain blood pH at about 7.4. Blood helps protect the body against invasion by disease-causing viruses and bacteria and against a potentially life-threatening loss of blood by clotting.

Plasma

Plasma contains a variety of inorganic and organic substances dissolved or suspended in water (Fig. 12.11*b*). Plasma proteins, which make up 7–8% of plasma, assist in transporting large organic molecules in blood. For example, **albumin** transports bilirubin, a breakdown product of hemoglobin. Lipoproteins transport cholesterol. Certain plasma proteins have specific functions. As discussed later in this section, fibrinogen is necessary to blood clotting, and immunoglobulins are antibodies that help fight infection. Plasma proteins also maintain blood volume because they are too large to leave capillaries. Therefore, blood in capillaries normally has a higher solute concentration than does tissue fluid, and water automatically diffuses into them.

The Red Blood Cells

Red blood cells (erythrocytes) are continuously manufactured in the red bone marrow of the skull, the ribs, the vertebrae, and the ends of the long bones. Normally, there are 4 to 6 million red blood cells per mm^3 of whole blood (see Fig. 12.11*d*).

Mature red blood cells don't have a nucleus and are biconcave disks (Fig. 12.12 *a, b*). Their shape increases their flexibility for moving through capillary beds and their surface area for diffusion of gases. Red blood cells carry oxygen because they contain **hemoglobin,** the respiratory pigment. Because hemoglobin is a red pigment, the cells are red. A hemoglobin molecule contains four polypeptide chains

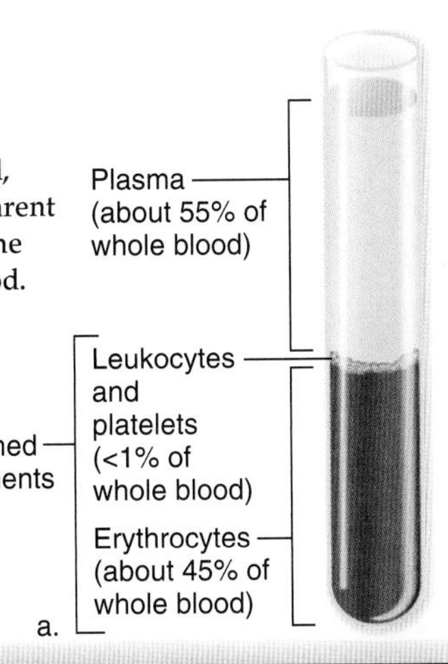

Figure 12.11
Composition of blood.
a. When blood is collected into a test tube containing an anticoagulant to prevent clotting and then centrifuged, it consists of three layers. The transparent straw-colored or yellow top layer is the plasma, the liquid portion of the blood. The thin, middle buffy coat layer consists of leukocytes and platelets. The bottom layer contains the erythrocytes. **b.** Breakdown of the components of plasma. **c.** Micrograph of the formed elements, which are also listed in (**d**).

Plasma		
Type	**Function**	**Source**
Water (90–92% of plasma)	Maintains blood volume; transports molecules	Absorbed from intestine
Plasma proteins (7–8% of plasma)	Maintain blood osmotic pressure and pH	
Albumin	Maintains blood volume and pressure, transport	Liver
Antibodies	Fight infection	B lymphocytes
Fibrinogen	Clotting	Liver
Salts (less than 1% of plasma)	Maintain blood osmotic pressure and pH; aid metabolism	Absorbed from intestine
Gases		
Oxygen	Cellular respiration	Lungs
Carbon dioxide	End product of metabolism	Tissues
Nutrients Lipids, Glucose, Amino acids	Food for cells	Absorbed from intestine
Nitrogenous wastes Urea, Uric acid	Excretion by kidneys	Liver
Other Hormones, vitamins, etc.	Aid metabolism	Varied

b.

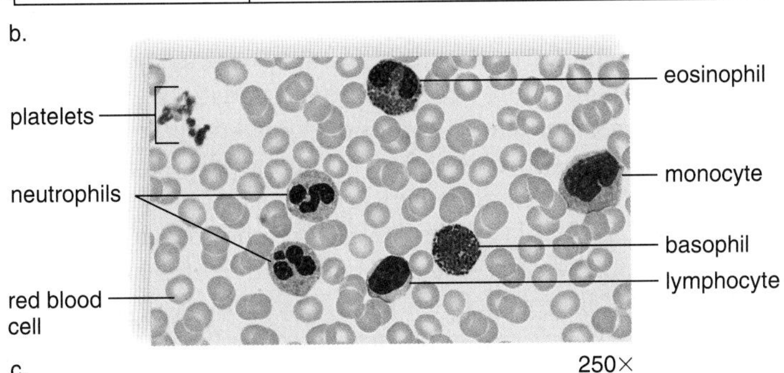

c.

Formed Elements		
Type	**Function and Description**	**Source**
Red blood cells (erythrocytes)	Transport O_2 and help transport CO_2	Red bone marrow
4 million–6 million per mm^3 blood	7–8 μm in diameter Bright-red to dark-purple biconcave disks without nuclei	
White blood cells (leukocytes) 5,000–11,000 per mm^3 blood	Fight infection	Red bone marrow
Granular leukocytes: Neutrophils * 40–70%	10–14 μm in diameter Spherical cells with multilobed nuclei; fine, pink granules in cytoplasm; phagocytize pathogens	
Granular leukocytes: Eosinophils * 1–4%	10–14 μm in diameter Spherical cells with bilobed nuclei; coarse, deep-red, uniformly sized granules in cytoplasm; phagocytize antigen-antibody complexes and allergens	
Granular leukocytes: Basophils * 0–1%	10–12 μm in diameter Spherical cells with lobed nuclei; large, irregularly shaped, deep-blue granules in cytoplasm; release histamine, which promotes blood flow to injured tissues	
Agranular leukocytes: Lymphocytes * 20–45%	5–17 μm in diameter (average 9–10 μm) Spherical cells with large, round nuclei; responsible for specific immunity	
Agranular leukocytes: Monocytes * 4–8%	10–24 μm in diameter Large spherical cells with kidney-shaped, round, or lobed nuclei; become macrophages that phagocytize pathogens and cellular debris	
Platelets (thrombocytes)	Aid clotting	Red bone marrow
150,000–300,000 per mm^3 blood	2–4 μm in diameter Disk-shaped cell fragments with no nuclei; purple granules in cytoplasm	

d.

* Appearance with Wright's stain.

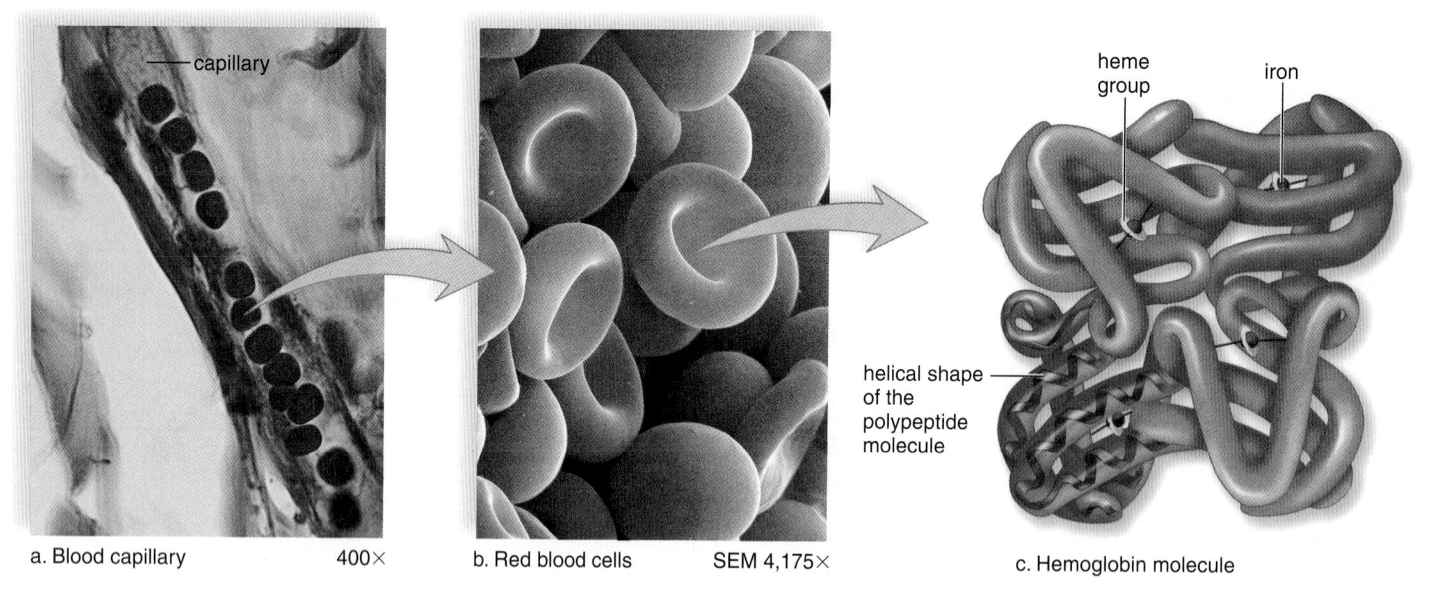

Figure 12.12 Physiology of red blood cells.
a. Red blood cells move in single file through the capillaries. **b.** Each red blood cell is a biconcave disk containing many molecules of hemoglobin, the respiratory pigment. **c.** Hemoglobin contains four polypeptide chains (purple). There is an iron-containing heme group in the center of each chain. Oxygen combines loosely with iron when hemoglobin is oxygenated. Oxyhemoglobin is bright red, and deoxyhemoglobin is a dark maroon color.

(Fig. 12.12*c*). Each chain is associated with heme, a complex iron-containing group. The iron portion of hemoglobin acquires oxygen in the lungs and gives it up in the tissues. Hemoglobin also helps to carry carbon dioxide from tissues back to the lungs.

Carbon monoxide is a colorless and odorless air pollutant that comes primarily from the incomplete combustion of natural gas and gasoline. Unfortunately, carbon monoxide combines with hemoglobin more readily than does oxygen, and it stays combined for several hours, making hemoglobin unavailable for oxygen transport. Carbon monoxide detectors in homes can help prevent accidental death due to a malfunctioning furnace or inadequate exhaust system.

Possibly because they lack nuclei, red blood cells live only about 120 days. They are destroyed chiefly in the liver and the spleen, where they are engulfed by large phagocytic cells. The iron is mostly salvaged and reused, while the heme portion undergoes chemical degradation, and the liver excretes it into the bile as bile pigments.

When the body has an insufficient number of red blood cells or the red blood cells do not contain enough hemoglobin, an individual suffers from **anemia** and has a tired, rundown feeling. There are three basic causes of anemia: (1) decreased production of red blood cells, (2) loss of red blood cells from the body, and (3) destruction of red blood cells within the body. In the most common type of anemia, iron-deficiency anemia, red blood cell production is decreased, most often due to a diet that does not contain enough iron.

Whenever arterial blood carries a reduced amount of oxygen, as in chronic anemia or when an individual first takes up residence at a high altitude, the kidneys increase their production of a hormone called **erythropoietin,** which speeds the maturation of red blood cells in the bone marrow.

The White Blood Cells

White blood cells (leukocytes) differ from red blood cells in that they are usually larger, have a nucleus, lack hemoglobin, and without staining appear translucent (see Fig. 12.11*c, d*). White blood cells are not as numerous as red blood cells, with only 5,000–11,000 cells per mm^3. White blood cells fight infection and play a role in the development of immunity, the ability to resist disease.

On the basis of structure, it is possible to divide white blood cells into **granular leukocytes** and **agranular leukocytes,** which are also called **mononuclear cells.** Granular leukocytes (neutrophils, basophils, and eosinophils) are filled with spheres that contain enzymes and proteins, which help white blood cells defend the body against microbes. **Neutrophils** are granular leukocytes with a multilobed nucleus joined by nuclear threads. They are the most abundant of the white blood cells and are able to phagocytize and digest bacteria. **Basophils** stain a deep blue and release histamine, which can be a problem for people with allergies. **Eosinophils** stain a deep red and are thought to fight parasitic worms, although they are also involved in some allergies.

Agranular leukocytes (monocytes and lymphocytes) typically have a kidney-shaped or spherical nucleus. **Monocytes** are the largest of the white blood cells, and they differentiate into phagocytic dendritic cells and macrophages. **Dendritic cells** are present in tissues that are in contact with the environment: skin, nose, lungs, and intestines. Once they have captured a microbe with their long, spiky arms, called dendrites, they stimulate other white blood cells to defend the body. **Macrophages,** well known as ferocious phagocytes (Fig. 12.13), play a similar role in other organs, such as the liver, kidney, and spleen. The **lymphocytes** are of two major types, B lymphocytes and T lymphocytes, and each type plays a specific role in immunity (see Chapter 13).

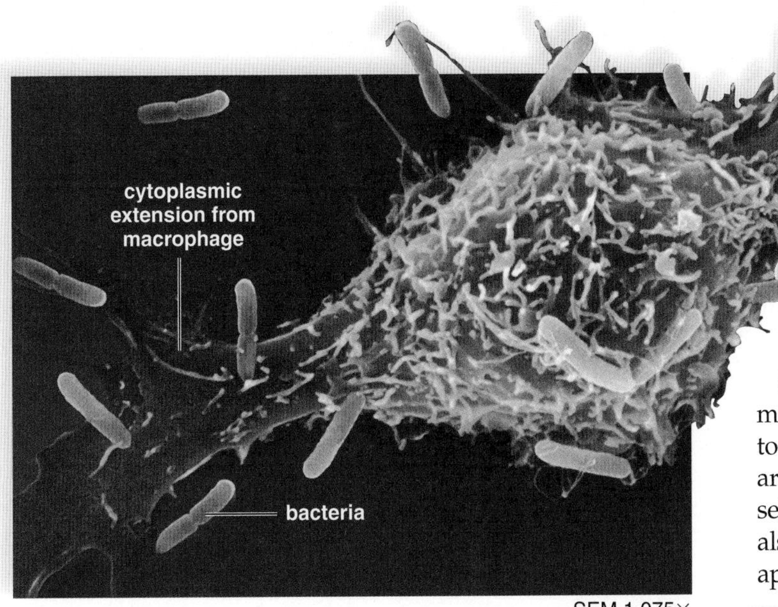

Figure 12.13 Macrophage (red) engulfing bacteria. Monocyte-derived macrophages are the body's scavengers. They engulf microbes and debris in the body's fluids and tissues, as illustrated in this colorized scanning electron micrograph.

If the total number of white blood cells increases or decreases beyond normal, disease may be present. Sometimes an increase or decrease of only one type of white blood cell is a sign of infection. A person with infectious mononucleosis, caused by the Epstein-Barr virus, often has an excessive number of lymphocytes of the B type. A person with AIDS, caused by an HIV infection, has an abnormally low number of T lymphocytes. Leukemia is a form of cancer characterized by uncontrolled production of abnormal white blood cells.

White blood cells live different lengths of time. Many live only a few days or may die combating invading pathogens. Others live for months or even years.

Check Your Progress

1. What are the major components of blood, and what are their functions?
2. What sort of information can be gained by checking the numbers and types of cells present in the blood of a person who is ill?

The Platelets and Blood Clotting

Platelets (also called thrombocytes) result from fragmentation of certain large cells, called **megakaryocytes,** in the red bone marrow. Platelets are produced at a rate of 200 billion a day, and the blood contains 150,000–300,000 per mm^3. These formed elements are involved in the process of blood **clotting,** or coagulation.

There are at least 12 clotting factors in the blood that participate with platelets in the formation of a blood clot. We will discuss the roles played by **fibrinogen** and **prothrombin,** which are proteins manufactured by the liver. Vitamin K, found in green vegetables and also formed by intestinal bacteria, is needed for prothrombin production. Vitamin K deficiency can result in clotting disorders.

Blood Clotting

When a blood vessel in the body is damaged, the process of clotting begins (Fig. 12.14*a*). First, platelets clump at the site of the puncture and partially seal the leak. Platelets and damaged tissue release **prothrombin activator,** which converts the plasma protein prothrombin to thrombin. This reaction requires calcium ions (Ca^{2+}). **Thrombin,** in turn, acts as an enzyme that severs two short amino acid chains from each fibrinogen molecule, activating it. The activated fragments then join end to end, forming long threads of **fibrin.** Fibrin threads wind around the platelet plug in the damaged area of the blood vessel and provide the framework for the clot. Red blood cells are also trapped within the fibrin threads. These cells make a clot appear red (Fig. 12.14*b*).

1. Blood vessel is punctured.

2. Platelets congregate and form a plug.

3. Platelets and damaged tissue cells release prothrombin activator, which initiates a cascade of enzymatic reactions.

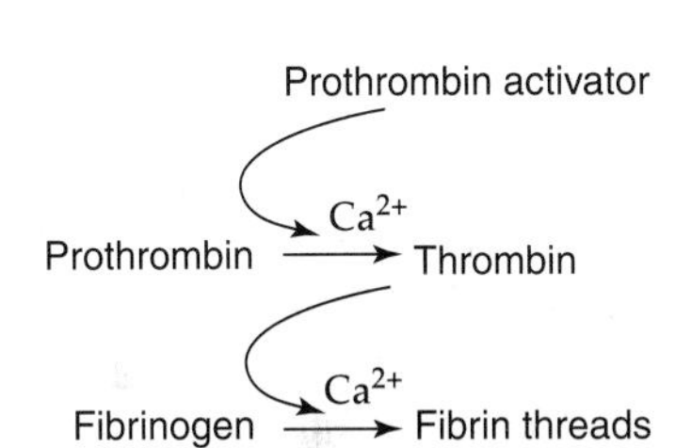

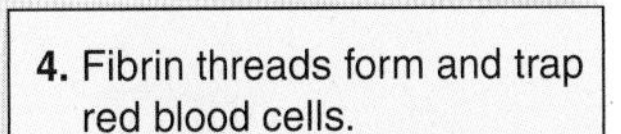

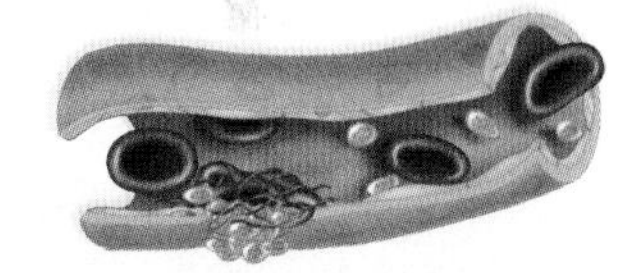

a. Blood-clotting process

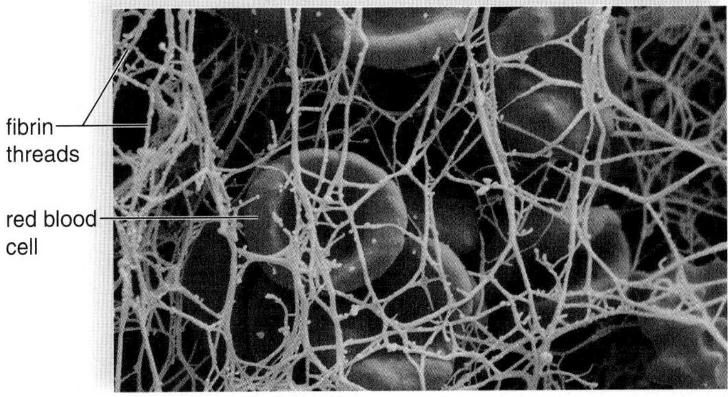

b. Blood clot 4,400×

Figure 12.14 Blood clotting.
a. Platelets and damaged tissue cells release prothrombin activator, which acts on prothrombin in the presence of Ca^{2+} (calcium ions) to produce thrombin. Thrombin acts on fibrinogen in the presence of Ca^{2+} to form fibrin threads. **b.** A scanning electron micrograph of a blood clot shows red blood cells caught in the fibrin threads.

A fibrin clot is only temporary. As soon as blood vessel repair is initiated, an enzyme called plasmin destroys the fibrin network and restores the fluidity of the plasma. Sometimes it is desirable to inhibit the clotting process. This can be done with medications such as heparin, or even with medicinal leeches (see the Health Focus on page 232).

If blood is allowed to clot in a test tube, a yellowish fluid develops above the clotted material. This fluid is called **serum,** and it contains all the components of plasma except fibrinogen. Table 12.1 reviews the body fluids related to blood.

TABLE 12.1 Body Fluids Related to Blood

Name	Composition
Blood	Formed elements and plasma
Plasma	Liquid portion of blood
Serum	Plasma minus fibrinogen
Tissue fluid	Plasma minus most proteins
Lymph	Tissue fluid within lymphatic vessels

Hemophilia

Hemophilia is an inherited clotting disorder caused by a deficiency in a clotting factor. The most common type, hemophila A, accounts for about 90% of all cases, and almost always occurs in males because the faulty gene is found on the X chromosome. (Females have two X's, therefore they have a backup copy of the gene.) The slightest bump to an affected person can cause bleeding into the joints. Cartilage degeneration in the joints and resorption of underlying bone can follow. Bleeding into muscles can lead to nerve damage and muscular atrophy. Death can result from bleeding into the brain with accompanying neurological damage. Hemophiliacs usually require frequent blood transfusions, although they may also be treated with injections of the specific clotting factor they are lacking.

Bone Marrow Stem Cells

A **stem cell** is a cell that is ever capable of dividing and producing new cells that go on to differentiate into particular types of cells. It's been known for some time that the bone marrow has multipotent stem cells, which have the potential to give rise to other stem cells for the various formed ele-

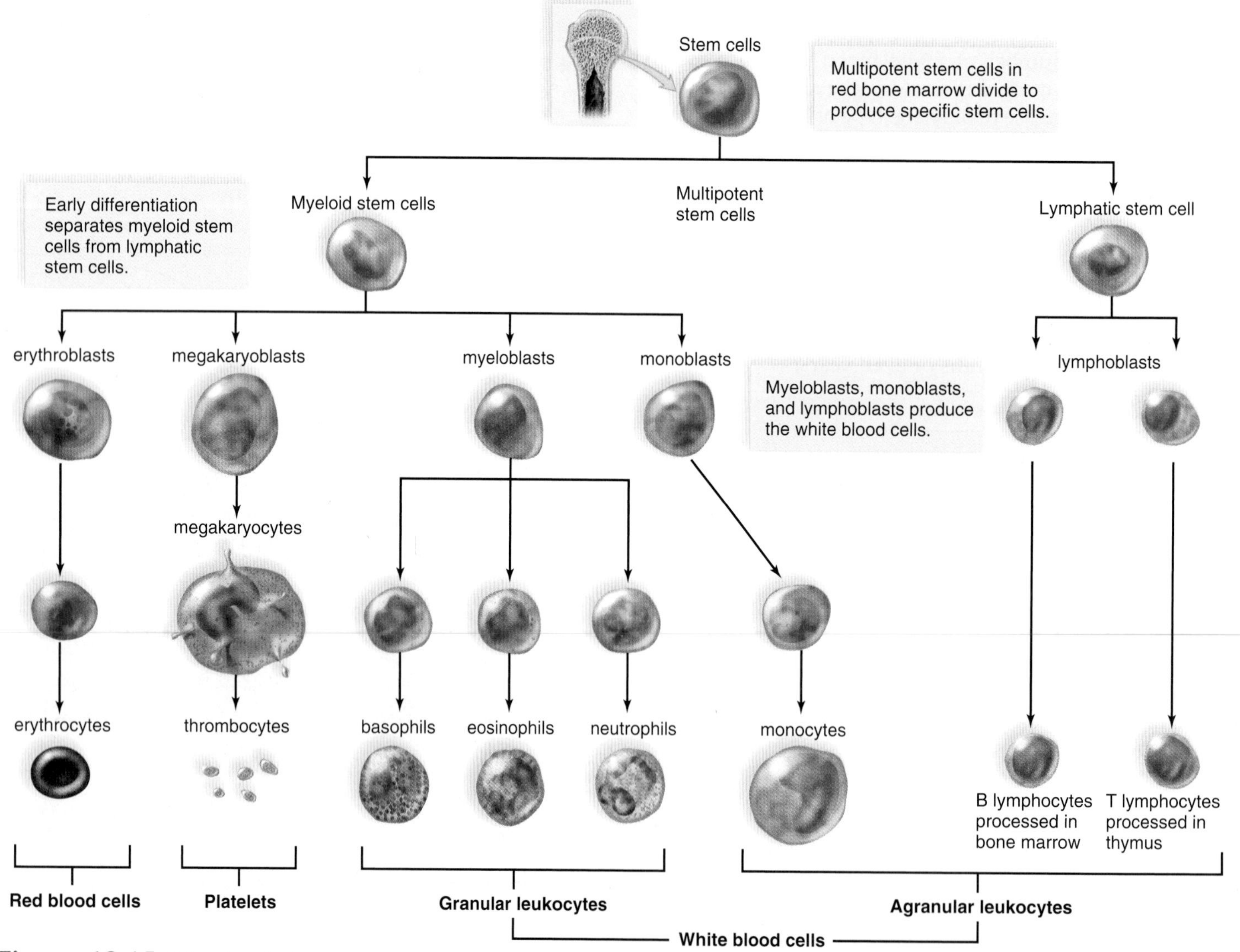

Figure 12.15 Blood cell formation in red bone marrow.
Multipotent stem cells give rise to two specialized types of stem cells. The myeloid stem cells give rise to still other cells, which become red blood cells, platelets, and all the white blood cells except lymphocytes. The lymphoid stem cells give rise to lymphoblasts, which become lymphocytes.

ments (Fig. 12.15). Recent research has shown that bone marrow stem cells are also able to differentiate into other types of cells, including liver, bone, fat, cartilage, heart, and even neurons. The possibility exists that a patient's own bone marrow stem cells could be used for curing certain conditions that might develop, such as diabetes, heart disease, liver disease, or even brain disorders (e.g., Alzheimer disease or Parkinson disease). In one study, researchers examined the brains of women who had received bone marrow stem cells from male donors as a part of their treatment for leukemia. All recipients had neurons in their brains that contained a Y chromosome! It seems that some of the bone marrow stem cells derived from male donors had traveled to the brain, where they differentiated into neurons.

The use of a person's own stem cells does away with the problem of possible rejection. Among all the various types of *adult stem cells* in the body, bone marrow stem cells are particularly attractive because they are the most accessible. Some researchers prefer to work with *embryonic stem cells,* thinking that they are more likely to become any type of cell. Embryonic stem cells are available because many early-stage embryos remain unused in fertility clinics, although this issue has become controversial.

Capillary Exchange

Two forces primarily control movement of fluid through the capillary wall: osmotic pressure, created by salts and plasma proteins, which tends to cause water to move from the tissue fluid to the blood, and blood pressure, which tends to cause water to move in the opposite direction. At the arterial end of a capillary, blood pressure is higher than the osmotic pressure of blood (Fig. 12.16), so water exits a capillary at this end.

Midway along the capillary, where blood pressure is lower, the two forces essentially cancel each other, and there is no net movement of water. Solutes now diffuse according to their concentration gradient—nutrients (glucose and amino acids) and oxygen diffuse out of the capillary, and wastes (carbon dioxide) diffuse into the capillary. In the pulmonary circuit, where the O_2 concentration is higher in the lung tissues, and the CO_2 concentration is lower, the movement of these gases is reversed.

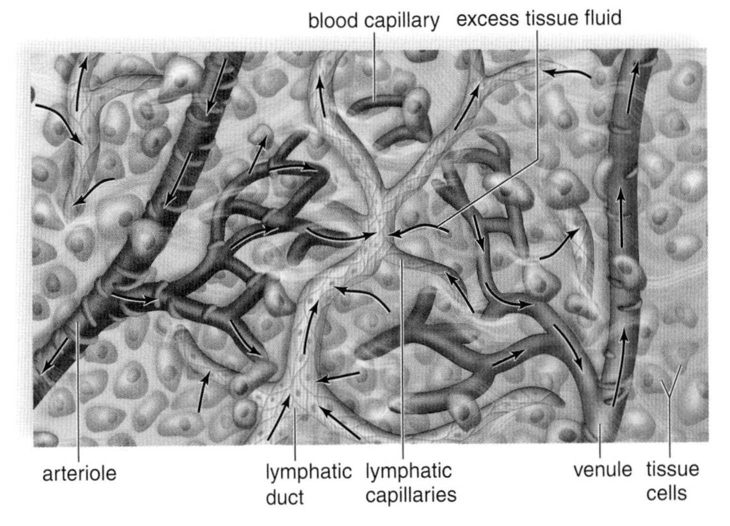

Figure 12.17 Lymphatic capillaries.
A lymphatic capillary bed (shown here in green) lies near a blood capillary bed. When lymphatic capillaries take up excess tissue fluid, it becomes lymph.

Red blood cells and almost all plasma proteins remain in the capillaries, but small substances leave, contributing to **tissue fluid,** the fluid between the body's cells. Tissue fluid tends to contain all the components of plasma but lesser amounts of protein, which generally stays in the capillaries.

At the venous end of a capillary, where blood pressure has fallen even more, osmotic pressure is greater than blood pressure, and water tends to move into the capillary. Almost the same amount of fluid that left the capillary returns to it, although some excess tissue fluid is always collected by the **lymphatic capillaries** (Fig. 12.17). Tissue fluid contained within lymphatic vessels is called **lymph.** Lymph is returned to the systemic venous blood when the major lymphatic vessels enter the subclavian veins in the shoulder region.

Check Your Progress

1. What major cellular and molecular events cause a blood clot?
2. What types of diseases would not be expected to be treatable with stem cell therapy?
3. What would be the effect on tissue fluid if the protein content of the blood was greatly reduced, i.e., by malnutrition?

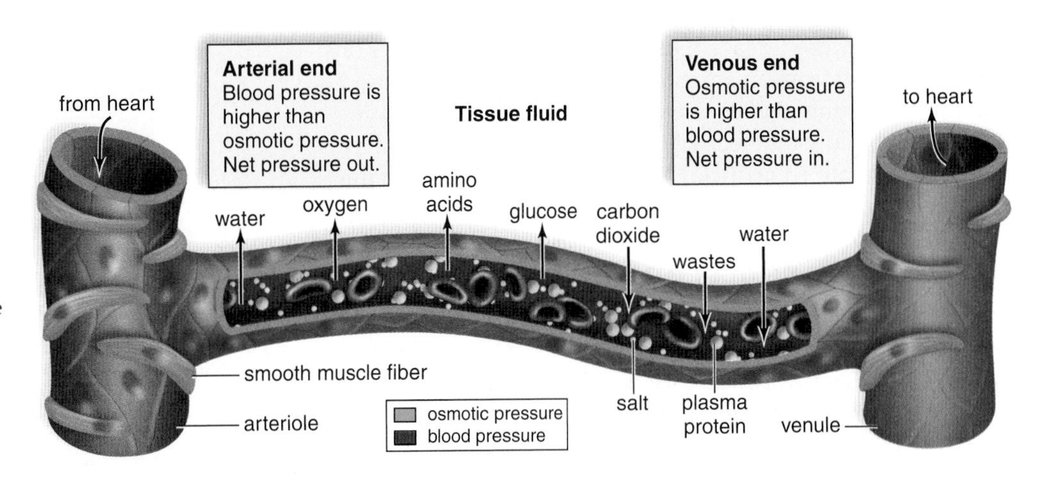

Figure 12.16 Capillary exchange in the systemic circuit.
At the arterial end of a capillary (*left*) the blood pressure is higher than the osmotic pressure; therefore, water tends to leave the bloodstream. In the midsection, molecules, including oxygen and carbon dioxide, follow their concentration gradients. At the venous end of a capillary (*right*), the osmotic pressure is higher than the blood pressure; therefore, water tends to enter the bloodstream. Notice that the red blood cells and the plasma proteins are too large to exit a capillary.

12.5 Cardiovascular Disorders

Learning Outcomes

1. Describe the major categories of cardiovascular disease that occur in the United States.
2. List three possible treatments for a blocked coronary artery.
3. Define hypertension and explain its most common causes.

Cardiovascular disease is the leading cause of untimely death in Western countries. Modern research efforts have resulted in improved diagnosis, treatment, and prevention. This section discusses some of the major advances that have been made in these areas. The Health Focus on page 230 describes ways to maintain cardiovascular health.

Atherosclerosis

Atherosclerosis is an accumulation of soft masses of fatty materials, particularly cholesterol, beneath the inner linings of arteries. Such deposits are called **plaque.** Plaque tends to protrude into the lumen of the vessel and interfere with the flow of blood (see Health Focus Fig. 12A). In most instances, atherosclerosis begins in early adulthood and develops progressively through middle age, but symptoms may not appear until an individual is 50 or older.

Plaque can cause platelets to adhere to the irregular arterial wall, forming a clot. As long as the clot remains stationary, it is called a **thrombus,** but if it dislodges and moves along with the blood, it is called an **embolus. Thromboembolism,** a clot that has been carried in the bloodstream but is now stationary, must be treated, or serious complications can arise.

Heart Valve Disease

Each year in the United States, around 90,000 people undergo surgery to have faulty heart valves repaired or replaced. In some patients, the heart valves are malformed at birth, but more commonly they degenerate due to age or disease, to the point where they no longer prevent the backflow of blood. A narrowing (stenosis) of the aortic valve opening is the most common, followed by mitral valve prolapse, in which abnormally thickened "leaflets" of the mitral valve protrude back into the left ventricle. In some cases, the faulty valves can be repaired during open-heart surgery, but more commonly they are replaced, using either artificial valves or valves removed from an animal (usually a pig) or a deceased human. Beginning in 2005, surgeons began developing procedures that allow a defective heart valve to be replaced by threading a compact artificial valve through an artery in the leg, thus avoiding open-heart surgery.

Stroke, Heart Attack, and Aneurysm

A cerebrovascular accident, also called a **stroke,** often results when an arteriole in the brain bursts or is blocked by an embolus. Lack of oxygen causes a portion of the brain to die, and paralysis or death can result. A person is sometimes forewarned of a stroke by a feeling of numbness in the hands or the face, difficulty in speaking, or temporary blindness in one eye. Cerebrovascular accidents become more common with age, but are certainly not limited to the elderly: in 2006, former Minnesota Twins baseball star Kirby Puckett died of a stroke at age 45.

If a coronary artery becomes partially blocked, the individual may suffer from **angina pectoris,** characterized by a squeezing or burning sensation in the chest. Nitroglycerin or related drugs dilate blood vessels and help relieve the pain. When a coronary artery is completely blocked, a portion of the heart muscle dies due to a lack of oxygen, and a myocardial infarction, or **heart attack,** occurs.

An **aneurysm** is the ballooning of a blood vessel, most often the abdominal aorta or the arteries leading to the brain. Atherosclerosis and high blood pressure can weaken the wall of an artery to the point that an aneurysm develops. If a major vessel such as the aorta should burst, about half of victims die before reaching a hospital. It is often possible to replace a damaged or diseased portion of a vessel, such as an artery, with a synthetic graft. At least two companies are developing wireless sensors that can be implanted along with these stents, to help doctors monitor changes in pressure that may signal an aneurism that is about to burst.

Coronary Bypass Operations

Since the late 1960s, coronary bypass surgery has been a common way to treat an obstructed coronary artery. During this operation, a surgeon usually takes a segment from another blood vessel and stitches one end to the aorta and the other end to a coronary artery past the point of obstruction. Figure 12.18 shows a heart in which three blocked coronary arteries have been bypassed using grafted arteries.

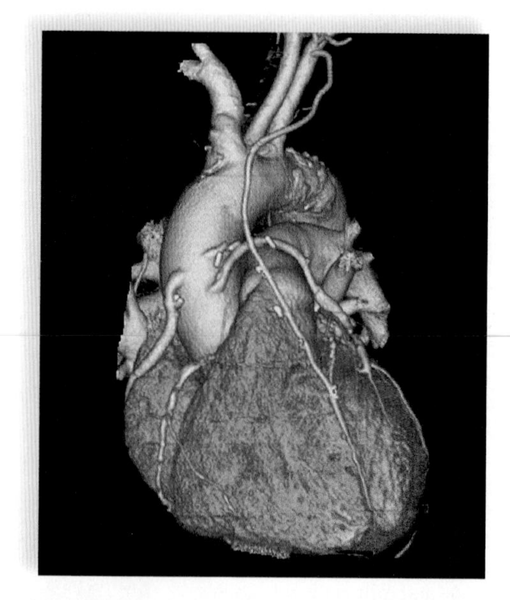

Figure 12.18 Bypassing blocked coronary arteries. This is a 3D scan of the heart of a patient who received a triple bypass operation. The surgeon has bypassed two blocked arteries using vessels removed from another part of the body, and used an existing artery that branches off of the left subclavian artery (see Figs. 12.3 and 12.4) to bypass a third blocked artery.

The results of some clinical trials have suggested that gene therapy may be another way to restore the blood supply to oxygen-starved cardiac muscle, without the need for major surgery. A harmless virus was engineered to contain a gene coding for a growth factor called VEGF (vascular endothelial growth factor), and then injected directly into affected areas of the heart. Although many patients appeared to show improvement, further testing will be required before such treatments can be judged safe enough for more widespread use.

Clearing Clogged Arteries

In **angioplasty,** a cardiologist threads a catheter into an artery in the groin or upper part of the arm and guides it through a major blood vessel toward the heart. When the tube arrives at the region of plaque in an artery, a balloon attached to the end of the tube is inflated, forcing the vessel open. However, the artery may not always remain open because the trauma can cause smooth muscle cells in the wall of the artery to proliferate and close it.

More often, the same operation utilizes a cylinder of expandable metal mesh called a **stent.** Once the stent is in place, a balloon inside the stent is inflated, expanding it, and locking it in place (Fig. 12.19). In recent years, the most commonly used stents were coated with a drug thought to inhibit scar tissue formation, but research published in 2007 suggested that patients with these stents may have an increased risk of dying compared with patients who received bare metal stents, or angioplasty plus medication. Since 2003, these drug-coated stents have been implanted in over 6 million people worldwide, so this will be an important issue to resolve!

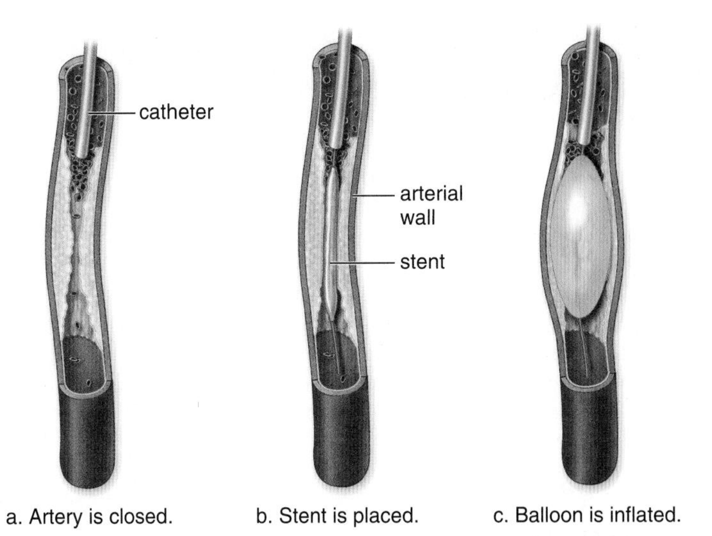

Figure 12.19 Angioplasty with stent placement.
a. A plastic tube (catheter) is inserted into the coronary artery until it reaches the clogged area. **b.** A metal stent with a balloon inside it is pushed out the end of the plastic tube into the clogged area. **c.** When the balloon is inflated, the vessel opens, and the stent is left in place to keep the vessel open.

Dissolving Blood Clots

Medical treatment for thromboembolism often includes the use of tissue plasminogen activator (tPA). This drug converts plasminogen into plasmin, an enzyme that dissolves blood clots, as mentioned in section 12.4. In fact, tPA is the body's own way of converting plasminogen to plasmin. tPA is also being used for thrombolytic stroke patients but with limited success because some patients experience life-threatening bleeding in the brain.

If a person has symptoms of angina or a stroke, aspirin may be prescribed. Aspirin reduces the stickiness of platelets, and thereby lowers the probability that a clot will form. Evidence indicates that aspirin protects against first heart attacks, but there is no clear support for taking aspirin every day to prevent strokes in symptom-free people. Even so, some physicians recommend taking a low dosage of aspirin every day.

CHECK YOUR PROGRESS

1. Define the following: thrombus, embolus, aneurysm.
2. What are some possible complications with stents placed in blocked arteries (i.e., what could go wrong)?

Heart Transplants and Artificial Hearts

Heart transplants are usually successful today, but unfortunately, the need for hearts to transplant is greater than the supply because of a shortage of human donors. Therefore, a left ventricular assist device (LVAD), implanted in the abdomen, may help patients waiting for a transplant. A tube passes blood from the left ventricle to the device, which pumps it to the aorta. A cable passes from the device through the skin to an external battery the patient totes around.

Only a few patients have ever received a so-called **total artificial heart (TAH).** One model now available is called an AbioCor (Fig. 12.20). An internal battery and controller regulate the pumping speed, and an external battery powers the device by passing electricity through the skin. A rotating centrifugal pump moves silicon hydraulic fluid between left and right sacs to force blood out of the heart into the pulmonary trunk and the aorta. The exterior is made mainly of titanium, but the valves and membranes inside the ventricles are made of plastic, which has held up to beating 100,000 times a day for years. So far, all recipients have been near death and most have lived for only a short time after getting an AbioCor.

Figure 12.20 A total artificial heart (TAH).
The AbioCor replacement heart is designed to be implanted within the chest cavity.

Health Focus

Prevention of Cardiovascular Disease

Many of us are predisposed to cardiovascular disease due to factors beyond our control. We can't change things like having a family history of heart attacks under age 55, or being a male, or our ethnicity (African Americans are at greater risk). However, many other risk factors for cardiovascular disease are influenced by our behavior. Because heart disease is such a common cause of death, everyone stands to benefit by following these recommendations:

The Don'ts

Smoking

When a person smokes, the drug nicotine, present in cigarette smoke, enters the bloodstream. Nicotine causes the arterioles to constrict and the blood pressure to rise. The heart has to pump harder to propel the blood through the lungs at a time when the blood's oxygen-carrying capacity is reduced by smoking.

Drug Abuse

Stimulants, such as cocaine and amphetamines, can cause an irregular heartbeat and lead to heart attacks even when using drugs for the first time. Intravenous drug use may also result in a cerebral blood clot and stroke.

Obesity

Hypertension (high blood pressure) is prevalent in persons who are more than 20% above the recommended weight for their height. More tissues require servicing, and the heart sends the extra blood out under greater pressure. Being overweight also increases the risk of type 2 diabetes, in which glucose damages blood vessels and makes them more prone to the development of plaque, as discussed next.

The Do's

Healthy Diet

Diet influences the amount of cholesterol in the blood. Cholesterol is ferried by two types of plasma proteins, called LDL (low-density lipoprotein) and HDL (high-density lipoprotein). LDL (called "bad" lipoprotein) takes cholesterol from the liver to the tissues, and HDL (called "good" lipoprotein) transports cholesterol out of the tissues to the liver. When the LDL level in blood is high or the HDL level is abnormally low, plaque, which interferes with circulation, accumulates on arterial walls (Fig. 12A).

Eating foods high in saturated fat (red meat, cream, and butter) or trans fats (most margarines, commercially baked goods, and deep-fried foods) raises the LDL-cholesterol level. Replacement of these harmful fats with monounsaturated fats (olive and canola oil) and polyunsaturated fats (corn, safflower, and soybean oil) is recommended. Cold-water fish (e.g., halibut, sardines, tuna, and salmon) also contain omega-3 polyunsaturated fatty acids that can reduce plaque. However, a study published in July 2008 suggests that the increasingly popular farm-raised tilapia contain low levels of these beneficial fatty acids and may, in fact, be detrimental to heart health (see Chapter 14).

Most nutritionists recommend consuming at least five servings of antioxidant-rich fruits and vegetables a day to protect against cardiovascular disease. Antioxidants protect the body from free radicals that oxidize cholesterol and damage the lining of an artery, leading to a blood clot that can block blood vessels. Taking antioxidant vitamins (A, E, and C) may have similar benefits.

Cholesterol Profile

Starting at age 20, all adults are advised to have their cholesterol levels tested at least every five years. Even in healthy individuals, an LDL level above 160 mg/100 ml and an HDL level below 40 mg/100 ml are matters of concern. If a person has heart disease or is at risk for heart disease, an LDL level below 100 mg/100 ml is now recommended. Cholesterol-lowering medications are available for those who do not meet these minimum guidelines.

Exercise

People who exercise are less apt to have cardiovascular disease. Exercise not only helps keep weight under control, but may also help minimize stress and reduce hypertension. And short bursts of exercise may be superior to longer sessions. In one study, as few as three 10-minute workout sessions a day reduced triglyceride levels in blood better than one 30-minute session.

Anxiety and Stress

Mental stress can increase the odds of a heart attack. Within an hour of a strong earthquake that struck near Los Angeles in 1994, 16 people died of sudden heart failure (compared to the average of about 4 per day). Over the next several days, the number of heart-related deaths declined, suggesting that emotional stress had triggered fatal complications in those who were already predisposed to them. Obviously, it is difficult to avoid earthquakes, but we can learn healthy ways to avoid and manage stress.

Can Alcohol Benefit the Heart?

Doctors usally advise patients with heart disease to avoid drinking alcohol. But a recent 21-year study of 12,000 people suggested that one or two drinks a day lowered the risk of heart attacks, presumably by increasing HDL cholesterol and reducing abnormal blood clots. Overall death rates were not affected, however, which perhaps should remind us of some of the potential downsides of alcohol consumption.

Discussion Questions

1. Considering the risk factors discussed in this reading, which, if any, could be affecting you right now?
2. Some scientists suspect that infections with certain types of bacteria and viruses may contribute to cardiovascular disease. What are some specific ways that infections might do this? If a connection could be proven, can you think of any new treatment strategies that should be tested?

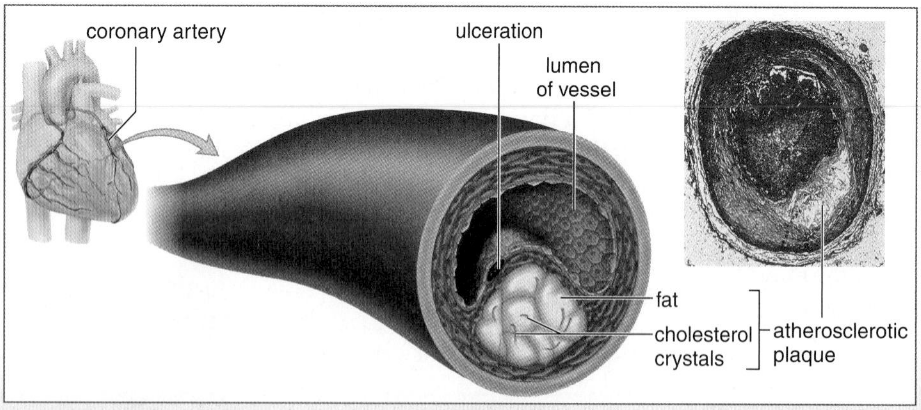

Figure 12A Coronary arteries and plaque.
Atherosclerotic plaque is an irregular accumulation of cholesterol and fat. When plaque is present in a coronary artery, a heart attack is more likely to occur because of restricted blood flow.

Health Focus

Medicinal Leeches: Medicine Meets *Fear Factor*

Although it may seem more like an episode of a popular TV show than a real-life medical treatment, the U.S. Food and Drug Administration has approved the use of leeches as medical "devices" for treating conditions involving poor blood supply to various tissues.

Leeches are bloodsucking, aquatic creatures, whose closest living relatives are earthworms (Fig. 12B). Prior to modern times, medical practitioners frequently applied leeches to patients, mainly in an attempt to remove the bad "humors" that they thought were responsible for many diseases. This practice was abandoned, thankfully, in the nineteenth century when it was realized that the "treatment" often harmed the patient.

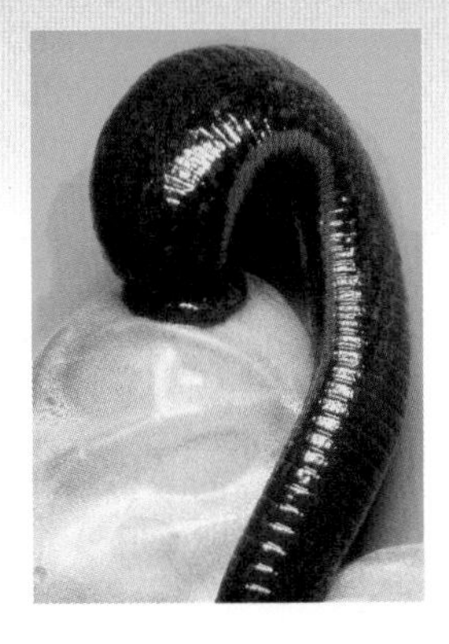

Figure 12B Leeches. Leeches can attach to human skin and suck blood out. Their use in medicine seems to be making a comeback.

True to their tenacious nature, however, leeches are making a comeback in twenty-first-century medicine. By applying leeches to tissues that have been injured by trauma or disease, blood supply can be improved. When reattaching a finger, for example, it is easier to suture together the thicker-walled arteries than the veins. Poorly draining blood from the veins can pool in the appendage and threaten its survival. It turns out that leech saliva contains chemicals that dilate blood vessels and prevent blood from clotting by blocking the activity of thrombin. These effects can improve the circulation to the body part. Another substance in leech saliva actually anesthetizes the bite wound. In a natural setting, this allows the leech to feast on the blood supply of its victim undetected, but in a medical setting, it makes the whole experience more tolerable, at least physically. Mentally, however, the application of leeches can still be a rather unsettling experience, and patient acceptance is a major factor limiting their more widespread use.

Discussion Questions

1. If you had an injury and your doctor said it might help, would you be willing to let leeches feast for a few minutes, say, on your hand? On your face?
2. Leeches were historically used for bloodletting, one of the oldest medical practices. Can you think of several reasons why, prior to the mid-nineteenth century, early physicians might have gotten the idea that removing blood could help cure various diseases?
3. The use of actual living organisms in medical treatments is sometimes called "biotherapy." Can you think of other situations in which living creatures might be used for human health benefits, either directly or indirectly?

Hypertension

Hypertension, or high blood pressure, affects about 20% of all Americans. It is sometimes called "the silent killer" because it may not be detected until a stroke or heart attack occurs. Normal blood pressure values vary somewhat among different age groups, body sizes, and levels of athletic conditioning, but according to guidelines issued in 2003 by the National Heart, Lung, and Blood Institute, prehypertension is present when the systolic pressure reading is between 120 and 139, or the diastolic pressure is 80 to 89. Prehypertension tends to get worse over time, and readings higher than 140/90 are considered to constitute hypertension for most individuals. Hypertension is most often due to a narrowing of the arteries. Consider trying to force a gallon of water through a straw, versus through a garden hose, in the same amount of time. You would need to apply much more force to get the water through the straw. In the same way, forcing blood through narrowed arteries over a prolonged period of time creates additional pressure on the cardiovascular system that can damage the blood vessels, heart, and other organs. Other factors that contribute to hypertension include obesity, smoking, chronic stress, and a high dietary salt intake (which tends to make the body retain more fluid, which in turn increases blood volume). Many medications can be used to treat high blood pressure, including diuretics (which reduce the blood volume by increasing urine output), vasodilators to dilate the blood vessels, and various drugs that improve heart function.

Check Your Progress

1. What are some treatment options, as well as potential limitations, for patients who need their heart replaced?
2. How does hypertension affect other organ systems? How can it be treated?

Applying the Concepts Revisited

TV personality Tim Russert was probably the most famous recent victim of cardiovascular disease, but it is very likely that everyone reading this knows someone who has suffered a similar fate. Thus, in terms of life and death applications, the cardiovascular system is in some ways the most important of all.

1. Was there anything else that Russert could have done to avoid this fatal outcome, or is it simply inevitable in some cases?
2. At what age should individuals begin to pay attention to the health of their cardiovascular system? What are the two or three most important steps they can take?

SUMMARIZING THE CONCEPTS

12.1 The Blood Vessels

There are three types of blood vessels:

- Arteries (and arterioles) take blood away from the heart.
- Capillaries, where exchange of substances with the tissues occurs.
- Veins (and venules) take blood to the heart.

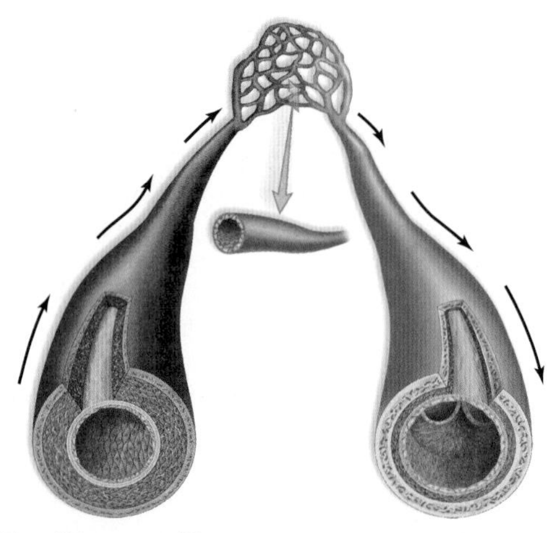

12.2 The Human Heart

The heart has a right and left side and four chambers:

- On the right side, the venae cavae deliver O_2-poor blood from the body into the right atrium, and the right ventricle pumps it via the pulmonary trunk into the pulmonary circuit.
- On the left side, the pulmonary veins bring O_2-rich blood from the lungs into the left atrium, and the left ventricle pumps it through the aorta into the systemic circuit.
- Blood from the atria passes through atrioventricular valves into the ventricles. Blood leaving the ventricles passes through semilunar valves.
- During the cardiac cycle, the SA node (cardiac pacemaker) initiates the heartbeat by causing the atria to contract. The AV node conveys the stimulus to the ventricles, causing them to contract.
- The heart sounds, "lub-dup," are due to the closing of the atrioventricular valves, followed by the closing of the semilunar valves.

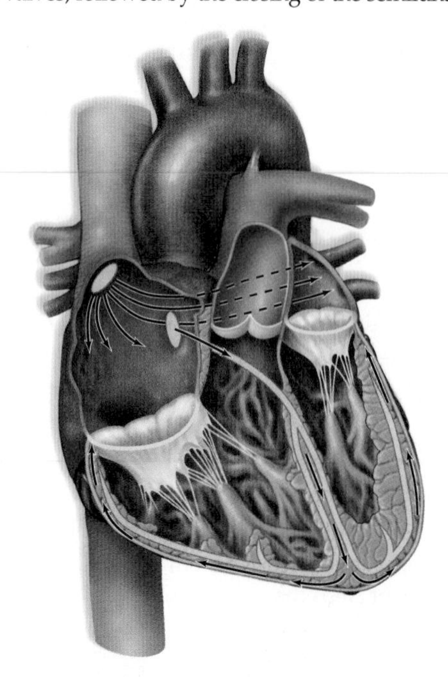

12.3 The Vascular Pathways

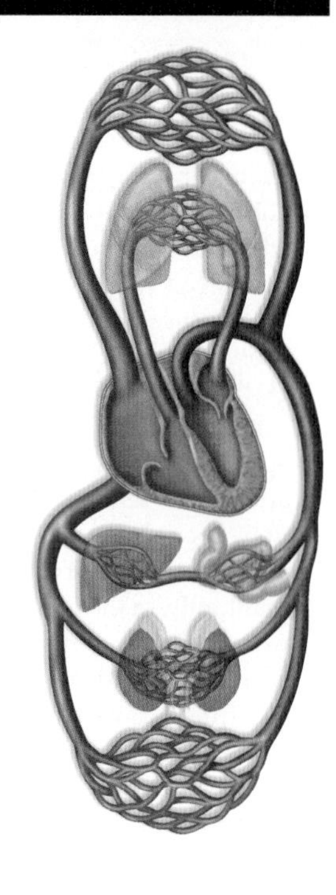

The cardiovascular system is divided into the pulmonary circuit and the systemic circuit:

- In the pulmonary circuit, the pulmonary trunk from the right ventricle of the heart and the two pulmonary arteries take O_2-poor blood to the lungs, and four pulmonary veins return O_2-rich blood to the left atrium of the heart.
- In the systemic circuit, O_2-rich blood is pumped from the left ventricle into the aorta, which branches off to form arteries going to specific organs. Eventually, arteries divide into arterioles and capillaries, and capillaries lead to venules, which join to form veins. The vein that carries O_2-poor blood to the vena cava most likely has the same name as the artery that delivered blood to the organ.
- Portal systems begin and end in capillaries. The hepatic portal vein originates from intestinal capillaries and finishes in hepatic capillaries.
- The left ventricle forces blood into the aorta under pressure. Systolic pressure is higher because it occurs during ventricular contraction; diastolic pressure occurs during relaxation. Blood pressure is higher in arteries than in veins, where muscular movement and the presence of one-way valves helps return venous blood to the heart.

12.4 Blood

Blood has two main parts: the plasma and the formed elements.

- Plasma contains mostly water and proteins, as well as nutrients and wastes. The plasma proteins prothrombin and fibrinogen are well known for their contribution to the clotting process.
- The formed elements are red blood cells, white blood cells, and platelets. Red blood cells contain hemoglobin and function in oxygen transport. White blood cells help defend against disease. Neutrophils and monocytes are phagocytic. Lymphocytes are involved in the development of adaptive immunity to disease. Platelets are involved in blood clotting.
- When blood reaches a capillary, water moves out at the arterial end due to blood pressure. At the venule end, water moves in due to osmotic pressure. In between, nutrients diffuse out and wastes diffuse in.

12.5 Cardiovascular Disorders

- Hypertension and atherosclerosis are two cardiovascular disorders that lead to stroke, heart attack, and aneurysm. Medical and surgical procedures are available to manage cardiovascular disease, but the best policy is prevention by following a heart-healthy diet, getting regular exercise, maintaining a proper weight, and not smoking.

TESTING YOURSELF

Choose the best answer for each question.

1. _____ lie between _____ and _____.
   a. Arteries, veins, capillaries
   b. Capillaries, arteries, veins
   c. Veins, arteries, capillaries
   d. None of these are correct.
2. Gas (oxygen and carbon dioxide) exchange occurs across the _____ of the _____.
   a. veins, lungs
   b. capillaries, tissues
   c. arteries, tissues
   d. All of these are correct.
3. The myocardium is made of
   a. muscle.
   b. epithelium.
   c. connective tissue.
   d. None of these are correct
4. The average adult heart rate is about _____ beats per minute.
   a. 120
   b. 100
   c. 45
   d. 70
5. The "lub," or first heart sound, is produced by closing of the
   a. aortic semilunar valve.
   b. pulmonary semilunar valve.
   c. tricuspid valve.
   d. bicuspid valve.
   e. both AV valves.
6. Label the following ECG wave chart.

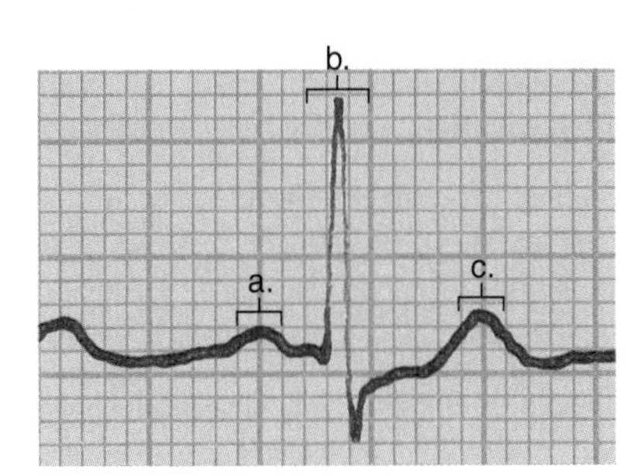

7. All arteries in the body contain O_2-rich blood with the exception of the
   a. aorta.
   b. pulmonary artery.
   c. renal artery.
   d. coronary arteries.
8. Which of the following assists in the return of venous blood to the heart?
   a. valves
   b. skeletal muscle contraction
   c. respiratory movements
   d. reduction in cross-sectional area from venules to veins
   e. All of these are correct.
9. The best explanation for the slow movement of blood in capillaries is
   a. skeletal muscles press on veins, not capillaries.
   b. capillaries have much thinner walls than arteries.
   c. there are many more capillaries than arterioles.
   d. venules are not prepared to receive so much blood from the capillaries.
   e. All of these are correct.
10. Label the following diagram of the systemic circuit.

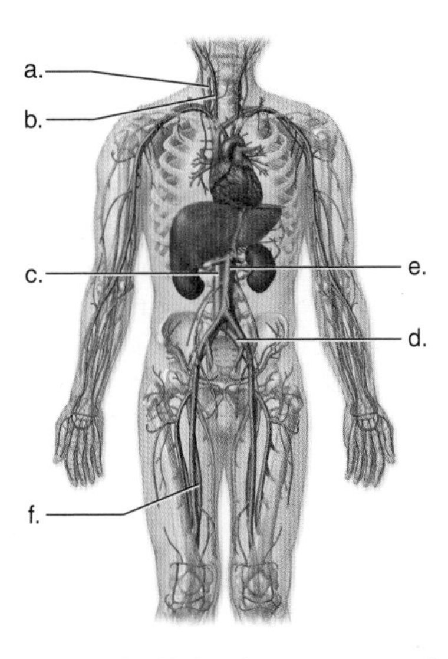

11. What is the type of cell that becomes any of the formed elements in blood?
    a. multipotent stem cells
    b. myeloid stem cells
    c. lymphoid stem cells
    d. All of these are correct.
12. Which association is incorrect?
    a. white blood cells—infection fighting
    b. red blood cells—blood clotting
    c. red blood cells—hemoglobin
    d. platelets—blood clotting
13. A decrease in lymphocytes would result in problems associated with
    a. clotting.
    b. immunity.
    c. oxygen transport.
    d. All of these are correct.
14. Anemia can result from
    a. an iron-deficient diet.
    b. a drop in red blood cell count.
    c. lack of hemoglobin production.
    d. All of these are correct.
15. In the tissues, the amount of carbon dioxide is _____ than in the capillaries.
    a. lower
    b. higher
    c. neither higher nor lower
    d. None of these are correct.
16. Lymph is formed from
    a. urine.
    b. whole plasma.
    c. excess extracellular tissue fluid.
    d. All of these are correct.
17. Water enters the venous side of capillaries because of
    a. active transport from tissue fluid.
    b. higher osmotic pressure than blood pressure.
    c. higher blood pressure on the venous side.
    d. higher blood pressure than osmotic pressure.
    e. higher red blood cell concentration on the venous side.
18. A stent would most likely be used to treat:
    a. hemophilia
    b. atherosclerosis
    c. mitral valve prolapse
    d. hypertension

UNDERSTANDING THE TERMS

THINKING CRITICALLY

1. A few specialized tissues do not contain any blood vessels, including capillaries. Can you think of two or three? How might these tissues survive without a direct blood supply?
2. Assume your heart rate is 70 beats per minute (bpm), and each minute your heart pumps 5.25 liters of blood to your body. Based on your age to the nearest day, about how many times has your heart beat so far, and what volume of blood has it pumped?
3. Examine the following abnormal ECG tracings. Compared with the normal tracing in Testing Yourself Question 6 or Figure 12.6*b*, what is your best guess as to what type of problem each patient may have? What type of device can be implanted in a patient's heart to correct some abnormalities of the heart, and how does it generally work?

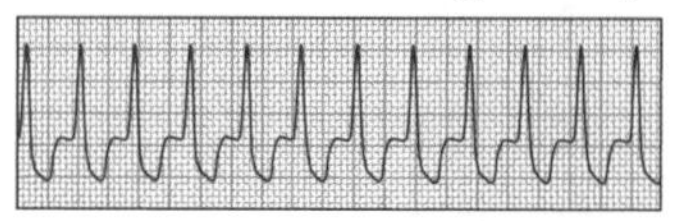

a.

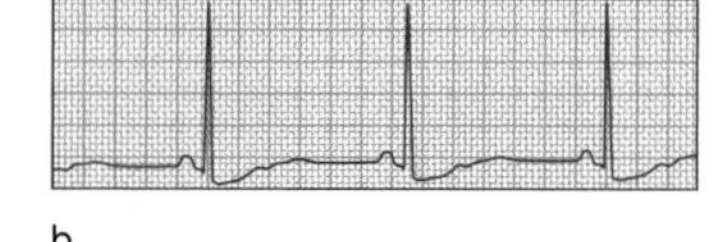

b.

4. There are not enough living hearts available to meet the need for heart transplants. What are some of the major problems a manufacturer would have to overcome when attempting to design an artificial heart that will last for years inside a patient's body?

INQUIRY INTO LIFE WEBSITE

The companion website for *Inquiry into Life* provides a wealth of information organized and integrated by chapter. You will find practice tests, animations, videos, and much more that will complement your learning and understanding of general biology.

http://www.mhhe.com/maderinquiry13

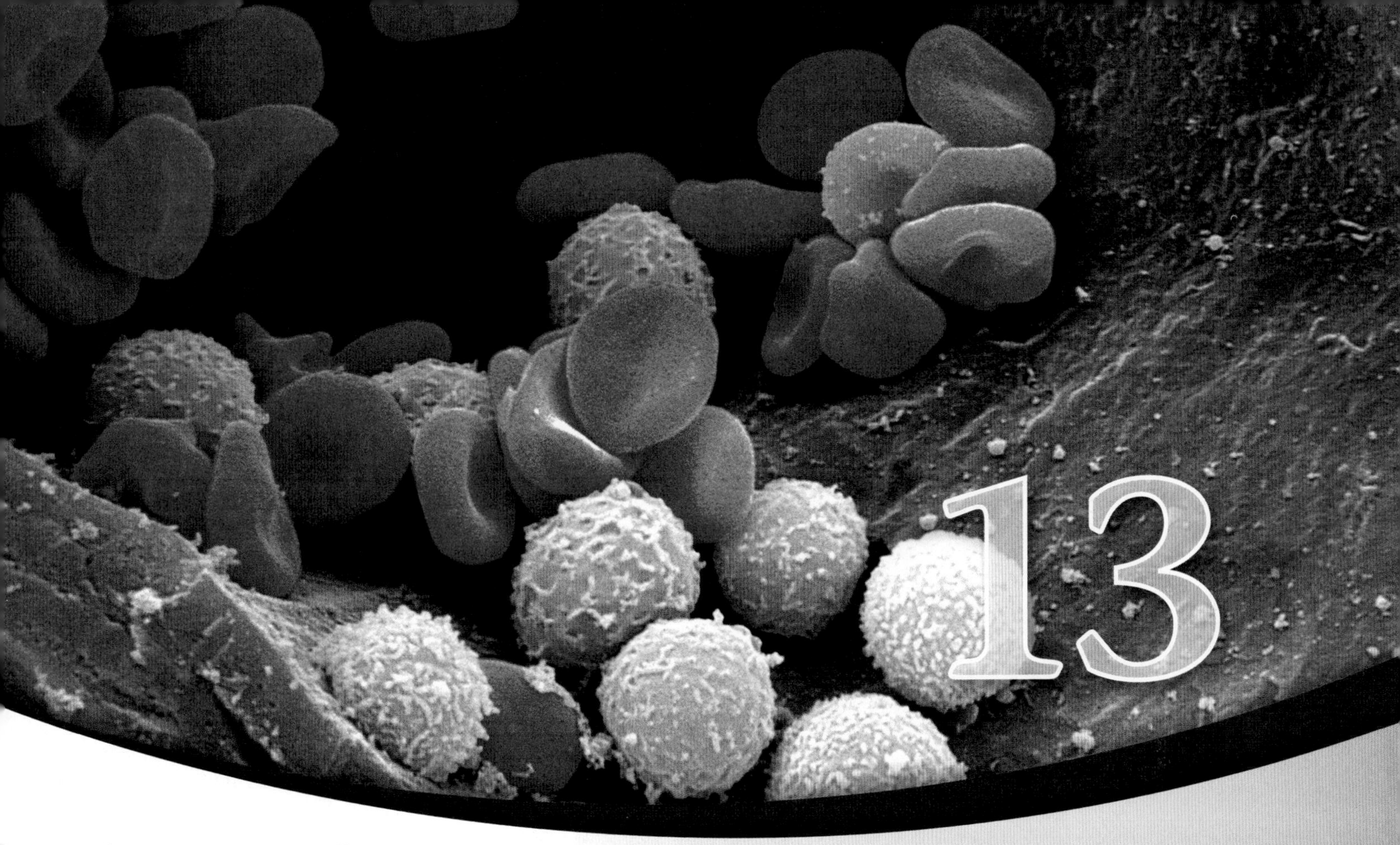

Applying the Concepts

Jason is an active three-year-old boy, but his parents are worried. They have been making frequent visits to the pediatrician's office since their son was about six months old. He seems to have had one ear infection after another, and they can't remember him going a month without having a cough or runny nose. These chronic infections led the doctor to suspect that Jason might be suffering from some kind of inherited defect in his immune system. After initial blood tests revealed that the boy had very low levels of antibodies in his blood, more specialized testing resulted in a diagnosis of X-linked agammaglobulinemia (XLA).

XLA is the most common of the inherited immunodeficiencies. Affected infants usually appear normal until about six months of age, when they start developing many infections. XLA is due to a mutated gene on the X chromosome that is needed for proper development of B lymphocytes, which are white blood cells that produce antibodies (the ruffled-looking cells in the photo are white blood cells adhering to a blood vessel wall). As we will see in this chapter, antibodies are an important component of the human immune system, and without them, serious or even fatal infections often occur. In this case, Jason will need to be treated with intravenous injections of human immunoglobulin every three to four weeks, for the rest of his life. His parents' hope, however, is that within their son's lifetime, doctors will be able to cure the genetic deficiency that caused his problem.

Lymphatic and Immune Systems

Chapter Outline

Photo: © Dr. Richard Kessel and Dr. Randy Kardon/Tissues and Organs/Visuals Unlimited

13.1 The Lymphatic System

Learning Outcomes

1. Describe three major functions of the lymphatic system.
2. Distinguish between primary and secondary lymphoid organs, and list the main functions of each.

The **lymphatic system** consists of lymphatic vessels and the lymphoid organs (Fig. 13.1). This system, which is closely associated with the cardiovascular system, has three main functions that contribute to homeostasis: (1) lymphatic capillaries take up excess tissue fluid and return it to the bloodstream; (2) small lymphatic capillaries absorb fats from the digestive tract and transport them to the bloodstream; and (3) lymphoid cells defend the body against disease. This last function is mainly carried out by the white blood cells present in lymphatic vessels and lymphoid organs, as well as in the blood.

Lymphatic Vessels

Lymphatic vessels form a one-way system that begins with lymphatic capillaries. Most regions of the body are richly supplied with lymphatic capillaries, tiny, closed-ended vessels whose walls consist of simple squamous epithelium (see Fig. 13.1). Lymphatic capillaries take up excess tissue fluid. Tissue fluid is mostly water, but it also contains solutes (i.e., nutrients, electrolytes, and oxygen) derived from plasma and cellular products (i.e., hormones, enzymes, and wastes) secreted by cells. The fluid inside lymphatic vessels is called **lymph.**

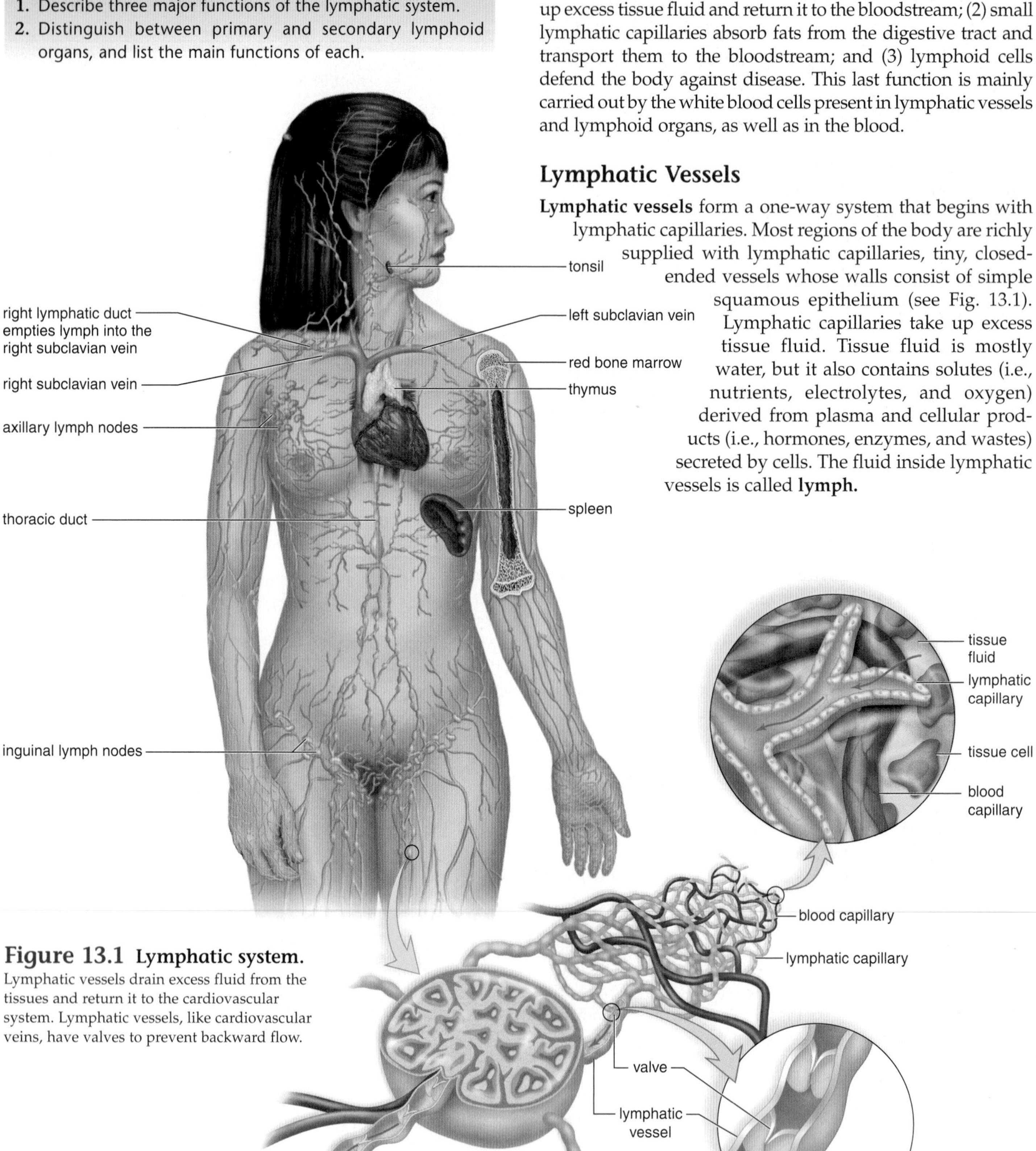

Figure 13.1 Lymphatic system. Lymphatic vessels drain excess fluid from the tissues and return it to the cardiovascular system. Lymphatic vessels, like cardiovascular veins, have valves to prevent backward flow.

The lymphatic capillaries join to form lymphatic vessels that merge before entering one of two ducts: the thoracic duct or the right lymphatic duct. The larger, thoracic duct returns lymph collected from the body below the thorax plus the left arm, the left side of the head, and the neck, into the left subclavian vein. The right lymphatic duct returns lymph from the right arm and right side of the head and neck into the right subclavian vein.

The structure of the larger lymphatic vessels is similar to that of cardiovascular veins. The movement of lymph within lymphatic capillaries is largely dependent upon skeletal muscle contraction. Lymph forced through lymphatic vessels as a result of muscular compression is prevented from flowing backward by one-way valves.

Edema is a localized swelling caused by the accumulation of tissue fluid that has not been collected by the lymphatic system. It occurs if too much tissue fluid is made and/or if not enough of it is drained away. Edema can lead to tissue damage and even death, illustrating the importance of the function of the lymphatic system. The fat absorption and defense functions of the lymphatic system are equally important. Unfortunately, cancer cells sometimes enter lymphatic vessels and move undetected to other regions of the body where they produce secondary tumors. In this way, the lymphatic system sometimes assists metastasis, the spread of cancer far from its place of origin.

Lymphoid Organs

The primary **lymphoid organs** are red bone marrow and the thymus. The secondary lymphoid organs are the spleen and lymph nodes (Fig. 13.2). Lymphoid organs contain large numbers of white blood cells known as lymphocytes. There are two main types of lymphocytes: B lymphocytes (B cells) and T lymphocytes (T cells). Lymphocytes develop and mature in the primary lymphoid organs, and some lymphocytes become activated in secondary lymphoid organs.

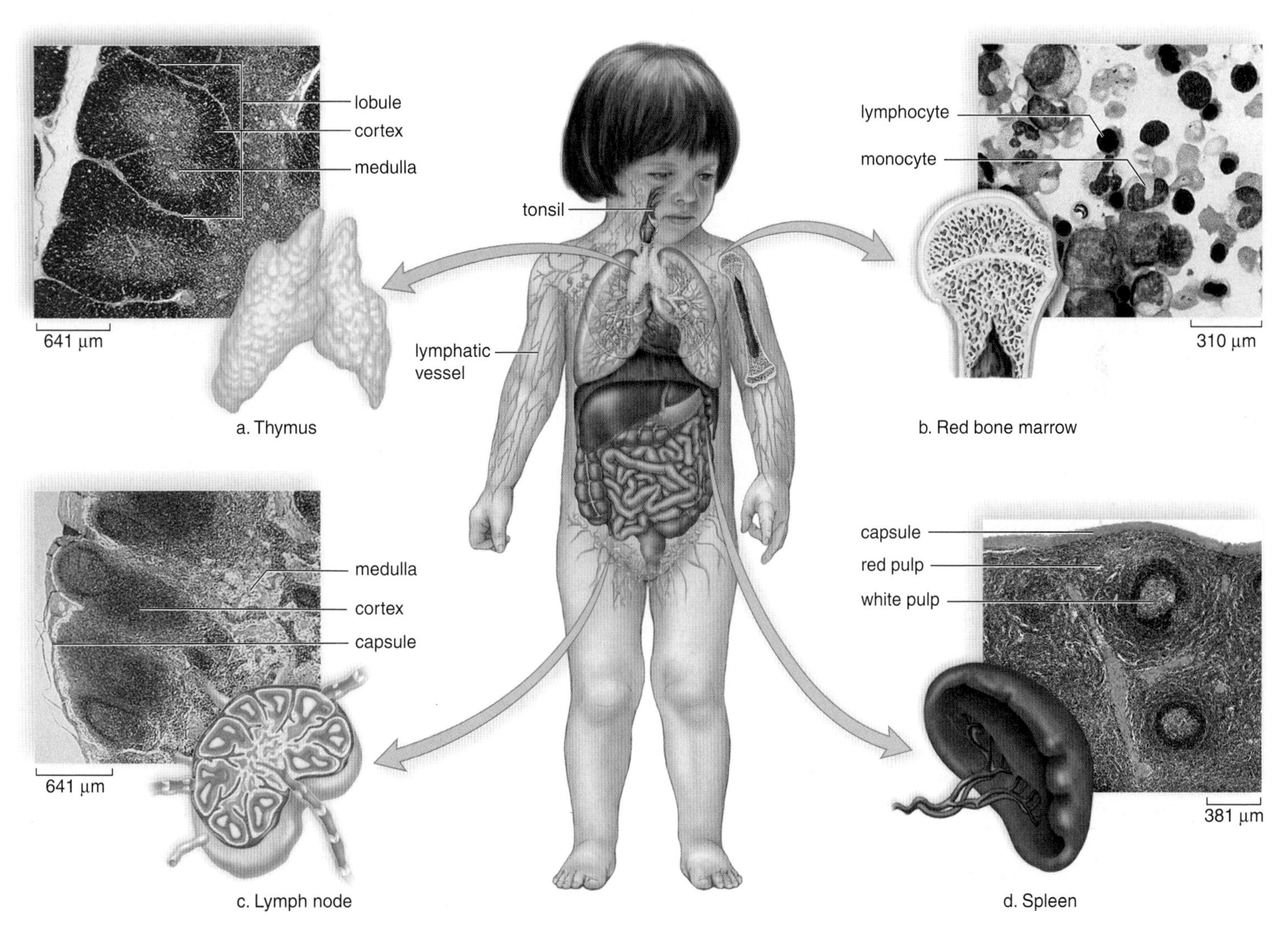

Figure 13.2 The lymphoid organs.
The thymus (**a**) and red bone marrow (**b**) are the primary lymphoid organs. Blood cells, including lymphocytes, are produced in red bone marrow. B cells mature in the bone marrow. T cells mature in the thymus. The lymph nodes (**c**) and the spleen (**d**) are secondary lymphoid organs. Lymph is cleansed in the nodes, and blood is cleansed in the spleen.

Primary Lymphoid Organs

Red bone marrow is the site of stem cells that are ever capable of dividing and producing blood cells. Some of these cells become the various types of white blood cells: neutrophils, eosinophils, basophils, lymphocytes, and monocytes (see Fig. 12.11).

In a child, most bones contain red bone marrow, but in an adult, it is present only in the bones of the skull, the sternum (breastbone), the ribs, the clavicle, the pelvic bones, and the vertebral column. The red bone marrow consists of a network of connective tissue fibers that support the stem cells and their progeny. The cells are packed around thin-walled sinuses filled with venous blood. Differentiated blood cells enter the bloodstream at these sinuses. Bone marrow is not only the source of B cells, but also the place where B cells mature. T cells mature in the thymus.

The soft, bilobed **thymus** is located in the thoracic cavity between the trachea and the sternum above the heart. The thymus varies in size, but it is largest in children and shrinks as we get older. Connective tissue divides the thymus into lobules, which are filled with T cells and supporting cells. Immature T cells migrate from the bone marrow through the bloodstream to the thymus, where they mature. Only about 5% of these cells ever leave the thymus. T cells that are capable of reacting to the body's own cells undergo apoptosis, which involves a cascade of specific cellular events leading to the death and destruction of the cell (see Chapter 5). Those that leave the thymus have the capability of reacting to foreign molecules or cells. These cells are said to be capable of distinguishing "self" from "nonself."

Secondary Lymphoid Organs

Lymphocytes frequently migrate from the bloodstream into the secondary lymphoid organs. Here, they may encounter foreign molecules or cells, after which they proliferate and become activated. These activated cells usually reenter the bloodstream, where they search for sites of infection or inflammation, like a squadron of highly trained military personnel seeking to destroy a specific enemy.

The **spleen** is located in the upper left side of the abdominal cavity behind the stomach. Most of the spleen is red pulp that filters the blood. Red pulp consists of blood vessels and sinuses where macrophages remove old and defective blood cells. Lymphocytes cleanse the blood of foreign particles. The spleen also has white pulp that is inside the red pulp and consists of small areas of lymphatic tissue.

The spleen's outer capsule is relatively thin, and an infection or trauma can cause the spleen to burst. Although the spleen's functions are largely replaced by other organs, a person without a spleen is slightly more susceptible to infections and may have to receive antibiotic therapy indefinitely.

Lymph nodes are small (from 1–25 mm in diameter), ovoid structures occurring along lymphatic vessels that cleanse lymph. Connective tissue divides the organ into nodules. Each of these is packed with B and T cells and contains a sinus. As lymph courses through the many sinuses, resident macrophages engulf **pathogens,** which are disease-causing agents such as viruses and bacteria, and also debris.

Sometimes incorrectly called "glands," lymph nodes are named for their location. For example, inguinal nodes are in the groin, and axillary nodes are in the armpits. Physicians often feel for the presence of swollen, tender lymph nodes as evidence that the body is fighting an infection. This is a noninvasive, preliminary way to help make such a diagnosis.

CHECK YOUR PROGRESS

1. How is lymph returned to the cardiovascular system?
2. Define edema. What can cause edema?
3. Name two primary and two secondary lymphoid organs. How does the function of the primary lymphoid organs differ from that of the secondary lymphoid organs?

13.2 Innate and Adaptive Immunity

LEARNING OUTCOMES

1. Define innate and adaptive immunity.
2. Describe the four mechanisms of innate immunity.
3. Compare B-cell versus T-cell responses.

Immunity is the body's capability of removing or killing foreign substances, pathogens, and cancer cells. Mechanisms of **innate immunity** are fully functional without previous exposure to these substances, while **adaptive immunity** is enhanced by exposure to specific antigens. An **antigen** is any molecule, usually a protein or carbohydrate, that stimulates an immune response. Mechanisms of innate immunity resemble a general police force, with members who are skilled at dealing with many different types of criminals. Adaptive immunity is carried out mainly by the lymphocytes, which more closely resemble bounty hunters trained to seek out a very specific perpetrator.

Innate Immunity

Mechanisms of innate immunity can be divided into at least four types—physical and chemical barriers, inflammation, phagocytes and natural killer cells, and protective proteins.

Physical and Chemical Barriers

Skin and the mucous membranes lining the respiratory, digestive, and urinary tracts serve as mechanical barriers to entry of pathogens. The upper respiratory tract is lined with ciliated cells that sweep mucus and trapped particles up into the throat, where they can be swallowed or expelled.

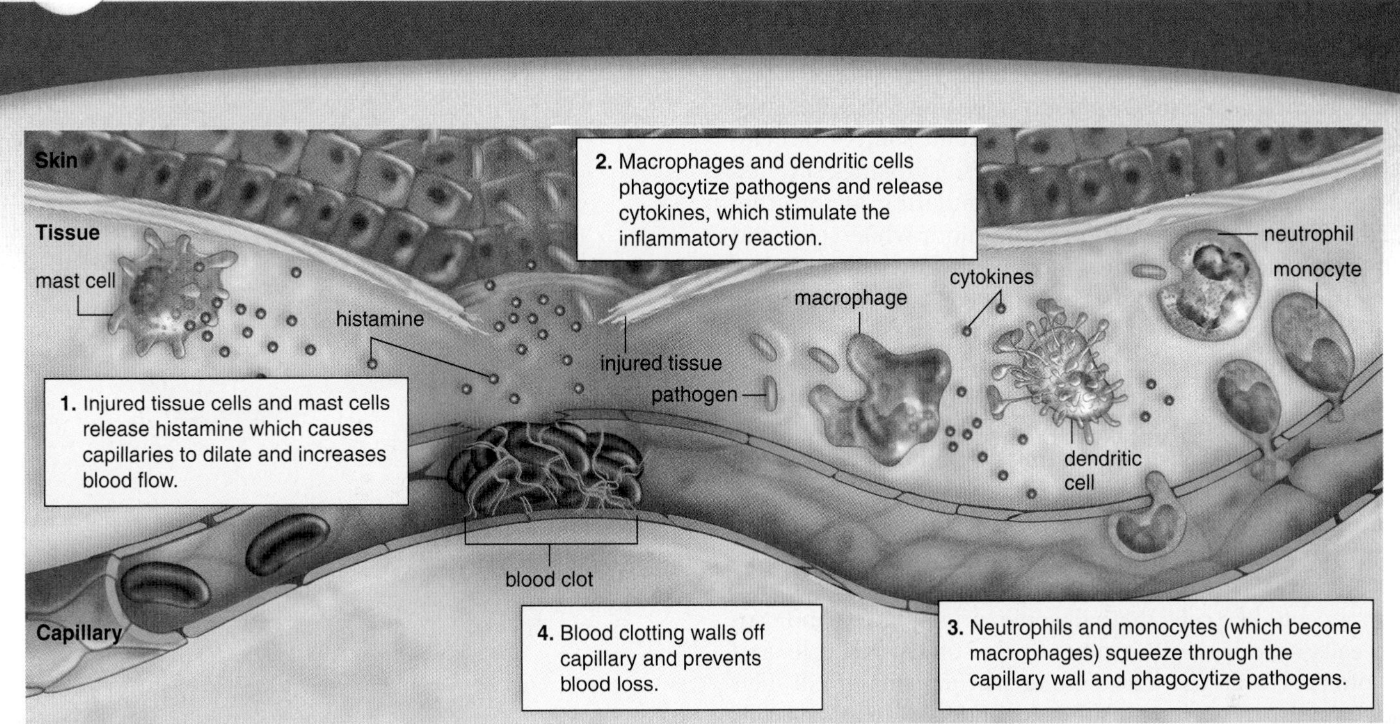

Figure 13.3 Inflammatory reaction.
Due to capillary changes in a damaged area and the release of chemical mediators, such as histamine by mast cells, an inflamed area exhibits redness, heat, swelling, and pain. The inflammatory reaction can be accompanied by other reactions to the injury. Macrophages and dendritic cells, present in the tissues, phagocytize pathogens, as do neutrophils, which squeeze through capillary walls from the blood. Macrophages and dendritic cells release cytokines, which stimulate the inflammatory and other immune reactions. A blood clot can form to seal a break in a blood vessel.

The secretions of oil glands in the skin contain chemicals that weaken or kill certain bacteria on the skin. The stomach has an acid pH, which kills many types of bacteria or inhibits their growth. The various bacteria that normally reside in the intestine and other areas, such as the vagina, take up nutrients and block binding sites that potentially could be used by pathogens.

Inflammation

Whenever tissue is damaged by physical or chemical agents or by pathogens, a series of events occurs that is known as the **inflammatory reaction.** Figure 13.3 illustrates the main participants in the inflammatory reaction. Inflammation tends to wall off infections and increase the exposure and access of the immune system to the inciting agent.

An inflamed area has four signs: redness, heat, swelling, and pain. Most of these signs are due to capillary changes in the damaged area. Chemicals such as **histamine,** released by damaged tissue cells as well as by **mast cells,** a type of immune cell found in tissues, cause nearby capillaries to dilate and become more permeable. This also results in increased blood flow, causing the skin to redden and feel warm. Increased capillary permeability allows fluids to escape into the tissues, resulting in swelling. The swollen area stimulates free nerve endings, causing the sensation of pain.

The inflammatory reaction can be accompanied by other responses to the injury. A blood clot can form to seal a break in a blood vessel. Phagocytes, discussed in the next section, enter inflamed tissues. Antigens, chemical mediators, dendritic cells, and macrophages move through the tissue fluid and lymph to the lymph nodes. Dendritic cells and macrophages cause T cells to mount a specific defense to the infection, as described later in this section.

If the cause of inflammation cannot be eliminated, as occurs in tuberculosis and some types of arthritis, the inflammatory reaction may persist and become harmful rather than helpful. In these cases, anti-inflammatory medications such as aspirin, ibuprofen, or cortisone may be used to minimize the detrimental effects of chronic inflammation.

Phagocytes and Natural Killer Cells

At sites of inflammation, at least two types of **phagocytes** which literally means "eating cell," migrate through the walls of dilated capillaries. **Neutrophils** are usually the first cells to arrive at the scene, and they may accumulate to form pus, which has a whitish appearance mainly due to their presence. If the inflammatory reaction continues, more **monocytes** will migrate from the blood to the tissues, where they are then called **macrophages** (meaning, literally, "large eaters"). Macrophages are also important sources of **cytokines,** which are chemical messengers. Cytokines secreted by macrophages include colony-stimulating factors that cause the bone marrow to produce more white cells, and others that act on the brain to induce a fever response. When a neutrophil or a macrophage encounters a pathogen, especially a bacterial cell, it will engulf the pathogen into an endocytic vesicle (see Fig. 4.11), which fuses with a lysosome inside the cell. The lysosome has an acid pH that activates hydrolytic enzymes and various reactive oxygen compounds, which can usually destroy the pathogen. Also present in various tissues are phagocytic **dendritic cells,** which play an important role in interacting with T cells during adaptive immune responses.

Natural killer (NK) cells are large, granular, lymphocyte-like cells that kill some virus-infected and cancer cells by cell-to-cell contact. Interestingly, NK cells and cytotoxic T cells use very similar mechanisms to kill their target cells, by inducing them to undergo apoptosis. NK cells, however, seek out and kill cells that lack a particular type of "self" molecule, called MHC-I (major histocompatibility class I), on their surface. Because some virus-infected and cancer cells may lack these MHC-I molecules, they often become susceptible to being killed by NK cells. However, because NK cells do not recognize specific viral or tumor antigens, and do not proliferate when exposed to a particular antigen, they are considered a part of the innate immune system.

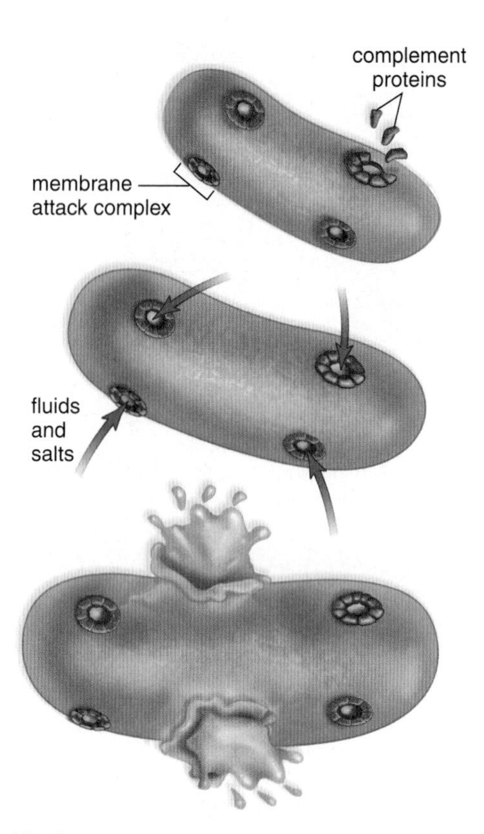

Figure 13.4 Action of the complement system against a bacterium.
When complement proteins in the blood plasma are activated in an immune response, they form a membrane attack complex. This complex makes holes in bacterial cell walls and plasma membranes, allowing fluids and salts to enter until the cell eventually bursts.

Protective Proteins

The **complement system,** often simply called complement, is composed of a number of blood plasma proteins designated by the letter C and a number (i.e., C3). The complement proteins "complement" certain immune responses, which accounts for their name. For example, they are involved in and amplify the inflammatory reaction because certain complement proteins can bind to mast cells and trigger histamine release, and others can attract phagocytes to the scene. Some complement proteins bind to the surface of pathogens, as described later in this section, which ensures that the pathogens will be phagocytized by a neutrophil, dendritic cell, or macrophage.

Certain other complement proteins join to form a membrane attack complex that produces holes in the surface of bacteria and some viruses. Fluids and salts then enter the bacterial cell or virus to the point that they burst (Fig. 13.4).

Interferons are proteins produced by virus-infected cells as a warning to noninfected cells in the area. Interferons can bind to receptors of noninfected cells, causing them to prepare for possible attack by producing substances that interfere with viral replication. Interferons may be administered to treat certain viral infections, such as hepatitis C.

Check Your Progress

1. Name a specific physical or chemical barrier.
2. Contrast the way that macrophages typically kill pathogens with the method used by natural killer cells.
3. What are three major functions of the complement system?

Adaptive Immunity

When innate defenses have failed to prevent an infection, adaptive defenses come into play. Because these defenses do not ordinarily react to our own normal cells, it is said that the immune system is able to distinguish "self" from "nonself." Adaptive defenses usually take five to seven days to become fully activated, and they may last for years. For example, once we recover from measles, we usually do not get measles a second time.

Adaptive defenses primarily depend on the action of B cells or T cells. B cells and T cells are capable of recognizing antigens because they have specific antigen receptors—plasma membrane receptor proteins whose shape allows them to combine with particular antigens. Each lymphocyte has only one type of receptor. It is often said that the

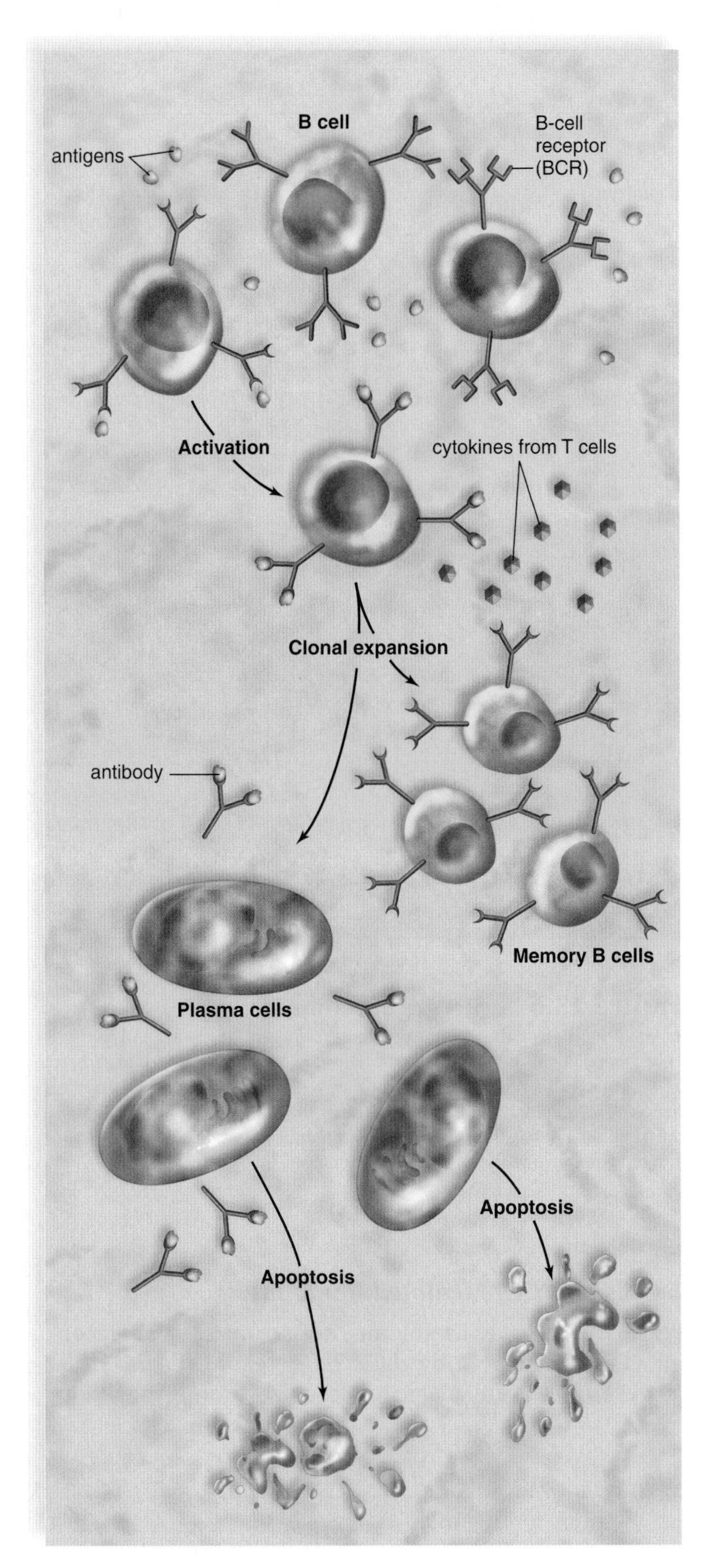

Figure 13.5 Clonal selection theory as it applies to B cells.
Each B cell has a B-cell receptor (BCR) designated by shape that will combine with a specific antigen. Activation of a B cell occurs when its BCR can combine with an antigen (colored green). In the presence of cytokines, the B cell undergoes clonal expansion, producing many plasma cells and memory B cells. These plasma cells secrete antibodies specific to the antigen, and memory B cells immediately recognize the antigen in the future. After the infection passes, plasma cells undergo apoptosis, also called programmed cell death.

receptor and the antigen fit together like a lock and a key. Because we may encounter millions of different antigens during our lifetime, we need a diversity of B cells and T cells to protect us against them. Remarkably, diversification occurs to such an extent during the maturation process that there are specific B cells and/or T cells for almost any possible antigen.

B cells give rise to plasma cells, which produce antibodies that are capable of combining with and neutralizing a particular antigen. In contrast, T cells do not produce antibodies. Instead, they differentiate into either helper T cells, which release chemicals to regulate the immune response, or cytotoxic T cells, which attack and kill virus-infected cells and tumor cells.

B Cells and Antibody-Mediated Immunity

The receptor for antigen on the surface of a B cell surface is called a **B-cell receptor (BCR).** B cells are activated in a lymph node or the spleen, when BCRs bind to specific antigens. Thereafter, the B cell divides by mitosis many times. In other words, it makes many copies of itself. Most of the resulting cells (clones) become plasma cells, which circulate in the blood and lymph. Plasma cells are larger than regular B cells because they have extensive rough endoplasmic reticulum for the mass production and secretion of antibodies to a specific antigen. Antibodies are the secreted form of the BCR of the B cell that was activated.

The **clonal selection theory** states that an antigen binds to the antigen receptor of only one type of B cell or T cell, and then this B cell or T cell divides, forming clones of itself. Note in Figure 13.5 that each B cell has a specific BCR represented by shape. Only the B cell with a BCR shape that fits the antigen (green circle) undergoes clonal expansion. During clonal expansion, cytokines secreted by helper T cells stimulate B cells to divide. Some cloned B cells become memory cells, which are the means by which long-term immunity is possible. If the same antigen enters the system again, memory B cells quickly divide and give rise to more plasma cells capable of rapidly producing the correct type of antibody.

Once the threat of an infection has passed, new plasma cells cease to develop, and most of those present undergo apoptosis (programmed cell death).

Defense by B cells is called **antibody-mediated immunity** because the various types of activated B cells become plasma cells that produce antibodies. It is also called humoral immunity because these antibodies are present in blood and lymph. (Historically, a humor is any fluid normally occurring in the body.)

Structure of an Antibody Antibodies are also called **immunoglobulins (Ig).** They are typically Y-shaped molecules with two arms. Each arm has a "heavy" (long) polypeptide chain and a "light" (short) polypeptide chain. These chains have constant (C) regions, where the sequence of amino acids is set, and variable (V) regions, where the sequence of amino acids varies between antibodies (Fig. 13.6). The constant regions are the same within a particular type or class of antibodies (see next section). The variable regions become hypervariable at their tips and form antigen-binding sites. It is the variable and hypervariable regions that allow the antibody to bind to a specific antigen. The antigen combines with the antibody at the antigen-binding site in a lock-and-key manner.

The antigen-antibody reaction can have several outcomes, but often the reaction produces complexes of antigens combined with antibodies. These complexes, also called immune complexes, mark the antigens for destruction. For example, an immune complex may be engulfed by neutrophils or macrophages, or it may activate complement. Antibodies may also "neutralize" viruses or toxins by preventing their binding to cells.

Types of Antibodies There are five major classes of antibodies in humans (Table 13.1). Immunoglobulin G (IgG) antibodies are the major type in blood, lymph, and tissue fluid. IgG antibodies bind to pathogens and their toxins and can activate complement. IgG is the only antibody class that can cross the placenta. Along with IgA, IgG is found in breast milk. IgM antibodies are pentamers, meaning that they contain five of the Y-shaped structures shown in Figure 13.6*a*. These antibodies appear in blood soon after a vaccination or infection and disappear relatively quickly. They are good complement activators. IgA antibodies are monomers in blood and lymph or dimers in tears, saliva, gastric juice, and mucous secretions. They are the main type of antibody found in body secretions. IgD molecules appear on the surface of B cells when they are ready to be activated. IgE antibodies are important in the immune system's response to parasites (e.g., parasitic worms). They are best known, however, for immediate allergic responses, including anaphylactic shock (discussed in section 13.4).

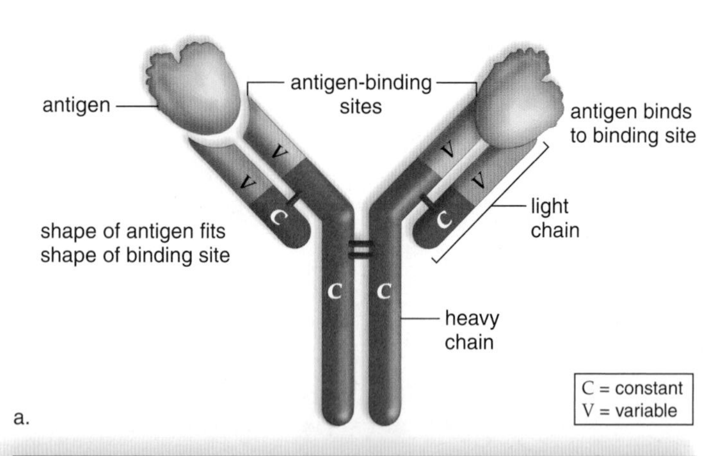

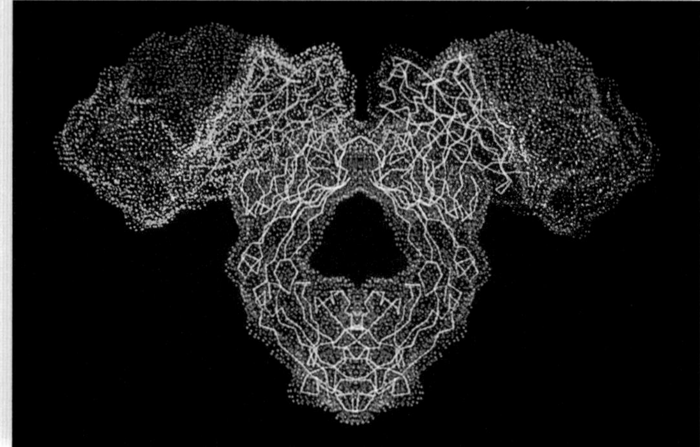

Figure 13.6 Structure of an antibody.
a. An antibody contains two heavy (long) polypeptide chains and two light (short) chains arranged so that there are two variable regions, where a particular antigen is capable of binding with an antibody (*V* = variable region, *C* = constant region). The shape of the antigen fits the shape of the binding site. **b.** Computer model of an antibody molecule. The antigen combines with the two side branches.

CHECK YOUR PROGRESS

1. Discuss the major tenets (principles) of the clonal selection theory.
2. Diagram the basic structure of an antibody molecule.
3. List the five classes of human antibodies and their main functions.

T Cells and Cell-Mediated Immunity

When a T cell leaves the thymus, it has a unique **T-cell receptor (TCR)** similar to the BCR on B cells. Unlike B cells, however, T cells are unable to recognize an antigen without

TABLE 13.1 Classes of Human Antibodies

Class	Distribution	Function
IgG	Main antibody type in circulation	Binds to pathogens, activates complement, and enhances phagocytosis
IgM	Antibody type found in circulation; largest antibody	Activates complement; clumps cells
IgA	Main antibody type in secretions such as saliva and milk	Prevents pathogens from attaching to epithelial cells in digestive and respiratory tract
IgD	Antibody type found on surface of immature B cells	Presence signifies readiness of B cell to respond to antigens
IgE	Antibody type found as antigen receptors on eosinophils in blood and on mast cells in tissues	Responsible for immediate allergic response and protection against certain parasitic worms

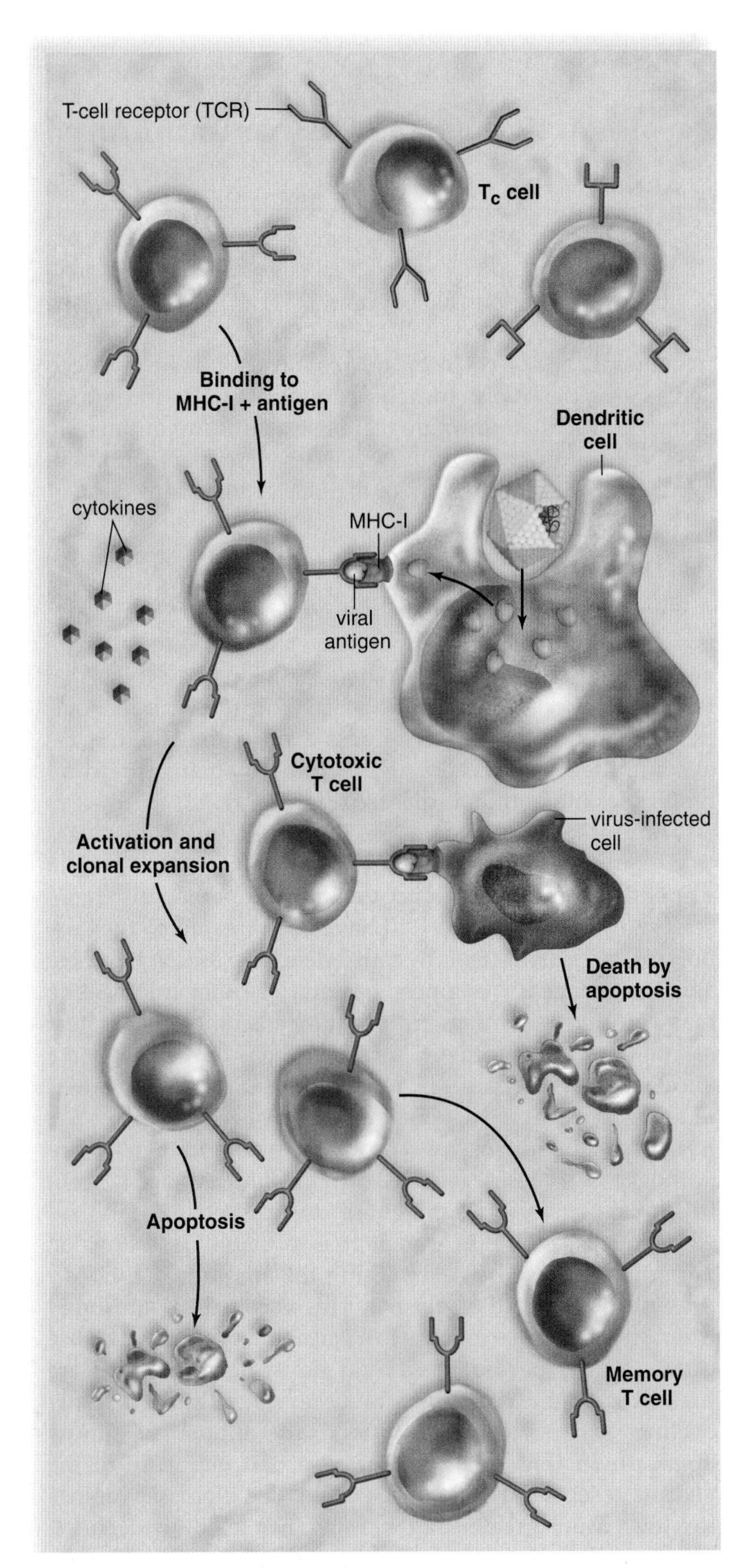

Figure 13.7 Clonal selection model as it applies to T_C cells.
Each T cell has a T-cell receptor (TCR) designated by a shape that will combine only with a specific antigen. Activation of a T cell occurs when its TCR can combine with an antigen. A dendritic cell presents the antigen (colored green) in the groove of an MHC class I (MHC-I) molecule. Thereafter, the T_C cell undergoes clonal expansion, and many copies of the same type of T cell are produced. After the immune response has been successful, the majority of T cells undergo apoptosis, but a small number are memory T cells. Memory T cells provide protection should the same antigen enter the body again at a future time.

help. The antigen must be displayed, or "presented," to the TCR by an **MHC (major histocompatibility complex)** protein on the surface of another cell.

There are two major types of T cells: **helper T cells** (T_H cells) and **cytotoxic T cells** (T_C cells, or CTLs). Each of these types has a TCR that can recognize an antigen fragment in combination with an MHC molecule. However, the major difference in antigen recognition by these two types of cells is that the T_H cells only recognize antigen presented by specialized **antigen-presenting cells (APCs)** with MHC class II molecules on their surface, while the T_C cells only recognize antigen presented by various cell types with MHC class I molecules on their surface (Fig. 13.7).

APCs such as dendritic cells or macrophages play an important role in initiating most adaptive immune responses. After ingesting and destroying a pathogen, APCs travel to a lymph node or the spleen, where T cells also congregate. After breaking the pathogen apart with lysosomal enzymes, the APC displays a piece of the pathogen on its surface, in the binding groove of an MHC class II molecule. After they are stimulated by APCs, T_H cells proliferate and secrete various cytokines. Cytokines produced by T_H cells help activate T_C cells, as well as B cells.

T_C cells have storage vacuoles that contain many molecules of perforin and enzymes called granzymes. After an activated T_C cell binds to a virus-infected or cancer cell that is presenting foreign antigen on its MHC class I molecules, the T_C cell releases perforin, which form pores in the plasma membrane of the abnormal cell. This allows the granzymes to enter the target cell, inducing it to undergo apoptosis and die (Fig. 13.8*a*). The T_C cell is then capable of moving on to kill another target cell (Fig. 13.8*b*).

Notice also in Figure 13.7 that, similar to B cells, some of the clonally expanded T cells become memory T cells. These may live for many years and can quickly jump-start an immune response to an antigen previously present in the body. Immunity mediated by T_H cells and T_C cells is sometimes referred to as **cell-mediated immunity.**

Because HIV, the virus that causes AIDS, infects helper T cells and other cells of the immune system, it suppresses many components of acquired immune responses and makes HIV-infected individuals susceptible to opportunistic infections (see the Health Focus, "Opportunistic Infections and HIV"). Infected macrophages and dendritic cells also serve as reservoirs for HIV.

Check Your Progress

1. How are T- and B-cell responses similar? How are they different?
2. What are the specific roles of APCs and MHC molecules in stimulating T cell responses?

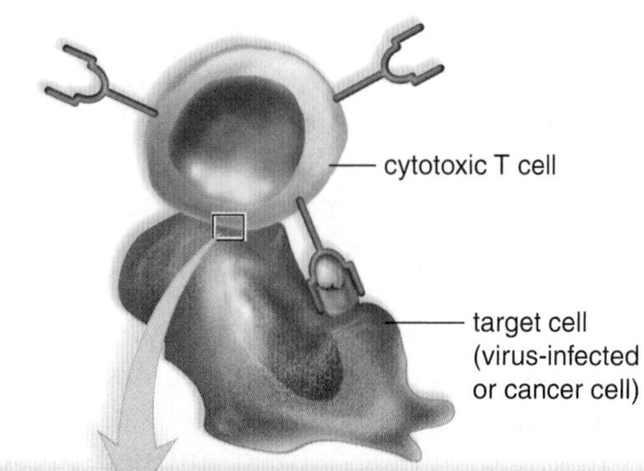

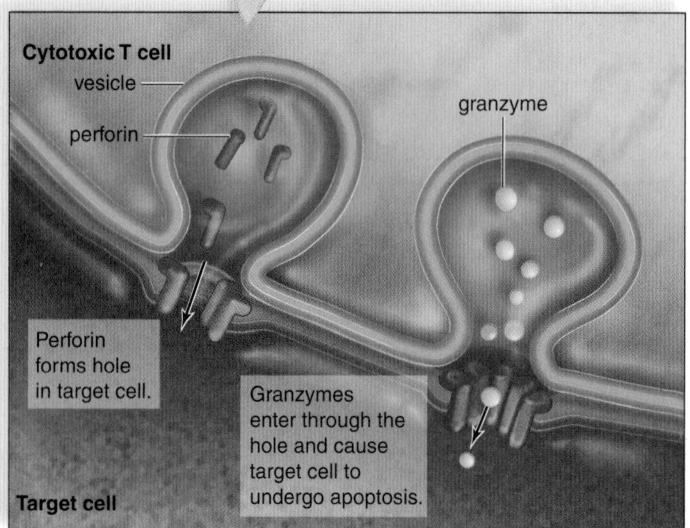

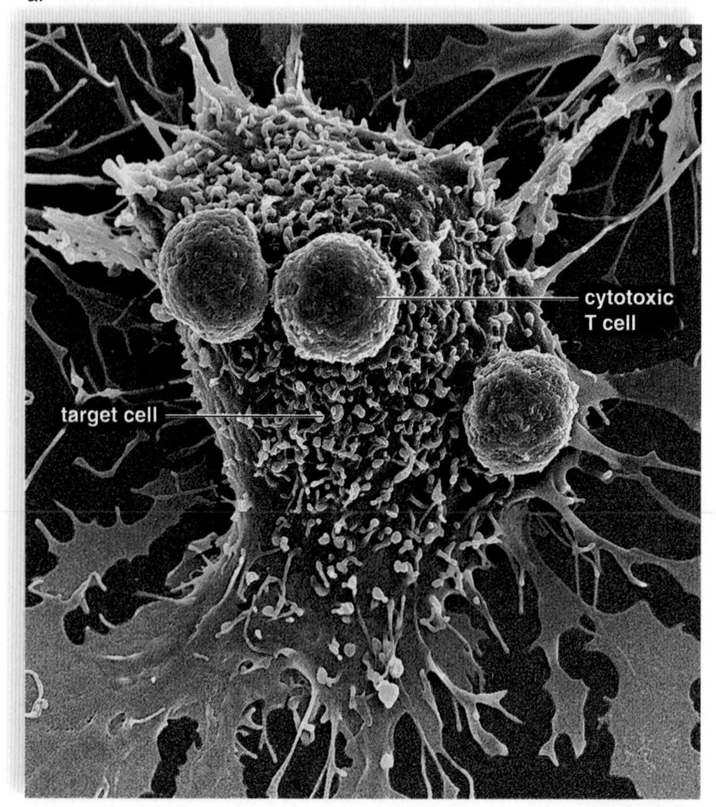

Figure 13.8 Cell-mediated immunity.
a. How a T_C cell destroys a virus-infected cell or cancer cell. **b.** The scanning electron micrograph shows cytotoxic T cells attacking and destroying a cancer cell (target cell).

13.3 Active Versus Passive Immunity

Learning Outcomes

1. Distinguish between active and passive immunity.
2. Describe some specific clinical applications of immune therapies and monoclonal antibodies.

In general, adaptive immune responses can be induced actively or passively. In **active immunity,** the individual alone produces an immune response against an antigen. In **passive immunity,** the individual is given prepared antibodies or cells, either naturally or artificially via an injection.

Active Immunity

Active immunity usually develops naturally after a person is infected with a pathogen. However, active immunity can also be induced artificially when a person is well, so that future infection will not take place. To prevent infections, people can be immunized against them. The United States is committed to immunizing all children against the common types of childhood disease (Fig. 13.9*a*).

Immunization involves the use of **vaccines,** substances that contain an antigen to which the immune system responds. Traditionally, vaccines are the pathogens themselves, or their products, that have been treated so they are no longer virulent (able to cause disease). Today, it is possible to genetically engineer bacteria or other microbes to mass-produce a protein from pathogens, and this protein can be used as a vaccine. This method has now produced a vaccine against hepatitis B, a viral disease, and is being used to develop a vaccine against malaria, a protozoal disease. It is also possible to use a carbohydrate component from a pathogen as a vaccine (such as the *H. influenzae* vaccine), but this must usually be "conjugated" to a protein.

After a vaccine is given, it is possible to follow an immune response by determining the amount of antibody present in a sample of plasma. This is called the **antibody titer.** After the first exposure to a vaccine, a primary response occurs. For a period of several days, no antibodies are present. Then the titer rises slowly, levels off, and gradually declines as the antibodies bind to the antigen or simply break down (Fig. 13.9*b*). After a second exposure to the vaccine, a secondary response occurs. The titer rises rapidly to a level much greater than before. Then it slowly declines. The second exposure is called a "booster" because it boosts the immune response to a high level. The high levels of antigen-specific T cells and antibodies prevent disease symptoms even if the individual is exposed to the disease-causing agent.

Active immunity is dependent upon the presence of memory B cells and memory T cells that are capable of responding quickly to an antigen. Active immunity is usually long-lasting, although a booster may be required periodically. In recent years, the safety of certain vaccines has come into question (see the Bioethical Focus at the end of this chapter).

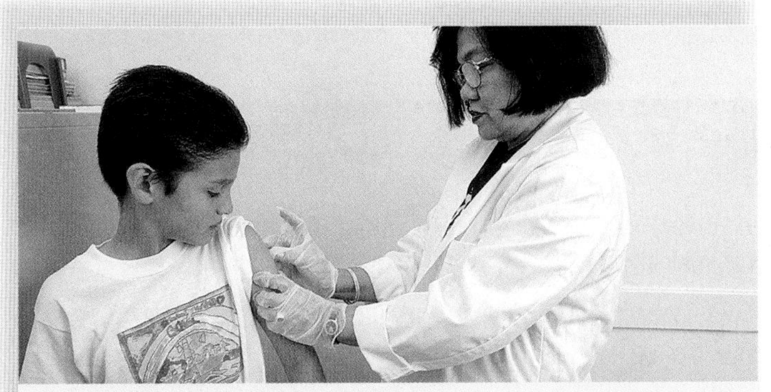

Suggested Immunization Schedule

Vaccine	Age (months)	Age (years)
Hepatitis B	Birth, 1–2, 6–18	
Diphtheria, tetanus, pertussis (DTP)	2, 4, 6, 15–18	4–6
Tetanus only		11–12, 13–18
Haemophilus influenzae, type b	2, 4, 6, 12–15	
Polio (IPV)	2, 4, 6–18	4–6
Pneumococcal	2, 4, 6, 12–15	
Measles, mumps, rubella (MMR)	12–15	4–6, 11–12
Varicella (chicken pox)	12–18	2–18
Hepatitis A (in selected areas)	12–18	2–18
Human papilloma-virus, types 6, 11, 16, 18	–	11–12

a.

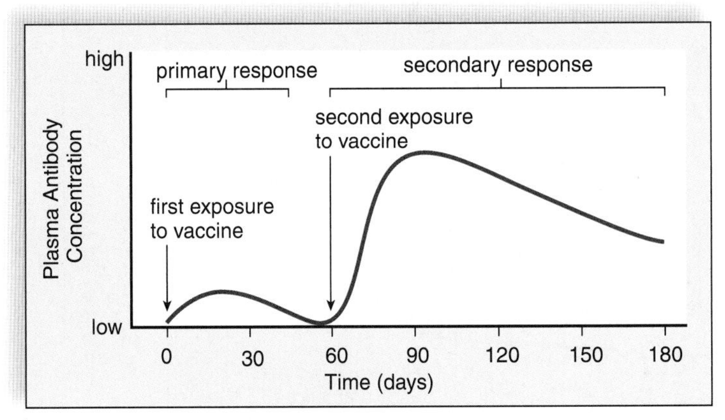

b.

Figure 13.9 Active immunity due to immunizations. **a.** Suggested immunization schedule for infants and children. **b.** During immunization, the primary response, after the first exposure to a vaccine, is minimal. The secondary response, which occurs after the second exposure, shows a dramatic rise in the amount of antibody present in plasma.

Passive Immunity

Passive immunity occurs when an individual receives another person's antibodies or immune cells. The passive transfer of antibodies is a common natural process. For example, newborn infants are passively immune to some diseases because IgG antibodies have crossed the placenta from the mother's blood (Fig. 13.10*a*). These antibodies soon disappear, however, so within a few months, infants become more susceptible to infections. Breast-feeding prolongs the natural passive immunity an infant receives from the mother because IgG and IgA antibodies are present in the mother's milk (Fig. 13.10*b*).

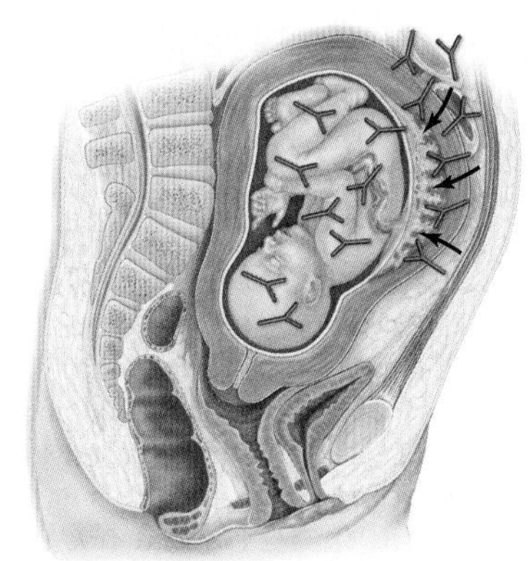

a. Antibodies (IgG) cross the placenta.

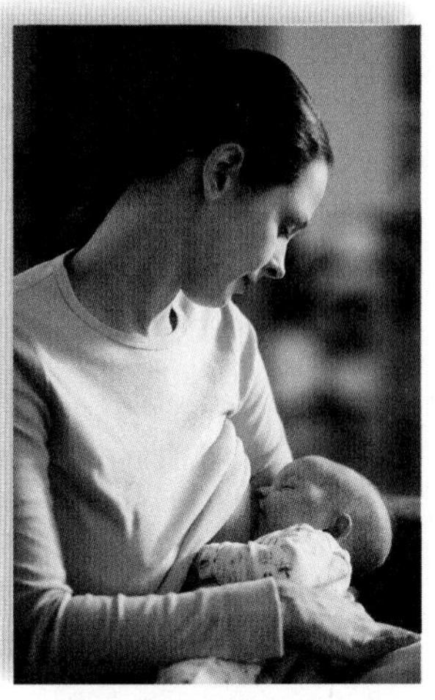

b. Antibodies (IgG, IgA) are secreted into breast milk.

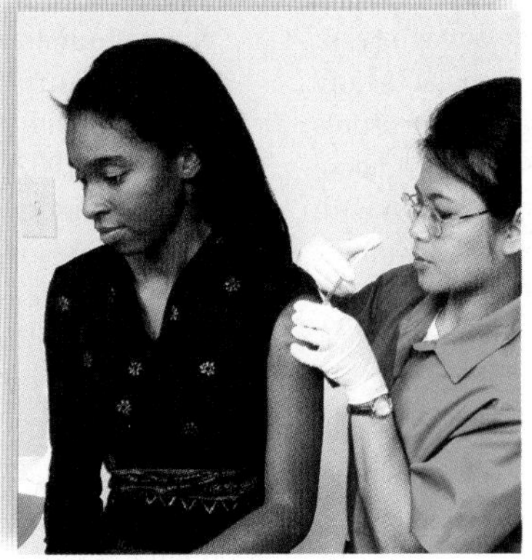

c. Antibodies can be injected by a physician.

Figure 13.10 Passive immunity. During passive immunity, antibodies are received by **a.** crossing the placenta, **b.** in breast milk, or **c.** by injection. Because the body is not producing the antibodies, passive immunity is short lived.

Even though passive immunity does not last, it is sometimes used to prevent illness in a patient who has been exposed to an infectious disease. Usually, the patient receives a gamma globulin injection (serum that contains antibodies), perhaps taken from individuals who have recovered from the illness (Fig. 13.10*c*). In the past, horses were immunized, and serum was taken from them to provide the needed antibodies against such diseases as diphtheria, botulism, and tetanus. Unfortunately, about 50% of patients who received these antibodies became ill, because their immune system recognized horse protein as foreign. This is called serum sickness. Passive immunity using horse antibodies against poisonous snake and spider venom is still used today, but human gamma globulin is used to prevent rabies and certain immunodeficiencies.

Instead of antibodies, cells of the immune system may be passively transferred into a patient, although this practice is less common. The best example is a bone marrow transplant, in which stem cells that produce blood cells are replenished after a cancer patient's own bone marrow has been intentionally destroyed by radiation or chemotherapy.

Check Your Progress

1. What are three different types of vaccines?
2. Why is the passive transfer of antibodies of great importance for the newborn?

Health Focus

Opportunistic Infections and HIV

AIDS (acquired immunodeficiency syndrome) is caused by the destruction of the immune system, especially the helper T cells, following an HIV (human immunodeficiency virus) infection. Then, the individual succumbs to many unusual types of infections that would not cause disease in a person with a healthy immune system. Such infections are known as opportunistic infections (OIs).

HIV not only kills helper T cells by directly infecting them, it also causes many uninfected T cells to die by a variety of mechanisms, including apoptosis. Many helper T cells are also killed by the person's own immune system as it tries to overcome the HIV infection. After initial infection with HIV, it may take up to ten years for an individual's helper T cells to become so depleted that the immune system can no longer organize a specific response to OIs (Fig. 13A). In a healthy individual, the number of helper T cells typically ranges from 800 to 1,000 cells per cubic millimeter (mm^3) of blood. The appearance of the following specific OIs can be associated with the helper T-cell count:

- Shingles. Painful infection with varicella-zoster (chicken pox) virus. Helper T-cell count of less than $500/mm^3$.
- Candidiasis. Yeast infection of the mouth, throat, or vagina. Helper T-cell count of about $350/mm^3$.
- *Pneumocystis* pneumonia. Fungal infection causing the lungs to fill with fluid and debris. Helper T-cell count of less than $200/mm^3$.
- Kaposi sarcoma. Cancer of blood vessels due to human herpesvirus 8; gives rise to reddish-purple, coin-sized spots and lesions on the skin. Helper T-cell count of less than $200/mm^3$.
- Toxoplasmic encephalitis. Protozoan infection characterized by severe headaches, fever, seizures, and coma. Helper T-cell count of less than $100/mm^3$.
- *Mycobacterium avium* complex (MAC). Bacterial infection resulting in persistent fever, night sweats, fatigue, weight loss, and anemia. Helper T-cell count of less than $75/mm^3$.
- Cytomegalovirus. Viral infection that leads to blindness, inflammation of the brain, throat ulcerations. Helper T-cell count of less than $50/mm^3$.

Due to development of powerful drug therapies that slow the progression of AIDS, people infected with HIV in the United States are suffering a lower incidence of OIs than was common in the 1980s and 1990s.

Discussion Questions

1. People who don't have HIV also sometimes get OIs, such as shingles or candidiasis. How would you expect the outcome of, for example, a yeast infection to be different in a person without HIV compared to a person with HIV?
2. Kaposi sarcoma occurs relatively rarely in people who are infected with human herpesvirus 8, but are HIV-negative. How might an HIV infection increase the odds of getting Kaposi sarcoma?
3. Why do you suppose that the amount of HIV in the blood increases rapidly during the later stages of AIDS, as shown in Figure 13A?

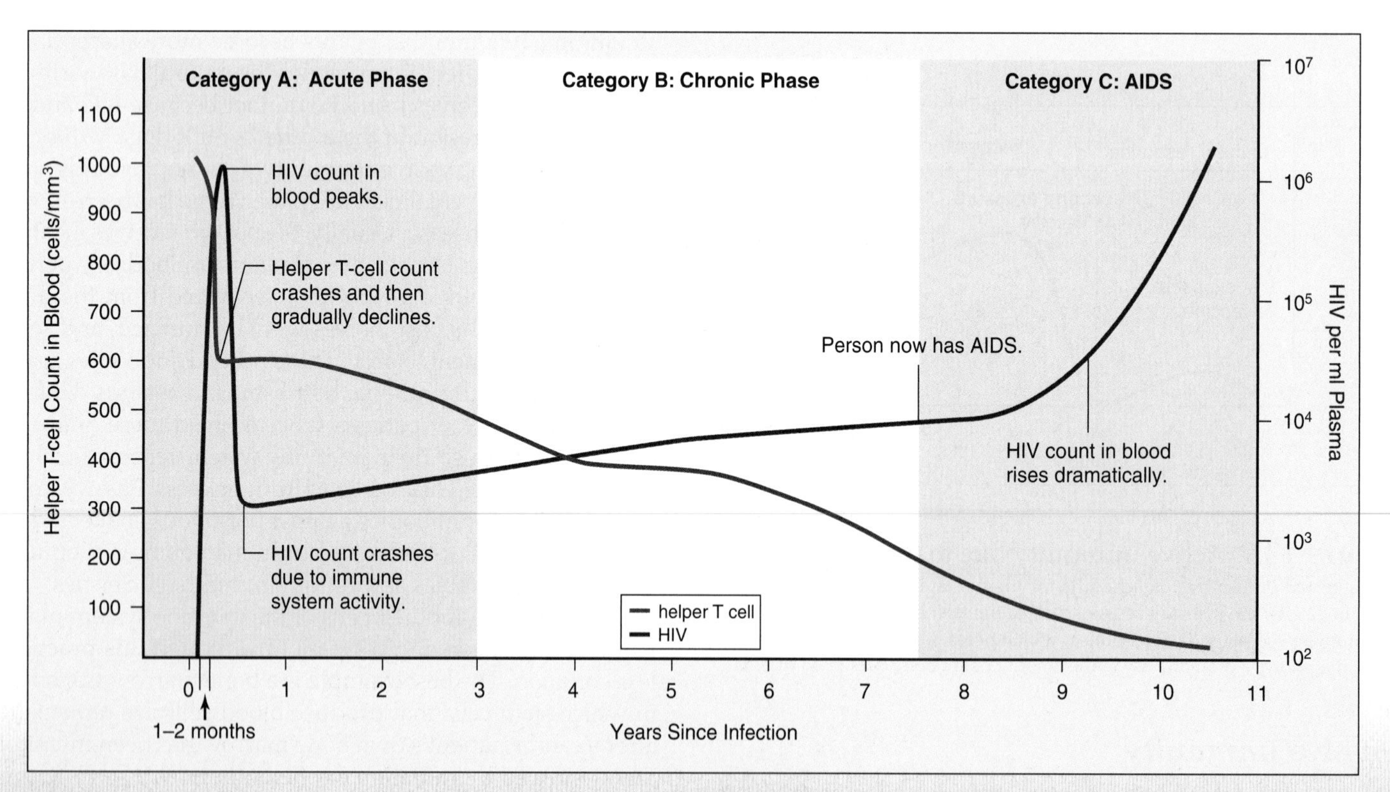

Figure 13A Progression of HIV infection during its three stages.
In category A, the individual may have no symptoms or very mild symptoms associated with the infection. By category B, opportunistic infections have begun to occur, such as candidiasis, shingles, and diarrhea. Category C is characterized by more severe opportunistic infections and is clinically described as AIDS.

Immune Therapies

Cytokines and Immunity

Cytokines are chemical messengers produced by T cells, macrophages, and other cells. Because cytokines regulate white blood cell formation and/or function, they are being investigated as possible adjunct therapy for cancer and AIDS. Both interferons and **interleukins,** which are cytokines produced by various white blood cells, have been used as immunotherapeutic drugs, particularly to enhance the ability of an individual's own T cells to fight cancer.

Because many cancer cells carry altered proteins on their cell surface, they should be attacked and destroyed by cytotoxic T cells. Whenever cancer develops, it is possible that cytotoxic T cells have not been activated. In that case, cytokines might awaken the immune system and lead to the destruction of the cancer. In one technique being investigated, researchers first withdraw T cells from the patient, present cancer cell antigens to them, and then activate the cells by culturing them in the presence of an interleukin. The T cells are reinjected into the patient, who is given doses of interleukin to maintain the killer activity of the T cells.

Scientists who are actively engaged in cytokine research believe that interleukins soon will be used as adjuncts for vaccines, for the treatment of chronic infectious disease, and perhaps for the treatment of cancer. Conversely, interleukin antagonists also may prove helpful in preventing skin and organ rejection, autoimmune diseases, and allergies.

Monoclonal Antibodies

Every plasma cell derived from the same B cell secretes antibodies against a specific antigen. These are **monoclonal antibodies** because all of them are the same type and because they are produced by plasma cells derived from the same B cell. One method of producing monoclonal antibodies in vitro (outside the body in glassware) is depicted in Figure 13.11. B cells are removed from an animal (usually mice are used) that has been exposed to a particular antigen. The resulting plasma cells are fused with myeloma cells (malignant plasma cells that live and divide indefinitely; they are immortal cells). The fused cells are called hybridomas—*hybrid-* because they result from the fusion of two different cells, and *-oma* because one of the cells is a cancer cell. Hybridomas producing the desired antibody are then isolated and purified.

Because they react very specifically with only one particular antigen, monoclonal antibodies can be used for quick and certain diagnosis of various conditions. For example, a particular hormone called hCG is present in the urine of a pregnant woman. A monoclonal antibody can be used to detect this hormone. If it is present, the woman is pregnant. Monoclonal antibodies are also used to identify infections. And because they can distinguish between cancerous and normal tissue cells, they are used to carry radioactive isotopes or toxic drugs to tumors, which can then be selectively destroyed. Herceptin is a monoclonal antibody used to treat breast cancer. Given intravenously, it binds to a protein receptor found on some breast cancer cells and prevents the cancer cells from dividing so fast. Antibodies that bind to cancer cells also can activate complement, increase phagocytosis by macrophages and neutrophils, and attract NK cells.

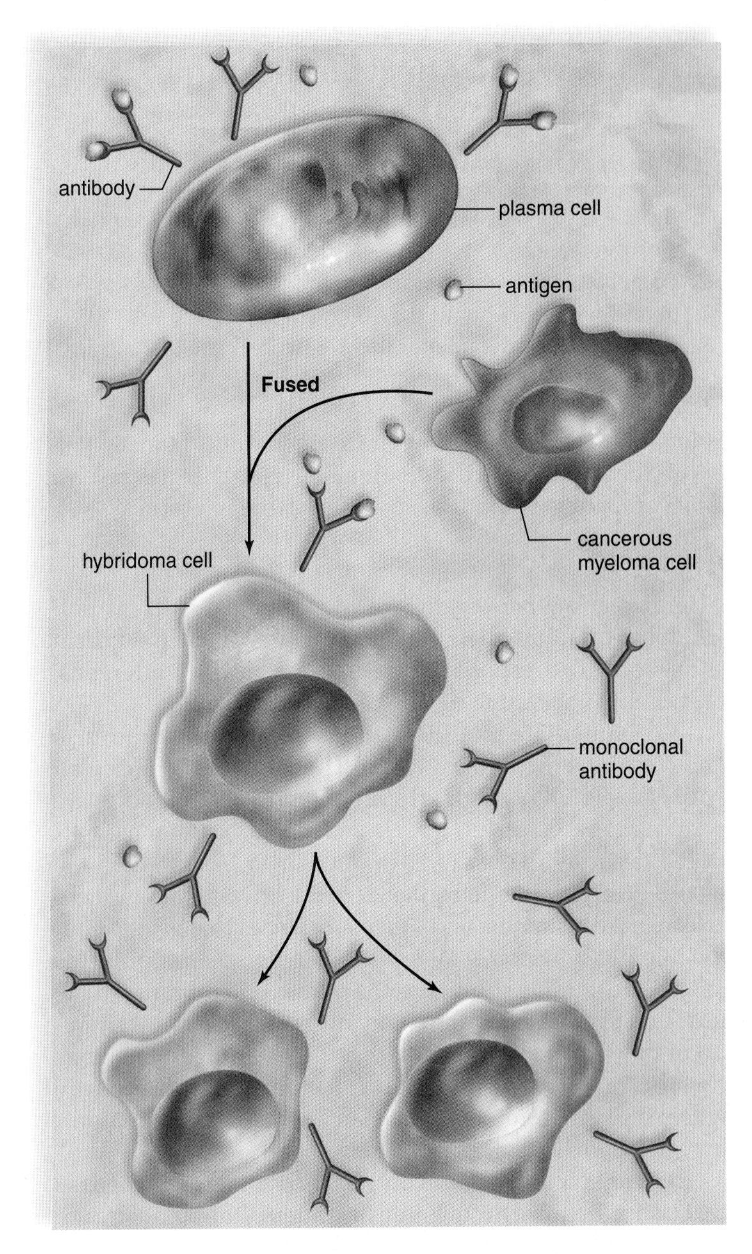

Figure 13.11 Production of monoclonal antibodies.
Plasma cells derived from mice immunized with a specific antigen are fused with myeloma (cancerous) cells, producing hybridoma cells that are "immortal." Hybridoma cells divide and continue to produce the same type of antibody, called monoclonal antibodies.

Check Your Progress

1. Cytokines interact with receptors on target cells. How could this interaction be blocked, or antagonized?
2. What is the main advantage of using a monoclonal antibody for diagnosis or treatment, versus using antibodies produced by injecting an animal with an antigen?

13.4 Adverse Effects of Immune Responses

Learning Outcomes

1. Discuss several situations where the immune response can have a harmful result.
2. Compare the adverse reactions involving the ABO blood system with those involving the Rh system.
3. Define xenotransplantation, and describe potential advantages and disadvantages of this procedure.

Sometimes the immune system responds in a manner that harms the body, as when individuals develop allergies, receive an incompatible blood type, or suffer tissue rejection.

Allergies

Allergies are hypersensitivities to substances, such as pollen, food, or animal hair, that ordinarily would do no harm to the body. The response to these antigens, called **allergens,** usually includes some degree of tissue damage.

An **immediate allergic response** can occur within seconds of contact with the antigen. The response is caused by antibodies known as IgE (see Table 13.1). IgE antibodies are attached to receptors on the plasma membrane of mast cells in the tissues and also to eosinophils and basophils in the blood. When an allergen attaches to the IgE antibodies, these cells release histamine and other substances that bring about the allergic symptoms. When an allergen such as pollen is inhaled, histamine stimulates the mucous membranes of the nose and eyes, causing the runny nose and watery eyes typical of **hay fever.** In **asthma,** the airways leading to the lungs constrict, resulting in difficult breathing accompanied by wheezing. When food contains an allergen, nausea, vomiting, and diarrhea often result.

Anaphylactic shock is an immediate allergic response that occurs because the allergen has entered the bloodstream. Bee stings and penicillin shots are known to cause this reaction. Anaphylactic shock is characterized by a sudden and life-threatening drop in blood pressure due to increased permeability of the capillaries caused by histamine. Injecting epinephrine can counteract this reaction until medical help is available.

Treating mild to moderate allergies usually involves antihistamines to relieve the symptoms. In more serious cases, injections of the allergen can be given so that the body will build up high quantities of IgG antibodies (see the Health Focus, "Immediate Allergic Responses"). The hope is that IgG antibodies will then combine with allergens before they have a chance to reach the IgE antibodies. A monoclonal antibody called Xolair is also available, which binds to and blocks the binding of IgE to the receptor found on inflammatory cells.

A **delayed allergic response** is initiated by memory T cells at the site of allergen contact in the body. The allergic response is regulated by the cytokines secreted by "sensitized" T cells at the site. A classic example of a delayed allergic response is the skin test for tuberculosis (TB). When the test result is positive, the tissue where the antigen was injected becomes red and hardened (Fig. 13.12), indicating prior exposure to tubercle bacilli, the cause of TB.

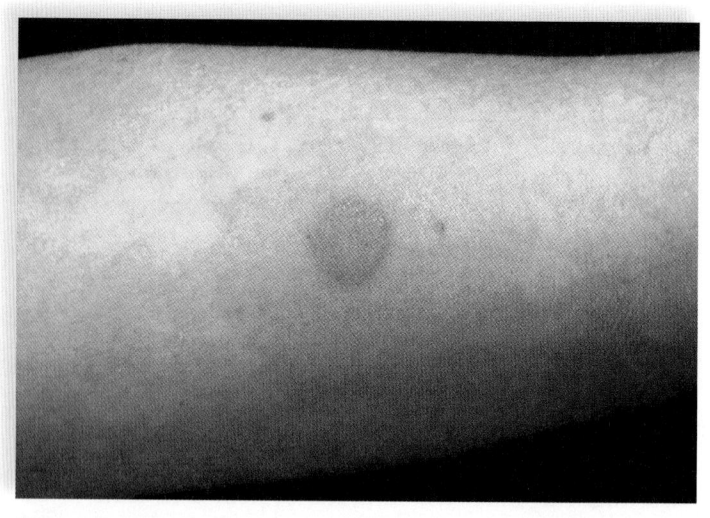

Figure 13.12 Skin testing for tuberculosis. The formation of a swollen, red area two to three days after injection of *Mycobacterium tuberculosis* antigens into the skin indicates previous exposure.

Blood-Type Reactions

Several blood typing systems are currently in use. The most important is the ABO system. People often carry information about their ABO type in their wallets in case an accident requires them to need blood.

ABO System

In the ABO system, the presence or absence of type A and type B antigens on red blood cells determines a person's blood type. For example, a person with type A blood has the A antigen on his or her red blood cells. Because it is considered "self," this molecule is not recognized as an antigen by this individual's body, although it can be an antigen to a recipient who does not have type A blood.

The ABO system identifies four types of blood: A, B, AB, and O. Because the A and B antigens are commonly found on microbes present in and on our bodies, a person's plasma contains antibodies to the antigens that are *not* present on the red blood cells. These antibodies are called anti-A and anti-B. The antibodies present in the plasma of each blood type are shown here:

Blood Type	Antigen on Red Blood Cells	Antibody in Plasma
A	A	Anti-B
B	B	Anti-A
AB	A, B	None
O	None	Anti-A and anti-B

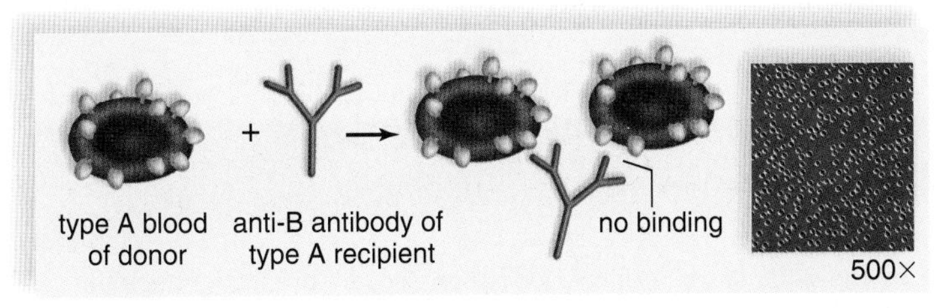

a. No agglutination

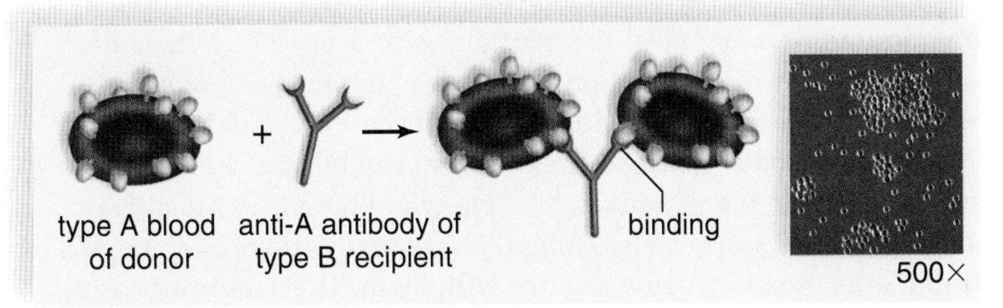

b. Agglutination

Figure 13.13 Blood transfusions.
No agglutination (**a**) versus agglutination (**b**) is determined by whether the recipient has antibodies in the plasma that can combine with antigens on the donor's red blood cells. The photos on the right show how the red blood cells appear under the microscope.

Because type A blood has anti-B and not anti-A antibodies in the plasma, a donor with type A blood can give blood to a recipient with type A blood (Fig. 13.13*a*). But giving type A blood to a type B recipient causes agglutination (Fig. 13.13*b*). **Agglutination,** or clumping of red blood cells, can stop blood from circulating in small blood vessels, leading to organ damage. It is also followed by *hemolysis,* or bursting of red blood cells, which if extensive can cause the individual to die.

Theoretically, which blood type would be accepted by all other blood types? Type O blood has no antigens on the red blood cells and is sometimes called the universal donor. Which blood type could receive blood from any other blood type? Type AB blood has no anti-A or anti-B antibodies in the plasma and is sometimes called the universal recipient. In practice, however, it is not safe to rely solely on the ABO system when matching blood, and instead, samples of the two types of blood are physically mixed, and the result is microscopically examined before blood transfusions are done.

Today, blood transfusions are a matter of concern not only because blood types should match, but also because each person wants to receive blood that is free of infectious agents. Blood is tested for the more serious agents, such as those that cause AIDS, hepatitis, and syphilis.

Rh System

Another important antigen in matching blood types is the Rh factor. Eighty-five percent of the U.S. population have this particular antigen on their red blood cells and are Rh-positive. Fifteen percent do not have this antigen and are Rh-negative. Rh-negative individuals normally do not have antibodies to the Rh factor, but they may make them when exposed to the Rh factor. The designation of blood type usually also includes whether the person has or does not have the Rh factor on the red blood cells, i.e., type A-positive.

During pregnancy, if the mother is Rh-negative and the father is Rh-positive, the fetus may be Rh-positive. The fetal Rh-positive red blood cells may leak across the placenta into the mother's cardiovascular system, especially as placental tissues normally break down during birth (Fig. 13.14*a*). Now, the mother produces anti-Rh antibodies. In a subsequent

a. Fetal Rh-positive red blood cells leak across placenta into mother's bloodstream.

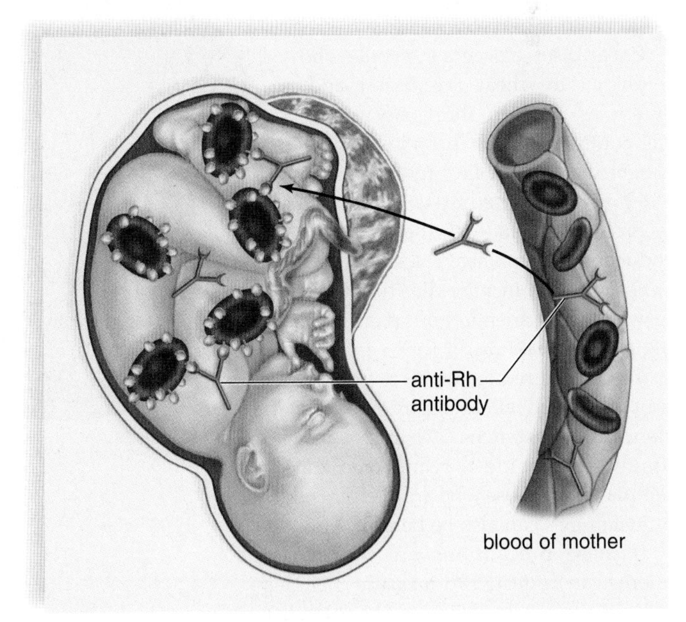

b. Mother forms anti-Rh antibodies that cross the placenta and attack fetal Rh-positive red blood cells.

Figure 13.14 Hemolytic disease of the newborn.
a. Due to a pregnancy in which the child is Rh positive, an Rh-negative mother can begin to produce antibodies against Rh-positive red blood cells. **b.** In the same, but more likely a subsequent pregnancy, these antibodies can cross the placenta and cause hemolysis of an Rh-positive child's red blood cells.

Health Focus

Immediate Allergic Responses

The runny nose and watery eyes of hay fever are often caused by an allergic reaction to the windblown pollen of trees, grasses, and ragweed, particularly in the spring and fall (Fig. 13B). Worse, if a person has asthma, the airways leading to the lungs constrict, resulting in difficult breathing characterized by wheezing. Most people can inhale pollen with no ill effects. But others have developed a hypersensitivity, meaning that their immune system responds in a deleterious manner. The problem stems from a type of antibody called immunoglobulin E (IgE) that causes the release of histamine from mast cells and also basophils whenever they are exposed to an allergen. Histamine causes the mucous membranes of the nose and eyes to release fluid as a defense against pathogen invasion. But in the case of allergies, copious fluid is released even though no real danger is present.

Many food allergies are also due to the presence of IgE antibodies, which usually bind to a protein in the food. The symptoms, such as nausea, vomiting, and diarrhea, are due to the mode of entry of the allergen. Skin symptoms may also occur, however. Adults are often allergic to shellfish, nuts, eggs, cows' milk, fish, and soybeans. Peanut allergy is a common food allergy in the United States, possibly because peanut butter is a staple in the diet. People seem to outgrow allergies to cows' milk and eggs more often than allergies to peanuts and soybeans.

Celiac disease occurs in people who are allergic to wheat, rye, barley, and sometimes oats—in short, any grain that contains gluten proteins. It is thought that the gluten proteins elicit a delayed cell-mediated immune response by T cells with the resultant production of cytokines. The symptoms of celiac disease include diarrhea, bloating, weight loss, anemia, bone pain, chronic fatigue, and weakness.

People can reduce their reactions to airborne and food allergens by avoiding the offending substances. Airlines are now required to provide a peanut-free zone in their planes for those who are allergic. Taking antihistamines can also be helpful.

If these precautions are inadequate, patients can be tested to measure their susceptibility to any number of possible allergens. A small quantity of a suspected allergen is injected just beneath the skin, and the strength of the subsequent reaction is noted. The appearance of a raised, red, so-called wheal-and-flare response at the skin prick site within a few minutes demonstrates that IgE antibodies attached to mast cells have reacted to an allergen. In an immunotherapy called hyposensitization, ever-increasing doses of the allergen are periodically injected subcutaneously with the hope that the body will produce an IgG response against the allergen. IgG, in contrast to IgE, does not cause the release of histamine after it combines with the allergen. If IgG combines first, the allergic response does not occur. Patients know they are cured when the allergic symptoms go away. Therapy may have to continue for as long as two to three years.

Allergic-type reactions can occur without involving the immune system. Wasp and bee stings contain substances that cause swelling, even in those whose immune system is not sensitized to substances in the sting. Also, jellyfish tentacles and certain foods (e.g., fish that is not fresh and strawberries) contain histamine or closely related substances that can cause a reaction. Furthermore, immunotherapy is not possible in people who are allergic to penicillin and bee stings. Upon the first exposure high sensitivity has built up so that when reexposed, anaphylactic shock can occur. Among its many effects, histamine causes increased permeability of the capillaries, the smallest blood vessels. When this reaction occurs throughout the body, these individuals experience a drastic decrease in blood pressure that can be fatal within a few minutes. People who know they are allergic to bee stings can obtain a syringe of epinephrine to carry with them. This medication can counteract the symptoms of anaphylactic shock until medical help is reached.

Discussion Questions

1. Based on what you have read in this essay and the chapter, what are some factors that might determine a person's tendency to develop an immediate allergic response to an allergen?
2. An increase in allergic diseases has been reported in developed countries, compared to lower rates in less-developed countries. What are some possible reasons for this difference?
3. What are the specific effects of epinephrine that can save the life of a person who is in anaphylactic shock?

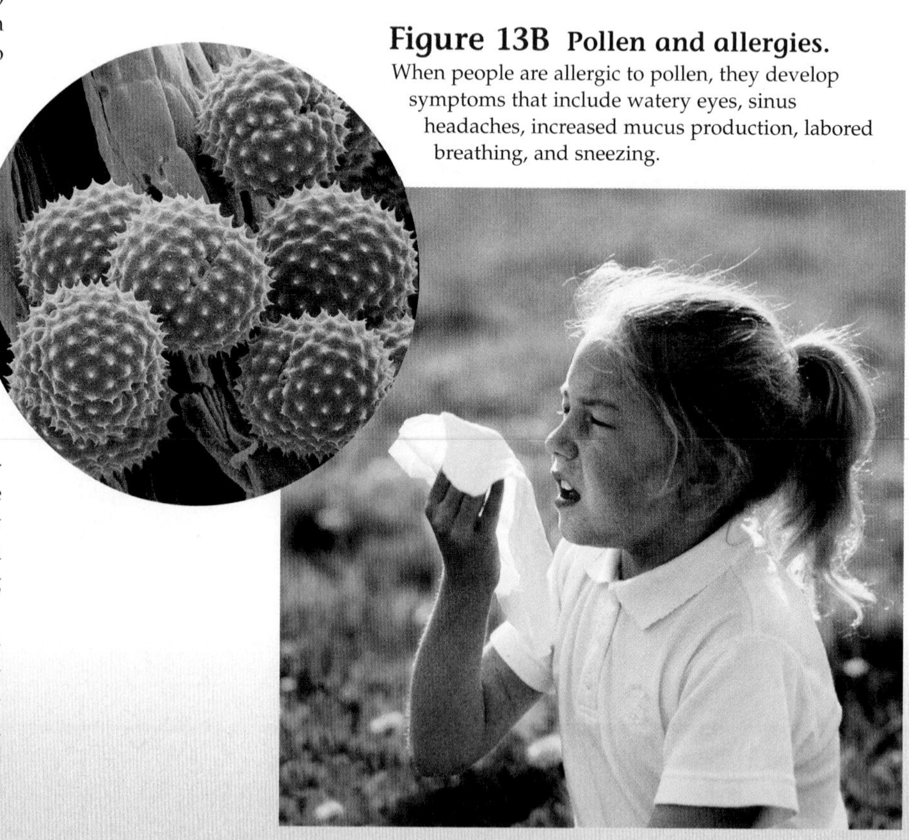

Figure 13B Pollen and allergies. When people are allergic to pollen, they develop symptoms that include watery eyes, sinus headaches, increased mucus production, labored breathing, and sneezing.

pregnancy with another Rh-positive child, these antibodies may cross the placenta and destroy the fetal red blood cells (Fig. 13.14*b*). The resulting condition, called hemolytic disease of the newborn, can lead to brain damage or even death due to anemia and excess bilirubin in the blood.

The Rh problem is prevented by giving Rh-negative women an Rh immunoglobulin injection, toward the end of the pregnancy and within 72 hours after giving birth to an Rh-positive child. This injection contains relatively low levels of anti-Rh antibodies that help destroy any baby's red blood cells in the mother's blood before her immune system produces high levels of anti-Rh antibodies.

Tissue Rejection

Certain organs, such as the skin, the heart, and the kidneys, would be relatively easy to transplant from one person to another if the recipient did not attempt to reject them. Unfortunately, in addition to their role in antigen presentation to T cells, MHC proteins are also involved in recognition of foreign tissues. Ideally, the transplanted organ would have exactly the same type of MHC antigens as those of the recipient. Because it is hard to find a perfect MHC match, the chance of organ rejection can also be diminished by administering **immunosuppressive** drugs. Two well-known immunosuppressive drugs, cyclosporine and tacrolimus, both act by inhibiting the production of certain cytokines by T cells.

Xenotransplantation is the use of animal organs instead of human organs in transplant patients. Scientists have chosen to use the pig because pigs are raised as a meat source and are prolific. Genetic engineering can make pig organs less antigenic by removing the MHC antigens. The ultimate goal is to make pig organs as widely accepted as type O blood. Although advances in xenotransplantation are encouraging, many researchers hope that tissue engineering, including the production of human organs that lack MHC antigens, will one day do away with the problem of rejection. Alternatively, scientists are learning how to grow organs from a patient's own tissues. In April of 2006, surgeons at Children's Hospital in Boston were able to use lab-grown urinary bladder tissue to rebuild defective bladders in human patients.

Check Your Progress

1. What is the immunological reason that immediate allergic responses can occur within seconds after being reexposed to an allergen, while delayed-type responses may take days?
2. Why is Rh incompatibility only a problem when a fetus is Rh-positive and the mother is Rh-negative, but not vice-versa?
3. Because the drugs cyclosporine and tacrolimus work by inhibiting cytokine production (rather than inhibiting specific responses to transplanted organs), what is a likely side effect of their use?
4. Considering the "normal" function of MHC molecules, what might be one problem if organs lacking these are transplanted into patients?

13.5 Disorders of the Immune System

Learning Outcomes

1. Define autoimmune disease, and provide some specific examples.
2. Differentiate between acquired and congenital immunodeficiencies.

When a person has an **autoimmune disease,** cytotoxic T cells or antibodies mistakenly attack the body's own cells as if they bear foreign antigens. Exactly what causes autoimmune diseases is not known. However, sometimes they occur after an individual has recovered from an infection.

In the autoimmune disease **myasthenia gravis,** antibodies attach to and interfere with neuromuscular junctions, and muscular weakness results. In **multiple sclerosis (MS),** T cells attack the myelin sheath of nerve fibers, and this causes various neuromuscular disorders. A person with **systemic lupus erythematosus (SLE)** has various symptoms prior to death due to kidney damage caused by the deposition of excessive antigen/antibody complexes. In **rheumatoid arthritis,** the joints are affected. Research suggests that heart damage following rheumatic fever and type 1 diabetes are also autoimmune illnesses. As yet, there are no cures for autoimmune diseases, but they can sometimes be controlled with immunosuppressive drugs.

When a person has an immunodeficiency, the immune system is unable to protect the body against disease. AIDS is an example of an acquired immunodeficiency (see the Health Focus, "Opportunistic Infections and HIV"). Immunodeficiencies may also be congenital (that is, inherited). Infrequently, a child may be born with an impaired immune system caused by a defect in lymphocyte development. In **severe combined immunodeficiency disease (SCID),** both antibody- and cell-mediated immunity are lacking or inadequate. Without treatment, even common infections can be fatal. Gene therapy has been successful in some SCID patients.

Check Your Progress

1. List four autoimmune diseases.
2. List two immunodeficiency diseases, and identify them as either congenital or acquired.

Applying the Concepts Revisited

Because the immune system is of critical importance in protecting an individual from infectious disease, disorders affecting the immune system can have serious or even fatal consequences. Inherited immune deficiencies can affect a very limited part of the immune system, or can more generally affect a major type of immune responses. In the story that opened this chapter, we learned about Jason, a three-year-old boy with an immunodeficiency called X-linked agammaglobulinemia (XLA), which mainly affects B-cell responses.

1. Why do you think that individuals born with XLA usually appear to be normal until they reach about six months of age?
2. Patients with XLA tend to suffer from extracellular infections caused by bacteria, more than viral infections. Based on what you have learned in this chapter, provide three specific reasons that might explain why this is the case.

Bioethical Focus

Vaccination Requirements

In July 1999, a vaccine called Rotashield was removed from the U.S. market. An advisory committee to the U.S. government decided that the vaccine, designed to protect against serious diarrheal disease caused by rotavirus, was unsafe. Since the introduction of the vaccine less than a year earlier, 15 cases of serious, life-threatening bowel obstructions had been reported within 1–2 weeks following vaccination.

Similar concerns have been raised about the safety of other vaccines. In the late 1990s, some scientific studies seemed to establish a link between the common childhood measles-mumps-rubella (MMR) vaccine and an increasing incidence of autism, even though subsequent studies have failed to confirm this link. When the U.S. Army announced that it would require all military personnel to be vaccinated against anthrax, many filed lawsuits because they objected to being forced to take the vaccine. This kind of skepticism about vaccines can be traced all the way back to the early 1800s in England, when news first spread about Edward Jenner's demonstration that smallpox could be prevented by intentionally exposing a child to cowpox (which we now know is caused by a related virus). Newspapers of the time mocked Jenner, and published cartoons showing cows growing out of people's bodies after receiving Jenner's vaccine. With such a long tradition of doubt about the safety of vaccines, it is not surprising that some parents remain hesitant to vaccinate their children today.

The history of medicine has demonstrated without a doubt that since vaccines have been introduced, previously feared diseases such as smallpox, measles, polio, diphtheria, whooping cough, and tetanus have been reduced to rare occurrences. Most parents of young children today have never seen a case of these diseases, leading some of them to wonder why it is necessary to take even a small risk of making their child sick from a vaccine. Ironically, even parents who don't vaccinate their children benefit from the fact that a large percentage of children are vaccinated, and thus are less likely to transmit infectious diseases to their unvaccinated child. This concept of "herd immunity" has been shown to be relevant in recent outbreaks of measles and whooping cough in communities where increasing numbers of parents are not vaccinating their children.

The concept that vaccinating a high percentage of children can protect entire communities has led to immunization requirements. Most states require that a parent provide written proof of a child's immunizations prior to registering for school. If the child is not immunized, he or she will not be allowed to attend public school, unless the parent can demonstrate a medical reason, or a religious objection, that explains why the child is not vaccinated. With the rising popularity of the Internet, there is no shortage of websites voicing objections to any vaccine requirements, and going so far as to question the benefit of all vaccinations. The challenge for all potential parents, and for future health-care workers in particular, is to remain as informed as possible about the benefits and risks of vaccination, so that they can make wise choices for their patients and for their own children.

Form Your Own Opinion

1. Do you think the benefits of vaccination outweigh the risks? What level of risk would seem acceptable if you were having your own child vaccinated against polio, for example?
2. Do you think it is appropriate to allow people to forgo the required vaccinations for their children because of religious objections?
3. Suppose you are a nurse or a pediatrician and parents tell you they don't think the government should force them to vaccinate their child, because they read on a website that all vaccines are ineffective and dangerous. How would you respond?

SUMMARIZING THE CONCEPTS

13.1 The Lymphatic System

The lymphatic system consists of lymphatic vessels and lymphoid organs:

- Lymphatic vessels absorb fats from the digestive tract and excess tissue fluid at blood capillaries and carry these to the bloodstream.
- Lymphoid organs are sites where lymphocytes are produced (primary organs) and where adaptive immune responses can be initiated (secondary organs). Primary lymphoid organs include the thymus and bone marrow. Secondary lymphoid organs include the lymph nodes and spleen.

13.2 Innate and Adaptive Immunity

Innate immunity is fully functional without previous exposure to antigens, while adaptive immunity is enhanced after exposure. Mechanisms of innate immunity include:

- Physical and chemical barriers, such as the ciliated cells and mucus of the respiratory tract, stomach acid, or the normal bacteria that inhabit the intestinal and genital tracts.
- Inflammation, which is the redness, heat, swelling, and pain that occur due to capillary changes in response to chemicals released by inflammatory cells.
- Phagocytes and natural killer cells, which either ingest and kill microorganisms (phagocytes), or kill virus-infected cells (NK cells).
- Protective proteins, such as complement and interferons.

Adaptive immunity requires B cells and T cells:

- B cells are responsible for antibody-mediated immunity. Upon exposure to an antigen that binds to its B-cell receptor, a B cell undergoes clonal selection, producing plasma cells and memory B cells. Plasma cells secrete antibodies and eventually undergo apoptosis. Memory B cells remain in the body and produce antibodies if the same antigen enters the body at a later date. The five main classes of human antibodies are IgG, IgM, IgA, IgE, and IgD. Antibodies are typically Y-shaped molecules with at least two binding sites for a specific antigen; IgA and IgM antibodies can also be dimeric and pentameric, respectively.
- T cells are responsible for cell-mediated immunity. Like B cells, each T cell bears antigen receptors. However, for a T cell to recognize an antigen, the antigen must be processed and presented on a cell surface by an MHC (major histocompatibility complex) molecule. The two main types of T cells are helper T (T_H) cells and cytotoxic T (T_C) cells. T_H cells respond to antigens presented by MHC class II molecules found on "antigen-presenting cells" such as dendritic cells or macrophages. The activated T_H cells produce cytokines that stimulate other immune cells. T_C cells kill virus-infected or cancer cells that bear a "nonself" protein on their MHC class I molecules. Thereafter, the activated T cell undergoes clonal expansion until the illness has been stemmed. Then, most of the activated T cells undergo apoptosis. A few cells remain, however, as memory T cells.

13.3 Active Versus Passive Immunity

Active and passive immunity are induced in different ways:

- Active immunity can be induced through infection or by vaccines. Vaccination is typically used when a person is well and in no immediate danger of contracting an infectious disease.
- Passive immunity is acquired from another individual, whether naturally as antibodies passed from a mother to child, or artificially via an injection.

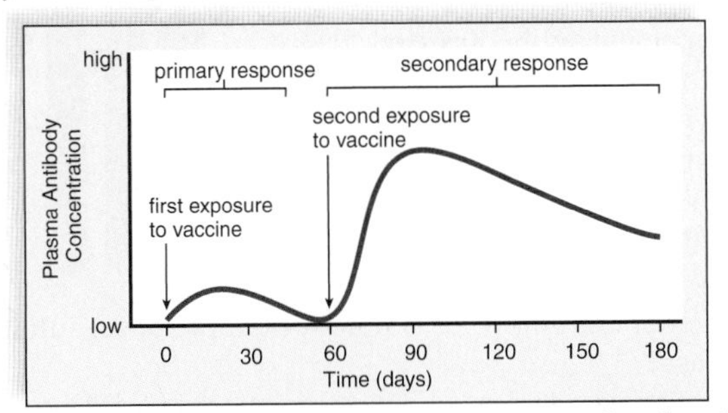

Immune therapies may include cytokines and monoclonal antibodies:

- Cytokines are chemical messengers that may be used as an adjunct to vaccination or to treat diseases.
- Monoclonal antibodies, which are produced by the same plasma cell, are used for various purposes, from detecting infections to treating cancer.

13.4 Adverse Effects of Immune Responses

Immune responses to harmless compounds (called allergens) are known as allergies or hypersensitivities:

- Immediate allergic responses occur due to an inappropriate production of IgE antibody, which results in an inflammatory response to subsequent exposures to the allergen. The most extreme form of this is anaphylaxis, which can be fatal.
- Delayed allergic responses, such as a positive TB skin test, are due to the activity of T cells.

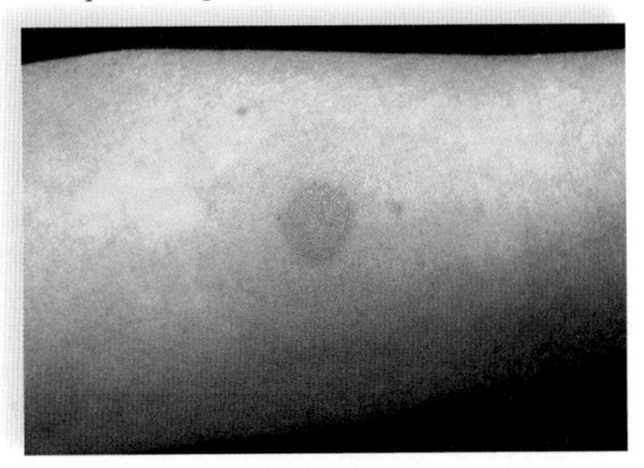

13.5 Disorders of the Immune System

In autoimmune diseases, a person's own T cells or antibodies target the body's own cells or molecules. Many types of immunodeficiencies also occur:

- Examples of autoimmune diseases include myasthenia gravis, multiple sclerosis, systemic lupus erythematosus, and rheumatoid arthritis.
- Immunodeficiency diseases can be acquired, as occurs with AIDS, or inherited, as is seen in children with SCID, who are born unable to produce their own antibody- and cell-mediated immunity. Depending on the severity, immunodeficiencies can be fatal, usually as a result of other, common infections.

TESTING YOURSELF

Choose the best answer for each question.

1. What is the term for localized swelling caused by fluid accumulation?
   a. node
   b. edema
   c. valve
   d. None of these are correct.
2. Label the four lymphoid organs, a–d. Which are primary, and which are secondary?

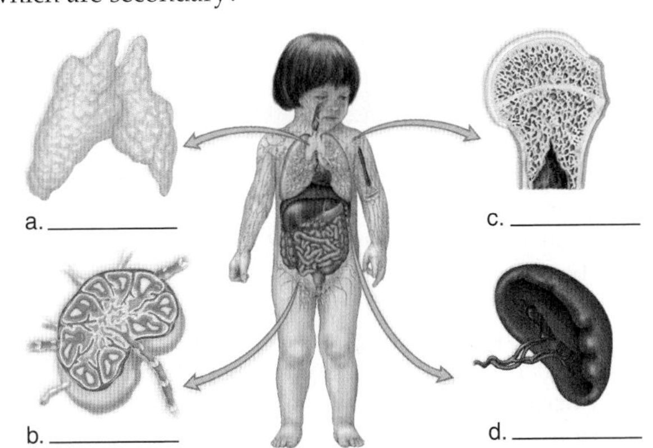

3. One-way flow of fluid in lymph vessels is aided by
   a. valves.
   b. skeletal muscle contractions.
   c. Both a and b are correct.
   d. None of these are correct.
4. Which of the following is not a function of the lymphatic system?
   a. produces red blood cells
   b. returns excess fluid to the blood
   c. transports lipids absorbed from the digestive system
   d. defends the body against pathogens
5. The ______ filters ______, and in its absence, problems associated with ______ may occur.
   a. spleen, lymph, oxygen transport
   b. liver, lymph, infection
   c. spleen, blood, infection
   d. spleen, blood, oxygen transport
6. Which cell(s) phagocytize pathogens?
   a. neutrophils
   b. macrophages
   c. lymphocytes
   d. Both a and b are correct.
7. ______ release histamines.
   a. Basophils
   b. Neutrophils
   c. Monocytes
   d. All of these are correct.
8. Complement
   a. is an innate defense mechanism.
   b. is involved in the inflammatory reaction.
   c. is a series of proteins present in the plasma.
   d. can enhance phagocytosis.
   e. All of these are correct.
9. Which of the following is an innate defense of the body?
   a. immunoglobulin
   b. B cell
   c. T cell
   d. vaccine
   e. inflammation
10. Antibody-mediated immunity is most directly associated with
    a. T cells.
    b. monocytes.
    c. basophils.
    d. B cells.
11. Plasma cells are
    a. the same as memory cells.
    b. formed from blood plasma.
    c. B cells that are actively secreting antibody.
    d. inactive T cells carried in the plasma.

12. Antibodies combine with antigens
    a. at variable regions.
    b. at constant regions.
    c. only if macrophages are present.
    d. Both a and c are correct.
13. Which applies to T cells?
    a. mature in bone marrow
    b. mature in thymus gland
    c. Both a and b are correct.
    d. None of these are correct.
14. An antigen-presenting cell (APC)
    a. presents antigens to T cells.
    b. secretes antibodies.
    c. marks each human cell as belonging to that particular person.
    d. secretes cytokines.
15. MHC antigens play a role in
    a. active immunity.
    b. presentation of antigens to T cells.
    c. tissue transplantation.
    d. All of these are correct.
16. Which of the following would not be a participant in cell-mediated immune responses?
    a. helper T cells
    b. macrophages
    c. cytokines
    d. cytotoxic T cells
    e. plasma cells
17. Type O blood has ____ antibodies in the plasma.
    a. anti-A and anti-B
    b. anti-O
    c. no
    d. All of these are correct.
18. Label a–c on this IgG molecule using these terms: antigen-binding sites, light chain, heavy chain. What do V and C stand for in the diagram?

19. Monoclonal antibodies are
    a. the same thing as vaccines.
    b. very specific in binding to a single antigen.
    c. usually isolated from the serum of immunized animals.
    d. All of these are correct.
20. Which of the following is not an example of an autoimmune disease?
    a. multiple sclerosis
    b. rheumatic fever
    c. hemolytic disease of the newborn
    d. systemic lupus erythematosus

UNDERSTANDING THE TERMS

active immunity 244
adaptive immunity 238
agglutination 249
allergen 248
allergy 248
anaphylactic shock 248
antibody-mediated immunity 241
antibody titer 244
antigen 238
antigen-presenting cell (APC) 243
asthma 248
autoimmune disease 251
B-cell receptor (BCR) 241
cell-mediated immunity 243
clonal selection theory 241
complement system 240
cytokine 240
cytotoxic T cell 243
delayed allergic response 248
dendritic cell 240
edema 237
hay fever 248
helper T cell 243
histamine 239
immediate allergic response 248
immunity 238
immunization 244
immunoglobulin(Ig) 241
immunosuppressive 251
inflammatory reaction 239
innate immunity 238
interferon 240
interleukin 247
lymph 236
lymphatic system 236
lymphatic vessel 236
lymph node 238
lymphoid organ 237
macrophage 240
major histocompatibility complex (MHC) 243
mast cell 239
monoclonal antibody 247
monocyte 240
multiple sclerosis (MS) 251
myasthenia gravis 251
natural killer (NK) cell 240
neutrophil 240
passive immunity 244
pathogen 238
phagocyte 239
red bone marrow 238
rheumatoid arthritis 251
severe combined immunodeficiency disease (SCID) 251
spleen 238
systemic lupus erythematosus (SLE) 251
T-cell receptor (TCR) 242
thymus 238
vaccine 244

THINKING CRITICALLY

1. Some primitive organisms, such as invertebrates, have no lymphocytes and thus lack an adaptive immune system, but they have some components of an innate immune system, including phagocytes and certain protective proteins. What are some general features of innate immunity that make it very valuable to organisms lacking more specific antibody- and cell-mediated responses? What are some disadvantages to having only an innate immune system?
2. Compare the ways in which natural killer cells and cytotoxic T cells recognize the cells they are going to kill. Why does it make good biological sense for NK cells and cytotoxic T cells to recognize, for example, virus-infected cells in these different ways?
3. Why is it that Rh incompatibility can be a serious problem when an Rh-negative mother is carrying an Rh-positive fetus, but ABO incompatibility between mother and fetus is usually no problem? That is, a type A mother can usually safely carry a type B fetus. (Hint: The antibodies produced by an Rh-negative mother against the Rh antigen are usually IgG, while the antibodies produced against the A or B antigens are IgM.) Because the Rh antigen obviously serves no vital function (most humans lack it), why do you think it hasn't been completely eliminated during human evolution?

INQUIRY INTO LIFE WEBSITE

The companion website for *Inquiry into Life* provides a wealth of information organized and integrated by chapter. You will find practice tests, animations, videos, and much more that will complement your learning and understanding of general biology.

http://www.mhhe.com/maderinquiry13

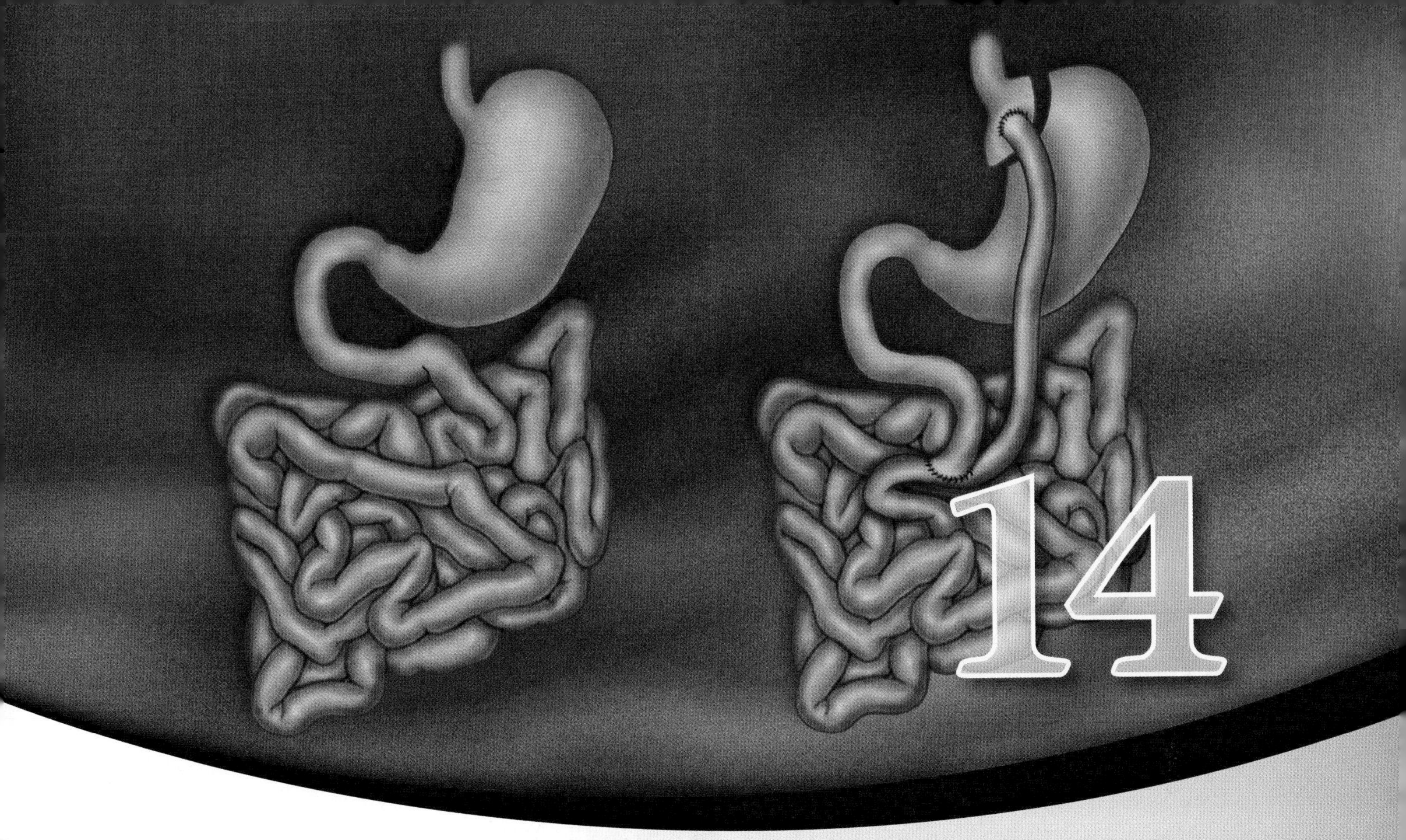

Applying the Concepts

Dennis had struggled with a weight problem since he was a teenager. At family gatherings, whenever he saw his overweight aunts, uncles, and grandparents, he was reminded that obesity was probably "in his genes," which only discouraged him from sticking with the various diets he had tried. And recent research has confirmed that, in fact, obesity may have a strong genetic component.

Tipping the scales at 450 lbs, Dennis had pretty much resigned himself to a lifetime of being out of breath while climbing stairs, requiring two seats on airplanes—and probably an early death from diabetes, heart disease, or some other condition that is far more common in obese people. Then, a co-worker whom Dennis always assumed was just "lucky" to be thin confided that she had undergone gastric bypass surgery.

After consulting with a surgeon, Dennis learned he was a good candidate for the Roux-en-Y procedure, in which a small pouch is created where the esophagus enters the stomach. This pouch is then connected about halfway down the small intestine, so that ingested food bypasses most of the stomach and upper small intestine. After the procedure, Dennis felt like his stomach was "full" even after a small meal, and he began dropping pounds. Although he had to put up with side effects such as abdominal pain and diarrhea after eating sweets, a year after the surgery, he had lost almost 200 lb. A few months later, at 220 lb, he went in for a second surgery—this time to remove the large amount of excess skin that had been stretched out during his years of being overweight. He now keeps only a few photos of himself from his "previous life," mostly to remind himself how much better he feels now.

Digestive System and Nutrition

Chapter Outline

14.1 The Digestive Tract

Learning Outcomes

1. Describe all the major components of the human digestive tract from the mouth to the anus.
2. Compare mechanical and chemical digestion with regard to where and how they occur.

Digestion takes place within a tube called the digestive tract, which begins with the mouth and ends with the anus (Fig. 14.1). The digestive system contributies to homeostasis by ingesting food, separating it into chemical nutrients that cells can use, absorbing those nutrients, and eliminating indigestible remains.

Digestion involves mechanical digestion and chemical digestion. Mechanical digestion begins with the chewing of food in the mouth and continues with the churning and mixing of food in the stomach. During chemical digestion, many different enzymes break down macromolecules to small organic molecules that can be absorbed.

The Mouth

The mouth, which receives food, is bounded externally by the lips and cheeks. The red portion of the lips is poorly keratinized, and this allows blood to show through.

Most people enjoy eating food largely because they like its texture and taste. Sensory receptors called taste buds occur primarily on the tongue, and when these are activated by the presence of food, nerve impulses travel by way of

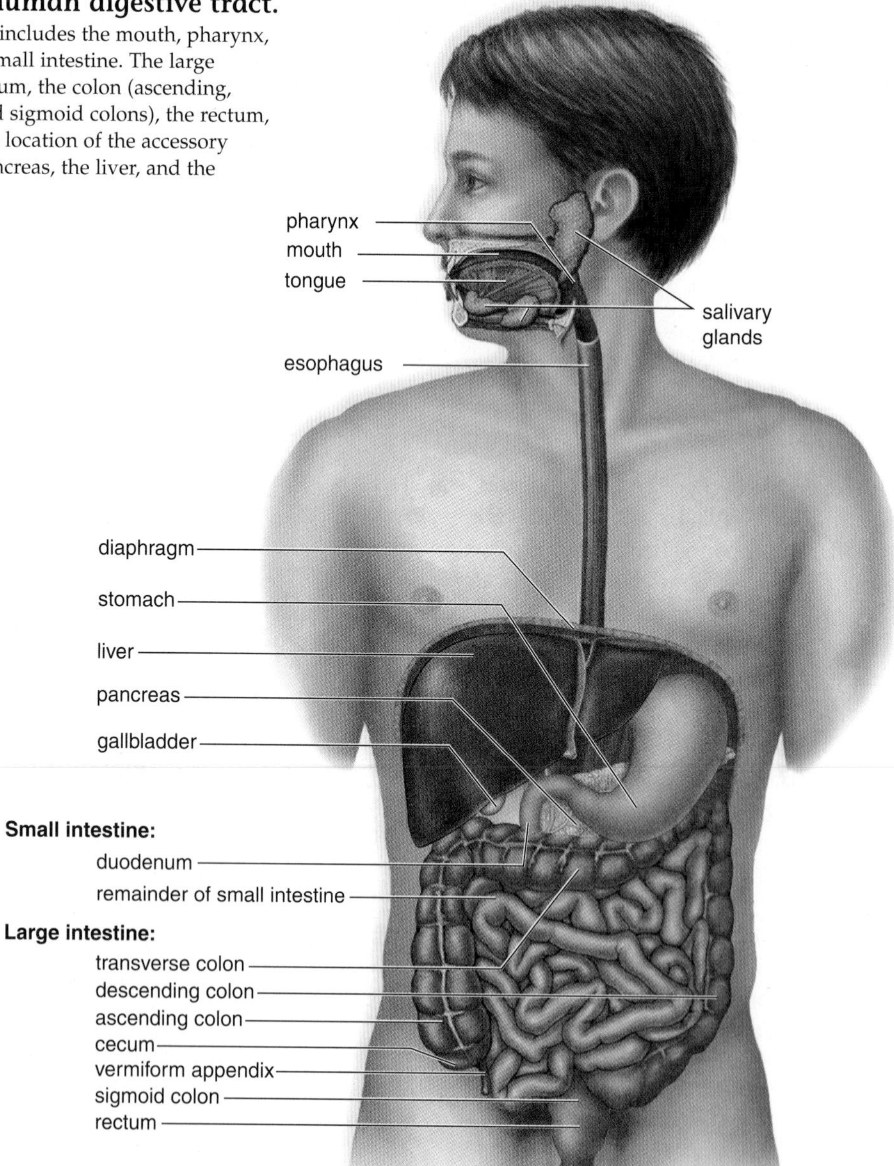

Figure 14.1 The human digestive tract. The upper part of the tract includes the mouth, pharynx, esophagus, stomach, and small intestine. The large intestine consists of the cecum, the colon (ascending, transverse, descending, and sigmoid colons), the rectum, and the anus. Note also the location of the accessory organs of digestion: the pancreas, the liver, and the gallbladder.

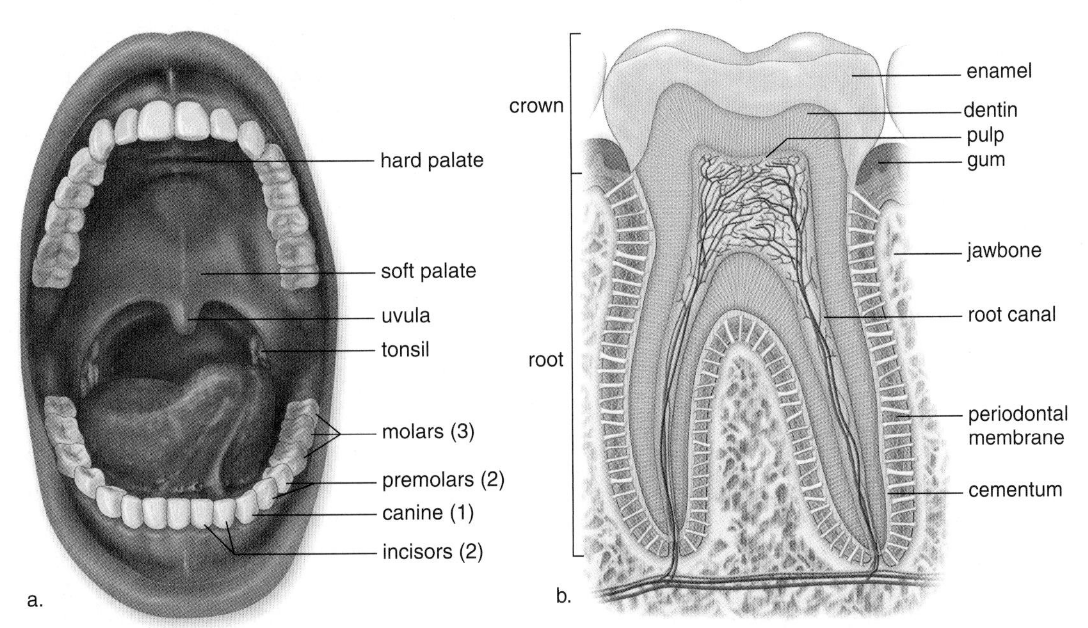

Figure 14.2 Adult mouth and teeth.
a. The chisel-shaped incisors bite; the pointed canines tear; the fairly flat premolars grind; and the flattened molars crush food. The last molars, called wisdom teeth, may fail to erupt, or if they do, they are sometimes crooked and useless. **b.** Longitudinal section of a tooth. The crown is the portion that projects above the gum line and can be replaced by a dentist if damaged. When a "root canal" is done, the nerves are removed. When the periodontal membrane is inflamed, the teeth can loosen.

cranial nerves to the brain. The tongue is composed of skeletal muscle that contracts to change the shape of the tongue. Muscles exterior to the tongue enable it to move about. A fold of mucous membrane on the underside of the tongue attaches it to the floor of the mouth.

The roof of the mouth separates the nasal cavities from the mouth. The roof has two parts: an anterior (toward the front) **hard palate** and a posterior (toward the back) **soft palate** (Fig. 14.2*a*). The hard palate contains several bones, but the soft palate is composed of muscle. The soft palate ends in a finger-shaped projection called the uvula. The tonsils are in the back of the mouth, on either side of the tongue, and in the nasopharynx, where they are called adenoids. The tonsils contain lymphoid tissue that helps protect the body against infections (see Chapter 13). If the tonsils become inflamed, the person has tonsillitis. If tonsillitis recurs repeatedly, the tonsils may be surgically removed (called a tonsillectomy).

Three pairs of **salivary glands** (see Fig. 14.1) produce saliva. Saliva keeps the mouth moist and contains an enzyme called **salivary amylase** that begins the process of digesting starch. One pair of salivary glands lies on either side of the face immediately below and in front of the ears. These glands swell when a person has the mumps, a disease caused by a viral infection. Another pair of salivary glands lies beneath the tongue, and still another pair lies beneath the floor of the mouth. The ducts from these salivary glands open under the tongue. You may be able to locate the openings if you use your tongue to feel for small flaps on the inside of your cheek and under your tongue.

The Teeth

With our teeth, we chew food into pieces convenient for swallowing. During the first two years of life, the smaller 20 deciduous, or baby, teeth appear. These are eventually replaced by 32 adult teeth (Fig. 14.2*a*). The third pair of molars, called the wisdom teeth, sometimes fail to erupt. If they push on the other teeth and/or cause pain, they can be removed by a dentist or oral surgeon.

Each tooth has two main divisions, a crown and a root (Fig. 14.2*b*). The crown has a layer of enamel, an extremely hard outer covering of calcium compounds; dentin, a thick layer of bonelike material; and an inner pulp, which contains the nerves and the blood vessels. Dentin and pulp are also found in the root. The gum tissue, or **gingiva,** surrounds the teeth and normally forms a tight seal around them.

Dental caries, or cavities, occur when bacteria adhering to the teeth metabolize sugar and give off acids, which may erode the enamel. Three measures can help to prevent tooth decay: limiting sugar intake, daily brushing and flossing, and regular visits to the dentist. Fluoride treatments, particularly in children, can make the enamel stronger and decay-resistant. Gum disease is more apt to occur with aging. Inflammation of the gums (gingivitis) can spread to the periodontal membrane, which lines the tooth socket. A person then has periodontitis, characterized by loss of bone and loosening of the teeth so that extensive dental work may be required. Stimulation of the gums in a manner advised by your dentist is helpful in controlling this condition.

Over time, and especially with exposure to cigarette smoke, soft drinks, or coffee, the enamel layer can become stained. In recent years, various tooth-whitening treatments, most of which contain a bleaching agent such as hydrogen peroxide, have become a popular way to remove these stains.

Check Your Progress

1. How does digestion specifically begin in the mouth?
2. What three steps can be taken to reduce dental caries?

TABLE 14.1 Path of Food

Organ	General Function(s)	Special Feature(s)	Function of Special Feature(s)
Mouth	Receives food; starts digestion of starch	Teeth Tongue	Chew food Forms bolus
Pharynx	Passageway	—	—
Esophagus	Passageway	—	—
Stomach	Storage of food; acidity kills bacteria; starts digestion of protein	Gastric glands	Release gastric juices
Small intestine	Digestion of all foods; absorption of nutrients	Intestinal glands Villi	Release intestinal juices Absorb nutrients
Large intestine	Absorption of water; storage of indigestible remains	—	—

The Pharynx

The **pharynx** is a region that receives air from the nasal cavities and food from the mouth (see Fig. 14.1).

Table 14.1 lists some specific functions of the digestive organs. From the mouth, food (the bolus) passes through the pharynx and esophagus to the stomach, small intestine, and large intestine. The food passage and air passage cross in the pharynx because the trachea (windpipe) is ventral to (in front of) the esophagus, which takes food to the stomach. Swallowing, a process that occurs in the pharynx (Fig. 14.3), is a **reflex action** performed automatically, without conscious thought. During swallowing, the soft palate moves back to close off the **nasopharynx,** and the trachea moves up under the **epiglottis** to cover the glottis. The **glottis** is the opening to the larynx (voice box), and therefore the air passage. The up-and-down movement of the Adam's apple, the front part of the larynx, is easy to observe when a person swallows. During swallowing, food normally enters the esophagus because the air passages are blocked. We do not breathe when we swallow.

Unfortunately, we have all had the unpleasant experience of having food "go the wrong way." The wrong way may be either into the nasal cavities or into the trachea. If it is the latter, coughing will most likely force the food up out of the trachea and into the pharynx again.

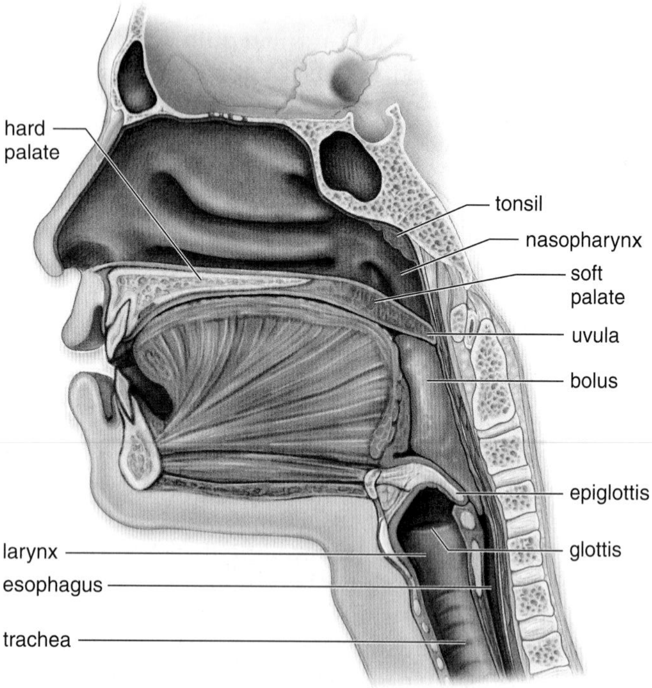

Figure 14.3 Swallowing.
When food is swallowed, the soft palate closes off the nasopharynx, and the epiglottis covers the glottis, forcing the bolus to pass down the esophagus. Therefore, a person does not breathe while swallowing.

The Esophagus

The **esophagus** is a long muscular tube that passes from the pharynx, through the thoracic cavity and diaphragm, and into the abdominal cavity, where it joins the stomach. The esophagus is ordinarily collapsed, but it opens and receives the bolus when swallowing occurs.

A rhythmic contraction called **peristalsis** pushes the food along the digestive tract. Peristalsis begins in the esophagus and continues in all the organs of the digestive tract. Occasionally, peristalsis begins even though there is no food in the esophagus. This produces the sensation of a lump in the throat.

The esophagus plays no role in the chemical digestion of food. Its sole purpose is to conduct the food bolus from the mouth to the stomach. **Sphincters** are muscles that encircle tubes and act as valves. The tubes close when sphincters contract, and they open when sphincters relax. The entrance of the esophagus to the stomach is marked by a constriction that is often called a sphincter, although the muscle is not as developed as it would be in a true sphincter. Relaxation of the sphincter allows the bolus to pass into the stomach, while contraction prevents the acidic contents of the stomach from backing up into the esophagus.

Heartburn, which feels like a burning pain rising into the throat, occurs when some of the stomach contents escape into the esophagus. Heartburn is a very common condition: two of the four best-selling drugs in the United States in 2007 targeted these symptoms by reducing acid production in the stomach. A more serious form of heartburn is GERD (gastroesophageal reflux disease), which is frequent or persistent reflux that may lead to more serious problems such as ulcers, difficulty swallowing, or even esophageal cancer due to chronic irritation of the esophageal wall by stomach acid.

Vomiting, which is the forceful expulsion of stomach contents out of the body through the mouth, can help protect against the ingestion of potentially harmful agents. In this scenario, certain cells in the intestinal tract, as well as in the brainstem, trigger the contraction of the abdominal muscles and diaphragm to propel the contents of the stomach upward through the esophagus.

The Wall of the Digestive Tract

The structure of the esophageal wall in the abdominal cavity is representative of that found in the stomach, small intestine, and large intestine. All are composed of four layers, as shown in Figure 14.4 and listed here:

Mucosa (mucous membrane layer) A layer of epithelium supported by connective tissue and smooth muscle lines the lumen (central cavity) and contains glandular epithelial cells that secrete digestive enzymes and goblet cells that secrete mucus.

Submucosa (submucosal layer) Beneath the mucosa lies the submucosa, a broad band of loose connective tissue that contains blood vessels. Lymph nodules, including some called Peyer's patches, are in the submucosa. Like the tonsils, they help protect us from disease.

Muscularis (smooth muscle layer) Two layers of smooth muscle make up this section. The inner, circular layer encircles the gut; the outer, longitudinal layer lies perpendicular to the circular layer. (The stomach also has oblique muscles.)

Serosa (serous membrane layer) Most of the digestive tract has a serosa, a very thin, outermost layer of squamous epithelium that secretes a serous fluid that keeps the outer surface of the intestines moist so that the organs of the abdominal cavity slide against one another.

Check Your Progress

1. What are the specific functions of the soft palate and the epiglottis during swallowing?
2. What is GERD, and why is it a potentially serious disease?
3. List the four layers that make up the wall of the esophagus, stomach, and intestines.

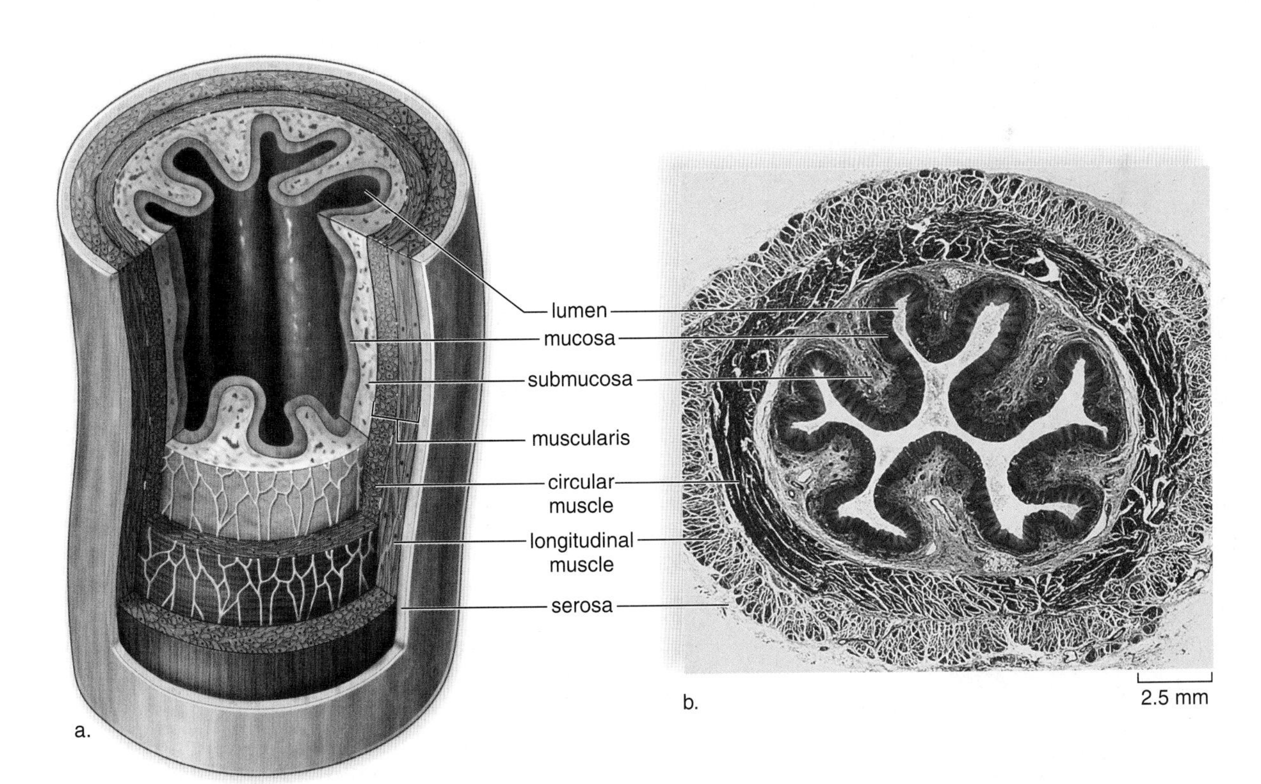

Figure 14.4 Wall of the digestive tract.
a. Several different types of tissues are found in the wall of the digestive tract. Note the placement of circular muscle inside longitudinal muscle.
b. Micrograph of the wall of the esophagus.

The Stomach

The **stomach** (Fig. 14.5*a*) is a thick-walled, J-shaped organ that lies on the left side of the abdominal cavity below the liver and diaphragm. It is continuous with the esophagus above and the duodenum of the small intestine below. The stomach receives food from the esophagus, starts the digestion of proteins, and moves food into the small intestine. The human stomach is about 25 cm (10 in.) long, regardless of the amount of food it holds, but the diameter varies, depending on how full it is. As the stomach expands, deep folds in its wall, called **rugae,** gradually disappear. When full, it can hold about 4 liters (1 gal).

The columnar epithelium lining the stomach has millions of gastric pits, which lead into **gastric glands** (Fig. 14.5*b*). (The term "gastric" always refers to the stomach.) The gastric glands produce gastric juice, which contains pepsinogen, hydrochloric acid (HCl), and mucus. Pepsinogen becomes the enzyme **pepsin** when exposed to HCl. The high acidity of the stomach (pH 1-2) is also beneficial because it kills most of the bacteria and other microbes present in food.

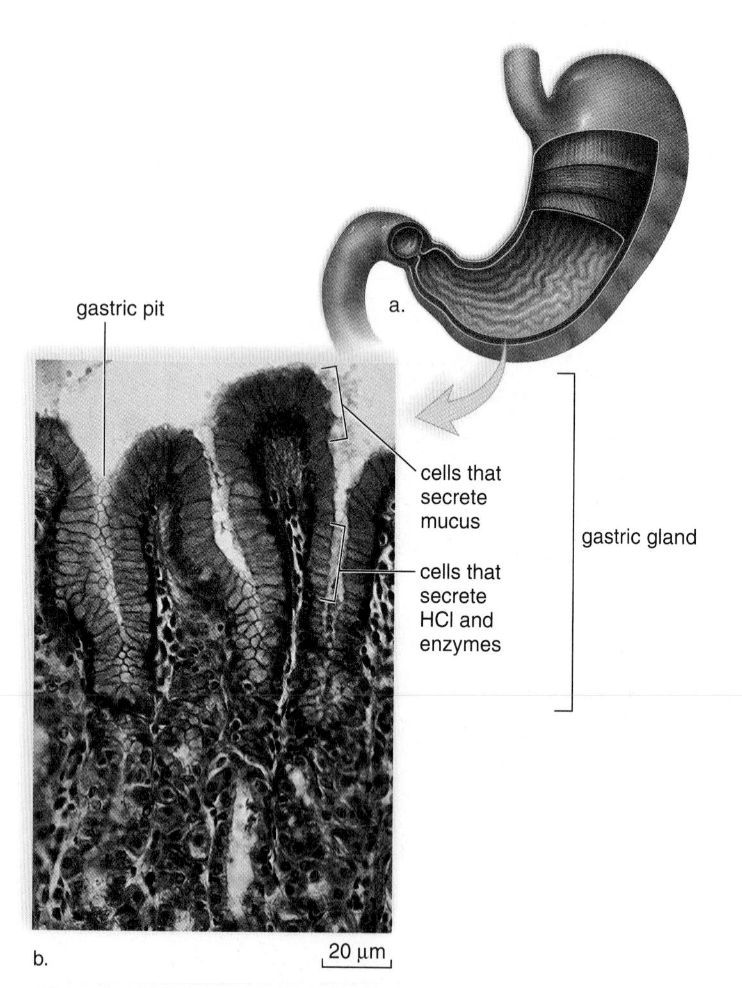

Figure 14.5 Anatomy of the stomach.
a. The stomach has a thick wall with folds that allow it to expand and fill with food. **b.** The mucosa contains gastric glands, which secrete gastric juice containing mucus and digestive enzymes.

The stomach acts both physically and chemically on food. Its wall contains three muscle layers: one layer is longitudinal and another is circular, as shown in Figure 14.4*a*; the third is obliquely arranged. This muscular wall not only moves the food along, but it also churns, mixing the food with gastric juice and breaking it down into small pieces.

Alcohol and other liquids are absorbed in the stomach, but food substances are not. Normally, the stomach empties in 2–6 hours. When food leaves the stomach, it is a thick, soupy liquid called **chyme.** Chyme enters the small intestine in squirts by way of the pyloric sphincter, which acts like a valve, repeatedly opening and closing.

The Small Intestine

The **small intestine** is named for its small diameter (compared to that of the large intestine), but perhaps it should be called the long intestine. It averages about 6 meters (almost 20 ft) in length, compared to the large intestine, which is about 1.5 meters (5 ft) in length.

The first 25 cm (1 ft) of the small intestine is called the **duodenum.** Ducts from the liver and pancreas join to form one common bile duct that enters the duodenum (see Fig. 14.10). The small intestine receives bile from the liver and pancreatic juice from the pancreas via this duct. **Bile** emulsifies fat, meaning that it causes fat droplets to disperse in water. The intestine has a slightly basic pH because pancreatic juice contains sodium bicarbonate ($NaHCO_3$), which neutralizes the acid in chyme. The enzymes in pancreatic juice and enzymes produced by the intestinal wall complete the process of food digestion.

The middle part of the small intestine is called the **jejunum,** and the remainder is the **ileum.** The submucosal layer of the ileum contains aggregations of lymphoid tissues called Peyer's patches, which are involved in generating immune responses to intestinal pathogens.

The surface area of the small intestine has been compared to that of a tennis court. What factors contribute to such a large surface area? The wall of the small intestine contains fingerlike projections called villi (sing. **villus**), which give the intestinal wall a soft, velvety appearance (Fig. 14.6). A villus has an outer layer of columnar epithelial cells, and each of these cells has thousands of microscopic extensions called microvilli. Microvilli greatly increase the surface area of the villus for the absorption of nutrients.

Nutrients are absorbed into the vessels of a villus. A villus contains blood capillaries and a small lymphatic capillary, called a **lacteal.** Glycerol and fatty acids (digested from fats) enter the epithelial cells of the villi, where they are joined and packaged as lipoprotein droplets that enter a lacteal. Lacteals are a part of the lymphatic system (see Chapter 13). The lymphatic system is an adjunct to the cardiovascular system; that is, its vessels carry a fluid called lymph to the cardiovascular veins. Sugars and amino acids directly enter the blood capillaries of a villus. After nutrients are absorbed, the bloodstream eventually carries them to all the cells of the body.

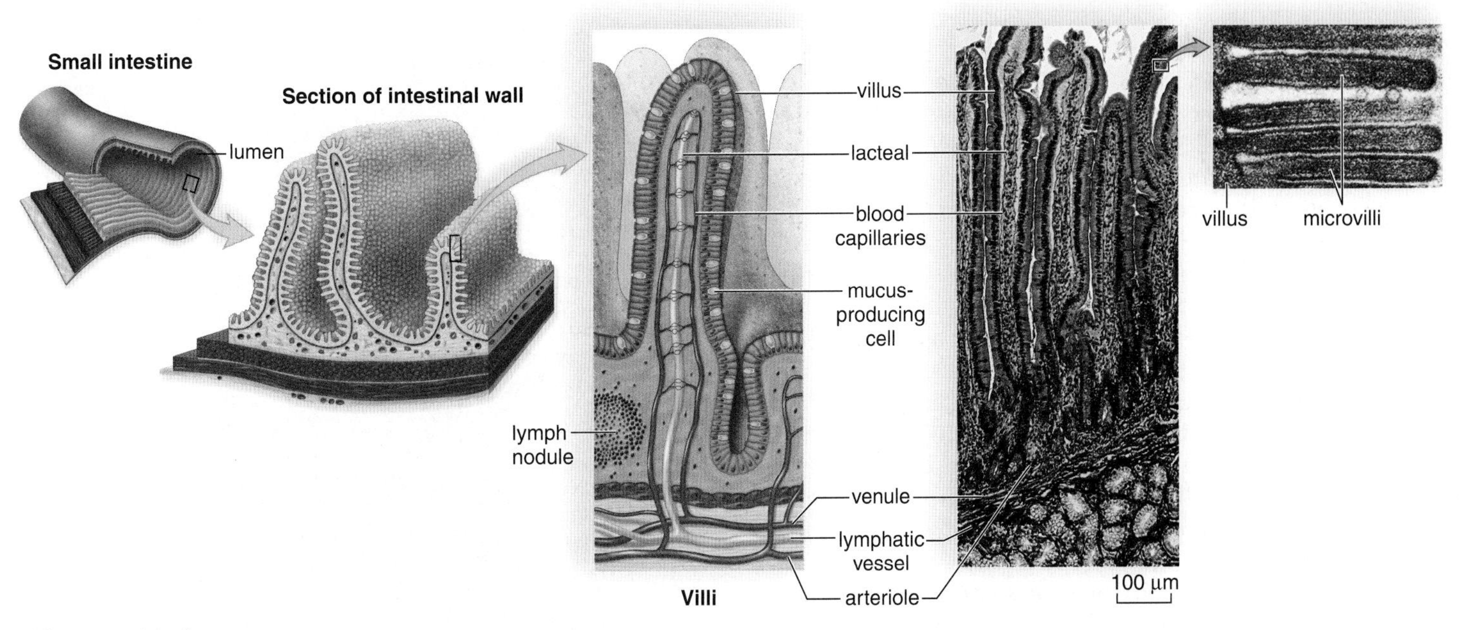

Figure 14.6 Anatomy of the small intestine.
The wall of the small intestine has folds that bear fingerlike projections called villi. Microvilli, which project from the villi, absorb the products of digestion into the blood capillaries and the lacteals of the villi.

Regulation of Digestive Secretions

The secretion of digestive juices is promoted by the nervous system and by hormones. A **hormone** is a substance produced by one set of cells that affects a different set of cells, the so-called target cells (see Chapter 20 for a more thorough discussion of the endocrine system). Hormones are usually transported by the bloodstream. For example, when a person has eaten a meal particularly rich in protein, the stomach produces the hormone gastrin. Gastrin enters the bloodstream, and soon the stomach is churning, and the secretory activity of gastric glands is increasing. A hormone produced by the duodenal wall, GIP (gastric inhibitory peptide), works opposite to gastrin: it inhibits gastric gland secretion.

Cells of the duodenal wall produce two other hormones that are of particular interest—secretin and CCK (cholecystokinin). Acid, especially hydrochloric acid (HCl) present in chyme, stimulates the release of secretin, while partially digested protein and fat stimulate the release of CCK. Soon after these hormones enter the bloodstream, the pancreas increases its output of pancreatic juice, which helps digest food, the liver increases its output of bile, and the gallbladder contracts to release bile. Figure 14.7 summarizes the actions of gastrin, secretin, and CCK.

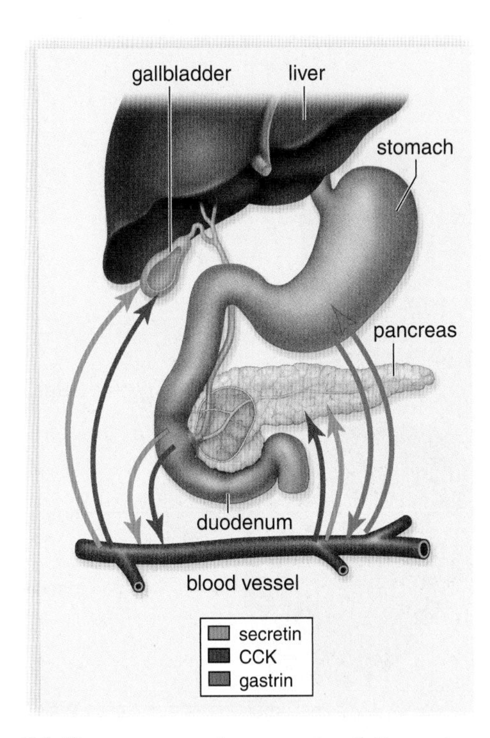

Figure 14.7 Hormonal control of digestive gland secretions.
Gastrin (blue), produced by the lower part of the stomach, enters the bloodstream and thereafter stimulates the stomach to produce more digestive juices. Secretin (green) and CCK (purple), produced by the duodenal wall, stimulate the pancreas to secrete its digestive juices and the gallbladder to release bile.

CHECK YOUR PROGRESS

1. Compare the functions of the stomach and small intestine in digestion. How are they similar? How are they different?
2. Might it be possible to help obese people lose weight by inhibiting the secretion and/or activity of certain hormones? What are some possible side effects of this approach?

The Large Intestine

The **large intestine** includes the cecum, colon, rectum, and anal canal. It is larger in diameter than the small intestine (6.5 cm compared to 2.5 cm), but shorter in length. The large intestine absorbs water, salts, and some vitamins. It also stores indigestible material until it is eliminated as feces.

The **cecum,** which lies below the junction with the small intestine, is a small pouch (6 cm long) that forms the first part of the large intestine. The human cecum has a small projection called the **vermiform appendix** (*vermiform* means wormlike) (Fig. 14.8). Like the tonsils, the appendix may play a role in fighting infection. This organ is subject to inflammation, a condition called appendicitis. If inflamed, the appendix should be removed before the appendix bursts, which could cause **peritonitis,** an inflammation of the lining of the abdominal cavity. Peritonitis can lead to death.

The **colon** includes the ascending colon, which goes up the right side of the body to the level of the liver; the transverse colon, which crosses the abdominal cavity just below the liver and the stomach; the descending colon, which passes down the left side of the body; and the sigmoid colon, which enters the **rectum,** the last 20 cm of the large intestine. The rectum opens at the **anus,** where **defecation,** the expulsion of feces, occurs. When feces are forced into the rectum by peristalsis, a defecation reflex occurs. The stretching of the rectal wall initiates nerve impulses to the spinal cord, and shortly thereafter, the rectal muscles contract, and the anal sphincters relax (Fig. 14.9). Ridding the body of indigestible remains is another way the digestive system helps maintain homeostasis. Feces are normally about three-quarters water and one-quarter solids. Bacteria, **fiber** (indigestible plant material), and other indigestible materials are in the solid portion. Bacterial action on undigested materials causes the odor of feces and also accounts for the presence of gas. A breakdown product of bilirubin (described in section 14.2) and the presence of oxidized iron cause the brown color of feces.

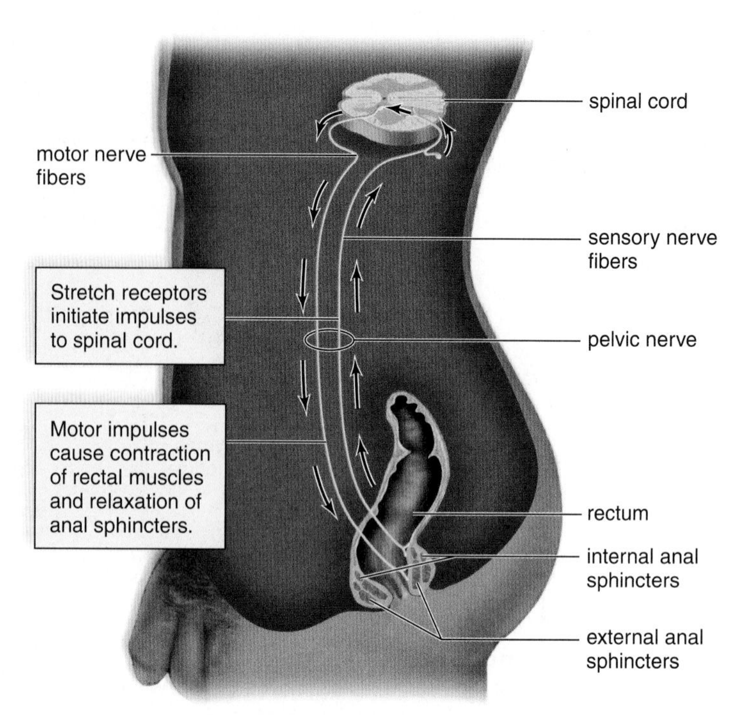

Figure 14.9 Defecation reflex.
The accumulation of feces in the rectum causes it to stretch, which initiates a reflex action resulting in rectal contraction and expulsion of the fecal material.

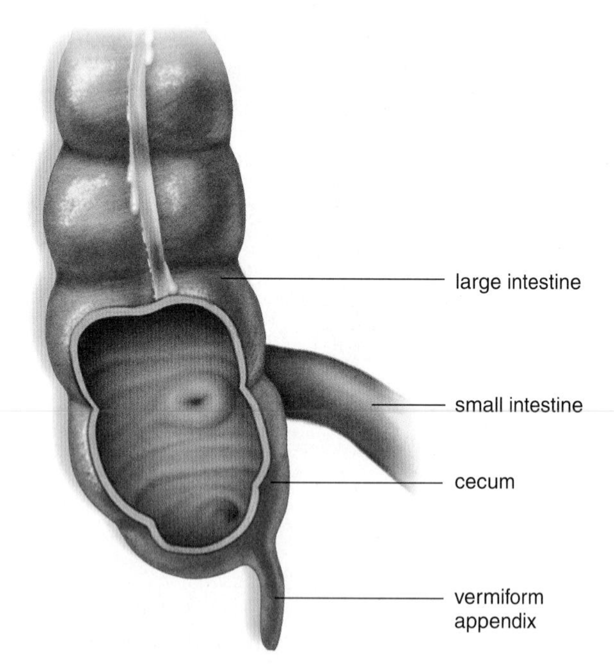

Figure 14.8 Junction of the small intestine and the large intestine.
The cecum is the blind end of the large intestine. The appendix is attached to the cecum.

About 40–50% of the fecal mass consists of bacteria and other microbes; there are about 100 billion bacteria per gram of feces! These include facultative bacteria (bacteria that can live with or without oxygen), such as *Escherichia coli,* as well as many obligate anaerobes (bacteria that die in the presence of oxygen). These bacteria break down some indigestible material, and produce some vitamins that our bodies can absorb and use. In this way, they perform a service for us, and we provide a good environment for them to live in.

Water is considered unsafe for drinking when the coliform (intestinal) bacterial count reaches a certain number. A high count indicates that a significant amount of feces has entered the water. The more feces present, the greater the possibility that disease-causing bacteria are also present.

Check Your Progress

1. What are the major functions of the large intestine?
2. What roles do bacteria play in colon function?

14.2 Accessory Organs of Digestion

Learning Outcomes

1. Summarize the major functions of the pancreas, the liver, and the gallbladder.
2. Summarize the structure and function of the hepatic portal.

The pancreas, liver, and gallbladder are accessory digestive organs. The salivary glands, discussed earlier, are also accessory organs. Figure 14.10*a* shows how the pancreatic duct from the pancreas and the common bile duct from the liver and gallbladder enter the duodenum.

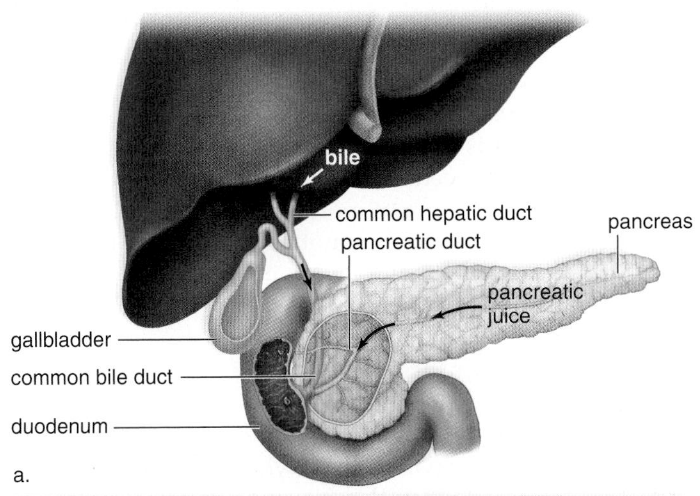

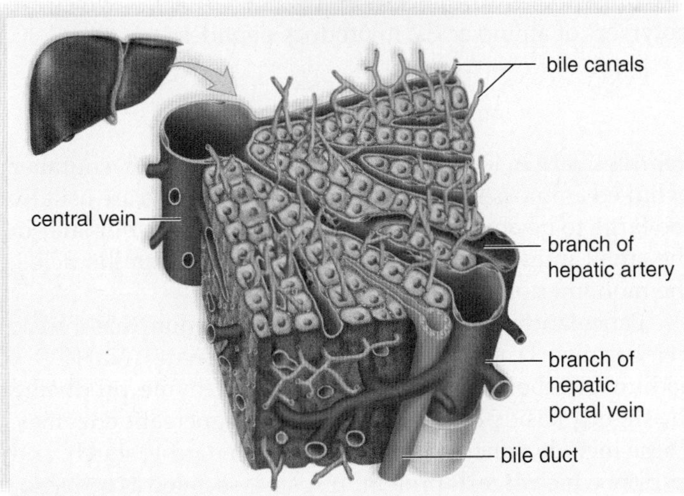

Figure 14.10 Liver, gallbladder, and pancreas.
a. The liver makes bile, which is stored in the gallbladder and sent (black arrow) to the small intestine by way of the common bile duct. The pancreas produces digestive enzymes that are sent (black arrows) to the small intestine by way of the pancreatic duct. **b.** A hepatic lobule. The liver contains over 100,000 lobules, each lobule composed of many cells that perform the various functions of the liver. They remove and add materials to the blood and deposit bile in a duct.

The Pancreas

The **pancreas** lies deep in the abdominal cavity, resting on the posterior abdominal wall. It is an elongated and somewhat flattened organ that has both an endocrine and an exocrine function. As an endocrine gland, it secretes insulin and glucagon, hormones that help keep the blood glucose level within normal limits. In this chapter, however, we are interested in its exocrine function. Most pancreatic cells produce pancreatic juice, which contains sodium bicarbonate ($NaHCO_3$) to help neutralize the stomach acid, and digestive enzymes for all types of food. **Pancreatic amylase** digests starch, **trypsin** digests protein, and **lipase** digests fat.

The Liver

The **liver,** which is the largest gland in the body, lies mainly in the upper right section of the abdominal cavity, under the diaphragm (see Fig. 14.1). The liver contains approximately 100,000 lobules that serve as its structural and functional units (Fig. 14.10*b*). Three structures are located between the lobules: a bile duct that takes bile away from the liver; a branch of the hepatic artery that brings O_2-rich blood to the liver; and a branch of the hepatic portal vein that transports nutrients from the intestines. Each lobule has a central vein that enters a hepatic vein. In Figure 14.11, trace the path of blood from the intestines to the liver via the hepatic portal vein and from the liver to the inferior vena cava via the hepatic veins.

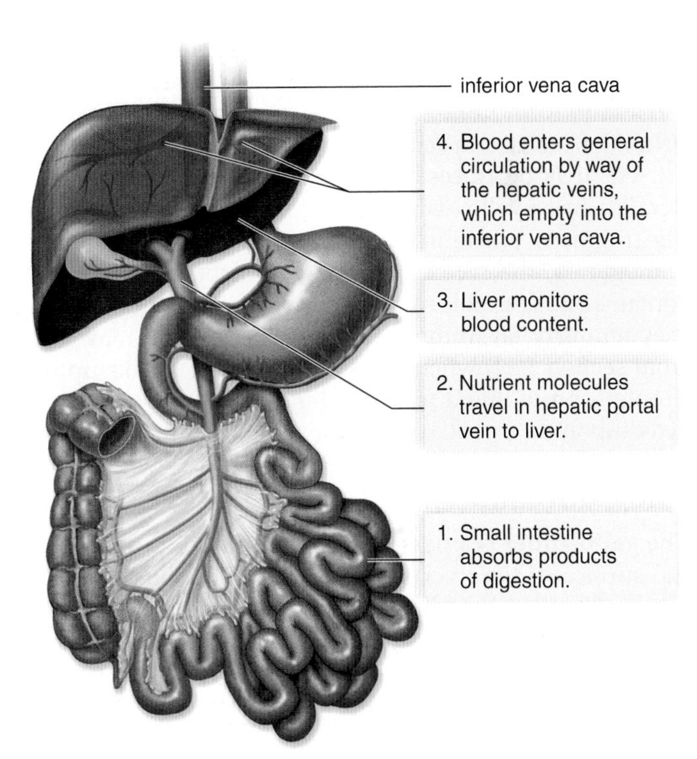

Figure 14.11 Hepatic portal system.
The hepatic portal vein takes the products of digestion from the digestive system to the liver, where they are processed before entering a hepatic vein.

TABLE 14.2 Functions of the Liver

Functions of the Liver
1. Detoxifies blood by removing and metabolizing poisonous substances
2. Stores iron (Fe^{2+}) and vitamins A, D, E, K, and B_{12}
3. Makes many plasma proteins, such as albumins and fibrinogen, from amino acids
4. Stores glucose as glycogen after a meal, and breaks down glycogen to glucose to maintain the glucose concentration of blood between eating periods
5. Produces urea after breaking down amino acids
6. Removes bilirubin, a breakdown product of hemoglobin from the blood, and excretes it in bile, a liver product
7. Helps regulate blood cholesterol level, converting some to bile salts

In some ways, the liver acts as the gatekeeper to the blood (Table 14.2). As blood from the hepatic portal vein passes through the liver, it removes poisonous substances and detoxifies them. The liver also removes and stores iron and vitamins A, D, E, K, and B_{12}. The liver makes many of the plasma proteins and helps regulate the quantity of cholesterol in the blood.

The liver maintains the blood glucose level at 50–80 mg/100 ml, even though a person eats intermittently. When insulin is present, any excess glucose in the blood is removed and stored by the liver as glycogen. Between meals, glycogen is broken down to glucose, which enters the hepatic veins, and in this way, the blood glucose level remains constant.

If the supply of glycogen is depleted, the liver converts glycerol (from fats) and amino acids to glucose molecules. The conversion of amino acids to glucose necessitates deamination, the removal of amino groups. By a complex metabolic pathway, the liver then combines ammonia with carbon dioxide to form urea. Urea is the usual nitrogenous waste product from amino acid breakdown in humans.

The liver produces bile, which is stored in the gallbladder. Bile has a yellowish-green color because it contains the bile pigment bilirubin, derived from the breakdown of hemoglobin, the red pigment in red blood cells. Bile also contains bile salts. Bile salts are derived from cholesterol, and they emulsify fat in the small intestine. As you may recall from section 14.1, when fat is emulsified, it breaks up into droplets, providing a much larger surface area that can be acted upon by a pancreatic lipase (see Section 14.3).

The Gallbladder

The **gallbladder** is a pear-shaped, muscular sac attached to the surface of the liver (see Fig. 14.1). The liver produces about 400–800 ml of bile each day, and any excess is stored in the gallbladder. Water is reabsorbed by the gallbladder so that bile becomes a thick, mucuslike material. When needed, bile leaves the gallbladder and proceeds to the duodenum via the common bile duct.

Check Your Progress

1. What does it mean to say that the pancreas has both endocrine and exocrine functions?
2. Why is the liver necessary for life, while the gallbladder can be removed with few consequences?

14.3 Digestive Enzymes

Learning Outcomes

1. Describe the overall function of digestive enzymes.
2. Compare the specific types of chemical digestion that take place in the mouth, stomach, and small intestine.

Digestive enzymes, like other enzymes, are proteins that speed up specific chemical reactions. Table 14.3 lists some of the major enzymes produced by the digestive tract. These enzymes help break down carbohydrates, proteins, nucleic acids, and fats, the major components of food.

Starch is a carbohydrate, and its digestion begins in the mouth. Saliva from the salivary glands has a neutral pH and contains salivary amylase, the first enzyme to act on starch:

$$\text{starch} + H_2O \xrightarrow{\text{salivary amylase}} \text{maltose}$$

In this equation, salivary amylase is written above the arrow to indicate that it is neither a reactant nor a product in the reaction. It merely speeds the reaction in which its substrate, starch, is digested to many molecules of maltose, a disaccharide. Maltose molecules cannot be absorbed by the intestine. Additional digestive action in the small intestine converts maltose to glucose, a monosaccharide. Glucose can be absorbed.

Protein digestion begins in the stomach. Gastric juice secreted by gastric glands has a very low pH of 1-2 because it contains hydrochloric acid (HCl). Pepsinogen, a precursor that is converted to the enzyme pepsin when exposed to HCl, is also present in gastric juice. Pepsin acts on proteins, which are polymers of amino acids, to produce peptides:

$$\text{protein} + H_2O \xrightarrow{\text{pepsin}} \text{peptides}$$

Peptides vary in length, but they always consist of a number of linked amino acids. Peptides from the stomach are usually too large to be absorbed by the intestinal lining, but later in the small intestine, they are broken down to amino acids, the monomer of a protein molecule.

Pancreatic juice, which enters the duodenum, has a basic pH because it contains sodium bicarbonate ($NaHCO_3$). Sodium bicarbonate neutralizes the acid in chyme, producing the slightly basic pH that is optimum for pancreatic enzymes. These include pancreatic amylase, which digests starch, and trypsin, which digests protein. Trypsin is secreted as trypsinogen, which is converted to trypsin in the duodenum.

Lipase, a third pancreatic enzyme, digests fat molecules in the fat droplets after they have been emulsified by bile salts:

$$\text{fat} \xrightarrow{\text{bile salts}} \text{fat droplets}$$

$$\text{fat droplets} + H_2O \xrightarrow{\text{lipase}} \text{glycerol} + 3 \text{ fatty acids}$$

Fats are triglycerides, which means that each one is composed of glycerol and three fatty acids. The end products of lipase digestion, glycerol and fatty acid molecules, are small enough

TABLE 14.3 Major Digestive Enzymes

Enzyme	Produced By	Site of Action	Optimum pH	Digestion
Salivary amylase	Salivary glands	Mouth	Neutral	Starch + H_2O $\longrightarrow$ maltose
Pancreatic amylase	Pancreas	Small intestine	Basic	Starch + H_2O $\longrightarrow$ maltose
Maltase	Small intestine	Small intestine	Basic	Maltose + H_2O $\longrightarrow$ glucose + glucose
Pepsin	Gastric glands	Stomach	Acidic	Protein + H_2O $\longrightarrow$ peptides
Trypsin	Pancreas	Small intestine	Basic	Protein + H_2O $\longrightarrow$ peptides
Peptidases	Small intestine	Small intestine	Basic	Peptide + H_2O $\longrightarrow$ amino acids
Nuclease	Pancreas	Small intestine	Basic	RNA and DNA + H_2O $\longrightarrow$ nucleotides
Nucleosidases	Small intestine	Small intestine	Basic	Nucleotide + H_2O $\longrightarrow$ base + sugar + phosphate
Lipase	Pancreas	Small intestine	Basic	Fat droplet + H_2O $\longrightarrow$ glycerol + fatty acids

to cross the cells of the intestinal villi, where absorption takes place. As mentioned in section 14.1, glycerol and fatty acids enter the cells of the villi, and within these cells, they are rejoined and packaged as lipoprotein droplets before entering the lacteals (see Fig. 14.6).

Peptidases and **maltase,** enzymes secreted by the surface cells of the small intestinal villi, complete the digestion of protein to amino acids and starch to glucose, respectively. These small molecules can then be absorbed by the cells of the villi. Peptides, which result from the first step in protein digestion, are digested to amino acids by peptidases:

$$\text{peptides} + H_2O \xrightarrow{\text{peptidases}} \text{amino acids}$$

Maltose, a disaccharide that results from the first step in starch digestion, is digested to glucose by maltase:

$$\text{maltose} + H_2O \xrightarrow{\text{maltase}} \text{glucose} + \text{glucose}$$

Other disaccharides, each of which has its own enzyme, are digested in the small intestine. The absence of any one of these enzymes can cause illness. The most common situation involves a deficiency of lactase, the enzyme that digests lactose, the sugar found in milk. Interestingly, the rate of this enzyme deficiency varies between human races. For example, only 25% of adult Caucasians, but 75–90% of adult African Americans, are lactase-deficient. Consuming dairy products often gives these individuals the symptoms of **lactose intolerance** (diarrhea, gas, cramps), caused by the fermentation of nondigested lactose by intestinal bacteria. However, lactose-reduced products are available for most kinds of dairy products, including milk, cheese, and ice cream. It is important to include foods that are high in calcium in the diet. Otherwise, osteoporosis, a condition characterized by weak and fragile bones, may develop (see section 19.6).

As mentioned in Chapter 6, enzymes function best at an optimum temperature and pH, which helps to maintain the proper shape to fit their substrate. Laboratory experiments can help to define these conditions. For example, the four test tubes shown in Figure 14.12 can be prepared and observed for the digestion of egg white, a protein digested in the stomach by the enzyme pepsin. After all tubes are placed in an incubator at body temperature for at least one hour, the results are as follows: Tube 1 is a control tube; no digestion has occurred (if a control gives a positive result, then the experiment is invalidated). Tube 2 shows limited or no digestion because HCl is missing, and therefore, the pH is too high for pepsin to be effective. Tube 3 shows no digestion because although HCl is present, the enzyme is missing. Tube 4 shows the best digestive action because the enzyme is present at an acidic pH.

CHECK YOUR PROGRESS

1. Where in the digestive tract does the chemical digestion of each of the following types of nutrients occur: carbohydrates, proteins, fats?
2. How could the experimental setup in Figure 14.12 be used to identify some other factors necessary for enzyme activity?

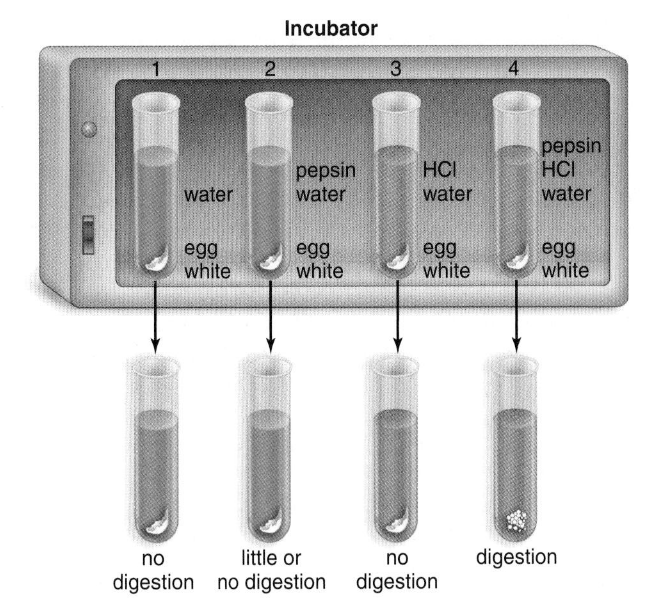

Figure 14.12 Digestion experiment.
The results of this experiment support the hypothesis that for digestion of egg white to occur, both the enzyme (pepsin) and acidic conditions must be present. Colors indicate pH of test tubes (blue = basic; pink = acidic.)

14.4 Nutrition

Learning Outcomes

1. List the six major classes of nutrients and their major functions in the body.
2. Compare your own typical daily food intake with the recommendations in this chapter for a heathy diet.
3. Describe three types of eating disorders, as well as some of the psychological and social factors that can lead to them.

The vigilance of your immune system, the strength of your muscles and bones, the ease with which your blood circulates—every aspect of your body's functioning depends on proper **nutrition,** the science of foods and nutrients. A **nutrient** is a component of food that performs a physiological function in the body. The six major classes of nutrients are carbohydrates, fats, proteins, vitamins, minerals, and water. The nutrients in our diet provide us with energy, promote growth and development, maintain the fluid balance and proper pH of blood, and regulate cellular metabolism. Carbohydrates and fats are the major sources of energy for the body. Proteins can be used for energy, but their primary function is to promote growth and development. Along with vitamins and minerals, proteins also help regulate metabolism, because nearly all enzymes are proteins. Water can also be considered an essential nutrient, because it serves many important roles in the body. The content of a cell is about 70–80% water.

Food choices to fulfill your nutrient needs constitute the diet. Nutritionists often present dietary recommendations in a pyramid form. In 2005, the U.S. Department of Agriculture (USDA) released new dietary recommendations (Fig. 14.13). The daily diet should include about 6 oz of whole-grain breads, crackers, pasta, cereals, or rice. Fruits and vegetables, collectively, should be consumed in even greater quantities than grains. The USDA recommends limiting solid fats such as butter, stick margarine, lard, and shortening. However, the USDA recommends that fat-free or low-fat milk or equivalent milk products be a part of the diet every day. Of course, individuals with lactose intolerance, other health conditions, or dietary preferences might need to modify this or other requirements. Interestingly, the USDA says people should fulfill their protein needs by consuming lean meat such as poultry, and limit their intake of red meat. Fish, beans, peas, nuts, and seeds are alternative sources of protein.

While food pyramids prepared by different sources may vary somewhat, most nutritionists can agree upon these simple guidelines:

1. To maintain a healthy weight, balance your energy from food intake against your energy output.
2. Eat a variety of foods, in order to acquire all needed nutrients.
3. A healthy diet
   - is moderate in total fat intake and low in saturated fat and cholesterol.
   - is rich in whole-grain products, legumes (e.g., beans and peas), and vegetables as sources of complex carbohydrates and fiber.
   - is low in refined carbohydrates, such as starches and sugars.
   - is low in salt and sodium intake.
   - contains adequate amounts of protein, largely from poultry, fish, and plant sources.
   - contains adequate amounts of minerals and vitamins, but avoids questionable food additives and supplements.

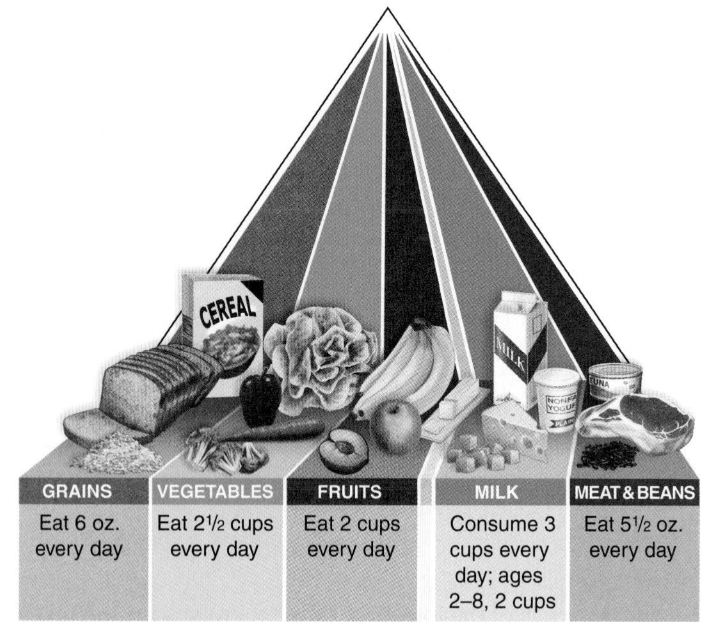

Figure 14.13 Food guide pyramid.
The U.S. Department of Agriculture (USDA) developed this pyramid as a guide to better health. The different widths of the food group bands suggest how much food a person should choose from each group. The six different colors illustrate that foods from all groups are needed each day for good health. The wider base is supposed to encourage the selection of foods with little or no solid fats or added sugars. The yellow band indicates fats and oils—eat only 1–2 tablespoons each day.

Carbohydrates

The quickest, most readily available source of energy for the body is glucose. Carbohydrates are digested to simple sugars, which can be converted to glucose. As mentioned in section 14.2, glucose is stored by the liver in the form of glycogen. Between eating periods, the blood glucose level is maintained at about 50–80 mg/100 ml of blood by the breakdown of glycogen or by the conversion of fat or amino acids to glucose. While body cells can utilize fatty acids as an energy source, brain cells require glucose.

Complex sources of carbohydrates, such as the whole-grain foods featured in Figure 14.14, are recommended because they are digested to sugars gradually and contain fiber. Insoluble fiber, such as that found in wheat bran, has a laxative effect and may possibly guard against colon cancer by limiting the amount of time cancer-causing substances are in contact with the intestinal wall. Soluble fiber, such as that found in oat bran, combines with bile acids and cholesterol in the intestine and

Figure 14.14 Complex carbohydrates.
To meet our energy needs, dietitians recommend consuming foods rich in complex carbohydrates, such as those shown here, rather than foods consisting of simple carbohydrates, such as candy and ice cream, or refined carbohydrates, such as white bread. Such carbohydrates provide monosaccharides, but few other types of nutrients.

prevents them from being absorbed. The liver then removes cholesterol from the blood and changes it to bile acids, replacing the bile acids that were lost. While the diet should have an adequate amount of fiber, some evidence suggests that a diet too high in fiber can be detrimental, possibly impairing the body's ability to absorb iron, zinc, and calcium.

Simple sugars in foods such as candy and ice cream immediately enter the bloodstream, as do those derived from the digestion of starch within white bread and potatoes. These foods are said to have a high **glycemic index (GI),** because the blood glucose response to these foods is high. When the blood glucose level rises rapidly, the pancreas produces an overload of insulin to bring the level under control. (Cells take up glucose in response to insulin.) Hunger quickly returns after a high glycemic meal. Investigators tell us that a chronically high insulin level also leads to insulin resistance and many harmful effects, such as increased fat deposition and a high blood fatty acid level. Over the years, the body's cells can become insulin resistant, and type 2 diabetes can develop. Even if insulin resistance does not develop, researchers warn that a high blood fatty acid level can lead to an increased risk of coronary heart disease, hypertension, liver disease, and several types of cancer.

Table 14.4 gives suggestions on how to reduce your intake of high-GI carbohydrates.

Check Your Progress

1. What does it mean to say a particular food has a high glycemic index? Does your own personal diet contain a high percentage of such foods?
2. Describe the benefits and risks of consuming a diet high in fiber.

TABLE 14.4 Reducing High Glycemic Index Carbohydrates

To reduce dietary sugar:
1. Eat fewer sweets, such as candy, soft drinks, ice cream, and pastry.
2. Eat fresh fruits or fruits canned without heavy syrup.
3. Use less sugar—white, brown, or raw—and less honey and syrups.
4. Avoid sweetened breakfast cereals.
5. Eat less jelly, jam, and preserves.
6. Drink pure fruit juices, not imitations.
7. Avoid potatoes and processed foods made from refined carbohydrates, such as white bread, rice, and pasta.

Proteins

Foods rich in protein include red meat, fish, poultry, dairy products, legumes (i.e., peas and beans), nuts, and cereals. Following digestion of protein, amino acids enter the bloodstream and are transported to the tissues. Ordinarily, amino acids are not used as an energy source. Most are incorporated into structural proteins found in muscles, skin, hair, and nails. Others are used to synthesize such proteins as hemoglobin, plasma proteins, enzymes, and hormones.

Adequate protein formation requires 20 different types of amino acids. Of these, eight are required from the diet in adults (nine in children) because the body is unable to produce them. These are termed the **essential amino acids.** The body is capable of producing the other amino acids by simply transforming one type into another type.

Some protein sources, such as meat, milk, and eggs, are complete; they provide all 20 types of amino acids in the approximate amounts needed by the human body. The amino acid composition of other dietary protein sources varies considerably. For example, corn is a poor source of the amino acids lysine and tryptophan. (However, high-lysine corn can now be produced through genetic modification of the plant.) Legumes (beans and peas), seeds, nuts, and grains can be good protein sources but some contain insufficient amounts of at least one essential amino acid. Therefore, vegetarians are often counseled to combine two or more incomplete types of plant products to acquire all the essential amino acids (see the Health Focus, page 268).

Amino acids are not stored in the body, thus a regular supply is needed. However, it does not take very much protein to meet the daily requirement. Two servings of meat a day (one serving is equal in size to a deck of cards) are usually plenty. While the body is harmed if the amount of protein in the diet is severely limited, it is also likely to be harmed by an overabundance of protein. Deamination of excess amino acids in the liver results in urea, our main nitrogenous excretion product. The water needed for excretion of urea can cause dehydration when a person is exercising and losing water by sweating. High-protein diets can also increase calcium loss in urine, which may increase the risk of kidney stones and osteoporosis.

Vegetarians: Where Do You Get Your Protein?

There are approximately 5 million adult vegetarians in the United States. Although definitions vary, most abstain from eating red meat, poultry, and fish. About a third of vegetarians are vegans, who avoid all animal products, while less stringent vegetarians may include eggs and/or dairy products as a part of their regular diet. Although vegetarianism is an important part of certain Eastern religions, most American vegetarians cite ethical or health reasons. Ethical concerns may include the treatment of animals raised as food, and/or the effects of animal agriculture on the environment. On the health side, most nutritional research seems to support the benefits of a low-fat, well-balanced vegetarian diet in reducing the incidence of cardiovascular disease, obesity, diabetes, and certain cancers, especially colon and prostate cancer.

There are certain health risks associated with vegetarian diets. Highly restricted vegetarian diets are probably not appropriate for pregnant or lactating women, or for very young children. Vegans, especially, have to be careful to obtain enough iron, vitamin D, vitamin B_{12}, and certain fatty acids. However, most American vegetarians would likely agree that the most common question they are asked about their dietary habits is, "Where do you get your protein?" Many people believe that humans require meat in their diet in order to obtain sufficient protein. Adding to the confusion, in 1971, Frances Moore Lappé published *Diet for a Small Planet,* which advocated vegetarianism as a more ecologically sustainable diet. That book also introduced the concept of protein complementation, the idea that foods with insufficient levels of one or more essential amino acids needed to be ingested at the same time as foods higher in those amino acids. This idea has influenced American nutritionists for several decades, but in a later edition of her famous book, Lappé noted that with the exception of a few extreme diets, "if people are getting enough calories, they are virtually certain of getting enough protein."

A position paper published in 2003 by the American Dietetic Association and Dietitians of Canada echoed these ideas, stating that "plant protein can meet requirements when a variety of plant foods is consumed and energy needs are met," as well as "complementary proteins do not need to be consumed at the same meal." Because proteins in plant food such as cereals may be harder to digest than animal proteins, vegetarians may need to include a higher percentage of protein in their diet. Beans, nuts, and legumes are particularly good protein sources, and according to the U.S. Food and Drug Administration, soybeans contain complete protein, meaning that like animal protein, soy protein contains sufficient amounts of all essential amino acids for human nutrition. Soy may have other benefits. In October 1999 the FDA gave food manufacturers permission to put labels on products high in soy protein indicating that they may help lower heart disease risk.

Figure 14A Vegetable proteins. For most adults, protein needs can be met by eating vegetables, such as soybeans, that contain high-quality proteins.

Discussion Questions

1. Consider animals such as cattle, horses, or even elephants. All are very large and powerful, while consuming a predominantly (or exclusively) vegetarian diet. In what way(s) is this a convincing argument that human vegetarians can obtain sufficient protein? Can you think of any potential flaws in this reasoning?
2. In general, proteins are not stored by the body in the way that carbohydrates or fats can be stored. Considering this idea, why do you think the idea of complementary proteins became so well-established?
3. Most proteins contain varying amounts of all 20 amino acids. During the process of protein synthesis, what would specifically happen if one required amino acid was missing? What specific step in translation would not occur?

Certain types of meat, especially red meat, are known to be high in saturated fats, while other sources, such as chicken, fish, and eggs, are more likely to be low in saturated fats. As discussed later in this section, a high intake of saturated fats is associated with a greater risk of cardiovascular disease. In one study, it was found that Hawaiians who follow the traditional native diet, which is rich in protein from plants, have a reduced incidence of cardiovascular disease and cancer, compared to those who follow a modern diet, which is rich in fat and animal protein (Fig. 14.15).

Check Your Progress

1. What is the main difference between an essential and a nonessential amino acid?
2. What are some specific consequences of consuming a diet that is deficient in protein? Of consuming too much protein?

Lipids

Fats, oils, and cholesterol are lipids. Saturated fats, which are solids at room temperature, usually have an animal origin. Two well-known exceptions are palm oil and coconut oil, which contain mostly saturated fats and come from the plants mentioned. Butter and meats, such as marbled red meats and bacon, contain saturated fats. Saturated fats are particularly associated with cardiovascular disease. The worst offenders are those that contain trans fatty acids, which arise when unsaturated fatty acids are hydrogenated to produce a solid fat. Found largely in commercially produced products, trans fatty acids (often called trans fats) may reduce the function of the plasma membrane receptors that clear cholesterol from the bloodstream.

Oils contain unsaturated fatty acids, which do not promote cardiovascular disease. Not only that, unsaturated oils,

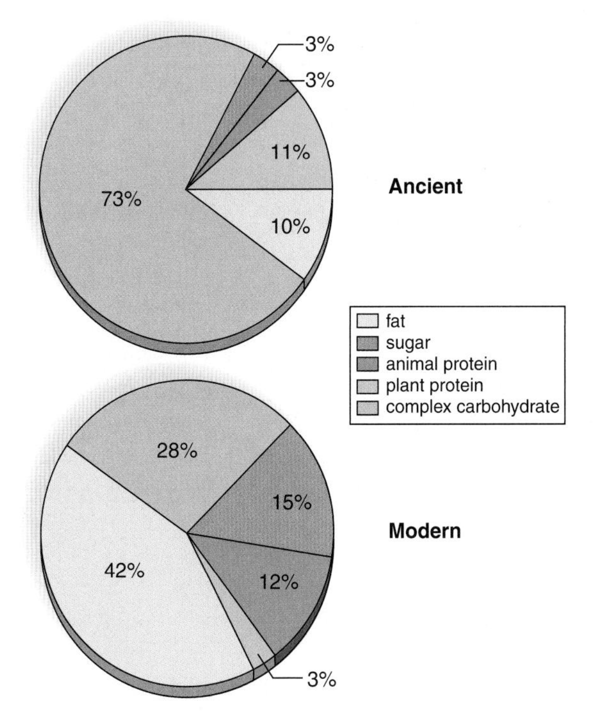

Figure 14.15 Ancient versus modern diet of native Hawaiians.

Among those Hawaiians who have switched back to an ancient diet that is rich in complex carbohydrates and plant protein and low in fat and sugar, the incidence of cardiovascular disease, cancer, and diabetes has dropped.

like whole grains and vegetables, have a low glycemic index and are more filling. Each type of oil has a percentage of monounsaturated and polyunsaturated fatty acids. Corn oil and safflower oil are high in polyunsaturated fatty acids. Polyunsaturated oils are nutritionally essential because they are the only type of fat that contains linoleic acid and linolenic acid, two fatty acids the body cannot make. The body needs these two polyunsaturated fatty acids to produce various hormones and the plasma membrane of cells. Because these fatty acids must be supplied by diet, they are called **essential fatty acids.**

Olive or canola oils contain a larger percentage of monounsaturated fatty acids than other types of cooking oils. Omega-3 fatty acids—characterized by a double bond in the third position—are believed to be especially protective against heart disease. Some cold-water fishes, such as salmon, sardines, and trout, are rich sources of omega-3, but they contain only about half as much as flaxseed oil, the best source from plants.

Fats That Cause Disease

Cardiovascular disease is often due to arteries blocked by **plaque,** which contains saturated fats and cholesterol. Cholesterol is carried in the blood by two types of lipoproteins: low-density lipoprotein (LDL) and high-density lipoprotein (HDL). LDL is thought of as "bad" because it carries cholesterol from the liver to the cells, while HDL is thought of as "good" because it carries cholesterol to the liver, which takes it up and converts it to bile salts. Saturated fats tend to raise LDL cholesterol levels, while unsaturated fats lower LDL cholesterol levels.

Controlled-feeding and statistical studies that have examined trans fat intake, in relation to the risk of heart disease and diabetes, indicate that trans fats are more prone to cause these diseases than saturated fats. Trans fatty acids are found in commercially packaged goods, such as cookies and crackers, commercially fried foods, such as french fries from some fast-food chains, and packaged snacks, such as microwave popcorn, as well as in vegetable shortening and some margarines. Any packaged goods that list partially hydrogenated vegetable oils or "shortening" on their label most likely contain trans fat.

In addition to avoiding trans fatty acids and saturated fats, everyone should use diet and exercise to keep their cholesterol level within normal limits, rather than relying on medications for this purpose. Table 14.5 suggests ways to reduce dietary saturated fat and cholesterol. It is not a good idea to rely on commercially produced low-fat foods or those that contain fake fat (e.g., olestra) to reduce total fat intake. In some products, carbohydrates, namely sugars, have replaced the fat, and in others, protein is used. We have already discussed the dangers of simple sugars in the diet, and while replacement of fat by protein is nutritionally more sound, protein still contributes calories to the diet. Nutritionists also express concern about the consumption of fake fat, because high levels of intake can lead to limited absorption of fat-soluble vitamins, diarrhea, and anal leakage. Products containing olestra often have extra fat-soluble vitamins added, however.

Check Your Progress

1. How are essential fatty acids and essential amino acids similar?
2. Describe the relationship between saturated fat intake, LDL, HDL, and heart disease. What other factors might influence your risk of having your arteries blocked by plaque?

TABLE 14.5 Reducing Certain Lipids

To reduce saturated fats and trans fats in the diet:
1. Choose poultry, fish, or dry beans and peas as a protein source.
2. Remove skin from poultry and trim fat from red meats before cooking and place on a rack so that fat drains off.
3. Broil, boil, or bake rather than fry.
4. Limit your intake of butter, cream, trans fats, shortenings, and tropical oils (coconut and palm oils).*
5. Use herbs and spices to season foods instead of butter, margarine, or sauces. Use lemon juice instead of salad dressing.
6. Drink skim milk instead of whole milk, and use skim milk in cooking and baking.
To reduce dietary cholesterol:
1. Limit cheese, egg yolks, liver, and certain shellfish (shrimp and lobster). Preferably, eat white fish and poultry.
2. Substitute egg whites for egg yolks in both cooking and eating.
3. Include soluble fiber in the diet. Oat bran, oatmeal, beans, corn, and fruits, such as apples, citrus fruits, and cranberries are high in soluble fiber.

*Although coconut and palm oils are from plant sources, they are mostly saturated fats.

TABLE 14.6 Fat-Soluble Vitamins

Vitamin	Functions	Food Sources	Conditions With	
			Too Little	*Too Much*
Vitamin A	Antioxidant synthesized from beta-carotene; needed for healthy eyes, skin, hair, and mucous membranes, and for proper bone growth	Deep yellow/orange and leafy, dark green vegetables, fruits, cheese, whole milk, butter, eggs	Night blindness, impaired growth of bones and teeth	Headache, dizziness, nausea, hair loss, abnormal development of fetus
Vitamin D	A group of steroids needed for development and maintenance of bones and teeth	Milk fortified with vitamin D, fish liver oil; also made in the skin when exposed to sunlight	Rickets, bone decalcification and weakening	Calcification of soft tissues, diarrhea, possible renal damage
Vitamin E	Antioxidant that prevents oxidation of vitamin A and polyunsaturated fatty acids	Leafy green vegetables, fruits, vegetable oils, nuts, whole-grain breads and cereals	Unknown	Diarrhea, nausea, headaches, fatigue, muscle weakness
Vitamin K	Needed for synthesis of substances active in clotting of blood	Leafy green vegetables, cabbage, cauliflower	Easy bruising and bleeding	Can interfere with anti-coagulant medication

Vitamins

Vitamins are organic compounds (other than carbohydrates, fats, and proteins) that the body needs for metabolic purposes but is unable to produce in adequate quantity. Many vitamins are portions of coenzymes, which are enzyme helpers. For example, niacin is part of the coenzyme NAD, and riboflavin is part of another important coenzyme, FAD. Our bodies need coenzymes in only small amounts because each can be used over and over again. Not all vitamins are coenzymes; vitamin A, for example, is a precursor for the visual pigment that prevents night blindness. If vitamins are lacking in the diet, various symptoms develop (Fig. 14.16). Altogether, there are 13 vitamins, which are divided into those that are fat soluble (Table 14.6) and those that are water soluble (Table 14.7).

Antioxidants

Over the past 20 years, numerous statistical studies have been done to determine whether a diet rich in fruits and vegetables can protect against cancer. Cellular metabolism generates free radicals, unstable molecules that carry an extra electron. The most common free radicals in cells are superoxide (O_2^-) and hydroxide (OH^-). In order to stabilize themselves, free radicals combine with DNA, proteins (including enzymes), or lipids, which are found in plasma membranes. Free radicals damage these cellular molecules and can thereby lead to disorders, perhaps even cancer.

Vitamins C, E, and A are believed to defend the body against free radicals, and therefore, they are termed antioxidants. These vitamins are especially abundant in fruits and

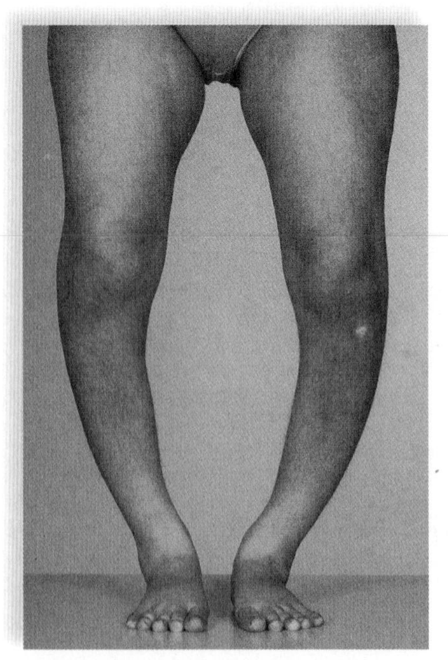
a.

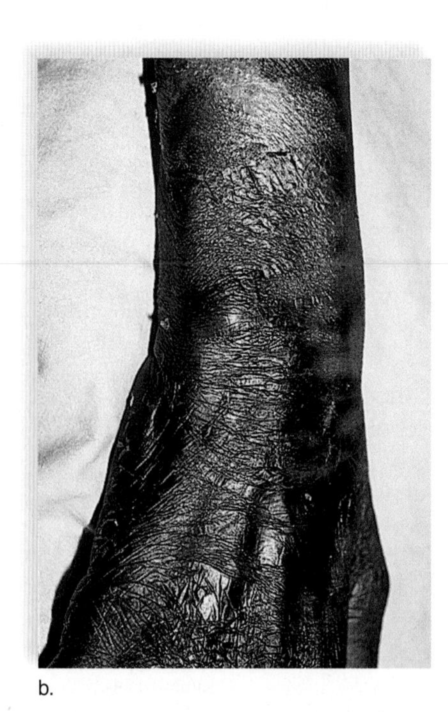
b.

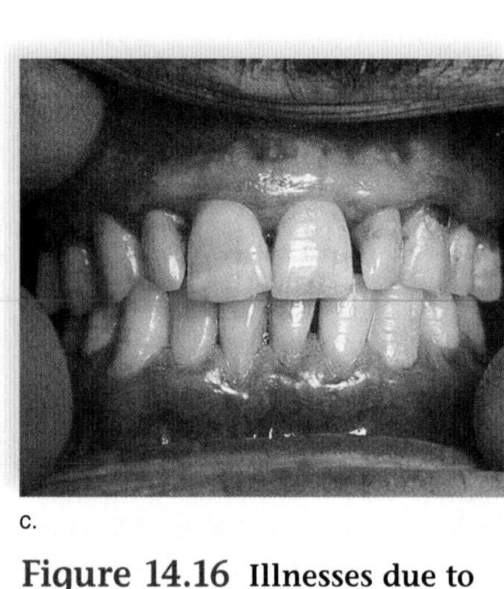
c.

Figure 14.16 Illnesses due to vitamin deficiency.
a. Bowing of bones (rickets) due to vitamin D deficiency. **b.** Dermatitis (pellagra) of areas exposed to light due to niacin (vitamin B_3) deficiency. **c.** Bleeding of gums (scurvy) due to vitamin C deficiency.

TABLE 14.7 Water-Soluble Vitamins

Vitamin	Functions	Food Sources	Conditions With	
			Too Little	*Too Much*
Vitamin C	Antioxidant; needed for forming collagen; helps maintain capillaries, bones, and teeth	Citrus fruits, leafy green vegetables, tomatoes, potatoes, cabbage	Scurvy, delayed wound healing, infections	Gout, kidney stones, diarrhea, decreased copper
Thiamine (vitamin B_1)	Part of coenzyme needed for cellular respiration; also promotes activity of the nervous system	Whole-grain cereals, dried beans and peas, sunflower seeds, nuts	Beriberi, muscular weakness, enlarged heart	Can interfere with absorption of other vitamins
Riboflavin (vitamin B_2)	Part of coenzymes, such as FAD; aids cellular respiration, including oxidation of protein and fat	Nuts, dairy products, whole-grain cereals, poultry, leafy green vegetables	Dermatitis, blurred vision, growth failure	Unknown
Niacin (nicotinic acid)	Part of coenzymes NAD and NADP; needed for cellular respiration, including oxidation of protein and fat	Peanuts, poultry, whole-grain cereals, leafy green vegetables, beans	Pellagra, diarrhea, mental disorders	High blood sugar and uric acid, vasodilation, etc.
Folacin (folic acid)	Coenzyme needed for production of hemoglobin and formation of DNA	Dark leafy green vegetables, nuts, beans, whole-grain cereals	Megaloblastic anemia, spina bifida	May mask B_{12} deficiency
Vitamin B_6	Coenzyme needed for synthesis of hormones and hemoglobin; CNS control	Whole-grain cereals, bananas, beans, poultry, nuts, leafy green vegetables	Rarely, convulsions, vomiting, seborrhea, muscular weakness	Insomnia, neuropathy
Pantothenic acid	Part of coenzyme A needed for oxidation of carbohydrates and fats; aids in the formation of hormones and certain neurotransmitters	Nuts, beans, dark green vegetables, poultry, fruits, milk	Rarely, loss of appetite, mental depression, numbness	Unknown
Vitamin B_{12}	Complex, cobalt-containing compound; part of the coenzyme needed for synthesis of nucleic acids and myelin	Dairy products, fish, poultry, eggs, fortified cereals	Pernicious anemia	Unknown
Biotin	Coenzyme needed for metabolism of amino acids and fatty acids	Generally in foods, especially eggs	Skin rash, nausea, fatigue	Unknown

vegetables. The dietary guidelines in Figure 14.13 suggest that we eat a minimum of four to five cups of fruits and vegetables a day. To achieve this goal, we should include salad greens, raw or cooked vegetables, dried fruit, and fruit juice, in addition to traditional apples and oranges.

Vitamin D (Calcitriol)

As discussed in Chapter 11, skin cells contain a precursor cholesterol molecule that is converted to vitamin D after UV exposure. Vitamin D leaves the skin and is modified first in the kidneys and then in the liver until finally it becomes calcitriol. Calcitriol promotes the absorption of calcium by the intestines. The lack of vitamin D leads to rickets in children (Fig. 14.16*a*). Rickets, characterized by bowing of the legs, is caused by defective mineralization of the skeleton. Since the 1930s, most milk has been fortified with vitamin D, which helps prevent the occurrence of rickets.

Dietary Supplements

Dietary supplements are nutrients and plant products (such as herbal teas) that are used to enhance health. The U.S. government does not require dietary supplements to undergo the same safety and effectiveness testing that new prescription drugs must complete before they are approved. Therefore, many of these products have not been tested scientifically to determine their benefits. People often think herbal products are safe because they are "natural," but many plants, such as lobelia, comfrey, and kava kava, can be poisonous.

Supplements can be useful to correct a deficiency, but most people in the United States take them to greatly increase the dietary intake of a particular nutrient. Everyone should be aware, however, that ingesting high levels of certain nutrients can cause harm. As mentioned earlier in this section, ingesting too much protein can be harmful, so most nutritionists do not recomment taking protein or amino acid supplements. Most fat-soluble vitamins are stored in the body and can accumulate to toxic levels, particularly vitamins A and D. Certain minerals can also be harmful, even deadly, when ingested in high doses.

Dietary supplements may provide a potential safeguard against cancer and cardiovascular disease, but nutritionists do not think people should take supplements instead of improving their intake of fruits and vegetables. There are many beneficial compounds in fruits that cannot be obtained from a vitamin pill.

CHECK YOUR PROGRESS

1. Define vitamin. How do the functions of vitamins A and D differ from those of most other vitamins?
2. What does it specifically mean to say a particular nutrient functions as an antioxidant? What are some examples?
3. What are some potential benefits and risks of taking dietary supplements?

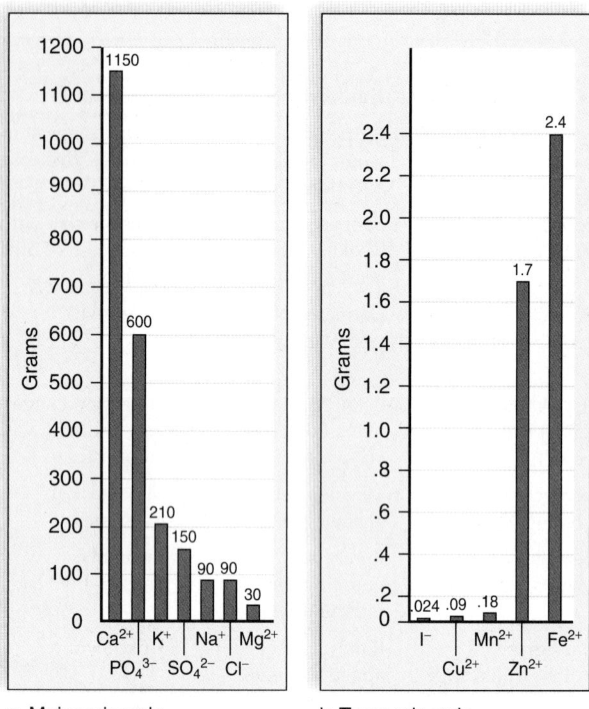

Figure 14.17 Minerals in the body.
These charts show the usual amount of certain minerals in a 60-kg (135 lb) person. The functions of minerals are given in Table 14.8. **a.** The major minerals are present in amounts larger than 5 g (about a teaspoon). **b.** Trace minerals are present in lesser amounts (notice the smaller scale).

Minerals

In addition to vitamins, the body requires various **minerals.** Minerals required as nutrients are divided into major minerals and trace minerals. The body contains more than 5 g of each major mineral and less than 5 g of each trace mineral (Fig. 14.17). The major minerals are constituents of cells and body fluids and are structural components of tissues. For example, calcium (present as Ca^{2+}) is needed for the construction of bones and teeth and for nerve conduction and muscle contraction. Phosphorus (present as PO_4^{3-}) is stored in the bones and teeth and is a part of phospholipids, ATP, and the nucleic acids. Potassium (K^+) is the major positive ion inside cells and is important in nerve conduction and muscle contraction, as is sodium (Na^+). Sodium also plays a major role in regulating the body's water balance, as does chloride (Cl^-). Magnesium (Mg^{2+}) is critical to the functioning of hundreds of enzymes.

The trace minerals are parts of larger molecules. For example, iron (Fe^{2+}) is present in hemoglobin, and iodine (I^-) is a part of thyroxine and triiodothyronine, hormones produced by the thyroid gland. Zinc (Zn^{2+}), copper (Cu^{2+}), and manganese (Mn^{2+}) are present in enzymes that catalyze a variety of reactions. Proteins called zinc-finger proteins, some of which contain zinc, bind to DNA when a particular gene is to be activated. As research continues, more and more elements are added to the list of trace minerals considered essential. During the past three decades, for example, very small amounts of selenium, molybdenum, chromium, nickel, vanadium, silicon, and even arsenic have been found to be essential to good health. Table 14.8 lists the functions of various minerals and gives their food sources and signs of deficiency and toxicity.

Occasionally, individuals do not receive enough iron, calcium, magnesium, or zinc in their diets. Adult females need more iron in the diet than males (18 mg compared to 10 mg) because they lose hemoglobin each month during menstruation. Stress can bring on a magnesium deficiency, and due to its high-fiber content, a vegetarian diet may make zinc less available to the body. However, a varied and complete diet usually supplies enough of each type of mineral.

Calcium

Many people take calcium supplements to counteract **osteoporosis,** a degenerative bone disease that afflicts an estimated one-fourth of older men and one-half of older women in the United States. Osteoporosis develops because bone-eating cells called osteoclasts become more active than bone-forming cells called osteoblasts. Therefore, the bones become porous, and they break easily because they lack sufficient calcium. Due to recent studies that show consuming more calcium does slow bone loss in elderly people, the guidelines have been revised. A calcium intake of 1,000 mg a day is recommended for men and for women who are premenopausal or who use estrogen replacement therapy; 1,300 mg a day is recommended for postmenopausal women who do not use estrogen replacement therapy. To achieve this amount, supplemental calcium is most likely necessary.

Because it promotes calcium absorption, vitamin D is an essential companion to calcium in preventing osteoporosis. Other nutrients may also be helpful. For example, magnesium has been found to suppress the cycle that leads to bone loss. Estrogen replacement therapy and exercise, in addition to adequate calcium and vitamin intake, also help prevent osteoporosis. A high daily caffeine intake and smoking are risk factors for osteoporosis. Medications are also available that slow bone loss while increasing skeletal mass. These are still being studied for their effectiveness and possible side effects.

Sodium

The recommended amount of sodium intake per day is less than 2,400 mg, although the average American takes in 4,000–4,700 mg every day. High sodium intake has been linked to hypertension (high blood pressure) in some people.

TABLE 14.8 Minerals

Mineral	Functions	Food Sources	Conditions With	
			Too Little	*Too Much*
Major minerals (more than 100 mg/day needed)				
Calcium (Ca^{2+})	Strong bones and teeth, nerve conduction, muscle contraction	Dairy products, leafy green vegetables	Stunted growth in children, low bone density in adults	Kidney stones; interferes with iron and zinc absorption
Phosphorus (PO_4^{3-})	Bone and soft tissue growth; part of phospholipids, ATP, and nucleic acids	Meat, dairy products, sunflower seeds, food additives	Weakness, confusion, pain in bones and joints	Low blood and bone calcium levels
Potassium (K^+)	Nerve conduction, muscle contraction	Many fruits and vegetables, bran	Paralysis, irregular heartbeat, eventual death	Vomiting, heart attack, death
Sodium (Na^+)	Nerve conduction, pH and water balance	Table salt	Lethargy, muscle cramps, loss of appetite	Edema, high blood pressure
Chloride (Cl^-)	Water balance	Table salt	Not likely	Vomiting, dehydration
Magnesium (Mg^{2+})	Part of various enzymes for nerve and muscle contraction, protein synthesis	Whole grains, leafy green vegetables	Muscle spasm, irregular heartbeat, convulsions, confusion, personality changes	Diarrhea
Trace minerals (less than 20 mg/day needed)				
Zinc (Zn^{2+})	Protein synthesis, wound healing, fetal development and growth, immune function	Meats, legumes, whole grains	Delayed wound healing, night blindness, diarrhea, mental lethargy	Anemia, diarrhea, vomiting, renal failure, abnormal cholesterol levels
Iron (Fe^{2+})	Hemoglobin synthesis	Whole grains, meats, prune juice	Anemia, physical and mental sluggishness	Iron toxicity disease, organ failure, eventual death
Copper (Cu^{2+})	Hemoglobin synthesis	Meat, nuts, legumes	Anemia, stunted growth in children	Damage to internal organs if not excreted
Iodine (I^-)	Thyroid hormone synthesis	Iodized table salt, seafood	Thyroid deficiency	Depressed thyroid function, anxiety
Selenium (SeO_4^{2-})	Part of antioxidant enzyme	Seafood, meats, eggs	Vascular collapse, possible cancer development	Hair and fingernail loss, discolored skin

TABLE 14.9 Reducing Dietary Sodium

To reduce dietary sodium:
1. Use spices instead of salt to flavor foods.
2. Add little or no salt to foods at the table, and add only small amounts of salt when you cook.
3. Eat unsalted crackers, pretzels, potato chips, nuts, and popcorn.
4. Avoid hot dogs, ham, bacon, luncheon meats, smoked salmon, sardines, and anchovies.
5. Avoid processed cheese and canned or dehydrated soups.
6. Avoid brine-soaked foods, such as pickles or olives.
7. Read labels to avoid high-salt products.

About one-third of the sodium we consume occurs naturally in foods; another one-third is added during commercial processing; and we add the last one-third either during home cooking or at the table in the form of table salt.

Clearly, it is possible for us to cut down on the amount of sodium in the diet. Table 14.9 gives recommendations for doing so.

Check Your Progress

1. List three major minerals and three trace minerals.
2. Use specific examples to explain why your body needs certain minerals in much larger amounts than it needs others.

Persons with obesity

- weigh 20% or more than is appropriate for their height.
- have more body fat than is consistent with optimal health.
- seldom exercise.

Figure 14.18 Recognizing obesity by its characteristics.

Eating Disorders

People with eating disorders have attitudes and behaviors toward food that are outside the norm. Anyone can develop an eating disorder, regardless of ethnicity, socioeconomic status, or intelligence level.

Eating disorders can be difficult to treat. A combination of medication and personal/family counseling is the most common form of treatment. If untreated, the most severe cases can be fatal.

Obesity

As indicated in Figure 14.18, **obesity** is most often defined as a body weight 20% or more above the ideal weight for a person's height. By this standard, approximately 30% of adults in the United States are obese. Moderate obesity is 41–100% above ideal weight, and severe obesity is 100% or more above ideal weight.

Obesity is most likely caused by a combination of hormonal, metabolic, and social factors. It is known that obese individuals have more fat cells than normal, and when they lose weight, the fat cells simply get smaller; they don't disappear. The inability to produce sufficient amounts of a satiety hormone, called leptin, might also be involved. As noted in the story that opened this chapter, it can be difficult to determine if a person has a genetic tendency to be obese or learns eating and exercise habits that lead to weight gain. People are certainly more sedentary than they used to be and they eat more processed foods, which can contain more calories than meals prepared from fresh foods at home.

The risk of heart disease is higher in obese individuals, and this alone tells us that excess body fat is not consistent with optimal health. The treatment depends on the degree of obesity. Surgery to remove body fat, or even to bypass part of the stomach (as in the story that opened this chapter), may be required for those who are severly obese. But for most people, a knowledge of good eating habits along with behavior modification may suffice, particularly if they combine a balanced diet with a sensible exercise program. A lifelong commitment to a healthy diet and exercise is the best way to prevent a cycle of weight gain followed by weight loss. Such a cycle is not conducive to good health.

Bulimia Nervosa

Bulimia nervosa can coexist with either obesity or anorexia nervosa, which is discussed next. According to the National Institute of Mental Health, 5–10% of adolescent and adult females, and 1% of males, suffer from these conditions. The desire for slenderness in females or a manly physique in males may lead to excessive dieting, which can be a

Persons with bulimia nervosa have

- recurrent episodes of binge eating: consuming a large amount of food in a short period and experiencing feelings of lack of control during the episode.
- obsession about body shape and weight.
- increase in fine body hair, halitosis, and gingivitis.

Body weight is regulated by

- a restrictive diet, excessive exercise.
- purging (self-induced vomiting or misuse of laxatives).

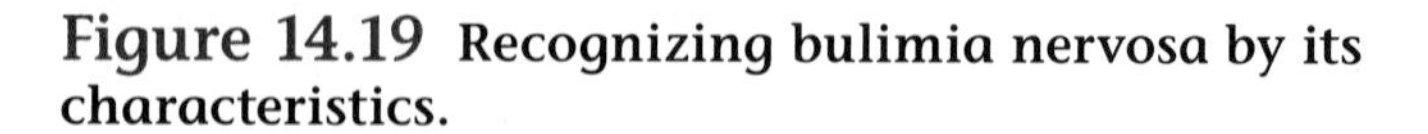

Figure 14.19 Recognizing bulimia nervosa by its characteristics.

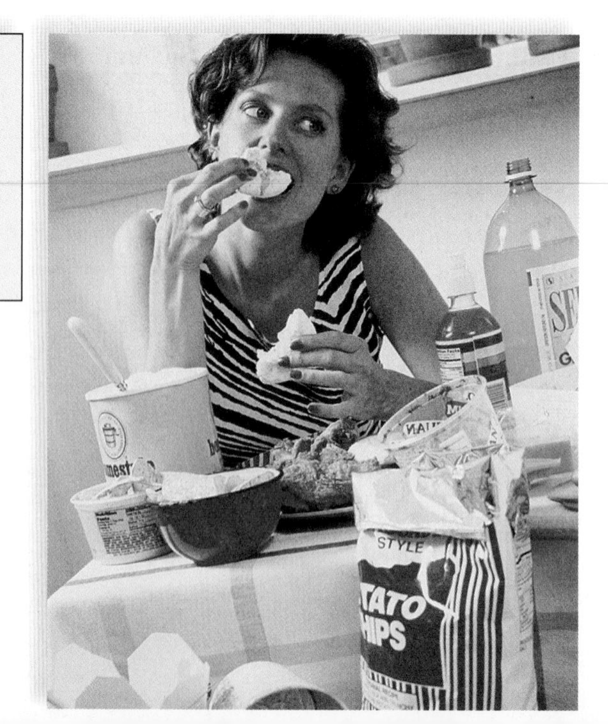

Persons with anorexia nervosa have	Body weight is kept too low by
• a morbid fear of gaining weight; body weight no more than 85% normal. • a distorted body image so that person feels fat even when emaciated. • in females, an absence of a menstrual cycle for at least three months.	• a restrictive diet, often with excessive exercise. • binge eating/purging (person engages in binge eating and then self-induces vomiting or misuses laxatives).

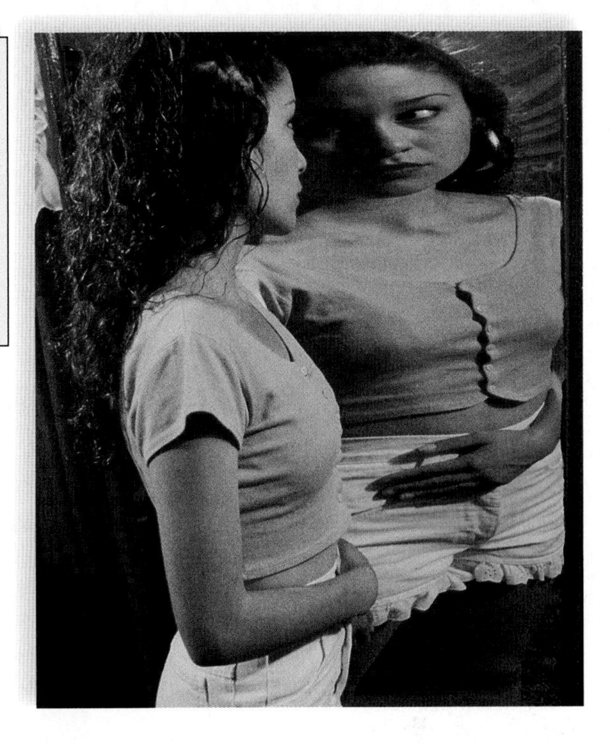

Figure 14.20 Recognizing anorexia nervosa by its characteristics.

precursor to bulimia nervosa and anorexia nervosa. The individual may come to think of weight control, weight gain, and rigid self-control of appetite as moral issues. People with bulimia have the habit of eating to excess (called binge eating) and then purging themselves by some artificial means, such as self-induced vomiting or use of a laxative. Bulimic individuals are overly concerned about their body shape and weight, and therefore, they may be on a very restrictive diet. A restrictive diet may bring on the desire to binge, and typically the person chooses to consume sweets, such as cakes, cookies, and ice cream (Fig. 14.19). The amount of food consumed is far beyond the normal number of calories for one meal, and the person keeps on eating until every bit is gone. Then, a feeling of guilt most likely brings on the next phase, which is a purging of all the calories that have been taken in.

Bulimia can be dangerous to a person's health. Blood composition is altered, leading to an abnormal heart rhythm, and damage to the kidneys can even result in death. At the very least, vomiting can lead to inflammation of the pharynx and esophagus, and stomach acids can cause the teeth to erode. The esophagus and stomach may even rupture and tear due to strong contractions during repeated vomiting.

The most important aspect of treatment is to get the patient on a sensible and consistent diet. Again, behavioral modification is helpful, and so perhaps is psychotherapy to help the patient understand the emotional causes of the behavior. Medications, including antidepressants, have sometimes helped to reduce the bulimic cycle and restore normal appetite.

Anorexia Nervosa

In **anorexia nervosa,** a morbid fear of gaining weight causes the person to be on a very restrictive diet. Athletes such as distance runners, wrestlers, and dancers are at risk of anorexia nervosa because they believe that being thin gives them a competitive edge. In addition to eating only low-calorie foods, the person may induce vomiting and use laxatives to bring about further weight loss. No matter how thin they have become, people with anorexia nervosa think they are overweight (Fig. 14.20). Such a distorted self-image may keep them from recognizing that they need medical help.

Actually, the person is starving and has all the symptoms of starvation, including low blood pressure, irregular heartbeat, constipation, and constant chilliness. Bone density decreases and stress fractures occur. The body begins to shut down; menstruation ceases in females; the internal organs, including the brain, don't function well; and the skin dries up. Impairment of the pancreas and digestive tract means that any food consumed does not provide nourishment. Death may be imminent. If so, the only recourse may be hospitalization and force-feeding. Eventually, it is necessary to use behavior therapy and psychotherapy to enlist the person's cooperation to eat properly. Family therapy may be necessary, because anorexia nervosa in children and teens is believed to be a way for them to gain some control over their lives.

Check Your Progress

1. What are some key differences that distinguish obesity, bulimia nervosa, and anorexia nervosa? Do they have any common causes?
2. From what you have learned about appetite control, what are some physiological factors that could result in an eating disorder?

A Bacterial Culprit for Stomach Ulcers

A stomach (gastric) ulcer is a gaping, sometimes bloody, indentation in the stomach lining (Fig. 14B). Australian physician Barry Marshall didn't believe the accepted explanation that the major causes of stomach ulcers were stress or spicy foods. Instead, he and his colleague, Dr. Robin Warren, thought most stomach ulcers were due to infection with a bacterial species called *Helicobacter pylori,* which they had discovered in 1982. At that time, most physicians disagreed with their hypothesis, because they didn't believe any bacteria could survive in the very acidic conditions of the stomach. However, Marshall and Warren, a pathologist, had observed that nearly every biopsy of stomach ulcer tissue contained the corkscrew-shaped *H. pylori* bacteria.

To get the data they needed, Dr. Marshall decided to use himself as a guinea pig. One day in July 1984, he drank a culture containing *H. pylori.* About a week later, he started vomiting and suffering the painful symptoms of an ulcer. Medical tests confirmed that his stomach lining was teeming with bacteria. In a few weeks, his ulcer had disappeared, but the experiment had served its purpose. In recognition of his efforts, in 2005, Marshall and Warren were awarded the Nobel Prize in Physiology or Medicine for their discovery of *H. pylori* and its role in ulcers and other stomach ailments.

Prior to Dr. Marshall's bold experiment, ulcers were usually treated mainly with drugs to reduce the acidity of the stomach. Patients got temporary relief, but the ulcers tended to return, and serious complications could arise. Some patients had bleeding ulcers and underwent an operation to remove a portion of the stomach. Thanks to the work of Marshall and Warren, in 1994 the National Institutes of Health endorsed antibiotics as a standard treatment for stomach ulcers. An international race is under way among scientists to develop a vaccine.

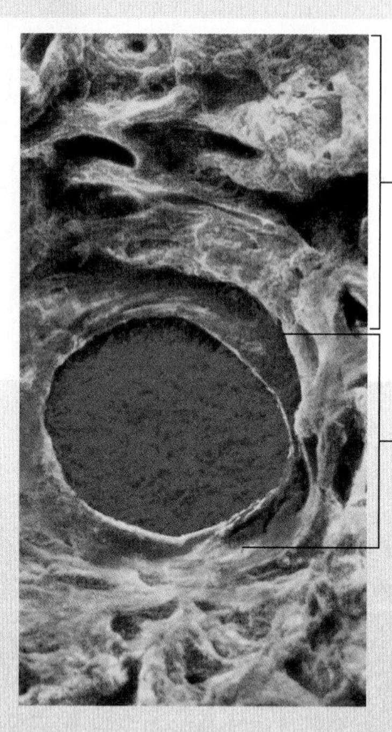

Figure 14B Stomach ulcer.
Drs. Barry Marshall and Robin Warren won a Nobel Prize in 2005 for discovering that most stomach ulcers are caused by a bacterial infection.

Discussion Questions

1. What kinds of adaptations would *H. pylori* need in order to survive in the acid pH of the stomach?
2. Why do you suppose that even after Dr. Marshall's experiment in 1984, it took ten years for his findings to be widely accepted by the medical community?
3. Do you think Dr. Marshall's experiment was ethical? If so, would it always be okay for scientists to experiment on themselves? If not, what would be a more ethical way to prove that *H. pylori* could cause ulcers?

14.5 Disorders of the Digestive System

LEARNING OUTCOMES

1. Describe a common disorder that affects each part of the digestive system.
2. Classify the digestive disorders by the type of cause (i.e., infectious, cancer, inflammatory, etc.).

Disorders of the digestive system can be grouped into two categories: disorders of the tract itself, and disorders of the accessory organs.

Disorders of the Digestive Tract

Stomach Ulcers

A thick layer of mucus normally protects the wall of the stomach. If this protective layer is broken down, the stomach wall can be damaged by the acidic pH of the stomach, resulting in an **ulcer,** or open sore in the wall caused by a gradual disintegration of the tissue. It now appears that most stomach ulcers are initiated by infection of the stomach by a bacterium called *Helicobacter pylori* (see the Health Focus above), which can impair the ability of the mucous cells to produce protective mucus. Thus, treatment of stomach ulcers now typically includes antibiotics to kill these bacteria, along with medication to reduce the amount of acid in the stomach. Other potential causes of stomach ulcers include certain viral infections, as well as the overuse of anti-inflammatory medications, which can have the side effect of damaging the stomach lining. Duodenal ulcers are also common because the duodenum receives acidic chyme from the stomach.

Intestinal Disorders

The most common problems associated with the small and large intestines are diarrhea and constipation. Diarrhea, or loose, watery feces, can be acute or chronic. Most cases of acute (sudden-onset) diarrhea are due to infections of the small or large intestines with any of several bacterial, viral, or protozoal agents. Collectively known as food poisoning, symptoms of these infections can range from a few hours of stomach and bowel discomfort, to life-threatening infections such as cholera, salmonellosis, and the more toxic forms of *E. coli.* In all cases, the intestinal wall becomes irritated by the infection, or by toxins produced by the organism, and peristalsis increases. Less water is absorbed, and the resulting diarrhea helps rid the body of the infectious organisms. Severe diarrhea can lead to dehydration because of water loss and to disturbances in the heart's contraction due to an imbalance of salts in the blood.

As opposed to intestinal infections that come and go, some people are afflicted with chronic diarrhea. About 7 in 100,000 people in the United States suffer from Crohn's disease, which is a persistent inflammation of the intestine that results in recurrent bouts of abdominal cramping (which may be severe) and bloody diarrhea. Usually diagnosed between ages 15 and 30, it seems to be caused by a misdirected immune response against one's own intestinal tissues, or perhaps against some of the normal bacteria that live in the gut. The incidence of Crohn's disease is higher among Caucasians compared to other races, and among wealthier, urban populations compared to poorer, rural ones. Genetic predisposition is certainly a factor, as are several environmental triggers, such as smoking and diet.

The opposite of diarrhea is constipation. When a person is constipated, the feces are dry and hard. One reason for this condition is that some people have learned to inhibit the defecation reflex to the point that the urge to defecate is ignored. Chronic constipation can lead to the development of hemorrhoids, which are enlarged and inflamed blood vessels of the anus. Increasing the amount of water and fiber in the diet can help prevent constipation. Laxatives and enemas can also be used, but frequent use is discouraged because they can be irritating to the colon.

Polyps and Colon Cancer

The colon is subject to the development of **polyps,** which are small growths arising from the epithelial lining. They may be benign or cancerous. Polyps are usually detected using a procedure known as colonoscopy, during which a long, flexible endoscope is inserted into the colon to visualize the wall of the large intestine. If colon cancer is detected and surgically removed while it is still confined to a polyp, a complete cure is expected.

Some investigators believe that dietary fat increases the likelihood of colon cancer because dietary fat causes an increase in bile secretion. Certain intestinal bacteria may convert bile salts to substances that promote the development of cancer. On the other hand, fiber in the diet seems to inhibit the development of colon cancer, possibly by diluting the concentration of bile salts and facilitating the movement of cancer-inducing substances through the intestine.

Disorders of the Accessory Organs

Disorders of the Pancreas

Pancreatitis is an inflammation of the pancreas. It can be caused by drinking too much alcohol, by gallstones that block the pancreatic duct, or by other unknown factors. In chronic pancreatitis, the digestive enzymes normally secreted by the pancreas can damage the pancreas and surrounding tissues. Eventually, chronic pancreatitis can decrease the ability of the pancreas to secrete insulin, potentially leading to diabetes.

Unfortunately, cancer of the pancreas is almost always fatal. Only about 20% of patients are alive one year after this diagnosis. Reasons for this include the essential functions of the pancreas, the resistance of many forms of the disease to treatment, and the tendency of the cancer cells to spread to other organs before any symptoms appear.

Disorders of the Liver and Gallbladder

The liver performs many functions that are essential to life; therefore, diseases that affect the liver can be life threatening. A person who has a liver ailment may develop **jaundice,** which is a yellowish coloring in the whites of the eyes, as well as in the skin of light-pigmented persons. Jaundice can be caused by an abnormally large amount of bilirubin in the blood. **Hepatitis,** or inflammation of the liver, is most commonly caused by one of several viruses. Hepatitis A virus is usually acquired by consuming water or food that has been contaminated by sewage. Hepatitis B virus is usually spread by sexual contact, but can also be acquired through blood transfusions or contaminated needles. The hepatitis B virus is more contagious than the AIDS virus, which is spread in similar ways. Thankfully, however, a vaccine is available to prevent hepatitis B. Hepatitis C virus is spread by the same routes as hepatitis B, and it can lead to chronic hepatitis, liver cancer, and death. Unfortunately, no vaccine is available for hepatitis C, although antiviral drugs can be effective.

Cirrhosis is another chronic disease of the liver. It is often seen in alcoholics due to malnutrition and to the toxic effects of consuming excessive amounts of alcohol, which acts as a toxin on the liver. In cirrhosis, the liver first becomes infiltrated with fat, and then the fatty liver tissue is replaced by fibrous scar tissue. Although the liver has amazing regenerative powers, if the rate of damage exceeds the rate of regeneration, there may not be enough time for the liver to repair itself. The preferred treatment in most cases of liver failure is liver transplantation, but the supply of livers currently cannot meet the demand. Artificial livers have been developed, but with very limited success so far.

In some individuals, the cholesterol present in bile can come out of solution and form crystals. If the crystals grow in size, they form gallstones. The passage of the stones from the gallbladder may block the common bile duct and cause obstructive jaundice. If this happens, the obstruction can sometimes be relieved; otherwise, the gallbladder must be removed. Interestingly, gallstone formation is particularly common in individuals who have lost a significant amount of weight in a short time, such as those who have undergone gastric bypass procedures.

Check Your Progress

1. What is now thought to be the cause of most stomach ulcers?
2. Why are the pancreas and liver such important organs to the health of an individual?

Applying the Concepts Revisited

The story that opened this chapter introduced Dennis, who struggled with obesity for most of his life until he had gastric bypass surgery.

1. Considering the specific roles that the stomach and duodenum normally play in digestion, why might it be a particularly effective weight-loss strategy to cause food to empty directly from the upper part of the stomach into the jejunum? (Hint: Where do pancreatic enzymes normally enter the digestive tract?)
2. In addition to where food enters the intestine, how might the control of digestion by the nervous system and by hormones (see Regulation of Digestive Secretions, section 14.1) be altered in an individual who has had gastric bypass surgery?

SUMMARIZING THE CONCEPTS

14.1 The Digestive Tract

The digestive tract is involved in the ingestion and digestion of food and elimination of indigestible material. It consists of the following parts:

- The mouth takes food into the body. Salivary glands send saliva into the mouth, where the teeth chew the food and the tongue forms a bolus for swallowing. Saliva contains salivary amylase, an enzyme that begins the digestion of starch.
- The pharynx is where the air passage and food passage cross. When a person swallows, the air passage is usually blocked off by the epiglottis, and food must enter the esophagus, where peristalsis begins.
- The esophagus is a muscular tube that conducts food from the pharynx to the stomach.
- The stomach expands and stores food. While food is in the stomach, the stomach churns, mixing food with the acidic gastric juice. Gastric juice contains pepsin, an enzyme that digests protein.
- The small intestine includes the duodenum, jejunum, and ileum. The duodenum receives bile from the liver and pancreatic juice from the pancreas. The walls of the small intestine have fingerlike projections called villi where small nutrient molecules are absorbed. Amino acids and glucose enter the blood vessels of a villus. Glycerol and fatty acids are joined and packaged as lipoproteins before entering lymphatic vessels called lacteals in a villus.
- The large intestine consists of the cecum, the colon (including the ascending, transverse, descending, and sigmoid colon), and the rectum, which ends at the anus. The large intestine absorbs water, salts, and some vitamins.

14.2 Accessory Organs of Digestion

Three accessory organs of digestion—the pancreas, liver, and gallbladder—send secretions to the duodenum via ducts:

- The pancreas produces pancreatic juice, which contains sodium bicarbonate and enzymes that chemically digest starch (pancreatic amylase), protein (trypsin), and fat (lipase).
- The liver has many important functions. It receives blood from the small intestine by way of the hepatic portal vein and helps to detoxify any harmful substances present. It monitors blood glucose concentrations and helps eliminate excess nitrogen from the body. It also produces many plasma proteins, as well as bile, which helps to emulsify (break apart) fats in the intestine.
- The gallbladder stores bile produced by the liver, until it is secreted into the duodenum through the common bile duct. Bile emulsifies fat and readies it for digestion by lipase.

14.3 Digestive Enzymes

Like all enzymes, digestive enzymes are specific to their substrate and speed up specific reactions at optimum body temperature and pH. They break down food into smaller molecules like glucose, amino acids, fatty acids, and glycerol, which can be absorbed. As food passes through the digestive tract:

- In the mouth, salivary amylase begins digestion of starch.
- In the stomach, pepsin begins digesting protein to peptides.
- In the small intestine, pancreatic amylase, trypsin, and lipase digest starch, protein, and fat. Enzymes produced by the small intestine finish the digestion of starch and protein.

14.4 Nutrition

The nutrients released by the digestive process should provide us with an adequate amount of energy, essential amino acids and fatty acids, and all necessary vitamins and minerals. However, some recommendations can be given, based on scientific studies:

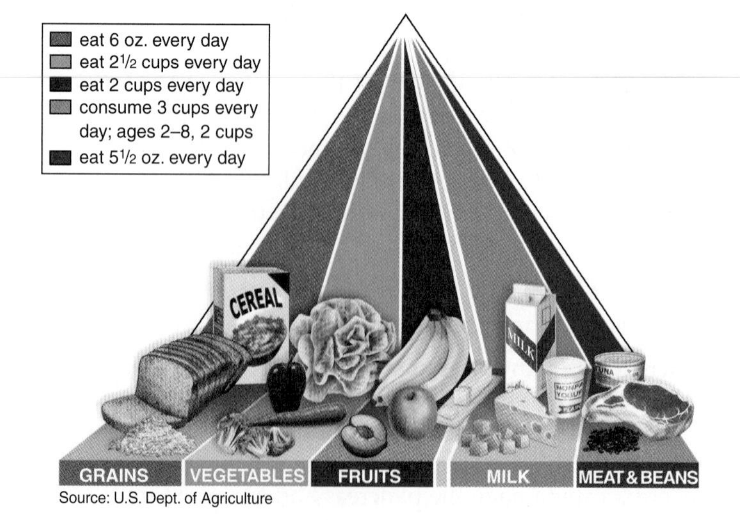

Source: U.S. Dept. of Agriculture

- Minimize your intake of high glycemic index carbohydrates (simple sugars and refined starches) because they cause a rapid release of insulin that can lead to health problems.
- Proteins supply us with essential amino acids, but it is wise to avoid meats that are fatty because fats from animal sources are saturated. While unsaturated fatty acids, particularly the omega-3 fatty acids, are protective against cardiovascular disease, the saturated fatty acids lead to plaque, which occludes blood vessels.
- Aside from carbohydrates, proteins, and fats, the body requires vitamins and minerals. The vitamins A, E, and C are antioxidants that protect cell contents from damage due to free radicals. The mineral calcium is needed for strong bones.

14.5 Disorders of the Digestive System

Various conditions can affect the function of the digestive tract, as well as the accessory organs.

- Most stomach ulcers are now thought to be caused by a bacterial infection and thus can be cured with antibiotics.
- Diarrhea is an increased frequency and looseness of the feces. Acute diarrhea is most frequently caused by intestinal infections. Chronic (longer-lasting) diarrhea may be due to an immune-mediated condition such as Crohn's disease.
- Disorders that affect the accessory digestive organs include pancreatitis, pancreatic cancer, hepatitis, and cirrhosis.

TESTING YOURSELF

Choose the best answer for each question.

1. The digestive system
   a. breaks down food into usable nutrients.
   b. absorbs nutrients.
   c. eliminates waste.
   d. All of these are correct.
2. Tooth decay is caused by bacteria metabolizing _____ and giving off _____.
   a. sugar, protein
   b. protein, sugar
   c. acids, sugar
   d. sugar, acids
3. Chemical digestion begins in the
   a. esophagus.
   b. stomach.
   c. pharynx.
   d. None of these are correct.
4. Which organ has both an exocrine and endocrine function?
   a. liver c. pancreas
   b. esophagus d. cecum
5. _____ is converted to _____ in the presence of _____.
   a. Pepsin, HCl, pepsinogen c. Amylase, maltose, HCl
   b. Amylase, HCl, pepsin d. Pepsinogen, pepsin, HCl
6. Which of these is not a function of the liver in adults?
   a. produces bile d. produces urea
   b. detoxifies alcohol e. makes red blood cells
   c. stores glucose
7. A good nonmeat source of protein is
   a. peas. c. peanuts.
   b. yogurt. d. All of these are correct.

For questions 8–12, match the ailment with the cause in the key.

Key:

a. too little calcium d. lack of iron
b. too much calcium e. too little vitamin A
c. artificial fat

8. Kidney stones
9. Anemia
10. Osteoporosis
11. Night blindness
12. Anal leakage
13. Bile
    a. is an important enzyme for the digestion of fats.
    b. cannot be stored.
    c. is made by the gallbladder.
    d. emulsifies fat.
    e. All of these are correct.
14. Label the following diagram of the large and small intestine junction.

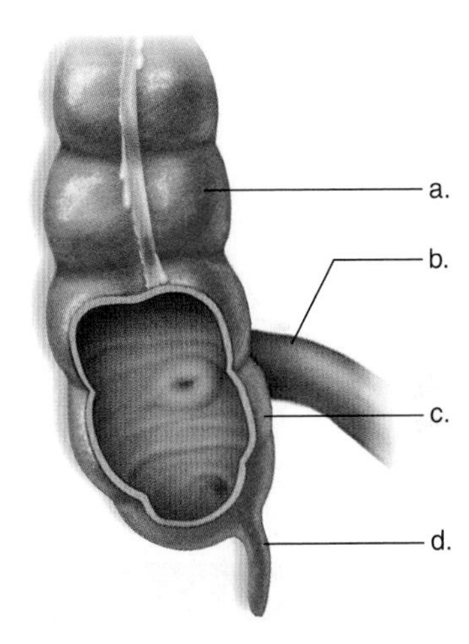

15. Label the following diagram of the hepatic lobule.

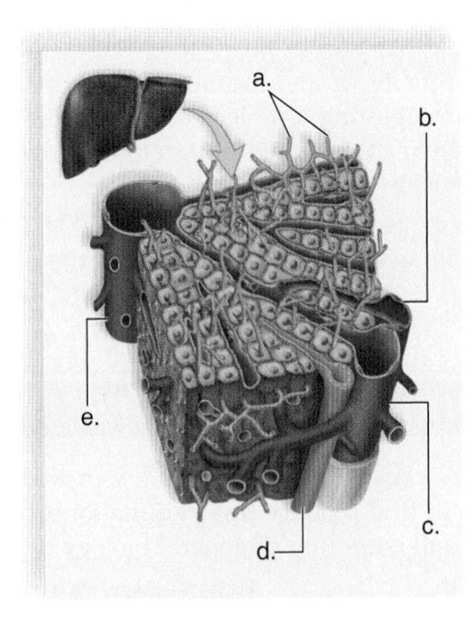

16 Most of the products of digestion are absorbed across the
  a. squamous epithelium of the esophagus.
  b. striated walls of the trachea.
  c. convoluted walls of the stomach.
  d. fingerlike villi of the small intestine.
  e. smooth wall of the large intestine.

17. The amino acids that must be consumed in the diet are called essential. Nonessential amino acids
  a. can be produced by the body.
  b. are needed only occasionally.
  c. are not needed for protein synthesis.
  d. are required in smaller amounts.

18. Which of the following are often organic portions of important coenzymes?
  a. minerals
  b. vitamins
  c. proteins
  d. carbohydrates

19. Bulimia nervosa is *not* characterized by
  a. a restrictive diet often with excessive exercise.
  b. binge eating followed by purging.
  c. an obsession with body shape and weight.
  d. a distorted body image, so that the person feels fat even when emaciated.
  e. a health risk due to this complex.

20. Pancreatic cancer is one of the most deadly forms of cancer because
  a. the pancreas is essential for life.
  b. the cancer usually spreads before it is detected.
  c. it is often resistant to treatment.
  d. All of these are correct.

UNDERSTANDING THE TERMS

anorexia nervosa 275
anus 262
bile 260
bulimia nervosa 274
cecum 262
chyme 260
cirrhosis 277
colon 262
defecation 262
dental caries 257
dietary supplement 271
duodenum 260
epiglottis 258
esophagus 258
essential amino acid 267
essential fatty acid 269
fiber 262
gallbladder 264
gastric gland 260
gingiva 257
glottis 258
glycemic index (GI) 267
hard palate 257
hepatitis 277
hormone 261
ileum 260
jaundice 277
jejunum 260
lacteal 260
lactose intolerance 265
large intestine 262
lipase 263
liver 263
maltase 265
mineral 272
nasopharynx 258
nutrient 266
nutrition 266
obesity 274
osteoporosis 272
pancreas 263
pancreatic amylase 263
pancreatitis 277
pepsin 260
peptidase 265
peristalsis 258
peritonitis 262
pharynx 258
plaque 269
polyp 277
rectum 262
reflex action 258
rugae 260
salivary amylase 257
salivary gland 257
small intestine 260
soft palate 257
sphincter 258
stomach 260
trypsin 263
ulcer 276
vermiform appendix 262
villus 260
vitamin 270

THINKING CRITICALLY

1. Suppose that, starting today, your body would lose the ability to perform either mechanical digestion or chemical digestion. Which type would you choose to keep and why?
2. Trace *all* of the types and locations of enzymes that might be involved in digesting a molecule of a complex carbohydrate such as starch, all the way through to the utilization of a glucose molecule by a body cell.
3. Pancreatic cancer is one of the most lethal cancers, with a survival rate of only about 5%. Why is the pancreas such a vital organ? Would you say it is more, less, or about equally as essential for human life as the liver and why?
4. Suppose you are taking large doses of creatine, an amino acid supplement that is advertised for its ability to enhance muscle growth. Because your muscles can grow only at a limited rate, what do you suppose happens to the excess creatine that is not used for synthesis of new muscle?

INQUIRY INTO LIFE WEBSITE

The companion website for *Inquiry into Life* provides a wealth of information organized and integrated by chapter. You will find practice tests, animations, videos, and much more that will complement your learning and understanding of general biology.

http://www.mhhe.com/maderinquiry13

Respiratory System

Chapter Outline

Applying the Concepts

Growing up living in a neighborhood next to a freeway in downtown Los Angeles, 13-year-old Juan had plenty of challenges to deal with. His parents wanted him to be the first in his family to attend college. That meant avoiding local gangs that he often heard about. Fortunately, Juan loved playing sports, and he knew if he stayed focused on playing for school teams, it was easier to stay out of trouble. Basketball was his favorite—but unfortunately, over the past few weeks, whenever he ran up and down the court, he would begin coughing and wheezing. Sometimes he couldn't catch his breath for several minutes. The school nurse diagnosed Juan with asthma.

Asthma is a disease in which the airways become constricted (narrowed) and inflamed (swollen), both of which can result in difficulty breathing. Over 20 million children and adults in the United States have asthma, and the incidence seems to be increasing. Experts offer various explanations for this. One hypothesis is that we may be "too clean," in the sense that we are not exposed to enough common bacteria, viruses, and even worm parasites as children. As a result, our immune system may react to harmless material we inhale. In Juan's case, living next to a freeway may be another significant factor. There are many harmful forms of air pollution, but scientists are becoming increasingly concerned about tiny "ultrafine" particles, which are produced at high levels by diesel engines. At no more than 0.1 micrometers (μm) across, these particles can bypass the normal defenses of the upper respiratory tract and end up lodging deep in the lungs, with damaging effects. Though incurable, asthma usually can be treated. By using an asthma inhaler and other medications, Juan was able to play the sports he enjoyed.

15.1 The Respiratory System

Learning Outcomes

1. Explain the major functions of the respiratory system.
2. Trace the path of air from the nose to the lungs.
3. Describe the layers of cells a molecule of oxygen must pass through when diffusing from a lung alveolus into a red blood cell, and from the red blood cell into the tissues.

The primary function of the respiratory system is to allow oxygen from the air to enter the blood and carbon dioxide from the blood to exit into the air. During **inspiration,** or inhalation (breathing in), and **expiration,** or exhalation (breathing out), air is conducted toward or away from the lungs by a series of cavities, tubes, and openings (Fig. 15.1). **Ventilation,** another term for breathing, encompasses both inspiration and expiration.

The respiratory system works with the cardiovascular system to accomplish these homeostatic functions:

1. External respiration, the exchange of gases (oxygen and carbon dioxide) between air and the blood.
2. Transport of gases to and from the lungs and the tissues.
3. Internal respiration, the exchange of gases between the blood and tissue fluid.

As described in Chapter 7, cellular respiration, which produces ATP, uses the oxygen and gives off the carbon dioxide that makes gas exchange with the environment necessary. Without a continuous supply of ATP, the cells cease to function.

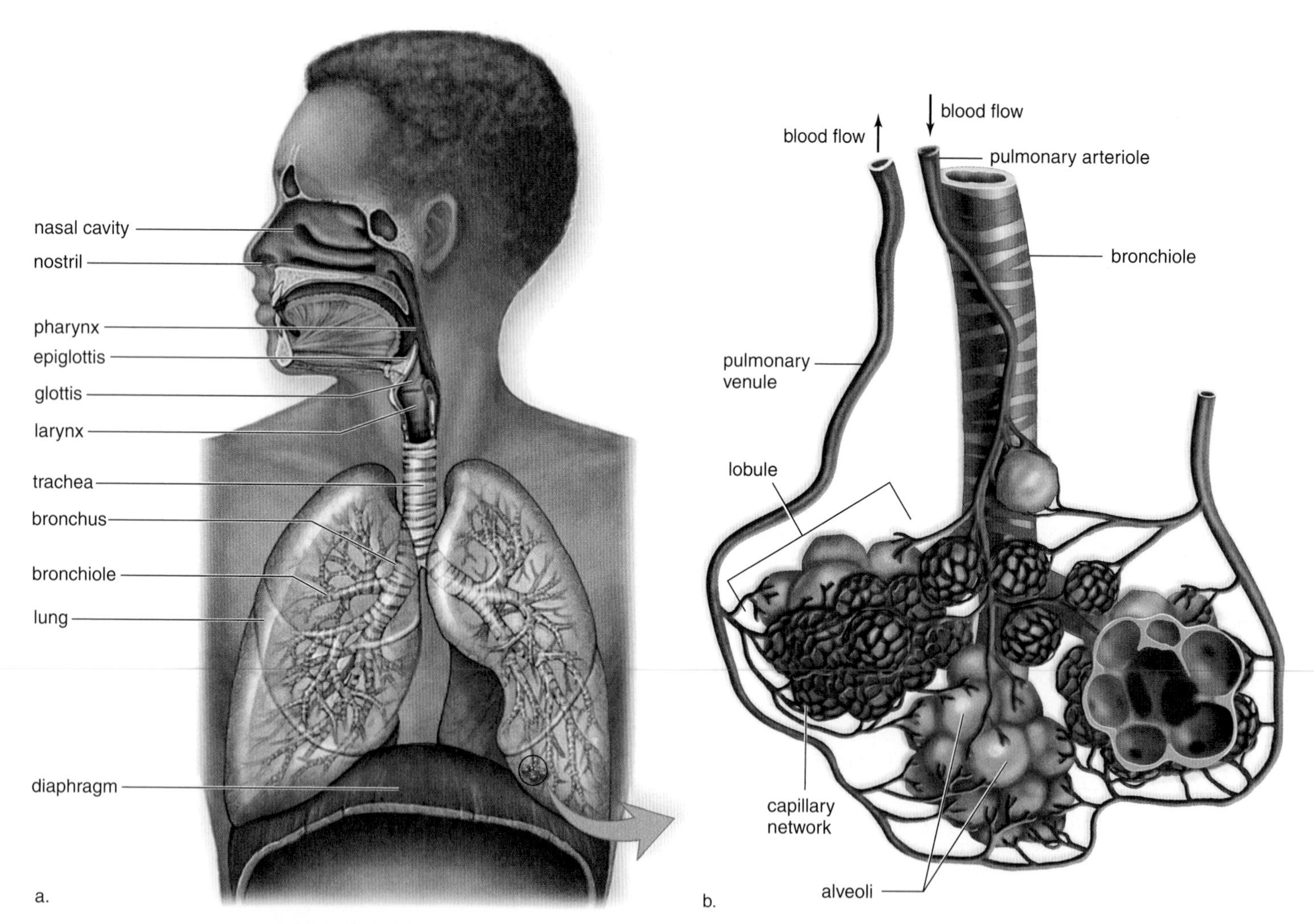

Figure 15.1 The respiratory tract.
a. The respiratory tract extends from the nose to the lungs, which are composed of air sacs called alveoli (arrow). **b.** Gas exchange occurs between air in the alveoli and blood within a capillary network that surrounds the alveoli. Notice that the pulmonary arteriole is colored blue—it carries O_2-poor blood away from the heart to the alveoli. Then carbon dioxide leaves the blood, and oxygen enters the blood. The pulmonary venule is colored red—it carries O_2-rich blood from alveoli toward the heart.

TABLE 15.1 Path of Air

Structure	Description	Function
Upper Respiratory Tract		
Nasal cavities	Hollow spaces in nose	Filter, warm, and moisten air
Pharynx	Chamber posterior to oral cavity; lies between nasal cavity and larynx	Connection to surrounding regions
Glottis	Opening into larynx	Passage of air into larynx
Larynx	Cartilaginous organ that houses the vocal cords; voice box	Sound production
Lower Respiratory Tract		
Trachea	Flexible tube that connects larynx with bronchi	Passage of air to bronchi
Bronchi	Paired tubes inferior to the trachea that enter the lungs	Passage of air to lungs
Bronchioles	Branched tubes that lead from bronchi to alveoli	Passage of air to each alveolus
Lungs	Soft, cone-shaped organs that occupy lateral portions of thoracic cavity	Contain alveoli and blood vessels
Alveoli	Thin-walled microscopic air sacs in lungs	Gas exchange between air and blood

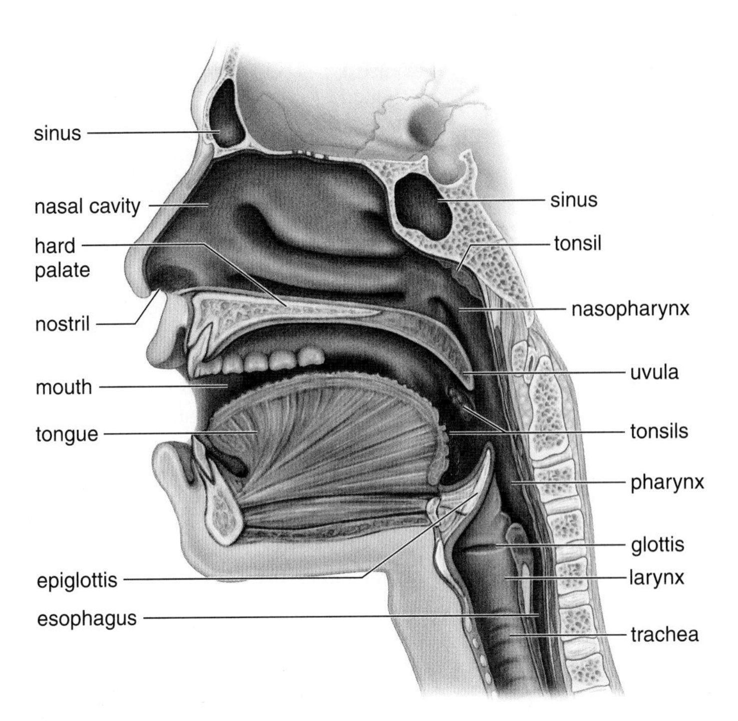

Figure 15.2 The path of air.
Air passes through the nasal cavities and mouth to and from the upper and lower respiratory tracts. The trachea is part of the lower respiratory tract; the other organs are in the upper respiratory tract.

The Respiratory Tract

Table 15.1 traces the path of air from the nose to the lungs. As air moves in along the airways, it is cleansed, warmed, and moistened. Cleansing is accomplished by coarse hairs just inside the nostrils and by cilia and mucus in the nasal cavities and the other airways of the respiratory tract. In the nose, the hairs and the cilia act as screening devices. In the trachea and other airways, the cilia beat upward, carrying mucus, dust, and occasional bits of food that "went down the wrong way" into the pharynx, where the accumulation can be swallowed or expelled. The air is warmed by heat given off by the blood vessels lying close to the surface of the lining of the airways, and it is moistened by the wet surface of these passages.

Conversely, as air moves out during expiration, it cools and loses its moisture. As the air cools, it deposits its moisture on the lining of the trachea and the nose, and the nose may even drip as a result of this condensation. The air still retains so much moisture, however, that upon expiration on a cold day, it condenses and can be seen as a small cloud.

The Nose

The nose, a prominent feature of the face, is the only external portion of the respiratory system. It is a part of the upper respiratory tract, which contains the nasal cavities, the pharynx, and the larynx (Fig. 15.2). Air enters the nose through external openings called **nostrils.** The nose contains two **nasal cavities,** which are narrow canals separated from one another by a septum composed of bone and cartilage. Mucous membranes line the nasal cavities. Bony ridges that project laterally into the nasal cavity increase the surface area for moistening and warming air during inhalation and for trapping water droplets during exhalation. Odor receptors are on the cilia of cells located high in the recesses of the nasal cavities.

The tear (lacrimal) glands drain into the nasal cavities by way of tear ducts. For this reason, crying produces a runny nose. The nasal cavities also communicate with **sinuses,** air-filled spaces that reduce the weight of the skull and act as resonating chambers for the voice. The nasal cavities are separated from the mouth by a partition called the palate, which has two portions. Anteriorly, the hard palate is supported by bone, while posteriorly the soft palate is not supported by bone.

The Pharynx

The **pharynx** is a funnel-shaped passageway that connects the nasal and oral cavities to the larynx. Consequently, the pharynx, commonly referred to as the "throat," has three parts: the nasopharynx, where the nasal cavities open posterior to the soft palate; the oropharynx, where the mouth opens; and the laryngopharynx, which opens into the larynx. The soft palate has a soft extension called the uvula projecting into the oropharynx, which you can see by looking into your throat using a mirror.

The tonsils form a protective ring at the junction of the mouth and the pharynx. The tonsils are lymphatic tissue containing lymphocytes that protect against invasion by inhaled bacteria and viruses.

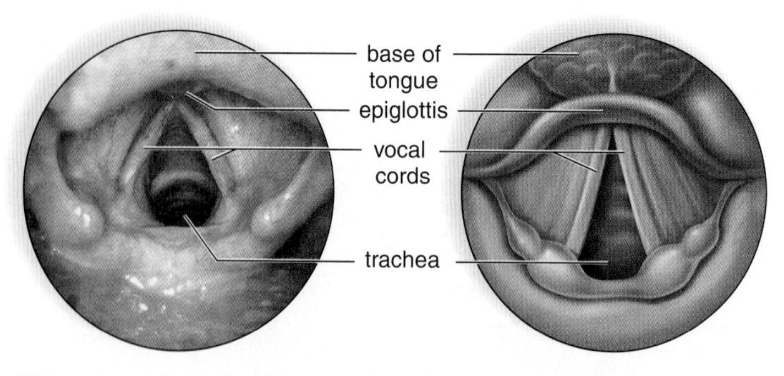

Figure 15.3 Placement of the vocal cords.
Viewed from above, the vocal cords stretch across the glottis, the opening to the trachea. When air is expelled through the glottis, the vocal cords vibrate, producing sound. The glottis is narrow when we make a high-pitched sound, and it widens as the pitch deepens.

In the pharynx, the air passage and the food passage cross because the larynx, which receives air, lies above and in front of the esophagus, which receives food. The larynx leads to the trachea. Both the larynx and the trachea are normally open, allowing air to pass, but the esophagus is normally closed and opens only when a person swallows.

The Larynx

The **larynx** is a cartilaginous structure that serves as a passageway for air between the pharynx and the trachea. The larynx can be pictured as a triangular box whose apex, the Adam's apple, is at the front of the neck. The larynx is called the voice box because it houses the vocal cords. The **vocal cords** are mucosal folds supported by elastic ligaments, and the slit between the vocal cords is an opening called the **glottis** (Fig. 15.3). When air is expelled past the vocal cords through the glottis, the vocal cords vibrate, producing sound. At the time of puberty, the growth of the larynx and the vocal cords is much more rapid and accentuated in the male than in the female, causing the male to develop a more prominent Adam's apple and a deeper voice. The voice "breaks" in the young male due to his inability to control the longer vocal cords.

The high or low pitch of the voice is regulated when speaking and singing by changing the tension on the vocal cords. The greater the tension, as when the glottis becomes more narrow, the higher the pitch. The loudness, or intensity, of the voice depends upon the amplitude of the vibrations—that is, the degree to which the vocal cords vibrate.

When food is swallowed, the larynx moves upward against the **epiglottis,** a flap of tissue that prevents food from passing into the larynx (Fig. 15.3). You can detect this movement by placing your hand gently on your larynx and swallowing.

The Trachea

The **trachea,** commonly called the windpipe, is a tube connecting the larynx to the primary bronchi. The trachea lies in front of the esophagus and is held open by C-shaped cartilaginous rings. The open part of the C-shaped rings faces the esophagus, and this allows the esophagus to expand when swallowing. The mucosa that lines the trachea has a layer of pseudostratified ciliated columnar epithelium (see Chapter 11). The cilia that project from the epithelium keep the lungs clean by sweeping mucus (produced by goblet cells) and debris toward the pharynx (Fig. 15.4). Smoking is known to destroy these cilia, and consequently, the toxins in cigarette smoke collect in the lungs. Smoking is discussed more fully in the Health Focus later in this chapter.

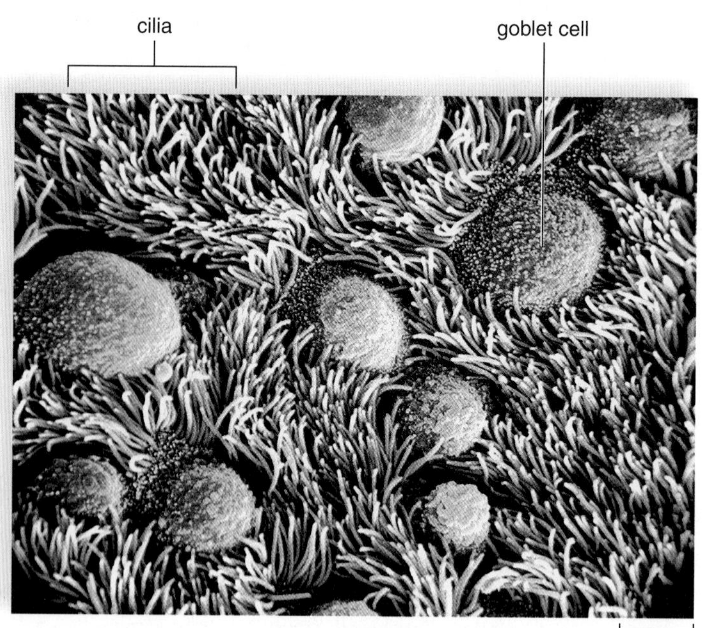

Figure 15.4 The surface of the trachea.
A scanning electron micrograph shows that the surface of the mucous membrane lining the trachea consists of goblet cells and ciliated cells. The cilia sweep mucus and the debris embedded in it toward the pharynx, where they are swallowed or expectorated. Smoking causes the cilia to disappear, allowing debris to enter the bronchi and lungs.

The Bronchial Tree

The trachea divides into right and left primary bronchi (sing., **bronchus**), which lead into the right and left lungs (see Fig. 15.1). The bronchi branch into a great number of secondary bronchi that eventually lead to **bronchioles.** The bronchi resemble the trachea in structure, but as the bronchial tubes divide and subdivide, their walls become thinner, and the small rings of cartilage are no longer present. Each bronchiole leads to an elongated space enclosed by a multitude of air pockets, or sacs, called alveoli (sing., **alveolus**). The components of the bronchial tree beyond the primary bronchi compose the lungs.

The Lungs

The **lungs** are paired, cone-shaped organs that occupy the thoracic cavity except for a central area that contains the trachea, the thymus, the heart, and the esophagus. The right lung has three lobes, and the left lung has two lobes, allowing room for the heart, whose apex points

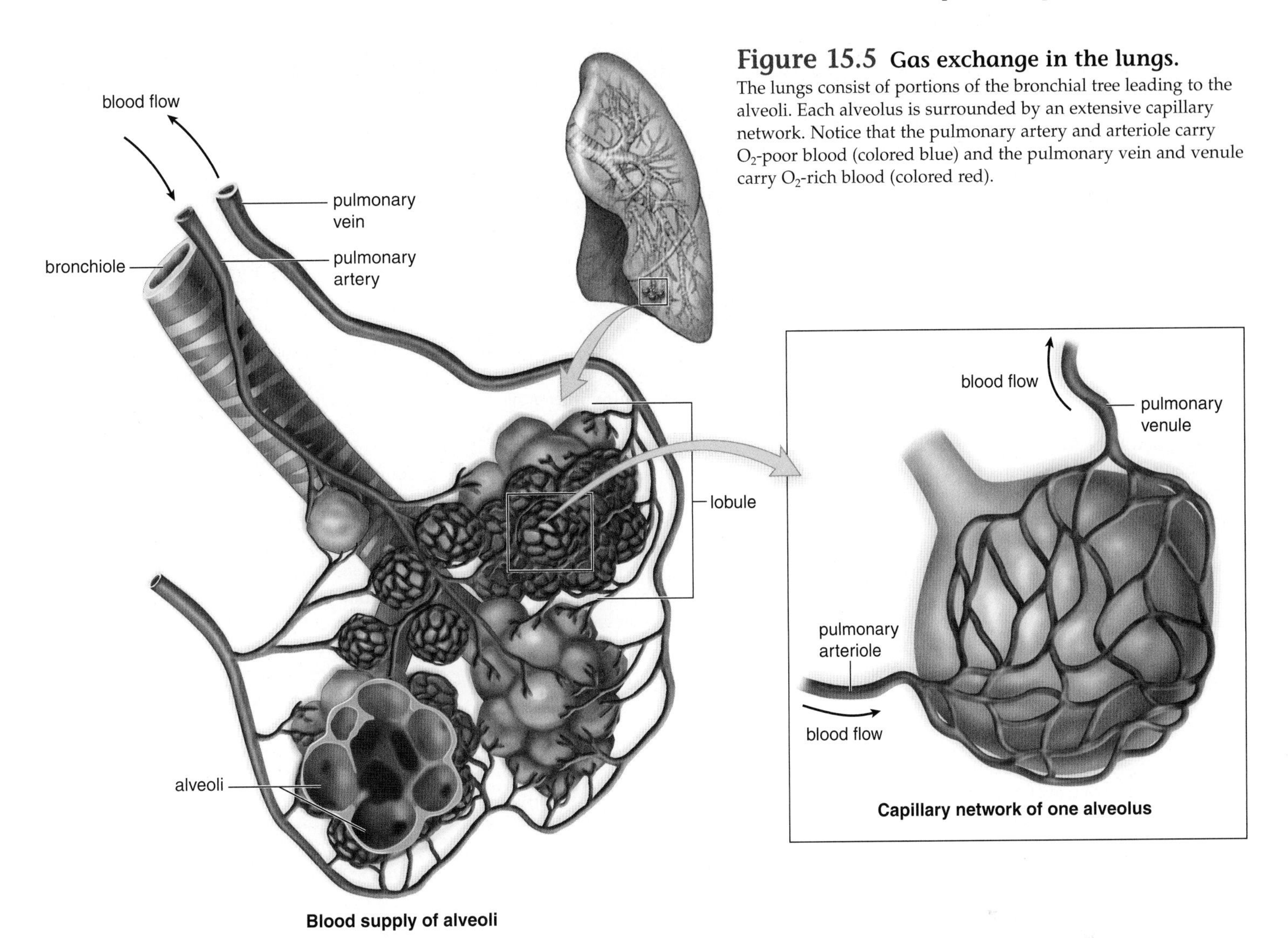

Figure 15.5 Gas exchange in the lungs. The lungs consist of portions of the bronchial tree leading to the alveoli. Each alveolus is surrounded by an extensive capillary network. Notice that the pulmonary artery and arteriole carry O_2-poor blood (colored blue) and the pulmonary vein and venule carry O_2-rich blood (colored red).

left. A lobe is further divided into lobules, and each lobule has a bronchiole serving many alveoli. The apex of a lung is narrow, while the base is broad and curves to fit the dome-shaped diaphragm, the muscle that separates the thoracic cavity from the abdominal cavity. The other surfaces of the lungs follow the contours of the ribs in the thoracic cavity.

Each lung is covered by a very thin serous membrane called a **pleura** (see Fig. 15.7*a*). Another pleura covers the internal chest wall and diaphragm. Both membranes produce a lubricating serous fluid that helps the pleurae slide freely against each other during inspiration and expiration. Surface tension is the tendency for water molecules to cling to each other due to hydrogen bonding between the molecules. Surface tension holds the two pleural layers together when the lungs recoil during expiration.

With each inhalation, air passes by way of the respiratory passageways to the alveoli. An alveolus is made up of simple squamous epithelium surrounded by blood capillaries. Gas exchange occurs between the air in an alveolus and the blood in the capillaries (Fig. 15.5). Oxygen diffuses across the alveolar and capillary walls to enter the bloodstream, while carbon dioxide diffuses from the blood across these walls to enter the alveoli.

If gas exchange is to occur, the alveoli must stay open to receive the inhaled air. Gas exchange takes place across moist cellular membranes, and yet the surface tension of water lining the alveoli is capable of causing them to close up. The alveoli are lined with **surfactant,** a film of lipoprotein that lowers the surface tension and prevents them from closing. The lungs collapse in some newborn babies, especially premature infants, who lack this film. The condition, called **infant respiratory distress syndrome,** is now treatable by surfactant replacement therapy.

Check Your Progress

1. Name the structures that comprise the upper and lower respiratory tracts.
2. Discuss the arrangement of the trachea, esophagus, larynx, and epiglottis, which normally prevents food from entering the trachea.
3. Trace the path of an oxygen molecule from the nose to the bloodstream.

Photochemical Smog Can Kill

Most industrialized cities have photochemical smog, at least occasionally. Photochemical smog arises when primary pollutants react with one another under the influence of sunlight to form a more deadly combination of chemicals. For example, two primary pollutants, nitrogen oxides (NO_x) and volatile organic compounds (VOCs) including hydrocarbons, as well as alcohols, aldehydes, and ethers, react with one another in the presence of sunlight to produce nitrogen dioxide (NO_2), ozone (O_3), and PAN (peroxy-acetylnitrate). Ozone and PAN are commonly referred to as oxidants. Breathing oxidants affects the respiratory and nervous systems, resulting in respiratory distress, headache, and exhaustion.

Cities with warm, sunny climates that are large and industrialized, such as Los Angeles, Denver, and Salt Lake City in the United States, Sydney in Australia, Mexico City in Mexico, and Buenos Aires in Argentina, are particularly susceptible to photochemical smog. If the city is surrounded by hills, a thermal inversion may aggravate the situation. Normally, warm air near the ground rises, so that pollutants are dispersed and carried away by air currents. But sometimes during a thermal inversion, smog gets trapped near the Earth by a blanket of warm air (Fig. 15A). This may occur when a cold front brings in cold air, which settles beneath a warm layer. The trapped pollutants cannot disperse, and the results are dangerous to a person's respiratory health. Even healthy adults experience a reduction in lung capacity when exposed to photochemical smog for long periods or during vigorous outdoor activities. Repeated exposures to high concentrations of ozone are associated with respiratory problems, such as an increased rate of lung infections and permanent lung damage. Children, the elderly, asthmatics, and individuals with emphysema or other similar disorders are particularly at risk.

Even though federal legislation is in place to bring air pollution under control, more than half the people in the United States live in cities polluted by too much smog. In the long run, pollution prevention is usually easier and cheaper than pollution cleanup. Some prevention suggestions are as follows:

- Encourage use of public transportation and burn fuels that do not produce pollutants.
- Increase recycling in order to reduce the amount of waste that is incinerated.
- Reduce energy use so that power plants need to provide less.
- Use renewable energy sources, such as solar, wind, or water power.
- Require industries to meet clean-air standards.

Discussion Questions

1. One of the most significant sources of the pollutants that contribute to photochemical smog is exhaust from automobiles. In June 2008, Honda introduced a car that runs on hydrogen and electricity and only emits water. How much extra would you be willing to pay for a car that doesn't emit any toxic exhaust fumes? $500? $5,000? Do you think that the government should provide strong incentives, such as tax breaks, for people who pay extra for these cars? Should automobile manufacturers and gas-station owners be required to convert to this technology?
2. Why do you think exposure to ozone causes more serious respiratory problems in the young and the elderly than in healthy adults?
3. How often do you use public transportation? Whether you live in a big city or not, would you be willing to use public transportation if it added, for example, an hour to the total time you spend each day traveling to school or work? Why or why not?

a. Ground-level ozone formation

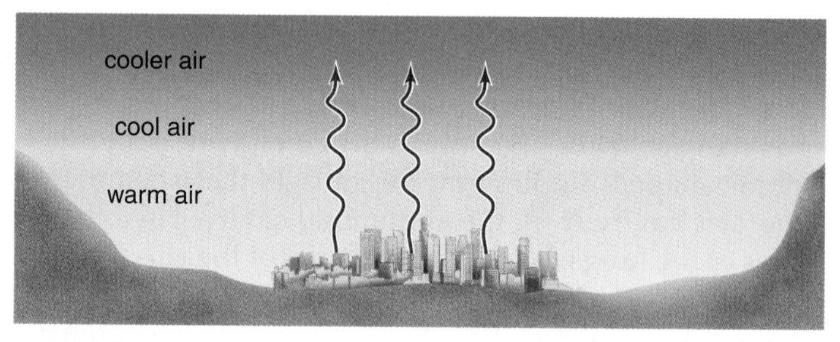

b. Normal pattern

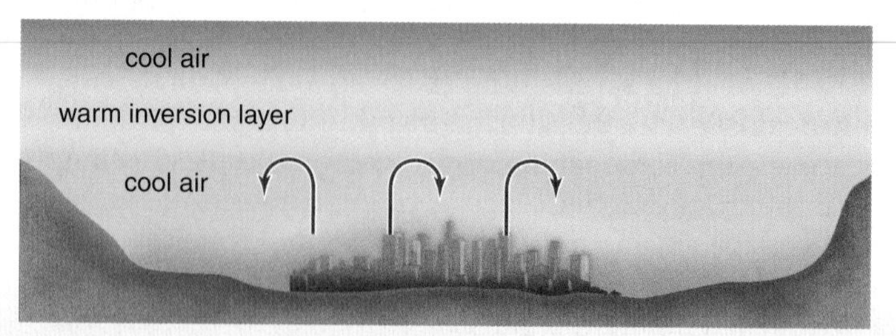

c. Thermal inversion

Figure 15A Thermal inversion.
a. Los Angeles is the "air pollution capital" of the world. Its millions of cars and thousands of factories make this city particularly susceptible to photochemical smog, which contains ozone due to the chemical reaction shown here. **b.** Normally, pollutants escape into the atmosphere when warm air rises. **c.** During a thermal inversion, a layer of warm air (warm inversion layer) overlies and traps pollutants in cool air below.

15.2 Mechanism of Breathing

Learning Outcomes

1. Identify the different respiratory volumes that can be measured with a spirometer.
2. Contrast the mechanisms of inspiration and expiration.
3. Indicate how ventilation is controlled by nervous and chemical mechanisms.

During ventilation (breathing), a free flow of air is vitally important. Therefore, a technique has been developed that allows physicians to detect any decrease in the ability of the lungs to fill with air and release it. An instrument called a spirometer records the volume of air exchanged during both normal and deep breathing. A spirogram shows the measurements recorded when a person breathes as directed by a technician (Fig. 15.6).

Respiratory Volumes

Normally when we are relaxed, only a small amount of air moves in and out with each breath. This amount of air, called the **tidal volume,** is only about 500 ml.

It is possible to increase the amount of air inhaled, and therefore the amount exhaled, by deep breathing. The maximum volume of air that can be moved in plus the maximum volume that can be moved out during a single breath is the **vital capacity.** It is called vital capacity because your life depends on breathing, and the more air you can move, the better off you are. A number of different illnesses discussed in section 15.4 can decrease vital capacity.

To increase your vital capacity, you must increase both the amount of air you breathe in and the amount you breathe out. You can increase inspiration (breathing in) by not only expanding the chest but also by lowering the diaphragm. Such forced inspiration usually increases the volume of inhaled air beyond the tidal volume by about 2,900 ml, and that amount is called the **inspiratory reserve volume.** You can increase the amount of air you exhale by contracting the abdominal and internal intercostal muscles. This so-called **expiratory reserve volume** is usually about 1,400 ml of air. You can see from Figure 15.6 that vital capacity is the sum of the tidal, inspiratory reserve, and expiratory reserve volumes.

In an average adult, only about 70% of the tidal volume actually reaches the alveoli; 30% remains in the airways. To ensure that a large portion of inhaled air reaches the lungs, it is better to breathe slowly and deeply. Also, note in Figure 15.6 that even after a very deep exhalation, some air (about 1,000 ml) remains in the lungs; this is called the **residual volume.** This air is not as useful for gas exchange because it has been depleted of oxygen. In some lung diseases, the residual volume builds up because the individual has difficulty emptying the lungs. This also means that the vital capacity is reduced. One environmental hazard that can lead to lung disease is photochemical smog (see the Ecology Focus).

Inspiration and Expiration

To understand ventilation, the manner in which air enters and exits the lungs, it is necessary to remember the following facts:

1. Normally, there is a continuous column of air from the pharynx to the alveoli of the lungs.
2. The lungs lie within the sealed-off thoracic cavity. The rib cage, consisting of the ribs joined to the vertebral column posteriorly and to the sternum anteriorly, forms the top and sides of the thoracic cavity. The intercostal muscles lie between the ribs. The diaphragm and connective tissue form the floor of the thoracic cavity.
3. The lungs adhere to the thoracic wall by way of the pleura. Normally, any space between the two pleurae is minimal due to the surface tension of the fluid between them.

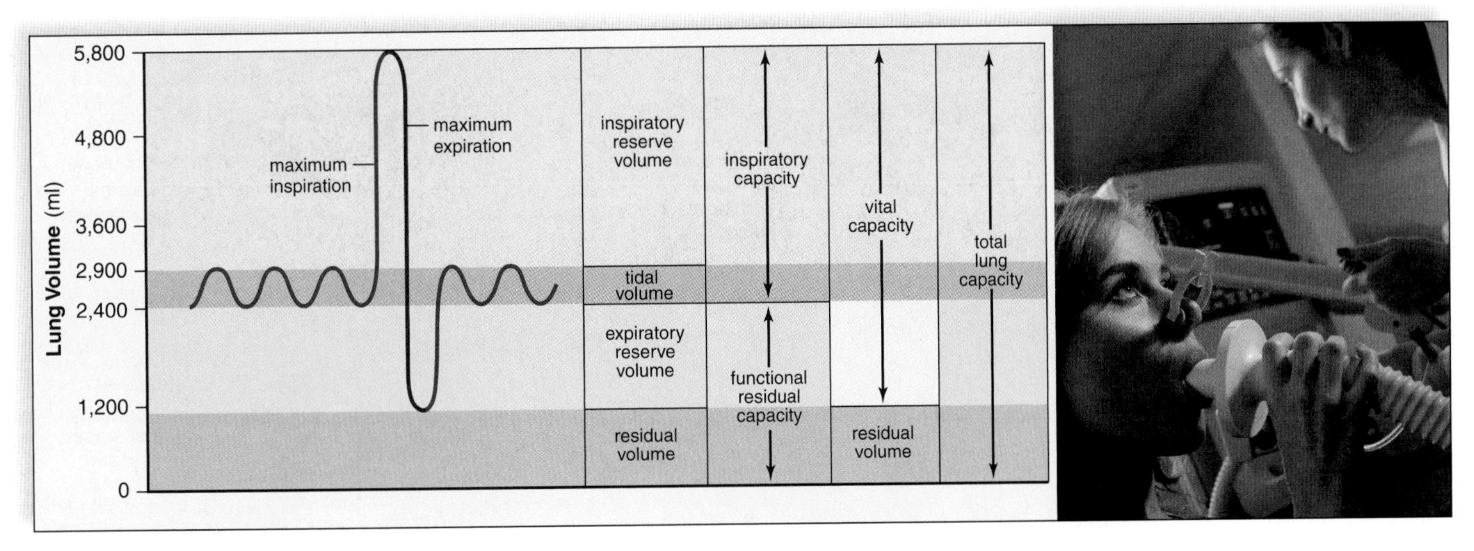

Figure 15.6 Measuring ventilation.
A spirometer measures the amount of air inhaled and exhaled with each breath. During inspiration, the pen moves up, and during expiration, the pen moves down. Vital capacity (red) is the maximum amount of air a person can exhale after taking the deepest inhalation possible.

Inspiration

Inspiration is the active phase of ventilation because this is the phase in which the diaphragm and the external intercostal muscles contract (Fig. 15.7*a*). In its relaxed state, the diaphragm is dome shaped. During deep inspiration, it contracts and lowers. Also, the external intercostal muscles contract, and the rib cage moves upward and outward.

Following this contraction, the volume of the thoracic cavity is larger than it was before. As the thoracic volume increases, the lungs expand. Now the air pressure within the alveoli decreases, creating a partial vacuum. Because alveolar pressure is now less than atmospheric pressure outside the lungs, air naturally flows from outside the body into the respiratory passages and into the alveoli.

Notice that air comes into the lungs because they have already opened up. Air does not force the lungs open. This is why it is sometimes said that *humans inhale by negative pressure.* While inspiration is the active phase of breathing, the actual flow of air into the alveoli is passive.

Expiration

Usually, expiration is the passive phase of breathing, and no effort is required to bring it about. During expiration, the elastic properties of the thoracic wall and lungs cause them to recoil. In addition, the lungs recoil because the surface tension of the fluid lining the alveoli tends to draw them closed. During expiration, the abdominal organs press up against the diaphragm, and the rib cage moves down and inward (Fig. 15.7*b*).

The diaphragm and external intercostal muscles are usually relaxed when expiration occurs. However, when breathing is deeper and/or more rapid, expiration can be active. Contraction of the internal intercostal muscles can force the rib cage to move downward and inward, and the increased pressure in the thoracic cavity helps expel air.

Control of Ventilation

Normally, adults have a breathing rate of 12 to 20 ventilations per minute. The rhythm of ventilation is controlled by a **respiratory center** located in the medulla oblongata of the brain.

The respiratory center causes inspiration to occur by automatically sending impulses to the diaphragm by way of the phrenic nerve, and to the intercostal muscles by way of the intercostal nerves (Fig. 15.8). When the respiratory center stops sending neuronal signals to the diaphragm and the rib cage, the diaphragm relaxes and resumes its dome shape. Now expiration occurs.

Although the respiratory center automatically controls the rate and depth of breathing, its activity can also be influenced by nervous input and chemical input. Following forced inhalation, stretch receptors in the alveolar walls initiate

Figure 15.7 Inspiration and expiration compared.
a. During inspiration, the thoracic cavity and lungs expand so that air is drawn in. **b.** During expiration, the thoracic cavity and lungs resume their original positions and pressures. Now, air is forced out.

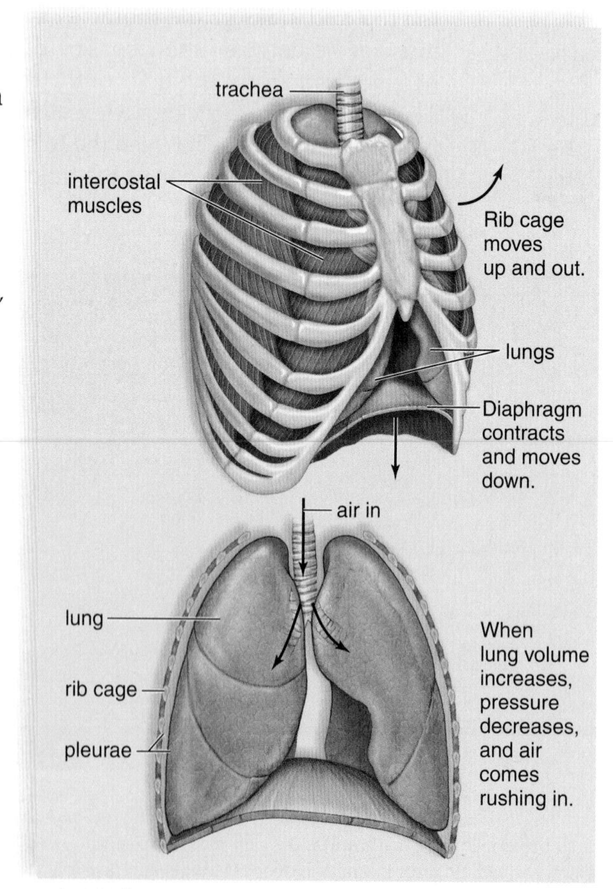

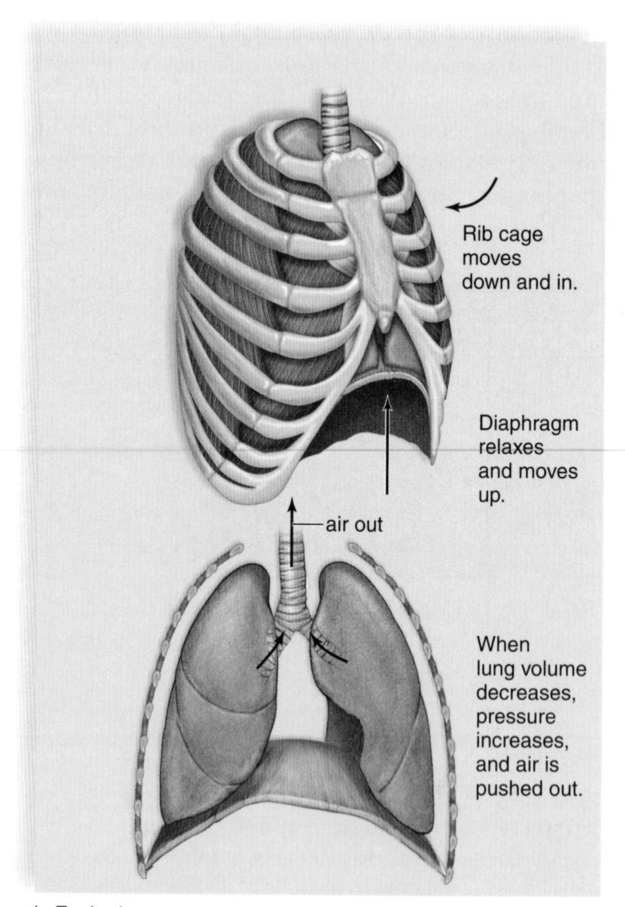

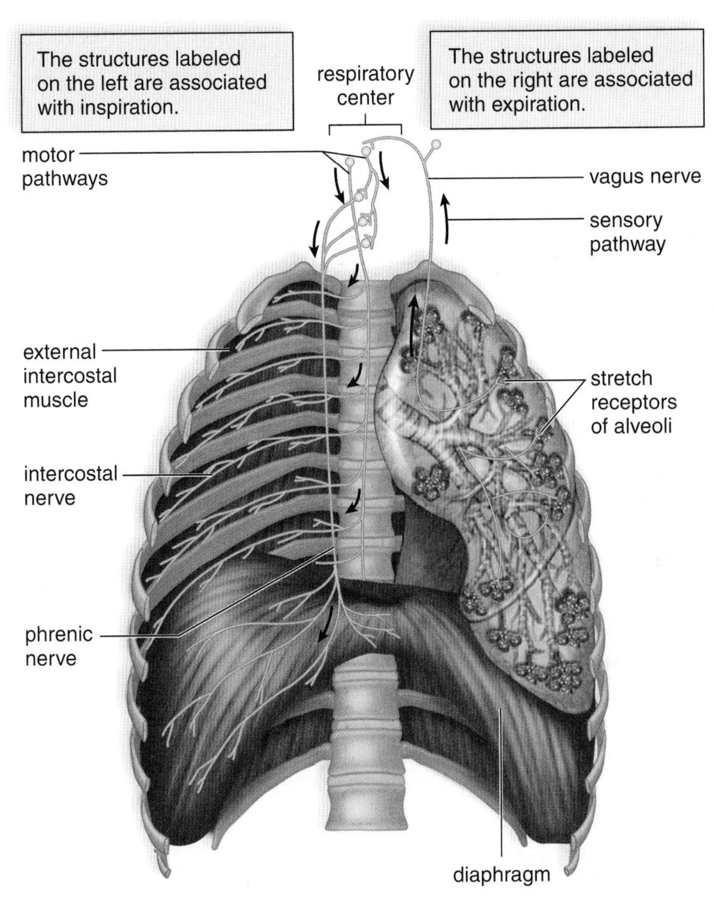

Figure 15.8 Nervous control of breathing.
The respiratory center automatically stimulates the external intercostal (rib) muscles and diaphragm to contract via the phrenic nerve. After forced inhalation, stretch receptors send inhibitory nerve impulses to the respiratory center via the vagus nerve. Usually, expiration automatically occurs due to lack of stimulation from the respiratory center to the diaphragm and intercostal muscles.

inhibitory nerve impulses that travel via the vagus nerve from the inflated lungs to the respiratory center. This stops the respiratory center from sending out nerve impulses.

Chemical Input The respiratory center is directly sensitive to the levels of carbon dioxide (CO_2) and hydrogen ions (H^+) in the blood. When they rise, the respiratory center increases the rate and depth of breathing. The center is not affected directly by low oxygen (O_2) levels. However, chemoreceptors in the *carotid bodies,* located in the carotid arteries, and in the *aortic bodies,* located in the aorta, are sensitive to the level of oxygen in the blood. When the concentration of oxygen decreases, these bodies communicate with the respiratory center, and the rate and depth of breathing increase.

Check Your Progress

1. Compare and contrast tidal volume, vital capacity, expiratory reserve volume, and residual volume.
2. Explain why inspiration is considered the active phase of ventilation and expiration the passive phase.
3. Discuss the roles of the following in controlling respiration: respiratory center, vagus nerve, and chemoreceptors.

15.3 Gas Exchanges in the Body

Learning Outcomes

1. Distinguish between external and internal respiration.
2. Interpret the chemical reaction discussed below to explain the effect of hyperventilation or hypoventilation on blood pH.
3. Review how the differences in P_{O_2} and P_{CO_2} in arterial and venous blood determine how gases are exchanged in the lungs versus the tissues.

The air passages of the respiratory system are simply conduits, and this system is absolutely essential to homeostasis, only because it allows gas exchange to occur. As mentioned previously, respiration includes not only the exchange of gases in the lungs (external respiration), but also the exchange of gases in the tissues (internal respiration) (Fig. 15.9). The principles of diffusion, alone, govern whether O_2 or CO_2 enters or leaves the blood in the lungs and in the tissues.

External Respiration

External respiration refers to the exchange of gases between air in the alveoli and blood in the pulmonary capillaries (see Fig. 15.5). Gases exert pressure, and the amount of pressure each gas exerts is called its partial pressure, symbolized as P_{O_2} and P_{CO_2}. Blood in the pulmonary capillaries has a higher P_{CO_2} than atmospheric air. Therefore, *CO_2 diffuses out of the plasma into the lungs.* Most of the CO_2 is carried as **bicarbonate ions** (HCO_3^-). As the little remaining free CO_2 begins to diffuse out, the following reaction is driven to the right:

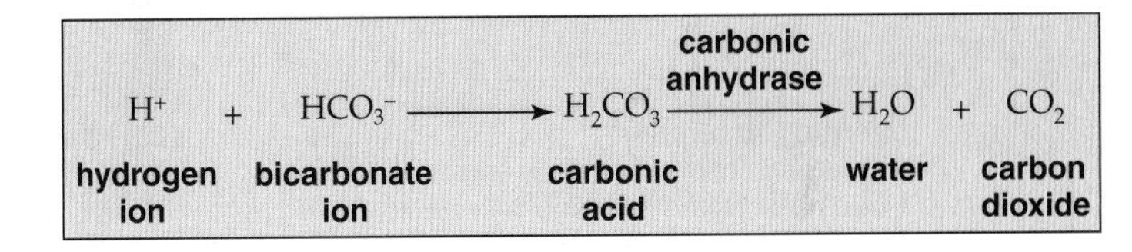

The enzyme **carbonic anhydrase,** present in red blood cells, speeds the breakdown of carbonic acid (H_2CO_3).

What happens if you hyperventilate (breathe at a high rate) and therefore push this reaction far to the right? The blood has fewer hydrogen ions, and alkalosis, a high blood pH, results. In that case, breathing will be inhibited, but in the meantime, you may suffer various symptoms, from dizziness to uncontrolled contractions of the muscles. What happens if you hypoventilate (breathe at a low rate) and this reaction does not occur? Hydrogen ions build up in the blood, and acidosis occurs. Buffers may compensate for the low pH, and breathing will most likely increase. Otherwise, you may become comatose and die.

The pressure pattern for O_2 during external respiration is the reverse of that for CO_2. Blood in the pulmonary capillaries is low in oxygen, and alveolar air contains a higher partial pressure of oxygen. Therefore, *O_2 diffuses into plasma and then into red blood cells in the lungs.* Hemoglobin takes up this oxygen and becomes **oxyhemoglobin** (HbO_2):

$$Hb + O_2 \longrightarrow HbO_2$$

hemoglobin — oxygen — oxyhemoglobin

Visual Focus

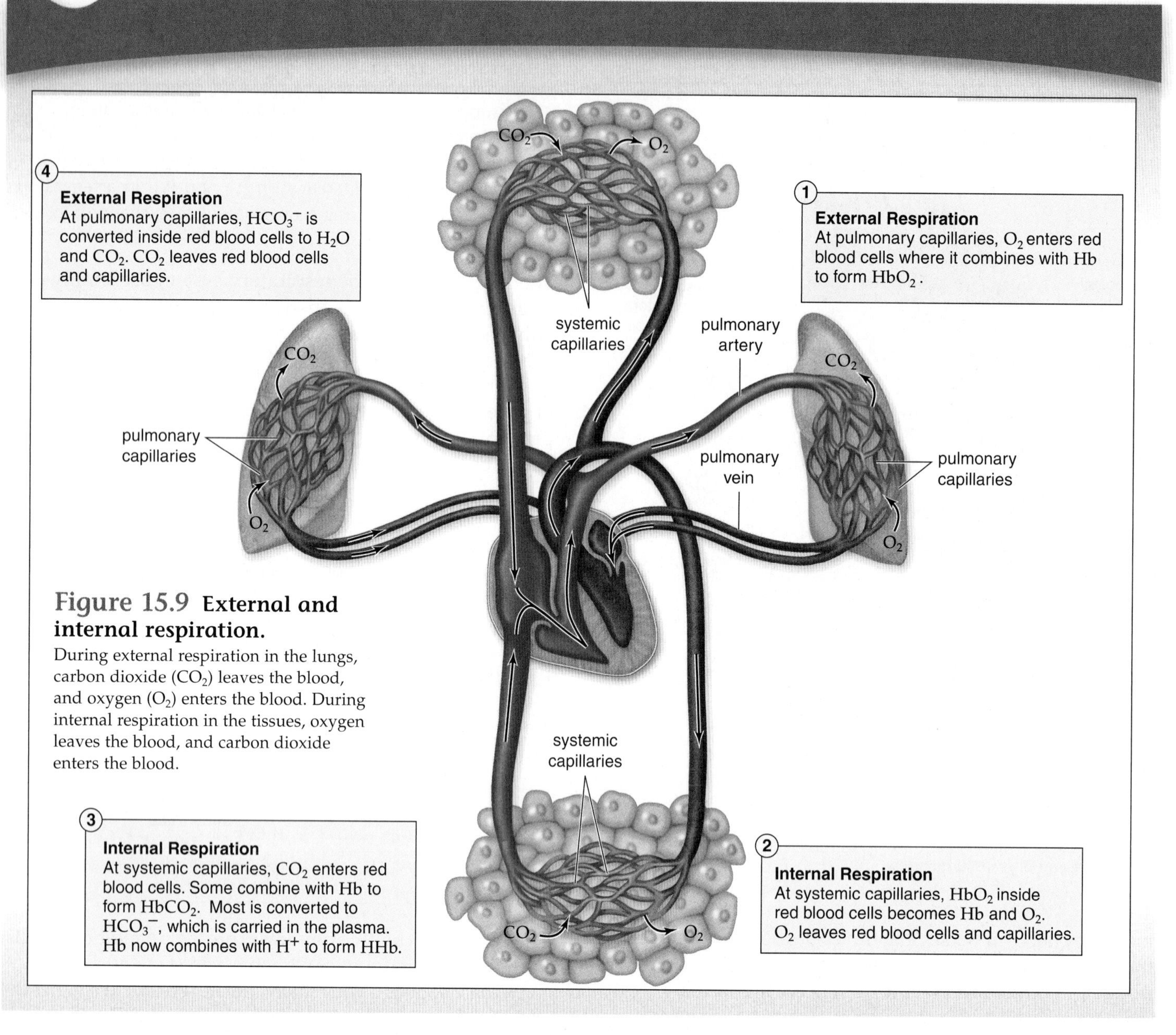

Figure 15.9 External and internal respiration. During external respiration in the lungs, carbon dioxide (CO_2) leaves the blood, and oxygen (O_2) enters the blood. During internal respiration in the tissues, oxygen leaves the blood, and carbon dioxide enters the blood.

Interestingly, only a relatively small percentage of the oxygen present in atmospheric air is utilized during normal external respiration. At sea level, air contains about 21% O_2, and exhaled air still contains 16–17% O_2. While this system may not seem very efficient, it does explain how a person whose heart and lungs have failed can sometimes be revived by mouth-to-mouth resuscitation, which involves filling the person's lungs with exhaled air.

Internal Respiration

Internal respiration refers to the exchange of gases between the blood in systemic capillaries and the tissue fluid. Therefore, internal respiration services tissue cells and without internal respiration, cells could not continue to produce the ATP that allows them to exist. Blood in the systemic capillaries is a bright red color because red blood cells contain oxyhemoglobin. Oxyhemoglobin gives up O_2, which diffuses out of the blood into the tissues:

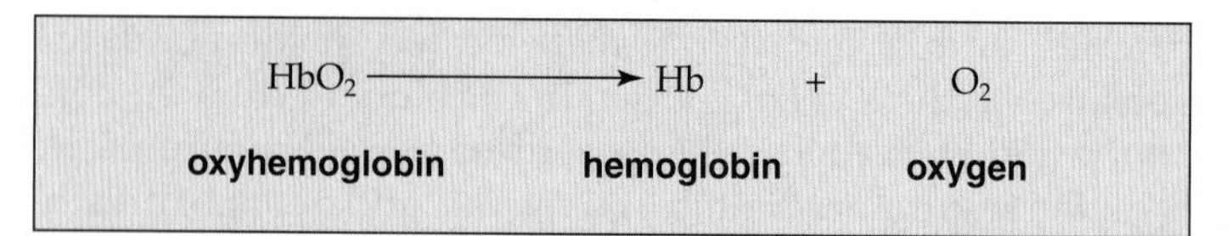

$$HbO_2 \longrightarrow Hb + O_2$$

oxyhemoglobin **hemoglobin** **oxygen**

Oxygen diffuses out of the blood into the tissues because the P_{O_2} of tissue fluid is lower than that of blood. The lower P_{O_2} is due to cells continuously using up oxygen in cellular respiration. *Carbon dioxide diffuses into the blood from the tissues* because the P_{CO_2} of tissue fluid is higher than that of blood. Carbon dioxide, produced continuously by cells, collects in tissue fluid.

After CO_2 diffuses into the blood, it enters the red blood cells, where a small amount is taken up by hemoglobin, forming **carbaminohemoglobin.** Most of the CO_2 combines with water, forming carbonic acid (H_2CO_3), which dissociates to hydrogen ions (H^+) and bicarbonate ions (HCO_3^-). The increased concentration of CO_2 in the blood drives the reaction to the right:

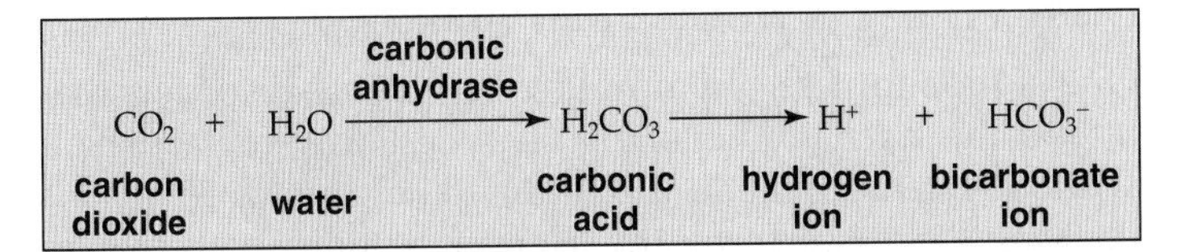

The enzyme carbonic anhydrase, mentioned previously, speeds up the reaction. Bicarbonate ions diffuse out of red blood cells and are carried in the plasma. The globin portion of hemoglobin combines with excess hydrogen ions produced by the overall reaction, and Hb becomes HHb, called **reduced hemoglobin.** In this way, the pH of blood remains fairly constant. Blood that leaves the systemic capillaries is a dark maroon color because red blood cells contain reduced hemoglobin (though veins may actually look blue when seen under the skin).

Check Your Progress

1. Explain the role of hemoglobin in carrying O_2, CO_2, and hydrogen ions. Which is hemoglobin's most essential function?
2. Why is arterial blood bright red in color, but venous blood is darker? This being the case, why does blood oozing from a cut always appear to be bright red?

15.4 Disorders of the Respiratory System

Learning Outcomes

1. List several disorders that affect the upper respiratory tract and several that occur in the lower respiratory tract.
2. Classify disorders of the respiratory system according to whether they are caused by allergies, infections, a genetic defect, or exposure to toxins.

The respiratory tract is constantly exposed to the air in our environment, and thus it is susceptible to various infectious agents, as well as to pollution and, in some individuals, tobacco smoke. Disorders of the upper respiratory tract will be discussed before disorders of the lower respiratory tract.

Disorders of the Upper Respiratory Tract

The upper respiratory tract consists of the nasal cavities, the pharynx, and the larynx. Because it is responsible for filtering out many of the pathogens and other materials that may be present in the air, the upper respiratory tract is susceptible to a variety of viral and bacterial infections. Upper respiratory infections can also spread from these areas to the middle ear or the sinuses.

The Common Cold

Most "colds" are relatively mild viral infections of the upper respiratory tract characterized by sneezing, a runny nose, and perhaps a mild fever. Many different viruses can cause colds, the most common being a group called the rhinoviruses (*rhin* is Greek for nose). Most colds last from a few days to a week, when the immune response is able to eliminate the virus. Because colds are caused by viruses, antibiotics do not help, although decongestants and anti-inflammatory medications can ease the symptoms.

Pharyngitis, Tonsillitis, and Laryngitis

Pharyngitis is an inflammation of the throat, usually because of an infection. If the tonsils have not been removed previously, they often become involved as well. What we call "strep throat" is a pharyngitis caused by the bacterium *Streptococcus pyogenes* that can lead to a generalized upper respiratory infection and even a systemic (affecting the body as a whole) infection. The symptoms of strep throat are severe sore throat, high fever, and white patches on a dark-red pharyngeal or tonsillar area (Fig. 15.10). Most cases of strep throat can be successfully treated with antibiotics.

Tonsillitis occurs when the **tonsils,** aggregates of lymphoid tissue in the pharynx, become inflamed and enlarged. The tonsils in the posterior wall of the nasopharynx are often called adenoids. If tonsillitis occurs frequently and the enlarged tonsils make breathing difficult, the tonsils can be removed surgically in a **tonsillectomy.** Fewer tonsillectomies are performed today than in the past because we now know that the tonsils help initiate immune responses to many of the pathogens that enter the pharynx. Therefore, they are an important component of the body's immune system.

Laryngitis is an inflammation of the larynx with accompanying hoarseness, often leading to the inability to talk in an audible voice. Usually, laryngitis disappears after resting the vocal cords and treating any infection present.

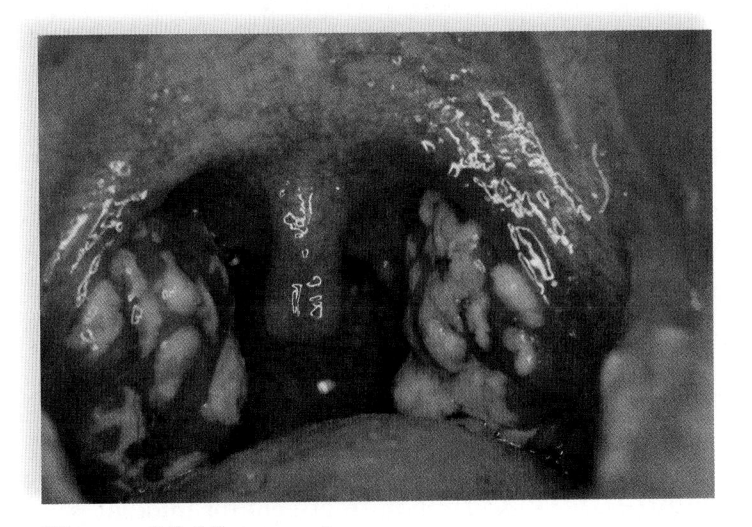

Figure 15.10 Strep throat.
Pharyngitis, often caused by the bacterium *S. pyogenes*, can lead to swollen, inflamed tonsils, sometimes with whitish patches.

Sinusitis

Sinusitis is an inflammation of the cranial sinuses, the cavities within the facial skeleton that drain into the nasal cavities. Sinusitis develops when nasal congestion blocks the tiny openings leading to the sinuses. Up to 10% of upper respiratory infections are accompanied by sinusitis, and allergies may also play a role. Symptoms may include postnasal discharge, headache, and facial pain that worsens when the patient bends forward. Successful treatment depends on addressing the cause of the inflammation, and restoring proper drainage of the sinuses. Rinsing the sinuses by instilling a warm saline solution into one nostril, and out the other, though somewhat uncomfortable, may help remove irritants and rinse out mucus.

Otitis Media

Otitis media is an inflammation of the middle ear. The middle ear is not a part of the respiratory tract, but this condition is considered here because, especially in children, nasal infections can spread to the ear by way of the **auditory (eustachian) tubes** that lead from the nasopharynx to the middle ear. Pain is the primary symptom of otitis media. A sense of fullness, hearing loss, vertigo (dizziness), and fever may also be present. If the cause is bacterial, antibiotic therapy is usually very effective. However, if viruses or allergies are the culprit, antibiotics are not effective. Tubes (called tympanostomy tubes) are sometimes surgically placed in the eardrums of children with multiple recurrences to help prevent the buildup of pressure in the middle ear and the possibility of hearing loss (Fig. 15.11). Normally, these tubes fall out with time.

Check Your Progress

1. List three common disorders of the upper respiratory tract, along with the most common cause of each.
2. Explain how infections can spread from the respiratory system to the middle ear.

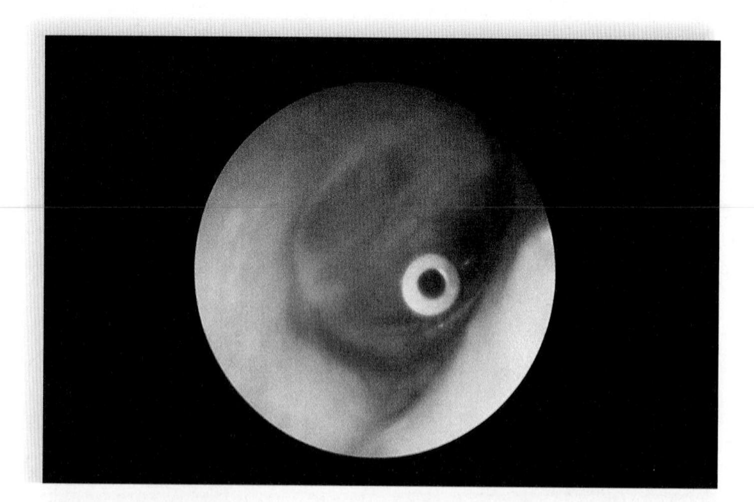

Figure 15.11 A common treatment for middle ear infections.
Tympanostomy tubes may be surgically implanted in the eardrums to drain fluid from the middle ears of children with chronic otitis media.

Disorders of the Lower Respiratory Tract

Several disorders of the lower respiratory tract cause problems by obstructing normal air flow. Their causes range from a foreign object lodged in the trachea to excessive mucus in the bronchi and bronchioles. Other conditions tend to restrict the normal elasticity of the lung tissue itself (Fig. 15.12).

Disorders of the Trachea and Bronchi

One of the simplest, but most life-threatening disorders that affects the trachea is choking. The best way for a person without extensive medical training to help someone who is choking is to perform the Heimlich maneuver, which involves grabbing the choking person around the waist from behind, and forcefully pulling both hands into their upper abdomen to expel whatever is lodged. Trained medical personnel may also insert a breathing tube by way of an incision made in the trachea. The operation is called a tracheotomy, and the opening is a **tracheostomy.** Sometimes people whose larynx or trachea has been damaged or destroyed, usually as a result of smoking, must have a permanent tracheostomy tube installed.

Acute bronchitis is an inflammation of the primary and secondary bronchi. Usually it is preceded by a viral infection that has led to a secondary bacterial infection. Most likely, a nonproductive cough has become a deep cough that produces more mucus and perhaps pus (Fig. 15.12*a*). Typically, the bacterial infection can be successfully treated with antibiotics.

In **chronic bronchitis,** the airways are inflamed and filled with mucus. A cough that brings up mucus is common. The bronchi have undergone degenerative changes, including the loss of cilia and their normal cleansing action. Under these conditions, an infection is more likely to occur. The most frequent cause of chronic bronchitis is smoking, although exposure to environmental pollutants can also be a contributing factor.

Asthma is a disease of the bronchi and bronchioles that is marked by wheezing, breathlessness, and sometimes a cough and expectoration of mucus. The airways are unusually sensitive to specific irritants, which can include a wide range of allergens such as pollen, animal dander, dust, cigarette smoke, and industrial fumes. Even cold air can be an irritant. When exposed to an irritant, the smooth muscle in the bronchioles undergoes spasms (Fig. 15.12*b*). Most asthma patients have some degree of bronchial inflammation that reduces the diameter of the airways and contributes to the seriousness of an attack. Asthma is not curable, but several types of drugs can prevent or treat asthma attacks. One group of drugs, called the beta-agonists, works by dilating the bronchioles. These drugs are usually administered using an asthma inhaler, as mentioned in the story that opened this chapter. Corticosteroids can help control the inflammation and hopefully prevent an attack.

Diseases of the Lungs

Pneumonia is an infection of the lungs in which the bronchi or alveoli fill with thick fluid (Fig. 15.12*c*). High fever and

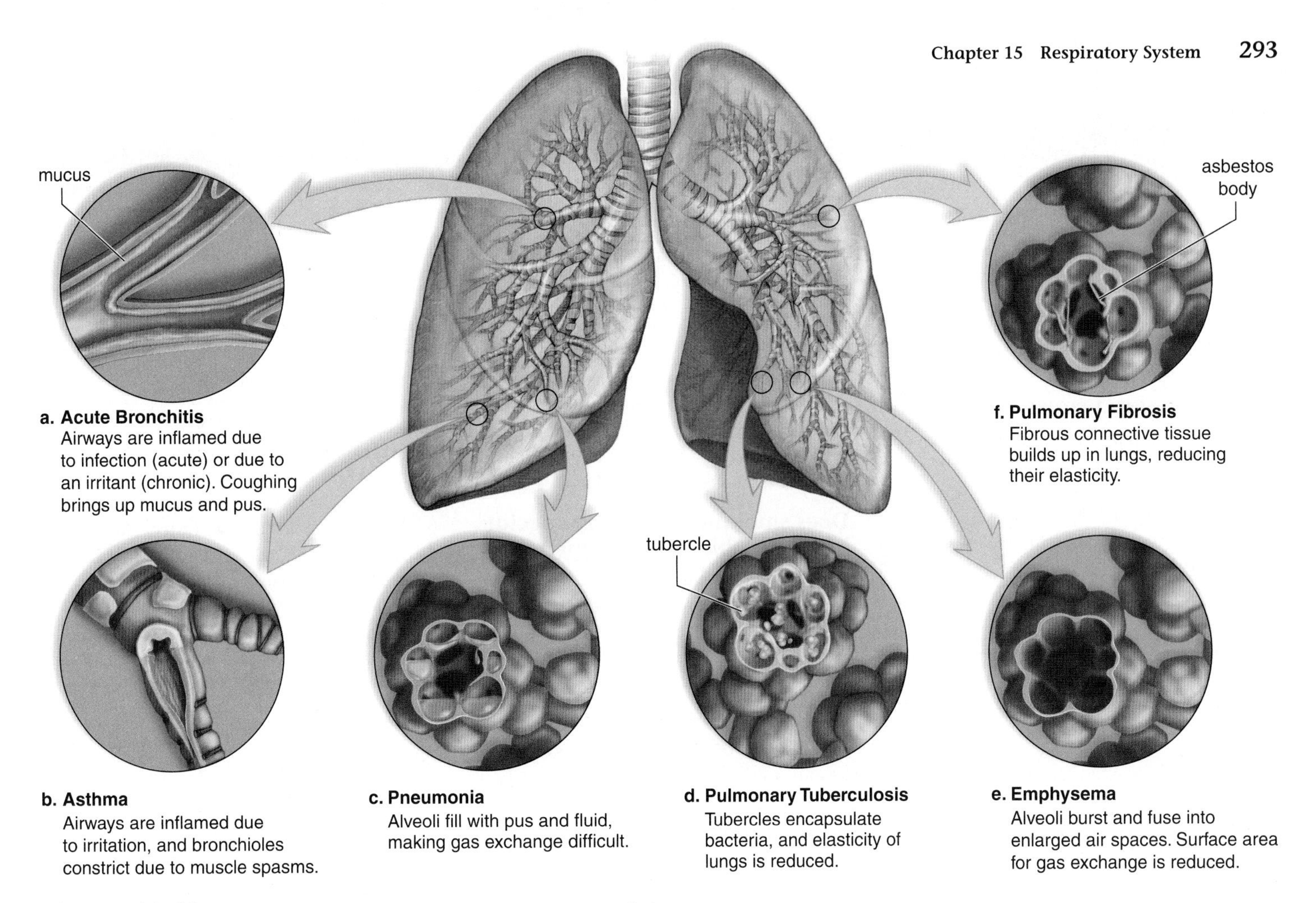

Figure 15.12 Common bronchial and pulmonary diseases.
Exposure to infectious pathogens and/or polluted air, including tobacco smoke, causes various diseases and disorders.

chills, with headache and chest pain, are symptoms of pneumonia. Rather than being a generalized lung infection, pneumonia may be localized in specific lobules of the lungs. Obviously, the more lobules involved, the more serious is the infection. Pneumonia can be caused by several types of bacteria, viruses, and other infectious agents. Certain types of pneumonia mainly strike individuals with reduced immunity. For example, AIDS patients are subject to a particularly rare form of pneumonia caused by the fungus *Pneumocystis carinii.*

Pulmonary tuberculosis is caused by the bacterium *Mycobacterium tuberculosis.* When *M. tuberculosis* invade the lung tissue, the cells accumulate around the invading bacteria, isolating them from the rest of the body. This accumulation of cells is called a tubercle (Fig. 15.12*d*). If the body's resistance is high, the imprisoned organisms die, but if the resistance is low, the organisms eventually escape and spread. If a chest X ray detects active tubercles, the individual is put on appropriate drug therapy to ensure the localization of the disease and the eventual destruction of any live bacteria. It is possible to tell if a person has ever been infected with tuberculosis bacteria with a TB skin test, in which a highly diluted extract of the bacteria is injected into the skin of the patient. A person who has never been exposed to *M. tuberculosis* shows no reaction, but one who has had or is fighting an infection shows an area of inflammation that peaks in about 48 hours (see Fig. 13.12).

Emphysema is a chronic and incurable disorder in which the alveoli are distended and their walls damaged so that the surface area available for gas exchange is reduced (Fig. 15.12*e*). Emphysema is often preceded by chronic bronchitis. Air trapped in the lungs leads to alveolar damage and a noticeable ballooning of the chest. The elastic recoil of the lungs is reduced, so the driving force behind expiration is also reduced. The victim often feels out of breath and may have a cough. Because the surface area for gas exchange is reduced, less oxygen reaches the heart and the brain. Because they are often seen in the same patient and tend to recur, emphysema, chronic bronchitis, and asthma are collectively called chronic obstructive pulmonary disease (COPD). COPD is the fourth leading cause of death in the United States, and is usually associated with smoking.

Cystic fibrosis (CF) is an example of a lung disease that is genetic, rather than infectious, although infections do play a role in the disease. One in 31 Americans carries the defective gene, but a child must inherit two copies of the faulty gene to have the disease. Still, CF is the most common inherited disease in the U.S. white population. The gene that is defective in CF codes for cystic fibrosis transmembrane

Health Focus

Frequently Asked Questions About Tobacco and Health

Is there a safe way to smoke?

No. All forms of tobacco can cause damage, and smoking even a small amount is dangerous. Tobacco is perhaps the only legal product whose advertised and intended use—that is, smoking it—will hurt the body.

Does smoking cause cancer?

Yes, and not only lung cancer. Besides lung cancer, smoking a pipe, cigarettes, or cigars is also a major cause of cancers of the mouth, larynx (voice box), and esophagus. In addition, smoking increases the risk of cancer of the bladder, kidney, pancreas, stomach, and uterine cervix.

What are the chances of being cured of lung cancer?

Very low; the combined five-year survival rate is only 15%. Fortunately, lung cancer is a largely preventable disease. In other words, by not smoking, it can probably be prevented.

Does smoking cause other lung diseases?

Yes. Smoking leads to chronic bronchitis, a disease in which the airways produce excess mucus, forcing the smoker to cough frequently. Smoking is also the major cause of emphysema, a disease that slowly destroys a person's ability to breathe.

Why do smokers have "smoker's cough"?

Normally, cilia (tiny hairlike formations that line the airways) beat outward and "sweep" harmful material out of the lungs. Smoke, however, decreases this sweeping action, so some of the poisons in the smoke remain in the lungs.

If you smoke but don't inhale, is there any danger?

Yes. Wherever smoke touches living cells, it does harm. So, even if smokers of pipes, cigarettes, and cigars don't inhale, they are at an increased risk for lip, mouth, and tongue cancer.

Does smoking affect the heart?

Yes. Smoking doubles the risk of heart disease, which is the United States' number-one killer. Smoking, high blood pressure, high cholesterol, and lack of exercise are all risk factors for heart disease.

Is there any risk for pregnant women and their babies?

Pregnant women who smoke endanger the health and lives of their unborn babies. When a pregnant woman smokes, she really is smoking for two because the nicotine, carbon monoxide, and other dangerous chemicals in smoke enter her bloodstream and then pass into the baby's body. Smoking mothers have more stillbirths and babies of low birth weight than nonsmoking mothers.

Does smoking cause any special health problems for women?

Yes. Women who smoke and use the birth control pill have an increased risk of stroke and blood clots in the legs. In addition, women who smoke increase their chances of getting cancer of the uterine cervix.

What are some of the short-term effects of smoking cigarettes?

Almost immediately, smoking can make it hard to breathe. Within a short time, it can also worsen asthma and allergies. Only seven seconds after a smoker takes a puff, nicotine reaches the brain, where it produces a morphinelike effect.

Are there any other risks to the smoker?

Yes, there are many more risks. Smoking is a cause of stroke, which is the third leading cause of death in the United States. Smokers are more likely to have and die from stomach ulcers than nonsmokers. Smokers have a higher incidence of cancer in general. If a person smokes and is exposed to radon or asbestos, the risk for lung cancer increases dramatically.

What are the dangers of passive smoking?

Passive smoking causes lung cancer in healthy nonsmokers. Children whose parents smoke are more likely to suffer from pneumonia or bronchitis in the first two years of life than children who come from smoke-free households. Passive smokers have a 30% greater risk of developing lung cancer than nonsmokers who live in a smoke-free house.

Are chewing tobacco and snuff safe alternatives to cigarette smoking?

No, they are not. Smokeless tobacco contains nicotine, the same addicting drug found in cigarettes and cigars. The juice from smokeless tobacco is absorbed through the lining of the mouth. There it can cause sores and white patches, which often lead to cancer of the mouth. Snuff dippers actually take in an average of over ten times more cancer-causing substances than cigarette smokers.

Lung Cancer Statistics

Lung cancer is the leading cause of death due to cancer for both men and women.

About 87% of lung cancer cases are linked to smoking. In the United States, an estimated 215,020 new cases and 161,840 deaths from lung cancer occurred in 2008.

This means that lung cancer will account for 29% of all cancer deaths. More people die of lung cancer than of colon, breast, and prostate cancers combined.

- Women who smoke are 1.5 times more likely to get lung cancer than are men who smoke.
- About 3,400 lung cancer deaths per year are attributed to secondhand smoke.
- Usually, 42% of people diagnosed with lung cancer die within the year.
- The five-year combined survival rate (the number of people alive after five years) after being diagnosed is only 15%. Even for those who are diagnosed early, the survival rate after five years is only about 49%.

Discussion Questions

1. Before reading this Health Focus, were you aware of all the risks associated with smoking? If you are a smoker, how does knowing these risks affect the chances that you will quit? If you are a nonsmoker, do you think you might ever start?
2. Smoking is clearly addictive, because many people who want to quit are unable to do so. Whether you are a smoker or not (and it might be interesting to compare the responses of these two groups), what do you think is the percentage of smoking addiction that is physical versus psychological?
3. Suppose you became addicted to smoking two packs a day. If you decided to quit, how would you go about it? All at once, or a little at a time? Would you use any aids, such as nicotine gum or patches, or drugs? Perhaps a device that administers a mild electrical shock every time you smoke a cigarette? (Such a device exists!)

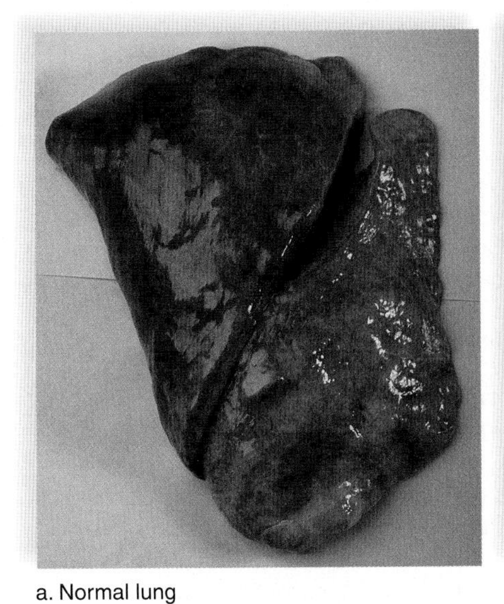
a. Normal lung

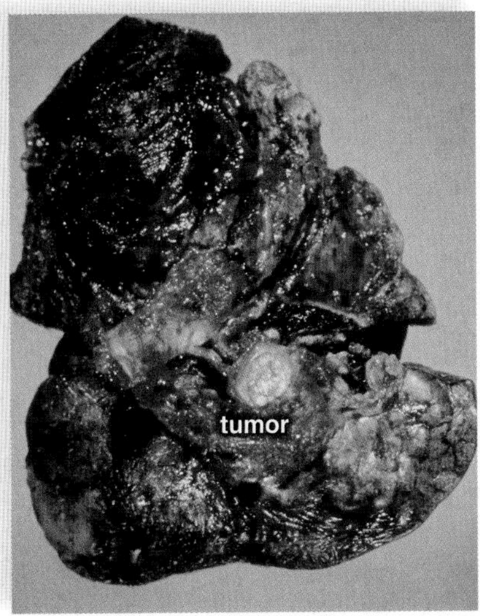

b. Lung cancer

Figure 15.13 Normal and cancerous lungs compared.
a. Normal lung. Note the healthy red color in both lobes. **b.** Lung of a heavy smoker. Notice that the lung is black except where cancerous tumors have formed.

regulator (CFTR), a protein needed for proper transport of Cl^- ions out of the epithelial cells of the lung. Because this also reduces the amount of water transported out of the lung cells, the mucus secretions become very sticky and can form plugs that interfere with breathing. Symptoms of CF include coughing and shortness of breath, and part of the treatment involves clearing mucus from the airways by vigorously slapping the patient on the back as well as by administering mucus-thinning drugs. None of these treatments is curative, however, and because the lungs can be severely affected, the median survival age for people with CF is only 30 years. Researchers are attempting to develop gene therapy strategies to replace the faulty *CFTR* gene.

Another common lung disease is **pulmonary fibrosis,** in which fibrous connective tissue builds up in the lungs, causing a loss of elasticity (Fig. 15.12*f*). This restricts the ability of the lungs to expand during inhalation, and so reduces the vital capacity and other lung volumes. Pulmonary fibrosis most commonly occurs in elderly persons, and the risk is increased by environmental exposure to silica (sand), various types of dust, and asbestos.

Lung cancer is more prevalent in men than in women, but it is the leading cause of cancer death in both genders. About 87% of lung cancers are associated with cigarette smoking. Autopsies on smokers have revealed the progressive steps by which the most common form of lung cancer develops. The first step appears to be thickening of the cells lining the bronchi. Then cilia are lost, making it impossible to prevent dust and dirt from settling in the lungs. Following this, cells with atypical nuclei appear, followed by a tumor consisting of disordered cells with atypical nuclei. A normal lung and a lung with cancerous tumors are shown in Figure 15.13. A final step occurs when some of these cells break loose and penetrate other tissues, a process called metastasis. Now the cancer has spread. The original tumor may grow until a bronchus is blocked, cutting off the supply of air to that lung. The entire lung then collapses, the secretions trapped in the lung spaces become infected, and pneumonia or a lung abscess (localized area of pus) results. The only treatment that offers a possibility of cure is **pneumonectomy,** in which a lobe or the whole lung is removed before metastasis has had time to occur. If the cancer has spread, chemotherapy and radiation are also required.

The Health Focus discusses the various illnesses, including cancer, that are apt to occur when a person smokes. The Bioethical Focus at the end of this chapter raises questions of fairness regarding attempts to cut back on passive smoking by issuing smoking bans. If a person stops voluntary smoking and avoids passive smoking, and if the body tissues are not already cancerous, they may return to normal over time.

Check Your Progress

1. Name one disorder of the lower respiratory tract that tends to cause a narrowing of the airways and one that restricts the lung's ability to expand normally.
2. In addition to the directly harmful effects of cigarette smoke, how might smoking predispose the lung to damage by other harmful substances?

Applying the Concepts [Revisited]

In the story that opened this chapter, 13-year-old Juan was having difficulty playing basketball due to the airway constriction and inflammation characteristic of asthma.

1. How would a narrowing and swelling of Juan's airways be expected to affect his respiratory volumes? On a spirogram, which of his specific respiratory volumes would be most affected?
2. Typical treatments for asthma involve drugs that reduce the symptoms. Why is it so difficult to develop a cure for asthma?

Bioethical Focus

Bans on Smoking

In 1964, the Surgeon General of the United States announced to the general public that smoking is hazardous to our health, and thereafter, a health warning was placed on packs of cigarettes. At that time, 40.4% of adults smoked, but by 2007, only 19.3% of adults smoked. In the meantime, however, the public became aware that passive smoking—that is, just being in the vicinity of someone who is smoking—can also lead to cancer and other health problems. By now, many state and local governments have passed legislation that bans smoking in public places such as restaurants, elevators, public meeting rooms, and the workplace.

Is legislation that restricts the freedom to smoke ethical? Or is such legislation akin to racism and creating a population of second-class citizens who are segregated from the majority on the basis of a habit? Are the desires of nonsmokers being allowed to infringe on the rights of smokers? Or is this legislation one way to help smokers become nonsmokers? One study showed that workplace bans on smoking reduce the daily consumption of cigarettes among smokers by 10%.

Is legislation that disallows smoking in family-style restaurants fair, especially if bars and restaurants associated with casinos are not included in the ban on smoking? The selling of tobacco, and even the increased need for health care it generates, helps the economy. One smoker writes, "Smoking causes people to drink more, eat more, and leave larger tips. Smoking also powers the economy of Wall Street." Is this a reason to allow smoking to continue? Or should we simply require all places of business to put in improved air filtration systems? Would that do away with the dangers of passive smoking?

Does legislation that bans smoking in certain areas represent government invasion of our privacy? If yes, is reducing the chance of cancer a good enough reason to allow the government to invade our privacy? Some people are more prone to cancer than others. Should we all be regulated by the same legislation? Are we our brothers' keepers, meaning that we have to look out for one another?

Form Your Own Opinion

1. Besides the Surgeon General's warning, are there any other possible explanations for why the percentage of smoking adults declined between 1964 and 2007?
2. If you are a smoker, do you smoke in your house? Your car? If not, why not? If you are a nonsmoker, how would you respond if a good friend or a relative wanted to smoke in your house?
3. If there are smoking bans in your area, how have you personally been affected? If your area has no bans but passed one tomorrow, how might you be affected? Why?

SUMMARIZING THE CONCEPTS

15.1 The Respiratory System

The major function of the respiratory system is to allow O_2 from the air to enter the blood, and CO_2 from the blood to exit the body. The respiratory tract can be divided into two major components:

- The upper respiratory tract, consisting of the nasal cavities (nose), the nasopharynx, the pharynx, and the larynx (which contains the vocal cords).
- The lower respiratory tract, made up of the trachea, the bronchi, the bronchioles, and the lungs. The lungs contain many alveoli, which are air sacs surrounded by a capillary network.

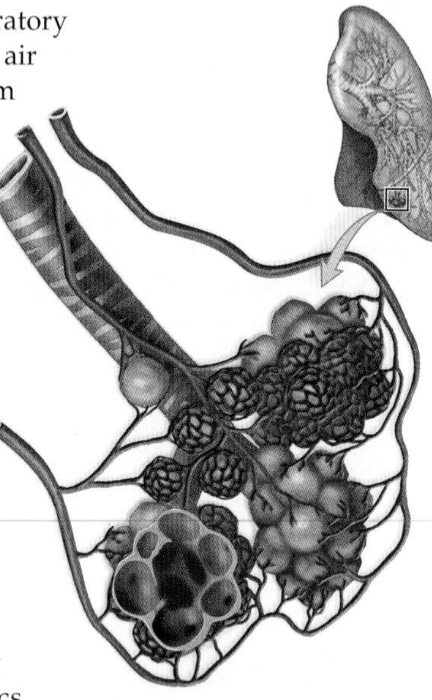

Blood supply of alveoli

15.2 Mechanism of Breathing

Ventilation includes both inspiration and expiration:

- Inspiration begins when the respiratory center in the medulla oblongata sends excitatory nerve impulses to the diaphragm and the muscles of the rib cage. As they contract, the diaphragm lowers, and the rib cage moves upward and outward; the lungs expand, creating a partial vacuum, which causes air to rush in (inspiration).
- Expiration occurs when the respiratory center stops sending impulses to the diaphragm and the muscles of the rib cage. As the diaphragm relaxes, it resumes its dome shape, and as the rib cage retracts, air is pushed out of the lungs (expiration).

15.3 Gas Exchanges in the Body

Oxygen and carbon dioxide are exchanged by diffusion during external respiration and internal respiration:

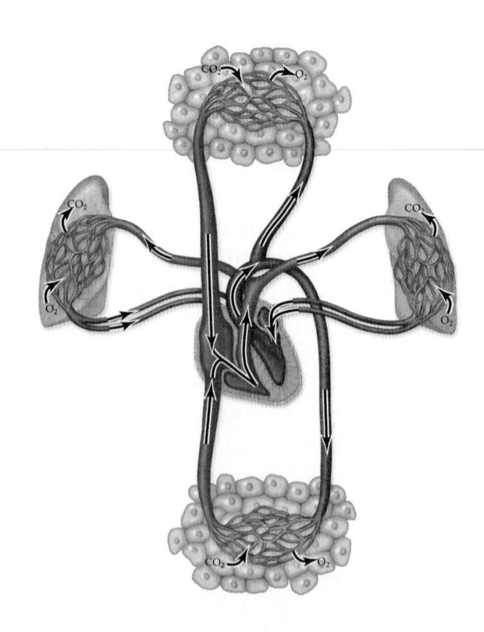

- External respiration occurs in the lungs when CO_2 leaves the blood via the alveoli and O_2 enters the blood from the alveoli. In the lungs, the P_{CO_2} is higher in pulmonary blood than in alveoli, so CO_2 diffuses from blood into the alveoli. Because carbon dioxide is present in the blood mainly as bicarbonate ion (HCO_3^-), carbonic acid first forms and is broken down to carbon dioxide and water. Because the P_{O_2} is higher in the alveoli than in blood, O_2 diffuses into the blood and is transported to the tissues in combination with hemoglobin as oxyhemoglobin (HbO_2).
- Internal respiration occurs in the tissues when O_2 leaves and CO_2 enters the blood. This occurs because the P_{O_2} is lower in the tissues compared to arterial blood, and the P_{CO_2} is higher. When carbon dioxide enters the blood, carbonic acid forms and is broken down to the bicarbonate ion (HCO_3^-) and hydrogen ions. Carbon dioxide is mainly carried to the lungs within the plasma as the bicarbonate ion. Hemoglobin combines with hydrogen ions and becomes reduced (HHb).

15.4 Disorders of the Respiratory System

Disorders of the respiratory system can be divided into those that affect the upper respiratory tract and those that affect the lower respiratory tract:

- In the upper respiratory tract, the common cold, pharyngitis, tonsillitis, and laryngitis are all conditions that most people experience at some time. In addition, these infections can spread into the sinuses and middle ears.
- The lower respiratory tract is subject to foreign bodies that can block the trachea and bronchi. In addition, infections can cause acute or chronic bronchitis, pneumonia, and pulmonary tuberculosis. Other common disorders include asthma, emphysema, cystic fibrosis, pulmonary fibrosis, and lung cancer. Smoking is a major risk factor associated with chronic bronchitis, emphysema, and lung cancer.

TESTING YOURSELF

Choose the best answer for each question.

1. Label this diagram of the human respiratory system.

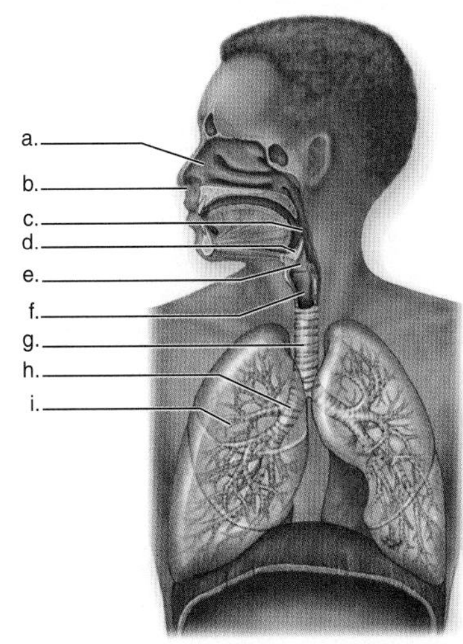

2. The pulmonary _____ are rich in oxygen.
   a. veins
   b. arteries
   c. Both a and b are correct.
   d. Neither a nor b is correct.
3. Food and air both travel through the
   a. lungs.
   b. pharynx.
   c. larynx.
   d. trachea.
4. Which of these statements is anatomically incorrect?
   a. The nose has two nasal cavities.
   b. The pharynx connects the nasal cavity and mouth to the larynx.
   c. The larynx contains the vocal cords.
   d. The trachea enters the lungs.
   e. The lungs contain many alveoli.
5. If the digestive and respiratory tracts were completely separate in humans, there would be no need for
   a. swallowing.
   b. a nose.
   c. an epiglottis.
   d. a diaphragm.
   e. All of these are correct.
6. Label the following diagram of the path of air.

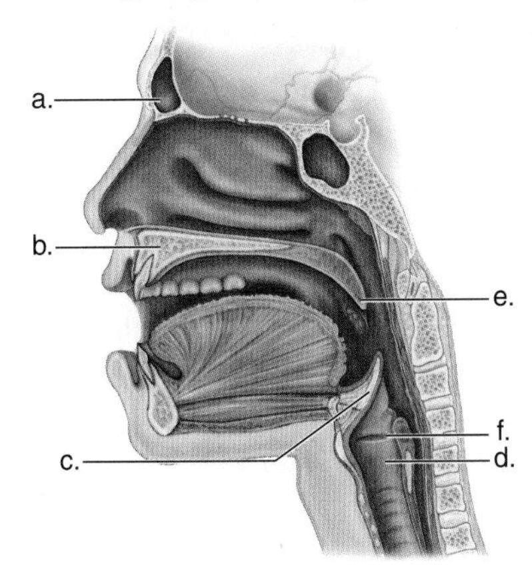

7. The amount of air moved into and out of the lungs during regular breathing is known as the
   a. vital capacity.
   b. tidal volume.
   c. residual volume.
   d. None of these are correct.
8. Label the respiratory volumes where indicated.

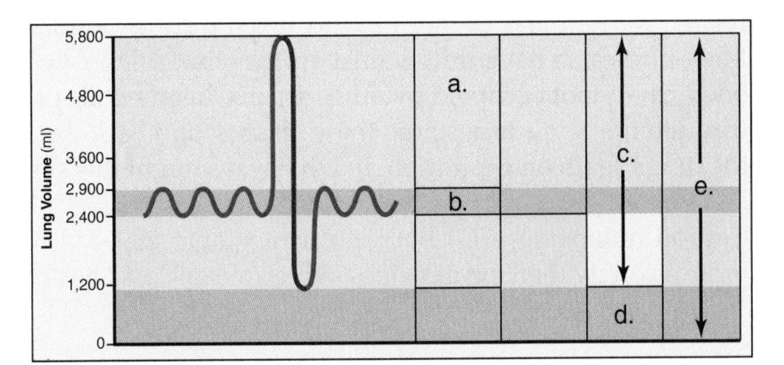

9. Which of these statements correctly goes with expiration rather than inspiration?
   a. Rib cage moves up and out.
   b. Diaphragm relaxes and moves up.
   c. Pressure in lungs decreases, and air comes rushing out.
   d. Diaphragm contracts and lowers.

10. The respiratory center is directly sensitive to
    a. carbon dioxide and hydrogen ions.
    b. oxygen and carbon dioxide.
    c. the amount of air in the alveoli.
    d. the breathing rate.
11. The phrenic nerve stimulates the ______.
    a. diaphragm
    b. lungs
    c. intestines
    d. All of these are correct.
12. Carbon dioxide is carried in the plasma
    a. in combination with hemoglobin.
    b. as the bicarbonate ion.
    c. combined with carbonic anhydrase.
    d. only as a part of tissue fluid.
    e. All of these are correct.
13. The presence of carbon dioxide in the blood
    a. affects the respiratory volumes.
    b. makes blood more acidic.
    c. makes respiratory exchanges more difficult.
    d. can increase the breathing rate.
    e. Both b and d are correct.
14. Internal respiration refers to
    a. the exchange of gases between the air and the blood in the lungs.
    b. the movement of air into the lungs.
    c. the exchange of gases between the blood and tissue fluid.
    d. cellular respiration, resulting in the production of ATP.
15. Reduced hemoglobin is carrying
    a. oxygen.
    b. CO_2.
    c. hydrogen ions.
    d. sodium.
16. Smoking is associated with an increased risk of:
    a. emphysema.
    b. several types of cancer.
    c. heart disease.
    d. All of these are correct.
17. Pulmonary fibrosis
    a. can be caused by inhaling asbestos.
    b. is most common in the elderly.
    c. leads to reduced vital capacity.
    d. All of these are correct.

UNDERSTANDING THE TERMS

acute bronchitis 292
alveolus 284
asthma 292
auditory (eustachian) tube 292
bicarbonate ion 289
bronchiole 284
bronchus 284
carbaminohemoglobin 291
carbonic anhydrase 289
chronic bronchitis 292
cystic fibrosis (CF) 293
emphysema 293
epiglottis 284
expiration 282
expiratory reserve volume 287
external respiration 289
glottis 284
infant respiratory distress syndrome 285
inspiration 282
inspiratory reserve volume 287
internal respiration 290
laryngitis 291
larynx 284
lung cancer 295
lungs 284
nasal cavity 283
nostril 283
otitis media 292
oxyhemoglobin 289
pharyngitis 291
pharynx 283
pleura 285
pneumonectomy 295
pneumonia 292
pulmonary fibrosis 295
pulmonary tuberculosis 293
reduced hemoglobin 291
residual volume 287
respiratory center 288
sinus 283
sinusitis 292
surfactant 285
tidal volume 287
tonsillectomy 291
tonsillitis 291
tonsils 291
trachea 284
tracheostomy 292
ventilation 282
vital capacity 287
vocal cord 284

THINKING CRITICALLY

1. The respiratory system of birds is quite different from that of mammals. Unlike mammalian lungs, the lungs of birds are relatively rigid and do not expand much during ventilation. Most birds also have thin-walled air sacs that fill most of the body cavity not occupied by other organs. Inspired air passes through the bird's lungs, into these air sacs, and back through the lungs again on expiration. In what ways might this system be more efficient than the mammalian lung?
2. Carbon monoxide (CO) binds to hemoglobin 230–270 times more strongly than oxygen does, which means less O_2 is being delivered to tissues. What would be the specific cause of death if too much hemoglobin became bound to CO instead of to O_2?
3. As mentioned in the chapter, sometimes it is necessary to install a permanent tracheostomy—for example, in smokers who develop laryngeal cancer. Besides interfering with the ability to speak, what sorts of health problems would you expect to see in an individual with a permanent tracheostomy, and why?

INQUIRY INTO LIFE WEBSITE

The companion website for *Inquiry into Life* provides a wealth of information organized and integrated by chapter. You will find practice tests, animations, videos, and much more that will complement your learning and understanding of general biology.

http://www.mhhe.com/maderinquiry13

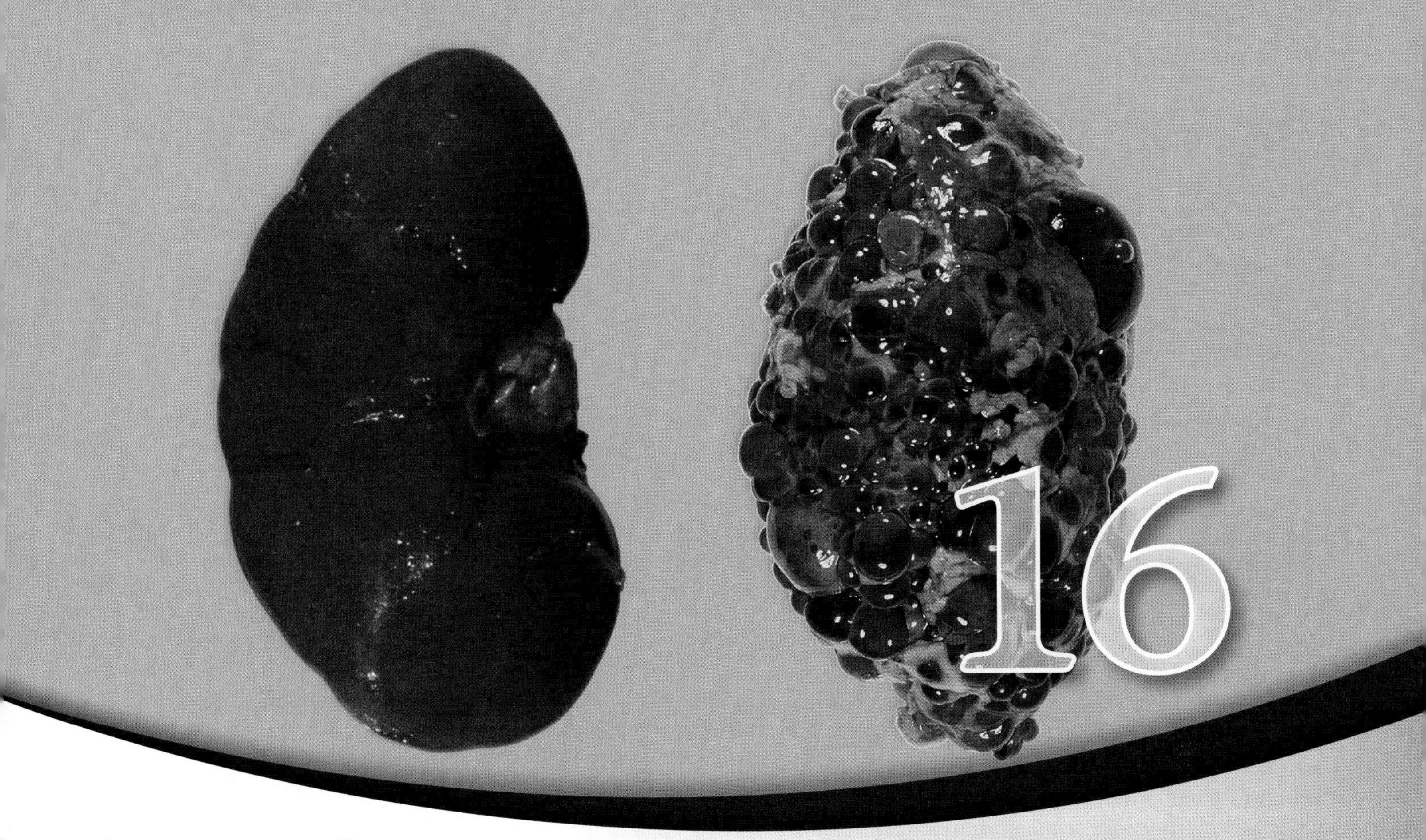

Urinary System and Excretion

Applying the Concepts

Michael and Jada were excited about the birth of their second baby. Married for three years, they already had a healthy daughter, and they were happy that Amber would have a younger sister to play with. After Aiesha was born, however, it soon became clear that something was wrong. She weighed only 5 lb, 4 oz at birth. The first time she urinated, there was an obvious tinge of blood in her urine. She also seemed to urinate much more frequently than normal. When her doctors used ultrasound to examine her abdominal organs, they found that Aiesha had signs of polycystic kidney disease (as shown by the kidney on the right, above). This meant that she might require hemodialysis, or even a kidney transplant, within a few years. Adding to their sadness, Michael and Jada now know that even though they are healthy, they are carriers of a faulty gene that they had unknowingly passed on to their new daughter.

There are two inherited forms of polycystic kidney disease (PKD): autosomal dominant and autosomal recessive. The autosomal dominant form is one of the most common inherited disorders in humans, and it is the most frequent genetic cause of kidney failure in the United States. However, the symptoms of this form of PKD usually don't show up until adulthood. Aiesha's doctors suspected that she had the autosomal recessive form, the symptoms of which can be much more severe in children, even while still in the womb. Although rare, this more severe form of PKD illustrates the critical functions of the urinary system, as described in this chapter.

Chapter Outline

16.1 Urinary System

Learning Outcomes

1. List four major functions of the urinary system.
2. Trace the path of urine from its formation to its exit from the body.
3. Compare and contrast the structure and functions of the male versus female urethra.

The kidneys are the primary organs of excretion. **Excretion** is the removal of metabolic wastes from the body. People sometimes confuse the terms excretion and defecation, but they do not refer to the same process. Defecation, the elimination of feces from the body, is a function of the digestive system. The undigested food and bacteria that make up feces have never been a part of the functioning of the body, while most substances excreted in urine were once metabolites in the body.

Functions of the Urinary System

The urinary system produces urine and conducts it to outside the body. As the kidneys produce urine, they carry out the following four functions that contribute to homeostasis:

1. *Excretion of Metabolic Wastes* The kidneys excrete metabolic wastes, notably nitrogenous wastes. Urea is the primary nitrogenous end product of metabolism in human beings, but we also excrete some ammonium, creatinine, and uric acid.

   Urea is a by-product of amino acid metabolism. The breakdown of amino acids in the liver releases ammonia, which the liver rapidly combines with carbon dioxide to produce urea. Ammonia is very toxic to cells, but urea is much less toxic. Some ammonia (NH_3) is also excreted as ammonium ion (NH_4^+).

   Creatine phosphate is a high-energy phosphate reserve molecule in muscles. The metabolic breakdown of creatine phosphate results in **creatinine.**

   The breakdown of nucleotides, such as those containing adenine and thymine, produces **uric acid.** Uric acid is rather insoluble. If too much uric acid is present in blood, crystals may form and precipitate out. Crystals of uric acid sometimes collect in the joints, producing a painful ailment called **gout.**

2. *Maintenance of Water-Salt Balance* A principal function of the kidneys is to maintain the appropriate balance of water and salt in the blood. Blood volume is intimately associated with the salt balance of the body. Salts, such as NaCl, have the ability to cause osmosis, the diffusion of water—in this case, into the blood. The more salts there are in the blood, the greater the blood volume and the greater the blood pressure. In this way, the kidneys are also involved in regulating blood pressure.

   The kidneys also help maintain the appropriate level of other ions, such as potassium (K^+), bicarbonate (HCO_3^-), and calcium (Ca^{2+}), in the blood.

3. *Maintenance of Acid-Base Balance* The kidneys regulate the acid-base balance of the blood. The kidneys monitor and help keep the blood pH at about 7.4, mainly by excreting hydrogen ions (H^+) and reabsorbing the bicarbonate ions (HCO_3^-) as needed. Human urine usually has a pH of 6 or lower because our diet often contains acidic foods.

4. *Secretion of Hormones* The kidneys assist the endocrine system in hormone secretion. The kidneys release renin, a substance that leads to the secretion of the hormone aldosterone from the adrenal cortex, the outer portion of the adrenal glands, which lie atop the kidneys. As described in section 16.3, aldosterone promotes the reabsorption of sodium ions (Na^+) by the kidneys.

   Whenever the oxygen-carrying capacity of the blood is reduced, or oxygen demand increases, the kidneys secrete the hormone **erythropoietin,** which stimulates red blood cell production.

   The kidneys also help activate vitamin D from the skin. Vitamin D is a hormone-like molecule that promotes calcium (Ca^{2+}) absorption from the digestive tract.

Organs of the Urinary System

The urinary system consists of the kidneys, ureters, urinary bladder, and urethra. Figure 16.1 shows these organs and also traces the path of urine.

Kidneys

The **kidneys** are paired organs located near the small of the back in the lumbar region on either side of the vertebral column. They lie in depressions dorsal to (behind) the peritoneum, where they receive some protection from the lower rib cage.

The kidneys are bean shaped and reddish-brown in color. The fist-sized organs are covered by a tough capsule of fibrous connective tissue called a renal capsule. The concave side of each kidney has a depression called the hilum where a **renal artery** enters and a **renal vein** and a ureter exit the kidney.

Ureters

The **ureters** conduct urine from the kidneys to the bladder. They are small, muscular tubes about 25 cm long and 5 mm in diameter. Each descends behind the peritoneum, from the hilum of a kidney, to enter the bladder at its dorsal surface.

The wall of each ureter has three layers: an inner mucosa, a smooth muscle layer, and an outer fibrous coat of connective tissue. Peristaltic contractions cause urine to enter the bladder in spurts, about one to five times per minute. Because a healthy adult human produces 1–2 liters of urine a day, an average of 40–80 ml per hour enters the bladder.

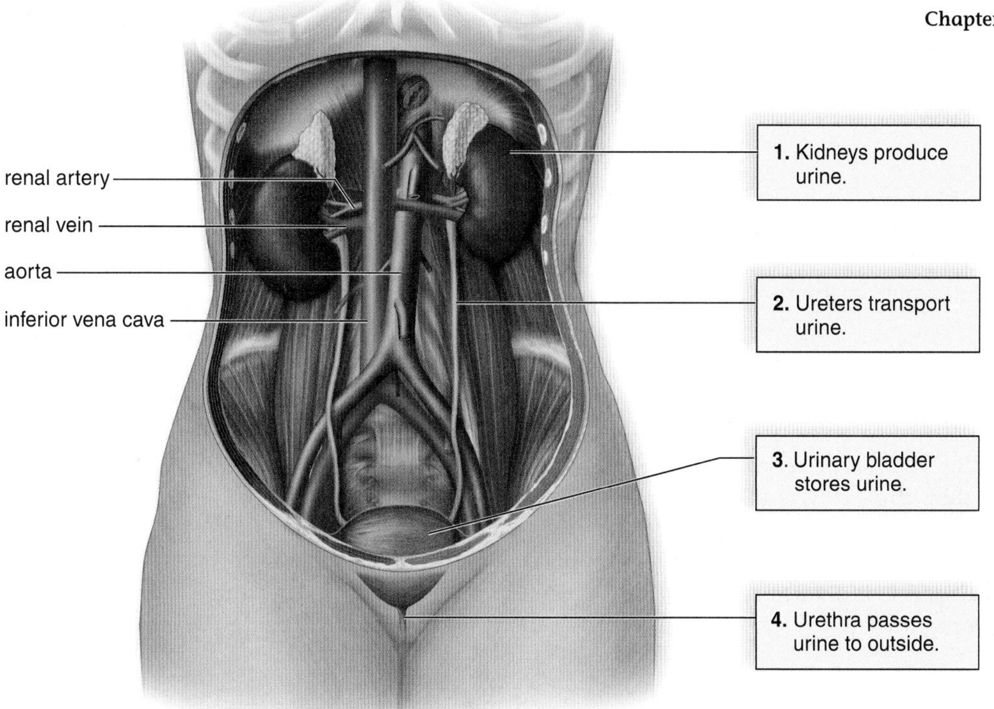

Figure 16.1 The urinary system.
Urine is found only within the kidneys, the ureters, the urinary bladder, and the urethra. The kidneys are important organs of homeostasis because they excrete metabolic wastes and adjust the water-salt balance and the acid-base balance of the blood.

Urinary Bladder

The **urinary bladder** stores urine until it is expelled from the body. The bladder is located in the pelvic cavity, behind the pubic symphysis and behind the peritoneum. The urinary bladder has three openings (orifices)—two for the ureters and one for the urethra, which drains the bladder (Fig. 16.2).

The bladder wall is expandable because it contains a middle layer of circular fiber and two layers of longitudinal muscle. The transitional epithelium of the mucosa becomes thinner, and folds in the mucosa called *rugae* disappear as the bladder enlarges.

After urine enters the bladder from a ureter, small folds of bladder mucosa act like a valve to prevent backward flow. Two sphincters lie in close proximity where the urethra exits the bladder. The internal sphincter occurs around the opening to the urethra. An external sphincter is composed of skeletal muscle that can be voluntarily controlled. Many people have some degree of **incontinence,** the involuntary loss of urine. At least one in ten people over age 65 have bladder control problems due to an age-related loss of muscle tone in the urinary bladder. Other causes of incontinence include enlarged prostate in men, pregnancy, and nervous system diseases that affect the control of urination.

Urethra

The **urethra** is a small tube that extends from the urinary bladder to an external opening. Therefore, its function is to remove urine from the body. In females, the urethra is only about 4 cm long, while in males, the urethra averages 20 cm when the penis is flaccid (nonerect). In females, the reproductive and urinary systems are not connected. In males, the urethra carries urine during urination and semen during ejaculation. As the urethra leaves the male urinary bladder, it is encircled by the prostate gland. In men over age 40, the prostate gland is subject to enlargement, as well as to prostate cancer. Either of these conditions can restrict urination, and may require treatment (see section 21.5).

Urination

When the urinary bladder fills to about 250 ml with urine, stretch receptors send sensory nerve impulses to the spinal cord. Subsequently, motor nerve impulses from the spinal cord cause the urinary bladder to contract and the sphincters to relax so that urination, also called **micturition,** is possible (Fig. 16.2). In older children and adults, the brain controls this reflex, delaying urination until a suitable time.

CHECK YOUR PROGRESS

1. Define excretion.
2. Of the four major functions of the urinary system, which is (are) exclusively performed by the kidneys, and which are shared with other body systems?

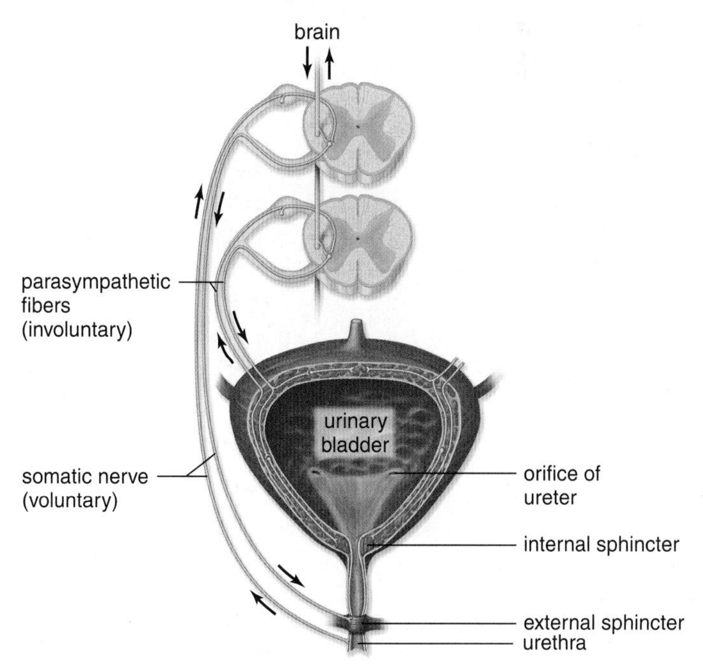

Figure 16.2 Urination.
As the bladder fills with urine, sensory impulses go to the spinal cord and then to the brain. When urination occurs, motor nerve impulses cause the bladder to contract and an internal sphincter to relax. Motor nerve impulses also cause an external sphincter to relax.

16.2 Anatomy of the Kidney and Excretion

Learning Outcomes

1. Illustrate and label the structure of a nephron.
2. Distinguish between glomerular filtration, tubular reabsorption, and tubular secretion.

A lengthwise section of a kidney shows that many branches of the renal artery and renal vein reach inside a kidney (Fig. 16.3*a*). A kidney has three regions (Fig. 16.3*c*). The **renal cortex** is an outer, granulated layer that dips down in between a radially striated inner layer called the renal medulla. The **renal medulla** contains cone-shaped tissue masses called renal pyramids. The **renal pelvis** is a central space, or cavity, that is continuous with the ureter.

Anatomy of a Nephron

Under higher magnification, the kidney is composed of over one million **nephrons,** sometimes called renal or kidney tubules (Fig. 16.3*b*). Each nephron has its own blood supply, including two capillary regions (Fig. 16.4). From the renal artery, an afferent arteriole leads to the **glomerulus,** a knot of capillaries inside the glomerular capsule. Blood leaving the glomerulus enters the efferent arteriole (the term afferent means going towards, while efferent means going away). The efferent arteriole takes blood to the **peritubular capillary network,** which surrounds the rest of the nephron. From there, the blood goes into a venule that joins the renal vein.

Parts of a Nephron

Each nephron is made up of several parts. First, the closed end of the nephron is pushed in on itself to form a cuplike structure called the **glomerular capsule** (Bowman capsule).

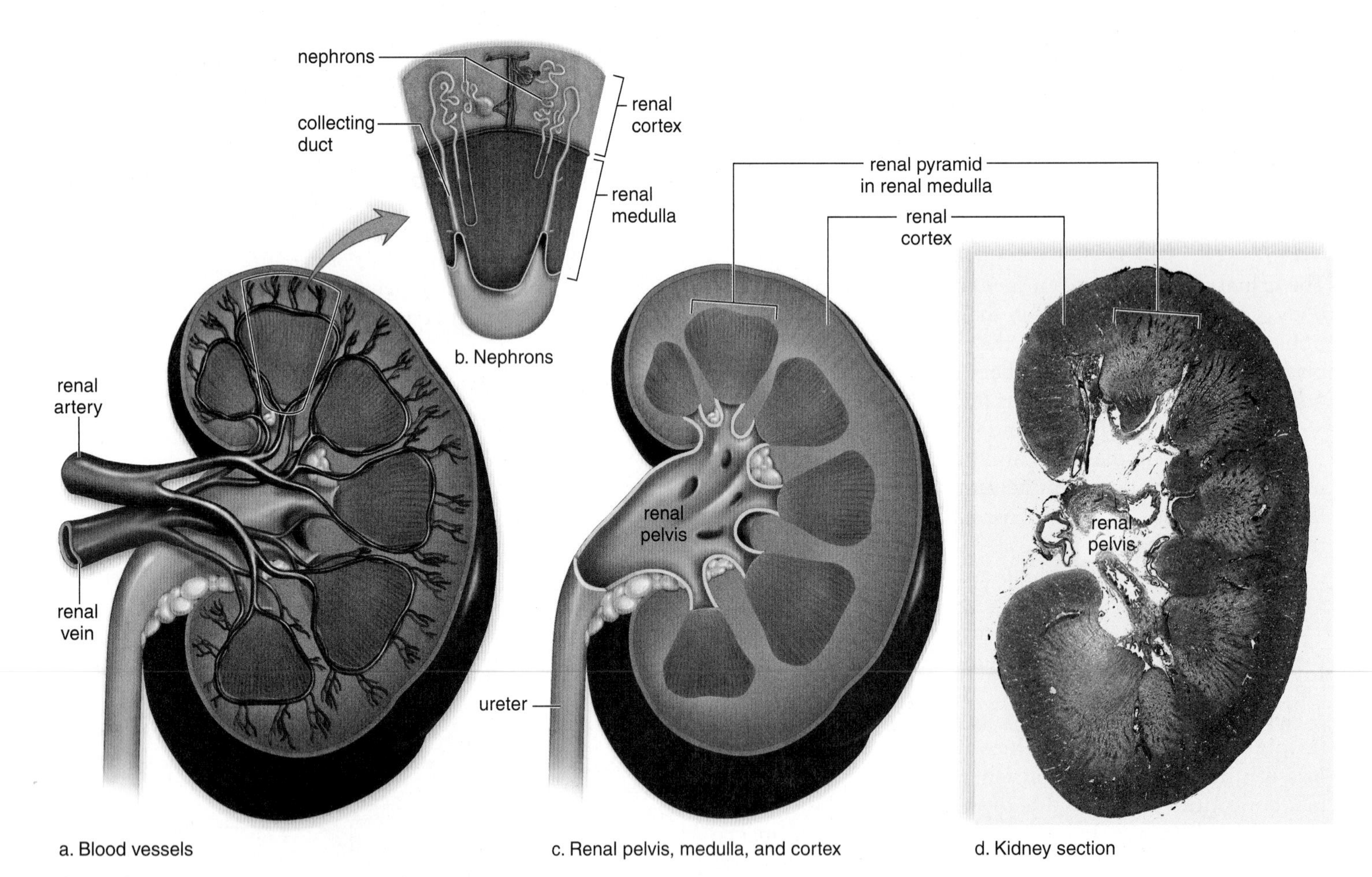

Figure 16.3 Anatomy of the kidney.
a. A lengthwise section of the kidney showing the blood supply. Note that the renal artery divides into smaller arteries, and these divide into arterioles. Venules join to form small veins, which form the renal vein. **b.** An enlargement showing the placement of nephrons. **c.** The same section as **(a)** without the blood supply. Now it is easier to distinguish the renal cortex, medulla, and pelvis, which connects with the ureter. **d.** A lengthwise section of an actual kidney.

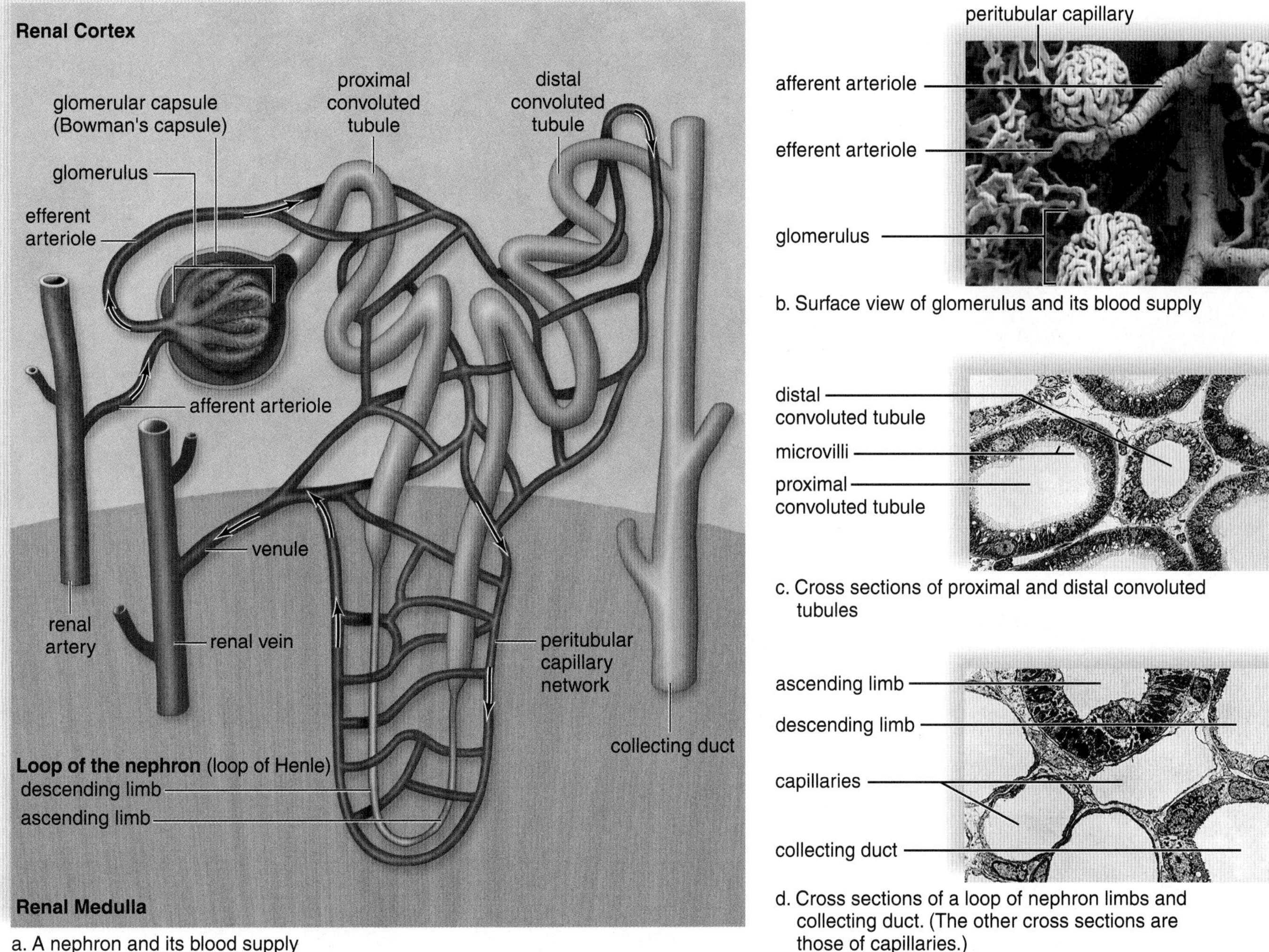

a. A nephron and its blood supply

b. Surface view of glomerulus and its blood supply

c. Cross sections of proximal and distal convoluted tubules

d. Cross sections of a loop of nephron limbs and collecting duct. (The other cross sections are those of capillaries.)

Figure 16.4 Nephron anatomy.

a. You can trace the path of blood through a nephron by following the black arrows. **b, c, d.** A nephron is made up of a glomerular capsule, the proximal convoluted tubule, the loop of the nephron, the distal convoluted tubule, and the collecting duct.

16.4b: © R. G. Kessel and R. H. Kardon, Tissues and Organs: A Text-Atlas of Scanning Electron Microscopy, *1979.*

The inner layer of the glomerular capsule is composed of *podocytes* (see Fig. 16.6) that have long cytoplasmic extensions. The podocytes cling to the capillary walls of the glomerulus and leave pores that allow easy passage of small molecules from the glomerulus to the inside of the glomerular capsule.

The glomerular capsule connects to a **proximal convoluted tubule (PCT).** The cuboidal epithelial cells lining this part of the nephron have numerous tightly packed microvilli that form a brush border, increasing the surface area for reabsorption. The tube then narrows and makes a U-turn called the **loop of the nephron** (loop of Henle), which is lined with simple squamous epithelium.

Following the loop, the tube becomes the **distal convoluted tubule (DCT),** which is composed of cuboidal epithelial cells that have numerous mitochondria but lack microvilli. This means the DCT is not specialized for reabsorption and is more likely to help move molecules from the blood into the tubule, a process called tubular secretion. The distal convoluted tubules of several nephrons enter one collecting duct. Many **collecting ducts** carry urine to the renal pelvis.

As shown in Figures 16.3*b* and 16.4*a*, the glomerular capsule and the convoluted tubules always lie within the renal cortex. The loop of the nephron dips down into the renal medulla; a few nephrons have a very long loop of the nephron, which penetrates deep into the renal medulla. Collecting ducts are also located in the renal medulla, and they help give the renal pyramids their lined appearance (see Fig. 16.3*d*).

Check Your Progress

1. Which parts of a nephron are found in the renal cortex? The renal medulla?
2. What difference in microscopic structure suggests that the PCT, but not the DCT, is specialized for reabsorption?

Visual Focus

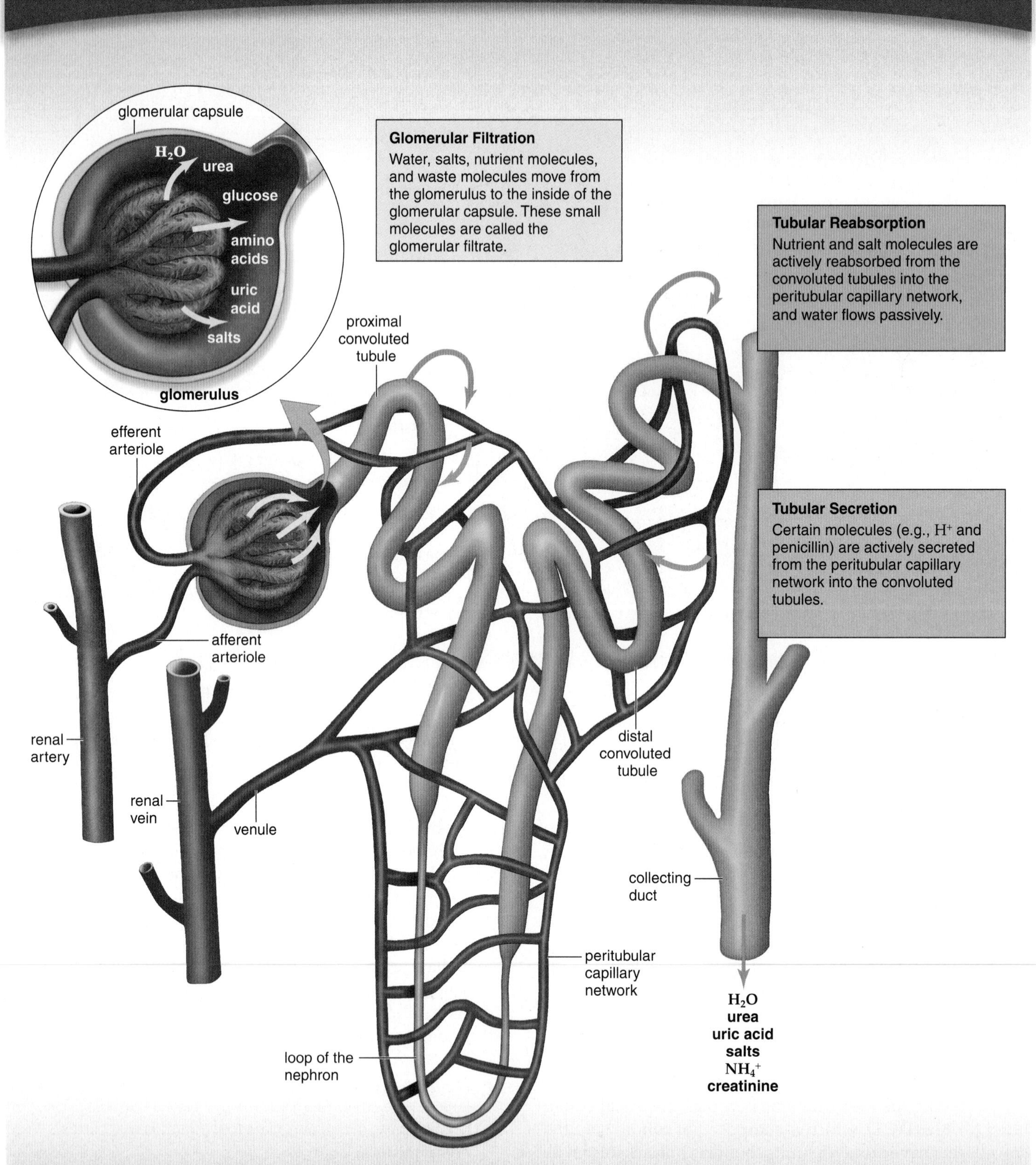

Figure 16.5 Processes in urine formation.
The three main processes in urine formation are described in boxes and color-coded to arrows that show the movement of molecules into or out of the nephron at specific locations. In the end, urine is composed of the substances within the collecting duct (see blue arrow).

Urine Formation

Figure 16.5 gives an overview of urine formation, which is divided into the following processes: glomerular filtration, tubular reabsorption, and tubular secretion.

Glomerular Filtration

Glomerular filtration occurs when whole blood enters the afferent arteriole and the glomerulus. Due to glomerular blood pressure, water and small molecules move from the glomerulus to the inside of the glomerular capsule. This is a filtration process because large molecules and formed elements are unable to pass through the capillary wall. In effect, then, blood in the glomerulus has two portions—the filterable components and the nonfilterable components:

Filterable Blood Components	Nonfilterable Blood Components
Water	Formed elements
Nitrogenous wastes	(blood cells and platelets)
Nutrients	Plasma proteins
Salts (ions)	

The **glomerular filtrate** contains small dissolved molecules in approximately the same concentration as plasma. Small molecules that escape being filtered and the nonfilterable components leave the glomerulus by way of the efferent arteriole.

As indicated in Table 16.1, nephrons in the kidneys filter 180 liters of water per day, along with a considerable amount of small molecules (such as glucose) and ions (such as sodium). If the composition of urine were the same as that of the glomerular filtrate, the body would continually lose water, salts, and nutrients. Therefore, we can conclude that the composition of the filtrate must be altered as this fluid passes through the remainder of the tubule.

Tubular Reabsorption

Tubular reabsorption occurs as molecules and ions are both passively and actively reabsorbed from the nephron into the blood of the peritubular capillary network. The osmolarity of the blood is maintained by the presence of both plasma proteins and salt. When sodium ions (Na^+) are actively reabsorbed, chloride ions (Cl^-) follow passively. The reabsorption of salt (NaCl) increases the osmolarity of the blood compared to the filtrate, and therefore water moves passively from the tubule into the blood. About 65% of Na^+ is reabsorbed at the proximal convoluted tubule.

Nutrients, such as glucose and amino acids, also return to the blood almost exclusively at the proximal convoluted tubule. This is a selective process because only molecules recognized by carrier proteins are actively reabsorbed. Glucose is an example of a molecule that ordinarily is completely reabsorbed because there is a plentiful supply of carrier proteins for it. However, every substance has a maximum rate of transport, and after all its carriers are in use, any excess in the filtrate will appear in the urine. For example, as reabsorbed levels of glucose approach 1.8–2 mg/ml of plasma, the rest appears in the urine.

In diabetes mellitus, excess glucose appears in the blood, because the liver and muscles have failed to store glucose as glycogen. The kidneys cannot reabsorb all the glucose in the filtrate, and it appears in the urine. The presence of glucose in the filtrate increases its osmolarity, and therefore, less water is reabsorbed into the peritubular capillary network. The frequent urination and increased thirst experienced by untreated diabetics are due to the fact that less water is being reabsorbed.

We have seen that the filtrate that enters the proximal convoluted tubule is divided into two portions—components that are reabsorbed from the tubule into blood and components that are not reabsorbed and continue to pass through the nephron to be further processed into urine:

Reabsorbed Filtrate Components	Nonreabsorbed Filtrate Components
Most water	Some water
Nutrients	Much nitrogenous waste
Required salts (ions)	Excess salts (ions)

The substances that are not reabsorbed become the tubular fluid, which enters the loop of the nephron.

Tubular Secretion

Tubular secretion is a second way by which substances are removed from blood in the peritubular network and added to the tubular fluid. Hydrogen ions, potassium ions, creatinine, and drugs, such as penicillin, are some of the substances that are removed by active transport from the blood. In the end, urine contains (1) substances that have undergone glomerular filtration but have not been reabsorbed and (2) substances that have undergone tubular secretion.

TABLE 16.1 Reabsorption from Nephrons

Substance	Amount Filtered (per day)	Amount Excreted (per day)	Reabsorption (%)
Water, L	180	1.8	99.0
Sodium, g	630	3.2	99.5
Glucose, g	180	0.0	100.0
Urea, g	54	30.0	44.0

L = liters, g = grams

Check Your Progress

1. Could certain drugs be filtered out of the blood into the glomerular filtrate? Why or why not?
2. What do the processes of diffusion, passive transport, and active transport have in common? How do they differ?
3. At what specific part of the nephron, and by what process, is glucose normally returned from the glomerular filtrate to the blood?

16.3 Regulatory Functions of the Kidneys

Learning Outcomes

1. Discuss where the reabsorption of salt occurs and how it is hormonally controlled.
2. Explain how a solute gradient is established in the renal medulla.
3. Describe two ways that the kidneys regulate blood pH.

The kidneys maintain the water-salt balance of the blood within normal limits. In this way, they also maintain the blood volume and blood pressure. Most of the water and salt (NaCl) present in the filtrate is reabsorbed across the wall of the proximal convoluted tubule.

Process of Water Reabsorption

The excretion of a hypertonic urine (one that is more concentrated than blood) is dependent upon the reabsorption of water from the loop of the nephron and the collecting duct. We can think of reabsorption of water as requiring (1) reabsorption of salt and (2) establishment of a solute gradient dependent on salt and urea before (3) water is reabsorbed.

Reabsorption of Salt

The kidneys regulate the salt balance in blood by controlling the excretion and the reabsorption of various ions. Sodium (Na^+) is an important ion in plasma that must be regulated, but the kidneys also excrete or reabsorb other ions, such as potassium ions (K^+), bicarbonate ions (HCO_3^-), and magnesium ions (Mg^{2+}), as needed.

Usually, more than 99% of sodium (Na^+) filtered at the glomerulus is returned to the blood. Most of this sodium (67%) is reabsorbed at the proximal tubule, and a sizable amount (25%) is reabsorbed by the ascending limb of the loop of the nephron. The rest is reabsorbed from the distal convoluted tubule and collecting duct.

Hormones regulate the reabsorption of sodium at the distal convoluted tubule. **Aldosterone,** a hormone secreted by the adrenal cortex, promotes the excretion of potassium ions (K^+) and the reabsorption of sodium ions (Na^+). The release of aldosterone is set in motion by the kidneys themselves. The **juxtaglomerular apparatus** is a region of contact between the afferent arteriole and the distal convoluted tubule (Fig. 16.6). When blood volume, and therefore blood pressure, is not sufficient to promote glomerular filtration, the juxtaglomerular apparatus secretes renin. **Renin** is an enzyme that changes angiotensinogen (a large plasma protein produced by the liver) into angiotensin I. Later, angiotensin I is converted to angiotensin II, a powerful vasoconstrictor that also stimulates the adrenal cortex to release aldosterone. The reabsorption of sodium ions is followed by the reabsorption of water. Therefore, blood volume and blood pressure increase.

Atrial natriuretic hormone (ANH) is a hormone secreted by the atria of the heart when cardiac cells are stretched due to increased blood volume. ANH inhibits the secretion of renin by the juxtaglomerular apparatus and the secretion of aldosterone by the adrenal cortex. Its effect, therefore, is to promote the excretion of Na^+, called natriuresis.

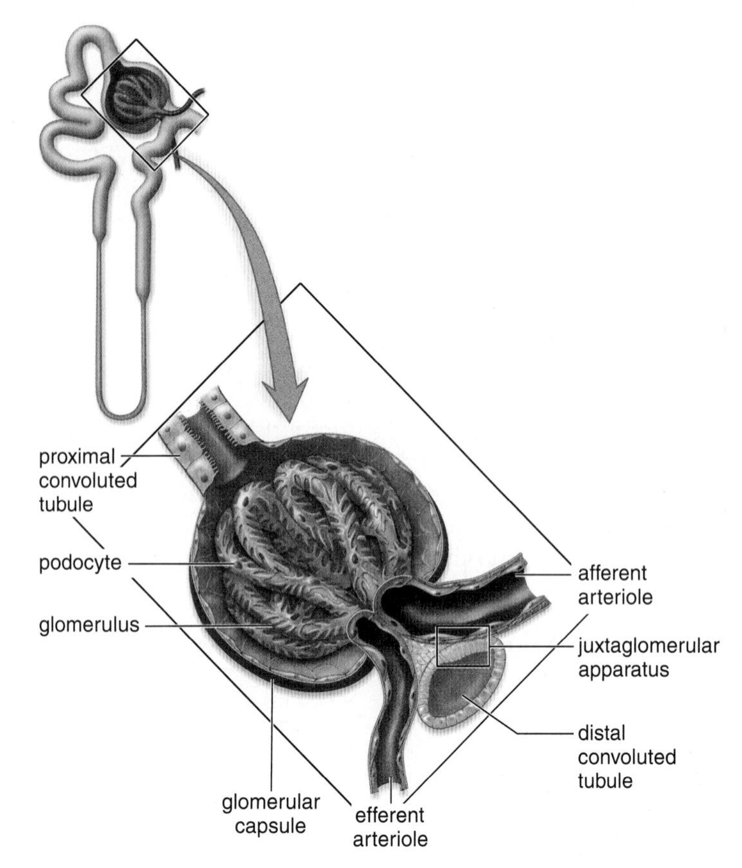

Figure 16.6 Juxtaglomerular apparatus.
The afferent arteriole and the distal convoluted tubule usually lie next to each other. The juxtaglomerular apparatus occurs where they touch. The juxtaglomerular apparatus secretes renin, a substance that leads to the release of aldosterone by the adrenal cortex. Reabsorption of sodium ions and water then occurs. Thereafter, blood volume and blood pressure increase.

Establishment of a Solute Gradient

The long loop of a nephron, which typically penetrates deep into the renal medulla, is made up of a descending limb and an ascending limb. Salt (NaCl) passively diffuses out of the lower portion of the ascending limb, but the upper, thick portion of the limb actively extrudes salt out into the tissue of the outer renal medulla (Fig. 16.7). Less and less salt is available for transport as fluid moves up the thick portion of the ascending limb. Because of these circumstances, there is an osmotic gradient within the tissues of the renal medulla: the concentration of salt is greater in the direction of the inner medulla. (Note that water cannot leave the ascending limb because the limb is impermeable to water.)

The large arrow at the far left in Figure 16.7 indicates that the innermost portion of the inner medulla has the highest concentration of solutes. This cannot be due to salt because active transport of salt does not start until fluid reaches the thick portion of the ascending limb. Urea is believed to leak from the lower portion of the collecting duct, and it is this molecule that contributes to the high solute concentration of the inner medulla.

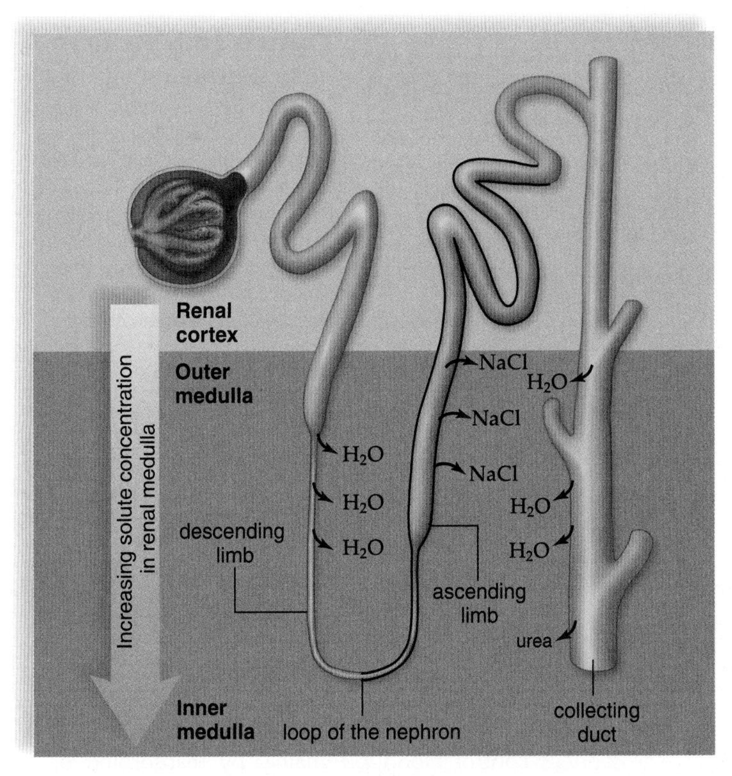

Figure 16.7 Reabsorption of water at the loop of the nephron and the collecting duct.
Salt (NaCl) diffuses and is actively transported out of the ascending limb of the loop of the nephron into the renal medulla. Also, urea is believed to leak from the collecting duct and to enter the tissues of the renal medulla. This creates a hypertonic environment, which draws water out of the descending limb and the collecting duct. This water is returned to the cardiovascular system. (The thick black outline of the ascending limb means that it is impermeable to water.)

Reabsorption of Water

Because of the osmotic gradient within the renal medulla, water leaves the descending limb along its entire length. This is a countercurrent mechanism: as water diffuses out of the descending limb, the remaining fluid within the limb encounters an even greater osmotic concentration of solute. Therefore, water continues to leave the descending limb from the top to the bottom.

Fluid enters the collecting duct from the distal convoluted tubule. This fluid is hypotonic to the renal cortex, so to this point, the active transport of salt out of the ascending limbs has decreased the osmolarity of the filtrate (see Fig. 16.7). This allows the production of urine that is hypotonic to the general body fluids—for example, when excess water needs to be excreted. When the urine needs to be hypertonic to body fluids (when the body is dehydrated, for example), the posterior pituitary gland releases **antidiuretic hormone (ADH).** In order to understand the action of this hormone, consider its name. *Diuresis* means an increased amount of urine, and *antidiuresis* means a decreased amount of urine. In the absence of ADH, the collecting duct is impermeable to water, and a dilute urine is produced. When ADH is present, the collecting duct becomes permeable to water. Because the filtrate within the collecting duct encounters the same osmotic gradient mentioned earlier, water can diffuse out of the collecting duct into the renal medulla. Usually, more ADH is produced at night, presumably so that we don't have to interrupt our sleep as often to urinate. This may also help to explain why the first urine of the day is more concentrated.

Diuretics

Diuretics are chemicals that increase the flow of urine. Drinking alcohol causes diuresis because it inhibits the secretion of ADH. The dehydration that follows is believed to contribute to the symptoms of a hangover. Caffeine is a diuretic because it increases the glomerular filtration rate and decreases the tubular reabsorption of Na^+. Diuretic drugs developed to counteract high blood pressure inhibit active transport of Na^+ at the loop of the nephron or at the distal convoluted tubule. A decrease in water reabsorption and a decrease in blood volume follow.

Besides their medical uses, diuretics can also be abused. Bodybuilders and other athletes may take them for a more "cut" look, or to lose weight quickly, in the form of water. Also, because it may be more difficult to detect drugs in dilute urine, drug users who are attempting to pass a urine drug test may use Lasix (furosemide) or other diuretics. While this may be effective, it is also risky. Electrolyte imbalances resulting from inappropriate use of Lasix or other diuretic drugs can result in dehydration, irregular heartbeat, and even death.

Check Your Progress

1. What three steps or processes are required for the excretion of a hypertonic urine?
2. Review the relationship between aldosterone, the juxtaglomerular apparatus, and renin.
3. What are some medical uses of diuretics? Why are they sometimes abused?

Acid-Base Balance

The normal pH for body fluids is about 7.4. This is the pH at which our proteins, such as cellular enzymes, function best. If the blood pH rises above 7.4, a person is said to have **alkalosis,** and if the blood pH falls below 7.4, a person is said to have **acidosis.** Alkalosis and acidosis are abnormal conditions that may need medical attention.

The foods we eat add basic or acidic substances to the blood, and so does metabolism. For example, cellular respiration adds carbon dioxide that combines with water to form carbonic acid, and fermentation adds lactic acid. The pH of body fluids stays at just about 7.4 via several mechanisms, primarily acid-base buffer systems, the respiratory center, and the kidneys.

Acid-Base Buffer Systems

The pH of the blood stays near 7.4 because the blood is buffered. A **buffer** is a chemical or a combination of chemicals that can take up excess hydrogen ions (H^+) or excess hydroxide ions (OH^-). One of the most important buffers in the blood is a combination of carbonic acid (H_2CO_3) and bicarbonate ions (HCO_3^-). When hydrogen ions (H^+) are added to blood, the following reaction occurs:

$$H^+ + HCO_3^- \longrightarrow H_2CO_3$$

When hydroxide ions (OH^-) are added to blood, this reaction occurs:

$$OH^- + H_2CO_3 \longrightarrow HCO_3^- + H_2O$$

These reactions temporarily prevent any significant change in blood pH. A blood buffer, however, can be overwhelmed unless some more permanent adjustment is made. The next adjustment to keep the pH of the blood constant occurs at pulmonary capillaries.

Respiratory Center

As discussed in Chapter 15, the respiratory center in the medulla oblongata increases the breathing rate if the hydrogen ion concentration of the blood rises. Increasing the breathing rate rids the body of hydrogen ions because the following reaction takes place in pulmonary capillaries:

$$H^+ + HCO_3^- \rightleftharpoons H_2CO_3 \rightleftharpoons H_2O + CO_2$$

In other words, when carbon dioxide is exhaled, this reaction shifts to the right, and the amount of hydrogen ions is reduced.

It is important to have the correct proportion of carbonic acid and bicarbonate ions in the blood. Breathing readjusts this proportion so that this particular acid-base buffer system can continue to absorb both H^+ and OH^- as needed.

The Kidneys

As powerful as the acid-base buffer and the respiratory center mechanisms are, only the kidneys can rid the body of a wide range of acidic and basic substances and otherwise adjust the pH. The kidneys are slower acting than the other two mechanisms, but they have a more powerful effect on pH. For the sake of simplicity, we can think of the kidneys as reabsorbing bicarbonate ions and excreting hydrogen ions as needed to maintain the normal pH of the blood (Fig. 16.8). If the blood is acidic, hydrogen ions are excreted, and bicarbonate ions are reabsorbed. If the blood is basic, hydrogen ions are not excreted, and bicarbonate ions are not reabsorbed. Because the urine is usually acidic, it follows that an excess of hydrogen ions is usually excreted. Ammonia (NH_3) provides another means of buffering and removing the hydrogen ions in urine: ($NH_3 + H^+ \longrightarrow NH_4^+$). Ammonia (whose presence is quite obvious in the diaper pail or kitty litter box) is produced in tubule cells by the deamination of amino acids. Phosphate provides another means of buffering hydrogen ions in urine.

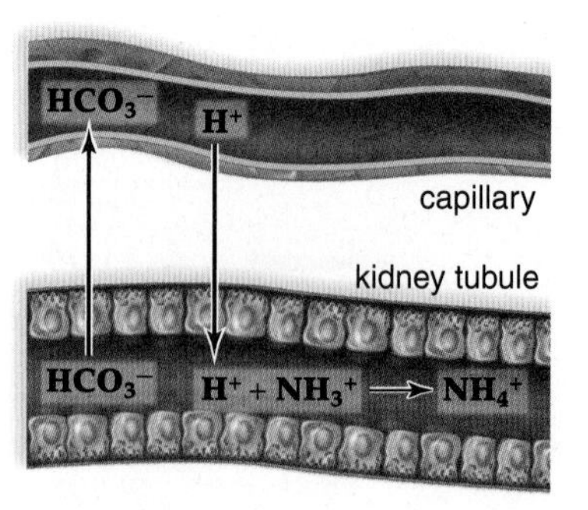

Figure 16.8 Acid-base balance.
In the kidneys, bicarbonate ions (HCO_3^-) are reabsorbed and hydrogen ions (H^+) are excreted as needed to maintain the pH of the blood. Excess hydrogen ions are buffered, for example, by ammonia (NH_3), which is produced in tubule cells by the deamination of amino acids.

The importance of the kidneys' ultimate control over the pH of the blood cannot be overemphasized. As mentioned, the enzymes of cells cannot continue to function if the internal environment does not have near-normal pH.

Check Your Progress

1. Hold your breath for as long as you can. What is happening to your blood pH? What chemical reaction is occurring? Once you start breathing again, what happens to your breathing rate and why?
2. The kidneys control blood pH mainly by reabsorbing or excreting what two ions?

16.4 Disorders of the Urinary System

Learning Outcomes

1. Indicate how damage to the glomeruli/nephrons can lead to uremia and edema.
2. List one advantage and one disadvantage of hemodialysis, continuous ambulatory peritoneal dialysis, and transplantation for kidney failure patients.
3. Describe three common disorders of the bladder and urethra.

Various disease conditions can affect the kidneys, ureters, urinary bladder, and urethra. The kidneys are essential for life, so if they are damaged sufficiently by disease, their function must somehow be restored if the patient is to survive.

Disorders of the Kidneys

Many major illnesses that affect other parts of the body, especially diabetes, hypertension, and certain autoimmune diseases, can also cause serious kidney disease. Most of these conditions tend to damage the glomeruli, resulting in a decreased glomerular filtration rate, and eventually kidney failure.

Infection of the kidneys is called **pyelonephritis.** Kidney infections usually result from infections of the urinary bladder that spread via the ureter(s) to the kidney(s). Kidney infections are usually curable with antibiotics if diagnosed in time. However, an infection called hemolytic uremic syn-

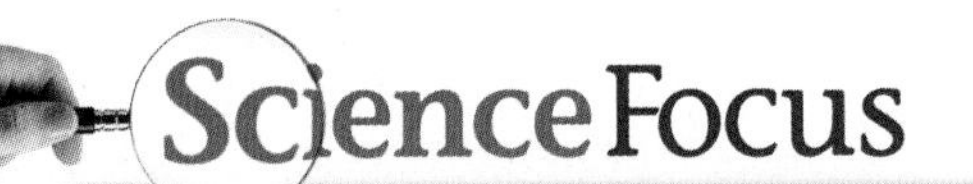

Matching Organs for Transplantation

As we have seen in this chapter, the kidneys are essential for life. Because humans normally have two kidneys, but can function quite normally with only one, the kidney was an obvious choice as the first complex organ to be successfully transplanted (Fig. 16A). This occurred in 1954 at Peter Bent Brigham Hospital in Boston, where one of 23-year-old Ronald Herrick's kidneys was transplanted into his twin, Richard, who was dying of kidney disease. The operation was successful enough to allow Richard to live another eight years, while his brother lived with one kidney for over 50 years.

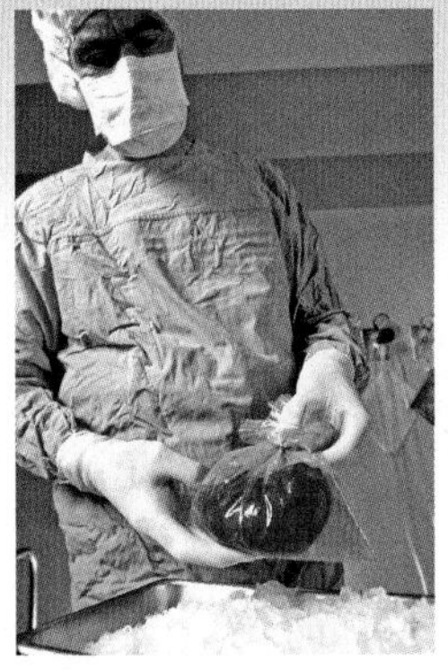

Figure 16A Preparing for transplantation. A donor kidney, kept moist and sterile inside a plastic bag, arrives in an operating room.

It was no accident, however, that the first successful transplant was performed between twins. Most cells and tissues of the body have molecules on their surface that can vary between individuals; that is, they are polymorphic. The major blood groups are a familiar example. Some individuals have only type A glycoproteins on their red blood cells, some have only type B, some have both (blood type AB), and some have neither (blood type O). Because an individual naturally has antibodies against the A or B glycoproteins their own cells lack, it is essential to make sure that donor blood is compatible before giving someone a blood transfusion. Moreover, because other cells besides red blood cells also have these glycoproteins, the first requirement when attempting to match a potential organ donor with a recipient is to make sure the blood types match.

Another step in matching organs for transplantation involves determining whether the donor and the recipient share the same proteins, called human leukocyte antigens (HLA) or major histocompatibility complex (MHC) proteins. As the latter name implies, these are the major molecules (antigens) that are recognized by the recipient's immune system when transplanted tissue is rejected. There are two types of MHC molecules: class I and class II. Because they are found on almost all cells of the body, the class I molecules probably play a greater role in transplant rejection reactions. Each individual inherits from each parent three genes coding for these MHC class I molecules, so most people have six different types of MHC-I molecules on their cells. Various types of tests can be done to determine what types of MHC-I molecules the donor cells and recipient cells have, but the donor organ is most compatible when all six match the recipient. This occurs 25% of the time between siblings who have the same mother and father, and also may occur by chance between unrelated individuals.

A final test that is usually performed prior to kidney transplantation is a cross-matching procedure in which the potential donor's cells are mixed with the blood (serum) of the recipient. This is to detect whether the recipient has any antibodies that could damage the cells of the donor kidney. Such antibodies could be present in patients who have been exposed to mismatched MHC molecules through a previous transplantation, blood transfusion, or even pregnancy. Even if the donor and recipient seem closely matched, if the recipient has preformed antibodies that could react with the donor organ, the organ would probably go to another recipient.

Because a perfect match (such as between two identical twins) is rare, transplantations would not have become as commonplace as they are today without the development of immunosuppressive drugs that help prevent transplant rejection. Unfortunately, at the present time all drugs used for this purpose, such as azathioprine and cyclosporine A, cause a general suppression of the immune system, leaving transplant patients more susceptible to infections. Because transplant patients usually need to take these drugs for the rest of their lives, there is a definite need for new drugs that more specifically target the cells responsible for the transplant rejection reaction, but leave the rest of the immune system intact.

Discussion Questions

1. As explained in Chapter 13, why do you suppose all humans with blood type A, for example, have antibodies against the B glycoprotein? Where have we all been exposed?
2. Even if two individuals are tested and found to have a perfect match of all of their MHC-I and MHC-II molecules, a kidney transplanted between them still has a higher chance of failing compared to a transplant between identical twins. Can you explain why?
3. Besides the matching criteria described here, what other medical issues should be taken into consideration before a decision is made to transplant an organ into a recipient?

drome, caused by various bacteria but most famously by a strain of *E. coli* known as O157:H7, can cause serious kidney damage in children even when antibiotics are administered.

Kidney stones are hard granules that can form in the renal pelvis. They may be composed of various materials, such as calcium, phosphate, uric acid, and protein. An estimated 5–15% of people will develop kidney stones at some time in their lives. Factors that contribute to the formation of these stones include too much animal protein in the diet, an imbalance in urinary pH, and/or urinary tract infection. Kidney stones may pass unnoticed in the urine flow. However, when a large kidney stone passes, strong contractions within a ureter can be excruciatingly painful. If a kidney stone grows large enough to block the renal pelvis or ureter, a reverse pressure builds up that may destroy the nephrons. To prevent this, it is sometimes necessary to break up a stone using a technique called lithotripsy, in which powerful ultrasound waves shatter the stone into smaller pieces that can be passed.

One of the first signs of kidney damage is the presence of albumin, white blood cells, and/or red blood cells in the urine. Once more than two-thirds of the nephrons have been destroyed by a disease process, urea and other waste products accumulate in the blood, a condition known as *uremia.* Although nitrogenous wastes can cause serious damage, the retention of water and salts by diseased kidneys is of even greater concern. The latter causes **edema,** the accumulation of fluid in body tissues. Imbalances in the ionic composition of body fluids can lead to loss of consciousness and heart failure.

Treatment Options for Kidney Failure

Patients whose kidneys are failing can undergo **hemodialysis,** continuous ambulatory peritoneal dialysis, or kidney transplantation. *Dialysis* is the diffusion of dissolved molecules through a semipermeable natural or synthetic membrane having pore sizes that allow only small molecules to pass through. In hemodialysis, the patient's blood is passed through an artificial kidney machine (Fig. 16.9). This machine contains a membranous tube, which is in contact with a dialysis solution, or **dialysate.** Substances more concentrated in the blood diffuse into the dialysate, and substances more common in the dialysate diffuse into the blood. The dialysate is continuously replaced to maintain favorable concentration gradients. In this way, the artificial kidney can be utilized either to extract substances from the blood, including waste products or toxic chemicals and drugs, or to add substances to the blood—for example, bicarbonate ions (HCO_3^-) if the blood is acidic. In the course of a three- to six-hour hemodialysis, 50–250 g of urea can be removed from a patient, which greatly exceeds the amount excreted by normal kidneys. Therefore, a patient needs to undergo treatment only about two or three times a week. Although about 98% of patients opt to be treated at dialysis centers, a growing number are purchasing portable units that can be used in their own homes.

Hemodialysis centers may not be available in rural areas, where continuous ambulatory peritoneal dialysis, or CAPD, is more commonly used. In CAPD, the peritoneum serves as the dialysis membrane. A fresh amount of dialysate is introduced directly into the abdominal cavity from a bag that is temporarily attached to a permanently implanted plastic tube. The dialysate flows into the peritoneal cavity by gravity. Waste and salt molecules pass from blood vessels in the abdominal wall into the dialysate before the fluid is collected several hours later. The solution is drained by gravity from the abdominal cavity into a bag, which is then discarded. One advantage of CAPD over an artificial kidney machine is that the individual can go about many of his or her normal activities during CAPD. A disadvantage, however, is an increased likelihood of infections.

Patients with renal failure may undergo a kidney transplant, which is the surgical replacement of a defective kidney with a functioning kidney from a donor. As with all organ transplants, organ rejection is a possibility (see the Science Focus). Receiving a kidney from a close relative decreases the chances of rejection. The current one-year survival rate following a kidney transplant is 95–98%, but donor organs are in short supply. In the future, it may be possible to use kidneys from pigs or kidneys created in the laboratory.

Disorders of the Urinary Bladder and Urethra

Infections are probably the most common cause of problems in the urinary bladder and urethra. Urine leaving the kidneys is normally free of bacteria. The distal parts of the urethra, however, are normally colonized with bacteria. Most of the time, the flow of urine should be one-way from the bladder to the urethra. However, sometimes harmful bacteria from the urethra are able to gain access to the bladder, especially in females, whose urethra is shorter and broader than that of males. This helps explain why females are relatively more prone to bladder infections (Fig. 16.10).

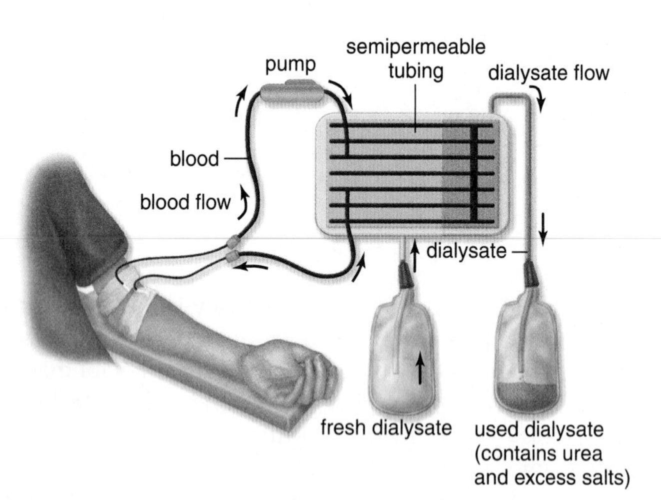

Figure 16.9 An artificial kidney machine.
As the patient's blood is pumped through dialysis tubing, the tubing is exposed to a dialysate (dialysis solution). Wastes exit the blood into the solution because of a preestablished concentration gradient. In this way, blood is not only cleansed, but its water-salt and acid-base balances are also adjusted.

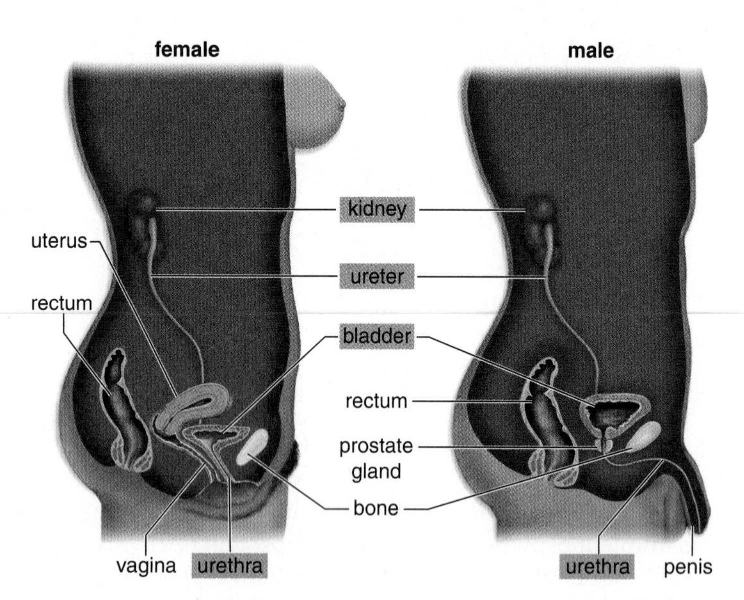

Figure 16.10 Female and male urinary tracts compared.
Females have a short urinary tract compared to that of males. This makes it easier for bacteria to invade the urethra and helps explain why females are more likely than males to get a urinary tract infection.

In the United States, approximately 25–40% of women aged 20–40 years have had a urinary tract infection. Infections of the bladder usually result in inflammation of the bladder, or **cystitis,** and infection of the urethra may result in **urethritis.** Almost all urinary tract infections can be cured with appropriate antibiotics.

Bladder stones (Fig. 16.11) can form in people of any age. They most commonly occur as a result of another condition that interferes with normal urine flow, such as a bladder infection with associated inflammation or prostate enlargement in men. They can also form from kidney stones that pass into the urinary bladder and become larger. The most common symptoms of bladder stones are pain, difficulty urinating or increased frequency of urination, and blood in the urine. Similar to kidney stones, bladder stones may be removed surgically or, preferably, by lithotripsy.

The most common type of cancer affecting the urinary system is cancer of the bladder. In the United States, bladder cancer is the fourth most common type of cancer in men and the tenth most common in women. Because several toxic by-products of tobacco are secreted in the urine, it is estimated that smoking quintuples the risk of bladder cancer. Certain types of bladder cancer are very malignant. In these cases, the bladder must be removed. With no bladder, the ureters must either be diverted into the large intestine or attached to an opening (or stoma) that is created in the skin near the navel. In the latter case, the patient typically must wear a removable plastic bag to collect the urine. Recently, however, researchers have had success in growing entire human bladders in the lab, and implanting them into a limited number of patients. Clinical trials of such replacement bladders may begin in 2009 (Fig. 16.12).

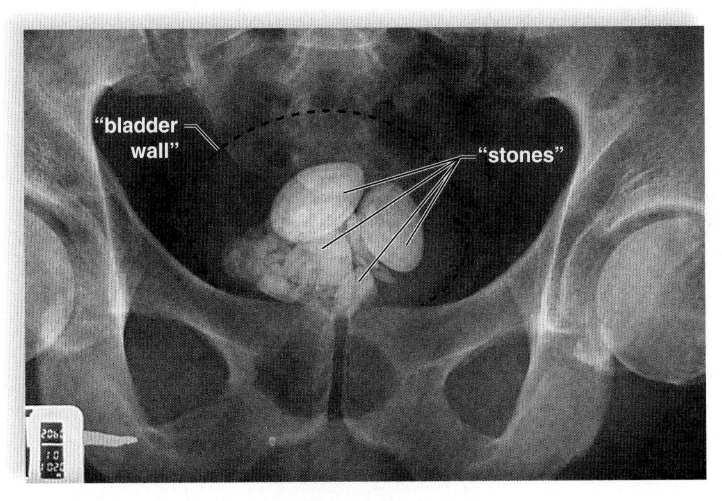

Figure 16.11 Bladder stones.
Hard granules such as those seen nearly filling the bladder in this X ray can form and lodge virtually anywhere in the urinary tract, including the kidneys, ureters, and urethra.

Figure 16.12 Organs grown in the lab.
Researchers at Children's Hospital, Boston, have cultured bladder cells on biodegradable bladder-shaped molds to produce an organ that can be implanted in individuals with diseased or defective bladders.

Check Your Progress

1. What are some potential specific causes of death in severe kidney failure?
2. What anatomical differences make females more prone to cystitis? What unique structure may cause difficult urination in males?

Applying the Concepts Revisited

In the story that opened this chapter, a little girl, Aiesha, was born with polycystic kidney disease. Take another look at the photo of a polycystic kidney on page 299, and answer the following:

1. Anatomically speaking, what specific parts of the kidney structure appear to be affected most severely by the fluid-filled cysts?
2. Polycystic kidney disease seems to cause more serious problems in African Americans, especially those who have sickle-cell disease. Because sickle-cell is mainly a disease of the red blood cells, what does this have to do with the kidneys?

SUMMARIZING THE CONCEPTS

16.1 Urinary System

The kidneys produce urine, which is conducted by the ureters to the bladder where it is stored before being released by way of the urethra. During this process the kidney serves three functions:

- Excretion of metabolic waste, especially nitrogenous waste.
- Maintenance of normal water-salt balance.
- Maintenance of the acid-base balance of the blood.

A fourth major function of the kidneys, which is unrelated to urine production, is the production of the following hormones:

- Erythropoietin, which stimulates red blood cell production.
- Renin, which leads to the secretion of aldosterone by the adrenal cortex.

The kidney also converts vitamin D to its biologically active form.

16.2 Anatomy of the Kidney and Excretion

Macroscopically, the kidneys are divided into the renal cortex, renal medulla, and renal pelvis. Microscopically, they contain the nephrons.

- Each nephron has its own blood supply. The afferent arteriole approaches the glomerular capsule and divides to become the glomerulus, a capillary tuft. The efferent arteriole leaves the capsule and immediately branches into the peritubular capillary network.
- A nephron has several parts: the glomerular capsule, the proximal convoluted tubule, the loop of the nephron, and the distal convoluted tubule. The spaces between the podocytes of the glomerular capsule allow small molecules to enter the capsule from the glomerulus. The cuboidal epithelial cells of the proximal convoluted tubule have many mitochondria and microvilli to carry out active transport (and allow passive transport) from the tubule to the blood. The cuboidal epithelial cells of the distal convoluted tubule have numerous mitochondria but lack microvilli. They carry out active transport from the blood to the tubule.

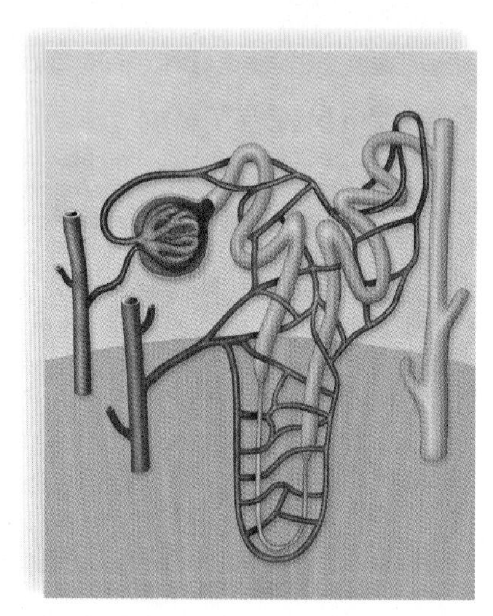

Urine is composed primarily of nitrogenous waste products and salts in water. The following steps occur during urine formation:

- Glomerular filtration, during which water and other small molecules consisting of both nutrients and wastes move from the blood to the inside of the glomerular capsule.
- Tubular reabsorption, during which water and nutrients move back into the blood, mostly at the proximal convoluted tubule.
- Tubular secretion, where certain substances (e.g., hydrogen ions) are transported from the blood into the distal convoluted tubule.

16.3 Regulatory Functions of the Kidneys

Reabsorption of water and the production of a hypertonic urine involves three steps:

- The reabsorption of salt increases blood volume and pressure because more water is also reabsorbed. Two hormones, aldosterone and ANH, control the kidneys' reabsorption of sodium (Na^+).
- Na^+ is actively transported out of the ascending limb of the loop of the nephron, establishing a solute gradient that increases toward the inner medulla.
- This gradient draws water from the descending limb of the loop of the nephron and also from the collecting duct. The permeability of the collecting duct is controlled by the hormone ADH.

The kidneys also keep blood pH within normal limits:

- They reabsorb HCO_3^- and excrete H^+ as needed to maintain the pH at about 7.4. Ammonia also buffers H^+ in the urine.

16.4 Disorders of the Urinary System

These disorders can be divided into those that affect the kidney itself and those that affect the rest of the urinary tract:

- Various types of medical conditions, including diabetes, kidney stones, and infections, can lead to kidney failure, which necessitates undergoing hemodialysis, CAPD, or kidney transplantation.
- The urinary bladder and urethra are susceptible to infections, bladder stones, and cancer.

TESTING YOURSELF

Choose the best answer for each question.

1. _____ is the elimination of metabolic waste, and _____ is the elimination of undigested food.
   a. Excretion, absorption
   b. Defecation, excretion
   c. Excretion, defecation
   d. None of these are correct.
2. Urea is a by-product of _____ metabolism.
   a. glucose
   b. salt
   c. fat
   d. amino acid
3. Breakdown of _____ produces _____, which can cause _____.
   a. proteins, fats, gout
   b. nucleotides, ammonia, high blood pressure
   c. nucleotides, uric acid, gout
   d. fats, acids, diabetes
4. Kidneys control blood pH by excreting
   a. HCO_3^-.
   b. H^+.
   c. Both a and b are correct.
   d. Neither a nor b is correct.
5. The function of erythropoietin is
   a. reabsorption of sodium ions.
   b. excretion of potassium ions.
   c. reabsorption of water.
   d. stimulation of red blood cell production.
   e. to increase blood pressure.
6. Label the following diagram of the urinary system and nearby blood vessels.

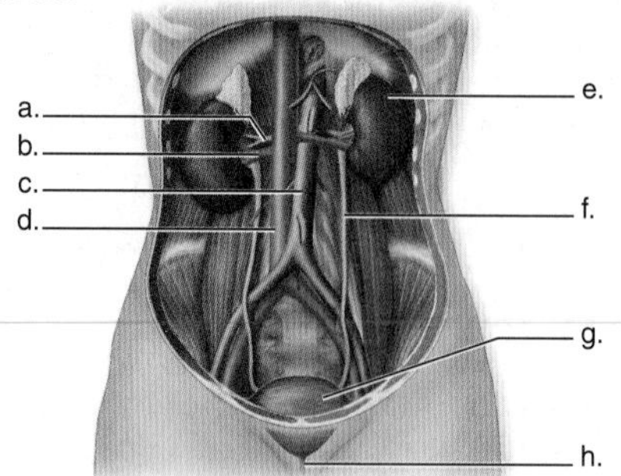

7. Label the following diagram of the kidneys.

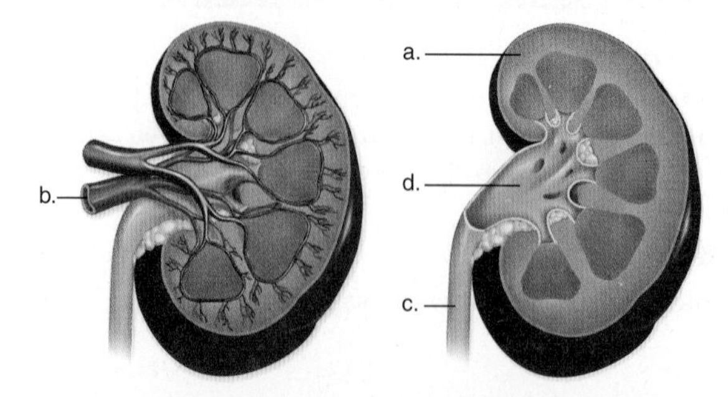

8. Which of the following is a difference between the urinary systems of males and females?
   a. Males have a longer urethra than females.
   b. In males, the urethra passes through the prostate.
   c. In males, the urethra serves both the urinary and reproductive systems.
   d. All of these are correct.
9. Put these parts of the nephron in correct order from start (nearest to glomerulus) to finish (nearest to the collecting duct).
   a. distal convoluted tubule
   b. proximal convoluted tubule
   c. glomerular capsule
   d. loop of the nephron
10. Filtration is associated with the
    a. glomerular capsule.
    b. distal convoluted tubule.
    c. collecting duct.
    d. All of these are correct.
11. An _____ artery exits the glomerulus.
    a. afferent
    b. efferent
    c. Both a and b are correct.
    d. Neither a nor b is correct.
12. Which of the following materials would not be filtered from the blood at the glomerulus?
    a. water
    b. urea
    c. protein
    d. glucose
    e. sodium ions
13. _____ reabsorption from the nephrons is, ordinarily, 100%.
    a. Sodium
    b. Glucose
    c. Water
    d. Urea
14. The purpose of the loop of the nephron in the process of urine formation is
    a. reabsorption of water.
    b. production of filtrate.
    c. reabsorption of solutes.
    d. secretion of solutes.
15. By what transport process are most molecules secreted from the blood into the tubule?
    a. osmosis
    b. diffusion
    c. active transport
    d. facilitated diffusion
16. The cells of the distal convoluted tubule
    a. have many mitochondria.
    b. lack microvilli.
    c. can move molecules from the blood to the tubules.
    d. All of these are correct.
17. Excretion of a hypertonic urine in humans is most associated with the
    a. glomerular capsule and the tubules.
    b. proximal convoluted tubule only.
    c. loop of the nephron and collecting duct.
    d. distal convoluted tubule and peritubular capillary.
18. Which of these associations is mismatched?
    a. atrial natriuretic hormone (ANH)—secreted by the heart
    b. renin—secreted by the anterior pituitary
    c. aldosterone—secreted by the adrenal cortex
    d. antidiuretic hormone (ADH)—secreted by the posterior pituitary
19. Which hormone causes a rise in blood pressure?
    a. aldosterone
    b. oxytocin
    c. erythropoietin
    d. atrial natriuretic hormone
20. Urinary tract infections are usually caused by
    a. viruses.
    b. bacteria.
    c. fungi.
    d. None of these are correct.
21. The use of an artificial kidney machine is known as
    a. filtration.
    b. excretion.
    c. hemodialysis.
    d. None of these are correct.

UNDERSTANDING THE TERMS

THINKING CRITICALLY

1. How do the afferent and efferent arterioles play important, and somewhat opposite, roles in determining the glomerular filtration rate (GFR)?
2. What would be the effect of a deficiency of antidiuretic hormone? Of too much aldosterone?
3. Suppose you developed kidney failure and needed hemodialysis, continuous ambulatory peritoneal dialysis, or a kidney transplant to survive. What would be some of the advantages and disadvantages to consider about each procedure?

INQUIRY INTO LIFE WEBSITE

The companion website for *Inquiry into Life* provides a wealth of information organized and integrated by chapter. You will find practice tests, animations, videos, and much more that will complement your learning and understanding of general biology.

http://www.mhhe.com/maderinquiry13

PART IV

Integration and Control of the Human Body

Nervous System

Chapter Outline

Applying the Concepts

As she sat in her car at the red light, Sarah noticed that the colors didn't seem quite right. "I must be stressed out," she thought to herself, but as she squinted at the traffic signal, the top light definitely seemed more orange than red. "Maybe there's something wrong with this light," she thought, but the next one was the same. At work later that day, she noticed that she was having trouble reading her e-mail, and by the end of the day she had a splitting headache. She kept telling herself that she had just been working too hard. But even as she tried to remain calm, deep down she had a bad feeling. Within a few weeks, she was almost completely blind in one eye, and the sensations in her feet felt muffled, like they were wrapped in gauze. Her doctors diagnosed multiple sclerosis (MS), and they were able to treat the inflammation in her optic nerves and spinal cord with high doses of immunosuppressive medications. Over the next few years, Sarah was able to keep her symptoms at bay by injecting herself daily with a drug called beta interferon, but she knew that someday she might need a wheelchair to get around.

MS is an inflammatory disease that affects the myelin sheaths, which wrap parts of some nerve cells like insulation around an electrical cord. As these sheaths deteriorate, the nerves no longer conduct impulses normally. For unknown reasons, MS often attacks the optic nerves first, before spreading to other areas of the brain. Most researchers think MS results from a misdirected attack on myelin by the body's immune system, although other factors may be involved. Like many diseases that affect the nervous system, there is no cure, so affected individuals must deal with their condition for the rest of their lives.

17.1 Nervous Tissue

Learning Outcomes

1. Describe the basic structure of a neuron, and compare the functions of the three classes of neurons.
2. Illustrate the changes in ion concentrations inside and outside of a neuron that result in an action potential.
3. Explain how a nerve impulse is transmitted across a chemical synapse.

Along with the endocrine system (see Chapter 20), the nervous system coordinates and regulates the functioning of the body's other systems. The nervous system has two major anatomical divisions. The **central nervous system (CNS)** consists of the brain and spinal cord, which are located in the midline of the body. The **peripheral nervous system (PNS)** consists of nerves that carry sensory messages to the CNS and motor commands from the CNS to the muscles and glands (Fig. 17.1). Although the CNS and PNS are anatomically and functionally distinct, the two systems work together and are connected to one another.

The nervous system contains two types of cells: neurons and neuroglia (neuroglial cells). **Neurons** are the cells that transmit nerve impulses between parts of the nervous system. **Neuroglia** support and nourish neurons, maintain homeostasis, form myelin, and may aid in signal transmission. This section discusses the structure and function of neurons, the myelin sheath that surrounds cell extensions called axons, and how nerve impulses are transmitted through neurons to effectors.

Types of Neurons and Neuron Structure

Although the structure of the nervous system is extremely complex, the principles of its operation are simple. There are three classes of neurons: sensory neurons, interneurons, and motor neurons (Fig. 17.2). Their functions are best described in relation to the CNS. A **sensory neuron** takes messages to the CNS. Sensory neurons may be equipped with specialized endings called sensory receptors that detect changes in the environment. An **interneuron** lies entirely within the CNS. Interneurons can receive input from sensory neurons and also from other interneurons in the CNS. Thereafter, they sum up all the messages received from these neurons before they communicate with motor neurons. A **motor neuron** takes messages away from the CNS to an effector (an organ, muscle fiber, or gland). Effectors carry out our responses to environmental changes.

Neurons vary in appearance, but most of them have just three parts: a cell body, dendrites, and an axon. The **cell body** contains the nucleus, as well as other organelles. **Dendrites** are extensions leading toward the cell body that receive signals from other neurons and send them on to the cell body. An **axon** conducts nerve impulses away from the cell body toward other neurons or effectors.

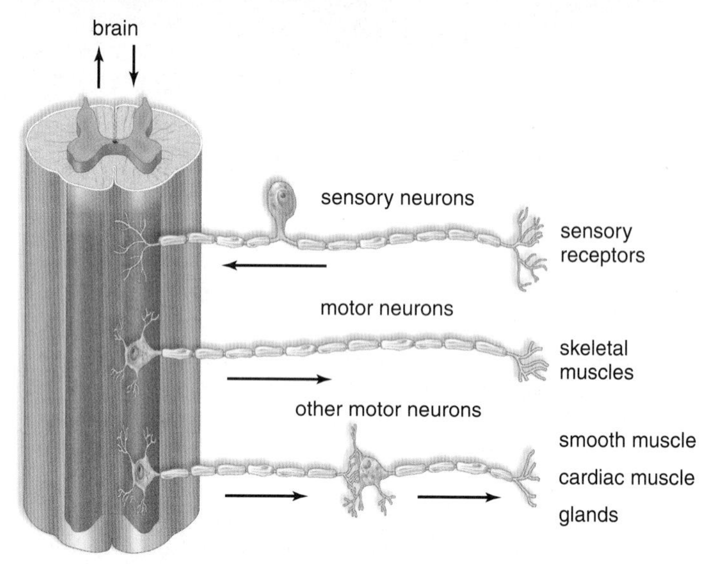

Figure 17.1 Organization of the nervous system. **a.** In paraplegics, messages no longer flow between the body's lower limbs and the central nervous system (the spinal cord and brain). **b.** The sensory neurons of the peripheral nervous system take nerve impulses from sensory receptors to the central nervous system (CNS), and motor neurons take nerve impulses from the CNS to muscles and glands.

Check Your Progress

1. List three functions of neuroglia.
2. Name three classes of neurons, and describe their relationship to each other.
3. Draw a diagram showing the three parts of every neuron.

Myelin Sheath

Some axons are covered by a protective **myelin sheath** (Fig. 17.3). In the PNS, this covering is formed by a type of neuroglia called **Schwann cells,** which contain the lipid substance myelin in their plasma membranes. The myelin sheath develops when Schwann cells wrap themselves around an axon as many as 100 times and in this way lay down many layers of plasma membrane. Because each Schwann cell myelinates only part of an axon, the myelin sheath is interrupted. The gaps where there is no myelin sheath are called **nodes of Ranvier**. Each Schwann cell only covers about one millimeter of an axon. Because a single axon may be several feet long (if it runs from the spinal cord to the foot, for example), several hundred or more Schwann cells may be required to myelinate a single axon.

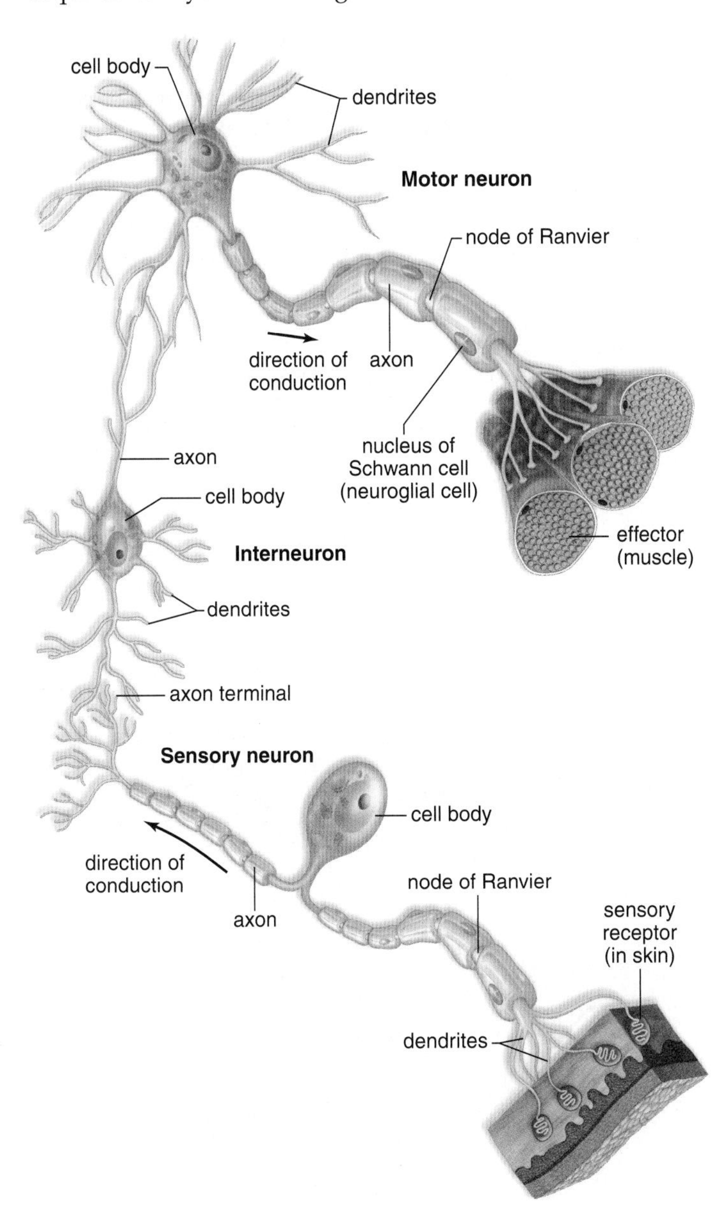

Figure 17.2 Types of neurons.
Sensory neurons, interneurons, and motor neurons have different arrangements in the body. How does each neuron's arrangement correlate with its function?

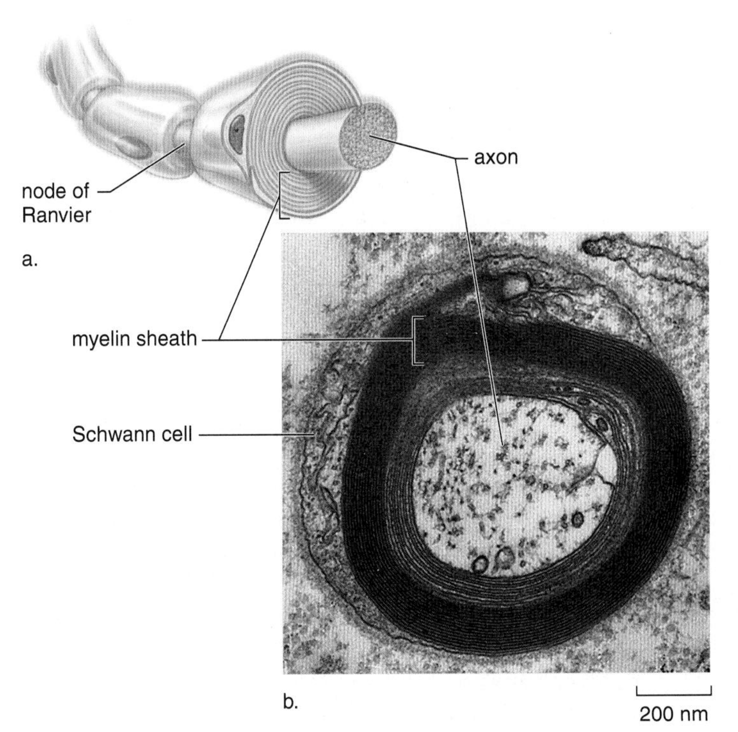

Figure 17.3 Myelin sheath.
a. In the PNS, a myelin sheath forms when Schwann cells wrap themselves around an axon. **b.** Electron micrograph of a cross section of an axon surrounded by a myelin sheath.

In the PNS, myelin gives nerve fibers their white, glistening appearance. The myelin sheath also plays an important role in nerve regeneration within the PNS. If an axon is accidentally severed, the myelin sheath may remain and serve as a passageway for new fiber growth. In the CNS, myelin is produced by the **oligodendroglial cells.** Unlike in the PNS, nerve regeneration does not seem to occur to any significant degree in the CNS.

The CNS is composed of two types of nervous tissue—gray matter and white matter. **Gray matter** is gray because it contains cell bodies and short, nonmyelinated fibers. **White matter** is white because it contains myelinated axons that run together in bundles called **tracts.** In the CNS, neurons with short axons make up the gray matter because no part of these neurons has a myelin sheath. However, some neurons in the CNS have long axons with a myelin sheath. These long axons, which make up the white matter of the brain, are carrying messages from one part of the CNS to another. The surface layer of the brain is gray matter, and the white matter lies deep within the gray matter. The central part of the spinal cord consists of gray matter, and the white matter surrounds the gray matter.

Check Your Progress

1. What cell types form the myelin in the PNS versus the CNS?
2. Review the structure of gray matter and white matter, and describe where each is found in the CNS and the PNS.
3. Formulate a reasonable hypothesis to explain why nerves can be regenerated in the PNS but not the CNS.

The Nerve Impulse

The nervous system uses the **nerve impulse** to convey information. The nature of a nerve impulse has been studied by using excised axons and a voltmeter called an oscilloscope. Voltage, which is expressed in millivolts (mV), is a measure of the electrical potential difference between two points. In the case of a neuron, the two points are the inside and the outside of the axon. Voltage is displayed on the voltmeter screen as a trace, or pattern, over time.

Resting Potential

In the experimental setup shown in Figure 17.4*a*, a voltmeter is wired to two electrodes: one electrode is placed inside an axon, and the other electrode is placed outside. When the axon is not conducting an impulse, the voltmeter records a potential difference across an axonal membrane (membrane of axon) equal to about –65 mV. This reading indicates that the inside of the axon is negative compared to the outside. This is called the **resting potential** because the axon is not conducting an impulse.

The existence of the polarity (charge difference) correlates with a difference in ion distribution on either side of the axonal membrane. As Figure 17.4*a* shows, the concentration of sodium ions (Na^+) is greater outside the axon than inside, and the concentration of potassium ions (K^+) is greater inside the axon than outside. The unequal distribution of these ions is maintained by the action of carrier proteins in the membrane called **sodium-potassium pumps** (see Fig. 4.9), which actively transport Na^+ out of and K^+ into the axon.

These pumps are always working because the membrane is somewhat permeable to these ions, and they tend to diffuse toward their lesser concentration. Because the membrane is more permeable to K^+ than to Na^+, there are always more positive ions outside the membrane than inside. This accounts for the polarity recorded by the voltmeter. Large, negatively charged organic ions in the axoplasm also contribute to the polarity across a resting axonal membrane.

Action Potential

An **action potential** is a rapid change in polarity across an axonal membrane as the nerve impulse occurs. An action potential is an all-or-none phenomenon. If a stimulus causes the axonal membrane to depolarize to a certain level, called **threshold,** an action potential occurs. The strength of an action potential does not change, but an intense stimulus can

Figure 17.4 Resting and action potentials of the axonal membrane.
a. Resting potential. A voltmeter that records voltage changes indicates the axonal membrane has a resting potential of –65 mV. There is a preponderance of Na^+ outside the axon and a preponderance of K^+ inside the axon. The permeability of the membrane to K^+ compared to Na^+ causes the inside to be negative compared to the outside. **b.** Action potential. Depolarization occurs when Na^+ gates open and Na^+ moves inside the axon. **c.** Repolarization occurs when K^+ gates open and K^+ moves outside the axon. **d.** Graph of the action potential.

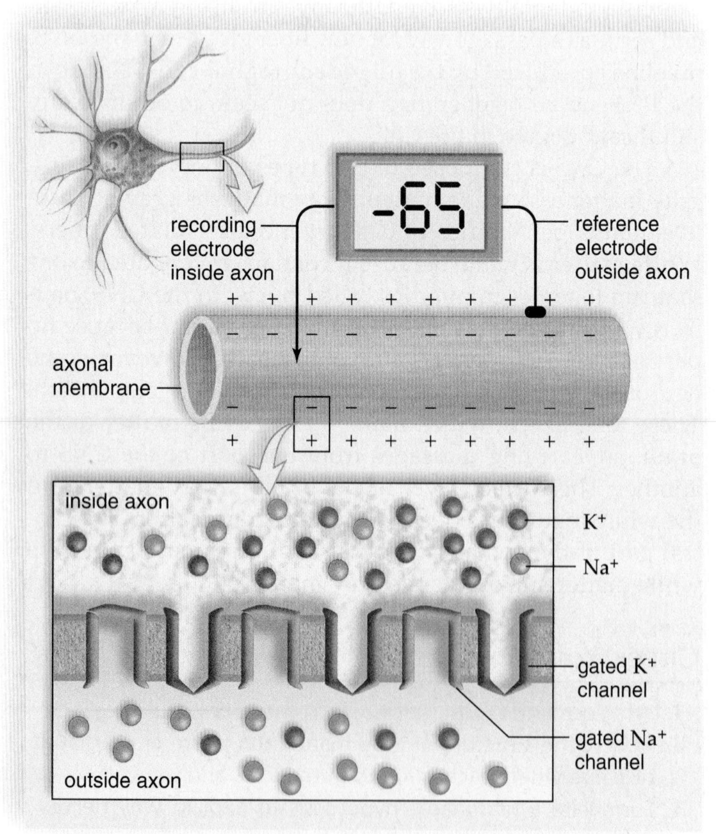

a. Resting potential: more Na^+ outside the axon and more K^+ inside the axon causes polarization.

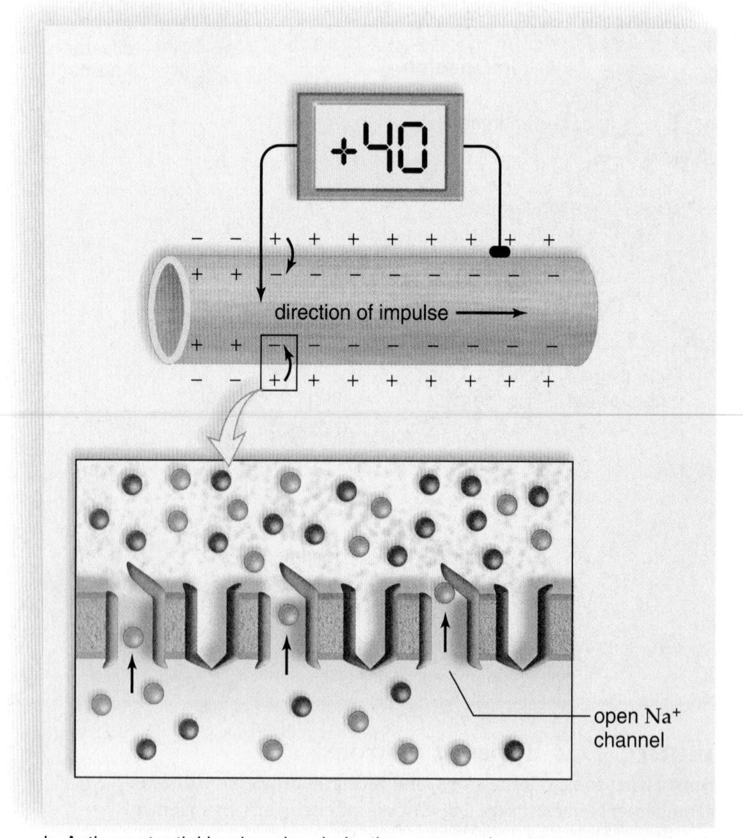

b. Action potential begins: depolarization occurs when Na^+ gates open and Na^+ moves inside the axon.

cause an axon to fire (start an axon potential) more often in a given time interval than a weak stimulus.

The action potential requires two types of gated channel proteins in the membrane. One gated channel protein opens to allow Na^+ to pass through the membrane to inside the cell, and the second channel opens to allow K^+ to pass through the membrane to outside the cell (Fig. 17.4*b*, *c*).

Sodium Gates Open When an action potential begins, the gates of sodium channels open first, and Na^+ flows down its concentration gradient into the axon. As Na^+ moves inside the axon, the membrane potential changes from –65 mV to +40 mV. This is called a *depolarization* because the charge inside the axon changes from negative to positive (Fig. 17.4*b*).

Potassium Gates Open Second, the gates of potassium channels open, and K^+ flows down its concentration gradient to outside the axon. As K^+ moves outside the axon, the action potential changes from +40 mV back to –65 mV. This is a *repolarization* because the inside of the axon resumes a negative charge as K^+ exits the axon (Fig. 17.4*c*). An action potential only takes 2 milliseconds (ms). In order to visualize such rapid fluctuations in voltage across the axonal membrane, researchers generally find it useful to plot the voltage changes over time (Fig. 17.4*d*).

Conduction of an Action Potential In nonmyelinated axons, the action potential travels down an axon one small section at a time. As soon as an action potential has moved on, the previous section undergoes a **refractory period,** during which the sodium gates are unable to open. Notice, therefore, that the action potential cannot move backward and instead always moves down an axon toward its terminals. When the refractory period is over, the sodium-potassium pump has restored the previous ion distribution by pumping Na^+ to outside the axon and K^+ to inside the axon.

In myelinated axons, the gated ion channels that produce an action potential are concentrated at the nodes of Ranvier. Because ion exchange occurs only at the nodes, the action potential travels faster than in nonmyelinated axons. This is called saltatory conduction, meaning that the action potential "jumps" from node to node. Speeds of 200 m per second (450 mph) have been recorded.

Check Your Progress

1. Describe the activity of the sodium-potassium pump present in neurons.
2. Explain the changes in Na^+ and K^+ ion concentrations that occur during an action potential. How are these changes associated with depolarization and repolarization?
3. Define node of Ranvier and saltatory conduction.

Transmission Across a Synapse

Every axon branches into many fine endings, each tipped by a small swelling called an **axon terminal.** Each terminal lies

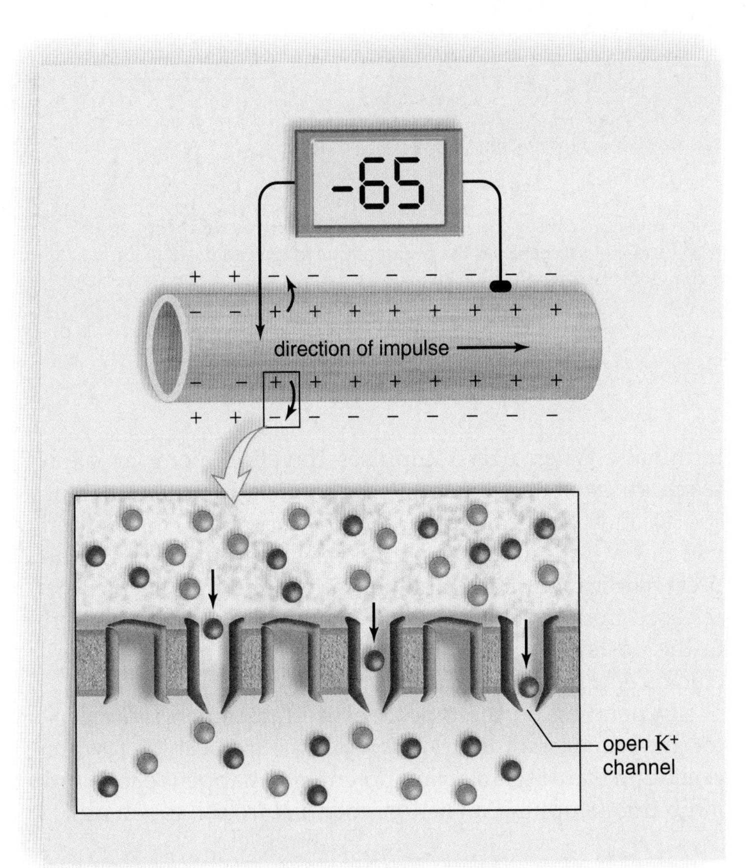

c. Action potential ends: repolarization occurs when K^+ gates open and K^+ moves outside the axon.

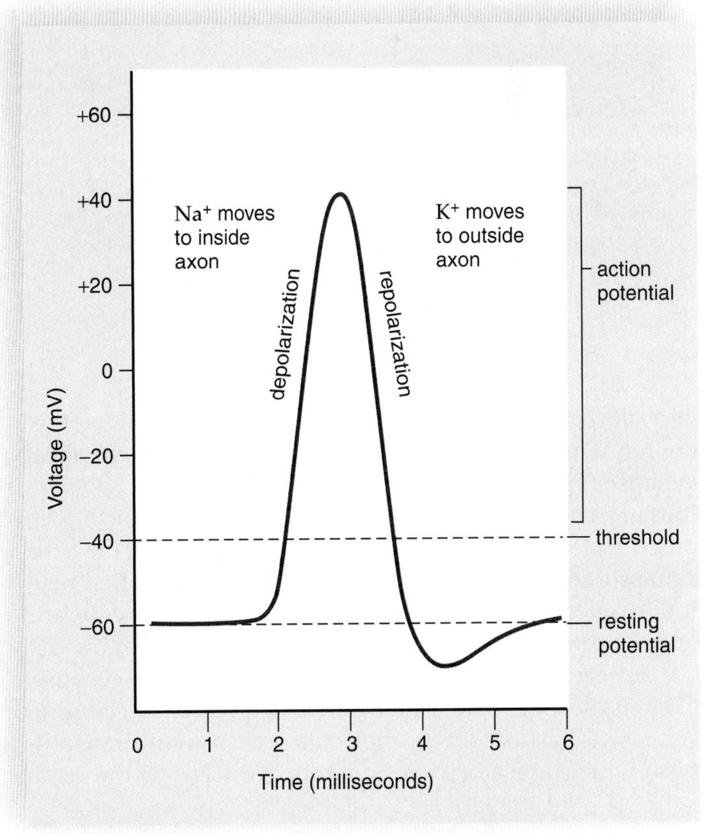

d. An action potential can be visualized if voltage changes are graphed over time.

Visual Focus

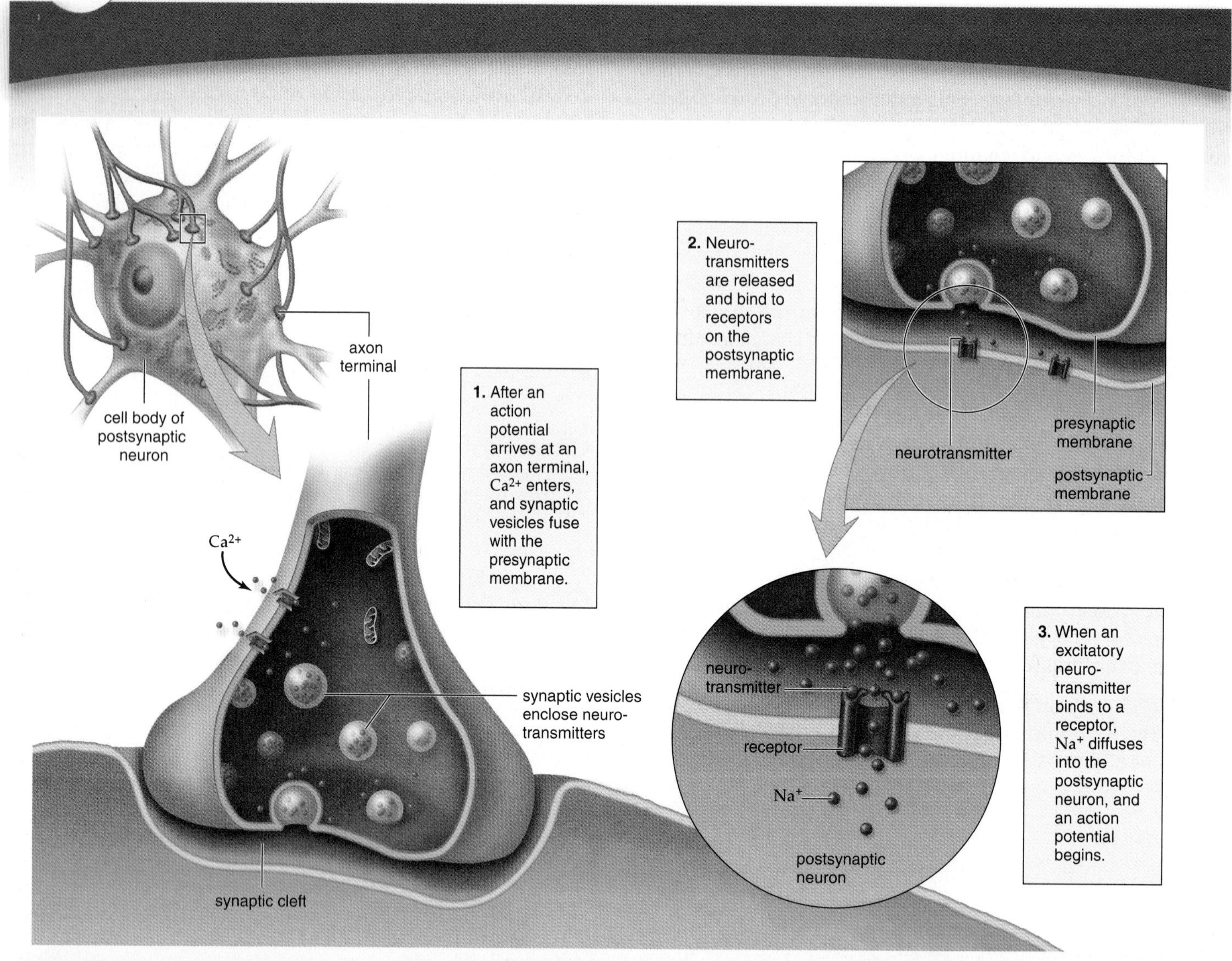

Figure 17.5 Chemical synapse structure and function.
Transmission across a synapse from one neuron to another occurs when an action potential causes a neurotransmitter to be released at the presynaptic membrane. The neurotransmitter diffuses across a synaptic cleft and binds to a receptor in the postsynaptic membrane. An action potential may begin in the postsynaptic membrane if the concentration of incoming Na^+ reaches threshold.

very close to either the dendrite or the cell body of another neuron. This region of close proximity is called a **chemical synapse** (Fig. 17.5). It is important to note that the two neurons at a chemical synapse don't ever physically touch each other. Instead, they are separated by a tiny gap called the **synaptic cleft.** At a synapse, the membrane of the first neuron is called the *pre*synaptic membrane, and the membrane of the next neuron is called the *post*synaptic membrane.

An action potential cannot cross a chemical synapse. Communication between the two neurons at a chemical synapse is carried out by molecules called **neurotransmitters,** which are stored in synaptic vesicles in the axon terminals. When nerve impulses traveling along an axon reach an axon terminal, gated channels for calcium ions (Ca^{2+}) open, and calcium enters the terminal. This sudden rise in Ca^{2+} stimulates synaptic vesicles to merge with the presynaptic membrane, and neurotransmitter molecules are released into the synaptic cleft. They diffuse across the cleft to the postsynaptic membrane, where they bind with specific receptor proteins.

Depending on the type of neurotransmitter and receptor, the response of the postsynaptic neuron can be toward excitation (causing an action potential to happen) or toward inhibition (stopping an action potential from happening).

Synaptic Integration

The dendrites and cell body of a neuron can have synapses with many other neurons, thus a single neuron may receive many excitatory and inhibitory signals. Excitatory signals have a depolarizing effect, causing the charge across the neuron membrane to move closer to the threshold needed to trigger an action potential. Inhibitory signals usually have a hyperpolarizing effect; that is, they increase the potential difference across the axonal membrane. **Synaptic integration** is the summing up of excitatory and inhibitory signals (Fig. 17.6). If a neuron receives many excitatory signals (either from different synapses or at a rapid rate from one synapse), the chances are that the neuron's axon will transmit a nerve impulse. On the other hand, if a neuron receives both inhibitory and excitatory signals, the summing up of these signals may prohibit the axon from firing.

Neurotransmitters

At least 25 different neurotransmitters have been identified, but two very well-known ones are **acetylcholine (ACh)** and **norepinephrine (NE).**

Once a neurotransmitter has been released into a synaptic cleft and has initiated a response, it is removed from the cleft. In some synapses, the postsynaptic membrane contains enzymes that rapidly inactivate the neurotransmitter. For example, the enzyme **acetylcholinesterase (AChE)** breaks down acetylcholine. In other synapses, the presynaptic membrane rapidly reabsorbs the neurotransmitter, possibly for repackaging in synaptic vesicles. The short existence of neurotransmitters at a synapse prevents continuous stimulation (or inhibition) of postsynaptic membranes.

It is of interest to note here that many drugs that affect the nervous system act by either interfering with or potentiating (enhancing) the action of neurotransmitters. As shown in Figure 17.18, drugs can enhance or block the release of a neurotransmitter, mimic the action of a neurotransmitter or block the receptor, or interfere with the removal of a neurotransmitter from a synaptic cleft. In addition, several naturally occurring venoms and poisons, organophosphate insecticides, and nerve agents such as sarin gas, all inhibit the AChE enzyme, thus prolonging the activity of ACh. In contrast, the bacterium responsible for botulism produces a toxin that inhibits ACh release, which can paralyze muscles. Botox, a highly diluted preparation of this toxin, is now being used to treat a variety of conditions, from eyelid spasms and back pain to the wrinkles that appear on our faces with age.

CHECK YOUR PROGRESS

1. Indicate two possible fates of the synaptic vesicles after they release their contents into the synaptic cleft.
2. Define synaptic integration.
3. The bite of a black widow spider injects a powerful AChE inhibitor. How could this explain these common symptoms: muscle cramps, salivation, fast heart rate, high blood pressure?

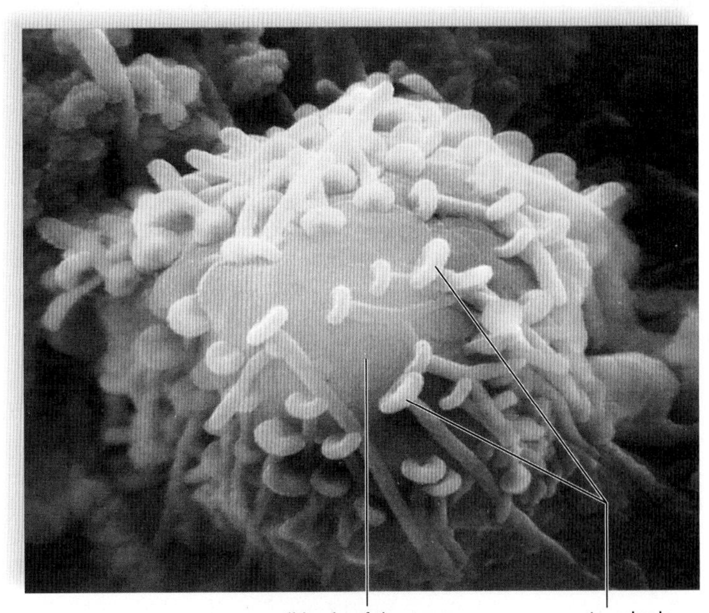

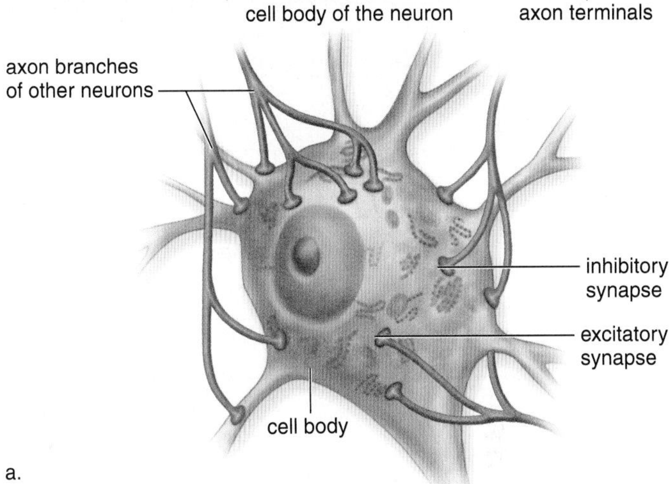

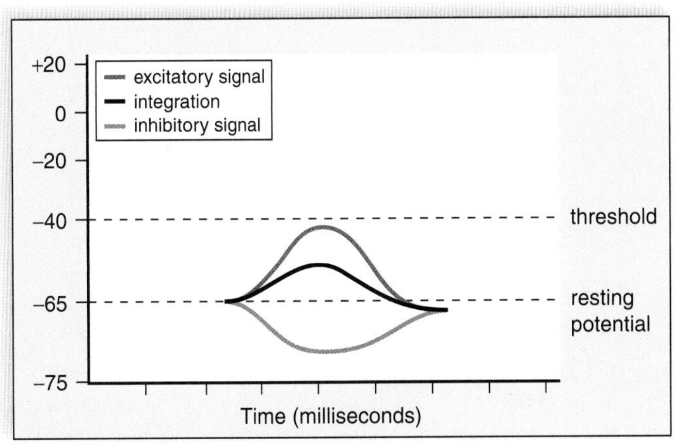

Figure 17.6 Synaptic integration.
a. Inhibitory signals and excitatory signals are summed up in the dendrites and cell body of the postsynaptic neuron. Only if the combined signals cause the membrane potential to rise above threshold does an action potential occur. **b.** In this example, threshold was not reached.

17.2 The Central Nervous System

Learning Outcomes

1. Draw and label the anatomy of the spinal cord.
2. List the four major regions of the brain and the main functions of each region.
3. Describe the functions and locations of these major parts of the brain: hypothalamus, thalamus, medulla oblongata.

The spinal cord and the brain make up the central nervous system (CNS), where sensory information is received and motor control is initiated. The brain controls or influences many bodily functions, such as breathing, heart rate, body temperature, and blood pressure. It is also the source of our emotions, as well as higher mental functions such as reasoning, memory, and creativity. Figure 17.7 illustrates how the CNS relates to the PNS. Both the spinal cord and the brain are protected by bone. The spinal cord is surrounded by vertebrae, and the brain is enclosed by the skull. Also, both the spinal cord and the brain are wrapped in protective membranes known as **meninges (sing., meninx).** The spaces between the meninges are filled with **cerebrospinal fluid (CSF),** which cushions and protects the CNS. A small amount of this fluid is sometimes withdrawn from around the cord for laboratory testing when a spinal tap (lumbar puncture) is performed to diagnose infections or other conditions that affect the brain and/or meninges.

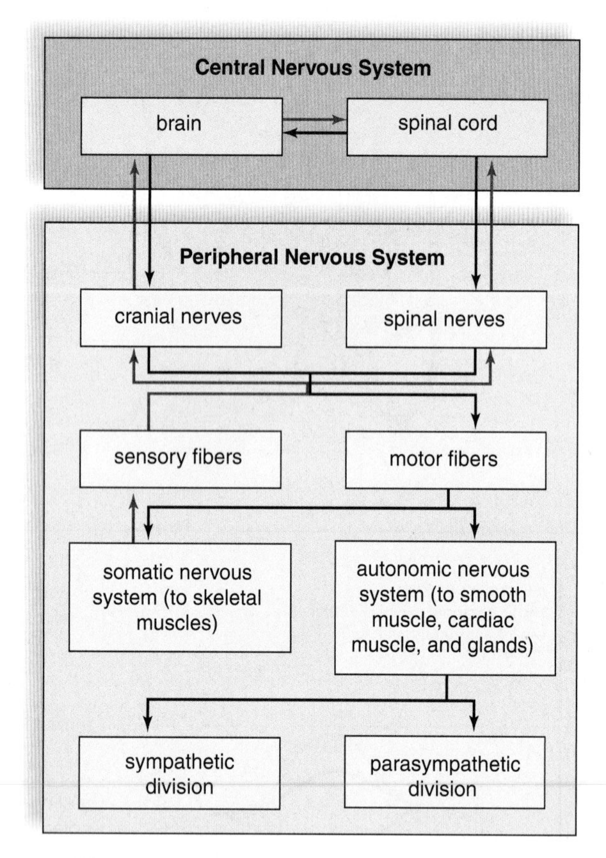

Figure 17.7 Organization of the nervous system.
The CNS is composed of the spinal cord and brain. The PNS is composed of the cranial and spinal nerves. Nerves contain both sensory and motor fibers. In the somatic system, nerves conduct impulses from sensory receptors to the CNS and motor impulses from the CNS to the skeletal muscles. In the autonomic system, consisting of the sympathetic and parasympathetic divisions, motor impulses travel to smooth muscle, cardiac muscle, and glands.

The brain has hollow interconnecting cavities termed **ventricles,** which also connect with the hollow **central canal** of the spinal cord. The ventricles produce and serve as a reservoir for cerebrospinal fluid as does the central canal. Normally, any excess cerebrospinal fluid drains away into the cardiovascular system. However, blockages can occur. In an infant, the brain and skull can enlarge due to cerebrospinal fluid accumulation, resulting in a condition called hydrocephalus ("water on the brain"). If cerebrospinal fluid collects in an adult, the brain cannot enlarge and instead is pushed against the skull. Regardless of age, the condition can cause brain damage if not corrected.

The Spinal Cord

The **spinal cord** extends from the base of the brain through a large opening in the skull called the foramen magnum and into the vertebral canal formed by openings in the vertebrae.

Structure of the Spinal Cord

Figure 17.8*a* shows how an individual vertebra protects the spinal cord. The spinal nerves project from the cord between the vertebrae that make up the vertebral column. Fluid-filled **intervertebral disks** cushion and separate the vertebrae. If a disk ruptures, the vertebrae press on the spinal cord and spinal nerves, causing pain and loss of motor function.

A cross section of the spinal cord shows a central canal, gray matter, and white matter (Fig. 17.8*b, c*). The central canal contains cerebrospinal fluid, as do the meninges that protect the spinal cord. The gray matter is centrally located and shaped like the letter H. Portions of sensory neurons and motor neurons are found there, as are interneurons that communicate with these two types of neurons. The dorsal root of a spinal nerve contains sensory fibers entering the gray matter, and the ventral root of a spinal nerve contains motor fibers exiting the gray matter. The dorsal and ventral roots join before the

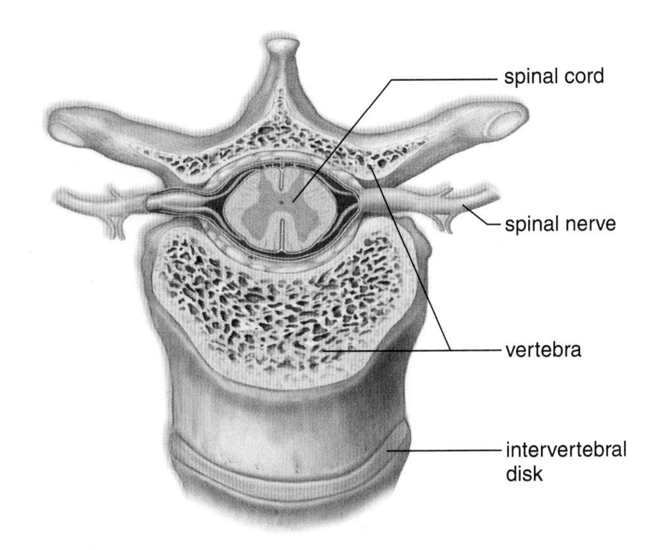

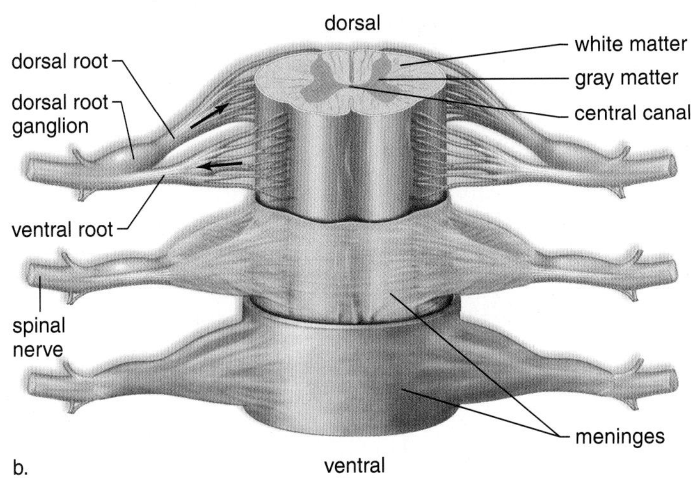

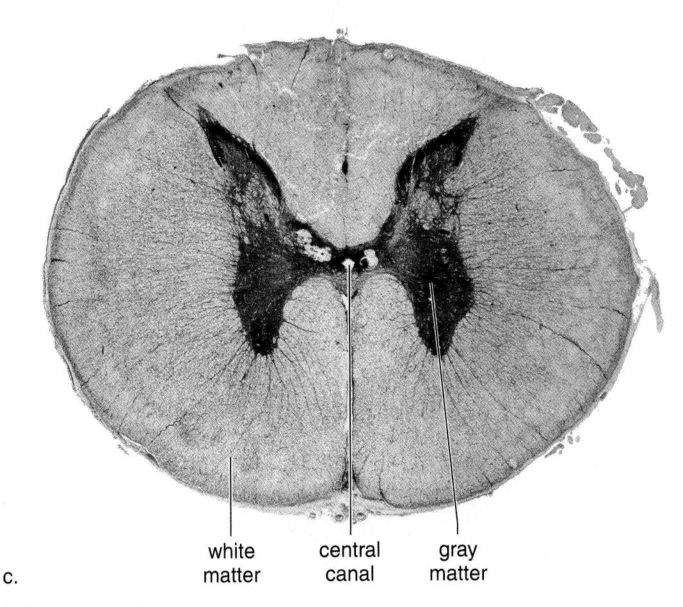

Figure 17.8 Spinal cord.
a. The spinal cord passes through the vertebral canal formed by the vertebrae. **b.** The spinal cord has a central canal filled with cerebrospinal fluid, gray matter in an H-shaped configuration, and white matter. The white matter contains tracts that take nerve impulses to and from the brain. **c.** Photomicrograph of a cross section of the spinal cord.

spinal nerve leaves the vertebral canal. Spinal nerves are a part of the PNS.

The white matter of the spinal cord surrounds the gray matter. The white matter contains ascending tracts taking information to the brain (primarily located dorsally) and descending tracts taking information from the brain (primarily located ventrally). Because the tracts cross after they enter and exit the CNS, the left side of the brain controls the right side of the body, and the right side of the brain controls the left side of the body.

Functions of the Spinal Cord

The spinal cord provides a means of communication between the brain and the peripheral nerves that leave the cord. When someone touches your hand, sensory receptors generate nerve impulses that pass through sensory fibers to the spinal cord and up ascending tracts to the brain. When we voluntarily move our limbs, motor impulses originating in the brain pass down descending tracts to the spinal cord and out to our muscles by way of motor fibers. Therefore, if the spinal cord is severed, we suffer a loss of sensation and a loss of voluntary control—that is, paralysis (see section 17.6).

In section 17.4 we will see that the spinal cord is also the center for thousands of reflex arcs, which allow the nerves and muscles to respond very quickly to potentially dangerous stimuli. A stimulus causes sensory receptors to generate nerve impulses that travel in sensory axons to the spinal cord. Interneurons integrate the incoming data and relay signals to motor neurons. The reflex is complete when motor axons cause skeletal muscles to contract or a gland or an organ to respond. Each interneuron in the spinal cord has synapses with many other neurons, and therefore, they send signals to several other interneurons and motor neurons.

The spinal cord plays a similar role for the internal organs. For example, when blood pressure falls, sensory receptors in the carotid arteries and aorta generate nerve impulses that pass through sensory fibers to the cord and then up an ascending tract to a cardiovascular center in the brain. Thereafter, nerve impulses pass down a descending tract to the spinal cord. Motor impulses then cause blood vessels to constrict so that the blood pressure rises.

Check Your Progress

1. Define meninges.
2. Describe the location and function of intervertebral disks.
3. Trace the path of a nerve impulse from a stimulus in an internal organ (such as food in the intestine stimulating peristalsis) to the brain and back.

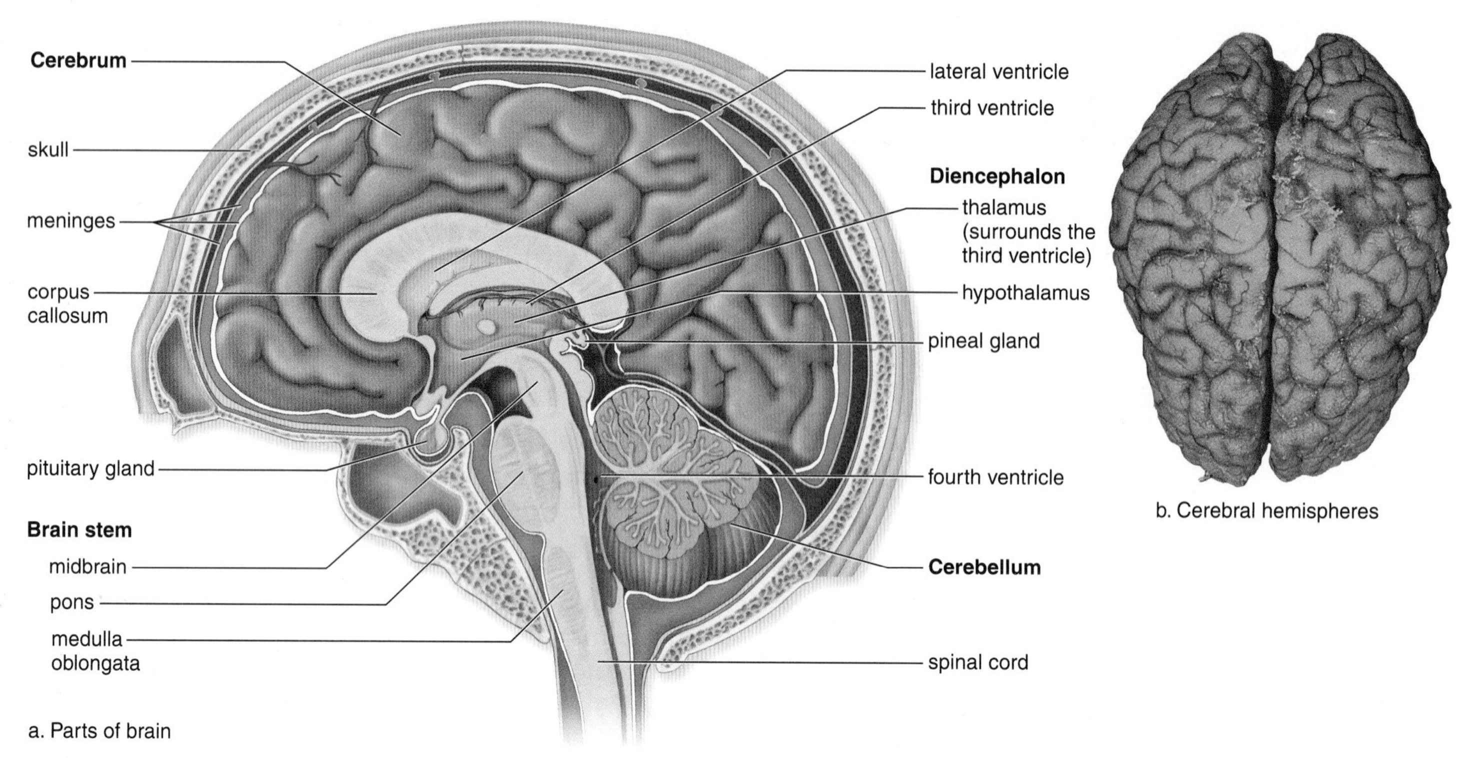

Figure 17.9 The human brain.
a. The cerebrum, seen here in longitudinal section, is the largest part of the brain in humans. The right cerebral hemisphere is shown here. **b.** Viewed from above, the cerebrum has left and right cerebral hemispheres. The hemispheres are connected by the corpus callosum.

The Brain

We will discuss the brain with reference to its four major parts: the cerebrum, the diencephalon, the cerebellum, and the brain stem. It will be helpful for you to associate these four regions with the brain's ventricles (fluid-filled spaces): the cerebrum with the two lateral ventricles, the diencephalon with the third ventricle, and the brain stem and the cerebellum with the fourth ventricle (Fig. 17.9*a*).

The Cerebrum

The **cerebrum** is the largest portion of the brain in humans. The cerebrum is the last center to receive sensory input and carry out integration before commanding voluntary motor responses. It communicates with and coordinates the activities of the other parts of the brain. The cerebrum carries out the higher thought processes required for learning and memory and for language and speech.

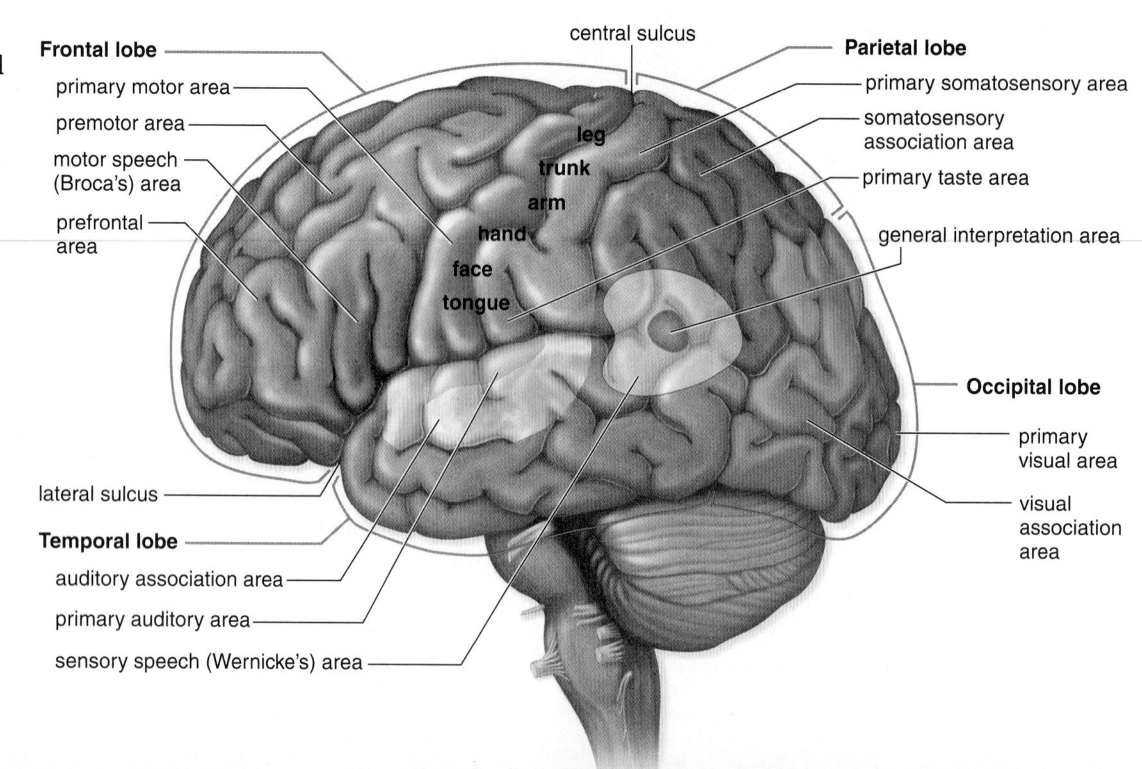

Figure 17.10 The lobes of a cerebral hemisphere.
Each cerebral hemisphere is divided into four lobes: frontal, parietal, temporal, and occipital. The frontal lobe contains centers for reasoning and movement, the parietal lobe for somatic sensing and taste, the temporal lobe for hearing, and the occipital lobe for vision.

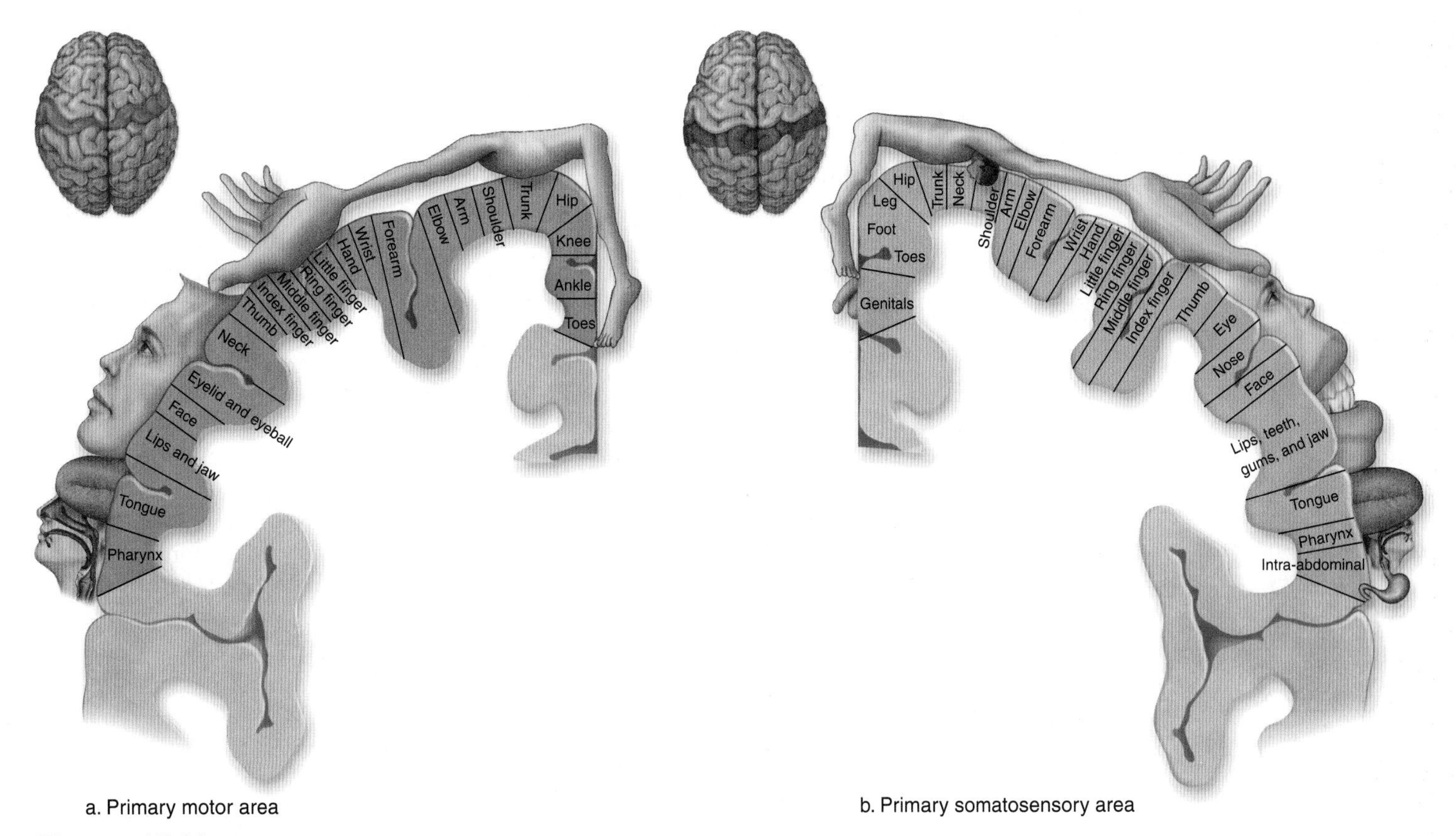

Figure 17.11 The primary motor and somatosensory areas.
In these drawings, the size of the body part reflects the amount of cerebral cortex devoted to that body part. For example, the amount of primary motor cortex **(a)** and somatosensory cortex **(b)** devoted to the thumb, fingers, and hand is greater than that for the foot and toes.

The Cerebral Hemispheres Just as the human body has two halves, so does the cerebrum. These halves are called the left and right **cerebral hemispheres** (Fig. 17.9*b*). A deep groove, the longitudinal *fissure,* divides the left and right cerebral hemispheres. Still, the two cerebral hemispheres are connected by a bridge of white matter within the **corpus callosum.**

Shallow grooves called *sulci* (sing., sulcus) divide each hemisphere into lobes (Fig. 17.10). The *frontal lobe* is the most ventral of the lobes (directly behind the forehead). The *parietal lobe* is dorsal to the frontal lobe. The *occipital lobe* is dorsal to the parietal lobe (at the rear of the head). The *temporal lobe* lies inferior to the frontal and parietal lobes (at the temple and the ear).

The **cerebral cortex** is a thin but highly convoluted outer layer of gray matter that covers the cerebral hemispheres. Each fold, or convolution, in the cortex is called a gyrus. The cerebral cortex contains over one billion cell bodies and is the region of the brain that accounts for sensation, voluntary movement, and all the thought processes we associate with consciousness.

Check Your Progress

1. Name the four major lobes of the brain.
2. Identify which of the four ventricles is associated with each major brain region.
3. Distinguish between the terms fissure, sulcus, and gyrus.

Primary Motor and Sensory Areas of the Cortex The **primary motor area** is in the frontal lobe just ventral to the central sulcus. Voluntary commands to skeletal muscles begin in the primary motor area, and each part of the body is controlled by a certain section. Control of the versatile human hand takes up an especially large portion of the primary motor area (Fig. 17.11).

The **primary somatosensory area** is just dorsal to the central sulcus in the parietal lobe. Sensory information from the skin and skeletal muscles arrives here, where each part of the body is sequentially represented. A primary visual area in the occipital lobe receives information from our eyes, and a primary auditory area in the temporal lobe receives information from our ears. A primary taste area in the parietal lobe accounts for taste sensations, and a primary olfactory area for smell is located in the temporal lobe.

Association Areas **Association areas** are places where integration occurs. Ventral to the primary motor area is a premotor area. The premotor area organizes motor functions for skilled motor activities, such as being able to walk and talk at the same time, and then the primary motor area sends signals to the cerebellum, which integrates them. Even a brief period of oxygen deprivation during birth can damage the motor areas of the cerebral cortex, which is one possible cause of cerebral palsy, a condition characterized by a spastic weakness of the arms and legs.

The somatosensory association area, located just dorsal to the primary somatosensory area, processes and analyzes sensory information from the skin and muscles. The visual association area associates new visual information with previously received visual information. It might "decide," for example, whether we have seen this face or scene or symbol in the past before. The auditory association area performs the same functions with regard to sounds.

Processing Centers Processing centers of the cortex receive information from the other association areas and perform higher-level analytical functions. The **prefrontal area,** a processing center in the frontal lobe, receives information from the other association areas and uses it to reason and plan our actions. Integration in this center accounts for our most cherished human abilities to think critically and to formulate appropriate behaviors.

The unique ability of humans to speak is partially dependent upon two processing centers found only in the left cerebral cortex: **Wernicke's area** in the dorsal part of the left temporal lobe and **Broca's area** in the left frontal lobe. Broca's area is located just ventral to the portion of the primary motor area for speech musculature (lips, tongue, larynx, and so forth) (see also Fig. 17.14). Wernicke's area helps us understand both the written and spoken word and sends the information to Broca's area. Broca's area adds grammatical refinements and directs the primary motor area to stimulate the appropriate muscles for speaking.

Central White Matter Much of the rest of the cerebrum beneath the cerebral cortex is composed of white matter. Tracts within the cerebrum take information between the different sensory, motor, and association areas pictured in Figures 17.10 and 17.11. The corpus callosum, previously mentioned, contains tracts that join the two cerebral hemispheres. Descending tracts from the primary motor area communicate with various parts of the brain, and ascending tracts from lower brain centers send sensory information up to the primary somatosensory area.

Basal Nuclei While the bulk of the cerebrum is composed of tracts, there are masses of gray matter located deep within the white matter. These so-called **basal nuclei** (formerly termed basal ganglia) integrate motor commands, ensuring that proper muscle groups are activated or inhibited.

The Diencephalon

The hypothalamus and the thalamus are in the **diencephalon,** a region that encircles the third ventricle (see Fig. 17.9*a*). The **hypothalamus** forms the floor of the third ventricle. The hypothalamus is an integrating center that helps maintain homeostasis by regulating hunger, sleep, thirst, body temperature, and water balance. The hypothalamus manufactures hormones and controls the pituitary gland. Therefore, the hypothalamus serves as a link between the nervous and endocrine systems.

The **thalamus** consists of two masses of gray matter that form the sides and roof of the third ventricle. Most of the sensory input from the visual, auditory, taste, and somatosensory systems arrives at the thalamus via the cranial nerves and tracts from the spinal cord. The thalamus integrates this information and sends it on to the appropriate portions of the cerebrum. Most information from the olfactory system, however, goes directly to the olfactory part of the cortex without passing through the thalamus, perhaps because olfaction is an evolutionarily older system compared to the other senses. The thalamus is involved in arousal of the cerebrum, and it also participates in higher mental functions, such as memory and emotions.

The pineal gland, which secretes the hormone melatonin and regulates our body's daily rhythms, is located in the diencephalon (see Fig. 17.9*a*).

The Cerebellum

The **cerebellum** is separated from the brain stem by the fourth ventricle (see Fig. 17.9*a*). The cerebellum has two portions that are joined by a narrow median portion. Each portion is primarily composed of white matter, which, in longitudinal section, has a treelike pattern. Overlying the white matter is a thin layer of gray matter that forms a series of complex folds.

The cerebellum receives sensory input from the joints, muscles, and other sensory pathways about the present position of body parts. It also receives motor output from the cerebral cortex about where these parts should be located. After integrating this information, the cerebellum sends motor impulses, by way of the brain stem, to the skeletal muscles. In this way, the cerebellum maintains posture and balance. It also ensures that all of the muscles work together to produce smooth, coordinated voluntary movements. In addition, the cerebellum assists the learning of new motor skills, such as playing the piano or hitting a baseball.

The Brain Stem

The **brain stem** contains the midbrain, the pons, and the medulla oblongata (see Fig. 17.9*a*). The **midbrain** acts as a relay station for tracts passing between the cerebrum and the spinal cord or cerebellum. It also has reflex centers for visual, auditory, and tactile responses. The word *pons* means bridge in Latin, and true to its name, the **pons** contains bundles of axons traveling between the cerebellum and the rest of the CNS. In addition, the pons functions with the medulla oblongata to regulate the breathing rate and has reflex centers concerned with head movements in response to visual and auditory stimuli.

The **medulla oblongata** contains a number of reflex centers for regulating heartbeat, breathing, and vasoconstriction (blood pressure). It also contains the reflex centers for vomiting, coughing, sneezing, hiccuping, and swallowing. The medulla oblongata lies just superior to the spinal cord.

Science Focus

Why Do We Sleep?

When was the last time you went without enough sleep? Studying for an exam, writing a paper, socializing with friends? How did you feel the next day? What do you think would happen if you only slept three or four hours a night for a month? But if you sleep eight hours a night, that's about 122 days a year. Couldn't you do something productive or fun with all that extra time?

Our bodies need sleep as much as they need food. Sleep deprivation affects us both mentally and physically. Without enough sleep we become irritable and have trouble concentrating. A 2008 study showed that adults deprived of sleep for 35 hours had increased activity in the amygdala, the brain's emotional center. Sleep-deprived bodies lose coordination and agility. Our reaction times are slower. Studies are also showing links between too little sleep and obesity, heart disease, and diabetes.

Virtually all mammals and birds sleep. Most fish and insects do too, or at least they enter a sleeplike state. Even dolphins, which need to keep surfacing to breathe, seem to sleep in one-half of their brain at a time! And yet, scientists are still debating exactly why we need to sleep at all. Considering that we spend around a third of our lives in this inactive, seemingly unproductive state, the essential function of sleep, as one sleep researcher noted, "may be the biggest open question in biology."

Most people probably believe that sleep is simply a time for the body to rest. However, no matter how quietly you may sit or lie down, watching TV or just meditating all day, you will still need to sleep. Another idea is that sleeping serves to protect various creatures by keeping them still and quiet at times when predators are more common. However, this does not explain the severe effects of sleep deprivation—and besides, a sleeping animal cannot run away or defend itself as well as an awake one.

Prior to 1953, most scientists believed that during sleep the brain was completely inactive. That was the year that a graduate student at the University of Chicago studied his sleeping son and noticed there were periods when his eyes darted rapidly back and forth. Subsequent research has shown that during this rapid eye movement (REM) sleep the brain is very active. REM sleep may be important for brain development; infants spend about 50% of their sleeping time in this state compared to 15–20% for adults. REM sleep may also be a time when the brain is building neurological connections to consolidate memories learned from experiences during that day. A 2004 study showed that subjects who were asked to work on a math problem that contained a "hidden solution" were about twice as likely to realize that solution when they returned after a night's sleep compared to subjects who spent an equal amount of time awake.

If a person is hooked up to an EEG during REM sleep, the brain appears to be awake, but the muscles are very still. If people are awakened during REM sleep, they are very likely to report that they have been dreaming, and dreams may also play a role in sorting through our experiences, saving some as memories and deleting others.

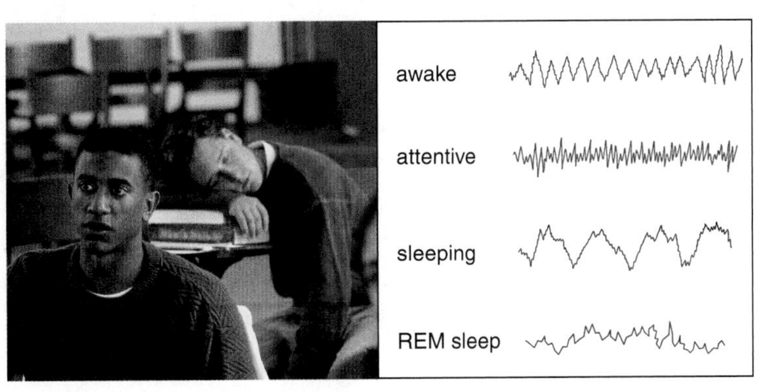

Figure 17A The sleeping brain.
In addition to an increased need for daytime naps , chronic sleep deprivation may lead to serious health problems. When we are awake, attentive, or in different stages of sleep, the brain produces characteristic types of electrical waves that can be measured with an EEG.

Another, often overlooked, task the brain must perform is forgetting. During the activities of an average day, the brain is taking in mountains of information by building new connections between neurons. Because maintaining these new connections requires energy, if some were not broken down, the brain would soon require more energy than the body could supply. A 2007 study suggested that this reduction of connections in the brain occurs during non-REM sleep. Perhaps sleep allows the brain time to "clean out" a large percentage of these memories, while solidifying others.

So if you have trouble sleeping, also known as insomnia, how can you get a better night's rest? Experts note that caffeine, nicotine, and even alcohol can disrupt sleep patterns. Exercising at night can raise metabolism when it needs to be slowing down for sleep. Taking naps longer than 15 minutes can make it harder to fall asleep later, as can exercising within three hours of bedtime. Regardless of the true purpose of sleep, all sleep researchers would agree that we can't rob ourselves of sleep for long, without serious consequences.

Discussion Questions

1. In one sentence, how would you define what sleep is?
2. Do you agree that the necessity of sleep for mammals and birds is one of the biggest mysteries in biology? What are some other big mysteries in biology?
3. Of the several ideas presented here regarding the purpose of sleep, which one makes the most sense to you? Which one seems the least likely?

It contains tracts that ascend or descend between the spinal cord and higher brain centers.

Electroencephalograms

The electrical activity of the brain can be recorded in the form of an **electroencephalogram (EEG).** Electrodes are taped to different parts of the scalp, and an instrument records the so-called brain waves. The EEG is a diagnostic tool. For example, an irregular pattern can signify epilepsy or a brain tumor. A flat EEG is often used as a clinical and legal criterion for brain death.

Check Your Progress

1. According to Figure 17.11, which parts of the body are represented by the largest amount of the primary motor area? The primary somatosensory area? Does the relative size of any of these areas come as a surprise to you?
2. Name the two major components of the diencephalon. What are their main functions?
3. What type of symptoms would you expect to see in a person who has sustained damage to their cerebellum? To their medulla oblongata?

17.3 The Limbic System and Higher Mental Functions

Learning Outcomes

1. Describe the anatomical and functional relationship between the limbic system and the cerebrum.
2. Distinguish between short-term, long-term, semantic, episodic, and skill memory.
3. Explain how certain diseases, accidents, and experiments have helped scientists understand some basic components of how memories are made.

Emotions and higher mental functions are associated with the limbic system in the brain. The limbic system blends primitive emotions (rage, fear, joy, sadness) and higher mental functions (reason, memory) into a united whole. It accounts for why activities such as sexual behavior and eating seem pleasurable and also for why, say, mental stress can cause high blood pressure.

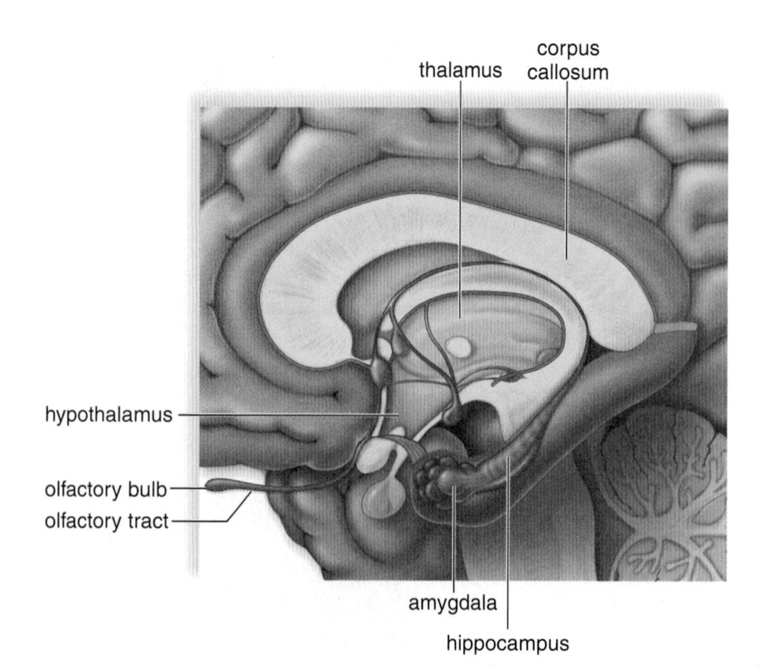

Figure 17.12 The limbic system.
Structures deep within the cerebral hemispheres and surrounding the diencephalon join higher mental functions such as reasoning with more primitive feelings such as fear and pleasure.

Anatomy of the Limbic System

The **limbic system** is a complex network of tracts and nuclei that incorporates portions of the cerebral lobes, the basal nuclei, and the diencephalon. Two significant structures within the limbic system are the hippocampus and the amygdala, which are essential for learning and memory (Fig. 17.12). The **hippocampus** is a seahorse-shaped structure deep in the temporal lobe that is well situated to communicate with the prefrontal area of the brain, which is also involved in learning and memory. The **amygdala** is an almond-shaped structure in the limbic system that allows us to respond to and display anger, avoidance, defensiveness, and fear. The amygdala prompts release of adrenaline and other hormones into the bloodstream. The inclusion of the frontal lobe in the limbic system means that we may be able to control strong feelings, despite the efforts of the amygdala to disrupt rational thought.

Higher Mental Functions

The well-developed human cerebrum is responsible for higher mental functions such as memory and learning, as well as language and speech. As in other areas of biological study, brain research has progressed due to technological breakthroughs. Neuroscientists now have a wide range of techniques at their disposal for studying the human brain, including modern technologies that allow us to record its functioning.

Memory and Learning

Just as the connecting tracts of the corpus callosum are evidence that the two cerebral hemispheres work together, so the limbic system indicates that cortical areas may work with lower centers to produce memory and learning. **Memory** is the ability to hold a thought in mind or to recall events from the past, ranging from a word we learned only yesterday to an early emotional experience that has shaped our lives. **Learning** takes place when we retain and utilize past memories.

Types of Memory The last time you decided to try a new pizza restaurant, you may have looked up the number and tried to remember it for a short period of time. If you said you were trying to keep it in the forefront of your brain, you were exactly correct. The prefrontal area, which is active during **short-term memory,** lies just behind our forehead! On the other hand, there are probably some telephone numbers that you have memorized. These have gone into **long-term memory**. Think of the phone number of someone close to you whom you call frequently. Can you bring the number to mind without also thinking about the place or person associated with that number? Most likely you cannot, because long-term memory is typically a mixture of **semantic memory** (of ideas, concepts, and meanings) and **episodic memory** (of specific facts, persons, events, etc.). Due to brain damage, some people lose one type of memory but not the other. For example, without a working episodic memory, they can carry on a conversation but have no recollection of recent

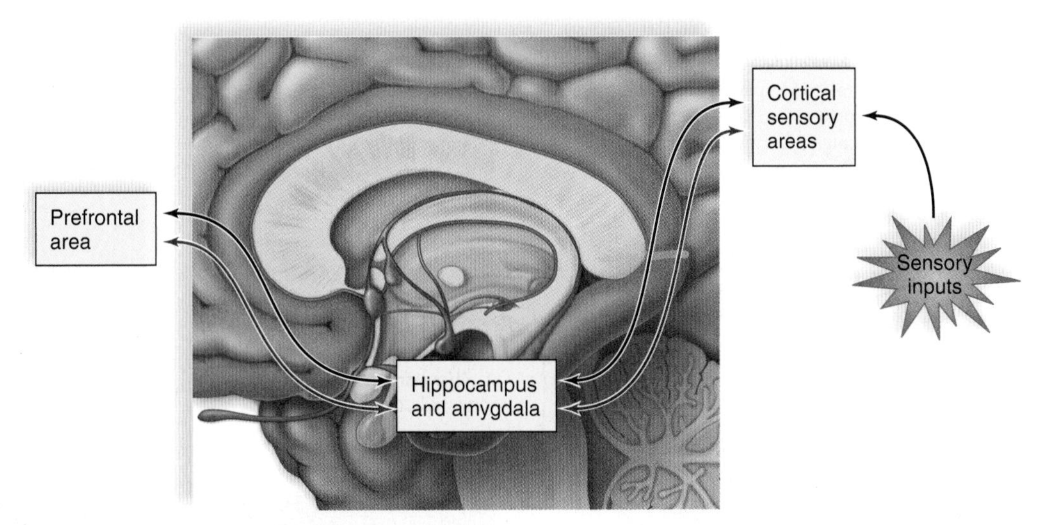

Figure 17.13 Long-term memory circuits.
The hippocampus and the amygdala are believed to be involved in the storage and retrieval of memories. Semantic memory (red arrows) and episodic memory (black arrows) are stored separately, and therefore, you can lose one without losing the other.

events. If you are talking to them and then leave the room, they don't remember you when you come back!

Skill memory is another type of memory that can exist independent of episodic memory. Skill memory is involved in performing motor activities, such as riding a bike, playing ice hockey, or typing a letter using a computer keyboard. When a person first learns a skill, more areas of the cerebral cortex are involved than after the skill is perfected. In other words, you have to think about what you are doing when you learn a skill, but later the actions become automatic. Skill memory involves all the motor areas of the cerebrum below the level of consciousness.

Long-Term Memory Storage and Retrieval The first step toward curing memory disorders is to know what parts of the brain are functioning when we remember something. Our long-term memories are stored in bits and pieces throughout the sensory association areas of the cerebral cortex. Visions are stored in the visual association area, sounds are stored in the auditory association area, and so forth. The hippocampus serves as a bridge between the sensory association areas and the prefrontal area. A classic case that demonstrated this involved a physicist named S.S., whose hippocampus was selectively destroyed by a herpesvirus infection. Even though he still had a high I.Q. and could remember childhood events and physics equations, S.S. forgot recent experiences within a few minutes. By observing S.S. and patients like him, we can surmise that the hippocampus must be involved in the conversion of short-term memories to long-term memories. As discussed in the Science Focus, different stages of sleep may be involved in the formation of memory connections in the brain.

So why are some memories so emotionally charged? The amygdala is responsible for fear conditioning and associating danger with sensory stimuli received from both the diencephalon and the cortical sensory areas. Figure 17.13 diagrammatically illustrates the long-term memory circuits of the brain.

In addition to studying what regions of the brain are involved in memory formation, neurobiologists want to know what exactly neurons are doing when we store memories and bring them back. They have discovered a chemical process called **long-term potentiation (LTP),** which is an enhanced response at synapses within the hippocampus. LTP seems to mainly involve glutamate, an amino acid that can function as a neurotransmitter. To prove this, investigators at the Massachusetts Institute of Technology produced mice that lacked the receptor for glutamate only in the hippocampus. Unlike control mice, the defective mice could not learn to run the maze, demonstrating an important role for this chemical in memory.

Check Your Progress

1. What would be some possible behavioral changes exhibited by a person who developed a tumor in his or her amygdala?
2. How did the unfortunate case of S.S. help prove the role of the hippocampus in long-term memory?
3. Summarize the evidence from the mouse study that showed the role of glutamate in memory formation.

Figure 17.14 Brain regions involved in language and speech.
a. The labeled areas are thought to be involved in speech comprehension and use. **b.** These PET images show the cortical pathway for reading words and then speaking them. Red indicates the most active areas of the brain, and blue indicates the least active areas.

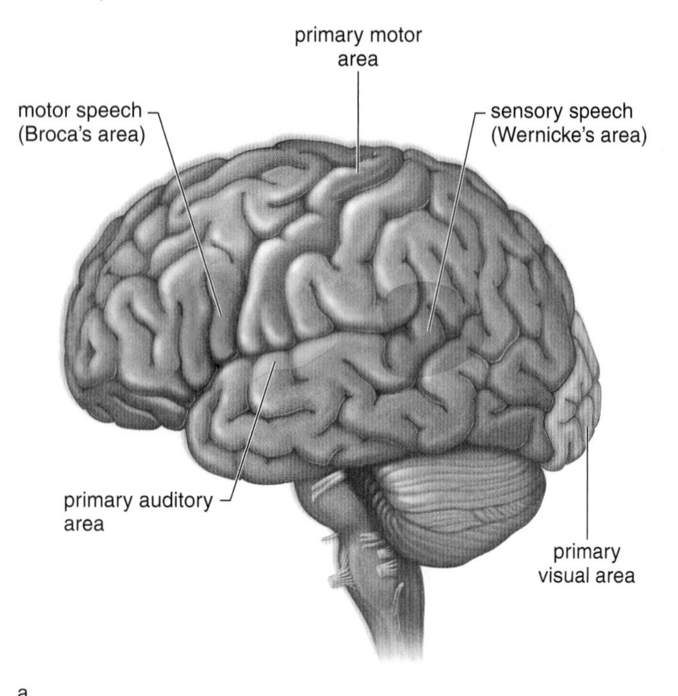

primary auditory cortex

visual cortex

Wernicke's area

1. The word is seen in the visual cortex.

2. Information concerning the word is interpreted in Wernicke's area.

primary motor cortex

Broca's area

3. Information from Wernicke's area is transferred to Broca's area.

4. Information is transferred from Broca's area to the primary motor area.

b.

Language and Speech

Language is obviously dependent upon semantic memory. Therefore, we would expect some of the same areas in the brain to be involved in both memory and language. Seeing words and hearing words depend on the primary visual cortex in the occipital lobe and the primary auditory cortex in the temporal lobe, respectively (Fig. 17.14*a*). And speaking words depends on motor centers in the frontal lobe. Functional imaging by a technique known as PET (positron emission tomography) supports these suppositions (Fig. 17.14*b*).

From studies of patients with speech disorders, it has been known for some time that damage to the motor speech (Broca's) area results in an inability to speak, while damage to the sensory speech (Wernicke's) area results in the inability to comprehend speech. Actually, any disruption of pathways between various centers of the brain can contribute to an inability to comprehend our environment and use speech correctly.

An additional observation pertaining to language and speech is the recognition that the left brain and right brain have different functions. Only the left hemisphere, not the right, contains a Broca's area and Wernicke's area. In an attempt to cure epilepsy in the early 1940s, the corpus callosum was surgically severed in some patients. Later studies showed that these split-brain patients could only name objects seen by the left hemisphere. If objects were viewed only by the right hemisphere, a split-brain patient could choose the proper object for a particular use but was unable to name it. Based on these and various other studies, the popular idea developed that the left brain can be contrasted with the right brain along these lines:

Left Hemisphere	Right Hemisphere
Verbal	Nonverbal, visuo-spatial
Logical, analytical	Intuitive
Rational	Creative

Further, researchers generally came to believe that one hemisphere was dominant in each person, accounting in part for personality traits. However, recent studies suggest that the hemispheres simply process the same information differently. The right hemisphere is more global in its approach, whereas the left hemisphere is more specific. For example, if only the left cerebral cortex is functional, a person has difficulty communicating the emotions involved with language. Therefore, both sides of the brain play a role in language use.

Check Your Progress

1. Describe the specific functions of Broca's and Wernicke's areas.
2. What is one disadvantage to having specialized areas of the brain on only one side, instead of duplicated?
3. Examine the table above and consider whether you are more of a left-brain-dominant or right-brain-dominant person.

17.4 The Peripheral Nervous System

Learning Outcomes

1. Describe the overall anatomy of the peripheral nervous system, including the cranial nerves and spinal nerves.
2. Explain how the somatic system differs from the autonomic system.
3. Contrast the overall functions of the sympathetic and parasympathetic divisions of the autonomic nervous system.

The peripheral nervous system (PNS) lies outside the central nervous system and is composed of nerves and ganglia. In the PNS, **nerves** are bundles of axons. The axons that occur in nerves are called nerve fibers.

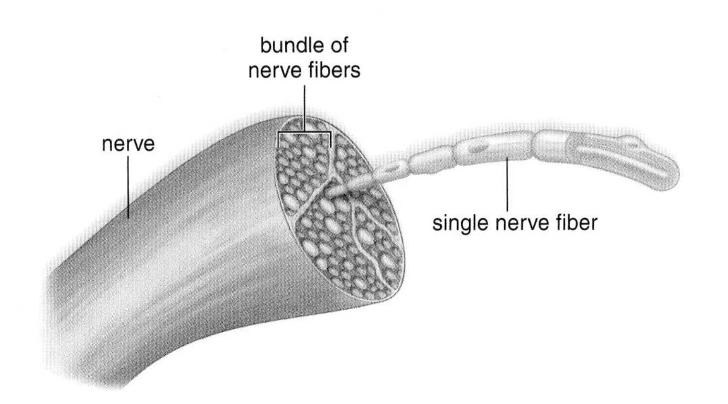

Sensory fibers carry information to the CNS, and motor fibers carry information away from the CNS. Ganglia (sing., **ganglion**) are swellings associated with nerves that contain collections of cell bodies.

Humans have 12 pairs of **cranial nerves** attached to the brain. By convention, the pairs of cranial nerves are referred to by Roman numerals (Fig. 17.15*a*). Some of the cranial nerves contain only sensory input fibers, while others contain only motor output fibers. The remaining are mixed nerves that contain both sensory and motor fibers. Cranial nerves are largely concerned with the head, neck, and facial regions of the body. However, the vagus nerve (X) has branches not only to the pharynx and larynx, but also to most of the internal organs.

The **spinal nerves** of humans emerge in 31 pairs from between openings in the vertebral column of the spinal cord. Each spinal nerve originates when two short branches, or roots, join together (Fig. 17.15*b*). The dorsal root (at the back) contains sensory fibers that conduct impulses inward (toward the spinal cord) from sensory receptors. The cell body of a sensory neuron is in a **dorsal root ganglion.** The ventral root (at the front) contains motor fibers that conduct impulses outward (away from the cord) to effectors. Notice, then, that all spinal nerves are mixed nerves that contain many sensory and motor fibers. Each spinal nerve serves the particular region of the body in which it is located. For example, the intercostal muscles of the rib cage are innervated by thoracic nerves.

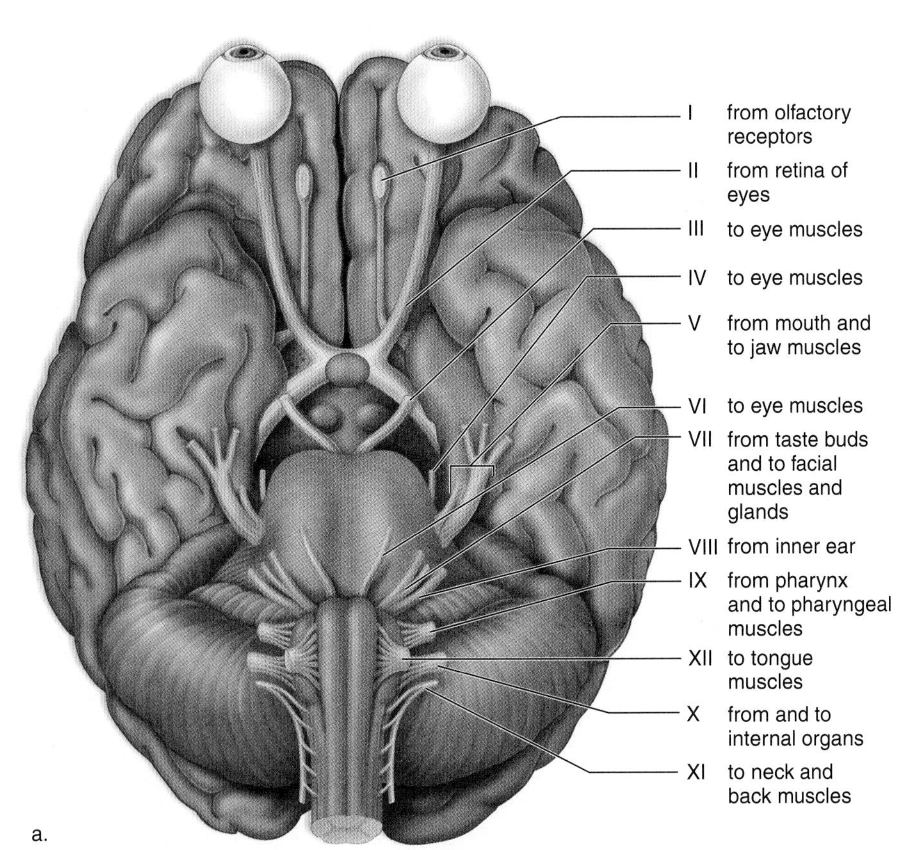

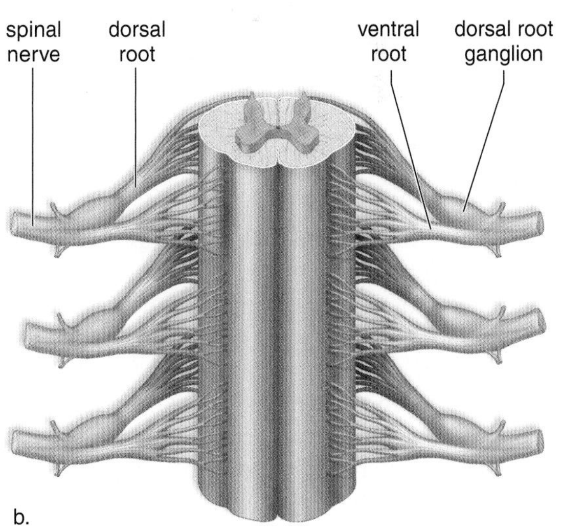

Figure 17.15 Cranial and spinal nerves.
a. Ventral surface of the brain showing the attachments of the 12 pairs of cranial nerves. **b.** Cross section of the spinal cord showing three pairs of spinal nerves. The human body has 31 pairs of spinal nerves altogether, and each spinal nerve has a dorsal root and a ventral root attached to the spinal cord.

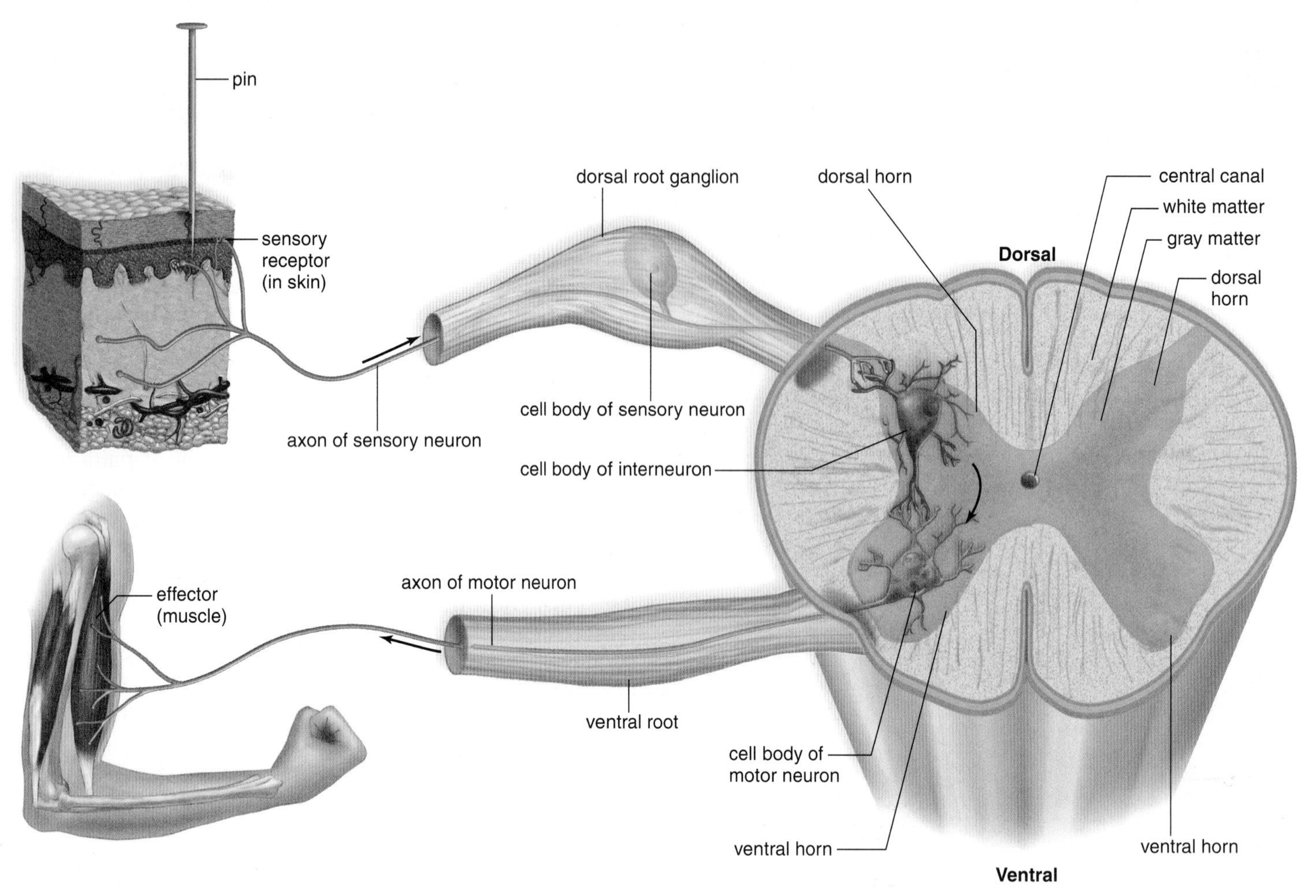

Figure 17.16 A somatic reflex arc showing the path of a spinal reflex.
A stimulus (e.g., sharp pin) causes sensory receptors in the skin to generate nerve impulses that travel in sensory axons to the spinal cord. Interneurons integrate data from sensory neurons and then relay signals to motor axons. Motor axons convey nerve impulses from the spinal cord to a sketetal muscle, which contracts. Movement of the hand away from the pin is the response to the stimulus.

Somatic System

The PNS is subdivided into the somatic system and the autonomic system. The **somatic system** serves the skin, skeletal muscles, and tendons. It includes nerves that take sensory information from external sensory receptors to the CNS and motor commands away from the CNS to the skeletal muscles. Some actions in the somatic system are due to **reflexes,** which are automatic responses to a stimulus. A reflex occurs quickly, without our even having to think about it. Other actions are voluntary, and these always originate in the cerebral cortex, as when we decide to move a limb.

The Reflex Arc

A reflex arc is a nerve pathway that carries out a reflex. Reflexes are programmed, built-in circuits that allow for protection and survival. They enable the body to react swiftly to stimuli that could disrupt homeostasis. They are present at birth and require no conscious thought to take place. Figure 17.16 illustrates the path of a withdrawal reflex that involves only the spinal cord. If your hand touches a sharp pin, sensory receptors in the skin generate nerve impulses that move along sensory fibers through the dorsal root ganglia toward the spinal cord. Sensory neurons that enter the cord dorsally pass signals on to many interneurons. Some of these interneurons synapse with motor neurons whose short dendrites and cell bodies are in the spinal cord. Nerve impulses travel along these motor fibers to an effector, which brings about a response to the stimulus. In this case, the effector is a muscle, which contracts so that you withdraw your hand from the pin. Various other reactions are also possible—you will most likely look at the pin, wince, and cry out in pain. This whole series of responses occurs because some of the interneurons involved carry nerve impulses to the brain. The brain makes you aware of the stimulus and directs these other reactions to it. You do not feel the pain until your brain receives the information and interprets it.

Check Your Progress

1. Define ganglion. What type of nerve cell body is in a dorsal root ganglion?
2. How does the vagus nerve differ from other cranial nerves?
3. If you touch a hot stove, which occurs first: withdrawing your hand or feeling the pain? Why?

Autonomic System

The **autonomic system** of the PNS regulates the activity of cardiac and smooth muscle and glands. The system is composed of the sympathetic and parasympathetic divisions (Fig. 17.17). These two divisions have several features in common: (1) they function automatically and usually in an involuntary manner; (2) they innervate all internal organs; and (3) for each signal, they utilize two motor neurons that synapse at a ganglion. The first neuron has a cell body within the CNS, and its axon is called the preganglionic fiber. The second neuron has a cell body within the ganglion, and its axon is termed the postganglionic fiber.

Reflex actions, such as those that regulate the blood pressure and breathing rate, are especially important to the maintenance of homeostasis. These reflexes begin when the sensory neurons in contact with internal organs send messages to the CNS. They are completed by motor neurons within the autonomic system.

Sympathetic Division

The cell bodies of preganglionic fibers in the **sympathetic division** are located in the middle, or thoracolumbar, portion of the spinal cord. The sympathetic division is especially important during emergency situations when you might be required to fight back or run away. It accelerates the heartbeat and dilates the bronchi. Active muscles, after all, require a ready supply of glucose and oxygen. On the other hand, the sympathetic division inhibits the digestive tract—digestion is not an immediate necessity if you are under attack.

One of the functions of the sympathetic division is to activate the adrenal medulla to secrete the hormones epinephrine (adrenaline) and norepinephrine (NE) into the blood (also see Chapter 20). In fact, one special preganglionic sympathetic neuron goes right to the adrenal gland without stopping at a ganglion. The adrenaline and NE released by the adrenal medulla bind to receptors on various cell types, adding to the "fight or flight" response. Interestingly, the neurotransmitter released by sympathetic postganglionic axons is also NE. A small amount of the NE released by these axons spills over into the blood. During times of high sympathetic nerve activity, the amount of NE entering the blood from these neurons increases significantly.

Because they cause the heart to beat stronger and faster and constrict certain blood vessels, adrenaline and NE tend to increase blood pressure. Certain drugs, called beta blockers, can be used in patients with hypertension to block the activity of adrenaline and NE, which slows the heart and decreases blood pressure.

Parasympathetic Division

The **parasympathetic division** includes a few cranial nerves (e.g., the vagus nerve), as well as fibers that arise from the sacral (bottom) portion of the spinal cord. Therefore, this division is often referred to as the craniosacral portion of the autonomic system. The parasympathetic division, sometimes called the housekeeper division, promotes all the internal responses we associate with "rest and digest." Opposite to the sympathetic system, it causes the pupil of the eye to contract, promotes digestion of food, and retards the heartbeat. The major neurotransmitter utilized by the parasympathetic division is acetylcholine (ACh).

Table 17.1 compares and contrasts the characteristics of the somatic and autonomic systems.

Check Your Progress

1. List several features of the autonomic system that are different from the somatic system.
2. Suppose you eat a big lunch, then go for a jog. If your stomach starts to ache, what could be the neurological explanation?

TABLE 17.1 Comparison of Somatic Motor and Autonomic Motor Pathways

Somatic Motor Pathway		Autonomic Motor Pathways	
		Sympathetic	*Parasympathetic*
Type of control	Voluntary/involuntary	Involuntary	Involuntary
Number of neurons per message	One	Two (preganglionic shorter than postganglionic)	Two (preganglionic longer than postganglionic)
Location of motor fiber	Most cranial nerves and all spinal nerves	Thoracolumbar spinal nerves	Cranial (e.g., vagus) and sacral spinal nerves
Neurotransmitter	Acetylcholine	Norepinephrine	Acetylcholine
Effectors	Skeletal muscles	Smooth and cardiac muscle, glands	Smooth and cardiac muscle, glands

Visual Focus

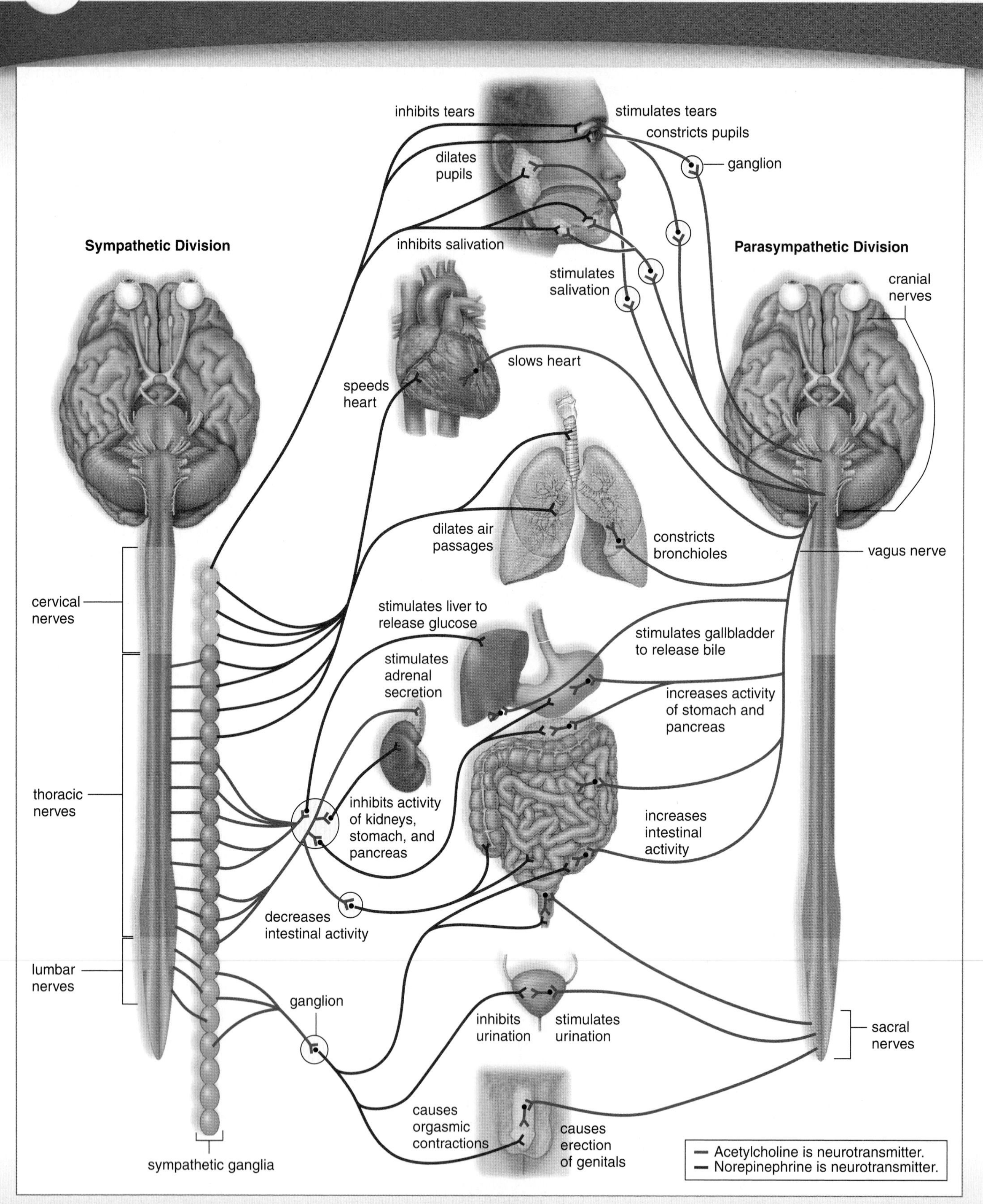

Figure 17.17 Autonomic system structure and function.
Sympathetic preganglionic fibers (*left*) arise from the cervical, thoracic, and lumbar portions of the spinal cord. Parasympathetic preganglionic fibers (*right*) arise from the cranial and sacral portions of the spinal cord. Each system innervates the same organs but has contrary effects.

17.5 Drug Abuse

Learning Outcomes

1. List three distinct ways that a drug could increase the effect of a neurotransmitter at a synapse.
2. Compare and contrast the specific mechanisms whereby the most common illicit drugs affect the brain.

A wide variety of drugs affect the nervous system and can alter the mood and/or emotional state. The first time someone tries a drug such as nicotine, alcohol, or cocaine, the person may experience strong feelings of pleasure. However, with continued use, larger amounts of the drug may be needed to create the same sensation. At some point, the users may become addicted, meaning they develop a physiological dependence on the drug and can't feel normal without it. At the extreme, drug addicts may prioritize getting "high" over food, hygiene, and interpersonal relationships. Even drug abusers who want to quit may face an enormous challenge from intense cravings and withdrawal symptoms when they stop taking a drug.

Most illicit drugs affect the action of a particular neurotransmitter at synapses in the brain (Fig. 17.18). Stimulants are drugs that increase the likelihood of neuron excitation, and depressants decrease the likelihood of excitation. Increasingly, researchers believe that dopamine is one of the main neurotransmitters in the brain that is responsible for mood. Cocaine is known to potentiate the effects of dopamine by interfering with its uptake from synaptic clefts (Fig. 17.19). Many of the new medications developed to counter drug dependence and mental illness affect the release, reception, or breakdown of dopamine.

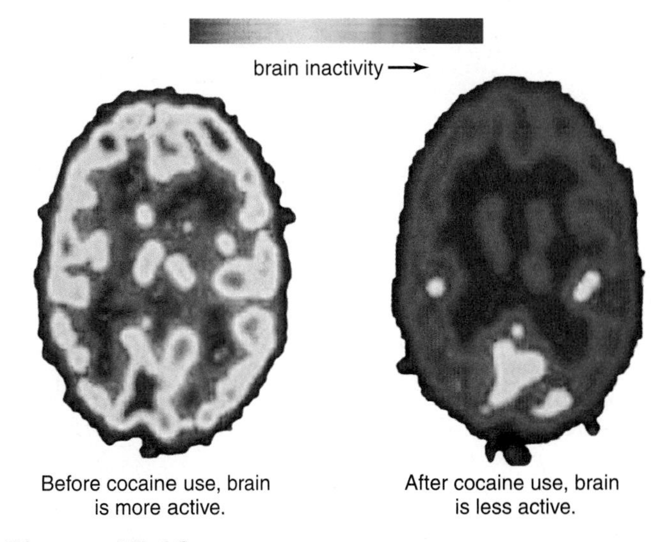

Figure 17.19 Drug abuse.
PET scans show that the usual activity of the brain is reduced and many areas become inactive after a person injects cocaine.

Some Specific Drugs of Abuse

Nicotine

In 2006, about 23% of U.S. high school students, and 8% of middle school students, were cigarette smokers. Young adults between ages 18 and 25 reported the highest usage of tobacco, at 45%. When a person smokes a cigarette or cigar, or chews tobacco, nicotine is quickly distributed throughout the body. This causes a release of epinephrine from the adrenal cortex, which increases blood sugar levels and causes an initial feeling of stimulation. Then, as blood sugar falls, depression and fatigue set in, leading the smoker to seek more nicotine. In the CNS, nicotine causes neurons to release excess dopamine, which has a reinforcing effect that leads to dependence on the drug. Nicotine induces both physiological and psychological dependence, and once they start, many tobacco users find it extremely difficult to give up the habit. Withdrawal symptoms include headache, stomach pain, irritability, and insomnia. Pregnant women who smoke are more likely to have stillborn or premature infants, or babies born with other health problems.

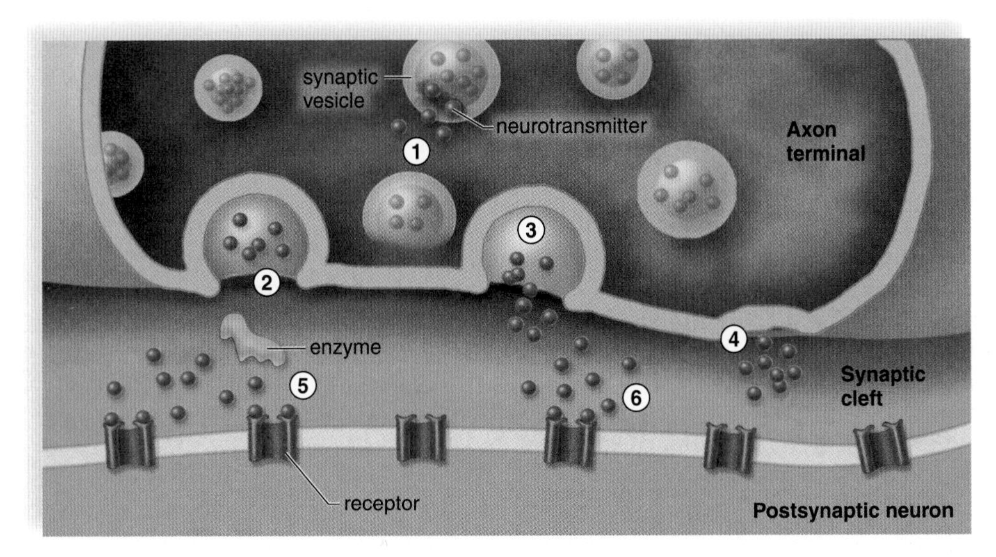

Figure 17.18
Drug actions at a synapse.
A drug can ① cause the neurotransmitter (NT) to leak out of a synaptic vesicle into the axon terminal; ② prevent release of NT into the synaptic cleft; ③ promote release of NT into the synaptic cleft; ④ prevent reuptake of NT by the presynaptic membrane; ⑤ block the enzyme that causes breakdown of the NT; or ⑥ bind to a receptor, mimicking the action of an NT.

Caffeine: Good or Bad for You?

Red Bull. Rockstar. Monster. Full Throttle. Walk down the beverage aisle of most U.S. grocery stores, and the colorful labels of these and other "energy drinks" compete for your attention (Fig. 17B). That's no surprise—the United States is the world's largest consumer of energy drinks, totaling roughly 290 million gallons in 2007.

Around 90% of American adults report using caffeine every day. Caffeine, or 1,3,7-trimethylxanthine, occurs naturally in over 60 plant species worldwide. In plants it is thought to function as a natural pesticide, killing certain insects that attempt to feed on the plant. It is naturally present in coffee, tea, and chocolate. It is also added to soft drinks and energy drinks, possibly to add flavor, though studies seem to prove that most people cannot taste the difference when caffeine is added to a caffeine-free beverage.

For humans, caffeine is a central nervous system stimulant. It readily crosses the blood-brain barrier, where it seems to block the effects of adenosine, which it resembles chemically. In the brain, adenosine normally seems to suppress brain activity, and by blocking this effect, caffeine can increase alertness. However, in response to continuing doses of caffeine, brain cells adjust by producing more adenosine receptors, so that increased doses of caffeine are needed for the same effect. If a regular user suddenly stops ingesting caffeine, withdrawal symptoms—such as headache, inability to concentrate, and insomnia—may ensue within 12 to 24 hours and last from one to five days.

Americans still get most of their caffeine from that tried-and-true source, the cup of coffee, which averages about 100 mg of caffeine. By comparison, a 12 oz can of Diet Coke contains about 38 mg (about 10 mg more than regular Coca-Cola), Mountain Dew has 55 mg, and Diet Pepsi Max has 69 mg. Energy drinks vary even more, with Red Bull at 80 mg, Full Throttle and Monster doubling that to around 160. At the extreme end of the spectrum, a 20 oz bottle of Fixx Energy Drink contains 500 mg of caffeine! In 2008, 100 scientists and physicians wrote a letter to the U.S. Food and Drug Administration asking for more regulation and better labeling of increasingly popular energy drinks because their high caffeine content puts young drinkers at possible risk for caffeine intoxication.

Figure 17B A sampling of energy drinks found in most U.S. grocery stores.
The caffeine content of these beverages varies widely.

As a stimulant, high doses of caffeine can have serious consequences. Adverse effects seen with high doses of caffeine include nervousness, irritability, tremors, insomnia, headaches, and high blood pressure. Caffeine reduces absorption of iron from the intestine, and because it acts as a diuretic (see Chapter 16), increases calcium loss in the urine. Some individuals, perhaps 20%, are more sensitive to lower levels of caffeine, so even a few milligrams may cause problems for them. Pregnant and nursing women should also be careful to limit their caffeine intake. Fortunately, caffeine-induced deaths are rare. The lethal dose for adults has been estimated at 5,000 mg.

So there is definitely a downside to caffeine use (or abuse). But might there be an upside? In 2008, researchers at Harvard published results of an 18-year study of thousands of men and women. They found that women who drank two to three cups of caffeinated coffee per day had a 25% lower risk of death from heart disease, and women who drank five to seven cups per week had a 7% lower risk of death from all causes than non-drinkers. For men, although the benefits were not statistically significant, the scientists found no health risk to drinking up to seven cups of coffee per day. Other studies have suggested a link between caffeine and increased athletic and academic performance, as well as a reduced risk of Parkinson disease, migraine headaches, and even diabetes. So although the debate over caffeine's risks versus benefits may never be completely resolved, for many people, it is possible that consumption of a moderate amount of caffeine may end up doing more good than harm.

Discussion Questions

1. Estimate how much caffeine you consume per day. What effects can you notice after consuming caffeine?
2. Would you agree that manufacturers should be required to display the caffeine content of soft drinks and energy drinks on the label? Why or why not?
3. Do you believe that there should be any restrictions on selling, for example, an energy drink containing 500 mg of caffeine? If so, what should the minimum age be?

Alcohol (Ethanol)

With the possible exception of caffeine (see the Science Focus, above), alcohol is the most used and abused drug among America's teenagers. According to a 2006 national survey, nearly one third of all high school students reported hazardous drinking (five-plus drinks in one setting) during the 30 days preceding the survey. Among adults, 61% had consumed alcohol in the past year, and 33% had more than five drinks on a single day.

When ingested, ethanol acts as a drug in the brain by influencing the action of gamma aminobutyrate (GABA), an inhibitory neurotransmitter, and other neurotransmitters like serotonin and dopamine. Alcohol is primarily metabolized in the liver, where it disrupts the normal workings of this organ so that fats cannot be broken down. Fat accumulation, the first stage of liver deterioration, begins after only a single night of heavy drinking. If heavy drinking continues, fibrous scar tissue is deposited during the next stage of liver damage. If heavy drinking stops at this point, the liver may still recover and become normal again. If not, the final and irrevocable stage, cirrhosis of the liver, occurs. Liver cells die, harden, and turn orange (cirrhosis means orange).

Alcohol is a carbohydrate and as such, can be converted into energy for the body to use. However, alcoholic beverages lack the vitamins, minerals, essential amino acids, and fatty acids the body needs. For this reason, many alcoholics are vitamin-deficient, undernourished, and prone to illness.

Based upon extensive research, the Surgeon General of the United States recommends that pregnant women drink no alcohol at all. Alcohol crosses the placenta freely and causes fetal alcohol syndrome in newborns, which is characterized by mental retardation and various physical defects.

Marijuana

Marijuana is the most commonly used illegal drug in the United States. Surveys vary, but in 2006 about 16% of young adults aged 18–25 reported using marijuana in the last month, and 40% of the U.S. population has tried it at least once. Marijuana, or "pot," refers to the dried flowering tops, leaves, and stems of the Indian hemp plant *Cannabis sativa,* which contain and are covered by a resin that is rich in THC (tetrahydrocannabinol). Researchers have found that THC binds to a receptor in the brain for anandamide, a naturally occurring neurotransmitter that is important for short-term memory processing, and perhaps for feelings of contentment. Usually, marijuana is smoked in a cigarette form called a "joint," or in pipes or other devices, but it can also be consumed. The occasional user experiences a mild euphoria along with alterations in vision and judgment, which result in distortions of space and time. Motor incoordination, including the inability to speak coherently, takes place. Heavy use can result in hallucinations, anxiety, depression, body image distortions, paranoid reactions, and similar psychotic symptoms. Some researchers believe that long-term marijuana use leads to brain impairment. Fetal cannabis syndrome, which resembles fetal alcohol syndrome, has also been reported.

Check Your Progress

1. Because most illicit drugs affect activity of neurotransmitters, examine Figure 17.18 and hypothesize why it might take increasing amounts of a drug to obtain the same effect over time.
2. Which drug do you believe causes the most harm to users: nicotine, alcohol, or marijuana? Why? Which does the most harm to society?

Cocaine and Crack

Cocaine is an alkaloid derived from the shrub *Erythroxylon coca.* It is sold in powder form and as crack, a more potent extract. Crack is cocaine that comes in a rocklike crystal that can be heated and its vapors inhaled. The term "crack" refers to the crackling sound heard when it is heated. Cocaine prevents the synaptic uptake of dopamine, and this causes the user to experience a rush sensation. The epinephrine-like effects of dopamine account for the state of arousal that lasts for several minutes after the rush experience.

A cocaine binge can go on for days, after which the user suffers a crash. During the binge period, the user is hyperactive and has little desire for food or sleep but an increased sex drive. During the crash period, the user is fatigued, depressed, and irritable, has memory and concentration problems, and displays no interest in sex. Indeed, men who use "coke" often become impotent.

Cocaine causes extreme physical dependence. With continued use, the body begins to make less dopamine to compensate for a seeming excess supply. The user subsequently experiences tolerance, withdrawal symptoms, and an intense craving for cocaine. Overdosing on cocaine can cause seizures and cardiac and respiratory arrest. It is possible that long-term cocaine abuse causes brain damage, and babies born to addicts suffer withdrawal symptoms and may have neurological and behavioral problems.

Heroin

Heroin is derived from morphine, an alkaloid of opium. Once heroin is injected into a vein, a feeling of euphoria, along with relief of any pain, occurs within three to six minutes. Heroin binds to receptors meant for the endorphins, naturally occurring neurotransmitters that kill pain and produce a feeling of tranquility. With repeated heroin use, the body's production of endorphins decreases. Tolerance develops so that the user needs to take more of the drug just to prevent withdrawal symptoms, and the original euphoria is no longer felt. Heroin withdrawal symptoms include perspiration, dilation of pupils, tremors, restlessness, abdominal cramps, vomiting, and increased blood pressure and respiratory rate. People who are excessively dependent may experience convulsions, respiratory failure, and death. Infants born to women who are physically dependent also experience those withdrawal symptoms.

"Club" Drugs

Ecstasy (MDMA), Rohypnol, and ketamine are examples of drugs that are abused by many teens and young adults who attend night-long dances called raves or trances. Of course, these drugs may also be abused by others as well. MDMA is chemically similar to methamphetamine. Many users say that "XTC" increases their feelings of well-being and friendliness, even love, for other people. However, it can cause many of the same side effects as other stimulants, such as increase in heart rate and blood pressure, muscle tension, and blurred vision. In high doses, MDMA can interfere with temperature regulation, leading to hyperthermia followed by damage to the liver, kidneys, and heart. Chronic MDMA use can also damage memory and lead to depression. Perhaps because of these side effects, several recent surveys have reported that MDMA use has been decreasing among youths aged 12–17.

Rohypnol is in the same class of drugs as Valium (diazepam), a popular sedative. When mixed with alcohol, Rohypnol, or "roofies," can render victims incapable of resisting, for example, sexual assault. It can also produce anterograde amnesia, which means victims may not remember events they experienced while under the influence of the drug. Ketamine, or "Special K," is an anesthetic that veterinarians use to perform surgery on animals. When administered to humans, it typically induces a dreamlike state, sometimes with hallucinations. Because it can render the user unable to move, ketamine has also been used as a date rape drug. At higher doses, ketamine can cause dangerous reductions in heart and respiratory functions.

Methamphetamine

Methamphetamine, also called "meth" or "crank," is a powerful CNS stimulant. Meth is often synthesized or "cooked" in makeshift home laboratories, usually starting with either ephedrine or pseudoephedrine, common ingredients in many cold and asthma medicines. Many states have passed laws making these medications more difficult to purchase. Meth is usually produced as a powder (speed) that can be snorted, or as crystals (crystal meth) that can be smoked. The most immediate effect is an initial "rush" of euphoria, due to high amounts of dopamine released in the brain. The user may also experience an increased energy level, alertness, and elevated mood. After the initial rush, a state of high agitation typically occurs that, in some individuals, leads to violent behavior. Chronic use can lead to paranoia, irritability, hallucinations, insomnia, tremors, hyperthermia, cardiovascular collapse, and death. Similar to cocaine, long-term meth abuse renders the brain less able to produce normal levels of dopamine. The user may have strong cravings for more meth, without being able to reach a satisfactory high.

Check Your Progress

1. Suppose a form of heroin could be synthesized that had only the desired effect (i.e., feelings of euphoria and pain relief) with none of the side effects. Could you see any harm in taking such a drug on a regular basis?
2. Do you think that any drugs that are legal should be illegal? Should any drugs that are illegal be legal? Why or why not?

17.6 Disorders of the Nervous System

Learning Outcomes

1. Describe two abnormalities seen in the brain of Alzheimer disease patients.
2. Compare the pathogenesis (mechanism causing disease) of Parkinson disease, MS, stroke, meningitis, and prion diseases.
3. Describe the symptoms seen in several disorders of the brain and spinal cord and discuss why it is difficult to find cures for these.

A myriad of abnormal conditions can affect the nervous system. We will consider only a few of the most serious, common, and/or interesting ones here.

Disorders of the Brain

Alzheimer disease (AD) is the most common cause of dementia, an impairment of brain function significant enough to interfere with the patient's ability to carry on daily activities. Signs of AD may appear before the age of 50, but the condition is usually seen in individuals past the age of 65. It is estimated that at least 40% of individuals older than 80 develop AD. One of the early symptoms is loss of memory, particularly for recent events. Characteristically, the person asks the same question repeatedly and becomes disoriented even in familiar places. Gradually, the person loses the ability to perform any type of daily activity and becomes bedridden. Death is often due to pneumonia resulting from the patient's debilitated state.

In AD patients, abnormal neurons are present throughout the brain but especially in the hippocampus and amygdala. These neurons have two abnormalities: (1) plaques, containing a protein called beta amyloid, envelop the axons, and (2) neurofibrillary tangles are in the axons (Fig. 17.20) and may extend

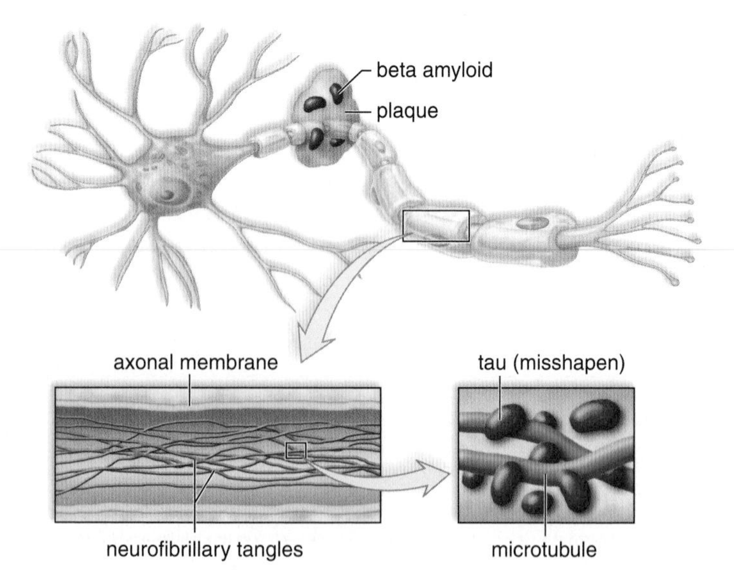

Figure 17.20 Alzheimer disease.
Some of the neurons of patients with AD have beta amyloid plaques and neurofibrillary tangles. AD neurons are present throughout the brain but concentrated in the hippocampus and amygdala.

upward to surround the nucleus. The neurofibrillary tangles arise because a protein called tau no longer has the correct shape. Tau holds microtubules in place so that they support the structure of neurons. When tau changes shape, it grabs onto other tau molecules, and the tangles result.

Researchers are working to discover the cause of beta amyloid plaques and neurofibrillary tangles. Several genes that predispose a person to AD have been identified, but it is unknown why inheritance of these genes leads to AD. Moreover, although several treatments have been tried to date, none has been shown to prevent or cure AD. A variety of medications can help ease symptoms such as memory loss, agitation, and depression, but brain cells still die. Alzhemed (tramiprosate), an experimental drug in clinical trials since 2004, has shown promise in reducing levels of beta amyloid in AD patients.

Parkinson disease (PD) is characterized by a gradual loss of motor control, typically beginning between the ages of 50 and 60. Eventually the person develops a wide-eyed, unblinking expression, involuntary tremors of the fingers and thumbs, muscular rigidity, and a shuffling gait. Speaking and performing ordinary daily tasks becomes laborious. In Parkinson patients, the basal nuclei (see section 17.2) function improperly because of a degeneration of the dopamine-releasing neurons in the brain. Without dopamine, the excessive excitatory signals from the motor cortex result in the symptoms of PD.

Unfortunately, it is not possible to give PD patients dopamine directly because of the impermeability of the capillaries serving the brain. However, symptoms can be alleviated by giving patients L-dopa, a chemical that can be changed to dopamine in the body, until too few cells are left to do the job. Then patients must turn to a number of controversial surgical procedures. In 1998 the actor Michael J. Fox, who had been diagnosed with PD seven years earlier at the age of 30, had a surgical procedure called a thalamotomy, which has been successful in reducing the symptoms of some PD patients. However, in his biography *Lucky Man: A Memoir*, Fox admits that after a temporary improvement, his PD symptoms returned in full force. In 2007, the first report was published on treating PD with gene therapy, in which a virus engineered to produce the required neurotransmitter was injected into the brains of 12 PD patients. In one year, motor function had improved in these patients from 25–65% of normal.

Multiple sclerosis (MS) is the most common neurological disease that afflicts young adults. Nearly 350,000 people in the United States have MS, with approximately 10,000 new cases diagnosed each year. MS affects the myelin sheath of neurons in the white matter of the brain. Although the initiating cause remains unknown, MS is considered an autoimmune disease in which the patient's own white blood cells attack the myelin, oligodendrocytes, and eventually, neurons in the CNS. Specifically, MS is characterized by an infiltration of lymphocytes and macrophages into the brain, brain stem, optic nerves, and spinal cord. The most common symptoms of MS include fatigue, vision problems, limb weakness and/or pain, and abnormal sensations such as numbness or tingling. These symptoms, however, can be similar to those of several other brain disorders, so a technique called magnetic resonance imaging (MRI) can be very helpful in both diagnosing and monitoring the disease (Fig. 17.21). MS can also take several clinical forms, from the milder relapsing-remitting form to more relentlessly progressive forms. Although MS is not generally considered a fatal disease, people with MS die an average of seven years earlier than the general population. No treatment has been shown to reverse the damage to the myelin of MS patients, but several medications are effective in slowing the progression of the disease, especially in the early stages.

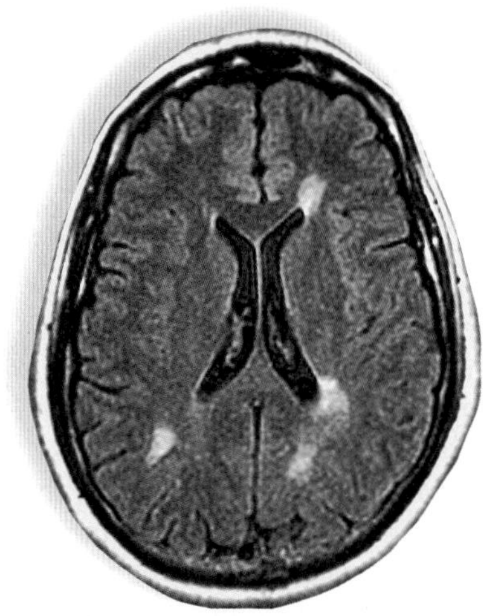

Figure 17.21
Multiple sclerosis.
This MRI scan shows the brain of an MS patient. The white areas, also known as MS plaques, are sites of active inflammation and myelin destruction.

CHECK YOUR PROGRESS

1. What are the major symptoms of Alzheimer disease? At what age can these symptoms appear?
2. How does L-dopa work to treat Parkinson disease?
3. What does it mean to say that MS is an autoimmune disease?

A **stroke** results in disruption of the blood supply to the brain. There are two major forms of stroke, hemorrhagic and ischemic. In hemorrhagic stroke, bleeding occurs into the brain due to leakage from small arteries, often after years of damage by high blood pressure. Ischemic stroke occurs when there is a sudden loss of blood supply to an area of the brain, usually due to a thrombus or thromboembolism (see Chapter 12). In both types of stroke, the symptoms depend on the amount and specific area of brain tissue affected. However, some degree of paralysis and aphasia (loss of speech) are common signs. The risk of stroke increases with age, and African Americans have a higher incidence than other races in the United States. Physicians treating stroke patients must be careful to determine which type they are dealing with. For example, tissue plasminogen activator (tPA) can be used to help dissolve clots when treating ischemic stroke, but could have disastrous consequences if administered to a hemorrhagic stroke patient!

Meningitis is an infection of the meninges that surround the brain and spinal cord. Most cases of meningitis are caused by bacteria or viruses that have usually gained access to the meninges after first infecting the blood, middle ear, or sinuses. Certain viruses are also capable of infecting the brain and meninges by infecting peripheral nerve cells and being transported back to the brain. Because cells of the immune system have somewhat limited access to the brain, the infection may spread into the brain tissue. Bacterial meningitis is especially serious, even life-threatening if untreated. Physicians may diagnose meningitis based on clinical signs such as a stiff neck, fever, and headache, and may confirm the diagnosis by examining a sample of cerebrospinal fluid, usually obtained by a lumbar puncture. Most cases of bacterial meningitis are caused by organisms called *Haemophilus influenzae* or *Neisseria meningitidis.* Outbreaks of *N. meningitidis,* also called meningococcus, have occurred recently among college students. Fortunately, vaccines are available to prevent infection with both organisms.

Several brain diseases of humans and animals are caused by prions, which are infectious agents that most researchers believe are composed of protein only (see Chapter 28). Examples of human diseases include kuru, Creutzfeldt-Jakob disease, and fatal familial insomnia. In animals, scrapie, mad cow disease, and chronic wasting disease are seen. Mad cow disease seems able to cross the species barrier and infect humans who consume infected tissues; over 100 human cases have been confirmed, mostly in the United Kingdom. All prion diseases are characterized by long incubation periods between initial exposure to the agent and onset of clinical signs, followed by rapid progress of the disease and, uniformly, death. Prion diseases are thought to result from the accumulation in the tissues of an abnormal form of a normal host (prion) protein. When this occurs in the brain, nerve cells die, forming areas that look like holes, which pathologists refer to as spongiform changes (Fig. 17.22). These diseases are also unique in that they can be inherited or can be spread by either eating infected tissues (as in mad cow disease) or through transplantation of certain organs or tissues (such as corneas) from an infected individual. Therefore, prion diseases are also called transmissible spongiform encephalopathies (TSEs). No treatment has yet been shown to delay the inevitably fatal progression of these diseases.

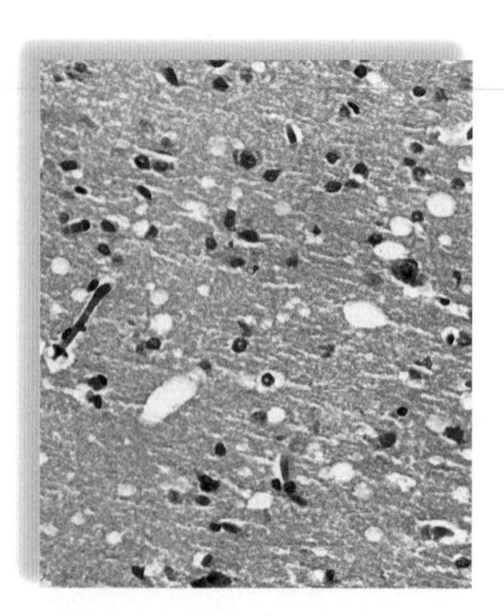

Figure 17.22
Effect of prions on the brain.
In this stained section of brain tissue from a patient with a prion disease, the spongiform changes (holes) in the tissue are due to accumulation of the abnormal prion protein.

Check Your Progress

1. In terms of treatment options, why is it important to determine whether a stroke is hemorrhagic or ischemic?
2. One cause of bacterial meningitis, *H. influenzae,* can also be found in the throat (pharynx) of healthy people. What factors might determine when *H. influenzae* causes disease?
3. What are some precautions you can take to reduce the risk of contracting a prion disease?

Disorders of the Spinal Cord

Spinal cord injuries may result from car accidents or other trauma. Because little or no nerve regeneration is possible in the CNS, any resulting disability is usually permanent. The cord may be completely cut across (called a transection) or only partially severed (a partial section). The location and extent of the damage produce a variety of effects, depending on the partial or complete stoppage of impulses passing up and down the spinal cord. If the spinal cord is completely transected, no sensations or somatic motor impulses are able to pass the point of injury. If the injury is between the first thoracic vertebra (T1) and the second lumbar vertebra (L2), paralysis of the lower body and legs occurs, a condition known as paraplegia (see Fig. 19.6 for the specific locations of these vertebrae). If the injury is in the neck region, the entire body below the neck is usually affected, a condition called quadriplegia. If the injury is above the second or third cervical vertebrae (C2 or C3), nervous control of respiration is also affected. In 1995, actor Christopher Reeve, best known for his movie role as "Superman," was thrown from his horse, crushing the top two cervical vertebrae. He immediately lost control of his arms and legs, and had to be placed on a respirator. Through an intense exercise and physical therapy program, Reeve regained the ability to move his left index finger and could take tiny steps while being held upright in a pool. Unfortunately, he died of cardiac arrest in 2004.

Amyotrophic lateral sclerosis (ALS), also known as Lou Gehrig's disease, is a devastating condition that affects the motor nerve cells of the spinal cord. ALS is incurable, and most people die within five years of being diagnosed, mainly due to failure of the respiratory muscles. Typically the first signs of ALS occur between the ages of 35 and 55 in the form of mild weakness or tremors in one or more limbs, starting in the fingers or toes and gradually involving the larger muscles nearer to the body. There may be multiple causes of nerve cell death in ALS. For example, decreased levels of a protein called insulin-like growth factor-1 have been reported in ALS patients, and autoimmune reactions may play a role. Other than Lou Gehrig, the most famous person with ALS is probably physicist Steven Hawking, who has

Figure 17.23 Physicist Steven Hawking.
Dr. Hawking has been a productive scientist and author for over 30 years, even though he has amyotrophic lateral sclerosis.

lived for over 30 years with the disease (Fig. 17.23). The fact that Hawking has remained an active scientist and author shows that ALS has very little effect on brain function.

Disorders of the Peripheral Nerves

Guillain-Barré syndrome (GBS) is an inflammatory disease that causes demyelination of peripheral nerve axons. It is thought to result from an abnormal immune reaction to one of several types of infectious agents. In GBS patients, antibodies (see Chapter 13) formed against these bacteria or viruses seem to cross-react with the myelin. Symptoms usually begin with weakness or unsteadiness two to four weeks after an infection or immunization. The muscle weakness typically starts in the lower limbs and ascends to the upper limbs over a period of days to weeks. The disease process may eventually affect the respiratory muscles, so that mechanical ventilation is required. In most cases, however, the paralysis begins to improve in a few weeks, and most patients make a full recovery within six to twelve months.

Myasthenia gravis (MG) is an autoimmune disorder in which antibodies are formed that react against the acetylcholine receptor (AchR) at the neuromuscular junction of the skeletal muscles. Normally, when the action potential arrives at the presynaptic terminal of peripheral motor nerves, acetylcholine (ACh) is released into the synaptic cleft and binds to the AchR on the muscle cell, stimulating contraction. In MG, antibodies bind to the AchR and block binding of ACh, preventing muscle stimulation.

Interestingly, the most common symptom of MG is weakness of the muscles of the eye or eyelid, although over time the weakness usually spreads to the face, trunk, and then the limbs. In severe cases, the respiratory muscles can be affected, leading to death if untreated. Even though there is no cure, MG patients often respond very well to treatment with immunosuppressive drugs and inhibitors of acetylcholinesterase (AChE), an enzyme that normally destroys the ACh after it is released. A technique called plasmapheresis can also be used in both GBS and MG to remove the harmful antibodies from the patient's blood.

Check Your Progress

1. Christopher Reeve was an advocate for stem cell research. How might stem cells be used to treat spinal cord injuries?
2. What advantage might be gained by bacteria or viruses that produce molecules that "mimic" the molecular structure of myelin?
3. Explain how each of the following treatments targets the cause of MG in different ways: immunosuppressive drugs, AChE inhibitors, plasmapheresis.

Applying the Concepts Revisited

MS is an autoimmune disease affecting the myelin. Years after her initial diagnosis, Sarah's condition continues to slowly deteriorate. Although she is still able to drive, she must walk with a cane and has occasional issues with her vision, plus fatigue, weakness, and other problems.

1. What is the function of myelin? Would this be more of a problem with the white matter or gray matter of the brain? What specific type of neurological process would be affected if myelin is damaged?
2. MS affects the CNS only, not the PNS. What cells synthesize myelin in these two systems, and how might this help to explain why MS affects myelin in the CNS only?

Summarizing the Concepts

17.1 Nervous Tissue

A neuron is composed of dendrites, a cell body, and an axon. Long axons are covered by a myelin sheath. There are three types of neurons:

- Sensory neurons take information from sensory receptors to the CNS.
- Interneurons occur within the CNS.
- Motor neurons take information from the CNS to effectors (muscles or glands).

In order to transmit a nerve impulse, an action potential must be generated, and the impulse must travel across synapses:

- When an axon is not conducting a nerve impulse, the inside of the axon is negative (−65 mV) compared to the outside. The sodium-potassium pump actively transports Na^+ out of an axon and K^+ to the inside of an axon. The resting potential is due to the leakage of K^+ to the outside of the neuron. When an axon is conducting a nerve impulse (action potential), Na^+ first moves into the axoplasm, and then K^+ moves out of the axoplasm.

Chemicals called neurotransmitters are involved in the transfer of messages between neurons, or between neurons and muscle cells:

- After being released by the presynaptic neuron, binding of the neurotransmitter to receptors in the postsynaptic membrane causes excitation or inhibition.
- Integration is the summing of excitatory and inhibitory signals.

17.2 The Central Nervous System

The CNS receives and integrates sensory input and formulates motor output. The CNS consists of the spinal cord and brain, which are both protected by bone:

- The spinal cord sends sensory information to the brain, receives motor output from the brain, and carries out reflex actions. The gray matter of the spinal cord contains neuron cell bodies. The white matter consists of myelinated axons that occur in bundles called tracts.
- The brain can be divided into several functional and anatomical regions. Sensation, reasoning, learning and memory, and language and speech take place in the cerebrum. The cerebrum has two cerebral hemispheres connected by the corpus callosum. The cerebral cortex is a thin layer of gray matter covering the cerebrum. The cerebral cortex of each cerebral hemisphere has four lobes: a frontal, parietal, occipital, and temporal lobe. The primary motor area, in the frontal lobe, sends out motor commands to lower brain centers, which pass them on to motor neurons. The primary somatosensory area, in the parietal lobe, receives sensory information from lower brain centers in communication with sensory neurons. The cortex also contains processing centers for reasoning and speech.
- Other regions of the brain include the hypothalamus, which controls homeostasis, and the thalamus, which specializes in sending sensory input on to the cerebrum. The cerebellum primarily coordinates skeletal muscle contractions, and the medulla oblongata and pons have centers for vital functions, such as breathing and the heartbeat.

17.3 The Limbic System and Higher Mental Functions

The limbic system is associated with emotions as well as higher brain functions:

- The limbic system contains the hippocampus, which acts as a conduit for sending messages to long-term memory and retrieving them once again, and the amygdala, which adds emotional overtones to memories In addition, it has various connections between the cerebral cortex and the hypothalamus, the thalamus, and the basal nuclei.

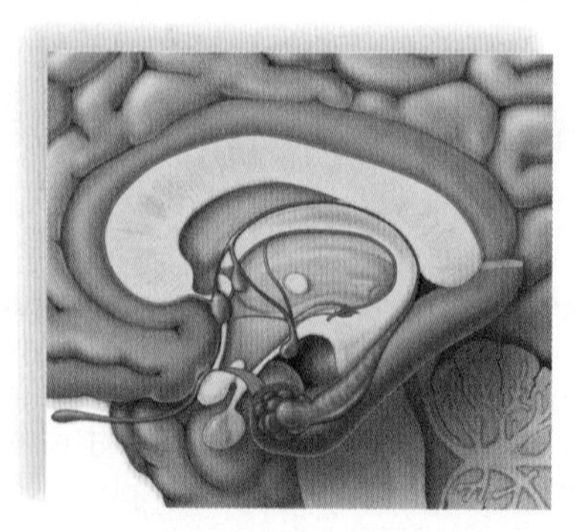

Higher brain functions include memory and learning, as well as language and speech:

- Two basic types of memory are short-term and long-term memory. The hippocampus is essential for formation of long-term memories. At the chemical level, this process involves glutamate and is known as long-term potentiation.
- Language and speech involve many areas of the brain, but Broca's area and Wernicke's area are indispensable. The right and left hemispheres play different, though overlapping, roles.

17.4 The Peripheral Nervous System

The PNS contains only nerves and ganglia. It can be subdivided into the somatic system and the autonomic system:

- The somatic system takes information from external sensory receptors to the CNS and voluntary motor commands from the cerebrum to skeletal muscles. It also contains many reflex arcs in which interneurons in the spine quickly send messages to motor nerves without CNS input.

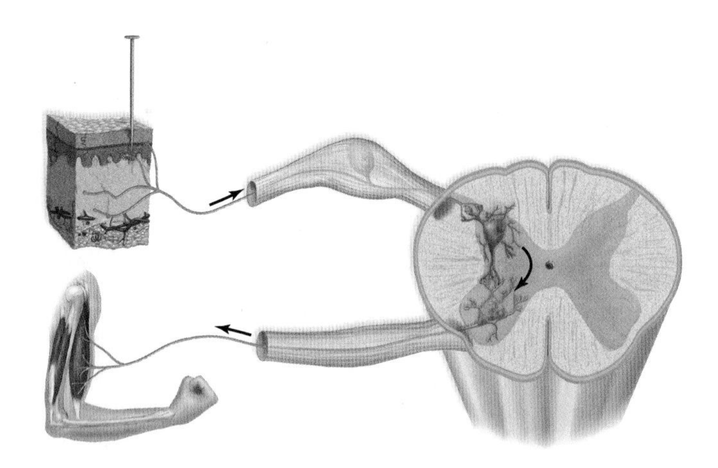

- The autonomic (involuntary) system controls the smooth muscle of the internal organs and glands. The sympathetic division is associated with responses that occur during times of stress, and the parasympathetic system is associated with responses that occur during times of relaxation.

17.5 Drug Abuse

Most illicit drugs affect neurotransmitters in the brain, either promoting or preventing the action of a particular neurotransmitter. Examples include:

- Nicotine, which increases dopamine release in the brain.
- Cocaine, which inhibits uptake of dopamine.
- Heroin, which binds to receptors for natural painkillers called endorphins.

17.6 Disorders of the Nervous System

The precise cause of many of the disorders affecting the CNS remains unclear, and many are untreatable. Examples of brain disorders include the following:

- Alzheimer disease seems to be due to an accumulation of abnormal proteins in the brain.
- Parkinson disease results from a degeneration of dopamine-secreting neurons.

- Multiple sclerosis is the result of autoimmune destruction of myelin and myelin-producing cells.
- Stroke, whether hemorrhagic or ischemic, occurs when the blood supply to the brain is disrupted.
- Infections of the brain include meningitis and prion diseases.

Similar to brain disorders, disorders of the spine and peripheral nerves can be serious and difficult to treat:

- Injuries to the spinal cord can cause a spectrum of problems, from weakness in one limb to complete paralysis.
- Amyotrophic lateral sclerosis is a fatal disease resulting from a loss of motor neurons in the spinal cord.
- Guillain-Barré syndrome and myasthenia gravis are both autoimmune disorders that attack the peripheral nerves.

TESTING YOURSELF

Choose the best answer for each question.

1. An interneuron can relay information
   a. from a sensory neuron to a motor neuron.
   b. only from a motor neuron to another motor neuron.
   c. only from a sensory neuron to another sensory neuron.
   d. None of these are correct.
2. In the PNS, myelin is formed by
   a. oligodendroglial cells. c. axons.
   b. Schwann cells. d. motor neurons.
3. Gray matter is found on the _____ of the brain and the _____ of the spinal cord.
   a. inside, inside c. outside, inside
   b. outside, outside d. inside, outside
4. Which of these correctly describes the distribution of ions on either side of an axon when it is not conducting a nerve impulse?
   a. more sodium ions (Na^+) outside and more potassium ions (K^+) inside
   b. more K^+ outside and less Na^+ inside
   c. charged protein outside and Na^+ and K^+ inside
   d. Na^+ and K^+ outside and water only inside
   e. chloride ions (Cl^-) outside and K^+ and Na^+ inside
5. Label this diagram of the action potential.

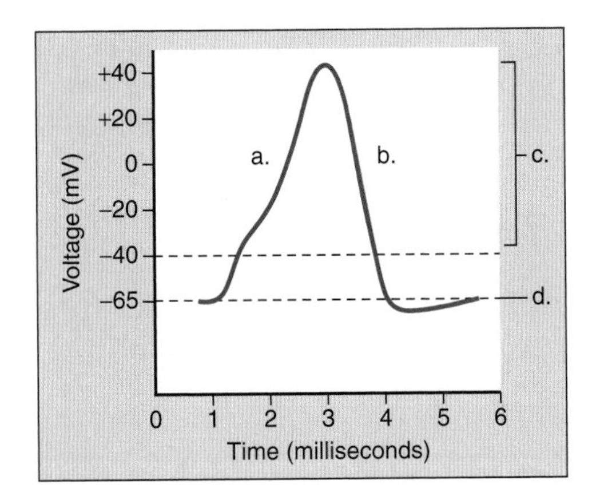

6. During the refractory period,
   a. sodium gates are open.
   b. sodium gates are closed.
   c. sodium and potassium gates are open.
   d. None of these are correct.
7. Transmission of an action potential across a synapse is accomplished by the
   a. movement of Na^+ and K^+.
   b. release of a neurotransmitter by a dendrite.
   c. release of a neurotransmitter by an axon.
   d. release of a neurotransmitter by a cell body.
   e. All of these are correct.
8. The summing up of inhibitory and excitatory signals is
   a. excitation.
   b. integration.
   c. depolarization.
   d. All of these are correct.
9. A spinal nerve takes nerve impulses
   a. to the CNS.
   b. away from the CNS.
   c. both to and away from the CNS.
   d. from the CNS to the spinal cord.
10. Label this diagram of the brain.

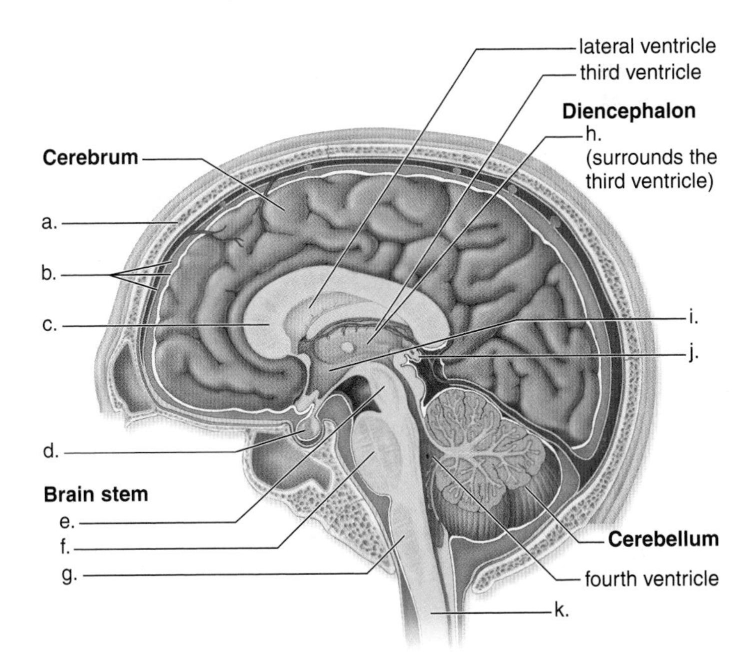

11. Homeostasis is dependent on which of these?
    a. hypothalamus c. parasympathetic division
    b. medulla oblongata d. All of these are correct.
12. The cerebellum
    a. coordinates skeletal muscle movements.
    b. receives sensory input from the joints and muscles.
    c. receives motor input from the cerebral cortex.
    d. All of these are correct.
13. The limbic system
    a. involves portions of the cerebral lobes, the basal nuclei, and the diencephalon.
    b. is responsible for our deepest emotions, including pleasure, rage, and fear.
    c. is a system necessary to memory storage.
    d. All of these are correct.

14. The sympathetic division of the autonomic system will
    a. increase heart rate and digestive activity.
    b. decrease heart rate and digestive activity.
    c. cause pupils to constrict.
    d. None of these are correct.
15. Somatic is to skeletal muscle as autonomic is to
    a. cardiac muscle.
    b. smooth muscle.
    c. gland.
    d. All of these are correct.
16. A damaged hippocampus can cause problems associated with
    a. long-term memory.
    b. episodic memory.
    c. consolidating short-term into long-term memory.
    d. All of these are correct.
17. Damage to the Wernicke's area of the brain could cause problems in
    a. vision.
    b. taste.
    c. smell.
    d. speech
18. Effects of drugs can include
    a. prevention of neurotransmitter release.
    b. prevention of reuptake by the presynaptic membrane.
    c. blockages to a receptor.
    d. All of these are correct.
    e. None of these are correct.
19. A stroke results from
    a. hemorrhage in the brain.
    b. a blockage of a blood vessel in the brain.
    c. a deficiency of tissue plasminogen activator.
    d. Can be either a or b.
20. What do Creutzfeldt-Jakob and mad cow disease have in common?
    a. Both are fatal diseases.
    b. Both are caused by prions.
    c. Both cause spongiform changes in the brain.
    d. All of these are correct.

UNDERSTANDING THE TERMS

acetylcholine (ACh) 321
acetylcholinesterase (AChE) 321
action potential 318
Alzheimer disease (AD) 338
amygdala 328
association area 325
autonomic system 333
axon 316
axon terminal 319
basal nuclei 326
brain stem 326
Broca's area 326
cell body 316
central canal 322
central nervous system (CNS) 316
cerebellum 326
cerebral cortex 325
cerebral hemisphere 325
cerebrospinal fluid (CSF) 322
cerebrum 324
chemical synapse 320
corpus callosum 325
cranial nerve 331
dendrite 316
diencephalon 326
dorsal root ganglion 331
electroencephalogram (EEG) 327
episodic memory 328
ganglion 331
gray matter 317
hippocampus 328
hypothalamus 326
interneuron 316
intervertebral disk 322
learning 328
limbic system 328
long-term memory 328
long-term potentiation (LTP) 329
medulla oblongata 326
memory 328
meninges (sing., meninx) 322
meningitis 340
midbrain 326
motor neuron 316
multiple sclerosis (MS) 339
myelin sheath 317
nerve 331
nerve impulse 318
neuroglia 316
neuron 316
neurotransmitter 320
node of Ranvier 317
norepinephrine (NE) 321
oligodendroglial cells 317
parasympathetic division 333
Parkinson disease (PD) 339
peripheral nervous system (PNS) 316
pons 326
prefrontal area 326
primary motor area 325
primary somatosensory area 325
reflex 332
refractory period 319
resting potential 318
Schwann cell 317
semantic memory 328
sensory neuron 316
short-term memory 328
skill memory 329
sodium-potassium pump 318
somatic system 332
spinal cord 322
spinal nerve 331
stroke 339
sympathetic division 333
synaptic cleft 320
synaptic integration 321
thalamus 326
threshold 318
tract 317
ventricle 322
Wernicke's area 326
white matter 317

THINKING CRITICALLY

1. How do you suppose scientists and physicians have determined which areas of the brain are responsible for different activities, as shown in Figures 17.10, 17.11, and 17.14?
2. Dopamine is a key neurotransmitter in the brain. Patients with Parkinson disease suffer because of a lack of this chemical, while abusers of drugs such as nicotine, cocaine, and methamphetamine enjoy an enhancement of dopamine activity. Would these drugs be a possible treatment, or even a cure, for Parkinson disease? Why or why not?
3. What are some of the factors that make brain diseases such as Alzheimer disease, Parkinson disease, and MS so difficult to treat?

INQUIRY INTO LIFE WEBSITE

The companion website for *Inquiry into Life* provides a wealth of information organized and integrated by chapter. You will find practice tests, animations, videos, and much more that will complement your learning and understanding of general biology.

http://www.mhhe.com/maderinquiry13

Senses

Chapter Outline

Applying the Concepts

Jacob and Marlene, both in their early thirties and childless, were married for almost a year when they learned Marlene was pregnant. Almost immediately, they began discussing what they would do if their baby was born with hearing problems. Marlene had been born deaf, due to a genetic condition in which her inner ear did not develop properly. She had learned sign language and had even become a teacher at a school for the deaf. However, because of technical advances since Marlene was a child, it was now possible to provide hearing for babies with conditions like Marlene's. Many in the deaf community, though, believe this is an ethical issue, because it may imply that something is "wrong" with deaf people, who need to be cured.

Cochlear implants are very different from hearing aids, which simply amplify sounds. A cochlear implant consists of an external device that sits behind the ear, and an internal device that is surgically implanted under the skin (see photo). The external part picks up sounds from the environment and converts them into electrical impulses, which are sent directly to different regions of the auditory nerve and then to the brain, as we will see in this chapter. As of 2006, about 23,000 adults and 15,000 children in the United States had received cochlear implants.

The sounds people hear with cochlear implants are different from what hearing people are familiar with. However, with training and experience, implant recipients can develop the ability to understand speech, and they function very much like a person with normal hearing. Several recent studies have shown that deaf babies who receive implants as young as six months of age develop language and speech skills very much like those of hearing children. It was not an easy decision, but after baby Tony was born with the same condition as his mother, they had cochlear implants placed in both his ears. Now at age three, Tony is learning to say his ABCs and sing nursery rhymes, but Marlene is making sure he learns sign language as well.

18.1 Sensory Receptors and Sensations

Learning Outcomes

1. Distinguish between interoceptors and exteroceptors.
2. Describe the four categories of sensory receptors.
3. Explain the significance of sensory transduction and sensory adaptation.

Sensory receptors are specialized cells that detect certain types of stimuli (sing., **stimulus**). Based on the source of the stimulus, sensory receptors can be classified as interoceptors or exteroceptors.

Interoceptors receive stimuli from inside the body, such as changes in blood pressure, blood volume, and the pH of the blood. Interoceptors are directly involved in homeostasis and are regulated by a negative feedback mechanism. For example, when blood pressure rises, pressoreceptors signal a regulatory center in the brain, which then sends out nerve impulses to the arterial walls, causing them to relax so that the blood pressure falls. Once the pressoreceptors are no longer stimulated, the system shuts down.

Exteroceptors are sensory receptors that detect stimuli from outside the body, such as those that result in taste, smell, vision, hearing, and equilibrium (Table 18.1). Their function is to send messages to the central nervous system to report changes in environmental conditions.

Types of Sensory Receptors

Based on the nature of the stimulus, interoceptors and exteroceptors can be further classified into four categories: chemoreceptors, photoreceptors, mechanoreceptors, and thermoreceptors.

Chemoreceptors respond to chemical substances in the immediate vicinity. As Table 18.1 indicates, taste and smell are dependent on this type of sensory receptor, but certain chemoreceptors in various other organs are sensitive to internal conditions. For example, chemoreceptors that monitor blood oxygen (O_2) levels are located in the carotid arteries and aorta. If O_2 levels fall, the breathing rate increases.

Photoreceptors respond to light energy. Our eyes contain photoreceptors that are sensitive to light rays, and thereby provide us with the sense of vision. Stimulation of the photoreceptors known as rod cells results in black-and-white vision, while stimulation of the photoreceptors known as cone cells results in color vision (see section 18.4).

Mechanoreceptors are stimulated by mechanical forces, which most often result in pressure of some sort. For example, when we hear, airborne sound waves are converted to fluid-borne pressure waves that can be detected by mechanoreceptors in the inner ear. Similarly, mechanoreceptors are responding to fluid-borne pressure waves when we detect changes in gravity and motion, helping us keep our balance. These receptors are in the vestibule and semicircular canals of the inner ear, respectively (see section 18.5).

The sense of touch is dependent on pressure receptors in the skin and tongue that are sensitive to either strong or slight pressures. Pressoreceptors located in certain arteries detect changes in blood pressure, and stretch receptors in the lungs detect the degree of lung inflation. Proprioceptors, which respond to the stretching of muscle fibers, tendons, joints, and ligaments, make us aware of the position of our limbs.

Thermoreceptors, located in the hypothalamus and skin, are stimulated by changes in temperature. Those that respond when temperatures rise are called warmth receptors, and those that respond when temperatures fall are called cold receptors.

How Sensation Occurs

Sensory receptors respond to environmental stimuli by generating nerve impulses. **Detection** occurs when environmental changes, such as pressure to the fingertips or light to the eye, stimulate sensory receptors. **Sensation** occurs when nerve impulses arrive at the cerebral cortex of the brain. **Perception** occurs when the brain interprets the meaning of stimuli.

As we discussed in Chapter 17, sensory receptors are the first element in a reflex arc. We are only aware of a reflex action when sensory information reaches the brain. At that time, the brain integrates this information with other information received from other sensory receptors. After all, if you burn yourself and quickly remove your hand from a hot stove, the brain receives information not only from your skin, but also from your eyes, nose, and all sorts of sensory receptors.

Some sensory receptors are free nerve endings or encapsulated nerve endings, while others are specialized cells closely associated with neurons. The plasma membrane of a sensory receptor contains receptor proteins that react to the

TABLE 18.1 Exteroceptors

Sensory Receptor	Stimulus	Category	Sense	Sensory Organ
Taste cells	Chemicals	Chemoreceptor	Taste	Taste buds
Olfactory cells	Chemicals	Chemoreceptor	Smell	Olfactory epithelium
Rod cells and cone cells in retina	Light rays	Photoreceptor	Vision	Eye
Hair cells in spiral organ	Sound waves	Mechanoreceptor	Hearing	Ear
Hair cells in semicircular canals	Motion	Mechanoreceptor	Rotational equilibrium	Ear
Hair cells in vestibule	Gravity	Mechanoreceptor	Gravitational equilibrium	Ear

stimulus. For example, the receptor proteins in the plasma membrane of chemoreceptors bind to certain molecules. When this happens, ion channels open, and ions flow across the plasma membrane. If the stimulus is sufficient, nerve impulses begin and are carried by a sensory nerve fiber within the PNS to the CNS (Fig. 18.1). In this process, known as *sensory transduction*, energy from the chemical or physical stimulus is transduced by the sensory receptors into an electrical signal. This electrical signal is in the form of nerve impulses that travel along the sensory nerve fiber. The stronger the stimulus, the greater the frequency of nerve impulses. Nerve impulses that reach the spinal cord first are conveyed to the brain by ascending tracts. If nerve impulses finally reach the cerebral cortex, sensation and perception occur.

All sensory receptors initiate nerve impulses; the sensation that results depends on the part of the brain receiving the nerve impulses. Nerve impulses that begin in the optic nerve reach the visual cortex, and thereafter, we see objects. Nerve impulses that begin in the auditory nerve reach the auditory cortex, and thereafter, we hear sounds.

Before sensory receptors initiate nerve impulses, they carry out **integration,** the summing up of signals. One type of integration is called **sensory adaptation,** a decrease in the response to a stimulus. For example, as you sit reading this, you are probably not aware of the feeling of the chair beneath you, until now as you are currently reminded. Some authorities believe that when sensory adaptation occurs, sensory receptors have stopped sending impulses to the brain. Others believe that the thalamus has filtered out the ongoing stimuli. As mentioned in section 17.2, most sensory information is conveyed from the brain stem through the thalamus to the cerebral cortex. The thalamus acts as a gatekeeper and only passes on information of immediate importance. Just as we gradually become unaware of particular environmental stimuli, we can suddenly become aware of stimuli that may have been present for some time. This can be attributed to the workings of the thalamus, which has synapses with all the major ascending sensory tracts.

The functioning of our sensory receptors makes a significant contribution to homeostasis. Without sensory input, we would not receive information about our internal and external environment. This information leads to appropriate reflex and voluntary actions to keep the internal environment constant.

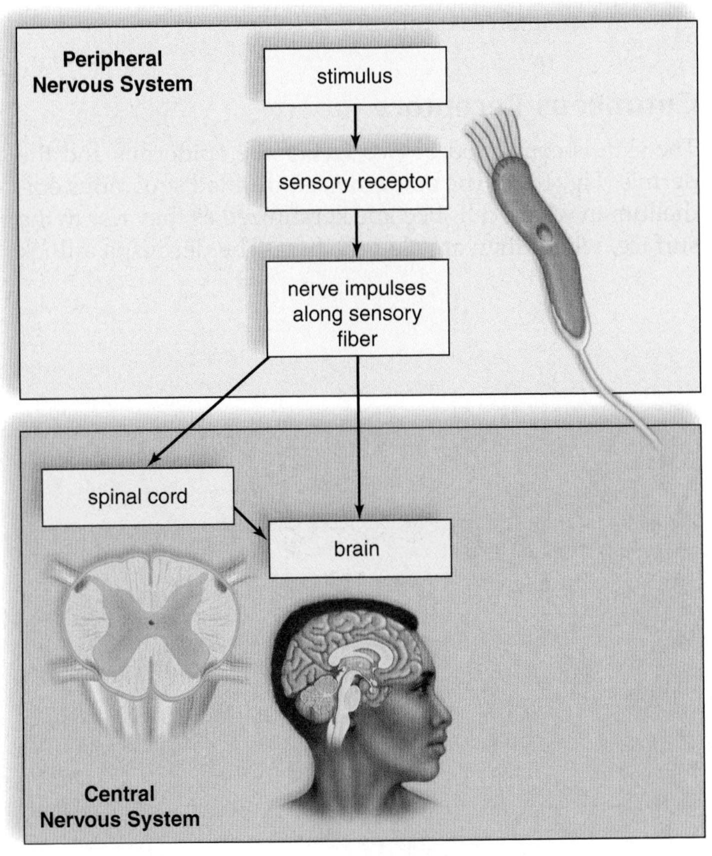

Figure 18.1 Sensation.
The stimulus is received by a sensory receptor, which generates nerve impulses (action potentials). Nerve impulses are conducted to the CNS by sensory nerve fibers within the PNS, and only those impulses that reach the cerebral cortex result in sensation and perception.

Check Your Progress

1. Describe how interoceptors are often involved in homeostasis.
2. List a specific example of each of the four types of sensory receptors.
3. Explain why sensory adaptation is important to an organism.

18.2 Somatic Senses

Learning Outcomes

1. Compare and contrast the functions of proprioceptors, cutaneous receptors, and pain receptors.
2. List specific types of cutaneous receptors that are sensitive to fine touch, pressure, and temperature.
3. Analyze the importance of pain receptors to the survival of an organism.

It is logical to consider those senses whose receptors are associated with the skin, muscles, joints, and viscera together as the somatic senses. Somatic sensory receptors can be categorized into three types: proprioceptors, cutaneous receptors, and pain receptors. Some of these are interoceptors, while others are exteroceptors. Regardless, these receptors send nerve impulses via the spinal cord to the somatosensory areas of the cerebral cortex (see Fig. 17.10).

Proprioceptors

Proprioceptors are mechanoreceptors involved in reflex actions that maintain muscle tone, and thereby the body's equilibrium and posture. They help us know the position of our limbs in space by detecting the degree of muscle relaxation, the stretch of tendons, and the movement of ligaments. Muscle spindles act to increase the degree of muscle contraction, and sensory receptors buried in the collagen

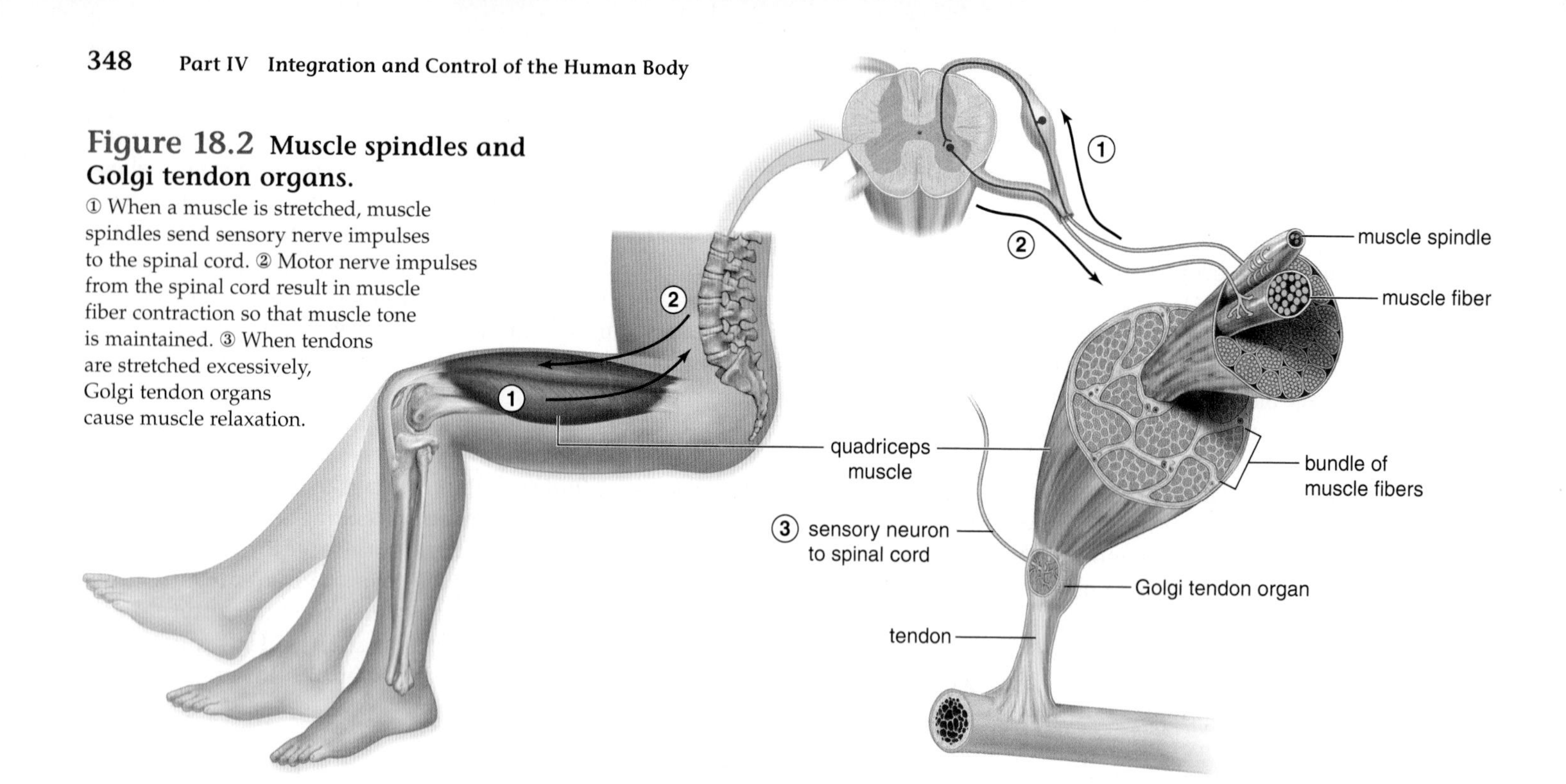

Figure 18.2 Muscle spindles and Golgi tendon organs.
① When a muscle is stretched, muscle spindles send sensory nerve impulses to the spinal cord. ② Motor nerve impulses from the spinal cord result in muscle fiber contraction so that muscle tone is maintained. ③ When tendons are stretched excessively, Golgi tendon organs cause muscle relaxation.

fibers of tendons, called Golgi tendon organs, act to decrease it. The result is a muscle that has the proper length and tension, or muscle tone.

Figure 18.2 illustrates the activity of a muscle spindle. In a muscle spindle, sensory nerve endings are wrapped around thin muscle cells within a connective tissue sheath. When the muscle relaxes and undue stretching of the muscle spindle occurs, nerve impulses are generated. The rapidity of the nerve impulses generated by the muscle spindle is proportional to the stretching of a muscle. A reflex action then occurs, which results in contraction of muscle fibers adjoining the muscle spindle. The knee-jerk reflex, which involves muscle spindles, offers an opportunity for physicians to test a reflex action. The information sent by muscle spindles to the CNS is used to maintain the body's equilibrium and posture, despite the force of gravity always acting upon the skeleton and muscles.

Cutaneous Receptors

The skin is composed of two layers: the epidermis and the dermis (Fig. 18.3). The epidermis is stratified squamous epithelium in which cells become keratinized as they rise to the surface, where they are sloughed off. The dermis is a thick

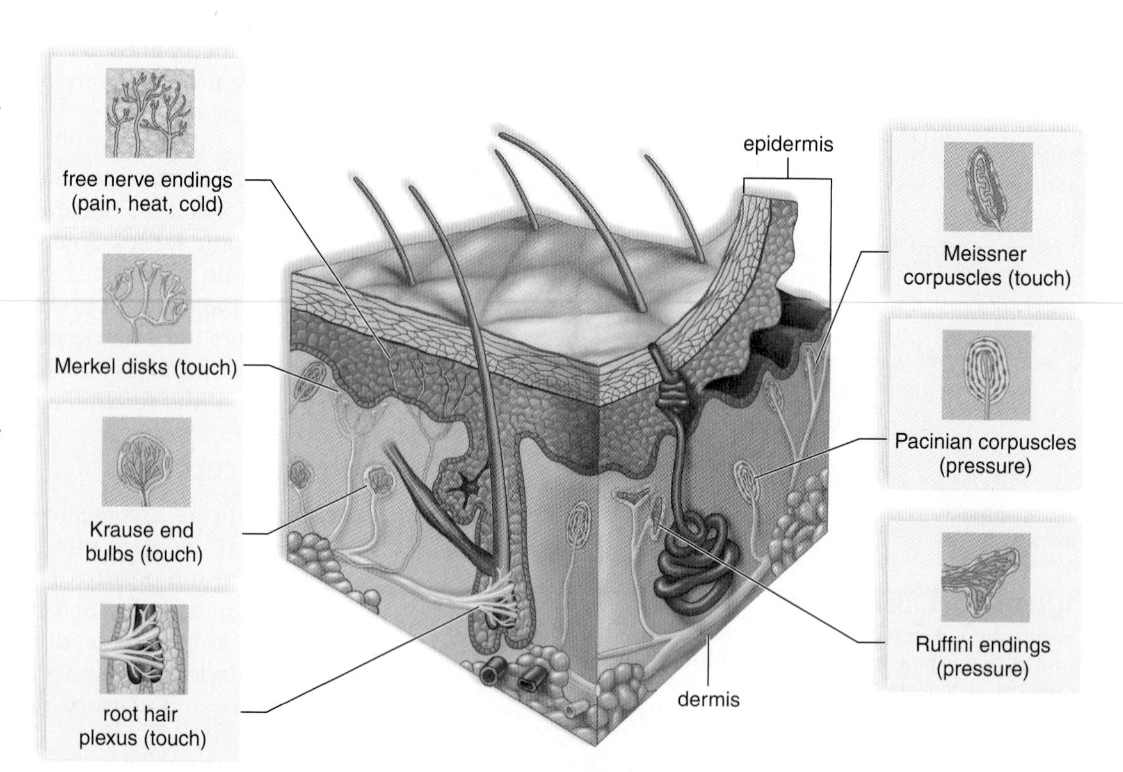

Figure 18.3 Cutaneous receptors in human skin.
The classical view is that each cutaneous receptor has the main function shown here. However, investigators report that matters are not so clear-cut. For example, microscopic examination of the skin of the ear shows only free nerve endings (pain receptors), and yet the skin of the ear is sensitive to all sensations. Therefore, it appears that the receptors of the skin are somewhat, but not completely, specialized.

connective tissue layer. The dermis contains **cutaneous receptors,** which make the skin sensitive to touch, pressure, pain, and temperature (warmth and cold). The dermis contains a mosaic of these tiny receptors, as you can determine by slowly passing a metal probe over your skin. At certain points, you will feel touch or pressure, and at others, you will feel heat or cold (depending on the probe's temperature).

Four types of cutaneous receptors are sensitive to fine touch. *Meissner corpuscles* and *Krause end bulbs* are concentrated in the fingertips, the palms, the lips, the tongue, the nipples, the penis, and the clitoris. *Merkel disks* are found where the epidermis meets the dermis. A free nerve ending called a *root hair plexus* winds around the base of a hair follicle and fires if the hair is touched.

Two different types of cutaneous receptors are sensitive to pressure. *Pacinian corpuscles* are onion-shaped sensory receptors that lie deep inside the dermis. *Ruffini endings* are encapsulated by sheaths of connective tissue and contain lacy networks of nerve fibers.

Temperature receptors are simply free nerve endings in the epidermis. Some free nerve endings are responsive to cold; others are responsive to warmth. Cold receptors are far more numerous than warmth receptors, but the two types have no known structural differences.

Pain Receptors

The skin and many internal organs and tissues have pain receptors (free nerve endings, also called *nociceptors*), which are sensitive to chemicals released by damaged cells. When damage occurs due to mechanical, thermal, or electrical stimulation, or a toxic substance, cells release chemicals that stimulate pain receptors. Aspirin and ibuprofen reduce pain by inhibiting the synthesis of one class of these chemicals.

Sometimes, stimulation of internal pain receptors is felt as pain from the skin, as well as the internal organs. This is called **referred pain.** Some internal organs have a referred pain relationship with areas located in the skin of the back, groin, and abdomen. For example, pain from the heart is typically felt in the left shoulder and arm. This most likely happens when nerve impulses from the pain receptors of internal organs travel to the spinal cord and synapse with neurons also receiving impulses from the skin. Pain receptors are protective because they alert us to possible danger. For example, without the pain of appendicitis, we might never seek the medical help needed to avoid a ruptured appendix.

Check Your Progress

1. Classify which somatic sensory receptors are interoceptors and which are exteroceptors.
2. What would be a potentially harmful outcome if an individual lacked muscle spindles? What if the person lacked pain receptors?

18.3 Senses of Taste and Smell

Learning Outcomes

1. List five types of taste receptors.
2. Compare and contrast how the brain receives information about taste versus smell.

Taste and smell are called chemical senses because their receptors are sensitive to molecules in the food we eat and the air we breathe. Taste cells and olfactory cells are classified as chemoreceptors.

Sense of Taste

The sensory receptors for the sense of taste, the taste cells, are located in **taste buds.** About 10,000 taste buds are embedded in epithelium, primarily on the tongue (Fig. 18.4). Many lie along the walls of the papillae, the small elevations on

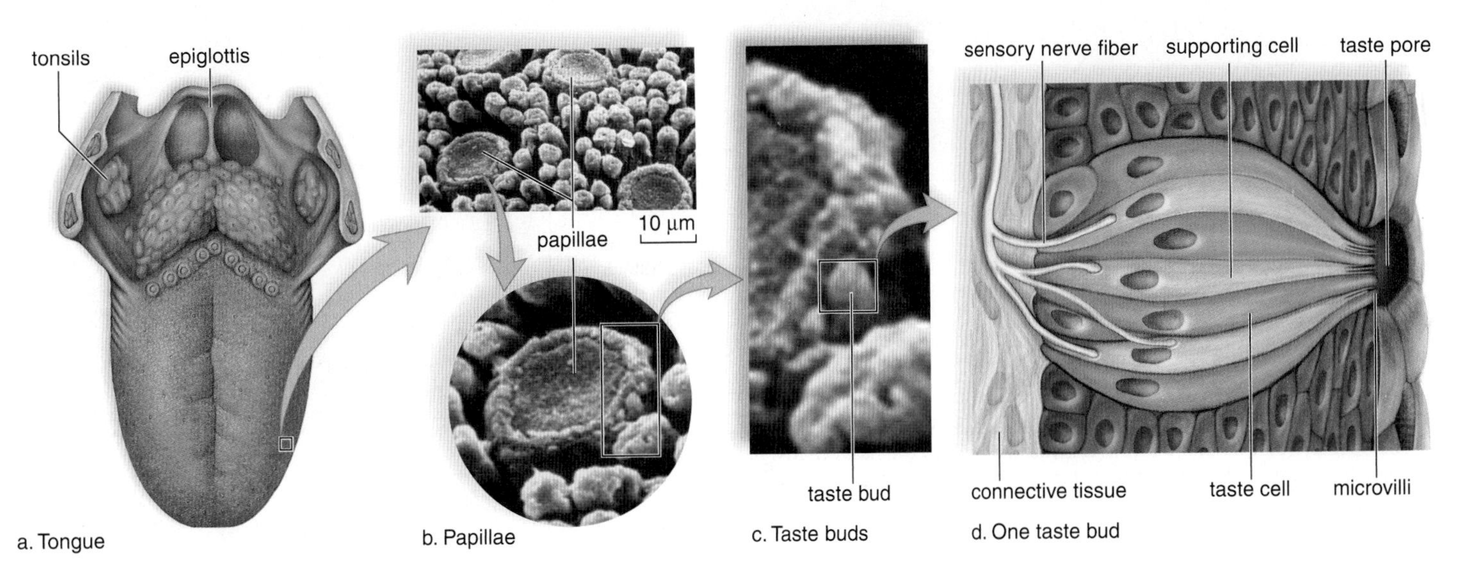

Figure 18.4 Taste buds in humans.
a, b. Papillae on the tongue contain taste buds that are sensitive to sweet, sour, salty, bitter, and perhaps umami. **c.** Taste buds occur along the walls of the papillae. **d.** Taste cells end in microvilli that bear receptor proteins for certain molecules. When molecules bind to the receptor proteins, nerve impulses are generated and go to the brain, where the sensation of taste occurs.

the tongue that are visible to the naked eye. Isolated taste buds are also present on the hard palate, the pharynx, and the epiglottis. Different taste cells can detect at least four primary types of taste: salty, sour, bitter, and sweet. However, there is evidence that more receptors do exist. For example, a receptor capable of detecting certain amino acids has been described. In particular, this receptor detects the amino acid glutamate, which is part of the flavor enhancer monosodium glutamate (MSG). These receptors have been called "umami" receptors, after the Japanese word meaning delicious. Taste buds containing sensory receptors for each type of taste are located throughout the tongue.

How the Brain Receives Taste Information

Taste buds open at a taste pore. They have supporting cells and a number of elongated taste cells that end in microvilli. When molecules bind to receptor proteins of the microvilli, nerve impulses are generated in sensory nerve fibers that go to the brain. When they reach the gustatory (taste) cortex, they are interpreted as particular tastes.

Because we can respond to a range of salty, sour, bitter, sweet, and perhaps other tastes, the brain appears to survey the overall pattern of incoming sensory impulses and to take a "weighted average" of their taste messages as the perceived taste. Note that even though our senses are dependent on sensory receptors, the cortex integrates the incoming information and gives us our sense perceptions.

Sense of Smell

Approximately 80–90% of what we perceive as "taste" is actually due to the sense of smell, which explains why food tastes dull when we have a head cold or a stuffed-up nose. Our sense of smell is dependent on **olfactory cells** located within olfactory epithelium high in the roof of the nasal cavity (Fig. 18.5). Olfactory cells are modified neurons. Each cell ends in a tuft of about five olfactory cilia, which bear receptor proteins for odor molecules.

How the Brain Receives Odor Information

Each olfactory cell has only one out of about 1,000 different types of receptor proteins. Nerve fibers from like olfactory cells lead to the same neuron in the olfactory bulb, an extension of the brain. An odor contains many odor molecules, which activate a characteristic combination of receptor proteins. For example, a rose might stimulate olfactory cells, designated by blue and green in Figure 18.5*b*, while a hyacinth might stimulate a different combination. An odor's signature in the olfactory bulb is determined by which

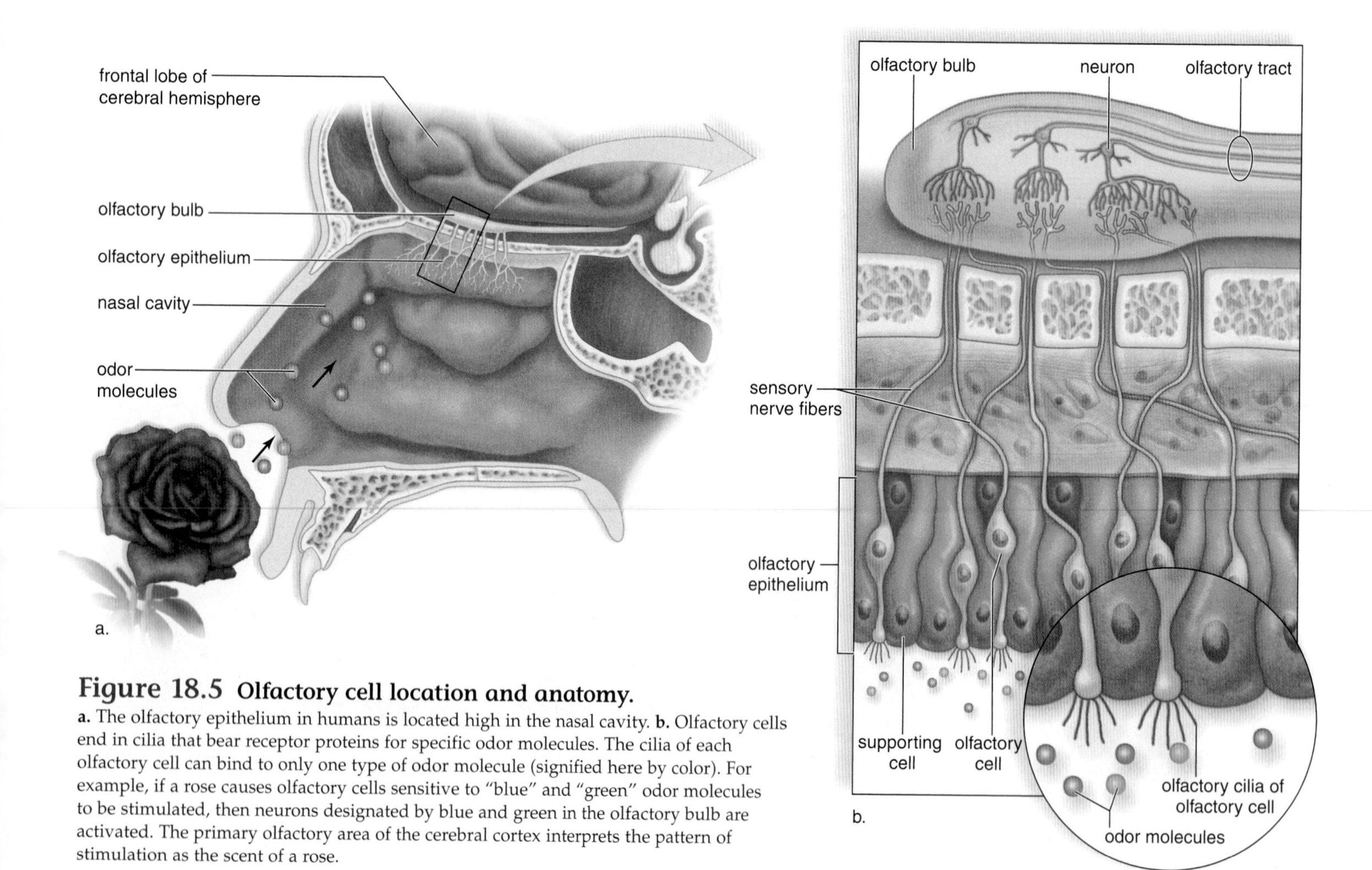

Figure 18.5 Olfactory cell location and anatomy.
a. The olfactory epithelium in humans is located high in the nasal cavity. **b.** Olfactory cells end in cilia that bear receptor proteins for specific odor molecules. The cilia of each olfactory cell can bind to only one type of odor molecule (signified here by color). For example, if a rose causes olfactory cells sensitive to "blue" and "green" odor molecules to be stimulated, then neurons designated by blue and green in the olfactory bulb are activated. The primary olfactory area of the cerebral cortex interprets the pattern of stimulation as the scent of a rose.

neurons are stimulated. The neurons communicate this information via the olfactory tract to the brain. When the information reaches the olfactory cortex, we know we have smelled a rose or a hyacinth.

Have you ever noticed that a certain aroma vividly brings to mind a certain person or place? A person's perfume may remind you of someone else, or the smell of boxwood may remind you of your grandfather's farm. The olfactory bulbs have direct connections to the limbic system and its centers for emotions and memory (see section 17.2). One investigator showed that when subjects smelled an orange while viewing a painting, they not only remembered the painting when asked about it later, but they also had many deep feelings about it.

Check Your Progress

1. What sort of structural variations could allow certain animals, such as dogs, to have a sense of smell thousands of times more sensitive than that of humans?
2. Anatomically speaking, what type of connections exist in the brain between smells and memories?

18.4 Sense of Vision

Learning Outcomes

1. List all tissues that light passes through from when it enters the eye until it is converted into a nerve impulse.
2. Discuss the role of rods and cones in transducing a light stimulus into a nerve impulse.
3. Trace the path of a nerve impulse from the retina to the visual cortex.

Vision requires the work of the eyes and the brain. Researchers estimate that at least a third of the cerebral cortex takes part in processing visual information.

Anatomy and Physiology of the Eye

The eyeball, an elongated sphere about 2.5 cm in diameter, has three layers, or coats: the sclera, the choroid, and the retina (Fig. 18.6). Only the retina contains photoreceptors for light energy. Table 18.2 lists the functions of the parts of the eye.

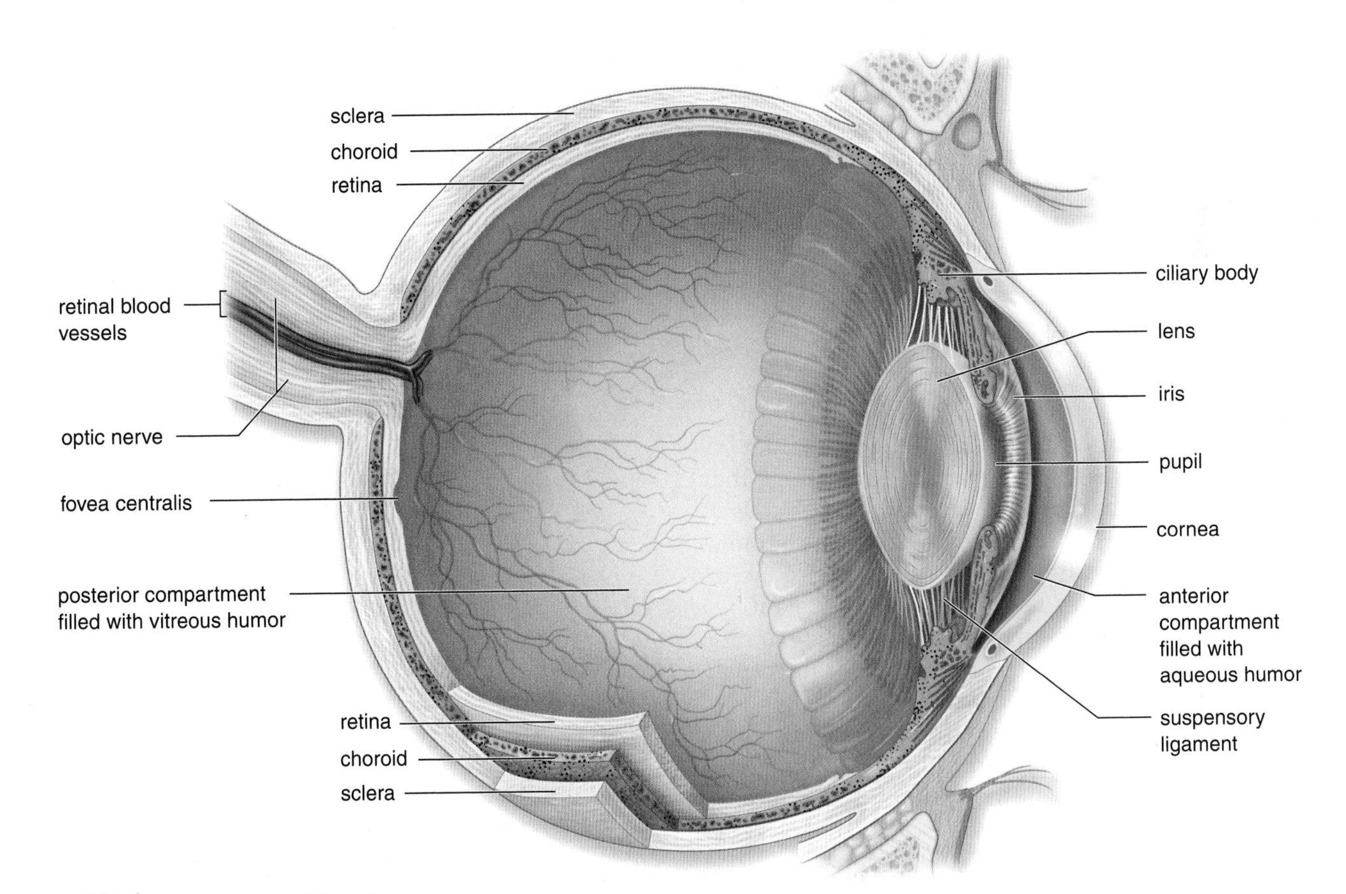

Figure 18.6 Anatomy of the human eye.
Notice that the sclera (the outer layer of the eye) becomes the cornea and that the choroid (the middle layer) is continuous with the ciliary body and the iris. The retina (the inner layer) contains the photoreceptors for vision. The fovea centralis is the region where vision is most acute.

TABLE 18.2 Functions of the Parts of the Eye

Part	Function
Sclera	Protects and supports eyeball
Cornea	Refracts light rays
Choroid	Absorbs stray light
Ciliary body	Holds lens in place; aids accommodation
Iris	Regulates light entrance
Pupil	Admits light
Retina	Contains sensory receptors for sight
Rods	Make black-and-white vision possible
Cones	Make color vision possible
Fovea centralis	Makes acute vision possible
Other	
Lens	Refracts and focuses light rays
Humors	Transmit light rays and support eyeball
Optic nerve	Transmits impulse to brain

The outer layer, called the **sclera,** is white and fibrous except for the **cornea,** which is made of transparent collagen fibers. The cornea is the window of the eye. The middle, thin, darkly pigmented layer, the **choroid,** is vascular and absorbs stray light rays that photoreceptors have not absorbed. Toward the front, the choroid becomes the donut-shaped **iris.** The iris contains smooth muscle that regulates the size of the **pupil,** a hole in the center of the iris through which light enters the eyeball. The color of the iris (and therefore, the color of your eyes) correlates with its pigmentation. Heavily pigmented eyes are brown, while lightly pigmented eyes are green or blue. Behind the iris, the choroid thickens and forms the circular ciliary body. The **ciliary body** contains the **ciliary muscle,** which controls the shape of the lens for near and far vision.

The **lens,** attached to the ciliary body by ligaments, divides the eye into two compartments. The one in front of the lens is the anterior compartment, and the one behind the lens is the posterior compartment. The anterior compartment is filled with a clear, watery fluid called the **aqueous humor.** A small amount of aqueous humor is continually produced each day. Normally, it leaves the anterior compartment by way of tiny ducts. When a person has **glaucoma,** these drainage ducts are blocked, and aqueous humor builds up, causing an increase in pressure (see section 18.7).

The third layer of the eye, the **retina,** is located in the posterior compartment, which is filled with a clear, gelatinous material called the **vitreous humor.** The retina contains photoreceptors called rod cells and cone cells. The rods are very sensitive to light, but they do not respond to color. Therefore, at night or in a darkened room, we see only shades of gray. The cones, which require bright light, are sensitive to different wavelengths of light, which give us the ability to distinguish colors. The retina has a special region called the **fovea centralis,** where cone cells are densely packed. Light is normally focused on the fovea when we look directly at an object. This is helpful because vision is most acute in the fovea centralis. Sensory nerve fibers from the retina form the **optic nerve,** which takes nerve impulses to the visual cortex.

Function of the Lens

The lens, assisted by the cornea and the humors, focuses images on the retina (Fig. 18.7). Focusing starts with the cornea and continues as the rays pass through the lens and the humors. The image produced is much smaller than the object because light rays are bent (refracted) when they are brought into **focus.** If the eyeball is too long or too short, the person may need corrective lenses to bring the image into focus.

Visual accommodation occurs for close vision. During visual accommodation, the lens rounds up in order to bring the image to focus on the retina. The shape of the lens is controlled by the ciliary muscle within the ciliary body. When we view a distant object, the ciliary muscle is relaxed, causing the suspensory ligaments attached to the ciliary body to be taut. Therefore, the lens remains relatively flat

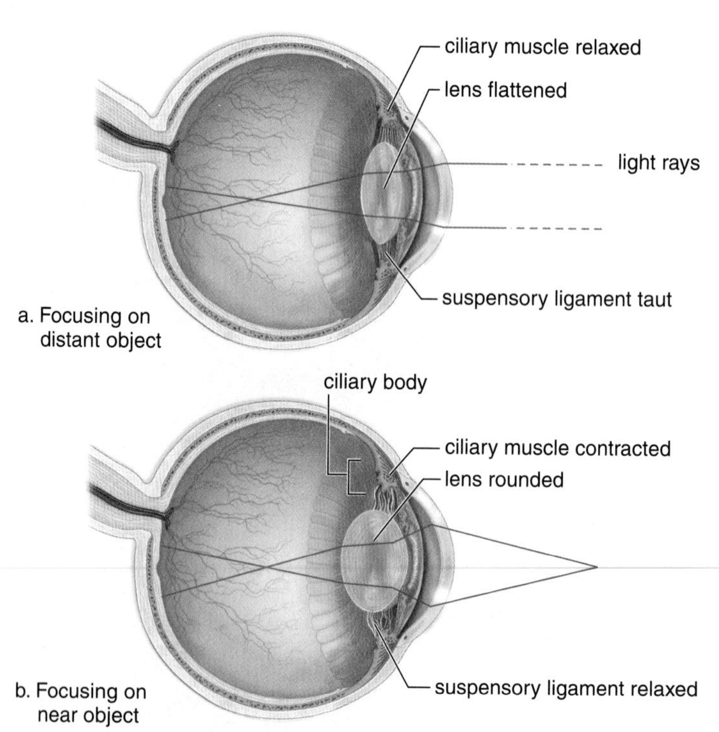

Figure 18.7 Focusing the human eye.
Light rays from each point on an object are bent by the cornea and the lens in such a way that an inverted and reversed image forms on the retina. **a.** When focusing on a distant object, the lens is flat because the ciliary muscle is relaxed and the suspensory ligament is taut. **b.** When focusing on a near object, the lens accommodates; that is, it becomes rounded because the ciliary muscle contracts, causing the suspensory ligament to relax.

(Fig. 18.7*a*). When we view a near object, the ciliary muscle contracts, releasing the tension on the suspensory ligaments, and the lens rounds up due to its natural elasticity (Fig. 18.7*b*). Now the image is focused on the retina. Because close work requires contraction of the ciliary muscle, it very often causes muscle fatigue, known as eyestrain. Usually after the age of 40, the lens loses some of its elasticity and is unable to accommodate. The use of reading glasses may then be necessary, or bifocal lenses for those who already have corrective lenses.

Check Your Progress

1. List the three main layers of the eye.
2. Define the following: pupil, iris, vitreous humor, fovea centralis, and optic nerve.
3. Describe how the ciliary muscle controls the shape of the lens.

Visual Pathway to the Brain

The pathway for vision begins once light has been focused on the photoreceptors in the retina. Some integration occurs in the retina, where nerve impulses begin before the optic nerve transmits them to the brain.

Function of Photoreceptors The photoreceptors in the eye are of two types—**rod cells** and **cone cells.** Figure 18.8 illustrates their structure. Both rods and cones have an outer segment joined to an inner segment by a stalk. Pigment molecules are embedded in the membrane of the many disks present in the outer segment. Synaptic vesicles are located at the synaptic endings of the inner segment.

The visual pigment in rods is a deep purple pigment called rhodopsin. **Rhodopsin** is a complex molecule made up of the protein opsin and a light-absorbing molecule called **retinal,** which is a derivative of vitamin A. When a rod absorbs light, rhodopsin splits into opsin and retinal, leading to a cascade of reactions and the closure of ion channels in the rod cell's plasma membrane. The release of inhibitory transmitter molecules from the rod's synaptic vesicles ceases. Thereafter, signals go to other neurons in the retina. Rods are very sensitive to light, and are therefore suited to night vision. (Because carrots are rich in vitamin A, it is true that eating carrots can improve your night vision.) With the exception of the fovea centralis, rod cells are plentiful throughout the entire retina. Therefore, they also provide us with peripheral vision and perception of motion.

The cones, on the other hand, are located primarily in the fovea centralis and are activated by bright light. They allow us to detect the fine detail and the color of an object. **Color vision** depends on three different kinds of cones, which contain pigments called the B (blue), G (green), and R (red) pigments. Each pigment is an iodopsin made up of retinal and an opsin. There is a slight difference in the opsin structure of each, which accounts for the ability of the pigments to absorb different wavelengths (colors) of visible light. Various combinations of cones are believed to be stimulated by in-between shades of color.

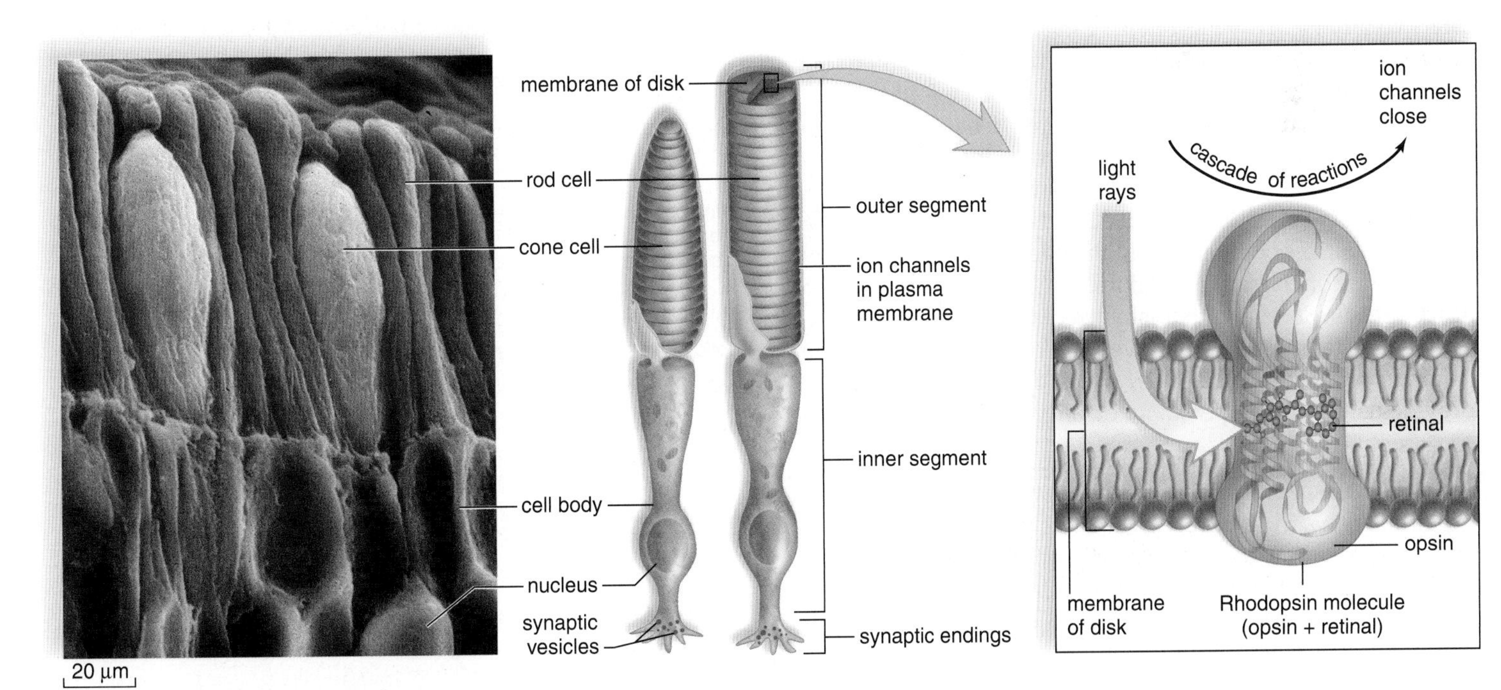

Figure 18.8 Photoreceptors in the eye.
The outer segment of rods and cones contains stacks of membranous disks that contain visual pigments. In rods, the membrane of each disk contains rhodopsin, a complex molecule composed of the protein opsin and the pigment retinal. When rhodopsin absorbs light energy, it splits, releasing opsin, which sets in motion a cascade of reactions that cause ion channels in the plasma membrane to close. Thereafter, nerve impulses go to the brain.

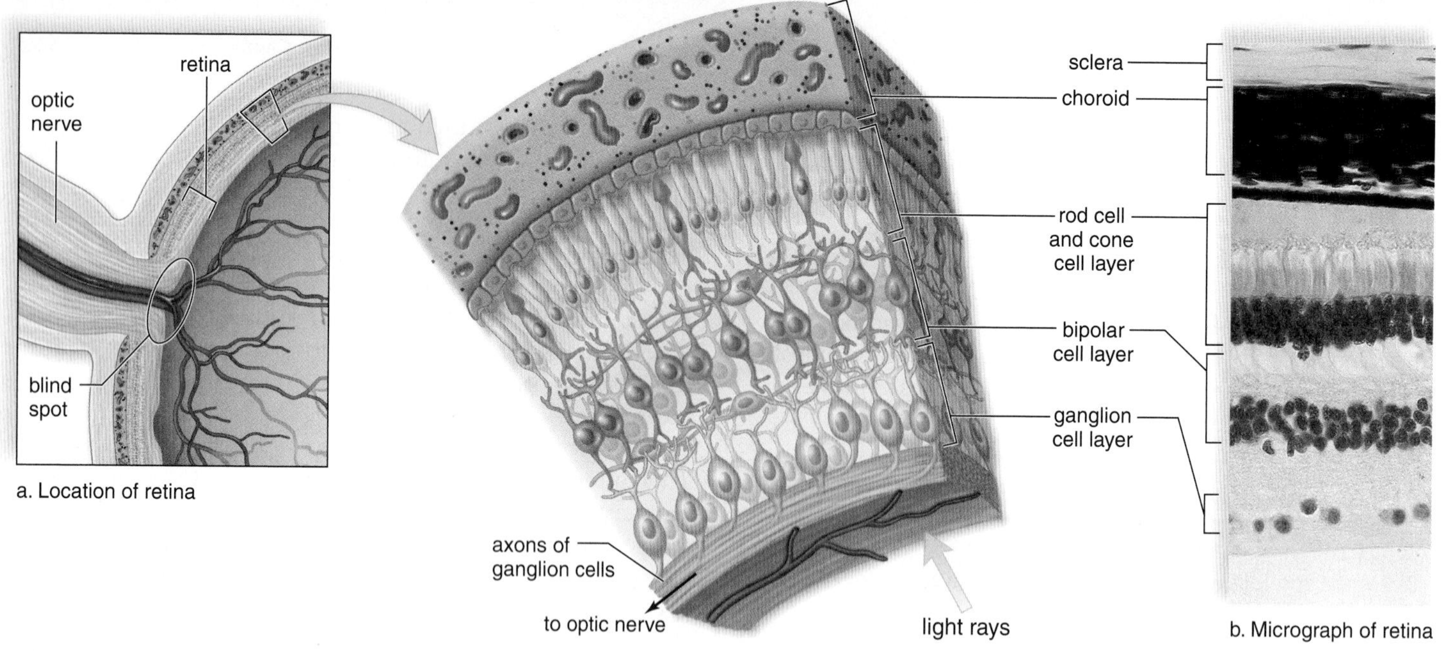

Figure 18.9 Structure and function of the retina.
a. The retina is the inner layer of the eyeball. Rod cells and cone cells, located at the back of the retina nearest the choroid, synapse with bipolar cells, which synapse with ganglion cells. Integration of signals occurs at these synapses. Therefore, much processing occurs in bipolar and ganglion cells. Notice also that many rod cells share one bipolar cell, but cone cells do not. Certain cone cells synapse with only one ganglion cell. Cone cells, in general, distinguish more detail than do rod cells. **b.** Micrograph shows that the sclera and choroid are relatively thin compared to the retina, which has several layers of cells.

Function of the Retina The retina has three layers of neurons (Fig. 18.9). The layer closest to the choroid (at the back of the eye) contains the rod cells and cone cells; the middle layer contains bipolar cells; and the innermost layer contains ganglion cells, whose sensory fibers become the optic nerve. Only the rod cells and the cone cells are sensitive to light, and therefore, light must penetrate to the back of the retina before they are stimulated.

The rod cells and the cone cells synapse with the bipolar cells, which in turn synapse with ganglion cells whose axons become the optic nerve. Notice in Figure 18.9 that there are many more rod cells and cone cells than ganglion cells. In fact, the retina has as many as 150 million rod cells and 6 million cone cells but only one million ganglion cells. The sensitivity of cones versus rods is due in part to how directly they connect to ganglion cells. As many as 150 rods, covering about one square millimeter of retina (about the size of a thumbtack hole) may synapse on the same ganglion cell. No wonder stimulation of rods results in vision that is blurred and indistinct. In contrast, some cone cells in the fovea centralis activate only one ganglion cell. This explains why cones, especially in the fovea, provide us with a sharper, more detailed image of an object.

Blind Spot Figure 18.9*a* provides an opportunity to point out that there are no rods and cones where the optic nerve exits the retina. Therefore, no vision is possible in this area. You can prove this to yourself by putting a dot to the right of center on a piece of paper. Use your right hand to move the paper slowly toward your right eye, while you look straight ahead. The dot will disappear at one point—this is your right eye's **blind spot.**

From the Retina to the Visual Cortex As stated previously, the axons of ganglion cells in the retina assemble to form the optic nerves. The optic nerves carry nerve impulses from the eyes to the optic chiasma. The **optic chiasma** has an X shape, formed by a crossing-over of optic nerve fibers. Fibers from the right half of each retina converge and continue on together in the *right optic tract,* and fibers from the left half of each retina converge and continue on together in the *left optic tract* (Fig. 18.10).

Figure 18.10 Optic chiasma.
Both eyes "see" the entire visual field. Because of the optic chiasma, data from the right half of each retina (red lines) go to the right visual cortex, and data from the left half of each retina (green lines) go to the left visual cortex. These data are then combined to allow us to see the entire visual field. Note that the visual pathway to the brain includes the thalamus, which has the ability to filter sensory stimuli.

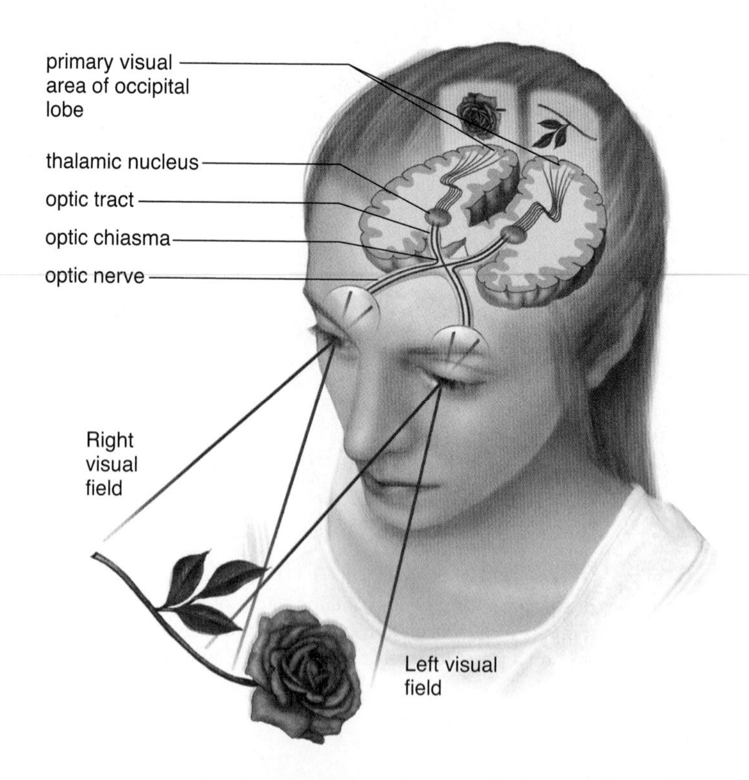

The optic tracts sweep around the hypothalamus, and most fibers synapse with neurons in nuclei (masses of neuron cell bodies) within the thalamus. Axons from the thalamic nuclei form optic radiations that take nerve impulses to the *visual area* within the occipital lobe. Notice in Figure 18.10 that the image arriving at the thalamus, and therefore the visual area, has been split because the left optic tract carries information about the right portion of the visual field (shown in green) and the right optic tract carries information about the left portion of the visual field (shown in red). Therefore, the right and left visual areas must communicate with each other for us to see the entire visual field. Also, because the image is inverted and reversed, it must be righted in the brain for us to correctly perceive the visual field.

The most surprising finding has been that the primary visual area acts like a post office, parceling out information regarding color, form, motion, and possibly other attributes to different portions of the adjoining visual association area. Therefore, the brain has taken the visual field apart, even though we see a unified visual field. The visual association areas are believed to rebuild the field and give us an understanding of it at the same time.

Check Your Progress

1. Examine the layers of cells that light must pass through before being detected by rod cells and cone cells. How is this the opposite of what might be expected?
2. Explain how the different number of rod or cone cells synapsing on a single ganglion cell helps explain why cones provide a clearer image.
3. Describe how the anatomy of the optic chiasma implies that the brain interprets visual information that has been "split" in half.

18.5 Sense of Hearing

Learning Outcomes

1. Distinguish the parts of the ear that make up the outer ear, middle ear, and inner ear.
2. Describe the mechanism by which sound waves in the outer ear are converted into nerve impulses in the inner ear.

The ear has two sensory functions: hearing and balance (equilibrium). The sensory receptors for both of these are located in the inner ear, and each consists of **hair cells** with stereocilia (long microvilli) that are sensitive to mechanical stimulation. They are mechanoreceptors.

Anatomy of the Ear

Figure 18.11 shows that the ear has three divisions: outer, middle, and inner. The **outer ear** consists of the **pinna** (external flap), which collects and funnels sound into the **auditory canal.** The opening of the auditory canal is lined with fine hairs and sweat glands. Modified sweat glands are located in the upper wall of the canal. They secrete earwax, a substance that helps guard the ear against the entrance of foreign materials, such as air pollutants and microorganisms.

The **middle ear** begins at the **tympanic membrane** (eardrum) and ends at a bony wall containing two small openings covered by membranes. These openings are called the **oval window** and the **round window.** Between the tympanic membrane and the oval window are three small bones, collectively called the **ossicles.** Individually, they are the **malleus** (hammer), the **incus** (anvil), and the **stapes** (stirrup) because their shapes resemble these objects. The malleus adheres to the tympanic membrane, and the stapes touches the oval window. An **auditory tube** (eustachian tube), which

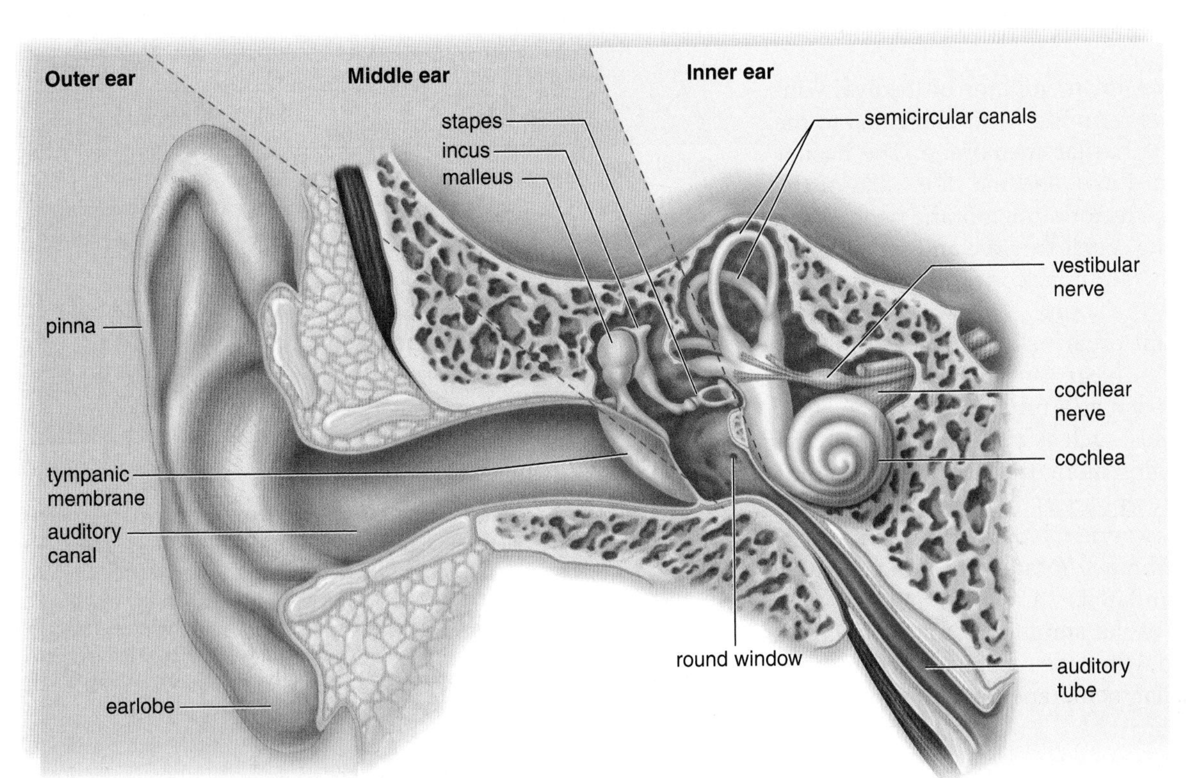

Figure 18.11 Anatomy of the human ear. In the middle ear, the malleus (hammer), the incus (anvil), and the stapes (stirrup) amplify sound waves. In the inner ear, the mechanoreceptors for equilibrium are in the semicircular canals and the vestibule, and the mechanoreceptors for hearing are in the cochlea.

extends from the middle ear to the nasopharynx, permits equalization of air pressure. Chewing gum, yawning, and swallowing in elevators and airplanes help move air through the auditory tubes upon ascent and descent. As this occurs, we often hear the ears "pop."

Whereas the outer ear and the middle ear contain air, the inner ear is filled with fluid. Anatomically speaking, the **inner ear** has three areas: the **semicircular canals** and the **vestibule** function in equilibrium; the **cochlea** functions in hearing. The cochlea resembles the shell of a snail because it spirals.

Auditory Pathway to the Brain

Hearing requires the ear, the cochlear nerve, and the auditory cortex of the brain. The sound pathway begins with the auditory canal.

Through the Auditory Canal and Middle Ear The process of hearing begins when sound waves enter the auditory canal. Just as ripples travel across the surface of a pond, sound waves travel by the successive vibrations of molecules. When these vibrations impinge upon the tympanic membrane, it begins to vibrate at the same frequency as the sound waves. The malleus then takes the pressure from the inner surface of the tympanic membrane and passes it, by means of the incus, to the stapes in such a way that the pressure is multiplied about 20 times as it moves. The stapes strikes the membrane of the oval window, causing it to vibrate, and in this way, the pressure is passed to the fluid within the cochlea.

From the Cochlea to the Auditory Cortex Examining part of the cochlea in cross section reveals that it has three canals: the vestibular canal, the **cochlear canal,** and the tympanic canal (Fig. 18.12). The sense organ for hearing, called the **spiral organ** (organ of Corti), is located in the cochlear canal. The spiral organ consists of little hair cells and a gelatinous material called the **tectorial membrane.** The hair cells sit on the basilar membrane, and their stereocilia are embedded in the tectorial membrane.

When the stapes strikes the membrane of the oval window, pressure waves move from the vestibular canal to the tympanic canal across the basilar membrane. The basilar membrane moves up and down, and the stereocilia of the hair cells embedded in the tectorial membrane bend. Then nerve impulses begin in the cochlear nerve and travel to the brain. When they reach the auditory cortex in the temporal lobe, they are interpreted as a sound.

Each part of the spiral organ is sensitive to different wave frequencies, or pitch. Near the tip, the spiral organ responds to low pitches, such as a tuba, and near the base, it responds to higher pitches, such as a bell or a whistle. The nerve fibers from each region along the length of the spiral organ lead to slightly different areas in the auditory cortex. The pitch sensation we experience depends upon which region of the basilar membrane vibrates and which area of the auditory cortex is stimulated.

Volume is a function of the amplitude of sound waves. Loud noises cause the fluid within the vestibular canal to exert more pressure and the basilar membrane to vibrate to a greater

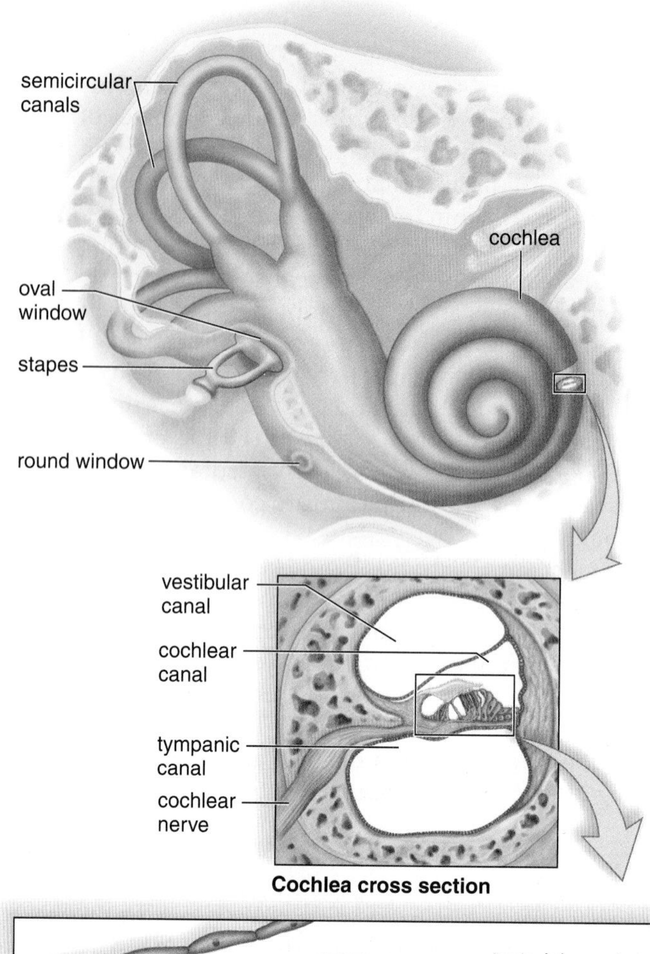

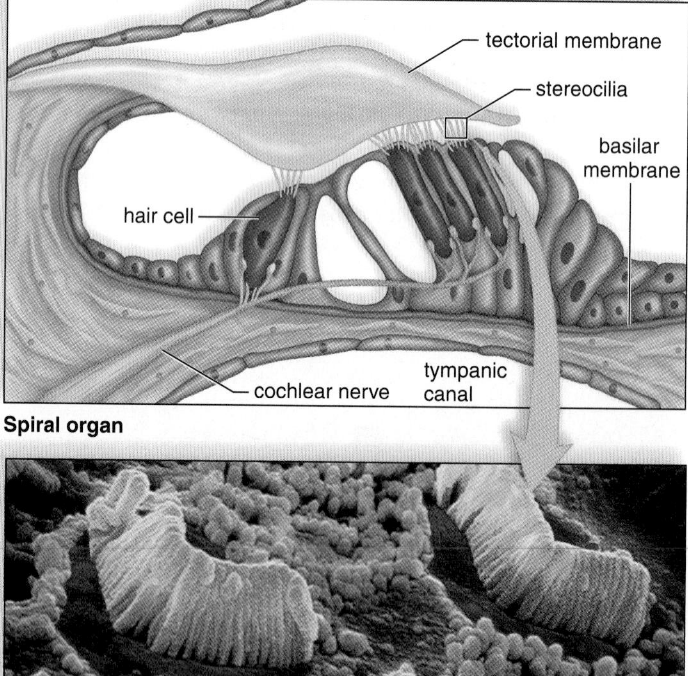

Figure 18.12 Mechanoreceptors for hearing.
The spiral organ is located within the cochlea. In the uncoiled cochlea, note that the spiral organ consists of hair cells resting on the basilar membrane, with the tectorial membrane above. Pressure waves move from the vestibular canal to the tympanic canal, causing the basilar membrane to vibrate. This causes the stereocilia (or at least a portion of the more than 20,000 hair cells) embedded in the tectorial membrane to bend. Nerve impulses traveling in the cochlear nerve to the auditory cortex result in hearing.

Health Focus

Preventing a Loss of Hearing

Some degree of age-associated hearing loss is very common (see section 18.7), but deafness due to stereocilia damage from exposure to loud noises is preventable. Hospitals are now aware that even the ears of newborns need to be protected from noise and are taking steps to make sure neonatal intensive care units and nurseries are as quiet as possible.

In today's society, exposure to the types of noises listed in Table 18A is common. Noise is measured in decibels (db), and any noise above a level of 80 db could result in damage to the hair cells of the spiral organ. Hearing loss occurs once 25–30% of these cells are damaged. Eventually, the stereocilia and then the hair cells disappear completely (Fig. 18A).

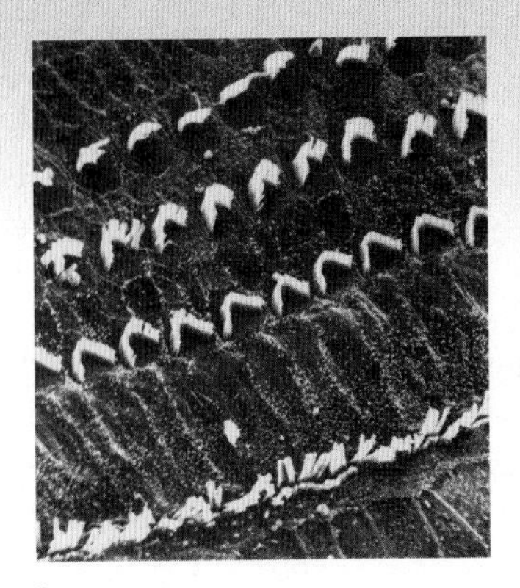
a.

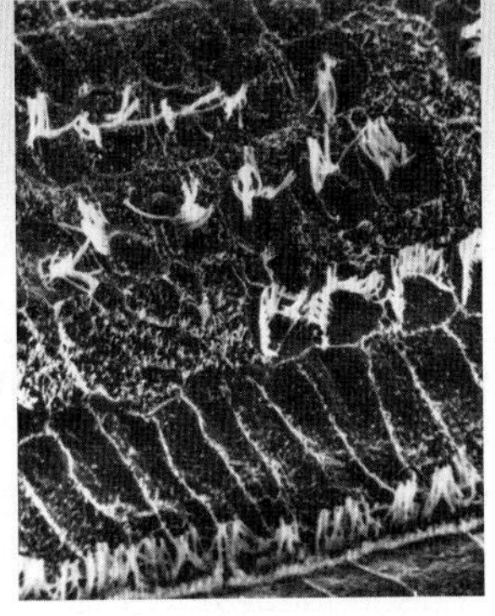
b.

Figure 18A Hearing loss. **a.** Normal hair cells in the spiral organ of a guinea pig. **b.** Damaged hair cells in the spiral organ of a guinea pig. This damage occurred after 24-hour exposure to a noise-level typical of a rock concert.

Legendary rock musician Pete Townsend of The Who recently admitted that long-term exposure to loud music has caused him to experience hearing loss. Digital music players may pose an additional risk because they can produce music as loud as 120 db, plus hold thousands of songs and play for hours without recharging. Like all the sensory systems, the ears can become desensitized to loud music, so that it seems less loud. The response of many music enthusiasts, of course, is to turn up the volume. In 2006, an iPod user filed a lawsuit against Apple, claiming the company did not warn users sufficiently about the dangers of loud music, and asking the company to offer a software update to limit the device's output to 100 db, as it has already done in France.

The first hint of danger could be temporary hearing loss, a "full" feeling in the ears, muffled hearing, or tinnitus (e.g., ringing in the ears). If you have any of these symptoms, modify your listening habits immediately to prevent further damage. If exposure to noise is unavoidable, specially designed noise-reduction earmuffs are available, and it is also possible to purchase earplugs made from a compressible, spongelike material at the drugstore or sporting-goods store. These earplugs are not the same as those worn for swimming, and they should not be used interchangeably.

In addition to loud music, noisy indoor or outdoor equipment can also be harmful to hearing if hearing protection is not used. Even motorcycles and recreational vehicles such as snowmobiles can contribute to a gradual loss of hearing. Exposure to intense sounds of short duration, such as a burst of gunfire, can result in an immediate hearing loss. Hunters may have a significant hearing reduction in the ear opposite the shoulder where the gun is held. (The butt of the rifle offers some protection to the ear nearest the gun when it is shot.)

Finally, people need to be aware that some medicines are ototoxic. Anticancer drugs, most notably cisplatin, and certain antibiotics (e.g., streptomycin, kanamycin, and gentamicin) make ears especially susceptible to a hearing loss. Anyone taking such medications needs to be especially careful to protect the ears from loud noises.

Discussion Questions

1. Do you think that manufacturers of digital music players should be required to restrict the output of these devices to 100 db in the United States?
2. If you listen to loud music or are exposed to other noise, are you concerned that the hair cells in the spiral organs of your inner ears may become damaged? Why or why not?
3. Assume that you suddenly became deaf today, and list three to five of your everyday activities that would be most affected. If your deafness could not be corrected medically, what specific steps could you take to help you communicate with hearing people? With other deaf people?

TABLE 18A Noises that Affect Hearing

Type of Noise	Sound Level (decibels)	Effect
Jet engine, shotgun, rock concert	Over 125	Beyond threshold of pain; potential for hearing loss high
Nightclub, "boom box," thunderclap	Over 120	Hearing loss likely
Chain saw, pneumatic drill, jackhammer, snowmobile, garbage truck, cement mixer	100–200	Regular exposure of more than 1 minute risks permanent hearing loss
Farm tractor, newspaper press, subway, motorcycle	90–100	Fifteen minutes of unprotected exposure potentially harmful
Lawn mower, food blender	85–90	Continuous daily exposure for more than eight hours can cause hearing damage
Diesel truck, average city traffic noise	80–85	Annoying; constant exposure may cause hearing damage

Source: National Institute on Deafness and Other Communication Disorders, National Institutes of Health.

extent. The brain interprets the resulting increased stimulation as volume. Researchers believe that the brain interprets the tone of a sound based on the distribution of the hair cells stimulated. Exposure to loud noises over a long period of time is one of the causes of hearing loss (see Health Focus).

Check Your Progress

1. How are sound waves transmitted from the tympanic membrane to the oval window?
2. As specifically as possible, where are the mechanoreceptors responsible for transducing sound waves into nerve impulses located?
3. How does the distribution of stereocilia along the spiral organ allow the brain to perceive different pitches? How are different volumes of sound distinguished?

18.6 Sense of Equilibrium

Learning Outcomes

1. Review the difference between rotational and gravitational equilibrium.
2. Compare the role of the fluid found in the semicircular canals with the otoliths present in the utricle and saccule.

Mechanoreceptors in the semicircular canals detect rotational and/or angular movement of the head (**rotational equilibrium**), while mechanoreceptors in the utricle and saccule detect movement of the head in the vertical or horizontal planes (**gravitational equilibrium**) (Fig. 18.13).

Through their communication with the brain, these mechanoreceptors help us achieve equilibrium, but other structures in the body are also involved. For example, we already mentioned that proprioceptors are necessary for maintaining our equilibrium. Vision, if available, also provides extremely helpful input the brain can act upon.

Rotational Equilibrium Pathway

Rotational equilibrium involves the three semicircular canals, which are arranged so that there is one in each dimension of space. The base of each of the three canals, called an **ampulla,** is slightly enlarged. Little hair cells, whose stereocilia are embedded within a gelatinous material called a cupula, are found within the ampullae. Because of the way the semicircular canals are arranged, each ampulla responds to head rotation in a different plane of space. As fluid within a semicircular canal flows over and displaces a cupula, the stereocilia of the hair cells bend, and the pattern of impulses carried by the vestibular nerve to the brain stem and cerebellum changes. The brain uses information from the hair cells within the ampullae of the semicircular canals to maintain rotational equilibrium through appropriate motor output to various skeletal muscles that can right our present position in space as need be.

Gravitational Equilibrium Pathway

Gravitational equilibrium depends on the **utricle** and the **saccule,** two membranous sacs located in the vestibule. Both of these sacs contain little hair cells, whose stereocilia are embedded within a gelatinous material called an otolithic membrane. Calcium carbonate ($CaCO_3$) granules, or **otoliths,** rest on this membrane. The utricle is especially sensitive to horizontal (back-and-forth) movements and the bending of the head, while the saccule responds best to vertical (up-and-down) movements.

When the body is still, the otoliths in the utricle and the saccule rest on the otolithic membrane above the hair cells. When the head bends or the body moves in the horizontal and vertical planes, the otoliths are displaced and the otolithic membrane sags, bending the stereocilia of the hair cells beneath. If the stereocilia move toward the largest stereocilium, called the kinocilium, nerve impulses increase in the vestibular nerve. If the stereocilia move away from the kinocilium, nerve impulses decrease in the vestibular nerve. If you are upside down, nerve impulses in the vestibular nerve cease. These data reach the vestibular cortex, which uses them to determine the direction of the movement of the head at the moment. The brain uses this information to maintain gravitational equilibrium through appropriate motor output to various skeletal muscles that can right our present position in space as need be.

Table 18.3 summarizes how the various parts of the ear function in both hearing and equilibrium.

Check Your Progress

1. Which of these are associated with the semicircular canals and which with the vestibule? Ampulla, cupula, otoliths, saccule, utricle.
2. How does having two separate systems for equilibrium provide more complete information than one system could?

TABLE 18.3 Functions of the Parts of the Ear

Part	Medium	Function	Mechanoreceptor
Outer Ear	*Air*		
Pinna		Collects sound waves	—
Auditory canal		Filters air	—
Middle Ear	*Air*		
Tympanic membrane and ossicles		Amplify sound waves	—
Auditory tube		Equalizes air pressure	—
Inner Ear	*Fluid*		
Semicircular canals		Rotational equilibrium	Stereocilia embedded in cupula
Vestibule (contains utricle and saccule)		Gravitational equilibrium	Stereocilia embedded in otolithic membrane
Cochlea (contains spiral organ)		Hearing	Stereocilia embedded in tectorial membrane

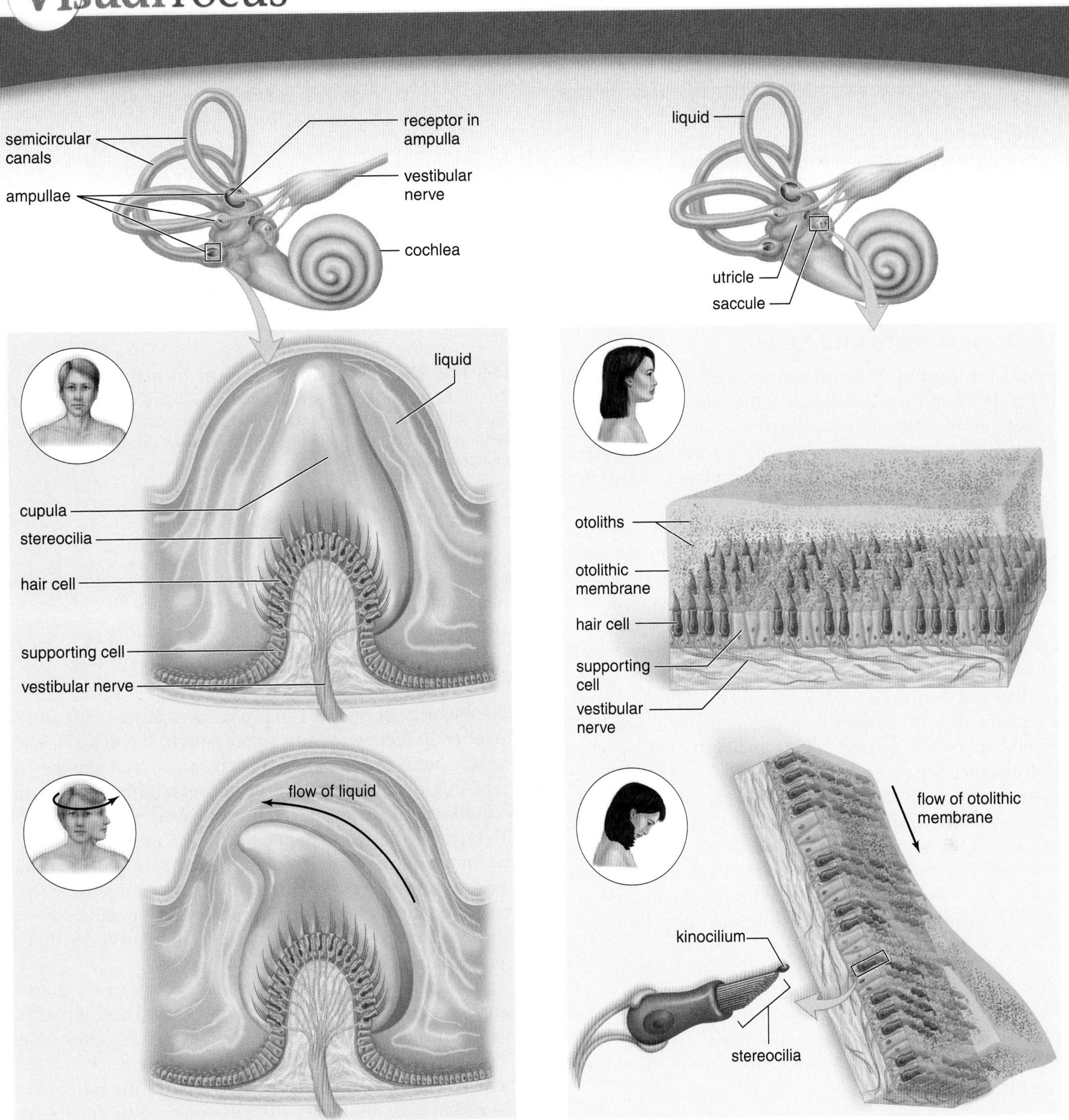

Figure 18.13 Mechanoreceptors for equilibrium.
a. Rotational equilibrium. The ampullae of the semicircular canals contain hair cells with stereocilia embedded in a cupula. When the head rotates, the cupula is displaced, bending the stereocilia. Thereafter, nerve impulses travel in the vestibular nerve to the brain. **b.** Gravitational equilibrium. The utricle and the saccule contain hair cells with stereocilia embedded in an otolithic membrane. When the head bends, otoliths are displaced, causing the membrane to sag and the stereocilia to bend. If the stereocilia bend toward the kinocilium, the longest of the stereocilia, nerve impulses increase in the vestibular nerve. If the stereocilia bend away from the kinocilium, nerve impulses decrease in the vestibular nerve. This difference tells the brain in which direction the head moved.

18.7 Disorders That Affect the Senses

Learning Outcomes

1. List three common disorders that can affect the senses of taste and smell.
2. Explain the anatomical abnormalities that cause color blindness, nearsightedness, farsightedness, and astigmatism.
3. Characterize the three most common causes of blindness in the United States.
4. Describe the specific mechanical changes behind the hearing loss experienced by many people as they age.

Disorders of Taste and Smell

It is hard for most of us to imagine our world without the ability to taste and smell. Disorders that affect these senses may not sound very serious, but these senses contribute substantially to our enjoyment of life. In addition, unpleasant smells and tastes can warn us about dangers such as fire, poisonous fumes, and spoiled food.

In most people, the sense of smell begins to decline after age 60, and a large proportion of elderly people lose some of their olfactory sense. Some people are born with a poor sense of smell or taste, and others completely lack a sense of smell, a condition known as anosmia. However, the most common conditions that affect our sense of taste and smell are upper respiratory infections, allergies, exposure to certain drugs and chemicals, and certain types of brain tumors. Brain injury due to trauma can also cause smell or taste problems, and smoking impairs the ability to identify odors and diminishes the sense of taste.

The extent of a loss of smell or taste can be tested using the lowest concentration of a chemical that a person can detect and recognize. Scientists have developed an easily administered "scratch-and-sniff" test to evaluate the sense of smell. For taste, patients are asked to evaluate different chemical concentrations, sometimes applied directly to specific areas of the tongue.

Disorders of the Eye

Color blindness and problems with visual focus are two common abnormalities of the eye. More serious disorders can result in blindness.

Color Blindness

Complete color blindness is extremely rare. Instead, in most instances a particular type of cone is lacking or deficient in number. The most common mutation is a lack of either red or green cones, resulting in various types of color blindness (Fig. 18.14). If the eye lacks red cones, the green colors are accentuated, and vice versa. Most genetic mutations affecting color vision are X-linked, recessive traits. Because females have two copies of the X chromosome, they are more likely to have a normal copy of the gene. Therefore, color blindness is much more common in males. Approximately 5–8% of males and 0.5% of females have some type of deficiency in color vision.

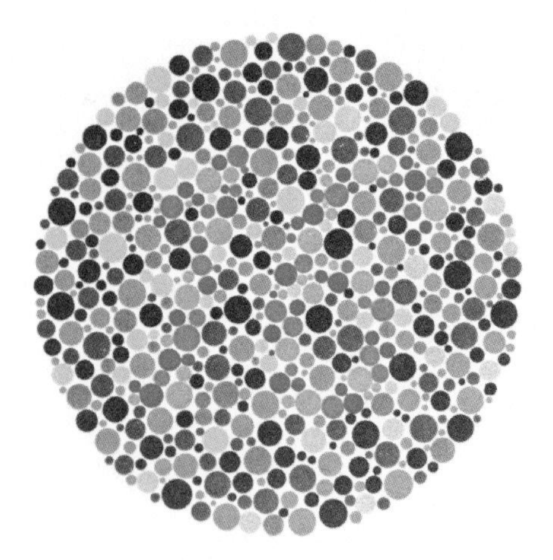

Figure 18.14 Testing for color blindness.
This is one type of test plate used to diagnose color blindness. There are many different types of color blindness, and thus many different tests are used. In this case, you should be able to see the number "16" in green among the dots.

Visual Focus

The majority of people can see what is designated as a size 20 letter 20 feet away, and so are said to have 20/20 vision. Persons who can see close objects but cannot see the letter from this distance are said to be **nearsighted.** These individuals usually have an elongated eyeball, so that when they attempt to look at a distant object, the lens of the eye brings the image into focus in front of the retina (Fig. 18.15*a*). They can see close objects because the lens can compensate for the long eyeball. To see distant objects, these people can wear concave lenses, which diverge the light rays so that the image focuses on the retina.

Persons who can easily see the optometrist's chart but cannot see close objects well are **farsighted.** These individuals can see distant objects better than they can see close objects. They have a shortened eyeball, and when they try to see close objects, the image is focused behind the retina. When the object is distant, the lens can compensate for the short eyeball. When the object is close, these persons must wear convex lenses to increase the bending of light rays so that the image can be focused on the retina (Fig. 18.15*b*). People over the age of 40 usually begin to have trouble seeing objects close up due to a condition called presbyopia, in which the lens of the eye begins to lose its elasticity and thus cannot focus on close objects.

When the cornea or lens is uneven, the image is fuzzy. The light rays cannot be evenly focused on the retina. This condition, known as **astigmatism,** can be corrected by wearing an unevenly ground lens to compensate for the uneven

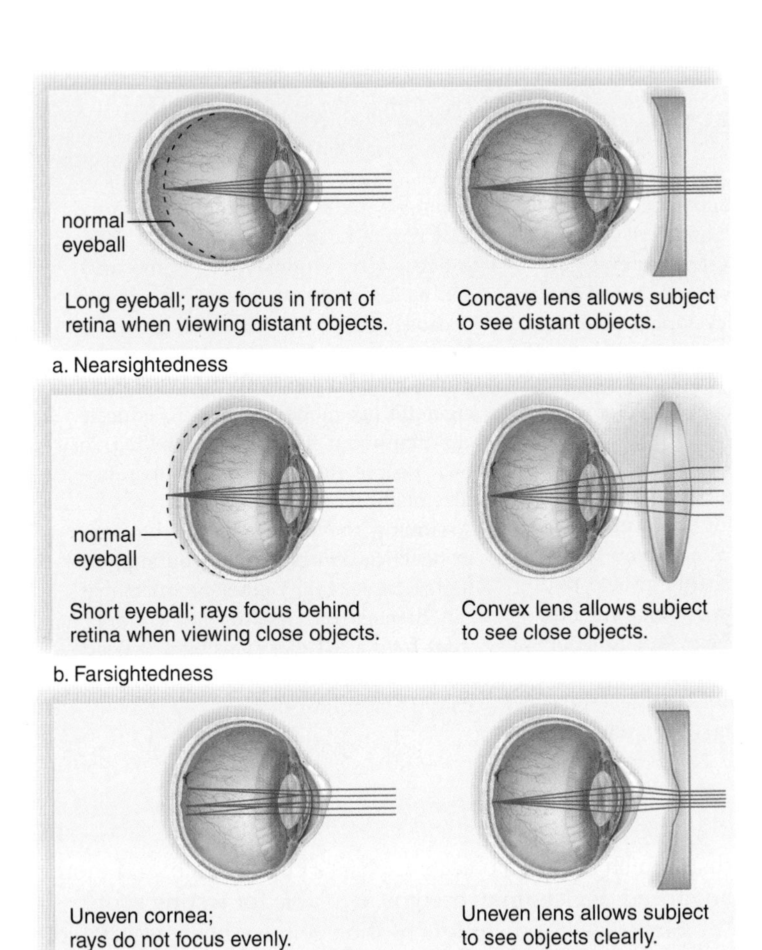

Figure 18.15 Common abnormalities of the eye and possible corrective lenses.
a. A concave lens in nearsighted persons focuses light rays on the retina.
b. A convex lens in farsighted persons focuses light rays on the retina.
c. An uneven lens in persons with astigmatism focuses light rays on the retina.

cornea (Fig. 18.15*c*). Rather than wearing glasses or contact lenses, many people with problems of visual focus are now choosing to undergo a laser surgery procedure called LASIK (laser-assisted in situ keratomileusis).

CHECK YOUR PROGRESS

1. Why is color blindness more common in males?
2. How can the shape of the eyeball result in nearsightedness or farsightedness? What causes astigmatism?

Common Causes of Blindness

The most frequent causes of blindness in adults are retinal disorders, glaucoma, and cataracts, in that order. Retinal disorders are varied. In diabetic retinopathy, the leading cause of blindness in the United States for people under the age of 65, capillaries to the retina become damaged, and hemorrhages and blocked vessels can occur. Surgical lasers can be used to seal the leaking blood vessels, but eventually scarring can lead to blindness. A common condition in people over 50 years of age is macular degeneration, in which the cones are destroyed because thickened choroid vessels no longer function as they should. The retina itself can also become torn or detached, often following trauma to the eye or head. If diagnosed quickly enough, the retina can usually be reattached with a laser.

Glaucoma occurs when the drainage system of the eyes fails, so fluid builds up and the resulting increase in pressure destroys the nerve fibers responsible for peripheral vision. If glaucoma is untreated, total blindness can result. Eye doctors usually measure the pressure inside the eye during routine exams, but it is advisable for everyone to be aware of the disorder because it can come on quickly, and damage to the nerves can be permanent. Those who have experienced acute glaucoma report that the eyeball feels as heavy as a stone. Rarely, the abnormalities that lead to glaucoma can be present at birth, even though the disease symptoms may not appear for a few months or years. Regular visits to an eye-care specialist, especially by the elderly, are a necessity in order to catch conditions such as glaucoma early enough to allow effective treatment.

With aging, the eye is increasingly subject to cataracts. Cataracts are cloudy spots on the lens of the eye that eventually pervade the whole lens, which takes on a milky-white to yellowish color and becomes incapable of transmitting light (Fig. 18.16). Factors that increase the risk of cataracts include exposure to ultraviolet or other types of radiation, diabetes, and heavy alcohol consumption. In addition, smoking can double the risk of cataracts, possibly by reducing the delivery of blood, and therefore nutrients, to the lens. In most cases vision can be restored by surgical removal of the unhealthy lens and replacement with a clear plastic artificial lens. Cataract surgery is the most frequently performed

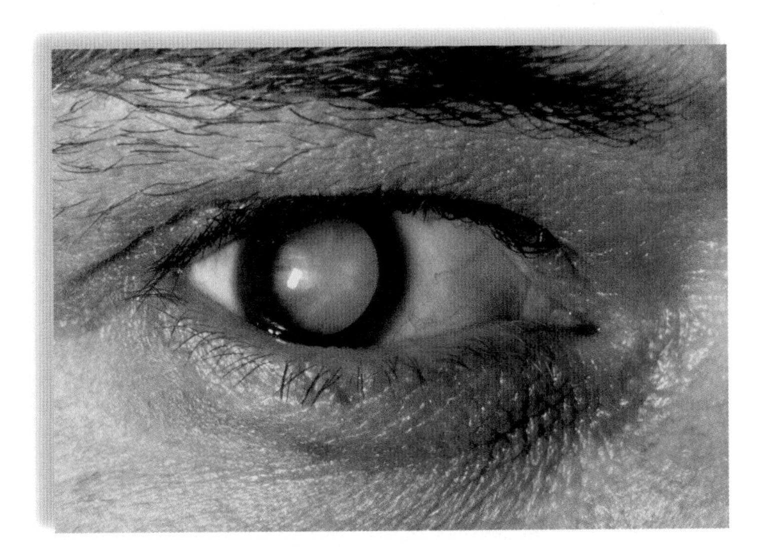

Figure 18.16 Cataract in a human eye.
A cataract is a lens that has become opaque or cloudy. Risk factors that increase the chances of cataract formation include UV exposure, diabetes, and smoking. Cataracts can usually be surgically removed and replaced with an artificial lens.

Bioethical Focus

Taking Personal Responsibility for Health

A cataract is a cloudiness of the lens that usually occurs in the elderly. A dense, centrally placed cataract causes severe blurring of vision. Certain factors have been identified as contributing to the chances of having a cataract:

- Smoking 20 or more cigarettes a day doubles the risk of cataracts in men, and smoking more than 30 cigarettes a day increases the chances of cataracts in women.
- Exposure to the ultraviolet radiation in sunlight can more than double the risk of cataracts.
- As many as one-third of cataracts may be caused by being overweight. Diet, rather than exercise, seems to reduce cataract formation.

Most cataract operations today are performed on an outpatient basis, with minimal postoperative discomfort and a high expectation of sight restoration. Should we simply rely on medical science to restore our eyesight? Or should we act responsibly and take all possible steps to keep from developing cataracts or other medical problems—such as refraining from smoking, watching our weight, wearing sunglasses, and turning down the volume?

Form Your Own Opinion

1. Suppose you lose your sense of hearing after years of listening to loud music on your MP3 player. Should you be eligible for insurance benefits to help pay for thousands of dollars of hearing aids or other devices to help you function normally?
2. Some state governments collect relatively large amounts of revenue by taxing cigarettes. For example, New York City currently adds $3 in taxes per pack. Interestingly, Virginia's tax is the lowest, at 2.5 cents per pack, presumably because much more tobacco is grown in Virginia than in New York. Which state's philosophy do you support? What if the cigarette tax money is used to educate children about the risks of smoking, to help people stop smoking, or to subsidize medical care for smokers?
3. Assuming the money would be used in beneficial ways, would you support higher taxes on any other products that can damage our health—for example, trans fatty acids that contribute to heart disease or digital music players that can damage your hearing?

surgery in the United States, with about 2.8 million cataract surgeries done in 2005. Also in 2005, surgeons began replacing cataracts with multifocal lenses that allow some patients to focus at different distances, even eliminating the need for glasses in some patients. (The Bioethical Focus explores the question of whether we should keep relying on surgery to solve this problem or take more preventive steps to keep cataracts from developing in the first place.)

In 2008, two groups of researchers published studies showing that a certain form of hereditary blindness could be successfully treated with gene therapy. Both groups injected a genetically modified virus into one eye of patients with Leber's congenital amaurosis (LCA), a rare condition in which retinal degeneration begins in childhood and eventually causes total blindness. Starting two weeks after the injections, vision in the injected eyes improved from detecting only hand movements to reading lines on an eye chart. Equally significant, the injected eyes showed no signs of inflammation in the retina or other toxic side effects. If these positive results are confirmed, gene therapy might be useful for other forms of blindness.

Although it may sound like science fiction, researchers are currently testing an artificial eye of sorts, which may allow blind people to see. For blind patients who still have intact connections between their retinas and brain, an array of electrodes can be surgically implanted in the retina. These electrodes receive input from a tiny digital camera mounted on eyeglass frames worn by the patient. This input is converted into a nerve impulse, which the individual's brain interprets as sight. Clinical trials have shown that by using such a device, some formerly blind patients can detect light or even distinguish between objects such as a cup or plate, and more advanced devices may become available for testing soon.

Since several disorders of the eye have been associated with exposure to ultraviolet irradiation, it is recommended that everyone, especially those who live in sunny climates or work outdoors, wear sunglasses that absorb ultraviolet light. The Sunglass Association of America has adopted the following system for categorizing sunglasses:

- Cosmetic lenses, which absorb at least 70% of UV-B (the more harmful type of ultraviolet radiation), 20% of UV-A, and 60% of visible light. Such lenses are worn for comfort or appearance rather than protection.
- General-purpose lenses that absorb at least 95% of UV-B, 60% of UV-A, and 60–92% of visible light. These lenses are good for outdoor activities in temperate regions.
- Special-purpose lenses, which block at least 99% of UV-B, 60% of UV-A, and 60–92% of visible light. They are good for bright sun combined with sand, snow, or water.

Disorders of Hearing and Equilibrium

Hearing loss can develop gradually or suddenly and has many potential causes. Especially when we are young, the middle ear is subject to infections that can lead to hearing impairment if not treated promptly by a physician (see section 15.4). Hearing loss is also one of the most common conditions of older adults, affecting one in three people over age 60. Age-associated hearing loss usually develops gradually. The first sign may be problems understanding conversations when there is noise in the background. Actually, in most individuals age-related hearing loss probably begins at around age 20, with the ability to hear high-pitched sounds affected first. With age, the mobility of ossicles decreases, and in otosclerosis, new filamentous bone grows over the stirrup, impeding its movement. Surgical treatment is the only remedy for this type of conduction deafness. Some types of hearing loss seem to run in families, but most can be attributed to the effect of years of frequent exposure to loud noise (see the Health Focus). As noted in the story that opened this chapter, some types of deafness can be treated with cochlear implants.

Sudden deafness, which may occur all at once or over a period of up to three days, most commonly affects people between the ages of 30 and 50, usually in only one ear. Potential causes include infections, head trauma, and the side effects of certain drugs. Depending on the cause, sudden deafness often resolves itself within a few days.

Deafness can also be present at birth. Several genetic disorders are associated with deafness at birth, as are infections with German measles (rubella) and mumps virus when contracted by a woman during her pregnancy. For this reason, every female should be immunized against these viruses before she reaches childbearing age.

Disorders of equilibrium often manifest as **vertigo,** the feeling that a person or the environment is moving when no motion is occurring. It is possible to simulate a feeling of vertigo by spinning your body rapidly and then stopping suddenly. Vertigo can be caused by problems in the brain as well as the inner ear. An estimated 20% of those who experience symptoms of vertigo have benign positional vertigo (BPV), which may result from the formation of particles in the semicircular canals. When individuals with BPV move their heads suddenly, especially when lying down, these particles shift like pebbles inside a tire. After the head movement, there is an initial delay, but when the movement stops, the particles tumble down with gravity, stimulating the stereocilia and resulting in the sensation of movement. Interestingly, most cases of BPV either resolve on their own or can be cured using a series of precise head movements designed to cause the particles to settle into a position where they can no longer stimulate the stereocilia.

Because the senses of hearing and equilibrium are anatomically linked, certain disorders can affect both. An example of this phenomenon is Meniere's disease, which is usually characterized by vertigo, a feeling of fullness in the affected ear(s), tinnitus (ringing in the ears), and hearing loss. In about 80% of patients, only one ear is affected, and the disease usually strikes between the ages of 20 and 50. The exact cause of Meniere's disease is unknown, but it seems related to an increased volume of fluid in the semicircular canals, vestibule, and/or cochlea. For this reason, it has been called "glaucoma of the ear." There is no cure for Meniere's disease, but the vertigo and feeling of pressure can often be managed by adhering to a low-salt diet, which is thought to decrease the volume of fluid produced. Any hearing loss that occurs, however, is usually permanent.

Check Your Progress

1. What does it mean to treat a disease condition with gene therapy?
2. What are some common causes of sudden hearing loss?
3. Define vertigo. What causes benign positional vertigo?
4. Why has Meniere's disease been called "glaucoma of the ear"?

Applying the Concepts Revisited

In the story that opened this chapter, a couple decided to have their baby, who was born deaf, treated with cochlear implants. Although somewhat controversial, this procedure can help some deaf children to speak and function much like a hearing child.

1. Some types of deafness are caused by abnormalities affecting the cochlea, while others affect only the auditory nerves. Which type(s) would be most amenable to treatment with cochlear implants, and why?
2. Cochlear implants work by converting sounds to electrical impulses, which are transmitted by electrodes implanted into the auditory nerve. Why is it very difficult to reproduce normal hearing with this technology? (Hint: An auditory nerve contains about 30,000 axons.)

Summarizing the Concepts

18.1 Sensory Receptors and Sensations

Sensory receptors can be classified in several ways:

- As interoceptors or exteroceptors, depending on whether they receive stimuli from inside or outside the body.
- As chemoreceptors, photoreceptors, mechanoreceptors, or thermoreceptors, according to the nature of the stimulus they receive.

Sensation, or conscious perception of stimuli, is the result of several processes: detection occurs when environmental changes stimulate sensory receptors, generating nerve impulses; sensation occurs when these nerve impulses reach the cerebral cortex; and perception occurs when the brain interprets the meaning of sensations.

18.2 Somatic Senses

The somatic sensory receptors include proprioceptors, sensory receptors, and pain receptors:

- Proprioceptors help maintain equilibrium and posture. Proprioception is illustrated by the action of muscle spindles that are stimulated when muscle fibers stretch and Golgi tendon organs that are stimulated when muscles contract.

- Cutaneous receptors in the skin detect touch, pressure, pain, and temperature (warmth and cold).
- Pain receptors are free nerve endings that respond to chemicals released by damaged cells. The pain of internal organs is sometimes felt in the skin and is called referred pain.

18.3 Senses of Taste and Smell

Taste and smell are due to the detection of molecules in food and in the air by chemoreceptors:

- After these molecules bind to plasma membrane receptor proteins on the microvilli of taste cells and the cilia of olfactory cells, nerve impulses eventually reach the cerebral cortex, which determines the taste and odor according to the pattern of stimulation. The gustatory (taste) cortex responds to a range of salty, sour, bitter, sweet, and perhaps "umami" tastes.

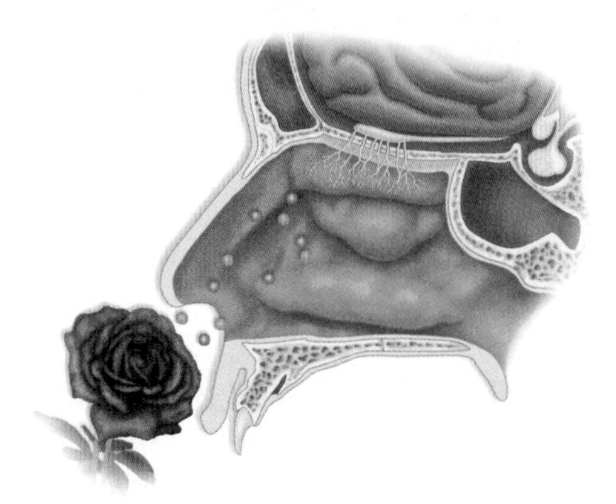

18.4 Sense of Vision

Vision requires the eye, the optic nerves, and the visual cortex in the occipital lobe:

- The eye has three layers. The outer layer, called the sclera, can be seen as the white of the eye. It also forms the transparent bulge in the front of the eye called the cornea. The middle pigmented layer, called the choroid, absorbs stray light rays. The rod cells (sensory receptors for dim light) and the cone cells (sensory receptors for bright light and color) are located in the retina, the inner layer of the eyeball. The cornea, the humors, and especially the lens bring the light rays to focus on the retina. To see a close object, accommodation occurs as the lens rounds up.
- Light is normally focused on the fovea centralis, an area of densely packed cone cells. The visual pathway begins when light strikes rhodopsin within the membranous disks of rod cells. A cascade of reactions leads to the closing of ion channels in a rod cell's plasma membrane, and signals are passed to other neurons in the retina. From the rod and cone cell layer, impulses pass to the bipolar cell layer and the ganglion cell layer. Many rods may synapse on the same ganglion cell, but some cone cells may activate only one ganglion cell.
- The axons of ganglion cells become the optic nerve, which takes nerve impulses to the brain. The visual field is taken apart by the optic chiasma before reaching the visual cortex in the occipital lobe. The primary visual area parcels out signals for color, form, and motion to the visual association area. Then the cortex rebuilds the field.

18.5 Sense of Hearing

Hearing is dependent on the ear, the cochlear nerve, and the auditory cortex of the brain:

- The ear is divided into three parts: outer, middle, and inner ear. The outer ear consists of the pinna and the auditory canal, which direct sound waves to the middle ear. The middle ear begins with the tympanic membrane and contains the ossicles (malleus, incus, and stapes). The malleus is attached to the tympanic membrane, and the stapes is attached to the oval window, which is covered by a membrane. The inner ear contains the cochlea and the semicircular canals, plus the utricle and the saccule.

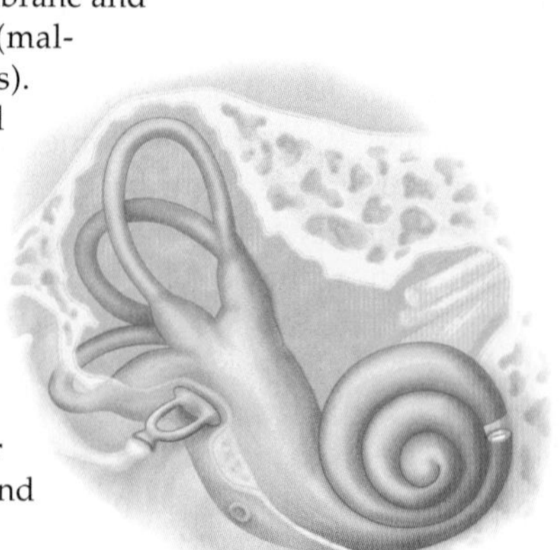

- The auditory pathway begins when the outer ear receives and the middle ear amplifies the sound waves, which then strike the oval window membrane. Its vibrations set up pressure waves across the cochlear canal, which contains the spiral organ, consisting of hair cells whose stereocilia are embedded within the tectorial membrane. When the basilar membrane vibrates, the stereocilia of the hair cells bend. This movement is transduced into nerve impulses that are carried by the cochlear nerve to the primary auditory area in the temporal lobe of the cerebral cortex.

18.6 Sense of Equilibrium

The ear also contains mechanoreceptors for our sense of equilibrium, or balance:

- Rotational equilibrium, which is the sensing of head movements, is dependent on the stimulation of hair cells within the ampullae of the semicircular canals.
- Gravitational equilibrium, which is the sensing of the position of the head in space, relies on the stimulation of hair cells within the utricle and the saccule by small granules called otoliths.

18.7 Disorders That Affect the Senses

Many conditions can affect the various sensory organs. Most are not life threatening, but all certainly can affect the quality of life:

- Disorders of taste and smell may result from infections, allergies, certain drugs, or tumors.
- Disorders of the eye include color blindness, nearsightedness and farsightedness, and blindness. The most common causes of blindness in the United States are retinal disorders, glaucoma, and cataracts.
- Disorders of hearing are very common, especially in the aging population. Possible causes include excessive noise exposure, infections, trauma, and certain drugs. Vertigo is an equilibrium disorder characterized by a feeling of motion or dizziness.

TESTING YOURSELF

Choose the best answer for each question.

1. Chemoreceptors are involved in
   a. hearing.
   b. taste.
   c. smell.
   d. vision.
   e. Both b and c are correct.
2. Free nerve endings sense temperature and
   a. pressure.
   b. pain.
   c. vision.
   d. hearing.
3. Tasting "sweet" versus "salty" is a result of
   a. activating different sensory receptors.
   b. activating many versus few sensory receptors.
   c. activating no sensory receptors.
   d. None of these are correct.
4. Our sense of smell
   a. is completely separate from our sense of taste.
   b. is dependent on olfactory cells, which are modified neurons.
   c. requires millions of different types of odor receptors.
   d. None of these are correct.
5. Label the following diagram of the human eye.

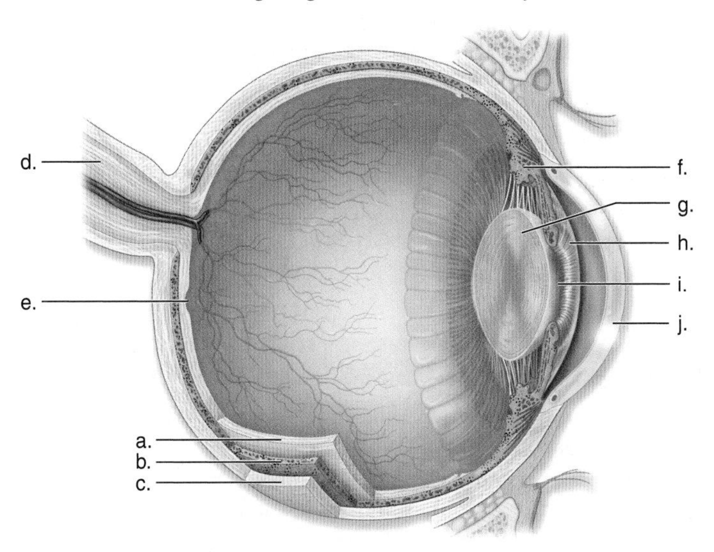

6. Which of the following structure-function associations is incorrect?
   a. lens—focusing
   b. cones—color vision
   c. iris—regulation of amount of light
   d. choroid—location of cones
   e. sclera—protection
7. Retinal is
   a. a derivative of vitamin A.
   b. sensitive to light energy.
   c. found in both rods and cones.
   d. a part of rhodopsin.
   e. All of these are correct.
8. Label the following diagram of the optic chiasma.

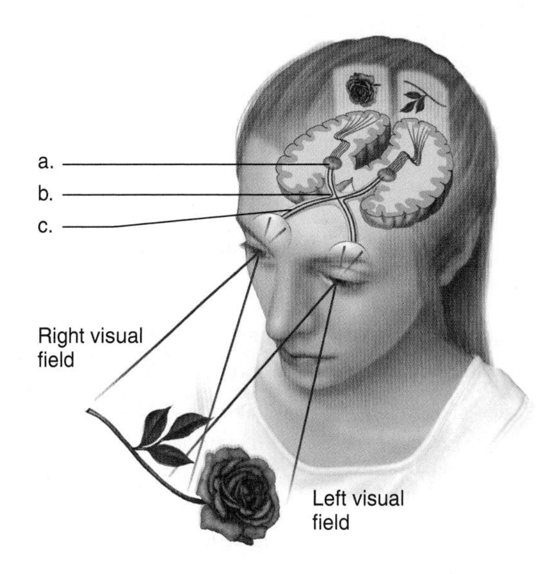

9. Label the following diagram of the human ear.

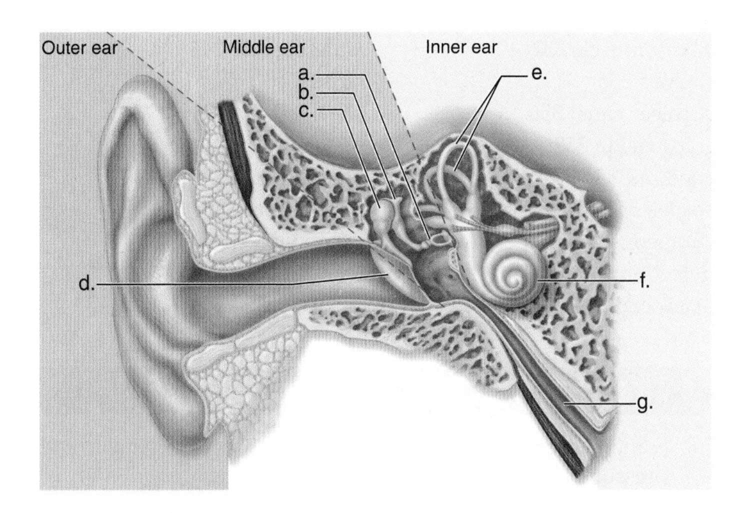

10. Which is the correct location of the spiral organ?
    a. between the tympanic membrane and the oval window in the inner ear
    b. in the utricle and saccule within the vestibule
    c. between the tectorial membrane and the basilar membrane in the cochlear canal
    d. between the nasal cavities and the throat
    e. between the outer and inner ear within the semicircular canals
11. Our perception of pitch is dependent upon the region of the _____ vibrated and the regions of the _____ stimulated.
    a. cochlea, spinal cord
    b. basilar membrane, spinal cord
    c. basilar membrane, auditory cortex
    d. cochlea, cerebellum

12. The ossicle that articulates with the tympanic membrane is the
    a. malleus.
    b. stapes.
    c. incus.
    d. All of these are correct.
13. Which of these associations is mismatched?
    a. semicircular canals—inner ear
    b. utricle and saccule—outer ear
    c. auditory canal—outer ear
    d. cochlea—inner ear
    e. ossicles—middle ear
14. Both olfactory receptors and sound receptors have cilia, and both
    a. are chemoreceptors.
    b. are a part of the brain.
    c. are mechanoreceptors.
    d. initiate nerve impulses.
    e. All of these are correct.
15. A color-blind person has (an) abnormal
    a. rods.
    b. cochlea.
    c. cornea.
    d. cones.
    e. None of these are correct.
16. Which pair of terms is properly matched?
    a. nearsightedness—light rays focus in front of the retina
    b. nearsightedness—light rays focus behind the retina
    c. farsightedness—light rays focus in front of the retina
    d. None of these are correct.

UNDERSTANDING THE TERMS

ampulla 358
aqueous humor 352
astigmatism 360
auditory canal 355
auditory tube 355
blind spot 354
chemoreceptor 346
choroid 352
ciliary body 352
ciliary muscle 352
cochlea 356
cochlear canal 356
color vision 353
cone cell 353
cornea 352
cutaneous receptor 349
detection 346
exteroceptor 346
farsighted 360
focus 352
fovea centralis 352
glaucoma 352
gravitational equilibrium 358
hair cell 355
incus 355
inner ear 356
integration 347
interoceptor 346
iris 352
lens 352
malleus 355
mechanoreceptor 346
middle ear 355
nearsighted 360
olfactory cell 350
optic chiasma 354
optic nerve 352
ossicle 355
otolith 358
outer ear 355
oval window 355
perception 346
photoreceptor 346
pinna 355
proprioceptor 347
pupil 352
referred pain 349
retina 352
retinal 353
rhodopsin 353
rod cell 353
rotational equilibrium 358
round window 355
saccule 358
sclera 352
semicircular canal 356
sensation 346
sensory adaptation 347
sensory receptor 346
spiral organ 356
stapes 355
stimulus 346
taste bud 349
tectorial membrane 356
thermoreceptor 346
tympanic membrane 355
utricle 358
vertigo 363
vestibule 356
visual accommodation 352
vitreous humor 352

THINKING CRITICALLY

1. Some sensory receptors, such as those for taste, smell, and pressure, readily undergo the process of sensory adaptation, or decreased response to a stimulus. In contrast, receptors for pain are less prone to adaptation. Why does this make good biological sense? What do you think happens to children who are born without the ability to feel pain normally?
2. Trace the path of a photon of light through all the layers of the retina discussed in this chapter. Then trace the nerve impulse from a photoreceptor located on the right side of the left eyeball through the visual cortex.
3. Suppose you find yourself in a room with very little light, and you are having trouble finding the light switch. You notice that you can see better with your peripheral vision than in the center of your visual field. Why is this the case?

INQUIRY INTO LIFE WEBSITE

The companion website for *Inquiry into Life* provides a wealth of information organized and integrated by chapter. You will find practice tests, animations, videos, and much more that will complement your learning and understanding of general biology.

http://www.mhhe.com/maderinquiry13

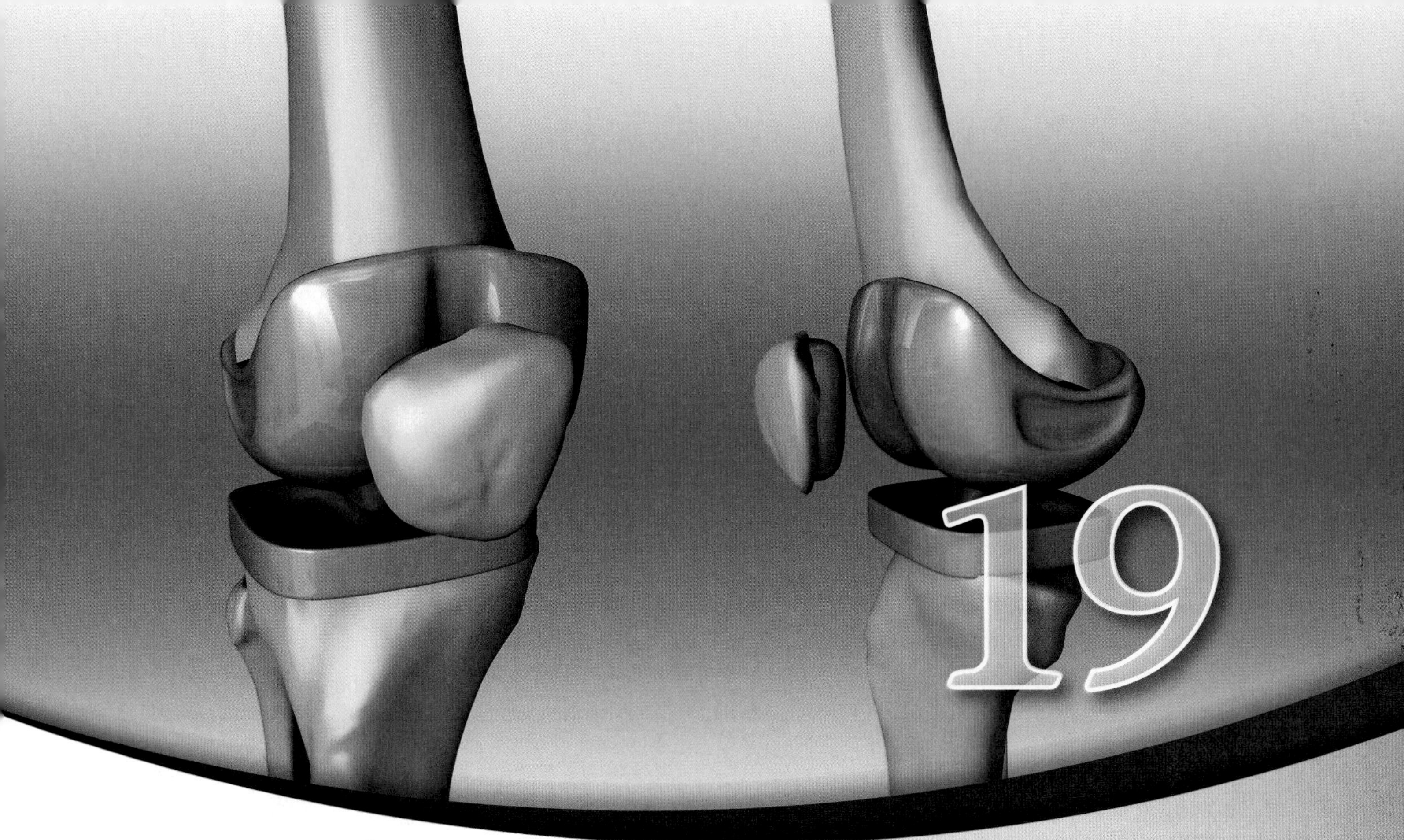

Musculoskeletal System

Chapter Outline

Applying the Concepts

Jackie was an outstanding athlete in high school, and even now, in her early fifties, she tried to stay in shape. But during her customary 3-mile jogs, she was having an increasingly hard time ignoring the pain in her left knee. She had torn ligaments in her knee playing intramural softball in college, and it had never quite felt the same. In her forties, she was able to control the pain by taking over-the-counter medications, but two years ago she had arthroscopic surgery on her knee to remove some torn cartilage and calcium deposits. Now that the pain was getting worse than before, she knew her best option might be a total knee replacement.

Although it sounds drastic, replacing old, arthritic joints with shiny new artificial ones is becoming increasingly routine. About 500,000 artificial knees were installed in U.S. patients in 2006, which represented a 65% increase from 2000. With the population aging, the demand for this procedure is expected to increase. During the procedure, a surgeon removes bone from the bottom of the femur and the top of the tibia, and replaces each with caps made of metal or ceramic, held in place with bone cement. A plastic plate is installed to allow the femur and tibia to move smoothly against each other, and a smaller plate is attached to the kneecap (patella) so it can function properly.

For the first painful month or so after having her knee replaced, Jackie wondered if she had made the right decision. Just walking down the hall or up stairs was excruciating at first. Within two months, however, she was walking and swimming. Her physical therapist attributed her rapid return to near-normalcy to her previous habits of staying in shape. But Jackie knows without twenty-first-century medicine, she might have had a very difficult time walking by the time she was 60. And she is reminded of the new metallic part of her every time she sets off an airport metal detector! In this chapter we will examine the anatomy and physiology of bones, joints, and muscles.

19.1 Anatomy and Physiology of Bone

LEARNING OUTCOMES

1. Describe the macroscopic and microscopic structure of bone.
2. Distinguish the structure and function of the three types of cartilage.
3. Explain how bones change during a person's lifetime.

The skeleton provides attachment sites for the muscles, whose contraction makes the bones move so that we can walk, play sports, hold this book, and perform all sorts of other activities. Together, the bones and muscles comprise the *musculoskeletal system.* In this chapter, we discuss the structures and functions of first the skeleton and then the muscles. Bones and the tissues associated with them are connective tissues, while muscles are, of course, composed of muscular tissue.

Organization of Tissues in the Skeleton

Bones are classified by their shape: long, short, flat, or irregular. Long bones do not have to be terribly long—the bones in your fingers are classified as long. But a long bone must be longer than it is wide. The bones of the arms and legs are long bones except for the kneecap and the wrist and ankle bones.

A long bone of the arm or leg can be used to illustrate the organization of tissues of the skeleton (Fig. 19.1). A bone is enclosed by a tough, fibrous, connective tissue covering called the **periosteum,** which is continuous with the ligaments that go across a joint. The periosteum contains blood vessels that enter the bone and give off branches that service various types of cells found in bone. Each expanded end of a long bone is termed an epiphysis; the portion between the epiphyses is called the diaphysis. Where the diaphysis of a long bone contacts (articulates with) another bone at a **joint,** it will be covered by a layer of hyaline cartilage, also called **articular cartilage.**

Structure of Bone and Associated Tissues

The primary connective tissues of the skeleton are bone, cartilage, and dense fibrous connective tissue. All connective tissues contain cells separated by a matrix that contains fibers. Bone tissue is particularly strong because the matrix contains collagen fibers plus mineral salts, notably calcium phosphate.

Bone

There are two types of bone tissue, compact bone and spongy bone. **Compact bone** is highly organized and composed of tubular units called **osteons.** Bone cells called **osteocytes** lie in lacunae, or tiny chambers, arranged in concentric circles around a central canal (see Fig. 19.1). Central canals contain blood vessels, lymphatic vessels, and nerves. Tiny canals called canaliculi run through the matrix, connecting the lacunae with each other and with the central canal. In this way, the canaliculi bring nutrients from the blood vessel in the central canal to the cells in the lacunae.

The epiphyses of a long bone are composed largely of spongy bone. Compared to compact bone, **spongy bone** has an unorganized appearance (see Fig. 19.1). Osteocytes are found in trabeculae, which are numerous thin plates separated by unequal spaces. Although the spaces make spongy bone lighter than compact bone, spongy bone is still designed for strength. Just as braces are used for support in buildings, the plates of spongy bone follow lines of stress.

Splitting a bone open, as in Figure 19.1, shows that the diaphysis is not solid but has a medullary cavity containing yellow bone marrow. Yellow bone marrow contains a large amount of fat.

The spaces of spongy bone are often filled with **red bone marrow,** a specialized tissue that produces all types of blood cells. In infants, red bone marrow is found in the cavities of most bones. In adults, red bone marrow occurs in a more limited number of bones (see section 19.2).

Cartilage

Cartilage is not as strong as bone, but it is more flexible because the matrix is gel-like and contains many collagenous and elastic fibers. The cells lie within lacunae, which are irregularly grouped. Cartilage has no blood vessels, and therefore, injured cartilage is slow to heal.

All three types of cartilage—hyaline cartilage, fibrocartilage, and elastic cartilage—are associated with bones. Hyaline cartilage is firm and somewhat flexible. The matrix appears uniform and glassy (Fig. 19.1), but actually it contains a generous supply of collagen fibers. Hyaline cartilage is found at the ends of long bones and in the nose, at the ends of the ribs, and in the larynx and trachea.

Fibrocartilage is stronger than hyaline cartilage because the matrix contains wide rows of thick collagen fibers. Fibrocartilage is able to withstand both tension and pressure. This type of cartilage is found where support is of prime importance—in the disks between the vertebrae and in the cartilaginous disks between the bones of the knee.

Elastic cartilage is more flexible than hyaline cartilage because the matrix contains mostly elastin fibers. This type of cartilage is found in the ear flaps and epiglottis.

Dense Fibrous Connective Tissue

Dense fibrous connective tissue contains rows of cells called fibroblasts separated by bundles of collagen fibers. Dense fibrous connective tissue forms the flared sides of the nose. **Ligaments,** which bind bone to bone, are dense fibrous connective tissue, and so are **tendons,** which connect muscle to bone at joints.

CHECK YOUR PROGRESS

1. In what type of bone would you find yellow versus red bone marrow? What is the function of red bone marrow?
2. Where in the body would you find hyaline cartilage, fibrocartilage, and elastic cartilage?
3. In what way are ligaments similar to tendons? How are they different?

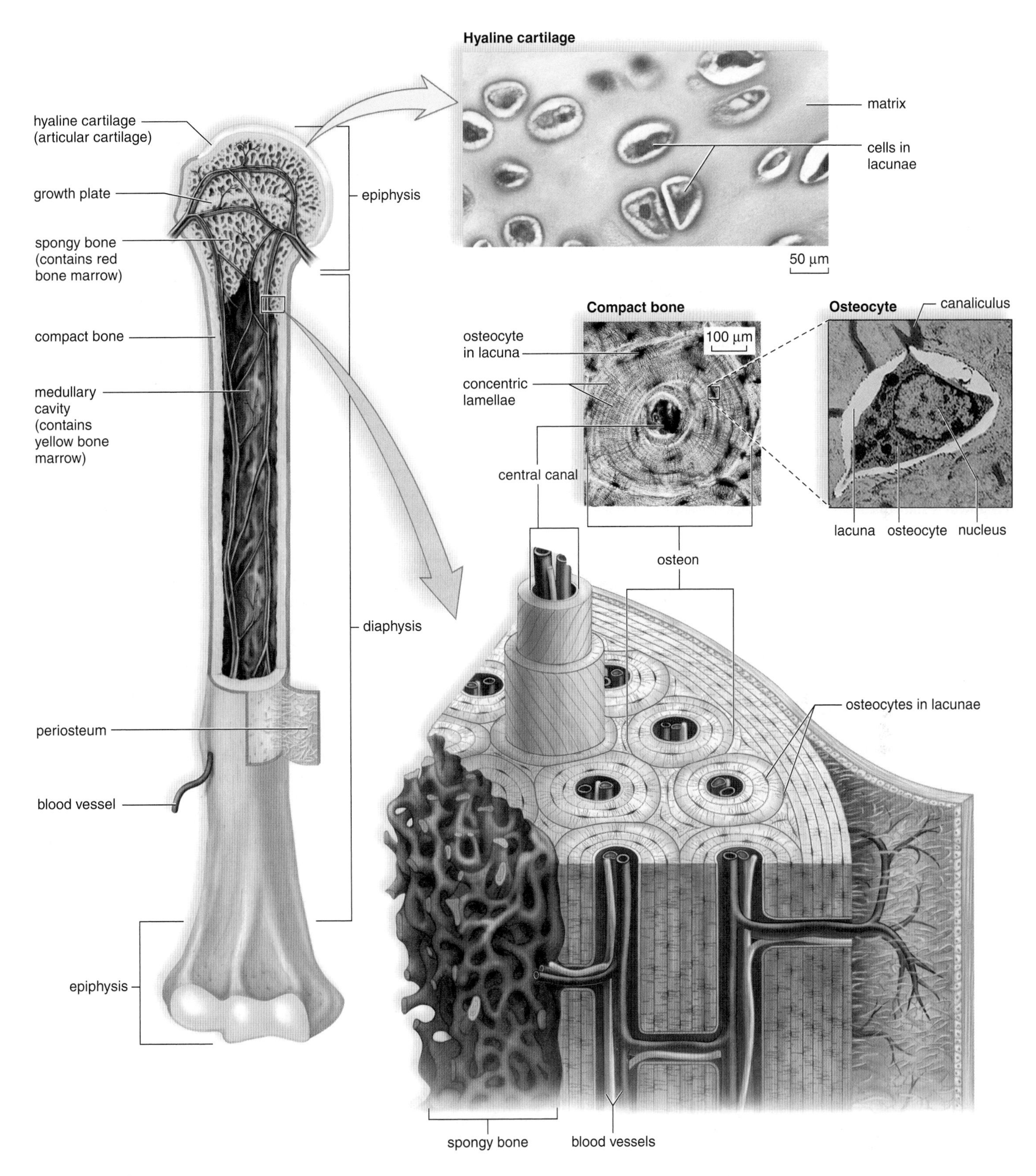

Figure 19.1 Anatomy of a bone from the macroscopic to the microscopic level.
A long bone is encased by fibrous membrane (periosteum) except where it is covered at the ends by hyaline cartilage, shown on the micrograph (*right, top*). Spongy bone, located beneath the cartilage, may contain red bone marrow. The central shaft contains yellow bone marrow and is bordered by compact bone, which is shown in the enlargement and micrograph (*right, center*).

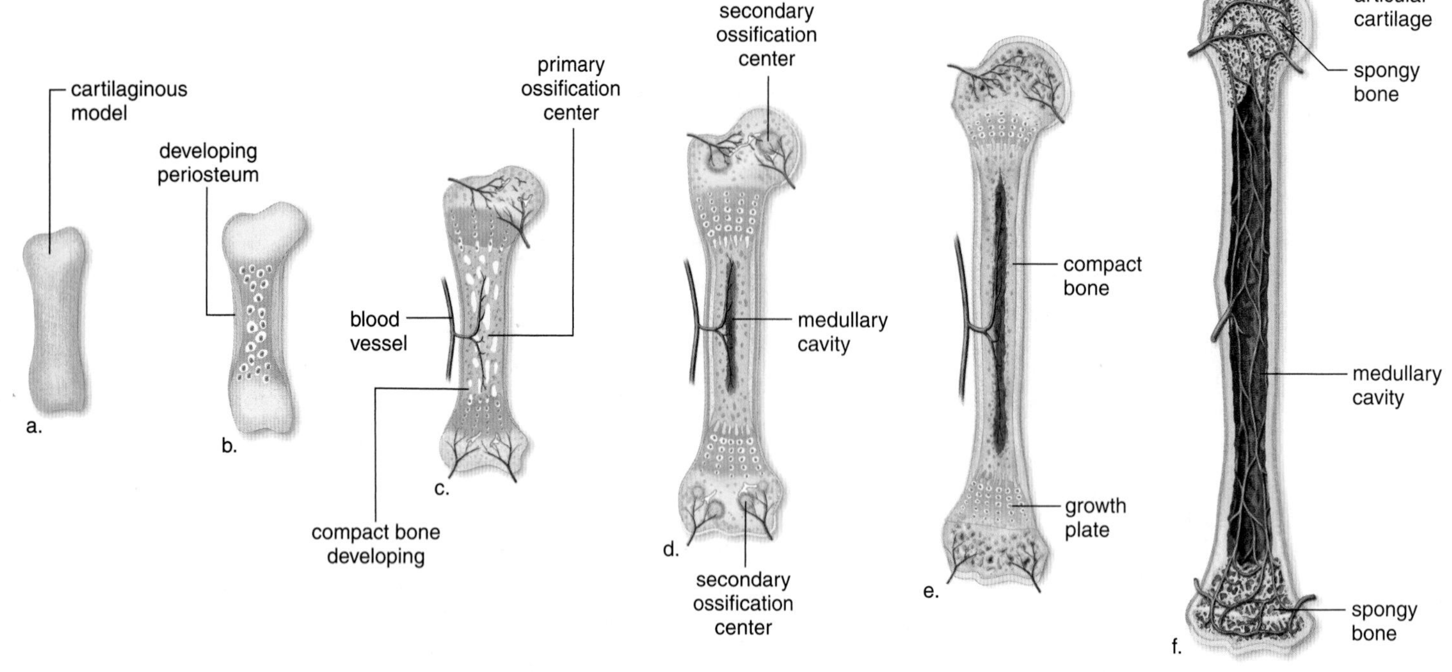

Figure 19.2 Endochondral ossification of a long bone.
a. A cartilaginous model develops during embryonic development. **b.** A periosteum develops. **c.** A primary ossification center contains spongy bone surrounded by compact bone. **d.** The medullary cavity forms in the shaft, and secondary ossification centers develop at the ends of the long bone. **e.** Growth is still possible as long as cartilage remains at the growth plate. **f.** When the bone is fully formed, the growth plate disappears.

Bone Growth and Remodeling

Bones are composed of living tissues, as exemplified by their ability to undergo growth and remodeling.

Bone Development and Growth

The bones of the human skeleton, except those of the skull, first appear during embryonic development as hyaline cartilage. These cartilaginous structures are then gradually replaced by bone, a process called endochondral ossification (Fig. 19.2).

During endochondral ossification, the cartilage begins to break down in the center of a long bone, which is now covered by a periosteum. **Osteoblasts** invade the region and begin to lay down spongy bone in what is called a primary ossification center. Other osteoblasts lay down compact bone beneath the periosteum. As the compact bone thickens, **osteoclasts,** which are large, multinucleated, macrophage-like cells, break down the spongy bone, and the cavity created becomes the medullary cavity.

The ends of developing bone continue to grow, but soon, secondary ossification centers appear in these regions. Here, spongy bone forms and does not break down. Also, a band of cartilage called a **growth plate** remains between the primary ossification center and each secondary center. The limbs keep increasing in length as long as the growth plates are present. The rate of growth is mainly controlled by growth hormone and sex hormones. Eventually, the growth plates become ossified, and the bone stops growing.

Remodeling of Bones

In the adult, bone is continually being broken down and built up again, a process called remodeling. Osteoclasts break down bone, resulting in increased calcium levels in the blood. Simultaneously, the bone is repaired by the work of osteoblasts. As they form new bone, osteoblasts take calcium from the blood. Eventually, some of these cells get caught in the matrix they secrete and are converted to osteocytes, the cells found within the lacunae of osteons. Because of continual remodeling, the thickness of bones can change due to excessive use or disuse. Hormones such as calcitonin and parathyroid hormone (see Chapter 20) can also affect the thickness of bones. When women reach menopause, the likelihood of a bone disease called osteoporosis increases, in part due to reduced estrogen levels (see section 19.6).

Bone remodeling also occurs when a fractured bone is healing. Bone healing can be divided into three major phases: 1) the reactive phase, which includes the body's inflammatory response to the injury; 2) the reparative phase, when the body generates a bone callus, which is a temporary connective tissue bridge that is gradually replaced in the last phase; 3) the remodeling phase, during which osteoclasts and osteoblasts gradually replace the temporary bone with compact bone. Depending on factors such as the type of fracture, the age of the patient, and medical intervention, some fractures are healed in as little as three or four weeks, while others may not be properly healed after a year or more.

Check Your Progress

1. Describe the role of osteoclasts and osteoblasts in remodeling bone.
2. What do you think might happen if a bone was fractured across a growth plate?
3. Besides changes in hormone levels, what other factors might contribute to osteoporosis?

19.2 Bones of the Skeleton

Learning Outcomes

1. Review the five major functions of the skeleton.
2. List the major bones that comprise the axial and appendicular skeletons.
3. Compare the structure and function of fibrous joints, cartilaginous joints, and synovial joints.

The functions of the skeleton pertain to particular bones:

The skeleton supports the body. The bones of the lower limbs (the femur in particular and also the tibia) support the entire body when we are standing, and the coxal bones of the pelvic girdle support the abdominal cavity.

The skeleton protects soft body parts. The bones of the skull protect the brain; the rib cage, composed of the ribs, thoracic vertebrae, and sternum, protects the heart and lungs.

The skeleton produces blood cells. All bones in the fetus have spongy bone containing red bone marrow that produces blood cells. In the adult, the flat bones of the skull, ribs, sternum, clavicles, and also the vertebrae and pelvis produce blood cells.

The skeleton stores minerals and fat. All bones have an extracellular substance that contains calcium phosphate. When bones are remodeled, osteoclasts break down bone and return calcium ions and phosphate ions to the bloodstream. Fat is stored in the yellow bone marrow (see Fig. 19.1).

The skeleton, along with the muscles, permits flexible body movement. While articulations (joints) occur between all the bones, we associate body movement, in particular, with the bones of the lower limbs (especially the femur and tibia) and the feet (tarsals, metatarsals, and phalanges) because we use them when walking.

Classification of the Bones

The approximately 206 bones of the skeleton are classified according to whether they occur in the axial skeleton or the appendicular skeleton. The **axial skeleton** is in the midline of the body, and the **appendicular skeleton** consists of the limbs along with their girdles (Fig. 19.3).

As mentioned, long bones, exemplified by the humerus and femur, are longer than they are wide. In contrast, short bones, such as the carpals and tarsals, are cube shaped—that is, their lengths and widths are about equal. Flat bones, such as those of the skull, are platelike, with broad surfaces. Irregular bones, such as the vertebrae and facial bones, have varied shapes that permit connections with other bones. Some authorities recognize another category, the round bones, which are circular in shape and usually embedded in a tendon. A good example of a round bone is the patella, or kneecap.

None of the bones of the skeleton are smooth. They have articulating depressions and protuberances at various joints. They also have projections, often called processes, where the muscles attach, as well as openings for nerves and/or blood vessels to pass through.

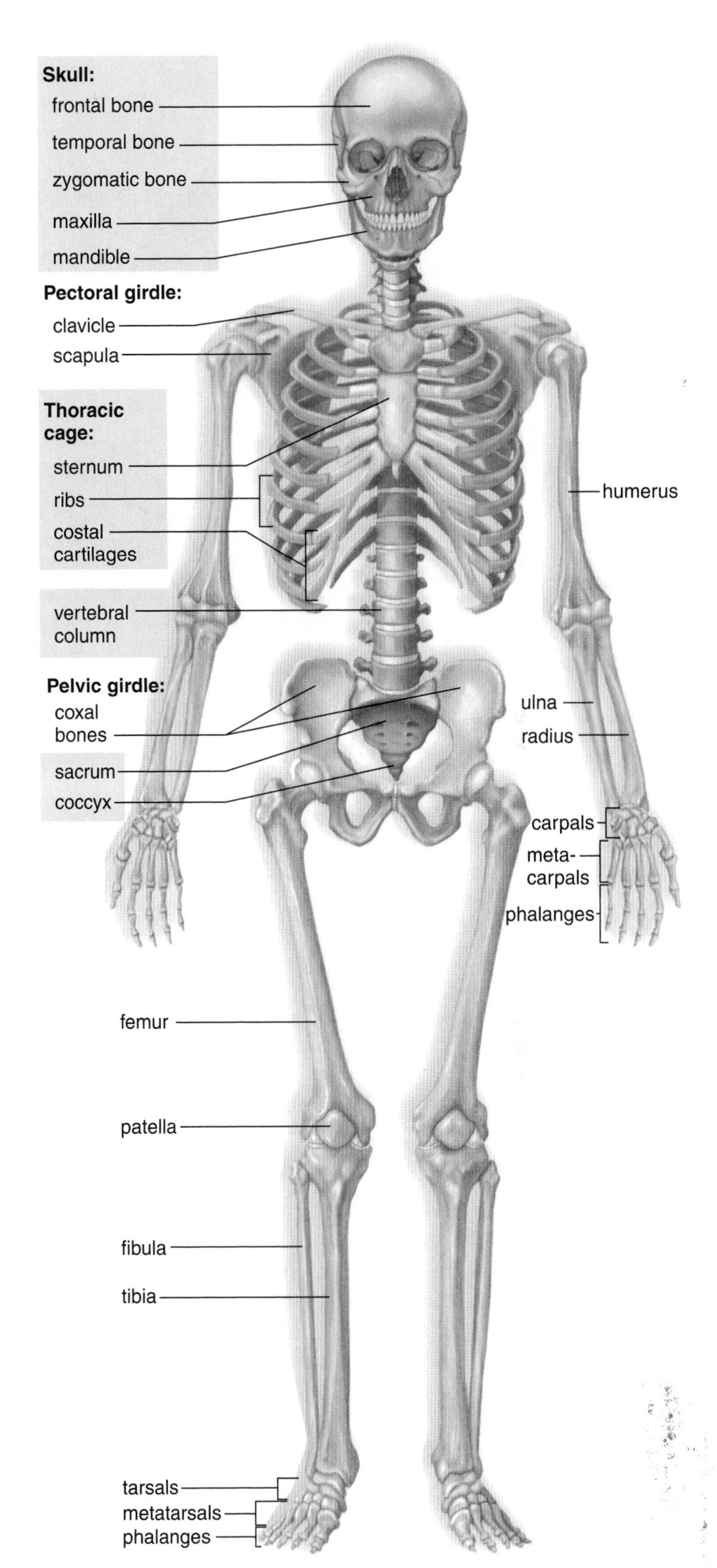

Figure 19.3 The skeleton.
The skeleton of a human adult contains bones that belong to the axial skeleton (shaded in blue) and those that belong to the appendicular skeleton (unshaded). The bones of the axial skeleton are located along the body's axis, and the bones of the appendicular skeleton are located in the girdles and appendages.

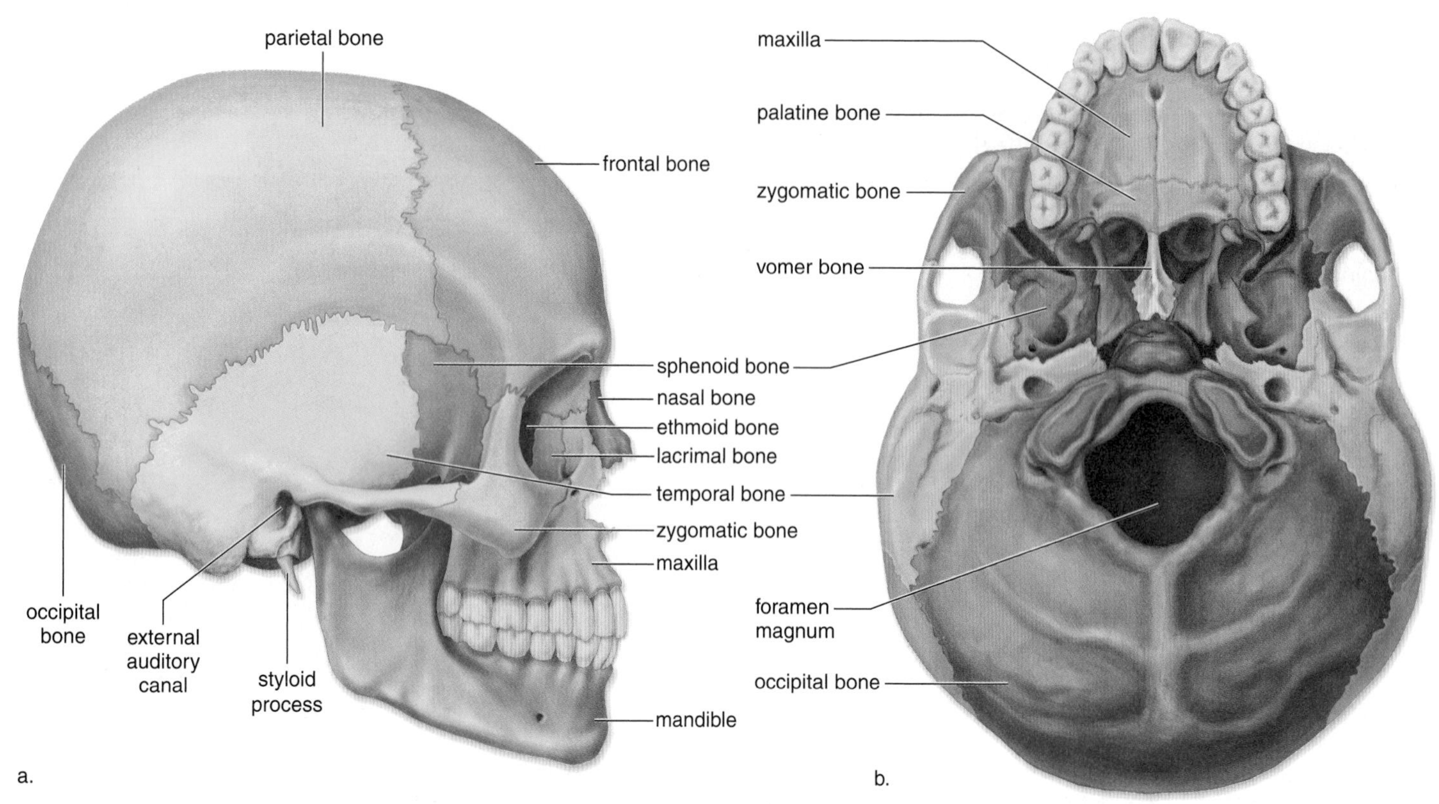

Figure 19.4 Bones of the skull.
a. Lateral view. **b.** Inferior view.

The Axial Skeleton

The axial skeleton lies in the midline of the body and consists of the skull, hyoid bone, vertebral column, rib cage, and ossicles.

The Skull

The **skull** is formed by the cranium (braincase) and the facial bones. It should be noted, however, that some cranial bones also help to form the face.

The Cranium The cranium protects the brain. In adults, it is composed of eight bones fitted tightly together. In newborns, certain cranial bones are joined by membranous regions called **fontanels,** which allows the cranuim to be compressed somewhat during birth. The fontanels usually close by the age of 24 months by the process of intramembranous ossification.

Some of the bones of the cranium contain the **sinuses,** air spaces lined by mucous membrane. The sinuses reduce the weight of the skull and give a resonant sound to the voice. Two sinuses, called the mastoid sinuses, drain into the middle ear. Infections of the upper respiratory tract can spread into various sinuses, sometimes resulting in chronic sinusitis that is difficult to treat (see section 15.4).

The major bones of the cranium have the same names as the lobes of the brain: frontal, parietal, occipital, and temporal. On the top of the cranium (Fig. 19.4*a*), the frontal bone forms the forehead, the parietal bones extend to the sides, and the occipital bone curves to form the base of the skull. At the base, there is a large opening, the **foramen magnum** (Fig. 19.4*b*), through which the spinal cord passes and connects with the brain stem. Below the much larger parietal bones, each temporal bone has an opening (external auditory canal) that leads to the middle ear.

The sphenoid bone, which is shaped like a bat with wings outstretched, extends across the floor of the cranium from one side to the other. The sphenoid is the keystone of the cranial bones because all the other bones articulate with it. The sphenoid completes the sides of the skull and also helps form the eye sockets. The eye sockets are called orbits because of our ability to rotate our eyes. The ethmoid bone, which lies in front of the sphenoid, also helps form the orbits and the nasal septum. The orbits are completed by various facial bones.

The Facial Bones The most prominent facial bones are the mandible, the maxillae (maxillary bones), the zygomatic bones, and the nasal bones.

The mandible, or lower jaw, is the only movable portion of the skull (Fig. 19.5*a*), and its action permits us to chew our food. It also forms the chin. Tooth sockets are located on the mandible and on the two maxillae, which form the upper jaw and also the anterior portion of the hard palate. The palatine bones make up the posterior portion of the hard palate and the floor of the nose (see Fig. 19.4*b*).

The lips and cheeks have a core of skeletal muscle. The zygomatic bones are the cheekbone prominences, and the nasal bones form the bridge of the nose. Other bones (e.g., ethmoid and vomer) are a part of the nasal septum, which divides the interior of the nose into two nasal cavities. The lacrimal bone

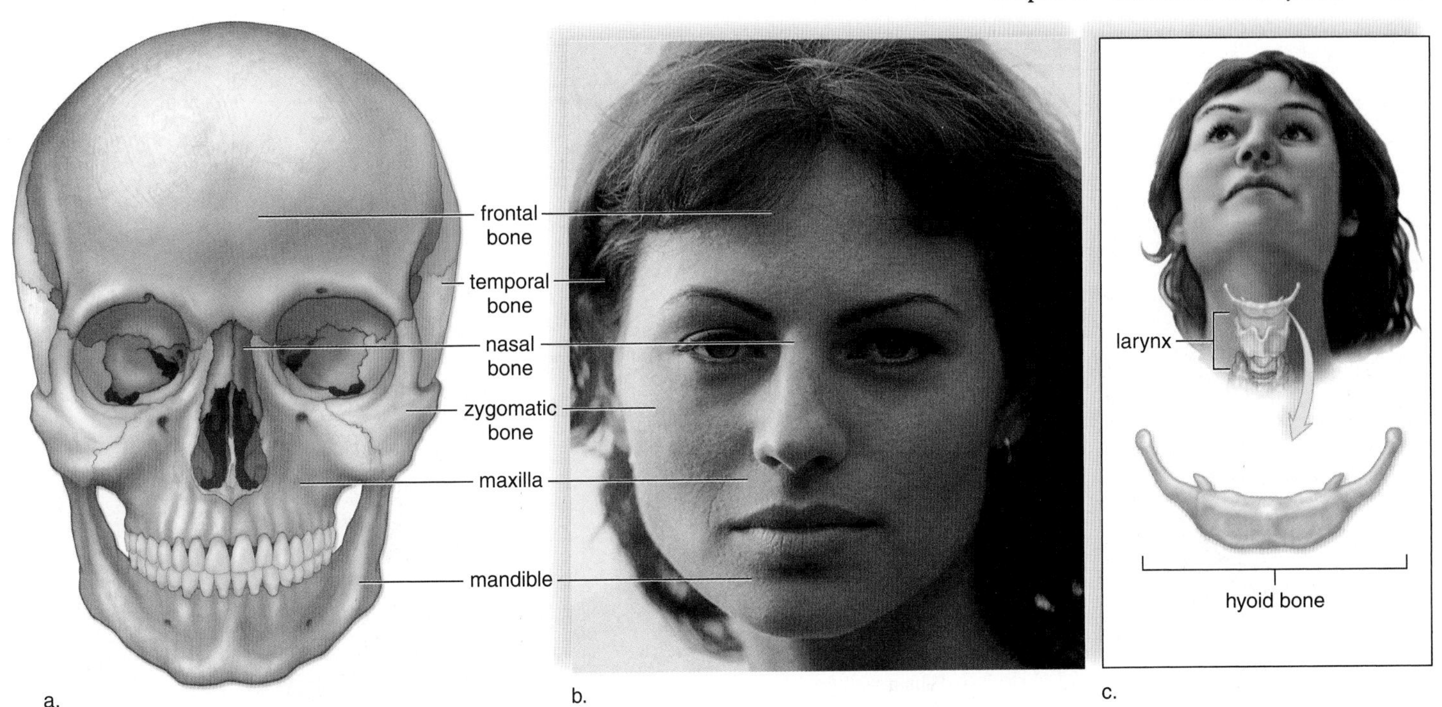

Figure 19.5 Bones of the face and the hyoid bone.
a. The frontal bone forms the forehead and eyebrow ridges, and the zygomatic bones form the cheekbones. The maxillae have numerous functions: assisting in the formation of the eye sockets and the nasal cavity; forming the upper jaw; and containing sockets for the upper teeth. The mandible is the lower jaw with sockets for the lower teeth. The mandible has a projection we call the chin. **b.** The maxillae, frontal bones, and nasal bones help form the external nose. **c.** The hyoid bone attaches to the larynx as shown.

(see Fig. 19.4*a*) contains the opening for the nasolacrimal canal, which brings tears from the eyes to the nose.

The temporal and frontal bones are cranial bones that contribute to the face. The temporal bones account for the flattened areas we call the temples. The frontal bone forms the forehead and has supraorbital ridges where the eyebrows are located. Glasses sit where the frontal bone joins the nasal bones (Fig. 19.5*b*).

While the ears are formed only by cartilage and not by bone, the nose is a mixture of bones, cartilages, and connective tissues. The cartilages complete the tip of the nose, and fibrous connective tissue forms the flared sides of the nose.

The Hyoid Bone

Although the hyoid bone is not part of the skull, we include it here because it is part of the axial skeleton. The hyoid is the only bone in the body that does not articulate with another bone. It is attached to processes of the temporal bones by muscles and ligaments and to the larynx by a membrane (Fig. 19.5*c*). The larynx is the voice box at the top of the trachea in the neck region. The hyoid bone anchors the tongue and serves as the site of attachment for the muscles associated with swallowing.

Check Your Progress

1. What are the four major bones of the cranium? What two bones form the floor of the cranium?
2. List the most prominent facial bones.
3. What is the function of the hyoid bone? What is most unique about its structure?

The Vertebral Column

The **vertebral column** consists of 33 vertebrae (Fig. 19.6). Normally, the vertebral column has four curvatures that provide resilience and strength for an upright posture. **Scoliosis** is an abnormal lateral (sideways) curvature of the spine. Two other well-known abnormal curvatures are kyphosis, an abnormal posterior curvature that often results in a hunchback, and lordosis, an abnormal anterior curvature resulting in a swayback.

The vertebral column forms when the vertebrae join. The spinal cord, which passes through the vertebral canal, gives off the spinal nerves at the intervertebral foramina. Among other functions, spinal nerves control skeletal muscle contraction. The spinous processes of the vertebrae can be felt as bony projections along the midline of the back. The spinous processes and also the transverse processes, which extend laterally, serve as attachment sites for the muscles that move the vertebral column.

Types of Vertebrae The various vertebrae are named according to their location in the vertebral column. The cervical vertebrae are in the neck. The first cervical vertebra, called the **atlas** holds up the head. It is so named because Atlas, of Greek mythology, held up the world. Movement of the atlas permits the "yes" motion of the head. It also allows the head to tilt from side to side. The second cervical vertebra is called the **axis** because it allows a degree of rotation, as when we shake the head "no." The thoracic vertebrae have long, thin, spinous processes and articular facets for the attachment of

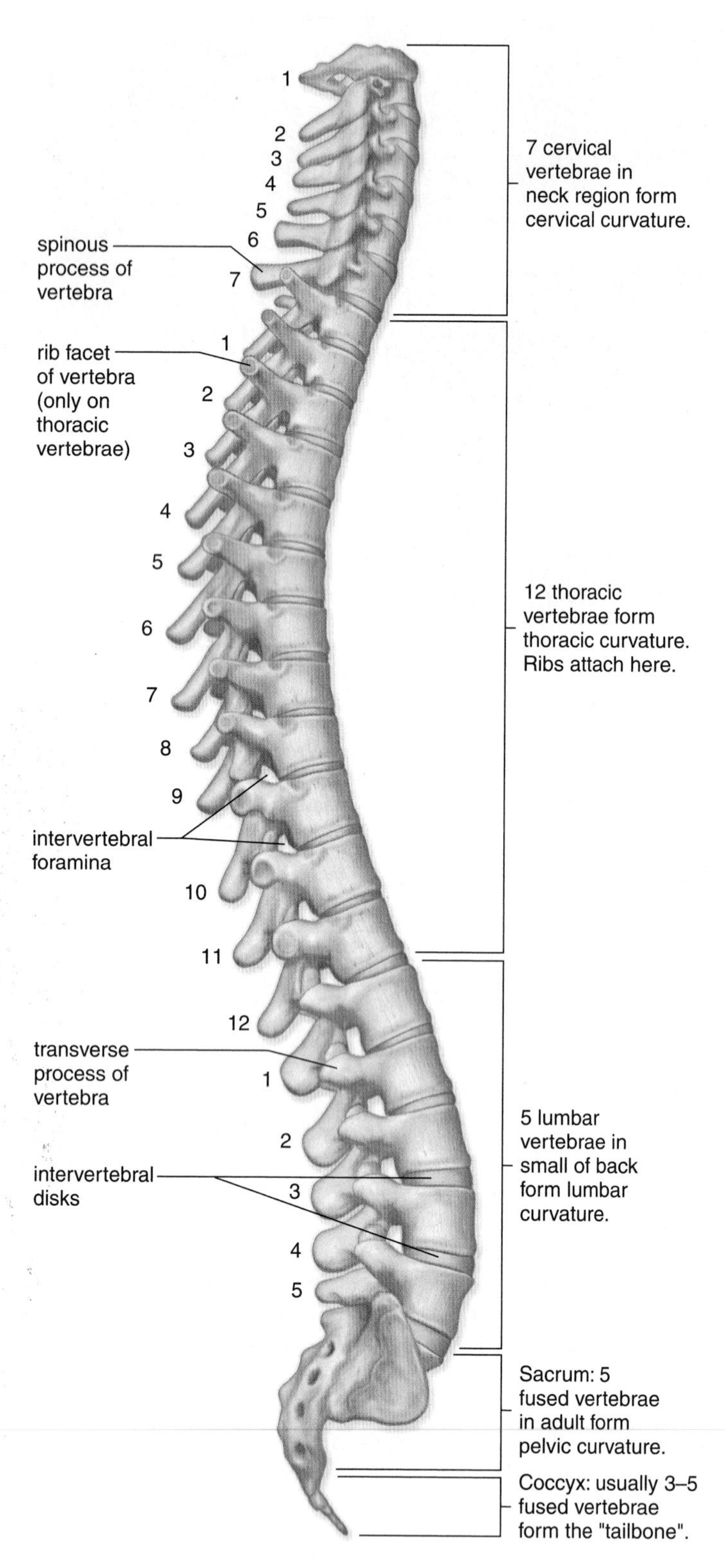

Figure 19.6 The vertebral column.
The vertebral column is flexible because the vertebrae are separated by intervertebral disks. The vertebrae are named for their location in the vertebral column. For example, the thoracic vertebrae are in the thorax. Note that humans have a coccyx, which is also called a tailbone.

the ribs (Fig. 19.7*a*). Lumbar vertebrae have a large body and thick processes. The five sacral vertebrae are fused together in the sacrum. The coccyx, or tailbone, is usually composed of three to five fused vertebrae.

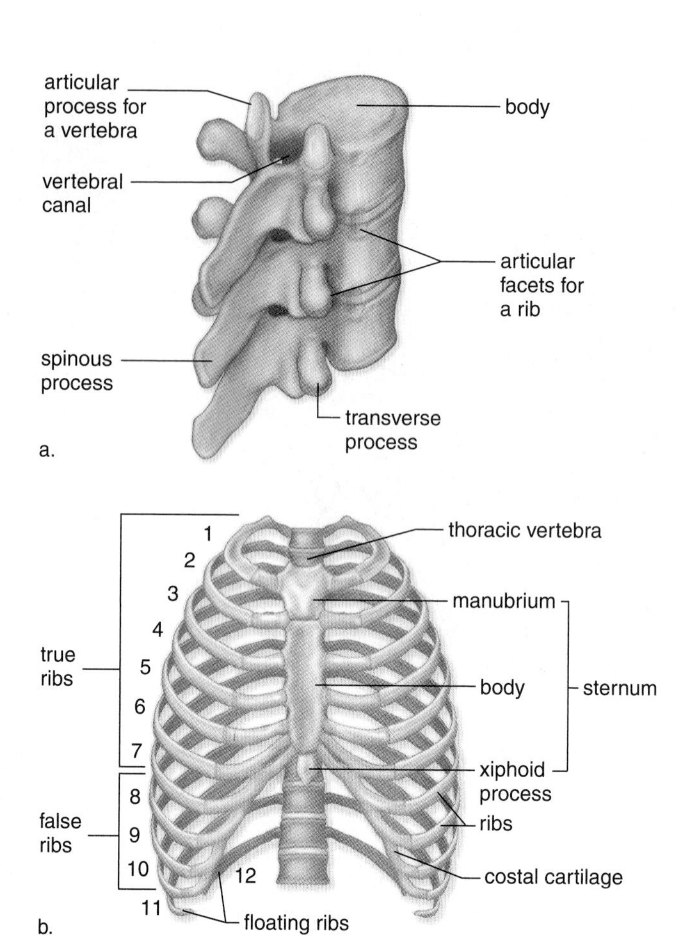

Figure 19.7 Thoracic vertebrae and the rib cage.
a. The thoracic vertebrae articulate with each other at the articular processes and with the ribs at the articular facets. A thoracic vertebra has two facets for articulation with a rib; one is on the body, and the other is on the transverse process. **b.** The rib cage consists of the 12 thoracic vertebrae, the 12 pairs of ribs, the costal cartilages, and the sternum. The rib cage protects the lungs and the heart.

Intervertebral Disks Between the vertebrae are **intervertebral disks** composed of fibrocartilage that act as a kind of padding. They prevent the vertebrae from grinding against one another and absorb shock caused by movements such as running, jumping, and even walking. Unfortunately, these disks become weakened with age and can even herniate and rupture. A herniated or "slipped" disk may press against the spinal cord and/or spinal nerves, resulting in pain, numbness, or even loss of function in the areas innervated by the spinal nerves at or below that region. Herniated disks are most common in the lumbar region; about 25% of people with lower back pain actually have a herneated disk. Although rest and treatment of symptoms usually results in much improvement, surgical treatment may be necessary, especially if there is chronic pain or loss of nerve function.

The Rib Cage

The rib cage, also called the thoracic cage, is composed of the thoracic vertebrae, the ribs and their associated cartilages, and the sternum (Fig. 19.7*b*). The rib cage is part of the axial skeleton.

The rib cage demonstrates how the skeleton is protective but also flexible. The rib cage protects the heart and lungs. Yet it swings outward and upward upon inspiration, and then downward and inward upon expiration. Because of their proximity to vital organs, however, the ribs also have the potential to cause harm. A fractured rib can puncture a lung.

The Ribs

There are twelve pairs of ribs, and all twelve connect directly to the thoracic vertebrae in the back. Each rib is a flattened bone that originates at a particular thoracic vertebra and proceeds toward the ventral thoracic wall. A rib articulates with the body and transverse process of its corresponding thoracic vertebra. It curves outward and then forward and downward.

The upper seven pairs of ribs connect directly to the sternum by means of costal cartilages. These are called the "true ribs." The next three pairs of ribs do not connect directly to the sternum, and they are called the "false ribs." They attach to the sternum by means of a common cartilage. The last two pairs are called "floating ribs" because they do not attach to the sternum at all.

The Sternum

The sternum, or breastbone, lies in the midline of the body. Along with the ribs, it helps protect the heart and lungs. The sternum is a flat bone that is shaped like a knife.

The sword-shaped sternum is composed of three bones that fuse during fetal development. These bones are the manubrium (the handle), the body (the blade), and the xiphoid process (the point of blade). The manubrium articulates with the clavicles of the appendicular skeleton and the first pair of ribs. The manubrium joins with the body of the sternum at an angle. This is an important anatomical landmark because it occurs at the level of the second rib, and therefore allows the ribs to be counted. Medical personnel sometimes count the ribs to locate the apex of a patient's heart, which is usually between the fifth and sixth ribs.

The xiphoid process is the third part of the sternum. Composed of hyaline cartilage in the child, it becomes ossified in the adult. The variably shaped xiphoid process serves as an attachment site for the diaphragm, which divides the thoracic cavity from the abdominal cavity.

Check Your Progress

1. How many total vertebrae does a normal human have? How many are cervical, thoracic, and lumbar vertebrae?
2. Define the difference between true ribs, false ribs, and floating ribs.
3. What are two functions of the sternum?

The Appendicular Skeleton

The appendicular skeleton consists of the bones within the pectoral and pelvic girdles and their attached limbs. The two (left and right) pectoral girdles and upper limbs are specialized for flexibility. The pelvic girdle and lower limbs are specialized for strength.

The Pectoral Girdle and Upper Limb

A **pectoral girdle** (shoulder girdle) consists of a scapula (shoulder blade) and a clavicle (collarbone) (Fig. 19.8). The clavicle extends across the top of the thorax. It articulates with (joins with) the sternum and the acromion process of

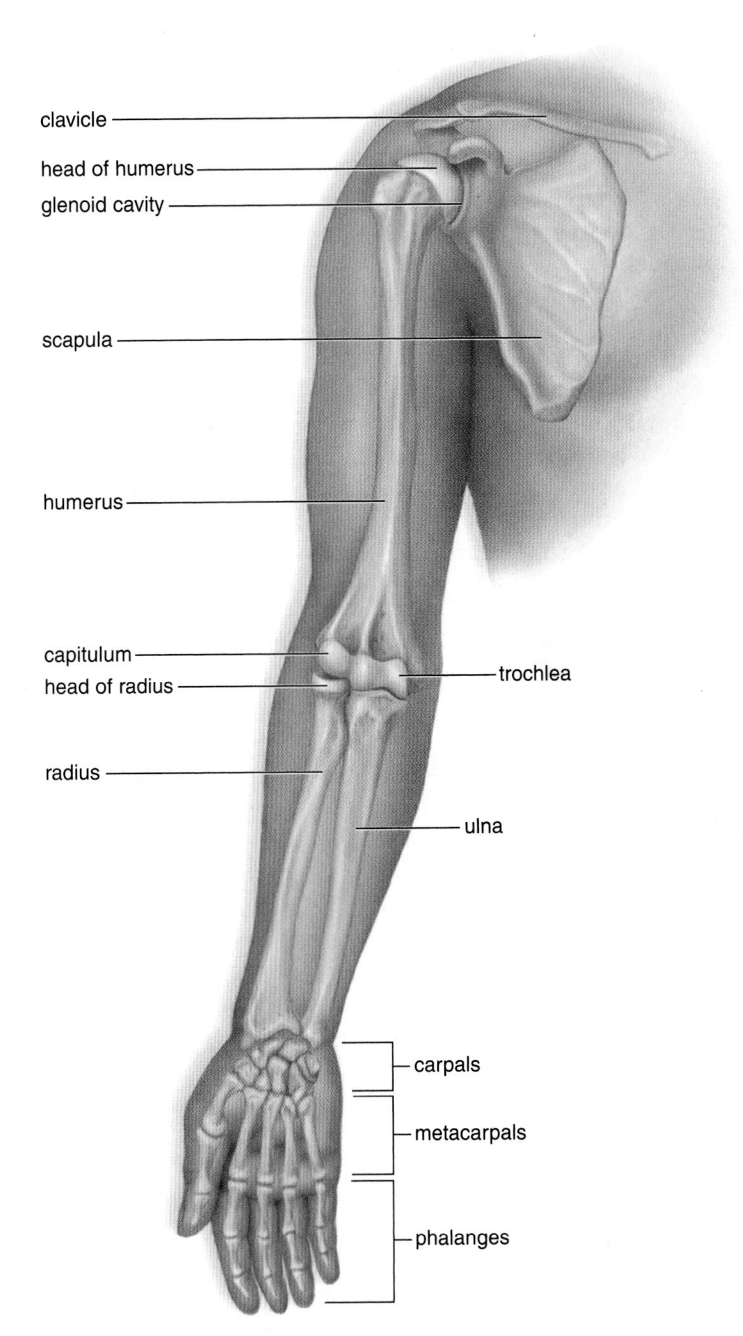

Figure 19.8 Bones of the pectoral girdle and upper limb (arm, forearm, and hand).

the scapula, a bone external to the ribs in the back. The muscles of the arm and chest attach to the coracoid process of the scapula. The glenoid cavity of the scapula articulates with the head of the humerus, the single long bone in the arm. This joint allows the arm to move in almost any direction, but has reduced stability and thus is the joint that is most apt to be dislocated. Tendons that encircle and help form a socket for the humerus are collectively called the rotator cuff. Vigorous circular movements of the arm (such as pitching a baseball) can lead to rotator cuff injuries.

The upper limb consists of the humerus in the arm and the radius and ulna in the forearm. The shaft of the humerus has a tuberosity (protuberance) where the deltoid, a prominent muscle of the shoulder, attaches. Enlargement of this tuberosity occurs in people who do a lot of heavy lifting.

The far end of the humerus has two protuberances, called the capitulum and the trochlea, which articulate respectively with the radius and the ulna at the elbow. The bump at the back of the elbow is the olecranon process of the ulna. This area is sometimes called the "funny bone" because the ulnar nerve passes behind it with only skin for protection—thus it is susceptible to being bumped, resulting in pain or numbness.

When the arm is held with the palm turned forward, the radius and ulna are about parallel to one another. When the arm is turned so that the palm faces backward, the radius crosses in front of the ulna, a feature that contributes to the easy twisting motion of the forearm.

The hand has many bones, and this increases its flexibility. The wrist has eight carpal bones, which look like small pebbles. From these, five metacarpal bones fan out to form a framework for the palm. The metacarpal bone that leads to the thumb is opposable to the other digits. (The term **digits** refers to either fingers or toes.) The knuckles are the enlarged distal ends of the metacarpals. Beyond the metacarpals are the phalanges, the bones of the fingers and the thumb. The phalanges of the hand are long, slender, and lightweight.

The Pelvic Girdle and Lower Limb

Figure 19.9 shows how the lower limb is attached to the pelvic girdle. The **pelvic girdle** consists of two heavy, large coxal bones (hipbones). The pelvis is a basin composed of the pelvic girdle, sacrum, and coccyx. The pelvis bears the weight of the body, protects the organs within the pelvic cavity, and serves as the site of attachment for the lower limbs.

Each coxal bone has three parts: the ilium, the ischium, and the pubis, which are fused in the adult (Fig. 19.9). The hip socket, called the acetabulum, occurs where these three bones meet. We sit on the ischium, which has a posterior spine called the ischial spine. The pubis, from which the term pubic hair is derived, is the anterior part of a coxal bone. The two pubic bones are joined together by a fibrocartilage disk at the pubic symphysis.

The male and female pelves differ from one another. In the female, the ilia are more flared, the pelvic cavity is broader, and the outlet is wider. These adaptations facilitate giving birth.

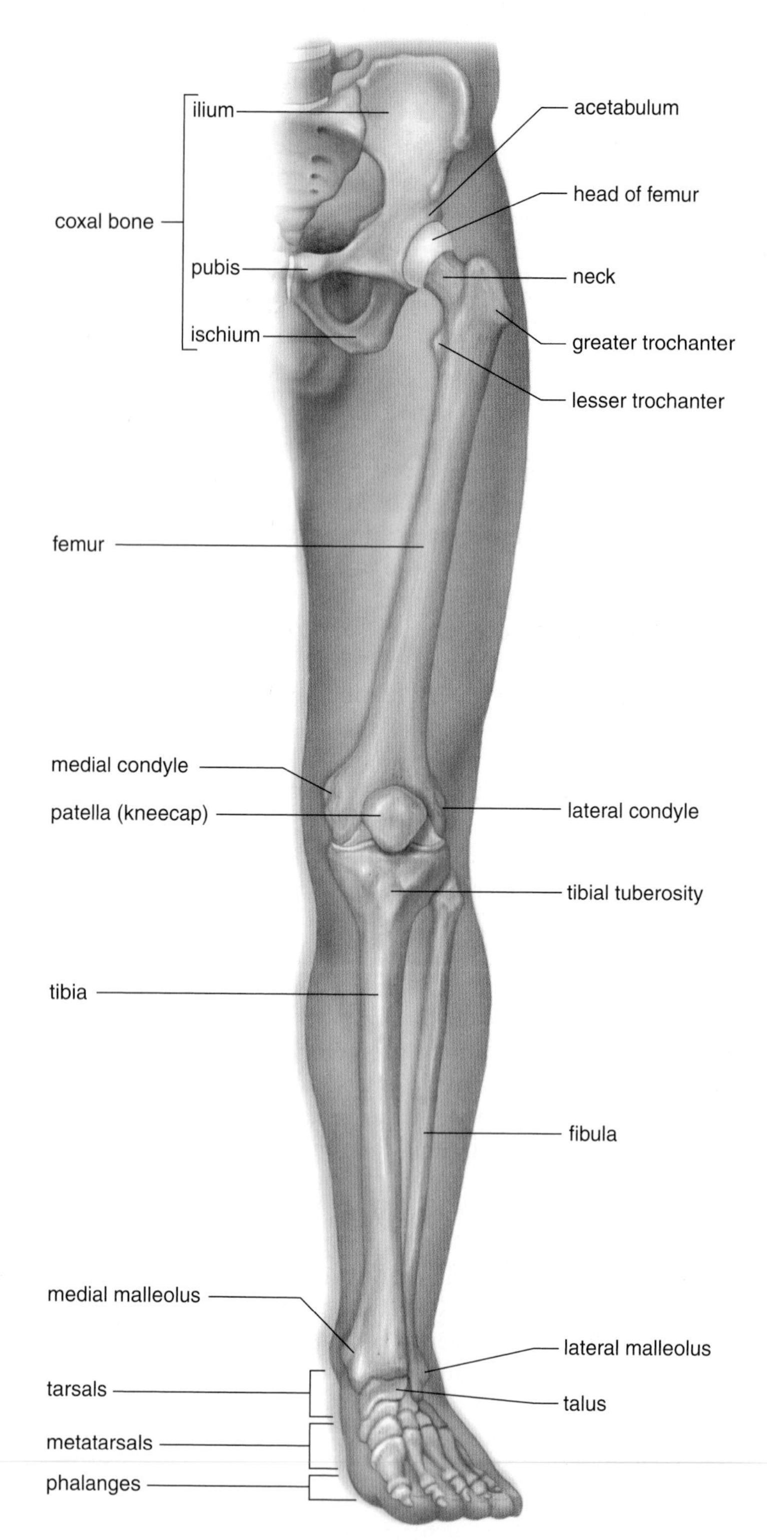

Figure 19.9 **Bones of the pelvic girdle and lower limb.**

The femur (thighbone) is the longest and strongest bone in the body. The head of the femur articulates with the coxal bones at the acetabulum, and the short neck better positions the legs for walking. The femur has two large processes, the greater and lesser trochanters, which are places of attachment for the thigh muscles and the muscles of the buttocks. At its distal end, the femur has medial and lateral condyles that articulate with the tibia. This is the region of the knee and the patella, or kneecap. The patella is held in place by the quadriceps tendon, which continues as a ligament that attaches to the tibial tuberosity. At the distal end, the medial malleolus of the tibia causes the inner bulge of the ankle. The fibula is the more slender bone in the leg. The fibula has a head that articulates with the tibia and a prominence (lateral malleolus) that forms the outer bulge of the ankle.

Each foot has an ankle, an instep, and five toes. The many bones of the foot give it considerable flexibility, especially on rough surfaces. The ankle contains seven tarsal bones, one of which (the talus) can move freely where it joins the tibia and fibula. Strange to say, the calcaneus, or heel bone, is also considered part of the ankle. The talus and calcaneus support the weight of the body.

The instep has five elongated metatarsal bones. The distal end of the metatarsals forms the ball of the foot. If the ligaments that bind the metatarsals together become weakened, flat feet are apt to result. The bones of the toes are called phalanges, just like those of the fingers, but in the foot, the phalanges are stout and extremely sturdy.

Joints

Bones are joined at the joints, which are classified as fibrous, cartilaginous, or synovial based on their structure and their ability to move. Most fibrous joints, such as the **sutures** between the cranial bones, are immovable. Cartilaginous joints are connected by hyaline cartilage, as in the costal cartilages that join the ribs to the sternum, or by fibrocartilage, as in the intervertebral disks. Cartilaginous joints tend to be slightly movable. Synovial joints are freely movable.

In **synovial joints,** the two bones are separated by a cavity. Ligaments hold the two bones in place as they form a capsule. Tendons also help stabilize the joint. The joint capsule is lined by a **synovial membrane,** which produces synovial fluid, a lubricant for the joint. Major synovial joints include the shoulder, elbow, hip, and knee. Aside from articular cartilage, the knee contains menisci (sing., **meniscus**), crescent-shaped pieces of hyaline cartilage between the bones (Fig. 19.10). These give added stability and act as shock absorbers. Unfortunately, athletes often suffer injury to the menisci, known as torn cartilage. The anterior and posterior ligaments, which help stabilize the knee by attaching the femur to the tibia, are also prone to being torn (sometimes called "ruptured"). The knee joint also contains 13 fluid-filled sacs called bursae (sing., **bursa**), which ease friction between the tendons and ligaments. Inflammation of the bursae is called **bursitis.** Tennis elbow is a form of bursitis.

There are different types of synovial joints. The knee and elbow joints are **hinge joints** because, like a hinged door, they largely permit movement in one direction only. The

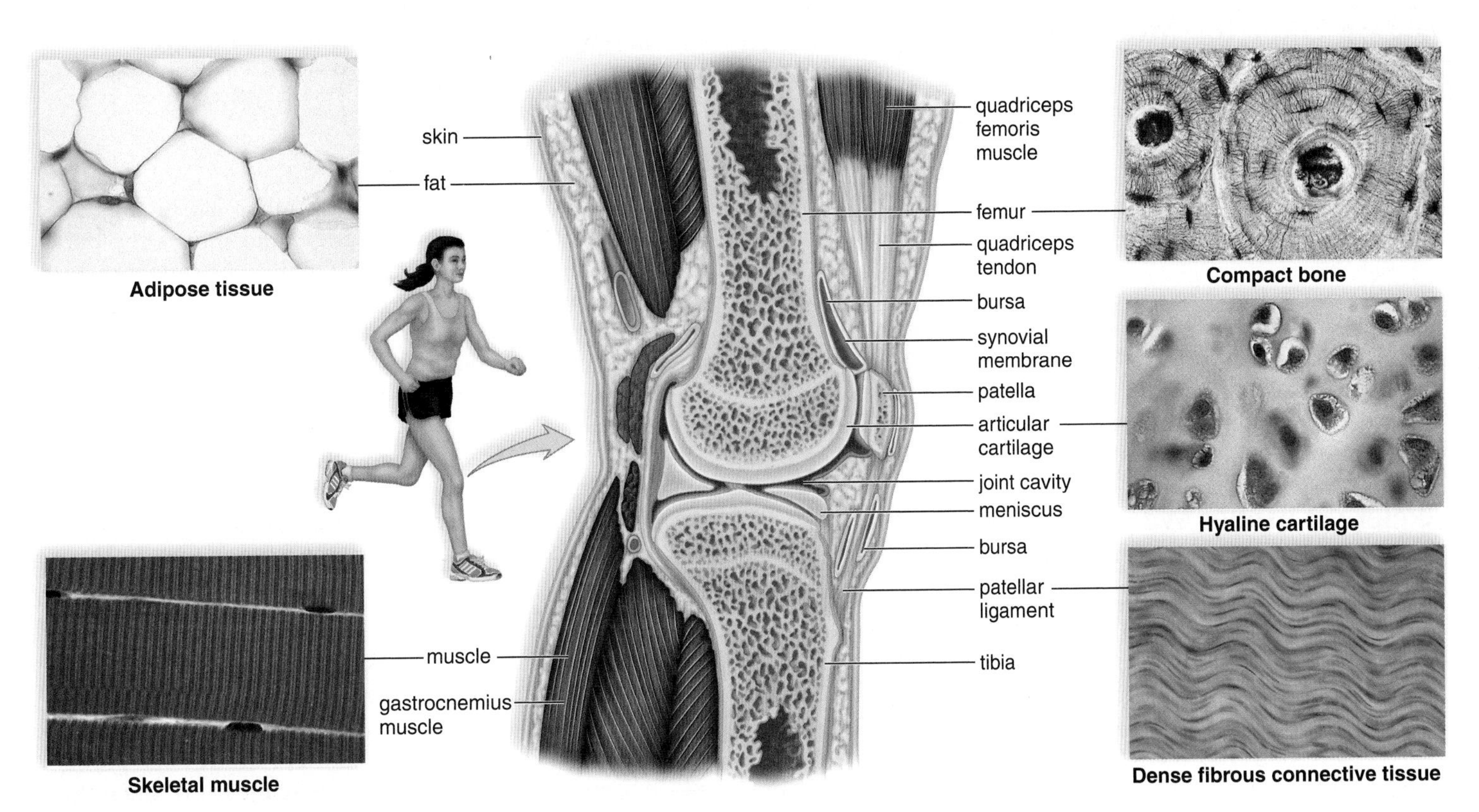

Figure 19.10 Knee joint.
The knee joint is a synovial joint. Notice the cavity between the bones, which is encased by ligaments and lined by synovial membrane. The patella (kneecap) serves to guide the quadriceps tendon over the joint when flexion or extension occurs.

TABLE 19.1 Examples of Movements at Synovial Joints

Type	Example
Flexion	Forearm toward the arm
Extension	Forearm away from the arm
Abduction	Arms sideways, away from body
Adduction	Arms back to the body
Rotation	Head to answer "no"

joint between the radius and ulna is a **pivot joint** in which only rotation is possible. More movable are the **ball-and-socket joints** which allow movement in all planes, even rotational movement. For example, the ball of the femur fits into a socket on the hipbone. Some of the movements at synovial joints are listed in Table 19.1.

Check Your Progress

1. Where would you find your carpal bones, metatarsal bones, and phalanges?
2. What are three differences between the male and female pelvis?
3. List one specific example each of a typical hinge joint, pivot joint, and ball-and-socket joint.

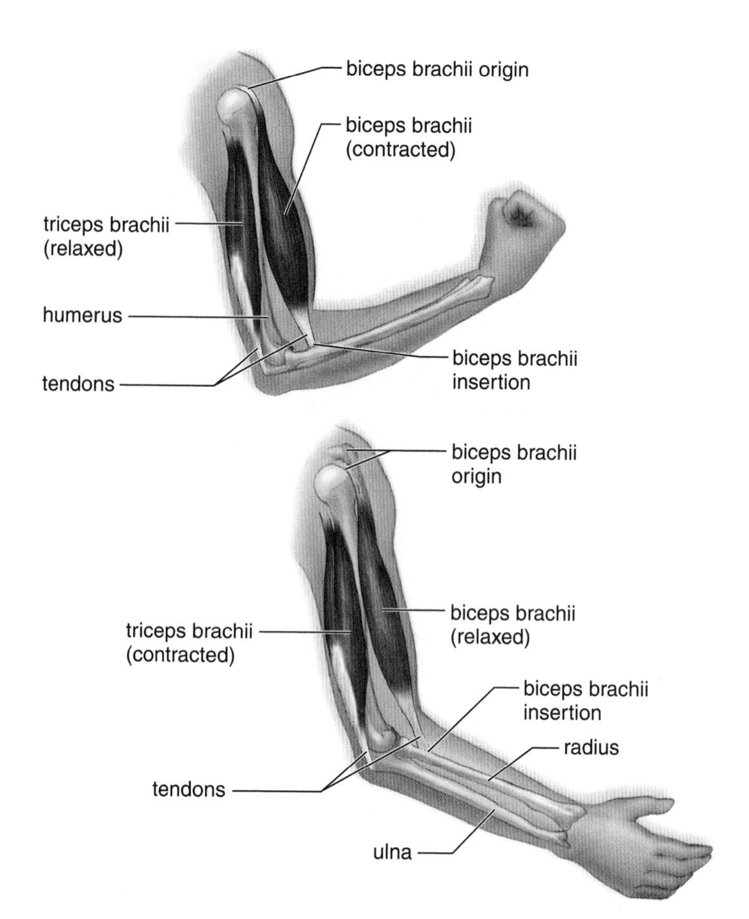

Figure 19.11 Antagonistic muscles.
Muscles can exert force only by shortening; therefore, they often work as antagonistic pairs. The biceps and triceps brachii exemplify an antagonistic pair of muscles that act opposite to one another. The biceps brachii raises the forearm, and the triceps brachii lowers the forearm.

19.3 Skeletal Muscles

Learning Outcomes

1. Explain how antagonistic muscle groups can move bones in different directions.
2. List the major muscles in the human body.
3. Review six characteristics that are used to name skeletal muscles.

Humans have three types of muscle tissue: smooth, cardiac, and skeletal (see Fig. 11.5). Skeletal muscle makes up the greatest percentage of muscle tissue in the body. Skeletal muscle is voluntary because its contraction can be consciously stimulated and controlled by the nervous system.

Skeletal Muscles Work in Pairs

Skeletal muscles are covered by several layers of fibrous connective tissue called fascia, which extends beyond the muscle to become its tendon. Tendons are attached to the skeleton, and the contraction of skeletal muscle causes the bones at a joint to move. When skeletal muscles contract, one bone remains fairly stationary, and the other one moves. The **origin** of a muscle is on the stationary bone, and the **insertion** of a muscle is on the bone that moves. Table 19.1 lists some terms that are used to describe these movements, and gives examples. Flexion means bending bones at a joint so that the angle between the bones decreases, and the involved bones move closer together, while extension increases that angle and moves the parts farther apart. Abduction at a joint moves the involved part (usually a limb) away from the middle of the body, while adduction brings it back toward the midline. Rotation moves the part around an axis, due to a twisting motion at the joint. Note that not all of these movements can occur at all joints.

Most muscles have antagonists, and antagonistic pairs bring about movement in opposite directions. For example, the biceps brachii and the triceps brachii are antagonists; one flexes the forearm, and the other extends the forearm (Fig. 19.11).

Major Skeletal Muscles

There are approximately 650 skeletal muscles in the human body. The major ones are shown in Figure 19.12. As listed in Tables 19.2 and 19.3, the muscles of the head are involved in facial expression and chewing; those of the neck participate in head movements; the muscles of the trunk and upper limb move the arm, the forearm, and the fingers; and the muscles of the buttocks and lower limb move the thigh, the leg, and the toes.

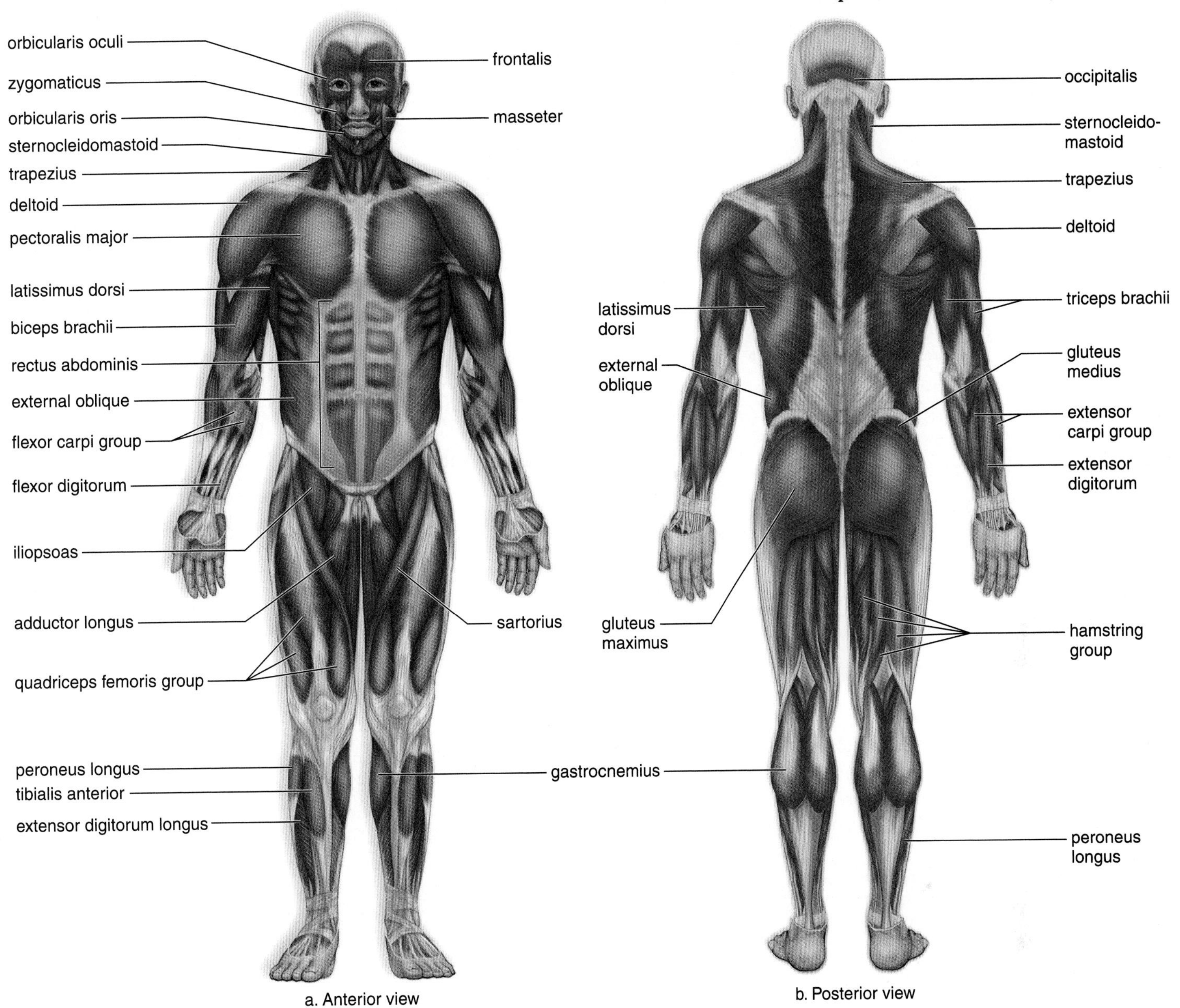

Figure 19.12 Human musculature.
Skeletal muscles located near the surface in **(a)** anterior and **(b)** posterior views.

Nomenclature

Skeletal muscles are named based on the following characteristics:

1. Size. The gluteus maximus that makes up the buttocks is the largest muscle (*maximus* means greatest). Other terms used to indicate size are vastus (huge) and longus (long).
2. Shape. The deltoid is shaped like a triangle. (The Greek letter delta has this appearance: Δ.) The trapezius is shaped like a trapezoid. Another term used to indicate shape is latissimus (wide).
3. Location. The frontalis overlies the frontal bone. The external obliques are located outside the internal obliques. Another term used to indicate location is pectoralis (chest).
4. Direction of muscle fibers. The rectus abdominis is a longitudinal muscle of the abdomen (*rectus* means straight). The orbicularis oculi is a circular muscle around the eye.
5. Number of attachments. The biceps brachii has two attachments, or origins (*bi-* means two).
6. Action. The extensor digitorum extends the fingers (digits). The adductor longus is a long muscle that adducts the thigh. Another term used to indicate action is masseter (to chew).

Check Your Progress

1. What are three types of muscle tissue?
2. Describe the difference between a muscle's origin and its insertion.
3. Define flexion, extension, abduction, adduction, and rotation.

TABLE 19.2 Muscles (Anterior View)

Name	Action
Head and neck	
Frontalis	Wrinkles forehead and lifts eyebrows
Orbicularis oculi	Closes eye (winking)
Zygomaticus	Raises corner of mouth (smiling)
Masseter	Closes jaw, chewing
Orbicularis oris	Closes and protrudes lips (kissing)
Upper limb and trunk	
External oblique	Compresses abdomen; rotates trunk
Rectus abdominis	Flexes spine; compresses abdomen
Pectoralis major	Flexes and adducts arm ventrally (pulls arm across chest)
Deltoid	Abducts and moves arm up and down in front
Biceps brachii	Flexes forearm and rotates hand outward
Lower limb	
Adductor longus	Adducts and flexes thigh
Iliopsoas	Flexes thigh at hip joint
Sartorius	Raises and rotates thigh and leg
Quadriceps femoris group	Extends leg at knee; flexes thigh
Peroneus longus	Everts foot
Tibialis anterior	Dorsiflexes and inverts foot
Flexor digitorum longus	Flexes toes
Extensor digitorum longus	Extends toes

TABLE 19.3 Muscles (Posterior View)

Name	Action
Head and neck	
Occipitalis	Moves scalp backward
Sternocleidomastoid	Turns head to side; flexes neck and head
Trapezius	Extends head; raises scapula as when shrugging shoulders
Upper limb and trunk	
Latissimus dorsi	Extends and adducts arm dorsally (pulls arm across back)
Deltoid	Abducts and moves arm up and down in front
External oblique	Rotates trunk
Triceps brachii	Extends forearm
Flexor carpi group	Flexes hand
Extensor carpi group	Extends hand
Flexor digitorum	Flexes fingers
Extensor digitorum	Extends fingers
Buttocks and lower limb	
Gluteus medius	Abducts thigh
Gluteus maximus	Extends thigh back (forms buttocks)
Hamstring group	Flexes leg and extends thigh
Gastrocnemius	Plantar flexes foot (tiptoeing)

19.4 Mechanism of Muscle Fiber Contraction

Learning Outcomes

1. Describe the microscopic structure of a muscle fiber.
2. Review the molecular mechanism of muscle contraction.
3. Indicate three ways that muscle cells can generate ATP.

We have already examined the structure of skeletal muscle as seen with the light microscope. Skeletal muscle tissue has alternating light and dark bands, giving it a striated appearance. These bands are due to the arrangement of myofilaments in a muscle fiber.

Muscle Fiber

A muscle fiber is a cell containing the usual cellular components, but special names have been assigned to some of these components (Table 19.4). Figure 19.13 shows the microscopic structure of a muscle fiber. The plasma membrane is called the **sarcolemma;** the cytoplasm is the sarcoplasm; and the endoplasmic reticulum is the **sarcoplasmic reticulum.** A muscle fiber also has some unique anatomical characteristics. One feature is its T (for transverse) system; the sarcolemma forms **T (transverse) tubules** that penetrate, or dip down, into the cell so that they come in contact—but do not fuse—with expanded portions of the sarcoplasmic reticulum. The expanded portions of the sarcoplasmic reticulum are calcium storage sites. Calcium ions (Ca^{2+}), as we shall see, are essential for muscle contraction.

The sarcoplasmic reticulum encases hundreds and sometimes even thousands of **myofibrils,** each about 1 μm in diameter, which are the contractile portions of the muscle fibers. Any other organelles, such as mitochondria, are located in the sarcoplasm between the myofibrils. The sarcoplasm

TABLE 19.4 Microscopic Anatomy of a Muscle

Name	Function
Sarcolemma	Plasma membrane of a muscle fiber that forms T tubules
Sarcoplasm	Cytoplasm of a muscle fiber that contains the organelles, including myofibrils
Glycogen	A polysaccharide that stores energy for muscle contraction
Myoglobin	A red pigment that stores oxygen for muscle contraction
T tubule	Extension of the sarcolemma that extends into the muscle fiber and conveys impulses that cause Ca^{2+} to be released from the sarcoplasmic reticulum
Sarcoplasmic reticulum	The smooth ER of a muscle fiber that stores Ca^{2+}
Myofibril	A bundle of myofilaments that contracts
Myofilament	Actin filaments and myosin filaments whose structure and functions account for muscle striations and contractions

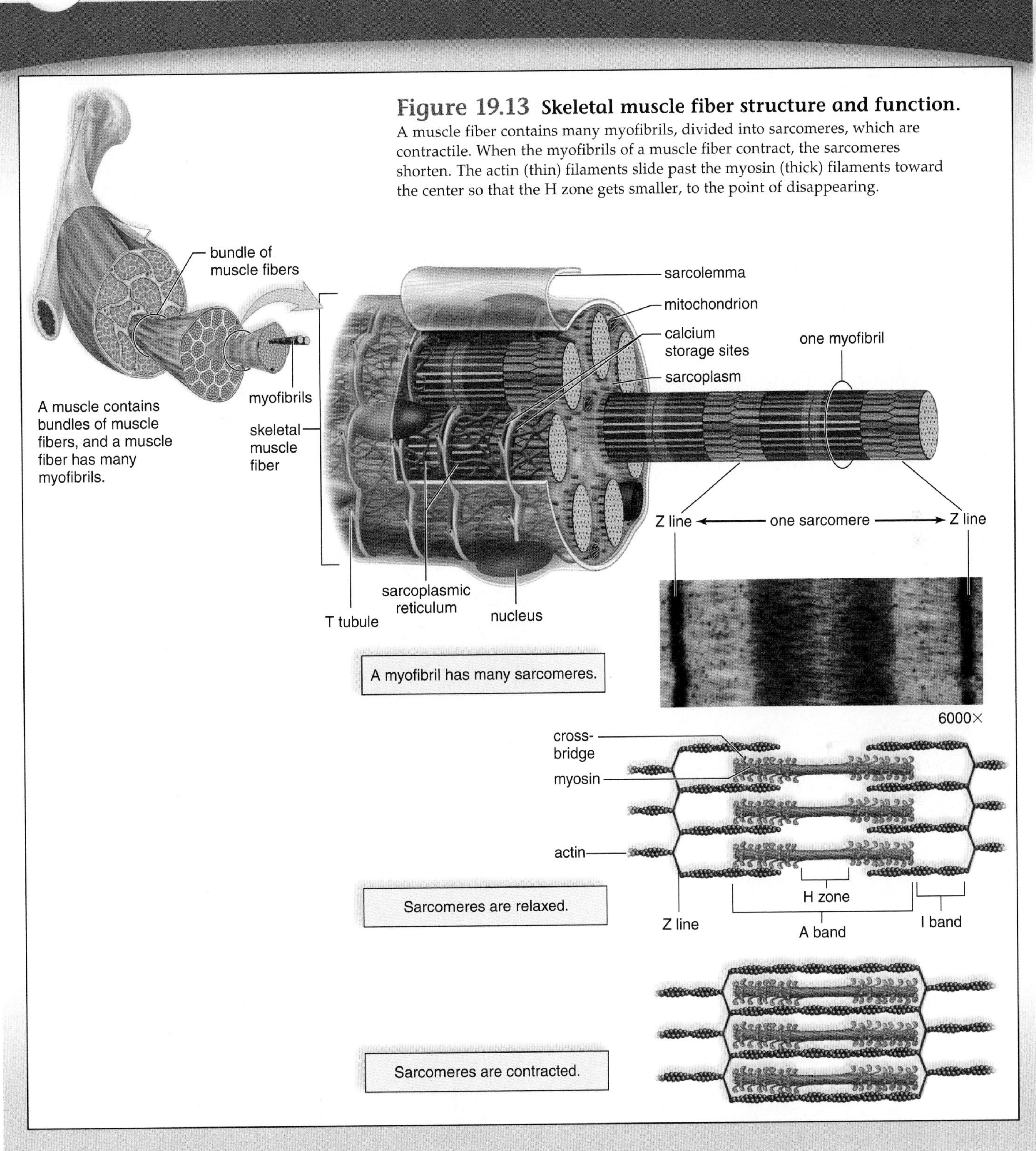

Figure 19.13 Skeletal muscle fiber structure and function. A muscle fiber contains many myofibrils, divided into sarcomeres, which are contractile. When the myofibrils of a muscle fiber contract, the sarcomeres shorten. The actin (thin) filaments slide past the myosin (thick) filaments toward the center so that the H zone gets smaller, to the point of disappearing.

also contains glycogen, which provides stored energy for muscle contraction, and the red pigment myoglobin, which binds oxygen until it is needed for muscle contraction.

Myofibrils and Sarcomeres

Myofibrils are cylindrical in shape and run the length of the muscle fiber. The light microscope shows that skeletal muscle fibers have light and dark bands called striations. The electron microscope shows that the striations of skeletal muscle fibers are formed by the placement of myofilaments within units of myofibrils called **sarcomeres.** A sarcomere extends between two dark lines, called the Z lines. A sarcomere contains two types of protein myofilaments. The thick filaments are made up of a protein called myosin, and the thin filaments are made up of a protein called actin. Other proteins are also present. The I band is light colored because it contains only actin filaments attached to a Z line. The dark regions of the A band contain overlapping actin and myosin filaments, and its H zone has only myosin filaments.

Myofilaments

The thick and thin filaments differ in the following ways:

Thick Filaments A thick filament is composed of several hundred molecules of the protein **myosin.** Each myosin molecule is shaped like a golf club, with the straight portion of the molecule ending in a double globular head. The myosin heads occur on each side of a sarcomere but not in the middle.

Thin Filaments A thin filament primarily consists of two intertwining strands of the protein **actin.** Two other proteins, called tropomyosin and troponin, also play a role, as we will discuss later in this section.

Sliding Filaments We will also see that when muscles are stimulated, impulses travel down a T tubule, and calcium is released from the sarcoplasmic reticulum. Calcium triggers muscle fiber contraction as the sarcomeres within the myofibrils shorten. When a sarcomere shortens, the actin (thin) filaments slide past the myosin (thick) filaments and approach one another. This causes the I band to shorten and the H zone to almost or completely disappear. The movement of actin filaments in relation to myosin filaments is called the **sliding filament theory** of muscle contraction. During the sliding process, the sarcomere shortens, even though the filaments themselves remain the same length. ATP supplies the energy for muscle contraction. Although the actin filaments slide past the myosin filaments, it is the myosin filaments that do the work. Myosin filaments break down ATP and form cross-bridges that pull the actin filaments toward the center of the sarcomere.

Skeletal Muscle Contraction

Muscle fibers are stimulated to contract by motor neurons whose axons are in nerves (Fig. 19.14). The axon of one motor neuron can stimulate from a few to several muscle fibers of a muscle

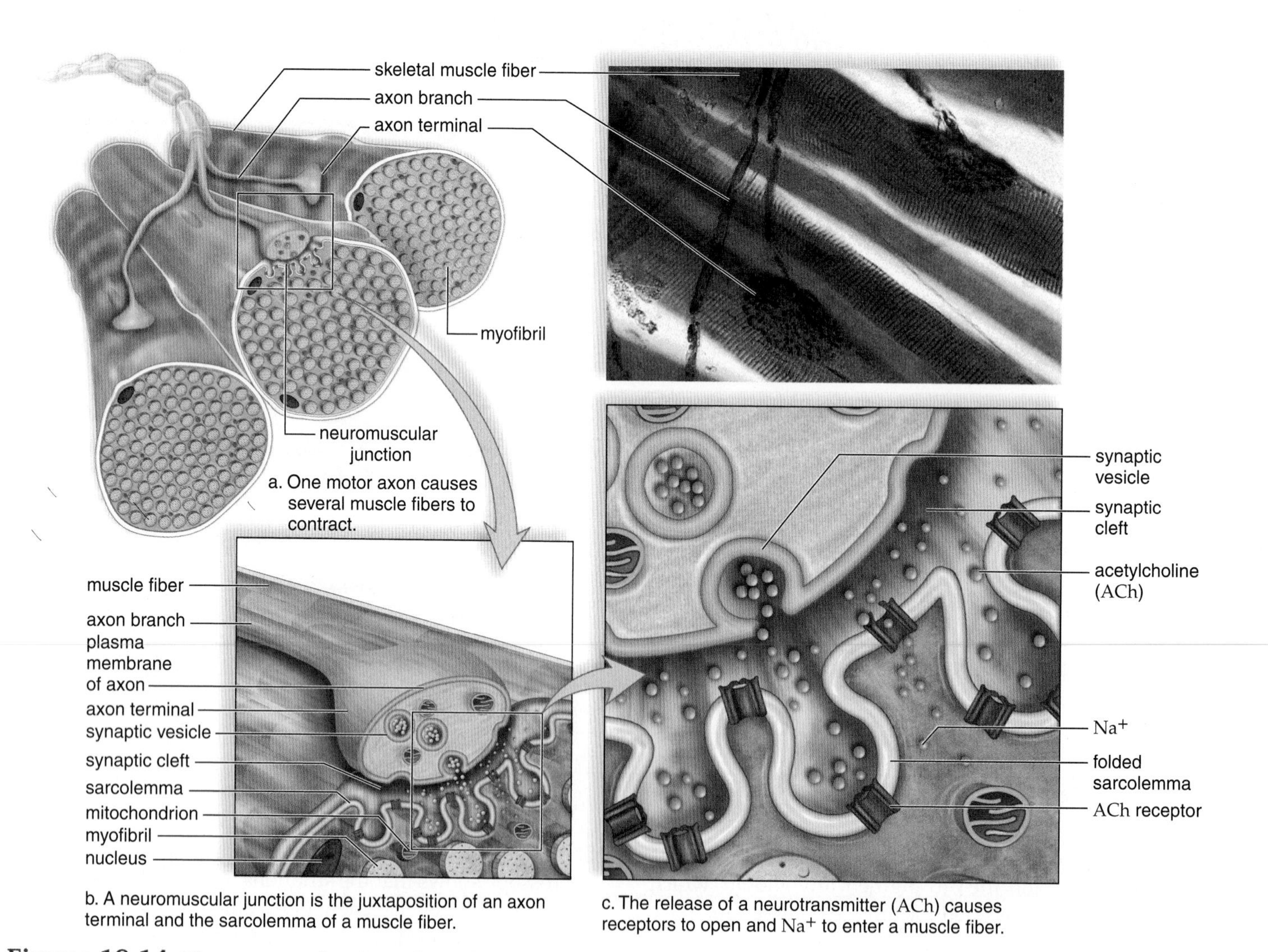

Figure 19.14 Neuromuscular junction.
a. The branch of a motor nerve fiber ends in an axon terminal that meets but does not touch a muscle fiber. **b.** A synaptic cleft separates the axon terminal from the sarcolemma of the muscle fiber. **c.** Nerve impulses traveling down a motor fiber cause synaptic vesicles to discharge a neurotransmitter (ACh) that diffuses across the synaptic cleft. When the neurotransmitter is received by the sarcolemma of a muscle fiber, impulses begin that lead to muscle fiber contraction.

because each axon has several branches. A branch of an axon ends in an axon terminal that lies in close proximity to the sarcolemma of a muscle fiber. A small gap, called a synaptic cleft, separates the axon terminal from the sarcolemma. This entire region is called a **neuromuscular junction** (see Fig. 19.14).

Axon terminals contain synaptic vesicles that are filled with the neurotransmitter acetylcholine (ACh). When nerve impulses traveling down a motor neuron arrive at an axon terminal, the synaptic vesicles release ACh into the synaptic cleft. ACh quickly diffuses across the cleft and binds to receptors in the sarcolemma. Now the sarcolemma generates impulses that spread over itself and down T tubules to the sarcoplasmic reticulum. The release of Ca^{2+} ions from the sarcoplasmic reticulum causes the filaments within the sarcomeres to slide past one another. Sarcomere contraction causes myofibril contraction, which in turn results in the contraction of a muscle fiber and, eventually, a whole muscle.

Botox is a trade name for botulinum toxin A, a neurotoxin produced by a bacterium. Botox, which can be injected by a physician to treat several medical conditions as well as to prevent wrinkling of the brow and skin around the eyes, prevents the release of ACh, and therefore, the contraction of muscles in these areas. Tetanus toxin, produced by a different bacterial species, interferes with muscle relaxation. Thus, it causes excessive contraction, which can be fatal if it affects the respiratory muscles. Fortunately, the tetanus vaccine is very effective at preventing this.

Check Your Progress

1. Name three microscopic components of muscle cells that are not found in other types of cells.
2. What is meant by the statement that "myosin filaments do the work" of muscle contraction?
3. How do neurons specifically control muscle contraction?

The Molecular Mechanism of Contraction

As shown in Figure 19.15, two other proteins are associated with actin filaments. Threads of *tropomyosin* wind about an actin filament, and *troponin* occurs at intervals along the

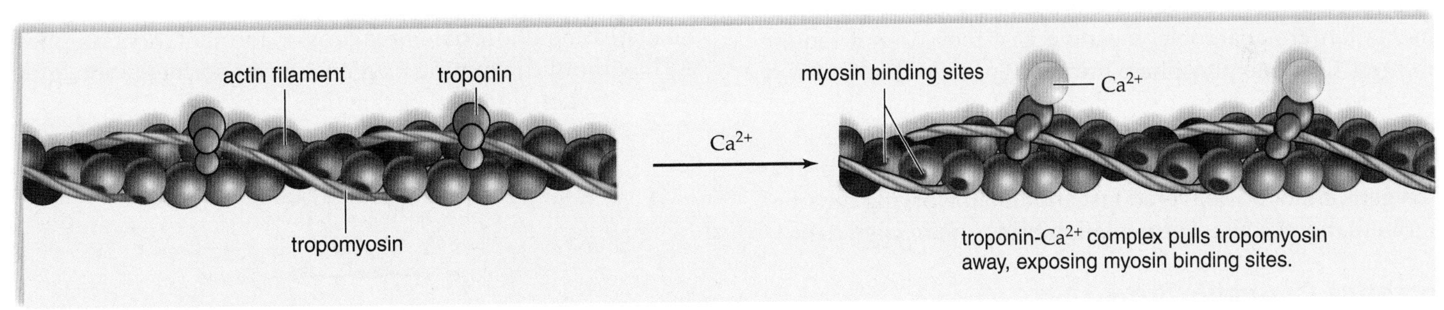

a. Function of Ca^{2+} ions in muscle contraction

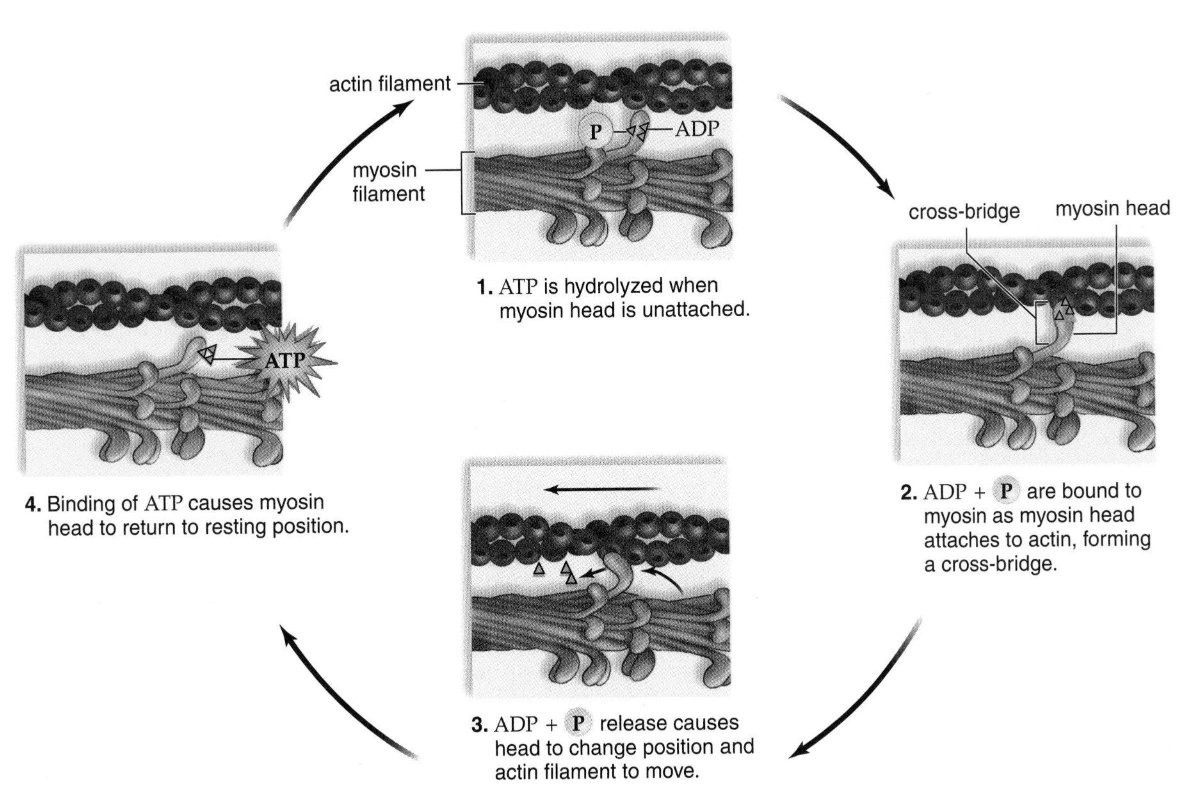

b. Function of cross-bridges in muscle contraction

Figure 19.15 The role of specific proteins and calcium in muscle contraction. **a.** Upon release, calcium binds to troponin, exposing myosin binding sites. **b.** After breaking down ATP (1), myosin heads bind to an actin filament (2), and later, a power stroke causes the actin filament to move (3). When another ATP binds to myosin, the head detaches from actin (4), and the cycle begins again. Although only one myosin head is featured, many heads are active at the same time.

threads. When Ca^{2+} ions are released from the sarcoplasmic reticulum, they combine with troponin, and this causes the tropomyosin threads to shift their position.

The double globular heads of a myosin filament have ATP binding sites, where an ATPase splits ATP into ADP and Ⓟ. The ADP and Ⓟ remain on the myosin heads until the heads attach to an actin filament, forming cross-bridges. Now, ADP and Ⓟ are released, and the cross-bridges change their positions. This is the power stroke that pulls the actin filament toward the center of the sarcomere. When ATP molecules again bind to the myosin heads, the cross-bridges are broken, and heads detach from the actin filament. Actin filaments move nearer the center of the sarcomere each time the cycle is repeated. When nerve impulses cease, the sarcoplasmic reticulum actively transports Ca^{2+} ions back into the sarcoplasmic reticulum, and the muscle relaxes.

Energy for Muscle Contraction

ATP produced previous to strenuous exercise lasts a few seconds, and then muscles acquire new ATP in three different ways: creatine phosphate breakdown, cellular respiration, and fermentation. Creatine phosphate breakdown and fermentation are anaerobic, meaning that they do not require oxygen. Creatine phosphate breakdown, used first, is a way to acquire ATP before oxygen starts entering mitochondria. Cellular respiration is aerobic and only takes place when oxygen is available. If exercise is vigorous to the point that oxygen cannot be delivered fast enough to working muscles, fermentation occurs. Fermentation brings on oxygen debt.

Creatine Phosphate Breakdown

Creatine phosphate is a high-energy compound built up when a muscle is resting. Creatine phosphate cannot participate directly in muscle contraction. Instead, it can regenerate ATP by the following reaction:

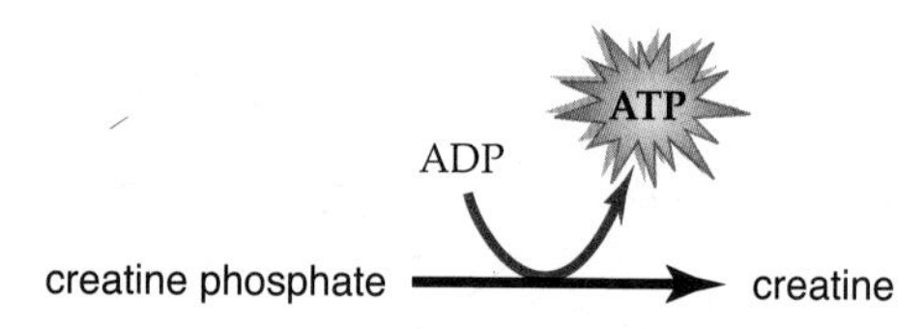

This reaction occurs in the midst of sliding filaments, and therefore is the speediest way to make ATP available to muscles. Creatine phosphate provides enough energy for only about eight seconds of intense activity, and then it is spent. Creatine phosphate is rebuilt when a muscle is resting by transferring a phosphate group from ATP to creatine. Creatine phosphate is also a very popular dietary supplement for training athletes, and there is evidence that consuming creatine in moderate doses is safe and may increase some types of athletic performance.

Cellular Respiration

Cellular respiration, completed in mitochondria, usually provides most of a muscle's ATP. Glycogen and fat are stored in muscle cells. Therefore, a muscle cell can use glucose (from glycogen) and fatty acids (from fat) as fuel to produce ATP if oxygen is available:

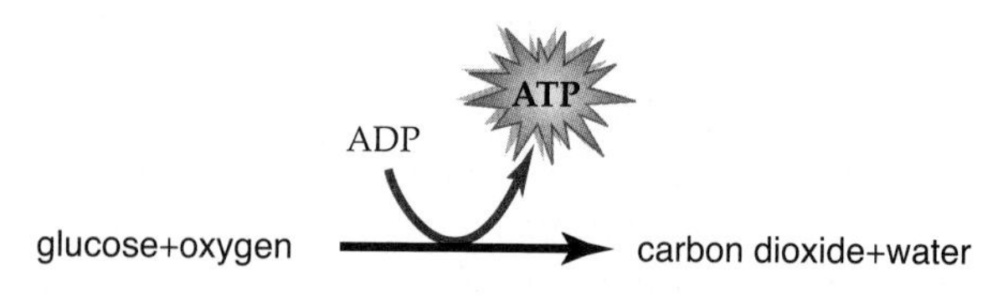

Myoglobin, an oxygen carrier similar to hemoglobin, is synthesized in muscle cells, and its presence accounts for the reddish-brown color of skeletal muscle fibers. Myoglobin has a higher affinity for oxygen than does hemoglobin. Therefore, myoglobin can pull oxygen out of blood and make it available to muscle mitochondria that are carrying on cellular respiration. Then, too, the ability of myoglobin to temporarily store oxygen reduces a muscle's immediate need for oxygen when cellular respiration begins. The end products (carbon dioxide and water) are usually no problem. Carbon dioxide leaves the body at the lungs, and water simply enters the extracellular space. The by-product, heat, keeps the entire body warm.

Fermentation

Like creatine phosphate breakdown, fermentation supplies ATP without consuming oxygen. During fermentation, glucose is broken down to lactate (lactic acid):

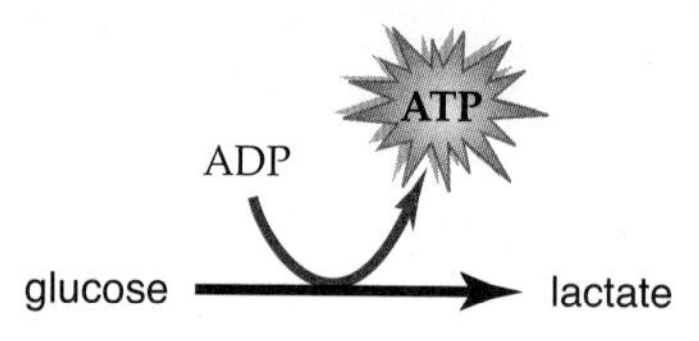

The accumulation of lactate in a muscle fiber makes the cytoplasm more acidic, and eventually enzymes cease to function well. If fermentation continues longer than two or three minutes, cramping and fatigue set in. Cramping seems to be caused by lack of the ATP needed to pump calcium ions back into the sarcoplasmic reticulum and to break the linkages between the actin and myosin filaments so that muscle fibers can relax.

Oxygen Debt

When a muscle uses creatine phosphate or fermentation to supply its energy needs, it incurs an **oxygen debt.** Oxygen debt is obvious when a person continues to breathe heavily after exercising. The ability to run up an oxygen debt is one of muscle tissue's greatest assets. Without oxygen, brain tissue cannot last nearly as long as muscles can.

In people who exercise regularly, the number of muscle mitochondria increases, and so fermentation is not needed to produce ATP. Their bodies' mitochondria can start consuming oxygen as soon as the ADP concentration begins to rise during muscle contraction. Because mitochondria can break down fatty acids instead of glucose, blood glucose is spared for the activity of the brain. (The brain, unlike other organs, can only utilize glucose to produce ATP.) Because less lactate is produced in people who train, the pH of the blood remains steady, and there is less of an oxygen debt.

Repaying an oxygen debt requires replenishing creatine phosphate supplies and disposing of lactate. Lactate can be changed back to pyruvate and metabolized completely in mitochondria, or pyruvate can be sent to the liver to reconstruct glycogen. A marathon runner who has just crossed the finish line is not usually exhausted due to oxygen debt. Instead, the runner has used up all the muscles', and probably the liver's, glycogen supply. It takes about two days to replace glycogen stores on a high-carbohydrate diet.

Check Your Progress

1. Describe the roles of tropomyosin and troponin in muscle contraction.
2. What is one specific advantage of each mechanism of generating ATP for muscle contraction?

19.5 Whole Muscle Contraction

Learning Outcomes

1. Explain what is happening to an isolated muscle fiber during summation, tetanus, and fatigue.
2. Describe how muscles generate different levels of contractive force.
3. Explain the types of exercise that require more fast-twitch versus slow-twitch muscle fibers.

Researchers sometimes study muscles in the laboratory in an effort to understand whole muscle contraction in the body.

In the Laboratory

When a muscle fiber is isolated, placed on a microscope slide, and provided with ATP plus various salts, it contracts completely along its entire length. This observation has resulted in the *all-or-none law,* which states that a muscle fiber contracts completely or not at all. In contrast, a whole muscle shows degrees of contraction. To study whole muscle contraction in the laboratory, an isolated muscle is stimulated electrically, and the mechanical force of contraction is recorded as a visual pattern called a **myogram.** When the strength of the stimulus is above a threshold level, the muscle contracts and then relaxes. This action—a single contraction that lasts only a fraction of a second—is called a **muscle twitch.**

Figure 19.16*a* is a myogram of a muscle twitch, which is customarily divided into three stages: the *latent period,* or the period of time between stimulation and initiation of contraction; the *contraction period,* when the muscle shortens; and the *relaxation period,* when the muscle returns to its former length. It's interesting to use our knowledge of muscle fiber contraction to understand these events. From our study thus far, we know that a muscle fiber in an intact muscle contracts when calcium leaves storage sacs and relaxes when calcium returns to storage sacs.

But unlike the contraction of a muscle fiber, a whole muscle has degrees of contraction, and a twitch can vary in height (strength) depending on the degree of stimulation. Why? Obviously, a stronger stimulation causes more individual fibers to contract than before.

If a whole muscle is given a rapid series of stimuli, it can respond to the next stimulus without relaxing completely. Summation is increased muscle contraction until maximal sustained contraction, called **tetanus,** is achieved (Fig. 19.16*b*). The myogram no longer shows individual twitches; rather, the twitches are fused and blended completely into a straight line. Tetanus continues until the muscle becomes fatigued due to depletion of energy reserves. Fatigue is apparent when a muscle relaxes even though stimulation continues.

In the Body

In the body, nerves cause muscles to contract. As mentioned, each axon within a nerve stimulates a number of muscle fibers. A nerve fiber, together with all of the muscle fibers it innervates, is called a **motor unit.** A motor unit obeys the all-or-none law. Why? Because all the muscle fibers in a motor unit are stimulated at once, and they all either contract or do not contract. A variable of interest is the number

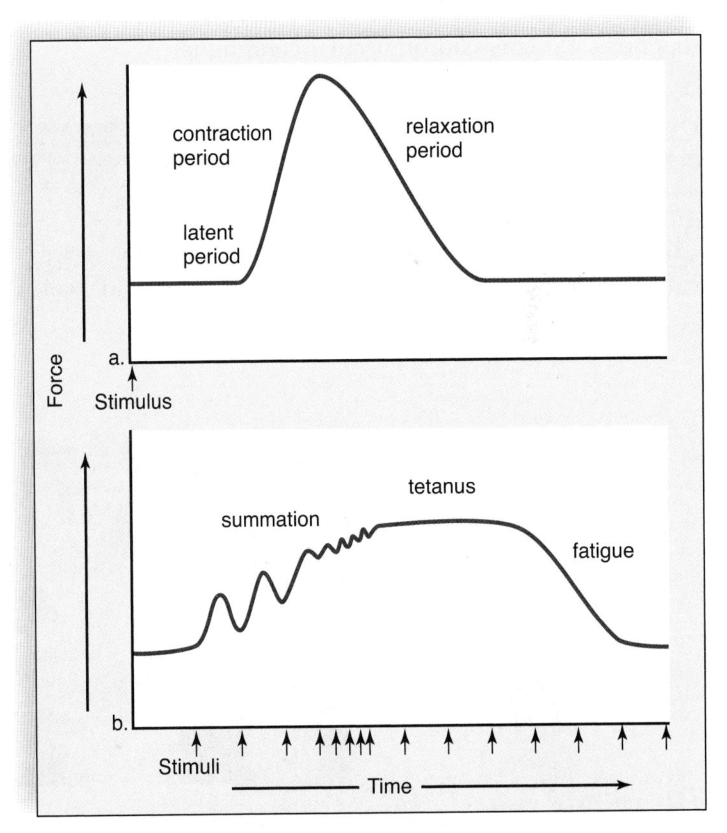

Figure 19.16 Physiology of skeletal muscle contraction.
Stimulation of a muscle dissected from a frog resulted in these myograms. **a.** A simple muscle twitch has three periods: latent, contraction, and relaxation. **b.** Summation and tetanus. When the muscle is not permitted to relax completely between stimuli, the contraction gradually increases in intensity. The muscle becomes maximally contracted until it fatigues.

of muscle fibers within a motor unit. For example, in the ocular muscles that move the eyes, the innervation ratio is one motor axon per 23 muscle fibers, while in the gastrocnemius muscle of the leg, the ratio is about one motor axon per 1,000 muscle fibers. Thus, moving the eyes allows finer control than moving the legs.

Tetanic contractions ordinarily occur in the body because, as the intensity of nervous stimulation increases, more and more motor units are activated. This phenomenon, known as recruitment, results in stronger and stronger muscle contractions. But while some muscle fibers are contracting, others are relaxing. Because of this, intact muscles rarely fatigue completely. Even when muscles appear to be at rest, they exhibit **muscle tone,** in which some of their fibers are always contracting. Muscle tone is particularly important in maintaining posture. If all the fibers within the muscles of the neck, trunk, and legs were to suddenly relax, the body would collapse.

Medical professionals can measure the electrical activity of muscles using a test called an electromyogram (EMG). Most commonly, needles are inserted at various points along a muscle suspected of functioning abnormally, and the patient is asked to smoothly contract the muscles. Because different conditions have varying effects on the electrical patterns generated by muscles, the EMG test can help in the diagnosis of many neuromuscular conditions.

Check Your Progress

1. What are the three typical stages of a muscle twitch, as seen in a myogram?
2. What would be the approximate ratio of motor nerve axons per muscle fiber in your index finger? In your gluteus maximus?

Athletics and Muscle Contraction

Athletes who excel in a particular sport, and much of the general public as well, are interested in staying fit by exercising. The Health Focus gives suggestions for exercise programs according to age.

Exercise and Size of Muscles Muscles that are not used or that are used for only very weak contractions decrease in size, or **atrophy.** Atrophy can occur when a limb is placed in a cast or when the nerve serving a muscle is damaged. If nerve stimulation is not restored, muscle fibers are gradually replaced by fat and fibrous tissue. Unfortunately, atrophy can cause muscle fibers to shorten progressively, leaving body parts contracted in contorted positions.

Forceful muscular activity over a prolonged period causes muscle to increase in size as the number of myofibrils within the muscle fibers increases. Increase in muscle size, called **hypertrophy,** occurs only if the muscle contracts to at least 75% of its maximum tension.

Slow-Twitch and Fast-Twitch Muscle Fibers We have seen that all muscle fibers metabolize both aerobically and anaerobically. Some muscle fibers, however, utilize one method more than the other to provide myofibrils with ATP. Slow-twitch fibers tend to be aerobic, and fast-twitch fibers tend to be anaerobic (Fig. 19.17).

Slow-twitch fibers have a steadier tug and more endurance, despite having motor units with a smaller number of fibers. These muscle fibers are most helpful in sports such as long-distance running, biking, jogging, and swimming. Because they produce most of their energy aerobically, they tire only when their fuel supply is gone. Slow-twitch fibers have many mitochondria and are dark in color because they contain myoglobin, the respiratory pigment found in muscles. They are also surrounded by dense capillary beds and draw more blood and oxygen than fast-twitch fibers. Slow-twitch fibers have a low maximum tension, which develops slowly, but these muscle fibers are highly resistant to fatigue. Because slow-twitch fibers

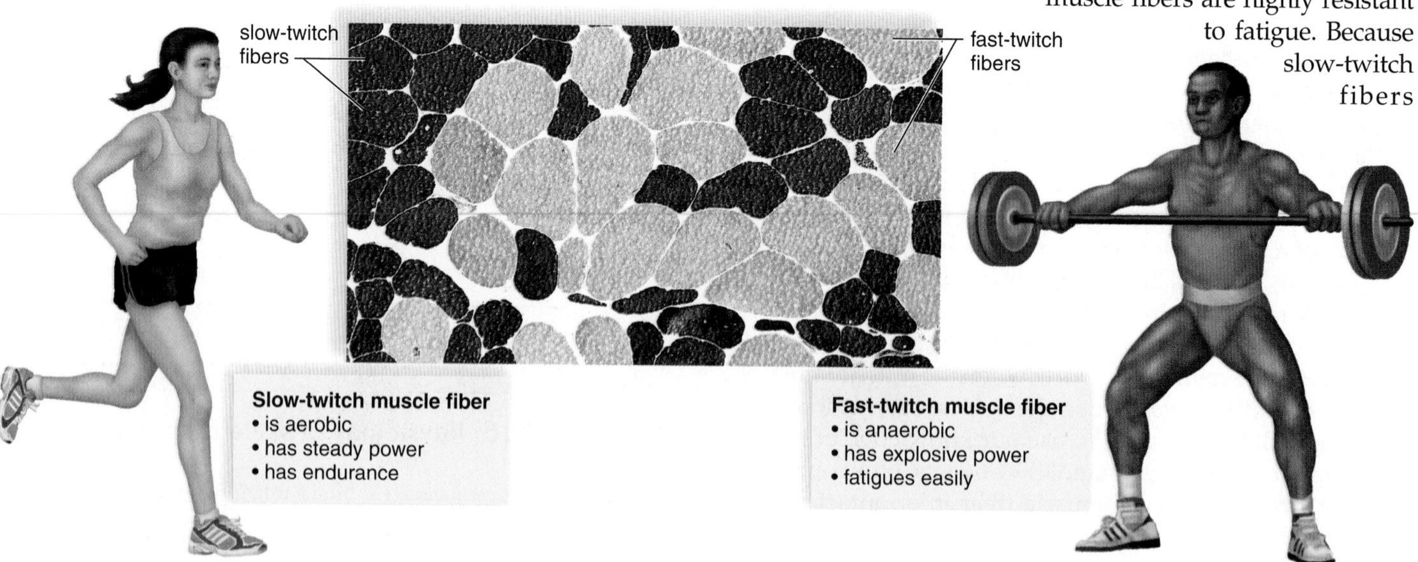

Figure 19.17 Slow- and fast-twitch muscle fibers.
If your muscles contain many slow-twitch fibers (dark color), you probably do better at a sport such as cross-country running. But if your muscles contain many fast-twitch fibers (light color), you probably do better at a sport such as weight lifting.

Health Focus

Exercise, Exercise, Exercise

Sensible exercise programs improve muscular strength, muscular endurance, and flexibility for people of all ages (Table 19A). Muscular strength is the force a muscle group (or muscle) can exert against a resistance in one maximal effort. Muscular endurance is judged by the ability of a muscle to contract repeatedly or sustain a contraction for an extended period. Flexibility is tested by observing the range of motion at a joint.

Exercise also improves cardiorespiratory endurance. The heart rate and capacity increase, and the air passages dilate so that the heart and lungs are able to support prolonged muscular activity. The blood level of high-density lipoprotein (HDL), the molecule that prevents the development of plaque in blood vessels, increases. Also, body composition, the proportion of protein to fat, changes favorably when you exercise.

Exercise reduces the risk of diabetes mellitus, especially type 2 or insulin-resistant diabetes, which is a large and growing health problem in many countries around the world. Being overweight and inactive increases the chances of developing type 2 diabetes. Studies have shown that healthy eating and regular exercise are more effective at reducing the risk of diabetes than medication alone. Exercise seems to help prevent certain kinds of cancer. Cancer prevention involves eating properly, not smoking, avoiding cancer-causing chemicals and radiation, undergoing appropriate medical screening tests, and knowing the early warning signs of cancer. However, studies show that people who exercise are less likely to develop colon, breast, cervical, uterine, and ovarian cancers.

Exercise promotes the activity of osteoblasts in young people as well as older people. Physical training with weights can improve bone density and also muscular strength and endurance in all adults, regardless of age. Even men and women in their eighties and nineties can make substantial gains in bone and muscle strength, which can help them lead more independent lives. Exercise also helps prevent osteoporosis, a condition in which the bones are weak and tend to break. The stronger the bones when a person is young, the lower the chance of osteoporosis as a person ages.

Exercise helps prevent weight gain, not only because the level of activity increases, but also because muscles metabolize faster than other tissues. As a person becomes more muscular, fat is less likely to accumulate.

Exercise relieves depression and enhances the mood. Some report that exercise actually makes them feel more energetic, and after exercising, particularly in the late afternoon, they sleep better that night. Self-esteem rises, not only because of improved appearance, but also due to other factors that are not well understood. However, it is known that vigorous exercise releases endorphins, hormone-like chemicals that alleviate pain and provide a feeling of tranquility.

A sensible exercise program is one that provides all the benefits without the detriments of a too strenuous program. Overexertion can actually be harmful to the body and might result in a lasting sports injury, such as a bad back or bad knees.

Dr. Arthur Leon at the University of Minnesota performed a study involving 12,000 men, and the results showed that only moderate exercise is needed to lower the risk of a heart attack by one-third. In another study conducted by the Institute for Aerobics Research in Dallas, Texas, which included 10,000 men and more than 3,000 women, even a little exercise was found to lower the risk of death from circulatory diseases and cancer. Simply increasing daily activity by walking to the corner store instead of driving and by taking the stairs instead of the elevator can improve your health.

Discussion Questions

1. What type(s) of muscle activity (strength, endurance, flexibility) are most important for the following sports?
   a. Football
   b. Golf
   c. Gymnastics
   d. Long-distance running
   e. Sprinting
2. Provide a specific, reasonable mechanism by which exercise could decrease your chances of developing the following: diabetes, hypertension, cancer.
3. Because there are so many documented heath benefits to exercising, what are some major factors that keep people from starting and sticking to a regular exercise program?

TABLE 19A A Checklist for Staying Fit

Children, 7–12	Teenagers, 13–18	Adults, 19–55	Seniors, 55 and up
Vigorous activity 1–2 hours daily	Vigorous activity 1 hour 3–5 days a week; otherwise, 1/2 hour daily moderate activity	Vigorous activity 1 hour 3 days a week; otherwise, 1/2 hour daily moderate activity	Moderate exercise 1 hour daily 3 days a week; otherwise, 1/2 hour daily moderate activity
Free play	Build muscle with calisthenics	Exercise to prevent lower back pain: aerobics, stretching, yoga	Plan a daily walk
Build motor skills through team sports, dance, swimming	Plan aerobic exercise to control buildup of fat cells	Take active vacations: hike, bicycle, cross-country ski	Daily stretching exercise
Encourage more exercise outside of physical education classes	Pursue tennis, swimming, horseback riding—sports that can be enjoyed for a lifetime	Find exercise partners: join a running club, bicycle club, outing group	Learn a new sport or activity: golf, fishing, ballroom dancing
Initiate family outings: bowling, boating, camping, hiking	Continue team sports, dancing, hiking, swimming		Try low-impact aerobics, boating. Before undertaking new exercises, consult your doctor

have a substantial reserve of glycogen and fat, their abundant mitochondria can maintain a steady, prolonged production of ATP when oxygen is available.

Fast-twitch fibers tend to be anaerobic and seem designed for strength because their motor units contain many fibers. They provide explosions of energy and are most helpful in sports activities such as sprinting, weight lifting, swinging a golf club, or throwing a shot. Fast-twitch fibers are light in color because they have fewer mitochondria, little or no myoglobin, and fewer blood vessels than slow-twitch fibers do. Fast-twitch fibers can develop maximum tension more rapidly than slow-twitch fibers can, and their maximum tension is greater. However, their dependence on anaerobic energy leaves them vulnerable to an accumulation of lactate that causes them to fatigue quickly.

Check Your Progress

1. Based on what you know about muscle tone, why would a muscle atrophy if the motor nerve supplying it is cut?
2. How can mitochondria specifically use glycogen and fat to produce ATP? (You may need to review section 6.4.)
3. Why would some domesticated birds, like turkeys and chickens, have mostly fast-twitch fibers in their breast (wing) muscles, while wild birds like geese and ducks have mostly slow-twitch fibers?

19.6 Disorders of the Musculoskeletal System

Learning Outcomes

1. List three common disorders that affect the bones and joints and three that affect the muscles.
2. Identify some of the most common steps that can be taken to help prevent osteoporosis.

Disorders of the Skeleton and Joints

Because bones are rigid, they are susceptible to being broken, or fractured. The most common cause of bone fractures is trauma, although pathological fractures can also occur during normal everyday activities if the bones are diseased. If the skin is punctured by the injury, the fracture is generally termed an "open" fracture; otherwise, it is "closed." Some common types of fractures include greenstick fractures, which do not pass all the way through the bone; comminuted fractures, which result in three or more bone fragments (Fig. 19.18); and stress fractures, one of the most common injuries in athletes. Stress fractures are small cracks in the bone that occur most commonly on the lower leg when muscles become fatigued and unable to absorb the stress of athletic activities. These small fractures usually heal with rest—that is, by avoiding the activity that produced them. Other types of uncomplicated fractures may be treated with casts or splints. Compound and comminuted fractures, however, often require surgery to realign bones or replace bone fragments. In children, the growth plate is the last part of the bone to ossify, leaving it more susceptible to fracture. Fracture of the growth plate may cause the affected bone to stop growing, resulting in limbs that are crooked or uneven in length.

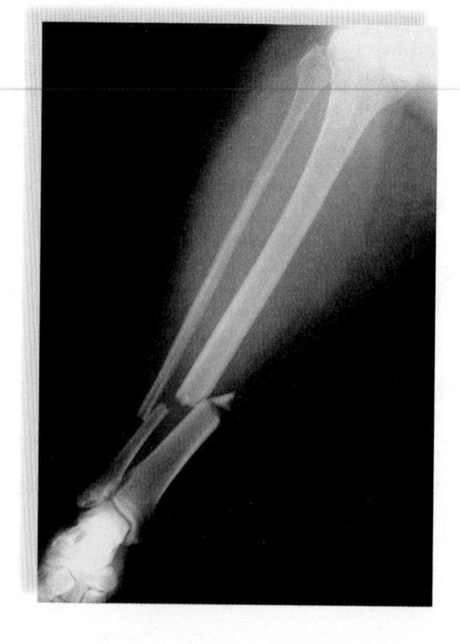

Figure 19.18 A comminuted fracture.
This X ray shows a tibia that has fractured into at least three pieces. The fibula (left) is also fractured.

Osteoporosis is a condition in which bone loses mass and mineral content. This leads to an increased risk of fractures, especially of the pelvis, vertebrae, and wrist. As the bone softens affected individuals often develop a characteristic curvature of the spine (Fig. 19.19). Eighty percent of those affected by osteoporosis are women. About one in four women will have an osteoporosis-related fracture in their lifetime. Women generally have about 30% less bone mass than men to begin with, and after menopause women typically lose about 2% of their bone mass per year. Estrogen replacement therapy (ERT) has been shown to increase bone mass and reduce fractures, but because it also can increase the risk of cardiovascular disease and certain types of cancer, patients must be well-informed before beginning ERT. Other factors that decrease the risk of osteoporosis for everyone include consuming 1,000–1,500 mg of calcium per day and engaging in regular physical exercise.

There are many different types of **arthritis,** or inflammation of the joints. The Arthritis Foundation estimates that nearly one in three adult Americans has some degree of chronic joint pain. Arthritis is second only to heart disease as a cause of work disability in the United States. The most common form of arthritis is **osteoarthritis (OA),** or degenerative joint disease, which results from the deterioration of the cartilage in one or more synovial joints. As the cartilage continues to soften, crack, and wear away, the resulting bone-on-bone contact causes pain and a decreased range of motion. Symptoms of OA, such as pain or stiffness in the joints after periods of inactivity, typically begin after age 40 and progress slowly.

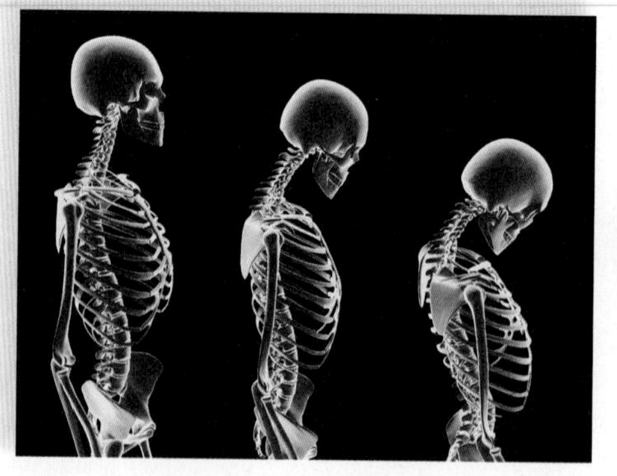

Figure 19.19 Osteoporosis.
Osteoporosis can eventually result in the slumped posture and loss of height characteristically seen in some elderly people.

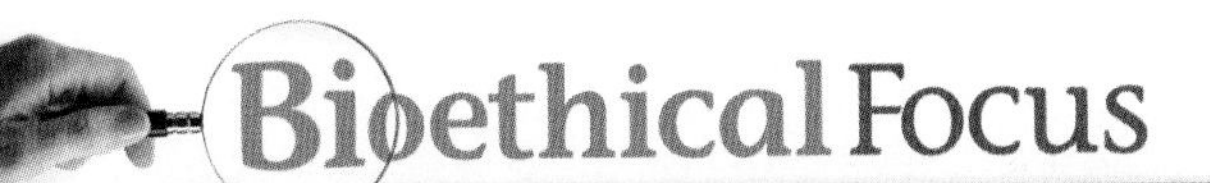

What Constitutes an Unfair Advantage in Sports?

Sports TV news shows are full of reports of the latest scandals involving professional athletes using performance-enhancing drugs. Baseball players either accused or proven to have used anabolic steroids or human growth hormone (HGH) may be banned from the game, or at least from the Hall of Fame. Floyd Landis, the American cyclist who won the Tour de France in 2006, had to give up his title when he tested positive for synthetic testosterone. This list seems to go on and on.

But even if there is general agreement that performance-enhancing drugs can give athletes an "unfair advantage," what about athletes with other types of advantages? We wouldn't tend to think that being born with malformed bones below the knee as an advantage, but for Oscar Pistorius, that may be exactly the case. Pistorius, who won a gold medal at the Athens Paralympics in 2004, was banned from the Beijing Olympics in 2008, because examinations showed that his blade-shaped prosthetic legs (Fig. 19A) are more efficient than normal human legs. Or consider the case of Casey Martin, who suffers from a painful leg defect, and who in 2001 sued the Professional Golf Association for the right to use a golf cart, which is normally banned in tournaments. Martin was granted the right to use a cart under the Americans with Disabilities Act, but did this give him an unfair advantage over able-bodied golfers?

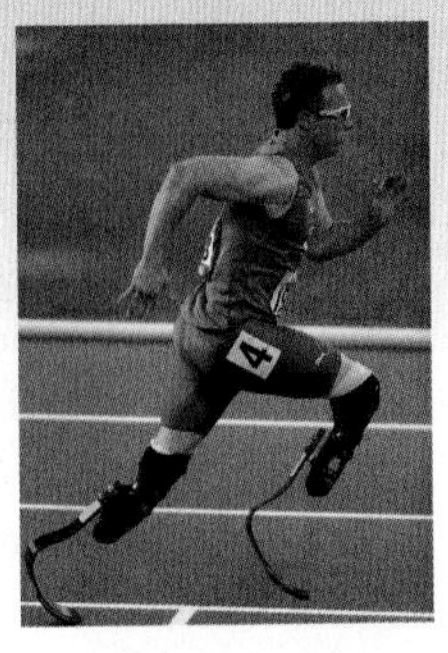

Figure 19A Oscar Pistorius was banned from competing in the 2008 Olympics.

One of the major reasons performance-enhancing drugs are banned is that they pose clear health risks to users. Even so, shouldn't individuals be allowed to take these risks if they want to? And on what basis can we outlaw adaptations that are actually designed to increase a person's health? In other words, how can you justify allowing some strategies that enhance performance and not others?

Form Your Own Opinion

1. Are there any similarities between taking anabolic steroids and taking legal drugs, such as aspirin, to reduce inflammation in the body?
2. Do you see a significant difference in the advantage gained by taking drugs, using a prosthesis, or otherwise compensating for a disability?
3. Suppose a new pay-per-view sports network allowed athletes to take any drug, or undergo any surgical modification they wanted, in order to see what the human body could achieve. Would you watch such a show? What factors would then limit the athlete's performance?

The hands, hips, knees, lower back, and neck are most often affected (Fig. 19.20). **Rheumatoid arthritis (RA)** is considered an autoimmune disease, in which the body's immune system attacks the joints as well as other tissues. RA affects more women than men, and often first strikes people in their twenties and thirties. The initiating cause of RA remains to be discovered, but genetic factors may play a major role. Over-the-counter anti-inflammatory drugs such as ibuprofen, or more powerful prescription drugs such as corticosteroids, can reduce the pain of most types of arthritis. Some studies have shown that dietary supplements containing glucosamine and chondroitin sulfate are also beneficial. As mentioned in the chapter-opening story, the damaged joint can also be surgically replaced with a prosthesis (artificial joint).

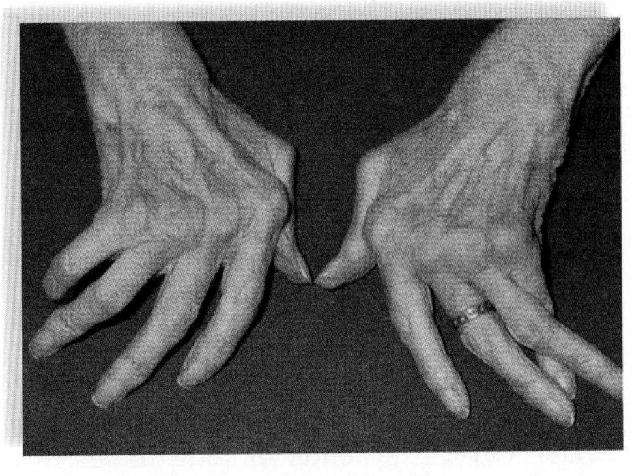

Figure 19.20 The effects of arthritis. Arthritis often affects the hands, resulting in swelling, pain, and loss of the ability to perform normal functions.

Disorders of the Muscles

Muscle cramps and twitches don't usually rise to the level of a "disorder," but almost everyone has experienced them. A true muscle cramp doesn't originate in the muscle itself, but rather is caused by abnormal nerve activity. This type of cramp is more common at night or after exercise, and can usually be ended by stretching the muscle. Most of us have also experienced occasional muscle twitches, or "fasciculations." These also result from abnormal nerve activity, but only involve a small part of a muscle. Twitches most commonly occur in the eyelids, between the thumb and forefinger, and in the feet. They may occur more frequently due to stress, caffeine, or lack of sleep.

Fibromyalgia is a disorder that causes chronic pain in the muscles and ligaments. It affects about 4 million Americans, mostly women in their mid-30s to late-50s. The pain is usually most severe in the neck, shoulders, back, and hips. Although the cause of fibromyalgia is unknown, it may be due to low levels of serotonin or other neurotransmitters involved in pain perception and sleep regulation. As a result, patients with fibromyalgia may be more sensitive to pain, and they usually are unable to get enough deep sleep. As a result, chronic fatigue is also a major symptom. There are no specific treatments, but most people can reduce their symptoms by combining medications to reduce pain and improve sleep with regular exercise and a healthy diet.

Muscular dystrophies (MD) are a group of genetic diseases that affect the muscles. Most people are probably most familiar with these diseases from the annual Jerry Lewis Labor Day telethon sponsored by the Muscular Dystrophy Association. These disorders vary considerably in their degree of severity. The more severe forms may cause problems beginning at a very young age, while other forms may not cause problems until middle age or later. The most common type of MD, called Duchenne MD, is due to an abnormality in the gene coding for a muscle-associated protein called dystrophin. Because it is an X-linked recessive disorder, Duchenne MD mainly affects boys.

Muscle weakness is usually noticeable by age 3 to 5. By age 12, most affected children are unable to walk, and will eventually need a respirator to breathe. Corticosteroids have been shown to slow the degenerative process of Duchenne MD, but these drugs often have side effects. Gene therapy, in which the correct form of the faulty gene is inserted into the patient's cells, is being investigated as a potential cure for MD.

Check Your Progress

1. What do you think would happen if a bone fracture was not treated?
2. Why would astronauts working in space be very prone to loss of bone density?
3. How does the cause of rheumatoid arthritis differ from the cause of osteoarthritis?

Applying the Concepts Revisited

In the story that opened this chapter, an athletic middle-aged woman had knee replacement surgery after years of worsening arthritis. A few months after surgery, she was able to do many activities that had been painful for many years.

1. The knee is by far the most common joint that is replaced, followed by the hip. What are two or three reasons why these joints would be replaced more often than, say, the elbow or shoulder?
2. A special bone cement is often used to hold artificial knees and hips in place. Because most artificial joints are designed to last 10–15 years, what would be some important properties for this cement to have?

SUMMARIZING THE CONCEPTS

19.1 Anatomy and Physiology of Bone

The skeletal system is comprised of bone, cartilage, and dense fibrous connective tissue:

- Bone is a connective tissue that contains mineral salts, which increase its strength. It is a living tissue that can grow and repair itself. In a long bone, the wall of the shaft is compact bone, which is highly organized. It is made up of units called osteons, which consist of a central canal surrounded by tiny canals called canaliculi and chambers called lacunae. The ends of a long bone contain spongy bone, which is less organized, although still very strong. The spaces in spongy bones often contain red bone marrow, which produces all types of blood cells.

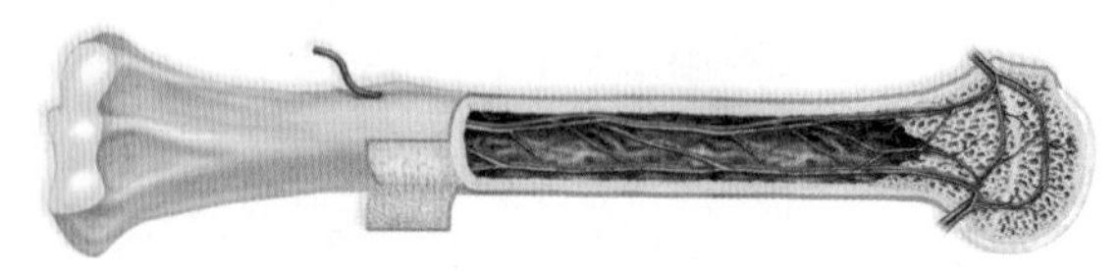

- Cartilage is not as strong as bone, but is more flexible because it contains many collagenous and elastic fibers. Hyaline cartilage covers the ends of long bones, while fibrocartilage is mainly used for support. Elastic cartilage is the most flexible type.
- Dense fibrous connective tissue, which contains fibroblasts and bundles of collagen fibers, forms the ligaments and tendons. Ligaments connect bone to bone, and tendons connect muscle to bone at joints.

19.2 Bones of the Skeleton

The skeleton supports and protects the body, produces blood cells, serves as a storehouse for fats and mineral salts, particularly calcium phosphate, and permits flexible movement. It can be divided into the axial skeleton and the appendicular skeleton, plus the associated joints:

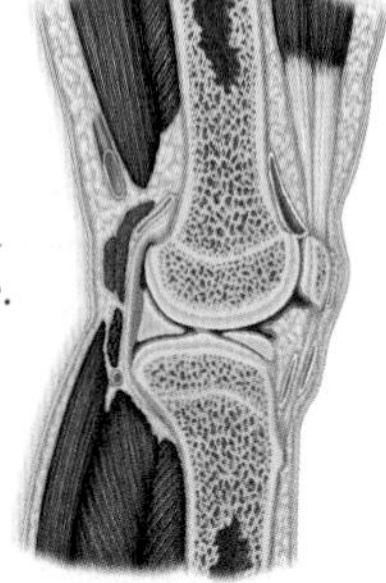

- The axial skeleton lies in the midline of the body and consists of the skull, the hyoid bone, the vertebral column, and the rib cage. The skull contains the cranium, which protects the brain, and the facial bones.
- The appendicular skeleton consists of the bones of the pectoral girdles, upper limbs, pelvic girdle, and lower limbs. The pectoral girdles and upper limbs are adapted for flexibility.
- There are three types of joints: fibrous joints, cartilaginous joints, and synovial joints.

19.3 Skeletal Muscles

Most skeletal (voluntary) muscles work in antagonistic pairs, causing bones to move at a joint:

- The origin of a muscle attaches to a stationary bone; the insertion is on the bone that moves.
- Terms used to describe the types of movements due to muscle contraction include flexion, extension, abduction, and adduction.
- Muscles are grouped by location and named for characteristics such as size, shape, location, and action.

19.4 Mechanism of Muscle Fiber Contraction

Understanding the microscopic anatomy of a muscle fiber helps explain the molecular mechanism of muscle contraction:

- The sarcolemma of a muscle fiber forms T tubules that extend into the fiber and almost touch the sarcoplasmic reticulum, which stores calcium ions. Each muscle fiber contains many myofibrils, which contain myofilaments made of actin and myosin. These are arranged in subuits called sarcomeres, which are the contractile units of a muscle fiber.
- At a neuromuscular junction, synaptic vesicles release acetylcholine (ACh), which binds to protein receptors on the sarcolemma, causing impulses to travel down T tubules and calcium to leave the sarcoplasmic reticulum. Calcium ions bind to troponin and cause tropomyosin threads to shift their position, revealing myosin binding sites on actin. Myosin is an ATPase, and once it breaks down ATP, the myosin head is ready to attach to actin. The release of ADP + Ⓟ causes the head to change its position. This is the power stroke that causes the actin filament to slide toward the center of a sarcomere.

A muscle fiber has three ways to acquire ATP after muscle contraction begins:

- Creatine phosphate, built up when a muscle is resting, can donate phosphates to ADP, forming ATP.
- Oxygen-dependent cellular respiration can occur within the mitochondria of the muscle cell. Myoglobin, an oxygen carrier

that has a higher affinity for oxygen than does hemoglobin, provides oxygen for this process.

- Fermentation of glucose, with the concomitant production of lactate, quickly produces ATP. Fermentation can result in oxygen debt, and lactate accumulation may cause fatigue and cramping.

19.5 Whole Muscle Contraction

The contraction of whole muscles can be studied in the laboratory as well as in the body:

- In the laboratory, a single all-or-none contraction of a muscle fiber is called a muscle twitch, and whole muscle contraction is described in terms of summation, tetanus, and fatigue.
- In the body, a motor unit consists of a single nerve fiber and all the muscle fibers it innervates. Muscles exhibit tone, in which tetanic contraction involving a number of muscle fibers is the rule. Two major types of muscle fibers can be identified, with respect to their predominant method of ATP production. Slow-twitch fibers are darker in color and have more mitochondria and myoglobin for aerobic cellular respiration. They seem to be specialized for activities that require endurance. Fast-twitch fibers have more muscle fibers in their motor units and rely more on an anaerobic means of acquiring ATP. These fibers provide quick bursts of power, but they fatigue quickly.

19.6 Disorders of the Musculoskeletal System

Disorders affecting the skeletal system can be grouped into those affecting the bones and joints, and those affecting the muscles:

- Disorders of the skeleton and joints include fractures, osteoporosis, and arthritis. Fractures can be open or closed, and some specific types are greenstick, comminuted, and stress fractures. Osteoporosis is a softening of bone due to loss of mass and mineral content, which affects women more commonly than men. Among the many types of arthritis are osteoarthritis, or degenerative joint disease, and rheumatoid arthritis, which is an autoimmune disease.
- Muscles can be affected by cramps and twitches, which are very common but usually just an annoyance. More serious disorders of the muscles include fibromyalgia, which results in intense pain and may be due to low levels of serotonin, and muscular dystrophy, a group of genetic disorders that affect muscle development.

TESTING YOURSELF

Choose the best answer for each question.

1. _____ are involved in the breakdown of bone.
   a. Monocytes
   b. Osteoblasts
   c. Lymphocytes
   d. Osteoclasts
2. _____ connect bone to bone, and _____ connect muscle to bone.
   a. Ligaments, ligaments
   b. Tendons, ligaments
   c. Ligaments, tendons
   d. None of these are correct.
3. The only bone that does *not* articulate with any other bone is the
   a. temporal bone.
   b. mandible.
   c. hyoid bone.
   d. maxilla.
   e. palatine bone.
4. The ribs articulate with the _____ vertebrae.
   a. cervical
   b. thoracic
   c. lumbar
   d. sacral
   e. All of these are correct.
5. Label the following diagram of the skull.

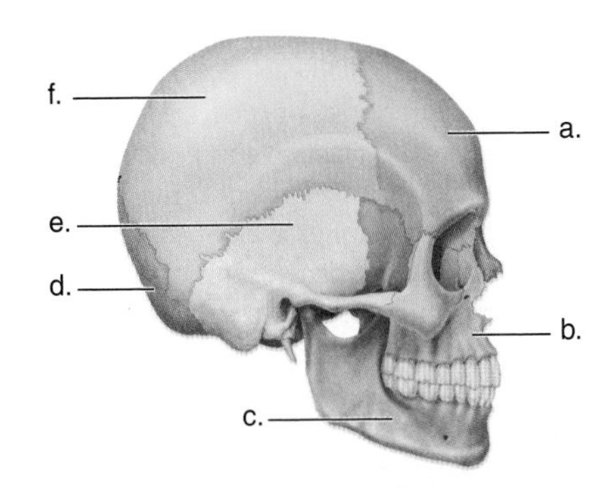

6. The knee joint contains
   a. cartilage.
   b. cruciate ligaments.
   c. synovial fluid.
   d. bursae.
   e. All of these are correct.
7. The _____ of the muscle attaches to the bone that is stationary during movement.
   a. origin
   b. insertion
   c. belly
   d. None of these are correct.
8. The biceps and triceps are considered
   a. synergists.
   b. antagonists.
   c. protagonists.
   d. None of these are correct.
9. Muscles can be named based on
   a. size.
   b. shape.
   c. location.
   d. action.
   e. All of these are correct.

For questions 10–14, match the names of the muscles in the key to the following diagram.

Key:

a. deltoid
b. sartorius
c. gastrocnemius
d. triceps brachii
e. biceps brachii

10.
11.
12.
13.
14.

15. Nervous stimulation of muscles
    a. occurs at a neuromuscular junction.
    b. results in an action potential that travels down the T system.
    c. causes calcium to be released from expanded regions of the sarcoplasmic reticulum.
    d. All of these are correct.
16. When muscles contract,
    a. sarcomeres increase in length.
    b. actin breaks down ATP.
    c. myosin slides past actin.
    d. the H zone disappears.
    e. calcium is taken up by the sarcoplasmic reticulum.

17. To increase the force of muscle contraction,
    a. individual muscle cells have to contract with greater force.
    b. motor units have to contract with greater force.
    c. more motor units need to be recruited.
    d. All of these are correct.
    e. None of these are correct.

For questions 18–23, label the diagram of a muscle fiber, using these terms: myofibril, Z line, T tubule, sarcomere, sarcolemma, sarcoplasmic reticulum.

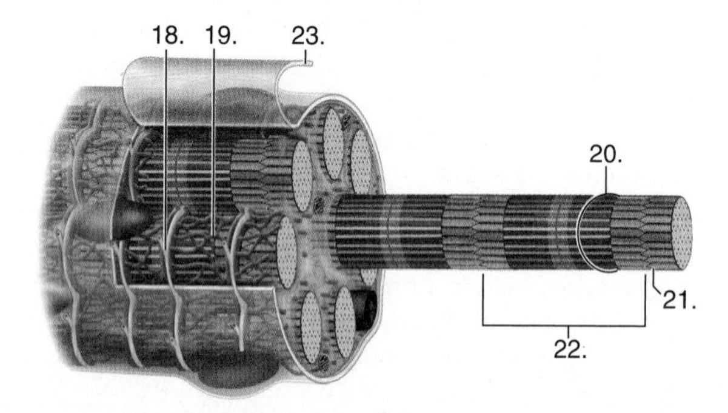

24. Myoglobin content is higher in _____-twitch fibers respiring _____.
    a. slow, aerobically
    b. fast, aerobically
    c. slow, anaerobically
    d. None of these are correct.
25. Rheumatoid arthritis
    a. affects one in three Americans.
    b. is mainly a disease of the elderly.
    c. affects men more than women.
    d. is caused by an autoimmune reaction against the joints.
    e. None of these are correct.

UNDERSTANDING THE TERMS

THINKING CRITICALLY

1. Determine whether each of the following joints is capable of these movements: flexion, extension, abduction, adduction, rotation. You may "cheat" by using your own body.
    a. elbow
    b. neck
    c. knee
    d. phalanges
    e. hip
2. Similar to the activity in our muscle cells, certain bacteria can utilize sugars by aerobic cellular respiration (which requires oxygen) or by fermentation (which does not). However, like muscle cells, these bacteria only use fermentation when oxygen is not available. Why is aerobic respiration preferable? What is the specific role of oxygen in this process? (You may want to refer back to Chapter 7.)
3. Most authorities emphasize the importance of calcium consumption for the prevention of osteoporosis. However, surveys have shown that osteoporosis is rare in some countries, such as China, where calcium intake is lower than in the United States. These studies often point to the higher level of animal protein consumed in the U.S. as having a detrimental effect on bone mass. If this were true, what physiological mechanism might explain it? Can you think of any other differences between the lifestyles in China and the U.S. that also might explain the difference?

INQUIRY INTO LIFE WEBSITE

The companion website for *Inquiry into Life* provides a wealth of information organized and integrated by chapter. You will find practice tests, animations, videos, and much more that will complement your learning and understanding of general biology.

http://www.mhhe.com/maderinquiry13

Endocrine System

Chapter Outline

Applying the Concepts

Hank is only in the second grade, but if you were around him for very long you would probably notice that he has some extra responsibilities compared to most other kids his age. Several times a day, Hank gets a small black kit out of his backpack, loads a plastic-covered needle onto an instrument, places it against his finger, and pushes a button. Once a drop of blood appears, he neatly and efficiently adds it to a strip, which he then inserts into a reader. If the number is too high, Hank knows it is time to go to the school nurse for an injection of insulin. If it is too low, he needs to eat a snack to restore his blood glucose level. Because Hank has juvenile-onset diabetes, he doesn't remember a time when he didn't have to perform these extra steps throughout his day.

Other advances in the care of diabetes, aside from those that allow a second-grader to monitor his blood sugar level, include the production of insulin. As we will see in this chapter, insulin is needed for the body to properly utilize glucose. Prior to the development of recombinant DNA technology, which now allows human insulin to be produced in large quantities, insulin was derived from the pancreases of pigs or cows. This required laborious purification, and because the animal insulins were not identical to the human form, sometimes immunological reactions occurred. Another recent advance is the insulin pump, a device a little bigger than a cell phone, which can deliver precise amounts of insulin via a small plastic catheter, more accurately mimicking the pancreas's natural release of the correct amount of insulin needed. Currently about 150,000 people in the United States are using an insulin pump. In the near future, it may be possible to implant a device into diabetes patients that will not only monitor the blood sugar level, but will also provide the appropriate doses of insulin. Until then, Hank and millions of other diabetics will keep checking and adjusting their blood glucose levels several times each day.

20.1 Endocrine Glands and Hormones

Learning Outcomes

1. List the major endocrine glands of the human body.
2. Compare the mechanism of action of peptide hormones and steroid hormones.
3. Differentiate between first messengers and second messengers.

The endocrine system consists of glands and tissues that secrete hormones. **Hormones** are chemicals that affect the behavior of other glands or tissues, which can be located far away from the sites of hormone production. The endocrine system is rivaled only by the nervous system with respect to the degree that both influence the body's other systems. Hormones influence the metabolism of cells, the growth and development of body parts, and homeostasis. Hormones are an essential component of the body's response to stress, and some (especially the reproductive hormones) act on the brain to influence behavior. Notice in Table 20.1 that hormones are involved in the function of various organs, including other endocrine organs, and that several hormones directly affect the blood glucose, calcium, and sodium levels. Hormones that promote cell division and mitosis are also sometimes called growth factors.

Not all hormones act between body parts. Local hormones are not carried in the bloodstream. Instead, they affect neighboring cells. As we shall see, prostaglandins, which mediate pain and inflammation, are good examples of local hormones.

Endocrine glands have no ducts. Instead, they usually secrete their hormones into tissue fluid. From there, hormones

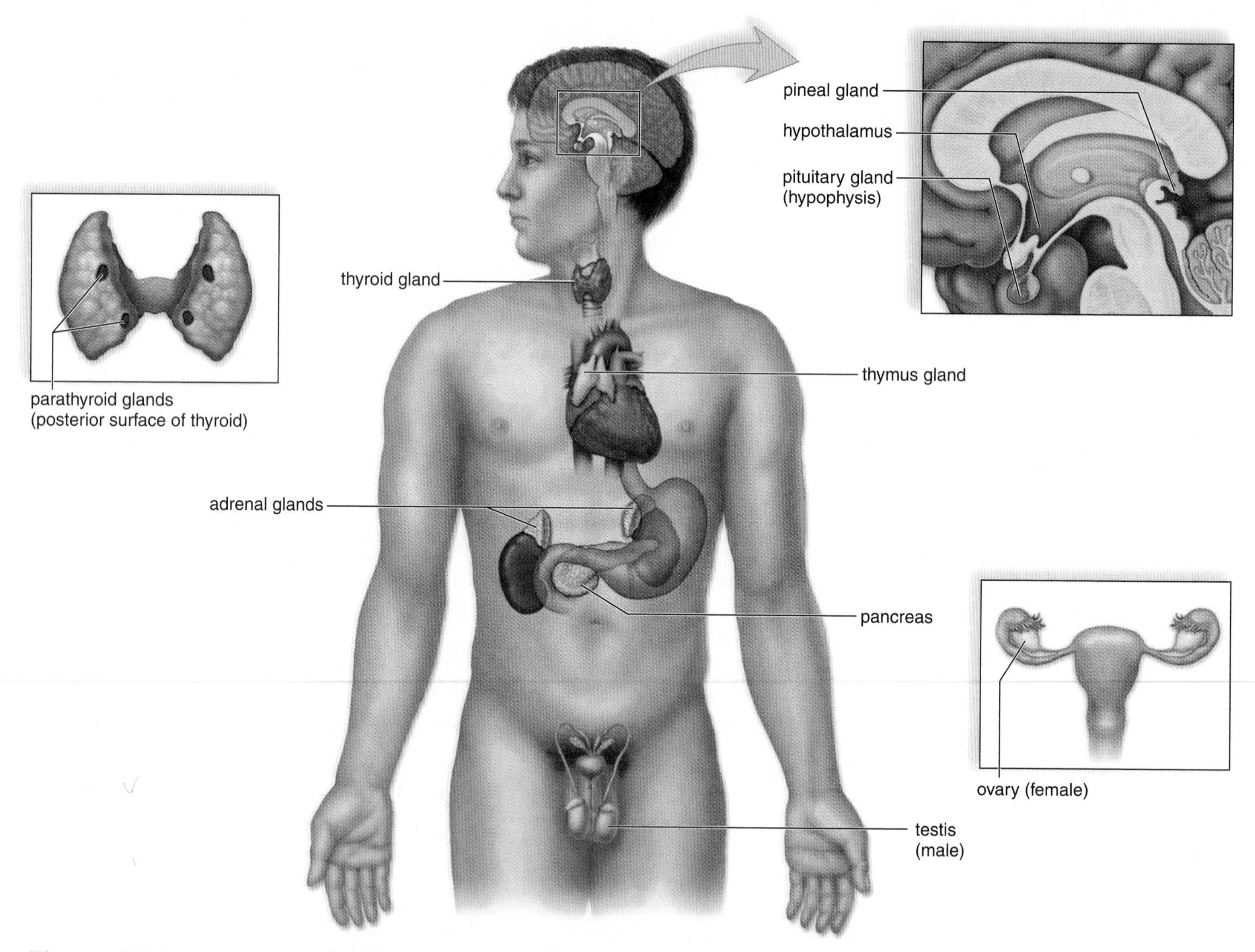

Figure 20.1 The endocrine system.
Anatomical location of major endocrine glands in the body. The hypothalamus and pituitary gland are in the brain, the thyroid and parathyroids are in the neck, and the adrenal glands and pancreas are in the abdominal cavity. The gonads include the ovaries in females, located in the pelvic cavity, and the testes in males, located outside this cavity in the scrotum. Also shown are the pineal gland, located in the brain, and the thymus gland, which lies ventral to the thoracic cavity.

TABLE 20.1 Principal Endocrine Glands and Hormones

Endocrine Gland	Hormone Released	Chemical Class	Target Tissues/Organs	Chief Function(s) of Hormone
Hypothalamus	Hypothalamic-releasing and -inhibiting hormones	Peptide	Anterior pituitary	Regulate anterior pituitary hormones
Pituitary gland				
Posterior pituitary	Antidiuretic (ADH)	Peptide	Kidneys	Stimulates water reabsorption by kidneys
	Oxytocin	Peptide	Uterus, mammary glands	Stimulates uterine muscle contraction, release of milk by mammary glands
Anterior pituitary	Thyroid-stimulating (TSH)	Glycoprotein	Thyroid	Stimulates thyroid
	Adrenocorticotropic (ACTH)	Peptide	Adrenal cortex	Stimulates adrenal cortex
	Gonadotropic (FSH, LH)	Glycoprotein	Gonads	Egg and sperm production; sex hormone production
	Prolactin (PRL)	Protein	Mammary glands	Milk production
	Growth (GH)	Protein	Soft tissues, bones	Cell division, protein synthesis, and bone growth
	Melanocyte-stimulating (MSH)	Peptide	Melanocytes in skin	Unknown function in humans; regulates skin color in lower vertebrates
Thyroid	Thyroxine (T_4) and triiodothyronine (T_3)	Iodinated amino acid	All tissues	Increases metabolic rate; regulates growth and development
	Calcitonin	Peptide	Bones, kidneys, intestine	Lowers blood calcium level
Parathyroids	Parathyroid (PTH)	Peptide	Bones, kidneys, intestine	Raises blood calcium level
Adrenal gland				
Adrenal cortex	Glucocorticoids (cortisol)	Steroid	All tissues	Raise blood glucose level; stimulate breakdown of protein
	Mineralocorticoids (aldosterone)	Steroid	Kidneys	Reabsorb sodium and excrete potassium
	Sex hormones	Steroid	Gonads, skin, muscles, bones	Stimulate reproductive organs and bring about sex characteristics
Adrenal medulla	Epinephrine and norepinephrine	Modified amino acid	Cardiac and other muscles	Released in emergency situations; raise blood glucose level
Pancreas	Insulin	Protein	Liver, muscles, adipose tissue	Lowers blood glucose level; promotes formation of glycogen
	Glucagon	Protein	Liver, muscles, adipose tissue	Raises blood glucose level
Gonads				
Testes	Androgens (testosterone)	Steroid	Gonads, skin, muscles, bones	Stimulate male sex characteristics
Ovaries	Estrogens and progesterone	Steroid	Gonads, skin, muscles, bones	Stimulate female sex characteristics
Thymus	Thymosins	Peptide	T lymphocytes	Stimulate production and maturation of T lymphocytes
Pineal gland	Melatonin	Modified amino acid	Brain	Controls circadian and circannual rhythms; possibly involved in maturation of sexual organs

diffuse into the bloodstream for distribution throughout the body. Figure 20.1 depicts the locations of the major endocrine glands in the body, and Table 20.1 lists the hormones they release. In contrast, exocrine glands secrete their products through ducts. For example, the salivary glands send saliva into the mouth by way of the salivary ducts.

Check Your Progress

1. What is the difference between a local hormone and a growth factor?
2. How could you distinguish an endocrine gland from an exocrine gland by its structure?

Hormones and Homeostasis

Like the nervous system, the endocrine system is intimately involved in homeostasis. However, while the nervous system is organized to respond rapidly to stimuli by transmitting nerve impulses, hormones secreted by the endocrine system must reach their target organs via the blood, resulting in a slower but often a more prolonged response. The blood concentration of a substance often prompts an endocrine gland to secrete its hormone. For example, the parathyroid glands secrete a hormone when the blood calcium level falls below normal. This hormone causes osteoclasts to slowly release calcium from bone. It takes time for the cells to respond, but the effect is longer lasting.

The production of hormones is usually controlled in two ways: by negative feedback, and by the action of other hormones. When controlled by **negative feedback,** an endocrine gland can be sensitive to either the condition it is regulating or the blood level of the hormone it is producing. For example, when the blood glucose level rises, the pancreas produces insulin. Insulin causes the liver to store glucose, and glucose is removed from the blood for liver storage. The stimulus for the production of insulin is, thereby, inhibited, and the pancreas stops producing insulin. Or, the anterior pituitary produces a thyroid-stimulating hormone that causes the thyroid to produce two hormones. When these hormones rise, the anterior pituitary stops producing thyroid-stimulating hormone.

The effect of a hormone also can be controlled by the release of an antagonistic hormone. The effect of insulin, for example, is offset by the production of glucagon by the pancreas. Insulin lowers the blood sugar level, while glucagon raises it. Similarly, a hormone produced by the thyroid gland lowers the blood calcium level, but another from the parathyroid gland increases blood calcium.

The Action of Hormones

Hormones have a wide range of effects on cells. Some of these effects induce a target cell to increase its uptake of particular substances (e.g., glucose) or ions (e.g., calcium). Other hormones bring about an alteration of the target cell's structure in some way.

Most hormones fall into two basic chemical classes: (1) **peptide hormones** are either peptides, proteins, glycoproteins, or modified amino acids; (2) **steroid hormones** always have the same complex of four carbon rings, but each specific hormone has different side chains. Because their effects are amplified through cellular mechanisms, both classes of hormones can function at extremely low concentrations. The chemical composition of a hormone also determines how it must be administered as a medication. Protein hormones, such as insulin, must be administered by injection. If these hormones were taken orally, they would be digested in the stomach and small intestine. Steroid hormones, such as those in birth control pills, can be taken orally.

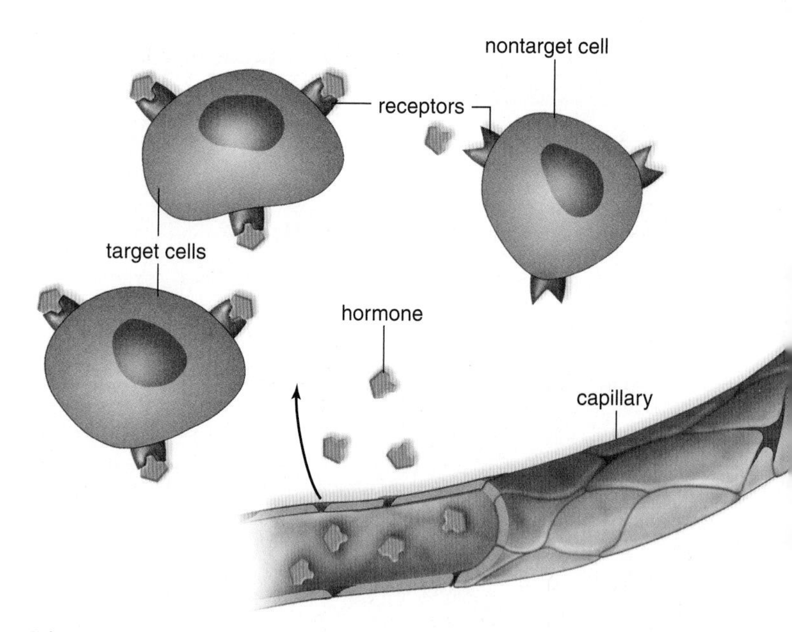

Figure 20.2 Target cell concept.
Most hormones are distributed by the bloodstream to target cells. Target cells have receptors for the hormone, and the hormone combines with the receptor as a key fits a lock.

Hormones are a type of chemical signal. **Chemical signals** are a means of communication between cells, between body parts, or even between individuals. They typically affect the metabolism of target cells that have receptors to receive them (Fig. 20.2). In the case of peptide hormones, these receptors are usually found on the cell surface (Fig. 20.3), while receptors for steroid hormones are usually in the nucleus but sometimes in the cytoplasm (Fig. 20.4). Epinephrine is a typical peptide hormone, which binds to a receptor on muscle cells. This leads to a conversion of ATP to cyclic AMP (adenosine monophosphate), or cAMP. In this chemical signaling process, the peptide hormone is called the **first messenger,** and cAMP is called the **second messenger.**

The second messenger typically sets in motion an enzymatic pathway sometimes termed an enzyme cascade, because each enzyme, in turn, activates another. Because enzymes work over and over, every step in an enzyme cascade leads to more reactions—the binding of even a single peptide hormone molecule can result in a thousandfold response. In muscle cells, the activation of these cellular signaling pathways by epinephrine leads to the breakdown of glycogen to glucose (see Fig. 20.3).

To better understand the terms "first messenger" and "second messenger," imagine that the adrenal medulla, which produces epinephrine, is the home office that sends out a courier (i.e., the first messenger, epinephrine) to a factory (the cell). The courier doesn't have clearance to enter

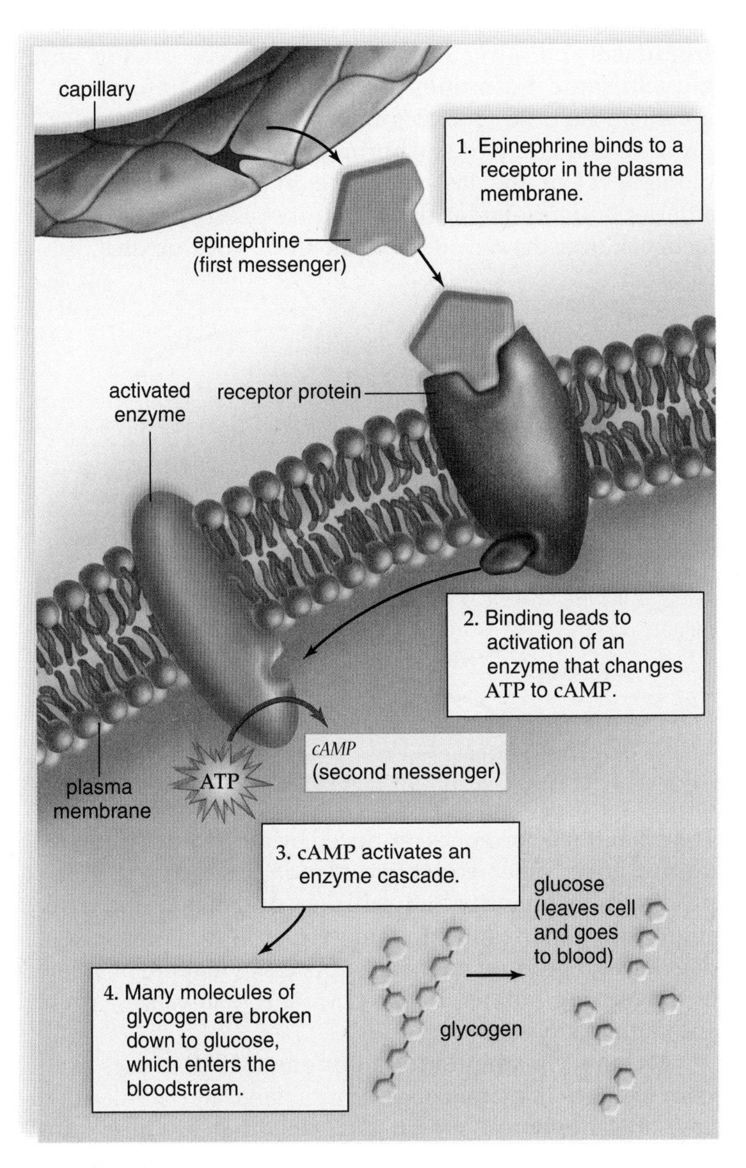

Figure 20.3 Action of epinephrine, a peptide hormone.
Epinephrine (first messenger) binds to a receptor in the plasma membrane of a muscle cell. Thereafter, a second messenger (cyclic AMP in this case) forms and activates an enzyme cascade.

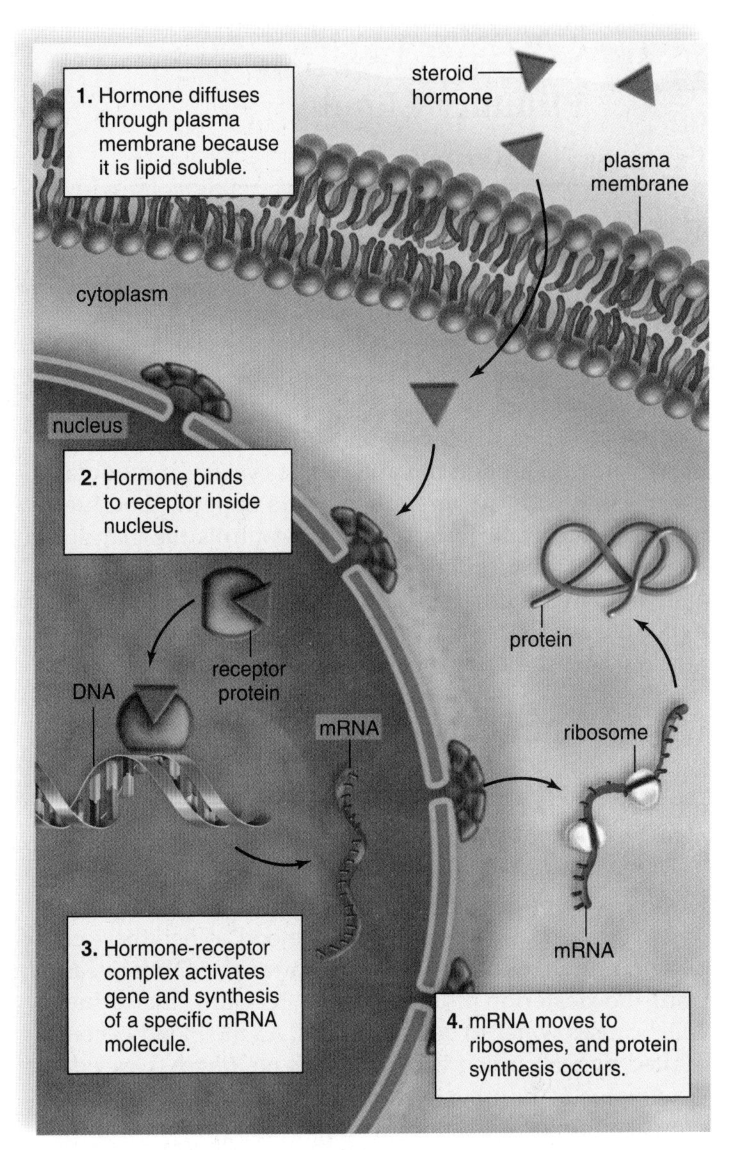

Figure 20.4 Steroid hormone action.
A steroid hormone passes directly through the target cell's plasma membrane before binding to a receptor in the nucleus or cytoplasm. The hormone-receptor complex binds to DNA, and gene expression follows.

the factory, so when he arrives there, he tells a supervisor through a screen door that the home office wants the factory to produce a particular product. The supervisor (i.e., cAMP, the second messenger) walks over and flips a switch (the enzymatic pathway), and a product is made.

Steroid hormones act in a somewhat different way. After diffusing through the plasma membrane, these hormones typically bind to receptors in the cytoplasm or nucleus (see Fig. 20.4). Either way, the hormone-receptor complex usually binds to DNA and activates transcription of certain genes. Translation of messenger RNA (mRNA) transcripts results in enzymes and other proteins that can carry out a response to the hormonal signal. Because it takes longer to synthesize new proteins than to activate enzymes already present in cells, steroid hormones generally act more slowly than peptide hormones. Their actions usually last longer, however.

Check Your Progress

1. Provide a specific example where secretion of a hormone is regulated by negative feedback.
2. Explain why steroid hormones usually take longer to have an effect compared to peptide hormones.
3. What is the specific role of cAMP in a cell's response to epinephrine?

20.2 Hypothalamus and Pituitary Gland

Learning Outcomes

1. Describe the relationship between the hypothalamus and the pituitary gland.
2. Name two hormones released by the posterior pituitary and six produced by the anterior pituitary.
3. Recognize which three pituitary hormones act on other endocrine glands.

The **hypothalamus** helps regulate the internal environment in two ways. Through the autonomic system, it modulates heartbeat, blood pressure, hunger and appetite, body temperature, and water balance. It also controls the glandular secretions of the **pituitary gland** (hypophysis). The pituitary, a small gland about 1 cm in diameter, is connected to the hypothalamus by a stalklike structure. The pituitary has two portions: the posterior pituitary and the anterior pituitary.

Posterior Pituitary

Neurons in the hypothalamus called neurosecretory cells produce the hormones **antidiuretic hormone (ADH)** and oxytocin (Fig. 20.5, *left*). These hormones pass through axons into the **posterior pituitary** where they are stored in axon terminals. Certain neurons in the hypothalamus are sensitive to the water-salt balance of the blood. When these cells determine that the blood is too concentrated, ADH is released from the posterior pituitary. Upon reaching the kidneys, ADH causes water to be reabsorbed. As the blood becomes diluted by reabsorbed water, ADH is no longer released.

Oxytocin, the other hormone made in the hypothalamus, causes uterine contraction during childbirth and milk letdown when a baby is nursing. The more the uterus contracts during labor, the more nerve impulses reach the hypothalamus, causing oxytocin to be released. Similarly, the more a baby suckles, the more oxytocin is released. In both instances, the release of oxytocin from the posterior pituitary is controlled by **positive feedback**—that is, the stimulus continues to bring about an effect that ever increases in intensity.

Anterior Pituitary

A portal system, consisting of two capillary networks connected by a vein, lies between the hypothalamus and the anterior pituitary (Fig. 20.5, *right*). The hypothalamus controls the anterior pituitary by producing **hypothalamic-releasing hormones** and, in some instances, **hypothalamic-inhibiting hormones.** For example, one hypothalamic-releasing hormone stimulates the anterior pituitary to secrete a thyroid-stimulating hormone, and a particular hypothalamic-inhibitory hormone prevents the anterior pituitary from secreting prolactin.

Three of the six hormones produced by the **anterior pituitary** have an effect on other glands: **thyroid-stimulating hormone (TSH)** stimulates the thyroid to produce the thyroid hormones; **adrenocorticotropic hormone (ACTH)** stimulates the adrenal cortex to produce cortisol; and **gonadotropic hormones** (FSH and LH) stimulate the gonads—the testes in males and the ovaries in females—to produce gametes and sex hormones. In each instance, the blood level of the last hormone in the hypothalamus-anterior pituitary-target gland control system exerts negative feedback over the secretions of the first two structures:

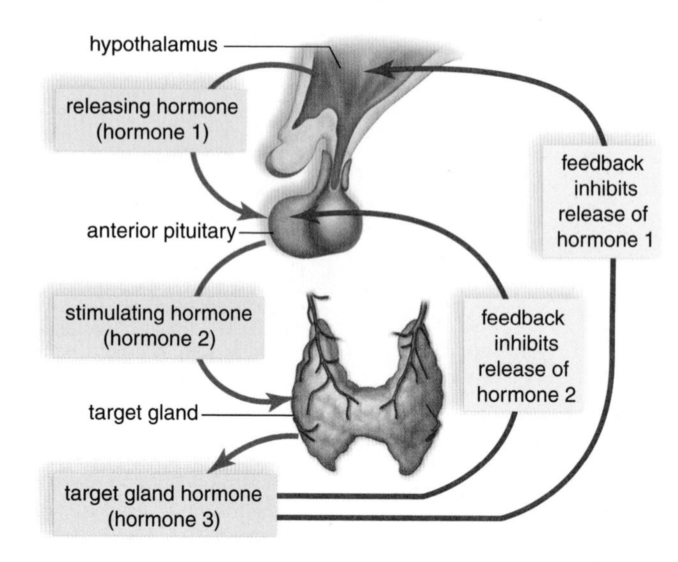

The other three hormones produced by the anterior pituitary do not affect other endocrine glands. **Prolactin (PRL)** is produced in quantity only during pregnancy and after childbirth. It causes the mammary glands in the breasts to develop and produce milk and suppress ovulation for a time in the nursing mother. It also plays a role in carbohydrate and fat metabolism.

Melanocyte-stimulating hormone (MSH) causes skin-color changes in many fishes, amphibians, and reptiles that have melanophores, special skin cells that produce color variations. The concentration of this hormone in humans is very low.

Growth hormone (GH), or somatotropic hormone, promotes skeletal and muscular growth. It increases the rate at which amino acids enter cells and causes increased protein synthesis. It also promotes fat metabolism as opposed to glucose metabolism.

The anterior pituitary normally produces higher levels of GH during childhood and adolescence, when most body growth is occurring. The amount of growth hormone produced during childhood is clearly one factor that influences height, although other factors, such as the genetic predisposition to attain a certain height, are also important (see section 20.8).

Check Your Progress

1. How do the posterior and anterior pituitary glands differ in the way that they are controlled by the hypothalamus?
2. What does it mean to say that oxytocin secretion is regulated by positive feedback?
3. Cancer of the pituitary gland can affect any of its normal functions. Why might these tumors be difficult to diagnose?

Visual Focus

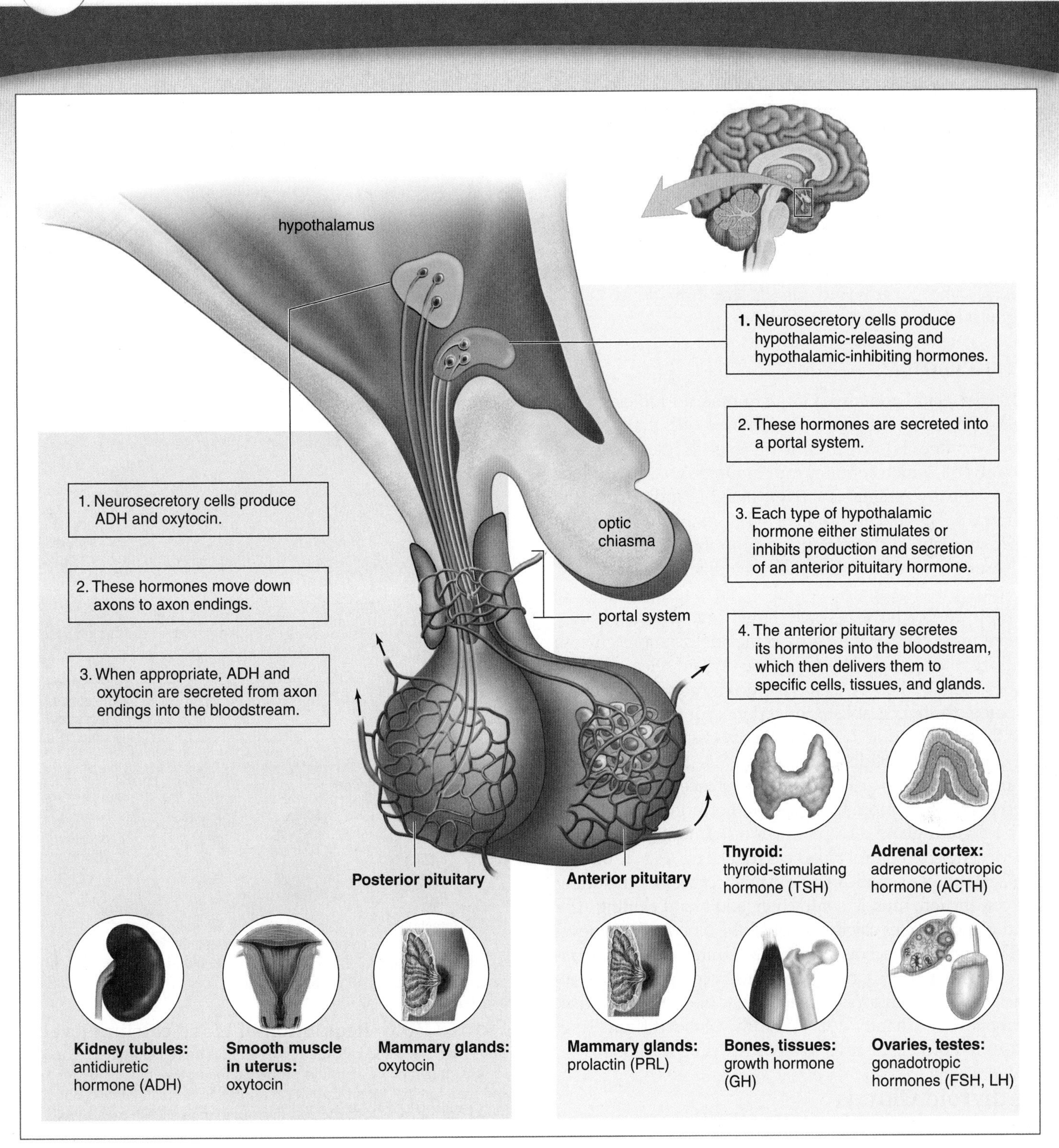

Figure 20.5 Hypothalamus and the pituitary.
Left: The hypothalamus produces two hormones, ADH and oxytocin, which are stored and secreted by the posterior pituitary. *Right:* The hypothalamus controls the secretions of the anterior pituitary, and the anterior pituitary controls the secretions of the thyroid gland, adrenal cortex, and gonads, which are also endocrine glands. These activities of the hypothalamus illustrate that the nervous system exerts control over the endocrine system and is intimately involved in maintaining the constancy of the internal environment.

20.3 Thyroid and Parathyroid Glands

Learning Outcomes

1. Identify the anatomical location of the thyroid and parathyroid glands.
2. Distinguish between the functions of T_3, T_4, calcitonin, and parathyroid hormone.

The **thyroid gland** is a large gland located in the neck, where it is attached to the trachea just below the larynx (see Fig. 20.1). The parathyroid glands are embedded in the posterior surface of the thyroid gland.

Thyroid Gland

The thyroid gland contains a large number of follicles, each a small spherical structure made of thyroid cells that produce triiodothyronine (T_3), which contains three iodine atoms, and **thyroxine (T_4),** which contains four iodine atoms. To produce T_3 and T_4, the thyroid gland actively acquires iodine from the bloodstream. The concentration of iodine in the thyroid gland is approximately 25 times that found in the blood. The primary source for iodine in our diets is iodized salt.

The thyroid hormones T_3 and T_4 increase the metabolic rate. They do not have a single target organ; instead, they stimulate most cells of the body to metabolize more glucose and utilize more energy. Although the thyroid releases about ten times more T_4 than T_3, T_3 has a much more potent activity. Fortunately, the liver is able to convert most of the T_4 produced by the thyroid gland to T_3. Interestingly, even though T_3 and T_4 are considered peptide hormones because they are synthesized from the amino acid tyrosine, their receptor is actually located inside cells, more like a steroid hormone receptor.

The thyroid gland also produces a hormone called **calcitonin,** which helps control blood calcium levels (Fig. 20.6). Calcium (Ca^{2+}) plays a significant role in many processes, including nerve conduction, muscle contraction, and blood clotting. The thyroid gland secretes calcitonin when the blood calcium level rises. The primary effect of calcitonin is to bring about the deposition of calcium in the bones. It does this by temporarily reducing the activity and number of osteoclasts, the cells that break down bone (see Chapter 19). When the blood calcium level decreases to normal, the thyroid stops releasing calcitonin.

Parathyroid Glands

Many years ago, the four parathyroid glands were sometimes mistakenly removed during thyroid surgery because they are so small. **Parathyroid hormone (PTH),** the hormone produced by the **parathyroid glands,** causes the blood phosphate (HPO_4^{2-}) level to decrease and the blood calcium level to increase.

PTH promotes the activity of osteoclasts and the release of calcium from the bones. PTH also promotes the reabsorption of calcium by the kidneys, where it helps activate vitamin D. Vitamin D, in turn, stimulates the absorption of calcium from the intestine. These effects bring the blood calcium level back to the normal range. As soon as blood calcium is back to normal, the parathyroid glands no longer secrete PTH.

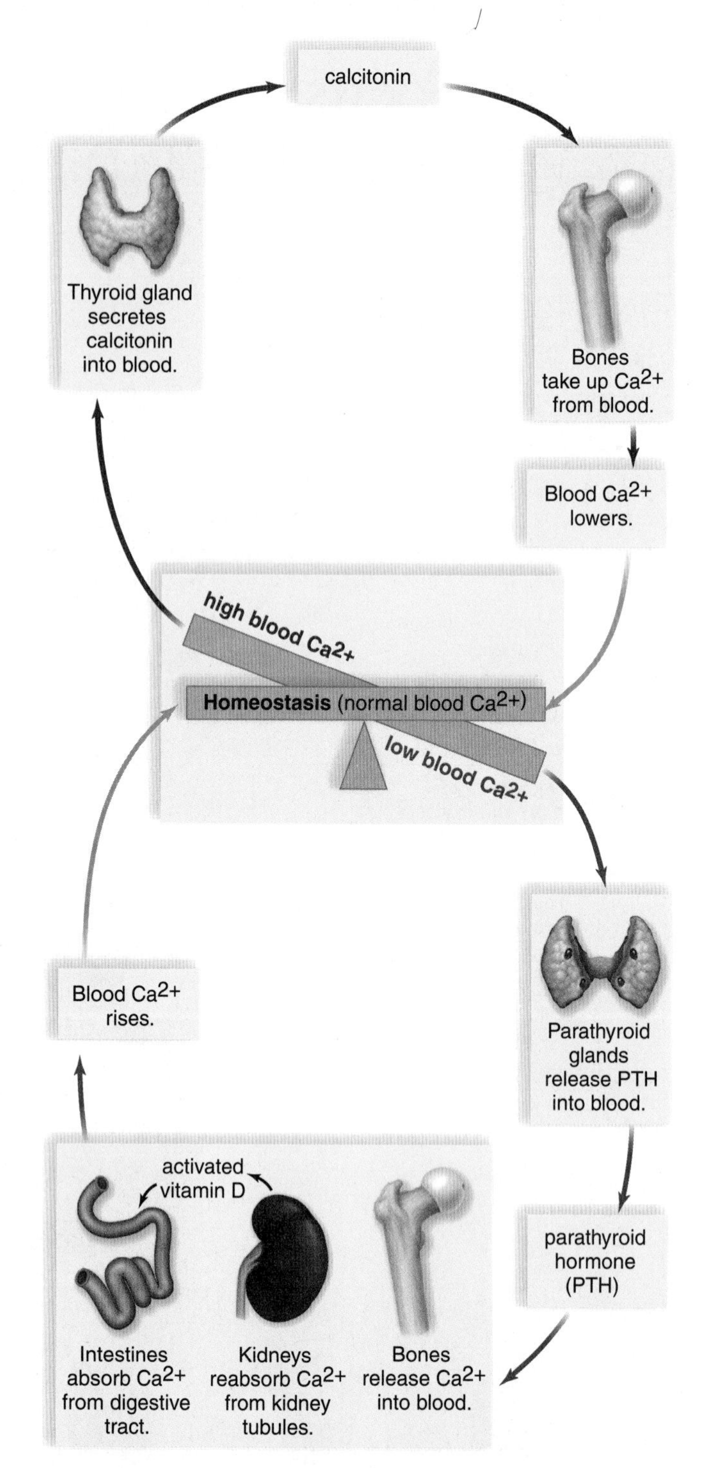

Figure 20.6 Regulation of blood calcium level.
Top: When the blood calcium (Ca^{2+}) level is high, the thyroid gland secretes calcitonin. Calcitonin promotes the uptake of Ca^{2+} by the bones, and therefore the blood Ca^{2+} level returns to normal. *Bottom:* When the blood Ca^{2+} level is low, the parathyroid glands release parathyroid hormone (PTH). PTH causes the bones to release Ca^{2+} and the kidneys to reabsorb Ca^{2+} and activate vitamin D. Thereafter, the intestines absorb Ca^{2+}. Therefore, the blood Ca^{2+} level returns to normal.

Check Your Progress

1. What controls the amount of T_3 and T_4 secreted by the thyroid gland? (Recall section 20.2.)
2. What symptoms might be expected in a person with a thyroid tumor that was producing excess calcitonin?
3. What would be the expected effect(s) of removal of the parathyroid glands?

20.4 Adrenal Glands

Learning Outcomes

1. Describe the location of the adrenal glands in the body.
2. List two hormones produced by the adrenal medulla and two produced by the adrenal cortex.
3. Review the role of the kidneys in regulating secretion of aldosterone.

The **adrenal glands** sit atop the kidneys (see Fig. 20.1). Each adrenal gland consists of an inner portion called the **adrenal medulla** and an outer portion called the **adrenal cortex.** These portions, like the anterior pituitary and the posterior pituitary, are two functionally distinct endocrine glands. The adrenal medulla is under nervous control, and a portion of the adrenal cortex is under the control of adrenocorticotropic hormone (ACTH), an anterior pituitary hormone. Stress of all types, including emotional and physical trauma, prompts the hypothalamus to stimulate a portion of the adrenal glands (Fig. 20.7).

Adrenal Medulla

The hypothalamus initiates nerve impulses that travel by way of the brain stem, spinal cord, and sympathetic nerve fibers to the adrenal medulla, which then secretes its hormones.

Epinephrine (adrenaline) and **norepinephrine** (NE) produced by the adrenal medulla rapidly bring about all the body changes that occur when an individual reacts to an emergency situation in a fight-or-flight manner (Fig. 20.7, *left*). The effects of these hormones provide a short-term response to stress.

Adrenal Cortex

The hormones produced by the adrenal cortex provide a longer-term response to stress (Fig. 20.7, *right*). The two major types of hormones produced by the adrenal cortex are the mineralocorticoids and the glucocorticoids. **Mineralocorticoids** regulate salt and water balance, leading to increases in blood volume and blood pressure. **Glucocorticoids** regulate carbohydrate, protein, and fat metabolism, leading to an increase in the blood glucose level. The adrenal cortex also secretes small amounts of male and female sex hormones in both sexes.

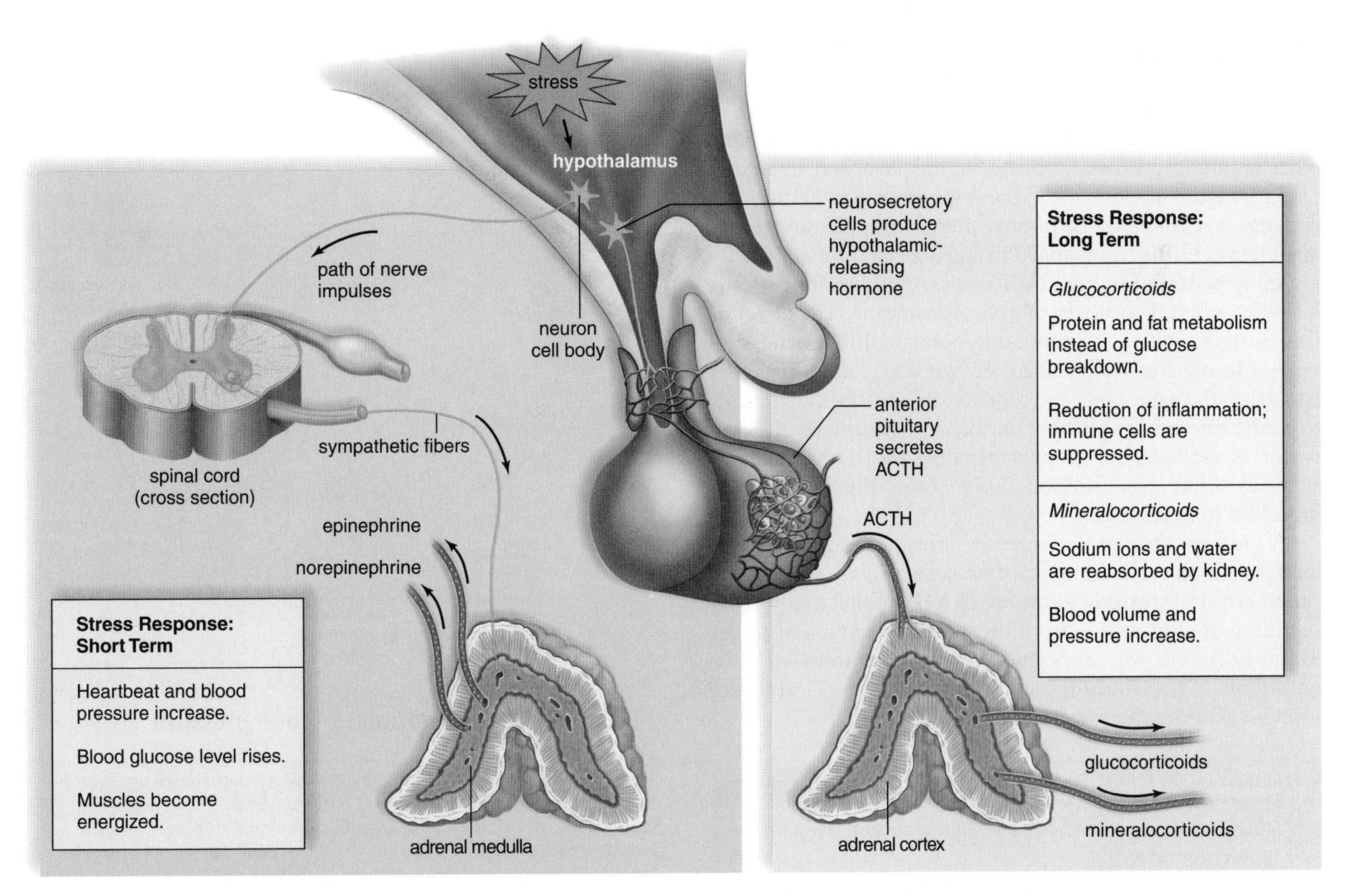

Figure 20.7 The stress response.
Both the adrenal cortex and the adrenal medulla are under the control of the hypothalamus when they help us respond to stress. *Left:* Nervous stimulation causes the adrenal medulla to provide a rapid, but short-term, stress response. *Right:* The adrenal cortex provides a slower, but long-term, stress response. ACTH causes the adrenal cortex to release glucocorticoids. Independently, the adrenal cortex releases mineralocorticoids.

Glucocorticoids

ACTH stimulates the portion of the adrenal cortex that secretes the glucocorticoids, of which **cortisol** is the most significant. Cortisol raises the blood glucose level in at least two ways: (1) it promotes the breakdown of muscle proteins to amino acids, which are taken up from the bloodstream and converted into glucose by the liver; (2) Cortisol promotes the metabolism of fatty acids, rather than carbohydrates, and this spares glucose. The rise in blood glucose is beneficial to a person under stress because glucose is the preferred energy source for neurons.

Cortisol also counteracts the inflammatory response that leads to the pain and swelling of joints in arthritis and bursitis. Cortisone, the medication often administered for inflammation of joints, is also a glucocorticoid. Very high levels of glucocorticoids in the blood can suppress the body's defense system, including the inflammatory response that occurs at infection sites. This can make a person more susceptible to injury and infection.

Mineralocorticoids

Aldosterone is the most important of the mineralocorticoids. Aldosterone primarily targets the kidney, where it promotes renal absorption of sodium (Na^+) and renal excretion of potassium (K^+), and thereby helps regulate blood volume and blood pressure (Fig. 20.8).

The secretion of mineralocorticoids is not controlled by the anterior pituitary. When the blood sodium level and, therefore, blood pressure are low, the kidneys secrete **renin.** Renin is an enzyme that converts the plasma protein angiotensinogen to angiotensin I, which is changed to angiotensin II by a converting enzyme found in lung capillaries. Angiotensin II stimulates the adrenal cortex to release aldosterone. The effect of this system, called the renin-angiotensin-aldosterone system, is to raise blood pressure in two ways: angiotensin II constricts the arterioles, and aldosterone causes the kidneys to reabsorb sodium. When the blood sodium level rises, water is reabsorbed, in part because the hypothalamus secretes ADH (see section 20.2). Then blood pressure increases to normal.

When the atria of the heart are stretched due to a great increase in blood volume, cardiac cells release a hormone called **atrial natriuretic hormone (ANH),** which inhibits the secretion of aldosterone from the adrenal cortex. The effect of this hormone is to cause the excretion of sodium—that is, *natriuresis.* When sodium is excreted, so is water, and therefore, blood pressure lowers to normal.

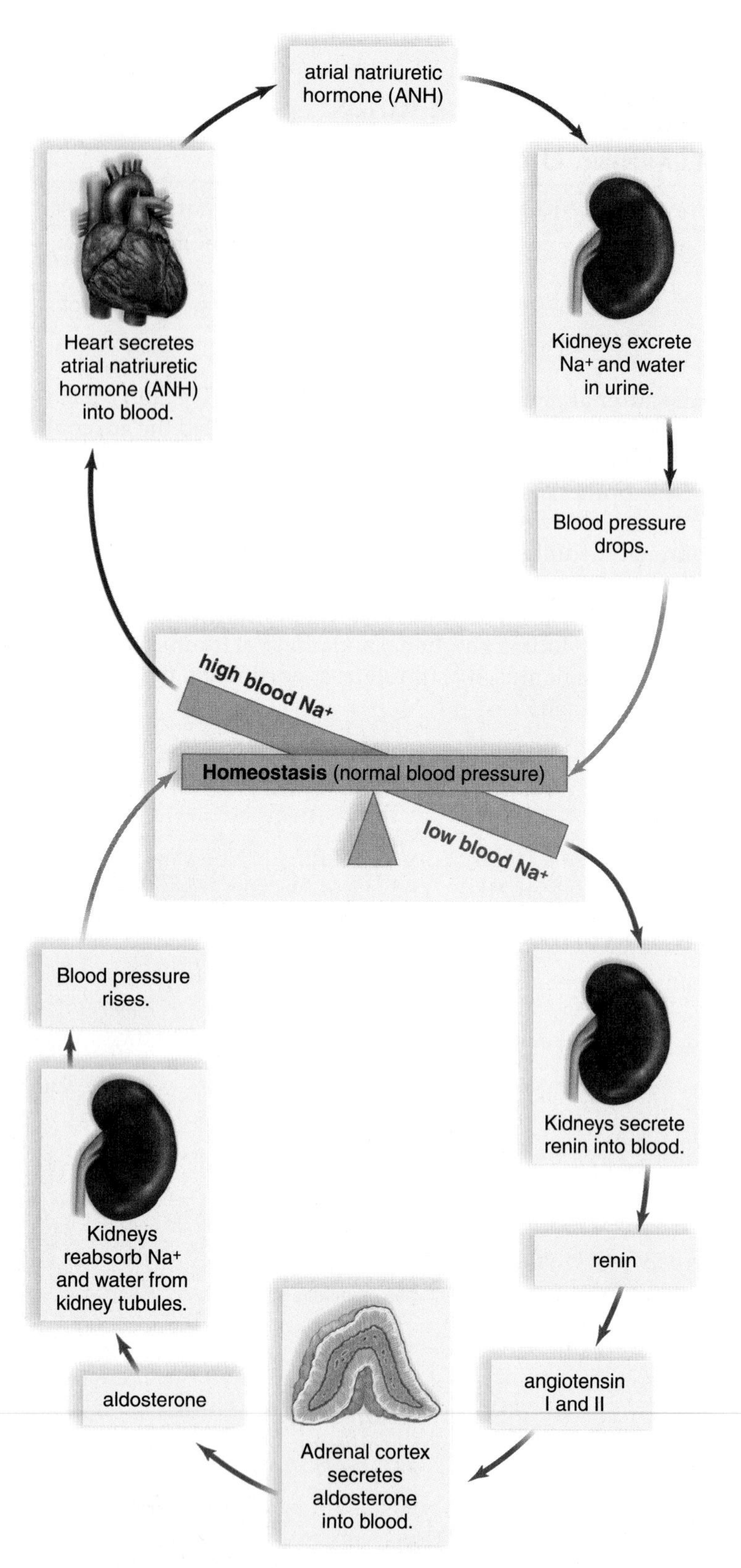

Figure 20.8 Regulation of blood pressure and volume.

Top: When the blood Na^+ is high, high blood volume causes the heart to secrete atrial natriuretic hormone (ANH). ANH causes the kidneys to excrete Na^+, and water follows. The blood volume and pressure then return to normal. *Bottom:* When the blood sodium (Na^+) level is low, low blood pressure causes the kidneys to secrete renin. Renin leads to the secretion of aldosterone from the adrenal cortex. Aldosterone causes the kidneys to reabsorb Na^+, and water follows, so that blood volume and pressure return to normal.

Check Your Progress

1. How is the secretion of hormones by the adrenal medulla and cortex controlled?
2. How do the hormones produced by the adrenal glands help the body respond to stress?
3. What is the specific effect of aldosterone on the kidney?

20.5 Pancreas

Learning Outcomes

1. Explain how the pancreas has both endocrine and exocrine functions.
2. Contrast the functions of insulin, glucagon, and somatostatin.

The **pancreas** is a long organ that lies in the abdomen between the kidneys and near the duodenum of the small intestine (see Fig. 20.1). It is composed of two types of tissue. Exocrine tissue produces and secretes digestive juices by way of ducts to the small intestine. The **pancreatic islets** (islets of Langerhans) are clusters of at least three types of endocrine cells: 1) alpha cells that produce **glucagon,** 2) beta cells that produce **insulin,** and 3) delta cells that produce **somatostatin.** Insulin and glucagon are important in regulating the blood glucose level (Fig. 20.9).

Insulin is secreted when the blood glucose level is high, which usually occurs just after eating. Insulin stimulates the uptake of glucose by cells, especially liver cells, muscle cells, and adipose tissue cells. In liver and muscle cells, glucose is then stored as glycogen. In muscle cells, the breakdown of glucose supplies energy for protein metabolism, and in fat cells, the breakdown of glucose supplies glycerol for the formation of fat. In these various ways, insulin lowers the blood glucose level.

Glucagon is secreted from the pancreas, usually before eating, when the blood glucose level is low. The major target tissues of glucagon are the liver and adipose tissue. Glucagon stimulates the liver to break down glycogen to glucose. Fat and protein are also used as energy sources by the liver, thus sparing glucose. Adipose tissue cells break down fat to glycerol and fatty acids. The liver takes these up and uses them as substrates for glucose formation. In these various ways, glucagon raises the blood glucose level.

Somatostatin is also known as growth hormone inhibiting hormone. Besides the pancreas, somatostatin is also produced by cells in the stomach and small intestine. Its main effects are to inhibit the release of growth hormone by the anterior pituitary, as well as to suppress the release of various hormones produced by the digestive system, including insulin and glucagon. Overall, somatostatin seems to generally decrease the absorption of nutrients from the digestive tract.

Check Your Progress

1. How is glucose normally taken up by cells? (See section 4.2.) How could insulin specifically affect this process?
2. What would determine whether a particular type of cell would be affected by insulin (a peptide hormone)?
3. What type of feedback mechanism does somatostatin exert on other hormones, such as insulin and glucagon?

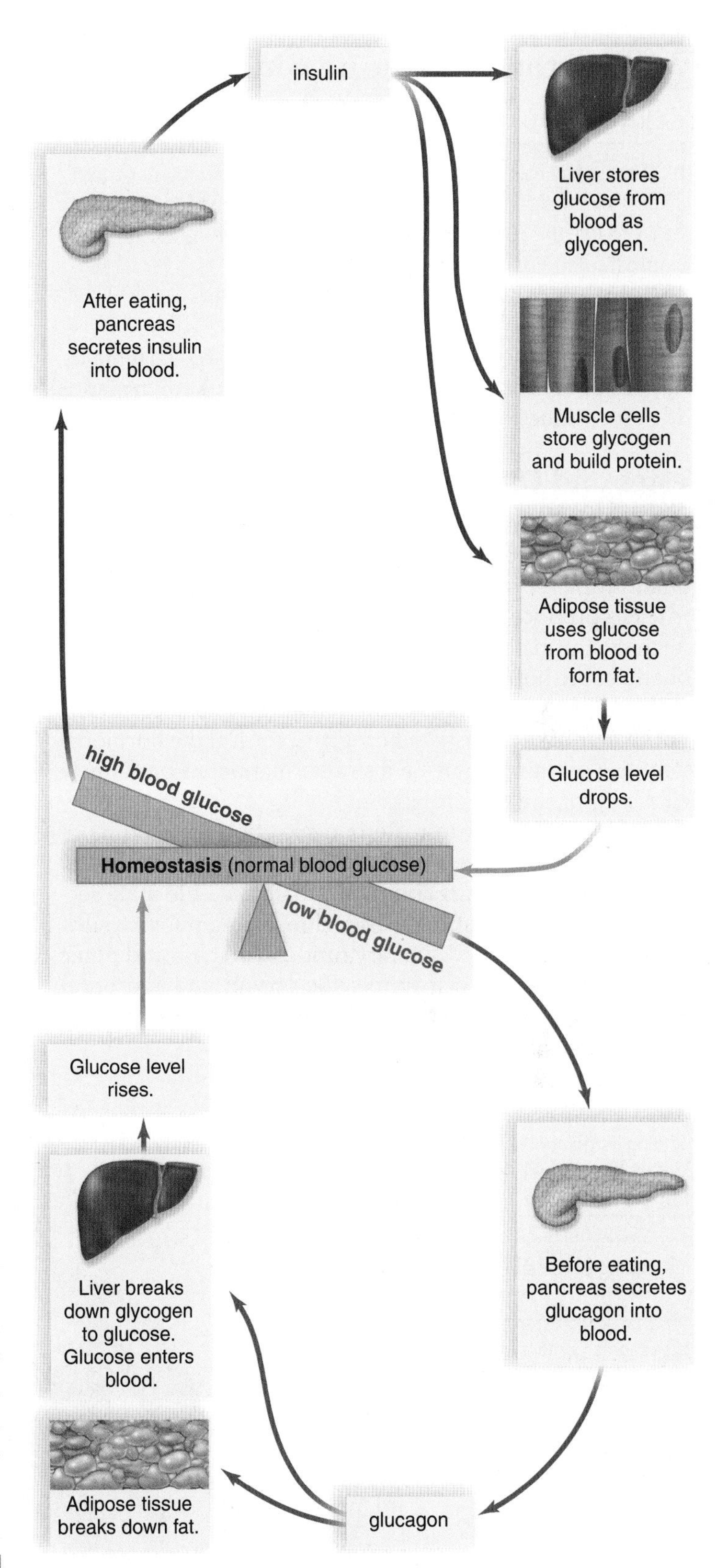

Figure 20.9 Regulation of blood glucose level. *Top:* When the blood glucose level is high, the pancreas secretes insulin. Insulin promotes the storage of glucose as glycogen and the synthesis of proteins and fats (as opposed to their use as energy sources). Therefore, insulin lowers the blood glucose level to normal. *Bottom:* When the blood glucose level is low, the pancreas secretes glucagon. Glucagon acts opposite to insulin. Therefore, glucagon raises the blood glucose level to normal.

20.6 Other Endocrine Glands

Learning Outcomes

1. Discuss the role of the testes and ovaries in the development of male and female secondary sex characteristics.
2. Describe the functions of thymosin, melatonin, leptin, and prostaglandins.
3. Define and list two examples of growth factors.

The **gonads** are the testes in males and the ovaries in females. The gonads are endocrine glands. Other lesser known glands and some tissues also produce hormones.

Testes and Ovaries

The **testes** are located in the scrotum, and the **ovaries** are located in the pelvic cavity. Under the influence of the gonadotropic hormones secreted by the anterior pituitary, the testes produce sperm and **androgens** (e.g., **testosterone**), which are the male sex hormones. Under the influence of the gonadotropic hormones, the ovaries produce eggs and also **estrogen** and **progesterone,** the female sex hormones. The hypothalamus and the pituitary gland control the hormonal secretions of these organs in a similar manner as previously described for the thyroid gland.

Greatly increased testosterone secretion at the time of puberty stimulates the growth of the penis and the testes. Testosterone also brings about and maintains the male secondary sex characteristics that develop during puberty, such as the growth of a beard, axillary (underarm) hair, and pubic hair. Testosterone also prompts the larynx and the vocal cords to enlarge, causing the voice to change. It is partially responsible for the muscular strength of males. This is the reason some athletes take supplemental amounts of illegal **anabolic steroids,** which are either testosterone or related chemicals. The side effects of taking anabolic steroids are listed in Figure 20.10. Testosterone also stimulates oil and sweat glands in the skin. Therefore, it is largely responsible for acne and body odor. Another side effect of testosterone is baldness. Genes for baldness are probably inherited by both sexes, but baldness occurs more often in males because of the presence of testosterone.

The female sex hormones, estrogen and progesterone, have many effects on the body. In particular, estrogen secreted at the time of puberty stimulates the growth of the uterus and the vagina. Estrogen is necessary for egg maturation and is largely responsible for the secondary sex characteristics in females, including female body hair and fat distribution. In general, females have a more rounded appearance than males because of a greater accumulation of fat beneath the skin. Also, the pelvic girdle is wider in females than in males, resulting in a larger pelvic cavity. Both estrogen and progesterone are required for breast development and regulation of the uterine cycle, which includes monthly menstruation (see Chapter 21).

Thymus Gland

The **thymus gland** is a lobular gland that lies just beneath the sternum (see Fig. 20.1). This organ reaches its largest size and is most active during childhood. With aging, the organ gets smaller and becomes fatty. Lymphocytes that originate in the bone marrow and then pass through the thymus differentiate into T lymphocytes. The lobules of the thymus are lined by epithelial cells that secrete hormones called

Figure 20.10 The side effects of anabolic steroid use.

Health Focus

Melatonin

The hormone melatonin is now sold as a nutritional supplement. The popular press promotes its use in pill form for sleep, aging, cancer treatment, sexuality, and more. At best, melatonin may have some benefits in certain sleep disorders. But most physicians do not yet recommend it for that use because so little is known about its dosage requirements and possible side effects.

Melatonin is produced by the pineal gland in greatest quantity at night and smallest quantity during the day. Notice in Figure 20A that melatonin's production cycle accompanies our natural sleep-wake cycle. Rhythms with a period of about 24 hours are called circadian ("around the day") rhythms. All circadian rhythms seem to be controlled by an internal biological clock because they are free-running—that is, they have a regular cycle even in the absence of environmental cues. In scientific experiments, humans have lived in underground bunkers where they never see the light of day. In a few people, the sleep-wake cycle drifts badly, but in most, the daily activity schedule is just about 25 hours.

An individual's internal biological clock is reset each day by the environmental day-night cycle. Characteristically, biological clocks that control circadian rhythms are reset by environmental cues, or else they drift out of phase with the environmental day-night cycle.

Recent research suggests that our biological clock lies in a cluster of neurons within the hypothalamus called the suprachiasmatic nucleus (SCN). The SCN undergoes spontaneous cyclical changes in activity, and therefore, it can act as a pacemaker for circadian rhythms. Neural connections between the retina and the SCN indicate that reception of light by the eyes most likely resets the SCN and keeps our biological rhythms on a 24-hour cycle. Some people suffer from seasonal affective disorder, or SAD. As the days get darker and darker during the fall and winter, they become depressed, sometimes severely. They find it difficult to keep up because their biological clock has fallen behind without early morning light to reset it. If so, a half-hour dose of simulated daylight from a portable light box first thing in the morning makes them feel operational again. The SCN also controls the secretion of melatonin by the pineal gland, and, in turn, melatonin may quiet the operation of the neurons in the SCN.

Research is still going forward to see if melatonin will be effective for circadian rhythm disorders such as SAD, jet lag, sleep phase problems, recurrent insomnia in the totally blind, and some other less common disorders. So-called jet lag occurs when you travel across several time zones and your biological clock is out of phase with local time. Jet-lag symptoms gradually disappear as your biological clock adjusts to the environmental signals of the local time zone.

Many young people have a sleep phase problem because their circadian cycle lasts 25 to 26 hours. As they lengthen their day, they get out of sync with normal times for sleep and activity.

Clinical trials revealed that melatonin could shift circadian rhythms and reset our biological clock. Melatonin given in the afternoon shifts rhythms earlier, while melatonin given in the morning shifts rhythms later. For most people, the process was gradual: the average rate of change was about an hour a day. Before you try melatonin, however, you might want to consider that melatonin is known to affect reproductive behavior in other mammals. It's a matter of deciding whether any potential side effect from melatonin use is worth the possible benefits.

Discussion Questions

1. Have you ever experienced any degree of seasonal affective disorder? If so, do you think it was at least partly psychological, or did you feel it was a "purely" chemical disorder over which you had no control? Is it possible that both explanations could be true?
2. Are you more of a "morning person" or a "night person"? Can you think of any way that variations in melatonin secretion could account for how different people feel about, for example, getting up early in the morning?
3. Currently melatonin is sold without a prescription in health food and drug stores in the United States. Do you agree with allowing this hormone to be so readily accessible to consumers? Why or why not?

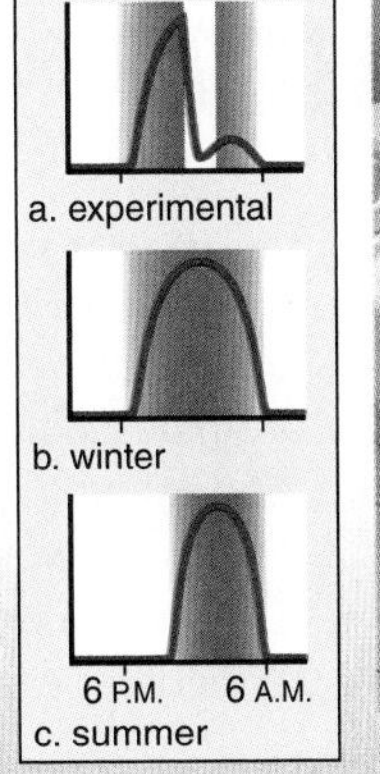

Figure 20A Melatonin production. Melatonin production is greatest at night when we are sleeping. Light suppresses melatonin production **(a)**, so its duration is longer in the winter **(b)** than in the summer **(c)**.

thymosins. These hormones aid in the differentiation of lymphocytes packed inside the lobules. There is some evidence that these hormones would enhance T-lymphocyte function if injected into AIDS or cancer patients.

Pineal Gland

The **pineal gland,** which is located in the brain (see Fig. 20.1), produces the hormone **melatonin,** primarily at night. Melatonin is involved in our daily sleep-wake cycle. Normally, we grow sleepy at night when melatonin levels increase and awaken once daylight returns and melatonin levels are low. Daily 24-hour cycles such as this are called **circadian rhythms,** and they are controlled by a biological clock located in the hypothalamus, as discussed in the Health Focus.

Check Your Progress

1. What are some of the most undesirable side effects of taking anabolic steroids?
2. What would be the main effect of removing someone's thymus gland?
3. Where in the brain is the pineal gland specifically located?

Hormones from Other Tissues

Some organs that are usually not considered endocrine glands can secrete hormones. We have already mentioned in this chapter that the heart produces atrial natriuretic hormone. Figure 14.7 showed that the stomach and small intestine produce peptide hormones that regulate digestive secretions. Other tissues that are usually not considered endocrine glands also secrete hormones.

Leptin

Leptin is a protein hormone produced by adipose tissue. Leptin acts on the hypothalamus, where it signals satiety—that the individual has had enough to eat. Paradoxically, the blood of obese individuals may be rich in leptin. It is possible that the leptin they produce is ineffective because of a genetic mutation, or else their hypothalamic cells lack a suitable number of receptors for leptin.

Growth Factors

A number of different types of organs and cells produce peptide **growth factors,** which stimulate cell division and mitosis. Like hormones, they act on cell types with specific receptors to receive them. Some, such as lymphokines, are released into the blood. Others diffuse to nearby cells. Growth factors of particular interest include:

Granulocyte-macrophage colony-stimulating factor is secreted by many different tissues. It causes bone marrow stem cells to form various types of white blood cells.

Platelet-derived growth factor is released from platelets and from many other cell types. It helps in wound healing and causes an increase in the number of fibroblasts, smooth muscle cells, and certain cells of the nervous system.

Epidermal growth factor and nerve growth factor stimulate the cells indicated by their names, as well as many others. These growth factors are also important in wound healing.

Prostaglandins

Prostaglandins are potent chemical signals produced within cells from arachidonate, a fatty acid. Prostaglandins are not distributed in the blood. Instead, they act locally, quite close to where they were produced. In the uterus, prostaglandins cause muscles to contract. Therefore, they are implicated in the pain and discomfort of menstruation in some women. Also, prostaglandins mediate the effects of pyrogens, chemicals that are believed to reset the temperature regulatory center in the brain. Aspirin reduces body temperature and controls pain because it inhibits the synthesis of prostaglandins.

CHECK YOUR PROGRESS

1. What is the normal function of leptin?
2. Recent studies have shown that injecting platelets directly into areas of injured muscles or tendons may speed healing—how might this work?
3. Why are prostaglandins considered to be local hormones?

20.7 Pheromones

LEARNING OUTCOMES

1. Define pheromone.
2. Provide a specific example that suggests human behavior may be influenced by pheromones.

Chemical signals that act between individuals of the same species are called **pheromones.** The effects of pheromones are better documented in other animals than in humans. For example, female moths release a sex attractant that is received by male moth antennae even several miles away. Humans do produce a rich supply of airborne chemicals from a variety of areas, including the scalp, oral cavity, armpit (axilla), genital areas, and feet. Studies suggest that women prefer the axillary odors of men who are a different MHC type from themselves. (MHC molecules are plasma membrane proteins involved in immunity.) Choosing a mate of a different MHC type could conceivably improve the immune response of offspring. Several studies indicate that axillary secretions can affect the menstrual cycle. Women who live in the same household often have menstrual cycles in synchrony. Researchers have found that a woman's axillary extract can alter another woman's cycle by a few days.

CHECK YOUR PROGRESS

1. The existence of human pheromones is still controversial. What type of study could you design to further investigate whether pheromones can affect human behavior?

20.8 Disorders of the Endocrine System

LEARNING OUTCOMES

1. Identify the cause of each of the following conditions: diabetes insipidus, pituitary dwarfism, acromegaly, Cushing syndrome, and Addison disease.
2. Describe two potential causes of hyperthyroidism.
3. Distinguish between the specific cause of type 1 versus type 2 diabetes mellitus.

The endocrine glands play a major role in regulating the development and function of many body systems. Therefore, an increase or decrease in the production of most hormones can cause significant disease. Increased production of a hormone is often due to cancer affecting the endocrine gland, while decreased production can be due to various conditions that result in destruction of the gland.

Disorders of the Pituitary Gland

Because it controls the secretions of several other endocrine glands, the pituitary has been called the "master gland" of the endocrine system. As such, disorders of the pituitary gland can have dramatic effects on the body. As discussed in

Figure 20.11 Effect of growth hormone on height.
Too much growth hormone can lead to giantism, while an insufficient amount results in limited stature and even pituitary dwarfism.

section 20.2, the posterior pituitary produces antidiuretic hormone (ADH), which reduces the amount of urine formed by increasing water reabsorption by the collecting ducts of the kidney (see section 16.3). If too little ADH is secreted, a condition known as **diabetes insipidus (DI)** results. Patients with DI are usually very thirsty, produce large volumes of urine, and can become severely dehydrated if the condition is untreated. Another type of DI occurs when the kidneys are unable to respond to ADH produced by the pituitary. Consumption of alcohol also inhibits ADH production, such that the symptoms of a hangover are, in part, caused by dehydration.

The anterior pituitary produces several important hormones, including growth hormone (GH) and adrenocorticotropic hormone (ACTH). If too little GH is produced during childhood, the individual will have **pituitary dwarfism,** characterized by normal proportions but small stature. Conversely, if too much GH is secreted during childhood, a person may become a giant (Fig. 20.11). Giants usually have poor health, primarily because GH has a secondary effect on the blood glucose-level, promoting diabetes mellitus (discussed later in this section).

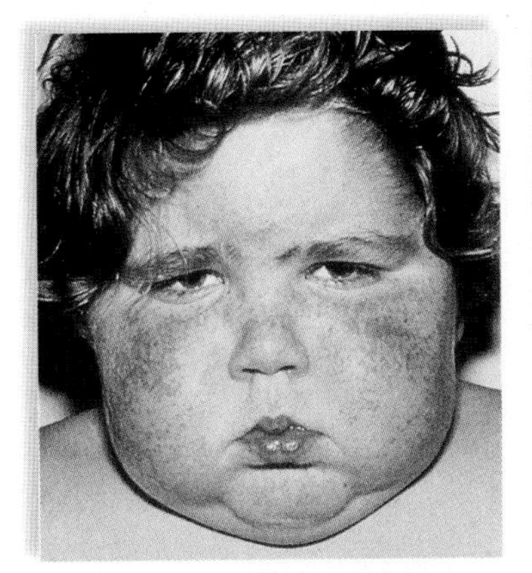

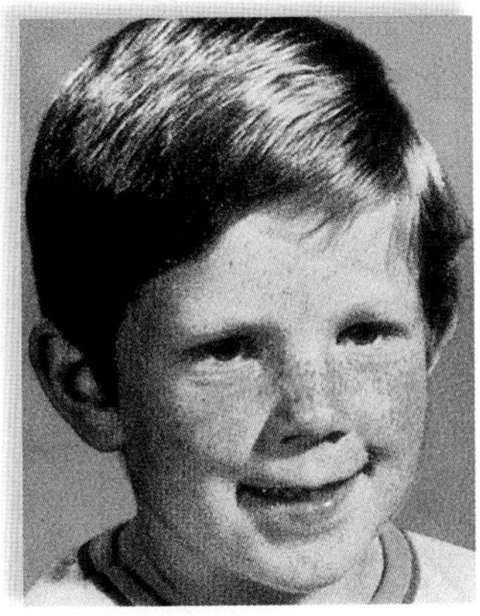

Figure 20.13 Cushing syndrome.
Cushing syndrome results from hypersecretion of hormones due to an adrenal cortex tumor. *Left:* Patient first diagnosed with Cushing syndrome. *Right:* Four months later, after therapy.

On occasion, GH is overproduced in the adult, and a condition known as **acromegaly** results. Because long bone growth is no longer possible in adults, only the feet, hands, and face (particularly the chin, nose, and eyebrow ridges) can respond, and these portions of the body become overly large (Fig. 20.12)

Some pituitary tumors produce large amounts of ACTH, which is a common cause of **Cushing syndrome.** The excess ACTH stimulates the adrenal cortex to overproduce cortisol. The increased amount of cortisol causes muscle protein to be metabolized and subcutaneous fat to be deposited in the midsection. The result is a swollen "moon" face and an obese trunk (Fig. 20.13). Depending on the cause and duration of Cushing syndrome, some people may experience more dramatic changes, including masculinization, high blood pressure, and weight gain. Cushing syndrome can also be caused by tumors of the adrenal gland or by taking high or prolonged doses of cortisone or similar drugs.

Age 9

Age 16

Age 33

Age 52

Figure 20.12 Acromegaly.
Acromegaly is caused by overproduction of GH in the adult. It is characterized by enlargement of the bones in the face, the hands, and the feet as a person ages.

a. Cretinism

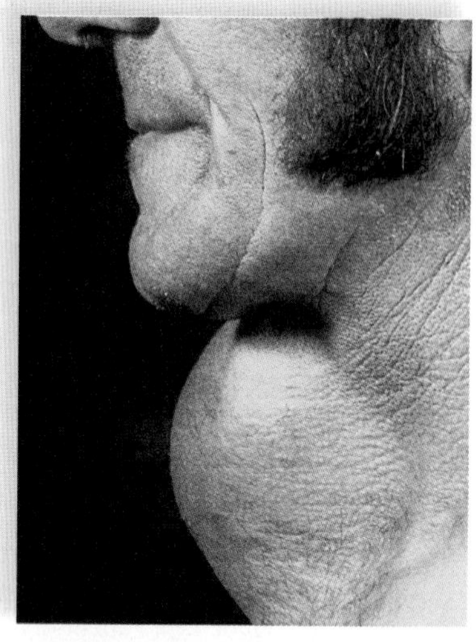
b. Simple goiter

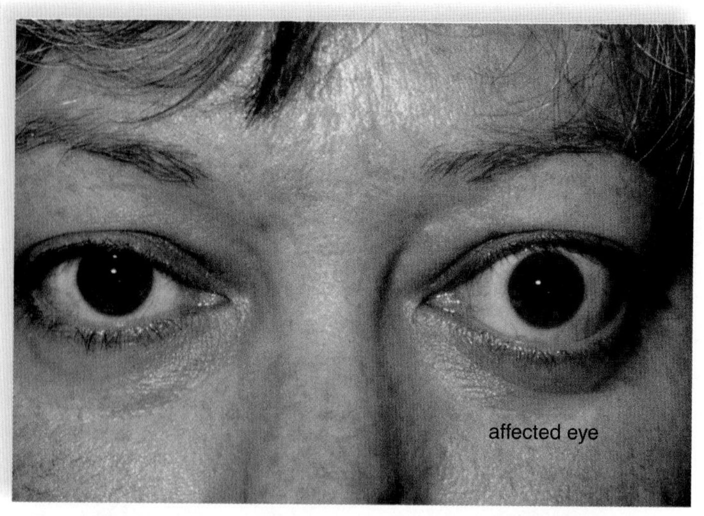

c. Exophthalmia

Figure 20.14 Abnormalities of the thyroid.
a. Individuals with cretinism have hypothyroidism. From infancy or childhood on, they do not grow and develop as others do. **b.** An enlarged thyroid gland is often caused by a lack of iodine in the diet. Without iodine, the thyroid is unable to produce its hormones, and continued anterior pituitary stimulation causes the gland to enlarge and form a simple goiter. **c.** In hyperthyroidism, accumulation of fluid behind the eyes can cause the eyes to protrude (exophthalmia). Only this individual's left eye was affected.

Disorders of the Thyroid, Parathyroid, and Adrenal Glands

If the thyroid gland does not produce enough thyroid hormones for any reason, **hypothyroidism** occurs. Failure of the thyroid to develop properly in childhood results in a condition called **cretinism** (Fig. 20.14*a*). Individuals with this condition are short and stocky. Thyroid hormone therapy can initiate growth, but unless treatment is begun within the first two months of life, mental retardation results. Hypothyroidism can also occur in adults, most often when the immune system produces antibodies that destroy the gland, a condition known as Hashimoto thyroiditis. Untreated hypothyroidism may produce a group of clinical symptoms called **myxedema,** which is characterized by lethargy, weight gain, hair loss, constipation, slow heart rate, and thickened or puffy skin. The administration of adequate doses of thyroid hormones restores normal body functions and appearance.

If iodine is lacking in the diet, the thyroid gland is unable to produce sufficient amounts of T_3 and T_4. In response to constant stimulation by TSH released by the anterior pituitary, the thyroid gland enlarges, resulting in a **simple goiter** (Fig. 20.14*b*). The incidence of simple goiters was much higher in the United States prior to the addition of iodine to most salt.

Hyperthyroidism results from the oversecretion of thyroid hormones. In Graves disease, antibodies are produced that react with the TSH receptor on thyroid follicular cells, mimicking the effect of TSH and causing the production of too much T_3 and T_4. One typical sign of Graves disease is **exophthalmia,** or excessive protrusion of the eyes (Fig. 20.14*c*). The eyes protrude because of the edema in eye socket tissues and swelling of the muscles that move the eyes. The patient with Graves disease usually becomes hyperactive, nervous, and irritable, and may suffer from insomnia, diarrhea, and abnormal heart rhythms. Drugs that inhibit the synthesis or release of thyroid hormones, or their effects on body tissues, are used to manage Graves disease. Hyperthyroidism can also be caused by a thyroid tumor, which is usually detected as a lump during physical examination. Removal or destruction of a portion of the thyroid gland, either by surgery or by administration of radioactive iodine, is often curative.

When insufficient parathyroid hormone production leads to a dramatic drop in the blood calcium level, tetany results. In **tetany,** the body shakes from continuous muscle contraction, brought about by increased excitability of the nerves.

As mentioned previously, an increased production of cortisol by the adrenal glands is known as Cushing syndrome. In contrast, adrenal gland insufficiency is called **Addison disease.** Perhaps the most famous person to suffer from Addison disease was President John F. Kennedy. The most common cause of Addison disease in the United States is destruction of the adrenal cortex by the immune system. Clinical symptoms do not appear until about 90% of both adrenal cortexes have been destroyed. These symptoms are fairly nonspecific and may include weakness, weight loss, abdominal pain, mood disturbances, and hyperpigmentation of the skin (Fig. 20.15). Because the decreased production of mineralocorticoids can affect the balance of sodium and potassium, which in turn can have dramatic effects on the heart, Addison disease can be fatal unless properly treated by replacing the missing hormones.

Check Your Progress

1. Drinking alcohol inhibits the secretion of ADH. What effect of alcohol consumption does this help to explain?
2. Why does a goiter develop in iodine deficiency?
3. How do antibodies play a different role in causing Graves disease versus Addison disease?

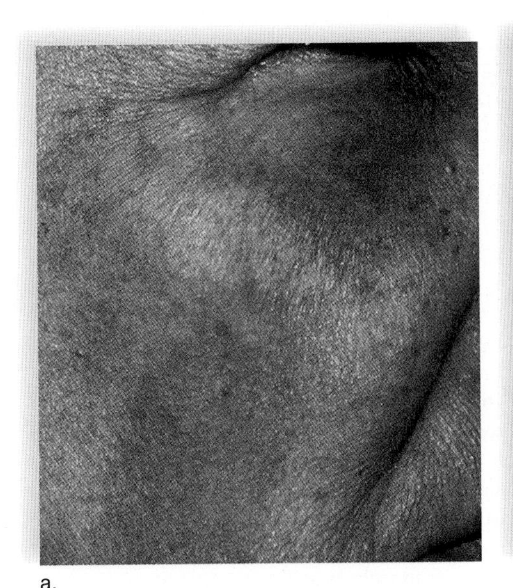

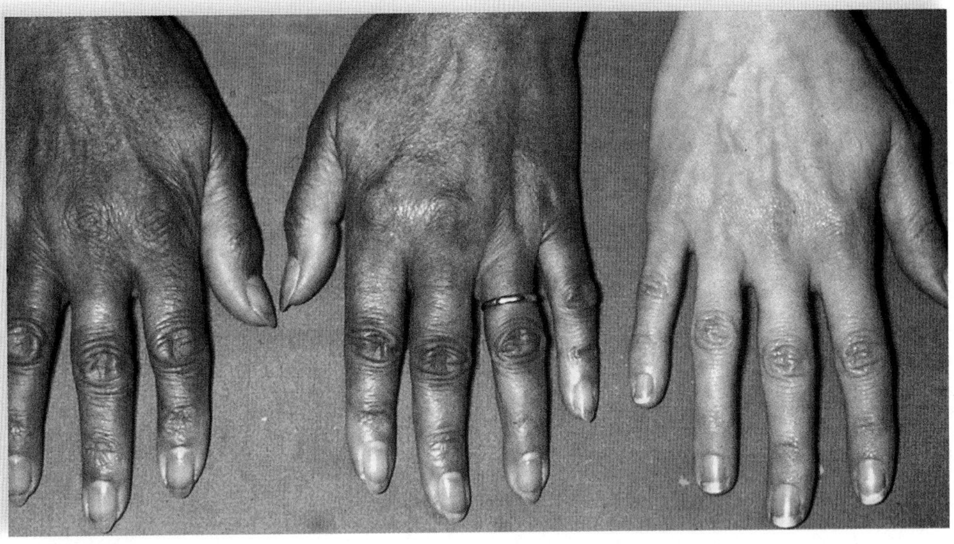

a. b.

Figure 20.15 Addison disease.
Addison disease is characterized by a peculiar bronzing of the skin, particularly noticeable in light-skinned individuals. Note the color of **(a)** the face and **(b)** the hands compared to the hand of an individual without the disease.

Diabetes Mellitus

As of 2007, it is estimated that about 24 million Americans, or 7.8% of the population, have **diabetes mellitus,** often referred to simply as diabetes, a condition that affects their ability to regulate their glucose metabolism. People with diabetes either do not produce enough insulin (type 1) or cannot properly use the insulin they produce (type 2). In either case, although blood glucose levels rise, cellular famine exists in the midst of plenty. Some of the excess glucose in the blood is excreted into the urine, and water follows, causing the volume of urine to increase. Because the glucose in the blood cannot be used, the body turns to the metabolism of fat, which leads to the buildup of ketones in the blood. In turn, these ketones are metabolized to form various acids, which can build up in the blood (acidosis) and lead to coma and death.

The glucose tolerance test is often used to diagnose diabetes mellitus. After a person ingests a known amount of glucose, the blood glucose concentration is measured at intervals. In a diabetic, the blood glucose level usually rises greatly and remains elevated for hours (Fig. 20.16). In the meantime, glucose appears in the urine. In a nondiabetic, the blood glucose level rises somewhat, and then returns to normal after about two hours.

About 10% of diabetics in the United States have type 1 diabetes. This condition usually begins in childhood or adolescence, and thus it is sometimes called juvenile-onset diabetes. However, type 1 diabetes can also occur later in life, following a viral infection, an autoimmune reaction, or an environmental agent that leads to destruction of the pancreatic islets. Because individuals with type 1 diabetes are suffering from an insulin shortage, treatment simply consists of providing the needed insulin through daily injections. These injections usually control the diabetic symptoms but if too much insulin is administered, or a meal is missed, the result can be hypoglycemia (low blood sugar). Symptoms of hypoglycemia include perspiration, pale skin, shallow breathing, and anxiety. Because the brain requires a constant supply of glucose, unconsciousness can result. The treatment is quite simple: immediate ingestion of a sugary snack or fruit juice can very quickly counteract hypoglycemia. Better control of glucose levels can sometimes be achieved with an insulin pump, a small device worn outside the body that is connected

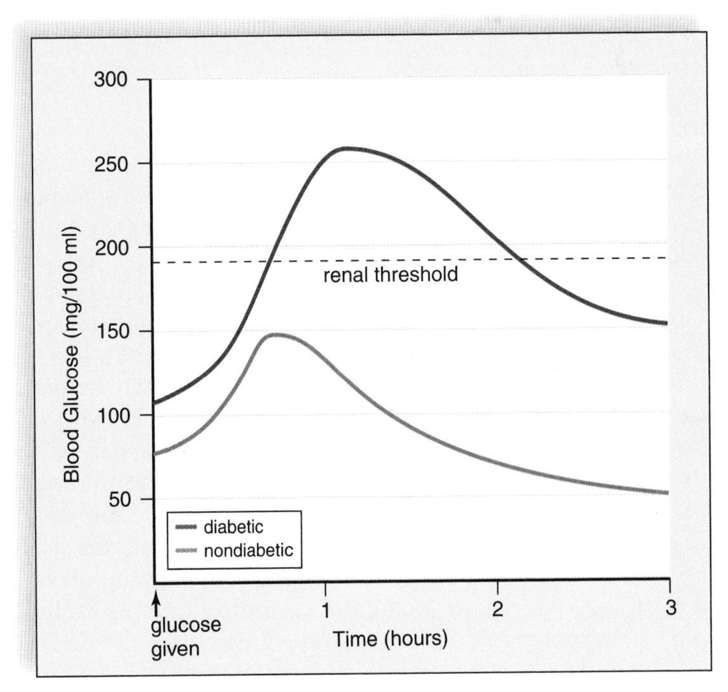

Figure 20.16 Glucose tolerance test.
Following the administration of 100 g of glucose, the blood glucose level rises dramatically in the diabetic and glucose appears in the urine. Also, the blood glucose level at 2 hours is equal to or more than 200 mg/100 ml.

Bioethical Focus

Human Growth Hormone: Does Rambo Know Something We Don't?

As discussed in the chapter, abnormally low levels of growth hormone production by the anterior pituitary can result in short stature. Suppose you have a son or daughter who is considerably shorter than most kids his or her age. The average adult American male is 5 ft, 9 in. tall. If you knew your son would only reach 5 ft, do you think it would be ethical to treat him with human growth hormone (hGH), which could easily add 3 or 4 in. to his height? And since hGH injections can cost between $10,000 and $30,000 per year, should that be covered by health insurance, even if it raises insurance rates for other people?

Now suppose you are 55 years old. You are finding that even though you run and lift weights three days a week, you are slowly losing muscle mass, and you don't feel nearly as energetic as you once felt, even in your forties. Then you hear about how some professional athletes and movie stars are swearing by hGH as a way to recapture their youth. You find a website touting "legal" hGH at affordable prices, and claiming hGH will help you lose weight, tone your muscles, and give you younger-looking skin in a matter of weeks. Assuming it was legal and you could afford it, would you give it a try?

Prior to the development of recombinant DNA techniques, treatment of children who failed to grow due to insufficient hGH production required isolation of hGH from the pituitary glands of human cadavers. In 1985, the FDA approved the first recombinant hGH, produced by inserting the hGH gene into *E. coli* bacteria. In 2003, the FDA approved a new use for hGH, in the treatment of children of short stature due to unknown causes. However, hGH must still be administered by a physician, and it is illegal to use it for fitness purposes.

Figure 20B
At age 61, actor Sylvester Stallone admitted to using human growth hormone and anabolic steroids to help him get in shape for filming his latest Rambo movie.

The production of recombinant hGH has been very beneficial in specific medical cases, but its use for other purposes remains controversial. Prior to filming his fourth Rambo movie, actor Sylvester Stallone admitted to using hGH as well as anabolic steroids to help tone his body (Fig. 20B). In 2007, Stallone was fined over $10,000 for attempting to smuggle 48 vials of hGH and 4 vials of testosterone into Australia. At age 61, he seems to be in incredible condition, but it is hard to say how much of that is due to hard workouts, genetics, or pharmaceuticals. Studies show that hGH can cut down on body fat and increase muscle mass, but no antiaging effects have been proven, and the list of scary side effects is long. Still, it's worth asking: Would you be willing to take the risk?

Form Your Own Opinion

1. Would you ever consider using hGH under any circumstances? Why or why not?
2. Do you believe that hGH should be illegal for personal use? If so, do you believe that taking hGH is worse than taking currently legal athletic supplements, such as creatine? Would it make a difference if someone was taking it to grow taller instead of for other perceived benefits?
3. Do you think that celebrities have an increased responsibility to avoid promoting behaviors that could be harmful to others?

to a plastic catheter inserted under the skin (Fig. 20.17). It is also possible to transplant a working pancreas, or even fetal pancreatic islet cells, into patients with type 1 diabetes. Some researchers believe that some form of artificial pancreas will become available in the early 2010s.

Most diabetics in the United States have type 2 diabetes. Often, the patient is overweight or obese, and adipose tissue may produce a substance that impairs insulin receptor function. Because type 2 diabetes usually doesn't occur before age 40, it was formerly called adult-onset diabetes. However, due to the increasing prevalence of obesity in children, type 2 diabetes is occurring at younger ages. Normally, the binding of insulin to its receptor on cell surfaces causes the number of glucose transporters to increase in the plasma membrane, but not in the type 2 diabetic. Treatment usually involves weight loss, which can sometimes control symptoms in the type 2 diabetic. However, many type 2 diabetics also have low insulin levels, so they may require injections of insulin, along with medications to increase the effectiveness of the insulin they produce.

Regardless of the type of diabetes they have, diabetics usually need to monitor their blood glucose levels several times a day. As described in the story that opened this chapter, this is usually done by poking a finger to obtain a drop of blood, which is tested with an external device. A U.S. company is currently testing a special tattoo ink that changes

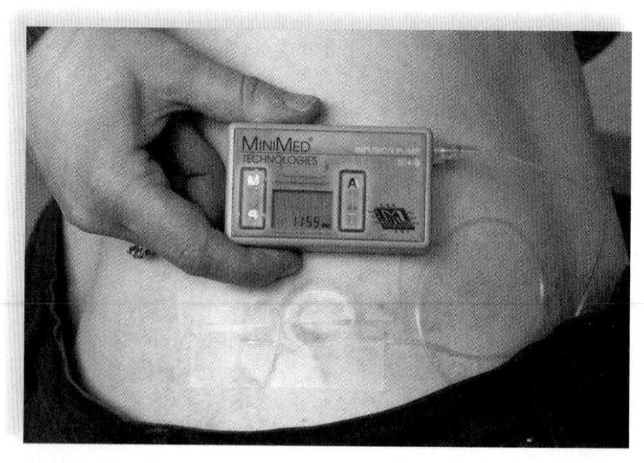

Figure 20.17 An insulin pump.
Insulin pumps administer preprogrammed small doses of insulin throughout the day via an implanted catheter. Most insulin pumps can be worn under clothing.

colors in response to glucose levels in the skin. The FDA has already approved a watchlike device that measures glucose levels in tiny amounts of fluid it painlessly extracts from the skin every 20 minutes. For insulin administration, insulin pumps are now replacing the need for injecting insulin with a needle and syringe (see Fig. 20.17).

Long-term complications of both types of diabetes are blindness, kidney disease, and cardiovascular disorders, including reduced circulation. The latter can lead to gangrene in the arms and legs. Diabetic women who become pregnant have an increased risk of diabetic coma, and the child of a diabetic is more likely to be stillborn or to die shortly after birth. These complications are uncommon, however, if the mother's blood glucose level is carefully regulated during pregnancy.

Check Your Progress

1. How do the causes of type 1 and type 2 diabetes differ?
2. What are the symptoms of hypoglycemia?
3. Why is a glucose tolerance test more valuable than simply measuring blood insulin levels in testing for diabetes?

Applying the Concepts [Revisited]

Hank is a second-grade boy with type 1 diabetes, meaning his pancreas doesn't produce enough insulin. He has adapted well to the routine of checking his blood sugar and receiving insulin injections, but recent medical advances may make life easier for him.

1. Besides the convenience of not needing several separate insulin injections a day, why might the use of insulin pumps help to delay some of the more serious long-term consequences of diabetes?
2. Some of the symptoms of hypoglycemia (low blood sugar) and hyperglycemia (high blood sugar) are similar. Both conditions are potentially dangerous—however, severe hypoglycemia is considered a more serious emergency. Why would this be the case?

SUMMARIZING THE CONCEPTS

20.1 Endocrine Glands and Hormones

Hormones are chemical signals that affect other tissues, sometimes at a distance. Along with the nervous system, the endocrine system maintains homeostasis. Negative feedback and antagonistic hormonal actions control the secretion of hormones.

- Hormones are either peptides or steroids. Binding of a peptide hormone to a receptor at the plasma membrane typically activates an enzyme cascade inside the cell. Steroid hormones combine with a receptor in the cell, and the complex attaches to and activates transcription of certain genes.

20.2 Hypothalamus and Pituitary Glands

The hypothalamus controls hormone secretion by the pituitary gland through two distinct mechanisms:

- The hypothalamus itself synthesizes antidiuretic hormone (ADH) and oxytocin, which are stored in the posterior pituitary until they are released.
- The hypothalamus communicates with the anterior pituitary via a portal system, through which hypothalamic-releasing and, in some instances, hypothalamic-inhibiting hormones, pass. The anterior pituitary produces at least six types of hormones, and some of these stimulate other hormonal glands to secrete hormones. In the latter case, the hypothalamus, anterior pituitary, and target gland are involved in a negative feedback control system.

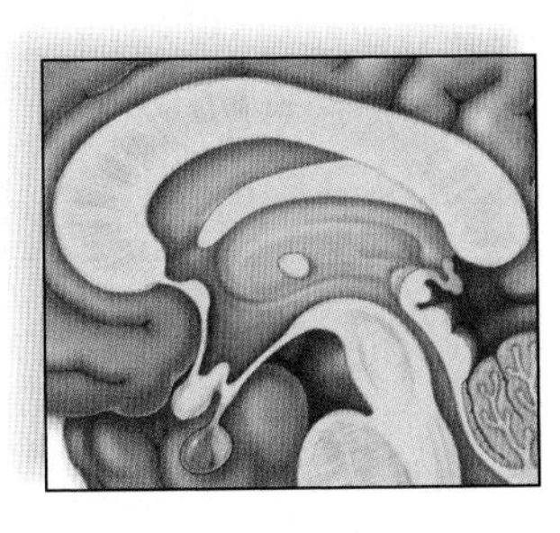

20.3 Thyroid and Parathyroid Glands

The thyroid gland produces three hormones:

- Thyroxine (T_4) and triiodothyronine (T_3) increase the metabolic rate of most cells in the body.
- Calcitonin helps lower the blood calcium level by reducing the activity of osteoclasts in the bone.

The parathyroid glands secrete parathyroid hormone, which raises the blood calcium and decreases the blood phosphate levels.

20.4 Adrenal Glands

The adrenal glands respond to stress. The adrenal medulla is controlled directly by the nervous system, allowing an almost immediate response. The adrenal cortex responds on a longer-term basis to ACTH released by the anterior pituitary:

- The adrenal medulla secretes epinephrine and norepinephrine, which bring about fight-or-flight responses we associate with emergency situations.

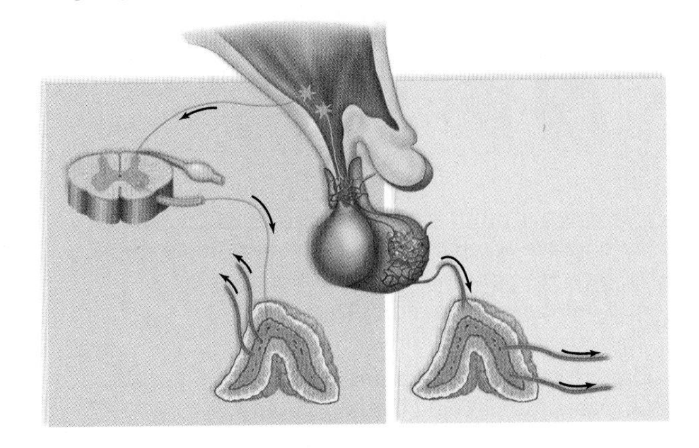

- The adrenal cortex produces the glucocorticoids (e.g., cortisol) and the mineralocorticoids (e.g., aldosterone). Cortisol stimulates the conversion of amino acids to glucose, raising the blood glucose level. It can also suppress inflammatory responses. Aldosterone causes the kidneys to reabsorb sodium ions (Na^+) and to excrete potassium ions (K^+), which raises the blood pressure.

20.5 Pancreas

The pancreas has both exocrine and endocrine functions. The pancreatic islets secrete insulin and glucagon:

- Insulin lowers the blood glucose level by stimulating the uptake of glucose by various types of cells.
- Glucagon raises the blood glucose level.

20.6 Other Endocrine Glands

Other organs and tissues that produce hormones include the following:

- The gonads produce the sex hormones.
- The thymus secretes thymosins, which stimulate T-lymphocyte production and maturation.
- The pineal gland produces melatonin, which may be involved in circadian rhythms and the development of the reproductive organs.
- Tissues also produce hormones. Adipose tissue produces leptin, which acts on the hypothalamus, and various tissues produce growth factors. Prostaglandins are produced and act locally.

20.7 Pheromones

Pheromones are chemical signals that act between individuals of the same species. There is some evidence for the activity of pheromone-like chemicals in humans.

20.8 Disorders of the Endocrine System

Increases or decreases in the amounts of hormones synthesized by endocrine glands cause several common disorders:

- Disorders caused by pituitary gland dysfunction include diabetes insipidus, growth hormone disorders, and Cushing syndrome.
- Insufficient thyroid gland function results in cretinism in childhood and myxedema in adults, while Graves disease is a type of hyperthyroidism.
- Addison disease is a serious condition in which the adrenal cortexes have been destroyed, often by an autoimmune process.
- Diabetes mellitus is a common condition affecting the body's ability to metabolize glucose. In type 1 diabetes, the pancreas doesn't produce enough insulin. In type 2 diabetes, which is more common, insulin receptors do not function properly. Insulin injections may be required for both types of diabetes, but type 2 diabetes can be more difficult to treat, resulting in long-term complications.

TESTING YOURSELF

Choose the best answer for each question.

1. Which hormones typically cross the plasma membrane?
   a. peptide hormones
   b. steroid hormones
   c. Both a and b are correct.
   d. Neither a nor b is correct.
2. Name the endocrine glands in the following diagram.

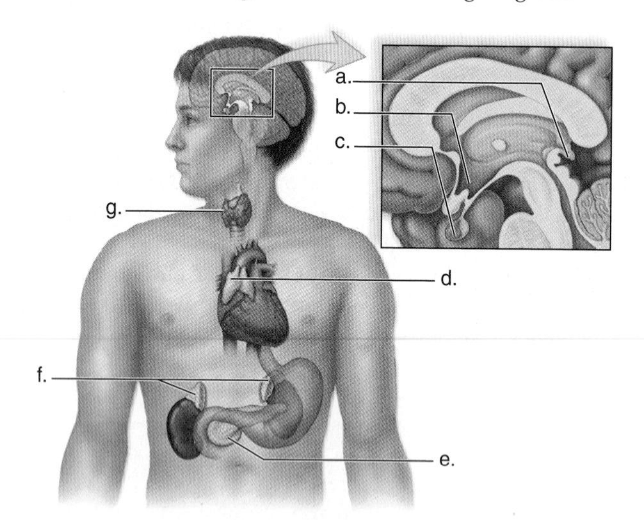

3. The anterior pituitary controls the secretion(s) of
   a. both the adrenal medulla and the adrenal cortex.
   b. both the pancreas and the adrenal cortex.
   c. both the ovaries and the testes.
   d. Both b and c are correct.
4. Growth hormone is produced by the
   a. posterior adrenal gland.
   b. posterior pituitary gland.
   c. anterior pituitary gland.
   d. kidneys.
   e. None of these are correct.
5. Complete the following diagram by filling in blanks a–e.

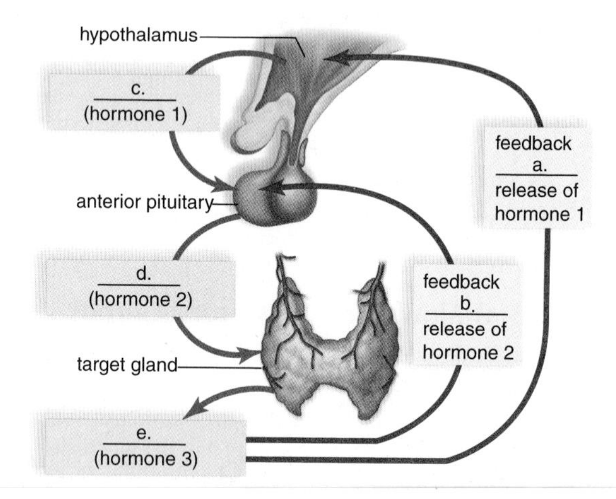

6. PTH causes the blood levels of ______ to increase and ______ to decrease.
   a. calcium, sodium
   b. calcium, phosphate
   c. phosphate, sodium
   d. phosphate, calcium
   e. None of these are correct.
7. The body's response to stress includes
   a. water reabsorption by the kidneys.
   b. blood pressure increase.
   c. increase in the blood glucose level.
   d. heart rate increase.
   e. All of these are correct.
8. Lack of aldosterone will cause a blood imbalance of
   a. sodium.
   b. potassium.
   c. water.
   d. All of these are correct.
   e. None of these are correct.

Match the hormones in questions 9–13 to the correct gland in the key.

Key:

a. pancreas
b. anterior pituitary
c. posterior pituitary
d. thyroid gland
e. adrenal medulla
f. adrenal cortex

9. Cortisol
10. Growth hormone (GH)
11. Oxytocin storage
12. Insulin
13. Epinephrine
14. Tropic hormones are hormones that affect other endocrine tissues. Which of the following would be considered a tropic hormone?
 a. calcitonin
 b. oxytocin
 c. glucagon
 d. melatonin
 e. follicle-stimulating hormone
15. Name the hormones in the diagram to the right.

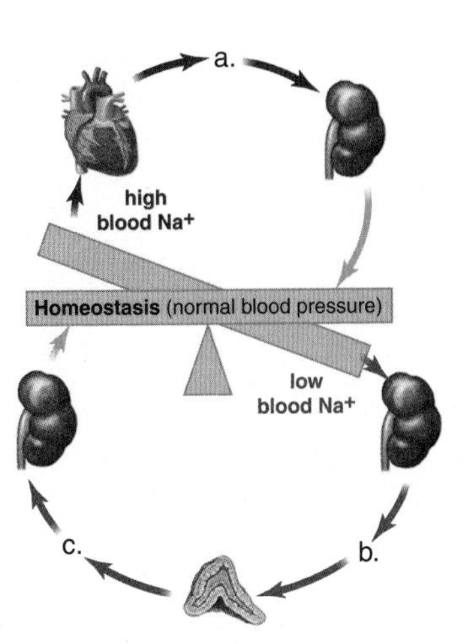

16. Which hormone and condition are mismatched?
 a. growth hormone—acromegaly
 b. thyroxine—goiter
 c. parathyroid hormone—tetany
 d. cortisol—myxedema
 e. insulin—diabetes
17. Long-term complications of diabetes include
 a. blindness.
 b. kidney disease.
 c. circulatory disorders.
 d. All of these are correct.

UNDERSTANDING THE TERMS

THINKING CRITICALLY

1. Because some of their functions overlap, why is it necessary to have both a nervous system and an endocrine system?
2. Certain endocrine disorders, such as Cushing syndrome, can be caused by excessive secretion of a hormone (in this case, ACTH) by the pituitary gland, or by a problem with the adrenal gland itself. If you were able to measure the ACTH levels of a Cushing patient, how could you tell the difference between a pituitary problem and a primary adrenal gland problem?
3. In animals, pheromones can affect many different behaviors. Because humans do produce a number of airborne chemicals, what potential human behaviors could be affected by these chemicals? How would these pheromones be received, and would we necessarily be conscious of their effects on us?

INQUIRY INTO LIFE WEBSITE

The companion website for *Inquiry into Life* provides a wealth of information organized and integrated by chapter. You will find practice tests, animations, videos, and much more that will complement your learning and understanding of general biology.

http://www.mhhe.com/maderinquiry13

PART V

Continuance of the Species

21

Applying the Concepts

The case of Nadya Suleman, who gave birth to eight babies in January 2009, made headlines around the world. Cable TV news organizations spent countless hours speculating about the ethics and motivation of this divorced single mother, also dubbed the "Octomom," who had six embryos (two of which became twins) implanted into her uterus at the same time. The embryos were left over from a previous in vitro fertilization (IVF) procedure using sperm from a donor. Complicating the situation was the fact that Ms. Suleman already had six children, she was unemployed and on government assistance, and she may have been interested in benefiting financially from the attention her story has received.

As will be described in this chapter, the reproductive system is a basic necessity for the survival of a species. However, the ethics of human reproduction can become quite complicated. In the great majority of cases, IVF and the other assisted reproductive technologies described in this chapter are used to help couples who are unable to conceive on their own. Guidelines from the American Society for Reproductive Medicine suggest that no more than three embryos should be implanted at a time, and therefore the procedure preformed on Ms. Suleman at an IVF clinic in Beverly Hills, California, is of questionable medical value, though not illegal. Some state legislatures are seeking to remedy that, however—soon after the Suleman case was publicized, legislatures in Missouri and Georgia began considering laws to limit the number of embryos that can be implanted by IVF. In addition to concerns about the babies' health, lawmakers expressed concern about the increased burden on taxpayers to care for children of families that may not even have health insurance. Many infertility doctors, however, argue that decisions on how many embryos to transfer should be left up to medical experts familiar with a patient's individual circumstances.

Reproductive System

Chapter Outline

21.1 Male Reproductive System

Learning Outcomes

1. Trace the path of sperm from the testes to the urethra.
2. List the organs that produce components of seminal fluid.
3. Analyze the role of gonadotropin-releasing hormone, luteinizing hormone, follicle-stimulating hormone, and testosterone in male sexual reproduction.

The male reproductive system includes the organs depicted in Figure 21.1 and listed in Table 21.1. The male gonads are paired **testes** (sing., testis), which are suspended within the sac-like **scrotum.**

Genital Tract

Sperm produced by the testes mature within the epididymides (sing., **epididymis**), which are tightly coiled ducts lying outside the testes. Maturation seems to be required in order for sperm to swim to the egg. When sperm leave an epididymis, they enter a **vas deferens** (pl., vasa deferentia), where they may also be stored for a time. Each vas deferens passes into the abdominal cavity, where it curves around the urinary bladder and empties into an ejaculatory duct. The ejaculatory ducts connect to the **urethra.**

TABLE 21.1 Male Reproductive Organs

Organ	Function
Testes	Produce sperm and sex hormones
Epididymides	Ducts where sperm mature and are stored
Vasa deferentia	Conduct and also store sperm
Seminal vesicles	Contribute nutrients and fluid to semen
Prostate gland	Contributes basic fluid to semen
Urethra	Conducts sperm
Bulbourethral glands	Contribute viscous fluid to semen
Penis	Organ of sexual intercourse

At the time of ejaculation, sperm leave the penis in a fluid called **semen** (seminal fluid). The seminal vesicles, the prostate gland, and the bulbourethral glands (Cowper glands) add secretions to semen. The pair of **seminal vesicles** lies at the base of the bladder, and each has a duct that joins with a vas deferens. The **prostate gland** is a single, donut-shaped gland that surrounds the upper portion of the urethra just below the urinary bladder. **Bulbourethral glands** are pea-sized organs that lie underneath the prostate on either side of the urethra.

Each component of semen seems to have a particular function. Sperm are more viable in a basic solution, and semen, which is milky in appearance, has a slightly basic

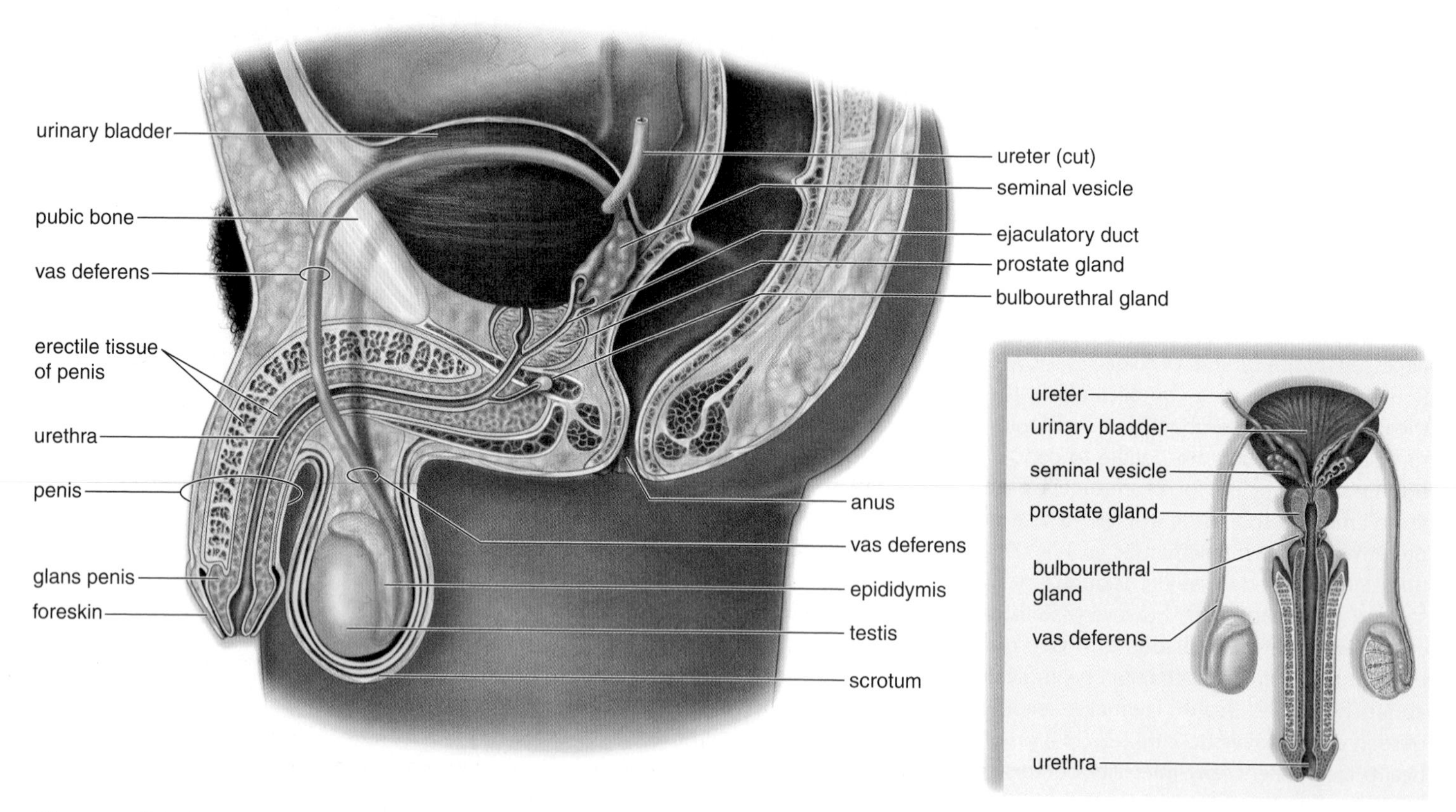

Figure 21.1 The male reproductive system.
The testes produce sperm. The seminal vesicles, the prostate gland, and the bulbourethral glands provide a fluid medium for the sperm, which move from the vasa deferentia through the ejaculatory ducts to the urethra in the penis. The foreskin (prepuce) is removed when a penis is circumcised.

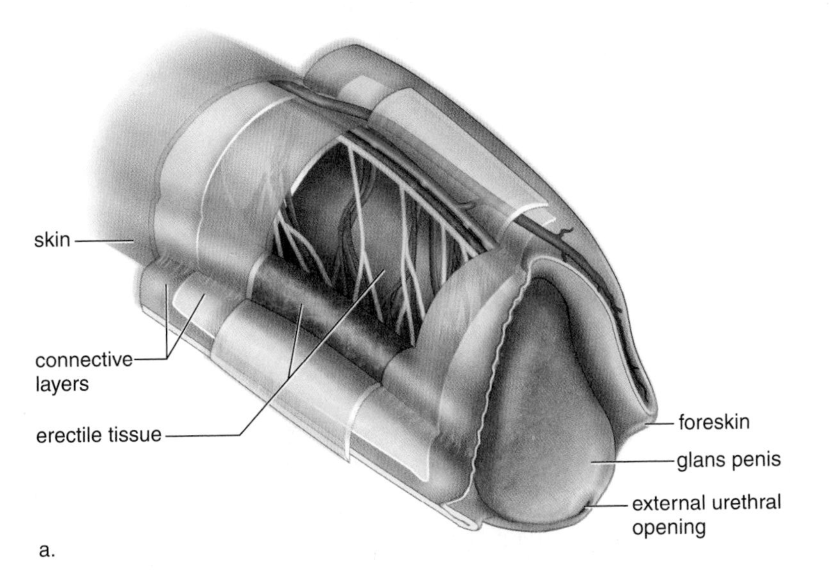

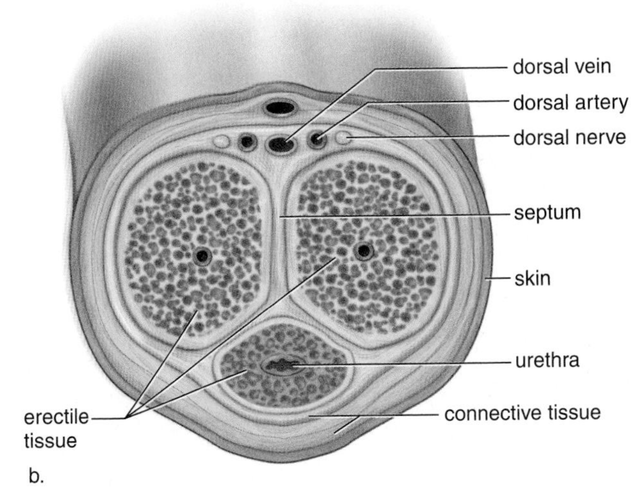

Figure 21.2 Penis anatomy.
a. Beneath the skin and the connective tissue lies the urethra, surrounded by erectile tissue. This tissue expands to form the glans penis, which in uncircumcised males is partially covered by the foreskin (prepuce). **b.** Two other columns of erectile tissue in the penis are located dorsally.

pH (about 7.5). Swimming sperm require energy, and semen contains the sugar fructose, which serves as an energy source. Semen also contains prostaglandins, local hormones that cause the uterus to contract. Some investigators believe that uterine contractions help propel the sperm toward the egg.

The **penis** (Fig. 21.2) is the male organ of sexual intercourse. The penis has a long shaft and an enlarged tip called the glans penis. At birth, the glans penis is covered by a layer of skin called the foreskin. Circumcision, the surgical removal of the foreskin, is usually done for religious reasons or perceived health benefits, including easier hygiene, decreased urinary tract infections, and penile cancer. A study of 3,000 Ugandan men, published in the *New England Journal of Medicine* in March of 2009, showed that circumcised men were 25 to 35 percent less likely to contract HIV, human papilloma virus, or genital herpes. However, opponents of the procedure focus on the risks, including pain and potential loss of sensitivity in the penis, and the American Academy of Pediatrics' position is that the benefits are not sufficient to recommend routine neonatal circumcision.

Erection and Orgasm in Males

Spongy, erectile tissue containing distensible blood spaces extends through the shaft of the penis. When a man is sexually excited, the arteries in the penis relax and widen. Increased blood flow causes the penis to enlarge and become erect. Also, the veins that normally carry blood away from the penis get compressed, and this maintains an erection.

When sexual stimulation intensifies, sperm enter the urethra from the vasa deferentia, and the accessory glands contribute secretions to the semen. Once semen is in the urethra, rhythmic muscle contractions cause it to be ejaculated from the penis in spurts. During ejaculation, a sphincter normally closes off the urinary bladder so that no urine enters the urethra, and no semen enters the bladder. (Notice that the urethra carries urine or semen at different times.)

The contractions that expel semen from the penis are a part of male orgasm, the physiological and psychological sensations that occur at the climax of sexual stimulation. The psychological sensation of pleasure is centered in the brain, but the physiological reactions involve the genital (reproductive) organs and associated muscles, as well as the entire body.

Following ejaculation and/or loss of sexual arousal, the penis returns to its normal flaccid state. After ejaculation, a male typically experiences a period of time, called the refractory period, during which stimulation does not bring about an erection. The length of the refractory period increases with age.

There may be in excess of 400 million sperm in the approximately 2–6 ml of semen expelled during ejaculation. However, the sperm count can be much lower than this, and fertilization of the egg by a sperm still can take place.

Check Your Progress

1. Compare the functions of the testes, epididymis, vasa deferentia, and urethra. Which also function(s) in urination?
2. The dilation of arteries in the penis is controlled by different enzymes than dilation of arteries elsewhere in the body. How do you think researchers took advantage of that fact in developing drugs like Viagra (sildenafil)?

Male Gonads, the Testes

The testes, which produce sperm and also the male sex hormones, lie outside the abdominal cavity of the human male, within the scrotum. The testes begin their development inside the abdominal cavity but descend into the scrotal sacs during the last two months of fetal development. If the testes fail to descend properly, and thus remain in the abdomen, male infertility may result. This is because the internal temperature of the body is too high to produce viable sperm. This condition can usually be surgically corrected, however. The scrotum helps regulate the temperature of the testes by holding them closer or farther away from the body. In fact, any activity that increases testicular temperature, such as taking hot baths, using a laptop computer, or even sitting for extended periods of time, may decrease sperm production.

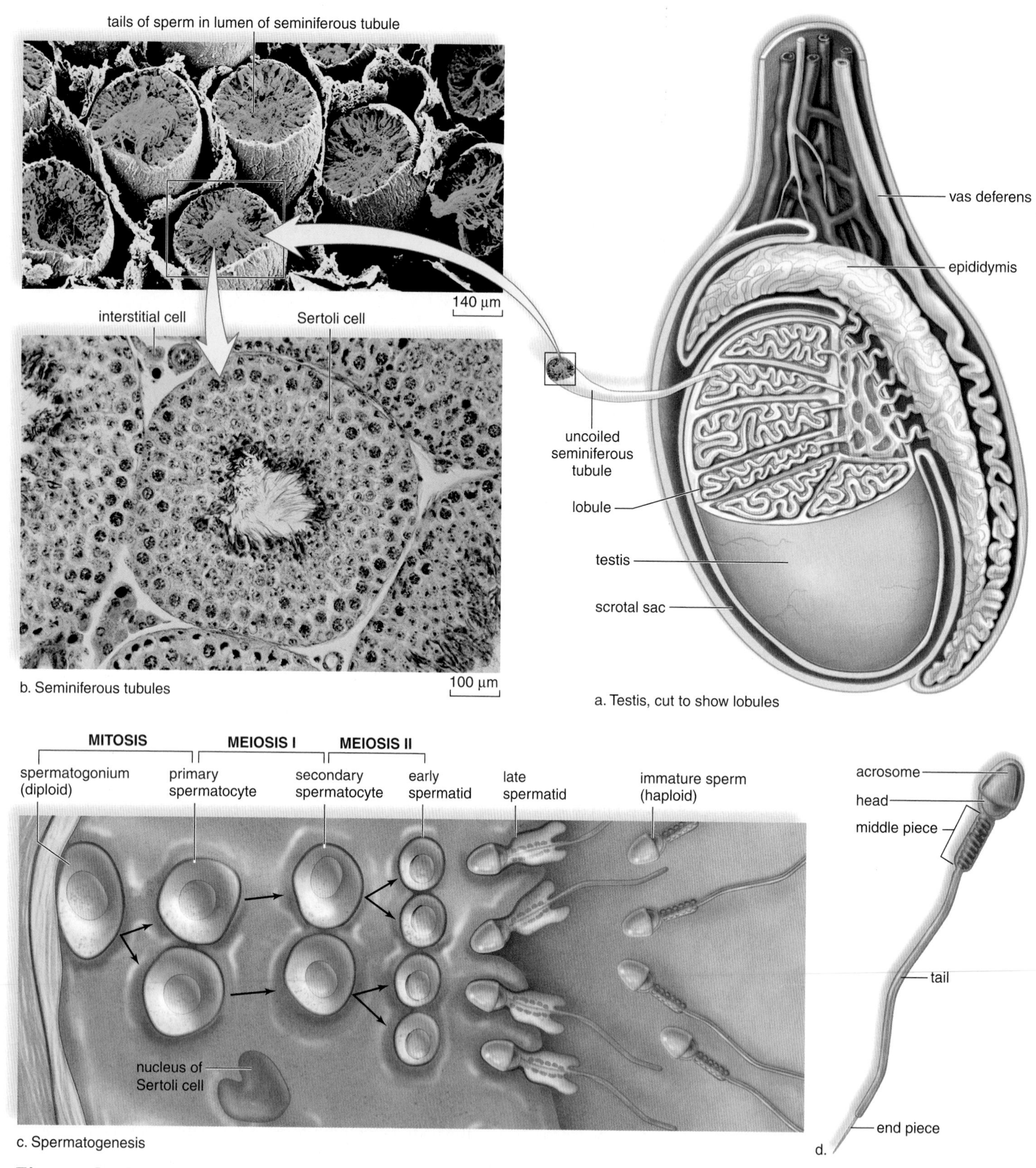

Figure 21.3 Testis and sperm.
a. The lobules of a testis contain seminiferous tubules. **b.** Light micrographs of a cross section of the seminiferous tubules, where spermatogenesis occurs. Note the location of interstitial cells in clumps among the seminiferous tubules. **c.** Diagrammatic representation of spermatogenesis, which occurs in the walls of the tubules. **d.** A sperm has a head, a middle piece, tail, and an end piece. The nucleus is in the head, which is capped by the enzyme-containing acrosome.

Seminiferous Tubules

A longitudinal section of a testis shows that it is composed of compartments called lobules, each of which contains one to three tightly coiled **seminiferous tubules** (Fig. 21.3*a*). Altogether, these tubules have a combined length of approximately 250 meters. A microscopic cross section of a seminiferous tubule reveals that it is packed with cells undergoing **spermatogenesis** (Fig. 21.3*b, c*), the production of sperm. Newly formed spermatogonia (sing., spermatogonium) move away from the outer wall and become primary spermatocytes that undergo meiosis I to produce secondary spermatocytes with 23 chromosomes. Secondary spermatocytes undergo meiosis II to produce four spermatids that are also haploid. Spermatids then differentiate into sperm. Also present are Sertoli cells (sustentacular cells), which support, nourish, and regulate the spermatogenic cells.

Mature **sperm,** or spermatozoa, have three main parts: a head, a middle piece, and a tail (Fig. 21.3*d*). Mitochondria in the middle piece provide energy for the movement of the tail, which has the structure of a flagellum (see Fig. 3.14). The head contains a nucleus covered by a cap called the **acrosome,** which stores enzymes needed to penetrate the egg. The ejaculated semen of a normal human male contains several hundred million sperm, but only one sperm normally enters an egg. Sperm usually do not live more than 48 hours in the female genital tract.

Interstitial Cells

The male sex hormones, the androgens, are secreted by cells that lie between the seminiferous tubules. Therefore, they are called **interstitial cells** (Fig. 21.3*b*). The most important of the androgens is testosterone, whose functions are discussed next.

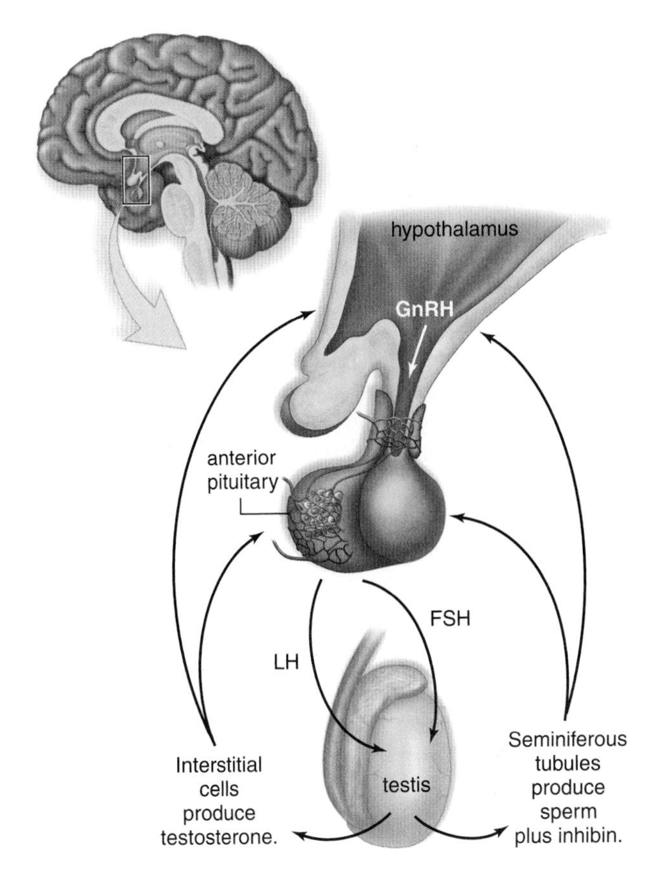

Figure 21.4 Hormonal control of testes.
GnRH stimulates the anterior pituitary to produce FSH and LH. FSH stimulates the testes to produce sperm, and LH stimulates the testes to produce testosterone. Testosterone from interstitial cells and inhibin from the seminiferous tubules exert negative feedback control over the hypothalamus and the anterior pituitary, and this ultimately regulates the level of testosterone in the blood.

Hormonal Regulation in Males

The hypothalamus has ultimate control of the testes' sexual function because it secretes a hormone called **gonadotropin-releasing hormone (GnRH)** that stimulates the anterior pituitary to secrete the gonadotropic hormones. There are two gonadotropic hormones—**follicle-stimulating hormone (FSH)** and **luteinizing hormone (LH)**—in both males and females. In males, FSH promotes the production of sperm in the seminiferous tubules, which also release the hormone inhibin. Inhibin, in turn, inhibits further FSH synthesis.

LH in males is sometimes given the name **interstitial cell-stimulating hormone (ICSH)** because it controls the production of testosterone by the interstitial cells. All these hormones are involved in a negative feedback relationship that maintains the fairly constant production of sperm and testosterone (Fig. 21.4).

Testosterone, the main sex hormone in males, is essential for the normal development and functioning of the organs listed in Table 21.1. Testosterone also brings about and maintains the male secondary sex characteristics that develop at the time of puberty. Males are generally taller than females and have broader shoulders and longer legs relative to trunk length. The deeper voices of males compared to those of females are due to a larger larynx with longer vocal cords. Because the so-called Adam's apple is a part of the larynx, it is usually more prominent in males than in females. Testosterone causes males to develop noticeable hair on the face, chest, and occasionally other regions of the body, such as the back. Testosterone also leads to the receding hairline and pattern baldness that occur in aging males.

Testosterone is responsible for the greater muscular development of males. Knowing this, both males and females sometimes take anabolic steroids, which are either testosterone or related steroid hormones resembling testosterone. Health problems involving the kidneys, the cardiovascular system, and hormonal imbalances can arise from such use. In males, the testes shrink in size, and feminization of other male traits occurs (see Fig. 20.10).

Check Your Progress

1. What is the role of the seminiferous tubules?
2. Why is meiosis necessary in the production of sperm?
3. What would be the predicted effects if interstitial cells did not develop and function normally?

21.2 Female Reproductive System

Learning Outcomes

1. Identify where an oocyte is produced and how it is transported to the uterus.
2. List the major components of the female external genitalia.
3. Describe the events that occur during a female orgasm.

The female reproductive system includes the organs depicted in Figure 21.5 and listed in Table 21.2. The female gonads are paired **ovaries** that lie in shallow depressions, one on each side of the upper pelvic cavity. **Oogenesis** is the production of an egg, or **oocyte,** the female gamete. The ovaries usually alternate in producing one oocyte per month. **Ovulation** is the process by which an oocyte bursts from an ovary and usually enters an oviduct.

The Genital Tract

The **oviducts,** also called uterine or fallopian tubes, extend from the uterus to the ovaries. However, the oviducts are not attached to the ovaries. Instead, they have fingerlike projections called fimbriae (sing., **fimbria**) that sweep over the ovaries. When an oocyte bursts from an ovary during ovulation, it usually is swept into an oviduct by the combined action of the fimbriae and the beating of cilia that line the oviducts.

TABLE 21.2 Female Reproductive Organs

Organ	Function
Ovaries	Produce oocyte and sex hormones
Oviducts (uterine or fallopian tubes)	Conduct oocyte; location of fertilization; transport early zygote
Uterus (womb)	Houses developing fetus
Cervix	Contains opening to uterus
Vagina	Receives penis during sexual intercourse; serves as birth canal and as the exit for menstrual flow

Once in the oviduct, the oocyte is propelled slowly by ciliary movement and tubular muscle contraction toward the uterus. An oocyte lives only approximately 6 to 24 hours unless fertilization occurs. Fertilization, and therefore formation of a **zygote,** usually takes place in the oviduct. The developing embryo normally arrives at the uterus after several days and then embeds, or **implants,** itself in the uterine lining, which has been prepared to receive it. Occasionally however, the embryo may implant in the oviduct or elsewhere, resulting in an ectopic pregnancy.

The **uterus** is a thick-walled, muscular organ about the size and shape of an inverted pear. The oviducts join the uterus at its upper end, while at its lower end, the **cervix** connects with the vagina nearly at a right angle (Fig. 21.5).

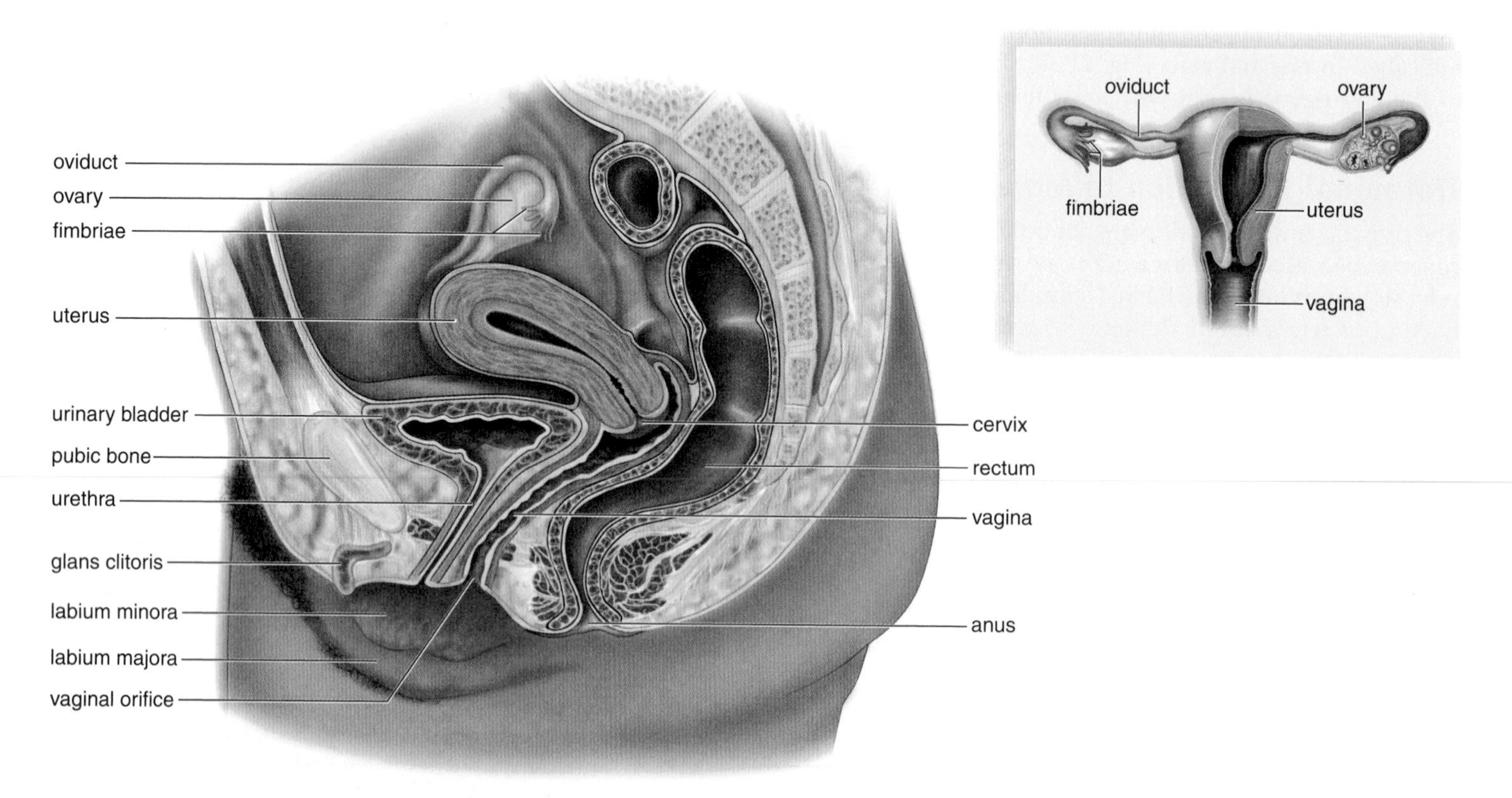

Figure 21.5 The female reproductive system.
The ovaries release one egg a month; fertilization occurs in the oviduct, and development occurs in the uterus. The vagina is the birth canal, as well as the organ of sexual intercourse and the outlet for menstrual flow.

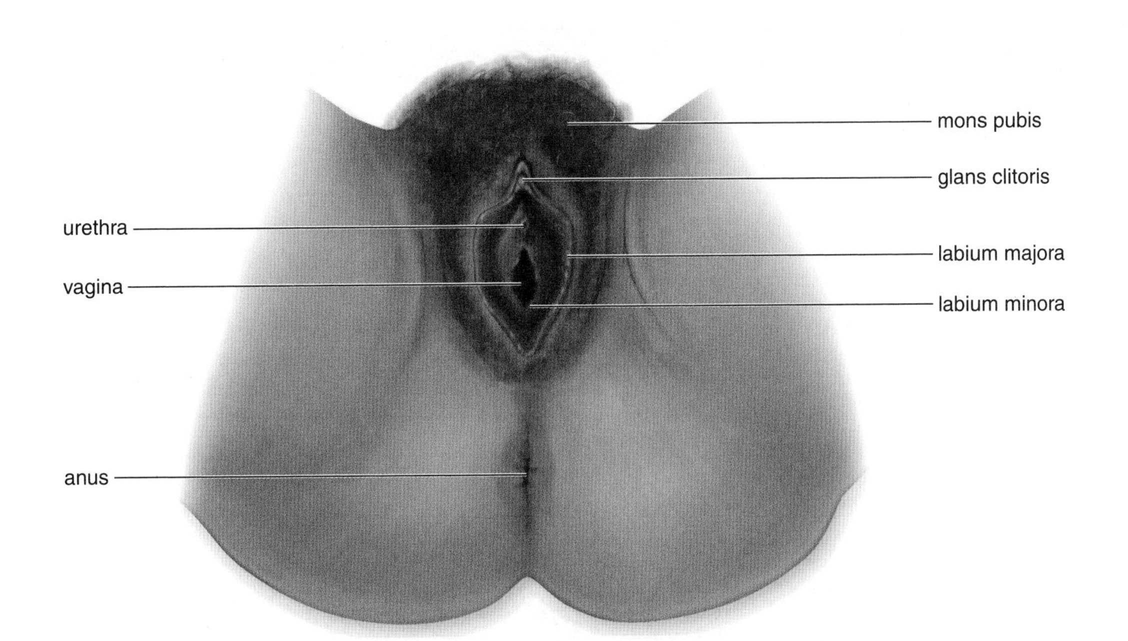

Figure 21.6 External genitals of the female. The opening of the vagina is partially blocked by a membrane called the hymen. Physical activities and sexual intercourse disrupt the hymen.

Development of the embryo normally takes place in the uterus. This organ, sometimes called the womb, is approximately 5 cm wide in its usual state but is capable of stretching to over 30 cm wide to accommodate a growing baby. The lining of the uterus, called the **endometrium,** participates in the formation of the placenta (see section 21.3), which supplies nutrients needed for embryonic and fetal development. In the nonpregnant female, the functional layer of the endometrium varies in thickness according to a monthly cycle of events called the uterine cycle.

A small opening in the cervix leads to the vaginal canal. The **vagina** is a tube that lies at a 45° angle to the small of the back. The mucosal lining of the vagina lies in folds and can extend. This is especially important when the vagina serves as the birth canal, and it facilitates sexual intercourse, when the vagina receives the penis. The vagina also acts as the exit for menstrual flow.

External Genitals

The external genital organs of the female are known collectively as the **vulva** (Fig. 21.6). The vulva includes two large, hair-covered folds of skin called the labia majora (sing., labium major). The labia majora extend backward from the mons pubis, a fatty prominence underlying the pubic hair. The labia minora are two small folds lying just inside the labia majora. They extend forward from the vaginal opening to encircle and form a foreskin for the glans clitoris. The glans clitoris is the organ of sexual arousal in females and, like the penis, contains a shaft of erectile tissue that becomes engorged with blood during sexual stimulation.

The cleft between the labia minora contains the openings of the urethra and the vagina. The vagina may be partially closed by a ring of tissue called the hymen. The hymen is ordinarily ruptured by sexual intercourse or by other types of physical activity. If remnants of the hymen persist after sexual intercourse, they can be surgically removed.

Notice that the urinary and reproductive systems in the female are entirely separate. The urethra carries only urine, and the vagina serves only as the birth canal and the organ for sexual intercourse.

Orgasm in Females

Upon sexual stimulation, the labia minora, the vaginal wall, and the clitoris become engorged with blood. The breasts also swell, and the nipples become erect. The labia majora enlarge, redden, and spread away from the vaginal opening.

The vagina expands and elongates. Blood vessels in the vaginal wall release small droplets of fluid that seep into the vagina and lubricate it. Mucus-secreting glands beneath the labia minora on either side of the vagina also provide lubrication for entry of the penis into the vagina. Although the vagina is the organ of sexual intercourse in females, the extremely sensitive clitoris plays a significant role in the female sexual response. The thrusting of the penis and the pressure of the pubic symphyses of the partners acts to stimulate the clitoris, which may swell to two or three times its usual size.

Orgasm occurs at the height of the sexual response. Blood pressure and pulse rate rise, breathing quickens, and the walls of the vagina, uterus, and oviducts contract rhythmically. A sensation of intense pleasure is followed by relaxation when organs return to their normal size. Females have little or no refractory period between orgasms, and multiple orgasms can occur during a single sexual experience.

Check Your Progress

1. What does it mean for a woman to "have her tubes tied"?
2. Explain how males and females differ in the degree of specialization of their genital tract for reproduction versus urination.
3. Which events occur during a female orgasm that may increase the chances of fertilization?

Visual Focus

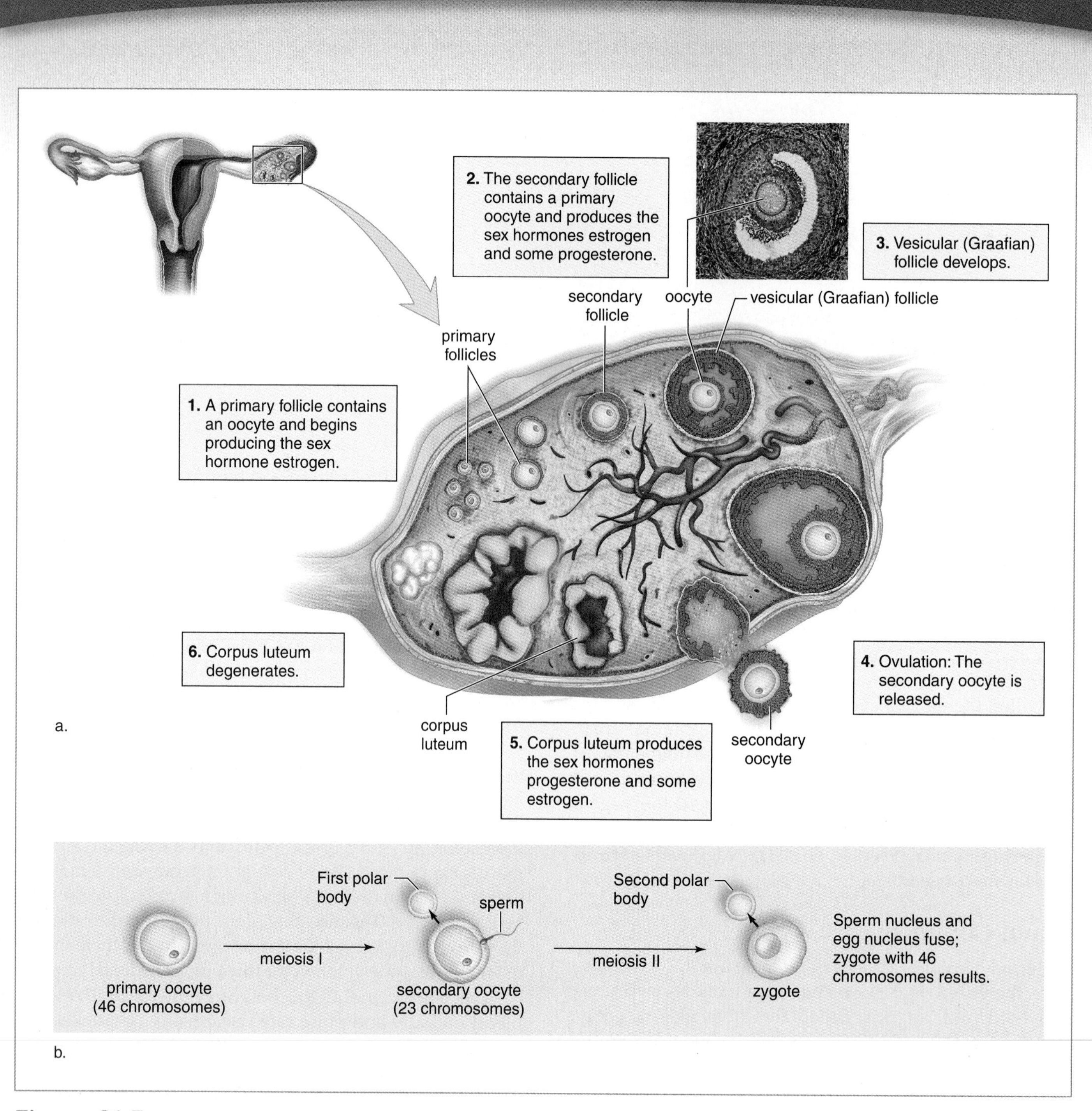

Figure 21.7 Ovarian cycle.
a. A single follicle goes through six stages in one place within the ovary. As a follicle matures, layers of follicle cells surround the developing oocyte. Eventually, the mature follicle ruptures, and the secondary oocyte is released. The follicle then becomes the corpus luteum, which eventually disintegrates. **b.** During oogenesis, the chromosome number is reduced from 46 to 23. Fertilization restores the full number of chromosomes.

21.3 Female Hormone Levels

Learning Outcomes

1. Explain the influence of hormones produced by the anterior pituitary on each phase of the ovarian cycle.
2. Indicate how estrogen and progesterone affect the uterus.
3. Describe changes that occur in the uterus during menstruation, pregnancy, and menopause.

Hormone levels cycle in the female on a monthly basis, and the ovarian cycle drives the uterine cycle.

The Ovarian Cycle

A longitudinal section through an ovary shows that it is made up of an outer cortex and an inner medulla (Fig. 21.7*a*). In the cortex are many **follicles,** each one containing an immature oocyte. A female is born with all the ovarian follicles she will ever have, an estimated 700,000. However, only about 400 of these follicles will ever mature because a female usually produces only one oocyte per month during her reproductive years. Since oocytes are present at birth, they age as the woman ages. This may be one reason older women are more likely to produce children with genetic defects.

The **ovarian cycle** occurs as a follicle changes from a primary to a secondary to a vesicular (Graafian) follicle (Fig. 21.7*a*). Epithelial cells of a primary follicle surround a primary oocyte. Pools of follicular fluid surround the oocyte in a secondary follicle. In a vesicular follicle, a fluid-filled cavity increases to the point that the follicle wall balloons out on the surface of the ovary.

As a follicle matures, oogenesis, depicted in Figure 21.7*b*, is initiated and continues. The primary oocyte divides, producing two haploid cells. One cell is a secondary oocyte, and the other is a polar body. The vesicular follicle bursts, releasing the secondary oocyte. This process is referred to as ovulation. Once a vesicular follicle has lost the secondary oocyte, it develops into a **corpus luteum,** a glandlike structure that produces progesterone.

The secondary oocyte enters an oviduct. If a sperm enters the secondary oocyte, fertilization occurs, and the secondary oocyte completes meiosis. An egg with 23 chromosomes and a second polar body result. When the sperm nucleus unites with the egg nucleus, a zygote with 46 chromosomes is produced. If zygote formation and pregnancy do not occur, the corpus luteum begins to degenerate after about ten days.

Phases of the Ovarian Cycle

The ovarian cycle is commonly divided into two phases. The first half of the cycle is called the follicular phase, and the second half is the luteal phase. During the *follicular phase,* follicle-stimulating hormone (FSH), produced by the anterior pituitary, promotes the development of a follicle in the ovary, which secretes estrogen and some progesterone

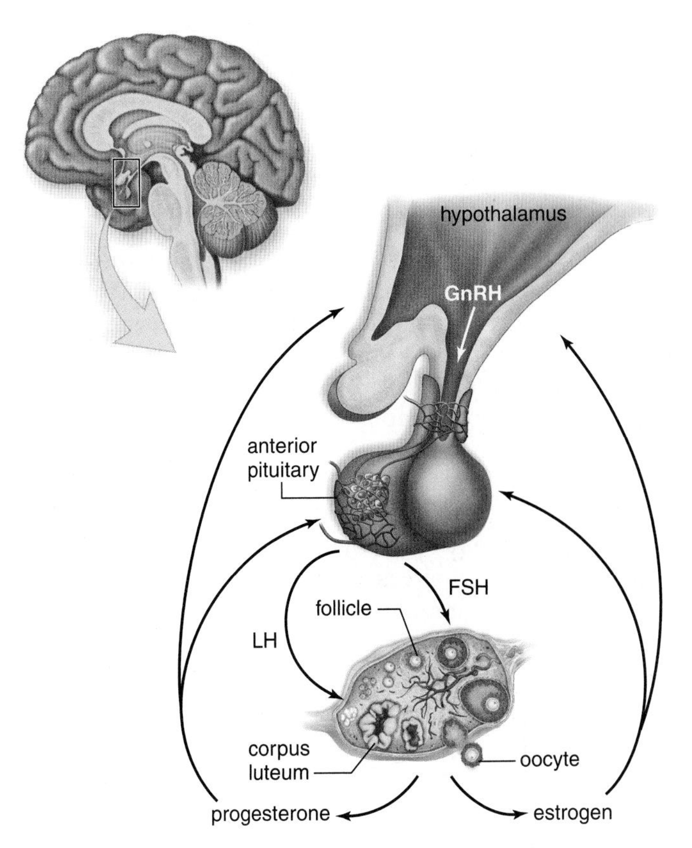

Figure 21.8 Hormonal control of ovaries.
The hypothalamus produces GnRH, which stimulates the anterior pituitary to produce FSH and LH. FSH stimulates the follicle to produce primarily estrogen, and LH stimulates the corpus luteum to produce primarily progesterone. Estrogen and progesterone maintain the sex organs (e.g., uterus) and the secondary sex characteristics, and they exert feedback control over the hypothalamus and the anterior pituitary. Feedback control regulates the relative amounts of estrogen and progesterone in the blood.

(Fig. 21.8). As the estrogen level in the blood rises, it exerts negative feedback control over the anterior pituitary secretion of FSH so that the follicular phase comes to an end.

Presumably, an estrogen spike causes a sudden secretion of a large amount of GnRH from the hypothalamus. This leads to a surge of luteinizing hormone (LH) from the anterior pituitary, which causes ovulation at about the 14th day of a 28-day cycle.

Now, the *luteal phase* begins. During this phase, LH promotes the development of the corpus luteum, which secretes progesterone and some estrogen. As the blood level of progesterone rises, it exerts feedback control over the anterior pituitary secretion of LH so that the corpus luteum in the ovary begins to degenerate. As the luteal phase comes to an end, the low levels of progesterone and estrogen in the body cause menstruation to begin, as discussed next.

Check Your Progress

1. Define corpus luteum and explain its function.
2. What events are occurring during the two phases of the ovarian cycle?

The Uterine Cycle

The female sex hormones, **estrogen** and **progesterone,** have numerous functions. One of their functions is to affect the endometrium, causing the uterus to undergo a cyclical series of events known as the **uterine cycle** (Fig. 21.9, *bottom*). Twenty-eight-day cycles are divided as follows:

During *days 1–5,* a low level of female sex hormones in the body causes the endometrium to disintegrate and its blood vessels to rupture. On day 1 of the cycle, a flow of blood and tissues passes out of the vagina during **menstruation,** also called the menstrual period.

During *days 6–13,* increased production of estrogen by a new ovarian follicle in the ovary causes the endometrium to thicken and become vascular and glandular. This is called the *proliferative phase* of the uterine cycle.

On *day 14* of a 28-day cycle, ovulation usually occurs.

During *days 15–28,* increased production of progesterone by the corpus luteum in the ovary causes the endometrium of the uterus to double or triple in thickness (from 1 mm to 2–3 mm) and the uterine glands to mature, producing a thick mucous secretion. This is called the *secretory phase* of the uterine cycle. The endometrium is now prepared to receive the developing embryo. If this does not occur, the corpus luteum in the ovary degenerates, and the low level of sex hormones in the female body results in the endometrium breaking down during menstruation.

Table 21.3 and Figure 21.9 compare the stages of the uterine cycle with those of the ovarian cycle.

Menstruation

During menstruation, arteries that supply the uterine lining constrict and the capillaries weaken. Blood spilling from the damaged vessels detaches layers of the lining, not all at once, but in random patches. Endometrium, mucus, and blood descend from the uterus, and through the vagina, creating menstrual flow. Fibrinolysin, an enzyme released by dying cells, prevents the blood from clotting. Normal menstruation lasts from 3–10 days. Some abdominal cramping, breast tenderness, and even moodiness are considered to be normal during the menstrual period. See section 21.5 for a discussion of abnormal menstruation.

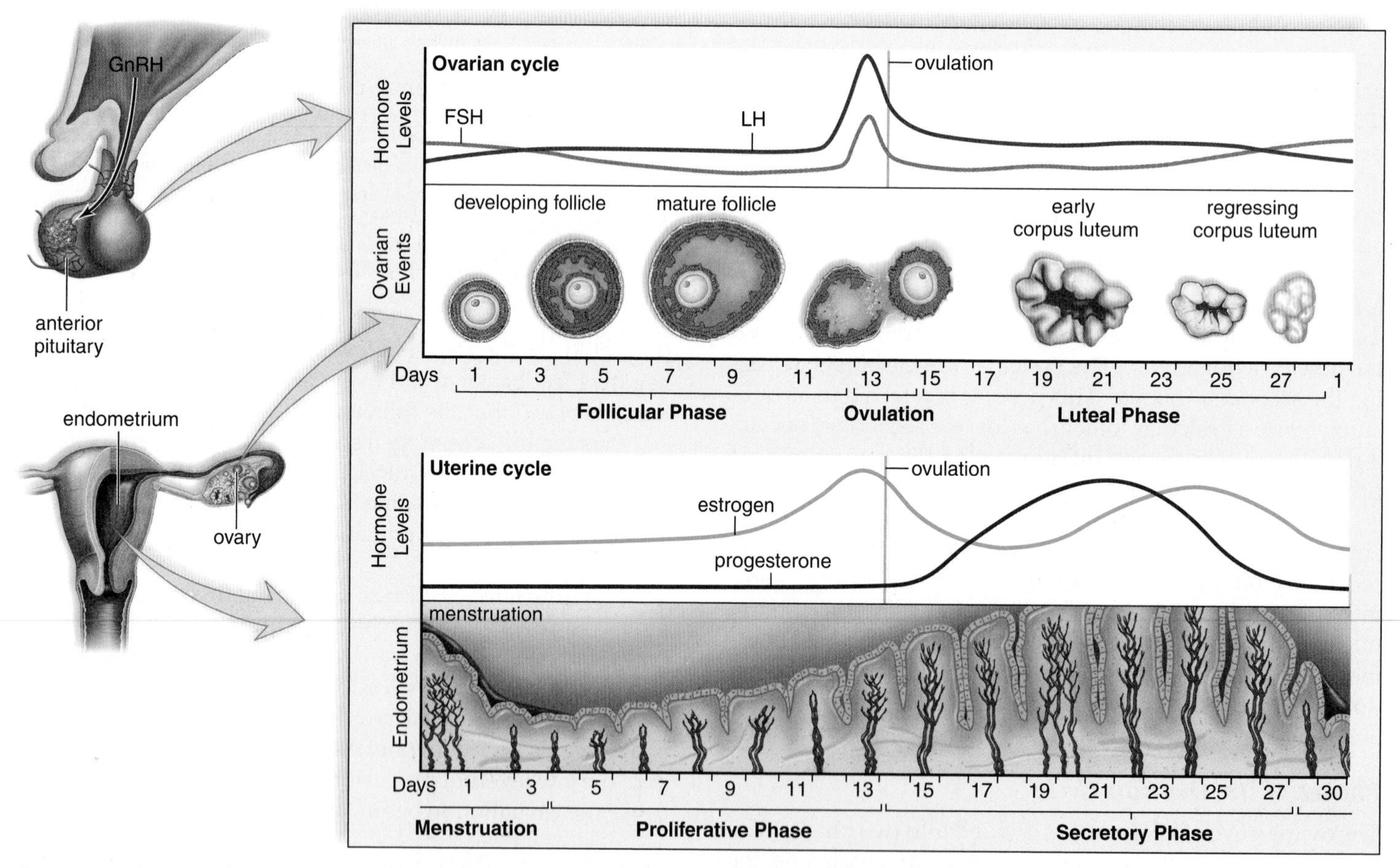

Figure 21.9 Female hormone levels during the ovarian and uterine cycles.
During the follicular phase of the ovarian cycle (*top*), FSH released by the anterior pituitary promotes the maturation of a follicle in the ovary. The ovarian follicle produces increasing levels of estrogen, which causes the endometrium to thicken during the proliferative phase of the uterine cycle (*bottom*). After ovulation and during the luteal phase of the ovarian cycle, LH promotes the development of the corpus luteum. This structure produces increasing levels of progesterone, which causes the endometrium to become secretory. Menstruation and the proliferative phase begin when progesterone production declines to a low level.

TABLE 21.3 Ovarian and Uterine Cycles

Ovarian Cycle	Events	Uterine Cycle	Events
Follicular phase—Days 1–13	FSH secretion begins.	Menstruation—Days 1–5	Endometrium breaks down.
	Follicle maturation occurs. Estrogen secretion is prominent.	Proliferative phase—Days 6–13	Endometrium rebuilds.
Ovulation—Day 14*	LH spike occurs.		
Luteal phase—Days 15–28*	LH secretion continues. Corpus luteum forms.	Secretory phase—Days 15–28	Endometrium thickens, and glands are secretory.

*Assuming a 28-day cycle.

Fertilization and Pregnancy

If fertilization does occur, an embryo begins development even as it travels down the oviduct to the uterus. The endometrium is now prepared to receive the developing embryo, which becomes implanted in the lining several days following fertilization (Fig. 21.10). Pregnancy has now begun.

The **placenta,** which sustains the developing embryo and later fetus, originates from both maternal and fetal tissues. It is the region of exchange of molecules between fetal and maternal blood, although the two rarely mix. At first, the placenta produces **human chorionic gonadotropin (hCG),** which maintains the corpus luteum in the ovary until the placenta begins its own production of progesterone and estrogen. A pregnancy test can usually detect the presence of hCG in the blood or urine by 10 days after fertilization.

Progesterone and estrogen produced by the placenta have two effects: they shut down the anterior pituitary so that no new follicle in the ovaries matures, and they maintain the endometrium so that the corpus luteum in the ovary is no longer needed. Usually, no menstruation occurs during pregnancy.

Estrogen and Progesterone

Estrogen and progesterone affect not only the uterus but other parts of the body as well. Estrogen is largely responsible for the secondary sex characteristics in females, including body hair and fat distribution. In general, females have a more rounded appearance than males because of a greater accumulation of fat beneath the skin. The pelvic girdle becomes wider and deeper in females, so the pelvic cavity usually has a larger relative size in women compared to men. Like males, females develop axillary and pubic hair during puberty. Both estrogen and progesterone are also required for breast development.

Menopause

Menopause, the period in a woman's life during which the ovarian and uterine cycles cease, usually occurs between ages 45 and 55. The ovaries become unresponsive to the gonadotropic hormones produced by the anterior pituitary, and they no longer secrete estrogen or progesterone. At the onset of menopause, the uterine cycle becomes irregular, but as long as menstruation occurs, it is still possible for a woman to conceive. Therefore, a woman is usually not considered to have completed menopause (and thus be infertile) until menstruation has been absent for a year.

The hormonal changes during menopause often produce physical symptoms, such as "hot flashes" (caused by circulatory irregularities), dizziness, headaches, insomnia, sleepiness, and depression. Until recently, many women took combined estrogen-progestin drugs to ease menopausal symptoms. However, accumulating data shows long-term use of the combined drugs by postmenopausal women may cause increases in breast cancer, heart attacks, strokes, and blood clots.

CHECK YOUR PROGRESS

1. Distinguish between the proliferative phase and the secretory phase of the uterine cycle and the hormones that promote each.
2. What are the functions of the hormones produced by the placenta?

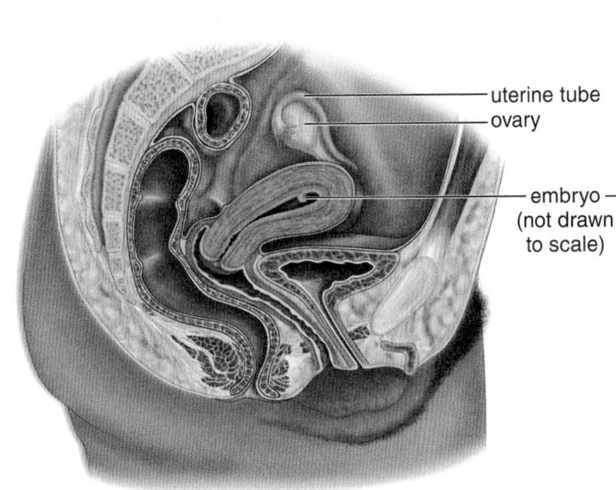

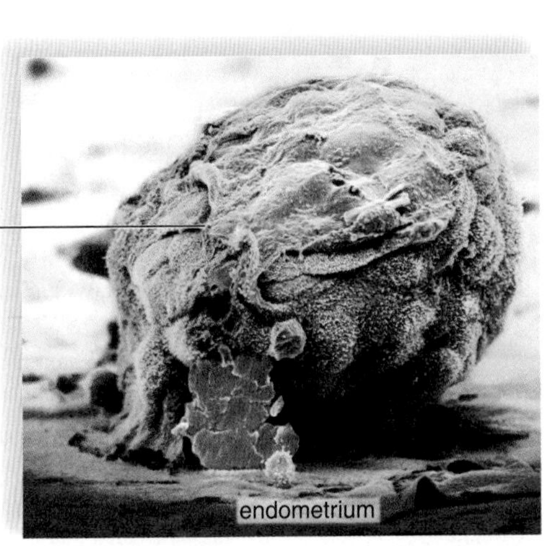

Figure 21.10 Implantation.
a. Site of implantation of an embryo in the uterine wall. **b.** A scanning electron micrograph showing an embryo implanted in the endometrium on day 12 following fertilization.

21.4 Control of Reproduction

Learning Outcomes

1. List five commonly available methods of birth control that are designed to prevent conception.
2. Note which birth control methods are effective at preventing the spread of sexually transmitted diseases.
3. Describe the mechanism of action of two types of so-called "morning-after pills."

Several means are available to dampen or enhance our reproductive potential. Birth control methods are used to regulate the number of children an individual or couple will have.

Birth Control Methods

The most reliable method of birth control is abstinence—that is, not engaging in sexual contact. This form of birth control has the added advantage of preventing the spread of sexually transmitted diseases (STDs). Table 21.4 lists other means of birth control used in the United States, and rates their effectiveness (when used correctly). For example, the birth control pill is nearly 100% effective, meaning that very few sexually active women will get pregnant while taking the pill. On the other hand, with natural family planning, one of the least effective methods, we expect that 70% will not get pregnant and 30% will get pregnant within the year.

Figure 21.11 features some of the most effective and commonly used means of birth control. **Contraceptives** are medications and devices that reduce the chance of pregnancy. Oral contraception **(birth control pills)** often involves taking a combination of estrogen and progesterone on a daily basis (Fig. 21.11*a*), which mimics the hormonal situation of pregnancy. The hormones in the birth control pill effectively shut down the pituitary production of both FSH and LH so that no follicle begins to develop in the ovary. Because ovulation does not occur, pregnancy cannot take place. Women taking birth control pills should see a physician regularly due to possible side effects.

An **intrauterine device (IUD)** is a small piece of molded plastic that is inserted into the uterus by a physician. IUDs are believed to alter the environment of the uterus and oviducts so that fertilization probably will not occur—but if fertilization should occur, implantation cannot take place. The type of IUD featured in Figure 21.11*b* has copper wire wrapped around the plastic.

TABLE 21.4 Common Birth Control Methods*

Name	Procedure	Methodology	Effectiveness	Risk
Abstinence	Refrain from sexual intercourse	No sperm in vagina	100%	None
Vasectomy	Vasa deferentia are cut and tied	No sperm in semen	Almost 100%	Irreversible sterility
Tubal ligation	Oviducts cut and tied	No oocytes in oviduct	Almost 100%	Irreversible sterility
Oral contraception	Hormone medication taken daily	Anterior pituitary does not release FSH and LH	Almost 100%	Increased risk of heart attack, stroke, and thromboembolism
Contraceptive implants	Tubes of progestin (form of progesterone) implanted under skin	Anterior pituitary does not release FSH and LH	More than 90%	Similar to other hormone treatments; rare problems with insertion/removal
Contraceptive injections/patch	Injections of hormones	Anterior pituitary does not release FSH and LH	About 99%	Similar to other hormone treatments
Vaginal contraceptive ring	Ring inserted into vagina each month	Anterior pituitary does not release FSH and LH	92% to 99%	May be somewhat lower than other hormone treatments
Intrauterine device (IUD)	Plastic coil inserted into uterus by physician	Prevents implantation	More than 90%	PID[1]; rare problems with insertion/removal
Diaphragm	Latex cup inserted into vagina to cover cervix before intercourse	Blocks entrance of sperm to uterus	With jelly, about 90%	Latex, spermicide allergy
Cervical cap	Latex cap held by suction over cervix	Delivers spermicide near cervix	Almost 85%	UTI[2]; latex or spermicide allergy
Male condom	Latex sheath fitted over erect penis	Traps sperm and prevents STDs	About 85%	Latex allergy
Female condom	Polyurethane liner fitted inside vagina	Blocks entrance of sperm to uterus and prevents STDs	About 85%	Latex allergy
Coitus interruptus	Penis withdrawn before ejaculation	Prevents sperm from entering	About 75%	Presently none known
Jellies, creams, foams	These spermicidal products are inserted before intercourse	Kill a large number of sperm	About 75%	UTI[2]; vaginitis, allergy
Natural family planning	Day of ovulation determined by record keeping, various methods of testing	Intercourse avoided on certain days of the month	About 70%	Presently none known
Douche	Vagina cleansed after intercourse	Washes out sperm	Less than 70%	Increased risk of infections and PID[1]

1 = pelvic inflammatory disease; 2 = urinary tract infection **Assuming a 28-day cycle.*

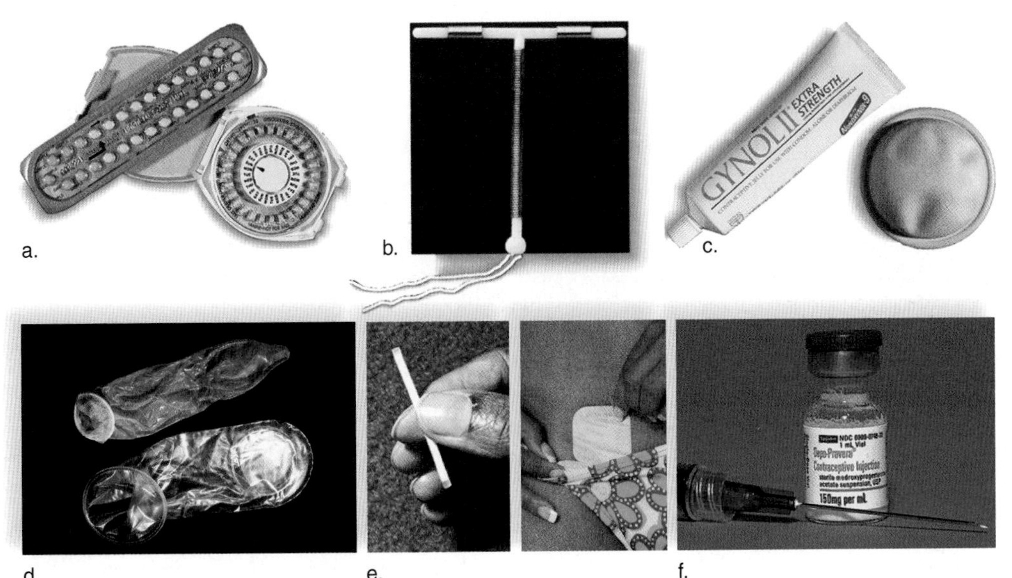

Figure 21.11 Various methods of birth control.
a. Oral contraception (birth control pills). **b.** Intrauterine device. **c.** Spermicidal jelly and diaphragm. **d.** Male and female condoms. **e.** Contraceptive implant and patch. **f.** Depo-Provera, a progesterone injection.

The **diaphragm** is a soft latex cup with a flexible rim that lodges behind the pubic bone and fits over the cervix (Fig. 21.11*c*). Each woman must be properly fitted by a physician, and the diaphragm can be inserted into the vagina no more than two hours before sexual relations. Also, it must be used with spermicidal jelly or cream and should be left in place for at least six hours after sexual relations. The cervical cap is a minidiaphragm.

In addition to helping prevent pregnancy, both male and female condoms provide some protection against sexually transmitted diseases (Fig. 21.11*d*). A **male condom** is most often a latex sheath that fits over the erect penis. The ejaculate is trapped inside the sheath, and thus does not enter the vagina. A **female condom** consists of a large polyurethane tube with a flexible ring that fits onto the cervix. The open end of the tube has a ring that covers the external genitals. Better protection is achieved when a condom is used in conjunction with a spermicide.

Contraceptive implants (Fig. 21.11*e*) utilize synthetic progesterone to prevent ovulation by disrupting the ovarian cycle. The newest version consists of a single capsule that remains effective for about three years. A **contraceptive patch** is also available. Known by the brand name Ortho Evra, the patch is applied to the skin once a week for three weeks, then no patch is worn during the menstrual period. Because women who use the patch are exposed to higher levels of estrogen than those who use oral contraceptives, the risk of side effects is somewhat higher with the patch.

Contraceptive injections are available as progesterone only (Fig. 21.11*f*) or as a combination of estrogen and progesterone. The length of time between injections can vary from three months to a few weeks.

One of the newer developments in hormonal contraceptives is the **vaginal contraceptive ring.** Also known by its brand name, NuvaRing, it is a soft, flexible, vinyl ring about two inches in diameter, that contains estrogen and progesterone. Currently available by prescription only, it is inserted into the vagina on about the fifth day after menstruation ceases, and left in place for 21 days. Used properly, the NuvaRing prevents ovuation for that month's cycle. It may also thicken the cervical mucus, thereby preventing the sperm from entering the uterus.

Contraceptive vaccines are now being developed. For example, a vaccine intended to immunize women against hCG, the hormone so necessary to maintaining the implantation of the embryo, was successful in a limited clinical trial. Because hCG is not normally present in the body, no autoimmune reaction is expected, but the immunization does wear off with time. It may also be possible to develop a safe antisperm vaccine for use in women.

Morning-After Pills

A morning-after pill, or emergency contraception, is a medication that can prevent pregnancy after unprotected intercourse. The expression "morning-after pill" is a misnomer in that a woman can begin taking the medication one to several days after unprotected intercourse.

One type, a kit called Preven, contains four synthetic progesterone pills; two are taken up to 72 hours after unprotected intercourse, and two more are taken 12 hours later. The medication upsets the normal uterine cycle, making it difficult for an embryo to implant itself in the endometrium. A recent study estimated that the medication was 85% effective in preventing unintended pregnancies.

Mifepristone, better known as RU-486, is a pill that prevents a fertilized embryo from implanting by blocking the progesterone receptor proteins of endometrial cells. Without functioning receptors for progesterone, the endometrium sloughs, carrying the embryo with it. When taken in conjunction with a prostaglandin to induce uterine contractions, RU-486 is 95% effective. In August 2006, the United States Food and Drug Administration approved another drug called Plan B, which is up to 89% effective in preventing pregnancy if taken within 72 hours after unprotected sex. It is available to women age 18 and older without a prescription.

Check Your Progress

1. Rank the following birth control methods regarding their effectiveness at preventing pregnancy: abstinence, birth control pill, condoms, natural family planning, and vasectomy.
2. Which of the birth control methods discussed in this section physically block sperm from entering the uterus?
3. Why is the "morning-after pill" more controversial for some people than other types of birth control?

Endocrine-Disrupting Contaminants

Rachel Carson's book *Silent Spring*, published in 1962, predicted that pesticides would have a deleterious effect on animal life. Soon thereafter, it was found that pesticides caused the eggshells of bald eagles to become so thin that their eggs broke and the chicks died. Additionally, populations of terns, gulls, cormorants, and lake trout declined after they ate fish contaminated by high levels of environmental toxins. The concern was so great that the U.S. Environmental Protection Agency (EPA) came into existence. This agency together with civilian environmental groups have brought about a reduction in pollution release and a cleaning up of emissions. Even so, we are now aware of more subtle effects of pollutants.

Hormones influence nearly all aspects of physiology and behavior in animals. Therefore, when wildlife in contaminated areas began to exhibit certain abnormalities, researchers began to think that certain pollutants could affect the endocrine system. In England, male fish exposed to sewage developed ovarian tissue and produced a metabolite normally found only in females during egg formation. In California, western gulls displayed abnormalities in gonad structure and nesting behaviors. Hatchling alligators in Florida possessed abnormal gonads and hormone concentrations linked to nesting.

At first, such effects seemed to involve only the female hormone estrogen, and researchers therefore called the contaminants ecoestrogens. Many of the contaminants interact with hormone receptors, and in that way cause developmental effects. Others bind directly with sex hormones, such as testosterone and estradiol (a potent estrogen). Still others alter the physiology of growth hormones and neurotransmitters responsible for brain development and behavior. Therefore, the preferred term today for these pollutants is endocrine-disrupting contaminants (EDCs).

Many EDCs are chemicals used as pesticides and herbicides in agriculture, and some are associated with the manufacture of various other organic molecules, such as PCBs (polychlorinated biphenyls). Some chemicals shown to influence hormones are found in plastics, food additives, and personal hygiene products. In mice, phthalate esters, which are plastic components, affect neonatal development when present in the part-per-trillion range. It is of great concern that EDCs have been found at levels comparable to functional hormone levels in the human body. Therefore, it is not surprising that EDCs are affecting the endocrine systems of a wide range of organisms (Fig. 21A).

Scientists and representatives of industrial manufacturers continue to debate whether EDCs pose a health risk to humans. Some suspect that EDCs lower sperm counts, reduce male and female fertility, and increase rates of certain cancers (breast, ovarian, testicular, and prostate). Additionally, some studies suggest that EDCs contribute to learning deficits and behavioral problems in children.

Laboratory and field research continues to identify chemicals that have the ability to influence the endocrine system. Millions of tons of potential EDCs are produced annually in the United States, and the EPA is under pressure to certify these compounds as safe. The European Economic Community has already restricted the use of certain EDCs and has banned the production of specific plastic components intended for use by children. Only through continued scientific research and the cooperation of industry can we identify the risks the EDCs pose to the environment, wildlife, and humans.

Discussion Questions

1. There are undeniable benefits to the use of pesticides, one of the major ones being more attractive, insect-free food. How would you personally weigh these benefits compared to the risks of ingesting endocrine-disrupting contaminants?
2. How might endocrine-disrupting chemicals specifically affect cells? (You may need to review how peptide and steroid hormones work in section 20.1.)
3. What would be some implications for agricultural and manufacturing industries if the use of all potential endocrine-disrupting chemicals were banned? How much extra would you be willing to pay for your monthly food (and other products) if their costs increased?

Figure 21A Exposure to endocrine-disrupting contaminants. Various types of wildlife, as well as humans, are exposed to endocrine-disrupting contaminants that can seriously affect their health and reproductive abilities.

21.5 Disorders of the Reproductive System

Learning Outcomes

1. List three STDs that are caused by viruses and three that are caused by bacteria.
2. Describe how HIV specifically affects the immune system and how this explains the three categories of HIV infection.
3. Discuss several common noninfectious conditions affecting the male and female reproductive systems.

In this section, we first discuss diseases that can be transmitted by sexual contact, such as AIDS and hepatitis, but that do not necessarily affect the reproductive organs themselves. Then we describe some of the most common conditions affecting the reproductive system.

Sexually Transmitted Diseases

Many diseases are transmitted by sexual contact and are therefore called sexually transmitted diseases (STDs). This discussion centers on those that are most prevalent. AIDS, genital herpes, genital warts, and hepatitis B are caused by viruses. Therefore, they do not respond to traditional antibiotics. Antiviral drugs have been developed to treat many of these diseases, but in many cases infection is lifelong. Chlamydia, gonorrhea, and syphilis are bacterial diseases that are usually curable with appropriate antibiotic therapy if diagnosed early enough. However, an increasing number of these bacteria are becoming resistant to antibiotics.

AIDS

Acquired immunodeficiency syndrome (AIDS) is caused by a virus known as HIV (human immunodeficiency virus). AIDS is considered a pandemic because the disease is prevalent in the entire human population around the globe. As of the end of 2007, the World Health Organization reports that AIDS has killed more than 25 million people worldwide, and about 33 million more are currently estimated to be infected with HIV. The epidemic has been particularly devastating in sub-Saharan Africa, where 30–40% of the population is infected in some countries. It is also estimated that AIDS could kill 31 million people in India and 18 million in China by the year 2025. In the United States, an estimated one million people are infected, and over 500,000 have died. Homosexual men make up the largest proportion of people with AIDS in the United States, but the fastest rate of increase is now seen among heterosexuals, and over half of all new HIV infections are occurring in people under the age of 25.

HIV is transmitted by sexual contact with an infected person, including vaginal or anal intercourse and oral/genital contact. Also, needle-sharing among intravenous drug users is a very efficient way to transmit HIV. The Health Focus on page 437 suggests ways to protect yourself against HIV and other STDs. Babies born to HIV-infected mothers may become infected before or during birth or through breast-feeding after birth. Cultural factors may also play a very important role in the transmission of HIV. For example, polygamy is common in many African countries, as are misconceptions about the effectiveness of birth control, and, indeed, about the fact that HIV is the cause of AIDS. Such practices and beliefs may significantly increase the risk of HIV infection.

Stages of an HIV Infection When a person initially becomes infected with HIV, the virus replicates at a high level in the macrophages and $CD4^+$ helper T cells, and spreads throughout the lymphoid tissues of the body (Fig. 21.12). Even though the person's blood contains a high "viral load" at this point, most experience only mild symptoms such as fever, chills, aches, and swollen lymph nodes, and soon these symptoms disappear. Even if a person had an HIV test at this time, it would probably be negative because it tests for the presence of anti-HIV antibodies, which are not detectable for an average of 25 days. This means that early in the infection, a person can be highly infectious, even though the HIV antibody test is negative.

Usually no other HIV-related symptoms occur for several years. Even so, the virus is beginning to cause a loss of $CD4^+$ T cells, which is one of the hallmarks of an HIV infection that is progressing to AIDS. HIV is thought to kill these cells directly by infecting them and also indirectly by inducing apoptosis, even of uninfected cells. Cytotoxic T cells are also thought to kill large numbers of HIV-infected cells. During this stage of infection, the body is able to maintain a reasonably healthy immune system by producing as many as one to two billion new $CD4^+$ helper T cells per day. As long as the body's production of new $CD4^+$ T cells is able to keep pace with the loss of $CD4^+$ T cells (much like leaving the water running in a sink with the drain open), the person stays relatively healthy. However, even during this asymptomatic stage, evidence indicates that the virus continues to replicate

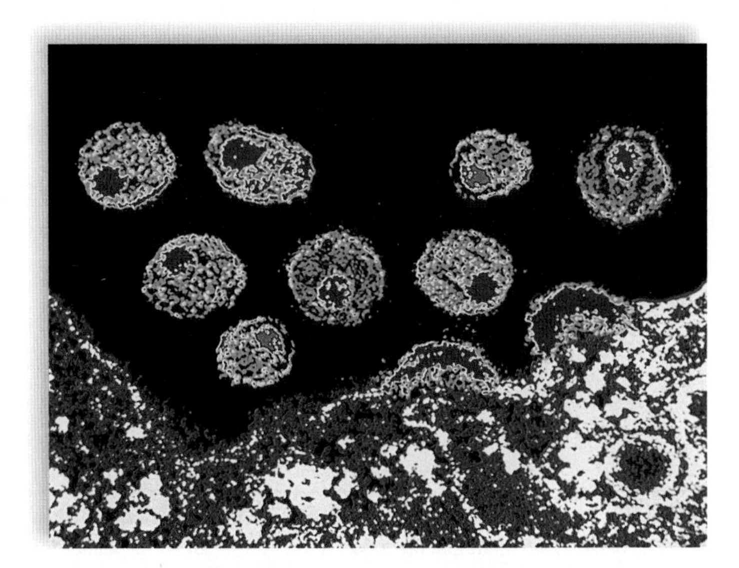

Figure 21.12 **HIV, the AIDS virus.**
False-colored micrographs show HIV particles budding from an infected helper T cell. These viruses can infect helper T cells and also macrophages, which work with helper T cells to stem the infection.

at a high rate. Eventually, the number of CD4$^+$ T cells falls so low that the patient begins to suffer from **opportunistic infections,** which have the *opportunity* to occur because the immune system is weakened.

In order to help physicians assess the prognosis of individuals at different stages of HIV infection, the U.S. Centers for Disease Control and Prevention have defined three categories of HIV infection, based mainly on the type of clinical disease seen in the patient (Fig. 21.13). Within each category, the number of CD4$^+$ T cells is also monitored. Patients in *category A* usually have CD4$^+$ T-cell counts of 500 or greater per mm^3 of blood, and either lack symptoms completely or have persistently enlarged lymph nodes. Patients in *category B* have symptoms that are indicative of an unhealthy immune system. Their CD4$^+$ T-cell count is usually between 200 and 499 per mm^3 of blood. Examples of category B conditions include thrush, a yeast infection of the oral cavity, and repeated episodes of shingles, a reactivation of a childhood chickenpox virus infection. Patients in *category C* usually have CD4$^+$ T-cell counts less than 200 per mm^3 of blood, and have one or more AIDS-indicator conditions, such as recurrent bacterial pneumonia, various fungal infections, Kaposi's sarcoma (caused by a human herpesvirus), mycobacterial infections, and toxoplasmosis of the brain (see the Health Focus on HIV in Chapter 13).

Although not composing an official category of HIV infection, another small, yet very interesting group of HIV-infected people remain healthy for at least ten years, even without any antiviral treatments. Often called long-term nonprogressors, these people have been very interesting for HIV researchers to study because their viral loads remain low and their CD4$^+$ T-cell counts usually remain normal. As of this writing, no single factor has been identified that explains how some people are able to remain relatively healthy in the face of HIV infection, but their existence continues to provide hope that effective host resistance to HIV is possible.

Treatments for HIV Infection Prior to the development of antiviral drugs, most HIV-infected individuals eventually progressed to category C or full-blown AIDS, which was almost inevitably fatal. There is still no cure for HIV infection, but since 1995 an increasing number of drugs have become available to help control the replication and spread of HIV, and the life span of many HIV-infected patients has increased to the point where HIV is no longer considered an automatic death sentence. Unfortunately, this is not the case in developing nations where most people cannot afford these drugs.

In order to understand how antiretroviral drugs work, a brief review of the life cycle of HIV is necessary (Fig. 21.14). HIV is a type of RNA virus called a retrovirus. HIV has a protein on its outer surface called gp120, which attaches to a CD4 molecule, plus another molecule called CCR5 or CXCR4 on a host T cell, macrophage, or dendritic cell. This attachment causes *fusion* of HIV with the host plasma membrane, releasing HIV's inner core inside the cell. This core contains the viral RNA, plus several viral enzymes. One such enzyme, called *reverse transcriptase,* makes a DNA copy of the viral RNA, called cDNA. Duplication occurs and the result is double-stranded DNA (dsDNA). During *integration,* the viral enzyme integrase inserts the viral dsDNA into host DNA, where it will remain for the life of the cell. Following integration, viral DNA serves as a template for production of more viral RNA. Some of the viral RNA proceeds to the ribosomes, where it brings about the synthesis of long polypeptides that need to be cleaved into smaller pieces before they can be assembled into new viruses. This cleavage process depends on a third viral enzyme, called *protease.* Finally, the new virus particles are assembled and then released from the host plasma membrane by a process called budding.

Currently, four classes of antiretroviral drugs are available: (1) reverse transcriptase inhibitors, which block the production of viral DNA from RNA; (2) protease inhibitors,

a. AIDS patient Tom Moran, at diagnosis

b. AIDS patient Tom Moran, 6 months later

c. AIDS patient Tom Moran, in category C of AIDS

Figure 21.13 The course of an AIDS infection.
These photos show the effect of an HIV infection in one individual who progressed through all the phases of HIV infection. At first, the body can fight off the effects of the virus, but as the infection takes over the number of CD4+ T cells drops from thousands to hundreds. Eventually, an AIDS patient will die from the opportunistic diseases that flourish because the immune system has been destroyed.

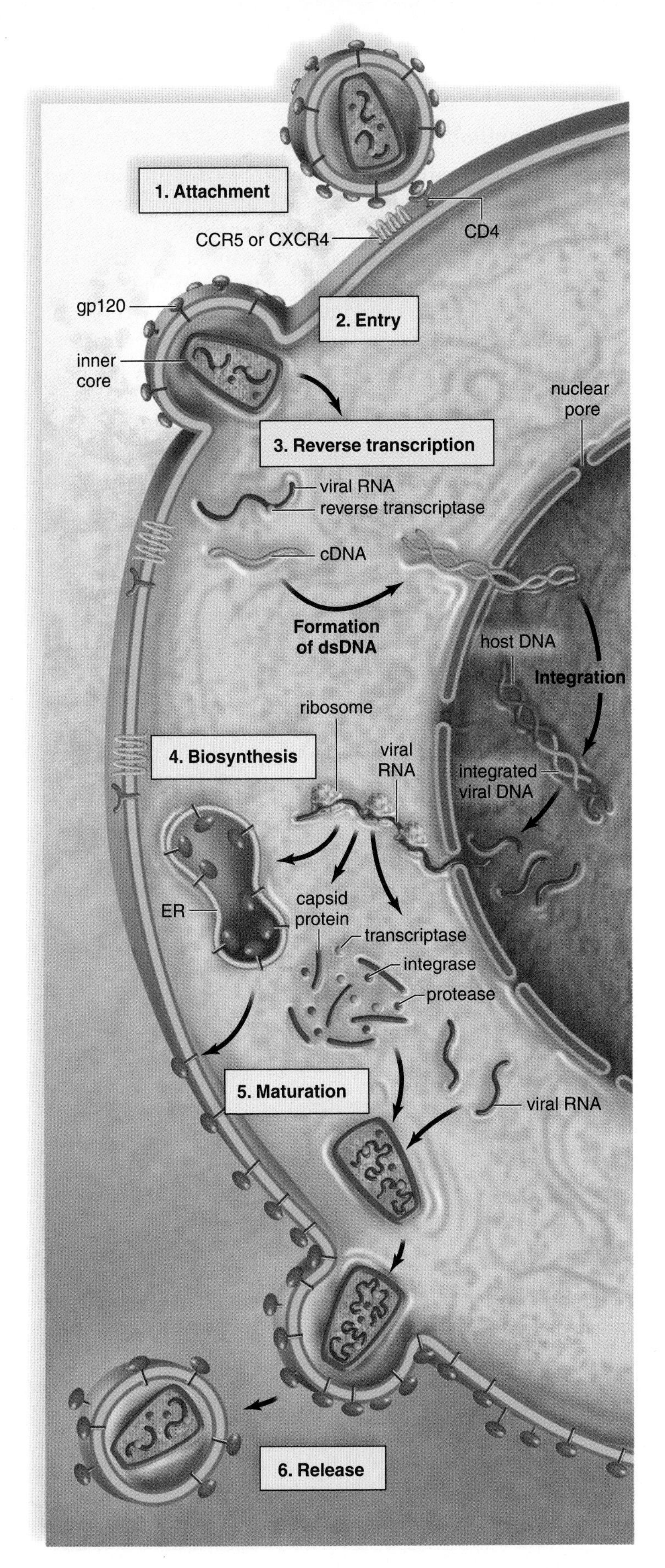

Figure 21.14 Reproduction of the retrovirus HIV. HIV attaches to CD4 plus CCR5 or CXCR4, then uses reverse transcription to produce a double-stranded DNA copy of viral RNA. The viral dsDNA integrates into the cell's chromosomes before the virus reproduces and buds from the cell.

which block the viral enzyme that cleaves long viral polypeptides prior to assembly; (3) fusion inhibitors, which interfere with HIV's ability to fuse with the host plasma membrane during initial infection; and (4) integrase inhibitors, which inhibit the insertion of viral DNA into host DNA. When the first antiretroviral drug, known as AZT, became available in 1995, it was initially very effective at reducing the viral load in many patients, but the virus tended to become resistant to the drug very quickly. This is because the HIV reverse transcriptase enzyme is very error-prone, which means the virus mutates at a very high rate, resulting in drug-resistant mutants. However, today HIV-infected patients are typically treated with three or four antiretroviral drugs at the same time, a combination called highly active antiretroviral therapy (HAART). Because it is unlikely that a particular mutant virus will be resistant to several drugs all at once, HAART is usually able to inhibit HIV replication to the point that no virus is detectable in the blood. It is important to note, however, that these patients presumably still have large numbers of HIV-infected cells, and thus they can still transmit the virus. Other potential problems with HAART include various side effects, which can be severe, the high cost, and the complexity of dosing schedules for the different medications.

Investigators have found that when HAART is discontinued, the virus rebounds. Therefore, therapy is usually continued indefinitely. An HIV-positive pregnant woman who takes reverse transcriptase inhibitors during her pregnancy reduces the chances of HIV transmission to her newborn. If possible, drug therapy should be delayed until the 10th to 12th weeks of her pregnancy to minimize any adverse effects of antiretroviral drugs on fetal development.

HIV Vaccines It is unlikely that the global HIV pandemic will ever be controlled unless an effective vaccine against HIV is developed. Some of the obstacles to developing an effective vaccine include the existence of many strains of HIV, the tendency of HIV to mutate its proteins, thus making it a "moving target" for the immune system, and the ability of HIV to "hide" inside cells. One type of vaccine that has been studied is a recombinant vaccine, in which a piece of DNA encoding an HIV protein is inserted into another organism (usually another type of virus), which can then infect cells and elicit a cytotoxic T-cell response, without any danger of causing an infection with HIV. Other types of vaccines focus more on inducing an antibody response, especially targeted against the gp120 protein (see Fig. 21.14). Unfortunately, most of these trials have yielded disappointing results, and most investigators now agree that a combination of various types of vaccines may be the best strategy.

Check Your Progress

1. Describe three ways that HIV may destroy $CD4^+$ T cells.
2. List four classes of antiretroviral drugs and the specific stage of the HIV life cycle that each class targets.
3. What are some of the major challenges faced by researchers attempting to develop an effective HIV vaccine?

Genital Herpes

Genital herpes is caused by the herpes simplex virus, of which there are two types: type 1, which usually causes cold sores and fever blisters, and type 2, which more often causes genital herpes. In the United States, it is estimated that approximately one in four women, and one in five men, have had a genital herpes infection, but 20% may never have symptoms. Even without symptoms, however, the virus can be transmitted. After becoming infected, most individuals experience a tingling or itching sensation before blisters appear at the infected site within 2–20 days (Fig. 21.15). Once the blisters rupture, they leave painful ulcers, which take from five days to three weeks to heal. These symptoms can be accompanied by fever, pain upon urination, and swollen lymph nodes. Several antiviral medications are available to treat herpesvirus infections. These can be used topically in ointment form, taken orally, or in severe cases, injected. All of these medications inhibit the replication of viral DNA.

After the blisters heal, the virus becomes latent (dormant), and blisters can recur repeatedly at variable intervals. Sunlight, sexual intercourse, menstruation, and stress seem to cause the symptoms of genital herpes to recur. While the virus is latent, it primarily resides in the ganglia of the sensory nerves associated with the affected skin. The ability of the virus to "hide" in this manner has made it more difficult to develop an effective vaccine.

Infection of the newborn can occur if the child comes in contact with a lesion in the birth canal. Within 1–3 weeks, the infant may be gravely ill and can become blind, have neurological disorders, including brain damage, or die. Birth by cesarean section prevents these adverse developments.

Human Papillomavirus

An estimated 20 million Americans are currently infected with the human papillomavirus (HPV). There are over 100 types of HPV. Most cause warts, and about 30 types cause **genital warts,** which are sexually transmitted. Genital warts may appear as flat or raised warts on the penis and/or foreskin of males, and on the vulva, vagina, and/or cervix of females. Note that if the wart(s) are only on the cervix, there may be no outward signs or symptoms of the disease. Newborns can also be infected with HPV during passage through the birth canal.

About ten types of HPV can cause cancer, especially cancer of the cervix, the second leading cause of cancer death in women in the United States (approximately 500,000 deaths per year). These HPV types produce a viral protein that inactivates a host protein called p53, which normally acts as a "brake" on cell division. Once p53 has been inactivated in a particular cell, that cell is more prone to the uncontrolled cell division characteristic of cancer. Early detection of cervical cancer is possible by means of a **Pap smear,** in which a few cells are removed from the region of the cervix for microscopic examination. If the cells are cancerous, a hysterectomy (removal of the uterus) may be recommended. In males, HPV can cause cancers of the penis, anus, and other areas. According to a 2008 study, HPV now causes as many cancers of the mouth and throat in U.S. males as does tobacco, perhaps due to an increase in oral sex as well as a decline in tobacco use.

Currently, there is no cure for an HPV infection, but the warts can be treated effectively by surgery, freezing, application of an acid, or laser burning. However, even after treatment, the virus can sometimes be transmitted. Therefore, once

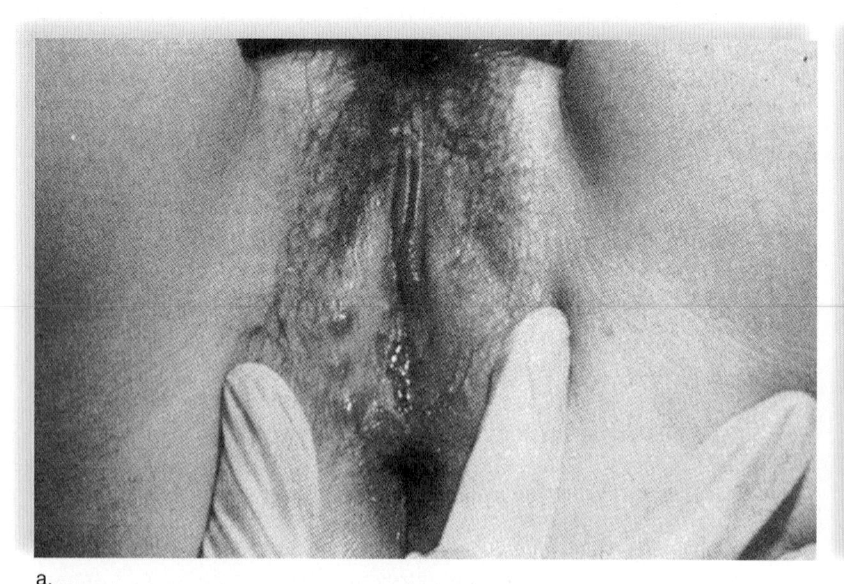
a.

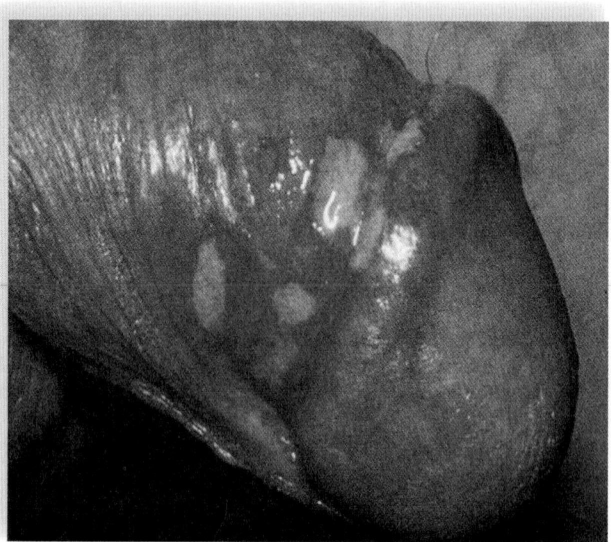
b.

Figure 21.15 Genital herpes.
There are several types of herpesviruses, and usually the type called herpes simplex 2 causes genital herpes. About 1 million persons become infected each year, most of them teens and young adults. Symptoms of genital herpes are due to an outbreak of blisters, which can be present on the labia of females **(a)** or on the penis of males **(b)**.

someone has been diagnosed with genital warts, abstinence or the use of a condom is recommended to prevent transmission of the virus.

In June 2006, the U.S. Food and Drug Administration licensed Gardasil, an HPV vaccine that is effective against the four most common types of HPV found in the United States, including the two types that cause about 70% of cervical cancers. Because the vaccine doesn't protect those who are already infected, ideally children should be vaccinated before they become sexually active. This vaccine has been controversial, however. For example, in February 2007, the governor of Texas issued an executive order requiring that all girls entering sixth grade be vaccinated for HPV. Within two months, however, the Texas legislature passed a bill blocking public officials from carrying out the mandatory vaccination.

Hepatitis B

Of the several types of viruses that can cause hepatitis, or inflammation of the liver, hepatitis B is the most likely to be transmitted sexually. Hepatitis B virus (HBV) is more contagious than HIV, and HBV is transmitted by similar routes. About 50% of persons infected with HBV develop flulike symptoms, including fatigue, fever, headache, nausea, vomiting, and muscle aches. As the virus replicates in the liver, a dull pain in the upper right half of the abdomen may occur, as may jaundice, a yellowish cast to the skin. Some persons have an acute infection that only lasts a few weeks, while others develop a chronic form of the disease that leads to liver failure and the need for a liver transplant. A safe and effective vaccine is available for the prevention of HBV infection. This vaccine is now on the list of recommended immunizations for children.

Check Your Progress

1. Why are viral diseases, in general, more difficult to treat than bacterial diseases?
2. What is the significance of latency in the ability of herpes simplex viruses to be transmitted, cause disease, and be resistant to vaccines?
3. How might inhibition of the normal function of p53 in a cell be beneficial to HPV?

Chlamydia

Chlamydia is named for the tiny bacterium that causes it, *Chlamydia trachomatis.* New chlamydial infections are more numerous than any other sexually transmitted disease. Some estimate that the actual incidence could be as high as 6 million new cases per year. For every reported case in men, more than five cases are detected in women. The low rates in men suggest that many of the sex partners of women with chlamydia are not diagnosed or reported.

Chlamydial infections of the lower reproductive tract are usually mild or asymptomatic. About 8–21 days after infection, men experience a mild burning sensation on urination and a mucous discharge. The infection can spread to the prostate gland and epididymides in males.

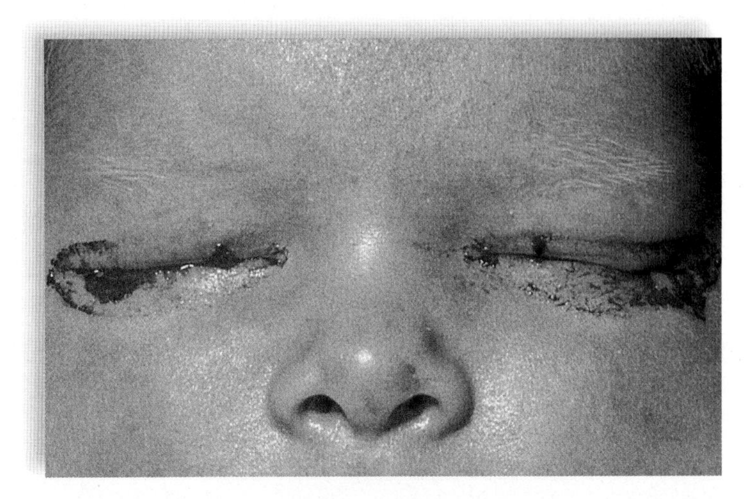

Figure 21.16 Chlamydial eye infection.
This newborn's eyes were infected after passing through the birth canal of an infected mother.

Women may have a vaginal discharge, along with the symptoms of a urinary tract infection. If not properly treated, the infection can eventually cause **pelvic inflammatory disease (PID).** This condition is characterized by inflamed oviducts that become partially or completely blocked by scar tissue. As a result, the woman may become sterile or subject to ectopic pregnancy, which can be a medical emergency. Some believe that, in any case, chlamydial infections increase the possibility of premature and stillborn births. If a newborn is exposed to chlamydia during delivery, pneumonia or inflammation of the eyes can result (Fig. 21.16).

Detection and Treatment of Chlamydia New and faster laboratory tests are now available for detecting a chlamydial infection. Their expense sometimes prevents public clinics from testing for chlamydia. Thus, the following criteria have been suggested to help physicians decide which women should be tested: no more than 24 years old; a new sex partner within the preceding two months; cervical discharge; bleeding during parts of the vaginal exam; and use of a nonbarrier method of contraception. Some doctors, however, are routinely prescribing additional antibiotics appropriate to treating chlamydia for anyone who has gonorrhea, because 40% of females and 20% of males with gonorrhea also have chlamydia.

Gonorrhea

Gonorrhea is caused by the bacterium *Neisseria gonorrhoeae.* Although as many as 20% of infected males are asymptomatic, usually they complain of pain at urination and have a thick, greenish-yellow urethral discharge three to five days after contact with an infected partner. Unfortunately, 60–80% of infected females are asymptomatic until they develop severe PID. Similarly, scarring of each vas deferens may occur in untreated males.

Gonorrhea proctitis, or infection of the anus, with symptoms that may include pain in the anus and blood or pus in the feces, is spread by anal intercourse. Oral sex can cause

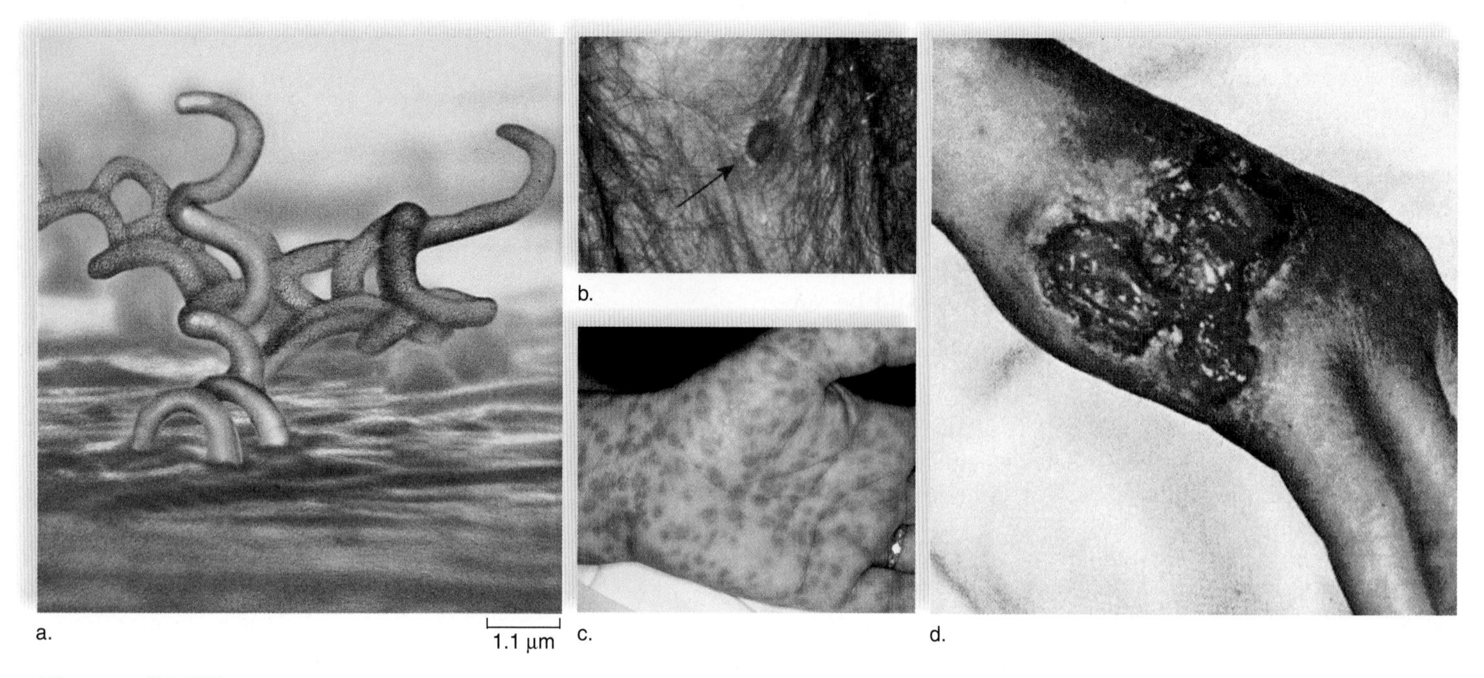

Figure 21.17 Syphilis.
a. Scanning electron micrograph of *Treponema pallidum,* the cause of syphilis. **b.** The primary stage of syphilis is a chancre at the site where the bacterium enters the body. **c.** The secondary stage is a body rash that occurs even on the palms of the hands and soles of the feet. **d.** In the tertiary stage, gummas may appear on the skin or internal organs.

infection of the throat and the tonsils. Gonorrhea can also spread to other parts of the body, causing heart damage or arthritis. If, by chance, the person touches infected genitals and then his or her eyes, a severe eye infection can result.

Eye infection leading to blindness can occur as a baby passes through the birth canal. Because of this, all vaginally delivered infants receive eyedrops containing antibacterial agents, such as silver nitrate, tetracycline, or penicillin, as a protective measure.

Reported gonorrhea rates declined steadily until the late 1990s, and since then, they have stayed fairly constant. In 2007, approximately 356,000 cases of gonorrhea were reported in the United States. About 75% of all reported cases are found in people between 15 and 29 years of age. Gonorrhea rates among African Americans are about 20 times greater than among Caucasians. Antibiotic-resistant strains of *N. gonorrhoeae* are also becoming more common. As with AIDS, condoms protect against gonorrheal and chlamydial infections.

Syphilis

Syphilis is caused by the bacterium *Treponema pallidum,* an actively motile corkscrewlike organism (Fig. 21.17*a*). Syphilis has three stages, which can be separated by intervening latent periods during which the bacteria are not multiplying. During the *primary stage,* a hard chancre (ulcerated sore with hard edges) indicates the site of infection (Fig. 21.17*b*). The chancre can go unnoticed, especially because it usually heals spontaneously, leaving little scarring. During the *secondary stage,* the individual breaks out in a rash, coinciding with the replication and spread of bacteria throughout the body. Curiously, the rash does not itch and is even seen on the palms of the hands and the soles of the feet (Fig. 21.17*c*). Hair loss can occur, and infectious gray patches may develop on the mucous membranes, including the mouth. These symptoms disappear of their own accord.

Not all cases of secondary syphilis go on to the tertiary stage. During the *tertiary stage,* which lasts until the patient dies, syphilis may affect the cardiovascular system; weakened arterial walls (aneurysms) develop, particularly in the aorta. In other instances, the disease may affect the nervous system so that the infected person shows psychological disturbances. In still another variety of the tertiary stage, gummas, large destructive ulcers, may develop on the skin or within the internal organs (Fig. 21.17*d*).

Congenital syphilis, which is present at birth, is caused by syphilitic bacteria crossing the placenta. The child is born blind and/or with numerous anatomical malformations.

As with many other bacterial diseases, penicillin has been effective in treating syphilis. Its control depends on prompt and adequate treatment of all new cases. Therefore, it is crucial for all sexual contacts to be traced so they can be treated. Diagnosis of syphilis can be made through blood tests or microscopic examination of fluids from lesions.

Check Your Progress

1. List three STDs that are caused by bacteria and thus can usually be treated with antibiotics.
2. Which STDs described in this chapter are most likely to cause infertility?
3. What are the major signs of primary, secondary, and tertiary syphilis?

Common Conditions Affecting the Male Reproductive System

Erectile Dysfunction

It is estimated that about 50% of men aged 40–75 have experienced some degree of **erectile dysfunction (ED),** or impotence, which is the inability to produce or maintain an erection sufficient to perform sexual intercourse. Beginning in 1998, several new drugs have been developed to treat ED. All of these drugs—like Viagra, Levitra, and Cialis—work by increasing blood flow to the penis during sexual arousal. Specifically, they inhibit an enzyme called PDE-5, found mainly in the penis, which normally breaks down some of the chemicals responsible for an erection. Surveys have shown these medications to be helpful in about 65% of men who are experiencing ED.

ED can have many causes. It is very likely that changes in the cardiovascular system associated with aging alone can also produce some degree of ED. A normal erection is dependent on blood flow, so almost any condition that affects the cardiovascular system, such as atherosclerosis, high blood pressure, or diabetes mellitus, can lead to ED. Certain medications can cause ED, as can smoking, alcohol, and some illicit drugs. Psychological factors, especially depression, can also be very important contributors to ED. Depending on the cause, other therapies besides the Viagra-type drugs may be indicated. For example, chemicals that dilate blood vessels can be injected directly into the penis, applied as a gel or a patch, or inserted into the urethra as a suppository. If chemical treatments do not work, it is possible to surgically implant a device into the penis that either causes it to remain semirigid all the time or that must be inflated with a small pump, which is usually implanted in the scrotum. Obviously, such devices are usually a last resort in men with ED for whom other approaches have been unsuccessful.

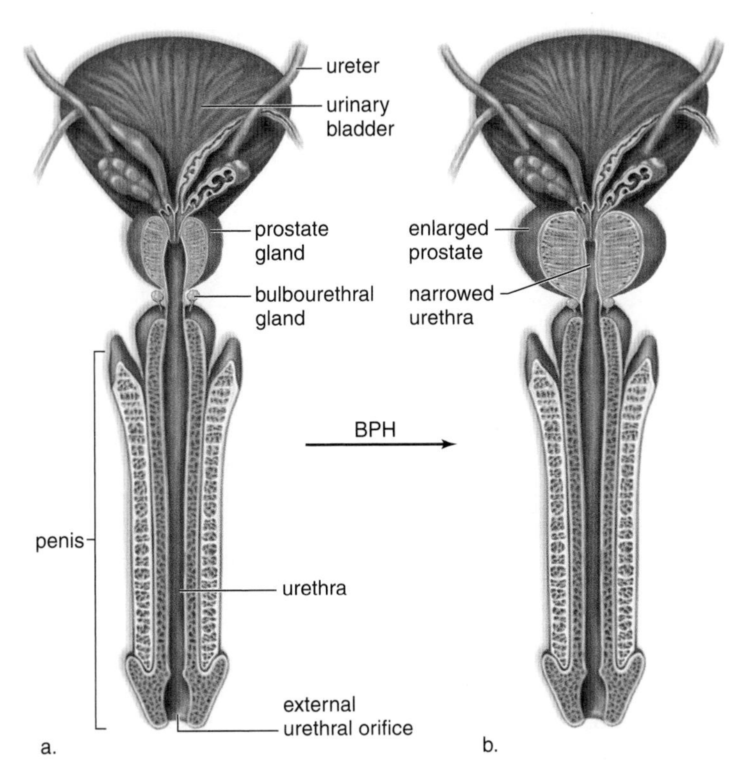

Figure 21.18 Benign prostatic hyperplasia (BPH). **a.** This longitudinal section shows how the prostate gland normally surrounds the urethra. **b.** When the prostate gland enlarges, it can constrict the urethra as it exits the bladder, making it difficult to fully empty the bladder.

Disorders of the Prostate

Beginning around age 40, most men have some enlargement of the prostate gland, which may eventually grow from its normal size of a walnut to that of a lime or even a lemon. This condition is called **benign prostatic hyperplasia (BPH).** Because the prostate gland wraps around the urethra, this enlargement initially may force the bladder muscle to work harder to empty the urethra, leading to irritation of the bladder and an urge to urinate more frequently (Fig. 21.18). As the prostate grows larger, this increased pressure may result in bladder or kidney damage, and may even completely block urination, a condition that is fatal if untreated.

Early signs of an enlarged prostate include a more frequent urge to urinate, a weak stream of urine, and/or a sense of not fully emptying the bladder after urination. The presence of an enlarged prostate can usually be confirmed by a simple digital exam of the prostate, which a physician performs by inserting the finger of a gloved hand into the rectum. Depending on the severity of the symptoms, treatments for BPH range from avoiding liquids in the evening (to reduce the need to urinate during the night), to taking medications that shrink the prostate and/or improve urine flow, to having surgery.

Prostate enlargement is often due to an enzyme (5-alpha reductase) that acts on the male sex hormone testosterone, converting it to a substance that promotes prostate growth. That growth is needed during puberty, but continued growth in the adult is undesirable. Several chemical substances have been shown to interfere with the action of this enzyme. Saw palmetto, which is sold in tablet form as an over-the-counter nutritional supplement, has been shown to be effective. Prescription drugs such as finasteride (Proscar) are more powerful inhibitors of the enzyme, but men taking these drugs may experience erectile dysfunction and loss of libido.

If drugs don't control symptoms or the symptoms are severe, surgery may be necessary to reduce the pressure on the urethra and/or reduce the size of the prostate. In the most common operation, called transurethral resection of the prostate, the surgeon scrapes away the innermost core of the gland using a small instrument inserted into the urethra. In a more limited operation, called transurethral incision of the prostate, the doctor simply makes one or two small cuts in the prostate to relieve pressure. Alternatively, a newer procedure called transurethral microwave thermotherapy uses microwaves to produce heat that destroys, and hopefully shrinks, the enlarged gland.

Cancer of the prostate is the most commonly diagnosed cancer in American males, with one in ten men expected to develop the disease in their lifetime. Many men are concerned that BPH may be associated with prostate cancer, but the two conditions are not necessarily related. If prostate cancer is detected early, usually by a blood test for prostate-specific antigen (PSA), it can usually be successfully treated. However, after lung cancer, prostate cancer is the second leading cancer-related cause of death in men in the United States, because it has a tendency to metastasize and spread throughout the body.

Testicular Cancer

Testicular cancer is the most common type of cancer in males aged 15–35. There are several types of testicular cancer, and some of them have a greater tendency to metastasize to other tissues. If discovered and treated early, testicular cancer is one of the most curable of all cancers. Therefore, experts recommend that men perform a monthly testicular self-examination after a hot shower or bath, when the scrotal skin is looser. Men should examine each testicle for any lumps or abnormal tenderness and for any change in the size of one testicle compared to the other. Any such findings should be reported to a doctor as soon as possible. If testicular cancer is confirmed, treatment usually includes surgery to remove the affected testicle, plus radiation or chemotherapy for types of cancer that have a high risk of spreading. However, if diagnosed and treated before it has spread, the cure rate for testicular cancer approaches 100%.

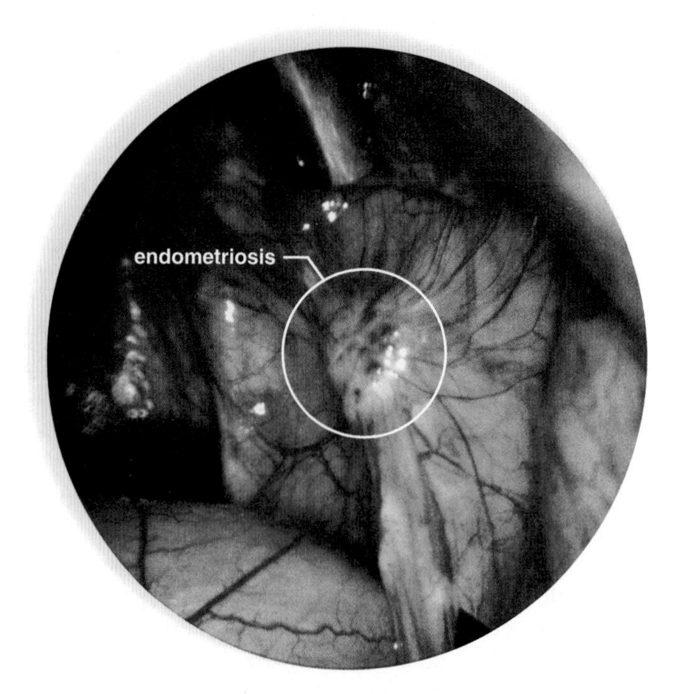

Figure 21.19 Endometriosis.
Endometriosis is a common cause of pelvic pain, abnormal menstruation, and infertility in women. The condition occurs when endometrial tissue, which is often darkly pigmented, grows outside the uterus (usually in the abdominal cavity).

Check Your Progress

1. How do Viagra and similar drugs work to enhance erections?
2. Describe two types of drugs and two types of surgery that can be used to treat BPH.
3. What are some factors that make testicular cancer so treatable?

Common Conditions Affecting the Female Reproductive System

Endometriosis

Endometriosis is the most common cause of chronic pelvic pain in women. Estimates indicate that between 10% and 33% of women aged 25–35 years will have some degree of this condition. **Endometriosis** is the presence of endometrial-like tissue at locations outside the uterine cavity, such as the ovaries, the oviducts, the outside of the uterus, or on virtually any other abdominal organ. The presence of this abnormal tissue can induce a chronic inflammatory reaction, which commonly results in pain, abnormal menstrual cycles, or infertility. The exact cause of endometriosis is unclear, but various theories have been suggested. During menstruation, it is possible that live cells shed from the endometrium actually flow backwards, through the oviducts and into the abdomen, where they are able to form endometrial tissue. Another hypothesis suggests that under certain hormonal conditions, the cells that normally line the abdominal organs can become endometrial tissue.

Endometriosis is often diagnosed when a surgeon sees the abnormal tissue during a surgical procedure to investigate the cause of abdominal pain (Fig. 21.19). The initial treatment usually focuses on controlling the pain, but may also involve drugs, such as oral contraceptives, which inhibit ovulation, and thus reduce the normal hormonally induced changes in the endometrial tissue. If these approaches do not work, surgery may be performed to either remove the abnormal endometrial tissue or to remove the uterus and ovaries in women who do not wish to have children.

Ovarian Cancer

Ovarian cancer accounts for about 4% of all cancers in the United States, translating into about 15,000 deaths in 2007. Gilda Radner, best known for being a member of the original cast of *Saturday Night Live,* died of the disease at age 42. Unlike testicular cancer, ovarian cancer typically occurs in individuals over the age of 40. However, it also has a worse prognosis than testicular cancer, with only 50% of women surviving five years after diagnosis. The main reason for this may be that ovarian cancer is often "silent," showing no obvious signs or symptoms until late in its development. The most common symptoms include abdominal pain, feeling bloated or full, and frequent urge to urinate.

Women over age 40 should have a cancer-related checkup every year. Early detection requires periodic, thorough pelvic examinations. The Pap smear, useful in detecting cervical cancer, does not reveal ovarian cancer. Testing for the level of tumor marker CA-125, a protein antigen, is helpful.

Health Focus

Preventing Transmission of STDs

Being aware of how STDs are spread (Fig. 21B) and then observing the following guidelines will greatly help prevent the transmission of STDs.

Sexual Activities Transmit STDs

Abstain from sexual intercourse or develop a long-term monogamous (always the same partner) *sexual relationship* with a partner who is free of STDs.

Refrain from multiple sex partners or having relations with someone who has multiple sex partners. If you have sex with two other people and each of these has sex with two people and so forth, the number of people having sexual contact is quite large.

Remember that, although the prevalence of AIDS is presently higher among homosexuals and bisexuals, the highest rate of increase is now occurring among heterosexuals. The lining of the uterus is only one cell thick, and thus it can allow infected cells from a sexual partner to enter the blood.

Be aware that having relations with an intravenous drug user is risky because the behavior of this group risks hepatitis and an HIV infection. Be aware that anyone who already has another sexually transmitted disease is more susceptible to an HIV infection.

Avoid anal-rectal intercourse (in which the penis is inserted into the rectum) because the lining of the rectum is thin and cells infected with HIV can easily enter the body there.

Unsafe Sexual Practices Transmit STDs

Always use a latex condom during sexual intercourse if you do not know for certain that your partner has been free of STDs for some time. Be sure to follow the directions supplied by the manufacturer.

Avoid fellatio (kissing and insertion of the penis into a partner's mouth) *and cunnilingus* (kissing and insertion of the tongue into the vagina) because they may be a means of transmission. The mouth and gums often have cuts and sores that facilitate the entrance of infected cells.

Be cautious about using alcohol or any drug that may prevent you from being able to control your behavior.

Drug Use Transmits Hepatitis and HIV

Stop, if necessary, or do not start the habit of injecting drugs into your veins. Be aware that hepatitis and HIV can be spread by blood-to-blood contact.

Always use a new, sterile needle for injection or one that has been cleaned in bleach if you are a drug user and cannot stop your behavior.

Discussion Questions

1. How can you be certain that your partner doesn't have any STDs? Do people necessarily know they have an STD?
2. What does it mean if a person tells you he or she just had an AIDS test, and it was negative? What does this test detect in the blood?
3. Which methods of birth control are most effective in preventing transmission of HIV? Which are least effective?

a.

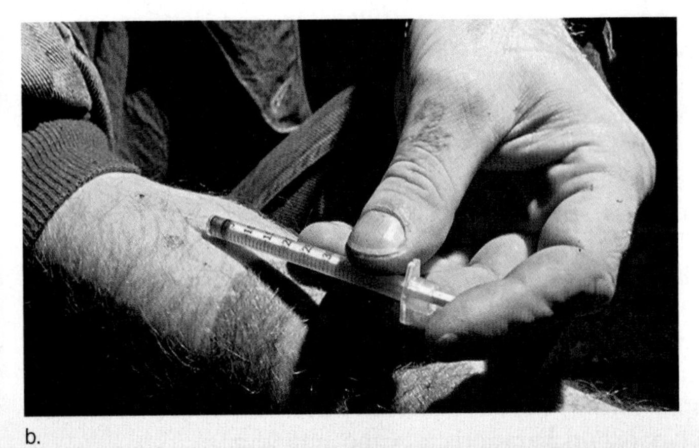

b.

Figure 21B Transmission of STDs.
a. Sexual activities transmit STDs. **b.** Sharing needles transmits STDs. In the United States, more than 65 million people are living with an incurable STD. An additional 15 million people become infected with one or more STDs each year. Approximately one-fourth of these new infections occur in teenagers. Despite the fact that STDs are quite widespread, many people remain unaware of the risks and the sometimes deadly consequences of becoming infected.

Surgery, radiation therapy, and drug therapy are treatment options. Surgery usually involves removing one or both ovaries, the uterus, and the oviducts.

Women who have never had children are twice as likely to develop ovarian cancer as those who have. Early age at first pregnancy, early menopause, and the use of oral contraceptives, which reduces ovulation frequency, can help protect against ovarian cancer. If a woman has had breast cancer, her chances of developing ovarian cancer double.

Disorders of Menstruation

Most menstrual cycles occur every 21–35 days, with 3–10 days of bleeding during which 30–40 ml of blood are lost. Abnormalities include too little bleeding (or a complete lack of menstruation) or too much bleeding. The former usually indicates a problem with the control of menstruation by the hypothalamus or pituitary gland. This may be seen, for example, in young women who are undergoing strenuous athletic training, and/or who have very low levels of body fat. Excessive bleeding may be due to a large number of other disorders, including various STDs, certain medications, or problems with blood clotting.

The most common disorder of menstruation is dysmenorrhea, or painful menstruation. About half of all women experience this condition at some point, which includes such symptoms as agonizing cramps, headaches, backaches, and nausea. Painful menstruation is the leading cause of lost time from school and work among young women. There are many potential causes of dysmenorrhea. One common factor is a high level of prostaglandins released by the endometrium during the luteal and menstrual phases of the cycle. Research has shown that prostaglandin levels are four or five times higher than normal in women who experience painful menstruation. Because prostaglandins are known to cause uterine contractions, investigators suspect that these chemicals are responsible for much of the pain these women experience.

Other potential causes of painful menstruation include endometriosis (discussed previously) and various types of ovarian cysts. These fluid-filled sacs can develop on the ovaries of females of all ages, but are more common in the childbearing years. Two major types of ovarian cysts are follicular cysts, usually caused by the failure of a developing follicle to burst, and corpus luteum cysts, which occur when the corpus luteum doesn't break down normally. On rare occasions, surgery may be needed to remove ovarian cysts.

Premenstrual syndrome (PMS) is a group of symptoms related to the menstrual cycle. Anywhere from two weeks to a few days before menstruation, a significant number of women experience such symptoms as irrational mood swings, headaches, joint pain, digestive upsets, and sore breasts. The mood swings can be particularly troublesome if they lead to behaviors uncharacteristic of the individual.

Self-help measures for painful menstruation include applying a heating pad to the lower abdomen, drinking warm beverages, and taking warm showers or baths. Over-the-counter anti-inflammatory medications may be helpful; if not, a physician can prescribe other medications. For PMS, avoiding salt, sugar, caffeine, and alcohol may also help, especially when symptoms are present.

CHECK YOUR PROGRESS

1. How could endometriosis cause infertility?
2. What are some factors that explain why ovarian cancer is more frequently fatal compared to testicular cancer?
3. Distinguish between dysmenorrhea and PMS.

Infertility

Infertility is the failure of a couple to achieve pregnancy after one year of regular, unprotected intercourse. The American Medical Association estimates that 15% of all couples are infertile. The cause of infertility can be attributed to the male (40%), the female (40%), or both (20%).

Causes of Infertility

The most frequent cause of infertility in males is low sperm count and/or a large proportion of abnormal sperm, which can have many causes. The public is particularly concerned about endocrine-disrupting contaminants, which are discussed in the Ecology Focus. Other possible causes of male infertility include various STDs, abnormal hormone levels, smoking, and alcohol. As noted in section 21.1, activities that increase testicular temperature can lower sperm count.

Common causes of infertility in females include blocked oviducts due to pelvic inflammatory disease and endometriosis, as discussed earlier in this section. Infertility problems also begin to increase sharply after age 35 in women, more so than in men. Several studies have also linked obesity and infertility in both genders, but the exact mechanism(s) involved remain unclear.

Sometimes the cause of infertility can be corrected by medical intervention so that couples can have children. It is possible to give females fertility drugs, which are gonadotropic hormones that stimulate the ovaries and bring about ovulation, although this may cause multiple births.

When reproduction does not occur in the usual manner, many couples adopt a child. Others sometimes try one of the assisted reproductive technologies discussed in the following paragraphs.

Assisted Reproductive Technologies

Assisted reproductive technologies (ART) are techniques that increase the chances of pregnancy.

Artificial Insemination by Donor (AID) During artificial insemination, sperm are placed in the vagina by a physician. Sometimes a woman is artificially inseminated by her partner's sperm. This is especially helpful if the partner has a low sperm count, because the sperm can be collected over a period of time and concentrated so that the sperm count is sufficient to result in fertilization. Often, however, a woman is inseminated by sperm acquired from a donor who is a complete stranger to her. At times, a combination of partner and donor sperm is used.

A variation of AID is *intrauterine insemination (IUI)*. In IUI, fertility drugs are given to stimulate the ovaries, and then the donor's sperm is placed in the uterus, rather than in the vagina.

If the prospective parents wish, sperm can be sorted into those that are believed to be X chromosome-bearing or Y chromosome-bearing to increase the chances of having a child of the desired sex.

Are Pharmacists Obligated to Fill All Prescriptions?

Suppose a married couple is engaging in sexual intercourse, and the condom breaks. It's rare, but it happens. Now imagine that they already have four kids and feel they cannot afford a fifth. The next day, the woman drives to the only drugstore in town to purchase the "morning-after pill," only to be told that the pharmacist refuses to sell it at that store. Situations like this are becoming increasingly frequent, as states battle over the importance of the personal beliefs of medical professionals versus the rights of patients to obtain legal services.

Especially regarding birth control, people have very different beliefs about what is appropriate behavior. Often this is related to their religious background. According to a pharmacist who was fired from a Kmart pharmacy in 2005 for refusing to fill birth control prescriptions, "What's been going on is the use of medication to stop human life. That violates the ideal of the Hippocratic oath that medical practitioners should do no harm." On the other hand, women's rights groups see these cases as a major threat to reproductive freedom, where medical professionals in white coats refuse to provide legal medications to those who need them.

Pharmacists can be fined or even lose their license for refusing to provide legal medications, but such actions are rarely taken by licensing boards. Still, several state legislatures have adopted "conscience clause" laws that allow pharmacists to avoid filling prescription that violate their beliefs. Large pharmacy chains such as Walgreens and CVS have instituted policies where pharmacists in this situation must notify their supervisor, who will make certain the customer has access to the medication by another means. The American Medical Association went a step further in 2005, adopting a resolution supporting pharmacists' right not to prescribe, but also recommending that if another source of medication is not available within a 30-mile radius, physicians should be allowed to directly provide medication to their patients.

Form Your Own Opinion

1. With whom do you tend to side in the situation where a married couple is denied access to a "morning-after pill"? Why?
2. Do you support the right of physicians to refuse to perform procedures, such as abortions, even if they are trained and qualified to do so? Should medical students be required to learn about these procedures even if they don't intend to perform them?
3. Can you imagine a "slippery slope" type of argument if nonmedical professionals are allowed to refuse services to others based on their own beliefs? For example, does a vegan have the right to work at a nonvegetarian restaurant but then refuse to serve steak to customers?

In Vitro Fertilization (IVF) During IVF, conception occurs in laboratory glassware. Ultrasound machines can now spot follicles in the ovaries that hold immature oocytes. Therefore, the latest method is to forgo the administration of fertility drugs and retrieve immature oocytes by using a needle. The immature oocytes are then brought to maturity in glassware before concentrated sperm are added. After about two to four days, the embryos are ready to be transferred to the uterus of the woman, who is now in the secretory phase of her uterine cycle. If desired, the embryos can be tested for a genetic disease, so that only those free of disease are used. If implantation is successful, development is normal and continues to term.

Gamete Intrafallopian Transfer (GIFT) The term **gamete** refers to a sex cell, either a sperm or an oocyte. Gamete intrafallopian transfer was devised to overcome the low success rate (15–20%) of in vitro fertilization. The method is exactly the same as in vitro fertilization, except the oocytes and the sperm are placed in the oviducts immediately after they have been brought together. GIFT has the advantage of being a one-step procedure for the woman—the oocytes are removed and reintroduced all in the same time period. A variation on this procedure is to fertilize the oocytes in the laboratory and then place the zygotes in the oviducts.

Surrogate Mothers In some instances, women are contracted and paid to have babies. These women are called surrogate mothers. The sperm and even the oocyte can be contributed by the contracting parents.

Intracytoplasmic Sperm Injection (ICSI) In this highly sophisticated procedure, a single sperm is injected into an oocyte. It is used effectively when a man has severe infertility problems.

Check Your Progress

1. Describe the most frequent causes of infertility in men and women.
2. Compare the degree of medical intervention required for artificial insemination (AID), in vitro fertilization (IVF), gamete intrafallopian transfer (GIFT), and intracytoplasmic sperm injection (ICSI).

Applying the Concepts [Revisited]

Almost anyone who owns a TV or reads the newspapers has heard of the case of Nadya Suleman, the "Octomom," who gave birth to octuplets in January 2009 following in vitro fertilization. Some of the issues raised by this case include the health of the babies, the financial burden on the taxpayers, the absence of a father, the mother's possible motivations, and the conduct of the IVF facility where the procedure was performed.

1. Rank these (or other) ethical issues raised by this case in their order of importance to you, and justify your rankings.
2. What specific physical factors limit the number of live children a mother can gestate and deliver?
3. Suppose these limitations could be overcome—for example, by incubating embryos in an artificial womb. Assuming a couple was able to afford it, should they be allowed to have, say, 100 children? Why or why not?

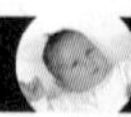

SUMMARIZING THE CONCEPTS

21.1 Male Reproductive System

The male genital tract consists of organs that produce the sperm and seminal fluid, plus the external genitalia:

- Spermatogenesis occurs in the seminiferous tubules of the testes, producing sperm that mature and are largely stored in the epididymides. Sperm enter the vasa deferentia before entering the urethra along with seminal fluid produced by the seminal vesicles, prostate gland, and bulbourethral glands. Sperm and these secretions are called semen.
- The external genitals of males are the penis, the organ of sexual intercourse, and the scrotum, which contains the testes. Erection of the penis occurs when expandable tissue in the penis fills with blood. Orgasm in males is a physical and emotional climax that results in ejaculation of semen from the penis.

The hypothalamus controls the male reproductive system through the secretion of hormones:

- GnRH from the hypothalamus causes the anterior pituitary to secrete FSH and LH. FSH promotes spermatogenesis in the seminiferous tubules of the testes. LH (ICSH) promotes testosterone production by the interstitial cells. Testosterone is required for proper functioning of the male sex organs, as well as for the development of male secondary sex characteristics.

21.2 Female Reproductive System

The female genital tract includes the internal organs that produce the oocyte and allow fertilization to occur, plus the external genitalia:

- Oogenesis is the production of an egg, or oocyte, by an ovary. After ovulation, or release of an oocyte, it is transported via an oviduct (fallopian tube) to the uterus, where a fertilized oocyte, or zygote, can become implanted in the endometrium and develop.
- The external genitals of the female include the vaginal opening, the clitoris, the labia minora, and the labia majora. The vagina is the organ of sexual intercourse, the birth canal, and the exit for menstrual fluids. The external genitals, especially the clitoris, play an active role in orgasm, which culminates in vaginal, uterine, and oviduct contractions.

21.3 Female Hormone Levels

The ovarian cycle is under the control of the hypothalamus and anterior pituitary. It is commonly divided into a follicular phase and a luteal phase:

- During the follicular phase, FSH causes maturation of a follicle that secretes estrogen and some progesterone.
- During the luteal phase, which occurs after ovulation and during the second half of the cycle, LH converts the follicle into the corpus luteum, which secretes progesterone and some estrogen.

The uterine cycle varies depending on whether fertilization occurs. In all cases, estrogen produced by a new ovarian follicle causes the endometrium to thicken. A surge of LH causes ovulation to occur, usually on day 14 of a 28-day cycle. Progesterone, produced by the corpus luteum, causes the endometrium to thicken and become secretory.

- If no pregnancy occurs, the low level of hormones causes the endometrium to break down, resulting in menstruation.
- If fertilization takes place, the embryo implants itself in the thickened endometrium. The corpus luteum in the ovary is maintained for a while because of hCG production by the placenta. Later, the placenta produces both estrogen and progesterone. Menstruation usually does not occur during pregnancy.

21.4 Control of Reproduction

Only abstinence and condoms provide protection against STDs. Several of the most important birth control methods are summarized here:

- Abstaining from sexual activity is the most reliable method of birth control.
- Oral contraceptive pills are nearly 100% effective in preventing ovulation.
- An intrauterine device (IUD) is inserted into the uterus to prevent either fertilization or implantation of a fertilized embryo.
- A vaginal contraceptive ring (NuvaRing) is inserted into the vagina every month to prevent ovulation.
- A diaphragm is inserted to cover the cervix, to prevent sperm from entering the uterus.
- Condoms, both male and female, form a barrier that prevents sperm from entering the uterus.
- Contraceptive implants, patches, and injections all utilize estrogen and/or progesterone to prevent ovulation.
- Morning-after pills, such as RU-486 and Plan B, can be taken after sexual activity (and fertilization) has occurred. They work by preventing implantation of the embryo.

21.5 Disorders of the Reproductive System

The reproductive system is affected by sexually transmitted diseases (STDs), along with several other common conditions:

- Common STDs include the viral diseases AIDS, a pandemic disease that devastates the immune system; hepatitis B, which can lead to liver failure; genital herpes, which flares up repeatedly; human papillomaviruses, the cause of genital warts and cervical cancer; and the bacterial diseases chlamydia and gonorrhea, which can lead to pelvic inflammatory disease, and syphilis, which may cause cardiovascular and neurological complications if untreated.

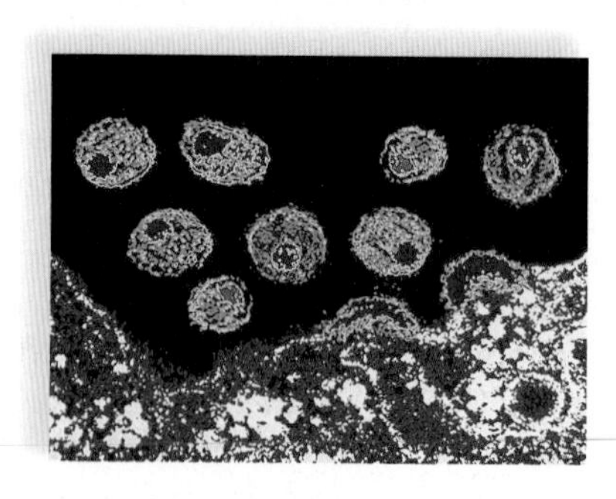

- Common disorders of the male reproductive tract include erectile dysfunction, prostate problems, and testicular cancer.
- Common disorders of the female reproductive tract include endometriosis, ovarian cancer, and various conditions that affect menstruation.
- Some couples are infertile, and if so, they may use assisted reproductive technologies in order to have a child. Such technologies include artificial insemination, in vitro fertilization, gamete intrafallopian transfer, and intracytoplasmic sperm injection.

TESTING YOURSELF

Choose the best answer for each question.

1. Which of these structures contributes fluid to semen?
   a. prostate
   b. seminal vesicles
   c. bulbourethral gland
   d. All of these are correct.
   e. None of these are correct.
2. Production of testosterone in the interstitial cells is controlled by
   a. LH.
   b. ICSH.
   c. GnRH.
   d. All of these are correct.
   e. Both a and b are correct.
3. Label the anatomical structures in this diagram of the male reproductive system.

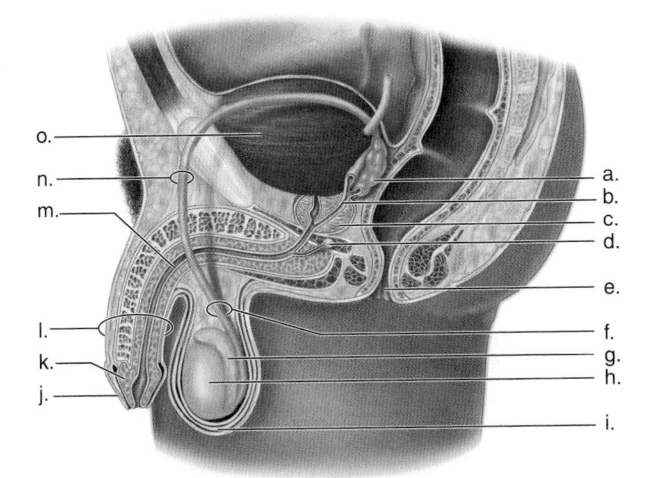

4. An oocyte is fertilized in the
   a. vagina.
   b. uterus.
   c. oviduct.
   d. ovary.
5. During pregnancy,
   a. the ovarian and uterine cycles occur more quickly than before.
   b. GnRH is produced at a higher level than before.
   c. the ovarian and uterine cycles do not occur.
   d. the female secondary sex characteristics are not maintained.
6. Home pregnancy tests are based on the presence of
   a. estrogen.
   b. progesterone.
   c. follicle-stimulating hormone.
   d. human chorionic gonadotropin.
7. Label the anatomical structures in this diagram of the female reproductive system.

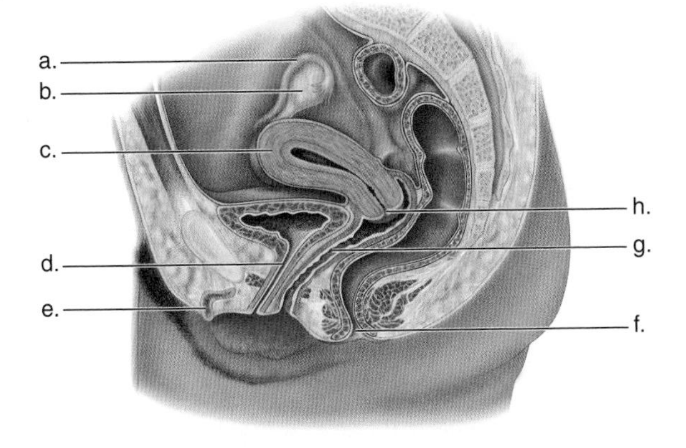

8. Female oral contraceptives prevent pregnancy because
   a. the pill inhibits the release of luteinizing hormone.
   b. oral contraceptives prevent the release of an egg.
   c. follicle-stimulating hormone is not released.
   d. All of these are correct.

For questions 9–15, match the terms in the key with the following graph of the female reproductive cycles.

Key:

a. estrogen
b. ovulation
c. menstruation
d. LH
e. follicular phase
f. luteal phase
g. progesterone

9.
10.
11.
12.
13.
14.
15.

16. In the category B stage of HIV infection, patients usually exhibit
    a. more T cells than category C, but certain other infections are common.
    b. a reduced amount of T cells, but often no symptoms of an HIV infection.
    c. opportunistic infections and likelihood of immediate death.
    d. a positive HIV blood test.
    e. Both a and d are correct.
17. Which of these diseases is caused by a virus?
    a. gonorrhea
    b. genital warts
    c. chlamydia
    d. syphilis
    e. None of these are correct.
18. Late detection of gonorrhea can cause
    a. chlamydia.
    b. hepatitis.
    c. PID.
    d. syphilis.
    e. All of these are correct.

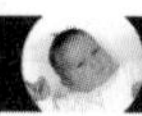

UNDERSTANDING THE TERMS

acquired immunodeficiency syndrome (AIDS) 429
acrosome 419
benign prostatic hyperplasia (BPH) 435
birth control pill 426
bulbourethral gland 416
cervix 420
chlamydia 433
contraceptive 426
contraceptive implant 427
contraceptive injection 427
contraceptive patch 427
contraceptive vaccine 427
corpus luteum 423
diaphragm 427
endometriosis 436
endometrium 421
epididymis 416
erectile dysfunction (ED) 435
estrogen 424
female condom 427
fimbria 420
follicle 423
follicle-stimulating hormone (FSH) 419
gamete 439
genital herpes 432
genital warts 432
gonadotropin-releasing hormone (GnRH) 419
gonorrhea 433
human chorionic gonadotropin (hCG) 425
implant 420
infertility 438
interstitial cell 419
interstitial cell-stimulating hormone (ICSH) 419
intrauterine device (IUD) 426
luteinizing hormone (LH) 419
male condom 427
menopause 425
menstruation 424
oocyte 420
oogenesis 420
opportunistic infection 430
ovarian cancer 436
ovarian cycle 423
ovary 420
oviduct 420
ovulation 420
Pap smear 432
pelvic inflammatory disease (PID) 433
penis 417
placenta 425
progesterone 424
prostate gland 416
scrotum 416
semen 416
seminal vesicle 416
seminiferous tubule 419
sperm 419
spermatogenesis 419
syphilis 434
testes 416
testicular cancer 436
testosterone 419
urethra 416
uterine cycle 424
uterus 420
vagina 421
vaginal contraceptive ring 427
vas deferens 416
vulva 421
zygote 420

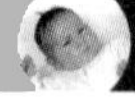

THINKING CRITICALLY

1. Compare the anatomy of the male and female reproductive tracts. Which organs serve similar functions? Which organs serve a unique function, with no analogous function in the other gender?
2. In the animal kingdom, only primates menstruate. Other mammals come into season, or "heat," during certain times of the year, while still others only ovulate after having sex. What would be some possible advantages of a monthly menstrual cycle?
3. The condition called benign prostatic hyperplasia is usually not life threatening, but prostate cancer can be. Since the prostate gland is typically enlarged in both conditions, why is one condition benign and the other potentially life threatening?

INQUIRY INTO LIFE WEBSITE

The companion website for *Inquiry into Life* provides a wealth of information organized and integrated by chapter. You will find practice tests, animations, videos, and much more that will complement your learning and understanding of general biology.

http://www.mhhe.com/maderinquiry13

Development and Aging

Chapter Outline

Applying the Concepts

Fifty is the new 30! In 2008, several famous women, including Madonna,Sharon Stone, and Michelle Pfeiffer turned 50. All of the media hype suggests that today's 50-year-old women are as healthy as 30-year-old women used to be. What health changes have occurred since 1958 when these women were born? In 1960, the life expectancy at birth in the United States was 69.7 years. In 2006, the last year with available statistics, the average life expectancy at birth was 78.1—a gain of 8.4 years. Many factors affect life expectancy, including reduction in infant mortality, improvement in living conditions and hygiene, and better nutrition. But changes in medicine and public health also play a role. For example, open heart and bypass surgery began in the 1960s. CAT (computerized axial tomography) scans and MRIs (magnetic resonance imaging) scans were invented in the 1970s. New antihypertension drugs were introduced in the 1980s. There has also been remarkable progress in cleaning up our environment and thereby reducing our health risks, due in part to the founding of the Environmental Protection Agency in 1970. A recent study showed that cleaner air from 1978 to the present has added approximately five months to our life expectancy. With today's emphasis on health and wellness, perhaps soon we will read headlines suggesting that 60 is the new 30!

22.1 Principles of Animal Development

Learning Outcomes

1. Outline the steps involved in fertilization.
2. Describe the cellular, tissue, and organ stages of development.
3. Explain how differentiation and morphogenesis alter the embryo.

Animal development begins with a single cell that multiplies and changes to form a complete organism.

Fertilization

Fertilization, which results in a zygote, requires that the sperm and oocyte interact. Recall that a sperm has three distinct parts: a head, a middle piece, and a tail. The head contains a nucleus and is capped by a membrane-bounded acrosome.

In humans, the plasma membrane of the oocyte is surrounded by an extracellular matrix termed the zona pellucida (Fig. 22.1). In turn, the zona pellucida is surrounded by a few layers of adhering follicle cells called the corona radiata. These cells nourished the oocyte when it was in a follicle of the ovary. During fertilization, a sperm travels from the interior of the oviduct to the inside of the oocyte. During this journey, (1) the sperm squeezes through the corona radiata; (2) the sperm releases acrosomal enzymes that allow it to penetrate the zona pellucida; (3) the plasma membrane of the sperm head fuses with the plasma membrane of the oocyte; (4) the sperm nucleus enters the oocyte and releases the chromatin; (5) cortical granules secrete enzymes that turn the zona pellucida into a fertilization membrane.

Several sperm attempt to make this journey, but only one completes it. It could be that the acrosomal enzymes of several sperm are needed to forge a pathway through the zona pellucida, and in this way the other sperm assist the one sperm that goes on to enter the oocyte. If more than one sperm enters an oocyte, an event called polyspermy, the

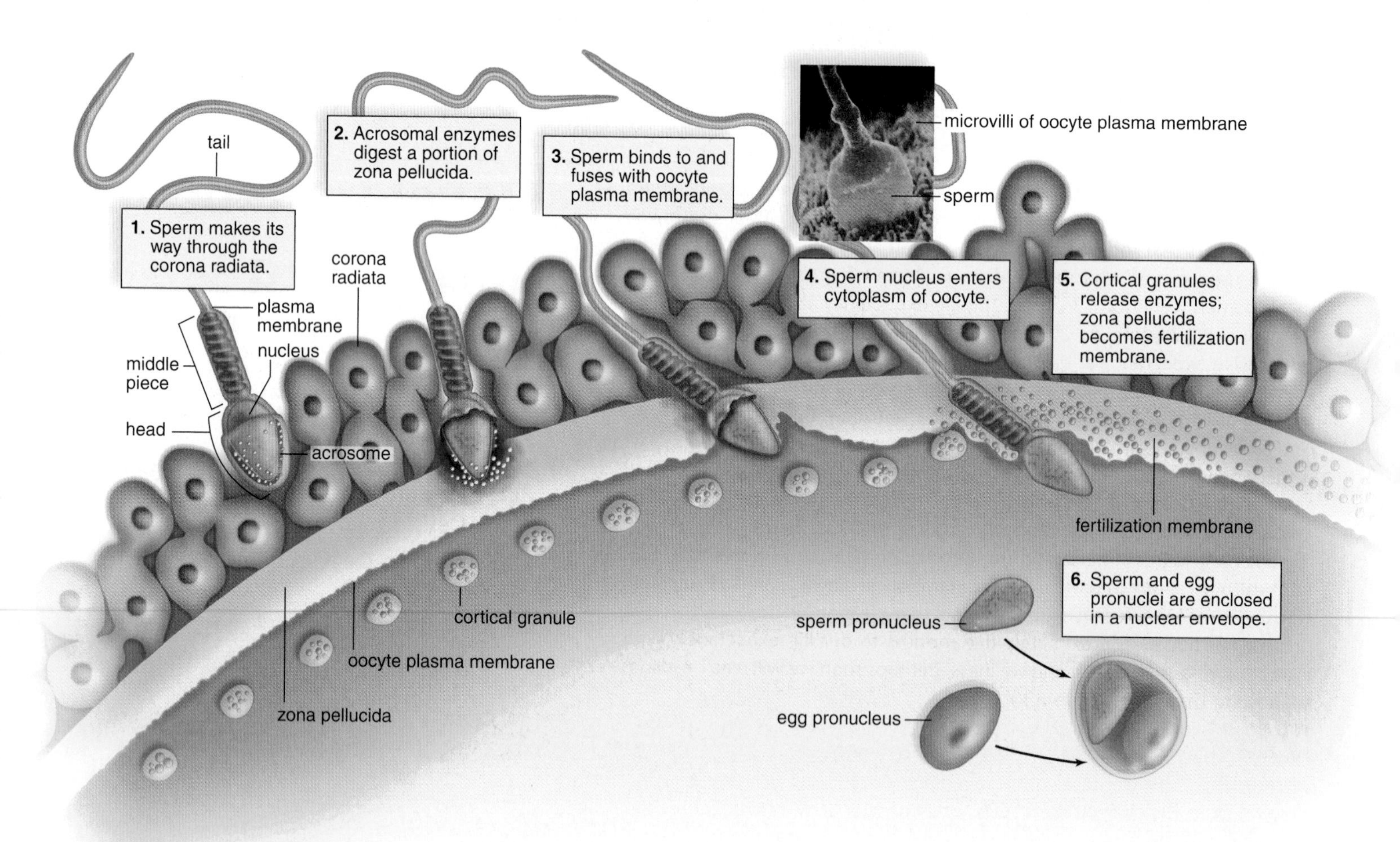

Figure 22.1 Fertilization.
During fertilization, a single sperm enters the oocyte. A sperm makes its way through the corona radiata surrounding the oocyte. The head of a sperm has a membrane-bounded acrosome filled with enzymes. When released, these enzymes digest a pathway for the sperm through the zona pellucida. After it binds to the plasma membrane of the oocyte, a sperm enters the oocyte. When the sperm pronucleus joins with the oocyte pronucleus, fertilization is complete.

zygote would have too many chromosomes, and development would be abnormal. Prevention of polyspermy depends on changes in the oocyte's plasma membrane and in the zona pellucida. As soon as a sperm touches an oocyte, the oocyte's plasma membrane depolarizes (from –65 mV to 10 mV), and this prevents the binding of any other sperm. Then the oocyte releases substances from cortical granules that lead to lifting of the zona pellucida away from the surface of the oocyte. Now, sperm cannot bind to the zona pellucida either.

The sperm chromatin reforms into a so-called sperm pronucleus. The secondary oocyte completes meiosis and its chromatin is contained with the egg pronucleus. A single nuclear envelope surrounds both pronuclei. Cell division begins almost immediately.

Check Your Progress

1. Outline the steps of fertilization beginning with sperm in the oviduct.
2. How does the fertilization membrane prevent polyspermy?

Early Stages of Animal Development

The early stages of animal development occur at the cellular, tissue, and organ levels of organization.

Cellular Stages of Development

The cellular stages of development are (1) cleavage resulting in a multicellular embryo, and (2) formation of the blastula. **Cleavage** is cell division without growth. DNA replication and mitotic cell division occur repeatedly, and the cells get smaller with each division. In other words, cleavage increases only the number of cells. It does not change the original volume of the egg cytoplasm.

As shown in Figure 22.2, cleavage in a lancelet, a primitive chordate, is equal (cells of uniform size) and results in a **morula,** which is a ball of cells. The next cellular stage in lancelet development is formation of a **blastula,** which is a hollow ball of cells having a fluid-filled cavity called a **blastocoel.** The blastocoel forms when the cells of the morula pump Na^+ into extracellular spaces and water follows by osmosis. The water collects in the center, and the result is a hollow ball of cells.

The zygotes of other animals, such as a frog, chick, or human, which are vertebrates, also undergo cleavage and form a morula. In frogs, cleavage is not equal because of the presence of yolk, a dense nutrient material. When yolk is present, the zygote and embryo are said to exhibit polarity because the embryo has an animal pole (smaller cells) and a vegetal pole (larger cells containing yolk). A chicken develops on land and lays a hard-shelled egg containing a large amount of yolk. In the developing chick, the yolk does not participate in cleavage. All vertebrates have a blastula stage, but the appearance of the blastula can be different from that of a lancelet. In the chick, the blastula forms as a layer of

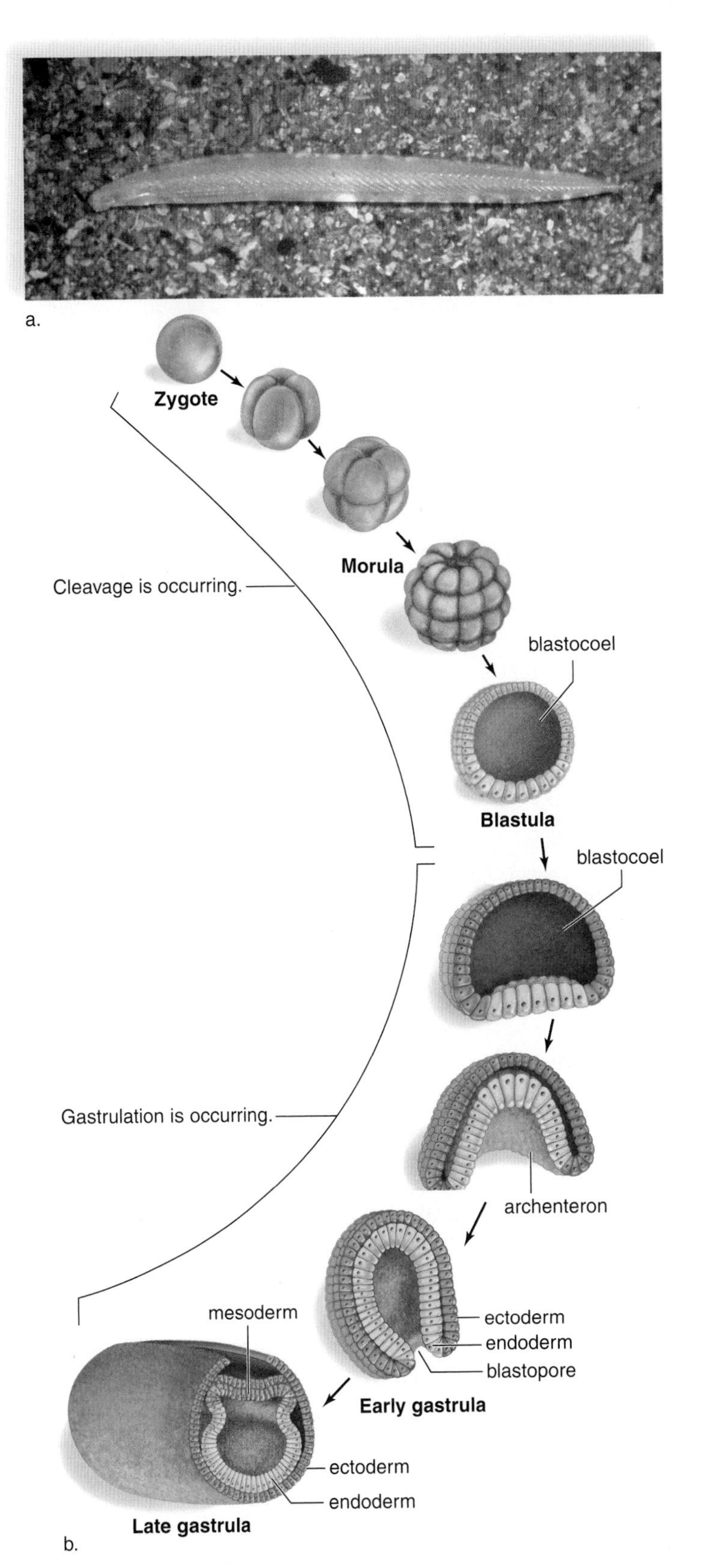

Figure 22.2 Lancelet early development.
a. A lancelet. **b.** The early stages of development are exemplified in the lancelet. Cleavage produces a number of cells that form a ball called a morula. The blastula then develops a cavity, the blastocoel. Invagination during gastrulation produces the germ layers ectoderm and endoderm. Then the mesoderm arises.

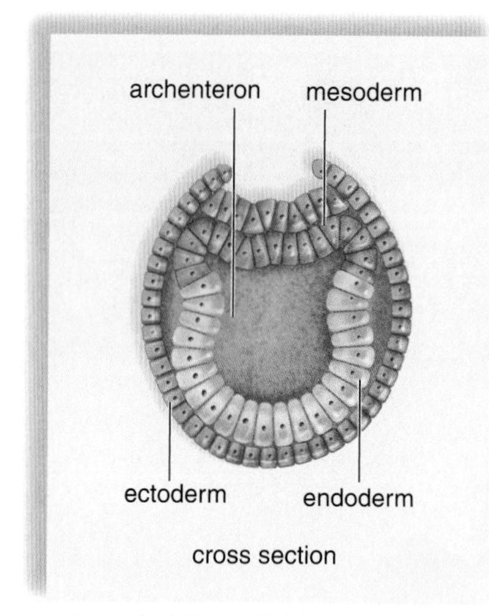

a. Lancelet late gastrula

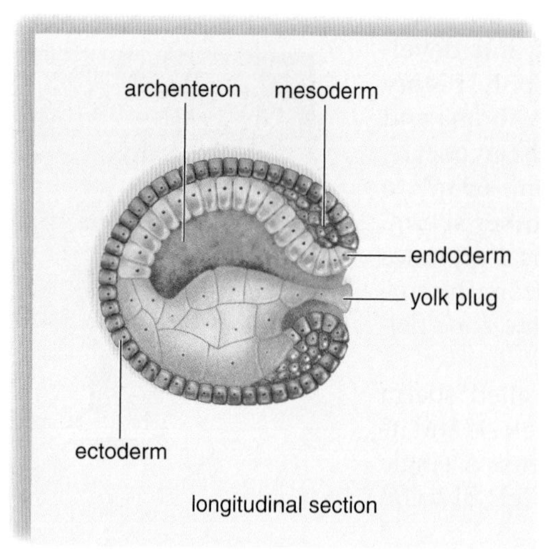

b. Frog late gastrula

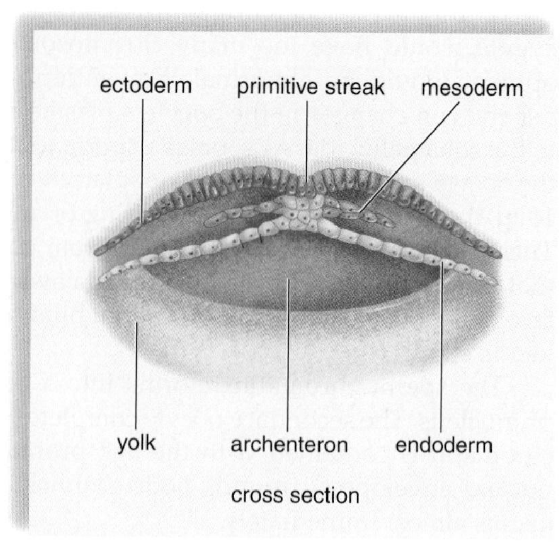

c. Chick late gastrula

Figure 22.3 Comparative development of mesoderm.
a. In the lancelet, mesoderm forms by an outpocketing of the archenteron. **b.** In the frog, mesoderm forms by migration of cells between the ectoderm and endoderm. **c.** In the chick, mesoderm also forms by invagination of cells.

cells that spreads out over the yolk. The blastocoel is a space that separates these cells from the yolk:

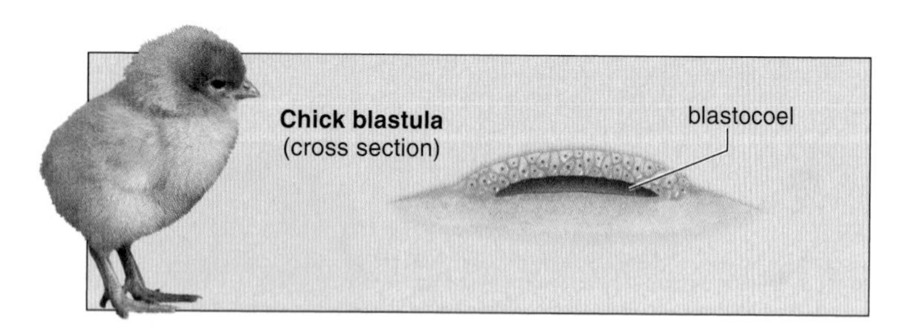

The blastula of humans resembles that of the chick embryo, but this resemblance cannot be related to the amount of yolk because the human egg contains little yolk. Rather, the evolutionary history of these two animals can explain this similarity. Both birds and mammals are related to reptiles, and all three groups develop similarly, despite a difference in the amount of yolk in their eggs.

Tissue Stages of Development

The tissue stages of development are: (1) the early gastrula and (2) the late gastrula. The early **gastrula** stage begins when certain cells begin to push, or invaginate, into the blastocoel, creating a double layer of cells (see Fig. 22.2). Cells migrate during this and other stages of development, sometimes traveling quite a distance before reaching a destination, where they continue developing. As cells migrate, they "feel their way" by changing their pattern of adhering to extracellular proteins and cytoskeletal elements.

An early gastrula has two layers of cells. The outer layer of cells is called the **ectoderm,** and the inner layer is called the **endoderm.** The endoderm borders the gut, which at this point is termed either the archenteron or the primitive gut. The pore, or hole, created by invagination is the **blastopore,** and in a lancelet, the blastopore eventually becomes the anus. **Gastrulation** is not complete until three layers of cells that will develop into adult organs are produced. In addition to ectoderm and endoderm, the late gastrula has a middle layer of cells called the **mesoderm.**

TABLE 22.1 Embryonic Germ Layers

Embryonic Germ Layer	Vertebrate Adult Structures
Ectoderm (outer layer)	Nervous system; epidermis of skin; epithelial lining of oral cavity and rectum
Mesoderm (middle layer)	Musculoskeletal system; dermis of skin; cardiovascular system; urinary system; reproductive system—including most epithelial linings; outer layers of respiratory and digestive systems
Endoderm (inner layer)	Epithelial lining of digestive tract and respiratory tract, associated glands of these systems, epithelial lining of urinary bladder

Figure 22.2 illustrates gastrulation in a lancelet, and Figure 22.3 compares the lancelet, frog, and chick late gastrula stages. In the lancelet, mesoderm formation begins as outpocketings from the archenteron (Fig. 22.3*a*). These outpocketings will grow in size until they meet and fuse, forming two layers of mesoderm. The space between them is the coelom. The coelom is a body cavity lined by mesoderm that contains internal organs. (In humans, the coelom becomes the thoracic and abdominal cavities.)

In the frog, the cells containing yolk do not participate in gastrulation, and therefore, they do not invaginate. Instead, a slitlike blastopore is formed when the animal pole cells begin to invaginate from above, forming endoderm. Animal pole cells also move down over the yolk, to invaginate from below. Some yolk cells, which remain temporarily in the region of the blastopore, are called the yolk plug. Mesoderm forms when cells migrate between the ectoderm and endoderm (Fig. 22.3*b*). Later, a splitting of the mesoderm creates the coelom.

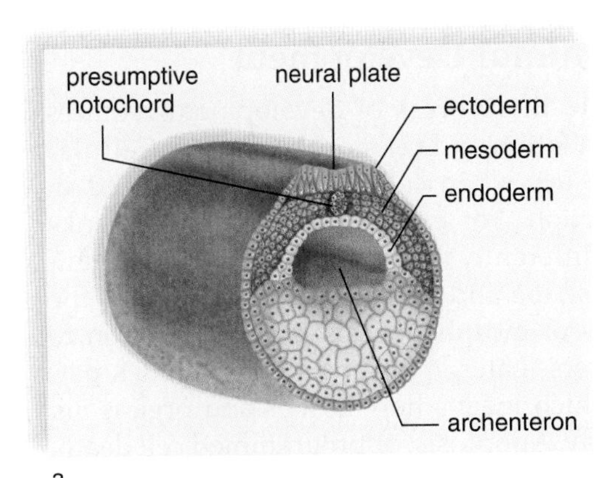

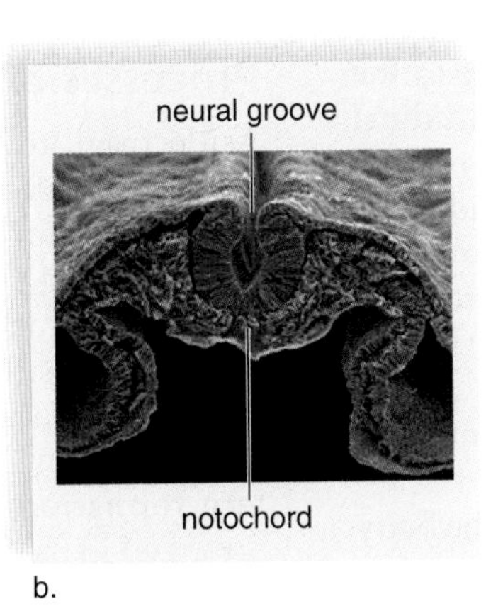

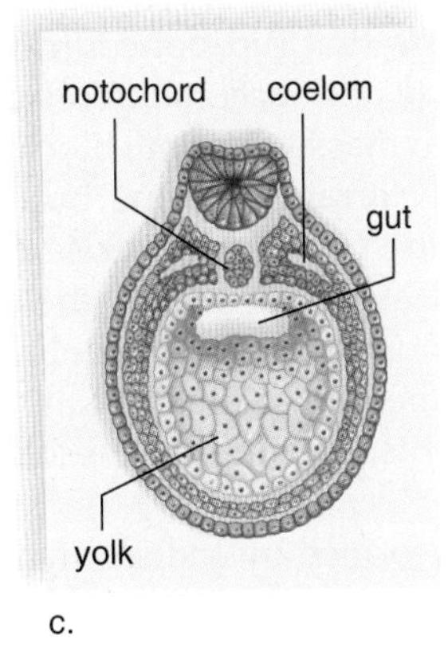

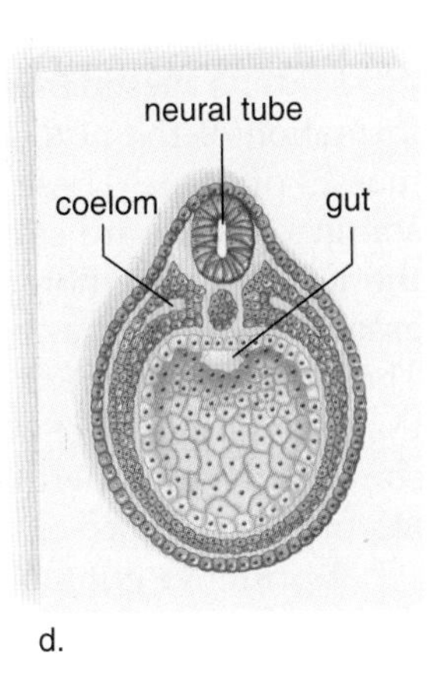

Figure 22.4 Development of neural tube and coelom in a frog embryo.
a. Ectodermal cells that lie above the future notochord (called presumptive notochord) thicken to form a neural plate. **b.** The neural groove and folds are noticeable as the neural tube begins to form. **c.** Splitting of the mesoderm produces a coelom, which is completely lined by mesoderm. **d.** A neural tube and a coelom have now developed.

The chicken egg contains so much yolk that endoderm formation does not occur by invagination. Instead, an upper layer of cells becomes ectoderm, and a lower layer becomes endoderm. Mesoderm arises by an invagination of cells along the edges of a longitudinal furrow in the midline of the embryo. Because of its appearance, this furrow is called the primitive streak (Fig. 22.3*c*). Later, the newly formed mesoderm splits to produce a coelomic cavity.

Ectoderm, mesoderm, and endoderm are called the embryonic **germ layers.** No matter how gastrulation takes place, the result is the same: three germ layers are formed. It is possible to relate the development of future organs to these germ layers (Table 22.1).

Check Your Progress

1. Compare and contrast the cellular stages of development and the tissue stages of development between the lancelet, the frog, and the chick.
2. How is human development similar to chick development at these stages?

Organ Stages of Development

The organs of an animal's body develop from the three embryonic germ layers. We will concentrate on how one organ system, the nervous system, develops.

The newly formed mesoderm cells lying along the main longitudinal axis of the animal coalesce to form a dorsal supporting rod called the **notochord.** The notochord persists in lancelets, but in frogs, chicks, and humans, it is later replaced by the vertebral column. Therefore, these animals are called vertebrates.

The nervous system develops from midline ectoderm located just above the notochord. At first, a thickening of cells, called the **neural plate,** is seen along the dorsal surface of the embryo. Then, neural folds develop on either side of a neural groove, which becomes the **neural tube** when these folds fuse. Figure 22.4 shows cross sections of frog development to illustrate the formation of the neural tube. At this point, the embryo is called a **neurula.** Later, the anterior end of the neural tube develops into the brain, and the rest becomes the spinal cord. In addition, the neural crest is a band of cells that develops where the neural tube pinches off from the ectoderm. Neural crest cells migrate to various locations, where they contribute to formation of skin and muscles, in addition to the adrenal medulla and the ganglia of the peripheral nervous system. Figure 22.5 will help you relate the formation of vertebrate structures and organs to the three embryonic layers of cells: the ectoderm, the mesoderm, and the endoderm.

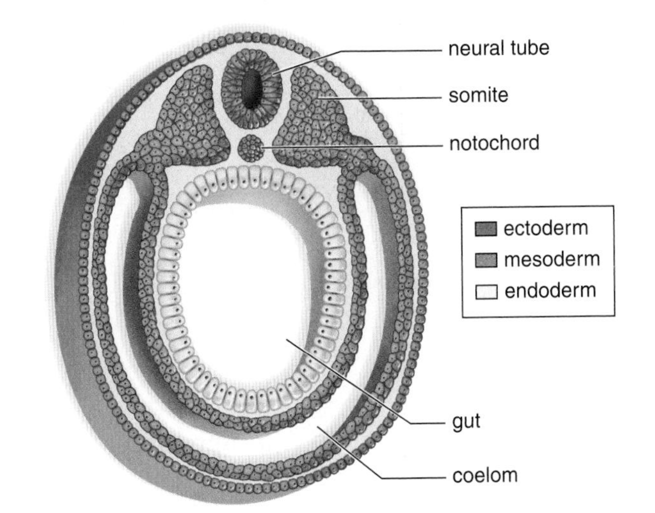

Figure 22.5 Vertebrate embryo, cross section.
At the neurula stage, each of the germ layers, indicated by color (see key), can be associated with the later development of particular parts. The somites give rise to the muscles of each segment and to the vertebrae, which replace the notochord in vertebrates.

Midline mesoderm cells that did not contribute to the formation of the notochord now become two longitudinal masses of tissue. These two masses become blocked off into somites, which are serially arranged along both sides for the length of the notochord. Somites give rise to muscles associated with the axial skeleton and to the vertebrae. The serial origin of axial muscles and the vertebrae testifies that vertebrates are segmented animals. Lateral to the somites, the mesoderm splits, forming the mesodermal lining of the coelom.

A primitive gut tube is formed by endoderm as the body itself folds into a tube. The heart, too, begins as a simple tubular pump. Organ formation continues until the germ layers have given rise to the specific organs listed in Table 22.1.

Check Your Progress

1. What type of tissue lines the gut? The coelom? How did these two cavities form?
2. At this stage of development, where is the notochord in relation to the neural tube?

Processes of Animal Development

Aside from growth, the process of development requires (1) cellular differentiation and (2) morphogenesis. **Cellular differentiation** occurs when cells become specialized in structure and function. For example, a muscle cell looks different and acts differently than a nerve cell. **Morphogenesis** produces the shape and form of the body. One of the earliest indications of morphogenesis is cell migration to form the germ layers. Later, morphogenesis includes **pattern formation,** which means how tissues and organs are arranged in the body. Apoptosis, or programmed cell death, which was first discussed in Chapter 5, is an important part of pattern formation.

Cellular Differentiation

At one time, investigators mistakenly believed that irreversible genetic changes must account for cellular differentiation. Perhaps the genes are parceled out as development occurs, they said, and that's why cells of the body have different structures and functions. However, the

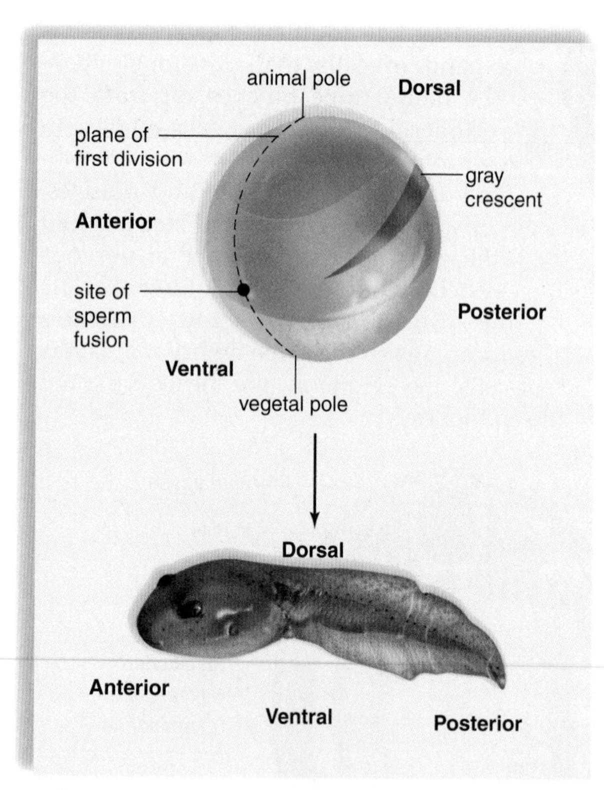

a. Frog's egg is polar and has axes.

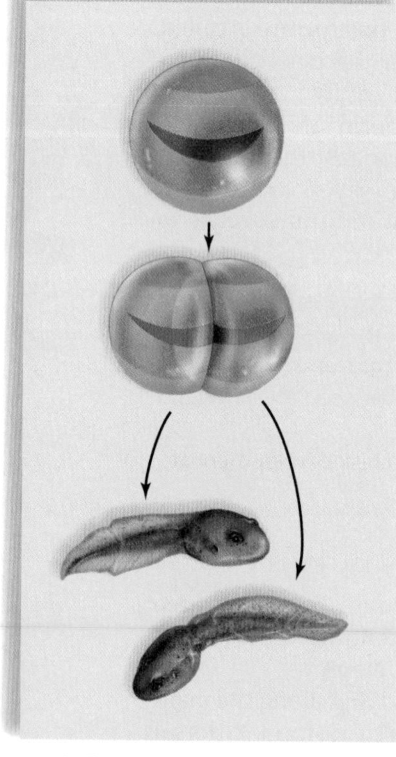

b. Each cell receives a part of the gray crescent.

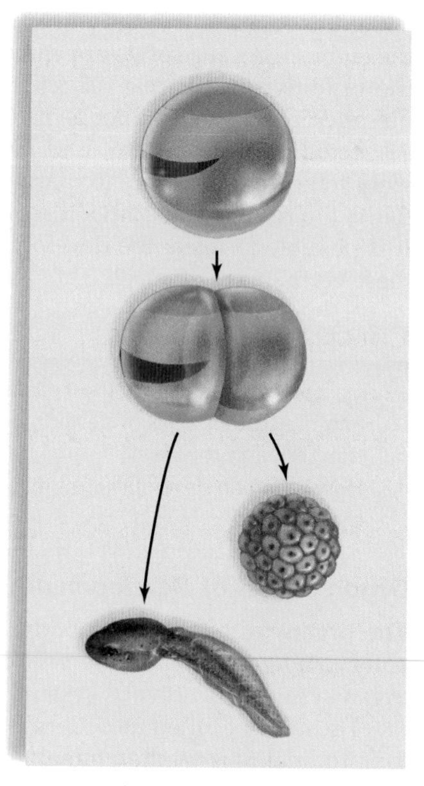

c. Only the cell on the left receives the gray crescent.

Figure 22.6 Experimental determination of cytoplasmic influence on development.
a. A frog's egg has anterior/posterior and dorsal/ventral axes that correlate with the position of the gray crescent. **b.** When researchers cut the fertilized egg such that the gray crescent is divided in half, each daughter cell is capable of developing into a complete tadpole. **c.** If the fertilized egg is cut such that only one daughter cell receives the gray crescent, then only that cell can become a complete embryo.

process of somatic cell cloning shows that all the nuclei of the body's cells are totipotent. Every cell in the body contains all the genes necessary to develop into an organism. In other words, all an organism's genes are in each cell of the body.

The process by which different tissues come about becomes clearer when we consider that specialized cells produce only certain proteins. In other words, we now know that specialization is not due to the parceling out of genes; rather, it is due to differential gene expression. Certain genes and not others are turned on in differentiated cells.

Therefore, in recent years, investigators have turned their attention to discovering the mechanisms that lead to differential gene expression during development. Differentiation must begin long before we can recognize specialized types of cells. Ectodermal, endodermal, and mesodermal cells in the gastrula look quite similar, but they must be different because they develop into different organs.

Investigations by early researchers showed that the cytoplasm of a frog's egg is not uniform. It is polar and has both an anterior/posterior axis and a dorsal/ventral axis, which can be correlated with the **gray crescent,** a gray area that appears after the sperm fertilizes the egg (Fig. 22.6*a*). When researchers in the lab divided the fertilized egg such that each half contained gray crescent, each experimentally separated daughter cell developed into a complete embryo (Fig. 22.6*b*). However, if the egg was divided so that only one daughter cell received the gray crescent, only that cell became a complete embryo (Fig. 22.6*c*). This experiment allows us to hypothesize that the gray crescent must contain particular chemical signals that are needed for development to proceed normally.

The oocyte is now known to contain substances called maternal determinants that influence the course of development. These determinants consist of RNAs and proteins synthesized from the maternal genome and stored in the egg. **Cytoplasmic segregation** is the parceling out of maternal determinants as mitosis occurs:

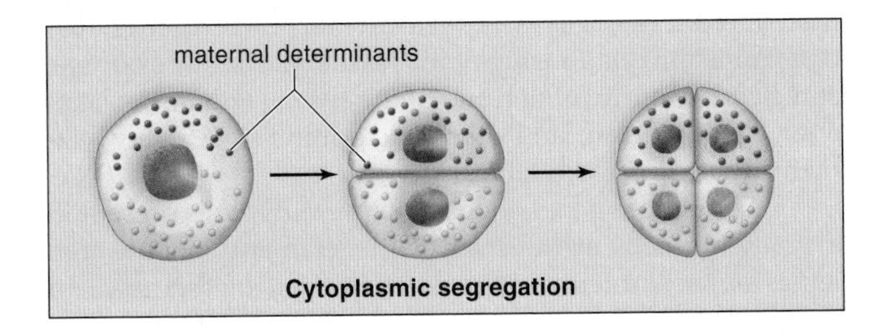

Cytoplasmic segregation helps determine how the various cells of the morula will later develop. As development proceeds, specialization of cells is influenced by maternal determinants and the signals given off by neighboring cells.

Induction and Frog Experiments **Induction** is the ability of one embryonic tissue to influence the development of another tissue by the use of signals called *inducers*. Inducers are chemical signals that alter the metabolism of the receiving cell and activate particular genes. A frog embryo's gray crescent becomes the dorsal lip of the blastopore, where gastrulation begins. Because this region is necessary for complete development, early investigators called the dorsal lip of the blastopore the primary organizer. The cells closest to the primary organizer become endoderm, those farther away become mesoderm, and those farthest away become ectoderm. This suggests that a molecular concentration gradient may act as a chemical signal (inducer) to cause germ layer differentiation.

The gray crescent of a frog's egg marks the dorsal side of the embryo where the mesoderm becomes notochord and ectoderm becomes nervous system. Experiments have shown that presumptive (potential) notochord tissue induces the formation of the nervous system. If presumptive nervous system tissue, located just above the presumptive notochord, is cut out and transplanted to another region of an embryo, nervous tissue does not develop in that region. The presumptive nervous system is missing some sort of signal to differentiate. On the other hand, if presumptive notochord tissue is cut out and transplanted beneath what would be ectoderm in another region, this other ectoderm does differentiate into nervous tissue. This experiment shows that signals from the presumptive notochord system cause ectoderm to differentiate into nervous tissue.

A well-known series of inductions accounts for the development of the vertebrate eye. An optic vesicle, which is a lateral outgrowth of a developing brain, induces the overlying ectoderm to thicken and become the lens of the eye. The developing lens, in turn, induces an optic vesicle to form an optic cup, where the retina forms.

Induction and Roundworm Experiments More recently, work with the roundworm *Caenorhabditis elegans* has also shown that induction is necessary to the process of differentiation. *C. elegans* is only 1 mm long, and vast numbers can be raised in the laboratory in either petri dishes or a liquid medium. The worm is hermaphroditic, and self-fertilization is the rule. Development of *C. elegans* takes three days, and the adult worm contains only 959 cells. Investigators have been able to watch the developmental process from beginning to end, especially because the worm is transparent. Its entire genome has been sequenced and many modern genetic studies have been done utilizing this worm. Individual genes have been altered and cloned, and their products have been injected into cells or extracellular fluid.

Fate maps, diagrams that trace the differentiation of developing cells, have been developed that show the destiny

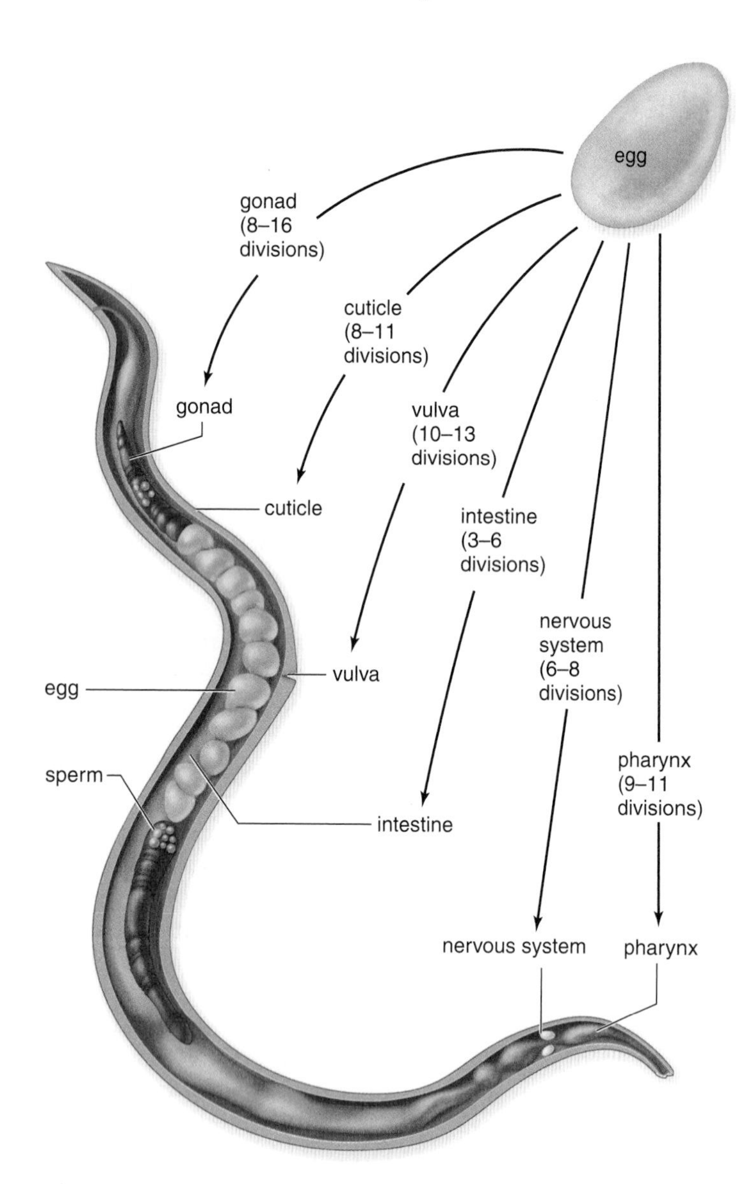

Figure 22.7 Development of *C. elegans*, a nematode. A fate map of the worm showing that as cells arise by cell division, they are destined to become particular structures.

of each cell as it arises following successive cell divisions (Fig. 22.7). Some investigators have studied in detail the development of the vulva, a pore through which eggs are laid. A cell called the anchor cell induces the vulva to form. The cell closest to the anchor cell receives the most inducer and becomes the inner vulva. This cell, in turn, produces another inducer, which acts on its two neighboring cells, and they become the outer vulva. Work with *C. elegans* has shown that a series of inductions occurs as development proceeds.

Morphogenesis

Pattern formation is the ultimate in morphogenesis. To understand the concept of pattern formation, think about the pattern of your own body. Your vertebral column runs along the main axis, and your arms and legs occur in certain locations. Pattern formation refers to how this arrangement of body parts comes about during development. Fruit fly experiments, in particular, have contributed to our knowledge of pattern formation. A few pairs of fruit flies can produce hundreds of offspring in a couple of weeks, all within a small bottle kept on a laboratory bench.

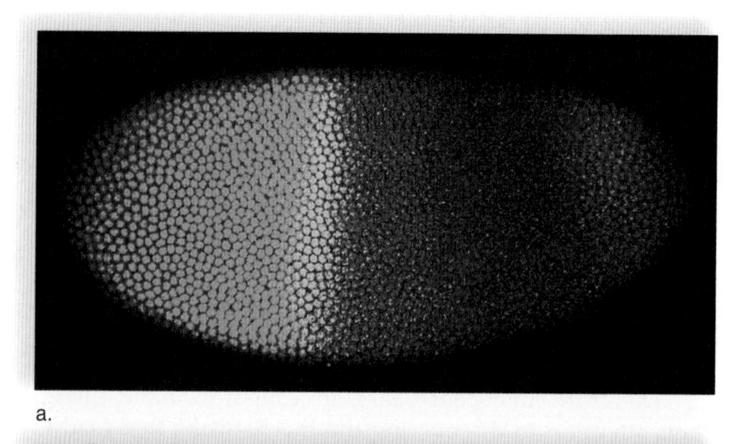

a.

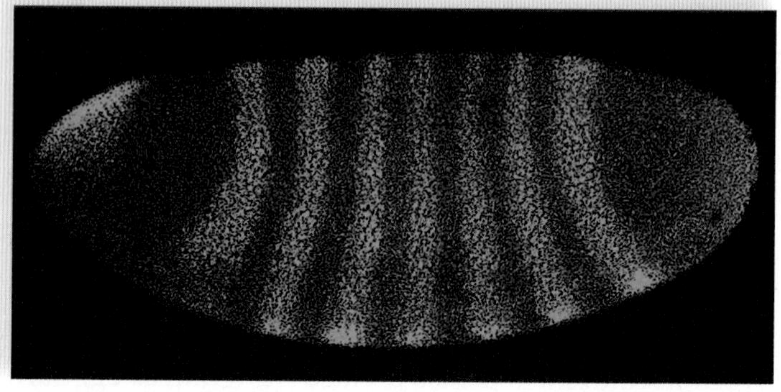

b.

Figure 22.8 Morphogen gradients in the fruit fly. Different morphogen gradients appear as development proceeds. **a.** This early gradient determines which end is the head and which is the tail. The different colors show the concentration gradients of two different proteins. **b.** Another gradient determines the number of segments.

The Fruit Fly Experiments In fruit flies, investigators have discovered certain genes, now called **morphogen genes,** that determine the relationship of individual parts. For example, some genes control which end of the animal will be the head and which the tail, and others determine how many segments the animal will have.

Each of these genes codes for a protein that is present in a gradient—cells at the start of the gradient contain high levels of the protein, and cells at the end of the gradient contain low levels of the protein. These gradients are called morphogen gradients because they determine the shape of the organism (Fig. 22.8). A morphogen gradient is efficient because it has a range of effects, depending on the particular concentration in a portion of the animal. Sequential sets of master genes code for morphogen gradients that activate the next set of master genes in turn. This is one reason development turns out to be so orderly.

Homeotic genes act by controlling the identity of each segment. They encode master regulatory proteins that

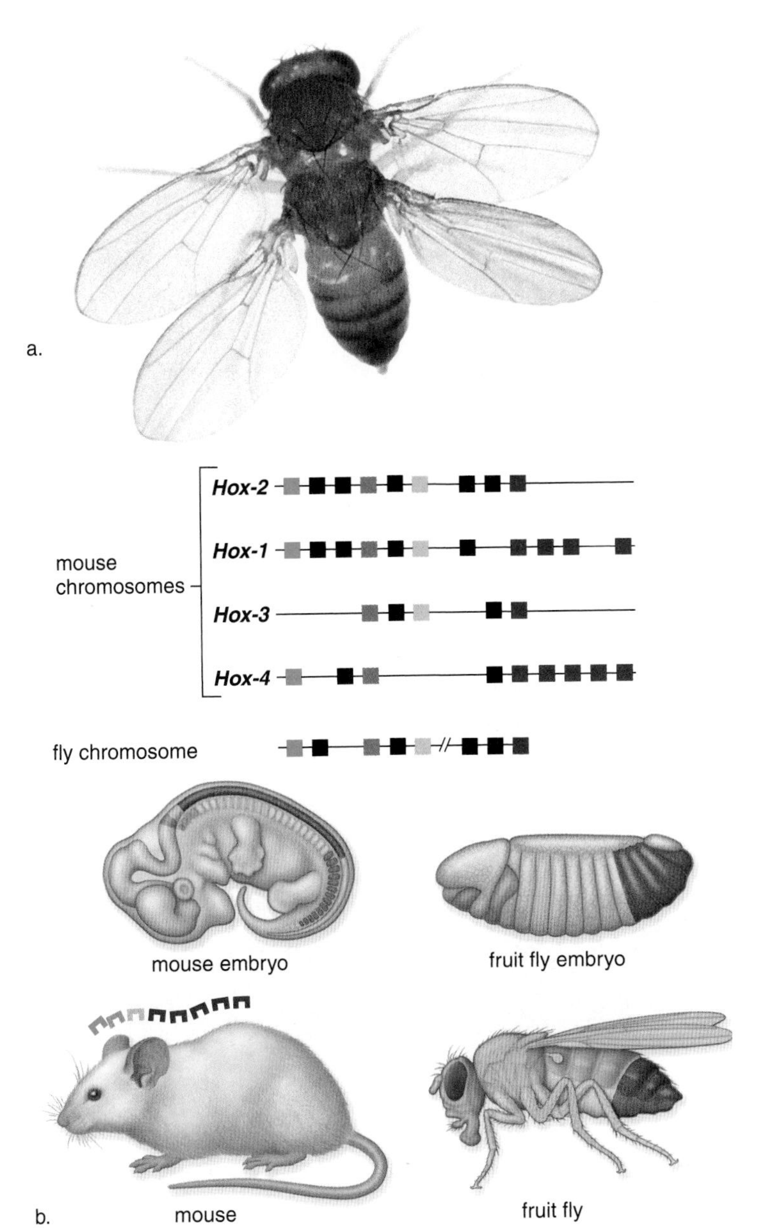

Figure 22.9 Pattern formation in *Drosophila*.
Homeotic genes control pattern formation, an aspect of morphogenesis. **a.** If homeotic genes are activated at inappropriate times, abnormalities such as a fly with four wings occur. **b.** The green, blue, yellow, and red colors show that homologous homeotic genes occur on four mouse chromosomes and on a fly chromosome in the same order. These genes are color-coded to the region of the embryo, and therefore the adult, where they regulate pattern formation. The black boxes are homeotic genes that are not identical between the two animals. In mammals, homeotic genes are called *Hox* genes.

control the expression of other genes, which in turn are responsible for the development of segment-specific structures. Mutations in homeotic genes cause particular segments to develop as if they were located elsewhere in the body. For example, the second thoracic segment of the fruit fly should develop a pair of wings. However, a mutation in the homeotic gene that controls the development of the third thoracic segment allows that segment to develop wings when it should not, resulting in a four-winged fly (Fig. 22.9*a*).

Homeotic genes have now been found in many other organisms, and surprisingly, they all contain the same particular sequence of nucleotides, called a **homeobox.** (Because homeotic genes in mammals contain a homeobox, they are called *Hox* genes.) The homeobox codes for a particular sequence of 60 amino acids called a homeodomain. A homeodomain protein binds to DNA and helps determine which genes are turned on:

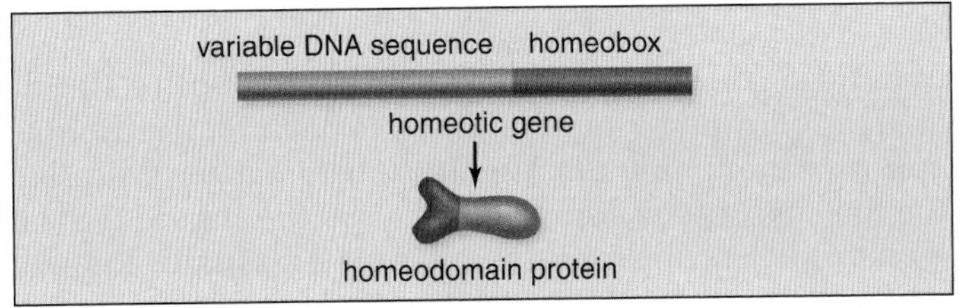

In *Drosophila,* homeotic genes are located on a single chromosome. In mice and also humans, the same four clusters of homeotic genes are located on four different chromosomes (Fig. 22.9*b*). In all three types of animals, homeotic genes are expressed from anterior to posterior in the same order. The first clusters determine the final development of anterior segments of the embryo, while those later in the sequence determine the final development of posterior segments of the embryo.

Because the homeotic genes of so many different organisms contain the same homeodomain, we know that this nucleotide sequence arose early in the history of life and that it has been largely conserved as evolution occurred. In general, it has been very surprising to learn that developmental genetics is similar in organisms ranging from yeasts to plants to a wide variety of animals. Certainly, the genetic mechanisms of development appear to be quite similar in all animals.

Apoptosis We have already discussed the importance of apoptosis (programmed cell death) in the normal day-to-day operation of the body (see Fig. 5.2). Apoptosis is also an important part of pattern formation in all organisms. In humans, for example, we know that apoptosis is necessary for the shaping of the hands and feet. If it does not occur, the child is born with webbing between the fingers and toes.

The fate maps of *C. elegans* indicate that apoptosis occurs in 131 cells as development takes place. When a cell-death signal is received, an inhibiting protein becomes inactive, allowing a cell-death cascade to proceed, which ends in enzymes destroying the cell.

Check Your Progress

1. What is the difference between cytoplasmic segregation and induction?
2. How does morphogenesis result in a segmented body plan?

22.2 Human Embryonic and Fetal Development

Learning Outcomes

1. Describe the membranes surrounding the embryo and their functions.
2. Chronologically list the events that occur during embryonic and fetal development.
3. Contrast the path of fetal blood with the path of blood in a newborn.
4. Outline the stages of birth.

In humans, the length of the time from conception (fertilization followed by implantation) to birth (parturition) is approximately nine months. It is customary to calculate the time of birth by adding 280 days to the start of the last menstrual period because this date is usually known, whereas the day of fertilization is usually unknown. Because the time of birth is influenced by so many variables, only about 5% of babies actually arrive on the forecasted date.

Human development before birth is often divided into embryonic development (months 1 and 2) and fetal development (months 3–9). **Embryonic development** consists of early formation of the major organs, and fetal development is the refinement of these structures. Today, developmental biology encompasses the study of all stages of the human life cycle, from the embryonic period through infancy, childhood, adolescence, and adulthood. During these stages, stem cells in particular continue to produce new cells, which undergo growth, differentiation, and morphogenesis.

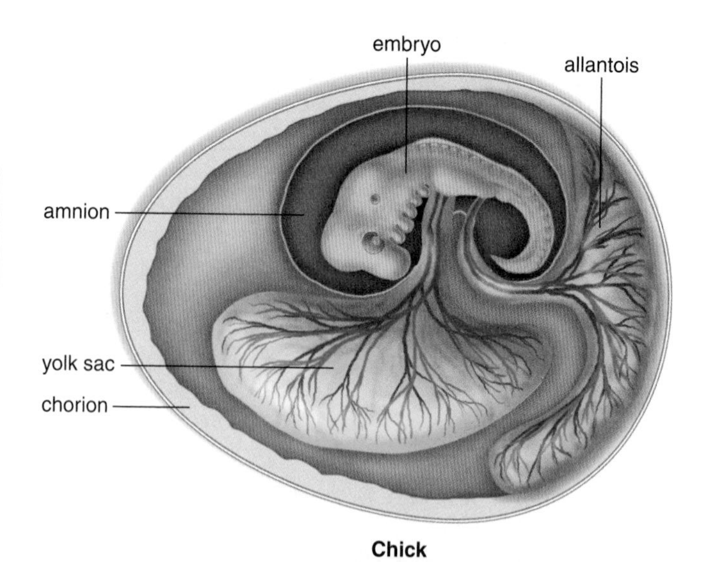

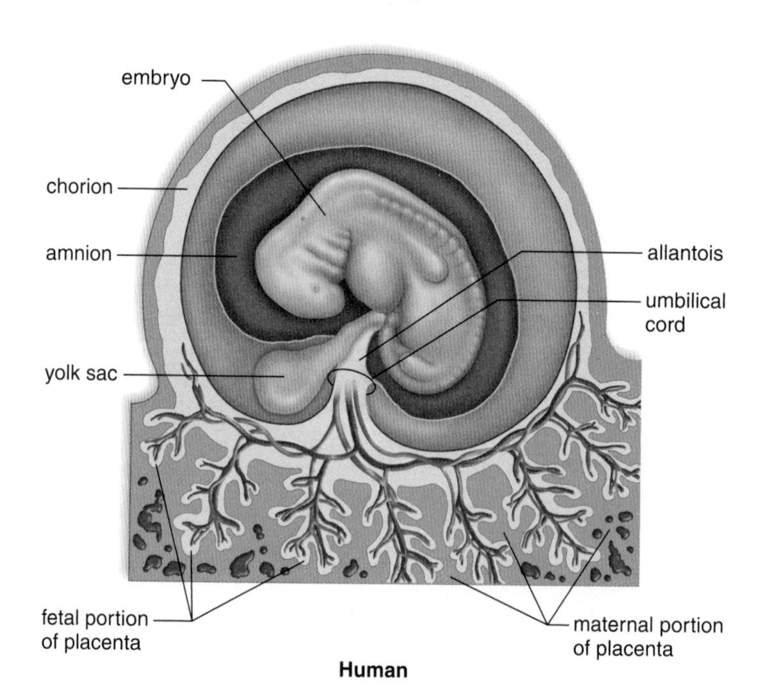

Figure 22.10 Extraembryonic membranes. Extraembryonic membranes, which are not part of the embryo, are found during the development of chicks and humans. Each has a specific function.

Extraembryonic Membranes

The **extraembryonic membranes** lie outside of the embryo. The names of the membranes are derived from their function in reptiles and birds. In reptiles, these membranes made development on land possible. If an embryo develops in the water, the water supplies oxygen for the embryo and takes away waste products. The surrounding water prevents desiccation, or drying out, and provides a protective cushion. For an embryo that develops on land, all these functions are performed by the extraembryonic membranes.

In the chick, the extraembryonic membranes develop from extensions of the germ layers, which spread out over the yolk. Figure 22.10 (*top*) shows the chick surrounded by the membranes. The **chorion** lies next to the shell and carries on gas exchange. The **amnion** contains the protective amniotic fluid, which bathes the developing embryo. The **allantois** collects nitrogenous wastes, and the **yolk sac** surrounds the remaining yolk, which provides nourishment.

Humans (and other mammals) also have these extraembryonic membranes (Fig. 22.10, *bottom*). The chorion develops into the fetal half of the placenta. The yolk sac, which has little yolk, is the first site of blood cell formation. The allantoic blood vessels become the umbilical blood vessels and the amnion contains fluid to cushion and protect the embryo, which develops into a fetus. Therefore, the functions of the extraembryonic membranes in humans have been modified to suit internal development, but their very presence indicates our relationship to birds and to reptiles. It is interesting to note that all chordate animals develop in water, either in bodies of water or within amniotic fluid.

Check Your Progress

1. Of the four extraembryonic membranes, which one maintains the most similar function between the chick and the human?

Embryonic Development

Embryonic development encompasses the first two months of development following fertilization.

The First Week

Fertilization occurs in the distal third (closest to the ovary) of an oviduct (Fig. 22.11), and cleavage begins even as the embryo passes down this duct to the uterus. By the time the embryo reaches the uterus on the third day, it is a morula. The morula is not much larger than the zygote because, even though multiple cell divisions have occurred, the newly formed cells do not grow. By about the fifth day, the morula is transformed into the blastocyst. The **blastocyst** has a fluid-filled cavity, a single layer of outer cells called the **trophoblast,** and an inner cell mass. Later, the trophoblast, reinforced by a layer of mesoderm, gives rise to the chorion, one of the extraembryonic membranes (see Fig. 22.10). The inner cell mass eventually becomes the embryo, which develops into a fetus. Embryonic stem cells are derived from this inner cell mass.

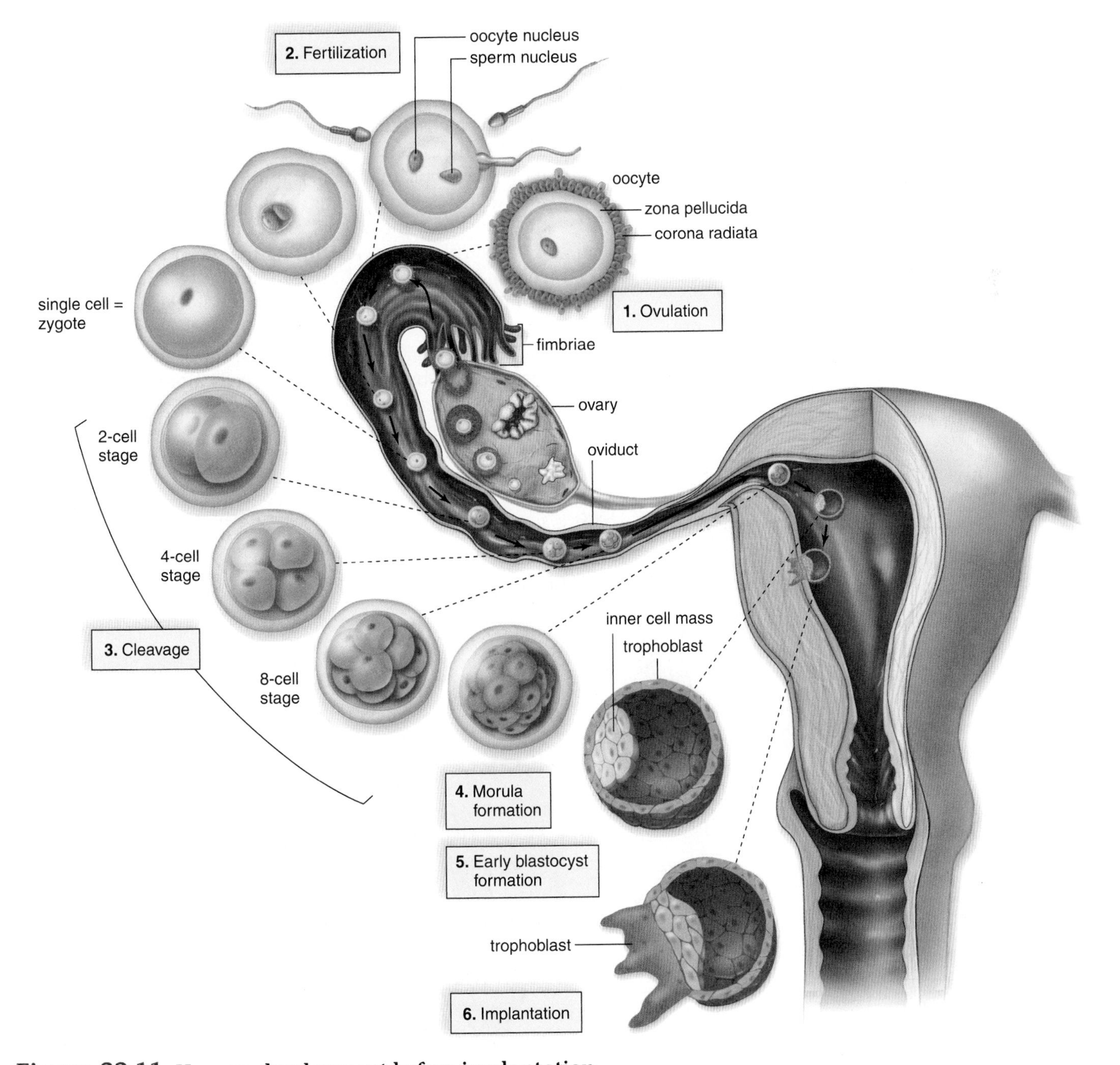

Figure 22.11 Human development before implantation.
Structures and events proceed counterclockwise. (1) At ovulation, the secondary oocyte leaves the ovary. A single sperm nucleus enters the oocyte and (2) fertilization occurs in the oviduct. As the zygote moves along the oviduct, it undergoes (3) cleavage to produce (4) a morula. (5) The blastocyst forms and (6) implants itself in the uterine lining.

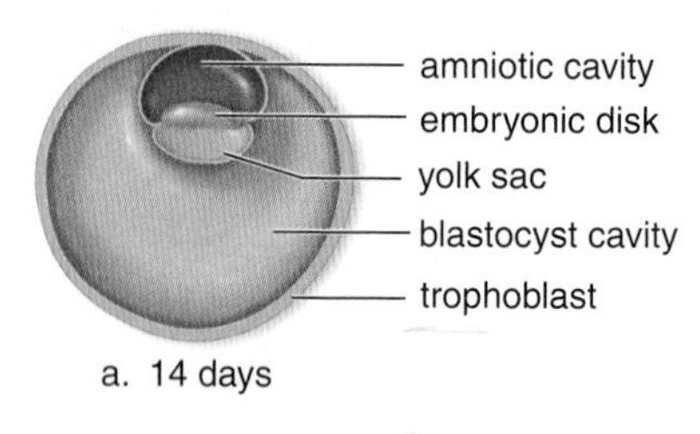

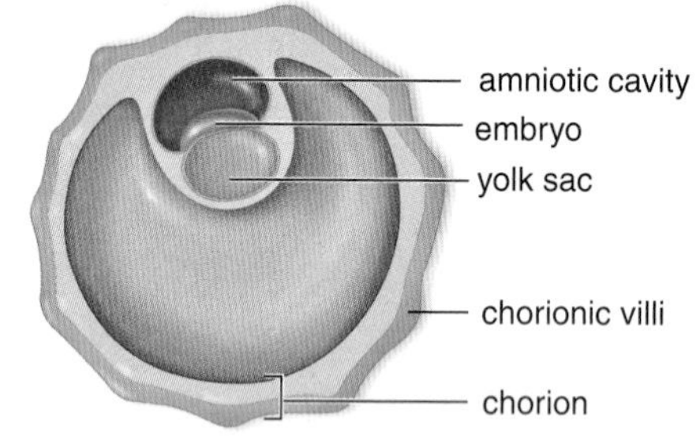

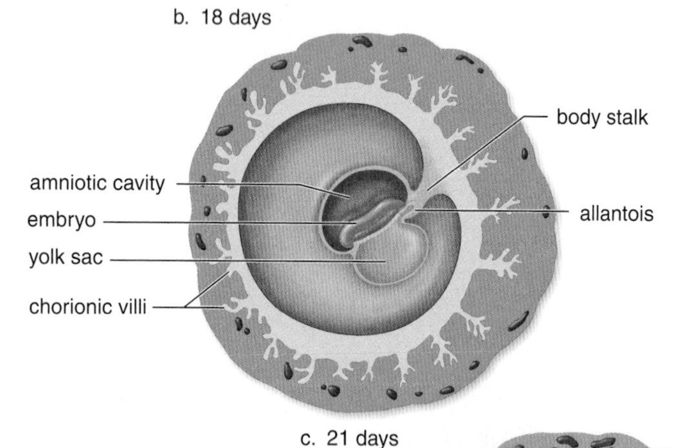

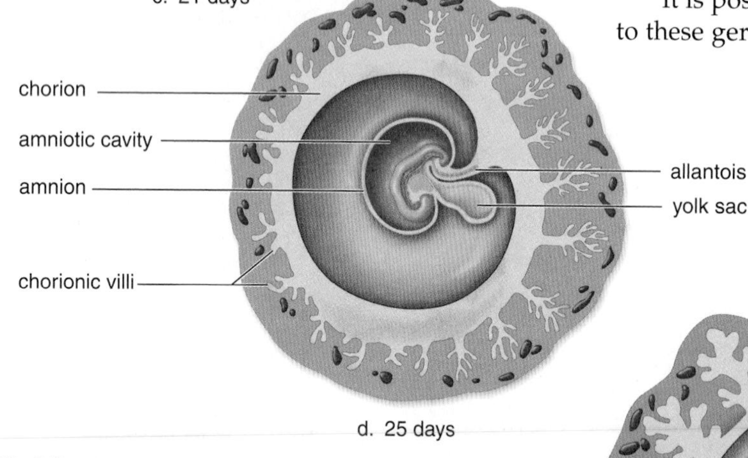

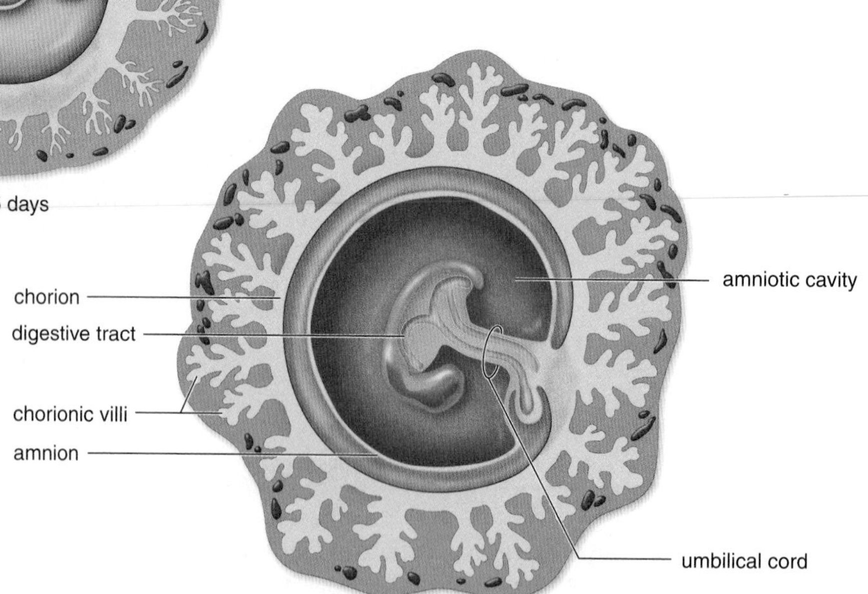

Figure 22.12 Human embryonic development.
a. At first, the embryo contains no organs, only tissues. The amniotic cavity is above the embryonic disk, and the yolk sac is below. **b.** The chorion develops villi, the structures so important to the exchange between mother and child. **c, d.** The allantois and yolk sac, two more extraembryonic membranes, are positioned inside the body stalk as it becomes the umbilical cord. **e.** At 35+ days, the embryo has a head region and a tail region. The umbilical cord takes blood vessels between the embryo and the chorion (placenta).

The Second Week

At the end of the first week, the embryo begins the process of implanting in the wall of the uterus. The trophoblast secretes enzymes to digest away some of the tissue and blood vessels of the endometrium of the uterus (see Fig. 22.11). The embryo is now about the size of the period at the end of this sentence. The trophoblast begins to secrete the hormone **human chorionic gonadotropin (hCG),** which is the basis for the pregnancy test and serves to maintain the corpus luteum past the time it normally disintegrates (see Chapter 21). Because of this, the endometrium is maintained and menstruation does not occur.

As the week progresses, the inner cell mass detaches itself from the trophoblast, and two more extraembryonic membranes form (Fig. 22.12*a*). The yolk sac, which forms below the embryonic disk, has no nutritive function as it does in chicks, but it is the first site of blood cell formation. The amnion has a cavity where the embryo (and then the fetus) develops. In humans, amniotic fluid acts as an insulator against cold and heat and also absorbs shock, such as that caused by the mother exercising.

Gastrulation occurs during the second week. The inner cell mass now has flattened into the **embryonic disk,** composed of two layers of cells: ectoderm above and endoderm below. Once the embryonic disk elongates to form the so-called primitive streak, the third germ layer, mesoderm, forms by invagination of cells along the streak. The trophoblast is reinforced by mesoderm and becomes the chorion (Fig. 22.12*b*).

It is possible to relate the development of future organs to these germ layers (see Table 22.1).

The Third Week

Two important organ systems appear during the third week. The nervous system is the first organ system to be visually evident. At first, a thickening appears along the entire dorsal length of the embryo, and then invagination occurs as neural folds appear. When the neural folds meet at the midline, the neural tube, which later develops into the brain and the nerve cord, is formed (see Fig. 22.4). After the notochord is replaced by the vertebral column, the nerve cord is called the spinal cord.

Development of the heart begins in the third week and continues into the fourth week. At first, there are right and left heart tubes. When these fuse, the heart begins pumping blood, even though its chambers are not fully formed. The veins enter posteriorly and the arteries exit anteriorly from this largely tubular heart, but later the heart twists so that all major blood vessels are located anteriorly.

The Fourth and Fifth Weeks

At four weeks, the embryo is barely larger than the height of this print. A bridge of mesoderm called the body stalk connects the caudal (tail) end of the embryo with the chorion, which has treelike projections called **chorionic villi** (Fig. 22.12*c, d*). The fourth extraembryonic membrane, the allantois, is contained within this stalk, and its blood vessels become the umbilical blood vessels. The head and the tail then lift up, and the body stalk moves anteriorly by constriction. Once this process is complete, the **umbilical cord,** which connects the developing embryo to the placenta, is fully formed (Fig. 22.12*e*).

Little flippers called limb buds appear (Fig. 22.13). Later, the arms and the legs develop from the limb buds, and even the hands and the feet become apparent. At the same time—during the fifth week—the head enlarges, and the sense organs become more prominent. It is possible to make out the developing eyes, ears, and even the nose.

The Sixth Through Eighth Weeks

A remarkable change in external appearance takes place during the sixth through eighth weeks of development. The embryo changes from a form that is difficult to recognize as human to one easily recognized as human. Concurrent with brain development, the head achieves its normal relationship with the body as a neck region develops. The nervous system is developed well enough to permit reflex actions, such as a startle response to touch. At the end of this period, the embryo is about 38 mm (1.5 inches) long and weighs no more than an aspirin tablet, even though all organ systems have been established.

Check Your Progress

1. Prepare a chart that lists the weeks of embryonic development and the changes that occur in the embryo and membranes.
2. What is the outcome of gastrulation?

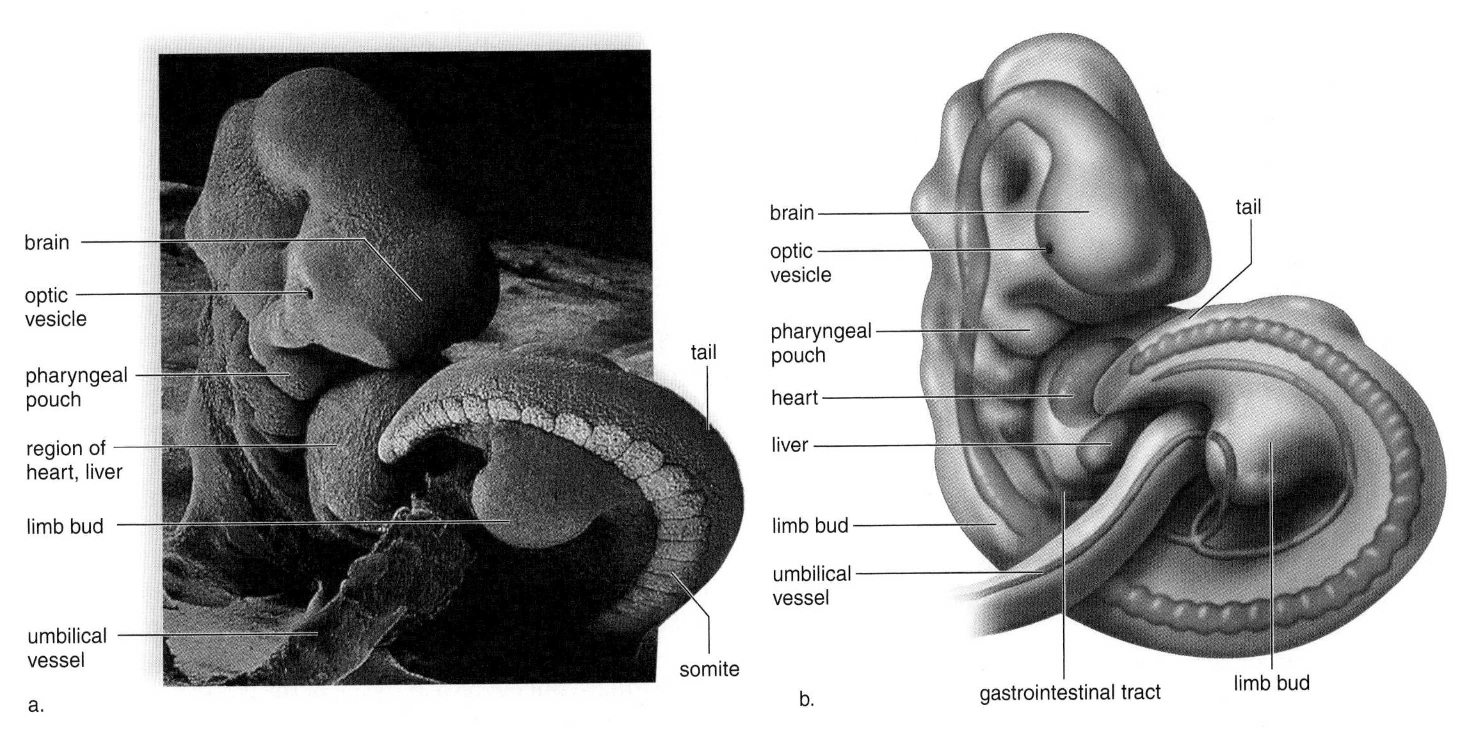

Figure 22.13 Human embryo at beginning of fifth week.
a. Scanning electron micrograph. **b.** The embryo is curled so that the head touches the heart and liver, the two organs whose development is farther along than the rest of the body. The organs of the gastrointestinal tract are forming, and the arms and the legs develop from the bulges that are called limb buds. The tail is an evolutionary remnant. Its bones regress and become those of the coccyx (tailbone). The pharyngeal pouches become functioning gills only in fishes and amphibian larvae. In humans, the first pair of pharyngeal pouches becomes the auditory tubes. The second pair becomes the tonsils, while the third and fourth pairs become the thymus gland and the parathyroid glands.

Fetal Development and Birth

Fetal development includes the third through ninth months of development. At this time, the fetus looks human (Fig. 22.14).

The Third and Fourth Months

At the beginning of the third month, the fetal head is still very large, the nose is flat, the eyes are far apart, and the ears are well formed. Head growth now begins to slow down as the rest of the body increases in length. Epidermal refinements, such as eyelashes, eyebrows, hair on head, fingernails, and nipples, appear.

Cartilage begins to be replaced by bone as ossification centers appear in most of the bones. Cartilage remains at the ends of the long bones, and ossification is not complete until age 18 or 20 years. The skull has six large membranous areas called **fontanels,** which permit a certain amount of flexibility as the head passes through the birth canal and allow rapid growth of the brain during infancy. Progressive fusion of the skull bones causes the fontanels to usually close by 16 months of age.

Sometime during the third month, it is possible to distinguish males from females. Researchers have discovered a series of genes on the X and Y chromosomes that cause the differentiation of gonads into testes or ovaries. Once these have differentiated, they produce the sex hormones that influence the differentiation of the genital tract.

At this time, either testes or ovaries are located within the abdominal cavity, but later, in the last trimester of fetal development, the testes descend into the scrotal sacs as discussed in Chapter 21. Sometimes the testes fail to descend, and in that case, an operation may be done later to place them in their proper location.

During the fourth month, the fetal heartbeat is loud enough to be heard when a physician applies a stethoscope to the mother's abdomen. By the end of this month, the fetus is about 152 mm (6 inches) in length and weighs about 171 grams (6 oz).

The Fifth Through Seventh Months

During the fifth through seventh months, the mother begins to feel movement. At first, there is only a fluttering sensation, but as the fetal legs grow and develop, kicks and jabs are felt. The fetus, though, is in the fetal position, with the head bent down and in contact with the flexed knees.

The wrinkled, translucent skin is covered by a fine down called **lanugo.** This, in turn, is coated with a white, greasy, cheeselike substance called **vernix caseosa,** which probably protects the delicate skin from the amniotic fluid. The eyelids are now fully open.

At the end of this period, the length has increased to about 300 mm (12 in.), and the weight is about 1,380 g (3 lb). It is possible that, if born now, the baby will survive.

Check Your Progress

1. How much does the fetus grow in length and weight from the end of the fourth month until the end of the seventh month?

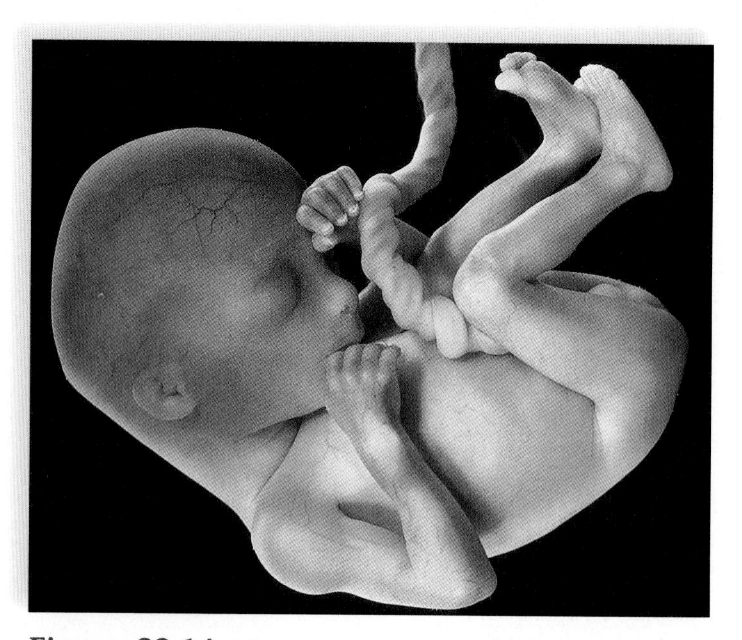

Figure 22.14 Three- to four-month-old fetus.
At this stage, the fetus looks human, with its face, hands, and fingers well defined.

Fetal Circulation

The fetus has circulatory features that are not present in the adult circulation (Fig. 22.15). All of these features are necessary because the fetus does not use its lungs for gas exchange. For example, much of the blood entering the right atrium is shunted into the left atrium through the **foramen ovale** between the two atria. Also, any blood that does enter the right ventricle and is pumped into the pulmonary trunk is shunted into the aorta by way of the **ductus arteriosus.**

Blood within the aorta travels to the various branches, including the iliac arteries, which connect to the **umbilical arteries** leading to the placenta. Exchange of gases and nutrients between maternal blood and fetal blood takes place at the placenta. The **umbilical vein** carries blood rich in nutrients and oxygen to the fetus. The umbilical vein enters the liver and then joins the **ductus venosus,** which merges with the inferior vena cava, a vessel that returns blood to the heart. It is interesting to note that the umbilical arteries and vein run alongside one another in the umbilical cord, which is cut at birth, leaving only the umbilicus (navel).

The most common of all cardiac defects in the newborn is the persistence of the foramen ovale. Once the umbilical cord has been tied and the lungs have expanded, blood enters the lungs in quantity. The return of this blood to the left side of the heart usually causes a flap to cover the opening. Incomplete closure occurs in nearly one out of four individuals, but even so, passage of the blood from the right atrium to the left atrium rarely occurs because either the opening is small or it closes when the atria contract. In a small number of cases, the passage of impure blood from the right side to the left side of the heart is sufficient to cause a "blue baby." Such a condition can now be corrected by open heart surgery.

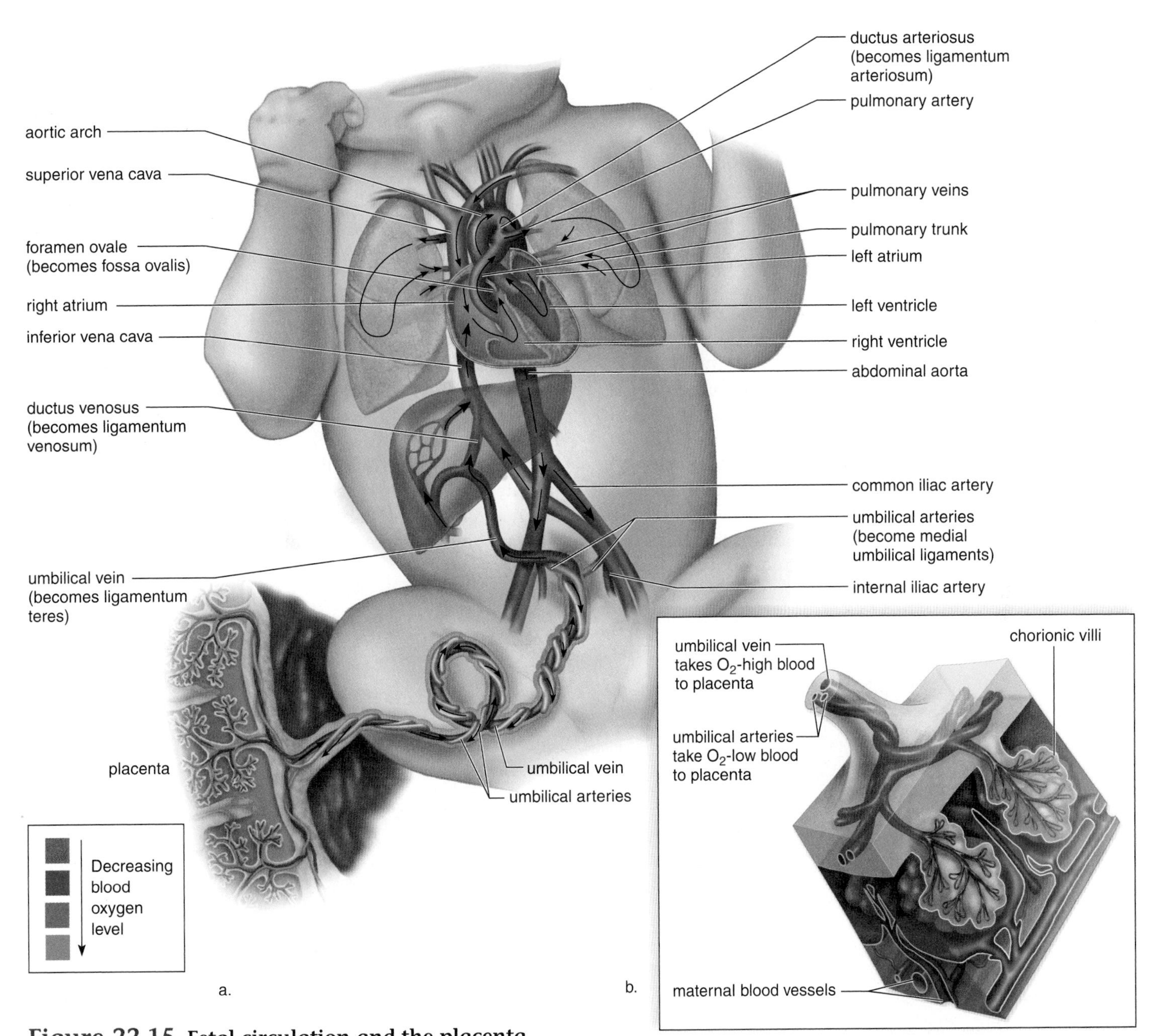

Figure 22.15 Fetal circulation and the placenta.
a. The lungs are not functional in the fetus, and the blood passes directly from the right atrium to the left atrium or from the right ventricle to the aorta. The umbilical arteries take fetal blood to the placenta, and the umbilical vein returns fetal blood from the placenta. **b.** At the placenta, an exchange of molecules between fetal and maternal blood takes place across the walls of the chorionic villi. Oxygen and nutrient molecules diffuse into the fetal blood, and carbon dioxide and urea diffuse out of the fetal blood.

The ductus arteriosus closes because endothelial cells divide and block off this duct. Remains of the ductus arteriosus and parts of the umbilical arteries and vein are later transformed into connective tissue.

Structure and Function of the Placenta Human beings belong to the group of mammals called placental mammals. The **placenta** is firmly attached to the uterine wall by the allantois and by fingerlike projections called the chorionic villi. The placenta is clearly a structure that functions only before birth. When the child is born, the placenta becomes part of the afterbirth, as described later in this section.

The placenta functions in gas, nutrient, and waste exchange between the embryonic (later fetal) and maternal circulatory systems. The placenta begins to form once the embryo is fully implanted. At first, the entire chorion has chorionic villi that project into the endometrium. Later, these disappear in all areas except where the placenta develops. By the tenth week, the placenta is fully formed and is producing progesterone and estrogen. These hormones have two effects: (1) due to their negative feedback control of the hypothalamus and the anterior pituitary, they prevent any new follicles from maturing, and (2) they maintain the lining of the uterus so that the corpus luteum is no longer needed. Usually no menstruation occurs during pregnancy.

The placenta has a fetal side contributed by the chorion and a maternal side consisting of uterine tissues. Notice in Figure 22.15 how the chorionic villi are surrounded by maternal blood. Yet, maternal and fetal blood never mix, because exchange always takes place across the walls of the chorionic villi. Carbon dioxide and other wastes move from the fetal side to the maternal side of the placenta, and nutrients and oxygen move from the maternal side to the fetal side.

The umbilical cord stretches between the placenta and the fetus. Although the umbilical cord may seem to travel from the placenta to the intestine, actually it is simply taking fetal blood to and from the placenta. The umbilical cord is the lifeline of the fetus because it contains the umbilical arteries and vein, which transport waste molecules (carbon dioxide and urea) to the placenta for disposal and take oxygen and nutrient molecules from the placenta to the rest of the fetal circulatory system.

As discussed in the Health Focus, harmful chemicals can also cross the placenta. This is of particular concern during the embryonic period, when various structures are first forming. Each organ or part seems to have a sensitive period during which a substance can alter its normal development. Even over-the-counter drugs should not be taken during pregnancy without permission from your physician.

Check Your Progress

1. Compare and contrast fetal circulation with adult circulation as shown in Figure 12.7.

Birth

The uterus has contractions during the last trimester of the pregnancy. At first, these are light, lasting about 20–30 seconds and occurring every 15–20 minutes. Near the end of pregnancy, the contractions may become stronger and more frequent. However, the onset of true labor is marked by uterine contractions that occur regularly every 10–15 minutes and last for 40 seconds or longer. These contractions are generally called "labor pains," although, when they first appear, they are not usually painful and only become painful at a later stage.

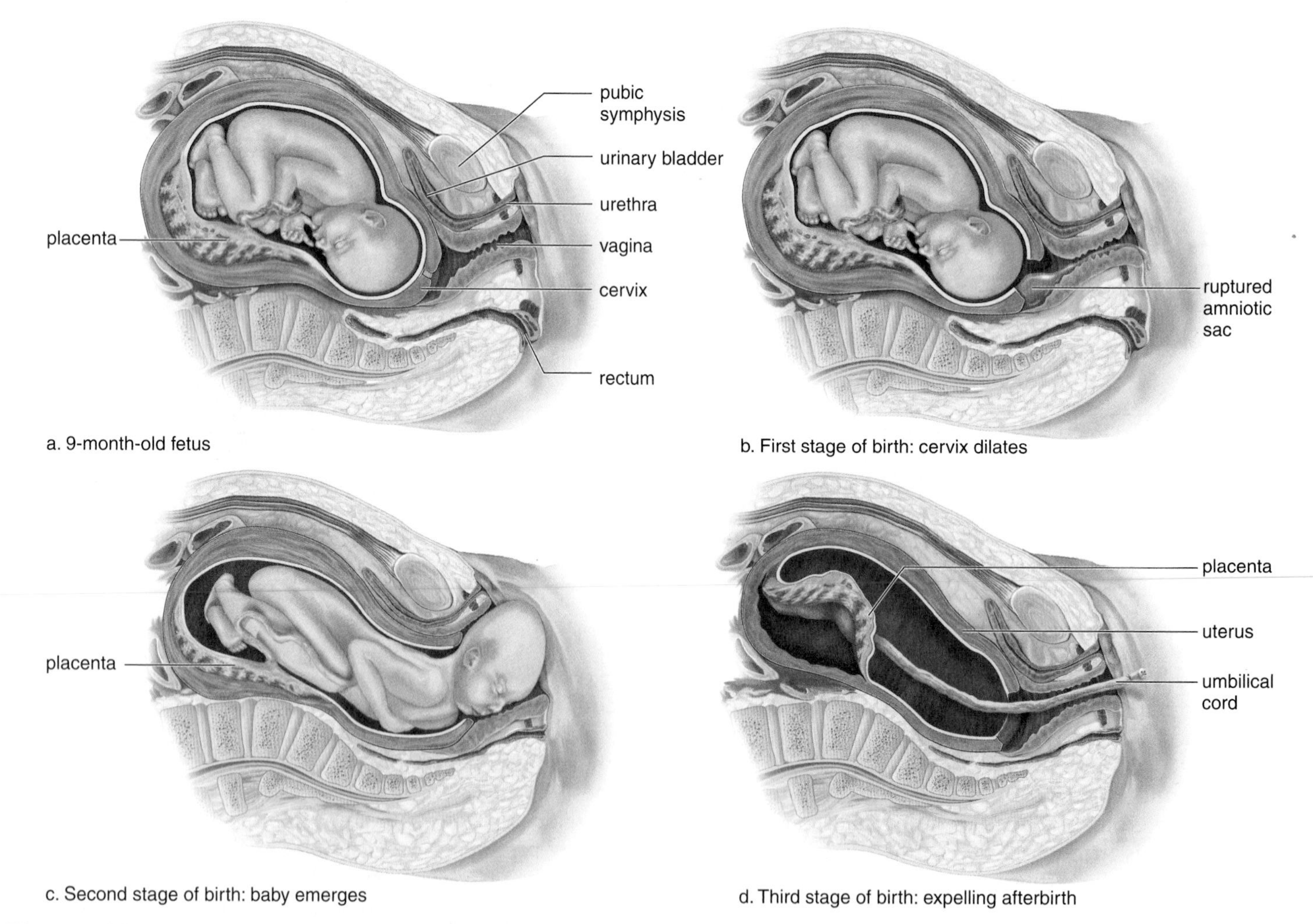

Figure 22.16 **Three stages of parturition (birth).**
a. Position of fetus just before birth begins. **b.** Dilation of cervix. **c.** Birth of baby. **d.** Expulsion of afterbirth.

A positive feedback mechanism regulates the onset and continuation of labor. Uterine contractions are induced by stretching of the cervix, which also brings about the release of oxytocin from the posterior pituitary. Oxytocin stimulates uterine contractions, which push the fetus downward, and the cervix stretches even more. This cycle keeps repeating itself until the baby is born. Three events, in any order, indicate that delivery will soon occur: (1) The uterine contractions are occurring about every five minutes and becoming stronger. (2) The amnion, which contains amniotic fluid, ruptures, causing water to flow out of the vagina. This event is sometimes referred to as "breaking water." (3) A plug of mucus from the cervical canal leaves the vagina. This plug prevents bacteria and sperm from entering the uterus during pregnancy. Its expulsion may be the least noticeable of the events, unless it is mixed with blood.

Stage 1

Parturition, the process of giving birth to an offspring, can be divided into three stages (Fig. 22.16). Prior to the first stage, there can be a "bloody show" caused by expulsion of the mucous plug. At first, the uterine contractions of labor occur in such a way that the cervical canal slowly disappears as the lower part of the uterus is pulled upward toward the baby's head (Fig. 22.16*b*). This process is called effacement, or "taking up the cervix." With further contractions, the baby's head acts as a wedge to assist cervical dilation. If the amniotic membrane has not already ruptured, it is apt to do so during this stage, releasing the amniotic fluid, which leaks out the vagina. The first stage of parturition ends once the cervix is dilated completely.

Stage 2

During the second stage of parturition, the uterine contractions occur every 1–2 minutes and last about one minute each. They are accompanied by a desire to push, or bear down. As the baby's head gradually descends into the vagina, the desire to push becomes greater. When the baby's head reaches the exterior, it turns so that the back of the head is uppermost (Fig. 22.16*c*). If the vaginal orifice does not expand enough to allow passage of the head, an **episiotomy** may be performed. This incision, which enlarges the opening, is sewn together later. As soon as the head is delivered, the baby's shoulders rotate so that the baby faces either to the right or the left. At this time, the physician may hold the head and guide it downward, while one shoulder and then the other emerges. The rest of the baby follows easily.

Once the baby is breathing normally, the umbilical cord is cut and tied, severing the child from the placenta. The stump of the cord shrivels and leaves a scar, which is the umbilicus.

Stage 3

The placenta, or **afterbirth,** is delivered during the third stage of parturition (Fig. 22.16*d*). About 15 minutes after delivery of the baby, uterine muscular contractions shrink the uterus and dislodge the placenta. The placenta is then expelled into the vagina. As soon as the placenta and its membranes are delivered, the third stage of parturition is complete.

Female Breast and Lactation

A female breast contains 15 to 25 lobules, each with a milk duct. Each milk duct begins at the nipple and divides into numerous other ducts that end in blind sacs called alveoli (Fig. 22.17).

During pregnancy, the breasts enlarge as the ducts and alveoli increase in number and size. The same hormones that affect the mother's breasts can also affect the child's. Some newborns, including males, even secrete a small amount of milk for a few days.

Usually, no milk is produced during pregnancy. The hormone prolactin is needed for lactation to begin, and the production of this hormone is suppressed because of the feedback control that the increased amount of estrogen and progesterone during pregnancy has on the pituitary. Once the baby is delivered, however, the pituitary begins secreting prolactin. It takes a couple of days for milk production to begin, and in the meantime, the breasts produce **colostrum,** a thin, yellow, milky fluid rich in protein, including antibodies.

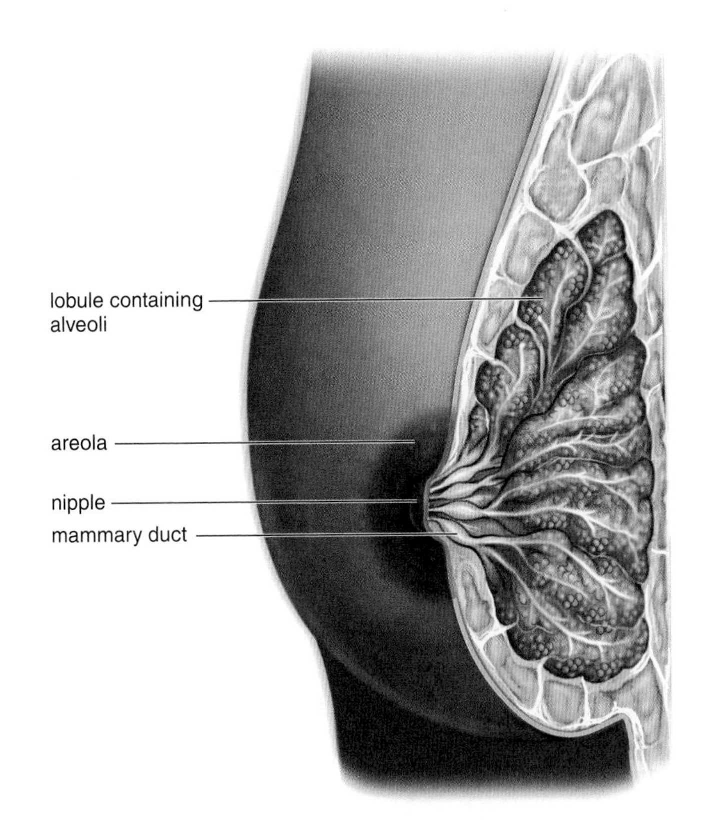

Figure 22.17 Female breast anatomy.
The female breast contains lobules consisting of ducts and alveoli. The alveoli are lined by milk-producing cells in the lactating (milk-producing) breast. Milk droplets congregate in the ducts that enter sinuses just behind the nipple. When a baby suckles, the sinuses are squeezed.

Health Focus

Preventing Birth Defects

While some birth defects are not preventable, others are. Therefore all females of childbearing age are advised to take everyday precautions to protect any future and/or presently developing embryos and fetuses from these defects. For example, it is best if a woman has a physical exam even before she becomes pregnant. At that time, it can be determined if she has been immunized against rubella (German measles). Depending on exactly when a pregnant woman has the disease, rubella can cause blindness, deafness, mental retardation, heart malformations, and other serious problems in an unborn child. A vaccine to prevent the disease can be given to a woman before she gets pregnant but not to a woman who is already pregnant because it contains live viruses. Also, the woman should be tested for the presence of HIV (the causative agent for AIDS) because preventive therapies are available to improve maternal and infant health.

Good health habits are a must during pregnancy, including proper nutrition, adequate rest, and exercise. Moderate exercise can usually continue throughout pregnancy and hopefully will contribute to ease of delivery. Basic nutrients are required in adequate amounts to meet the demands of both fetus and mother. A growing number of studies confirm that small, thin newborns are more likely to develop certain chronic diseases, such as diabetes and high blood pressure, when they become adults than are babies who are born heavier.

An increased amount of minerals, such as calcium for bone growth and iron for red blood cell formation, and certain vitamins, such as vitamin B_6 for proper metabolism and folate (folic acid), are required. A pregnant woman needs more folate per day to meet an increased rate of cell division and DNA synthesis in her own body and in that of the developing child. A maternal deficiency of folate has been linked to development of neural tube defects in the fetus, which include spina bifida (spinal cord or spinal fluid bulge through the back) and anencephaly (absence of a brain). Perhaps as many as 75% of these defects could be avoided by adequate folate intake even before pregnancy occurs. Consuming fortified breakfast cereals is a good way to meet folate needs, because they contain a more absorbable form of folate.

TABLE 22A Behaviors Harmful to the Unborn

Drinking alcohol
Smoking cigarettes
Taking illegal drugs
Taking any medication not approved by a physician
Exposure to environmental toxins and radiation

As shown in Figure 22.15, the placenta is the region for exchange between the fetal and maternal circulatory systems. Good health habits include avoiding the substances listed in Table 22A, which can cross the placenta and harm the fetus. Cigarette smoke poses a serious threat to the health of a fetus because it contains not only carbon monoxide but also other fetotoxic chemicals. Children born to smoking mothers have a greater chance of a cleft lip or palate, increased incidence of respiratory diseases, and later on, more reading disorders than those born to mothers who did not smoke during pregnancy.

Alcohol easily crosses the placenta, and only one drink a day appears to increase the chance of a spontaneous abortion. The more alcohol consumed, the greater the chance of physical abnormalities if the pregnancy continues. Heavy consumption of alcohol puts a fetus at risk of a mental defect because alcohol enters the brain of the fetus. Babies born to heavy drinkers are apt to undergo delirium tremens after birth—shaking, vomiting, and extreme irritability—and to have fetal alcohol syndrome (FAS). Children with FAS have decreased weight, height, and head size, with malformation of the head and face (Fig. 22A). Later, mental retardation is common, as are numerous other physical malformations.

Certainly, pregnant women should completely avoid illegal drugs, such as marijuana, cocaine, and heroin. "Cocaine babies" now make up 60% of drug-affected babies. Severe fluctuations in blood pressure produced by the use of cocaine temporarily deprive the developing brain of oxygen. Cocaine babies have visual problems, lack coordination, and are mentally retarded.

Children born to women who received X-ray treatment during pregnancy for, say, cancer are apt to have birth defects and/or to develop leukemia later. Even low X-ray levels are apt to cause mutations in the developing embryo or fetus. Dental and other diagnostic X rays that result in only a small amount of radiation are probably safe. Still, a woman should be sure her physician knows that she is or may be pregnant. Similarly, toxic chemicals, such as pesticides, and many organic industrial chemicals, such as vinyl chloride, formaldehyde, asbestos, and benzenes, are mutagenic and can cross the placenta, resulting in abnormalities. Lead circulating in a pregnant woman's blood can cause a child to be mentally retarded.

A woman has to be very careful about taking medications while pregnant. Excessive vitamin A, sometimes used to treat acne, may damage an embryo. In the 1950s and 1960s, DES (diethylstilbestrol), a synthetic hormone related to the natural female hormone estrogen, was given to pregnant women to prevent cramps, bleeding, and threatened miscarriage. But in the 1970s and 1980s, some adolescent girls and young women whose mothers had been treated with DES showed various abnormalities of the reproductive

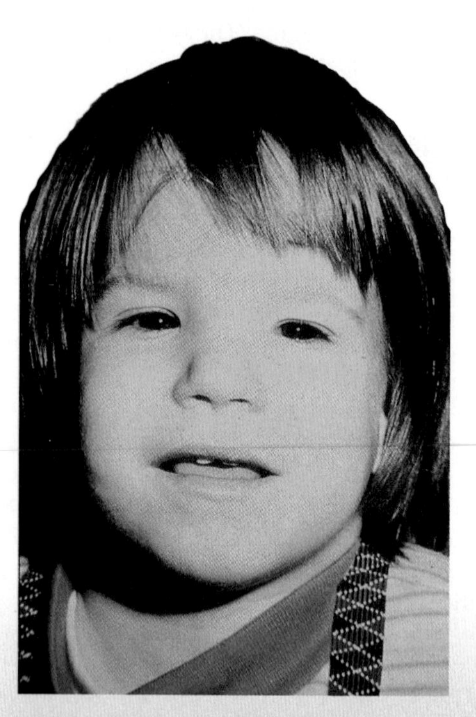

Figure 22A A child with fetal alcohol syndrome.

organs and an increased tendency toward cervical cancer. Other sex hormones, including birth control pills, can possibly cause abnormal fetal development, including abnormalities of the sex organs.

The drug thalidomide was a popular tranquilizer during the 1950s and 1960s in many European countries and to a degree in the United States. The drug, which was taken to prevent nausea in pregnant women, arrested the development of arms and legs in some children and also damaged heart and blood vessels, ears, and the digestive tract. Some mothers of affected children report that they took the drug for only a few days. Because of such experiences, physicians are generally very cautious about prescribing drugs during pregnancy, and no pregnant woman should take any drug—even an ordinary cold remedy—without first checking with her physician.

Unfortunately, immunization for sexually transmitted diseases is not possible. The AIDS virus can cross the placenta and cause mental retardation. As mentioned, proper medication can greatly reduce the chance of this happening. When a mother has herpes, gonorrhea, or chlamydia, newborns can become infected as they pass through the birth canal. Blindness and other physical and mental defects may develop. Birth by cesarean section could prevent these occurrences. Rh-negative women who have given birth to Rh-positive children should receive an Rh immunoglobulin injection within 72 hours to prevent the mother's body from producing Rh antibodies. She will start producing these antibodies when some of the child's Rh-positive red blood cells enter her bloodstream, possibly before but particularly at birth. Rh antibodies can cause nervous system and heart defects in a fetus. The first Rh-positive baby is not usually affected. But in subsequent pregnancies, antibodies created at the time of the first birth cross the placenta and begin to destroy the blood cells of the fetus, causing anemia and other complications.

The birth defects we have been discussing are particularly preventable because they are not due to inheritance of an abnormal number of chromosomes or any other genetic abnormality. According to the National Center for Health Statistics, between 1980 and 2004 the number of women giving birth at age 30 doubled. In the same time period, women giving birth after age 40 has almost quadrupled. The chance of an older woman bearing a child with a birth defect unrelated to genetic inheritance is no greater than that of a younger woman. However, as discussed in Chapter 24, there is a greater risk of an older woman having a child with a chromosomal abnormality, leading to premature delivery, cesarean section, low birth weight, and certain syndromes. Chapters 24 and 26 discuss how to detect chromosomal and other genetic defects in utero so that therapy for these disorders can begin as soon as possible.

Now that physicians and laypeople are aware of the various ways birth defects can be prevented, it is hoped that the incidence of birth defects will decrease in the future (Fig. 22B).

Discussion Questions

1. Should a woman be prosecuted for causing damage to her unborn child by drinking or illegal drug use?
2. Why does good medical care during pregnancy only reduce the *risk* of birth defects but not guarantee a healthy child?
3. Because folate, which has been shown to reduce the risk of neural tube defects, is needed by the embryo during the early weeks of development, when should a woman begin to take extra folate?

Figure 22B Child care.
Proper care of a child begins before birth in order to prevent birth defects.

The continued production of milk requires a suckling child. When a breast is suckled, the nerve endings in the areola are stimulated, and a nerve impulse travels along neural pathways from the nipples to the hypothalamus, which directs the pituitary gland to release the hormone oxytocin. When this hormone arrives at the breast, it causes contraction of the lobules so that milk flows into the ducts (called milk letdown), where it may be drawn out of the nipple by the suckling child.

Whether to breast-feed or not is a private decision based in part on a woman's particular circumstances. However, it is well known that breast milk contains antibodies produced by the mother that can help a baby survive. Babies have immature immune systems, less stomach acid to destroy foreign antigens, and also unsanitary habits. Breast-fed babies are less likely to develop stomach and intestinal illnesses, including diarrhea, during the first 13 weeks of life. Breast-feeding also has physiological benefits for the mother. Suckling causes uterine contractions that can help the uterus return to its normal size, and breast-feeding uses up calories and thus can help a woman return to her normal weight.

Check Your Progress

1. At what stage of parturition does the baby arrive?
2. How is milk production under hormonal control?

22.3 Human Development After Birth

Learning Outcomes

1. Understand how development continues after birth.
2. Describe three hypotheses about what causes aging and outline the changes that occur in the body due to aging.

Development does not cease once birth has occurred but continues throughout the stages of life: infancy, childhood, adolescence, and adulthood. Infancy, the toddler years, and preschool years are times of remarkable growth. During the birth to 5-year-old stage, humans acquire gross motor and fine motor skills. These include the ability to sit up and then to walk, as well as being able to hold a spoon and manipulate small objects. Language usage begins during this time and will become increasingly sophisticated throughout childhood. As infants and toddlers explore their environment, their senses—vision, taste, hearing, smell, and touch—mature dramatically. Socialization is very important, as a child forms emotional ties with its caregivers and learns to separate self from others. Babies do not all develop at the same rate, and there is a large variation in what is considered normal.

The preadolescent years, from 6 to 12 years of age, are a time of continued rapid growth and learning. Preadolescents form identities apart from parents, and peer approval becomes very important. Adolescence begins with the onset of puberty, as the young person achieves sexual maturity. For girls, puberty begins between 10 and 14 years of age, whereas for boys it generally occurs between ages 12 and 16. During this time, the sex-specific hormones cause the secondary sexual characteristics to appear. Profound social and psychological changes are also associated with the transition from childhood to adulthood.

Aging encompasses these progressive changes, which contribute to an increased risk of infirmity, disease, and death (Fig. 22.18). Today, there is great interest in **gerontology,** the study of aging, because our society includes more older individuals than ever before, and the number is expected to rise even more. In the next half-century, the number of people over age 65 will increase by 147%. The human life span is judged to be a maximum of 120–125 years. The present goal of gerontology is not necessarily to increase the life span, but to increase the health span, the number of years an individual enjoys the full functions of all body parts and processes.

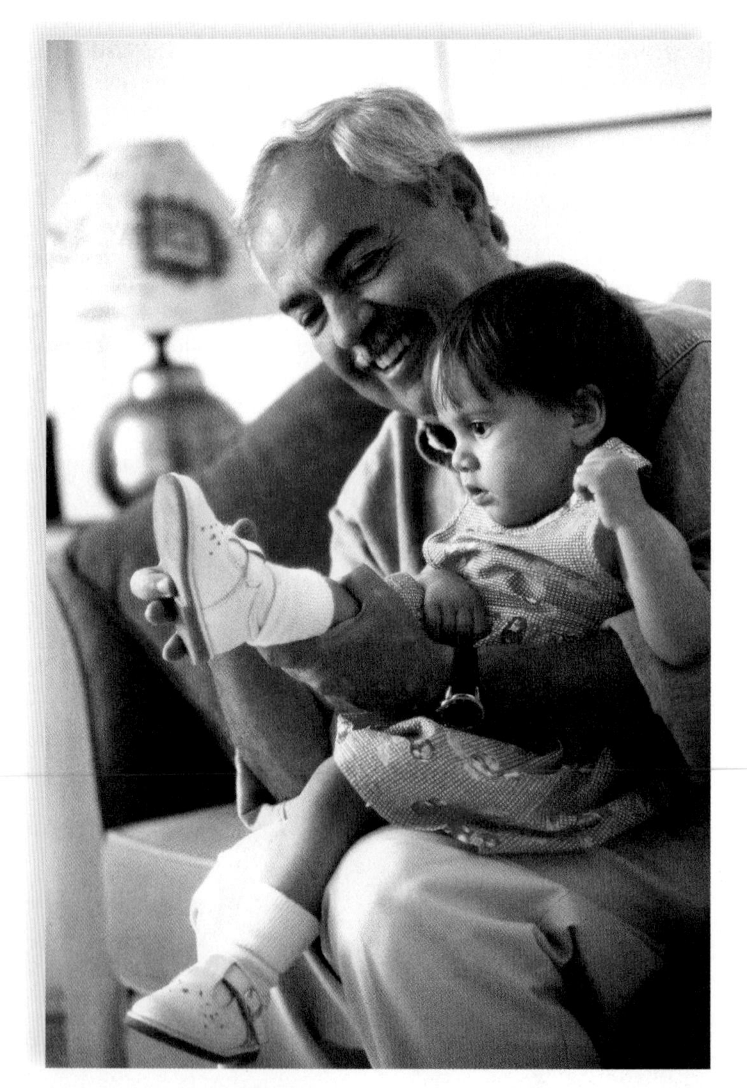

Figure 22.18 Aging.
Aging is a slow process during which the body undergoes changes that eventually bring about death, even if no marked disease or disorder is present. Medical science is trying to extend the human life span as well as the health span, the length of time the body functions normally.

Hypotheses About Aging

Of the many hypotheses about what causes aging, three are considered here.

Genetic in Origin

Several lines of evidence indicate that aging has a genetic basis: (1) The number of times a cell divides is species-specific. Perhaps as we grow older, more and more cells are unable to divide, and instead, they undergo degenerative changes and die. (2) Some cell lines may become nonfunctional long before the maximum number of divisions has occurred. Whenever DNA replicates, mutations can occur, and this can lead to the production of nonfunctional proteins. Eventually, the number of inadequately functioning cells can build up, which contributes to the aging process. (3) The children of long-lived parents tend to live longer than those of short-lived parents. Recent work suggests that when an animal produces fewer free radicals, it lives longer. Free radicals are unstable molecules that have an unpaired electron. In order to stabilize themselves, free radicals react with another molecule, such as DNA or proteins (e.g., enzymes) or lipids found in plasma membranes. Eventually, these molecules are unable to function, and the cell is destroyed. Certain genes code for antioxidant enzymes that detoxify free radicals. Research suggests that animals with particular forms of these genes—and therefore more efficient antioxidant enzymes—live longer.

Whole-Body Process

A decline in the hormonal system can affect many different organs of the body. For example, type 2 diabetes is common in older individuals. The pancreas makes insulin, but the cells lack the receptors that enable them to respond. Menopause in women occurs for a similar reason. There is plenty of follicle-stimulating hormone in the bloodstream, but the ovaries do not respond. Perhaps aging results from the loss of hormonal activities and a decline in the functions they control.

The immune system, too, no longer performs as it once did, and this can affect the body as a whole. The thymus gland gradually decreases in size, and eventually most of it is replaced by fat and connective tissue. The incidence of cancer increases among the elderly, which may signify that the immune system is no longer functioning as it should. This idea is also substantiated by the increased incidence of autoimmune diseases in older individuals.

It is possible, though, that aging is not due to the failure of a particular system that can affect the body as a whole, but to a specific type of tissue change that affects all of the organs and even the genes. It has been noticed for some time that proteins—such as the collagen fibers present in many support tissues—become increasingly cross-linked as people age. Undoubtedly, this cross-linking contributes to the stiffening and loss of elasticity characteristic of aging tendons and ligaments. It may also account for the inability of such organs as the blood vessels, the heart, and the lungs to function as they once did. Some researchers have now found that glucose has the tendency to attach to any type of protein, which is the first step in a cross-linking process. They are presently experimenting with drugs that can prevent cross-linking.

Extrinsic Factors

The current data about the effects of aging are often based on comparisons of the elderly to younger age groups. But perhaps today's elderly were not as aware when they were younger of the importance, for example, of diet and exercise to general health. It is possible, then, that much of what we attribute to aging is instead due to years of poor health habits.

Consider, for example, osteoporosis. This condition is associated with a progressive decline in bone density in both males and females so that fractures are more likely to occur after only minor trauma. Osteoporosis is common in the elderly—by age 65, one-third of women will have vertebral fractures, and by age 81, one-third of women and one-sixth of men will have suffered a hip fracture. While there is no denying that a decline in bone mass occurs as a result of aging, certain extrinsic factors are also important. The occurrence of osteoporosis itself is associated with cigarette smoking, heavy alcohol intake, and inadequate calcium intake. Not only is it possible to eliminate these negative factors by personal choice, but it is also possible to add a positive factor. A moderate exercise program has been found to slow the progressive loss of bone mass.

Even more important, a sensible exercise program and a proper diet that includes at least five servings of fruits and vegetables a day will most likely help eliminate cardiovascular disease. Experts no longer believe that the cardiovascular system necessarily suffers a large decrease in functioning ability with age. Persons 65 years of age and older can have well-functioning hearts and open coronary arteries if their health habits are good and they continue to exercise regularly.

Check Your Progress

1. List the changes that occur from infancy to adulthood.
2. How do the three hypotheses of aging influence each other? Explain your answer.

Effect of Age on Body Systems

Data about how aging affects body systems are necessarily based on past events. It is possible that, in the future, age will not have these effects or at least not to the same degree as those described here.

Skin

As aging occurs, skin becomes thinner and less elastic because the number of elastic fibers decreases and the collagen fibers undergo cross-linking, as discussed previously. Also, there is less adipose tissue in the subcutaneous layer; therefore, older people are more likely to feel cold. The loss of thickness partially accounts for sagging and wrinkling of the skin.

Homeostatic adjustment to heat is also limited because there are fewer sweat glands for sweating to occur. There are fewer hair follicles, so the hair on the scalp and the limbs thins out. The number of oil (sebaceous) glands is reduced, and the skin tends to crack. Older people also experience a decrease in the number of melanocytes, making their hair gray and their skin pale. In contrast, some of the remaining pigment cells are larger, and pigmented blotches appear on the skin.

Processing and Transporting

Cardiovascular disorders are the leading cause of death today. The heart shrinks because of a reduction in cardiac muscle cell size. This leads to loss of cardiac muscle strength and reduced cardiac output. Still, in the absence of disease, the heart is able to meet the demands of increased activity. It can double its rate or triple the amount of blood pumped each minute even though the maximum possible output declines.

Because the middle layer of arteries contains elastic fibers, which are most likely subject to cross-linking, the arteries become more rigid with time. The internal diameter of arteries is further reduced by plaque, a buildup of fatty material. Therefore, blood pressure readings gradually rise with age. Such changes are common in individuals living in Western industrialized countries but not in agricultural societies. A diet low in cholesterol and saturated fatty acid has been suggested as a way to control degenerative changes in the cardiovascular system.

Blood flow to the liver is reduced, and this organ does not metabolize drugs as efficiently as before. This means that, as a person gets older, less medication is needed to maintain the same level in the bloodstream.

Cardiovascular problems are often accompanied by respiratory disorders, and vice versa. Growing inelasticity of lung tissue means that ventilation is reduced. Because we rarely use the entire vital capacity, these effects are not noticed unless the demand for oxygen increases.

Blood supply to the kidneys is also reduced. The kidneys become smaller and less efficient at filtering wastes. Salt and water balance are difficult to maintain, and the elderly dehydrate faster than young people. Difficulties involving urination include incontinence (lack of bladder control) and the inability to urinate. In men, the prostate gland may enlarge and reduce the diameter of the urethra, making urination so difficult that surgery may be needed.

The loss of teeth, which is frequently seen in elderly people, is more apt to be the result of long-term neglect than aging. The digestive tract loses tone, and secretion of saliva and gastric juice is reduced, but there is no indication of reduced absorption. Therefore, an adequate diet, rather than vitamin and mineral supplements, is recommended. Elderly people commonly complain of constipation, increased gas, and heartburn. Gastritis, ulcers, and cancer can also occur.

Integration and Coordination

While most tissues of the body regularly replace their cells, some at a faster rate than others, the brain and the muscles ordinarily do not. However, contrary to previous opinion, recent studies show that few neural cells of the cerebral cortex are lost during the normal aging process. This means that cognitive skills remain unchanged even though a loss in short-term memory characteristically occurs. Although the elderly learn more slowly than the young, they can acquire and remember new material. It has been noted that when more time is given for the subject to respond, age differences in learning decrease.

Neurons are extremely sensitive to oxygen deficiency, and if neuron death does occur, it may be due not to aging itself but to reduced blood flow in narrowed blood vessels. Specific disorders, such as depression, Parkinson disease, and Alzheimer disease, are sometimes seen, but they are not common. Reaction time, however, does slow, and more stimulation is needed for hearing, taste, and smell receptors to function as before. After age 50, the ability to hear tones at higher frequencies decreases gradually, and this can make it difficult to identify individual voices and to understand conversation in a group. The lens of the eye does not accommodate as well and also may develop a cataract. Glaucoma, the buildup of pressure due to increased fluid, is more likely to develop because of a reduction in the size of the anterior cavity of the eye.

Loss of skeletal muscle mass is not uncommon, but it can be controlled by a regular exercise program. The capacity to do heavy labor decreases, but routine physical work should be no problem. A decrease in the strength of the respiratory muscles and inflexibility of the rib cage contribute to the inability of the lungs to expand as before, and reduced muscularity of the urinary bladder contributes to an inability to empty the bladder completely, and therefore to the occurrence of urinary infections.

As noted before, aging is accompanied by a decline in bone density. Osteoporosis, characterized by a loss of calcium and mineral from bone, is not uncommon, but evidence indicates that proper health habits can prevent its occurrence. Arthritis, which causes pain upon movement of a joint, is also seen.

Weight gain occurs because the basal metabolism decreases and inactivity increases. Muscle mass is replaced by stored fat and retained water.

The Reproductive System

Females undergo menopause, and thereafter, the level of female sex hormones in the blood falls markedly. The uterus and the cervix are reduced in size, and the walls of the oviducts and the vagina become thinner. The external genitals become less pronounced. In males, the level of androgens falls gradually over the age span of 50–90, but sperm production continues until death. Sexual activity need not decline with age, however, due to hormone therapy and other treatments.

It is of interest that, as a group, females live longer than males. Although their health habits may be better, it is also possible that the female sex hormone estrogen offers women some protection against cardiovascular disorders when they

Bioethical Focus

End-of-life Decisions

In 2006, 12.4% of the U.S. population was over 65 years of age. According to U.S. Census Bureau projections, as the "baby boomers" age, the over-65 age group is expected to make up 20% of the population by 2030. As the population of the United States ages, more and more people will be dealing with "end-of-life" issues. These are the practical, legal, medical, and even spiritual decisions that must be made when a person is expected to die within approximately six months. You may be involved in making such decisions with your parents. Legal decisions involve financial matters, and a will should be made to protect these decisions. Medical decisions relate to the degree of family involvement, the type of care desired, and whether this care will be delivered in the dying person's own home, in the home of a family member, or in an institutional care setting. Often, family members or physicians end up making these medical end-of-life decisions because the patients have waited until they are too sick to communicate their desires. For some, this is due to the discomfort of discussing the possibility of their death with family members. Often the medical terminology is difficult to comprehend, and a patient may not understand the choices available. For many, cultural or religious beliefs influence the type of medical treatment they want or do not want, including the use of feeding tubes, ventilators, chemotherapy, and drastic lifesaving procedures.

People can plan their end-of-life medical treatment in what is termed an *advance directive.* The two types of advance directives are a living will and a durable power of attorney for health care. A living will expresses the patient's desires concerning the type and degree of medical intervention. A durable power of attorney names another person who will make only health-care decisions for the person. A durable power of attorney is especially critical if the patient wants a nonfamily member to make these decisions, as in the case of an unmarried couple or same-sex partners. The American Psychological Association suggests that patients plan ahead while they are not ill and have time to find answers to all of their questions, write down their choices, complete the legal paperwork, and share their decisions with their family members and physicians.

Form Your Own Opinion

1. Should all Americans be required to complete an advance directive for their medical care? If so, by what age?
2. In the absence of an advance directive, how should decisions be made concerning medical treatment for an incapacitated patient if family members disagree?
3. How would you feel if asked to discuss these issues with your parents while they are still healthy?

are younger. Males suffer a marked increase in heart disease in their forties, but an increase is not noted in females until after menopause, when women lead men in the incidence of stroke. Men are still more likely than women to have a heart attack, however.

Figure 22.19 Remaining active.
The aim of gerontology is to allow seniors to enjoy living.

Conclusion

We have listed many adverse effects of aging, but it is important to emphasize that such effects are not inevitable (Fig. 22.19). We must discover any extrinsic factors that precipitate these adverse effects and guard against them. Just as it is wise to make the proper preparations to remain financially independent when older, it is also wise to realize that biologically successful old age begins with the health habits developed when we are younger. The Bioethical Focus suggests other ways that aging people can plan ahead.

Check Your Progress

1. List the changes that occur as the body ages. Are these changes inevitable?

Applying the Concepts [Revisited]

Although there have been many advances made in medicine and public health, much of the responsibility for good health as we age rests with choices made by the individual. Madonna is a "fitness fanatic" who exercises daily and eats a macrobiotic diet. Sharon Stone exercises and abstains from all caffeine and alcohol.

1. What good health habits have you incorporated into your routine that will extend your health span?
2. What bad habits should you try to eliminate from your life now, before they cause damage that will age you?

SUMMARIZING THE CONCEPTS

22.1 Principles of Animal Development

- Development begins at fertilization. Only one sperm actually enters the oocyte, and this sperm's nucleus fuses with the oocyte nucleus.
- The early developmental stages in animals proceed from the cellular level to the tissue level to the organ level. During the cellular stages, cleavage (cell division) occurs, but there is no overall growth. The result is a morula, which becomes the blastula when an internal cavity (the blastocoel) appears.
- During the tissue stages, gastrulation (invagination of cells into the blastocoel) results in formation of the germ layers: ectoderm, mesoderm, and endoderm. Both the cellular and tissue stages can be affected by the amount of yolk.
- Organ formation can be related to germ layers. For example, during neurulation, the nervous system develops from midline ectoderm, just above the notochord.

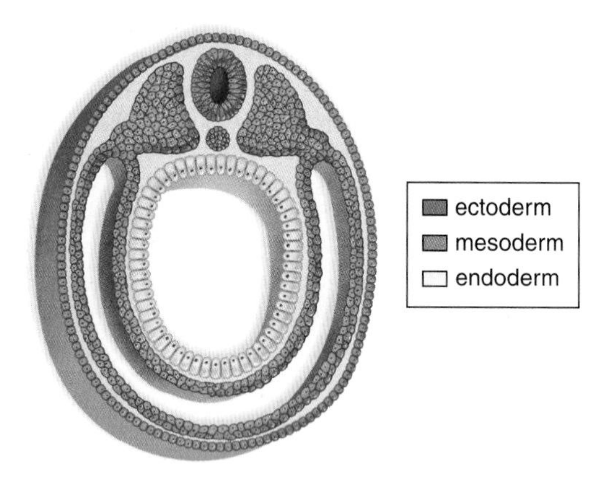

- Cellular differentiation begins with cytoplasmic segregation in the egg. Induction is also part of cellular differentiation. In *C. elegans,* investigators have shown that induction is an ongoing process in which one tissue after the other regulates the development of another tissue through chemical signals coded for by particular genes.
- Work with *Drosophila* has allowed researchers to identify morphogen genes that determine the shape and form of the body. During development, sequential sets of morphogen genes code for morphogen gradients that activate the next set of morphogen genes, in turn. Homeotic genes code for proteins that contain a homeodomain, a particular sequence of 60 amino acids. Homologous homeotic genes have been found in a wide variety of organisms, and therefore, they must have arisen early in the history of life and been conserved.

22.2 Human Embryonic and Fetal Development

- Human development before birth can be divided into embryonic development (months 1 and 2) and fetal development (months 3–9).
- The extraembryonic membranes appear early in human development. The trophoblast of the blastocyst is the first sign of the chorion, which goes on to become the fetal part of the placenta. Exchange of gases, nutrients, and wastes occurs between fetal and maternal blood at the placenta. The amnion contains amniotic fluid, which cushions and protects the embryo. The yolk sac and allantois are also present.

- Fertilization occurs in the oviduct, and cleavage occurs as the embryo moves toward the uterus. The morula becomes the blastocyst before implanting in the endometrium of the uterus.
- Organ development begins with neural tube and heart formation. There follows a steady progression of organ formation during embryonic development. During fetal development, features are refined, and the fetus adds weight.
- Birth, which requires three stages, occurs about 280 days after the start of the mother's last menstruation.

22.3 Human Development After Birth

- Development after birth consists of infancy, childhood, adolescence, and adulthood.
- Aging may be due to cellular repair changes, which are genetic in origin. Other factors that may affect aging are changes in body processes and certain extrinsic factors.

TESTING YOURSELF

Choose the best answer for each question.

1. Which occurs in the egg to prevent fertilization by multiple sperm cells?
   a. Plasma membrane depolarizes.
   b. Zona pellucida moves away from the egg.
   c. Killer enzymes are secreted.
   d. Both a and b are correct.
2. The inner lining of the digestive tract forms from which germ layer?
   a. endoderm
   b. ectoderm
   c. mesoderm
   d. blastula
3. In some animals, the notochord is replaced by the
   a. neural plate.
   b. vertebral column.
   c. neurula.
   d. None of these are correct.

4. The umbilical vein carries blood rich in
   a. nutrients.
   b. oxygen.
   c. Both a and b are correct.
   d. Neither a nor b is correct.
5. The archenteron will develop into the
   a. heart.
   b. vertebral column.
   c. gut.
   d. None of these are correct.
6. The outer cell layer in the blastocyst is called the
   a. ectoderm.
   b. endoderm.
   c. trophoblast.
   d. embryo.
   e. None of these are correct.
7. Implantation occurs at the _____ stage.
   a. gastrula
   b. blastula
   c. morula
   d. None of these are correct.
8. The cavity in the gastrula is the
   a. blastocoel.
   b. gastrocoel.
   c. archenteron.
   d. embryo cavity.
   e. None of these are correct.
9. The ability to experimentally cause a second neural plate to develop above notochord donor tissue is due to
   a. segregation.
   b. induction.
   c. reduction.
   d. gastrulation.
   e. None of these are correct.
10. Identify the stages of animal development in the following diagram.

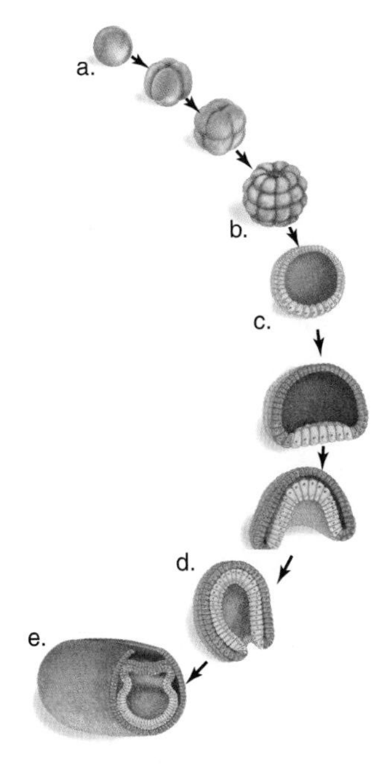

11. Identify the regions of a cross section of an animal embryo in the following diagram.

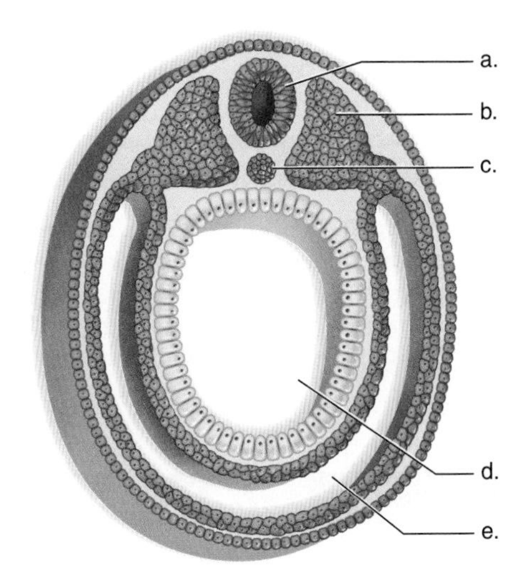

12. In human development, which part of the blastocyst develops into a fetus?
    a. morula
    b. trophoblast
    c. inner cell mass
    d. chorion
    e. yolk sac
13. Genes that control the structural pattern of an animal are
    a. homeotic genes.
    b. morphogen genes.
    c. induction genes.
    d. All of these are correct.
    e. Both a and b are correct.
14. Morphogenesis is associated with
    a. protein gradients.
    b. pattern formation.
    c. homeotic genes.
    d. All of these are correct.
15. Which of the germ layers is best associated with development of the heart?
    a. ectoderm
    b. mesoderm
    c. endoderm
    d. neurula
    e. All of these are correct.

UNDERSTANDING THE TERMS

afterbirth 459
aging 462
allantois 452
amnion 452
blastocoel 445
blastocyst 453
blastopore 446
blastula 445
cellular differentiation 448
chorion 452
chorionic villi 455
cleavage 445
colostrum 459
cytoplasmic segregation 449
ductus arteriosus 456
ductus venosus 456
ectoderm 446
embryonic development 452
embryonic disk 454
endoderm 446
episiotomy 459
extraembryonic membrane 452
fate map 449
fertilization 444
fontanel 456
foramen ovale 456
gastrula 446
gastrulation 446
germ layer 447
gerontology 462
gray crescent 449
homeobox 451
homeotic gene 450
human chorionic gonadotropin (hCG) 454

induction 449
lanugo 456
mesoderm 446
morphogenesis 448
morphogen gene 450
morula 445
neural plate 447
neural tube 447
neurula 447
notochord 447
parturition 459
pattern formation 448
placenta 457
trophoblast 453
umbilical artery 456
umbilical cord 455
umbilical vein 456
vernix caseosa 456
yolk sac 452

THINKING CRITICALLY

1. Mitochondria contain their own genetic material. Mitochondrial mutations are inherited from mother to child, without any contribution from the father. Why?
2. In one form of prenatal testing for fetal genetic abnormalities, chorionic villi samples are taken and analyzed. Why is this possible?
3. The Health Focus in this chapter suggests that women of childbearing age begin taking precautions to prevent birth defects (good health habits, rubella vaccination, etc.). Why should these precautions begin before a woman is pregnant?

INQUIRY INTO LIFE WEBSITE

The companion website for *Inquiry into Life* provides a wealth of information organized and integrated by chapter. You will find practice tests, animations, videos, and much more that will complement your learning and understanding of general biology.

http://www.mhhe.com/maderinquiry13

23 Patterns of Gene Inheritance

Applying the Concepts

Flo Hyman was considered the best volleyball player of her time. Born in 1954, Hyman was athletically talented and very tall—over 6 ft tall in junior high school. Her final height was 6 ft 5 in. At the 1981 World Cup in Tokyo, she was named to the six-member All-World Cup Team. Hyman went on to lead the U.S. Olympic team to a silver medal in the 1984 Summer Olympic Games in Los Angeles, California. She earned the nickname "Flying Clutchman" and could serve a volleyball at speeds of up to 100 miles per hour. After the Olympics, Hyman lobbied for the Civil Rights Restoration Act and testified before Congress in 1985 on behalf of Title IX legislation, which prohibits sex discrimination in sports by schools receiving federal funds. In 1982, Hyman returned to Japan to play professional women's volleyball. Her height, however, was a symptom of the condition that would kill her. Hyman collapsed during a match in January 1986 and died later that evening of a ruptured aorta, a complication of Marfan syndrome. As described in this chapter, Marfan syndrome is an autosomal dominant genetic disorder that results in a defect in an elastic connective tissue protein called fibrillin. Connective tissue is abundant throughout the body and fibrillin plays a special role in strengthening the wall of the aorta. Hyman did not suffer from other abnormalities common to those with Marfan syndrome, including curvature of the spine and breastbone, and this is probably why she went undiagnosed. To honor her athleticism, the Women's Sports Foundation instituted an annual Flo Hyman Award given to the female athlete who best exemplifies Hyman's "dignity, spirit, and commitment to excellence."

Chapter Outline

23.1 Mendel's Laws

Learning Outcomes

1. State Mendel's law of segregation and law of independent assortment, and relate them to the movements of chromosomes during gamete formation and fertilization.
2. Define allele and explain what it means for an allele to be dominant or recessive.
3. Contrast genotype and phenotype, and use appropriate terms for describing genotypes.
4. Predict outcome ratios and probabilities for one- and two-trait crosses; design and interpret testcrosses.

Today, most people know that DNA is the genetic material, and they may have heard that scientists have completed the sequence of the human genome—all the bases in the DNA of human cells. In contrast, they may never have heard of Gregor Mendel, an Austrian monk who, in 1860, developed certain laws of heredity after doing crosses between garden pea plants (Fig. 23.1). Mendel investigated genetics at the organism level, and this is still the level that intrigues most of us on a daily basis. We observe, for example, that facial and other features run in families, and we would like some convenient way of explaining this observation. And so it is appropriate to begin our study of genetics at the organism level and learn to use Mendel's laws of heredity.

Gregor Mendel

Mendel's parents were farmers, so he no doubt acquired the practical experience he needed to grow pea plants during childhood. Mendel was also a mathematician. He kept careful and complete records, even though he crossed and catalogued some 24,034 plants through several generations. He concluded that the plants transmitted distinct factors (now called genes found on chromosomes) to offspring. In pea plants and in humans, the chromosomes come in pairs, called *homologous pairs* of chromosomes. One member of the pair is inherited from the mother, while the other member is inherited from the father. We can recognize a homologous pair because both of its members have the same length and centromere location. They also carry genes for the same traits in the same order. For example, one trait in humans is finger length. One member of the homologous pair of chromosomes could contain the gene for long fingers, while the other member could contain the gene for short fingers. Alternate forms of a gene for the same trait are called **alleles.** Alleles are always at the same spot, or **locus,** on each member of the homologous pair. In Figure 23.2, the letters on the homologous chromosomes stand for alleles. *G* is an allele of *g*, and vice versa; *R* is an allele of *r*, and vice versa. *G* could never be an allele of *R* because *G* and *R* are at different loci and govern different traits.

Figure 23.1 Gregor Mendel, 1822–84.
Mendel grew and tended the pea plants he used for his experiments. For each experiment, he observed as many offspring as possible. For a cross that required him to count the number of round seeds to wrinkled seeds, he observed and counted a total of 7,324 peas!

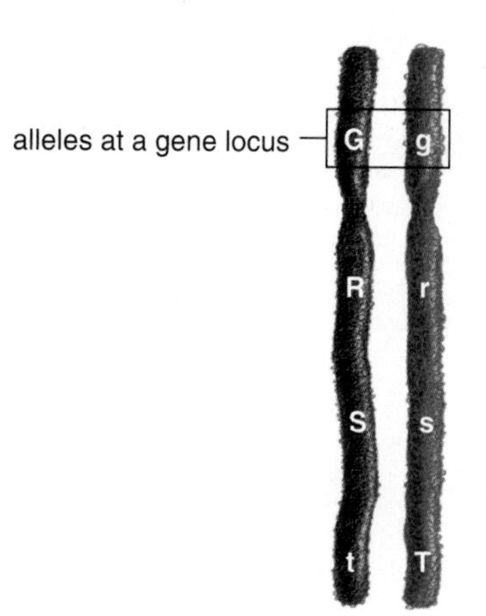

Figure 23.2 Homologous chromosomes.
The letters represent alleles—that is, alternate forms of a gene. Each allelic pair, such as *Gg* or *Tt*, is located on homologous pairs of chromosomes at a particular gene locus. The different colors represent the maternal or paternal origin of the chromosome.

Mendel knew nothing about homologous pairs of chromosomes, but his work with pea plants (described in the Science Focus) caused him to decide that pea plants have two factors for every trait, such as stem length. He observed that one of the factors controlling the same trait can be dominant over the other, which is recessive. Therefore, a tall pea plant could have one factor for long stem and another for short stem. While most of the offspring from two such tall pea plants were tall, some were short. Mendel reasoned that this would be possible if the gametes (i.e., sperm and oocyte) contained only one factor for each trait. Only in that way could an offspring receive two factors for short stem. On the basis of such studies, Mendel formulated his first law, the **law of segregation.** The law of segregation states the following:

- Each individual has two factors for each trait.
- The factors segregate (separate) during the formation of the gametes.
- Each gamete contains only one factor from each pair of factors.
- Fertilization gives each new individual two factors for each trait.

Check Your Progress

1. How are homologous chromosomes alike? How are they different?
2. State Mendel's law of segregation.

a. Widow's peak

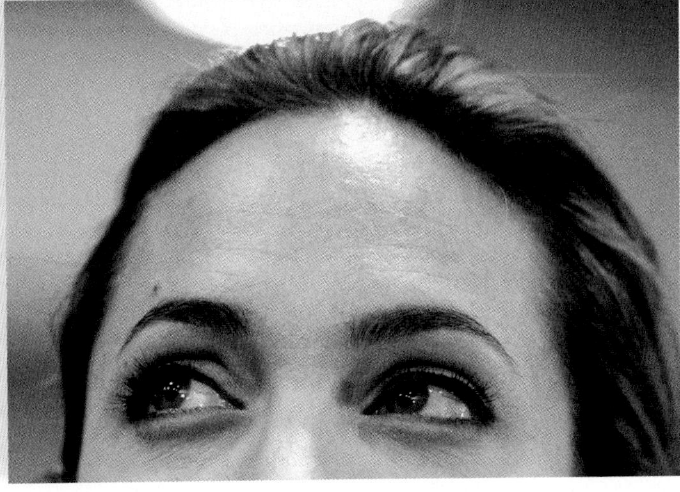

b. Straight hairline

Figure 23.3 Widow's peak.
In humans, **(a)** widow's peak is dominant over **(b)** straight hairline. Therefore, the man in (a) could have the genotype *Ww* or *WW*. The woman's genotype has to be *ww* because only this genotype can result in straight hairline. Suppose this man reproduces with this woman. If the man's genotype is *Ww*, then they could have a child with a straight hairline.

The Inheritance of a Single Trait

The **phenotype** of an organism refers to the individual's actual appearance. Phenotype may describe either an individual's physical characteristics, or microscopic and metabolic characteristics. For example, a widow's peak and a straight hairline are considered phenotypes (Fig. 23.3). The **genotype** refers to the alleles the chromosomes carry that are responsible for that trait. Because each chromosome is found in a homologous pair in diploid organisms, there are two alleles for each trait, one on each member of the homologous pair. Mendel suggested using letters to indicate these factors, which we now call alleles. A capital letter indicates a **dominant allele** and a lowercase letter indicates a **recessive allele.** The word *dominant* is not meant to imply that the dominant allele is better or stronger than the recessive allele. Dominant means that this allele will mask the expression of the recessive allele when they are together in the same organism.

When working with the alleles that determine hairline, we will use this key:

W = Widow's peak (dominant allele)
w = Straight hairline (recessive allele)

The key tells us which letter of the alphabet to use for the gene in a particular problem. It also tells which allele is dominant, because a capital letter signifies dominance.

In the case of a single trait, such as hairline, there are three possible combinations of the two alleles: *WW*, *Ww* and *ww*. These are considered the possible genotypes for this trait. If the two members of the allelic pair are the same (*homo*), the organism is said to be **homozygous.** Both *WW* and *ww* are homozygous, with *WW* homozygous dominant, and *ww* homozygous recessive. If the two members of the allelic pair are different (*hetero*), the organism is said to be **heterozygous.** Only *Ww* is a heterozygous genotype. But what about the phenotype? When an organism has a *W*, it exhibits the dominant phenotype regardless of whether the other allele is *W* or *w*. Both *WW* and *Ww* exhibit the dominant phenotype, a widow's peak. The recessive allele *w* is masked by the expression of the dominant allele *W* in the heterozygous genotype. The only time the recessive phenotype, straight hairline, is expressed, is when the genotype is homozygous recessive, or *ww*. The phenotypes and genotypes for this trait are summarized in Table 23.1.

Gamete Formation

The genotype has two alleles for each trait, whereas the gametes (i.e., sperm and oocyte) have only one allele for each trait in accordance with Mendel's law of segregation. During gamete formation, the homologous chromosomes separate, and there is only one member of each pair of chromosomes in each gamete. Therefore, there is only one allele for each trait, such as type of hairline, in each gamete. (You may want to review meiosis in Chapter 5 at this time.)

In the simplest terms, *no two letters in a gamete can be the same letter of the alphabet.* If the genotype of an individual is *Ww*, the gametes for this individual will contain either a *W* or a *w* but not both. If the genotype is *WwLl*, all combinations of any two different letters can be present.

TABLE 23.1 Genotype Related to Phenotype

Genotype	Genotype	Phenotype
WW	Homozygous dominant	Widow's peak
Ww	Heterozygous	Widow's peak
ww	Homozygous recessive	Straight hairline

The Investigations of Gregor Mendel

Virtually every culture in history has attempted to explain inheritance patterns. An understanding of these patterns has always been important to agriculture and animal husbandry, the science of breeding animals. However, it was not until the 1860s that the Austrian monk Gregor Mendel developed the fundamental laws of heredity after performing a series of ingenious experiments. Previously, he had studied science and mathematics at the University of Vienna. When Mendel began his work, most plant and animal breeders acknowledged that both sexes contribute equally to a new individual. They thought that parents of contrasting appearance always produced offspring of intermediate appearance. This concept, called the *blending concept of inheritance*, meant that a cross between plants with red flowers and plants with white flowers would yield only plants with pink flowers. When red and white flowers reappeared in future generations, the breeders mistakenly attributed this to instability in the genetic material.

Mendel chose to work with the garden pea, *Pisum sativum* (Fig. 23A). The garden pea was a good choice for several reasons. The plants were easy to cultivate and had a short generation time. Although peas normally self-pollinate (pollen only goes to the same flower), they could be cross-pollinated by hand by transferring pollen from an anther to a stigma. Many varieties of peas were available, and Mendel chose 22 for his experiments. When these varieties self-pollinated, they were *true-breeding*—meaning that the

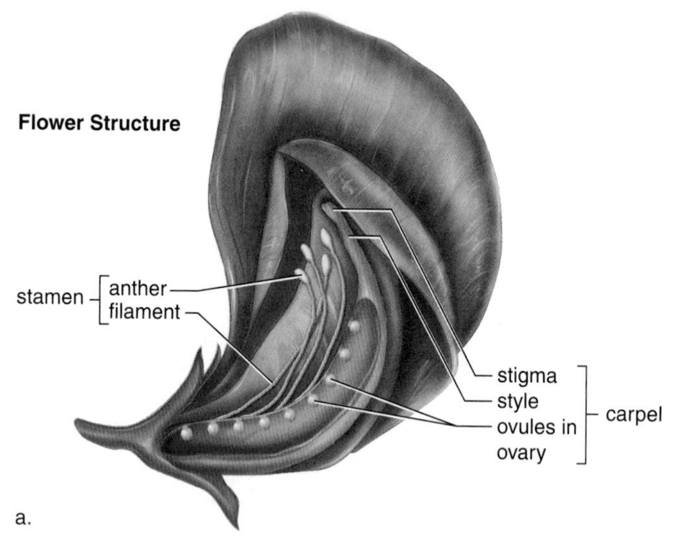

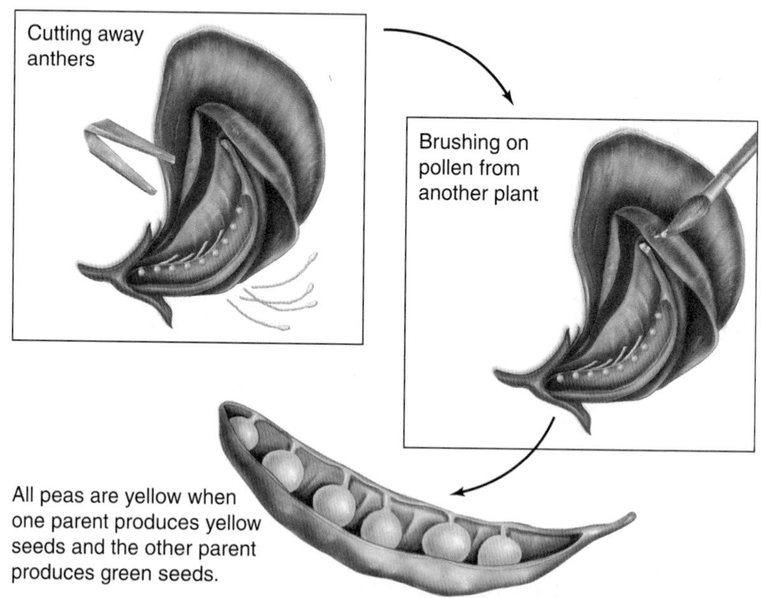

Figure 23A Garden pea anatomy and a few traits.
a. In the garden pea, *Pisum sativum*, pollen grains produced in the anther contain sperm, and ovules in the ovary contain oocytes. When Mendel performed crosses, he brushed pollen from one plant onto the stigma of another plant. After sperm fertilized oocytes, the ovules developed into seeds (peas). The open pod shows the results of a cross between plants with yellow seeds and plants with green seeds. **b.** Mendel selected traits like these for study. He made sure his parent (P generation) plants bred true, and then he cross-pollinated the plants. The offspring called F1 (first filial) generation always resembled the parent with the dominant characteristic (*left*). Mendel then allowed the F1 plants to self-pollinate. In the F2 (second filial) generation, he always achieved a 3:1 (dominant to recessive) phenotypic ratio.

Practice Problems 1*

1. For each of the following genotypes, give all possible gametes.
   a. *WW* b. *WWSs* c. *Tt*
   d. *Ttgg* e. *AaBb*
2. For each of the following, state whether a genotype or a gamete is represented.
   a. *D* b. *Ll*
   c. *Pw* d. *LlGg*

**Answers to Practice Problems appear in Appendix A.*

One-Trait Crosses

It is now possible for us to consider a particular cross. If a homozygous man with a widow's peak (Fig. 23.3*a*) reproduces with a woman with a straight hairline (Fig. 23.3*b*), what kind of hairline will their children have? To solve this problem, (1) use the key already established (Table 23.1) to indicate the genotype of each parent; (2) determine the possible gametes from each parent; (3) combine all possible gametes; and (4) determine the genotypes and the phenotypes of all the offspring. (When writing a heterozygous genotype, *always put the capital letter first* to avoid confusion.)

offspring were like the parent plants and like each other. In contrast to his predecessors, Mendel studied the inheritance of relatively simple and discrete traits, such as seed shape, seed color, and flower color.

Mendel studied both one-trait and two-trait crosses. He cross-pollinated the plants for the F_1 (first filial) generation. None of the offspring of this generation were intermediate between the parents. All of the offspring exhibited the trait of one or the other parent, which Mendel termed the dominant trait. The trait that disappeared in the offspring he called the recessive trait. He then allowed the F_1 plants to self-pollinate to obtain an F_2 generation. He counted many plants and always found approximately the same ratios of plants exhibiting dominant and recessive traits in this generation.

Most likely, Mendel's background in mathematics prompted him to use a statistical basis for his breeding experiments. As Mendel followed the inheritance of individual traits, he kept careful records, and he used his understanding of the mathematical laws of probability to interpret his results. He arrived at a theory that is now called a *particulate theory of inheritance* because it is based on the existence of minute particles or hereditary units, which we call genes. Inheritance involves reshuffling the same genes from generation to generation.

Mendel achieved his success in genetics by studying large numbers of offspring, keeping careful records, and treating his data quantitatively. However, a certain amount of luck was involved in the traits Mendel chose. None of the traits he analyzed exhibited the more complicated patterns of inheritance.

Discussion Questions

1. Would the genotype of a true-breeding plant be homozygous or heterozygous? Why?
2. Besides the reasons listed above, what other reasons would make the pea plant a better object for genetic study than human beings?
3. What "more complicated patterns of inheritance" could have confused Mendel's results?

Trait	Characteristics		F_2 Results*	
	Dominant	Recessive	Dominant	Recessive
Stem length	Tall	Short	787	277
Seed shape	Round	Wrinkled	5,474	1,850
Seed color	Yellow	Green	6,022	2,001
Flower color	Purple	White	705	224

*All of these produce approximately a 3:1 ratio. For example, $\frac{787}{277} \cong \frac{3}{1}$.

b.

In the following diagram, the letters in the first row are the genotypes of the parents. Each parent has only one type of gamete in regard to hairline, and therefore all the offspring have similar genotypes and phenotypes. The children are heterozygous (*Ww*) and have a widow's peak:

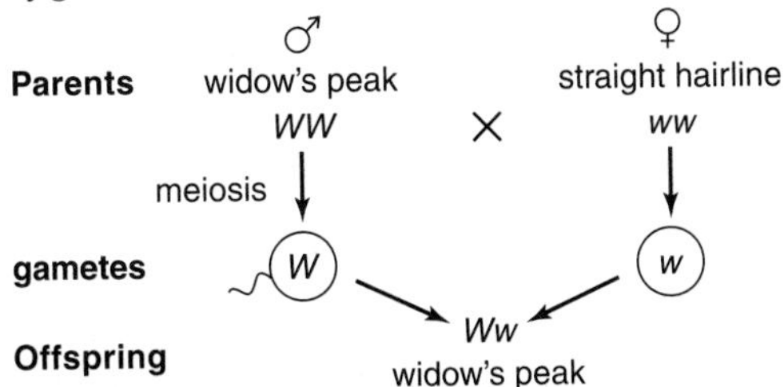

These children are **monohybrids;** that is, they are heterozygous for only one pair of alleles. If they reproduce with someone else of the same genotype, what type of hairline will their children have? In this problem (*Ww* × *Ww*), each parent has two possible types of gametes (*W* or *w*), and we must ensure that all types of sperm have an equal chance to fertilize all possible types of oocytes. One way to do this is to use a **Punnett square** (Fig. 23.4), in which all possible types of sperm are lined up vertically and all possible types of oocytes are lined up horizontally (or vice versa), and every possible combination of gametes occurs within the squares.

After we determine the genotypes and the phenotypes of the offspring, we can determine the genotypic and the phenotypic ratios. The genotypic ratio is 1 *WW*: 2 *Ww*: 1 *ww*, or simply 1:2:1, but the phenotypic ratio is 3:1. Why? Because three individuals have a widow's peak and one has a straight hairline. This 3:1 phenotypic ratio is always expected for a

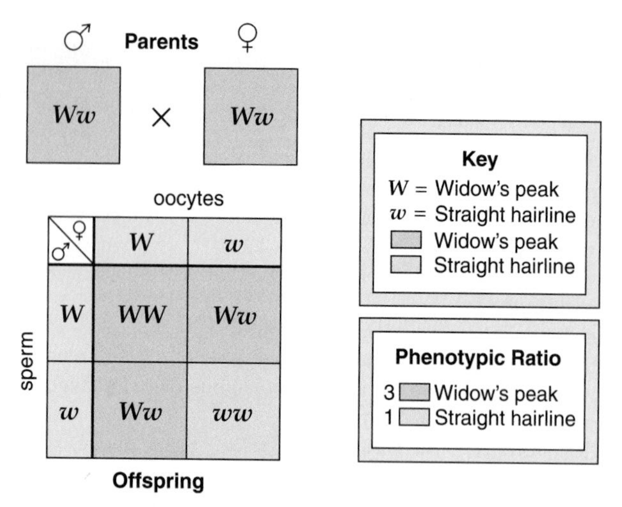

Figure 23.4 Monohybrid cross.
A Punnett square diagrams the results of a cross. When the parents are heterozygous, each child has a 75% chance of having the dominant phenotype and a 25% chance of having the recessive phenotype.

monohybrid cross when one allele is completely dominant over the other. The exact ratio is more likely to be observed if a large number of matings take place and if a large number of offspring result. Only then do all possible kinds of sperm have an equal chance of fertilizing all possible kinds of oocytes. Naturally, in humans, we do not routinely observe hundreds of offspring from a single type of cross. The best interpretation of Figure 23.4 in humans is to say that each child has three chances out of four to have a widow's peak, or one chance out of four to have a straight hairline. It is important to realize that *chance has no memory;* for example, if two heterozygous parents already have three children with a widow's peak and are expecting a fourth child, this child still has a 75% chance of having a widow's peak and a 25% chance of having a straight hairline.

One-Trait Crosses and Probability

Another method of calculating the expected ratios uses the rules of probability or chance. First, we must know the *product rule of probability,* which states that the chance of two or more independent events occurring together is the product (multiplication) of their chance of occurring separately. For example, if you flip a coin, the chance of getting "heads" is ½ (one head out of two possibilities, or ½). If you flip two coins, the chance of getting two heads is ½ × ½, or ¼. You must multiply together the probability of each head occurring separately (½).

In the cross just considered (*Ww* × *Ww*), what is the chance of obtaining either a *W* or a *w* from a parent?

The chance of *W* = ½, and the chance of *w* = ½.

Therefore, the probability of having these genotypes is as follows:

1. The chance of *WW* = ½ × ½ = ¼
2. The chance of *Ww* = ½ × ½ = ¼
3. The chance of *wW* = ½ × ½ = ¼
4. The chance of *ww* = ½ × ½ = ¼

Now we must consider the *sum rule of probability,* which states that the chance of an event that can occur in more than one way is the sum (addition) of the individual chances. Therefore, to calculate the chance of an offspring having a widow's peak, add the chances of *WW*, *Ww*, or *wW* from the preceding list to arrive at ¾, or 75%. The chance of offspring with straight hairline (only *ww* from the preceding) is ¼, or 25%.

The One-Trait Testcross

It is not possible to tell by observation if an individual expressing a dominant allele is homozygous dominant or heterozygous. Therefore, breeders sometimes do a so-called **testcross**—they cross an organism showing the dominant phenotype with one showing the recessive phenotype. The recessive phenotype is used because it has a *known* genotype. The results of a testcross should indicate the genotype of an organism with the dominant phenotype (Fig. 23.5).

The results of crosses between people also tell us the genotype of the parents. For example, Figure 23.5 shows two possible results when a man with a widow's peak reproduces with a woman who has a straight hairline. If the man is homozygous dominant, all his children will have a widow's peak. If the man is heterozygous, each child has a 50% chance of having a straight hairline. The birth of just one child with a straight hairline indicates that the man is heterozygous.

CHECK YOUR PROGRESS

1. Define homozygous and heterozygous.
2. Why is a Punnett square used for crosses?
3. What is the purpose of a testcross?
4. What type of cross results in a 3:1 ratio? A 1:1 ratio?

Practice Problems 2*

1. Both a man and a woman are heterozygous for freckles. Freckles (*F*) are dominant over no freckles (*f*). What is the chance that their child will have freckles?
2. Both you and your sister or brother have attached earlobes, but your parents have unattached earlobes. Unattached earlobes (*E*) are dominant over attached earlobes (*e*). What are the genotypes of your parents?
3. A father has dimples, the mother does not have dimples, and all five of their children have dimples. Dimples (*D*) are dominant over no dimples (*d*). Give the probable genotypes of all persons concerned.

**Answers to Practice Problems appear in Appendix A.*

The Inheritance of Two Traits

Figure 23.6 represents two pairs of homologues undergoing meiosis. The homologues are distinguished by length—that is, one pair of homologues is short and the other is long. (The color signifies that we inherit chromosomes from our parents; one homologue of each pair is the "paternal" chromosome, and the other is the "maternal" chromosome.) When the homologues separate (segregate) during meiosis, each

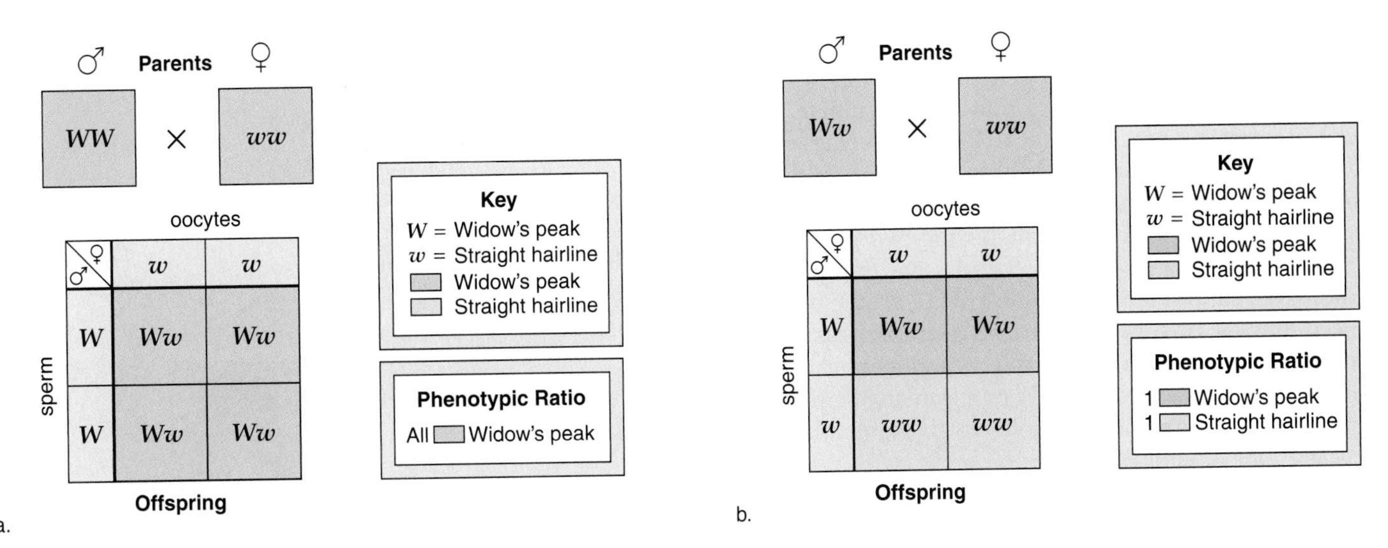

Figure 23.5 One-trait testcross.
A testcross determines if an individual with the dominant phenotype is homozygous or heterozygous. **a.** Because all offspring show the dominant characteristic, the individual is most likely homozygous, as shown. **b.** Because the offspring show a 1:1 phenotypic ratio, the individual is heterozygous, as shown.

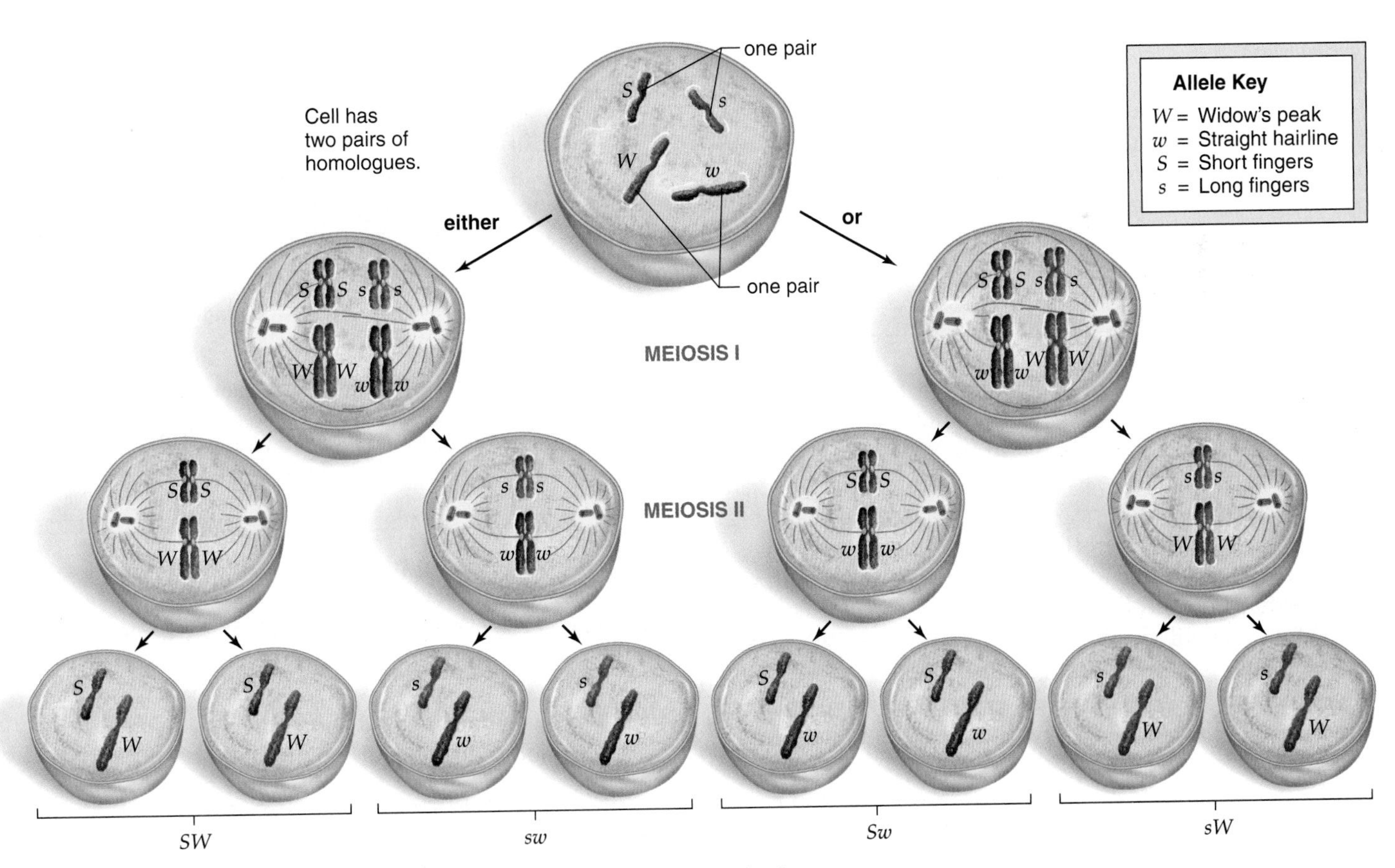

Figure 23.6 Homologous pairs of chromosomes during meiosis.
A cell has two pairs of homologous chromosomes (homologues), a short pair and a long pair. The short pair of homologues carries alleles for finger length, and the long pair of homologues carries alleles for type of hairline. The homologues, and the alleles they carry, segregate independently during meiosis I. This means that either a red or blue chromosome from each pair can go into either of the daughter cells. Therefore, all possible combinations of chromosomes and alleles occur in the gametes. The last line of the illustration lists the possible allele combinations for this particular cross.

gamete receives one member from each pair of homologues. The homologues separate independently. It does not matter which member of a pair goes into which gamete. In the simplest terms, a gamete in Figure 23.6 will *receive one short and one long chromosome of either color.* Therefore, all possible combinations of chromosomes are in the gametes.

We will assume that the alleles for two genes are on these homologues. The alleles *S* and *s* are on one pair of homologues, and the alleles *W* and *w* are on the other pair of homologues. Because there are no restrictions as to which homologue goes into which gamete, a gamete can receive either an *S* or an *s* and either a *W* or a *w* in any combination. In the end, the gametes will collectively have all possible combinations of alleles.

Independent Assortment

Mendel had no knowledge of meiosis, but he could see that his results were attainable only if the sperm and eggs contain every possible combination of factors. This caused him to formulate his second law, the **law of independent assortment.** The law of independent assortment states the following:

- Each pair of factors separates independently (without regard to how the others separate).
- All possible combinations of factors can occur in the gametes.

Figure 23.6 illustrates the law of segregation and the law of independent assortment. Because it does not matter which homologue of a pair faces which spindle pole, a daughter cell can receive either homologue and either allele.

Two-Trait Crosses

In the two-trait cross depicted in Figure 23.7, a person homozygous for widow's peak and short fingers (*WWSS*) reproduces with one who has a straight hairline and long fingers (*wwss*). The law of segregation tells us that the gametes for the *WWSS* parent have to be *WS* and the gametes for the *wwss* parent have to be *ws*. Therefore, their offspring will all have the genotype *WwSs* and the same phenotype (widow's peak with short fingers). This genotype is called a **dihybrid** because the individual is heterozygous in two regards: hairline and fingers.

When the dihybrid *WwSs* reproduces with another dihybrid that is *WwSs*, what gametes are possible? The law of segregation tells us that each gamete can have only one letter of each kind, and the law of independent assortment tells us that all combinations are possible. Therefore, these are the gametes for both dihybrids: *WS*, *Ws*, *wS*, and *ws*.

A Punnett square represents all possible sperm fertilizing all possible oocytes, so following are the expected phenotypic results for a dihybrid cross:

9 widow's peak and short fingers
3 widow's peak and long fingers

P generation ♂ *WWSS* × *wwss* ♀

P gametes *WS* *ws*

F_1 generation *WwSs*

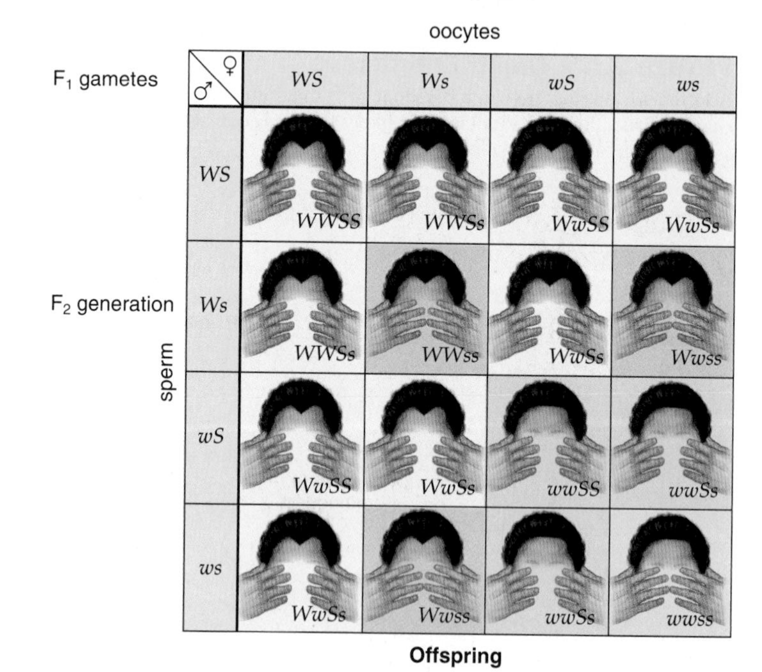

Allele Key
W = Widow's peak
w = Straight hairline
S = Short fingers
s = Long fingers

Phenotypic Ratio
9 Widow's peak, short fingers
3 Widow's peak, long fingers
3 Straight hairline, short fingers
1 Straight hairline, long fingers

Figure 23.7 Dihybrid cross.
Each dihybrid can form four possible types of gametes, so four different phenotypes occur among the offspring in the proportions shown.

3 straight hairline and short fingers
1 straight hairline and long fingers

This 9:3:3:1 phenotypic ratio is always expected for a dihybrid cross when simple dominance is present. We can use this expected ratio to predict the chances of each child receiving a certain phenotype. For example, the chance of getting the two dominant phenotypes together is 9 out of 16, and the chance of getting the two recessive phenotypes together is 1 out of 16.

Two-Trait Crosses and Probability

Instead of using a Punnett square, it is also possible to use the product rule and the sum rule of probability to predict the results of a dihybrid cross. For example, we know that the probable results for two separate monohybrid crosses are as follows:

Probability of widow's peak = $3/4$
Probability of short fingers = $3/4$
Probability of straight hairline = $1/4$
Probability of long fingers = $1/4$

Using the product rule, we can calculate the probable outcome of a dihybrid cross as follows:

Probability of widow's peak and short fingers =
$3/4 \times 3/4 = 9/16$

Probability of widow's peak and long fingers =
$3/4 \times 1/4 = 3/16$

Probability of straight hairline and short fingers =
$1/4 \times 3/4 = 3/16$

Probability of straight hairline and long fingers =
$1/4 \times 1/4 = 1/16$

In this way, the rules of probability tell us that the expected phenotypic ratio when all possible sperm fertilize all possible oocytes is 9:3:3:1.

The Two-Trait Testcross

It is impossible to tell by inspection whether an individual expressing the dominant allele for two traits is homozygous dominant or heterozygous in regard to these traits. A two-trait testcross occurs when this individual is crossed with a homozygous recessive for both traits. The recessive phenotype is used because it has a *known* phenotype.

For example, if a man who is homozygous dominant for widow's peak and short fingers reproduces with a woman who is homozygous recessive for both traits, then all his children will have the dominant phenotypes. However, if a man is heterozygous for both traits, then each child has a 25% chance of showing either one or both recessive traits. A Punnett square (Fig. 23.8) shows that the expected ratio is 1 widow's peak with short fingers: 1 widow's peak with long fingers: 1 straight hairline with short fingers: 1 straight hairline with long fingers, or 1:1:1:1.

Parents
♂ *WwSs* × *wwss* ♀

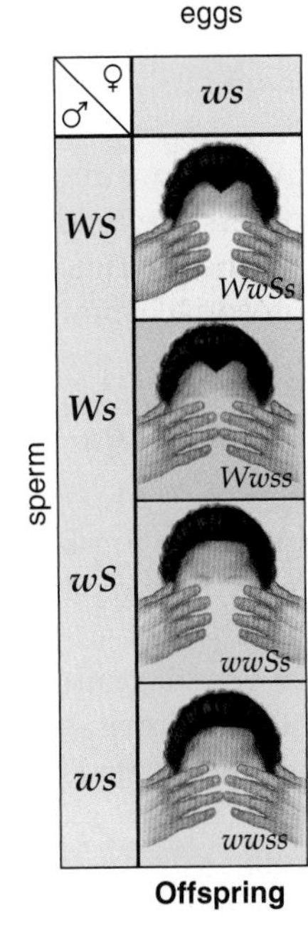

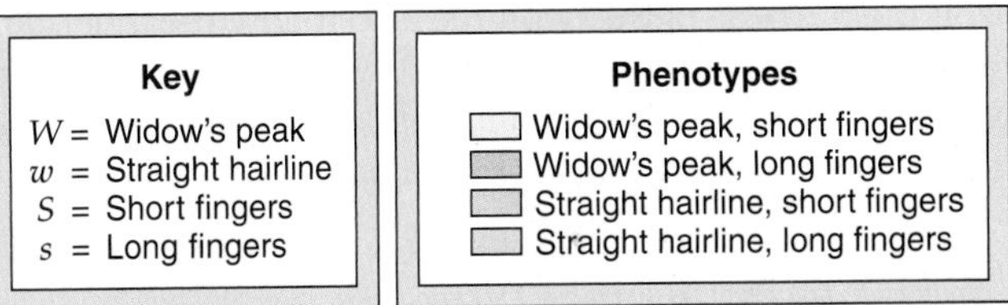

Figure 23.8 Two-trait testcross.
A testcross determines if the individual with a dominant phenotype is homozygous or heterozygous. If the individual is heterozygous as shown, there is a 25% chance for each possible phenotype. If the individual in question was *WWSs*, could any of the children have a straight hairline? If the individual in question was *WwSS*, could any of the children have long fingers?

Practice Problems 3*

1. Attached earlobes are recessive. What genotype do children have if one parent is homozygous recessive for attached earlobes and homozygous dominant for widow's peak, and the other is homozygous dominant for unattached earlobes and homozygous recessive for straight hairline?
2. If an individual from this cross reproduces with another of the same genotype, what are the chances that they will have a child with a straight hairline and attached earlobes?
3. A child who does not have dimples or freckles is born to a man who has dimples and freckles (both dominant) and a woman who does not. What are the genotypes of all persons concerned?

**Answers to Practice Problems appear in Appendix A.*

Check Your Progress

1. How many possible gametes form in a dihybrid cross?
2. State Mendel's law of independent assortment.
3. What type of cross results in a 9:3:3:1 ratio? A 1:1:1:1 ratio?

23.2 Pedigree Analysis and Genetic Disorders

Learning Outcomes

1. Recognize autosomal dominant and autosomal recessive pattern of inheritance when examining a pedigree.
2. Discuss certain genetic disorders and do problems that involve these disorders.

It is now apparent that many human disorders are genetic in origin. Genetic disorders are medical conditions caused by alleles inherited from the parents. Some of these conditions are controlled by autosomal dominant or recessive alleles.

Patterns of Inheritance

In humans, it is unethical or impossible to carry out some of the crosses that can be done in plants, such as a testcross or self-pollination. Instead, human inheritance is diagrammed in a pedigree. A **pedigree** is a chart of a family's history with regard to a particular genetic trait. Pedigrees can be done for humans as well as other animals, such as dogs and livestock. In the chart, males are designated by squares and females by circles. A line between a square and a circle represents a mating. A vertical line going downward leads directly to a single child. If there are more children, they are placed off a horizontal line. If the purpose of the pedigree is to follow the inheritance of a particular trait, those individuals with the trait, called the affected individuals, are usually shaded. A pedigree is often used to determine the inheritance of a genetic disease.

Pedigrees for Autosomal Disorders

It is possible to decide if an inherited condition is due to an autosomal dominant or an autosomal recessive allele by studying a pedigree. Does the following pattern of inheritance represent an autosomal dominant or an autosomal recessive characteristic?

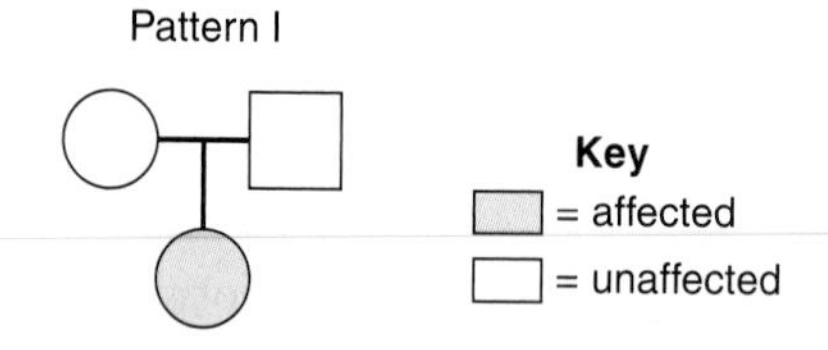

In this pattern, the child is affected, but neither parent is. This can happen only if the disorder is recessive and the parents are heterozygotes. Notice that the parents are carriers because they are unaffected but are capable of having a child with the genetic disorder. If the family pedigree suggests that the parents are carriers for an autosomal recessive disorder, it would be possible to do prenatal testing for the genetic disorder, as described in Chapter 26.

Notice that in the pedigree in Figure 23.9, cousins are the parents of three children, two of whom have the disorder. Aside from illustrating that reproduction between cousins is more likely to bring out recessive traits, this pedigree also shows that "chance has no memory." Each child born to heterozygous parents has a 25% chance of having the disorder. In other words, it is possible that if a heterozygous couple has four children, each child might have the condition.

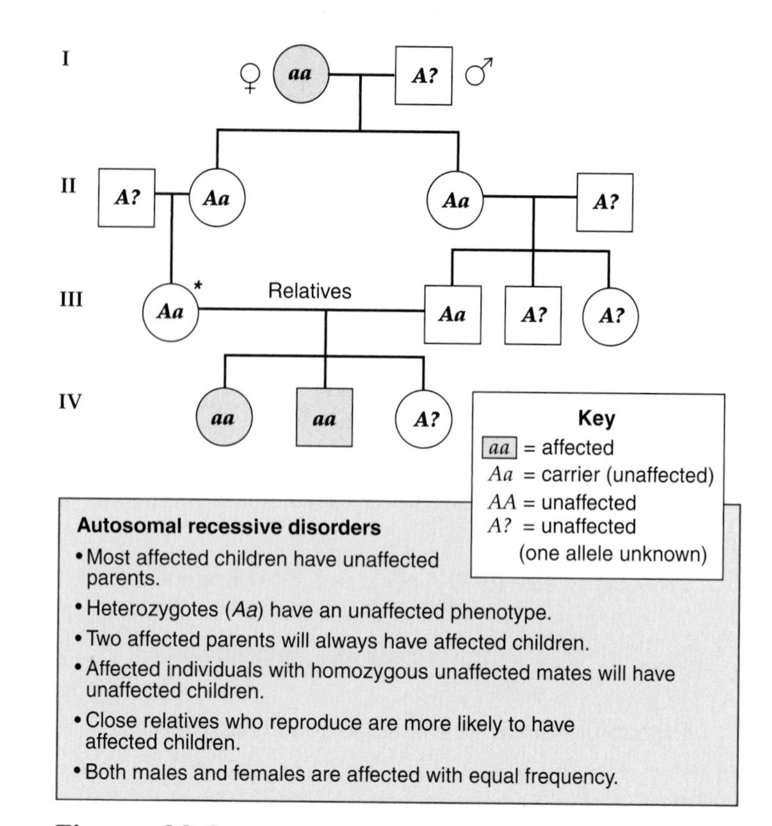

Figure 23.9 Autosomal recessive pedigree. The list gives ways to recognize an autosomal recessive disorder. How would you know the individual at the asterisk is heterozygous?

*See Appendix A for answers.

Now consider this pattern of inheritance:

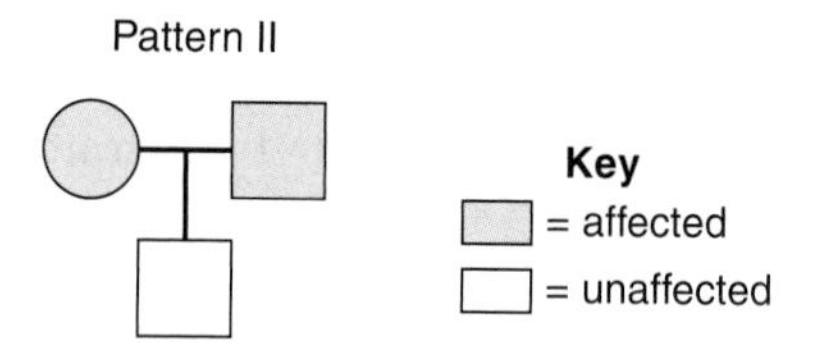

In this pattern, the child is unaffected, but the parents are both affected. This can happen if the condition is autosomal dominant and the parents are heterozygotes. Figure 23.10 lists other ways to recognize an autosomal dominant pattern of inheritance. This pedigree also illustrates that when both parents are unaffected, all their children are unaffected. Why? Because neither parent has a dominant gene that passes the condition on.

Autosomal Recessive Disorders

Inheritance of two recessive alleles is required before an autosomal recessive disorder will appear.

Tay-Sachs disease is a well-known autosomal recessive disorder that usually occurs among Jewish people in the

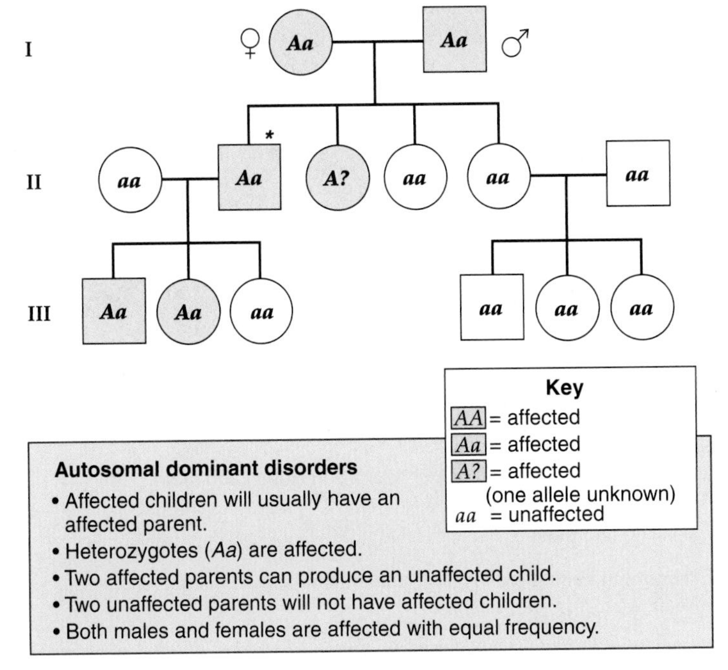

Figure 23.10 Autosomal dominant pedigree.
The list gives ways to recognize an autosomal dominant disorder. How would you know the individual at the asterisk is heterozygous?

* See Appendix A for answers.

United States, most of whom are of central and eastern European descent. Tay-Sachs disease results from a lack of the enzyme hexosaminidase A (Hex A) and the subsequent storage of its substrate, a glycosphingolipid, in lysosomes. Lysosomes build up in many body cells, but the primary sites of storage are the cells of the brain, which accounts for the onset of symptoms and the progressive deterioration of psychomotor functions (Fig. 23.11*a*).

At first, it is not apparent that a baby has Tay-Sachs disease. However, development begins to slow down between four and eight months of age, and neurological impairment and psychomotor difficulties then become apparent. The child gradually becomes blind and helpless, develops uncontrollable seizures, and eventually becomes paralyzed.

Cystic fibrosis (CF) is an autosomal recessive disorder that occurs among all ethnic groups, but it is the most common lethal genetic disorder among Caucasians in the United States. Research has demonstrated that chloride ions (Cl^-) fail to pass through a plasma membrane channel protein in the cells of these patients (Fig. 23.11*b*). Ordinarily, after chloride ions have passed through the membrane, sodium ions (Na^+) and water follow. It is believed that lack of water is the cause of abnormally thick mucus in bronchial tubes and pancreatic ducts. In these children, this mucus is so thick and viscous that it interferes with the function of the lungs and pancreas. To ease breathing, the thick mucus in the lungs must be manually loosened periodically, but the lungs still become infected frequently. Clogged pancreatic ducts prevent digestive enzymes from reaching the small intestine, and to improve digestion, CF patients take digestive enzymes before every meal.

Phenylketonuria (PKU) is an autosomal recessive metabolic disorder that affects nervous system development. Affected individuals lack the enzyme needed for the normal metabolism of the amino acid phenylalanine, and therefore, it appears in the urine and the blood. Newborns are routinely tested in the hospital for elevated levels of phenylalanine in the blood. If an elevated level is detected, the newborn will develop normally if placed on a diet low in phenylalanine, which must be continued until the brain is fully developed, around the age of seven, or else severe mental retardation will develop. Some doctors recommend that the diet continue for life, but in any case, a pregnant woman with phenylketonuria must be on the diet in order to protect her unborn child from harm.

Sickle cell disease is an autosomal recessive disorder in which the red blood cells are not biconcave disks like normal red blood cells; rather, they are irregular, and often sickle

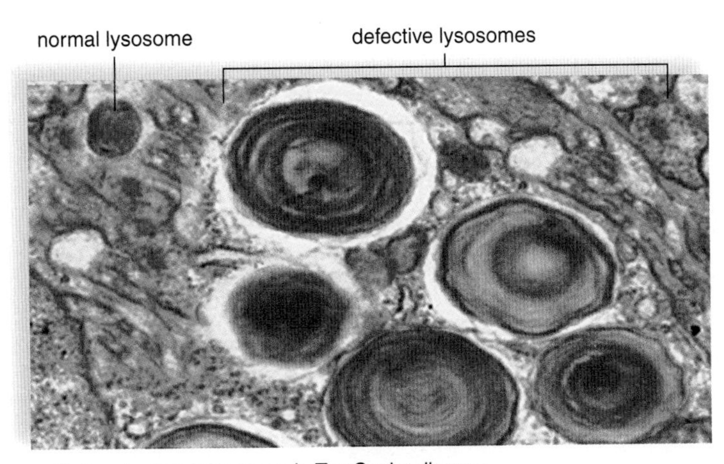

a. Malfunctioning lysosomes in Tay-Sachs disease.

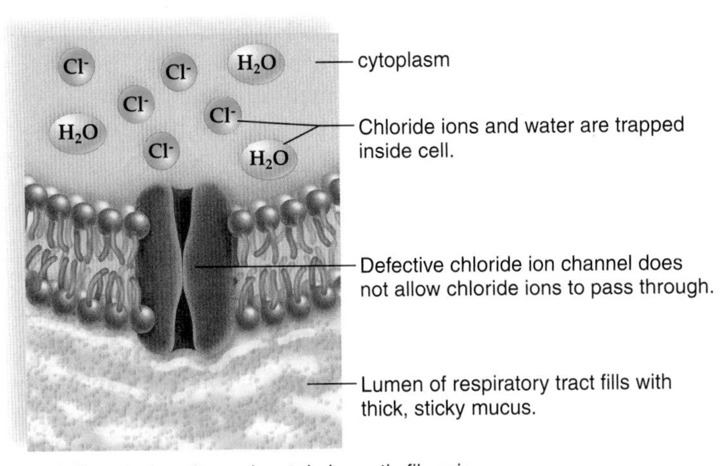

b. Malfunctioning channel protein in cystic fibrosis.

Figure 23.11 Genetic disorders.
a. A neuron in Tay-Sachs disease contains many defective lysosomes because of a deficient lysosomal enzyme. **b.** Cystic fibrosis is due to a faulty protein that is supposed to regulate the flow of chloride ions into and out of cells through a channel protein.

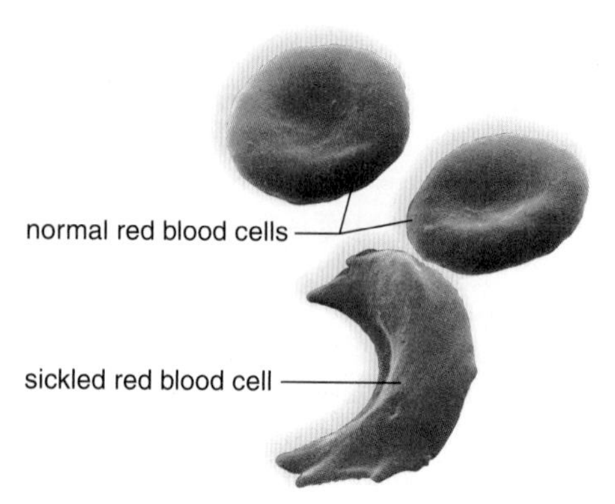

Abnormally shaped red blood cells in sickle cell disease.
a.

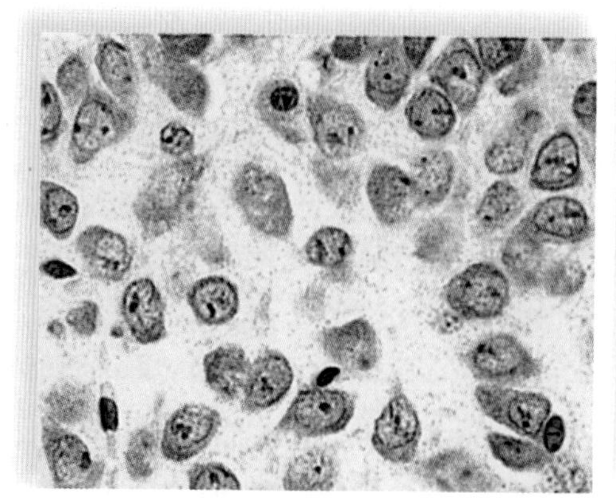
Many neurons in normal brain.
b.

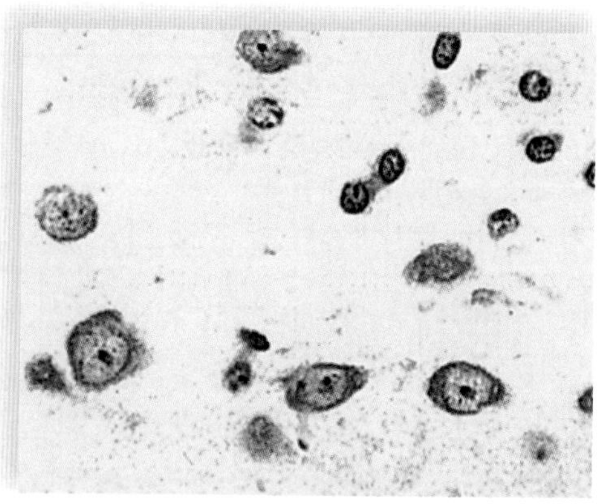
Loss of neurons in Huntington brain.

Figure 23.12 Genetic disorders.
a. Persons with sickle cell disease have sickle-shaped red blood cells because of an abnormal hemoglobin A molecule. **b.** Huntington disease is characterized by increasingly serious psychomotor and mental disturbances because of the loss of nerve cells.

shaped (Fig. 23.12*a*). This defect is caused by an abnormal type of hemoglobin that differs from normal hemoglobin by one amino acid in the protein globin. The single amino acid change causes hemoglobin molecules to stack up and form insoluble rods, and the red blood cells become sickle shaped.

Because sickle-shaped cells can't pass along narrow capillary passageways as disk-shaped cells can, they clog the vessels and break down. This is why persons with sickle cell disease experience poor circulation, anemia, and low resistance to infection. Internal hemorrhaging leads to further complications, such as jaundice, episodic pain in the abdomen and joints, and damage to internal organs.

Sickle cell heterozygotes have sickle cell traits in which the blood cells are normal unless they experience dehydration or mild oxygen deprivation. Still, at present, most experts believe that persons with the sickle cell trait do not need to restrict their physical activity.

Autosomal Dominant Disorders

Inheritance of only one dominant allele is necessary for an autosomal dominant genetic disorder to appear.

Marfan syndrome, an autosomal dominant disorder, is caused by a defect in an elastic connective tissue protein called fibrillin. This protein is normally abundant in the lens of the eye, the bones of the limbs, fingers, and ribs, and also in the wall of the aorta. The affected person often has a dislocated lens, long limbs and fingers, and a caved-in chest. The aorta wall is weak and may burst without warning. A tissue graft can strengthen the aorta, but people with Marfan syndrome still should not overexert themselves.

Huntington disease is a neurological disorder that leads to progressive degeneration of brain cells (Fig. 23.12*b*). The disease is caused by a mutated copy of the gene for a protein called huntingtin. Most patients appear normal until they are of middle age and have already had children, who may later also be stricken. Occasionally, the first sign of the disease appears during the teen years or even earlier. There is no effective treatment, and death comes 10 to 15 years after the onset of symptoms.

Several years ago, researchers found that the gene for Huntington disease was located on chromosome 4. They developed a test to detect the presence of the gene. However, few people want to know they have inherited the gene because there is no cure. At least now we know that the disease stems from a mutation that causes the huntingtin protein to have too many copies of the amino acid glutamine. The normal version of huntingtin has stretches of between 10 and 25 glutamines. If huntingtin has more than 36 glutamines, it changes shape and forms large clumps inside neurons. Even worse, it attracts and causes other proteins to clump with it. One of these proteins, called CBP, which helps nerve cells survive is inactivated when it clumps with huntingtin. Researchers hope to combat the disease by boosting CBP levels.

Check Your Progress

1. List the characteristics of a pedigree for an autosomal dominant and an autosomal recessive trait.
2. Describe two autosomal recessive and two autosomal dominant genetic disorders.

23.3 Beyond Simple Inheritance Patterns

Learning Outcomes

1. Explain how the phenotypes for incompletely dominant and codominant traits differ from dominant/recessive trait phenotypes.
2. Understand how ABO blood types are an example of multiple allele inheritance.
3. Describe polygenic inheritance.

Certain traits, such as those studied in sections 23.1 and 23.2, are controlled by one set of alleles that follows a simple dominant or recessive inheritance. But we now know of many other types of inheritance patterns.

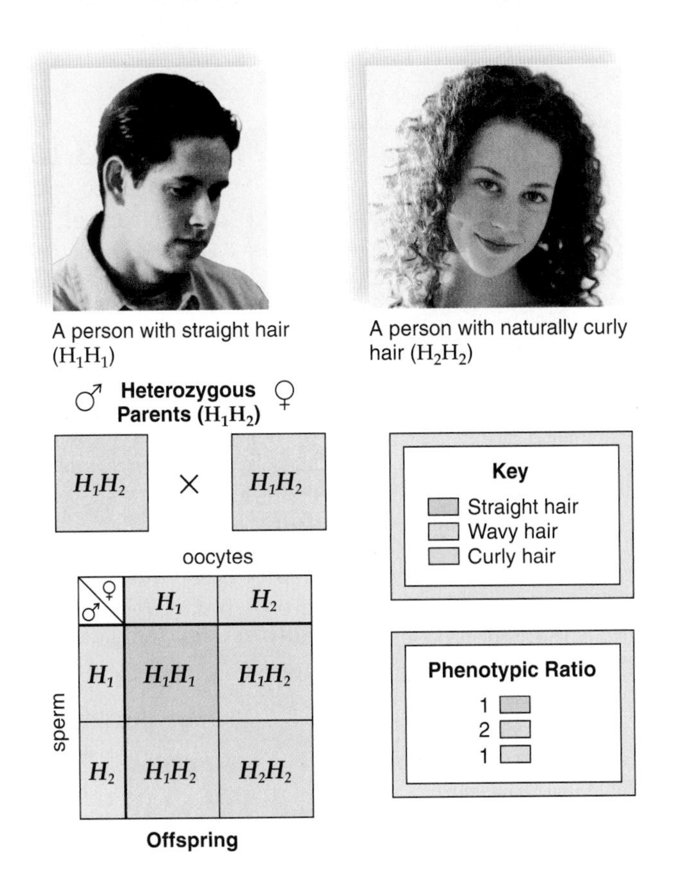

Figure 23.13 Incomplete dominance.
Among Caucasians, neither straight nor curly hair is dominant. When two wavy-haired individuals reproduce, each offspring has a 25% chance of having either straight or curly hair and a 50% chance of having wavy hair, the intermediate phenotype.

Incomplete Dominance and Codominance

Incomplete dominance occurs when the heterozygote is intermediate between the two homozygotes. For example, when a curly-haired Caucasian reproduces with a straight-haired Caucasian, their children have wavy hair. When two wavy-haired persons reproduce, the expected phenotypic ratio among the offspring is 1:2:1—that is, one curly-haired child to two with wavy hair to one with straight hair (Fig. 23.13). We can explain incomplete dominance by assuming that only one allele codes for a product and the single dose of the product gives the intermediate result. Notice in Figure 23.13 that we do not use the same capital and lower case letters as we used with dominant/recessive traits.

Codominance occurs when alleles are equally expressed in a heterozygote. A familiar example is the human blood type AB, in which the red blood cells have the characteristics of both type A and type B blood. We can explain codominance by assuming that both genes code for a product, and we observe the results of both products being present. Blood type inheritance is also said to be an example of multiple alleles.

Incompletely Dominant Disorders

The severity of symptoms exhibited in **familial hypercholesterolemia (FH)** parallels the number of LDL-cholesterol receptor proteins in the plasma membrane. A person with two mutated alleles lacks LDL-cholesterol receptors; a person with only one mutated allele has half the normal number of receptors; and a person with two normal alleles has the usual number of receptors. The presence of excessive cholesterol in the blood causes cardiovascular disease. Therefore, people with no receptors die of cardiovascular disease before 30 years of age. These individuals develop cholesterol deposits in the skin, tendons, and cornea. Individuals with one-half the number of receptors may die while still young or after they have reached middle age. People with the full number of cholesterol receptors do not have familial hypercholesterolemia (Fig. 23.14).

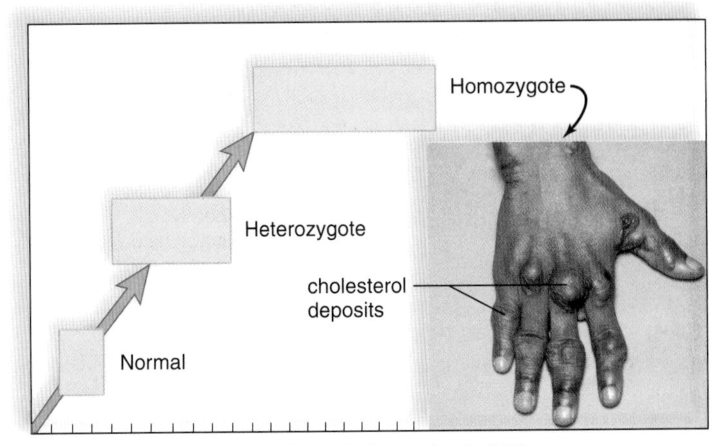

Figure 23.14 Incompletely dominant genetic disorder.
Familial hypercholesterolemia is incompletely dominant. Persons with one mutated allele (heterozygotes) have an abnormally high level of cholesterol in the blood, and those with two mutated alleles (homozygotes) have a higher level still. In homozygotes, the cholesterol level is sometimes so high that the excess forms visible deposits in the skin.

Multiple Allele Inheritance

When a trait is controlled by **multiple alleles,** the gene exists in several allelic forms. But each person has only two of the possible alleles.

ABO Blood Types

Three alleles for the same gene control the inheritance of ABO blood types. These alleles determine the presence or absence of antigens on red blood cells:

I^A = A antigen on red blood cells
I^B = B antigen on red blood cells
i = Neither A nor B antigen on red blood cells

Each person has only two of the three possible alleles, and both I^A and I^B are dominant over i. Therefore, there are two possible genotypes for type A blood and two possible genotypes for type B blood. On the other hand, I^A and I^B are fully expressed in the presence of the other. Therefore, if a person inherits one of each of these alleles, that person will have type AB blood. Type O blood can result only from the inheritance of two i alleles.

The possible genotypes and phenotypes for blood type are as follows:

Phenotype	Genotype
A	I^AI^A, I^Ai
B	I^BI^B, I^Bi
AB	I^AI^B
O	ii

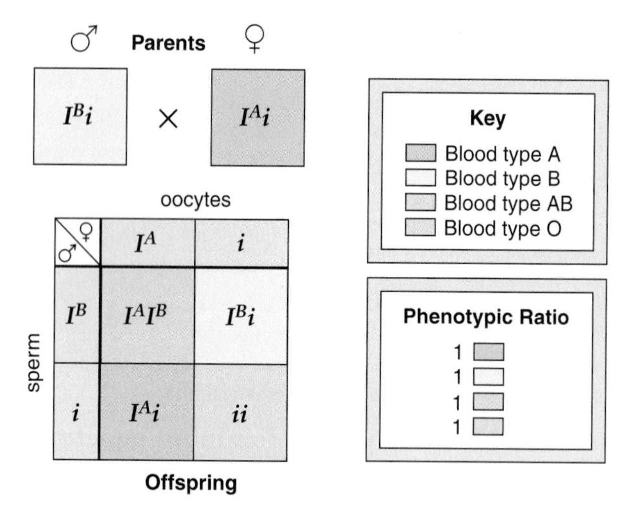

Figure 23.15 Inheritance of blood types.
Blood type exemplifies multiple allele inheritance. A mating between individuals with type A blood and type B blood can result in any one of the four blood types. Why? Because the parents are I^Ai and I^Bi. If both parents had type AB blood, predict the blood types of their potential offspring.

In the past, blood typing was used in paternity suits. However, the results of a blood test can only suggest that the tested individual *might* be the father, not that he definitely *is* the father. For example, it is possible, but not definite, that a man with type A blood (genotype I^Ai) is the father of a child with type O blood. On the other hand, a blood test can sometimes definitely prove that a man is *not* the father. For example, a man with type AB blood cannot possibly be the father of a child with type O blood. Therefore, blood tests can be used in legal cases only to exclude a man from possible paternity.

As a point of interest, the Rh factor is inherited separately from A, B, AB, or O blood types. When you are Rh positive, your red blood cells have a particular antigen, and when you are Rh negative, that antigen is absent. There are multiple recessive alleles for Rh^-, but they are all recessive to Rh^+.

Figure 23.15 shows that matings between certain genotypes can have surprising results in terms of blood type.

Polygenic Inheritance

Many traits, such as size or height, shape, weight, skin color, metabolic rate, and behavior, are governed by several genes (each with sets of alleles). **Polygenic inheritance** occurs when a trait is governed by two or more genes (sets of alleles). The individual has a copy of all allelic pairs, possibly located on many different pairs of chromosomes. Each dominant allele codes for a product, and therefore the dominant alleles have a quantitative effect on the phenotype, and these effects are additive. The result is a *continuous variation* of phenotypes, resulting in a distribution of these phenotypes that resembles a bell-shaped curve. The more genes involved, the more continuous are the variations and distribution of the phenotypes.

Skin Color

Skin color is an example of a polygenic trait that is likely controlled by many pairs of alleles. Even so, we will use the simplest model and assume that skin has only two pairs of alleles (*Aa* and *Bb*) and that each capital letter contributes pigment to the skin. When a very dark person reproduces with a very light person, the children have medium-brown skin. When two people with the genotype *AaBb* reproduce with one another, the children may range in skin color from very dark to very light:

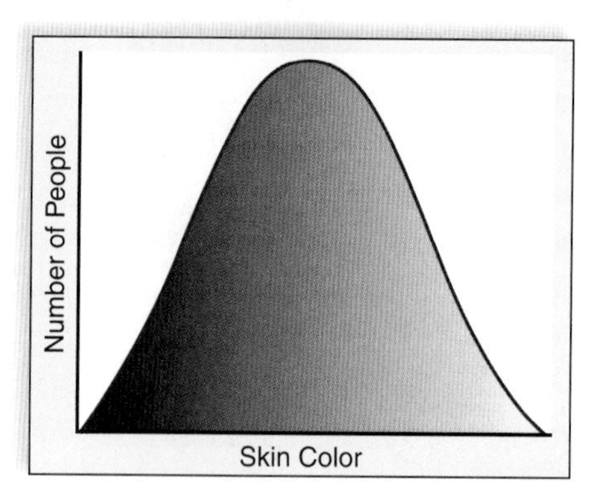

Figure 23.16 Polygenic inheritance.
Skin color is controlled by many pairs of alleles, which results in a range of phenotypes. The vast majority of people have skin colors in the middle range, while fewer people have skin colors in the extreme range.

Genotypes	Phenotypes
AABB	Very dark
AABb or *AaBB*	Dark
AaBb or *AAbb* or *aaBB*	Medium brown
Aabb or *aaBb*	Light
aabb	Very light

Notice again that a range of phenotypes exists and several possible phenotypes fall between the two extremes. Therefore, the distribution of these phenotypes is expected to follow a bell-shaped curve, meaning that few people have the extreme phenotypes and most people have the phenotype that lies in the middle (Fig. 23.16).

CHECK YOUR PROGRESS

1. If Mendel had used an incompletely dominant trait, how might his results have been different?
2. Why are the symbols I^A, I^B, and *i* used for the blood types?
3. Regardless of how many potential alleles there are, why does each person only get two?

Practice Problems 4*

1. If a person with straight hair marries someone with wavy hair, can they have a child with curly hair?
2. A child with type O blood is born to a mother with type A blood. What is the genotype of the child? The mother? What are the possible genotypes of the father?
3. From the following blood types, determine which baby belongs to which parents:

Baby 1, type O	Mrs. Doe, type A
Baby 2, type B	Mr. Doe, type A
	Mrs. Jones, type A
	Mr. Jones, type AB

4. A certain polygenic trait is controlled by three pairs of alleles: ***A*** versus ***a***, ***B*** versus ***b***, and ***C*** versus ***c***. What are the two extreme genotypes for this trait?

Answers to Practice Problems appear in Appendix A.

23.4 Environmental Influences

Learning Outcomes

1. Describe ways in which the environment can influence genetic traits.
2. Understand how scientists determine the effect of nature versus nurture.

Environmental factors, such as nutrition or temperature, can also influence the expression of genetic traits. Polygenic traits seem to be particularly influenced by the environment. For example, in the case of height, differences in nutrition are one of the factors that bring about a bell-shaped curve (Fig. 23.17).

Temperature can also affect the phenotypes of plants and animals. For example, primroses have white flowers when grown above 32°C and red flowers when grown at 24°C. The coats of Siamese cats and Himalayan rabbits are darker in color at the ears, nose, paws, and tail. Himalayan rabbits are known to be homozygous for the allele *ch*, which is involved in the production of melanin. Experimental evidence suggests that the enzyme coded for by this gene is active only at low temperatures, and therefore black fur occurs only at the extremities, where body heat is lost to the environment (Fig. 23.18).

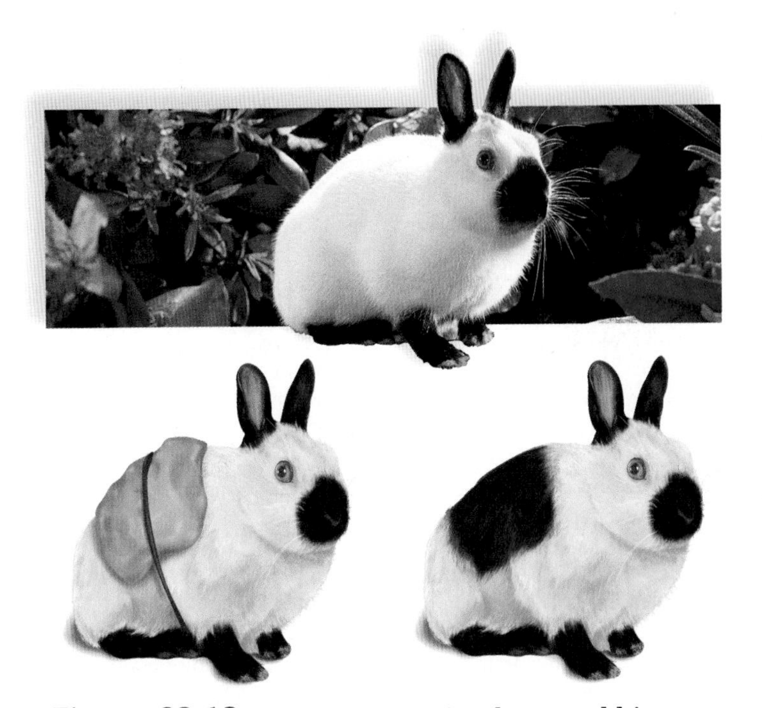

Figure 23.18 Coat color in Himalayan rabbits. The influence of the environment on the phenotype has been demonstrated by plucking out the fur from one area and applying an ice pack. The new fur that grew in was black instead of white, showing that the enzyme that produces melanin (a dark pigment) in the rabbit is active only at low temperatures.

Investigators try to determine what percentage of various human traits is due to nature (inheritance) and what percentage is due to nurture (the environment).

Some studies use identical and fraternal twins who have been separated since birth and thus have been raised in different environments. The supposition is that if identical twins in different environments share the same trait, that trait is most likely inherited. These studies have found that identical twins are more similar in their intellectual talents, personality traits, and levels of lifelong happiness than are fraternal twins when both groups have been separated since

Figure 23.17 Polygenic inheritance and environmental effects. When you record the heights of a large group of people chosen at random, the values follow a bell-shaped curve. Such a continuous distribution is due to control of a trait by several sets of alleles. Environmental effects are also involved.

Science Focus

But It Tastes Bad!

People differ in how they experience the taste of food. As children, we all refused to eat certain foods. Sometimes we were just being stubborn, but other times it was because that food tasted horrible! Researchers are looking at the genetics of taste. On the tongue are approximately 10,000 taste buds and within each taste bud, there are 50 to 100 taste receptor cells. There are five different types of taste receptors: sweet, sour, salty, bitter, and umami (savory). When a food molecule binds to a taste receptor on the outside of the cell, the cell triggers a nerve impulse that travels to the brain where it is interpreted as a particular taste. There is no taste receptor for spicy food. Spicy food causes pain and is registered by the brain as that!

Humans contain about 30 different genes for bitter taste receptors. Natural plant toxins are generally bitter so it makes sense that humans would evolve an aversion to bitter-tasting compounds. In one study, children who possess two positive alleles (+ +) for a particular bitter taste receptor were very sensitive to any bitterness in their food. Interestingly, the mothers of those children who were also + + did not share that same degree of sensitivity. Researchers think that as we age our sense of taste may lessen and so the mothers' tastes were not quite as sensitive. So a child who refuses to eat a particularly bitter-tasting vegetable may find it much more offensive than mom does.

A common genetics laboratory experiment is to taste PTC (phenylthiocarbamide), a compound not found in nature. A single gene with at least three different alleles is responsible for the ability to taste PTC. A PTC taster finds low concentrations extremely bitter, while a nontaster cannot taste anything even at higher concentrations. For those who can taste PTC, the bitter taste is unforgettable! Researchers have found that there is a correlation between the ability to taste PTC and food preferences. Those who taste PTC tend to dislike bitter foods such as dark chocolate and black coffee. And broccoli. Or that could just be stubbornness.

Discussion Questions

1. Are taste receptors on the tongue the only thing involved in how we perceive taste?
2. If something tastes very bitter, it is more likely that a human or animal will not try to eat it (at least after the first try). Explain why it is advantageous for plants to produce bitter chemicals.

birth. Biologists conclude that all behavioral traits are partly inheritable and that genes exert their effects by acting together in complex combinations susceptible to environmental influences.

Scientists have also learned how to manipulate the environment in order to prevent the damaging effects of some genetic disorders. For example, the bodies of people who have the genetic disorder phenylketonuria (PKU) lack an enzyme that breaks down the amino acid phenylalanine. The accumulation of phenylalanine leads to neurological damage in these patients. But scientists have discovered that if these patients modify their diet so that they consume very little protein containing phenylalanine, they can avoid many of the damaging effects of this disease.

Check Your Progress

1. How do scientists study the effect of nature versus nurture?
2. Why are polygenic traits particularly affected by the environment?

Applying the Concepts [Revisited]

Flo Hyman suffered from an autosomal dominant disorder known as Marfan syndrome. It is necessary to inherit only one allele to suffer from an autosomal dominant disorder. Marfan syndrome sufferers are very tall and often have weakened artery walls; heart valve problems; slender, tapering fingers (often called spider fingers); long arms and legs; curvature of the spine; and eye problems. In Hyman's case, she did not exhibit all the various symptoms of the disease but did exhibit excessive height.

1. What are the possible genotypes of Hyman's parents?
2. Marfan syndrome is a connective tissue disorder. How does this account for the symptoms of the disease?

SUMMARIZING THE CONCEPTS

23.1 Mendel's Laws

- Genes are on the chromosomes and each gene has a minimum of two alternative forms, called alleles.
- It is customary to use letters to represent the genotypes of individuals. Homozygous dominant is indicated by two capital letters, and homozygous recessive is indicated by two lowercase letters. Heterozygous is indicated by a capital letter and a lowercase letter.
- An individual has two alleles for each trait, whereas the gametes have one allele for each trait. Therefore, in keeping with

Mendel's law of segregation, each heterozygous individual can form two types of gametes.

- When performing an actual cross, it is assumed that all possible types of sperm fertilize all possible types of oocytes. The Punnett square is based on this assumption. The results may be expressed as a probable phenotypic ratio. It is also possible to state the chance of an offspring having a particular phenotype. When a heterozygote reproduces with a heterozygote (called a monohybrid cross), the phenotypic ratio is 3:1; that is, each child has a 75% chance of having the dominant phenotype and a 25% chance of having the recessive phenotype.

oocytes

♂ \ ♀	*W*	*w*
W	*WW*	*Ww*
w	*Ww*	*ww*

sperm

Offspring

- In keeping with Mendel's law of independent assortment, an individual heterozygous for two traits can form four types of gametes. Therefore, the phenotypic ratio for a dihybrid cross is 9:3:3:1; in other words, $\frac{9}{16}$ of the offspring have the two dominant traits, $\frac{3}{16}$ have one of the dominant traits with one of the recessive traits, $\frac{3}{16}$ have the other dominant trait with the other recessive trait, and $\frac{1}{16}$ have both recessive traits.
- Testcrosses are used to determine if an individual with the dominant phenotype is homozygous or heterozygous.

23.2 Pedigree Analysis and Genetic Disorders

- A pedigree is a chart of a family's history with regard to a particular genetic trait. It may be possible to decide if an inherited condition is due to an autosomal dominant or an autosomal recessive allele by studying a pedigree.
- Tay-Sachs disease (a lysosomal storage disease), cystic fibrosis (faulty regulator of chloride channel), phenylketonuria (inability to metabolize phenylalanine), and sickle cell disease (sickle-shaped red blood cells) are all autosomal recessive disorders.
- Marfan syndrome (defective elastic connective tissue) and Huntington disease (abnormal huntingtin protein) are both autosomal dominant disorders.

23.3 Beyond Simple Inheritance Patterns

- There are many exceptions to Mendel's laws, including incomplete dominance (curly hair), codominance, and multiple alleles (ABO blood types).
- Familial hypercholesterolemia (liver cells lack cholesterol receptors) is an incompletely dominant autosomal disorder.
- For polygenic traits such as skin color and height, several genes each contribute to the overall phenotype in equal, small degrees, resulting in a bell-shaped curve.

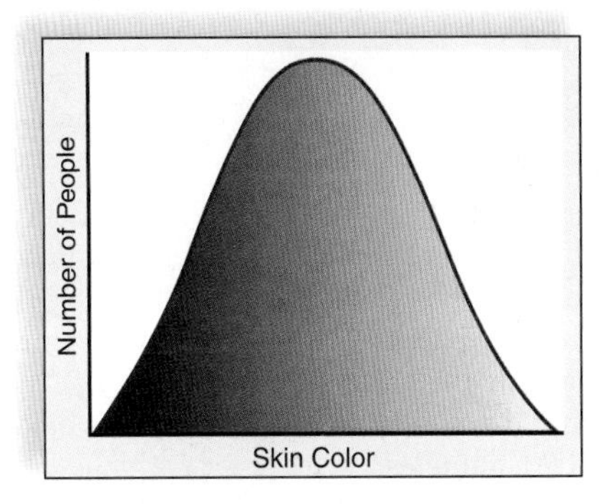

23.4 Environmental Influences

- Environmental factors, such as nutrition and temperature, affect the expression of certain traits. In humans, nutrition helps determine height. In Siamese cats and Himalayan rabbits, the effect of temperature is apparent in the distribution of fur color.
- Humans can manipulate the environment to lessen the damaging effects of genetic defects.

TESTING YOURSELF

Choose the best answer for each question.

1. According to the law of segregation,
   a. each gamete contains two copies of each factor for each trait.
   b. each individual has one copy of each factor for each trait.
   c. fertilization gives each new individual one factor for each trait.
   d. All of these are correct.
   e. None of these are correct.
2. Peas are good for genetic studies because they
   a. cannot self-pollinate.
   b. have a long generation time.
   c. are easy to grow.
   d. are more genetically variable than most plants.
3. Which of the following genotypes indicates a heterozygous individual?
   a. *AB*
   b. *AA*
   c. *Aa*
   d. *aa*
4. List the gametes produced by *AaBb*.
   a. *Aa*, *Bb*
   b. *A*, *a*, *B*, *b*
   c. *AB*, *ab*
   d. *AB*, *Ab*, *aB*, *ab*
5. Homologous chromosomes contain identical
   a. genes.
   b. alleles.
   c. DNA sequences.
   d. All of these are correct.
6. The genotypic ratio from the F_2 of a monohybrid cross is
   a. 1:1.
   b. 3:1.
   c. 1:2:1.
   d. 9:3:3:1.
   e. 1:1:1:1.
7. If two parents with short fingers (dominant) have a child with long fingers, what is the chance their next child will have long fingers?
   a. There is no chance.
   b. $\frac{1}{2}$
   c. $\frac{1}{4}$
   d. $\frac{1}{16}$
   e. $\frac{3}{16}$

8. Using the product rule, determine the probability that an *Aa* individual will be produced from an *Aa* × *Aa* cross.
   a. 50%
   b. 25%
   c. 75%
   d. 100%
   e. 0%
9. Using the following diagram, show how four types of gametes (*SW*, *sw*, *Sw*, *sW*) are produced from the dihybrid by attaching the correct letter to each chromosome.

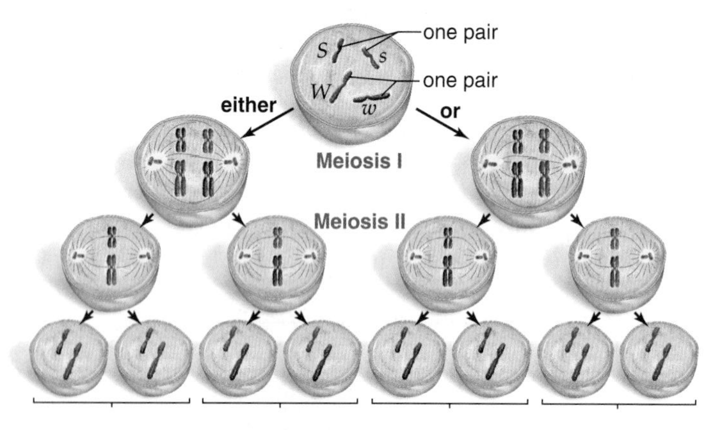

10. According to the law of independent assortment,
    a. all possible combinations of factors can occur in the gametes.
    b. only the parental combinations of gametes can occur in the gametes.
    c. only the nonparental combinations of gametes can occur in the gametes.
11. Which of the following is *not* a feature of polygenic inheritance?
    a. Effects of dominant alleles are additive.
    b. Genes affecting the trait may be on multiple chromosomes.
    c. Environment influences phenotype.
    d. Recessive alleles are harmful.
12. The ABO blood system exhibits
    a. codominance.
    b. dominance.
    c. multiple alleles.
    d. All of these are correct.
    e. None of these are correct.
13. A woman with blood type A and a man with blood type B have a baby. The blood type of the baby could be
    a. A or B
    b. A, B, or O
    c. A, B, AB, or O
    d. A only
14. Parents who do not have Tay-Sachs disease produce a child who has Tay-Sachs. What are the chances that each child born to this couple will have Tay-Sachs disease?
    a. 100%
    b. 75%
    c. 25%
    d. 0%
    e. None of these are correct.
15. Two affected parents have an unaffected child. The trait involved is
    a. autosomal recessive.
    b. incompletely dominant.
    c. controlled by multiple alleles.
    d. autosomal dominant.
16. Which is a late-onset neuromuscular genetic disorder?
    a. cystic fibrosis
    b. Huntington disease
    c. phenylketonuria
    d. Tay-Sachs disease

UNDERSTANDING THE TERMS

allele 470
codominance 481
cystic fibrosis (CF) 479
dihybrid 476
dominant allele 471
familial hypercholesterolemia (FH) 481
genotype 471
heterozygous 471
homozygous 471
Huntington disease 480
incomplete dominance 481
law of independent assortment 476
law of segregation 470
locus 470
Marfan syndrome 480
monohybrid 473
multiple alleles 481
pedigree 478
phenotype 471
phenylketonuria (PKU) 479
polygenic inheritance 482
Punnett square 473
recessive allele 471
sickle cell disease 479
Tay-Sachs disease 478
testcross 474

THINKING CRITICALLY

1. Explain at the molecular level why blood types A and B are dominant to blood type O.
2. Explain what is meant by "chance has no memory."
3. If identical twins have exactly the same genetic makeup, why are there any differences at all between them?

INQUIRY INTO LIFE WEBSITE

The companion website for *Inquiry into Life* provides a wealth of information organized and integrated by chapter. You will find practice tests, animations, videos, and much more that will complement your learning and understanding of general biology.

http://www.mhhe.com/maderinquiry13

Chromosomal Basis of Inheritance

Applying the Concepts

Sarah looked at her youngest son, Brandon, sitting at the kitchen counter watching her. Brandon broke into a grin when he caught his mother's eye. Sarah was 44 when she gave birth to Brandon and knew he had Down syndrome before he was born. Brandon had just come home from school and was waiting for a cookie. As soon as she set the plate in front of him, Brandon started telling her about his exciting adventures at school that day.

Down syndrome, which occurs when an individual inherits an extra chromosome 21, is the most common chromosomal anomaly in humans. People with a syndrome have several of the same symptoms, and in the case of Down syndrome these symptoms include an upward slant to the eyes, low muscle tone, a deep crease across the center of the palm, and an increased risk of cardiac defects among others. There is some degree of intellectual disability, but most individuals with Down syndrome have IQs in the mild to moderate disability range.

The National Down Syndrome Society tells us that individuals with Down syndrome can be integrated into all aspects of society. They attend school, work, and participate in community activities. Some individuals graduate from high school with a normal diploma and attend college. Some socialize, date, and even marry. The National Down Syndrome Society is a strong advocate for the value, acceptance, and inclusion of people with Down syndrome.

In this chapter, we will continue our discussion of genetic diseases and also include various syndromes that arise due to changes in chromosomal number and structure.

Chapter Outline

24.1 Sex-Linked Inheritance

Learning Outcomes

1. Explain X-linked genetic inheritance.
2. Solve crosses involving sex-linked traits.
3. Describe four X-linked recessive disorders.

Normally, both males and females have 23 pairs of chromosomes; 22 pairs are called **autosomes**, and one pair is the sex chromosomes. The **sex chromosomes** are so named because they differ between the sexes. In humans, males have the sex chromosomes X and Y, and females have two X chromosomes.

Traits controlled by genes on the sex chromosomes are said to be **sex-linked**; an allele on an X chromosome is **X-linked**, and an allele on the Y chromosome is Y-linked. Most sex-linked genes are only on the X chromosomes, and the Y chromosome is blank for these. Very few alleles have been found on the much smaller Y chromosome.

It would be logical to suppose that a sex-linked trait is passed from father to son or from mother to daughter, but this is not the case. A male always receives an X-linked allele from his mother, from whom he inherited an X chromosome. *The Y chromosome from the father does not carry an allele for the trait.* Usually a sex-linked genetic disorder is recessive. Therefore, a female must receive two alleles, one from each parent, before she has the condition.

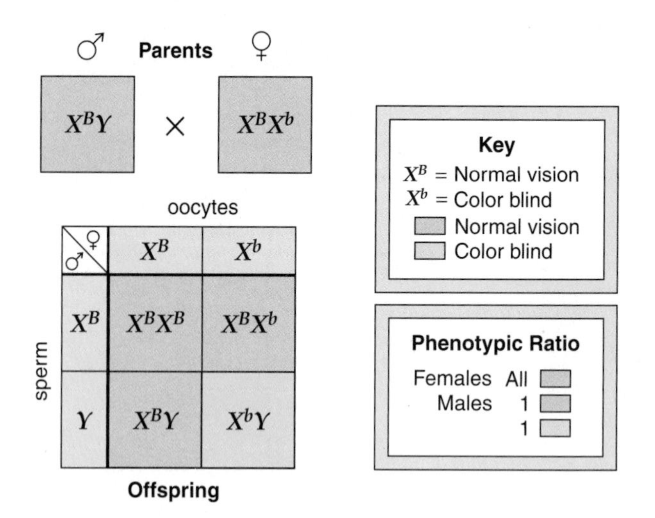

Figure 24.1 Cross involving an X-linked allele. The male parent is normal, but the female parent is a carrier—that is, an allele for color blindness is located on one of her X chromosomes. Therefore, each son has a 50% chance of being color blind. The daughters will have normal color vision, but each one has a 50% chance of being a carrier.

Sex-Linked Alleles

When considering X-linked traits, the allele on the X chromosome is shown as a letter attached to the X chromosome. For example, following is the key for red-green color blindness, a well-known X-linked recessive disorder:

$$X^B = \text{normal vision}$$

$$X^b = \text{color blindness}$$

The possible genotypes and phenotypes in both males and females are:

Genotypes	Phenotypes
X^BX^B	Female who has normal color vision
X^BX^b	Carrier female who has normal color vision
X^bX^b	Female who is color blind
X^BY	Male who has normal vision
X^bY	Male who is color blind

The second genotype is a **carrier** female because, although a female with this genotype has normal color vision, she is capable of passing on an allele for color blindness. Color-blind females are rare because they must receive the allele from both parents. Color-blind males are more common because they need only one recessive allele to be color blind. The allele for color blindness has to be inherited from their mother because it is on the X chromosome. Males only inherit the Y chromosome from their father.

Now let us consider a mating between a man with normal vision and a heterozygous woman (Fig. 24.1). What is the chance that this couple will have a color-blind daughter? A color-blind son? All daughters will have normal color vision because they all receive an X^B from their father. The sons, however, have a 50% chance of being color blind, depending on whether they receive an X^B or an X^b from their mother. The inheritance of a Y chromosome from their father cannot offset the inheritance of an X^b from their mother. Because the Y chromosome doesn't have an allele for the trait, it can't possibly prevent color blindness in a son. Note in Figure 24.1 that the phenotypic results for sex-linked traits are given separately for males and females.

Check Your Progress

1. Describe the sex-linked inheritance, and explain why it almost always involves the X chromosome.
2. What is a carrier?

Pedigree for X-Linked Disorders

Like color blindness, most sex-linked disorders are carried on the X chromosome. Figure 24.2 gives a pedigree for an *X-linked recessive disorder.* More males than females have the disorder because recessive alleles on the X chromosome are always expressed in males. The Y chromosome lacks an allele for the disorder. X-linked recessive conditions often pass from grandfather to grandson because the daughters of a male with the disorder are carriers. Figure 24.2 lists various ways to recognize a recessive X-linked disorder.

Only a few known traits are *X-linked dominant.* If a disorder is X-linked dominant, affected males pass the trait *only* to daughters, who have a 100% change of having the condition. Females can pass an X-linked dominant allele to both

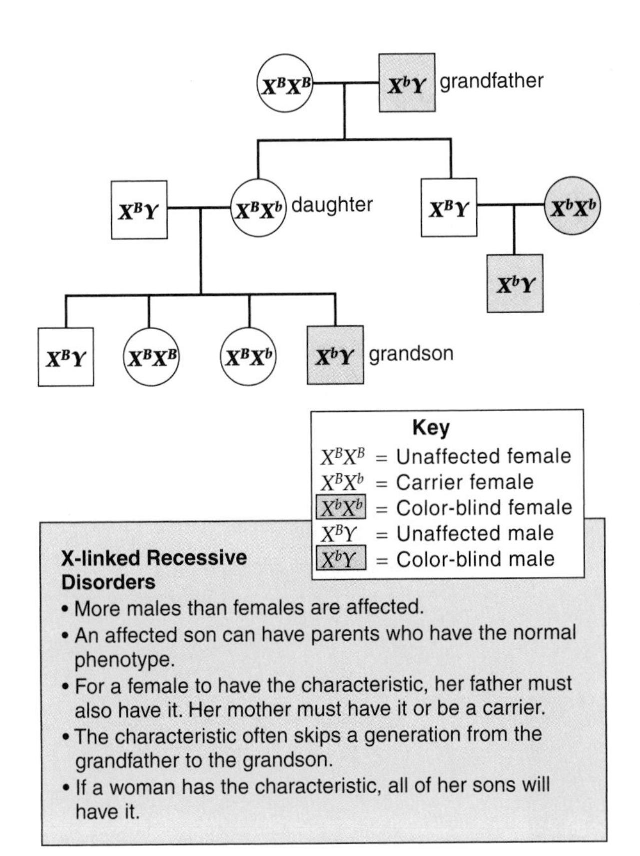

Figure 24.2 X-linked recessive pedigree.
This pedigree for color blindness exemplifies the inheritance pattern of an X-linked recessive disorder. The list gives various ways of recognizing the X-linked recessive pattern of inheritance.

sons and daughters. If a female is heterozygous and her partner is normal, each child has a 50% chance of escaping an X-linked dominant disorder. This depends on which maternal X chromosome is inherited.

X-Linked Recessive Disorders of Interest

Color blindness, an X-linked recessive disorder, does not prevent males from leading a normal life. About 8% of Caucasian men have red-green color blindness. Most of them see brighter greens as tans, olive greens as browns, and reds as reddish browns. A few cannot tell reds from greens at all. They see only yellows, blues, blacks, whites, and grays.

Duchenne muscular dystrophy is an X-linked recessive disorder characterized by wasting away of the muscles. Symptoms, which include a waddling gait, toe walking, frequent falls, and difficulty in rising, may appear as soon as the child starts to walk. Muscle weakness intensifies until the individual is confined to a wheelchair. Death usually occurs by age 20; therefore, affected males are rarely fathers. The recessive allele remains in the population by passing from carrier mother to carrier daughter.

The absence of a protein, now called dystrophin, is the cause of Duchenne muscular dystrophy. Much investigative work determined that the lack of dystrophin causes calcium to leak into the cell, which promotes the action of an enzyme that dissolves muscle fibers. When the body attempts to repair the tissue, fibrous tissue forms (Fig. 24.3), and this cuts off the blood supply so that more and more cells die. Immature muscle cells can be injected into muscles, but it takes 100,000 cells for dystrophin production to increase by 30–40%.

Fragile X syndrome (see the Health Focus box) is the most common cause of inherited mental impairment. These impairments can range from mild learning disabilities to more severe intellectual disabilities. It is also the most common known cause of *autism*, a class of social, behavioral, and communication disorders. Males with full symptoms of the condition have characteristic physical abnormalities. A long face, prominent jaw, and large ears are facial features often seen in fragile X males. Excessively flexible joints and genital abnormalities are common as well. Females with the condition have variable symptoms. Most of the same traits seen in males with fragile X have been reported in females as well, but often the symptoms are milder in females and present with lower frequency.

Hemophilia, of which there are two common types, is another X-linked recessive disorder. Hemophilia A is due to the absence or minimal presence of a clotting factor known as factor VIII, and hemophilia B is due to the absence of clotting factor IX. Hemophilia is called the bleeder's disease because the affected person's blood either does not clot or clots very slowly. Although hemophiliacs bleed externally after an injury, they also bleed internally, particularly around joints. Hemorrhages can be stopped with transfusions of fresh blood (or plasma) or concentrates of the missing clotting protein. Factors VIII and IX are also now available as biotechnology products.

At the turn of the century, hemophilia was prevalent among the royal families of Europe, and all of the affected males could trace their ancestry to Queen Victoria of England

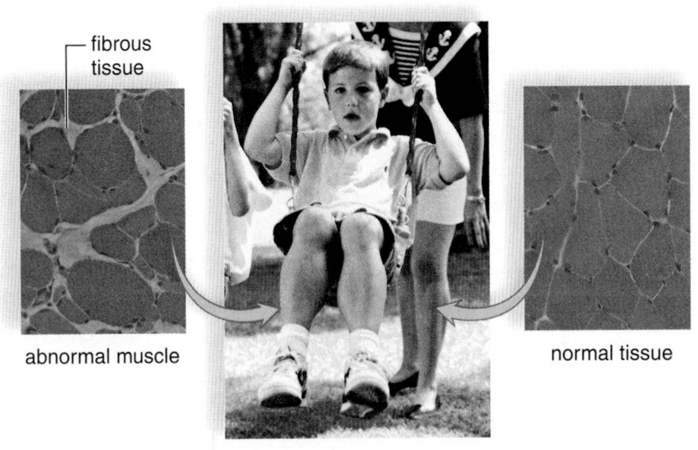

Figure 24.3 X-linked genetic disorder.
In Duchenne muscular dystrophy, an X-linked recessive disorder, the calves of the legs enlarge because fibrous tissue develops as muscles waste away due to lack of the protein dystrophin.

(Fig. 24.4). Of Queen Victoria's 26 grandchildren, four grandsons had hemophilia, and four granddaughters were carriers. Because none of Queen Victoria's ancestors were affected, it seems that the faulty allele she carried arose by mutation, either in Victoria or in one of her parents. Her carrier daughters, Alice and Beatrice, introduced the allele into the ruling houses of Russia and Spain, respectively. Alexei, the last heir to the Russian throne before the Russian Revolution, was a hemophiliac. There are no hemophiliacs in the present British royal family because Victoria's eldest son, King Edward VII, did not receive the allele.

Check Your Progress

1. Why are more males than females color blind?
2. What phenotypic ratio is expected for a cross in which both parents have one X-linked recessive allele?
3. Describe the symptoms of fragile X syndrome.

Practice Problems 1*

1. What phenotypic ratio is expected for a cross in which both parents have one X-linked dominant allele?
2. Both the mother and the father of a son with hemophilia appear to be normal. From whom did the son inherit the allele for hemophilia? What are the genotypes of the mother, the father, and the son?
3. A woman is color blind. What are the chances that her sons will be color blind? If she is married to a man with normal vision, what are the chances that her daughters will be color blind? Will be carriers?
4. Both the husband and wife have normal vision. The wife gives birth to a color-blind daughter. Is it more likely the father had normal vision or was color blind? What does this lead you to deduce about the girl's parentage?

Answers to Practice Problems appear in Appendix A.

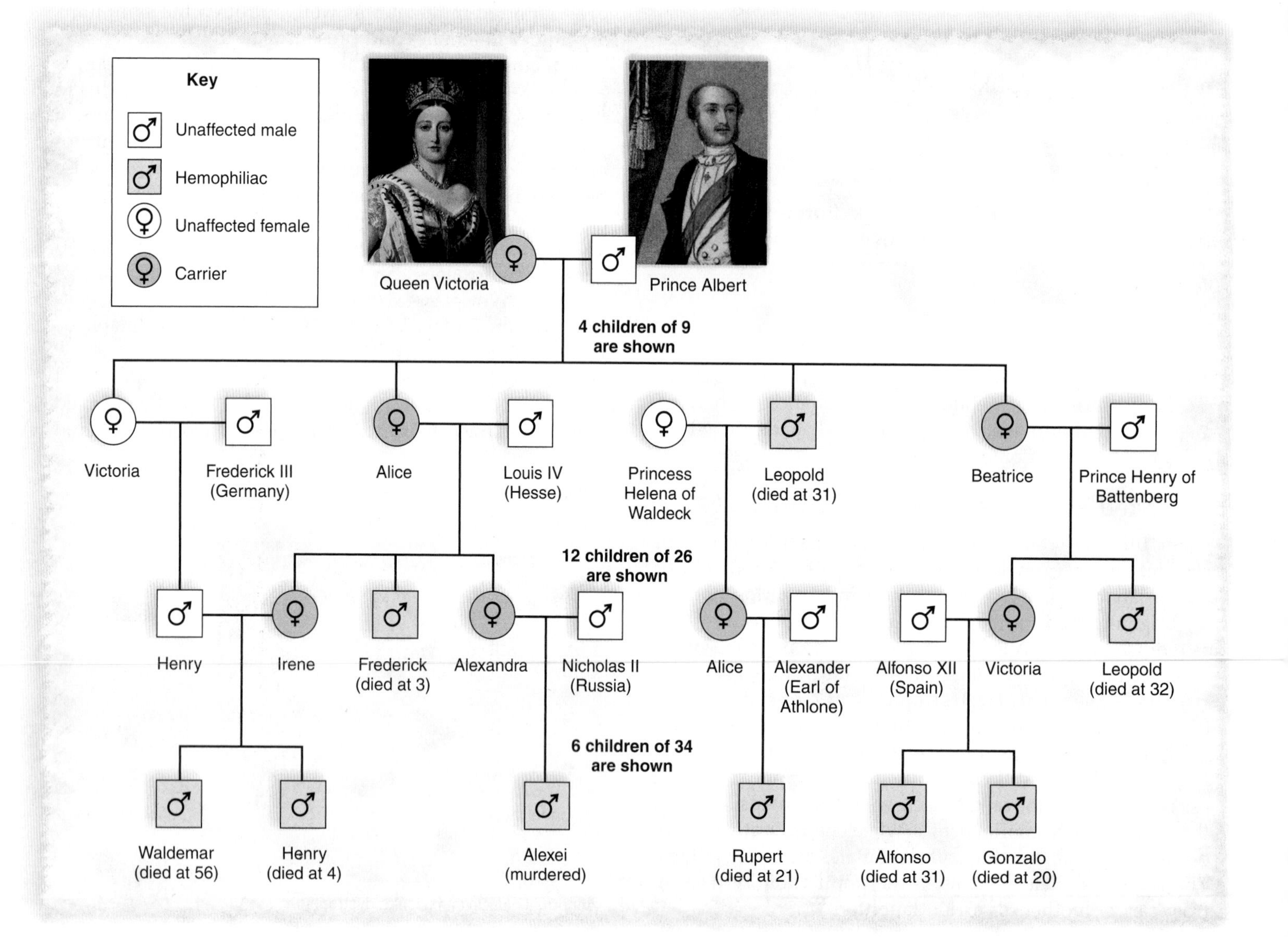

Figure 24.4 A simplified pedigree showing X-linked inheritance of hemophilia in European royal families. Because Queen Victoria was a carrier, each of her sons had a 50% chance of having the disorder, and each of her daughters had a 50% chance of being a carrier. This pedigree shows only the affected descendants. Many others are unaffected, including the members of the present British royal family.

24.2 Gene Linkage

Learning Outcomes

1. Diagram a linkage group.
2. Predict which alleles will cross over more frequently.

The study of sex linkage allowed investigators to realize that many different types of alleles are on a single chromosome. A chromosome doesn't contain just one or two alleles; it contains a long series of alleles in a definite sequence. The sequence is fixed because each allele has its own particular locus on a chromosome. All the alleles on one chromosome form a **linkage group** because they tend to be inherited together. (The term sex linkage refers to a sex chromosome while linked genes refer to two or more genes on the same chromosome.) When we do two trait crosses, we are assuming that the alleles are on nonhomologous chromosomes and therefore are not linked (see Fig. 23.7). Alleles that are linked do not show independent assortment. Figure 24.5*a* shows the results of a cross when linkage is complete: a dihybrid produces only two types of gametes in equal proportion.

During meiosis, crossing-over sometimes occurs between the nonsister chromatids of a tetrad (see Fig. 5.11). During crossing-over, the chromatids exchange genetic material, and therefore genes. If crossing-over occurs between the two alleles of interest, a dihybrid produces four types of gametes instead of two (Fig. 24.5*b*). Recombinant gametes occur in reduced number because crossing-over is infrequent.

The occurrence of crossing-over can help tell the sequence of genes on a chromosome because crossing-over occurs more often between distant genes than between genes that are close together on a chromosome. For example, consider these homologous chromosomes:

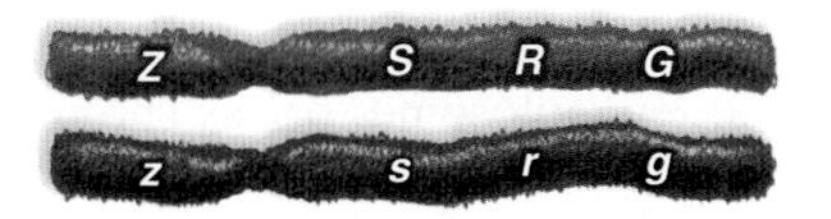

pair of homologous chromosomes

We expect recombinant gametes to include *G* and *z* more often than *R* and *s*. In keeping with this observation, investigators formerly used recombination frequencies to map the chromosomes of, say, fruit flies, but also humans to a degree. Each 1% of crossing-over is equivalent to one map unit between genes.

Check Your Progress

1. Explain why linked alleles do not show independent assortment.
2. How many types of gametes are produced when linkage is complete and the diploid number is 2?

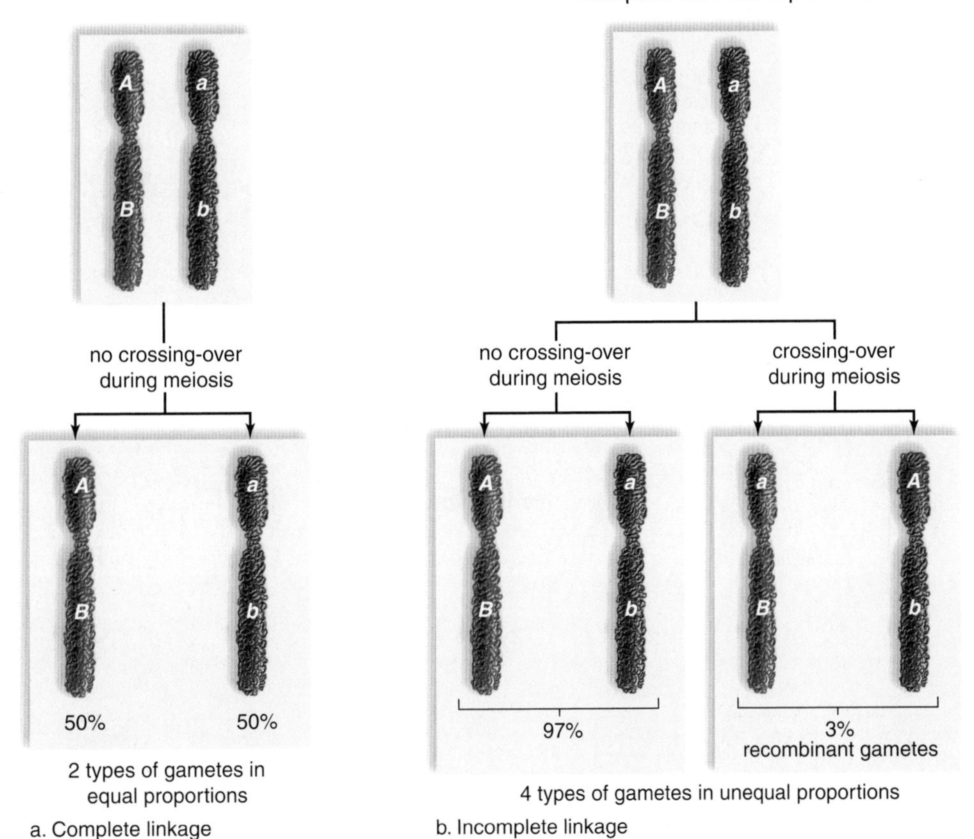

Figure 24.5 Linkage group.
In this individual, alleles *A* and *B* are on one member of a homologous pair, and alleles *a* and *b* are on the other member. Therefore, the individual is a dihybrid. **a.** When linkage is complete, this dihybrid produces only two types of gametes in equal proportion. **b.** When linkage is incomplete, this dihybrid produces four types of gametes because crossing-over has occurred. The recombinant gametes occur in reduced proportion because crossing-over occurs infrequently.

24.3 Changes in Chromosome Number

Learning Outcomes

1. Describe how chromosome number disorders arise.
2. Give several examples of chromosome number disorders.

Normally, a human receives 22 pairs of autosomes and two sex chromosomes. Sometimes individuals are born with either too many or too few autosomes or sex chromosomes, most likely due to nondisjunction during meiosis. **Nondisjunction** occurs during meiosis I, when both members of a homologous pair go into the same daughter cell, or during meiosis II, when the sister chromatids fail to separate and both daughter chromosomes go into the same gamete. Figure 24.6 assumes that nondisjunction has occurred during oogenesis. Some abnormal eggs have 24 chromosomes, while others have only 22 chromosomes. If an egg with 24 chromosomes is fertilized with a normal sperm, the result is a **trisomy,** so called because one type of chromosome is present in three copies. If an egg with 22 chromosomes is fertilized with a normal sperm, the result is a **monosomy,** so called because one type of chromosome is present in a single copy.

Normal development depends on the presence of exactly two of each kind of chromosome. Too many chromosomes are tolerated better than a deficiency of chromosomes, and several trisomies are known to occur in humans. Among autosomal trisomies, only trisomy 21 (Down syndrome) has a reasonable chance of survival after birth. This is probably due to the fact that chromosome 21 is one of the smallest chromosomes.

The chances of survival are greater when trisomy or monosomy involves the sex chromosomes. In normal XX females, one of the X chromosomes becomes a darkly stained mass of chromatin called a **Barr body** (after the person who discovered it). A Barr body is an inactive X chromosome. Therefore, we now know that the cells of females function with a single X chromosome just as those of males do. This is most likely the reason that a zygote with one X chromosome (Turner syndrome) can survive. Then, too, all extra X chromosomes beyond a single one become Barr bodies, and this explains why poly-X females and XXY males are seen fairly frequently. An extra Y chromosome, called Jacobs syndrome, is tolerated in humans, most likely because the Y chromosome carries few genes. Jacobs syndrome (XYY) is due to nondisjunction during meiosis II of spermatogenesis. We know this because two Ys are present only during meiosis II in males.

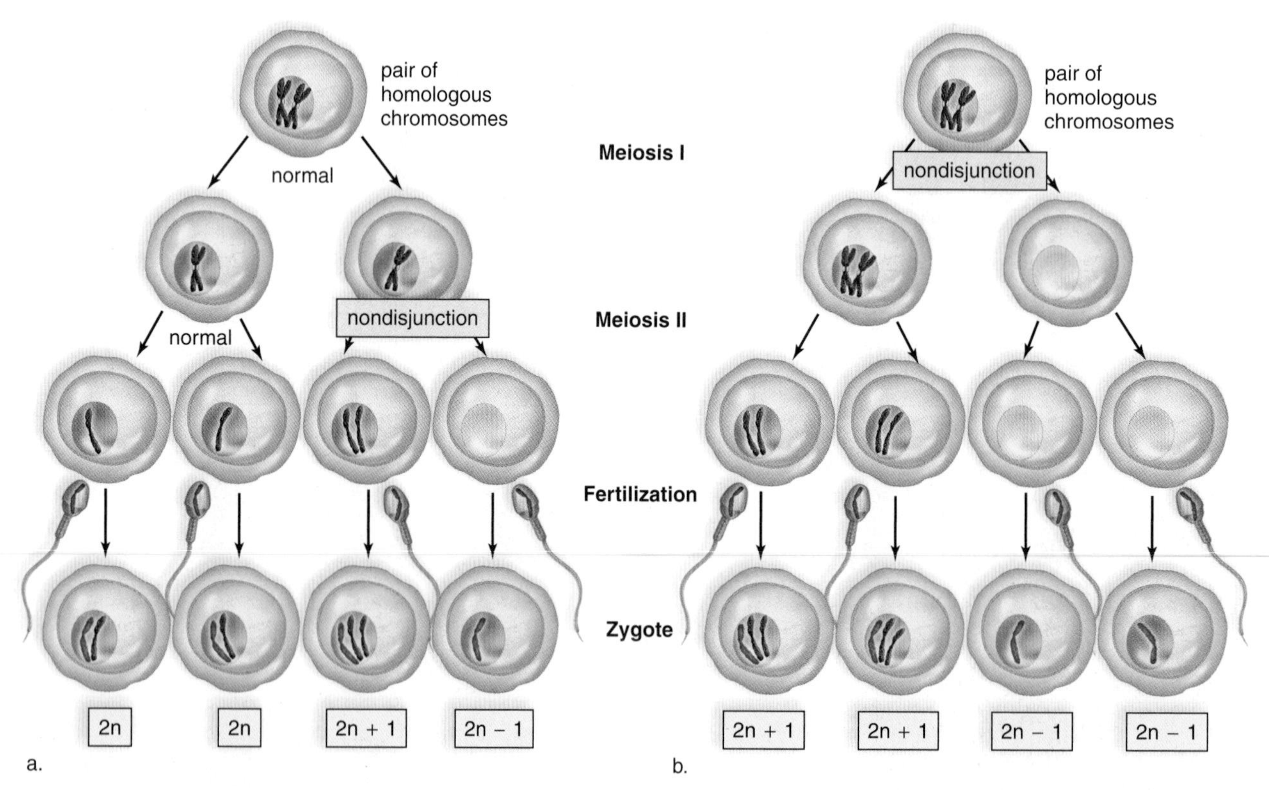

Figure 24.6 Nondisjunction of chromosomes during oogenesis.
a. Nondisjunction can occur during meiosis II if the sister chromatids separate but the resulting chromosomes go into the same daughter cell. Then the egg will have one more or one less than the usual number of chromosomes. Fertilization of these abnormal eggs with normal sperm produces an abnormal zygote with abnormal chromosome numbers. **b.** Nondisjunction can also occur during meiosis I and result in abnormal eggs that also have one more or one less than the normal number of chromosomes. Fertilization of these abnormal eggs with normal sperm results in a zygote with an abnormal chromosome number. In humans (2n), the normal diploid number of chromosomes is 46. An individual who is 2n + 1 would have 47 chromosomes, while an individual who is 2n – 1 would have 45 chromosomes.

Down Syndrome

The most common autosomal trisomy seen among humans is trisomy 21, also called Down syndrome (Fig. 24.7). This syndrome is easily recognized by the following characteristics: short stature; an eyelid fold; a flat face; stubby fingers; a wide gap between the first and second toes; a large, fissured tongue; a round head; a palm crease (the so-called simian line); and mental impairment, which may range from mild to severe.

Persons with Down syndrome usually have three copies of chromosome 21 because the egg had two copies instead of one. (In 23% of the cases studied, however, the sperm had the extra chromosome 21.) The chances of a woman having a Down syndrome child increase rapidly with age, starting at about age 40, and the reasons for this are still being determined.

Although an older woman is more likely to have a Down syndrome child, most babies with Down syndrome are born to women younger than age 40 because this is the age group having the most babies. Karyotyping can detect a child with Down syndrome. However, young women are not routinely encouraged to undergo the procedures necessary to get a sample of fetal cells (i.e., amniocentesis or chorionic villus sampling) because the risk of complications is greater than the risk of having a Down syndrome child. Fortunately, a test based on substances in maternal blood can help identify fetuses that may need to be karyotyped. The Science Focus in this chapter describes karyotyping, and amniocentesis and chorionic villus sampling.

The genes that cause Down syndrome are located on the long arm of chromosome 21 (Fig. 24.7*b*), and extensive investigative work has been directed toward discovering the specific genes responsible for the characteristics of the syndrome. Thus far, investigators have discovered several genes that may account for various conditions seen in persons with Down syndrome. For example, they have located genes most likely responsible for the increased tendency toward leukemia, cataracts, and accelerated rate of aging. Researchers have also discovered that an extra copy of the *Gart* gene causes an increased level of purines in the blood, a finding associated with mental impairment. One day, it may be possible to control the expression of the *Gart* gene even before birth so that at least this symptom of Down syndrome does not appear.

Changes in Sex Chromosome Number

An abnormal sex chromosome number is the result of inheriting too many or too few X or Y chromosomes. Figure 24.6 can be used to illustrate nondisjunction of the sex chromosomes during oogenesis, if you assume that the chromosomes shown represent X chromosomes. Nondisjunction during oogenesis or spermatogenesis can result in gametes that have too few or too many X or Y chromosomes. After fertilization, the syndromes listed in Table 24A (other than Down syndrome, which is autosomal) are possibilities.

A person with only one sex chromosome, an X (Turner syndrome), is a female, and a person with more than one X chromosome plus a Y (Klinefelter syndrome) is a male. This shows that in humans, the presence of a Y chromosome, not

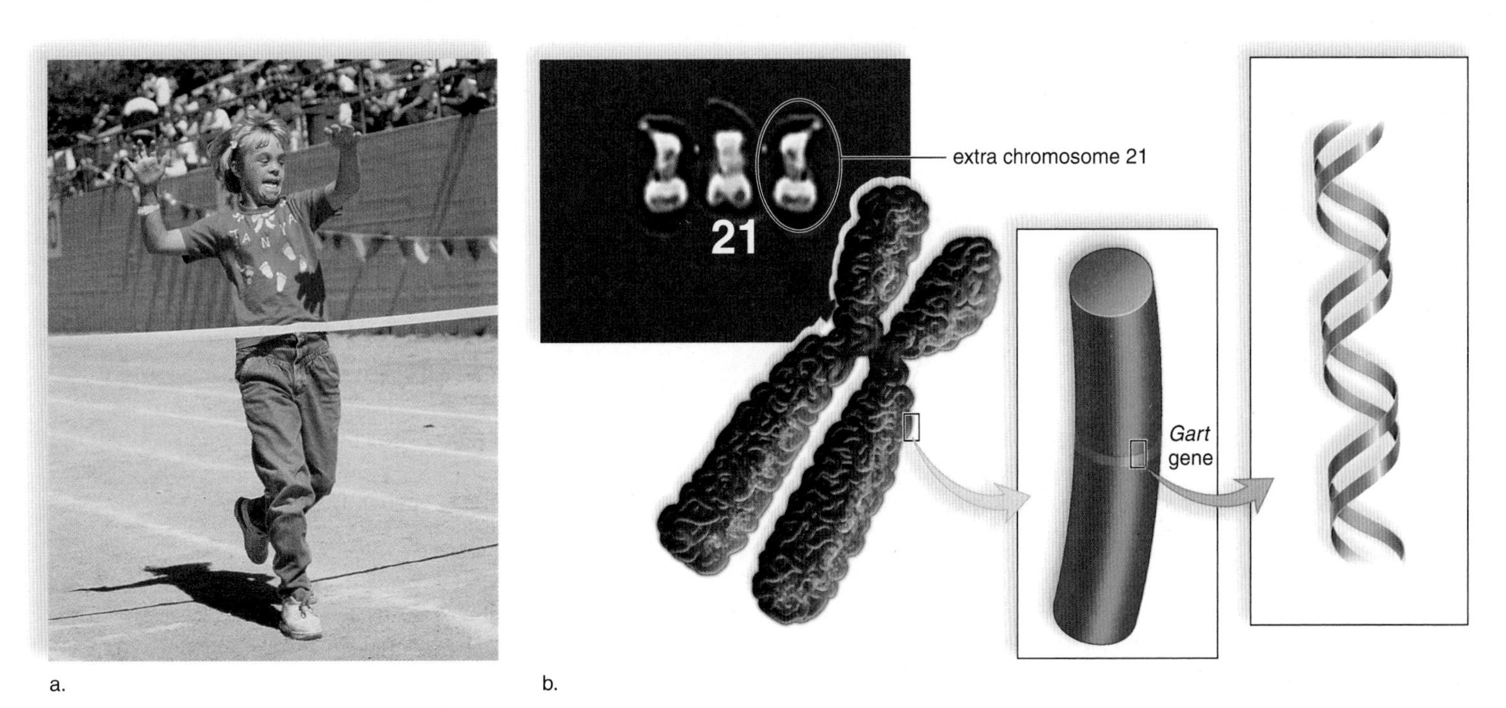

Figure 24.7 Abnormal autosomal chromosome number.
Persons with Down syndrome have an extra chromosome 21. **a.** Common characteristics of the syndrome include a wide, rounded face and a fold on the upper eyelids. **b.** Karyotype of an individual with Down syndrome shows an extra chromosome 21. More sophisticated technologies allow investigators to pinpoint the location of specific genes associated with the syndrome. An extra copy of the *Gart* gene, which leads to a high level of purines in the blood, may account for the mental impairment seen in persons with Down syndrome.

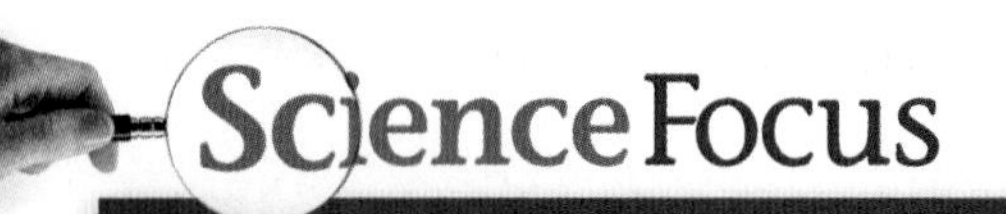

Viewing the Chromosomes

A **karyotype** is a visual display of the chromosomes, arranged by size, shape, and banding pattern. Usually, both human males and females have 23 pairs of chromosomes: 22 pairs are autosomes, and 1 pair is the sex chromosomes. Males have the sex chromosomes X and Y, and females have two X chromosomes.

Various human disorders result from abnormal chromosome number or structure. Such disorders often cause a **syndrome,** which is a group of symptoms that always occur together. Table 24A lists several syndromes that are due to an abnormal chromosome number. Doing a karyotype will reveal such abnormalities.

Any cell in the body, except red blood cells, which lack a nucleus, can be a source of chromosomes for karyotyping. In adults, it is easiest to use white blood cells separated from a blood sample for this purpose. For fetuses, cells can be obtained by either amniocentesis or chorionic villus sampling. These cells can be used for karyotype analysis or for DNA analysis, as described in Chapter 26.

Amniocentesis

Amniocentesis is a procedure for obtaining a sample of amniotic fluid from the uterus of a pregnant woman. Blood tests and the mother's age are used to determine whether the procedure should be done. The risk of spontaneous abortion increases by about 0.25–0.5% due to amniocentesis, and doctors use the procedure only if it is medically warranted.

Amniocentesis is not usually performed until after the fifteenth week of pregnancy. A long needle is passed through the abdominal and uterine walls to withdraw a small amount of fluid, which also contains fetal cells (Fig. 24A*a*). The amniotic fluid is then tested. Karyotyping of the chromosomes may be delayed as long as four weeks so that the cells can be cultured to increase their number. DNA tests for other genetic disorders can also be done.

Chorionic Villus Sampling

Chorionic villus sampling (CVS) is a procedure for obtaining chorionic cells in the region where the placenta will develop. This procedure is usually done between the tenth and fourteenth weeks of pregnancy. A long, thin suction tube is inserted through the vagina into the uterus (Fig. 24.A*b*). Ultrasound, which gives a picture of the uterine contents, is used to place the tube between the uterine lining and the chorionic villi. Then a sampling of chorionic cells is obtained by suction. Results are available within one to two weeks. But testing amniotic fluid is not possible because no amniotic fluid is collected. Also, CVS carries a greater risk of spontaneous abortion than amniocentesis—0.5–1% compared to 0.25–0.5%. The advantage of CVS is getting the results of karyotyping at an earlier date.

TABLE 24A Syndromes from Abnormal Chromosome Numbers

Syndrome	Sex	Chromosomes	Chromosome Number	Frequency	
				Spontaneous Abortions	*Live Births*
Down	M or F	Trisomy 21	47	1/40	1/733
Poly-X	F	XXX (or XXXX)	47 or 48	0	1/1,500
Klinefelter	M	XXY (or XXXY)	47 or 48	1/300	1/800
Jacobs	M	XYY	47	?	1/1,000
Turner	F	X	45	1/18	1/2,500

the number of X chromosomes, determines maleness. The *SRY* (sex-determining *r*egion of *Y*) gene, on the short arm of the Y chromosome, produces a hormone called testis-determining factor, which plays a critical role in the development of male genitals.

Turner Syndrome Females with Turner syndrome have only one sex chromosome, an X. They are usually short and may have malformed features, such as a webbed neck, high palate, and small jaw. Many have congenital heart and kidney defects. Most have ovarian failure and do not undergo puberty or menstruate without sex hormone replacement therapy. However, pregnancy has been achieved through in vitro fertilization using donor eggs. Women with Turner syndrome have a normal range of intelligence but often have a nonverbal learning disability. They can lead very successful, fulfilling lives if they receive appropriate care.

Klinefelter Syndrome About 1 in 650 males is born with two X chromosomes and one Y chromosome. The symptoms of this condition (referred to as "47, XXY") are often so subtle that only 25% are ever diagnosed, and those are usually not diagnosed until after age 15. Earlier diagnosis opens the possibility for educational accommodations and other interventions that can help mitigate common symptoms, which include speech and language delays. Those 47, XXY males who develop more severe symptoms as adults are referred to as having "Klinefelter syndrome." All 47, XXY adults will require assisted reproduction

Karyotyping

After a cell sample has been obtained, the cells are stimulated to divide in a culture medium. A chemical is used to stop mitosis during metaphase when chromosomes are the most highly compacted and condensed. The cells are then killed, spread on a microscope slide, and dried. Stains are applied to the slides, and the cells are photographed. Staining causes the chromosomes to have dark and light cross-bands of varying widths, and these can be used, in addition to size and shape, to pair up the chromosomes. Today, a computer is used to arrange the chromosomes in pairs (Fig. 24.A*c*). The karyotype of a person who has Down syndrome usually has three chromosomes 21, instead of the usual two (Fig. 24.A*d*, *e*).

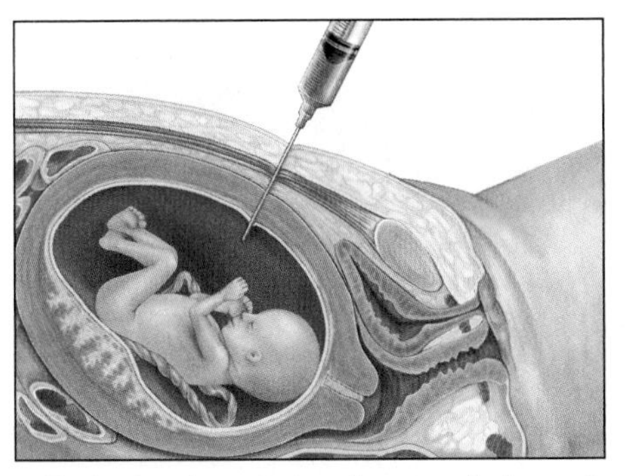

a. During amniocentesis, a long needle is used to withdraw amniotic fluid containing fetal cells.

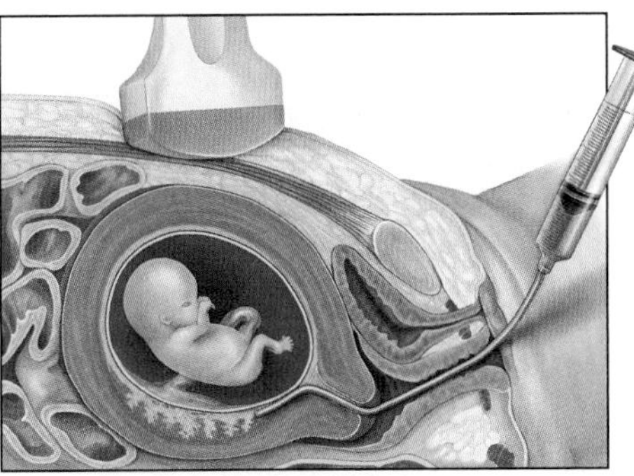

b. During chorionic villus sampling, a suction tube is used to remove cells from the chorion, where the placenta will develop.

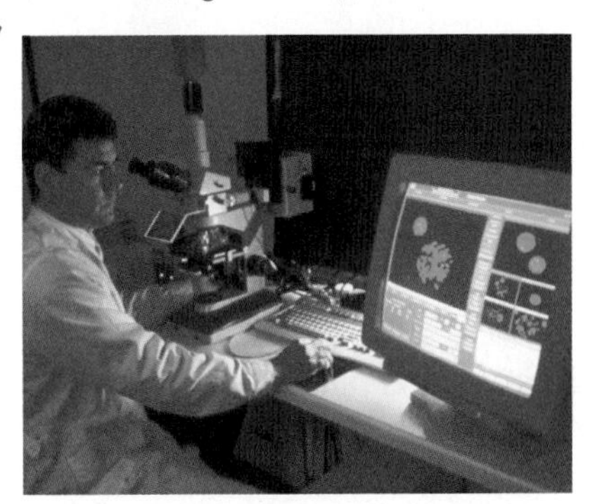

c. Cells are microscopically examined and photographed. A computer is used to arrange chromosomes by pairs.

Figure 24A Human karyotype preparation.

A karyotype is an arrangement of an individual's chromosomes into numbered pairs according to their size, shape, and banding pattern. **a.** Amniocentesis and **(b)** chorionic villus sampling provides cells for karyotyping to determine if the unborn child has a chromosomal abnormality. **c.** After cells are treated, a computer constructs the karyotype. **d.** Karyotype of a normal male. **e.** Karyotype of a male with Down syndrome, showing three chromosomes 21.

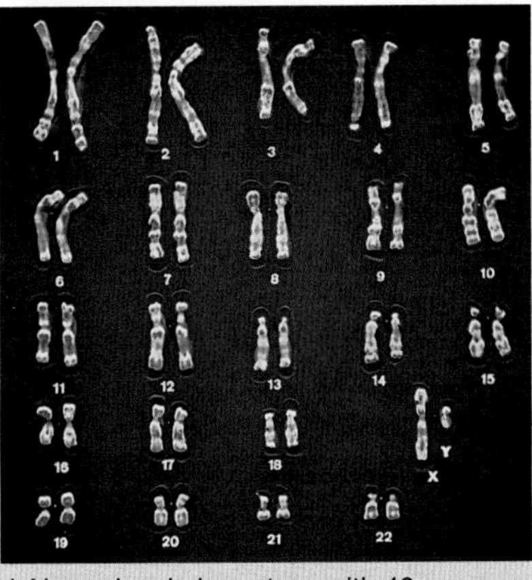

d. Normal male karyotype with 46 chromosomes

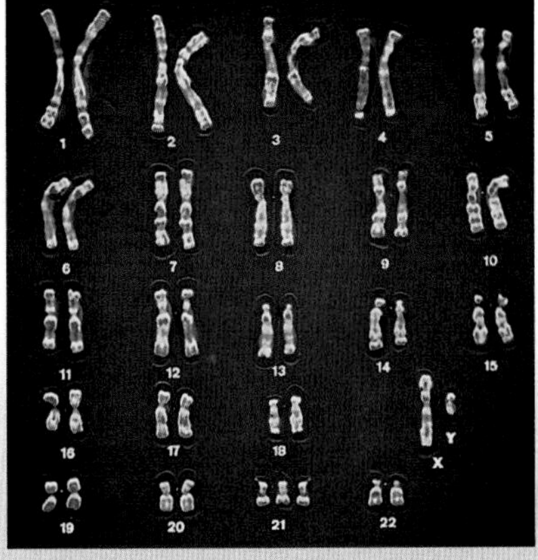

e. Down syndrome karyotype with an extra chromosome 21

in order to father children. Affected individuals commonly receive testosterone supplementation beginning at puberty.

Poly-X Females A poly-X female has more than two X chromosomes and thus extra Barr bodies in the nucleus. Females with three X chromosomes have no distinctive phenotype, aside from a tendency to be tall and thin. Although some have delayed motor and language development, most poly-X females are not mentally impaired. Some may have menstrual difficulties, but many menstruate regularly and are fertile. Their children usually have a normal karyotype.

Females with more than three X chromosomes occur rarely. Unlike XXX females, XXXX females are tall and more likely to be mentally impaired. They exhibit various physical abnormalities, but may menstruate normally.

Jacobs Syndrome XYY males, who have Jacobs syndrome, can only result from nondisjunction during spermatogenesis. Affected males are usually taller than average, suffer from persistent acne, and tend to have speech and reading problems. At one time, it was suggested that these men were likely to be criminally aggressive, but more recent evidence indicates that the incidence of such behavior among them may be no greater than among XY males.

Check Your Progress

1. Why are the chances of survival greater for a trisomy or monosomy of the sex chromosomes versus autosomes?
2. What determines whether a fetus develops into a male?

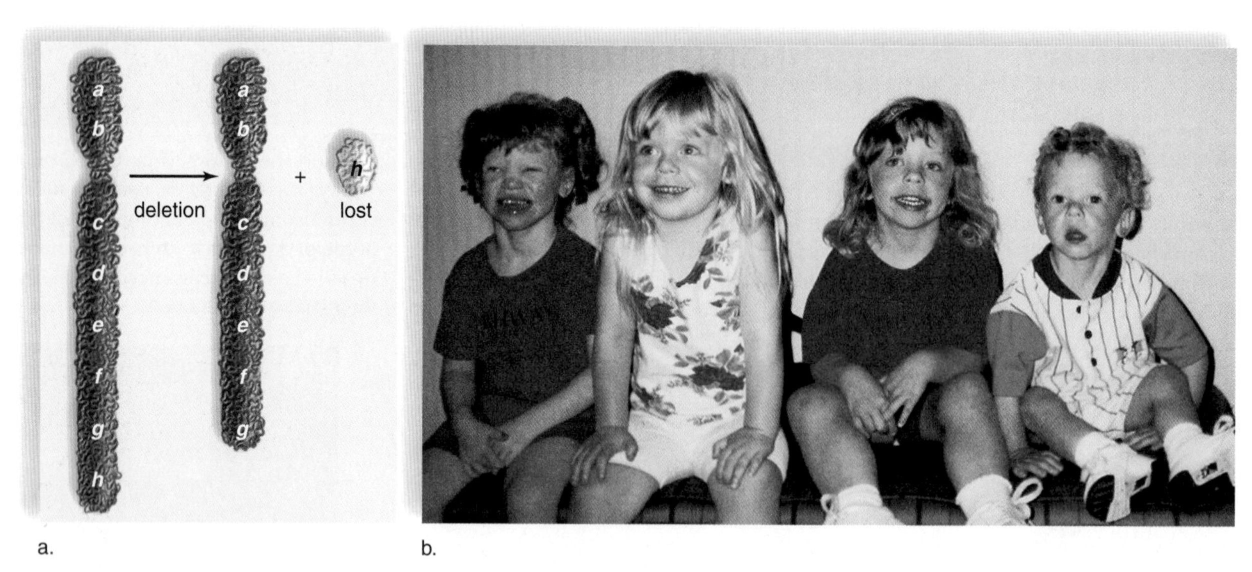

Figure 24.8 Deletion.
a. When chromosome 7 loses an end piece, the result is Williams syndrome. **b.** These children, although unrelated, have similar appearance, health, and behavioral problems.

24.4 Changes in Chromosome Structure

Learning Outcomes

1. Describe a chromosomal deletion, duplication, translocation, and inversion.
2. List disorders caused by changes in chromosome structure.

Chromosomal mutations occur when chromosomes break. Various environmental agents—radiation, certain organic chemicals, or even viruses—can cause chromosomes to break apart. Ordinarily, when breaks occur, the segments reunite to give the same sequence of genes. But their failure to reunite correctly can result in one of several types of mutations: deletion, duplication, translocation, or inversion. Chromosomal mutations can occur during meiosis, and if the offspring inherits the abnormal chromosome, a syndrome may develop.

Deletions and Duplications

A **deletion** occurs when a single break causes a chromosome to lose an end piece or when two simultaneous breaks lead to the loss of an internal chromosomal segment. An individual who inherits a normal chromosome from one parent and a chromosome with a deletion from the other parent no longer has a pair of alleles for each trait, and a syndrome can result.

Williams syndrome occurs when chromosome 7 loses a tiny end piece (Fig. 24.8). Children who have this syndrome look like pixies because they have turned-up noses, wide mouths, small chins and large ears. Although their academic skills are poor, they exhibit excellent verbal and musical abilities. The gene that governs the production of the protein elastin is missing, which affects the health of the cardiovascular system and causes their skin to age prematurely. Such individuals are very friendly but need an ordered life, perhaps because of the loss of a gene for a protein that is normally active in the brain.

Cri du chat (cat's cry) syndrome is seen when chromosome 5 is missing an end piece. The affected individual has a small head, is mentally impaired, and has facial abnormalities. Abnormal development of the glottis and larynx results in the most characteristic symptom—the infant's cry resembles that of a cat.

In a **duplication,** a chromosomal segment is repeated in the same chromosome or in a nonhomologous chromosome. In any case, the individual has more than two alleles for certain traits. An inverted duplication is known to occur in chromosome 15. **Inversion** means that a segment joins in the direction opposite from normal. Children with this syndrome, called inv dup 15 syndrome, have poor muscle tone, mental impairment, seizures, a curved spine, and autistic characteristics, including poor speech, hand flapping, and lack of eye contact (Fig. 24.9).

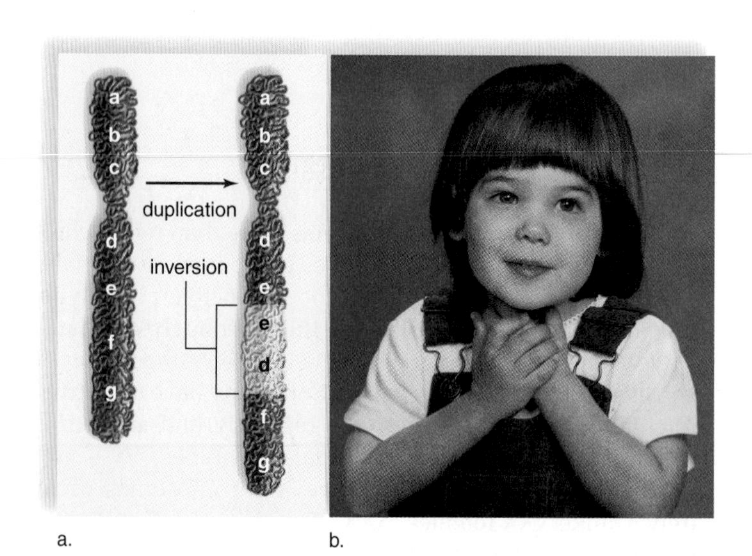

Figure 24.9 Duplication.
a. When a piece of chromosome 15 is duplicated and inverted, **(b)** a syndrome results that is characterized by poor muscle tone and autistic characteristics.

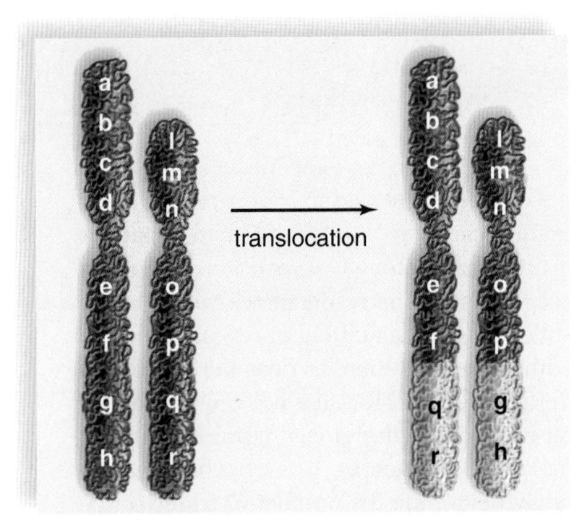

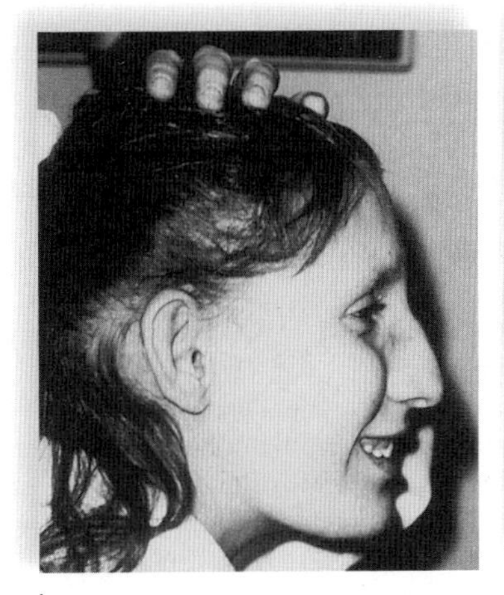

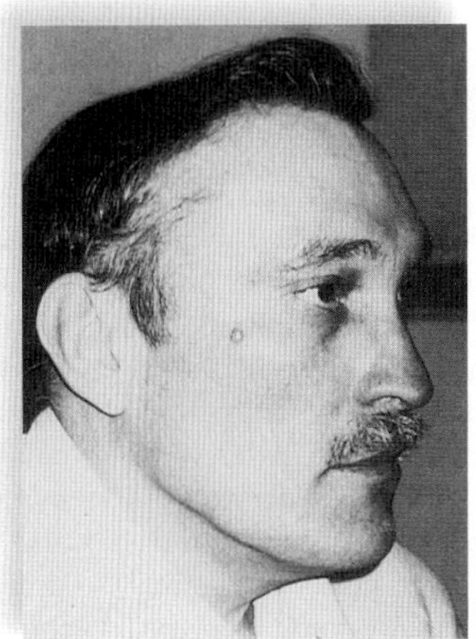

Figure 24.10 Translocation.
a. When chromosomes 2 and 20 exchange segments, (**b**) Alagille syndrome, characterized by distinctive facial features, sometimes results because the translocation disrupts an allele on chromosome 20.

The Health Focus describes a type of duplication called a trinucleotide repeat expansion. In the case of Huntington disease, the trinucleotide repeat consists of the sequence CAG repeated over and over again. In the case of fragile X syndrome, the trinucleotide repeat is CGG. Persons with duplications over a particular number, which is specific for each disease, exhibit symptoms of the disease. In the case of Huntington disease, sufferers have over 40 copies of the CAG sequence within the huntingtin protein gene.

Translocation

A **translocation** is the exchange of chromosomal segments between two nonhomologous chromosomes. A person who has both of the involved chromosomes has the normal amount of genetic material and is healthy, unless the chromosome exchange breaks an allele into two pieces. The person who inherits only one of the translocated chromosomes will have only one copy of certain alleles and three copies of certain other alleles.

In 5% of cases, a translocation that occurred in a previous generation between chromosomes 21 and 14 is the cause of one type of Down syndrome. The affected person inherits two normal chromosomes 21 and an abnormal chromosome 14 that contains a segment of chromosome 21. In these cases, Down syndrome is not related to the age of the mother, but instead tends to run in the family of either the father or the mother.

Figure 24.10 shows a daughter and father who have a translocation between chromosomes 2 and 20. Although they have the normal amount of genetic material, they have the distinctive face, abnormalities of the eyes and internal organs, and severe itching characteristic of Alagille syndrome. People with this syndrome ordinarily have a deletion on chromosome 20. Therefore, it can be deduced that the translocation disrupted an allele on chromosome 20 in the father. The symptoms of Alagille syndrome range from mild to severe, so some people may not be aware they have the syndrome. This father did not realize it until he had a child with the syndrome.

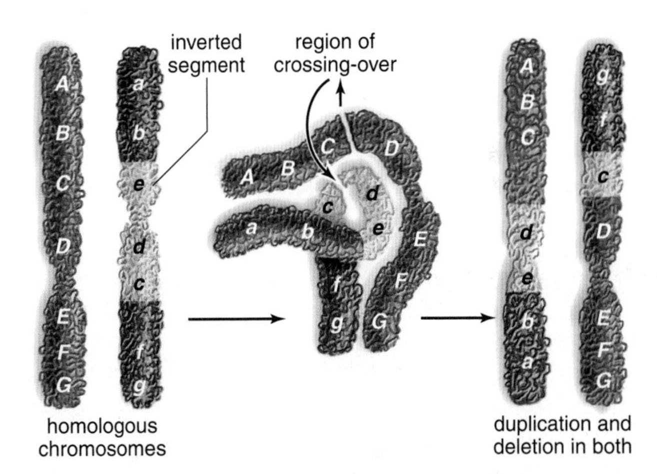

Figure 24.11 Inversion.
Left: A segment of one homologue is inverted. Notice that in the inverted segment *edc* occurs instead of *cde*. *Middle:* The two homologues can pair only when the inverted sequence forms an internal loop. After crossing-over, a duplication and a deletion can occur. *Right:* The homologue on the left has *AB* and *ba* sequences and neither *gf* nor *FG* genes. The homologue on the right has *gf* and *FG* sequences and neither *AB* nor *ba* genes.

Inversion

An inversion occurs when a segment of a chromosome is turned 180 degrees. You might think this is not a problem because the same genes are present, but the reverse sequence of alleles can lead to altered gene activity.

Crossing-over between an inverted chromosome and the noninverted homologue can lead to recombinant chromosomes that have both duplicated and deleted segments. This happens because alignment between the two homologues is only possible when the inverted chromosome forms a loop (Fig. 24.11).

Check Your Progress

1. Why does the absence of a single protein, such as elastin in Williams syndrome, lead to multiple health problems?
2. How is it possible for a person with a translocation to be phenotypically normal?

Health Focus

Trinucleotide Repeat Expansion Disorders (TRED)

Several of the genetic diseases we have studied result from a single change in the DNA that results in a single change in an amino acid in a particular protein. For example, the most common form of sickle cell disease results from a one-nucleotide change in the DNA that causes a one-amino-acid difference in the protein hemoglobin. However, in at least 15 human diseases, including Huntington disease and fragile X syndrome, the problem lies in a DNA sequence that is repeated too many times, called a *trinucleotide repeat.*

As described in Chapter 23, Huntington disease is a neurological disorder in which the huntingtin protein is altered. The trinucleotide sequence CAG normally appears several times at the beginning of the gene that encodes the huntingtin protein. The presence of multiple CAGs results in multiple glutamines in the protein. People with 10–35 copies of the CAG display no signs of Huntington disease. However, people with more than 40 copies have the disease.

Fragile X syndrome is one of the most common genetic causes of mental impairment. As children, fragile X individuals may be hyperactive or autistic. Their speech is delayed in development and often repetitive in nature. As adults, males have large testes and big, usually protruding ears. They are short in stature, but the jaw is prominent, and the face is long and narrow (Fig. 24B*a*). The term *fragile X syndrome* originated because diagnosis used to be dependent upon observation of an X chromosome whose tip is attached to the rest of the chromosome by only a thin thread (Fig. 24B*b*). This disease results from repeats of the trinucleotide CGG. People with 6–50 copies appear normal, while those with over 230 copies have the disease. Those with between 50 and 230 copies are carriers of fragile X, with either no symptoms or very mild symptoms.

People who have the intermediate number of repeats are said to have a premutation. Premutations can lead to full-blown mutations in future generations by expansion of the number of repeats. This is sometimes called genetic anticipation. The number of repeats increases with each generation, as does the severity of the disease (Fig. 24B*c*). It is not known why the repeats expand in number. Using the current molecular technology, it is possible to accurately determine the number of trinucleotide repeats present in an individual.

Discussion Questions

1. The repeat expansion appears to occur during meiosis but not mitosis. Why might this be true?
2. If a male has fragile X syndrome, is his mother or father more likely to be a carrier for the disease? Why?
3. Why is fragile X but not Huntington disease considered a syndrome?

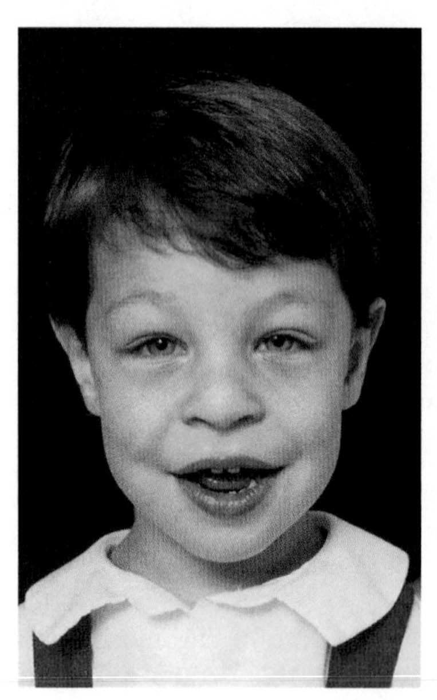

a. Young male with fragile X syndrome

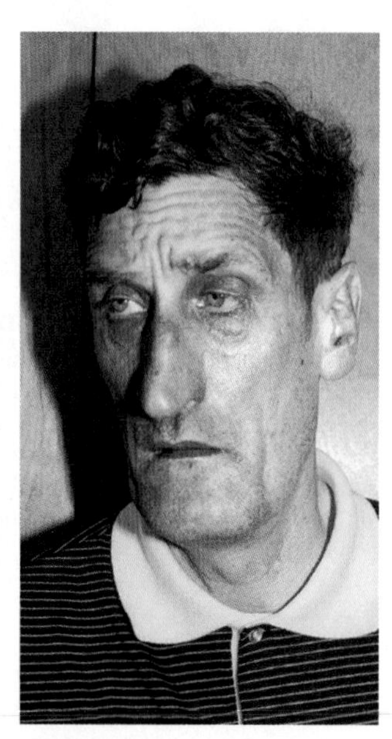

Same individual when mature

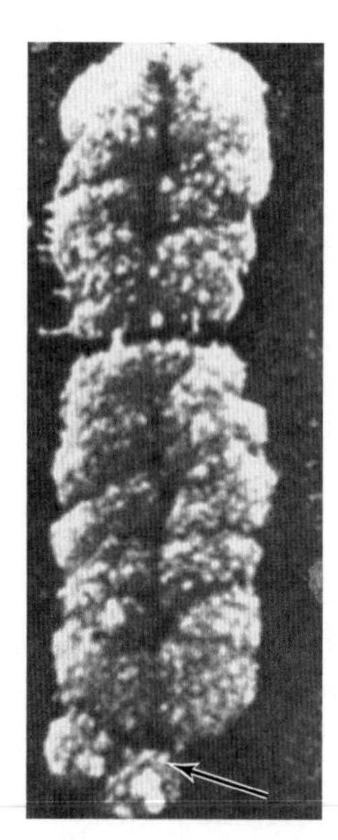

b. Fragile X chromosome

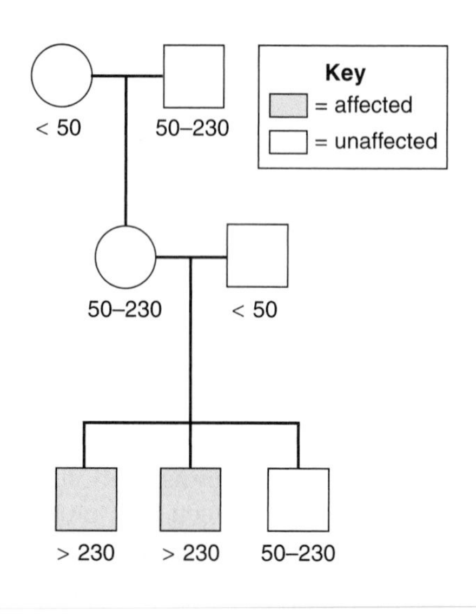

c. Inheritance pattern for fragile X syndrome

Figure 24B TRED in fragile X syndrome.
a. A young male with fragile X syndrome appears normal but with age develops an elongated face with a prominent jaw and ears that noticeably protrude. **b.** An arrow points out the fragile site of this fragile X chromosome. **c.** The number of base-pair repeats at the fragile site are given for each person, showing that the incidence and severity increase with each succeeding generation. A grandfather who has a premutation with 50–230 repeats has no symptoms but transmits the condition to his grandsons through his daughter. Two grandsons have full-blown mutations with more than 230 base repeats.

Applying the Concepts [Revisited]

Brandon suffers from Down syndrome. Ninety-five percent of Down syndrome cases are a result of nondisjunction of chromosome 21 resulting in three copies of this chromosome.

1. Why might Sarah have had prenatal testing during her pregnancy with Brandon?
2. Why does an additional chromosome result in multiple difficulties, often termed a syndrome?

SUMMARIZING THE CONCEPTS

24.1 Sex-Linked Inheritance

Some traits are sex-linked, meaning that although they do not determine gender, they are carried on the sex chromosomes. Most of the alleles for these traits are carried on the X chromosome, while the Y does not bear alleles for those same traits.

- The phenotypic results of sex-linked crosses are given for females and males separately.
- An X-linked recessive pedigree shows why more males than females have these disorders. Females are carriers of the disease.

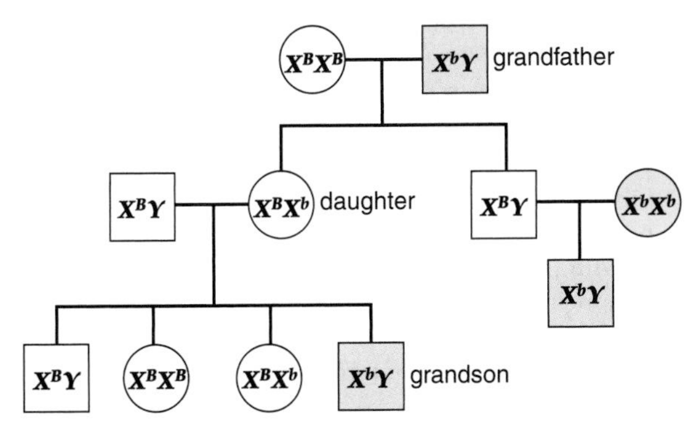

- Color blindness, Duchenne muscular dystrophy, fragile X syndrome, and hemophilia are all X-linked recessive disorders.

24.2 Gene Linkage

- All the genes on one chromosome form a linkage group, which is broken only when crossing-over occurs. Genes that are linked tend to go together into the same gamete.
- If crossing-over occurs, a dihybrid cross yields all possible phenotypes among the offspring, but the expected ratio is greatly changed because recombinant phenotypes are reduced in number.

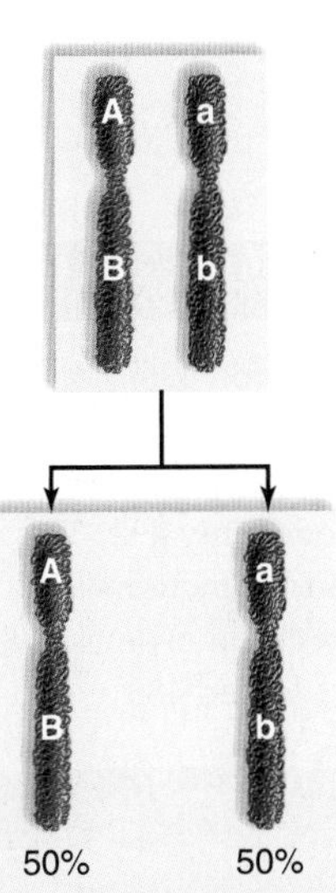

24.3 Changes in Chromosome Number

- Nondisjunction during meiosis can result in an abnormal number of autosomes or sex chromosomes in the gametes.
- Down syndrome results when an individual inherits three copies of chromosome 21.
- Females who are XO have Turner syndrome, and those who are XXX are poly-X females. Males with Klinefelter syndrome are XXY. Males who are XYY have Jacobs syndrome.

24.4 Changes in Chromosome Structure

- Changes in chromosomal structure also affect the phenotype. Chromosomal mutations include deletions, duplications, translocations, and inversions.
- Translocations do not necessarily cause any difficulties if the person has inherited both translocated chromosomes. However, the translocation can disrupt a particular allele, and then a syndrome will follow.
- An inversion can lead to chromosomes that have a deletion and a duplication when the inverted piece loops back to align with the noninverted homologue, and crossing-over follows between the nonsister chromatids.

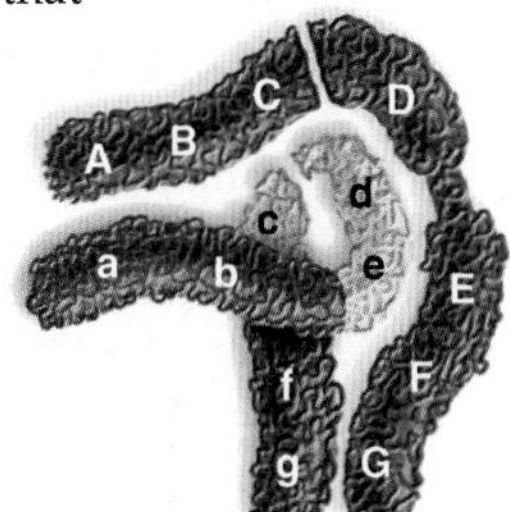

TESTING YOURSELF

1. The following pedigree pertains to color blindness. Using the letter *B* for the normal allele, the genotype of the individual with the asterisk is
   a. $X^B X^B$
   b. $X^B X^b$
   c. $X^b X^b$

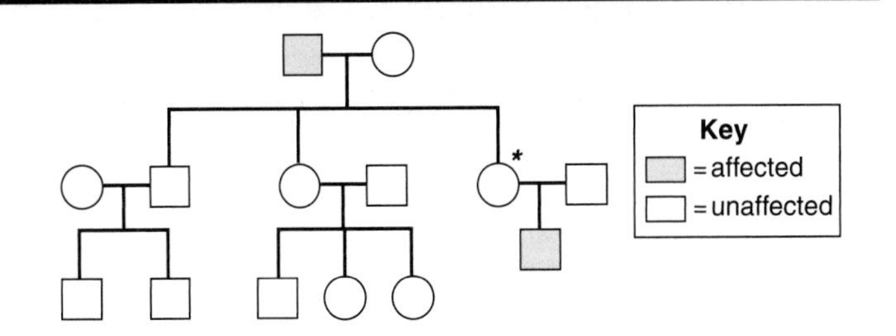

2. All the genes on one chromosome are said to form a
   a. chromosomal group.
   b. recombination group.
   c. linkage group.
   d. crossing-over group.
3. If alleles *R* and *s* are linked on one chromosome and *r* and *S* are linked on the homologous chromosome, what gametes will be produced? (Assume that no crossing-over occurs.)
   a. *RS, Rs, rS, rs*
   b. *RS, rs*
   c. *Rs, rS*
   d. *R, S, r, s*
4. An individual can have too many or too few chromosomes as a result of
   a. nondisjunction.
   b. Barr bodies.
   c. trisomy.
   d. amniocentesis.
   e. monosomy.
5. Assume that two parents with normal vision have a color-blind son. Which parent is responsible for the son's color blindness?
   a. the mother
   b. the father
   c. either parent
   d. None of these are correct because two normal parents cannot have a color-blind son.

For questions 6–10, match the chromosome disorder to its description in the key.

Key:

a. female with undeveloped ovaries and uterus, unable to undergo puberty, normal intelligence, can live normally with hormone replacement
b. XXY male, can inherit more than two X chromosomes
c. male or female, mentally impaired, short stature, flat face, stubby fingers, large tongue, simian palm crease
d. XXX or XXXX female
e. caused by nondisjunction during spermatogenesis

6. Klinefelter syndrome
7. Poly-X female
8. Down syndrome
9. Turner syndrome
10. Jacobs syndrome

For questions 11–14, match the chromosome mutation to its description in the key.

Key:

a. turned-up nose, wide mouth, small chin, large ears, poor academic skills, excellent verbal and musical abilities, prematurely aging cardiovascular system
b. deletion in chromosome 5
c. poor muscle tone, mental impairment, seizures, curved spine, autistic characteristics, poor speech, hand flapping, lack of eye contact
d. translocation between chromosomes 2 and 20

11. Alagille syndrome
12. Inv dup 15 syndrome
13. Williams syndrome
14. Cri du chat syndrome
15. A karyotype is prepared
    a. for an unborn chld through amniocentesis or chorionic villi sampling.
    b. by arranging chromosomes into pairs by size, shape, and banding pattern.
    c. as a means of diagnosing an abnormal chromosomal disorder.
    d. using a photograph of a cell sample arrested during metaphase.
    e. All of these are correct.

UNDERSTANDING THE TERMS

amniocentesis 494
autosome 488
Barr body 492
carrier 488
chorionic villus sampling (CVS) 494
chromosomal mutation 496
color blindness 489
deletion 496
Duchenne muscular dystrophy 489
duplication 496
fragile X syndrome 489
hemophilia 489
inversion 496
karyotype 494
linkage group 491
monosomy 492
nondisjunction 492
sex chromosomes 488
sex-linked 488
syndrome 494
translocation 497
trisomy 492
X-linked 488

THINKING CRITICALLY

1. Why might expectant parents want to undergo fetal genetic testing if they know that, regardless of the findings, they plan to have the baby?
2. Why do you think there are no viable trisomies of chromosome 1?
3. Why can a person carrying a translocation be normal except for the inability to have children?

INQUIRY INTO LIFE WEBSITE

The companion website for *Inquiry into Life* provides a wealth of information organized and integrated by chapter. You will find practice tests, animations, videos, and much more that will complement your learning and understanding of general biology.

http://www.mhhe.com/maderinquiry13

DNA Structure and Control of Gene Expression

Applying the Concepts

Sometimes they are called "children of the night" because they cannot play outside in the sunlight. They even have a summer camp named Camp Sundown in which they do all the things that other children do at summer camp such as hiking, horseback riding, and swimming, except they do it after sundown. Whenever these children are outside during daylight, they wear protective clothing and eyewear, use sunblock, and travel in cars with tinted windows. These are children that suffer from a genetic disease called xeroderma pigmentosum (XP). This is a genetic disease in which the enzymes that are needed to repair DNA damage due to ultraviolet (UV) light are defective. Therefore, those who suffer with XP cannot be exposed to UV light. XP is very rare, only about 1 in a million individuals in the United States have XP, although those with Japanese descent have a higher incidence. There is no cure. DNA damage cannot be repaired and mutations accumulate throughout the lifetime of the patient. These individuals have a 1,000-fold higher risk of skin cancer than those with normal DNA repair enzymes. Exposure to UV light results in blistering and freckling, and individuals suffer from premature aging of the skin along with eye tumors. Some individuals with more severe XP also suffer progressive neurological complications such as mental retardation and hearing loss. Less than 40% of individuals with XP survive beyond the age of 20, although those with milder cases may survive into middle age. The most common cause of death is skin cancer, either squamous cell carcinoma or metastatic melanoma. This chapter describes the structure of DNA and how mutations could affect the function of RNA and proteins. It also describes the progression of cancer from a single mutation to a metastatic tumor.

Chapter Outline

25.1 DNA Structure and Replication

Learning Outcomes

1. Review the experiment that enabled scientists to determine that DNA made up the genetic material.
2. Diagram the structure of DNA and identify the components.
3. Explain how DNA is copied or replicated.

We now know that **DNA (deoxyribonucleic acid)** makes up our genes, but that is a relatively recent discovery. A little more than 50 years ago, scientists were not sure whether the genetic material was composed of DNA or protein. In 1952, Alfred Hershey and Martha Chase used experiments with viruses to answer this question. Hershey and Chase thought that if they could determine which part of the virus—the DNA in the viral core, or the protein in the viral coat, or capsid—enters a bacterium and directs the production of more viruses, they would know whether genes were made of DNA or protein.

Hershey and Chase performed two experiments (Fig. 25.1). In the first experiment, viral DNA was labeled with radioactive ^{32}P (represented in yellow). The viruses were allowed to attach to and inject their genetic material into the bacterium *Escherichia coli* (*E. coli*). Then the culture was agitated in a kitchen blender to remove whatever remained of the viruses on the outside of the bacteria. Finally, the culture was centrifuged (spun at high speed) so that the bacteria collected as a pellet at the bottom of the centrifuge tube. In this experiment investigators found most of the ^{32}P-labeled DNA (yellow) in the bacteria and not in the liquid medium. Why? Because the DNA had entered the bacteria.

In the second experiment, viral protein in capsids was labeled with radioactive ^{35}S (yellow). The viruses were allowed to attach to and inject their genetic material into *E. coli,* then the same procedure was followed to remove viruses and collect a pellet of bacteria. In this experiment scientists found ^{35}S-labeled protein (yellow) in the liquid medium and not in the bacteria. Why? Because the radioactive capsids remained on the outside of the bacteria.

These results indicated that the DNA of a virus (not a protein) enters the host, where viral reproduction takes place. Therefore, DNA is the genetic material and transmits the genetic information needed to make new viruses.

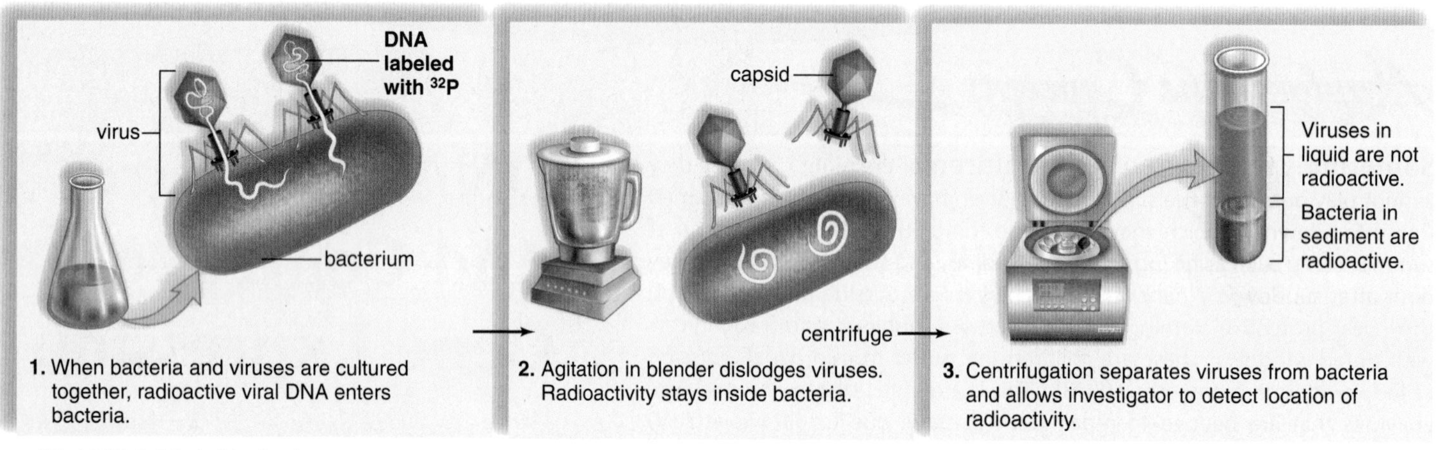

a. Viral DNA is labeled (yellow).

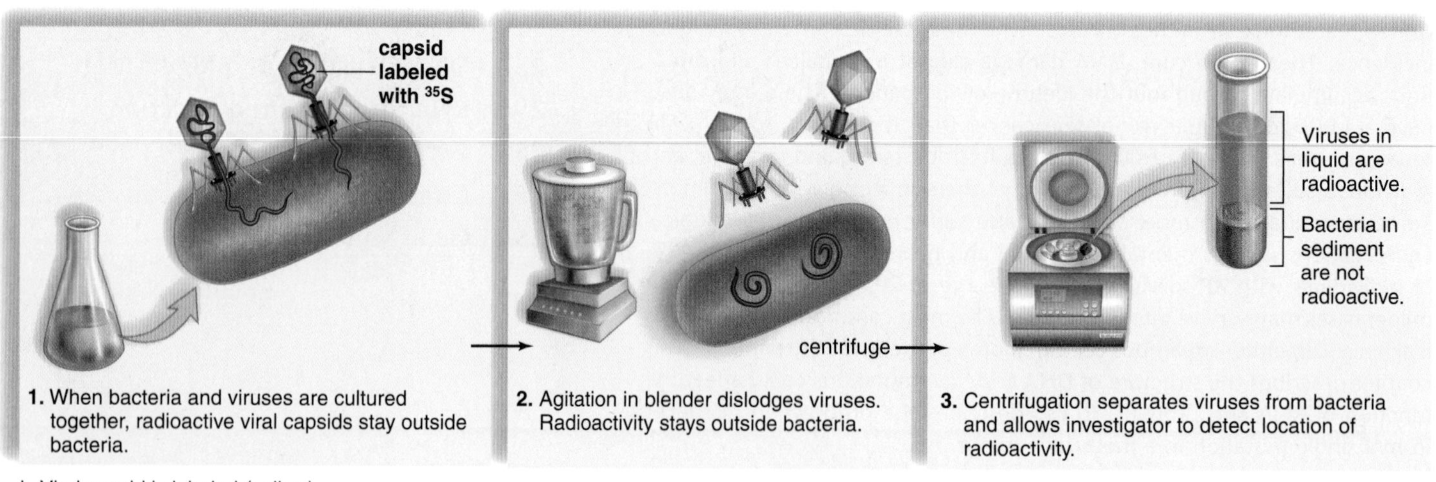

b. Viral capsid is labeled (yellow).

Figure 25.1 Hershey-Chase experiments.
These experiments concluded that viral DNA, not protein, was responsible for directing the production of new viruses.

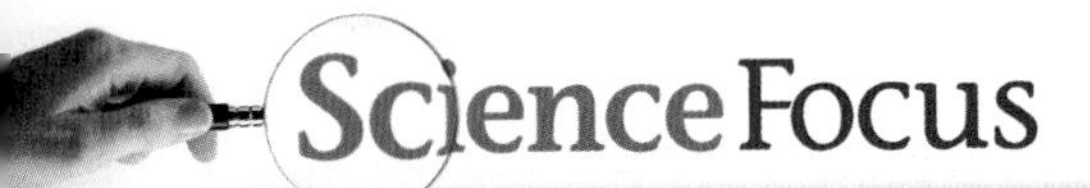

Science Focus

Finding the Structure of DNA

In 1953, James Watson, an American biologist, began an internship at the University of Cambridge, England. There he met Francis Crick, a British physicist, who was interested in molecular structures. Together, they set out to determine the structure of DNA and to build a model that would explain how DNA, the genetic material, can vary from species to species and even from individual to individual. They also discovered how DNA replicates (makes a copy of itself) so that daughter cells receive an identical copy.

The bits and pieces of data available to Watson and Crick were like puzzle pieces they had to fit together. *This is what they knew from the research of others:*

1. DNA is a polymer of nucleotides, each one having a phosphate group, the sugar deoxyribose, and a nitrogen-containing base. There are four types of nucleotides because there are four different bases: adenine (A) and guanine (G) are purines, while cytosine (C) and thymine (T) are pyrimidines.
2. A chemist, Erwin Chargaff, had determined in the late 1940s that regardless of the species under consideration, the number of purines in DNA always equals the number of pyrimidines. Further, the amount of adenine equals the amount of thymine (A = T), and the amount of guanine equals the amount of cytosine (G = C). These findings came to be known as Chargaff's rules.
3. Rosalind Franklin (Fig. 25A*a*), working with Maurice Wilkins at King's College, London, had just prepared an X-ray diffraction photograph of DNA. It showed that DNA is a double helix of constant diameter and that the bases are regularly stacked on top of one another (Fig. 25A*b*).

Using these data, Watson and Crick deduced that DNA has a twisted, ladderlike structure. The sugar-phosphate molecules make up the sides of the ladder, and the bases make up the rungs. Further, they determined that if A is normally hydrogen-bonded with T, and G is normally hydrogen-bonded with C (in keeping with Chargaff's rules), then the rungs always have a constant width, consistent with the X-ray photograph.

Watson and Crick built an actual model of DNA out of wire and tin. This double-helix model does indeed allow for differences in DNA structure between species because the base pairs can be in any order. Also, the model suggests that complementary base pairing plays a role in the replication of DNA. As Watson and Crick pointed out in their original paper, "It has not escaped our notice that the specific pairing we have postulated immediately suggests a possible copying mechanism for the genetic material."

When the Nobel Prize for the discovery of the double helix was awarded in 1962, the honor went to James Watson, Francis Crick, and Maurice Wilkins. Rosalind Franklin was not listed as one of the recipients. Tragically, Franklin developed ovarian cancer in 1956, and she died in 1958 at the age of 37. The Nobel Prize is not awarded posthumously, and so Franklin was ineligible.

Discussion Questions

1. Based on Chargaff's rules, if a segment of DNA is composed of 20% adenine (A) bases, what is the percentage of guanine (G)?
2. Watson and Crick's discovery of DNA is clearly one of the most important biological discoveries in the last century. What advances in medicine and science can you think of that are built on knowing the structure of DNA?
3. Describe why the structure of DNA led Watson and Crick to point out "a possible copying mechanism for the genetic material."

a.

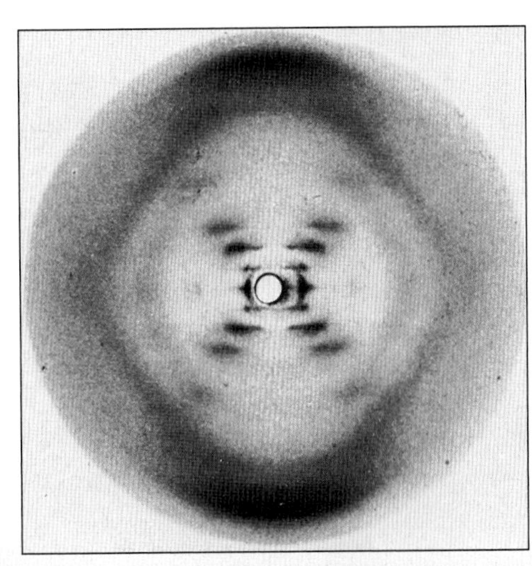

b.

Figure 25A X-ray diffraction of DNA.
a. Rosalind Franklin, 1920–1958. **b.** The diffraction pattern of DNA produced by Rosalind Franklin. The crossed (X) pattern in the center told investigators that DNA is a helix, and the dark portions at the top and the bottom told them that some feature is repeated over and over. Watson and Crick determined that this feature was the hydrogen-bonded bases.

Structure of DNA

The structure of DNA was determined by James Watson and Francis Crick in the early 1950s. The data they used and how they interpreted the data to deduce DNA's structure are reviewed in the Science Focus.

DNA is a chain of nucleotides. Each nucleotide is a complex of three subunits—phosphoric acid (phosphate), a pentose sugar (deoxyribose), and a nitrogen-containing base. There are four possible bases: two are **purines** with a double ring, and two are **pyrimidines** with a single ring. Adenine (A) and guanine (G) are purines; thymine (T) and cytosine (C) are pyrimidines.

A DNA polynucleotide *strand* has a backbone made up of alternating phosphate and sugar molecules. The bases are attached to the sugar but project to one side. DNA has two such strands, and the two strands twist about one another in the form of a **double helix** (Fig. 25.2*a*). The strands are held together by hydrogen bonding between the bases: A always pairs with T by forming two hydrogen bonds, and G always pairs with C by forming three hydrogen bonds. Notice that a purine is always bonded to a pyrimidine. This is called **complementary base pairing.** When the DNA helix unwinds, it resembles a ladder (Fig. 25.2*b*). The sides of the ladder are the phosphate-sugar backbones, and the rungs of the ladder are the complementary paired bases.

The two DNA strands are antiparallel—that is, they run in opposite directions, which you can verify by noticing that the sugar molecules are oriented differently. The carbon atoms in a sugar molecule are numbered, and the fifth carbon atom (5′) is uppermost in the strand on the left, while the third carbon atom (3′) is uppermost in the strand on the right (Fig. 25.2*c*).

Replication of DNA

When the body grows or heals itself, cells divide. Each new cell requires an exact copy of the DNA contained in the chromosomes. The process of copying one DNA double helix into two identical double helices is called **DNA replication.** Each original strand serves as a template for the formation of a complementary new strand. DNA replication is termed *semiconservative* because a new double helix has one conserved old strand and one new strand. Replication results in two DNA helices that are identical to each other and to the original molecule (Fig. 25.3).

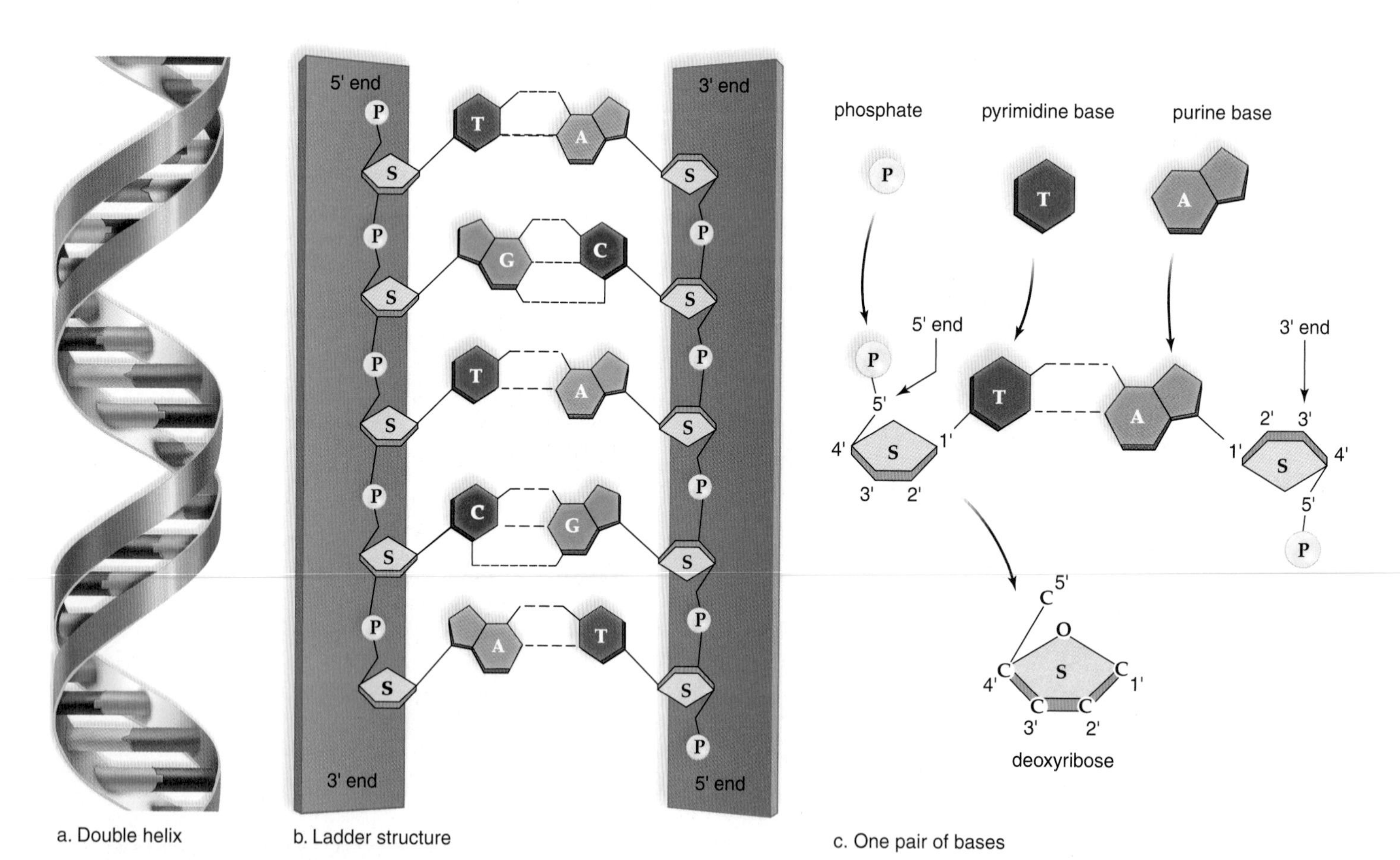

Figure 25.2 Overview of DNA structure.
a. DNA double helix. **b.** Unwinding the helix reveals a ladder configuration in which the uprights are composed of sugar and phosphate molecules and the rungs are complementary bases. The bases in DNA pair in such a way that the phosphate-sugar backbones are oriented in different directions. **c.** Notice that 3′ and 5′ are part of the system for numbering the carbon atoms that make up the sugar.

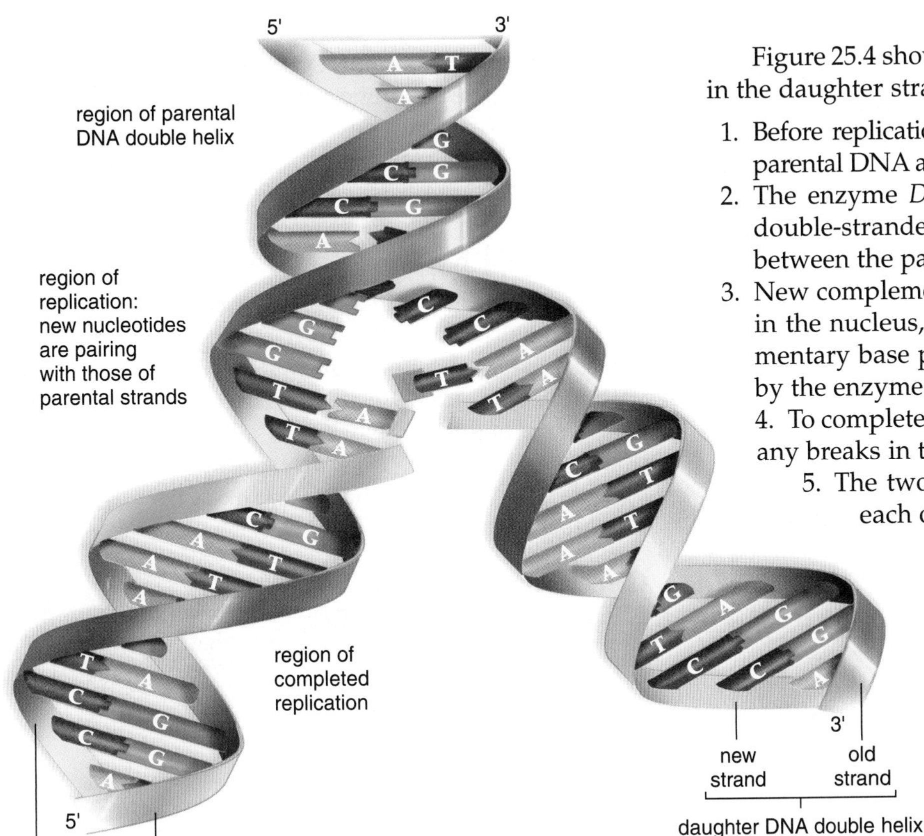

Figure 25.3 Overview of DNA replication.
Replication is called semiconservative because each new double helix is composed of an old (parental) strand and a new (daughter) strand.

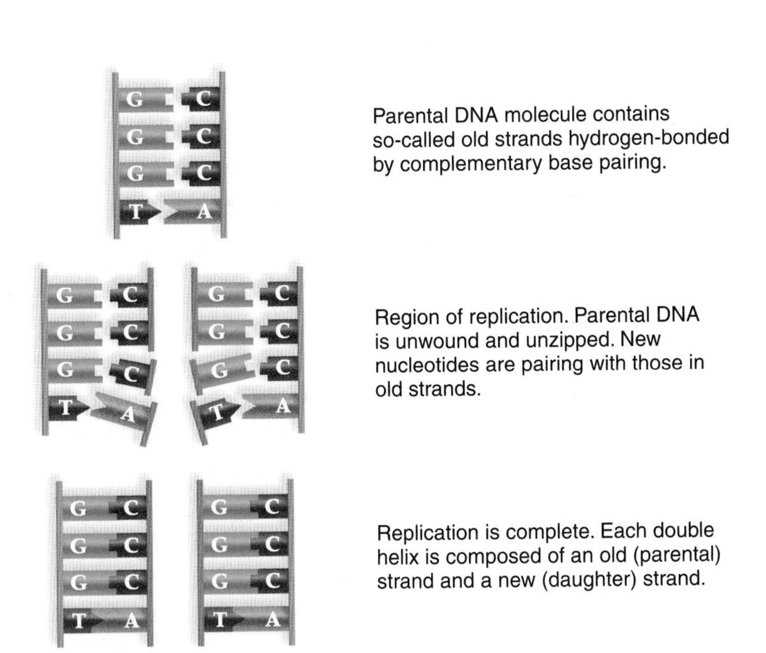

Figure 25.4 Ladder configuration and DNA replication.
The ladder configuration best illustrates how complementary nucleotides, available in the cell, pair with those of each old strand before they are joined together to form a daughter strand.

Figure 25.4 shows how complementary nucleotides pair in the daughter strand.

1. Before replication begins, the two strands that make up parental DNA are hydrogen-bonded to one another.
2. The enzyme *DNA helicase* unwinds and "unzips" the double-stranded DNA (i.e., the weak hydrogen bonds between the paired bases break).
3. New complementary DNA nucleotides, always present in the nucleus, fit into place by the process of complementary base pairing. These are positioned and joined by the enzyme *DNA polymerase.*
4. To complete replication, the enzyme *DNA ligase* seals any breaks in the sugar-phosphate backbone.
5. The two double helix molecules are identical to each other and to the original DNA molecule.

Chemotherapeutic drugs for cancer treatment stop replication and, therefore, cell division. Some chemotherapeutic drugs are analogs that have a similar, but not identical, structure to one of the four nucleotides in DNA. When these are mistakenly used by the cancer cells to synthesize DNA, replication stops, and the cancer cells die off.

Check Your Progress

1. At the completion of their experiment, where did Hershey and Chase find radioactive DNA and why was this significant?
2. Explain the "ladder" structure of DNA.
3. During DNA replication what process allows bases to be properly sequenced?

25.2 RNA Structure and Function

Learning Objectives

1. Describe the structure of RNA and tell how it differs from DNA.
2. Name the three major types of RNA and tell how they function in protein synthesis.

RNA (ribonucleic acid) is made up of nucleotides containing the sugar ribose, thus accounting for its name. The four nucleotides that make up an RNA molecule have the following bases: adenine (A), **uracil (U),** cytosine (C), and guanine (G). Notice that in RNA, the base uracil replaces the base thymine (Fig. 25.5).

RNA, unlike DNA, is single-stranded, but the single RNA strand sometimes doubles back on itself, allowing complementary base pairing to occur. Similarities and differences between these two nucleic acid molecules are listed in Table 25.1.

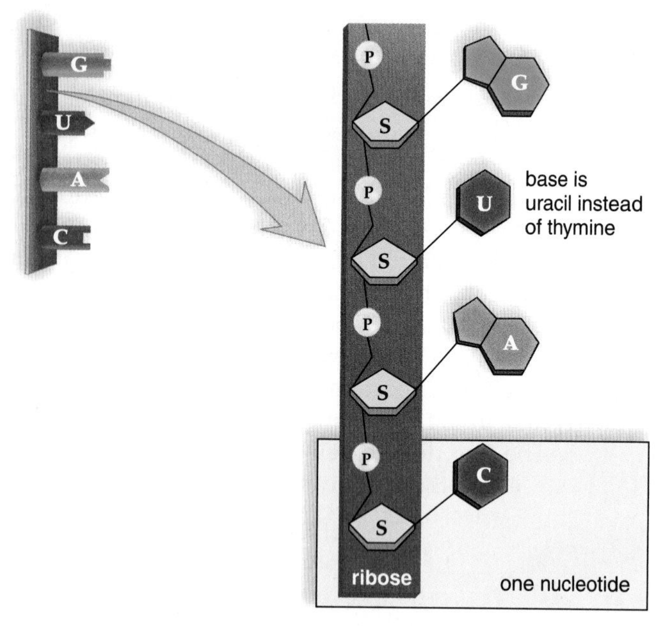

Figure 25.5 Structure of RNA.
Like DNA, RNA is a polymer of nucleotides. In an RNA nucleotide, the sugar ribose is attached to a phosphate molecule and to a base, either G, U, A, or C. Notice that in RNA, the base uracil replaces thymine as one of the pyrimidine bases. RNA is generally single stranded, whereas DNA is double stranded.

TABLE 25.1 Comparison of DNA and RNA

	DNA	RNA
Sugar	Deoxyribose	Ribose
Bases	Adenine, guanine, thymine, cytosine	Adenine, guanine, uracil, cytosine
Strands	Double stranded	Single stranded
Helix	Yes	No

In general, RNA is a helper to DNA, allowing protein synthesis to occur according to the genetic information that DNA provides. There are three major types of RNA, each with a specific function in protein synthesis.

Messenger RNA

Messenger RNA (mRNA) is produced in the nucleus, or in the nucleoid of prokaryotes. DNA serves as a template for its formation during a process called transcription. Which genes are transcribed is tightly controlled in each type of cell and accounts for why some cells are nerve cells and others are muscle cells, for example. Once formed mRNA carries genetic information from DNA to the ribosomes in the cytoplasm, where protein synthesis occurs.

Transfer RNA

Transfer RNA (tRNA) is also produced in the nucleus, and a portion of DNA also serves as a template for its production. Because DNA serves as a template for both rRNA and tRNA, it is obvious that not all DNA specifies protein synthesis.

Appropriate to its name, tRNA transfers amino acids to the ribosomes, where the amino acids are joined, forming a protein. There are 20 different types of amino acids in proteins. Therefore, at least 20 different tRNAs must be functioning in the cell. Each type of tRNA carries only one type of amino acid.

Ribosomal RNA

In eukaryotic cells, **ribosomal RNA (rRNA)** is produced in the nucleolus of a nucleus, where a portion of DNA also serves as a template for its formation. Ribosomal RNA joins with proteins made in the cytoplasm to form the subunits of ribosomes, one large and one small. Each subunit has its own mix of proteins and rRNA. The subunits leave the nucleus and come together in the cytoplasm when protein synthesis is about to begin.

Check Your Progress

1. Contrast the structure of RNA with that of DNA.
2. Contrast the function of messenger RNA with that of transfer RNA.

25.3 Gene Expression

Learning Outcomes

1. State and explain the two steps of gene expression.
2. Use messenger RNA codons to translate a mRNA into a sequence of amino acids.
3. Describe mRNA processing.

The process of using a gene sequence to synthesize a protein is called gene expression. Specifically, gene expression requires two processes called transcription and translation. In eukaryotes, transcription takes place in the nucleus and translation takes place in the cytoplasm. During **transcription,** a portion of DNA serves as a template for mRNA formation. During **translation,** the sequence of mRNA bases (which are complementary to those in the template DNA) determines the sequence of amino acids in a polypeptide. So, in effect, genetic information lies in the sequence of the bases in DNA, which through mRNA determines the sequence of amino acids in a protein. Transfer RNA assists mRNA during protein synthesis by bringing amino acids to the ribosomes. Proteins differ from one another by the sequence of their amino acids, and proteins determine the structure and function of cells and the phenotype of the organism.

Transcription

During transcription, a segment of the DNA called a **gene** serves as a template for the production of an RNA molecule. Previously, molecular genetics considered a gene to be a nucleic acid sequence that codes for the sequence of amino acids in a protein. In contrast to this definition, we have known for some time all three types of RNA are transcribed from DNA and that these RNAs are useful products. We also know that protein-coding regions can be interrupted by regions that do not code for a protein. In recognition of these new findings, Mark Gerstein and associates in 2007 suggested a new definition for a gene: "A gene is a genomic sequence (either DNA or RNA) directly encoding functional products, either RNA or protein." Although all three classes of RNA are formed by transcription, we will focus on transcription to form messenger RNA (mRNA), the first step in protein synthesis.

Messenger RNA

Transcription begins when the enzyme **RNA polymerase** binds tightly to a **promoter,** a region of DNA that contains a special sequence of nucleotides. This enzyme opens up the DNA helix just in front of it so that complementary base pairing can occur in the same way as in DNA replication. Then, RNA polymerase joins the RNA nucleotides, and an mRNA molecule results. When mRNA forms, it has a sequence of bases complementary to that of the DNA; wherever A, T, G, or C are present in the DNA template, U, A, C, or G, respectively, are incorporated into the mRNA molecule (Fig. 25.6). Now, mRNA is a faithful copy of the sequence of bases in DNA.

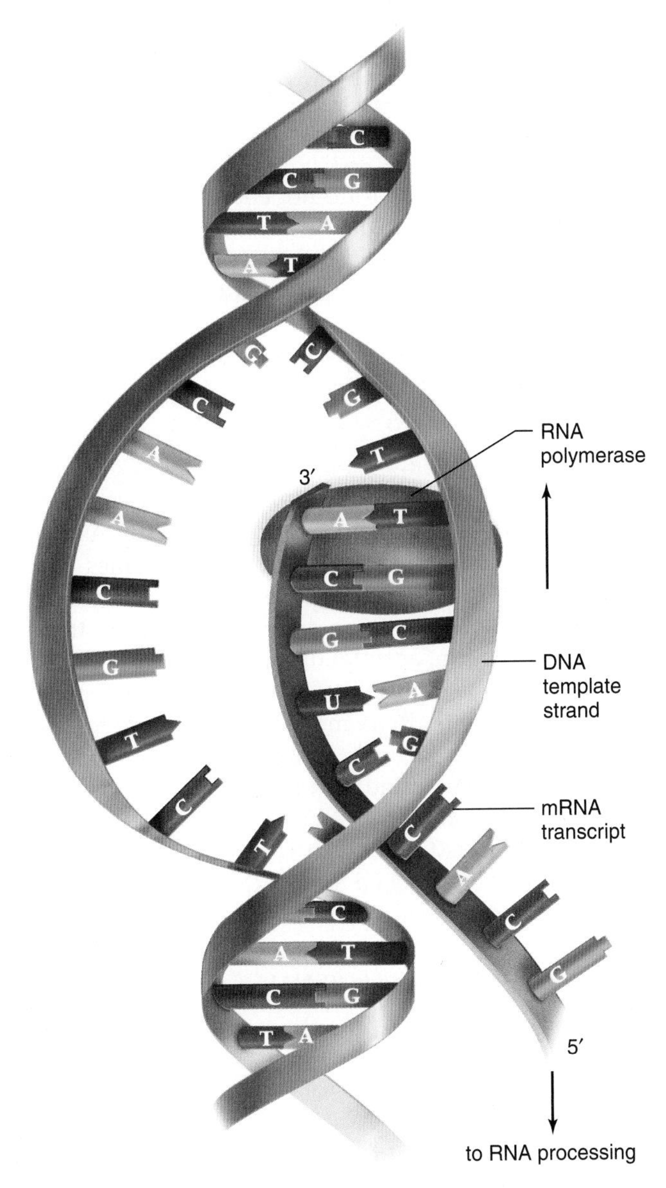

Figure 25.6 Transcription of DNA to form mRNA.
During transcription, complementary RNA is made from a DNA template. At the point of attachment of RNA polymerase, the DNA helix unwinds and unzips, and complementary RNA nucleotides are joined together. After RNA polymerase has passed by, the DNA strands rejoin and the mRNA transcript is released.

Processing of mRNA After the mRNA is transcribed in eukaryotic cells, it must be processed before entering the cytoplasm.

The newly synthesized **primary mRNA** molecule becomes a *mature mRNA* molecule after processing. Most genes in humans are interrupted by segments of DNA that are not part of the gene. These portions are called **introns** because they are intragene segments. The other portions of the gene are called **exons** because they are ultimately expressed. Only exons result in a protein product.

Primary mRNA contains bases that are complementary to both exons and introns, but during processing, (1) one end of the mRNA is modified by the addition of a cap, composed of an altered guanine nucleotide, and the other end is modified by the addition of a poly-A tail, a series of adenosine nucleotides. (2) The introns are removed, and the exons are joined to form a mature mRNA molecule consisting of continuous exons (Fig. 25.7). Ordinarily, processing brings together all the exons of a gene. In some instances, cells use only certain exons to form a mature RNA transcript. The result can be a different protein product in each cell. Alternate mRNA splicing may account for white blood cells' ability to produce a specific antibody for each type of bacteria and virus we encounter on a daily basis.

Translation

Translation is the second process by which gene expression leads to protein synthesis. Translation requires several enzymes, mRNA, and two other types of RNA: transfer RNA and ribosomal RNA.

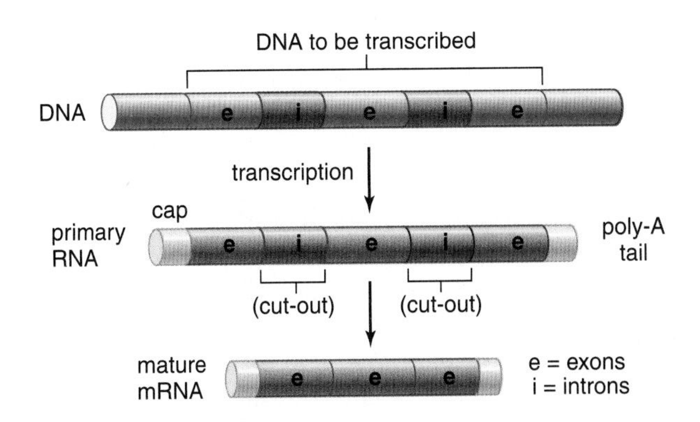

Figure 25.7 mRNA processing.
During processing, a cap and tail are added to mRNA, and the introns (i) are removed so that only exons (e) remain.

First Base	Second Base				Third Base
	U	C	A	G	
U	UUU phenylalanine	UCU serine	UAU tyrosine	UGU cysteine	U
	UUC phenylalanine	UCC serine	UAC tyrosine	UGC cysteine	C
	UUA leucine	UCA serine	UAA *stop*	UGA *stop*	A
	UUG leucine	UCG serine	UAG *stop*	UGG tryptophan	G
C	CUU leucine	CCU proline	CAU histidine	CGU arginine	U
	CUC leucine	CCC proline	CAC histidine	CGC arginine	C
	CUA leucine	CCA proline	CAA glutamine	CGA arginine	A
	CUG leucine	CCG proline	CAG glutamine	CGG arginine	G
A	AUU isoleucine	ACU threonine	AAU asparagine	AGU serine	U
	AUC isoleucine	ACC threonine	AAC asparagine	AGC serine	C
	AUA isoleucine	ACA threonine	AAA lysine	AGA arginine	A
	AUG (*start*) methionine	ACG threonine	AAG lysine	AGG arginine	G
G	GUU valine	GCU alanine	GAU aspartate	GGU glycine	U
	GUC valine	GCC alanine	GAC aspartate	GGC glycine	C
	GUA valine	GCA alanine	GAA glutamate	GGA glycine	A
	GUG valine	GCG alanine	GAG glutamate	GGG glycine	G

Figure 25.8 Messenger RNA codons.
Notice that in this chart, each of the codons (in boxes) is composed of three letters representing the first base, second base, and third base. For example, find the box where C for the first base and A for the second base intersect. You will see that U, C, A, or G can be the third base. The bases CAU and CAC are codons for histidine; the bases CAA and CAG are codons for glutamine.

The Genetic Code

The sequence of bases in DNA is transcribed into mRNA, which ultimately codes for a particular sequence of amino acids to form a polypeptide. Can four mRNA bases (A, C, G, U) provide enough combinations to code for 20 amino acids? If only one base stood for an amino acid (i.e., a "singlet code"), then only four amino acids would be possible. If two bases stood for one amino acid, there would only be 16 possible combinations (4×4). If the code is a triplet, then there are 64 possible triplets of the four bases ($4 \times 4 \times 4$). It should come as no surprise then to learn that the genetic code is a **triplet code** (each triplet is also called a **codon**) and that this code is *redundant.* That is, notice in Figure 25.8 that most amino acids are coded for by more than one codon. For example, leucine has six codons and serine has four codons. This redundancy offers some protection against possibly harmful mutations that change the sequence of bases. Of the 64 codons, 61 code for amino acids, while the remaining three are *stop codons* (UAA, UGA, UAG), codons that do not code for amino acids but instead signal polypeptide termination.

The genetic code is just about universal in all living things. This means that a codon in a fruit fly codes for the same amino acid as in a bird, a fern, or a human. It also suggests that the code dates back to the very first organisms on Earth and that all living things are related.

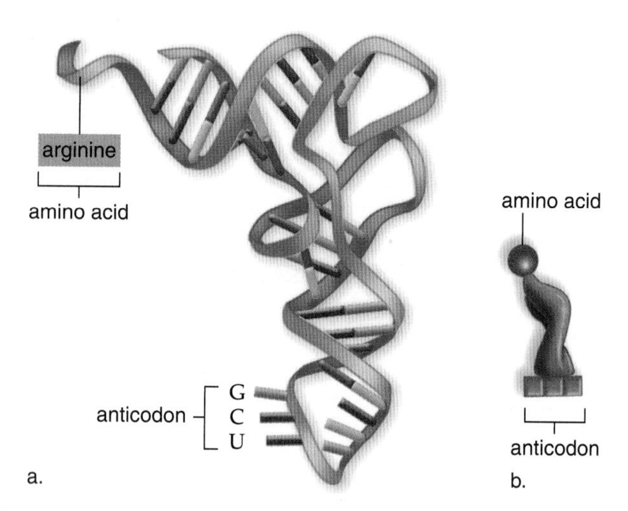

Figure 25.9 Transfer RNA: amino acid carrier.
a. A tRNA is a polynucleotide that folds into a bootlike shape because of complementary base pairing. At one end of the molecule is its specific anticodon—in this case, GCU (which hybridizes to the codon CGA). At the other end, an amino acid attaches that corresponds to this anticodon—in this case, arginine. **b.** tRNA is represented like this in the illustrations that follow.

Transfer RNA

Transfer RNA (tRNA) molecules bring amino acids to the ribosomes, the site of protein synthesis. Each tRNA molecule is a single-stranded polynucleotide that doubles back on itself such that complementary base pairing creates a bootlike shape. On one end is an amino acid, and on the other end is an **anticodon,** a triplet of three bases complementary to a specific codon of mRNA (Fig. 25.9). Because there are 64 codons, at least one tRNA molecule is possible for each of the 20 amino acids found in proteins.

When a tRNA–amino acid complex comes to the ribosome, its anticodon pairs with an mRNA codon. For example, if the codon is CGG, what is the anticodon, and what amino acid will be attached to the tRNA molecule? Based on Figure 25.8, the answer to this question is as follows:

Codon (mRNA)	Anticodon (tRNA)	Amino Acid (protein)
CGG	GCC	Arginine

The order of the codons of the mRNA determines the order that tRNA–amino acids come to a ribosome and, therefore, the final sequence of amino acids in a protein.

Figure 25.10 shows gene expression that results in a protein product. During transcription, the base sequence in DNA is copied into a sequence of bases in mRNA. During translation, tRNAs bring amino acids to the ribosomes in the order dictated by the base sequence of mRNA. The sequence of amino acids forms a polypeptide chain, or complete protein.

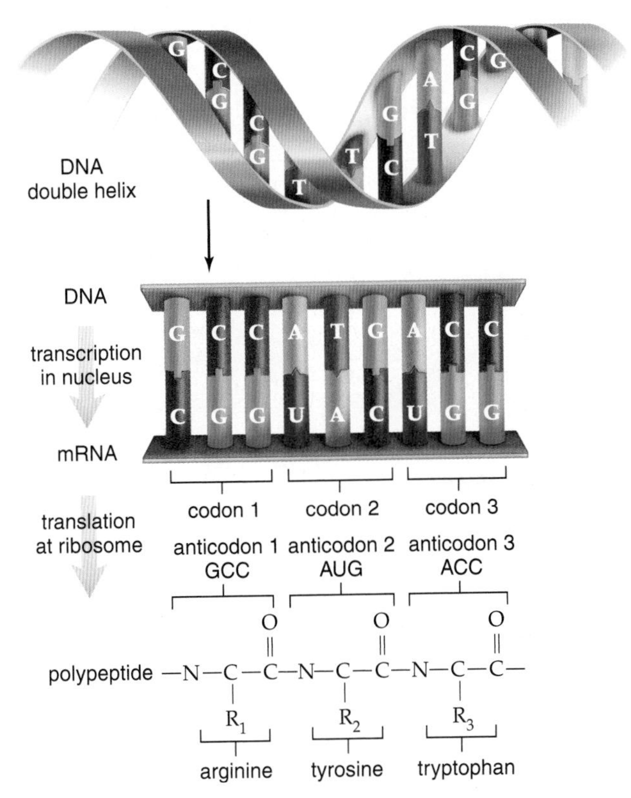

Figure 25.10 Overview of gene expression. One strand of DNA acts as a template for mRNA synthesis, and the sequence of bases in mRNA determines the sequence of amino acids in a polypeptide.

Ribosomes and Ribosomal RNA

Ribosomes are small structural bodies found in the cytoplasm and on the endoplasmic reticulum where translation occurs. Ribosomes are composed of many proteins and several ribosomal RNAs (rRNAs). In eukaryotic cells, rRNA is produced in a nucleolus within the nucleus. Then the rRNA joins with proteins manufactured in and imported from the cytoplasm to form two ribosomal subunits, one large and one small. The subunits leave the nucleus and join together in the cytoplasm to form a ribosome just as protein synthesis begins.

A ribosome has a binding site for mRNA as well as binding sites for three tRNA molecules. These binding sites facilitate complementary base pairing between tRNA anticodons and mRNA codons. As the ribosome moves down the mRNA molecule, new tRNAs arrive, and a polypeptide forms and grows longer. Translation terminates once the polypeptide is fully formed and an mRNA stop codon is reached. The ribosome then dissociates into its two subunits and falls off the mRNA molecule.

As soon as the initial portion of mRNA has been translated by one ribosome and the ribosome has begun to move down the mRNA, another ribosome attaches to the same mRNA. Therefore, several ribosomes are often attached to and translating a single mRNA, thus forming several copies of a polypeptide simultaneously. The entire complex is called a **polyribosome** (Fig. 25.11).

CHECK YOUR PROGRESS

1. Describe how transcription of DNA results in an mRNA.
2. If the bases in the DNA are CGGAAA, what are the mRNA codons, the tRNA anticodons, and the resulting amino acids in the dipeptide?
3. What are the two steps required for processing of an mRNA?

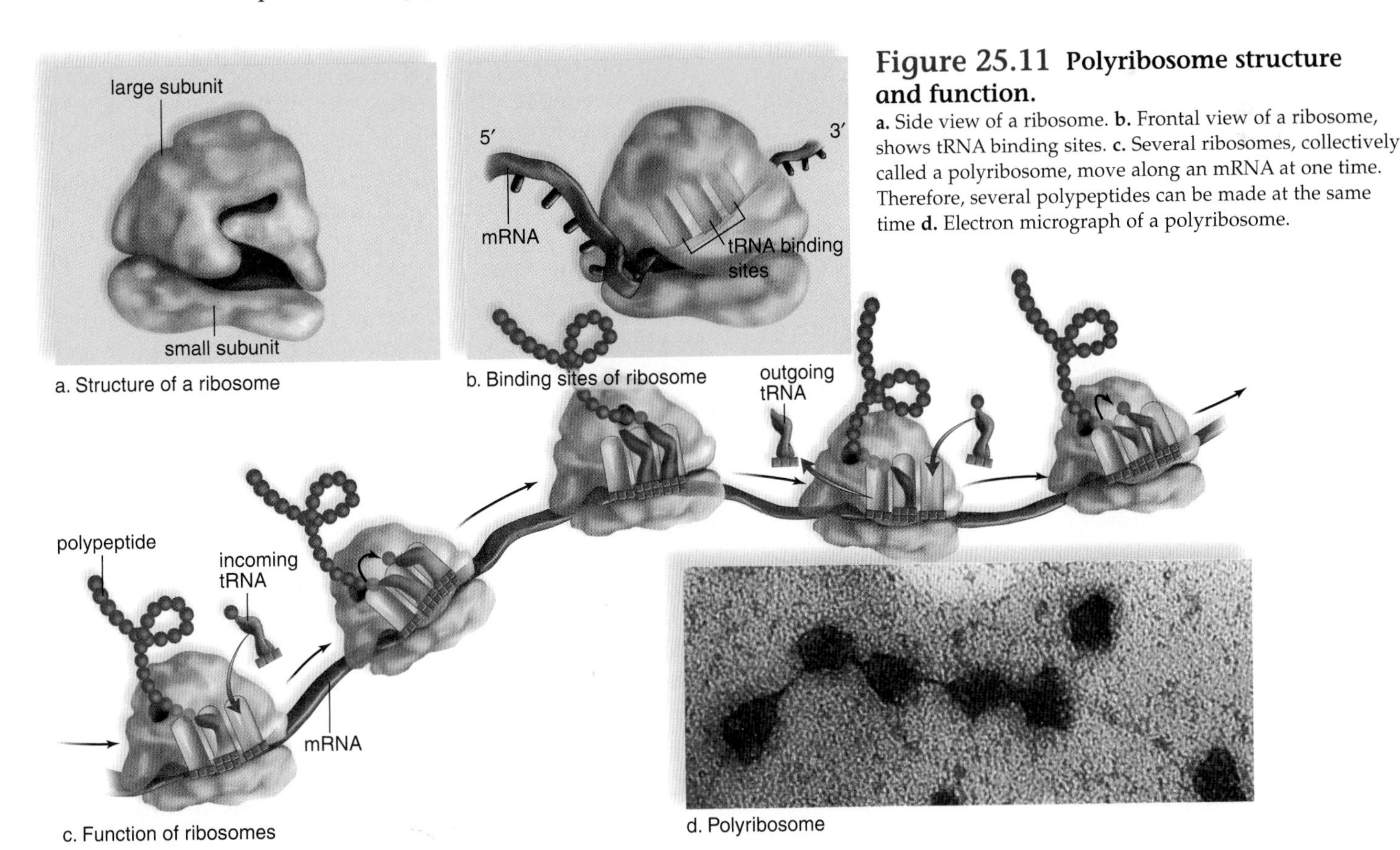

Figure 25.11 Polyribosome structure and function. **a.** Side view of a ribosome. **b.** Frontal view of a ribosome, shows tRNA binding sites. **c.** Several ribosomes, collectively called a polyribosome, move along an mRNA at one time. Therefore, several polypeptides can be made at the same time **d.** Electron micrograph of a polyribosome.

Translation Requires Three Steps

During translation, the codons of an mRNA base pair with the anticodons of tRNA molecules carrying specific amino acids. The order of the codons determines the order of the tRNA molecules at a ribosome and the sequence of amino acids in a polypeptide. The process of translation must be extremely orderly so that the amino acids of a polypeptide are sequenced correctly.

Protein synthesis involves three steps: initiation, elongation, and termination. Enzymes are required for each of the three steps to function properly. The first two steps, initiation and elongation, require energy.

Initiation

Initiation is the step that brings all the translation components together. Proteins called initiation factors are required to assemble the small ribosomal subunit, mRNA, initiator tRNA, and the large ribosomal subunit for the start of protein synthesis.

Initiation is shown in Figure 25.12. In prokaryotes, a small ribosomal subunit attaches to the mRNA in the vicinity of the *start codon* (AUG). The first or initiator tRNA pairs with this codon because its anticodon is UAC. Then, a large ribosomal subunit joins to the small subunit (Fig. 25.12). Although similar in many ways, initiation in eukaryotes is much more complicated and will not be discussed here.

A ribosome has three binding sites for tRNAs. One of these is called the P (for peptide) site, and the other is the A (for amino acid) site. The tRNA exits at the E site. The initiator tRNA happens to be capable of binding to the P site, even though it carries only the amino acid methionine (see Fig. 25.8). The A site is for tRNA carrying the next amino acid.

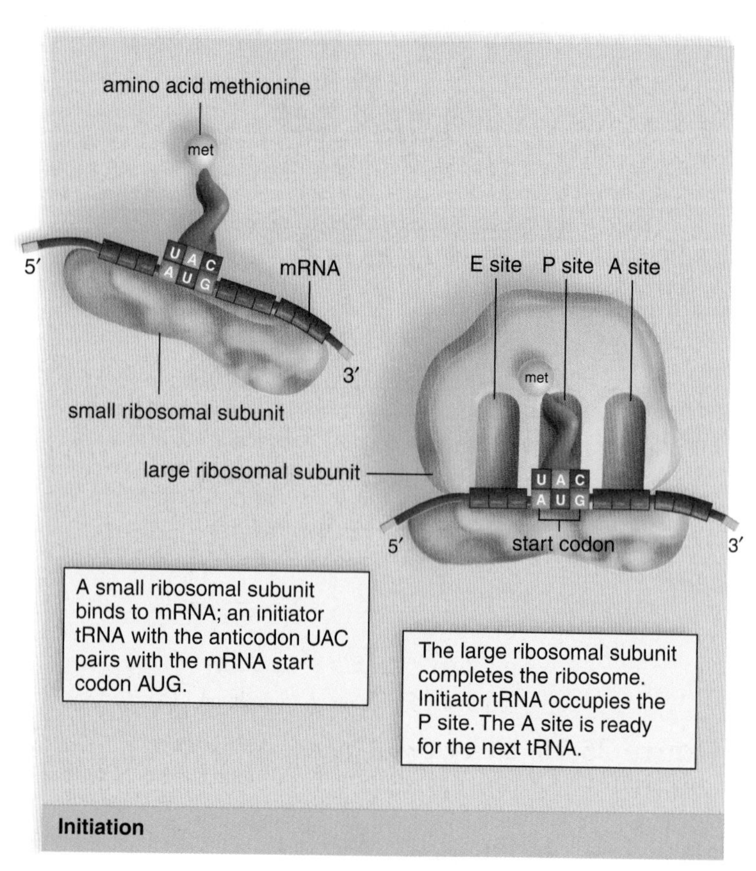

Figure 25.12 Initiation.
During initiation, participants in the translation process assemble as shown. The start codon, AUG, also codes for the first amino acid methionine.

Elongation

Elongation is the protein synthesis step in which a polypeptide increases in length one amino acid at a time. In addition to the participation of tRNAs, elongation requires elongation factors, which facilitate the binding of tRNA anticodons to mRNA codons at a ribosome.

Elongation is shown in Figure 25.13 as a series of four steps:

1. A tRNA with an attached peptide is already at the P site, and a tRNA carrying the next amino acid in the chain is just arriving at the A site.
2. Once the next tRNA is in place at the A site, the peptide chain will be transferred to this tRNA.
3. Energy and part of the ribosomal subunit are needed to bring about this transfer. The energy contributes to peptide bond formation, which makes the peptide one amino acid longer by adding the peptide from the A site.
4. Next, translocation occurs—the mRNA moves forward one codon length, and the peptide-bearing tRNA is now at the ribosome P site. The "spent" tRNA now exits. The new codon is at the A site and is ready to receive the next complementary tRNA.

The complete cycle in steps 1–4 is repeated at a rapid rate (e.g., about 15 times each second in *E. coli*).

Termination

Termination is the final step in protein synthesis. During termination, as shown in Figure 25.14, the polypeptide and the assembled components that carried out protein synthesis are separated from one another.

Termination of polypeptide synthesis occurs at a stop codon. Termination requires a protein called a release factor, which cleaves the polypeptide from the last tRNA. After this occurs, the polypeptide is set free and begins to take on its three-dimensional shape. The ribosome dissociates into its two subunits.

Properly functioning proteins are of paramount importance to the cell and to the organism. For example, if an organism inherits a faulty gene, the result can be a genetic disorder (such as Huntington disease) caused by a malfunctioning protein or a propensity toward cancer. Proteins are the link between genotype and phenotype. The DNA sequence underlying these proteins distinguishes

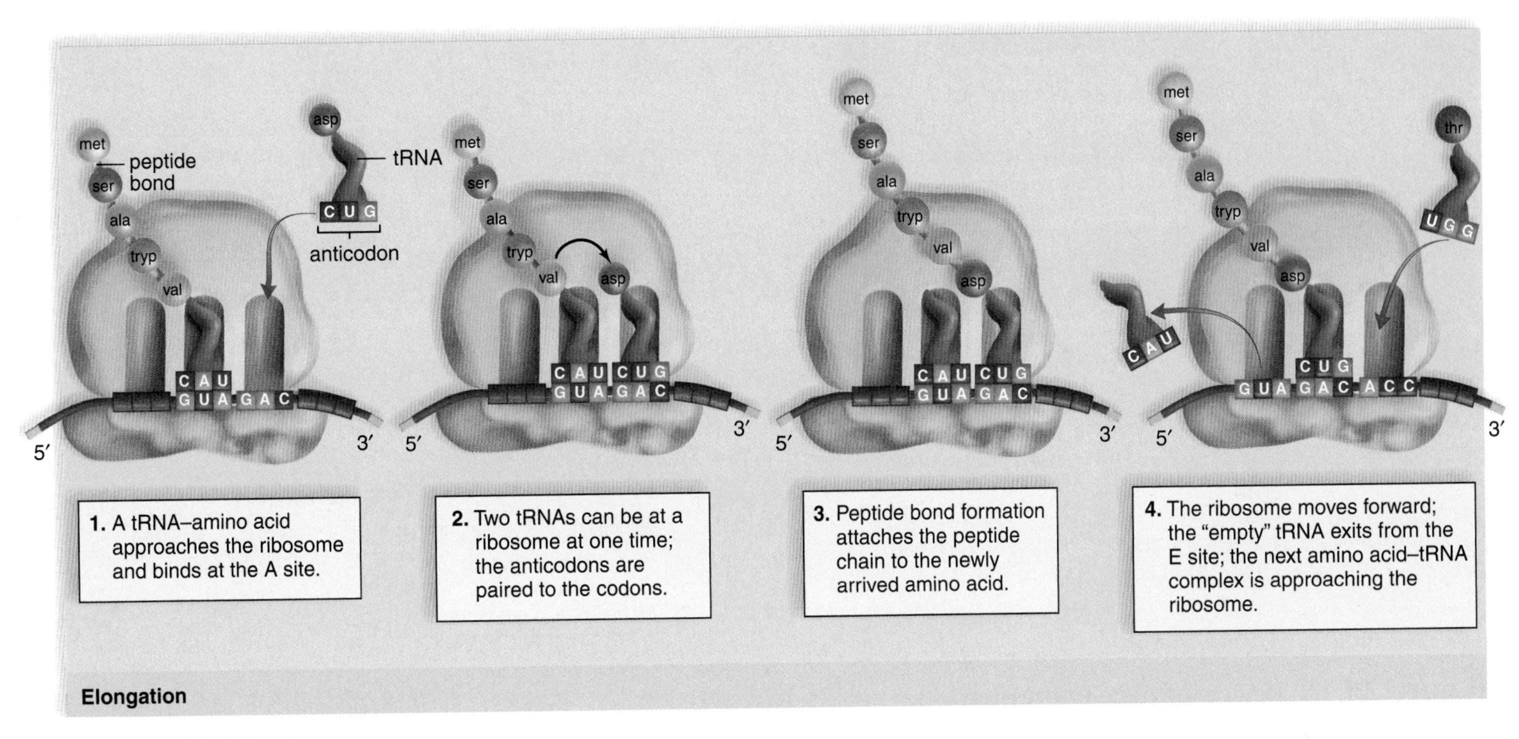

Figure 25.13 Elongation.
Note that a polypeptide is already at the P site. During elongation, polypeptide synthesis occurs as amino acids are added one at a time to the growing chain.

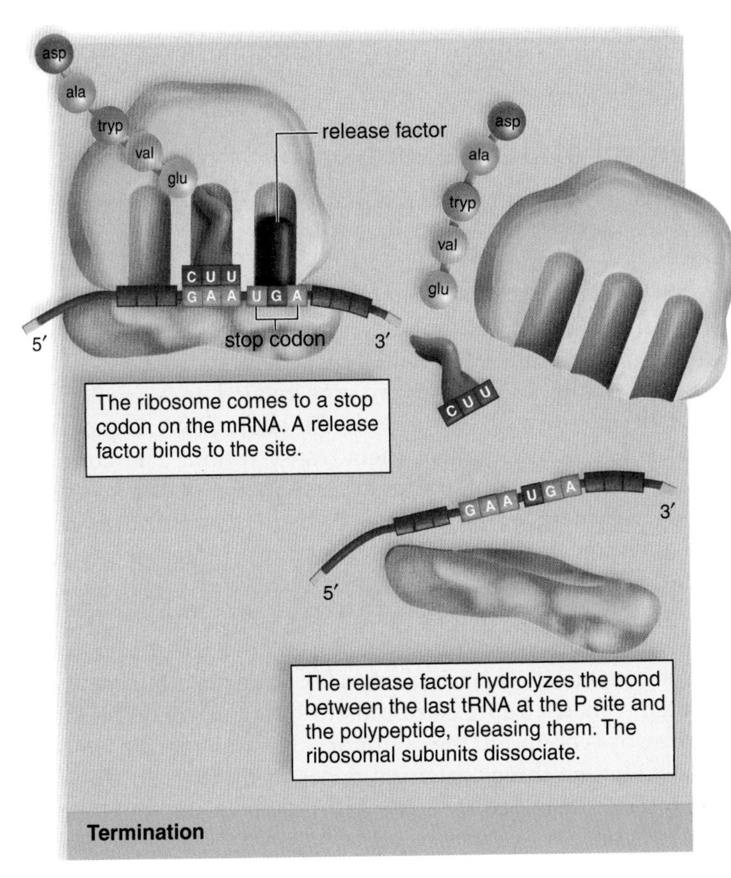

Figure 25.14 Termination.
During termination, the finished polypeptide is released, as are the mRNA and the last tRNA.

different types of organisms. In addition to accounting for the difference between cell types, proteins account for the differences between organisms.

Review of Gene Expression

A gene is expressed when its protein product has been synthesized. Protein synthesis requires the process of transcription and translation (Fig. 25.15). During transcription, a segment of a DNA strand serves as a template for the formation of messenger RNA (mRNA). The bases in mRNA are complementary to those in DNA. Every three mRNA bases is a *codon* (a triplet code) for a certain amino acid. Messenger RNA is processed before it leaves the nucleus, during which time the introns are removed and the ends are modified. Messenger RNA carries a sequence of codons to the *ribosomes*. During translation, tRNAs bring attached amino acids to the ribosomes. Because tRNA anticodons pair with codons, the amino acids become sequenced in the order originally specified by DNA. The genes we receive from our parents determine the proteins in our cells and these proteins are responsible for our inherited traits!

Check Your Progress

1. What is the start codon for the initiation of translation? What amino acid is on the tRNA that pairs with this codon?
2. Explain the process of translocation.
3. What are the stop codons? Do any tRNAs pair with these codons?

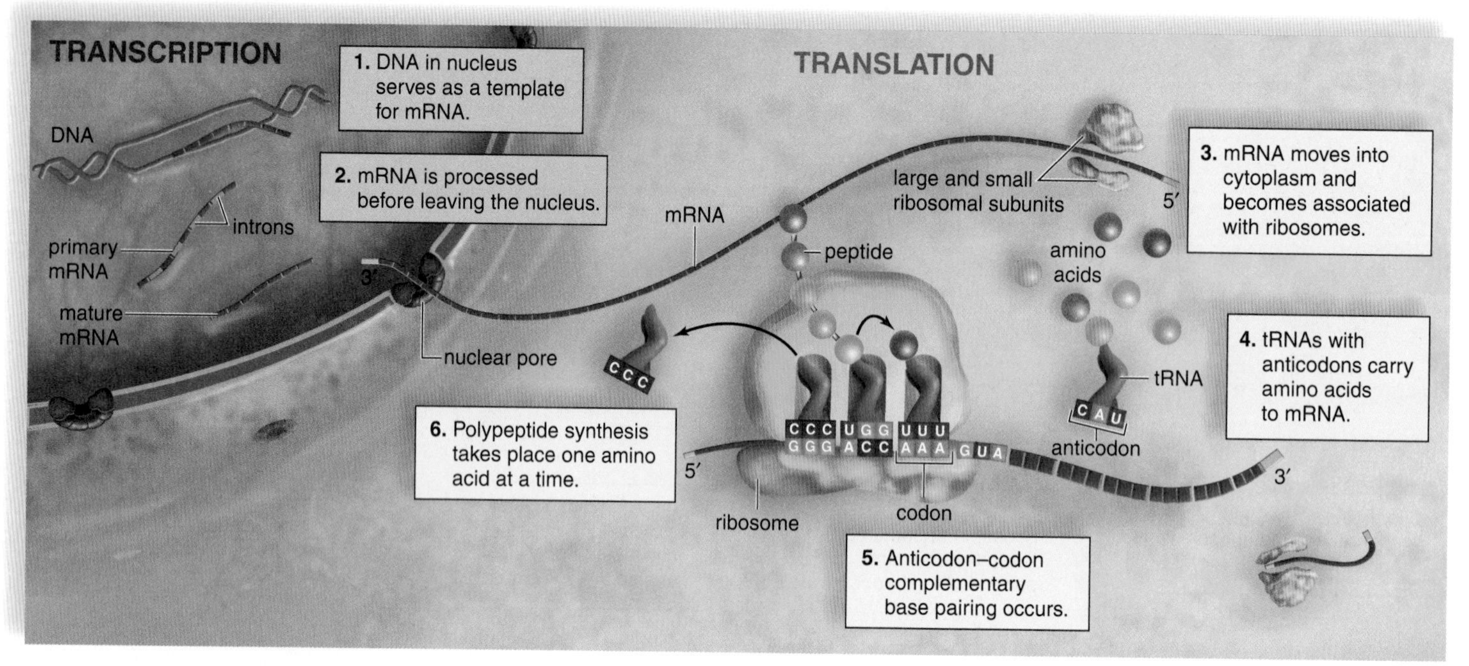

Figure 25.15 Review of gene expression.
Messenger RNA is produced and processed in the nucleus during transcription, and protein synthesis occurs in the cytoplasm at the ribosomes during translation.

25.4 Control of Gene Expression

Learning Outcomes

1. Understand the relationship between gene regulation and gene activity in a cell.
2. Compare regulation in prokaryotes to regulation in eukaryotes.
3. Discuss the many ways genes are regulated in eukaryotes.

The human body contains many types of cells that differ in structure and function. Each cell type must contain its own mix of proteins that make it different from all other cell types. Therefore, only certain genes are active in cells that perform specialized functions, such as nerve, muscle, gland, and blood cells.

Some of these active genes are called housekeeping genes because they govern functions that are common to many types of cells, such as glucose metabolism. But otherwise, the activity of selected genes accounts for the specialization of cells:

Cell type Red blood Muscle Pancreatic

Gene type
Housekeeping
Hemoglobin
Insulin
Myosin

In other words, gene expression is controlled in a cell, and this control accounts for its specialization. Let's begin by examining a simpler system—the control of transcription in prokaryotes.

Control of Gene Expression in Prokaryotes

The bacterium *Escherichia coli* that lives in your intestine can use various sugars as a source of energy and carbon. This organism can quickly adjust its gene expression to match your diet. The enzymes that are needed to break down lactose, the sugar present in milk, are found encoded together in an operon in the bacterial DNA. An **operon** is a cluster of genes usually coding for proteins related to a particular metabolic pathway, along with the short DNA sequences that coordinately control their transcription. The control sequences consist of a promoter, a sequence of DNA where RNA polymerase first attaches to begin transcription, and an operator, a sequence of DNA where a repressor protein binds (Fig. 25.16).

In the *lac* operon, the structural genes for three enzymes that are needed for lactose metabolism are under the control of one promoter/operator complex. If lactose is absent, a protein called a **repressor** binds to the operator. When the repressor is bound to the operator, RNA polymerase cannot transcribe the three structural genes of the operon. The *lac* repressor is encoded by a **regulator gene** located outside of the operon.

When lactose is present, the lactose binds with the *lac* repressor so that the repressor is unable to bind to the operator. Then RNA polymerase is able to transcribe the structural genes into a single mRNA, which is then translated into the three

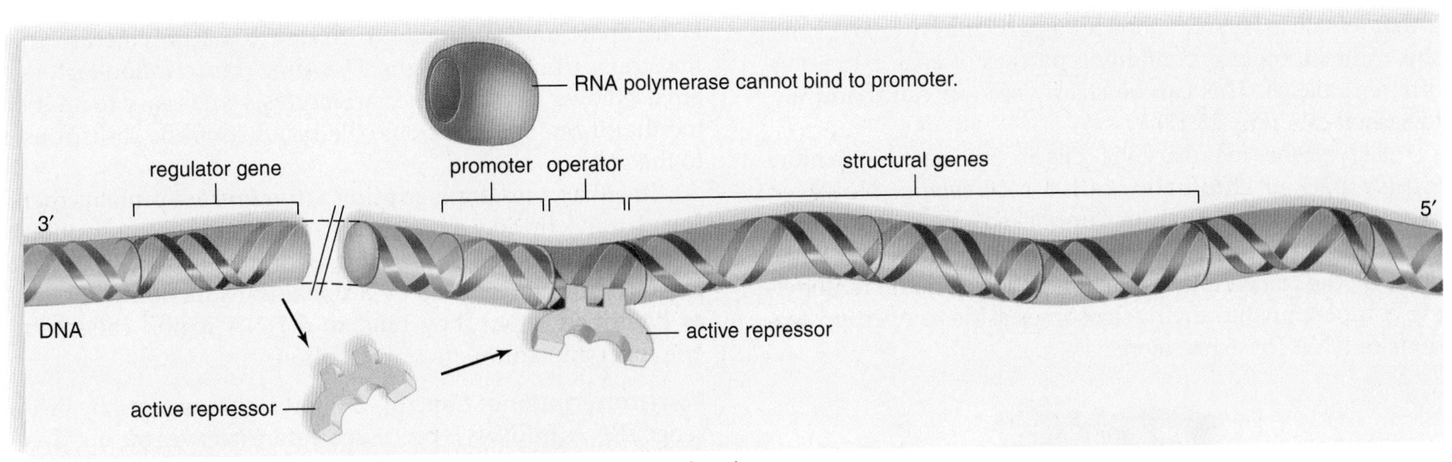

a. **Lactose absent.** Enzymes needed to metabolize lactose are not produced.

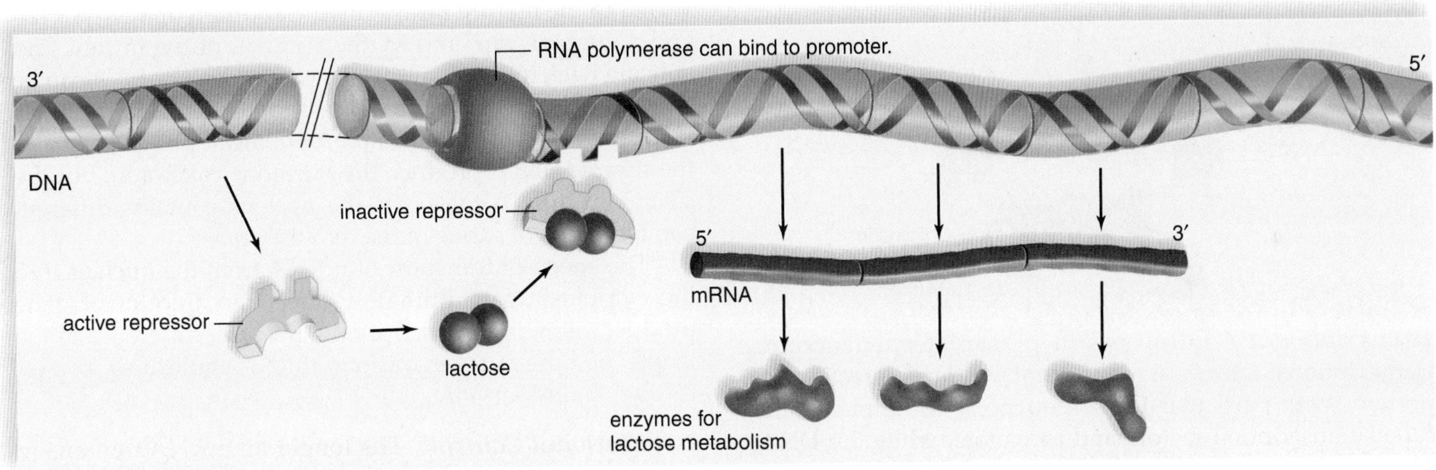

b. **Lactose present.** Enzymes needed to metabolize lactose are produced only when lactose is present.

Figure 25.16 The *lac* operon.
a. When lactose is absent, the regulator gene codes for a repressor that is normally active. When it binds to the operator, RNA polymerase cannot attach to the promoter, and structural genes are not expressed. **b.** When lactose is present, it binds to the repressor, changing its shape so that it is inactive and cannot bind to the operator. Now, RNA polymerase binds to the promoter, and the structural genes are expressed.

different enzymes. In this way, the enzymes needed to break down lactose are only synthesized when lactose is present.

The *lac* operon is considered an inducible operon. It is only activated when lactose induces its expression. Other bacterial operons are repressible operons. They are usually active until a repressor turns them off.

Control of Gene Expression in Eukaryotes

In bacteria, a single promoter serves several genes that make up a transcription unit, while in eukaryotes, each gene has its own promoter where RNA polymerase binds. Bacteria rely mostly on transcriptional control, but eukaryotes employ a variety of mechanisms to regulate gene expression. These mechanisms affect whether the gene is expressed, the speed with which it is expressed, and how long it is expressed.

Levels of Gene Control

Eukaryotic genes exhibit control of gene expression at five different levels: (1) pretranscriptional control, (2) transcriptional control, (3) posttranscriptional control, (4) translational control, and (5) posttranslational control.

Pretranscriptional Control Eukaryotes use DNA methylation and chromatin packing as a way to keep genes turned off. Genes within darkly staining, highly condensed portions of chromatin, called *heterochromatin*, are inactive. A dramatic example of this occurs with the X chromosome in mammalian females. Females have two X chromosomes, while males have only one X chromosome. This inequality is balanced by the inactivation of one X chromosome in every cell of the female body. Each inactivated X chromosome, called a Barr body in honor of its discoverer, can be seen as a small, darkly staining mass of condensed chromatin along the inner edge of the nuclear envelope. During early prenatal development, one or the other X chromosome in a cell is randomly shut off by chromatin condensation into heterochromatin. All of the cells that form from division of that cell will have the same X chromosome inactivated. Therefore, females have patches of tissue that differ in which X chromosome is being expressed.

If that female is heterozygous for a particular X-linked gene, she will be a mosaic, containing patches of cells expressing different alleles. This can be clearly seen in calico and tortoiseshell cats (Fig. 25.17).

Active genes in eukaryotic cells are associated with more loosely packed chromatin, called *euchromatin*. However, even euchromatin must be "unpacked" before it can be transcribed. The presence of nucleosomes limits access to the DNA by the transcription machinery. A chromatin remodeling complex pushes the nucleosomes aside to open up sections of DNA for expression:

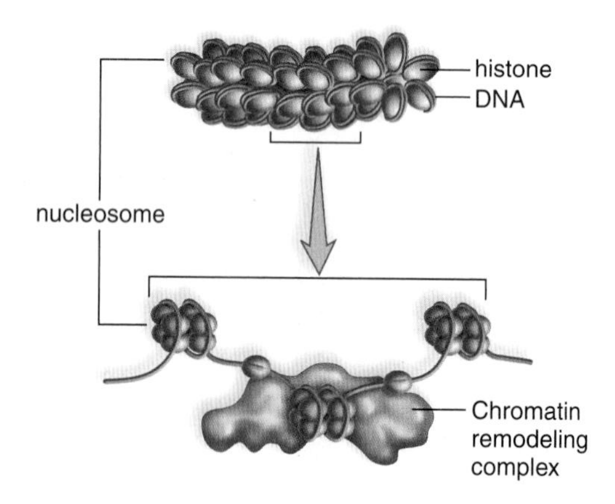

Transcriptional Control As in prokaryotes, eukaryotic transcriptional control is dependent on the interaction of proteins with particular DNA sequences. The proteins are called transcription factors and activators, while the DNA sequences are called **enhancers** and promoters. In eukaryotes, each gene has its own promoter. **Transcription factors** are proteins that help RNA polymerase bind to a promoter. Several transcription factors per gene form a *transcription initiation complex* that also helps pull double-stranded DNA apart and even acts to release RNA polymerase so that transcription can begin. The same transcription factors are used over again at other promoters, so it is easy to imagine that if one malfunctions, the result could be disastrous to the cell.

In eukaryotes, **transcription activators** are proteins that speed transcription dramatically. In general, transcription activators are themselves activated in response to a signal received and transmitted by a signal transduction pathway as described later. They bind to a DNA region called an **enhancer** and stimulate transcription.

Posttranscriptional Control Following transcription, messenger RNA (mRNA) is processed before it leaves the nucleus and passes into the cytoplasm. The primary mRNA is converted to the mature mRNA by the addition of a poly-A tail and a guanine cap, and by the removal of the introns and splicing back together of the exons. The same mRNA can be spliced in different ways to make slightly different products in different tissues. For example, both the hypothalamus and the thyroid gland produce the hormone calcitonin, but the calcitonin mRNA that exits the nucleus contains different combinations of exons in the two tissues.

The speed of transport of mRNA from the nucleus into the cytoplasm can ultimately affect the amount of gene product following transcription. There is a difference in the length of time it takes various mRNA molecules to pass through a nuclear pore.

Translational Control The longer an mRNA remains in the cytoplasm before it is broken down, the more gene product can be translated. Differences in the poly-A tail or the guanine cap can determine how long a particular transcript remains active before it is destroyed by a ribonuclease associated with ribosomes. Hormones can cause

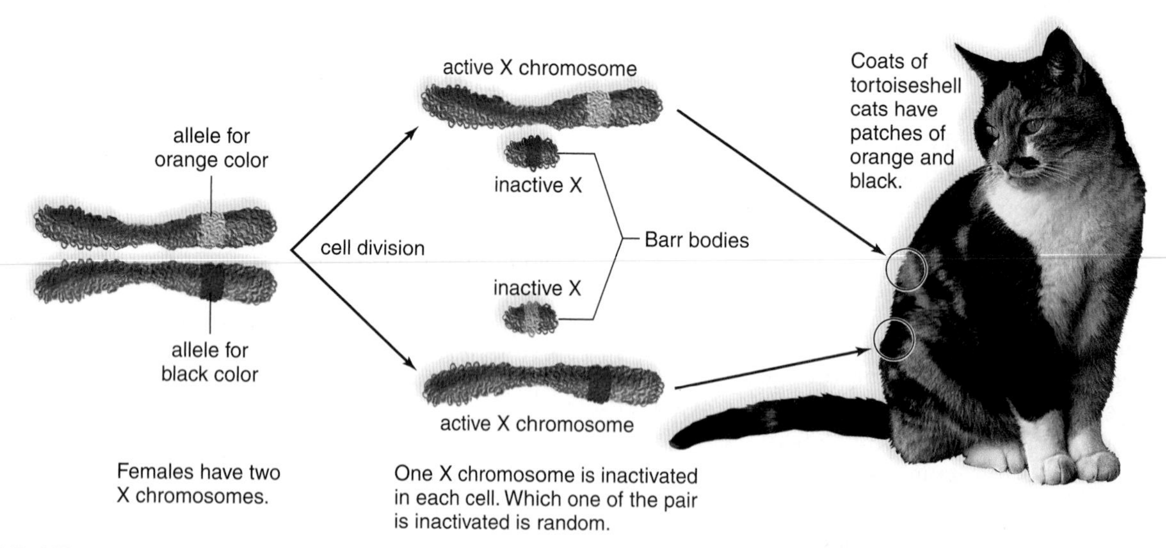

Figure 25.17 X-inactivation.
In cats, the alleles for black or orange coat color are carried on the X chromosome. Random X-inactivation occurs in females. Therefore, in heterozygous females, some of the cells express the allele for black coat color, while other cells express the allele for orange coat color. If the X-inactivation happened early in development, resulting in large patches, a calico pattern is found. If it occurred later in development, producing smaller patches of color, a tortoiseshell pattern is found. The white color on calico cats is provided by another gene.

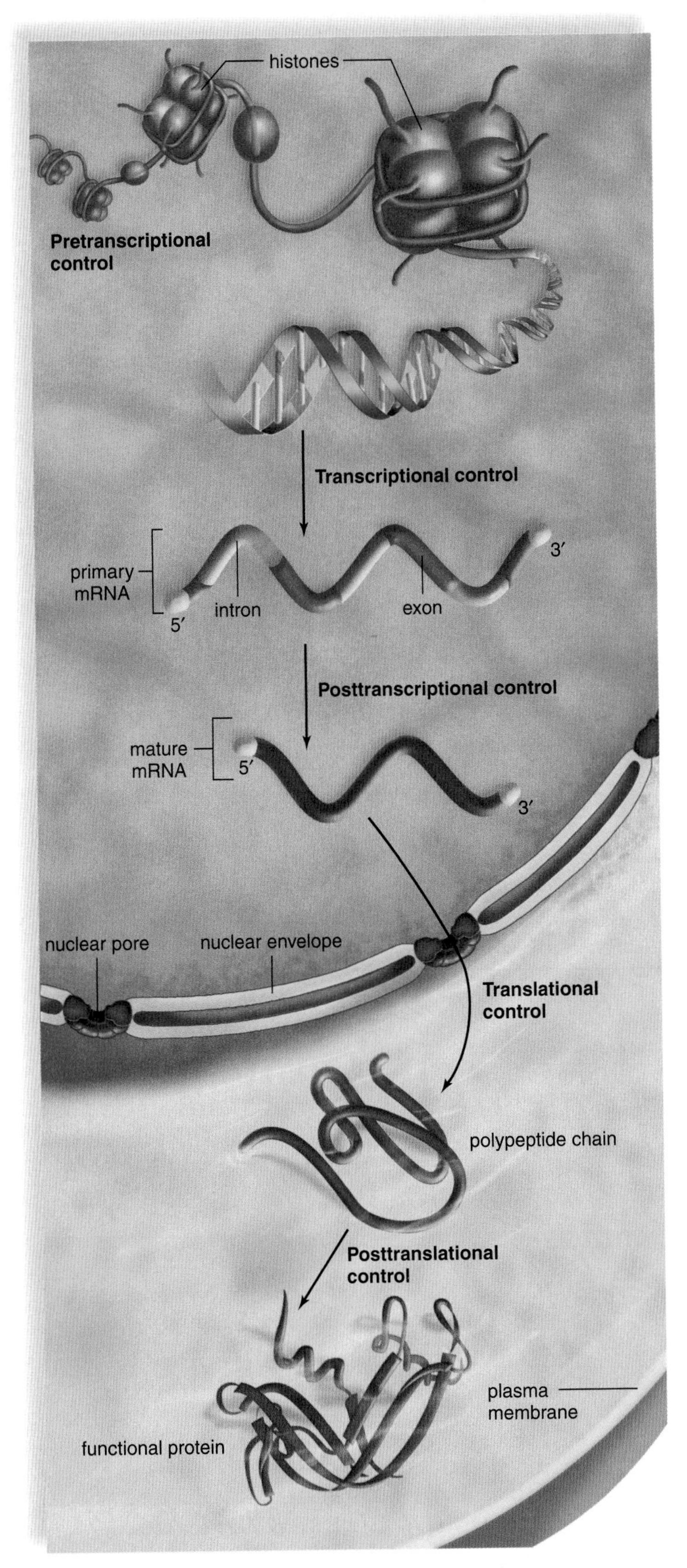

Figure 25.18 Levels at which control of gene expression occurs in eukaryotic cells.
The five levels of control are (1) pretranscriptional control, (2) transcriptional control, and (3) posttranscriptional control, which occur in the nucleus; and (4) translational and (5) posttranslational control, which occur in the cytoplasm.

the stabilization of certain mRNA transcripts. For example, the mRNA for vitelline, an egg membrane protein, can persist for three weeks if it is exposed to estrogen, as opposed to 15 hours without estrogen.

Posttranslational Control Some proteins are not active immediately after synthesis. After translation, insulin is folded into a three-dimensional structure that is inactive. Then a sequence of about 30 amino acids is enzymatically removed from the middle of the molecule, leaving two polypeptide chains that are bonded together by disulfide (S—S) bonds. This activates the protein. Other modifications, such as phosphorylation, also affect the activity of a protein. Many proteins only function a short time before they are degraded or destroyed by the cell.

The various levels of gene control are summarized in Figure 25.18.

Check Your Progress

1. What is the relationship between cell specificity and gene regulation?
2. What type of gene regulation is present in both prokaryotes and eukaryotes? How do they differ?
3. Name the different levels of gene regulation in eukaryotes.

25.5 Gene Mutations

Learning Outcomes

1. Define a gene mutation, and explain what causes mutations.
2. Describe why cancer is a failure of genetic control.
3. List the characteristics of cancer cells.

A **gene mutation** is a permanent change in the sequence of bases in DNA. The effect of a DNA base sequence change on protein activity can range from no effect to complete inactivity. Germ-line mutations are those that originally occurred in sex cells and can be passed to subsequent generations. We now know that some germ-line mutations can result in cancer. Also, somatic mutations that are not passed on to future generations can sometimes lead to the development of cancer.

Causes of Mutations

Three causes of mutations are errors in replication, mutagens, and transposons.

Errors in Replication DNA replication errors are a rare source of mutations. DNA polymerase, the enzyme that carries out replication, proofreads the new strand against the old strand. Usually mismatched pairs are then replaced with the correct nucleotides. In the end, there is typically only one mistake for every 1 billion nucleotide pairs replicated.

Mutagens Environmental influences called **mutagens** cause mutations in humans. Mutagens include radiation (e.g.,

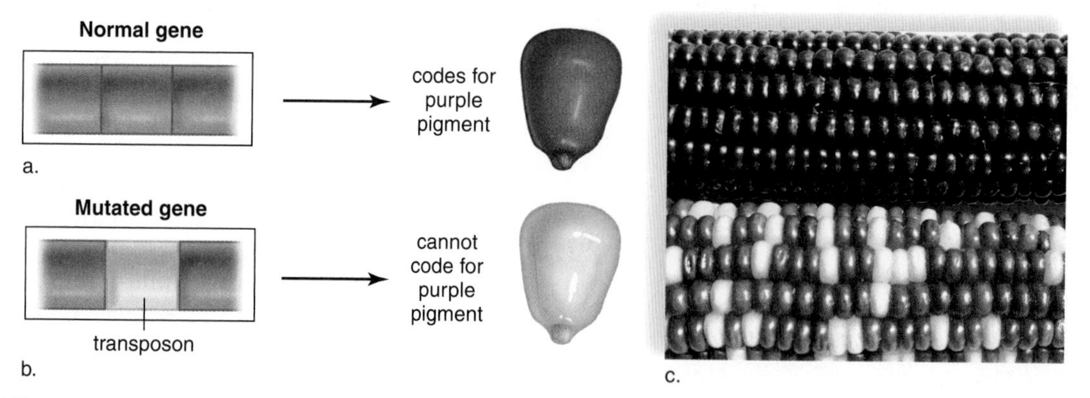

Figure 25.19 Transposon.
a. A purple coding gene ordinarily codes for a purple pigment. **b.** A transposon "jumps" into the purple-coding gene. This mutated gene is unable to code for purple pigment and a white kernel results. **c.** Indian corn displays a variety of colors and patterns due to transposon activity.

radioactive elements, X rays, ultraviolet [UV] radiation) and certain organic chemicals (e.g., chemicals in cigarette smoke and certain pesticides). The rate of mutations resulting from mutagens is generally low because DNA repair enzymes constantly monitor and repair any irregularities.

Transposons **Transposons** are specific DNA sequences that have the remarkable ability to move within and between chromosomes. Their movement to a new location sometimes alters neighboring genes, particularly by increasing or decreasing their expression. Although "movable elements" in corn were described over 50 years ago by Barbara McClintock, their significance was only realized recently. So-called jumping genes have now been discovered in bacteria, fruit flies, and humans, and it is likely that all organisms have such elements. McClintock described how the presence of white kernels in corn is due to a transposon located within a gene coding for a pigment-producing enzyme (Fig. 25.19*a*, *b*). So-called "Indian corn" displays a variety of colors and patterns because of transposons (Fig. 25.19*c*). In a rare human neurological disorder called Charcot-Marie-Tooth disease, a transposon called Mariner causes the muscles and nerves of the legs and feet to gradually wither away.

Effect of Mutations on Protein Activity

Point mutations involve a change in a single DNA nucleotide and, therefore, a possible change in a specific amino acid. The base change in the second row of Figure 25.20*a* has no effect on the resulting amino acid in hemoglobin. The change in the third row, however, codes for the amino acid glutamic acid instead of valine. This base change accounts for the genetic disorder sickle-cell disease because the incorporation of valine, instead of glutamic acid, causes hemoglobin molecules to form semirigid rods, and the red blood cells become sickle shaped. (Compare Figure 25.20*b* to Figure 25.20*c*.) Sickle-shaped cells clog blood vessels and die off more quickly than normal-shaped cells. The base change in the fourth row of Figure 25.20*a* may also have drastic results because the DNA now codes for a stop codon.

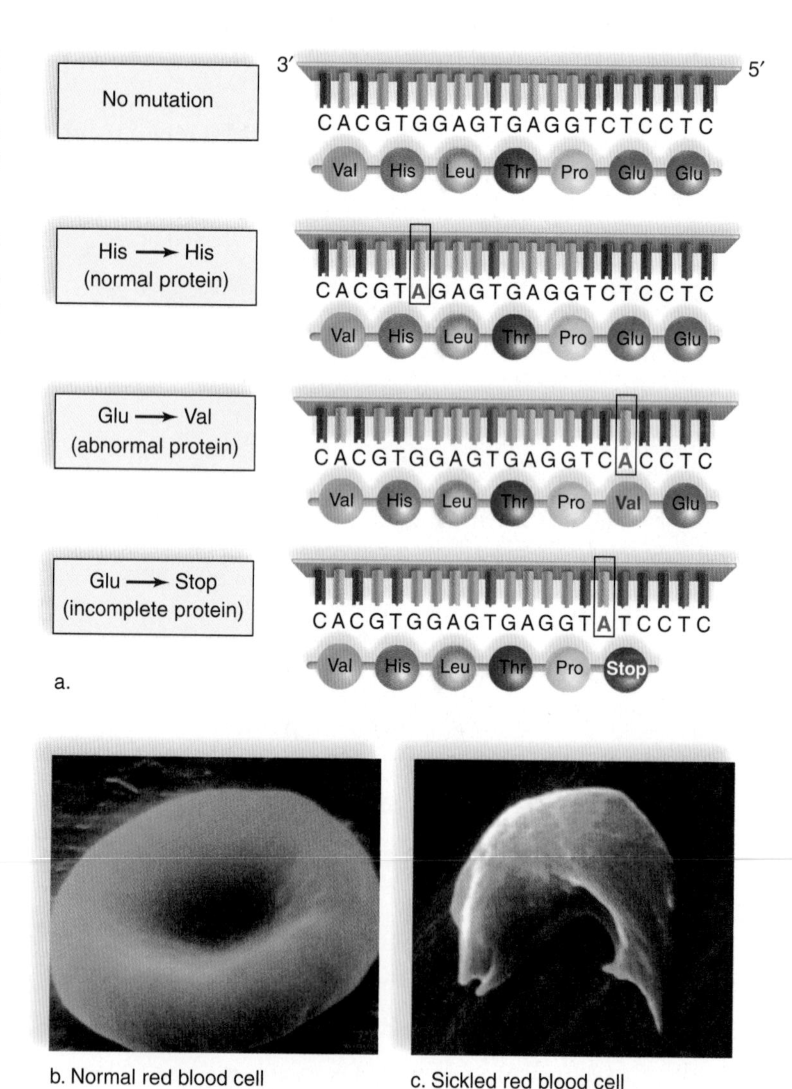

Figure 25.20 Point mutations in hemoglobin.
The effect of a point mutation can vary. **a.** Starting at the *top*: Normal sequence of bases in hemoglobin; next, the base change has no effect; next, due to base change, DNA now codes for valine instead of glutamic acid, and the result is that normal red blood cells **(b)** become sickle shaped **(c)**; next, base change will cause DNA to code for termination and the protein will be incomplete.

Frameshift mutations occur most often because one or more nucleotides are either inserted or deleted from DNA. The result of a frameshift mutation can be a completely new sequence of codons and nonfunctional protein. Here is how this occurs: The sequence of codons is read from a specific starting point, as in this sentence, THE CAT ATE THE RAT. If the letter C is deleted from this sentence and the reading frame is shifted, we read THE ATA TET HER AT—something that doesn't make sense.

Nonfunctional Proteins

A single nonfunctioning protein can have a dramatic effect on the phenotype, because enzymes are often a part of metabolic pathways. One particular metabolic pathway in cells is as follows:

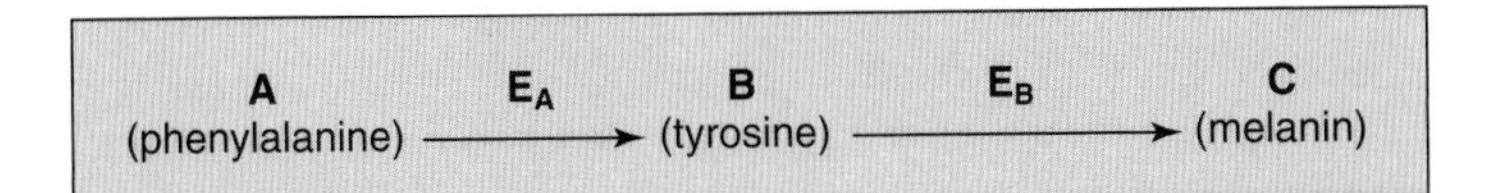

If a faulty code for enzyme E_A is inherited, a person is unable to convert the molecule A to B. Phenylalanine builds up in the system, and the excess causes mental impairment and the other symptoms of the genetic disorder phenylketonuria (PKU). In the same pathway, if a person inherits a faculty code for enzyme E_B, then B cannot be converted to C, and the individual is an albino.

A rare condition called androgen insensitivity is due to a faulty receptor for androgens, which are male sex hormones such as testosterone. Although there is plenty of testosterone in the blood, the cells are unable to respond to it. Female instead of male external genitals form, and female instead of male secondary sex characteristics occur. The individual, who appears to be a normal female, may be prompted to seek medical advice when menstruation never occurs. The karyotype is that of a male rather than a female, and the individual does not have the internal sexual organs of a female.

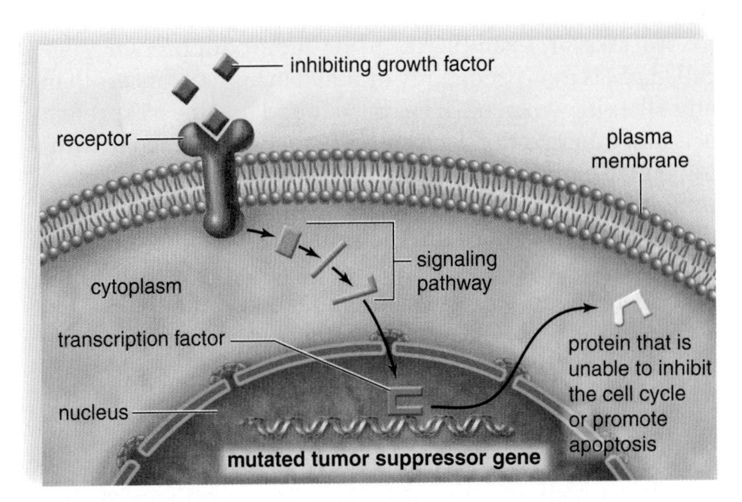

Figure 25.21 Cell signaling pathway that no longer inhibits a mutated tumor suppressor gene. A mutated tumor suppressor gene codes for a product that directly or indirectly no longer inhibits the cell cycle.

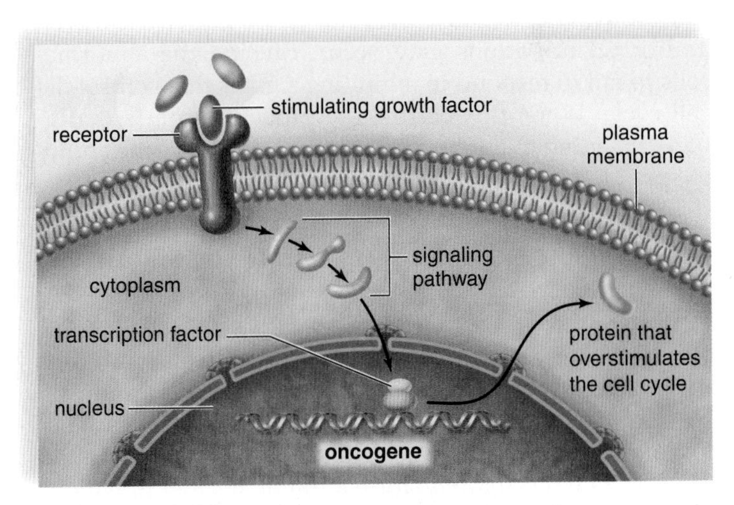

Figure 25.22 Cell signaling pathway that stimulates an oncogene. An oncogene codes for a product that either directly or indirectly overstimulates the cell cycle.

Mutations Can Cause Cancer

It is estimated that one-third of the children born in 1999 will develop cancer at some time in their lives. Of these affected individuals, one-third of the females and one-fourth of the males will die due to cancer. In the United States, the three deadliest forms of cancer are lung cancer, colon and rectal cancer, and breast cancer.

The development of cancer involves a series of accumulating mutations that can be different for each type of cancer. As discussed in Chapter 5, tumor suppressor genes ordinarily act as brakes on cell division, especially when it begins to occur abnormally. Proto-oncogenes stimulate cell division but are usually turned off in fully differentiated nondividing cells. When proto-oncogenes mutate, they become oncogenes that are active all the time. Carcinogenesis begins with the loss of tumor suppressor gene activity and/or the gain of oncogene activity. When tumor suppressor genes are inactive and oncogenes are active, cell division occurs uncontrollably because a cell signaling pathway that reaches from the plasma membrane to the nucleus no longer functions as it should (Fig. 25.21 and Fig. 25.22).

It often happens that tumor suppressor genes and proto-oncogenes code for transcription factors or proteins that control transcription factors. As we have seen, transcription factors are a part of the rich and diverse types of mechanisms that control gene expression in cells. They are of fundamental importance to DNA replication and repair, cell growth and division, control of apoptosis, and cellular differentiation. Therefore, it is not surprising that inherited or acquired defects in transcription factor structure and function contribute to the development of cancer.

To take an example, a major tumor suppressor gene called *p53* is more frequently mutated in human cancers than any other known gene. It has been found that the *p53* protein acts as a transcription factor, and as such is involved in turning on the expression of genes whose products are cell cycle inhibitors. *p53* also promotes apoptosis (programmed cell death) when it is needed. The retinoblastoma protein (RB) controls the activity of a transcription factor for cyclin D and other genes whose products promote entry into the S stage of the cell cycle. When the tumor suppressor gene *p16* mutates, the RB protein is always available, and the result is too much active cyclin D in the cell.

Mutations in many other genes also contribute to the development of cancer. Several proto-oncogenes code for ras proteins, which are needed for cells to grow, to make new DNA, and to not grow out of control. A point mutation is sufficient to turn a normally functioning *ras* proto-oncogene into an oncogene. Abnormal growth results.

Although cancers vary greatly, they usually follow a common multistep progression (Fig. 25.23). Most cancers begin as an abnormal cell growth that is **benign,** or not cancerous, and usually does not grow larger. However, additional mutations may occur, causing the abnormal cells to fail to respond to inhibiting signals that control the cell cycle. When this occurs, the growth becomes **malignant,** meaning that it is cancerous and possesses the ability to spread.

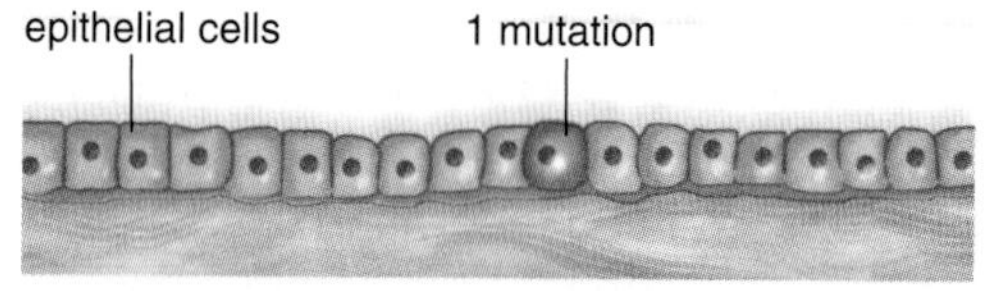

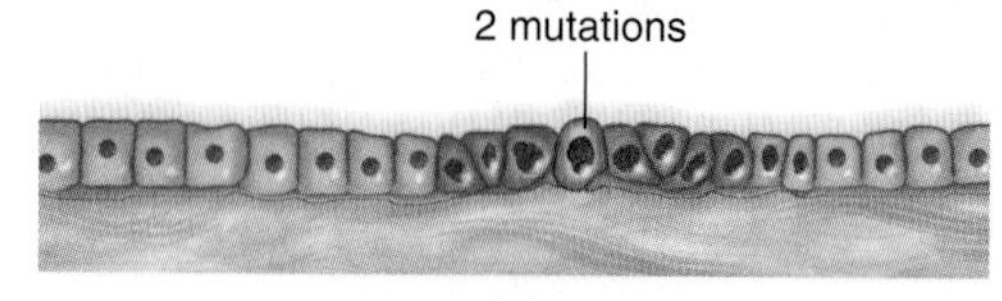

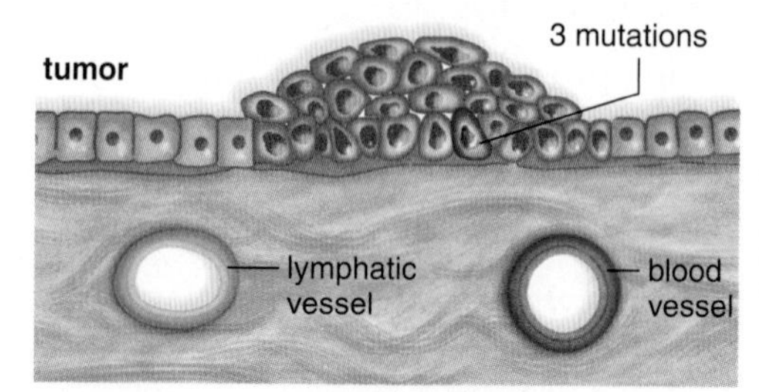

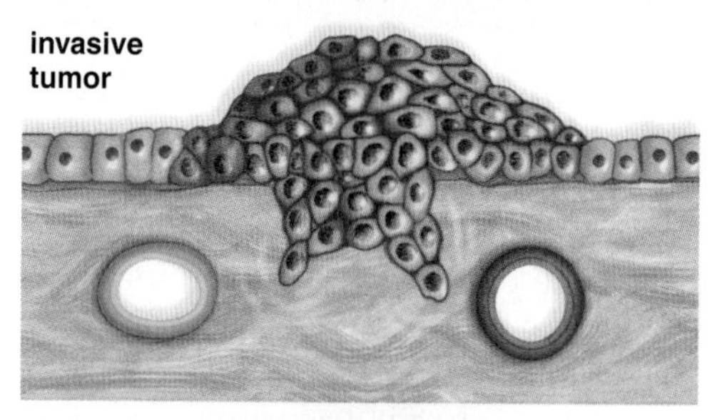

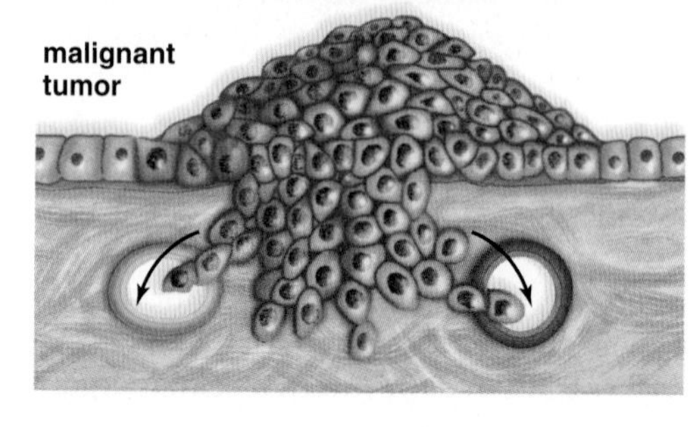

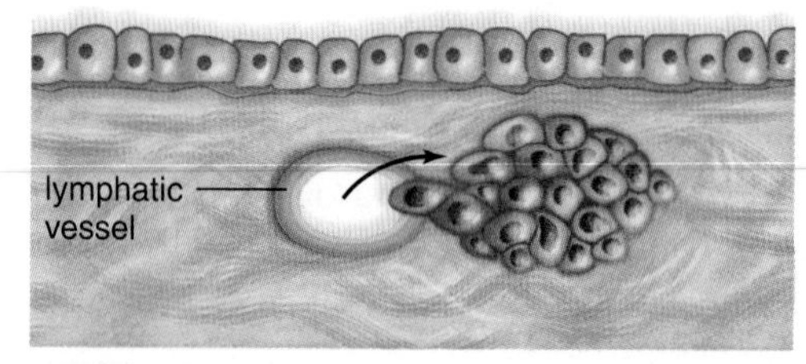

Figure 25.23 Development of cancer.
Several mutations contribute to the development of a tumor. A single abnormal cell begins the process, and the most aggressive cell, thereafter, becomes the one that divides the most and forms the tumor. Eventually, cancer cells gain the ability to invade underlying tissue and travel to other parts of the body, where they develop new tumors.

Characteristics of Cancer Cells

The primary characteristics of cancer cells are as follows:

Cancer cells are genetically unstable. Generation of cancer cells appears to be linked to mutagenesis. A cell acquires a mutation that allows it to continue to divide. Eventually one of the progeny cells will acquire another mutation and gain the ability to form a tumor. Further mutations occur, and the most aggressive cell becomes the dominant cell of the tumor. Tumor cells undergo multiple mutations and also tend to have chromosomal aberrations and rearrangements.

Cancer cells do not correctly regulate the cell cycle. Cancer cells continue to cycle through the cell cycle. The normal controls of the cell cycle do not operate to stop the cycle and allow the cells to differentiate. Because of that, cancer cells tend to be nonspecialized. Both the rate of cell division and the number of cells increase.

Cancer cells escape the signals for cell death. A cell that has genetic damage or problems with the cell cycle will initiate apoptosis, or programmed cell death. However, cancer cells do not respond to internal signals to die, and they continue to divide even with genetic damage. Cells from the immune system, when they detect an

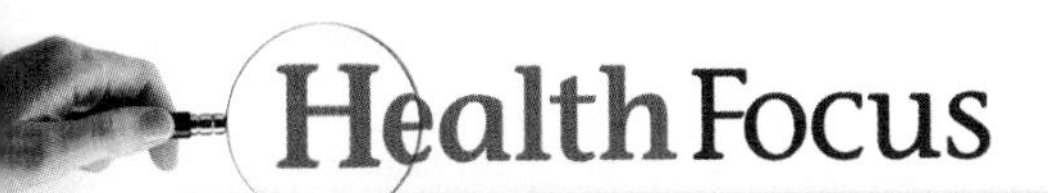

Health Focus

Prevention of Cancer

Protective Behaviors

Be tested for cancer.

Do a shower test for breast cancer or testicular cancer. Women should get a Pap smear for cervical cancer annually if they are over 21 or are sexually active. Have a friend or physician check your skin annually for any unusual moles. Have other exams done regularly by a physician.

Be aware of occupational hazards.

Exposure to several different industrial agents (nickel, chromate, asbestos, vinyl chloride, etc.) and/or radiation increases the risk of various cancers. Your employer should notify you of potential hazards in your workplace. The risk of several of these cancers is increased when combined with cigarette smoking.

Use sunscreen.

Almost all cases of basal-cell and squamous-cell skin cancers are sun-related. Use a sunscreen of at least SPF 15 (Fig. 25B) and wear protective clothing if you are going to be out during the brightest part of the day. Don't sunbathe on the beach or in a tanning salon.

Check your home for radon.

Excessive radon exposure in homes increases the risk of lung cancer, especially in cigarette smokers. It is best to test your home and take the proper remedial actions.

Avoid unnecessary X rays.

Even though most medical and dental X rays are adjusted to deliver the lowest dose possible, unnecessary X rays should be avoided. Sensitive areas of the body that are not being X-rayed should be protected with lead screens.

Practice safe sex.

Human papillomaviruses (HPVs) cause genital warts and are associated with cervical cancer, as well as tumors of the vulva, vagina, anus, penis, and mouth. HPVs may be involved in 90–95% of all cases of cervical cancer, and 20 million people in the United States have an infection that can be transmitted to others.

Carefully consider hormone therapy.

Estrogen therapy to control menopausal symptoms increases the risk of endometrial cancer. However, including progesterone in estrogen replacement therapy helps minimize this risk.

Figure 25B UV Protection. Use of sunscreen with SPF of 15 or higher can help prevent skin cancer, as can limiting midday exposure.

Maintain a healthy weight.

The risk of cancer (especially colon, breast, and uterine cancers) is 55% greater among obese women and 33% greater among obese men, compared to people of normal weight. Eating a low-fat, healthy diet helps maintain your weight as well as reducing your risk of cancer.

Exercise regularly.

Regular activity can reduce the incidence of certain cancers, especially colon and breast cancer.

Increase consumption of foods that are rich in vitamins A and C.

Beta-carotene, a precursor of vitamin A, is found in dark-green, leafy vegetables, carrots, and various fruits. Vitamin C is present in citrus fruits. These vitamins are called antioxidants because they prevent the formation of chemicals called free radicals that can damage the DNA in the cell. Vitamin C also prevents the conversion of nitrates and nitrites into carcinogenic nitrosamines in the digestive tract.

Include vegetables from the cabbage family in your diet.

The cabbage family includes cabbage, broccoli, Brussels sprouts, kohlrabi, and cauliflower. These vegetables may reduce the risk of gastrointestinal and respiratory tract cancers.

Limit consumption of salt-cured, smoked, or nitrite-cured foods.

Salt-cured or pickled foods may increase the risk of stomach and esophageal cancer. Smoked foods, such as ham and sausage, contain chemical carcinogens similar to those in tobacco smoke. Nitrites are sometimes added to processed meats (e.g., hot dogs and cold cuts) and other foods to protect them from spoilage. These are converted to carcinogenic nitrosamines in the digestive tract.

Be moderate in the consumption of alcohol.

Cancers of the mouth, throat, esophagus, larynx, and liver occur more frequently among heavy drinkers, especially when accompanied by smoking cigarettes or chewing tobacco.

Don't smoke.

Cigarette smoking accounts for about 30% of all cancer deaths. Smoking is responsible for 90% of lung cancer cases among men and 79% among women—about 87% altogether. People who smoke two or more packs of cigarettes a day have lung cancer mortality rates 13 to 23 times greater than those of nonsmokers. Cigars and smokeless tobacco (chewing tobacco or snuff) increase the risk of cancers of the mouth, larynx, throat, and esophagus.

Discussion Questions

1. Which of these cancer-preventive measures are you currently practicing?
2. Why do you think tobacco use increases the risk of other types of cancer besides lung cancer?
3. Why does the use of tanning beds also increase the incidence of skin cancer?

Bioethical Focus

Genetic Testing for Cancer Genes

Over the past decade, genetic tests have become available for certain cancer genes. If women test positive for defective *BRCA1* and *BRCA2* genes, they have an increased risk for early-onset breast and ovarian cancer. If individuals test positive for the *APC* gene, they are at greater risk for the development of colon cancer. Other genetic tests exist for rare cancers, including retinoblastoma and Wilms tumor.

Advocates of genetic testing say that it can alert those who test positive for these mutated genes to undergo more frequent mammograms or colonoscopies. Early detection of cancer clearly offers the best chance for successful treatment.

Others feel that genetic testing is unnecessary because nothing can presently be done to prevent the disease. Perhaps it is enough for those who have a family history of cancer to schedule more frequent check-ups, beginning at a younger age.

People opposed to genetic testing worry that being predisposed to cancer might threaten one's job or health insurance. They suggest that genetic testing be confined to a research setting, especially because it is not known which particular mutations in the genes predispose a person to cancer. They are afraid, for example, that a woman with a defective *BRCA1* or *BRCA2* gene might make the unnecessary decision to have a bilateral mastectomy. The lack of proper counseling also concerns many. In a study of 177 patients who underwent *APC* gene testing for susceptibility to colon cancer, less than 20% received counseling before the test. Moreover, physicians misinterpreted the test results in nearly one-third of the cases.

Another concern is that testing negative for a particular genetic mutation may give people the false impression that they are not at risk for cancer. Such a false sense of security can prevent them from having routine cancer screening. Regular testing and avoiding known causes of cancer—such as smoking, a high-fat diet, or too much sunlight—are important for everyone.

Form Your Own Opinion

1. Should genetic testing for cancer be available for everyone, or should genetic testing be confined to a research setting? Explain.
2. If genetic testing for cancer were offered to you, would you take advantage of it? Why or why not?
3. Do you feel that everyone should do all they can to avoid having cancer, such as by not smoking, or is it the individual's choice to take that risk? Explain.

abnormal cell, will send signals to that cell, inducing apoptosis. Cancer cells also ignore these signals.

Most normal cells have a built-in limit to the number of times they can divide before they die. One of the reasons normal cells stop entering the cell cycle is that the telomeres become shortened. **Telomeres** are sequences at the ends of the chromosomes that keep them from fusing with each other. With each cell division, the telomeres shorten, eventually becoming short enough to signal apoptosis. Cancer cells turn on the gene that encodes the enzyme telomerase, which is capable of rebuilding and lengthening the telomeres. Cancer cells thus show characteristics of "immortality" in that they can enter the cell cycle repeatedly.

Cancer cells can survive and proliferate elsewhere in the body. Many of the changes that must occur for cancer cells to form tumors elsewhere in the body are not understood. The cells apparently disrupt the normal adhesive mechanism and move to another place within the body. They travel through the blood and lymphatic vessels and then invade new tissues, where they form tumors. This process is known as **metastasis**. As a tumor grows, it must increase its blood supply by forming new blood vessels, a process called **angiogenesis.** Tumor cells switch on genes that code for the production of growth factors encouraging blood vessel formation. These new blood vessels supply the tumor cells with the nutrients and oxygen they require for rapid growth, but they also rob normal tissues of nutrients and oxygen.

Check Your Progress

1. What are some common causes of mutations?
2. Give an example of a mutation that may not affect protein activity and one that does affect protein activity.
3. What types of mutations can cause cancer?

Applying the Concepts [Revisited]

Children who suffer from xeroderma pigmentosum are missing the enzymes that repair DNA after exposure to UV light. These enzymes constantly search the DNA for abnormalities and then either directly repair the damage, or remove a section of the DNA with the abnormality and resynthesize a new strand of DNA.

1. Besides UV light, what other types of radiation or chemicals can cause damage to DNA? Should XP sufferers also avoid these?
2. Why would skin and eyes be most susceptible to DNA damage?

SUMMARIZING THE CONCEPTS

25.1 DNA Structure and Replication

- In two separate experiments, investigators labeled the protein coat of a bacterial virus with ^{35}S and the DNA with ^{32}P. They then showed that the radioactive P alone is largely taken up by the bacterial host and is followed by reproduction of viruses. These are convincing data that DNA is the genetic material.
- DNA is a double helix with two sugar-phosphate backbones and paired nitrogen-containing bases. A (adenine) is paired with T (thymine), and G (guanine) is paired with C (cytosine).
- During replication, DNA "unzips," and then a complementary strand forms opposite to each original strand. This means that replication is semiconservative because each double helix contains one old strand and one new strand.

25.2 RNA Structure and Function

- RNA is a single-stranded nucleic acid, complementary to a DNA segment, in which U (uracil) replaces T.
- Functional RNAs include messenger RNA (mRNA), ribosomal RNA (rRNA), and transfer RNA (tRNA). Each functions in a specific manner in protein synthesis.

25.3 Gene Expression

- Gene expression involves two processes: transcription and translation.
- Transcription produces an RNA sequence complementary to one of the DNA strands. Messenger RNA produced in eukaryotes must be processed by the addition of a cap and tail, and removal of introns, before being transported to the cytoplasm where it can be translated.
- DNA specifies the synthesis of proteins using a triplet code: every three bases transcribed into an mRNA codon codes for one amino acid.
- During translation, tRNA molecules, attached to their own particular amino acid, travel to a ribosome, and through complementary base pairing between anticodons of tRNA and codons of mRNA, the tRNAs, and the amino acids they carry, are sequenced in a predetermined way to form a polypeptide chain.
- A ribosome has a binding site for two tRNAs at a time. The tRNA at the P site passes a peptide to a newly arrived tRNA–amino acid at the A site. Then translocation occurs: the ribosome moves, and the tRNA with the peptide is at the P site. In this way, an amino acid grows one amino acid at a time until a polypeptide is formed.

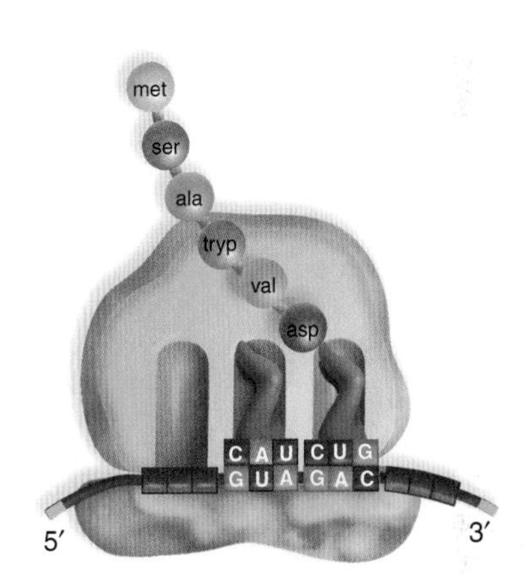

25.4 Control of Gene Expression

- Each cell type in the body contains its own mix of proteins that make it different from all other cell types. Only certain genes are active in cells that perform specialized functions. Other active genes are called housekeeping genes because they govern functions that are common to many types of cells.
- In bacteria, genes are clustered into operons. An operon contains a group of genes, usually coding for proteins related to a particular metabolic pathway, along with the promoter, a DNA sequence where RNA polymerase binds to begin transcription, and an operator, where a repressor protein binds. In the case of the *lac* operon, the structural genes code for enzymes that metabolize lactose. When lactose is absent, the repressor binds to the operator and shuts off the operon. When lactose is present, it binds to the repressor so that it can no longer bind to the operator, and this allows RNA polymerase to transcribe the structural genes. Some bacterial operons are repressible instead of inducible.

- Control of gene expression is more complicated in eukaryotes. Eukaryotes have five levels of control: pretranscriptional control, transcriptional control, posttranscriptional control, translational control, and posttranslational control.
- Genes that are found within highly condensed heterochromatin are inactive. A dramatic example of this is inactivated X chromosomes, called Barr bodies. Even genes that are found in looser-packed DNA, called euchromatin, must have the nucleosomes shifted in order to be expressed.
- The transcriptional level of control includes transcription factors that assist the RNA polymerase and transcriptional activators which greatly increases the rate of transcription.
- In posttranscriptional control, differences in mRNA processing affect gene expression.
- Translational controls occur in the cytoplasm and involve the length of time the mRNA is functional.
- Posttranslational controls also occur in the cytoplasm and involve the length of time the protein is functional. Proteins can be activated by a variety of methods.

25.5 Gene Mutations

- A gene mutation is a permanent change in the sequence of bases in DNA. The effect of a DNA base sequence change on protein activity can range from no effect to complete inactivity.
- Gene mutations result in a change in the sequence of nucleotides in the DNA and can be caused by errors in replication, mutagens, and transposons.
- Point mutations involve a change in a single DNA nucleotide and, therefore, a possible change in a specific amino acid. Frameshift mutations occur most often because one or more nucleotides are either inserted or deleted from DNA.
- The development of cancer involves a series of accumulating mutations that can be different for each type of cancer. Carcinogenesis begins with the loss of tumor suppressor gene activity and/or the gain of oncogene activity. Mutations in many other genes also contribute to the development of cancer. Although cancers vary greatly, they usually follow a common multistep progression. Cancer cells defy the normal regulation of the cell cycle and can invade and colonize other areas of the body. Cancer cells do not exhibit contact inhibition and thus form tumors. When tumor cells gain the ability to invade surrounding tissues, they are said to be malignant.

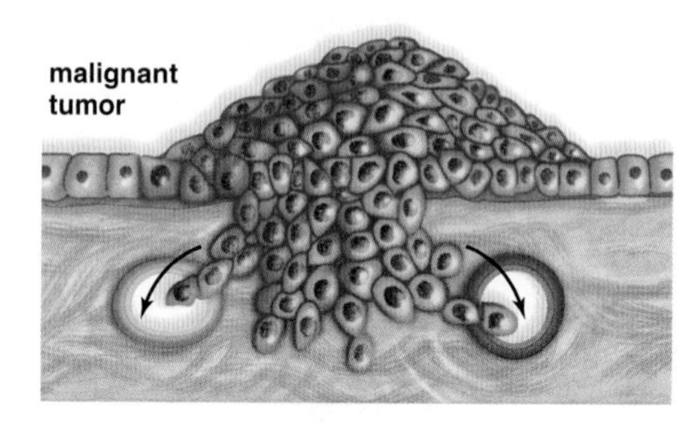

- In terms of their primary characteristics, cancer cells are genetically unstable, do not correctly regulate the cell cycle, escape the signals for cell death, and can survive and proliferate elsewhere in the body, a process called metastasis. Multiple mutations and chromosomal aberrations are present in cancer cells. They continue to cycle through the cell cycle and tend to be nonspecialized. They do not respond to either internal or external signals for apoptosis, and they induce the expression of telomerase to repair their telomeres.

TESTING YOURSELF

Choose the best answer for each question.

1. In the Hershey-Chase experiments, radioactive phosphorus was found
   a. outside the bacterial cells.
   b. inside the bacterial cells.
   c. both inside and outside the bacterial cells.
2. According to Chargaff's rules, in DNA, the amount of
   a. A = T and G = C.
   b. A = G and T = C.
   c. A = C and T = G.
   d. A = T = C = G.
3. During DNA replication, the parental strand ATTGGC would code for the daughter strand
   a. ATTGGC.
   b. CGGTTA.
   c. TAACCG.
   d. GCCAAT.

For questions 4–6, match the function with the type of RNA in the key.

Key:

a. messenger RNA
b. ribosomal RNA
c. transfer RNA

4. Carries amino acids to ribosomes
5. Is a structural component of the organelle responsible for protein synthesis
6. Carries information from genes to the translation machinery
7. In contrast to DNA, RNA
   a. is a helix.
   b. contains uracil.
   c. is double stranded.
   d. contains deoxyribose sugar.
8. Using the information in Figure 25.8, determine the sequence of amino acids that would be produced following transcription and translation of the following DNA template strand:

   TACGTTCCAACT

   a. tyrosine - valine - proline - threonine
   b. methionine - glutamine - glycine
   c. isoleucine - valine - proline
   d. methionine - glutamine - glycine - threonine
9. Mutation rates are typically low due to
   a. proofreading by DNA polymerase.
   b. DNA repair enzymes.
   c. silent mutations.
   d. None of these are correct.
   e. All of these are correct.

10. The process of converting the information contained in the nucleotide sequence of RNA into a sequence of amino acids is called
    a. transcription.
    b. translation.
    c. translocation.
    d. replication.
11. During protein synthesis, an anticodon of a transfer RNA (tRNA) pairs with
    a. amino acids in the polypeptide.
    b. DNA nucleotide bases.
    c. ribosomal RNA (rRNA) nucleotide bases.
    d. messenger RNA (mRNA) nucleotide bases.
    e. other tRNA nucleotide bases.
12. An operon is a short sequence of DNA
    a. that prevents RNA polymerase from binding to the promoter.
    b. that prevents transcription from occurring.
    c. and the sequences that control its transcription.
    d. that codes for the repressor protein.
    e. that functions to prevent the repressor from binding to the operator.
13. How is transcription directly controlled in eukaryotic cells?
    a. through the use of phosphorylation
    b. by means of apoptosis
    c. using transcription factors and activators
    d. when chromatin is packed to keep genes turned on
    e. None of these are correct.
14. Which of these is a characteristic of cancer cells?
    a. They do not exhibit contact inhibition.
    b. They lack specialization.
    c. They have abnormal chromosomes.
    d. They fail to undergo apoptosis.
    e. All of these are correct.
15. Which of these is not a possible cause of cancer?
    a. radiation
    b. metastasis
    c. viruses
    d. genes
    e. oncogenes
16. Which of these is not a possible behavior that could help prevent cancer?
    a. maintaining a healthy weight
    b. eating more dark-green, leafy vegetables, carrots, and various fruits
    c. not smoking
    d. consuming large amounts of smoked and pickled foods
    e. consuming alcohol only in moderation
17. If the original DNA sequence is ACGCGT, which of the following represents a point mutation?
    a. ACGGGT
    b. TGCCGT
    c. GGGCCC
    d. ACGCGT

UNDERSTANDING THE TERMS

angiogenesis 520
anticodon 508
benign 518
codon 508
complementary base pairing 504
DNA (deoxyribonucleic acid) 502
DNA replication 504
double helix 504
elongation 510
enhancer 514
exon 507
gene 506
gene mutation 515
initiation 510
intron 507
malignant 518
messenger RNA (mRNA) 506
metastasis 520
mutagen 515
operon 512
polyribosome 509
primary mRNA 507
promoter 507
purine 504
pyrimidine 504
regulator gene 512
repressor 512
ribosomal RNA (rRNA) 506
RNA (ribonucleic acid) 505
RNA polymerase 507
telomere 520
termination 510
transcription 506
transcription activator 514
transcription factor 514
transfer RNA (tRNA) 506
translation 506
transposon 516
triplet code 508
uracil 505

THINKING CRITICALLY

1. What kind of genes do you think would be included in the category of "housekeeping" genes?
2. When the enzyme telomerase was first discovered, some people thought it might be "the fountain of youth" because it could immortalize cells. Why was this not found to be true?
3. Why is it only the *risk* for cancer that is inherited?

INQUIRY INTO LIFE WEBSITE

The companion website for *Inquiry into Life* provides a wealth of information organized and integrated by chapter. You will find practice tests, animations, videos, and much more that will complement your learning and understanding of general biology.

http://www.mhhe.com/maderinquiry13

26

Applying the Concepts

Carena and Esteban have always wanted a baby boy. They now have four girls and are considering having a fifth child. When they went to the doctor, much to their surprise, Carena and Esteban found out there are now techniques to sort sperm and thereby choose the gender of a child with a relatively high degree of certainty. Although biotechnology is at the point where we can choose the sex of our children, we cannot ensure that they will not have debilitating or even lethal genetic disorders or diseases.

Now imagine the day that we can detect these diseases in the parents and correct them in vitro using gene therapy. That day might not be too far off in the distant future. The human genome sequence is now complete, and as a result, we are developing increasingly effective methods to detect genetic disorders. Although gene therapy is in its infancy and has met with limited success, we now have the technology to deliver functional genes to cells with dysfunctional or mutated genes in affected patients. One day, these technological advances may be able to be applied to developing fetuses.

Surely recent developments in biotechnology will be tremendously beneficial for human health in the future. But what about the ethical implications? Some might consider correcting a genetic disorder in a fetus, while others may find this to be unnatural. What if people could choose their child's eye color? Hair color? Height? Intelligence? As we continue in the twenty-first century, biotechnology will likely make great strides in treating and even curing many human diseases. But, as a society, we will also face many ethical dilemmas.

Biotechnology and Genomics

Chapter Outline

26.1 DNA Cloning

Learning Outcomes

1. Illustrate the steps in forming recombinant DNA.
2. Discuss how the polymerase chain reaction works.
3. Explain how DNA is analyzed in a forensics case.

Knowledge of DNA biology has led to an ability to manipulate the genes of organisms. We can clone genes and then use them to alter the **genome** (the complete genetic makeup of an organism) of viruses and cells, whether bacterial, plant, or animal cells. This practice, called **genetic engineering,** has innumerable uses, from producing a product to treating cancer and genetic disorders.

The Cloning of a Gene

We often think of **cloning** as the production of identical copies of an organism through some asexual means. The members of a bacterial colony on a petri dish are clones because they all came from the division of the same original cell. Human identical twins are also considered clones, because the first two cells of the embryo separated and each became a complete individual.

Another major biological application of cloning is **gene cloning,** which is the production of many identical copies of a single gene. Biologists clone genes for a number of reasons. They might want to produce large quantities of the gene's protein product, such as human insulin, learn how a cloned gene codes for a particular protein, or use the genes to alter the phenotypes of other organisms in a beneficial way. When cloned genes are used to modify a human, the process is called gene therapy. Otherwise, organisms with foreign DNA or genes inserted into them are called **transgenic organisms,** which are frequently used to produce a product desired by humans. Recombinant DNA technology and the polymerase chain reaction (PCR) are two techniques that scientists use to clone DNA.

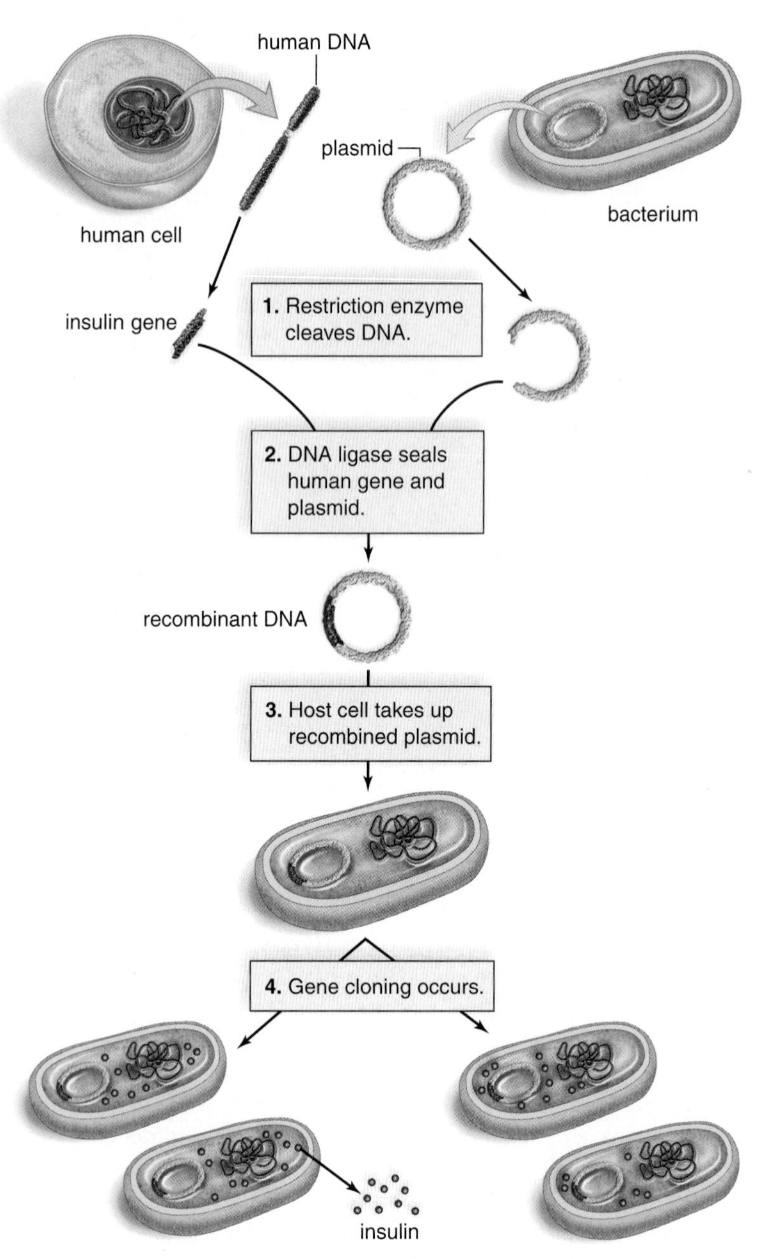

Figure 26.1 Cloning a human gene.
Human DNA and plasmid DNA are cleaved by a specific type of restriction enzyme. For example, human DNA containing the insulin gene is spliced into a plasmid by the enzyme DNA ligase. Gene cloning is achieved after a bacterium takes up the plasmid. If the gene functions normally as expected, the product (e.g., insulin) may also be retrieved.

Recombinant DNA Technology

Recombinant DNA (rDNA) contains DNA from two or more different sources, such as the human cell and the bacterial cell in Figure 26.1. To make rDNA, a researcher needs a **vector,** a piece of DNA that can be manipulated such that foreign DNA can be added to it. One common vector is a plasmid. **Plasmids** are small accessory rings of DNA from bacteria that are not part of the bacterial chromosome and are capable of self-replicating. Plasmids were discovered by investigators studying the bacterium *E. coli.*

Two enzymes are needed to introduce foreign DNA into vector DNA: (1) a **restriction enzyme** to cleave the vector DNA and (2) **DNA ligase** to seal DNA into an opening created by the restriction enzyme. Hundreds of restriction enzymes occur naturally in bacteria, where they act as a primitive immune system by cutting up any viral DNA that enters the cell. They are called restriction enzymes because they *restrict* the growth of viruses, but they can also be used as molecular scissors to cut double-stranded DNA at a specific site. For example, the restriction enzyme called *Eco*RI always recognizes and cuts double-stranded

DNA in the following manner when DNA has the sequence of bases GAATTC:

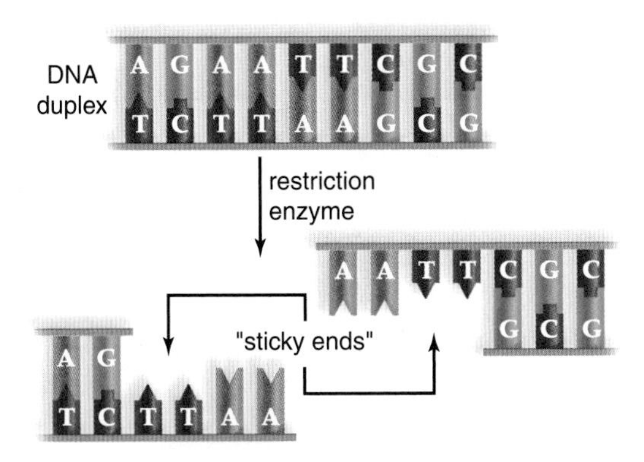

Notice that a gap now exists into which a piece of foreign DNA can be placed if it ends in bases complementary to those exposed by the restriction enzyme. To ensure this, it is only necessary to cleave the foreign DNA with the same type of restriction enzyme. The single-stranded, but complementary, ends of the two DNA molecules are called "sticky ends" because they can bind a piece of foreign DNA by complementary base pairing. Sticky ends facilitate the insertion of foreign DNA into vector DNA.

DNA ligase, an enzyme that functions in DNA replication, is then used to seal the foreign piece of DNA into the vector. Bacterial cells take up recombinant plasmids, especially if the cells are treated to make them more permeable. Thereafter, as the plasmid replicates, so does the foreign DNA and thus the gene is cloned.

The Polymerase Chain Reaction

The **polymerase chain reaction (PCR)** can create billions of copies of a segment of DNA in a test tube in a matter of hours. PCR is very specific—it *amplifies* (makes copies of) a targeted DNA sequence, usually a few hundred bases in length. PCR requires the use of DNA polymerase, the enzyme that carries out DNA replication, and a supply of nucleotides for the new DNA strands. PCR involves three basic steps that occur repeatedly, usually for about 35–40 cycles: (1) a denaturation step at 95°C, where DNA is heated to become single stranded; (2) an annealing step at a temperature usually between 50–60°C, where an oligonucleotide primer hybridizes to each of the single DNA strands; and (3) an extension step at 72°C, where an engineered DNA polymerase adds complementary bases to each of the single DNA strands, creating double-stranded DNA.

PCR is a chain reaction because the targeted DNA is repeatedly replicated, much in the same way natural DNA replication occurs, as long as the process continues. Figure 26.2 uses color to distinguish the old strand from the new DNA strand. Notice that the amount of DNA doubles with each replication cycle.

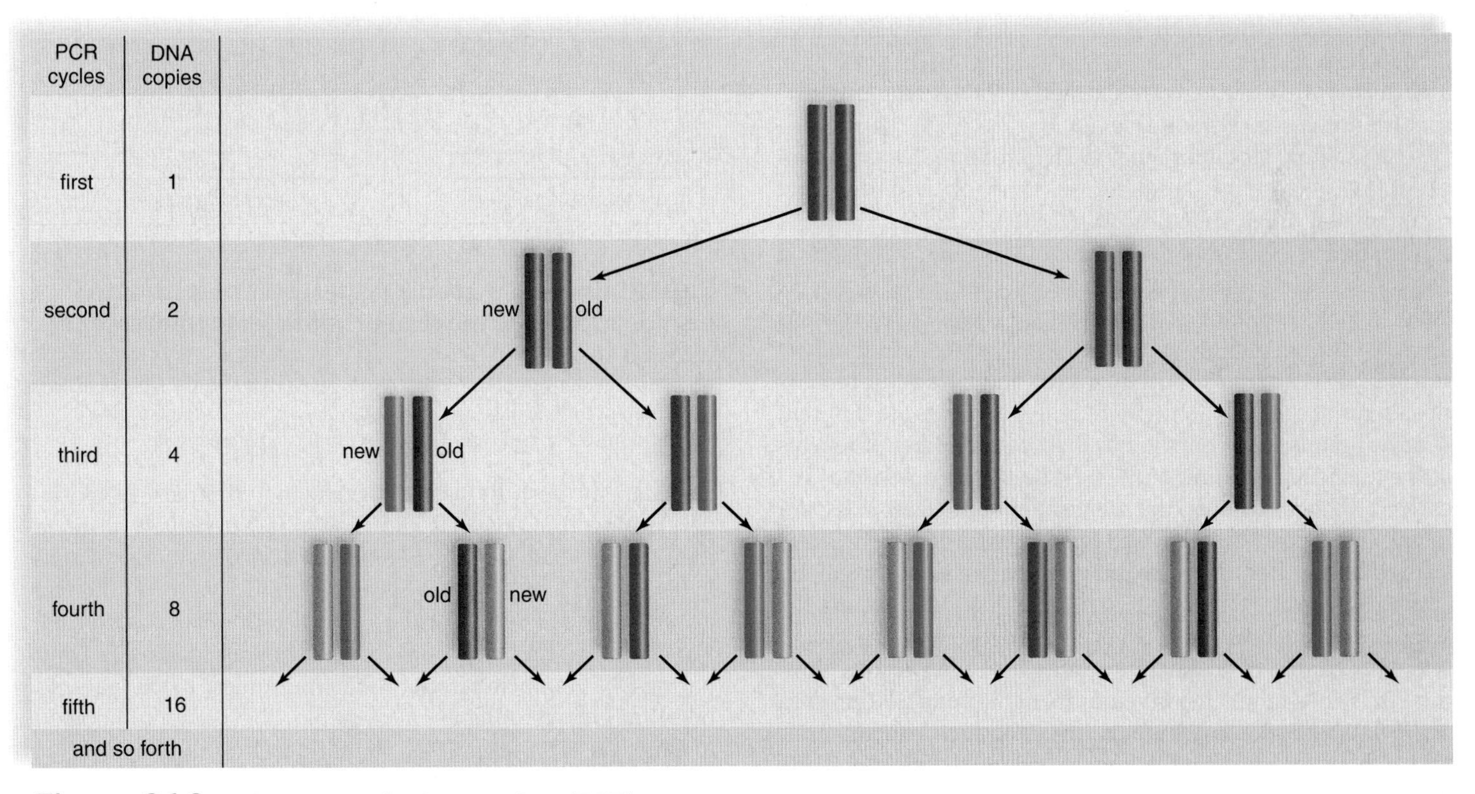

Figure 26.2 Polymerase chain reaction (PCR).
PCR allows the production of many identical copies of DNA in a laboratory setting. Assuming you start with only one copy, you can create millions of copies with only 20–25 cycles. For this reason, only a tiny DNA sample is necessary for forensic genetics.

Thus, assuming you start with only one copy of DNA, after one cycle, you will have two copies, after two cycles four copies, and so on. PCR has been in use since its development in 1985 by Kary Banks Mullis, and now almost every laboratory has automated PCR machines to carry out the procedure. Automation became possible after a temperature-insensitive (thermostable) DNA polymerase was extracted from the bacterium *Thermus aquaticus,* which lives in hot springs. The enzyme can withstand the high temperature used to denature double-stranded DNA. Therefore, replication does not have to be interrupted by the need to add more enzyme.

DNA amplified by PCR is often analyzed for various purposes. For example, mitochondrial DNA base sequences have been used to decipher the evolutionary history of human populations. Since so little DNA is required for PCR to be effective, it is commonly used as a forensic method for analyzing DNA found at crime scenes—only a drop of semen, a flake of skin, or the root of a single hair is necessary!

DNA Analysis

Analysis of DNA following PCR has undergone improvements over the years. At first, the entire genome was treated with restriction enzymes, and because each person has their own restriction enzyme sites, they would have a unique collection of DNA fragment sizes. During a process called gel electrophoresis, whereby an electrical current is used to force DNA though a porous gel material, these fragments are separated according to their size. Smaller fragments move farther through the gel than larger fragments, and result in a pattern of distinctive bands, called a **"DNA fingerprint,"** that identified the person (Fig. 26.3).

Now, **short tandem repeat (STR)** profiling is the method of choice. STRs are the same short sequence of DNA bases that recur several times, as in GATAGATAGATA. STR profiling is advantageous because it doesn't require the use of restriction enzymes. Instead, PCR is used to amplify target sequences of DNA, which are fluorescently labeled. These PCR products are run through an automated DNA sequencer, and the fluorescent labels are picked up by a laser and a detector records the length of each DNA fragment. The fragments are different lengths because each person has their own number of repeats at the particular location of the STR on the chromosome (i.e., each STR locus). That is, the greater the number of STRs at a locus, the longer the DNA fragment amplified by PCR. If individuals are homozygotes, they will have a single fragment, and heterozygotes will have two fragments of different lengths (Fig. 26.4). The more STR loci employed, the more confident scientists can be of distinctive results for each person. Now, the FBI is collecting a massive number of human STR profiles in its databases for current and future crime scene analyses, and genetic profiles may someday be on people's driver's licenses in the United States. The Bioethical Focus in this chapter explores some of the pros and cons associated with DNA forensics that may result from these trends.

Applications of PCR are limited only by our imaginations. When the DNA matches that of a virus or mutated gene, it is known that a viral infection, genetic disorder, or cancer is present. DNA fingerprints from blood or tissues at a crime scene has been successfully used in convicting criminals. DNA fingerprinting through STR profiling was extensively used to identify the victims of the September 11, 2001 terrorist attacks in the United States. Relatives can be found, paternity suits can be settled (see Fig. 26.3), and genetic disorders can be detected. PCR has also shed new light on evolutionary studies by comparing DNA extracted from human mummies 76,000 years old or animal fossils millions of years old. The National Football League has even used DNA to mark each of the Super Bowl footballs to be able to authenticate them.

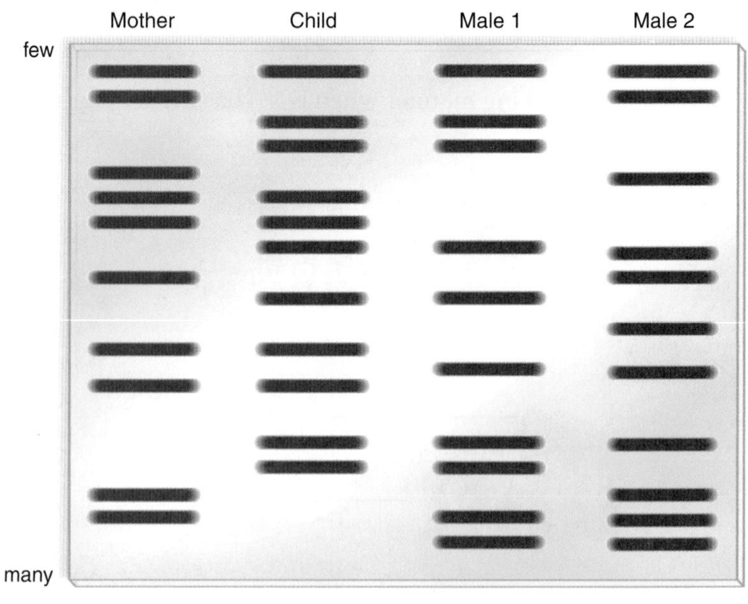

Figure 26.3 The use of DNA fingerprints to establish paternity.
In this method, DNA fragments resulting from restriction enzyme cuts are separated by gel electrophoresis. Male 1 is the father.

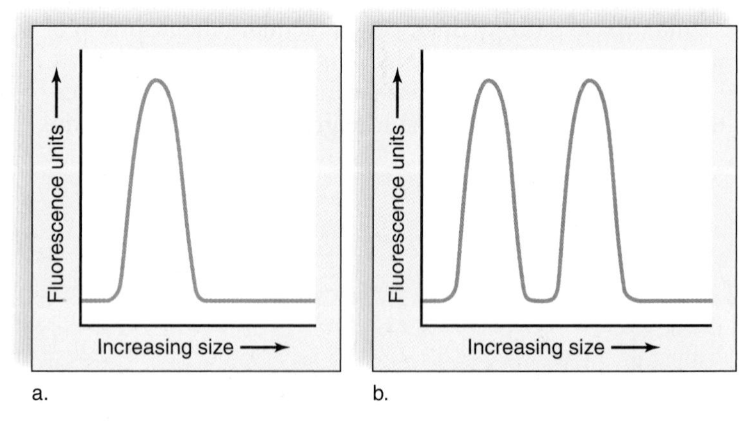

Figure 26.4 A person's fingerprint pattern can also be read by a machine that detects fluorescence using a laser.
Above are two electropherograms (laser graphs) of a single STR locus in two different individuals. **a.** This individual is a homozygote (note the single fluorescence peak). **b.** This individual is a heterozygote.

Check Your Progress

1. List and explain the two required steps for producing recombinant DNA.
2. What is a DNA fingerprint? Give an example of how DNA fingerprints are used.

26.2 Biotechnology Products

Learning Outcomes

1. Compare and contrast production of transgenic bacteria, plants, and animals.
2. Discuss the various uses for genetically modified bacteria, plants, and animals.

Today, transgenic bacteria, plants, and animals are often called genetically modified organisms (GMOs), and the products they produce are called **biotechnology** products (Fig. 26.5).

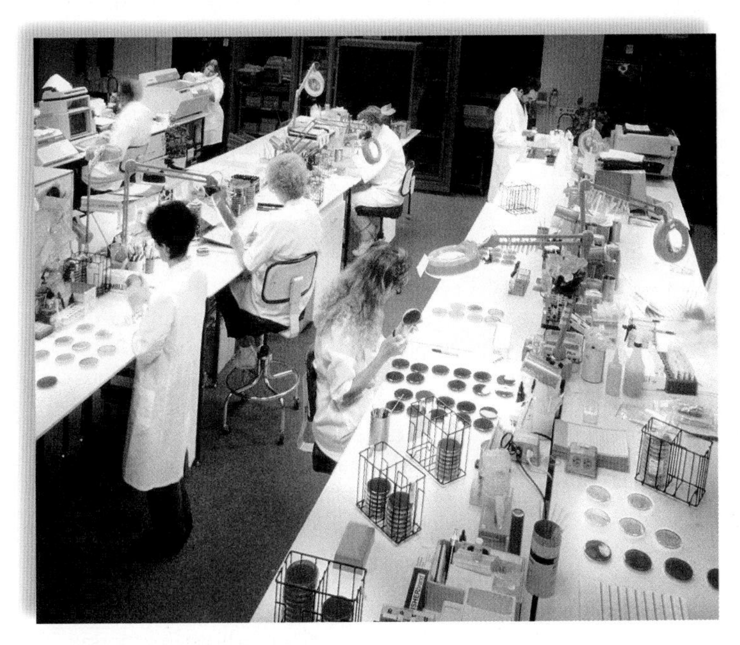

Figure 26.5 Biotechnology products.
Products such as clotting factor VIII, which is administered to hemophiliacs, can be made by transgenic bacteria, plants, or animals. After being processed and packaged, they are sold as commercial products.

Transgenic Bacteria

Recombinant DNA technology is used to produce transgenic bacteria, which are grown in huge vats called bioreactors. The bacteria express the cloned gene, and the gene product is usually collected from the medium in which the bacteria are grown. Biotechnology products produced by bacteria include insulin, human growth hormone, tPA (tissue plasminogen activator), and hepatitis B vaccine.

Transgenic bacteria have many other uses as well. Some have been produced to promote the health of plants. For example, bacteria that normally live on plants and encourage the formation of ice crystals have been changed from frost-plus to frost-minus bacteria. As a result, new crops such as frost-resistant strawberries are being developed. Also, a bacterium that normally colonizes the roots of corn plants has now been endowed with genes (from another bacterium) that code for an insect toxin. The toxin protects the roots from insects.

Bacteria can be selected for their ability to degrade a particular substance, and this ability can then be enhanced by bioengineering. For instance, naturally occurring bacteria that eat oil can be genetically engineered to do an even better job of cleaning up beaches after oil spills (Fig. 26.6). Bacteria can also remove sulfur from coal before it is burned, resulting in cleaner emissions. One bacterial strain was given genes that allowed it to clean up levels of toxins that would have killed other bacterial strains. Further, these bacteria were given "suicide" genes that caused them to self-destruct when their job was done.

Organic chemicals are often synthesized by having catalysts act on precursor molecules or by using bacteria to carry out the synthesis. Today, it is possible to go one step further and manipulate the genes that code for these enzymes. For instance, biochemists discovered a strain of bacteria that is especially good at producing phenylalanine, an organic chemical needed to make aspartame, better known as NutraSweet. They isolated, altered, and cloned the appropriate genes so that various bacteria could be genetically engineered to produce phenylalanine.

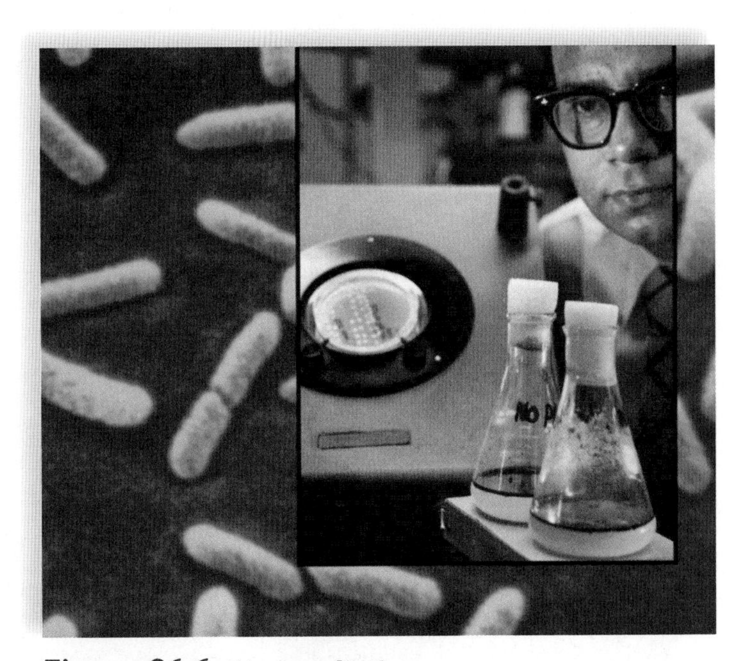

Figure 26.6 Bioremediation.
These bacteria, which are capable of decomposing oil, were engineered and patented by the investigator Dr. Chakrabarty. The flask toward the rear contains oil and no bacteria. The flask toward the front contains the engineered bacteria and is almost clear of oil.

Health Focus

Are Genetically Engineered Foods Safe?

A series of focus groups conducted by the Food and Drug Administration (FDA) in 2000 showed that although most participants believed that genetically engineered foods, now also called GMOs, might offer benefits, they also feared possible unknown long-term health consequences. In Canada, Conrad G. Brunk, a bioethicist at the University of Waterloo in Ontario, has said, "When it comes to human and environmental safety, there should be clear evidence of the absence of risks. The mere absence of evidence is not enough."

The discovery by activists that a type of genetically engineered corn called StarLink had inadvertently made it into the food supply triggered the recall of taco shells, tortillas, and many other corn-based foodstuffs from supermarkets. Further, the makers of StarLink were forced to buy back StarLink from farmers and to compensate food producers at an estimated cost of several hundred million dollars in late 2000.

StarLink is a type of "BT" corn. It contains a foreign gene taken from a common soil organism, *Bacillus thuringiensis,* which makes a protein that is toxic to many insect pests. About a dozen BT varieties, including corn, potato, and even a tomato, have now been approved for human consumption. These strains contain a gene for an insecticidal protein called CryIA. Instead, StarLink contained a gene for a related protein called Cry9C, which researchers thought might slow down the chances of pest resistance to BT corn. In order to get FDA approval for use in foods, the makers of StarLink performed the required tests. Like the other now-approved strains, StarLink wasn't poisonous to rodents, and its biochemical structure is not similar to those of most chemicals in food that commonly cause allergic reactions in humans (called allergens). But the Cry9C protein resisted digestion longer than the other BT proteins when it was put in simulated stomach acid and subjected to heat. Because most food allergens resist digestion in a similar fashion, StarLink was not approved for human consumption.

The scientific community is now trying to devise more tests for allergens because it has not been possible to determine conclusively whether Cry9C is or is not an allergen. Also, at this point, it is unclear how resistant to digestion a protein must be in order to be an allergen, and it is also unclear what degree of amino acid sequence similarity a potential allergen must have to a known allergen to raise concern. Dean D. Metcalfe, chief of the Laboratory of Allergic Diseases at the National Institute of Allergy and Infectious Diseases, said, "We need to understand thresholds for sensitization to food allergens and thresholds for elicitation of a reaction with food allergens."

Other scientists are concerned about the following potential drawbacks to the planting of BT corn: (1) resistance among populations of the target pest, (2) exchange of genetic material between the transgenic crop and related plant species, and (3) BT crops' impact on nontarget species. They feel that many more studies are needed before stating for certain that BT corn has no ecological drawbacks.

Despite controversies, the planting of genetically engineered corn increased from 2001–2007. The USDA reports that U.S. farmers planted genetically engineered corn on 45% of all corn acres, up from 26% in 2001. In all, U.S. farmers planted at least 100 million acres with mostly genetically engineered corn, soybeans, and cotton (Fig. 26A). Some groups advocate that GMOs should be labeled as such, but this may not be easy to accomplish because, for example, most cornmeal is derived from both conventional and genetically engineered corn. So far, there has been no attempt to sort out one type of food product from the other. However, at many health food stores, foods that *do not* contain GMOs are now labeled.

Discussion Questions

1. Do you think genetically modified organisms (GMOs) should be labeled? Construct an argument both for and against labeling GMOs.
2. Some people are strongly advocating the complete removal of GMOs from the market. A few people, called ecoterrorists, are taking such drastic actions as burning crops or even setting biotechnology labs on fire. Do you think GMOs should be removed from the market? Why or why not? What further information would you need to make your decision?
3. Rice is a staple in the diet of millions of people worldwide, many of them living in less-developed countries. In some of those same countries, vitamin A deficiency is a major cause of blindness in small, malnourished children. Scientists have developed a new form of rice, called "golden rice," which is genetically modified to assist with the metabolism of vitamin A and might prevent millions of cases of blindness. The scientists who created golden rice claim it is safe for consumption, although critics say the health effects are not yet fully understood. Given its nutritional potential, should golden rice be planted and distributed on a wide scale? What information would you need to make your decision?

a.

b.

Figure 26A Genetically engineered crops.

Genetically engineered (**a**) corn, (**b**) soybeans, and (**c**) cotton crops are increasingly being planted by today's farmers.

c.

Transgenic Plants

Techniques have been developed to introduce foreign genes into immature plant embryos or into plant cells called **protoplasts** that have had their cell wall removed. It is possible to treat protoplasts with an electric current while they are suspended in a liquid containing foreign DNA. The electric current makes tiny, self-sealing holes in the plasma membrane through which the desired genetic material can enter. Protoplasts go on to develop into mature plants containing and expressing the foreign DNA.

One altered plant known as the pomato is the result of these technologies. This plant produces potatoes belowground and tomatoes aboveground. Foreign genes transferred to cotton, corn, and potato strains have made these plants resistant to pests because their cells now produce an insect toxin. Similarly, soybeans have been made resistant to a common herbicide that is sprayed to kill weeds that compete with soybean growth. Some corn and cotton plants are both pest- and herbicide-resistant. These and other genetically engineered crops that are expected to have increased yields are now commonly sold commercially. However, as discussed in the Health Focus, the public is concerned about the possible effect of genetically modified organisms, also called GMOs, on human health, as well as the environment.

Like bacteria, plants are also being engineered to produce human proteins, such as hormones, clotting factors, and antibodies, in their seeds. One type of antibody made by corn can deliver radioisotopes to tumor cells, and another made by soybeans can be used to treat genital herpes.

Transgenic Animals

Techniques have been developed to insert genes into the eggs of animals. It is possible to microinject foreign genes into eggs by hand, but another method uses vortex mixing. The eggs are placed in an agitator with DNA and silicon-carbide needles. The needles make tiny holes in the eggs through which the DNA can enter. When these eggs are fertilized, the resulting offspring are transgenic animals. Using this technique, many types of animal eggs have acquired the gene for bovine growth hormone (bGH). The procedure has been used to produce larger fishes, cows, pigs, rabbits, and sheep.

Gene pharming, the use of transgenic farm animals to produce pharmaceuticals, is being pursued by a number of firms. Genes that code for therapeutic and diagnostic proteins are incorporated into an animal's DNA, and the proteins appear in the animal's milk. Plans are under way to produce drugs for the treatment of cystic fibrosis, cancer, blood diseases, and other disorders by this method. Figure 26.7 outlines the procedure for producing transgenic mammals: DNA containing the gene of interest is injected into donor eggs. Following in vitro fertilization, the zygotes are placed in host females, where they develop. After female offspring mature, the product is secreted in their milk.

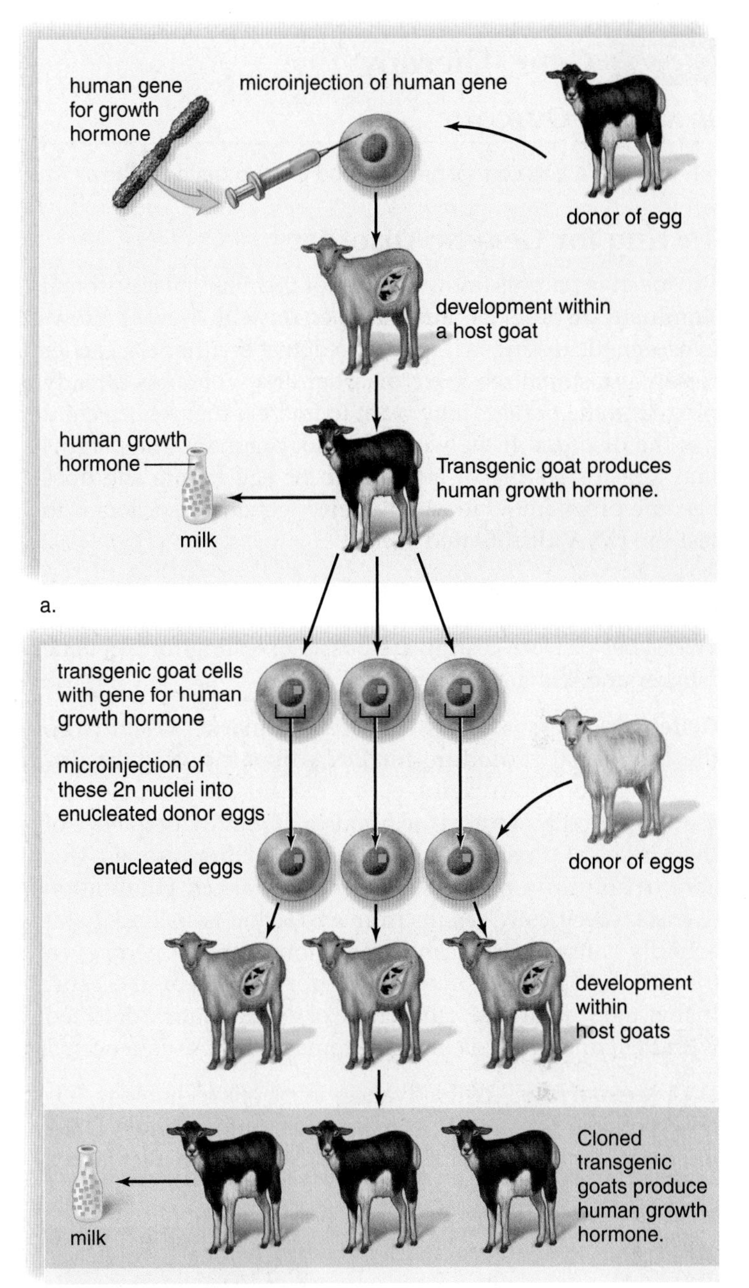

Figure 26.7 Transgenic animals.
a. A genetically engineered egg develops in a host to create a transgenic goat that produces a biotechnology product in its milk. **b.** Nuclei from the transgenic goat are transferred into donor eggs, which develop into cloned transgenic goats.

Check Your Progress

1. What difficulties are there in creating transgenic animals versus transgenic bacteria?
2. What are three practical uses of transgenic organisms?

26.3 Gene Therapy

Learning Outcome

1. Compare and contrast in vivo and ex vivo gene therapy.

Testing for Genetic Disorders

Prospective parents know if either of them has an autosomal dominant disorder because the person will show it. However, genetic testing is required to detect if either is a carrier for an autosomal recessive disorder. If a woman is already pregnant, the parents may want to know if the unborn child has the disorder. If the woman is not pregnant, the parents may opt for testing of an embryo or egg before she does become pregnant. One way to detect genetic disorders is to test the DNA for mutated genes.

Testing the DNA

Two types of DNA testing are possible: testing for a genetic marker and using a DNA probe.

Genetic Markers Testing for a genetic marker is similar to the traditional procedure for DNA fingerprinting, as discussed earlier. As an example, consider that individuals with Huntington disease have an abnormality in the sequence of their bases at a particular location on a chromosome. This abnormality in sequence is a **genetic marker.** Huntington disease, specifically, results from a STR that is so long that it actually causes a frameshift mutation within a gene even though the STR itself occurs outside, but nearby the gene. In this and similar cases, the length of the STR can be detected with PCR and analysis on an automated DNA sequencer.

DNA Microarrays With advances in robotic technology, it is now possible to spot all human genes onto a single **DNA microarray** (or "gene chip") (Fig 26.8). A mutation microarray, the most common type, can be used to generate a person's genetic profile. The DNA microarray contains hundreds to thousands of known disease-associated mutant gene alleles. Genomic DNA from the individual to be tested is labeled with a fluorescent dye, and then added to the microarray. The spots on the microarray fluoresce if the individual's DNA binds to the mutant genes on the chip, indicating that the individual may have a particular disorder or is at risk for developing it later in life. This technique can generate a genetic profile much more quickly and inexpensively than older methods involving DNA sequencing. **Genetic profiling** might become a regular part of a medical checkup in the future.

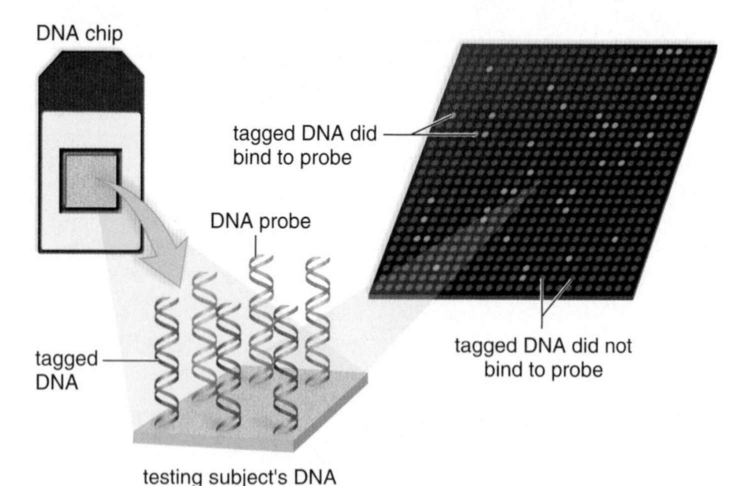

Figure 26.8 Use of a DNA chip to test for mutated genes.
This DNA chip contains rows of DNA sequences for mutations that indicate the presence of particular genetic disorders. If DNA fragments derived from an individual's DNA bind to a sequence representing a mutation on the DNA chip, that sequence fluoresces, and the individual has the mutation.

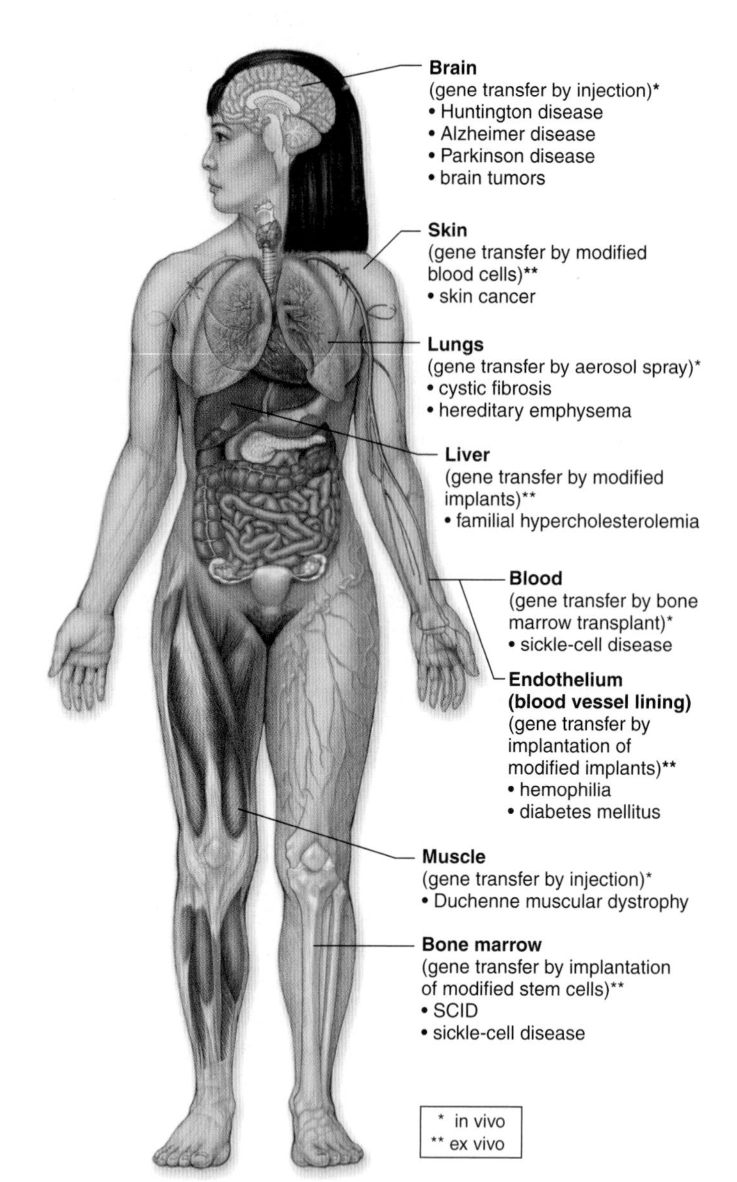

Figure 26.9 Gene therapy.
Sites of ex vivo and in vivo gene therapy to cure the conditions noted.

Gene Therapy

Once a genetic disorder is detected, gene therapy is a potential course of treatment. **Gene therapy** is the insertion of genetic material into human cells for the treatment of genetic disorders and various other human illnesses, such as cardiovascular disease and cancer. Figure 26.9 shows regions of the body that have received copies of normal genes by various methods of gene transfer. Viruses genetically modified to be

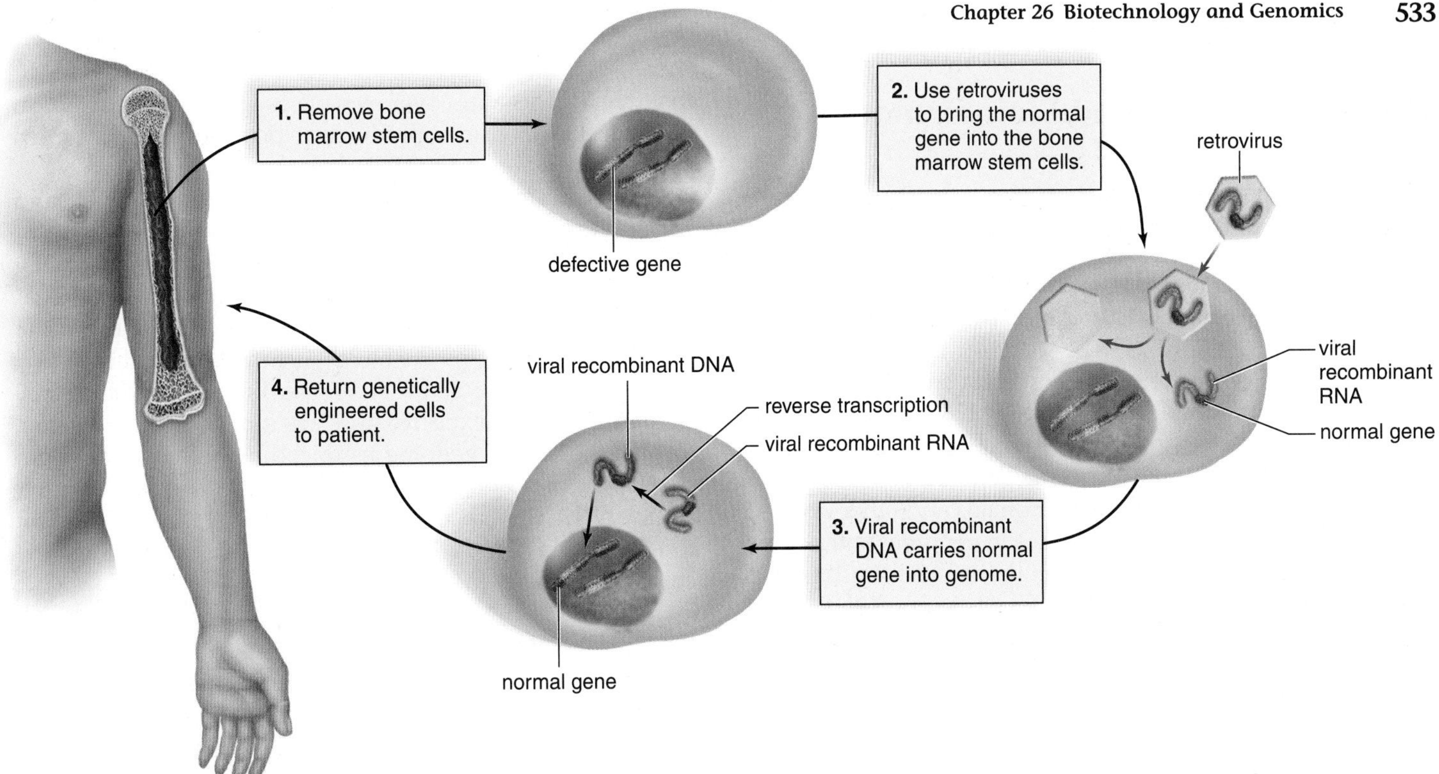

Figure 26.10 Ex vivo gene therapy in humans.
Bone marrow stem cells are withdrawn from the body, an RNA retrovirus is used to insert a normal gene into them, and they are then returned to the body.

safe can be used to ferry a normal gene into the body, and so can liposomes, which are microscopic globules of lipids specially prepared to enclose the normal gene (i.e., ex vivo gene therapy). On the other hand, sometimes the gene is injected directly into a particular region of the body (i.e., in vivo gene therapy).

Note that despite its promise for treating disorders, gene therapy is still in its infancy and has had detrimental side effects for many patients, such as causing leukemia. Nonetheless, some strides are being made with particular diseases, such as those described in the following sections.

Ex Vivo Gene Therapy

Figure 26.10 describes an ex vivo methodology for treating children who have SCID (severe combined immunodeficiency). These children lack the enzyme ADA (adenosine deaminase), which is involved in the maturation of T and B cells. Therefore, these children are prone to constant infections and may die without treatment. To carry out gene therapy, bone marrow stem cells are removed from the bone marrow of the patient and infected with a virus that carries a normal gene for the enzyme into their DNA. Then the cells are returned to the patient, where it is hoped they will divide to produce more blood cells with the same genes. Patients who have undergone this procedure show significantly improved immune function associated with a sustained rise in the level of ADA enzyme activity in the blood.

Another example of ex vivo gene therapy has been used to treat familial hypercholesterolemia, a condition that develops when liver cells lack a receptor protein for removing cholesterol from the blood. The high levels of blood cholesterol make the patient subject to fatal heart attacks at a young age. A small portion of the liver is surgically excised and then infected with a virus containing a normal gene for the receptor before being returned to the patient. Patients are expected to experience lowered serum cholesterol levels following this procedure.

In Vivo Gene Therapy

Cystic fibrosis patients lack a gene that codes for the transmembrane carrier of the chloride ion (see section 23.2). They often die due to numerous infections of the respiratory tract because a thick mucus forms in the lungs and attracts bacteria and other antigens. In gene therapy trials, the gene needed to cure cystic fibrosis is sprayed into the nose or delivered to the lower respiratory tract by an adenovirus vector or by using liposomes. So far, these treatments have met with limited success, but investigators are trying to improve uptake by using a combination of different vectors.

Gene therapy is increasingly relied upon as a part of cancer treatment. Genes are being used to make healthy cells more tolerant of chemotherapy, while making tumor cells more sensitive. Knowing that the tumor suppressor gene *p53* brings about apoptosis (cell death), researchers are interested in finding a way to selectively introduce *p53* into cancer cells, and in that way, kill them.

Check Your Progress

1. What methods are being used to introduce genes into human beings for gene therapy?
2. Give an example of ex vivo and of in vivo gene therapy.

26.4 Genomics and Bioinformatics

Learning Outcomes

1. Discuss the implications of knowing the human genome sequence.
2. Describe a major insight from comparative genomics.
3. Compare and contrast functional genomics and proteomics.

In the twentieth century, researchers discovered the structure of DNA, how DNA replicates, and how DNA and RNA are involved in the process of protein synthesis. Genetics in the twenty-first century largely concerns **genomics,** the study of the complete genetic sequences of humans and other organisms. Knowing the sequence of bases in genomes is the first step, and mapping their location on the chromosomes is the next step. The enormity of the task can be appreciated by knowing not only that we have over 20,000 genes that code for proteins, but also that nearly 99% of the 3.2 billion bases of our genome is noncoding and contains many repetitive sequences of unknown function. Many other organisms have even a larger number of protein-coding genes but fewer noncoding regions when compared to the human genome.

Sequencing the Genome

We now know the sequence of the roughly 3.2 billion pairs of DNA bases in our genome. Stretched out, the DNA in each of our cells is six feet long, and the nucleotides would make a book over one half million pages if printed as text. This feat, which has been compared to completion of the periodic table of the elements in chemistry, was accomplished by the **Human Genome Project (HGP),** a 13-year effort that involved both university and private laboratories around the world. How did they do it? First, investigators developed a laboratory procedure that would allow them to decipher a short sequence of base pairs, and then instruments became available that could carry out sequencing automatically. Over the 13-year span, DNA sequencers were constantly improved, and today's instruments can automatically analyze up to 120 million base pairs of DNA in a 24-hour period. So, new genomes are being sequenced all the time, and at a much faster rate than the human genome. For example, the genome of the African clawed frog, *Xenopus laevis,* which is roughly the same size as the human genome, was sequenced in under a year.

Completion of the human genome sequence has opened up great possibilities for biomedical research and treatment. These methods have been used to screen for particular diseases, as discussed earlier in this chapter. The HGP also led to the discovery of many small regions of DNA that vary among individuals (polymorphisms). Most of these are single nucleotide polymorphisms (SNPs) (a difference of only one nucleotide). Many SNPs have no effect. Others may contribute to protein-coding differences affecting the phenotype. It's possible that certain SNP patterns change an individual's susceptibility to disease and alter their response to medical treatments. These discoveries have now led pharmaceutical companies to consider producing "designer drugs," which are tailored for an individual's genotype.

Determining that humans have between 20,000 and 25,000 genes required a number of techniques, many of which relied on identifying RNAs in cells and then working backward to find the DNA that can pair with that RNA. Structural genomics—knowing the sequence of the bases and how many genes we have—is now being followed by functional genomics. Most of the known 20,000–25,000 human genes are expected to code for proteins. However, most of the human genome is noncoding because it does not specify the order of amino acids in a polypeptide. This noncoding DNA, once dismissed as "junk DNA," is now thought to serve many important functions, and has piqued the curiosity of many investigators.

Genome Architecture

As mentioned above, researchers were somewhat surprised that nearly 99% of the human genome is DNA that does not directly code for amino acid sequences. Some of the DNA that does not specify polypeptides is transcribed into ribosomal RNA and transfer RNA, both structural molecules involved in protein assembly. The rest of the genome is comprised of transposable elements (or transposons), repetitive DNA elements, and sequences with unknown function. Transposable elements make up approximately 45% of the human genome. Transposable elements, originally discovered by Barbara McClintock in 1950 (who later won a Nobel Prize for this work), are short sequences of DNA that are able to jump from one location on a chromosome to another. Their movement to a new location sometimes alters neighboring genes, particularly decreasing their expression. In other words, a transposon sometimes acts like a regulator gene. The movement of transposons throughout the genome is thought to be a driving force in the evolution of living things.

Nearly half of the human genome is made up of repetitive elements, which occur when the same sequence of two or more nucleotides (e.g., CACACA) are repeated many times along the length of one or more chromosomes. Although many scientists still dismiss them as having no function, others point out that the centromeres and telomeres of chromosomes are composed of repetitive elements and, therefore, repetitive DNA elements may not be as useless as once thought. Telomeres are repetitive DNA sequences found near the ends of chromosomes and are thought to help maintain their structural stability. In addition, perhaps repetitive sequences in centromeres may help with segregating sister chromatids during cell division.

What Is a Gene?

Knowledge of the human genome sequences has changed the way researchers think about the concept of a "gene." Historically, a gene was thought of as a particular location (locus) of a chromosome.

While prokaryotes typically possess a single circular chromosome with genes that are tightly packed together, eukaryotic chromosomes are much more complex. The genes are seemingly randomly distributed along the length of a chromosome and are fragmented into exons, with intervening sequences called introns scattered throughout the length of the gene. In fact, 95% or more of most human genes is composed of introns. Recall that after transcription, introns need to be spliced out and exons joined together to form a functional mRNA transcript that will next be translated into a protein. Once regarded as merely intervening sequences, introns are now attracting attention as regulators of gene expression. The presence of introns allows exons to be put together in various sequences so that different mRNAs and proteins can result from a single gene. It could also be that introns function to regulate gene expression and help determine which genes are to be expressed and how they are to be spliced. In fact, entire genes have been found embedded within the introns of other genes.

Thus, perhaps the modern definition of a gene should take the emphasis away from the chromosome and place it on the results of transcription. Previously, molecular genetics considered a gene to be a nucleic acid sequence that codes for the sequence of amino acids in a protein. In contrast to this definition, we have known for some time all three types of RNA (rRNA, mRNA, and tRNA) are transcribed from DNA and that these RNAs are useful products. We also know that protein-coding regions can be interrupted by regions that do not code for a protein but do produce RNAs with various functions. In light of these new findings, Mark Gerstein and associates suggested a new definition in 2007: "A gene is a genomic sequence (either DNA or RNA) directly encoding functional products, either RNA or protein." This definition takes into account three new things we have learned by investigating genomic sequences: 1) a gene product may not necessarily be a protein; 2) a gene may not be found at a particular locus on a chromsome; and, 3) the genetic material need not be only DNA—some prokaryotes have RNA genes.

Functional and Comparative Genomics

In addition to the human genome, the genomes of many other organisms, including a common bacterium, a yeast, and a mouse, are also complete (Table 26.1). Because we now know the nucleotide sequence of many genomes, we can now focus on comparative and functional genomics.

Using **comparative genomics,** researchers have identified many similarities between the sequence of human bases and those of other organisms. Model organisms (e.g., those found in Table 26.1) can be used in these types of genetic analyses because they share many mechanisms and cellular pathways with other organisms, including humans. For example, scientists inserted a human gene associated with Parkinson disease into the fruit fly, *Drosophila melanogaster*, and the flies showed symptoms similar to those seen in humans with the disorder. These studies confirmed that the suspected gene is involved in Parkinson disease and suggested we might be able to use fruit flies to test potential therapies.

Comparative genomics also offers a way to study changes in the genome through time because some diseases, as well as model organisms, have a shorter generation time than humans. In this way, we have been able to track the evolution of the human immunodeficiency virus (HIV), the virus that causes AIDS, in individual patients. Tracking the genome sequence of the virus through time has allowed scientists and doctors to understand how the virus responds to different drug therapy regimens and, in some cases, modify treatment to improve a patient's life span.

Comparing genomes will also help us understand the evolutionary relationships among organisms. One surprising discovery is that the genomes of all vertebrates are similar. Researchers were not surprised to find that the genomes of humans and chimpanzees were approximately 98% alike, but they did not expect to find that the human and mouse sequence were 85% similar. Genomic comparisons will likely yield improved ability to reconstruct evolutionary relationships among organisms, as discussed in Chapter 27.

TABLE 26.1 Comparison of Sequenced Genomes

Organism	*Homo sapiens* (human)	*Mus musculus* (mouse)	*Drosophila melanogaster* (fruit fly)	*Arabidopsis thaliana* (flowering plant)	*Caenorhabditis elegans* (roundworm)	*Saccharomyces cerevisiae* (yeast)
Estimated Size	~3,000 million bases	2,500 million bases	180 million bases	125 million bases	97 million bases	12 million bases
Estimated Number of Genes	~20,500	~30,000	13,600	25,500	19,100	6,300
Chromosome Number	46	40	8	10	12	32

Comparative genomics also allows researchers to infer function of unknown genes in one species through amino acid similarity of those genes in another species.

The aim of **functional genomics** is to understand the function of the various genes discovered within each genomic sequence and how these genes interact. In fact, functional genomics has utilized comparative genomics to assess similarities between the human genes and genes of other organisms to help deduce the probable function of many of our 20,000+ genes. Functional genomics also uses DNA microarrays to monitor the expression of thousands of genes simultaneously (see Fig. 26.8). The use of a microarray can tell what genes are turned on in a specific cell or tissue type in a particular organism at a particular point in time and under certain environmental circumstances. For example, we could compare gene expression of a patient in different stages of cancer growth to assist with treatment. As discussed earlier in this chapter, DNA microarrays can also be used to identify various mutations in a human's genome. This is called the person's **genetic profile,** which can be used to determine if various genetic diseases are likely, as well as to suggest which drug therapy may be most appropriate based on the individual's genotype.

Figure 26.11 Bioinformatics.
New computer programs are being developed to make sense out of the raw data generated by genomics and proteomics. Bioinformatics allows researchers to study both functional and comparative genomics in a meaningful way.

Proteomics

Now that entire genomes are being published for different species, there is a race to sequence their proteomes, or a species' entire collection of proteins. **Proteomics** is the study of the structure, function, and interaction of cellular proteins, which differ depending on each cell type. Each cell produces hundreds of different proteins that can vary between cells and within the same cell, depending on conditions. Therefore, the goal of proteomics is an overwhelming endeavor.

Computer modeling of the three-dimensional shape of these proteins is an important part of proteomics. The study of protein shape and function is essential to the discovery of better drugs so that their chemical structure can match that of different protein shapes. One day, it may be possible to correlate drug treatment to the particular genome of the individual to increase efficiency and decrease side effects.

Bioinformatics

Bioinformatics is the application of computer technologies, specially developed software, and statistical techniques to the study of biological information, particularly databases that contain much genomic and proteomic information (Fig. 26.11). The new data produced by structural genomics and proteomics have produced literally terabytes of raw data stored in databases that are readily available to research scientists. It is called raw data because billions of base pairs of DNA nucleotide sequence have little meaning by themselves. Functional genomics and proteomics are dependent on computer analysis to find significant patterns in the raw data. For example, BLAST, which stands for *b*asic *l*ocal *a*lignment *s*earch *t*ool, is a computer program that can identify homologous genes among the genomic sequences of model organisms. Homologous genes are genes that code for the same proteins, although the base sequence may be slightly different. Finding these differences can help identify the putative function of genes as new organisms' genomes are sequenced, and also help to trace the history of evolution among a group of organisms. For example, researchers found the function of the protein that causes cystic fibrosis by using the computer to search for genes in model organisms that have the same sequence. Because they knew the function of this same gene in model organisms, they could deduce the function in humans. This was a necessary step toward possibly developing specific treatments for cystic fibrosis.

Bioinformatics has various applications in human genetics. The human genome has 3 billion known base pairs, and without the computer it would be almost impossible to make sense of these data. For example, it is now known that an individual's genome often contains multiple copies of a gene. But individuals may differ as to the number of copies—called copy number variations. Now it seems that the number of copies of a gene in a genome can be associated with specific diseases. The computer can help make correlations between genomic differences among large numbers of people and certain diseases.

It is safe to say that without bioinformatics, our progress in assembling DNA sequences into genomes; determining the function of DNA sequences; mapping genes on chromosomes; comparing our genome to model organisms; knowing how genes and proteins interact in cells; and so forth, would be extremely slow. Instead, with the help of bioinformatics, progress should proceed rapidly in these and other areas.

Check Your Progress

1. Discuss the benefits of whole genome sequencing.
2. How can comparative genomics give us insights into gene function?
3. Once a complete genome is sequenced, discuss the contributions of functional genomics and proteomics.

Bioethical Focus

DNA Forensics

Traditional fingerprinting has been used for years to identify criminals and to exonerate those wrongly accused of crimes. However, sometimes fingerprints are not left at the crime scene, while at other times fingerprints may be present but not connected to the particular crime at all. Forensic DNA analysis, on the other hand, does not depend on fingerprints. It can be used to identify suspects based on the unique DNA sequences in their genes, as obtained from biological samples found at the crime scene. One advantage of DNA profiling is that only a tiny sample is needed because PCR is used. So, the sample can come from a drop of blood, semen from a rape victim, or even a single hair root!

In DNA profiling, target sequences of DNA simple tandem repeats (also called **"microsatellites"**) are amplified using PCR and separated using automated equipment that can determine the length of the fragments because the DNA is labeled with fluorescent dyes that are picked up by a laser beam as they move through a gel contained in capillaries (see Fig. 26.3). A person could have a single fragment at an STR locus, indicating he or she is a homozygote, or two fragments, indicating he or she is a heterozygote (see Fig. 26.4). The suspect's genotype is then compared to that found at the crime scene. A mismatch at one STR locus is enough to exclude a suspect. However, a match at a single locus is not enough to prove that the DNA sample at the crime scene belongs to a particular suspect. Why? The particular genotype is compared to the frequency of that genotype in the FBI database. Let's say that a particular genotype is present in 10% of the population of U.S. males, and that a suspect in a rape case has DNA that matches that in semen collected from the crime scene. This means there is a 10% chance of *accidental match.* Is this guilt beyond a reasonable doubt? Surely not. Therefore, the microsatellite genotyping procedure is done over several different locations in the genome (usually 7–8) to ensure that the probability of accidental match is incredibly low. Thus, advocates of DNA profiling claim that identification is "beyond a reasonable doubt."

Opponents of this technology, however, point out that it is not without its problems. Police or laboratory negligence can invalidate the evidence. For example, during the O.J. Simpson trial, the defense claimed that the DNA evidence was inadmissible because it could not be proven that the police had not "planted" O.J.'s blood at the crime scene. Problems involving sloppy laboratory procedures and the credibility of forensic experts have also been reported.

In addition to identifying criminals, DNA fingerprinting can be used to establish paternity, determine nationality for immigration purposes, and identify victims of a national disaster, such as a tsunami or earthquake.

Now that genomic sequencing is continually becoming cheaper and faster, it may not be in the too distant future that the genome of all humans can be sequenced inexpensively. Considering the usefulness of DNA fingerprints, perhaps everyone should be required to contribute blood to create a national DNA fingerprint databank. To this end, some have even argued for having a complete genome sequence of each person on their driver's license. Others say, however, this would constitute "search without cause," which is unconstitutional.

Form Your Own Opinion

1. Would you be willing to provide your DNA for a national DNA databank? Why or why not?
2. If not everyone, do you think that convicted felons, at least, should be required to provide DNA for a databank?
3. Should all defendants have access to DNA fingerprinting (at government expense) to prove they didn't commit a crime? Should this include those already convicted of crimes who want to reopen their cases using new DNA evidence?

Applying the Concepts [Revisited]

Through reading this chapter, it should be clear that advances in genomics and biotechnology will allow us to better understand the causes of human diseases, as well as to develop potential treatments and cures. One of the great challenges of the twenty-first century will be to learn how to assemble and interpret the massive quantities of data generated through genome sequencing projects, comparative genomics, and proteomics. An even greater challenge to society is to use these data and the technologies that result in an ethically responsible manner. As you think about these ethical dilemmas, answer the following questions.

1. In this chapter, you learned about gene cloning. Now, as you have likely heard in the news, we can clone whole organisms—most recently, dogs, cats, and mules. Soon, we may even be able to clone human beings! Should cloning of pets be allowed (for a cost of about $30,000 to the consumer)? What about humans? What are the scientific arguments for and against cloning?
2. Currently, couples who are expecting a baby often perform routine tests on the fetus, such as the test for Down syndrome. In the near future, we may be able to test the whole genome—potentially for thousands of genetic disorders. If you detect a genetic disorder in your unborn child, would you elect to correct it using gene therapy? What if we knew the child would be short as an adult? Would you elect to make him or her taller? Discuss what you believe are the ethical limits to the use of such technology.

SUMMARIZING THE CONCEPTS

26.1 DNA Cloning

- DNA cloning can be used to isolate a gene and produce many copies of it. The gene can be studied in the laboratory or inserted into a bacterium, plant, or animal, and produced in mass quantities, as a commercial product or a medicine.
- Two methods are currently available for making copies of DNA: recombinant DNA technology and the polymerase chain reaction (PCR). Recombinant DNA contains DNA from two different sources. A restriction enzyme cleaves both vector (plasmid) DNA and foreign DNA. The resulting "sticky ends" facilitate

insertion of foreign DNA into vector DNA. The foreign gene is sealed into the vector DNA by DNA ligase.

- PCR uses a heat-resistant DNA polymerase to quickly make multiple copies of a specific piece (target) of DNA. PCR is a chain reaction because the targeted DNA is replicated over and over again. Analysis of DNA segments following PCR has multiple uses from assisting genomic research to DNA forensics studies.

26.2 Biotechnology Products

- Transgenic organisms have had a foreign gene inserted into them.
- Genetically modified bacteria, agricultural plants, and farm animals now produce commercial products of interest to humans, such as hormones and vaccines. Bacteria usually secrete the product, but the seeds of plants and the milk of animals contain the product.
- Transgenic bacteria have also been engineered to promote the health of plants, extract minerals, and produce medically important chemicals.
- Transgenic crops, engineered to resist herbicides and pests, are commercially available.
- Transgenic animals have been given various genes, in particular the one for bovine growth hormone (bGH). Cloning of whole animals is now possible.

26.3 Gene Therapy

- Testing the DNA can reveal the presence of certain genetic disorders. To test the DNA for mutations, it is possible to use genetic markers or DNA chips.

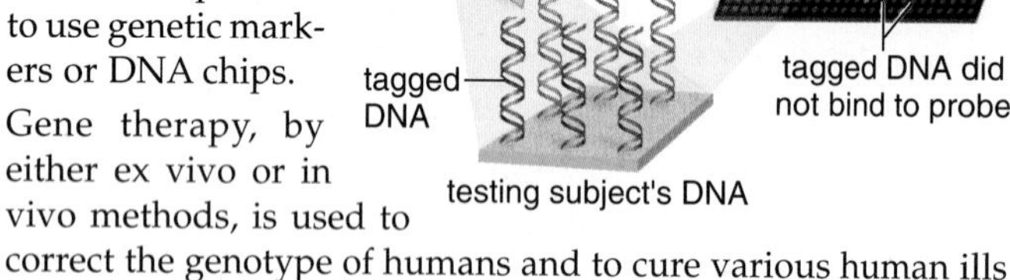

- Gene therapy, by either ex vivo or in vivo methods, is used to correct the genotype of humans and to cure various human ills by giving the patient a foreign gene.
- During ex vivo therapy, cells are removed from the patient, treated, and returned to the patient. Ex vivo gene therapy has apparently helped children with SCID lead normal lives.

1. Remove bone marrow stem cells.
2. Use retroviruses to bring the normal gene into the bone marrow stem cells.
3. Viral recombinant DNA carries normal gene into genome.
4. Return genetically engineered cells to patient.

- In vivo therapy consists of directly giving the patient a foreign gene that will improve his or her health. Although it has limited success for treating cystic fibrosis, a number of in vivo therapies are being employed in the war against cancer and other human illnesses, such as cardiovascular disease.

26.4 Genomics and Bioinformatics

- Researchers now know the sequence of all the base pairs of the human genome. So far, between 20,000 and 25,000 genes that code for proteins have been found; the rest of our DNA consists of noncoding regions.
- Noncoding regions include: introns, transposable elements, and repetitive DNA.
- Noncoding DNA, once dismissed as "junk," is now thought to have important functions that may include maintaining structural integrity of chromosomes and gene regulation.
- Currently, researchers are placing an emphasis on functional and comparative genomics.
- Functional genomics aims to understand the function of protein-coding regions and noncoding regions of our genome. To that end, researchers are utilizing new tools such as DNA microarrays.
- Microarrays can also be used to create an individual's genetic profile, which can be helpful in predicting illnesses and how a person will react to particular medications.
- Comparative genomics has revealed little difference between the DNA sequence of our bases and those of many other organisms. Genome comparisons have revolutionized our understanding of evolutionary relations by revealing previously unknown relationships between organisms.
- Proteomics is the study of which genes are active in producing proteins in which cells and under which circumstances.
- Bioinformatics is the use of computers to assist with analysis of data from proteomics and functional and comparative genomics.

TESTING YOURSELF

Choose the best answer for each question.

1. Restriction enzymes found in bacterial cells are ordinarily used
   a. during DNA replication.
   b. to degrade the bacterial cell's DNA.
   c. to degrade viral DNA that enters the cell.
   d. to attach pieces of DNA together.
2. Which of the following enzymes are needed to introduce foreign DNA into a vector?
   a. DNA gyrase and DNA ligase
   b. DNA ligase and DNA polymerase
   c. DNA gyrase and DNA polymerase
   d. restriction enzyme and DNA gyrase
   e. restriction enzyme and DNA ligase
3. A genetic profile can
   a. assist in maintaining good health.
   b. be accomplished utilizing bioinformatics.
   c. show how many genes are normal.
   d. be accomplished utilizing a microarray.
   e. Both a and d are correct.
4. The polymerase chain reaction
   a. uses RNA polymerase.
   b. takes place in huge bioreactors.
   c. uses a temperature-insensitive enzyme.
   d. makes many nonidentical copies of DNA.
   e. All of these are correct.
5. During the PCR reaction, the DNA sample is heated in order to
   a. separate it into single strands.
   b. allow primers to bind.
   c. allow DNA polymerase to work.
   d. synthesize RNA.

6. Gel electrophoresis separates DNA fragments according to
   a. the sequence of their bases.
   b. differences in electrical charge.
   c. their length.
   d. the number of mutations they carry.
7. Because of the Human Genome Project, we know
   a. humans have large portions of our genome with no known function.
   b. humans have approximately 20,500 genes.
   c. our genome size.
   d. All of these are correct.
   e. Only a and c are correct.
8. Gene therapy
   a. is sometimes used in medicine today.
   b. is always successful.
   c. is used only to cure genetic disorders, such as SCID and cystic fibrosis.
   d. makes use of viruses to carry foreign genes into human cells.
   e. Both a and d are correct.
9. Because of the Human Genome Project, we now know
   a. the sequence of the base pairs of our DNA.
   b. the sequence of all genes along the human chromosomes.
   c. all the mutations that lead to genetic disorders.
   d. All of these are correct.
   e. Only a and c are correct.
10. Bioinformatics can
    a. assist genomics and proteomics.
    b. compare our genome to that of a monkey.
    c. depend on computer technology.
    d. match up genes with proteins.
    e. All of these are correct.
11. Proteomics is used to discover
    a. which genes are active in which cells.
    b. which proteins are active in which cells.
    c. the structure and function of proteins.
    d. how proteins interact.
    e. All but a are correct.
12. Which of the following statements is incorrect?
    a. Bacteria usually secrete the biotechnology product into the medium.
    b. Plants are being engineered to have human proteins in their seeds.
    c. Animals are engineered to have a human protein in their milk.
    d. Animals can be cloned, but plants and bacteria cannot.
13. Which of these is a true statement?
    a. Plasmids can serve as vectors.
    b. Plasmids can carry recombinant DNA, but viruses cannot.
    c. Vectors carry only the foreign gene into the host cell.
    d. Only gene therapy uses vectors.
    e. Both a and d are correct.
14. Comparative genomics
    a. is the application of computer technologies to the study of the genome.
    b. is the study of the structure, function, and interaction of cellular proteins.
    c. can be used to understand human gene function by investigating genes in other species.
    d. involves studying all the genes that occur in a cell.
    e. is the study of a person's complete genotype, or genetic profile.
15. What is ex vivo gene therapy?
    a. when a person's genes are cloned using PCR
    b. when a person is infected with a virus that delivers a functional gene into a chromosome with a dysfunctional gene
    c. when cells containing a dysfunctional gene are removed from the patient, treated, and then returned to the patient
    d. when a person's gene is cloned into a plant, which then produces the correct gene product in its seeds
    e. None of these are correct.

UNDERSTANDING THE TERMS

bioinformatics 536
biotechnology 529
cloning 526
comparative genomics 535
DNA fingerprint 528
DNA ligase 526
DNA microarray 532
functional genomics 536
gene cloning 526
gene therapy 532
genetic engineering 526
genetic marker 532
genetic profile 536
genetic profiling 532
genome 526
genomics 534
Human Genome Project (HGP) 534
microsatellites 537
plasmid 526
polymerase chain reaction (PCR) 527
proteomics 536
protoplasts 531
recombinant DNA (rDNA) 526
restriction enzyme 526
short tandem repeat (STR) 528
transgenic organism 526
vector 526

THINKING CRITICALLY

1. We can use transgenic viruses to infect humans and help treat genetic disorders. That is, the viruses are genetically modified to contain "normal" human genes to try to replace nonfunctional human genes. Using this type of gene therapy, a person is infected with a particular virus, which then delivers the "normal" human gene to cells by infecting them. What are some pros and cons of viral gene therapy?
2. In a genomic comparison between humans and yeast, what genes would you expect to be similar?

INQUIRY INTO LIFE WEBSITE

The companion website for *Inquiry into Life* provides a wealth of information organized and integrated by chapter. You will find practice tests, animations, videos, and much more that will complement your learning and understanding of general biology.

http://www.mhhe.com/maderinquiry13

PART VI

Evolution and Diversity

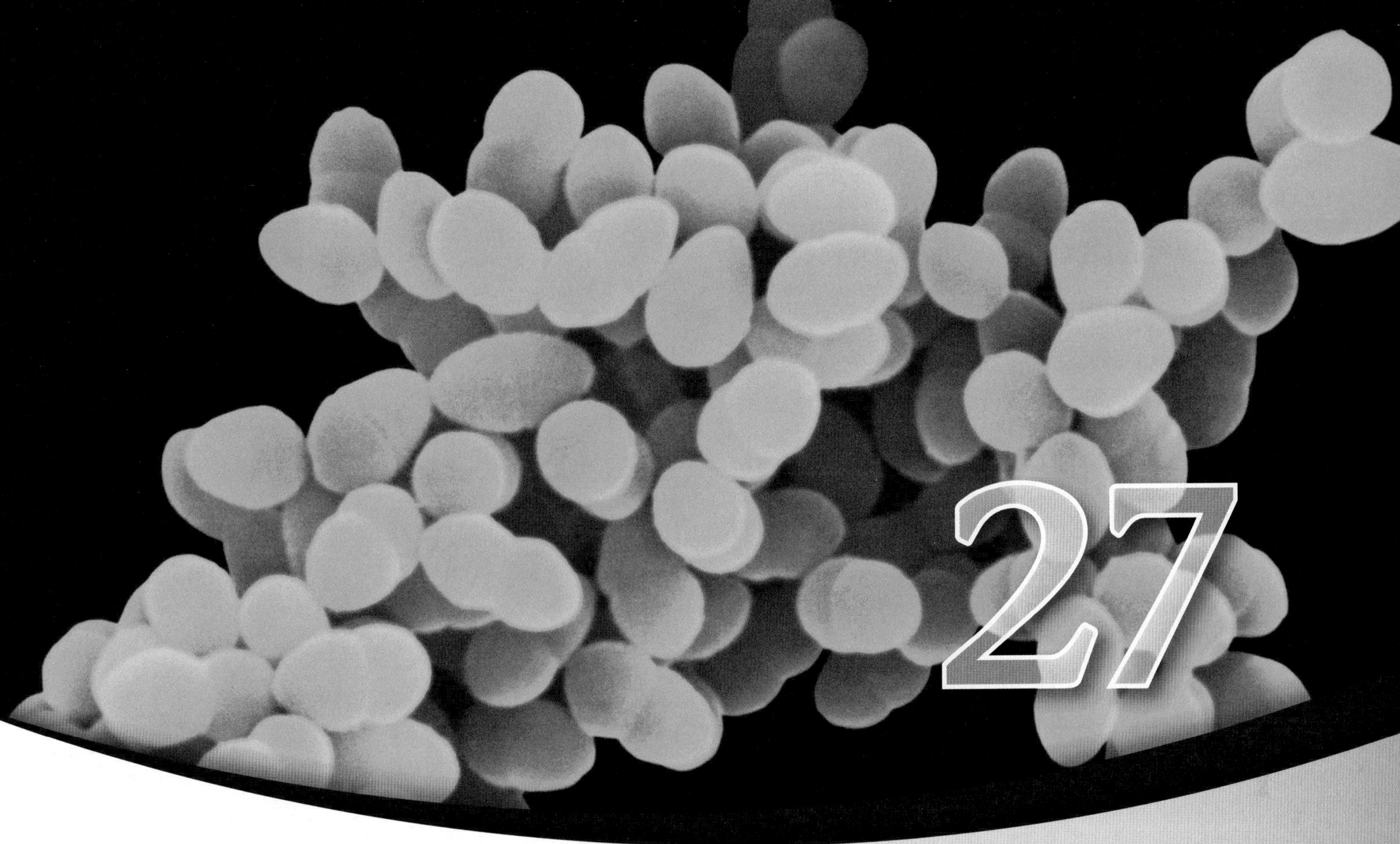

Evolution of Life

Chapter Outline

Applying the Concepts

The death of a Virginia high school student in October 2007 sparked a wave of concern about MRSA.

MRSA (methicillin-resistant *Staphylococcus aureus*) is a strain of a common bacterium that can be deadly because of its resistance to several types of antibiotics. The bacterium is often found on the skin and in the noses of healthy people, but *S. aureus* can cause "staph" infections in scrapes, cuts, and open wounds. Until recently, severe staph infections were found primarily in hospitals. Now, MRSA has escaped and is spreading through prisons and schools, especially in gymnasiums and locker rooms. The 2007 death resulted in the closure of all 21 Bedford County, Virginia, school buildings after student protests of unsanitary conditions. Regular hand washing with soap and warm water is recommended to prevent staph infections, particularly after you have gone to the gym.

Use and overuse of antibiotics has resulted in evolution of resistant bacterial strains, such as MRSA. Although we tend to think of evolution as happening over long timescales, human activities can accelerate the process of evolution quite rapidly. In fact, evolution of resistance to the antibiotic methicillin occurred in just one year! Antibiotic-resistant strains of bacteria are generally hard to treat, and treatment of infected patients can cost thousands of dollars.

Some scientists believe that "superbugs," or bacteria that have evolved antibiotic resistance, will be a far bigger threat to human health than emerging diseases such as H1N1 flu and AIDS. The good news is that our understanding of evolutionary biology has helped change human behavior to deal with superbugs. For example, doctors no longer prescribe antibiotics unless they are relatively certain a patient has a bacterial infection. Antibiotic resistance is an example why evolution is important in people's everyday lives. In this chapter, you will learn about evidence that indicates evolution has occurred and about how the evolutionary process works.

27.1 Origin of Life

Learning Outcomes

1. Describe the steps that may have led to the formation of the first cells.
2. Explain how the Miller-Urey experiment supports our current understanding of the origin of small organic molecules.
3. Relate three theories about the origin of macromolecules to the origin of the protocell and eventually the true cell.

The common ancestor for all living things was the first cell or cells. The planet Earth existed for a long time before the first cell arose—over a billion years, in fact. Earth is 4.6 billion years old, and the earliest fossils of prokaryotes are 3.5 billion years old. A billion years is about 13 million human life spans, assuming humans live 75 years. The origin of the first cell is an event of low probability because a complex series of events would have had to occur—but this length of time is long enough for such an event to have occurred. Today we do not believe that life arises spontaneously from nonlife, and we say that "life comes only from life." However, the very first living thing had to have come from nonliving chemicals. Under the conditions of early Earth, it is possible that a chemical reaction produced the first cell(s). A particular mix of inorganic chemicals could have reacted to produce small organic molecules such as glucose, amino acids, and nucleotides. Then these would have polymerized into macromolecules. Once a plasma membrane formed through other means, a structure called a **protocell** could have come into existence. Once these protocells could self-replicate, they were then called true cells (Fig. 27.1).

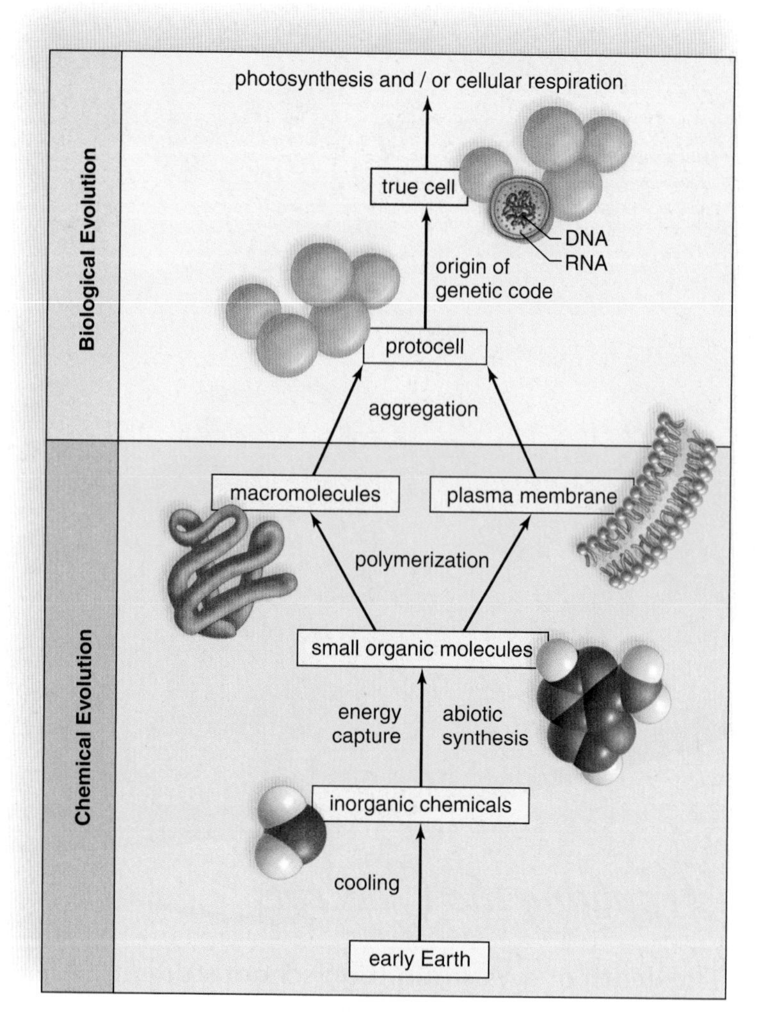

Figure 27.1 Origin of the first cell(s).
A chemical evolution may have produced the first cell. First, inorganic chemicals present on early Earth reacted to produce small organic molecules, which polymerized to form macromolecules. With the origination of the plasma membrane, the first primitive cell (a protocell) evolved. Once this protocell acquired genes and could replicate, a true cell had evolved.

Evolution of Small Organic Molecules

Most chemical reactions take place in water, and the first protocell undoubtedly arose somewhere in the ocean. Two hypotheses are that these molecules arose either on the surface of the ocean or in mid-oceanic ridges. Protocells could have formed on the surface of the seas and in seaside pools where much energy was available. Ultraviolet radiation was intense in these areas because there was no ozone shield and surface waters offer little protection from ultraviolet radiation. The ozone shield arose later when photosynthesis added oxygen to the atmosphere.To test the hypothesis that small organic molecules formed at the ocean's surface, Stanley Miller and Harold Urey performed an experiment (known as the Miller-Urey experiment) in 1953. In the early Earth, volcanoes erupted constantly, and the first atmospheric gases would have consequently contained methane (CH_4), ammonia (NH_3), and hydrogen (H_2). These gases could then have been washed into the ocean by the first rains. Fierce lightning and unabated ultraviolet radiation would have allowed them to react and produce the first organic molecules.

To simulate the Earth's early environment, Miller placed the inorganic chemicals mentioned in a closed system, heated the mixture, and circulated it past an electric spark (simulating lightning). After a week, the solution contained a variety of amino acids and organic acids (Fig. 27.2). This and other similar experiments support the hypothesis that inorganic chemicals in the absence of oxygen (O_2) and in the presence of a strong energy source can result in organic molecules.

In contrast, is also possible that small organic molecules could have arisen amidst mid-oceanic ridges within the depths of the sea. Hydrothermal (hot water) vents (openings) occur in the region of mid-oceanic ridges. A vent can be huge, measuring 10–15 m wide with sides about 15–20 m high. Hot water spewing out of these vents contains a mix of iron-nickel sulfides. Amazingly, scientists have discovered communities of organisms, including tube

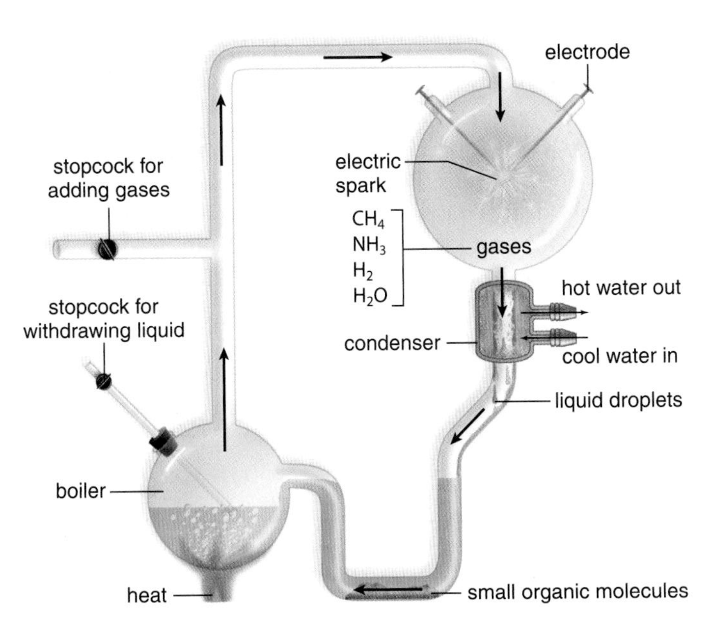

Figure 27.2 Miller and Urey's apparatus and experiment.
Gases that were thought to be present in the early Earth's atmosphere were admitted to the apparatus, circulated past an energy source (electric spark), and cooled to produce a liquid that could be withdrawn. Upon chemical analysis, the liquid was found to contain various small organic molecules.

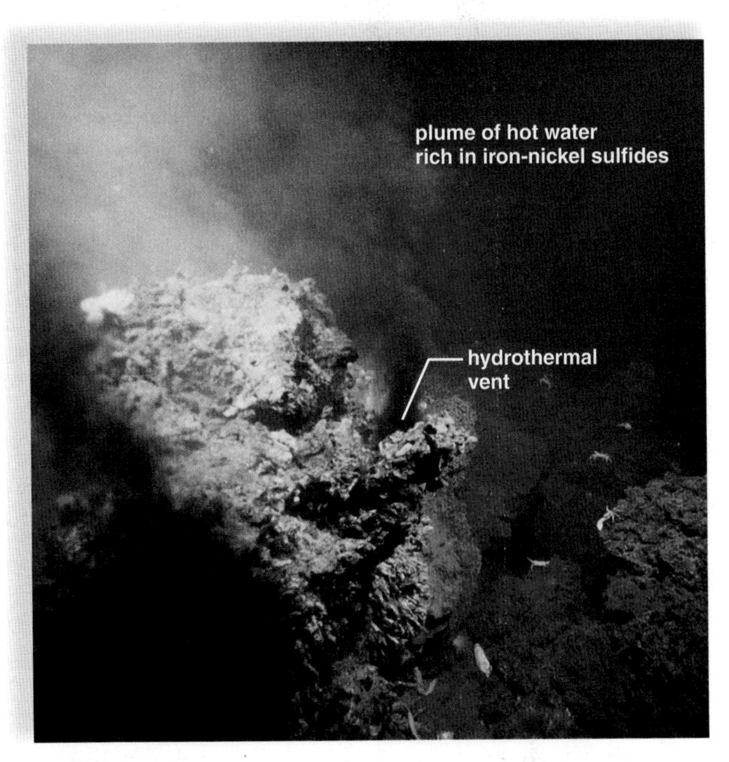

Figure 27.3 Chemical evolution at hydrothermal vents.
Minerals that form at deep-sea hydrothermal vents like this one can catalyze the formation of ammonia and even organic molecules.

worms and giant clams, living in the regions of hydrothermal vents. It is possible that in the past, the right combination of conditions occurred at these vents to initiate life (Fig. 27.3).

Macromolecules

Once formed, the first small organic molecules gave rise to still larger molecules and then macromolecules. There are three primary hypotheses concerning this stage in the origin of life. One is the **RNA-first hypothesis,** which suggests that only the macromolecule RNA (ribonucleic acid) was needed at this time to progress toward formation of the first cell or cells. Scientists formulated this hypothesis after discovering that RNA can sometimes be both a genetic substrate and an enzyme. Such RNA molecules are called ribozymes. The first genes and enzymes could thus both have been composed of RNA, as we now know that ribozymes exist. Scientists who support this hypothesis say it was an "RNA world" some 4 billion years ago.

Another hypothesis is termed the **protein-first hypothesis.** Sidney Fox has shown that amino acids polymerize abiotically (without life) when exposed to dry heat. He suggests that amino acids collected in shallow puddles along the rocky shore, and the heat of the sun caused them to form **proteinoids,** small polypeptides that have some catalytic properties. When proteinoids are returned to water, they form **microspheres,** structures composed only of protein that have many of the properties of a cell. Some of these proteins could have had enzymatic properties. This hypothesis assumes that DNA genes came after protein enzymes arose.

A combination of these two hypotheses is put forth by Graham Cairns-Smith. He thinks that clay was especially helpful in causing the polymerization of both proteins and nucleic acids at the same time. Clay attracts small organic molecules and contains iron and zinc, which may have served as inorganic catalysts for polypeptide formation. In addition, clay tends to collect energy from radioactive decay and then discharges it when the temperature or humidity changes, possibly providing a source of energy for polymerization. Cairns-Smith suggests that RNA nucleotides and amino acids became associated in such a way that polypeptides were ordered by, and helped synthesize, RNA.

Check Your Progress

1. Describe how the Miller-Urey experiment supports the hypothesis that small organic molecules were produced at the ocean's surface.
2. Compare and contrast the RNA-first hypothesis, the protein-first hypothesis, and the combination hypothesis for the origin of macromolecules.

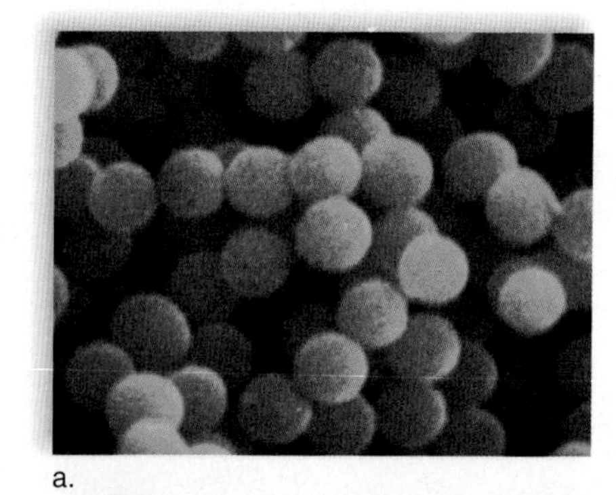

a.

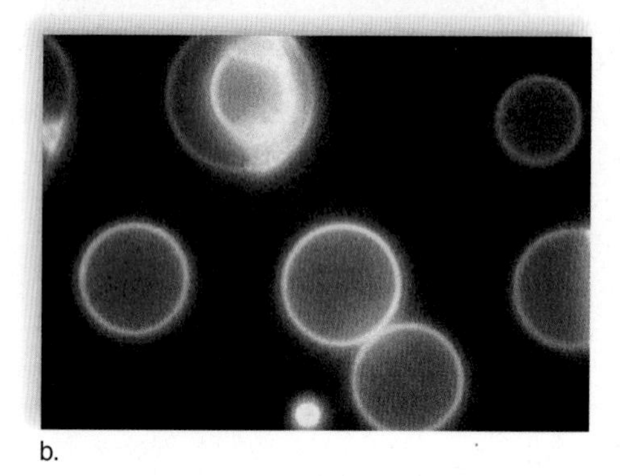

b.

Figure 27.4 Protocell components.
a. Microspheres, which are composed only of protein, have a number of cellular characteristics and could have evolved into the protocell.
b. Liposomes form automatically when phospholipid molecules are put into water. Plasma membranes may have evolved similarly.

The Protocell

After macromolecules formed, something akin to a modern plasma membrane was needed to separate them from the environment. Thus, before the first true cell arose, there would likely have been a protocell, a structure that had a lipid-protein membrane and carried on energy metabolism. Fox has shown that if lipids are made available to microspheres, the two tend to become associated, producing a lipid-protein membrane (Fig. 27.4*a*).

Aleksandr Oparin, a Soviet biochemist, showed that under appropriate conditions of temperature, ionic composition, and pH, concentrated mixtures of macromolecules can give rise to complex units called coacervate droplets. Coacervate droplets have a tendency to absorb and incorporate various substances from the surrounding solution. Eventually, a semipermeable boundary may form around the droplet. In a liquid environment, phospholipid molecules automatically form droplets called **liposomes** (Fig. 27.4*b*). Perhaps the first plasma-like membrane formed in this manner. However it happened, development of the plasma membrane was key because it separated the genetic material from the outside environment.

The Heterotroph Hypothesis

It has been suggested that the protocell likely was a **heterotroph,** an organism that takes in preformed food. During the early evolution of life, the ocean contained abundant nutrition in the form of small organic molecules. This suggests that heterotrophs preceded **autotrophs,** organisms that make their own food.

At first, the protocell may have used preformed ATP, but as this supply dwindled, cells that could extract energy from carbohydrates to transform ADP to ATP were favored. Glycolysis is a common metabolic pathway in living things, and this testifies to its early evolution in the history of life. Since there was no free oxygen, we can assume that the protocell carried on a form of fermentation using enzymes. It seems logical that the protocell at first had limited ability to break down organic molecules and that it took millions of years for glycolysis to evolve completely.

The True Cell

A true cell is a membrane-bounded structure that can carry on protein synthesis to produce the enzymes that allow DNA to replicate. The central concept of genetics states that DNA directs protein synthesis and that information flows from DNA to RNA to protein. It is possible that this sequence developed in stages.

According to the RNA-first hypothesis, RNA would have been the first genetic material to evolve, and the first true cell would have had RNA genes. These genes would have directed and enzymatically carried out protein synthesis, as in ribozymes. Also, today we know that some viruses have RNA as their genetic material. These viruses have a protein enzyme called reverse transcriptase that uses RNA as a template to form DNA, which then initiates and directs protein formation. Perhaps, with time, reverse transcription gave rise to DNA genes. Once DNA genes existed, they may have specified proteins.

A second possibility is the protein-first hypothesis, which suggests that proteins, or at least polypeptides, were the first of the three molecules (i.e., DNA, RNA, and protein) to arise. Only after the protocell developed sophisticated enzymes did it have the ability to synthesize DNA and RNA from small molecules provided by the ocean. Researchers point out that because nucleic acids are very complicated molecules, the likelihood that RNA arose on its own is minimal.

A third alternative is proposed by Cairns-Smith, who suggests that polypeptides and RNA evolved simultaneously. Therefore, the first true cell would have contained RNA genes that could have replicated because of the presence of proteins. This eliminates the baffling chicken-and-egg paradox: which came first, proteins or RNA? It does mean, however, that two unlikely events would have had to happen at the same time.

Once the protocells acquired genes that could replicate, they became cells capable of reproducing, and biological evolution began.

Check Your Progress

1. State two attributes of the protocell.
2. How do the three hypotheses concerning the origin of macromolecules pertain to the origin of a true cell?

27.2 Evidence of Evolution

Learning Outcomes

1. Define biological evolution.
2. Discuss four lines of evidence for evolution.

Once a true cell formed, biological evolution began. Biological evolution, or simply **evolution** is all the changes that have occurred in living things since the beginning of life due to differential reproductive success. That is, some individuals reproduce more than others because they are better suited to their environment. Table 27.1 indicates that Earth is about 4.6 billion years old and that prokaryotes, probably the first living organisms, evolved about 3.5 billion years ago. The eukaryotic cell arose about 2.1 billion years ago, but multicellularity didn't begin until perhaps 700 million years ago. This means that only unicellular organisms were present for 80% of the time that life has existed on Earth. Consequently, most evolutionary events we will be discussing in the next few chapters occurred in less than 20% of the history of life!

Because of descent with modification, all living things share the same fundamental characteristics: they are made of cells, take chemicals and energy from the environment, respond to external stimuli, and reproduce. Living things are diverse because organisms are adapted to different environments and the features that enable them to survive in those environments vary tremendously.

Many fields of biology provide evidence that evolution through descent with modification occurred in the past and is still occurring. Let us look at the various types of evidence for evolution.

Fossil Evidence

Fossils are the remains and traces of past life or any other direct evidence of past life. Most fossils consist only of hard parts of organisms, such as shells, bones, or teeth, because these are usually preserved after death. The soft parts of a dead organism are often consumed by scavengers or decomposed by bacteria. Occasionally, however, an organism is buried quickly and in such a way that decomposition is never completed or is completed so slowly that the soft parts leave an imprint of their structure. Traces include trails, footprints, burrows, worm casts, or even preserved droppings.

The great majority of fossils are found embedded in sedimentary rock. Sedimentation, a process that has been going on since Earth formed, can take place on land or in bodies of water. The weathering and erosion of rocks produces particles that vary in size and are called sediment. As such particles accumulate, sediment becomes a stratum (pl., strata), a recognizable layer of rock. Any given stratum is older than the one above it and younger than the one immediately below it, so that the relative age of fossils can be determined based on their depth.

Paleontologists are biologists who study the fossil record and from it draw conclusions about the history of life. When fossils are arranged from oldest to youngest, they can provide evidence of evolutionary change through time. A particularly good example of this is the horse, which evolved from a dog-sized, forest dwelling tree browser with forward looking eyes and teeth geared toward chewing leaves, to the animal we recognize as a horse today. Modern horses are adapted for a open field-type habitat, with eye sockets more to the sides of their head to allow for better peripheral vision and thereby detection of possible predators, as well as teeth geared toward grinding grasses.

Particularly interesting are the fossils that serve as **transitional links** between groups. A famous example are the fossils of *Archaeopteryx* that lived about 165 million years ago (Fig. 27.5). The fossil clearly seems to be an intermediate form between dinosaurs and birds. *Archaeopteryx* had dinosaur-like features including jaws with teeth and a long, jointed tail, but it also also had feathers and wings similar to those of modern birds. The fossils of *Archaeopteryx* are similar to other transitional fossils in that they have some traits like their ancestors and others like their descendants, rather than expressing intermediate traits. Other transitional links among fossil vertebrates suggest that fishes evolved before amphibians, which evolved before reptiles, which evolved before mammals in the history of life.

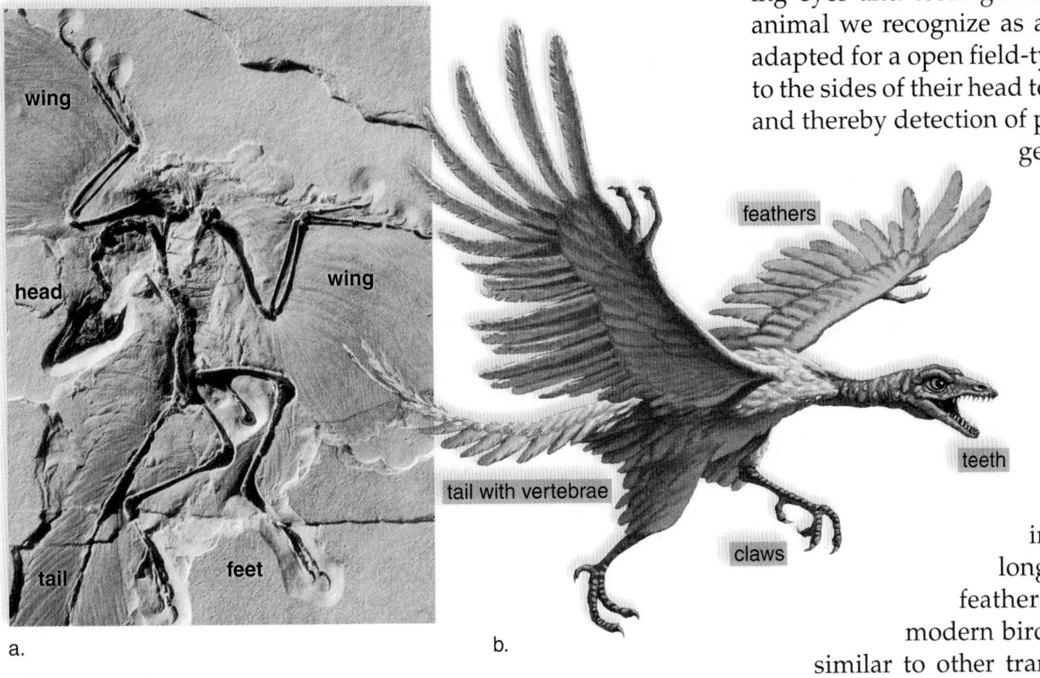

Figure 27.5 Transitional fossils.
a. A fossil of *Archaeopteryx,* now considered the first bird, has features of both birds and dinosaurs. Fossils indicate it had feathers and wing claws. Most likely, it was a poor flier. Perhaps it ran over the ground on strong legs and climbed into trees with the assistance of these claws. **b.** *Archaeopteryx* also had a feather-covered, reptile-like tail that shows up well in this artist's representation.

TABLE 27.1 The Geological Timescale: Major Divisions of Geological Time and Some of the Major Evolutionary Events of Each Time Period

Era	Period	Epoch	Millions of Years Ago	Plant Life	Animal Life
		Holocene	(0.01–0)	Human influence on plant life	Age of *Homo sapiens*
		Significant Mammalian Extinction			
	Quaternary	Pleistocene	(1.8–0.01)	Herbaceous plants spread and diversify.	Presence of ice age mammals. Modern humans appear.
		Pliocene	(5.33–1.8)	Herbaceous angiosperms flourish.	First hominids appear.
		Miocene	(23.03–5.33)	Grasslands spread as forests contract.	Apelike mammals and grazing mammals flourish; insects flourish.
*Cenozoic**		Oligocene	(33.9–23.03)	Many modern families of flowering plants evolve; appearance of grasses.	Browsing mammals and monkeylike primates appear.
	Tertiary	Eocene	(55.8–33.9)	Subtropical forests with heavy rainfall thrive.	All modern orders of mammals are represented.
		Paleocene	(65.5–55.8)	Flowering plants continue to diversify.	Ancestral primates, herbivores, carnivores, and insectivores appear.
		Mass Extinction: Dinosaurs and Most Reptiles			
	Cretaceous		(145.5–65.5)	Flowering plants spread; conifers persist.	Placental mammals appear; modern insect groups appear.
Mesozoic	Jurassic		(199.6–145.5)	Flowering plants appear.	Dinosaurs flourish; birds appear.
		Mass Extinction			
	Triassic		(251–199.6)	Forests of conifers and cycads dominate.	First mammals appear; first dinosaurs appear; corals and molluscs dominate seas.
		Mass Extinction			
	Permian		(299–251)	Gymnosperms diversify.	Reptiles diversify; amphibians decline.
	Carboniferous		(359.2–299)	Age of great coal-forming forests: ferns, club mosses, and horsetails flourish.	Amphibians diversify; first reptiles appear; first great radiation of insects.
		Mass Extinction			
Paleozoic	Devonian		(416–359.2)	First seed plants appear. Seedless vascular plants diversify.	First insects and first amphibians appear on land.
	Silurian		(443.7–416)	Seedless vascular plants appear.	Jawed fishes diversify and dominate the seas.
		Mass Extinction			
	Ordovician		(488.3–443.7)	Nonvascular land plants appear.	Invertebrates spread and diversify; first jawless and then jawed fishes appear.
	Cambrian		(542–488.3)	Marine algae flourish.	All invertebrate phyla present; first chordates appear.
			630	Soft-bodied invertebrates	
			1,000	Protists diversify.	
			2,100	First eukaryotic cells	
Precambrian Time			2,700	O_2 accumulates in atmosphere.	
			3,500	First prokaryotic cells.	
			4,570	Earth forms.	

*Many authorities divide the Cenozoic era into the Paleogene period (contains the Paleocene, Eocene, and Oligocene epochs) and the Neogene period (contains the Miocene, Pliocene, Pleistocene, and Holocene epochs).

Geological Timescale

As a result of studying strata, scientists have divided Earth's history into eras, and then periods and epochs (see Table 27.1). The fossil record has helped determine the dates given in the table. There are two ways to age fossils. The relative dating method determines the relative order of fossils and strata depending on the layer of rock in which they were found, but it does not determine the actual date they were formed.

The absolute dating method relies on radioactive dating techniques to assign an actual date to a fossil. All radioactive isotopes have a particular half-life, the length of time it takes for half of the radioactive isotope to change into another stable element. Carbon 14 (^{14}C) is the only radioactive isotope in organic matter. Assuming a fossil contains organic matter, half of the ^{14}C will have changed to nitrogen 14 (^{14}N) in 5,730 years. For example, if a fossil has one-fourth the amount of radioactive ^{14}C as a modern sample, then the fossil is approximately 11,460 years old (2 half-lives).

Biogeographical Evidence

Another type of evidence that supports evolution through descent with modification is found in the field of **biogeography,** the study of the range and distribution of plants and animals in different places throughout the world. Such distributions are consistent with the hypothesis that, when forms are related, they evolved in one locale and then spread to accessible regions. Therefore, a different mix of plants and animals would be expected whenever geography separates continents, islands, seas, and so on.

Many of these barriers arose through a process called **continental drift.** That is, the continents have never been fixed. Rather, their positions and the positions of the oceans have changed over time (Fig. 27.6). During the Permian period, all the present landmasses belonged to one continent and then later drifted apart. As evidence of this, fossils of one species of seed fern (*Glossopteris*) have been found on all the southern continents separated by oceans. This species' presence on Antarctica is evidence that this continent was not always frozen. In contrast, many Australian species are restricted to that continent, including the majority of marsupials (pouched mammals such as the kangaroo). What is the explanation for these distributions? Some organisms must have evolved and spread out before the continents broke up. Then they became extinct.

The world's six biogeographical regions each have their own distinctive mix of living things. Darwin noted that South America lacked rabbits, even though the environment was quite suitable to them. He concluded there are no rabbits in South America because rabbits evolved somewhere else and had no means of reaching South America. To take another example, both cacti and spurges (*Euphorbia*) are plants adapted to a hot, dry environment—both are succulent, spiny, flowering plants. Why do cacti grow in the American deserts and most *Euphorbia* grow in African deserts when each would do well on the other continent? It seems obvious that they just happened to evolve on their respective continents. What is the best explanation for this phenomenon? Different mammals and flowering plants evolved separately in each biogeographical region, and

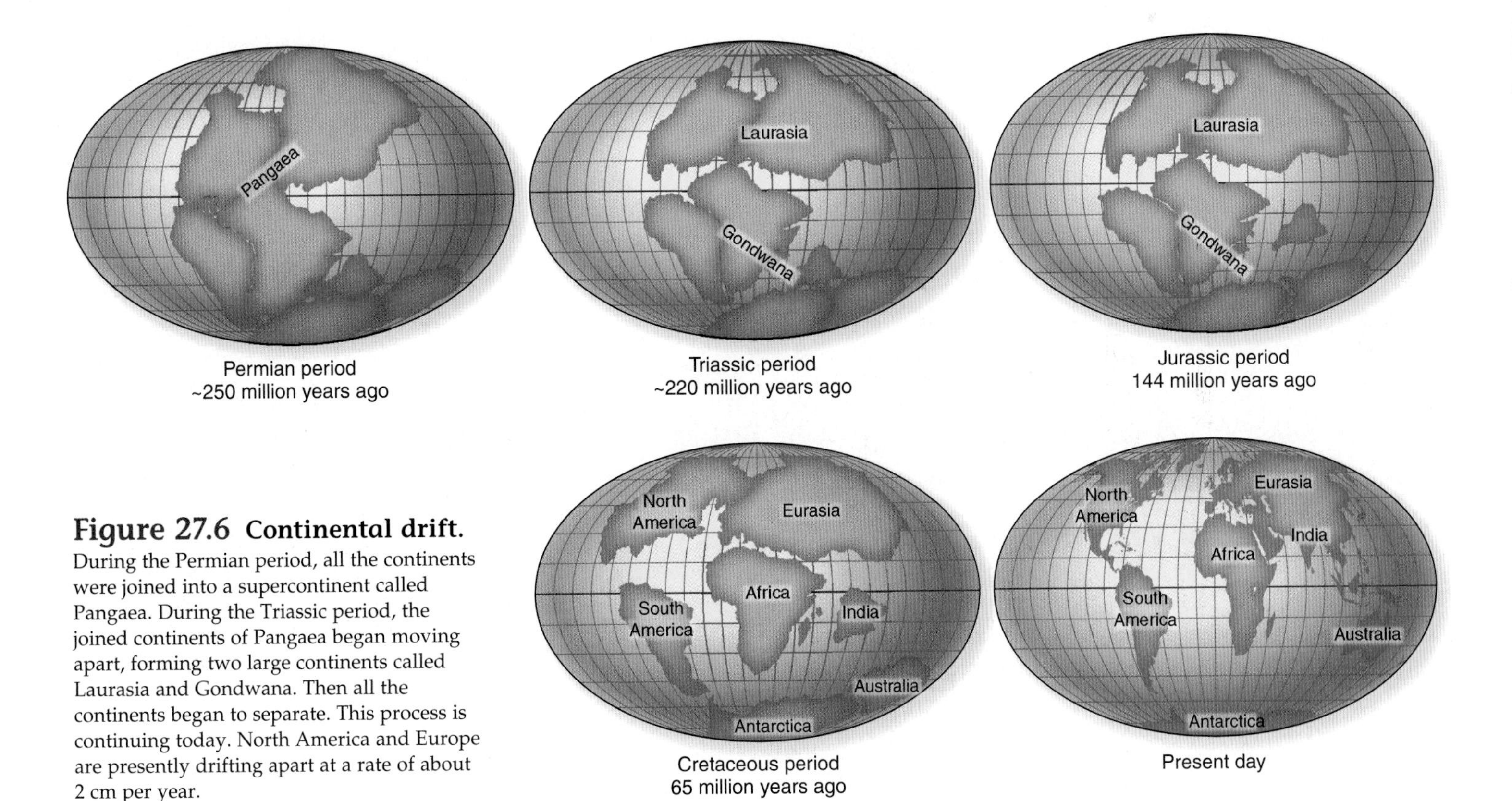

Figure 27.6 Continental drift. During the Permian period, all the continents were joined into a supercontinent called Pangaea. During the Triassic period, the joined continents of Pangaea began moving apart, forming two large continents called Laurasia and Gondwana. Then all the continents began to separate. This process is continuing today. North America and Europe are presently drifting apart at a rate of about 2 cm per year.

barriers such as mountain ranges and oceans prevented them from migrating to other regions.

Mass Extinctions

Extinction is the death of every member of a species. During mass extinctions, a large percentage of species become extinct within a relatively short period of time. So far, there have been five major mass extinctions. These occurred at the ends of the Ordovician, Devonian, Permian, Triassic, and Cretaceous periods (see Table 27.1), and a sixth is likely occurring now, probably as a result of human activities (discussed in Chapter 36). Following mass extinctions, the remaining groups of organisms are likely to spread out and fill the habitats vacated by those that have become extinct.

It was proposed in 1977 that the Cretaceous extinction (or "Cretaceous crisis") was due to an asteroid that exploded, producing meteorites that fell to Earth. A large meteorite striking Earth could have produced a cloud of dust that mushroomed into the atmosphere, blocking out the sun and causing plants to freeze and die. A huge crater that could have been caused by a meteorite involved in the Cretaceous extinction was found in the Caribbean–Gulf of Mexico region on the Yucatán Peninsula. During the Cretaceous period, great herds of dinosaurs roamed the plains, as did *Parasaurolophus walkeri* and *Triceratops* (Fig. 27.7), but all dinosaur species went extinct near the end of the Cretaceous period.

Certainly, continental drift contributed to the Ordovician extinction. This extinction occurred after Gondwana arrived at the South Pole. Immense glaciers, which drew water from the oceans, chilled even once-tropical land. Marine invertebrates and coral reefs, which were especially hard hit, didn't recover until Gondwana drifted away from the pole and warmth returned. The mass extinction at the end of the Devonian period saw an end to 70% of marine invertebrates. Other scientists believe that this mass extinction could have been due to movement of Gondwana back to the South Pole.

Check Your Progress

1. How do the geological timescale and fossils offer evidence of evolution?
2. How does biogeography help explain why certain animals and plants are found in some areas but not others?
3. How does continental drift help explain the uniqueness of biodiversity in the world's different biogeographical regions?

Anatomical Evidence

The concept of common descent offers a plausible explanation for anatomical similarities among organisms. Vertebrate forelimbs are used for flight (birds and bats), orientation during swimming (whales and seals), running (horses), climbing (arboreal lizards), or swinging from tree branches (monkeys). Yet all vertebrate forelimbs contain the same sets of bones organized in similar ways, despite their dissimilar functions. The most plausible explanation for this unity is

Figure 27.7 Dinosaurs of the late Cretaceous period.
Parasaurolophus walkeri, although not as large as other dinosaurs, was one of the largest plant-eaters of the late Cretaceous period. The crest atop its head was about 2 m long and was used to make booming calls. Also living at this time were the rhinolike dinosaurs represented here by *Triceratops* (left), another herbivore.

that the basic forelimb plan belonged to a common ancestor, and then the plan was modified in the succeeding groups as each continued along its own evolutionary pathway. Structures that are anatomically similar because they are inherited from a common ancestor are called **homologous structures** (Fig. 27.8). In contrast, **analogous structures** serve the same function, but are not constructed similarly, nor do they share a common ancestry. The wings of birds and insects and the eyes of octopi and humans are analogous structures and are similar due to a common environment, not common ancestry. The presence of homology, not analogy, is evidence that organisms are related.

Vestigial structures are anatomical features that are fully developed in one group of organisms but that are reduced and may have no function in similar groups. Most birds, for example, have well-developed wings for flight. However, some bird species (e.g., ostrich) have greatly reduced wings and do not fly. Similarly, snakes have no use for hindlimbs, and yet some have remnants of hindlimbs in a pelvic girdle and legs. The presence of vestigial structures can be explained by common descent. Vestigial structures occur because organisms inherit their anatomy from their ancestors. They are traces of an organism's evolutionary history.

The homology shared by vertebrates extends to their embryological development (Fig. 27.9). At some time during development, all vertebrates have a postanal tail and exhibit paired pharyngeal pouches. In fishes and amphibian larvae, these pouches develop into functioning gills. In humans, the first pair of pouches becomes the cavity of the middle ear and the auditory tube. The second pair becomes the tonsils, while the third and fourth pairs become the thymus and parathyroid glands. Why should terrestrial vertebrates develop and then modify structures like pharyngeal pouches that have lost their original function? The most likely explanation is that fishes are ancestral to other vertebrate groups.

In 1859, Charles Darwin (see the Science Focus, page 554) speculated that whales evolved from a land mammal. His hypothesis has now been substantiated. In recent years the fossil record has yielded an incredible parade of fossils that link modern whales and dolphins to land ancestors

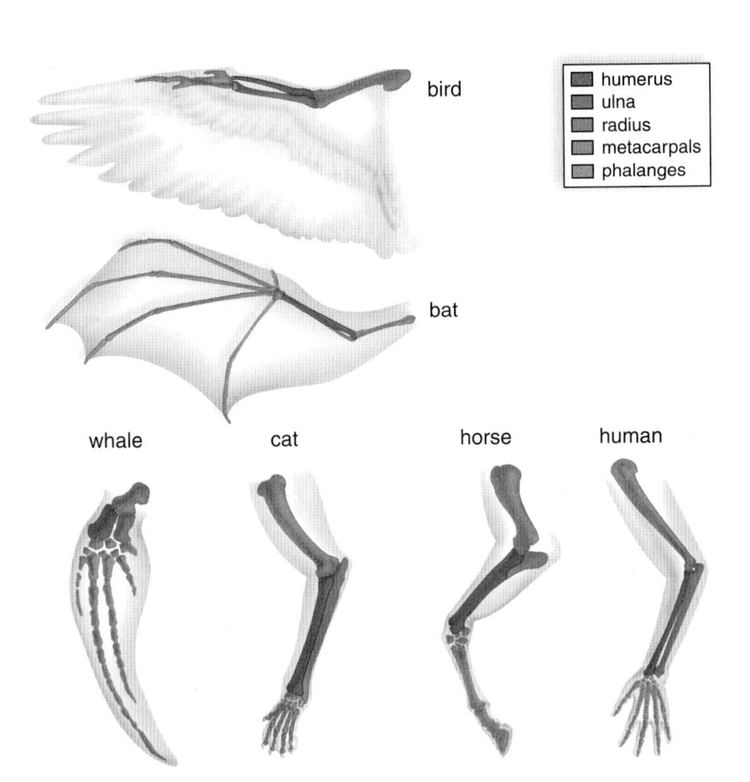

Figure 27.8 Significance of homologous structures.
Although the specific design details of vertebrate forelimbs are different, the same bones are present (note color-coding). Homologous structures provide evidence of a common ancestor.

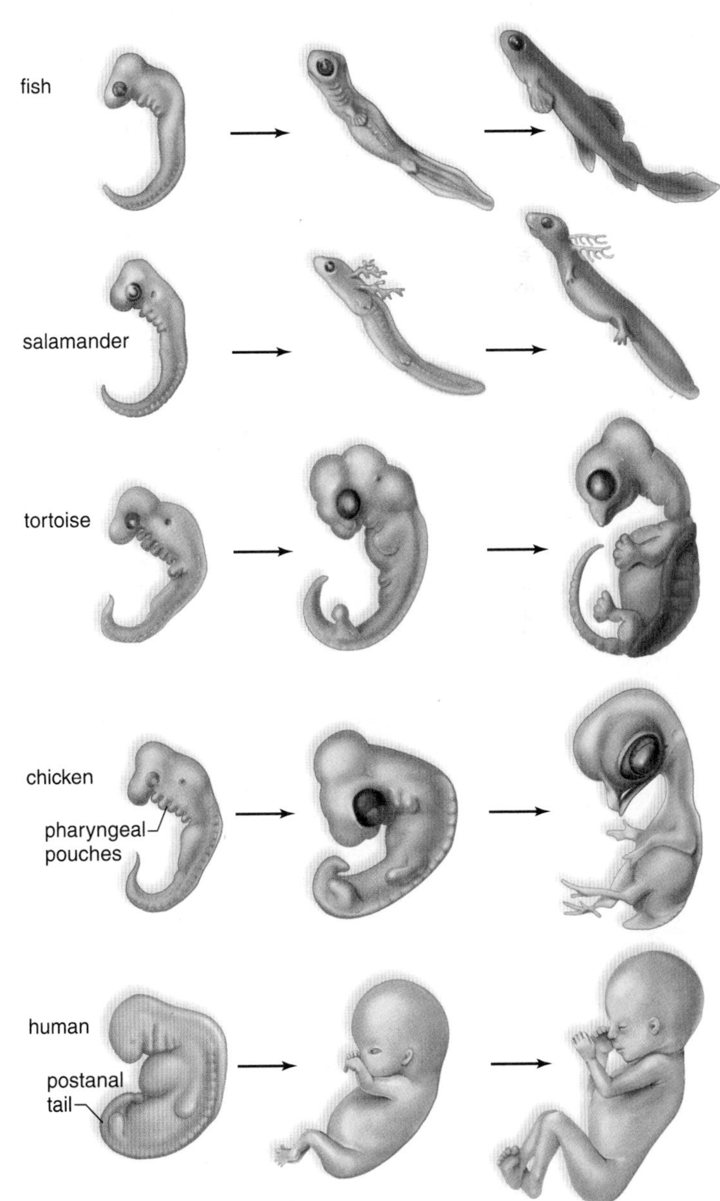

Figure 27.9 Significance of developmental similarities.
At these comparable developmental stages, vertebrate embryos have many features in common, which suggests they evolved from a common ancestor. (These embryos are not drawn to scale.)

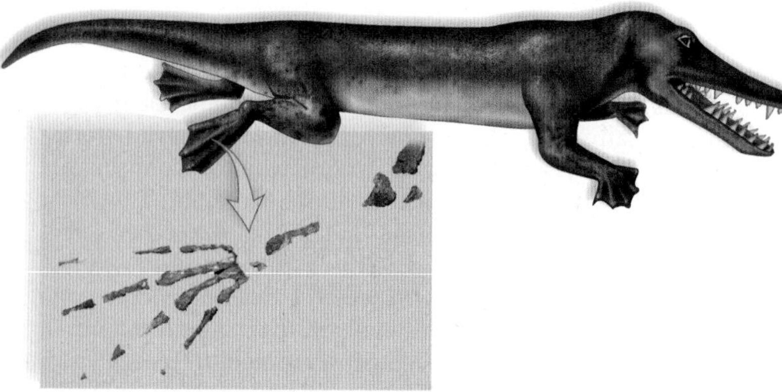

Figure 27.10 Ancestor to whales.
Ambulocetus, an ancestor to modern whales, dated 50 MYA. The presence of limbs is evidence that land-based mammals gave rise to whales.

(Fig. 27.10). The presence of a vestigial pelvic girdle and legs in modern whales is also significant evidence that ancestors of these creatures once walked on land. However, that these structures are severely reduced in modern whales is consistent with the fact that they are exclusively aquatic today.

Biochemical Evidence

Almost all living organisms use the same basic biochemical molecules, including DNA, ATP (adenosine triphosphate), and many identical or nearly identical enzymes. Further, organisms use the same DNA triplet code for the same 20 amino acids in their proteins. Because the sequences of DNA bases in the genomes of many organisms are now known, it has become clear that humans share a large number of genes with much simpler organisms. It appears that life's vast diversity has come about by only a slight difference in many of the same genes and regulatory genes often found in introns and other regions of the genome. The result has been widely divergent types of bodies.

When the degree of similarity in DNA nucleotide sequences or the degree of similarity in amino acid sequences of proteins is examined, the more similar the DNA sequences are, generally the more closely related the organisms are. For example, humans and chimpanzees are about 97% similar! Cytochrome *c* is a molecule that is used in the electron transport chain of all the organisms appearing in Figure 27.11. Data regarding differences in the amino acid sequence of cytochrome *c* show that the sequence in a human differs from that in a monkey by only one amino acid, from that in a duck by 11 amino acids, and from that in a yeast by 51 amino acids. These data are consistent with other data regarding the anatomical similarities of these organisms and, therefore, how closely they are related.

Evolution is no longer considered a hypothesis. It is one of the great unifying theories of biology. In science, the word *theory* is reserved for those conceptual schemes that are supported by a large number of observations and scientific experiments. The theory of evolution has the same status in biology that the theory of gravity has in physics.

Check Your Progress

1. How do homologous structures give evidence for evolution?
2. Which would have have fewer protein and DNA base sequence differences with humans—chimpanzees or cows? Explain.

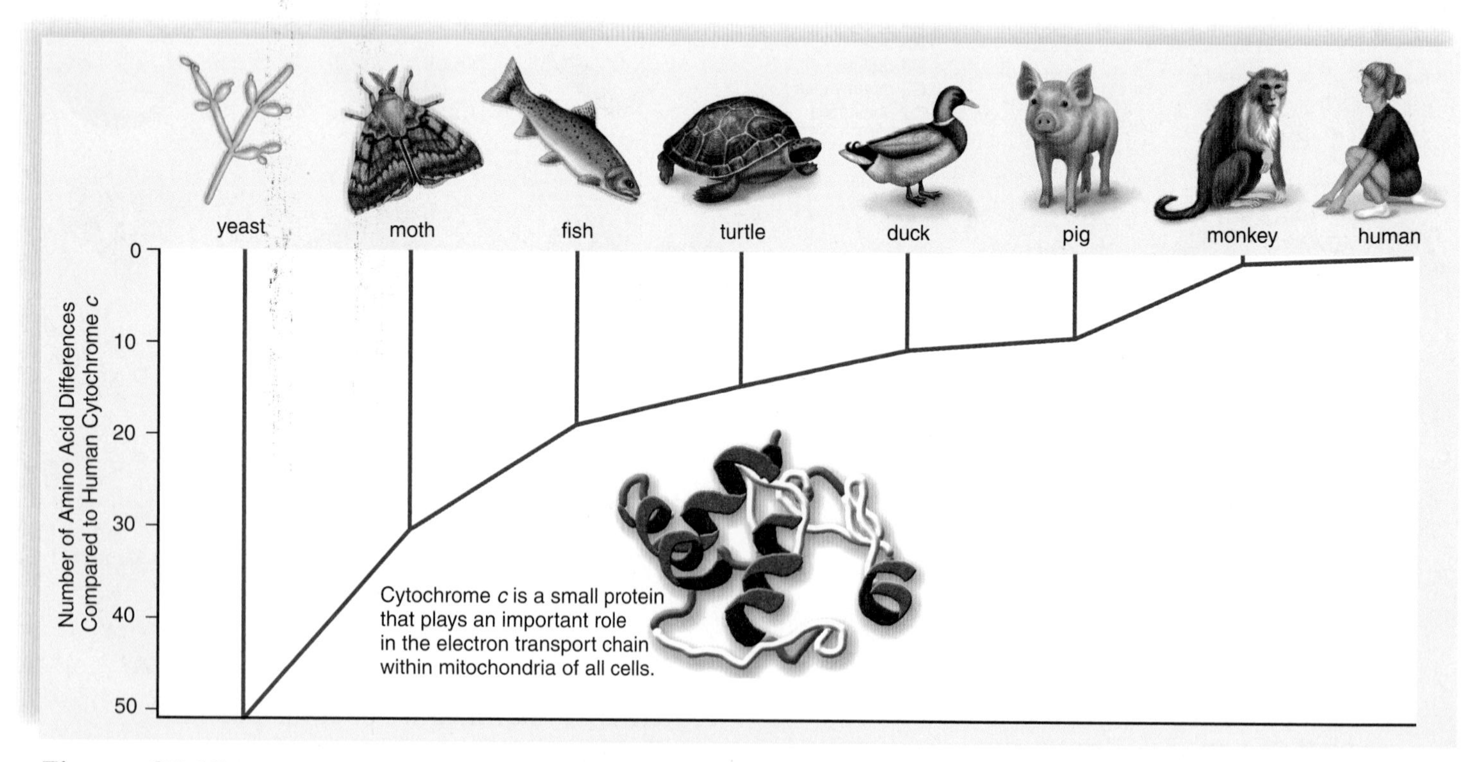

Figure 27.11 Significance of biochemical differences.
The branch points in this diagram indicate the number of amino acids that differ between human cytochrome *c* and the organisms depicted. These biochemical data are consistent with those provided by a study of the fossil record and comparative anatomy.

27.3 The Process of Evolution

Learning Outcomes

1. List the five conditions necessary to maintain Hardy-Weinberg equilibrium.
2. Describe the agents of evolutionary change.
3. Compare and contrast the three types of natural selection.

Some people have the misconception that individuals evolve. However, evolution occurs at the population level. As evolution takes place, genetic changes occur within a population and, over generations, these lead to phenotypic changes that are commonly seen in that population. In this section we will consider a change in gene frequencies within a population over time, defined as **microevolution.** Microevolution can be studied using population genetics, which investigates changes in gene frequencies.

Population Genetics

A **population** is all the members of a single species that occupy a particular area at the same time and that interbreed and exchange genes. A population could be all the green frogs in a frog pond, all the field mice on a farm, or all the English daisies on a hill. The members of a population reproduce with one another to produce the next generation.

Each member of a population is assumed to be free to reproduce with any other member, and when reproduction occurs, the genes of one generation are passed on in the manner described by Mendel's laws. Therefore, in this so-called Mendelian population (as discussed in Chapter 23) of sexually reproducing individuals, the total number of alleles at all the gene loci in all the members make up a **gene pool** for the population. It is customary to describe this gene pool in terms of allele frequencies for the various genes. Using this methodology, two investigators, G. H. Hardy, an English mathematician, and W. Weinberg, a German physician, discovered a principle that now bears their names.

Hardy and Weinberg decided to use the binomial equation $p^2 + 2pq + q^2$ to calculate the genotype and allele frequencies of a population. Figure 27.12 shows how this is done. Once you know the allele frequencies, you can calculate the ratio of genotypes in the next generation using a Punnett square. The data from Figure 27.12 reveal that the next generation will have exactly the same ratio of genotypes as before:

		eggs	
		0.20 *D*	0.80 *d*
sperm	0.20 *D*	0.04 *DD*	0.16 *Dd*
	0.80 *d*	0.16 *Dd*	0.64 *dd*

Genotype frequencies:
0.04 *DD* + 0.32 *Dd* + 0.64 *dd* = 1
or
$D^2 + 2Dd + d^2$

It is important to realize that the sperm and eggs represented in this Punnett square are actually the frequencies of alleles *D* and *d* in an entire population, not gametes produced by individuals.

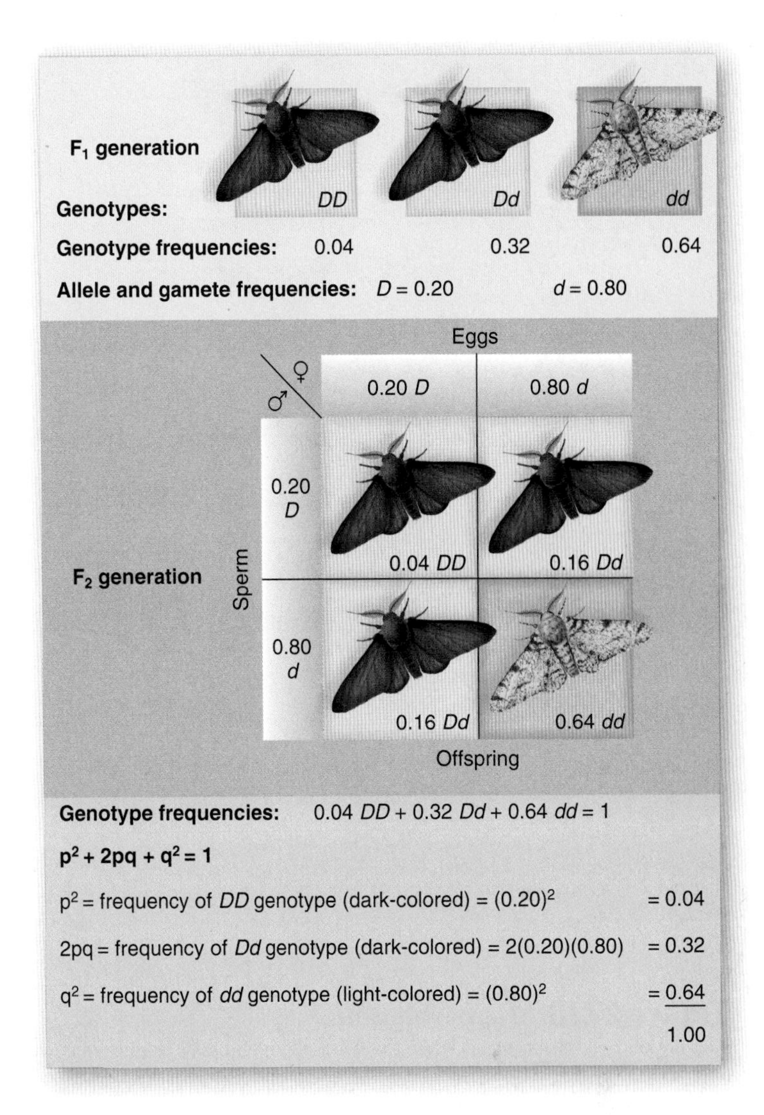

Figure 27.12 Hardy-Weinberg equilibrium. Using the gamete frequencies in a population, it is possible to employ a Punnett square to calculate the genotype frequencies of the next generation. When this is done, it can be shown that sexual reproduction alone does not alter a Hardy-Weinberg equilibrium; the genotype and therefore allele frequencies remain the same. Hardy-Weinberg used the binomial expression to calculate the genotype frequencies of a population.

The Hardy-Weinberg Principle

The Hardy-Weinberg principle states that allele frequencies in a gene pool will remain at equilibrium, and thus constant, after one generation of random mating in a large, sexually reproducing population as long as five conditions are met:

1. *No mutations.* Genetic mutations are an alteration in an allele, due to a change in DNA composition. Under Hardy-Weinberg assumptions, allele changes do not occur, or changes in one direction are balanced by changes in the opposite direction.

Figure 27.13 Microevolution.
Microevolution has occurred when there is a change in gene pool frequencies—in this case, due to natural selection. (*Left*) Light-colored moths are more frequent in the population because birds that eat moths are less likely to see light-colored peppered moths against light vegetation. (*Right*) Dark-colored moths are more frequent in the population because birds are less likely to see dark-colored moths against dark vegetation.

2. *No genetic drift.* Genetic drift is random changes in allele frequencies by chance. If a population is very large, changes in allele frequencies due to chance alone are insignificant.
3. *No gene flow.* Gene flow is the sharing of alleles between two populations through interbreeding. If there is no gene flow, migration of individuals, and therefore their genes, into or out of the population does not occur.
4. *Random mating.* Random mating occurs when individuals pair by chance, not according to their genotypes or phenotypes.
5. *No selection.* Often, the environment selects certain phenotypes to reproduce and have more offspring than other phenotypes. If selection does not occur, no phenotype is favored over another to reproduce.

In real life, these conditions are rarely, if ever, met, and allele frequencies in the gene pool of a population do change from one generation to the next. Because a change in allele frequencies is our definition of *micro*evolution, then evolution has occurred. A significance of the Hardy-Weinberg principle is that microevolution can be detected by noting deviations from a Hardy-Weinberg equilibrium of allele frequencies in the gene pool of a population.

Such deviations suggest that one or more of the five conditions is occurring in a population. Figure 27.13 gives an example of microevolution due to selection in a population of peppered moths. Peppered moths can be dark colored or light colored, and the percentage of each in the population can vary. Predatory birds are the selective agent that causes the makeup of the population to vary. When dark-colored moths rest on light trunks in a nonpolluted area, they are seen and eaten by these birds. With pollution, the trunks of trees darken, so light-colored moths stand out and are eaten more than dark-colored moths. We know that evolution has occurred in Figure 27.13 because the population changes from 36% dark-colored phenotype to 64% dark-colored phenotype over time. In this example, evolution has occurred because a selective force (predatory birds) favored one genotype over another.

Check Your Progress

1. What are the five conditions necessary to maintain a Hardy-Weinberg equilibrium in a population?
2. Give an example of what happens to allele frequencies in a population if these predictions are not met.

Five Agents of Evolutionary Change

The list of conditions for genetic equilibrium stated previously implies that the opposite conditions can cause evolutionary change. These conditions are mutations, genetic drift, gene flow, nonrandom mating, and natural selection.

Mutations

The Hardy-Weinberg principle recognizes new mutations as a force that can cause the allele frequencies to change in a gene pool and cause microevolution to occur. **Mutations**, which are permanent genetic changes, are the raw material for evolutionary change because without mutations, there could be no heritable genetic diversity among members of a population. It is important to realize that evolution is not directed, meaning that no mutation arises because the organism "needs" one. For example, a mutation that causes some bacteria to be resistant was already present before antibiotics appeared in the environment.

Mutations are random and are most often thought to result in no change or a negative effect on an individual's reproductive success. For example, a mutation can result in a nucleotide change in a gene. A mutation can be "silent" if the nucleotide change does not result in an amino acid change or if it is recessive and masked by a dominant allele in a diploid organism. The importance of recessive alleles increases if the environment is changing; it's possible that a particular homozygous recessive genotype could be helpful in a new environment, if not the present one. Mutations that cause a malfunctioning protein can be harmful to an organism.

In contrast to diploid organisms, mutations are the primary source of genetic differences among prokaryotes that reproduce asexually. Generation time is so short that many mutations can occur quickly, even though the rate is low, and since these organisms are haploid, any mutation that results in a phenotypic change is immediately tested by the environment.

Genetic Drift

Genetic drift refers to changes in the allele frequencies of a gene pool due to chance, as illustrated by the green and brown frogs in Figure 27.14. As you can imagine, genetic drift has greater effects in smaller populations. For example, the chance death of one individual in a population of a million will not have an appreciable effect on allele frequencies, but the chance death of one individual in a population of ten could change the frequency of an allele by 10% or even cause its loss altogether (if that individual was the only one with that allele). In nature, two situations, called founder effect and bottleneck effect, lead to small populations whereby genetic drift can drastically affect allele frequencies in a gene pool.

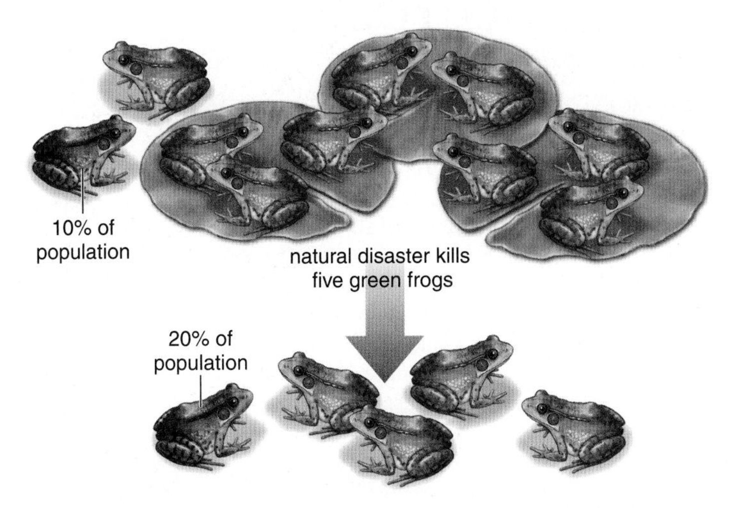

Figure 27.14 Genetic drift.
Genetic drift occurs when, by chance, only certain members of a population (in this case, green frogs) reproduce and pass on their alleles to the next generation. A natural disaster can cause the allele frequencies of the next generation's gene pool to be markedly different from those of the previous generation.

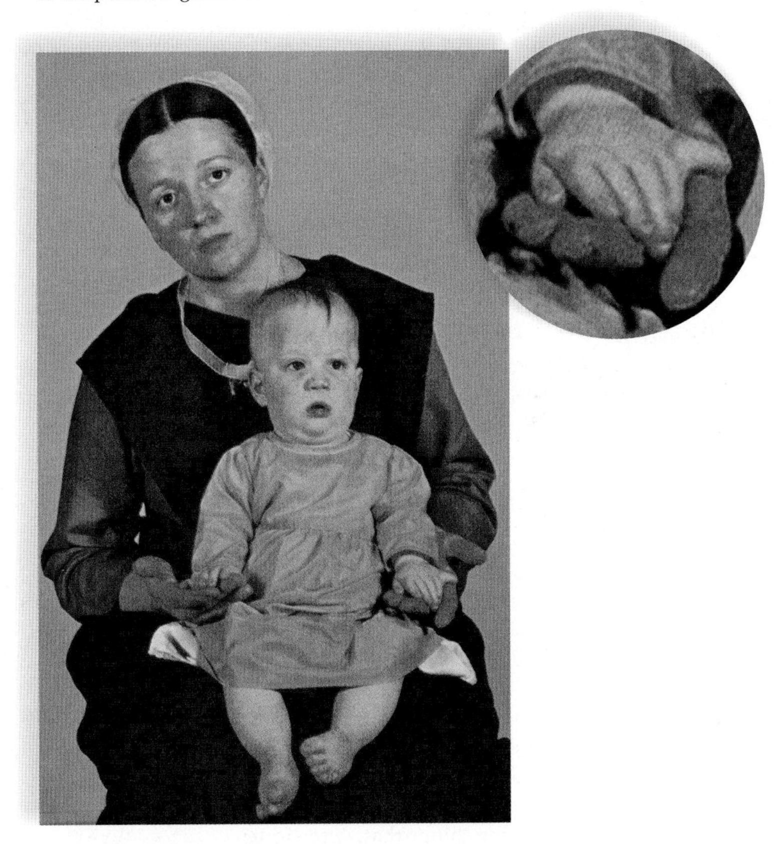

Figure 27.15 Founder effect.
A member of the founding population of Amish in Pennsylvania had a recessive allele for a rare kind of dwarfism linked with polydactylism. The percentage of the Amish population now carrying this allele is much higher compared to that of the general population.

The **founder effect** occurs when a few individuals form a new colony, and only a fraction of the total genetic diversity of the original gene pool is represented in these individuals. The particular alleles carried by the founders are dictated by chance alone. The Amish population of Lancaster, Pennsylvania, is an isolated religious sect descended from a few German founders. Today, as many as one in 14 individuals in this group carries a recessive allele that causes an unusual form of dwarfism (it affects only the lower arms and legs) and polydactylism (extra fingers) (Fig. 27.15). Genetic drift has caused this proportion to be much higher in the Amish than the non-Amish, where the allele is found in only one in 1,000 people.

Science Focus

Charles Darwin's Theory of Natural Selection

Although Charles Darwin is often credited as the first to propose descent with modification, biologists before him had slowly begun to accept the idea of evolution. Jean-Baptiste de Lamarck (1744–1829), a predecessor of Darwin, concluded after studying the succession of life-forms in geological strata, that more complex organisms are descended from less complex organisms. To explain the process of adaptation to the environment, Lamarck proposed the theory of inheritance of acquired characteristics—that the environment can bring about inherited change. One example he gave—and the one for which he is most famous—is that the long neck of a giraffe developed over time because animals stretched their necks to reach food high in trees and then passed gradually longer necks to their offspring (Fig. 27A).

This hypothesis for the inheritance of acquired characteristics has never been substantiated. The molecular mechanism of inheritance explains why. Phenotypic changes acquired during an organism's lifetime do not result in genetic changes that can be passed to subsequent generations. As an example, consider tail cropping in Doberman pinschers. All Doberman puppies are born with tails, even though their parents' tails are most often cropped. That is, tail cropping is a phenotypic change that is not inherited in the DNA. We now know that Lamarck's ideas, although important for advancing ideas about evolution, were incorrect.

Charles Darwin (1809–1882) came to a different conclusion from Lamarck's after going on a five-year trip as a naturalist aboard the ship the HMS *Beagle*. He read a book by Charles Lyell, a geologist who suggested the world is very old and has been undergoing gradual changes for a great many years. This meant that there was time for evolution to occur.

Because the ship sailed in the tropics of the Southern Hemisphere, Darwin encountered different living things that were more abundant and varied than those found in his native England. When Darwin compared the animals of Africa to those of South America, he noted that the African ostrich and the South American rhea, although similar in appearance, were actually different animals. He reasoned that they had a different line of descent because they were on different continents. When he arrived at the Galápagos Islands, he began to study the diversity of finches (see Fig. 27.20), whose adaptations could best be explained by assuming they had diverged from a common ancestor. He found such a hypothetical ancestor on the mainland of South America, supporting his theory. With this type of evidence, Darwin concluded that species evolve (change) with time.

Early giraffes probably had short necks that they stretched to reach food.

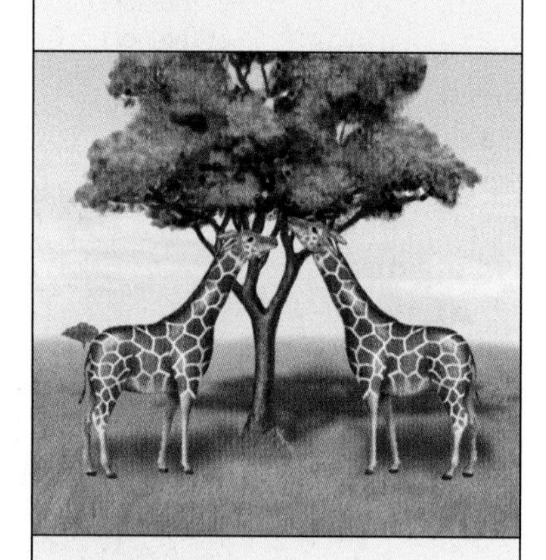

Their offspring had longer necks that they stretched to reach food.

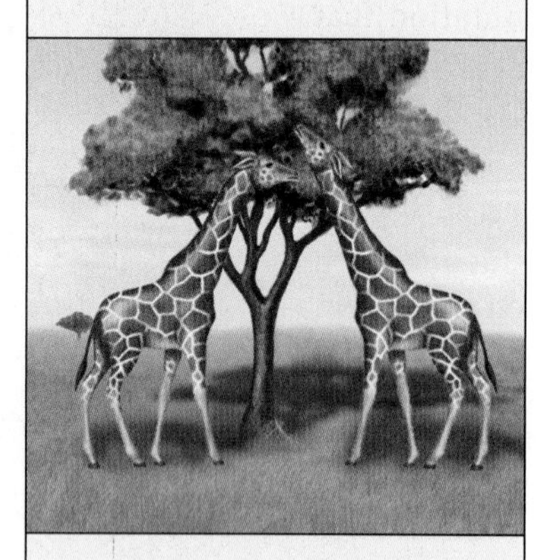

Eventually, the continued stretching of the neck resulted in today's giraffe.

Figure 27A Jean-Baptiste de Lamarck's proposal of acquired characteristics.

When Darwin returned home, he spent the next 20 years gathering data to support the principle of biological evolution. His most significant contribution was his theory of natural selection, which explains how populations of a species become adapted to their environment. This theory is explained in Figure 27B. Before formulating the theory, Darwin read an essay on human population growth written by Thomas Malthus. Malthus observed that although humans have a great reproductive potential, many environmental variables, such as availability of food and living space, tend to keep the human population in check with factors such as disease and famine. Darwin applied these ideas to all populations of organisms. A population is all the members of a species living in one particular place. Darwin calculated that a single pair of elephants could have 19 million descendants in 750 years. He realized that other organisms have an even greater reproductive potential than this pair of elephants, yet usually population sizes remain about the same. Darwin decided there is a constant struggle for existence, whereby only certain members of a population survive to reproduce. Those individuals best adapted to their environment produce the greatest number of offspring, and it is their traits that increase in frequency in a population in successive generations. This so-called "survival of the fittest" causes the next generation to be better adapted to the environment than the previous generation.

Darwin's theory of natural selection was nonteleological, meaning that there is no design or purpose in the works or processes of nature. However, rather than believing that organisms strive to adapt themselves to the environment, Darwin concluded that the *environment acts on individual phenotypes to select those individuals that are best adapted.* These individuals have been "naturally selected" to pass on their characteristics to the next generation. In contrast, the Lamarckian explanation for the long neck of the giraffe was incorrect because ancestors of the modern giraffe were "trying" to reach into the trees to browse on high-growing vegetation. Lamarck's proposal is teleological because, according to him, the outcome (longer necks) is predetermined. Darwin's theory of evolution, rather than being progressive or "forward looking" implies that the changing environment does not move toward any predetermined outcome.

The critical elements of Darwin's theory are as follows:

- **Variations.** Individual members of a population vary in physical characteristics. To be affected by natural selection, physical variations must be inherited from generation to generation by reproduction rather than being environmentally induced. If there is no variation in a trait in a population, natural selection cannot act.
- **Overproduction and struggle for existence.** The members of all populations compete with each other for limited resources. Certain members are able to capture or utilize these resources better than others.
- **Survival of the fittest.** Just as humans carry on artificial breeding programs to select which plants and animals will reproduce, natural selection by the environment determines which members of a population survive and reproduce. While Darwin emphasized the importance of survival, modern evolutionists emphasize the importance of unequal reproduction. That is, certain members of the population produce more offspring than others simply because they happen to have a variation or variations that make them better suited to the environment. In a biological sense, *fitness* is the number of fertile offspring an individual produces throughout its lifetime.
- **Adaptation.** The result of natural selection is that populations come to resemble the "best types"—those individuals that produce the most offspring because they are better adapted to the environment.

Early giraffes probably had necks of various lengths.

Natural selection due to competition led to survival of the longer-necked giraffes and their offspring.

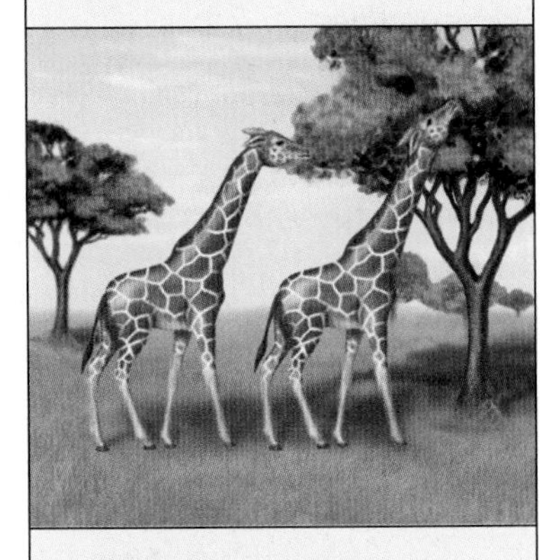

Eventually, only long-necked giraffes survived the competition.

Figure 27B Charles Darwin's theory of natural selection.

Darwin was prompted to publish his findings only after he received a letter from another naturalist, Alfred Russel Wallace, who had come to the same conclusions about evolution. Although both scientists subsequently presented their ideas at the same meeting of the famed Royal Society in London in 1858, only Darwin had outlined his reasoning for the theory in a draft of *The Origin of Species by Means of Natural Selection,* which he had completed 16 years earlier and eventually published in 1859. This book is still studied by many biologists today.

Can natural selection account for the origin of new species and for the great diversity of life? Yes, Darwinian selection is the only accepted scientific theory for the diversity of life.

Discussion Questions

1. Currently, a debate is in progress regarding the teaching of intelligent design alongside evolution as a theory for the diversity of life. Explain why the intelligent design idea is teleological.
2. Explain why variation in a trait must be present in order for natural selection to operate.
3. Explain why Lamarck's idea of "inheritance of acquired characteristics" is incorrect.

Sometimes a population is subjected to near extinction because of a natural disaster (e.g., earthquake or fire) or because of human interference. The disaster acts as a bottleneck, preventing the majority of genotypes from breeding to form the next generation. For example, the large genetic similarity found in cheetahs is hypothesized to be due to a **bottleneck effect.** In a skin grafting study, most cheetahs failed to reject skin grafts from unrelated cheetahs because they were so genetically similar.

Gene Flow

Gene flow is the movement of alleles among populations, as occurs when individuals migrate from one population to another and breed in that new population. For example, adult plants are not able to migrate, but their gametes are often either blown by the wind or carried by insects. The wind, in particular, can carry pollen for long distances and can therefore be a factor in gene flow among plant populations.

Gene flow among populations keeps their gene pools similar. It also prevents close adaptation to a local environment.

Nonrandom Mating

Nonrandom mating occurs when individuals pair up, not by chance, but according to their genotypes or phenotypes. Inbreeding, or mating between relatives to a greater extent than by chance, is an example of nonrandom mating. Inbreeding decreases the proportion of heterozygotes and increases the proportions of homozygotes at all gene loci. In a human population, inbreeding increases the frequency of recessive abnormalities (see Fig. 27.15).

Natural Selection

Natural selection is the process by which some individuals produce more offspring than others. The Science Focus outlines how Charles Darwin explained evolution by natural selection. Here, we restate these steps in the context of modern evolutionary theory. Evolution by natural selection requires:

1. **Individual variation.** The members of a population differ from one another.
2. **Inheritance.** Many of these differences are heritable genetic differences.
3. **Overproduction.** Individuals in a population are engaged in a struggle for existence because breeding individuals in a population tend to produce more offspring than the environment can support.
4. **Differential reproductive success.** Individuals that are better adapted to their environment produce more offspring than those that are not as well adapted, and consequently, their fertile offspring will make up a greater proportion of the next generation.

In biology, the **fitness** of an individual is measured by the number of fertile offspring produced throughout its lifetime. Gene mutations are the ultimate source of variation because they provide new alleles. Sometimes, these mutations result in positive fitness consequences for an organism perhaps providing variation for adaptation to a new environmental change. However, in sexually reproducing organisms, genetic variation can also result from crossing-over and independent assortment of chromosomes during meiosis and also fertilization when gametes are combined. A different combination of alleles can lead to a new and different phenotype.

In this context, consider that most of the traits on which natural selection acts are polygenic and thus controlled by more than one gene. Such traits have a range of phenotypes that often follow a bell-shaped curve.

The three main types of natural selection are stabilizing selection, directional selection, and disruptive selection.

Stabilizing selection With stabilizing selection, extreme phenotypes are selected against, and individuals near the average are favored. Stabilizing selection can improve adaptation of the population to those aspects of the environment that remain constant. As an example, consider that when Swiss starlings lay four to five eggs, more young survive than when the female lays more or less than this number (Fig. 27.16). Genes determining physiological characteristics, such as the production of yolk, and behavioral characteristics, such as how long the female will mate, are involved in determining clutch size.

Through the years, hospital data have shown that human infants born with an intermediate birth weight (3–4 kg) have a better chance of survival than those born with a birth weight either higher or lower than that range. Thus, stabilizing selection serves to reduce the variability in birth weight in human populations.

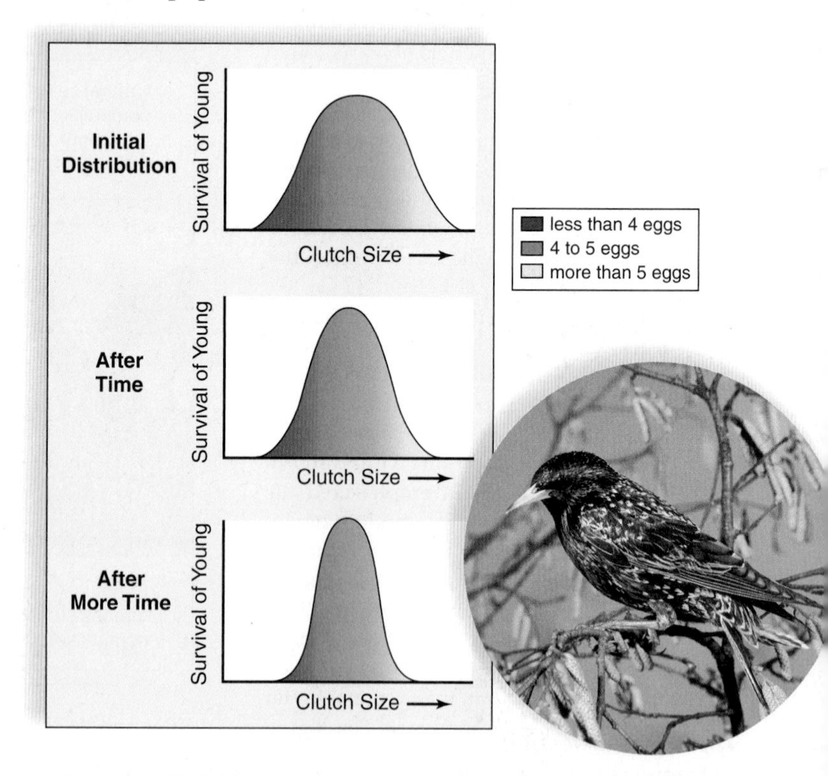

Figure 27.16 Stabilizing selection.
Stabilizing selection occurs when natural selection favors the intermediate phenotype over the extremes. For example, Swiss starlings that lay four to five eggs (usual clutch size) have more surviving young than birds that lay fewer than four eggs or more than five eggs.

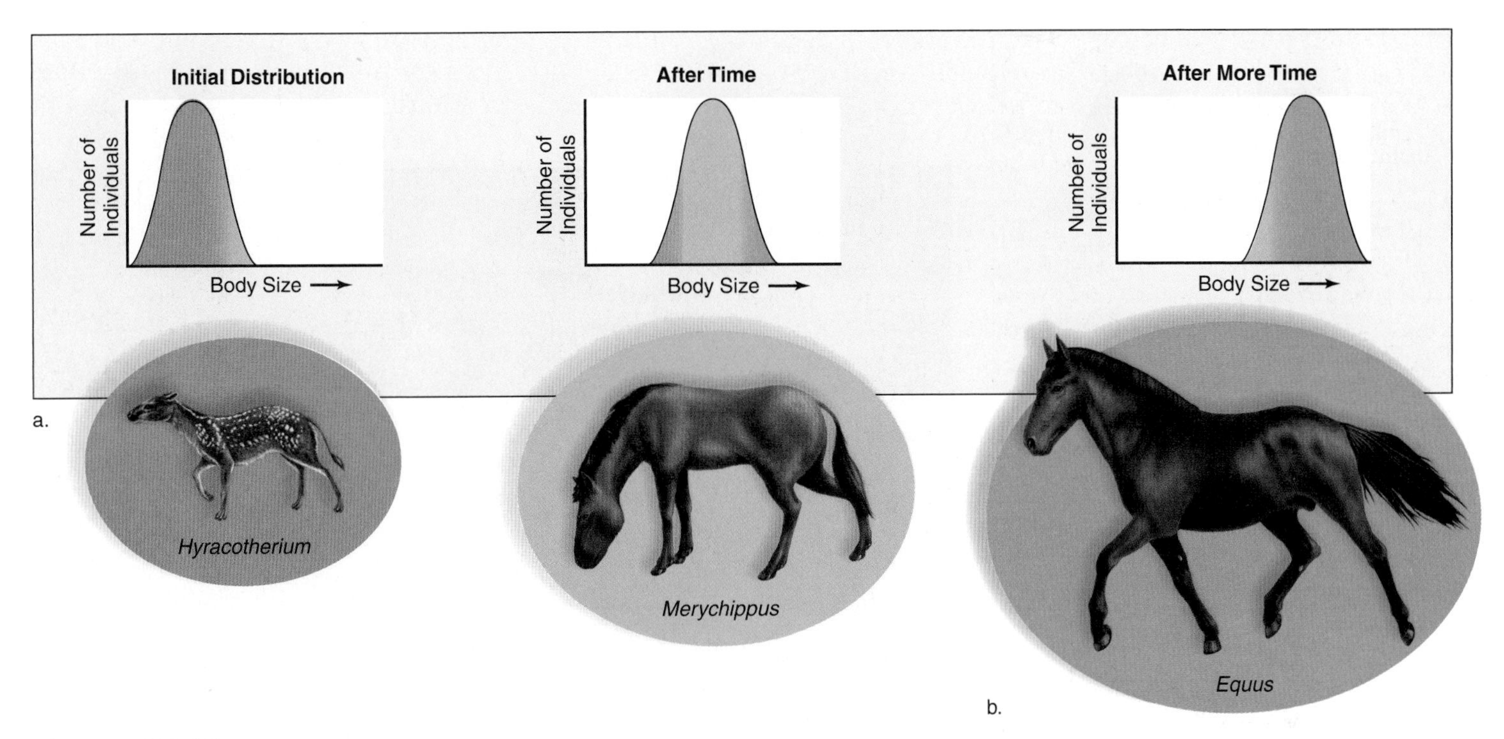

Figure 27.17 Directional selection.
a. Directional selection occurs when natural selection favors one extreme phenotype, resulting in a shift in the distribution curve. **b.** For example, *Equus,* the modern-day horse, which is adapted to a grassland habitat, is much larger than its ancestor, *Hyracotherium,* which was adapted to a forest habitat.

Directional Selection **Directional selection** occurs when an extreme phenotype is favored and the distribution curve shifts in that direction (Fig. 27.17). This changes the average phenotype in a population. Such a shift can occur when a population is adapting to a changing environment. For example, the gradual increase in the size of the modern horse, *Equus,* can be correlated with a change in the environment from forest conditions to grassland conditions. *Hyracotherium,* the ancestor of the modern horse, was about the size of a dog and was adapted to the forestlike environment of the Eocene epoch of the Paleocene period. This animal could have hidden among the trees for protection, and its low-crowned teeth would have been appropriate for browsing on leaves. Later, in the Miocene and Pliocene epochs, grasslands began to replace the forests. Then the ancestors of *Equus* were subject to selective pressure for the development of strength, intelligence, speed, and durable grinding teeth. A larger size provided the strength needed for combat, elongated legs ending in hooves gave speed for escaping from enemies, and the durable grinding teeth enabled the animals to feed efficiently on grasses. Nevertheless, the evolution of the horse should not be viewed as a straight line of descent; there were many side branches that became extinct.

Disruptive Selection In **disruptive selection,** two or more extreme phenotypes are favored over any intermediate phenotype (Fig. 27.18). For example, British land snails (*Cepaea nemoralis*) have a wide habitat range that includes grass fields and hedgerows and forested areas. In areas with low-lying vegetation, thrushes feed mainly on snails with dark shells that lack light bands, and in forested areas, they feed mainly on snails with light-banded shells. Therefore, the two different habitats have resulted in two different phenotypes in the population.

CHECK YOUR PROGRESS

1. Why would you expect natural selection and not genetic drift to result in adaptation to the environment?
2. Which four conditions must be met for natural selection to operate?
3. Compare and contrast the end result of stabilizing, directional, and disruptive selection.

Maintenance of Variation

A population nearly always shows some genotypic variation. The maintenance of variation is beneficial because populations with limited variation may not be able to adapt to new conditions should the environment change and may become extinct. How can variation be maintained in spite of selection constantly working to reduce it? First, we must remember that the forces that promote variation are still at work: mutation still generates new alleles, recombination and independent assortment still shuffle the alleles during gametogenesis, and fertilization still creates new combinations of alleles from those present in the gene pool. Second, gene flow might still occur. If the receiving population is small and mostly homozygous, gene flow can be a significant source of new alleles. Finally, natural selection favors certain phenotypes, but the other types may still remain in reduced

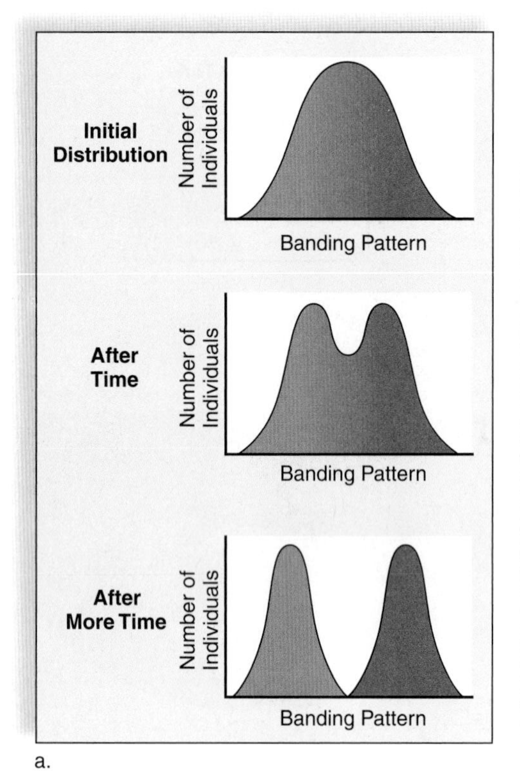

Figure 27.18 Disruptive selection.
a. Disruptive selection favors two or more extreme phenotypes. **b.** Today, British land snails comprise mainly two different phenotypes, each adapted to a different habitat. Snails with dark shells are more prevalent in forested areas, and light-banded snails are more prevalent in areas with low-lying vegetation.

frequency. Disruptive selection even promotes polymorphism in a population. Genetic variation can be maintained in diploid organisms in heterozygote individuals that can maintain recessive alleles in the gene pool.

Diploidy and the Heterozygote

Only alleles that are expressed (cause a phenotypic difference) are subject to natural selection. In diploid organisms, this fact makes the heterozygote a potential protector of recessive alleles that might otherwise be weeded out of the gene pool by natural selection. Because the heterozygote remains in a population, so does the possibility of the recessive phenotype, which might have greater fitness in a changing environment. When natural selection favors the ratio of two or more phenotypes in generation after generation, it is called balanced polymorphism. Sickle cell disease offers an example of balanced polymorphism.

Individuals with sickle cell disease have the genotype Hb^SHb^S (Hb = hemoglobin, the oxygen-carrying protein in red blood cells; s = Sickle cell) and tend to die at an early age due to hemorrhaging and organ destruction. Those who are heterozygous and have sickle cell trait (Hb^AHb^S; A = normal) are better off because their red blood cells usually become sickle-shaped only when the oxygen content of the environment is low. Ordinarily, those with a normal genotype (Hb^AHb^A) are the most fit.

Geneticists studying the distribution of sickle cell disease in Africa have found that the recessive allele (Hb^S) has a higher frequency (0.2 to as high as 0.4 in a few areas) in regions where malaria is also prevalent. What is the connection between higher frequency of the recessive allele and malaria? Malaria is caused by a parasite that lives in and destroys the red blood cells of the normal homozygote (Hb^AHb^A). However, the parasite is unable to live in the red blood cells of the heterozygote (Hb^AHb^S) because the infection causes the red blood cells to become sickle-shaped. Sickle-shaped red blood cells lose potassium, and this causes the parasite to die. Thus, in an environment where malaria is prevalent, the heterozygote is favored. Each of the homozygotes is selected against but is maintained because the heterozygote is favored in parts of Africa subject to malaria. Table 27.2 summarizes the effects of the three possible genotypes.

TABLE 27.2 Sickle Cell Disease

Genotype	Phenotype	Result
Hb^AHb^A	Normal	Dies due to malarial infection
Hb^AHb^S	Sickle cell trait	Lives due to protection from both
Hb^SHb^S	Sickle cell disease	Dies due to sickle cell disease

CHECK YOUR PROGRESS

1. List forces that help maintain genetic variability in a population.
2. Explain the higher incidence of sickle cell disease in populations subject to malaria.

27.4 Speciation

Learning Outcomes

1. Give examples to illustrate the process of speciation.
2. Explain how adaptive radiation can lead to speciation.
3. Compare and contrast phyletic gradualism with punctuated equilibrium.

Usually, a species occupies a certain geographical range, within which several subpopulations exist. For our present discussion, **species** is defined as a group of subpopulations that are capable of interbreeding and are isolated reproductively from other species. The subpopulations of the same species can exchange genes, but different species do not exchange genes. Reproductive isolation of similar species is accomplished by the isolating mechanisms listed in Table 27.3. **Prezygotic isolating mechanisms** are in place before fertilization, and thus reproduction is never attempted. **Postzygotic isolating mechanisms** are in place after fertilization, so reproduction may take place, but it does not produce fertile offspring.

The Process of Speciation

Speciation has occurred when one species gives rise to two species, each of which continues on its own evolutionary pathway. How can we recognize speciation? Whenever reproductive isolation develops between two formerly interbreeding groups of populations, speciation has occurred. One type of speciation, called **allopatric speciation,** usually occurs when populations become separated by a geographic barrier and gene flow is no longer possible. Figure 27.19 illustrates an example of allopatric speciation that has been extensively studied in California. Apparently, members of an ancestral population of *Ensatina* salamanders existing in the Pacific Northwest migrated southward, establishing a series of populations. Each population was exposed to its own selective pressures along the coastal and Sierra Nevada Mountains. Due

TABLE 27.3 Reproductive Isolating Mechanisms

Isolating Mechanism	Example
Prezygotic	
Habitat isolation	Species at same locale occupy different habitats
Temporal isolation	Species reproduce at different seasons or different times of day
Behavioral isolation	In animals, courtship behavior differs, or they respond to different songs, calls, pheromones, or other signals
Mechanical isolation	Genitalia unsuitable for one another
Postzygotic	
Gamete isolation	Sperm cannot reach or fertilize egg
Zygote mortality	Fertilization occurs, but zygote does not survive
Hybrid sterility	Hybrid survives but is sterile and cannot reproduce
F_2 fitness	Hybrid is fertile, but F_2 hybrid has reduced fitness

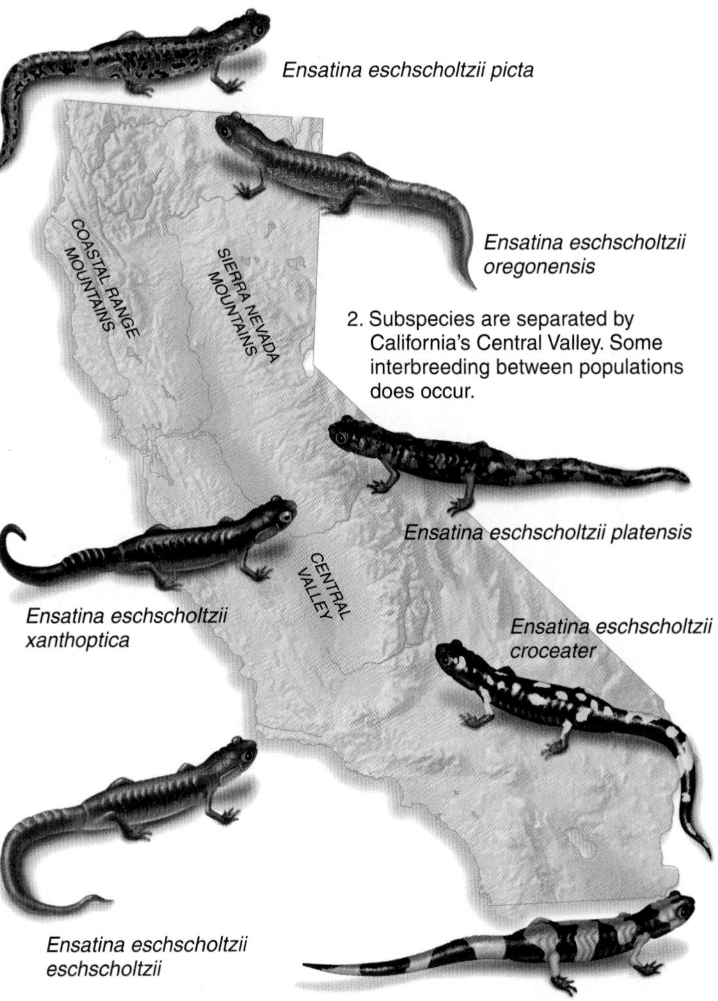

Figure 27.19 Allopatric speciation.
In this example of allopatric speciation, the Central Valley of California is separating a range of populations descended from the same northern ancestral species. Those to the west along the coastal mountains and those to the east along the Sierra Nevada Mountains experience gene flow, but gene flow is limited between the eastern populations and the western populations. Members of the most southerly eastern and western populations are quite different in color pattern.

to the presence of the Central Valley of California, which is largely dry and thus unsuitable habitat for amphibians, gene flow rarely occurs between eastern and western populations of *Ensatina*. Genetic differences also increased from north to south, resulting in two distinct forms of *Ensatina* salamanders in Southern California that differ dramatically in color.

It is also possible that a single population could suddenly divide into two reproductively isolated groups without being geographically isolated. The best evidence for this type of speciation, called **sympatric speciation,** is found among plants, where multiplication of the chromosome number in one plant prevents it from successfully reproducing with others of its kind. Self-reproduction can maintain such a new plant species.

Check Your Progress

1. Compare and contrast allopatric and sympatric speciation.

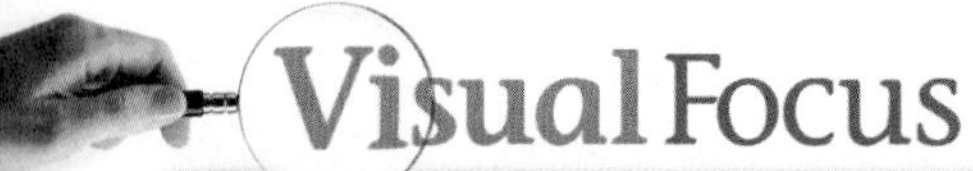

Figure 27.20 The Galápagos finches.
Each of these finches is adapted to gathering and eating a different type of food. Note the different sizes and shapes of the beaks in the different species. Tree finches have beaks largely adapted to eating insects and, at times, plants. The woodpecker finch, a tool-user, uses a cactus spine or twig to probe in the bark of a tree for insects. Ground finches have beaks adapted to eating prickly-pear cactus or different-sized seeds.

Adaptive Radiation

One of the best examples of "allopatric" speciation is provided by the finches on the Galápagos Islands, located 600 miles west of Ecuador, South America. The 13 species of finches found there are often called Darwin's finches because Darwin first realized their significance as an example of how evolution works. These species appear to be descended from mainland finches that migrated to one of the islands. It is likely that after the original species on a single island increased, some individuals dispersed to other islands.

The islands are ecologically different enough to have promoted divergent feeding habits. This is apparent because, although the birds physically resemble each other in many respects, they have different beaks, each adapted to gathering and eating a different type of food (Fig. 27.20). There are seed-eating ground finches, cactus-eating ground finches, insect-eating tree finches, also with different-sized beaks; and a warbler-type tree finch, with a beak adapted to eating insects and gathering nectar. Among the tree finches is a woodpecker type, which lacks the long tongue of a true woodpecker but makes up for this by using a cactus spine or a twig to ferret out insects. Remarkably, each of these types is found on islands where its beak matches the abundant food type. Therefore, Darwin's finches are an example of **adaptive radiation,** or the proliferation of a species by adaptation to different ways of life.

The Pace of Speciation

Currently, there are two hypotheses about the pace of speciation and, therefore, evolution. One hypothesis is called the phyletic gradualism model, and the other is called the punctuated equilibrium model. Each model gives a different answer to the question of why so few transitional links are found in the fossil record.

Traditionally, evolutionists have supported a model called **phyletic gradualism,** which states that change is very slow but steady within a lineage before and after a divergence (splitting of the line of descent) (Fig. 27.21*a*). Therefore, it is not surprising that few transitional links such as *Archaeopteryx* (see Fig. 27.5) have been found. Indeed, the fossil record, even if it were complete, might be unable to show when speciation has occurred. Why? Because a new species comes about after reproductive isolation, and reproductive isolation cannot be detected in the fossil record! Only when a new species evolves and displaces the existing species is the new species likely to show up in the fossil record.

A model of evolution called **punctuated equilibrium** has also been proposed (Fig. 27.21*b*). It says that long periods of stasis, or no visible change, are followed by rapid periods of speciation. With reference to the length of the fossil record (about 3.5 billion years), speciation occurs relatively rapidly, and this can explain why few transitional links are found. Mass extinction events are often followed by rapid (relative to the age of the Earth) periods of speciation.

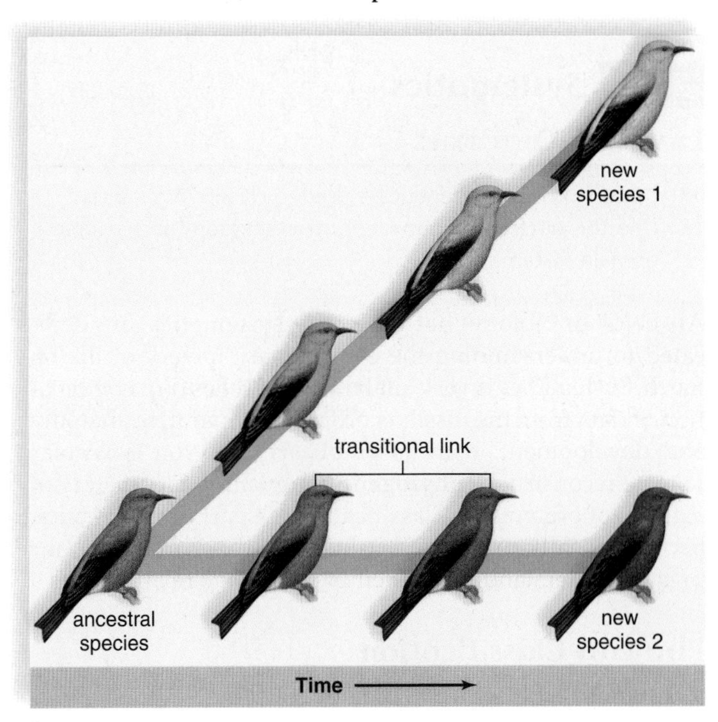

a.

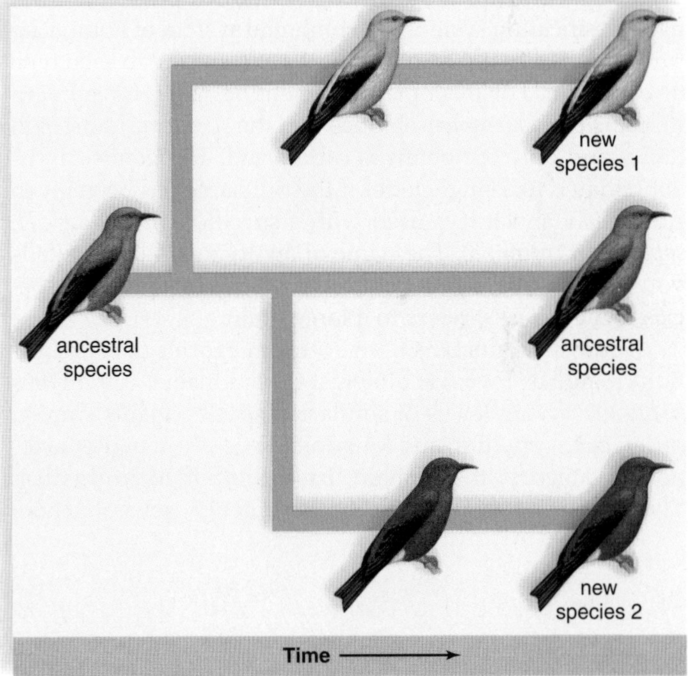

b.

Figure 27.21 Phyletic gradualism compared to punctuated equilibrium.
a. Supporters of the phyletic gradualism model hypothesize that speciation takes place gradually and many transitional links occur. **b.** Supporters of the more recent, punctuated equilibrium model hypothesize that speciation occurs rapidly, with no transitional links.

Check Your Progress

1. Define adaptive radiation.
2. Compare models of how speciation can occur rapidly versus slowly.

27.5 Systematics

Learning Outcomes

1. Discuss how phylogenetics is used to classify organisms.
2. Cite the differences between a five-kingdom and a three-domain system.

All fields of biology, but especially **systematics**, are dedicated to understanding the evolutionary history of life on Earth. Systematics is very analytical and relies on a combination of data from the fossil record and comparative anatomy and development, with an emphasis today on molecular data, to reconstruct a **phylogeny,** the evolutionary history of a group of organisms. Classification is a part of systematics, because ideally organisms are classified according to our present understanding of their evolutionary relationships.

Linnean Classification

Taxonomy is the branch of biology concerned with identifying, naming, and classifying organisms. A taxon (pl., taxa) is an organism or a group of organisms at a particular level in a classification system. The binomial system of nomenclature assigns a two-part name (*genus* and *species*) to each type of organism. For example, the scientific name for modern humans is *Homo sapiens*. Notice that the scientific name is in italics and only the genus is capitalized. The genus can be abbreviated to a single letter if the full name has been given previously and if it is used with a specific epithet (e.g., *H. sapiens* for humans). The name of an organism usually tells you something about the organism. In this instance, the species name, *sapiens,* refers to a large brain.

Today, taxonomists use several categories of classification created by Swedish biologist Carl Linnaeus in the 1700s to show varying levels of similarity: species, genus, family, order, class, phylum, and kingdom. Recently, a higher taxonomic category, the domain, has been added to this list. There can be several species within a genus, several genera within a family, and so forth. In this hierarchy, the higher the category, the more inclusive it is. Therefore species in the same domain have general traits in common, while those in the same genus have quite specific traits in common.

In most cases, each category of classification can be subdivided into three additional categories, as in superorder, order, suborder, and infraorder. Considering these, there are more than 30 categories of classification.

Phylogenetics

Phylogenetics is the modern way in which organisms are classified and arranged in evolutionary trees. Phylogeneticists arrange species and higher classification categories into clades. **Clades** may be represented on a diagram called a **cladogram.** A clade contains a most recent common ancestor and all its descendant species—the common ancestor is presumed and not identified. Figure 27.22 depicts a cladogram for seven groups of vertebrates. Only the lamprey, the so-called "out-group," lacks jaws, but the other six groups of vertebrates are in the same clade because they all have jaws, a shared derived characteristic (one that occurred more recently) relative to their ancestors. On the other hand, the vertebrates beyond the shark are all in the same clade because they have lungs, and so forth.

Figure 27.22 is somewhat misleading because, although single traits are noted on the tree, phylogeneticists use much more data to arrange groups of organisms into clades. Phylogeneticists are aided in their endeavor by the computer and any and all available data, including morphological data and DNA sequences. It is important to note that phylogeneticists arrange groups hierarchically based on homologous structures or DNA sequences. Deciphering homology is sometimes difficult because of convergent evolution. **Convergent evolution** is the acquisition of the same or similar traits in distantly related lines of descent. Similarity due to convergence results in analogous structures such as insect and bat wings. Analogous structures have the same function in different groups, but organisms with these structures do not share a recent

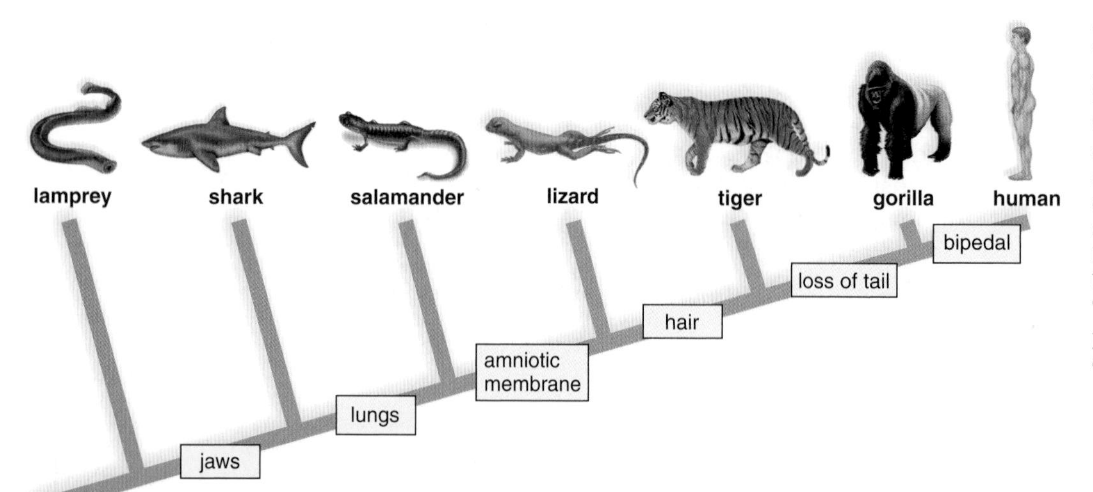

Figure 27.22 Cladogram. A cladogram gives comparative information about relationships. Organisms in the same clade share the same derived characteristics. Humans and all the other vertebrates shown, except lampreys, are in the same clade as sharks because they all have jaws. However, humans are alone on a branch because only they are bipedal. This cladogram is simplified because phylogeneticists actually use a great deal more data to construct cladograms.

common ancestor. Instead, analogous structures arise because of adaptations to similar environments.

In making decisions, phylogeneticists were traditionally guided by the principle of parsimony, which states that the pattern that requires the fewest evolutionary changes is the most likely. More modern approaches use complicated computer simulations (e.g., Maximum likelihood and Bayesian analyses), to determine the most likely phylogenetic reconstruction based on the data that they have accumulated. Today, these data are most commonly DNA or rRNA sequences. As a result, modern phylogenetic studies are based on the assumption that the more closely species are related, the fewer changes will be found in molecular nucleotide sequences (Fig. 27.23). Computer software breakthroughs have made it possible to analyze nucleotide sequences quickly and accurately. Phylogenetic analyses are now possible for anyone doing comparative studies because DNA sequences are archived on publically accessible websites, so each investigator doesn't necessarily have to generate the raw nucleotide sequence data themselves. The combination of computation speed and availability of vast amounts of data, even entire genomes, has made molecular systematics a standard way to study the evolutionary relationships among organisms today.

Linnaean Classification Versus Phylogenetics

Figure 27.24 illustrates the types of problems that arise when trying to reconcile Linnaean classification with the principles of phylogenetics. Figure 27.24, which is based on phylogenetics, shows that birds are in a clade with crocodiles because they share a recent common ancestor. This ancestor had a gizzard. An examination of the skulls of crocodiles and birds would show other derived traits that they share. Birds have scaly skin and share this ancestral characteristic with other reptiles as well. Yet Linnaean classification places birds in their own group separate from crocodiles and separate from reptiles in general. In many other instances, also, Linnaean classification is not consistent with new understandings about phylogenetic relationships. Therefore some phylogeneticists have proposed a different system of classification called the International Code of Phylogenetic Nomenclature, or PhyloCode, which sets forth rules to follow in naming of clades. Other biologists are hoping to modify Linnaean classification to be consistent with the principles of modern phylogenetic systematics.

Two major problems may be unsolvable, however: (1) Clades are hierarchically arranged as are Linnaean categories. However, many more clades may exist than do Linnaean taxonomic categories and it is therefore difficult to equate clades with individual taxa or their respective groups. (2) Taxonomic groups are not necessarily equivalent in the Linnaean system. To take an example, the family taxon within Kingdom Plantae may not be equivalent to the

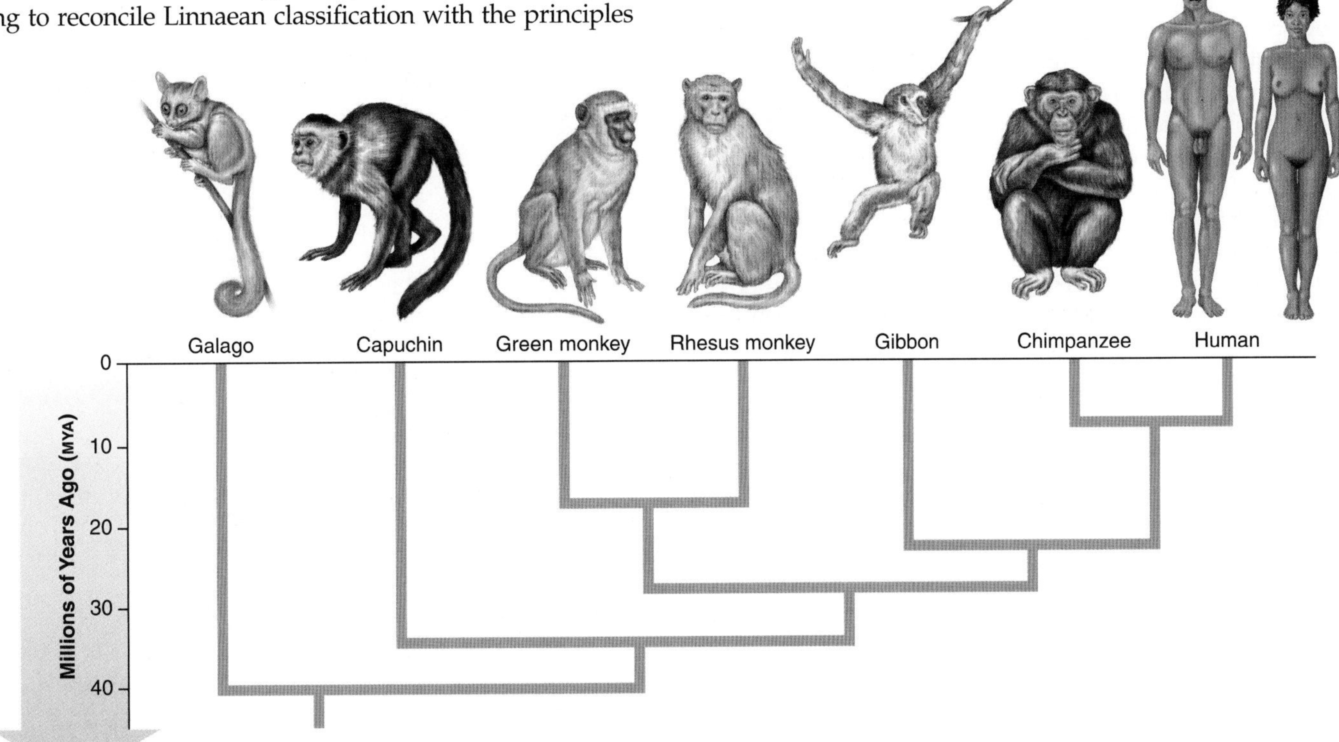

Figure 27.23 Molecular data.
The relationships of certain primate species are based on a study of their genomes. The length of the branches indicates the relative number of DNA base-pair differences between the groups. These data, along with knowledge of the fossil record for one divergence, make it possible to suggest a date for the other divergences in the tree.

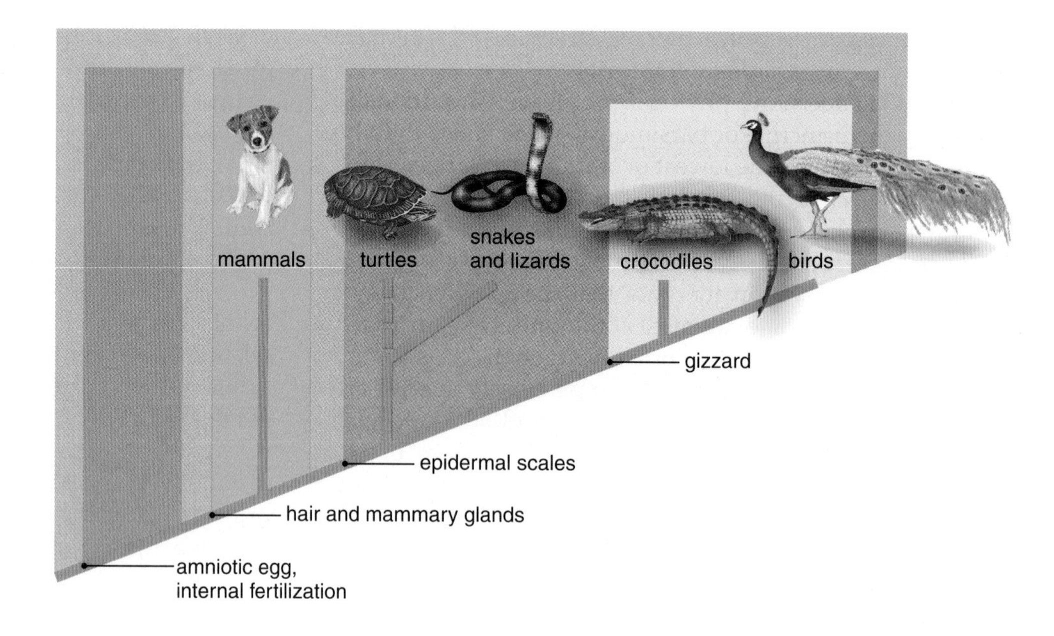

Figure 27.24 Cladistic classification.
Taxonomic designations are based upon evolutionary history. Each taxon includes a common ancestor and all of its descendants.

family taxon in Kingdom Animalia. Because of such problems, some phylogeneticists recommend abandoning Linnaeus' system altogether.

Three-Domain Classification System

Classification systems change over time. Historically, most biologists utilized the five-kingdom system of classification, which contains kingdoms for the plants, animals, fungi, protists, and monerans. Organisms were placed into these kingdoms based on type of cell (prokaryotic or eukaryotic), level of organization (unicellular or multicellular), and type of nutrition. In the five-kingdom system, the monerans were distinguished by their structure—they were prokaryotic (lack a membrane-bound nucleus)—whereas the organisms in the other kingdoms were eukaryotic (have a membrane-bound nucleus). The kingdom Monera contained all prokaryotes, which according to the fossil record evolved first.

Sequencing the genes for rRNA provided new information that called into question the five-kingdom system of classification. Aside from molecular data, cellular data also suggests that there are two groups of prokaryotes, named the Bacteria and the Archaea. These two groups are so fundamentally different from each other they have been assigned to separate domains, a category of classification that is higher than the kingdom category. The Bacteria arose first, followed by the Archaea and then the Eukarya (Fig. 27.25). The Archaea and Eukarya are more closely related to each other than either is to the Bacteria. Systematists, using the **three-domain system** of classification, are in the process of sorting out what kingdoms belong within **domain Bacteria** and **domain Archaea**. **Domain Eukarya** contains kingdoms for

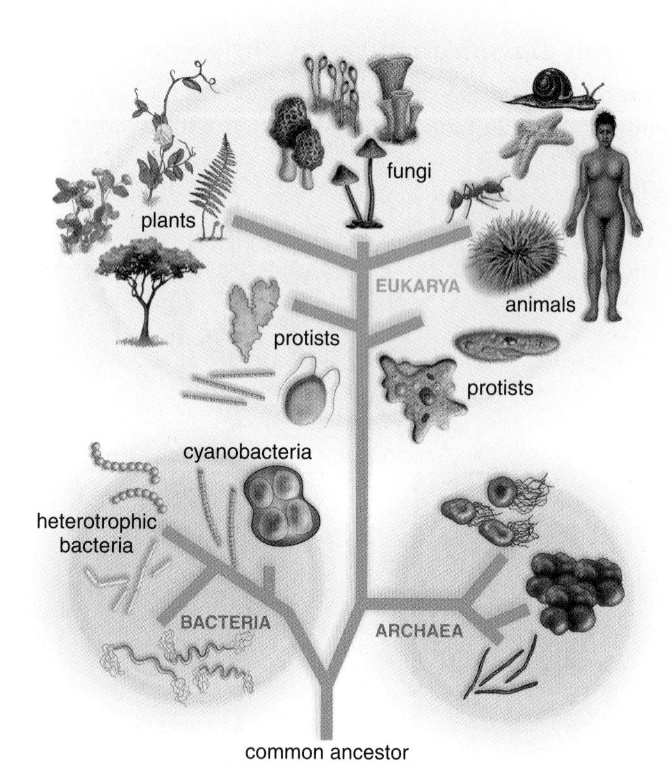

Figure 27.25 The three-domain system of classification.
Representatives of each domain are depicted. The phylogenetic tree shows that domain Archaea is more closely related to domain Eukarya than either is to domain Bacteria.

Evolution of Pesticide and Antibiotic Resistance

Through Darwin's theory of natural selection, we have come to understand why insects become resistant to the pesticides we spray on our crops. Some people refer to the use of pesticides as "artificial selection," because humans are involved. Nonetheless, the process is the same as natural selection—pesticides kill individuals that are susceptible, but insect populations within a single field are usually so large that some resistant individuals are likely to survive and reproduce. Through time, the frequency of resistant individuals increases in the population to the point that a certain pesticide may no longer be effective. Overall, pests cause over a quarter of a trillion dollars per year in damage to crops across the globe. Keep in mind that resistant insects may require several pesticides for effective control.

However, recent concern about the effects of pesticide use on the environment and human health has led farmers to seek alternatives. An evolutionary approach to dealing with insect damage to crops is called, "integrated pest management." Instead of using the strongest pesticide year after year, farmers will use natural pests, such as spiders, to control insect populations. Genetic engineering is also used to insert genes into plants so that they produce toxins that are poisonous to most species of insects or resistant to herbicides. Crop rotation is a part of integrated pest management as well.

In the same way that we select pests resistant to insecticides, we also artificially select for resistant bacterial strains using antibiotics. In fact, resistance has often evolved soon after the introduction of new antibiotics (Table 27A). This is the type of "accelerated evolution" that creates superbugs that you learned about in the beginning of this chapter. Bacterial resistance creates the need for even newer antibiotics. The development of a single new antibiotic is estimated to cost between $400 and $500 million. Antibiotic resistance adds over $30 billion to annual medical costs in the United States alone!

Our understanding of evolutionary biology has helped doctors treat patients appropriately. Doctors generally no longer prescribe antibiotics for colds or flu because they are viral infections. Antibiotics only kill bacteria, not viruses. In the case of TB, the strain infecting a patient is tested for antibiotic resistance and the patient is treated accordingly. The antibiotic isoniazid is used to treat patients with nonresistant strains, whereas a four-drug regimen is recommended for treatment of resistant strains.

Discussion Questions

1. Your crops are infested with pests that are resistant to the insecticides you are using. Using what you know about evolutionary biology, describe your approach to this problem.
2. In New York City, state health officials have the power to quarantine TB patients who do not take their medicine. That is, they can essentially lock them up for as long as needed (often up to 18 months) to treat their illness. Note also that some of the medications can have serious side effects. What do you think about this policy?
3. Many people in less-developed countries die from TB (tuberculosis), not because their disease is incurable, but simply because they do not have health insurance and cannot afford the medications. Should we in the United States pay more for our medications so that pharmaceutical companies can provide them to lower-income people at a reduced cost or for free?

TABLE 27A Dates of Antibiotic Discovery and Resistance

Antibiotic	Discovery/ Introduction	Resistance
Penicillin	1928/1943	1946
Sulfonamides	1930s	1940s
Streptomycin	1943/1945	1959
Chloramphenicol	1947	1959
Tetracycline	1948	1953
Erythromycin	1952	1988
Vancomycin	1956	1988/1993
Methicillin	1960	1961
Ampicillin	1961	1972
Cefotaxime/ ceftazidime	1981/1985	1983/1984/1988

all eukaryotes, including protists, animals, fungi, and plants. Later in this text, we will study the individual kingdoms that occur within the domain Eukarya. The protists themselves do not share one common ancestor, and some suggest that the kingdom should be divided into many different kingdoms. The number of kingdoms is still being determined among systematists, illustrating that classification is changing as new data become available.

Check Your Progress

1. Describe the binomial nomenclature system of naming species.
2. What is a clade?
3. Describe how a three-domain classification system differs from a five-kingdom system.

Applying the Concepts [Revisited]

In this chapter, you have learned about both macroevolution and microevolution. Although people tend to think of evolution occurring over long timescales, it can occur quickly, particularly under human influence. The evolution of antibiotic resistance and pesticide resistance are examples of how such rapid evolution can be important in your everyday life. The evolution of antibiotic-resistant strains of bacteria such as methicillin-resistant *Staphylococcus aureus* is an increasing health problem, costing billions in increased medical costs and even resulting in deaths of some affected patients. By the same token, pesticide-resistant insects are costing billions in crop damage and decreasing yields. An evolutionary approach to treating bacterial infections involves careful use of antibiotics, thereby slowing the rate at which newly resistant strains evolve.

1. In the cases of evolution of insecticide and pesticide resistance, what type of selection is operating? What do you expect to happen to gene frequencies in the population (of insects or bacteria) that has evolved resistance?
2. Using what you have learned in this chapter, discuss some strategies doctors might use to help reduce the rate at which antibiotic-resistant bacteria evolve.

SUMMARIZING THE CONCEPTS

27.1 Origin of Life

- Chemical reactions are hypothesized to have led to the formation of the first true cell(s) in the following steps:

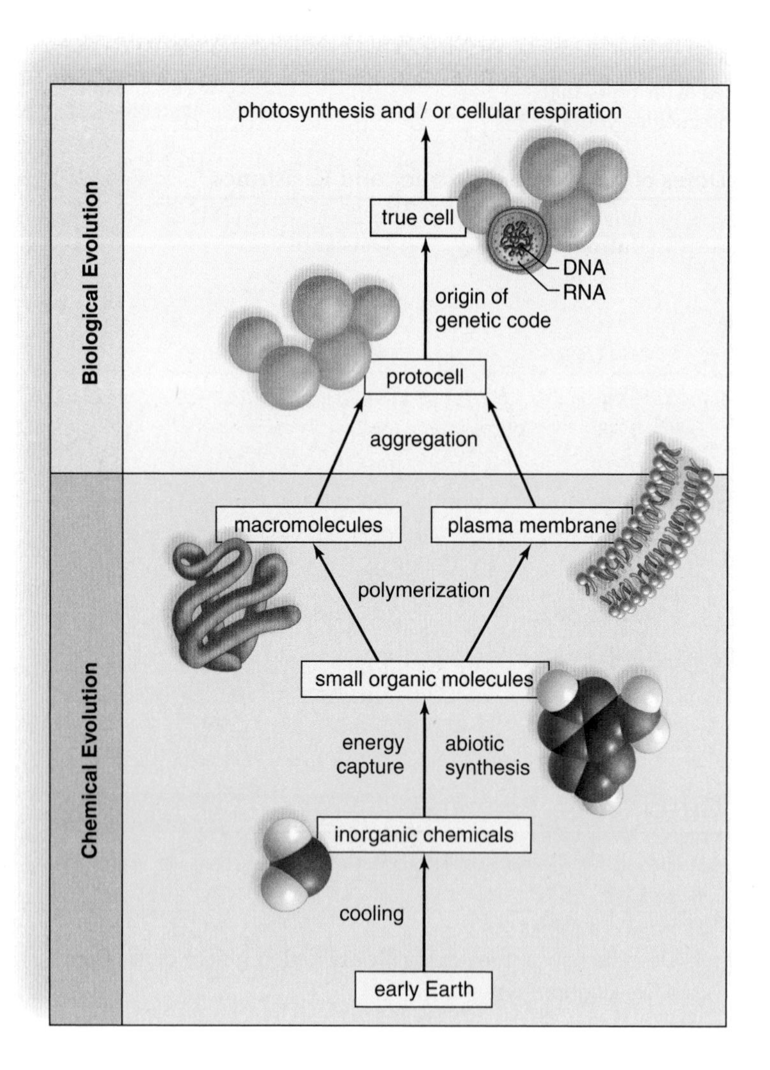

1. Inorganic chemicals, probably derived from the primitive atmosphere, reacted in the ocean to form small organic molecules.
2. Small organic molecules then polymerized to form macromolecules. Amino acids joined to form proteins, and nucleotides joined to form nucleic acids.
3. There are competing theories as to whether RNA was the first nucleic acid or proteins formed first.
4. Once a plasma membrane developed, the protocell came into being.
5. Eventually, the DNA ⟶ RNA ⟶ protein system evolved, and a true cell came into being.

27.2 Evidence of Evolution

The fossil record and biogeography, as well as studies of comparative anatomy and development, and biochemistry, all provide evidence of evolution:

- The fossil record gives clues about the history of life in general and allows us to trace the descent of a particular group.
- Biogeography shows that the distribution of organisms on Earth can be influenced by a combination of evolutionary and geological processes.
- Comparing the anatomy and the development of organisms reveals homologous structures among those that share common ancestry.
- All organisms have certain biochemical molecules in common, and these chemical similarities indicate the degree of relatedness.

27.3 The Process of Evolution

- Microevolution is a process that involves a change in allele frequencies within the gene pool of a sexually reproducing population.
- The Hardy-Weinberg principle states that gene pool frequencies arrive at an equilibrium that is maintained generation after generation unless disrupted by mutations, genetic drift, gene flow, nonrandom mating, or natural selection.
- Any change from the initial allele frequencies in the gene pool of a population signifies that evolution has occurred.

27.4 Speciation

- Speciation is the origin of new species. This usually requires geographic isolation, followed by reproductive isolation. The evolution of several species of finches on the Galápagos Islands is an example of speciation caused by adaptive radiation because each one has a different way of life.
- Currently, there are two models about the pace of speciation. Phyletic gradualism is slow, steady change leading to speciation. In contrast, punctuated equilibrium proposes that long periods of stasis are interrupted by rapid speciation.

27.5 Systematics

- Systematics involves assigning species to a hierarchy of categories: domain, kingdom, phylum, class, order, family, genus, and species.
- Phylogeneticists classify and diagram the evolutionary relationships among organisms. They use as many characteristics as possible to put species in clades, which are portions of a diagram called a cladogram. Each clade contains a most recent common ancestor and all its descendant species, which share the same derived characteristics relative to their ancestors.
- Modern phylogenetic methods most commonly analyze similarities and differences among nucleotide sequences, aided by computers, to reconstruct phylogenies.
- The three-domain system (Bacteria, Archaea, and Eukarya), based on molecular data, is currently preferred to the formerly used five-kingdom system (Monera, Protista, Fungi, Plantae, Animalia). Both bacteria and archaea are prokaryotes. Members of the kingdoms Protista, Fungi, Plantae, and Animalia are eukaryotes.

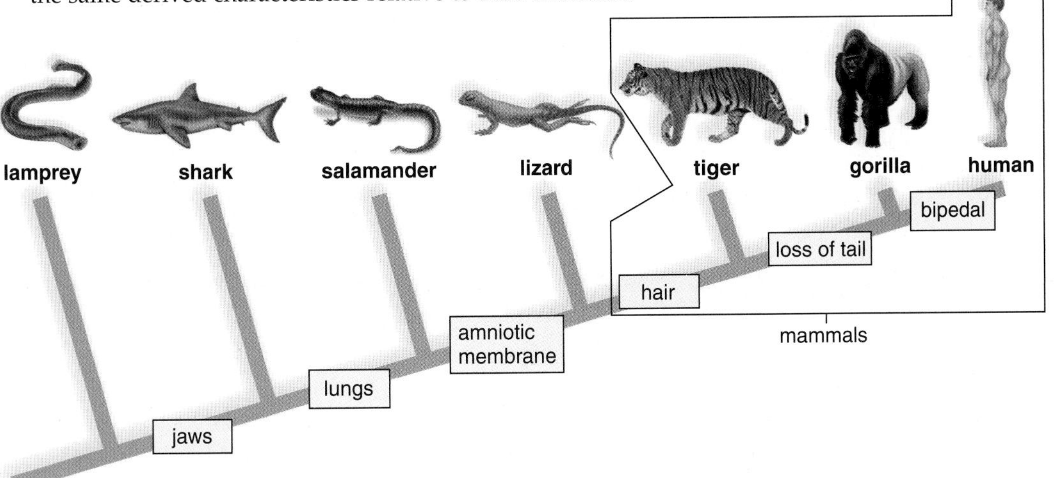

TESTING YOURSELF

Choose the best answer for each question.

1. The RNA-first hypothesis for the origin of cells is supported by the discovery of
   a. ribozymes.
   b. proteinoids.
   c. polypeptides.
   d. nucleic acid polymerization.
2. Protocells probably obtained energy as
   a. photosynthetic autotrophs.
   b. chemoautotrophs.
   c. heterotrophs.
   d. None of these are correct.
3. All true cells are able to
   a. replicate DNA.
   b. synthesize sugars.
   c. absorb nutrients.
   d. export minerals.
4. Fossils that serve as transitional links allow scientists to
   a. determine how prehistoric animals interacted with each other.
   b. deduce the order in which various groups of animals arose.
   c. relate climate change to evolutionary trends.
   d. determine why evolutionary changes occur.
5. Carbon dating cannot be used to determine the age of dinosaur fossils because
   a. levels of atmospheric carbon were very low when dinosaurs were alive.
   b. dinosaurs contained very low levels of carbon.
   c. dinosaur fossils contain very low levels of carbon.
   d. the half-life of radioactive carbon is too short.
6. The flipper of a dolphin and the fin of a tuna are
   a. homologous structures.
   b. homogeneous structures.
   c. analogous structures.
   d. reciprocal structures.
7. The frequency of a rare disorder expressed as an autosomal recessive trait is 0.0064. Using the Hardy-Weinberg principle, determine the frequency of carriers for the disease in this population.
   a. 0.147
   b. 0.080
   c. 0.920
   d. 0.020
   e. 0.846

For questions 8–12, match the description with the appropriate term in the key.

Key:

a. mutation
b. natural selection
c. founder effect
d. bottleneck
e. gene flow
f. nonrandom mating

8. The Northern elephant seal went through a severe population decline as a result of hunting in the late 1800s. As a result of a hunting ban, the population has rebounded but is now homozygous for nearly every gene studied.
9. A small, reproductively isolated religious sect called the Dunkers was established by 27 families that came to the United States from Germany over 200 years ago. The frequencies for blood group alleles in this population differ significantly from those in the general U.S. population.
10. Turtles on a small island tend to mate with relatives more often than turtles on the mainland.

11. Within a population, plants that produce an insect toxin are more likely to survive and reproduce than plants that do not produce the toxin.
12. The gene pool of a population of bighorn sheep in the southwest U.S. is altered when several animals cross over a mountain pass and join the population.
13. People who are heterozygous for the cystic fibrosis gene are more likely than others to survive a cholera epidemic. This heterozygote advantage seems to explain why homozygotes are maintained in the human population, and is an example of
    a. disruptive selection.
    b. balanced polymorphism.
    c. high mutation rate.
    d. nonrandom mating.
14. The creation of new species due to geographic barriers is called
    a. isolation speciation.
    b. allopatric speciation.
    c. allelomorphic speciation.
    d. sympatric speciation.
    e. symbiotic speciation.
15. Which of the following models is supported by the observation that few transitional links are found in the fossil record?
    a. phyletic gradualism
    b. punctuated equilibrium
    c. Both a and b are correct.
    d. None of these are correct.
16. The three-domain classification system has recently been developed based on
    a. mitochondrial biochemistry and plasma membrane structure.
    b. cellular and rRNA sequence data.
    c. plasma membrane and cell wall structure.
    d. rRNA sequence data and plasma membrane structure.
    e. nuclear and mitochondrial biochemistry.

UNDERSTANDING THE TERMS

adaptive radiation 561
allopatric speciation 559
analogous structure 549
autotroph 544
biogeography 547
bottleneck effect 556
clade 562
cladogram 562
continental drift 547
convergent evolution 562
directional selection 557
disruptive selection 557
domain Archaea 564
domain Bacteria 564
domain Eukarya 564
evolution 545
fitness 556
fossil 545
founder effect 553
gene flow 556
gene pool 551
genetic drift 553
heterotroph 544
homologous structure 549
liposome 544
microevolution 551
microsphere 543
mutation 553
natural selection 556
nonrandom mating 556
phyletic gradualism 561
phylogenetics 562
phylogeny 562
population 551
postzygotic isolating mechanism 559
prezygotic isolating mechanism 559
protein-first hypothesis 543
proteinoid 543
protocell 542
punctuated equilibrium 561
RNA-first hypothesis 543
speciation 559
species 559
sympatric speciation 559
systematics 562
taxonomy 562
three-domain system 564
transitional link 545
vestigial structure 549

THINKING CRITICALLY

1. Viruses such as HIV are rapidly replicated and have very high mutation rates. Thus, evolution of the virus can be observed in a single infected person. Using what you have learned in this chapter, explain why HIV is so hard to treat, even though multiple drugs to treat HIV have been developed.
2. You observe a wasting disease in cattle that you know is genetically caused and thus heritable. The disease is fatal in young cattle. The allele frequency for the gene that causes the disease is 0.05 in the United States, but 0.35 in South America. Explain why such a difference in allele frequencies might exist.
3. Why are homologous structures, as opposed to analogous ones, used to determine the evolutionary relationships of species and to reconstruct phylogenies?

INQUIRY INTO LIFE WEBSITE

The companion website for *Inquiry into Life* provides a wealth of information organized and integrated by chapter. You will find practice tests, animations, videos, and much more that will complement your learning and understanding of general biology.

http://www.mhhe.com/maderinquiry13

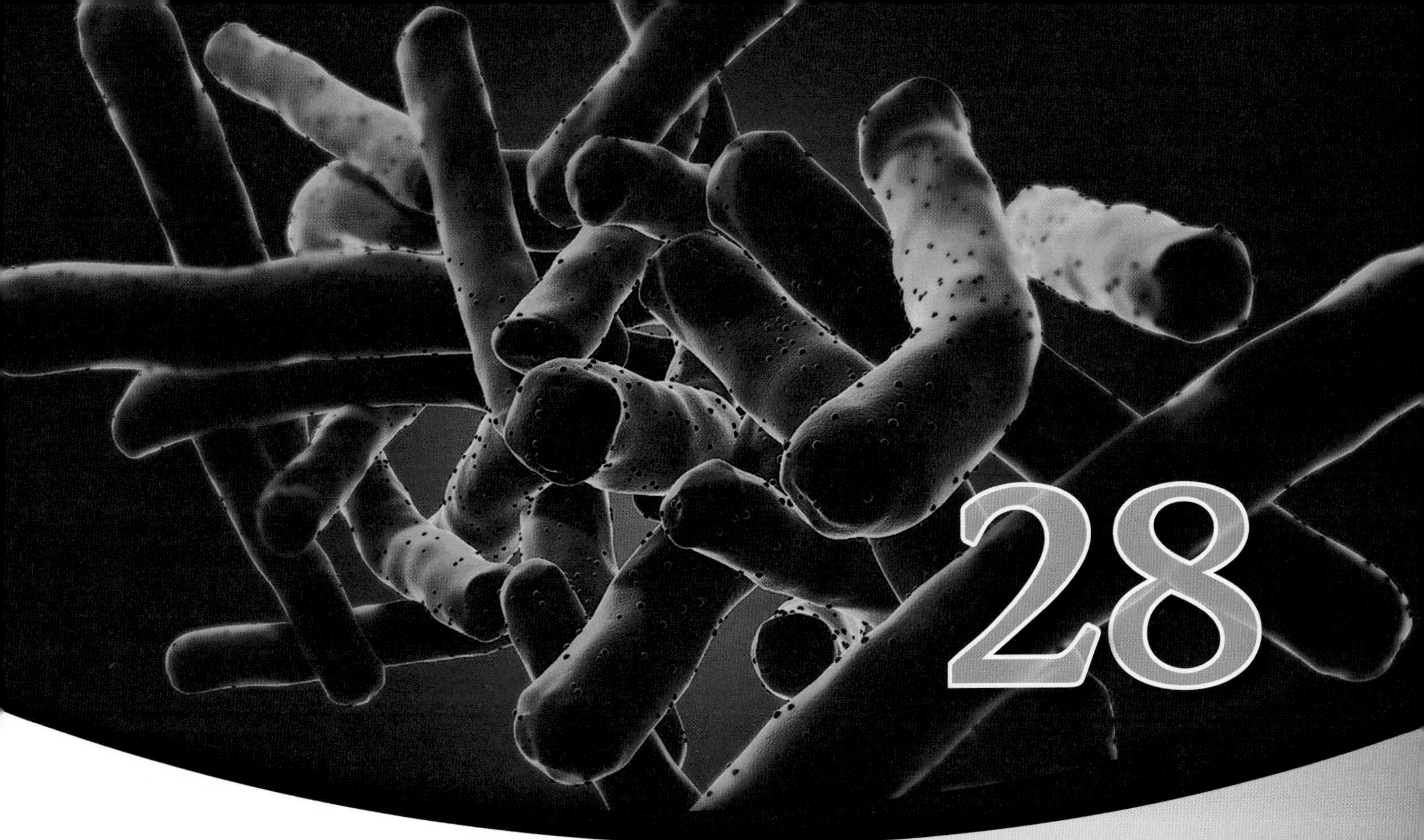

Microbiology

Chapter Outline

Applying the Concepts

Howard Hughes was the world's richest man in the 1930s and 1940s. He built airplanes, produced movies, and romanced movie stars. But he was also terrified of "germs," and throughout his life he became increasingly fearful of contact with other people, whom he considered to be the source of most harmful microbes. According to the 2004 movie *The Aviator*, Hughes' mother instilled this fear of germs in him beginning when he was a small boy. But in his later years, Hughes' germ phobia became so extreme that he wasted away in a darkened hotel room he considered to be "germ-free."

Even those of us who are not terrified of germs tend to become more aware of microbes when we use the bathroom. However, because they aren't cleaned regularly, desks, not toilets, are among the most germ-filled indoor environments—with teachers' desks being the worst! A 2006 study estimated that the average office worker's hands contact 10 million bacteria a day. Almost anything that people touch frequently, from ATM keypads to cell phones, is teeming with microbial life. Paper money is another common source of bacteria. Presumably Mr. Hughes was unaware that his billions of dollars could serve as temporary homes for even more billions of thriving germs!

Fortunately, humans have always lived in close contact with bacteria, viruses, and other microbes, so there is no need to don surgical masks and sequester ourselves in sterilized hotel rooms. In fact, many experts suggest that our children are receiving too little exposure to common, harmless microbes early in life, leading to decreased immunity to harmful varieties and, perhaps, to an increasing incidence of allergies and autoimmune disease. Since we cannot avoid microbes no matter how hard we try, and because exposure to the harmless forms may actually be healthy for us, it seems that while taking care to wash our hands or use a hand sanitizer a little more often, we might also want to embrace, at least figuratively, the microscopic life-forms with which we share the planet.

28.1 The Microbial World

Learning Outcomes

1. Describe the contributions of Leeuwenhoek and Pasteur to the science of microbiology.
2. List the major groups of microbes studied in microbiology.
3. Provide several specific examples of the beneficial effects of microbes.

Antonie van Leeuwenhoek (1632–1723) was a Dutch tradesman and scientist. He was apparently very skilled at working with glass lenses, which enabled him to greatly improve the microscopes that already existed in the 1600s. Using these instruments, he was among the first to view microscopic life forms in a drop of water, which he called "animalcules" and described in this way:

> *[They] were incredibly small, nay so small, in my sight, that I judged that even if 100 of these very wee animals lay stretched out one against another, they could not reach to the length of a grain of coarse sand.*

Leeuwenhoek and others after him believed that the "wee animals" he had observed could arise spontaneously from inanimate matter. For about 200 years, scientists carried out various experiments to determine the origin of microscopic organisms in laboratory cultures. Finally, in about 1859, Louis Pasteur devised the experiment shown in Figure 28.1. In the first experiment, when flasks containing sterilized broth were exposed to either outdoor or indoor air, they often became contaminated with microbial growth. In the second experiment, however, if the neck of the flask was curved so that microbes could not enter the broth from the air, no growth occurred. Thus, the bacteria were not arising spontaneously in the broth, but rather were coming from the air. In 1884, Pasteur also suggested that something even smaller than a bacterium was the cause of rabies, and it was he who chose the word *virus* from a Latin word meaning poison. Pasteur went on to develop a vaccine for rabies, which he used to save the life of a young French boy who had been bitten by a rabid dog.

Microbiology is the study of microbes, a term that includes the bacteria, archaea, protists, fungi, viruses, viroids, and prions. Most, but not all, of these organisms are so small that they require a microscope to be seen. Today we know that bacteria and other microbes are incredibly numerous in air, water, soil, and on objects. A single spoonful of soil can contain 10^{10} bacteria, and the total number of bacteria on Earth has been estimated at around 10^{30}, which clearly exceeds the numbers of any other type of organism on Earth (with the exception of viruses that infect all those bacteria!). It is a good thing bacteria are microscopic—if they were the size of beetles, the Earth would be covered in a layer of bacteria several miles deep!

We need only to turn on the nightly TV news to hear about diseases caused by microbes. Mad cow disease, H1N1 flu, and diseases caused by antibiotic-resistant strains of the bacteria *Staphylococcus aureus* and *Mycobacterium tuberculosis*, are recent examples of "emerging" infectious diseases. However, many microbes have beneficial—and sometimes essential—roles to play in human health as well as in the proper functioning of the biosphere. For example, you have approximately 10^{14} bacteria living on or in your body right now, even though your body is composed of only about 10^{13} human cells! In other words, you have ten times as many bacterial cells as human cells. (Of course, the bacterial cells are much smaller). Most of these microbes, also known as the **normal microflora,** have beneficial effects.

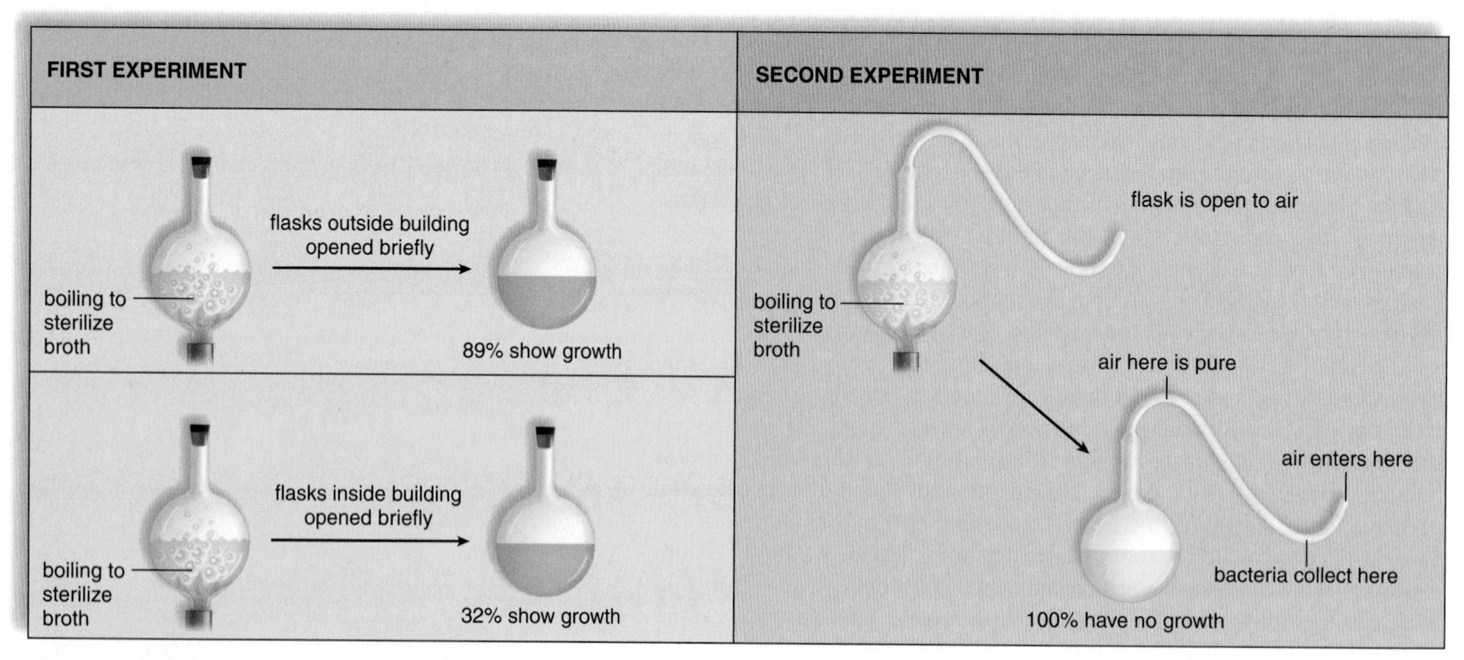

Figure 28.1 Pasteur's experiments.
Pasteur disproved the theory of spontaneous generation of microbes by performing these types of experiments.

Bacteria that live on your skin help crowd out harmful microbes that might grow in those areas, and bacteria in the intestines aid in digestion and synthesize vitamin K and vitamin B_{12}. Our bodies are actually a finely balanced ecosystem for microbes.

Bacteria and fungi are the **decomposers** that play essential roles in various nutrient cycles on Earth by breaking down organic and inorganic materials that can be reused by plants and animals. Photosynthetic algae and other protists are *primary producers* that capture energy from the sun or inorganic material, providing nutrients for more complex organisms. Some bacteria also perform photosynthesis, often in ways very different from algae or plants. We have also learned to use bacteria to clean up our environment. Wastewater that leaves your bathroom and kitchen is probably cleaned at a treatment plant by processes that rely on the activity of bacteria. It seems that bacteria can consume virtually any substance found on Earth. Soil contaminated by oil spills or just about any other toxic compounds can be cleaned by encouraging the growth of bacteria that eat these compounds.

Not only do many microbes play essential roles in the environment, but they are also valuable for industrial processes, particularly food processing. Also, most antibiotics known today were first discovered in soil bacteria or fungi. Genetic techniques can be used to alter the products generated by bacterial cultures. A variety of valuable products can be made in this way, including insulin and vaccines. Although it might not seem so to an individual suffering from tuberculosis, AIDS, or another serious infectious disease, most microbes are far more beneficial than harmful—in fact, we could not live without them!

Check Your Progress

1. What specific experiment did Pasteur use to disprove the idea of spontaneous generation of microbes?
2. List five major groups of microbes.
3. Define the term decomposers. If all decomposers vanished tomorrow, what would be the consequences?

28.2 Bacteria and Archaea

Learning Outcomes

1. Describe bacterial shapes, structures, reproduction, and metabolism.
2. Name several major bacterial diseases of humans and describe how they are treated.
3. Compare and contrast the biology of bacteria and archaea.

Both bacteria (domain Bacteria) and archaea (domain Archaea) are prokaryotes, but each is placed in its own domain because of molecular and cellular differences. Prokaryotes do not have nuclei or the membrane-bound cytoplasmic organelles found in eukaryotic cells.

Biology of Bacteria

Most **bacteria** are between 0.2–10 μm in size. A few, however, are quite large, including one species that is about the same size as the period at the end of this sentence. Bacteria have three basic shapes: rod (bacillus, pl., bacilli); spherical (coccus, pl., cocci); and spiral-shaped or helical (Fig. 28.2). A bacillus or coccus can occur singly or may occur in particular arrangements. For example, when cocci form a cluster, they are called staphylococci, while cocci that form chains are called streptococci.

The typical structure of a bacterium is shown in Figure 28.3. All bacterial cells have a plasma membrane, which is a lipid bilayer, similar to the plasma membrane in plant and animal cells. Most bacterial cells are further protected by a cell wall that contains the unique molecule peptidoglycan. Bacteria can be classified by differences in their cell walls, which are detected using a staining procedure devised more than 100 years ago by Hans Christian Gram. When you go to the doctor for a bacterial infection, one of the most common tests performed is the **Gram stain,** because different antibiotics tend to be more effective against Gram-positive or Gram-negative bacteria. Cell walls that have a thick layer of peptidoglycan outside the plasma membrane stain purple with the Gram stain procedure, and are called Gram-positive bacteria. If the

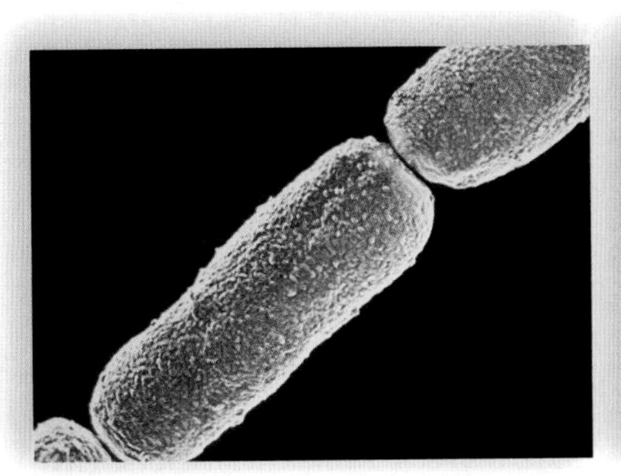

a. Bacilli (rod): *Bacillus anthracis* SEM 35,000×

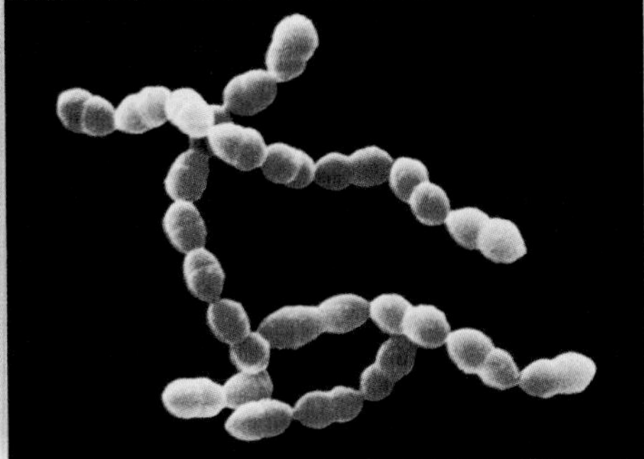

b. Cocci (spherical): *Streptococcus thermophilus* SEM 3,520×

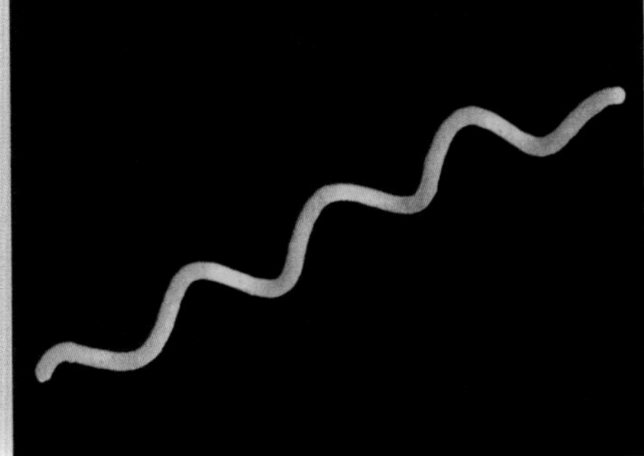

c. Spirillum (curved): *Spirillum volutans* SEM 6,250×

Figure 28.2 Typical shapes of bacteria.
a. Bacilli, rod-shaped bacteria. **b.** Cocci, round bacteria. **c.** Spirillum, a spiral-shaped bacterium.

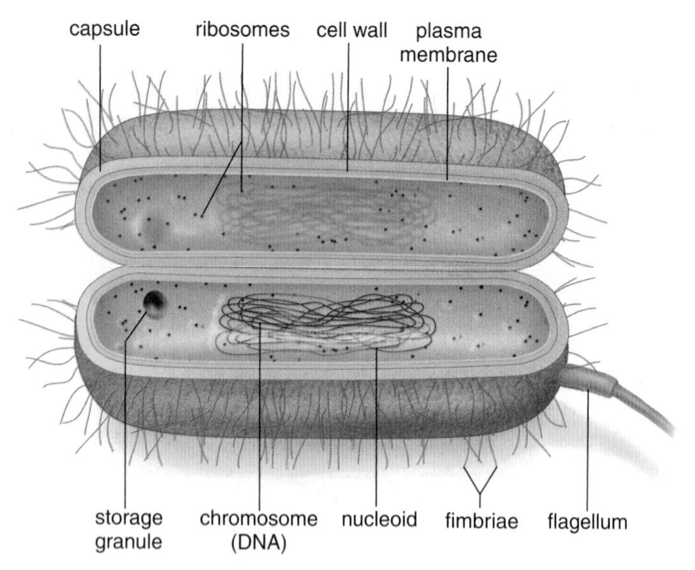

Figure 28.3 Typical bacterial cell.
Bacteria are prokaryotes and thus have no membrane-bounded nuclei or organelles.

peptidoglycan layer is either thin or lacking altogether, the cells stain pink and are considered Gram-negative. In addition to their plasma membrane, Gram-negative bacteria have an outer membrane that contains **lipopolysaccharide (LPS)** molecules. When these Gram-negative cells are killed by your immune system, LPS molecules are released, stimulating inflammation and fever. Beyond the cell wall, some bacteria have a slimy polysaccharide capsule that can protect the cell from dehydration and the immune system.

Motile bacteria have **flagella** for locomotion but never cilia. The flagellum is a stiff, curved filament that rotates like a propeller. Bacterial flagella are structurally distinct from eukaryotic flagella. Some bacteria have fimbriae that bind to various surfaces. For example, bacteria that cause urinary tract infections can bind to urinary tract cells. Drinking cranberry juice seems to inhibit this binding.

Most bacteria have a single circular chromosome, which is located in a **nucleoid** region, rather than in a membrane-bounded nucleus. Many bacteria also harbor accessory rings of DNA called plasmids that can carry genes for antibiotic resistance, among other things, and are commonly used to carry foreign DNA into other bacteria during genetic engineering (see Chapter 26). Bacterial cells also contain abundant ribosomes, as well as various types of granules that store nutrients such as glycogen and lipids. Notice that bacteria do not have an endoplasmic reticulum, a Golgi apparatus, mitochondria, or chloroplasts. This diagram summarizes the major structural features of bacteria:

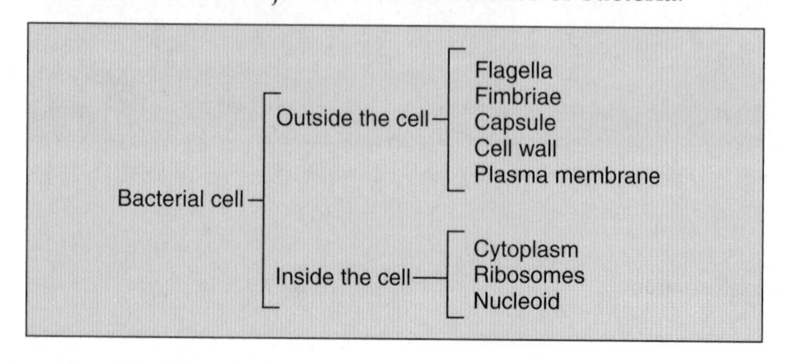

Bacterial Reproduction and Gene Transfer

Bacteria reproduce asexually. After a period of sufficient growth, the bacterial cell simply divides into two new cells, with each cell getting a copy of the genome and approximately half of the cytoplasm. This process is known as **binary fission** (Fig. 28.4). Each daughter cell is a *clone,* or exact copy, of the parent cell. When bacteria are spread out and grown on agar plates, each individual cell can give rise to a colony of millions of cells. In the absence of genetic mutations, each of these cells is a clone of the initial cell. Some bacteria need only 20 minutes to reproduce, while others grow more slowly, with generation times of a day or more.

Under unfavorable conditions, some bacteria (such as *Clostridium tetani,* the cause of tetanus) can produce resistant structures called **endospores,** thick-walled, dehydrated structures capable of surviving the harshest conditions, perhaps even for thousands of years. Endospores are *not for reproduction,* but are simply a way to survive unfavorable conditions.

Sexual reproduction does not occur among prokaryotes, but at least three means of gene transfer have been observed. **Conjugation** takes place when a donor cell passes DNA to a recipient cell by way of a *sex pilus.* **Transformation** occurs when a bacterium takes up DNA released into the medium by dead bacteria. During **transduction,** viruses carry portions of bacterial DNA from one bacterium to another.

Bacterial Metabolism

Although most bacteria are structurally similar, they demonstrate a remarkable range of metabolic abilities. Most bacteria are **heterotrophic** and require an outside source of organic compounds in the same way that animals do. Some heterotrophic bacteria are anaerobic and cannot use oxygen to capture electrons at the end of their electron transport chain (see Chapter 7). Instead, they use a variety of substances as final electron acceptors. Sulfate-reducing bacteria transfer the electrons to sulfate, producing hydrogen sulfide, which smells like rotten eggs. Denitrifying bacteria use nitrate, while others use minerals, such as iron or manganese, as electron receivers.

Other bacteria are **chemoautotrophs.** They reduce carbon dioxide to an organic compound by using energetic electrons derived from chemicals, such as ammonia, hydrogen gas, and hydrogen sulfide. Electrons can also be extracted from certain minerals, such as iron.

Some bacteria are photosynthesizers that use solar energy to produce their own food. **Cyanobacteria** are believed to have arisen some 3.8 billion years ago and to have produced much of the oxygen in the atmosphere at that time. Sometimes erroneously called blue-green algae, these organisms contain green chlorophyll and have other pigments that give them a bluish-green color (Fig. 28.5). Other bacterial photosynthesizers split hydrogen sulfide, instead of water, in anaerobic environments. Therefore, they produce sulfur rather than oxygen as a by-product of photosynthesis.

Aside from producing oxygen, cyanobacteria are often the first colonizers of rocks. Many are capable of both carbon fixation and nitrogen fixation, needing only minerals, air,

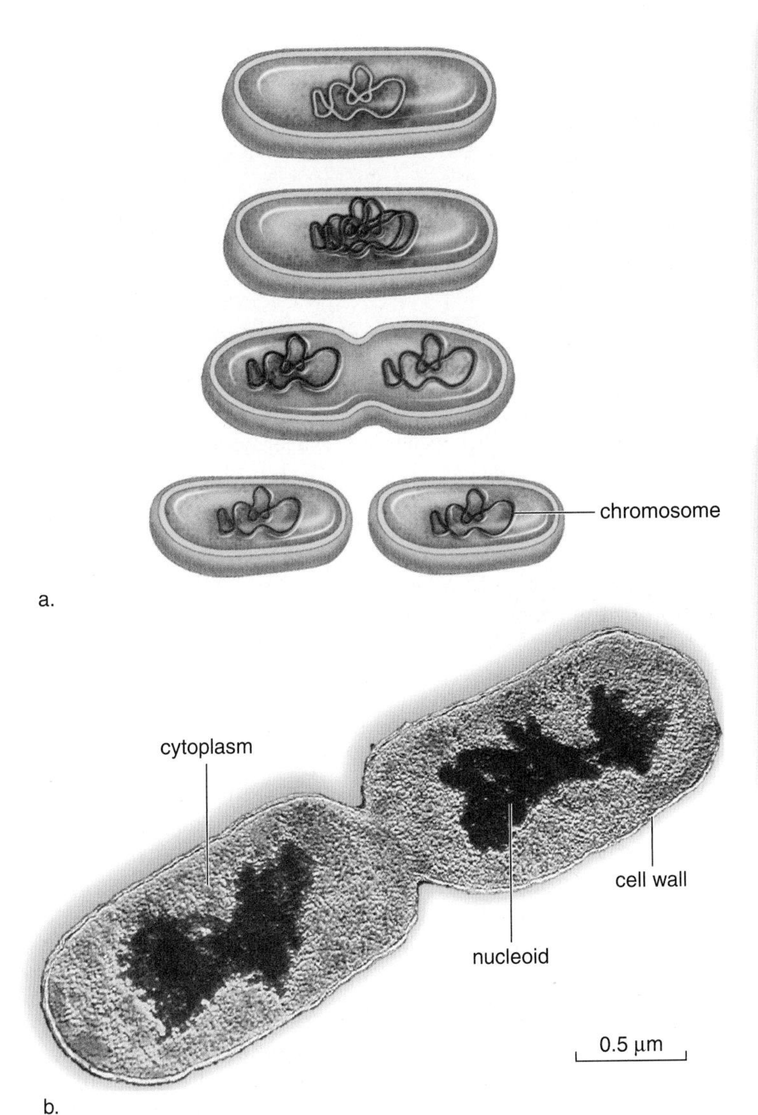

Figure 28.4 Binary fission.
a. A diagram and **(b)** a micrograph illustrate binary fission, a process that occurs when a bacterium reproduces. Bacteria grow and the chromosome is duplicated before binary fission occurs.

Figure 28.5 Cyanobacteria.
Cyanobacteria are photosynthetic bacteria that contain chlorophyll.
a. In *Chroococcus*, single cells are grouped in a common gelatinous sheath. **b.** Filaments of cells occur in *Oscillatoria*, a common inhabitant in freshwater environments.

sunlight, and water for growth. Cyanobacteria are also notorious for forming toxic "blue-green algae" blooms in waters enriched by nutrients, which may require that some lakes be closed to human use. Cyanobacteria (and in some cases eukaryotic algae) form a symbiotic relationship with fungi in a lichen. The fungi provide a place for the cyanobacteria to grow and obtain water and mineral nutrients. Some cyanobacteria even appear to live in a symbiotic relationship in the fur of sloths, providing the animals with a form of camouflage in their dense green forest home!

Bacterial Diseases in Humans

Most types of bacteria do not cause disease, but a significant number do. Why does one microbe cause disease, while another, closely related species is completely harmless? Pathogenic microbes often carry genes that code for specific virulence factors that determine the type and extent of illness they are capable of causing. For example, right now you have billions of *E. coli* living in your large intestine, without causing any problems. However, some strains of *E. coli* have acquired genes that make them dangerous invasive pathogens. The strain of *E. coli* called O157:H7 has the ability to generate a toxin that damages the lining of the intestine, resulting in a bloody diarrhea. It has also obtained other virulence factors that help it stick to the lining of the gut more efficiently. In 2006, around 200 people in the United States became ill and at least three died from *E. coli* O157:H7 contamination in bagged spinach.

By acquiring genes coding for virulence factors (i.e., via conjugation, transformation, and/or transduction) harmless bacteria may be converted into pathogens. Antibiotic resistance genes may pass between organisms in the same ways, creating antibiotic-resistant pathogens.

Here we discuss a few of the most common types of bacterial diseases in humans.

Streptococcal Infections More different types of human disease are caused by bacteria from the genus *Streptococcus* than by any other type of bacterium. *Streptococcus pneumoniae* can cause pneumonia, meningitis, and middle ear infections. Up to 70% of apparently healthy adults are carriers of this potentially harmful bacterium, which causes disease mainly in children and the elderly. *Streptococcus mutans* is found on the teeth and contributes to tooth decay and the formation of dental caries.

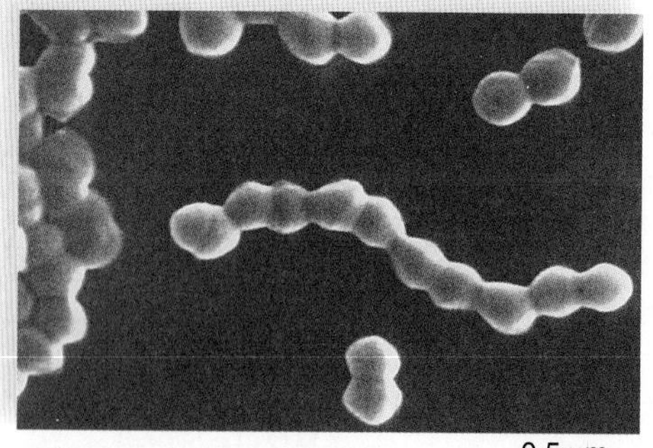

a. *Streptococcus pyogenes*

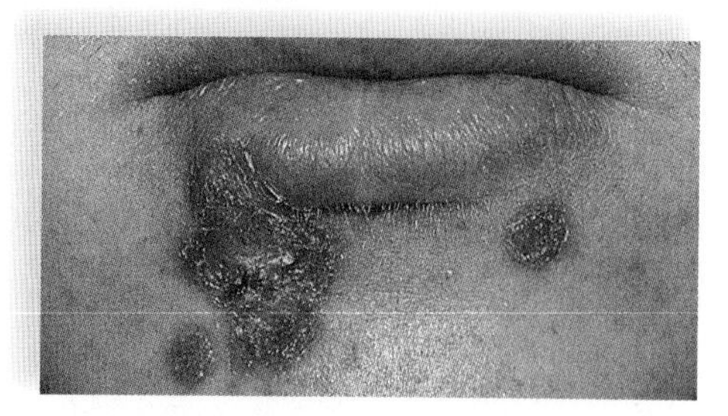
b. Impetigo

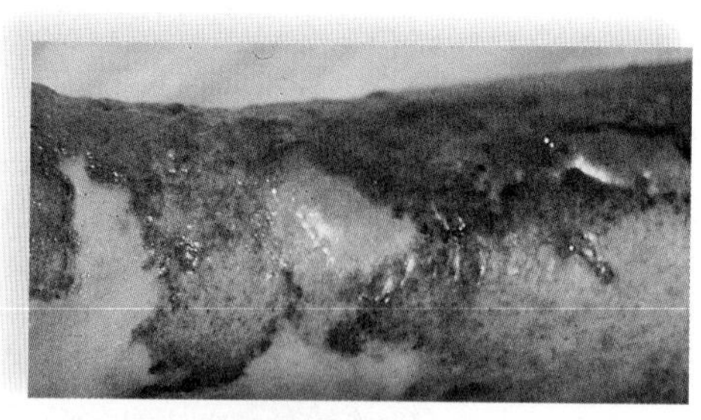
c. Flesh-eating disease

Figure 28.6 ***Streptococcus pyogenes.***
a. *Streptococcus pyogenes,* shown here in an electron micrograph, often forms chains of cells, as seen here. **b.** Impetigo is a common mild skin infection. **c.** Flesh-eating disease is rare but life-threatening.

The most common illness caused by streptococci is pharyngitis, commonly called strep throat, which is usually due to infection by *Streptococcus pyogenes* (Fig. 28.6*a*). *Streptococcus pyogenes* also causes relatively mild skin diseases, such as impetigo in infants (Fig. 28.6*b*). A number of more serious and invasive diseases can occur after or during infections with *Streptococcus pyogenes.* The nature and severity of these diseases depend on which virulence factors the pathogen carries. Scarlet fever is a strep infection caused by a *Streptococcus* strain that produces a toxin causing a red rash. An intense immune response to the infection can lead to rheumatic fever, which is characterized by a high temperature, swollen joints that form nodules, and heart damage. Although rarely seen in the United States today, rheumatic fever killed more school-age children than all other diseases combined in the early twentieth century.

By releasing enzymes that destroy connective and muscle tissues and kill cells, *Streptococcus pyogenes* infection can lead to large-scale cell lysis and tissue destruction. So-called "flesh-eating" bacteria cause necrotizing fasciitis (Fig. 28.6*c*), in which rapid tissue damage can require amputation of infected limbs or, in 40% of cases, rapid death.

***Staphylococcus aureus* and MRSA** *Staphylococcus aureus* is perhaps the bacterial species that has received the most media attention recently. About 20% of people are carriers of this bacterium (mostly on their skin or in their nostrils) without any symptoms. When *Staphylococcus aureus* does cause disease, it is usually limited to skin infections. However, in people who are very young, very old, or immunocompromised for any other reason, it can invade the body and cause life-threatening disease. Moreover, a strain of *Staphylococcus aureus* that is resistant to methicillin, called MRSA, is killing an increasing number of young, otherwise healthy individuals (about 19,000 in 2007). Besides being resistant to many antibiotics, MRSA strains often possess genes coding for toxins not found in other *Staphylococcus aureus* strains. Some of these toxins can be very damaging to tissues. In February 2009, a Texas jury awarded $17.5 million to a man who lost both arms and legs to a hospital-acquired MRSA infection, although the award was later reduced.

Tuberculosis

Tuberculosis (TB) is one of the leading worldwide causes of death due to infectious disease. When first described in 1882 by the father of medical microbiology, Robert Koch, TB caused one of every seven deaths in Europe, and one-third of all deaths of young adults. Today, it is estimated that one-third of the world's population is infected with the TB bacterium, causing approximately 2 million deaths each year. Tuberculosis is a chronic disease caused by *Mycobacterium tuberculosis,* a pathogen closely related to *Mycobacterium leprae,* the causative agent of leprosy. Generally, *M. tuberculosis* infection occurs in the lungs, but it can occur elsewhere as well.

M. tuberculosis is very slow growing, and the extent of the disease is determined by host susceptibility. An immune response results in inflammation of the lungs. Active lesions can produce dense structures called **tubercles** (Fig. 28.7). These can persist for years, causing symptoms such as coughing and spreading the bacteria to other areas of the lungs and body as well as to other individuals. Eventually, the damaged lung tissue hardens and calcifies, leaving characteristic spots that can be observed with chest X rays. The number of TB cases in the United States surged in the late 1980s, mainly due to increased rates of infection with HIV, which damages helper T cells that are needed to fight the TB bacterium. Since 1992, however, the prevalence of TB in the United States has steadily declined, to a rate of 4.4 cases per 100,000 people in 2007. This reduction is mainly attributed to better testing and treatment of infected persons.This trend is also occurring worldwide, although numbers of TB cases continue to increase in Africa and Southeast Asia.

The 2007 story of a 31-year-old Atlanta lawyer with TB shows how easily an infectious agent can spread around the world. Despite knowing that he was infected with *M. tuberculosis,* he flew from Atlanta to Paris to get married, traveling through five countries in the process. Making matters worse, he was infected with a strain of *M. tuberculosis* called XDR, which is resistant to nearly all drugs available to treat TB. Upon returning to the United States, he was taken into forced quarantine for four weeks, the first time an American had been forcibly isolated by the Centers for Disease Control and Prevention (CDC) in several decades. He was released after it was determined that he was no longer infectious. It should be noted, however, that there is no way to know how many people infected with *M. tuberculosis* are traveling on airplanes every day.

Food Poisoning Whether the source was mishandled food at a salad bar, or potato salad that sat in the sun too long at a picnic, all of us have probably experienced the intestinal discomfort known as food poisoning. Two basic types of bacteria cause food poisoning: those that produce toxins while they are growing in food, and those that cause infections once they are in the intestine.

Several species of bacteria can produce toxins in foods, especially those containing dairy, eggs, or meat products. The symptoms, which consist mainly of vomiting and diarrhea, tend to appear suddenly within a few hours of ingestion, and are usually self-limiting. In contrast, *Clostridium botulinum,* the causative agent of *botulism,* produces one of the most toxic substances on Earth. When people are canning and don't heat foods to a high enough temperature, *Clostridium* can produce endospores that survive the canning process. These endospores then germinate in the airless environment of the can or bottle and become toxin-producing cells. If untreated, about 25 percent of people who ingest botulism toxin die from respiratory paralysis.

Salmonella is a classic example of a bacterial food poisoning agent that does not produce gastroenteritis until it has reproduced in the intestines for several days. A recent large outbreak of salmonellosis in the United States occurred between late summer of 2008 and January 2009, when FDA recalled every product made from peanuts processed by a Georgia plant, which filed for bankruptcy in February 2009. Over 700 people became ill, and at least nine died after ingesting contaminated products.

Drug Control of Bacterial Diseases

Most **antibiotics** kill or inhibit bacteria by interfering with their unique metabolic pathways. Therefore, they are not expected to harm human cells. For example, erythromycin and tetracyclines inhibit bacterial protein synthesis by binding to bacterial ribosomes, while penicillins and cephalosporins inhibit bacterial cell wall synthesis.

There are problems associated with antibiotic therapy. Some individuals are allergic to certain antibiotics, and the reaction may even be fatal. Antibiotics not only kill off disease-causing bacteria, but they may also reduce the number of beneficial bacteria in the intestinal tract and vagina. This may allow the overgrowth of certain harmful bacteria or yeast, especially in the intestine, vagina, or mouth. See the Health Focus for a discussion of how probiotics can be used to replenish these beneficial organisms.

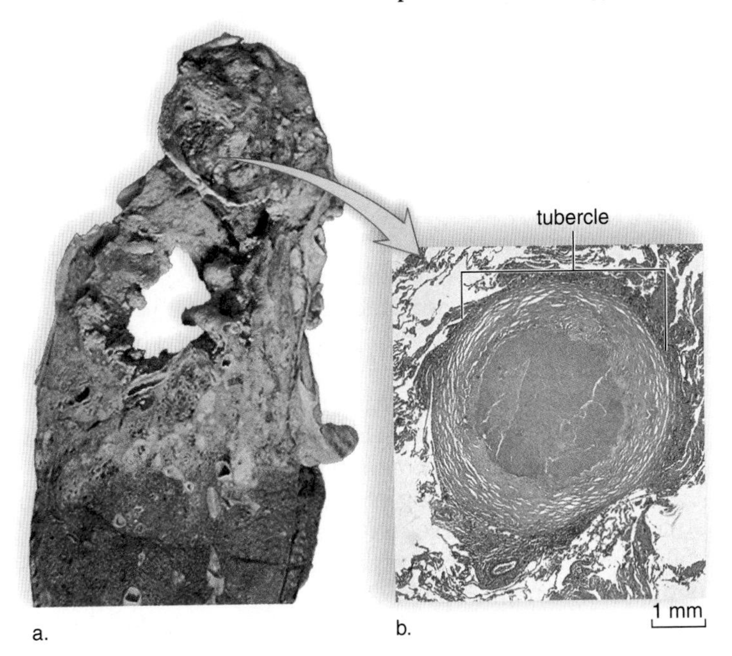

Figure 28.7 Tuberculosis.
Tuberculosis, an infection caused by the bacterium *Mycobacterium tuberculosis,* usually settles in the lungs. **a.** Lung tissue with a large cavity surrounded by tubercles, which are hard, calcified nodules where the bacteria are trapped. **b.** Photomicrograph shows a cross section of one large tubercle, with a diseased center.

Most important perhaps is the growing resistance of bacteria to antibiotics, as discussed in the story that opened Chapter 27. Antibiotics were introduced in the 1940s, and for several decades they worked so well it appeared that infectious diseases had been brought under control. However, we now know that bacterial strains can mutate and become resistant to a particular antibiotic. Worse yet, when bacteria exchange genetic material, resistance can pass between different bacteria. Penicillin and tetracycline now have a failure rate of more than 22% against *Neisseria gonorrhoeae,* which causes gonorrhea. Bacteria with multiple drug resistances, such as MRSA and multidrug-resistant *M. tuberculosis,* are an especially serious threat in hospitals, prisons, and long-term care facilities. With the rise of antibiotic resistance, many companies are now working to develop innovative kinds of antibacterial therapies. The need for new, potent antibiotics is especially critical since the U.S. anthrax attacks of October 2001 raised the potential of bioterrorist activity (see the Bioethical Focus at the end of the chapter).

Check Your Progress

1. What are the three basic bacterial shapes?
2. How does bacterial conjugation differ from transformation and transduction?
3. What is a virulence factor? Name specific virulence factors produced by *E. coli* O157:H7 and MRSA.
4. What are three human diseases caused by bacteria of the genus *Streptococcus?*
5. Consider how antibiotics work, and then describe two specific mechanisms bacteria could use to become resistant.

Biology of Archaea

Archaea and bacteria are not close relatives, even though both are prokaryotes. Therefore, they are placed in separate domains. Based on a number of criteria, including nucleic acid similarities, archaea may be more closely related to eukarya, even though the prokaryotic cell is much different from the eukaryotic cell (Table 28.1).

Many of the **archaea** studied to date live in so-called "extreme" environments that feature very salty, hot, acidic, and/or anaerobic conditions (Fig. 28.8). Since these conditions may resemble those on early Earth, it has been suggested that archaea may be the most ancient forms of life. However, some bacteria are also found in extreme environments, and the question of which group evolved first remains unresolved. Archaea have remarkable strategies for thriving under these conditions, and these abilities are reflected in the major groups of archaea: *thermoacidophiles* (high temperature and low pH environments); *methanogens* (anaerobic environments); and *halophiles* (salty environments). Nevertheless, in recent years, it has become evident that archaea are also found in large numbers in less extreme environments. Archaea have been isolated from the human colon. However, no archaea have been proven to be associated with any human disease.

TABLE 28.1 Comparison of Domains Archaea and Eukarya

	DOMAIN	
Feature	**Archaea**	**Eukarya**
Nucleus	No	Yes
Organelles	No	Yes
Introns	Sometimes	Yes
Histones	Yes	Yes
RNA polymerase	Several types	Several types
Methionine is at start of protein synthesis	Yes	Yes

Archaeal Structure

Archaea usually range from 0.1–15 μm in size. Their genome is a single, closed, circular DNA molecule, often smaller than a bacterial genome. Archaea that stain Gram-positive have a thick polysaccharide cell wall. Those that stain Gram-negative have a protein or glycoprotein surface layer. The plasma membrane of archaea differs markedly from those of bacteria and eukaryotes. Rather than a lipid bilayer, archaea often have a monolayer of lipids with branched side chains. Their chemical characteristics make archaea tolerant to acid and heat. Like bacteria, archaea reproduce asexually by binary fission.

Archaeal Metabolism

Some archaea are heterotrophs, while others are autotrophs. Many halophiles perform a unique type of photosynthesis by utilizing a pigment similar to retinal in human eyes. When the pigment absorbs solar energy, it moves a hydrogen ion to outside the plasma membrane. This establishes an H^+ gradient that promotes ATP synthesis.

Among archaea, the **methanogens** are mostly chemoautotrophs that can use carbon dioxide and hydrogen as energy sources, producing methane as a byproduct. Methanogens are found in anaerobic environments such as swamps, lake sediments, rice paddies, and the intestines of animals. Cows, which have large populations of methanogens in their digestive tracts, release a significant amount of methane into the environment. Since methane is a greenhouse gas that may contribute to global warming, some scientists are suggesting that the amount of methane produced by cattle should be reduced by changing the animals' diets, adding substances to their feed to alter the microbial populations, or developing a vaccine to specifically inhibit the growth of methanogens. However, the relative contribution of cattle, compared to other potential sources of methane, remains controversial.

CHECK YOUR PROGRESS

1. How do the plasma membranes of archaea differ from those of bacteria and eukaryotes?
2. What are some specific adaptations that might allow archaea to survive in extremely hot or salty environments?
3. Why might a dedicated environmentalist give up eating beef?

a.

b.

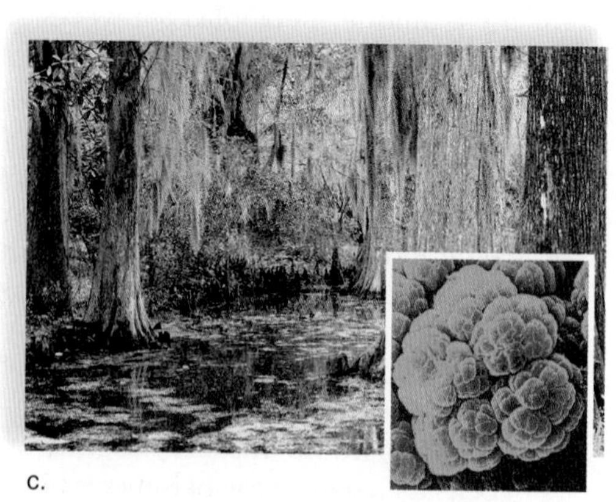

c.

Figure 28.8 Extreme habitats.

Many archaea thrive in unusual environmental conditions. a. Halophilic archaea can live in salt lakes. **b.** Thermophilic archaea can live in the hot springs of Yellowstone National Park. **c.** Methanogens live in anaerobic swamps and in the guts of animals.

28.3 Protists

Learning Outcomes

1. Discuss the evolution of the first unicellular protist.
2. Decribe a protist in general and the shared structural components of each group of protists.
3. Analyze the unique and specific features of each group of protists, including how they reproduce and obtain food.

Protists (domain Eukarya, kingdom Protista) are eukaryotes, thus they have membranous organelles. For the most part, protists are unicellular and microscopic, which sets them apart from other eukaryotes. They are such a diverse group, however, that perhaps the best definition of a protist is any eukaryotic organism that is not a plant, animal, or fungus.

Protists offer us a glimpse into the past because they are most likely related to the first eukaryotic cell to have evolved. According to the endosymbiotic theory, mitochondria may have resulted when a nucleated cell engulfed aerobic bacteria, and chloroplasts may have originated when a nucleated cell with mitochondria engulfed cyanobacteria (see Fig. 3.15). Protists also bridge the gap between the first eukaryotic cells and multicellular organisms—the other three types of eukaryotes (fungi, plants, and animals) all trace their ancestry to a protist. As further proof of this, some protists are not only multicellular, but also have differentiated tissues.

Biology of Protists

Protists can be unicellular, colonial, or multicellular. They are structurally diverse, having many complex shapes. Some have an outer cell wall, while others have external shells or tests made of minerals, cellulose, or even glass! Some protists have more than one nucleus. In some cases they have organelles that are not found in other eukaryotes. For example, **contractile vacuoles** are responsible for osmoregulation, particularly in freshwater environments. They pump out the excess water that tends to enter the cell. Some motile, photosynthetic protists have an **eyespot apparatus** that allows them to sense light intensity and move to areas for optimum photosynthesis.

The variability of protists also extends to their lifestyles and modes of reproduction. Most carry out asexual reproduction, with sexual reproduction as an option, particularly under unfavorable conditions. Sexual reproduction often results in a spore or cyst that can survive until environmental conditions improve. Spore formation may also promote dispersal to areas better suited for growth. Encystation involves the development of a cell with a thick cell wall and low metabolic rate, analogous to the endospores produced by some bacteria. Some protists, especially parasitic forms such as *Plasmodium* that cause malaria, have complicated life cycles in their different hosts.

A **protozoan** is a usually motile, eukaryotic, unicellular protist. One common way to divide protozoans is based on mode of locomotion. One group is nonmotile, and the other groups use flagella, cilia, or pseudopods for getting about. Protozoans are distributed in great number in many habitats. For example, in aquatic environments, they are part of the **zooplankton,** microscopic suspended organisms that feed on other organisms. A number of protozoans are parasitic, often causing diseases of the blood. In many cases, their complex life cycles hinder the development of suitable treatments.

Diversity of Protists

Many different classification schemes have been used to define relationships between the protists. As described in the previous section, protists can have a combination of characteristics not seen in other eukaryotic groups, which makes them difficult to classify. They have significant variations in cellular organization, structure, nutrition, locomotion, and reproduction. Traditionally, protists have been classified by their source of energy and nutrients: the algae are photosynthetic (but use a variety of pigments), the protozoans are heterotrophic by ingestion, and the water molds and slime molds are heterotrophic by absorption. Newer data, especially genetic sequence information, has resulted in some controversy about how to classify the protists: some experts lump them into as few as five major groups; others split them into as many as 15 separate phyla. Undoubtedly as newer molecular techniques are used to learn more about these organisms, a more uniform classification system will emerge. As shown in Figure 28.9, for this discussion we will group the protists into six groups according to some of their major shared characteristics: (1) photosynthetic protists, (2) flagellates, (3) ciliates, (4) amoeboids, (5) sporozoans, and (6) water molds and slime molds.

Photosynthetic Protists

Algae are photosynthetic eukaryotes that can be unicellular or form colonies or filaments. The unicellular and colonial forms may have flagella. Some types of algae are multicellular

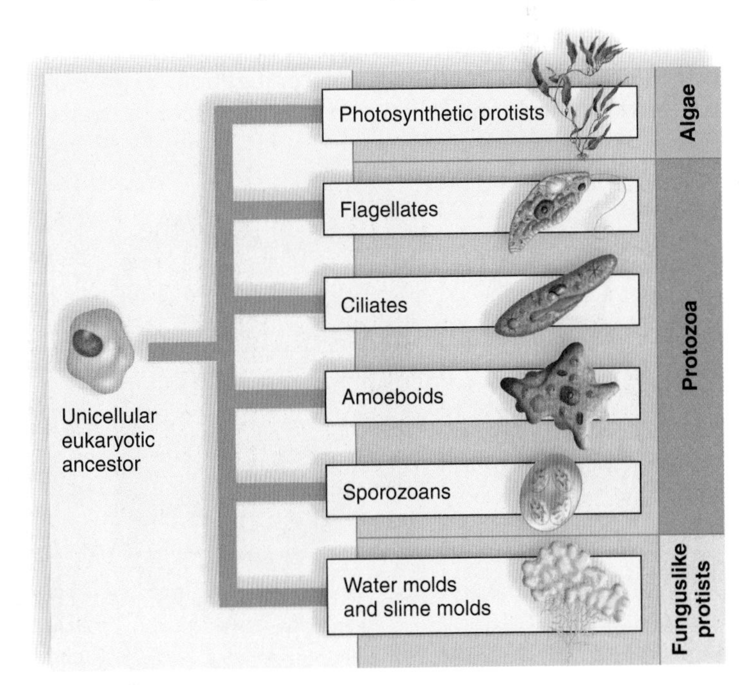

Figure 28.9 Major groups of protists.
Because the precise evolutionary relationships between these groups are not yet known, they are grouped here (and discussed in the text) by major shared characteristics.

seaweeds, such as kelp. Unicellular types can be as small as bacteria (about 1 μm), while multicellular forms can grow as large as hundreds of feet. In aqueous environments, small algae are often a component of suspended, photosynthesizing organisms termed **phytoplankton.** Algae are important in terrestrial ecosystems as well, being found in soils, on rocks, and on trees. In addition, algae contribute to the formation of coral reefs and partner with fungi in lichens (see Fig. 28.32).

Algae have chloroplasts that contain chlorophyll and sometimes other pigments. Algae perform plantlike photosynthesis. Oxygen is released from water, and electrons energized by the sun are used to reduce carbon dioxide to carbohydrate. Thus, algae are important primary producers of food for heterotrophs and provide oxygen to the atmosphere. **Pyrenoids** are organelles found in algae that are active in starch storage and metabolism.

Algae generally have a rigid cell wall, sometimes made of cellulose. Beyond the cell wall, many algae produce a slime layer that can be harvested and used in foods. Some algae also produce oils that can be burned, which has stimulated much interest in algae as an alternative to petroleum fuel.

Green algae are believed to be closely related to the first plants, because both groups have some of the same characteristics: (1) a cell wall that contains cellulose, (2) chlorophylls *a* and *b,* and (3) storage of reserve food as starch inside the chloroplast.

The many different types of green algae include those that are unicellular, colonial, or multicellular (Fig. 28.10). *Chlamydomonas* is a unicellular green alga with a cup-shaped chloroplast and two whiplike flagella. *Volvox* is a colony (loose association of cells) in which thousands of flagellated cells are arranged in a single layer surrounding a watery interior. Each cell resembles a *Chlamydomonas. Spirogyra* is filamentous, consisting of end-to-end chains of cells. A ribbonlike chloroplast is arranged in a spiral within each cell. *Spirogyra* can reproduce by conjugation, a process whereby physical contact between cells allows for the transmission of genetic material. Multicellular *Ulva* is commonly called sea lettuce because of its leafy appearance.

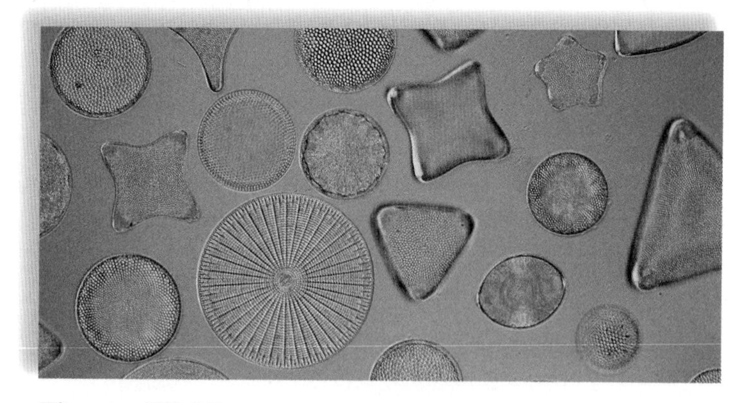

Figure 28.11 Assorted fossilized diatoms.
Diatoms have overlapping shells made of silica. Scientists use their delicate markings to identify the particular species.

Diatoms are the most numerous unicellular algae in the oceans, and they are also common in fresh water. In addition to chlorophyll, they utilize a brown pigment to capture solar energy. Diatoms constitute a large fraction of the phytoplankton and are important primary producers at the base of marine food chains. Diatoms are easy to recognize under the microscope because they have a wide variety of elaborate shells made of silica (Fig. 28.11). The two overlapping shells have intricately shaped depressions, pores, and passageways that bring the diatom's plasma membrane in contact with the environment. The fossilized remains of diatoms, also called diatomaceous earth, accumulate on the ocean floor and can be mined for use as filtering agents and abrasives.

Dinoflagellates are best known for the **red tide** they cause when they greatly increase in number, an event called an algal bloom. *Gonyaulax,* one species implicated in red tides, produces a very potent toxin (Fig. 28.12). This can be harmful by itself, but it also accumulates in shellfish. The shellfish are not damaged, but people can become quite ill when they eat them.

Despite some harmful effects, dinoflagellates are important members of the phytoplankton in marine and freshwater ecosystems. Generally, dinoflagellates are photosynthetic,

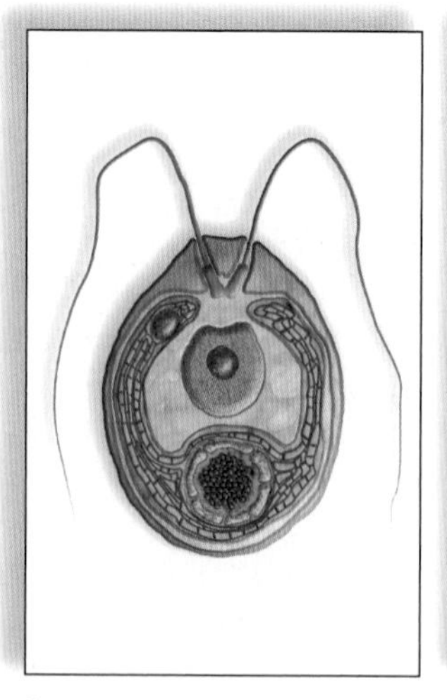
a.

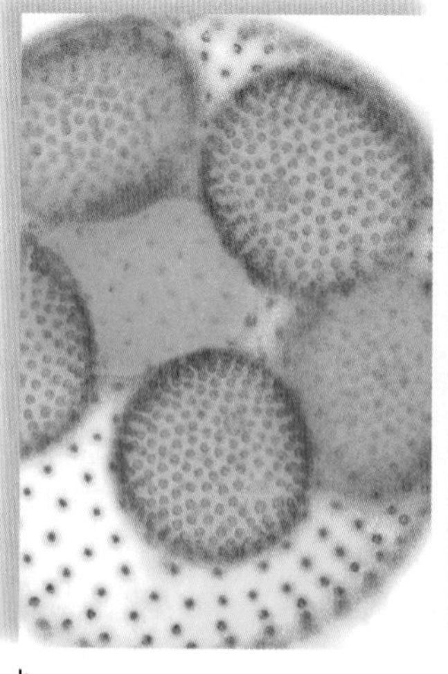
b.

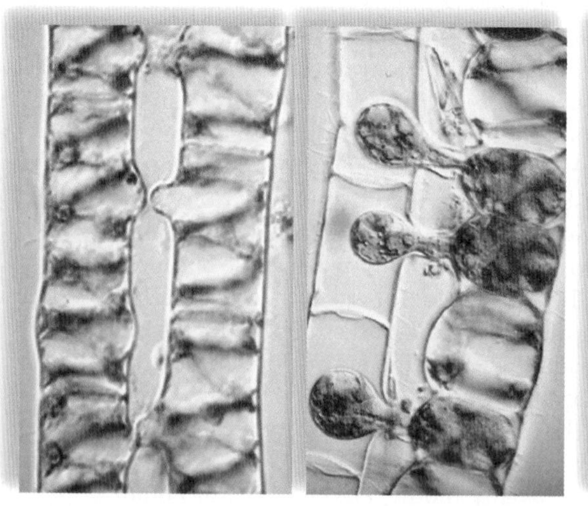
c.

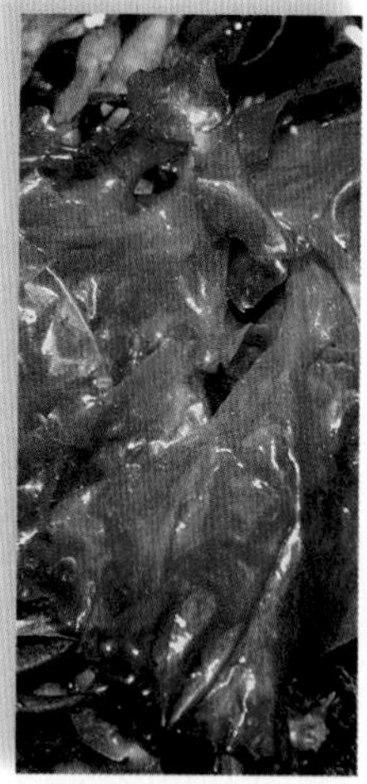
d.

Figure 28.10 Representative green algae.
Green algae all contain the pigment chlorophyll. a. *Chlamydomonas* is a motile, unicellular green alga. **b.** *Volvox* is a colony of flagellated cells. **c.** *Spirogyra* is a filamentous green alga in which each cell has a ribbonlike chloroplast. During conjugation, the cell contents of one filament enter the cells of another filament. Zygote formation follows. **d.** *Ulva* is a multicellular green alga known as sea lettuce.

although colorless heterotrophic forms are known to live as symbionts inside other organisms. Corals, which build coral reefs, contain large numbers of symbiotic dinoflagellates.

Many dinoflagellates have protective cellulose plates that become encrusted with silica, turning them into hard shells. Dinoflagellates have two flagella. One is located in a groove that encircles the protist, and the other is in a longitudinal groove and has a free end. The arrangement of the flagella makes a dinoflagellate whirl as it moves. In fact, the name dinoflagellate is derived from the Greek *dino-*, for whirling.

It is interesting to note that luminous dinoflagellates produce a twinkling light that can give seas a phosphorescent glow at night, particularly in the tropics.

Most **red algae** are multicellular, ranging from simple filaments to leafy structures. As seaweeds, red algae often resemble the brown algae seaweeds, although red algae can be more delicate (Fig. 28.13).

In addition to chlorophyll, the red algae contain red and blue pigments that give them characteristic colors. Coralline algae have calcium carbonate in their cell walls and contribute to the formation of coral reefs.

Red algae produce a number of useful gelling agents. Agar is commonly used in microbiology laboratories to solidify culture media and commercially to encapsulate vitamins or drugs. Agar is also a gelatin substitute for vegetarians, an antidrying agent in baked goods, and an additive in cosmetics, jellies, and desserts. Carrageenan is a related product used in cosmetics and chocolate manufacture. *Porphyra*, a red seaweed, is popular as a sushi wrap in Japan.

Brown algae are the conspicuous multicellular seaweeds that dominate rocky shores along cold and temperate coasts. The color of brown algae is due to accessory pigments that actually range from pale beige to yellow-brown to almost black. These pigments allow the brown algae to extend their range down into deeper waters because the pigments are more efficient than green chlorophyll in absorbing the sunlight away from the ocean surface. The alga produces a slimy matrix that retains water when the tide is out and the seaweed is exposed. This gelatinous material, algin, is used in ice cream, cream cheese, and some cosmetics.

The Sargasso Sea in the North Atlantic Ocean commonly has large floating mats of brown algae called sargassum. The most familiar brown algae are kelp (genus *Laminaria*) in coastal regions and giant marine kelp (genus *Macrocystis*) in deeper waters (Fig. 28.14). Tissue differentiation in kelp results in blades, stalks, and holdfasts, which are analogous to the leaves, stems, and roots of plants.

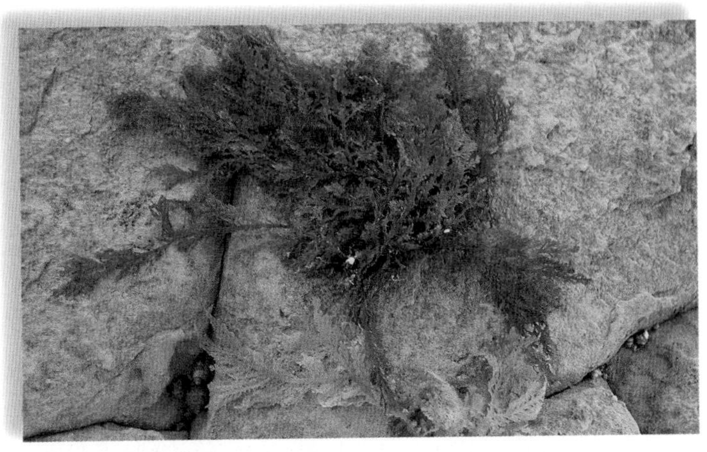

Figure 28.13 Red alga.
Red algae contain accessory pigments in addition to chlorophyll. Represented here by *Chondrus crispus*, they are smaller and more delicate than brown algae.

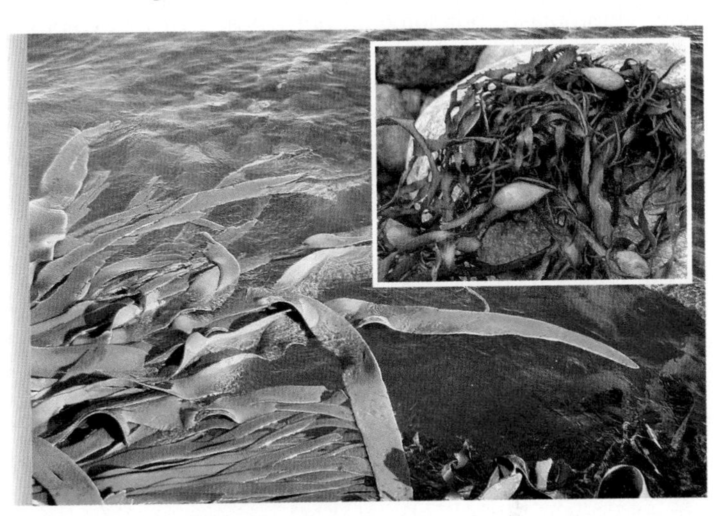

Figure 28.14 Brown alga.
Brown algae have accessory pigments that allow them to obtain their light energy while living in deeper waters compared to other algae. This is a type of brown algae known as bull kelp, *Nereocystis luetkeana*. The photo on the upper right shows flotation bladders that brown algae form to keep their blades close to the surface.

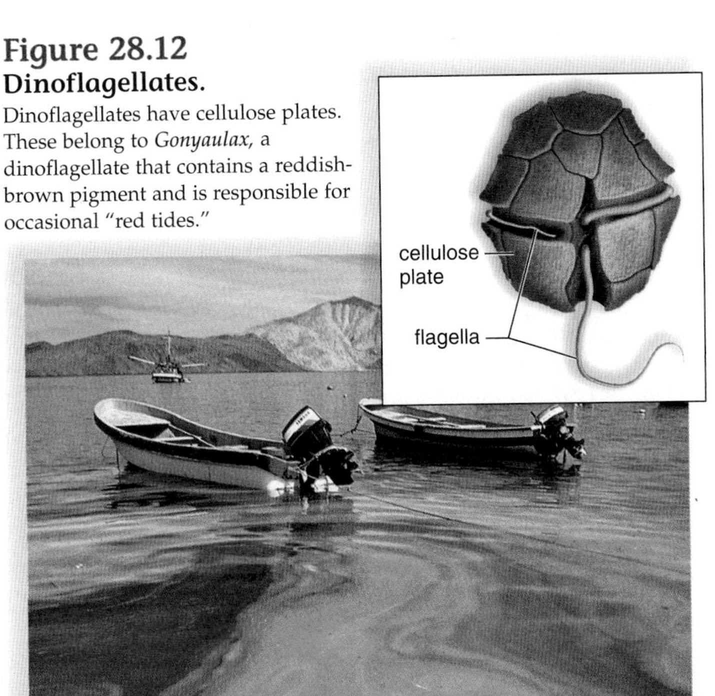

Figure 28.12 Dinoflagellates.
Dinoflagellates have cellulose plates. These belong to *Gonyaulax*, a dinoflagellate that contains a reddish-brown pigment and is responsible for occasional "red tides."

Check Your Progress

1. Describe the endosymbiotic theory.
2. Under what circumstances do some protists produce spores or cysts? What benefit does this provide for the organism?
3. Name three specific similarities between green algae and plants.
4. Describe the unique features of a diatom's shell.

Flagellates

Flagellates are heterotrophic protozoans that propel themselves using one or more flagella. Most are symbiotic, and many are parasitic. *Trypanosoma brucei,* the cause of African sleeping sickness, is transmitted by the tsetse fly (Fig. 28.15). It attacks the patient's blood, causing inflammation that decreases oxygen flow to the brain. Chagas disease, caused by *T. cruzi,* is transmitted by the kissing bug, so called because it tends to bite lips. *Giardia lamblia* is well known as a human pathogen that can cause epidemics of diarrhea, known as giardiasis. The cysts of *Giardia* can pass through water treatment plants because they are not well retained by sand filters and are impervious to chlorine disinfection. *Giardia* is common in environments contaminated with fecal materials, such as day-care centers, and can be found in natural surface waters.

Euglenoids are freshwater unicellular organisms that typify the problem of classifying protists. Many euglenoids have chloroplasts, but some do not. Those that lack chloroplasts ingest or absorb their food. Those that have chloroplasts are believed to have originally acquired them by ingestion and subsequent endosymbiosis of a green algal cell. Three, rather than two, membranes surround these chloroplasts. The outermost membrane is believed to represent the plasma membrane of an original host cell that engulfed a green alga.

Euglenoids have two flagella, one of which is typically much longer than the other and projects out of the anterior, vase-shaped invagination (Fig. 28.16). Near the base of this flagellum is an eyespot apparatus, which is a photoreceptive organelle for detecting light. Because euglenoids are bounded by a flexible pellicle composed of protein strips lying side by side, they can assume different shapes as the underlying cytoplasm undulates and contracts.

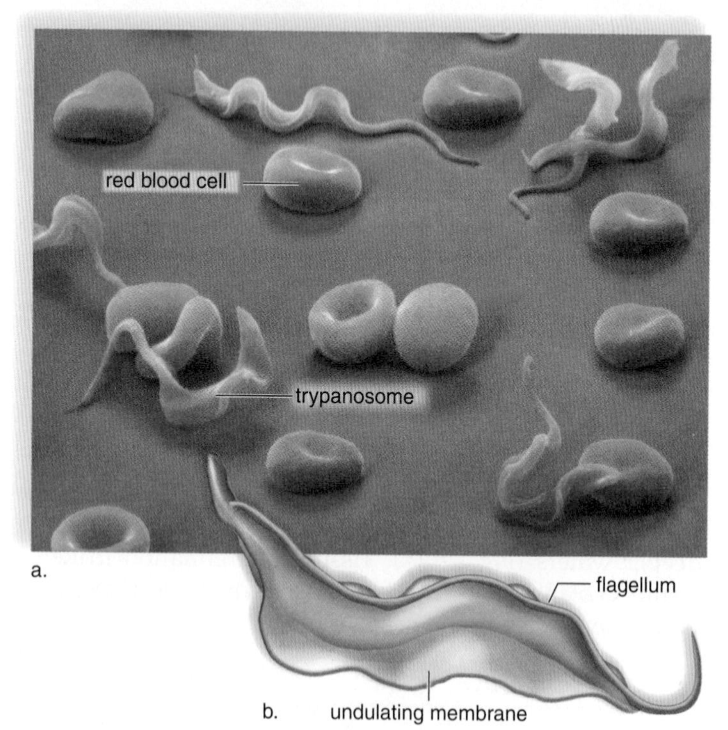

Figure 28.15 Flagellates.
Flagellates are motile by means of one or more flagella. a. Micrograph showing *Trypanosoma brucei,* a causal agent of African sleeping sickness, among red blood cells. **b.** The drawing shows the general structure of *T. brucei.*

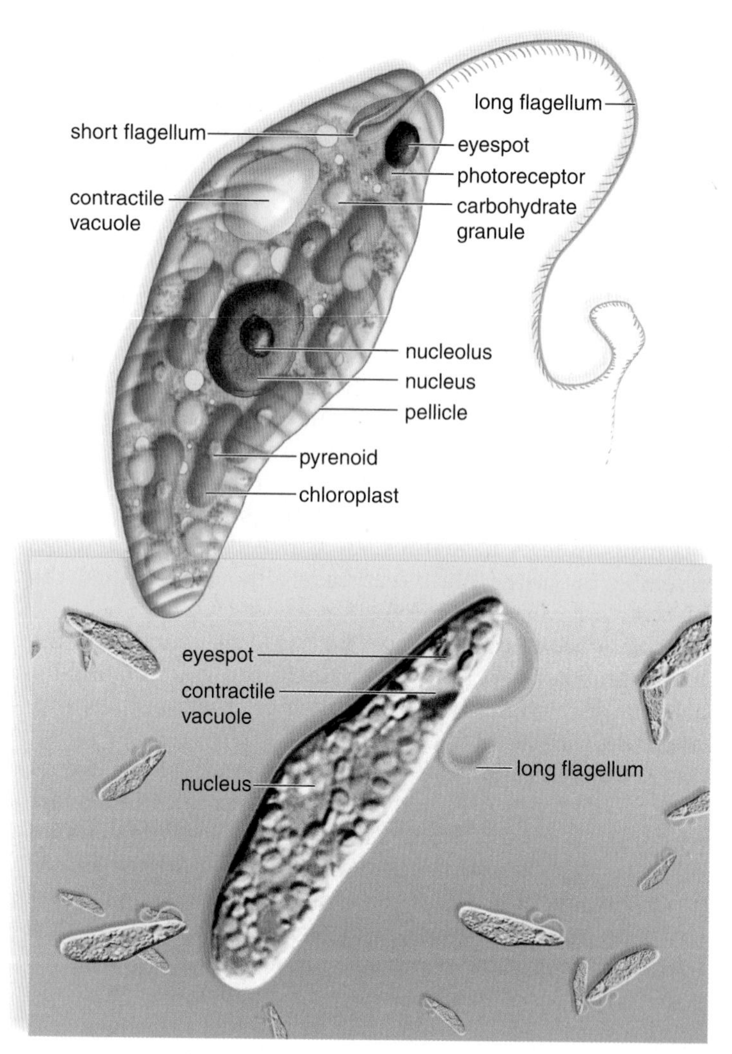

Figure 28.16 *Euglena.* LM 200×
Euglenoids have a flexible pellicle and a long flagellum that propels the body.

Ciliates

Ciliates are the largest group of protozoans. All of them have cilia, hairlike structures that rhythmically beat, moving the cell forward or in reverse. Typically, the cell rotates as it moves. Cilia also help capture prey and particles and then move them toward the mouthparts. After phagocytic vacuoles engulf food, they combine with lysosomes, which supply the enzymes needed for digestion.

Some ciliates are up to 3 mm long, which is large enough to be seen with the naked eye. Most are freely motile, but some can be anchored like members of the genus *Stentor,* which form a stalk, attaching to a surface and collecting food with cilia (Fig. 28.17*a*). *Paramecium* is the most widely known ciliate, and it is commonly used for research and teaching. It is shaped like a slipper and has visible contractile vacuoles (Fig. 28.17*b*). Members of the genus *Paramecium* have a large macronucleus and a small micronucleus. The macronucleus produces mRNA and directs metabolic functions. The micronucleus is important during sexual reproduction. *Paramecium* has been important

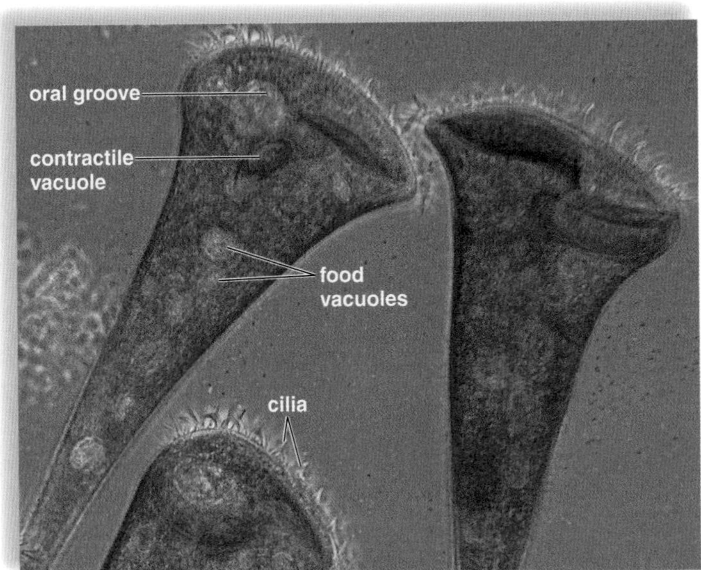

a. *Stentor*

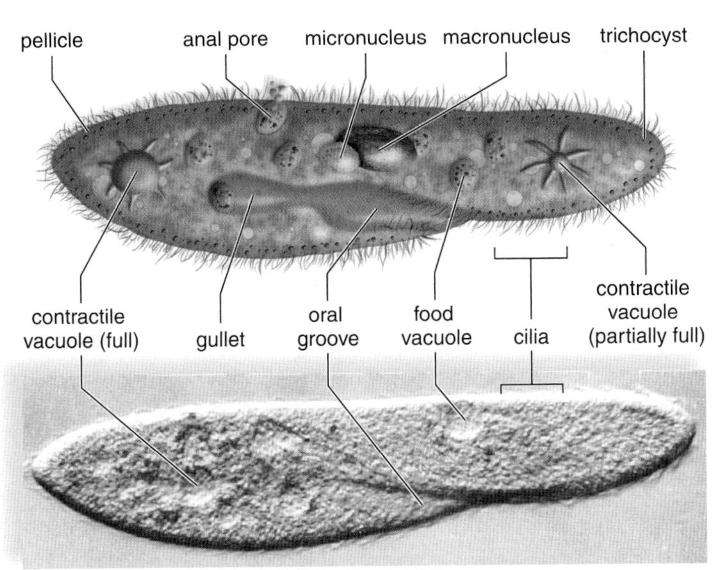

b. *Paramecium*

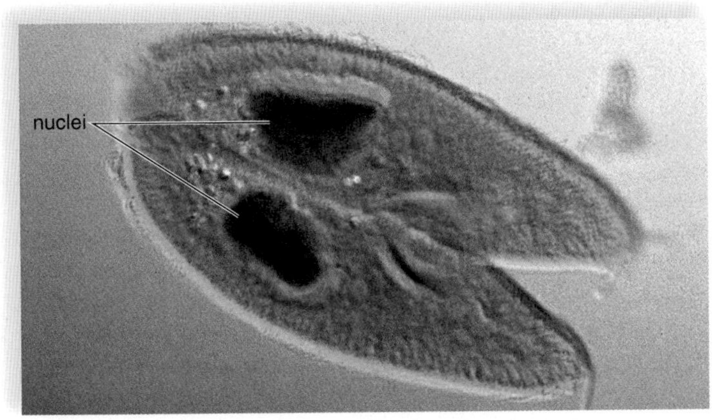

c. During conjugation, two paramecia first unite at oral areas.

Figure 28.17 Ciliates.
Ciliates are motile by means of cliia. **a.** *Stentor,* a large, vase-shaped, freshwater ciliate. **b.** Structure of *Paramecium,* adjacent to an electron micrograph. Note the oral groove and the gullet and anal pore. **c.** A form of sexual reproduction called conjugation occurs periodically in paramecia.

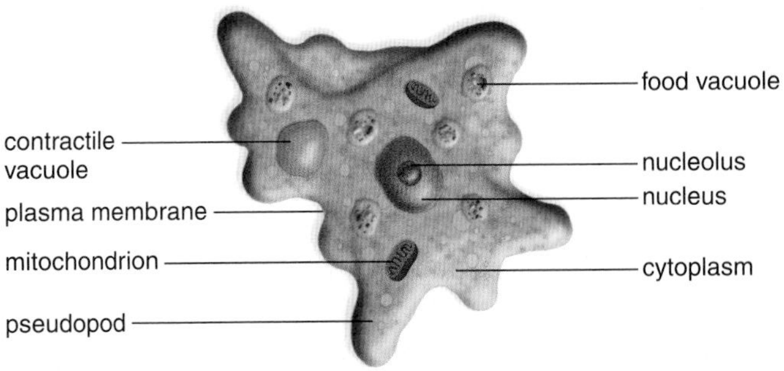

a. Amoeba, *Amoeba proteus*

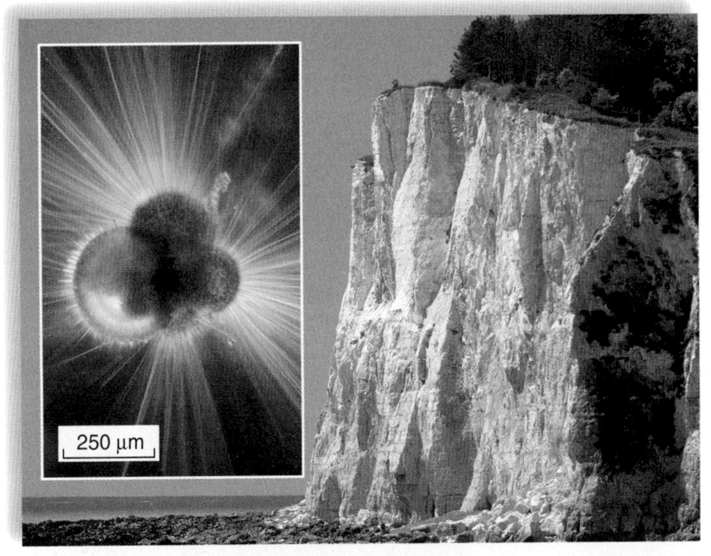

b. Foraminiferan, *Globigerina,* and the White Cliffs of Dover, England

c. Radiolarian tests SEM 200×

Figure 28.18 Amoeboids.
Amoeboids are motile by way of pseudopods. a. Structure of *Amoeba proteus,* an amoeboid common in freshwater ponds. **b.** Pseudopods of a live foraminiferan project through holes in the calcium carbonate shell. Fossilized shells were so numerous they became a large part of the White Cliffs of Dover when a geologic upheaval occurred. **c.** Skeletal tests of radiolarians.

for studying ciliate sexual reproduction, which involves *conjugation,* with interactions between micronuclei and macronuclei (Fig. 28.17*c*).

Amoeboids

Amoeboids move by pseudopods, processes that form when cytoplasm streams forward in a particular direction (Fig. 28.18*a*). They usually live in aquatic environments, such as oceans and freshwater lakes and ponds, where they are a part of the zooplankton. When amoeboids feed, their pseudopods surround and engulf their prey, which may be algae, bacteria, or other protists. Digestion then occurs within a food vacuole.

Parasitic amoeboids in the genus *Entamoeba* cause amoebic dysentery. Complications arise when the parasite invades the intestinal lining and reproduces there. If the parasite enters the body proper, liver and brain involvement can be fatal.

Foraminiferans and radiolarians are related amoeboid groups that have an external skeleton (test). In **foraminiferans,** pseudopods extend through openings in the test, which covers the plasma membrane. Because each geological time period has a distinctive form of foraminiferan, they can be used to date sedimentary rock. Depositions over millions of years followed by geological upheaval formed the White Cliffs of Dover along the southern coast of England (Fig. 28.18*b*). Also, the great Egyptian pyramids are built of foraminiferous limestone. In radiolarians, the test is internal (Fig. 28.18*c*). The tests of dead foraminiferans and radiolarians form a deep layer of sediment on the ocean floor. Their presence is used as an indicator of oil deposits on land or sea.

Sporozoans

Members of the phylum Apicomplexa are commonly called **sporozoans** because they produce spores. All phases of the life cycle are generally nonmotile, with the exception of male gametes and zygotes. Sporozoans are either intercellular or extracellular parasites. An apical complex composed of fibrils, microtubules, and organelles is found at one end of the single cell. Enzymes for attacking host cells are secreted through a pore at the end of the apical complex.

Malaria The most widespread and dangerous sporozoan disease is **malaria.** It is endemic in approximately 50% of the habitable surface of Earth. In 2006, an estimated 247 million people were infected with malaria worldwide, and 881,000 of these people died. The malaria parasite, one of several *Plasmodium* species, has a complex life cycle that involves transmission by a mosquito vector (Fig. 28.19).

The sexual reproduction phase in the mosquito ends with the migration of a developmental stage called sporozoites into the salivary glands. These sporozoites are then transmitted by the mosquito's bite to a human, where they reproduce asexually in the liver and red blood cells, forming merozoites. The infected red blood cells frequently rupture, releasing merozoites and toxins, and causing the person to experience the chills and fever that are characteristic of malaria. Some of these merozoites become gametocytes, which, if ingested by another mosquito, can start the sexual reproduction phase again. People who survive malaria have only a limited immunity to reinfection. Interestingly, the sickle cell trait found in up to one-third of sub-Saharan Africans provides some protection against malaria. Efforts to control mosquito populations have

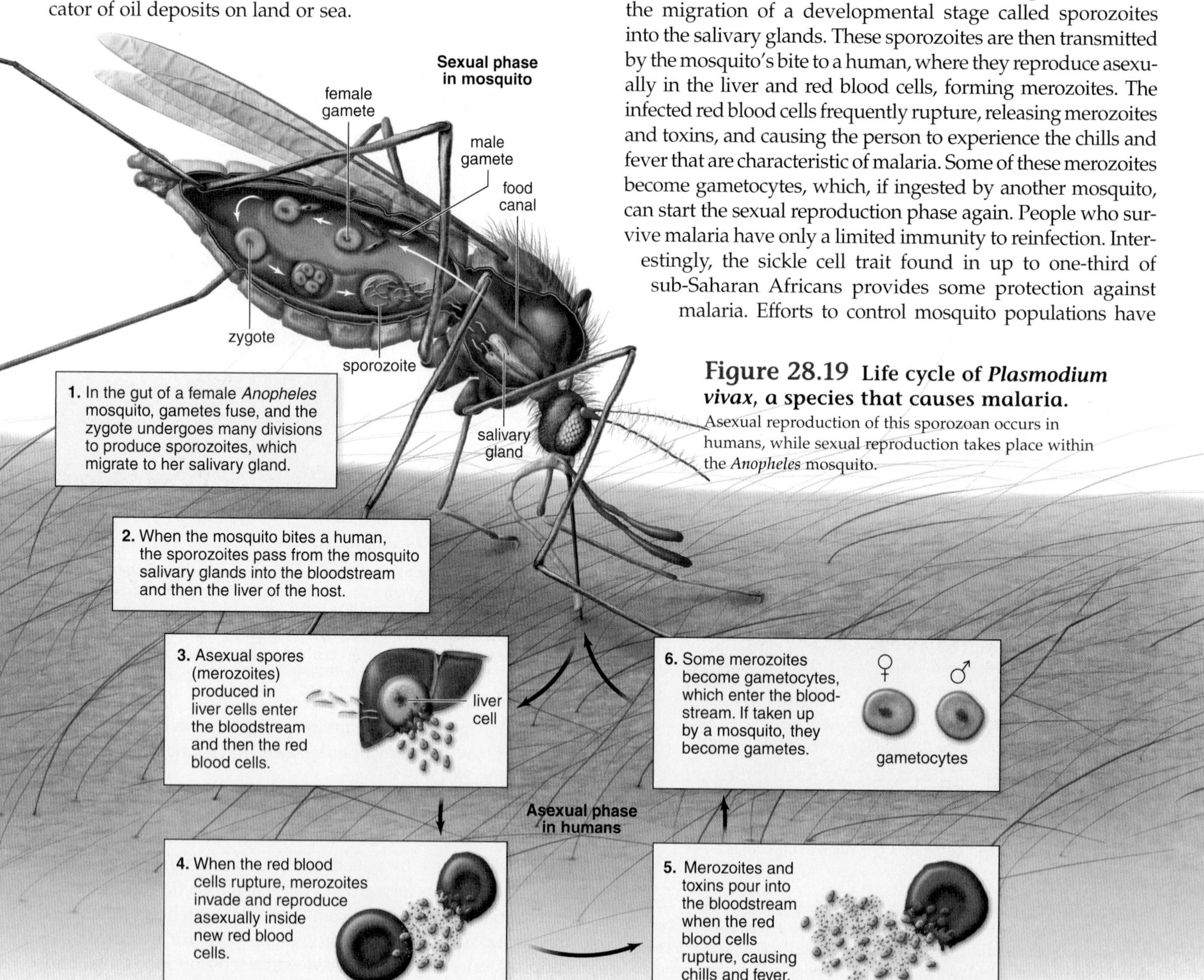

Figure 28.19 Life cycle of *Plasmodium vivax*, a species that causes malaria. Asexual reproduction of this sporozoan occurs in humans, while sexual reproduction takes place within the *Anopheles* mosquito.

been successful in the United States, where malaria is rarely seen. The control of mosquitoes worldwide has now been hampered by the rise of resistance to pesticides like DDT. Scientists are working to find new ways to inhibit mosquitoes, such as genetically engineering them to be resistant to *Plasmodium* infection. A Seattle company is even designing a laser system that could protect villages by zapping mosquitoes in flight!

Attempts to produce an effective vaccine against malaria have been largely unsuccessful, but two studies published in late 2008 showed significant protection of African infants (65% protected) and young children (53% protected) against malaria. This research has been funded largely by the Bill and Melinda Gates Foundation, which has contributed about $1.2 billion to malaria research since 2000. The vaccine is being tested in sub-Saharan Africa in 2009, and may be available for widespread use by 2012.

Other Sporozoan Diseases A related protozoan, *Toxoplasma*, is commonly transmitted by cat feces. The disease toxoplasmosis generally causes no appreciable symptoms, but the parasite can be harmful to a developing fetus. Thus, pregnant women are advised not to empty cat litter boxes and to avoid working in gardens or other locations where cats may defecate. Another related protozoal disease is caused by *Cryptosporidium*. The organism and its cysts are common in surface waters and in the feces of animals and birds, and can pass through sand filters in drinking-water plants, while being unaffected by chlorine treatment. *Cryptosporidium* infection usually causes a self-limiting gastroenteritis, but can lead to a fatal watery diarrhea in some cases.

Water Molds and Slime Molds

We tend to think of molds as fungi, but the water molds and slime molds are classified as protists.

Water Molds Most **water molds** are saprotrophic, meaning that they feed on dead organic matter. They usually live in water, where they decompose remains and form furry growths when they parasitize fish. In spite of their common name, some water molds live on land and parasitize insects and plants. The water mold *Phytophthora* was responsible for the 1840s potato famine in Ireland.

Water molds have a filamentous body, as do fungi, but the cell walls of water molds are largely composed of cellulose, whereas fungi have cell walls of chitin. During asexual reproduction, water molds produce flagellated spores. During sexual reproduction, they produce eggs and sperm. Their phylum name, Oomycota, refers to the enlarged tips (oogonia), where eggs are produced.

Slime Molds In forests and woodlands, slime molds feed on dead plant material. They also feed on bacteria, keeping their population under control. The vegetative cells of slime molds are mobile and amoeboid. They ingest their food by phagocytosis.

Usually, **plasmodial (acellular) slime molds** exist as a plasmodium, a diploid, multinucleated, cytoplasmic mass enveloped by a slime sheath, that creeps along, phagocytizing decaying plant material in a forest or agricultural field (Fig. 28.20). At times unfavorable to growth,

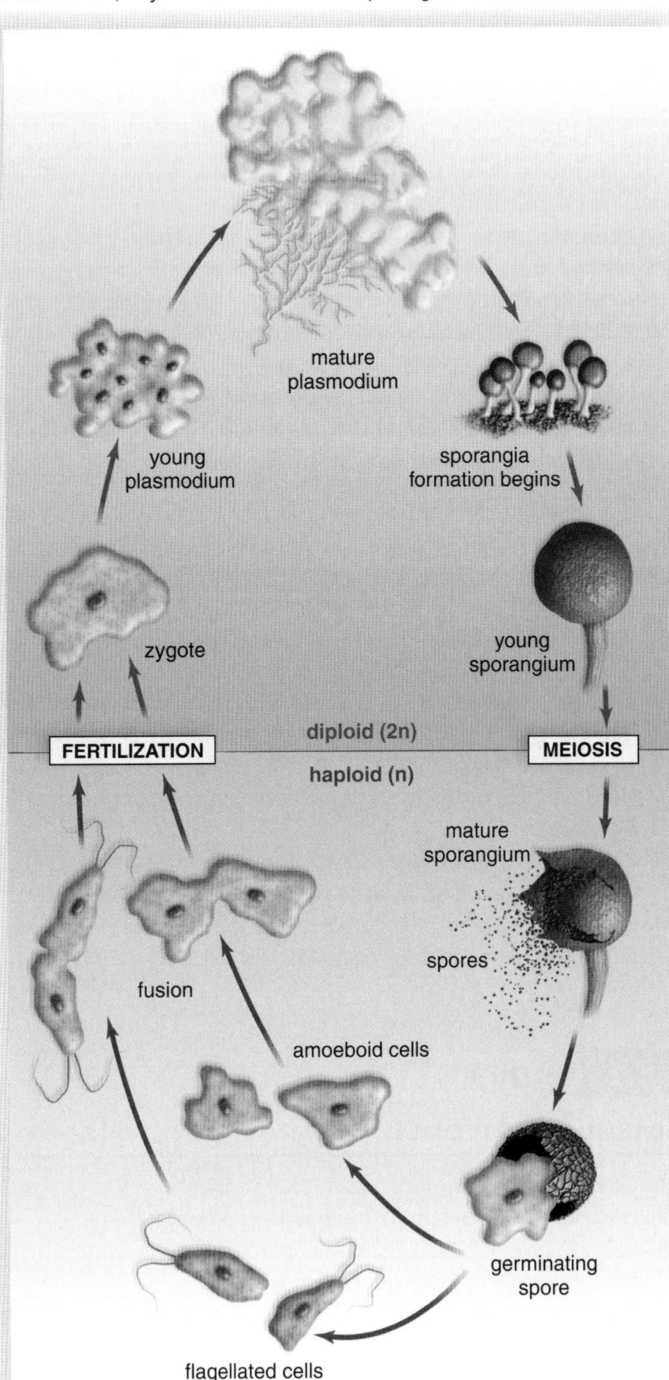

Figure 28.20 Plasmodial slime molds.
The diploid adult forms sporangia during sexual reproduction, when conditions are unfavorable to growth. Haploid spores germinate, releasing haploid amoeboid or flagellated cells that fuse.

such as during a drought, the plasmodium develops many sporangia, reproductive structures that produce spores. The spores produced by a sporangium can survive until moisture is sufficient for them to germinate. In plasmodial slime molds, spores release a haploid flagellated cell or an amoeboid cell. Eventually, two of these fuse to form a zygote that feeds and grows, producing a multinucleated plasmodium once again.

Cellular Slime Molds **Cellular slime molds** exist as individual amoeboid cells. They are common in soil, where they feed on bacteria and yeasts. *Dictyostelium* is the commonly researched example. As food supplies dwindle, the cells release a chemical that causes them to aggregate into a sluglike pseudoplasmodium, which eventually gives rise to a fruiting body that produces spores. When favorable conditions return, the spores germinate, releasing haploid amoeboid cells, and this cycle, which is asexual, begins again. A sexual cycle is known to occur under very moist conditions.

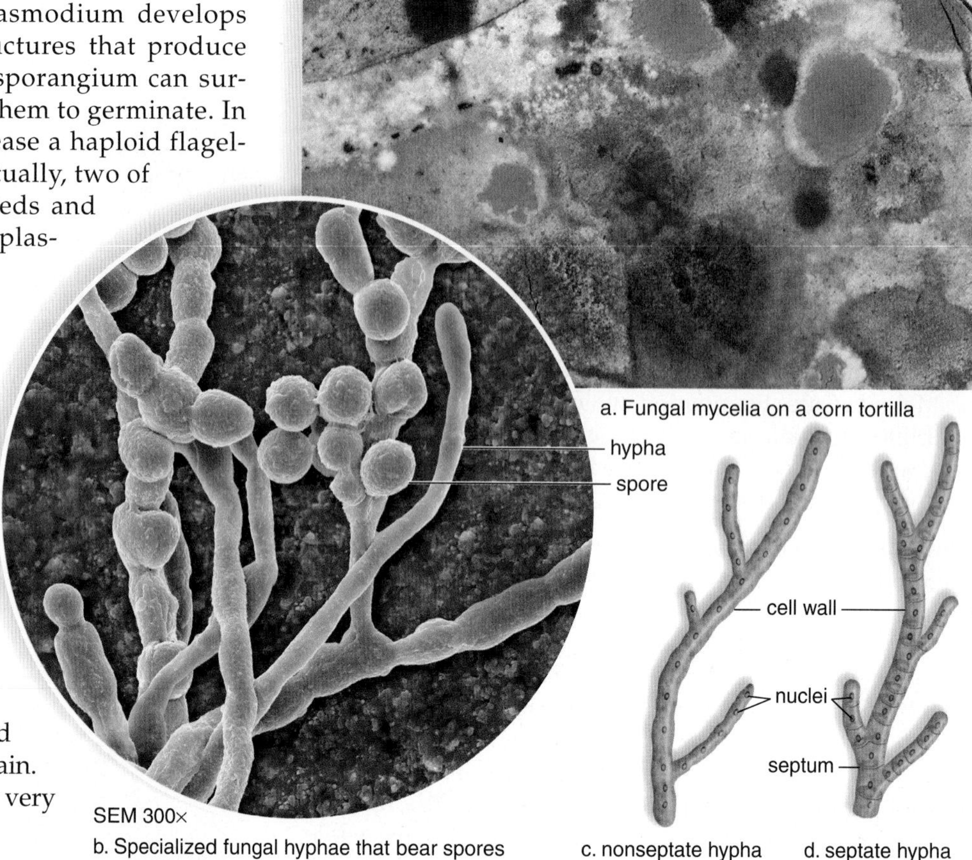

Figure 28.21 Fungal mycelia and hyphae.
a. A colorful variety of fungal mycelia growing on a corn tortilla.
b. Scanning electron micrograph of specialized aerial fungal hyphae that bear spores. Hyphae are either **(c)** nonseptate (do not have cross walls) or **(d)** septate (have cross walls).

Check Your Progress

1. What type of protist causes African sleeping sickness? How is it transmitted?
2. What characteristics of euglenoids illustrate the difficulty in classifying protists?
3. The White Cliffs of Dover and the Egyptian pyramids have what structural component in common?
4. How do the sporozoans differ from the other major groups of protozoans?
5. What are some possible reasons (both scientific and social) why it has been so difficult to develop a vaccine for malaria?
6. Compare the types of nutrients used, and the mechanism of feeding, of the water molds, slime molds, and cellular slime molds.

28.4 Fungi

Learning Outcomes

1. Explain how most fungi obtain their nutrients.
2. Describe the structural components of a fungal body.
3. Identify several fungal diseases of humans, and indicate why fungal diseases can be difficult to treat.
4. Compare the reproductive strategies of the major groups of fungi.

Fungi (domain Eukarya, kingdom Fungi) are a structurally diverse group of eukaryotes that are strict heterotrophs. Unlike animals, fungi (sing., fungus) release digestive enzymes into their external environment and digest their food outside the body, while animals ingest their food and digest it internally. A few fungi are parasitic, but most are saprotrophs that decompose dead plants, animals, and microbes. Along with bacteria, fungi play an important role in ecosystems by breaking down complex organic molecules and returning inorganic nutrients to producers of food—that is, photosynthesizers. Fungi can degrade even cellulose and lignin in the woody parts of trees. It is common to see fungi (brown rot or white rot) on the trunks of fallen trees. The body of a fungus can become large enough to cover acres of land.

Biology of Fungi

The body of a fungus is composed of a mass of individual filaments called hyphae (sing., **hypha**). Collectively, the mass of filaments is called a **mycelium** (pl., mycelia) (Fig. 28.21). Some fungi have cross walls that divide a hypha into a chain of cells. These hyphae are termed septate. Septa have pores that allow cytoplasm and even organelles to pass from one cell to the other along the length of the hypha. Nonseptate fungi have no cross walls, and their hyphae are multinucleated.

a.

b.

Figure 28.22 Dispersal of spores.
Fungi reproduce by spore formation. **a.** As with this earthstar fungus, hordes of spores are released into the air. **b.** The microscopic appearance of the spores.

Source image: From C.Y. Shih and R.G. Kessel, Living Images. *Sciences Books International, 1982, Boston.*

Hyphae give the mycelium quite a large surface area per volume of cytoplasm, which facilitates absorption of nutrients into the body of a fungus. Hyphae grow at their tips, and the mycelium absorbs and then passes nutrients on to the growing tips. Individual hyphae can grow quite rapidly, as much as 18 feet per day. Altogether, a single mycelium may add as much as a kilometer of new hyphae in a single day!

Fungal cells are quite different from plant cells, not only because they lack chloroplasts, but also because their cell walls contain chitin, not cellulose. Chitin, like cellulose, is a polymer of glucose, but in chitin, each glucose molecule has a nitrogen-containing amino group attached to it. Chitin is also the major structural component of the exoskeleton of arthropods, such as insects, lobsters, and crabs. Unlike plants, the energy reserve of fungi is not starch, but glycogen, as in animals. Fungi are nonmotile and most do not have flagella at any stage in their life cycle. They move toward a food source by growing toward it.

Although a few fungal species are aquatic, most have adapted to life on land by producing windblown spores during both asexual and sexual reproduction (Fig. 28.22). A **spore** is a haploid reproductive cell that develops into a new organism without the need to fuse with another reproductive cell. In fungi, spores germinate into new mycelia. Sexual reproduction in fungi involves conjugation of hyphae from two different mating types (usually designated + and –). Often, the haploid nuclei from the two hyphae do not immediately fuse to form a zygote. The hyphae contain + and – nuclei for long periods of time. Eventually, the nuclei fuse to form a zygote that undergoes meiosis, followed by spore formation.

Fungal Diseases of Humans

Mycoses, fungal diseases of humans, vary in levels of seriousness. Fungi called dermatophytes cause infections of the skin called **tineas.** Athlete's foot is a tinea characterized by itching and peeling of the skin between the toes (Fig. 28.23*a*). In ringworm, which is not caused by a worm but instead by several different fungi, the fungus releases enzymes that degrade keratin and collagen in skin. The area of infection becomes red and inflamed, and the fungal colony grows outward, forming a ring of inflammation. As the center of the lesion begins to heal, ringworm acquires its characteristic appearance, a red ring surrounding an area of healed skin (Fig. 28.23*b*). Some people can harbor and spread the fungus, but show no signs of disease. At any one time, an estimated 3–8% of the U.S. population is infected with scalp ringworm.

Also known as black mold, *Stachybotrys chartarum* grows well on building materials—especially those containing cellulose—that become moist. It prefers dark, unventilated locations. *S. chartarum* is thought to play a

Figure 28.23 Human fungal diseases.
a. Athlete's foot and **(b)** ringworm are caused by fungi called *dermatophytes*. **c.** Thrush, or oral candidiasis, is characterized by the formation of white patches on the tongue.

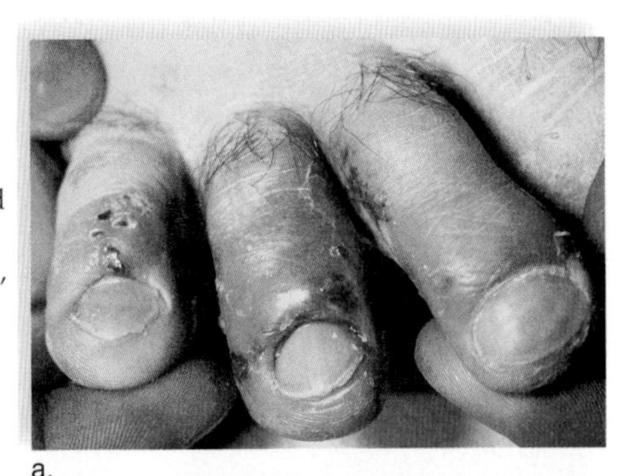

a.

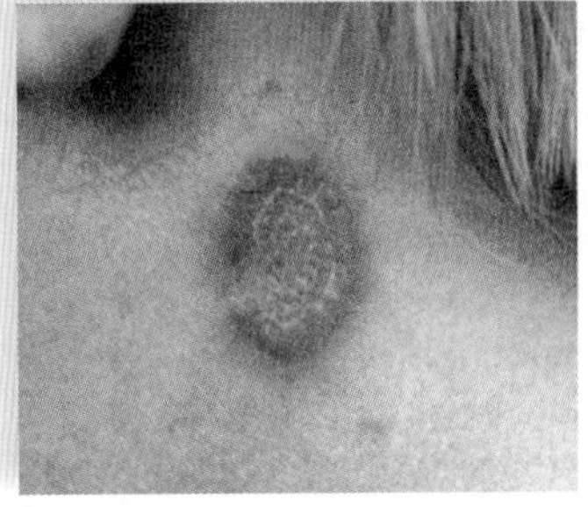

b.

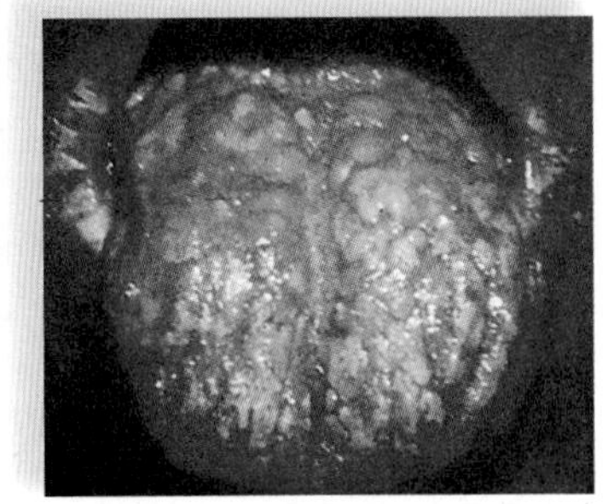

c.

role in "sick building syndrome," in which individuals exposed to toxins produced by the fungus may experience allergies, flulike symptoms, headaches, fatigue, and dermatitis.

The majority of people living in the Midwest and eastern United States have been infected with *Histoplasma capsulatum*, a common soil fungus often associated with bird droppings. Only about 5% of infected individuals notice any symptoms, but about 3,000 a year develop a lung disease called histoplasmosis, which resembles tuberculosis. People with cancer, AIDS, or other forms of immunosuppression may develop a disseminated form, which is usually fatal if not treated.

Candida albicans causes a wide variety of fungal infections. Candida infections tend to occur when the normal microflora balance in an organ is disturbed, or the immune system is suppressed. For example, the bacteria of the genus *Lactobacillus* normally produces organic acids that lower the pH of the vagina, and this inhibits *Candida*, a normal inhabitant of the vagina, from proliferating. When lactobacilli are killed off by antibiotics, *Candida* proliferates, resulting in inflammation, itching, and discharge. Oral thrush, a *Candida* infection of the mouth, is common in newborns and AIDS patients (Fig. 28.23*c*). In immunosuppressed individuals, *Candida* may cause an invasive infection that can damage the heart, the brain, and other organs.

Control of Fungi

As eukaryotes, fungal cells contain a nucleus, organelles, and ribosomes that are like those of human cells. These similarities make it a challenge to design antimicrobials against fungi that do not also harm humans. It is generally safer to treat fungal skin infections with a topical medication, which is not absorbed. For systemic fungal infections, medication must be taken into the body, and thus side effects are more of a problem. To minimize this, researchers exploit any biochemical differences they can discover. The biosynthesis of membrane sterols differs somewhat between fungi and humans. Therefore, a variety of fungicides are directed against sterol biosynthesis.

Check Your Progress

1. Define the following terms: saprotroph, hyphae, mycelium, septate, chitin, and spore.
2. How do fungal cell walls differ from plant cell walls?
3. What is a tinea?
4. Why is it more difficult to develop safe and effective antifungal drugs compared to antibacterial drugs?

Diversity of Fungi

Fungi are traditionally classified based on their mode of sexual reproduction. Figure 28.24 is an evolutionary tree showing the how the five major groups of fungi that will be discussed here are hypothesized to be related. Major fungal groups (phyla) include the chytrids, zygospore fungi, sac fungi, club fungi, and AM fungi.

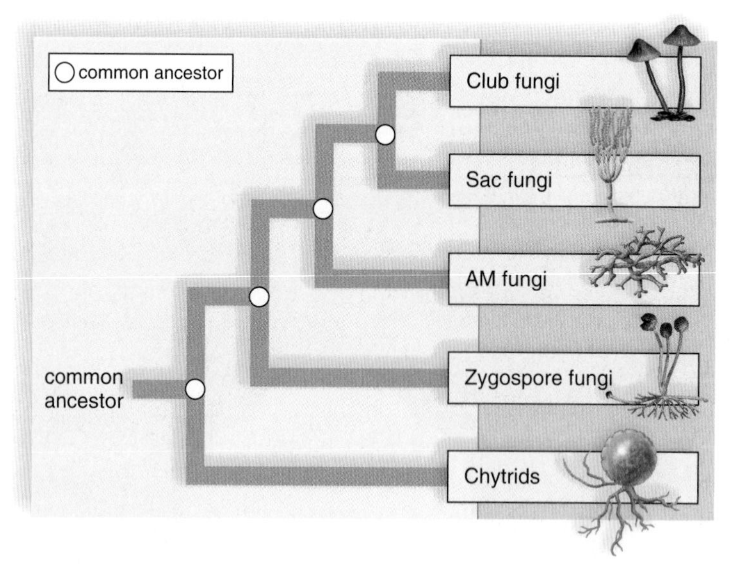

Figure 28.24 Evolutionary relationships among the fungi.
The common ancestor for the fungi was a flagellated saprotroph with chitin in the cell walls. All but chytrids lost the flagella at some point. They also became multicellular.

Chytrid Fungi

The **chytrid fungi** (phylum Chytridomycota) may have been the first type of fungi to evolve. The are unique among the fungi because (1) they are aquatic, and (2) they produce flagellated reproductive cells. Most chytrids reproduce asexually though the production of zoospores, which grow into new chytrids. Many chytrids play a role in the decay and digestion of dead aquatic organisms, but some are parasitic on plants, animals, and protists (Fig. 28.25).

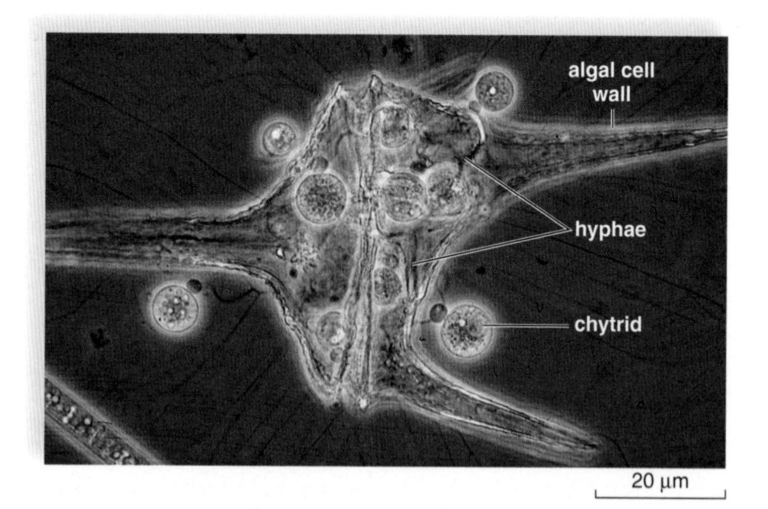

Figure 28.25 Chytrids parasitizing a protist.
These aquatic chytrids *(Chytriomyces hyalinus)* have penetrated the cell walls of this dinoflagellate and are absorbing nutrients meant for their host. They will produce flagellated zoospores that will go on to parasitize other protists.

Zygospore Fungi

The **zygospore fungi** (phylum Zygomycota) are mainly saprotrophs, but some are parasites of small soil protists or worms and even insects, such as the housefly. *Rhizopus stolonifer* (Fig. 28.26) is well known to many of us as the mold that appears on old bread, even when it has been refrigerated. In *Rhizopus,* the hyphae are specialized: some are horizontal and exist on the surface of the bread; others grow into the bread to anchor the mycelium and carry out digestion; and still others are stalks that bear sporangia. A **sporangium** is a capsule that produces spores.

When *Rhizopus* reproduces sexually, the ends of + and – hyphae join, haploid nuclei fuse, and a thick-walled zygospore results. The zygospore undergoes a period of dormancy before meiosis and germination take place. Following germination, aerial hyphae, with sporangia at their tips, produce many spores. The spores are dispersed by air currents and give rise to new mycelia.

Sac Fungi

Approximately 75% of all known fungi are **sac fungi** (phylum Ascomycota), named for their characteristic cuplike sexual reproductive structure called an ascocarp. Many sac fungi reproduce by producing chains of sexual spores called conidia (sing., **conidium**). Cup fungi, morels, and truffles have conspicuous ascocarps (Fig. 28.27). Truffles, underground symbionts of hazelnut and oak trees, are highly prized as gourmet delights. Pigs and dogs are trained to sniff out truffles in the woods, but they are also cultivated on the roots of seedlings.

Most fungal plant pathogens are sac fungi. Examples of these pathogenic fungi include powdery mildews that grow on plant leaves, as well as chestnut blight and Dutch elm disease that destroy trees. *Ergot,* a parasitic sac fungus that infects rye, produces hallucinogenic compounds similar to LSD. Psychoses resulting from ingesting grain contaminated with rye ergot may have led to the Salem, Massachusetts, witch trials in 1692.

Some of the most familiar fungi were formerly considered "imperfect fungi," because their means of sexual reproduction is unknown. However, based on DNA sequencing and other recent information, these fungi are now classified as sac fungi. Examples include *Penicillium,* the original source of penicillin, a breakthrough antibiotic that led to the important class of "cillin" antibiotics that have saved millions of lives. Other species of *Penicillium*—*P. roquefortii* and *P. camemberti*—are necessary to the production of blue

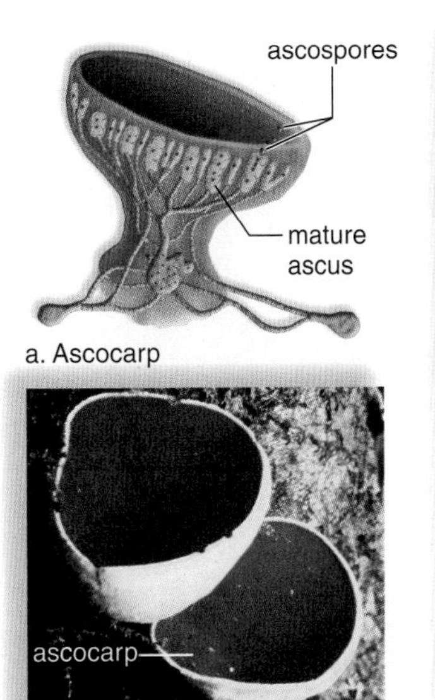

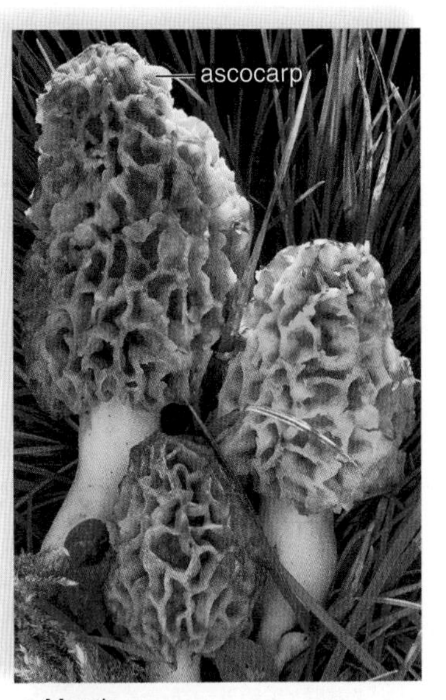

Figure 28.27 Sexual reproduction in sac fungi.
Some sac fungi are known to us by their ascocarp, the structure that produces spores as a part of sexual reproduction. **a.** Diagram of an ascocarp containing asci, where spores are produced. **b.** The ascocarps of cup fungi. **c.** The ascocarp of a morel, which is a prized delicacy.

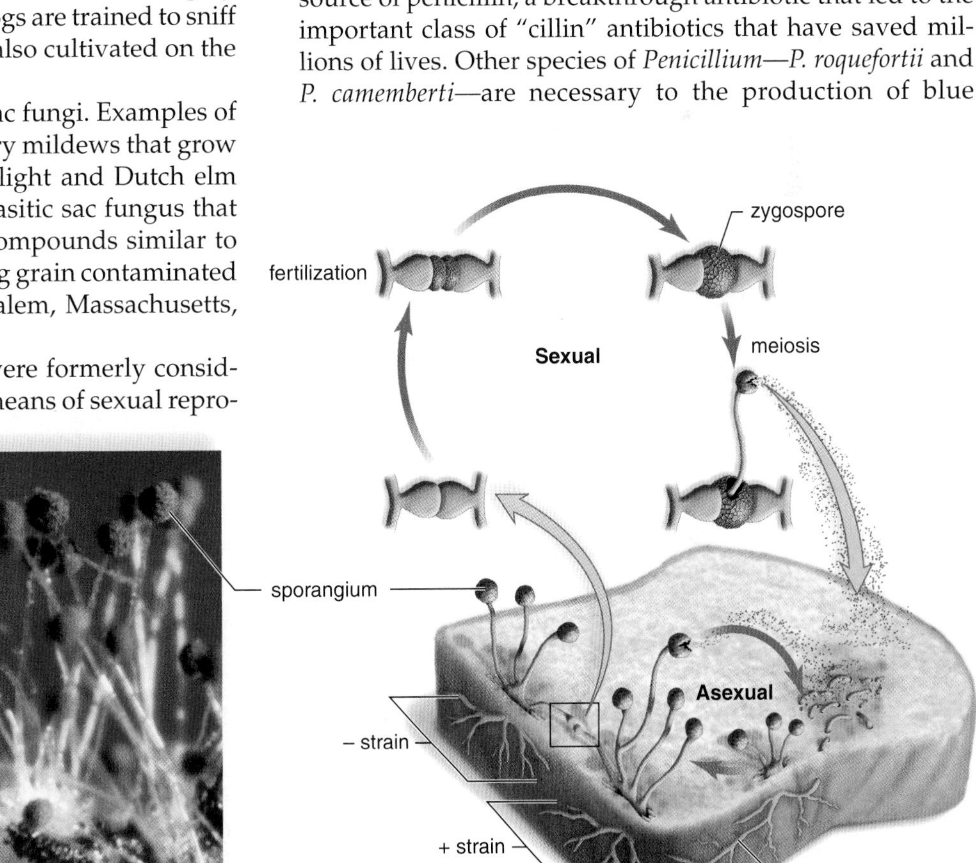

Figure 28.26 Black bread mold, *Rhizopus stolonifer.*
The mycelium of this mold utilizes sporangia to produce windblown spores. A zygospore forms during the sexual life cycle.

Figure 28.28 Blue cheese.
The blue color of this aromatic cheese is due to the presence of fungal conidia.

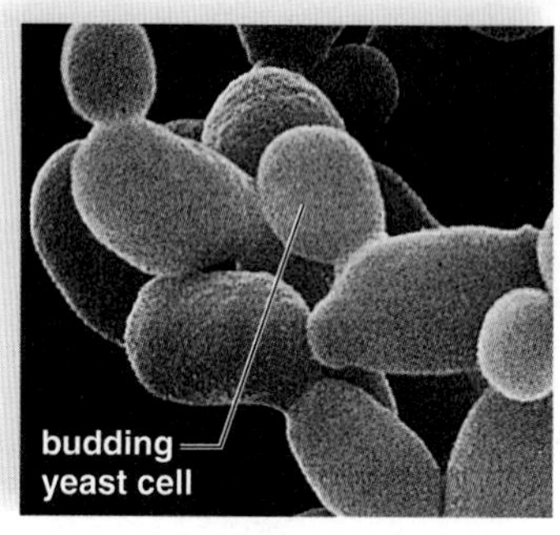

a.

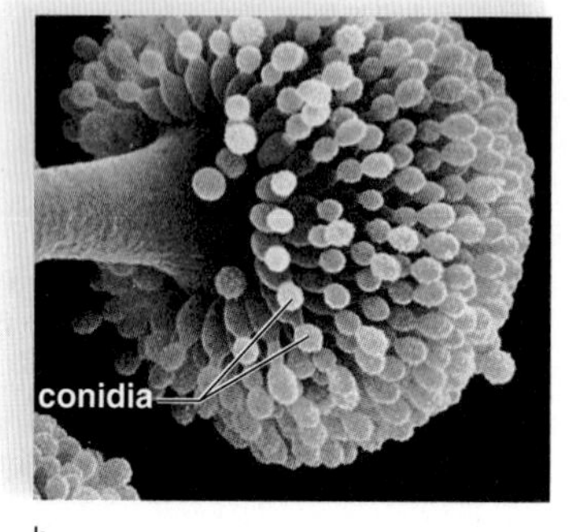

b.

Figure 28.29 Asexual reproduction in sac fungi.
a. Yeasts, unique among fungi, reproduce by budding. **b.** The sac fungi usually reproduce asexually by producing spores called conidia.

cheeses (Fig. 28.28). *Aspergillus* is used widely for its ability to produce citric acid and various enzymes. Certain species also cause serious infections in people with compromised immune systems.

Yeasts The term **yeasts** is generally applied to unicellular fungi, and many of these organisms are sac fungi. *Saccharomyces cerevisiae,* brewer's yeast, is representative of budding yeasts. When unequal binary fission occurs, a small cell gets pinched off and then grows to full size. Asexual reproduction occurs when the food supply runs out, producing spores (Fig. 28.29).

When some yeasts ferment, they produce ethanol and carbon dioxide. In the wild, yeasts grow on fruits, and historically, the yeasts already present on grapes were used to produce wine. Today, selected yeasts are added to relatively sterile grape juice in order to make wine. Also, yeasts are added to grains to make beer and liquor. Both the ethanol and carbon dioxide are retained in beers and sparkling wines, while carbon dioxide is released from wines. In breadmaking, the carbon dioxide produced by yeasts causes the dough to rise, and the ethanol quickly evaporates. The gas pockets are preserved as the bread bakes.

Club Fungi

Club fungi (phylum Basidiomycota) are named for their characteristic sexual reproductive structure called a basidium. Basidia are enclosed within a basidiocarp, which develops after + and – hyphae join. The union of + and – nuclei occurs in the basidium, which produces spores by meiosis. We recognize most club fungi by their basidiocarp (Fig. 28.30). When you eat a mushroom, you are consuming a basidiocarp. Certain mushrooms are poisonous, especially those of the genus *Amanitas*. *A. phalloides*, also known as the death cap, causes 90% of fatalities related to mushroom poisoning, due to its production of a toxin that interferes with gene transcription by inhibiting RNA polymerase. Other club fungi, mainly of the genus *Psilocybe,* produce a hallucinogenic chemical called psilocybin that is a structural analog of LSD. These mushrooms have been used in religious ceremonies since ancient times, though their recreational use is currently illegal in the United States.

Shelf or bracket fungi found on dead trees are also basidiocarps. Less well known are puffballs and stinkhorns. In

nuclei in basidium
fusion
meiosis
spores
gill of mushroom
basidiocarp
–
+
a. Sexual reproduction

Figure 28.30 Sexual reproduction in club fungi.
Most club fungi are known to us by their basidiocarp, the structure that produces spores as a part of sexual reproduction. **a.** Sexual life cycle of a basidiocarp and a basidium (pl., basidia) where spores are produced. Also shown are basidiocarps of **(b)** scarlet hood mushroom, **(c)** chicken-of-the-woods shelf fungi, and **(d)** a giant puffball. Puffballs contain many spores, and their basidiocarps tend to be rounded. When they are mature, any pressure from outside, such as a raindrop or the kick of a shoe, ejects the spores through a hole as a cloud of dust.

b. Mushroom

c. Shelf fungi

d. Giant puffball

puffballs, spores are produced inside parchmentlike membranes, and the spores are released through a pore or when the membrane breaks down. Stinkhorns resemble a mushroom, but they emit a very disagreeable odor. Flies are attracted by the odor. When they linger to feed on the sweet jelly, they pick up spores and later distribute them.

Smuts and rusts are club fungi that parasitize cereal crops, such as corn, wheat, oats, and rye. They are of great economic importance because of the crops they destroy each year. The corn smut mycelium grows between corn kernels and secretes substances that cause tumor-like swellings to develop (Fig. 28.31*a*). The life cycle of rusts often requires two different plant host species to complete the cycle, and one way to keep them in check is to eradicate the alternate host. Wheat rust (Fig. 28.31*b*) can also be controlled by producing new and resistant strains of wheat.

AM Fungi

AM fungi (phylum Glomeromycota) are a recently recognized group whose name stands for *a*rbuscular *m*ycorrhial fungi. Arbuscules are branching invaginations the fungus makes when it invades plant roots. AM fungi are one type of mycorrhizae, fungi that form mutually beneficial relationships with the roots of plants (see below).

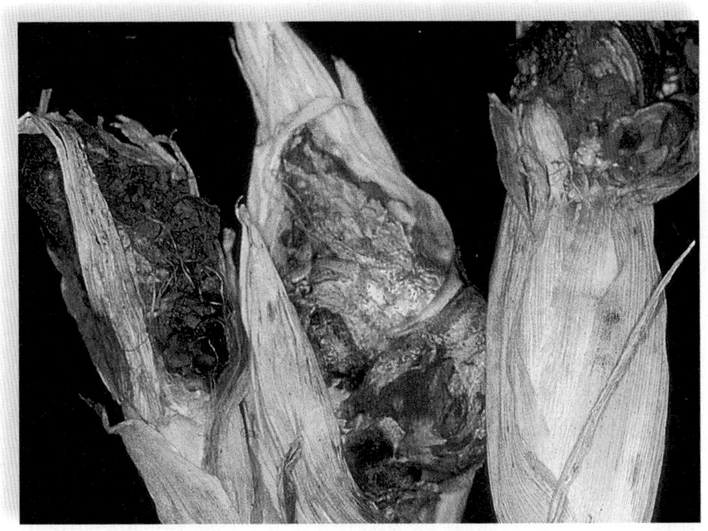

a. Corn smut, *Ustilago*

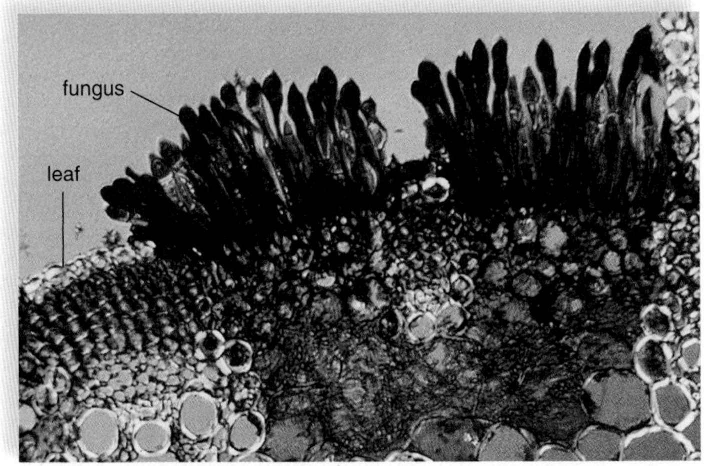

b. Wheat rust, *Puccinia*

Figure 28.31 Smuts and rusts.
a. Corn smut. **b.** Wheat rust. Both are important fungal diseases affecting agricultural crops in the United States and other countries.

Symbiotic Relationships of Fungi

Fungi often have symbiotic relationships with other organisms.

Fungi and Photosynthesizers

Lichens are associations between fungi and cyanobacteria or green algae. The different lichen species are identified according to the fungal partner. Lichens are efficient at acquiring nutrients and moisture, and therefore, they can survive in poor soils, as well as on rocks with no soil. These primary colonizers produce organic matter and create new soil, allowing plants to invade the area. Lichens exhibit three structures: compact *crustose* lichens, often seen on bare rocks or tree bark; shrublike *fruticose* lichens; and leaflike *foliose* lichens (Fig. 28.32).

The body of a lichen has three layers. The fungus forms a thin, tough upper layer and a loosely packed lower layer. These shield the photosynthetic cells in the middle layer. Specialized fungal hyphae, which penetrate or envelop the photosynthetic cells, transfer organic nutrients directly to the fungus. The fungus protects the algae from predation and desiccation and provides them with minerals and water. Lichens can reproduce asexually by releasing fragments that contain hyphae and an algal cell. As with many symbioses, the relationship between fungi and algae was likely a pathogen-and-host interaction originally, but became mutually beneficial over evolutionary time.

Mycorrhizal fungi form mutualistic relationships with the roots of most plants, helping them grow more successfully in dry or poor soils, particularly those deficient in phosphates. It is important to encourage the growth of mycorrhizal fungi when restoring lands damaged by strip mining or chemical pollution. The relationship is very ancient, as it is seen in early plant fossils. Perhaps it helped plants adapt to life on dry land. Mycorrhizal fungi generally go unnoticed, except for the truffle, discussed previously.

Mycorrhizal fungi may live on the outside of roots, enter the cortex of roots, or penetrate root cells. The fungus and plant cells can easily exchange nutrients, with the plant providing organic nutrients to the fungus and the fungus bringing water and minerals to the plant. The fungal hyphae greatly increase the surface area from which the plant can absorb water and nutrients.

Check Your Progress

1. In what two ways do chytrid fungi differ significantly from all other fungal groups?
2. List one specific fungus belonging to the zygospore fungi, sac fungi, and club fungi.
3. What specific role do fungi play in the production of bread and alcoholic beverages?
4. What benefit(s) does each partner derive in a lichen?
5. What type of symbiotic relationship exists in mycorrhiza?

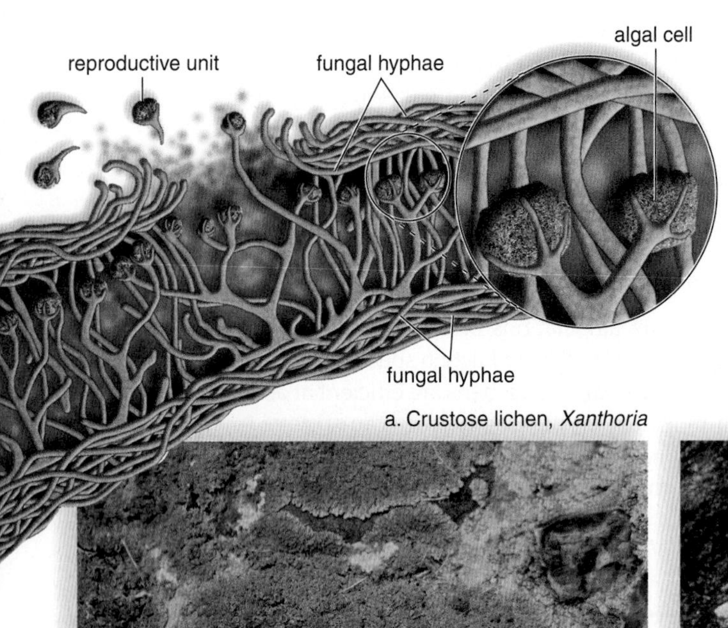

a. Crustose lichen, *Xanthoria*

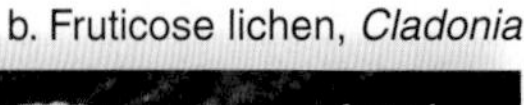

b. Fruticose lichen, *Cladonia*

c. Foliose lichen, *Xanthoparmelia*

Figure 28.32 Lichen morphology.
a. A section of a compact crustose lichen shows the placement of the algal cells and the fungal hyphae, which encircle and penetrate the algal cells. **b.** Fruticose lichens are shrublike. **c.** Foliose lichens are leaflike.

28.5 Viruses, Viroids, and Prions

Learning Outcomes

1. Review the major structural features of enveloped and nonenveloped viruses.
2. List the six steps in a typical viral reproductive cycle, and explain how some viruses become latent.
3. Describe several viral diseases of humans, and explain why it is difficult to produce vaccines against some viruses.
4. Distinguish between viruses, viroids, and prions.

As we learned in Chapter 1, all living things are composed of cells. **Viruses** are not composed of cells. They are acellular. Also, viruses are obligate parasites, meaning that they can only reproduce inside a living cell (called the host cell) by utilizing at least some of the machinery (ribosomes, certain enzymes, etc.) of that cell. Therefore, the question arises, "Are viruses alive?"

Scientists and philosophers have long argued this question. Some say that viruses are not alive. After all, not only are they acellular, but some have been synthesized in the lab from chemicals! Moreover, when viruses are outside a host cell, they are totally quiescent, exhibiting no metabolic activity. Others argue that viruses are alive because they have a genome that directs their reproduction when they are inside the host cell.

If viruses are not considered alive, then certainly the simpler viroids and prions are not alive either. Viroids are strands of RNA that can reproduce inside a cell, and prions are protein molecules that cause other proteins to become prions.

Biology of Viruses

The structure of viruses and the precise way they reproduce varies considerably for different viruses.

Viral Structure

Most viruses are much smaller than bacteria. Viruses typically measure between 0.03–0.2 μm, while most bacteria measure at least 0.5 μm. Interestingly, the largest virus discovered so far, called the Mimivirus, is about 0.4 μm in size, which is larger than the smallest known bacterium.

Viruses come in a variety of shapes, including helical, spheres, polyhedrons, and more complex forms. A virus always has at least two parts—an outer **capsid** composed of protein subunits, which protects an inner core of nucleic acid (Fig. 28.33). The viral genome can be single- or double-stranded DNA, or single- or double-stranded RNA. This diversity in genetic material is different from all cellular

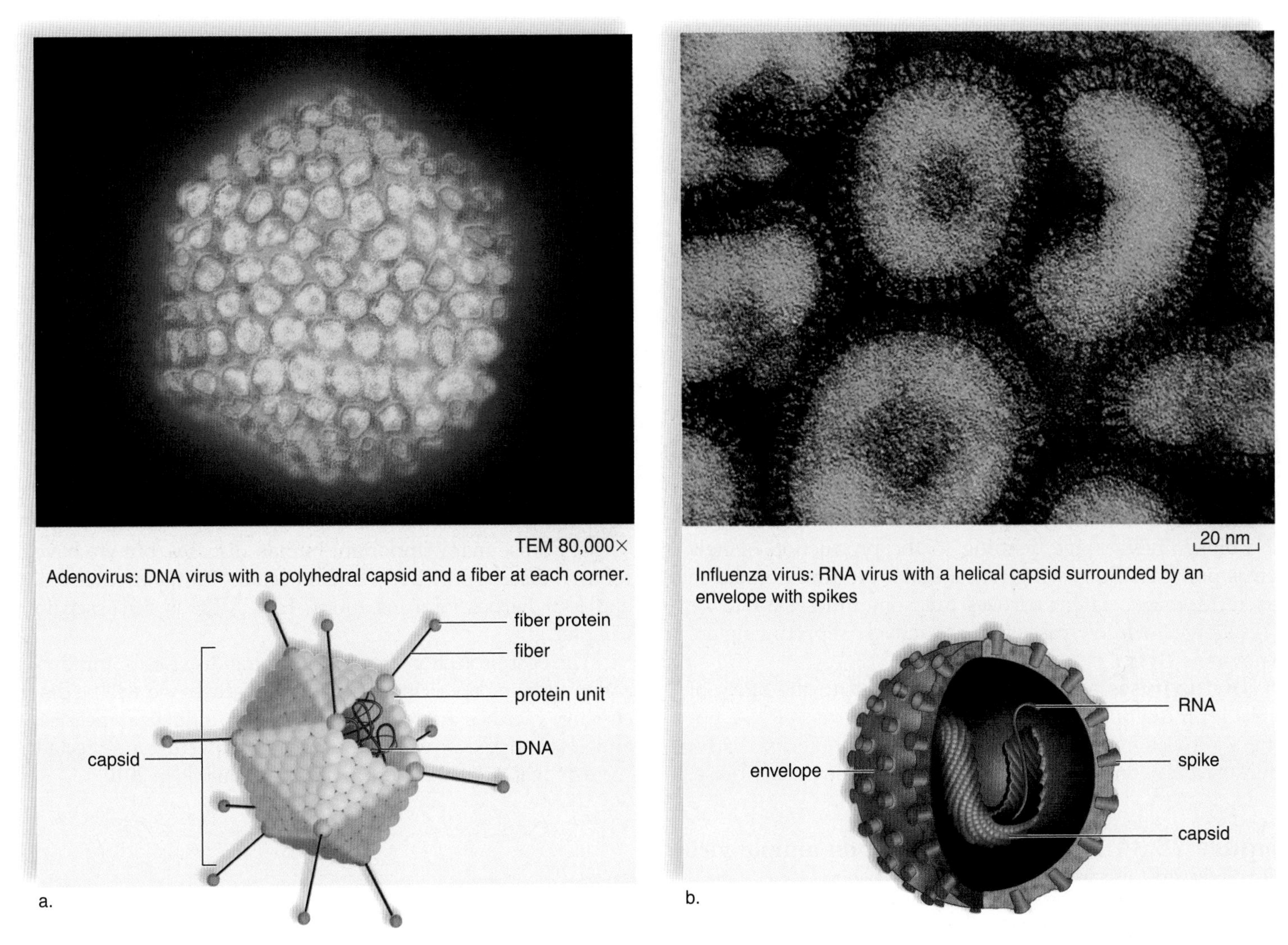

Figure 28.33 Viruses.
a. Despite their diversity, all viruses have an outer capsid composed of protein subunits and a nucleic acid core that is composed of either DNA or RNA. **b.** Some types of viruses also have a membranous envelope.

organisms, which always have a double-stranded DNA genome. Special viral enzymes that help a virus reproduce can also be inside the capsid.

In some viruses, especially those that infect animals, the capsid is surrounded by a membrane called an **envelope.** This envelope is made of lipid, and is usually derived from the host cell's plasma membrane when the virus buds from the host cell. Viral glycoproteins called spikes often extend from the envelope. These spikes are critical to viral infection because they help the virus bind to the surface of the host cell before entering it.

Viral Reproduction

Viruses infect almost every type of organism on earth. Viruses called bacteriophages infect bacterial cells, usually just a particular group or species of bacteria. Some viruses only infect plants, while others only infect animals. Some viruses can infect and cause disease only in humans. The specificity of a virus for a host occurs because the spike of a virus and a receptor molecule on a cell's plasma membrane fit together like a hand fits a glove, which triggers events allowing viral entry into the cell.

Figure 28.34 illustrates the reproductive cycle of a typical enveloped animal RNA virus. This reproductive cycle has six steps:

1. During *attachment,* the spikes of the virus bind to a specific receptor molecule on the surface of a host cell. The host cell normally uses this receptor for another purpose, but it is effectively "hijacked" by the virus for its own purposes.
2. During *entry,* also called penetration, the viral envelope fuses with the host's plasma membrane, and the rest of the virus (capsid and viral genome) enters the cell. The genome is freed when cellular enzymes remove the capsid, a process called uncoating.
3. *Replication* occurs when a viral enzyme makes complementary copies of the genome, which is RNA in this case.

4. During *biosynthesis,* some of these RNA molecules serve as mRNA for the production of more capsid and spike proteins, using host ribosomes.
5. At *assembly,* a mature capsid forms around a copy of the viral genome.
6. During *budding,* new viruses are released from the cell surface. During this process, they acquire a portion of the host cell's plasma membrane and spikes, which were specified by viral genes during biosynthesis. The enveloped viruses are now free to spread the infection to other cells.

Latency Some animal viruses can become latent (hidden) inside the host cell. Herpesviruses and retroviruses are well-known for using this strategy, which helps them avoid detection by the host immune system. During latency, new viruses are not produced, but the viral genome is reproduced along with the host cell. Environmental stresses, such as ultraviolet radiation, can induce the latent virus to enter the biosynthesis stage, leading to the production of new virus particles. Borrowing terms that were first defined in bacteriophages, latent viruses are sometimes said to be "lysogenic," while viruses that are actively reproducing are known as "lytic."

Retroviruses add an interesting twist to the story of viral reproduction. The genome of a retrovirus is RNA, but these viruses are able to convert their genome into DNA because they contain an enzyme called reverse transcriptase. This enzyme, which is not found in host cells, is so named because it catalyzes a process that is the reverse of normal transcription, which goes from DNA to RNA. The DNA copy (cDNA) of the retroviral genome can be integrated into the host DNA, where it is called a **provirus** (see Fig. 21.14). Not only is the integrated provirus resistant to antiviral medications taken by the host, but it is also able to escape detection by the host immune system.

Check Your Progress

1. Draw the basic structure of a typical enveloped virus, including capsid, nucleic acid, envelope, and spikes.
2. What is the difference between viral entry and uncoating? Between replication and biosynthesis?
3. Define viral latency. How do retroviruses accomplish this?

Viral Diseases in Humans

Viruses cause many important human diseases, but we have room to discuss only a few of them: colds, influenza, measles, and four herpes viral infections. HIV/AIDS is discussed in Chapter 21.

The best protection against most viral diseases is immunization utilizing a vaccine. Of the viral diseases we will be discussing, vaccines are readily available for influenza, measles, and chickenpox. A vaccine aimed at protecting uninfected women against genital herpes may be available in 2010.

Figure 28.34 Reproductive cycle of an animal virus. The rubella virus, like many others that infect animal cells, has an RNA genome. Notice also that the mode of entry requires uncoating and that the virus acquires an envelope and spikes when it buds from the host cell.

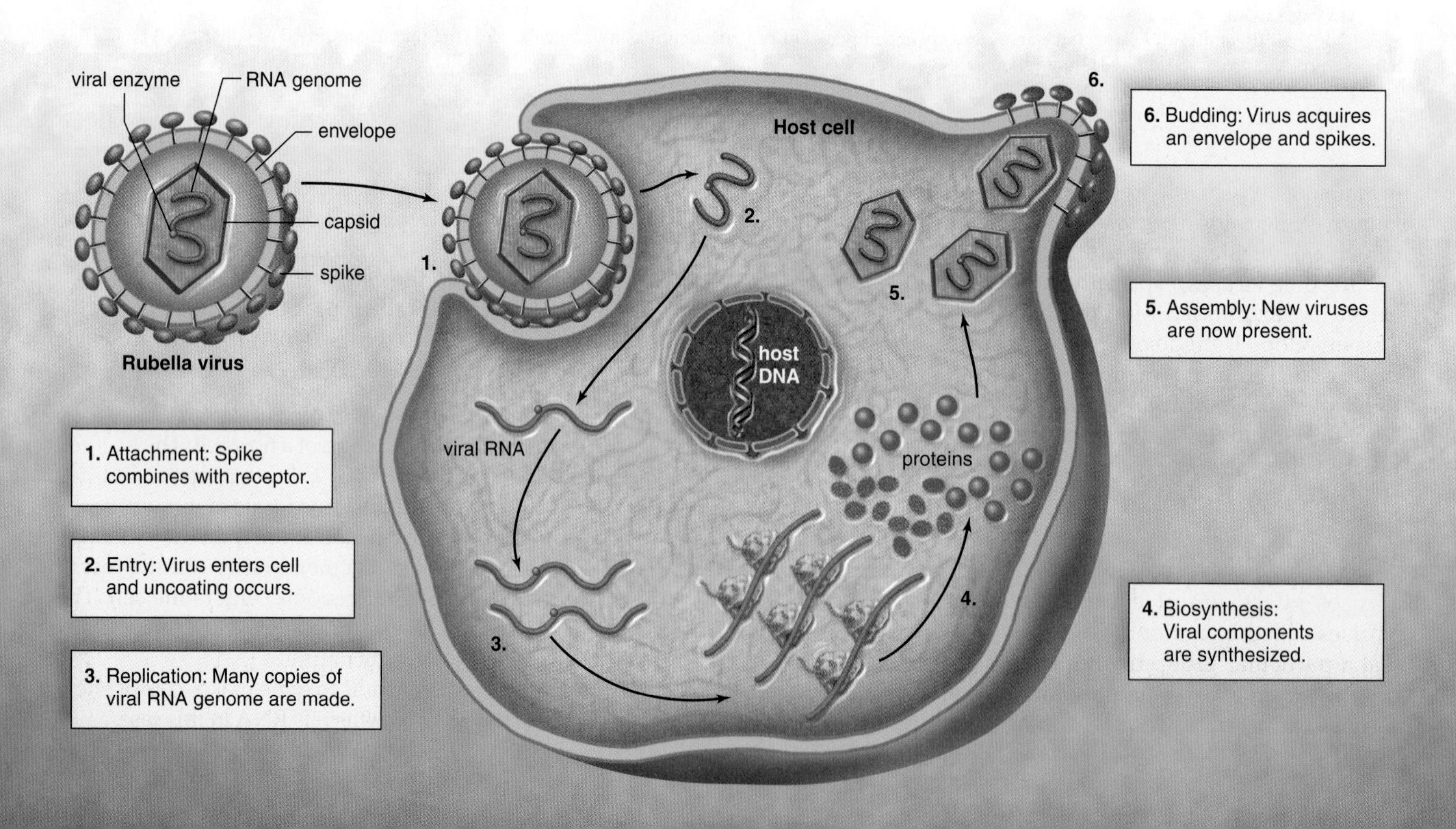

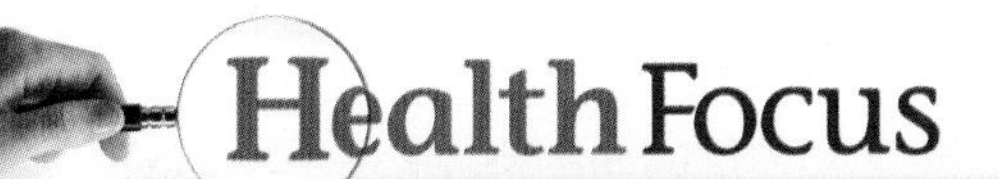

Health Focus

Antibiotics and Probiotics

In 1928 at a London lab, Scottish scientist Alexander Fleming was researching substances that might inhibit bacterial growth when he took a two-week vacation. Upon returning, he noticed some petri dishes that had bacteria growing on them were contaminated with *Penicillium* fungus. In one of the most famous moments in medical history, Fleming realized that the fungus was producing a chemical that inhibited the growth of the bacteria (Fig. 28A). Of course that substance turned out to be penicillin, and the practice of medicine was changed forever. Many bacterial infections that had previously meant serious illness or even death were now treatable.

As often happens, however, the widespread use of Fleming's new drug, and many others like it, had unintended consequences. Especially in patients who take antibiotics frequently or for a prolonged period, normal microbial populations in the body can be disrupted, resulting in diarrhea, yeast infections, or worse. A bacterium called *Clostridium difficile*, which is normally present in the intestines at relatively low levels, tends to overgrow in these patients, and may cause a severe, or even fatal, inflammation of the colon. Treatment for this disease is usually more antibiotics, but some authorities are suggesting that probiotics may be helpful in preventing this and many similar diseases.

Grocery store shelves and Internet health sites are full of products claiming to contain "probiotics," usually containing bacteria of the genera *Lactobacillus* or *Bifidobacterium*. According to the U.N. Food and Agricultural Organization, probiotics are "live microorganisms, which, when administered in adequate amounts, confer a health benefit on the host." By this definition, foods like yogurt, cheeses, and other fermented dairy products, are probiotics. But because of the increased interest in the health benefits of "friendly" bacteria, live bacteria are being added to an increasing variety of products, and Americans spent three times more on probiotics in 2003 than in 1994 (Fig. 28A).

How do probiotics work? As already mentioned in this chapter, the normal microflora of the body can produce vitamins and aid digestion. They also inhibit harmful microbes by simply taking up space and nutrients, and in some cases, by secreting chemicals that directly kill pathogens. But the beneficial effects of normal microflora go beyond competing with other microbes. Body surfaces such as the intestinal wall are fortified with large populations of immune cells (see Chapter 13) that guard against invasion of the tissues by pathogens. An ongoing interaction between these cells and the normal microflora is necessary to maintain a healthy immune system. Experimentally, animals that are raised under gnotobiotic (germ-free) conditions have poorly developed immune systems, and replenishing their bodies with "good" bacteria restores healthy immune function, sort of like priming a pump. That is one reason why probiotics are also being tested for treatment of inflammatory conditions such as irritable bowel syndrome, ulcerative colitis, and Crohn disease. Future applications may also include treating urinary tract and vaginal infections, inhibiting food allergies, preventing tooth decay, and improving the effectiveness of vaccines!

Despite their potential, however, one aspect of probiotics to be aware of is quality control. Currently there is very little regulation of which bacterial strains, or even the number of beneficial organisms, a product must contain to be labeled as probiotic. And of course, it is extremely unlikely that all the potential benefits of probiotics will pan out. In a period of less than 100 years, however, we have gone from discovering a new way to inhibit or kill harmful bacteria that invade our bodies, to realizing how closely our health is linked to our relationship with harmless microflora.

Discussion Questions

1. If no antibiotics existed, how might your life be different?
2. Besides taking antibiotics, what other factors might influence the numbers and types of microflora in your body?
3. High numbers of "good" bacteria are found in the intestine and on the skin. The immune system needs to protect these areas from invading microbes, but cannot respond as strongly to the normal microflora without causing problems. What are some possible ways that immune cells could distinguish "good" from "bad" bacteria?

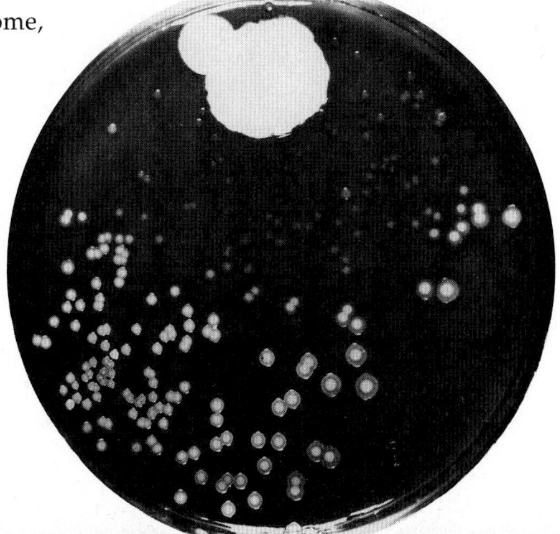

Figure 28A Changing attitudes about microbes.
When Fleming discovered the inhibition of bacterial growth by *Penicillium* (a fungal contaminant) in 1928, few would have predicted the increasing awareness of beneficial bacteria (probiotics) less than 100 years later.

The Common Cold and Influenza Colds are most commonly caused by rhinoviruses, and the symptoms usually include a runny nose, mild fever, and fatigue. The flu, caused by the influenza virus, is characterized by more severe symptoms, such as a high fever, body aches, and severe fatigue. Cold symptoms tend to subside within a week, but the flu may last for two or three weeks, and can result in death, especially in elderly patients or those with weakened immune systems.

While most common cold viruses are endemic (always present) in the human population, flu outbreaks tend to be epidemic, affecting large numbers of people in more limited geographic areas at any one time. Why can you get several colds or the flu year after year? There are over 100 different strains of rhinoviruses, plus several other viruses that cause very similar symptoms. In the case of colds, you become immune only to those strains you have contracted before. However, the reason you may get the flu each year is that the influenza virus can change rapidly. Small changes called **antigenic drift,** especially those affecting the surface spikes, may be enough to make the virus capable of temporarily evading the immune response of individuals who were immune to the original virus (Fig. 28.35*a*). Therefore, manufacturers of influenza vaccines try to incorporate the strains that are predicted to cause the highest number of cases each year. Still, antigenic drift can lead to local epidemics.

The genome of the influenza virus is composed of eight segments of RNA. When two different influenza viruses infect the same cell, these RNA segments can get mixed up as the viruses reproduce. When new virus particles are assembled that contain RNA from both original viruses, this *reassortment* event, called an **antigenic shift,** may lead to new combinations of surface spikes (Fig. 28.35*b*). Because the human population has not previously been exposed to this combination of antigens, the result can be a pandemic, or worldwide epidemic. The "avian flu" virus, also called H5N1, which arose in southern China in 2005, contains genes from a bird virus, as well as a human virus, and thus may have resulted from a reassortment event. Fortunately, the H5N1 virus doesn't seem to be efficiently transmitted to humans. However, the most recent "swine flu" outbreak of 2009 serves to remind us it is impossible to predict what future reassortment events may occur, so vaccine manufacturers must play catch-up whenever a new virus evolves.

Measles Measles is one of the most contagious human diseases. An unvaccinated person can become infected simply by breathing viral particles in the air of a room where an infected person has been, even two hours later. After exposure, there is a seven- to twelve-day incubation period before the onset of flulike symptoms, including fever. A red rash

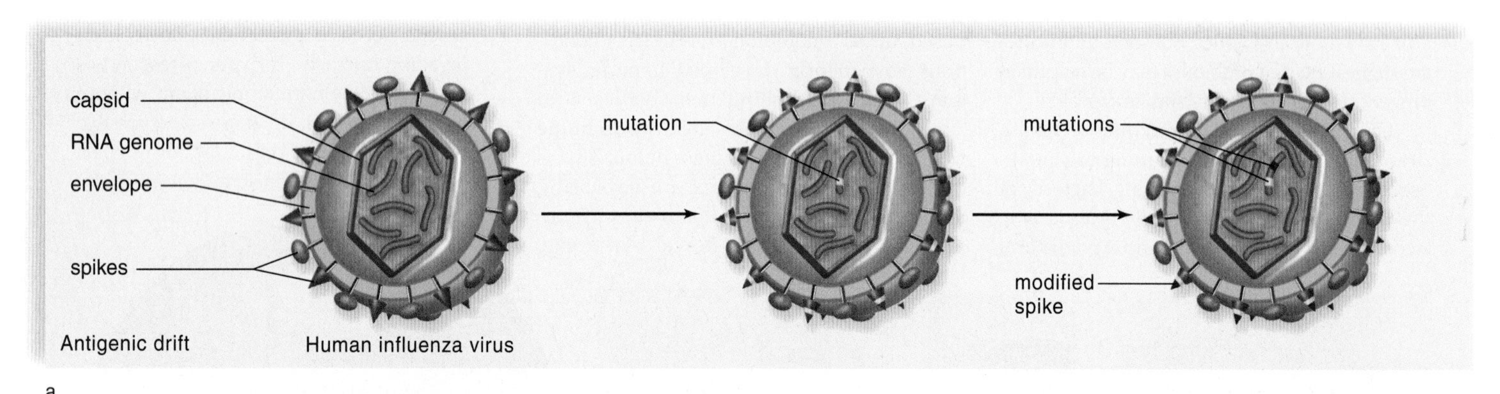

Figure 28.35 Antigenic drift and shift.
a. In antigenic drift, small mutations gradually change surface antigens so that antibodies to the original virus become less effective. As time goes by, people are more likely to become ill upon exposure to the virus. **b.** In antigenic shift, even more serious major changes take place on surface antigens as genome segments are reassorted between two influenza viruses that infect the same cell. Now, antibodies are more likely to be ineffective, and most people will become ill when they are exposed to the virus.

develops on the face and moves to the trunk and limbs. The rash lasts for several days. In the more-developed nations, the fatality rate is about 1 in 3,000. However, in the less-developed countries, the fatality rate is 10–15%. The number of deaths is decreasing in many areas, however, due to efforts to vaccinate children, who still make up the majority of measles fatalities. In the United States, the number of measles cases dropped from 5 million each year to a few thousand after the introduction of the measles vaccine in 1963. Between 2000 and 2007, measles deaths worldwide fell by 74%, from an estimated 750,000 to 197,000. The measles vaccine is usually given to infants in combination with vaccines against mumps and rubella, collectively called the MMR vaccine (see Fig. 13.9).

Herpesviruses Herpesviruses cause chronic infections that remain latent for much of the time. Eight herpesviruses capable of infecting humans have been described. Some of these infect epithelial cells and neurons, while others infect blood cells. Some herpesviruses do not seem to cause any pathology, but we will discuss four herpesviruses that cause disease in humans.

Herpes simplex virus type 1 (HSV-1) is usually associated with cold sores and fever blisters around the mouth. It is estimated that 70–90% of the adult population is infected with HSV-1. Herpes infections of the genitals are generally caused by HSV-2 and are transmitted through sexual contact. Painful blisters fill with clear fluid that is very infectious (see Fig. 21.15). Genital herpes infections are widespread in the United States, affecting 20–25% of the adult population. Most individuals infected with genital herpes will experience recurrences of symptoms induced by various stresses.

Another herpesvirus that infects humans is the varicella-zoster virus, which causes chickenpox (Fig. 28.36). Later in life, usually after age 60, the latent virus can reemerge as a related disease called shingles. Painful blisters form in an area innervated by a single sensory neuron, often on the upper chest or face. The symptoms may last for several weeks or, in some cases, months. The FDA approved a vaccine for the prevention of chickenpox in 1995, and many states are now requiring that children receive this vaccine. In May 2006, the FDA approved the first vaccine for prevention of shingles. It is for use in people over age 60.

Epstein-Barr virus (EBV) is the herpesvirus associated with infectious mononucleosis, or "mono." Like HSV-1, EBV is very common—as many as 95% of adults aged 35 to 40 have been infected. When infected with EBV as children, most people have no symptoms. In contrast, those infected as teenagers and young adults often develop infectious mononucleosis. The disease is named after the tendency of the cells infected by the virus, which are also known as mononuclear cells, to become more numerous in the blood. The symptoms of mono usually include fever, fatigue, and swollen lymph nodes. Like all herpesviruses, EBV establishes a lifelong latent infection. EBV has also been associated with other conditions, such as chronic fatigue syndrome and certain rare cancers.

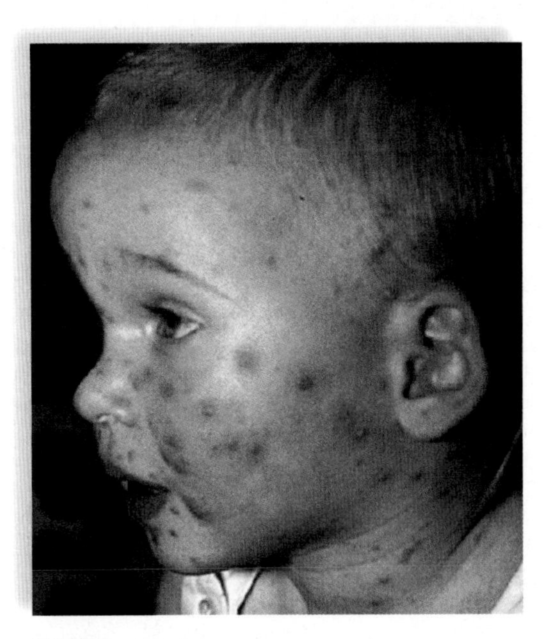

Figure 28.36 Chickenpox.
The lesions characteristic of chickenpox are pus-filled blisters that break and crust. The varicella-zoster virus remains latent in nerve cells and can produce shingles decades later.

Antiviral Drugs

Because viruses use the machinery of host cells for viral replication, it is difficult to develop drugs that affect viral replication without harming host cells. As noted earlier, however, many viruses use their own enzymes to copy their genetic material, and many antiviral drugs inhibit these enzymes. As discussed in Chapter 21, a variety of antiretroviral agents have been developed to inhibit the reverse transcriptase and protease enzymes used by HIV, which can delay the onset of AIDS. Other drugs, such as acyclovir and valacyclovir (or Valtrex), are commonly used to control recurrences of HSV-1 and -2 infections. These compounds mimic nucleotides, and thus can inhibit viral genome synthesis. Other antiviral drugs may affect virus attachment, entry, or assembly. If no effective antiviral drugs are available for a particular viral infection, patients are often told to simply let the virus run its course, because antibiotics are not effective against viral infections.

Check Your Progress

1. How do influenza viruses undergo antigenic shifts?
2. Why did measles deaths decline dramatically in the United States after 1963?

Viroids and Prions

Viroids and prions are also acellular pathogens. A **viroid** consists of a circular piece of naked RNA. The viroid RNA is ten times shorter than that of most viral genomes, and apparently it does not code for any proteins. Viroid replication causes diseases in plants, the only known hosts. The mechanism of viroid diseases, such as potato spindle tuber and apple scar skin, are not known.

Bioterrorism

The events of September 11, 2001, caused scientists and ethicists to consider the possibility of bioterrorism—the use of a biological agent, such as a microbe, to kill thousands of people at once. For a few weeks after 9/11, this possibility became even more real as several people died when anthrax spores were sent through the U.S. mail. Unfortunately, the person(s) responsible for those deaths have not been identified, although a leading suspect, microbiologist Bruce Ivins, committed suicide in July 2008 as the government was preparing to charge him for the attacks.

Bioterrorism has been employed throughout the history of the civilized world, so much so, that the United States, the United Kingdom, and the USSR signed the 1972 Biological Weapons Convention banning research and the production of offensive biological agents. However, the possibility of using microbes and inexpensive nerve agents as weapons is particularly attractive to countries that lack the means to produce conventional weapons of war. When a group of Japanese extremists used sarin, a nerve agent, to kill many innocent civilians in the Tokyo subway system in 1995, it became apparent that individuals, as well as hostile countries, can obtain and use these agents as weapons.

Numerous ethical questions surround the process of researching biological weapons. For example, should scientists be allowed to perfect and publish their findings regarding such agents? Many biologists say not publishing these findings would be detrimental to the progress of science. "How do doctors talk about research if we don't publish it?" asks Dr. Ariella Rosengard of the University of Pennsylvania. In 2002, her laboratory published a paper on the variola virus, which causes smallpox, showing that a viral protein is 100 times more effective at inhibiting immune system proteins called complement (see section 13.2) than is a similar protein from a related virus. This knowledge may help to explain why variola virus is so lethal to humans, but it could also be useful to bioterrorists attempting to produce even more lethal forms of the virus. While it may seem that potentially harmful information like this shouldn't be published, such research is important for developing suitable vaccines.

Other scientists feel the need to be more socially responsible. "When you wield power equivalent to nuclear weapons, you need some oversight," says John Steinbruner of the Center for International and Security Studies in Maryland. This organization and others are attempting to define what types of research should be allowed, and what types should possibly be banned. The main difficulty is that a better understanding of what makes these biological agents so deadly could save lives, as well as destroy them.

Form Your Own Opinion

1. Should the United States maintain an active research program to develop biological weapons? Why or why not? Is the use of these kinds of weapons more objectionable than conventional warfare? Than nuclear warfare?
2. Many experts believe that Russia still possesses large stockpiles of smallpox virus produced during the Cold War. The United States also retains some smallpox virus, reportedly for research purposes. Ideally, if both countries could agree, would it be best to destroy all samples of this deadly virus?
3. After September 11, 2001, the U.S. government has increased funding for research into infectious agents that bioterrorists could use. However, this also means that more scientists will be working with these agents and developing knowledge that could be used for detrimental purposes. Should scientists working, for example, with the bacterium that causes anthrax, be required to submit to extensive background checks? To periodic questioning by the FBI? If so, might this discourage people with good intentions from doing this work?

Prions are proteinaceous infectious particles that cause degenerative diseases of the nervous system in humans and other animals. They are derived from normal proteins of unknown function in the brains of healthy individuals. Disease occurs when the normal proteins change into the abnormal, prion shape. This forms more prion proteins, which go on to convert other normal proteins into the wrong shape. These abnormal proteins seem to build up in the brain, causing a loss of neurons and what appear to be holes in the tissue (see Fig. 17.22).

The first prion disease to be described was scrapie, which occurs in sheep. It is called scrapie because, as the disease affects the animal's brain, a sheep begins to scrape off much of its wool. Scrapie is not capable of causing disease in humans, but another prion disease in cattle, called bovine spongiform encephalopathy, or mad cow disease, does appear capable of infecting humans and has caused over 150 human deaths in Great Britain and a few other countries. This disease is now called variant Creutzfeldt-Jakob disease (vCJD), to distinguish it from the "classic" form of CJD that occurs spontaneously in a very low percentage of people. Most cases of vCJD have occurred in people who ate meat from cows that had mad cow disease. As of 2008, only three cases of vCJD have been documented in the United States, and all were thought to have originated in other countries. Chronic wasting disease is another prion disease that occurs in deer, elk, and moose, but there have been no documented cases of transmission to humans. One prion disease that passes directly from human to human is kuru, which was transmitted among a cannibalistic tribe in New Guinea due to their traditional practice of consuming their dead relatives. Even though prion diseases are as frightening as they are fascinating, their incidence in humans remains very low.

Check Your Progress

1. How do viroids differ from viruses?
2. How do prions differ from all other microbes?
3. What is unique about how prions "replicate?"

Applying the Concepts [Revisited]

By all accounts, Howard Hughes was terrified of "germs" for most of his life. During his later years he lived as a recluse, because he failed to understand some of the most basic facts about the microbial world.

1. Because the word *germs* really has no precise scientific meaning, what do you think people mean by it?
2. Why is it difficult, if not impossible, for someone to live a microbe-free life?
3. Suppose you could actually remove all microbes from your body, as well as your surroundings. What do you think would be the immediate effects? The long-term effects?

SUMMARIZING THE CONCEPTS

28.1 The Microbial World

- Microbiology is primarily the study of bacteria, archaea, protists, fungi, and viruses. Most microbes are microscopic, and some are extremely numerous.
- Two of the most significant contributors to the early discipline of microbiology were Antonie van Leeuwenhoek, who was the first to visualize microbes using his microscopes, and Louis Pasteur, whose many accomplishments include disproving the idea that microbes could arise spontaneously.
- Microbes are perhaps best known for causing infectious diseases, which can range from the merely annoying to the swiftly fatal. However, most microorganisms are far more beneficial than harmful, not only to humans, but to Earth's ecosystem.

28.2 Bacteria and Archaea

Bacteria and archaea are unicellular prokaryotic organisms that lack nuclei and membrane-bounded organelles. There are fundamental differences between domain Bacteria and domain Archaea, however. Several important structural, reproductive, and metabolic features of bacteria can be noted:

- Most bacteria have a cell wall that contains peptidoglycan, and stain either Gram-postitive or Gram-negative. All have a DNA genome, usually found on a single, circular chromosome.

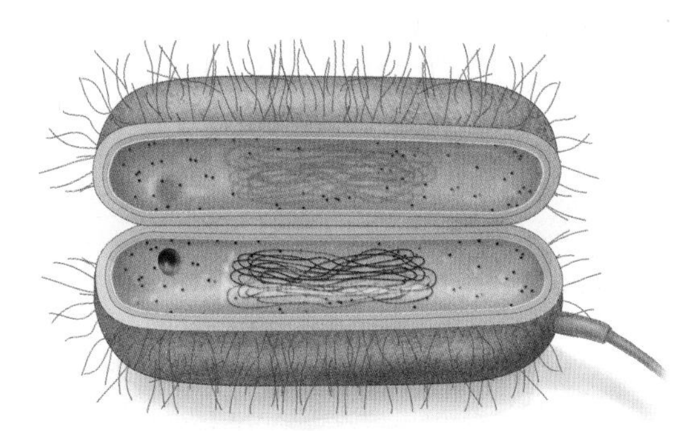

- Bacteria reproduce asexually by binary fission. However, bacteria can share DNA by conjugation, transformation, and transduction.
- Most bacteria are heterotrophs, requiring carbon in the form of organic molecules. Some are autotrophs, which acquire carbon from carbon dioxide and energy from chemicals or the sun. Cyanobacteria are an important group of photosynthetic bacteria.
- Bacterial species that cause important diseases of humans include members of the genera *Staphylococcus*, *Streptococcus*, *Mycobacterium*, and *Salmonella*.
- Archaea are often associated with extreme habitats, but actually are widespread in the environment. They are distinguished from bacteria by nucleic acid sequence and biochemical differences. The archaeal plasma membrane has unique lipids that help some to survive in extreme environments. Some perform photosynthesis using a unique pigment. Others are methanogenic, producing methane.

28.3 Protists

Protists are a diverse group of mostly unicellular eukaryotes that are widespread in aquatic and moist environments: Protists are quite complex because they (1) have a combination of characteristics not seen in the other eukaryotic kingdoms, (2) have complicated life cycles that include the ability to withstand hostile environments, and (3) sometimes have organelles not seen in other types of eukaryotic cells. Although classification schemes for protists are still being debated, they can be placed into six groups:

- The photosynthetic protists derive energy from the sun, and reduce carbon dioxide to carbohydrates. Green algae possess chlorophyll, store reserve food as starch, and have cell walls of cellulose, as do plants. Green algae include those that are unicellular (*Chlamydomonas*), filamentous (*Spirogyra*), multicellular (*Ulva*), and colonial (*Volvox*). Diatoms, which have an outer layer of silica, are extremely numerous in both marine and freshwater ecosystems. Dinoflagellates usually have cellulose plates and two flagella, one at a right angle to the other. They are extremely numerous in the ocean and, on occasion, produce a neurotoxin when they form red tides. The large red and brown algae, called seaweeds, are economically important. Red algae are often more delicate than brown algae and are usually found in warmer waters. Giant kelp grows under the surface of the sea in dense groves.
- Flagellates, which move by flagella, are often symbiotic. Trypanosomes are flagellates that cause African sleeping sickness in humans. Euglenoids are flagellated protists with a pellicle instead of a cell wall. Their chloroplasts are most likely derived from a green alga, through endosymbiosis.

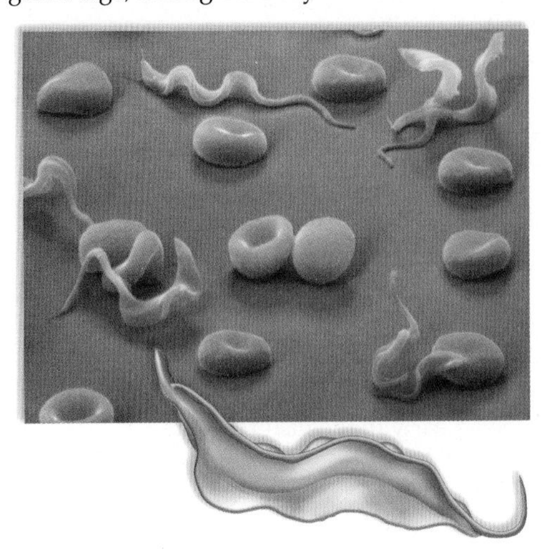

- Ciliates move by means of their many cilia. They are remarkably diverse in form—and as exemplified by *Paramecium,* they show how complex a protist can be, despite being a single cell.
- Amoeboids move and feed by forming pseudopods. Parasitic amoeboids, in the genus *Entamoeba,* cause amoebic dysentery.
- Sporozoans are nonmotile parasites that form spores. *Plasmodium* causes malaria. *Toxoplasma* and *Cryptosporidium* are also important pathogens.
- Water molds and slime molds are actually classified as protists. Water molds, which ordinarily are saprotrophs, are sometimes parasitic. Slime molds are amoeboid and ingest food by phagocytosis. Cellular slime molds exist as individual amoeboid cells until nutrients become limited. The cells then fuse into a pseudoplasmodium that produces spores.

28.4 Fungi

Fungi are unicellular or multicellular eukaryotes that are strict heterotrophs:

- After external digestion, fungi absorb the resulting nutrient molecules. Most fungi are saprotrophs that aid the cycling of chemicals in ecosystems by decomposing dead remains.
- The body of a fungus is composed of thin filaments, called hyphae, which form masses termed mycelia. Nonseptate hyphae have no cross walls, forming multinucleate cells. Septate hyphae have cross walls, but pores allow for exchange of cytoplasm. The cell wall contains chitin. Most fungi are nonmotile throughout their life cycle.
- Fungi produce nonmotile, and often windblown, spores during both asexual and sexual reproduction. During sexual reproduction, hyphae tips fuse so that hyphae containing both + and – nuclei usually result. Following nuclear fusion, meiosis occurs during the production of the sexual spores.
- Fungal diseases of humans include skin infections called tineas, toxin-based diseases, as well as systemic infections like histoplasmosis and candidiasis.
- Fungi are divided into the chytrids, which are aquatic and have flagella at some stages; zygospore fungi, which form zygospores; sac fungi, which form ascocarps; club fungi, which form basidiocarps, and AM fungi, which are mycorrhizae.
- Fungi often have symbiotic relationships with other organisms, and some are parasites. Lichens are symbiotic associations between fungi and cyanobacteria, or green algae. The algae provide organic nutrients to the fungi, while the fungi provide protection and enhanced absorption of water and minerals. Mycorrhizal fungi enjoy a mutualistic relationship with plant roots, such that the fungus helps the plant absorb minerals and water, while the plant supplies the fungus with organic nutrients.

28.5 Viruses, Viroids, and Prions

Viruses are minute, acellular pathogens that reproduce as obligate intracellular parasites:

- Structural features shared by all viruses include an outer capsid composed of protein and an inner core of nucleic acid. The viral genome can be single- or double-stranded DNA or RNA. Many animal viruses also have an envelope with spikes.
- The reproductive cycle of an animal virus typically has six steps: attachment, entry, genome replication, biosynthesis, assembly, and budding. During budding, enveloped viruses usually acquire a portion of the host cell's plasma membrane. Retroviruses (e.g., HIV) have RNA genomes that are converted to DNA by reverse transcriptase, before integrating into the host cell's chromosome as a provirus.
- Viruses infect almost all types of organisms and cause many important diseases. Viral diseases of humans include the common cold, influenza, measles, and chickenpox. Antiviral drugs usually inhibit viral enzymes.

Like viruses, viroids and prions are acellular pathogens:

- Viroids are obligate, intracellular plant pathogens that are autonomously replicating short RNA molecules. They contain no protein.
- Prions contain no nucleic acid. They are proteinaceous, infectious particles that convert a normal cellular protein into the abnormal form, which accumulates and damages brain tissue, inducing dementia. Diseases caused by prions include scrapie, kuru, and mad cow disease.

TESTING YOURSELF

Choose the best answer for each question.

1. Decomposers
   a. break down dead organic matter in the environment by secreting digestive enzymes.
   b. break down living organic matter by secreting digestive enzymes.
   c. destroy living cells and then break them down with digestive enzymes.
   d. live in close association with another species.
2. Which is not true of prokaryotes? They
   a. are living cells.
   b. lack a nucleus.
   c. all are parasitic.
   d. include both archaea and bacteria.
   e. evolved early in the history of life.
3. Which bacterium is spherical in shape?
   a. bacillus
   b. coccus
   c. spirillum
4. Bacterial endospores function in
   a. reproduction.
   b. survival.
   c. protein synthesis.
   d. storage.
5. Chemoautotrophic bacteria acquire energy
   a. by parasitizing other organisms.
   b. from the sun.
   c. by oxidizing inorganic compounds.
   d. by anaerobic photosynthesis.
   e. Both c and d are correct.

6. Scarlet fever and rheumatic fever are both caused by
   a. *Escherichia coli* O157:H7.
   b. *Mycobacterium tuberculosis.*
   c. *Staphylococcus aureus.*
   d. *Streptococcus pyogenes.*
7. Archaea differ from bacteria in that they
   a. have a nucleus.
   b. have membrane-bound organelles.
   c. have peptidoglycan in their cell walls.
   d. are often photosynthetic.
   e. None of these are correct.
8. Dinoflagellates
   a. usually reproduce sexually.
   b. have protective cellulose plates.
   c. are insignificant producers of food and oxygen.
   d. have cilia instead of flagella.
   e. tend to be larger than brown algae.
9. Ciliates
   a. can move by pseudopods.
   b. are not as varied as other protists.
   c. have a gullet for food gathering.
   d. are closely related to the radiolarians.
10. Which is found in slime molds but not in fungi?
    a. nonmotile spores
    b. amoeboid vegetative cells
    c. zygote formation
    d. photosynthesis
    e. All of these are correct.
11. Which feature is best associated with hyphae?
    a. strong, impermeable walls
    b. rapid growth
    c. large surface area
    d. pigmented cells
    e. Both b and c are correct.
12. Fungi are classified according to
    a. sexual reproductive structures.
    b. shape of their hyphae.
    c. mode of nutrition.
    d. type of cell wall.
    e. All of these are correct.
13. Conidia are formed
    a. asexually at the tips of special hyphae.
    b. during sexual reproduction.
    c. by all types of fungi except water molds.
    d. only when it is windy and dry.
    e. as a way to survive a harsh environment.
14. Lichens
    a. cannot reproduce.
    b. need a nitrogen source to live.
    c. are parasitic on trees.
    d. are able to live in extreme environments.
15. Mycorrhizal fungi
    a. are a type of lichen.
    b. are in a mutualistic relationship.
    c. help plants gather solar energy.
    d. help plants gather inorganic nutrients.
    e. Both b and d are correct.
16. Which of the following diseases is caused by *Candida*?
    a. oral thrush
    b. tinea
    c. histoplasmosis
    d. ringworm
17. Label the reproductive cycle of an RNA animal virus in the following diagram.

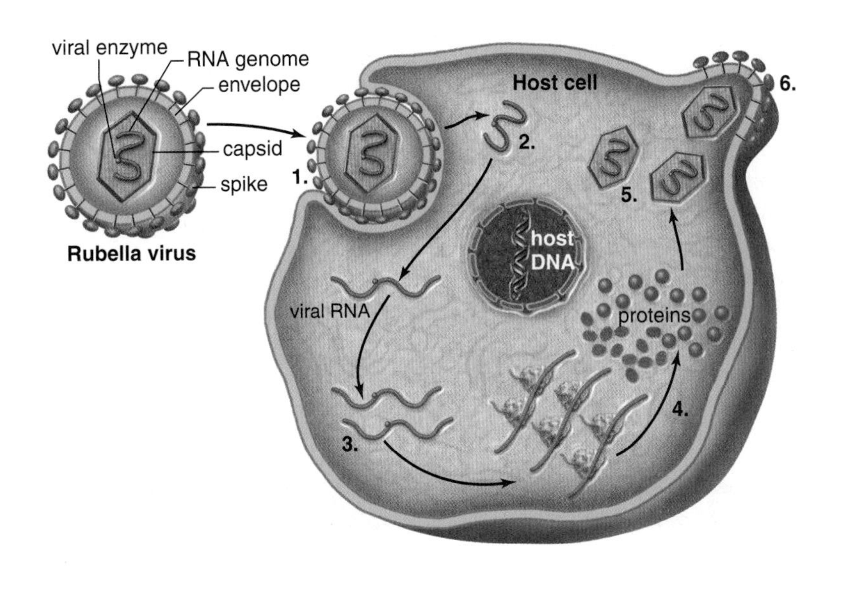

18. The envelope of an animal virus is usually derived from the _____ of its host cell.
    a. cell wall
    b. plasma membrane
    c. capsule
    d. receptors
19. Capsid proteins are synthesized during which phase of viral replication?
    a. replication
    b. biosynthesis
    c. assembly
    d. proteination
    e. All of these are correct.
20. Retroviruses have a unique enzyme that
    a. disintegrates host DNA.
    b. polymerizes host DNA.
    c. transcribes viral RNA to cDNA.
    d. translates host DNA.
    e. repairs viral DNA.
21. Small changes in influenza surface antigens lead to
    a. antigenic drift.
    b. antigenic shift.
    c. antigenic schism.
    d. antigenic sift.
22. Shingles is a disease observed mainly in the elderly that is a reactivation of which of the following?
    a. herpes
    b. polio
    c. chickenpox
    d. AIDS
23. Prions contain
    a. DNA only.
    b. protein only.
    c. RNA only.
    d. DNA, RNA, and protein.

 UNDERSTANDING THE TERMS

algae 577
AM fungi 589
amoeboid 581
antibiotic 575
antigenic drift 594
antigenic shift 594
archaea 576
bacteria 571
binary fission 572
brown alga 579
capsid 590
cellular slime mold 584
chemoautotroph 572
chytrid fungi 586
ciliate 580
club fungi 588
conidium (pl., conidia) 587
conjugation 572
contractile vacuole 577
cyanobacteria 572
decomposer 571
diatom 578
dinoflagellate 578
endospore 572
envelope 591
euglenoid 580
eyespot apparatus 577
flagella 572
flagellate 580
fungi 584
Gram stain 571
green algae 578
heterotrophic 572
hypha 584
lichen 589
lipopolysaccharide (LPS) 572
malaria 582
methanogen 576
mycelium 584
mycorrhizal fungi 589
mycoses 585
normal microflora 570
nucleoid 572
phytoplankton 578
plasmodial (acellular) slime mold 583
prion 596
protist 577
protozoan 577
provirus 592
pyrenoid 578
red alga 579
red tide 578
retrovirus 592
sac fungi 587
sporangium 587
spore 585
sporozoan 582
tinea 585
transduction 572
transformation 572
tubercle 574
viroid 595
virus 590
water mold 583
yeast 588
zooplankton 577
zygospore fungi 587

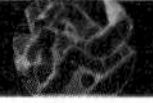 THINKING CRITICALLY

1. Sickle cell disease occurs mostly in people of African ancestry. When a person inherits two copies of the sickle cell gene, his or her hemoglobin is abnormal, causing red blood cells to be more fragile and sticky than usual. This leads to anemia and circulatory problems. About 70,000 Americans have sickle cell disease, and about 2 million carry one copy of the gene. Even though it can cause a serious disease, the sickle cell gene is thought to have been favored by natural selection, because it gives some protection against malaria. Considering the type of cells infected by *Plasmodium* species, how might this protection work? What do you think may happen to the prevalence of the sickle cell disease gene if an effective vaccine against malaria is ever developed and widely used?
2. Many viruses contain their own enzymes for replicating their genetic material, while others use the host cell's enzymes. Would DNA viruses or RNA viruses be more likely to produce their own enzyme(s) for this purpose, and why?
3. Why is it easier to create drugs to treat bacterial infections than protozoal or fungal infections? Specifically, why do drugs such as penicillin have a very low incidence of side effects?

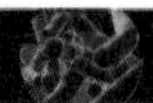 *INQUIRY INTO LIFE* WEBSITE

The companion website for *Inquiry into Life* provides a wealth of information organized and integrated by chapter. You will find practice tests, animations, videos, and much more that will complement your learning and understanding of general biology.

http://www.mhhe.com/maderinquiry13

Plants

Applying the Concepts

A student asks his professor, "What happens if plants disappear?" His teacher responds, "If all the plants on Earth were suddenly to die, all animals, including humans, would soon perish. Yet, if animals were to die, many plants would still survive and flourish." Plants provide important food and nesting sites for many animals, trees provide oxygen to the air, and even many medicines come directly from plants or plant products. Plants have been around for a long time. The fossil record indicates that plants performed the incredible feat of moving onto dry land from an aquatic environment before animals. The first plants to conquer land were the bryophytes (mosses and liverworts), which evolved around 450 million years ago (MYA). Their lack of vascular tissue and the inability to conduct water and nutrients keeps the bryophytes to a small size. The seedless vascular plants, of which ferns are the most notable, evolved about 350 MYA. Vascular tissue allowed ferns and their allies to grow very tall. If you could travel back to this time, you would see steamy jungles and swamps dominated by large, seedless vascular plants. Gymnosperms were the first seed plants to evolve during the Devonian period around 365 MYA. The ability to produce seeds is a highly successful trait that increases reproductive success by protecting the embryo and allowing dispersal. Gymnosperms, such as cycads and evergreen trees, were quite large and dominated the land during the Mesozoic era when the dinosaurs ruled. In fact, botanists refer to the Mesozoic as the "age of the cycads" instead of the "age of the dinosaurs" because the cycads were much larger and more numerous than they are today. Later, between 200 and 135 MYA, angiosperms evolved. Today, angiosperms are the most widespread of the plants because of their ability to produce flowers and seeds enclosed by fruit. Overall, plants display an astonishing diversity on every continent, as well as amazing abilities to adapt to varied environments.

Chapter Outline

29.1 Evolutionary History of Plants

Learning Outcomes

1. List the similarities between charophytes and land plants.
2. List the five major events in the evolutionary history of plants.
3. Describe alternation of generations in plants.

Plants (domain Eukarya, kingdom Plantae) are vital to human survival. Our dependence on them is nothing less than absolute (see the Ecology Focus at the end of this chapter). Although a land environment does offer advantages such as plentiful light, it also has certain challenges such as the constant threat of desiccation (drying out). Most important, all stages of reproduction—gametes, zygote, and embryo—must be protected from the drying effects of air. To keep the internal environment of cells moist, a land plant must acquire water, and transport it to all parts of the body, while keeping the body in an erect position. We will see how plants have adapted to these problems by evolving an internal vascular system.

The evolutionary history of plants begins in the water. Most likely land plants evolved from a form of freshwater green algae some 450 MYA (million years ago): both green algae and plants (1) contain chlorophylls *a* and *b* and various accessory pigments, (2) store excess carbohydrates as starch, and (3) have cellulose in their cell walls. A comparison of DNA and RNA base sequences suggest that land plants are most closely related to a group of freshwater green algae known as charophytes. Although the common ancestor of modern charophytes and land plants no longer exists, if it did, it would have features resembling those of the *Chara* and *Coleochaete*.

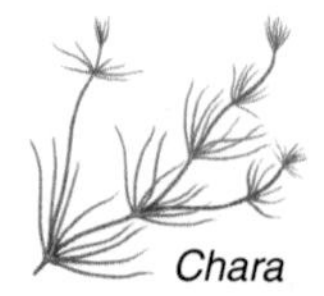

Let's take a look at these filamentous green algae. *Chara* are commonly known as stoneworts because they are encrusted with calcium carbonate deposits. The body consists of a single file of very long cells anchored in mud by thin filaments. Whorls of branches occur at regions called nodes, located between the cells of the main axis. Male and female reproductive structures grow at the nodes. A *Coleochaete* looks flat like a pancake, but the body is actually composed of long, branched filaments of cells. Most important to the evolution of plants, charophytes protect the zygote. Land plants not only protect the zygote, they also protect and nourish the resulting embryo—this may be the first derived feature that separates land plants from green algae.

To illustrate the evolution of land plants, we will be discussing five groups. It is possible to associate each group with one of five major evolutionary events, each one an adaptation to a land existence.

Mosses are low-lying plants that lack vascular tissue and therefore have no means of transporting water, but they do protect the body of the plant from drying out and protect the embryo within a special structure. The lycophytes, which evolved around 420 MYA, are among the first plants to have a vascular system that transports water and solutes from the roots to the leaves of land plants. Plants with **vascular tissue** have true roots, stems, and leaves. The leaves of lycophytes are very narrow. Ferns are well-known plants with large leaves called **megaphylls.** The evolution of branching and megaphylls allows a plant to increase the amount of photosynthesis and carbohydrate produced. Without an adequate production of food, a plant cannot increase in size.

The next evolutionary event was the evolution of **seeds.** A seed contains an embryo and stored organic nutrients within a protective coat. Seeds are highly resistant structures well suited to protect a plant embryo from drying out until conditions are favorable for germination. The gymnosperms were the first seed plants to appear about 365 MYA. The final evolutionary event of interest to us is the evolution of the flower, a reproductive structure found in angiosperms. Flowers attract pollinators such as insects, and they also give rise to fruits that cover seeds. Plants with flowers evolved somewhere between 200 and 135 MYA. Figure 29.1 traces the evolutionary history of land plants and will serve as a backdrop as we discuss major groups of plants in the pages that follow.

Alternation of Generations

All plants have a life cycle that includes **alternation of generations.** In this life cycle, two multicellular individuals alternate, each producing the other (Fig. 29.2). The two individuals are (1) a sporophyte, which represents the diploid generation, and (2) a gametophyte, which represents the haploid generation.

The **sporophyte** (2n) is so named for its production of spores by meiosis. A **spore** is a haploid reproductive cell that develops into a new organism without needing to fuse with another reproductive cell. In the plant life cycle, a spore undergoes mitosis and becomes a gametophyte.

The **gametophyte** (n) is so named for its production of gametes. In plants, eggs and sperm are produced by mitotic cell division. A sperm and egg fuse, forming a diploid zygote that undergoes mitosis and becomes the sporophyte.

Two observations are in order. First, meiosis produces haploid spores. This is consistent with the realization that the sporophyte is the diploid generation and spores are haploid reproductive cells. Second, mitosis occurs as a spore becomes a gametophyte, and mitosis also occurs as a zygote becomes a sporophyte. Indeed, it is the occurrence of mitosis at these times that results in two generations.

Plants differ as to which generation is dominant—that is, more conspicuous. In nonvascular plants, the gametophyte is dominant, but in the other three groups of plants, the sporophyte is dominant. In the history of plants, only the sporophyte evolves vascular tissue. Therefore, the shift to sporophyte dominance is an adaptation to life on land. Notice that as the sporophyte gains in dominance, the

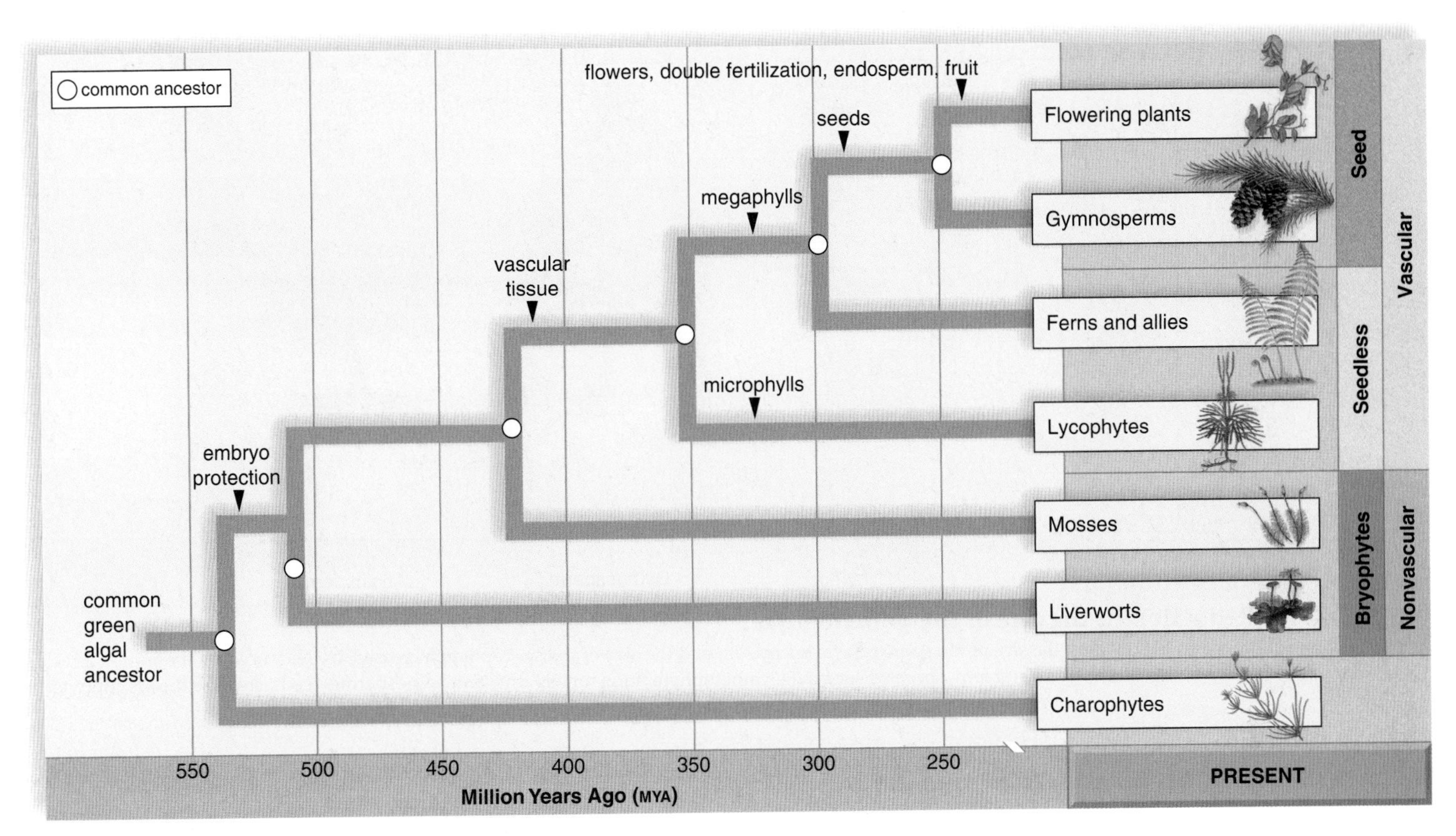

Figure 29.1 Evolutionary history of plants.
The evolution of plants involves five significant innovations: (1) Protection of a multicellular embryo was seen in the first plants to live on land. (2) The evolution of vascular tissue was another important adaptation for life on land. (3) The evolution of megaphylls aided dispersion of nutrients into leaves by veination. (4) The evolution of the seed increased the chance of survival for the next generation. (5) The evolution of the flower fostered the use of animals as pollinators and the use of fruits to aid in the dispersal of seeds.

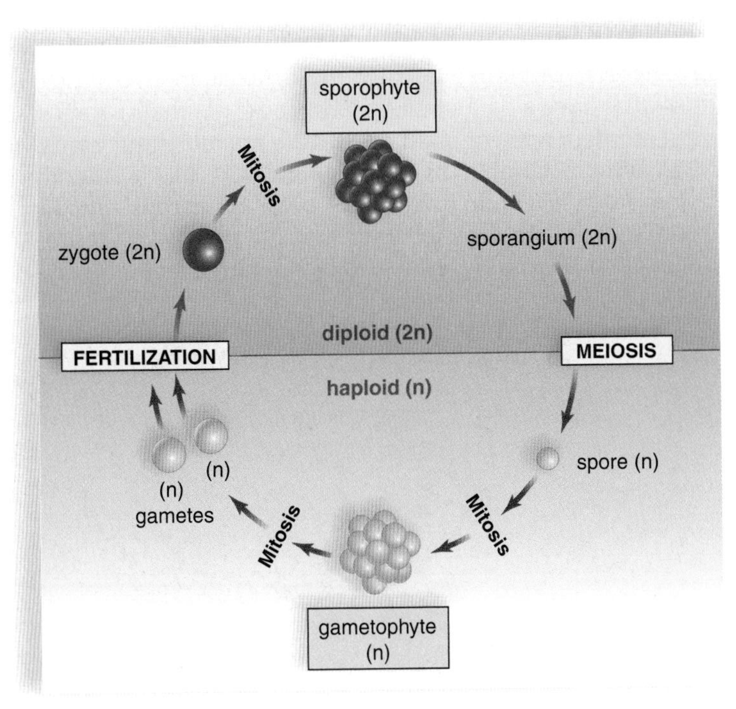

Figure 29.2 Alternation of generations.

gametophyte becomes microscopic (Fig. 29.3). It also becomes dependent upon the sporophyte.

All plants have an alternation of generations life cycle, but the appearance of the generations varies widely. In ferns, a gametophyte is a small, independent, heart-shaped structure. Archegonia are female reproductive organs that produce eggs. Antheridia are male reproductive organs that produce motile sperm. The eggs are fertilized by flagellated sperm, which swim to the archegonia in a film of water. In contrast, the female gametophyte of an angiosperm, called the embryo sac, is retained within the body of the sporophyte plant and consists of seven cells within a structure called an ovule. Following fertilization, the ovule becomes a seed. In seed plants, pollen grains are mature, sperm-bearing male gametophytes. Pollen grains are transported by wind, insects, bats, or birds, and therefore, they do not need free water to reach the egg. In seed plants, reproductive cells are protected from drying out in the land environment.

Check Your Progress

1. What are the five main events that marked the evolution of plants?
2. Describe alternation of generations in plants.

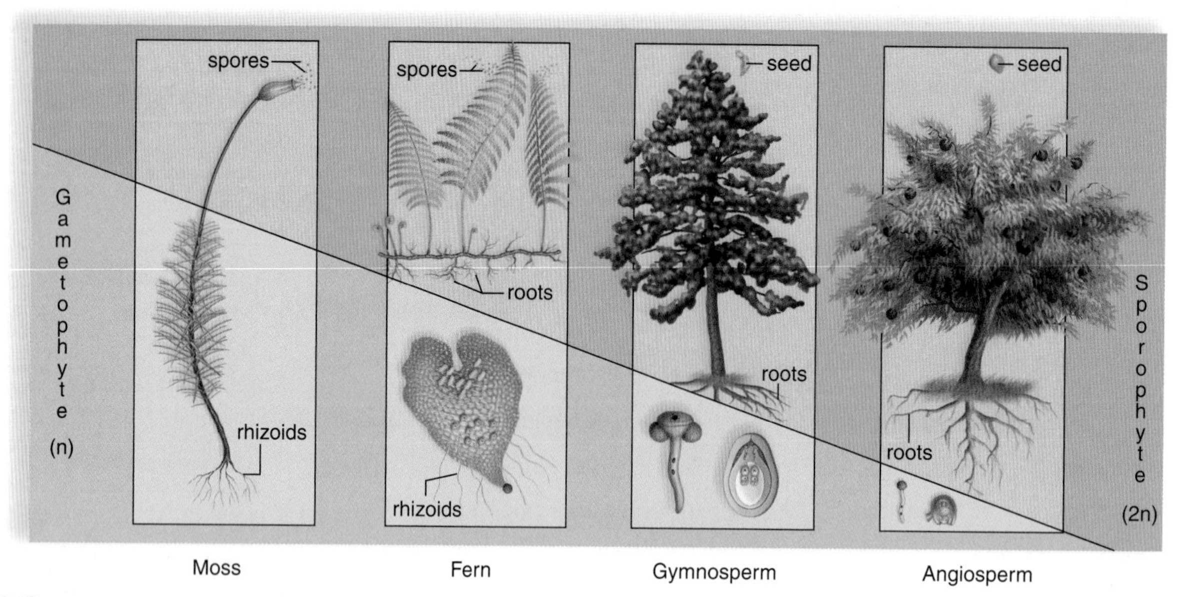

Figure 29.3 Reduction in the size of the gametophyte.
As plants became adapted to life on land, the size of the gametophyte decreased, and the size of the sporophyte increased, as evidenced by these representatives of today's plants. In the moss and fern, spores disperse the gametophyte. In gymnosperms and angiosperms, seeds disperse the sporophyte.

29.2 Nonvascular Plants

Learning Outcomes

1. Discuss the characteristics of nonvascular plants including the ways the moss life cycle shows nonvascular plants are/are not adapted to the land environment.
2. Describe the characteristics of liverworts and mosses.

The **nonvascular plants** lack vascular tissue. Although the nonvascular plants often have a "leafy" appearance, they do not have true roots, stems, and leaves—which, by definition, must contain true vascular tissue. Therefore, the nonvascular plants are said to have rootlike, stemlike, and leaflike structures.

The gametophyte is the dominant generation in nonvascular plants—that is, it is the generation we recognize as the plant. Further, flagellated sperm swim in a continuous film of water to the vicinity of the egg. The sporophyte, which develops from the zygote, is attached to, and derives its nourishment from, the gametophyte shoot.

The nonvascular plants consist of three divisions—one each for hornworts, liverworts, and mosses , which together are considered bryophytes. The placement of these plants in separate divisions reflects current thinking that they are not closely related. In this section, we discuss the liverworts and mosses.

Liverworts

Liverworts (phylum Hepaticophyta) exist in two types: those that have a flat, lobed thallus (body; pl., thalli) are the most familiar, while those that are leafy are more numerous. *Marchantia,* a familiar liverwort (Fig. 29.4), has a smooth upper surface. The lower surface bears numerous *rhizoids* (rootlike hairs) that project into the soil. *Marchantia* reproduces both asexually and sexually. Gemmae cups on the

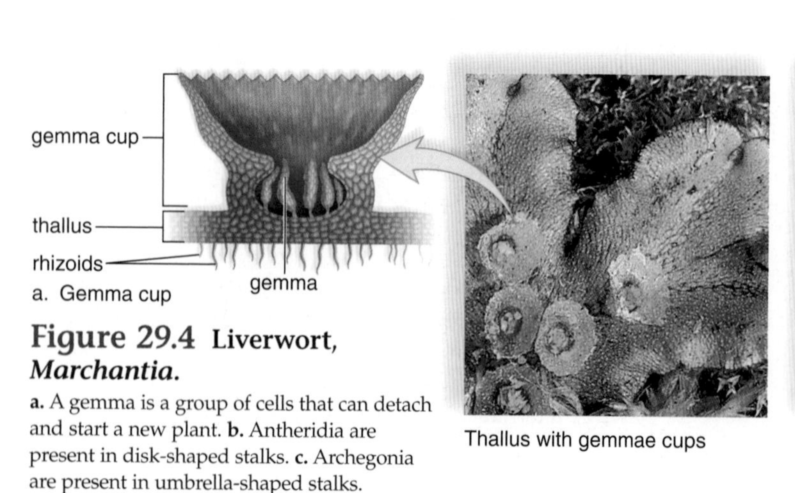

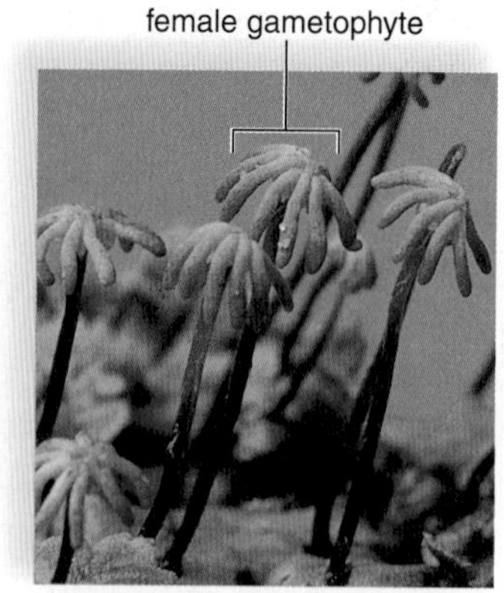

Figure 29.4 Liverwort, *Marchantia*.
a. A gemma is a group of cells that can detach and start a new plant. **b.** Antheridia are present in disk-shaped stalks. **c.** Archegonia are present in umbrella-shaped stalks.

upper surface of the thallus contain gemmae, groups of cells that detach from the thallus and can asexually start a new plant. Sexual reproduction depends on disk-headed stalks that bear antheridia, where flagellated sperm are produced, and on umbrella-headed stalks that bear archegonia, where eggs are produced. Following fertilization, tiny sporophytes composed of a foot, a short stalk, and a capsule begin growing within archegonia. Windblown spores are produced within the capsule.

Mosses

Mosses (phylum Bryophyta) can be found from the Arctic through the tropics to parts of the Antarctic. Although most live in damp, shaded locations in the temperate zone, some survive in deserts, while others inhabit bogs and streams. In forests, mosses frequently form a mat that covers the ground or the surfaces of rotting logs. Mosses can store large quantities of water in their cells, but if a dry spell continues for long, they become dormant until it rains.

Most mosses can reproduce asexually by fragmentation. Just about any part of the plant is able to grow and eventually produce leaflike thalli. Figure 29.5 describes the life cycle of a typical temperate-zone moss. The gametophyte of mosses has two stages. First, there is the algalike *protonema*, a branching filament of cells. Then, after about three days of favorable growing conditions, upright leafy thalli appear at

Figure 29.5 Moss life cycle, ***Polytrichum*** sp.

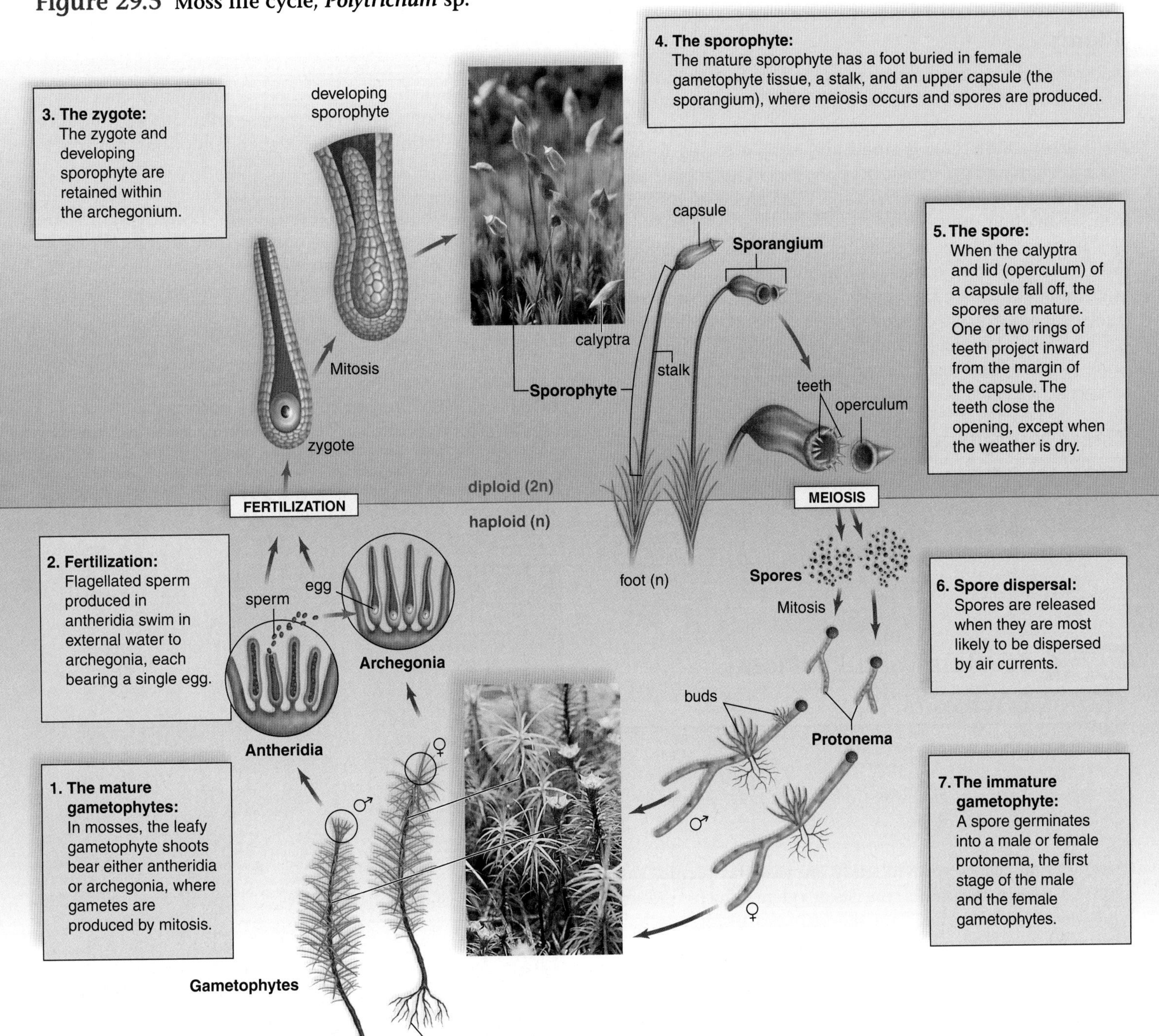

intervals along the protonema. Rhizoids anchor the thalli, which bear antheridia and archegonia. An **antheridium** consists of a short stalk, an outer layer of sterile cells, and an inner mass of cells that become the flagellated sperm. An **archegonium,** which looks like a vase with a long neck, has a single egg located inside its base.

The dependent sporophyte consists of a *foot,* which grows down into the gametophyte tissue; a *stalk;* and an upper *capsule,* or *sporangium,* where windblown spores are produced. At first, the sporophyte is green and photosynthetic. At maturity, it is brown and nonphotosynthetic. Because the gametophyte is the dominant generation, it seems consistent for spores to be dispersal agents—that is, when the haploid spores germinate, the gametophyte is in a new location.

Adaptations and Uses of Nonvascular Plants

Although mosses are nonvascular, they are better than flowering plants at living on stone walls and on fences and even in the shady cracks of hot, exposed rocks. For these particular microhabitats, being small and simple seems to be a selective advantage. When bryophytes colonize bare rock, they help convert the rocks to soil that can be used for the growth of other organisms.

In areas such as bogs, where the ground is wet and acidic, dead mosses, especially *Sphagnum,* do not decay. The accumulated moss, called peat or bog moss, can be used as fuel. Peat moss is also commercially important because it has special nonliving cells that can absorb moisture. Thus, peat moss is often used in gardens to improve the water-holding capacity of the soil.

Check Your Progress

1. (a) Name two features limiting the adaptation of nonvascular plants and (b) two features that indicate they are adapted to the land environment.
2. How would you know whether you are looking at a liverwort or a moss?

29.3 Seedless Vascular Plants

Learning Outcomes

1. Give two reasons why ferns are the more plentiful group of seedless vascular plants.
2. Describe the similarities and differences between the fern and the moss life cycle.

All the other plants we will study are **vascular plants.** Vascular tissue in these plants consists of **xylem,** which conducts water and minerals up from the soil, and **phloem,** which transports organic nutrients from one part of the plant to another. The vascular plants usually have true roots, stems, and leaves. The roots absorb water from the soil, and the stem conducts water to the leaves. Xylem, with its strong-walled cells, supports the body of the plant against the pull of gravity. The leaves are fully covered by a waxy cuticle except where it is interrupted by stomata, little pores whose size can be regulated to control water loss.

The sporophyte is the dominant generation in vascular plants. Seedless vascular plants include two groups: lycophytes and ferns and their allies. Both of these groups disperse their offspring by producing wind-blown spores. This is advantageous because the sporophyte is the generation with vascular tissue. Another advantage of having a dominant sporophyte relates to its being diploid. If a faulty gene is present, it can be masked by a functional gene. Then, too, the greater the amount of genetic material, the greater the possibility of mutations that will lead to increased variety and complexity. Indeed, vascular plants are complex, extremely varied, and widely distributed.

Some vascular plants do not produce seeds. Seedless vascular plants include whisk ferns, club mosses, horsetails, and ferns, which disperse their offspring by producing windblown spores. When the spores germinate, they produce a small gametophyte that is independent of the sporophyte for its nutrition. In these plants, antheridia release flagellated sperm, which swim in a film of external water to the archegonia, where fertilization occurs. Because spores are dispersal agents and the nonvascular gametophyte is independent of the sporophyte, these plants cannot wholly benefit from the adaptations of the sporophyte to a terrestrial environment.

Figure 29.6 The Carboniferous period. Growing in the swamp forests of the Carboniferous period were treelike club mosses (*left*), treelike horsetails (*right*), and shorter, fernlike foliage (*lower left*). When the trees fell, they were covered by water and did not decompose well. Sediment built up and turned to rock, and the resulting compression caused the organic material to become coal, a fossil fuel that helps run our industrialized society today.

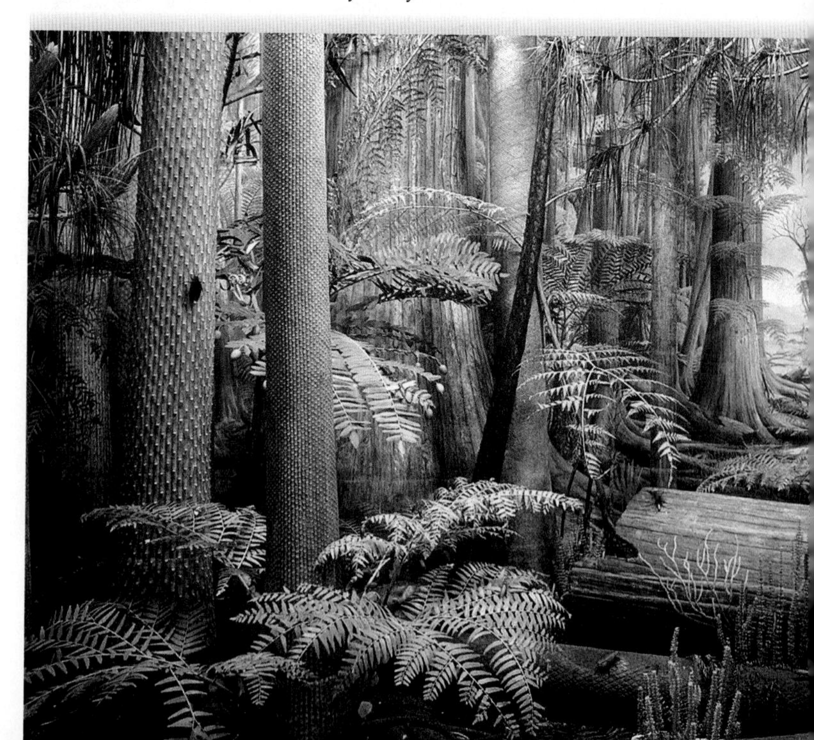

The seedless vascular plants formed the great swamp forests of the Carboniferous period (Fig. 29.6). A large number of these plants died but did not decompose completely. Instead, they were compressed to form the coal that we still mine and burn today. (Oil has a similar origin, but it most likely formed in marine sedimentary rocks and included animal remains.)

Lycophytes

Lycophytes (phylum Lycophyta) are also called **club mosses** and were among the first land plants to have vascular tissue. Unlike true mosses (bryophytes), the lycophytes have well-developed vascular tissue in roots, stems, and leaves (Fig. 29.7). Typically, a fleshy underground and horizontal stem, called a rhizome, sends up upright arial stems. Tightly packed, scalelike leaves cover the stems and branches, giving the plant a mossy look. The small leaves, termed microphylls, each have a single vein composed of xylem and phloem. The sporangia are borne on terminal clusters of leaves, called strobili, which are club-shaped. The spores are sometimes harvested and sold as lycopodium powder, or vegetable sulfur, for use in pharmaceuticals and in fireworks because it is highly flammable. *Lycopodium* featured in Figure 29.7 are common in moist woodlands in temperate climates, where they are called ground pines; they are also abundant in the tropics and subtropics.

Ferns and Their Allies

In this section, we will discuss the characteristics of ferns, whisk ferns, and horsetails.

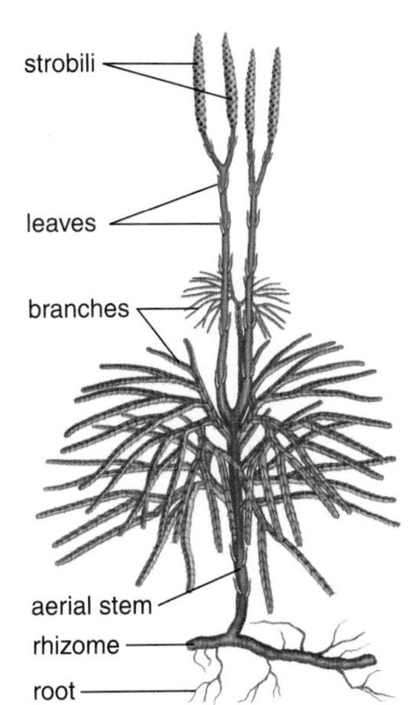

Figure 29.7 Ground pine, *Lycopodium*.
Lycophytes such as *Lycopodium* have vascular tissue and thus true roots, stems, and leaves. The *Lycopodium* sporophyte develops an underground rhizome system. A rhizome is an underground stem; this rhizome produces roots along its length.

Ferns

Ferns (phylum Polypodiophyta) are the largest group of plants other than the flowering plants, and they display great diversity in form and habitat. Ferns are most abundant in warm, moist, tropical regions, but they are also found in northern regions and in relatively dry, rocky, places. They range in size from low-growing plants resembling mosses to tall trees. The *fronds* (leaves) that grow from a rhizome can vary. The royal fern has fronds that stand 6 ft tall; those of the maidenhair fern are branched, with broad leaflets; and those of the hart's tongue fern are straplike and leathery. Figure 29.8 gives examples of fern diversity.

Cinnamon fern, *Osmunda cinnamomea*

Hart's tongue fern, *Campyloneurum scolopendrium*

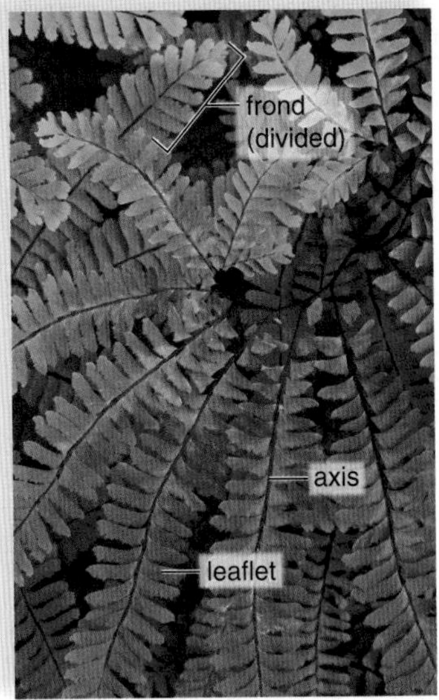

Maidenhair fern, *Adiantum pedatum*

Figure 29.8 Fern diversity.
The many different types of ferns all grow in places that offer moisture. These photos show the dominant sporophyte. The separate, delicate, and almost microscopic gametophyte is water dependent and produces sperm that require moisture to swim to the egg.

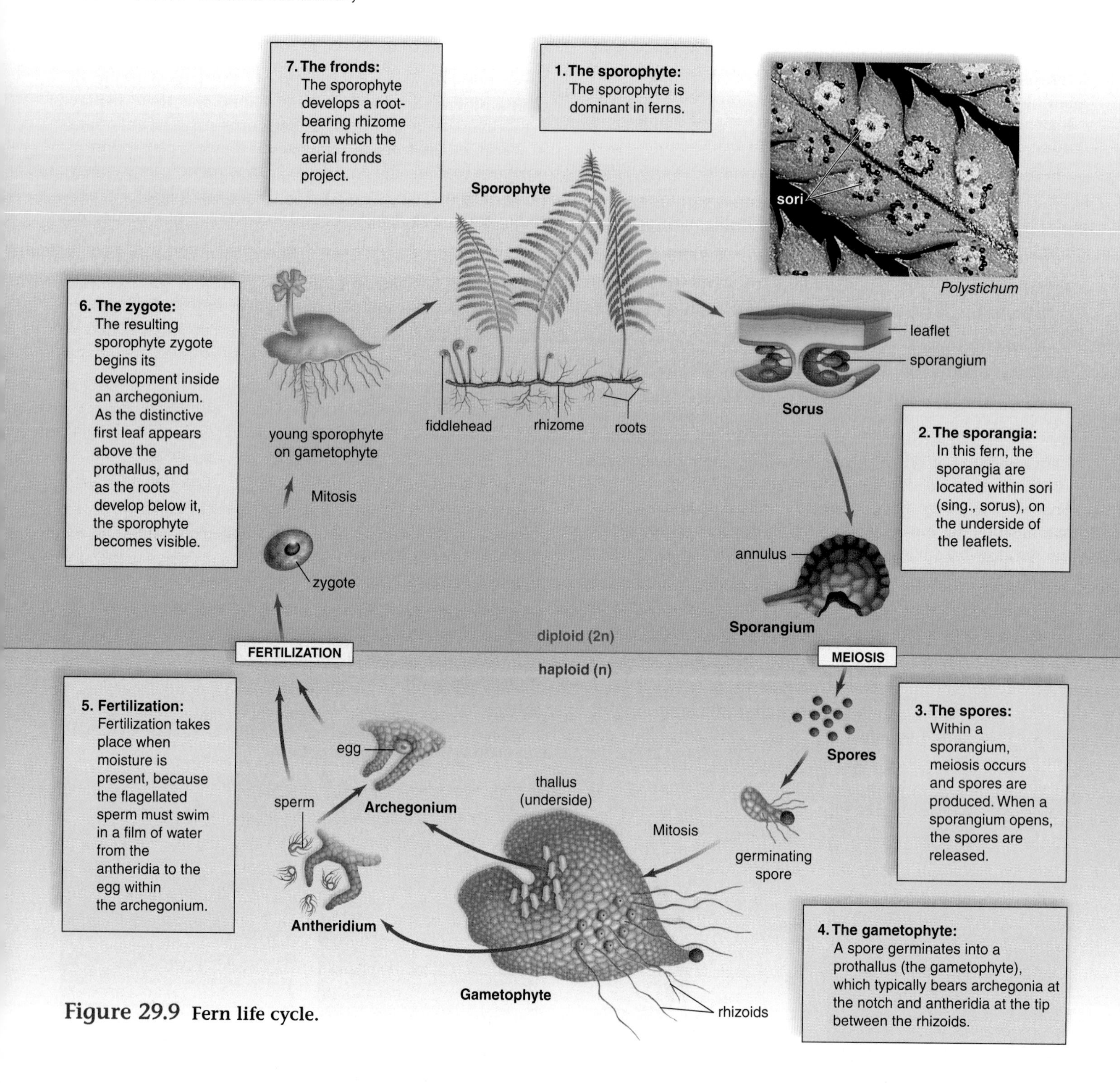

Figure 29.9 Fern life cycle.

Life Cycle, Adaptations, and Uses of Ferns Ferns have true roots, stems, and leaves; that is, they contain vascular tissue. The well-developed leaves fan out, capture solar energy, and photosynthesize. The life cycle of a typical temperate fern is shown in Figure 29.9. The dominant sporophyte produces windblown spores. When the spores germinate, a tiny green and independent gametophyte develops. The gametophyte is water dependent because it lacks vascular tissue. Also, flagellated sperm produced within antheridia require an outside source of moisture to swim to the eggs in the archegonia. Upon fertilization the zygote develops into a sporophyte. In nearly all ferns, the leaves of the sporophyte

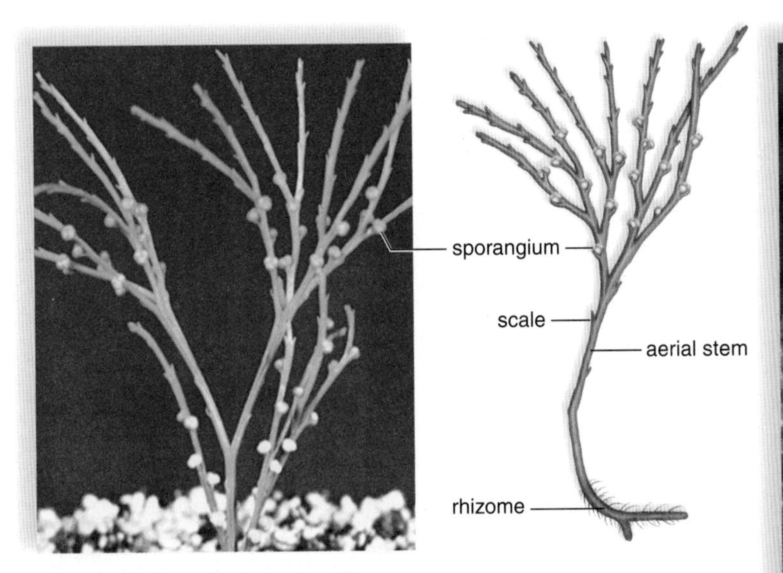

Figure 29.10 Whisk fern, *Psilotum*.
Whisk ferns have no roots or leaves—the branches carry on photosynthesis. The sporangia are yellow.

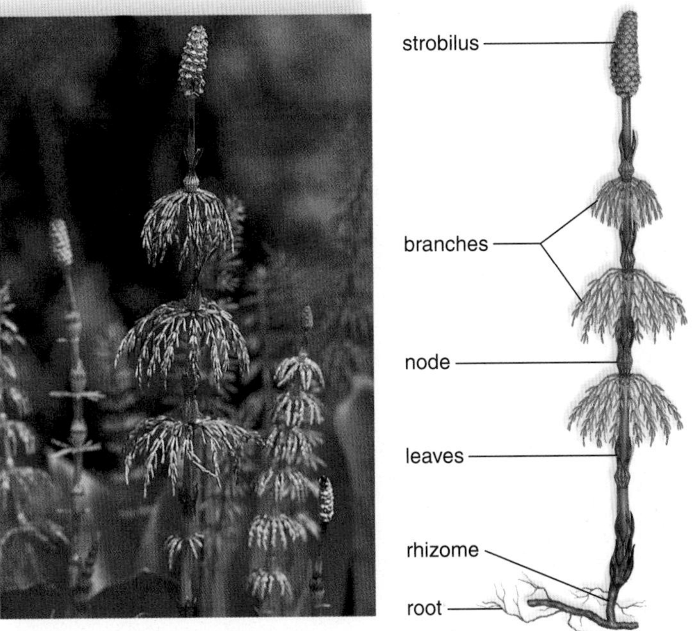

Figure 29.11 Horsetail, *Equisetum*.
Whorls of branches and tiny leaves are at the nodes of the stem. Spore-producing sporangia are borne in strobili (cones).

first appear in a curled-up form called a fiddlehead, which unrolls as it grows. Once established, some ferns, such as the bracken fern *Pteridium aquilinum,* can spread into drier areas by means of vegetative (asexual) reproduction. For example, ferns can spread when their rhizomes grow horizontally in the soil, producing fiddleheads that become new fronds.

In terms of economic value, people use ferns in decorative bouquets and as ornamental plants in the home and garden. Wood from tropical tree ferns often serves as a building material because it resists decay, as well as damage by termites. Ferns, especially the ostrich fern, are eaten as food, and in the Northeast, many restaurants feature fiddleheads as a special treat, although studies have shown that some are carcinogenic.

Whisk Ferns and Horsetails

Whisk ferns (phylum Psilotophyta), named for their resemblance to whisk brooms, are found in Arizona, Texas, Louisiana, and Florida, as well as Hawaii and Puerto Rico. *Psilotum* (Fig. 29.10) looks like a rhyniophyte, an ancient vascular plant that is known only from the fossil record. Its aerial stem forks repeatedly and is attached to a *rhizome,* a fleshy horizontal stem that lies underground. Whisk ferns have no leaves, so the branches carry on photosynthesis. Sporangia, located at the ends of short branches, produce spores that are dispersal agents.

The *Psilotum* gametophyte is independent, small (less than a centimeter), and found underground associated with mutualistic mycorrhizal fungi. The resemblance of its life cycle to that of a fern suggests that *Psilotum* is actually a fern devoid of leaves and roots. If so, it could be said to be a vestigial fern.

Horsetails (phylum Equisetophyta), which thrive in moist habitats around the globe, are represented by *Equisetum,* the only genus in existence today (Fig. 29.11). A perennial rhizome produces roots and photosynthetic aerial stems that can stand about 1.3 m high. In some species, the whorls of slender green side branches at the joints (nodes) of the stem bear a fanciful resemblance to a horse's tail. The small, scalelike leaves are whorled at the nodes and are nonphotosynthetic. Many horsetails have strobili at the tips of branch-bearing stems. Others send up special buff-colored, naked stems that bear the strobili.

The stems of horsetails are tough and rigid because of silica deposited in their cell walls. Early Americans, in particular, used horsetails to scour pots and called them "scouring rushes." Today, they are still used as ingredients in a few abrasive powders. *Equisetum* is sometimes considered a weed in the garden, and it is difficult to control because any portion of the rhizome left in the soil will regenerate.

Check Your Progress

1. What is the significance of lycophytes in plant evolution?
2. Name two significant differences between the fern and moss life cycle.
3. Why are ferns the more plentiful group of seedless vascular plants?

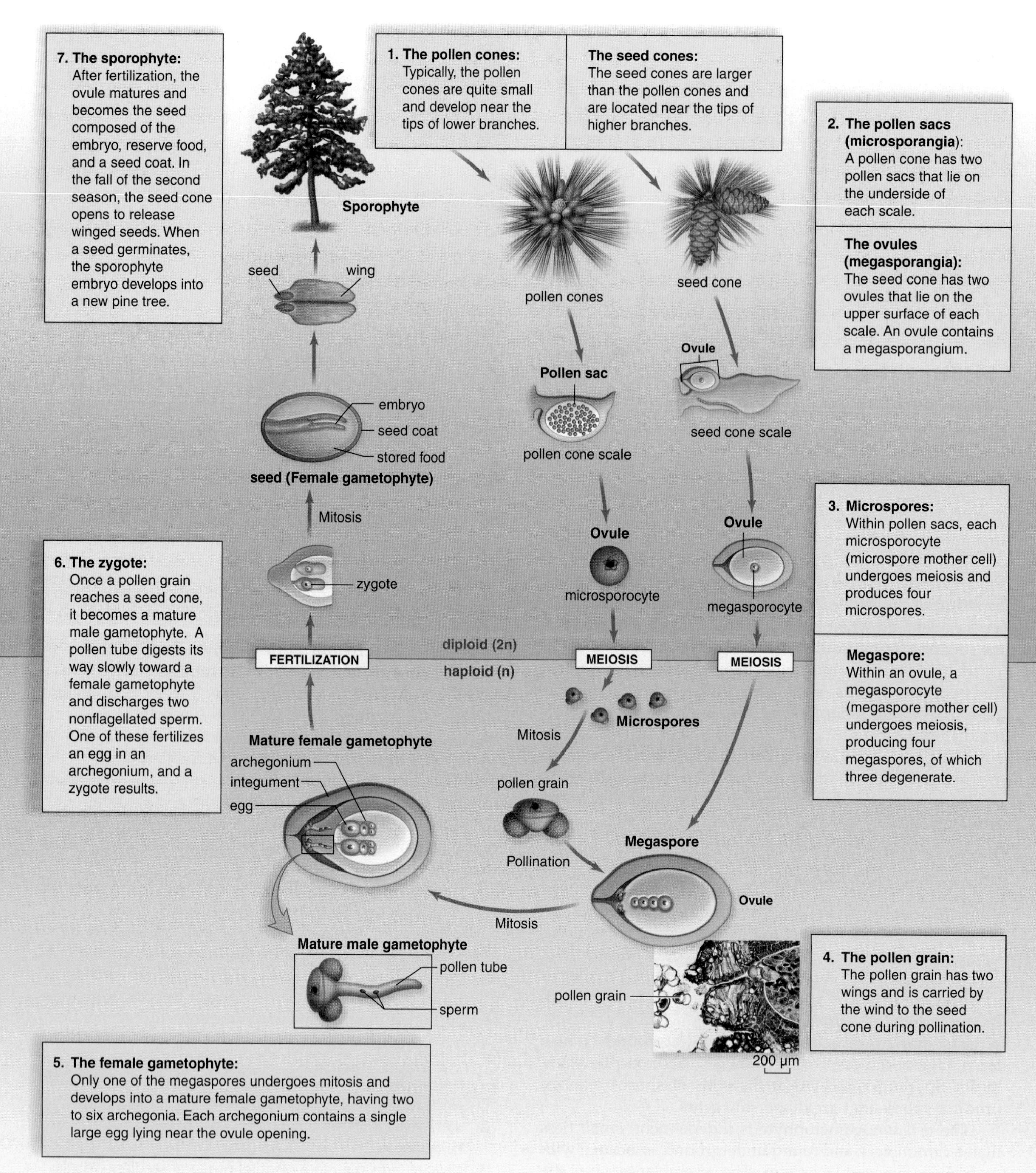

Figure 29.12 Pine life cycle.
The sporophyte is dominant, and its sporangia are borne in cones. There are two types of cones: pollen cones (microstrobili) and seed cones (megastrobili).

29.4 Seed Plants

Learning Outcomes

1. Discuss with reference to their life cycle why seed plants are fully adapted to life on land.
2. Compare and contrast gymnosperms with angiosperms, explaining why angiosperms are more species-rich today.

The gymnosperms and angiosperms are seed plants, the most plentiful plants in the biosphere today. Seeds contain a sporophyte embryo and stored food within a protective seed coat. The seed coat and stored food allow an embryo to survive harsh conditions during long periods of dormancy (arrested state) until environmental conditions become favorable for growth. Seeds can even remain dormant for hundreds of years. When a seed germinates, the stored food is a source of nutrients for the growing seedling. The survival value of seeds largely accounts for the dominance of seed plants today.

In the life cycle of the pine, a gymnosperm (Fig. 29.12), seed plants are *heterosporous*, meaning that they have two types of spores, and they produce two kinds of gametophytes—male and female. **Pollen grains,** which resist drying out, become the multicellular male gametophyte. **Pollination** occurs when a pollen grain is brought to the vicinity of a female gametophyte by wind or, in the case of flowering plants, by either wind or a pollinator. A **pollinator** is an animal, usually an insect, that carries pollen from flower to flower. After a pollen grain germinates, sperm move toward the female gametophyte through a growing **pollen tube.** *No external water is needed to accomplish fertilization.* The female gametophyte develops within an **ovule,** which eventually becomes a seed. The seed disperses offspring.

a. A northern coniferous forest

b. Seed cones and pollen cones of lodgepole pine

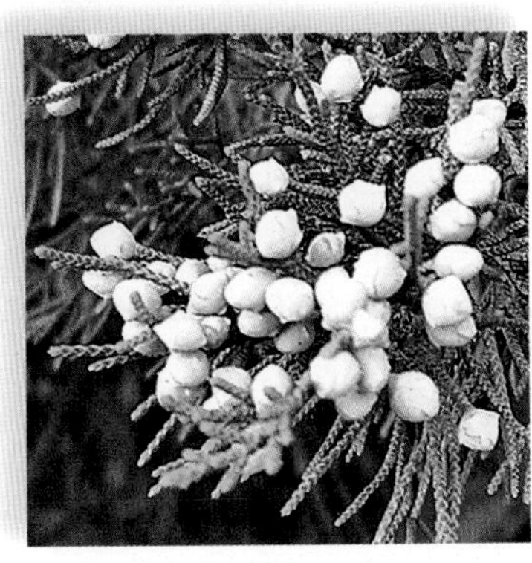

c. Juniper seed cones are fleshy

Figure 29.13 Conifers.
a. Conifers are adapted to living in cold climates. **b.** In pine trees, the seed cones become woody, but the pollen cones do not. **c.** The seed cones of a juniper have fleshy scales that fuse into a berrylike structure.

Gymnosperms

Most of the approximately 800 species of **gymnosperms** are cone-bearing plants. On the surfaces of their cone scales are ovules, which later become seeds. These ovules are said to be naked because they are not completely enclosed by diploid tissue (as the ovules in angiosperms are). This characteristic gives the gymnosperms their name; *gymnos* is a Greek word meaning naked, and *sperma* means seed. Botanists now divide the gymnosperms into four groups: conifers, cycads, ginkgoes, and gnetophytes.

Conifers

The better-known gymnosperms are evergreen, cone-bearing trees called **conifers** (phylum Pinophyta). The 575 species of conifers include the familiar pine, spruce, fir, cedar, hemlock, redwood, and cypress. Perhaps the oldest and largest trees in the world are conifers. Bristlecone pines in the Nevada mountains are known to be more than 4,900 years old, and a number of redwood trees in California are 2,000 years old and more than 100 meters tall.

Adaptations and Uses of Conifers Conifers are adapted to cold, dry weather. Thus, vast areas of northern temperate regions are covered in coniferous forests (Fig. 29.13*a*). The tough, needlelike leaves of pines conserve water because they have a thick cuticle and recessed stomata.

Note in the life cycle of the pine (see Fig. 29.12) that the dominant sporophyte produces two kinds of cones. **Pollen cones** (microstrobili) produce windblown pollen, and **seed cones** (megastrobili) produce windblown seeds (Fig. 29.13*b, c*). These cones are adaptations to a land environment.

Conifers supply much of the wood used to construct buildings and to manufacture paper. They also produce many valuable chemicals, such as those extracted from resin (pitch), a substance that protects conifers from attack by fungi and insects, and is used for waterproofing and in varnish and paints.

Other Gymnosperms

Cycads (phylum Cycadophyta, approximately 130 species) have large, finely divided leaves growing in clusters at the

a.

b.

c.

Figure 29.14 Other gymnosperms.
a. Cycad cones. **b.** The female Ginkgo trees are known for their malodorous seeds. **c.** *Welwitschia* is known for its enormous straplike leaves.

top of the stem. Therefore, they resemble palms or ferns, depending on their height. Pollen cones or seed cones, which grow at the top of the stem surrounded by the leaves, can be huge—more than a meter long and weighing 40 kg (Fig. 29.14*a*). Cycads were very plentiful in the Mesozoic era at the time of the dinosaurs, and it's likely that dinosaurs fed on cycad seeds. Now, cycads are in danger of extinction because they grow very slowly, a distinct disadvantage.

Ginkgoes (phylum Ginkgophyta), although plentiful in the fossil record, are represented today by only one surviving species, *Ginkgo biloba,* the maidenhair tree, so named because its leaves resemble those of the southern maidenhair fern, *Adiantum capillus-veneris.* Female trees produce fleshy seeds, which ripen in the fall, and give off such a foul odor that male trees are usually preferred for planting (Fig. 29.14*b*). Ginkgo trees are resistant to pollution and do well along city streets and in city parks.

The three living genera of **gnetophytes** don't resemble one another. *Gnetum,* which occurs in the tropics, consists of trees or climbing vines with broad, leathery leaves arranged in pairs. *Ephedra,* occurring only in southwestern North America and Southeast Asia, is a shrub with small, scalelike leaves. *Welwitschia,* living in the deserts of southwestern Africa, has only two enormous, straplike leaves, which split lengthwise as the plant ages (Fig. 29.14*c*).

TABLE 29.1 Monocots and Eudicots

Monocots	Eudicots
One cotyledon	Two cotyledons
Flower parts in threes or multiples of three	Flower parts in fours or fives or multiples of four or five
Usually herbaceous	Woody or herbaceous
Usually parallel venation	Usually net venation
Scattered bundles in stem	Vascular bundles in a ring
Fibrous root system	Taproot system

Angiosperms

Angiosperms, the flowering plants (phylum Magnoliophyta), are an exceptionally large and successful group of plants. There are 240,000 known species of angiosperms—six times the number of species of all other plant groups combined. Angiosperms live in all sorts of habitats, from fresh water to desert, and from the frigid north to the torrid tropics. They range in size from the tiny, almost microscopic duck weed to *Eucalyptus* trees over 100 m tall. It would be impossible to exaggerate the importance of angiosperms in our everyday lives. As discussed in the Ecology Focus (page 616), they provide us with clothing, food, medicines, and other commercially valuable products.

The name angiosperm is derived from the Greek words *angio,* meaning vessel, and *sperma,* meaning seed. The seed develops from an ovule within an ovary, which becomes a fruit. Therefore, angiosperms produce covered seeds (in contrast to the exposed seeds of gymnosperms). Although the oldest fossils of angiosperms date back no more than about 135 million years, the angiosperms probably arose much earlier. Indirect evidence suggests the possible ancestors of angiosperms may have originated as long as 200 million years ago. Scientists hypothesize that angiosperms coevolved with the insects that act as pollinators. As angiosperms became diverse, flying insects also enjoyed an adaptive radiation increasing in morphological and ecological diversity.

Monocots and Eudicots

Most flowering plants belong to one of two classes. The **Monocotyledones** (Liliopsida), often shortened to **monocots,** are composed of about 65,000 species. The **Eudicotyledones,** shortened to **eudicots** (formerly known as dicots), are composed of about 175,000 species. Table 29.1 lists the differences between monocots and eudicots.

Monocots are so called because they have only one cotyledon in their seeds, whereas eudicots have two cotyledons. Cotyledons are the seed leaves—they contain nutrients that nourish the developing embryo.

The Flower

Although flowers vary widely in appearance (Fig. 29.15), most have certain structures in common. The flower stalk expands slightly at the tip into a **receptacle,** which bears

Beavertail cactus, *Opuntia basilaris*

Water lily, *Nymphaea odorata*

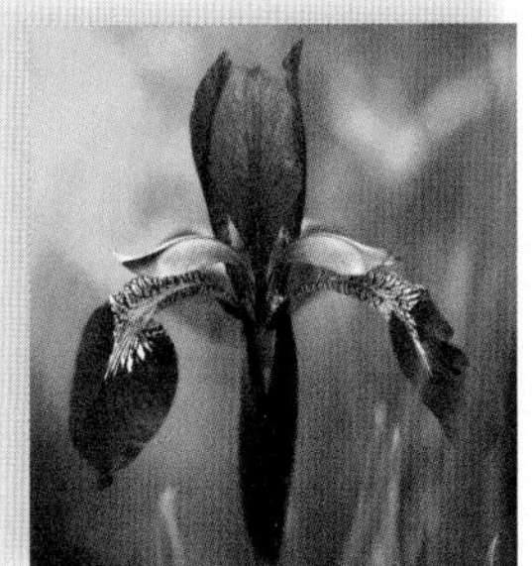
Blue flag iris, *Iris versicolor*

Snow trillium, *Trillium nivale*

Apple blossom, *Malus domestica*

Butterfly weed, *Asclepias tuberosa*

Figure 29.15 Flower diversity.
Regardless of size and shape, flowers share certain features.

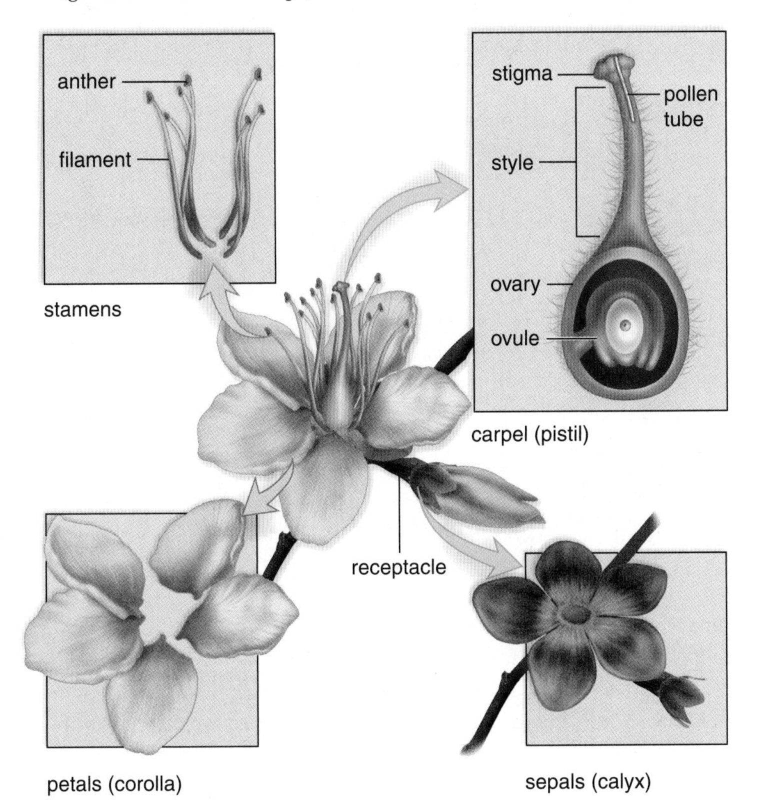

Figure 29.16 Generalized flower.
A flower has four main parts: sepals, petals, stamens, and carpels. Each stamen has an anther and a filament. A carpel has a stigma, a style, and an ovary. The ovary contains ovules.

the other flower parts. These parts, called **sepals, petals, stamens,** and **carpels,** are attached to the receptacle in whorls (circles) (Fig. 29.16). The sepals, collectively called the **calyx,** protect the flower bud before it opens. The sepals may drop off or may be colored like the petals. Usually, however, sepals are green and remain attached to the receptacle. The petals, collectively called the **corolla,** are quite diverse in size, shape, and color. The corolla often attracts a particular pollinator.

Each stamen consists of a saclike **anther,** where pollen is produced, and a stalk called a **filament.** In most flowers, the anther is positioned where the pollen can be carried away by wind or a pollinator. One or more carpels (sometimes called pistils) is at the center of a flower. A carpel has three major regions: ovary, style, and stigma. The swollen base is the **ovary,** which contains from one to hundreds of ovules. The **style** elevates the **stigma,** which is sticky or otherwise adapted to receive pollen grains. Glands in the region of the ovary produce nectar, a nutrient that pollinators gather as they go from flower to flower. The flower explains the success of the angiosperms.

Life Cycle, Adaptations, and Uses of Flowering Plants Figure 29.17 depicts the life cycle of a typical flowering plant. Sexual reproduction in flowering plants is dependent on the flower, which produces both pollen and seeds. A gymnosperm, such as a pine tree, you will recall, depends wholly on the wind to disperse both pollen and seeds. In contrast, among angiosperms, some species have windblown pollen, and others rely on a pollinator to carry pollen to another member of the same species. Bees, wasps, flies, butterflies, moths, beetles, and even bats are pollinators. The flower provides the pollinator with nectar, a sugary substance that, in the case of bees, will become honey in the hive. Several species of bees (such as honeybees and bumblebees) also have two "pollen baskets," one on each of their hind legs. When bees make a brief stop on a flower, they suck up nectar with specialized mouthparts and gather pollen grains, which are stored in their pollen baskets. Honey and pollen are the main diets of immature bees (called larvae). When a bee visits another flower, some of the pollen grains are rubbed off, and in this way the bee delivers pollen from flower to flower. Other types of pollinators also carry pollen between flowers of

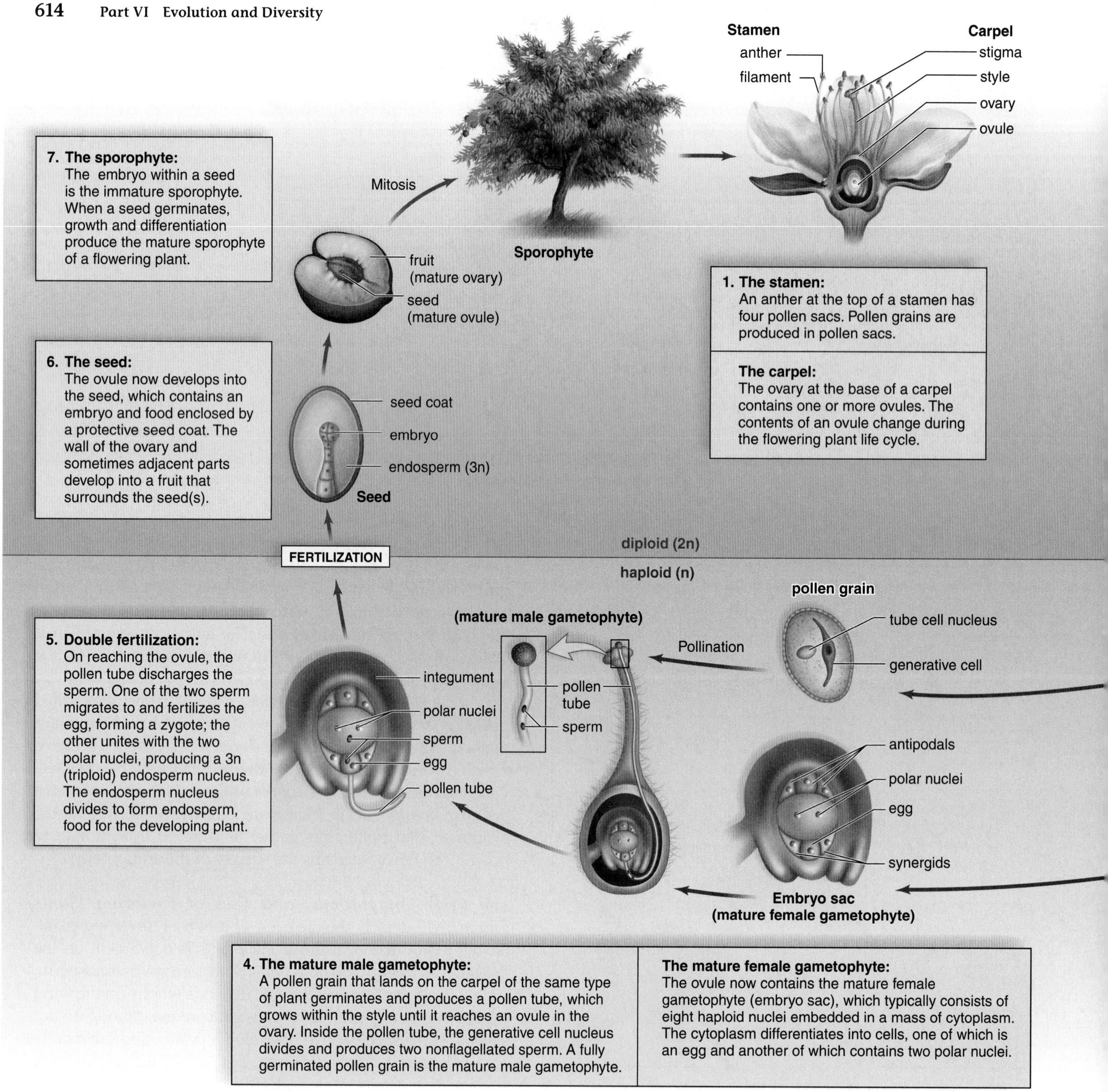

Stamen
anther
filament
Carpel
stigma
style
ovary
ovule

1. The stamen:
An anther at the top of a stamen has four pollen sacs. Pollen grains are produced in pollen sacs.

The carpel:
The ovary at the base of a carpel contains one or more ovules. The contents of an ovule change during the flowering plant life cycle.

pollen grain
tube cell nucleus
generative cell
Pollination
(mature male gametophyte)
pollen tube
sperm
antipodals
polar nuclei
egg
synergids
Embryo sac
(mature female gametophyte)

4. The mature male gametophyte:
A pollen grain that lands on the carpel of the same type of plant germinates and produces a pollen tube, which grows within the style until it reaches an ovule in the ovary. Inside the pollen tube, the generative cell nucleus divides and produces two nonflagellated sperm. A fully germinated pollen grain is the mature male gametophyte.

The mature female gametophyte:
The ovule now contains the mature female gametophyte (embryo sac), which typically consists of eight haploid nuclei embedded in a mass of cytoplasm. The cytoplasm differentiates into cells, one of which is an egg and another of which contains two polar nuclei.

integument
polar nuclei
sperm
egg
pollen tube

5. Double fertilization:
On reaching the ovule, the pollen tube discharges the sperm. One of the two sperm migrates to and fertilizes the egg, forming a zygote; the other unites with the two polar nuclei, producing a 3n (triploid) endosperm nucleus. The endosperm nucleus divides to form endosperm, food for the developing plant.

FERTILIZATION
diploid (2n)
haploid (n)

seed coat
embryo
endosperm (3n)
Seed

6. The seed:
The ovule now develops into the seed, which contains an embryo and food enclosed by a protective seed coat. The wall of the ovary and sometimes adjacent parts develop into a fruit that surrounds the seed(s).

fruit
(mature ovary)
seed
(mature ovule)
Mitosis
Sporophyte

7. The sporophyte:
The embryo within a seed is the immature sporophyte. When a seed germinates, growth and differentiation produce the mature sporophyte of a flowering plant.

Figure 29.17
Flowering plant life cycle.
The parts of the flower involved in reproduction are the stamens and the carpel. Reproduction has been divided into significant stages: female gametophyte development, male gametophyte development, and sporophyte development.

the same species, because pollinators and flowers are adapted to one another.

Flowering plants that have windblown pollen are usually not showy, whereas insect-pollinated flowers and bird-pollinated flowers are often colorful. Many flowers are adapted to attract quite specific pollinators. For example, bee-pollinated flowers are usually blue or yellow and have ultraviolet shadings that lead the pollinator to the location of nectar at the base of the flower. Night-blooming flowers are usually aromatic and white or cream-colored, so that their smell alone, rather than color, can attract nocturnal pollinators such as bats.

2. The microsporangia: Pollen sacs of the anther are microsporangia, where each microsporocyte undergoes meiosis to produce four microspores.	**The megasporangium:** First, an ovule within an ovary contains a megasporangium, where a megasporocyte undergoes meiosis to produce four megaspores.

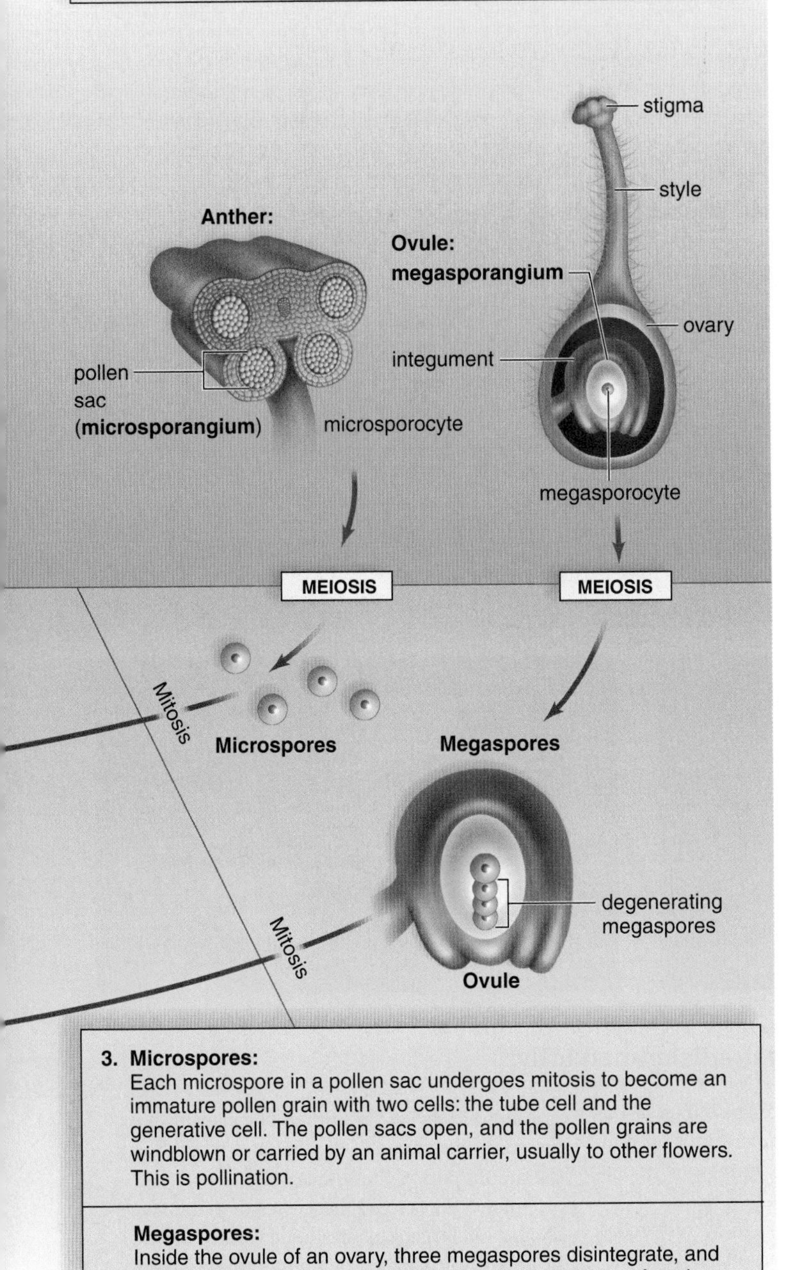

3. Microspores:
Each microspore in a pollen sac undergoes mitosis to become an immature pollen grain with two cells: the tube cell and the generative cell. The pollen sacs open, and the pollen grains are windblown or carried by an animal carrier, usually to other flowers. This is pollination.

Megaspores:
Inside the ovule of an ovary, three megaspores disintegrate, and only the remaining one undergoes mitosis to become a female gametophyte.

Fruits, the final product of a flower, aid in the dispersal of seeds (Fig. 29.18). The production of fruit is another advantage angiosperms have over gymnosperms. Dry fruits, such as pods, assist the dispersal of windblown seeds. Mature pods sometimes explode and shoot out the seeds. Other pods, such as those of peas and beans, simply break open and then the seeds are scattered. The infamous dandelion produces a one-seeded fruit with an umbrella-like crown of intricately branched hairs. The slightest gust of wind catches the hairs, raising and propelling the fruit into the air like a parachute.

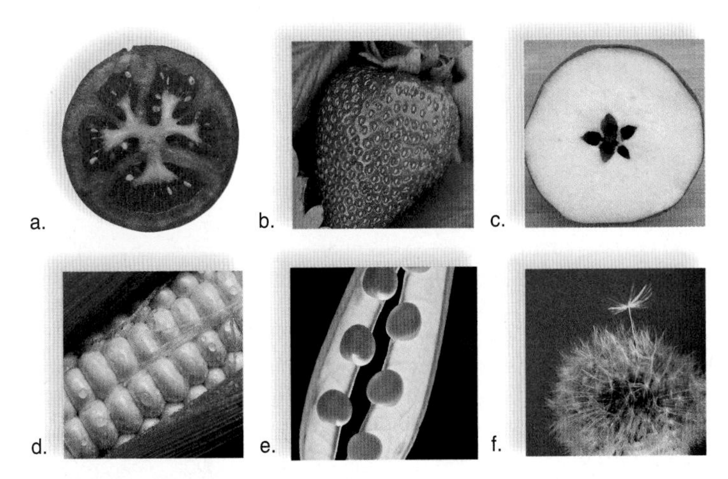

Figure 29.18 Fruits.
Angiosperms have diverse fruits. **a.** A tomato is technically a berry. **b.** A strawberry is an aggregate fruit formed from many carpels within a single flower. **c.** Apples have seeds inside a fleshy fruit. **d.** A corn cob carries many fruits (kernels). **e.** Peas have seeds in pods. **f.** Dandelion fruits are carried by wind.

Fruits also help disperse angiosperm seeds in two other ways. One means is exemplified by the coconut, a fruit that can drift for thousands of kilometers across seas and oceans. A second method of seed dispersal occurs because the seeds of some fruits have a fleshy covering that animals eat as a source of food. Only the fleshy part is digested, so the stones or seeds pass through the animal's digestive system and are deposited perhaps far away from the parent plant. Still other fruits, like those of the cocklebur, have seeds with hooks that catch on the fur of animals and are carried away.

Check Your Progress

1. What two types of spores occur in the seed plant life cycle, where do they develop, and what role do they play in the life cycle?
2. What features of the flowering plant life cycle are not found in any other group of plants?

Applying the Concepts Revisited

If we suddenly lost all of the plants in the world, the animals would soon follow. Plants, and particularly angiosperms, provide us important products, such as food, clothing, and medicines. Land plants also provide shelter for many animals, oxygen to the air, and help buffer noise in cities. People also often appreciate the aesthetic value of land plants, such as that provided by the seclusion and peacefulness of walking on a trail through a forest.

1. Why would animals have evolved to live on land after plants did?
2. Name three major problems solved by plants during their evolution to be adapted to a land existence.
3. How has symbiosis evolved in plants and their pollinators to the benefit of both?

Ecology Focus

Plants: Could We Do Without Them?

Humans derive most of their nourishment from three flowering plants: wheat, corn, and rice. All three of these plants are members of the grass family and are collectively called grains (Fig. 29A). Wheat is commonly used to produce flour and bread. It was first cultivated in the Middle East about 8000 B.C., and is thought to be one of the earliest cultivated plants. Early settlers brought wheat to North America, where many varieties of corn, more properly called maize, were already in existence. Rice originated in southeastern Asia several thousand years ago, where it grew in swamps. Many foods of both plant and animal origin are considered bland or tasteless without seasonings or spices. The Americas were discovered when Columbus was seeking a new route to the Far East to acquire spices, such as nutmeg, oregano, rosemary, and sage. In addition, our primaty sweetener, sugar, comes almost exclusively from two plants—sugarcane (grown in South America, Africa, Asia, and the Caribbean) and sugar beets (grown mostly in Europe and North America). Our most popular drinks—coffee, tea, and cola—also come from flowering plants. Coffee originated in Ethiopia, whereas tea is thought to have been first used somewhere in central Asia. Cotton and rubber are two plants that still have many uses today (Fig. 29B). Until a few decades ago, cotton and other natural fibers were our only source of clothing. Interestingly, when Levi Strauss wanted to make a tough pair of jeans, he needed a stronger fiber than cotton, so he used hemp. Hemp comes from a plant that is closely related to marijuana, but unlike marijuana, has extremely small amounts of the chemical THC that causes the hallucinogenic effect when marijuana is smoked. Today, hemp is increasingly used to make clothing due to its toughness and wearability. Rubber had its origin in Brazil from the thick, white sap of the rubber tree. To produce a stronger rubber, such as that in tires, sulfur is added, and the sap is heated in a process called vulcanization. This produces a flexible material less sensitive to temperature changes. However, at present, most rubber is synthetically produced.

Plants have also been used for centuries for a number of important household items, including the house itself. We are most familiar with lumber as the major structural portion in buildings. This wood comes mainly from a variety of conifers: pine, fir, and spruce, among others. In the tropics, trees and even herbs provide important components for houses. In rural parts of Central and South America, palm leaves are preferable to tin for roofs, because they last as long as ten years and are quieter during rainstorms. In the Middle East, numerous houses along rivers are made entirely of reeds.

Today, plants are increasingly researched for their use in medicinal products. Currently, about 50% of all pharmaceutical drugs originate from plants or are derived from plant products. Some cancers are even treatable with medicinal plants, such as the rosy periwinkle, extracts from which has shown some success in treating childhood leukemia. Indeed, the National Cancer Institute and most pharmaceutical companies have spent millions of dollars to send botanists out to collect and test plant samples from around the world. Tribal medicine men, or shamans, of South America and Africa have already been of great importance in developing numerous drugs. However, plant extracts also continue to be misused for their hallucinogenic or other effects on the human body; examples are coca, the source of cocaine and crack, and the opium poppy, the source of heroin.

In addition to all these uses of plants, we should not forget their aesthetic value. Flowers brighten any yard, ornamental plants accent landscaping, and trees provide cooling shade during the summer as well as shelter from harsh winds during the winter.

Wheat plants, *Triticum*

Rice plants, *Oryza*

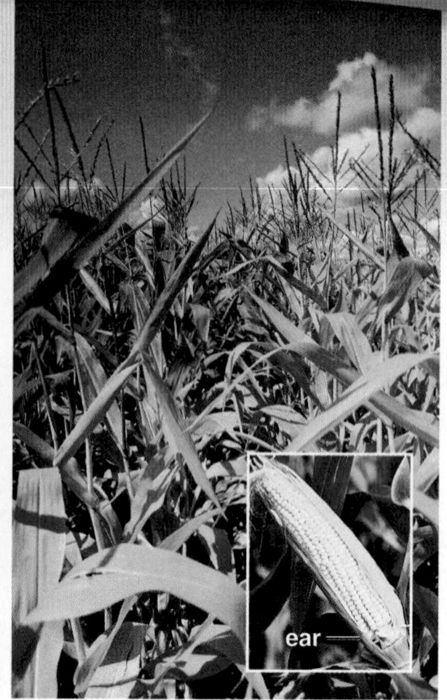

Corn plants, *Zea*

Figure 29A Plants used for food.

Rubber tree, *Hevea*

Cotton, *Gossypium*

Figure 29B Plants used commercially.

Discussion Questions

1. Why do you think plants are such a good source of drugs for human use? What advantage does this give the plants?
2. With over 250,000 species of angiosperms alone, why do you think we get most of our nourishment from only three plants (wheat, corn, and rice)?
3. Can you think of any uses of plants other than those discussed above?

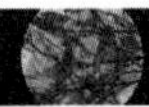

SUMMARIZING THE CONCEPTS

29.1 Evolutionary History of Plants

- Plants are multicellular, photosynthetic organisms adapted to a land existence that probably evolved from a multicellular, freshwater green alga about 500 MYA.
- Five significant events associated with plant adaptation to a land existence are: (1) embryo protection, (2) vascular tissue, (3) megaphylls, (4) seeds, and (5) the flower.
- The life cycle of plants is characterized by alternation of generations—some have a dominant gametophyte (haploid generation) and others have a dominant sporophyte (diploid generation).
- Land plants are divided into nonvascular plants, seedless vascular plants, and seed plants based on presence/absence of vascular tissue and their dispersal mechanism.

29.2 Nonvascular Plants

- Nonvascular plants, including liverworts and mosses, lack vascular tissue—and therefore, they do not have true roots, stems, or leaves.
- The gametophyte is dominant. Antheridia produce swimming sperm that use external water to reach eggs in the archegonia. Following fertilization, the dependent moss sporophyte consists of a foot, a stalk, and a capsule within which windblown spores are produced by meiosis.
- Use of windblown spores to disperse the gametophyte is nonvascular plants' chief adaptation for reproducing on land. Each spore germinates to produce a gametophyte.

29.3 Seedless Vascular Plants

- Vascular plants have vascular tissue—xylem and phloem. Xylem transports water and minerals and also supports plants against the pull of gravity. Phloem transports organic nutrients from one part of the plant to another.
- Lycophytes have true roots, stems, and leaves because they contain vascular tissue. Lycophytes have narrow leaves called microphylls.
- Lycophytes and ferns and their allies (whisk ferns and horsetails) were prominent in swamp forests of the Carboniferous period.
- The sporophyte is dominant; in the life cycle of seedless vascular plants, windborne spores disperse the gametophyte, and the separate gametophyte produces flagellated sperm within antheridia and eggs within archegonia.
- Vegetative (asexual) reproduction is sometimes used to disperse ferns in dry habitats.

29.4 Seed Plants

- Seed plants have a life cycle that is fully adapted to existence on land.
- The spores develop inside the body of the sporophyte and are of two types. The male gametophyte is the pollen grain, which produces nonflagellated sperm. The pollen grain is windblown or carried by an animal in the case of angiosperms, whereas flagellated sperm of seedless vascular plants require an outside source of moisture to reach an egg. The egg is located within an ovule and produced by the female gametophyte.
- Gymnosperms, which include the conifers, produce seeds that are uncovered (naked). Male gametophytes develop in pollen cones. The female gametophyte develops within an ovule located on the scales of seed cones. Following pollination and fertilization, the ovule becomes a winged seed dispersed by wind. Seeds contain the next sporophyte generation.
- Angiosperms, also called the flowering plants, produce seeds covered by fruits. The petals of flowers attract pollinators, and the ovary develops into a fruit, which aids seed dispersal. Angiosperms provide most of the food that sustains terrestrial animals, and they are the source of many products used by humans.

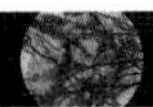

TESTING YOURSELF

Choose the best answer for each question.

1. The spores produced by a plant are
   a. haploid and genetically different from each other.
   b. haploid and genetically identical to each other.
   c. diploid and genetically different from each other.
   d. diploid and genetically identical to each other.
2. The gametophyte is the dominant generation in
   a. ferns. c. gymnosperms.
   b. mosses. d. angiosperms.
3. Label the stages in the following diagram of the plant life cycle.

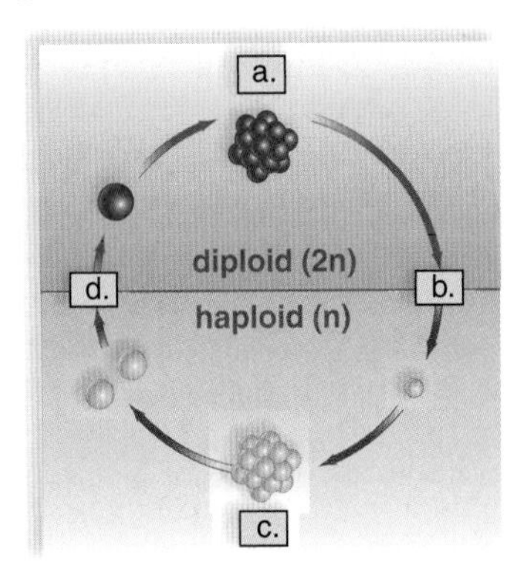

4. In mosses, meiosis occurs in
   a. antheridia.
   b. archegonia.
   c. capsules.
   d. protonema.

For questions 5–10, match the description with the appropriate plant group in the key. Some answers can be used more than once. Some questions will have more than one answer.

Key:

a. nonvascular plants
b. seedless vascular plants
c. gymnosperms
d. angiosperms

5. Contain xylem and phloem
6. Produce seeds
7. Produce flagellated sperm
8. Gametophyte is separate from sporophyte
9. Produce seeds on scales
10. Produce flowers

11. In contrast to seed plants, seedless vascular plants
    a. are not heterosporous.
    b. produce gametophytes.
    c. have male gametophytes that travel to the female gametophytes.
    d. produce ovules.
12. Which statement about the conifer life cycle is false?
    a. Meiosis produces spores.
    b. The gametophyte is the dominant generation.
    c. The seed is a dispersal agent.
    d. Male gametophytes are carried by the wind.
13. Cycads are in danger of extinction because
    a. they occupy very specialized niches.
    b. they have unusual mineral nutrient requirements.
    c. they grow very slowly.
    d. their natural habitats are becoming too warm.
14. Label the parts of the flower in the following diagram.

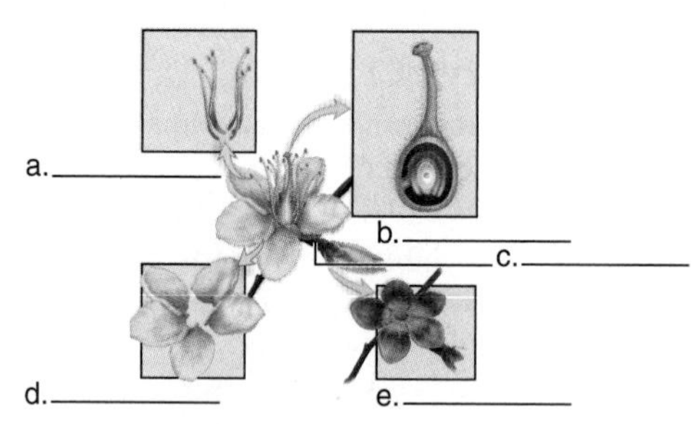

15. Unlike eudicots, monocots have
    a. woody tissue.
    b. two seed leaves.
    c. scattered vascular bundles in their stems.
    d. flower parts in multiples of fours or fives.

UNDERSTANDING THE TERMS

alternation of generations 602
angiosperm 612
anther 613
antheridium 606
archegonium 606
calyx 613
carpel 613
club moss 607
conifer 611
corolla 613
cycad 611
eudicot (Eudicotyledones) 612
fern 607
filament 613
fruit 615
gametophyte 602
ginkgo 612
gnetophyte 612
gymnosperm 611
horsetail 609
liverwort 604
megaphylls 602
monocot (Monocotyledones) 612
moss 605
nonvascular plant 604
ovary 613
ovule 611
petal 613
phloem 606
pollen cone 611
pollen grain 611
pollen tube 611
pollination 611
pollinator 611
receptacle 612
seed 602
seed cone 611
sepal 613
spore 602
sporophyte 602
stamen 613
stigma 613
style 613
vascular plant 606
vascular tissue 602
xylem 606

THINKING CRITICALLY

1. What characteristics of angiosperms have allowed this group of plants to dominate Earth?
2. Compare and contrast the plant alternation of generation life cycle with the human life cycle.

INQUIRY INTO LIFE WEBSITE

The companion website for *Inquiry into Life* provides a wealth of information organized and integrated by chapter. You will find practice tests, animations, videos, and much more that will complement your learning and understanding of general biology.

http://www.mhhe.com/maderinquiry13

Animals: Part I

Applying the Concepts

Insects are the most diverse of the invertebrates, numbering over 1 million species. Although insects are often viewed as pests, the vast majority of species are not pests and the lifestyles of many are fascinating.

Take, for example, the termite, a colonial insect. Although termites are generally thought of as wood-eating nuisances, there are over 3,000 species, most of which are not pests! As a group, termites are over 200 million years old, and some species are capable of building mounds that are the strongest and tallest (20 ft high, equivalent to a 2,000-ft building) nonhuman structures on Earth. These mounds are often found away from human dwellings and thus do not interfere with human activites.

There are typically three social classes of termites (all born from the same queen, which can lay 3 million eggs per year): workers build, soldiers defend, and harvesters help to feed the colony. Harvesters can recycle nutrients to improve soil fertility. Within the mound, harvesters actually cultivate fungi, which serve as food for the nest and may also have antibacterial properties that prevent infection. Workers build ventilation shafts that function like air-conditioning units, keeping the mound at a constant temperature. In the meantime, soldiers will vigorously defend the mound against any intruders. In all, there are often more than 1 million termites in a mound, all working together to keep things running smoothly.

Termites are an example of the fascinating diversity of invertebrates throughout the world. In this chapter, you will learn about the lifestyles and features of some of the major phyla of invertebrates.

Chapter Outline

30.1 Evolutionary Trends Among Animals

Learning Outcomes

1. List the five general characteristics of animals.
2. Describe the differences between protostomes and deuterostomes.

Animals are members of the domain Eukarya and the kingdom Animalia. Animals are extremely diverse (Fig. 30.1), and can be contrasted with plants and fungi. All three groups are multicellular eukaryotes, but unlike plants, which make their food through photosynthesis, animals are heterotrophs, and must acquire nutrients from an external source. Fungi digest their food externally and absorb the breakdown products, whereas animals ingest (eat) whole food and digest it internally. In general, animals share the following characteristics:

1. Typically have the power of movement or locomotion by means of muscle fibers
2. Multicellular; most have specialized cells that form tissues and organs
3. Have a life cycle in which the adult is typically diploid
4. Usually undergo sexual reproduction and produce an embryo that goes through developmental stages
5. Heterotrophic; usually acquire food by ingestion followed by digestion

The more than 35 animal phyla are hypothesized to have evolved from a protistan ancestor approximately 700 million years ago. This chapter will consider only groups of **invertebrates** which are animals without an endoskeleton of bone or cartilage. The invertebrates which evolved first far outnumber the vertebrates, which do have such an endoskeleton. There is no adequate fossil record to trace the early evolution of animals. Therefore, the evolutionary tree shown in Figure 30.2, is based on molecular and anatomical data. It is assumed that the more closely related two organisms are, the more nucleotide sequences they will have in common. The anatomical data are discussed next and outlined in Table 30.1.

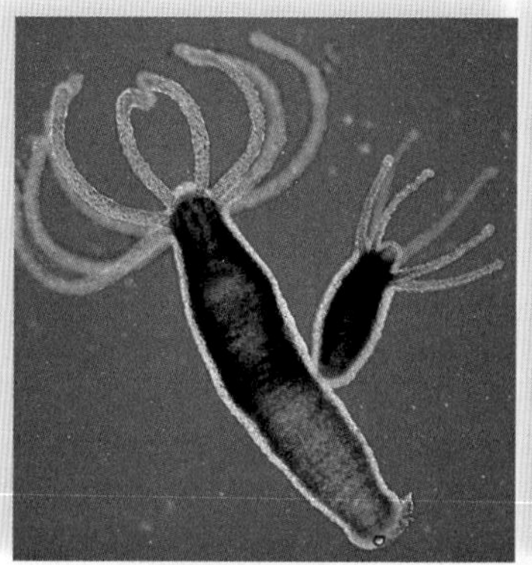

Hydra, *Hydra*

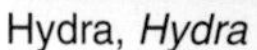

Green crayfish, *Barbi*

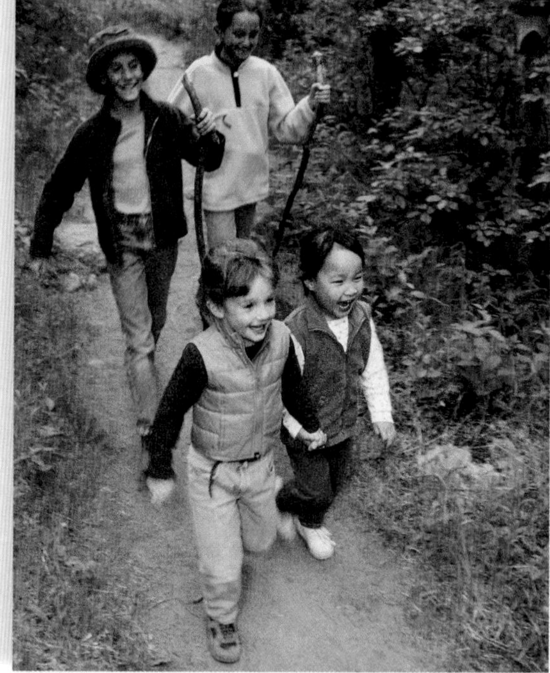

Human being, *Homo*

Figure 30.1 Animal diversity.
Hydras, crayfish, and humans are all multicellular heterotrophic animals, but they differ in certain fundamental ways. In terms of level of organization, hydras have only tissues, but crayfish and humans have organ systems. Hydras do not have an identifiable head, whereas crayfish and humans do. Hydras do not have a complete digestive system, whereas crayfish and humans have a complete digestive tract surrounded by a body wall. Crayfish and humans share other features, even though one is an invertebrate and the other is a vertebrate. They both have a skeleton with jointed appendages, but the crayfish skeleton is external, while the human skeleton is internal.

Anatomical Data

Refer to Figure 30.2 as we discuss the anatomical characteristics substantiating the molecular data used to construct the tree.

Type of Symmetry

Animals can be asymmetrical, radially symmetrical, or bilaterally symmetrical. **Asymmetrical** animals have no particular symmetry, such as some species of sponges. **Radial symmetry** means that the animal is organized circularly, similar to a wheel, so that no matter how the animal is sliced longitudinally, mirror images are obtained. **Bilateral symmetry** means that the animal has definite right and left halves; only a longitudinal cut down the center of the animal will produce a mirror image.

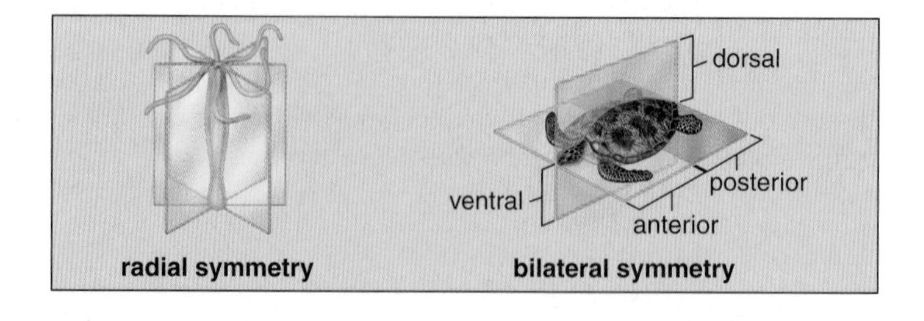

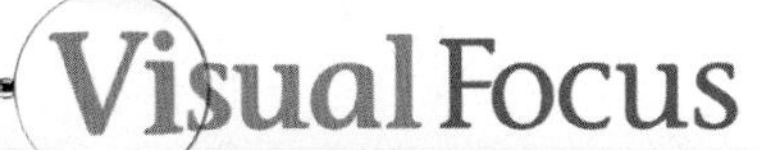

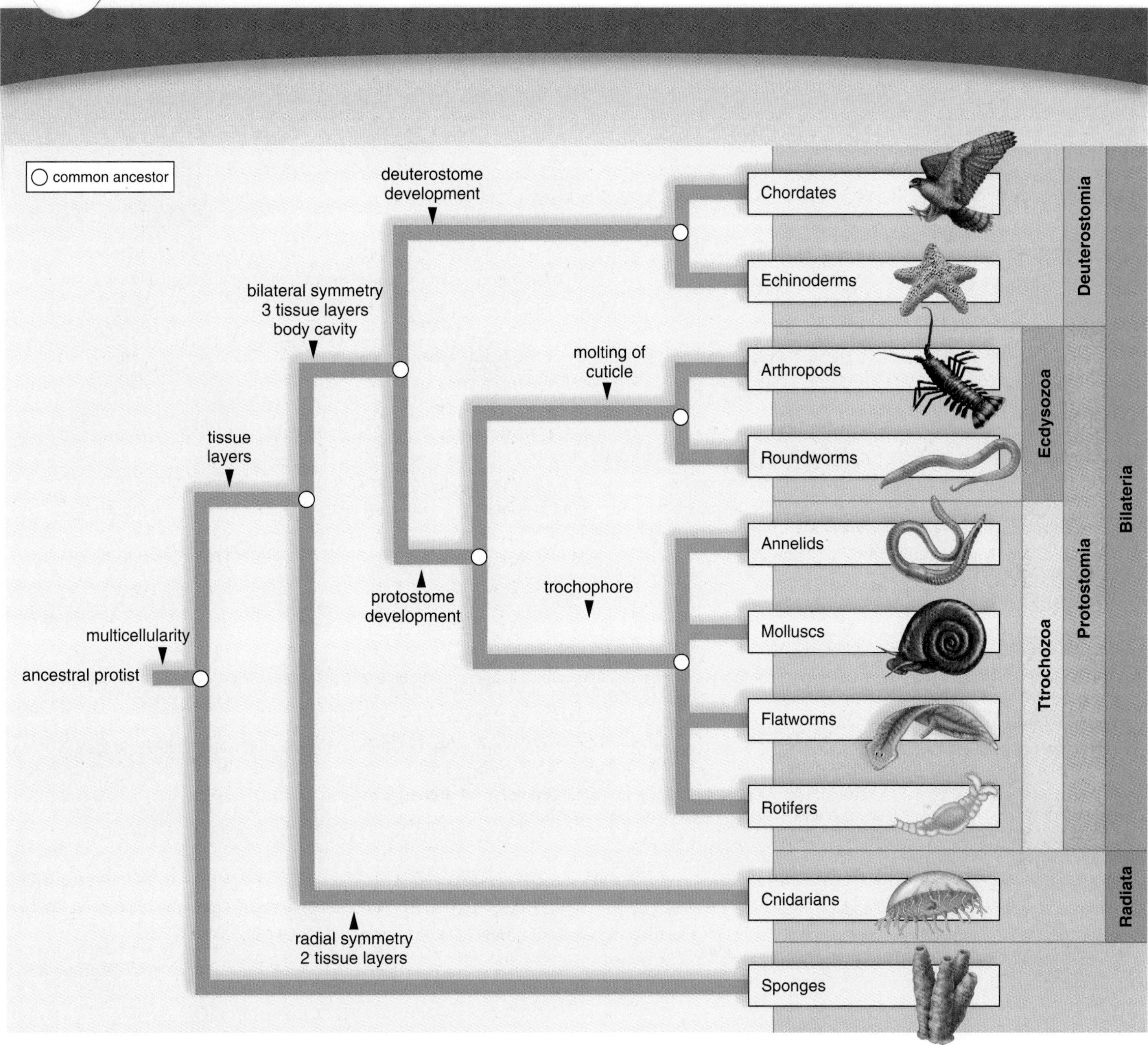

Figure 30.2 Phylogenetic tree of animals.
All animal phyla living today are most likely descended from an ancestral protist living about 700 million years ago. This evolutionary tree is based on molecular data (nucleotide sequences) and anatomical data used to indicate which phyla are most closely related to one another.

TABLE 30.1 Animal Characteristics

DOMAIN: Eukarya
KINGDOM: Animals

CHARACTERISTICS
Multicellular, usually with specialized tissues; ingest or absorb food; diploid life cycle

INVERTEBRATES

Sponges (bony, glass, spongin):* Asymmetrical, saclike body perforated by pores; internal cavity lined by choanocytes; spicules serve as internal skeleton. 5,150+

Radial Symmetry

Cnidarians (hydra, jellyfish, corals, sea anemones): Radially symmetrical with two tissue layers; incomplete digestive tract; tentacles with nematocysts. 10,000+

Protostomes (Trochozoa)

Flatworms (planarians, tapeworms, flukes): Bilateral symmetry with cephalization; three tissue layers and organ systems; acoelomate with incomplete digestive tract that can be lost in parasites; hermaphroditic. 20,000+

Rotifers (wheel animals): Microscopic animals with a corona (crown of cilia) that looks like a spinning wheel when in motion. 2,000+

Molluscs (chitons, clams, snails, squids): Coelom, all have a foot, mantle, and visceral mass; foot is variously modified; in many, the mantle secretes a calcium carbonate shell as an exoskeleton; true coelom and all organ systems. 110,000+

Annelids (polychaetes, earthworms, leeches): Segmented with body rings and setae; cephalization in some polychaetes; hydroskeleton; closed circulatory system. 16,000+

Protostomes (Ecdysozoa)

Roundworms (*Ascaris*, pinworms, hookworms, filarial worms): Pseudocoelom and hydroskeleton; complete digestive tract; free-living forms in soil and water; parasites common. 25,000+

Arthropods (crustaceans, spiders, scorpions, centipedes, millipedes, insects): Chitinous exoskeleton with jointed appendages undergoes molting; insects—most have wings—are most numerous of all animals. 1,000,000+

Deuterostomes

Echinoderms (sea stars, sea urchins, sand dollars, sea cucumbers): Radial symmetry as adults; unique water-vascular system and tube feet; endoskeleton of calcium plates. 7,000+

Chordates (tunicates, lancelets, vertebrates): All have notochord, dorsal tubular nerve cord, pharyngeal pouches, and postanal tail at some time; contains mostly vertebrates in which notochord is replaced by vertebral column. 56,000+

VERTEBRATES

Fishes (jawless, cartilaginous, bony): Endoskeleton, jaws, and paired appendages in most; internal gills; single-loop circulation; usually scales. 28,000+

Amphibians (frogs, toads, salamanders): Most have jointed limbs; lungs; three-chambered heart with double-loop circulation; moist, thin skin. 5,383+

Reptiles (snakes, turtles, crocodiles): Amniotic egg; rib cage in addition to lungs; three- or four-chambered heart typical; scaly, dry skin; copulatory organ in males and internal fertilization. 8,000+

Birds (songbirds, waterfowl, parrots, ostriches): Endothermy, feathers, and skeletal modifications for flying; lungs with air sacs; four-chambered heart. 10,000+

Mammals (monotremes, marsupials, placental): Hair and mammary glands. 4,800+

*After a character is listed, it is present in the rest, unless stated otherwise.
+Number of species.

Radially symmetrical animals have the advantage of being able to reach out in all directions from one center. Some radially symmetrical animals are attached to a substrate—that is, they are **sessile.** Others are free floating, such as jellyfish. Bilaterally symmetrical animals tend to be more active, with movement directed toward an anterior end. During the evolution of animals, bilateral symmetry is accompanied by **cephalization,** localization of a brain and specialized sensory organs at the anterior end.

Embryonic Development

Sponges are multicellular like all animals, but unlike other animals, they do not have true tissues. Therefore, sponges have the *cellular level of organization. True tissues* appear in the other animals as they undergo embryological development. The first three tissue layers are often called germ layers because they give rise to the organs and organ systems of complex animals. Animals such as the cnidarians, which have only two tissue layers (ectoderm and endoderm) as embryos, are diploblastic with the tissue level of organization. Those animals that develop further and have all three tissue layers (ectoderm, mesoderm, and endoderm) as embryos are triploblastic and have the organ level of organization. Notice in the phylogenetic tree that the animals with three tissue layers are either **protostomes** [Gk. *proto*, first; *stoma*, mouth] or **deuterostomes** [Gk. *deuter*, second; *stoma*, mouth].

Figure 30.3 shows that protostome and deuterostome development are differentiated by three major events:

1. Cleavage, the first event of development, is cell division without cell growth. In protostomes, spiral cleavage occurs, and daughter cells sit in grooves formed by the previous cleavages. The fate of these cells is fixed and determinate in protostomes; each can contribute to development in only one particular way. In deuterostomes, radial cleavage occurs, and the daughter cells sit right on top of the previous cells. The fate of these cells is indeterminate—that is, if they are separated from one another, each cell can go on to become a complete organism.
2. As development proceeds, a hollow sphere of cells, or blastula, forms and the indentation that follows produces an opening called the blastopore. In protostomes, the mouth appears at or near the blastopore, hence the origin of their name. In deuterostomes, the anus appears at or near the blastopore, and only later does a second opening form the mouth, hence the origin of their name.
3. Certain members of the protostomes and all deuterostomes have a body cavity completely lined by mesoderm, called a true coelom. However, a true coelom develops differently in the two groups. In protostomes, the mesoderm arises from cells located near the embryonic blastopore, and a splitting occurs that produces the coelom. In deuterostomes, the coelom arises as a pair of mesodermal pouches from the wall of the primitive gut. The pouches enlarge until they meet and fuse.

The deuterostomes include the echinoderms and the chordates, two groups of animals that will be examined in detail in Chapter 31. The protostomes are divided into the two groups: the **ecdysozoa** and the **trochozoa.** The ecdysozoans include the roundworms and arthropods. Both of these types of animals molt; they shed their outer covering as they grow. Ecdysozoa means molting animals. The trochozoa contain **trochophores,** which either have presently or their ancestors had a trochophore larva. A trochophore larva has many cilia to allow free-swimming in the planktonic marine environment.

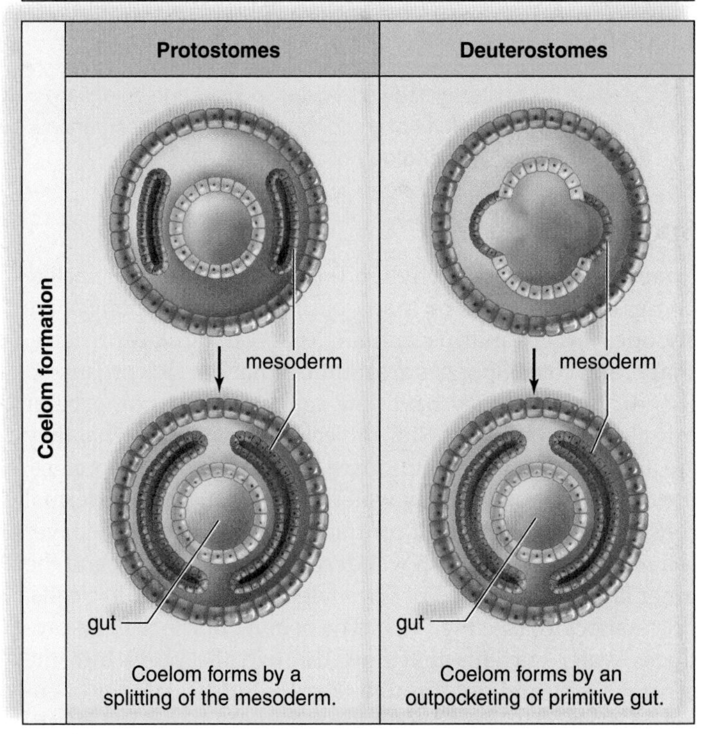

Figure 30.3 Protostomes compared to deuterostomes.

Left: In the embryo of protostomes, cleavage is spiral—new cells are at an angle to old cells—and each cell has limited potential and cannot develop into a complete embryo; the blastopore is associated with the mouth; and the coelom, if present, develops by a splitting of the mesoderm. *Right:* In deuterostomes, cleavage is radial—new cells sit on top of old cells—and each one can develop into a complete embryo; the blastopore is associated with the anus; and the coelom, if present, develops by an outpocketing of the primitive gut.

Check Your Progress

1. List five characteristics that animals have in common.
2. With reference to figure 30.2, list the characteristics that pertain to arthropods.

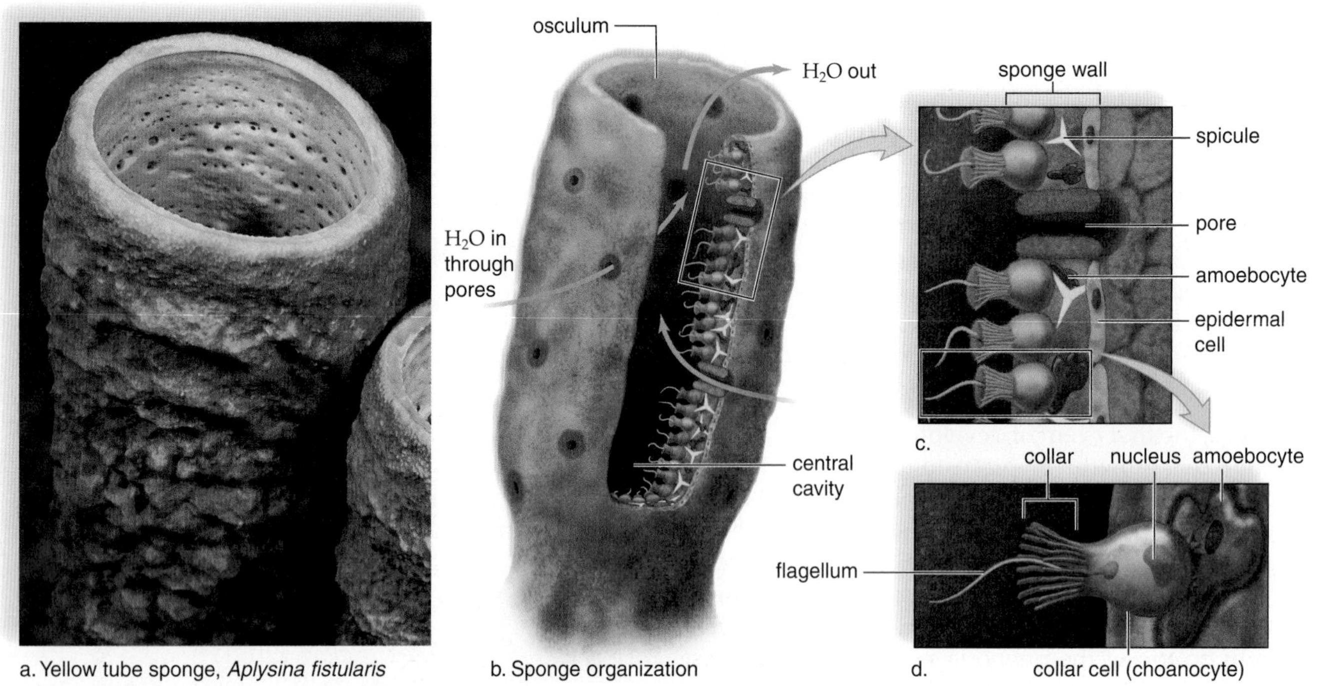

Figure 30.4 Simple sponge anatomy.
a. A simple sponge. **b.** The wall of a sponge contains two layers of cells, the outer epidermal cells and the inner collar cells as shown in **(c).** The collar cells (**d,** enlarged) have flagella that beat, moving the water through pores as indicated by the blue arrows in **(b).** Food particles in the water are trapped by the collar cells and digested within their food vacuoles. Amoebocytes transport nutrients from cell to cell. Spicules form an internal skeleton in some sponges.

30.2 Introducing the Invertebrates

Learning Outcomes

1. Describe the anatomy and life cycle of sponges and cnidarians.
2. Contrast the way of life of a sponge with that of a cnidarian.
3. Recall the basic morphology of a hydra.

Sponges

Sponges are placed in phylum Porifera because their saclike bodies are perforated by many pores (Fig. 30.4). Sponges are aquatic, largely marine animals that vary greatly in size, shape, and color. Sponges are multicellular but lack organized tissues. Actually, they have few cell types and no nerve or muscle cells to speak of. Still, molecular data suggest that they are at the base of the evolutionary tree of animals. In sponges, the outer layer of the body wall contains flattened epidermal cells, some of which have contractile fibers; the middle layer is a semifluid matrix with wandering amoeboid cells; and the inner layer is composed of flagellated cells called **collar cells,** or choanocytes (see Fig. 30.4). The beating of the flagella produces water currents that flow through the pores into the central cavity and out through the osculum, the upper opening of the body. Even a simple sponge only 10 cm tall is estimated to filter as much as 100 liters of water each day. It takes this much water to supply the needs of the sponge. A sponge is a sessile **filter feeder,** an organism that filters its food from the water by means of a straining device—in this case, the pores of the walls and the microvilli making up the collar of collar cells. Microscopic food particles that pass between the microvilli are engulfed by the collar cells and digested by them in food vacuoles or are passed to the amoeboid cells for digestion. The amoeboid cells also act as a circulatory device to transport nutrients from cell to cell, and they produce the sex cells (the egg and the sperm) and spicules.

Sponges can reproduce both asexually and sexually. They reproduce asexually by fragmentation followed by regeneration, gemmule formation, or budding. A gemmule is like a spore and is resistant to drying out, freezing, or the lack of oxygen. Gemmule formation occurs in freshwater sponges, and when conditions are favorable, a gemmule gives rise to an adult sponge. During sexual reproduction, sperm are released through the osculum and drawn into other sponges through the pores, where they fertilize eggs within the body. Most sponges are **hermaphroditic,** meaning they possess both male and female sex organs. However, they usually do not self-fertilize. After fertilization, the zygote develops into a flagellated larva that may swim to a new location. Sponges are capable of regeneration—if the cells of a sponge are separated, they are capable of reassembling and regenerating into a complete and functioning organism!

Sponges are classified on the basis of their skeleton. Some sponges have an internal skeleton composed of **spicules,** small needle-shaped structures with one to six rays. Chalk sponges have spicules made of calcium carbonate; glass sponges have spicules that contain silica. Most sponges have a skeleton composed of fibers of spongin, a soft protein. A bath sponge is the dried spongin skeleton from which all living tissue has been removed. Today, however, commercial "sponges" are usually synthetic. Note that the popular "loofah" is not really a sponge; it is actually a spongy plant related to cucumbers.

Check Your Progress

1. Describe three main characteristics of sponges.
2. Describe the sponge's way of life.

Cnidarians

These animals have true tissues and, as embryos, they have the two germ layers ectoderm and endoderm. They are radially symmetrical as adults, a type of symmetry with the advantages discussed on page 620.

Cnidarians (phylum Cnidaria) are tubular or bell-shaped animals that reside mainly in shallow coastal waters. However, there are some freshwater, brackish, and oceanic forms. The term cnidaria is derived from the presence of specialized stinging cells called **cnidocytes**. Each cnidocyte has a toxin-filled capsule called a **nematocyst** that contains a long, spirally coiled hollow thread. When the trigger of the cnidocyte is touched, the nematocyst is discharged. Some threads merely trap prey, and others have spines that penetrate and inject paralyzing venom.

The body of a cnidarian is a two-layered sac. The outer tissue layer is a protective epidermis derived from ectoderm. The inner tissue layer, which is derived from endoderm, secretes digestive juices into the internal cavity, called the **gastrovascular cavity** because it serves for digestion of food and circulation of nutrients. The fluid-filled gastrovascular cavity also serves as a supportive hydrostatic skeleton, so called because it offers some resistance to the contraction of muscle but permits flexibility. The two tissue layers are separated by mesoglea.

Two basic body forms are seen among cnidarians (Fig. 30.5*a*). The mouth of a **polyp** is directed upward, while the mouth of a jellyfish, or medusa, is directed downward. The bell-shaped medusa has more mesoglea than a polyp, and the tentacles are concentrated on the margin of the bell. At one time, both body forms may have been a part of the life cycle of all cnidarians. When both are present, the animal is dimorphic: the sessile polyp stage produces medusae by asexual budding, and the motile **medusan** stage produces egg and sperm. In some cnidarians, one stage is dominant and the other is reduced; in other species, one form is absent altogether.

Cnidarian Diversity

Cnidarians are quite diverse (Fig. 30.5*b–e*). Sea anemones (Fig. 30.5*b*) are sessile polyps that live attached to submerged rocks, timbers, or other substrate. Most sea anemones range in size from 0.5–20 cm in length and 0.5–10 cm in diameter and are often colorful. Their upward-turned oral disk that contains the mouth is surrounded by a large number of hollow tentacles containing nematocysts.

Corals (Fig. 30.5*c*) resemble sea anemones encased in a calcium carbonate (limestone) house. The coral polyp can extend into the water to feed on microorganisms and retreat into the house for safety. Some corals are solitary, but the vast majority live in colonies that vary in shape from rounded to branching. Many corals exhibit elaborate geometric designs and stunning colors and are responsible for the building of coral reefs. The slow accumulation of limestone can result in massive structures, such as the Great Barrier Reef along the eastern coast of Australia. Coral reef ecosystems are very productive, and an extremely diverse group of marine life call the reef home.

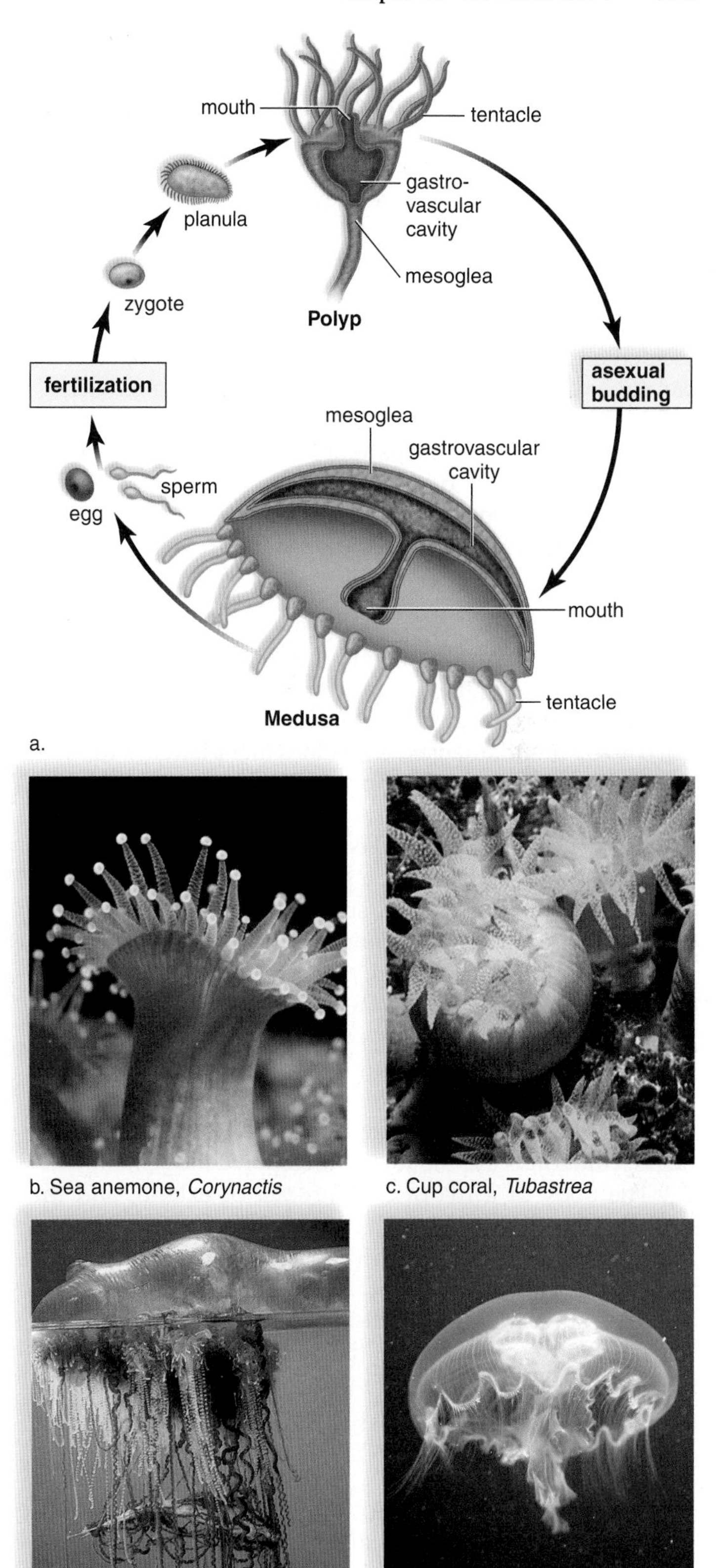

Figure 30.5 Cnidarian diversity.
a. The life cycle of a cnidarian. Some cnidarians have both a polyp stage and a medusa stage. In others, one stage may be dominant or absent altogether. **b.** The anemone, which is sometimes called the flower of the sea, is a solitary polyp. **c.** Corals are colonial polyps residing in a calcium carbonate or proteinaceous skeleton. **d.** The Portuguese man-of-war is a colony of modified polyps and medusae. **e.** True jellyfishes undergo the complete life cycle. This is the medusa stage. The polyp is small.

The hydrozoans have a dominant polyp. Hydra (see Fig. 30.6) is a hydrozoan, and so is a Portuguese man-of war (Fig. 30.5*d*). You might think the Portuguese man-of-war is an odd-shaped medusa, but actually it is a colony of polyps. The original polyp becomes a gas-filled float that provides buoyancy, keeping the colony afloat. Other polyps, which bud from this one, are specialized for feeding or for reproduction. A long, single tentacle armed with numerous nematocysts arises from the base of each feeding polyp. Swimmers who accidentally come upon a Portuguese man-of-war can receive painful, even serious, injuries from these stinging tentacles.

In true jellyfishes (Fig. 30.5*e*), the medusa is the primary stage, and the polyp remains small. Jellyfishes are zooplankton and depend on tides and currents for their primary means of movement. They feed on a variety of invertebrates and fishes and are themselves food for marine animals.

Hydra

The body of a hydra is a small tubular polyp about one-quarter inch in length. Hydras are often studied in biology classes and labs as an example of a cnidarian. Hydras are likely to be found attached to underwater plants or rocks in most lakes and ponds. The only opening (the mouth) is in a raised area surrounded by four to six tentacles that contain a large number of nematocysts.

The outer tissue layer is a protective epidermis derived from ectoderm. The inner tissue layer, derived from endoderm, is called a gastrodermis. The two tissue layers are separated by mesoglea. There are both circular and longitudinal muscle fibers. Nerve cells located below the epidermis, near the mesoglea, interconnect and form a nerve net that communicates with sensory cells throughout the body. The nerve net allows transmission of impulses in several directions at once. Because they have both muscle fibers and nerve fibers, cnidarians are capable of directional movement.

The body of a hydra can contract or extend, and the tentacles that ring the mouth can reach out and grasp prey and discharge nematocysts (Fig. 30.6). Digestion begins within the central cavity but is completed within the food vacuoles of gastrodermal cells. Nutrient molecules pass by diffusion to the other cells of the body. The large gastrovascular cavity allows digestion and gastrodermal cells to exchange gases directly with a watery medium.

Although hydras exist only as polyps (there are no medusae), they can reproduce either sexually or asexually. When sexual reproduction is going to occur, an ovary or a testis develops in the body wall. Like the sponges, cnidarians have great regenerative powers, and hydras can grow an entire organism from a small piece. When conditions are favorable, hydras reproduce asexually by making small outgrowths, or buds, that pinch off and begin to live independently.

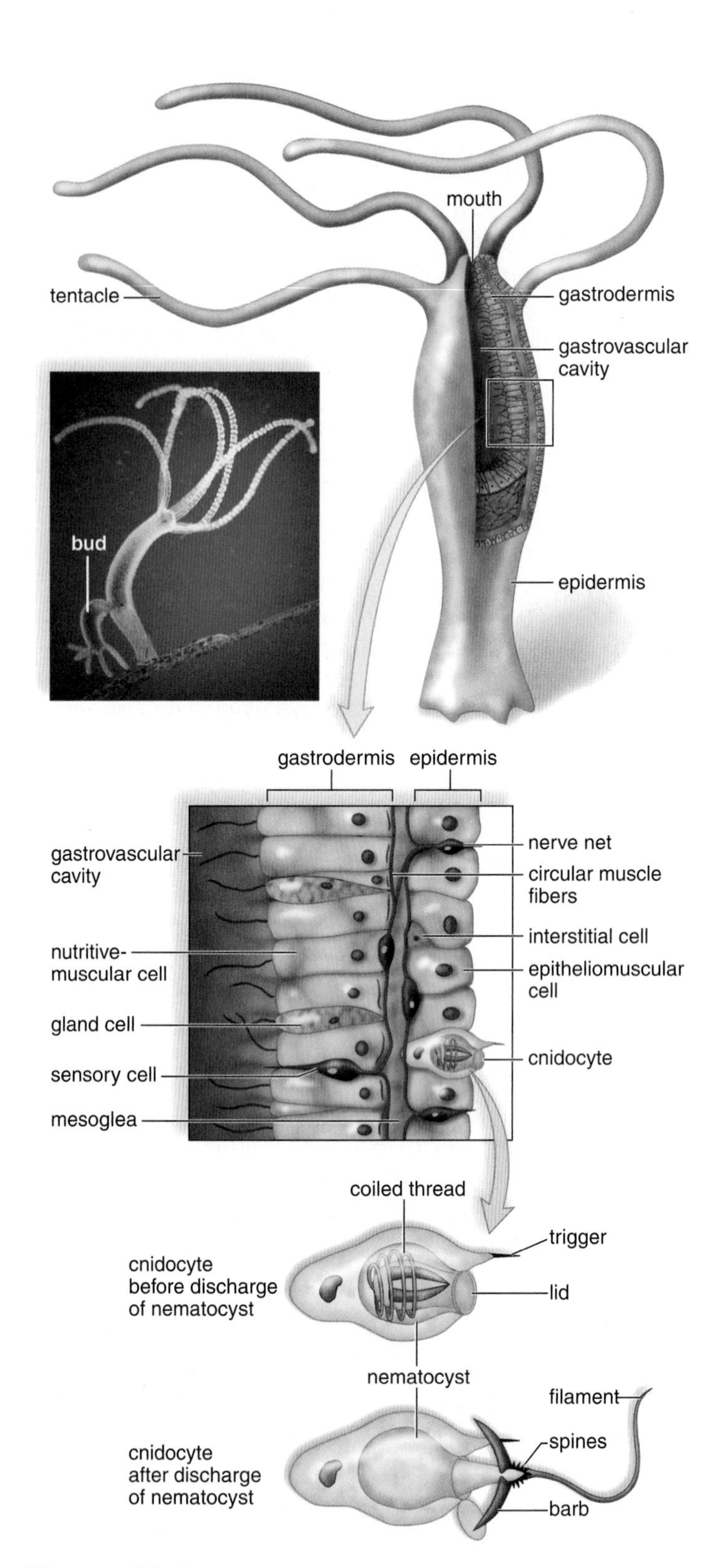

Figure 30.6 Anatomy of *Hydra*.
Top: The body of *Hydra* is a small, tubular polyp whose wall contains two tissue layers. *Hydra* reproduces asexually by forming outgrowths called buds (see photo) that develop into a complete animal. *Middle:* Various types of cells in the body wall. *Bottom:* Cnidocytes are cells that contain nematocysts.

Check Your Progress

1. How are cnidarians more anatomically complex than sponges?
2. Compare and contrast the two life cycle stages of cnidarians.
3. What are the basic anatomical characteristics of a hydra?

30.3 The Trochozoa

Learning Outcomes

1. Contrast the main anatomical features of free-living flatworms with parasitic flatworms.
2. List the unique features of molluscs and the specific features of gastropods, cephalopods, and bivalves.
3. Recall the similarities of and the differences between the polychaetes (e.g., earthworm) and oligochaetes (e.g., clam worm).

The trochozoa are bilaterally symmetrical at least in some stage of their development. As embryos, they have three germ layers, and as adults, they have the organ level of organization. Trochozoans are protostomes and include the trochophores (flatworms, rotifers, molluscs, and annelids). The trochophores either have a trochophore larva today (molluscs and annelids), or an ancestor had one in the past (flatworms and rotifers).

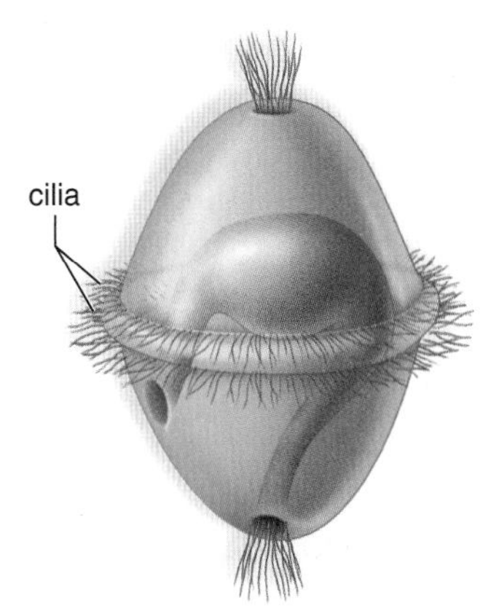

trochophore larva

Flatworms

Flatworms (phylum Platyhelminthes) are aptly named because they have an extremely flat body. Like the cnidarians, flatworms have an incomplete digestive tract and only one opening, the mouth. When one opening is present, the digestive tract is said to be *incomplete*, and when two openings are present, the digestive tract is *complete*. Also, flatworms have no body cavity, and instead the third germ layer, mesoderm, fills the space between their organs.

Among flatworms, planarians are free-living; flukes and tapeworms are parasitic. Free-living flatworms have muscles and excretory, reproductive, and digestive systems. The worms lack respiratory and circulatory systems, however. Because the body is flat and thin, diffusion alone can pass needed oxygen and other substances from cell to cell.

Free-Living Flatworms

Freshwater planarians, shown in Figure 30.7, are small (several millimeters to several centimeters) worms that live in lakes, ponds, streams, and springs. Some tend to be colorless; others have brown or black pigmentation. They feed on small living or dead organisms, such as worms and crustaceans.

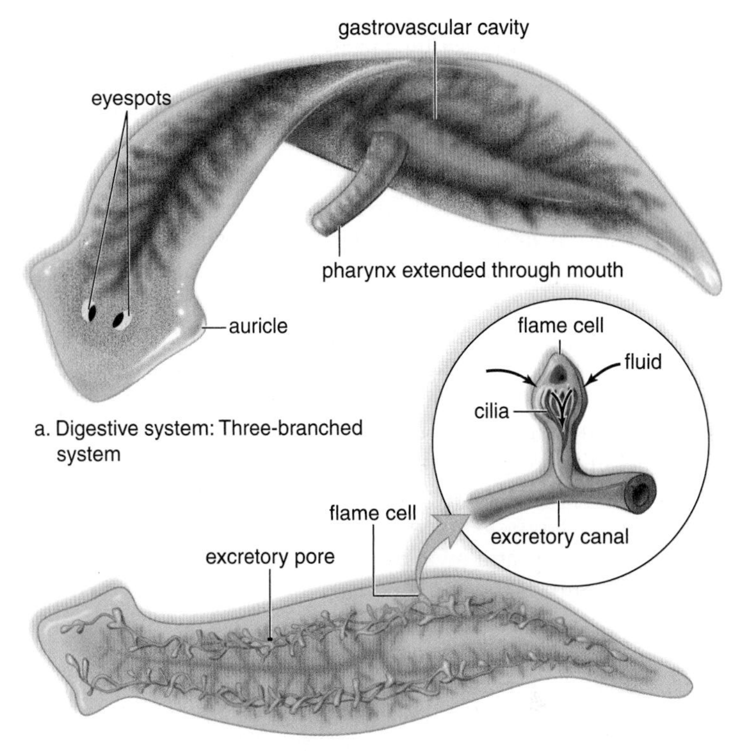

a. Digestive system: Three-branched system

b. Excretory system: Flame-cell system

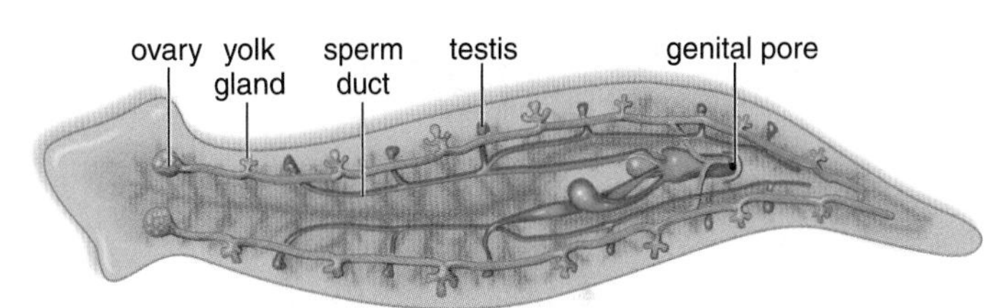

c. Reproductive system: Hermaphroditic system

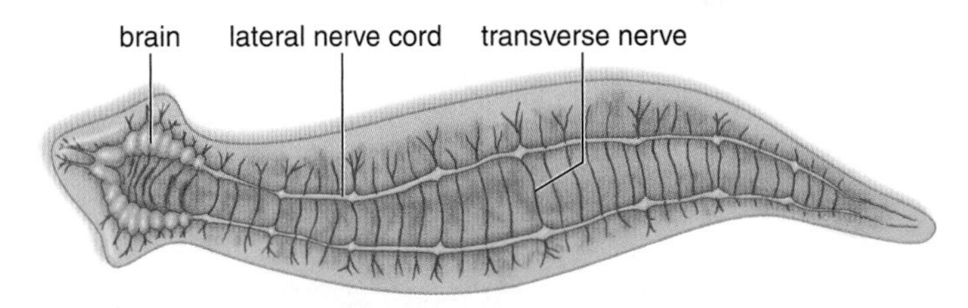

d. Nervous system: Ladder-style system

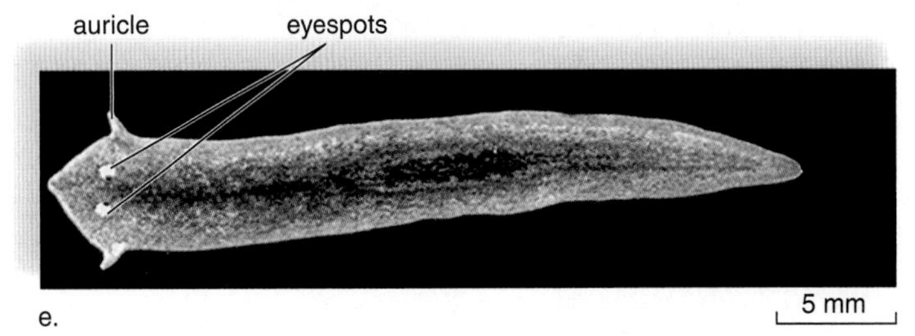

Figure 30.7 Planarian anatomy.
a. A planarian extends its pharynx to suck food into a gastrovascular cavity that branches throughout the body. **b.** The excretory system includes flame cells. **c.** The reproductive system (shown in pink and blue) has both male and female organs. **d.** The nervous system has a ladderlike appearance. **e.** A flatworm, such as *Dugesia*, is bilaterally symmetrical and has a head region with eyespots.

Planarians have an excretory system that consists of a network of interconnecting canals extending through much of the body. The beating of cilia in the **flame cells** (so named because the beating of the cilia reminded early investigators of the flickering of a flame) keeps the water moving toward the excretory pores.

Planarians have a ladderlike nervous system. A small anterior brain and two lateral nerve cords are joined by cross-branches. Planarians exhibit cephalization; in addition to a brain, they have light-sensitive organs (the eyespots) and chemosensitive organs located on the auricles. Their three muscle layers—outer circular, inner longitudinal, and diagonal—allow for varied movement. A ciliated epidermis allows planarians to glide along a film of mucus.

The animal captures food by wrapping itself around the prey, entangling it in slime, and pinning it down. Then the planarian extends a muscular pharynx, and by a sucking motion, tears up and swallows the food. The pharynx leads into a three-branched gastrovascular cavity within which digestion is completed. The digestive tract is considered incomplete because it has only one opening.

Planarians can reproduce both sexually and asexually. Although *hermaphroditic,* the worms practice cross-fertilization, in which the penis of one is inserted into the genital pore of the other (and vice versa), and reciprocal transfer of sperm takes place. Fertilized eggs hatch in 2–3 weeks as tiny worms. Asexual reproduction occurs by regeneration—if you slice a planarian in half, two new planarians will grow. You can even make a two-headed planarian by slicing the head in half!

Parasitic Flatworms

The parasitic flatworms belong to two classes: the tapeworms (class Cestoda) and the flukes (class Trematoda).

Tapeworms As adults, tapeworms are endoparasites (internal parasites) of various vertebrates, including humans. They vary in length from a few millimeters to nearly 20 m.

Tapeworms have a tough integument, a specialized body covering resistant to the host's digestive juices. Their excretory, muscular, and nervous systems are similar to those of other flatworms. Tapeworms have a well-developed anterior region, called the **scolex,** that bears hooks for attachment to the intestinal wall of the host and suckers for feeding. Behind the scolex are **proglottids,** a series of reproductive units with a full set of male and female sex organs. Each proglottid fertilizes its own eggs, which number in the thousands. Immature proglottids are located closer to the scolex, while mature (or gravid) proglottids are farther away. The gravids, which contain fertilized eggs, break away and are eliminated in the host's feces. A secondary host must ingest the eggs for the life cycle to continue. Figure 30.8 illustrates the life cycle of the pork tapeworm, *Taenia solium,* where the human is the primary host and the pig is the secondary host. The larvae burrow through the intestinal wall and travel in the bloodstream to finally lodge and encyst in muscle. The cyst is a small, hard-walled structure that contains a larva called a bladder worm. When humans eat infected meat that has not been thoroughly cooked, the bladder worm breaks out of the cyst, attaches itself to the intestinal wall, and grows to adulthood. Then the life cycle begins again.

Flukes Flukes are all endoparasites of various vertebrates. Their flattened and oval-to-elongated body is covered by a nonciliated integument. At the anterior end of these animals, an oral sucker is surrounded by sensory papillae, and there is at least one other sucker used for attachment to the host. Although the digestive system is reduced compared to that of free-living flatworms, the alimentary canal is well developed. The excretory

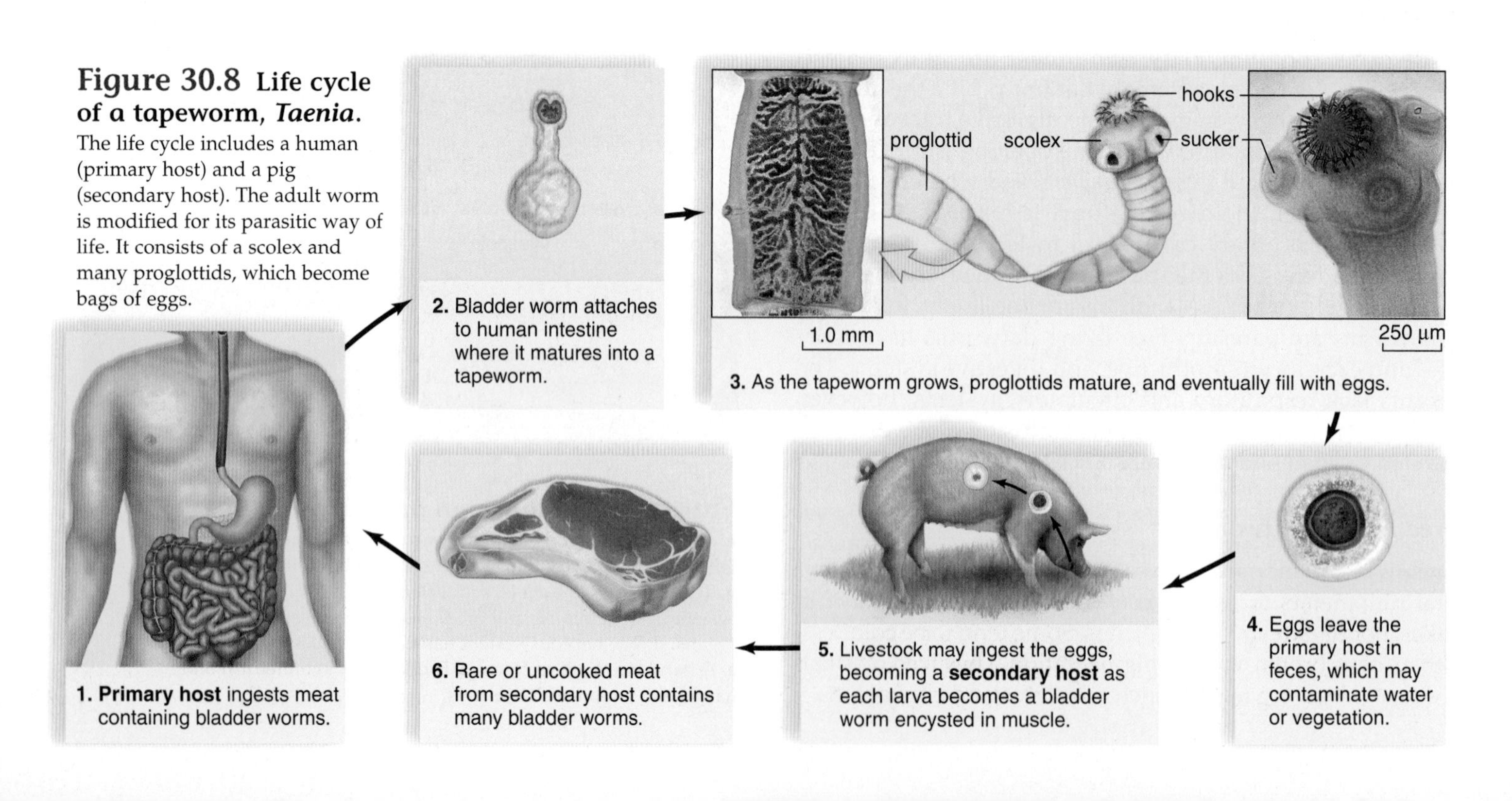

Figure 30.8 Life cycle of a tapeworm, *Taenia*. The life cycle includes a human (primary host) and a pig (secondary host). The adult worm is modified for its parasitic way of life. It consists of a scolex and many proglottids, which become bags of eggs.

and muscular systems are similar to those of free-living flatworms, but the nervous system is reduced, with poorly developed sense organs. Most flukes are hermaphroditic.

Flukes are usually named for the type of vertebrate organ they inhabit; for example, there are blood, liver, and lung flukes. The blood fluke (*Schistosoma* spp.) occurs predominantly in the Middle East, Asia, and Africa. Nearly 800,000 infected persons die each year from schistosomiasis. Adult flukes are small (approximately 2.5 cm long) and may live for years in their human hosts. The Chinese liver fluke, *Clonorchis sinensis,* is a major parasite of humans, cats, dogs, and pigs. This 20-mm-long fluke is commonly found in many regions of the Orient. It requires two intermediate hosts, a snail and a fish. Eggs are shed into the water in feces of the human or other mammal host and enter the body of a snail, where they undergo development. Larvae escape into the water and bore into the muscles of a fish. When humans eat infected fish, the juveniles migrate into the bile duct, where they mature. A heavy infection can cause destruction of the liver and death.

Check Your Progress

1. What are the basic characteristics of a free-living flatworm such as a planarian?
2. What two forms of life illustrate that some flatworms are parasitic?

Rotifers

Rotifers (phylum Rotifera) are related to the flatworms and both are trochozoans. Anton von Leeuwenhoek viewed rotifers through his microscope and called them the "wheel animalcules." Rotifers have a crown of cilia, known as the corona, on their heads (Fig. 30.9). When in motion, the corona, which looks like a spinning wheel, serves as an organ of locomotion and also directs food into the mouth. The approximately 2,000 species primarily live in fresh water; however, some marine and terrestrial forms exist. The majority of rotifers are transparent, but some are very colorful. Many species of rotifers can desiccate during harsh conditions and remain dormant for lengthy periods of time. This characteristic has earned them the title "resurrection animalcules."

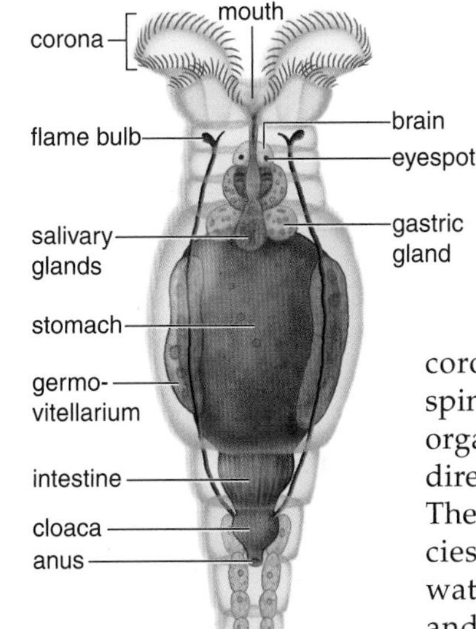

Figure 30.9 Rotifer. Rotifers are microscopic animals only 0.1–3 mm in length. The beating of cilia on two lobes at the anterior end of the animal gives the impression of a pair of spinning wheels.

Molluscs

The **molluscs** (phylum Mollusca) are the second most numerous group of animals, numbering over 110,000 species. They inhabit a variety of environments, including marine, freshwater, and terrestrial habitats. Although almost everyone enjoys looking at the intricate patterns and beauty of seashells, few people are aware of the tremendous diversity found in the phylum Mollusca (Fig. 30.10). Molluscs

a. Chiton, *Tonicella*

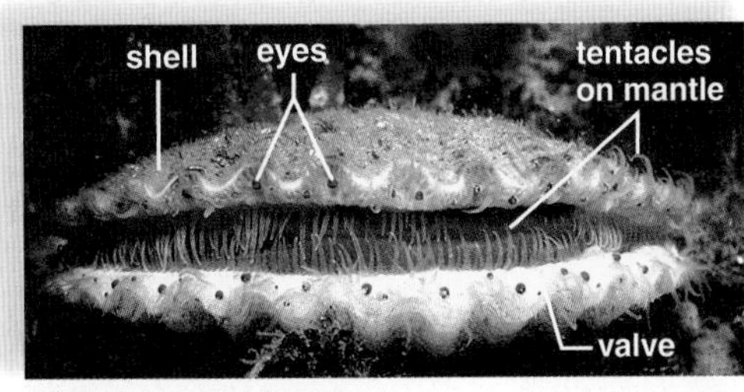

b. Scallop, *Pecten* sp.

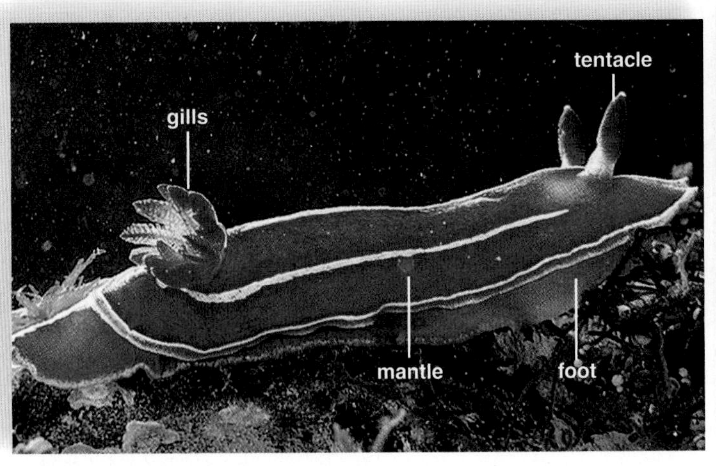

c. Nudibranch, *Glossodoris macfarlandi*

d. Two-spotted octopus, *Octopus bimaculoides*

Figure 30.10 Diversity of molluscs. **a.** A chiton is believed to be most like the ancestral mollusc that gave rise to all the other types. Recent evidence suggests that ancestral molluscs such as the chiton were segmented, but that this segmentation was lost throughout much of the phylum. **b.** Like clams, a scallop is a bivalve. **c.** Like snails, a nudibranch is a gastropod. **d.** Like squids, an octopus is a cephalopod.

include chitons, limpets, slugs, snails, abalones, conchs, nudibranchs, clams, scallops, squid, and octopuses. Molluscs have a true coelom, and all coelomates have bilateral symmetry, three germ layers, the organ level of organization, and a complete digestive tract. The advantages of having a coelom include freer body movements, space for development of complex organs, greater surface area for absorption of nutrients, and protection of internal organs from damage.

The Unique Characteristics of Molluscs

Despite being a very large and diversified group, all molluscs have a body composed of at least three distinct parts:

1. The **visceral mass** is the soft-bodied portion that contains internal organs, including a highly specialized digestive tract, paired kidneys, and reproductive organs. The nervous system of a mollusc consists of several ganglia connected by nerve cords. The amount of cephalization and sensory organs varies from nonexistent in clams to complex in squid and octopuses.
2. The foot is a strong, muscular portion used for locomotion. Molluscan groups can be distinguished by modifications of the foot. Molluscs exhibit varying amounts of mobility. Oysters are sessile; snails are extremely slow moving; and squid are fast-moving, active predators.
3. The **mantle,** a membranous or sometimes muscular covering, envelops but does not completely enclose the visceral mass. The *mantle cavity* is the space between the two folds of the mantle. The mantle may secrete a *shell,* which is an exoskeleton.

Another feature often present is a rasping, tongue-like *radula,* an organ that bears many rows of teeth and is used to obtain food.

Gastropods

The **gastropods** (class Gastropoda) include nudibranchs, conchs, and snails. In gastropods, whose name means "stomach-footed," the foot is ventrally flattened, and the animal moves by muscle contractions that pass along the foot. Many are herbivores that use their radulas to scrape

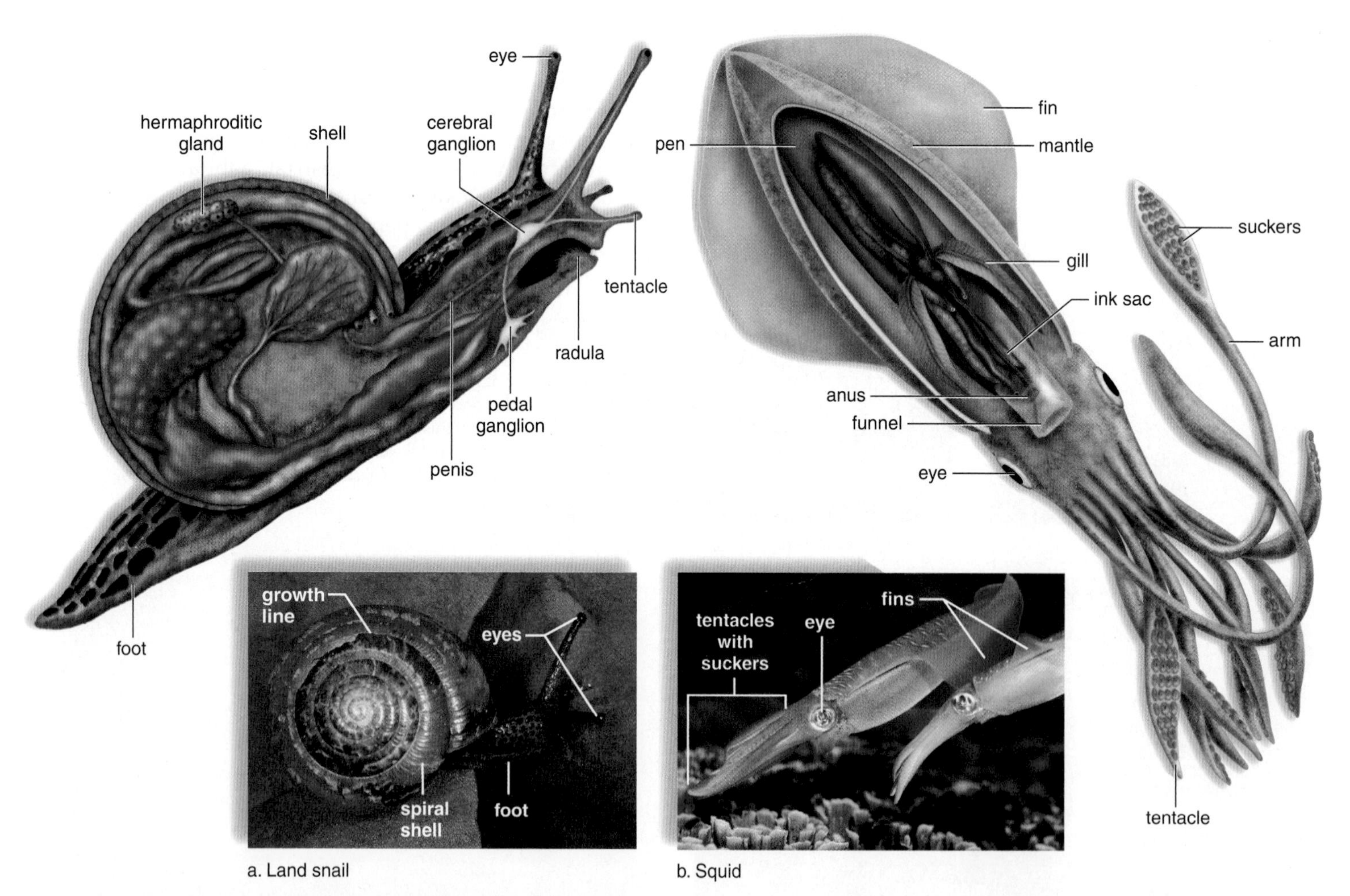

Figure 30.11 Gastropod and cephalopod anatomy.
a. Snails have a long muscular foot that allows them to creep along slowly by way of muscular contraction. Note the absence of lungs in this land snail. **b.** Squids are torpedo-shaped and adapted for fast swimming by jet propulsion.

food from surfaces. Others are carnivores, using their radulas to bore through surfaces, such as bivalve shells, to obtain food.

While nudibranchs, also called sea slugs, lack a shell, conchs and snails have a univalve coiled shell in which the visceral mass spirals. Land snails, such as *Helix aspera*, have a head with two pairs of tentacles; one pair bears eyes at the tips (Fig. 30.11*a*). The shell not only offers protection, but also prevents desiccation (drying out). While aquatic gastropods have gills, land snails have a mantle that is richly supplied with blood vessels and functions as a lung. Reproduction is also adapted to a land existence. Land snails are hermaphroditic. When two snails meet, each inserts its penis into the vagina of the other, and following fertilization and the deposit of eggs externally, development proceeds directly without formation of swimming larvae.

Cephalopods

In **cephalopods** (class Cephalopoda, meaning head-footed), including octopuses, squid, and nautiluses, the foot has evolved into a funnel or siphon about the head (Fig. 30.11*b*). Aside from the tentacles, which seize prey, cephalopods have a powerful beak and a radula (toothy tongue) to tear prey apart. Cephalization is apparent. The eyes are superficially similar to those of vertebrates—they have a lens and a retina with photoreceptors. However, the eye is constructed so differently from the vertebrate eye that the so-called camera-type eye must have evolved independently in both the molluscs and in the vertebrates. In cephalopods, the brain is formed from a fusion of ganglia, and nerves leaving the brain supply various parts of the body. An especially large pair of nerves controls the rapid contraction of the mantle, allowing these animals to move quickly by a jet propulsion of water. Rapid movement and the secretion of a dark ink help cephalopods escape their enemies. Octopuses have no shell, and squid have only a remnant of a shell concealed beneath the skin. Octopuses, like some other species of cephalopods, are thought to be among the most intelligent invertebrates and are even capable of learning!

Bivalves

Clams, mussels, oysters, and scallops are called **bivalves** (class Bivalvia) because their shells have two parts. In a clam, the shell, which is secreted by the mantle, is composed of protein and calcium carbonate, with an inner layer of mother-of-pearl. If a foreign body is placed between the mantle and the shell, pearls form as concentric layers of shell are deposited about the particle.

Figure 30.12 shows the internal anatomy of the freshwater clam, *Anodonta*. The adductor muscles hold the valves of the shell together. Within the mantle cavity, the gills, which are the organs for gas exchange in aquatic forms, hang down on either side of the visceral mass, which lies above the foot.

The clam is a filter feeder. Food particles and water enter the mantle cavity by way of the incurrent siphon, a posterior opening between the two valves. Mucous secretions cause smaller particles to adhere to the gills, and ciliary action sweeps them toward the mouth.

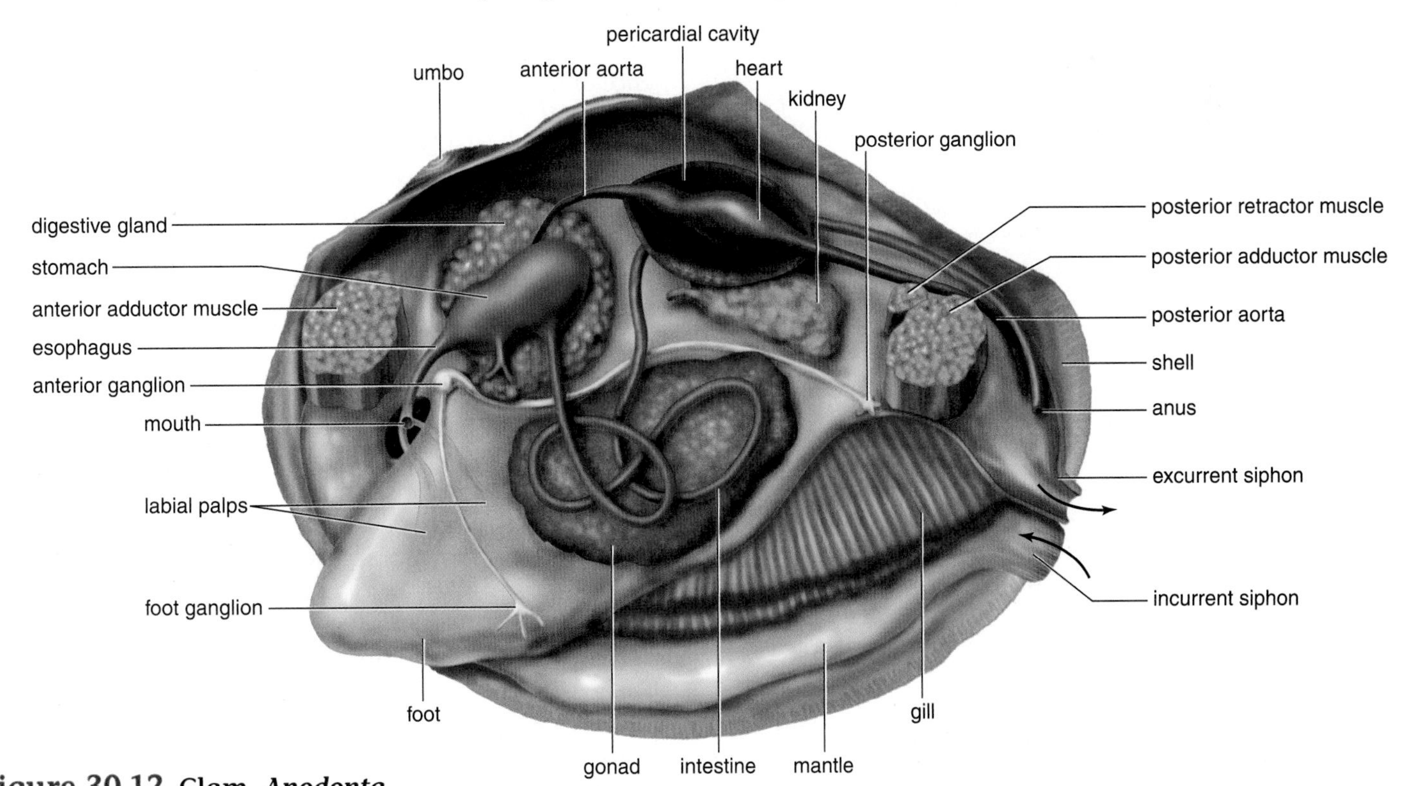

Figure 30.12 **Clam, *Anodonta*.**
In this drawing of a clam, one of the bivalve shells and the mantle have been removed from one side. Follow the path of food from the incurrent siphon to the gills, the mouth, the stomach, the intestine, the anus, and the excurrent siphon. Locate the three ganglia: anterior, foot, and posterior. The heart lies in the reduced coelom.

The Visceral Mass The heart of a clam lies just below the hump of the shell within the pericardial cavity, the only remains of the reduced coelom. The heart pumps blood into a dorsal aorta that leads to the various organs of the body. Within the organs, however, blood flows through spaces, or sinuses, rather than through vessels. This is called an **open circulatory system** because the blood is not entirely contained within blood vessels. An open circulatory system would be inefficient in a fast-moving animal.

The nervous system of a clam is composed of three pairs of ganglia (anterior, foot, and posterior), which are all connected by nerves. Clams lack cephalization. The "hatchet" foot projects anteriorly from the shell, and by expanding the tip of the foot with blood and pulling the body after it, the clam moves forward.

The digestive system of the clam includes a mouth with labial palps, an esophagus, a stomach, and an intestine, which coils about in the visceral mass and then is surrounded by the heart as it extends to the anus. Two excretory kidneys remove waste from the pericardial cavity for excretion into the mantle cavity.

The sexes are usually separate. The gonad (i.e., ovary or testis) is located around the coils of the intestine. All clams have some type of larval stage, and marine clams have a trochophore larva.

Comparison

Table 30.2 compares the clam and land snail, which are adapted to less active lifestyles, with the squid, a more active animal.

TABLE 30.2 Comparison of Clam, Squid, and Land Snail

Feature	Clam	Squid	Land Snail
Food gathering	Filter feeder	Active predator	Herbivore
Skeleton	Heavy shell for protection	No external skeleton	Shell protects visceral mass
Circulation	Open	Closed	Open
Cephalization	None	Marked	Present
Locomotion	"Hatchet" foot	Jet propulsion	Flat foot
Nervous system	3 separate ganglia	Brain and nerves	Cerebral ganglion

CHECK YOUR PROGRESS

1. What are the three characteristics of all molluscs?
2. Name the different groups of molluscs and give an example from each group.

Annelids

Annelids (phylum Annelida, about 12,000 species) are *segmented,* as evidenced by the rings encircling the outside of their bodies. Segmentation is also seen in arthropods and chordates, although annelids are the only trochozoan with segmentation and a well-developed coelom. Internally, the segments of annelids are partitioned by septa. Worms, in general, do not have an internal or external skeleton, but they do have a **hydrostatic skeleton,** a fluid-filled interior that supports muscle contraction and enhances flexibility. Along with the partitioning of the fluid-filled coelom, this hydroskeleton permits each body segment to move independently. Locomotion occurs by contraction and expansion of each body segment, propelling the animal forward. Thus, a terrestrial annelid is capable of crawling on the surface in addition to burrowing in the mud. Although the most familiar members of this phylum are leeches and earthworms, the majority of annelids are marine. Annelids vary in size from microscopic to tropical earthworms as much as 4 m long.

The body plan in annelids has led to specialization of the digestive tract. For example, the digestive system may include a pharynx, esophagus, crop, gizzard, intestine, and accessory glands. Annelids have an extensive **closed circulatory system** with blood vessels that run the length of the body and branch to every segment. The nervous system consists of a brain connected to a **ventral solid nerve cord,** with ganglia in each segment. The excretory system consists of nephridia in most segments. A **nephridium** (pl., nephridia) is a tubule that collects waste material and excretes it through an opening in the body wall.

Polychaetes

Marine annelids are the Polychaeta, which refers to the presence of many setae. **Setae** are bristles that anchor the worm or help it move. The setae are in bundles on parapodia, which are paddlelike appendages found on most segments. These are used in swimming, but also as respiratory organs. Clam worms, such as *Nereis* (Fig. 30.13*a*), prey on crustaceans and other small animals, which they capture using a pair of strong, chitinous jaws that extend with the pharynx. Associated with its way of life, *Nereis* has a well-defined head region with eyes and other sense organs (Fig.30.13*b*).

Other polychaetes are sessile tube worms, with tentacles that form a funnel-shaped fan (Fig. 30.13*c*). Water currents, created by the action of cilia, trap food particles that are directed toward the mouth of these filter feeders.

Polychaetes have breeding seasons, and only during these times do they possess functional sex organs. In *Nereis,* many worms concurrently shed a portion of their bodies containing either eggs or sperm, and these float to the surface, where fertilization takes place. The zygote rapidly develops into a trochophore larva, just as in marine clams. The existence of this larval form in both the annelids and molluscs is evidence that these two groups of animals are evolutionarily related.

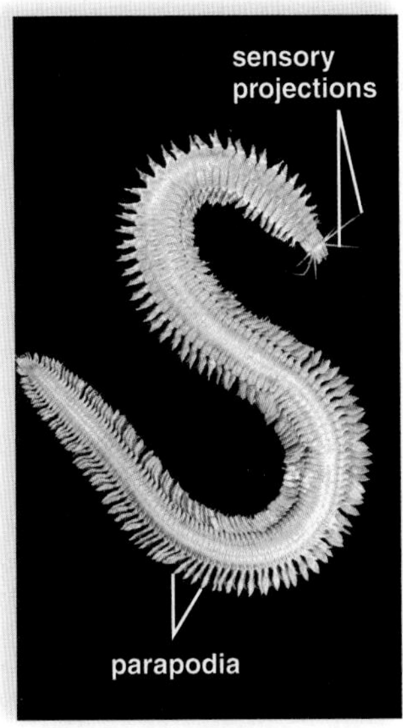

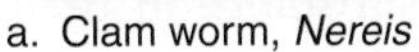
a. Clam worm, *Nereis*

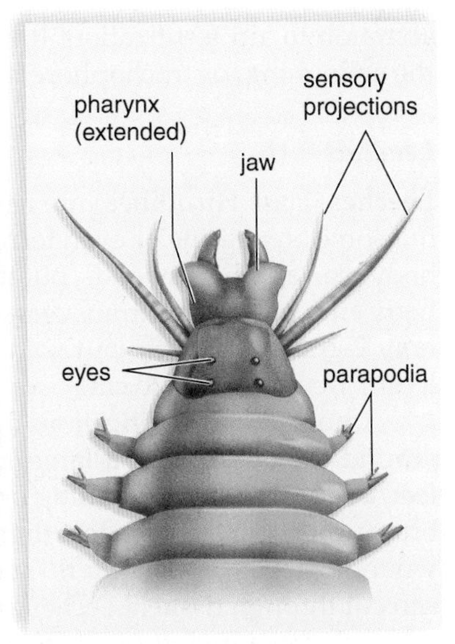

b. Head region of *Nereis*

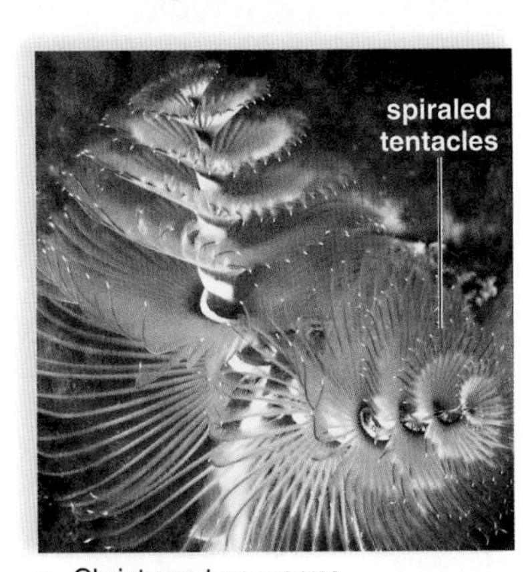

c. Christmas tree worms, *Spirobranchus*

Figure 30.13
Polychaete diversity.
a. Clam worms are predatory polychaetes that have a well-defined head region. **b.** Note also the parapodia, which are used for swimming and as respiratory organs. **c.** Christmas tree worms (a type of tube worm) are sessile filter feeders whose ciliated tentacles spiral.

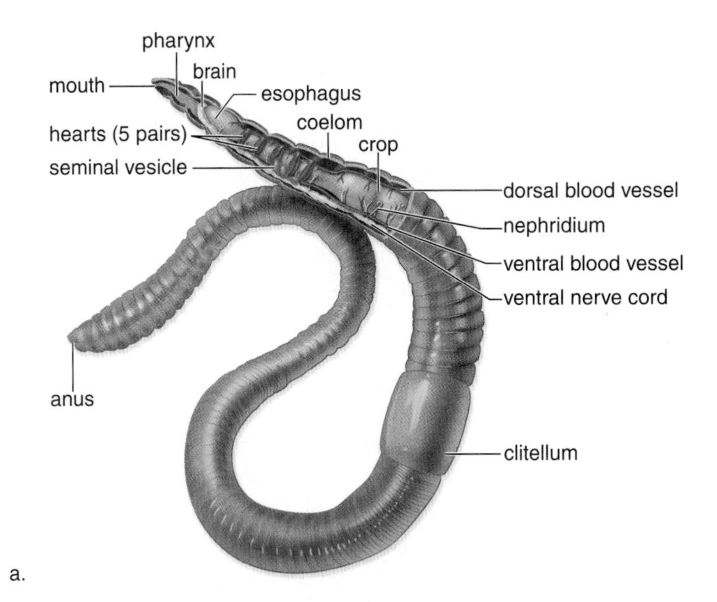

a.

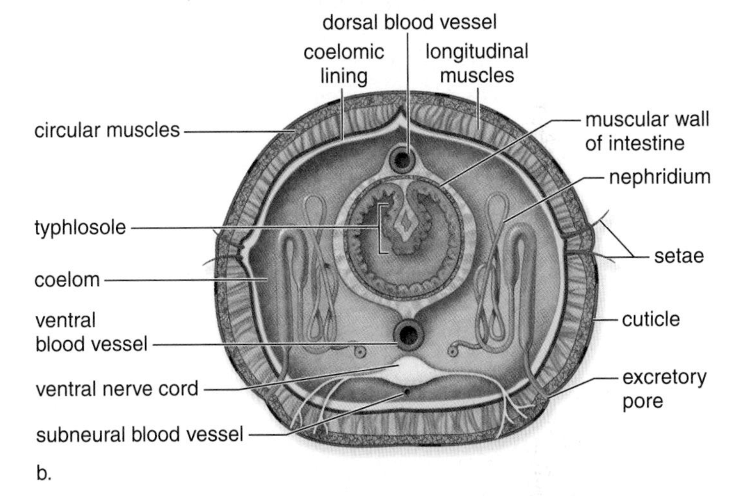

b.

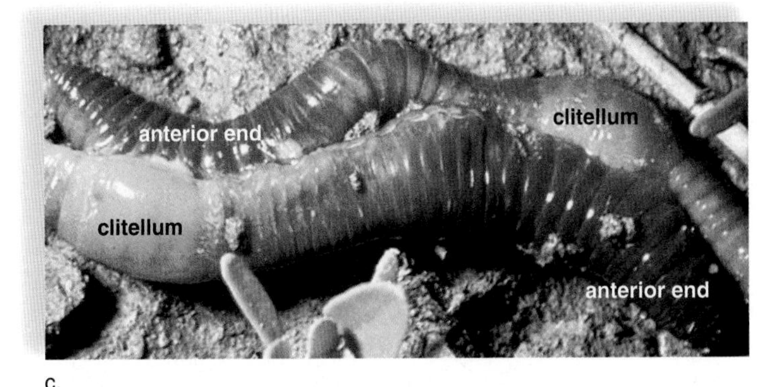

c.

Figure 30.14 Earthworm, *Lumbricus*.
a. Internal anatomy of the anterior part of an earthworm. Notice that each body segment bears a pair of setae and that internal septa divide the coelom into compartments. **b.** Cross section of an earthworm. **c.** When earthworms mate, they are held in place by a mucus secreted by the clitellum. The worms are hermaphroditic, and when mating, sperm pass from the seminal vesicles of each to the seminal receptacles of the other.

Oligochaetes

The oligochaetes (class Oligochaeta), which include earthworms, have few setae per segment. Earthworms (e.g., *Lumbricus*) (Fig. 30.14) do not have a well-developed head or parapodia. Their setae protrude in pairs directly from the surface of the body. Locomotion, which is accomplished section by section, utilizes muscle contraction and the setae. When longitudinal muscles contract, segments bulge, and their setae protrude into the soil. Then, when circular muscles contract, the setae are withdrawn, and these segments move forward in sequence, pushing the whole animal forward.

Earthworms reside in soil where there is adequate moisture to keep the body wall moist for gas exchange. They are scavengers that do not have an obvious head and feed on leaves or any other organic matter, living or dead, which they can conveniently take into their mouths along with dirt. Food drawn into the mouth by the action of the muscular pharynx is stored in a crop and ground up in a thick, muscular gizzard. Digestion and absorption occur in a long intestine, whose dorsal surface has an expanded region, called a typhlosole, that increases the surface for absorption.

Segmentation Earthworm segmentation, which is obvious externally, is also internally evidenced by the presence of septa. The long, ventral solid nerve cord leading from the brain has ganglionic swellings and lateral nerves in each segment. The paired nephridia in most segments have two openings: one is a ciliated funnel that collects coelomic fluid, and the other is an exit in the body wall. Between the two openings is a convoluted region where waste material is removed from the blood vessels about the tubule. Red blood moves anteriorly in the dorsal blood vessel, which connects to the ventral blood vessel by five pairs of connectives called "hearts." Pulsations of the dorsal blood vessel and the five pairs of hearts are responsible for blood flow. As the ventral vessel takes the blood toward the posterior regions of the worm's body, it gives off branches in every segment. Altogether, segmentation is evidenced by:

- body rings
- coelom divided by septa
- setae on most segments
- ganglia and lateral nerves in each segment
- nephridia in most segments
- branch blood vessels in each segment

Reproduction Earthworms are hermaphroditic: the male organs are the testes, the seminal vesicles, and the sperm ducts, and the female organs are the ovaries, the oviducts, and the seminal receptacles. When mating, two worms lie parallel to each other facing in opposite directions. The fused midbody segment, called a clitellum, secretes mucus that protects the sperm from drying out as they pass between the worms. After the worms separate, the clitellum of each produces a slime tube, which is moved along over the anterior end by muscular contractions. As it passes, eggs and the sperm received earlier are deposited, and fertilization occurs. The slime tube then forms a cocoon to protect the worms as they develop. There is no larval stage.

Comparison with Clam Worms Comparing the anatomy of marine clam worms (class Polychaeta) to that of terrestrial earthworms (class Oligochaeta) highlights the manner in which earthworms are adapted to life on land. A lack of cephalization is seen in the nonpredatory earthworms that extract organic remains from the soil they eat. Their lack of parapodia helps reduce the possibility of water loss and facilitates burrowing in soil. The clam worm makes use of external water, while the earthworm provides a mucous secretion to aid fertilization. It is the aquatic form that has the swimming, or trochophore larva, not the land form.

Leeches

Leeches (class Hirudinea) are usually found in fresh water, but some are marine or even terrestrial. They have the same body plan as other annelids, but they have no setae, and each body ring has several transverse grooves. Most leeches are only 2–6 cm in length, but some, including the medicinal leech, are as long as 20 cm.

Among their modifications are two suckers, a small one around the mouth and a large, posterior one. While some leeches are free-living, feeding on plant material or invertebrates, most are fluid feeders that attach themselves to open wounds. Some bloodsuckers, such as the medicinal leech, can cut through tissue. Leeches are able to keep blood flowing and prevent clotting by means of a substance in their saliva known as hirudin, a powerful anticoagulant.

Check Your Progress

1. Compare a clam to an earthworm with regard to body cavity and circulatory system.
2. Compare the mating behavior of the earthworm and the clam worm.
3. Name two features of leeches that adapt them to a parasitic way of life.

30.4 The Ecdysozoa

Learning Outcomes

1. Describe the major anatomical features of a roundworm.
2. Explain the characteristics contributing to the success and diversity of arthropods.
3. List the major features that distinguish the crustaceans, insects, and arachnids.
4. Compare the anatomy of a grasshopper and a crayfish.

The ecdysozoans are protostomes, as are the trochozoa. The term *ecdysis* means molting, and both roundworms and arthropods, which belong to this group, periodically shed their outer covering.

Roundworms

Roundworms (phylum Nematoda, about 90,000 species) are nonsegmented worms that are prevalent in almost any environment. Generally, nematodes are colorless and range in size from microscopic to exceeding 1 m in length. The internal organs, including the tubular reproductive organs, lie within the pseudocoelom. A **pseudocoelom** is a body cavity that is incompletely lined by mesoderm. In other words, mesoderm occurs inside the body wall but not around the digestive cavity (gut). The fluid-filled pseudocoelom provides space for the development of organs, substitutes for a circulatory system by allowing easy passage of molecules, and provides a type of skeleton.

Nematodes have developed a variety of lifestyles from free-living to parasitic. One species, *Caenorhabditis elegans*, a free-living nematode, is a model animal used in genetics and developmental biology as it was one of the first species to have its genome sequenced.

Ascaris

In the roundworm, *Ascaris lumbricoides* (Fig. 30.15*a*), females tend to be larger (20–35 cm in length) than males. Both sexes move by a characteristic whiplike motion because they have only longitudinal muscles and no circular muscles next to the body wall (Fig. 30.15*b*). *Ascaris* species are most commonly parasites of humans and pigs.

A female *Ascaris* is very reproductively prolific, producing over 200,000 eggs daily. The eggs are passed with host feces and, under the right conditions, can develop into a worm within two weeks. The eggs then enter a new host's body via uncooked vegetables, soiled fingers, or ingested fecal material and hatch in the intestines. The juveniles make their way into the veins and lymphatic vessels and are carried to the heart and lungs. From the lungs, the larvae travel up the trachea, where they are swallowed and eventually reach the intestines. There, the larvae mature and begin feeding on intestinal contents. The symptoms of an *Ascaris* infection depend upon the site and the stage of infection.

Other Roundworms

Trichinosis is a fairly serious infection caused by *Trichinella spiralis*, a roundworm that rarely infects humans in the United States. Humans contract the disease when they eat pork that is not fully cooked and contains encysted larvae. After maturation, the female adult burrows into the wall of the host's small intestine, where she deposits live larvae that are then carried by the bloodstream and encyst in the skeletal muscles (Fig. 30.15*c*). When the adults are in the small intestine, digestive disorders, fatigue, and fever occur. When the larvae encyst, the symptoms include aching joints, muscle pain, and itchy skin.

Elephantiasis is caused by a roundworm called the filarial worm, which utilizes the mosquito as an intermediate host. When a mosquito bites an infected person, it transports larvae to a new host. Because the adult worms reside in lymphatic vessels, fluid return is impeded, and the limbs of an infected human can swell to an enormous size, even resembling those of an elephant.

Other roundworm infections are more common in the United States. Children frequently acquire a pinworm infection, and hookworm is seen in the southern states, as well as worldwide. Good hygiene, proper disposal of sewage, and cooking meat thoroughly usually protect people from parasitic roundworms.

Check Your Progress

1. What type of symmetry and digestive tract characterizes the roundworms?
2. Name two parasitic roundworms and the symptoms they cause.

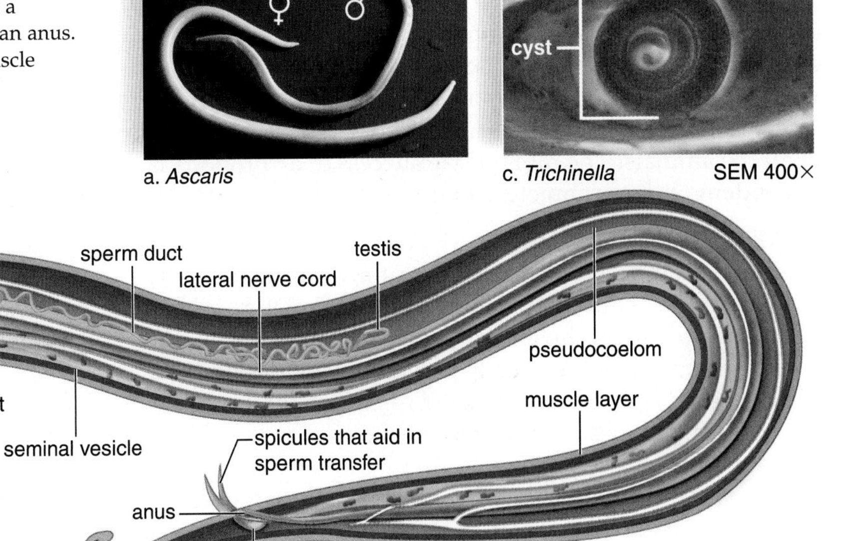

Figure 30.15 Roundworm anatomy.
a. The roundworm *Ascaris*. **b.** Roundworms such as *Ascaris* have a pseudocoelom and a complete digestive tract with a mouth and an anus. **c.** The larvae of the roundworm *Trichinella* penetrate striated muscle fibers, where they coil in a cyst formed from the muscle fiber.

a. Flat-backed millipede, *Sigmoria*

b. Tarantula, *Aphonopelma*

e. Stone centipede, *Lithobius*

c. Dungeness crab, *Cancer*

d. Paper wasp, *Polistes*

Figure 30.16 Arthropod diversity.
a. A millipede has only one pair of antennae, and the head is followed by a series of segments, each with two pairs of appendages. **b.** The hairy tarantulas of the genus *Aphonopelma* are dark in color and move carefully and steadily. Their bite is harmless to people. **c.** A crab is a crustacean with a calcified exoskeleton, one pair of claws, and four other pairs of walking legs. **d.** A wasp is an insect with two pairs of wings, both used for flying, and three pairs of walking legs. **e.** A centipede has only one pair of antennae, and the head is followed by a series of segments, each with a single pair of appendages.

Arthropods

Arthropods (phylum Arthropoda) are extremely diverse (Fig. 30.16), ranging in size from less than 0.1 mm (mites) to 4 m (Japanese crab) in length. Over 1 million species have been discovered and described, but some experts suggest that as many as 30 million arthropods could exist—most of them insects. Arthropods, which also occupy every type of habitat, are considered the most successful group of all the animals. The remarkable success of arthropods is dependent on five characteristics:

1. *A rigid but jointed exoskeleton* (Fig. 30.17*a, b*). The exoskeleton is composed primarily of chitin, a strong, flexible, nitrogenous polysaccharide. The exoskeleton serves many functions, including protection, attachment for muscles, locomotion, and prevention of desiccation. However, because it is hard and nonexpandable, arthropods must molt, or shed, the exoskeleton as they grow larger (Fig. 30.17*c*).
2. *Segmentation.* Segmentation is readily apparent because each segment has a pair of jointed appendages, even though certain segments are fused into a head, thorax, and abdomen. The jointed appendages of arthropods are basically hollow tubes moved by muscles. Typically, the appendages are highly adapted for a particular function, such as food gathering, reproduction, and locomotion. In addition, many appendages are associated with sensory structures and used for tactile purposes.
3. *Well-developed nervous system.* Arthropods have a brain and a ventral nerve cord. The head bears various types of sense organs, including eyes of two types—simple and compound. The compound eye is composed of many complete visual units, each of which operates independently (Fig. 30.17*d*). The lens of each visual unit focuses an image on a small number of photoreceptors within that unit. The simple eye, like that of vertebrates, has a single lens that brings the image to focus onto many receptors, each of which receives only a portion of the image. In addition to sight, many arthropods have well-developed touch, smell, taste, balance, and hearing. Arthropods, such as the termite discussed at the beginning of this chapter, display many complex behaviors and methods of communication.
4. *Variety of respiratory organs.* Marine forms use gills, which are vascularized, highly convoluted, thin-walled tissue specialized for gas exchange. Terrestrial forms have book lungs (e.g., spiders) or air tubes called tracheae. Tracheae serve as a rapid way to transport oxygen directly to the cells.
5. *Reduced competition through metamorphosis.* Many arthropods undergo a drastic change in form and physiology that occurs as an immature stage, called a larva, becomes an adult. Among arthropods, the larva eats different

food and lives in a different environment than the adult. For example, larval crabs live among and feed on plankton, while adult crabs are bottom dwellers that catch live prey or scavenge dead organic matter. Among insects, such as butterflies, the caterpillar feeds on leafy vegetation, while the adult feeds on nectar.

Crustaceans

Crustaceans (subphylum Crustacea) are a group of largely marine arthropods that include barnacles, shrimps, lobsters, and crabs. There are also some freshwater crustaceans, including the crayfish, and some terrestrial ones, including the sowbug, or pillbug.

Crustaceans are named for their hard shells. The exoskeleton is calcified to a greater degree in some forms than in others. Although crustacean anatomy is extremely diverse, the head usually bears a pair of compound eyes and five pairs of appendages. The first two pairs, called antennae, lie in front of the mouth and have sensory functions. The other three pairs are mouthparts used in feeding.

In crayfish, such as *Cambarus*, the thorax bears five pairs of walking legs. The first walking leg is a pinching claw (Fig. 30.18). The gills are situated above the walking legs. The head and thorax are fused into a cephalothorax,

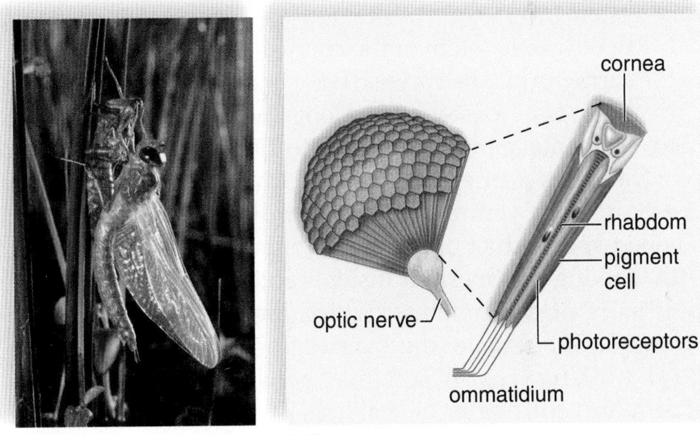

Figure 30.17 Arthropod skeleton and eye.
a. The joint in an arthropod skeleton is a region where the cuticle is thinner and not as hard as the rest of the cuticle. The direction of movement is toward the flexor muscle or the extensor muscle, whichever one has contracted. **b.** The exoskeleton is secreted by the epidermis and consists of the endocuticle; the exocuticle, hardened by the deposition of calcium carbonate; and the epicuticle, a waxy layer. Chitin makes up the bulk of the exo- and endocuticles. **c.** Because the exoskeleton is nonliving, it must be shed through a process called molting for the arthropod to grow. **d.** Arthropods have a compound eye that contains many individual units, each with its own lens and photoreceptors.

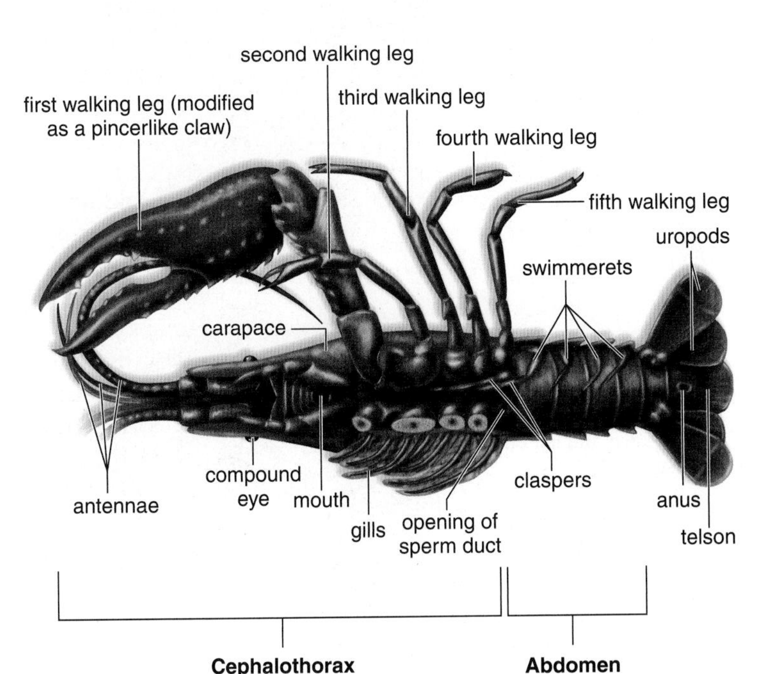

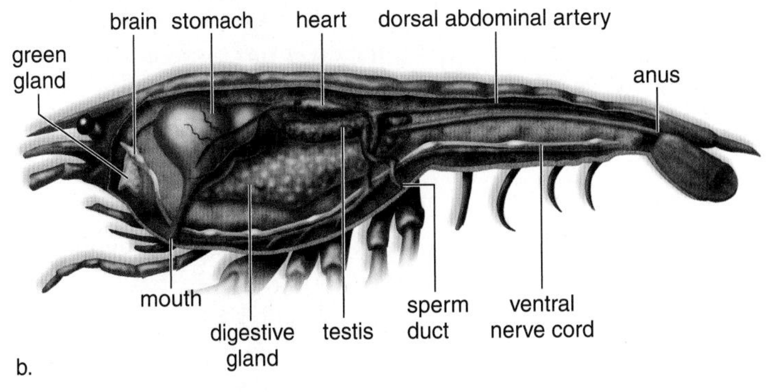

Figure 30.18 Male crayfish, *Cambarus*.
a. Externally, it is possible to observe the crayfish's jointed appendages, including the swimmerets, and the walking legs, which include claws. These appendages, plus a portion of the carapace, have been removed from the right side so that the gills are visible. **b.** An internal view shows the parts of the digestive and circulatory systems. Note the ventral nerve cord.

which is covered on the top and sides by a nonsegmented carapace. The abdominal segments, which contain much musculature, are equipped with swimmerets, small paddle-like structures. The last two segments bear the uropods and the telson, which make up a fan-shaped tail to propel the crayfish backward.

Internal Organs The digestive system includes a stomach, which is divided into two main regions. The anterior portion consists of a gastric mill, a special grinding apparatus equipped with chitinous teeth. The posterior region acts as a filter to prevent coarse particles from entering the digestive glands where absorption takes place. **Green glands** lying in the head region, anterior to the esophagus, excrete metabolic wastes through a duct that opens externally at the base of the antennae. The coelom is reduced to a space around the reproductive system. A heart within a pericardial cavity pumps blood containing the respiratory pigment hemocyanin into a **hemocoel** consisting of sinuses (open spaces), where the hemolymph flows about the organs. (Whereas hemoglobin is a red, iron-containing pigment, hemocyanin is a blue, copper-containing pigment.) Thus, the blood is not contained within blood vessels, constituting an open circulatory system.

The nervous system of the crayfish is very similar to that of the earthworm. The crayfish has a brain and a ventral nerve cord that passes posteriorly. Along the length of the nerve cord, segmental ganglia give off 8 to 19 paired lateral nerves.

The sexes are separate in the crayfish, and the gonads are located just ventral to the pericardial cavity. In the male, a coiled sperm duct opens to the outside at the base of the fifth walking leg. Sperm transfer is accomplished by the first two pairs of swimmerets, which are enlarged and quite strong. In the female, the ovaries open at the bases of the third walking legs. A stiff fold between the bases of the fourth and fifth pairs of walking legs serves as a seminal receptacle. Following fertilization, the eggs are attached to the swimmerets of the female.

Check Your Progress

1. What features account for the great success of arthropods?
2. Briefly describe the major anatomical characteristics of a crustacean.

Insects

Insects (subphylum Uniramia) are so numerous (likely over 1 million species) and so diverse (Fig. 30.19) that the study of this one group is a major specialty in biology called entomology. Some insects show remarkable behavioral adaptations, as exemplified by the social systems of bees, ants, termites, and other colonial insects.

Insects have certain features in common. The body is divided into a head, a thorax, and an abdomen. The head usually bears a pair of sensory antennae, a pair of compound eyes, and several simple eyes. The mouthparts are adapted to the specific insect's way of life. For example, a grasshopper

a. Housefly, *Musca*

b. Walking stick, *Diapheromera*

c. Bee, *Apis*

d. Dragonfly, *Aeshna*

e. Birdwind butterfly, *Troides*

Figure 30.19
Insect diversity.
a. Flies have a single pair of wings and lapping mouthparts. **b.** Walking sticks are herbivorous, with biting and chewing mouthparts. **c.** Bees have four translucent wings and a thorax separated from the abdomen by a narrow waist. **d.** Dragonflies have two pairs of similar wings. They catch and eat other insects while flying. **e.** Butterflies have forewings larger than their hindwings. Their mouthparts form a long tube for siphoning up nectar from flowers.

has mouthparts that chew, and a butterfly has a long tube for siphoning the nectar of flowers. The abdomen contains most of the internal organs. The thorax bears three pairs of legs and the wings—either one or two pairs or none. Wings, if present, enhance an insect's ability to survive by providing a way of escaping enemies, finding food, facilitating mating, and dispersing the offspring. The exoskeleton of an insect is lighter and contains less chitin than that of many other arthropods.

Here, we will discuss the features of a grasshopper as a representative example of insect form and function. In the grasshopper (Fig. 30.20), the third pair of legs is suited to jumping. There are two pairs of wings. The forewings are tough and leathery, and when folded back at rest, they protect the broad, thin hindwings. On the lateral surface, the first abdominal segment bears a large tympanum on each side for the reception of sound waves.

Internal Organs The digestive system of a grasshopper is suitable for a herbivorous diet. The mouthparts chew the food, which is temporarily stored in the crop before passing into a gastric mill, where it is finely ground before digestion is completed in the stomach. Nutrients are absorbed into the hemocoel from outpockets called gastric ceca. The stomach leads into an intestine and a rectum, which empties by way of an anus. The excretory system consists of **Malpighian tubules,** which extend into the hemocoel and collect nitrogenous wastes that are excreted into the digestive tract. The formation of a solid nitrogenous waste, namely uric acid, conserves water.

The respiratory system begins with openings in the exoskeleton called spiracles. From here, the air enters small tubules called **tracheae** (Fig. 30.20*a*). The tracheae branch and rebranch until they end intracellularly, where the actual exchange of gases takes place. The movement of air through this complex of tubules is not passive. Rather, air is actively pumped by alternate contraction and relaxation of the body wall. Breathing by tracheae is suitable to the small size of insects (most are less than 60 mm in length), as the tracheae would be crushed by any significant amount of weight.

The circulatory system contains a slender, tubular heart that lies against the dorsal wall of the abdominal exoskeleton and pumps hemolymph into an aorta that leads to a hemocoel, where it circulates before returning to the heart again. The hemolymph is colorless and lacks a respiratory pigment because the tracheal system is responsible for gas exchange.

Reproduction and Development Grasshopper reproduction is adapted to life on land. The male grasshopper has a penis, and sperm passed to the female are stored in a seminal receptacle. Internal fertilization protects both gametes and zygotes from drying out. The female deposits the fertilized eggs in the ground with her ovipositor.

Metamorphosis is a change in form and physiology that occurs as an immature stage, called a larva, becomes an adult. Grasshoppers undergo gradual metamorphosis as they mature. The immature grasshopper, called a nymph, looks like an adult grasshopper, even though it differs somewhat in shape and form. Other insects, such as butterflies, undergo complete metamorphosis, involving drastic changes in form. At first, the animal is a wormlike larva (caterpillar) with chewing mouthparts. It then forms a case, or cocoon, about itself and becomes a pupa. During this stage, the body parts are completely reorganized. The adult then emerges from the cocoon. This life cycle allows the larvae and adults to use different food sources. Most eating by insects occurs during the larval stage. The field of forensic science has found a way to put our knowledge of insect life cycle stages to good use (see the Science Focus).

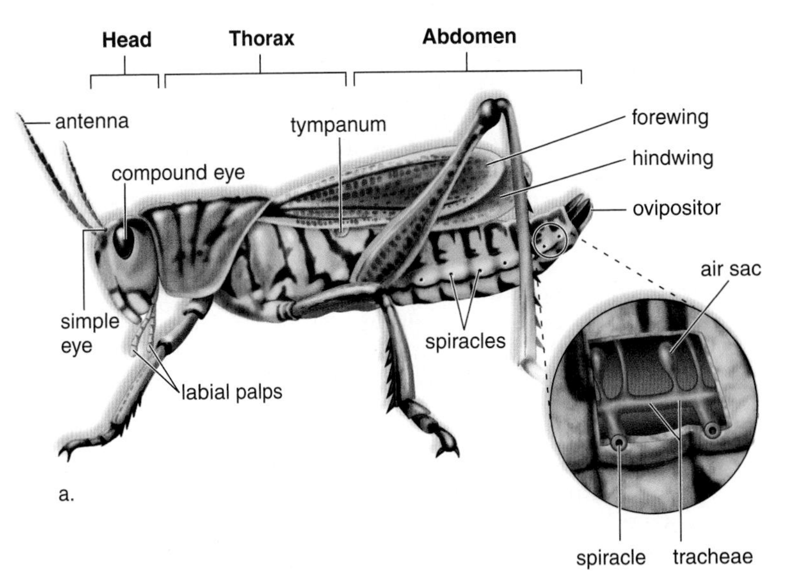

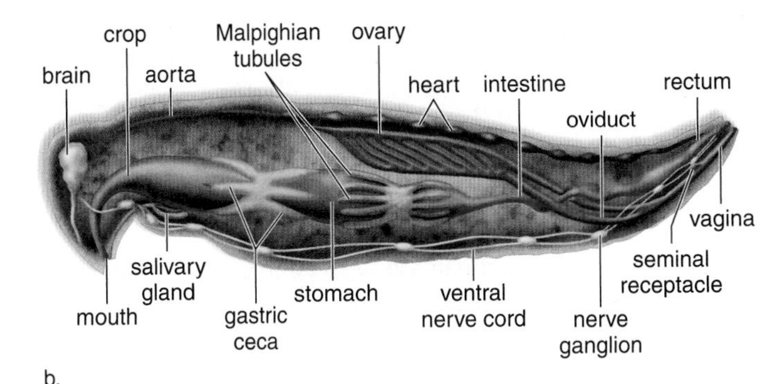

Figure 30.20 Female grasshopper, *Romalea*.
a. Externally, the body of a grasshopper is divided into three sections and has three pairs of legs. The tympanum receives sound waves, and the jumping legs and the wings are for locomotion. **b.** Internally, the digestive system is specialized. The Malpighian tubules excrete a solid nitrogenous waste (uric acid). A seminal receptacle receives sperm from the male, which has a penis.

Comparison with Crayfish The grasshopper and the crayfish share a common ancestor, but they have diverged in their morphology to adapt to aquatic versus terrestrial environments. In crayfish, gills take up oxygen from water, while

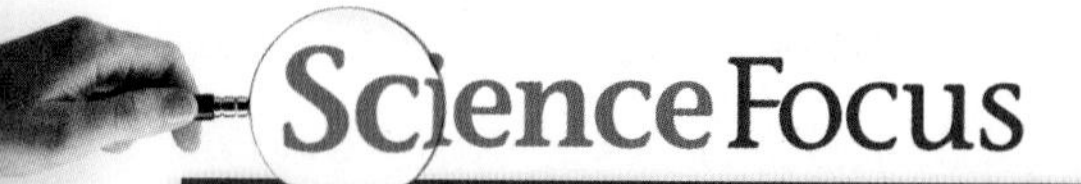

Maggots: A Surprising Tool for Crime Scene Investigation

A fresh human corpse is lying on the floor. Within 10–15 minutes, blowflies with blue and green metallic bodies arrive. An animal corpse is a perfect site for egg production. The flies lay eggs in the mouth, nostrils, ears, eyes, and any open wounds. Depending on the temperature, the eggs will hatch in the next 12–24 hours. The small hatchling larval flies, more commonly known as maggots, feed on fat and collagen in the corpse in order to grow. Their development occurs in three stages called instars, during which the larvae shed skin and grow. When the third instar grows, it leaves the corpse and forms a pupa that takes about 14 days (depending on the temperature) to emerge as an adult blowfly. The adult has to wait a day or two until it can fly because the wings need to expand and the body has to harden. Once capable of flight, the adult leaves to find a mate as well as another corpse in which to lay eggs, thereby completing the life cycle.

What do blowflies have to do with crime scene investigation? If the corpse involved is a human, the timing of the various stages of the life cycle of the flies (e.g., newly hatched eggs, maggot size, maggot weight, or pupa remains) can help forensic scientists determine the postmortem interval (or PMI) since the time of death. Forensic entomology uses insects such as blowflies to determine the time of death. A forensic entomologist can even determine if a human corpse has been moved based on the fact that different species of blowflies prefer to lay eggs in different environments. For example, some species lay their eggs in shade, while others lay eggs in sunny habitats; similarly, some species inhabit urban areas, while others inhabit rural areas. Thus, the combination of life stage(s) and species of blowfly found on a corpse can help a forensic entomologist determine the time and possible location of a person's death.

Discussion Questions

1. Given that the hatching of eggs and the growth rate of blowfly larvae depend on temperature, how sure do you think forensic entomologists can be about time of death?
2. As a juror, how would you feel about forensic entomology evidence entered into court?
3. When might forensic entomology evidence be useful in a murder case?

in the grasshopper, tracheae allow oxygen-laden air to enter the body. Appropriately, the crayfish has an oxygen-carrying pigment, but a grasshopper has no such pigment in its blood. The crayfish excretes a liquid nitrogenous waste (ammonia), while the grasshopper excretes a solid nitrogenous waste (uric acid). Only grasshoppers have (1) a tympanum for the reception of sound waves, and (2) a penis in males for passing sperm to females without possible desiccation, and an ovipositor in females for laying eggs in soil. Crayfish utilize their uropods when they swim; a grasshopper has legs for hopping and wings for flying.

Arachnids

The **arachnids** (subphylum Chelicerata) include scorpions, spiders, ticks, and mites (Fig. 30.21). In this group, the cephalothorax bears six pairs of appendages: the chelicerae and the pedipalps and four pairs of walking legs. The cephalothorax is followed by an abdomen that contains internal organs.

Scorpions are the oldest terrestrial arthropods. Today, they are widely distributed in the tropics, subtropics, and temperate regions, with a notable absence from New Zealand. They are nocturnal and spend most of the day hidden under a log or a rock. In scorpions, the pedipalps are large pincers, and the long abdomen ends with a stinger that contains venom.

Ticks and mites are parasites. Ticks suck the blood of vertebrates and sometimes transmit diseases, such as Rocky Mountain spotted fever or Lyme disease. Chiggers, the larvae of certain mites, feed on the skin of vertebrates.

Spiders, the most familiar arachnids, have a narrow waist that separates the cephalothorax from the abdomen. Each chelicera consists of a basal segment and a fang that delivers venom to paralyze or kill prey. Spiders use silk threads for all sorts of purposes, from lining their nests to catching prey. Biologists have used web-building behavior to discover how spider families are related.

The internal organs of spiders also show how they are adapted to a terrestrial way of life. Malpighian tubules work in conjunction with rectal glands to reabsorb ions and water before a relatively dry nitrogenous waste (uric acid) is excreted. Invaginations of the inner body wall form the lamellae ("pages") of spiders' so-called book lungs. Air flowing into the folded lamellae on one side exchanges gases with blood flowing in the opposite direction on the other side.

Check Your Progress

1. What anatomical features would allow you to recognize an insect?
2. Name two ways a crayfish is adapted to life in the water and a grasshopper is adapted to life on land.
3. Why is a spider classified as an arachnid and not as an insect?

Bioethical Focus

Conservation of Coral Reefs

Coral reefs are among the most biologically diverse and productive communities on Earth. Coral reefs tend to be found in warm, clear, and shallow tropical waters worldwide and are typically formed by reef-building corals, which are cnidarians. Aside from being beautiful and giving shelter to many colorful species of fishes, coral reefs help generate economic income from tourism, protect ocean shores from erosion, and may serve as the source of medicines derived from antimicrobial compounds that reef-dwelling organisms produce. However, coral reefs around the globe are being destroyed for a variety of reasons, most of them linked to human development. Deforestation, for example, causes tons of soil to settle on the top of coral reefs. This sediment prevents photosynthesis of symbiotic algae that provide food for the corals. When the algae die, so do the corals, which then turn white. This so-called coral "bleaching" has been seen in the Pacific Ocean and the Caribbean Sea. Recent evidence suggests that global warming could be linked to coral bleaching and death because corals can tolerate only a narrow range of temperatures. As global temperatures rise, so do water temperatures, and corals can die as a result. Global warming also contributes to favorable conditions for various pathogens that can kill corals, such as those similar to pathogens that cause cholera in humans. Increases in aquatic nutrients from fertilizers that wash into the ocean also make corals more susceptible to diseases, which can also kill them. Marine scientist Edgardo Gomez estimates that 90% of coral reefs in the Philippines are dead or deteriorating due to human activities such as pollution and, especially, overfishing. Fishing methods that employ dynamite or cyanide to kill or stun the fish for food or the pet trade can easily kill corals. Paleobiologist Jeremy Jackson of the Smithsonian Tropical Research Institute in Panama estimates that we may lose 60% of the world's coral reefs by the year 2050.

Figure 30A **Stony coral polyps, *Tubastraea aurea*.**

Form Your Own Opinion

1. What features of coral reefs help explain why they are so biologically diverse?
2. Considering what is causing the loss of coral reefs, is it possible to save them? How?

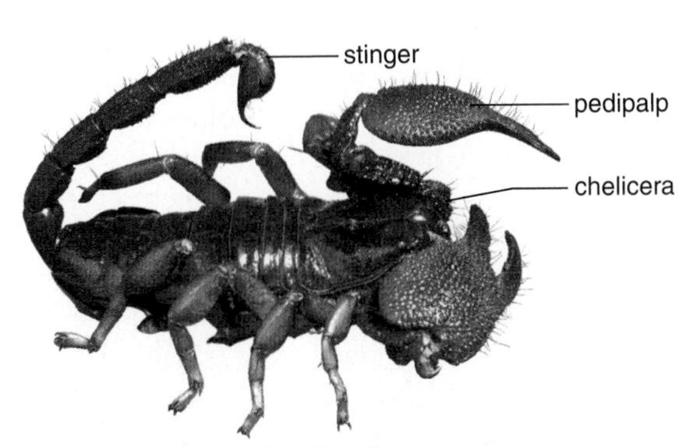

a. Kenyan giant scorpion, *Pandinus*

b. Goldenrod spider, *Misumena*

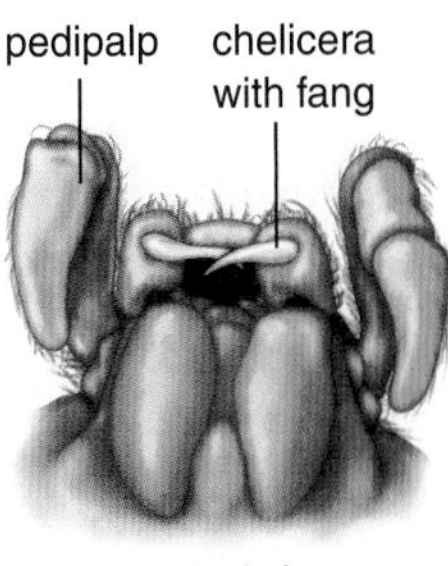

ventral view of mouthparts

c. Wood ticks, *Ixodes*

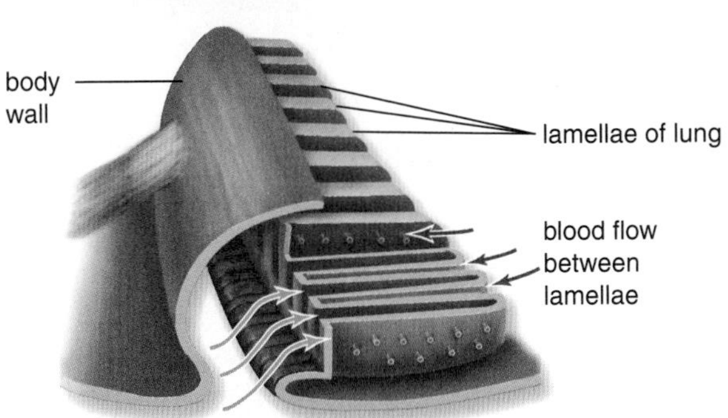

d. Book lung anatomy

Figure 30.21 **Arachnid diversity.**
a. A scorpion has pincerlike chelicerae and pedipalps. Its long abdomen ends with a stinger that contains venom. **b.** Goldenrod spiders can change colors from yellow to white to camouflage themselves with the color of the flower they are found on. **c.** In the western United States, the wood tick carries a debilitating disease called Rocky Mountain spotted fever. **d.** Arachnids breathe by means of book lungs, in which the "pages" are double sheets of thin tissue (lamellae).

Applying the Concepts Revisited

Although often viewed as small or insignificant, insects are the most diverse invertebrates, numbering about 1 million species. The great success of insects can be attributed to a light, flexible exoskeleton and remarkable behavioral adaptations that vary among species. The termites, for example, are all born to the same queen, yet soldiers, workers, and harvesters are all morphologically distinct and have specific roles within the mound. Soldiers defend the colony, workers build, and harvesters cultivate food. As with ants and bees, the cooperation of thousands to millions of individuals is what keeps the colony going.

1. In what way did anatomical complexity increase during the evolutionary history of the invertebrates we discussed in this chapter?
2. How might the ability to fly help explain the diversity of insect species?

SUMMARIZING THE CONCEPTS

30.1 Evolutionary Trends Among Animals

- Animals are motile, multicellular heterotrophs that ingest their food.
- They have the diploid life cycle.
- A phylogenetic tree depicts evolutionary trends among the animals.

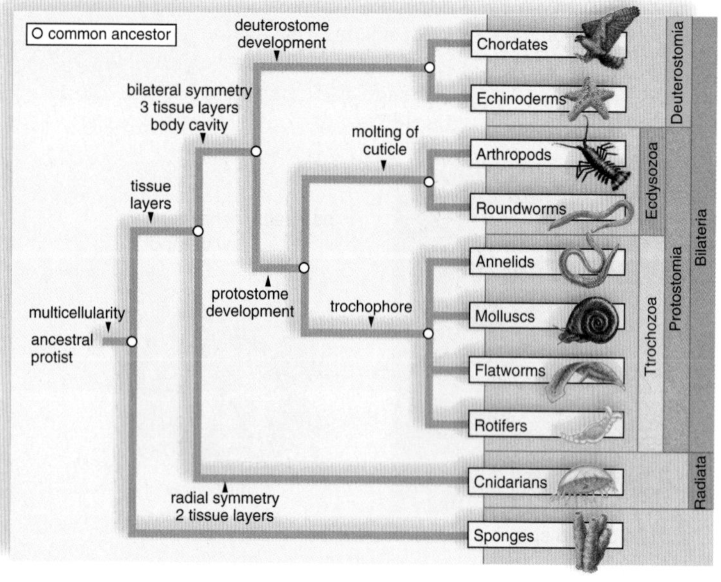

30.2 Introducing the Invertebrates

- Table 30.3 contrasts selected anatomical features used to help classify animals.

- Sponges have the cellular level of organization, lack true tissues, and are typically asymmetrical.
- Cnidarians are radially symmetrical and have two tissue layers derived from the germ layers ectoderm and endoderm.
- Cnidarians exist as either polyps or medusae, or they can alternate between the two.

30.3 The Trochozoa

- Flatworms have three tissue layers and no coelom.
- Planarians have cephalization, muscles, and a ladder-type nervous system.
- Rotifers are microscopic and have a corona that resembles a spinning wheel when in motion.
- The body of a mollusc typically contains a visceral mass, a mantle, and a foot. Many molluscs also have a head and a radula.
- Molluscs often have a reduced coelom and an open circulatory system.

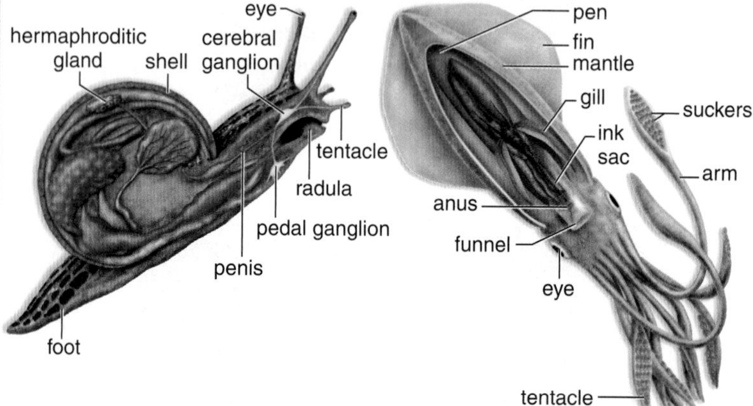

TABLE 30.3 Comparison of Invertebrates

	Sponges	Cnidarians	Flatworms	Roundworms	Other Phyla*
Level of organization	Cell	Tissue	Organs	Organs	Organs
Type of body plan	—	Incomplete digestive system	Incomplete digestive system	Complete digestive system	Complete digestive system
Type of symmetry	Radial or none	Radial	Bilateral	Bilateral	Bilateral except echinoderms
Type of body cavity	—	—	None	Pseudocoelom	Coelom
Segmentation	—	—	—	—	Only annelids, arthropods, and chordates
Jointed appendages	—	—	—	—	Only arthropods and chordates

*Molluscs, annelids, arthropods, echinoderms, chordates

- Predaceous squid display marked cephalization, move rapidly by jet propulsion, and have a closed circulatory system.
- Bivalves (e.g., clams), which have a hatchet foot, are filter feeders.
- Annelids are segmented both externally and internally.
- Polychaetes are marine worms that have parapodia.
- A clam worm is a predaceous marine worm with a defined head region.
- Earthworms are oligochaetes without a well-defined head region that scavenge for food in the soil.

30.4 The Ecdysozoa

- Roundworms are usually small, very diverse, and are present almost everywhere in great numbers. Roundworms have a pseudocoelom.
- Arthropods are the most diverse group of animals. Their success is largely attributable to a flexible exoskeleton, segmentation, a well-developed nervous system, respiratory organs, and metamorphosis.
- Like many other arthropods, crustaceans have a head that bears compound eyes, antennae, and mouthparts.
- Like many other insects, grasshoppers have wings and three pairs of legs attached to the thorax. Grasshoppers also have several adaptations to a terrestrial life, including respiration by tracheae.
- Spiders are arachnids with chelicerae, pedipalps, and four pairs of walking legs attached to a cephalothorax. Spiders, too, are adapted to life on land, and they spin silk that is used in various ways.
- The crayfish, a crustacean, also has an open circulatory system, respiration by gills, and a ventral solid nerve cord.

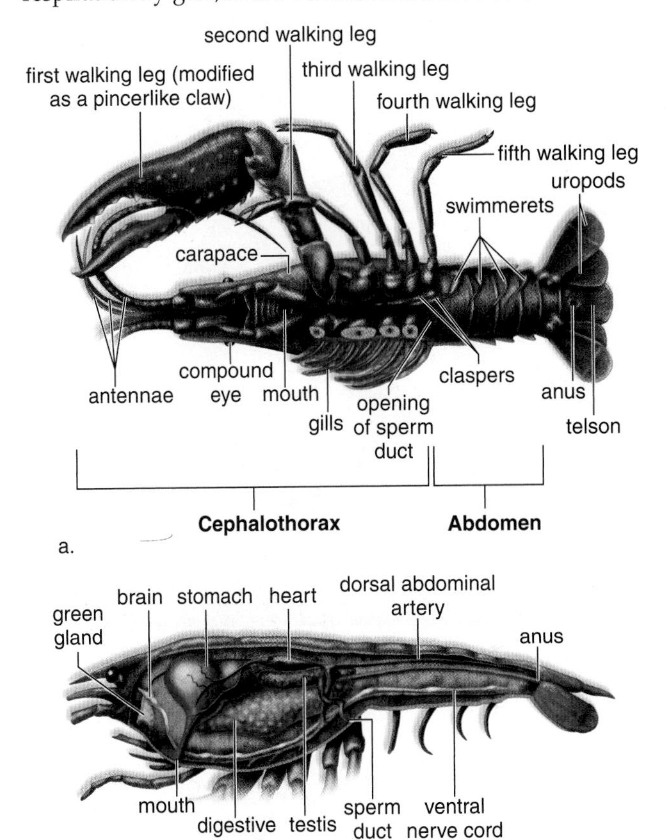

TESTING YOURSELF

Choose the best answer for each question.

1. Animals with a cellular level of organization produce
   a. cells only.
   b. cells and tissues.
   c. cells and organs.
   d. cells, tissues, and organs.
2. Which of the following has an incomplete digestive tract?
   a. earthworm
   b. hydra
   c. leech
   d. ant
   e. mosquito

For problems 3–7, indicate the type of symmetry exhibited by the animal listed. Each answer can be used more than once.

Key:

a. asymmetry
b. radial symmetry
c. bilateral symmetry

3. Lobster
4. Jellyfish
5. Sponge
6. Planarian
7. Octopus

For problems 8–10, match the feature with the type of flatworm. Each answer may be used more than once. Each problem may have more than one answer.

Key:

a. planarian
b. tapeworm
c. fluke

8. Endoparasite
9. Light-sensitive organs
10. Digestive tract lacks an anus
11. Compared to an animal species that has a coelom, one that lacks a coelom
    a. is more flexible.
    b. has more complex organs.
    c. is more likely to tolerate temperature variations.
    d. is less able to store excretory wastes.
12. Which of the following is not a feature of most coelomates?
    a. radial symmetry
    b. three germ layers
    c. complete digestive tract
    d. organ level of organization

13. A feature of annelids is
    a. segmented body.
    b. acoelomate.
    c. incomplete digestive tract.
    d. radial symmetry.
14. Label the parts of an earthworm in the following illustration.

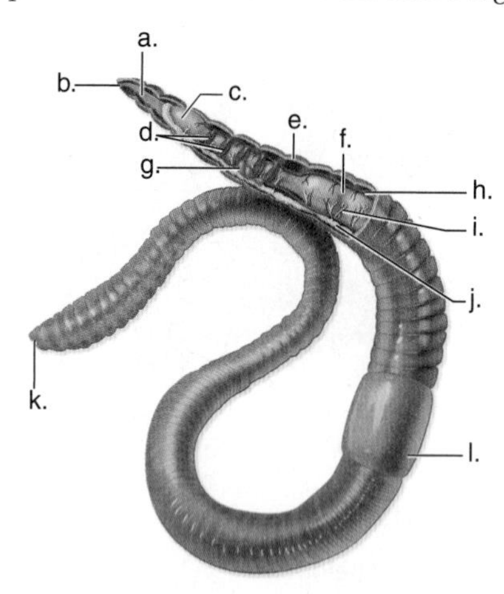

15. Which of the following is not a feature that has led to the success of the arthropods?
    a. They have two types of eyes.
    b. They have jointed appendages.
    c. They produce an exoskeleton.
    d. They have ears that can hear a wide range of sounds.
    e. They have well-developed eyes.

UNDERSTANDING THE TERMS

annelid 632
arachnid 640
arthropod 636
asymmetrical 620
bilateral symmetry 620
bivalve 631
cephalization 622
cephalopod 631
closed circulatory system 632
cnidocytes 625
collar cells 624
crustacean 637
deuterostome 623
ecdysozoa 623
elephantiasis 635
filter feeder 624
flame cells 628
flatworm 627
gastropod 630
gastrovascular cavity 625
green glands 638
hemocoel 638
hermaphroditic 624
hydrostatic skeleton 632
insect 638
invertebrate 620
leech 634
Malpighian tubule 639
mantle 630
medusan 625
metamorphosis 639
mollusc 629
nematocyst 625
nephridium (pl., nephridia) 632
open circulatory system 632
polyp 625
proglottids 628
protostome 623
pseudocoelom 635
radial symmetry 620
roundworm 634
scolex 628
sessile 622
setae 632
spicules 624
sponge 624
tracheae 639
trichinosis 635
trochophore 623
trochozoa 623
ventral solid nerve cord 632
visceral mass 630

THINKING CRITICALLY

1. Cnidarians are radially symmetrical, while flatworms, roundworms, molluscs, annelids, and arthropods are bilaterally symmetrical. How does the type of symmetry relate to the lifestyle of each type of animal?
2. What features of arthropods make them the most diverse animal phylum on Earth?
3. Why do you think coelomates far outnumber acoelomates or pseudocoelomates in terms of species diversity?

INQUIRY INTO LIFE WEBSITE

The companion website for *Inquiry into Life* provides a wealth of information organized and integrated by chapter. You will find practice tests, animations, videos, and much more that will complement your learning and understanding of general biology.

http://www.mhhe.com/maderinquiry13

Animals: Part II

Chapter Outline

Applying the Concepts

Thought to be extinct for 60 years, the ivory-billed woodpecker was likely rediscovered in 2005 in the Cache River National Wildlife Refuge in Arkansas. The ivory-billed is the largest woodpecker in the United States and the second largest in the world. After reports that biologists had heard the bird's call, Dr. John Fitzpatrick, Director of the Cornell Lab of Ornithology, and Scott Simon, Arkansas State Director of The Nature Conservancy, spent a year leading a team that searched for the bird. Its existence has been confirmed by video footage and seven eyewitness sightings. The Nature Conservancy and the Cornell Laboratory of Ornithology have since led a multiorganization effort to form the Big Woods Conservation Partnership to protect the bird's habitat. Their 10-year goal is to protect more than 200,000 acres in the Big Woods of Arkansas. Thus far, the Nature Conservancy has worked to protect more than 18,000 acres near the Cache River and White River National Wildlife Refuges.

In Chapter 30, we learned about invertebrates, new species of which are being discovered all the time. We are also discovering and rediscovering vertebrates such as the ivory-billed woodpecker. However, approximately 20% of all mammals, 12% of birds, and 32% of amphibians are threatened with extinction. The future of many vertebrates may, in large part, depend on our ability to protect them. Amidst such discouraging messages, the ivory-billed woodpecker brings with it a ray of hope. The large-scale cooperative effort to conserve the ivory-billed woodpecker's habitat and help to protect other species that inhabit the area is an example of how people can work together to protect biodiversity.

31.1 Echinoderms

Learning Outcomes

1. Describe the similarities between echinoderms and chordates.
2. Outline the basic morphological characteristics of echinoderms.

In Chapter 30, we introduced the invertebrates, the large group of animals characterized by their lack of an endoskeleton of bone or cartilage. This chapter will focus mainly on the echinoderms and the chordates, which include the vertebrates. Table 31.1 previews the animal groups we will discuss in this chapter.

Among animals, chordates are most closely related to the **echinoderms** (phylum Echinodermata) as witnessed by their similar embryological development, even though echinoderms are invertebrates. Both echinoderms and chordates are deuterostomes, in which the second embryonic opening becomes the mouth, and the coelom forms by outpocketing the primitive gut. Recall from Chapter 30 that protostomes, in contrast, are characterized by a mouth that forms from the first embryonic opening (the blastopore).

Characteristics of Echinoderms

Echinoderms are a diverse group of marine animals. There are no terrestrial echinoderms. They have an endoskeleton (internal skeleton) consisting of spine-bearing, calcium-rich plates. The spines, which stick out through their delicate skin, account for their name.

It may seem surprising that echinoderms, although related to chordates, lack those features we associate with vertebrates such as human beings. For example, echinoderms are often radially, not bilaterally, symmetrical. Their larva is a free-swimming filter feeder with bilateral symmetry, but it typically metamorphoses into a radially symmetrical adult. Recall that with radial symmetry, it is possible to obtain two mirror images, no matter how the animal is sliced longitudinally, whereas in bilaterally symmetrical animals, only a longitudinal cut gives two mirror images.

Echinoderm Diversity

Echinoderms are quite diverse (Fig. 31.1). They include sea lilies, motile feather stars, brittle stars, and sea cucumbers. Sea cucumbers actually look like a cucumber, except they have feeding tentacles surrounding their mouth. You may

a. Brittle star, *Ophiothrix*

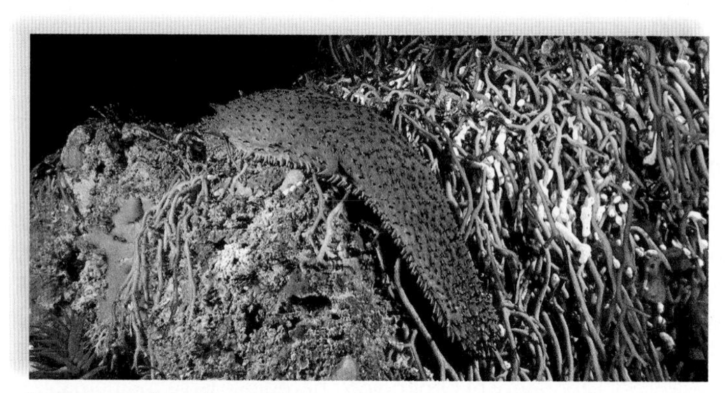

b. Sea cucumber, *Cucumaria*

c. Sea urchin, *Strongylocentrotus*

d. Feather stars, *Antedon*

Figure 31.1 Echinoderm diversity.
a. Brittle stars have long, slender arms, which make them the most mobile of echinoderms. **b.** Sea cucumbers look like a cucumber. They lack arms but have tentacle-like tube feet with suckers around the mouth. **c.** Sea urchins have large, colored, external spines for protection. **d.** A feather star extends its arms to filter suspended food particles from the sea.

TABLE 31.1 Animal Phyla, Part II

Phylum Echinodermata: echinoderms
Radially symmetrical as adults; endoskeleton of calcium-rich plates; spiny skin; unique water vascular system with tube feet; able to regenerate lost body parts
Phylum Chordata: chordates
Segmented, with a notochord; dorsal tubular nerve cord; pharyngeal pouches; postanal tail at some time during the life cycle
Subphylum Urochordata
Sessile adult; encased in a tunic; lacks notochord and nerve cord. Example: tunicates
Subphylum Cephalochordata
Retains the four chordate characteristics as adults; obvious segmentation throughout body. Example: lancelets
Subphylum Vertebrata: vertebrates
The notochord is replaced by the vertebral column. Examples: fishes, amphibians, reptiles (including birds), mammals

be more familiar with sea urchins and sand dollars in the class Echinoidea, which have spines for locomotion, defense, and burrowing. Sea urchins are food for sea otters off the coast of California. We will use the sea star as an example of echinoderm characteristics.

Sea Stars

Sea stars (starfish) are commonly found along rocky coasts, where they feed on clams, oysters, and other bivalve molluscs. The five-rayed body has an oral, or mouth, side (the underside) and an aboral, or anus, side (the upper side) (Fig. 31.2). Various structures project through the body wall: (1) spines from the endoskeletal plates offer some protection; (2) pincerlike structures called pedicellariae keep the surface free of small particles; and (3) skin gills, tiny fingerlike extensions of the skin, are used for gas exchange. On the oral surface, each arm has a groove lined by small tube feet.

To feed, a sea star positions itself over a bivalve such as a clam and attaches some of its tube feet to each side of the shell (Fig. 31.2*a*). By working its tube feet in alternation, it pulls the shell open. A very small crack is enough for the sea star to evert its cardiac stomach and push it through the crack to contact the soft parts of the bivalve. The stomach secretes enzymes, and digestion begins even while the bivalve is attempting to close its shell. Later, partly digested food is taken into the sea star's body, where digestion continues in the pyloric stomach, using enzymes from digestive glands found in each arm. A short intestine opens at the anus on the aboral side.

In each arm, the well-developed coelomic cavity contains not only a pair of digestive glands, but also gonads (either male or female), which open on the aboral surface by very small pores. The nervous system consists of a central nerve ring from which radial nerves extend into each arm. A light-sensitive eyespot is at the tip of each arm. Sea stars are capable of coordinated but slow responses and body movements.

Locomotion depends on their water vascular system. Water enters this system through a structure on the aboral

a.

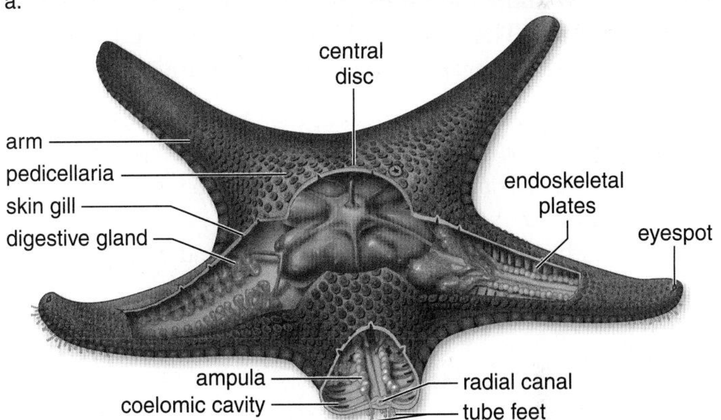

b. Aboral side showing ray cross-section

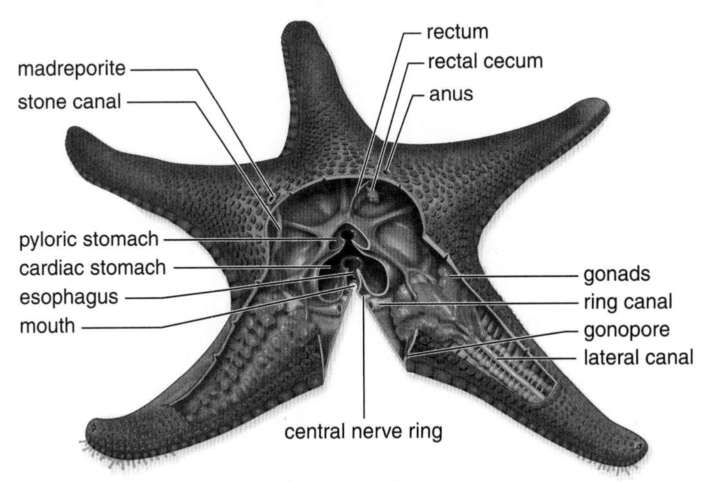

c. Aboral side showing internal cross-section

Figure 31.2 Sea star anatomy and behavior.
a. A sea star uses the suction of its tube feet to open a bivalve mollusc, its primary source of food. **b.** Each arm of a sea star contains digestive glands, gonads, and portions of the water vascular system. This system (colored orange) terminates in tube feet. **c.** A different cross-section shows the orientation of mouth, digestive system, and anus.

side called the madreporite, or sieve plate. From there it passes down a stone canal into a ring canal, which surrounds the mouth. A radial canal in each arm branches off from the central ring canal. For locomotion, water enters the ampullae from the radial canals. Contraction of ampullae forces water into the tube foot, expanding it. When the foot touches a surface, the center is withdrawn, giving it suction so that it can adhere to the surface. By alternating the expansion and contraction of the tube feet, a sea star moves slowly along.

Echinoderms don't have a respiratory, excretory, or circulatory system. Fluids within the coelomic cavity and the water vascular system carry out many of these functions. For example, gas exchange occurs across the skin gills and the tube feet. Nitrogenous wastes diffuse through the coelomic fluid and the body wall. Cilia on the peritoneum lining the coelom keep the coelomic fluid moving.

Sea stars reproduce both asexually and sexually. If the body is fragmented, each fragment can regenerate a whole animal as long as the fragment contains part of the central disk. Fishermen who try to get rid of sea stars by cutting them up and tossing them overboard are merely propagating more sea stars! Sea stars also spawn, releasing either eggs or sperm. The larva is bilaterally symmetrical and metamorphoses to become the radially symmetrical adult.

Check Your Progress

1. Why are echinoderms more closely related to chordates than invertebrates?
2. Draw and label a basic diagram of an echinoderm.

31.2 Chordates

Learning Outcomes

1. List the four morphological characteristics unique to the chordates.
2. Explain why lancelets and tunicates are closely related to vertebrates.

To be classified as a **chordate** (phylum Chordata, about 45,000 species), an animal must have, at some time during its life history, the characteristics depicted in Figure 31.3 and listed here:

1. A **dorsal supporting rod** called a **notochord.** The notochord is located just below the nerve cord toward the back (i.e., dorsal). Vertebrates have an embryonic notochord that is replaced by the vertebral column during development.
2. A **dorsal tubular nerve cord.** Tubular means that the cord contains a canal filled with fluid. In vertebrates, the nerve cord is protected by the vertebrae. Therefore, it is called the spinal cord because the vertebrae form the spine.
3. **Pharyngeal pouches.** Most vertebrates have pharyngeal pouches only during embryonic development. In the nonvertebrate chordates, the fishes, and some amphibian larvae, the pharyngeal pouches become functioning gills. Water passing into the mouth and the pharynx goes through the gill slits, which are supported by gill arches. In terrestrial vertebrates that breathe by lungs, the pouches are modified for various purposes. In humans, the first pair of pouches becomes the auditory tubes. The second pair becomes the tonsils, while the third and fourth pairs become the thymus gland and the parathyroids.
4. **Postanal tail.** The tail extends beyond the anus, hence the term postanal.

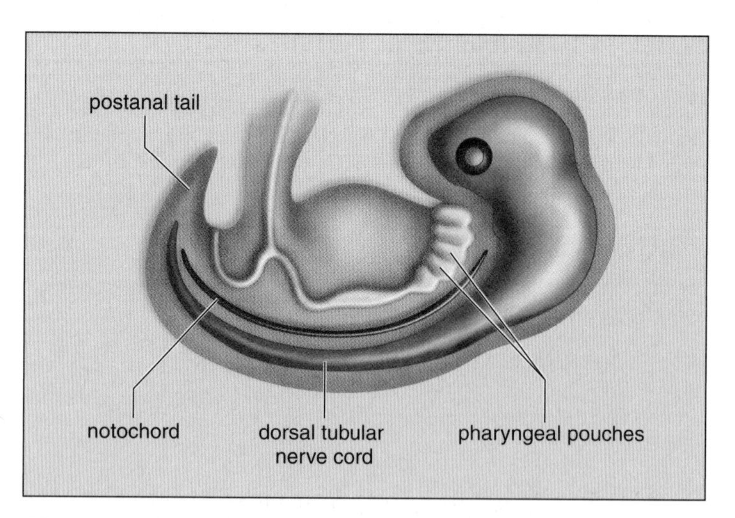

Figure 31.3 Chordate characteristics.

Evolutionary Trends Among the Chordates

Figure 31.4 depicts the phylogenetic tree of the chordates, The figure also lists at least one main evolutionary trend that distinguishes each group of animals from the preceding one.

The tunicates and lancelets are nonvertebrate chordates, meaning that they do not have vertebrae. The vertebrates are the fishes, amphibians, reptiles, birds, and mammals. The cartilaginous fishes were the first to have jaws, and some early bony fishes had lungs. However, not all bony fishes have lungs. Amphibians are the first group to clearly have jointed appendages and to invade land. However, the fleshy appendages of lobe-finned fishes from the Devonian era contained bones homologous to those of terrestrial vertebrates. These fishes are believed to be ancestral to the amphibians. Reptiles and mammals have a means of reproduction suitable for land. During development, an amnion and other extraembryonic membranes are present. These membranes support an embryo and prevent it from drying out as it develops into a particular species' offspring.

Check Your Progress

1. What are the four features that distinguish chordates from other animal phyla?
2. With reference to Figure 31.4, list three innovations that distinguish mammals from cartilaginous fishes.

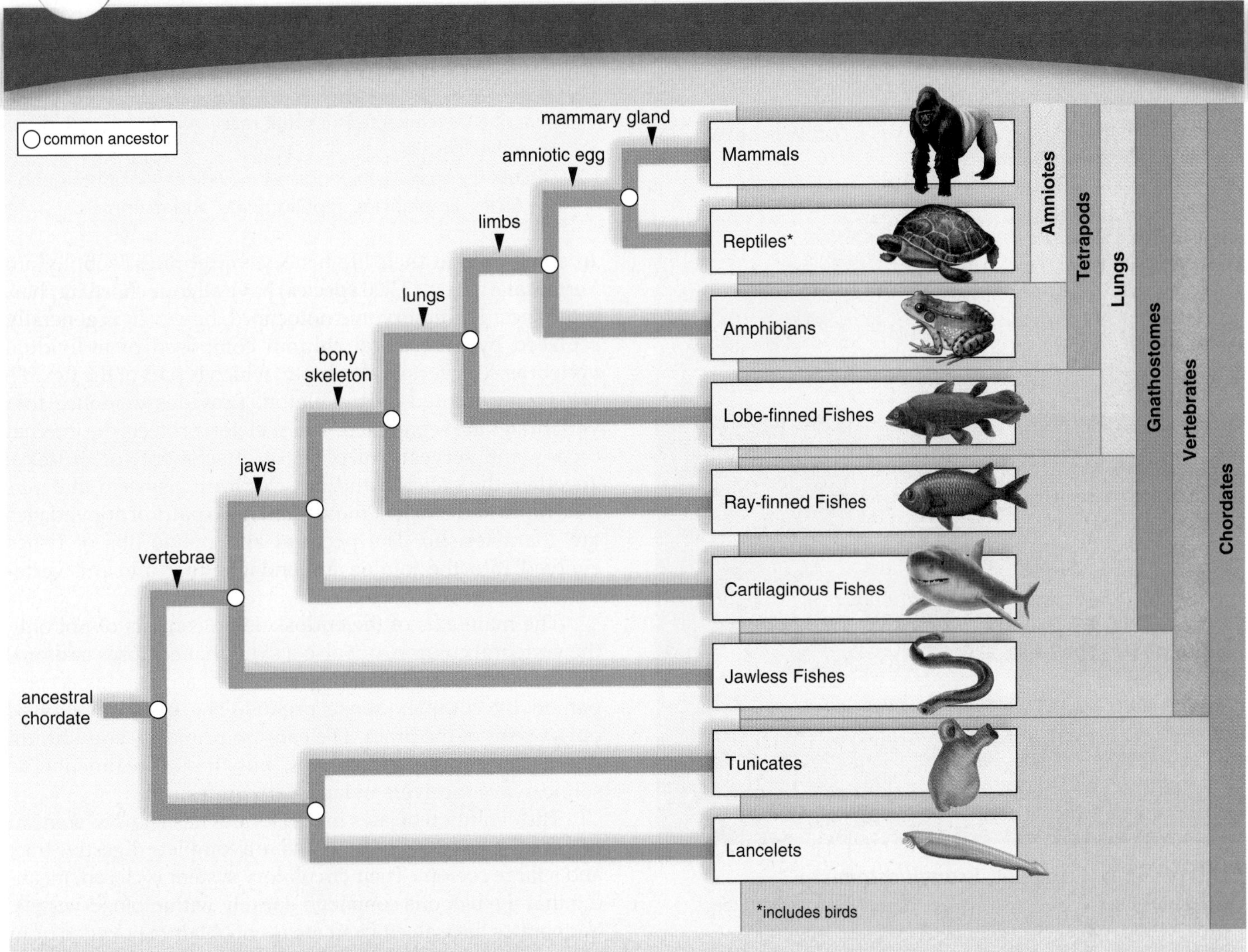

Figure 31.4 Phylogenetic tree of chordates.
Each of the innovations listed is an evolutionary trend shared by the classes beyond the branch point.

Nonvertebrate Chordates

Nonvertebrate chordates are characterized by the fact that the notochord never becomes a vertebral column.

Tunicates (subphylum Urochordata, about 1,250 species) live on the ocean floor and take their name from a tunic that makes the adults look like thick-walled, squat sacs (Fig. 31.5). They are also called sea squirts because they squirt water from one of their siphons when disturbed. The tunicate larva is bilaterally symmetrical and has the four chordate characteristics. Metamorphosis produces the sessile adult, which has incurrent and excurrent siphons.

The pharynx is lined by numerous cilia. Beating of these cilia creates a current of water that moves into the pharynx and out the numerous gill slits, the only chordate characteristic that remains in the adult. Microscopic particles adhere to a mucous secretion and are digested.

Some biologists have suggested that tunicates are directly related to vertebrates. They hypothesize that a larva with the

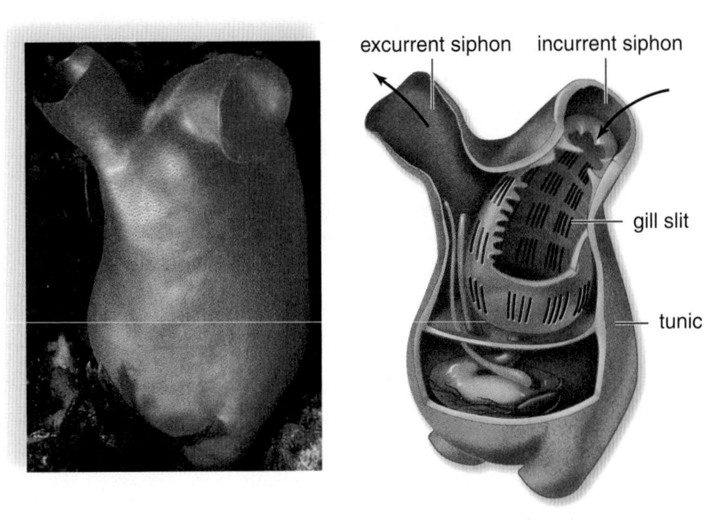

Figure 31.5 Sea squirt, *Halocynthia*.
The only chordate characteristic remaining in the adult is gill slits.

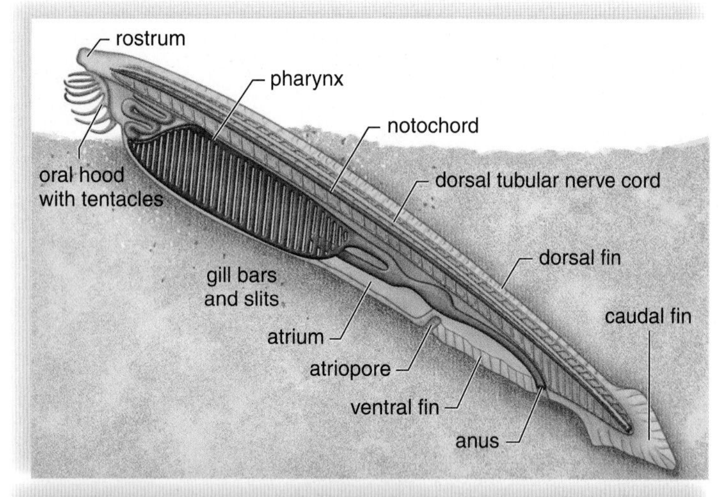

Figure 31.6 Lancelet, *Branchiostoma*.
Lancelets are filter feeders. Water enters the mouth and exits at the atriopore after passing through the gill slits.

four chordate characteristics may have become sexually mature, and evolution thereafter produced a fishlike vertebrate.

Lancelets, formerly referred to as amphioxus (subphylum Cephalochordata, about 23 species), are marine chordates only a few centimeters long. They are named for their resemblance to a lancet—a small, two-edged surgical knife (Fig. 31.6). Lancelets are found in shallow water along most coasts, where they usually lie partly buried in sandy or muddy substrates with only their anterior mouth and gill apparatus exposed. They feed on microscopic particles filtered out of a constant stream of water that enters the mouth and exits through the gill slits.

Lancelets retain the four chordate characteristics as adults. In addition, segmentation is present, as evidenced by the segmental arrangement of the muscles and the periodic branches of the dorsal tubular nerve cord.

Check Your Progress

1. Describe the basic life cycle of a tunicate.
2. Why are tunicates considered by some to be most closely related to the vertebrates?
3. Highlight the morphological differences between a tunicate and a lancelet.

31.3 Vertebrates

Learning Outcomes

1. Describe the characteristics that make vertebrates suited to an active lifestyle.
2. Discuss the major evolutionary innovations that distinguish the fishes, amphibians, reptiles, birds, and mammals.

At some time in their life history, **vertebrates** (subphylum Vertebrata, about 43,700 species) have all four chordate characteristics. The embryonic notochord, however, is generally replaced by a vertebral column composed of individual vertebrae. The vertebral column, which is part of the flexible but strong jointed endoskeleton, provides evidence that vertebrates are segmented. The skeleton protects the internal organs and serves as a place of attachment for muscles. Together, the skeleton and muscles form a system that permits rapid and efficient movement. Two pairs of appendages are characteristic. The pectoral and pelvic fins of fishes evolved into the jointed appendages that allowed vertebrates to move onto land.

The main axis of the endoskeleton consists of not only the vertebral column, but also a skull that encloses and protects the brain. The high degree of cephalization is accompanied by complex sense organs. The eyes develop as outgrowths of the brain. The ears are primarily equilibrium devices in aquatic vertebrates, but they also function as sound-wave receivers in land vertebrates.

The evolution of jaws in vertebrates has allowed some to become predators. Vertebrates have a complete digestive tract and a large coelom. Their circulatory system is closed, meaning that the blood is contained entirely within blood vessels. Vertebrates have an efficient means of obtaining oxygen from water or air, as appropriate. The kidneys are important excretory and water-regulating organs. The sexes are generally separate, and reproduction is usually sexual. The evolution of the amnion allowed reproduction to take place on land. Many species of reptiles and a few species of mammals lay a shelled egg. In placental mammals, development takes place within the uterus of the female. Although use of vertebrates, as well as other animals in laboratory research has been very beneficial in many ways, it also generates controversy for some. The Bioethical Focus later in this chapter discusses issues related to use of animals in scientific research.

A strong, jointed endoskeleton, a vertebral column composed of vertebrae, a closed circulatory system, efficient respiration and excretion, and a high degree of cephalization are characteristics demonstrating vertebrates are adapted to an active lifestyle. Figure 31.7 shows major milestones in the history of vertebrates: the evolution of jaws, limbs, and the amnion, an extraembryonic membrane that is first seen in the shelled **amniotic egg** of reptiles.

Fishes: First Jaws, Then Lungs

The first vertebrates were jawless fishes that wiggled through the water and sucked up food from the ocean floor.

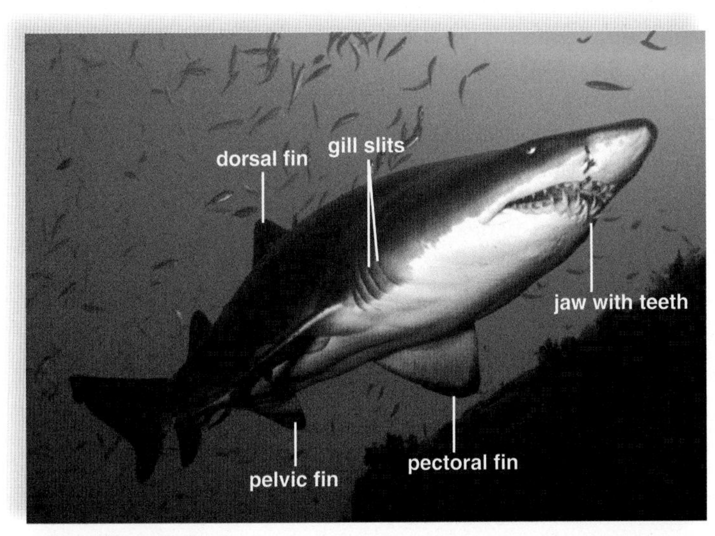

a. Sand tiger shark

b. Spotted salamander, *Ambystoma*

c. Veiled chameleons, *Chamaeleo*

Figure 31.7 Milestones in vertebrate evolution.
a. The evolution of jaws in fishes allows animals to be predators and to feed off other animals. **b.** The evolution of limbs in most amphibians is adaptive for locomotion on land. **c.** The evolution of an amnion and a shelled egg in reptiles is adaptive for reproduction on land.

Today, there are three living groups of fishes: jawless fishes, cartilaginous fishes, and bony fishes. The two latter groups have *jaws,* tooth-bearing structures in the head. Jaws evolved from the first pair of gill arches, structures that ordinarily support gills. The second pair of gill arches became support structures for the jaws, instead of gills. The presence of jaws permits a predatory way of life in many species of fishes.

Fishes are adapted to life in the water. Usually, they shed their sperm and eggs into the water, where fertilization occurs. The zygote develops into a swimming larva, which must fend for itself until it develops into the adult form.

Jawless Fishes

Living representatives of the **jawless fishes** (superclass Agnatha, about 63 species) are cylindrical and up to a meter long. They have smooth, scaleless skin and no jaws or paired fins. The two groups of living jawless fishes are *hagfishes* and *lampreys.* The hagfishes are scavengers, feeding mainly on dead fishes, while some lampreys are parasitic. When parasitic, the round mouth of the lamprey serves as a sucker. The lamprey attaches itself to another fish and taps into its circulatory system. Unlike other fishes, the lamprey cannot take in water through its mouth. Instead, water moves in and out through the gill openings.

Cartilaginous Fishes

Cartilaginous fishes (class Chondrichthyes, about 850 species) are the sharks (see Fig. 31.7*a*), the rays (Fig. 31.8*a*), and the skates. This group of fishes is so named because they have skeletons of cartilage instead of bone. The small dogfish shark is often dissected in biology laboratories. One of the most dangerous sharks inhabiting both tropical and temperate waters is the hammerhead shark, although people are rarely attacked by sharks. Each year more people are injured or killed by dog attacks than shark attacks. The largest sharks, the whale sharks, feed on small fishes and marine invertebrates and do not attack humans. Skates and rays are rather flat fishes that live partly buried in the sand and feed on mussels and clams.

Three well-developed senses enable sharks and rays to detect their prey: (1) They have the ability to sense electric currents in water—even those generated by the muscle movements of animals. (2) They have a lateral line system, a series of pressure-sensitive cells along both sides of the body that can sense pressure caused by nearby movement in the water. (3) They have a keen sense of smell. The part of the brain associated with this sense is enlarged relative to the other parts. Sharks can detect about one drop of blood in 25 gallons of water.

Bony Fishes

Bony fishes (class Osteichthyes, about 20,000 species) are by far the most numerous and diverse of all the vertebrates.

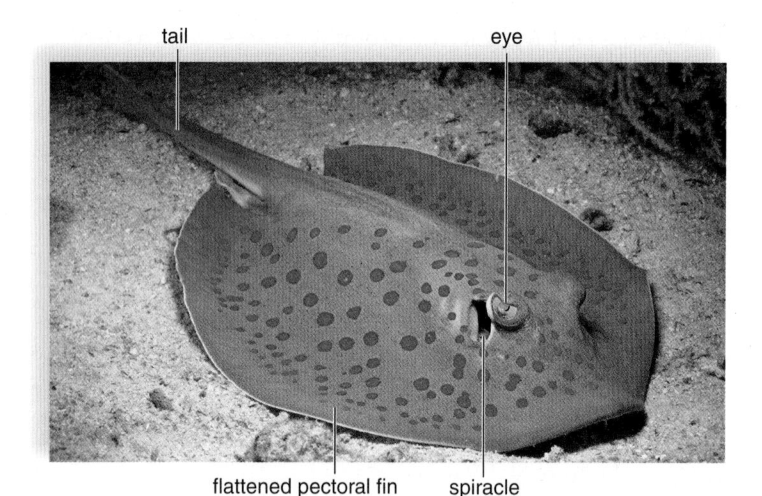

a. Blue-spotted stingray, *Taeniura lymma*

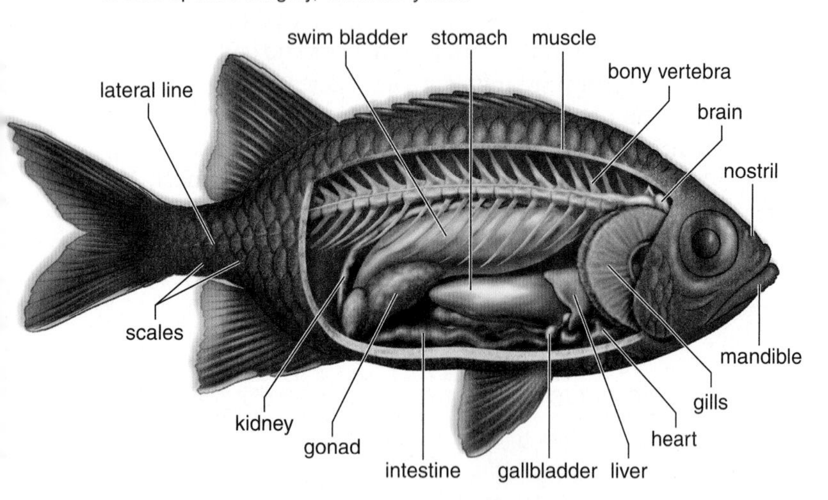

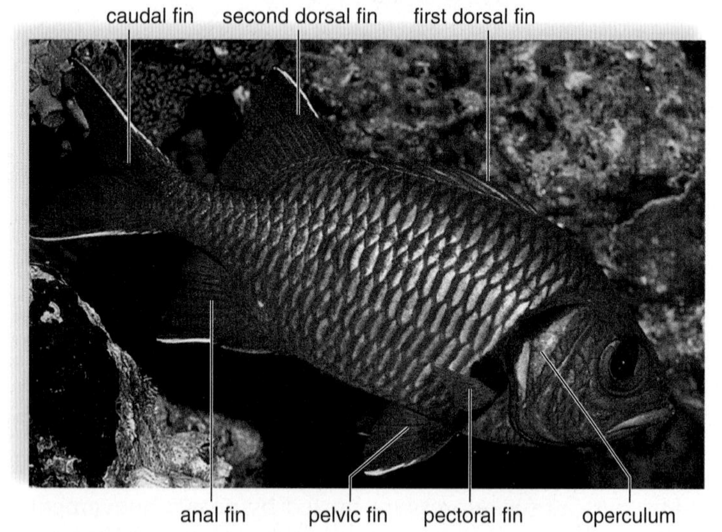

b. Soldierfish, *Myripristis jacobus*

Figure 31.8 Jawed fishes.
a. Rays, such as this blue-spotted stingray, are cartilaginous fishes. **b.** A soldierfish has the typical appearance and anatomy of a ray-finned fish, the most common type of bony fish.

Most of the bony fishes we eat, such as cod, trout, salmon, and haddock, are **ray-finned fishes.** Their paired fins, which they use to balance and propel the body, are thin and supported by bony rays. Ray-finned fishes have various ways of life. Some, such as herring, are filter feeders; others, such as trout, are opportunists; and still others are predaceous carnivores, such as piranhas and barracudas.

Ray-finned fishes have a swim bladder, which usually regulates buoyancy (Fig. 31.8*b*). By secreting gases into or absorbing gases from the bladder, these fishes can change their density and thus go up or down in the water, respectively. The streamlined shape, fins, and muscle action of ray-finned fishes are all suited for locomotion in the water. Their skin is covered by bony scales that protect the body but do not prevent water loss. When ray-finned fishes respire, their gills are kept continuously moist by the passage of water through the mouth and out the gill slits. As the water passes over the gills, oxygen is absorbed by the blood, and carbon dioxide is given off. Ray-finned fishes have a single-circuit circulatory system. The heart is a simple pump, and the blood flows through the chambers, including a nondivided atrium and ventricle, to the gills. Oxygenated blood leaves the gills and is circulated throughout the body by the heart (see Fig. 31.12*c*).

Another type of bony fish, called the **lobe-finned fishes,** evolved into the amphibians. These fishes not only had fleshy appendages that could be adapted to land locomotion, but most also had a **lung** that was used for respiration. One type of lobe-finned fish, the coelacanth, was thought to have gone extinct about 20,000 years ago. However, some coelacanths were recently discovered off the coasts of Eastern Africa and Indonesia, making them the only "living fossil" among these fishes.

Check Your Progress

1. Which features allow vertebrates to be relatively active?
2. List characteristics that distinguish bony fishes from other vertebrates.
3. How are ray-finned fishes different from lobe-finned fishes?

Amphibians: Jointed Appendages

Amphibians (class Amphibia, about 6,600 species), whose class name means living on both land and in the water, are represented today by frogs, toads, newts, salamanders, and caecilians, which are fossorial, wormlike amphibians that spend most of their life underground. Aside from *jointed appendages* (which have been lost in the limbless caecilians), amphibians have other features not seen in bony fishes: they usually have four limbs, eyelids for keeping their eyes moist, ears (a tympanum) for picking up sound waves, and a voice-producing larynx. Relative to body size, the brain is larger than that of a fish. Adult amphibians usually have small

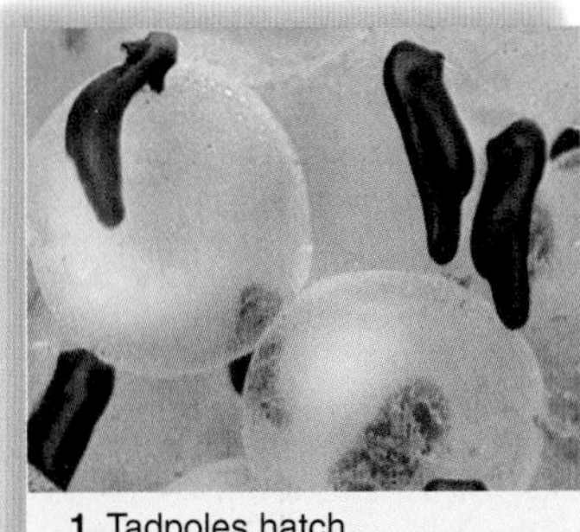
1. Tadpoles hatch.

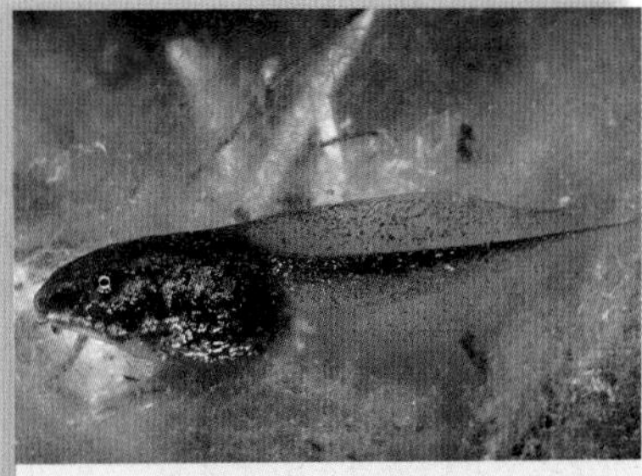
2. Tadpole respires with gills.

3. Front and hind legs are present.

4. Frog respires with lungs.

Figure 31.9 Frog metamorphosis.
During metamorphosis, the animal changes from an aquatic to a terrestrial organism.

lungs, but some species respire entirely through their skin. Air enters the mouth by way of nostrils, and when the floor of the mouth is raised, air is forced into the relatively small lungs. Respiration is supplemented by gas exchange through the smooth, moist, and glandular skin. The amphibian heart has a divided atrium but a single ventricle. The right atrium receives de-oxygenated blood from the body, and the left atrium receives oxygenated blood from the lungs. These two types of blood are partially mixed in the single ventricle (see Fig. 31.12*c*). Mixed blood is then sent to all parts of the body. Some is sent to the skin, where it is further oxygenated.

Most members of this group lead an amphibious life—that is, the larval stage lives in the water, and the adult stage lives on land. However, the adult usually returns to the water to reproduce. Figure 31.9 illustrates how a frog tadpole undergoes metamorphosis into a tetrapod (four-limbed) adult before taking up life on land. Currently, there is widespread concern about the future of amphibians because over 40% of all amphibian species are threatened with extinction. Because of their permeable skin and a life cycle that often depends on both water and land, they are thought to be "indicator species" of environmental quality. That is, they are the first to respond to environmental degradation, and disappearances of populations and species thus generate concern for other species—even humans.

Check Your Progress

1. List three features that distinguish amphibians.
2. Which features, in particular, allow most amphibians to have life stages in water and on land?

Reptiles: Amniotic Egg

Reptiles (class Reptilia, about 15,000 species) diversified and were most abundant between 245 and 65 million years ago. These reptiles included the mammal-like reptiles, the ancestors of today's living reptiles, and the dinosaurs, which became extinct. Some dinosaurs are remembered for their great size. *Brachiosaurus,* an herbivore, was about 23 m (75 ft) long and about 17 m (56 ft) tall. *Tyrannosaurus rex,* a carnivore, was 5 m (16 ft) tall when standing on its hind legs. The bipedal stance of some reptiles preceded the evolution of wings in birds. In fact, some say birds are actually living dinosaurs. New molecular, morphological, and behavioral studies support this hypothesis.

The reptiles living today are mainly turtles, alligators, snakes, lizards, and birds. The body is covered with hard, keratinized scales, which protect the animal from desiccation and from predators. Another adaptation for a land existence is the manner in which snakes typically use their tongue as a sense organ (Fig. 31.10). Reptiles have well-developed lungs enclosed by a protective rib cage. When the rib cage expands, a partial vacuum establishes a negative pressure, which causes air to rush into the lungs. The atrium of the heart is always separated into right and left chambers, but division of the ventricle varies. An interventricular septum is incomplete in certain species. Therefore, some oxygenated and deoxygenated blood is mixed between the ventricles.

Figure 31.10 The tongue as a sense organ.
Snakes wave a forked tongue in the air to collect chemical molecules, which are then brought back into the mouth and delivered to an organ in the roof of the mouth. Analyzed chemicals help the snake trail a prey animal, recognize a predator, or find a mate.

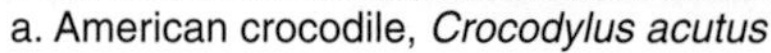
a. American crocodile, *Crocodylus acutus*

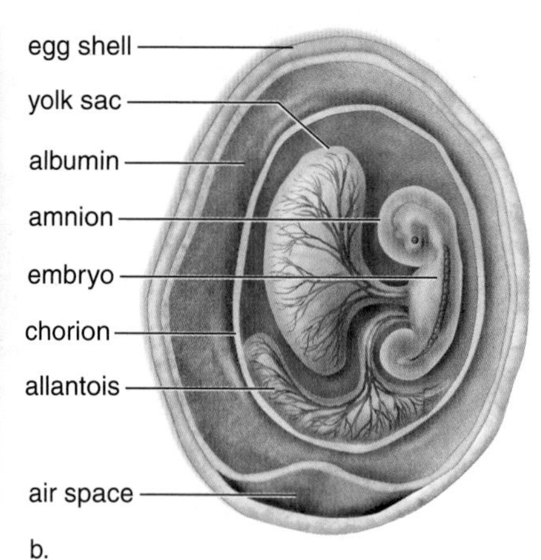

b.

Figure 31.11
The reptilian egg allows reproduction on land.
a. Baby American crocodile, *Crocodylus*, hatching out of its shell. Note that the shell is leathery and flexible, not brittle like birds' eggs. **b.** Inside the egg, the embryo is surrounded by extraembryonic membranes. The chorion aids gas exchange, the yolk sac provides nutrients, the allantois stores waste, and the amnion encloses a fluid that prevents drying out and provides protection.

Perhaps the most outstanding evolutionary innovation of reptiles is that their means of reproduction is suitable to a land existence. The penis of the male passes sperm directly to the female. Fertilization is internal, and the female typically possesses leathery, flexible, shelled eggs. Some snakes lay eggs, while other snakes actually give birth to live young. This *amniotic egg* made development on land possible and eliminated the need for a swimming-larval stage during development. This egg provides the developing embryo with atmospheric oxygen, food, and water; it removes nitrogenous wastes; and it protects the embryo from drying out and from mechanical injury. This is accomplished by the presence of extraembryonic membranes such as the chorion (Fig. 31.11).

Fishes, amphibians, and living reptiles other than birds are **ectothermic,** meaning that their body temperature matches the temperature of the external environment. If it is cold externally, they are cold internally; if it is hot externally, they are hot internally. Reptiles try to regulate their body temperatures by basking in the sun if they need warmth or by hiding in the shadows if they need cooling off.

Check Your Progress

1. List three features that distinguish reptiles.
2. Which features, in particular, make reptiles suited to an existence on land?

Feathered Reptiles

To many people, the **birds** are the most conspicuous, melodic, beautiful, and fascinating group of vertebrates. Over 9,000 species of birds have been described. Birds range in size from the tiny "bee" hummingbird at 1.8 g to the ostrich at a maximum weight of 160 kg and a height of 2.7 m. The fossil record of birds is unraveling evolutionary mysteries because *Archaeopteryx* and newly described fossils provide evidence that birds evolved from reptiles. Combined with recent molecular evidence, birds are now considered part of the reptilia. Their legs have scales and their feathers (Fig. 31.12*a*) are modified reptilian scales. However, birds typically lay a hard-shelled amniotic egg, rather than the leathery egg of reptiles.

Diversity of Birds

The majority of birds, including eagles, geese, and mockingbirds, have the ability to fly. However, some birds, such as emus, penguins, and ostriches, are flightless. Traditionally, birds have been classified based on beak and foot type (Fig. 31.13) and, to some extent, on their habitat and behavior. The various orders include birds of prey with notched beaks and sharp talons; shorebirds with long, slender, probing beaks and long, stiltlike legs; woodpeckers with sharp, chisel-like beaks and grasping feet; waterfowl with broad beaks and webbed toes; penguins with wings modified as paddles; and songbirds with perching feet.

Anatomy and Physiology of Birds

Nearly every anatomical feature of a bird can be related to its ability to fly (see Fig. 31.12*a*). The forelimbs are modified as wings. The hollow, very light bones are laced with air cavities. A horny beak has replaced jaws equipped with teeth, and a slender neck connects the head to a rounded, compact torso. The sternum is enlarged and has a keel, to which strong muscles are attached for flying. Respiration is efficient because the lobular lungs form anterior and posterior air sacs. The presence of these sacs means that the air circulates one way through the lungs and that gases are continuously exchanged across respiratory tissues (Fig. 31.12*b*). Another benefit of air sacs is that they lighten the body and thereby aid flying.

Birds have a four-chambered heart that completely separates O_2-rich blood from O_2-poor blood. The left ventricle pumps O_2-rich blood under pressure to the muscles (Fig. 31.12*c*). Birds are **endothermic.** Like mammals, their internal temperature is constant because they generate and maintain metabolic heat. This may be associated with their efficient nervous, respiratory, and circulatory systems. Also, their feathers provide insulation. Birds have no bladder and excrete uric acid in a semi-dry state.

Flight requires acute sense organs and an intricate nervous system. Birds have particularly good vision and well-developed brains. Their muscle reflexes are excellent. An

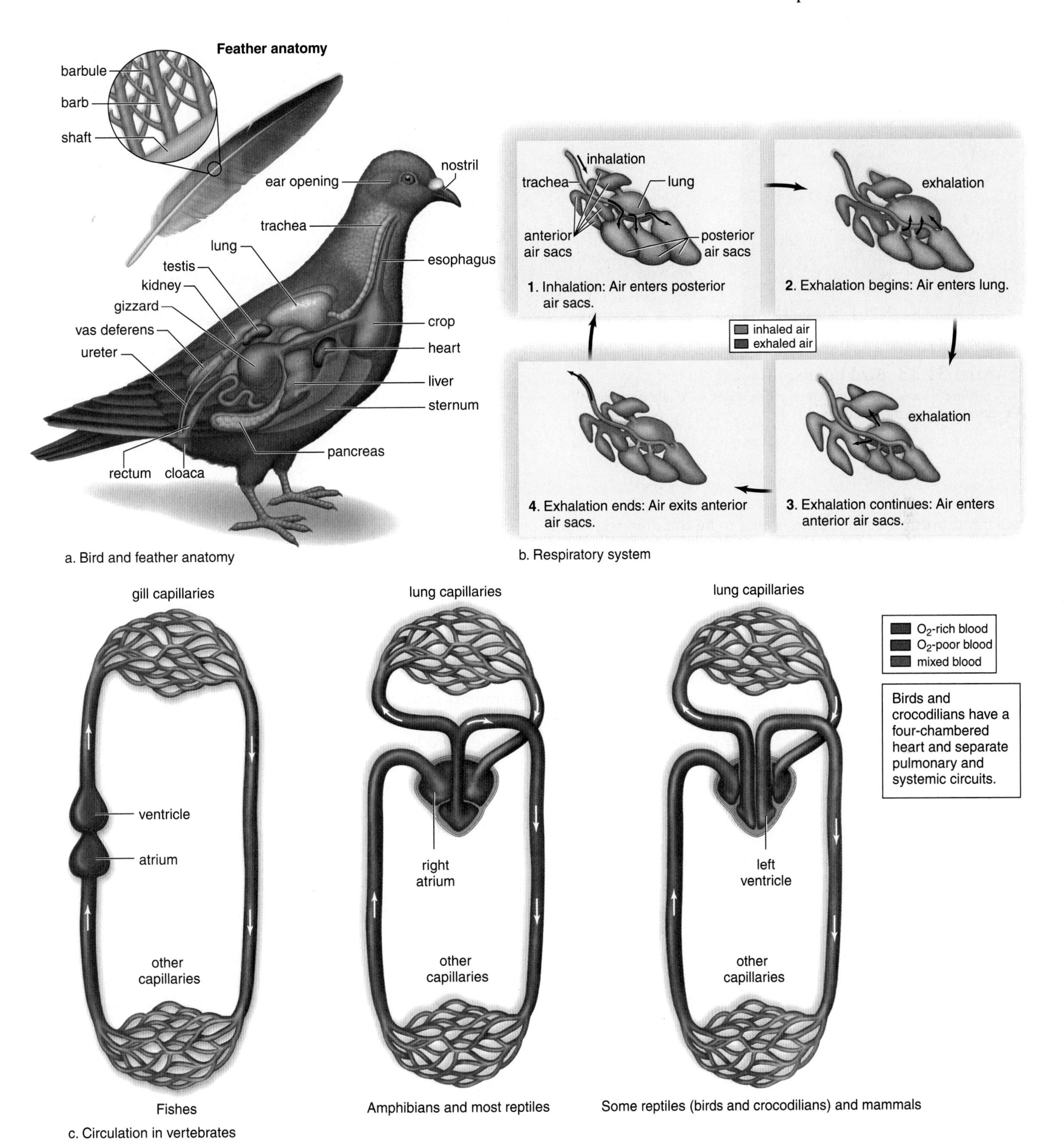

Figure 31.12 Bird anatomy and physiology.
a. The anatomy of a pigeon is representative of bird anatomy. **b.** In birds, air passes one way through the lungs. **c.** In fishes, a single-loop system utilizes a two-chambered heart. In amphibians and most reptiles, the heart has three chambers, and some mixing of O_2-rich and O_2-poor blood takes place. In birds, mammals, and some reptiles, the heart has four chambers and sends only O_2-poor blood to the lungs and O_2-rich blood to the body.

a. Cardinal, *Cardinalis cardinalis*

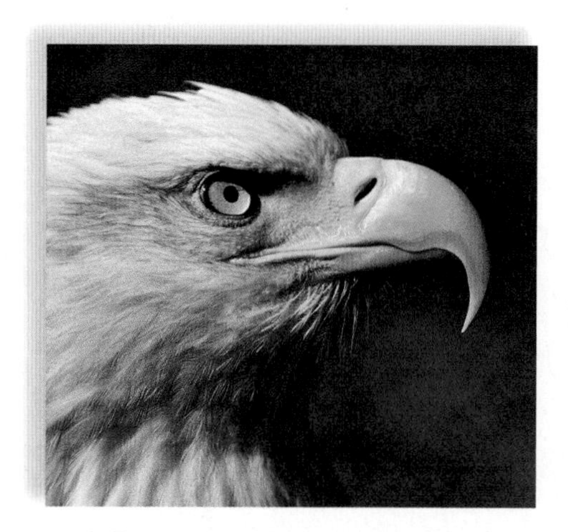

b. Bald eagle, *Haliaetus leucocephalus*

c. Flamingo, *Phoenicopterus ruber*

Figure 31.13 Bird beaks.
a. A cardinal's beak allows it to crack tough seeds. **b.** A bald eagle's beak allows it to tear prey apart. **c.** A flamingo's beak strains food from the water with bristles that fringe the mandibles.

enlarged portion of the brain seems to be the area responsible for instinctive behavior. A ritualized courtship often precedes mating. Many newly hatched birds require parental care before they are able to fly away and seek food for themselves. A remarkable aspect of bird behavior is the seasonal migration of many species over very long distances. Birds navigate by day and night, whether it's sunny or cloudy, by using the sun and stars and even Earth's magnetic field to guide them.

Check Your Progress

1. List the features that distinguish birds from other groups.
2. Which morphological features adapt birds for flight?

Mammals: Hair and Mammary Glands

Mammals (class Mammalia, about 4,600 species) also evolved from the reptiles. The first mammals were small, about the size of mice. During all the time the dinosaurs flourished (165–65 million years ago), mammals remained small in size and changed little evolutionarily. Some of the earliest mammalian groups are still represented today by the monotremes and marsupials, but they are not abundant. The placental mammals that evolved later went on to live in many habitats, including air, land, and sea.

The chief characteristics of mammals are body hair and milk-producing mammary glands. Almost all mammals are endothermic and maintain a constant internal temperature. Many of the adaptations of mammals are related to temperature control. *Hair,* for example, provides insulation against heat loss and allows mammals to be active, even in cold weather. Like birds, mammals have efficient respiratory and circulatory systems, which ensure a ready supply of oxygen to muscles whose contraction produces body heat. Also, like birds, mammals have double-loop circulation and a four-chambered heart (see Fig. 31.12*c*).

Mammary glands enable females to feed (nurse) their young without leaving them to find food. Nursing also creates a bond between mother and offspring that helps ensure parental care while the young are helpless. In all mammals (except monotremes), the young are born alive after a period of development in the uterus, a part of the female reproductive tract. Internal development shelters the young and allows the female to move actively about while the young are maturing. Mammals are classified as monotremes, marsupials, or placental mammals, according to how they reproduce.

Monotremes

Monotremes are mammals that, like birds, have a *cloaca,* a terminal region of the digestive tract that serves as a common chamber for feces, excretory wastes, and sex cells. They also lay hard-shelled amniotic eggs. They are represented by only two species: the spiny anteater and the duckbill platypus, both of which are found in Australia (Fig. 31.14*a*). The female duckbill platypus lays her eggs in a burrow in the ground. She incubates the eggs, and after hatching, the young lick up milk that seeps from modified sweat glands on the abdomens of both males and females. The spiny anteater has a pouch on the belly side formed by swollen mammary glands and longitudinal muscle. The egg moves from the cloaca to this pouch, where hatching takes place and the young remain for about 53 days. Then they stay in a burrow, where the mother periodically visits and nurses them.

a. Duckbill platypus, *Ornithorhynchus anatinus*

b. Virginia opossum, *Didelphis virginiana*

c. Koala, *Phascolarctos cinereus*

Figure 31.14 Monotremes and marsupials.
a. The duckbill platypus is a monotreme that inhabits Australian streams. **b.** The opossum is the only marsupial in the Americas. The Virginia opossum is found in a variety of habitats. **c.** The koala is an Australian marsupial that lives in trees.

Marsupials

The young of **marsupials** begin their development inside the female's body, but they are born in a very immature condition. Newborns crawl up into a pouch on their mother's abdomen. Inside the pouch, they attach to the nipples of mammary glands and continue to develop. Frequently, more are born than can be accommodated by the number of nipples, and it's "first come, first served."

Marsupial mammals are found mainly in Australia, where they underwent adaptive radiation for several million years without competition from placental mammals. The Virginia opossum (*Didelphis virginiana*) is the only marsupial that occurs north of Mexico (Fig. 31.14*b*), and the remainder occur in Central and South America. Among the herbivorous marsupials in Australia, koalas are tree-climbing browsers (Fig. 31.14*c*), and kangaroos are grazers. The Tasmanian devil, a carnivorous marsupial about the size of a small to medium-sized dog, is now threatened with extinction due to a transmissible tumor disease.

Placental Mammals

The vast majority of living mammals are **placental mammals** (Fig. 31.15). In these mammals, the extraembryonic membranes of the reptilian egg (see Fig. 31.11) have been modified for internal development within the uterus of the female. The chorion contributes to the fetal portion of the placenta, while a part of the uterine wall contributes to the maternal portion. Here, nutrients, oxygen, and waste are exchanged between fetal and maternal blood.

Mammals, with the exception of marine mammals, are adapted to life on land and have limbs that allow them to move rapidly. In fact, an evaluation of mammalian features leads us to the obvious conclusion that they lead active lives. The lungs are expanded not only by the action of the rib cage, but also by the contraction of the diaphragm, a horizontal muscle that divides the thoracic cavity from the abdominal cavity. The heart has four chambers. The internal temperature is constant, and hair, when abundant, helps insulate the body.

The mammalian brain is well developed and enlarged due to the expansion of the cerebral hemispheres, which control the rest of the brain. The brain is not fully developed until after birth, and the young learn to take care of themselves during a period of dependency on their parents.

Mammals have differentiated teeth. Typically, in the front, the incisors and canine teeth have cutting edges for capturing and killing prey. On the sides, the premolars and molars chew food. The specific shape and size of the teeth may be associated with whether the mammal is an herbivore (eats vegetation), a carnivore (eats meat), or an omnivore (eats both meat and vegetation). For example, mice (order Rodentia) have continuously growing incisors; horses have large, grinding molars; and dogs (order Carnivora) have long canine teeth. Placental mammals are classified based on their methods of obtaining food and their mode of locomotion. For example, bats (order Chiroptera) have membranous wings supported by digits; horses (order Perissodactyla) have long, hoofed legs; and whales (order Cetacea) have paddlelike forelimbs.

Check Your Progress

1. How do mammals maintain a constant body temperature?
2. Which features distinguish placental mammals from marsupials?

Science Focus

Uses of Animals for Human Medicine

Hundreds of pharmaceutical products come from animals, and even animals that produce poisons and toxins give us beneficial medicines. The Thailand cobra paralyzes its victim's nerves and muscles with a potent venom that eventually leads to respiratory arrest. However, that venom is also the source of the drug Immunokine, which has been used for over ten years to treat multiple sclerosis patients. Immunokine, which is almost without side effects, actually protects patients' nerve cells from destruction by their immune system.

A compound known as ABT-594, derived from the skin of the poison-dart frog (Fig. 31A*a*), is approximately 50 times more powerful than morphine in relieving chronic and acute pain without the addictive properties. Recently, a study showed that certain natural proteins contained in frog skin (called antimicrobial peptides) are very effective in killing the HIV virus and preventing it from leaving dendritic cells, where the virus first "hides out" in the early stages of infection. The copperhead snake, the fer-de-lance viper, and the giant Israeli scorpion are also some unlikely animals that directly provide, or serve as a model for the development of, pharmaceuticals such as painkillers, antibiotics, and anticancer drugs.

Other animals produce proteins similar enough to human proteins to be used for medical treatment. Until 1978, when recombinant DNA human insulin was produced, diabetics injected insulin purified directly from pigs. Currently, the flu vaccine is produced in fertilized chicken eggs. However, the production of vaccines and animal-derived drugs is often time-consuming, labor intensive, and expensive. In 2003, pharmaceutical companies used 90 million chicken eggs and took nine months to produce the flu vaccine. In 2004, several million doses of the flu vaccine were contaminated, creating a shortage.

Powerful applications of genetic engineering can be found in the development of treatments for human diseases. In fact, this advent in biotechnology has actually led to a whole new industry: animal pharming. Animal pharming uses genetically altered animals, such as mice, sheep, goats, cows, pigs, and chickens, to produce medically useful pharmaceutical products. A human gene for a particular useful product is inserted into the embryo of the animal. That embryo is implanted into a foster mother, which gives birth to a transgenic animal, so called because it contains genes from two sources. An adult transgenic animal produces large quantities of the pharmed product in its blood, eggs, or milk, which is then harvested and purified. Alpha-1-antitrypsin for the treatment of emphysema and cystic fibrosis is an example of such a product that is now undergoing clinical trials.

Some species of vertebrates have amazing abilities to regenerate body parts. For example, salamanders can regenerate limbs, upper and lower jaws, and the retina of the eye, while zebrafish can regenerate fins, the spinal cord, and part of the heart. Scientists hypothesize that because regeneration exists throughout the animal kingdom, humans may already have the genes to regenerate body parts, but these genes are turned off. Recent research suggests that a gene associated with regeneration in zebrafish also exists in humans. Perhaps drugs could be developed to turn these genes on. Or, it is possible that these genes or their products could be utilized for human medicine by animal pharming or other recombinant DNA methods.

Because there is an alarming shortage of human donor organs, xenotransplantation, the transplantation of animal tissues and organs into human beings, is another benefit of genetically altered animals. For example, in the late 1990s, two patients were kept alive using a pig liver outside of their body to filter their blood until a human organ was available for transplantation. Although baboons are genetically closer to humans than pigs, pigs are generally healthier, produce more offspring in a shorter time, and are already farmed for food (Fig. 31A*b*). Currently, pig heart valves and skin are routinely used to treat humans (Fig. 31A*c*). Miniature pigs, whose heart size is similar to that of humans, are being genetically engineered to make their tissues less foreign to the human body, in order to avoid rejection.

Discussion Questions

1. Is it ethical to change the genetic makeup of animals to use them as drug or organ factories?
2. Given the potential health effects and concern of transfer of diseases from animals to humans, what guidelines should there be with regard to xenotransplantation?
3. Using the information you learned from this reading, develop an argument for the conservation of biodiversity.

a. Poison-dart frogs, source of a medicine

b. Pigs, source of organs

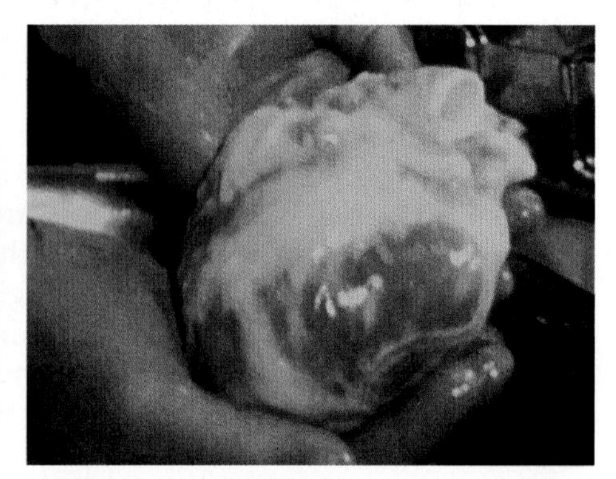
c. Heart for transplantation

Figure 31A Medical uses of animals.
a. The poison-dart frog is the source of a pain medication. **b.** Pigs are now being genetically altered to provide a supply of hearts for **(c)** heart transplant operations.

a. White-tailed deer, *Odocoileus virginianus*

b. African lioness, *Panthera leo*

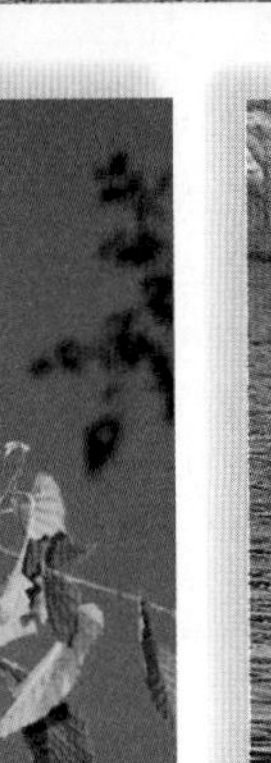

c. Squirrel monkey, *Saimiri sciureus*

d. Killer whale, *Orcinus orca*

Figure 31.15
Placental mammals.
Placental mammals have adapted to various ways of life. **a.** Deer are herbivores that live in forests. **b.** Lions are carnivores on the African plain. **c.** Monkeys inhabit tropical forests. **d.** Whales are sea-dwelling placental mammals.

Primates

Primata are members of the order Primates. In contrast to the other orders of placental mammals, most primates are adapted to an arboreal life—that is, for living in trees. Primate limbs are mobile, and the hands and feet both have five digits each. Many primates have a big toe and a thumb that are both opposable, meaning the big toe or thumb can touch each of the other toes or fingers. (Humans don't have an opposable big toe, but the thumb is opposable, and this results in a powerful and precise grip.) The opposable thumb allows a primate to easily reach out and bring food such as fruit to the mouth. When locomoting, primates grasp and release tree limbs freely because they have nails instead of the claws of their evolutionary ancestors.

In primates, the snout is shortened considerably, allowing the eyes to move to the front of the head. The stereoscopic vision (or depth perception) that results permits primates to make accurate judgments about the distance and position of adjoining tree limbs. Some primates that are active during the day, such as humans, have color vision and strong visual acuity because the retina contains cone cells in addition to rod cells. Cone cells require bright light, but the image is sharp and in color. The lens of the eye focuses light directly on the fovea, a region of the retina where cone cells are concentrated.

The evolutionary trend among primates is toward a larger and more complex brain. The brain size is smallest in prosimians and largest in modern humans. Increases in size of the cerebral cortex, which is associated with learning, memory, thought, and awareness, is primarily responsible for the trend toward increased brain size. The portion of the brain devoted to smell gets smaller, and the portions devoted to sight increase in size and complexity during primate evolution. Also, more and more of the brain is involved in controlling and processing information received from the hands and the thumb. The result is good hand-eye coordination in humans.

It is difficult to care for several offspring while moving from limb to limb, and so one birth at a time is the norm in primates. The juvenile period of dependency on parental care is extended, and there is an emphasis on learned behavior and complex social interactions.

The order Primata has two suborders: the **prosimians** (lemurs, tarsiers, and lorises) and the **anthropoids** (monkeys, apes, and humans). Thus, humans are more closely related to the monkeys and apes than they are to the prosimians.

Check Your Progress

1. Which characteristics separate primates from the other mammals?
2. What are the outcomes of increased brain size in more recently evolved primates?

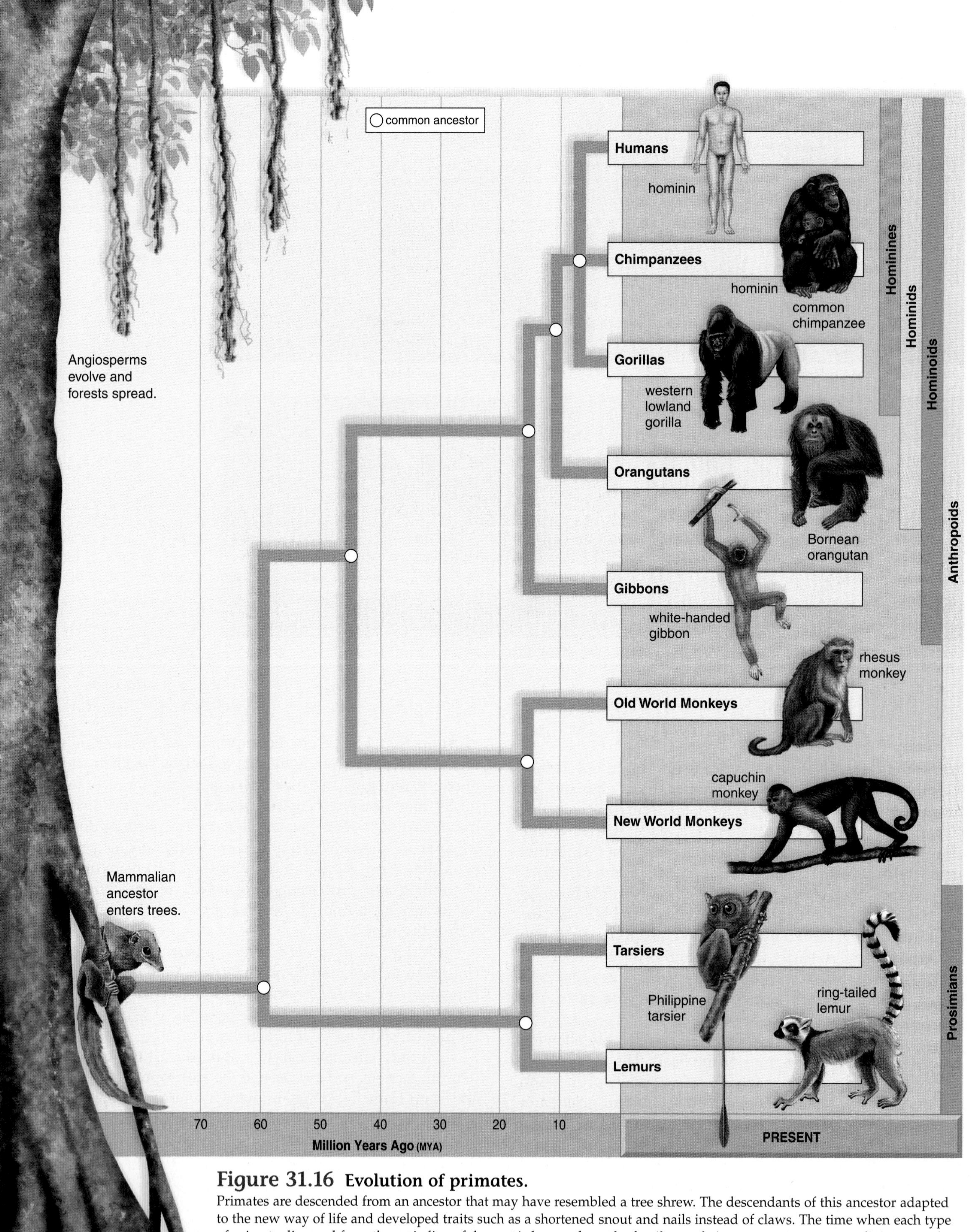

Figure 31.16 Evolution of primates.
Primates are descended from an ancestor that may have resembled a tree shrew. The descendants of this ancestor adapted to the new way of life and developed traits such as a shortened snout and nails instead of claws. The time when each type of primate diverged from the main line of descent is known from the fossil record. A common ancestor was living at each point of divergence—for example, there was a common ancestor for hominines about 7 MYA, for the hominoids about 15 MYA, and one for anthropoids about 45 MYA.

31.4 Human Evolution

Learning Outcomes

1. Recall the sequence of primate evolution.
2. Discuss the course of hominid evolution.
3. Outline the differences between the out-of-Africa versus the multiregional continuity hypotheses of the evolution of modern humans.

The evolutionary tree in Figure 31.16 shows that all primates share one common ancestor and that the other types of primates diverged from the human line of descent (called a **lineage**) over time. Notice that prosimians, represented by lemurs and tarsiers, were the first types of primates to diverge from the human line of descent, and African apes were the last group to diverge from our line of descent. Also note that the tree suggests that humans are most closely related to African apes. One of the most unfortunate misconceptions concerning human evolution is the belief that Darwin and others suggested that humans evolved from apes. On the contrary, humans and apes shared a common apelike ancestor. Today's apes are our distant cousins, and we couldn't have evolved from our cousins because we are contemporaries—living on Earth at the same time.

Molecular data have been used to estimate the date of the split between the human lineage and that of apes. When two lines of descent first diverge from a common ancestor, the genes of the two lineages are nearly identical. But as time goes by, each lineage accumulates genetic changes. Many genetic changes are neutral (not tied to adaptation) and accumulate at a fairly constant rate. Such changes can be used as a kind of **molecular clock** to infer how closely two groups are related and the point at which they diverged. Molecular data suggest that the split between the ape and human lineages occurred about 6–8 million years ago (MYA). Currently, humans and chimpanzees are probably the most closely related, sharing greater than 96% of our DNA. Hominins (the designation that includes chimpanzees, humans, and species very closely related to humans) first evolved about 5 MYA. Molecular data show that hominins and gorillas are closely related and these two groups must have shared a common ancestor sometime during the Miocene. Hominins and gorillas are now grouped together as hominines. The hominids include the hominines and the orangutan. The hominoids include the gibbon and the hominids. The hominoid common ancestor first evolved at the beginning of the Miocene about 23 MYA.

Evolution of Humanlike Hominins

The classification of humans is given in the box at the right. Within the hominins are the tribe hominini. The biggest derived characteristic that separates modern humans is an anatomy suitable for standing erect and walking on two feet (called **bipedalism**).

CLASSIFICATION

ORDER: Primates

- Adapted to an arboreal life
- Prosimians, Anthropoids

FAMILY: Hominidae (hominids)
SUBFAMILY: Homininae (hominines)
TRIBE*: Hominini (hominins)
Early Humanlike Hominins → *Sahelanthropus*, ardipithecines
Later Humanlike Hominins → australopithecines

GENUS: *Homo* (humans) → *Homo habilis*, *Homo rudolfensis*, *Homo ergaster*, *Homo erectus*
Early *Homo*
Brain size greater than 600 cc; tool use and culture

Later *Homo* → *Homo heidelbergensis*, *Homo neandertalensis*, *Homo sapiens*
Brain size greater than 1,000 cc; tool use and culture

A new taxonomic level that lies between subfamily and genus.

Until recently, many scientists thought that hominids evolved because of a dramatic change in climate that caused forests to be replaced by grassland about 4 MYA. It is thought that hominids began to stand upright to keep the sun off their backs and to have better vision for detecting predators or prey in grasslands. However, this is currently under debate because recent fossil findings suggest hominids may have begun to walk upright as early as 7 MYA, while they still lived in forests. While living in trees, the first hominids may have walked upright on large branches as they collected fruit from overhead. Then, when they began to forage on the ground, an upright stance would have made it easier for them to travel more easily from woodland to woodland and/or easier to forage among bushes.

Since 2000, scientists have found several fossils that may provide clues to early hominid evolution. One such skull fossil, with the species name *Sahelanthropus tchadensis,* is estimated to date from approximately 7 MYA, but researchers are debating whether it is a hominid or an ape. Only the braincase has been found, and it is very apelike. However, a point at the back of the skull where the neck muscles attach suggests bipedalism. Also, the canines are smaller and the tooth enamel thicker than in the teeth of an ape. Another series of leg fossils (genus *Orrorin*), which are about 6 million years old, may suggest upright walking because the femur appears strong and able to support more weight than that of the apes. An area called Middle Awash, Ethiopia, seems to be a "hotbed" of recent primate fossil findings. There, remains of two chimplike creatures were found. A more recent discovery, dated between 5.8 and 5.2 MYA, may have been the first creature to walk erect and thus, possibly, the first hominid. The fossil's name, *Ardipithecus ramidus kadabba,* is derived from the Afar language spoken in Ethiopia. *Ardi* means ground or floor; *ramid* means root; and *kadabba* means family ancestor.

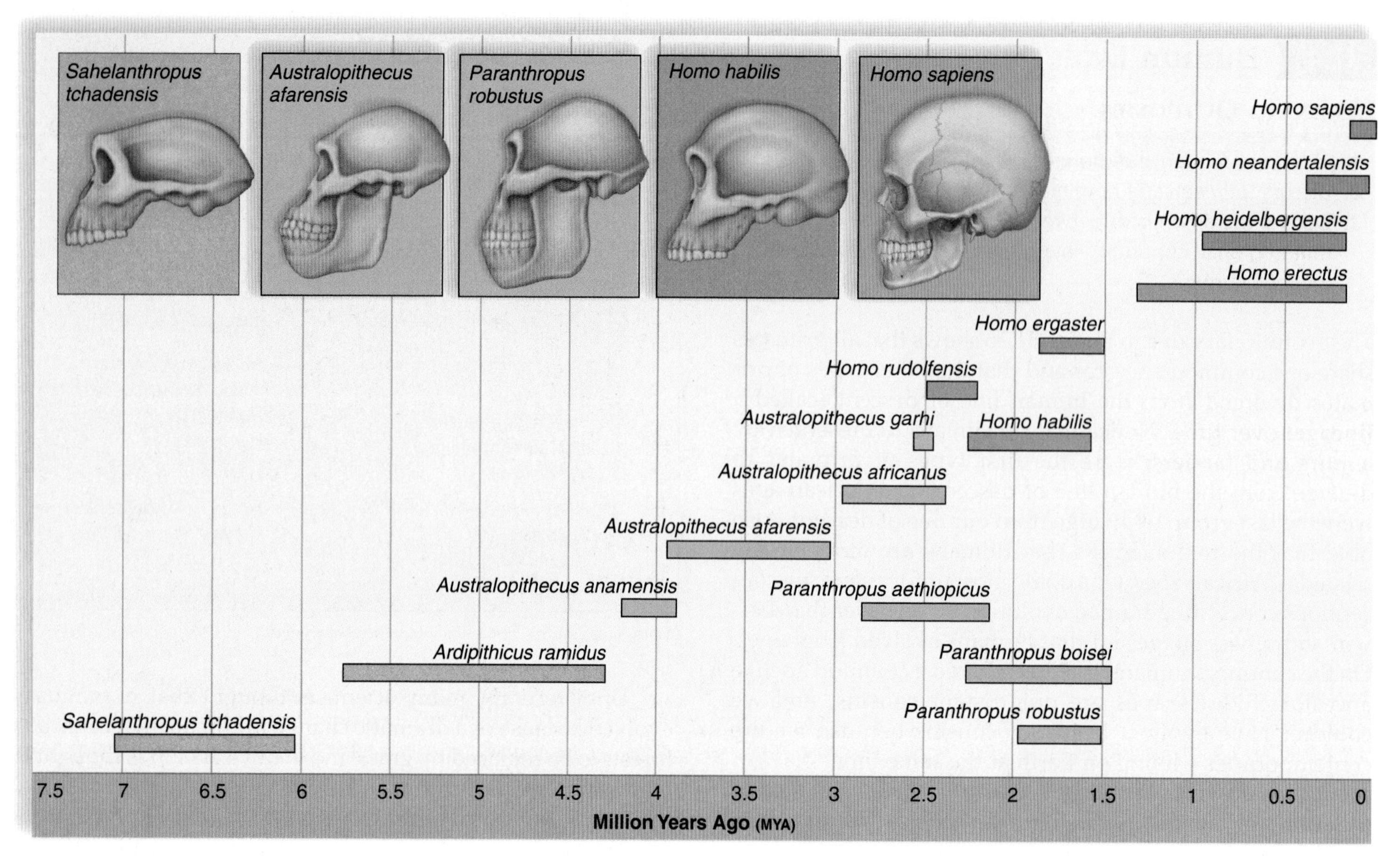

Figure 31.17 Human evolution.
Several groups of extinct hominins preceded the evolution of modern humans. These groups have been divided into the early humanlike hominins (orange), later humanlike hominins (green), early *Homo* species (lavender), and finally the later *Homo* species (blue). Only modern humans are classified as *Homo sapiens.*

Figure 31.17 is a timeline of human evolution in which these potential early hominids are represented by orange-colored bars.

Australopithecines

Most recently, scientists have discovered a fossil, which they have named *Australopithecus anamensis,* in Middle Awash, where they have found eight hominid-like species so far. This fossil may be the missing evolutionary link between *Ardipithecus* and the **australopithecines** (represented by green-colored bars in Fig. 31.17), the possible direct hominid ancestors for humans (genus *Homo*). It is thought that the australopithecines evolved and diversified in Africa 4 MYA and went extinct about 1 MYA. Australopithecines had a small brain (an apelike characteristic) and walked erect (a humanlike characteristic). Thus, it seems that humanlike characteristics did not all evolve at the same time. Australopithecines give evidence of **mosaic evolution,** meaning that different body parts changed at different rates and therefore at different times.

Australopithecines stood about 100–115 cm high and had a brain slightly larger than that of a chimpanzee (about 370–515 cc). The forehead was low, and the face projected forward. There is no evidence that they used tools. Some are thought to have been slight of frame, or "gracile" (slender), while others were robust (powerful), with strong upper bodies and massive jaws powered by chewing muscles attached to a prominent bony crest on top of the skull. The gracile types probably fed on soft fruits and leaves, while the robust types likely had a more fibrous diet that may have included hard nuts.

Some fossil remains of australopithecines have been found in southern Africa, while others have been found in eastern Africa. The exact evolutionary relationship between these two groups is currently not known. The first australopithecine was discovered in southern Africa by Raymond Dart in the 1920s. This hominid, named *Australopithecus africanus,* is a gracile type. A second southern African fossil specimen, *A. robustus,* is a robust type. Both *A. africanus* and *A. robustus* had a brain size of about 500 cc. These hominids clearly walked upright, but the proportions of their limbs are apelike (i.e., longer forelimbs than hindlimbs). Therefore, most scientists do not believe that *A. africanus* is ancestral to early *Homo.*

In 1973, a team led by Donald Johanson unearthed nearly 250 fossils of a hominid called *Australopithecus afarensis* in eastern Africa. A now-famous female skeleton discovered in this group is known worldwide by its field name, "Lucy." This skeleton is about 3.2 million years old. Although her brain was quite small (400 cc), the shapes and relative proportions of her limbs indicate that she stood up and walked bipedally (Fig. 31.18*a*). Even better evidence of bipedal locomotion comes from a trail of footprints in Laetoli from about 3.7 MYA (Fig. 31.18*b*). For these reasons,

Figure 31.18
Australopithecus afarensis.
a. A reconstruction of Lucy on display at the St. Louis Zoo. **b.** These fossilized footprints occur in ash from a volcanic eruption some 3.7 MYA. The larger footprints are double, and a third, smaller individual was walking to the side. (A female holding the hand of a youngster may have been walking in the footprints of a male.) The footprints suggest that *A. afarensis* walked bipedally.

A. afarensis is usually favored as being more directly related to *Homo* than *A. africanus. A. afarensis,* a gracile type, is believed to be ancestral to the robust types found in eastern Africa such as *A. boisei,* which had a powerful upper body and the largest molars of any humanlike hominin.

In 2000, a team of scientists from the Max Planck Institute unearthed the fossilized remains of a 3.3-million-year-old juvenile *A. afarensis* just 4 km from where Lucy had been discovered. Dubbed Salem by her discoverer, she is often called "Lucy's baby," even though she is tens of thousands of years older than Lucy. Not only is this fossil exceptional because the remains of infants and juveniles rarely fossilize, but it represents the most complete *A. afarensis* fossil to date. An earlier find called *A. garhi* may be the transitional link between the australopiths and the next group of fossils we will be discussing, namely, the early *Homo* species, represented in Figure 31.17 by lavender-colored bars. *A. garhi* is an australopith, but it made tools.

CHECK YOUR PROGRESS

1. When and why might bipedalism have evolved?
2. Which feature distinguishes the first hominids?

Evolution of Early Homo

Fossils designated as early *Homo* species are represented in Figure 31.17 by a lavender colored bar. These fossils appear in the fossil record approximately 2 MYA. They all have a brain size that is 600 cc or greater, their jaw and teeth resemble those of humans, and tool use is in evidence.

Homo habilis and *Homo rudolfensis*

Homo habilis and *Homo rudolfensis* are closely related and will be considered together. *Homo habilis* literally means handyman, and these two species are credited by some as being the first peoples to use stone tools. Most experts agree that while they were socially organized, they were probably scavengers rather than hunters. The cheek teeth of these hominins tend to be smaller than even those of the gracile australopiths. This is also evidence that they were omnivorous and ate meat, in addition to plant material. Compared to australopiths, the protrusion of the face was less, and the brain was larger. Although the height of *H. rudolfensis* did not exceed that of the australopiths, some of this species' fossils have a brain size as large as 800 cc, which is considerably larger than that of *A. afarensis.*

Homo ergaster and *Homo erectus*

Homo ergaster evolved in Africa perhaps from *H. rudolfensis.* Similar fossils found in Asia are different enough to be classified as *Homo erectus,* literally meaning "upright man." These fossils span the dates between 1.9 and 0.3 MYA, and many other fossils belonging to both species have been found in Africa and Asia. Compared to *H. habilis, H. ergaster* had a larger brain (about 1,000 cc) and a flatter face, with a projecting nose. This type of nose is adaptive for a hot, dry climate because it permits water to be absorbed before air leaves the body. The recovery of an almost complete skeleton of a 10-year-old boy indicates that *H. ergaster* was much taller than the hominids discussed thus far (Fig. 31.19). Males were 1.8 m tall, and females were 1.55 m tall. Indeed, these hominids stood erect and most likely had a *striding gait* similar to modern humans. The robust and most likely heavily muscled skeleton still retained some australopithecine features. Even so, the size of the birth canal indicates that infants were born in an immature state that required an extended period of parental care.

H. ergaster first appeared in Africa but then migrated into Europe and Asia sometime between 1 and 2 MYA. Most likely, *H. erectus* evolved from *H. ergaster* after *H. ergaster* arrived in Asia. In any case, such an extensive population movement is a first in the history of humankind and a tribute to the intellectual and physical skills of these peoples. They also had a knowledge of fire and may have been the first to cook meat.

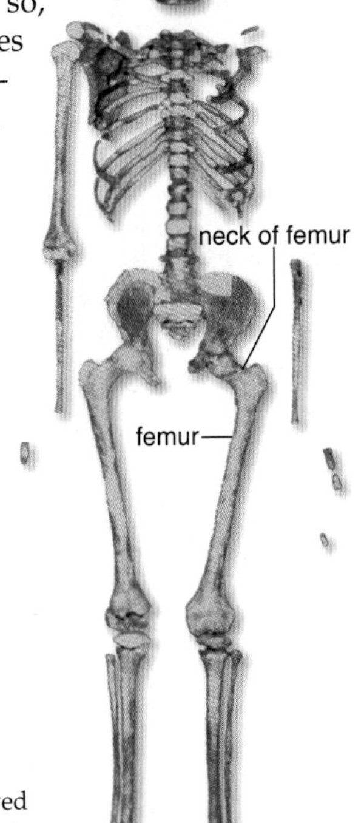

Figure 31.19 ***Homo ergaster.***
This skeleton of a 10-year-old boy who lived 1.6 MYA in eastern Africa shows femurs that are angled because the neck of the femur is quite long.

Homo floresiensis In 2004, scientists announced the discovery of the fossil remains of *Homo floresiensis*. The 18,000-year-old fossil of a 1 m tall, 25 kg adult female was discovered on the island of Flores in the South Pacific. This important finding suggests that a species of *Homo* coexisted with modern *Homo sapiens* much more recently than Neandertals, which went extinct about 28,000 years ago. The specimen was the size of a three-year-old *Homo sapiens sapiens* but possessed a braincase only one-third the size of that of a modern human. Apparently, *H. floresiensis* used tools and fire. A 2007 study supports the hypothesis that this diminutive hominin and her peers evolved from normal-sized, island hopping *Homo erectus* populations that reached Flores about 840,000 years ago. Some scientists think that its small size is due to island dwarfing, the reduction in size of large animals when their gene pool is limited to a small environment. This phenomenon is seen in other island populations.

Check Your Progress

1. Which characteristics describe the early *Homo* species?
2. When did *Homo ergaster* evolve?
3. What are *Homo ergaster's* distinguishing characteristics?

Evolution of Modern Humans

The later species of *Homo* are represented in Figure 31.17 by blue-colored bars. Most researchers think that modern humans *Homo sapiens* evolved from *H. ergaster*, but they differ as to the details. Many disparate early *Homo* species in Europe are now classified as *Homo heidelbergensis*. Just as *H. erectus* is hypothesized to have evolved from *H. ergaster* in Asia, so *H. heidelbergensis* is hypothesized to have evolved from *H. ergaster* in Europe. Further, for the sake of discussion, *H. ergaster* in Africa, *H. erectus* in Asia, and *H. heidelbergensis* (and *H. neandertalensis*) in Europe can be grouped together as archaic humans who lived as long as a million years ago. The presence of archaic humans at these different locations suggests to some that modern humans evolved from archaic humans separately in all three places (Fig. 31.20*a*). The most widely accepted hypothesis for the evolution of modern humans from archaic humans is referred to as the replacement model or **out-of-Africa** hypothesis, which proposes that modern humans evolved from archaic humans only in Africa, and then modern humans migrated to Asia and Europe, where it replaced the archaic species about 100,000 years bp (before the present) (Fig. 31.20*b*). The replacement model is supported by the fossil record. The earliest remains of modern humans (Cro-Magnon), dating

Millions of Years Ago (MYA)
0.1
1
2
ASIA
AFRICA
EUROPE
Homo sapiens
Homo sapiens
Homo sapiens
Homo ergaster evolves into modern humans in Asia, Africa, and Europe.
interbreeding
interbreeding
migration of Homo ergaster
migration of Homo ergaster
Homo ergaster

a. Multiregional continuity

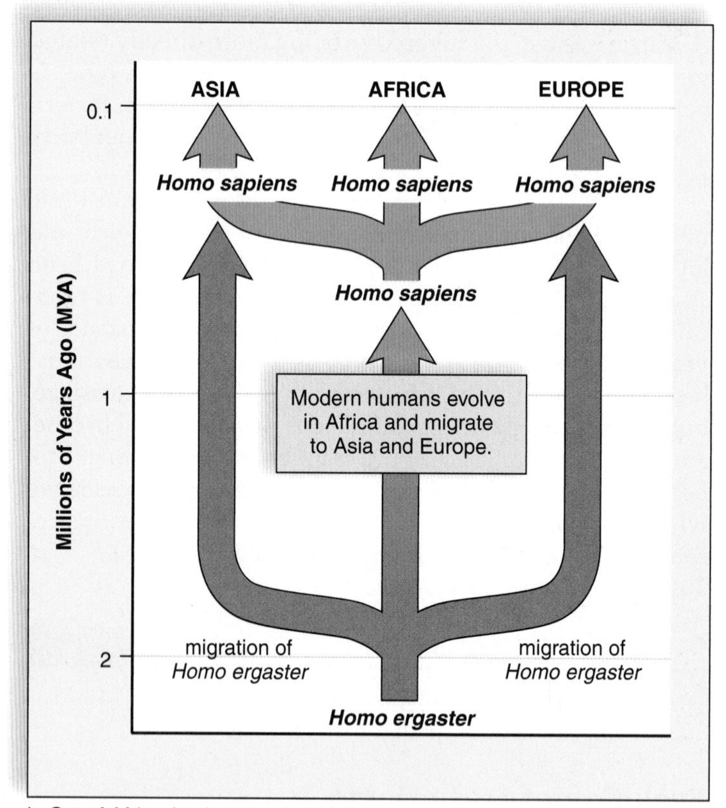

b. Out of Africa (replacement model)

Figure 31.20 Evolution of modern humans.
a. The multiregional continuity hypothesis proposes that *Homo sapiens* evolved separately in at least three different places: Asia, Africa, and Europe. Therefore, continuity of genotypes and phenotypes is expected in each region but not between regions. **b.** The out-of-Africa hypothesis proposes that *Homo sapiens* evolved only in Africa and then migrated out of Africa, as *H. ergaster* did many thousands of years before. *H. sapiens* would have supplanted populations of ancestral *Homo* in Asia and Europe about 100,000 years ago.

at least 130,000 years bp, have been found only in Africa. Modern humans are not found in Asia until 100,000 years bp and not in Europe until 60,000 years bp. Until earlier modern human fossils are found in Asia and Europe, the replacement model is supported. The replacement model is also supported by DNA data. Several years ago, a study showed that the mitochondrial DNA of Africans is more diverse than the DNA of the people in Europe (and the world). This is significant because if mitochondrial DNA has a constant rate of mutation, Africans should show the greatest diversity, since modern humans have existed the longest in Africa. Similarly, Europeans are quite similar genetically, consistent with recent colonization of Europe under the out-of-Africa model.

An opposing hypothesis to the out-of-Africa hypothesis does exist. This hypothesis, called the multiregional continuity hypothesis, proposes that modern humans arose from archaic humans in essentially the same manner in Africa, Asia, and Europe. The hypothesis is multiregional because it applies equally to Africa, Asia, and Europe, and it supposes that in these regions, genetic continuity will be found between modern populations and archaic populations. This contrasting hypothesis has sparked many innovative studies to test which hypothesis is correct.

Neandertals

The **Neandertals,** sometimes classified as *Homo neandertalensis,* are a species of archaic humans that lived between 200,000 and 28,000 years ago. Neandertal fossils have been found from the Middle East throughout Europe. Neandertals take their name from Germany's Neander Valley, where one of the first Neandertal skeletons, estimated to be about 200,000 years old, was discovered.

According to the replacement model, Neandertals were supplanted by modern humans. Surprisingly, however, the Neandertal brain was, on the average, slightly larger than that of *Homo sapiens* (1,400 cc, compared with 1,360 cc in most modern humans). The Neandertals had massive brow ridges and wide, flat noses. They also had a forward-sloping forehead and a receding lower jaw. Their nose, jaws, and teeth protruded far forward. Physically, the Neandertals were powerful and heavily muscled, especially in the shoulders and neck (Fig. 31.21). The bones of Neandertals were shorter and thicker than those of modern humans. New fossils show that the pubic bone was long compared to that of modern humans. The Neandertals lived in Europe and Asia during the last Ice Age, and their sturdy build could have helped conserve heat.

Archaeological evidence suggests that Neandertals were culturally advanced. Some Neandertals lived in caves. However, others probably constructed shelters. They manufactured a variety of stone tools, including spear points, which they could have used for hunting, and scrapers and knives, which would have helped in food preparation. They most likely successfully hunted bears, woolly mammoths, rhinoceroses, reindeer, and other contemporary animals. They used and could control fire, which probably helped in cooking frozen meat and in keeping warm. They even buried their dead with flowers and tools and may have had a religion.

Cro-Magnons

Cro-Magnons are the oldest fossils to be designated *Homo sapiens sapiens.* In keeping with the out-of-Africa hypothesis, the Cro-Magnons, who are named after a fossil location in France, were the modern humans who entered Asia and Europe from Africa 100,000 years ago or even earlier. They probably reached western Europe about 40,000 years ago. Cro-Magnons had a thoroughly modern appearance (Fig. 31.22). They had lighter bones, flat high foreheads, domed skulls housing brains of 1,350 cc, small teeth, and a distinct chin. They made advanced stone tools, including compound tools, such as by fitting stone flakes to a wooden handle. They may have been the first to make knifelike blades and to throw spears,

Figure 31.21 Neandertals.
This drawing shows that the nose and the mouth of the Neandertals protruded from their faces, and their muscles were massive. They made stone tools and were most likely excellent hunters.

Figure 31.22 Cro-Magnons.
Cro-Magnon people are the first to be designated *Homo sapiens sapiens.* Their tool-making ability and other cultural attributes, such as their artistic talents, are legendary.

Bioethical Focus

Animals in the Laboratory

Some people suggest that animals should be protected in every way and should not be used in laboratory research. As a result, there is a growing recognition of what is generally referred to as animal rights. Psychologists with Ph.D.s earned in the 1990s are half as likely to express strong support for animal research as those who earned their Ph.D.s before 1970.

Those who approve of laboratory research involving animals point out that even today it is difficult to develop new vaccines and medicines against infectious diseases, new surgical techniques for saving human lives, or new treatments for spinal cord injuries without using animals. Even so, most scientists today are in favor of what are now called the "three Rs": replacement of animals by in vitro, or test-tube, methods whenever possible; reduction of the numbers of animals used in experiments; and refinement of experiments to cause less suffering to animals. In the Netherlands, all scientists engaging in research that involves animals are well trained in the three Rs. After designing an experiment that uses animals, they are asked to find ways to answer the same questions without using animals. In the United States, the Animal Welfare Act regulates all federally funded institutions to minimize pain and suffering in vertebrate animal research.

F. Barbara Orlans of the Kennedy Institute of Ethics at Georgetown University says, "It is possible to be both pro research and pro reform." She feels animal rights activists need to accept that sometimes animal research is beneficial to humans, and all scientists need to consider the ethical dilemmas that arise when animals are used for laboratory research.

Form Your Own Opinion

1. Are you opposed to the use of animals in laboratory experiments? Always or under certain circumstances?
2. Are we redefining the relationship between animals and humans to the detriment of both? Why or why not?
3. Do you feel that it would be possible for animal activists and scientists to find a compromise they can both accept?

enabling them to kill animals from a distance. They were such accomplished hunters that some researchers hypothesize they may have been responsible for the extinction of many larger mammals, such as the giant sloth, the mammoth, the saber-toothed tiger, and the giant ox, during the late Pleistocene epoch. This event is known as the Pleistocene overkill.

Cro-Magnons hunted cooperatively, and perhaps they were the first to have language. They are believed to have lived in small groups, with the men hunting by day while the women remained at home with the children. It's quite possible that this hunting way of life among prehistoric people influences our behavior even today. The Cro-Magnon culture included art. They sculpted small figurines out of reindeer bones and antlers. They also painted beautiful drawings of animals on cave walls in Spain and France (see Fig. 31.22).

Check Your Progress

1. Discuss the differences between the multiregional continuity hypothesis and the out-of-Africa hypothesis.
2. Compare and contrast Neandertals and Cro-Magnons.

Applying the Concepts [Revisited]

At the beginning of this chapter, you learned about the apparent rediscovery of the ivory-billed woodpecker, a majestic component of the biodiversity in the southeastern United States. Conservation efforts are now underway to protect the once-extinct bird's habitat. Throughout the chapter, you have learned the major animal groups, their characteristics, and their evolutionary histories.

1. Throughout animal history, there have been five mass extinction events that have, in part, led to the animal diversity we see today. Most scientists argue that we are now in a sixth mass extinction, with the current biodiversity threats caused by humans rather than by natural events as in the past. What should we do about it? Counteract the argument that humans should not care because in the past, animal diversity increases after a mass extinction event.
2. Modern humans, *Homo sapiens sapiens* likely coexisted with *Homo neandertalensis* and *Homo floresiensis.* Why do you think these lineages went extinct while we survived?

SUMMARIZING THE CONCEPTS

31.1 Echinoderms

- Echinoderms and chordates are deuterostomes, whereby the second embryonic opening becomes the mouth and the coelom forms by outpocketing of the primitive gut.
- Echinoderms (e.g., sea stars, sea urchins, sea cucumbers, sea lilies) have radial symmetry as adults (not as larvae) and spines from endoskeletal plates.
- Typical of echinoderms, sea stars have tiny skin gills, a central nerve ring with branches, and a water vascular system for locomotion. Each arm of a sea star contains branches from the nervous, digestive, and reproductive systems.

31.2 Chordates

- Chordates (tunicates, lancelets, and vertebrates) have a notochord, a dorsal tubular nerve cord, pharyngeal pouches, and a postanal tail at some time in their life history.

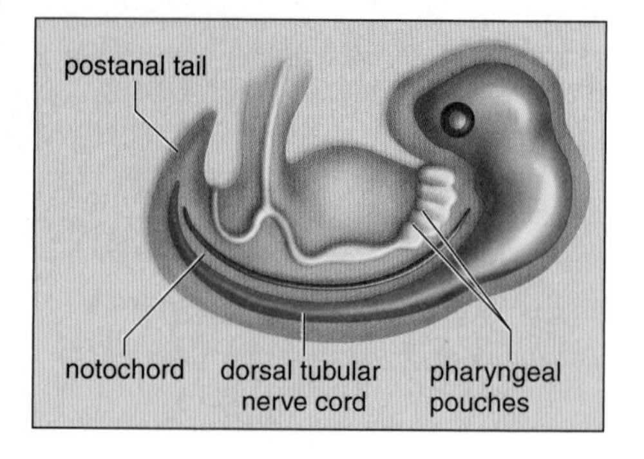

- Tunicates and lancelets are the nonvertebrate chordates. Adult tunicates lack chordate characteristics except gill slits, but adult lancelets have the four chordate characteristics.

31.3 Vertebrates

- Vertebrates have the four chordate characteristics as embryos, but the notochord is replaced by the vertebral column.
- Internal organs are well-developed, and cephalization is apparent.
- The vertebrate classes trace their evolutionary history as follows:
  - The first vertebrates, represented by hagfishes and lampreys, lacked jaws and fins.
  - Bony fishes have jaws and two pairs of fins. The bony fishes include those that are ray-finned and a few that are lobe-finned (some of which have lungs).
  - Amphibians (e.g., frogs and salamanders) evolved from lobe-finned fishes and have two pairs of limbs. Most amphibian species return to the water to reproduce, and their tadpoles or larvae then metamorphose into terrestrial adults.
  - Reptiles most often possess shelled eggs, which contain extraembryonic membranes, including an amnion that allows them to reproduce on land.
  - Birds are feathered reptiles, which helps maintain a constant body temperature. Hollow bones, lungs with air sacs that penetrate bones and allow one-way ventilation, and a keeled breastbone adapt birds for flight.
  - Mammals have hair to help them maintain a constant body temperature, and mammary glands for nursing of young. Monotremes lay eggs; marsupials have a pouch in which the newborn matures; and placental mammals, which are far more varied and numerous, retain offspring inside the uterus until birth.

31.4 Human Evolution

- Primates are mammals adapted to living in trees.
- Prosimians (tarsiers and lemurs) diverged first, followed by the monkeys and then the apes.
- Molecular evidence suggests humans are most closely related to African apes, whose ancestry split from ours about 6–10 MYA.
- Hominid evolution continued in eastern Africa with the rise of ardipithecines and the australopithecines. The most famous australopithecine is Lucy (3.2 MYA), who walked bipedally and had a small brain.
- *Homo habilis,* present about 2 MYA, is certain to have made tools.
- *Homo erectus,* with a brain capacity of 1,000 cc and a striding gait, was the first to migrate out of Africa.
- Two contradicting hypotheses concern the origin of modern humans. The multiregional continuity hypothesis posits that modern humans originated separately in Asia, Europe, and Africa. The out-of-Africa hypothesis suggests that modern humans originated in Africa and, after migrating into Europe and Asia, replaced the other *Homo* species (including Neandertals) found there. Most evidence to date supports the latter hypothesis.
- The recent discovery of the fossil, *Homo floresiensis,* however, suggests a species of *Homo* coexisted with modern humans more recently than the Neandertals.

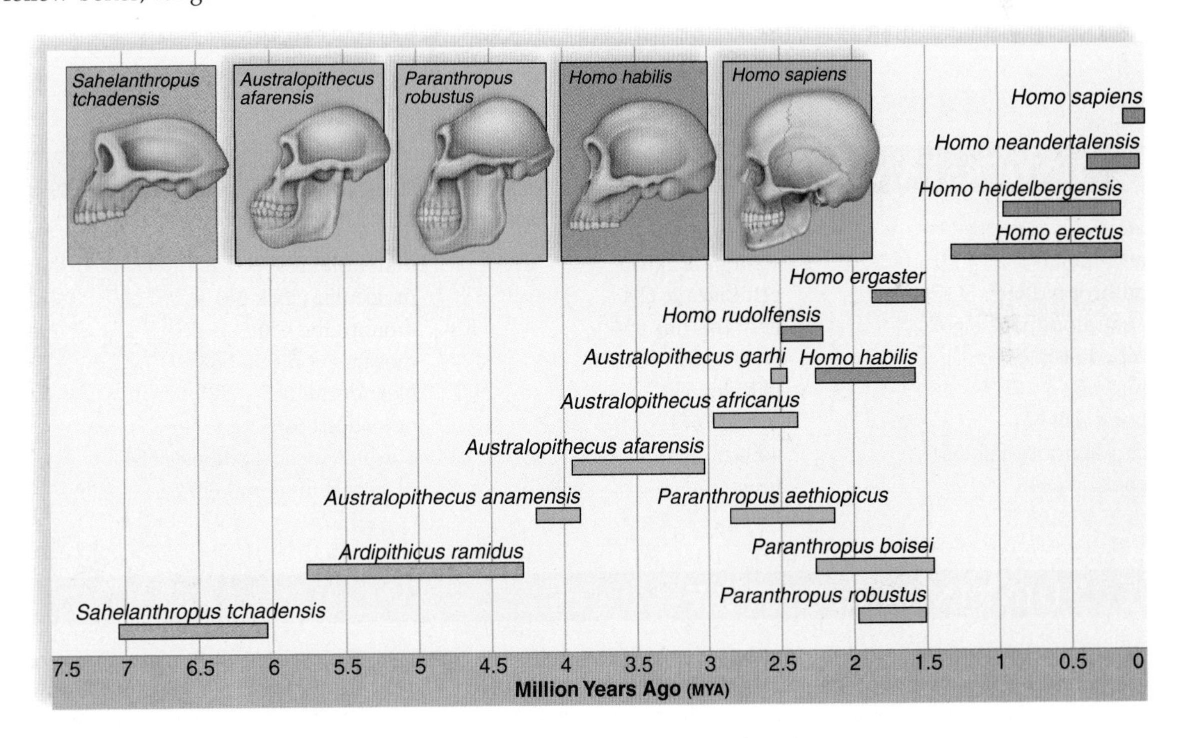

TESTING YOURSELF

Choose the best answer for each question.

For questions 1–5, match the feature with the animal group in the key. Each answer may be used more than once. Each question may have more than one answer.

Key:

a. echinoderms
b. nonvertebrate chordates
c. vertebrate chordates

1. Closed circulatory system
2. Notochord
3. Vertebral column
4. Lancelets and tunicates
5. Paired appendages

For questions 6–9, match the feature with the animal group in the key. Each answer may be used more than once. Each question may have more than one answer.

Key:

a. fishes
b. amphibians
c. reptiles
d. birds

6. Internal fertilization
7. Jointed appendages
8. Internal skeleton
9. Ectothermic
10. Label the parts of a bony fish in the following diagram.

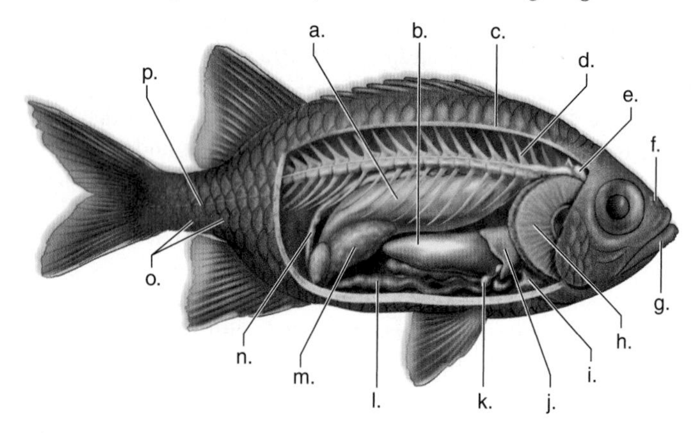

11. Which of the following is not an adaptation for flight in birds?
    a. air sacs
    b. modified forelimbs
    c. bones with air cavities
    d. enlarged sternum
    e. well-developed bladder
12. There are more marsupials in Australia than in North America because
    a. they did not need to compete with placental mammals until recently.
    b. their environment is more favorable for their survival in Australia.
    c. volcanoes led to the mass extinction of marsupials in North America.
    d. food sources are more diverse in Australia.
13. Stereoscopic vision is possible in primates due to
    a. the presence of cone cells.
    b. an enlarged brain.
    c. a shortened snout.
    d. None of these are correct.
14. The first humanlike feature to evolve in the hominids was
    a. a large brain.
    b. massive jaws.
    c. a slender body.
    d. bipedal locomotion.
15. In *H. habilis,* enlargement of the portions of the brain associated with speech probably led to
    a. cooperative hunting.
    b. the sharing of food.
    c. the development of culture.
    d. All of these are correct.

UNDERSTANDING THE TERMS

amniotic egg 650
amphibian 652
anthropoid 659
australopithecine 662
bipedalism 661
bird 654
bony fish 651
cartilaginous fish 651
chordate 648
Cro-Magnon 665
echinoderm 646
ectothermic 654
endothermic 654
jawless fish 651
lancelet 650
lineage 661
lobe-finned fish 652
lung 652
mammal 656
marsupial 657
molecular clock 661
monotreme 656
mosaic evolution 662
Neandertal 665
notochord 648
out-of-Africa hypothesis 664
placental mammal 657
primata 659
prosimian 659
ray-finned fish 652
reptile 653
tunicate 649
vertebrate 650

THINKING CRITICALLY

1. Researchers are now sequencing the genome of several vertebrate species, including the platypus and the African clawed frog. What information might these studies give us that could be applied to the development of new medical treatments for humans?
2. Bipedalism has many selective advantages, including the increased ability to spot predators and prey. However, bipedalism has one particular disadvantage—upright posture leads to a smaller pelvic opening, which makes giving birth to an offspring with a large head very difficult. This situation results in a higher percentage of deaths (of both mother and child) during birth in humans compared to other primates. How can you explain the selection for a trait, such as bipedalism, that has both positive and negative consequences for fitness?
3. In studying recent fossils of the genus *Homo,* such as Cro-Magnon, biologists have determined that modern humans have not undergone much biological evolution in the past 50,000 years. Rather, cultural anthropologists argue that cultural evolution has been far more important than biological evolution in the recent history of modern humans. What do they mean by this? Support your argument with some examples.

INQUIRY INTO LIFE WEBSITE

The companion website for *Inquiry into Life* provides a wealth of information organized and integrated by chapter. You will find practice tests, animations, videos, and much more that will complement your learning and understanding of general biology.

http://www.mhhe.com/maderinquiry13

PART VII

Behavior and Ecology

32 Behavioral Ecology

Applying the Concepts

Emperor penguins tend to their young. These penguins live in Antarctica and breed on glaciers, where temperatures can reach 100° below zero! As you can imagine, they have developed many adaptive behaviors in order to survive. The recent popular movie "March of the Penguins" revealed their secrets to survival.

Emperor penguins breed in winter, which begins in March in Antarctica. After an extended courtship with a male, a female emperor penguin lays a single egg. She then passes this egg—very carefully—to the male so that she can return to the water to feed. She has lost much of her body weight in the 50-mile migration to the breeding grounds. If the egg touches the ground—even for a second—it will crack open and the chick will die. The male must then keep the egg on top of his feet and under a pouch of skin in order to keep it warm—for about two months, during which time he does not feed.

The female returns and feeds the newly hatched young. The male, who has not eaten in over four months, then returns 50 miles to the ocean and feeds. Meanwhile, the female keeps the young in her brood pouch for another two months. The non-feeding penguins huddle together in large groups in order to survive the cold winter temperatures. This social behavior is necessary for survival, and penguins cooperate, taking turns between the outside and the middle of the huddle where it's warm.

In this chapter you will learn about animal behavior in general and gain a better understanding of the complexity of the emperor penguin lifestyle. First, we will discuss the basis of behavior—that is, whether behaviors have a genetic or environmental basis. Then we will examine mating behaviors, social behaviors, and communication, necessary components of all animals', including humans', struggle to survive.

Chapter Outline

32.1 Nature Versus Nurture: Genetic Influences

LEARNING OUTCOME

1. Describe experiments that show behavior can be genetically based.

The "nature versus nurture" question asks to what degree our genes (nature) and environmental influences (nurture) affect our behavior. **Behavior** encompasses any action that can be observed and described. Because both the anatomy and physiology of animals, including humans, determine what types of behavior are possible, we immediately know that the genes, to a degree, control behavior. Many experiments have been conducted to discover the degree to which genetics controls behavior.

Experiments with Lovebirds, Snakes, and Snails

Lovebirds are small, green and pink African parrots that nest in tree hollows. There are several closely related species of lovebirds in the genus *Agapornis,* whose behavior differs by the way they build nests. Fischer lovebirds, *Agapornis fischeri,* pick up a large leaf (or in the laboratory, a piece of paper) with their bills, perforate it with a series of longitudinal bites, and then tear out long strips. They carry the strips in their bills to the nest (Fig. 32.1*a*) and weave them with others to make a deep cup. Peach-faced lovebirds, *Agapornis roseicollis,* cut somewhat shorter strips in a similar manner, but then carry them to the nest in a very unusual way. They pick up the strips in their bills and insert them deep into their rump feathers (Fig. 32.1*b*). In this way, they can carry several short strips during each trip to the nest, whereas Fischer lovebirds can carry only one longer strip at a time.

Researchers hypothesized that if the behavior for obtaining and carrying nesting material is inherited, then hybrids might show intermediate behavior. When the two species of birds were crossed, the hybrid offspring had difficulty carrying nesting materials. They cut strips of intermediate length and then attempted to tuck the strips into their rump feathers. They did not push the strips far enough into the feathers, however, and when they walked or flew, the strips always fell out. Hybrid birds eventually (after about three years) learned to carry the cut strips in their bills, but still briefly turned their heads toward their rumps before flying off. The intermediate behavior exhibited by the hybrids supports the hypothesis that behavior has a genetic basis.

a. Fischer lovebird with nesting material in its beak.

b. Peach-faced lovebird with nesting material in its rump feathers.

Figure 32.1 Nest building behavior in lovebirds. **a.** Fischer lovebirds carry strips of nesting material in their bills, as do most other birds. **b.** Peach-faced lovebirds tuck strips of nesting material into their rump feathers before flying back to the nest.

Several experiments have been conducted using the garter snake, *Thamnophis elegans,* to determine if food preference has a genetic basis. In California, inland garter snake populations are aquatic and commonly feed underwater on frogs and fish, whereas coastal populations are terrestrial and feed mainly on slugs. In the laboratory, inland adult snakes refused to eat slugs, while coastal snakes readily did so. Newborns resulting from matings between snakes from the two populations (inland and coastal) have an intermediate preference for slugs, suggesting a genetic basis for food choice.

Differences between slug acceptors and slug rejecters appear to be inherited, but what physiological difference is there between the two populations? A clever experiment

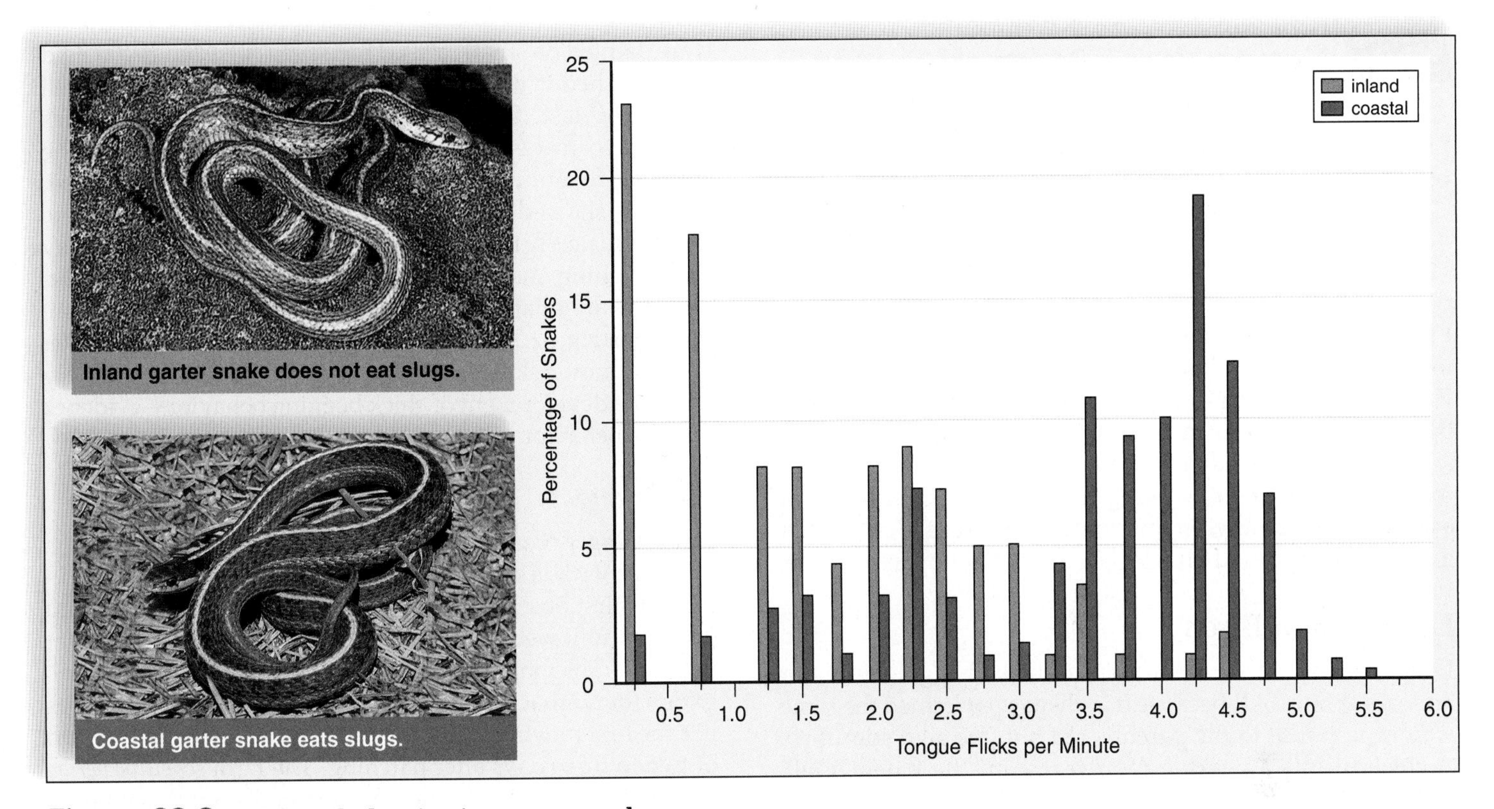

Figure 32.2 Feeding behavior in garter snakes.
The number of tongue flicks by inland and coastal garter snakes is measured in terms of their response to slug extract on cotton swabs. Coastal snakes tongue-flicked more than inland snakes, indicating that coastal snakes are more sensitive to the smell of slugs than inland snakes.

answered this question. Snakes use tongue flicks to recognize their prey, and their tongues carry chemicals to an odor receptor in their mouth. Newborns will even flick their tongues at cotton swabs dipped in fluids of their prey. Swabs were dipped in slug extract, and the number of tongue flicks were counted for newborn inland and coastal snakes. Coastal snakes had a higher number of tongue flicks (indicating more interest in the smell) than inland snakes (Fig. 32.2). Thus, inland snakes apparently do not eat slugs because they are not sensitive to their smell. A genetic difference between the two populations of snakes has resulted in a physiological difference in their nervous systems. Although hybrids showed a great deal of variation in the number of tongue flicks, they were generally intermediate, as predicted by the genetic hypothesis.

The nervous and endocrine systems are both responsible for the coordination of body systems. Is the endocrine system also involved in behavior? Experiments support endocrine system involvement. For example, the egg-laying behavior in the marine snail *Aplysia* involves a set sequence of movements. Following copulation, the animal extrudes long strings of more than a million eggs per egg case. It then takes the egg case string in its mouth, covers it with mucus, waves its head back and forth to wind the string into an irregular mass, and attaches the mass to a solid object, such as a rock. Several years ago, scientists isolated and analyzed an egg-laying hormone (ELH) that causes the snail to lay eggs even if it has not mated. ELH was found to be a small protein of 36 amino acids that diffuses into the circulatory system and causes the smooth muscle cells of the reproductive duct to contract and expel the egg string. Using recombinant DNA techniques, the investigators isolated the entire ELH gene. The gene's product turned out to be a protein with 271 amino acids. The protein can be cleaved into as many as 11 possible products, and the ELH hormone is one of these. The hormone alone, or in conjunction with the gene's other products, is thought to control all the components of egg-laying behavior in *Aplysia*.

Experiments with Humans

Human twins, on occasion, have been separated at birth and raised under different environmental conditions. Studies of separated twins show that they have similar food preferences and activity patterns, and even select mates with similar characteristics. These twin studies lend support to the hypothesis that at least certain types of behavior are primarily influenced by nature (i.e., genes).

Check Your Progress

1. How can studies of identical twins illustrate that behavior can be genetically based?

32.2 Nature Versus Nurture: Environmental Influences

Learning Outcomes

1. Illustrate an experiment that shows behavior can be environmentally influenced.
2. Depict three different types of learned behavior.

Even though genetic inheritance serves as a basis for behavior, environmental influences (nurture) also affect behavior. For example, behaviorists originally believed that some behaviors were **fixed action patterns (FAPs)** elicited by a sign stimulus. But then they found that many behaviors formerly thought to be FAPs improve with practice—for example, by learning. In this context, **learning** is defined as a durable change in behavior brought about by experience.

Learning in Birds

Laughing gull *(Leucophaeus atricilla)* chicks' begging behavior appears to be a FAP, because it is always performed the same way in response to the parent's red bill (the sign stimulus). A chick directs a pecking motion toward the parent's bill, grasps it, and strokes it downward (Fig. 32.3*a*). Parents bring about the begging behavior by swinging their bill gently from side to side. After the chick responds, the parent regurgitates food onto the nest floor. If need be, the parent then encourages the chick to eat. This interaction between the chicks and their parents suggests that the begging behavior involves learning. To test this hypothesis, diagrammatic pictures of gull heads were painted on small cards. Then, eggs collected in the field were hatched in a dark incubator to eliminate visual stimuli before the test. On the day of hatching, each chick was allowed to make about a dozen pecks at the model. The chicks were returned to the nest, and then each was retested. The tests showed that on the average, only one-third of the pecks by a newly hatched chick strike the model. But one day after hatching, more than half of the pecks are accurate, and two days after hatching, the accuracy reaches a level of more than 75% (Fig. 32.3*b*, *c*). Investigators concluded that improvement in motor skills, as well as visual experience, strongly affect the development of chick begging behavior—evidence that the behavior is, in part, learned.

Imprinting

Imprinting is considered a very simple form of learning, although it has a strong genetic component as well. Imprinting was first observed in birds when chicks, ducklings, and goslings followed the first moving object they saw after hatching. This object is ordinarily their mother, but investigators found that birds can imprint on any object, as long as it is the first moving object they see during a sensitive period of two to three days after hatching. The term *sensitive period* means that the behavior develops only during this time.

A chick imprinted on a red ball follows it around and chirps whenever the ball is moved out of sight. Social interactions between parent and offspring during the sensitive period seem key to normal imprinting. For example, female mallards cluck during the entire time imprinting is occurring, and it could be that vocalization before and after hatching is necessary for normal imprinting.

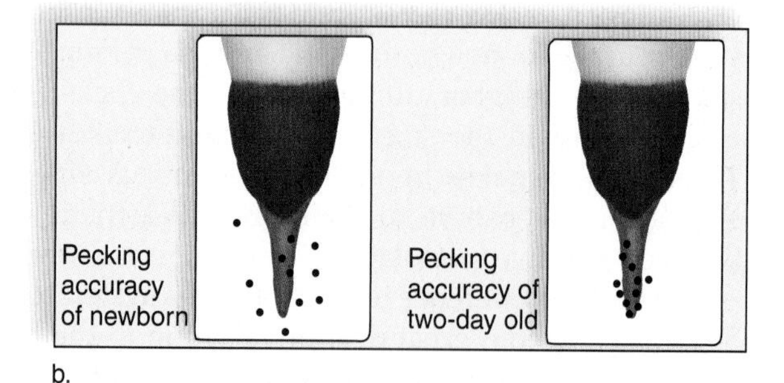

Figure 32.3 Begging behavior in laughing gull chicks. **a.** At about three days of age, a laughing gull chick grasps the red bill of a parent, stroking it downward, and the parent then regurgitates food. **b.** The accuracy of a chick striking a test probe, painted red. **c.** Chick-pecking accuracy illustrated graphically. Note from these diagrams that a chick markedly improves its ability (within only two days) to peck a bill, a behavior that normally causes a parent to regurgitate food.

Song Learning

White-crowned sparrows sing a species-specific song, but males of a particular region have their own dialect. Birds were caged to test the hypothesis that young white-crowned sparrows learn how to sing from older members of their species.

Three groups of birds were tested. Birds in the first group *heard no songs at all.* When grown, these birds sang a song, but it was not fully developed. Birds in the second group *heard tapes of white-crowns singing.* When grown, they sang in that dialect, as long as the tapes had been played during a sensitive period from about age 10 to 50 days. White-crowned sparrows' dialects (or other species' songs) played before or after this sensitive period had no effect on their song. Birds in a third group did not hear tapes and instead were *given an adult tutor of a different species.* These birds sang the song of a still different species—no matter when the tutoring began—showing that social interactions apparently assist learning in birds.

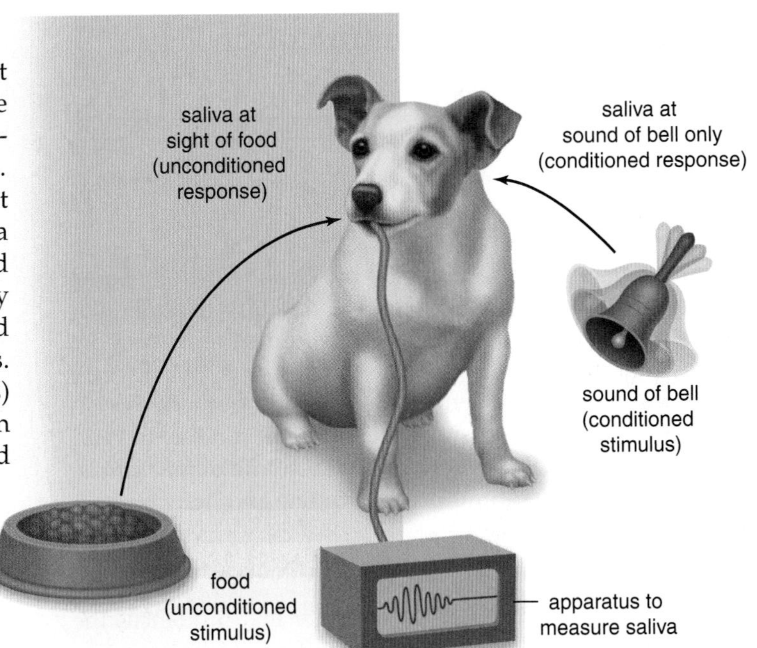

Figure 32.4 Classical conditioning.
Ivan Pavlov discovered classical conditioning by performing this experiment with dogs. A bell is rung when a dog is fed. Salivation is noted. Eventually, the dog salivates when the bell is rung even though no food is presented. Food is an unconditioned stimulus, and the sound of the bell is a conditioned stimulus that brings about the response—that is, salivation.

Associative Learning

A change in behavior that involves an association between two events is termed **associative learning.** For example, birds that get sick after eating a monarch butterfly no longer prey on monarch butterflies, even though they may be readily available. Or the smell of fresh-baked bread may entice you to eat even though you may have just eaten. Because you really enjoyed bread in the past, you associate the smell with those past experiences, which makes you hungry. Both classical conditioning and operant conditioning are examples of associative learning.

Classical Conditioning

In **classical conditioning,** the presentation of two different types of stimuli at the same time causes an animal to form an association between them. The best-known laboratory example of classical conditioning is an experiment done by the Russian psychologist Ivan Pavlov. First, Pavlov observed that dogs salivate when presented with food. Then he rang a bell whenever the dogs were fed. Eventually, the dogs would salivate whenever the bell was rung, regardless of whether food was present (Fig. 32.4).

Classical conditioning suggests that an organism can be trained—that is, conditioned—to associate any response to any stimulus. Unconditioned responses are those that occur naturally, as when salivation follows the presentation of food. Conditioned responses are those that are learned, as when a dog learns to salivate when it hears a bell. Advertisements attempt to use classical conditioning. For example, commercials pair attractive people with a particular product in the hope that viewers will associate attractiveness with that product. This pleasant association may cause them to buy the product.

In a similar way, it's been suggested that holding children on your lap when reading to them facilitates an interest in reading. Why? Because they will associate a pleasant feeling with reading.

Operant Conditioning

During **operant conditioning,** a stimulus-response connection is strengthened. Most people know that it is helpful to give an animal a reward, such as food or affection, when teaching them a trick. When we go to an animal show, it is quite obvious that trainers use operant conditioning. They present a stimulus, such as a hoop, and then give a reward (food) for the proper response (jumping through the hoop).

B. F. Skinner is well known for studying this type of learning in the laboratory. In Skinner's simplest type of experiment, a caged rat happens to press a lever and is rewarded with sugar pellets, which it avidly consumes. Thereafter, the rat regularly presses the lever whenever it is hungry. In more sophisticated experiments, Skinner even taught pigeons to play ping-pong by rewarding desired responses to stimuli.

When operant conditioning is applied to child rearing, it's been suggested that parents who give a positive reinforcement for good behavior will be more successful than parents who punish behaviors they believe are undesirable.

Check Your Progress

1. Teaching a dog to sit by rewarding him with treats is an example of which type of learning?
2. What type of learning is illustrated by a cat that comes every time a pull-top can is opened in the kitchen?

32.3 Animal Communication

LEARNING OUTCOMES

1. Define communication.
2. Compare and contrast chemical, auditory, visual, and tactile communication.

Animals exhibit a wide diversity of social behaviors. Some animal species are largely solitary and join with a member of the opposite sex only for the purpose of reproduction. Others pair, bond, and cooperate in raising offspring. Still others form a **society** in which members of species are organized in a cooperative manner, extending beyond sexual and parental behavior. Social behavior in these and other animals requires that they communicate with one another.

Communicative Behavior

Communication is a signal by a sender that influences the behavior of a receiver. The communication can be purposeful, but does not have to be. Bats send out a series of sound pulses and listen for the corresponding echoes in order to find their way through dark caves and locate food at night. Some moths have an ability to hear these sound pulses, and they begin evasive tactics when they sense that a bat is near. Are the bats purposefully communicating with the moths? No, bat sounds are simply a cue to the moths that danger is near.

The types of communication signals used by animals include chemical, auditory, visual, and tactile.

Chemical Communication

Chemical signals have the advantage of being effective both night and day. The term **pheromone** designates chemical signals in low concentration that are passed between members of the same species. For example, female moths secrete chemicals from special abdominal glands, which are detected downwind by receptors on male antennae. The antennae are especially sensitive, and this ensures that only male moths of the correct species (not predators) will be able to detect them.

Cheetahs and other cats mark their territories by depositing urine, feces, and anal gland secretions at the boundaries (Fig. 32.5). Klipspringers (small antelope) use secretions from a gland below the eye to mark twigs and grasses of their territory.

To what degree do pheromones, in addition to hormones, affect the behavior of mammals, even humans, determining whether they carry out parental care, become agressive, or engage in courtship behavior? Some researchers maintain that human behavior is influenced by undetectable pheromones wafting through the air. They have discovered that like the mouse, humans have an organ in the nose, called the vomeronasal organ (VNO), that can detect not only odors, but also pheromones. The neurons from this organ lead to the hypothalamus, the part of the brain that controls the release of many hormones in the body. Perhaps this organ is even involved in how people choose their mates (see Science Focus, "Mate Choice and Smelly T-shirts").

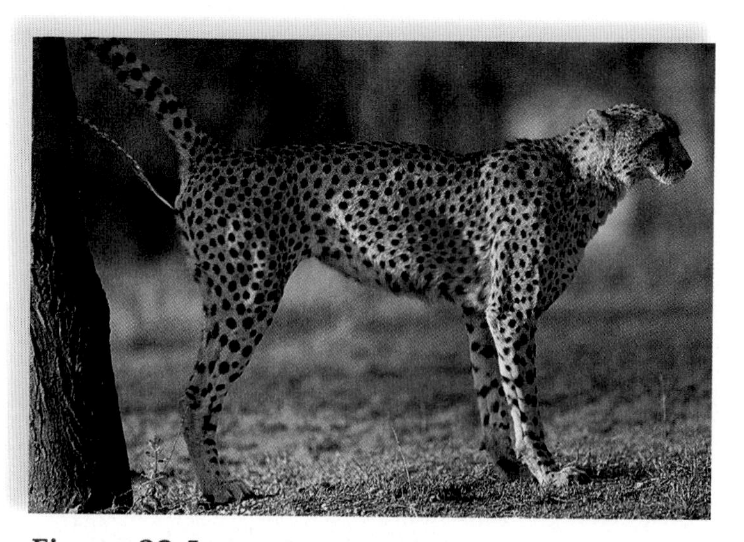

Figure 32.5 Use of a pheromone.
This male cheetah is spraying urine onto a tree to mark its territory.

Auditory Communication

Auditory (sound) communication has some advantages over other kinds of communication. It is faster than chemical communication, and it too is effective both night and day. Further, auditory communication can be modified not only by loudness but also by pattern, duration, and repetition. In an experiment with rats, a researcher discovered that an intruder can avoid attack by increasing the frequency with which it makes an appeasement sound.

Male birds have songs for a number of different occasions, such as one song for distress, another for courting, and still another for marking territories. Sailors have long heard the songs of humpback whales transmitted through the hull of a ship. However, only recently has it been shown that the song has six basic themes, each with its own phrases, that can vary in length, be interspersed with sundry cries and chirps, and be heard for many miles. Interestingly, humpbacks songs change from year to year, but all the humpbacks in an area learn and converge on singing the same song. The purpose of the song is probably sexual, serving to advertise the availability of the singer.

Language is the ultimate auditory communication. Only humans have the biological ability to produce a large number of different sounds and to put them together in many different ways. Nonhuman primates have different vocalizations, each having a definite meaning, such as when vervet monkeys give alarm calls (Fig. 32.6). Although chimpanzees can be taught to use an artificial language, they never progress beyond the capability level of a two-year-old child. It also has been difficult to prove that chimps understand the concept of grammar or can use their language to reason. As such, most anthropologists argue that humans possess a communication ability unparalleled by other animals.

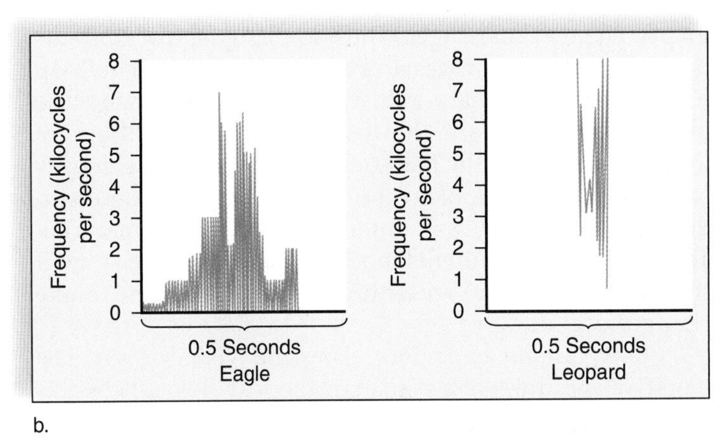

Figure 32.6 Auditory communication.
a. Vervet monkeys, *Cercopithecus aethiops*, are responding to an alarm call. Vervet monkeys can give different alarm calls according to whether a troop member sights an eagle or a leopard, for example. **b.** The frequency per second of the sound differs for each type of call.

Visual Communication

Visual signals are most often used by species active during the day. Contests between males make use of threat postures and possibly prevent outright fighting, a behavior that might result in reduced fitness. A male baboon displaying full threat is an awesome sight that establishes his dominance and keeps peace within the baboon troop (Fig. 32.7). Hippopotamuses perform territorial displays that include mouth opening.

Many animals use complex courtship behaviors and displays. The plumage of a male Raggiana Bird of Paradise allows him to put on a spectacular courtship dance to attract a female, giving her a basis on which to select a mate. Defense and courtship displays are exaggerated and always performed in the same way so that their meaning is clear. Fireflies use a flash pattern to signal females of the same species (Fig. 32.8).

Figure 32.7 A male olive baboon displaying full threat.
In olive baboons, males are larger than females and have enlarged canines. Competition between males establishes a dominance hierarchy for the distribution of resources.

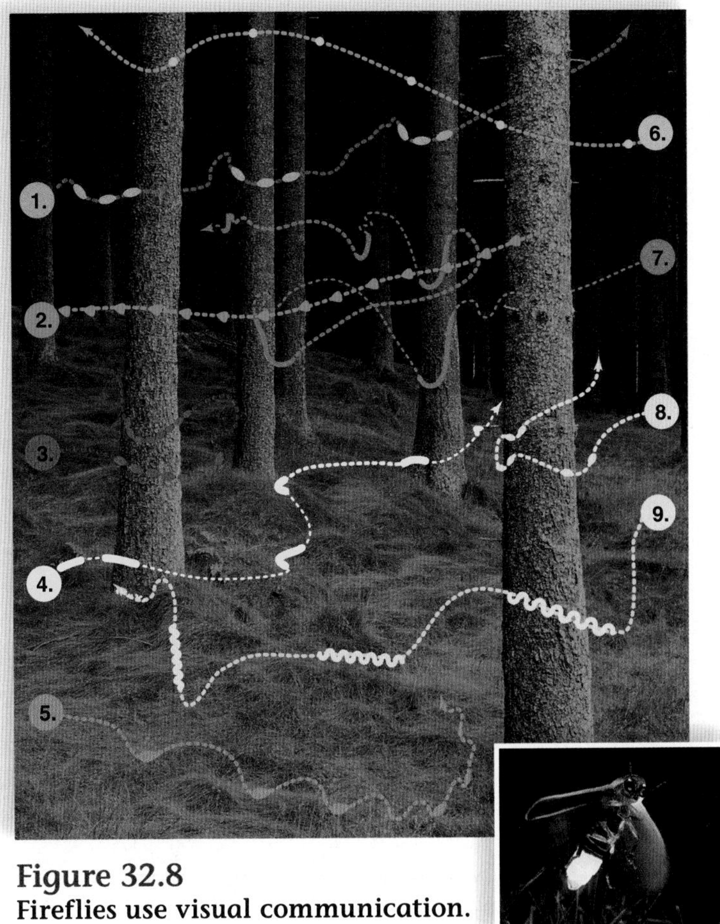

Figure 32.8
Fireflies use visual communication.
Each number represents the male flash pattern of a different species. The patterns are a behavioral reproductive isolation mechanism.

Visual communication allows animals to signal others of their intentions without the need to provide any auditory or chemical messages. The body language of students during a lecture provides an example. Some students lean forward in their seats and make eye contact with the instructor. They want the instructor to know they are interested and find the material of value. Others lean back in their chairs and look at the floor or doodle. These students indicate they are not interested in the material. Teachers can use students' body language to determine if they are effectively presenting the material and make changes accordingly.

Other human behaviors also send visual clues. The hairstyle and dress of a person or the way he or she walks and talks are ways to send messages to others. Psychologists have long tried to understand how visual clues can be used to better understand human emotions and behavior. Some studies have suggested that women are apt to dress in an appealing manner and be sexually inviting when they are ovulating. People who dress in black, move slowly, fail to make eye contact, and sit alone may be telling others that they are unhappy. Similarly, body language in animals is being used to suggest that they too have emotions, as discussed in the Science Focus, "Do Animals Have Emotions?"

Tactile Communication

Tactile communication occurs when one animal touches another. For example, laughing gull chicks peck at the parent's beak to induce the parent to feed them (see Fig. 32.3). A male leopard nuzzles the female's neck to calm her and stimulate her willingness to mate. In primates, grooming—one animal cleaning the coat and skin of another—helps cement social bonds within a group.

Honeybees use a combination of communication methods, including tactile ones, to convey information about their environment. When a foraging bee returns to the hive, it performs a waggle dance that communicates the distance and direction to the food source (Fig. 32.9). Inside the hive, other foragers crowd around her and touch the forager with their antennae, and as the bee moves between the two loops of a figure 8, it buzzes noisily and shakes its entire body in so-called waggles. The angle of the straight run of the figure 8 to that of the direction of gravity is the same as that of the angle between the sun and the food source. In other words, a 40° angle to the left of vertical means that food is 40° to the left of the sun. Bees can use the sun as a compass because they have a biological clock that allows them to compensate for the movement of the sun in the sky. (A biological clock is an internal means of telling time—for example, darkness outside stimulates many animals to sleep, including humans.) Outside the hive, the dance is done on a horizontal surface, and the straight-run part of the figure 8 indicates the direction of the food.

Check Your Progress

1. Name two ways in which communication is meant to affect the behavior of a receiver.
2. Which types of communication would be most likely to be used in dense forest? Why?

a.

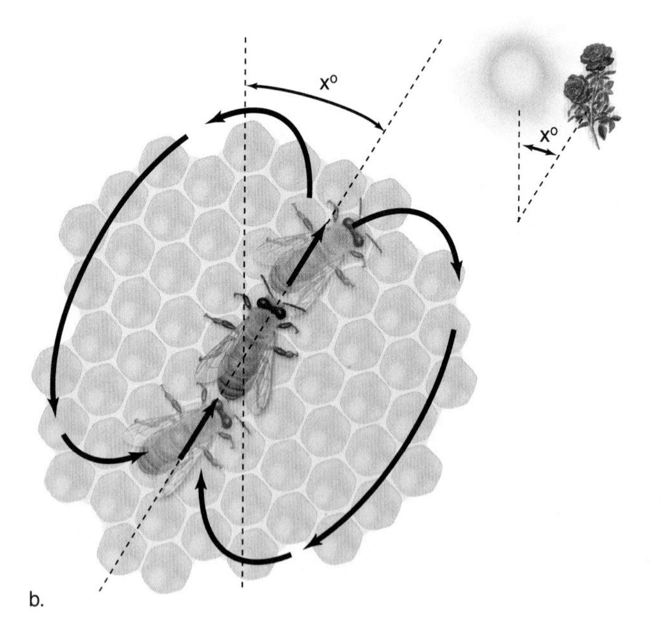

b.

Figure 32.9 Communication among bees.
a. Honeybees do a waggle dance to indicate the direction of food. **b.** If the dance is done outside the hive on a horizontal surface, the straight run of the dance will point to the food source. If the dance is done inside the hive on a vertical surface, the angle of the straightaway to that of the direction of gravity is the same as the angle of the food source to the sun.

Do Animals Have Emotions?

Recently, investigators have become interested in determining whether animals have emotions. The body language of animals suggests that animals do have feelings (Fig. 32A). When wolves reunite, they wag their tails to and fro, whine, and jump up and down. Elephants vocalize—emit their "greeting rumble"—flap their ears, and spin about. Many young animals play with one another, or even by themselves, as when dogs chase their own tails. On the other hand, upon the death of a friend or parent, chimps are apt to sulk, stop eating, and even die. It seems reasonable to hypothesize that animals are "happy" when they reunite, "enjoy" themselves when they play, and are "depressed" over the loss of a troop member or relative. Even people who rarely observe animals usually agree about what the animal appears to be feeling.

a.

b.

Figure 32A Emotions in animals.
a. Do chimpanzees feel motherly love, as this photograph seems to suggest? **b.** Does the hare feel fear as the lynx closes in?

In the past, scientists found it expedient to collect data only about observable behavior and to ignore the possible mental state of the animal. Why? Because emotions are personal and no one can ever know exactly how another animal is feeling. B. F. Skinner, whose research method is described in section 32.2, regarded animals as robots that become conditioned to respond automatically to various stimuli. He and others never considered that animals might have feelings. But now, some scientists argue that sufficient data exist to suggest that at least other vertebrates and/or mammals do have feelings, including fear, joy, embarrassment, jealousy, anger, love, sadness, and grief. And they believe that those who hypothesize otherwise should have to present the opposing data.

Perhaps it would be reasonable to consider the suggestion of Charles Darwin, the father of evolutionary biology, who said that animals are different in degree rather than in kind. This means that all animals can, say, feel love, but perhaps not to the degree that humans can. B. Würsig watched the courtship of two baleen whales. They touched, caressed, rolled side-by-side, and eventually swam off together. He wondered if their behavior indicated they felt love for one another. When you think about it, it seems unlikely that emotions first appeared in humans with no evolutionary homologies in animals.

Neurobiological data support the hypothesis that animals other than humans are capable of enjoying themselves when they perform an activity, such as playing. Researchers have found a high level of dopamine in the brain when rats play, and the dopamine level increases even when rats anticipate the opportunity to play. Certainly even the staunchest critic is aware that many different species of animals have limbic systems and are capable of fight-or-flight responses in dangerous situations. Can we go further and suggest that animals feel fear even when no physiological response has yet occurred?

Laboratory animals may be too stressed to provide convincing data on emotions. This makes field research more useful. Then, too, we have to consider that emotions evolved under an animal's normal environmental conditions. It is possible to fit animals with devices that transmit information on heart rate, body temperature, and eye movements as they go about their daily routine. Such information will help researchers learn how animal emotions might correlate with their behavior. In humans, emotions influence behavior, and the same may be true of other animals. One possible definition of emotion is a psychological phenomenon that helps animals direct and manage their behavior.

M. Bekoff, who is prominent in studying the basis of animal behavior, encourages us to be open to the possibility that animals have emotions. He states:

> By remaining open to the idea that many animals have rich emotional lives, even if we are wrong in some cases, little truly is lost. By closing the door on the possibility that many animals have rich emotional lives, even if they are very different from our own or from those of animals with whom we are most familiar, we will lose great opportunities to learn about the lives of animals with whom we share this wondrous planet.[1]

Discussion Questions

1. Do you believe animals have emotions? Why or why not? If not, what experiment(s) would convince you that animals do have emotions?
2. Pet psychology, an emerging field, is based on the premise that pets have feelings and emotions. Would you spend \$50–100 per hour to take your pet to a psychologist?
3. Does knowing of existing evidence that animals may have emotions change your opinion about how animal research should be conducted, as was discussed in Chapter 31?

[1]*Bekoff, M.* Animal emotions: Exploring passionate natures. *October 2000.* Bioscience *50:10, page 869.*

32.4 Behaviors That Affect Fitness

Learning Outcomes

1. Discuss why territorality evolved in certain animal groups.
2. Recall why females should be choosier than males in selecting a mate.
3. Describe the costs and benefits of living in a society.
4. Create an argument to explain the evolution of altruism.

Behavioral ecology assumes that most behavior is subject to natural selection. We have established that behaviors can have a genetic basis, and we would expect that certain behaviors more than others will lead to increased survival and number of offspring. Therefore, much of the behavior of organisms we can observe today must have adaptive value.

Figure 32.10 Male and female gibbons.
Siamang gibbons, *Hylobates syndactylus*, are monogamous, and they both share the task of raising offspring. They also share the task of marking their territory by singing. As is often the case in monogamous relationships, the sexes are similar in appearance. Male is above and female is below.

Territoriality and Fitness

In order to gather food, animals often have a particular home range where they can be found during the course of the day. Animals may actively defend a portion of their home range for their exclusive use as a food source or as a mating area. This portion of the home range is called their territory and the behavior is called territoriality.

Territoriality is more likely to occur during times of reproduction. For example, gibbons live in the tropical rain forest of South and Southeast Asia, and territories are maintained by loud singing (Fig. 32.10). Males sing just before sunrise, and mated pairs sing duets during the morning. Males, but not females, show evidence of fighting to defend their territory in the form of broken teeth and scars. Obviously, defense of a territory has a certain cost; it takes energy to sing and fight off others. Also, you might get hurt. So, what is the adaptive value of being territorial? Chief among the benefits of territoriality are to ensure a source of food, exclusive rights to one or more females, and to have a place to rear young and possibly to protect yourself from predators.

The territory has to be the right size for the animal. Too large a territory cannot be defended, and too small a territory may not contain enough food. Cheetahs require a large territory to hunt for their prey and, therefore, they use urine (the scent lasts awhile) to mark their territory (see Fig. 32.5). Hummingbirds are known to defend a very small territory because they depend on only a small patch of flowers as their food source.

Foraging for Food

Animals need to ingest food that will provide more energy than the effort expended acquiring the food. In one study, it was shown that shore crabs eat intermediate-sized mussels because the net energy gain was more than if they ate larger-sized mussels (Fig. 32.11). The large mussels take too much energy to open per the amount of energy they provide. The optimal foraging model states that it is adaptive for foraging behavior (i.e., searching for food) and food choice to be as energetically efficient as possible. Even though it can be shown that animals that take in more energy are the ones that are likely to have more offspring, animals often have to consider other factors such as escaping from predation. If an animal is killed and eaten, it has no chance at all of having offspring in the future. Thus, animals may have to avoid foraging when the threat of predation is high. Animals often face these types of trade-offs that cause them to modify their behavior. As another example, many animal species stop eating during peak reproductive periods because competition for mates is often fierce.

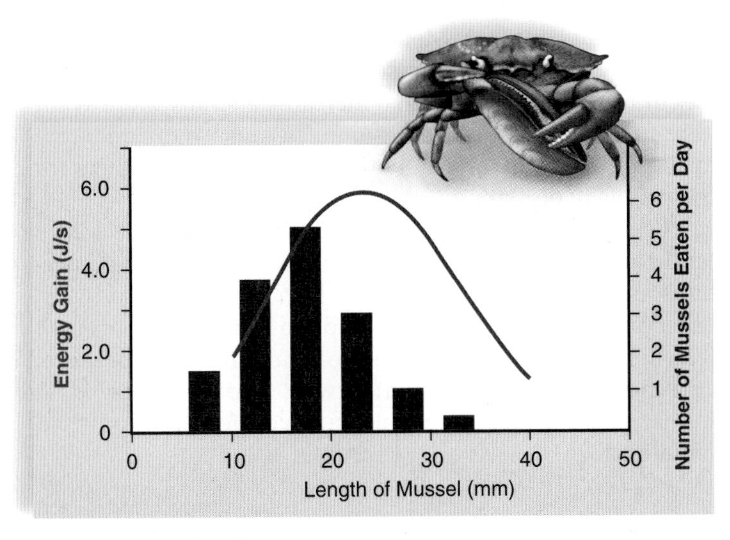

Figure 32.11 Foraging for food.
When offered a choice of an equal number of each size of mussel, the shore crab, *Carcinus maenas*, prefers the intermediate size. This size provides the highest rate of net energy return. Net energy is determined by the energetic yield of the crab minus the energetic costs of breaking open the shell and digesting the crab.

Figure 32.12 Hamadryas baboons.
Among Hamadryas baboons, *Papio hamadryas*, a male, which is silver-white and twice the size of a female, keeps and guards a harem of females with whom he mates exclusively.

Reproductive Strategies and Fitness

Usually, primates are polygamous, and males monopolize multiple females. Because of gestation and lactation, females invest more in their offspring than do males and may not always be available for mating. Under these circumstances, it is adaptive for females to be concerned with a good food source. When food sources are clumped, females congregate in small groups. Because only a few females are expected to be receptive at a time, males will likely be able to defend these few from other males. Males are expected to compete with other males for the limited number of receptive females available (Fig. 32.12).

A limited number of primates are polyandrus. Tamarins are squirrel-sized New World monkeys that live in Central or South America. Tamarins live together in groups of one or more families in which one female mates with more than one male. The female normally gives birth to twins of such a large size that the father, and not the mother, carry it about. This may be the reason these animals are polyanthrus. Polyandry also occurs when the environment does not have sufficient resources to support several young at a time.

Gibbons, as previously mentioned, are monogamous, which means that they pair bond. Subsequently, both the male and female help with the rearing of the young. Males are active fathers, frequently grooming and handling infants. Monogamy is relatively rare in primates, which includes prosimians, monkeys, and apes (only about 18% are monogamous). In primates, monogamy occurs when males have limited mating opportunities, territoriality exists, and the male is fairly certain the offspring are his. In gibbons, females are evenly distributed in the environment, most likely because they are aggressive to one another. Studies have shown that females do attack a speaker when it plays female sounds in their territory.

Sexual Selection

Sexual selection is a form of natural selection that favors features that increase an animal's chances of mating. In other words, these features are adaptive in the sense that they lead to increased fitness. Sexual selection often results in female choice and male competition. Because females produce a limited number of eggs in their lifetime and are generally the sex that provides more parental care, it is adaptive for them to be choosy about their mate. If they choose a mate that passes on features to a male offspring that will cause him to be chosen by females, their fitness has increased. Whether females choose features that are adaptive to the environment is in question. For example, peahens are likely to choose peacocks that have the most elaborate tails. Such a fancy tail could otherwise be detrimental to the male and make him more likely to be captured by a predator. In one study, an extra ornament was attached to a father zebra finch and the daughters of this bird underwent the process of imprinting. Now, these females were more likely to choose a mate that also had the same artificial ornament.

While females can always be sure an offspring is theirs, males do not have this certainty. However, males produce a plentiful supply of sperm. The best strategy for males to increase their fitness, therefore, is to have as many offspring as possible. Competition may be required for them to gain access to females, and ornaments, such as antlers,

Figure 32.13 Competition.
During the mating season, bull elk, *Cervus elaphus*, males may find it necessary to engage in antler wrestling in order to have sole access to females in a territory.

can enhance a male's ability to fight (Fig. 32.13). When bull elk compete, they issue a loud number of screams that gives way to a series of grunts. If still necessary, the two bulls walk in parallel to show each other their physique. If this doesn't convince one or the other to back off, the pair resorts to ramming each other with their antlers. Rarely is either bull actually hurt. Whereas a peacock cannot shed his tail, elk shed their antlers as soon as mating season is over.

Check Your Progress

1. How is territoriality related to foraging for food?
2. Why should females be choosier than males in selecting a mate?

Mating in Humans

A study of human mating behavior shows that the concepts of female choice and male competition apply to humans as well as to other animals. That is, mate choice behavior in men and women in most human cultures seems to be influenced by its fitness consequences. Of course, applying the principles of evolutionary biology to human behavior is not without controversy because we can't be reduced to being "preprogrammed" and still have the ability to make conscious choices. That said, understanding the evolutionary basis of animal behavior can provide some interesting insights into why people behave the way they do.

Human Males Compete

Consider that women, by nature, must invest more in having a child than men. After all, it takes nine months to have a child, and pregnancy is followed by lactation when a woman may nurse her infant. Men, on the other hand, need only contribute sperm during a sex act that may require only a few minutes. The result is that men are generally more available for mating than are women. Because more men are available, they necessarily have to compete with others for the privilege of mating.

Like many other animals, humans are dimorphic. Males tend to be larger and more aggressive than females, perhaps as a result of past sexual selection by females. As in other animals, males may pay a price due to high levels of testosterone, the energetic costs of male-male competition, and increased stress associated with finding mates. Male humans live on average four to seven fewer years than females do.

Females Choose

David Buss, an evolutionary psychologist at the University of Texas, conducted studies of female preference of a male mate across cultures in over 20 countries. His research, although somewhat controversial, suggests that the number-one trait females prefer in a male mate is his ability to obtain (financial) resources. Financial success means that men are more likely to provide females with the resources they need to raise their children, and thus increase the fitness (reproductive success) of the female.

Other recent studies have shown that symmetry in facial and body features is also important in female mate choice, possibly related to the "good genes" hypothesis. That is, females may choose males with "good genes" so that their offspring have good genes and a better chance at survival. As an example, symmetry in other animals is often a sign of good health and a strong immune system (e.g., animals with high parasite loads are often asymmetrical). Females may thus choose symmetrical males to pass these traits on to their offspring.

Men Also Have a Choice

Just as women choose men who can provide resources, men prefer youthfulness and attractiveness in females, signs that their partner can provide them with children. In one controversial study, men ages 8 to 80 across four continents preferred women with a waist-to-hip ratio (WHR) of 0.7, regardless of weight. Research showed that a WHR of 0.7 is optimal for conception; that is, for each increase of 10% in WHR, the odds of conception during each ovulation event decrease by 30%. Men responding to questionnaires prefer physical attributes in their female mates that biologists associate with a strong immune system and good health (e.g., symmetry), high estrogen levels (indicating fertility), and especially youthfulness. On average, men marry women 2.5 years younger than they are, but as men age, they tend to prefer women who are many years younger. Men are capable of reproduction for many more years than women. Therefore, by choosing younger women, older men can increase their fitness

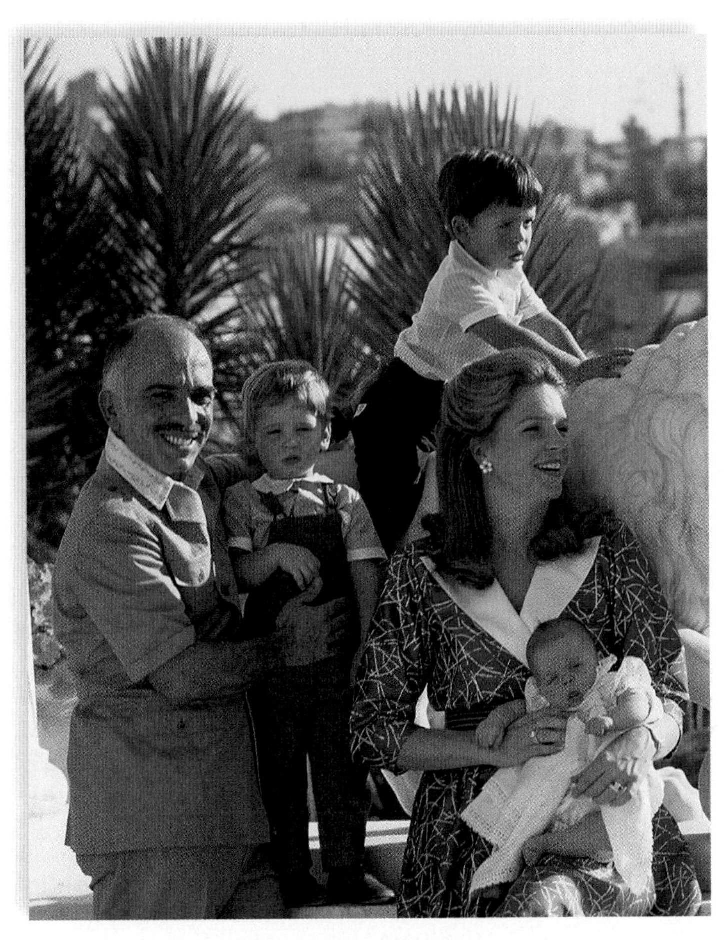

Figure 32.14 King Hussein and family.
The tendency of men to mate with fertile younger women is exemplified by King Hussein of Jordan, who was about 16 years older than his wife, Queen Noor. One of King Hussein's children from his second wife (not pictured here) is now King Abdullah of Jordan.

(Fig. 32.14). Other factors are also involved in human mate choice, as explained in the Science Focus, "Mate Choice and Smelly T-Shirts."

Societies and Fitness

The principles of evolutionary biology can be applied to the study of social behavior in animals. Sociobiologists hypothesize that societies form when living in a society has a greater reproductive benefit than reproductive cost. A **cost-benefit analysis** can help determine if this hypothesis is supported. Group living does have its benefits. It can help an animal avoid predators, rear offspring, and find food. A group of impalas is more likely to hear an approaching predator than a solitary one. Many fish moving rapidly in different directions might distract a would-be predator. Weaver birds form giant colonies that help protect them from predators, but the birds may also share information about food sources. Primate members of the same baboon troop signal to one another when they have found an especially bountiful fruit tree. Lions working together are able to capture larger prey, such as zebra and buffalo, than they could alone.

Group living also has its disadvantages. When animals are crowded together into a small area, disputes can arise over access to the best feeding places and sleeping sites. Dominance hierarchies are one way to apportion resources, but this puts subordinates at a disadvantage. Among red deer, sons are preferable because, as a harem master, sons will result in a greater number of grandchildren. However, sons, being larger than daughters, need to be nursed more frequently and for a longer period of time. Subordinate females do not have access to enough food resources to adequately nurse sons and, therefore, they tend to rear daughters, not sons. Still, like the subordinate males in a baboon troop, subordinate females in a red deer harem may be better off in terms of fitness if they stay with a group, despite the cost involved.

Living in close quarters exposes individuals to illness and parasites that can easily pass from one animal to another. Social behavior helps to offset some of the proximity disadvantages. For example, baboons and other types of social primates invest much time in grooming one another, and this most likely helps them remain healthy. Humans use extensive medical care to help offset the health problems that arise from living in the densely populated cities of the world.

Sociobiology and Human Culture

Humans today rely on living in organized societies. Clearly, the benefits of social living must outweigh the costs because we have organized governments, with laws that tend to increase the potential for human survival. The **culture** of a human society involves a wide spectrum of customs, ranging from how to dress to forms of entertainment, marriage rituals, and types of food. Language and the use of tools are essential to human culture. Language is used to socialize children, teach them how to use tools, educate them, and train them in skills that may help them maintain a household or begin a career. Technological skills, such as knowing how to use a computer or how to fix plumbing, are taught from an early age. Certainly cultural evolution has surpassed biological evolution in the past few centuries. For example, medical advances, which are passed on through language, have increased the human life expectancy from 40–45 years a century ago to nearly 80 years today.

Why did human societies originate? Perhaps the earliest organized societies were composed of "hunter-gatherers." Which of our ancestors first became hunters is debated among paleontologists (biologists who study fossils) and sociologists. Nonetheless, scientists speculate that a predatory lifestyle may have encouraged the evolution of intelligence and the development of language. That is, cooperation, communication, and tools (such as weapons) are necessary for hunting large animals. In hunter-gatherer societies, men tended to hunt, while women utilized wild and cultivated

Mate Choice and Smelly T-Shirts

Mate choice has been studied extensively in humans, largely because of our curiosity about how evolutionary biology and behavior in other animals may give us insight into human behavior. And, of course, mate choice is one of the more fascinating human behaviors. In one unusual and recent study, scientists had several men wear T-shirts for a few nights while they slept, thereby infusing the T-shirts with their unique smell. Then they asked women to rate the attractiveness of the male without seeing the person—that is, based only on the smell of the T-shirt. It turns out that women tended to choose men whose major histocompatability complex (MHC) alleles were different from their own. Recall from Chapter 13 that the MHC is associated with recognition of foreign antigens (e.g., viruses or bacteria) and enhances the immune system's ability to fight off infections. In choosing men with different (complementary) MHC alleles, females are maximizing the chance that their offspring have highly heterozygous MHC and potentially a more robust immune system. A more heterozygous MHC is associated with a better innate ability to recognize a more diverse group of antigens. In fact, females in this study said unattractive shirt smells reminded them of their fathers, who, of course, have very similar MHC alleles. Where did scientists get the idea for such a strange experiment? It has already been well documented that mice preferentially choose mates with complementary MHC alleles based on smell.

Of course, while evolutionary and behavioral biology can provide insights into human mate choice, the process is clearly complex. Although males prefer young faithful females with physical attributes that indicate fertility, they also look for symmetry, intelligence, and a sense of humor. And while females prefer males with financial resources, they also list such qualities as dependability and emotional stability (indicators of good parenting), physical attractiveness (indicated by symmetry and testosterone markers, such as above-average strength and height, possibly to defend resources), sense of humor, and intelligence. The way a person smells also could play a factor, as indicated by the T-shirt study. Then there are unknowns—the context under which people meet, their professions, their likes/dislikes, etc.

Discussion Questions

1. Scientists have found a biological basis for infanticide. In lions, for example, males who kill the offspring of other males when they take over a pride then induce females to come into estrus, so they can mate with them. This increases male fitness. Should a biological basis for infanticide be used as a defense in human court cases where, say, a stepfather kills a stepchild?
2. As discussed in section 32.4, some studies suggest that men prefer women with a waist-to-hip ratio of 0.7. Women of different weights can have this ratio because it is based on bone structure, rather than weight. Nonetheless, women may misinterpret this information, which can contribute to the epidemic of eating disorders, such as bulimia and anorexia. How can we better educate people about the scientific information on mate choice and the dangers of eating disorders?
3. Studies of human mate choice are generally complex, and often controversial. Should federal agencies fund these studies? Why or why not?

plants as a source of food. Sometimes, when animal food was scarce, hunter-gatherer societies relied on plant food cultivated by women. In this sense, cooperation ensured that people were fed.

Altruism Versus Self-Interest

Altruism can be generally described as a self-sacrificing behavior for the good of another member of the society. In an evolutionary sense, altruism may compromise the fitness of the altruist, while benefiting the fitness of the recipient. Are animals truly altruists, given that natural selection should eventually rid populations of altruists due to their decreased reproductive success? In humans, we can clearly think of examples—a volunteer firefighter dying while trying to save a stranger, a soldier losing his life for his country, a woman diving in front of a car to save an elderly person from being hit. But what about other animals?

In general, altruistic behavior in animals is explained by the concept of **kin selection.** In other words, because close relatives share many of your genes, it may make sense to self-sacrifice to save them. For example, your siblings share 50% of your genes, so sacrificing your life for two brothers or sisters makes sense evolutionarily. **Inclusive fitness** refers to an individual's personal reproductive success, as well as that of his or her relatives, and thus to an individual's total genetic contribution to the next generation. The concepts of kin selection and inclusive fitness are well supported by research in complex animal societies. For example, in a colony of army ants, the queen is inseminated only during her nuptial flight. Thereafter, she spends the rest of her life reproducing constantly, laying up to 30,000 eggs per day! The eggs hatch into three different sizes of sterile female workers whose jobs contribute to the colony, which can include over 1 million individuals. The smallest workers (3 mm), called nurses, take care of the queen and the larvae

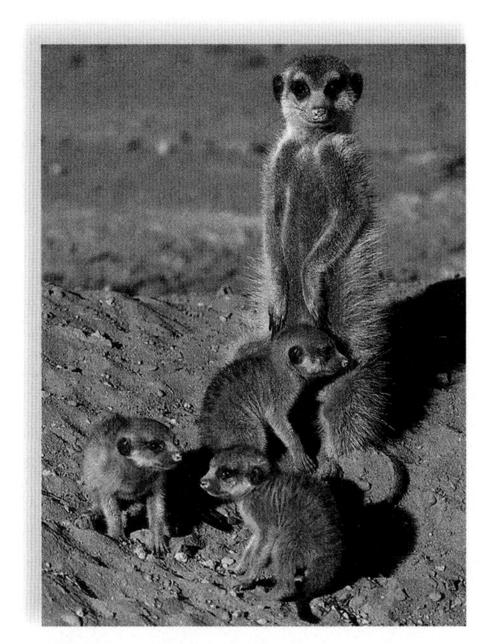

Figure 32.15 Inclusive fitness.
A meerkat is acting as a babysitter for its young sisters and brothers while their mother is away. Researchers point out that the helpful behavior of the older meerkat can lead to increased inclusive fitness.

by feeding and cleaning them. The intermediate-sized workers (3–12 mm), which constitute most of the population, go out on raids to collect food. The soldiers (14 mm), which have relatively huge heads and powerful jaws, surround raiding parties to protect them and the colony from attack by intruders. Why do the worker ants engage in this apparently altruistic behavior?

The queen ant is diploid (2n), but her mate is haploid. Thus, assuming the queen has had only one mate, her sister workers are more closely related to each other than normal siblings (75% versus 50%). Therefore, a worker can achieve higher inclusive fitness by helping her mother (the queen) produce additional sisters than by producing her own offspring, which would only share 50% of her genes. Thus, this behavior is not really altruistic. Rather, it is adaptive because it is likely due to the increase in inclusive fitness of the helpers relative to those that hypothetically might defect and breed on their own. Similar social patterns are observed in certain species of bees and wasps.

Reciprocal Altruism

In some bird species, offspring from a previous clutch of eggs may stay at the nest to help parents rear the next batch of offspring. In a study of Florida scrub jays, the number of fledglings produced by an adult pair doubled when they had helpers. Mammalian offspring are also observed to help their parents (Fig. 32.15). Among jackals in Africa, solitary pairs managed to rear an average of 1.4 pups, whereas pairs with helpers reared 3.6 pups. What are the benefits of staying behind to help? First, a helper is contributing to the survival of its own kin. Therefore, the helper actually gains a fitness benefit. Second, a helper is more likely than a nonhelper to inherit a parental territory—including other helpers. Helping, then, involves making a minimal, short-term reproductive sacrifice in order to maximize future reproductive potential. Therefore, helpers at the nest are also practicing a form of reciprocal altruism. Reciprocal altruism also occurs in animals that are not necessarily closely related. In this event, an animal helps or cooperates with another animal with no immediate benefit. However, the animal that was helped will repay the debt at some later time. Reciprocal altruism usually occurs in groups of animals that are mutually dependent. Cheaters in reciprocal altruism are recognized and not reciprocated in future events. Reciprocal altruism occurs in vampire bats that live in the tropics. Bats returning to the roost after a feeding activity share their blood meal with other bats in the roost. If a bat fails to share blood with one that had previously shared blood with it, the cheater bat will be excluded from future blood sharing.

CHECK YOUR PROGRESS

1. Describe the advantages and disadvantages of social living.
2. Use your understanding of the concepts kin selection and inclusive fitness to explain the maintenance of altruism in spite of its fitness costs.

Applying the Concepts [Revisited]

At the beginning of this chapter, you learned a bit about the complexity of emperor penguin behaviors that enable them to survive in Antarctica, one of the harshest of the world's environments. Their lifestyle incorporates complex mating behavior, intense amounts of parental care by both parents, communication among parents and offspring, and social behavior to ensure survival.

1. Describe one emperor penguin behavior that you think is learned and one that you think is genetically controlled. Justify your choices.
2. Apply what you have learned about mate choice to the emperor penguin system. Should males be choosy as well as females? Why or why not?
3. Why do you think social living has evolved in emperor penguins?

SUMMARIZING THE CONCEPTS

32.1 Nature Versus Nurture: Genetic Influences

- Investigators have long been interested in the degree to which nature (genetics) or nurture (environment) influence behavior.
- Hybrid studies with lovebirds and human twin studies are consistent with the hypothesis that behavior has a genetic basis. Garter snake experiments indicate that the nervous system controls behavior. Egg-laying behavior in *Aplysia* has endocrine system involvement.

32.2 Nature Versus Nurture: Environmental Influences

- Even some behaviors formerly thought to be fixed action patterns (FAPs) can sometimes be modified by learning. A red spot on the bill of laughing gulls initiates chick begging behavior. However, with experience, chick begging accuracy improves.
- Other studies also support the involvement of learning in behaviors. Imprinting during a sensitive period causes birds to follow the first moving object they see.
- Song learning in birds involves various elements—including learning during the sensitive period, as well as the effect of social interactions outside the period.
- Associative learning includes classical conditioning and operant conditioning. In classical conditioning, the pairing of two different types of stimuli causes an animal to form an association between them (e.g., dogs salivating at the sound of a bell). In operant conditioning, animals learn behaviors because they are rewarded when they perform them.

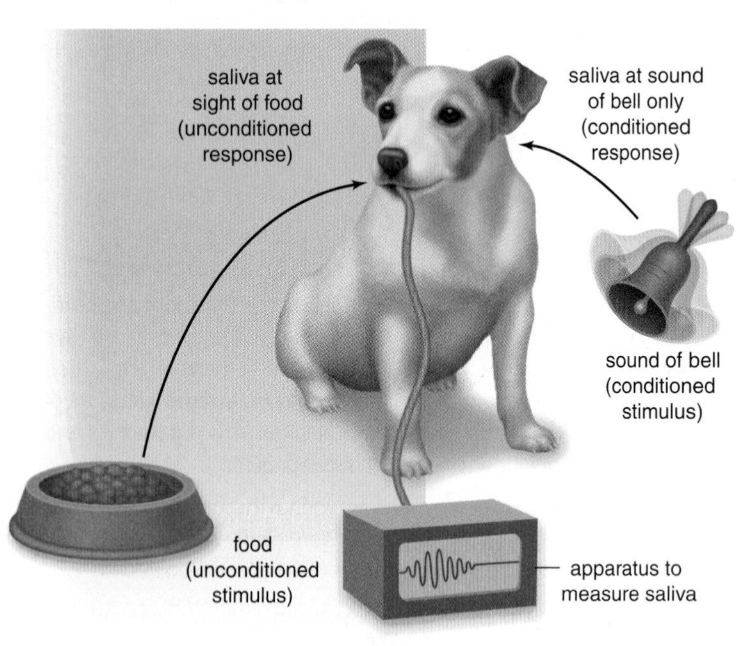

32.3 Animal Communication

- Communication is an action by a sender that affects the behavior of a receiver.

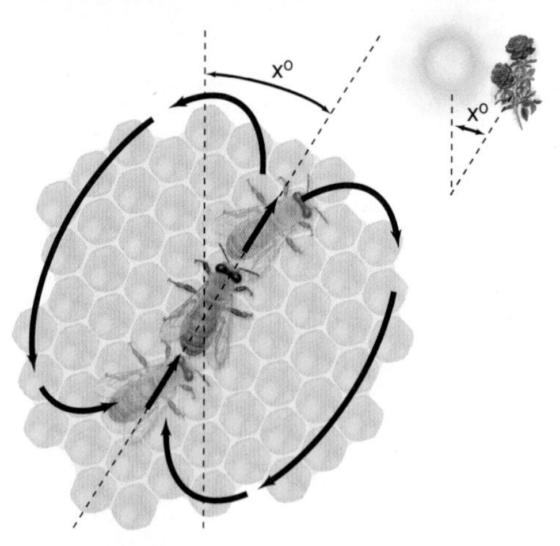

- Pheromones are chemical signals passed between members of the same species.
- Auditory communication includes language, which occurs in humans and other animals.
- Visual communication allows signaling without auditory or chemical messages.
- Tactile communication is especially associated with social behavior.

32.4 Behaviors That Affect Fitness

- Traits that promote reproductive success are expected to have fitness advantages that outweigh their disadvantages.
- Some animals are territorial and defend a territory that may have high quality food and/or nesting sites.
- Sexual selection is a form of natural selection that selects for traits that increase an animal's fitness.
- Males, who produce many sperm, are expected to compete to inseminate as many females as possible.
- Females, who produce few eggs, are expected to be selective about their mates. Females may choose mates based on increased survival of offspring or on traits that make their sons more attractive to other females.
- It is likely that mating behavior and mate choice in humans have also been shaped by evolutionary forces.
- Living in a social group can have its advantages (e.g., avoiding predators, raising young, and finding food). It also has disadvantages (e.g., competition between members, spread of illness and parasites, and reduced reproductive potential).
- When animals live in groups, the fitness (number of fertile offspring produced) benefits must outweigh the costs, or social behavior would not exist.
- Sometimes animals perform what are apparently altruistic acts. However, it is necessary to consider inclusive fitness, which includes personal reproductive success, as well as the reproductive success of relatives. For example, social insects help their mother reproduce, but this behavior seems reasonable when we consider that siblings share 75% of their genes. Parental helpers in mammals and birds often inherit the parent's territory.

TESTING YOURSELF

Choose the best answer for each question.

1. In lovebirds, if carrying strips in the bill is controlled by a single dominant gene, then hybrids between peach-faced lovebirds and Fischer lovebirds would
   a. carry strips in their bills.
   b. carry strips in their rump feathers.
   c. not carry strips.
   d. carry strips in both their bills and rump feathers.
   e. have decreased fitness.
2. Which of the following is not an example of a genetically based behavior?
   a. Inland garter snakes do not eat slugs, while coastal populations do.
   b. One species of lovebird carries nesting strips one at a time, while another carries several.
   c. One species of warbler migrates, while another one does not.
   d. Snails lay eggs in response to egg-laying hormone.
   e. Wild foxes raised in captivity are not capable of hunting for food.
3. How would the following graph differ if begging behavior in laughing gulls was a fixed action pattern?

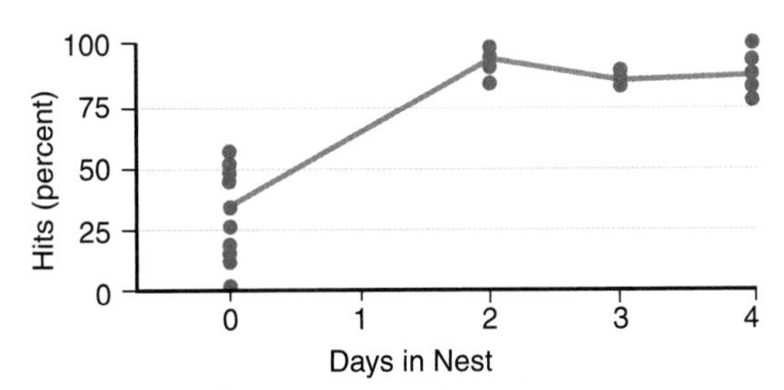

   a. It would be a diagonal line with an upward incline.
   b. It would be a diagonal line with a downward incline.
   c. It would be a horizontal line.
   d. It would be a vertical line.
   e. None of these are correct.
4. Using treats to train a dog to do a trick is an example of
   a. imprinting.
   b. tutoring.
   c. vocalization.
   d. operant conditioning.
5. In white-crowned sparrows, social experience exhibits a very strong influence over the development of singing patterns. What observation led to this conclusion?
   a. Birds learned to sing only when they were trained by other birds.
   b. The window in which birds learn from other birds is wider than that in which birds learn from tape recordings.
   c. Birds could learn different dialects only from other birds.
   d. Birds that learned to sing from a tape recorder could change their song when they listened to another bird.
6. Which reaction best provides physiological evidence that animals have emotions?
   a. Adrenaline levels increase when a deer senses danger.
   b. Dopamine levels in the brains of rats increase when they play.
   c. Chimpanzees experience decreases in acetylcholine levels when they are content.
   d. The blood pressure of wolves drops when they return to the pack.
7. Which of the following best describes classical conditioning?
   a. the gradual strengthening of stimulus-response connections that are seemingly unrelated
   b. a type of associative learning in which there is no contingency between response and reinforcer
   c. the learning behavior in which an organism follows the first moving object it encounters
   d. the learning behavior in which an organism exhibits a fixed action pattern from the time of birth
8. Bees that do a waggle dance are teaching other bees
   a. how to dance.
   b. where to find food.
   c. how to find and use the sun for navigation.
   d. how to use auditory communication.
9. The benefits of imprinting generally outweigh the costs because
   a. an animal that has been imprinted on the wrong object can be reimprinted on the mother.
   b. imprinting behavior never lasts more than a few months.
   c. animals in the wild rarely imprint on anything other than their mother.
   d. animals that imprint on the wrong object generally die before they pass on their genes.
10. Which of the following are costs that a dominant male baboon must pay in order to gain a reproductive benefit?
    a. He requires more food and must travel longer distances.
    b. He requires more food and must care for his young.
    c. He is more prone to injury and requires more food.
    d. He is more prone to injury and must care for his young.
    e. He must care for his young and travel longer distances.
11. A red deer harem master typically dies earlier than other males because he is
    a. likely to get expelled from the herd and cannot survive alone.
    b. more prone to disease because he interacts with so many animals.
    c. likely to starve to death.
    d. apt to place himself between a predator and the herd to protect the herd.

For problems 12–15, match the type of communication in the key with its description. Answers may be used more than once.

Key:

a. chemical communication
b. auditory communication
c. visual communication
d. tactile communication

12. Aphids (insects) release an alarm pheromone when they sense they are in danger.
13. Male peacocks exhibit an elaborate display of feathers to attract females.
14. Ground squirrels give an alarm call to warn others of the approach of a predator.
15. Male silk moths are attracted to females by a sex attractant released by the female moth.

UNDERSTANDING THE TERMS

altruism 684
associative learning 675
behavior 672
classical conditioning 675
communication 676
cost-benefit analysis 683
culture 683
fixed action pattern (FAP) 674
imprinting 674
inclusive fitness 684
kin selection 684
learning 674
operant conditioning 675
pheromone 676
sexual selection 681
society 676

THINKING CRITICALLY

1. You are studying two populations of rats—one in New York and the other in Florida. Adult New York rats seem to prefer swiss cheese, while Florida rats seem to prefer cheddar. Design an experiment to determine the extent to which this behavior may be genetically controlled versus environmentally influenced.
2. Until about 15 years ago, it was long thought that most bird species were monogamous—that is, at least within a breeding season, males and females paired and mated only with each other. However, with advances in DNA technology, researchers began to find out that offspring in a nest were often sired by males in neighboring territories. When females mate with neighboring males, it is called "extra pair copulation" or EPC. Why do you think EPC behavior might have evolved in many bird species?
3. If altruistic behavior has usually evolved to improve the altruist's inclusive fitness by helping relatives, then why do you think humans perform altruistic acts for completely unrelated individuals (e.g., jumping into a pool to save a drowning person)?

INQUIRY INTO LIFE WEBSITE

The companion website for *Inquiry into Life* provides a wealth of information organized and integrated by chapter. You will find practice tests, animations, videos, and much more that will complement your learning and understanding of general biology.

http://www.mhhe.com/maderinquiry13

33

Population Ecology

Applying the Concepts

Overfishing threatens the world's fish populations, such as that of the swordfish. Global fisheries' production has increased steadily over the past 50 years, from about 18 million tons per year in 1950 to over 100 million tons in 2007. The United Nations Food and Agricultural Organization estimates that we will need to catch an additional 37 tons per year to feed the growing human population by 2030. These numbers lead some to estimate a global fishery collapse by 2048. Overfishing occurs when the catch of a particular fish species exceeds reproduction, resulting in declining numbers. In other words, the death rate exceeds the birthrate. Normally, fish populations exist at the carrying capacity of the environment (the numbers the environment can support). However, if humans reduce fish populations well below the carrying capacity, a lag in population growth occurs. If fishing continues at a high rate during this lag phase, it can drive fish populations to extinction.

Population ecologists study the distribution and abundance of organisms, such as fish species that are commercially exploited. Our knowledge of population growth and regulation has led to controls on fishing, including size limits and catch limits, to try to sustain the world's commercially valuable fish populations. Thus, the study of population ecology is necessary for the conservation of species and the maintenance of viable populations for commercial use.

Chapter Outline

33.1 The Scope of Ecology

Learning Outcome

1. Describe the levels of ecological study.

Ecology is the study of the interactions of organisms with each other and with their physical environment. Ecology, like so many biological disciplines, is wide-ranging and involves several levels of study (Fig. 33.1).

At the most basic level, ecologists study how organisms are adapted to their environment. For example, they might study why a particular species of fish in a coral reef lives only within a narrow temperature range in warm tropical waters. Most organisms do not exist singly. Rather, they are part of a **population,** which is defined as all the organisms of the same species interacting with the environment in a particular area. At the population level of study, ecologists describe the size of populations over time. For example, an ecologist might study the relative sizes of parrotfishes' populations within a coral reef over time. A **community** consists of all the various populations at a particular locale. A coral reef community contains numerous populations of fish species, crustaceans, corals, and so forth. At the community level, ecologists study how various extrinsic factors (e.g., weather) and intrinsic factors (e.g., species' competition for resources) affect the size of these populations. An **ecosystem** encompasses a community of populations, as well as the nonliving environment. For example, energy flow and chemical cycling in a coral reef can affect the success of the organisms that inhabit it. Finally, the **biosphere** is that portion of the entire Earth's surface—air, water, land—where living things exist. Knowing the composition and diversity of the coral reef ecosystem is important to the dynamics of the biosphere. Table 33.1 defines and summarizes the levels commonly studied in ecology.

Modern ecology is not just descriptive; it primarily focuses on developing testable hypotheses. A central goal of modern ecology is to develop models that explain and predict the distribution and abundance of populations and species. Ultimately, ecology considers not just one particular area, but species' distributions throughout the biosphere. In this chapter, we particularly concentrate on the patterns of population growth, and how population growth is regulated. Then, at the community level, we examine how interacting populations change through time during the process of ecological succession.

TABLE 33.1 Ecological Terms

Term	Definition
Ecology	Study of the interactions of organisms with each other and with the physical environment
Population	All the members of the same species that inhabit a particular area
Community	All the populations found in a particular area
Ecosystem	A community and its physical environment, including both nonliving (abiotic) and living (biotic) components
Biosphere	All of the ecosystems on Earth, including their biotic communities

Check Your Progress

1. What is a major goal of modern ecology?

Figure 33.1 Levels of ecological organization. The study of ecology encompasses levels of organization that proceed from the individual organism to the population, to the community, and to an ecosystem.

33.2 Patterns of Population Growth

Learning Outcomes

1. Calculate the per capita rate of increase in a population.
2. Compare exponential and logistic population growth curves.
3. Explain why human population growth in less-developed countries exceeds that of more-developed countries.

Each population has particular patterns of growth. The population size can stay the same, increase, or decrease, according to a *per capita rate of increase* or growth rate. Suppose, for example, a human population presently has a size of 1,000 individuals, the birthrate is 30 per year, and the death rate is 10 per year. The growth rate per year will be:

$$30 - 10 \ / \ 1{,}000 = 0.02 = 2.0\% \text{ per year}$$

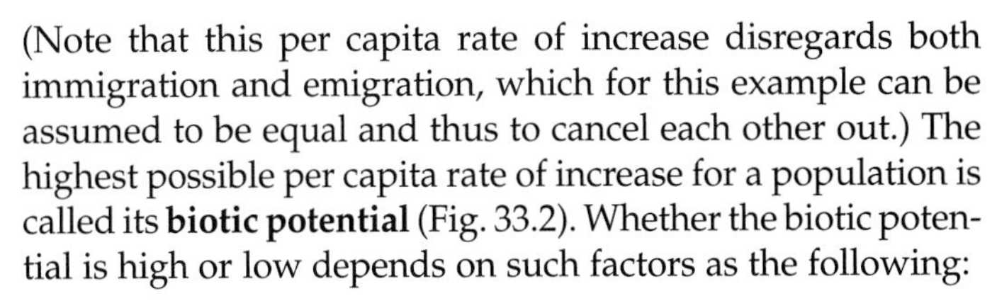

(Note that this per capita rate of increase disregards both immigration and emigration, which for this example can be assumed to be equal and thus to cancel each other out.) The highest possible per capita rate of increase for a population is called its **biotic potential** (Fig. 33.2). Whether the biotic potential is high or low depends on such factors as the following:

1. Average number of offspring per reproduction
2. Chances of survival until age of reproduction
3. Age at first reproduction
4. How often each individual reproduces

Suppose we are studying the growth of a population of insects capable of infesting and taking over an area. Under these circumstances, **exponential growth** is expected. An exponential pattern of population growth results in a J-shaped curve (Fig. 33.3*a*). This pattern of population growth can be likened to compound interest at the bank: the more your money increases, the more interest you will get.

a.

b.

Figure 33.2 Biotic potential.
A population's maximum growth rate under ideal conditions—that is, its biotic potential—is greatly influenced by the number of offspring produced in each reproductive event. **a.** Pigs, which produce many offspring that quickly mature to produce more offspring, have a much higher biotic potential than **(b)** the rhinoceros, which produces only one or two offspring during infrequent reproductive events.

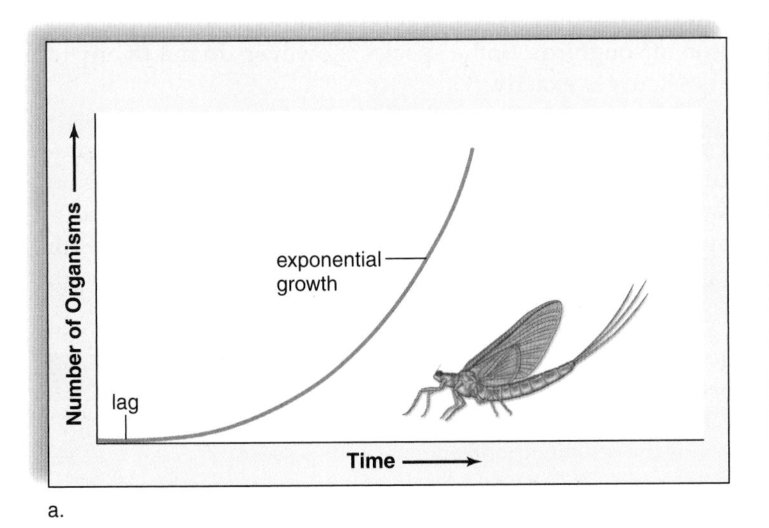

a.

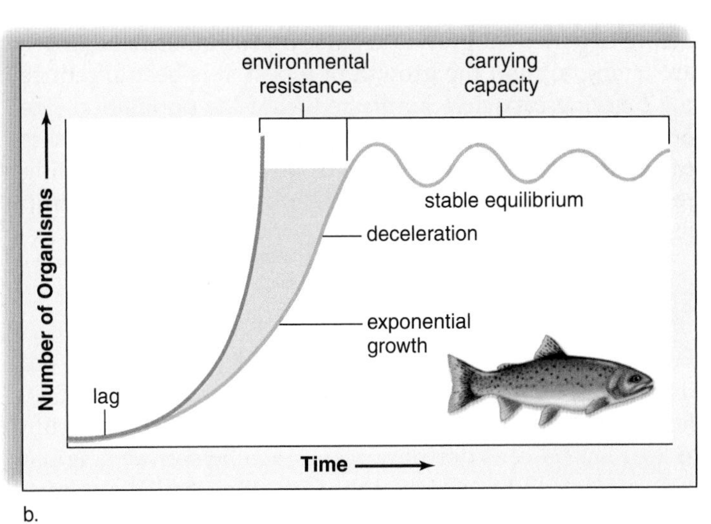

b.

Figure 33.3 Patterns of population growth.
a. Exponential growth results in a J-shaped growth curve because the growth rate is positive, and increasing, as in insects. **b.** Logistic growth results in an S-shaped growth curve because environmental resistance causes the population size to level off and be in a steady state at the carrying capacity of the environment, as in fish.

If the insect population has 2,000 individuals and the per capita rate of increase is 20% per month, there will be 2,400 insects after one month, 2,880 after two months, 3,456 after three months, and so forth.

Notice that a J-shaped curve has these phases:

Lag phase. Growth is slow because the population is small.
Exponential growth phase. Growth is accelerating, and the population is exhibiting its biotic potential.

Usually, exponential growth cannot continue for long because of environmental resistance. **Environmental resistance** is all those environmental conditions—such as limited food supply, accumulation of waste products, increased competition, or predation—that prevent populations from achieving their biotic potential. The intensity of environmental resistance increases as the population grows larger, until population growth levels off, resulting in an S-shaped pattern called **logistic growth** (Fig. 33.3*b*).

Notice that an S-shaped curve has these phases:

Lag phase. Growth is slow because the population is small.
Exponential growth phase. Growth is accelerating, due to biotic potential.
Deceleration phase. The rate of population growth slows down.
Stable equilibrium phase. Little if any growth takes place because births and deaths are about equal.

The stable equilibrium phase is said to occur at the **carrying capacity** of the environment. The carrying capacity is the number of individuals of a species that a particular environment can support.

Our knowledge of logistic growth has practical implications. The model predicts that exponential growth occurs only when population size is much lower than the carrying capacity. So, for example, if humans are using a fish population as a continuous food source, it would be best to maintain that population size in the exponential phase of growth when the birthrate is the highest. If we overfish, the fish population will sink into the lag phase, and it will be years before exponential growth recurs. On the other hand, if we are trying to limit the growth of a pest, it is best to reduce the carrying capacity. Simply reducing the population size only encourages exponential growth to begin again. Farmers can reduce the carrying capacity for a pest by alternating rows of different crops rather than growing one type of crop throughout the entire field.

Survivorship

Population growth patterns assume that populations are made up of identically aged individuals. In real populations, however, individuals are in different stages of their life spans. Let us consider how many members of an original group of individuals born at the same time, called a **cohort,** are still alive after certain intervals of time. The result is a survivorship curve.

For the sake of discussion, three types of idealized survivorship curves are recognized (Fig. 33.4). The type I curve is characteristic of a population of humans in which most individuals survive well past the midpoint, and death comes near the end of the maximum life span. On the other hand, the type III curve is typical for a population of oysters in which most individuals die very young. In the type II curve, survivorship decreases at a constant rate throughout the life span. This has been found typical of a population of songbirds. Some species, however, do not fit any of these curves exactly.

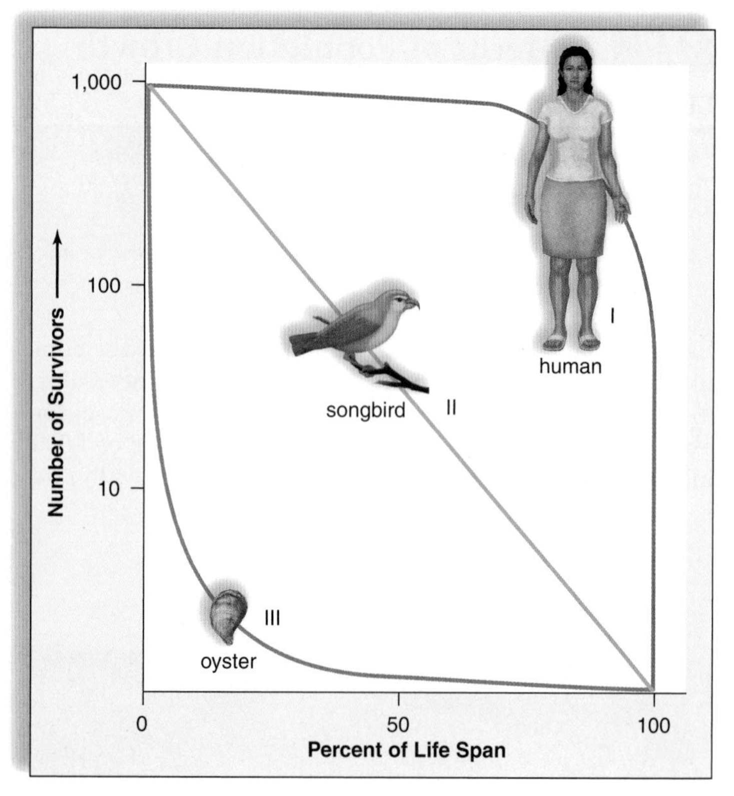

Figure 33.4 Survivorship curves.
Human beings have a type I survivorship curve: the individual usually lives a normal life span, and then death is increasingly expected. Songbirds have a type II curve: the chances of surviving are the same for any particular age. Oysters have a type III curve: most deaths occur during the free-swimming larva stage, but oysters that survive to adulthood usually live a normal life span.

Much can be learned about the life history of a species by studying its survivorship curve. Would you predict that most or only a few members of a population with a type III survivorship curve are contributing offspring to the next generation? Obviously, because death comes early for most members, only a few live long enough to reproduce.

Check Your Progress

1. Why don't populations grow to their biotic potential? What is the resulting population growth curve?
2. Describe the three types of idealized population survivorship curves.
3. Assuming an annual birth rate of 100 per 1,000 individuals and a death rate of 50 per 1,000 individuals in a population, what is the per capita rate of increase?

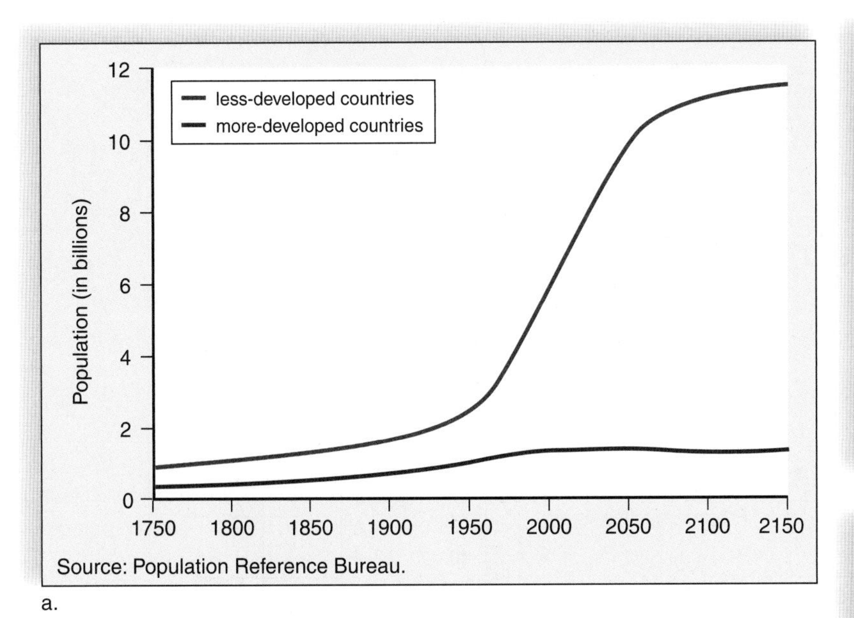

a.

b.

c.

Figure 33.5 World population growth.
a. The graph shows world population size in the past, with estimates to 2150. People in **(b)** more-developed countries have a high standard of living and will contribute the least to world population growth, while people in **(c)** less-developed countries have a low standard of living and will contribute the most to world population growth.

Human Population Growth

Figure 33.5*a* illustrates human population growth. Growth in *less-developed countries* is still in the exponential phase and is not expected to level off until about 2040, while growth in *more-developed countries* has leveled off. The equivalent of a medium-sized city (roughly 220,000) is added to the world's population every day, and 79 million (equivalent to the combined populations of Argentina, Ecuador, and Peru) are added every year.

The rapid growth of the human population can be appreciated by considering the **doubling time,** the length of time it takes for the population size to double. Currently, the doubling time is estimated to be 52 years. Such an increase in population size will put extreme demands on our ability to produce and distribute resources. In 52 years, the world will need to double the amount of food, jobs, water, energy, and so on just to maintain the present standard of living.

Many people are gravely concerned that the amount of time needed to add each additional billion persons to the world population has become shorter. The first billion was reached in 1800; the second billion in 1930; the third billion in 1960; and today there are over 6.7 billion. By contrast, **zero population growth,** in which the birthrate equals the death rate and population size remains steady, can only be achieved if the per capita rate of increase declines. The world's population may level off at 10, 12, or 14.2 billion, depending on the speed with which the per capita rate of increase declines.

More-Developed Versus Less-Developed Countries

The **more-developed countries (MDCs),** typified by countries in North America and Europe, are those in which population growth is low and people enjoy a good standard of living (Fig. 33.5*b*). The **less-developed countries (LDCs),** such as countries in Latin America, Africa, and Asia, are those in which population growth is expanding rapidly and the majority of people live in poverty (Fig. 33.5*c*).

The MDCs doubled their populations between 1850 and 1950 due to a decline in the death rate, the development of modern medicine, and improved socioeconomic conditions. However, the transition from an agricultural society to an urban society reduced the incentive for people to have large families (e.g., more people to help on the farm) because maintaining large families in urban areas is more costly than in rural areas, causing birthrates to decline. These factors, combined with the decreased death rate, caused populations in MDCs to experience only modest growth between 1950 and 1975. This sequence of events (i.e., decreased death rate followed by decreased birthrate) is termed a **demographic transition.**

Yearly growth of the MDCs as a whole has now stabilized at about 0.1%. The populations of several of the MDCs, including Germany, Greece, Italy, Hungary, and Sweden, are not growing or are actually decreasing in size. In contrast, there is no leveling off in U.S. population growth. Although yearly growth of the United States is only about 0.8%, many people immigrate to the United States each year. In addition, an unusually large number of babies were born

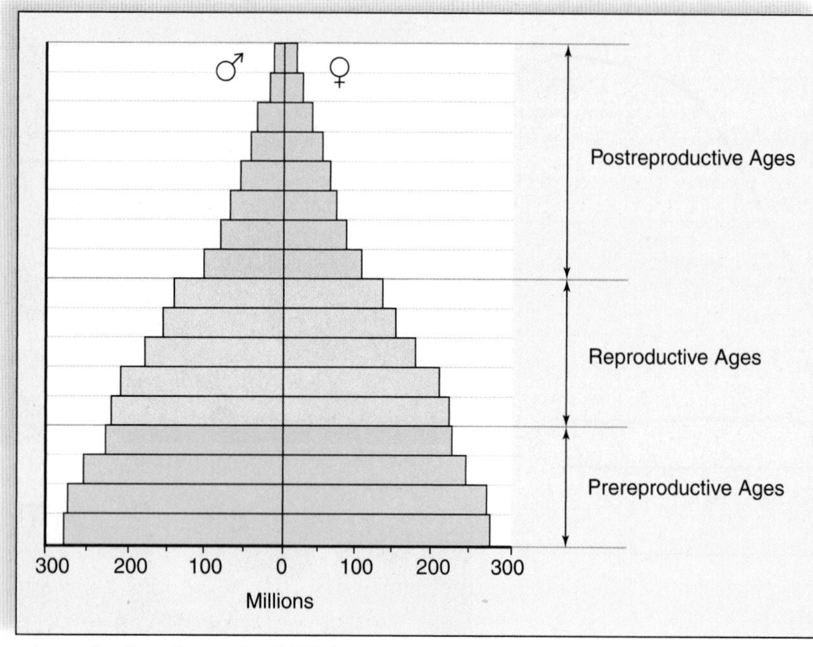

a. Less-developed countries (LDCs)

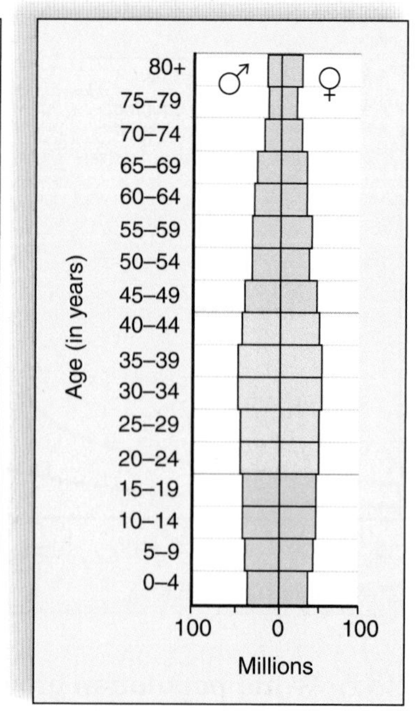

b. More-developed countries (MDCs)

Figure 33.6 Age structure diagrams.
The diagrams illustrate that **(a)** the LCDs will expand rapidly, resulting in unstable age structure, whereas **(b)** the MDCs are approaching stabilization.

between 1947 and 1964 (called a baby boom). Therefore, a large number of women are still of reproductive age.

Although death rates began to decline steeply in the LDCs following World War II with the importation of modern medicine from the MDCs, birthrates remained high. The yearly growth of the LDCs peaked at 2.5% between 1960 and 1965. Since that time, a demographic transition has occurred: the decline in the death rates slowed, and birthrates fell. Yearly growth now averages 1.9%. Still, because of exponential growth, the population of LDCs may explode from 4.9 billion today to 11 billion by 2100. Most of this growth will occur in Africa, Asia, and Latin America. Suggestions for greatly reducing the expected increase include the following:

1. Establish and/or strengthen family planning programs. A decline in growth is seen in countries with good family planning programs supported by community leaders. Currently, 25% of women in sub-Saharan Africa say they would like to delay or stop childbearing, and yet they lack access to birth control. Similarly, 15% of women in Asia and Latin America have an unmet need for birth control.
2. Use social progress to reduce the desire for large families. Many couples in LDCs presently desire as many as four to six children. But providing education, raising the status of women, reducing child mortality, and improving economic stability are desirable social improvements that could help decrease population growth.
3. Delay the onset of childbearing. A delay in the onset of childbearing and wider spacing of births could cause a temporary decline in the birthrate and reduce the present reproductive rate.

Age Distributions

An **age-structure diagram** divides the population into three age groups: prereproductive, reproductive, and postreproductive (Fig. 33.6). The LDCs are experiencing a population momentum because they have more women entering the reproductive years than older women leaving them.

It is a misconception that if each couple has two children, referred to as **replacement reproduction,** that zero population growth will take place immediately. Actually, the population will continue to grow as long as there are more young women entering their reproductive years than there are older women leaving them. This is true of most LDCs, where a large proportion of the population is younger than 15, thereby causing population expansion even after replacement reproduction is attained. This type of population growth results in *unstable age structure* (Fig. 33.6*a*). In LDCs, the more quickly replacement reproduction is achieved, the sooner zero population growth will occur. The Bioethical Focus at the end of this chapter discusses population growth in LDCs.

In MDCs, on the other hand, it is more typical that the number of people in the prereproductive class roughly equals the number in the reproductive age class. This causes a *stable age structure,* and population numbers will remain about the same for the foreseeable future (Fig. 33.6*b*).

Check Your Progress

1. Describe a demographic transition.
2. How can differences in age structure explain differences in population growth rates between LDCs and MDCs?

33.3 Regulation of Population Growth

Learning Outcomes

1. Know the difference between density dependence and density independence.
2. Describe how competition and predation can limit population growth.
3. List the three types of symbiosis and discuss the characteristics of each.

Ecologists wish to determine the factors that regulate population growth. In general, ecologists have described two types of life history patterns characterized by how long it takes to reach reproductive maturity and how great is the reproductive output: an *opportunistic pattern* and an *equilibrium pattern* (Fig. 33.7). Members of opportunistic populations are small in size, mature early, and have a short life span. They tend to produce many relatively small offspring and to forego parental care in favor of a greater number of offspring. The more offspring, the more likely it is that some of them will survive a population crash. Classic examples of opportunistic species are many insects and weeds.

In contrast, the size of equilibrium populations remains pretty much at the carrying capacity. Resources such as food and shelter are relatively scarce for these individuals, and those who are best able to compete will have the largest number of offspring. These organisms allocate more energy than opportunists to their own growth and survival and to the growth and survival of their offspring. Therefore, they are fairly large, are slow to mature, and have a fairly long life span. They are specialists rather than colonizers and tend to become extinct when their normal way of life is destroyed. The best examples of equilibrium species are found among birds and mammals. For example, the Florida panther, the largest animal in the Florida Everglades, requires a very large range, and produces few offspring, which must be cared for. Currently, the Florida panther is unable to compensate for a reduction in its range due to human destruction of its habitat and is therefore on the verge of extinction.

For some time, ecologists have recognized that the environment contains both abiotic and biotic components. Abiotic factors, such as weather and natural disasters, are **density-independent,** meaning that the effects of the factor are the same for all sizes of populations. For example, fires don't necessarily kill a larger percentage of individuals as the population increases in size. On the other hand, biotic factors, such as competition, predation, and parasitism, are called **density-dependent.** The effects of a density-dependent factor depend on the size of the population. The denser a population is, the faster a disease might spread, for example. Populations that have the opportunistic life history pattern tend to be regulated by density-independent factors, and those that have the equilibrium life history pattern tend to be regulated by density-dependent factors.

Check Your Progress

1. Which is more likely to experience opportunistic population growth—a population of elephants or mosquitoes? Why?
2. A fire wipes out a population of American robins in a forest. Which type of regulatory factor does this represent?

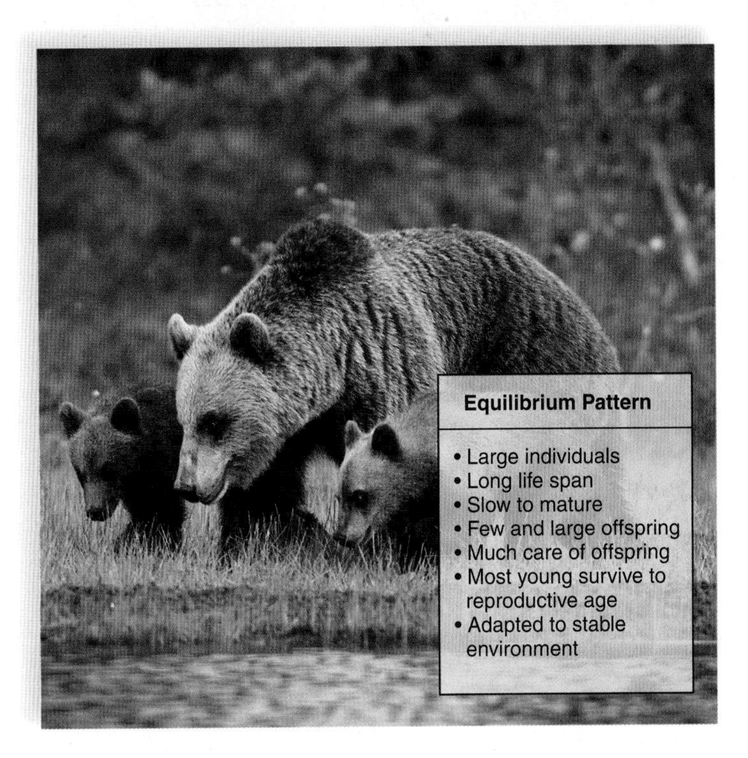

Figure 33.7 Life history patterns.
Dandelions are an opportunistic species with the characteristics noted, and bears are an equilibrium species with the characteristics noted. Often the distinctions between these two possible life history patterns are not as clear-cut as they may seem.

Competition

Competition, a density-dependent factor, occurs when members of different species try to utilize a resource (such as light, space, or nutrients) that is in limited supply. Understanding the effects of competition first requires a grasp of the concept of ecological niche. The **ecological niche** is the role a species plays in the community, including the **habitat** it requires and its interactions with other organisms. The niche is essentially a species' total way of life, and thus includes the resources needed to meet its energy, nutrient, survival, and reproductive demands.

According to the **competitive exclusion principle,** no two species can occupy the same ecological niche at the same time if resources are limiting. This principle is exemplified by an experiment whereby two species of paramecia grown separately survive. But when they are grown together in one test tube, resources are limiting, and only one species survives (Fig. 33.8). In another laboratory experiment, two species of paramecia continue to occupy the same tube when one species feeds on bacteria at the bottom of the tube and the other feeds on bacteria suspended in solution. The division of feeding niches, called **resource partitioning,** decreases competition between the two species and allows occupancy of different niches and therefore survival. As another example, consider that swallows, swifts, and martins all eat flying insects and parachuting spiders. These birds even frequently fly in mixed flocks. But each species of bird has different nesting sites and migrates at a slightly different time of year. Therefore, they are not competing for the same food source when they are feeding their young. Thus, while it may seem as if several populations living in the same area are occupying the same niche, it is usually possible to find slight differences.

On the Scottish coast, a small barnacle (*Chthamalus stellatus*) lives on the high part of the intertidal zone, and a large barnacle (*Balanus balanoides*) lives on the lower part (Fig. 33.9). Free-swimming larvae of both species attach themselves to rocks at any point in the intertidal zone, where they develop into the sessile adult forms. In the lower zone, the large *Balanus* barnacles seem to either force the smaller *Chthamalus* individuals off the rocks or grow over them. Competition is therefore restricting the range of *Chthamalus* on the rocks. *Chthamalus* is more resistant to drying out than is *Balanus*. Therefore, it has an advantage that permits it to grow in the upper intertidal zone.

Check Your Progress

1. What is meant by competitive exclusion?
2. How can resource partitioning allow coexistence of species with very similar niches?

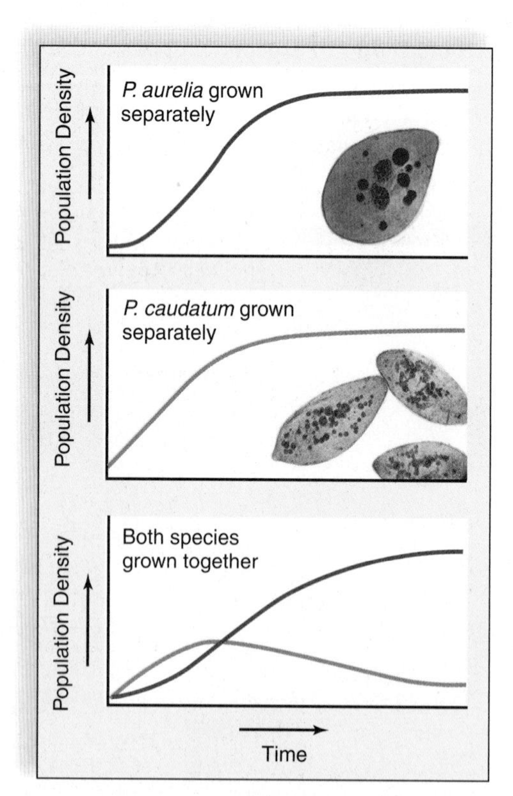

Figure 33.8 Competition between two laboratory populations of paramecia.
When grown alone in pure culture (top two graphs), *Paramecium caudatum* and *Paramecium aurelia* exhibit logistic growth. When the two species are grown together in mixed culture (bottom graph), *P. aurelia* is the better competitor, and *P. caudatum* dies out. This experiment illustrates the competitive exclusion principle.

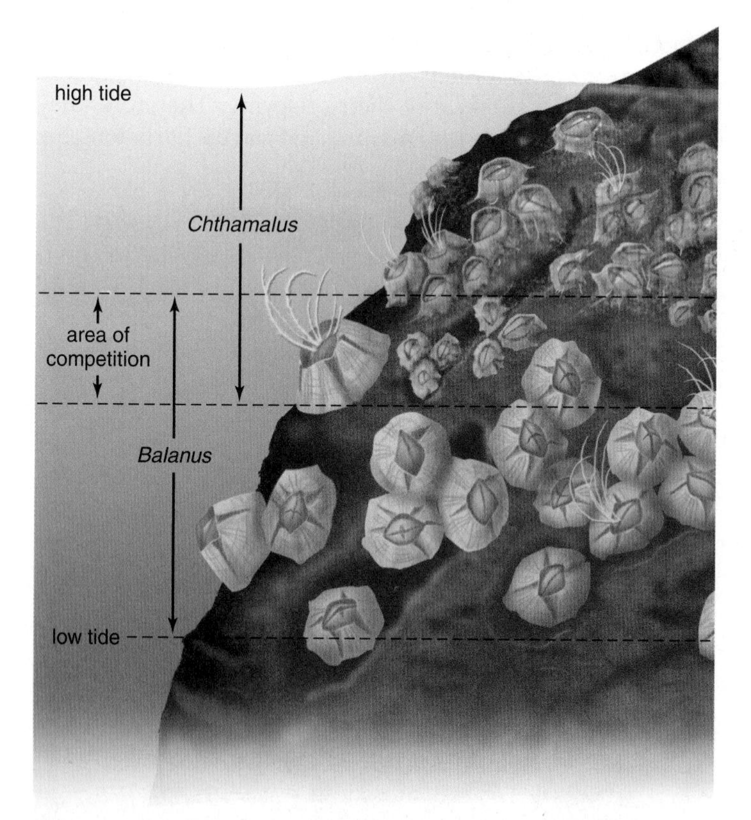

Figure 33.9 Competition between two species of barnacles.
Competition prevents two species of barnacles from occupying as much of the intertidal zone as possible. Both exist in the area of competition between *Chthamalus* and *Balanus*. Above this area, only *Chthamalus* survives, and below it only *Balanus* survives.

Predation

Predation occurs when one organism, called the **predator,** feeds on another, called the **prey.** In the broadest sense, predators include not only animals such as lions that kill zebras, but also filter-feeding blue whales that strain krill from ocean waters, and even parasitic ticks that suck blood from their victims.

Predator-Prey Population Dynamics

Predators reduce the population density of prey, as shown by a laboratory study in which the protozoans *Paramecium caudatum* (prey) and *Didinium nasutum* (predator) were grown together in a culture medium. *Didinium* ate all the *Paramecium* and then died of starvation. In nature, we can find a similar example. When a gardener brought prickly-pear cactus to Australia from South America, the cactus spread out of control until millions of acres were covered with nothing but cacti. The cacti were brought under control when a moth from South America, whose caterpillar feeds only on the cactus, was introduced. Now, both cactus and moth are found at greatly reduced densities in Australia.

Mathematical formulas predict that predator and prey populations may sometimes cycle instead of maintaining a steady state. Cycling could occur if (1) the predator population overkills the prey and then the predator population also declines in number, and (2) the prey population overshoots the carrying capacity and suffers a crash, and then the predator population follows suit because of a lack of food. In either case, the result would be a series of peaks and valleys in population density of both species, with the predator population lagging slightly behind the prey.

A famous case of predator-prey cycles occurs between the snowshoe hare and the Canadian lynx, a type of small, predatory cat (Fig. 33.10). The snowshoe hare is a common herbivore in the coniferous forests of North America, where it feeds on terminal twigs of various shrubs and small trees. The Canadian lynx feeds on snowshoe hares but also on ruffed grouse and spruce grouse, two types of birds. Investigators at first assumed that the lynx had brought about the decline of the hare population. But others noted that the decline in snowshoe hare abundance was accompanied by low growth and reproductive rates that could be signs of a food shortage. It appears that both explanations apply to the data. In other words, both a predator-hare cycle and a hare-food cycle have combined to produce an overall effect, which is observed in Figure 33.10. Although not shown on the graph, densities of

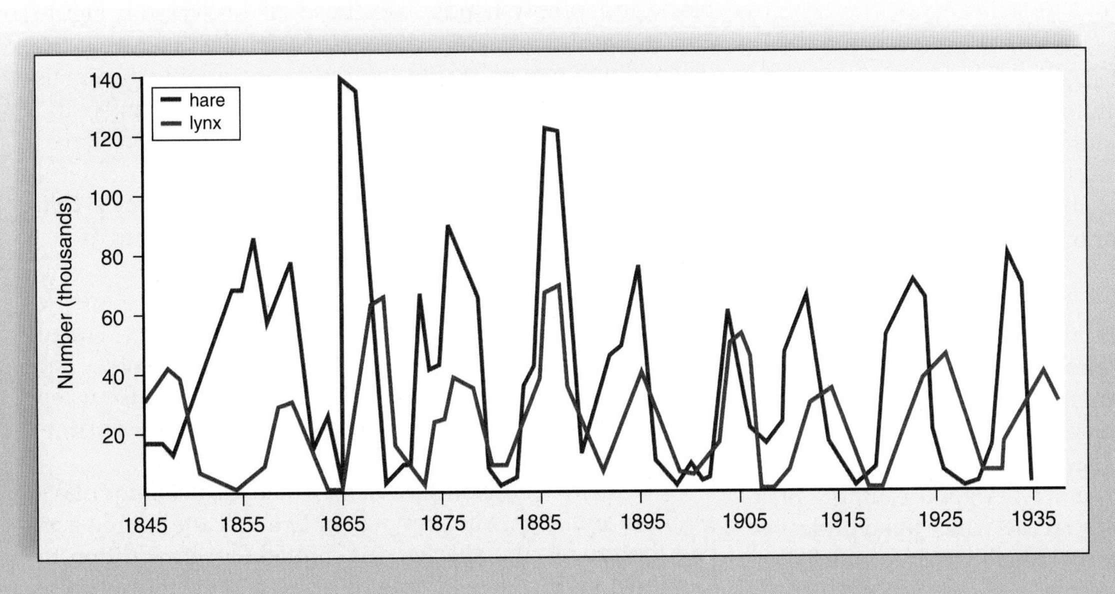

Figure 33.10 Predator-prey cycling of a lynx and a snowshoe hare. The number of pelts received yearly by the Hudson Bay Company for almost 100 years shows a pattern of ten-year cycles in population densities. The snowshoe hare (prey) population reaches a peak before that of the lynx (predator) by a year or more.

Data from D. A. MacLulich, *Fluctuations in the Numbers of the Varying Hare* (Lepus americanus), University of Toronto Press, Toronto, 1937; reprinted 1974, p. 543.

a.

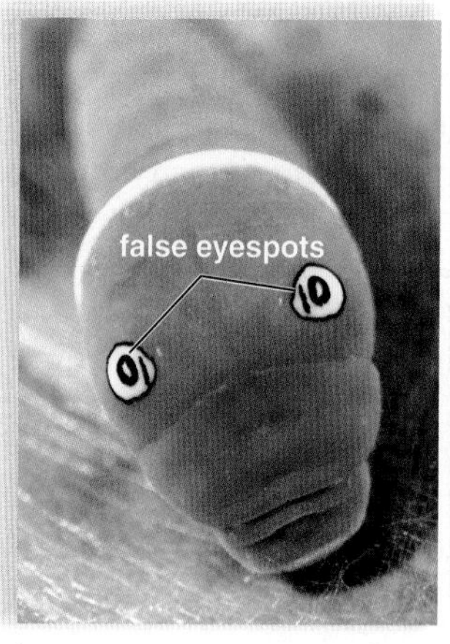

b.

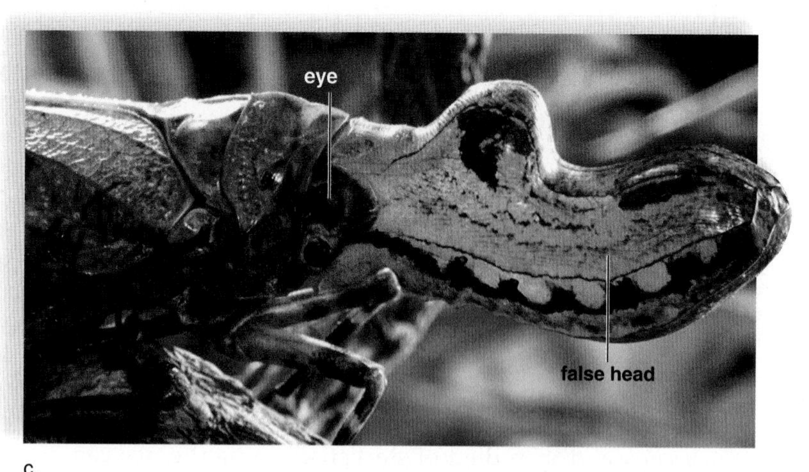

c.

Figure 33.11 Antipredator defenses.
a. The skin secretions of poison-dart frogs are so poisonous that they were used by natives to make their arrows instant lethal weapons. The bright coloration of these frogs is called "warning coloration" because brightly colored species are generally toxic. **b.** The caterpillar of the eastern tiger swallowtail butterfly has false eyespots used to confuse a predator. **c.** The South American lantern fly has a large false head that resembles that of an alligator. This may frighten a predator into thinking it is facing a dangerous animal.

the grouse populations also cycle, perhaps because the lynx switches to this food source when the hare population declines. Predators and prey do not normally exist as simple, two-species systems, and therefore, abundance patterns should be viewed with the complete community in mind.

Antipredator Defenses

While predators have evolved strategies to secure the maximum amount of food with minimal expenditure of energy, prey organisms have evolved strategies to escape predation. **Coevolution** is present when each of two species adapts in response to selective pressure imposed by the other. This phenomenon applies to predation, as well as to symbiotic interactions (discussed next).

In plants, the sharp spines of the cactus, the pointed leaves of holly, and the tough, leathery leaves of the oak tree all discourage predation by insects. Plants even produce poisonous chemical compounds. Animals have varied antipredator defenses. Some effective defenses are poisonous secretions, concealment, fright, flocking together, and warning coloration (Fig. 33.11), as well as mimicry. A good example of concealment occurs with caddisfly larvae, which build protective cases out of sticks and sand, making themselves look like the stream bottom where they reside.

Mimicry **Mimicry** occurs when one species resembles another species that has evolved to defend against predators or resembles an object in the environment to deceive prey. Mimicry can help a predator capture food or a prey avoid capture. For example, angler fishes have lures that resemble worms for the purpose of bringing fish within reach. To avoid capture, some inchworms resemble twigs, and some caterpillars can transform themselves into shapes resembling snakes.

Batesian mimicry (named for Henry Bates, who discovered it) occurs when a prey species that is not harmful mimics another species that has a successful antipredator defense. Many examples of Batesian mimicry involve warning coloration. For example, stinging insects all tend to have black and yellow bands, like those of the wasp in Figure 33.12*a*. Once a predator has experienced the defense of the wasp, it will remember the coloration and avoid all animals that look similar, such as the nonstinging flower fly and longhorn beetle in Figure 33.12*b, c*, which can be considered Batesian mimics.

Another type of mimicry occurs when species that resemble each other all have successful defenses. For example, many coral snake species have brilliant red, black, and yellow body rings and are venomous. Mimics that share the same protective defense are called Müllerian mimics, named after Fritz Müller, who discovered the phenomenon. The bumblebee in Figure 33.12*d* is a Müllerian mimic of the wasp in Figure 33.12*a*—that is, both insects have the same appearance and mode of defense.

Just as with other antipredator defenses, behavior plays a role in mimicry. Mimicry works better if the mimic acts like the model (the species it is mimicking) in addition to looking like it. For example, beetles that phenotypically resemble a wasp actively fly from place to place and spend most of their time in the same habitat as the wasp model.

Check Your Progress

1. How can interactions between predators and prey result in population cycles?
2. What are some of the ways by which prey can escape predation?

a.

b.

c.

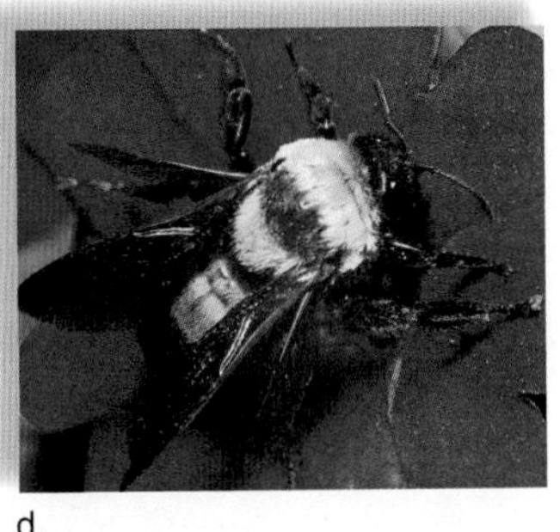
d.

Figure 33.12 Mimicry among insects.
a. A yellow jacket wasp uses stinging as its defense. **b.** A flower fly and **(c)** a longhorn beetle are Batesian mimics because they are incapable of stinging another animal, and yet they have the same appearance as the yellow jacket wasp. **d.** A bumblebee and the yellow jacket wasp are Müllerian mimics because they have a similar appearance, and both use stinging as a defense.

Symbiosis

Symbiosis refers to close interactions between members of two species. Three types of symbiotic relationships have traditionally been defined—parasitism, commensalism, and mutualism. Table 33.2 lists these three categories in terms of their benefits to one species or the other. Of the three types, only mutualism may increase the population size of both species. However, some ecologists now consider it difficult to try to classify symbiotic relationships into these categories, because the amount of harm or good the individuals of two species do to one another depends on what the investigator chooses to measure. Although the following discussion describes the traditional classification system, bear in mind that symbiotic relationships do not always fall neatly into these three categories.

Parasitism

Parasitism is a symbiotic relationship in which the *parasite* derives nourishment from another organism, called the *host*. Therefore, the parasite benefits and the host is harmed. Parasites occur in all kingdoms of life. Bacteria (e.g., strep infection), protists (e.g., malaria), fungi (e.g., athlete's foot), plants (e.g., mistletoe), and animals (e.g., tapeworm) all include parasitic species. The effects of parasites on the health of the host can range from slightly weakening them to actually killing them over time.

In addition to providing nourishment, some host organisms also provide their parasites with a place to live and reproduce, as well as a mechanism for dispersing offspring to new hosts. Many parasites have both a primary and secondary host. The secondary host may be a vector that transmits the parasite to the next primary host. As an example, consider the deer ticks *Ixodes dammini* and *I. ricinus* in the eastern and western United States, respectively. Deer ticks are arthropods that go through a number of stages (egg, larva, nymph, adult). They are so named because adults feed and mate on white-tailed deer in the fall. The female lays her eggs on the ground, and when the eggs hatch in the spring, they become larvae that feed primarily on white-footed mice. If a mouse is infected with the bacterium *Borrelia burgdorferi*, the larvae become infected also. The fed larvae overwinter and molt the next spring to become nymphs that can, by chance, take a blood meal from a human. At this time, the tick may pass the bacterium on to a human, who subsequently comes down with Lyme disease, characterized by arthritic-like symptoms and often a "bulls-eye" rash around the site of the tick bite. The fed nymphs develop into adults, and the cycle begins again.

TABLE 33.2 Symbiotic Relationships

	Species 1	Species 2
Parasitism*	Benefited	Harmed
Commensalism	Benefited	No effect
Mutualism	Benefited	Benefited

*Can be considered a type of predation.

Commensalism

Commensalism is a symbiotic relationship between two species in which one species is benefited and the other is neither benefited nor harmed. Often one species provides a home and/or transportation for the other species, as when barnacles attach themselves to the backs of whales.

Clownfishes live within the waving mass of tentacles of sea anemones. Because most fishes avoid the venomous tentacles of the anemones, clownfishes are protected from predators. If clownfishes attract other fishes on which the anemone can feed, this relationship borders on mutualism. Other examples of commensalism may also be considered mutualistic. For example, birds called cattle egrets benefit from feeding near cattle because the cattle flush insects and other animals from the vegetation as they graze (Fig. 33.13).

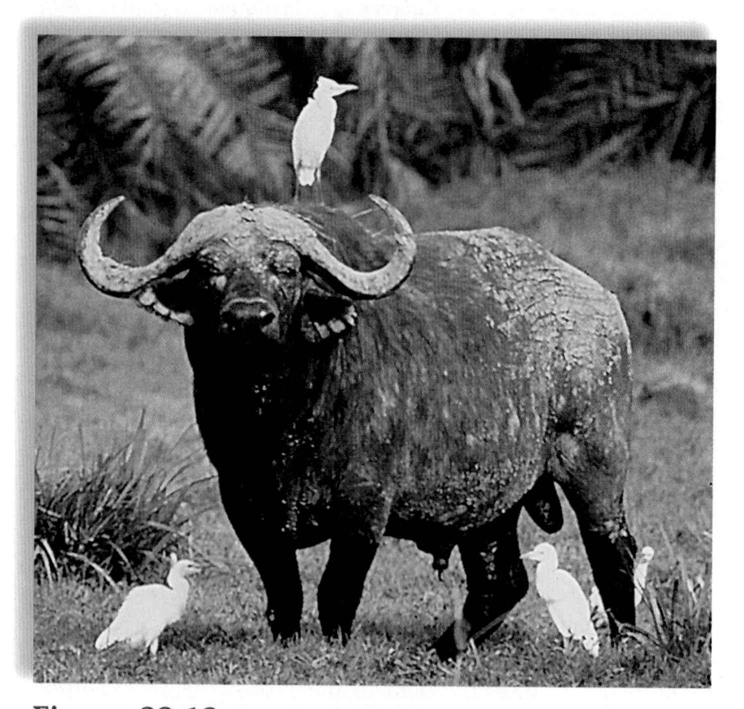

Figure 33.13 Egret symbiosis.
Cattle egrets eat insects off and around various animals, such as this African cape buffalo.

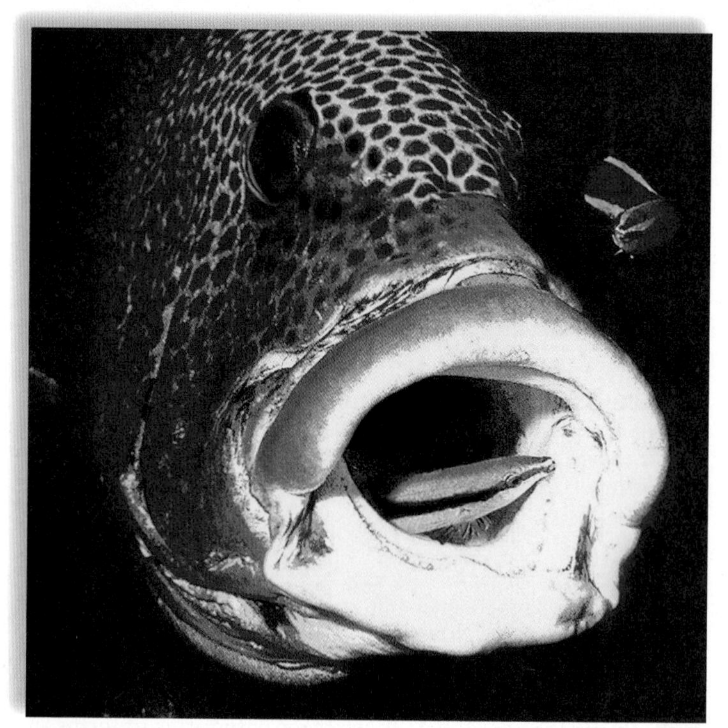

Figure 33.14 Cleaning symbiosis.
A cleaner wrasse, *Labroides dimidiatus*, in the mouth of a spotted sweetlip, *Plectorhincus chaetodontoides*, is feeding off parasites. Does this association improve the health of the sweetlip, or is the sweetlip being exploited? Investigation is under way.

Mutualism

Mutualism is a symbiotic relationship in which both members of the association benefit, though not necessarily equally. An example is the relationship between plants and their animal pollinators, which began when herbivores, such as insects, feasted on pollen. The provision of nectar by plants may have spared the pollen and at the same time allowed the herbivore to become an instrument of pollination.

Ants form mutualistic relationships with both plants and insects. In tropical America, the bullhorn acacia tree is adapted to provide a home for ants of the species *Pseudomyrmex ferruginea.* Unlike other acacias, this species has swollen thorns with a hollow interior, where ant larvae can grow and develop. In addition to housing the ants, the acacias provide them with food. The ants constantly protect the plant from the caterpillars of moths and butterflies by swarming and stinging them, and from other plants that might shade the plant. When the ants on experimental trees were poisoned, the trees died.

Cleaning symbiosis is a relationship in which the individual being cleaned is often a vertebrate. Crustaceans, fish, and birds act as cleaners and are associated with a variety of vertebrate clients. Large fish in coral reefs line up at cleaning stations and wait their turn to be cleaned by small fish that even enter the mouths of the large fish (Fig. 33.14). It's been suggested that cleaners may be exploiting the relationship by feeding on host tissues as well as on ectoparasites. On the other hand, cleaning could ultimately lead to net gains in client fitness by ridding them of parasites.

Check Your Progress

1. Give an example of mutualism.
2. How can parasitism regulate population size?

33.4 Ecological Succession

Learning Outcomes

1. Discuss the differences between the inhibition, tolerance, and facilitation models of succession.
2. Discuss the reasons why a community might not always reach a successional climax.

Like populations, communities, which are composed of all of the interacting populations in an area, can also be regulated by biotic interactions and ecological disturbances. **Ecological succession** is a change in a community's composition that is directional and follows a continuous pattern of extinction and colonization by new species. On land, *primary succession* is the establishment of a plant community (usually first by early colonizers such as lichens) in a newly formed area where there is no soil formation. Primary succession is typically instigated by a major disturbance such as a volcanic eruption or a glacial retreat. *Secondary succession* is the return of a community to its natural vegetation following a disturbance, as when a cultivated cornfield returns to its natural state.

The first species to begin the process of secondary succession are plants that are colonizers of disturbed habitats, called *pioneer species.* Then succession progresses through a series of stages, as shown in Figure 33.15. Note that, in this case, the process began with grasses followed by shrubs and then to a mixture of shrubs and trees, until finally there were only trees.

Population Growth in Less-Developed Countries

Population experts have discovered that when women in the less-developed countries have rights equal to those of men they tend to have fewer children. Also, family planning leads to healthier women, and healthier women have healthier children, and the cycle continues. Under these circumstances, women do not have to have many babies for a few to survive. More education is also helpful because better-educated people are more interested in postponing childbearing and in promoting the status of women.

"There isn't any place where women who have the choice haven't chosen to have fewer children," says Beverly Winikoff at the Population Council in New York City. Bangladesh is a case in point. Bangladesh is one of the densest and poorest countries in the world. In 1990, the birthrate was 4.9 children per woman, and now it is 2.9. This achievement was due in part to the Dhaka-based Grameen Bank, which lends small amounts of money, mostly to destitute women, to start a business. The bank discovered that when women start making decisions about their lives, they also start making decisions about the size of their families. Family planning within Grameen families is twice as common as the national average. In fact, those women who get a loan promise to keep their families small! Also helpful has been the network of village clinics that counsel women who want to use contraceptives. The expression "contraceptives are the best contraceptives" refers to the fact that it is not necessary to wait for social changes to get people to use contraceptives—the two can occur simultaneously and reinforce each other.

Form Your Own Opinion

1. Do you think less-developed countries should simply make contraceptives available, or should more persuasive methods be employed? Explain.
2. Do you think that more-developed countries should be concerned about population growth in the less-developed countries? Why or why not?
3. Are you in favor of providing foreign financial aid to help countries develop family planning programs? Why or why not?

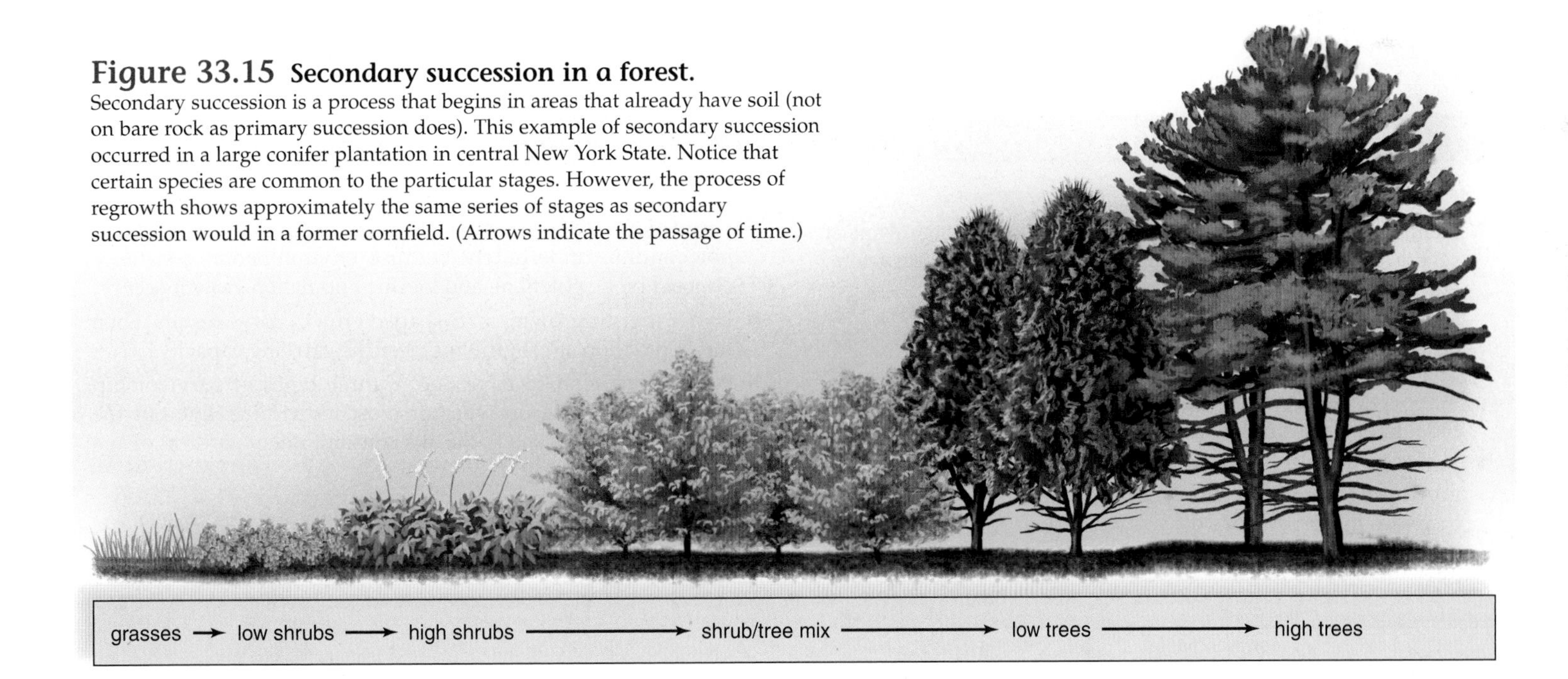

Figure 33.15 Secondary succession in a forest. Secondary succession is a process that begins in areas that already have soil (not on bare rock as primary succession does). This example of secondary succession occurred in a large conifer plantation in central New York State. Notice that certain species are common to the particular stages. However, the process of regrowth shows approximately the same series of stages as secondary succession would in a former cornfield. (Arrows indicate the passage of time.)

Succession can also occur in aquatic communities, as when lakes and ponds sometimes undergo stages whereby they become filled in and eventually disappear.

Models of Succession

Ecologists have developed various hypotheses to explain succession and predict future events. The *climax-pattern model* of succession says that particular areas will always lead to the same type of community, called a **climax community,** that usually exists for a long time relative to its predecessors. This model is based on the observation that climate, in particular, determines whether a desert, a grassland, or a particular type of forest results. Therefore, coniferous forests are expected to occur in northern latitudes, deciduous forests in temperate zones, and tropical rain forests in the tropics. The climax-pattern model of succession now recognizes that the exact composition of a community need not always be the same. That is, while we might expect to see a coniferous forest as opposed to a tropical rain forest in northern latitudes, the exact mix of plants and animals after each stage of succession can vary.

Does each stage of succession facilitate (help) or inhibit the next stage? To support a *facilitation model,* it can be observed as shown in the example in Figure 33.15 that shrubs can't grow on sand dunes until dune grass has caused enough soil to develop. Thus, it's possible that each successive community prepares the way for the next, and this is why grass-shrub-forest development occurs sequentially.

On the other hand, the *inhibition model* says that colonists hold on to their space and inhibit the growth of other plants until the colonists die or are damaged. Still another possible model, the *tolerance model*, predicts that different types of plants can colonize an area at the same time. Sheer chance determines which seeds arrive first, and successional stages may simply reflect the length of time it takes individuals to mature. This alone could account for the grass-shrub-forest development seen in Figure 33.15. In reality, the models we have mentioned are not mutually exclusive, and succession is recognized as a complex process.

Check Your Progress

1. What type of community has been stable for a long time in a given area?
2. Which model of succession is represented by mosses that help weather rocks into soil so that it can later be colonized by grasses?

Applying the Concepts Revisited

At the beginning of this chapter, you learned that overfishing may cause a complete collapse of the global fishery by the year 2048. One of the primary reasons for the decline in the stability of many fish populations is that they are fished to the point of being in the lag phase of population growth. In addition, constant removal of the largest, most reproductively fit individuals from populations limits the number of individuals in the reproductive portion of the population's age structure.

1. Given what you have learned about population regulation in this chapter, what are some strategies to help create sustainable fisheries?
2. The increased demand for fish stems from human population growth—perhaps our biggest environmental concern. The more people there are, the more of the world's resources we must use to feed, shelter, and provide energy for ourselves. Population growth has slowed in MDCs due to a demographic transition. However, more population growth is predicted to occur in LDCs. Describe strategies to help stimulate a demographic transition in LDCs.

Summarizing the Concepts

33.1 The Scope of Ecology

- Ecologists study biotic and abiotic interactions of organisms at several levels: individual, population, community, ecosystem, and biosphere.

33.2 Patterns of Population Growth

- The per capita rate of increase is calculated by subtracting the number of deaths from the number of births and dividing by the number of individuals in the population.
- Two patterns of population growth are possible. Exponential growth results in a J-shaped curve because, as the population increases in size, so does the number of new members.
- Under ideal circumstances, exponential growth occurs due to a population's biotic potential. However, exponential growth cannot continue indefinitely because environmental resistance opposes biotic potential, and logistic population growth occurs.
- Under logistic growth, an S-shaped growth curve results when the population stops growing near the carrying capacity.
- Populations tend to have one of three types of survivorship curves, depending on whether most individuals live out the normal life span (type I), die at a constant rate regardless of age (type II), or die early (type III).

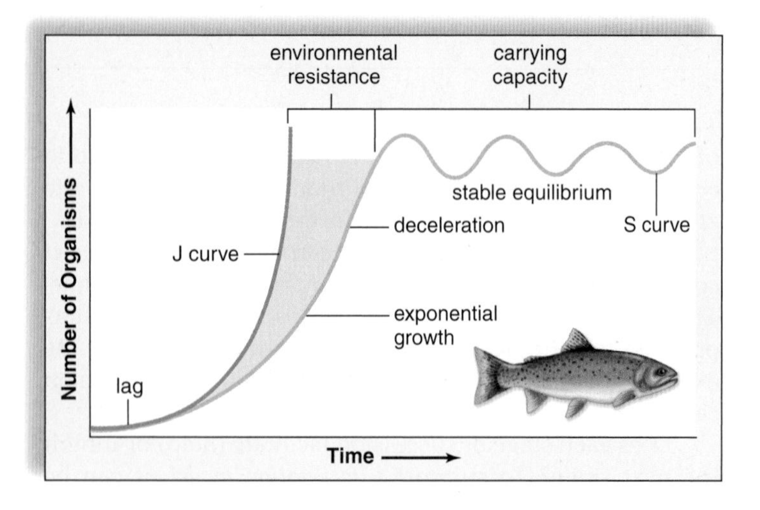

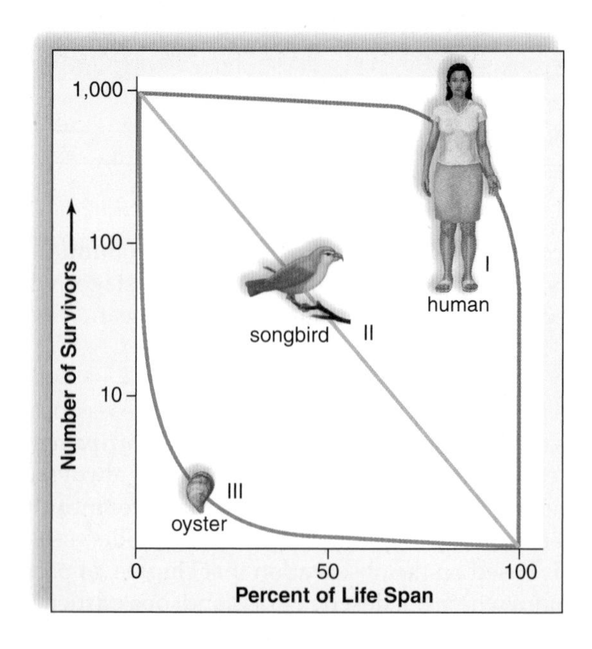

- The human population is expanding exponentially, mostly within less-developed countries in Africa, Asia, and Latin America. Support for family planning, human development, and delayed childbearing could help lessen the expected increase.

33.3 Regulation of Population Growth

- Opportunistic species rapidly produce many young, generally lack parental care, and rely on rapid dispersal to new, unoccupied environments. Their population size is regulated by density-independent factors.
- Equilibrium species produce few young, which often require parental care, and their population size is typically regulated by density-dependent factors, such as competition and predation.
- The competitive exclusion principle states that no two species can occupy the same ecological niche at the same time when resources are limiting. With limited resources, either resource partitioning occurs or one species goes locally extinct.
- Predator-prey interactions can cause prey populations to decline and remain at relatively low densities, decline of predator populations, or cycling of both predator and prey population densities.
- Prey defenses take many forms, such as concealment, use of fright, and warning coloration. Batesian mimicry occurs when one species has the warning coloration but lacks the defense of another species. Müllerian mimicry occurs when two species with the same warning coloration have a similar defense.
- In a parasitic relationship, the parasite benefits and the host is harmed. Parasites often utilize more than one host. In a commensalistic relationship, neither party is harmed, and one species often provides a home and/or transportation for another species. And in a mutualistic relationship, both partners benefit. Examples of mutualistic relationships include flowers and their pollinators, ants and acacia trees, and cleaning symbiosis.

33.4 Ecological Succession

- A change in community composition over time is called ecological succession.
- The climax-pattern model of succession says that particular areas will always lead to the same type of community.
- The facilitation model implies that some community member prepare the landscape for new species during succession. The inhibition model suggests that some species prevent colonization, and the tolerance model suggests that species colonize simultaneously.
- A climax community is stable and associated with a particular geographic area.

TESTING YOURSELF

Choose the best answer for each question.

1. Place the following levels of organization in order, from lowest to highest.
   a. population, organism, community, ecosystem
   b. community, ecosystem, population, organism
   c. organism, community, population, ecosystem
   d. population, ecosystem, organism, community
   e. organism, population, community, ecosystem
2. Assume that a deer population contains 500 animals. The birthrate is 105 animals per year, and the death rate is 100 animals per year. The per capita rate of increase per year is
   a. 5%.
   b. 4%.
   c. 3%.
   d. 2%.
   e. 1%.
3. What type of survivorship curve would you expect for a plant species in which most seedlings die?
   a. type I
   b. type II
   c. type III
4. Label this S-shaped growth curve:

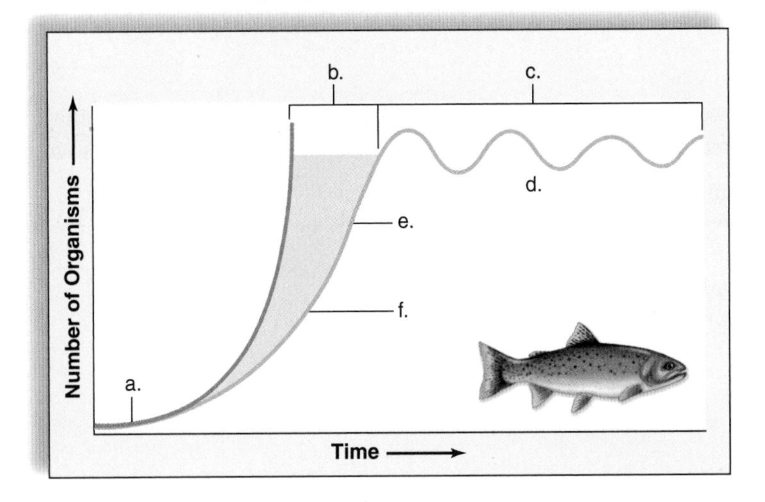

5. If an age-structure diagram indicates that there are more people in the prereproductive stage of life than in the reproductive stage, then replacement reproduction, over the long run, will result in
   a. zero population growth.
   b. an increase in population size.
   c. a decrease in population size.
6. An ecological niche includes an organism's
   a. interactions with other organisms.
   b. habitat.
   c. resources.
   d. Both a and b are correct.
   e. All of these are correct.
7. A colorful, nonpoisonous frog that mimics a poison-dart frog is exhibiting
   a. Batesian mimicry.
   b. Müllerian mimicry.

For problems 8–10, indicate the type of symbiosis illustrated by each example. Some answers may be used more than once.

Key:

a. parasitism
b. commensalism
c. mutualism

8. Unicellular algae live in the tissues of coral animals. The algae provide food for the coral, while the coral provides a stable home for the algae.
9. Flowering plants reward visiting insects with nectar. The insects, in turn, carry pollen to other flowers.
10. Small wasps lay eggs on other insects. The eggs hatch into larvae that feed on the insects and kill them.

11. If a population has a type I survivorship curve (most of its members live the entire life span), which of these would you also expect?
    a. a single reproductive event per adult
    b. most individuals reproduce
    c. sporadic reproductive events
    d. reproduction occurring near the end of the life span
    e. None of these are correct.
12. Which of these is a density-independent regulating factor?
    a. competition
    b. predation
    c. size of population
    d. weather
    e. resource availability
13. Six species of monkeys are found in a tropical forest. Most likely, they
    a. occupy the same ecological niche.
    b. eat different foods and occupy different ranges.
    c. spend much time fighting each other.
    d. are from different stages of succession.
    e. All of these are correct.
14. Mosses growing on bare rock will eventually help to create soil. These mosses are involved in _____ succession.
    a. primary
    b. secondary
    c. tertiary
15. Assume that a farm field is allowed to return to its natural state. By chance, the field is first colonized by native grasses, which begin the succession process. This is an example of which model of succession?
    a. climax pattern
    b. tolerance
    c. facilitation
    d. inhibition

UNDERSTANDING THE TERMS

age-structure diagram 694
biosphere 690
biotic potential 691
carrying capacity 692
climax community 701
coevolution 698
cohort 692
commensalism 699
community 690
competition 696
competitive exclusion principle 696
demographic transition 693
density-dependent 695
density-independent 695
doubling time 693
ecological niche 696
ecological succession 700
ecology 690
ecosystem 690
environmental resistance 692
exponential growth 691
habitat 696
less-developed country (LDC) 693
logistic growth 692
mimicry 698
more-developed country (MDC) 693
mutualism 700
parasitism 699
population 690
predation 697
predator 697
prey 697
replacement reproduction 694
resource partitioning 696
symbiosis 699
zero population growth 693

THINKING CRITICALLY

1. You are a farmer fighting off a crop insect pest. You know from evolutionary biology and natural selection that using the same pesticide causes resistance to evolve in the insect population. From what you have learned about population growth and regulation, what are some strategies you might use to control the insect population?
2. You find two species of insects, both using bright coloration and similar color patterns as an antipredator defense. Design an experiment to tell whether this is an example of Müllerian or Batesian mimicry.
3. If predators tend to reduce the population densities of prey, what prevents predators from reducing prey populations to such low levels that they drive themselves extinct?

INQUIRY INTO LIFE WEBSITE

The companion website for *Inquiry into Life* provides a wealth of information organized and integrated by chapter. You will find practice tests, animations, videos, and much more that will complement your learning and understanding of general biology.

http://www.mhhe.com/maderinquiry13

34 Nature of Ecosystems

Applying the Concepts

It is the year 2100—the future Earth. Pristine ecosystems such as the one pictured may be a thing of the past. People in the northeastern United States no longer have shovels or snowblowers as winters barely get below freezing temperatures. Corn is now being grown in the middle latitudes of Canada, as opposed to in Iowa. New Orleans is completely under water. Skin cancer has reached an all-time high and affects half of all people in the world. Nearly all water bodies are overgrown with algae and pond scum, and freshwater fish are disappearing. Fresh water is limited and costs over $50 per bottle. Hurricanes are more severe than ever before. . . .

Although extreme, this scenario is possible given the influence of modern humans on ecosystems and chemical cycling. An ecosystem is characterized by the interactions of its biotic (living) and abiotic (nonliving) components. The biotic community depends on cycling of chemicals, such as carbon, nitrogen, and phosphorus. As we continue to burn carbon-rich fossil fuels such as oil, carbon dioxide is released into the atmosphere and acts like a blanket, trapping in solar radiation. Steady atmospheric increases in carbon dioxide contribute to global warming, which can cause climate shifts. In turn, sea levels are predicted to rise and extreme weather events, like hurricanes and cyclones, are predicted to increase in frequency. Use of detergents with phosphorus and nitrogen-containing fertilizers add excess amounts of these nutrients to water bodies, such as ponds and lakes. The result is cultural eutrophication, which causes massive algal blooms. When these blooms decompose, oxygen levels in water bodies plummet, causing widespread fish kills.

In this chapter, you will learn about how energy flows through the members of biotic communities and how abiotic chemicals and nutrients cycle through ecosystems. This knowledge will help you understand why phenomena such as global warming are a major concern.

Chapter Outline

34.1 The Biotic Components of Ecosystems

Learning Outcomes

1. Name three biotic components of an ecosystem and the ecological roles of each.
2. Discuss what happens to energy flow and how nutrients cycle in ecosystems.

An **ecosystem** possesses both abiotic and biotic components. The abiotic components include resources, such as sunlight, inorganic nutrients, and water availability, and conditions, such as type of soil, prevailing temperature, and wind speed. The biotic components of an ecosystem are the various populations of species that form a community.

Populations Within an Ecosystem

The populations within an ecosystem are categorized according to their food source. **Autotrophs** (i.e., "self-feeders") require only inorganic nutrients and an outside energy source to produce organic nutrients for their own use as food. Because they produce food for themselves and other members of the community, they are called **producers** (Fig. 34.1*a*). Photosynthetic organisms such as algae and plants produce most of the organic nutrients for the biosphere.

Some bacteria are chemoautotrophs. They reduce carbon dioxide to an organic compound by using energetic electrons derived from inorganic compounds, such as hydrogen sulfides, hydrogen gas, or ammonium ions. Chemoautotrophs have been found to support communities at hydrothermal vents along deep-sea oceanic ridges.

Heterotrophs (i.e., "other feeders") need an outside source of organic nutrients. Because they consume food, they are called **consumers. Herbivores** are animals that feed directly on plants or algae (Fig. 34.1*b*). **Carnivores** feed on other animals. Birds that feed on insects are carnivores, and so are hawks that feed on birds (Fig. 34.1*c*). In this example, herbivorous insects are primary consumers, birds are secondary consumers, and hawks are tertiary consumers. Sometimes tertiary consumers are called top predators. **Omnivores** are animals that feed on both plants and animals, such as most humans. **Decomposers** are heterotrophic

a. Producers

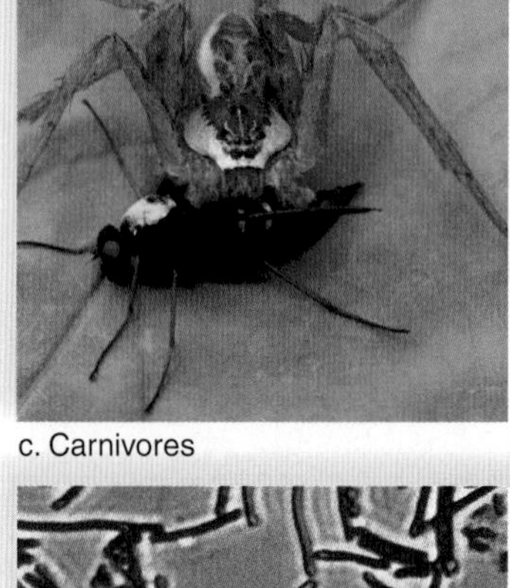

c. Carnivores

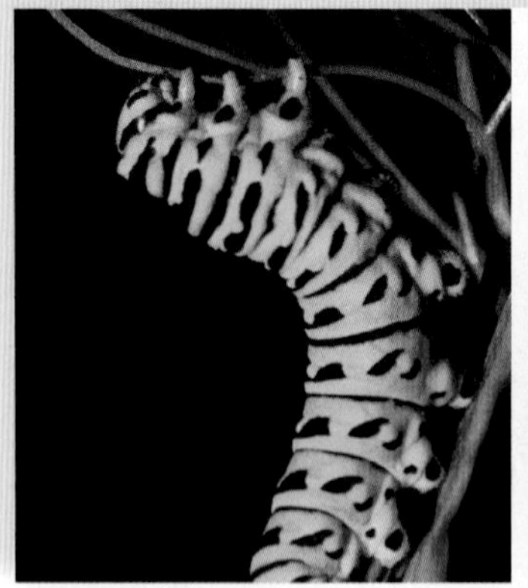

b. Herbivores

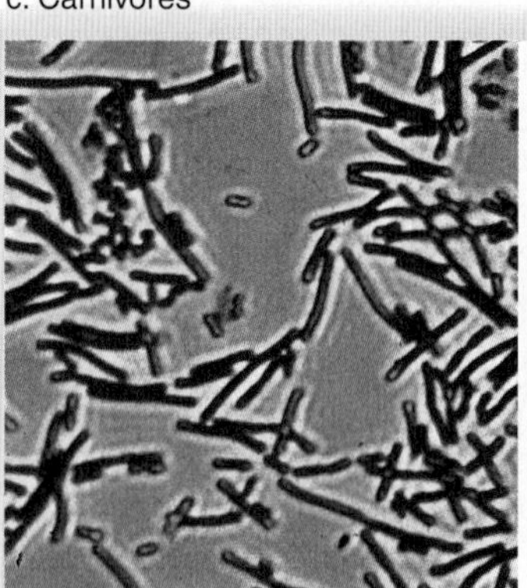

d. Decomposers

Figure 34.1 Biotic components.

a. Diatoms, a type of algae, and green plants are producers. **b.** Caterpillars and rabbits are herbivores. **c.** Spiders and ospreys are carnivores. **d.** Bacteria and mushrooms are decomposers.

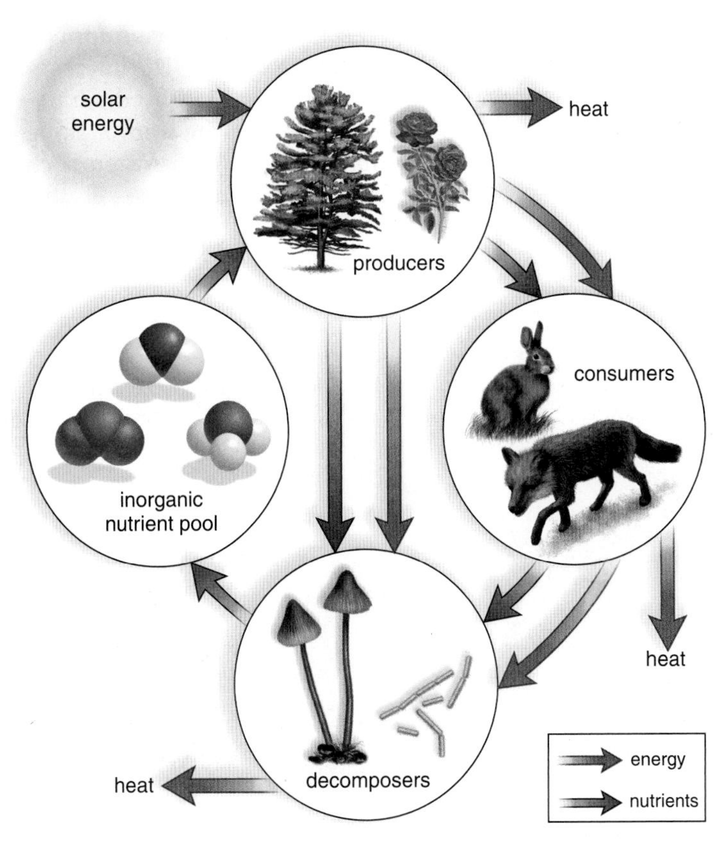

Figure 34.2 Energy flow and nutrient cycling.
Nutrients cycle, but energy flows through an ecosystem. As energy transformations repeatedly occur, all the energy derived from the sun eventually dissipates as heat.

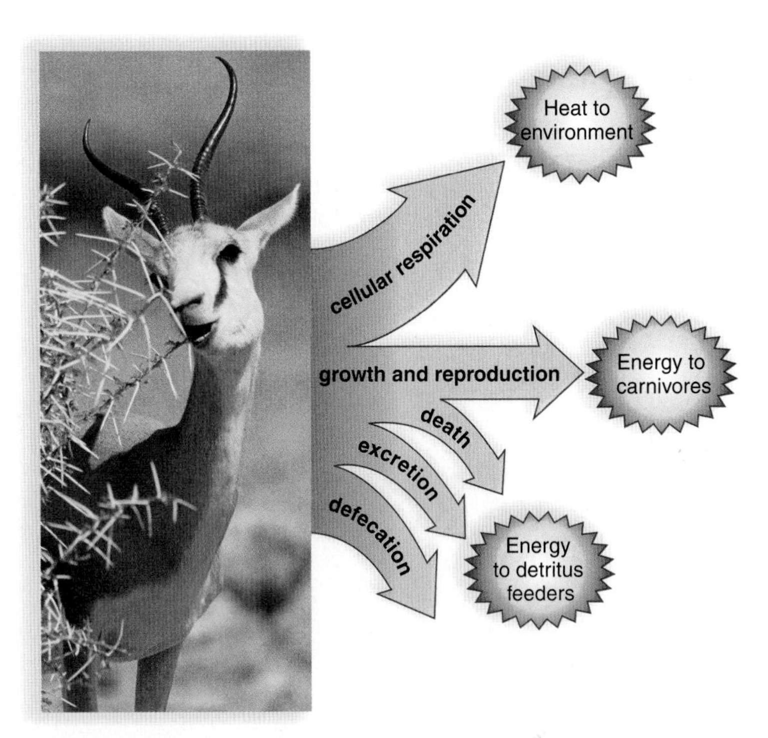

Figure 34.3 Energy balances.
Only about 10% of the food energy taken in by a herbivore is passed on to carnivores and much of the rest is lost as heat to the environment. A large portion goes to detritivores via defecation, excretion, and death, and another large portion is used for cellular respiration.

bacteria and fungi, such as molds and mushrooms, that break down nonliving organic matter, including animal wastes (Fig. 34.1*d*). They perform a very valuable service because they release inorganic nutrients that are taken up by plants back into the environment. **Detritus** is partially decomposed matter in the water or soil. Detritivores are found in all ecosystems. Earthworms and some beetles, termites, and maggots are soil detritivores.

Energy Flow and Chemical Cycling

Every ecosystem is characterized by two fundamental phenomena: energy flow and chemical cycling. Energy flow begins when producers absorb solar energy, and chemical cycling begins when producers take in inorganic nutrients from the physical environment (Fig. 34.2). Thereafter, producers make organic nutrients (food) directly for themselves and indirectly for the other populations of the ecosystem. Most ecosystems cannot exist without a continual supply of solar energy. Chemicals cycle when inorganic nutrients are returned to the producers from the atmosphere or soil, as appropriate.

Only a portion of the organic nutrients made by autotrophs is passed on to heterotrophs because plants use organic molecules to fuel their own cellular respiration. Similarly, only a small percentage of nutrients taken in by heterotrophs is available to higher-level consumers. Figure 34.3 shows why. A certain amount of the food eaten by a herbivore is never digested and is eliminated as feces, which decomposers recycle. Metabolic wastes are excreted as urine. Of the assimilated energy, a large portion is utilized during cellular respiration to provide energy to the organism and thereafter becomes heat. Only the remaining food, which is converted into increased body weight (or additional offspring), becomes available to carnivores.

In Chapter 6, you learned that energy flow is described by the laws of thermodynamics: (1) energy cannot be created or destroyed; and (2) in every energy transformation, some energy is lost as heat. These principles explain why ecosystems are dependent on a continual outside source of energy, such as sunlight, and why only a part of the original energy from producers is available to consumers.

Check Your Progress

1. What is the difference between producers, consumers, and decomposers?
2. Why does most energy fail to be converted to a usable form when one organism eats another?

34.2 Energy Flow

Learning Outcomes

1. Describe the difference between a food chain and a food web.
2. Illustrate what is meant by an ecological pyramid.

The principles we have been discussing can now be applied to an actual ecosystem—a forest, for example. The interconnecting paths of energy flow are represented by diagramming a **food web.** Figure 34.4*a* is a **grazing food web.** Grazing food webs do not necessarily have to begin with grazers such as cattle; they just have to begin with a producer—in this

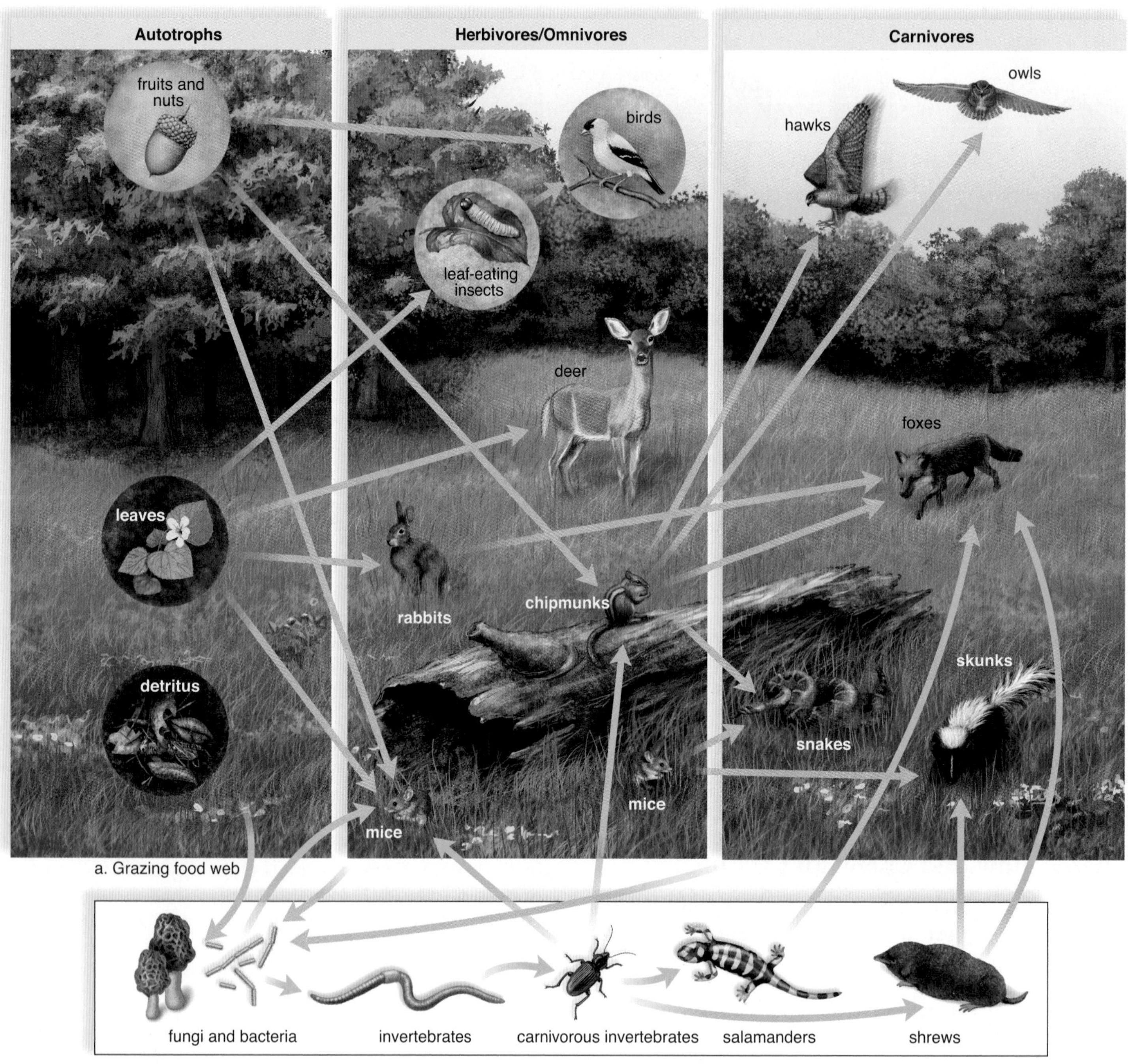

Figure 34.4 Grazing and detrital food webs.
Food webs are descriptions of who eats whom. **a.** Tan arrows illustrate possible grazing food webs. For example, birds, which feed on nuts, may be eaten by a hawk. Autotrophs such as the tree are producers (first trophic level); the first series of animals are primary consumers (second trophic level); and the next group of animals are secondary consumers (third trophic level). **b.** Green arrows illustrate possible detrital food webs, which begin with detritus, the waste products and remains of dead organisms. A large portion of these remains are from the grazing food web illustrated in **(a).** The organisms in the detrital food web are sometimes consumed by animals in the grazing food web, as when robins feed on earthworms. Thus, the grazing food web and the detrital food web are connected.

case, the oak tree. Caterpillars feed on leaves in the trees, and mice, rabbits, and deer feed on leaves at or near the ground. Birds, chipmunks, and mice eat fruits and nuts, but are omnivores because they also feed on caterpillars. These herbivores and omnivores all provide food for a number of different carnivores.

Figure 34.4*b* is a **detrital food web,** which begins with detritus. Detritus is food for soil organisms such as earthworms and beetles. These animals are, in turn, fed on by salamanders and shrews. Because the members of the detrital food web may become food for aboveground carnivores, the detrital and grazing food webs are joined.

In this particular forest, the organic matter lying on the forest floor and mixed into the soil contains over twice as much energy as the leaves of living trees. Therefore, more energy in a forest may be funneling through the detrital food web than through the grazing food web.

Trophic Levels

You can see that Figure 34.4*a* would allow us to link organisms one to another in a straight line, according to who eats whom. Diagrams that show a single path of energy flow such as this are called **food chains.** For example, in the grazing food web, we could find this **grazing food chain:**

leaves ⟶ caterpillars ⟶ tree birds ⟶ hawks

And in the detrital food web (Fig. 34.4*b*), we could find this **detrital food chain:**

detritus ⟶ earthworms ⟶ shrews

A **trophic level** is composed of all the organisms that feed at a particular link in a food chain. In the grazing food web in Figure 34.4*a*, going from left to right, the trees are primary producers (first trophic level). The first series of animals, the herbivores, are primary consumers (second trophic level). The group of animals at the far right, the carnivores, are secondary consumers (third trophic level).

Ecological Pyramids

The shortness of food chains can be attributed to the loss of energy between trophic levels. In general, only about 10% of the energy of one trophic level is available to the next trophic level. Therefore, if a herbivore population consumes 1,000 kg of plant material, only about 100 kg is converted to herbivore tissue, 10 kg to first-level carnivores, and 1 kg to second-level carnivores. The so-called 10% rule of thumb explains why few carnivores can be supported in a food web. The large energy losses that occur between successive trophic levels are sometimes depicted as an **ecological pyramid** (Fig. 34.5).

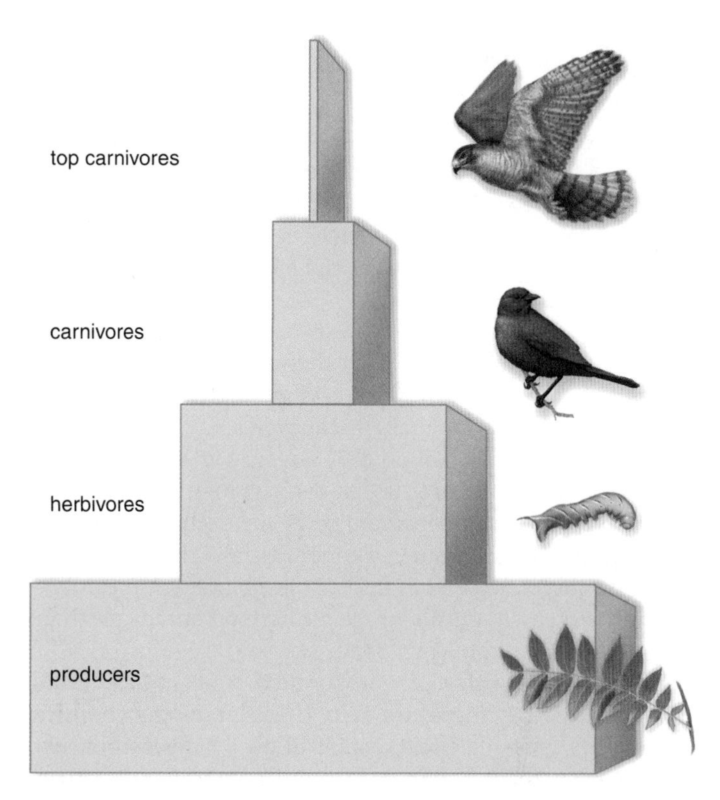

Figure 34.5 Ecological pyramid.
An ecological pyramid reflects the loss of energy from one trophic level to the next. Energy is lost as as herbivores feed on producers, carnivores feed on herbivores, and top carnivores feed on carnivores. This loss of energy results in decreasing biomass and numbers of organisms at higher trophic levels relative to lower trophic levels (levels not to scale).

Biomass is the number of organisms multiplied by their weight. Thinking in terms of biomass rather than simply the number of organisms can be useful, since in terms of only numbers, one oak tree may have hundreds of caterpillars. Although it appears that there are more herbivores than autotrophs, the biomass of autotrophs is much greater. Similarly, the biomass of herbivores is greater than that of carnivores.

However, in aquatic ecosystems, such as lakes and open seas where algae are the main producers, the herbivores may have a greater biomass than the producers. The reason is that, over time, the algae reproduce rapidly, but they are also consumed at a high rate. Any pyramids like this one, which have more herbivores than producers, are called inverted pyramids:

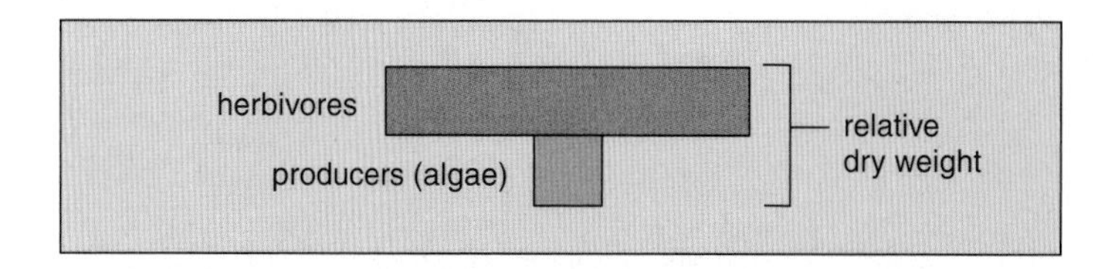

Ecologists are hesitant to use pyramids to describe ecological relationships for all situations. Detrital food chains are rarely included in pyramids, but they may represent a large portion of energy in many ecosystems, making conclusions based on aboveground pyramids potentially inaccurate.

Check Your Progress

1. Why do ecosystems generally support few carnivores?
2. What is meant by the term "inverted ecological pyramid?"

34.3 Global Biogeochemical Cycles

Learning Outcomes

1. Define biogeochemical cycles.
2. Recall the steps of the water cycle, the phosphorus cycle, the nitrogen cycle, and the carbon cycle.
3. Discuss how human activities can alter each of the biogeochemical cycles.

Because the pathways by which chemicals circulate through ecosystems involve both biotic and geological components, they are known as **biogeochemical cycles.** For each element, chemical cycling may involve (1) a reservoir—a source normally unavailable to producers, such as fossilized remains, rocks, and deep-sea sediments; (2) an exchange pool—a source from which organisms generally take chemicals, such as the atmosphere or soil; and (3) the biotic community—through which chemicals move along food chains, perhaps never entering a pool (Fig. 34.6).

With the exception of water, which exists in gas, liquid, and solid forms, there are two types of biogeochemical cycles. In a gaseous cycle, exemplified by the carbon and nitrogen cycles, the element is withdrawn from and returns to the atmosphere as a gas. In a sedimentary cycle, exemplified by the phosphorus cycle, the element is absorbed from soil by plant roots, passed to heterotrophs, and eventually returned to the soil by decomposers, usually nearby.

The diagrams on the next few pages show how nutrients flow between terrestrial and aquatic ecosystems. In the nitrogen and phosphorus cycles, these nutrients run off from a terrestrial to an aquatic ecosystem and in that way enrich the aquatic ecosystem. However, too much of these nutrients can be harmful—for example, by causing plant overgrowth. Decaying organic matter in aquatic ecosystems can be a source of nutrients for intertidal inhabitants such as fiddler crabs. Seabirds feed on fish but deposit guano (droppings) on land, and in that way phosphorus from the water is deposited on land. Anything put into the environment in one ecosystem can find its way to another ecosystem. As evidence of this, scientists find the soot from urban areas and pesticides from agricultural fields in the snow and animals of the Arctic.

The Water Cycle

The **water cycle** (or hydrologic cycle) is described in Figure 34.7. During the water cycle, fresh water is distilled from salt water. First, evaporation occurs. During **evaporation,** a liquid (in this case, water) changes from a liquid state to a gaseous state. The sun's rays cause fresh water to evaporate from sea-water, and the salts are left behind. Next, condensation occurs. During **condensation,** a gas is changed to a liquid. For example, vaporized fresh water rises into the atmosphere, is stored in clouds, cools, and falls as rain over the oceans and the land.

Water also evaporates from land, from plants (transpiration), and from bodies of fresh water. Because land lies above

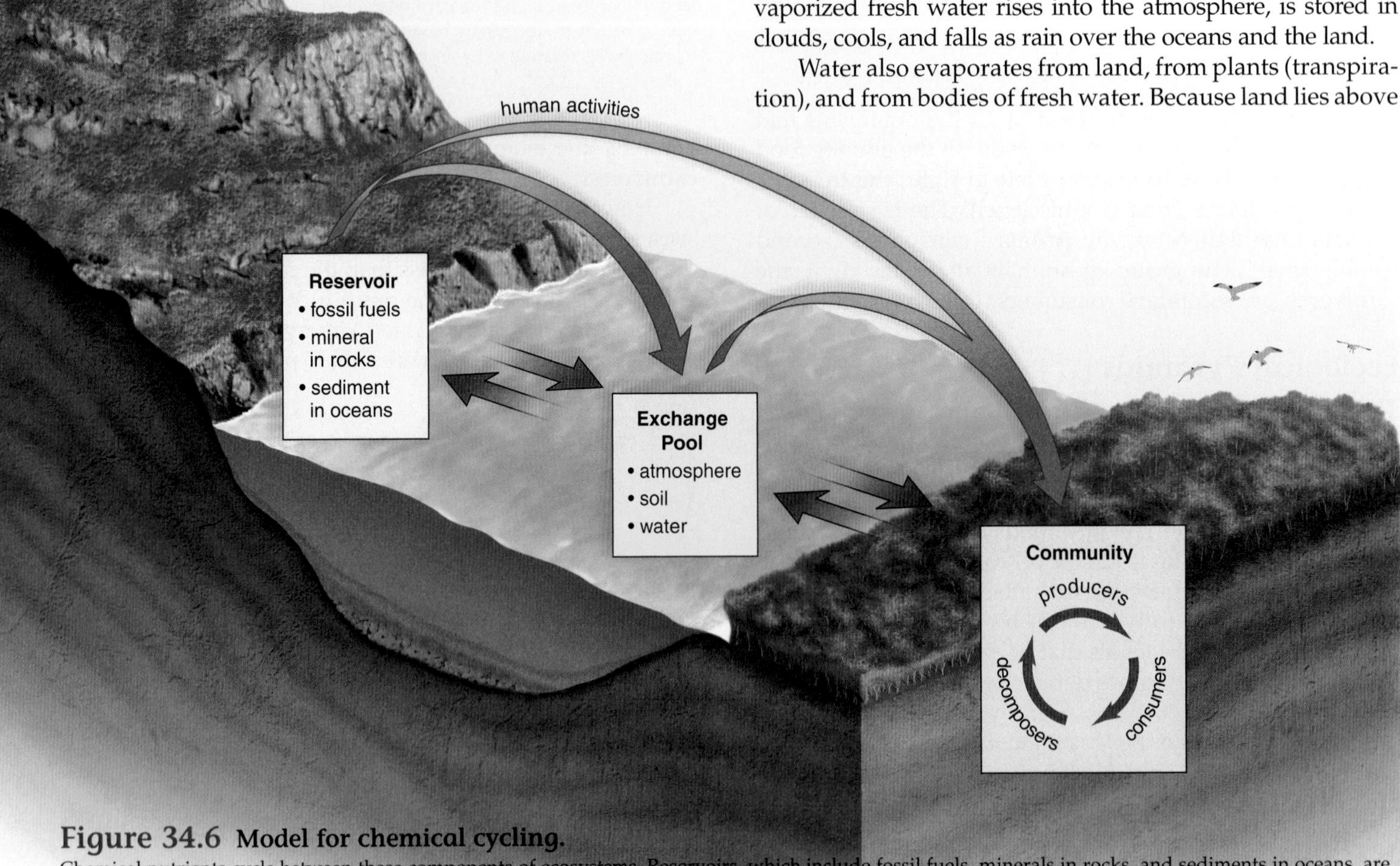

Figure 34.6 Model for chemical cycling.
Chemical nutrients cycle between these components of ecosystems. Reservoirs, which include fossil fuels, minerals in rocks, and sediments in oceans, are normally relatively unavailable sources. However, exchange pools, such as those in the atmosphere, soil, and water, are available sources of chemicals for the biotic community. When human activities (purple arrows) remove chemicals from reservoirs and pools and make them available to the biotic community, pollution can result.

sea level, gravity eventually returns all fresh water to the sea. In the meantime, water is contained within standing waters (lakes and ponds), flowing water (streams and rivers), and groundwater.

Some of the water from **precipitation** (e.g., rain, snow, sleet, hail, and fog) sinks, or percolates, into the ground and saturates the earth to a certain level. The top of the saturation zone is called the groundwater table, or simply, the water table. Because water infiltrates through the soil and rock layers, sometimes groundwater is also located in **aquifers,** rock layers that contain water and release it in appreciable quantities to wells or springs. Aquifers are recharged when rainfall and melted snow percolate into the soil.

Human Activities

In some parts of the United States, especially the arid West and southern Florida, withdrawals from aquifers exceed any possibility of recharge. This is called "groundwater mining." In these locations, the groundwater level is dropping, and residents may run out of groundwater, at least for irrigation purposes, within a few years.

Fresh water, which makes up only about 3% of the world's supply of water, is called a renewable resource because a new supply is always being produced as a result of the water cycle. But it is possible to run out of fresh water when the available supply is not adequate or has become polluted so that it is not usable. Notice that the price of bottled water has gone up steadily over the past several years.

CHECK YOUR PROGRESS

1. Discuss the steps of the water cycle.

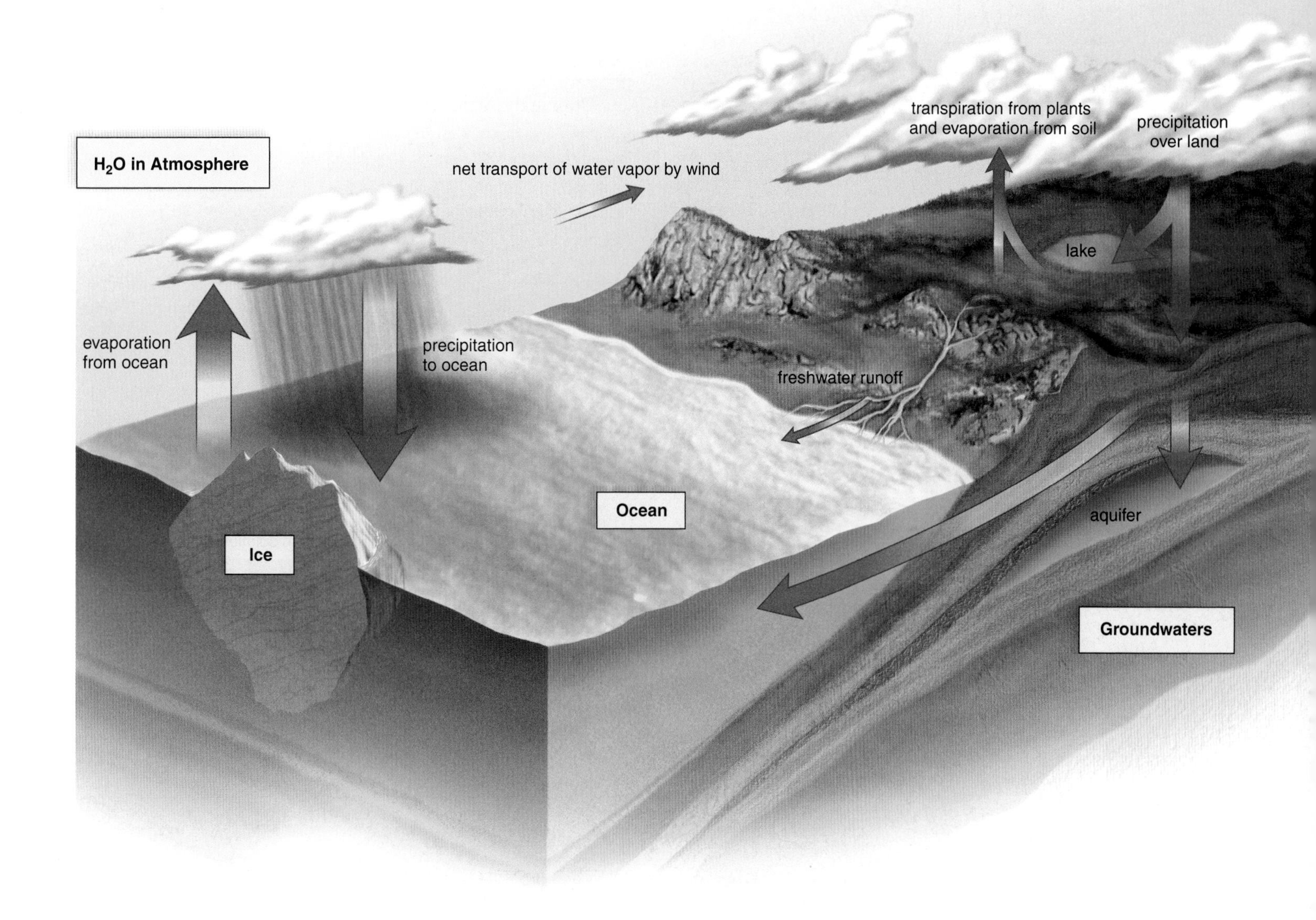

Figure 34.7 The water cycle.
Evaporation from the ocean exceeds precipitation, so there is a net movement of water vapor onto land, where precipitation results in the eventual flow of surface water and groundwater back to the sea. On land, transpiration by plants contributes to evaporation.

The Phosphorus Cycle

In the phosphorus cycle, phosphorus moves from rocks on land to the oceans, where it gets trapped in sediments. Then phosphorus moves back onto land following a geological upheaval. You can verify this by following the appropriate arrows in Figure 34.8. However, on land, the very slow weathering of rocks makes phosphate ions (PO_4^{3-} and HPO_4^{2-}) available to plants, which take up phosphate from the soil.

Producers use phosphate to form a variety of molecules, including phospholipids, ATP, and the nucleotides that become a part of DNA and RNA. Animals incorporate some of the phosphate into teeth, bones, and shells that take many years to decompose. Eventually, however, phosphate ions become available to producers once again. Because the available amount of phosphate is already being utilized within food chains, phosphate is usually a limiting inorganic nutrient for plants—that is, their growth is limited by the amount of available phosphorus.

Some phosphate runs off into aquatic ecosystems, where algae acquire phosphate before it becomes trapped in sediments. Phosphate in marine sediments does not become available to producers on land again until a geological upheaval exposes sedimentary rocks to weathering once more. Phosphorus does not enter the atmosphere, therefore, the phosphorus cycle is called a sedimentary cycle.

Human Activities

Human beings boost the amount of available phosphate by mining phosphate ores and using them to make fertilizers, animal feed supplements, and detergents. Fertilizers usually contain three basic ingredients: nitrogen, phosphorus, and potassium. Most laundry detergents contain approximately 35–75% sodium triphosphate. The amount of phosphate in animal feed varies.

Animal wastes from livestock feedlots, fertilizers from lawns and cropland, and untreated and treated sewage discharged from cities all add excess phosphate to nearby waters. The end result is **cultural eutrophication** (overenrichment), which can lead to an algal bloom, as indicated by green scum floating on the water or excessive mats of filamentous algae. When the algae die off, decomposers use up all available oxygen for cellular respiration. The result can be a massive fish kill.

CHECK YOUR PROGRESS

1. Outline the steps of the phosphorus cycle.

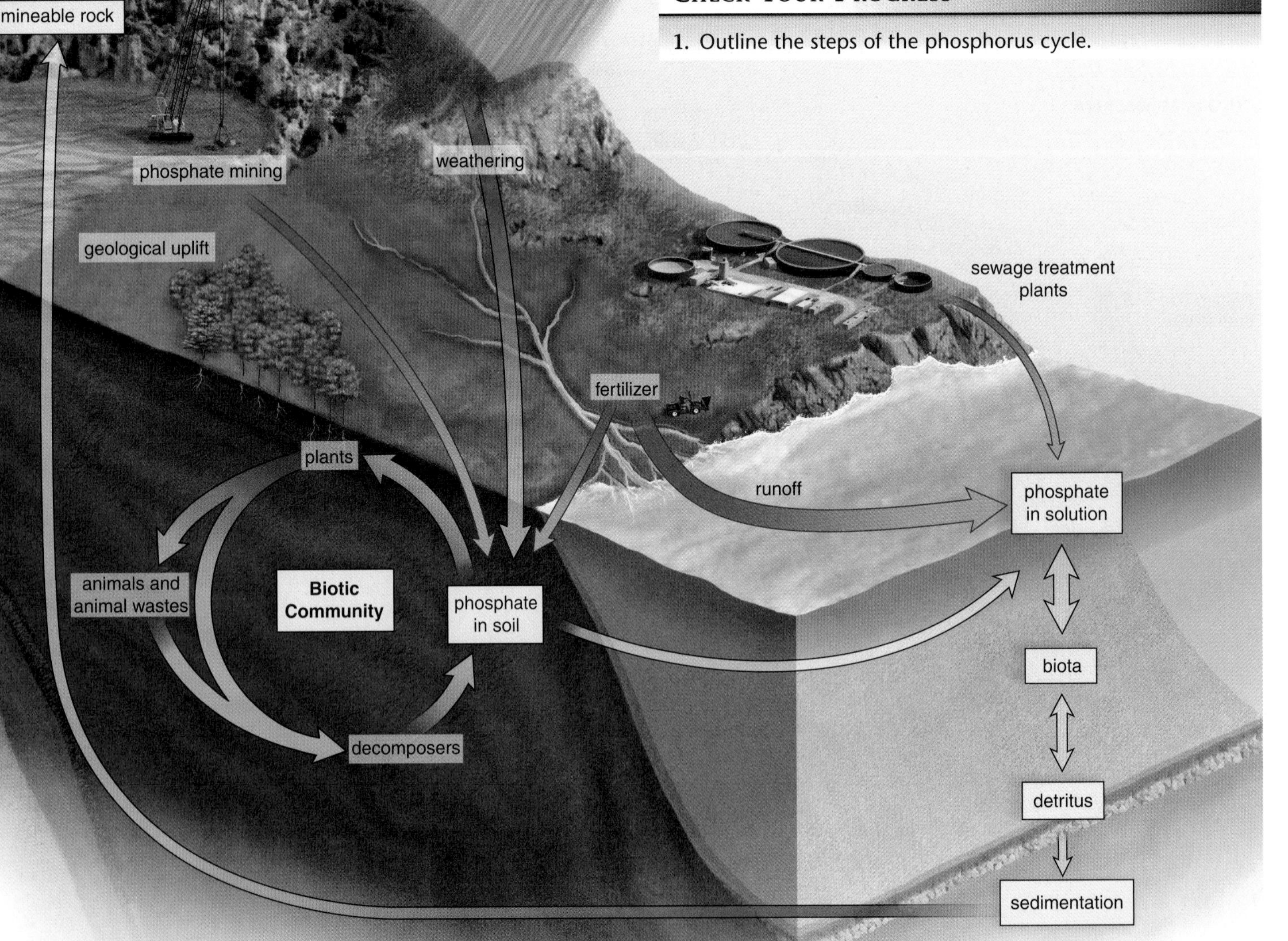

Figure 34.8 The phosphorus cycle.
Purple arrows represent human activities; gray arrows represent natural events.

The Nitrogen Cycle

Nitrogen gas (N_2) makes up about 78% of the atmosphere, but plants cannot use nitrogen gas. Therefore, nitrogen is also a limiting inorganic nutrient for plants.

Nitrogen Fixation

In the nitrogen cycle, **nitrogen fixation** occurs when nitrogen gas (N_2) is converted to ammonium ions (NH_4^+), a form plants can use (Fig. 34.9). Some cyanobacteria in aquatic ecosystems and some free-living bacteria in soil are able to fix atmospheric nitrogen in this way. Other nitrogen-fixing bacteria live in nodules on the roots of legumes (see Chapter 9). They make organic compounds containing nitrogen available to the host plants so that the plant can form proteins and nucleic acids.

Nitrification

Plants can also use nitrates as a source of nitrogen. The production of nitrates during the nitrogen cycle is called **nitrification.** Nitrification can occur in two ways: (1) Nitrogen gas is converted to nitrate (NO_3^-) in the atmosphere when cosmic radiation, meteor trails, and lightning provide the high energy needed for nitrogen to react with oxygen. (2) Ammonium ions in the soil from decomposition are converted to nitrate by soil bacteria in a two-step process. First, nitrite-producing bacteria convert ammonium to nitrite (NO_2^-), and then nitrate-producing bacteria convert nitrite to nitrate. These two groups of bacteria, called the nitrifying bacteria, are chemoautotrophs.

Denitrification

Denitrification is the conversion of nitrate back to nitrogen gas, which enters the atmosphere. Denitrifying bacteria

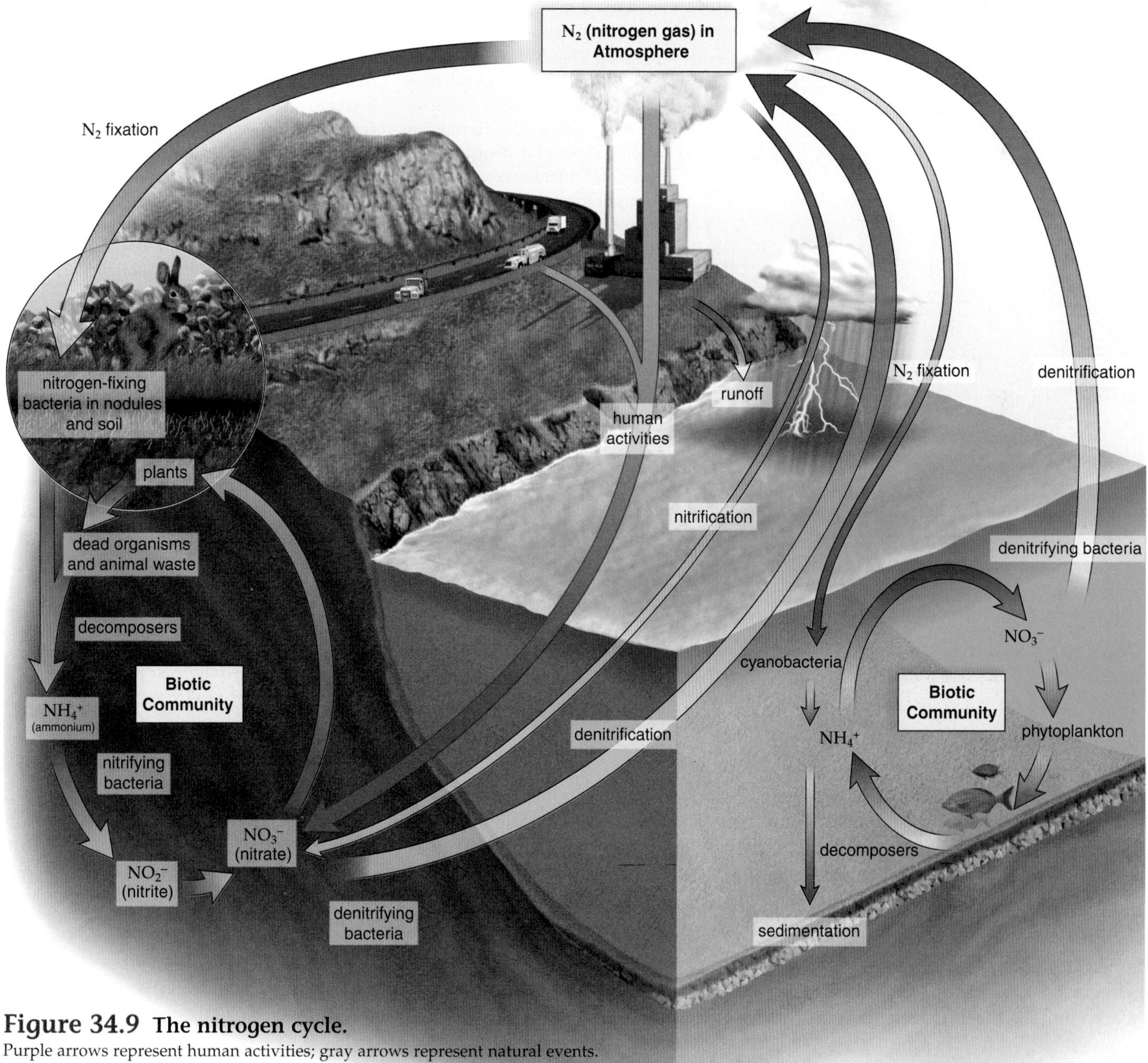

Figure 34.9 The nitrogen cycle.
Purple arrows represent human activities; gray arrows represent natural events.

Ecology Focus

Ozone Shield Depletion

Ozone is a gas that forms in Earth's atmosphere when ultraviolet radiation from the sun splits oxygen molecules (O_2), and then the single oxygen atoms (O) combine with other oxygen molecules to form ozone (O_3). When ozone is present in the troposphere, the layer of the atmosphere closest to the ground, it is considered a pollutant because it adversely affects plant growth and our ability to breathe oxygen. But in the stratosphere, some 50 kilometers (km) above Earth, ozone forms the **ozone shield,** a layer that absorbs solar ultraviolet radiation.

The absorption of UV radiation by the ozone shield is critical for living things. In humans, UV radiation causes mutations that can lead to skin cancer and can make the lens of the eye develop cataracts. In addition, it adversely affects the immune system and our ability to resist infectious diseases. UV radiation also impairs crop and tree growth, and kills off algae (phytoplankton) and tiny shrimplike animals (krill) that sustain oceanic life. Without an adequate ozone shield, therefore, our health and food sources are threatened.

It became apparent in the 1980s that ozone levels had become depleted worldwide and that the depletion was most severe above the Antarctic every spring. This so-called **ozone hole** is equivalent in area to two and a half times the size of Europe (Fig. 34A). It exposed not only Antarctica but also the southern tip of South America and vast areas of the Pacific and Atlantic oceans to harmful UV rays. Of even greater concern, an ozone hole has now appeared above the Arctic as well, and ozone holes have also been detected within northern and southern latitudes where many people live. In Queensland, Australia, an area of the world where the ozone layer is the thinnest, one in two people contract some form of skin cancer during their lives. A United Nations Environmental Program report predicts a 26% rise in cataracts and nonmelanoma skin cancers for every 10% drop in the ozone level. A 26% increase translates into 1.75 million additional cases of cataracts and 300,000 more skin cancer cases every year, worldwide.

Scientists found that the cause of ozone depletion is chlorine atoms (Cl), which can destroy up to 100,000 molecules of ozone before settling to Earth's surface as chloride years later. These chlorine atoms come primarily from the breakdown of **chlorofluorocarbons (CFCs).** The best-known CFC is Freon, a coolant found in refrigerators and air conditioners. CFCs are also used as a foaming agent during the production of Styrofoam coffee cups, egg cartons, insulation, and padding. Formerly, CFCs were used as propellants in spray cans, but this application is now banned in the United States and several European countries.

Most countries in the world have stopped using CFCs, and the United States halted production in 1995. Since that time, satellite measurements indicate that the amount of harmful chlorine pollution in the stratosphere has started to decline. It is clear, however, that recovery of the ozone shield may take many more years and involve other pollution-fighting approaches, aside from lowering chlorine pollution. For example, as greenhouse gases increase global warming (see the Bioethical Focus), the stratosphere becomes colder because less heat is reflected from Earth's surface. Ozone depletion increases when the stratosphere is very cold.

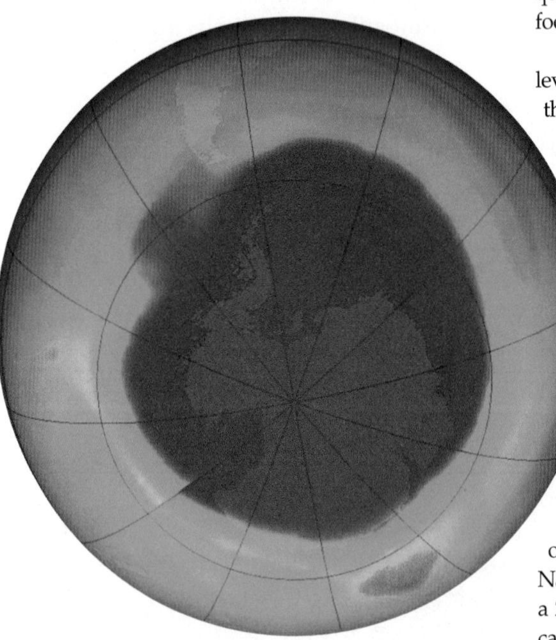

Figure 34A Ozone shield depletion.
Map of ozone levels in the atmosphere of the Southern Hemisphere, September 13, 2007. The ozone depletion is demarcated in the purple and blue colors. The areas in purple indicate the thinnest layer, and are often called an ozone hole. This area is much larger than the size of Europe.

Discussion Questions

1. Should we care about ozone depletion, especially when the most serious ozone depletion is over the Antarctic, where almost no humans live?
2. Ozone depletion and global warming are connected in that warming of Earth's surface causes less heat to radiate in the atmosphere. Should we decrease our burning of fossil fuels, which reduces the amount of greenhouse gases released into the atmosphere—even though this action might decrease profits for U.S. corporations such as oil companies and automobile manufacturers?
3. Some chemicals, such as industrial solvents, still release harmful chlorine into the atmosphere. As the owner of a factory, would you support using alternative "environmentally friendly" solvents even if they would cost more? Why or why not?

living in the anaerobic mud of lakes, bogs, and estuaries carry out this process as a part of their own metabolism. In the nitrogen cycle, denitrification would counterbalance nitrogen fixation except for human activities.

Human Activities

Human activities significantly alter the transfer rates in the nitrogen cycle by producing fertilizers from N_2—in fact, they nearly double the nitrogen fixation rate. Fertilizer, which also contains phosphate, runs off into lakes and rivers and results in algal overgrowth and fish kill.

Fertilizer use also results in the release of nitrous oxide (N_2O), a greenhouse gas, component of acid rain, and contributor to ozone shield depletion (see the Ecology Focus).

Check Your Progress

1. Illustrate the process of nitrogen fixation in the nitrogen cycle.
2. What is the difference between nitrification and denitrification?

The Carbon Cycle

In the carbon cycle, plants in both terrestrial and aquatic ecosystems take up carbon dioxide (CO_2) from the air, and through photosynthesis, they incorporate carbon into food that is used by autotrophs and heterotrophs alike (Fig. 34.10). When all organisms, including plants, respire, a portion of this carbon is returned to the atmosphere as carbon dioxide.

In aquatic ecosystems, the exchange of carbon dioxide with the atmosphere is primarily indirect. However, there is a small quantity of free carbon dioxide in the water. Carbon dioxide from the air combines with water to produce bicarbonate ions (HCO_3^-), a source of carbon for algae that produce food for themselves and for heterotrophs. Similarly, when aquatic organisms respire, the carbon dioxide they give off becomes bicarbonate ion. The amount of bicarbonate in the water is in equilibrium with the amount of carbon dioxide in the air.

Reservoirs Hold Carbon

Living and dead organisms contain organic carbon and serve as one of the reservoirs for the carbon cycle. The world's biotic components, particularly trees, contain 800 billion tons of organic carbon, and an additional 1,000–3,000 billion metric tons are estimated to be held in the remains of plants and animals in the soil. Before decomposition can occur, some of these remains are subjected to physical processes that transform them into coal, oil, and natural gas. We call these materials the **fossil fuels.** Most of the fossil fuels were formed during the Carboniferous period, 286–360 million years ago, when an exceptionally large amount of organic matter was buried before decomposing. Another reservoir is the inorganic carbonate that accumulates in limestone and in calcium carbonate shells of many marine organisms.

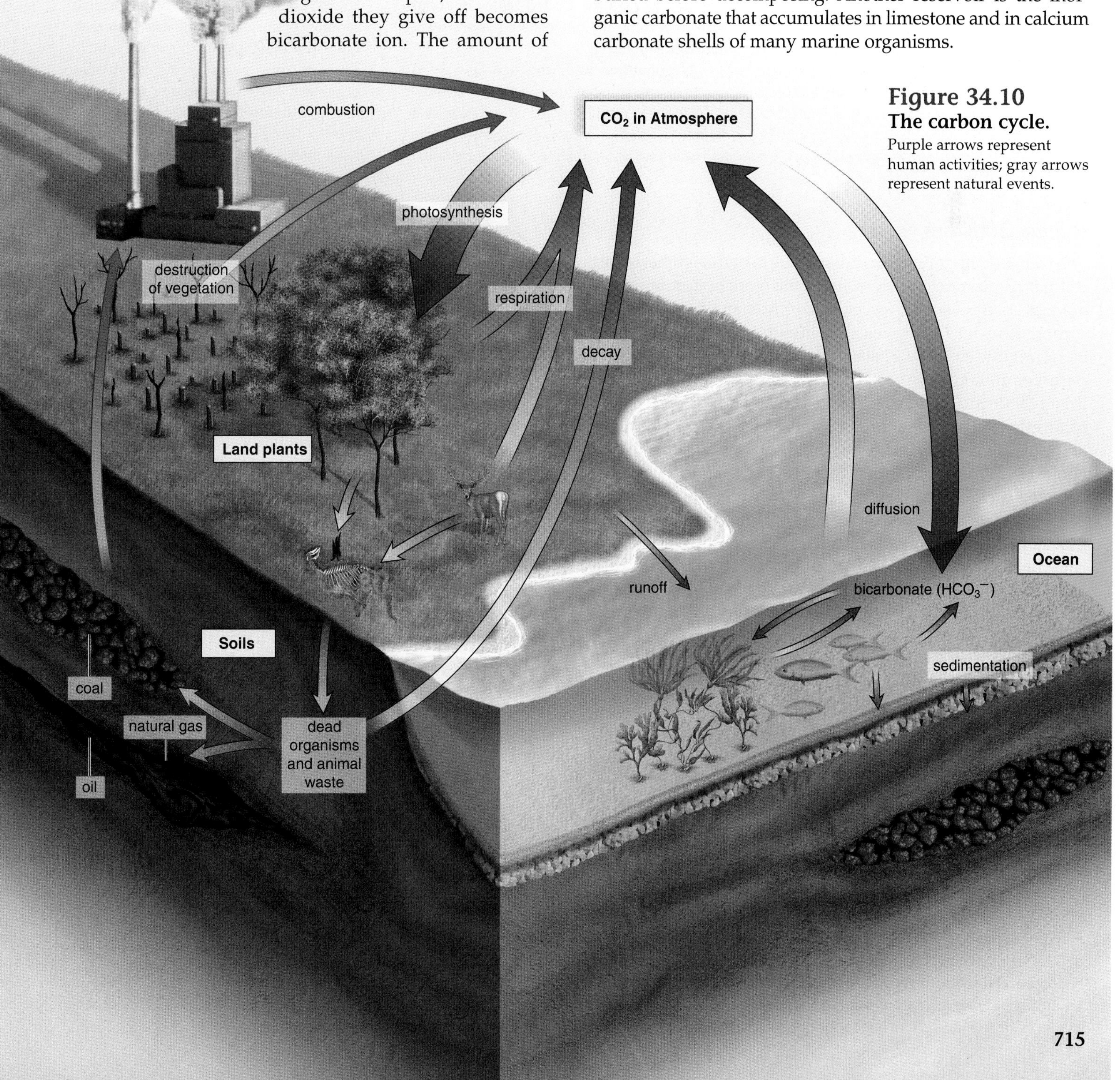

Figure 34.10
The carbon cycle.
Purple arrows represent human activities; gray arrows represent natural events.

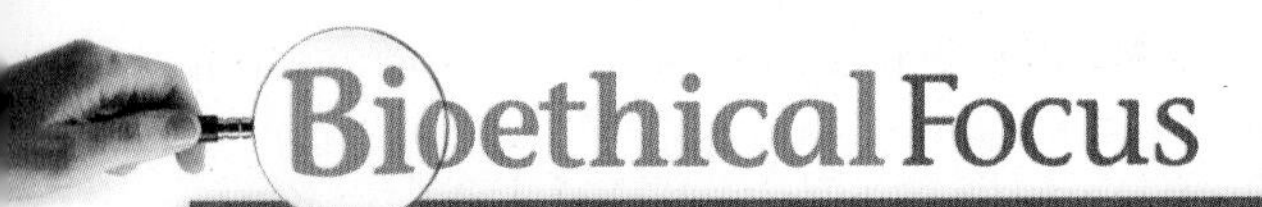

Bioethical Focus

Curtailing Greenhouse Gases

Global warming is predicted to upset normal weather cycles, which will most likely lead to outbreaks of such diseases as malaria, dengue and yellow fevers, filariasis, encephalitis, schistosomiasis, and cholera. That is, increased temperatures will lead to the spread of disease-carrying insects, such as mosquitoes that carry malaria, from tropical to temperate areas. Clearly, use of greenhouse gases should be curtailed.

In December 1997, 159 countries met in Kyoto, Japan, to work out a protocol that would reduce greenhouse gas emissions worldwide. Greenhouse gases are those that are produced in large part through the burning of fossil fuels. Carbon dioxide and methane are two of the most common ones. Such gases allow the sun's rays to pass through but then trap the heat from escaping. It is believed that the emission of greenhouse gases, especially from power plants, will cause Earth's temperature to rise 1.5–4.5°C on average by 2060. Today, over 180 countries have ratified the Kyoto protocol, with the United States a notable exception. The U.S. Senate did not want to ratify the agreement because it did not include a binding emissions commitment from the less-developed countries. While the United States presently emits a large proportion of the world's greenhouse gases, China is expected to pass the United States in about 2020 to become the biggest source of greenhouse emissions.

Negotiations with the less-developed countries are still going on, and some creative ideas have been put forward. Why not have a trading program that allows companies to buy and sell emission credits across international boundaries? Greenhouse reduction techniques would also be for sale. Presumably, if it became monetarily worth their while, companies in less-developed countries would have an incentive to use such techniques to reduce greenhouse emissions.

Form Your Own Opinion

1. Should all countries be expected to reduce their greenhouse emissions even though only a few countries, such as the United States and China, are responsible for most of the greenhouse gases released into the atmosphere? Why or why not?
2. Do you approve of giving companies monetary incentives to reduce greenhouse emissions? Why or why not?
3. Should the United States agree to the Kyoto protocol? Why or why not?

Human Activities

The transfer rates of carbon dioxide due to photosynthesis and cellular respiration, which includes the work of decomposers, are just about even. However, due to human activities, more carbon dioxide is being released into the atmosphere than is being removed. In 1850, atmospheric CO_2 was at about 280 parts per million (ppm); today, it is over 380 ppm. This increase is largely due to the burning of fossil fuels and the destruction of forests to make way for farmland and pasture. Today, the amount of carbon dioxide released into the atmosphere is about twice the amount that remains in the atmosphere. It's believed that most of this dissolves in the ocean.

The increased amount of carbon dioxide (and other gases) in the atmosphere is causing a rise in temperature called **global warming.** These gases allow the sun's rays to pass through, but they absorb and radiate heat back to Earth, a phenomenon called the **greenhouse effect.** Global warming is expected to cause dire consequences, as discussed in Chapter 36. The Bioethical Focus discusses worldwide efforts to cut back on greenhouse gases.

Check Your Progress

1. How can human activities disrupt the carbon cycle to contribute to global warming?

Applying the Concepts Revisited

At the beginning of this chapter, you were presented with a scenario of what the Earth may be like in about 90 years. Human disruption of biogeochemical cycles can cause global warming, eutrophication, smog, and acid rain.

1. Based on what you know about food chains, what do you expect to happen to the concentration of pollutants in animals as you move from lower to higher trophic levels?
2. By the year 2100, the world population has surpassed 13 billion people. Advances in technology have made it possible to create an artificial atmosphere on the moon that traps appropriate levels of oxygen and allows ample sunlight for life to persist. You are a member of a team of ecologists assigned the task of making the moon habitable. You realize that much more than a suitable atmosphere is necessary for survival. Construct an ecosystem that will support human and other life as we move people to the moon. Justify your choices of organisms used in this ecosystem.
3. How do we alter the nitrogen cycle, and what are some of the environmental consequences?

SUMMARIZING THE CONCEPTS

34.1 The Biotic Components of Ecosystems

- An ecosystem is composed of populations of organisms plus their physical environment.
- Producers are autotrophs that transform solar energy into food for themselves and all consumers. Consumers are heterotrophs that take in organic food. As herbivores feed on plants or algae and carnivores feed on herbivores, energy is converted to heat. Feces, urine, and dead bodies become food for decomposers. Decomposers return some proportion of inorganic nutrients to autotrophs, and other portions are imported or exported among ecosystems in global cycles.

- Ecosystems are characterized by energy flow and chemical cycling. Energy is lost from the biosphere, but inorganic nutrients are not. They recycle within and among ecosystems. Eventually, all the solar energy that enters an ecosystem is converted to heat, and thus ecosystems require a continual supply of solar energy.

34.2 Energy Flow

- Ecosystems contain food webs in which the various organisms are connected by trophic relationships. In a grazing food web, food chains begin with a producer. In a detrital food web, food chains begin with detritus. The two food webs are joined when the same consumer links both a grazing and a detrital food chain.
- A trophic level contains all the organisms that feed at a particular link in a food chain. Ecological pyramids show trophic levels stacked one on top of the other like building blocks. They are shaped like pyramids because most energy is lost from one trophic level to the next, and thus the number of species that can be sustained decreases.

34.3 Global Biogeochemical Cycles

- Biogeochemical cycles consist of reservoirs, exchange pools, and biotic communities. A reservoir pool contains elements available on a limited basis to living things, such as fossil fuels, sediments, and rocks. An exchange pool, such as the atmosphere, soil, and water, is a ready source of nutrients for living things.
- In the water cycle, evaporation over the ocean is not compensated for by rainfall. Evaporation from terrestrial ecosystems includes transpiration from plants. Rainfall over land results in bodies of fresh water plus groundwater, including aquifers. Eventually, all water returns to the oceans.
- In the phosphorus cycle, the biotic community recycles phosphorus back to the producers, and only limited quantities are made available by the weathering of rocks. Phosphates are mined for fertilizer production. When phosphates and nitrates enter lakes, ponds, and eventually the ocean, pollution and overenrichment occurs.
- In the nitrogen cycle, certain bacteria in water, soil, and on root nodules can fix atmospheric nitrogen. Other bacteria return nitrogen to the atmosphere. Human activities convert atmospheric nitrogen to fertilizer, which is broken down by soil bacteria. Humans also burn fossil fuels. In this way, nitrogen oxides can contribute to the formation of smog and acid rain, both of which are detrimental to animal and plant life.
- In the carbon cycle, organisms add as much carbon dioxide to the atmosphere as they remove. Shells in ocean sediments, organic compounds in living and dead organisms, and fossil fuels are reservoirs for carbon. Human activities, such as burning fossil fuels and trees, add carbon dioxide and other gases to the atmosphere. A buildup of these "greenhouse gases" is leading to global warming. A rise in sea level and a change in climate patterns is expected to follow.

TESTING YOURSELF

Choose the best answer for each question.

1. Which of the following would be a primary consumer in a vegetable garden?
   a. aphid sucking sap from cucumber leaves
   b. lady beetle eating aphids
   c. songbird eating lady beetles
   d. fox eating songbirds
   e. All of these are correct.
2. Label the populations in this diagram of an ecosystem.

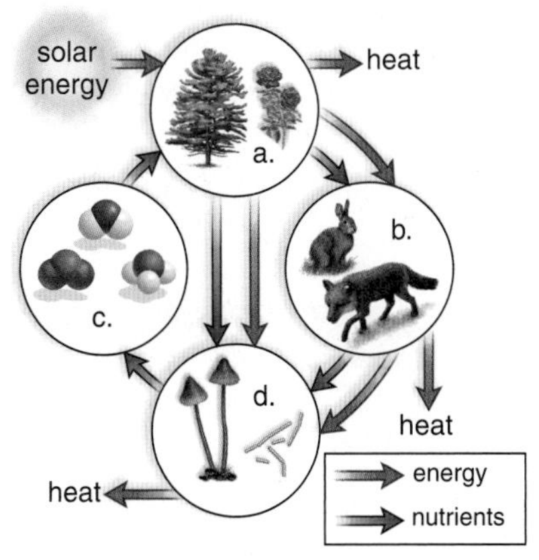

3. During chemical cycling, inorganic nutrients are typically returned to the soil by
   a. autotrophs.
   b. detritivores.
   c. decomposers.
   d. tertiary consumers.
4. When a heterotroph takes in food, only a small percentage of the energy in that food is used for growth. The remainder is
   a. not digested and eliminated as feces.
   b. excreted as urine.
   c. given off as heat.
   d. All of these are correct.
   e. None of these are correct.
5. The first trophic level on a food web is composed of the
   a. producers.
   b. primary consumers.
   c. secondary consumers.
   d. tertiary consumers.
6. Which of the following is a grazing food chain?
   a. leaves ⟶ detritivores ⟶ deer ⟶ owls
   b. birds ⟶ mice ⟶ snakes
   c. nuts ⟶ leaf-eating insects ⟶ chipmunks ⟶ hawks
   d. leaves ⟶ leaf-eating insects ⟶ mice ⟶ snakes
7. In this diagram, label the trophic levels (blanks a–d), using two of these terms for each level: producers, top carnivores, secondary consumers, autotrophs, primary consumers, tertiary consumers, carnivores, herbivores.

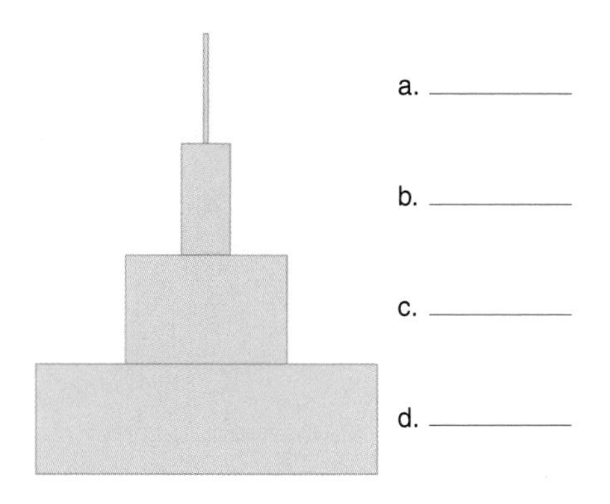

8. In a grazing food web, carnivores that eat herbivores are
   a. producers.
   b. primary consumers.
   c. secondary consumers.
   d. tertiary consumers.
9. Few carnivores can be supported in a food web because
   a. mineral nutrients do not cycle quickly enough.
   b. they produce only a few young per generation.
   c. their generation time is very long.
   d. a large amount of energy is lost at each trophic level.
10. Which of the following is a sedimentary biogeochemical cycle?
    a. carbon
    b. nitrogen
    c. phosphorus
    d. water
11. Underground oil is an example of a carbon
    a. cycle.
    b. pathway.
    c. reservoir.
    d. exchange pool.
12. Which of the following is not a component of the water cycle?
    a. evaporation
    b. filling of aquifers
    c. runoff
    d. photosynthesis
    e. condensation

For questions 13–15, match each type of nitrogen conversion with a process in the key.

Key:

a. nitrogen fixation
b. nitrification
c. denitrification

13. Nitrate to nitrogen gas
14. Nitrogen gas to nitrate
15. Nitrogen gas to ammonium
16. Choose the statement that is true concerning this food chain: grass ⟶ rabbits ⟶ snakes ⟶ hawks
    a. Each predator population has a greater biomass than its prey population.
    b. Each prey population has a greater biomass than its predator population.
    c. Each population is omnivorous.
    d. Each population returns inorganic nutrients and energy to the producer.
    e. Both a and c are correct.

UNDERSTANDING THE TERMS

THINKING CRITICALLY

1. A large forest has been removed by timber harvest (clear-cutting) and the land has not been replanted. After several years, humidity and rainfall in the area seem to have decreased. Use your knowledge of the water cycle to explain these observations.
2. In mountainous regions of the western United States, large predators (e.g., wolves in Yellowstone National Park), which were previously driven out of the area, are being reintroduced. People living in these areas are concerned that livestock may be lost if there is not enough wild food for the predators. What types of data are needed concerning food webs and ecological pyramids to ensure the successful reintroduction of these predators and minimal impact on livestock?
3. Based on your knowledge of the carbon cycle, why are fossil fuels such as oil so limited?

INQUIRY INTO LIFE WEBSITE

The companion website for *Inquiry into Life* provides a wealth of information organized and integrated by chapter. You will find practice tests, animations, videos, and much more that will complement your learning and understanding of general biology.

http://www.mhhe.com/maderinquiry13

Major Ecosystems of the Biosphere

Applying the Concepts

From space, the Earth looks like a pristine aqua globe, hovering in a vast backdrop of darkness. As you move closer, Earth's churning atmosphere and immense water systems come into view. Not until the planet's surface is visible, as from an airplane, can we make out the vast differences in land formations and physical features. Rainfall and temperature largely account for the great terrestrial biomes of the world—from the freezing, snow-covered Arctic tundra to the arid, scorched deserts, to the lush tropical rain forests. From the window of an airplane, we can also make out signs of human development—such as farms, towns, and cities—all of which have reduced or fragmented Earth's ecosystems to a mere fraction of their original size. Pollution does not stay localized, reaching the far corners of the biosphere—even the most remote areas. These major environmental changes have altered the world's climate, the biomes around the globe, and the functioning of ecosystems within them.

In this chapter, we see how climate determines the characteristics of major biomes, each of which has its own mix of species adapted to living under the particular environmental conditions characteristic of that biome. By understanding these environmental characteristics, we can learn the value of ecosystem services and, in turn, help conserve species—even our own.

Chapter Outline

35.1 Climate and the Biosphere

Learning Outcomes

1. Explain the relationship between climate and latitude.
2. Discuss how the three major types of global wind patterns are determined.
3. Give an example of how topography affects climate.

Climate refers to the prevailing weather conditions in a particular region. Climate is dictated by temperature and rainfall, which in turn are influenced by the following factors: (1) variations in the distribution of solar radiation due to the tilt of the spherical Earth as it orbits about the sun; and (2) other effects, such as topography and proximity to water bodies.

Effect of Solar Radiation

Because the Earth is spherically shaped, the sun's rays are more direct at the equator and more spread out progressing toward the poles. Therefore, the tropical regions nearest the equator are warmer than the temperate regions farther away from the equator. In addition, Earth does not face the sun directly. Rather, it is on a slight tilt (about 23°) away from the sun. As the Earth orbits around the sun throughout the year, different parts of the planet are tilted toward or away from the sun, which determines the seasons. For example, during winter, the Northern Hemisphere is tilted away from the sun (Fig. 35.1). At the same time, the Southern Hemisphere is tilted toward the sun and it is summer.

Because the Earth completes one rotation on its axis per day and its surface consists of continents and oceans, the flows of warm and cold air form three large circulation patterns in each hemisphere. The direction in which the air rises and cools determines the direction of the wind (Fig. 35.2). At the equator, the sun heats the air and evaporates water. The warm, moist air rises and loses most of its moisture as rain, resulting in the greatest amounts of rainfall occurring nearest to the equator. The rising air flows toward the poles, but at about 30° north and south latitude, it sinks toward Earth's surface and reheats. As the dry air descends and warms once again, areas of high pressure are generated, which results in low rainfall. As a result, the great deserts of Africa, Australia, and the Americas occur at these latitudes. At about 60° north and south latitude, the warm air rises and cools as it does near the equator, producing a low-pressure area. Low pressure results in zones of high rainfall. Between 30° and 60° latitude, the strong wind patterns known as the westerlies occur in both the Northern and Southern Hemispheres. The westerlies move from west to east. The west coasts of the continents at these latitudes are wet, as in the U.S. Pacific Northwest, where a temperate (evergreen) rain forest is located. Weaker winds, called the polar easterlies, blow from east to west at latitudes higher than 60° in both hemispheres.

The direction of wind patterns, such as the easterlies and westerlies, is affected by the spinning of the Earth about its axis. That is, in the Northern Hemisphere, large-scale winds generally move clockwise, while in the Southern Hemisphere, they move counterclockwise. This explains, for example, why the northeast trade winds blow from the northeast toward the southwest and the southeast trade winds blow from the southeast toward the northwest (see Fig. 35.2). Trade winds are so called because early sailors depended on them to power the movement of sailing ships.

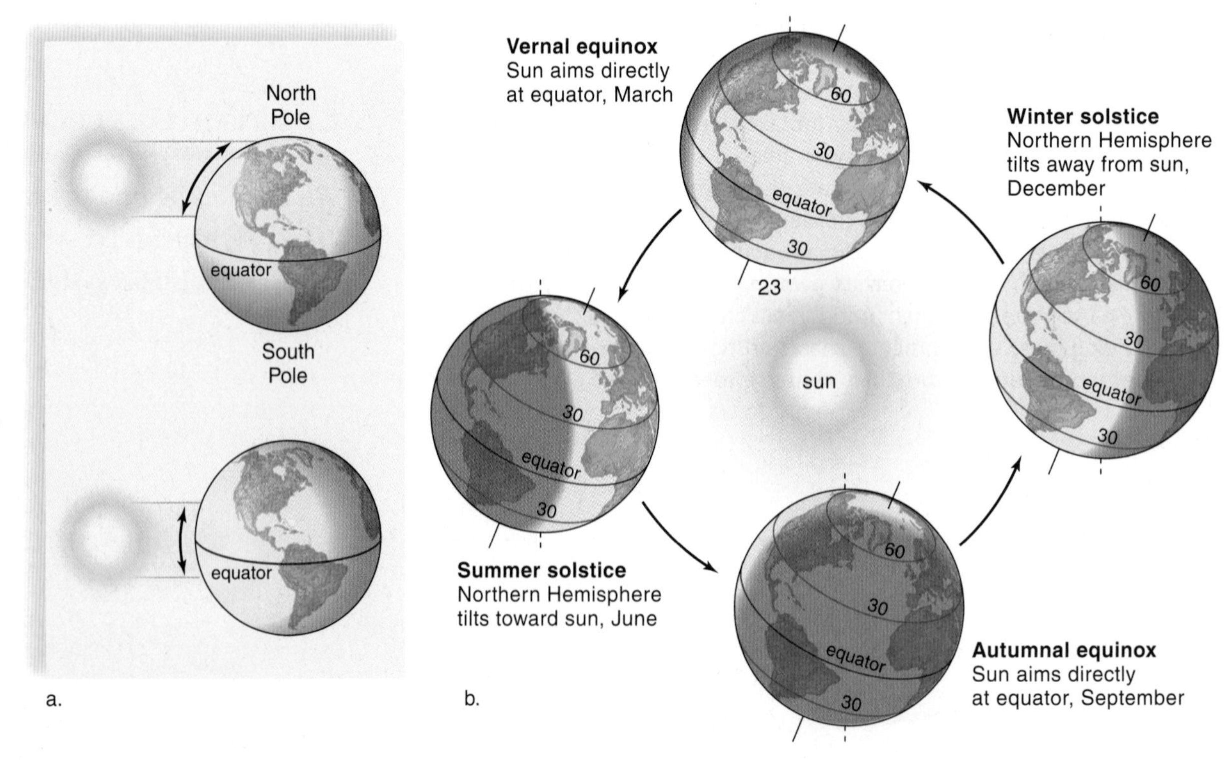

Figure 35.1 Distribution of solar energy. **a.** Because Earth is a sphere, beams of solar energy striking the planet near one of the poles are spread over a wider area than similar beams striking Earth at the equator. **b.** The seasons of the Northern and Southern Hemispheres are due to the tilt of Earth on its axis as it rotates around the sun.

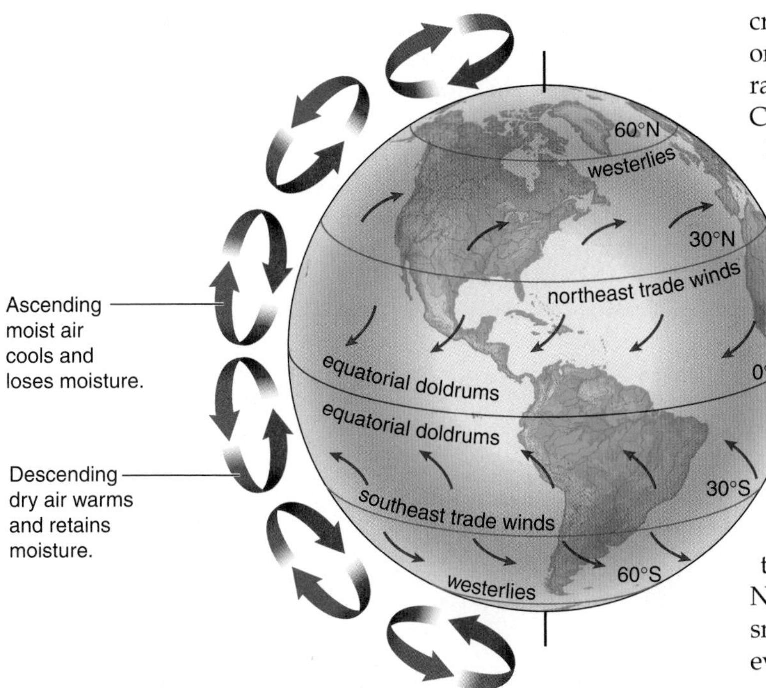

Figure 35.2 Global wind circulation.
Air ascends and descends as shown because Earth rotates on its axis. Also, the trade winds blow from the northeast to the west in the Northern Hemisphere, and blow the southeast to the west in the Southern Hemisphere. The westerlies blow toward the east.

Other Effects

The term topography refers to the physical features, or "the lay," of the land. One physical feature that affects climate is the presence of mountains. As air blows up and over a mountain range, it rises and cools. One side of the mountain, called the windward side, receives more rainfall than the other side, called the leeward side. On the leeward side, the air descends, picks up moisture, and generally produces clear weather (Fig. 35.3). The difference between the windward side and the leeward side can be quite dramatic. In the Hawaiian Islands, for example, the windward side of the mountains typically receives more than 750 cm of rain a year while the leeward side, which is in a **rain shadow,** gets an average of only 50 cm of rain and is generally sunny. In the United States, the western side of the Sierra Nevada mountain range is lush, while the eastern side is a semidesert.

In contrast to landmasses, the oceans are slower to change temperature. This causes coasts to have a unique weather pattern that is not seen inland. During the day, the land warms more quickly than the ocean, and the air above the land rises. Then a cool sea breeze blows in from the ocean. At night, the reverse happens. The breeze blows from the land to the sea.

In India and some other countries in southern Asia, the land heats more rapidly than the waters of the Indian Ocean during spring. The difference in temperature between the land and the ocean causes a gigantic circulation of air: warm air rises over the land, and cooler air comes in off the ocean to replace it. As the warm air rises, it loses its moisture, creating a **monsoon** climate in which wet ocean winds blow onshore for almost half the year. During the monsoon season, rainfall is particularly heavy on the windward side of hills. Cherrapunji in northern India receives an annual average of 1,090 cm of rain a year because of its high altitude. The chief crop of India is rice, which starts to grow when the monsoon rains begin. The weather pattern has reversed by November, when the land has become cooler than the ocean. Therefore, dry winds blow from the Asian continent across the Indian Ocean. In the winter, the air over the land is dry, the skies cloudless, and temperatures pleasant.

Other large bodies of water create major weather patterns. For example, in the United States, people often speak of the "lake effect," meaning that in the winter, arctic winds blowing over the Great Lakes become warm and moisture-laden. When these winds rise and lose their moisture, snow begins to fall. Places such as Buffalo, New York, get heavy snowfalls due to the lake effect, and snow is on the ground there for an average of 90–140 days every year.

Check Your Progress

1. What determines the different seasons?
2. What determines the direction of wind patterns in different latitudes?
3. How does topography affect rainfall?

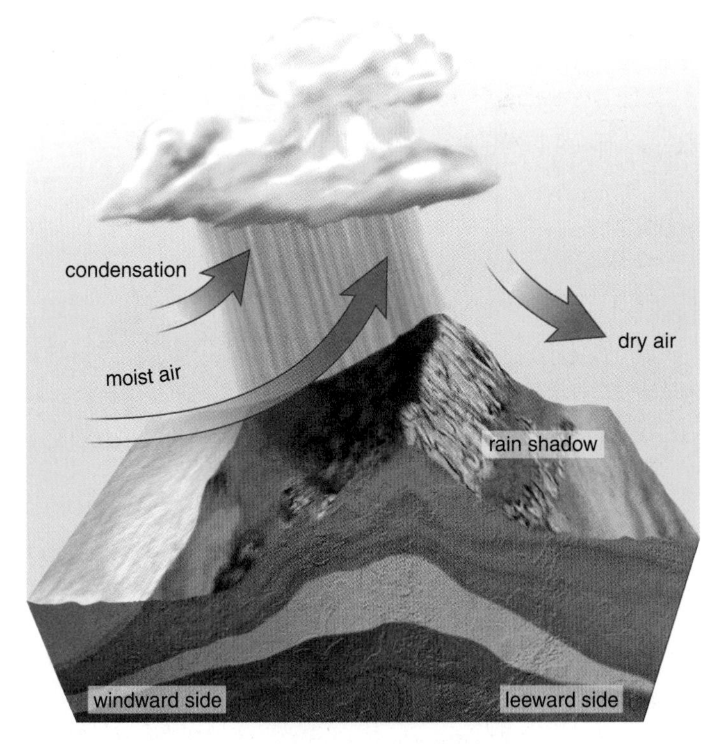

Figure 35.3 Formation of a rain shadow.
When winds from the sea cross a coastal mountain range, they rise and release their moisture as they cool the windward side of the mountain. The leeward side of a mountain is warmer and receives relatively little rain. Therefore, it is said to lie in a "rain shadow."

35.2 Terrestrial Ecosystems

Learning Outcomes

1. List the environmental conditions that determine the locations of the various biomes around the world.
2. Briefly describe the characteristics of the seven major types of terrestrial biomes.

A major type of terrestrial ecosystem is called a **biome.** Depending on where they are found in terms of latitude and longitude and the amount of precipitation they receive, terrestrial ecosystems have a particular mix of plants and animals adapted to living there. When terrestrial biomes are plotted according to their mean annual temperature and mean annual precipitation, a particular pattern results (Fig. 35.4*a*). The distribution of biomes is shown in Figure 35.4*b*. Even though

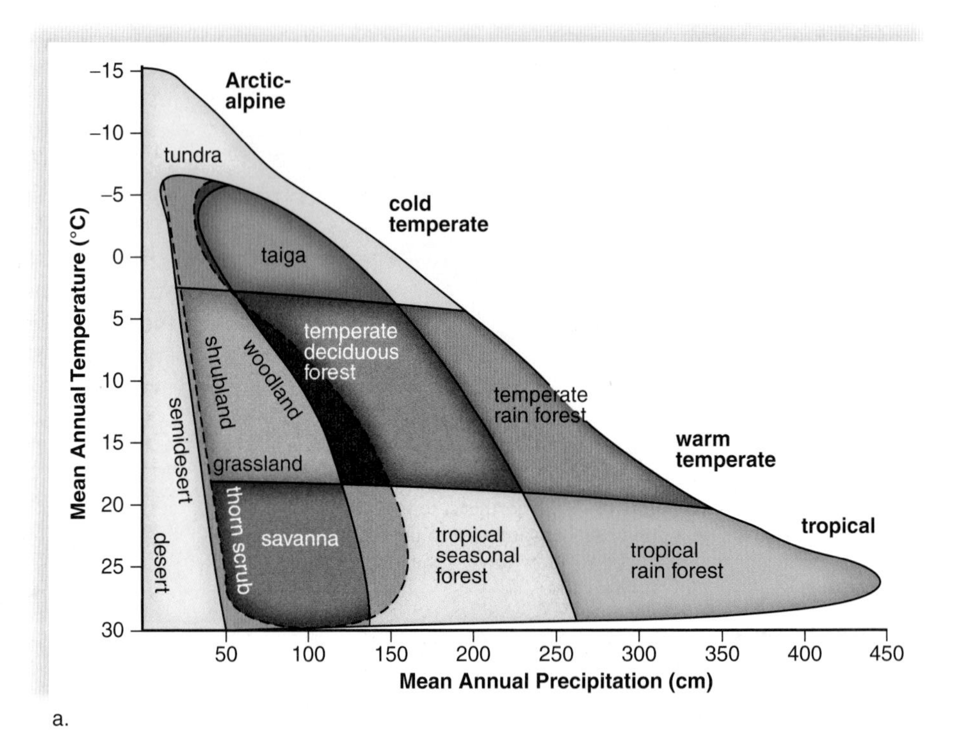

a.

polar ice
tundra
taiga
mountain zone
temperate deciduous forest
temperate rain forest
tropical deciduous forest
tropical seasonal forest
tropical rain forest
shrubland
temperate grassland
savanna
semidesert
desert

b.

Figure 35.4 Pattern of biome distribution.
a. Pattern of world biomes in relation to temperature and moisture. The dashed line encloses a wide range of environments in which either grasses or woody plants can dominate the area, depending on the soil type. **b.** The same type of biome can occur in different regions of the world, as shown on this global map.

Figure 35.4 shows what appear to be clear demarcations, note that biomes gradually change from one type to another. To simplify our discussion in this chapter, we will group the biomes into the seven general categories listed in Table 35.1. Although we will be discussing each type of biome separately, keep in mind that all biomes are linked to form the biosphere, with each biome having inputs from and outputs to all the other terrestrial and aquatic ecosystems.

The distribution of the biomes and their corresponding communities of organisms is determined principally by differences in climate due to the distribution of solar radiation, water, and defining topographical features. Both latitude and altitude are responsible for temperature gradients. When traveling from the equator to the North Pole, it would be possible to observe first a tropical rain forest, followed by a temperate deciduous forest, a coniferous forest, and tundra. This sequence can also be seen when ascending a mountain (Fig. 35.5). The coniferous forest of a mountain is called a *montane coniferous forest*, and the tundra near the peak of a mountain is called an *alpine tundra*. However, when going from the equator to the South Pole, one would not reach a region corresponding to the coniferous forest and tundra of the Northern Hemisphere because the majority of the landmasses are shifted toward the north.

TABLE 35.1 Selected Biomes

Name	Characteristics
Tundra (arctic tundra, alpine tundra)	Around North Pole; average annual temperature is −12°C to −6°C; low annual precipitation (less than 25 cm); permafrost (permanent ice) year-round within a meter of surface
Taiga (coniferous forests, montane forests)	Large northern biome that circles just below the Arctic Circle; temperature is below freezing for half the year; moderate annual precipitation (30–85 cm); long nights in winter and long days in summer
Temperate (deciduous forests)	Eastern half of United States, Canada, Europe, and parts of Russia; four seasons of the year with hot summers and cold winters; relatively high annual precipitation (75–150 cm)
Tropical forests (tropical rain, tropical deciduous)	Located near the equator in Latin America, Southeast Asia, and West Africa; warm (20–25°C) and wet (190 cm/year); has wet/dry season
Shrublands	In the western U.S., western South America, and central Asia; contains shrubs that are adapted to arid conditions with small, thick leaves; low to moderate annual precipitation (20–100 cm/year)
Grasslands (savanna prairie, steppe)	Called prairies in North America, savannas in Africa, pampas in South America, steppes in Europe; hot in summer and cold in winter (United States); moderate annual precipitation; good soil for agriculture
Deserts	Northern and Southern Hemispheres at 30° latitude; hot (38°C) days and cold (7°C) nights; low annual precipitation (less than 25 cm)

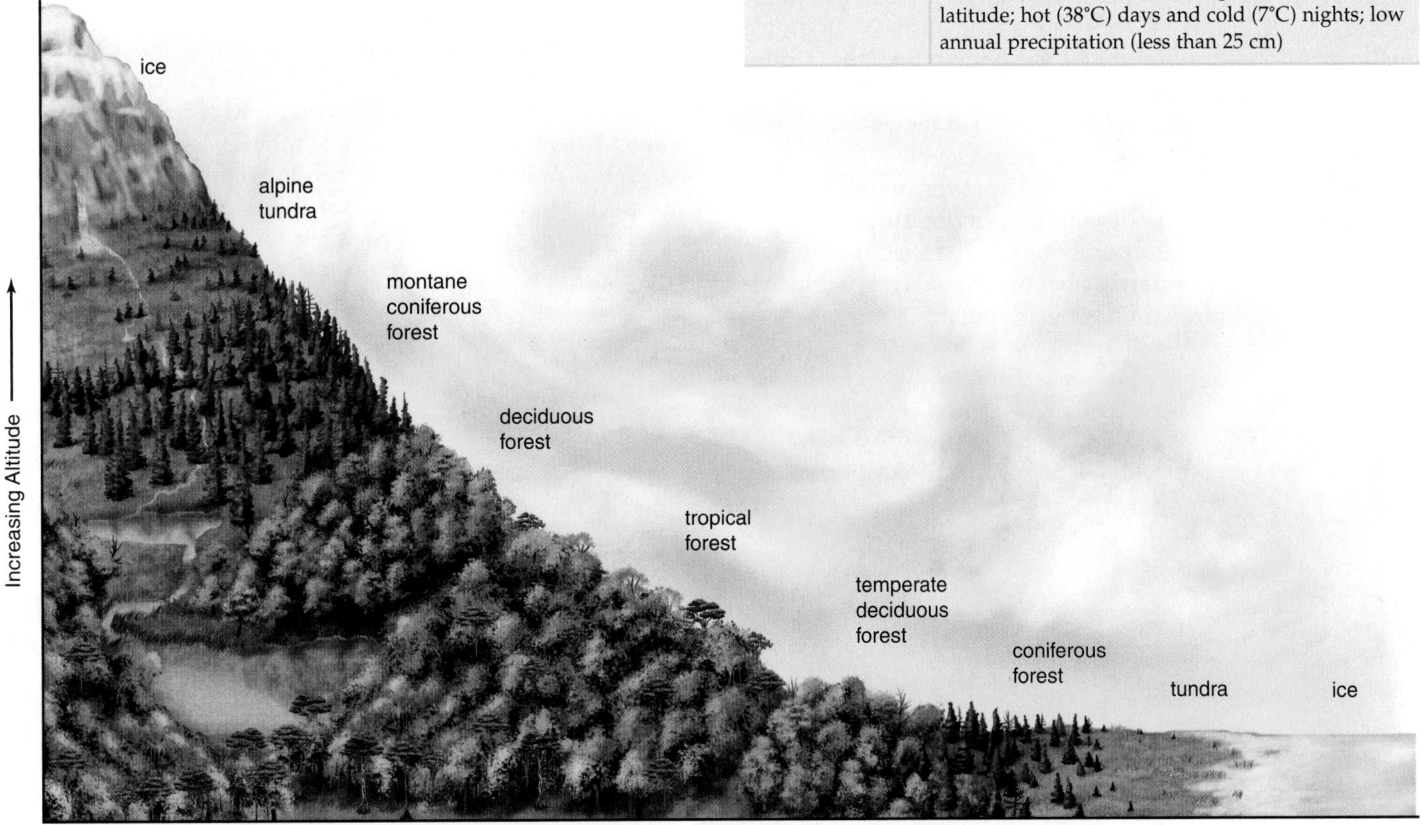

Figure 35.5 Climate and terrestrial biomes.
Biomes change with altitude just as they do with latitude because vegetation is partly determined by temperature. Precipitation also plays a significant role, which is one reason grasslands, instead of tropical or deciduous forests, are sometimes found at the base of mountains.

a. Tundra

b. Bull caribou

Figure 35.6 Tundra.
a. Vegetation consists principally of lichens, mosses, grasses, and low-growing shrubs. **b.** Caribou feed on the vegetation in the summer.

Tundra

The **tundra** biome, which encircles the arctic region just south of the ice-covered polar seas in the Northern Hemisphere, covers about 20% of Earth's land surface. As previously mentioned, a similar ecosystem, called the alpine tundra, occurs above the timberline on mountain ranges. The *arctic tundra* is cold and dark much of the year. Because rainfall amounts to only about 20 cm a year, the tundra could possibly be considered a desert. However, melting snow creates a landscape of pools and bogs in the summer, especially because so little evaporates. Only the topmost layer of earth thaws. The **permafrost** beneath this layer is always frozen, and therefore, drainage is minimal.

Trees are not found in the tundra because the growing season is too short, their roots cannot penetrate the permafrost, and they cannot become anchored in the shallow, boggy soil of summer. Instead, the ground is covered with short grasses and sedges, as well as numerous patches of lichens and mosses in summer (Fig. 35.6*a*). Dwarf woody shrubs, such as dwarf birch, flower and seed quickly while there is plentiful sun for photosynthesis.

A few animals live in the tundra year-round, but nearly all species have adaptations for living in extreme cold and short growing seasons. In winter, for example, the ptarmigan (a grouse) burrows in the snow during storms, and the musk ox conserves heat because of its thick coat and short, squat body. Other species, such as caribou (Fig. 35.6*b*) and

a. Taiga

Figure 35.7 Coniferous forest.
a. The taiga, which means swampland, is a coniferous forest that spans northern Europe, Asia, and North America. **b.** The appellation "spruce-moose" refers to the dominant presence of spruce trees and moose, which frequent the ponds.

b. Bull moose

reindeer also migrate to and from the tundra, as do the wolves that prey upon them. In the summer, the tundra is alive with numerous insects and birds, particularly shorebirds and waterfowl that migrate inland.

Coniferous Forests

Coniferous forests are found in three locations: in the **taiga,** which extends around the world in the northern part of North America and Eurasia; near mountaintops (where it is called a montane coniferous forest); and along the Pacific coast of North America, as far south as northern California.

Taiga typifies the coniferous forest with its cone-bearing trees, such as spruce, fir, and pine (Fig. 35.7*a*). These trees are well adapted to the cold because both the leaves (reduced to needles) and bark have thick coverings. Also, the needlelike leaves can withstand the weight of heavy snow. There is a limited understory of plants, but the floor is covered by low-lying mosses and lichens beneath a layer of needles. Birds harvest the seeds of the conifers, and bears, deer, moose, beaver, and muskrat live around the cool lakes and along the streams (Fig. 35.7*b*). Wolves prey on these larger mammals. A montane coniferous forest also harbors the wolverine and the mountain lion.

The coniferous forest that runs along the west coast of Canada and the United States is sometimes called a *temperate rain forest.* The plentiful rainfall and rich soil have produced some of the tallest conifer trees ever known, including the coastal redwoods. Small sections of this forest are considered old-growth forest because trees average over 150 years old, with some trees even as old as 800 years. It truly is an evergreen forest because mosses, ferns, and other plants often grow on tree trunks. Whether the limited portion of old-growth forest that remains should be protected from logging is an important and controversial conservation issue.

Temperate Deciduous Forests

Temperate deciduous forests are found south of the taiga in eastern North America, eastern Asia, and much of Europe. The climate in these areas is moderate, with relatively high rainfall (75–150 cm per year). The seasons are well defined, and the growing season ranges between 140 and 300 days. The trees, such as oak, beech, and maple, have broad leaves and are deciduous because they lose their leaves in fall and regrow them in spring.

The tallest trees form a canopy, an upper layer of leaves that are the first to receive sunlight and thereby create shade below (Fig. 35.8). Even so, enough sunlight penetrates to provide energy for another layer of trees, called understory trees. Beneath these trees are shrubs and herbaceous

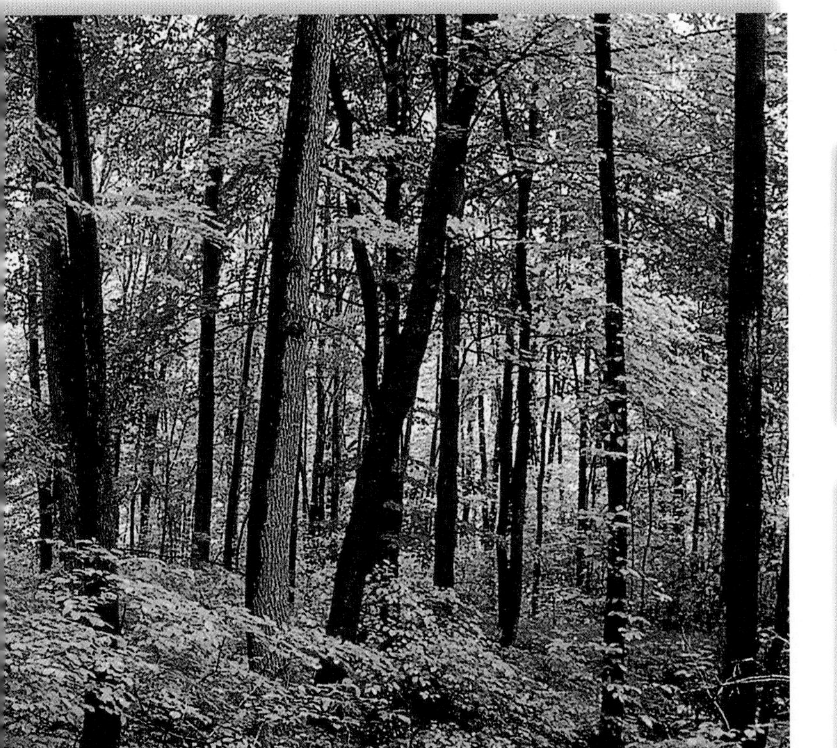

Figure 35.8 Temperate deciduous forest.
A temperate deciduous forest is home to many varied plants and animals. Millipedes can be found among leaf litter, chipmunks feed on acorns, and bobcats prey on these and other small mammals.

millipede

eastern chipmunk

bobcat

plants that may flower in the spring before the trees have put forth their leaves. Mosses, lichens, and ferns then reside beneath the shrub layer. This stratification provides a variety of habitats for insects and birds. Ground life is also plentiful. Squirrels, cottontail rabbits, shrews, skunks, woodchucks, and chipmunks are small herbivores. These and ground birds such as turkeys, pheasants, and grouse are preyed on by red foxes. White-tail deer and black bears have recently increased in number. In contrast to the taiga, a greater diversity of amphibians and reptiles occur in this biome because the winters are not as cold. Frogs and turtles generally prefer an aquatic existence, as do the beaver and muskrat, which are mammals.

Autumn fruits, nuts, and berries provide a supply of food for the winter, and the leaves, after turning brilliant colors and falling to the ground, contribute to the rich layer of humus after decomposition. The minerals within the rich soil are washed far into the ground by spring rains, but the deep tree roots capture these nutrients and cycle them back through the forest system.

Tropical Forests

The most common type of **tropical forest** is the *tropical rain forest,* which is found in areas of South America, Africa, and the Indo-Malayan region near the equator. The weather in a tropical rain forest is always warm (between 20° and 25°C), and rainfall is plentiful (a minimum of 190 cm per year). As a result of these favorable climate conditions, this is likely the richest biome in terms of both number of species and their abundance. The diversity of species is enormous—a 10-km^2 area of tropical rain forest may contain 750 species of trees and 1,500 species of flowering plants.

A tropical rain forest has a complex structure, with many levels of life. Some broadleaf evergreen trees grow to heights of 15–50 m or more. These tall trees often have trunks buttressed at ground level to prevent them from toppling over. Lianas, or woody vines, which often encircle rainforest trees as they grow, also help strengthen the trunk.

Although some animals live on the ground (e.g., pacas, agoutis, peccaries, and armadillos), many also live in the trees (Fig. 35.9). Insect life is so abundant that the majority of species have not yet been identified. Termites play a vital role in the decomposition of woody plant material, and ants are found everywhere. The various birds, such as hummingbirds, parakeets, parrots, and toucans, are often beautifully colored. Amphibians and reptiles are well represented by many types of frogs, snakes, and lizards. Lemurs, sloths, and monkeys are well-known primates that feed on the fruits of the trees. The largest carnivores are the big cats—the jaguars in South America and the leopards in Africa and Asia.

Many animals spend their entire life in the canopy, as do some plants. **Epiphytes** are plants that grow on other plants but usually have roots of their own that absorb

Figure 35.9 Representative animals of the tropical rain forests of the world.

poison-dart frog

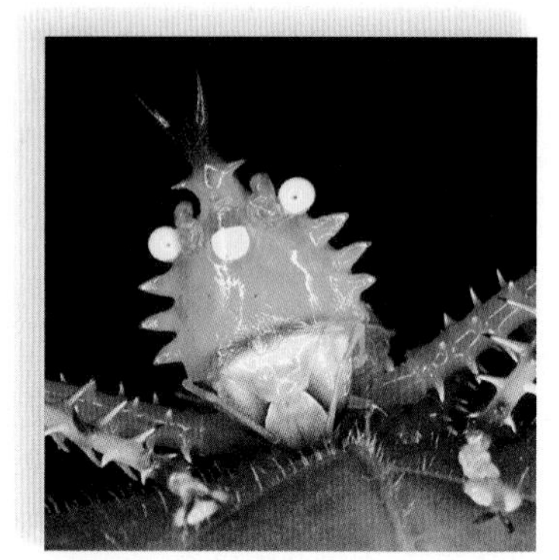
spike-headed katydid

ocelot

blue and gold macaw

blue morpho butterfly

lemur

chameleon

a. California chaparral

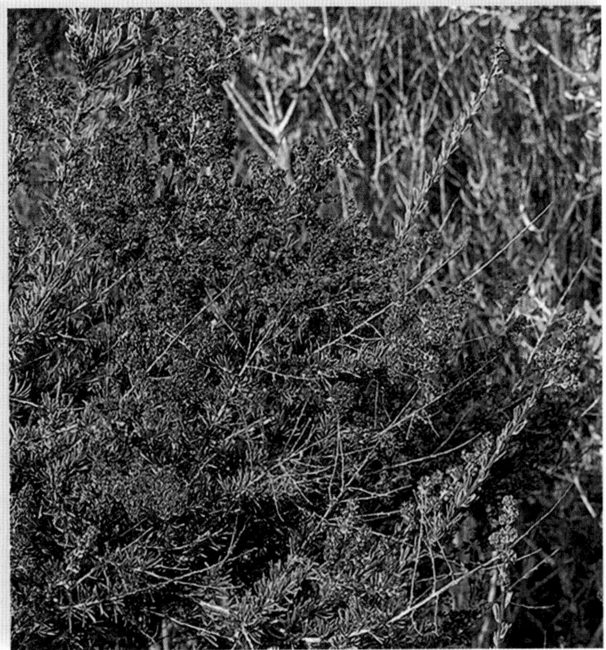

b. Chemise

Figure 35.10 Shrublands.
Shrublands, such as the chaparral in California **(a)**, are subject to raging fires, but the shrubs, such as chemise in **(b)**, are adapted to quickly regrow.

moisture and minerals leached from the canopy. Others catch rain and debris in hollows produced by overlapping leaf bases. The most common epiphytes are related to pineapples, orchids, and ferns.

Whereas the soil of a temperate deciduous forest biome is rich enough for agricultural purposes, the soil of a tropical rain forest biome is not. Nutrients are cycled directly from the litter to the plants again. Productivity is high because of warm and consistent temperatures, a year-long growing season, and the rapid recycling of nutrients from litter decomposition. To make up for the soil's low fertility, trees are sometimes cut and burned, so that the resulting ashes can provide enough nutrients for several harvests. This practice, called swidden agriculture or slash-and-burn agriculture, has been successful yet destructive in the tropics. Once the crops are harvested, the soil nutrients become depleted, consequently removing nutrients from the system. Furthermore, soil erodes due to lack of tree roots, washing away additional nutrients with it.

While we usually think of tropical forests as nonseasonal rain forests, *tropical deciduous forests* that have wet and dry seasons are found in India, Southeast Asia, West Africa, South and Central America, the West Indies, and northern Australia. They also have deciduous trees, and some of these forests also contain elephants, tigers, and hippopotamuses in addition to the animals mentioned previously.

Check Your Progress

1. Why do you think a tropical rain forest has higher biodiversity than the taiga?
2. Contrast the vegetation of the tropical rain forest with that of a temperate deciduous forest.

Shrublands

Shrublands tend to occur along coasts that have dry summers and receive most of their rainfall in winter. They are characterized by shrubs that have small but thick evergreen leaves often coated with a waxy material that prevents loss of moisture. Shrubs are adapted to withstand arid conditions and can also quickly regrow after a fire. In fact, the seeds of many species require the heat and scarring action of fire to induce germination. Other shrubs sprout from the roots after a fire. Typical shrubland species include coyotes, jackrabbits, gophers, and other rodents, as well as fire-adapted plant species such as manzanita. The dense shrubland that occurs in California is known as **chaparral** (Fig. 35.10). This type of shrubland, called Mediterranean, lacks an understory and ground litter, and is highly flammable.

Grasslands

Grasslands occur where rainfall is greater than 25 cm but generally insufficient to support trees. For example, it is too dry for forests and too wet for deserts to form in temperate areas where rainfall is between 10 and 30 inches. Natural grasslands once covered more than 40% of Earth's land surface, but most areas that were once grasslands are now used to grow crops such as wheat and corn.

Grasses are well adapted to a changing environment and can tolerate some grazing, as well as flooding, drought, and sometimes fire. Where rainfall is high, large tall grasses that reach more than 2 m in height (e.g., pampas grass) can flourish. In drier areas, shorter grasses between 5 and 10 cm are dominant (e.g., grama grass). The growth of grasses is also seasonal. As a result, some grassland animals such as bison migrate and ground squirrels hibernate, when there is little grass for them to eat.

The temperate grasslands include the Russian *steppes,* the South American *pampas,* and the North American *prairies* (Fig. 35.11). Large herds of bison—estimated at hundreds of thousands—once roamed the prairies, as did herds of pronghorn antelope. Now, small mammals, such as mice, prairie dogs, and rabbits, typically live below ground but usually feed above ground. Hawks, snakes, badgers, coyotes, and foxes feed on these mammals.

Savannas

Savannas, which are grasslands that contain some trees, occur in regions where a relatively cool dry season is followed by a hot, rainy season. One tree that can survive the severe dry season is the flat-topped acacia, which sheds its leaves during a drought. The African savanna supports the greatest variety and number of large herbivores of all the biomes (Fig. 35.12). Elephants and giraffes are browsers that feed on tree vegetation. Antelopes, zebras, wildebeests, water buffalo, and rhinoceroses are grazers that feed on grasses. Any plant litter that is not consumed by grazers is attacked by a variety of small organisms, among them termites. Termites build towering mounds in which they tend fungal gardens that they use for food. The herbivores support a large population of carnivores. Lions and hyenas sometimes hunt in packs, cheetahs hunt singly by day, and leopards hunt singly by night.

Deserts

As discussed in section 35.1, **deserts** are usually found at latitudes of about 30°, in both the Northern and Southern Hemispheres. The winds that descend in these regions lack moisture, and annual rainfall is less than 25 cm. Days are hot because a lack of cloud cover allows the sun's rays to penetrate easily, but nights are cold because heat escapes easily into the atmosphere.

Tall-grass prairie

American bison

Figure 35.11 The prairie.
Tall-grass prairies are seas of grasses dotted by pines and junipers. Bison, once abundant, are now being reintroduced into certain areas.

cheetah

giraffe

wildebeest

zebra (in foreground)

Figure 35.12 The savanna.
The African savanna varies from grassland to widely spaced shrubs and trees. This biome supports a large assemblage of herbivores, including zebras, wildebeests, giraffes, and gazelles. Carnivores, such as the cheetah, prey on these.

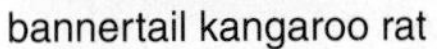

Figure 35.13 **Deserts.**
Plants and animals that live in a desert are adapted to arid conditions. The plants are either succulents, which retain moisture, or shrubs with woody stems and small leaves, which lose little moisture. The kangaroo rat feeds on seeds and other vegetation. The roadrunner preys on insects, lizards, and snakes. The kit fox is a desert carnivore.

The Sahara—which stretches all the way from the Atlantic coast of Africa to the Arabian Peninsula—and a few other deserts have little or no vegetation. But most deserts have a variety of plants (Fig. 35.13). The best-known desert perennials in North America are the succulent, spiny-leafed cacti, which have stems that store water and carry on photosynthesis. Also common are nonsucculent shrubs, such as the many-branched sagebrush and the spiny-branched ocotillo that produces leaves during wet periods and sheds them during dry periods.

Some animals are adapted to the desert environment. For example, many desert animals are nocturnal. They are active at night when it is cooler. Reptiles and insects have waterproof outer coverings that conserve water. A desert has numerous insects, which pass through the stages of development from pupa to pupa again while there is rain. Reptiles, especially lizards and snakes, are a characteristic group of vertebrates found in deserts, but running birds (e.g., the roadrunner) and rodents (e.g., the kangaroo rat) are also well known (see Fig. 35.13). Coyotes and hawks prey on the rodents.

Check Your Progress

1. Which biome is most likely to have the fewest species? Why?

35.3 Aquatic Ecosystems

Learning Outcomes

1. Describe the key features of a freshwater ecosystem.
2. Explain why coastal ecosystems are so productive.
3. Discuss the differences between the pelagic and benthic zones in an ocean ecosystem.

Aquatic ecosystems are classified as either of two types: freshwater (inland) or saltwater (usually marine). Brackish water, however, is a mixture of fresh and salt water. In this section, we describe lakes as examples of freshwater ecosystems and oceans as examples of saltwater ecosystems. Coastal ecosystems will represent areas of brackish water.

Lakes

Lakes are bodies of fresh water often classified by their nutrient abundance. Oligotrophic (nutrient-poor) lakes are characterized by a small amount of organic matter and low

a.

b.

Figure 35.14 Types of lakes.
Lakes can be classified according to whether they are **(a)** oligotrophic (nutrient-poor) or **(b)** eutrophic (nutrient-rich). Eutrophic lakes tend to have large populations of algae and rooted plants, resulting in a large population of decomposers that use up much of the oxygen and leave little oxygen for fishes.

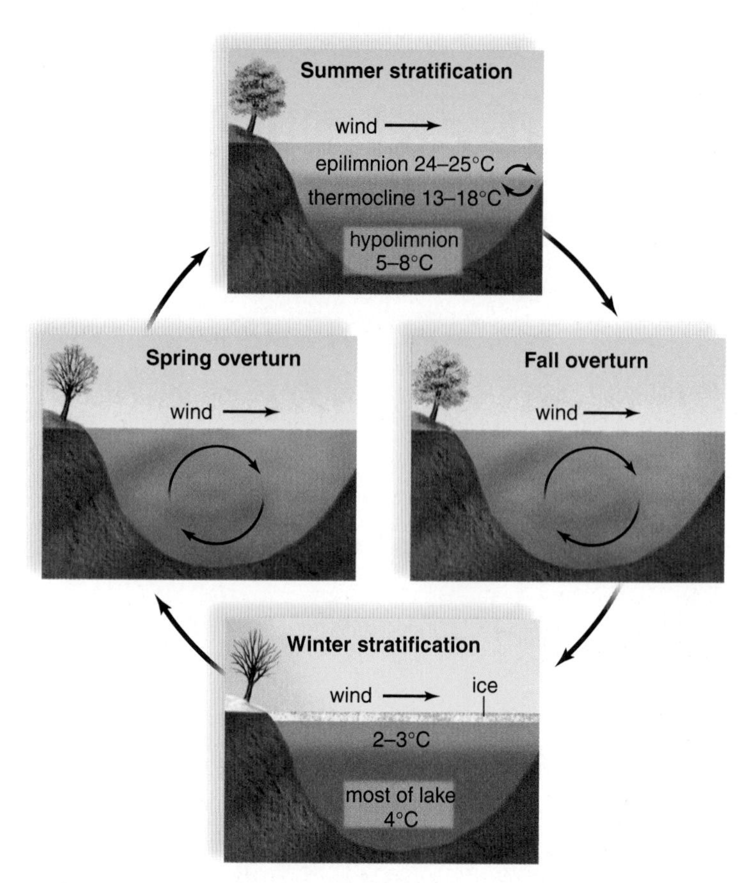

Figure 35.15 Lake stratification in a temperate region.
Temperature profiles of a large oligotrophic lake in a temperate region vary with the season. During the spring and fall overturns, the deep waters receive oxygen from surface waters, and surface waters receive inorganic nutrients from deep waters.

productivity (Fig. 35.14*a*). Eutrophic (nutrient-rich) lakes typically have plentiful organic matter and high productivity (Fig. 35.14*b*). Such lakes are usually situated in naturally nutrient-rich regions or are enriched by agricultural or urban and suburban runoff. Oligotrophic lakes can become eutrophic through large inputs of nutrients, a process called **eutrophication.** Excess nutrient inputs, such as nitrogen fertilizer, can cause eutrophication and fish kills. See the Bioethical Focus in this chapter on page 737 for a discussion of freshwater pollution legislation.

In the temperate zone, deep lakes are stratified in the summer and winter. In summer, lakes in the temperate zone have three layers of water that differ in temperature (Fig. 35.15). The surface layer, the epilimnion, is warm due to solar radiation; the middle thermocline layer experiences an abrupt drop in temperature; and the lowest layer, the hypolimnion, is cold. These differences in temperature prevent mixing. The warmer, less dense water of the epilimnion "floats" on top of the colder, more dense water of the hypolimnion.

As the season progresses, the epilimnion becomes nutrient-poor, while the hypolimnion begins to be depleted of oxygen. Phytoplankton (see Chapter 28) found in the sunlit epilimnion use up nutrients as they photosynthesize. Photosynthesis releases oxygen, giving this layer a ready supply. Detritus naturally falls by gravity to the bottom of the lake, and here oxygen is used up as decomposition occurs. Decomposition releases nutrients, however.

In the fall, as the epilimnion cools, and in the spring, as it warms, an overturn occurs. In the fall, the upper epilimnion waters become cooler than the hypolimnion waters. This causes the surface water to sink and the deep water to rise. The **fall overturn** continues until the temperature is uniform throughout the lake. At this point, wind helps circulate the water so that mixing occurs.

As winter approaches, the water cools further. Ice formation begins at the top, and the ice floats because ice is less dense than cold water. Ice has an insulating effect, preventing further cooling of the water below. This permits aquatic organisms to live through the winter in the water beneath the surface of the ice.

In the spring, as the ice melts, the cooler water on top sinks below the warmer water below it. The **spring overturn** continues until the temperature is uniform through the lake. At this point, wind aids in the circulation of water as before. When the surface waters absorb solar radiation, thermal stratification occurs once more.

This vertical stratification and seasonal change of temperatures in a lake basin influence the seasonal distribution of fish and other aquatic life. For example, coldwater fish move to the deeper water in summer and inhabit the upper water in winter. In the fall and spring just after mixing occurs, phytoplankton growth at the surface is most abundant.

Life Zones

In both fresh and salt water, free-drifting microscopic **plankton,** which includes phytoplankton and also zooplankton (see Chapter 28), are important components of the ecosystem. Lakes and ponds can be divided into several life zones, as shown in Figure 35.16. Aquatic plants are rooted in the shallow littoral zone of a lake, and various microscopic organisms cling to these plants and to rocks. Some organisms, such as the water strider, live at the water-air interface and can literally walk on water. In the limnetic zone, small fishes, such as minnows and killifish, feed on plankton and also serve as food for large fishes. In the profundal zone, zooplankton and fishes, such as whitefish, feed on debris that falls from above. Pike species are an example of "lurking predators." They wait among vegetation around the margins of lakes and surge out to catch passing prey.

A few insect larvae are in the limnetic zone, but they are far more abundant in both the littoral and profundal zones. Midge larvae and ghost worms are common members of the benthos, animals that live on the bottom in the benthic zone. In a lake, the benthos includes crayfishes, snails, clams, and various types of worms and insect larvae.

Check Your Progress

1. Is an oligotrophic or eutrophic lake more likely to support higher biodiversity? Why?
2. How do fall and spring overturn affect lake stratification?

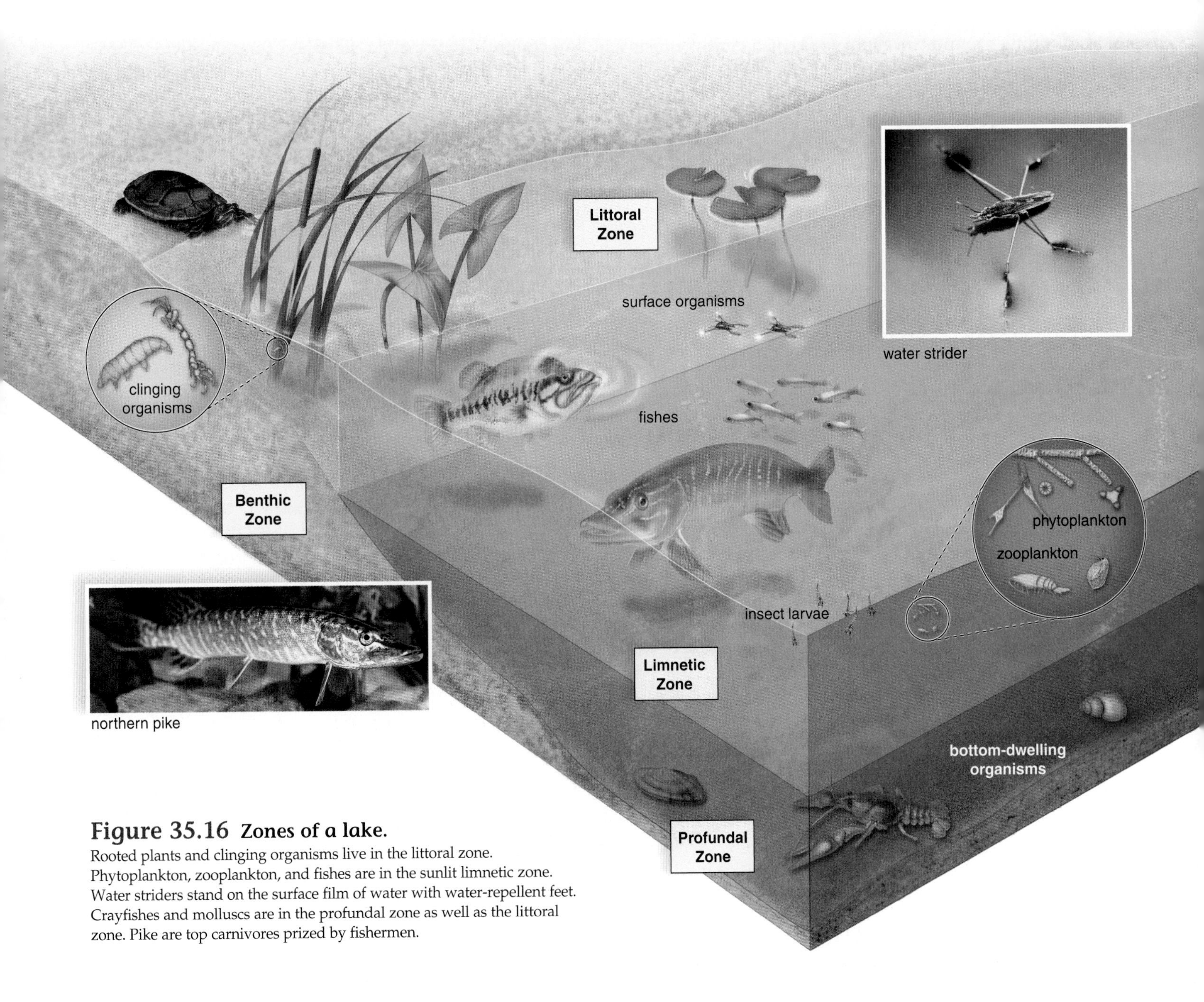

Figure 35.16 Zones of a lake.
Rooted plants and clinging organisms live in the littoral zone. Phytoplankton, zooplankton, and fishes are in the sunlit limnetic zone. Water striders stand on the surface film of water with water-repellent feet. Crayfishes and molluscs are in the profundal zone as well as the littoral zone. Pike are top carnivores prized by fishermen.

Coastal Ecosystems

Estuaries

An **estuary** is a partially enclosed body of water where fresh water and salt water meet and mix (Fig. 35.17 and see Fig. 35.18 for examples). Coastal bays, tidal marshes, fjords (an inlet of water between high cliffs), some deltas (a triangular-shaped area of land at the mouth of a river), and lagoons (a body of water separated from the sea by a narrow strip of land) are all examples of estuaries.

Organisms living in an estuary must be able to withstand constant mixing of waters and rapid changes in salinity. Not many organisms are adapted to this environment, but those that are suited find an abundance of nutrients. An estuary acts as a nutrient trap because the sea prevents the rapid escape of nutrients brought by a river. It has been estimated that well over half of all marine fishes develop in the protective environment of an estuary, which explains why estuaries are called the nurseries of the sea. Estuaries are also feeding grounds for many birds, fish, and shellfish because they offer a ready supply of food.

Salt marshes, dominated by salt marsh cordgrass, are often associated with estuaries. So are mudflats and mangrove swamps, where sediment and nutrients from the land collect (Fig. 35.18).

Seashores

Seashores, which may be rocky or sandy, are constantly bombarded by the sea as the tides roll in and out (Fig. 35.19). The **littoral zone** lies between the high- and low-tide marks. The littoral zone of a rocky beach is divided into subzones. In the upper portion of the littoral zone, barnacles are glued so tightly to stone by their own secretions that their calcareous outer plates remain in place even after the enclosed shrimplike animal dies. In the midportion of the littoral zone, brown algae known as rockweed may overlie the barnacles. In the lower portions of the littoral zone, oysters and mussels attach themselves to the rocks by filaments called *byssal threads*. Also present are snails called limpets and periwinkles. Periwinkles have a coiled shell and secure themselves by hiding in crevices or under seaweeds, while limpets press their single, flattened cone tightly to a rock. Below the littoral zone, macroscopic seaweeds, which are the main photosynthesizers, anchor themselves to the rocks by holdfasts.

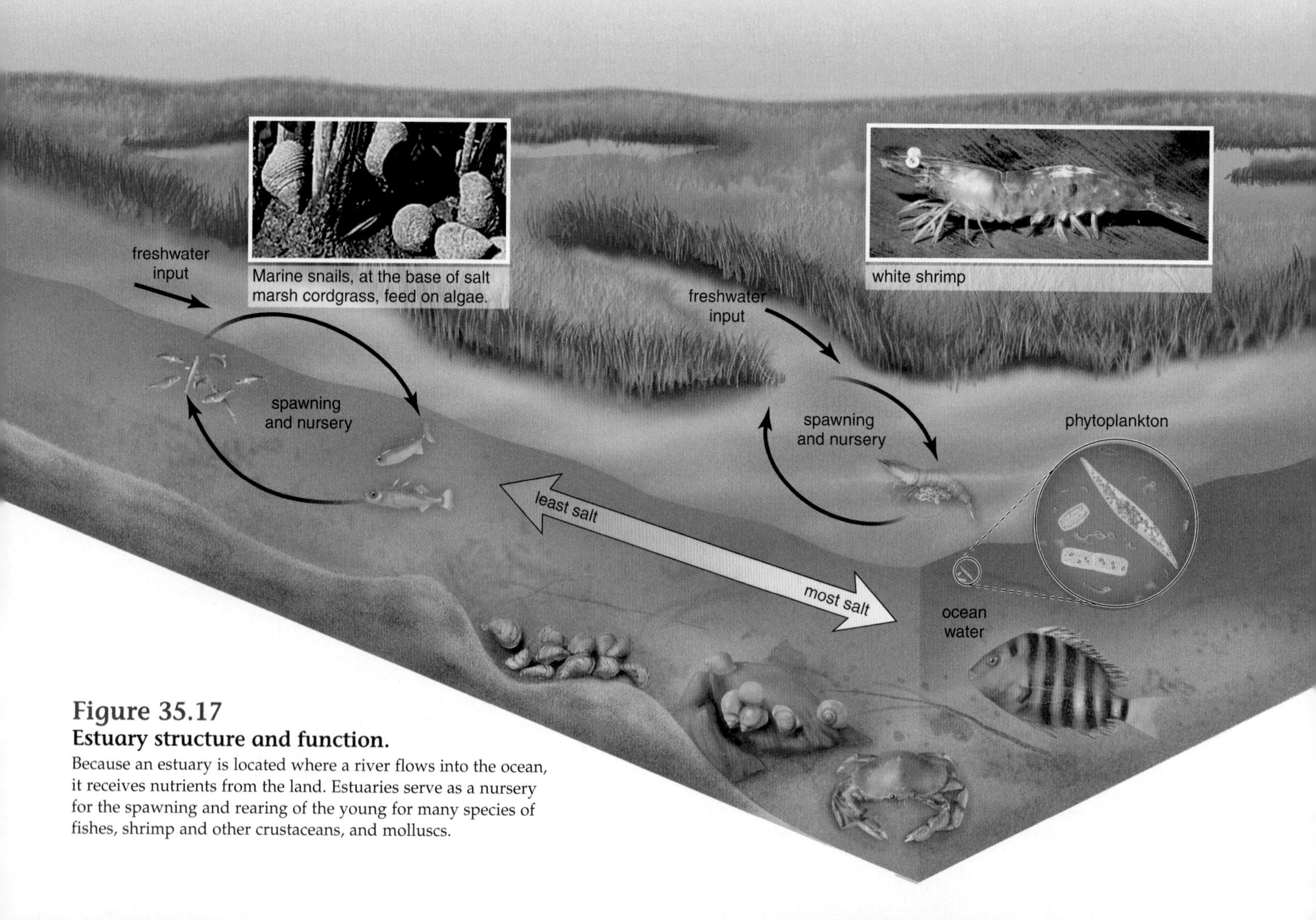

Figure 35.17
Estuary structure and function.
Because an estuary is located where a river flows into the ocean, it receives nutrients from the land. Estuaries serve as a nursery for the spawning and rearing of the young for many species of fishes, shrimp and other crustaceans, and molluscs.

a.

b.

Figure 35.18 Types of estuaries.
Types of estuaries include (**a**) mudflats, which form at the mouth of rivers in temperate zones and are frequented by migrant birds, and (**b**) mangrove swamps, which skirt the coastlines of many tropical and subtropical lands. The tangled roots of mangrove trees trap sediments and nutrients that sustain many immature forms of sea life. The salt marsh depicted in Figure 35.17 is yet another type of estuary-associated system.

a. Sea stars at low tide

b. Shorebirds

Figure 35.19 Seashores.
a. Some organisms of a rocky coast live in tidal pools. **b.** A sandy shore looks devoid of life except for the birds that feed there.

Organisms cannot attach themselves to shifting, unstable sands on a sandy beach. Therefore, nearly all the permanent residents dwell underground. They either burrow during the day and surface to feed at night, or they remain permanently within their burrows and tubes. Ghost crabs and sandhoppers (amphipods) burrow themselves above the high-tide mark and feed at night when the tide is out. Sandworms and sand (ghost) shrimp remain within their burrows in the littoral zone and feed on detritus whenever possible. Still lower in the sand, clams, cockles, and sand dollars are found.

Check Your Progress

1. Why are coastal ecosystems so productive and thus called the nurseries of the sea?

Oceans

Oceans cover approximately three-quarters of our planet. Climate is driven by the sun, but the oceans play a major role in redistributing heat in the biosphere. Water tends to be warm at the equator and much cooler at the poles because of the distribution of the sun's rays, as discussed earlier in this chapter (see Fig. 35.1*a*). Air takes on the temperature of the water below, and then warm air moves from the equator to the poles. In other words, the oceans have a major influence on wind patterns. (The landmasses also play a role, but water holds heat longer and remains cool longer during periods of changing temperature than land.)

When the wind blows strongly and steadily across a great expanse of ocean for a long time, friction from the moving air begins to drag the water along with it. Once the water has been set in motion, its momentum, aided by the wind, keeps it moving in a steady flow we call a current. Because the ocean currents eventually strike land, they move in a circular path—clockwise in the Northern Hemisphere and counterclockwise in the Southern Hemisphere. As the currents flow, they take warm water from the equator to the poles. One such current, called the Gulf Stream, brings tropical Caribbean water to the east coast of North America and the higher latitudes of western Europe. Without the Gulf Stream, Great Britain, which has a relatively warm temperature, would be as cold as Greenland. In the Southern Hemisphere, another major ocean current warms the eastern coast of South America.

Also, in the Southern Hemisphere, the Humboldt Current flows toward the equator. The Humboldt Current carries phosphorus-rich cold water northward along the west coast of South America. During a process called **upwelling,** cold offshore winds cause cold nutrient-rich waters to rise and take the place of warm nutrient-poor waters. In South America, the enriched waters nourish an abundance of marine life that supports the fisheries of Peru and northern Chile. When the Humboldt Current is not as cool as usual, upwelling does not occur, stagnation results, the fisheries decline, and climate patterns change globally. This phenomenon is discussed later in the Ecology Focus, "El Niño–Southern Oscillation."

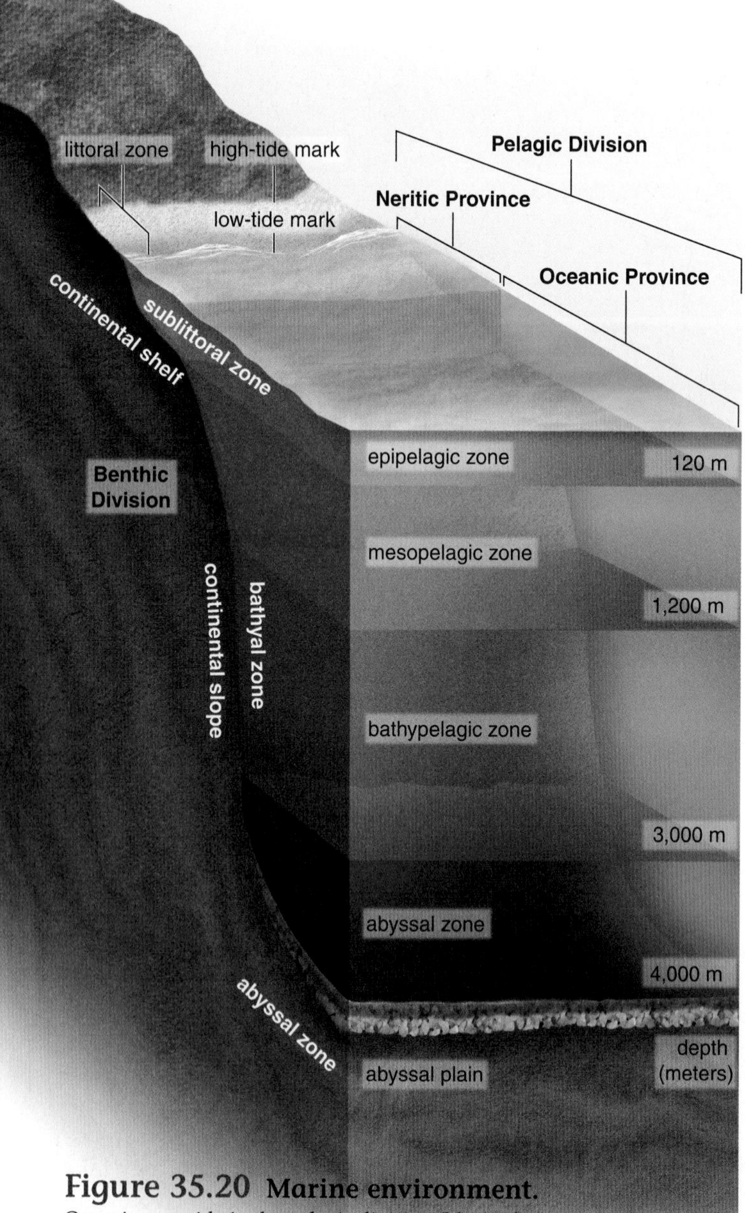

Figure 35.20 Marine environment.
Organisms reside in the pelagic division (blue) where waters are divided as indicated. Organisms also reside in the benthic division (brown) in the surfaces and zones indicated.

Pelagic Division

It is customary to place the organisms of the oceans into either the pelagic division (open waters) or the benthic division (ocean floor). The geographic areas and zones of an ocean are shown in Figure 35.20, and the animals that inhabit the different zones are shown in Figure 35.21. The **pelagic division** includes the neritic province and the oceanic province.

Neritic Province The neritic province hosts a greater concentration of organisms than the oceanic province because the neritic province is sunlit and also has a supply of inorganic nutrients for photosynthesizers. Phytoplankton, consisting of suspended algae, is food not only for zooplankton but also for small fishes. These small fishes, in turn, are food for commercially valuable fishes.

Coral reefs are areas of high biological abundance and productivity found just below the surface of shallow, warm, tropical waters. Their chief constituents are stony corals, animals that have a calcium carbonate (limestone) exoskeleton, and calcareous red and green algae. Corals do not usually occur individually. Rather, they form colonies derived from individual corals that have reproduced by means of budding. Corals provide a home for a microscopic alga called zooxanthella. The corals, which feed at night, and the algae, which photosynthesize during the day, are mutualistic and share materials and nutrients. The close relationship of corals and zooxanthellae is likely the reason coral reefs form only in shallow sunlit water—the algae need sunlight for photosynthesis.

A reef is densely populated with life. The large number of crevices and caves provide shelter for filter feeders (e.g., sponges, sea squirts, and fan worms) and for scavengers (e.g., crabs and sea urchins). There are many types of small, beautifully colored fishes. Parrotfishes feed directly on corals, and others feed on plankton or detritus. Small fishes become food for larger fishes, including snappers and other fish that are caught for human consumption. The barracuda, moray eel, and various shark species are top predators in coral reefs.

Oceanic Province The oceanic province lacks the inorganic nutrients of the neritic province and, therefore, does not have as high a concentration of phytoplankton, even though the epipelagic zone is sunlit. Still, the photosynthesizers are food for a large assembly of zooplankton, which then become food for herrings and bluefishes. These, in turn, are eaten by larger mackerels, tunas, and sharks. Flying fishes, which glide above the surface, are preyed upon by dolphins. Whales are other mammals found in the epipelagic zone (see Fig. 35.21). Baleen whales strain krill (small crustaceans) from the water, and toothed sperm whales feed primarily on the common squid.

Animals in the mesopelagic zone are carnivores, are adapted to the absence of light, and tend to be translucent, red in color, or even luminescent. There are luminescent shrimps, squids, and fishes, such as lantern and hatchet fishes.

The bathypelagic zone is in complete darkness except for an occasional flash of bioluminescent light. Carnivores and scavengers are found in this zone. Strange-looking fishes with distensible mouths and abdomens and small, tubular eyes feed on infrequent prey.

Benthic Division

The **benthic division** includes organisms that live on or in the soil of the continental shelf (sublittoral zone), the continental slope (bathyal zone), and the abyssal plain (abyssal zone) (see Fig. 35.20).

Seaweed grows in the sublittoral zone, and it can be found in batches on outcroppings as the water gets deeper. More diversity of life exists in the sublittoral and bathyal zones than in the abyssal zone. In the first two zones, clams, worms, and sea urchins are preyed upon by starfishes, lobsters, crabs, and brittle stars. Photosynthesizing algae occur in the sunlit sublittoral zone, but benthic organisms in the bathyal zone are dependent on the slow rain of detritus from the waters above.

The abyssal zone is inhabited by animals that live at the soil-water interface of the abyssal plain (see Fig. 35.21). It once was thought that few animals exist in this zone because of the intense pressure and the extreme cold. Yet many invertebrates live there by feeding on debris floating down from the mesopelagic zone. Sea lilies rise above the seafloor; sea cucumbers and sea urchins crawl around on the sea bottom; and tube worms burrow in the mud.

The flat abyssal plain is interrupted by enormous underwater mountain chains called oceanic ridges. Along the axes of the ridges, crustal plates spread apart, and molten magma from Earth's core rises to fill the gap. At **hydrothermal vents,** seawater percolates through cracks and is heated to about 350°C, causing sulfate to react with water and form hydrogen sulfide (H_2S). Free-living or mutualistic chemoautotrophic bacteria use energetic electrons derived from hydrogen sulfide to reduce bicarbonate to organic compounds. These compounds support an ecosystem that includes huge tube worms and clams. It was a surprise to find such communities of organisms living so deep in the ocean, where light never penetrates. Life is possible so deep because unlike photosynthesizers, chemoautotrophs are producers that do not require solar energy.

Check Your Progress

1. What are the two major oceanic divisions? Explain the major characteristics of each.

Figure 35.21 Ocean inhabitants.
Different organisms are characteristic of the epipelagic, mesopelagic, and bathypelagic zones of the pelagic division compared to the abyssal zone of the benthic division.

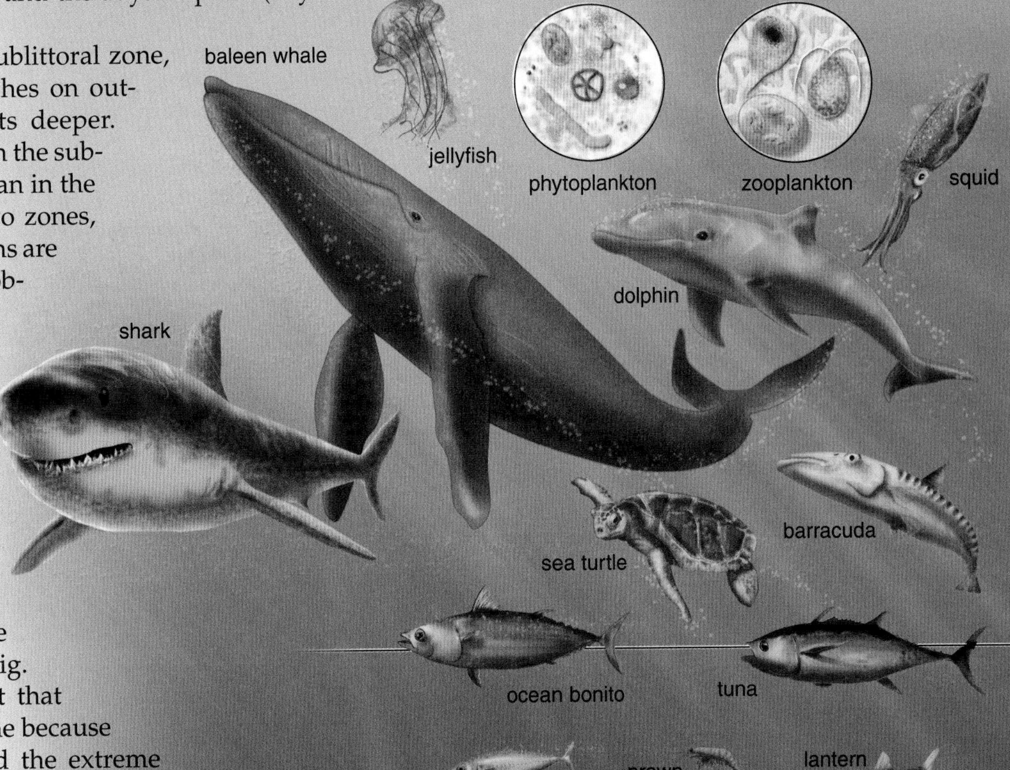

El Niño–Southern Oscillation

Climate largely determines the distribution of life on Earth. Short-term variations in climate, which we call weather, also have a pronounced effect on living things. There is no better example than an El Niño. Originally, El Niño referred to a warming of the seas off the coast of Peru at Christmastime—hence, the name El Niño, "the boy child," after Jesus.

Now scientists prefer the term **El Niño–Southern Oscillation (ENSO)** to denote a severe weather change brought on by an interaction between the atmosphere and ocean currents. Ordinarily, the southeast trade winds move along the coast of South America and turn west because of Earth's daily rotation on its axis. As the winds drag warm ocean waters from east to west, there is an upwelling of nutrient rich cold water from the ocean's depths, resulting in a bountiful Peruvian harvest of anchovies. When the warm ocean waters reach their western destination, the monsoons bring rain to India and Indonesia. Scientists have noted that these events correlate with a difference in the barometric pressure over the Indian Ocean and the southeastern Pacific—that is, the barometric pressure is low over the Indian Ocean and high over the southeastern Pacific. But when a "southern oscillation" occurs and the barometric pressures switch, an El Niño begins.

During an El Niño, both the northeast and the southeast trade winds slacken. Upwelling no longer occurs, and the anchovy catch off the coast of Peru plummets. During a severe El Niño, waters from the east never reach the west, and the winds lose their moisture in the middle of the Pacific instead of over the Indian Ocean. The monsoons fail, and drought occurs in India, Indonesia, Africa, and Australia. Harvests decline, cattle must be slaughtered, and famine is likely in highly populated India and Africa, where funds to import replacement supplies of food are limited. In 2009, severe drought caused fires that killed hundreds of people and caused billions of dollars in damage in the state of Victoria, Australia.

A backward movement of winds and ocean currents may even occur so that the waters warm to more than 14° above normal along the west coast of the Americas. This is a sign that a severe El Niño has occurred, and the weather changes are dramatic in the Americas also. Southern California is hit by storms and even hurricanes, and the deserts of Peru and Chile receive so much rain that flooding occurs. A jet stream (strong wind currents) can carry moisture into Texas, Louisiana, and Florida, with flooding a near certainty. Or the winds can turn northward and deposit snow in the mountains along the west coast so that flooding occurs in the spring. Some parts of the United States, however, benefit from an El Niño. The Northeast is warmer than usual, few if any hurricanes hit the east coast, and there is a lull in tornadoes throughout the Midwest. Altogether, a severe El Niño affects the weather over three-quarters of the globe.

Eventually, an El Niño dies out, and normal conditions return. The normal cold-water state off the coast of Peru is known as La Niña (the girl). Figure 35A contrasts the weather conditions of a La Niña with those of an El Niño. Since 1991, the sea surface has been almost continuously warm, and two record-breaking El Niños have occurred. What could be causing more of the El Niño state than the La Niña state? Some scientists argue that this environmental change is related to global warming, a rise in environmental temperature due to greenhouse gases in the atmosphere as discussed in Chapter 34. Greenhouse gases, including carbon dioxide, are released by humans in mass quantities into the atmosphere by, for example, burning fossil fuels. Like the glass of a greenhouse, these gases allow the sun's rays to pass through, but they trap the heat.

Discussion Questions

1. Recent research has suggested that global warming is contributing to the increased frequency and severity of El Niño events. This can cause severe drought in India and Africa, as well as greatly reduce fish catch in parts of the Southern Hemisphere. Should the responsibility for decreasing greenhouse gas emissions be greater for more-developed countries (MDCs), because they produce a greater amount of greenhouse gases per capita than less-developed countries (LDCs), even though El Niños have more severe consequences for LDCs?
2. Aside from the recent increases in gasoline prices, would you consider buying a more fuel-efficient car or spend more on plane flights to purchase "carbon credits" to do your part in decreasing the emission of greenhouse gases such as carbon dioxide? Why or why not?

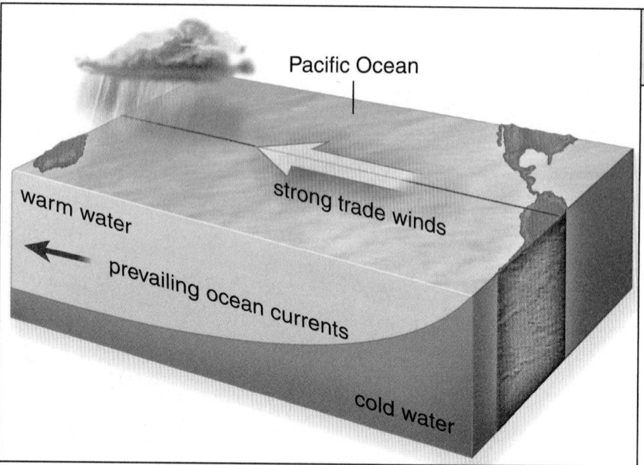

La Niña

- Upwelling off the west coast of South America brings cold waters to the surface.
- Barometric pressure is high over the southeastern Pacific.
- Monsoons associated with the Indian Ocean occur.
- Hurricanes occur off the east coast of the United States.

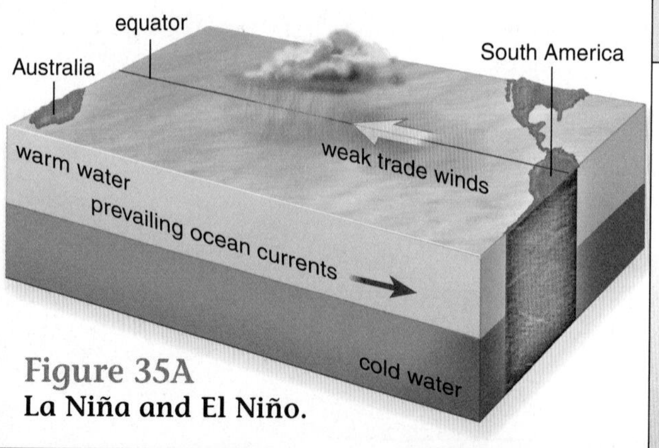

El Niño

- Great ocean warming occurs off the west coast of the Americas.
- Barometric pressure is low over the southeastern Pacific.
- Monsoons associated with the Indian Ocean fail.
- Hurricanes occur off the west coast of the United States.

Figure 35A
La Niña and El Niño.

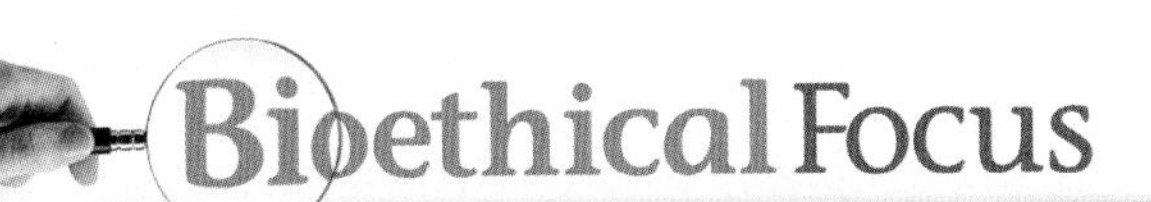

Pollution Legislation

Agricultural fertilizers are the chief cause of nitrate contamination of groundwater that eventually ends up in drinking-water wells. Excessive nitrates in a baby's bloodstream can lead to a slow suffocation known as blue-baby syndrome. In addition, agricultural herbicides are suspected carcinogens in the tap water of scattered communities coast to coast. What can be done?

Some farmers are already using irrigation methods that deliver water directly to plant roots, no-till agriculture that reduces the loss of topsoil and cuts back on herbicide use, and integrated pest management that relies heavily on good bugs to kill bad bugs. Perhaps more should do so. Encouraged—in some cases, compelled—by state and federal agents, dairy farmers have built sheds, concrete containments, and underground liquid storage tanks to hold the wastes from rainy days. Then, the manure can be trucked to fields and spread as fertilizer.

Like golf courses and ski resorts, home owners also contribute to the problem. Manicuring lawns, using motor vehicles, and constructing and using roads and buildings all add contaminants to streams, lakes, and aquifers. Citizens around Grand Traverse Bay on the eastern shore of Lake Michigan have gotten the message, especially because they want to keep on enjoying water-dependent activities such as boating, swimming, and fishing. James Haverman, a concerned member of the Traverse Bay Watershed Initiative, says aptly, "If we can't change the way people live their everyday lives, we are not going to be able to make a difference." Builders in Traverse County are already required to control soil erosion with filter fences, steer rainwater away from exposed soil, build sediment basins, and plant protective buffers. Home owners must have a 25-ft setback from wetlands and a 50-ft setback from lakes and creeks. They are also encouraged to pump out their septic systems every two years.

Form Your Own Opinion

1. Do you approve of legislation that requires farmers, businesses, and home owners to protect freshwater supplies? Why or why not?
2. Should landscapers and gardeners be required to follow restrictions on their use of fertilizers, herbicides, and pesticides? Why or why not?

Applying the Concepts [Revisited]

In the beginning of this chapter, we took a look at Earth as we approached from space. As our planet came into view, we saw the landmasses and the oceans. Now, we have learned how to distinguish the different types of terrestrial and aquatic biomes distributed throughout the biosphere.

1. Highlight the major differences between aquatic and terrestrial biomes.
2. Explain why coral reefs are some of the most diverse, yet most fragile, ecosystems on Earth. Hypothesize why nearby logging that causes deposits of silt (dirt runoff) in coastal areas is thought to be a major cause of coral reef decline.

SUMMARIZING THE CONCEPTS

35.1 Climate and the Biosphere

- The sun's rays are spread out over a larger area at the poles than the vertical rays at the equator because Earth is spherical. Thus, surface temperatures decrease from the equator toward each pole.
- The planet is tilted on its axis, and the seasons change depending on which hemisphere is closest to the sun as Earth completes its annual rotation.
- Warm air rises near the equator, loses its moisture, and then descends at about 30° north and south latitude, and continuing toward the poles. When the air descends, it retains moisture, and therefore, the great deserts of the world are formed at 30° latitudes.
- Because Earth rotates on its axis daily, the winds blow in opposite directions above and below the equator.
- Air rising over coastal ranges loses its moisture on the windward side, making the leeward side drier.

35.2 Terrestrial Ecosystems

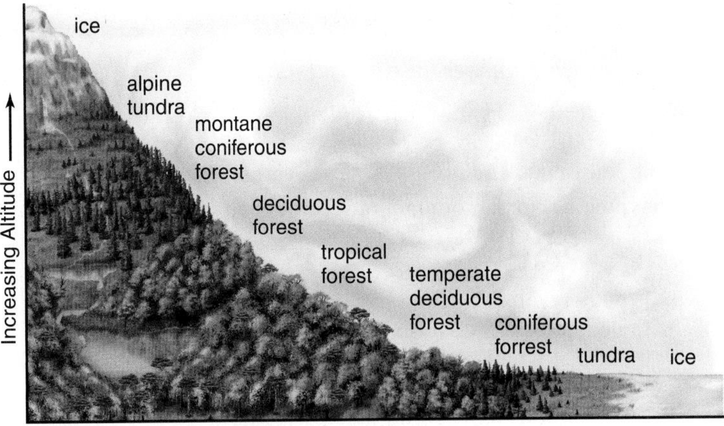

- A biome is a major type of terrestrial ecosystem.
- Temperature and rainfall influence the distribution of biomes around the world.
- The tundra is the northernmost biome and consists largely of short grasses and sedges and dwarf woody plants. Most of the water in the soil is frozen year-round (i.e., permafrost) because of cold winters and short summers.
- The taiga, a coniferous forest, has less rainfall than other types of forests.
- The temperate deciduous forest has trees that gain and lose their leaves during spring and fall, respectively.
- Tropical forests are the most complex and productive of all biomes.
- Shrublands usually occur along coasts that have dry summers and receive most of their rainfall in the winter.
- Among grasslands, the savanna, a tropical grassland, supports the greatest number of different types of large herbivores. Prairies, such as those in the mid United States, have a limited variety of vegetation and animal life.
- Deserts are characterized by rainfall less than 25 cm a year. Plants, such as cacti, are succulents, and others are shrubs with thick leaves—both of which are adaptations to conserve water.

35.3 Aquatic Ecosystems

- In deep temperate zone lakes, water depth determines the temperature and the concentrations of nutrients and gases. During spring and fall turnover, nutrients from the bottom lake layers are redistributed while the entire body of water is cycled.
- Lakes and ponds have three life zones. Rooted plants and clinging organisms live in the littoral zone, plankton and fishes live in the sunlit limnetic zone, and bottom-dwelling organisms, such as crayfishes and molluscs, live in the profundal zone.
- Marine ecosystems are divided into coastal ecosystems and the oceans.
- The coastal ecosystems, especially estuaries, are more productive than the oceans. Estuaries (e.g., salt marshes, mudflats, and mangrove forests) are near the mouth of a river and are called the nurseries of the sea.
- An ocean is divided into the pelagic division and the benthic division. The pelagic division (open waters) has three zones. The epipelagic zone receives adequate sunlight and supports the most life. The mesopelagic and bathypelagic zones contain organisms adapted to minimum light and no light, respectively. The benthic division (ocean floor) includes organisms living on the continental shelf in the sublittoral zone, the continental slope in the bathyal zone, and the abyssal plain in the abyssal zone.

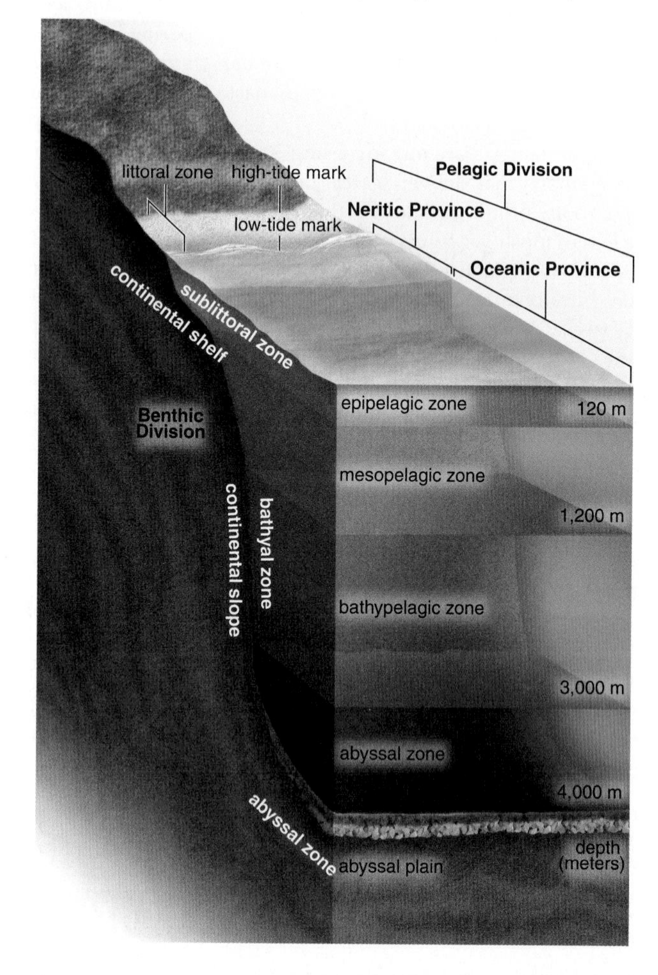

TESTING YOURSELF

Choose the best answer for each question.

1. The Northern Hemisphere is tilted toward the sun during the
   a. winter solstice.
   b. autumnal equinox.
   c. summer solstice.
   d. vernal equinox.
2. Which side of a mountain range lies in a rain shadow?
   a. windward
   b. leeward
3. Whether you are ascending a mountain or traveling from the equator to the North Pole, you would observe terrestrial biomes in which order?
   a. coniferous forest, tropical rain forest, temperate deciduous forest, tundra
   b. tropical rain forest, coniferous forest, temperate deciduous forest, tundra
   c. tropical rain forest, temperate deciduous forest, coniferous forest, tundra
   d. tropical rain forest, coniferous forest, tundra, temperate deciduous forest
   e. tundra, temperate deciduous forest, tropical rain forest, coniferous forest

For questions 4–10, match the description to the biome in the key.

Key:

a. tundra
b. taiga
c. temperate deciduous forests
d. tropical forests
e. shrublands
f. grasslands
g. deserts

4. Dominated by coniferous trees
5. Chaparral in California
6. Contains permafrost
7. Typically found at latitudes of about 30°
8. Has a complex structure, with many levels of life, from the canopy to the floor
9. Moderate climate, with relatively high rainfall and well-defined seasons
10. Includes savannas
11. Compared to the hypolimnion layer, the epilimnion layer of a lake in late summer is
    a. nutrient-rich and oxygen-rich.
    b. nutrient-rich and oxygen-poor.
    c. nutrient-poor and oxygen-rich.
    d. nutrient-poor and oxygen-poor.

For questions 12–15, indicate the life zone that each lake organism is most likely to inhabit. A question may have more than one answer.

Key:

a. littoral
b. limnetic
c. profundal
d. benthic

12. Predatory fish
13. Water lily
14. Clams
15. Unicellular algae

UNDERSTANDING THE TERMS

benthic division 735
biome 722
chaparral 727
climate 720
coniferous forest 725
coral reef 734
deserts 728
El Niño–Southern Oscillation (ENSO) 736
epiphyte 726
estuary 732
eutrophication 730
fall overturn 730
grasslands 727
hydrothermal vent 735
lake 729
littoral zone 732
monsoon 721
pelagic division 734
permafrost 724
plankton 731
rain shadow 721
shrubland 727
spring overturn 730
taiga 725
temperate deciduous forest 725
tropical forest 726
tundra 724
upwelling 734

THINKING CRITICALLY

1. Global warming is predicted to have the least severe consequences in terms of temperature change nearest the equator and more severe consequences progressing into the more temperate zones. Why do you think this is?
2. Scientists have documented a phenomenon called "biodiversity gradients," which shows that the greatest amounts of species diversity exist nearest the equator and that diversity tends to diminish moving through increasing latitudes. Based on your knowledge of climate, topography, and general characteristics of terrestrial biomes, explain the phenomenon of biodiversity gradients.
3. Based on what you know about tropical forests versus temperate forests, why do you think temperate forest habitats are more suitable for long-term agriculture, despite the fact that they have a shorter growing season than tropical forests?

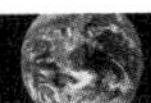

INQUIRY INTO LIFE WEBSITE

The companion website for *Inquiry into Life* provides a wealth of information organized and integrated by chapter. You will find practice tests, animations, videos, and much more that will complement your learning and understanding of general biology.

http://www.mhhe.com/maderinquiry13

Applying the Concepts

The demands made on the Earth's natural resources by the growth of the human population have created a biodiversity crisis through habitat destruction, introduction of exotic species, pollution, overexploitation, and spread of disease.

Biodiversity has both direct and indirect value to humans. Billions of dollars worth of medicines have come from plant and animal products. Plants and animals also provide valuable resources, such as timber and food, as well as help to control pests that threaten our crops. Indirect values of biodiversity, often referred to as "ecosystem services," are estimated to be more valuable than the total gross national product of the entire world! For example, preserving biogeochemical cycles, such as the water cycle, provides clean drinking water, and conserving topsoil provides billions worth of indirect value for crop growth and erosion control.

Conservation biology is directed at protecting biodiversity for the good of all living things, including humans. As an illustration, consider the past plight of the bald eagle. In the twentieth century, the pesticide DDT entered food chains after being used worldwide to control mosquitoes and other insect pests. The concentration of DDT became magnified up the food chain. As a result, bald eagles, which are tertiary consumers, were driven to near extinction. DDT in their food caused their eggs to be extremely thin and easily crushed by incubating parents. The Environmental Protection Agency banned the use of DDT in the United States in 1972, and since then, bald eagle populations have recovered to the point that, although they are still protected, they were removed from the endangered species list in 2007.

Conservation Biology

Chapter Outline

36.1 Conservation Biology and Biodiversity

Learning Outcomes

1. Define conservation biology.
2. Explain how biodiversity is classified into four levels.

Conservation biology is an interdisciplinary discipline with the explicit goal of protecting biodiversity and Earth's natural resources. Conservation biology is considered *interdisciplinary* because it relies on many subdisciplines of biology for the development of basic scientific concepts to describe biodiversity, as well as the application of these concepts such that biodiversity is sustainably managed and conserved for future generations.

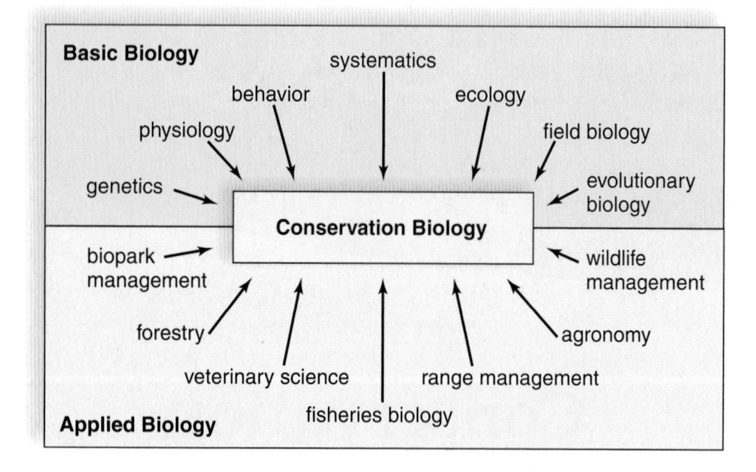

The application of basic scientific principles for conservation and management means that it is often necessary for conservation biologists to work with government officials at both the local and federal levels. Social scientists and economists are also involved because many conservation decisions have socio-economic impacts in addition to biological ones. Public education is also critical for the success of conservation biology—educating people facilitates better-informed consumer decisions, such as product choices that minimize environmental impact.

Conservation biology is unique among the life sciences in that it embodies the following ethical principles: (1) biodiversity is desirable for the biosphere and therefore for humans; (2) human-induced extinctions are therefore undesirable; (3) complex interactions within ecosystems and communities support biodiversity and the maintenance of such interactions is therefore desirable; and (4) biodiversity generated by evolutionary change has intrinsic value, regardless of any practical benefit. Because species within communities often interact in subtle and complex ways, disruption to the ecosystems in which they live can have unpredictable and potentially dire consequences. For example, loss of a single, ecologically important species could result in large numbers of secondary extinctions. Thus, disruption of ecosystems or ecosystem functioning should be avoided.

Conservation biology has often been called a "crisis discipline" or a "discipline with a deadline" because we are facing a sixth mass extinction. Unlike the five previous mass extinctions that have occurred naturally throughout Earth's history, the current mass extinction is much more rapid. It is estimated that as many as 10–20% of all species may be extinct in the next 30–50 years, and some researchers suggest that we may lose as many as 50% of all species by the year 2100. Thus, it is urgent that we take steps to ensure that all people understand the concept of biodiversity, the value of biodiversity, the causes of biodiversity loss, and the potential consequences of reduced biodiversity.

Biodiversity

At its simplest level, **biodiversity** is the variety of life on Earth. Commonly, biodiversity is described in terms of the number of species found in a given area or ecosystem. To date, approximately 1.9 million species have been described across the globe (Fig. 36.1). This may only be a small fraction of Earth's species, however. Estimates are that there may be between 10 and 50 million species in all, leaving many species yet to be discovered and described.

Of those that are described, nearly 2,000 species in the United States and 75,000 species worldwide are threatened with extinction. An **endangered species** is one that faces immediate extinction throughout all or most of its range. Examples of endangered species include the black lace cactus, armored snail, hawksbill sea turtle, California condor, West Indian manatee, and snow leopard. Threatened species are likely to become endangered in the foreseeable future. Examples of **threatened species** include the Navaho sedge, puritan tiger beetle, gopher tortoise, bald eagle, gray wolf, and Louisiana black bear.

To develop a meaningful understanding of life on Earth, we need to know more about species than their total number. Although we most often think at the species level, ecologists

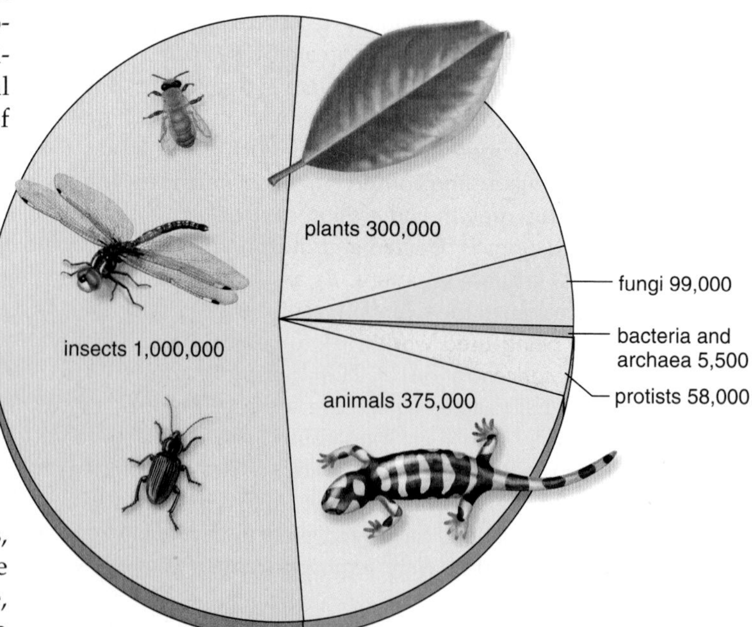

Figure 36.1 Number of described species.
There are only about 1.9 million described species; insects are far more prevalent than organisms in other groups. Undescribed species probably number far more than those species that have been described.

and conservation biologists necessarily describe biodiversity at three other levels of biological organization: genetic diversity, ecosystem diversity, and landscape diversity.

Genetic diversity includes the number of different alleles, as well as the relative frequencies of those alleles in populations and species. It is genetic diversity that underlies the evolutionary potential of populations, or their capacity to adapt to future environmental change. Thus, populations with high genetic diversity are more likely to have some individuals that can survive changes in their community or ecosystem. Low genetic diversity made the 1846 potato blight in Ireland and the 1984 outbreak of citrus canker in Florida worse than they would have been if the populations were more genetically diverse. Higher genetic diversity would have likely meant that there would have been a higher proportion of individuals that, by chance, were resistant to the pathogens that decimated these crops.

Ecosystem diversity is dependent on the interactions among species and with their abiotic environment in a particular area. An ecosystem may have one or several types of communities within it. The composition of species within a community may differ dramatically among different community types. Consider the species you would find in a temperate rain forest versus a savannah, for example. Variation in community composition, therefore, increases the levels of biodiversity in the biosphere. Although past conservation efforts frequently focused on saving particular charismatic species, such as the California condor, the black-footed ferret, or the spotted owl, such species-specific methods may not be the most beneficial for all species in the ecosystem. Instead, a more effective approach is to conserve species with critical roles to play in an ecosystem, such as those whose loss may result in many secondary extinctions.

As another example, reintroduction of wolves in Yellowstone National Park has had dramatic effects. Without wolves, elk are numerous and alter aspen groves, their favorite food. Wolves have helped reduce the number of elk, thereby increasing the number of aspen and other trees found in floodplains that, in turn, increase the habitat for songbirds and beavers. Beavers have then built dams that have flooded areas, making them suitable for muskrats, ducks, otters, and amphibians. Further, elk carcasses that are left by wolves become food for grizzly bears and eagles. Disrupting a community can also threaten the existence of more than one species. Opossum shrimp, *Mysis relicta,* were introduced into Flathead Lake in Montana and its tributaries as food for salmon. The shrimp ate so much zooplankton that, in the end, there was far less food for the fish and ultimately for the grizzly bears and bald eagles (Fig. 36.2).

Landscape diversity involves a group of interacting ecosystems within a single landscape. For example, plains, mountains and rivers may be fragmented to the point that they are connected only by patches (or corridors) that allow organisms to move from one ecosystem to the other. Fragmentation of the landscape may reduce reproductive capacity by, for example, increasing predation risk during dispersal between suitable habitat patches or creating difficulty in finding mates.

Check Your Progress

1. What is conservation biology?
2. Why might ecosystem-level conservation be more important than species-level conservation?

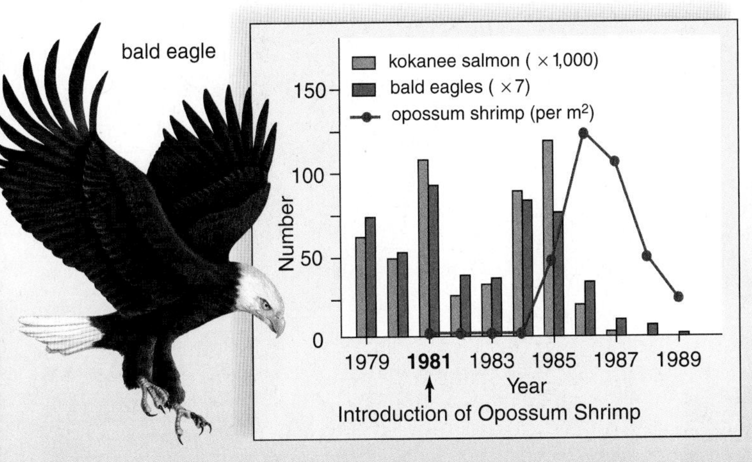

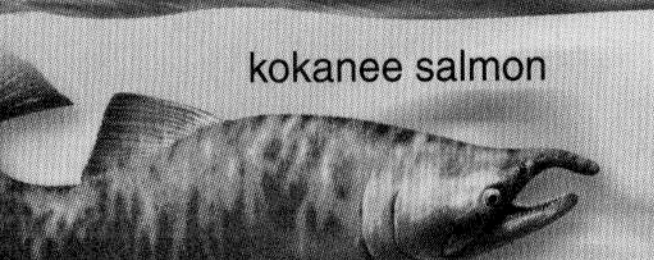

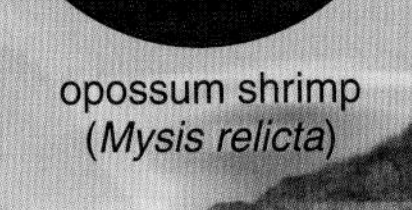

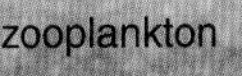

Figure 36.2 Eagles and bears feed on spawning salmon. Humans introduced the opossum shrimp as prey for salmon. Instead, the shrimp competed with salmon for zooplankton as a food source. The salmon, eagle, and bear populations subsequently declined.

36.2 Value of Biodiversity

Learning Outcomes

1. Argue the value of biodiversity.
2. Explain the difference between direct and indirect values of biodiversity.

Conservation biology strives to reverse the trend toward the possible extinction of thousands of plants and animals. To achieve this goal, it is necessary to make all people aware that biodiversity is a resource of immense value.

Direct Value

Various individual species perform services for human beings and contribute greatly to the value we should place on biodiversity. Figure 36.3 gives examples of the direct value of wildlife.

Medicinal Value

Most of the prescription drugs used in the United States, valued at over $200 billion, were originally derived from living organisms. The rosy periwinkle from Madagascar is an excellent example of a tropical plant that has provided us with useful medicines. Potent chemicals from this plant are now used to treat two forms of cancer: leukemia and Hodgkin disease. Because of these drugs, the survival rate for childhood leukemia has improved from 10–90%, and Hodgkin disease is usually curable. Although the value of saving a life cannot be calculated, it is still sometimes easier for us to appreciate the worth of a resource if it is explained in monetary terms. Thus, researchers estimate that, judging from the success rate in the past, an additional 328 types of drugs are yet to be found in tropical rain forests, and the value of this resource to society is probably $147 billion.

You may already know that the antibiotic penicillin is derived from a fungus and that certain species of bacteria produce the antibiotics tetracycline and streptomycin. These drugs have proven to be indispensable in the treatment of diseases, including sexually transmitted diseases (STDs) such as chlamydia, the most common STD in the United States and a leading cause of sterility in women.

Leprosy is among those diseases for which there is as yet no cure. The bacterium that causes leprosy will not grow in the laboratory, but scientists discovered that it grows naturally in the nine-banded armadillo. Having a source of

Figure 36.3 Direct value of wildlife. The direct services of wild species benefit human beings immensely. It is sometimes possible to calculate the monetary value, which is always surprisingly large.

Wild species, like the rosy periwinkle, are sources of many medicines.

Wild species, like many marine species, provide us with food.

Wild species, like the nine-banded armadillo, play a role in medical research.

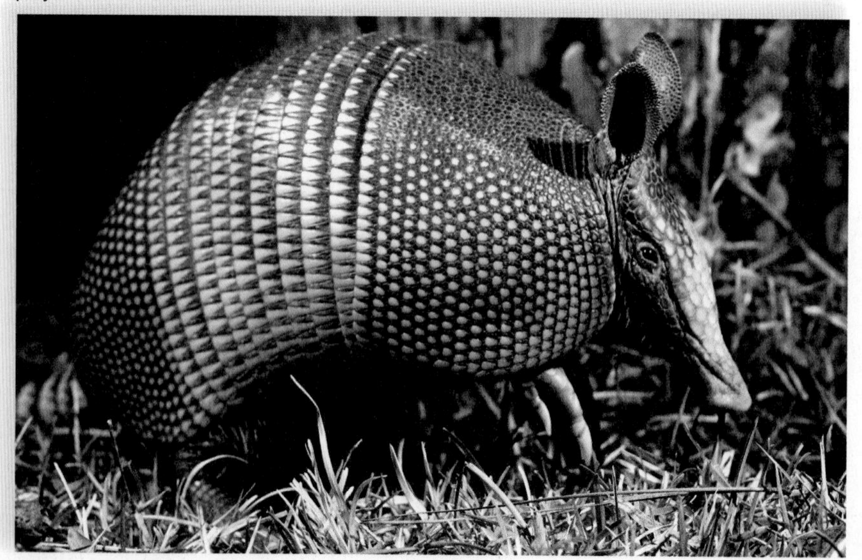

the bacterium may make it possible to find a cure for leprosy. The blood of horseshoe crabs contains a substance called limulus amoebocyte lysate, which is used to ensure that medical devices, such as pacemakers, surgical implants, and prosthetic devices, are free of bacteria. Blood is taken from 250,000 crabs a year, which are then returned to the sea unharmed.

Agricultural Value

Crops such as wheat, corn, and rice are derived from wild plants that have been modified to be high producers. The same high-yield, genetically similar strains tend to be grown worldwide. When rice crops in Africa were being devastated by a virus, researchers grew wild rice plants from thousands of seed samples until they found one that contained a gene for resistance to the virus. These wild plants were then used in a breeding program to transfer the gene into high-yield rice plants. If this variety of wild rice had become extinct before it could be discovered, rice cultivation in Africa might have collapsed.

Biological pest controls—natural predators and parasites—are often preferable to using chemical pesticides. When a rice pest called the brown planthopper became resistant to pesticides, farmers began to use natural brown planthopper enemies instead. The economic savings were calculated at well over $1 billion. Similarly, cotton growers in Cañete Valley, Peru, found that pesticides were no longer working against the cotton aphid because resistance had evolved. Research identified natural predators that are now being used to an ever-greater degree by cotton farmers. Again, savings have been enormous.

Most flowering plants are pollinated by animals such as bees, wasps, butterflies, beetles, birds, and bats. The honeybee, *Apis mellifera,* has been domesticated, and it pollinates almost $10 billion worth of food crops annually in the United States. The danger of this dependency on a single species is exemplified by the fact that mites have now wiped out more than 20% of the commercial U.S. honeybee population. Where can we get resistant bees? From the wild, of course. The value of wild pollinators to the U.S. agricultural economy has been calculated at $4.1 to $6.7 billion a year.

Consumptive Use Value

We have experienced much success in cultivating crops, keeping domesticated animals, growing trees in plantations, and so forth. But so far, aquaculture, the growing of fish and

Wild species, like the lesser long-nosed bat, are pollinators of agricultural and other plants.

Wild species, like ladybugs, play a role in biological control of agricultural pests.

Wild species, like rubber trees, can provide a product indefinitely if the forest is not destroyed.

shellfish for human consumption, has contributed only minimally to human welfare. Instead, most freshwater and marine harvests depend on wild animals, such as fishes (e.g., trout, cod, and tuna) and crustaceans (e.g., lobsters, shrimps, and crabs). Obviously, these aquatic organisms are an invaluable biodiversity resource.

The environment provides all sorts of other products that are sold in the marketplace worldwide, including wild fruits and vegetables, skins, fibers, beeswax, and seaweed. Also, some people obtain their meat directly from the wild. In one study, researchers calculated that the economic value of wild pig in the diet of native hunters in Sarawak, East Malaysia, was approximately $40 million per year.

Similarly, many trees are still felled for their wood. Researchers have calculated that a species-rich forest in the Peruvian Amazon region is worth far more if the forest is used for fruit and rubber production than for timber production. Whereas fruit and the latex needed to produce rubber can be brought to market indefinitely, no more timber or other resulting products can be produced when trees are cut down.

Check Your Progress

1. What is the difference between consumptive use value and agricultural value of biodiversity?

Indirect Value

The wild species we have been discussing live in ecosystems. If we want to preserve them, many scientists argue it is more economical to save the ecosystems than the individual species. Ecosystems perform many services for modern humans, who increasingly live in cities. These services are said to be indirect because they are pervasive and not easily discernible (Fig. 36.4). Even so, our survival depends on the functions that ecosystems perform for us. Recent studies suggest the indirect value of ecosystem services surpasses that of the global gross national product—about $33 trillion per year!

Biogeochemical Cycles

You'll recall from Chapter 34 that ecosystems are characterized by energy flow and chemical cycling. The biodiversity within ecosystems contributes to the workings of the water, carbon, nitrogen, phosphorus, and other biogeochemical cycles. We are dependent on these cycles for fresh water, removal of carbon dioxide from the atmosphere, uptake of excess soil nitrogen, and provision of phosphate. When human activities upset the usual workings of biogeochemical cycles, the dire environmental consequences include the release of excess pollutants that are

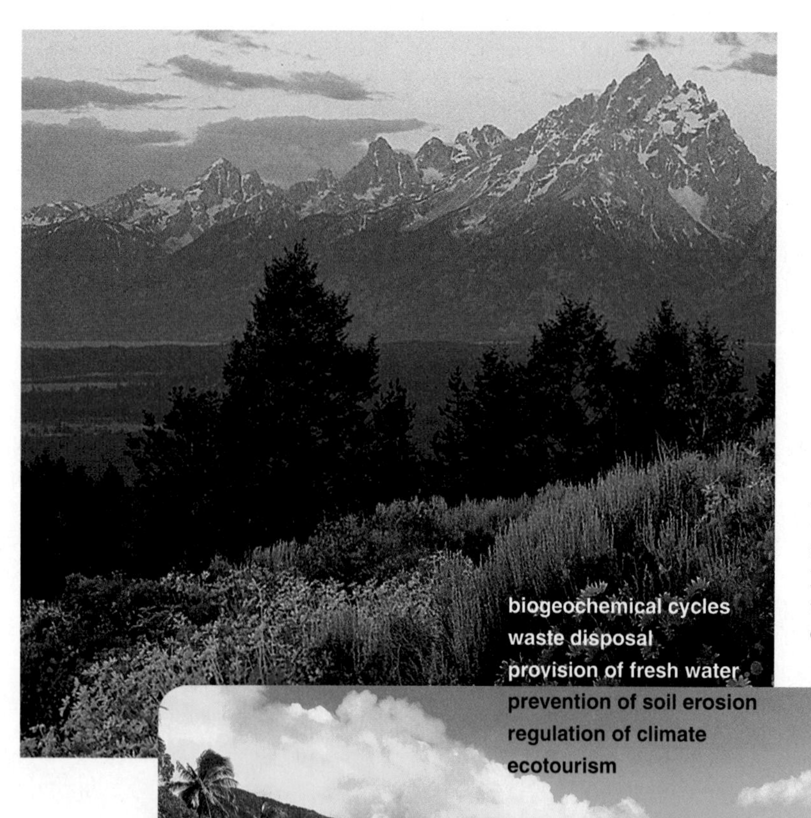

Figure 36.4 Indirect value of ecosystem. Forests and the oceans perform many of the functions listed as the indirect value of ecosystems.

harmful to us. Unfortunately, technology is unable to artificially contribute to or recreate any of the biogeochemical cycles.

Waste Disposal

Decomposers break down dead organic matter and other types of wastes to inorganic nutrients that are used by the producers within ecosystems. This function aids humans immensely because we dump millions of tons of waste material into natural ecosystems each year. If it were not for decomposition, waste would soon cover the entire surface of our planet. We can build sewage treatment plants, but they are expensive. In addition, few of them break down solid wastes completely to inorganic nutrients. It is less expensive and more efficient to water plants and trees with partially treated wastewater and let soil bacteria cleanse it completely.

Biological communities are also capable of breaking down and immobilizing pollutants, such as heavy metals and pesticides, that humans release into the environment. A review of wetland functions in Canada assigned a value of \$50,000 per hectare (2.5 acres or 10,000 m^2) per year to the ability of natural areas to purify water and take up pollutants.

Provision of Fresh Water

Few terrestrial organisms are adapted to living in a salty environment—they need fresh water. The water cycle continually supplies fresh water to terrestrial ecosystems. Humans use fresh water in innumerable ways, including for drinking and for irrigating their crops. Freshwater ecosystems, such as rivers and lakes, also provide us with fish and other types of organisms for food. Unlike other commodities, there is no substitute for fresh water. We can remove salt from seawater to obtain fresh water, but the cost of desalinization is about four to eight times the average cost of fresh water acquired via the water cycle.

Flood Prevention

Forests and other natural ecosystems exert a "sponge effect," thereby reducing flooding. They soak up water and then release it at a regular rate. When rain falls in a natural area, plant foliage and dead leaves lessen its impact, and the soil slowly absorbs it, especially if the soil has been aerated by organisms. The water-holding capacity of forests reduces the possibility of flooding. The value of a marshland outside Boston, Massachusetts, has been estimated at \$72,000 per hectare per year based solely on its ability to reduce floods. Flooding in New Orleans by Hurricane Katrina in 2005 would likely have been much less severe if the natural wetlands and marshlands around the Gulf of Mexico had still been intact.

Prevention of Soil Erosion

Intact ecosystems naturally retain soil and prevent soil erosion. The importance of this ecosystem attribute is especially observed following deforestation. In Pakistan, the world's largest dam, the Tarbela Dam, is losing its storage capacity of 12 billion cubic meters (m^3) many years sooner than expected because silt is building up behind the dam due to deforestation. At one time, the Philippines exported \$100 million worth of oysters, mussels, and clams each year. Now, silt carried down rivers following deforestation is smothering the mangrove ecosystem that serves as a nursery for the sea. In general, most coastal ecosystems are not as bountiful as they once were because of deforestation, coastal development, and aquatic pollution.

Regulation of Climate

At the local level, trees provide shade, reduce the need for fans and air conditioners during the summer, and provide buffers from noise. In tropical rain forests, trees maintain regional rainfall, and without them the forests become arid.

Globally, forests ameliorate the climate because they take up carbon dioxide and release oxygen. The leaves of trees use carbon dioxide when they photosynthesize, and the bodies of the trees store carbon. When trees are cut and burned, carbon dioxide is released into the atmosphere. Carbon dioxide makes a significant contribution to global warming, which is expected to be stressful for many plants and animals.

Ecotourism

Many people prefer to vacation in natural areas. In the United States, nearly 100 million people spend nearly \$5 billion each year on fees, travel, lodging, and food associated with ecotourism. Tourists are often interested in activities such as sport fishing, whale watching, boat riding, hiking, birdwatching, and the like. Some merely want to immerse themselves in the beauty and serenity of nature. Many underdeveloped countries in tropical regions, such as Belize and Costa Rica, are taking advantage of this by offering "ecotours" of the local biodiversity. Providing guided tours of natural ecosystems, such as tropical rainforests, is often more profitable than destroying them. Ecotourism also provides indirect value for wildlife. As an example, the tusks of an elephant are worth around \$10,000 for the ivory, but the Kenyan government estimates that a single elephant is worth as much as \$1.2 million in tourist dollars.

CHECK YOUR PROGRESS

1. People going to see elephants in Africa represent which type of value of biodiversity?
2. What is the difference between direct and indirect values of biodiversity?

36.3 Threats to Biodiversity

Learning Outcomes

1. Describe the five causes of species' extinctions.
2. Consider how these causes can work together to reduce biodiversity.

We are presently in a biodiversity crisis—the number of extinctions (loss of species) expected to occur in the near future is unparalleled in Earth's history. To identify the role that humans are playing in modern extinctions, researchers examined the records of 1,880 threatened and endangered wild species in the United States. Habitat loss was the most important, and it was involved in 85% of the cases (Fig. 36.5*a*). Exotic species had a hand in nearly 50%, pollution was a factor in 24%, overexploitation in 17%, and disease in 3%. Note that the percentages add up to more than 100% because most of these species are imperiled for more than one reason. These five causes reflect the relative importance of the five leading causes for species' extinctions worldwide. Macaws illustrate that a combination of factors can lead to a species decline (Fig. 36.5*b*). Not only has their habitat been reduced by encroaching timber and mining companies, but macaws are also hunted for food and collected for the pet trade. DNA technology may now help enforce laws to protect wildlife collected for the pet trade, as described in the Ecology Focus.

Habitat Loss

Habitat loss has occurred in all ecosystems, and human disruption of natural habitats is the most important cause of biodiversity loss. Concern has now centered on tropical rain forests and coral reefs because they are particularly rich in species diversity. A sequence of events in Brazil offers a fairly typical example of the manner in which rain forest is converted to land uninhabitable for wildlife. The construction of a major highway into the forest first provided a way to reach the interior of the forest (see Fig. 36.5*c*). Small towns and industries sprang up along the highway, and roads branching off the main highway gave rise to even more roads. The result was fragmentation of the once immense forest. The government offered subsidies to anyone willing to take up residence in the forest, and the people who came

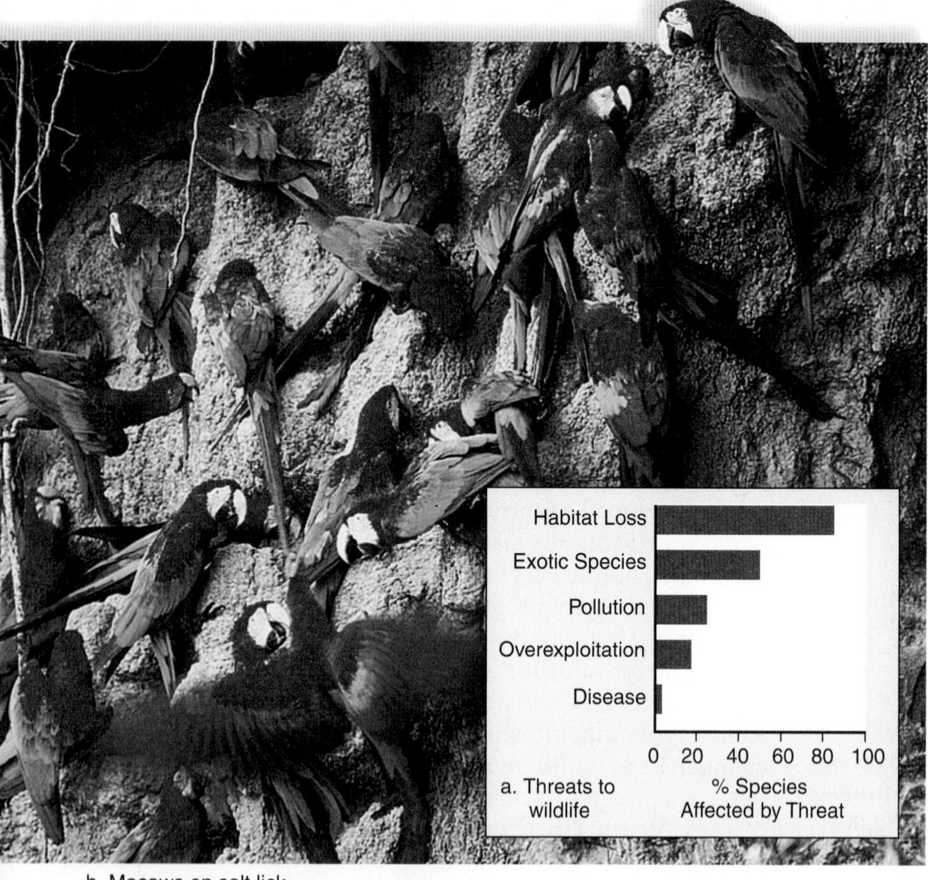

b. Macaws on salt lick

Roads cut through forest

Forest occurs in patches

Destroyed areas

c. Wildlife habitat is reduced

Figure 36.5 Habitat loss.
a. In a study that examined records of imperiled U.S. plants and animals, habitat loss emerged as the greatest threat to wildlife. **b.** Macaws that reside in South American tropical rain forests are endangered for some of the reasons listed in the graph in **(a)**. **c.** Habitat loss due to road construction in Brazil. (*top*) Road construction opened up the rain forest and subjected it to fragmentation. (*middle*) The result was patches of forest and degraded land. (*bottom*) Wildlife could not live in destroyed portions of the forest.

Wildlife Conservation and DNA

After DNA analysis, scientists were amazed to find that some 60% of loggerhead turtles drowning in the nets and hooks of fisheries in the Mediterranean Sea were from beaches in the southeastern United States. Since the unlucky creatures were a good representative sample of the turtles in the area, that meant more than half of the young turtles living in the Mediterranean Sea had hatched from nests on beaches in Florida, Georgia, and South Carolina (Fig. 36A*a*). Some 20,000–50,000 loggerheads die each year due to the Mediterranean fisheries, which may partly explain the decline in loggerheads nesting on southeastern U.S. beaches for the last 25 years.

The sequencing of DNA from Alaskan brown bears allowed Sandra Talbot (a graduate student at the University of Alaska's Institute of Arctic Biology) and wildlife geneticist Gerald Shields to conclude that there are two types of brown bears in Alaska. One type resides only on southeastern Alaska's Admiralty, Baranof, and Chichagof islands, known as the ABC Islands. The other brown bear in Alaska is found throughout the rest of the state, as well as in Siberia and western Asia (Fig. 36A*b*).

A third distinct type of brown bear, known as the Montana grizzly, resides in other parts of North America. These three types comprise all of the known brown bears in the New World.

The ABC bears' uniqueness may be bad news for the timber industry, which has expressed interest in logging parts of the ABC Islands. Says Shields, "Studies show that when roads are built and the habitat is fragmented, the population of brown bears declines. Our genetic observations suggest they are truly unique, and we should consider their heritage. They could never be replaced by transplants."

In what will become a classic example of how DNA analysis might be used to protect endangered species from future ruin, scientists from the United States and New Zealand carried out discreet experiments in a Japanese hotel room on whale sushi bought in local markets. Sushi, a staple of the Japanese diet, is rice and meat wrapped in seaweed. Armed with a miniature DNA sampling machine, the scientists found that, of the 16 pieces of whale sushi they examined, many were from whales that are endangered or protected under an international moratorium on whaling. "Their findings demonstrated the true power of DNA studies," says David Woodruff, a conservation biologist at the University of California, San Diego.

One sample was from an endangered humpback, four were from fin whales, one was from a northern minke, and another from a beaked whale. Stephen Palumbi of Harvard University says the technique could be used for monitoring and verifying catches. Until then, he says, "no species of whale can be considered safe."

Meanwhile, Ken Goddard, director of the unique U.S. Fish and Wildlife Service Forensics Laboratory in Ashland, Oregon, is already on the watch for wildlife crimes in the United States and 122 other countries that send samples to him for analysis. "DNA is one of the most powerful tools we've got," says Goddard, a former California police crime-lab director.

The lab has blood samples, for example, for all of the wolves being released into Yellowstone Park—"for the obvious reason that we can match those samples to a crime scene," says Goddard. The lab has many cases currently pending in court that he cannot discuss. But he likes to tell the story of the lab's first DNA-matching case. Shortly after the lab opened in 1989, California wildlife authorities contacted Goddard. They had seized the carcass of a trophy-sized deer from a hunter. They believed the deer had been shot illegally on a 3,000-acre preserve owned by actor Clint Eastwood. The agents found a gut pile on the property but had no way to match it to the carcass. The hunter had two witnesses to deny the deer had been shot on the preserve.

Goddard's lab analysis made a perfect match between tissue from the gut pile and tissue from the carcass. Says Goddard: "We now have a cardboard cutout of Clint Eastwood at the lab saying 'Go ahead: Make my DNA.'"

Discussion Questions

1. Recently, distinctiveness in terms of genetic DNA sequence has been used to set policy. That is, if researchers find a population of a species, such as ABC brown bears, to be genetically distinct from other populations of the same species, then that population may be afforded special protection, such as threatened or endangered status under the U.S. Endangered Species Act. This policy is then used to make major environmental decisions, such as halting logging in that area. Should DNA distinctness be used to make such major policy decisions? Why or why not?
2. Forensic DNA evidence has recently been used to prosecute hunters who illegally hunt animals, such as deer, in areas where they do not have a proper hunting license. Occasionally, such hunters face harsh sentences that may even include jail time. Should people who hunt illegally be put in jail or simply fined?
3. DNA studies, such as those conducted in Japan, opened our eyes to the fact that illegal whale hunting still occurs. Some people have recommended we screen the DNA of much of the imported fish meat coming from different countries. However, this can be quite expensive. Do you support genetic testing of imported fish meat, even if it increases the cost of the meat to the consumer?

a. Loggerhead turtle

b. Brown bears

Figure 36A DNA studies.
a. Many loggerhead turtles found in the Mediterranean Sea are from the southeastern United States. **b.** These two brown bears appear similar, but one type, known as an ABC bear, resides only on southeastern Alaska's Admiralty, Baranof, and Chichagof islands.

cut and burned trees in patches to facilitate grazing (see Fig. 36.5*c*). Tropical soils contain limited nutrients, and burning of trees releases additional nutrients that support cattle grazing for only about three years. Once the land was degraded, the farmers moved on to another portion of the forest to start over again.

Loss of habitat also affects freshwater and marine biodiversity. Coastal degradation is mainly due to the large concentration of people living on or near the coast. At least 40% of the world's population lives within 100 km (60 mi) of a coastline, and this number is expected to increase. In the United States, over one-half of the population lives within 80 km (50 mi) of the coasts (including the Great Lakes). Coastal habitation leads to beach erosion and direct inputs of pollutants into freshwater or marine systems. Already, 60% of coral reefs have been destroyed or nearly so, with siltation resulting from beach erosion as a leading cause. It is possible that all coral reefs may disappear during the next 40 years unless our behaviors drastically change. Mangrove forest destruction is also a problem. Indonesia, with the most mangrove acreage, has lost 45% of its mangroves, and the percentage is even higher for other tropical countries. Wetland areas, estuaries, and seagrass beds are also being rapidly destroyed by the actions of humans.

Exotic Species

Exotic species, sometimes also called alien species, are nonnative members of a community. Communities around the world are characterized by unique assemblages of species that have evolved together in an area. Introduction of exotic species can disrupt this balance by changing the interactions of species in a food web, as shown in Figure 36.2. In this example, opossum shrimp introduced in a lake in Montana added a trophic level that, in the end, meant less food for bald eagles and grizzly bears. Often, exotic species can directly compete with or prey upon native species, thereby reducing their abundance or causing localized extinction. For these reasons, exotics are the second most important reason for biodiversity loss. Humans have introduced exotic species into new ecosystems by the following means:

Colonization. Europeans, in particular, brought various familiar species with them when they colonized new places. For example, the pilgrims brought the dandelion to the United States as a familiar salad green. The British brought foxes and stoats (a type of weasel) to New Zealand, and these predators have resulted in the loss of nearly 40% of all of New Zealand's bird species. These birds are easy prey because they evolved without mammalian predators, which were completely absent from New Zealand.

Horticulture and agriculture. Some exotics now taking over vast tracts of land have escaped from cultivated areas. Kudzu is a vine from Japan that the U.S. Department of Agriculture thought would help prevent soil erosion. The plant now covers much of the landscape in the South, including even walnut, magnolia, and gum trees (Fig. 36.6*a*). Similarly, the water hyacinth was introduced to the United States from South America because of its beautiful flowers. Today, it clogs up waterways and diminishes natural diversity.

Accidental transport. Global trade and travel accidentally bring many new species from one country to another. The zebra mussel from the Caspian Sea was accidentally introduced into the Great Lakes in 1988. It now forms dense beds that reduce biodiversity by outcompeting native mussels. Zebra mussels also decrease the amount of food available for higher levels in the food chain because they are such effective filter feeders. Zebra mussels cause millions in damage per year by clogging sewage and other pipes. Other exotics introduced into the United States include the Argentinian fire ant and the nutria, a large rodent found throughout the Southeast.

Exotics on Islands

Islands are particularly susceptible to environmental discord caused by the introduction of exotic species. Islands have unique assemblages of native species that are closely adapted to one another and cannot compete well against exotics. Myrtle trees, *Myrica faya,* introduced into the Hawaiian Islands from the Canary Islands, are symbiotic with a type of bacterium that is capable of nitrogen fixation. This feature allows the species to establish itself on nutrient-poor volcanic soil such as that found in Hawaii. Once established, myrtle trees halt normal ecological succession by outcompeting native plants on volcanic soil.

The brown tree snake has been introduced onto a number of islands in the Pacific Ocean. The snake eats eggs, nestlings, and adult birds. On Guam, it has reduced ten native bird species to the point of extinction. Mongooses introduced into the Hawaiian Islands to control rats have increased dramatically in abundance and also prey on native birds, causing some population declines (Fig. 36.6*b*).

CHECK YOUR PROGRESS

1. Discuss two ways in which exotic species are introduced into new areas.

a.

b.

Figure 36.6 Exotic species.
a. Kudzu, a vine from Japan, was introduced in several southern states to control erosion. Today, kudzu has taken over and displaced many native plants. **b.** Mongooses were introduced into Hawaii to control rats, but they also prey on native birds.

Pollution

In the present context, **pollution** can be defined as any environmental change that adversely affects the lives and health of living things. Pollution has been identified as the third main cause of extinction. Pollution can also weaken organisms and lead to disease, the fifth leading cause of extinction. Biodiversity is particularly threatened by the following types of environmental pollution:

Acid deposition. Sulfur dioxide from power plants and nitrogen oxides from auto exhaust are converted to acids when combined with atmospheric water vapor. These acids return to Earth during precipitation (rain or snow) or via dry deposition. This acid deposition decimates forests because it causes trees to weaken and increases their susceptibility to disease and insects. In addition, because acid deposition can make the pH of water too low for organisms to survive, many lakes in northern states are now lifeless.

Eutrophication. Lakes are also under stress due to overenrichment. When lakes receive excess nutrients due to runoff from agricultural fields and wastewater from sewage treatment, algae begin to grow in abundance. Death of algae leads to an abundance of decomposers that, in turn, decrease available oxygen and often lead to fish kills.

Ozone depletion. The Ecology Focus in Chapter 34 tells how the ozone shield protects Earth from harmful ultraviolet (UV) radiation. The release of chlorofluorocarbons (CFCs), in particular, into the atmosphere causes the shield to break down, leading to impaired growth of crops and trees and the death of plankton that sustain oceanic life. The immune systems and ability of all organisms to resist disease will most likely be weakened. Ozone depletion has also led to dramatic increases in skin cancer.

Organic chemicals. Our modern society uses organic chemicals in all sorts of ways. Organic chemicals called nonylphenols are used in products ranging from pesticides to dishwashing detergents, cosmetics, plastics, and spermicides. Many of these chemicals can mimic the effects of hormones, and if so, they are called endocrine-disruptors. For example, investigators exposed young salmon to nonylphenol. Although these fish are born in fresh water and mature in salt water, 20–30% of fish exposed to nonylphenol were unable to make the transition from fresh to salt water. Nonylphenols cause the pituitary to produce prolactin, a hormone that may prevent saltwater adaptation. Other endocrine-disrupting contaminants can possibly affect the endocrine system and thereby the reproductive potential of other animals, including humans.

Global warming. Global warming is an increase in Earth's temperature due to the increase of greenhouse gases, such as carbon dioxide and methane, in Earth's atmosphere. Global warming is expected to have many detrimental effects, including the destruction of coastal wetlands due to a rise in sea levels, a shift in suitable temperatures to locations where various species cannot live, and the death of coral reefs (Fig. 36.7). An upward shift in temperatures could influence everything from growing seasons in plants to migratory patterns in animals. As temperatures rise, regions of suitable climate for various terrestrial species may shift toward the poles and higher elevations. Extinctions are expected because the present assemblages of species in ecosystems will be disrupted as some species migrate northward (or southward in the Southern Hemisphere) to track the environmental changes, whereas other species with limited mobility may not be able to migrate quickly enough. For example, to remain in a favorable habitat, it's been calculated that the rate of beech tree migration would have to be 40 times faster than has ever been observed.

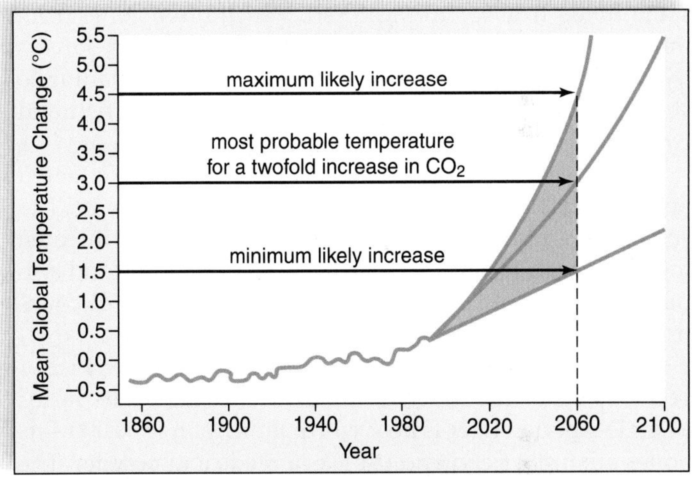

a.

b.

Figure 36.7 Global warming.
a. Mean global temperature is expected to rise due to the introduction of greenhouse gases into the atmosphere. **b.** Global warming has the potential to significantly affect the world's biodiversity distribution. A temperature rise of only a few degrees causes coral reefs to "bleach" and become lifeless. If, in the meantime, migration occurs, coral reefs could move northward.

Overexploitation

Overexploitation occurs when the number of individuals taken from a wild population is so great that the population becomes severely reduced in numbers. A positive feedback cycle explains overexploitation: the smaller the population, the more commercially valuable its members, and the greater the incentive to capture the few remaining organisms. Poachers are very active in the collecting and sale of endangered and threatened species because it has become so lucrative. The overall international value of trading wildlife species is $20 billion, of which $8 billion is attributed to the illegal sale of rare species.

Markets for rare plants and exotic pets support both legal and illegal trade in wild species. Rustlers dig up rare cacti, such as the crested saguaros, and sell them to gardeners for as much as $15,000 each. Parrots are among birds taken from the wild for sale to pet owners. For every bird delivered alive, many more have died in the process. The same holds true for tropical fish, which often come from the coral reefs of Indonesia and the Philippines. Divers dynamite reefs or use plastic squeeze-bottles of cyanide to stun the fish within them. In the process, many fish and corals die.

The Convention on International Trade in Endangered Species (CITES) was an agreement established in 1973 to ensure that international trade of species does not threaten their survival. Today, over 30,000 species of plants and animals receive some level of protection from over 172 countries worldwide.

Poachers still hunt for hides, claws, tusks, horns, or bones of many endangered mammals. Because of its rarity, a single Siberian tiger is now worth more than $500,000—its bones are pulverized and used as a medicinal powder. The horns of rhinoceroses become ornate carved daggers, and their bones are also ground up to sell as a medicine. The ivory of an elephant's tusk is used to make art objects, jewelry, or piano keys. The fur of a Bengal tiger sells for as much as $100,000 in Tokyo.

The Food and Agricultural Organization of the United Nations tells us that we have now overexploited 11 of 15 major oceanic fishing areas. Fish are a renewable resource only if harvesting does not exceed the ability of the fish to reproduce. Our society uses larger and more efficient fishing fleets to decimate fishing stocks. Pelagic species such as tuna are captured by purseseine fishing, in which a very large net surrounds a school of fish, and then the net is closed in the same manner as a drawstring purse. Up to thousands of dolphins that swim above schools of tuna are often captured and then killed in this type of net. However, many tuna suppliers advertise their product as "dolphin safe." Other fishing boats drag huge trawling nets, large enough to accommodate 12 jumbo jets, along the seafloor to capture bottom-dwelling fish (Fig. 36.8*a*). Only large fish are kept; undesirable small fish and sea turtles are discarded, dying, back into the ocean. Trawling has been called the marine equivalent of clear-cutting trees because after the net goes by, the sea bottom is devastated (Fig. 36.8*b*). Today's fishing practices don't allow fisheries to recover. Cod and haddock, once the most abundant bottom-dwelling fish along the northeast coast of the United States, are now often outnumbered by dogfish and skate.

Entire marine ecosystems can be disrupted by overfishing, as exemplified on the U.S. west coast. When sea otters began to decline in numbers, investigators found that they were being eaten by orcas (killer whales). Usually orcas prefer seals and sea lions to sea otters, but they began eating sea otters when few seals and sea lions could be found. What caused a decline in seals and sea lions? Their preferred food sources—perch and herring—were no longer in abundance due to overfishing. Ordinarily, sea otters keep the population

a. Fishing by use of a drag net

b. Result of drag net fishing

Figure 36.8 Trawling.
a. These Alaskan pollock were caught by dragging a net along the seafloor. **b.** Appearance of the seafloor after the net passed.

Figure 36.9 Amphibian at risk.
The harlequin toad is nearly extinct due to infections by a fungal pathogen. Diseases are implicated in global population declines and extinctions of amphibians.

of sea urchins, which feed on kelp, under control. But with fewer sea otters around, the sea urchin population exploded and decimated the kelp beds. Thus, overfishing set in motion a chain of events that detrimentally altered the food web of an ecosystem.

Disease

Wildlife is subject to emerging diseases just as humans are. Due to the encroachment of humans on their habitat, wildlife have been exposed to new pathogens from domestic animals living nearby, and due to human intervention, they have been exposed to animals they would ordinarily not come in contact with. For example, canine distemper was spread from domesticated dogs to lions in the African Serengeti, causing population declines. Avian influenza likely emerged from domesticated fowl (e.g., chicken) populations and has led to the deaths of millions of wild birds.

The significant effect of diseases on biodiversity is underscored by National Wildlife Health Center findings that almost half of sea otter deaths along the coast of California are due to infectious diseases. Scientists tell us the number of pathogens that cause disease is on the rise, and just as human health is threatened, so is that of wildlife. Extinctions due simply to disease may occur, and are currently implicated in the worldwide decline of amphibians (Fig. 36.9).

Check Your Progress

1. What are the five main causes of extinction?
2. Give two reasons why pollution can impact biodiversity.

36.4 Habitat Conservation and Restoration

Learning Outcomes

1. Describe how habitat is prioritized for conservation.
2. Synthesize the components of nature reserve design.
3. Outline the goals of habitat restoration.

Habitat Conservation

Because habitat loss is the leading cause of species' extinctions, conservation of habitat is of primary concern. One way to prioritize which habitats to conserve is to focus on those that harbor the highest levels of biodiversity. Generally, biodiversity is highest at the tropics, and it declines toward each pole. Tropical rain forests and coral reefs are examples of ecosystems known for maintaining high levels of biodiversity.

Some regions of the world are called **biodiversity hotspots** because they contain unusually large concentrations of species. Biodiversity in these hotspots accounts for about 44% of all known higher plant species and 35% of all terrestrial vertebrate species but cover only about half of 1.4% of Earth's land area. Hotspots may also include large numbers of endemic species, or those not found anywhere else. An example of such a hotspot is the tropical forests of Madagascar, within which 93% of the primate species, 99% of the amphibian species, and over 80% of the plant species are endemic. The Cape region of South Africa, Indonesia, the coast of California, and the Great Barrier Reef of Australia are also considered biodiversity hotspots.

We can also focus our efforts on conserving habitat for **keystone species,** or those whose loss would result in a great number of secondary extinctions. The "keystone" analogy comes from the stone in a top of a doorway that supports all the other stones in the doorway—removal of the stone would result in the doorway's collapse. Keystone species need not be the most abundant. Although they are relatively low in numbers, wolves can be considered a keystone species in Yellowstone National Park, as discussed earlier. Bats have also been designated keystone species in Old World tropical forests. Bats are extremely important pollinators and dispersers of tree seeds. The loss of bats results in failure of many tree species to reproduce and can lead to a loss of biodiversity. Other keystone species include grizzly bears (Fig. 36.10*a*), beavers in wetlands, and elephants in grasslands and forests.

Keystone species should not be confused with **flagship species,** or those that evoke an emotional response from humans. Flagship species are considered charismatic and valued for their beauty, regal nature, or similarity to people's pets. Flagship species such as lions, tigers, dolphins, and the giant panda can motivate the public to conserve biodiversity.

Landscape Conservation and Reserve Design

Conservation often has to occur at the landscape level because sufficient habitat may not be available in a single place to sustain a viable population of a particular species due to widespread human development. Grizzly bears, once numbering between 50,000–100,000 south of Canada, are now estimated at about 1,000 individuals in six small, subdivided populations. Grizzly bear conservation entails maintenance of a number of different types of ecosystems, including plains, mountains, and rivers. Other species also rely on multiple ecosystem types, such as amphibians, that rely on wetlands and terrestrial habitat. Conserving a single one of these ecosystems would not be sufficient, nor would conservation of several ecosystems in isolation of one another. It is thus necessary to establish **conservation corridors** that allow animals to move safely between habitats that would otherwise be isolated. Corridors are often necessary because landscapes are subdivided due to urbanization, agriculture, and other aspects of human development. A corridor may be as small as an overpass for wildlife to cross a highway or a strip of forested habitat that is maintained along a stream after timber harvest. A corridor may be as small as an overpass for wildlife to cross a highway or as large as a patch of land several kilometers wide to allow movement of large mammals.

a. Grizzly bear, *Ursus arctos horribilis*

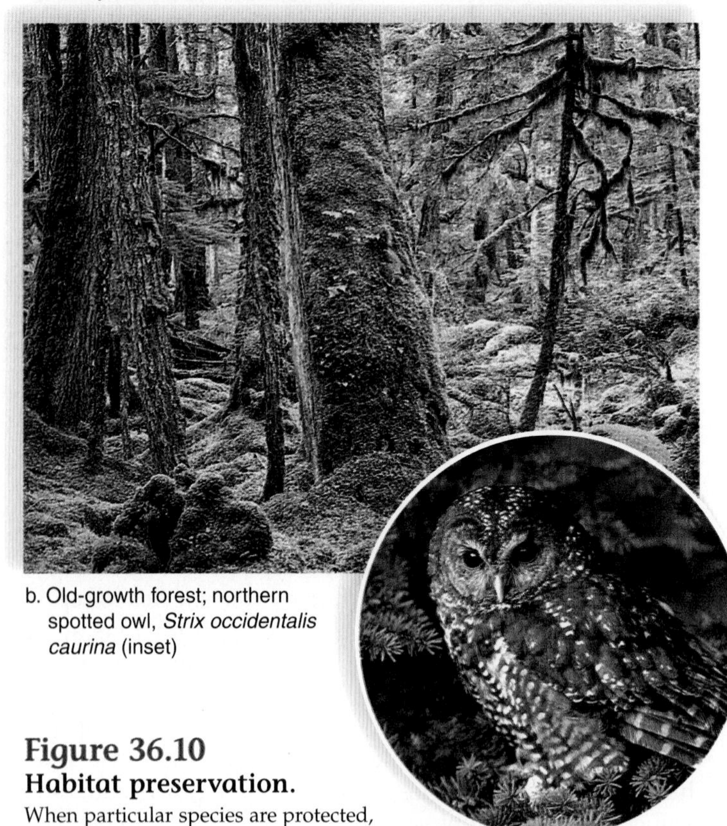

b. Old-growth forest; northern spotted owl, *Strix occidentalis caurina* (inset)

Figure 36.10
Habitat preservation.
When particular species are protected, other wildlife benefits. **a.** The Greater Yellowstone Ecosystem has been delineated in an effort to save grizzly bears, which need a very large habitat. **b.** Currently, the remaining portions of old-growth forests in the Pacific Northwest are not being logged in order to save the northern spotted owl (inset).

Landscape conservation for a single species is often beneficial for other species that share the same habitats. For example, conservation of the northern spotted owl (see Fig. 36.10*b*) helps protect many other species that inhabit temperate old growth forests. The last of the contiguous 48 states' harlequin ducks, westslope cutthroat trout, lynx, pine martins, wolverines, and great gray owls are found in areas occupied by grizzly bears. The geographic range of grizzly bears also overlaps with 40% of Montana's vascular plants of conservation concern.

Edge Effects

When conserving landscapes, it is necessary to consider **edge effects.** An edge reduces the amount of habitat typical of an ecosystem because the edges around a patch have a habitat slightly different from the interior of the patch. For example, forest edges are brighter, warmer, drier, and windier, with more vines, shrubs, and weeds than the forest interior. Also, Figure 36.11*a* shows that a small and a large patch of habitat have the same amount of edge. Therefore, the effective habitat shrinks as a patch gets smaller.

Many popular game animals, such as turkeys and white-tailed deer, are more plentiful in the edge region of a particular area. However, today it is known that creating edges can be detrimental to wildlife because of habitat fragmentation.

Edge effects can also have a serious impact on population size. Songbird populations west of the Mississippi have been declining of late, and ornithologists have noticed that the nesting success of songbirds is quite low at the edge of a forest. The cause turns out to be the brown-headed cowbird, a social parasite of songbirds. Adult cowbirds prefer to feed in open agricultural areas, and they only briefly enter the forest when searching for a host nest in which to lay their eggs (Fig. 36.11*b*). Cowbirds are therefore benefited, while songbirds are disadvantaged, by the edge effect.

Reserve Design

Conservation reserves are those areas that are set aside with the primary goal of protecting biodiversity within them. As such, reserves should be largely protected from human

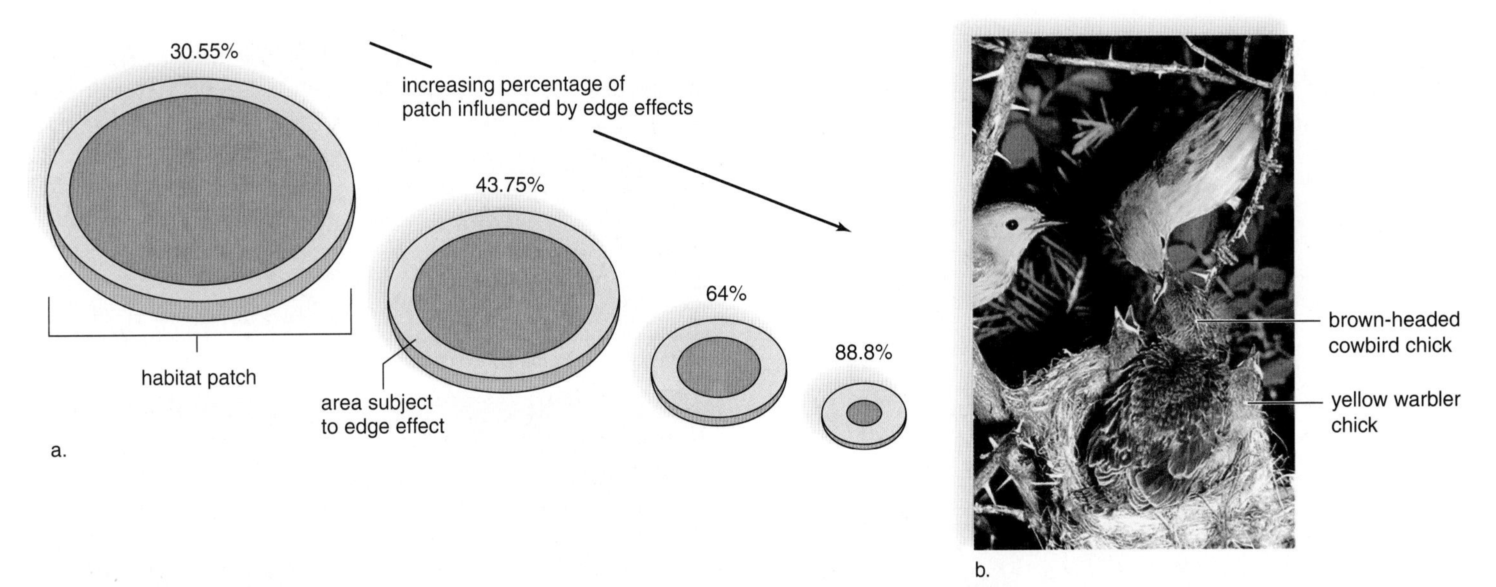

Figure 36.11 Edge effect.
a. The smaller the patch, the greater the proportion that is subject to the edge effect. **b.** Cowbirds lay their eggs in the nests of songbirds (yellow warblers). A cowbird is bigger than a warbler nestling and will be able to acquire most of the food brought by the warbler parent.

activities, except perhaps ecotourism or limited scientific research. When considering the arguments above, reserves should contain sufficient amounts of habitat to sustain the biodiversity within them. This space should include multiple ecosystems that are connected by corridors to allow movement of animals and dispersal of plant species between habitat types. This space should also include enough area to account for edge effects, meaning that the absolute amount of land included in each reserve is likely higher than the usable habitat within it.

Globally, a network of more than 480 biosphere reserves has been designated by the United Nations. The maintenance of biosphere reserves is in the hands of the countries and territories in which the reserves are located, and compliance is voluntary. Because the goals of biosphere reserves include preservation of local cultural values, as well as maintenance of biodiversity and the ability to foster sustainable human development, there has been more cooperation than was initially expected.

Each biosphere reserve is divided into three areas. First, there is the central core reserve, which allows only research, light tourism, and limited sustainable use for cultural purposes. Next, there is a buffer zone that surrounds the core where only low-impact human activities are allowed. A transition area then surrounds the buffer zone. This area is meant to support sustainable human development, tourism, and agriculture.

Although there has been widespread cooperation with the establishment of biosphere reserves, there are few that strictly follow the United Nations' model. People are often unwilling to modify their lifestyles to the extent necessary to achieve sustainability, and funding for biosphere maintenance is often limited. For example, funding levels may be insufficient for compensating landowners for adhering to sustainable development practices in buffer zones or transition areas.

Check Your Progress

1. Name two ways to prioritize habitats for conservation.
2. How should edge effects affect nature reserve design?

Habitat Restoration

In cases where habitat has already been modified in an area to the extent that conservation and reserve formation may not be viable, or to reverse existing damage, habitat restoration is an alternative. **Restoration ecology** is a new subdiscipline of conservation biology that seeks scientific ways to return ecosystems to their state prior to habitat degradation. Although habitat restoration is perceived as beneficial, there is some concern that the restored areas may not be functionally equivalent to the natural regions that were once there. Nonetheless, habitat restoration is increasing, and in the process, three principles have emerged thus far. First, it is best to begin as soon as possible before remaining fragments of the original habitat are lost. These fragments are sources of wildlife and seeds from which to restock the restored habitat. Second, once the natural histories of the species in the habitat are understood, it is best to use biological techniques that mimic natural processes to bring about restoration. This might take the form of using controlled burns to bring back grassland habitats, biological pest controls to rid the area of exotic species, or bioremediation techniques to clean up pollutants. Third, the goal is sustainable development, the ability of an ecosystem to maintain itself while providing services to human beings. The Everglades ecosystem is used here to illustrate these principles.

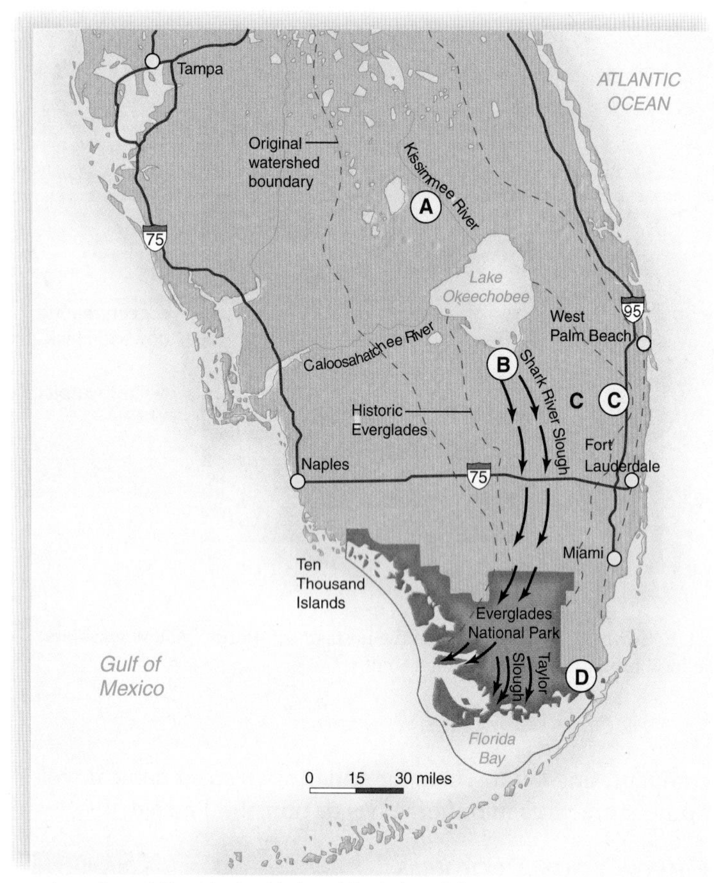

a. Location of Everglades National Park (purple)

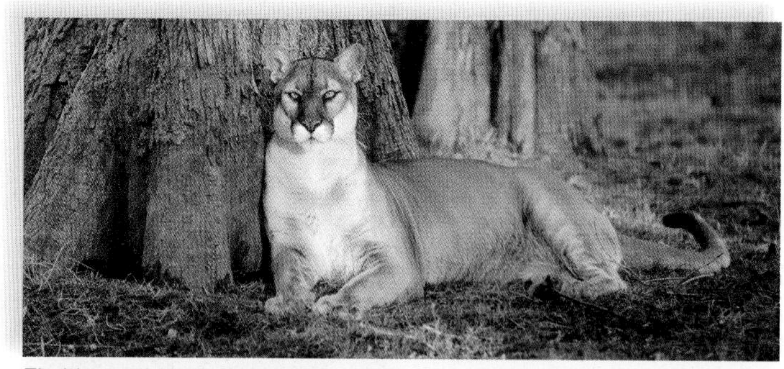

Florida panther, *Puma concolor coryi*

American alligator, *Alligator mississippiensis*

White ibis, *Eudocimus albus*

Roseate spoonbill, *Ajaia ajaja*

Wood stork, *Mycteria americana*

b. Wildlife in Everglades

Figure 36.12 Restoration of the Everglades.
a. Restoration plans call for adding curves and habitat back to the Kissimmee River Ⓐ; creating large marshes and making the Shark River Slough free-flowing again Ⓑ; creating a buffer zone of wetlands between urban development along Florida's eastern coast Ⓒ; and reducing salinity by letting fresh water flow into and through Taylor Slough Ⓓ. **b.** Wildlife of the Florida Everglades.

The Everglades

Originally, the Everglades encompassed the whole of southern Florida from Lake Okeechobee down to Florida Bay (Fig. 36.12). This ecosystem is a vast sawgrass prairie, interrupted occasionally by a cypress dome or hardwood tree island. Within these islands, both temperate and tropical evergreen trees grow among dense and tangled vegetation. Mangroves are found along sloughs (creeks) and at the shoreline. The prop roots of red mangroves protect over 40 different types of juvenile fishes as they grow to maturity. During the wet season, from May to November, animals disperse throughout the region, but in the dry season, from December to April, they congregate wherever pools of water are found. Alligators are famous for making "gator holes," where water collects and fishes, shrimp, crabs, birds, and a host of living things survive until the rains come again. The Everglades once supported millions of large and colorful birds, including herons, egrets, the white ibis, and the roseate spoonbill (see Fig. 36.12).

At the turn of the century, settlers began to drain the land just south of Lake Okeechobee to grow crops in the newly established Everglades Agricultural Area (EAA). To provide flood protection for urban development, a large dike now rings Lake Okeechobee and water is shunted through the St. Lucie Canal to the Atlantic Ocean or through the canalized Caloosahatchee River to the Gulf of Mexico.

The Central and Southern Florida (C&SF) Flood Control Project included the construction of over 2,250 km of canals, 125 water control stations, and 18 large pumping stations. Now the Everglades National Park receives water only when it is discharged artificially from a conservation area, and the discharge is according to the convenience of the C&SF Flood Control Project rather than according to the natural wet/dry season of southern Florida. Because of the huge disruptions to the natural water flows, the Everglades are now dying, as witnessed by declining bird populations. The birds now number in the thousands instead of the millions.

Restoration Plan

A restoration plan has been developed that will sustain the Everglades ecosystem while maintaining the services society requires. The U.S. Army Corps of Engineers is to redesign the Flood Control Project so that the Everglades receive a more natural flow of water from Lake Okeechobee. This will require flooding the EAA and growing only crops such as sugarcane and rice that can tolerate these wetter conditions. This has the benefit of stopping the loss of topsoil and preventing possible residential development in the area. There will also be an extended buffer zone between an expanded Everglades and the urban areas on Florida's east coast. The buffer zone will contain a contiguous system of interconnected marsh areas, detention reservoirs, seepage barriers, and water treatment areas. This plan is expected to stop the decline of the Everglades and its biodiversity, while still allowing agriculture to continue and providing water and flood control to the eastern coast. Sustainable development will maintain the ecosystem indefinitely and still meet human needs.

Check Your Progress

1. What are three principles of habitat restoration?

36.5 Working Toward a Sustainable Society

Learning Outcomes

1. Compare renewable versus nonrenewable resources.
2. Outline two ways to transition to renewable energy resources.
3. Discuss how modern agriculture can be changed to minimize environmental impacts.

Ultimately, biodiversity loss results from human consumption of resources (Fig. 36.13). Some resources are nonrenewable, and some are renewable. **Nonrenewable resources** are limited in supply. For example, the amount of land, fossil fuels, and minerals is finite and can be exhausted. Even though better extraction methods can make more fossil fuels and minerals available and efficient use and recycling can make the supply last longer, eventually these resources will run out. **Renewable resources** are not limited in supply. For example, certain forms of energy, such as solar and wind energy, can be replenished.

The following characteristics indicate that human society today is most likely not sustainable (Fig. 36.14*a*):

1. As discussed previously, a considerable proportion of land, and therefore of natural ecosystems, is being altered for human purposes.

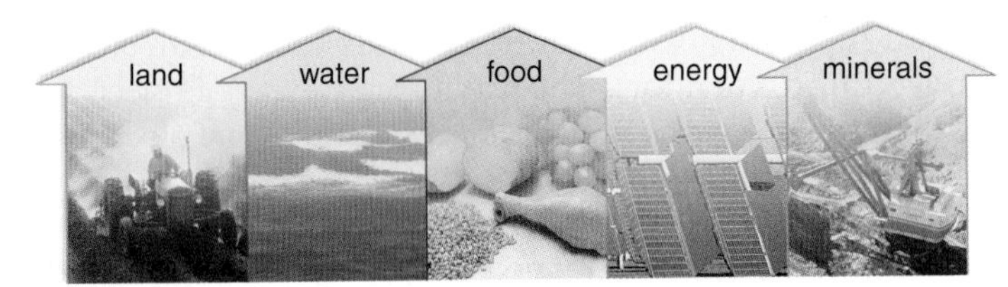

Figure 36.13 Resources.
Human beings use land, water, food, energy, and minerals to meet their basic needs, such as a place to live, food to eat, and products that make their lives easier.

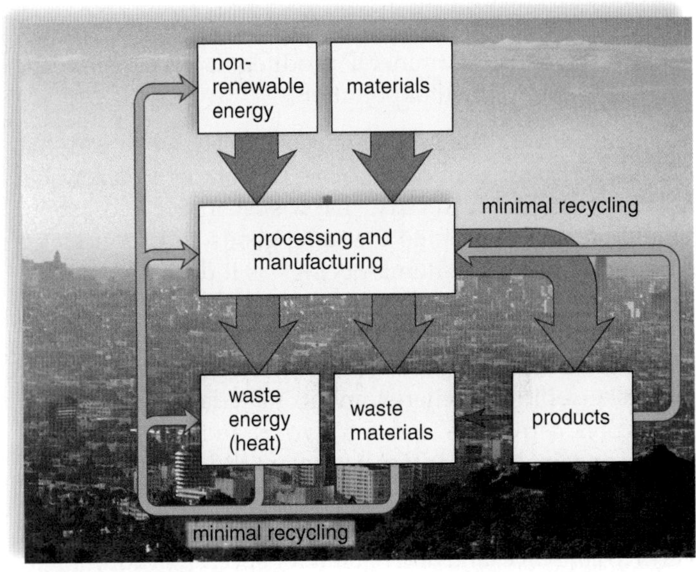

a. Human society at present

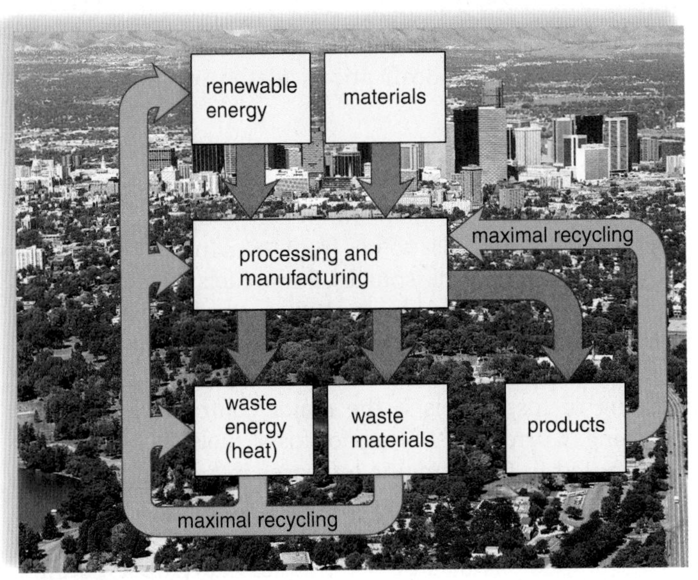

b. Sustainable society

Figure 36.14 Current human society versus a sustainable society.
a. At present, our "throwaway" society is characterized by high input of energy and raw materials, large output of waste materials and energy in the form of heat (red arrows), and minimal recycling (purple arrows). **b.** A sustainable society would be characterized by the use of only renewable energy sources, reuse of heat and waste materials, and maximal recycling of products (purple arrows).

2. Our society primarily utilizes nonrenewable fossil fuel energy, which leads to global warming, acid precipitation, and smog.
3. Even though fresh water is a renewable resource, we are using it faster than it can be replenished in aquifers and other sources.
4. Agriculture requires large inputs of nonrenewable fossil fuel energy, fertilizer, and pesticides, which create much pollution.
5. At least half of the agricultural yield in the United States goes toward feeding animals; it takes 10 lb of grain to produce 1 lb of meat.
6. Minerals are nonrenewable, and the mining, manufacture, and use of mineral products is responsible for much environmental pollution.

To move toward a more sustainable society, we should shift our efforts from use of nonrenewable resources to renewable ones. A **sustainable** society, like a sustainable ecosystem, would be able to provide the same goods and services for future generations of human beings as it does now. At the same time, biodiversity would be conserved.

A natural ecosystem can offer clues as to what a sustainable human society would be like. A natural ecosystem uses only renewable solar energy, and its materials cycle through herbivores, carnivores, and detritivores, and back to producers once again. It is clear that if we want to develop a sustainable society, we too should use renewable energy sources and recycle materials (Fig. 36.14*b*). Section 36.4 discusses ways to conserve land and habitat for species. In addition to recycling and reuse of many of Earth's minerals (e.g., aluminum, copper, iron, and gold), which are nonrenewable, we should also consider alternative energy use, water conservation, and modifications to the way we conduct modern agriculture.

Energy

Presently, about 6% of the world's energy supply comes from nuclear power, and 75% comes from fossil fuels; both of these are finite, nonrenewable sources. Fossil fuels (oil, natural gas, and coal) are so named because they are derived from the compressed remains of plants and animals that died millions of years ago, and as has been discussed previously, burning of fossil fuels is the major contributor to global warming. Currently, shortage of fossil fuels such as oil has contributed to the dramatic increases in heating and gasoline costs. Comparatively speaking, each person in a more-developed country (MDC) uses approximately as much energy in one day as a person in a less-developed country (LDC) uses in one year. Increasing our reliance on renewable energy resources is a major step toward becoming a sustainable society.

Renewable Energy Sources

Renewable types of energy include hydropower, geothermal energy, wind power, and solar energy. Hydroelectric plants convert the energy of falling water into electricity (Fig. 36.15).

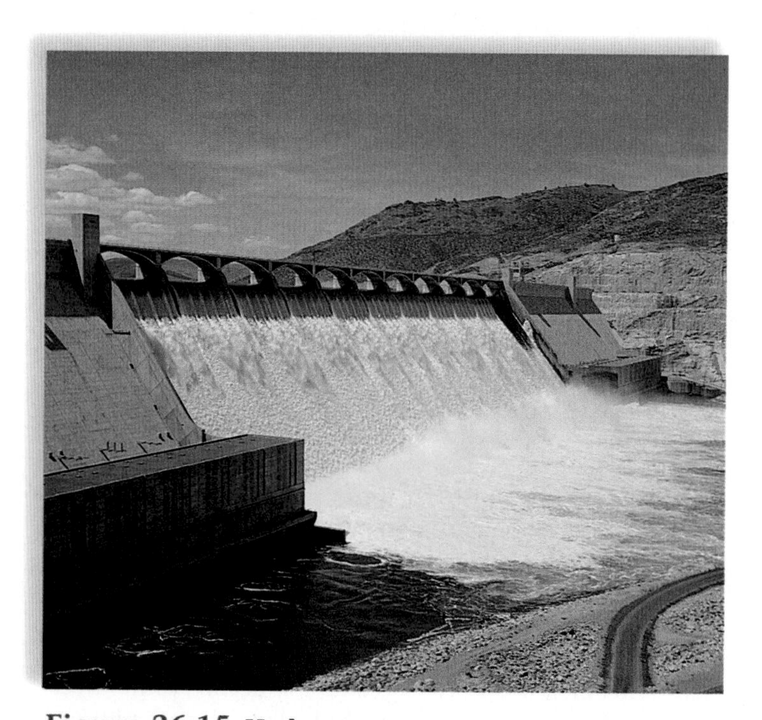

Figure 36.15 Hydropower.
Hydropower dams provide a clean form of energy but can be ecologically disastrous in other ways.

Worldwide, hydropower presently generates 20% of all electricity utilized. In the United States, hydropower accounts for approximately 6% of total energy produced and 67% of the renewable energy used. Brazil, New Zealand, and Switzerland produce at least 75% of their electricity with hydropower, but Canada is the world's leading hydropower producer, accounting for 97% of its renewable energy generation. One way to reduce our reliance on nonrenewable resources such as fossil fuels is to increase our reliance on hydropower, which is expected to rise. However, much of the recent hydropower development has been due to construction of enormous dams that have detrimental environmental effects. Small-scale dams that generate less power per dam, but do not have the same environmental impact, are believed to be the more environmentally responsible choice.

Geothermal energy is produced because Earth has an internal source of heat. Elements such as uranium, thorium, radium, and plutonium undergo radioactive decay underground and then heat the surrounding rocks to hundreds of degrees Celsius. When the rocks are in contact with underground streams or lakes, huge amounts of steam and hot water are produced. This steam can be piped up to the surface to supply hot water for home heating or to run steam-driven turbogenerators. The California Geysers Project, for example, is one of the world's largest geothermal electricity generating complexes.

Wind power is expected to account for a significant percentage of our energy needs in the future (Fig. 36.16*a*). Despite the common belief that a huge amount of land is required for "wind farms" that produce commercial electricity, the actual amount of space for a wind farm compares favorably to the amount of land required by a coal-fired

a.

b.

c.

Figure 36.16 Other renewable energy sources.
a. Wind power requires land on which to place enough windmills to generate energy. **b.** Photovoltaic cells on rooftops and **(c)** the use of hydrogen cars will reduce air pollution and dependence on fossil fuels.

power plant or a solar thermal energy system. A community that generates its own electricity by using wind power can solve the problem of uneven energy production by selling electricity to a local public utility when an excess is available and buying electricity from the same facility when wind power is in short supply.

Solar energy is diffuse energy that must be (1) collected, (2) converted to another form, and (3) stored if it is to compete with other available forms of energy. Passive solar heating of a house is successful when the windows of the house face the sun, the building is well insulated, and heat can be stored in water tanks, rocks, bricks, or some other suitable material.

In a **photovoltaic (solar) cell,** a wafer of an electron-emitting metal is in contact with another metal that collects the electrons and passes them along into wires in a steady stream. Spurred by the oil shocks of the 1970s and most recently in 2008, the U.S. government has been supporting the development of photovoltaic cells. As a result, the price of buying one has dropped from about $100 per watt to around $4. The photovoltaic cells placed on roofs, for example, generate electricity that can be used inside a building and/or sold back to a power company (Fig. 36.16*b*).

Traditional cars have internal combustion engines that run on gasoline, but hybrid cars that are increasing in popularity due to rising fuel costs, run on both gasoline and electricity, increasing mileage per gallon. Now that we have better ways to capture solar energy, scientists are investigating the possibility of using solar energy to extract hydrogen from water via electrolysis. The hydrogen can then be used as a clean-burning fuel. When it burns, water is produced.

Increasingly, vehicles are powered by fuel cells that use hydrogen to produce electricity. Fuel cells are now powering buses in Vancouver and Chicago. Hydrogen cars are already in limited production (Fig. 36.16*c*). Hydrogen fuel can be produced locally or in central locations, using energy from photovoltaic cells. If in central locations, hydrogen can be piped to filling stations using the natural gas pipes already plentiful in the United States. The advantages of such a solar-hydrogen revolution would be at least twofold: (1) the world would no longer be dependent on its limited oil reserves, and (2) environmental problems, such as global warming, acid rain, and smog, would begin to lessen.

Water

In some areas of the world, people do not have ready access to drinking water, and if they do, the water may be impure. It's considered a human right for people to have clean drinking water, but actually, most fresh water is utilized by industry and agriculture. Worldwide, 70% of all fresh water is used to irrigate crops! Domestically in MDCs, more water is usually used for bathing, flushing toilets, and watering lawns than for drinking and cooking.

Although the needs of the human population overall do not exceed the renewable supply of water, this is not the case in certain regions of the United States and the world. When needed, humans increase the supply of fresh water by damming rivers and withdrawing water from aquifers. Dams have drawbacks: (1) Reservoirs behind the dam lose water due to evaporation and seepage into underlying rock beds. The amount of water lost sometimes equals the amount dams make available! (2) The salt left behind by evaporation and agricultural runoff increases salinity and can make a river's water unusable farther downstream. (3) Over time, dams hold back less water because of sediment buildup. Sometimes a reservoir becomes so full of silt that it is no longer useful for storing water. (4) The reduced water below the dam has a negative impact on the native wildlife.

To meet their freshwater needs, people are pumping vast amounts of water from **aquifers,** which are reservoirs just below or as much as 1 km below the ground's surface. Aquifers hold about 1,000 times the amount of water that falls on land as precipitation each year. In the past 50 years, groundwater depletion has become a problem in many areas of the world. Removal of water is causing land **subsidence,** settling of the soil as it dries out. In California's San Joaquin valley, an area of more than 13,000 km^2 has subsided at least 30 cm due to groundwater depletion, and in the worst spot, the surface of the ground has dropped more than 9 m! Subsidence damages canals, buildings, and underground pipes.

Conservation of Water

By 2025, two-thirds of the world's population may be facing serious water shortages. Some solutions for expanding water supplies have been suggested. Planting drought- and salt-tolerant crops would help a lot. Development of many such crops is already underway due to genetic engineering,

a. Polyculture

b. Biological pest control

c. Contour farming

Figure 36.17 Agricultural conservation methods.
a. Polyculture reduces the potential damaging effects of parasites and lessens the need for herbicides. Here, alfalfa has been planted between strips of corn. Alfalfa, a legume, has root nodules that contain nitrogen-fixing bacteria, thus helping to replenish the nitrogen content of the soil and reducing the amount of fertilizers needed. **b.** Instead of pesticides, it is sometimes possible to use a natural predator. Here, ladybird beetles are feeding on cottony-cushion scale insects on citrus trees. **c.** Contour farming conserves topsoil in gently sloping areas because water has less tendency to run off.

as discussed in Chapter 26. Using drip irrigation delivers more water to crops and saves about 50% over traditional methods while increasing crop yields as well. Although the first drip systems were developed in 1960, they are currently used on less than 1% of irrigated land. Most governments subsidize irrigation so heavily that farmers have little incentive to invest in drip systems or other water-saving methods.

Reusing water and adopting conservation measures could help the world's industries cut their water demands by more than half. For example, recycling washing machine water, shower water, or water used to wash cars through a filter before it is discarded as sewage could significantly reduce domestic water usage. Home yard irrigation should occur during dusk and dawn hours, as opposed to in the middle of the day when evaporation is at its highest. Purchasing and using dual-flush toilets can also save millions of gallons of water per year.

Agriculture

In 1950, the global human population numbered 2.5 billion, and at that time, there was only enough food to provide less than 2,000 calories per person per day. Today, current agricultural practices provide enough food to provide everyone on Earth a healthy diet consisting of 2,500 calories per day. However, one-sixth of the world's population (over 1 billion people) are currently considered malnourished due to lack of proper distribution and the fact that much of the grain that is produced in MDCs is used to feed livestock rather than humans.

In addition, modern farming methods are environmentally destructive in the following ways. First, planting only a few genetic varieties, or monocultures (a genetically identical crop), means that a single destructive parasite or pathogen can cause huge crop losses. Second, modern farming relies on heavy use of fertilizers, pesticides, and herbicides. Pesticides reduce soil fertility because they kill off beneficial soil organisms as well as pests. Pesticides also select for artificial resistance in insects, increasing eradication costs via increased pesticide use. In addition, fertilizers, pesticides, and herbicides all contribute to pollution. Third, as discussed above, modern agriculture uses significant amounts of fresh water through irrigation. Fourth, modern agriculture uses large amounts of fuels. Fertilizer production is energy intensive, irrigation pumps require energy to remove water from aquifers, large tractors and even airplanes are used to spread fertilizers, pesticides, and herbicides, and large machines are often used to harvest crops.

Several alternatives exist to employing modern agricultural methods. Polyculture, or planting of several varieties of a crop such as wheat, can reduce the susceptibility of crops to pests or diseases (Fig. 36.17*a*). Polyculture also reduces the amount of herbicides necessary to kill competing weeds and can be used to replenish nutrients to topsoil. Crop rotation—where, for example, nitrogen-fixing crops, such as legumes, are alternated across harvest years with crops that utilize soil nitrogen, such as corn—can help reduce the use of nitrogen-containing fertilizers. Such multiuse farming techniques generally help increase the amount of organic matter and nutrients in the soil.

Organic farms are also increasing in number. Organic farms, as mandated by the U.S. Department of Agriculture, are those in which synthetic pesticides and herbicides cannot be used. Organic farming has become increasingly profitable in recent years because people are more willing to purchase organic produce, despite the increased cost relative to nonorganic produce. Health concerns surrounding pesticide content in nonorganic produce, as well as better tasting food, have helped this trend.

One way that organic farmers have eliminated the need for pesticides is by using integrated pest management, which encourages the growth of competitive beneficial insects and uses biological pest control methods (also called "biocontrol"). Biocontrol has also helped reduce pesticide use in traditional farms as well. Such methods include using natural predators, such as spiders and ladybird beetles, to reduce the numbers of pests (Fig. 36.17*b*). Use of natural or engineered pathogens has also been suggested for biocontrol of pests, but there is concern that such diseases will escape crop areas and affect natural plants or wildlife.

Several types of farming now exist that reduce erosion and help to minimize topsoil loss, such as contour farming (Fig. 36.17*c*). Terrace farming involves converting steep slopes into steplike hills to minimize erosion, and some farmers plant "natural fences," such as rows of trees, around crops to prevent topsoil loss due to wind or other factors. These trees can also be used as other products; mature rubber trees provide us with rubber, and tagua nuts are an excellent substitute for ivory, for example. Cover crops, which are often a mixture of legumes and grasses, also help stabilize soil between rows of cash crops. Finally, avoiding farming on steep slopes also helps reduce erosion. Soil nutrients can be increased through composting, organic farming techniques, or other self-renewable methods. In general, we should consider using precision farming (PF) techniques that rely on accumulated knowledge to reduce habitat destruction, while improving crop yields.

Urban Growth

More and more people are moving to cities. Growth of cities increases pollution via many sources, including automobile exhaust, runoff of pollutants on impervious surfaces (e.g., roads) and into waterways, noise pollution, and industrial and domestic wastes. Thus, the growth of cities involves careful planning to serve the needs of new arrivals, without overexpansion of the city.

Energy use in cities can be curtailed by providing a good public transportation system, preferably one that is energy-efficient. Many U.S. cities are now encouraging carpooling by having HOV (high-occupancy vehicle) lanes on highways, as well as reduced toll costs for cars occupied by more than one person. Portland, Oregon, even has short-term electric car rental stations around the city, whereby people can rent cars for an evening and easily drop them off when they are finished. Maintaining a network of safe bicycle lanes also encourages people to ride their bikes to work.

A way to curtail urban sprawl, or extensive expansion of cities outward, is to build them upward. Several cities, such as Vancouver, Canada, are building many high-rise apartment buildings to accommodate more people in a smaller area. As new buildings are built, they should be as "green" as possible—by using solar or geothermal energy for heating, as well as being constructed out of sustainable or recycled materials. Space on top of buildings could be used to make "green roofs," whereby a wild garden of grasses, herbs, and vegetables is planted to assist temperature control, supply food, reduce rainwater runoff, and be visually appealing. As more people move into cities, city officials can focus on renovating older parts of the city, as opposed to spreading out further.

Modern cities can also have more planned green spaces, including plentiful walking and bicycle paths. In city parks, native species that attract bees and butterflies and require less water and fertilizers could be planted, as opposed to traditional grasses.

Sustainable cities can also improve storm-water management by using sediment traps for storm drains, artificial wetlands, and holding ponds. As new development occurs, cities can increase the use of porous surfaces for walking paths, parking lots, and roads. These surfaces reflect less heat, while soaking up rainwater runoff.

Check Your Progress

1. How is nonrenewable energy use environmentally harmful?
2. How can we conserve water to minimize negative environmental impacts?
3. Discuss two alternative agricultural practices that are environmentally friendly.

Applying the Concepts [Revisited]

At the beginning of the chapter, the conservation success of the bald eagle was highlighted. We have learned that biodiversity is valuable for many reasons, from being charismatic to providing services that help ecosystems function. Currently, we are in a biodiversity crisis that is largely caused by humans, and to survive this crisis and save many of Earth's species, people will have to modify their way of life.

1. Why should we care about conserving biodiversity, even if doing so may increase costs and threaten people's livelihoods?
2. List some ways that you, personally, can contribute to moving humans toward a more sustainable society.

SUMMARIZING THE CONCEPTS

36.1 Conservation Biology and Biodiversity

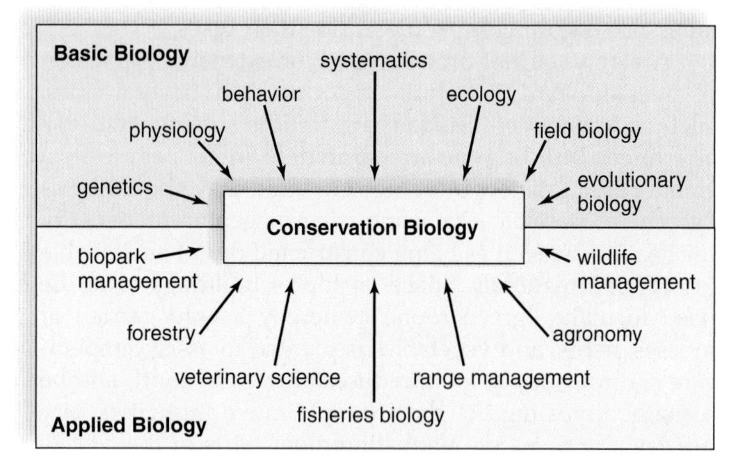

- Conservation is an interdisciplinary science with the goal of conserving Earth's biodiversity for the good of all species.
- Conservation biology often entails complex decisions and thus requires input from people among several subdisciplines of biology, as well as economists, social scientists, and politicians.
- Although biodiversity is often described as the total number of species, it is also considered at the genetic, ecosystem, and landscape levels.
- Genetic diversity helps ensure that populations can evolve in the face of future environmental change.
- Ecosystem diversity considers the functioning of communities within ecosystems and the interconnectedness of species within such communities.
- Landscape diversity includes all of the ecosystems that may interact in a particular area.

36.2 Value of Biodiversity

- A direct value of biodiversity includes the use of wild species for our medical needs.
- Wild species are also directly valuable for their agricultural value as domesticated animals, use for biological controls, and as animal pollinators of crops.
- Biodiversity is also directly valuable for consumptive purposes, such as uses of hardwood trees for lumber.
- Biodiversity also provides immense indirect value by helping to maintain ecosystem functioning.
- Ecosystem services include the workings of biogeochemical cycles, waste disposal, provision of fresh water, prevention of soil erosion, and regulation of climate.
- Ecotourism is an indirect value of biodiversity that is increasing as an important source of income for many LDCs.

36.3 Threats to Biodiversity

- The five major causes of extinction, in decreasing order of importance, are habitat loss, introduction of exotic species, pollution, overexploitation, and disease.
- Habitat loss has occurred in all parts of the biosphere, but concern has now centered on tropical rain forests and coral reefs, where biodiversity is especially high.
- Exotic species have been introduced into foreign ecosystems because of colonization, horticulture or agriculture, and accidental transport.
- Among the various causes of pollution (acid rain, eutrophication, ozone depletion, and organic chemicals), global warming is expected to cause the most instances of extinction.
- Overexploitation is exemplified by commercial fishing, which is so efficient that fisheries of the world are collapsing.
- Wildlife species are increasingly threatened by disease-causing pathogens.

36.4 Habitat Conservation and Restoration

- Biodiversity hotspots, such as tropical rain forests and coral reefs, are often high-priority conservation areas because they contain unusually large concentrations of species.
- Keystone species are often a target for conservation because their loss would result in many secondary extinctions.
- Habitat conservation should be considered at the landscape level, because large species often require several different types of ecosystems to survive.
- Conservation corridors should be maintained or constructed so that species can travel safely between ecosystem types.
- Edge effects can often reduce the amount of suitable habitat for some species. Reserve design should consider edge effects.
- Biosphere reserves are found globally. Each reserve should consist of a core area with nearly no disturbance, a surrounding buffer area where only low-impact human activities are allowed, and an outer transition area meant to support sustainable human development, tourism, and agriculture.
- Restoration ecology has recently emerged with the goal of returning ecosystems to their state prior to habitat degradation.
- Restoration ecology has three basic principles: (1) begin as soon as possible, (2) use biological techniques that mimic natural processes to bring about restoration, and (3) maintain ability of an ecosystem to maintain itself while providing services to human beings.

36.5 Working Toward a Sustainable Society

- The current biodiversity crisis is ultimately caused by human use of Earth's resources.
- Today's society is considered unsustainable because we rely heavily on nonrenewable resources to maintain our existence.
- A shift of focus toward renewable resources and sustainable use of water and agricultural practices will help reduce human impacts in the future.
- Renewable energy sources include wind power, solar power, geothermal power, and hydropower.
- Water conservation methods include drip irrigation, planting drought- and salt-tolerant crops, and household water reuse and recycling methods.
- Modern-day agriculture can employ new methods such as biocontrol to reduce reliance on pesticides and contour farming or similar methods that reduce topsoil erosion.

TESTING YOURSELF

Choose the best answer for each question.

1. Which type of resource is renewable and readily available?
   a. oil
   b. coal
   c. solar energy
   d. underground water supplies
   e. copper
2. Most fresh water is used for
   a. domestic purposes, such as bathing, flushing toilets, and watering lawns.
   b. domestic purposes, such as cooking and drinking.
   c. agriculture.
   d. industry.
3. Removal of groundwater from aquifers may cause
   a. pollution.
   b. subsidence.
   c. mineral depletion.
   d. soil erosion.
   e. All of these are correct.
4. The most significant cause of the loss of biodiversity is
   a. habitat loss.
   b. pollution.
   c. exotic species.
   d. disease.
   e. overexploitation.
5. Exotic species are introduced into new areas by which method?
   a. colonization
   b. agriculture
   c. accidental transport
   d. All of these are correct.
6. Edge effects
   a. result from species crowding at edge of habitats.
   b. mean that no plants can grow well at the edge of a mountain.
   c. may increase wind and temperatures around the outside of habitats.
   d. increase competition between species at the edge of their territories.
   e. All of these are correct.
7. Agricultural runoff of nutrients can result in a fish kill due to
   a. toxic effects of the high nutrient levels on the fish.
   b. toxic effects of the high nutrient levels of the food sources of the fish.
   c. low oxygen levels in the water as decomposers break down algae.
   d. the cooling of the water as algae shade the bottom.
8. A transition to hydrogen fuel technology will
   a. be long in coming and probably not of major significance.
   b. lessen many current environmental problems.
   c. not be likely since it will always be expensive and as polluting as natural gas.
   d. be of major consequence, but resource limitations for obtaining hydrogen will hinder its progress.
9. Global warming does not bring about
   a. the inability of species to migrate to cooler climates as environmental temperatures rise.
   b. the bleaching and drowning of coral reefs.
   c. a rise in sea levels and loss of wetlands.
   d. preservation of species because cold weather causes hardships.
   e. All of these results are expected.
10. Biodiversity hotspots
    a. have few populations because the temperature is too hot.
    b. contain about 20% of Earth's species even though their area is small.
    c. are always found in tropical rain forests and coral reefs.
    d. are sources of species for the ecosystems of the world.
    e. All except a are correct.
11. Consumptive use value
    a. means we should think of conservation in terms of the long run.
    b. means we are placing too much emphasis on living things that are useful to us.
    c. means some organisms, other than crops and farm animals, are valuable as products.
    d. is a type of direct value.
    e. Both c and d are correct.
12. Which of these is not an indirect value of species?
    a. participates in biogeochemical cycles
    b. participates in waste disposal
    c. helps provide fresh water
    d. prevents soil erosion
    e. All of these are indirect values.
13. The services provided to us by ecosystems are unseen. This means
    a. they are not valuable.
    b. they are noticed particularly when the service is disrupted.
    c. biodiversity is not needed for ecosystems to keep functioning as before.
    d. we should be knowledgeable about them and protect them.
    e. Both b and d are correct.
14. Eagles and bears feed on spawning salmon. If shrimp are introduced that compete with salmon for food,
    a. the salmon population will decline.
    b. the eagle and bear populations will decline.
    c. only the shrimp population will decline.
    d. all populations will increase in size.
    e. Both a and b are correct.
15. Which of the following should not be considered when designing nature reserves?
    a. edge effects that may require an increase in size
    b. corridors to facilitate wildlife movement among habitat patches
    c. primarily the needs of a single, important species
    d. inclusion of multiple ecosystem types

UNDERSTANDING THE TERMS

aquifer 759
biodiversity 742
biodiversity hotspots 753
conservation biology 742
conservation corridors 754
conservation reserves 754
ecosystem diversity 743
edge effects 754
endangered species 742
exotic species 750
flagship species 753
genetic diversity 743
keystone species 753
landscape diversity 743
nonrenewable resource 757
overexploitation 752
photovoltaic (solar) cell 759
pollution 751
renewable resource 757
restoration ecology 755
subsidence 759
sustainable 758
threatened species 742

THINKING CRITICALLY

1. What are some ways we can reduce our reliance on nonrenewable resources and move toward increased reliance on renewable resources?
2. Why should people be generally more concerned about pollutants in our meat than in our vegetables?
3. Discuss how removal of a keystone species can disrupt ecosystem functioning. Give an example.

INQUIRY INTO LIFE WEBSITE

The companion website for *Inquiry into Life* provides a wealth of information organized and integrated by chapter. You will find practice tests, animations, videos, and much more that will complement your learning and understanding of general biology.

http://www.mhhe.com/maderinquiry13

Appendix A

Answer Key

This appendix contains the answers to the Check Your Progress and Testing Yourself questions for each chapter, and the Practice Problems and Additional Genetics Problems, which appear in Chapters 23 and 24. Answers to the Applying the Concepts Revisited and Thinking Critically questions that appear at the end of each chapter can be found on the book's website at www.mhhe.com/maderinquiry13.

Chapter 1

Check Your Progress

p. 5: 1. 1) organization; 2) ways to acquire materials and energy; 3) reproduction; 4) responses to stimuli; 5) homeostasis; 6) growth and development; and, 7) capacity for adaptation to their environment. **2.** Organisms share these characteristics because at some point in the very early history of life, all species share a common ancestor which was the first cell or cells. Evolution over approximately the last 3.5 billion years has since shaped the diversity of species we see today. **p. 7: 1.** Domains are the largest, most inclusive classification category. The three domains are Archaea, Bacteria, and Eukarya. The first two are composed of species with prokaryotic cells (no membrane bound nucleus), and the third has organisms with eukaryotic cells. Archaea live in extreme aquatic environments, and eukarya and bacteria live almost everywhere else. Four kingdoms have been classified within the Eukarya, Protista, Fungi, Plantae, and Animalia. **2.** A hierarchical classification system allows increasing specificity about which organisms are under study, and the binomial nomenclature (or 2 name naming system) of the Latin names of genus and species allow researchers to speak a common language about the organisms that they study. **p. 9: 1.** Ecosystems are characterized as the interaction of all the biological communities in an area with their abiotic environment. **2.** Human activities can disrupt basic ecosystem processes by changing climate (i.e., global warming), since climate primarily determines which ecosystems are found around the globe. We also disrupt ecosystems through human development that results in habitat destruction or pollution. **p. 11: 1.** A scientific hypothesis is a tentative explanation for an observed phenomenon that is presented as a falsifiable statement. **2.** The steps of the scientific method are described in Fig. 1.8 and include: 1) make observations or gather data; 2) formulate a hypothesis based on the observations or data; 3) conduct an experiment to test the hypothesis; 4) formulate a conclusion as to whether your experiment supports or refutes the hypothesis; 5) if a hypothesis *is not* supported by the experiment (i.e., rejected) then researchers may formulate new hypotheses or design new experiments to further test the existing hypothesis; 6) if the hypothesis *is* supported by repeated experiments, then it can be used to formulate a theory. **p. 14: 1.** Scientists predicted that rotation with peas, which are legumes and help fix nitrogen in the soil, would increase yield of winter wheat with increased soil nitrogen. They tested this prediction by planting winter wheat in test pots in the laboratory and having control pots (no added nitrogen); pots with equivalent of 45 kg/ha added nitrogen; pots with equivalent of 90 kg/ha added nitrogen; and pots rotated between winter wheat and peas. They found the highest biomass (as a measure of wheat production) in the 90 kg of fertilizer treatment in year 1, and the pea/wheat rotation in years 2–3. These results support their prediction that pigeon pea rotation with winter wheat is more beneficial for wheat growth than simply adding fertilizer. **2.** A control must be a group that is identical to the experimental group, but does not receive the treatment of interest. A common test of this is with drug trials. Test patients either receive a placebo (a pill with nothing in it) or an actual drug being tested. They do not know whether they receive the placebo or real drug, and researchers then record how they respond to their treatment and determine at the end of the experiment whether there is a difference between the control and experimental groups. The null hypothesis is usually no difference. In this case, no difference would suggest the drug is ineffectual and a significant improvement in the treatment group relative to the control group would support that the drug may be appropriate for treatment. **p. 15: 1.** Students could come up with just about any example, just as long as they discuss the pros and cons. They could discuss the DDT example as mentioned in the chapter. A student could do more research on the use of malathion for mosquito control to help control West Nile Virus in the United States (a pro) and come up with cons such as adverse effects on amphibians, fish, and other wildlife.

Testing Yourself

1. c; **2.** d; **3.** b; **4.** a; **5.** b; **6.** b; **7.** d; **8.** a; **9.** c; **10.** c; **11.** a; **12.** e; **13.** d; **14.** c; **15.** b

Chapter 2

Check Your Progress

p. 21: 1. Earth's crust has a different function than living organisms. It is not a carbon-based life form. **2.** The drawing would be similar to the drawings in Fig. 2.2, with twenty protons and twenty neutrons in the nucleus, two electrons in the first orbital, eight electrons in the second orbital, eight electrons in the third orbital, and two electrons in the fourth orbital. **3.** eight. **p. 23: 1.** Since radiation is used to kill things, there are times when that is beneficial (sterilization of food) and times when that is harmful (killing of normal cells). **2.** Carbon only has six electrons, two in the first orbital, and four in the second orbital. It does not need a third orbital. **3.** It is more likely to gain one electron. **p. 26: 1.** Negative, because it needs one electron to fill its outer orbital. **2.** It has four electrons in the outer shell so it can share four pairs of electrons with another atom. **3.** Because hydrogen is such a small atom (it only has one orbital), any bond it forms (except with itself) will be with a larger atom. **4.** Since the electrons are shared, but shared unequally, the charge difference is only partial. The atom has not lost or gained electrons which would result in a positive or negative charge. **p. 29: 1.** Heat capacity is the amount of energy needed to raise the temperature of one gram of water one degree Celsius while heat of vaporization is the amount of energy needed to convert one gram of the hottest water to a gas. **2.** Charged molecules tend to be hydrophilic since they can form hydrogen bonds with water. **3.** A solution at pH 6 contains 1×10^{-6} H^+ ions while a solution at pH 8 contains 1×10^{-8} H^+ ions. 1×10^{-6} is a larger number than 1×10^{-8}. **4.** Because it minimally dissociates and can reform easily. **p. 31: 1.** Because they contain carbon and hydrogen. **2.** The addition or removal of a molecule of water across a chemical bond. **p. 33: 1.** A ring structure with each carbon bonded to a hydrogen and an OH group. **2.** Because our bodies are able to break the bonds between the glucose molecules in starch but not between the glucose molecules in cellulose. **p. 35: 1.** They both have a hydrophobic end and a hydrophilic end. **2.** A double bond tends to make a fatty acid liquid since it introduces a "kink" in the chain that prevents the fatty acids from packing as tightly together. **p. 39: 1.** Enzymes, structural proteins, hormones. **2.** Each one has an amino group and an acid group. **3.** Peptide bonds, hydrogen bonds, covalent bonds. **p. 41: 1.** In the sequence of the nucleotide bases A, C, T, and G. **2.** In the chemical bonds of the phosphate groups.

Testing Yourself

1. d; **2.** d; **3.** d; **4.** d; **5.** c; **6.** d; **7.** a; **8.** a; **9.** a; **10.** d; **11.** b; **12.** c; **13.** a; **14.** a; **15.** a; **16.** d; **17.** a; **18.** d

Chapter 3

Check Your Progress

p. 47: 1. No, it is too large. **2.** Eukaryotic. **p. 48: 1.** To form a barrier between inside and outside and regulate what crosses that barrier. **2.** See Fig. 3.3 for the answer. **3.** A bacterium contains all the necessary structures to live an independent lifestyle—it can move, regulate what comes in and out, make proteins, protect itself from danger, etc. **p. 49: 1.** Protection itself. **2.** Any well-defined subcellular structure that performs a particular function. **p. 56: 1.** Storage of genetic information, synthesis of DNA and RNA. **2.** Protein synthesis. **3.** To get rid of waste, due to a specialized function. **4.** Different processes require different chemical environments and/or different structures. **p. 59: 1.** They are both subcellular organelles, bounded by a double membrane, have an interior compartment, divide by splitting in two, etc. **2.** Sequential, the chloroplast uses the energy of the sun and inorganic molecules to make organic molecules that the mitochondria break down. **p. 61: 1.** All three are part of the cytoskeleton and are dynamic, intermediate filaments play a structural role while both actin filaments and microtubules have a structural and a movement role, actin filaments are composed of actin monomers, microtubules are composed of tubulin monomers, and intermediate filaments are composed of various types of subunits. **2.** Cilia and flagella have a 9 + 2 pattern of microtubules while centrioles have a 9 + 0 pattern. **3.** The dynein side arms slide past each other using the energy of ATP. **p. 63: 1.** They are similar to bacteria in size and structure, they are bounded by a double membrane, they contain a limited amount of genetic material and divide by splitting, they have their own ribosomes (similar to prokaryotic ribosomes) and do produce some proteins, the ribosomal RNA sequences are similar to prokaryotic rRNA sequences.

Testing Yourself

1. a; **2.** d; **3.** c; **4.** d; **5.** c; **6.** c; **7.** d; **8.** c; **9.** a; **10.** b; **11.** e; **12.** d; **13.** d; **14.** a; **15.** **a.** RER, folding, processing of proteins, **b.** nucleus, information in the form of DNA to make the proteins, **c.** nucleolus, where ribosomes are assembled, rRNA made, **d.** Golgi apparatus, proteins are transported from here; **16.** c; **17.** c; **18.** d

Chapter 4

Check Your Progress

p. 69: 1. To separate inside from outside so can control what gets into and out of the cell, maintain different chemical environments, etc. **2.** Channel proteins are not involved in the passage of molecules through the membrane as carrier proteins are. Channel proteins do not combine with the molecule or help it cross the membrane. **p. 71: 1.** Size, charge, chemical composition. **p. 74: 1.** Molecules move from an area of high concentration to an area of low concentration. **2.** In osmosis, the molecules are crossing a semipermeable membrane whereas in diffusion there is no membrane. **3.** This is a hypertonic solution to red blood cells, so they would shrivel. **p. 76: 1.** Both move molecules across the plasma membrane. Both require a carrier molecule. Facilitated transport does not use energy while active transport does. Facilitated transport moves items down their concentration gradient while active transport moves against the concentration gradient. **2.** It is charged and not lipid soluble. **p. 78: 1.** The Golgi apparatus membrane, the plasma membrane. **2.** All three forms bring substances into the cell via vesicle formation. Phagocytosis brings in large molecules, while pinocytosis brings in liquids or very small molecules. Receptor-mediated endocytosis is a form of pinocytosis which uses a receptor protein.

Testing Yourself

1. c; **2.** d; **3.** c; **4.** b; **5.** d; **6.** a; **7.** b; **8.** b; **9.** c; **10.** a; **11.** e; **12.** d; **13.** b; **14.** a; **15.** c; **16.** b

Chapter 5

Check Your Progress

p. 83: 1. See Fig. 5.1. **2.** It removes unwanted tissue and abnormal cells. **p. 84: 1.** Cells stop at the G_1 checkpoint if conditions are not favorable for cell division and/or there is DNA damage. Cells stop at the G_2 checkpoint if the DNA has not finished replicating and/or there is DNA damage. Cells stop at the M checkpoint if the cells are not lined up correctly at the metaphase plate. **2.** Oncogenes encode proteins that promote the cell cycle so a promoting protein that is inappropriately expressed can lead to cancer. Tumor suppressor genes stop the cell cycle so a stopping protein that is inactive leads to a continuation of the cell cycle and possibly cancer. **p. 88: 1.** One, two. **2.** See Fig. 5.5. **3.** Both result in two daughter cells with identical genetic material. Both are divided into four phases. Animal cells have centrioles while plant cells do not. Animal cells pinch in two, while plant cells build a cell plate between the daughter cells. **p. 92: 1.** In order to divide the DNA content in half. **2.** Because the DNA was already replicated before meiosis began. **3.** See Figs. 5.10 and 5.13. **4.** Crossing-over introduces material to one chromosome from the homologous chromosome while independent assortment mixes whole chromosomes donated by both parents. The end result of both is cells that do not have genetic material identical to what was found in either parent. **p. 94: 1.** See Table 5.1. **2.** See Table 5.2. **p. 96: 1.** Sperm and egg cells, sexual reproduction (production of the next generation). **2.** Four, one.

Testing Yourself

1. a; **2.** b; **3.** b; **4.** a; **5.** d; **6.** c; **7.** a; **8.** d; **9.** a; **10.** f; **11.** c; **12.** e; **13.** d; **14.** d; **15.** e; **16.** d; **17.** a; **18.** b; **19.** b; **20.** e

Chapter 6

Check Your Progress

p. 101: 1. Once energy is released as heat it cannot be recaptured. **2.** The energy from the sun which is lost as heat. **p. 103: 1.** Endergonic. **2.** It is available as small packets of energy that are spent to perform work. **p. 105: 1.** So that the energy can be captured and used to perform work instead of being lost as heat, so the reaction can be regulated by the cell. **2.** The enzyme actually does change shape slightly when the substrate binds. **p. 107: 1.** Heat, pH. **2.** The pathway is inhibited at the very first step so energy is not wasted going through the remaining steps. **p. 110: 1.** They both involve the gain and loss of hydrogen atoms. In photosynthesis, water loses hydrogen atoms and carbon dioxide gains them. In cellular respiration, glucose loses hydrogen atoms and oxygen gains them. **2.** See Fig. 6.11 and legend. **3.** They take in carbohydrates and oxygen produced by plants, and produce carbon dioxide and water for consumption by plants. **4.** Carbohydrates, fat, proteins.

Testing Yourself

1. b; **2.** c; **3.** b; **4.** c; **5.** a; **6.** d; **7.** b; **8.** d; **9.** d; **10.** a; **11.** a; **12.** b; **13.** c; **14.** e; **15.** e

Chapter 7

Check Your Progress

p. 115: 1. A hydrogen ion and an electron ($H^+ + e^-$). **2.** Glycolysis occurs in the cytoplasm, preparatory reaction, citric acid cycle, and electron transport chain occur in the mitochondria **3.** Electron transport chain. **p. 116: 1.** The loss of hydrogen atoms which results in a high-energy phosphate group, NAD^+. **2.** Four ATP, two ATP were required to activate glucose in step 1. **p. 119: 1.** Inputs: pyruvate, CoA, NAD^+, outputs: acetyl-CoA, NADH + H^+, carbon dioxide. **2.** Inputs: acetyl-CoA, NAD^+, FAD, outputs: NADH + H^+, CoA, carbon dioxide, $FADH_2$. **3.** They pick up electrons and deliver them to the electron transport chain, then are recycled. **p. 121: 1.** Hydrogen ions are pumped from the matrix into the intermembrane space by the electron transport chain. **2.** Hydrogen ions are trapped in the intermembrane space. When the ATP synthase complex opens, hydrogen ions rush through and generate energy. **3.** Because $FADH_2$ delivers its electrons lower in the electron transport chain than does NADH. **p. 122: 1.** It would be unable to produce any energy and would die. **2.** Lactate.

Testing Yourself

1. b; **2.** d; **3.** a; **4.** c; **5.** e; **6.** b; **7.** a; **8.** b; **9.** c; **10.** b; **11.** b; **12.** e; **13.** b; **14.** b; **15.** d; **16.** d; **17.** c; **18.** a

Chapter 8

Check Your Progress

p. 129: 1. The sun. **2.** The chloroplasts. **p. 130: 1.** $6\ CO_2 + 6\ H_2O \longrightarrow C_6H_{12}O_6 + 6\ O_2$. **2.** Light reactions, Calvin cycle reactions. **p. 133: 1.** Violet-blue-green visible light. **2.** Because they reflect green visible light. **3.** In the noncyclic electron pathway, the excited electron from photosystem II passes down an electron transport chain to photosystem I producing ATP. The excited electron from photosystem I is passed to $NADP^+$ making NADPH. Both ATP and NADPH are used in the Calvin cycle reactions. **4.** In the cyclic electron pathway, the excited electron from photosystem I is returned to photosystem I via an electron transport chain, making ATP. This ATP is used in the Calvin cycle and other enzymatic reactions. **p. 135: 1.** The thylakoid membrane. **2.** Because the movement of the H^+ through the membrane via the ATP synthase complex generates energy (ATP). **p. 136: 1.** Carbon dioxide is attached to RuBP by the enzyme RuBP carboxylase, and then converted to G3P in a two step reaction involving ATP as the energy source and NADPH + H^+ as the source of the electrons for the reduction step. **2.** Six, it takes three turns to generate one G3P and two G3P make one glucose. **3.** G3P can be converted into glucose, sucrose, starch, cellulose, fatty acids, or amino acids. **p. 138: 1.** Sugarcane, corn, succulent plants such as cacti. **2.** It allows the stomata to close during the day so there is not much loss of water without affecting the plant's ability to fix carbon dioxide. **p. 140: 1.** Because the chemical pathways are not the reverse of each other, there are many steps represented in that overall chemical reaction and those are not opposite for photosynthesis and cellular respiration.

Testing Yourself

1. d; **2.** c; **3.** d; **4.** d; **5.** b; **6.** a; **7.** a; **8.** b; **9.** a; **10.** d; **11.** c; **12.** c; **13.** e; **14. a.** thylakoid, **b.** oxygen, **c.** stroma, **d.** Calvin cycle, **e.** granum; **15.** a; **16.** a; **17.** e; **18.** e

Chapter 9

Check Your Progress

p. 145: 1. Because all stems have nodes and internodes while not all stems are aboveground, tubers are underground and yet are stems. **2.** Roots: the root hairs and surface area of the roots allow for uptake of water and minerals, the branching of the roots allow them to stabilize the aboveground portion of the plant, stems: stems allow for continued growth and contain xylem and phloem for transportation of water and nutrients through the plant, leaves: leaves are thin so that sunlight can reach the chloroplasts and gases can diffuse easily, they are carried on a petiole that allows for maximum exposure to the sun. **p. 148: 1.** Meristematic tissue. **2.** Epidermal tissue: protection, conservation of water, ground tissue: photosynthesis and carbohydrate storage, vascular tissue: transport of water and nutrients. **3.** Ground tissue: parenchyma (progenitor cells), collenchyma (flexible support), sclerenchyma (support, some transport water), vascular tissue: xylem (transport water and minerals from roots to leaves), phloem (transports sugar and other organic compounds throughout the plant). **p. 149: 1.** Root: in a ring, stem: in scattered bundles, leaves: in parallel veins. **p. 151: 1.** Via a Casparian strip, requiring water and minerals to pass through the cells to enter the xylem. **2.** The monocot root has a vascular cylinder surrounding a central pith, while the eudicot root has the phloem and xylem in the center of the root. **p. 153: 1.** Because some plants store carbohydrates in their roots. **2.** Adventitious roots (stabilizing the shoot system), epiphytes (absorption of water), haustoria (absorption of water and nutrients), root nodules (obtaining nutrients), mycorrhizae (absorption of water and nutrients). **p. 155: 1.** Protoderm gives rise to the epidermis, ground meristem gives rise to pith and cortex, procambium gives rise to xylem and phloem. **2.** Both contain xylem and phloem, are covered with epidermis, and support the leaves. The eudicot stem contains a central pith surrounded by vascular bundles, while the monocot has vascular bundles scattered throughout the ground tissue. **p. 158: 1.** Due to the growth rate of the tree, spring wood has sufficient water and so contains wide vessels while summer wood has less water and contains fewer vessels. **2.** The sapwood (See Fig. 9.19). **3.** It contains nodes and internodes. **p. 162: 1.** Palisade mesophyll (elongated cells packed tightly) and spongy mesophyll (irregular cells bounded by air spaces). **2.** Shade plants tend to have broad, wide leaves, desert plants tend to have reduced leaves with sunken stomata, leaves tend to be thin and flat to allow maximum sunlight absorption. **p. 165: 1.** Cohesion-water molecules tend to cling together, adhesion-water molecules interact with the walls of the xylem vessels, both are required to keep the column of water together as it ascends the xylem channel. **2.** Evaporation of water from a leaf, removing water from the leaf of the plant causes the whole channel of water to move upward drawing in more water from the roots. **3.** Active transport of potassium into the guard cells surrounding the stomata causes water to flow into the cell by osmosis, this results in the stomata opening, loss of potassium has the opposite effect. **p. 167: 1.** After sugar is actively transported into the phloem sieve tubes, water follows by osmosis, this creates a pressure that causes the phloem contents to flow from the source to the sink. **2.** Not always, phloem transports nutrients from the source to the sink, recently formed leaves could be a sink and then phloem would transport nutrients to them.

Testing Yourself

1. c; **2.** c; **3.** c; **4.** c; **5.** c; **6.** d; **7.** c; **8.** d; **9.** d; **10.** b; **11.** a; **12.** d; **13.** b; **14.** a; **15.** c

Chapter 10

Check Your Progress

p. 173: 1. The sporophyte generation produces microspores and megaspores. A microspore undergoes mitosis and becomes a pollen grain (male gametophyte), and the megaspore undergone mitosis to become a microscopic embryo sac (female gametophyte). **2.** Anther: development of pollen grains. Filament: holds up anther. Stamen: male part of flower consisting of anther and filament. Pistil (or carpel): female part of flower, consisting of stigma, style, ovary and ovule. Stigma: sticky knob for holding pollen. Style: holds up stigma. Ovary: enlarged base that holds one or more ovules. Ovule: house megaspores and when fertilized, is chamber where seed develops. **p. 175: 1.** Because they produce both microspores and megaspores. **2.** A megaspore mother cell undergoes meiosis to form 4 megaspores. Three are discarded and one forms an embryo sac. Within the embryo sac, meiosis forms one egg cell and three polar bodies. After pollination, one pollen grain will fertilize an egg cell in the embryo sac and the other joins with the polar nuclei to form endosperm. The embryo sac then develops into a seed. **3.** Once pollination occurs, one sperm nucleus unites with the egg nucleus, forming a 2n zygote, and the other sperm nucleus migrates and unites with the polar nuclei of the central cell, forming a 3n endosperm cell. **p. 177: 1.** The hypocotyls and cotyledons contribute to above ground growth and the radicle contributes to below ground growth. **2.** In monocots, the endosperm is used to store food for the cotyledon. In eudocots, the endosperm usually is absorbed and the food is stored directly in the cotyledons. **p. 179: 1.** Seeds with structures such as woolly hairs, plumes, and wings can facilitate dispersal by wind. **2.** Some seeds stay dormant for a period of time until it is favorable by growth. **3.** In eudicots, the epicotyl bears young leaves called a plumule, the hypocotyls becomes part of the stem, and the radicle develops into the roots. In monocots, the plumule and radicle are enclosed in protective sheaths and the radicle bursts through these coverings during germination. **p. 183: 1.** Plants can produce outgrowths such as stolons (e.g., strawberries) or rhizomes (violets). Potatoes produce "eyes" or buds. **2.** Plant cells are totipotent, which means that an entire plant can be produced from most plant cells. Plant cell walls can be removed, resulting in protoplasts, and these are grown in a nutrient medium under ideal growth conditions. **3.** Traditionally, plants have been genetically engineered by hybridization. Now, they are also genetically engineered by inserting genes from other plants (i.e., they are transgenic) to make them taste better, to make them more resistant to pesticides or herbicides, or to be toxic to insects. **p. 186: 1.** Plant hormones are small organic molecules produced by the plant that serve as chemical signals between cells and tissues. **2.** Auxin causes the shoot to grow from the top. Gibberelins cause stem elongation. Cytokinins promote cell division. Abscicic acid initiates and maintains seed and bud dormancy and closes stomata when a plant is water stressed. Ethylene is involved in abscicion and fruit ripening.

Testing Yourself

1. b; **2.** c; **3.** b; **4.** c; **5.** e; **6.** a; **7.** d; **8.** c; **9.** d; **10.** d; **11.** *a.* stigma, *b.* style, *c.* ovary, *d.* ovule, *e.* receptacle, *f.* peduncle, *g.* sepal, *h.* petal, *i.* anther, *j.* filament ; **12.** *a.* diploid, *b.* anther, *c.* ovule, *d.* ovary, *e.* haploid, *f.* megaspore, *g.* male gametophyte, *h.* female gametophyte, *i.* sperm, *j.* seed; **13.** b; **14.** b; **15.** c

Chapter 11

Check Your Progress

p. 196: 1. The main functions of epithelial tissue are protection, secretion, absorption, excretion, and filtration. **2.** Epithelial tissues are classified by shape of cells and number of cell layers. **3.** Tight junctions form barriers, gap junctions add strength but allow passage of small molecules, adhesion junctions act like spot-welds to add strength. **p. 197: 1.** Loose and dense fibrous connective tissue both contain collagen and elastic fibers, as well as fibroblasts. In dense fibrous connective tissue, the collagen fibers are packed together more closely. **2.** Hyaline cartilage is found in joints, elastic cartilage is found in the outer ear, and fibrocartilage is found in the pads between the vertebrae. **3.** Bone is the most rigid connective tissue because it contains a hard matrix of inorganic salts, especially calcium. **p. 198: 1.** The main components of blood are the plasma plus the formed elements. **2.** Red blood cells transport oxygen (and CO_2), white blood cells fight infection, and platelets are involved in blood clotting. **3.** Blood contains cells separated by a matrix, but it is a liquid. **p. 199: 1.** The three types of muscle tissue are skeletal, smooth, and cardiac. **2.** Muscle tissue is required for heart contraction, as well as the function of other organs such as the stomach, intestine, and blood vessels. **p. 200: 1.** Three parts of a neuron are the cell body, dendrites, and axon. A nerve fiber is an axon plus its myelin sheath. **2.** Three functions of the nervous system are sensory input, integration of data, and motor output. **3.** Four types of neuroglial cells are microglia, which engulf bacteria and debris, astrocytes, which produce nutrients and hormones, oligodendroglial cells, which form myelin, and ependymal cells, which line the ventricles. **p. 201: 1.** The diaphragm separates the thoracic and abdominal (abdominopelvic) cavities. **2.** The fluids produced by body membranes include mucus, which helps protect against infection, and serous fluid that is mainly involved in lubrication. **p. 203: 1.** There are many possible cells unique to each system, but examples include: hair follicle cells in the integumentary system; heart muscle cells in the cardiovascular system; white blood cells in the lymphatic system; liver cells in the digestive system; lung cells involved in gas exchange (called alveolar cells, see chapter 15) in the respiratory system; kidney cells in the urinary system; bone cells in the skeletal system; skeletal muscle cells in the muscular system; nerve cells (neurons) in the nervous system; hormone-producing cells in the endocrine system; and sperm cells and oocytes in the reproductive system. **2.** Body systems that work together include the respiratory and digestive systems, which help the body acquire essential oxygen and nutrients, the cardiovascular and respiratory systems, which distribute oxygen to tissues, the cardiovascular and lymphatic systems, which circulate immune cells throughout the body, the respiratory, digestive, and urinary systems, which are all involved in removing wastes from the body, the nervous and endocrine systems, which control the other systems, and the nervous, skeletal, and muscular systems, which carry out body movement. **p. 205: 1.** Compared to mucous and serous membranes, skin has a thicker epithelium called the epidermis, with an outer layer of cells containing keratin. Under the epidermis, skin has a thick dermis that contains collagen, blood vessels, and many specialized sensory receptors. **2.** Three accessory organs of skin include nails, which aid dexterity, oil glands that lubricate the hair and skin, and sweat glands that help to cool the body. **p. 208: 1.** The primary homeostatic mechanism for most body systems is negative feedback. **2.** In negative feedback, a sensor detects a change in internal conditions, resulting in a response that brings the conditions back to normal. Positive feedback involves an ever-greater change in the same direction. **p. 209: 1.** The respiratory, digestive, and urinary systems interact to take in O_2 and nutrients needed by the body, and eliminate unused or waste products. **2.** Failure of the immune system results in various infectious diseases, failure of the cardiovascular system results in heart failure or strokes, failure of endocrine organs can result in diabetes or hypothyroidism.

Testing Yourself

1. c; **2.** d; **3.** d; **4.** b; **5.** a; **6.** d; **7.** b; **8. a.** ventral cavity, **b.** thoracic cavity, **c.** abdominal cavity, **d.** pelvic cavity, **e.** dorsal cavity, **f.** cranial cavity, **g.** vertebral canal, **h.** diaphragm; **9.** e; **10.** c; **11.** c; **12.** c; **13. a.** hair shaft, **b.** epidermis, **c.** dermis, **d.** subcutaneous layer, **e.** adipose tissue, **f.** sweat gland, **g.** hair follicle, **h.** arrector pili muscle, **i.** oil gland, **j.** sensory receptor, **k.** basal cells, **l.** sweat pore; **14.** a; **15.** b; **16.** d

Chapter 12

Check Your Progress

p. 215: 1. Blood flow is controlled in arteries by contraction of smooth muscle in artery walls. Capillary blood flow is affected by arterial supply, plus the contraction of precapillary sphincters. Venous blood flow is affected by arterial and capillary blood flow, and valves prevent blood from flowing backward. **2.** O_2 and CO_2 move across capillary walls by diffusion. **p. 217: 1.** Blood enters the right atrium and right ventricle, then travels though the pulmonary arteries to the lungs, and through the pulmonary veins back to the left atrium and left ventricle. **2.** If the left ventricle is not working properly, blood would back up into the lung, increasing blood pressure in the lung, resulting in increased fluid leakage from blood vessels in the lung. **p. 219: 1.** The heartbeat sounds are caused by the closure of the atrioventricular valves, then the semilunar valves. **2.** Intrinsic control of the heartbeat is needed to maintain regular contractions, while extrinsic control allows the heartbeat to change in respond to different situations. **p. 222: 1.** The pulmonary arteries carry O_2-poor blood. **2.** Blood flows to the lungs via the pulmonary arteries, and returns to the heart via the pulmonary veins. Blood flows to the body via the aorta, and returns via the superior and inferior vena cava. **3.** Blood flows in arteries due to the pumping of the heart, and in veins mainly due to muscle contraction. **4.** Systolic pressure occurs during heart contraction; diastolic pressure occurs during heart relaxation. **p. 225: 1.** The major components of blood are plasma, the liquid portion that contains many inorganic and organic nutrients, plus clotting factors and antibodies; and the formed elements, which are the red blood cells (O_2 transport), white blood cells (fight infection), and platelets (clotting). **2.** An increased number of certain types of white blood cells can indicate infection, or even leukemia. A decreased number of red blood cells signifies anemia. **p. 227: 1.** A blood clot begins when platelets clump together and release prothrombin activator, which converts prothrombin to thrombin. Thrombin activates fibrinogen to form fibrin threads that adhere to the platelets and form the rest of the clot. **2.** Diseases such as cancer, which results from too much cell replication, or most infectious diseases, would not be expected to be treatable with stem cells. 3. Since blood proteins are a major factor that determines osmotic pressure, a significant decrease in blood proteins would cause more fluid to leave the blood vessel and enter the tissues, causing edema. **p. 229: 1.** A thrombus is a blood clot that remains stationary, an embolus is a clot that moves with the blood, and an aneurysm is the ballooning of a weakened blood vessel. **2.** Stents could directly activate the blood clotting system, or irritate blood vessels in which they are placed, causing inflammation. **p. 231: 1.** Failing hearts can be replaced with a transplant from a human donor, or with a total artificial heart. A left ventricular assist device can help a failing heart function until it can be replaced. Potential problems include rejection of a transplanted heart by the recipient's immune system, and wear-and-tear on an artificial pump that must beat millions of times. **2.** Hypertension damages blood vessels due to the increased pressure, and also causes the heart to work harder. Hypertension can be treated by addressing the causes (stopping smoking, losing weight), by reducing fluid in the body (through a low-salt diet or medications), and/or by taking medication to dilate the blood vessels.

Testing Yourself

1. b; **2.** b; **3.** a; **4.** d; **5.** e; **6.** *a.* P wave, *b.* QRS complex, *c.* T wave; **7.** b; **8.** e; **9.** c; **10.** *a.* internal jugular vein, *b.* common carotid artery, *c.* inferior vena cava, *d.* common iliac artery, *e.* aorta, *f.* femoral artery; **11.** a; **12.** b; **13.** b; **14.** d; **15.** b; **16.** c; **17.** b; **18.** b

Chapter 13

Check Your Progress

p. 238: 1. Lymph is returned to the cardiovascular system via the lymphatic vessels, which eventually merge into either the thoracic duct or the right lymphatic duct, both of which return lymph to the blood. **2.** Edema is the accumulation of excess tissue fluid, caused by too much fluid produced (i.e., by a tumor), or not enough drained away (i.e., because of low blood protein content or blockage in a lymphatic vessel). **3.** Primary lymphoid organs such as the bone marrow and thymus are sites of lymphocyte development; secondary lymphoid organs such as lymph nodes and the spleen are sites of lymphocyte responses to antigens. **p. 240: 1.** The skin and mucous membranes are physical barriers; a specific chemical barrier is the acid of the stomach. **2.** Macrophages kill pathogens by engulfing them into a vesicle that has an acid pH, hydrolytic enzymes, and reactive oxygen compounds. NK cells induce cells that lack self-MHC-I molecules to undergo apoptosis (cell suicide). **3.** The complement system enhances phagocytosis of pathogens, activates inflammation, and kills pathogens by forming a membrane attack complex. **p. 242: 1.** The clonal selection theory states that B cells and T cells have cell surface receptors for only one specific antigen. When the cell contacts that specific antigen, it is selected to undergo clonal expansion (divide) and differentiate into either memory cells or cells that actively fight infection. **2.** A basic antibody molecule has two heavy chains bound to each other. One (smaller) light chain is attached to each heavy chain, and each combination of heavy and light chains forms an antigen-binding site. **3.** The five classes of human antibodies are IgG (activates complement, crosses the placenta), IgM (early response, activate complement), IgA (found in body secretions), IgD (found only on immature B cells), and IgE (protect against parasites). **p. 243: 1.** T cell and B cell responses are similar in that they both result in activation and proliferation of cells that react to one very specific antigen, but they are different in that activated T cells produce cytokines or kill virus-infected cells, whereas activated B cells secrete antibodies. **2.** APCs can break antigens apart and present pieces on the MHC class II molecules found on their surface, where they can be recognized by helper T cells. **p. 245: 1.** Three types of vaccines include attenuated (live but nonvirulent) vaccines, genetically engineered vaccines (also called subunit vaccines), and conjugated vaccines. **2.** The immune system of the newborn is immature, plus the newborn has not been exposed to any infectious agents, so it would be very susceptible to infectious diseases if it did not receive passive protection from its mother. **p. 247: 1.** The interaction of cytokines with their receptors could be blocked using antibodies that react with the cytokine, or possibly with the receptor. **2.** Monoclonal antibodies are extremely specific, i.e., they react with only one antigen. **p. 251 (left column): 1.** Immediate allergic responses are caused by IgE antibodies that have already been produced against the offending allergen; delayed-type responses take longer because they require sensitized T cells to secrete cytokines that cause inflammation. **2.** If a mother is Rh-positive, she will not produce anti-Rh antibodies, because it is a self-antigen. **3.** Drugs that inhibit cytokine production may inhibit certain desirable immune responses. **4.** Normally MHC molecules are involved in presentation of antigens to T cells, so there may be some inhibition of normal immune responses. **p. 251 (right column): 1.** Autoimmune diseases include myasthenia gravis, multiple sclerosis, systemic lupus, and rheumatoid arthritis. **2.** Congenital immunodeficiency diseases include severe combine immunodeficiency (SCID). AIDS is an acquired immunodeficiency disease.

Testing Yourself

1. b; **2. a.** thymus (primary), **b.** lymph node (secondary), **c.** bone marrow (primary), **d.** spleen (secondary); **3.** c; **4.** a; **5.** c; **6.** d; **7.** a; **8.** e; **9.** e; **10.** d; **11.** c; **12.** a; **13.** b; **14.** a; **15.** d; **16.** e; **17.** a; **18. a.** Ag-binding sites, **b.** light chain, **c.** heavy chain, V = variable, C = constant; **19.** b; **20.** c

Chapter 14

Check Your Progress

p. 257: 1. Mechanically by chewing, chemically by salivary amylase. **2.** Limit sugar intake, brush and floss daily, regular dental visits. **p. 259: 1.** The soft palate prevents food from entering the nasal cavity, the epiglottis covers the trachea. **2.** Gastroesophageal reflux disease can lead to ulcers, difficulty swallowing, or esophageal cancer. **3.** The four layers that make up these organs are mucosa, submucosa, muscularis, and serosa. **p. 261: 1.** The stomach is involved in both mechanical and chemical digestion, while the small intestine is more specialize for chemical digestion. **2.** Hormones that increase digestive activity include gastrin, secretin, and CCK. Inhibiting these might cause difficulties in digesting food, leading to excess gas or constipation. **p. 262: 1.** The large intestine absorbs water, salts, and some vitamins, and eliminates indigestible material as feces. **2.** Bacteria help to break down some indigestible material and also produce some vitamins. **p. 264: 1.** The pancreas secretes hormones such as insulin and glucagon into the blood, and also secretes digestive enzymes into the small intestine by way of a duct. **2.** The liver performs many functions that are essential for life, such as removing toxins from the blood, processing some wastes, and synthesizing some essential plasma proteins such as albumin. The main function of the gallbladder is to store bile, a function that can be dispensed with. **p. 265: 1.** Carbohydrate digestion begins in the mouth, with salivary amylase, and continues in the small intestine, due to maltase and pancreatic amylase. Protein digestion begins in the stomach, where pepsin breaks proteins into peptides, and continues with trypsin and peptidases in the small intestine. Fats are mainly digested by pancreatic lipase in the small intestine. **2.** In the experimental setup shown in Figure 14.12, the tubes could also be kept at different temperatures, or various concentrations of salt or other chemicals could be included in the tubes to determine the effect of these factors on digestion. **p. 267: 1.** The glycemic index (GI) of foods is an indicator of whether the blood glucose response to their ingestion is high or low. The advantage of consuming foods with a high GI is that they can provide a quick burst of energy, but that energy is usually short-lived and over time, consuming too many high-GI foods can lead to type 2 diabetes, heart disease, and other illnesses. Students should assess their personal diet. **2.** A high-fiber diet helps prevent colon cancer and prevents some cholesterol from being absorbed from the intestine, but extremely high-fiber diets can interfere with the absorption of certain minerals. **p. 268: 1.** Essential amino acids must be consumed in the diet, while non-essential amino acids can be synthesized by the body. **2.** Protein-deficient diets can lead to an inability to synthesize essential body components like muscles tissue, antibodies, serum albumin, and many others, but a diet with too much protein may lead to chronic dehydration, kidney stones, and osteoporosis. **p. 269: 1.** Both essential fatty acids and essential amino acids must be consumed in the diet. **2.** Saturated fats tend to raise LDL cholesterol levels. LDL carries cholesterol to the cells, raising the risk of heart disease, while HDL carries cholesterol to the liver to be eliminated, lowering the risk. Consuming trans fats may increase the risk even more, along with a lack of exercise. **p. 271: 1.** A vitamin is an organic substance (other than essential amino acids and fatty acids) needed by the body that can't be synthesized. Unlike most vitamins, vitamins A and D do not act as coenzymes. **2.** Antioxidants such as vitamins C, E, and A, destroy or inactivate free radicals like superoxide and hydroxide. **3.** Dietary supplements may help correct a deficiency, but taking in abnormally large amounts of almost any nutrient may cause harm. **p. 273: 1.** Major minerals required in the diet include calcium, phosphorus, sodium, and magnesium. Trace minerals needed in lower amounts include iron, iodine, zinc, copper, and manganese. **2.** Calcium and phosphorus are needed in large amounts because they are part of bones, and sodium and chloride regulate the body's water balance. **p. 275: 1.** Obesity is having a body weight greater than 20% above the ideal, bulimia nervosa is a desire to be thin that involves binge eating followed by purging (vomiting), and anorexia nervosa is a morbid fear of gaining weight that leads to a very restrictive diet. All three conditions involve an abnormal psychological relationship with food. **2.** Abnormally increased or decreased levels of hormones that influence appetite, such as gastrin, gastric inhibitory peptide (GIP), or leptin, could predispose a person to these conditions. **p. 277: 1.** Most stomach ulcers are caused by the bacterium *Helicobacter pylori*. **2.** The pancreas is important digestive organ, but its most essential function is the secretion of insulin and glucagon. The liver synthesizes bile and many essential proteins, regulate glucose and cholesterol levels, metabolizes many waste products, and stores iron and many vitamins.

Testing Yourself

1. d; **2.** d; **3.** d; **4.** c; **5.** d; **6.** e; **7.** d; **8.** b; **9.** d; **10.** a; **11.** e; **12.** c; **13.** d; **14. a.** large intestine, **b.** small intestine (ileum), **c.** cecum, **d.** vermiform appendix; **15.** a. bile canals, **b.** hepatic artery, **c.** portal vein, **d.** bile duct, **e.** central vein; **16.** d; **17.** a; **18.** b; **19.** b; **20.** d

Chapter 15

Check Your Progress

p. 285: 1. The upper respiratory tract consists of the nasal cavities, pharynx, and larynx. The lower respiratory tract consists of trachea, bronchial tree, and lungs. **2.** The trachea lies in front of (ventral to) the esophagus. When food is swallowed, the epiglottis covers the tracheal opening (glottis) so the food slides over the epiglottis and into the esophagus. **3.** O_2 travels through the nasal cavity, pharynx, larynx, trachea, bronchial tree, and lungs, where it diffuses across the alveolar and capillary endothelial cells. **p. 289: 1.** Tidal volume is the amount of air that normally moves in and out of the lungs with each breath; vital capacity is the maximum volume of air that can be moved in and out during a single breath; expiratory reserve volume is the air that can be forcibly exhaled beyond the tidal volume; residual volume is the air left in the lungs after a forced exhalation. **2.** Inspiration is considered the active phase of respiration because it requires the contraction of the diaphragm and external intercostal muscles; expiration is the passive phase because it requires no muscle contractions, just the elastic recoil of the thoracic wall and lungs. **3.** The respiratory center in the brain automatically sends nerve impulses to the diaphragm and intercostal muscles; the vagus nerve carries inhibitory impulses from the lungs to the brain to stop the lungs from overstretching; and chemoreceptors monitor levels of CO_2, H^+, and O_2 in the blood. **p. 291: 1.** Hemoglobin's most essential function is to carry O_2 (as oxyhemoglobin) to the tissues, but it also carries a small amount of CO_2 (as carbaminohemoglobin) back to the lungs. Excess H^+ ions can be taken up by hemoglobin, which then becomes reduced hemoglobin. **2.** Arterial blood is brighter red than venous blood because the red blood cells in arterial blood contain oxyhemoglobin, which is bright red, compared to the red blood cells in venous blood that contain reduced hemoglobin, which is darker. Blood from a cut appears bright red because hemoglobin becomes oxyhemoglobin upon exposure to O_2 in the air. **p. 292: 1.** Three common disorders of the upper respiratory tract are the common cold, usually caused by rhinoviruses; strep throat, caused by the bacterium *Streptococcus pyogenes*; and otitis media (middle ear infections), caused by various bacteria, viruses, and allergies. **2.** Infections can easily spread from the respiratory system (pharynx) to the middle ear via the auditory (eustachian) tubes. **p. 295: 1.** Asthma is a disorder that causes a narrowing of the bronchi and bronchioles, mostly because of an allergic-type reaction to a variety of irritants. Emphysema, in contrast, restricts the ability of the lungs to expand normally, due to damage to the alveolar walls. **2.** Smoking can predispose the lungs to harm by damaging the cilia that normally help remove mucus, toxins, microbes and debris from the trachea.

Testing Yourself

1. a. nasal cavity, **b.** nostril, **c.** pharynx, **d.** epiglottis, **e.** glottis, **f.** larynx, **g.** trachea, **h.** bronchus, **i.** bronchiole; **2.** a; **3.** b; **4.** d; **5.** c; **6. a.** sinus, **b.** hard palate, **c.** epiglottis, **d.** larynx, **e.** uvula, **f.** glottis; **7.** b; **8. a.** inspiratory reserve, **b.** tidal volume, **c.** vital capacity, **d.** residual volume, **e.** total lung capacity; **9.** b; **10.** a; **11.** a; **12.** b; **13.** e; **14.** c; **15.** c; **16.** d; **17.** d

Chapter 16

Check Your Progress

p. 301: 1. Excretion is the removal of metabolic wastes from the body. **2.** The four major functions are excretion, maintenance of water-salt balance, maintenance of acid-base balance, and secretion of hormones. Excretion is mainly performed by the kidneys. Water-salt balance is also affected by the digestive system as well as the integumentary system (sweating). Acid-base balance is affected by the digestive system and the respiratory system. Hormones are mainly produced by the endocrine system. **p. 303: 1.** The renal cortex contains the glomeruli, proximal convoluted tubules, and distal convoluted tubules. The renal medulla contains the loops of the nephron (loops of Henle). **2.** The epithelium of the PCT has many microvilli (brush border); the DCT doesn't. **p. 305: 1.** Yes, as long the drug molecules are small enough. **2.** All these processes involve the movement of a substance across biological membranes. Diffusion and passive transport are dependent on concentration; active transport is not (but requires energy). Active and passive transport usually require carrier molecules. **3.** Glucose is normally returned from the glomerular filtrate to the blood by reabsorption at the proximal convoluted tubule. **p. 307: 1.** Reabsorption of salt, establishment of a solute gradient, reabsorption of water. **2.** When blood volume/pressure is low, the JG apparatus secretes renin, which converts angiotensin I, which is then converted to angiotensin II that causes aldosterone secretion. **3.** Diuretics can be used medically to reduce blood volume and lower blood pressure. They are used illicitly to lose weight quickly or to dilute urine, thus making it more difficult to detect drugs. **p. 308: 1.** As you hold your breath, your blood pH is decreasing as CO_2 builds up and combines with water, forming carbonic acid (H_2CO_3), which breaks down to form bicarbonate (HCO_3^-) ions and H^+ ions. When you start breathing again, the rate is rapid, mainly because chemoreceptors are sensing the high levels of CO_2 and H^+ ions. **2.** The kidneys control blood pH mainly by reabsorbing HCO_3^- ions and excreting H^+ ions. **p. 311: 1.** Accumulation of toxic metabolic wastes, retention of water and salts, excessive loss of protein and other essential components. **2.** Females have a shorter, broader urethra that is closer to the anal opening, this more easily contaminated with bacteria. An enlarged prostate may cause difficult urination.

Testing Yourself

1. c; **2.** d; **3.** c; **4.** b; **5.** d; **6. a.** renal artery, **b.** renal vein, **c.** aorta, **d.** inferior vena cava, **e.** kidneys, **f.** ureters, **g.** urinary bladder, **h.** urethra; **7. a.** renal cortex, **b.** renal vein, **c.** ureter, **d.** renal pelvis; **8.** d; **9.** c, b, d, a; **10.** a; **11.** b; **12.** c; **13.** b; **14.** a; **15.** c; **16.** d; **17.** c; **18.** b; **19.** a; **20.** b; **21.** c

Chapter 17

Check Your Progress

p. 316: 1. Three functions of neuroglia include nourishing neurons, formation of myelin, and aiding signal transduction. **2.** The three classes of neurons are sensory neurons, which take messages to the CNS, interneurons, which sum up messages from sensory neurons and other interneurons and communicate with motor neurons, which take messages away from the CNS to effector organs, muscles, or glands. **3.** Most neurons contain dendrites, a cell body, and an axon (see Fig. 17.2). **p. 317: 1.** Myelin is formed by Schwann cells in the PNS, and by oligodendroglial cells in the CNS. **2.** Gray matter contains nonmyelinated nerve fibers; in white matter the fibers are myelinated. The brain has gray matter on the surface and white matter in deeper tissue; that pattern is reversed in the spinal cord. **3.** One reasonable hypothesis would be that the type of myelin found in the PNS differs from that in the CNS in such a way that myelin can guide regeneration in PNS but not in the CNS. **p. 319: 1.** The sodium-potassium pumps in neurons are always transporting Na^+ to the outside, and K^+ to the inside of the cell. **2.** During an action potential, the Na^+ gates open and Na^+ enters the cell, causing a depolarization. Then the Na^+ gates close and the K^+ gates open, causing a repolarization (even a slight hyperpolarization, see Fig. 17.4d). **3.** A node of Ranvier is a gap between Schwann cells that make up the myelin sheath of an axon of a PNS neuron. Saltatory conduction is the "jumping" of action potentials as they spread from node to node. **p. 321: 1.** After they release their contents, synaptic vesicles could remain fused with the presynaptic membrane and become a part of it, or they could be recycled and refilled with neurotransmitter. **2.** Synaptic integration is the summing up of all incoming excitatory and inhibitory messages by a neuron. **3.** AChE normally breaks down acetylcholine in the synaptic cleft. The listed effects of black widow spider venom suggest that acetylcholine normally transmits impulses that relax certain muscles and decrease salivation, heart rate, and blood pressure. **p. 323: 1.** The meninges are tough protective membranes that surround the brain and spinal cord. **2.** Intervertebral discs are fluid-filled pads that separate and cushion the vertebrae. **3.** The presence of food in the intestine generates impulses that travel via sensory nerves to the spinal cord, where they synapse on interneurons in a dorsal root. Impulses travel up an ascending tract (in white matter) to a digestive center in the brain, which returns motor impulses down the spinal cord (in white matter) that exit the spinal cord at a ventral root and travel back to the intestine to simulate peristalsis. **p. 325: 1.** The four major lobes of the brain are the cerebrum, the diencephalon, the cerebellum, and the brain stem. **2.** The cerebrum has two lateral ventricles, the diencephalon contains the third ventricle, and the cerebellum and brain stem are associated with the fourth ventricle. **3.** The cerebrum is divided by a major fissure, or cleft, into right and left hemispheres. Sulci are shallow grooves that divide the hemispheres into lobes, and the lobes are subdivided into folds or convolutions called gyri. **p. 327: 1.** The largest amount of both the primary motor and somatosensory areas is devoted to the face, hands, forearms, and tongue. **2.** The two major components of the diencephalon are the hypothalamus, which regulates most basic body functions and links the nervous and endocrine systems by producing many hormones, and the thalamus, which acts as a relay center by receiving information from the senses and distributes it to appropriate brain areas. **3.** The cerebellum integrates sensory information related to body position, and integrates motor impulses to produce smooth coordinated movements. Damage to this area would produce jerky, uncontrolled movements. The medulla oblongata in the brain stem regulates essential functions such as heartbeat, breathing, and blood pressure, so damage to this area would usually produce death. **p. 329: 1.** Because the amygdala influences responses like anger and fear, a tumor in this location could cause a person to be excessively angry or fearful. **2.** S.S. had his hippocampus destroyed by a viral infection, and he couldn't convert short-term to long-term memories. **3.** Mice that lack a glutamate receptor in their hippocampus were unable to learn to run mazes. **p. 330: 1.** Wernicke's area is involved in the comprehension of speech; Broca's area is involved in the actual motor function of speech. **2.** Since Broca's and Wernicke's areas are normally only found in the left hemisphere, there are no redundant areas on the right side to take over their functions if the left side is damaged. **3.** Some scientists believe that "left-brained" persons are more talkative and analytical, and "right-brained" persons are more creative and artistic. **p. 333 (left column): 1.** A ganglion is a collection of nerve cell bodies. Dorsal root ganglia in the spinal cord contain cell bodies of sensory nerves. **2.** The vagus nerve differs from other cranial nerves because it sends branches to many internal organs. **3.** When you touch a hot stove your hand withdraws before you feel the pain, because the nerve pathway for a reflex arc travels directly through the spinal cord, without input from the brain. **p. 333 (right column): 1.** While the somatic system usually requires conscious input from the brain, the autonomic system usually does not. The autonomic system is also divided into sympathetic and parasympathetic divisions; the somatic system is not. The somatic system is mostly involved with body sensation and movement; the autonomic system regulates the heart, internal organs, and glands. **2.** The neurological explanation for a stomachache after jogging might be that your brain directed more activity to the sympathetic system, and less to the parasympathetic. This could decrease peristalsis and direct blood away from the digestive system, leading to some discomfort. **p. 337: 1.** Continued exposure to a drug over time might lead to a compensatory response at synaptic clefts that results in a decreased sensitivity to the drug. For example, if the drug prevents release of NT (see Fig. 17.18 part 2), the cell might synthesize more of the NT. **2.** Students might also want to consider that nicotine and alcohol are more commonly used than marijuana, but only marijuana is illegal. Should all three be illegal? Should marijuana be legalized? **p. 338: 1.** This is an opinion question, but students might first need to define what the "desired" vs. "undesired" effects. **2.** Currently legal drugs include alcohol, nicotine, and caffeine. The discussion about legal drugs could also be broadened to include various prescription drugs with which the students are familiar. **p. 339: 1.** Major symptoms of AD include loss of short-term memory, progressing to an ability to carry out basic functions. These symptoms can begin before age 50, but usually not before 65. **2.** L-dopa is converted to dopamine, which helps to compensate for the loss of dopamine-producing cells in the brain. **3.** Autoimmune diseases occur when the immune system attacks "self" cells, tissues, or organs. In MS, the immune system attacks myelin, oligodendroglial cells, and neurons. **p. 340: 1.** The treatments for different types of stroke are very different. If the stroke is caused by a blood clot, anti-clotting or clot-dissolving drugs (such as tPA) will be administered. These drugs, however, could make a stroke worse if it is caused by excessive bleeding in the brain. **2.** *H. influenzae* and similar microbes are more likely to cause disease if the microbes somehow gain access to the blood or CNS, or in people who are immunocompromised (for example, by very young or old age, AIDS, or chemotherapy). **3.** The risk of acquiring a prion disease can be reduced by not consuming potentially infected tissues (especially brains) from species that are susceptible to prion diseases, as well as by avoiding other exposure to potentially infected tissues (i.e., corneal transplants). **p. 341: 1.** Since stem cells can theoretically differentiate into any type of tissues, they could become CNS neurons after being injected into a damaged area of spinal cord. **2.** Molecular mimicry may allow bacteria or viruses to hide from the host immune response, since the host tends not to react to self-molecules. **3.** Immunosuppressive drugs help to treat MG by inhibiting the production of antibodies that react against the AChR; AChE inhibitors prolong the presence of ACh in neuromuscular synapses; plasmapheresis removes antibodies from the MG patient's blood.

Testing Yourself

1. a; **2.** b; **3.** c; **4.** a; **5. a.** depolarization, **b.** repolarization, **c.** action potential, **d.** resting potential; **6.** b; **7.** c; **8.** b; **9.** c; **10. a.** skull, **b.** meninges, **c.** corpus callosum, **d.** pituitary gland, **e.** midbrain, **f.** pons, **g.** medulla oblongata, **h.** thalamus, **i.** hypothalamus, **j.** pineal gland, **k.** spinal cord; **11.** d; **12.** d; **13.** d; **14.** d; **15.** d; **16.** d; **17.** d; **18.** d; **19.** d; **20.** d

Chapter 18

Check Your Progress

p. 347: 1. Interoceptors are involved in regulating critical body functions like blood pressure, blood volume, and blood pH. **2.** Chemoreceptors—taste and smell; photoreceptors—vision; mechanoreceptors—hearing and balance; thermoreceptors—heat and cold. **3.** Without sensory adaptation, we would constantly be stimulated by all sorts of conditions, which would overwhelm the brain. **p. 349: 1.** Somatic sensory receptors that are interoceptors include proprioceptors (such as muscle spindles and Golgi tendon organs), and pain receptors in the skin and internal organs. Somatic sensory receptors that are exteroceptors include cutaneous receptors like Meissner corpuscles, Krause end bulbs, Merkel disks, Pacinian corpuscles, Ruffini endings, and free nerve endings sensitive to temperature. **2.** Someone who lacks muscle spindles would be susceptible to injuring their muscles and joints due to a lack of awareness of limb position and degree of muscle contraction. A person who lacked pain receptors would be very prone to injury, without the benefit of the protective function of pain. **p. 351: 1.** Adaptations that allow some animals to have more sensitive sense of smell would include having a greater number of olfactory receptors (and increased space in the nasal cavity for these receptors) and having more sensory nerve axons to receive signals from these receptors. **2.** The olfactory bulbs in the brain have direct connections to centers for memory in the limbic system. **p. 353: 1.** The three main layers of the eye are the sclera, the choroid, and the retina. **2.** The pupil is the opening in the center of the iris; the iris is a ring of smooth muscle that regulates the amount of light entering the eye; the vitreous humor is the clear, gelatinous material that fills the eye's posterior compartment; the fovea centralis is the area of the retina where light normally focuses (which is packed with cone cells); and the optic nerve is a bundle of sensory nerve fibers that travel from the retina to the visual cortex. **3.** When the ciliary muscle is relaxed, the suspensory ligaments that attach to the lens are taut, which causes the lens to become flatter. When the ciliary muscle contracts, the suspensory ligaments are relaxed, allow the lens to become more round (i.e., when viewing an object up close). **p. 355: 1.** As light enters the eye, it first passes through the cornea and lens. When light reaches the back of the eye it must first pass through the ganglion cells, then the bipolar cells, and finally the rod cells and cone cells. The rods and cones would be expected to be on the innermost layer, so light would not have to pass through the other cell layers. **2.** As many as 150 rod cells, but perhaps only one cone cell, synapse(s) on an individual ganglion cell, so that more information can be sent to the brain for an individual cone cell vs. a rod cell. **3.** The left optic tract carries information about the right part of the visual field, and vice versa. This information must be synthesized by the visual area of the brain so we can perceive a unified visual field. **p. 358 (left column): 1.** Vibrations of the tympanic membrane are transmitted via the ossicles (malleus, incus, and stapes) to the oval window. **2.** The mechanoreceptors responsible for transducing sound waves into nerve impulses are located on hair cells on the basilar membrane of the spiral organ in the cochlea of the inner ear. **3.** Sound waves transmitted via the oval window cause the basilar membrane to vibrate. Near the tip of the spiral organ, the basilar membrane vibrates more in response to higher pitched sound, and at the base it responds to lower pitches. Different sensory nerves supply each part of the basilar membrane, and take information to different parts of the auditory cortex, where the pitch is interpreted. **p. 358 (right column): 1.** The ampulla and cupula are associated with the semicircular canals; the otoliths, saccule, and utricle are associated with the vestibule. **2.** Having two systems for equilibrium allows the brain to receive information about rotational movement from the semicircular canals, and about the body's position at rest from the vestibule. Although it might be possible to distinguish between these with just one system, it might not be as efficient or effective. **p. 361: 1.** Color blindness is more common in males because it is usually carried on the X-chromosome. Since female cells have two × chromosomes, they are less likely to express the defect. **2.** Nearsightedness results when the eyeball is elongated; hence the lens focuses in front of the retina; farsightedness results when the eyeball is too short. Astigmatism is caused by a misshapen cornea or lens. **p. 363: 1.** In gene therapy, the affected cells of a patient with a genetic deficiency are treated with DNA containing the missing or defective gene. **2.** Causes of sudden hearing loss include infections, head trauma, and certain drugs. **3.** Vertigo is the feeling that the environment is moving when no motion is occurring. In benign positional vertigo, abnormal particles form in the semicircular canals and stimulate a sensation of movement. **4.** Meniere's disease is sometimes compared to glaucoma because the loss of hearing seems related to increased fluid in the inner ear.

Testing Yourself

1. e; **2.** b; **3.** a; **4.** b; **5. a.** retina, **b.** choroid, **c.** sclera, **d.** optic nerve, **e.** fovea centralis, **f.** ciliary body, **g.** lens, **h.** iris, **i.** pupil, **j.** cornea; **6.** d; **7.** e; **8. a.** thalamic nucleus, **b.** optic chiasma, **c.** optic nerve; **9. a.** stapes, **b.** incus, **c.** malleus, **d.** tympanic membrane, **e.** semicircular canals, **f.** cochlea, **g.** auditory tube; **10.** c; **11.** c; **12.** a; **13.** b; **14.** d; **15.** d; **16.** a

Chapter 19

Check Your Progress

p. 368: 1. Yellow bone marrow is found in the medullary cavity of long bones; red bone marrow often fills the spaces of spongy bones. Red bone marrow produces all types of blood cells. **2.** Hyaline cartilage is found at the ends of long bones (in joints), in the nose, ribs, larynx, and trachea. The disks between the vertebrae and in the knee are made of fibrocartilage, and elastic cartilage is found in the ear and epiglottis. **3.** Both ligaments and tendons are composed of dense fibrous connective tissue, but ligaments connect bone to bone, while tendons connect muscle to bone. **p. 370: 1.** During bone remodeling, osteoclasts break down bone, and osteoblasts rebuild it. **2.** Sometimes when a bone is fractured across a growth plate, the plate "closes" and the bone stops growing at that end. **3.** Besides hormone levels, diet, exercise, and genetic predisposition are all important factors in osteoporosis. **p. 373: 1.** The four major bones of the cranium are the frontal, parietal, occipital, and temporal bones. **2.** The most prominent facial bones are the mandible, maxillae, zygomatic bones, and nasal bones. **3.** The function of the hyoid bone is to anchor the tongue and serve as site of attachment of muscles involved in swallowing. It is unique among bones because it does not articulate with any other bone. **p. 375: 1.** A normal human has 33 vertebrae. Seven of these are cervical, 12 are thoracic, and five are lumbar vertebrae. **2.** True ribs are attached to the sternum by costal cartilage, false ribs connect to the sternum via a common cartilage, and floating ribs do not attach to the sternum. **3.** The sternum protects the heart and lungs, and provides attachment sites for the true and false ribs. **p. 378: 1.** Carpal bones are found in the wrist, metatarsal bones are found in the foot, and phalanges are found in the fingers and toes.
2. Compared to the male pelvis, in the female pelvis the ilia are more flared, the pelvic cavity is broader, and the pelvic outlet is wider. **3.** The knee is a typical hinge joint, the elbow is a pivot joint, and the hip is a ball-and-socket joint. **p. 379: 1.** The three types of muscle tissue are smooth, cardiac, and skeletal. **2.** The origin of a muscle is its bone attachment that remains stationary, while the insertion is on the bone that moves. **3.** Flexion involves movement that decreases the angle between bones; extension is the opposite of flexion; abduction moves a bone away from the body; adduction brings a limb towards the midline; rotation moves the part around an axis. **p. 383: 1.** Three unique components of muscle cells are the T tubules, the myofibrils, and myoglobin. **2.** During muscle contraction, myosin filaments use energy from ATP to pull on actin filaments towards the center of the sarcomere. **3.** Neurons control muscle contraction by releasing neurotransmitters at the neuromuscular junction. These chemicals bind to receptors on the sarcolemma, stimulating impulses that spread via the T tubules. **p. 385: 1.** Tropomyosin and troponin are involved in the regulation of muscle contraction by calcium. Threads of tropomyosin wrap around actin filaments, and troponin occurs at intervals along the threads. Calcium released from the sarcoplasmic reticulum binds to troponin, causing the tropomyosin to switch positions, exposing sites where myosin heads can bind. **2.** Creatine phosphate is the fastest way to provide ATP for muscle contraction but it only supplies enough energy for a few seconds of intense activity; cellular respiration provides most ATP used by a muscle cell by breaking down glycogen and fatty acids; fermentation can supply ATP without requiring oxygen. **p. 386: 1.** Three stages of a muscle twitch are the latent, contraction, and relaxation periods. **2.** The ratio of motor nerve axons per muscle fiber in the index finger would probably be similar to that in the eye muscles (1:23), whereas the ratio would be much higher in your gluteus maximus muscle (around 1:1000). **p. 388: 1.** Muscle tone is maintained by having some fibers of a muscle always contracting. This may help to maintain muscle size as well, and explains why muscle atrophy occurs when a nerve is cut. **2.** Mitochondria produce ATP by breaking down glycogen to glucose, which is further broken down in glycolysis. Fat can be broken down to acetyl Co-A, which enters Krebs cycle. Glycolysis and Krebs cycle produced reduced NADH that donates electrons to the electron transport chain in the mitochondrial inner membrane, where most ATP is produced. **3.** Domesticated birds no longer migrate by flying long distances, so they have no need for the endurance associated with slow-twitch fibers. However, there may still be some benefit for them to have the strength and bursts of energy associated with fast-twitch fibers. **p. 390: 1.** Some untreated bone fractures do heal, especially if the individual is able to prevent the bone from moving, or it occurs in a bone that doesn't move a lot (i.e., a rib). However, if two broken ends of a bone continue to move, usually an exaggerated bone callus forms, and the fracture may not heal. **2.** Astronauts would be prone to loss of bone density because if the bones are not bearing weight, they would be remodeled over time by overactivity of the osteoclasts. **3.** Rheumatoid arthritis is an autoimmune disease in which the immune system attacks and damages the joints; osteoarthritis is usually secondary to traumatic damage to a joint (either from an injury or over the lifetime of an older person).

Testing Yourself

1. d; **2.** c; **3.** c; **4.** b; **5. a.** frontal bone, **b.** maxilla, **c.** mandible, **d.** occipital bone, **e.** temporal bone, **f.** parietal bone; **6.** e; **7.** a; **8.** b; **9.** e; **10.** b; **11.** e; **12.** a; **13.** d; **14.** c. **15.** d; **16.** d; **17.** c; **18.** T tubule; **19.** sarcoplasmic reticulum; **20.** myofibril; **21.** Z line; **22.** sarcomere; **23.** sarcolemma; **24.** a; **25.** d

Chapter 20

Check Your Progress

p. 395: 1. Local hormones are chemicals that affect other glands or tissues, but unlike "regular" hormones, local hormones are not carried in the bloodstream. Growth factors are hormones that promote cell division.
2. Exocrine glands have ducts; endocrine glands lack them. **p. 397: 1.** The secretion of insulin is regulated by negative feedback. The pancreas secretes insulin in response to high blood sugar levels, and when these levels fall, insulin secretion is inhibited. **2.** Compared to peptide hormones, it usually takes longer for steroid hormones to have an effect because they usually turn on transcription of certain genes, while peptide hormones usually activate enzymes already present in cells. **3.** In a cell's response to epinephrine, cAMP acts as a second messenger, which means that it helps to convert the binding of a hormone to its receptor into some chemical action inside of cells. **p. 398: 1.** Hormones released by the posterior pituitary are first synthesized in the hypothalamus and stored in axon terminals that extend into the posterior pituitary. The hypothalamus controls the anterior pituitary by producing chemicals that travel via a portal system to act on the anterior pituitary.
2. Oxytocin is regulated by positive feedback because it stimulates the uterus to contract, and the uterine contractions stimulate more oxytocin release. **3.** Since the pituitary gland affects so many other tissues, a cancer of the pituitary could be manifested as effects on any of these tissues. **p. 400: 1.** Levels of T_3 and T_4 are controlled by negative feedback—when levels of T_3 and T_4 rise, the anterior posterior stops producing thyroid-stimulating hormone. **2.** Since calcitonin decreases blood calcium by increasing calcium deposition in bones, a person with a calcitonin-producing thyroid tumor would be expected to have low blood calcium levels and increased bone density. **3.** Parathyroid hormone increases the activity of osteoclasts, reabsorption of calcium by the kidneys, and activates vitamin D, so removal of the parathyroid glands would tend to cause decreased blood calcium and increased blood phosphate. **p. 402: 1.** Secretion of hormones by the adrenal medulla is under direct nervous control; secretions of the adrenal cortex are controlled either by release of ACTH from the anterior pituitary (for glucocorticoids) or by the renin-angiotensin system (for mineralcorticoids). **2.** Epinephrine from the adrenal medulla helps the body deal with stress by activating the fight-or-flight response (increased heart rate, blood pressure, and energy level); glucocorticoids from the adrenal cortex raise blood glucose levels and inhibit the body's inflammatory response. **3.** Aldosterone causes the kidneys to absorb sodium and excrete potassium, which generally raises blood pressure. **p. 403: 1.** Glucose is normally taken up by cells by a transporter protein in the plasma membrane (see section 4.2). Insulin could affect this process by increasing the number of glucose transporters on the cell surface, or by increasing the activity (speed) at which these transporters work. **2.** A cell could only respond to insulin if it has insulin receptors for this peptide hormone on its surface.
3. Somatostatin inhibits glucagon and insulin secretion by negative feedback mechanism. **p. 405: 1.** Some of the worst side effects from taking anabolic steroids are liver dysfunction and cancer, kidney disease, heart damage, stunted growth, infertility, and impotency. **2.** Depending on age, removal of the thymus gland may cause a T cell deficiency. **3.** The pineal gland is located in the midbrain, directly behind the thalamus (see Fig. 17.9).
p. 406 (left column): 1. The function of leptin is to signal satiety, a sense of not being hungry. **2.** Platelets contain growth factors that may stimulate wound healing.
3. Prostaglandins are not distributed in the blood, but instead act close to where they are produced. **p. 406 (right column): 1.** One possible study to investigate the possible influence of pheromones on human behavior would involve exposing females to the T-shirts of men they have not seen, to see which odors they might prefer (or find less offensive), then testing whether those same women tend to be sexually attracted to the men whose odor they preferred. **p. 408: 1.** Because ADH cause more water to be reabsorbed in the kidneys, inhibition of ADH secretion by alcohol helps to explain the increased amount and frequency of urination. **2.** In iodine deficiency, the thyroid gland cannot synthesize enough T_3 and T_4 (which both contain iodine). The anterior pituitary responds by releasing large amounts of TSH, which stimulates the thyroid gland to enlarge.
3. In Addison disease, antibodies play a role in destruction of the adrenal cortex by the immune system. In Graves disease, antibodies bind to the TSH receptor, stimulating production of T_3 and T_4 rather than destroying the gland. **p. 411: 1.** Type 1 diabetes is due to an insulin shortage, whereas type 2 is an insensitivity to insulin, often specifically due to insufficient numbers of insulin receptors on cells. **2.** Symptoms of hypoglycemia include perspiration, shallow breathing, anxiety, and loss of consciousness. **3.** Testing the ability of the body to reduce blood glucose levels is more valuable than measuring blood insulin, because insulin levels may be normal in a person with type 2 diabetes.

Testing Yourself

1. b; **2. a.** pineal gland, **b.** hypothalamus, **c.** pituitary gland, **d.** thymus, **e.** pancreas, **f.** adrenal glands, **g.** thyroid gland; **3.** c; **4.** c; **5. a.** inhibits, **b.** inhibits, **c.** releasing hormone, **d.** stimulating hormone, **e.** target gland hormone; **6.** b; **7.** e; **8.** d; **9.** f; **10.** b; **11.** c; **12.** a; **13.** e; **14.** e; **15. a.** ANH, **b.** renin, **c.** aldosterone; **16.** d; **17.** d

Chapter 21

Check Your Progress

p. 417: 1. The testes produce sperm and testosterone. The epididymides store the sperm, and the vasa deferentia and urethra carry the sperm to the female.

The urethra also functions in urination. **2.** Viagra (sildenafil) is a drug that binds to and inhibits a specific enzyme involved in reducing the dilation of arteries in the penis. **p. 419: 1.** The seminiferous tubules are sites of sperm maturation. **2.** Meiosis is necessary to produce sperm that are haploid, i.e., only contain a single set of chromosomes (as opposed to pairs). For example, each sperm will only contain an "X" chromosome or a "Y" chromosome, not both. **3.** If interstitial cells did not develop and function normally, male hormones such as testosterone would not be produced. When the affected male reached puberty, he would not develop normal secondary sex characteristics (body hair, muscle and skeletal density, genital development, etc.). **p. 421: 1.** The "tubes" that are tied are the fallopian tubes, which may be actually severed and a section removed rather than just being tied off. **2.** In males, the urethra and external genitalia (the penis) function in both reproduction and urination. In contrast, these two systems are entirely separate in females. **3.** During female orgasm, contractions of the uterus may help to move sperm from the uterus to the fallopian tubes. **p. 423: 1.** The corpus luteum is a glandlike structure that develops at the site where a secondary oocyte has been released from a follicle during ovulation. It produces progesterone and some estrogen, which act by negative feedback to decrease the secretion of LH by the anterior pituitary. **2.** During the follicular phase, FSH from the anterior pituitary promotes the development of an ovarian follicle. Ovulation signals the end of the follicular phase. The major feature of the luteal phase is secretion of progesterone by the corpus luteum, which causes the uterine lining to thicken in order to be ready to receive a fertilized oocyte. **p. 425: 1.** Progesterone stimulates the proliferative phase of the uterine cycle, during which the endometrium begins to thicken. The secretory phase of the uterine cycle occurs once the endometrium has thickened, and uterine glands produce a thick mucus. **2.** The placenta produces human chorionic gonadotropin, which maintains progesterone production by the corpus luteum until the placenta can produce progesterone and estrogen, which inhibit the ovarian cycle and maintain the endometrium. **p. 427: 1.** Abstinence is the most effective method for preventing pregnancy, followed by vasectomy, birth control pill, condoms, and natural family planning. **2.** The following methods physically block sperm from entering the uterus: condoms (male and female), diaphragm, and cervical cap. **3.** Morning after pills are controversial because they prevent the implantation of a fertilized human embryo. **p. 431: 1.** HIV may kill CD4+ T cells by directly infecting them, or by inducing apoptosis, or by the action of cytotoxic T cells on the HIV-infected cells. **2.** Reverse transcriptase inhibitors block production of viral DNA from RNA, protease inhibitors prevent cleavage of long viral proteins, fusion inhibitors interfere with HIV entry into cells, and integrase inhibitors prevent insertion of viral DNA into host DNA. **3.** Difficulties faced by HIV vaccine researchers include various strains of HIV, the ability of the virus to mutate, and the tendency of HIV to hide within host cellular DNA. **p. 433: 1.** Viral diseases are generally more difficult to treat than bacterial diseases because there are many differences between bacteria (as prokaryotes) and human cells. In contrast, it is more difficult to kill viruses without harming the cells they infect. **2.** Herpes simplex viruses are less likely to be transmitted and cause disease when they are in the latent stage, but latent viruses are also more resistant to the host immune response (to vaccines). **3.** The cellular protein p53 normally acts as a "brake" on cell division, so when p53 is inhibited by HPV, infected cells will divide more rapidly, which can help the virus to replicate and spread. **p. 434: 1.** Three STDs that are caused by bacteria include chlamydia, gonorrhea, and syphilis. **2.** STDs most likely to cause infertility include chlamydia and gonorrhea. **3.** The major sign of primary syphilis is the formation of a chancre at the site of infection. During secondary syphilis, a rash is accompanied by hair loss and gray patches at nucous membranes. The tertiary stage is marked by cardiovascular and nervous system effects. **p. 436: 1.** Viagra and similar drugs inhibit an enzyme called PDE-5, which normally breaks down chemicals responsible for an erection. **2.** Benign prostate hypertrophy can be treated by drugs such as saw palmetto and finasteride, which inhibit a hormone that encourages prostate growth. BPH can also be treated surgically by transurethral resection, or microwave thermotherapy. **3.** Testicular cancer is relatively treatable because the cancer can be detected early by testicular exam, plus the affected testicle can be removed. **p. 438: 1.** Endometriosis, or inflammation of the uterine lining, can cause infertility by preventing implantation of a fertilized oocyte. **2.** Ovarian cancer is more frequently fatal than testicular cancer mainly because the ovaries are inside a woman's body, therefore it is usually detected much later, after it has spread to other organs. **3.** Dysmenorrhea is any painful menstruation, including symptoms like cramps, headaches, backaches, and nausea. Premenstrual syndrome is a group of symptoms that includes mood swings, joint pain, digestive upsets, and sore breasts. **p. 439: 1.** The most frequent causes of infertility in men are low sperm count, various STDs, abnormal hormone levels, smoking, and alcohol. In females, age and pelvic inflammatory disease are also significant factors. **2.** Artificial insemination by donor requires the least medical intervention, since sperm is simply collected, concentrated, and placed into the vagina (or uterus) by a physician. Gamete intrafallopian transfer requires more intervention, since the oocytes and sperm are brought together in vitro, then introduced into the woman's oviduct. Intracytoplasmic sperm injection requires the most intervention, as a single sperm must be injected into an oocyte.

Testing Yourself

1. d; **2.** e; **3. a.** seminal vesicle, **b.** ejaculatory duct, **c.** prostate gland, **d.** bulbourethral gland, **e.** anus, **f.** vas deferens, **g.** epididymis, **h.** testis, **i.** scrotum, **j.** foreskin, **k.** glans penis, **l.** penis, **m.** urethra, **n.** vas deferens, **o.** urinary bladder; **4.** c; **5.** c; **6.** d; **7. a.** oviduct, **b.** ovary, **c.** uterus, **d.** urethra, **e.** clitoris, **f.** anus, **g.** vagina, **h.** cervix; **8.** d. **9.** d; **10.** e; **11.** b; **12.** f; **13.** a. **14.** g; **15.** c; **16.** e; **17.** b; **18.** c

Chapter 22

Check Your Progress

p. 445: 1. The sperm travels from the oviduct to the egg and squeezes through the corona radiata. Its plasma membrane fuses with the plasma membrane of the oocyte releasing the sperm nucleus into the oocyte. **2.** The membrane depolarizes and lifts off of the oocyte, preventing any further sperm from binding. **p. 447: 1.** Cellular—All three have cleavage resulting in the formation of a multicellular embryo and the formation of a blastula. In the lancelet, the cleavage produces a ball of cells of equal size. In the frog, the cells are not of equal size, and in the chicken, cleavage is restricted to a layer of cells over the yolk. Tissue—All develop three germ layers: ectoderm, mesoderm, and endoderm. In the lancelet, mesoderm forms by outpocketing of the archenteron, while in the frog and chicken, mesoderm forms by migration of cells between the ectoderm and endoderm. **2.** The blastula of the human is similar to the blastula of the chick, even though this difference is not due to the presence of yolk in the chicken egg. Gastrulation is similar between the chick and human. **p. 448: 1.** Endoderm, mesoderm, the gut forms during gastrulation while the coelom forms later from a splitting of the mesoderm. **2.** Ventral (or underneath it toward the gut). **p. 451: 1.** Cytoplasmic segregation is the parceling out of maternal determinants as mitosis occurs while induction is the ability of one embryonic tissue to influence the development of another tissue by the use of signals called inducers. **2.** Morphogenesis determines which sections will develop into particular body parts depending on where that section is relative to other sections in the pattern of the body axis. **p. 452: 1.** The amnion. **p. 455: 1.** Week 1—fertilization, cleavage, blastocyst formation; Week 2—implantation, yolk sac and amnion form, gastrulation occurs; Week 3—the nervous system appears, development of the heart begins; Weeks 4–5—development of the chorion and allantois, umbilical cord forms, limb buds appear, head enlarges, developing eyes, ears, and nose; Weeks 6–8—head assumes normal position relative to body, all organ systems are developed. **2.** An embryo with three germ layers. **p. 456: 1.** The length increases from 152 mm to 300 mm and the weigh increases from 171 g to 1,380 g. **p. 458: 1.** Both have a heart, arteries, veins, capillaries, etc. Many other aspects of the circulatory system are the same except that blood from the right atrium is shunted into the left atrium, and blood that does make it into the right ventricle is shunted to the aorta by the ductus arteriosius. Also blood within the aorta travels through the various branches and then connects with the umbilical arteries which lead to the placenta. The umbilical vein then carries blood from the placenta to the liver where it joins the ductus venosus which merges with the inferior vena cava. **p. 462: 1.** Stage 2. **2.** Prolactin is needed for milk production. During pregnancy the hormone is suppressed due to increased amounts of estrogen and progesterone inhibiting the pituitary. When the baby is delivered, the pituitary begins to secrete prolactin and milk production begins. **p. 463: 1.** Physical, social, and psychological changes, growth, puberty, development of language, sensory development, skill development, learning, etc. **2.** Genetic changes may control how the body responds to extrinsic factors. An ineffective immune system may change how the body responds to extrinsic factors and increase the number of genetic mutations. Extrinsic factors may stress the body such that mutations increase and the whole body responds poorly. **p. 465: 1.** Skin becomes thinner and less elastic, there is less adipose tissue in the subcutaneous layer, the heart shrinks, arteries become more rigid, blood flow to the liver is reduced, loss of skeletal muscle mass, decline in bone density, women undergo menopause, etc.

Testing Yourself

1. d; **2.** a; **3.** b; **4.** c; **5.** c; **6.** c; **7.** b; **8.** c; **9.** b; **10. a.** zygote, **b.** morula, **c.** blastula, **d.** early gastrula, **e.** late gastrula; **11. a.** neural tube, **b.** somite, **c.** notochord, **d.** gut, **e.** coelom; **12.** c; **13.** d; **14.** d; **15.** b

Chapter 23

Check Your Progress

p. 470: 1. They have the same genes in the same order, the DNA sequences do not have to be identical. **2.** Each individual has two factors for each trait and these factors segregate during the formation of gametes such that each gamete contains only one factor from each pair of factors. Fertilization restores the number of factors. **p. 474: 1.** Homozygous—if two members of the allelic pair are the same, heterozygous—if the two members of the allelic pair are different. **2.** To ensure that all sperm/egg combinations are accounted for. **3.** To determine if an individual expressing a dominant allele is homozygous dominant or heterozygous. **4.** A monohybrid cross, a one trait testcross if the one parent is heterozygous. **p. 477: 1.** Four. **2.** Each pair of factors separates independently such that all possible combinations of factors can occur in the gametes. **3.** A dihybrid cross, a two trait testcross if the one parent is heterozygous. **p. 480: 1.** Autosomal dominant—affected children usually have an affected parent, heterozygotes are affected, two affected parents can produce an unaffected child, two unaffected parents will not have affected children, both males and females are affected with equal frequency. Autosomal recessive-most affected children have unaffected parents, heterozygotes are unaffected, two affected parents always have affected children, affected individuals with homozygous unaffected mates have unaffected children, close relatives who reproduce are more likely to have affected children, both males and females are affected with equal frequency. **2.** Autosomal recessive—Tay-Sachs Disease (usually occurs among Jewish people of central/eastern European descent, lack of hexosaminidase A, lysosomal storage disease), cystic fibrosis (most common lethal genetic disorder among U.S. Caucasians, chloride ions fail to pass through a plasma membrane channel protein, thick mucus in lungs and pancreas). Autosomal dominant—Marfan syndrome (defect in fibrillin, connective tissue disease, dislocated lens of eye, long limbs and fingers, caved-in chest, at risk for aorta wall rupture), Huntington disease (neurological disorder that leads to progressive degeneration of brain cells, symptoms usually appear during middle age). **p. 482: 1.** His factors would not have been dominant and recessive. It would have appeared that there was a blending of the traits in the next generation. **2.** Since A and B are codominant alleles, they are given capital letter symbols (I^A and I^B), since both are dominant to O, O is given a lower case symbol (*i*). **3.** Because they receive one from their mother and one from their father, humans are diploid. **p. 484: 1.** Often with twin studies, twins who have been raised in different environments. **2.** Because the phenotypes are not discrete, they tend to be additive.

Practice Problems

p. 472: 1. a. W, **b.** WS, Ws, **c.** T, t, **d.** Tg, tg, **e.** AB, Ab, aB, ab. **2. a.** gamete, **b.** genotype, **c.** gamete, **d.** genotype. **p. 474: 1.** 75%. **2.** Both are Ee. **3.** father DD, mother dd, children Dd. **p. 477: 1.** WwEe. **2.** 1/16 or 6.25%. **3.** father DdFf, mother ddff, child ddff. **Fig. 23.9:** Because they do not have the disorder, yet they must have passed on an "a" allele to their first two children. **Fig. 23.10:** Because they have the disorder, yet they must have passed on an "a" allele to their third child. **p. 482: 1.** No. **2.** Child *ii*, mother I^Ai, father *ii*, I^Ai, I^Bi. **3.** Baby 1 belongs to Mr. and Mrs. Doe, Baby 2 belongs to Mr. and Mrs. Jones. **4.** AABBCC and aabbcc.

Testing Yourself

1. e; **2.** c; **3.** c; **4.** d; **5.** a; **6.** c; **7.** c; **8.** a; **9.** See Fig. 23.6 for answer; **10.** a; **11.** d; **12.** d; **13.** c; **14.** c; **15.** d; **16.** b

Chapter 24

Check Your Progress

p. 488: 1. Because a father passes a Y chromosome to his son. **2.** A person who does not exhibit the trait phenotypically, but can pass the particular allele on to their offspring. **p. 490: 1.** Because females have two × chromosomes, thus they have a normal copy of the gene to mask the mutated copy. **2.** 50% of the offspring (both male and female) will be affected, all of the unaffected females will be carriers. **3.** Long face, prominent jaw, large ears, joint laxity, genital abnormalities, mental impairment, autism. **p. 491: 1.** Because the factors (alleles) "travel together" on the same chromosome, they do not segregate. **2.** Two. **p. 495: 1.** Because all X chromosomes but one in a female are inactivated, because the Y chromosome is so small it carries little genetic information. **2.** The sex-determining region of Y (SRY) gene. **p. 497: 1.** Because elastin is present in many different places and tissues in the body affecting the function of all of them. **2.** As long as all the information is present and the translocation did not break an allele into two pieces, the person can be phenotypically normal.

Practice Problems

p. 490: 1. Females all have the dominant trait, half of the males have the dominant trait, half of the males have the recessive trait. **2.** Mother, mother X^HX^h, father X^HY, son X^hY. **3.** 100%, 0%, 100%. **4.** The father must be color blind, unless there is a genetic abnormality in the daughter (i.e., she has only one X chromosome), the husband is not the father.

Testing Yourself

1. c; **2.** c; **3.** c; **4.** a; **5.** a; **6.** b; **7.** d; **8.** c; **9.** a; **10.** e; **11.** d; **12.** c; **13.** a; **14.** b; **15.** e

Chapter 25

Check Your Progress

p. 505: 1. The radioactive DNA was in the bacterial cells, it showed that the DNA was what entered the cells to allow reproduction of the viruses, therefore, DNA is the genetic material. **2.** When DNA is unwound and drawn flat, the sugar-phosphate backbone forms the sides of the ladder and the bases form the rungs. **3.** Base pairing of A with T and C with G. **p. 506: 1.** RNA is single stranded while DNA is double stranded, RNA has U instead of T (found in DNA), RNA contains ribose while DNA contains deoxyribose, RNA is not wound into a helix while DNA is. **2.** mRNA carries sequence information from the DNA in the nucleus to the cytoplasm, tRNA carries amino acids to the ribosomes. **p. 509: 1.** The double stranded DNA is unwound, an RNA polymerase copies one strand of the DNA into an mRNA molecule, the DNA closes back up. **2.** mRNA codons—GCC/UUU, tRNA anticodons—CGG/AAA, amino acids—alanine/phenylalanine. **3.** Add a cap and tail, remove introns and splice exons back together. **p. 511: 1.** AUG, methionine. **2.** The mRNA moves forward one codon length positioning the peptide-bearing tRNA in the P site. **3.** UAA, UAG, UGA, no. **p. 515: 1.** The specificity of a cell is related to how the genes are regulated, turned on and off, etc. **2.** Transcriptional, genes are clustered in operons which are coordinately transcribed in prokaryotes, genes are individually transcribed in eukaryotes. **3.** Pretranscriptional, transcriptional, posttranscriptional, translational, posttranslational. **p. 520: 1.** Errors in replication, mutagens, transposons. **2.** No effect—a point mutation that is in the third position of a codon that does not change the amino acid assignment, effect—an addition of a nucleotide within a coding sequence resulting in a frameshift and a nonfunctional protein. **3.** Often mutations in tumor-suppressor genes and proto-oncogenes.

Testing Yourself

1. b; **2.** a; **3.** c; **4.** c; **5.** b; **6.** a; **7.** b; **8.** b; **9.** e; **10.** b; **11.** d; **12.** c; **13.** c; **14.** e; **15.** b; **16.** d; **17.** a

Chapter 26

Check Your Progress

p. 528: 1. Using a restriction enzyme to cut open plasmid or other DNA followed by use of DNA ligase to seal the foreign gene into the plasmid. **2.** A DNA fingerprint is a unique genetic signature belonging to each individual. DNA fingerprints can be visualized using gel electrophoresis or STRs. DNA fingerprints can be used for criminology or studying paternity. **p. 531: 1.** Transgenic animals are harder to produce, because it is done individually. Transgenic bacteria can be grown in huge numbers in vats called bioreactors. **2.** Improved agriculture, production of pharmaceuticals, production of hormones, cleanup of environmental pollution. **p. 533: 1.** In vivo and ex vivo methods using viral vectors to infect cells and introduce corrected genes. They also infect people with viruses that help fight off cancer. **2.** An example of in vivo gene therapy is use of an adenovirus vector as an inhalant to treat cystic fibrosis patients. An example of ex vivo gene therapy is removal of dysfunctional genes from an SCID patient, using a virus to "inject" the correct gene into these cells, and re-inoculating the patient with the corrected cells. **p. 536: 1.** Whole genome sequencing can allow us to know all of the genes in an organism for possible health treatments. **2.** Comparative genomics can allow us to better understand gene function because unknown function in a focal organism that are close in genetic sequence to known genes in related organisms can often have a similar function. **3.** The raw genome sequence, in and of itself, needs to be interpreted, especially because there are large portions of the genome (i.e., "gene deserts") with no known function. Functional genomics helps understand the function of focal genes and proteomics helps researchers understand the proteins that they produce, which are the end product of most gene function.

Testing Yourself

1. c; **2.** e; **3.** e; **4.** c; **5.** a; **6.** c; **7.** d; **8.** e; **9.** a; **10.** e; **11.** e; **12.** d; **13.** a; **14.** c; **15.** c

Chapter 27

Check Your Progress

p. 543: 1. Miller and Urey (in 1953) set up a laboratory apparatus that simulated the Earth's environment about 3 billion years ago. Their apparatus contained methane, ammonia and hydrogen and used electric sparks to simulate lightning. After a week, the solution in the apparatus contained several amino acids and organic acids. **2.** The RNA-first hypothesis suggests that RNA was the first macromolecule formed on the way to forming the first cells. This hypothesis is supported by the discovery of ribozymes, the fact that RNA can be both a genetic molecule, as well as an enzyme. The protein-first hypothesis argues that the heat of the sun in the ancient Earth's environment heated up amino acids in shallow puddles and caused them to polymerize. Upon return to water, these structures form microspheres, which have similar properties to a cell. The combination hypothesis suggests that clay in the Earth's environment was conducive to polymerization of both amino acids and proteins. **p. 544: 1.** A protocell likely had a lipid-protein membrane and carried on energy metabolism. **2.** Under the RNA-first hypothesis, RNA would have been the first genetic material to evolve, and the first true cell would have had RNA genes. The protein-first hypothesis suggests that proteins, or at least polypeptides, arose before DNA or RNA. The combination hypothesis suggests that polypeptides and RNA evolved simultaneously. **p. 548: 1.** Fossils show evolutionary links between modern organisms and ancient organisms. They also show transitions, such as *Archaeopteryx*, that is an evolutionary link between reptiles and birds. **2.** Plants and animals, over long periods of time, need to adapt to their environmental conditions or go extinct. Because the environment varies, for example, between the equator and more temperate climates, only certain species that survive in the cooler temperate climates are found there. **3.** Biogeography incorporates the fact that the Earth's continents have been constantly moving over millions of years. This so called continental drift has separated certain land masses for long periods of time, allowing plants and animals to evolve independently to the different conditions found on each land mass. **p. 550: 1.** Homologous structures are those shared through common ancestry, such as hair in mammals. The more closely species are related evolutionarily, the more homologous structures they will share. **2.** Chimpanzees—because they are more closely related evolutionarily. **p. 552: 1.** No mutations, selection, genetic drift or gene flow. The population must also have random mating. **2.** Allele frequencies will change if these predictions are not met. As an example, directional selection will decrease the frequency of the allele that is not favored. **p. 557: 1.** Natural selection results in individuals that are better adapted to the environment having more offspring than those that are less well-adapted. Genetic drift results in random changes in allele frequencies in small populations. **2.** There must be variation in phenotypic traits. This variation must be heritable. There must be over-reproduction and a struggle for survival due to limited resources. There must be differential reproductive success, where some individuals produce more offspring than others. **3.** Stabilizing selection favors intermediate traits of a phenotype in a population. Directional selection favors extreme traits at one end of the spectrum or the other, while disruptive selection favors traits at both extremes simultaneously. **p. 558: 1.** Gene flow, genetic drift, stabilizing selection. **2.** Heterozygotes for sickle cell disease are also resistant to malaria. As such, this creates a balanced polymorphism whereby the sickle cell allele is maintained in relatively high frequency in populations subject to malaria. **p. 559: 1.** Allopatric speciation occurs when population evolve independently because of the presence or emergence of a geographic barrier to dispersal. Sympatric speciation occurs when populations diverge and evolve independently when they are in the same geographic area without a dispersal barrier. **p. 561: 1.** Adaptive radiation is characterized by rapid evolutionary change in morphological and or ecological diversity of a diversifying lineage (usually a species). A classic example is Darwin's finches, with 14 different species, each found on different islands in the Galapagos. **2.** There are two major models that incorporate the rate of speciation. The first is the phyletic gradualism model, which argues that evolution is a slow, gradual process. In contrast is the punctuated equilibrium model, which states that there are long periods of evolutionary stasis followed by rapid speciation events that occur over relatively short periods (in geological time). **p. 565: 1.** The binomial nomenclature is a "two name naming system" that uses a Latin genus and species name to name each of the nearly two million described species on Earth. **2.** A clade is a phylogenetic grouping that includes an ancestor and all of its descendents. **3.** The three domain system classifies life into the Archaea, Bacteria and Eukaryote domains. The prokaryotes are separated into two groups. The five kingdom system recognizes: Monera, Protista, Plantae, Fungi and Animalia. Note here that the eukaryotes are separated into four kingdoms, and the prokaryotes are in their own kingdom (Monera).

Testing Yourself

1. a; **2.** c; **3.** a; **4.** b; **5.** d; **6.** c; **7.** b; **8.** d; **9.** c; **10.** f; **11.** a; **12.** e; **13.** b; **14.** b; **15.** b; **16.** b

Chapter 28

Check Your Progress

p. 571: 1. Pasteur used an S-shaped flask to prove that sterile broth did not spontaneously generate microbes, as long as the flask had a curve in its neck to prevent microbes in the air from contaminating the broth. **2.** Types of microbes include bacteria, archaea, protists, fungi, and viruses. **3.** Decomposers break down and thereby recycle organic and inorganic material. Without decomposers, all dead organisms and organic material would accumulate, and eventually the supply of many nutrients would run out. **p. 575: 1.** The three basic bacterial shapes are bacillus, coccus, and spiral-shaped. **2.** In bacterial conjugation, actual physical contact (via a sex pilus) is required for two bacteria to share DNA. Transformation is the uptake of DNA by a bacterial cell, and in transduction DNA is transferred from one bacterium to another by viruses. **3.** Virulence factors are gene products that increase the ability of microbes to cause disease. For example, *E. coli* O157:H7 has genes coding for toxins, and MRSA has genes coding for antibiotic resistance. **4.** Human diseases caused by bacteria of the genus *Streptococcus* include dental caries, "strep throat," scarlet fever, rheumatic fever, and necrotizing fasciitis, or "flesh-eating" disease. **5.** Since many antibiotics work by inhibiting normal bacterial metabolism, bacteria can become resistant to antibiotics if a mutation occurs that alters the target of the drug (often an enzyme). Alternatively, some bacteria may acquire the ability to alter the antibiotic itself, so the drug is destroyed or inactivated. **p. 576: 1.** The plasma membranes of archaea have a monolayer of lipids with branched side chains, compared to the lipid bilayer of bacterial membranes, which contributes to the ability of many archaea to grow at very high temperatures. **2.** In addition to the membrane structural variations described above, archaeal enzymes must be able to function at high temperatures and/or salt concentrations that would inactivate the enzymes of most other organisms. **3.** Archaea living in the digestive tracts of cattle produce methane that may contribute to global warming, so those concerned about climate change might choose to consider a vegetarian diet keeping in mind that the production of fruits, grains, and vegetables also relies on the use of gasoline or diesel engines that themselves may contribute to global warming!) **p. 579: 1.** The endosymbiotic theory states that mitochondria evolved when an aerobic bacterium was engulfed by a nucleated cell and survived, and that chloroplasts were similarly derived from a photosynthetic bacterium. **2.** Under unfavorable environmental conditions, many protists produce spores or cysts that can survive for long periods, allowing the organism to survive until environmental conditions improve. **3.** Both green algae and plants have cellulose in their cell walls, perform photosynthesis using chlorophylls a and b, and store their carbohydrate reserves as starch. **4.** The two overlapping outer shells of diatoms are made of silica, and contain many intricate depressions and passageways that allow contact between the plasma membrane and the environment. **p. 584: 1.** African sleeping sickness is caused by *Trypanosoma brucei*, a flagellated protozoan. It is transmitted to humans by the bite of the tsetse fly. **2.** Some protists of the genus Euglena are photosynthetic and motile, while some are heterotrophic, suggesting they could be classified as either algae or flagellates. **3.** The White Cliffs of Dover and the Egyptian pyramids are both largely composed of the tests of dead foraminiferans. **4.** The Apicomplexa produce spores, and most phases of their life cycle are nonmotile. Most are also parasitic. **5.** Members of the genus Plasmodium carry out their complex life cycles in various host cells, making it difficult for the immune system to eliminate the parasite. Malaria also occurs mostly in developing countries, where there is less financial incentive for companies to invest in vaccines or treatments. **6.** Most water molds are saprophytes, and some are parasitic. Plasmodial slime molds feed on dead plant material and bacteria by phagocytosis. Cellular slime molds are also phagocytic, feeding on bacteria and yeasts. **p. 586: 1.** A saprotroph obtains its nutrition from dead organic material; hyphae are the filaments that make up a fungal body or mycelium; septate fungi have hyphae that are divided into chains of cells; chitin is the major chemical component of fungal cell walls, and fungal spores are haploid reproductive cells. **2.** Fungal cell walls contain chitin; plant cell walls contain cellulose. **3.** A tinea is a fungal skin infection. **4.** Fungi are eukaryotes, thus their cellular structures and metabolic pathways are very similar to those of their host. This means it is more challenging to find drugs that can selectively inhibit fungal pathogens compared to inhibiting bacteria, which are prokaryotes. **p. 589: 1.** Unlike other fungi, chytrid fungi are aquatic and have a motile stage. **2.** Rhizopus is a zygospore fungus, Penicillium is a sac fungus, and mushrooms are club fungi. **3.** Some yeasts produce carbon dioxide when they ferment sugars, and this causes bread dough to rise. Yeast fermentation also produces ethanol. **4.** A lichen is an association between a fungus and either cyanobacteria or green algae. The fungus receives organic nutrients from the photosynthetic partner, and the algae or cyanobacteria are physically protected and receive minerals and water. **5.** Mycorrhizae are mutualistic relationships (both partners benefit) between fungi and the roots of plants. **p. 592: 1.** A typical enveloped virus has a protein capsid that contains its nucleic acid (DNA or RNA) and is surrounded by a lipid envelope in which protein spikes are embedded. **2.** Viral entry refers to penetration of the viral capsid into the host cells, while uncoating occurs when the viral nucleic acid is freed from the capsid inside the cell. Replication refers to copying of the viral nucleic acid, prior to biosynthesis of capsid and spike protiens. **3.** During viral latency viruses are hidden inside cells, expressing none or just a few of their genes. Retroviruses accomplish this by integrating into the host DNA as a provirus. **p. 595: 1.** Influenza viruses may undergo antigenic shifts when two viruses infect the same cell. When new viruses are assembled, some may contain RNA segments from both original viruses, leading to new combinations. **2.** Measles deaths declined dramatically in the United States after the introduction of the measles vaccine. **p. 596: 1.** A viroid is a circular piece of naked RNA. Unlike viruses, viroids lack a protein capsid, and the viroid's RNA doesn't code for any proteins. **2.** Unlike all other microbes, prions are thought to contain no nucleic acid, just protein. **3.** Prions are unique in their ability to act as a template to convert proteins of normal shape into the abnormal, disease-producing form.

Testing Yourself

1. a; **2.** c; **3.** b; **4.** b; **5.** c; **6.** d; **7.** e; **8.** b; **9.** c; **10.** b; **11.** e; **12.** a; **13.** a; **14.** d; **15.** e; **16.** a; **17.** 1. attachment, 2. entry, 3. replication, 4. biosynthesis, 5. assembly, 6. budding; **18.** b; **19.** b; **20** c; **21.** a; **22.** c; **23.** b

Chapter 29

Check Your Progress

p. 603: 1. Five significant events associated plant evolution are: (1) embryo protection, (2) vascular tissue, (3) megaphylls, (4) seeds, and (5) the flower. **2.** There is an alternation between the diploid sporophyte generation and the haploid gametophyte generation. The sporophyte produces haploid spores that are reproductive cells and can develop into a new organism without fusing with another reproductive cell. The spore undergoes mitosis and becomes a gametophyte. The gametophyte then produces gametes by mitosis. **p. 606: 1. a.** Nonvascular plants are limited due to swimming sperm, and they lack vascular tissue, thereby limiting the transport of nutrients and water. **b.** Nonvascular plants are adapted to a land environment with wind-blown spores and the fact that they have embryonic protection. **2.** A liverwort has a flat lobed thallus, whereas a moss has leafy green shoots. **p. 609: 1.** Lycophytes were the first plants to have vascular tissue and true roots, stems, and leaves. **2.** The sporophyte is dominant in ferns (gametophyte is dominant in mosses); the gametophyte is separate from the sporophyte in ferns, whereas the two life stages are attached in mosses. **3.** Ferns have megaphylls. They can proliferate in dry areas by vegetative propagation. **p. 615: 1.** Microspores develop in pollen sacs and they become pollen grains (male gametophye). Megaspores develop in ovule and one out of four becomes an emboryo sac (female gametophyte). **2.** Double fertilization results in presence of endosperm in addition to a zygote. Presence of an ovary leads to production of seeds enclosed by a fruit. Animals are often used as pollinators.

Testing Yourself

1. b; **2.** b; **3.** *a.* sporophyte, *b.* meiosis, *c.* gametophyte, *d.* fertilization; **4.** b; **5.** b, c, d; **6.** c, d; **7.** a; **8.** b; **9.** c; **10.** d; **11.** c; **12.** d; **13.** c; **14.** *a.* stamen, *b.* pistil, *c.* petals, *d.* sepals; **15.** c

Chapter 30

Check Your Progress

p. 623: 1. Five characteristics animals have in common are: 1. movement or locomotion via muscle fibers; 2. multicellularity; 3. typically diploid adults; 4. usually sexual reproduction; 5. usually heterotrophic. **2.** Arthropods have: bilateral symmetry, three tissue layers, a body cavity, protostome development and molt their cuticle. **p. 624: 1.** Sponges: 1) are aquatic; 2) have many pores; 3) are multicellular but lack tissues; 4) are asymmetrical; 5) are filter feeders; 6) are usually hermaphroditic; 7) can reproduce both sexually and asexually; 8) have a skeleton composed of spicules. **2.** A sponge is a sessile filter feeder that uses flagellated collar cells to cause water flow in its inner cavity for feeding. Most sponges are hermaphroditic, although they usually do not self fertilize. They can reproduce asexually or sexually. **p. 626: 1.** Cniderians have true tissues, whereas sponges only have the cellular level of organization. **2.** Cniderians have a polyp and medusa stage. Polyps are sessile, have an upward directed mouth, and produce medusae by asexual budding. The medusan stage has a downward directed mouth, is motile, and produces eggs and sperm. **3.** A hydra is a small tubular polyp about one quarter inch in length with an incomplete digestive tract. The single mouth is surrounded by four to six tentacles that contain a large number of nematocysts. It has two tissue layers separated by mesoglea. **p. 629: 1.** Flat, incomplete digestive tract, single opening (mouth), tend to be colorless, flame cells for excretion and a ladder-like nervous system. **2.** Flukes and tapeworms. **p. 632: 1.** They have a visceral mass, a strong, muscular foot and a mantle. **2.** Gastropods—nudibranchs, conchs and snails; Cephalopods—octopuses, nautilus and squid; Bivalves—clams, mussels, oysters and scallops. **p. 634: 1.** The body cavity is reduced in a clam but is well developed in earthworm. A clam has an open circulatory system, whereas an earthworm has a closed system. **2.** Polychaetes, such as the clam worm possess functional sex organs only during a breeding season. During this season, many clam worms concurrently shed a portion of their bodies containing either eggs or sperm, and these float to the surface, where fertilization takes place. In contrast, earthworms are hermaphroditic, and when mating, two worms lie parallel to each other facing in opposite directions. They both fertilize one another simultaneously. **3.** Two suckers and presence of the anticoagulant, hirudin, in their saliva. **p. 635: 1.** Roundworms have bilateral symmetry and have a complete digestive tract with both a mouth and an anus. **2.** *Ascaris* are lung and intestinal parasites in humans and pigs. The symptoms vary depending on severity of infection. *Trichinella* cause trichinosis, a severe infection caused by eating pork that is not fully cooked. Encystment of the larvae in skeletal muscle can cause aching joints, fatigue and muscle pain. Adult presence in the intestine causes digestive disorders, fatigue and fever. Elephantiasis is caused by a filarial worm and is characterized by swelling of limbs of infected humans, often to enormous sizes. **p. 638: 1.** Rigid, but jointed exoskeleton, segmentation, well developed nervous system, wide variety of respiratory organs and reduced competition through metamorphosis. **2.** Crustaceans have hard, calcified exoskeletons and typically have heads with compound eyes and five pairs of appendages. **p. 640: 1.** The body of an insect is divided into a head, a thorax, and an abdomen. The head usually bears a pair of sensory antennae, a pair of compound eyes, and several simple eyes. **2.** In crayfish, gills take up oxygen from water, and oxygen carrying pigment. In contrast, the grasshopper has tracheae to allow to enter the body from the air but lacks an oxygen carrying pigment. The crayfish excretes a liquid nitrogenous waste (ammonia), while the grasshopper excretes a solid nitrogenous waste (uric acid). **3.** Spiders have a fused cephalothorax, whereas insects have a separate head and thorax. The cephalothorax bears six pairs of appendages: the chelicerae and the pedipalps and four pairs of walking legs, but insects typically have three pairs of legs.

Testing Yourself

1. a; **2.** b; **3.** c; **4.** b; **5.** a; **6.** c; **7.** b; **8.** b, c; **9.** a; **10.** a; **11.** d; **12.** a; **13.** b; **14.** *a.* pharynx, *b.* mouth, *c.* esophagus, *d.* hearts, *e.* coelom, *f.* crop, *g.* seminal vesicle, *h.* dorsal blood vessel, *i.* nephridium, *j.* ventral nerve cord, *k.* anus, *l.* clitellum; **15.** d

Chapter 31

Check Your Progress

p. 648 (left column): 1. They share the deuterstome form of embryonic development with chordates, whereas the rest of the invertebrates are protostomes. **2.** Check the student's diagram of an echinoderm. **p. 648 (right column): 1.** Chordates have a notochord, a dorsal tubular nerve cord, pharyngeal pouches and a postanal tail at some point in development. **2.** Relative to cartilaginous fishes, mammals have: limbs, amniotic eggs, mammary glands, and hair. **p. 650: 1.** Tunicates have a bilaterally symmetrical larva that metamorphose into sessile, filter feeding adults that live on the ocean floor. **2.** The larvae have all four chordate characteristics, and upon sexual maturity, may have given rise to fishes later. **3.** Tunicates have four chordate characteristics as larvae, while lancelets have them as adults. Also, lancelets are segmented and elongate, while tunicates are asymmetrical and non-segmented. **p. 652: 1.** Vertebrates have a vertebral column, which is part of a flexible and strong jointed endoskeleton. Together, the skeleton and attached muscles form a system that permits rapid and efficient movement. **2.** Bony fishes have a swim bladder that regulates buoyancy, skin covered by bony scales, and a single circuit circulatory system. **3.** Ray-finned fishes respire exclusively in water via gills, whereas lobe-finned fishes have lungs and can obtain oxygen from air as well as water. **p. 653: 1.** Most amphibians have jointed appendages. They typically have a life cycle that includes both aquatic and terrestrial phases. Relative to bony fishes, they also usually have four limbs, eyelids for keeping their eyes moist, a tympanum for picking up sound waves, and a voice producing larynx. **2.** They undergo metamorphosis with gills as tadpoles and lungs as adults. **p. 654: 1.** Relative to amphibians, reptiles have an amniotic egg. They also have scales. Snakes have tongues as a sensory organ. Reptiles also have a well-developed rib cage that protects the lungs. Most have a three-chambered heart. **2.** Probably the most important features are: internal fertilization, amniotic egg and scales, all of which protect the animals and their offspring from water loss. **p. 656: 1.** They have feathers, hollow, light bones, beaks and a sternum with attached keel. **2.** Birds have feathers and hollow, light bones. They also have excellent vision and muscle coordination. **p. 657: 1.** Mammals have hair which provides insulation and efficient respiratory and circulatory systems. **2.** Early development of the young occurs in the pouch of the female in marsupials. In placental mammals, young are born later and receive nutrition while still inside the mother via diffusion through the placenta. **p. 659: 1.** Primate limbs are mobile, and the hands and feet both have five digits each. Many primates have a big toe and a thumb that are both opposable. The snout is shortened considerably, with the eyes toward the front of the head. This allows the stereoscopic vision (or depth perception). **2.** Increased brain size is associated with learning, memory, thought, awareness and good hand eye coordination. **p. 663: 1.** Bipedalism evolved either 4 or 7 million years ago. A dramatic change in climate caused forests to be replaced by grassland about 4 MYA. It is thought that hominids began to stand upright to keep the sun off their backs and to have better vision for detecting predators or prey in grasslands. However, recent fossil findings suggest bipedalism might have evolved as early as 7 MYA, while hominids still lived in forests. The first hominids may have walked upright on large branches as they collected fruit from overhead. The upright stance later made it easier for traveling on the ground and foraging among bushes. **2.** Bipedalism. **p. 664: 1.** They all have a brain size that is 600 cc or greater, their jaw and teeth resemble those of humans, and tool use is in evidence. **2.** *Homo ergaster* likely evolved around 1.9 million years ago. **3.** *H. ergaster* had a larger brain (about 1,000 cc) and a flatter face, with a projecting nose. They were also taller than previous hominids. They had a robust and likely heavily muscled skeleton. **p. 666: 1.** The replacement model or out-of-Africa hypothesis, proposes that modern humans evolved from archaic humans only in Africa, and then modern humans migrated to Asia and Europe, where it replaced the archaic species about 100,000 years bp. In contrast, the multiregional continuity hypothesis, proposes that modern humans arose from archaic humans in Africa, Asia, and Europe at roughly the same time. **2.** Neandertals lived between 200,000 and 28,000 years ago and were likely supplanted by modern humans (Cro-Magnons at the time). Neandertal brains were slightly larger (1400 cc). They had massive brow ridges, a forward-sloping forehead, a receding lower jaw, and wide, flat noses. The bones of Neandertals were shorter and thicker than those of Cro-Magnons. Cro-Magnons had had lighter bones, flat high foreheads, domed skulls housing brains of 1,350 cc, small teeth, and a distinct chin.

Testing Yourself

1. a, b; **2.** b, c; **3.** c; **4.** b; **5.** c; **6.** c, d; **7.** b, c, d; **8.** a, b, c, d; **9.** a, b, c; **10.** *a.* swim bladder, *b.* stomach, *c.* muscle, *d.* vertebra(e), *e.* brain, *f.* nostril, *g.* mandible, *h.* gills, *i.* heart, *j.* liver, *k.* gallbladder, *l.* intestine, *m.* gonad, *n.* kidney, *o.* scales, *p.* lateral line; **11.** e; **12.** a; **13.** c; **14.** d; **15.** c

Chapter 32

Check Your Progress

p. 673: 1. Identical twins that are separated at birth and exposed to different environments growing up will show similarities, simply because they are genetically identical. Similar food preferences or other behaviors suggest that there is a genetic basis, since their environments were mostly different. **p. 675: 1.** Operant conditioning. **2.** Classical conditioning. **p. 678: 1.** Communication could: 1) warn a receiver; 2) show receptivity to mating; 3) mark a territory; 4) tell a receiver where food is, etc. **2.** Auditory communication. Visual would be hard to see. **p. 682: 1.** Animals may defend a portion of their home range that has particularly good food resources. **2.** Females produce a limited number of eggs in their lifetime, whereas males produce essentially limitless amounts of sperm. Females are also usually the sex that provides more parental care than males, so they should be choosy to ensure the best quality males in terms of fitness and/or parental cares that the males may provide for the offspring. **p. 685: 1.** Advantages: avoiding predators, finding food, cooperative offspring rearing. Disadvantages: higher competition for mates and food, spread of diseases. **2.** Kin selection involves consideration of the fitness of related individuals in addition to yourself. Since your siblings are 50% related to you genetically, if you save 2 siblings, you are essentially breaking even. If you sacrifice yourself to save 3, you are actually benefitting in terms of your inclusive fitness (although your individual fitness is lost).

Testing Yourself

1. a; **2.** e; **3.** c; **4.** d; **5.** b; **6.** b; **7.** a; **8.** b; **9.** c; **10.** c; **11.** c; **12.** a; **13.** c; **14.** b; **15.** a

Chapter 33

Check Your Progress

p. 690: 1. Modern ecology seeks to describe the distribution and abundance of organisms across the Earth. **p. 692: 1.** Populations don't grow to their biotic potential because the environment is always limiting in terms of resources such as food, nesting sites, etc. The resulting population growth curve is S-shaped, whereby population growth slows as it reaches the carrying capacity of the environment. **2.** Type I is representative of long-lived species like humans. Survivorship is high until late life, where it rapidly declines as individuals approach the maximum life span. Type II is representative of species with constant mortality rates, such as some species of birds. Type III is representative of species with very high reproductive rates (e.g., clams) where very few young survive and then survivorship gradually decreases as individuals reach their maximum life span. **3.** (100–50)/1000 = 5% per year. **p. 694: 1.** A demographic transition is decreased birth rate followed by decreased death rate. The transition from an agricultural society to an urban society reduced the incentive for people to have large families. Urban living also resulted in a decreased death rate. **2.** Because more people are pre-reproductive age in LDC than are post-reproductive age, population growth will increase when these people reach reproductive age. In MDCs, more people are post-reproductive age than pre-reproductive age due to lower overall mortality rates. As such, population growth rates will decrease. **p. 695: 1.** Mosquitoes because they produce large numbers of offspring and are short-lived. **2.** This is a density-independent factor. **p. 696: 1.** No two species can occupy the same niche, so when two species are in the same area competing for the same resources, one species will go locally extinct. **2.** Species with very similar niches can specialize on different food sources, for example, thereby allowing coexistence. **p. 698: 1.** Predators can increase in abundance after prey increase in abundance. This predator increase, however, decreases the number of prey available, which, in turn, decreases the number of predators. They prey then eventually become more abundant and the process begins again. **2.** Prey can escape predation by: having warning coloration, being toxic, looking like other prey species that are toxic (i.e., mimicry), and presenting "alarm" signals when threatened, such as eyespots. **p. 700: 1.** A classic example is plants and pollinators, whereby plants provide the pollinators with nectar and pollinators help the plants fertilize one another. **2.** As populations grow large and more dense, parasites and diseases spread more easily. As parasites reduce fitness, population sizes may consequently decline. **p. 702: 1.** A climax community. **2.** Primary succession, as well as facilitation succession.

Testing Yourself

1. e; **2.** e; **3.** c; **4.** *a.* lag, *b.* environmental resistance, *c.* carrying capacity, *d.* stable equilibrium, *e.* deceleration, *f.* exponential growth; **5.** b; **6.** e; **7.** a; **8.** c; **9.** c; **10.** a; **11.** b; **12.** d; **13.** b; **14.** a; **15.** c

Chapter 34

Check Your Progress

p. 707: 1. Producers are at the base of the food chain and are plants and thus autotrophs. Consumers need to eat other organisms to survive and are heterotrophs. Decomposers break down dead organic matter. **2.** Most energy (90%) is lost as heat to the environment due to inefficiencies in consumption and digestion. **p. 709: 1.** Because most energy is lost as heat when one organism eats another, little biomass remains at the top trophic levels in an ecosystem. **2.** Sometimes in aquatic ecosystems such as lakes, herbivores can have higher biomass than producers. **p. 711: 1.** The water cycle consists of evaporation whereby water on the Earth's surface enters the environment, condensation, whereby environmental water accumulates in clouds, and precipitation, where water is returned from the environment to the Earth's surface. **p. 712: 1.** In the phosphorus cycle, phosphorus moves from rocks on land to the oceans, where it gets trapped in sediments. Then phosphorus moves back onto land following a geological upheaval. Phosphorus also cycles through the biological community. **p. 714: 1.** Nitrogen fixation occurs when nitrogen gas (N_2) is converted to ammonium ions (NH_4^+), a form plants can use by nitrogen fixing bacteria in soil. **2.** Nitrification is the conversion of nitrogen gas into nitrate. In contrast, denitrification is the conversion of nitrate back to atmospheric nitrogen gas. **p. 716: 1.** Human activities burn fossil fuels, releasing carbon dioxide into the atmosphere. Large amounts of this gas allow the sun's rays to pass through, but they absorb and radiate heat back to Earth, thus causing rises in global temperatures.

Testing Yourself

1. a; **2.** **a.** producers, **b.** consumers, **c.** decomposers, **d.** inorganic nutrient pool; **3.** c; **4.** c; **5.** a; **6.** d; **7.** **a.** tertiary consumers, **b.** secondary consumers,

c. primary consumers, **d.** producers; **8.** c; **9.** d; **10.** c; **11.** c; **12.** d; **13.** c; **14.** b; **15.** a; **16.** b

Chapter 35

Check Your Progress

p. 721: 1. The proximity of the portion of the Earth to the sun determines the different seasons. Because the Earth is tilted on its axis, during part of the year the southern hemisphere is further from the sun than the northern hemisphere and vice versa. **2.** The direction of wind patterns, such as the easterlies and westerlies, is affected by the spinning of the Earth about its axis. **3.** Mountains affect rainfall. As air blows up and over a mountain range, it rises and cools. One side of the mountain is called the windward side, and this side is windier and rainier than the leeward side. On the leeward side, the air descends, picks up moisture, and generally produces clear weather. **p. 727: 1.** A tropical rainforest has a more stable climate than the taiga, which has extreme winter and a long period of frost. **2.** Deciduous forests are characterized by trees that shed their leaves during fall. Tropical rainforests do not experience as cold weather and hence are characterized by trees that do not shed leaves. Tropical rainforests also are typically more lush and have higher biodiversity than temperate deciduous forests. **p. 729: 1.** Tundra—because it experiences permafrost. **p. 731: 1.** Oligotrophic lakes are nutrient poor, whereas eutrophic lakes are nutrient rich. Thus, eutrophic lakes are likely to have more biodiversity because they can provide more nutrients for the food webs within them. **2.** Both fall and spring overturn cause cold water to sink and warm water to rise, eventually causing uniform temperatures in the lake and temporarily removing stratification (of temperatures). **p. 733: 1.** An estuary acts as a nutrient trap because the sea prevents the rapid escape of nutrients brought by a river. Many marine fishes develop in the protective environment of an estuary. **p. 735: 1.** The two major oceanic divisions are the pelagic division and the benthic division. The pelagic division is characterized by organisms that live in the open waters, whereas the benthic division is characterized by organisms that live on the ocean floor.

Testing Yourself

1. c; **2.** b; **3.** c; **4.** b; **5.** e; **6.** a; **7.** g; **8.** d; **9.** c; **10.** f; **11.** c; **12.** b, c; **13.** a; **14.** c, d; **15.** b

Chapter 36

Check Your Progress

p. 743: 1. Conservation biology is an interdisciplinary discipline with the explicit goal of protecting biodiversity and Earth's natural resources. **2.** Ecosystem-level conservation will protect habitat for many communities and thus species, whereas species-level conservation may protect one species while not accounting for the needs of others. **p. 746: 1.** Consumptive use value describes the direct value of a natural product, such as timber. Agricultural use value is the value of those products produced by humans for agricultural purposes. **p. 747: 1.** Ecotourism. **2.** Direct values are the values of a product such as consumptive use value or agricultural value. Indirect values include possible development of medicines, value for ecotourism, etc. **p. 750: 1.** Exotic species can be introduced by humans when they colonize a new area, via accidental transport (e.g., zebra mussels on ship ballasts), or via horticulture and agriculture (e.g., yard ornamental plants accidentally spreading). **p. 753: 1.** The five main causes of extinction are: habitat loss, introduction of exotic species, pollution, overexploitation and diseases. **2.** Pollution could weaken immunity and lead to increased susceptibility to disease. Pollution leads to global warming, which can have negative impacts on species' geographic ranges. **p. 755: 1.** One way to conserve many species is to focus on protecting habitat in biodiversity hotspots because they contain large numbers of species. A second way to prioritize is to focus on conservation of keystone species, the loss of which can lead to many secondary extinctions. **2.** Because edge effects essentially remove the amount of suitable habitat in the interior, nature reserves should account for this and generally be designed to be larger. **p. 757: 1.** Three principles of habitat restoration are: 1) begin as soon as possible to protect remaining habitat; 2) use biological techniques to mimic natural processes; and, 3) focus on sustainable development, allowing the ecosystem to maintain itself while providing services to humans. **p. 761: 1.** Most non-renewable energy use is concentrated on fossil fuel burning, which releases greenhouse gases into the environment and contributes to global warming. **2.** We can: plant salt and drought tolerant crops, use drip irrigation, recycle home use water, such as laundry water, limit yard irrigation to dawn and dusk hours, and use dual flush toilets. **3.** Crop rotation helps maintain nutrients in the soil (e.g., by alternating nitrogen fixing crops, such as legumes). Both organic farming and biological pest control remove or reduce the use of pesticides and herbicides. Contour farming, terrace farming, planting cover crops and natural fences all help reduce erosion.

Testing Yourself

1. c; **2.** c; **3.** c; **4.** a; **5.** d; **6.** c; **7.** c; **8.** b; **9.** d; **10.** b; **11.** e; **12.** e; **13.** e; **14.** e; **15.** c

Appendix B

Classification of Organisms

Domain Bacteria

Prokaryotic, unicellular organisms that lack a membrane-bounded nucleus and reproduce asexually. Metabolically diverse being heterotrophic by absorption; autotrophic by chemosynthesis or by photosynthesis. Motile forms move by flagella consisting of a single filament. (571)

Domain Archaea

Prokaryotic, unicellular organisms that lack a membrane-bounded nucleus and reproduce asexually. Many are autotrophic by chemosynthesis; some are heterotrophic by absorption. Many live in extreme or anaerobic environments. Major groups include thermoacidophiles, methanogens, and halophiles. Archaea are distinguishable from bacteria by their unique rRNA base sequence and their distinctive plasma membrane and cell wall chemistry. (576)

Domain Eukarya

Eukaryotic, unicellular to multicellular organisms that have a membrane-bounded nucleus containing several chromosomes. Sexual reproduction is common. Phenotypes and nutrition are diverse; each kingdom has specializations that distinguish it from the other kingdoms. Flagella, if present, have a 9 + 2 organization. (577)

Kingdom Protista

Eukaryotic, unicellular organisms and their immediate multicellular descendants. Asexual reproduction is common, but sexual reproduction as a part of various life cycles does occur. Phenotypically and nutritionally diverse, being either photosynthetic or heterotrophic by various means. Locomotion, if present, utilizes flagella, cilia, or pseudopods. (577)

Algae (photosynthetic protists)*

- Phylum Chlorophyta: green algae (578)
- Phylum Chrysophyta: diatoms, golden-brown algae (578)
- Phylum Pyrrophyta: dinoflagellates (578)
- Phylum Rhodophyta: red algae (579)
- Phylum Phaeophyta: brown algae (579)

Protozoans*

- Phylum Zoomastigophora: flagellates (580)
- Phylum Ciliophora: ciliates (580)
- Phylum Rhizopoda: amoeboids (581)
- Phylum Apicomplexa: sporozoans (582)

Water Molds*

- Phylum Oomycota: water molds (583)

Slime Molds*

- Phylum Myxomycota: plasmodial slime molds (583)
- Phylum Acrasiomycota: cellular slime molds (584)

Kingdom Fungi

Structurally diverse, heterotrophic eukaryotes that are usually saprotrophic. Most are multicellular; some have unicellular yeast forms. Most produce nonmotile spores during both asexual and sexual reproduction. The only multicellular forms of life to be heterotrophic by absorption. (584)

- Phylum Chytridomycota: chytrid fungi (586)
- Phylum Zygomycota: zygospore fungi (587)
- Phylum Ascomycota: sac fungi (587)
- Phylum Basidiomycota: club fungi (588)
- Phylum Glomeromycota: AM fungi (589)

*Not a classification category, but added for clarity

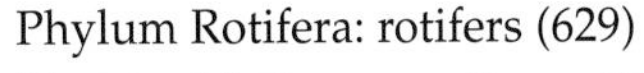

Kingdom Plantae

Multicellular, primarily terrestrial eukaryotes with well-developed tissues. Plants have an alternation of generations life cycle and are usually photosynthetic. Like green algae, they contain chlorophylls a and b, carotenoids; store starch in chloroplasts; and have a cell wall that contains cellulose. (602)

Nonvascular Plants*

- Phylum Hepaticophyta: liverworts (604)
- Phylum Bryophyta: mosses (605)

Seedless Vascular Plants*

- Phylum Lycophyta: club mosses, spike mosses, quillworts (607)
- Phylum Polypodiophyta: ferns (607)
- Phylum Psilotophyta: whisk ferns (609)
- Phylum Equisetophyta: horsetails (609)

Gymnosperms*

- Phylum Pinophyta: conifers, such as pines, firs, yews, redwoods, spruces (611)
- Phylum Cycadophyta: cycads (611)
- Phylum Ginkgophyta: maidenhair tree (612)
- Phylum Gnetophyta: gnetophytes (612)

Angiosperms*

- Phylum Magnoliophyta: flowering plants (612)
  - Class Monocotyledones: monocots (612)
  - Class Eudicotyledones: eudicots (612)

Kingdom Animalia

Multicellular organisms with well-developed tissues that have the diplontic life cycle. Animals tend to be mobile and are heterotrophic by ingestion, generally in a digestive cavity. Complexity varies; the more complex forms have well-developed organ systems. More than a million species have been described. (620)

Invertebrates*

- Phylum Porifera: sponges (624)
- Phylum Cnidaria: hydras, jellyfishes (625)

Trochozoa*

- Phylum Platyhelminthes: flatworms (627)
  - Class Cestoda: tapeworms (628)
  - Class Trematoda: flukes (628)
- Phylum Rotifera: rotifers (629)
- Phylum Mollusca: molluscs (629)
  - Class Gastropoda: snails, slugs, nudibranchs (630)
  - Class Cephalopoda: squid, chambered nautilus, octopus (631)
  - Class Bivalvia: clams, scallops, oysters, mussels (631)
- Phylum Annelida: annelids (632)
  - Class Polychaeta: clam worms, tube worms (632)
  - Class Oligochaeta: earthworms (633)
  - Class Hirudinea: leeches (634)

Ecdysozoa*

- Phylum Nematoda: roundworms (634)
- Phylum Arthropoda: arthropods (636)
  - Subphylum Crustacea: crustaceans (shrimps, crabs, obsters, barnacles) (637)
  - Subphylum Uniramia: insects, millipedes, centipedes (638)
  - Subphylum Chelicerata: spiders, scorpions, horseshoe crabs (640)
- Phylum Echinodermata: echinoderms (646)
- Phylum Chordata: chordates (648)
  - Subphylum Urochordata: tunicates (649)
  - Subphylum Cephalochordata: lancelets (650)

Vertebrates*

- Subphylum Vertebrata: Vertebrates (650)
  - Superclass Agnatha: jawless fishes (hagfishes, lampreys) (651)
    - Class Chondrichthyes: cartilaginous fishes (sharks, skates) (651)
    - Class Osteichthyes: bony fishes (lobe-finned fishes and ray-finned fishes) (651)
    - Class Amphibia: amphibians (frogs, toads) (652)
    - Class Reptilia: reptiles (turtles, snakes, lizards, crocodiles, birds) (653)
    - Class Mammalia: mammals (monotremes—duckbill platypus, spiny anteater; marsupials—opposums, kangaroos, koalas; placental mammals—shrews, whales, rats, rabbits, dogs, cats) (656)
      - Order Primata: primates (prosimians, monkeys, apes) (659)
        - Family hominids (661)
          - Genus *Homo:* humans (661)

*Not a classification category, but added for clarity.

Appendix C

Metric System

Unit and Abbreviation	Metric Equivalent	Approximate English-to-Metric Equivalents
Length		
nanometer (nm)	= 10^{-9} m (10^{-3} μm)	
micrometer (μm)	= 10^{-6} m (10^{-3} mm)	
millimeter (mm)	= 0.001 (10^{-3}) m	
centimeter (cm)	= 0.01 (10^{-2}) m	1 inch = 2.54 cm 1 foot = 30.5 cm
meter (m)	= 100 (10^{2}) cm = 1,000 mm	1 foot = 0.30 m 1 yard = 0.91 m
kilometer (km)	= 1,000 (10^{3}) m	1 mi = 1.6 km
Weight (mass)		
nanogram (ng)	= 10^{-9} g	
microgram (μg)	= 10^{-6} g	
milligram (mg)	= 10^{-3} g	
gram (g)	= 1,000 mg	1 ounce = 28.3 g 1 pound = 454 g
kilogram (kg)	= 1,000 (10^{3}) g	= 0.45 kg
metric ton (t)	= 1,000 kg	1 ton = 0.91 t
Volume		
microliter (μl)	= 10^{-6} l (10^{-3} ml)	
milliliter (ml)	= 10^{-3} liter = 1 cm^{3} (cc) = 1,000 mm^{3}	1 tsp = 5 ml 1 fl oz = 30 ml
liter (l)	= 1,000 ml	1 pint = 0.47 liter 1 quart = 0.95 liter 1 gallon = 3.79 liter
kiloliter (kl)	= 1,000 liter	

°C	°F	
100	212	Water boils at standard temperature and pressure.
71	160	Flash pasteurization of milk
57	134	Highest recorded temperature in the United States, Death Valley, July 10, 1913
41	105.8	Average body temperature of a marathon runner in hot weather
37	98.6	Human body temperature
13.7	56.66	Human survival is still possible at this temperature.
0	32.0	Water freezes at standard temperature and pressure.

Units of Temperature

To convert temperature scales:

$$°C = \frac{(°F - 32)}{1.8}$$

$$°F = 1.8\,(°C) + 32$$

Appendix D

Periodic Table of the Elements

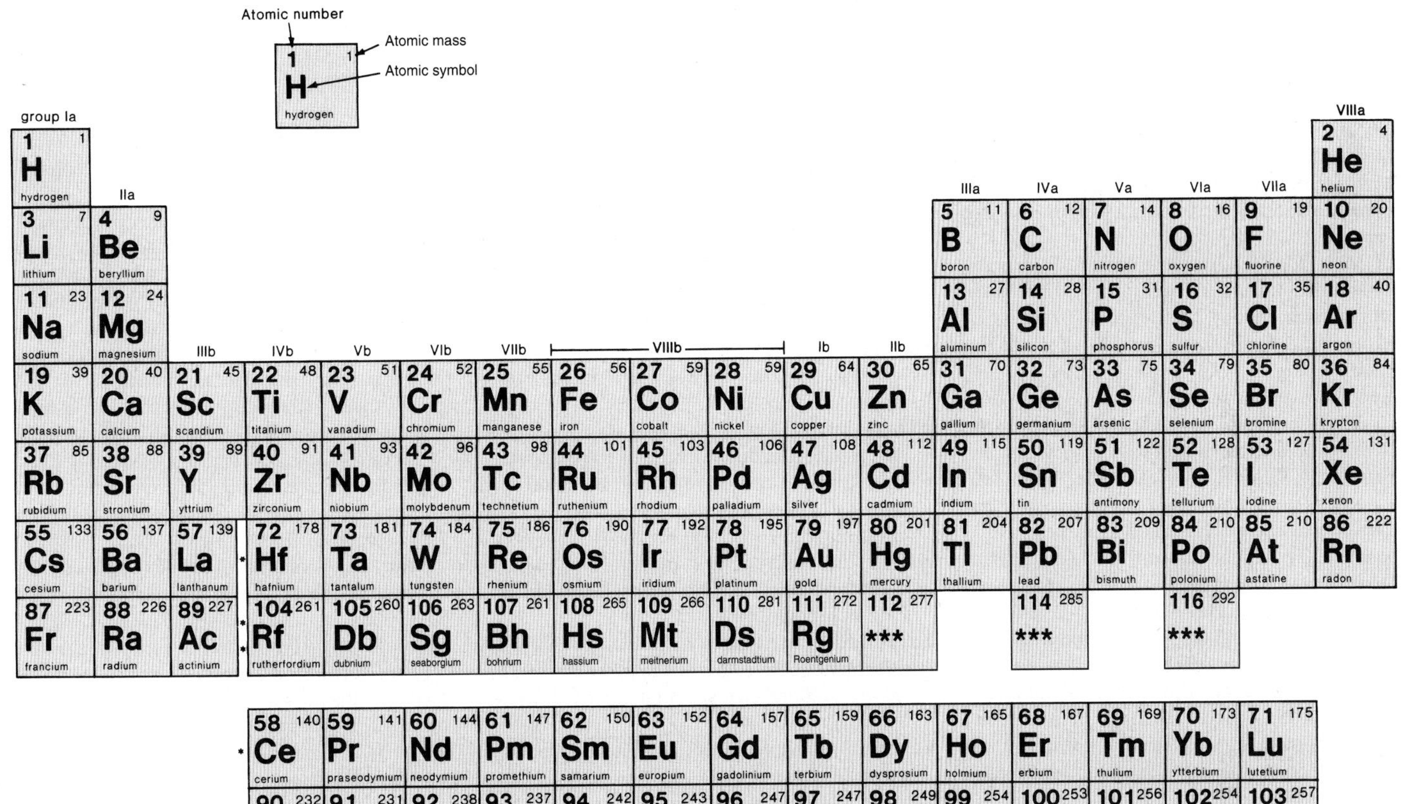

Glossary

A

abscisic acid (ABA) Plant hormone that causes stomata to close and initiates and maintains dormancy. 186

abscission Dropping of leaves, fruit, or flowers from a plant. 186

acetylcholine (ACh) Neurotransmitter active in both the peripheral and central nervous systems. 321

acetylcholinesterase (AChE) Enzyme that breaks down acetylcholine within a synapse. 321

acid Molecules tending to raise the hydrogen ion concentration in a solution and to lower its pH numerically. 28

acidosis Excessive accumulation of acids in body fluids. 307

acquired immunodeficiency syndrome (AIDS) Disease caused by the human immunodeficiency virus (HIV). 429

acromegaly Condition resulting from an increase in growth hormone production after adult height has been achieved. 407

acrosome Cap at the anterior end of a sperm that contains enzymes that help the sperm penetrate the oocyte. 419

actin Muscle protein making up the thin filaments in a sarcomere. 382

actin filaments Long, thin fibers of the cytoskeleton composed of actin monomers. 59

action potential Electrochemical changes that take place across the axon membrane; the nerve impulse. 318

active immunity Resistance to disease due to an individual's own immune system's response to a microorganism or a vaccine. 244

active site Region on the surface of an enzyme where the substrate binds and where the reaction occurs. 105

active transport Use of a plasma membrane carrier protein to move a substance into or out of a cell from lower to higher concentration. 75

acute bronchitis Inflammation of the bronchi that develops relatively quickly, usually due to an infectious agent. 292

acute disease An abnormal condition that occurs suddenly and tends to last a short time. 209

adaptation A feature of an organism, in terms of its structure, function, or behavior, that is suitable to the environment and increases the organism's reproductive success. 5

adaptive immunity The branch of the immune system that is enhanced by exposure to specific antigens. 238

adaptive radiation Evolution of several species from a common ancestor into new ecological or geographical zones. 561

Addison disease Condition resulting from a deficiency of adrenal cortex hormones. 408

adenine (A) One of four nitrogen-containing bases in nucleotides composing the structure of DNA and RNA. 39

adhesion junction Junction between cells in which the adjacent plasma membranes do not touch but are held together by intercellular filaments attached to buttonlike thickenings. 196

adipose tissue Connective tissue in which fat is stored. 196

ADP (adenosine diphosphate) Nucleotide with two phosphate groups that can accept another phosphate group and become ATP. 41, 102

adrenal cortex Outer portion of the adrenal gland; secretes mineralocorticoids and glucocorticoids. 401

adrenal gland Gland that lies atop a kidney; the adrenal medulla produces the hormones epinephrine and norepinephrine, and the adrenal cortex produces the glucocorticoid and mineralocorticoid hormones. 401

adrenal medulla Inner portion of the adrenal gland; secretes the hormones epinephrine and norepinephrine. 401

adrenocorticotropic hormone (ACTH) Hormone secreted by the anterior lobe of the pituitary gland that stimulates activity in the adrenal cortex. 398

adventitious roots Fibrous roots that develop from stems or leaves. 152

aerobic With oxygen. 115

afterbirth The placenta and the extraembryonic membranes that are delivered (expelled) during the third stage of childbirth. 459

age-structure diagram In demographics, a display of the age groups of a population. 694

aging Progressive changes over time, leading to loss of physiological function and eventual death. 462

agglutination Clumping of cells due to a reaction between antigens on cell plasma membranes and antibodies. 249

agranular leukocyte White blood cell that does not contain distinctive granules; also called mononuclear cells. 224

albumin Plasma protein of the blood having transport and osmotic functions. 222

aldosterone Hormone secreted by the adrenal cortex that decreases sodium and increases potassium excretion; raises blood volume and pressure. 306, 402

algae (sing., alga) Type of protist that carries on photosynthesis. 577

alkalosis Excessive accumulation of bases in body fluids. 307

allantois Extraembryonic membrane that accumulates nitrogenous wastes in birds and reptiles and contributes to the formation of umbilical blood vessels in mammals, including humans. 452

allele Alternative form of a gene. 470

allergen Foreign substance capable of stimulating an allergic response. 248

allergy Immune response to substances that usually are not dangerous to the body. 248

allopatric speciation Origin of new species in populations that are separated geographically. 559

alternation of generations Life cycle, typical of plants, in which a diploid sporophyte alternates with a haploid gametophyte. 172, 602

altruism Social interaction that has the potential to decrease the lifetime reproductive success of the member exhibiting the behavior. 684

alveolus (pl., alveoli) Air sac of a lung. 284

Alzheimer disease (AD) Brain disorder characterized by a general loss of mental abilities. 338

AM Fungus Arbuscular mycorrhial fungi that form beneficial associations with plant roots. 589

amino acid Monomer of a protein; takes its name from the fact that it contains an amino group (—NH_2) and an acid group (—COOH). 37

amniocentesis Procedure for removing amniotic fluid surrounding the developing fetus in order to test the fluid or cells within the fluid. 494

amnion Extraembryonic membrane of reptiles, birds, and mammals that forms an enclosing, fluid-filled sac. 452

amniotic egg Egg that has an amnion, as seen during the development of reptiles, birds, and mammals. 650

amoeboid Cell that moves and engulfs debris with pseudopods. 581

amphibian Member of a class of vertebrates that includes frogs, toads, and salamanders. 652

ampulla Base of a semicircular canal in the inner ear. 358

amygdala Portion of the limbic system that adds emotional overtones to memories. 328

anabolic steroid Synthetic steroid that mimics the effect of testosterone. 404

anabolism Metabolic process by which larger molecules are synthesized from smaller ones. 102

analogous structure Structure that has a similar function in separate lineages but differs in anatomy and ancestry. 549

anaphase Mitotic phase during which daughter chromosomes move toward the poles of the spindle. 86

anaphylactic shock Severe systemic form of an allergic reaction involving bronchiolar constriction, impaired breathing, vasodilation, and a rapid drop in blood pressure with a threat of circulatory failure. 248

androgen Male sex hormone (e.g., testosterone). 404

anemia Inefficiency in the oxygen-carrying ability of blood due to a shortage of red blood cells or hemoglobin. 224

aneurysm Ballooning of a blood vessel. 228

angina pectoris Condition characterized by thoracic pain resulting from occluded coronary arteries. 228

angiogenesis Formation of new blood vessels. 520

angioplasty Surgical procedure for treating clogged arteries. 229

angiosperm Flowering plant that produces seeds within an ovary, which eventually develops into a fruit. 144, 612

animal Multicellular, heterotrophic organism belonging to the animal kingdom. 7

annelid Member of a phylum of invertebrates that contains segmented worms, such as the earthworm. 632

annual ring Layer of wood (secondary xylem) usually produced during one growing season. 157

anorexia nervosa Eating disorder characterized by a morbid fear of gaining weight. 275

anterior pituitary Portion of the pituitary gland that is controlled by the hypothalamus and produces six types of hormones. 398

anther In flowering plants, pollen-bearing portion of the stamen. 172, 613

antheridium Sperm-producing structure, as in the moss life cycle. 606

anthropoid A suborder of Primates containing apes, monkeys, and humans. 659

antibiotic Microbial product or its derivative that kills susceptible microorganisms or inhibits their growth. 575

antibody-mediated immunity Specific mechanism of defense in which plasma cells derived from B cells produce antibodies that combine with antigens. 241

antibody titer The amount of antibody present in a sample of plasma. 244

anticodon Three-base sequence in a transfer RNA molecule that pairs with a complementary codon in mRNA. 508

antidiuretic hormone (ADH) Hormone secreted by the posterior pituitary that increases the permeability of the collecting ducts in a kidney. 307, 398

antigen Foreign substance, usually a protein or a polysaccharide, that stimulates the immune system to react. 238

antigenic drift Small change in the antigenic character of an organism that allows it to avoid attack by the immune system. 594

antigenic shift Major change in the antigenic character of an organism that alters it to an antigenic strain unrecognized by host immune mechanisms. 594

antigen-presenting cell (APC) Cell that displays fragments of an antigen to T cells so they can defend the body against that particular antigen. 243

anus Outlet of the digestive tract. 262

aorta Major systemic artery that receives blood from the left ventricle. 220

apical dominance Influence of a terminal bud in suppressing the growth of lateral buds. 184

apical meristem In vascular plants, masses of cells in the root and shoot that reproduce and elongate as primary growth occurs. 146

apoptosis Programmed cell death involving a cascade of specific cellular events leading to death and destruction of the cell. 82

appendicular skeleton Portion of the skeleton forming the pectoral girdle and upper limbs, and the pelvic girdle and lower limbs. 371

aqueous humor Clear, watery fluid between the cornea and the lens of the eye. 352

aquifer Rock layers that contain water and release it in appreciable quantities to wells or springs. 711, 759

arachnid Group of arthropods that includes spiders and scorpions. 640

Archaea One of the three domains of life; contains prokaryotic cells, called archaea, that often live in extreme habitats and have unique genetic, biochemical, and physiological characteristics. 7, 576

archegonium (pl., archegonia) Egg-producing structure, as in the moss life cycle. 606

arteriole Vessel that takes blood from an artery to capillaries. 214

artery Vessel that takes blood away from the heart to arterioles. 214

arthritis Inflammation of one or more joints. 388

arthropod Member of a phylum of invertebrates containing, among other groups, crustaceans and insects that have an exoskeleton and jointed appendages. 636

articular cartilage Hyaline cartilaginous covering over the articulating surface of the bones of synovial joints. 368

association area Region of the cerebral cortex related to memory, reasoning, judgment, and emotional feelings. 325

associative learning Acquired ability to associate two stimuli or a stimulus and a response. 675

aster Short, radiating fiber produced by the centrosomes in animal cells. 85

asthma Condition in which bronchioles constrict and cause difficulty in breathing. 248, 292

astigmatism Blurred vision due to an irregular curvature of the cornea or the lens. 360

asymmetrical Dissimilarity in corresponding parts or organs on opposite sides of the body that are normally alike. 620

atherosclerosis Condition in which fatty substances accumulate abnormally beneath the inner linings of the arteries. 228

atlas First cervical vertebrae, holds up the head. 373

atom Smallest particle of an element that displays the properties of the element. 20

atomic mass Mass of an atom equal to the number of protons plus the number of neutrons within the nucleus. 21

atomic number Number of protons within the nucleus of an atom. 21

ATP (adenosine triphosphate) Nucleotide with three phosphate groups. The breakdown of ATP into ADP + P makes energy available for energy-requiring processes in cells. 40, 102

ATP synthase Enzyme complex that functions in the production of ATP. 133

atrial natriuretic hormone (ANH) Hormone secreted by the heart that increases sodium excretion and therefore lowers blood volume and pressure. 306, 402

atrioventricular bundle Group of specialized fibers that conducts impulses from the atrioventricular node to the ventricles of the heart; also called the AV bundle. 218

atrioventricular valve Valve located between the atrium and the ventricle. 216

atrium One of the upper chambers of the heart that receives blood. 216

atrophy Wasting away or decrease in size of a tissue or body part, usually due to disuse or injury. 386

auditory canal Tube in the outer ear that leads to the tympanic membrane. 355

auditory tube Extension from the middle ear to the nasopharynx that equalizes air pressure on the eardrum; sometimes called the eustachian tube. 292, 355

australopithecine One of several species of Australopithecus, a genus that contains the first generally recognized hominids. 662

autoimmune disease Disease that results when the immune system mistakenly attacks the body's own tissues. 251

autonomic system Branch of the peripheral nervous system that has control over the internal organs; consists of the sympathetic and parasympathetic divisions. 333

autosome Any chromosome other than the sex chromosomes. 488

autotroph Organism that can capture energy and synthesize organic molecules from inorganic nutrients. 128, 544, 706

auxin Plant hormone regulating growth, particularly cell elongation; most often called indoleacetic acid (IAA). 184

AV (atrioventricular) node Small region of neuromuscular tissue that transmits impulses received from the SA node to the ventricular walls. 218

axial skeleton Portion of the skeleton that supports and protects the organs of the head, the neck, and the trunk. 371

axillary bud Bud located in the axil of a leaf. 145

axis Second cervical vertebrae, allows a degree of rotation for the head. 373

axon Fiber of a neuron that conducts nerve impulses away from the cell body. 316

axon terminal Small swelling at the tip of one of many endings of the axon. 319

B

bacteria One of the three domains of life; contains prokaryotic cells called bacteria that differ from archaea. 7, 571

ball-and-socket joint The most freely movable type of joint; e.g., the shoulder or hip joint. 378

bark External part of a tree, containing cork, cork cambium, and phloem. 156

Barr body Dark-staining body in the nuclei of female mammals that contains a condensed, inactive X chromosome. 492

basal nuclei Subcortical nuclei deep within the white matter that serve as relay stations for motor impulses and produce dopamine to help control skeletal muscle activities. 326

base Molecules tending to lower the hydrogen ion concentration in a solution and raise the pH numerically. 28

basement membrane Layer of nonliving material that anchors epithelial tissue to underlying connective tissue. 194

basophil White blood cell with a granular cytoplasm; able to be stained with a basic dye. 224

B-cell receptor (BCR) A molecule on the surface of a B cell that binds to a specific antigen. 241

behavior Observable, coordinated responses to environmental stimuli. 4, 672

benign Abnormal cell growth that is not cancerous. 518

benign prostatic hyperplasia (BPH) Enlargement of the prostate gland. 435

benthic division Ocean or lake floor, extending from the high-tide mark to the deepest depths. 735

bicarbonate ion Ion that participates in buffering the blood, and the form in which carbon dioxide is transported in the bloodstream. 289

bilateral symmetry Body plan having two corresponding or complementary halves. 620

bile Secretion of the liver that is temporarily stored and concentrated in the gallbladder before being released into the small intestine, where it emulsifies fat. 260

binary fission Bacterial reproduction resulting in two daughter cells. 572

biodiversity Total number of species, the variability of their genes, and the communities in which they live. 9, 742

biodiversity hotspots Regions of the world that contain unusually large concentrations of species. 753

biogeochemical cycle Circulating pathway of elements such as carbon and nitrogen involving exchange pools, storage areas, and biotic communities. 710

biogeography Study of the geographical distribution of organisms. 547

bioinformatics Computer technologies used to study the genome. 536

biology Scientific study of life. 10

biomass The number of organisms multiplied by their weight. 709

biome One of the biosphere's major communities, characterized in particular by certain climatic conditions and particular types of plants. 722

biosphere That portion of Earth's surface (air, water, and land) where living things exist. 8, 690

biotechnology Term that encompasses genetic engineering and other techniques that make use of natural biological systems to create a product or achieve a particular result desired by humans. 529

biotic potential Maximum population growth rate under ideal conditions. 691

bipedalism Walking erect on two feet. 661

bird Endothermic vertebrate that has feathers and wings, is often adapted for flight, and lays hard-shelled eggs. 654

birth control pill Oral contraception containing estrogen and progesterone. 426

bivalve Type of mollusc with a shell composed of two valves; includes clams, oysters, and scallops. 631

blade Broad, expanded portion of a plant leaf. 145

blastocoel Fluid-filled cavity of a blastula. 445

blastocyst Early stage of human embryonic development that consists of a hollow, fluid-filled ball of cells. 453

blastopore Opening into the primitive gut formed at gastrulation. 446

blastula Hollow, fluid-filled ball of cells occurring during animal development prior to gastrula formation. 445

blind spot Region of the retina, lacking rods or cones, where the optic nerve leaves the eye. 354

blood Type of connective tissue in which cells are separated by a liquid called plasma. 198

blood pressure Force of blood pushing against the inside wall of blood vessels. 221

bone Connective tissue having protein fibers and a hard matrix of inorganic salts, notably calcium salts. 197

bony fish Member of a class of vertebrates (class Osteichthyes) containing numerous diverse fishes that have a bony rather than a cartilaginous skeleton. 651

bottleneck effect Type of genetic drift; occurs when a majority of genotypes are prevented from participating in the production of the next generation as a result of a natural disaster or human interference. 556

brain stem Portion of the brain consisting of the medulla oblongata, pons, and midbrain. 326

Broca's area Region of the frontal lobe that coordinates complex muscular actions of the mouth, tongue, and larynx, making speech possible. 326

bronchioles Smaller air passages in the lungs that begin at the bronchi and terminate in alveoli. 284

bronchus (pl., bronchi) One of two major divisions of the trachea leading to the lungs. 284

brown algae Multicellular sea weeds that dominate rocky shores along cold and temperate coasts. 579

buffer Substance or group of substances that tend to resist pH changes in a solution. 29, 307

bulbourethral gland Either of two small structures located below the prostate gland in males; adds secretions to semen. 416

bulimia nervosa Eating disorder characterized by binge eating followed by purging via self-induced vomiting or use of a laxative. 274

bursa Saclike, fluid-filled structure, lined with synovial membrane, that occurs near a synovial joint. 377

bursitis Inflammation of any of the friction-easing sacs called bursae, which are found near joints. 377

C

C3 plant Plant that fixes carbon dioxide via the Calvin cycle; the first stable product of C_3 photosynthesis is a 3-carbon compound. 137

C4 plant Plant that fixes carbon dioxide to produce a C_4 molecule that releases carbon dioxide to the Calvin cycle. 137

calcitonin Hormone secreted by the thyroid gland that increases the blood calcium level. 400

calorie Amount of heat energy required to raise the temperature of water 1°C. 27

Calvin cycle reaction Pathway of photosynthesis that converts carbon dioxide to carbohydrate. 130

calyx The sepals collectively; the outermost flower whorl. 613

CAM pathway Alternative pathway for photosynthesis in which the components are separated by time, carbon dioxide is fixed into a 4-carbon molecule at night. 138

capillary Microscopic vessel connecting arterioles to venules; exchange of substances between blood and tissue fluid occurs across their thin walls. 194, 214

capsid Protein-containing layer that surrounds and protects the genetic material of a virus. 590

capsule Gelatinous layer surrounding the cells of blue-green algae and certain bacteria. 47

carbaminohemoglobin Hemoglobin-carrying carbon dioxide. 291

carbohydrate Class of organic compounds that includes monosaccharides, disaccharides, and polysaccharides. 32

carbon dioxide (CO_2) fixation Photosynthetic reaction in which carbon dioxide is attached to an organic compound. 136

carbonic anhydrase Enzyme in red blood cells that speeds the formation of carbonic acid from water and carbon dioxide. 289

carcinoma Cancer arising in epithelial tissue. 194

cardiac cycle One complete cycle of systole and diastole for all heart chambers. 218

cardiac muscle Striated, involuntary muscle tissue found only in the heart. 199

cardiac pacemaker Keeps the heartbeat regular; another name for the SA node. 218

cardiovascular system Organ system in which blood vessels distribute blood under the pumping action of the heart. 202

carnivore Secondary or higher consumer in a food chain that eats other animals. 706

carotenoid Yellow or orange pigment that serves as an accessory to chlorophyll in photosynthesis. 128

carpel Ovule-bearing unit that is a part of a pistil. 173, 613

carrier Heterozygous individual who has no apparent abnormality but can pass on an allele for a recessively inherited genetic disorder. 488

carrier protein Protein that combines with and transports a molecule or ion across the plasma membrane. 69

carrying capacity The limit of the number of individuals of a particular species that the environment can support in a particular area. 692

cartilage Connective tissue in which the cells lie within lacunae embedded in a flexible, proteinaceous matrix. 196

cartilaginous fish Member of a class of vertebrates (class Chondrichthyes) with a cartilaginous rather than a bony skeleton; includes sharks, rays, and skates. 651

Casparian strip Impermeable layer bordering four sides of root endodermal cells; prevents water and solute transport between adjacent cells. 151

catabolism Metabolic process that breaks down large molecules into smaller ones. 102

cecum Small pouch. In humans, a cecum lies below the entrance of the small intestine and is the blind end of the large intestine. 262

cell Structural and functional unit of an organism; the smallest structure capable of performing all the functions necessary for life. 3

cell body Portion of a neuron that contains a nucleus and from which dendrites and an axon extend. 316

cell cycle Repeating sequence of events in eukaryotes that involves cell growth and nuclear division. 82

cell-mediated immunity Specific mechanism of defense in which T cells destroy antigen-bearing cells. 243

cell plate Structure that precedes the formation of the cell wall as a part of cytokinesis in plants. 88

cell recognition protein Glycoprotein that helps the body defend itself against pathogens. 69

cell theory One of the major theories of biology; states that all organisms are made up of cells and that cells are capable of self-reproduction and come only from preexisting cells. 46

cellular differentiation Process and developmental stages by which a cell becomes specialized for a particular function. 448

cellular respiration Metabolic reactions that use the energy from carbohydrate, fatty acid, or amino acid breakdown to produce ATP molecules. 57, 114

cellular slime mold Free-living protist that exists as individual amoeboid cells. 584

cellulose Polysaccharide that is the major complex carbohydrate in plant cell walls. 33

cell wall Structure that surrounds a plant, protistan, fungal, or bacterial cell and maintains the cell's shape and rigidity. 47

central canal Tube within the spinal cord that is continuous with the ventricle of the brain and contains cerebrospinal fluid. 322

central nervous system (CNS) Portion of the nervous system consisting of the brain and spinal cord. 316

centriole Cell organelle, existing in pairs, that occurs in the centrosome and may help organize a mitotic spindle for chromosome movement during animal cell division. 61, 85

centromere Constricted region of a chromosome where sister chromatids are attached to one another and where the chromosome attaches to a spindle fiber. 85

centrosome Central microtubule organizing center of cells. 61

cephalization Having a well-recognized anterior head with a brain and sensory receptors. 622

cephalopod Type of mollusc in which a modified foot develops into the head region; includes squids, cuttlefish, octopuses, and nautiluses. 631

cerebellum Part of the brain located posterior to the medulla oblongata and pons that coordinates skeletal muscles to produce smooth, graceful motions. 326

cerebral cortex Outer layer of the cerebral hemispheres; receives sensory information and controls motor activities. 325

cerebral hemisphere One of the large, paired structures that together constitute the cerebrum of the brain. 325

cerebrospinal fluid (CSF) Fluid found in the ventricles of the brain, in the central canal of the spinal cord, and in association with the meninges. 322

cerebrum Main part of the brain consisting of two large masses, or cerebral hemispheres. 324

cervix Narrow end of the uterus, which leads into the vagina. 420

channel protein Forms a channel to allow a particular molecule or ion to cross the plasma membrane. 69

chaparral Biome characterized by broad-leafed evergreen shrubs forming dense thickets. 727

chemical energy Energy associated with the interaction of atoms in a molecule. 100

chemical signal Molecule(s) released by cells, or even by organisms, that affect the metabolism of target cells or even the behavior of other individuals. 396

chemical synapse Junction between neurons consisting of the presynaptic (axon) membrane, the synaptic cleft, and the postsynaptic (usually dendrite) membrane. 320

chemiosmosis Ability of certain membranes to use a hydrogen ion gradient to drive ATP formation. 120

chemoautotroph Organism able to synthesize organic molecules by using carbon dioxide as the carbon source and the oxidation of an inorganic substance (such as hydrogen sulfide) as the energy source. 572

chemoreceptor Sensory receptor that is sensitive to chemical stimulation; e.g., receptors for taste and smell. 346

chitin Strong but flexible nitrogenous polysaccharide found in fungal cell walls and the exoskeleton of arthropods. 641

chlamydia Sexually transmitted disease caused by the bacterium *Chlamydia trachomatis*. 433

chlorofluorocarbons (CFCs) Organic compounds containing carbon, chlorine, and fluorine atoms. CFCs such as Freon can deplete the ozone shield by releasing chlorine atoms in the upper atmosphere. 714

chlorophyll Green pigment that captures solar energy during photosynthesis. 128

chloroplast Membranous organelle that contains chlorophyll and is the site of photosynthesis. 57

chordae tendineae Tough bands of connective tissue that attach the papillary muscles to the atrioventricular valves within the heart. 216

chordate Member of the phylum Chordata, which includes lancelets, tunicates, fishes, amphibians, reptiles, birds, and mammals; characterized by a notochord, dorsal tubular nerve cord, pharyngeal gill pouches, and postanal tail at some point in the life cycle. 648

chorion Extraembryonic membrane functioning for respiratory exchange in birds and reptiles; contributes to placenta formation in mammals. 452

chorionic villi Treelike extensions of the chorion, an extraembryonic membrane, projecting into the maternal tissues at the placenta. 455

chorionic villus sampling (CVS) Prenatal test in which a sample of chorionic villi cells is removed for diagnostic purposes. 494

choroid Vascular, pigmented middle layer of the eyeball. 352

chromatin Network of fibrils consisting of DNA and associated proteins within a nucleus. 52, 84

chromosomal mutation Variation in regard to the normal number of chromosomes inherited or in regard to the normal sequence of alleles on a chromosome. 496

chromosome Composed of chromatin and contains the hereditary units, or genes. 52, 84

chronic bronchitis Obstructive pulmonary disorder that tends to recur; marked by inflamed airways filled with mucus and degenerative changes in the bronchi. 292

chronic disease Disease that persists over a long duration. 209

chyme Thick, semiliquid food material that passes from the stomach to the small intestine. 260

chytrid fungi Aquatic fungi that may have been the first type of fungi to evolve. 586

ciliary body In the eye, the structure that contains the ciliary muscle, which controls the shape of the lens. 352

ciliary muscle Muscle that controls the curvature of the eye. 352

ciliate Complex unicellular protist that moves by means of cilia and digests food in food vacuoles. 580

cilium (pl., cilia) Motile, short, hairlike extensions on the exposed surfaces of cells. 61

circadian rhythm Regular physiological or behavioral activity that occurs on an approximately 24-hour cycle. 405

cirrhosis Chronic, irreversible injury to liver tissue. 277

citric acid cycle Cyclic metabolic pathway found in the matrix of mitochondria that participates in cellular respiration. 114

clade Taxon or group of taxa consisting of an ancestral species and all of its descendants, forming a distinct branch on a phylogenetic tree. 562

cladogram A branching diagram that shows the relationship among species in regard to their shared, derived characters. 562

class One of the categories, or taxa, used by taxonomists to group species. 6

classical conditioning Type of learning whereby an unconditioned stimulus that elicits a specific response is paired with a neutral stimulus so that the response becomes conditioned. 675

cleavage Cell division without cytoplasmic addition or enlargement. 445

cleavage furrow Indentation that begins the process of cleavage, by which animal cells undergo cytokinesis. 88

climate Prevailing weather conditions (e.g., temperature, precipitation) in a region. 720

climax community In ecology, the community that results when succession has come to an end. 701

clonal selection theory The concept that an antigen selects which lymphocyte will undergo clonal expansion and produce more lymphocytes bearing the same type of antigen receptor. 241

cloning Production of identical copies. In organisms, the production of organisms with the same genes; in genetic engineering, the production of many identical copies of a gene. 526

closed circulatory system A circulatory system in which blood is completely contained within vessels. 632

clotting Process of blood coagulation, usually after an injury. 225

club fungi Members of the phylum Basidiomycota. 588, 607

club moss Type of seedless vascular plants that are also called ground pines. 607

cnidocytes Specialized stinging cells of cnidarians. 625

cochlea Portion of the inner ear that resembles a snail's shell and contains the spiral organ, the sense organ for hearing. 356

cochlear canal Canal within the cochlea that bears the spiral organ. 356

codominance Pattern of inheritance in which both alleles of a gene are equally expressed. 481

codon Three-base sequence of messenger RNA that causes the insertion of a particular amino acid into a protein or termination of translation. 508

coelom Embryonic body cavity lying between the digestive tract and the body wall that is completely lined by mesoderm. 201

coenzyme Nonprotein organic molecule that aids the action of the enzyme to which it is loosely bound. 107

coevolution Interaction of two species such that each influences the evolution of the other species. 698

cofactor Nonprotein adjunct required by an enzyme in order to function. 107

cohesion-tension model Explanation for upward transport of water in xylem based upon transpiration-created tension and the cohesive properties of water molecules. 163

collagen fiber White fiber in the matrix of connective tissue, giving flexibility and strength. 196

collar cells The inner layer of cells in sponges composed of flagellated cells. 624

collecting duct Duct within the kidney that receives fluid from several nephrons; the reabsorption of water occurs here. 303

collenchyma Plant tissue composed of cells with unevenly thickened walls; supports growth of stems and petioles. 147

colon The major portion of the large intestine, consisting of the ascending colon, the transverse colon, and the descending colon. 262

color blindness Deficiency in one or more of the three kinds of cones responsible for color vision. 489

color vision Perception of and ability to distinguish colors. 353

colostrum Thin, milky fluid rich in proteins, including antibodies, that is secreted by the mammary glands a few days prior to or after delivery before true milk is secreted. 459

columnar epithelium Type of epithelial tissue with cylindrical cells. 194, 498

commensalism Symbiotic relationship in which one species is benefited, and the other is neither harmed nor benefited. 699

communication Signal by a sender that influences the behavior of a receiver. 676

community Assemblage of populations interacting with one another within the same environment. 8, 690

compact bone Type of bone that contains osteons consisting of concentric layers of matrix and osteocytes in lacunae. 197, 368

companion cell Cell associated with sieve-tube members in phloem of vascular plants. 148

comparative genomics A subfield within genomics which studies the similarities and differences between genomes of organisms. 535

competition Interaction between two organisms in which both require the same limited resource, which results in harm to both. 696

competitive exclusion principle The conclusion that no two species can occupy the same niche at the same time. 696

complementary base pairing Hydrogen bonding between particular bases. In DNA, thymine (T) pairs with adenine (A), and guanine (G) pairs with cytosine (C). 504

complement system Group of plasma proteins that form a nonspecific defense mechanism, often by puncturing microbes. 240

compound Substance having two or more different elements united chemically in a fixed ratio. 24

concentration gradient Gradual change in chemical concentration from one point to another. 71

conclusion Statement made following an experiment as to whether the results support the hypothesis. 11

condensation The process by which water vapor turns into liquid. 710

cone cell Photoreceptor in the retina of eye that responds to bright light; detects color and provides visual acuity. 353

conidium (pl., conidia) Structure that asexually produces fungal spores. 587

conifer Cone-bearing gymnosperm plants that include pine, cedar, and spruce trees. 611

coniferous forest Forested area generally found in areas with temperate climate and characterized by cone-bearing trees that are mostly evergreens with needle-shaped or scale-like leaves. 725

conjugation Transfer of genetic material from one cell to another. 572

connective tissue Type of tissue characterized by cells separated by a matrix that often contains fibers. 196

conservation biology An interdisciplinary discipline with the explicit goal of protecting biodiversity and Earth's natural resources. 742

conservation corridors Strips of land that allow animals to move safely between habitats that would otherwise be isolated. 754

conservation reserves Areas that are set aside with the primary goal of protecting biodiversity within them. 754

consumer Organism that feeds on another organism in a food chain; primary consumers eat plants, and secondary (or higher) consumers eat animals. 706

continental drift Movement of continents with respect to one another over Earth's surfaces. 547

contraceptive Medication or device used to reduce the chance of pregnancy. 426

contraceptive implant Birth control method utilizing synthetic progesterone; prevents ovulation by disrupting the ovarian cycle. 427

contraceptive injection Birth control method utilizing progesterone or estrogen and progesterone; prevents ovulation by disrupting the ovarian cycle. 427

contraceptive patch Birth control method that involves a hormone patch applied to the skin. 427

contraceptive vaccine Birth control method that immunizes against the hormone hCG, crucial to maintaining implantation of the embryo. 427

contractile vacuole Organelle that pumps water out of the cell. 577

control Sample that goes through all the steps of an experiment but does not contain the variable being tested; a standard against which the results of an experiment are checked. 11

convergent evolution The acquisition of the same or similar traits in distantly related lines of evolutionary descent. 562

coral reef Structure found in tropical waters that is formed by the buildup of coral skeletons and is home to many and various types of organisms. 734

cork Outer covering of the bark of trees; made of dead cells that may be sloughed off. 147

cork cambium Lateral meristem that produces cork. 156

cornea Transparent, anterior portion of the outer layer of the eyeball. 352

corolla The petals, collectively; usually the conspicuously colored flower whorl. 613

coronary artery Artery that supplies blood to the wall of the heart. 220

corpus callosum Mass of white matter within the brain, composed of nerve fibers connecting the right and left cerebral hemispheres. 325

corpus luteum Yellow body that forms in the ovary from a follicle that has discharged its secondary oocyte; it secretes progesterone and some estrogen. 423

cortex In plants, ground tissue bounded by the epidermis and vascular tissue in stems and roots; in animals, outer layer of an organ, such as the cortex of the kidney or adrenal gland. 151

cortisol Glucocorticoid secreted by the adrenal cortex that responds to stress on a long-term basis; reduces inflammation and promotes protein and fat metabolism. 402

cost-benefit analysis A weighing-out of the costs and benefits (in terms of contributions to reproductive success) of a particular strategy or behavior. 683

cotyledon Seed leaf for the embryo of a flowering plant; provides nutrient molecules for the developing plant before photosynthesis begins. 149, 176

coupled reactions Reactions that occur simultaneously; one is an exergonic reaction that releases energy, and the other is an endergonic reaction that requires an input of energy in order to occur. 103

covalent bond Chemical bond in which atoms share one pair of electrons. 25

cranial nerve Nerve that arises from the brain. 331

creatine phosphate Compound unique to muscles that contains a high-energy phosphate bond. 384

creatinine Nitrogenous waste; the end product of creatine phosphate metabolism. 300

crenation The shriveling of an animal cell in a hypertonic solution due to osmosis. 74

cretinism Condition resulting from improper development of the thyroid in an infant; characterized by stunted growth and mental retardation. 408

cristae (sing., crista) Short, fingerlike projections formed by the folding of the inner membrane of mitochondria. 59, 118

Cro-Magnon Common name for the first fossils to be designated *Homo sapiens sapiens*. 665

crossing-over Exchange of corresponding segments of genetic material between nonsister chromatids of homologous chromosomes during synapsis of meiosis. 90

crustacean Member of a group of marine arthropods that contains, among others, shrimps, crabs, crayfish, and lobsters. 637

cuboidal epithelium Type of epithelial tissue characterized by cube-shaped cells. 194

cultural eutrophication Overenrichment caused by human activities leading to excessive bacterial growth and oxygen depletion of a body of water. 712

culture Total pattern of human behavior; includes technology and the arts, and is dependent upon the capacity to speak and transmit knowledge. 683

Cushing syndrome Condition resulting from hypersecretion of glucocorticoids; often characterized by thin arms and legs and a "moon face," and accompanied by high blood glucose and sodium levels. 407

cutaneous receptor Sensory receptors for pressure and touch found in the dermis of the skin. 349

cuticle Waxy layer covering the epidermis of plants that protects the plant against water loss and disease-causing organisms. 146

cyanobacterium Photosynthetic bacterium that contains chlorophyll and releases oxygen; formerly called a blue-green alga. 572

cycad Type of gymnosperm with palmate leaves and massive cones. 611

cyclic electron pathway Portion of the light reaction that involves only photosystem I and generates ATP. 133

cyclin Protein that cycles in quantity as the cell cycle progresses; combines with and activates the kinases that function to promote the events of the cycle. 83

cystic fibrosis (CF) A generalized, autosomal recessive disorder of infants and children characterized by widespread dysfunction of the exocrine glands and accumulation of thick mucus in the lungs. 293, 479

cystitis Inflammation of the urinary bladder. 311

cytochrome Any of several iron-containing protein molecules that serve as electron carriers in photosynthesis and cellular respiration. 119

cytokine Chemical messenger secreted by cells of the immune system that stimulates other cells to perform their various functions. 240

cytokinesis Division of the cytoplasm following mitosis and meiosis. 83

cytokinin Plant hormone that promotes cell division. 185

cytolysis Disruption or bursting of a cell; can be in response to osmosis in a hypotonic solution. 73

cytoplasm Contents of a cell between the nucleus and the plasma membrane that contains the organelles. 47

cytoplasmic segregation In development, the parceling of signals present in the egg cytoplasm to the cells that result from cleavage. 449

cytosine (C) One of four nitrogen-containing bases in the nucleotides composing the structure of DNA and RNA. 39

cytoskeleton Internal framework of the cell, consisting of microtubules, actin filaments, and intermediate filaments. 59

cytotoxic T cell Type of T cell that attacks and kills antigen-bearing cells. 243

D

data Facts or pieces of information collected through observation and/or experimentation. 11

daughter chromosome Separate chromatids become daughter chromosomes during anaphase of mitosis and anaphase II of meiosis. 85

day-neutral plant Plant whose flowering is not dependent on day length. 187

deamination Removal of an amino group ($-NH_2$) from an amino acid or other organic compound. 110

deciduous Plants that lose their leaves annually. 145

decomposer Organism, usually a bacterium or fungus, that breaks down organic matter into inorganic nutrients that can be recycled in the environment. 571, 706

deductive reasoning Process of logic and reasoning using "if . . . then" statements. 10

defecation Discharge of feces from the rectum through the anus. 262

dehydration reaction Chemical reaction resulting in a covalent bond with the accompanying loss of a water molecule. 31

delayed allergic response Allergic response initiated at the site of the allergen by sensitized T cells, involving macrophages and regulated by cytokines. 248

deletion Change in chromosome structure in which the end of a chromosome breaks off, or two simultaneous breaks lead to the loss of an internal segment. 496

demographic transition Decline in the birthrate following reduction in the death rate. 693

denaturated (denaturation) Loss of normal shape by an enzyme so that it no longer functions. 39, 106

dendrite Part of a neuron that sends signals toward the cell body. 316

dendritic cell Type of white blood cell that presents fragments of antigens to T cells. 224, 240

denitrification Conversion of nitrate or nitrite to nitrogen gas by bacteria in soil. 713

dense fibrous connective tissue Type of connective tissue containing many collagen fibers packed together. 196

density-dependent Biotic factor, such as disease or competition, that affects population size according to the population's density. 695

density-independent Abiotic factor, such as fire or flood, that affects population size independent of the population's density. 695

dental caries Tooth decay that occurs when bacteria within the mouth metabolize sugar and give off acids that erode teeth; a cavity. 257

dermis Region of the skin that lies beneath the epidermis. 204

desert Ecological biome characterized by a limited amount of rainfall; deserts have hot days and cool nights. 728

detection Conversion of environmental signals, such as light or sound, to nerve impulses by the various sensory systems. 346

detrital food chain Straight-line linking of organisms according to who eats whom, beginning with detritus. 709

detrital food web Complex pattern of interlocking and crisscrossing food chains, beginning with detritus. 709

deuterostome Group of coelomate animals in which the second embryonic opening is associated with the mouth; the first embryonic opening, the blastopore, is associated with the anus. 623

development Series of stages by which a zygote becomes an organism or by which an organism changes during its life span. 5

diabetes insipidus (DI) A disorder caused by insufficient secretion of antidiuretic hormone (ADH) by the posterior pituitary gland, causing production of large amounts of dilute urine and excessive thirst. 407

diabetes mellitus Condition characterized by a high blood glucose level and the appearance of glucose in the urine due to a deficiency of insulin production or inability of cells to respond normally to insulin. 409

dialysate Material that passes through the membrane in dialysis. 309

diaphragm A birth control device consisting of a soft rubber or latex cup that fits over the cervix. 427

diastole Relaxation period of a heart chamber during the cardiac cycle. 218

diastolic pressure Arterial blood pressure during the diastolic phase of the cardiac cycle. 221

diatom Golden-brown alga with a cell wall in two parts, or valves; significant part of phytoplankton. 578

diencephalon Portion of the brain in the region of the third ventricle that includes the thalamus and hypothalamus. 326

dietary supplement Product added to a person's diet that is not considered food; usually contains vitamins, minerals, amino acids, or a concentrated energy source. 271

differentially permeable Ability of plasma membranes to regulate the passage of substances into and out of the cell, allowing some to pass through and preventing the passage of others; sometimes called selectively permeable. 71

diffusion Movement of molecules or ions from a region of higher concentration to one of lower concentration; it requires no energy and stops when the distribution is equal. 72

digestive system Organ system that includes the mouth, esophagus, stomach, small intestine, and large intestine (colon), which receives food and digests it into nutrient molecules. Also has associated organs: teeth, tongue, salivary glands, liver, gallbladder, and pancreas. 202

digit Referring to fingers or toes. 376

dihybrid Individual that is heterozygous for two traits. 476

dinoflagellate Photosynthetic unicellular protist with two flagella, one a whiplash and the other located within a groove between protective cellulose plates; significant part of phytoplankton. 578

diploid (2n) number Twice the number of chromosomes found in the gametes. 84

directional selection Natural selection in which an extreme phenotype is favored, usually in a changing environment. 557

disaccharide Sugar that contains two units of a monosaccharide. 32

disease Illness; state of homeostatic imbalance in which part or all of the body does not function properly. 209

disruptive selection Natural selection in which extreme phenotypes are favored over the average phenotype, leading to more than one distinct form. 557

distal convoluted tubule (DCT) Final portion of a nephron that joins with a collecting duct; associated with tubular secretion. 303

diuretic Drug used to counteract hypertension by causing the excretion of water. 307

DNA (deoxyribonucleic acid) Nucleic acid found in cells; the genetic material that specifies protein synthesis in cells. 39, 502

DNA fingerprint DNA fragment lengths resulting from restriction enzyme cleavage to identify particular individuals. 528

DNA ligase Enzyme that links DNA fragments; used during production of recombinant DNA to join foreign DNA to vector DNA. 526

DNA microarray Rows of DNA sequences on a small chip for use in molecular genetic testing; DNA chip. 532

DNA replication Synthesis of a new DNA double helix prior to mitosis and meiosis in eukaryotic cells and during prokaryotic fission in prokaryotic cells. 504

domain Largest of the categories, or taxa, used by taxonomists to group species; the three domains are Archaea, Bacteria, and Eukarya. 6

domain Archaea One of three domains; contains prokaryotic organisms. 564

domain Bacteria One of the three domains; contains prokaryotic organisms. 564

domain Eukarya One of three domains; contains eukaryotic organisms. 564

dominant allele Allele that exerts its phenotypic effect in the heterozygote; it masks the expression of the recessive allele. 471

dormancy In plants, a cessation of growth under conditions that seem appropriate for growth. 179

dorsal root ganglion Mass of sensory neuron cell bodies located in the dorsal root of a spinal nerve. 331

double fertilization In flowering plants, one sperm fuses with the egg to produce a zygote and a second sperm fuses with the polar nuclei within the embryo sac to produce a 3n endosperm nucleus. 175

double helix Double spiral; describes the three-dimensional shape of DNA. 40, 504

doubling time Number of years it takes for a population to double in size. 693

ductus arteriosus Part of fetal circulation that shunts blood from the pulmonary trunk into the aorta. 456

ductus venosus Part of fetal circulation that connects the umbilical vein with the inferior vena cava. 456

Duchenne muscular dystrophy Chronic progressive disease affecting the shoulder and pelvic girdles, commencing in early childhood; transmitted as an X-linked trait, and affected individuals, predominantly males, rarely survive to maturity. 489

duodenum First part of the small intestine, where chyme enters from the stomach. 260

duplication Change in chromosome structure in which a particular segment is present more than once. 496

E

ecdysozoa Division of the protostome animals that includes roundworms and arthropods (molting animals). 623

echinoderm Phylum of marine animals that includes sea stars, sea urchins, and sand dollars. 646

ecological niche Role an organism plays in its community, including its habitat and its interactions with other organisms. 696

ecological pyramid Pictorial graph based on the biomass, number of organisms, or energy content of various trophic levels in a food web. 709

ecological succession Gradual replacement of communities in an area following a disturbance (secondary succession) or the creation of new soil (primary succession). 700

ecology Study of the interactions of organisms with other organisms and the physical environment. 690

ecosystem Biological community together with the associated abiotic environment; characterized by energy flow and chemical cycling. 8, 692, 706

ecosystem diversity Considers the functioning of communities within ecosystems and the interconnectedness of species within such communities. 743

ectoderm Outermost germ layer of the embryonic gastrula; it gives rise to the nervous system and skin. 446

ectothermic Having a body temperature that varies according to the environmental temperature. 654

edema Swelling due to tissue fluid accumulation in the intercellular spaces. 237, 310

edge effect The reduction of effective habitat due to fragmentation of ecosystems. 754

egg A gamete; the final product of oogenesis. 96

elastic cartilage Type of cartilage composed of elastic fibers, allowing greater flexibility. 197

elastic fiber Yellow fiber in the matrix of connective tissue, providing flexibility. 196

electrocardiogram (ECG) Recording of the electrical activity associated with the heartbeat. 219

electroencephalogram (EEG) Graphic recording of the brain's electrical activity. 327

electron Subatomic particle that has almost no weight and carries a negative charge; orbits the nucleus of an atom in an orbital. 20

electron transport chain Chain of electron carriers in the cristae of mitochondria and the thylakoid membrane of chloroplasts. As the electrons pass from one carrier to the next, energy is released and used to establish a hydrogen ion gradient that is used to produce ATP. 115

element Substance that cannot be broken down into substances with different properties; composed of only one type of atom. 20

elephantiasis Disease caused by a roundworm called the filarial worm. 635

El Niño–Southern Oscillation (ENSO) Warming of water in the Eastern Pacific equatorial region such that the Humboldt Current is displaced. 736

elongation During DNA replication or the formation of mRNA during transcription, elongation is the step whereby the bases are added to the new or "daughter" strands of DNA or the mRNA transcript strand is lengthened, respectively. 510

embolus Moving blood clot that is carried through the bloodstream. 228

embryo Stage of a multicellular organism that develops from a zygote before it becomes free-living; in seed plants, the embryo is part of the seed. 175

embryonic development Months 1 and 2 of human development during which early development of major organs occurs. 452

embryonic disk During human development, flattened area during gastrulation from which the embryo arises. 454

embryo sac Female gametophyte of flowering plants that produces an egg cell. 175

emphysema Degenerative disorder in which the bursting of alveolar walls reduces the total surface area for gas exchange. 293

emulsification Breaking up of fat globules into smaller droplets by the action of bile salts or any other emulsifier. 34

endangered species A species that faces immediate extinction throughout all or most of its range. 742

endergonic reaction Chemical reaction that requires an input of energy. 102

endocrine gland Ductless organ that secretes hormone(s) into the bloodstream. 394

endocrine system Organ system involved in the coordination of body activities; uses hormones as chemical signals secreted into the bloodstream. 203

endocytosis Process by which substances are moved into the cell from the environment by phagocytosis (cellular eating) or pinocytosis (cellular drinking). 77

endoderm Inner germ layer of the embryonic gastrula that becomes the lining of the digestive and respiratory tracts and associated organs. 446

endodermis Plant root tissue that forms a boundary between the cortex and the vascular cylinder. 151

endometriosis The presence of endometrial tissue in places other than the uterus, usually within the abdominal cavity, that often results in severe pain and infertility. 436

endometrium Lining of the uterus; becomes thickened and vascular during the uterine cycle. 421

endoplasmic reticulum (ER) Membranous system of tubules, vesicles, and sacs in cells, sometimes having attached ribosomes. Rough ER has ribosomes; smooth ER does not. 54

endosperm In angiosperms, the 3n tissue that nourishes the embryo and seedling and is formed as a result of a sperm joining with two polar nuclei. 175

endospore Thick-walled, resilient structure formed within certain bacteria. 572

endosymbiotic theory Possible explanation for the evolution of eukaryotic organelles by phagocytosis of prokaryotes. 63

endothelium A single layer of simple squamous epithelium that forms the innermost lining of blood vessels, lymphatic vessels, and the heart. 214

endothermic Maintenance of a constant body temperature independent of the environmental temperature. 654

energy Capacity to do work and bring about change; occurs in a variety of forms. 4, 100

energy of activation (E_a) Energy that must be added to cause molecules to react with one another. 104

enhancer Element that regulates transcription from nearby genes; functions by acting as a binding site for transcription factors. 514

entropy Measure of disorder or randomness. 101

envelope Lipid layer around some viruses. 591

environmental resistance Total of factors in the environment that limit the numerical increase of a population in a particular region. 692

enzymatic protein Protein that catalyzes a specific reaction. 69

enzyme Organic catalyst, usually a protein, that speeds up a reaction in cells due to its particular shape. 37, 104

enzyme inhibition Means by which cells regulate enzyme activity; may be competitive or noncompetitive. 107

eosinophil White blood cell containing cytoplasmic granules that stain with acidic dye. 224

epidermal tissue Exterior tissue, usually one cell thick, of leaves, young stems, roots, and other parts of plants. 146

epidermis In plants, tissue that covers roots, leaves, and stems of a nonwoody organism; in mammals, the outer protective region of the skin. 146, 204

epididymis (pl., epididymides) Coiled tubule next to the testes where sperm mature and may be stored for a short time. 416

epiglottis Structure that covers the glottis, the opening to the air tract, during the process of swallowing. 258, 284

epinephrine Hormone secreted by the adrenal medulla in times of stress; also called adrenaline. 401

epiphyte Plant that takes its nourishment from the air because its placement in other plants gives it an aerial position. 152, 726

episiotomy Surgical procedure performed during childbirth in which the opening of the vagina is enlarged to avoid tearing. 459

episodic memory Capacity of the brain to store and retrieve information about persons and events. 328

epithelial tissue Type of tissue that lines hollow organs and covers surfaces; also called epithelium. 194

erectile dysfunction (ED) Failure of the penis to achieve erection. 435

erythropoietin Hormone, produced by the kidneys, that speeds red blood cell formation. 224, 300

esophagus Muscular tube for moving swallowed food from the pharynx to the stomach. 258

essential amino acids Amino acids required in the human diet because the body cannot make them. 267

essential fatty acid Fatty acid that is necessary in the diet because the body is unable to manufacture it. 269

estrogen Female sex hormone that helps maintain sexual organs and secondary sex characteristics. 404, 424

estuary Portion of the ocean located where a river enters and fresh water mixes with salt water. 732

ethylene Plant hormone that causes ripening of fruit and is also involved in abscission. 186

eudicot (Eudicotyledones) A member of the class of flowering plants called Eudicotyledones. These plants have several common characteristics, including two embryonic leaves (cotyledons) in the seed, net-veined leaves, cylindrical arrangement of vascular bundles, and flower parts in fours or fives or multiples of four or five. 149, 612

euglenoid Flagellated and flexible freshwater unicellular protist that usually contains chloroplasts and has a semirigid cell wall. 580

Eukarya One of the three domains of life, consisting of eukaryotes, which are organisms in the kingdoms Protista, Fungi, Plantae, and Animalia. 7, 570

eukaryote An organism composed of eukaryotic cell(s). 7

eukaryotic cell Type of cell that has a membrane-bounded nucleus and organelles. 49

eutrophication Enrichment of water by inorganic nutrients used by phytoplankton. 730

evaporation The process by which liquid water turns into gaseous form (water vapor). 710

evergreen A tree that does not lose its leaves when seasons change. 145

evolution Changes that occur in the members of a species with the passage of time, often resulting in increased adaptation of organisms to the environment. 5, 545

excretion Removal of metabolic wastes from the body. 300

exergonic reaction Chemical reaction that releases energy; opposite of endergonic reaction. 102

exocytosis Process in which an intracellular vesicle fuses with the plasma membrane so that the vesicle's contents are released outside the cell. 77

exon A segment of a gene that codes for a protein. 507

exophthalmia An abnormal protrusion of the eyes, often caused by hyperactivity of the thyroid gland. 408

exotic species Nonnative species that migrate or are introduced by humans into a new ecosystem. 750

experimental design A test that controls the conditions under which experimental observations are made. The experimental design should control all factors except the one that researchers are interested in. 10

experimental variable A treatment that is varied in an experiment to test a hypothesis. 12

expiration Act of expelling air from the lungs; exhalation. 282

expiratory reserve volume Volume of air that can be forcibly exhaled after normal exhalation. 287

exponential growth Growth, particularly of a population, in which the increase occurs in the same manner as compound interest. 691

external respiration The exchange of gases between air in the lung alveoli and blood in the pulmonary capillaries. 289

exteroceptor Sensory receptor that detects stimuli from outside the body; e.g., taste, smell, vision, hearing, and equilibrium. 346

extraembryonic membrane Membrane that is not a part of the embryo but is necessary to the continued existence and health of the embryo. 452

eye spot apparatus Allows some photosynthetic protists to sense light intensity. 577

F

facilitated transport Use of a plasma membrane carrier to move a substance into or out of a cell from a higher to a lower concentration; no energy required. 75

FAD (flavin adenine dinucleotide) Coenzyme of oxidation-reduction that becomes $FADH_2$ as oxidation of substrates occurs and then delivers electrons to the electron transport system in mitochondria during cellular respiration. 114

fall overturn Mixing process that occurs in fall in stratified lakes whereby oxygen-rich top waters mix with nutrient-rich bottom waters. 730

familial hypercholesterolemia (FH) Inability to remove cholesterol from the bloodstream; predisposes individual to heart attack. 481

family One of the categories, or taxa, used by taxonomists to group species. 6

farsighted Abnormal vision due to a shortened eyeball from front to back; light rays focus in back of the retina when viewing close objects. 360

fat Organic molecule that contains glycerol and fatty acids and is found in adipose tissue. 34

fate map Diagram that traces the differentiation of cells during development, from their origin to their final structure and function. 449

fatty acid Molecule that contains a hydrocarbon chain and ends with an acid group. 34

female condom Birth control method that blocks the entrance of sperm to the uterus; also prevents STDs. 427

female gametophyte In seed plants, the gametophyte that produces an egg; in flowering plants, an embryo sac. 174

fermentation Anaerobic breakdown of glucose that results in a gain of two ATP and end products such as alcohol and lactate. 115

fern Member of a group of plants that have large fronds. 607

fertilization Union of a sperm nucleus and an oocyte nucleus to create a zygote with the diploid number of chromosomes. 89, 444

fiber Structure resembling a thread; also, plant material that is nondigestible. 262

fibrin Insoluble protein threads formed from fibrinogen during blood clotting. 225

fibrinogen Plasma protein that is converted into fibrin threads during blood clotting. 225

fibroblast Cell in connective tissues that produces fibers and other substances. 196

fibrocartilage Cartilage with a matrix of strong collagenous fibers. 197

fibrous root system In most monocots, a mass of similarly sized roots that cling to the soil. 152

filament End-to-end chains of cells that form as cell division occurs in only one plane; in plants, the elongated stalk of a stamen. 173, 615

filter feeder Method of obtaining nourishment by certain animals that strain minute organic particles from the water in a way that deposits them in the digestive tract. 624

fimbria (pl., fimbriae) Fingerlike extension from the oviduct near the ovary. 47, 420

first messenger An extracellular chemical, such as a hormone, that binds to a cellular receptor and initiates intracellular activity. 396

fitness Lifetime reproductive success; the relative ability of an individual to produce fertile offspring as measured against other individuals of the same species in the same environment. 556

fixed action pattern (FAP) Innate behavior pattern that is stereotyped, spontaneous, genetically encoded, and independent of individual learning. 674

flagellate Heterotrophic protozoans that propel themselves using one or more flagella. 580

flagellum (pl., flagella) Long, slender extension used for locomotion by some bacteria, protozoans, and sperm. 47, 572

flagship species Species that evoke an emotional response from humans and encourage conservation. 753

flame cells Part of the excretory system of Planaria containing cilia. 628

flatworm Unsegmented worm lacking a body cavity; phylum Platyhelminthes. 627

flower Reproductive organ of plants that contains the structures for the production of pollen grains and covered seeds. 172

fluid-mosaic model Model of the plasma membrane based on the changing location and pattern of protein molecules in a fluid phospholipid bilayer. 68

focus Bending of light rays by the cornea, lens, and humors so that they converge and create an image on the retina. 352

follicle Structure in the ovary that produces the oocyte and, in particular, the female sex hormones estrogen and progesterone. 423

follicle-stimulating hormone (FSH) Hormone secreted by the anterior pituitary gland that stimulates the development of an ovarian follicle in a female or the production of sperm in a male. 419

fontanel Membranous region located between certain cranial bones in the skull of a fetus or infant. 372, 456

food chain The order in which one population feeds on another in an ecosystem. 709

food web Complex pattern of energy flow in an ecosystem represented by interlocking and crisscrossing food chains. 708

foramen magnum Opening in the occipital bone of the vertebrate skull through which the spinal cord passes. 372

foramen ovale Part of fetal circulation that shunts blood from the right atrium to the left atrium. 456

formed element Constituent of blood that is cellular (red blood cells and white blood cells) or at least cellular in origin (platelets). 222

fossil Any past evidence of an organism that has been preserved in Earth's crust. 545

fossil fuel Remains of once-living organisms that are burned to release energy, such as coal, oil, and natural gas. 715

founder effect Type of genetic drift in which only a fraction of the total genetic diversity of the original gene pool is represented as a result of a few individuals founding a colony. 556

fovea centralis Region of the retina, consisting of densely packed cones, that is responsible for the greatest visual acuity. 352

Fragile X syndrome Genetic disease that results in a constellation of abnormalities due to a trinucleotide repeat expansion. 489

free energy Energy in a system that is capable of performing work. 102

fruit In a flowering plant, the structure that forms from an ovary and associated tissues and encloses seeds. 172, 614

functional genomics Study of the function of the various genes discovered within each genomic sequence and how these genes interact. 536

functional group Specific cluster of atoms attached to the carbon skeleton of organic molecules that enters into reactions and behaves in a predictable way. 31

fungus (pl., fungi) Saprotrophic decomposer; the body is made up of filaments called hyphae that form a mass called a mycelium; e.g., mushrooms and molds. 7, 584

G

G3P (glyceraldehyde 3-phosphate) Significant metabolite in both the glycolytic pathway and the Calvin cycle; G3P is oxidized during glycolysis and reduced during the Calvin cycle. 116

gallbladder Organ attached to the liver that stores and concentrates bile. 264

gamete Haploid sex cell; an oocyte or a sperm that joins during fertilization to form a zygote. 89, 439

gametophyte Haploid generation of the alternation of generations life cycle of a plant; produces gametes that unite to form a diploid zygote. 172, 602

ganglion Collection or bundle of neuron cell bodies usually outside the central nervous system. 331

gap junction Region between cells formed by the joining of two adjacent plasma membranes. 196

gastric gland Gland within the stomach wall that secretes gastric juice. 260

gastropod Mollusc with a broad flat foot for crawling. 630

gastrovascular cavity Blind digestive cavity that also serves a circulatory (transport) function in animals lacking a circulatory system. 625

gastrula In animal development, the embryonic stage that follows formation of the germ layers. 446

gastrulation Process during animal development by which the germ layers form. 446

gene Unit of heredity existing as alleles on the chromosomes. 4, 506

gene cloning Production of one or more copies of the same gene. 526

gene flow Sharing of genes between two populations through interbreeding. 556

gene mutation Alteration in a gene due to a change in DNA composition. 515

gene pool Total of all the genes of all the individuals in a population. 551

gene therapy Correction of a detrimental mutation by the addition of new DNA and its insertion in a genome. 532

genetically modified plant A plant that carries the genes of another organism as a result of DNA technology; also, a transgenic plant. 182

genetic diversity The number of different alleles, as well as the relative frequencies of those alleles in populations and species. 743

genetic drift Mechanism of evolution due to random changes in the allelic frequencies of a population; more likely to occur in small populations or when only a few individuals of a large population reproduce. 553

genetic engineering Alteration of genomes for medical or industrial purposes. 526

genetic marker Abnormality in the sequence of a base at a particular location on a chromosome, signifying a disorder. 532

genetic profile A compilation of the list of mutations within a person's genome based on microarray studies. 536

genetic profiling An individual's complete genotype, including any possible mutations. 532

genetic variation Due to crossing-over and independent assortment during meiosis, offspring do not have the same genetic makeup as either parent. 90

genital herpes Sexually transmitted disease caused by herpes simplex virus. 432

genital warts Warts on or near the genitals that are caused by human papillomaviruses. 432

genome Full set of genetic information of a species or a virus. 526

genomics The study of all the nucleotide sequences, including structural genes, regulatory sequences, and noncoding DNA segments, in the chromosomes of an organism. 534

genotype Genes of an individual for a particular trait or traits; often designated by letters, such as BB or Aa. 471

genus One of the categories, or taxa, used by taxonomists to group species; contains those species that are most closely related through evolution. 6

germinate Beginning of the growth of a seed, spore, or zygote, especially after a period of dormancy. 178

germ layer Developmental layer of the body—that is, ectoderm, mesoderm, or endoderm. 447

gerontology Study of aging. 462

gibberellin Plant hormone promoting increased stem growth; also involved in flowering and seed germination. 184

gingiva Gum tissue. 257

ginkgo One of four phyla of gymnosperms. 612

girdling Removal of a strip of bark from around a tree. 165

gland Epithelial cell or group of epithelial cells that are specialized to secrete a substance. 194

glaucoma Increasing loss of the field of vision caused by blockage of the ducts that drain the aqueous humor, creating pressure buildup and nerve damage. 352

global warming Predicted increase in Earth's temperature, due to the greenhouse effect, which will lead to the melting of polar ice and a rise in sea levels. 716

glomerular capsule Double-walled cup that surrounds the glomerulus at the beginning of the nephron. 302

glomerular filtrate Filtered portion of blood contained within the glomerular capsule. 305

glomerular filtration Movement of small molecules from the glomerulus into the glomerular capsule due to blood pressure. 305

glomerulus Capillary network within the glomerular capsule of a nephron, where glomerular filtration takes place. 302

glottis Opening for airflow in the larynx. 258, 284

glucagon Hormone secreted by the pancreas that causes the liver to break down glycogen and raises the blood glucose level. 403

glucocorticoid Type of hormone secreted by the adrenal cortex that influences carbohydrate, fat, and protein metabolism. 401

glucose Six-carbon sugar that organisms degrade as a source of energy during cellular respiration. 32

glycemic index (GI) The blood glucose response of a given food compared to, for example, white bread. 267

glycogen Storage polysaccharide, found in animals, that is composed of glucose molecules joined in a linear fashion but having numerous branches. 32

glycolipid Lipid in plasma membranes that bears a carbohydrate chain attached to a hydrophobic tail. 69

glycolysis Anaerobic metabolic pathway found in the cytoplasm that participates in cellular respiration and fermentation; it converts glucose to two molecules of pyruvate. 114

glycoprotein Protein in plasma membranes that bears a carbohydrate chain. 69

gnetophyte Division of plants composed of three related woody gymnosperms. 612

Golgi apparatus Organelle, consisting of flattened saccules and also vesicles, that processes, packages, and distributes molecules within or from the cell. 54

gonad Organ that produces gametes; the ovary produces oocytes, and the testis produces sperm. 404

gonadotropic hormone Substance secreted by the anterior pituitary that regulates the activity of the ovaries and testes; principally, follicle-stimulating hormone (FSH) and luteinizing hormone (LH). 398

gonadotropin-releasing hormone (GnRH) Hormone secreted by the hypothalamus that stimulates the anterior pituitary to secrete follicle-stimulating hormone and luteinizing hormone. 419

gonorrhea Sexually transmitted disease caused by the bacterium *Neisseria gonorrhoeae*. 433

gout Joint inflammation caused by accumulation of uric acid. 300

Gram stain Common laboratory test performed to help identify bacteria. 571

granular leukocyte White blood cell with prominent granules in the cytoplasm. 224

granum (pl., grana) Stack of chlorophyll-containing thylakoids in a chloroplast. 58

grassland A biome characterized by grasses and small plants, but no large trees. 727

gravitational equilbrium The ability to detect movement of the head in the vertical or horizontal planes. 358

gravitropism Growth response of roots and stems of plants to Earth's gravity. 184

gray crescent Gray area that appears in an amphibian egg after being fertilized by the sperm; thought to contain chemical signals that turn on the genes that control development. 449

gray matter Nonmyelinated nerve fibers and cell bodies in the central nervous system. 317

grazing food chain A flow of energy to a straight-line linking of organisms according to who eats whom. 709

grazing food web Complex pattern of interlocking and crisscrossing food chains that begins with populations of autotrophs serving as a producers. 708

green alga Member of a diverse group of photosynthetic protists that contains chlorophylls a and b and has other biochemical characteristics like those of plants. 578

green glands Excrete metabolic waste in crustaceans. 638

greenhouse effect Reradiation of solar heat toward Earth caused by gases in the atmosphere. 716

ground tissue Tissue that constitutes most of the body of a plant; consists of parenchyma, collenchyma, and sclerenchyma cells that function in storage, basic metabolism, and support. 146

growth factor Chemical signal that regulates mitosis and differentiation of cells that have receptors for it. 406, 518

growth hormone (GH) Substance secreted by the anterior pituitary; controls size of individual by promoting cell division, protein synthesis, and bone growth. 398

growth plate The area of a long bone between the epiphysis and the diaphysis where growth in length occurs. 370

guanine (G) One of four nitrogen-containing bases in nucleotides composing the structure of DNA and RNA. 39

guard cell One of two cells that surround a leaf stoma. 147

gymnosperm Type of woody seed plant in which the seeds are not enclosed by fruit and are usually borne in cones, such as those of the conifers. 611

H

habitat Place where an organism lives and is able to survive and reproduce. 696

hair cell Cell with stereocilia (long microvilli) that is sensitive to mechanical stimulation; mechanoreceptor for hearing and equilibrium in the inner ear. 355

hair follicle Tubelike depression in the skin in which a hair develops. 205

haploid (n) number Half the diploid number; the number characteristic of gametes that contain only one set of chromosomes. 85

hard palate Bony, anterior portion of the roof of the mouth. 257

hay fever Seasonal variety of allergic reaction to a specific allergen. 248

heart Muscular organ located in the thoracic cavity whose rhythmic contractions maintain blood circulation. 216

heart attack Damage to the myocardium due to blocked circulation in the coronary arteries; also called a myocardial infarction (MI). 228

helper T cell Type of T cell that releases cytokines and stimulates certain other immune cells to perform their respective functions. 243

hemocoel Residual coelom found in arthropods, which is filled with hemolymph. 638

hemodialysis Cleansing of blood by using an artificial membrane that causes substances to diffuse from blood into a dialysis fluid. 310

hemoglobin Iron-containing pigment in red blood cells that combines with and transports oxygen. 222

hemophilia X-linked, recessive genetic disease in which one or more clotting factors are missing. 489

hepatic portal system Portal system that begins at the villi of the small intestine and ends at the liver. 221

hepatic portal vein A vein formed by capillaries in the digestive tract that carries nutrients to the liver. 221

hepatic vein Vein that runs between the liver and the inferior vena cava. 221

hepatitis Inflammation of the liver. 277

herbaceous stem Nonwoody stem. 155

herbivore Primary consumer in a grazing food chain; a plant eater. 706

hermaphroditic Animals having both male and female sex organs. 624

heterotroph Organism that cannot synthesize organic molecules from inorganic nutrients and therefore must take in organic nutrients (food). 128, 544, 706

heterotrophic Requires an outside source of organic compounds. 572

heterozygous Having two different alleles (e.g., Aa) for a given trait. 471

hexose Six-carbon sugar. 32

hinge joint Type of joint that allows movement as a hinge does. 377

hippocampus Part of the cerebral cortex where memories form. 328

histamine Substance, produced by basophils and mast cells in connective tissue, that causes capillaries to dilate. 239

homeobox Nucleotide sequence located in all homeotic genes and serving to identify portions of the genome in many different types of organisms that are active in pattern formation. 451

homeostasis Maintenance of normal internal conditions in a cell or an organism by means of self-regulating mechanisms. 4, 207

homeotic gene Gene that controls the overall body plan by controlling the fate of groups of cells during development. 450

homologous chromosomes (homologues) Similarly constructed chromosomes that have the same shape and contain genes for the same traits. 89

homologous structure Structure that is similar in two or more species because of common ancestry. 549

homozygous Having identical alleles (e.g., AA or aa) for a given trait; pure breeding. 471

hormone Chemical signal produced in one part of the body that controls the activity of other parts. 261, 394

horsetail Member of a seedless vascular plant phylum that has only one genus (Equisetum) in existence today. 609

human chorionic gonadotropin (hCG) Hormone produced by the chorion that functions to maintain the uterine lining. 425, 454

Human Genome Project Initiative to determine the complete sequence of the human genome and to analyze this information. 534

Huntington disease Genetic disease marked by progressive deterioration of the nervous system due to deficiency of a neurotransmitter. 480

hyaline cartilage Cartilage whose cells lie in lacunae separated by a white translucent matrix containing very fine collagen fibers. 196

hybridization Crossing of different species. 182

hydrogen bond Weak bond that arises between a slightly positive hydrogen atom of one molecule and a slightly negative atom of another molecule or between parts of the same molecule. 26

hydrolysis reaction Splitting of a compound by the addition of water, with the H^+ being incorporated in one fragment and the OH^- in the other. 31

hydrophilic Type of molecule that interacts with water by dissolving in water and/or by forming hydrogen bonds with water molecules. 27

hydrophobic Type of molecule that does not interact with water because it is nonpolar. 27

hydrostatic skeleton Fluid-filled body compartment that provides support for muscle contraction resulting in movement. 632

hydrothermal vent Hot springs in the seafloor along ocean ridges where heated seawater and sulfate react to produce hydrogen sulfide. 735

hyperthyroidism A disorder in which the thyroid gland produces excess thyroid hormone(s). 408

hypertension Elevated blood pressure, particularly the diastolic pressure. 231

hypertonic solution Higher solute concentration (less water) than the cell; causes cell to lose water by osmosis. 74

hypertrophy Increase in size of a tissue or body part. 386

hypha Filament of the vegetative body of a fungus. 584

hypothalamus Part of the brain located below the thalamus that helps regulate the internal environment of the body and produces releasing factors that control the anterior pituitary. 326, 398

hypothesis Supposition formulated after making an observation. 10

hypothyroidism A disorder in which the thyroid gland produces insufficient thyroid hormone(s). 408

hypothalamic-inhibitinghormone Hormone released by the hypothalamus to inhibit the anterior pituitary. 398

hypothalamic-releasing hormone Hormone released by the hypothalamus to control the anterior pituitary. 398

hypotonic solution Lower solute (more water) concentration than the cytosol of a cell; causes cell to gain water by osmosis. 73

I

immediate allergic response Allergic response that occurs within seconds of contact with an antigen, due to IgE. 248

ileum The final portion of the small intestine, contains Peyer's patches. 260

immune system All the cells and molecules in the body that protect the body against foreign organisms and substances, and also against cancerous cells. 202

immunity Ability of the body to protect itself from foreign substances and cells, including disease-causing agents. 238

immunization Use of a vaccine to protect the body against specific disease-causing agents. 244

immunoglobulin (Ig) Globular plasma protein that functions as an antibody. 241

immunosuppressive Any condition that inhibits the activity of the immune system. 251

implant Attachment and penetration of the embryo into the lining of the uterus (endometrium). 420

imprinting Learning to make a particular response to only one type of animal or object. 674

inclusive fitness Fitness that results from personal reproduction and from helping nondescendant relatives reproduce. 684

incomplete dominance Occurs when the heterozygote is intermediate in phenotype between the two homozygotes. 481

incontinence Inability to control excretory functions, especially urination. 301

incus The middle of the three ossicles of the ear that conduct vibrations from the tympanic membrane to the oval window of the inner ear. 355

independent assortment Alleles of unlinked genes segregate independently of each other during meiosis so that the gametes contain all possible combinations of alleles. 91

induced fit model Change in the shape of an enzyme's active site that enhances the fit between the active site and its substrate(s). 105

induction Ability of a chemical or a tissue to influence the development of another tissue. 449

inductive reasoning Using specific observations and the process of logic and reasoning to arrive at a hypothesis. 10

infant respiratory distress syndrome Breathing disorder in premature infants due to the lack of surfactant. 285

inferior vena cava Large vein that enters the right atrium from below and carries blood from the trunk and lower extremities. 220

infertility Inability to have as many children as desired. 438

inflammatory reaction Tissue response to injury, characterized by redness, swelling, pain, and heat. 239

initiation During DNA replication or transcription, initiation is the step whereby DNA replication begins or transcription begins. 510

innate immunity Protective responses of the body that are fully functional without previous exposure to a disease agent. 238

inner ear Portion of the ear consisting of a vestibule, semicircular canals, and the cochlea, where equilibrium is maintained and sound is transmitted. 356

inorganic molecule Type of molecule that is not derived from a living organism. 30

insect Member of a group of arthropods having a head with antennae, compound eyes, and simple eyes; a thorax with three pairs of legs and often wings; and an abdomen containing internal organs. 638

insertion End of a muscle that is attached to a movable bone. 378

inspiration Taking air into the lungs; inhalation. 282

inspiratory reserve volume Volume of air that can be forcibly inhaled after normal inhalation. 287

insulin Hormone secreted by the pancreas that lowers the blood glucose level by promoting the uptake of glucose by cells and the conversion of glucose to glycogen by the liver and skeletal muscles. 403

integration The summing of incoming signals by sensory receptors. 347

integumentary system Organ system consisting of skin and various organs, such as hair, that are found in skin. 202

intercalated disk Region that holds adjacent cardiac muscle cells together and appears as dense bands at right angles to the muscle striations. 199

interferon Protein formed by a cell infected with a virus that blocks the infection of another cell. 240

interkinesis Period of time between meiosis I and meiosis II during which no DNA replication takes place. 91

interleukin Cytokine produced by immune cells, such as macrophages and T cells, that functions as a regulator of the immune response. 247

intermediate filaments Group of fibrous polypeptides of various types that make up part of the cytoskeleton. 59

intermembrane space The space between the inner and outer membrane of the mitochondrial membrane. 120

internal respiration Exchange of oxygen and carbon dioxide between blood and tissue fluid. 290

interneuron Neuron located within the central nervous system that conveys messages between parts of the central nervous system. 316

internode In vascular plants, the region of a stem between two successive nodes. 145

interoceptor Sensory receptor that detects stimuli from inside the body; e.g., pressoreceptors, osmoreceptors, chemoreceptors. 346

interphase Stages of the cell cycle (G1, S, G2) during which growth and DNA synthesis occur when the nucleus is not actively dividing. 82

interstitial cell Hormone-secreting cell located between the seminiferous tubules of the testes. 419

interstitial cell-stimulating hormone (ICSH) Name sometimes given to luteinizing hormone in males; controls the production of testosterone by interstitial cells. 419

intervertebral disk Layer of cartilage located between adjacent vertebrae. 322, 374

intrauterine device (IUD) Birth control device consisting of a small piece of molded plastic inserted into the uterus. 426

inversion Change in chromosome structure in which a segment of a chromosome is turned around 180°. 496

invertebrate Referring to an animal without a serial arrangement of vertebrae. 620

ion Particle that carries a negative or positive charge. 24

ionic bond Chemical bond in which ions are attracted to one another by opposite charges. 24

iris Muscular ring that surrounds the pupil and regulates the passage of light through its opening. 352

isotonic solution Solution that is equal in solute concentration to that of the cytoplasm of a cell; causes cell to neither lose nor gain water by osmosis. 73

isotope Atom of the same element having the same atomic number but a different mass number due to the number of neutrons. 22

J

jaundice Yellowish tint to the skin caused by an abnormal amount of bilirubin (bile pigment) in the blood, indicating liver malfunction. 277

jawless fish Type of fish that has no jaws; includes today's hagfishes and lampreys. 651

jejunum The middle third of the small intestine. 260

joint Union of two or more bones; an articulation. 368

juxtaglomerular apparatus Structure located in the walls of arterioles near the glomerulus that regulates renal blood flow. 306

K

karyotype Chromosomes arranged by pairs according to their size, shape, and general appearance in mitotic metaphase. 494

keystone species Those species whose loss would result in a great number of secondary extinctions. 753

kidney One of a pair of organs in the urinary system that produce and excrete urine. 300

kidney stones Solid mass of crystals within the kidney. 309

kinetic energy Energy associated with motion. 100

kingdom The taxonomic grouping above phylum. 6

kin selection An explanation for altruistic behavior based on the concept that relatives share many genes. 684

L

lacteal Lymphatic vessel in an intestinal villus; aids in the absorption of lipids. 260

lactose intolerance Inability to digest lactose because of an enzyme deficiency. 265

lacuna Small pit or hollow cavity, as in bone or cartilage, where a cell or cells are located. 196

lake Body of fresh water. 729

lancelet Invertebrate chordate with a body that resembles a lancet and has the four chordate characteristics as an adult. 650

landscape diversity Includes all of the ecosystems that may interact in a particular area. 743

lanugo Short, fine hair that is present during the later portion of fetal development. 456

large intestine Last major portion of the digestive tract, extending from the small intestine to the anus and consisting of the cecum, the colon, the rectum, and the anus. 262

laryngitis Inflammation of the larynx with accompanying hoarseness. 291

larynx Cartilaginous organ located between the pharynx and the trachea that contains the vocal cords; also called the voice box. 284

law Theory that is generally accepted by an overwhelming number of scientists. 11

law of independent assortment Alleles of unlinked genes assort independently of each other during meiosis so that the gametes contain all possible combinations of alleles. 476

law of segregation Separation of alleles from each other during meiosis so that the gametes contain one from each pair. 470

leaf Lateral appendage of a stem, highly variable in structure, often containing cells that carry out photosynthesis. 145

leaf vein Vascular tissue within a leaf. 148

learning Relatively permanent change in an animal's behavior that results from practice and experience. 328, 674

leech Blood-sucking annelid, usually found in fresh water, with a sucker at each end of a segmented body. 634

lens Clear, membranelike structure found in the vertebrate eye behind the iris; brings objects into focus. 352

lenticel Permits gas exchange between the interior of a plant and the external atmosphere. 147

leptin Hormone produced by adipose tissue that acts on the hypothalamus to signal satiety. 406

less-developed country (LDC) Country in which population growth is expanding rapidly and the majority of people live in poverty. 693

lichen Fungi and algae coexisting in a symbiotic relationship. 589

ligament Tough connective tissue that usually connects bone to bone. 196, 368

light reactions Portion of photosynthesis that captures solar energy and takes place in thylakoid membranes. 130

lignin Chemical that hardens the cell walls of plants. 147

limbic system Association of various brain centers, including the amygdala and hippocampus; governs learning and memory and various emotions such as pleasure, fear, and happiness. 328

lineage An evolutionary group held together by common ancestry. 661

linkage group Alleles of different genes that are located on the same chromosome and tend to be inherited together. 491

lipase Fat-digesting enzyme secreted by the pancreas. 263, 484

lipid Organic compound that is insoluble in water; notably fats, oils, and steroids. 34

lipopolysaccharide (LPS) Molecule containing both lipid and polysaccharide, which is important in the outer membrane of the gram-negative cell wall. 572

liposome A microscopic artificial sac composed of fatty substances and used in experimental research on the cell. 544

littoral zone Shore zone between high-tide mark and low-tide mark; also, shallow water of a lake where light penetrates to the bottom. 732

liver Large, dark-red internal organ that produces urea and bile, detoxifies the blood, stores glycogen, and produces the plasma proteins, among other functions. 263

liverwort Type of bryophyte. 604

lobe-finned fish Type of fish with limblike fins. 652

localized Condition that affects only a limited area of the body. 209

locus Particular site where a gene is found on a chromosome. 470

logistic growth An "S-shaped" population growth curve. 692

long-day plant Plant that flowers when day length is longer than a critical length. 187

long-term memory Retention of information that lasts longer than a few minutes. 328

long-term potentiation (LTP) Increase in synaptic strength following frequent stimulation. 329

loop of the nephron Portion of a nephron between the proximal and distal convoluted tubules; functions in water reabsorption. 303

loose fibrous connective tissue Tissue composed mainly of fibroblasts widely separated by a matrix containing collagen and elastic fibers. 196

lung cancer Malignant tumors in the lungs that are highly associated with smoking. 295

lungs In humans, paired, cone-shaped organs within the thoracic cavity; function in internal respiration and contain moist surfaces for gas exchange. 284, 652

luteinizing hormone (LH) Hormone produced by the anterior pituitary gland that stimulates the development of the corpus luteum in females and the production of testosterone in males. 419

lymph Fluid, derived from tissue fluid, that is carried in lymphatic vessels. 227, 236

lymphatic capillaries Small vessels that form a blind pouch at one end and collect lymphatic fluid in various organs and tissues. 227

lymphatic system Organ system consisting of lymphatic vessels and lymphoid organs; transports lymph and lipids, and aids the immune system. 202, 236

lymphatic vessel Vessel that carries lymph. 236

lymph node Mass of lymphoid tissue located along the course of a lymphatic vessel. 238

lymphocyte Specialized white blood cell that functions in acquired immunity; occurs in two forms—T lymphocytes and B lymphocytes. 224

lymphoid (lymphatic) organ Collections of lymphocytes and other immune cells; divided into primary lymphoid organs where lymphocytes develop and secondary lymphoid organs where most lymphocyte responses occur. 237

lysosome Membrane-bounded vesicle that contains hydrolytic enzymes for digesting macromolecules. 56

M

macrophage Large phagocytic cell, derived from a monocyte, that ingests microbes and debris. 224, 240

major histocompatibility complex (MHC) Cell surface proteins that bind an antigen for presentation to a T cell. 243

malaria Serious infectious illness caused by the parasitic protozoan *Plasmodium*. 582

male condom Sheath used to cover the penis during sexual intercourse; used as a contraceptive and, if latex, to minimize the risk of transmitting infection. 427

male gametophyte In seed plants, the gametophyte that produces sperm; a pollen grain. Sometimes called a microgametophyte. 174

malignant Tumors with the ability to invade surrounding tissue and metastasize. 518

malleus The first of the three ossicles of the ear that conduct vibrations from the tympanic membrane to the oval window of the inner ear. 355

Malpighian tubule Blind, threadlike excretory tubule attached to the gut of an insect. 639

maltase Enzyme produced in the small intestine that breaks down maltose to two glucose molecules. 265

mammal Homeothermic vertebrate characterized especially by the presence of hair and mammary glands. 656

mantle A membranous or sometimes muscular covering that envelops but does not completely enclose the visceral mass in molluscs. 630

Marfan syndrome Congenital disorder of connective tissue characterized by abnormal length of the limbs. 480

marsupial Member of a group of mammals bearing immature young nursed in a marsupium, or pouch. 657

mast cell Cell to which IgE antibodies, which form in response to allergens, attach, causing the cell to release allergy mediators that cause symptoms. 239

matrix Unstructured semifluid substance that fills the space between cells in connective tissues or inside organelles. 59, 118, 196

matter Anything that takes up space and has mass. 20

mechanical energy A type of kinetic energy, such as walking or running. 100

mechanoreceptor Sensory receptor that is sensitive to mechanical stimulation, such as pressure, sound waves, or gravity. 346

medulla oblongata Part of the brain stem that is continuous with the spinal cord; controls heartbeat, blood pressure, breathing, and other vital functions. 326

medusan Pertaining to the motile stage of cnidarians in which the mouth is pointed downward. 625

megakaryocyte Large bone marrow cell that gives rise to blood platelets. 225

megaspore One of the two types of spores produced by seed plants; develops into a female gametophyte (embryo sac). 172

megasporocyte Megaspore mother cell; produces megaspores by meiosis. 174

meiosis Type of nuclear division that occurs as part of sexual reproduction in which the daughter cells receive the haploid number of chromosomes in varied combinations. 89

melanocyte Specialized cell in the epidermis that produces melanin, the pigment responsible for skin color. 204

melanocyte-stimulating hormone (MSH) Substance that causes melanocytes to secrete melanin in lower vertebrates. 398

melatonin Hormone, secreted by the pineal gland, that is involved in biorhythms. 405

memory The brain function of recalling something that has been learned. 328

meninges Protective membranous coverings of the central nervous system. 201, 322

meningitis Inflammation of the meninges, the membranes that surround the brain and spinal cord. 340

menisci (sing., meniscus) Cartilaginous wedges that separate the surfaces of bones in synovial joints. 377

menopause Termination of the ovarian and uterine cycles in older women. 425

menstruation Loss of blood and tissue from the uterus at the end of a uterine cycle. 424

meristematic tissue Undifferentiated embryonic tissue in the active growth regions of plants. 146

mesoderm Middle germ layer of embryonic gastrula; gives rise to muscles, the connective tissue, and the circulatory system. 446

mesophyll Inner, thickest layer of a leaf consisting of palisade and spongy mesophyll; the site of most photosynthesis. 160, 602

messenger RNA (mRNA) Ribonucleic acid whose sequence of codons specifies the sequence of amino acids during protein synthesis. 506

metabolic pathway Series of linked reactions, beginning with a particular reactant and terminating with an end product. 104

metabolism All of the chemical changes that occur within a cell. 102

metamorphosis Change in shape and form that some animals, such as amphibians and insects, undergo during development. 639

metaphase Mitotic phase during which chromosomes are aligned at the metaphase plate. 86

metaphase plate A disk formed during metaphase in which all of a cell's chromosomes lie in a single plane at right angles to the spindle fibers. 86

metastasis The ability of malignant tumor cells to invade surrounding tissue and spread throughout the body. 520

methanogen Any of the various species of archaea that produce methane gas as a metabolic by-product. 576

microevolution Change in gene frequencies between populations of a species over time. 551

microsatellite Tandem repeat of DNA nucleotides that varies in length among individuals. 537

microsphere Structure composed only of protein that has many of the properties of a cell. 543

microspore One of two types of spores produced by seed plants; develops into a male gametophyte (pollen grain). 172

microsporocyte Microspore mother cell; produces microspores by meiosis. 174

microtubules Small, hollow cylinders of the cytoskeleton composed of tubulin, involved in the structure and movement of cells. 61

microvillus Cylindrical process that extends from an epithelial cell of a villus of the intestinal wall and serves to increase the surface area of the cell. 194

micturition Emptying of the bladder; urination. 301

midbrain Part of the brain located below the thalamus and above the pons; contains reflex centers and tracts. 326

middle ear Portion of the ear consisting of the tympanic membrane, the oval and round windows, and the ossicles, where sound is amplified. 355

mimicry Superficial resemblance of two or more species; mechanism by which an organism avoids predation by appearing to be noxious. 698

mineral Naturally occurring inorganic substance containing two or more elements. 272, 753

mineralocorticoids Hormones secreted by the adrenal cortex that regulate salt and water balance, leading to increases in blood volume and blood pressure. 401

mitochondrion Membrane-bounded organelle in which ATP molecules are produced during the process of cellular respiration. 57

mitosis Type of cell division in which daughter cells receive the exact chromosomal and genetic makeup of the parental cell; occurs during growth and repair. 83

model In an experiment, an organism that serves as a representative for the organism of interest. 11

molecular clock Idea that the rate at which mutational changes accumulate in certain genes is constant over time and is not involved in adaptation to the environment. 661

molecule Union of two or more atoms of the same element; also, the smallest part of a compound that retains the properties of the compound. 24

mollusc Member of the phylum Mollusca, which includes squid, clams, snails, and chitons; characterized by a visceral mass, a mantle, and a foot. 629

monoclonal antibody One of many antibodies produced by a clone of hybridoma cells that all bind to the same antigen. 247

monocot (Monocotyledones) A member of the class of flowering plants called Monocotyledones. Monocots have several common characteristics, including one cotyledon in the seed, scattered vascular bundles in the stem, and flower parts in threes or multiples of three. 149, 612

monocyte Type of agranular leukocyte that functions as a phagocyte, particularly after it becomes a macrophage, which is also an antigen-presenting cell. 224, 240

monohybrid Individual that is heterozygous for one trait. 473

mononuclear cell Another name for agranular leukocytes. 224

monosomy One less chromosome than usual. 492

monotreme Egg-laying mammal. 656

monsoon Climate in India and southern Asia caused by wet ocean winds that blow onshore for almost half the year. 721

more-developed country (MDC) Country in which population growth is low and the people enjoy a good standard of living. 693

morphogenesis Emergence of shape in tissues, organs, or an entire embryo during development. 448

morphogen gene Unit of inheritance that controls a gradient that influences morphogenesis. 450

morula Spherical mass of cells resulting from cleavage during animal development prior to the blastula stage. 445

mosaic evolution Concept that human characteristics did not evolve at the same rate; e.g., some body parts are more humanlike than others in early hominids. 662

moss Type of bryophyte. 605

motor molecule Proteins that attach to cytoskeletal elements and allow for cell/organelle movement. 59

motor neuron Nerve cell that conducts nerve impulses away from the central nervous system and innervates effectors (muscles and glands). 316

motor unit Motor neuron and all the muscle fibers it innervates. 385

mucous membrane Membrane that lines a cavity or tube that opens to the outside of the body; also called mucosa. 201

multiple allele Inheritance pattern in which there are more than two alleles for a particular trait. 481

multiple sclerosis (MS) Disease in which the outer, myelin layer of nerve fiber insulation becomes damaged, interfering with normal conduction of nerve impulses. 251, 339

muscle tone Contraction of some fibers in skeletal muscle at any given time. 386

muscle twitch Contraction of a whole muscle in response to a single stimulus. 385

muscular system System of muscles that produces both movement within the body and movement of its limbs; principal components are skeletal muscle, smooth muscle, and cardiac muscle. 203

muscular tissue Type of tissue composed of fibers that can shorten and thicken. 199

mutagen An environmental agent, such as ultraviolet radiation, that causes mutations in DNA. 515

mutation Alteration in chromosome structure or number; also, an alteration in a gene due to a change in DNA composition. 553

mutualism Symbiotic relationship in which both species benefit. 700

myasthenia gravis Chronic disease characterized by muscles that are weak and easily fatigued. 251

mycelium Mass of hyphae that makes up the body of a fungus. 584

mycorrhiza (pl., mycorrhizae) Mutually beneficial symbiotic relationship between a fungus and the roots of vascular plants. 153

mycorrhizal fungi Fungi that grow on the roots of plants. 589

mycoses Any disease caused by a fungus. 585

myelin sheath White, fatty material—derived from the membrane of Schwann cells—that forms a covering for nerve fibers. 317

myocardium Cardiac muscle in the wall of the heart. 216

myofibril Specific muscle cell organelle containing a linear arrangement of sarcomeres, which shorten to produce muscle contraction. 380

myogram Recording of a muscular contraction. 385

myosin Muscle protein making up the thick filaments in a sarcomere; it pulls actin to shorten the sarcomere, yielding muscle contraction. 382

myxedema Condition resulting from a deficiency of thyroid hormone in an adult. 408

N

NAD$^+$ (nicotinamide adenine dinucleotide) Coenzyme of oxidation-reduction that accepts electrons and hydrogen ions to become NADH + H$^+$ as oxidation of substrates occurs. During cellular respiration, NADH carries electrons to the electron transport system in mitochondria. 114

NADP$^+$ (nicotinamide adenine dinucleotide phosphate) Coenzyme of oxidation-reduction that accepts electrons and hydrogen ions to become NADPH + H$^+$. During photosynthesis, NADPH participates in the reduction of carbon dioxide to glucose. 130

nail A protective covering of the distal part of the digits. 205

nasal cavity One of two canals in the nose, separated by a septum. 283

nasopharynx Region of the pharynx associated with the nasal cavity. 258

natural killer (NK) cell Lymphocyte-like cell that causes certain virus-infected or cancerous cells to undergo apoptosis. 240

natural selection Mechanism of evolution caused by environmental selection of organisms most fit to reproduce; results in adaptation to the environment. 5, 556

Neandertal Hominid with a sturdy build that lived during the last Ice Age in Europe and the Middle East. 665

nearsighted Vision abnormality due to an elongated eyeball from front to back; light rays focus in front of the retina when viewing distant objects. 360

negative feedback Mechanism of homeostatic response by which the output of a system suppresses or inhibits the activity of the system. 207, 396

nematocyst In cnidarians, a capsule that contains a threadlike fiber whose release aids in the capture of prey. 625

nephridium (pl. nephridia) Segmentally arranged, paired excretory tubules of many invertebrates, as in the earthworm. 632

nephron Anatomical and functional unit of the kidney; a kidney tubule. 302

nerve Bundle of nerve fibers outside the central nervous system. 200, 331

nerve impulse Action potential (electrochemical change) traveling along a neuron. 318

nervous system Organ system consisting of the brain, spinal cord, and associated nerves that coordinates the other organ systems of the body. 203

nervous tissue Tissue that contains nerve cells (neurons), which conduct impulses, and neuroglial cells, which support, protect, and provide nutrients to neurons. 200

neural plate Region of the dorsal surface of the chordate embryo that marks the future location of the neural tube. 447

neural tube Tube formed by closure of the neural groove during development. In vertebrates, the neural tube develops into the spinal cord and brain. 447

neuroglia Nonconducting nerve cells that are intimately associated with neurons and function in a supportive capacity. 200, 316

neuromuscular junction Region where an axon bulb approaches a muscle fiber; contains a presynaptic membrane, a synaptic cleft, and a postsynaptic membrane. 383

neuron Nerve cell that characteristically has three parts: a cell body, dendrites, and an axon. 200, 316

neurotransmitter Chemical stored at the ends of axons that is responsible for transmission across a synapse. 320

neurula The early embryo during the development of the neural tube from the neural plate, marking the first appearance of the nervous system. 447

neutrophil Granular leukocyte that is the most abundant of the white blood cells. 224, 240

nitrification Process by which nitrogen in ammonia and organic compounds is oxidized to nitrites and nitrates by soil bacteria. 713

nitrogen fixation Process whereby free atmospheric nitrogen is converted into compounds, such as ammonium and nitrates, usually by bacteria. 713

node In plants, the place where one or more leaves attach to a stem. 145

node of Ranvier Gap in the myelin sheath around a nerve fiber; increases the speed of nerve impulse conduction. 317

noncyclic electron pathway Portion of the light-dependent reaction of photosynthesis that involves both photosystem I and photosystem II. It generates both ATP and NADPH. 132

nondisjunction Failure of homologous chromosomes or daughter chromosomes to separate during meiosis I and meiosis II, respectively. 492

nonrandom mating Mating among individuals on the basis of their phenotypic similarities or differences, rather than mating on a random basis. 556

nonrenewable resource Minerals, fossil fuels, and other materials present in essentially fixed amounts (within the human timescale) in our environment. 757

nonvascular plant Bryophytes such as mosses and liverworts that have no vascular tissue and either occur in moist locations or have special adaptations for living in dry locations. 604

norepinephrine (NE) Neurotransmitter of the postganglionic fibers in the sympathetic division of the autonomic system; also, a hormone produced by the adrenal medulla. 321, 401

normal microflora Microbes, mostly bacteria, that live on and in the body and usually have beneficial effects. 570

nostril One of the two openings in the nose that allow air to enter the nasal cavities. 283

notochord Dorsal supporting rod that exists in all chordates at some time in their life history; replaced by the vertebral column in vertebrates. 447, 648

nuclear envelope Double membrane that surrounds the nucleus and is continuous with the endoplasmic reticulum. 52

nuclear pore Opening in the nuclear envelope that permits the passage of proteins into the nucleus and ribosomal subunits out of the nucleus. 52

nucleoid Region of prokaryotic cells where DNA is located; it is not bounded by a nuclear envelope. 48, 572

nucleolus (pl., nucleoli) Dark-staining, spherical body in the nucleus that produces ribosomal subunits. 52

nucleoplasm Semifluid medium of the nucleus containing chromatin. 52

nucleotide Monomer of DNA and RNA consisting of a 5-carbon sugar bonded to a nitrogenous base and a phosphate group. 39

nucleus Membrane-bounded organelle within a eukaryotic cell that contains chromosomes and controls the structure and function of the cell. 52

nutrients Chemical substances in foods that are essential to the diet and contribute to good health. 266

nutrition The process by which living organisms assimilate food. 266

O

obesity Excess adipose tissue; exceeding ideal weight by more than 20%. 274

observation Step in the scientific method by which data are collected before a conclusion is drawn. 10

oil Triglyceride, usually of plant origin, that is composed of glycerol and three fatty acids and is liquid in consistency. 34

oil gland Secretes sebum to lubricate the hair and skin. 205

olfactory cell Modified neuron that is a sensory receptor for the sense of smell. 350

oligodendroglial cells Cells that form the myelin sheaths surrounding axons in the central nervous system. 317

omnivore Organism in a food chain that feeds on both plants and animals. 706

oncogene Cancer-causing gene. In its normal state, involved in regulation of the growth of cells; when mutated, allows cells to grow out of control. 84, 518

oocyte The female gamete, or egg, prior to fertilization. 420

oogenesis Production of oocytes in females by the process of meiosis and maturation. 95, 420

open circulatory system Circulatory system in which blood is not entirely contained within blood vessels. 632

operant conditioning Learning that results from rewarding or reinforcing a particular behavior. 675

operon In bacteria, a group of structural and regulatory genes that acts as a unit. 512

opportunistic infection Disease that arises in the presence of an impaired immune system. 430

optic chiasma X-shaped structure on the underside of the brain formed by partial crossing-over of optic nerve fibers. 354

optic nerve Cranial nerve that carries nerve impulses from the retina of the eye to the brain, thereby contributing to the sense of sight. 352

order One of the categories, or taxa, used by taxonomists to group species. 6

organ Combination of two or more different tissues performing a common function. 3, 144

organelle Small, often membranous structure in the cytoplasm having a specific structure and function. 46

organic molecule Molecule that always contains carbon and hydrogen, and often contains oxygen as well. 30

organism Individual living thing. 3

organ system Group of related organs working together. 3

origin End of a muscle that is attached to a relatively immovable bone. 378

osmosis Diffusion of water through a differentially permeable membrane. 73

osmotic pressure Measure of the tendency of water to move across a differentially permeable membrane. 73

ossicle One of the small bones of the vertebrate middle ear—malleus, incus, or stapes. 355

osteoarthritis (OA) Disintegration of the cartilage between bones at a synovial joint. 388

osteoblast Bone-forming cell. 370

osteoclast Cell that causes erosion of bone. 370

osteocyte Mature bone cell located within the lacunae of bone. 368

osteon Cylindrical-shaped unit containing bone cells that surround an osteonic canal; also called a Haversian system. 368

osteoporosis Condition in which bones break easily because calcium is removed from them faster than it is replaced. 272, 388

otitis media Inflammation of the middle ear, often caused by a bacterial infection. 292

otolith Calcium carbonate granule associated with ciliated cells in the utricle and saccule. 358

outer ear Portion of the ear consisting of the pinna and the auditory canal. 355

out-of-Africa hypothesis Proposal that modern humans originated only in Africa; then they migrated and supplanted populations of Homo in Asia and Europe about 100,000 years ago. 664

oval window The oval-shaped opening between the middle ear and the vestibule to which the stapes bone attaches. 355

ovarian cancer Malignant tumor of the ovaries in females. 436

ovarian cycle Monthly follicle changes occurring in the ovary that control the level of sex hormones in the blood and the uterine cycle. 423

ovary In animals, the female gonad, the organ that produces oocytes, estrogen, and progesterone; in flowering plants, the base of the pistil that protects ovules and, along with associated tissues, becomes a fruit. 173, 404, 420, 613

overexploitation When the number of individuals taken from a wild population is so great that the population becomes severely reduced in numbers. 752

ovule In seed plants, a structure where the female megaspore becomes an egg-producing female gametophyte that develops into a seed following fertilization. 173, 611

oxidation Loss of one or more electrons from an atom or molecule; in biological systems, generally the loss of hydrogen atoms. 108

oxygen debt Amount of oxygen needed to metabolize lactate, a compound that accumulates during vigorous exercise. 122, 384

oxyhemoglobin Compound formed when oxygen combines with hemoglobin. 289

oxytocin Hormone released by the posterior pituitary that causes contraction of the uterus and milk letdown. 398

ozone hole Seasonal thinning of the ozone shield in the lower stratosphere at the North and South Poles. 714

ozone shield Accumulation of O_3, formed from oxygen in the upper atmosphere; a filtering layer that protects Earth from ultraviolet radiation. 714

P

palisade mesophyll In a plant leaf, the layer of mesophyll containing elongated cells with many chloroplasts. 160

pancreas Large gland that lies behind the stomach in humans; has both digestive and endocrine functions. 263, 403

pancreatic amylase Enzyme in the pancreas that digests starch to maltose. 263

pancreatic islet (of Langerhans) Distinctive group of cells within the pancreas that secretes insulin and glucagon. 403

pancreatitis Inflammation of the pancreas. 277

Pap smear Sample of cells removed from the tip of the cervix and then stained and examined microscopically. 432

parasitism Symbiotic relationship in which one species (the parasite) benefits in terms of growth and reproduction to the detriment of the other species (the host). 699

parasympathetic division That part of the autonomic system that is active under normal conditions. 333

parathyroid gland One of four glands embedded in the posterior surface of the thyroid gland; produces parathyroid hormone. 400

parathyroid hormone (PTH) Hormone secreted by the four parathyroid glands that increases the blood calcium level and decreases the phosphate level. 400

parenchyma Thin-walled, minimally differentiated cell that photosynthesizes or stores the products of photosynthesis. 147

Parkinson disease (PD) Progressive deterioration of the central nervous system due to a deficiency in the neurotransmitter dopamine. 339

parturition Processes that lead to and include birth and the expulsion of the afterbirth. 459

passive immunity Protection against infection acquired by transfer of antibodies to a susceptible individual. 244

pathogen Disease-causing agent such as a virus, parasitic bacterium, fungus, or animal. 238

pattern formation During development, how tissues and organs are arranged in the body. 448

pectoral girdle Portion of the skeleton that provides support and attachment for the upper limbs. 375

pedigree Chart showing the relationships of relatives and their particular traits. 478

pelagic division Open portion of the sea. 734

pelvic girdle Portion of the skeleton to which the lower limbs are attached. 376

pelvic inflammatory disease (PID) Latent infection of gonorrhea or chlamydia in the vasa deferentia or uterine tubes. 433

penis External organ in males through which the urethra passes; also serves as the organ of sexual intercourse. 417

pentose Five-carbon sugar. 32

pepsin Enzyme secreted by gastric glands that digests proteins to peptides. 260

peptidase Intestinal enzyme that breaks down short chains of amino acids to individual amino acids that are absorbed across the intestinal wall. 265

peptide bond Covalent bond that joins two amino acids. 37

peptide hormone Type of hormone that is a protein, a peptide, or derived from an amino acid. 396

perception Mental awareness of sensory stimulation. 346

pericardium Protective serous membrane that surrounds the heart. 216

pericycle Layer of cells surrounding the vascular tissue of roots; produces branch roots. 151

periderm Protective tissue that replaces the epidermis of a stem and includes the cork and cork cambium. 147

periosteum Fibrous connective tissue covering the surface of bone. 368

peripheral nervous system (PNS) Nerves and ganglia that lie outside the central nervous system. 316

peristalsis Wavelike contractions that propel substances along a tubular structure such as the esophagus. 258

peritonitis Generalized infection of the lining of the abdominal cavity. 262

peritubular capillary network Capillary network that surrounds a nephron and functions in reabsorption during urine formation. 302

permafrost Permanently frozen ground, usually in the tundra, a biome of arctic regions. 724

peroxisome Enzyme-filled vesicle in which fatty acids and amino acids are metabolized to hydrogen peroxide that is then broken down to harmless products. 56

petal A flower part just inside the sepals; often conspicuously colored to attract pollinators. 172, 613

petiole The part of a plant leaf that connects the blade to the stem. 145

phagocyte Designation for neutrophils and macrophages that engulf antigens. 239

phagocytosis Process by which amoeboid cells engulf large substances, forming an intracellular vacuole. 77

pharyngitis Inflammation of the pharynx. 291

pharynx Portion of the digestive tract between the mouth and the esophagus that serves as a passageway for food on its way to the stomach and air on its way to the trachea. 258, 283

phenomenon Something that a scientist observes in nature, such as the law of gravity. 10

phenotype Outward appearance of an organism caused by the genotype and environmental influences. 471

phenylketonuria (PKU) Condition caused by the accumulation of phenylalanine. 479

pheromone Chemical signal that works at a distance and alters the behavior of another member of the same species. 406, 676

phloem Vascular tissue that conducts organic solutes in plants. 145, 606

phospholipid Molecule that forms the bilayer of the cell's membranes. 35

photoperiod Relative lengths of daylight and darkness that affect the physiology and behavior of an organism. 187

photoreceptor Sensory receptor in the retina that responds to light stimuli. 346

photosynthesis Process by which plants and algae make their own food using the energy of the sun. 57, 128

photosystem Cluster of light-absorbing pigment molecules within thylakoid membranes. 132

phototropism Growth response of plant stems to light. 184

photovoltaic (solar) cell An energy-conversion device that captures solar energy and directly converts it to electrical current. 759

pH scale Measurement scale for hydrogen ion concentration; ranges from 0 (acid) to 14 (basic), with 7 neutral. 29

phyletic gradualism Evolutionary model that proposes that evolutionary change resulting in a new species can occur gradually in an unbranched lineage. 561

phylogenetics The study of how organisms are evolutionarily related; used to build evolutionary trees to describe these relationships. 562

phylogeny Evolutionary history of a group of organisms. 562

phylum One of the categories, or taxa, used by taxonomists to group species. 6

phytochrome Photoreversible plant pigment involved in photoperiodism and other responses of plants such as etiolation. 188

phytoplankton Part of plankton containing organisms that photosynthesize, releasing oxygen into the atmosphere and serving as food producers in aquatic ecosystems. 578

pineal gland Endocrine gland located in the third ventricle of the brain that produces melatonin. 405

pinna Outer, funnel-like structure of the ear that picks up sound waves. 355

pinocytosis Process by which vesicle formation brings molecules into the cell. 77

pith Parenchyma tissue in the center of some stems and roots. 151

pituitary dwarfism Condition characterized by normal proportions but small stature; caused by inadequate growth hormone. 407

pituitary gland Endocrine gland that lies just inferior to the hypothalamus; consists of the anterior pituitary and the posterior pituitary. 398

pivot joint Type of joint, such as that between the radius and ulna, that only allows rotation. 378

placenta Structure that forms from the chorion and the uterine wall and allows the embryo, and then the fetus, to acquire nutrients and rid itself of wastes. 425, 457

placental mammal A group of species that rely on internal development whereby the fetus exchanges nutrients and wastes with its mother via a placenta. 657

plankton Freshwater and marine organisms suspended on or near the surface of the water; includes phytoplankton and zooplankton. 731

plant Multicellular, usually photosynthetic organism belonging to the plant kingdom. 7

plant hormone Chemical signal that is produced by various plant tissues and coordinates the activities of plant cells. 184

plaque Accumulation of soft masses of fatty material, particularly cholesterol, beneath the inner linings of the arteries. 228, 269

plasma Liquid portion of blood. 198, 222

plasma membrane Membrane surrounding the cytoplasm that consists of a phospholipid bilayer with embedded proteins; regulates the entrance and exit of molecules from the cell. 47

plasmid Self-duplicating ring of accessory DNA in the cytoplasm of bacteria. 48, 526

plasmodial (acellular) slime mold Free-living mass of cytoplasm that moves by pseudopods on a forest floor or in a field, feeding on decaying plant material by phagocytosis. 583

plasmodesmata Cytoplasmic channels that cross plant cell walls, aid in communication and transport between cells. 148

plasmolysis Contraction of the cell contents due to the loss of water. 74

platelet Cell fragment that is necessary to blood clotting; also called a thrombocyte. 198, 225

pleura Serous membrane that encloses the lungs. 285

plumule In flowering plants, the embryonic plant shoot that bears young leaves. 179

pneumonectomy Surgical removal of all or part of a lung. 295

pneumonia Infection of the lungs that causes alveoli to fill with mucus and pus. 292

polar body Nonfunctioning daughter cell, formed during oogenesis, that has little cytoplasm. 96

pollen cone One of two types of cones produced by gymnosperms; contains windblown pollen (male gametophyte). 611

pollen grain In seed plants, the sperm-producing male gametophyte. 174, 611

pollen tube In seed plants, a tube that forms when a pollen grain lands on the stigma and germinates. The tube grows to reach the egg inside an ovule, where fertilization occurs. 611

pollination In seed plants, the delivery of pollen to the vicinity of the egg-producing female gametophyte. 175, 611

pollinator An animal, usually an insect, that carries pollen from flower to flower. 611

pollution Any environmental change that adversely affects the lives and health of living things. 751

polygenic inheritance Inheritance pattern in which a trait is controlled by several allelic pairs. 482

polymer Macromolecule consisting of covalently bonded monomers. 31

polymerase chain reaction (PCR) Technique that uses the enzyme DNA polymerase to produce copies of a particular piece of DNA within a test tube. 527

polyp Small, abnormal growth that arises from the epithelial lining; a body form of cnidarians with the mouth pointing upward. 277, 625

polypeptide Polymer of many amino acids linked by peptide bonds. 37

polyribosome String of ribosomes simultaneously translating regions of the same mRNA strand during protein synthesis. 54, 509

polysaccharide Polymer made from sugar monomers. 32

pons Portion of the brain stem above the medulla oblongata and below the midbrain; assists the medulla oblongata in regulating the breathing rate. 326

population Group of organisms of the same species occupying a certain area and sharing a common gene pool. 8, 551, 690

positive feedback Mechanism of homeostatic response in which the output intensifies and increases the likelihood of response instead of countering and canceling it. 208, 398

posterior pituitary Portion of the pituitary gland that stores and secretes oxytocin and antidiuretic hormone, which are produced by the hypothalamus. 398

postzygotic isolating mechanism A mechanism that prevents an offspring from developing or becoming sexually mature after fertilization has taken place. 559

potential energy Stored energy as a result of location or spatial arrangement. 100

precipitation The process by which water leaves the atmosphere and falls to the ground as water, ice, or snow. 711

predation Interaction in which one organism uses another, called the prey, as a food source. 697

predator Any nonherbivorous species that is not a producer. 697

prediction A prediction results from a hypothesis and is usually stated in an "if . . . then" way. 10

prefrontal area Association area in the frontal lobe that receives information from other association areas and uses it to reason and plan actions. 326

preparatory (prep) reaction Reaction that converts pyruvate from glycolysis into acetyl CoA for entrance into the citric acid cycle. 114

pressure-flow model Explanation for phloem transport; osmotic pressure following active transport of sugar into phloem brings a flow of sap from a source to a sink. 166

prey Organism that provides nourishment for a predator. 697

prezygotic isolating mechanism A mechanism that prevents fertilization. 559

primary motor area Area in the frontal lobe where voluntary commands begin. 325

primary mRNA After transcription, primary mRNA is formed. This mRNA is processed before translation can occur. 507

primary root Original root that grows straight down and remains the dominant root of the plant. 152

primary somatosensory area Area of the brain in the parietal lobe where sensory information from the skin and skeletal muscles arrive. 325

primata Member of the order Primate; includes prosimians, monkeys, apes, and hominids. 659

principle Theory that is generally accepted by an overwhelming number of scientists; also called a law. 11

prion Infectious particle, consisting of protein only and no nucleic acid, that is believed to be linked to several diseases of the central nervous system; stands for proteinlike infectious agent. 596

producer Photosynthetic organism at the start of a grazing food chain that makes its own food. 706

product Substance that forms as a result of a reaction. 102

progesterone Female hormone that helps maintain sexual organs and secondary sex characteristics. 404, 424

proglottids A series of reproductive units with a full set of male and female sex organs in parasitic flatworms. 628

prokaryote Lacking a membrane-bounded nucleus and organelles; a prokaryotic cell is the cell type within the domains Bacteria and Archaea. 7, 47

prolactin (PRL) Hormone secreted by the anterior pituitary that stimulates the production of milk from the mammary glands. 398

promoter In an operon, a sequence of DNA where RNA polymerase binds prior to transcription. 507

prophase Mitotic phase during which chromatin condenses so that chromosomes appear. 86

proprioceptor Sensory receptor that assists the brain in knowing the position of the limbs. 347

prosimian Group of primates that includes lemurs and tarsiers, and may resemble the first primates to have evolved. 659

prostaglandin (PG) Hormone that has various and powerful local effects. 406

prostate gland Gland located around the male urethra below the urinary bladder; adds secretions to semen. 416

protein Organic macromolecule composed of one or several polypeptides. 37

protein-first hypothesis In chemical evolution, the proposal that protein originated before other macromolecules and made possible the formation of protocells. 543

proteinoid Abiotically polymerized amino acids that, when exposed to water, become microspheres. 543

proteomics The study of all the proteins in an organism. 536

prothrombin Plasma protein converted to thrombin during the steps of blood clotting. 225

prothrombin activator Enzyme that catalyzes the transformation of the precursor prothrombin to the active enzyme thrombin. 225

protist Member of the kingdom Protista, one of the six kingdoms in the classification of organisms. 7, 577

protocell In biological evolution, a possible cell forerunner that became a cell once it could reproduce. 542

proton Positive subatomic particle, located in the nucleus and having a weight of approximately one atomic mass unit. 20

proto-oncogene A gene that encodes a protein that promotes the cell cycle and prevents apoptosis. 84

protoplast Plant cell from which the cell wall has been removed. 181, 531

protostome Group of coelomate animals in which the first embryonic opening (the blastopore) is associated with the mouth. 623

protozoan Heterotrophic unicellular protist that moves by flagella, cilia, or pseudopodia, or is immobile. 577

provirus Latent form of a retrovirus in which the viral DNA is incorporated into the chromosome of the host. 592

proximal convoluted tubule (PCT) Highly coiled region of a nephron near the glomerular capsule, where tubular reabsorption takes place. 303

pseudocoelom Body cavity lying between the digestive tract and the body wall that is incompletely lined by mesoderm. 635

pulmonary artery Blood vessel that takes blood away from the heart to the lungs. 217

pulmonary circuit Circulatory pathway that consists of the pulmonary trunk, the pulmonary arteries, and the pulmonary veins; takes O_2-poor blood from the heart to the lungs and O_2-rich blood from the lungs to the heart. 220

pulmonary fibrosis Accumulation of fibrous connective tissue in the lungs; usually caused by inhaling irritating particles. 295

pulmonary tuberculosis Tuberculosis of the lungs, caused by the bacterium *Mycobacterium tuberculosis*. 293

pulmonary vein Blood vessel that takes blood to the heart from the lungs. 217

pulse Rhythmic expansion and recoil of arteries, due to contraction of the heart, that can be felt in certain areas at the body surface. 217

punctuated equilibrium Evolutionary model that proposes that periods of rapid change dependent on speciation are followed by long periods of stasis. 561

Punnett square Grid created to calculate the expected results of simple genetic crosses. 473

pupil The opening in the iris of the eye through which light passes. 352

purine Type of nitrogen-containing base, such as adenine or guanine, having a double-ring structure. 504

Purkinje fibers Specialized muscle fibers that conduct the cardiac impulse from the AV bundle into the ventricles. 218

pyelonephritis Inflammation of the renal pelvis and kidney tissue. 308

pyrenoid Spherical protein body embedded in an algal chloroplast. 578

pyrimidine Type of nitrogen-containing base, such as cytosine, thymine, or uracil, having a single-ring structure. 504

pyruvate End product of glycolysis; its further fate, involving fermentation or entry into a mitochondrion, depends on oxygen availability. 115

R

radial symmetry Body plan in which similar parts are arranged around a central axis, like the spokes of a wheel. 620

radioactive isotope Unstable form of an atom that spontaneously emits radiation in the form of radioactive particles or radiant energy. 22

rain shadow A drier climate area, usually found on the leeward (nonwindy) side of a mountain range. 721

ray-finned fish Group of bony fishes having fins supported by parallel bony rays connected by webs of thin tissue. 652

reactant Substance that participates in a reaction. 102

receptacle Area where a flower attaches to a floral stalk. 612

receptor-mediated endocytosis Selective uptake of molecules into a cell by vacuole formation after they bind to specific receptor proteins in the plasma membrane. 77

receptor protein Protein in the plasma membrane or within the cell that binds to a substance that alters some metabolic aspect of the cell. 69

recessive allele Hereditary factor that expresses itself in the phenotype only when the genotype is homozygous. 471

recombinant DNA (rDNA) DNA that contains genes from more than one source. 526

rectum Terminal end of the digestive tube between the sigmoid colon and the anus. 262

red algae Marine photosynthetic protist with a notable abundance of phycobilin pigments. 579

red blood cell (erythrocyte) Formed element that contains hemoglobin and carries oxygen from the lungs to the tissues. 198, 222

red bone marrow Vascularized modified connective tissue sometimes found in the cavities of spongy bone; site of blood cell formation. 238, 368

redox reaction Oxidation-reduction reaction; one molecule loses electrons (oxidation), while another molecule simultaneously gains electrons (reduction). 108

red tide Occurs frequently in coastal areas and is often associated with population blooms of dinoflagellates. 578

reduced hemoglobin Hemoglobin that is carrying hydrogen ions. 291

reduction Chemical reaction that results in the addition of one or more electrons to an atom, ion, or compound. 108

referred pain Pain perceived as having come from a site other than that of its actual origin. 349

reflex Automatic, involuntary response of an organism to a stimulus. 332

reflex action An action performed automatically, without conscious thought; e.g., swallowing. 258

refractory period Time following an action potential when a neuron is unable to conduct another nerve impulse. 319

regulator gene In an operon, a gene coding for a protein that regulates the expression of other genes. 512

renal artery Vessel that originates from the aorta and delivers blood to the kidney. 300

renal cortex Outer portion of the kidney that appears granular. 302

renal medulla Inner portion of the kidney that consists of renal pyramids. 302

renal pelvis Hollow chamber in the kidney that receives freshly prepared urine from the collecting ducts. 302

renal vein Vessel that takes blood from the kidney to the interior vena cava. 300

renewable resource Resource normally replaced or replenished by natural processes and not depleted by moderate use. 757

renin Enzyme released by the kidneys that leads to the secretion of aldosterone and a rise in blood pressure. 306, 402

replacement reproduction The number of offspring necessary to keep the population at a constant size through time. 694

repressor In an operon, protein molecule that binds to an operator, preventing transcription of structural genes. 512

reproduction To make more of itself. 4

reproductive system Organ system that contains male or female organs and specializes in the production of offspring. 203

reptile Member of a class of terrestrial vertebrates with internal fertilization, scaly skin, and an egg with a leathery shell. 653

residual volume Amount of air remaining in the lungs after a forceful expiration. 287

resource partitioning Mechanism that increases the number of niches by apportioning the supply of a resource, such as food or living space, between species. 696

respiratory center Group of nerve cells in the medulla oblongata that send out rhythmic nerve impulses, resulting in involuntary inspiration on an ongoing basis. 288

respiratory system Organ system consisting of the lungs and tubes that bring oxygen into the lungs and take carbon dioxide out. 203

response variable A variable that is measured in an experiment due to a treatment. 12

resting potential Polarity across the plasma membrane of a resting neuron due to an unequal distribution of ions. 318

restoration ecology Seeks scientific ways to return ecosystems to their state prior to habitat degradation. 755

restriction enzyme Bacterial enzyme that stops viral reproduction by cleaving viral DNA; used to cut DNA at specific points during the production of recombinant DNA. 526

reticular connective tissue Composed of reticular cells and reticular fibers; found in bone marrow and lymphoid organs, and in lesser amounts elsewhere. 196

reticular fiber Very thin collagen fibers in the matrix of connective tissue, highly branched and forming delicate supporting networks. 196

retina Innermost layer of the vertebrate eyeball; contains the rod cells and the cone cells. 352

retinal Light-absorbing molecule that is a derivative of vitamin A and a component of rhodopsin. 353

retrovirus RNA virus containing the enzyme reverse transcriptase that synthesizes DNA from RNA. 592

rheumatoid arthritis (RA) Persistent inflammation of synovial joints due to an autoimmune reaction. 251, 389

rhizome Rootlike underground stem. 158

rhodopsin Visual pigment found in the rods whose activation by light energy leads to vision. 353

ribosomal RNA (rRNA) Type of RNA found in ribosomes where protein synthesis occurs. 506

ribosome RNA and protein in two subunits; site of protein synthesis in the cytoplasm. 48, 497

RNA (ribonucleic acid) Nucleic acid produced by covalent bonding of nucleotide monomers that contain the sugar ribose; occurs in three forms: messenger RNA, ribosomal RNA, and transfer RNA. 39, 505

RNA-first hypothesis In chemical evolution, the proposal that RNA originated before other macromolecules and allowed the formation of the first cell(s). 543

RNA polymerase During transcription, an enzyme that joins nucleotides complementary to a portion of DNA. 507

rod cell Photoreceptor in the retina of the eye that responds to dim light. 353

root cap Protective covering of the root tip, whose cells are constantly replaced as they are ground off when the root pushes through rough soil particles. 150

root hair Extension of a root epidermal cell that increases the surface area for the absorption of water and minerals. 146

root nodule Structure on plant root that contains nitrogen-fixing bacteria. 153

root system Includes the main root and any and all of its lateral (side) branches. 144

rotational equilibrium The ability to detect rotational and/or angular movement of the head. 358

round window Membrane-covered opening between the inner ear and the middle ear. 355

roundworm Member of the phylum Nematoda, having a cylindrical body, a complete digestive tract, and a pseudocoelom. Some forms are free-living in water and soil; many are parasitic. 634

RuBP (ribulose-1, 5-bisphosphate) The starting material of the Calvin cycle. 136

rugae Deep folds, as in the wall of the stomach. 260

S

saccule Saclike cavity in the vestibule of the inner ear; contains sensory receptors for gravitational equilibrium. 358

sac fungi Members of the phylum Ascomycota. 587

salivary amylase Secreted by the salivary glands; the first enzyme to act on starch. 257

salivary gland Sends salvia by way of ducts to the mouth. 257

SA (sinoatrial) node Small region of neuromuscular tissue that initiates the heartbeat; also called the pacemaker. 218

sarcolemma The plasma membrane enclosing a muscle fiber. 380

sarcomere One of many units, arranged linearly within a myofibril, whose contraction produces muscle contraction. 381

sarcoplasmic reticulum Smooth endoplasmic reticulum of skeletal muscle cells; surrounds the myofibrils and stores calcium ions. 380

saturated fatty acid Molecule that lacks double bonds between the carbons of its hydrocarbon chain. 34

Schwann cell Cell that surrounds a peripheral nerve fiber and forms the myelin sheath. 317

scientific theory Concept supported by a broad range of observations, experiments, and conclusions. 11

sclera White, fibrous, outer layer of the eyeball. 352

sclerenchyma Plant tissue composed of cells with heavily lignified cell walls; functions in support. 147

scolex Anterior region of parasitic flatworms that bear hooks for attachment. 628

scoliosis Abnormal lateral (side-to-side) curvature of the vertebral column. 373

scrotum Pouch of skin that encloses the testes. 416

secondary oocyte In oogenesis, the functional product of meiosis I. 96

secondary spermatocyte In spermatogenesis, the functional product of meiosis I. 96

second messenger Chemical produced within a cell in response to the binding of a messenger to a membrane receptor; triggers a metabolic reaction in the cell. 396

secretion Release of a substance by exocytosis from a cell that may be a gland or part of a gland. 55

seed Mature ovule that contains a sporophyte embryo with stored food enclosed by a protective coat. 172, 602

seed cone One of two types of cones produced by gymnosperms; contains windblown seeds (female gametophyte). 611

semantic memory Capacity of the brain to store and retrieve information with regard to words or numbers. 328

semen Thick, whitish fluid consisting of sperm and secretions from several glands of the male reproductive tract. 416

semicircular canal One of three tubular structures within the inner ear that contain the sensory receptors responsible for the sense of rotational equilibrium. 356

semilunar valve Valve resembling a half moon located between the ventricles and their attached vessels. 216

seminal vesicle Convoluted, saclike structure attached to the vas deferens near the base of the urinary bladder in males; adds secretions to semen. 416

seminiferous tubule Highly coiled duct within the male testes that produces and transports sperm. 419

senescence Sum of processes involving the aging, decline, and eventual death of a plant or plant part. 185

sensation Conscious awareness of a stimulus due to a nerve impulse sent to the brain from a sensory receptor by way of sensory neurons. 346

sensory adaptation Phenomenon in which a sensation becomes less noticeable once it has been recognized by constant repeated stimulation. 347

sensory neuron Nerve cell that transmits nerve impulses to the central nervous system after a sensory receptor has been stimulated. 316

sensory receptor Structure that receives either external or internal environmental stimuli and is a part of a sensory neuron or transmits signals to a sensory neuron. 346

sepal Outermost, sterile, leaflike covering of the flower; usually green in color. 172, 613

serous membrane Membrane that covers internal organs and lines cavities without an opening to the outside of the body; also called the serosa. 201

serum Light yellow liquid left after clotting of blood. 226

sessile Animal that tends to stay in one place. 622

setae Bristles that anchor or help move annelid worms. 632

severe combined immunodeficiency disease (SCID) Congenital illness in which both antibody- and cell-mediated immunity are lacking. 251

sex chromosome Chromosome that determines the sex of an individual; in humans, females have two X chromosomes, and males have an X and a Y chromosome. 488

sex-linked Allele that occurs on the sex chromosomes but may control a trait having nothing to do with the sexual characteristics of an individual. 488

sexual selection Changes in males and females, often due to male competition and female selectivity, leading to increased fitness. 681

shoot system Aboveground portion of a plant consisting of the stem, leaves, and flowers. 144

short-day plant Plant that flowers when day length is shorter than a critical length. 187

short tandem repeat (STR) The same short sequence of DNA bases that recur several times. 528

short-term memory Retention of information for only a few minutes. 328

shrubland Arid terrestrial biome characterized by shrubs and tending to occur along coasts that have dry summers and receive most of their rainfall in the winter. 727

sickle cell disease Hereditary disease due to a mutation in the hemoglobin gene, in which red blood cells are narrow and curved. 479

sieve-tube member Member that joins with others in the phloem tissue of plants as a means of transport for nutrient sap. 148

simple goiter Condition in which an enlarged thyroid produces low levels of thyroxine. 408

sink In the pressure-flow model of phloem transport, the location (roots) from which sugar is constantly being removed. 167

sinus Cavity or hollow space in an organ such as the skull. 283, 372

sinusitis Inflammation of the sinuses, usually caused by an infection, which leads to blockage of the openings to the sinuses, and is characterized by postnasal discharge and facial pain. 292

sister chromatid One of two genetically identical chromosomal units that are the result of DNA replication and are attached to each other at the centromere. 85

skeletal muscle Striated, voluntary muscle tissue that comprises skeletal muscles; also called striated muscle. 199

skeletal system System of bones, cartilage, and ligaments that works with the muscular system to protect the body and provide support for locomotion and movement. 203

skill memory Capacity of the brain to store and retrieve information necessary to perform motor activities, such as riding a bike. 329

skin Outer covering of the body; part of the integumentary system, which also includes organs such as the sense organs. 204

skull Bony framework of the head, composed of cranial bones and the bones of the face. 372

sliding filament theory An explanation for muscle contraction based on the movement of actin filaments in relation to myosin filaments. 382

slime layer Gelatinous sheath surrounding the cell walls of certain bacteria. 47

small intestine Long, tubelike chamber of the digestive tract between the stomach and the large intestine. 260

smooth muscle Nonstriated, involuntary muscle tissue found in the walls of internal organs; also called visceral muscle. 199

society Organization of the members of species in a cooperative manner, extending beyond sexual and parental behavior. 676

sodium-potassium pump Carrier protein in the plasma membrane that moves sodium ions out of and potassium ions into cells; important in nerve and muscle cells. 75, 318

soft palate Entirely muscular posterior portion of the roof of the mouth. 257

solute Substance that is dissolved in a solvent, forming a solution. 27

solvent Fluid, such as water, that dissolves solutes. 72

somatic cell A body cell; excludes cells that undergo meiosis and become a sperm or an oocyte. 82

somatic embryo Plant cell embryo that is asexually produced through tissue culture techniques. 181

somatic system The portion of the peripheral nervous system containing motor neurons that control skeletal muscles. 332

somatostatin Growth hormone inhibiting hormone produced by the pancreas, stomach, and small intestine. 403

source In the pressure-flow model of phloem transport, the location (leaves) of sugar production. 166

speciation When one species gives rise to two species, each of which continues on its own evolutionary pathway. 559

species Group of similarly constructed organisms capable of interbreeding and producing fertile offspring; organisms that share a common gene pool. 559

sperm Male sex cell with three distinct parts at maturity; head, middle piece, and tail. 96, 419

spermatid In spermatogenesis, the functional product of meiosis II; becomes sperm. 96

spermatogenesis Production of sperm in males by the process of meiosis and maturation. 95, 419

spicules Small needle-shaped structures within the internal skeleton of sponges. 624

sphincter Muscle that surrounds a tube and closes or opens the tube by contracting and relaxing. 258

spinal cord Part of the central nervous system; the nerve cord that is continuous with the base of the brain and housed within the vertebral column. 322

spinal nerve Nerve that arises from the spinal cord. 331

spindle Microtubule structure that brings about chromosomal movement during nuclear division. 85

spiral organ Organ in the cochlear duct of the inner ear that is responsible for hearing; also called the organ of Corti. 356

spleen Large organ located in the upper left region of the abdomen; stores and purifies blood. 238

sponge Invertebrate animal of the phylum Porifera; pore-bearing filter feeder whose inner body wall is lined by collar cells. 624

spongy bone Type of bone that has an irregular, meshlike arrangement of thin plates of bone; often contains red bone marrow. 197, 368

spongy mesophyll Layer of tissue in a plant leaf containing loosely packed cells, increasing the amount of surface area for gas exchange. 161

sporangium Structure that produces spores. 587

spore Haploid reproductive cell, sometimes resistant to unfavorable environmental conditions, which is capable of producing a new individual that is also haploid. 172, 585, 602

sporophyte Diploid generation of the alternation of generations life cycle of a plant; produces haploid spores that develop into the haploid generation. 172, 602

sporozoan Spore-forming protist that has no means of locomotion and is typically a parasite with a complex life cycle having both sexual and asexual phases. 582

spring overturn Mixing process that occurs in spring in stratified lakes whereby the oxygen-rich top waters mix with the nutrient-rich bottom waters. 730

squamous epithelium Type of epithelial tissue that contains flat cells. 194

stamen In flowering plants, the portion of the flower that consists of a filament and an anther containing pollen sacs where pollen is produced. 172, 613

stapes The last of the three ossicles of the ear that conduct vibrations from the tympanic membrane to the oval window membrane of the inner ear. 355

starch Storage polysaccharide found in plants; composed of glucose molecules joined in a linear fashion with side chains. 32

stem Usually the upright, vertical portion of a plant that transports substances to and from the leaves. 145

stem cell Any cell that can divide and differentiate into more functionally specific cell types. 226

stent Metal or plastic tube inserted into a blood vessel to help keep it open; often used in the treatment of vascular disease due to atherosclerosis. 229

steroid Type of lipid molecule having a complex of four carbon rings. 35

steroid hormone Type of hormone that has a complex of four carbon rings but different side chains from those of other steroid hormones. 396

stigma In flowering plants, portion of the pistil where pollen grains adhere and germinate before fertilization can occur. 173, 613

stimulus Change in the internal or external environment that a sensory receptor can detect, leading to nerve impulses in sensory neurons. 346

stolon Stem that grows horizontally along the ground and may give rise to new plants where it contacts the soil. 158

stoma (pl., stomata) Microscopic opening bordered by guard cells in the leaves of plants through which gas exchange takes place. 129

stomach Muscular sac that mixes food with gastric juices to form chyme, which enters the small intestine. 260

striated Having bands; in cardiac and skeletal muscle, alternating light and dark crossbands produced by the distribution of contractile proteins. 199

stroke Condition resulting when an arteriole in the brain bursts or becomes blocked by an embolism; also called a cerebrovascular accident. 228, 339

stroma Fluid within a chloroplast that contains enzymes involved in the synthesis of carbohydrates during photosynthesis. 58

style Elongated, central portion of the pistil between the ovary and the stigma. 173, 613

subcutaneous layer A sheet that lies just beneath the skin and consists of loose connective and adipose tissue. 205

subsidence Occurs when a portion of Earth's surface gradually settles downward. 759

substrate Reactant in a reaction controlled by an enzyme. 104

substrate-level ATP synthesis Process in which ATP is formed by transferring a phosphate from a metabolic substrate to ADP. 116

superior vena cava Large vein that enters the right atrium from above and carries blood from the head, thorax, and upper limbs to the heart. 220

surfactant Agent that reduces the surface tension of water; in the lungs, a surfactant prevents the alveoli from collapsing. 285

sustainable Ability of a society or ecosystem to maintain itself while also providing services to human beings. 758

suture Type of immovable joint articulation found between the bones of the skull. 377

sweat gland Skin gland that secretes a fluid substance for evaporative cooling; also called a sudoriferous gland. 205

symbiosis Relationship that occurs when two different species live together in a unique way; it may be beneficial, neutral, or detrimental to one and/or the other species. 699

sympathetic division The part of the autonomic system that usually promotes activities associated with emergency (fight-or-flight) situations; uses norepinephrine as a neurotransmitter. 333

sympatric speciation Origin of new species in populations that overlap geographically. 559

synapsis Pairing of homologous chromosomes during prophase I of meiosis. 89

synaptic cleft Small gap between the presynaptic and postsynaptic membranes of a synapse. 320

synaptic integration Summing up of excitatory and inhibitory signals by a neuron or by some part of the brain. 321

syndrome Group of symptoms that appear together and tend to indicate the presence of a particular disorder. 494

synovial joint Freely moving joint in which two bones are separated by a cavity. 377

synovial membrane Membrane that forms the inner lining of the capsule of a freely movable joint. 201, 377

syphilis Sexually transmitted disease caused by the bacterium *Treponema pallidum*. 434

systematics The discipline of identifying and classifying organisms according to their evolutionary relationships. 6, 562

systemic Occurring throughout the body; invading many compartments and organs via circulation. 209

systemic circuit That part of the circulatory system that serves body parts other than the gas-exchanging surfaces in the lungs. 220

systemic lupus erythematosis (SLE) Autoimmune disease that causes a wide variety of symptoms. 251

systole Contraction period of the heart during the cardiac cycle. 218

systolic pressure Arterial blood pressure during the systolic phase of the cardiac cycle. 221

T

taiga A coniferous forest extending in a broad belt across northern Eurasia and North America. 725

taproot Main axis of a root that penetrates deeply and is used by certain plants, such as carrots, for food storage. 152

taste bud Sense organ containing the receptors associated with the sense of taste. 349

taxonomy The assignment of a binomial name to each species. 7, 562

Tay-Sachs disease Lethal genetic disease in which the newborn has a faulty lysosomal digestive enzyme. 478

T-cell receptor (TCR) A molecule on the surface of a T cell that can bind to a specific antigen fragment in combination with an MHC molecule. 242

technology The science or study of the practical or industrial arts. 14

tectorial membrane Membrane that lies above and makes contact with the hair cells in the spiral organ. 356

telomere The end of a chromosome; shortens with every round of replication and therefore helps determine the number of times a cell can divide. 520

telophase Mitotic phase during which daughter cells are located at each pole. 86

temperate deciduous forest Forest found south of the taiga, characterized by the following: deciduous trees; moderate climate; relatively high rainfall; stratified plant growth; and plentiful ground life. 725

tendon Strap of fibrous connective tissue that connects skeletal muscle to bone. 196, 368

terminal bud Bud that develops at the apex of a shoot. 154

termination The final step in protein synthesis when the polypeptide and the assembled components that carried out protein synthesis are separated from one another. 510

testcross Cross between an individual with the dominant phenotype and an individual with the recessive phenotype. The resulting phenotypic ratio indicates whether the dominant phenotype is homozygous or heterozygous. 474

testes Male gonads that produce sperm and the male sex hormones. 404, 416

testicular cancer Malignant tumor of the testicles in males. 436

testosterone Male sex hormone that helps maintain sexual organs and secondary sex characteristics. 404, 419

tetanus Sustained muscle contraction without relaxation. 385

tetany Severe twitching caused by involuntary contraction of the skeletal muscles due to a calcium imbalance. 408

thalamus Part of the brain located in the lateral walls of the third ventricle that serves as the integrating center for sensory input; plays a role in arousing the cerebral cortex. 326

thermoreceptor Sensory receptor that is sensitive to changes in temperature. 346

threatened species Species that are likely to become endangered in the foreseeable future. 742

three-domain system Method of classification in which the largest grouping is the domain; The three domains are Bacteria, Archaea, and Eukarya. 564

threshold Electrical potential level (voltage) at which an action potential or nerve impulse is produced. 318

thrombin Enzyme that converts fibrinogen to fibrin threads during blood clotting. 225

thromboembolism Obstruction of a blood vessel by a thrombus that has dislodged from the site of its formation. 228

thrombus Blood clot that remains in the blood vessel where it formed. 228

thylakoid Flattened sac within a granum whose membrane contains chlorophyll and where the light-dependent reactions of photosynthesis occur. 48

thymine (T) One of four nitrogen-containing bases in nucleotides composing the structure of DNA. 39

thymus gland Lymphoid organ, located along the trachea behind the sternum, involved in the maturation of T cells in the thymus gland. 238, 404

thyroid gland Endocrine gland in the neck that produces several important hormones, including thyroxine, triiodothyronine, and calcitonin. 400

thyroid-stimulating hormone (TSH) Substance produced by the anterior pituitary that causes the thyroid to secrete thyroxine and triiodothyronine. 398

thyroxine (T_4) Hormone secreted by the thyroid gland that promotes growth and development; in general, it increases the metabolic rate of cells. 400

tidal volume Amount of air normally moved in the human body during an inspiration or an expiration. 287

tight junction Region between cells where adjacent plasma membrane proteins join to form an impermeable barrier. 194

tinea Name for many different kinds of superficial fungal infections of the skin, nails, and hair. 585

tissue Group of similar cells that perform a common function. 3, 194

tissue culture Process of growing tissue artificially, usually in a liquid medium in laboratory glassware. 181

tissue fluid Fluid that surrounds the body's cells; consists of dissolved substances that leave the blood capillaries by filtration and diffusion. 227

tonsillectomy Surgical removal of the tonsils. 291

tonsillitis Infection of the tonsils that causes inflammation and can spread to the middle ears. 291

tonsils Partially encapsulated lymph nodules located in the pharynx. 291

total artificial heart (TAH) A mechanical replacement for the heart as opposed to replacement with a natural human heart. 229

totipotent Cell that has the full genetic potential of the organism and the potential to develop into a complete organism. 181

trachea Windpipe from the larynx to the bronchi; in insects, tracheae are tubes that transport air to the tissues. 284, 639

tracheid In flowering plants, type of cell in xylem that has tapered ends and pits through which water and minerals flow. 148

tracheostomy Creation of an artificial airway by incision of the trachea and insertion of a tube. 292

tract Bundle of myelinated axons in the central nervous system. 317

transcription Process resulting in the production of a strand of mRNA that is complementary to a segment of DNA. 506

transcription activator Protein that speeds transcription. 514

transcription factor Protein that initiates transcription by RNA polymerase and thereby starts the process that results in gene expression. 514

transduction The transfer of genetic material between bacteria by viruses. 572

transfer RNA (tRNA) Type of RNA that transfers a particular amino acid to a ribosome during protein synthesis; at one end, it binds to the amino acid, and at the other end, it has an anticodon that binds to an mRNA codon. 506

transformation The taking up of extraneous genetic material from the environment by bacteria. 572

transgenic organism Free-living organism in the environment that has a foreign gene in its cells. 526

transgenic plant A plant that carries the genes of another organism as a result of DNA technology; also, a genetically modified plant. 182

transitional link A form, usually extinct, that bears a resemblance to two groups that in the present day are classified separately. 545

translation Process by which the sequence of codons in mRNA dictates the sequence of amino acids in a polypeptide. 506

translocation Movement of a chromosomal segment from one chromosome to another, nonhomologous chromosome, leading to abnormalities. 497

transpiration Plant's loss of water to the atmosphere, mainly through evaporation at leaf stomata. 164

transposon DNA sequence capable of randomly moving from one site to another in the genome. 516

trichinosis Infection caused by the roundworm *Trichinella spiralis*. 635

trichome In plants, a specialized outgrowth of the epidermis, such as root hairs. 146

triglyceride Neutral fat composed of glycerol and three fatty acids. 34

trisomy One more chromosome than usual. 492

trochophore A form of larva with many cilia to allow free-swimming in the planktonic marine environment. 623

trochozoa Division of protostome animals with trochophore larva. 623

trophic level Feeding level of one or more populations in a food web. 709

trophoblast Outer membrane surrounding the embryo in mammals; when thickened by a layer of mesoderm, it becomes the chorion, an extraembryonic membrane. 453

tropical forest Can be wet or dry. Tropical rain forests and tropical deciduous forests are characterized by high annual precipitation, lush vegetation, an annual mean temperature of 20–25°C, high humidity and characterized by both a wet and dry season. The difference between the two forest types is in the length of the dry season. Tropical deciduous forests have a longer dry season and consequently, trees lose their leaves. 726

tropism In plants, a growth response toward or away from a directional stimulus. 187

trypsin Protein-digesting enzyme secreted by the pancreas. 263

T (transverse) tubules Extensions of the sarcolemma that protrude into the muscle cell. 380

tubercle Small, rounded nodular lesion produced by *Mycobacterium tuberculosis*. 574

tubular reabsorption Movement of primarily nutrient molecules and water from the contents of the nephron into the blood at the proximal convoluted tubule. 305

tubular secretion Movement of certain molecules from the blood into the distal convoluted tubule of a nephron so that they are added to urine. 305

tumor suppressor gene Gene coding for a protein that ordinarily suppresses cell division; inactivity can lead to a tumor. 84

tundra Biome characterized by permanently frozen subsoil found between the ice cap and the tree line of regions of the Arctic, just south of the ice-covered polar seas in the Northern Hemisphere; also known as the arctic tundra. Alpine tundra, having similar characteristics, is found near the peak of a mountain. 724

tunicate Type of primitive invertebrate chordate. 649

turgor pressure In plant cells, pressure of the cell contents against the cell wall when the central vacuole is full. 73

tympanic membrane Membranous region that receives air vibrations in an auditory organ; in humans, the eardrum. 355

U

ulcer Open sore in the lining of the stomach; frequently caused by bacterial infection. 276

umbilical artery Fetal blood vessel that travels to and from the placenta. 456

umbilical cord Cord connecting the fetus to the placenta through which blood vessels pass. 455

umbilical vein Fetal blood vessel that travels to and from the placenta. 456

unsaturated fatty acid Fatty acid molecule that has one or more double bonds between the atoms of its carbon chain. 34

upwelling Upward movement of deep, nutrient-rich water along coasts; it replaces surface waters that move away from shore when the direction of prevailing wind shifts. 734

uracil (U) One of the four nitrogen-containing bases in nucleotides composing the structure of RNA. 39

urea Primary nitrogenous waste of humans derived from amino acid breakdown. 300

ureter One of two tubes that take urine from the kidneys to the urinary bladder. 300

urethra Tubular structure that receives urine from the bladder and carries it to the outside of the body. 301, 416

urethritis Inflammation of the urethra. 311

uric acid Waste product of nucleotide metabolism. 300

urinary bladder Organ where urine is stored before being discharged by way of the urethra. 301

urinary system Organ system consisting of the kidneys and urinary bladder; rids the body of nitrogenous wastes and helps regulate the water-salt balance of the blood. 203

uterine cycle Monthly-occurring changes in the characteristics of the uterine lining (endometrium). 424

uterus Organ located in the female pelvis where the fetus develops; also called the womb. 420

utricle Saclike cavity in the vestibule of the inner ear that contains sensory receptors for gravitational equilibrium. 358

V

vaccine Antigens prepared in such a way that they can promote active immunity without causing disease. 244

vacuole Membrane-bounded sac that holds fluid and a variety of other substances. 56

vagina Organ that leads from the uterus to the vestibule and serves as the birth canal and organ of sexual intercourse in females. 421

vaginal contraceptive ring A soft, flexible, vinyl ring containing estrogen and progesterone that prevents ovulation. 427

valve Membranous extension of a vessel of the heart wall that opens and closes, ensuring one-way flow. 215

varicose veins Irregular dilation of superficial veins, seen particularly in the lower legs, due to weakened valves within the veins. 215

vascular bundle In plants, primary phloem and primary xylem enclosed by a bundle sheath. 148

vascular cambium In plants, lateral meristem that produces secondary phloem and secondary xylem. 154

vascular cylinder In dicot roots, a core of tissues bounded by the endodermis, consisting of vascular tissues and pericycle. 148

vascular plant Plant that has vascular tissue (xylem and phloem); includes seedless vascular plants (e.g., ferns) and seed plants (gymnosperms and angiosperms). 606

vascular tissue Transport tissue in plants, consisting of xylem and phloem. 146, 602

vas deferens (pl., vasa deferentia) Tube that leads from the epididymis to the urethra in males. 416

vector In genetic engineering, a means of transferring foreign genetic material into a cell; e.g., a plasmid. 526

vein Vessel usually having nonelastic walls that takes blood to the heart from venules; in plants, veins consist of vascular bundles that branch into the leaves. 214

ventilation Process of moving air into and out of the lungs; breathing. 282

ventral solid nerve cord Part of an invertebrate's nervous system located ventrally. 632

ventricle Cavity in an organ, such as a lower chamber of the heart or the ventricles of the brain. 216, 322

venule Vessel that takes blood from capillaries to a vein. 215

vermiform appendix Small, tubular appendage that extends outward from the cecum of the large intestine. 262

vernix caseosa Cheeselike substance covering the skin of the fetus. 456

vertebral column Portion of the vertebrate endoskeleton that houses the spinal cord; consists of many vertebrae separated by intervertebral disks. 373

vertebrate Chordate in which the notochord is replaced by a vertebral column. 650

vertigo Symptom associated with disorders of the inner ear in which the patient experiences a sense of movement or dizziness. 363

vesicle Small, membrane-bounded sac that stores substances within a cell. 54

vessel element Cell that joins with others to form a major conducting tube found in xylem. 148

vestibule Space or cavity at the entrance to a canal, such as the cavity that lies between the semicircular canals and the cochlea. 356

vestigial structure Underdeveloped structure that was functional in some ancestor but is no longer functional in a particular organism. 549

villus Fingerlike projection from the wall of the small intestine that functions in absorption. 260

viroid Infectious strand of RNA devoid of a capsid and much smaller than a virus. 595

virus Noncellular obligate parasite of living cells consisting of an inner core of nucleic acid surrounded by a protein capsid. Some viruses have an additional lipid layer called an envelope. 590

visceral mass The soft-bodied portion of molloscs that contains internal organs. 630

visual accommodation Ability of the eye to focus at different distances by changing the curvature of the lens. 352

vital capacity Maximum amount of air moved in or out of the human body with each breathing cycle. 287

vitamin Essential organic requirement in the diet, needed in small amounts. Vitamins are often part of coenzymes. 107, 270

vitreous humor Clear, gelatinous material between the lens of the eye and the retina. 352

vocal cord Fold of tissue within the larynx; creates vocal sounds when it vibrates. 284

vulva External genitals of the female that surround the opening of the vagina. 421

W

wart Small areas of skin proliferation caused by the human papillomavirus. 205

water (hydrologic) cycle Interdependent and continuous circulation of water from the ocean to the atmosphere, to the land, and back to the ocean. 710

water mold Filamentous organism having cell walls made of cellulose; typically a decomposer of dead freshwater organisms, but some are parasites of aquatic or terrestrial organisms. 583

Wernicke's area Brain area involved in language comprehension. 326

white blood cell (leukocyte) One of several types of cells in the blood, each having a specific function in protecting the body from invasion by foreign substances and organisms. 198, 224

white matter Myelinated axons in the central nervous system. 317

wood Secondary xylem that builds up year after year in woody plants and becomes the annual rings. 156

X

X-linked Allele that is located on an X chromosome but may control a trait that has nothing to do with the sex characteristics of an individual. 488

xylem Vascular tissue that transports water and mineral solutes upward through the plant body; contains vessel elements and tracheids. 145, 606

Y

yeast Unicellular fungus that has a single nucleus and reproduces asexually by budding or fission, or sexually through spore formation. 588

yolk sac Extraembryonic membrane that encloses the yolk of birds; in humans, it is the first site of blood cell formation. 452

Z

zero population growth No growth in population size. 693

zooplankton Part of plankton containing protozoans and other types of microscopic animals. 577

zygospore fungi Members of the phylum Zygomycota. 587

zygote Diploid cell formed by the union of sperm and oocyte; the product of fertilization. 95, 420

Credits

History of Biology

Leeuwenhoek, Darwin, Pasteur, Koch, Pavlov, Lorenz, Pauling: © Bettman/Corbis; McClintock: © AP Images; Franklin: © Photo Researchers, Inc.

Chapter 1

Opener: © Brand X Pictures/PunchStock RF; 1.1(sequoia): © Robert Glusic/Getty RF; 1.1(mushroom): © IT Stock/Age Fotostock RF; 1.1(peacock): © Brand X Pictures/PunchStock RF; 1.1(humans): © Heath Korvola/UpperCut Images/Getty RF; 1.1(giraffes): © Dr. Sylvia S. Mader; 1.1(butterfly): © Creatas/PunchStock RF; 1.1(Earth): © Ingram Publishing/Alamy RF; 1.3: © John Cancalosi/Peter Arnold; 1.4(mature tree): © PhotoDisc Website RF; 1.4(sapling): Courtesy Paul Wray, Iowa State University; 1.4(seedling): © Herman Eisenbeiss/Photo Researchers, Inc.; 1.5(bacteria): © A.B. Dowsett/SPL/Photo Researchers, Inc.; 1.5(archaean): © Ralph Robinson/Visuals Unlimited; 1.8: Courtesy Leica Microsystems, Inc.; 1.9: © Dr. Jeremy Burgess/Photo Researchers, Inc.; 1.10(all): Courtesy Jim Bidlack.

Chapter 2

Opener: © image100/Corbis RF; 2.1: © Gunter Ziesler/Peter Arnold; 2.4a: © Biomed Commun./Custom Medical Stock Photo; 2.4b(left): © Mazzlota et al./Photo Researchers, Inc; 2.4b(right): © Hank Morgan/Rainbow; 2.5a: © Tony Freeman/PhotoEdit; 2.5b: © Geoff Tompkinson/SPL/Photo Researchers, Inc.; 2.7(both): © The McGraw-Hill Companies, Inc./ Evelyn Jo Johnson photographer; 2.10a: © PNC/Brand X/ Corbis RF; 2.10b: © Simone Mueller/The Image Bank/Getty Images; 2A(aquariums): Courtesy of the U.S. Geological Survey; 2.13: © The McGraw Hill Companies, Inc./John Thoeming, photographer; 2.17: © Jeremy Burgess/SPL/Photo Researchers, Inc.; 2.18: © Don W. Fawcett/Photo Researchers, Inc.; 2.19: © Science Source/J.D. Litvay/Visuals Unlimited; 2B (line art): U.S. Department of Agriculture; 2.26a: © Radius Images/Alamy RF. Agriculture.

Chapter 3

Opener: © Yorgos Nikas/Stone/Getty Images; 3.3: © Ralph A. Slepecky/Visuals Unlimited; 3.4: © Dr. Dennis Kunkel/Visuals Unlimited; 3.5: © E.H. Newcomb/W.P. Wergin/Biological Photo Service; 3.6(top): Courtesy Ron Milligan/Scripps Research Institute; 3.6(close up): Courtesy E.G. Pollock; 3Aa(amoeba): © Stephen Durr; 3Ab(mitochondrion): © Dr. Don W. Fawcett/Visuals Unlimited; 3Ac(dinoflagellate): © Biophoto Assoc./Photo Researchers, Inc.; 3.7: © R. Bolender & D. Fawcett/Visuals Unlimited; 3.9: © EM Research Services, Newcastle University; 3.10: Courtesy Herbert W. Israel, Cornell University; 3.11: Courtesy Dr. Keith Porter; 3.12a(actin): © M. Schliwa/Visuals Unlimited; 3.12a(*Chara*): © The McGraw-Hill Companies, Inc. /Dennis Strete and Darrell Vodopich, photographers; 3.12b(intermediate): © K.G. Murti/Visuals Unlimited; 3.12b(humans): © Amos Morgan/Getty RF; 3.12c(microtubules): © K.G. Murti/Visuals Unlimited; 3.12c(chameleon): © Photodisc/Vol. 6/Getty RF; 3.14(cilia): © Dr. G. Moscoso/Photo Researchers, Inc.; 3.14(flagellum): © William L. Dentler/Biological Photo Service; 3.14(sperm): © David M. Phillips/Photo Researchers, Inc.

Chapter 4

Opener: © Tooga Productions, Inc./Getty Images; 4.7(all top): © David M. Phillips/Photo Researchers, Inc.; 4.7(bottom left, center): © Dwight Kuhn; 4.7(bottom right): © Ed Reschke/Peter Arnold; 4.11a(right): © Eric Grave/Phototake; 4.11b(right): © Don W. Fawcett/Photo Researchers, Inc.; 4.11c(both): Courtesy Mark Bretscher.

Chapter 5

Opener: © Kevin Mazur, Contributor/WireImage/Getty Images; 5.2: Courtesy Douglas R. Green/LaJolla Institute for Allergy and Immunology; 5.5: (early prophase, prophase, metaphase, anaphase, telophase)): © Ed Reschke; 5.5(late prophase): © Michael Abbey/Photo Researchers, Inc.; 5.6:(prophase, metaphase, anaphase): © R. Calentine/Visuals Unlimited; 5.6(telophase): © Jack M. Bostrack/Visuals Unlimited; 5.7(both): © R.G. Kessel and C.Y. Shih, "Scanning Electron Microscopy in Biology: A Student's Atlas on Biological Organization," 1974 Springer-Verlag, New York. All rights reserved; 5.8: © Katherine Esau; permission courtesy of the Botanical Society of America.

Chapter 6

Opener: © Michael Fogden/Animals Animals; 6.3: © Darwin Dale/Photo Researchers, Inc.; 6.8b: © Brand X Pictures/PunchStock RF; 6.8c: © Digital Vision/PunchStock RF; 6.11(leaves): © Comstock/PunchStock RF; 6.11(runner): © PhotoDisc/Getty RF; 6.12: © Getty Images/SW Productions RF.

Chapter 7

Opener: © PCN Photography/PCN/Corbis.

Chapter 8

Opener: © The McGraw-Hill Companies, Inc./ Evelyn Jo Johnson photographer; 8.1(*Oscillatoria*): © Tom Adams/Visuals Unlimited/Getty Images; 8.1(kelp): © Chuck Davis/Stone/Getty Images; 8.1(sequoia): © Thomas Wiewandt/ChromoSohm Media Inc./Photo Researchers, Inc.; 8.2: © Dr. George Chapman/Visuals Unlimited; 8.10a: © The McGraw-Hill Companies, Inc./Evelyn Jo Johnson, photographer; 8.10b: © Corbis RF; 8.10c: © S. Alden/PhotoLink/Getty RF; 8A: © Doable/A.collection/Getty Images.

Chapter 9

Opener: © The McGraw-Hill Companies Inc./Evelyn Jo Johnson, photographer; 9.2(all): © The McGraw-Hill Companies Inc./Evelyn Jo Johnson, photographer; 9.3a: © Runk/Schoenberger/Grant Heilman Photography; 9.3b: © J. Robert Waaland/Biological Photo Service; 9.3c: © Steven P. Lynch; 9.4a: © Dr. Ken Wagner/Visuals Unlimited; 9.4b: © George Wilder/Visuals Unlimited; 9.4c: © Biophoto Assoc./Photo Researchers, Inc.; 9.5a: © N.C. Brown Center for Ultrastructure Studies, SUNY, College of Environmental Science & Forestry, Syracuse, NY; 9.5b: © Dr. Michael Clayton, University of Wisconsin, Madison, Dept. of Botany; 9.7a: Courtesy Ray F. Evert/University of Wisconsin Madison; 9.7b: © CABISCO/Phototake; 9.8: © Steven P. Lynch; 9.9a: © John D. Cunningham/Visuals Unlimited; 9.9b: Courtesy George Ellmore, Tufts University; 9.10a: © Dr. Robert Calentine/Visuals Unlimited; 9.10b: © The McGraw-Hill Companies Inc./Evelyn Jo Johnson, photographer; 9.10c: © Brad Mogen/Visuals Unlimited/Getty Images; 9.10d: © Tim Laman/National Geographic/Getty RF; 9.10e(dodder): © Pat Pendarvis; 9.10e(dodder micrograph): © BioPhot; 9.11a: © Dwight Kuhn; 9.11b: © E.H. Newcomb & S.R. Tardon/Biological Photo Service; 9.12(top): © Science VU/R. Roncadori/Visuals Unlimited; 9.12(bottom): © Dana Richter/Visuals Unlimited; 9.15(top): © Ed Reschke; 9.15(bottom): Courtesy Ray F. Evert/University of Wisconsin Madison; 9.16(top): © CABISCO/Phototake; 9.16(bottom): © Kingsley Stern; 9.18: © Ed Reschke/Peter Arnold; 9.19a: © Ardea London Limited; 9A: © Inga Spence/Visuals Unlimited/Getty Images; 9.20a: © The McGraw-Hill Companies Inc./Evelyn Jo Johnson, photographer; 9.20b: © Wally Eberhart/Visuals Unlimited; 9.20c, d: © The McGraw Hill Companies, Inc./Carlyn Iverson, photographer; 9.21: © Jeremy Burgess/SPL/Photo Researchers, Inc.; 9B: Courtesy of the National Park Service (NPS); 9.23a: © Corbis RF; 9.23b: © Gerald & Buff Corsi/Visuals Unlimited; 9.23c: © Steven P. Lynch; 9.25(both): © Jeremy Burgess/SPL/Photo Researchers, Inc.; 9.26a: © M. H. Zimmermann, Courtesy Dr. P. B. Tomlinson, Harvard University; 9.26b: © Steven P. Lynch; photo, p. 168(tree trunk): © Ardea London Limited.

Chapter 10

Opener: © Mitch Hrdlicka/Getty RF; 10.3a: © Farley Bridges; 10.3b: © Pat Pendarvis; 10.4a: © Arthur C. Smith, III/Grant Heilman Photography; 10.4b: © Larry Lefever/Grant Heilman Photography; 10.5(top): Courtesy Graham Kent; 10.5(bottom): © Ed Reschke; 10.6a: © George Bernard/Animals Animals; 10.6b: © Simko/Visuals Unlimited; 10.6c: © Dwight Kuhn; 10.8a: © James Mauseth; 10.8b: © Edward S. Ross; 10.8c: © Runk/Schoenberger/Grant Heilman Photography; 10.8d: © Joe Solem/Riser/Getty Images; 10.9: © BJ Miller/Biological Photo Service; 10.10b(left): © James Mauseth; 10.10b(right): © Barry L. Runk/Grant Heilman Photography; 10Aa: © Steven P. Lynch; 10Ab: © IT Stock Free/Alamy RF; 10.11(all): Courtesy Prof. Dr. Hans-Ulrich Koop, from Plant Cell Reports, 17:601-604; 10.12(both): Courtesy Monsanto; 10.13b: Courtesy Eduardo Blumwald; 10.15(both): Courtesy Prof. Malcolm B. Wilkins; 10.16: © Sylvan Wittwer/Visuals Unlimited; 10.18a, b: © Kingsley Stern; 10.18c: © Kent Knudson/PhotoLink/Getty RF; 10.19a: © Runk/Schoenberger/Grant Heilman Photography; 10.19b: © Kingsley Stern; 10.21a, b(both): © Nigel Cattlin/Visuals Unlimited/Getty Images.

Chapter 11

Opener: © Punchstock/Digital Vision RF; 11.1(all): © Ed Reschke; 11.2a: © David M. Phillips/Visuals Unlimited; 11.2b: Courtesy Camillo Peracchia, M.D.; 11.2c: © Kelly, 1966. Originally published in *The Journal of Cell Biology*, 28: 51–72; 11.3a: © Ed Reschke; 11.3b, e: © The McGraw-Hill Companies, Inc./Dennis Strete, photographer; 11.3c: © The McGraw Hill Companies, Inc./Al Telser, photographer; 11.3d: © Dr. Fred Hossler/Visuals Unlimited/Getty Images; 11.5a, c: © Ed Reschke; 11.5b: © The McGraw-Hill Companies, Inc./Dennis Strete, photographer; 11.6b: © Ed Reschke; 11Aa: © Science Source/Photo Researchers, Inc.; 11Ab: © James Stevenson/SPL/Photo Researchers, Inc.

Chapter 12

Opener: © Mark Sullivan/WireImage/Getty Images; 12.1d, 12.6 (both), 12.11c: © Ed Reschke; 12.12a: © Ed Reschke/Peter Arnold; 12.12b: © Andrew Syred/Photo Researchers, Inc.; 12.13: © Dennis Kunkel/Phototake; 12.14b: © Eye of Science/Photo Researchers, Inc.; 12.18: © CCN/Phototake; 12.20: © ABIOMED, Inc.; 12A: © Biophoto Associates/Photo Researchers, Inc.; 12B: © Astrid & Hanns-Frieder Michler/Photo Researchers, Inc.; p. 233(ECG): © Ed Reschke.

Chapter 13

Opener: © Dr. Richard Kessel & Dr. Randy Kardon/Tissues & Organs/Visuals Unlimited; 13.2a: © The McGraw-Hill Companies, Inc./Dennis Strete, photographer; 13.2b: © R. Valentine/Visuals Unlimited; 13.2c: © Fred E. Hossler/Visuals Unlimited; 13.2d: © The McGraw Hill Companies, Inc./Al Telser, photographer; 13.6b: Courtesy Dr. Arthur J. Olson, Scripps Institute; 13.8b: © Steve Gschmeissner/Photo Researchers, Inc.; 13.9a(photo): © Michael Newman/PhotoEdit; 13.9a (line art): Approved by the Advisory Committee on Immunization Practices (ACIP), the American Academy of Pediatrics (AAP), and the American Academy of Family Physicians; 13.10b: © Digital Vision/Getty RF; 13.10c: © Aaron Haupt/Photo Researchers, Inc.; 13.12: © Bart's Medical Library/Phototake;

13.13(both): © J. C. Revy/Phototake; 13B(right): © Damien Lovegrove/SPL/Photo Researchers, Inc.; 13B(left): © David Scharf/SPL/Photo Researchers, Inc.; p. 233: © Bart's Medical Library/Phototake.

Chapter 14

Figure 14.4b: © Biophoto Associates/Photo Researchers, Inc.; 14.5b: © Ed Reschke/Peter Arnold; 14.6(villi): © Manfred Kage/Peter Arnold; 14.6(microvilli): Photo published in Medical Cell Biology, Charles Flickinger, photo by Susumu Ito, Copyright Elsevier, 1979.; 14.13 (line art): U.S. Department of Agriculture; 14.14: © Digital Vision/Getty RF; 14.15: Data from T.T. Shintani, *Eat More, Weigh Less*™ Diet, 1983; 14A: © The McGraw Hill Companies, Inc./Andrew Resek, photographer; 14.16a, c: © Biophoto Associates/Photo Researchers, Inc.; 14.16b: © Ken Greer/Visuals Unlimited; 14.18: © Karen Kasmauski/Corbis; 14.19: © Donna Day/Stone/Getty Images; 14.20: © Tony Freeman/PhotoEdit; 14B: © J. James/Photo Researchers, Inc.

Chapter 15

Opener: © Michael Keller/Corbis; 15.3(left): © CNRI/Phototake; 15.4: © Dr. Kessel & Dr. Kardon/Tissues & Organs/Visuals Unlimited; 15Aa: © Bill Aron/PhotoEdit; 15.6: © Yoav Levy/Phototake; 15.10: © Dr. P. Marazzi/Photo Researchers, Inc.; 15.11: © Clinica Claros/Phototake; Heath Focus (text), p. 294: Adapted from The Most Often Asked Questions About Smoking Tobacco and the Answers, July 1993, The American Cancer Society; 15.13a: © Matt Meadows/Peter Arnold; 15.13b: © Biophoto Associates/Photo Researchers, Inc.

Chapter 16

Opener (polycystic): © Biophoto Associates/Photo Researchers, Inc.; opener (normal): © SIU/Visuals Unlimited/Getty Images; 16.3d: © Manfred Kage/Peter Arnold; 16.4b: © R.G. Kessel and R H. Kardon, Tissues and Organs: A Text-Atlas of Biological Organization, 1979. W.H. Freeman. All rights reserved; 16.4c,d:Reprinted from *Journal of Ultrastructure Research*, Vol. 15, A.B. Maunsbach, pages 242–282, copyright 1966, with permission of Elsevier.; 16A: © Will & Deni McIntyre/Photo Researchers, Inc.; 16.11: Image reprinted with permission from eMedicine.com, 2008. Available at http://emedicine.medscape.com/article/440657-overview; 16.12: © AP Images/Brian Walker.

Chapter 17

Opener: © Corbis RF; 17.1a: © Gerhard Gscheidle/Peter Arnold; 17.3b: © The McGraw Hill Companies, Inc./Dr. Dennis Emery, Dept. of Zoology and Genetics, Iowa State University, photographer; 17.6a: Courtesy Dr. E.R. Lewis, University of California Berkeley; 17.8c: © Karl E. Deckart/Phototake; 17.9b: © Colin Chumbley/Science Source/Photo Researchers, Inc.; 17A(student): © Photodisc/Alamy RF; 17A(brain waves): © L. Birmington/Custom Medical Stock Photo; 17.14(all): © Marcus Raichle; 17.19(both): © Science VU/Visuals Unlimited; 17B: © The McGraw Hill Companies, Inc./Mark A.S. Dierker, photographer; 17.21: © ISM/Phototake; 17.22: © Ralph C. Eagle, Jr./Photo Researchers, Inc.; 17.23: © Topham/The Image Works.

Chapter 18

Openers (both): © AP Images/Gene J. Puskar; 18.4b(all): © Omikron/SPL/Photo Researchers, Inc.; 18.8: © Lennart Nilsson, from "The Incredible Machine"; 18.9b: © Biophoto Associates/Photo Researchers, Inc.; 18.12: © P. Motta/SPL/Photo Researchers, Inc.; Table 18.A (text): National Institute on Deafness and Other Communication Disorders, National Institute of Health; 18A(both):Robert S. Preston and Joseph E. Hawkins, Kresge Hearing Research Institute, University of Michigan; 18.16: © Sue Ford/Photo Researchers, Inc.

Chapter 19

Openers (both): © 3D4Medical.com/Getty Images; 19.1(hyaline, bone): © Ed Reschke; 19.1(osteocyte): © Biophoto Associates/Photo Researchers, Inc.; 19.5b: © Corbis RF; 19.10(adipose, bone, hyaline, muscle): © Ed Reschke; 19.10(connective tissue): © The McGraw-Hill Companies, Inc./Dennis Strete, photographer; 19.13: © EM Research Services, Newcastle University; 19.14a: © Victor B. Eichler, Ph.D.; 19.17: © G.W. Willis/Visuals Unlimited; 19.18: © Scott Camazine/Phototake; 19.19(left): © Alfred Pasieka/Photo Researchers, Inc.; 19.19(right): © Yoav Levy/Phototake USA.com; 19.20: © Dr. P. Marazzi/Photo Researchers, Inc.; 19A: © AFP/Getty Images.

Chapter 20

Opener: © Michael Krasowitz/Getty Images; 20A: © James Darell/Stone/Getty Images; Heath Focus (text), p. 405: Adapted from Robert L. Sack, Alfred J. Lewy, and Mary L. Blood, *Journal of Visual Impairment and Blindness*, 92: 145-161, March 1998; 20.11: © General Photographic Agency/Getty Images; 20.12(all): Reprinted from Clinical Pathological Conference, *American Journal of Medicine*, Vol. 20, page 133, "Acromegaly, Diabetes, Hypermetabolism, Proteinura and Heart Failure", copyright 1956, with permission from Elsevier.; 20.13(both):"Atlas of Pediatric Physical Diagnosis," Second Edition by Zitelli & Davis, 1992. Mosby-Wolfe Europe Limited, London, UK; 20.14a: © John Paul Kay/Peter Arnold; 20.14b: © Biophoto Associates/Photo Researchers, Inc.; 20.14c: © Dr. P. Marazzi/SPL/Photo Researchers, Inc.; 20.15a: © Custom Medical Stock Photo; 20.15b: © NMSB/Custom Medical Stock Photos; 20.17: © BSIP/Phototake; 20B: © Ramey Photo Agency.

Chapter 21

Opener: © Stockbyte/Getty RF; 21.3b(top): © CNRI/Photo Researchers, Inc.; 21.3b(bottom): © Ed Reschke; 21.7(oocyte): © Ed Reschke/Peter Arnold; 21.10b: © Bettman/Corbis; 21.11a, c, f: © The McGraw-Hill Companies, Inc./Bob Coyle, photographer; 21.11b: © The McGraw-Hill Companies, Inc./Vincent Ho, photographer; 21.11d: © Scott Camazone/Photo Researchers, Inc.; 21.11e(left): Courtesy Population Council, Karen-Tweedy Holmes.; 21.11e(right): © Bob Pardue/Alamy; 21A(alligators): © Index Stock; 21A(woman): © Steven Peters/Stone/Getty Images; 21A(owls): © Art Wolfe/Stone/Getty Images; 21A(fish): © Georgette Douwma/Getty RF; 21.12: © Scott Camazine/Photo Researchers, Inc.; Ecology Focus (text), p. 428: Courtesy of Dr. John M. Matter, Juaniata College; 21.13(all): © Nicholas Nixon; 21.15aboth): Courtesy of the Centers for Disease Control, Atlanta, GA. (Crooks, R. and Baur, K. "Our Sexuality, 8/e", Wadsworth 200d 17.4, pg. 489; 21.16: Courtesy Seattle STD Training Center/HIV Prevention, Seattle, WA; 21.17a: © Science VU/CDC/Visuals Unlimited; 21.17b: © Courtesy of the Centers for Disease Control and Prevention, Dr. N.J. Fiumara, Dr. Gavin Hart; 21.17c: © Collection CNRI/Phototake; 21.17d: © Science VU/Visuals Unlimited; 21.19: © Laparoscopic Appearance of Endometriosis (copyright 1991), www.memfert.com, Dan Martin, M. D.; 21Ba(left): © Vol. 161/Corbis; 21Ba(right): © Vol. 178/Corbis; 21Bb: © Argus Fotoarchiv/Peter Arnold; p. 440(birth control pills): © The McGraw-Hill Companies, Inc./Bob Coyle, photographer; p. 440(HIV): © Scott Camazine/Photo Researchers, Inc.

Chapter 22

Opener: © WireImage/Getty Images; 22.1: © David M. Phillips/Visuals Unlimited; 22.2a: © William Jorgensen/Visuals Unlimited; © Photodisc/Getty Images; p. 446: © Photodisc/Getty Images; 22.4b: Courtesy Kathryn Tosney; p. 447(frog): © Photodisc/Getty Images; 22.8(both): Courtesy Steve Paddock, Howard Hughes Medical Research Institute; 22.9a: Courtesy E.B. Lewis; 22.13a: © Lennart Nilsson "A Child is Born" Dell Publishing; 22.14: © John Watney/Photo Researchers, Inc.; 22A: Courtesy Dr. Ann Streissguth, University of Washington School of Medicine, Fetal Alcohol and Drug Unit; 22B: © PhotoDisc Vol. 170/Getty CD; 22.18: © Jose Luis Pelaez, Inc./Corbis; 22.19: © Corbis RF.

Chapter 23

Opener: © Sports Illustrated/Getty Images; 23.1: © Ned M. Seidler/National Geographic Image Collection; 23.3a: © Rufus F. Flokks/Corbis; 23.3b: © Larry Downing/Reuters/Corbis; 23.11a: Courtesy Robert D. Terry/University of California, San Diego School of Medicine; 23.12a: © Eye of Science/Photo Researchers, Inc.; 23.12b, c: Courtesy Dr. Hemachandra Reddy, The Neurological Science Institute, Oregon Health & Science University; 23.13(man): © Vol. 88/PhotoDisc; 23.13(woman): © Larry Williams/Corbis; 23.14: © Mediscan/Medical-On-Line; 23.17: Courtesy University of Connecticut, Peter Morenus, photographer; 23.18: © Jane Burton/Bruce Coleman, Inc.

Chapter 24

Opener: © moodboard/Alamy; 24.3(left, right): Courtesy Dr. Rabi Tawil, Director, Neuromuscular Pathology Laboratory, University of Rochester Medical Center; 24.3(center): Courtesy Muscular Dystrophy; 24.4(queen): © Stapleton Collection/Corbis; 24.4(prince): © Huton Archive/Getty Images; 24.7a: © Jose Carrilo/PhotoEdit; 24.7b: © CNRI/SPL/Photo Researchers, Inc.; 24Ac: © James King-Holmes/SPL/Photo Researchers, Inc.; 24Ad, e: © CNRI/SPL/Photo Researchers, Inc.; 24.8: Courtesy The Williams Syndrome Association; 24.9: Courtesy Kathy Wise; 24Ba(both):From R. Simensen and R. Curtis Rogers, "Fragile X Syndrome," American Family Physician 39(5):186, May 1989. © American Academy of Family Physicians; 24Bb: © David M. Phillips/Visuals Unlimited; 24.10b(both): From N.B. Spinner et al., *American Journal of Human Genetics* 55 (1994): p. 239. The University of Chicago Press.

Chapter 25

Opener: © AP Images/Jim McKnight; 25Aa, b: © Science Source/Photo Researchers, Inc.; 25.11: Courtesy Alexander Rich; 25.17: © Photodisc/Getty RF; 25.19c: © Mondae Leigh Baker; 25.20(both): © Stan Flegler/Visuals Unlimited; 25B: © Jeff Maloney/Getty Images.

Chapter 26

Opener: © Blend Images Photography/Veer; 26.5: © Lester Lefkowitz/Corbis; 26.6(both): Courtesy General Electric Research & Development.; 26A(corn): © Larry Lefever/Grant Heilman Photography; 26A (soybeans): © Norm Thomas/Photo Researchers, Inc.; 26A(cotton): Courtesy Carolyn Merchant/ UC Berkeley; 26.1(human): © Image Source/JupiterImages; 26.1(mouse): © David A. Northcott/Corbis; 26.1(fly): © Herman Eisenbeiss/Photo Researchers, Inc.; 26.1(plant): © Brad Mogen/Visuals Unlimited; 26.1(roundworm): © 2007 J.L. Carson/Custom Medical Stock Photo; 26.1(yeast): © David Scharf/Photo Researchers, Inc.; 26.11: Courtesy Dr. Michael Thompson.

Chapter 27

Opener: © Dr. Dennis Kunkel/Getty Images; 27.3: © Ralph White/Corbis; 27.4a: © Science VU/Visuals Unlimited; 27.4b: Courtesy Dr. David Deamer; 27.5a: © Jean-Claude Carton/Bruce Coleman, Inc.; 27.5b: © Joe Tucciarone; 27.7: © Chase Studio/Photo Researchers, Inc.; 27.10: © J.G.M. Thewissen, http://darla.neoucom.edu/DEPTS/ANAT/Thewissen/; 27.13(both): © Michael Wilmer Forbes Tweedie/Photo Researchers, Inc.; 27.15: Courtesy Victor McKusick; 27.16: © Robert Maier/Animals Animals; 27.18b: © Bob Evans/Peter Arnold; 27.20(med ground, Mangrove, vegetarian finches): © Greg Lasley Nature Photography; 27.20(small tree and large cactus finches): © Rob and Ann Simpson/Visuals Unlimited; 27.20(medium tree finch): © David Hosking/Alamy; 27.20(small ground and sharp-beaked ground finches): © Gerald & Buff Corsi/Visuals Unlimited; 27.20(cactus finch): © Fritz Polking/Visuals Unlimited; 27.20(large tree finch): © Steve Bird, Birdseekers Ltd.; 27.20(woodpecker finch): © R. Koster/OSF/Animals Animals; 27.20(warbler finch): © Joe McDonald/Animals Animals; 27.20(large ground finch): © Adrienne Gibson/Animals Animals.

Chapter 28

Opener: © 3D4Medical.com/Getty Images; 28.2a: © Gary Gaugler/Visuals Unlimited; 28.2b: © SciMAT/Photo Researchers, Inc.; 28.2c: © Dr. Richard Kessel & Dr. Gene Shih/Visuals Unlimited; 28.4b: © CNRI/SPL/Photo Researchers, Inc.; 28.5a: © R. Knauft/Biology Media/Photo Researchers, Inc.; 28.5b: © Eric Grave/Photo Researchers, Inc.; 28.6a: © Fred Hossler/Visuals Unlimited; 28.6b: © SPL/Photo Researchers, Inc.; 28.6c: © Kenneth E. Greer/Visuals Unlimited; 28.7(both): © Leonard V. Crowley, *An Introduction to Human Disease: Pathology and Pathophysiology Correlations* 5/e, page 383 right and page 385 left; 2001: Jones and Bartlett Publishers, Sudbury, MA. Reprinted with permission; 28.8a(main): © John Sohlden/Visuals Unlimited; 28.8a(inset): From J.T. Staley, et al., *Bergey's Manual of Systematic Bacteriology*, Vol. 13, © 1989 Williams & Wilkins Co., Baltimore. Prepared by A.L. Usted; Photography by Dept. of Biophysics, Norwegian Institue of Technology; 28.8b(main): © Jeff Lepore/Photo Researchers, Inc.; 28.8b(inset): Courtesy Dennis W. Grogan, Univ. of Cincinnati; 28.8c(main): © Susan Rosenthal/Corbis; 28.8c(inset): © Ralph Robinson/Visuals Unlimited; 28.10b: © Stephen Durr; 28.10c(both): © Biophoto Associates/

Photo Researchers, Inc.; 28.10d: © Dr. John D. Cunningham/Visuals Unlimited; 28.11: © Darlyne A. Murawski/Getty Images; 28.12(top): © Kevin Schafer/Peter Arnold; 28.13, 28.14 (both): © Dan Ippolito; 28.15a: © Eye of Science/Photo Researchers, Inc.; 28.16: © Michael Abbey/Visuals Unlimited; 28.17a: © Eric Grave/Photo Researchers, Inc.; 28.17b: © CABISCO/Phototake; 28.17c, 28.18b (inset): © Manfred Kage/Peter Arnold; 28.18b(cliffs): © Rick Ergenbright/Corbis; 28.18c: © Dr. Richard Kessel & Dr. Gene Shih/Visuals Unlimited; 28.20(plasmodium): © CABISCO/Visuals Unlimited; 28.20(sporangia): © V. Duran/Visuals Unlimited; 28.21a: © Gary R. Robinson/Visuals Unlimited; 28.21b: © Dennis Kunkel/Visuals Unlimited; 28.22a: © Jeffrey Lepore/Photo Researchers, Inc.; 28.22b:From C.Y. Shih and R.G. Kessel, *Living Images* Science Books International, Boston, 1982; 28.23a: © P. Marazzi/SPL/Photo Researchers, Inc.; 28.23b: © John Hadfield/SPL/Photo Researchers, Inc.; 28.23c: © Everett S. Beneke/Visuals Unlimited; 28.25:Reproduced with permission of the Freshwater Biological Association on behalf of The Estate of Dr Hilda Canter-Lund. (c) The Freshwater Biological Association; 28.26: © Runk/Schoenberger/Grant Heilman Photography; 28.27b: © Robert Calentine/Visuals Unlimited; 28.27c: © Michael Viard/Peter Arnold; 28.28: © Mondae Leigh Baker; 28.29(both): © David Philips/Visuals Unlimited; 28.30b: © Biophoto Assoc./Photo Researchers, Inc.; 28.30c: © Inga Spence/Photo Researchers, Inc.; 28.30d: © L. West/Photo Researchers, Inc.; 28.31a: © Steven P. Lynch; 28.31b: © Arthur M. Siegelman/Visuals Unlimited; 28.32a: © Dr. Jeremy Burgess/SPL/Photo Researchers, Inc.; 28.32b: © Stephen Sharnoff/Visuals Unlimited; 28.32c: © Kerry T. Givens; 28.33a: © Dr. Hans Gelderblom/Visuals Unlimited; 28.33b: © K.G. Murti/Visuals Unlimited; 28A(foods): © The McGraw-Hill Companies, Inc./ Mark Dierker, photographer; 28A(petri dish): © Bettmann/Corbis; 28.36: © John D. Cunningham/Visuals Unlimited; photo, p. 601: © Eye of Science/Photo Researchers, Inc.

Chapter 29

Opener: © David Dilcher and Ge Sun; 29.4a: © Ed Reschke/Peter Arnold; 29.4b: © J.M. Conrarder/Nat'l Audubon Society/Photo Researchers, Inc.; 29.4c: © R. Calentine/Visuals Unlimited; 29.5(sporophyte): © Heather Angel/Natural Visions; 29.5(gametophyte): © Bruce Iverson; 29.6: © The Field Museum; 29.7: © Steve Solum/Bruce Coleman, Inc.; 29.8(cinnamon): © James Strawser/Grant Heilman Photography; 29.8(hart's tongue): © Walter H. Hodge/Peter Arnold; 29.8(maidenhair): © Larry Lefever/Jane Grushow/Grant Heilman Photography; 29.9: © Steven P. Lynch; 29.10: © CABISCO/Phototake; 29.11: © Robert P. Carr/Bruce Coleman, Inc.; 29.12(pollen grain): © Phototake; 29.13a: © Grant Heilman/Grant Heilman Photography; 29.13b: © Walt Anderson/Visuals Unlimited; 29.13c: © The McGraw-Hill Companies, Inc./Evelyn Jo Johnson, photographer; 29.14a(both): © Steven P. Lynch; 29.14b: © Runk/Schoenberger/Grant Heilman Photography; 29.14c: Courtesy Fiona Norris; 29.15(cactus): © Christi Carter/Grant Heilman Photography; 29.15(waterlily): © Pat Pendarvis; 29.15(iris): © David Cavanaugh/Peter Arnold; 29.15(trillium): © Adam Jones/Photo Researchers, Inc.; 29.15(apple blossoms): © Inga Spence/Photo Researchers, Inc.; 29.15(butterfly weed): © Courtesy Evelyn Jo Johnson; 29.18a: © Stockbyte/PunchStock; 29.18b, c: © Mondae Leigh Baker; 29.18d: © Photolink/Getty RF; 29.18e: © Corbis RF; 29.18f: © Brand X Pictures RF; 29A(wheat plants): © Pixtal /age fotostock; 29A(corn plants): © Corbis RF; 29A(corn ear): © Nigel Cattlin/Photo Researchers, Inc.; 29A(rice plants): © Corbis RF; 29A(rice grain heads): © Dex Image/Getty RF; 29B(rubber): © Steven King/Peter Arnold; 29B(cotton): © Dale Jackson/Visuals Unlimited; Ecology Focus (text), pg. 616: Charles N. Horn.

Chapter 30

Opener: © John William Banagan/Getty Images; 30.1(hydra): © Biophoto Associates/Photo Researchers, Inc.; 30.1(crayfish): © David M. Dennis; 30.1(humans): © Stock Connection Distribution/Alamy; 30.4a: © Andrew J. Martinez/Photo Researchers, Inc.; 30.5b: © Azure Computer & Photo Services/Animals Animals; 30.5c: © Ron & Valerie Taylor/Bruce Coleman, Inc.; 30.5d: © Runk/Schoenberger/Grant Heilman Photography; 30.5e: © © Amos Nachoum/Corbis; 30.6: © CABISCO/Visuals Unlimited; 30.7: © Tom E. Adams/Peter Arnold; 30.8(proglottid): © John D. Cunningham/Visuals Unlimited; 30.8(scolex): © James Webb/Phototake; 30.10a: © Fred Bavendam/Peter Arnold; 30.10b: Courtesy Larry S. Roberts; 30.10c: © Kenneth W. Fink/Bruce Coleman, Inc.; 30.10d: © Ken Lucas/Visuals Unlimited; 30.11a: © Farley Bridges; 30.11b: © Georgette Douwma/Photo Researchers, Inc.; 30.13a: © Heather Angel/Natural Visions; 30.13b: © Diane R. Nelson; 30.14: © Roger K. Burnard/Biological Photo Service; 30.15a: © Lauritz Jensen/Visuals Unlimited; 30.15c: © James Solliday/Biological Photo Service; 30.16a: © John MacGregor/Peter Arnold; 30.16b: © G.C. Kelley/Photo Researchers, Inc.; 30.16c: © Robert Evans/Peter Arnold; 30.16d: © James Robinson/Photo Researchers, Inc.; 30.16e: © Dwight Kuhn; 30.17c: © OSF/London Scientific Films/Animals Animals; 30.19a: © Joel Sartore/Getty RF; 30.19b: © Creatas Images/PictureQuest; 30.19c: © MedioImages/Photodisc/Getty RF; 30.19d: © Photos.comSelect/Index Stock Imagery; 30.19e: © Darlyne A. Murawski/Getty Images; 30A: © Vioila's Photo Visions Inc./Animals Animals; 30.21a: © Tom McHugh/Photo Researchers, Inc.; 30.21b: © Robert Lubeck/Animals Animals; 30.21c: © Scott Camazine/Photo Researchers, Inc.

Chapter 31

Opener (inset): © Tom McHugh/Photo Researchers, Inc.; opener(main): © AP Images/Mike Wintroath; 31.1a: © Diane R. Nelson; 31.1b, c: © David Wrobel/Getty Images; 31.1d: © Philippe Bourseiller/Getty Images; 31.2: © Randy Morse, GoldenStateImages.com; 31.5: © Rick Harbo; 31.6: © Heather Angel/Natural Visions; 31.7a: © James Watt/Animals Animals; 31.7b: © Dwight Kuhn; 31.7c: © Zig Leszczynski/Animals Animals; 31.8a: © Comstock Images/PictureQuest; 31.8b: © Ron & Valerie Taylor/Bruce Coleman, Inc.; 31.9(1): © Jane Burton/Bruce Coleman, Inc.; 31.9(2, 3): © Michael Redmer/Visuals Unlimited; 31.9(4): © Joe McDonald/Visuals Unlimited; 31.10: © Zig Leszczynski/Animals Animals; 31.11a: © Martin Harvey/Gallo Images/Corbis; 31.13a: © Gary W. Carter/Corbis; 31.13b: © Dale DeGabriele/Getty Images; 31.13c: © Medford Taylor/Getty Images; 31.14a: © Tom McHugh/Photo Researchers, Inc.; 31.14b: © Leonard Lee Rue/Photo Researchers, Inc.; 31.14c: © Fritz Prenzel/Animals Animals; 31.15a: © Stephen J. Krasemann/Photo Researchers, Inc.; 31.15b: © John Downer/Getty Images; 31.15c: © Gerald Lacz/Animals Animals; 31.15d: © Mike Bacon; 31Aa: © MedioImages/SuperStock RF; 31Ab: © Allan Friedlander/SuperStock; 31Ac: © Account Phototake/Phototake; 31.18a: © Dan Dreyfus and Associates; 31.18b: © John Reader/Photo Researchers, Inc.; 31.19: © National Museum of Kenya; 31.21:(C) The Field Museum #A102513c; 31.22: Courtesy Dept. of Library Services, American Museum of Natural History.

Chapter 32

Opener: © Frans Lanting/Corbis; 32.1a: © Joe McDonald; 32.1b: Courtesy Jeff and Wendy Martin, Refuge for Saving the Wildlife, Inc.; 32.2(coastal): © John Sullivan/Monica Rua/Ribbitt Photography; 32.2(inland): © R. Andrew Odum/Peter Arnold; 32.2 (line art): Data from S.J. Arnold, "The Microevolution of Feeding Behavior" in *Foraging Behavior: Ecology, Ethological, and Psychological Approaches*, edited by A. Kamil and T. Sargent, 1980, Garland Publishing Company, New York, NY. p. 824; 32.5: © Gregory G. Dimijian/Photo Researchers, Inc.; 32.6(main): © Arco Images/GmbH/Alamy; 32.6(inset): © Fritz Polking/Visuals Unlimited; 32.7: © Barbara Gerlach/Visuals Unlimited; 32.8(trees): © PhotoLink RF; 32.8(firefly): © Phil Degginger/Alamy; 32.9a: © OSF/Animals Animals; 32Aa: © Tom McHugh/Photo Researchers, Inc.; 32Ab: © Alan Carey/Photo Researchers, Inc.; 32.10: © Nicole Duplaix/Peter Arnold; 32.12: © Thomas Dobner 2006/Alamy RF; 32.13: © D. Robert & Lorri Franz/Corbis; 32.14: © Jodi Cobb/National Geographic Image Collection; 32.15: © J & B Photo/Animals Animals.

Chapter 33

Opener: © Masa Ushioda/SeaPics.com; 33.1: © David Hall/Photo Researchers, Inc.; 33.2a: © age fotostock/SuperStock; 33.2b: © Tracey Thompson/Corbis; 33.5b: © Image State/Alamy; 33.5c: © Ben Osborne/OSF/Animals Animals; 33.6 (line art): United Nations Population Division, 2002, p. 850; 33.7(dandelions): © Ted Levin/Animals Animals; 33.7(bears): © Winfried Wisniewski/Getty Images; 33.8: Data from G.F. Gause, *The Struggle for Existence*, 1934, Williams & Wilkens Company, Baltimore, MD, 33.10: © Alan Carey/Photo Researchers, Inc.; 33.10: Data from D.A. MacLulich, Fluctuations in the Numbers of the Varying Hare (Lepus americanus), University of Toronto Press, Toronto, 1937; reprinted 1974, p. 543; 33.11a: © Michael Fogden/Animals Animals; 33.11b: © Scott Camazine/Photo Researchers, Inc.; 33.11c: © National Audubon Society/A. Cosmos Blank/Photo Researchers, Inc.; 33.12a, b, c: © Edward S. Ross; 33.12d: © James H. Robinson/Photo Researchers, Inc.; 33.13: © Gunter Ziesler/Peter Arnold; 33.14: © Bill Wood/Bruce Coleman, Inc. p. 864.

Chapter 34

Opener: © Jeremy Woodhouse 2008/Getty Images; 34.1a(diatom): © Ed Reschke/Peter Arnold; 34.1a(tree): © © Herman Eisenbeiss/Photo Researchers, Inc.; 34.1b(caterpillar): © Corbis RF; 34.1b(rabbit): © Gerald C. Kelley/Photo Researchers, Inc.; 34.1c(spider): © Bill Beatty/Visuals Unlimited; 34.1c(osprey): © Joe McDonald/Visuals Unlimited; 34.1d(bacteria): © David M. Phillips/Visuals Unlimited; 34.1d(mushrooms): © Michael Beug; 34.3: © George D. Lepp/Photo Researchers, Inc.; 34A: Courtesy NASA.

Chapter 35

Opener: Courtesy NASA; 35.6a: © John Shaw/Tom Stack & Assoc.; 35.6b: © John Shaw/Bruce Coleman, Inc.; 35.7a: © Mack Henly/Visuals Unlimited; 35.7b: © Bill Silliker, Jr./Animals Animals; 35.8(forest): © E. R. Degginger/Color Pic/Animals Animals; 35.8(bobcat): © Tom McHugh/Photo Researchers, Inc.; 35.8(millipede): © OSF/Animals Animals; 35.8(chipmunk): © Zig Lesczynski/Animals Animals; 35.9(frog): © Art Wolfe/Photo Researchers, Inc.; 35.9(katydid): © M. Fogden/OSF/Animals Animals; 35.9(ocelot): © Martin Wendler/Peter Arnold; 35.9(macaw): © PhotoLink/Getty RF; 35.9(butterfly): © Kevin Schafer/Getty Images; 35.9(lemur): © Brand X Pictures RF; 35.9(lizard): © J.H. Pete Carmichael/Getty Images; 35.10a: © Bruce Iverson; 35.10b: © Kathy Merrifield/Photo Researchers, Inc.; 35.11(prairie): © Jim Steinberg/Photo Researchers, Inc.; 35.11(bison): © Steven Fuller/Animals Animals; 35.12(cheetah): © Digital Vision/Getty RF; 35.12(giraffes): © George W. Cox; 35.12(wildebeests, zebras): © Darla G. Cox; 35.13(desert): © John Shaw/Bruce Coleman, Inc.; 35.13(rat): © Bob Calhoun/Bruce Coleman, Inc.; 35.13(roadrunner): © Jack Wilburn/Animals Animals; 35.13(fox): © Jeri Gleiter/Peter Arnold; 35.14a: © Roger Evans/Photo Researchers, Inc.; 35.14b: © Michael Gadomski/Animals Animals; 35.16(pike): © Robert Maier/Animals Animals; 35.16(water strider): © Frank Greenaway/Getty Images; 35.17(shrimp): © Ken Lucas/Ardea London Limited; 35.17(snails): © Fred Whitehead/Animals Animals; 35.18a: © John Eastcott/Yva Momatiuk/Animals Animals; 35.18b: © Mark Lewis/Getty Images; 35.19a: © Brandon Cole/Visuals Unlimited; 35.19b: © Jeff Greenburg/Photo Researchers, Inc.

Chapter 36

Opener: © Bennett, Darren/Animals Animals; 36.3(periwinkle © Kevin Schafer/Peter Arnold, Inc.; 36.3(armadillo): © John Cancalosi/Peter Arnold, Inc.; 36.3(fishing): © Herve Donnezan/Photo Researchers, Inc.; 36.3(bat): © Merlin D. Tuttle/Bat Conservation International; 36.3(ladybug): © D. Hurst/Alamy RF; 36.3(man): © Bryn Campbell/Stone/Getty Images; 36.4(forest): © William Smithey Jr./Getty Images; 36.4(seashore): © Vol. 121/Corbis RF; 36.5b: © Gunter Ziesler/Peter Arnold; 36.5c(top): Courtesy Woods Hole Research Center; 36.5c(center): Courtesy R.O. Bierregaard, Jr.; 36.5c(bottom): Courtesy Thomas Stone, Woods Hole Research Center; 36Aa: © Porterfield/Chickering/Photo Researchers, Inc.; 36Ab: © Michio Hoshino/Minden Pictures; 36.6a: © Chuck Pratt/Bruce Coleman, Inc.; 36.6b: © Chris Johns/National Geographic Image Collection; 36.7b: Courtesy Walter C. Jaap/Florida Fish & Wildlife Conservation Commission; 36.8a: © Shane Moore/Animals Animals; 36.8b: © Peter Auster/University of Connecticut; 36.9: © Dr. Paul A. Zahl/Photo Researchers, Inc; 36.10a: © Gerard Lacz/Peter Arnold; 36.10b(forest): © Art Wolfe/Artwolfe.com; 36.10b(owl): © Pat & Tom Leeson/Photo Researchers, Inc.; 36.11b: © Jeff Foott Productions; 36.12(panther): © Tom & Pat Leeson/Photo Researchers, Inc.; 36.12(alligator): © Fritz Polking/Visuals Unlimited; 36.12(ibis): © Stephen G. Maka; 36.12(spoonbill): © Kim Heacox/Peter Arnold; 36.12(stork): © Millard H. Sharp/Photo Researchers, Inc.; 36.13(land): © Vol. 39/PhotoDisc/Getty Images; 36.13(water): © Courtesy Evelyn Jo Johnson; 36.13(food): © The McGraw-Hill Companies, Inc./John Thoeming photographer; 36.13(energy): © Gerald and Buff Corsi/Visuals Unlimited; 36.13(minerals): © James P. Blair/National Geographic Image Collection; 36.14a: © Kent Knudson/PhotoLink/Getty Images RF; 36.14b: © Scenics of America/PhotoLink/Getty RF; 36.15: © David L. Pearson/Visuals Unlimited; 36.16a: © S.K. Patrick/Visuals Unlimited; 36.16b: © Argus Fotoarchiv/Peter Arnold; 36.16c: © 2009 The Associated Press; 36.17a: © Laish Bristol/Visuals Unlimited; 36.17b:V. Jane Windsor, Division of Plant Industry, Florida Department of Agriculture & Consumer Services.

Index

Note: Page numbers in *italics* indicate material presented in figures and tables.

A

B

C

F

H

I

J

K

L

M

Q

R

S

U

V

W

X

Y

Z

History of Biology

Year	Name	Country	Contribution
1628	William Harvey	Britain	Demonstrates that the blood circulates and the heart is a pump.
1665	Robert Hooke	Britain	Uses the word *cell* to describe compartments he sees in cork under the microscope.
1668	Francesco Redi	Italy	Shows that decaying meat protected from flies does not spontaneously produce maggots.
1673	Antonie van Leeuwenhoek	Holland	Uses microscope to view living microorganisms.
1735	Carolus Linnaeus	Sweden	Initiates the binomial system of naming organisms.
1809	Jean B. Lamarck	France	Supports the idea of evolution but thinks there is inheritance of acquired characteristics.
1825	Georges Cuvier	France	Founds the science of paleontology and shows that fossils are related to living forms.
1828	Karl E. von Baer	Germany	Establishes the germ layer theory of development.
1838	Matthias Schleiden	Germany	States that plants are multicellular organisms.
1839	Theodor Schwann	Germany	States that animals are multicellular organisms.
1851	Claude Bernard	France	Concludes that a relatively constant internal environment allows organisms to survive under varying conditions.
1858	Rudolf Virchow	Germany	States that cells come only from preexisting cells.
1858	Charles Darwin	Britain	Presents evidence that natural selection guides the evolutionary process.
1858	Alfred R. Wallace	Britain	Independently comes to same conclusions as Darwin.
1865	Louis Pasteur	France	Disproves the theory of spontaneous generation for bacteria; shows that infections are caused by bacteria, and develops vaccines against rabies and anthrax.
1866	Gregor Mendel	Austria	Proposes basic laws of genetics based on his experiments with garden peas.
1882	Robert Koch	Germany	Establishes the germ theory of disease and develops many techniques used in bacteriology.
1900	Walter Reed	United States	Discovers that the yellow fever virus is transmitted by a mosquito.
1902	Walter S. Sutton Theodor Boveri	United States Germany	Suggest that genes are on the chromosomes, after noting the similar behavior of genes and chromosomes.
1903	Karl Landsteiner	Austria	Discovers ABO blood types.
1904	Ivan Pavlov	Russia	Shows that conditioned reflexes affect behavior, based on experiments with dogs.
1910	Thomas H. Morgan	United States	States that each gene has a locus on a particular chromosome, based on experiments with *Drosophila*.
1922	Sir Frederick Banting Charles Best	Canada	Isolate insulin from the pancreas.
1924	Hans Spemann Hilde Mangold	Germany	Show that induction occurs during development, based on experiments with frog embryos.
1927	Hermann J. Muller	United States	Proves that X rays cause mutations.
1929	Sir Alexander Fleming	Britain	Discovers the toxic effect of a mold product he called penicillin on certain bacteria.

Antonie van Leeuwenhoek

Charles Darwin

Louis Pasteur

Robert Koch

Ivan Pavlov